# GRAY'S ANATOMY FOR STUDENTS

# Make the most of your Gray's experience with extensive online content

## eChapter 9 Neuroanatomy by Jennifer M. McBride, PhD

## Gray's Systemic Anatomy

1. Cardiovascular System
2. Respiratory System
3. Gastrointestinal System
4. Urogenital System
5. Lymphatic System
6. Nervous System

## Gray's Surface Anatomy Interactive Tool

## PT Cases by Jennifer L. Artz, DPT

Updated for this edition, 40 PT cases provide cases with explanations of physical exam findings and treatments

## Clinical Cases

Over 100 Clinical Cases

## Online Anatomy and Embryology Self-Study Course

Provides 77 modules with self-check activities to support learning

## Short Questions

Updated for this edition, 70 write-in questions requiring short answers

## Self-Assessment Questions

Over 150 National Board style multiple-choice questions

**See inside front cover for your access instructions**

# GRAY'S ANATOMY FOR STUDENTS

Fourth Edition

**Richard L. Drake, PhD, FAAA**
Director of Anatomy
Professor of Surgery
Cleveland Clinic Lerner College of Medicine
Case Western Reserve University
Cleveland, Ohio

**A. Wayne Vogl, PhD, FAAA**
Professor of Anatomy and Cell Biology
Department of Cellular and Physiological Sciences
Faculty of Medicine
University of British Columbia
Vancouver, British Columbia, Canada

**Adam W. M. Mitchell, MB BS, FRCS, FRCR**
Consultant Radiologist
Director of Radiology
Fortius Clinic
London, United Kingdom

*Illustrations by*
**Richard Tibbitts** and **Paul Richardson**

*Photographs by*
**Ansell Horn**

ELSEVIER

GRAY'S ANATOMY FOR STUDENTS, FOURTH EDITION

ISBN: 978-0-323-39304-1
IE ISBN: 978-0-323-61104-6



**Notices**

Knowledge and best practice in this field are constantly changing. As new research and experience broaden our understanding, changes in research methods, professional practices, or medical treatment may become necessary. Practitioners and researchers must always rely on their own experience and knowledge in evaluating and using any information, methods, compounds, or experiments described herein. In using such information or methods they should be mindful of their own safety and the safety of others, including parties for whom they have a professional responsibility. With respect to any drug or pharmaceutical products identified, readers are advised to check the most current information provided (i) on procedures featured or (ii) by the manufacturer of each product to be administered, to verify the recommended dose or formula, the method and duration of administration, and contraindications. It is the responsibility of practitioners, relying on their own experience and knowledge of their patients, to make diagnoses, to determine dosages and the best treatment for each individual patient, and to take all appropriate safety precautions. To the fullest extent of the law, neither the Publisher nor the authors, contributors, or editors, assume any liability for any injury and/or damage to persons or property as a matter of products liability, negligence or otherwise, or from any use or operation of any methods, products, instructions, or ideas contained in the material herein.

***The Publisher***

Library of Congress Control Number: 2018952008

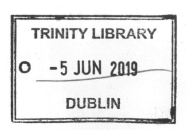

*Senior Content Strategist:* Jeremy Bowes
*Director, Content Development:* Rebecca Gruliow
*Publishing Services Manager:* Catherine Jackson
*Senior Project Manager:* John Casey
*Senior Book Designer:* Amy Buxton

Printed in Canada

9 8 7 6 5 4 3 2 1

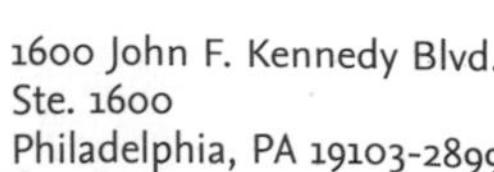
1600 John F. Kennedy Blvd.
Ste. 1600
Philadelphia, PA 19103-2899

To my wife, Cheryl, who has supported me; and my parents,
who have guided me.
***RLD***

To my family, to my professional colleagues and role models,
and to my students—this book is for you.
***AWV***

To Max, Elsa, and Cathy.
And the man that made us all,
Timothy Ianthor Mitchell!
***AWMM***

# Acknowledgments

First, we would like to collectively thank those who have reviewed drafts of various editions of this book—anatomists, educators, and student members of the editorial review board from around the world. Your input was invaluable.

We'd also like to thank Richard Tibbitts and Paul Richardson for their skill in turning our visual ideas into a reality that is not only a foundation for the acquisition of anatomical knowledge, but also is beautiful.

Thanks must also go to Madelene Hyde, Jeremy Bowes, Bill Schmitt, Rebecca Gruliow, John Casey, and all the team at Elsevier for guiding us through the preparation of this book.

We'd also like to thank Professor Richard A. Buckingham of the Abraham Lincoln School of Medicine, University of Illinois for the provision of Fig. 8.114B. Finally, because we worked separately, distanced by, in some cases, thousands of miles, there are various people who gave local support, whom we would like to make mention of individually. We've gratefully listed them here:

Dr. Leonard Epp, Dr. Carl Morgan, Dr. Robert Shellhamer, and Dr. Robert Cardell who profoundly influenced my career as a scientist and an educator.

***Richard L. Drake***

Dr. Sydney Friedman, Dr. Elio Raviola, and Dr. Charles Slonecker, for their inspiration and support and for instilling in me a passion for the discipline of Anatomy.

Dr. Murray Morrison, Dr. Joanne Matsubara, Dr. Brian Westerberg, Laura Hall, and Jing Cui, for contributing images for the chapter on the head and neck.

Dr. Bruce Crawford and Logan Lee, for help with images for the surface anatomy of the upper limb.

Professor Elizabeth Akesson and Dr. Donna Ford, for their enthusiastic support and valuable critiques.

Dr. Sam Wiseman, for contributing surgical and other images in the abdomen and head and neck chapters.

Dr. Rosemary Basson for writing the 'Erectile dysfunction - In the clinic' in the pelvis and perineum chapter.

***A. Wayne Vogl***

Anatomy changes! We see it through new "glasses"—X-rays, CT, MRI, and now AI ... what's next (I do not know, but it's exciting). We need new eyes and new blood, and I am delighted to have been helped by two inspirational colleagues, Dr. Monika Rowe and Dr. Rajat Choudhury with the clinical. They are stars of the future.

Dr. Justin Lee and Dr. Gajan Rajeswaran are the greatest colleagues who challenge me daily with MSK anatomy.

Many thanks for constant support from Mr. Andrew Williams, Prof. James Calder, and Lucy Ball. They are all awesome!

***Adam W. M. Mitchell***

# Preface

The 4th edition of *Gray's Anatomy for Students* maintains the goals and objectives of the 1st, 2nd, and 3rd editions while at the same time continuing to incorporate input from our readers and adjust the content to align with the evolving educational environment.

One of the major focuses of our attention as we prepared the 4th edition was adjusting print content to accommodate the evolving capacity to move material onto e-learning platforms. We have expanded the number of "In the Clinic" boxes in the printed book but have moved some of the "Clinical Cases" in earlier editions of the book from the print version to the online platform. Moving some of the clinical cases to the e-learning platform has allowed us to add new material to the print version without expanding the size of the book or compromising the basic goals and objectives of the work. New material added has included new imaging to reflect recent advances in the field of radiology and, in response to reader feedback, we have added simple summary line diagrams ("Quick draw sketches") to some of the key figures that can be easily replicated by students.

A new feature in the 4th edition is an accompanying online e-book organized systemically and referred to as *Gray's Systemic Anatomy*. This e-book has chapters on the cardiovascular, respiratory, gastrointestinal, nervous, urogenital, and lymphatic systems. We feel students may find this material useful as many medical school curriculums now use a systems-based, integrated approach.

A second new feature to the 4th edition is the inclusion of a neuroanatomy chapter. We hope that this eChapter will assist students as they advance their knowledge of neuroanatomy.

As was the case in previous editions, review materials/study aids are available on Student Consult as an online resource with the appropriate review materials for each chapter listed at the beginning of that chapter. This information includes an online anatomy and embryology self-study course, medical clinical cases, physical therapy clinical cases, self-assessment questions, an interactive surface anatomy module, and short answer questions. We believe that with these changes the 4th edition of *Gray's Anatomy for Students* is an improved version of the 3rd edition and hope that the book will continue to be a valuable learning resource for students.

**Richard L. Drake**
**A. Wayne Vogl**
**Adam W. M. Mitchell**
*October 2018*

# About the Book

## The idea

In the past 20 years or so, there have been many changes that have shaped how students learn human anatomy in medical and dental schools and in allied health programs, with curricula becoming either more integrated or more systems based. In addition, instructional methods focus on the use of small group activities with the goals of increasing the amount of self-directed learning and acquiring the skills for the life-long acquisition of knowledge. An explosion of information in every discipline has also been a force in driving curricular change as it increases the amount to be learned without necessarily increasing the time available. With these changes, we felt it was time for a new text to be written that would allow students to learn anatomy within the context of many different curricular designs, and within ever-increasing time constraints.

We began in the fall of 2001 by considering the various approaches and formats that we might adopt, eventually deciding upon a regional approach to anatomy with each chapter having four sections. From the beginning, we wanted the book to be designed with multiple entry points, to be targeted at introductory level students in a broad spectrum of fields, and to be a student-oriented companion text for *Gray's Anatomy*, which is aimed at a more professional audience. We wrote the text first and subsequently constructed all the artwork and illustrations to complement and augment the words. Preliminary drafts of chapters, when complete, were distributed to an international editorial board of anatomists, educators, and anatomy students for review. Their comments were then considered carefully in the preparation of the final book.

The text is not meant to be exhaustive in coverage, but to present enough anatomy to provide students with a structural and functional context in which to add further detail as they progress through their careers. *Gray's Anatomy* was used as the major reference, both for the text and for the illustrations, during the preparation of this book, and it is the recommended source for acquiring additional detail.

## The book

*Gray's Anatomy for Students* is a clinically oriented, student-friendly textbook of human anatomy. It has been prepared primarily for students in a variety of professional programs (e.g., medical, dental, chiropractic, and physical therapy programs). It can be used by students in traditional, systemic, combined traditional/systemic, and problem-based curricula and will be particularly useful to students when lectures and laboratories in gross anatomy are minimal.



### ORGANIZATION

Using a regional approach, *Gray's Anatomy for Students* progresses through the body in a logical fashion, building on the body's complexities as the reader becomes more comfortable with the subject matter. Each chapter can be used as an independent learning module, and varying the sequence will not affect the quality of the educational experience. The sequence we have chosen to follow is back, thorax, abdomen, pelvis and perineum, lower limb, upper limb, and head and neck.

We begin with "The body," which contains an overview of the discipline of gross anatomy and an introduction to imaging modalities and general body systems. We follow this with the back because it is often the first area dissected by students. The thorax is next because of its central location and its contents (i.e., the heart, the great vessels, and the lungs). This also begins a progression through the body's cavities. The abdomen and pelvis and perineum follow logically in sequence from the thorax. Continuing downward toward the feet, the lower limb is next, followed by the upper limb. The last region discussed is the head and neck. This region contains some of the most difficult anatomy in the body. Covering all other regions first gives the student the opportunity to build a strong foundation from which to understand this complex region.

### CONTENT

Each regional anatomy chapter consists of four consecutive sections: conceptual overview, regional anatomy, surface anatomy, and clinical cases.

The conceptual overview provides the basis on which information in the later sections is built. This section can be read independently of the rest of the text by students who require only a basic level of understanding and can also be read as a summary of important concepts after the regional anatomy has been mastered.

The regional anatomy section provides more detailed anatomy along with a substantial amount of relevant clinical correlations. It is not an exhaustive discussion but instead provides information to a level that we feel is necessary for understanding the organization of the region. Throughout this section, two levels of clinical material are provided. Clinical hooks are fully integrated with the main anatomical text and function to relate ("hook") the anatomy discussed directly to a clinical application without taking students out of their train of thought and without disrupting the flow of the text. Although fully integrated with the anatomical text, these passages are differentiated from it by the use of green highlighting. "In the clinic" summaries provide students with useful and relevant clinical information demonstrating how applying anatomical knowledge facilitates the solving of clinical problems. These are spread throughout the text close to the most relevant anatomical discussion.

Surface anatomy assists students in visualizing the relationship between anatomical structures and surface landmarks. This section also provides students with practical applications of the anatomical information, combining visual inspection with functional assessment, as occurs during any type of patient examination.

The final section of each chapter consists of clinical cases. These cases represent the third level of clinical material in the book. In these cases the clinical problem is described, and a step-by-step process of questions and answers leads the reader to the resolution of the case. The inclusion of these cases in each chapter provides students with the opportunity to apply an understanding of anatomy to the resolution of a clinical problem.

Illustrations are an integral part of any anatomy text. They must present the reader with a visual image that brings the text to life and presents views that will assist in the understanding and comprehension of the anatomy. The artwork in this text accomplishes all of these goals. The illustrations are original and vibrant, and many views are unique. They have been designed to integrate with the text, present the anatomy in new ways, deal with the issues that students find particularly difficult, and provide a conceptual framework for building further understanding. To ensure that the illustrations of the book work together and to enable students to cross-refer from one illustration to another, we have used standard colors throughout the book, except where indicated otherwise.

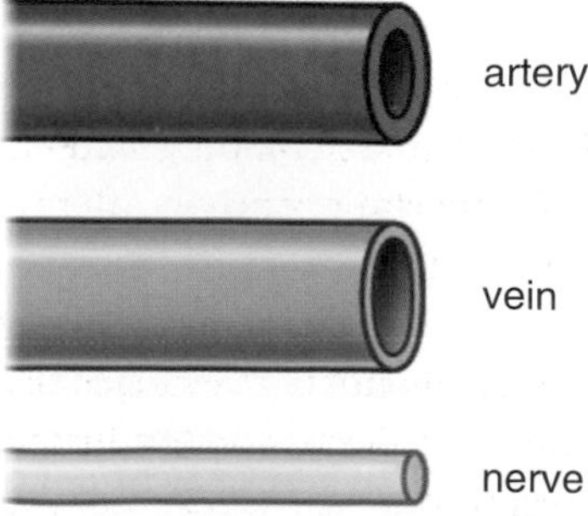


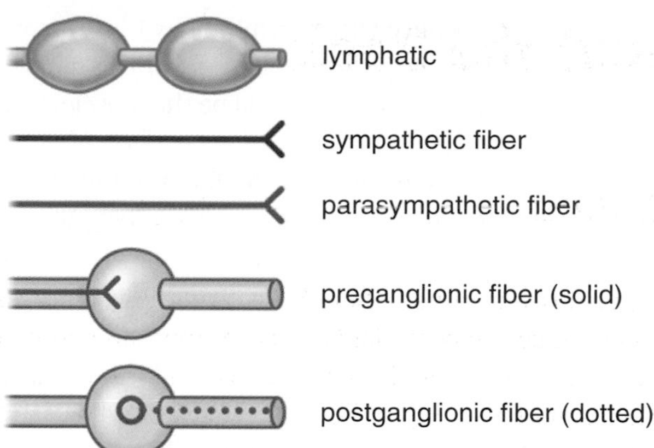


The position and size of the artwork was one of the parameters considered in the overall design of each page of the book.

Clinical images are also an important tool in understanding anatomy and are abundant throughout the text. Examples of state-of-the-art medical imaging, including MRIs, CTs, PETs, and ultrasound, as well as high-quality radiographs, provide students with additional tools to increase their ability to visualize anatomy in vivo and, thus, increase their understanding.

### What the book does not contain

*Gray's Anatomy for Students* focuses on gross anatomy. While many curricula around the world are being presented in a more integrated format combining anatomy, physiology, histology, and embryology, we have focused this textbook on understanding only the anatomy and its application to clinical problems. Except for some brief references to embryology where necessary for a better understanding of the anatomy, material from other disciplines is not included. We felt that there are many outstanding textbooks covering these subject areas, and that trying to cover everything in a single book would produce a text of questionable quality and usefulness, not to mention enormous size!

## Terminology

In any anatomical text or atlas, terminology is always an interesting issue. In 1989, the Federative Committee on Anatomical Terminology (FCAT) was formed and was charged with developing the official terminology of the anatomical sciences. The *Terminologia Anatomica* (2nd edition, Thieme, Stuttgart/New York, 2011) was a joint publication by this group and the 56 member associations of the International Federation of Associations of Anatomists (IFAA). We have chosen to use the terminology presented in this publication in the interest of uniformity. Other terminology is not incorrect; we just felt that using terminology from this single, internationally recognized source would be the most logical and straightforward approach.

Although we use anatomical terms for orientation as much as possible, we also use terms such as "behind" or "in front of" occasionally to make the text more readable. In these cases, the context clarifies the meaning.

## Anatomical use of adverbs

During the writing of this book, we had many long discussions about how we were going to describe anatomical relationships as clearly as possible but maintain the readability of the text. One issue that arose continually in our discussions was the correct use of the "-ly" adverb form of anatomical orientation terms, such as anterior, posterior, superior, inferior, lateral, and medial. We reached the following consensus:

***-ly adverbs*** *e.g., anteriorly, posteriorly*, have been used to modify (describe) verbs in passages where an action or direction is mentioned. For example, "The trachea passes inferiorly through the thorax."

***circumstantial adverbs***, *e.g., anterior, posterior*, have been used to indicate the fixed location of an anatomical feature. For example, "The trachea is anterior to the esophagus."

Furthermore, both usages may occur in the same passage. For example, "The trachea passes inferiorly through the thorax, anterior to the esophagus."

We have very much enjoyed the process of putting this book together. We hope that you enjoy using it to the same degree.

**Richard L. Drake**
**A. Wayne Vogl**
**Adam W. M. Mitchell**

# Index of Clinical Content

## 1 The body

### In the Clinic



### Clinical Case



## 2 Back

### In the Clinic



### Clinical Cases



### Available online only*



*All Clinical Cases are available online at StudentConsult.com.

## 3 Thorax

### In the Clinic

### Clinical Cases



### Available online only*



*All Clinical Cases are available online at StudentConsult.com.

## 4 *Abdomen*

### In the Clinic



### Clinical Cases

*All Clinical Cases are available online at StudentConsult.com.

## 5 Pelvis and perineum



*All Clinical Cases are available online at StudentConsult.com.

## 6 Lower limb

**Clinical Cases**



**Available online only***



*All Clinical Cases are available online at StudentConsult.com.

## 7 Upper limb

**In the Clinic**



**Clinical Cases**



**Available online only***

*All Clinical Cases are available online at StudentConsult.com.

## 8 *Head and neck*

### In the Clinic



### Clinical Cases



### Available online only*



*All Clinical Cases are available online at StudentConsult.com.

# Contents

## 3 Thorax



## 4 Abdomen

# 6 Lower limb



# 7 Upper limb

1

# The Body

**ADDITIONAL LEARNING RESOURCES for Chapter 1, The Body, on STUDENT CONSULT (www.studentconsult.com)**

- Short Questions—these are questions requiring short responses, Chapter 1

# What is anatomy?

Anatomy includes those structures that can be seen grossly (without the aid of magnification) and microscopically (with the aid of magnification). Typically, when used by itself, the term *anatomy* tends to mean gross or macroscopic anatomy—that is, the study of structures that can be seen without using a microscopic. Microscopic anatomy, also called histology, is the study of cells and tissues using a microscope.

Anatomy forms the basis for the practice of medicine. Anatomy leads the physician toward an understanding of a patient's disease, whether he or she is carrying out a physical examination or using the most advanced imaging techniques. Anatomy is also important for dentists, chiropractors, physical therapists, and all others involved in any aspect of patient treatment that begins with an analysis of clinical signs. The ability to interpret a clinical observation correctly is therefore the endpoint of a sound anatomical understanding.

Observation and visualization are the primary techniques a student should use to learn anatomy. Anatomy is much more than just memorization of lists of names. Although the language of anatomy is important, the network of information needed to visualize the position of physical structures in a patient goes far beyond simple memorization. Knowing the names of the various branches of the external carotid artery is not the same as being able to visualize the course of the lingual artery from its origin in the neck to its termination in the tongue. Similarly, understanding the organization of the soft palate, how it is related to the oral and nasal cavities, and how it moves during swallowing is very different from being able to recite the names of its individual muscles and nerves. An understanding of anatomy requires an understanding of the context in which the terminology can be remembered.

## How can gross anatomy be studied?

The term *anatomy* is derived from the Greek word *temnein*, meaning "to cut." Clearly, therefore, the study of anatomy is linked, at its root, to dissection, although dissection of cadavers by students is now augmented, or even in some cases replaced, by viewing prosected (previously dissected) material and plastic models, or using computer teaching modules and other learning aids.

Anatomy can be studied following either a regional or a systemic approach.

- With a **regional approach**, each *region* of the body is studied separately and all aspects of that region are studied at the same time. For example, if the thorax is to be studied, all of its structures are examined. This includes the vasculature, the nerves, the bones, the muscles, and all other structures and organs located in the region of the body defined as the thorax. After studying this region, the other regions of the body (i.e., the abdomen, pelvis, lower limb, upper limb, back, head, and neck) are studied in a similar fashion.
- In contrast, in a **systemic approach**, each *system* of the body is studied and followed throughout the entire body. For example, a study of the cardiovascular system looks at the heart and all of the blood vessels in the body. When this is completed, the nervous system (brain, spinal cord, and all the nerves) might be examined in detail. This approach continues for the whole body until every system, including the nervous, skeletal, muscular, gastrointestinal, respiratory, lymphatic, and reproductive systems, has been studied.

Each of these approaches has benefits and deficiencies. The regional approach works very well if the anatomy course involves cadaver dissection but falls short when it comes to understanding the continuity of an entire system throughout the body. Similarly, the systemic approach fosters an understanding of an entire system throughout the body, but it is very difficult to coordinate this directly with a cadaver dissection or to acquire sufficient detail.

## Important anatomical terms

### The anatomical position

The anatomical position is the standard reference position of the body used to describe the location of structures (Fig. 1.1). The body is in the anatomical position when standing upright with feet together, hands by the side and face looking forward. The mouth is closed and the facial expression is neutral. The rim of bone under the eyes is in the same horizontal plane as the top of the opening to the ear, and the eyes are open and focused on something in the distance. The palms of the hands face forward with the fingers straight and together and with the pad of the thumb turned 90° to the pads of the fingers. The toes point forward.

### Anatomical planes

Three major groups of planes pass through the body in the anatomical position (Fig. 1.1).

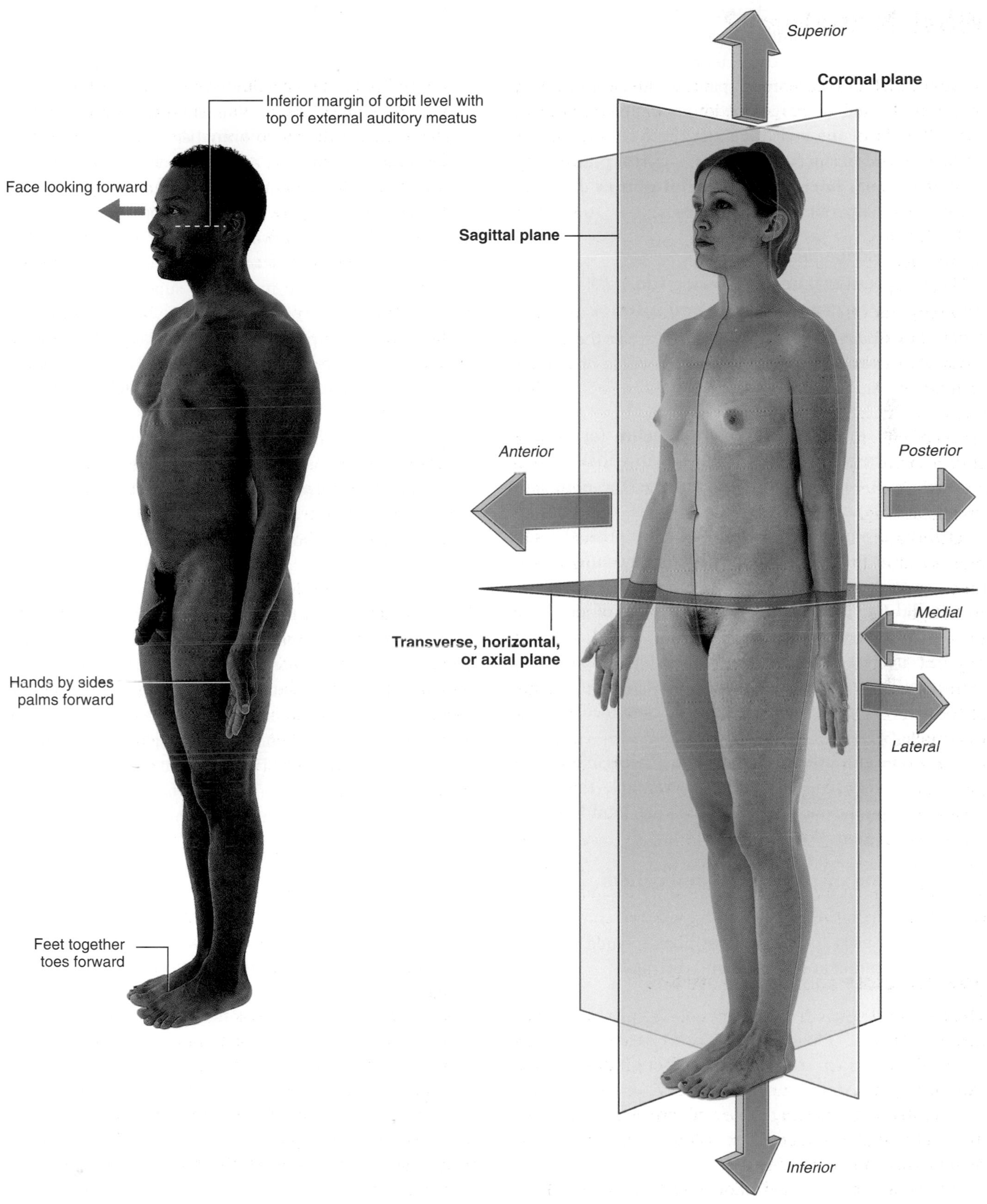


**Fig. 1.1** The anatomical position, planes, and terms of location and orientation.

- **Coronal planes** are oriented vertically and divide the body into anterior and posterior parts.
- **Sagittal planes** also are oriented vertically but are at right angles to the coronal planes and divide the body into right and left parts. The plane that passes through the center of the body dividing it into equal right and left halves is termed the **median sagittal plane**.
- **Transverse, horizontal**, or **axial planes** divide the body into superior and inferior parts.

## Terms to describe location

### Anterior (ventral) and posterior (dorsal), medial and lateral, superior and inferior

Three major pairs of terms are used to describe the location of structures relative to the body as a whole or to other structures (Fig. 1.1).

- **Anterior** (or **ventral**) and **posterior** (or **dorsal**) describe the position of structures relative to the "front" and "back" of the body. For example, the nose is an anterior (ventral) structure, whereas the vertebral column is a posterior (dorsal) structure. Also, the nose is anterior to the ears and the vertebral column is posterior to the sternum.
- **Medial** and **lateral** describe the position of structures relative to the median sagittal plane and the sides of the body. For example, the thumb is lateral to the little finger. The nose is in the median sagittal plane and is medial to the eyes, which are in turn medial to the external ears.
- **Superior** and **inferior** describe structures in reference to the vertical axis of the body. For example, the head is superior to the shoulders and the knee joint is inferior to the hip joint.

### Proximal and distal, cranial and caudal, and rostral

Other terms used to describe positions include proximal and distal, cranial and caudal, and rostral.

- **Proximal** and **distal** are used with reference to being closer to or farther from a structure's origin, particularly in the limbs. For example, the hand is distal to the elbow joint. The glenohumeral joint is proximal to the elbow joint. These terms are also used to describe the relative positions of branches along the course of linear structures, such as airways, vessels, and nerves. For example, distal branches occur farther away toward the ends of the system, whereas proximal branches occur closer to and toward the origin of the system.
- **Cranial** (toward the head) and **caudal** (toward the tail) are sometimes used instead of superior and inferior, respectively.
- **Rostral** is used, particularly in the head, to describe the position of a structure with reference to the nose. For example, the forebrain is rostral to the hindbrain.

### Superficial and deep

Two other terms used to describe the position of structures in the body are **superficial** and **deep**. These terms are used to describe the relative positions of two structures with respect to the surface of the body. For example, the sternum is superficial to the heart, and the stomach is deep to the abdominal wall.

Superficial and deep can also be used in a more absolute fashion to define two major regions of the body. The superficial region of the body is external to the outer layer of deep fascia. Deep structures are enclosed by this layer. Structures in the superficial region of the body include the skin, superficial fascia, and mammary glands. Deep structures include most skeletal muscles and viscera. Superficial wounds are external to the outer layer of deep fascia, whereas deep wounds penetrate through it.

# Imaging

## Diagnostic imaging techniques

In 1895 Wilhelm Roentgen used the X-rays from a cathode ray tube to expose a photographic plate and produce the first radiographic exposure of his wife's hand. Over the past 35 years there has been a revolution in body imaging, which has been paralleled by developments in computer technology.

### Plain radiography

X-rays are photons (a type of electromagnetic radiation) and are generated from a complex X-ray tube, which is a type of cathode ray tube (Fig. 1.2). The X-rays are then collimated (i.e., directed through lead-lined shutters to stop them from fanning out) to the appropriate area of the body. As the X-rays pass through the body they are attenuated (reduced in energy) by the tissues. Those X-rays that pass through the tissues interact with the photographic film.

In the body:

- air attenuates X-rays a little;
- fat attenuates X-rays more than air but less than water; and
- bone attenuates X-rays the most.

These differences in attenuation result in differences in the level of exposure of the film. When the photographic film is developed, bone appears white on the film because this region of the film has been exposed to the least amount of X-rays. Air appears dark on the film because these regions were exposed to the greatest number of X-rays.

Modifications to this X-ray technique allow a continuous stream of X-rays to be produced from the X-ray tube and collected on an input screen to allow real-time visualization of moving anatomical structures, barium studies, angiography, and fluoroscopy (Fig. 1.3).

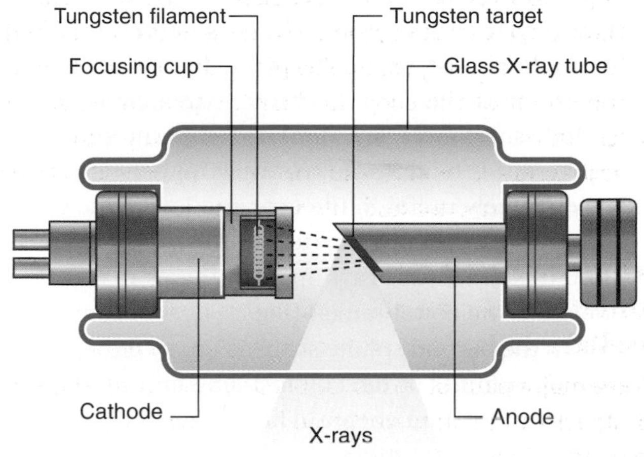


**Fig. 1.2** Cathode ray tube for the production of X-rays.

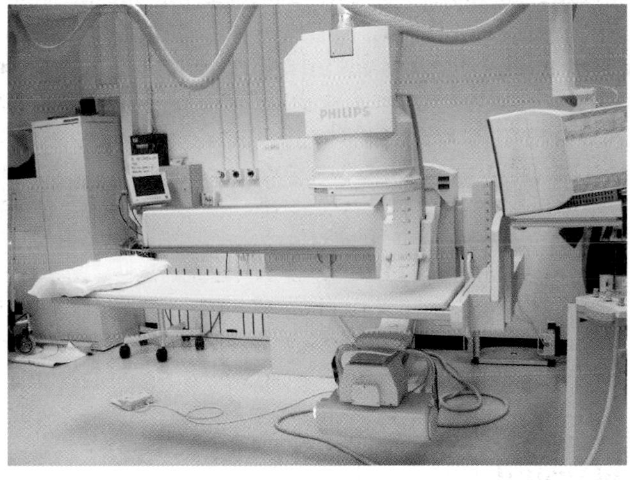

**Fig. 1.3** Fluoroscopy unit.

### Contrast agents

To demonstrate specific structures, such as bowel loops or arteries, it may be necessary to fill these structures with a substance that attenuates X-rays more than bowel loops or arteries do normally. It is, however, extremely important that these substances are nontoxic. Barium sulfate, an insoluble salt, is a nontoxic, relatively high-density agent that is extremely useful in the examination of the gastrointestinal tract. When a **barium sulfate suspension** is ingested it attenuates X-rays and can therefore be used to demonstrate the bowel lumen (Fig. 1.4). It is common to add air to the barium sulfate suspension, by either ingesting "fizzy" granules or directly instilling air into the body cavity, as in a barium enema. This is known as a double-contrast (air/barium) study.

For some patients it is necessary to inject contrast agents directly into arteries or veins. In this case, iodine-based molecules are suitable contrast agents. **Iodine** is chosen because it has a relatively high atomic mass and so markedly attenuates X-rays, but also, importantly, it is naturally excreted via the urinary system. Intra-arterial and intravenous contrast agents are extremely safe and are well tolerated by most patients. Rarely, some patients have an anaphylactic reaction to intra-arterial or intravenous injections, so the necessary precautions must be taken. Intra-arterial and intravenous contrast agents not only help in visualizing the arteries and veins but because they are excreted by the urinary system, can also be used to visualize the kidneys, ureter, and bladder in a process known as **intravenous urography.**

### Subtraction angiography

During angiography it is often difficult to appreciate the contrast agent in the vessels through the overlying bony structures. To circumvent this, the technique of subtraction angiography has been developed. Simply, one or two images are obtained before the injection of contrast media. These images are inverted (such that a negative is created from the positive image). After injection of the contrast media into the vessels, a further series of images are obtained, demonstrating the passage of the contrast through the arteries into the veins and around the circulation. By adding the "negative precontrast image" to the positive postcontrast images, the bones and soft tissues are subtracted to produce a solitary image of contrast only. Before the advent of digital imaging this was a challenge, but now the use of computers has made this technique relatively straightforward and instantaneous (Fig. 1.5).

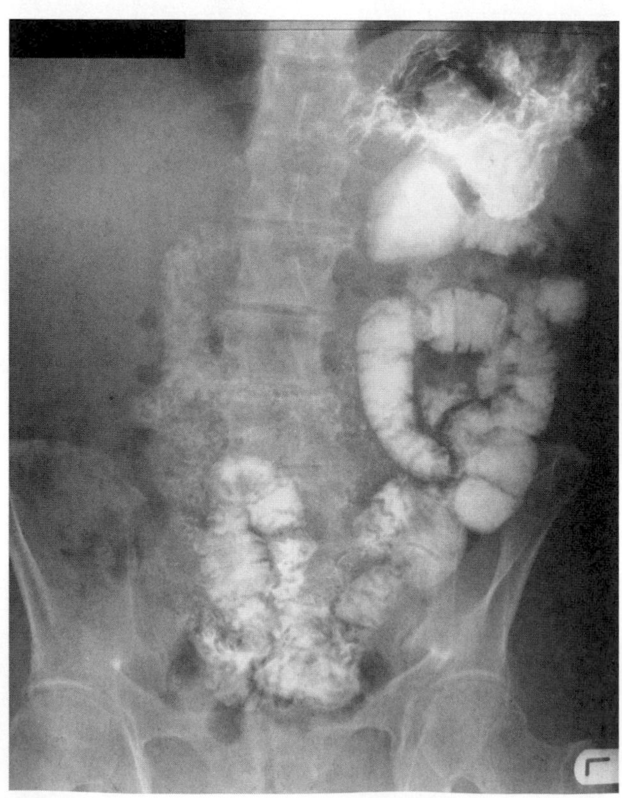

**Fig. 1.4** Barium sulfate follow-through.

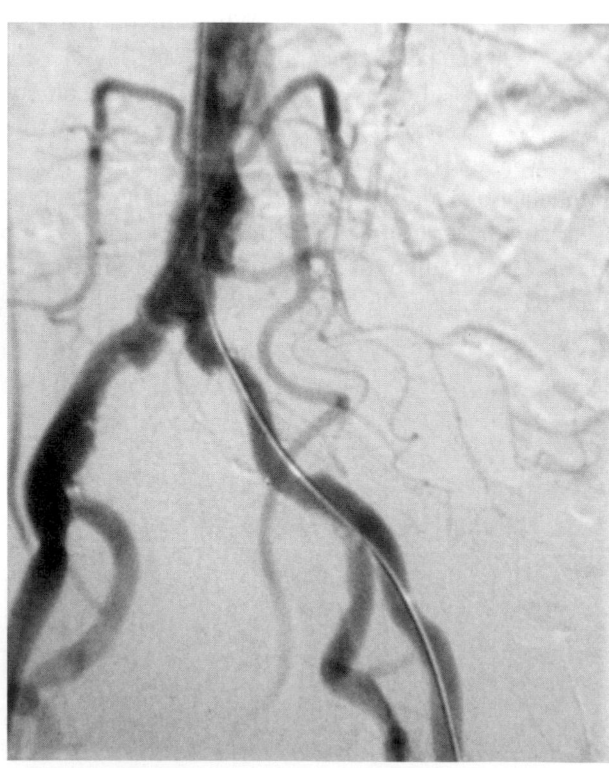

**Fig. 1.5** Digital subtraction angiogram.

## Ultrasound

Ultrasonography of the body is widely used for all aspects of medicine.

Ultrasound is a very high frequency sound wave (not electromagnetic radiation) generated by piezoelectric materials, such that a series of sound waves is produced. Importantly, the piezoelectric material can also receive the sound waves that bounce back from the internal organs. The sound waves are then interpreted by a powerful computer, and a real-time image is produced on the display panel.

Developments in ultrasound technology, including the size of the probes and the frequency range, mean that a broad range of areas can now be scanned.

Traditionally ultrasound is used for assessing the abdomen (Fig. 1.6) and the fetus in pregnant women. Ultrasound is also widely used to assess the eyes, neck, soft tissues, and peripheral musculoskeletal system. Probes have been placed on endoscopes, and endoluminal ultrasound of the esophagus, stomach, and duodenum is now routine. Endocavity ultrasound is carried out most commonly to assess the genital tract in women using a transvaginal or transrectal route. In men, transrectal ultrasound is the imaging method of choice to assess the prostate in those with suspected prostate hypertrophy or malignancy.

### Doppler ultrasound

Doppler ultrasound enables determination of flow, its direction, and its velocity within a vessel using simple ultrasound techniques. Sound waves bounce off moving structures and are returned. The degree of frequency shift determines whether the object is moving away from or toward the probe and the speed at which it is traveling. Precise measurements of blood flow and blood velocity can therefore be obtained, which in turn can indicate sites of blockage in blood vessels.

## Computed tomography

Computed tomography (CT) was invented in the 1970s by Sir Godfrey Hounsfield, who was awarded the Nobel Prize in Medicine in 1979. Since this inspired invention there have been many generations of CT scanners.

A CT scanner obtains a series of images of the body (slices) in the axial plane. The patient lies on a bed, an X-ray tube passes around the body (Fig. 1.7), and a series of images are obtained. A computer carries out a complex mathematical transformation on the multitude of images to produce the final image (Fig. 1.8).

## Magnetic resonance imaging

Nuclear magnetic resonance imaging was first described in 1946 and used to determine the structure of complex

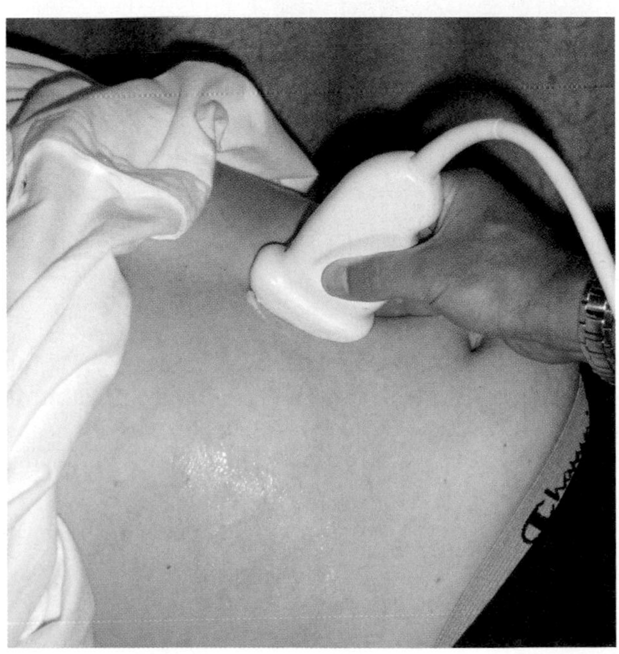

**Fig. 1.6** Ultrasound examination of the abdomen.

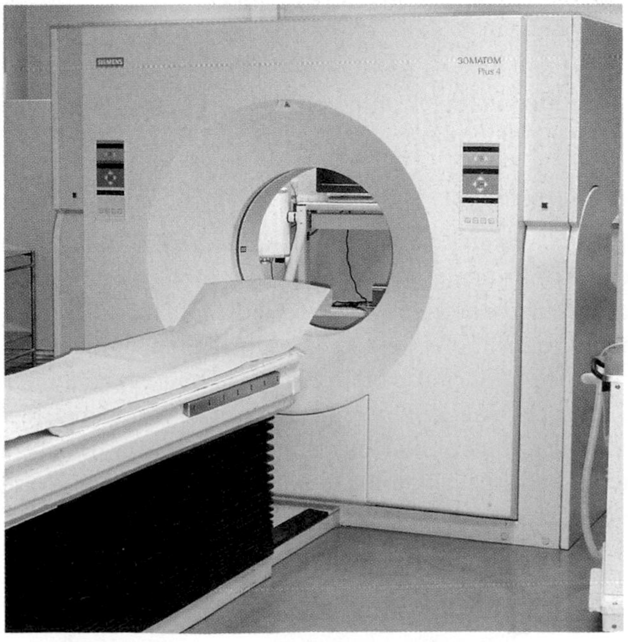


**Fig. 1.7** Computed tomography scanner.

molecules. The process of magnetic resonance imaging (MRI) is dependent on the free protons in the hydrogen nuclei in molecules of water ($H_2O$). Because water is present in almost all biological tissues, the hydrogen proton is ideal. The protons within a patient's hydrogen nuclei can be regarded as small bar magnets, which are randomly oriented in space. The patient is placed in a strong magnetic field, which aligns the bar magnets. When a pulse of radio waves is passed through the patient the magnets are deflected, and as they return to their aligned position they emit small radio pulses. The strength and frequency of the emitted pulses and the time it takes for the protons to return to their pre-excited state produce a signal. These signals are analyzed by a powerful computer, and an image is created (Fig. 1.9).

By altering the sequence of pulses to which the protons are subjected, different properties of the protons can be assessed. These properties are referred to as the "weighting" of the scan. By altering the pulse sequence and the scanning parameters, T1-weighted images (Fig. 1.10A) and T2-weighted images (Fig. 1.10B) can be obtained. These two types of imaging sequences provide differences in image contrast, which accentuate and optimize different tissue characteristics.

From the clinical point of view:

- Most T1-weighted images show dark fluid and bright fat—for example, within the brain the cerebrospinal fluid (CSF) is dark.
- T2-weighted images demonstrate a bright signal from fluid and an intermediate signal from fat—for example, in the brain the CSF appears white.

MRI can also be used to assess flow within vessels and to produce complex angiograms of the peripheral and cerebral circulation.

### Diffusion-weighted imaging

Diffusion-weighted imaging provides information on the degree of Brownian motion of water molecules in various tissues. There is relatively free diffusion in extracellular spaces and more restricted diffusion in intracellular spaces. In tumors and infarcted tissue, there is an increase in intracellular fluid water molecules compared with the extracellular fluid environment resulting in overall increased restricted diffusion, and therefore identification of abnormal from normal tissue.

## Nuclear medicine imaging

Nuclear medicine involves imaging using gamma rays, which are another type of electromagnetic radiation.

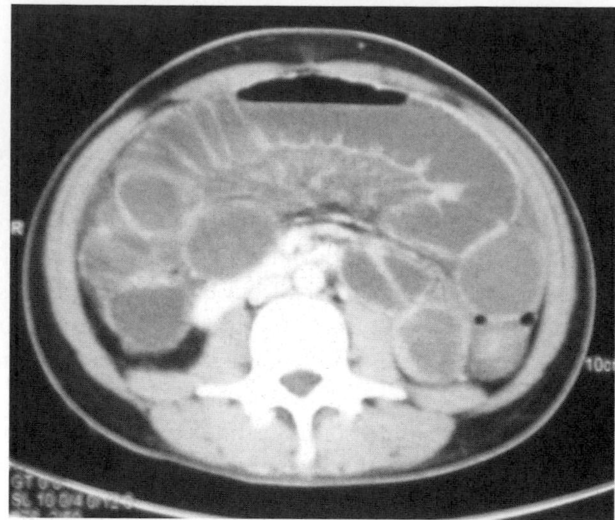


**Fig. 1.8** Computed tomography scan of the abdomen at vertebral level L2.

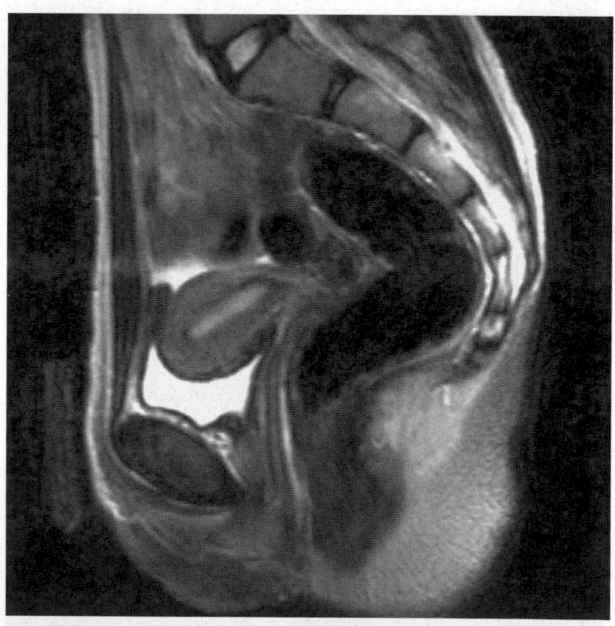

**Fig. 1.9** A T2-weighted MR image in the sagittal plane of the pelvic viscera in a woman.

The important difference between gamma rays and X-rays is that gamma rays are produced from within the nucleus of an atom when an unstable nucleus decays, whereas X-rays are produced by bombarding an atom with electrons.

For an area to be visualized, the patient must receive a gamma ray emitter, which must have a number of properties to be useful, including:

- a reasonable half-life (e.g., 6 to 24 hours),
- an easily measurable gamma ray, and

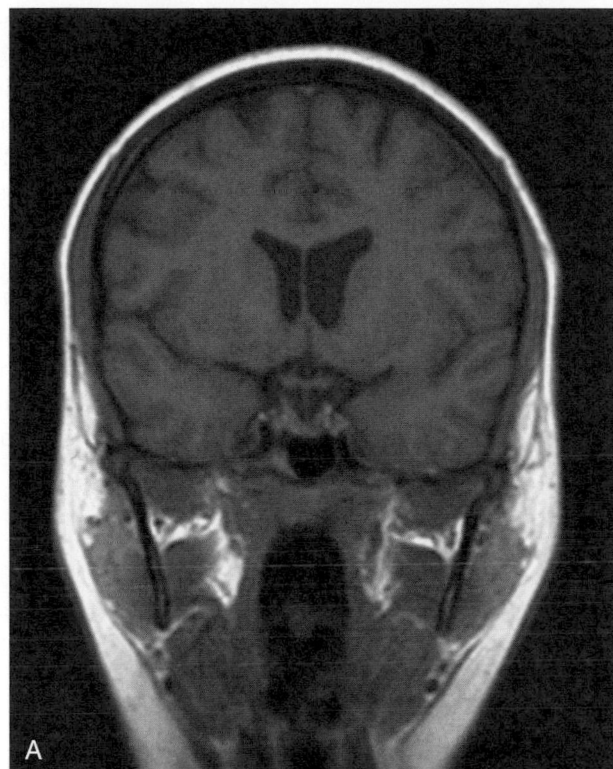


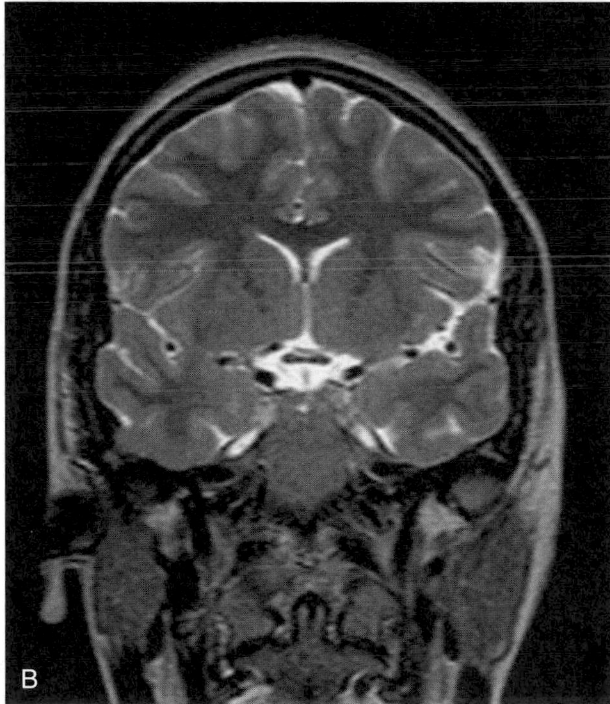


**Fig. 1.10** T1-weighted (**A**) and T2-weighted (**B**) MR images of the brain in the coronal plane.

- energy deposition in as low a dose as possible in the patient's tissues.

The most commonly used radionuclide (radioisotope) is technetium-99m. This may be injected as a technetium salt or combined with other complex molecules. For example, by combining technetium-99m with methylene diphosphonate (MDP), a radiopharmaceutical is produced. When injected into the body this radiopharmaceutical specifically binds to bone, allowing assessment of the skeleton. Similarly, combining technetium-99m with other compounds permits assessment of other parts of the body, for example the urinary tract and cerebral blood flow.

Depending on how the radiopharmaceutical is absorbed, distributed, metabolized, and excreted by the body after injection, images are obtained using a gamma camera (Fig. 1.11).

### Positron emission tomography

Positron emission tomography (PET) is an imaging modality for detecting positron-emitting radionuclides. A positron is an anti-electron, which is a positively charged particle of antimatter. Positrons are emitted from the decay of proton-rich radionuclides. Most of these radionuclides are made in a cyclotron and have extremely short half-lives.

The most commonly used PET radionuclide is fluorodeoxyglucose (FDG) labeled with fluorine-18 (a positron

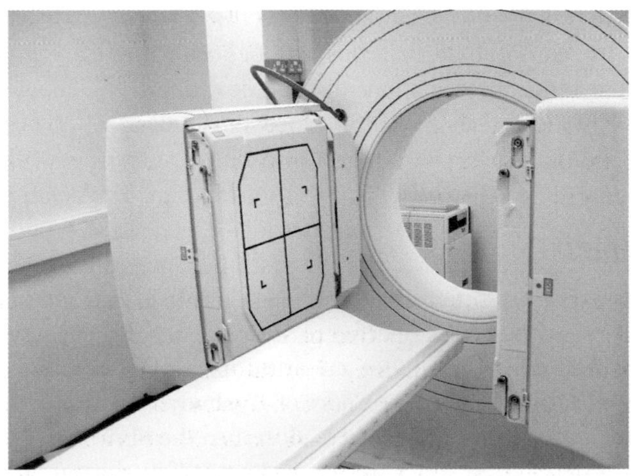

**Fig. 1.11** A gamma camera.

emitter). Tissues that are actively metabolizing glucose take up this compound, and the resulting localized high concentration of this molecule compared to background emission is detected as a "hot spot."

PET has become an important imaging modality in the detection of cancer and the assessment of its treatment and recurrence.

#### Single photon emission computed tomography

Single photon emission computed tomography (SPECT) is an imaging modality for detecting gamma rays emitted from the decay of injected radionuclides such as technetium-99m, iodine-123, or iodine-131. The rays are detected by a 360-degree rotating camera, which allows the construction of 3D images. SPECT can be used to diagnose a wide range of disease conditions such as coronary artery disease and bone fractures.

## IMAGE INTERPRETATION

Imaging is necessary in most clinical specialties to diagnose pathological changes to tissues. It is paramount to appreciate what is normal and what is abnormal. An appreciation of how the image is obtained, what the normal variations are, and what technical considerations are necessary to obtain a radiological diagnosis. Without understanding the anatomy of the region imaged, it is impossible to comment on the abnormal.

### Plain radiography

Plain radiographs are undoubtedly the most common form of image obtained in a hospital or local practice. Before interpretation, it is important to know about the imaging technique and the views obtained as standard.

In most instances (apart from chest radiography) the X-ray tube is 1 m away from the X-ray film. The object in question, for example a hand or a foot, is placed upon the film. When describing subject placement for radiography, the part closest to the X-ray tube is referred to first and that closest to the film is referred to second. For example, when positioning a patient for an anteroposterior (AP) radiograph, the more anterior part of the body is closest to the tube and the posterior part is closest to the film.

When X-rays are viewed on a viewing box, the right side of the patient is placed to the observer's left; therefore, the observer views the radiograph as though looking at a patient in the anatomical position.

#### Chest radiograph

The chest radiograph is one of the most commonly requested plain radiographs. An image is taken with the patient erect and placed posteroanteriorly (PA chest radiograph; that is, with the patient's back closest to the X-ray tube.).

Occasionally, when patients are too unwell to stand erect, films are obtained on the bed in an anteroposterior (AP) position. These films are less standardized than PA films, and caution should always be taken when interpreting AP radiographs.

The plain chest radiograph should always be checked for quality. Film markers should be placed on the appropriate side. (Occasionally patients have dextrocardia, which may be misinterpreted if the film marker is placed inappropriately.) A good-quality chest radiograph will demonstrate the lungs, cardiomediastinal contour, diaphragm, ribs, and peripheral soft tissues.

#### Abdominal radiograph

Plain abdominal radiographs are obtained in the AP supine position. From time to time an erect plain abdominal radiograph is obtained when small bowel obstruction is suspected.

#### Gastrointestinal contrast examinations

High-density contrast medium is ingested to opacify the esophagus, stomach, small bowel, and large bowel. As described previously (p. 6), the bowel is insufflated with air (or carbon dioxide) to provide a double-contrast study. In many countries, endoscopy has superseded upper gastrointestinal imaging, but the mainstay of imaging the large bowel is the double-contrast barium enema. Typically the patient needs to undergo bowel preparation, in which powerful cathartics are used to empty the bowel. At the time of the examination a small tube is placed into the rectum and a barium suspension is run into the large bowel. The patient undergoes a series of twists and turns so that the contrast passes through the entire large bowel. The contrast is emptied and air is passed through the same tube to insufflate the large bowel. A thin layer of barium coats the normal mucosa, allowing mucosal detail to be visualized (see Fig. 1.4).

#### Urological contrast studies

Intravenous urography is the standard investigation for assessing the urinary tract. Intravenous contrast medium is injected, and images are obtained as the medium is excreted through the kidneys. A series of films are obtained during this period from immediately after the injection up to approximately 20 minutes later, when the bladder is full of contrast medium.

This series of radiographs demonstrates the kidneys, ureters, and bladder and enables assessment of the retroperitoneum and other structures that may press on the urinary tract.

## Computed tomography

Computed tomography is the preferred terminology rather than computerized tomography, though both terms are used interchangeably by physicians.

It is important for the student to understand the presentation of images. Most images are acquired in the axial plane and viewed such that the observer looks from below and upward toward the head (from the foot of the bed). By implication:

- the right side of the patient is on the left side of the image, and
- the uppermost border of the image is anterior.

Many patients are given oral and intravenous contrast media to differentiate bowel loops from other abdominal organs and to assess the vascularity of normal anatomical structures. When intravenous contrast is given, the earlier the images are obtained, the greater the likelihood of arterial enhancement. As the time is delayed between injection and image acquisition, a venous phase and an equilibrium phase are also obtained.

The great advantage of CT scanning is the ability to extend and compress the gray scale to visualize the bones, soft tissues, and visceral organs. Altering the window settings and window centering provides the physician with specific information about these structures.

## Magnetic resonance imaging

There is no doubt that MRI has revolutionized the understanding and interpretation of the brain and its coverings. Furthermore, it has significantly altered the practice of musculoskeletal medicine and surgery. Images can be obtained in any plane and in most sequences. Typically the images are viewed using the same principles as CT. Intravenous contrast agents are also used to further enhance tissue contrast. Typically, MRI contrast agents contain paramagnetic substances (e.g., gadolinium and manganese).

## Nuclear medicine imaging

Most nuclear medicine images are functional studies. Images are usually interpreted directly from a computer, and a series of representative films are obtained for clinical use.

# SAFETY IN IMAGING

Whenever a patient undergoes an X-ray or nuclear medicine investigation, a dose of radiation is given (Table 1.1). As a general principle it is expected that the dose given is as low as reasonably possible for a diagnostic image to be obtained. Numerous laws govern the amount of radiation exposure that a patient can undergo for a variety of procedures, and these are monitored to prevent any excess or additional dosage. Whenever a radiograph is booked, the clinician ordering the procedure must appreciate its necessity and understand the dose given to the patient to ensure that the benefits significantly outweigh the risks.

Imaging modalities such as ultrasound and MRI are ideal because they do not impart significant risk to the patient. Moreover, ultrasound imaging is the modality of choice for assessing the fetus.

Any imaging device is expensive, and consequently the more complex the imaging technique (e.g., MRI) the more expensive the investigation. Investigations must be carried out judiciously, based on a sound clinical history and examination, for which an understanding of anatomy is vital.

**Table 1.1** The approximate dosage of radiation exposure as an order of magnitude

| Examination | Typical effective dose (mSv) | Equivalent duration of background exposure |
|---|---|---|
| Chest radiograph | 0.02 | 3 days |
| Abdomen | 1.00 | 6 months |
| Intravenous urography | 2.50 | 14 months |
| CT scan of head | 2.30 | 1 year |
| CT scan of abdomen and pelvis | 10.00 | 4.5 years |

# Body systems

## SKELETAL SYSTEM

The skeleton can be divided into two subgroups, the axial skeleton and the appendicular skeleton. The axial skeleton consists of the bones of the skull (cranium), vertebral column, ribs, and sternum, whereas the appendicular skeleton consists of the bones of the upper and lower limbs (Fig. 1.12).

The skeletal system consists of cartilage and bone.

### Cartilage

Cartilage is an avascular form of connective tissue consisting of extracellular fibers embedded in a matrix that contains cells localized in small cavities. The amount and kind of extracellular fibers in the matrix varies depending on the type of cartilage. In heavy weightbearing areas or areas prone to pulling forces, the amount of collagen is greatly increased and the cartilage is almost inextensible. In contrast, in areas where weightbearing demands and stress are less, cartilage containing elastic fibers and fewer collagen fibers is common. The functions of cartilage are to:

- support soft tissues,
- provide a smooth, gliding surface for bone articulations at joints, and
- enable the development and growth of long bones.

There are three types of cartilage:

- hyaline—most common; matrix contains a moderate amount of collagen fibers (e.g., articular surfaces of bones);
- elastic—matrix contains collagen fibers along with a large number of elastic fibers (e.g., external ear);
- fibrocartilage—matrix contains a limited number of cells and ground substance amidst a substantial amount of collagen fibers (e.g., intervertebral discs).

Cartilage is nourished by diffusion and has no blood vessels, lymphatics, or nerves.

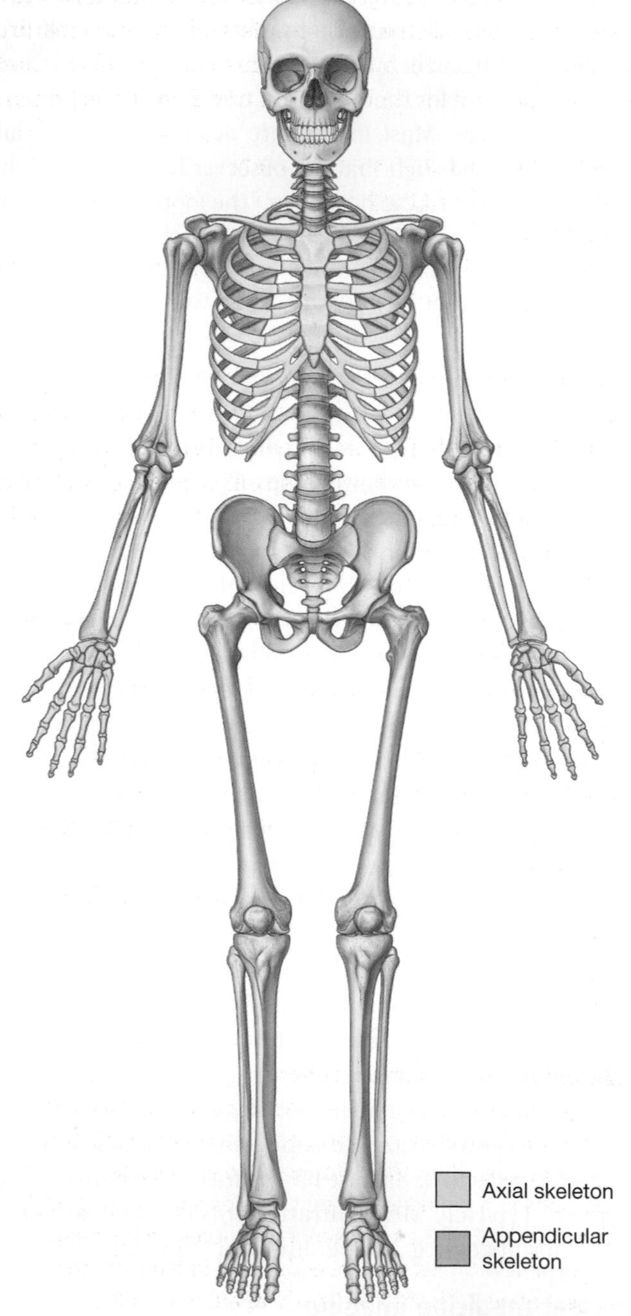


**Fig. 1.12** The axial skeleton and the appendicular skeleton.

## Bone

Bone is a calcified, living, connective tissue that forms the majority of the skeleton. It consists of an intercellular calcified matrix, which also contains collagen fibers, and several types of cells within the matrix. Bones function as:

- supportive structures for the body,
- protectors of vital organs,
- reservoirs of calcium and phosphorus,
- levers on which muscles act to produce movement, and
- containers for blood-producing cells.

There are two types of bone, compact and spongy (trabecular or cancellous). Compact bone is dense bone that forms the outer shell of all bones and surrounds spongy bone. Spongy bone consists of spicules of bone enclosing cavities containing blood-forming cells (marrow). Classification of bones is by shape.

- Long bones are tubular (e.g., humerus in upper limb; femur in lower limb).
- Short bones are cuboidal (e.g., bones of the wrist and ankle).
- Flat bones consist of two compact bone plates separated by spongy bone (e.g., skull).
- Irregular bones are bones with various shapes (e.g., bones of the face).
- Sesamoid bones are round or oval bones that develop in tendons.

### In the clinic

**Accessory and sesamoid bones**

These are extra bones that are not usually found as part of the normal skeleton, but can exist as a normal variant in many people. They are typically found in multiple locations in the wrist and hands, ankles and feet (Fig. 1.13). These should not be mistaken for fractures on imaging.

Sesamoid bones are embedded within tendons, the largest of which is the patella. There are many other sesamoids in the body particularly in tendons of the hands and feet, and most frequently in flexor tendons of the thumb and big toe.

Degenerative and inflammatory changes of, as well as mechanical stresses on, the accessory bones and sesamoids can cause pain, which can be treated with physiotherapy and targeted steroid injections, but in some severe cases it may be necessary to surgically remove the bone.

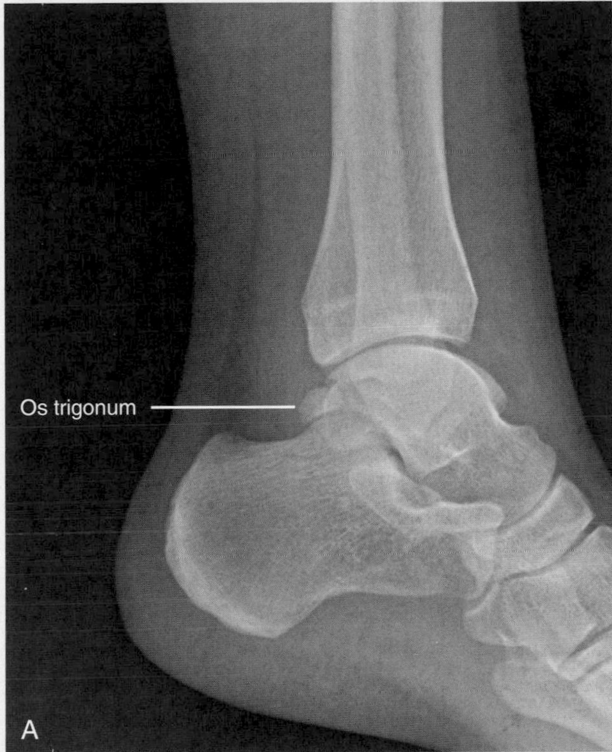


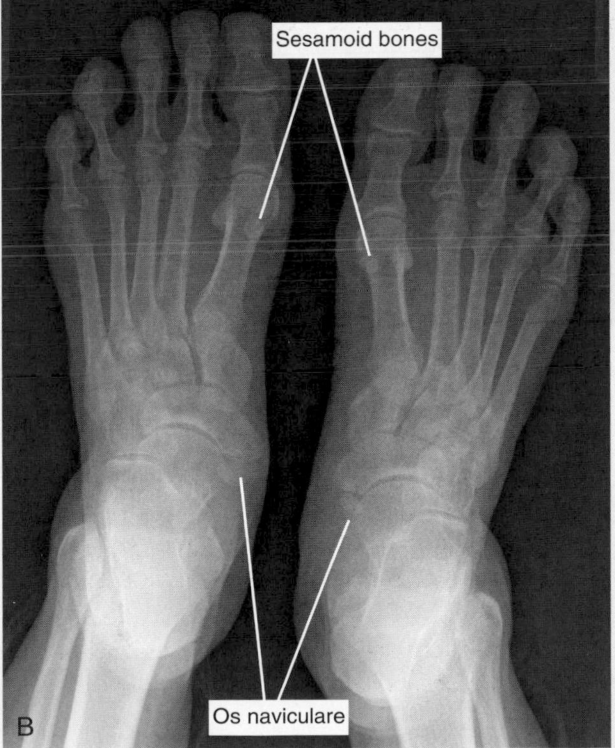


**Fig. 1.13** Accessory and sesamoid bones. **A.** Radiograph of the ankle region showing an accessory bone (os trigonum). **B.** Radiograph of the feet showing numerous sesamoid bones and an accessory bone (os naviculare).

Bones are vascular and are innervated. Generally, an adjacent artery gives off a nutrient artery, usually one per bone, that directly enters the internal cavity of the bone and supplies the marrow, spongy bone, and inner layers of compact bone. In addition, all bones are covered externally, except in the area of a joint where articular cartilage is present, by a fibrous connective tissue membrane called the periosteum, which has the unique capability of forming new bone. This membrane receives blood vessels whose branches supply the outer layers of compact bone. A bone stripped of its periosteum will not survive. Nerves accompany the vessels that supply the bone and the periosteum. Most of the nerves passing into the internal cavity with the nutrient artery are vasomotor fibers that regulate blood flow. Bone itself has few sensory nerve fibers. On the other hand, the periosteum is supplied with numerous sensory nerve fibers and is very sensitive to any type of injury.

Developmentally, all bones come from mesenchyme by either intramembranous ossification, in which mesenchymal models of bones undergo ossification, or endochondral ossification, in which cartilaginous models of bones form from mesenchyme and undergo ossification.

## In the clinic

### Determination of skeletal age

Throughout life the bones develop in a predictable way to form the skeletally mature adult at the end of puberty. In western countries skeletal maturity tends to occur between the ages of 20 and 25 years. However, this may well vary according to geography and socioeconomic conditions. Skeletal maturity will also be determined by genetic factors and disease states.

Up until the age of skeletal maturity, bony growth and development follows a typically predictable ordered state, which can be measured through either ultrasound, plain radiographs, or MRI scanning. Typically, the nondominant (left) hand is radiographed, and the radiograph is compared to a series of standard radiographs. From these images the bone age can be determined (Fig. 1.14).

In certain disease states, such as malnutrition and hypothyroidism, bony maturity may be slow. If the skeletal bone age is significantly reduced from the patient's true age, treatment may be required.

In the healthy individual the bone age accurately represents the true age of the patient. This is important in determining the true age of the subject. This may also have medicolegal importance.

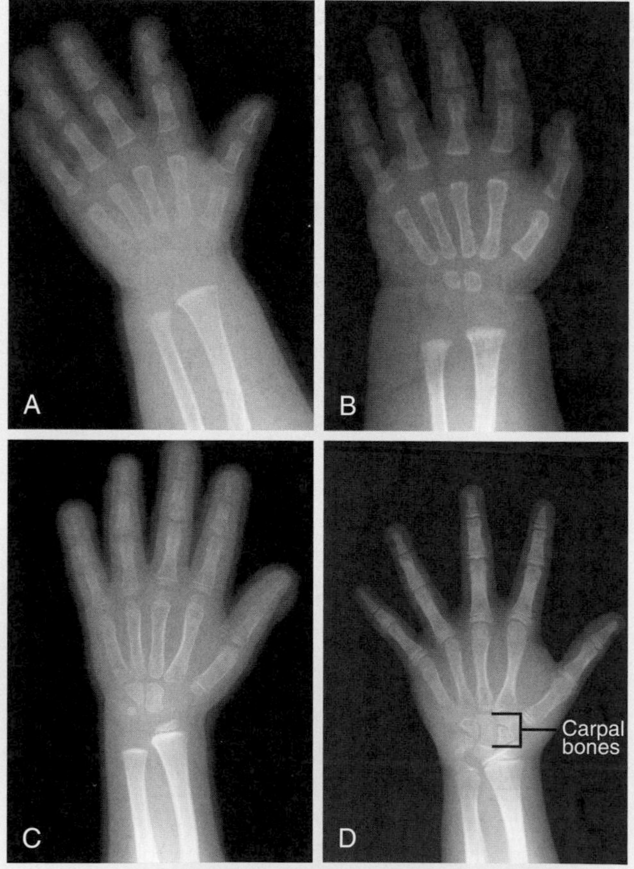


**Fig. 1.14** A developmental series of radiographs showing the progressive ossification of carpal (wrist) bones from 3 (**A**) to 10 (**D**) years of age.

## In the clinic

### Bone marrow transplants

The bone marrow serves an important function. There are two types of bone marrow, red marrow (otherwise known as myeloid tissue) and yellow marrow. Red blood cells, platelets, and most white blood cells arise from within the red marrow. In the yellow marrow a few white cells are made; however, this marrow is dominated by large fat globules (producing its yellow appearance) (Fig. 1.15).

From birth most of the body's marrow is red; however, as the subject ages, more red marrow is converted into yellow marrow within the medulla of the long and flat bones.

Bone marrow contains two types of stem cells. Hemopoietic stem cells give rise to the white blood cells, red blood cells, and platelets. Mesenchymal stem cells differentiate into structures that form bone, cartilage, and muscle.

There are a number of diseases that may involve the bone marrow, including infection and malignancy. In patients who develop a bone marrow malignancy (e.g., leukemia) it may be possible to harvest nonmalignant cells from the patient's bone marrow or cells from another person's bone marrow. The patient's own marrow can be destroyed with chemotherapy or radiation and the new cells infused. This treatment is bone marrow transplantation.

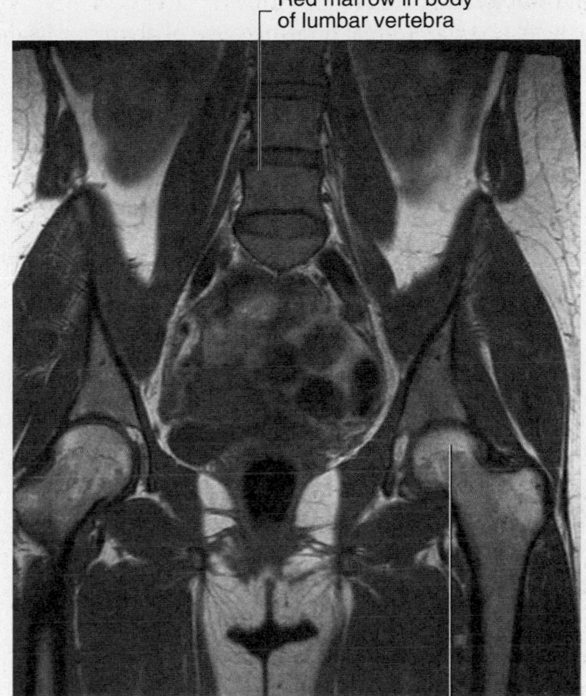


**Fig. 1.15** T1-weighted image in the coronal plane, demonstrating the relatively high signal intensity returned from the femoral heads and proximal femoral necks, consistent with yellow marrow. In this young patient, the vertebral bodies return an intermediate darker signal that represents red marrow. There is relatively little fat in these vertebrae; hence the lower signal return.

## In the clinic

### Bone fractures

Fractures occur in normal bone because of abnormal load or stress, in which the bone gives way (Fig. 1.16A). Fractures may also occur in bone that is of poor quality (osteoporosis); in such cases a normal stress is placed upon a bone that is not of sufficient quality to withstand this force and subsequently fractures.

In children whose bones are still developing, fractures may occur across the growth plate or across the shaft. These shaft fractures typically involve partial cortical disruption, similar to breaking a branch of a young tree; hence they are termed "greenstick" fractures.

After a fracture has occurred, the natural response is to heal the fracture. Between the fracture margins a blood clot is formed into which new vessels grow. A jelly-like matrix is formed, and further migration of collagen-producing cells occurs. On this soft tissue framework, calcium hydroxyapatite is produced by osteoblasts and forms insoluble crystals, and then bone matrix is laid down. As more bone is produced, a callus can be demonstrated forming across the fracture site.

Treatment of fractures requires a fracture line reduction. If this cannot be maintained in a plaster of Paris cast, it may require internal or external fixation with screws and metal rods (Fig. 1.16B).

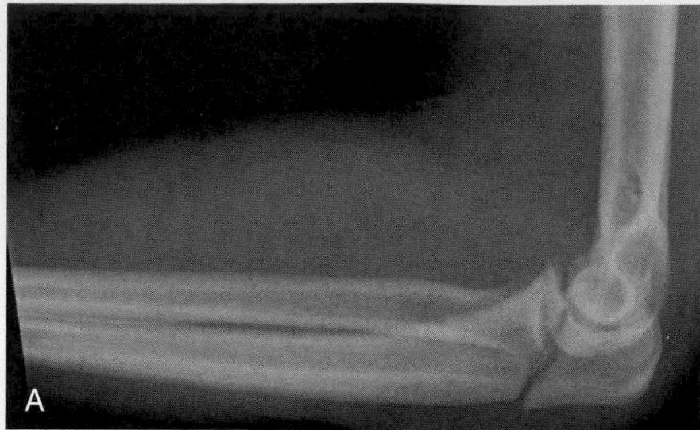


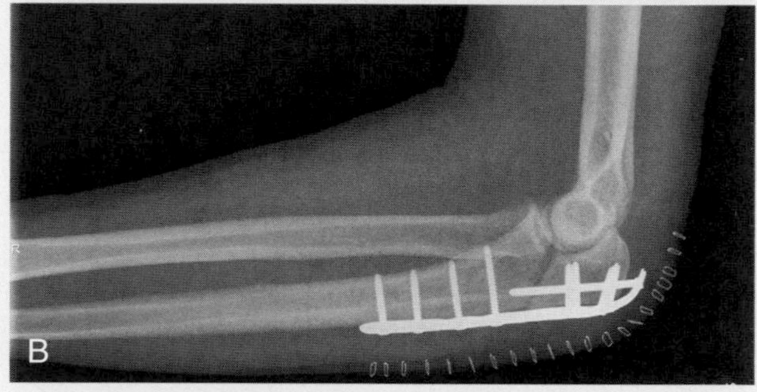


**Fig. 1.16** Radiograph, lateral view, showing fracture of the ulna at the elbow joint **(A)** and repair of this fracture **(B)** using internal fixation with a plate and multiple screws.

## In the clinic

### Avascular necrosis

Avascular necrosis is cellular death of bone resulting from a temporary or permanent loss of blood supply to that bone. Avascular necrosis may occur in a variety of medical conditions, some of which have an etiology that is less than clear. A typical site for avascular necrosis is a fracture across the femoral neck in an elderly patient. In these patients there is loss of continuity of the cortical medullary blood flow with loss of blood flow deep to the retinacular fibers. This essentially renders the femoral head bloodless; it subsequently undergoes necrosis and collapses (Fig. 1.17). In these patients it is necessary to replace the femoral head with a prosthesis.

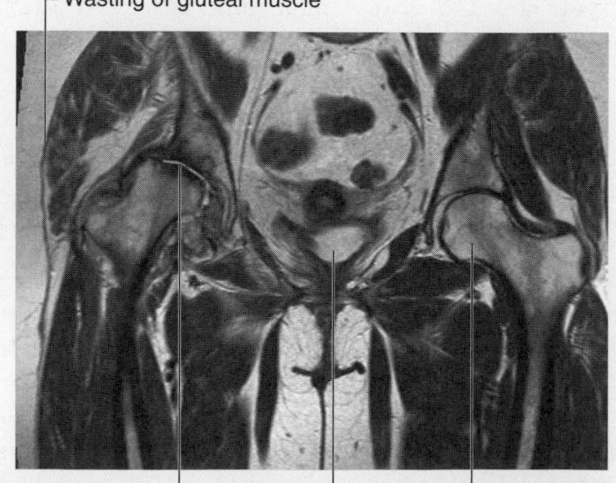


**Fig. 1.17** Image of the hip joints demonstrating loss of height of the right femoral head with juxta-articular bony sclerosis and subchondral cyst formation secondary to avascular necrosis. There is also significant wasting of the muscles supporting the hip, which is secondary to disuse and pain.

## In the clinic

### Epiphyseal fractures

As the skeleton develops, there are stages of intense growth typically around the ages of 7 to 10 years and later in puberty. These growth spurts are associated with increased cellular activity around the growth plate between the head and shaft of a bone. This increase in activity renders the growth plates more vulnerable to injuries, which may occur from dislocation across a growth plate or fracture through a growth plate. Occasionally an injury may result in growth plate compression, destroying that region of the growth plate, which may result in asymmetrical growth across that joint region. All fractures across the growth plate must be treated with care and expediency, requiring fracture reduction.

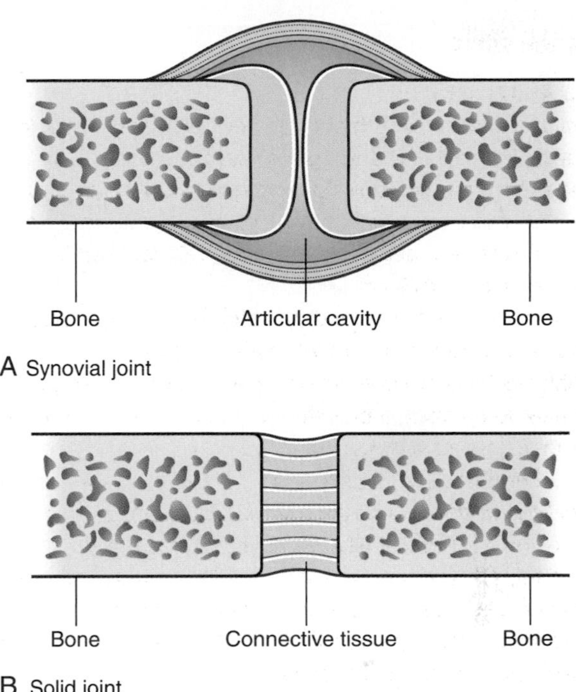


**Fig. 1.18** Joints. **A.** Synovial joint. **B.** Solid joint.

## Joints

The sites where two skeletal elements come together are termed joints. The two general categories of joints (Fig. 1.18) are those in which:

- the skeletal elements are separated by a cavity (i.e., **synovial joints**), and
- there is no cavity and the components are held together by connective tissue (i.e., **solid joints**).

Blood vessels that cross over a joint and nerves that innervate muscles acting on a joint usually contribute articular branches to that joint.

### Synovial joints

Synovial joints are connections between skeletal components where the elements involved are separated by a narrow articular cavity (Fig. 1.19). In addition to containing an articular cavity, these joints have a number of characteristic features.

First, a layer of cartilage, usually **hyaline cartilage**, covers the articulating surfaces of the skeletal elements. In other words, bony surfaces do not normally contact one another directly. As a consequence, when these joints are viewed in normal radiographs, a wide gap seems to separate the adjacent bones because the cartilage that covers the articulating surfaces is more transparent to X-rays than bone.

A second characteristic feature of synovial joints is the presence of a **joint capsule** consisting of an inner **synovial membrane** and an outer **fibrous membrane**.

- The synovial membrane attaches to the margins of the joint surfaces at the interface between the cartilage and bone and encloses the articular cavity. The synovial membrane is highly vascular and produces synovial fluid, which percolates into the articular cavity and lubricates the articulating surfaces. Closed sacs of synovial membrane also occur outside joints, where they form synovial bursae or tendon sheaths. Bursae often intervene between structures, such as tendons and bone, tendons and joints, or skin and bone, and reduce the friction of one structure moving over the other. Tendon sheaths surround tendons and also reduce friction.
- The **fibrous membrane** is formed by dense connective tissue and surrounds and stabilizes the joint. Parts of the fibrous membrane may thicken to form ligaments, which further stabilize the joint. Ligaments outside the capsule usually provide additional reinforcement.

Another common but not universal feature of synovial joints is the presence of additional structures within the area enclosed by the capsule or synovial membrane, such as **articular discs** (usually composed of fibrocartilage), **fat pads**, and **tendons**. Articular discs absorb compression forces, adjust to changes in the contours of joint surfaces during movements, and increase the range of movements that can occur at joints. Fat pads usually occur between the synovial membrane and the capsule and move

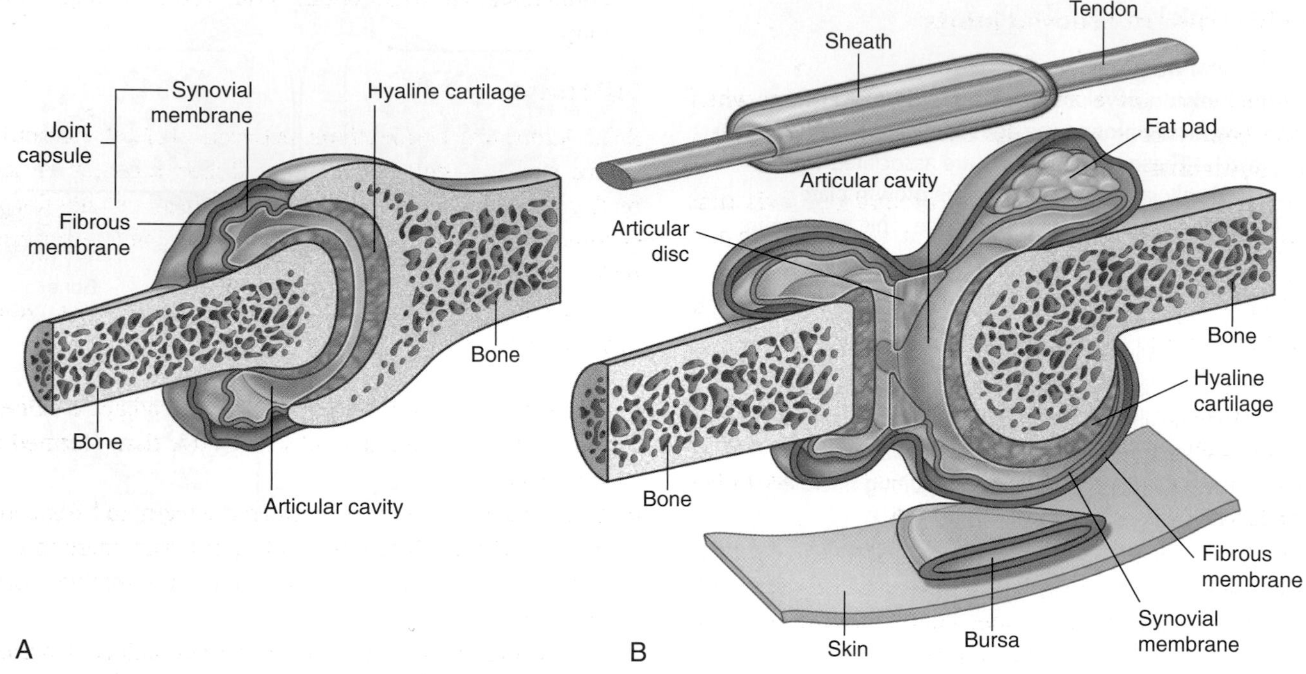


**Fig. 1.19** Synovial joints. **A.** Major features of a synovial joint. **B.** Accessory structures associated with synovial joints.

into and out of regions as joint contours change during movement. Redundant regions of the synovial membrane and fibrous membrane allow for large movements at joints.

### Descriptions of synovial joints based on shape and movement

Synovial joints are described based on shape and movement:

- based on the shape of their articular surfaces, synovial joints are described as plane (flat), hinge, pivot, bicondylar (two sets of contact points), condylar (ellipsoid), saddle, and ball and socket;
- based on movement, synovial joints are described as uniaxial (movement in one plane), biaxial (movement in two planes), and multiaxial (movement in three planes).

Hinge joints are uniaxial, whereas ball and socket joints are multiaxial.

### Specific types of synovial joints (Fig. 1.20)

- Plane joints—allow sliding or gliding movements when one bone moves across the surface of another (e.g., acromioclavicular joint)
- Hinge joints—allow movement around one axis that passes transversely through the joint; permit flexion and extension (e.g., elbow [humero-ulnar] joint)
- Pivot joints—allow movement around one axis that passes longitudinally along the shaft of the bone; permit rotation (e.g., atlanto-axial joint)
- Bicondylar joints—allow movement mostly in one axis with limited rotation around a second axis; formed by two convex condyles that articulate with concave or flat surfaces (e.g., knee joint)
- Condylar (ellipsoid) joints—allow movement around two axes that are at right angles to each other; permit flexion, extension, abduction, adduction, and circumduction (limited) (e.g., wrist joint)
- Saddle joints—allow movement around two axes that are at right angles to each other; the articular surfaces are saddle shaped; permit flexion, extension, abduction, adduction, and circumduction (e.g., carpometacarpal joint of the thumb)
- Ball and socket joints—allow movement around multiple axes; permit flexion, extension, abduction, adduction, circumduction, and rotation (e.g., hip joint)

### Solid joints

Solid joints are connections between skeletal elements where the adjacent surfaces are linked together either by fibrous connective tissue or by cartilage, usually fibrocartilage (Fig. 1.21). Movements at these joints are more restricted than at synovial joints.

**Fibrous joints** include sutures, gomphoses, and syndesmoses.

- **Sutures** occur only in the skull where adjacent bones are linked by a thin layer of connective tissue termed a *sutural ligament.*
- **Gomphoses** occur only between the teeth and adjacent bone. In these joints, short collagen tissue fibers in the periodontal ligament run between the root of the tooth and the bony socket.
- **Syndesmoses** are joints in which two adjacent bones are linked by a ligament. Examples are the ligamentum flavum, which connects adjacent vertebral laminae, and an interosseous membrane, which links, for example, the radius and ulna in the forearm.

**Cartilaginous joints** include synchondroses and symphyses.

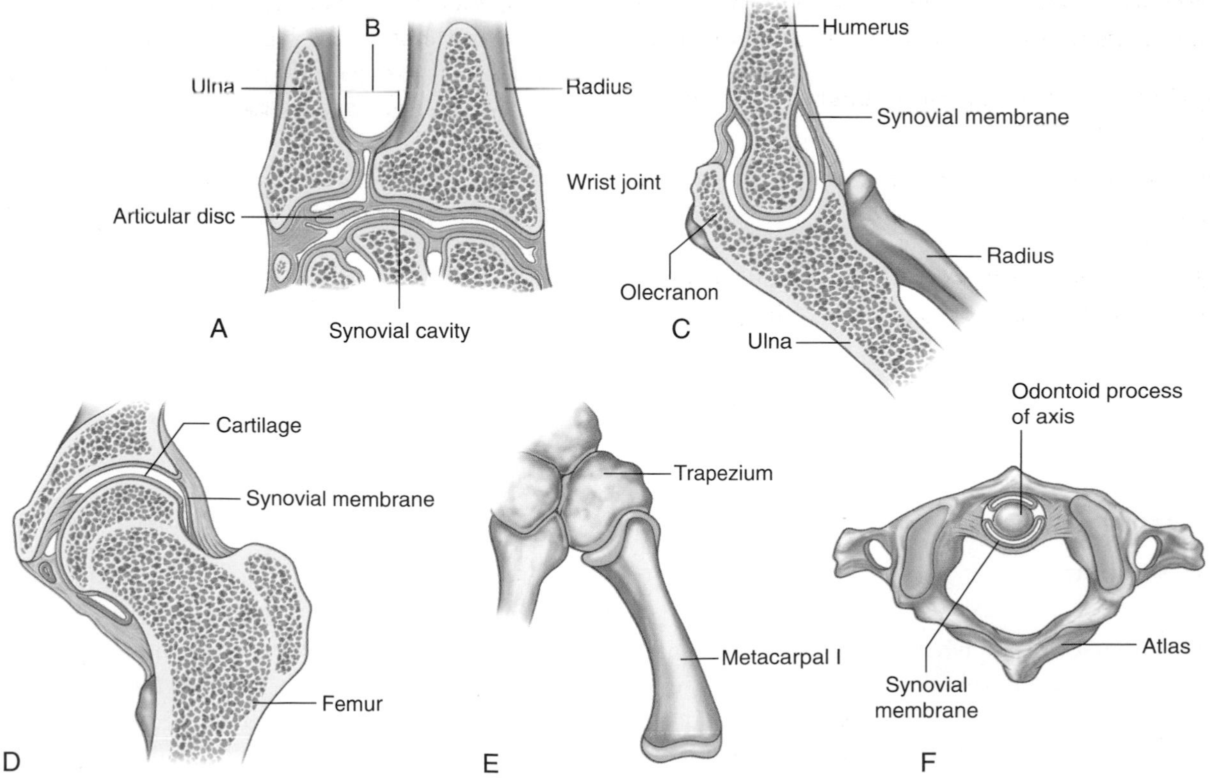


**Fig. 1.20** Various types of synovial joints. **A.** Condylar (wrist). **B.** Gliding (radio-ulnar). **C.** Hinge (elbow). **D.** Ball and socket (hip). **E.** Saddle (carpometacarpal of thumb). **F.** Pivot (atlanto-axial).

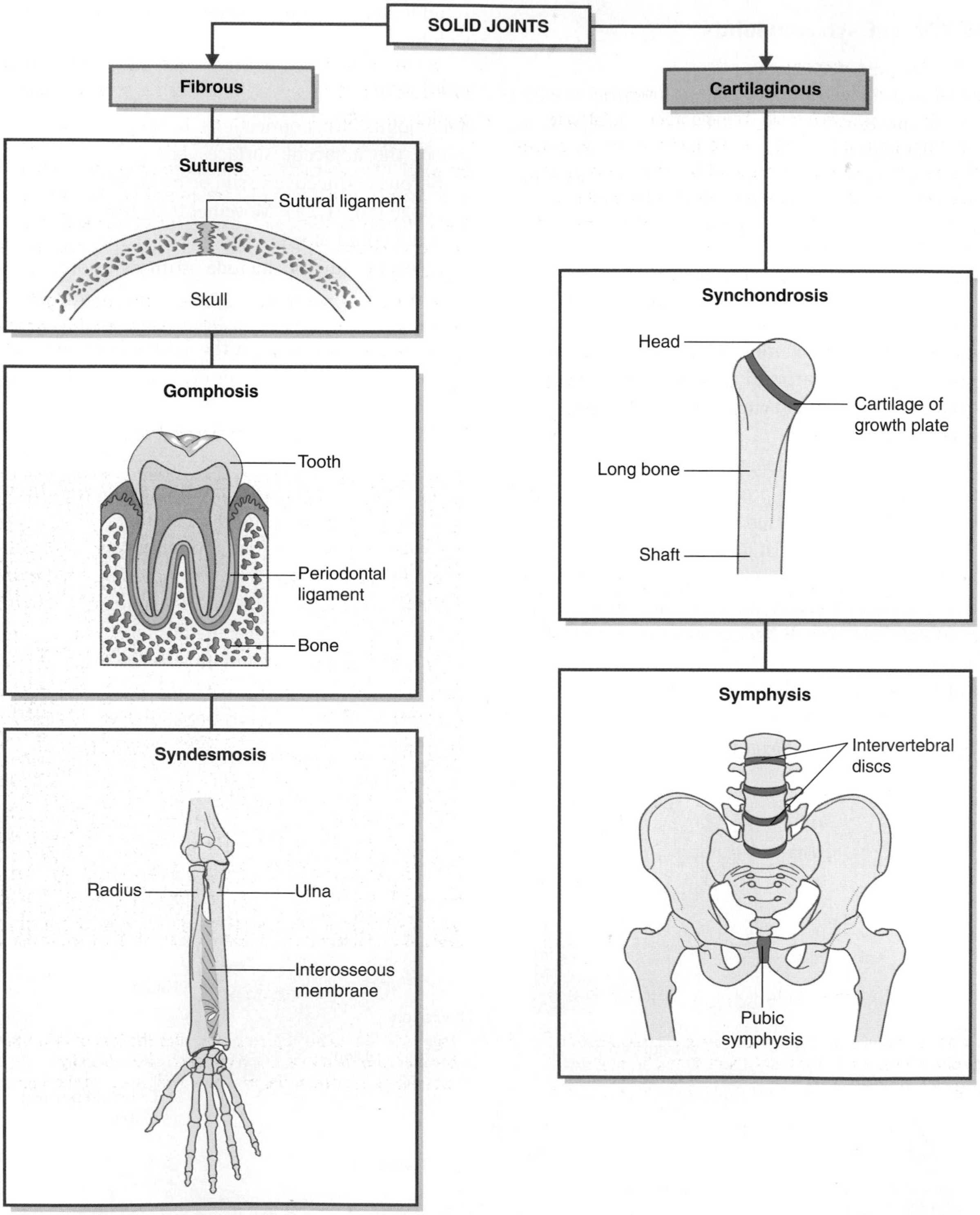

**Fig. 1.21** Solid joints.

- **Synchondroses** occur where two ossification centers in a developing bone remain separated by a layer of cartilage, for example, the growth plate that occurs between the head and shaft of developing long bones. These joints allow bone growth and eventually become completely ossified.
- **Symphyses** occur where two separate bones are interconnected by cartilage. Most of these types of joints occur in the midline and include the pubic symphysis between the two pelvic bones, and intervertebral discs between adjacent vertebrae.

## In the clinic

### Degenerative joint disease

Degenerative joint disease is commonly known as osteoarthritis or osteoarthrosis. The disorder is related to aging but not caused by aging. Typically there are decreases in water and proteoglycan content within the cartilage. The cartilage becomes more fragile and more susceptible to mechanical disruption (Fig. 1.22). As the cartilage wears, the underlying bone becomes fissured and also thickens. Synovial fluid may be forced into small cracks that appear in the bone's surface, which produces large cysts. Furthermore, reactive juxta-articular bony nodules are formed (osteophytes) (Fig. 1.23). As these processes occur, there is slight deformation, which alters the biomechanical forces through the joint. This in turn creates abnormal stresses, which further disrupt the joint.

In the United States, osteoarthritis accounts for up to one-quarter of primary health care visits and is regarded as a significant problem.

The etiology of osteoarthritis is not clear; however, osteoarthritis can occur secondary to other joint diseases, such as rheumatoid arthritis and infection. Overuse of joints and abnormal strains, such as those experienced by people who play sports, often cause one to be more susceptible to chronic joint osteoarthritis.

Various treatments are available, including weight reduction, proper exercise, anti-inflammatory drug treatment, and joint replacement (Fig. 1.24).

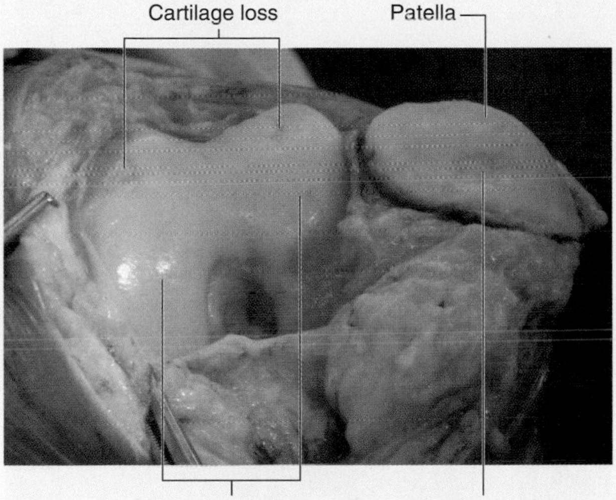

Cartilage loss
Patella
Femoral condyles
Cartilage loss

**Fig. 1.22** This operative photograph demonstrates the focal areas of cartilage loss in the patella and femoral condyles throughout the knee joint.

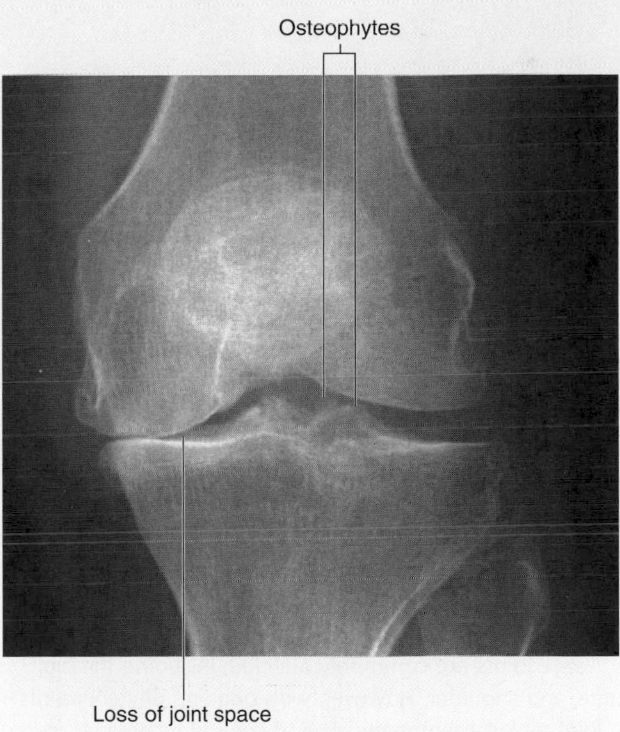


**Fig. 1.23** This radiograph demonstrates the loss of joint space in the medial compartment and presence of small spiky osteophytic regions at the medial lateral aspect of the joint.

## In the clinic—cont'd

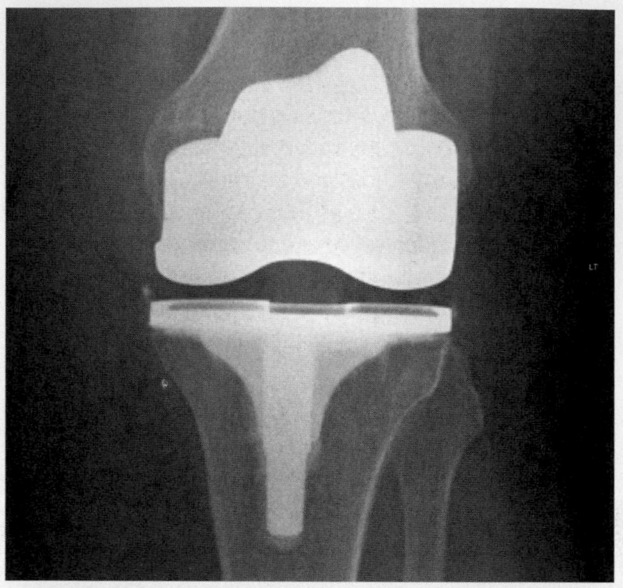

**Fig. 1.24** After knee replacement. This radiograph shows the position of the prosthesis.

### *Arthroscopy*

Arthroscopy is a technique of visualizing the inside of a joint using a small telescope placed through a tiny incision in the skin. Arthroscopy can be performed in most joints. However, it is most commonly performed in the knee, shoulder, ankle, and hip joints.

Arthroscopy allows the surgeon to view the inside of the joint and its contents. Notably, in the knee, the menisci and the ligaments are easily seen, and it is possible using separate puncture sites and specific instruments to remove the menisci and replace the cruciate ligaments. The advantages of arthroscopy are that it is performed through small incisions, it enables patients to quickly recover and return to normal activity, and it only requires either a light anesthetic or regional anesthesia during the procedure.

## In the clinic

### Joint replacement

Joint replacement is undertaken for a variety of reasons. These predominantly include degenerative joint disease and joint destruction. Joints that have severely degenerated or lack their normal function are painful. In some patients, the pain may be so severe that it prevents them from leaving the house and undertaking even the smallest of activities without discomfort.

Large joints are commonly affected, including the hip, knee, and shoulder. However, with ongoing developments in joint replacement materials and surgical techniques, even small joints of the fingers can be replaced.

Typically, both sides of the joint are replaced; in the hip joint the acetabulum will be reamed, and a plastic or metal cup will be introduced. The femoral component will be fitted precisely to the femur and cemented in place (Fig. 1.25).

Most patients derive significant benefit from joint replacement and continue to lead an active life afterward. In a minority of patients who have been fitted with a metal acetabular cup and metal femoral component, an aseptic lymphocyte-dominated vasculitis-associated lesion (ALVAL) may develop, possibly caused by a hypersensitivity response to the release of metal ions in adjacent tissues. These patients often have chronic pain and might need additional surgery to replace these joint replacements with safer models.

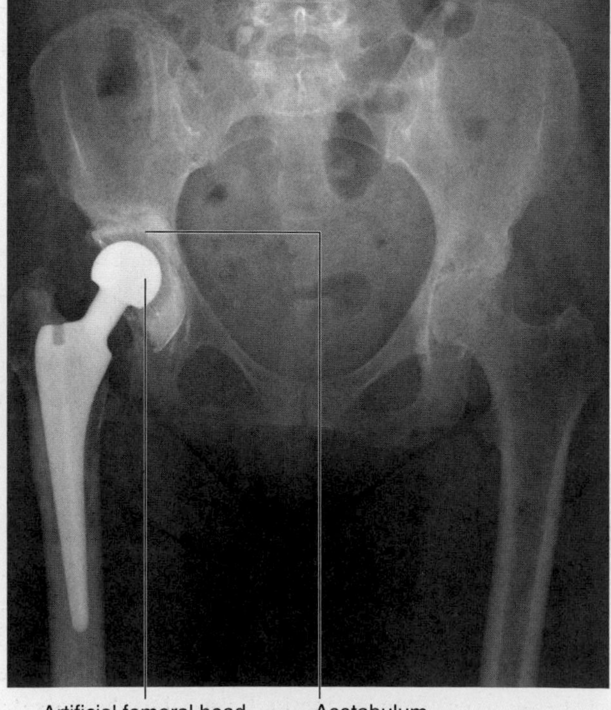


**Fig. 1.25** This is a radiograph, anteroposterior view, of the pelvis after a right total hip replacement. There are additional significant degenerative changes in the left hip joint, which will also need to be replaced.

## SKIN AND FASCIAS

### Skin

The skin is the largest organ of the body. It consists of the epidermis and the dermis. The epidermis is the outer cellular layer of stratified squamous epithelium, which is avascular and varies in thickness. The dermis is a dense bed of vascular connective tissue.

The skin functions as a mechanical and permeability barrier, and as a sensory and thermoregulatory organ. It also can initiate primary immune responses.

### Fascia

Fascia is connective tissue containing varying amounts of fat that separate, support, and interconnect organs and structures, enable movement of one structure relative to another, and allow the transit of vessels and nerves from one area to another. There are two general categories of fascia: superficial and deep.

- Superficial (subcutaneous) fascia lies just deep to and is attached to the dermis of the skin. It is made up of loose connective tissue usually containing a large amount of fat. The thickness of the superficial fascia (subcutaneous tissue) varies considerably, both from one area of the body to another and from one individual to another. The superficial fascia allows movement of the skin over deeper areas of the body, acts as a conduit for vessels and nerves coursing to and from the skin, and serves as an energy (fat) reservoir.
- Deep fascia usually consists of dense, organized connective tissue. The outer layer of deep fascia is attached to the deep surface of the superficial fascia and forms a thin fibrous covering over most of the deeper region of the body. Inward extensions of this fascial layer form intermuscular septa that compartmentalize groups of muscles with similar functions and innervations. Other extensions surround individual muscles and groups of vessels and nerves, forming an investing fascia. Near some joints the deep fascia thickens, forming retinacula. These fascial retinacula hold tendons in place and prevent them from bowing during movements at the joints. Finally, there is a layer of deep fascia separating the membrane lining the abdominal cavity (the parietal peritoneum) from the fascia covering the deep surface of the muscles of the abdominal wall (the transversalis fascia). This layer is referred to as **extraperitoneal fascia**. A similar layer of fascia in the thorax is termed the **endothoracic fascia**.

#### In the clinic

**The importance of fascias**

A fascia is a thin band of tissue that surrounds muscles, bones, organs, nerves, and blood vessels and often remains uninterrupted as a 3D structure between tissues. It provides important support for tissues and can provide a boundary between structures.

Clinically, fascias are extremely important because they often limit the spread of infection and malignant disease. When infections or malignant diseases cross a fascial plain, a primary surgical clearance may require a far more extensive dissection to render the area free of tumor or infection.

A typical example of the clinical importance of a fascial layer would be of that covering the psoas muscle. Infection within an intervertebral body secondary to tuberculosis can pass laterally into the psoas muscle. Pus fills the psoas muscle but is limited from further spread by the psoas fascia, which surrounds the muscle and extends inferiorly into the groin pointing below the inguinal ligament.

#### In the clinic

**Placement of skin incisions and scarring**

Surgical skin incisions are ideally placed along or parallel to Langer's lines, which are lines of skin tension that correspond to the orientation of the dermal collagen fibers. They tend to run in the same direction as the underlying muscle fibers and incisions that are made along these lines tend to heal better with less scarring. In contrast, incisions made perpendicular to Langer's lines are more likely to heal with a prominent scar and in some severe cases can lead to raised, firm, hypertrophic, or keloid, scars.

## MUSCULAR SYSTEM

The muscular system is generally regarded as consisting of one type of muscle found in the body—skeletal muscle. However, there are two other types of muscle tissue found in the body, smooth muscle and cardiac muscle, that are important components of other systems. These three types of muscle can be characterized by whether they are controlled voluntarily or involuntarily, whether they appear striated (striped) or smooth, and whether they are associated with the body wall (somatic) or with organs and blood vessels (visceral).

- Skeletal muscle forms the majority of the muscle tissue in the body. It consists of parallel bundles of long,

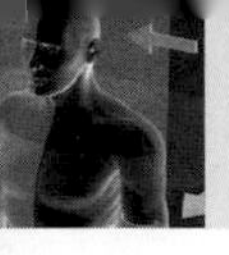



multinucleated fibers with transverse stripes, is capable of powerful contractions, and is innervated by somatic and branchial motor nerves. This muscle is used to move bones and other structures, and provides support and gives form to the body. Individual skeletal muscles are often named on the basis of shape (e.g., rhomboid major muscle), attachments (e.g., sternohyoid muscle), function (e.g., flexor pollicis longus muscle), position (e.g., palmar interosseous muscle), or fiber orientation (e.g., external oblique muscle).

- Cardiac muscle is striated muscle found only in the walls of the heart (myocardium) and in some of the large vessels close to where they join the heart. It consists of a branching network of individual cells linked electrically and mechanically to work as a unit. Its contractions are less powerful than those of skeletal muscle and it is resistant to fatigue. Cardiac muscle is innervated by visceral motor nerves.
- Smooth muscle (absence of stripes) consists of elongated or spindle-shaped fibers capable of slow and sustained contractions. It is found in the walls of blood vessels (tunica media), associated with hair follicles in the skin, located in the eyeball, and found in the walls of various structures associated with the gastrointestinal, respiratory, genitourinary, and urogenital systems. Smooth muscle is innervated by visceral motor nerves.

## In the clinic

### Muscle paralysis

Muscle paralysis is the inability to move a specific muscle or muscle group and may be associated with other neurological abnormalities, including loss of sensation. Major causes include stroke, trauma, poliomyelitis, and iatrogenic factors. Paralysis may be due to abnormalities in the brain, the spinal cord, and the nerves supplying the muscles.

In the long term, muscle paralysis will produce secondary muscle wasting and overall atrophy of the region due to disuse.

## In the clinic

### Muscle atrophy

Muscle atrophy is a wasting disorder of muscle. It can be produced by a variety of causes, which include nerve damage to the muscle and disuse.

Muscle atrophy is an important problem in patients who have undergone long-term rest or disuse, requiring extensive rehabilitation and muscle building exercises to maintain normal activities of daily living.

## In the clinic

### Muscle injuries and strains

Muscle injuries and strains tend to occur in specific muscle groups and usually are related to a sudden exertion and muscle disruption. They typically occur in athletes.

Muscle tears may involve a small interstitial injury up to a complete muscle disruption (Fig. 1.26). It is important to identify which muscle groups are affected and the extent of the tear to facilitate treatment and obtain a prognosis, which will determine the length of rehabilitation necessary to return to normal activity.

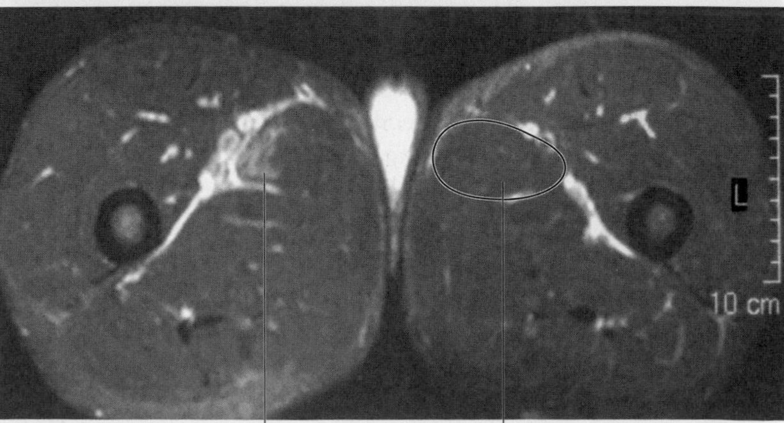


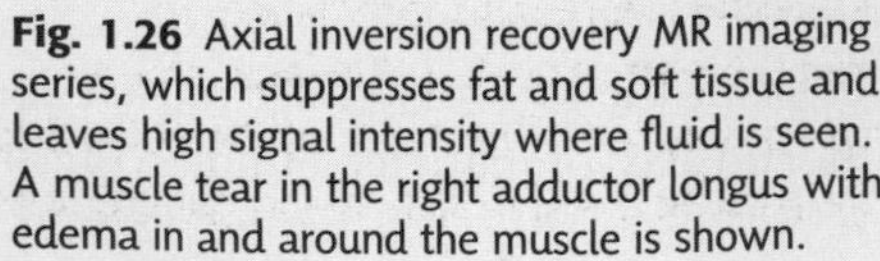

**Fig. 1.26** Axial inversion recovery MR imaging series, which suppresses fat and soft tissue and leaves high signal intensity where fluid is seen. A muscle tear in the right adductor longus with edema in and around the muscle is shown.

## CARDIOVASCULAR SYSTEM

The cardiovascular system consists of the heart, which pumps blood throughout the body, and the blood vessels, which are a closed network of tubes that transport the blood. There are three types of blood vessels:

- arteries, which transport blood away from the heart;
- veins, which transport blood toward the heart;
- capillaries, which connect the arteries and veins, are the smallest of the blood vessels and are where oxygen, nutrients, and wastes are exchanged within the tissues.

The walls of the blood vessels of the cardiovascular system usually consist of three layers or tunics:

- tunica externa (adventitia)—the outer connective tissue layer,
- tunica media—the middle smooth muscle layer (may also contain varying amounts of elastic fibers in medium and large arteries), and
- tunica intima—the inner endothelial lining of the blood vessels.

Arteries are usually further subdivided into three classes, according to the variable amounts of smooth muscle and elastic fibers contributing to the thickness of the tunica media, the overall size of the vessel, and its function.

- Large elastic arteries contain substantial amounts of elastic fibers in the tunica media, allowing expansion and recoil during the normal cardiac cycle. This helps maintain a constant flow of blood during diastole. Examples of large elastic arteries are the aorta, the brachiocephalic trunk, the left common carotid artery, the left subclavian artery, and the pulmonary trunk.
- Medium muscular arteries are composed of a tunica media that contains mostly smooth muscle fibers. This characteristic allows these vessels to regulate their diameter and control the flow of blood to different parts of the body. Examples of medium muscular arteries are most of the named arteries, including the femoral, axillary, and radial arteries.
- Small arteries and arterioles control the filling of the capillaries and directly contribute to the arterial pressure in the vascular system.

Veins also are subdivided into three classes.

- Large veins contain some smooth muscle in the tunica media, but the thickest layer is the tunica externa. Examples of large veins are the superior vena cava, the inferior vena cava, and the portal vein.
- Small and medium veins contain small amounts of smooth muscle, and the thickest layer is the tunica externa. Examples of small and medium veins are superficial veins in the upper and lower limbs and deeper veins of the leg and forearm.
- Venules are the smallest veins and drain the capillaries.

Although veins are similar in general structure to arteries, they have a number of distinguishing features.

- The walls of veins, specifically the tunica media, are thin.
- The luminal diameters of veins are large.
- There often are multiple veins (venae comitantes) closely associated with arteries in peripheral regions.
- Valves often are present in veins, particularly in peripheral vessels inferior to the level of the heart. These are usually paired cusps that facilitate blood flow toward the heart.

More specific information about the cardiovascular system and how it relates to the circulation of blood throughout the body will be discussed, where appropriate, in each of the succeeding chapters of the text.

### In the clinic

**Atherosclerosis**

Atherosclerosis is a disease that affects arteries. There is a chronic inflammatory reaction in the walls of the arteries, with deposition of cholesterol and fatty proteins. This may in turn lead to secondary calcification, with reduction in the diameter of the vessels impeding distal flow. The plaque itself may be a site for attraction of platelets that may "fall off" (embolize) distally. Plaque fissuring may occur, which allows fresh clots to form and occlude the vessel.

The importance of atherosclerosis and its effects depend upon which vessel is affected. If atherosclerosis occurs in the carotid artery, small emboli may form and produce a stroke. In the heart, plaque fissuring may produce an acute vessel thrombosis, producing a myocardial infarction (heart attack). In the legs, chronic narrowing of vessels may limit the ability of the patient to walk and ultimately cause distal ischemia and gangrene of the toes.

## In the clinic

### Varicose veins

Varicose veins are tortuous dilated veins that typically occur in the legs, although they may occur in the superficial veins of the arm and in other organs.

In normal individuals the movement of adjacent leg muscles pumps the blood in the veins to the heart. Blood is also pumped from the superficial veins through the investing layer of fascia of the leg into the deep veins. Valves in these perforating veins may become damaged, allowing blood to pass in the opposite direction. This increased volume and pressure produces dilatation and tortuosity of the superficial veins (Fig. 1.27). Apart from the unsightliness of larger veins, the skin may become pigmented and atrophic with a poor response to tissue trauma. In some patients even small trauma may produce skin ulceration, which requires elevation of the limb and application of pressure bandages to heal.

Treatment of varicose veins depends on their location, size, and severity. Typically the superficial varicose veins can be excised and stripped, allowing blood only to drain into the deep system.

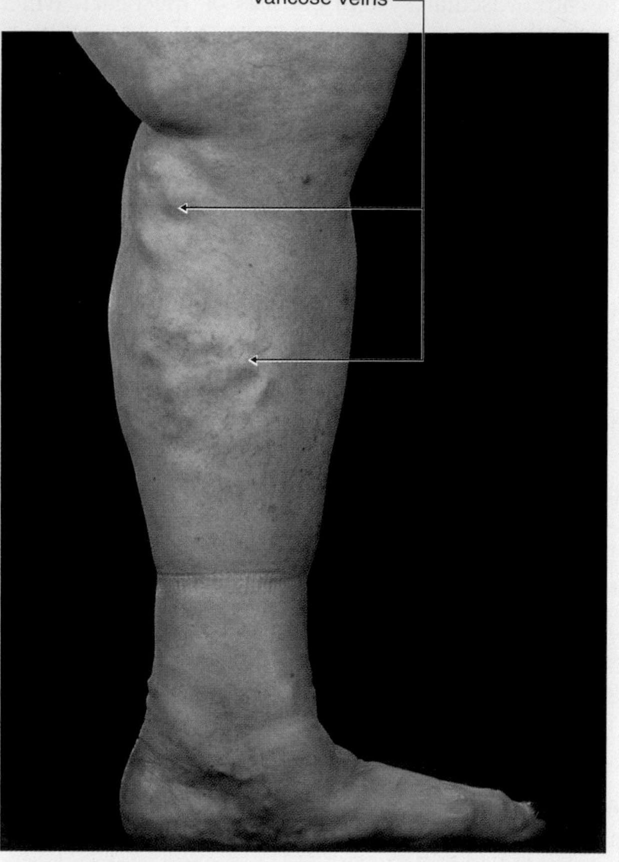


**Fig. 1.27** Photograph demonstrating varicose veins.

## In the clinic

### Anastomoses and collateral circulation

All organs require a blood supply from the arteries and drainage by veins. Within most organs there are multiple ways of perfusing the tissue such that if the main vessel feeding the organ or vein draining the organ is blocked, a series of smaller vessels (collateral vessels) continue to supply and drain the organ.

In certain circumstances, organs have more than one vessel perfusing them, such as the hand, which is supplied by the radial and ulnar arteries. Loss of either the radial or the ulnar artery may not produce any symptoms of reduced perfusion to the hand.

There are circumstances in which loss of a vein produces significant venous collateralization. Some of these venous collaterals become susceptible to bleeding. This is a considerable problem in patients who have undergone portal vein thrombosis or occlusion, where venous drainage from the gut bypasses the liver through collateral veins to return to the systemic circulation.

Normal vascular anastomoses associated with an organ are important. Some organs, such as the duodenum, have a dual blood supply arising from the branches of the celiac trunk and also from the branches of the superior mesenteric artery. Should either of these vessels be damaged, blood supply will be maintained to the organ. The brain has multiple vessels supplying it, dominated by the carotid arteries and the vertebral arteries. Vessels within the brain are end arteries and have a poor collateral circulation; hence any occlusion will produce long-term cerebral damage.

## LYMPHATIC SYSTEM

### Lymphatic vessels

Lymphatic vessels form an extensive and complex interconnected network of channels, which begin as "porous" blind-ended lymphatic capillaries in tissues of the body and converge to form a number of larger vessels, which ultimately connect with large veins in the root of the neck.

Lymphatic vessels mainly collect fluid lost from vascular capillary beds during nutrient exchange processes and deliver it back to the venous side of the vascular system (Fig. 1.28). Also included in this interstitial fluid that drains into the lymphatic capillaries are pathogens, cells of the lymphocytic system, cell products (such as hormones), and cell debris.

In the small intestine, certain fats absorbed and processed by the intestinal epithelium are packaged into protein-coated lipid droplets (**chylomicrons**), which are released from the epithelial cells and enter the interstitial compartment. Together with other components of the interstitial fluid, the chylomicrons drain into lymphatic capillaries (known as **lacteals** in the small intestine) and are ultimately delivered to the venous system in the neck. The lymphatic system is therefore also a major route of transport for fat absorbed by the gut.

The fluid in most lymphatic vessels is clear and colorless and is known as **lymph**. That carried by lymphatic vessels from the small intestine is opaque and milky because of the presence of chylomicrons and is termed **chyle**.

There are lymphatic vessels in most areas of the body, including those associated with the central nervous system (Louveau A et al., *Nature* 2015; 523:337-41; Aspelund A et al., *J Exp Med* 2015; 212:991-9). Exceptions include bone marrow and avascular tissues such as epithelia and cartilage.

The movement of lymph through the lymphatic vessels is generated mainly by the indirect action of adjacent structures, particularly by contraction of skeletal muscles and pulses in arteries. Unidirectional flow is maintained by the presence of valves in the vessels.

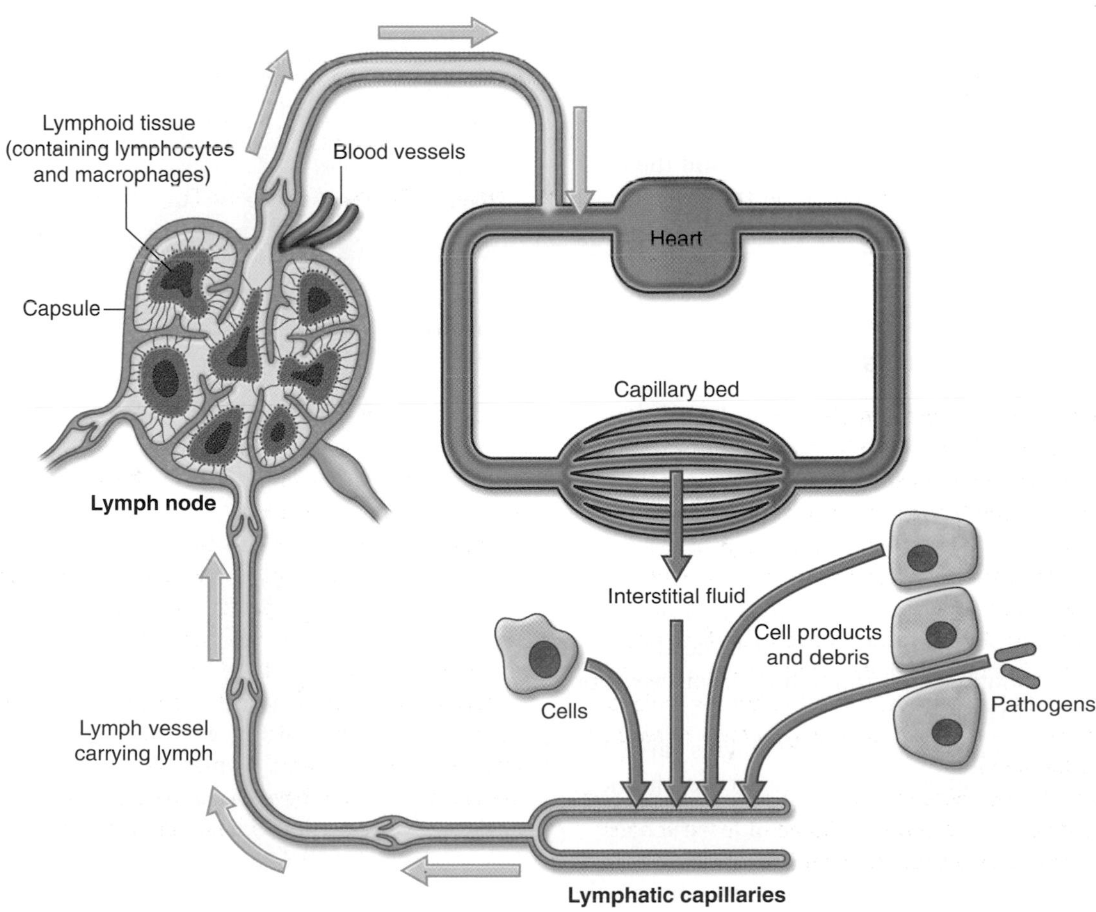


**Fig. 1.28** Lymphatic vessels mainly collect fluid lost from vascular capillary beds during nutrient exchange processes and deliver it back to the venous side of the vascular system.

## Lymph nodes

Lymph nodes are small (0.1–2.5 cm long) encapsulated structures that interrupt the course of lymphatic vessels and contain elements of the body's defense system, such as clusters of lymphocytes and macrophages. They act as elaborate filters that trap and phagocytose particulate matter in the lymph that percolates through them. In addition, they detect and defend against foreign antigens that are also carried in the lymph (Fig. 1.28).

Because lymph nodes are efficient filters and flow through them is slow, cells that metastasize from (migrate away from) primary tumors and enter lymphatic vessels often lodge and grow as secondary tumors in lymph nodes. Lymph nodes that drain regions that are infected or contain other forms of disease can enlarge or undergo certain physical changes, such as becoming "hard" or "tender." These changes can be used by clinicians to detect pathologic changes or to track spread of disease.

A number of regions in the body are associated with clusters or a particular abundance of lymph nodes (Fig. 1.29). Not surprisingly, nodes in many of these regions drain the body's surface, the digestive system, or the respiratory system. All three of these areas are high-risk sites for the entry of foreign pathogens.

Lymph nodes are abundant and accessible to palpation in the axilla, the groin and femoral region, and the neck. Deep sites that are not palpable include those associated with the trachea and bronchi in the thorax, and with the aorta and its branches in the abdomen.

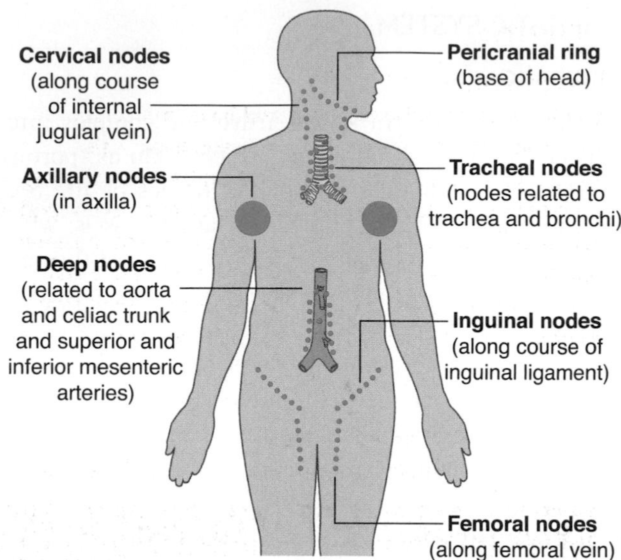


**Fig. 1.29** Regions associated with clusters or a particular abundance of lymph nodes.

## Lymphatic trunks and ducts

All lymphatic vessels coalesce to form larger trunks or ducts, which drain into the venous system at sites in the neck where the internal jugular veins join the subclavian veins to form the brachiocephalic veins (Fig. 1.30):

- Lymph from the right side of the head and neck, the right upper limb, and the right side of the thorax is carried by lymphatic vessels that connect with veins on the right side of the neck.
- Lymph from all other regions of the body is carried by lymphatic vessels that drain into veins on the left side of the neck.

Specific information about the organization of the lymphatic system in each region of the body is discussed in the appropriate chapter.

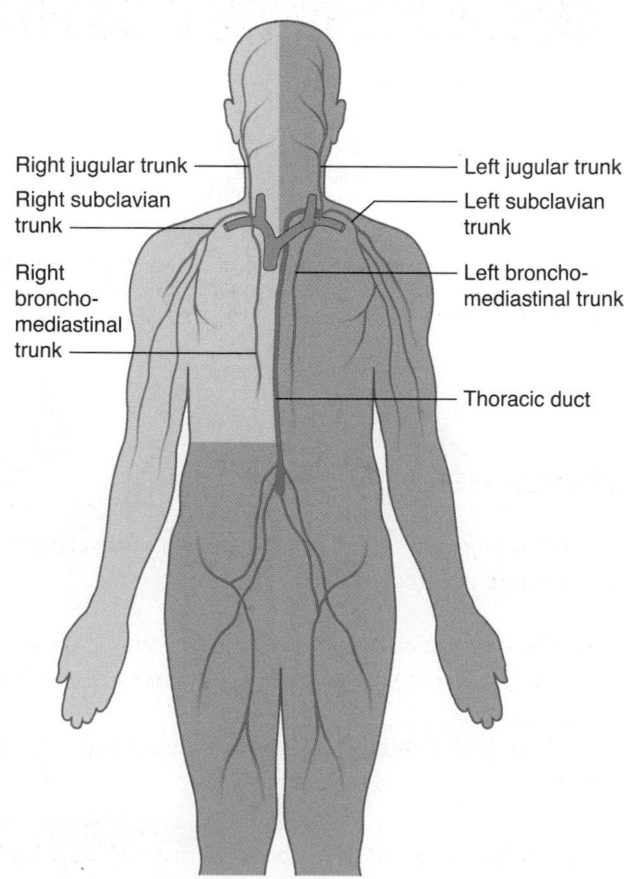


**Fig. 1.30** Major lymphatic vessels that drain into large veins in the neck.

## In the clinic

### Lymph nodes

Lymph nodes are efficient filters and have an internal honeycomb of reticular connective tissue filled with lymphocytes. These lymphocytes act on bacteria, viruses, and other bodily cells to destroy them. Lymph nodes tend to drain specific areas, and if infection occurs within a drainage area, the lymph node will become active. The rapid cell turnover and production of local inflammatory mediators may cause the node to enlarge and become tender.

Similarly, in patients with malignancy the lymphatics may drain metastasizing cells to the lymph nodes. These can become enlarged and inflamed and will need to be removed if clinically symptomatic.

Lymph nodes may become diffusely enlarged in certain systemic illnesses (e.g., viral infection), or local groups may become enlarged with primary lymph node malignancies, such as lymphoma (Fig. 1.31).

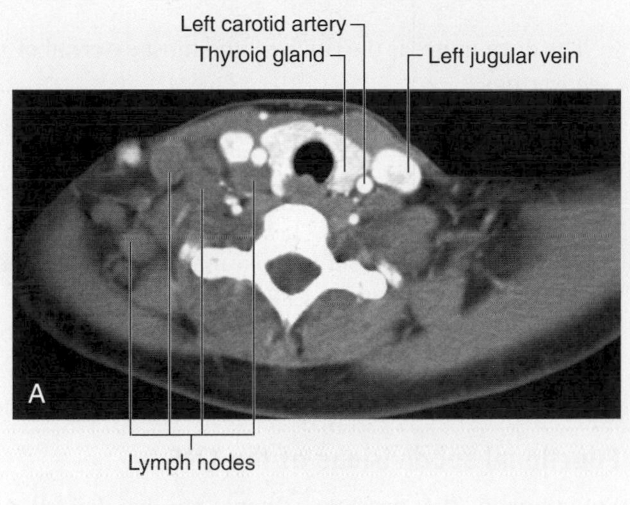


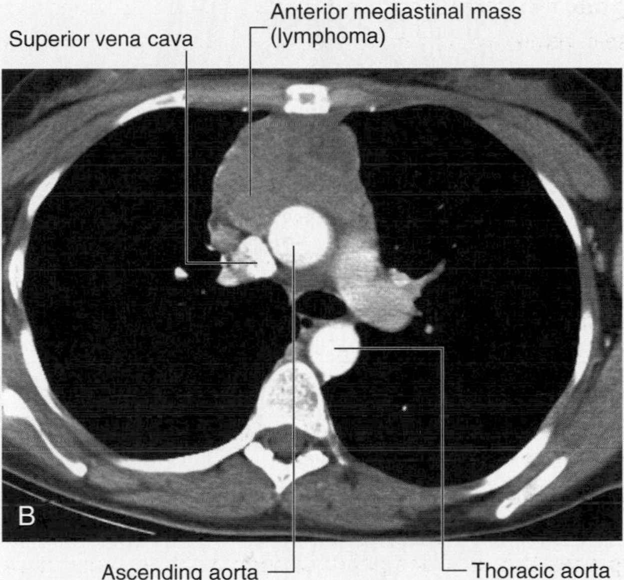


**Fig. 1.31** **A.** This computed tomogram with contrast, in the axial plane, demonstrates the normal common carotid arteries and internal jugular veins with numerous other nonenhancing nodules that represent lymph nodes in a patient with lymphoma. **B.** This computed tomogram with contrast, in the axial plane, demonstrates a large anterior soft tissue mediastinal mass that represents a lymphoma.

## NERVOUS SYSTEM

The nervous system can be separated into parts based on structure and on function:

- structurally, it can be divided into the central nervous system (CNS) and the peripheral nervous system (PNS) (Fig. 1.32);
- functionally, it can be divided into somatic and visceral parts.

The CNS is composed of the brain and spinal cord, both of which develop from the neural tube in the embryo.

The PNS is composed of all nervous structures outside the CNS that connect the CNS to the body. Elements of this system develop from neural crest cells and as outgrowths of the CNS. The PNS consists of the spinal and cranial nerves, visceral nerves and plexuses, and the enteric system. The detailed anatomy of a typical spinal nerve is described in Chapter 2, as is the way spinal nerves are numbered. Cranial nerves are described in Chapter 8. The details of nerve plexuses are described in chapters dealing with the specific regions in which the plexuses are located.

### Central nervous system

#### Brain

The parts of the brain are the cerebral hemispheres, the cerebellum, and the brainstem. The cerebral hemispheres

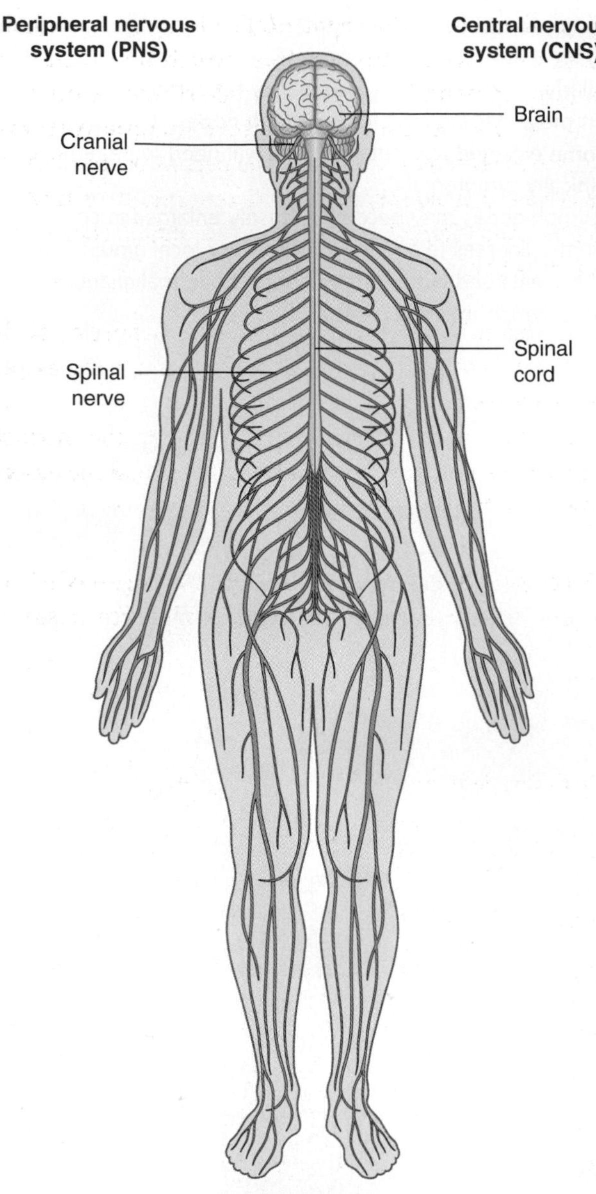


**Fig. 1.32** CNS and PNS.

consist of an outer portion, or the **gray matter**, containing cell bodies; an inner portion, or the **white matter**, made up of axons forming tracts or pathways; and the **ventricles**, which are spaces filled with CSF.

The cerebellum has two lateral lobes and a midline portion. The components of the brainstem are classically defined as the diencephalon, midbrain, pons, and medulla. However, in common usage today, the term "brainstem" usually refers to the midbrain, pons, and medulla.

A further discussion of the brain can be found in Chapter 8.

### Spinal cord

The spinal cord is the part of the CNS in the superior two thirds of the vertebral canal. It is roughly cylindrical in shape, and is circular to oval in cross section with a central canal. A further discussion of the spinal cord can be found in Chapter 2.

### Meninges

The meninges (Fig. 1.33) are three connective tissue coverings that surround, protect, and suspend the brain and spinal cord within the cranial cavity and vertebral canal, respectively:

- The dura mater is the thickest and most external of the coverings.
- The arachnoid mater is against the internal surface of the dura mater.
- The pia mater is adherent to the brain and spinal cord.

Between the arachnoid and pia mater is the subarachnoid space, which contains CSF.

A further discussion of the cranial meninges can be found in Chapter 8 and of the spinal meninges in Chapter 2.

## Functional subdivisions of the CNS

Functionally, the nervous system can be divided into somatic and visceral parts.

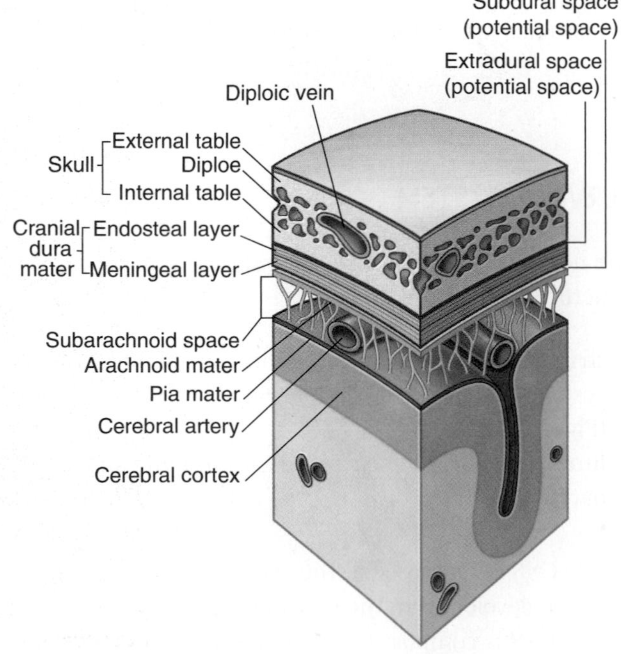


**Fig. 1.33** Arrangement of meninges in the cranial cavity.

- The **somatic part** (*soma*, from the Greek for "body") innervates structures (skin and most skeletal muscle) derived from somites in the embryo, and is mainly involved with receiving and responding to information from the external environment.
- The **visceral part** (*viscera*, from the Greek for "guts") innervates organ systems in the body and other visceral elements, such as smooth muscle and glands, in peripheral regions of the body. It is concerned mainly with detecting and responding to information from the internal environment.

### Somatic part of the nervous system

The somatic part of the nervous system consists of:

- nerves that carry conscious sensations from peripheral regions back to the CNS, and
- nerves that innervate voluntary muscles.

Somatic nerves arise segmentally along the developing CNS in association with **somites**, which are themselves arranged segmentally along each side of the neural tube (Fig. 1.34). Part of each somite (the **dermatomyotome**) gives rise to skeletal muscle and the dermis of the skin. As cells of the dermatomyotome differentiate, they migrate into posterior (dorsal) and anterior (ventral) areas of the developing body:

- Cells that migrate anteriorly give rise to muscles of the limbs and trunk (**hypaxial muscles**) and to the associated dermis.
- Cells that migrate posteriorly give rise to the intrinsic muscles of the back (**epaxial muscles**) and the associated dermis.

Developing nerve cells within anterior regions of the neural tube extend processes peripherally into posterior

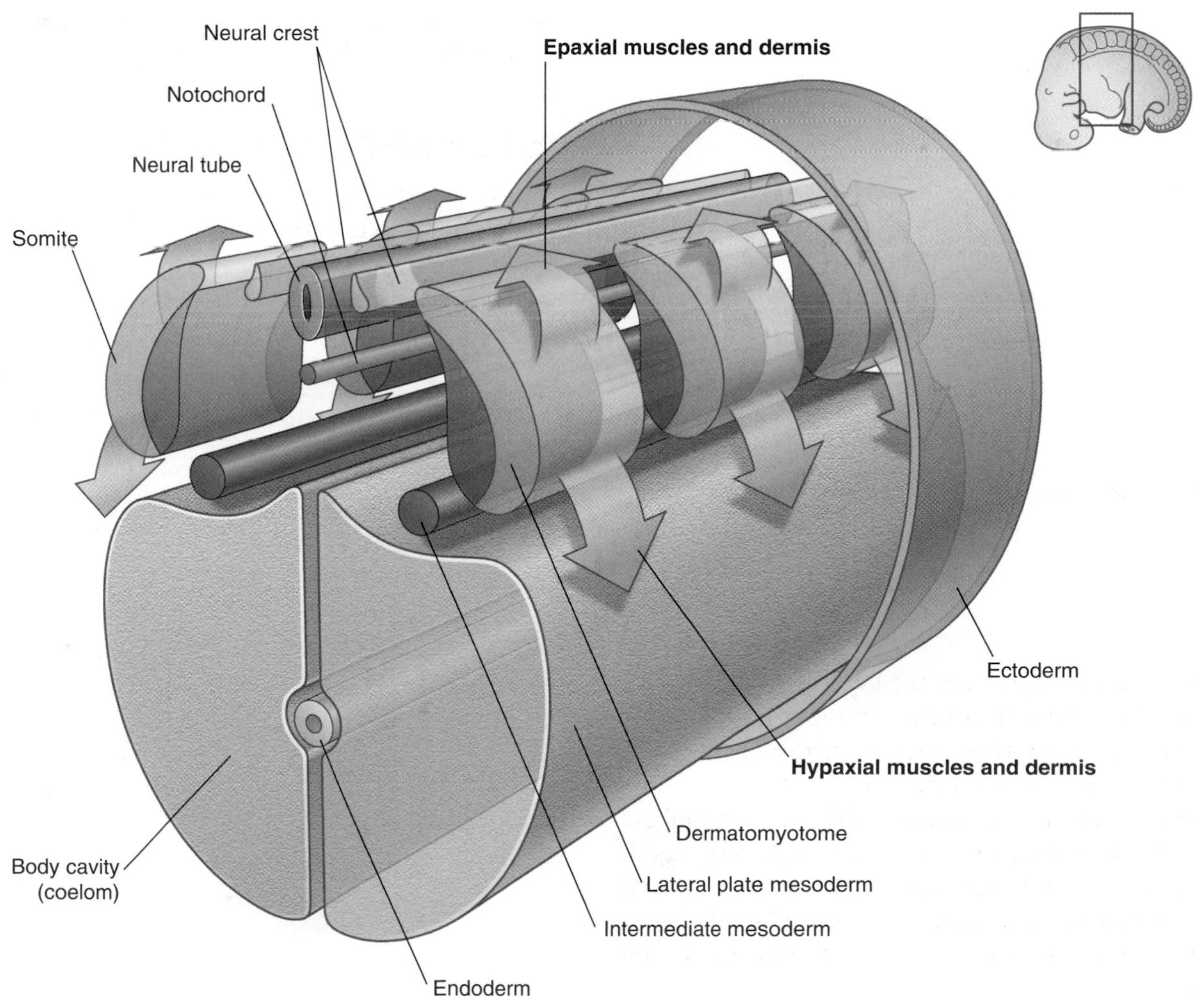


**Fig. 1.34** Differentiation of somites in a "tubular" embryo.

and anterior regions of the differentiating dermatomyotome of each somite.

Simultaneously, derivatives of neural crest cells (cells derived from neural folds during formation of the neural tube) differentiate into neurons on each side of the neural tube and extend processes both medially and laterally (Fig. 1.35):

- Medial processes pass into the posterior aspect of the neural tube.
- Lateral processes pass into the differentiating regions of the adjacent dermatomyotome.

Neurons that develop from cells within the spinal cord are **motor neurons** and those that develop from neural crest cells are **sensory neurons**.

Somatic sensory and somatic motor fibers that are organized segmentally along the neural tube become parts of all spinal nerves and some cranial nerves.

The clusters of sensory nerve cell bodies derived from neural crest cells and located outside the CNS form sensory ganglia.

Generally, all sensory information passes into the posterior aspect of the spinal cord, and all motor fibers leave anteriorly.

**Somatic sensory neurons** carry information from the periphery into the CNS and are also called **somatic sensory afferents** or **general somatic afferents (GSAs)**. The modalities carried by these nerves include temperature, pain, touch, and proprioception. Proprioception is the sense of determining the position and movement of the musculoskeletal system detected by special receptors in muscles and tendons.

**Somatic motor fibers** carry information away from the CNS to skeletal muscles and are also called **somatic motor efferents** or **general somatic efferents** (**GSEs**). Like somatic sensory fibers that come from the periphery, somatic motor fibers can be very long. They extend from cell bodies in the spinal cord to the muscle cells they innervate.

### Dermatomes

Because cells from a specific somite develop into the dermis of the skin in a precise location, somatic sensory fibers originally associated with that somite enter the posterior

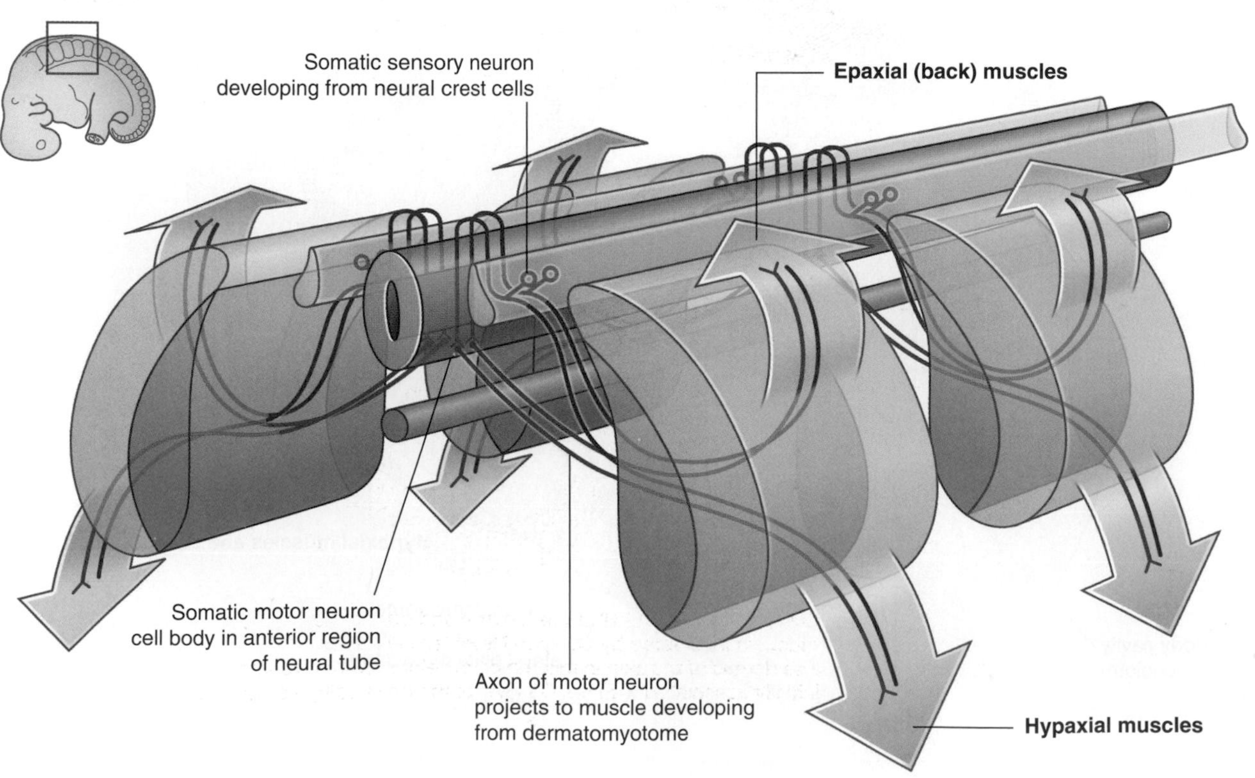


**Fig. 1.35** Somatic sensory and motor neurons. Blue lines indicate motor nerves and red lines indicate sensory nerves.

region of the spinal cord at a specific level and become part of one specific spinal nerve (Fig. 1.36). Each spinal nerve therefore carries somatic sensory information from a specific area of skin on the surface of the body. A **dermatome** is that area of skin supplied by a single spinal cord level, or on one side, by a single spinal nerve.

There is overlap in the distribution of dermatomes, but usually a specific region within each dermatome can be identified as an area supplied by a single spinal cord level. Testing touch in these autonomous zones in a conscious patient can be used to localize lesions to a specific spinal nerve or to a specific level in the spinal cord.

### Myotomes

Somatic motor nerves that were originally associated with a specific somite emerge from the anterior region of the spinal cord and, together with sensory nerves from the same level, become part of one spinal nerve. Therefore each spinal nerve carries somatic motor fibers to muscles that originally developed from the related somite. A **myotome** is that portion of a skeletal muscle innervated by a single spinal cord level or, on one side, by a single spinal nerve.

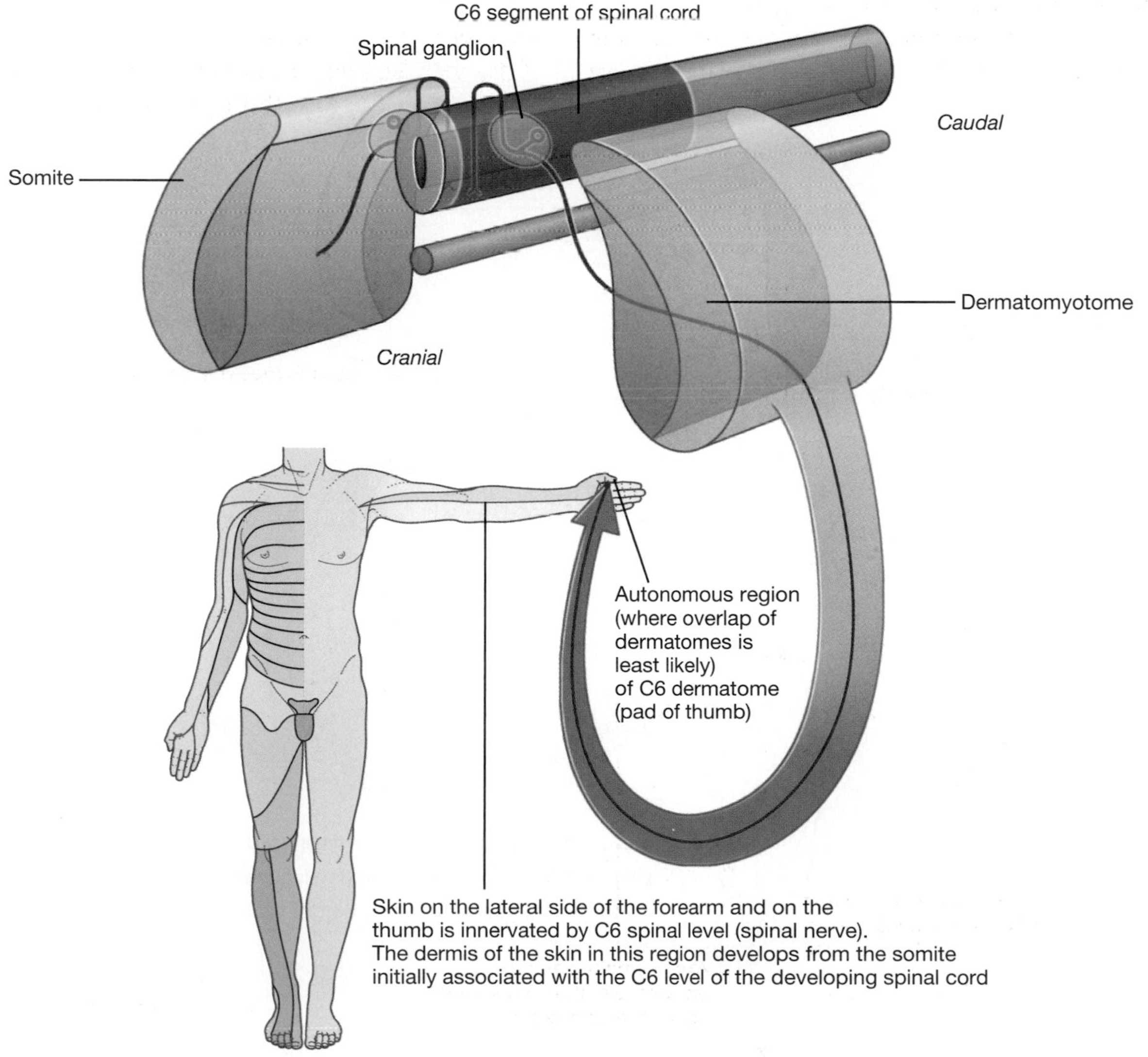
C6 segment of spinal cord
Spinal ganglion
Caudal
Somite
Dermatomyotome
Cranial
Autonomous region (where overlap of dermatomes is least likely) of C6 dermatome (pad of thumb)
Skin on the lateral side of the forearm and on the thumb is innervated by C6 spinal level (spinal nerve).
The dermis of the skin in this region develops from the somite initially associated with the C6 level of the developing spinal cord

**Fig. 1.36** Dermatomes.

Myotomes are generally more difficult to test than dermatomes because each skeletal muscle in the body often develops from more than one somite and is therefore innervated by nerves derived from more than one spinal cord level (Fig. 1.37).

Testing movements at successive joints can help in localizing lesions to specific nerves or to a specific spinal cord level. For example:

- Muscles that move the shoulder joint are innervated mainly by spinal nerves from spinal cord levels C5 and C6.
- Muscles that move the elbow are innervated mainly by spinal nerves from spinal cord levels C6 and C7.
- Muscles in the hand are innervated mainly by spinal nerves from spinal cord levels C8 and T1.

### Visceral part of the nervous system

The visceral part of the nervous system, as in the somatic part, consists of motor and sensory components:

- Sensory nerves monitor changes in the viscera.
- Motor nerves mainly innervate smooth muscle, cardiac muscle, and glands.

The visceral motor component is commonly referred to as the **autonomic division of the PNS** and is subdivided into **sympathetic** and **parasympathetic** parts.

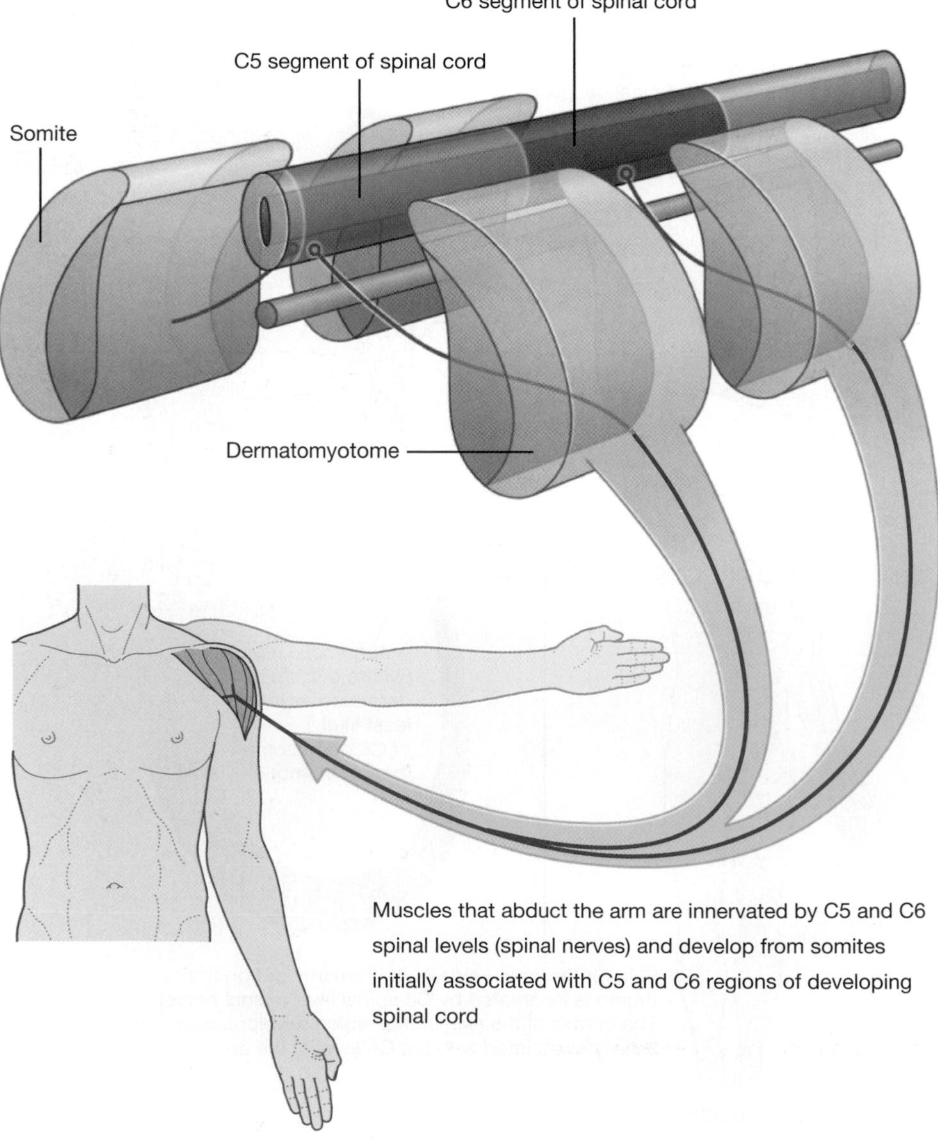
C6 segment of spinal cord
C5 segment of spinal cord
Somite
Dermatomyotome
Muscles that abduct the arm are innervated by C5 and C6 spinal levels (spinal nerves) and develop from somites initially associated with C5 and C6 regions of developing spinal cord

**Fig. 1.37** Myotomes.

## In the clinic

### Dermatomes and myotomes

A knowledge of dermatomes and myotomes is absolutely fundamental to carrying out a neurological examination. A typical dermatome map is shown in Fig. 1.38.

Clinically, a dermatome is that area of skin supplied by a single spinal nerve or spinal cord level. A myotome is that region of skeletal muscle innervated by a single spinal nerve or spinal cord level. Most individual muscles of the body are innervated by more than one spinal cord level, so the evaluation of myotomes is usually accomplished by testing movements of joints or muscle groups.

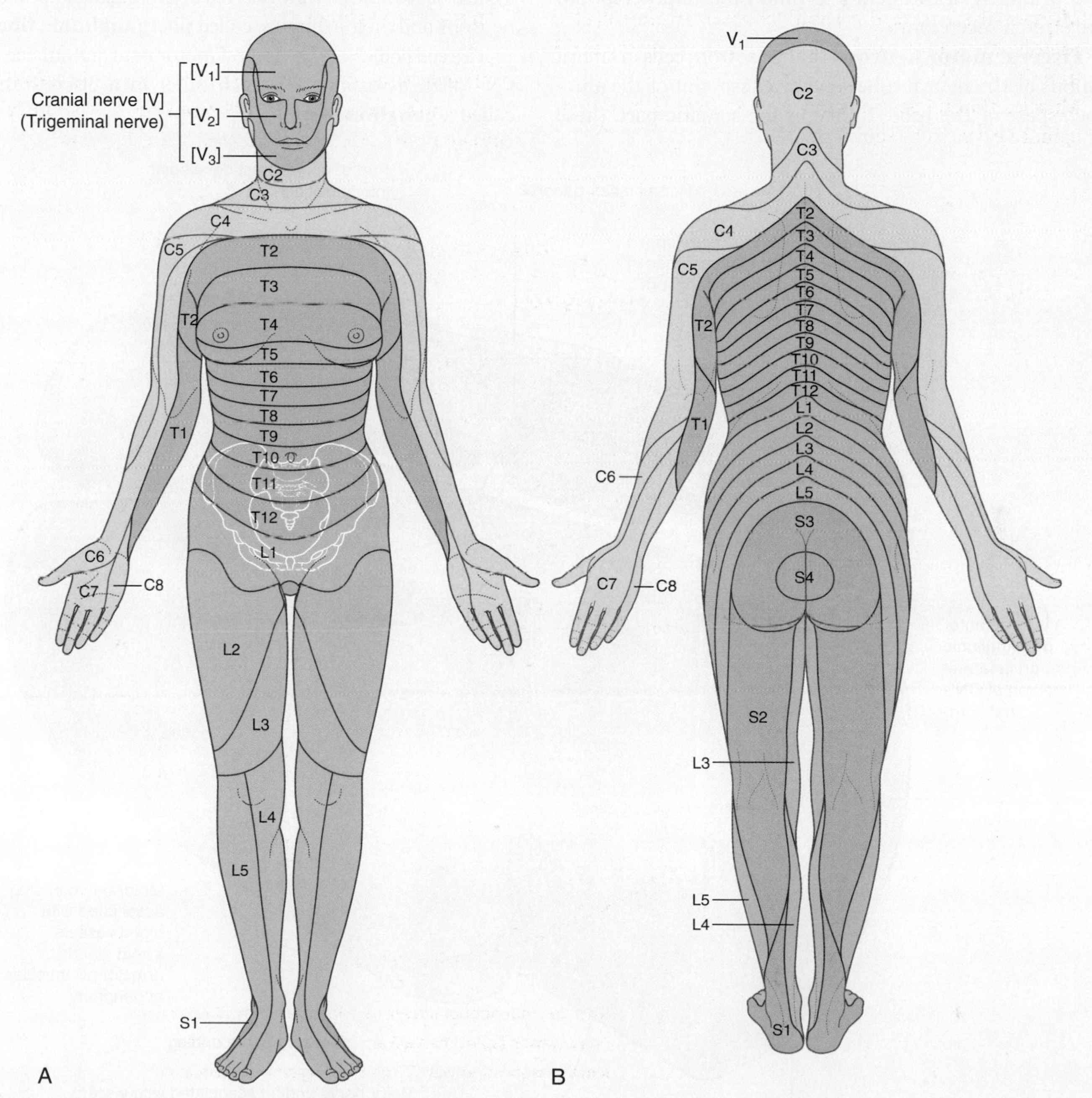


**Fig. 1.38** Dermatomes. **A.** Anterior view. **B.** Posterior view.

Like the somatic part of the nervous system, the visceral part is segmentally arranged and develops in a parallel fashion (Fig. 1.39).

**Visceral sensory neurons** that arise from neural crest cells send processes medially into the adjacent neural tube and laterally into regions associated with the developing body. These sensory neurons and their processes, referred to as **general visceral afferent fibers** (**GVAs**), are associated primarily with chemoreception, mechanoreception, and stretch reception.

**Visceral motor neurons** that arise from cells in lateral regions of the neural tube send processes out of the anterior aspect of the tube. Unlike in the somatic part, these processes, containing **general visceral efferent fibers (GVEs)**, synapse with other cells, usually other visceral motor neurons, that develop outside the CNS from neural crest cells that migrate away from their original positions close to the developing neural tube.

The visceral motor neurons located in the spinal cord are referred to as preganglionic motor neurons and their axons are called **preganglionic fibers**; the visceral motor neurons located outside the CNS are referred to as postganglionic motor neurons and their axons are called **postganglionic fibers**.

The cell bodies of the visceral motor neurons outside the CNS often associate with each other in a discrete mass called a **ganglion**.

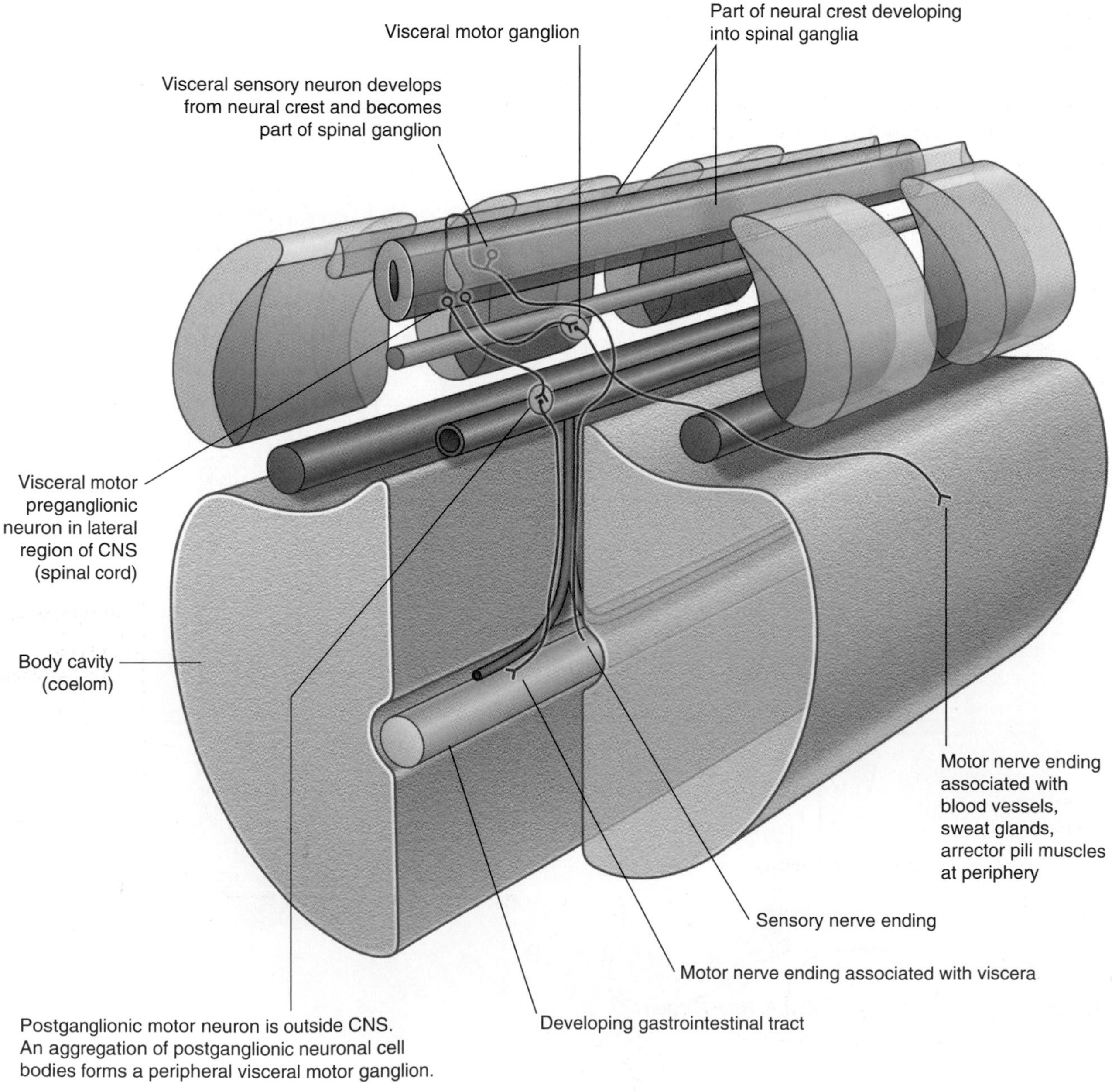

**Fig. 1.39** Development of the visceral part of the nervous system.

Visceral sensory and motor fibers enter and leave the CNS with their somatic equivalents (Fig. 1.40). Visceral sensory fibers enter the spinal cord together with somatic sensory fibers through posterior roots of spinal nerves. Preganglionic fibers of visceral motor neurons exit the spinal cord in the anterior roots of spinal nerves, along with fibers from somatic motor neurons.

Postganglionic fibers traveling to visceral elements in the periphery are found in the posterior and anterior rami (branches) of spinal nerves.

Visceral motor and sensory fibers that travel to and from viscera form named visceral branches that are separate from the somatic branches. These nerves generally form plexuses from which arise branches to the viscera.

Visceral motor and sensory fibers do not enter and leave the CNS at all levels (Fig. 1.41):

- In the cranial region, visceral components are associated with four of the twelve cranial nerves (CN III, VII, IX, and X).
- In the spinal cord, visceral components are associated mainly with spinal cord levels T1 to L2 and S2 to S4.

Visceral motor components associated with spinal levels T1 to L2 are termed **sympathetic**. Those visceral motor components in cranial and sacral regions, on either side of the sympathetic region, are termed **parasympathetic**:

- The sympathetic system innervates structures in peripheral regions of the body and viscera.
- The parasympathetic system is more restricted to innervation of the viscera only.

## Terminology

Spinal sympathetic and spinal parasympathetic neurons share certain developmental and phenotypic features that are different from those of cranial parasympathetic neurons. Based on this, some researchers have suggested reclassifying all spinal visceral motor neurons as sympathetic (Espinosa-Medina I et al. Science 2016;354:893-897). Others are against reclassification, arguing that the results only indicate that the neurons are spinal in origin

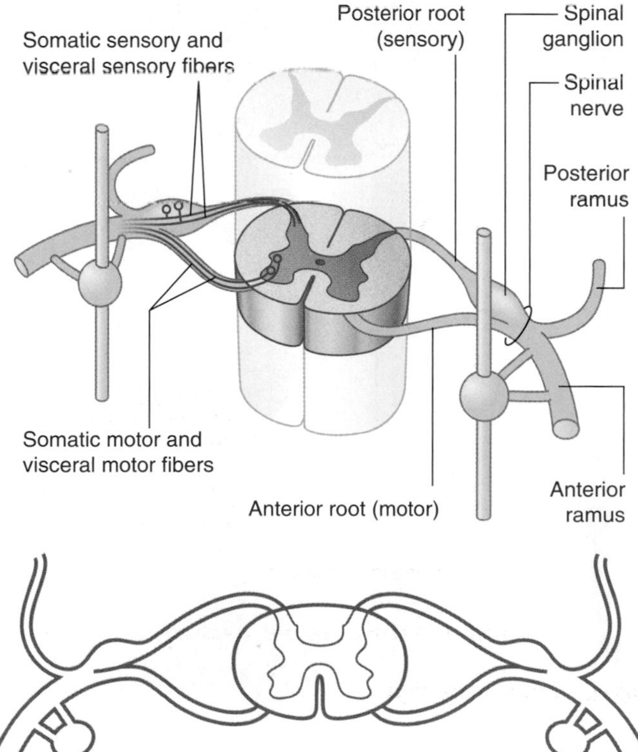


**Fig. 1.40** Basic anatomy of a thoracic spinal nerve.

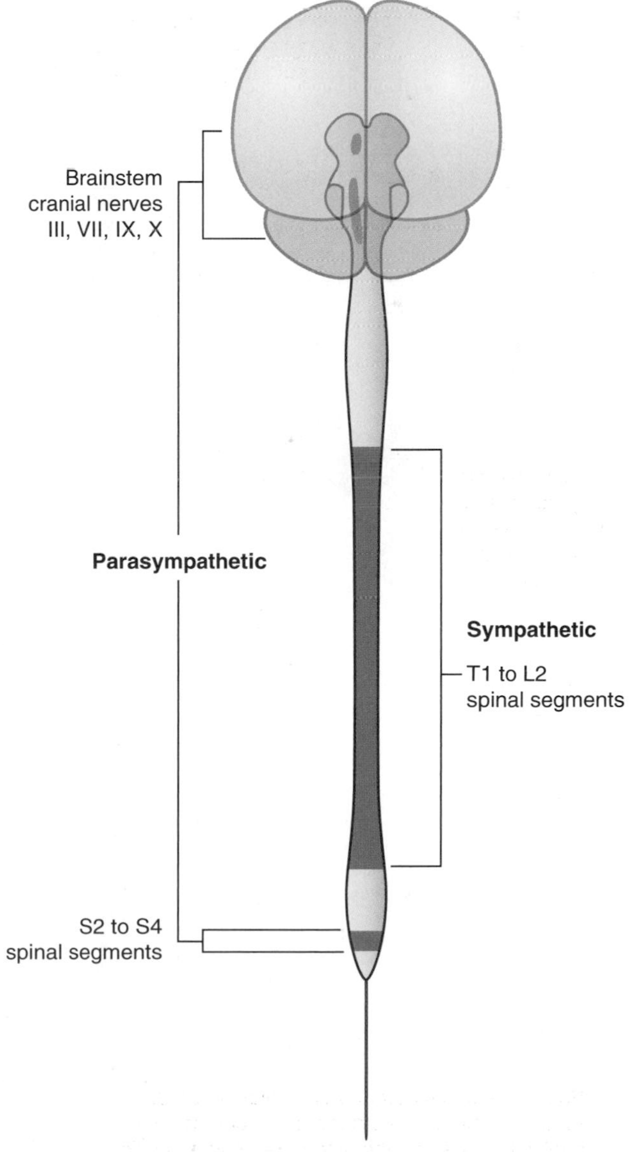


**Fig. 1.41** Parts of the CNS associated with visceral motor components.

(Neuhuber W et al. Anat Rec 2017;300:1369-1370). In addition, sacral nerves do not enter the sympathetic trunk, nor do they have postganglionic fibers that travel to the periphery on spinal nerves, as do T1-L2 visceral motor fibers. We have chosen to retain the classification of S2,3,4 visceral motor neurons as parasympathetic. "Parasympathetic" simply means on either side of the "sympathetic," which correctly describes their anatomy.

### Sympathetic system

The sympathetic part of the autonomic division of the PNS leaves thoracolumbar regions of the spinal cord with the somatic components of spinal nerves T1 to L2 (Fig. 1.42). On each side, a paravertebral sympathetic trunk extends from the base of the skull to the inferior end of the vertebral column where the two trunks converge anteriorly to the coccyx at the ganglion impar. Each trunk is attached

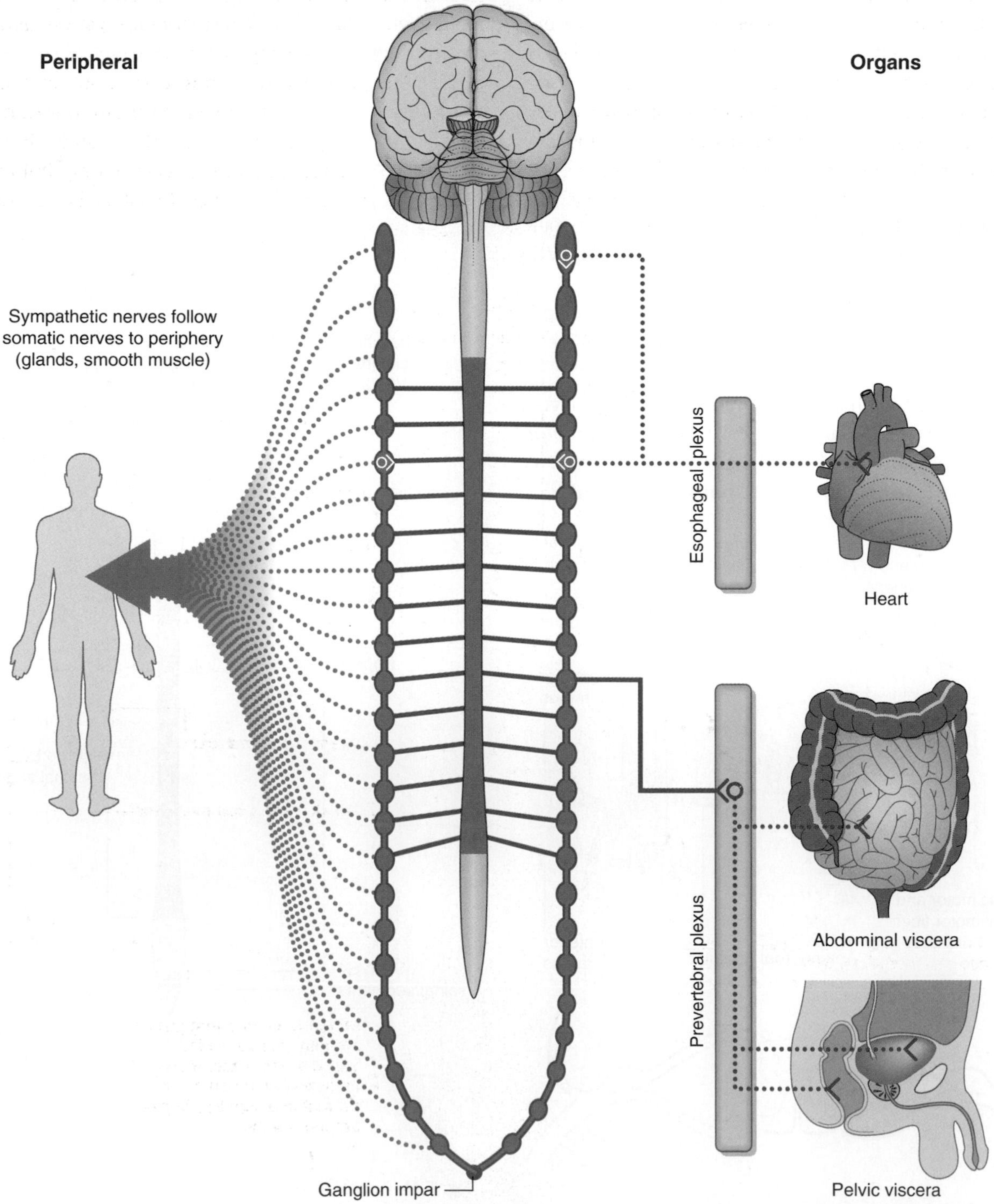

**Fig. 1.42** Sympathetic part of the autonomic division of the PNS.

to the anterior rami of spinal nerves and becomes the route by which sympathetics are distributed to the periphery and all viscera.

Visceral motor preganglionic fibers leave the T1 to L2 part of the spinal cord in anterior roots. The fibers then enter the spinal nerves, pass through the anterior rami and into the sympathetic trunks. One trunk is located on each side of the vertebral column (paravertebral) and positioned anterior to the anterior rami. Along the trunk is a series of segmentally arranged ganglia formed from collections of postganglionic neuronal cell bodies where the preganglionic neurons synapse with postganglionic neurons. Anterior rami of T1 to L2 are connected to the sympathetic trunk or to a ganglion by a **white ramus communicans,** which carries preganglionic sympathetic fibers and appears white because the fibers it contains are myelinated.

Preganglionic sympathetic fibers that enter a paravertebral ganglion or the sympathetic trunk through a white ramus communicans may take the following four pathways to target tissues:

### 1. Peripheral sympathetic innervation at the level of origin of the preganglionic fiber

Preganglionic sympathetic fibers may synapse with postganglionic motor neurons in ganglia associated with the sympathetic trunk, after which postganglionic fibers enter the same anterior ramus and are distributed with peripheral branches of the posterior and anterior rami of that spinal nerve (Fig. 1.43). The fibers innervate structures at the periphery of the body in regions supplied by the spinal nerve. The **gray ramus communicans** connects the sympathetic trunk or a ganglion to the anterior ramus and contains the postganglionic sympathetic fibers. It appears gray because postganglionic fibers are nonmyelinated. The gray ramus communicans is positioned medial to the white ramus communicans.

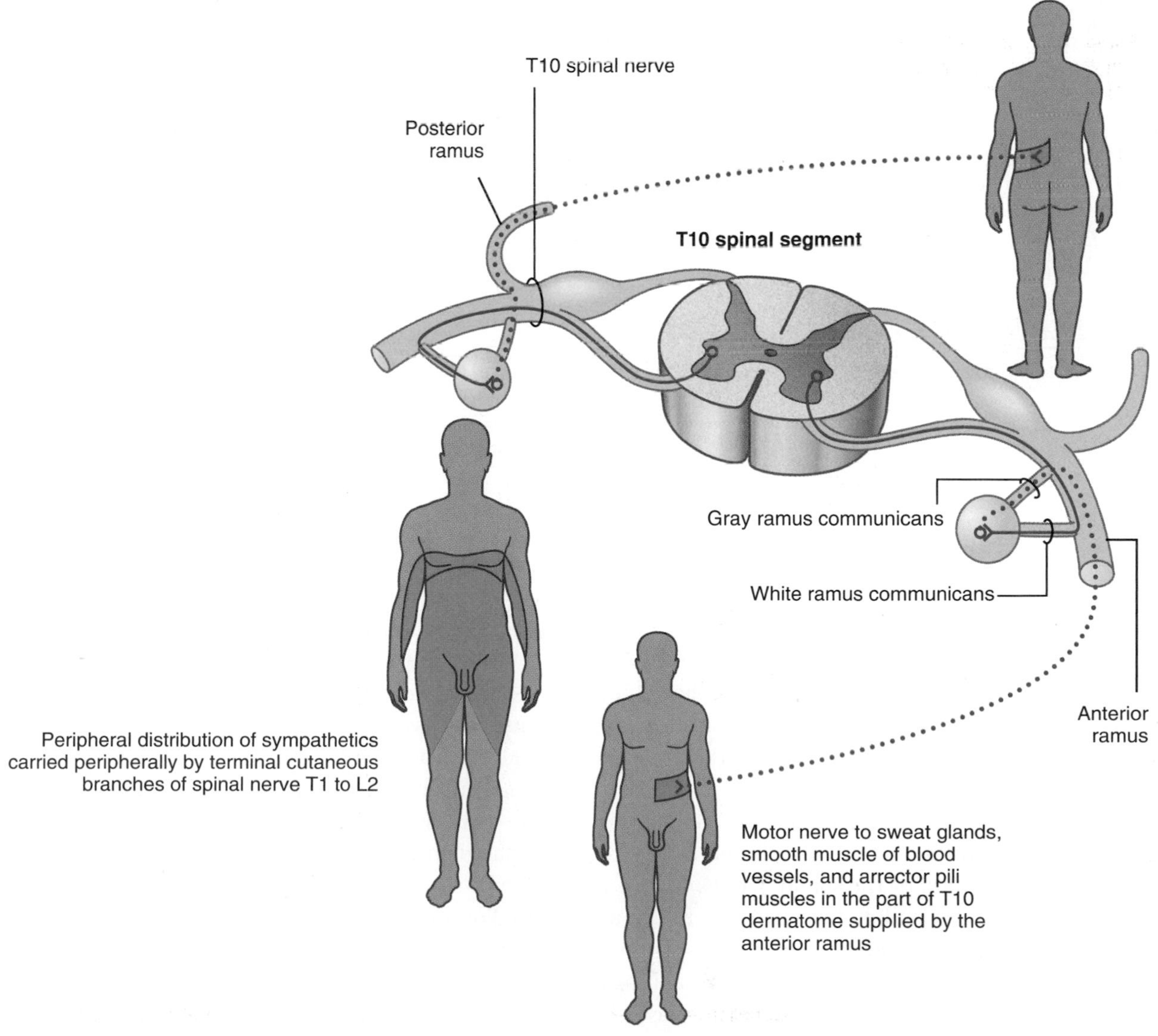


**Fig. 1.43** Course of sympathetic fibers that travel to the periphery in the same spinal nerves in which they travel out of the spinal cord.

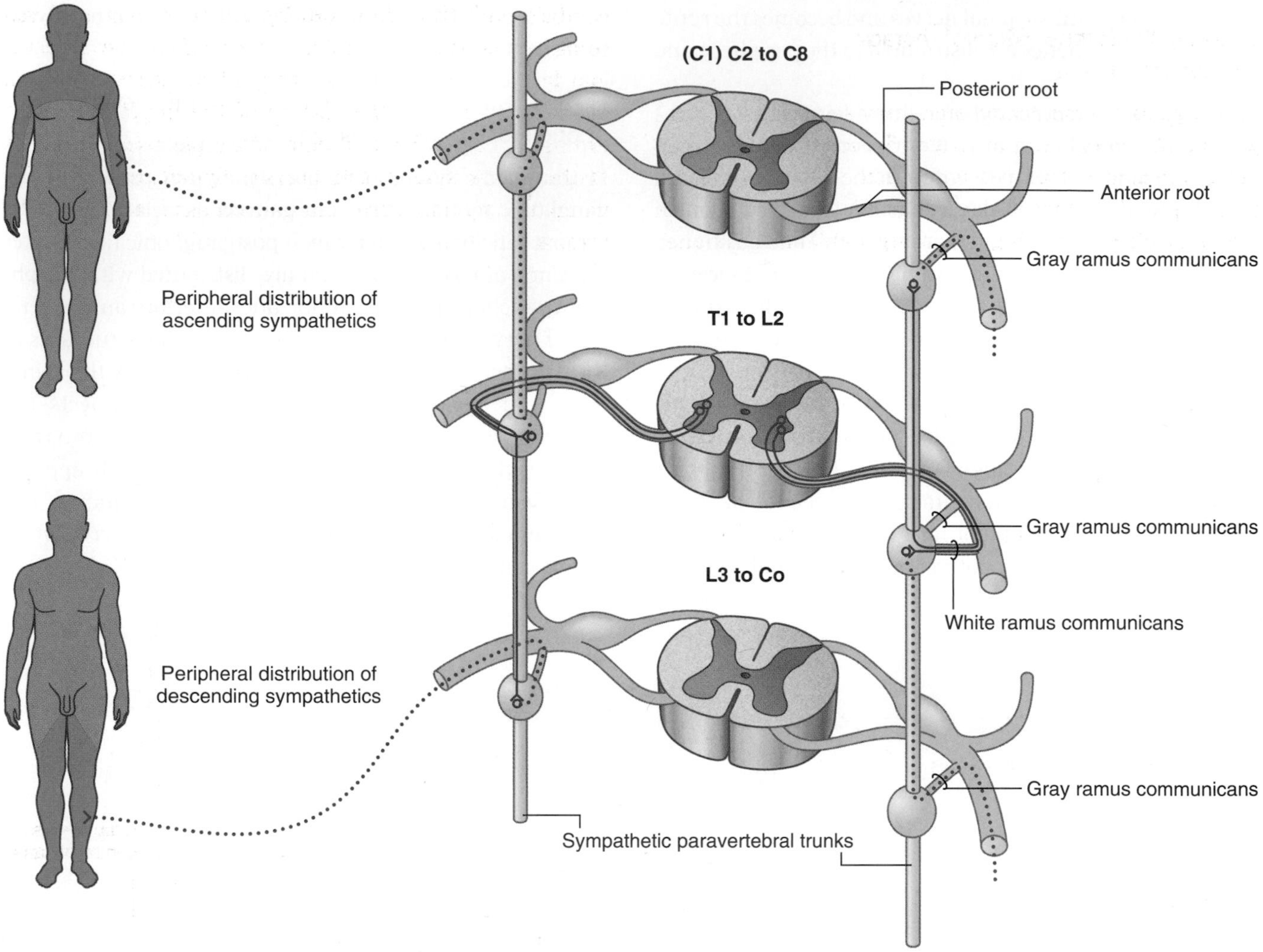

**Fig. 1.44** Course of sympathetic nerves that travel to the periphery in spinal nerves that are not the ones through which they left the spinal cord.

### 2. Peripheral sympathetic innervation above or below the level of origin of the preganglionic fiber

Preganglionic sympathetic fibers may ascend or descend to other vertebral levels where they synapse in ganglia associated with spinal nerves that may or may not have visceral motor input directly from the spinal cord (i.e., those nerves other than T1 to L2) (Fig. 1.44).

The postganglionic fibers leave the distant ganglia via gray rami communicantes and are distributed along the posterior and anterior rami of the spinal nerves.

The ascending and descending fibers, together with all the ganglia, form the **paravertebral sympathetic trunk**, which extends the entire length of the vertebral column. The formation of this trunk, on each side, enables visceral motor fibers of the sympathetic part of the autonomic division of the PNS, which ultimately emerge from only a small region of the spinal cord (T1 to L2), to be distributed to peripheral regions innervated by all spinal nerves.

White rami communicantes only occur in association with spinal nerves T1 to L2, whereas gray rami communicantes are associated with all spinal nerves.

Fibers from spinal cord levels T1 to T5 pass predominantly superiorly, whereas fibers from T5 to L2 pass inferiorly. All sympathetics passing into the head have preganglionic fibers that emerge from spinal cord level T1 and ascend in the sympathetic trunks to the highest ganglion in the neck (the **superior cervical ganglion**), where they synapse. Postganglionic fibers then travel along blood vessels to target tissues in the head, including blood vessels, sweat glands, small smooth muscles associated with the upper eyelids, and the dilator of the pupil.

### 3. Sympathetic innervation of thoracic and cervical viscera

Preganglionic sympathetic fibers may synapse with postganglionic motor neurons in ganglia and then leave the ganglia medially to innervate thoracic or cervical viscera (Fig. 1.45). They may ascend in the trunk before synapsing, and after synapsing the postganglionic fibers may combine with those from other levels to form named visceral nerves, such as cardiac nerves. Often, these nerves join branches from the parasympathetic system to form plexuses on or near the surface of the target organ, for example, the cardiac and pulmonary plexuses. Branches of the plexus innervate the organ. Spinal cord levels T1 to T5 mainly innervate cranial, cervical, and thoracic viscera.

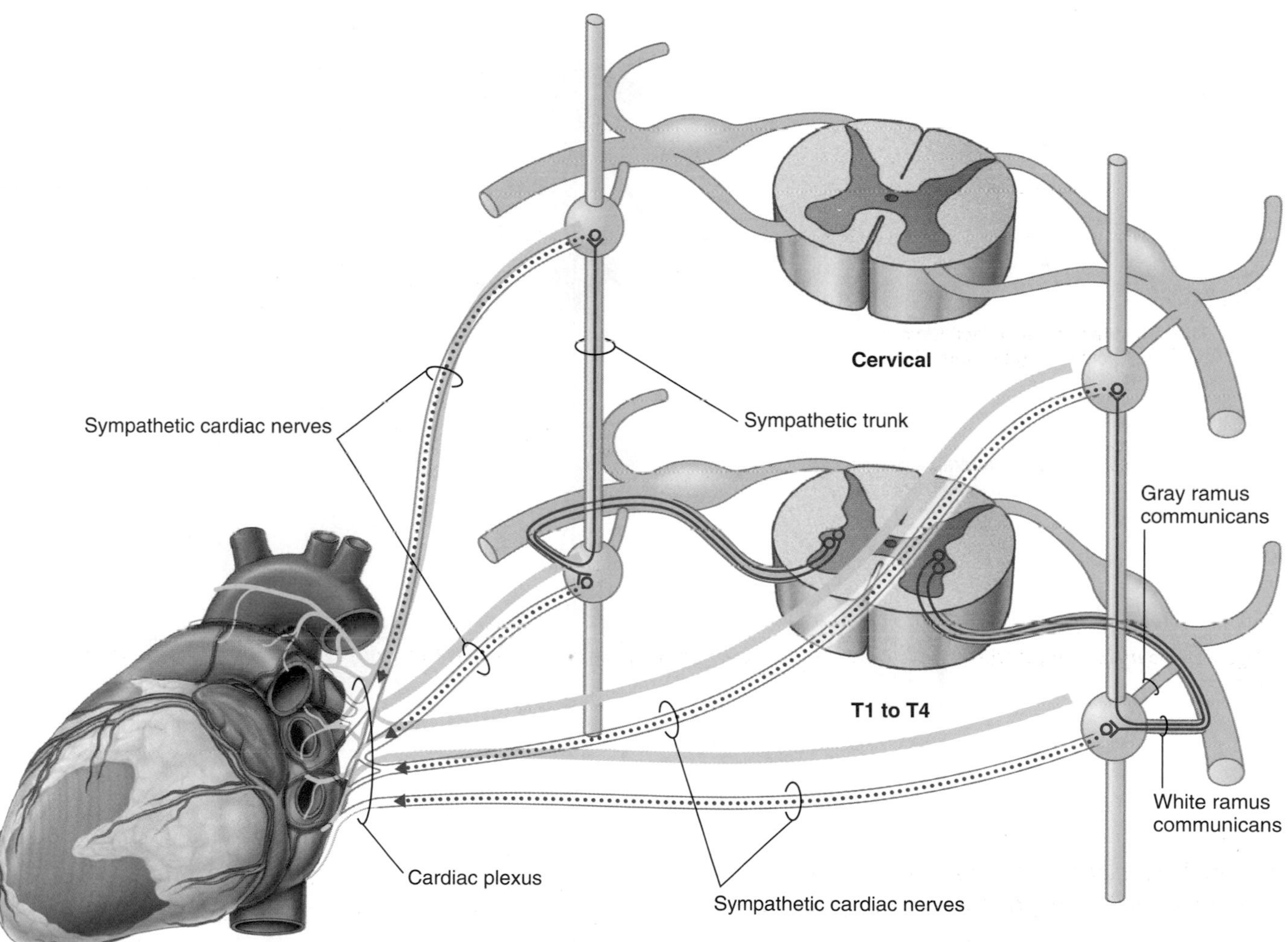


**Fig. 1.45** Course of sympathetic nerves traveling to the heart.

### 4. Sympathetic innervation of the abdomen and pelvic regions and the adrenals

Preganglionic sympathetic fibers may pass through the sympathetic trunk and paravertebral ganglia without synapsing and, together with similar fibers from other levels, form **splanchnic nerves** (**greater, lesser, least, lumbar**, and **sacral**), which pass into the abdomen and pelvic regions (Fig. 1.46). The preganglionic fibers in these nerves are derived from spinal cord levels T5 to L2.

The splanchnic nerves generally connect with sympathetic ganglia around the roots of major arteries that branch from the abdominal aorta. These ganglia are part of a large prevertebral plexus that also has input from the

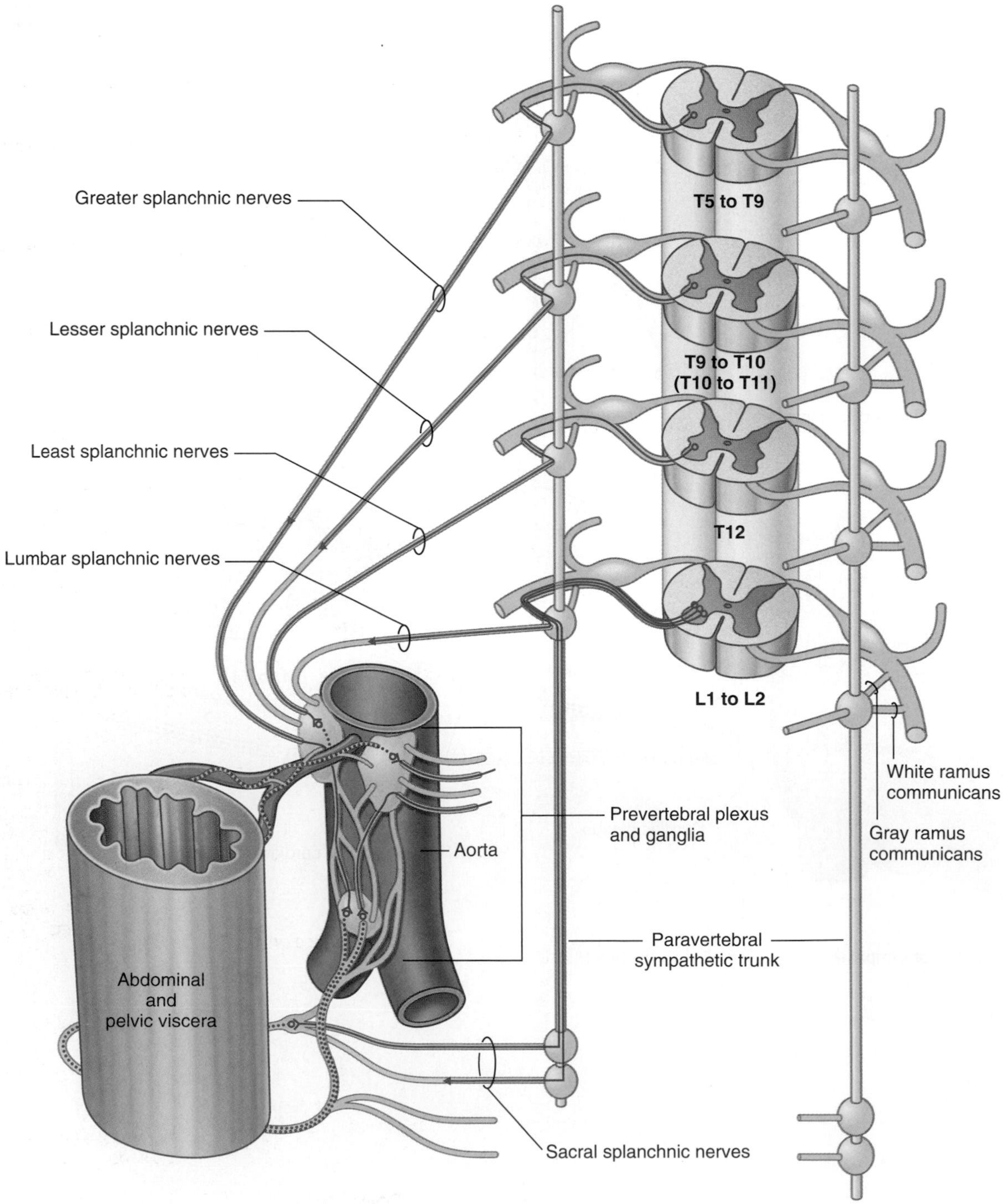

**Fig. 1.46** Course of sympathetic nerves traveling to abdominal and pelvic viscera.

parasympathetic part of the autonomic division of the PNS. Postganglionic sympathetic fibers are distributed in extensions of this plexus, predominantly along arteries, to viscera in the abdomen and pelvis.

Some of the preganglionic fibers in the prevertebral plexus do not synapse in the sympathetic ganglia of the plexus but pass through the system to the adrenal gland, where they synapse directly with cells of the adrenal medulla. These cells are homologues of sympathetic postganglionic neurons and secrete adrenaline and noradrenaline into the vascular system.

### Parasympathetic system

The parasympathetic part of the autonomic division of the PNS (Fig. 1.47) leaves cranial and sacral regions of the CNS in association with:

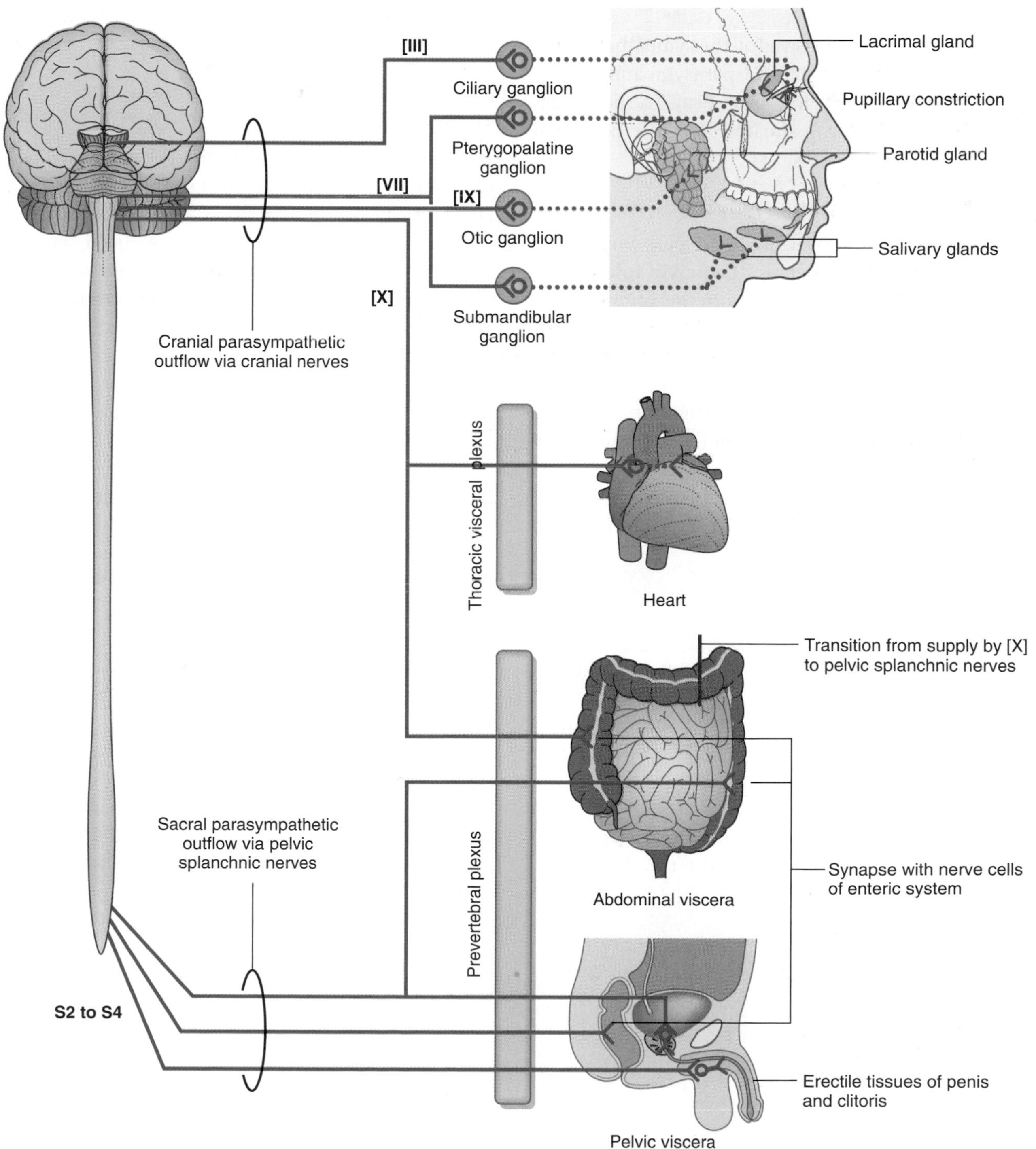


**Fig. 1.47** Parasympathetic part of the autonomic division of the PNS.

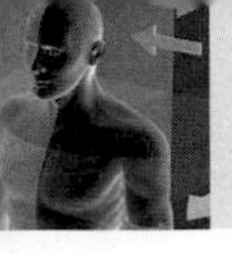



- cranial nerves III, VII, IX, and X: III, VII, and IX carry parasympathetic fibers to structures within the head and neck only, whereas X (the vagus nerve) also innervates thoracic and most abdominal viscera; and
- spinal nerves S2 to S4: sacral parasympathetic fibers innervate inferior abdominal viscera, pelvic viscera, and the arteries associated with erectile tissues of the perineum.

Like the visceral motor nerves of the sympathetic part, the visceral motor nerves of the parasympathetic part generally have two neurons in the pathway. The preganglionic neurons are in the CNS, and fibers leave in the cranial nerves.

### Sacral preganglionic parasympathetic fibers

In the sacral region, the preganglionic parasympathetic fibers form special visceral nerves (the **pelvic splanchnic nerves**), which originate from the anterior rami of S2 to S4 and enter pelvic extensions of the large prevertebral plexus formed around the abdominal aorta. These fibers are distributed to pelvic and abdominal viscera mainly along blood vessels. The postganglionic motor neurons are in the walls of the viscera. In organs of the gastrointestinal system, preganglionic fibers do not have a postganglionic parasympathetic motor neuron in the pathway; instead, preganglionic fibers synapse directly on neurons in the ganglia of the enteric system.

### Cranial nerve preganglionic parasympathetic fibers

The preganglionic parasympathetic motor fibers in CN III, VII, and IX separate from the nerves and connect with one of four distinct ganglia, which house postganglionic motor neurons. These four ganglia are near major branches of CN V. Postganglionic fibers leave the ganglia, join the branches of CN V, and are carried to target tissues (salivary, mucous, and lacrimal glands; constrictor muscle of the pupil; and ciliary muscle in the eye) with these branches.

The vagus nerve [X] gives rise to visceral branches along its course. These branches contribute to plexuses associated with thoracic viscera or to the large prevertebral plexus in the abdomen and pelvis. Many of these plexuses also contain sympathetic fibers.

When present, postganglionic parasympathetic neurons are in the walls of the target viscera.

### Visceral sensory innervation (visceral afferents)

Visceral sensory fibers generally accompany visceral motor fibers.

### Visceral sensory fibers accompany sympathetic fibers

Visceral sensory fibers follow the course of sympathetic fibers entering the spinal cord at similar spinal cord levels. However, visceral sensory fibers may also enter the spinal cord at levels other than those associated with motor output. For example, visceral sensory fibers from the heart may enter at levels higher than spinal cord level T1. Visceral sensory fibers that accompany sympathetic fibers are mainly concerned with detecting pain.

### Visceral sensory fibers accompany parasympathetic fibers

Visceral sensory fibers accompanying parasympathetic fibers are carried mainly in IX and X and in spinal nerves S2 to S4.

Visceral sensory fibers in IX carry information from chemoreceptors and baroreceptors associated with the walls of major arteries in the neck, and from receptors in the pharynx.

Visceral sensory fibers in X include those from cervical viscera, and major vessels and viscera in the thorax and abdomen.

Visceral sensory fibers from pelvic viscera and the distal parts of the colon are carried in S2 to S4.

Visceral sensory fibers associated with parasympathetic fibers primarily relay information to the CNS about the status of normal physiological processes and reflex activities.

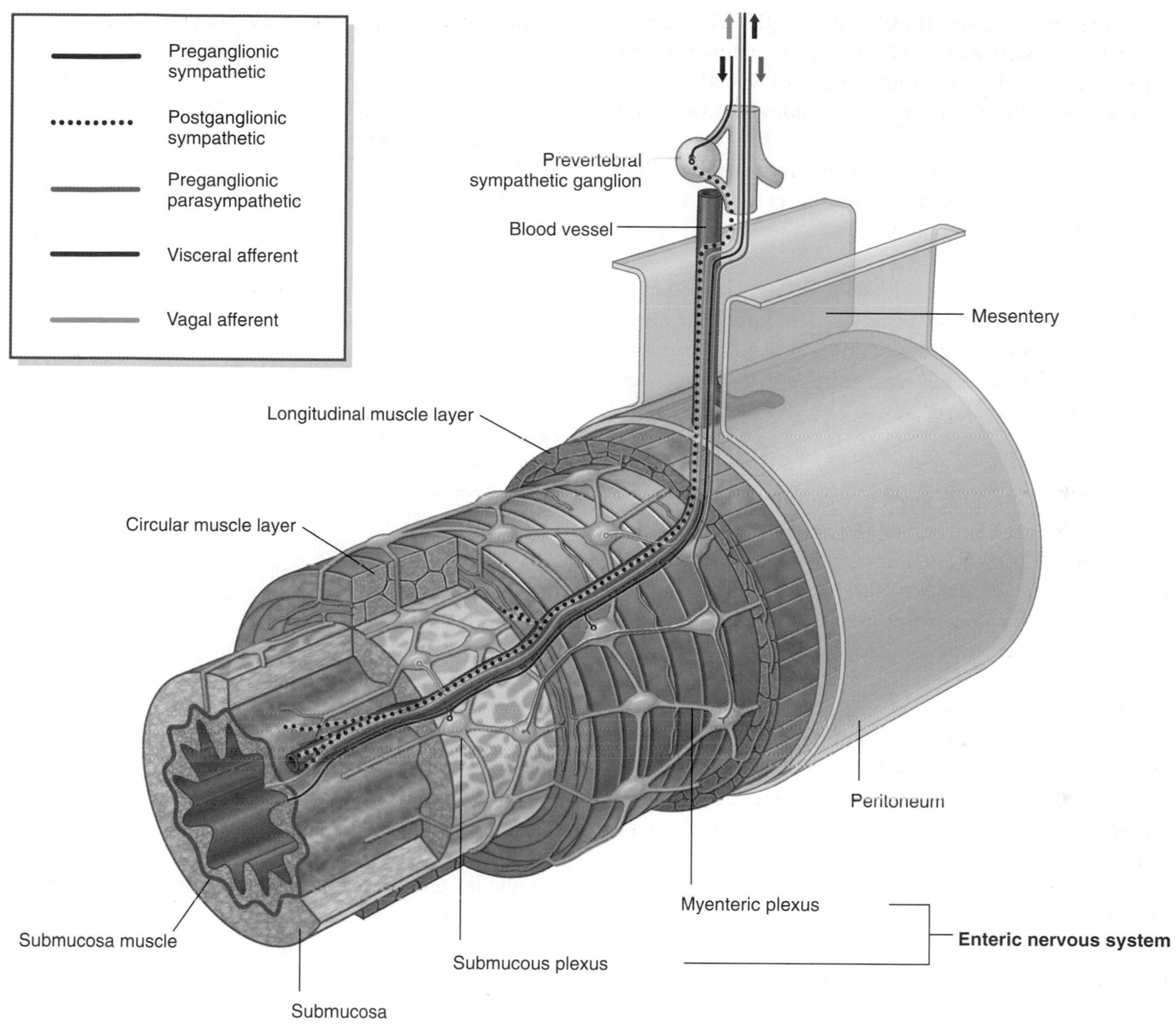


**Fig. 1.48** Enteric part of the nervous system.

## The enteric system

The enteric nervous system consists of motor and sensory neurons and their support cells, which form two interconnected plexuses, the **myenteric** and **submucous nerve plexuses**, within the walls of the gastrointestinal tract (Fig. 1.48). Each of these plexuses is formed by:

- ganglia, which house the nerve cell bodies and associated cells, and
- bundles of nerve fibers, which pass between ganglia and from the ganglia into surrounding tissues.

Neurons in the enteric system are derived from neural crest cells originally associated with occipitocervical and sacral regions. Interestingly, more neurons are reported to be in the enteric system than in the spinal cord itself.

Sensory and motor neurons within the enteric system control reflex activity within and between parts of the

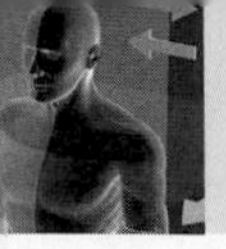



gastrointestinal system. These reflexes regulate peristalsis, secretomotor activity, and vascular tone. These activities can occur independently of the brain and spinal cord, but can also be modified by input from preganglionic parasympathetic and postganglionic sympathetic fibers.

Sensory information from the enteric system is carried back to the CNS by visceral sensory fibers.

## Nerve plexuses

Nerve plexuses are either somatic or visceral and combine fibers from different sources or levels to form new nerves with specific targets or destinations (Fig. 1.49). Plexuses of the enteric system also generate reflex activity independent of the CNS.

### Somatic plexuses

Major somatic plexuses formed from the anterior rami of spinal nerves are the cervical (C1 to C4), brachial (C5 to T1), lumbar (L1 to L4), sacral (L4 to S4), and coccygeal (S5 to Co) plexuses. Except for spinal nerve T1, the anterior rami of thoracic spinal nerves remain independent and do not participate in plexuses.

### Visceral plexuses

Visceral nerve plexuses are formed in association with viscera and generally contain efferent (sympathetic and parasympathetic) and afferent components (Fig. 1.49). These plexuses include cardiac and pulmonary plexuses in the thorax and a large prevertebral plexus in the abdomen anterior to the aorta, which extends inferiorly onto the lateral walls of the pelvis. The massive prevertebral plexus supplies input to and receives output from all abdominal and pelvic viscera.

## In the clinic

**Referred pain**

Referred pain occurs when sensory information comes to the spinal cord from one location but is interpreted by the CNS as coming from another location innervated by the same spinal cord level. Usually, this happens when the pain information comes from a region, such as the gut, which has a low amount of sensory output. These afferents converge on neurons at the same spinal cord level that receive information from the skin, which is an area with a high amount of sensory output. As a result, pain from the normally low output region is interpreted as coming from the normally high output region.

Pain is most often referred from a region innervated by the visceral part of the nervous system to a region innervated, at the same spinal cord level, by the somatic side of the nervous system.

Pain can also be referred from one somatic region to another. For example, irritation of the peritoneum on the inferior surface of the diaphragm, which is innervated by the phrenic nerve, can be referred to the skin on the top of the shoulder, which is innervated by other somatic nerves arising at the same spinal cord level.

## OTHER SYSTEMS

Specific information about the organization and components of the respiratory, gastrointestinal, and urogenital systems will be discussed in each of the succeeding chapters of this text.

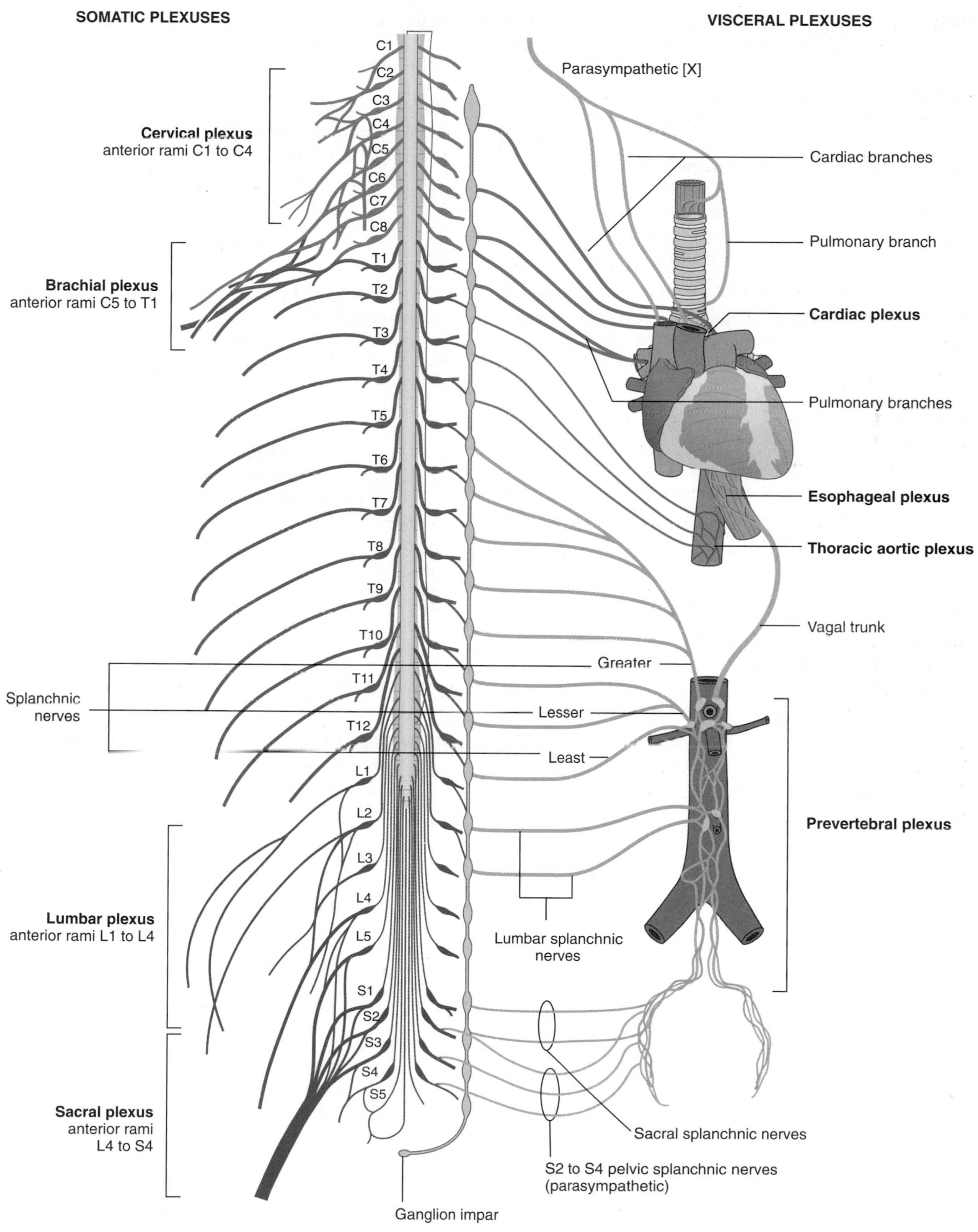
SOMATIC PLEXUSES
VISCERAL PLEXUSES
C1
C2
C3
C4
C5
C6
C7
C8
T1
T2
T3
T4
T5
T6
T7
T8
T9
T10
T11
T12
L1
L2
L3
L4
L5
S1
S2
S3
S4
S5
Cervical plexus
anterior rami C1 to C4
Brachial plexus
anterior rami C5 to T1
Splanchnic nerves
Lumbar plexus
anterior rami L1 to L4
Sacral plexus
anterior rami
L4 to S4
Parasympathetic [X]
Cardiac branches
Pulmonary branch
Cardiac plexus
Pulmonary branches
Esophageal plexus
Thoracic aortic plexus
Vagal trunk
Greater
Lesser
Least
Prevertebral plexus
Lumbar splanchnic nerves
Sacral splanchnic nerves
S2 to S4 pelvic splanchnic nerves
(parasympathetic)
Ganglion impar

**Fig. 1.49** Nerve plexuses.

# Clinical cases

## Case 1

APPENDICITIS

**A young man sought medical care because of central abdominal pain that was diffuse and colicky. After some hours, the pain began to localize in the right iliac fossa and became constant. He was referred to an abdominal surgeon, who removed a grossly inflamed appendix. The patient made an uneventful recovery.**

When the appendix becomes inflamed, the visceral sensory fibers are stimulated. These fibers enter the spinal cord with the sympathetic fibers at spinal cord level T10. The pain is referred to the dermatome of T10, which is in the umbilical region (Fig. 1.50). The pain is diffuse, not focal; every time a peristaltic wave passes through the ileocecal region, the pain recurs. This intermittent type of pain is referred to as colic.

In the later stages of the disease, the appendix contacts and irritates the parietal peritoneum in the right iliac fossa, which is innervated by somatic sensory nerves. This produces a constant focal pain, which predominates over the colicky pain that the patient felt some hours previously. The patient no longer interprets the referred pain from the T10 dermatome.

Although this is a typical history for appendicitis, it should always be borne in mind that the patient's symptoms and signs may vary. The appendix is situated in a retrocecal position in approximately 70% of patients; therefore it may never contact the parietal peritoneum anteriorly in the right iliac fossa. It is also possible that the appendix is long and may directly contact other structures. As a consequence, the patient may have other symptoms (e.g., the appendix may contact the ureter, and the patient may then develop urological symptoms).

Although appendicitis is common, other disorders, for example of the bowel and pelvis, may produce similar symptoms.

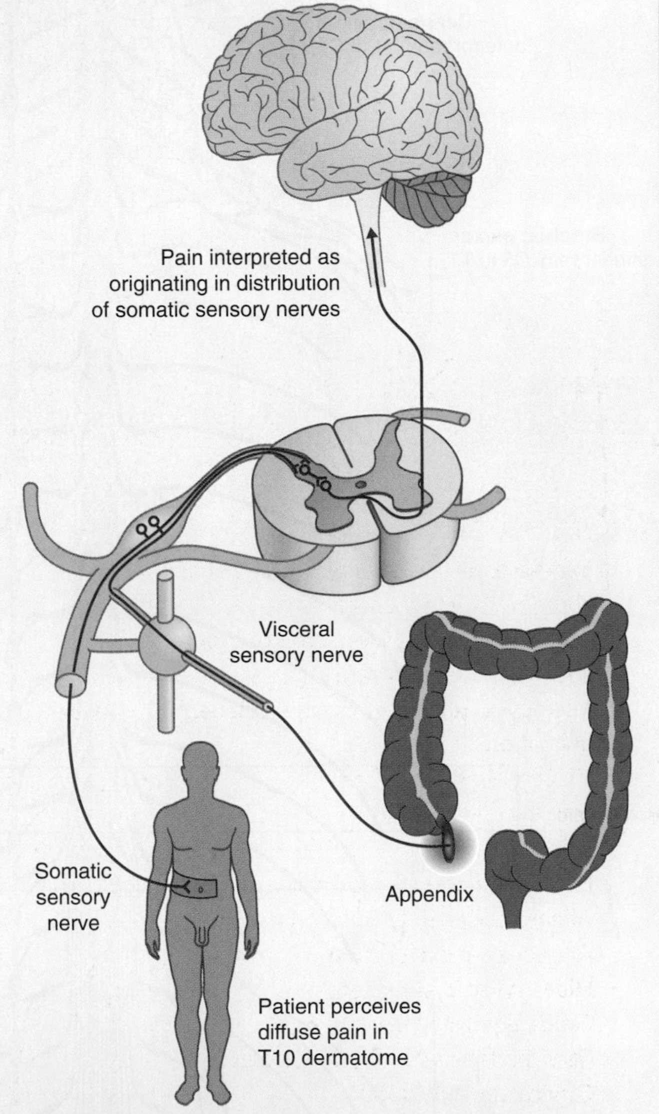


**Fig. 1.50** Mechanism for referred pain from an inflamed appendix to the T10 dermatome.

# 2

# Back



**ADDITIONAL LEARNING RESOURCES for Chapter 2, Back, on STUDENT CONSULT (www.studentconsult.com):**

- Self-Assessment—National Board style multiple-choice questions, Chapter 2
- Short Questions—these are questions requiring short responses, Chapter 2
- Interactive Surface Anatomy—interactive surface animations, Chapter 2
- PT Case Studies, Chapter 2
  Low back instability
  Stenosis
  Herniated nucleus pulposus
  Atlanto-occipital dysfunction
  Atlanto-axial dysfunction
  Mid-cervical dysfunction
  Cauda equina syndrome
  Cervical degenerative joint disease
  Cervical radiculopathy
- Clinical Cases, Chapter 2
  Ankylosing spondylitis
  Atlas fracture
  Cervical facet syndrome
  Dissecting thoracic aneurysm
  Lumbar puncture
  Psoas abscess
  Sacral tumor
  Spinal cord infarction

**FREE Online Anatomy and Embryology Self-Study Course**

Anatomy modules 23 through 25

# Conceptual overview

## GENERAL DESCRIPTION

The back consists of the posterior aspect of the body and provides the musculoskeletal axis of support for the trunk. Bony elements consist mainly of the vertebrae, although proximal elements of the ribs, superior aspects of the pelvic bones, and posterior basal regions of the skull contribute to the back's skeletal framework (Fig. 2.1).

Associated muscles interconnect the vertebrae and ribs with each other and with the pelvis and skull. The back contains the spinal cord and proximal parts of the spinal nerves, which send and receive information to and from most of the body.

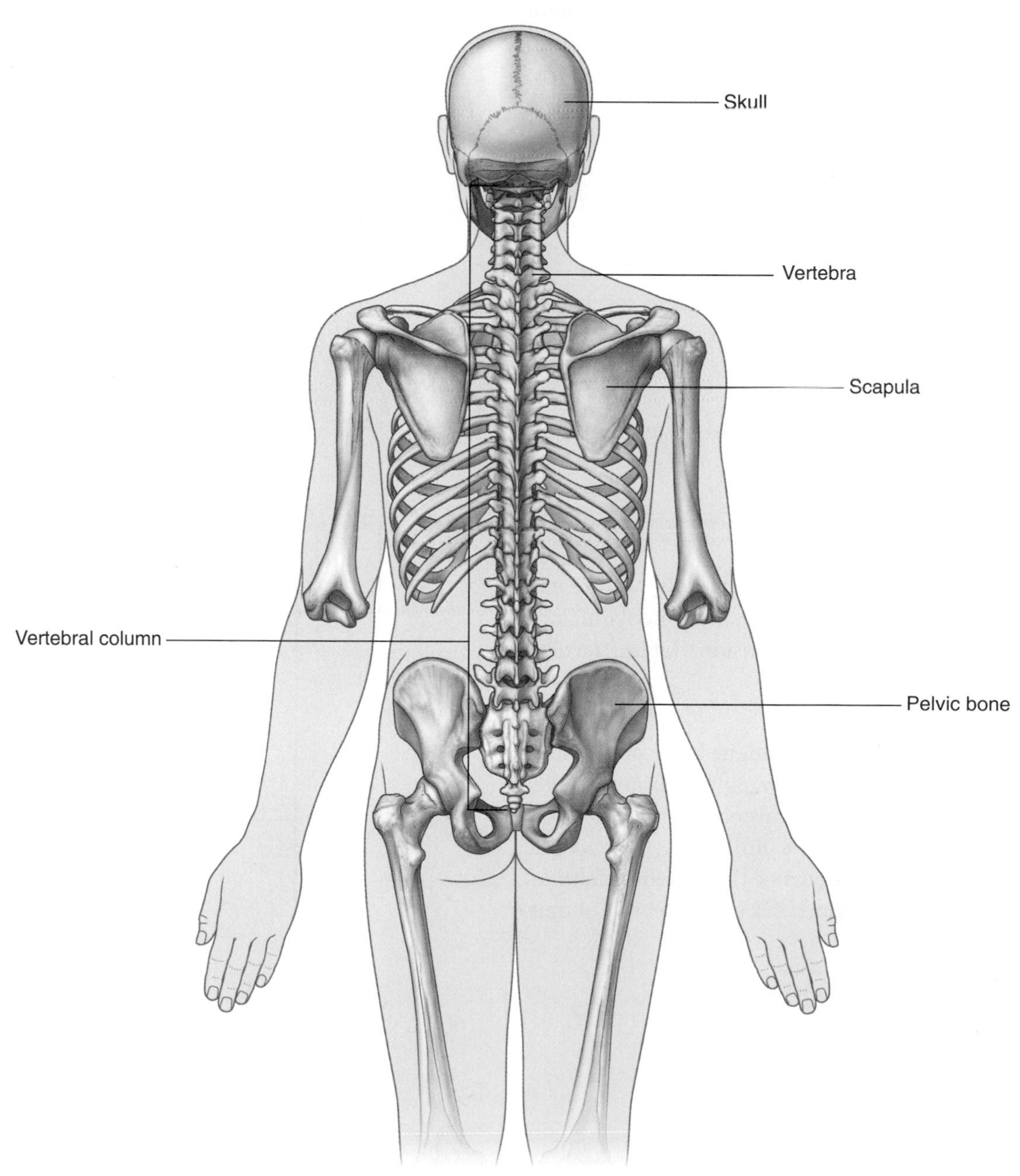


**Fig. 2.1** Skeletal framework of the back.

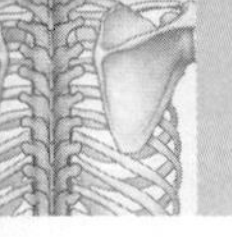



## FUNCTIONS

### Support

The skeletal and muscular elements of the back support the body's weight, transmit forces through the pelvis to the lower limbs, carry and position the head, and brace and help maneuver the upper limbs. The vertebral column is positioned posteriorly in the body at the midline. When viewed laterally, it has a number of curvatures (Fig. 2.2):

- The primary curvature of the vertebral column is concave anteriorly, reflecting the original shape of the embryo, and is retained in the thoracic and sacral regions in adults.
- Secondary curvatures, which are concave posteriorly, form in the cervical and lumbar regions and bring the center of gravity into a vertical line, which allows the body's weight to be balanced on the vertebral column in a way that expends the least amount of muscular energy to maintain an upright bipedal stance.

As stresses on the back increase from the cervical to lumbar regions, lower back problems are common.

### Movement

Muscles of the back consist of extrinsic and intrinsic groups:

- The extrinsic muscles of the back move the upper limbs and the ribs.
- The intrinsic muscles of the back maintain posture and move the vertebral column; these movements include flexion (anterior bending), extension, lateral flexion, and rotation (Fig. 2.3).

Although the amount of movement between any two vertebrae is limited, the effects between vertebrae are additive along the length of the vertebral column. Also, freedom of movement and extension are limited in the thoracic region relative to the lumbar part of the vertebral column. Muscles in more anterior regions flex the vertebral column.

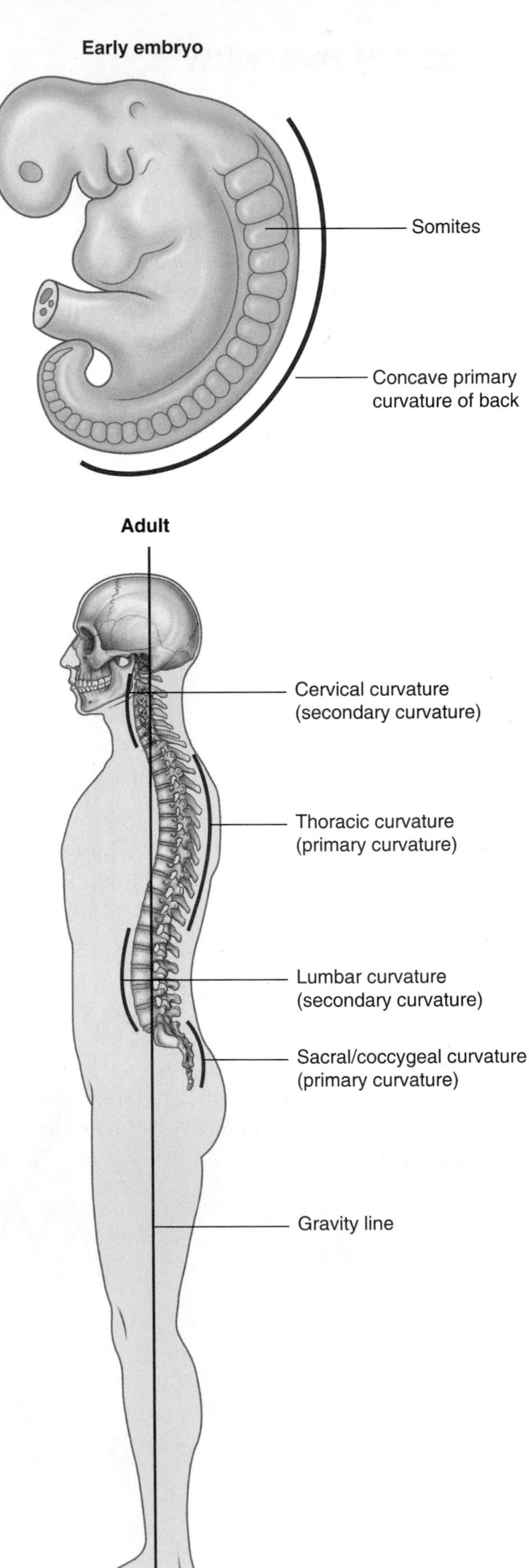


**Fig. 2.2** Curvatures of the vertebral column.

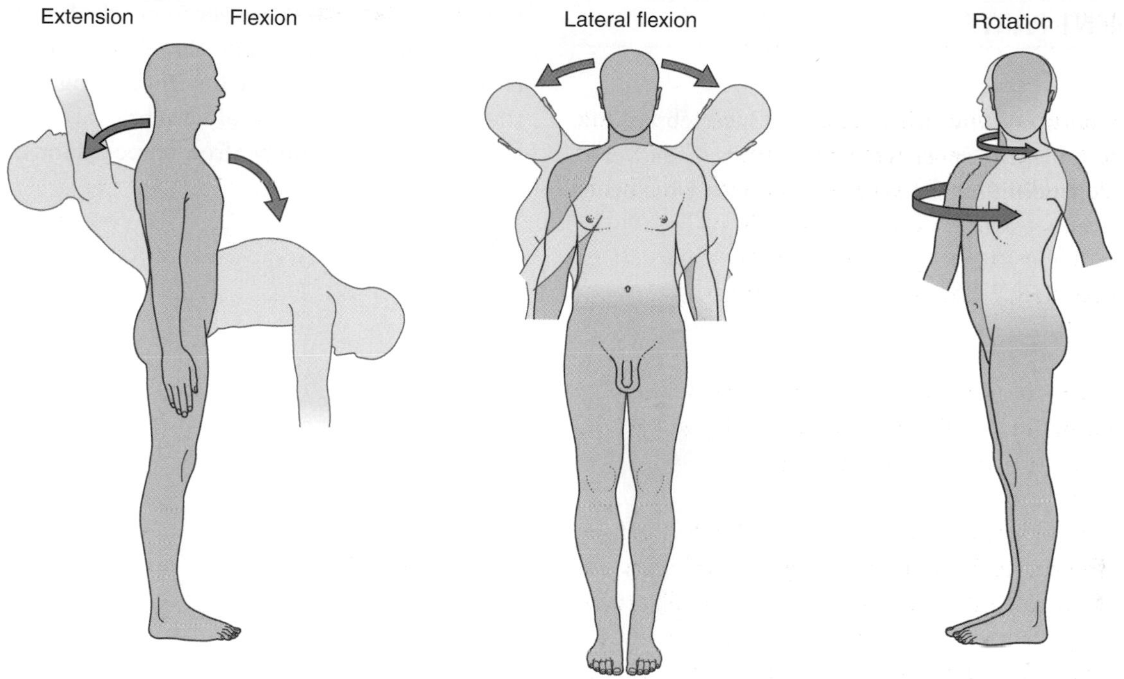


**Fig. 2.3** Back movements.

In the cervical region, the first two vertebrae and associated muscles are specifically modified to support and position the head. The head flexes and extends, in the nodding motion, on vertebra CI, and rotation of the head occurs as vertebra CI moves on vertebra CII (Fig. 2.3).

## Protection of the nervous system

The vertebral column and associated soft tissues of the back contain the spinal cord and proximal parts of the spinal nerves (Fig. 2.4). The more distal parts of the spinal nerves pass into all other regions of the body, including certain regions of the head.

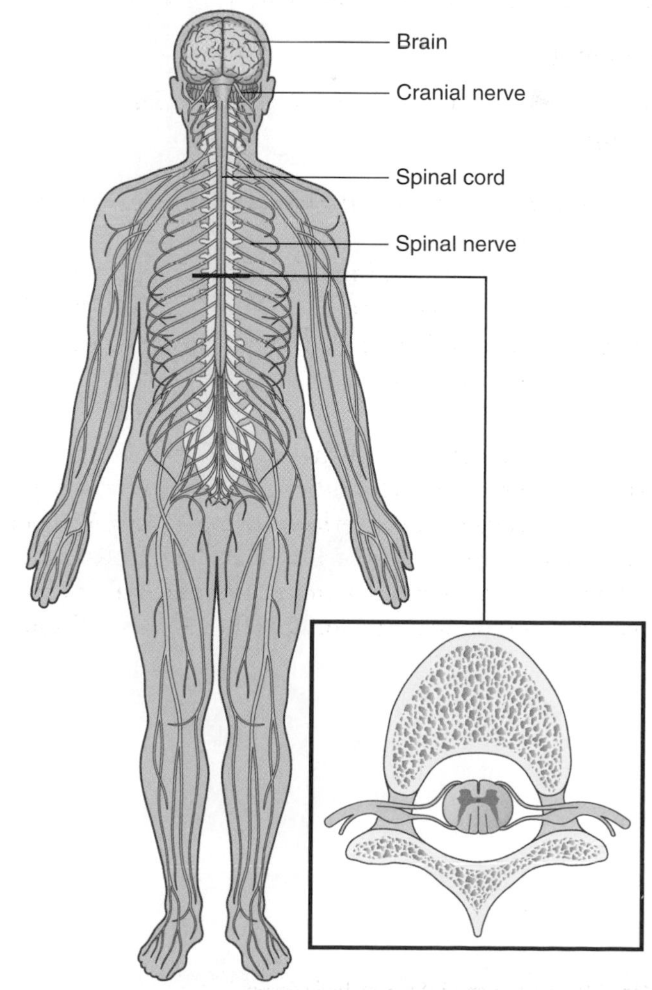


**Fig. 2.4** Nervous system.

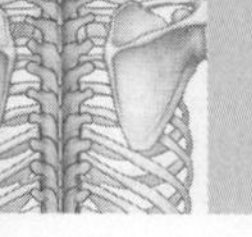



## COMPONENT PARTS

### Bones

The major bones of the back are the 33 vertebrae (Fig. 2.5). The number and specific characteristics of the vertebrae vary depending on the body region with which they are associated. There are seven cervical, twelve thoracic, five lumbar, five sacral, and three to four coccygeal vertebrae. The sacral vertebrae fuse into a single bony element, the sacrum. The coccygeal vertebrae are rudimentary in structure, vary in number from three to four, and often fuse into a single coccyx.

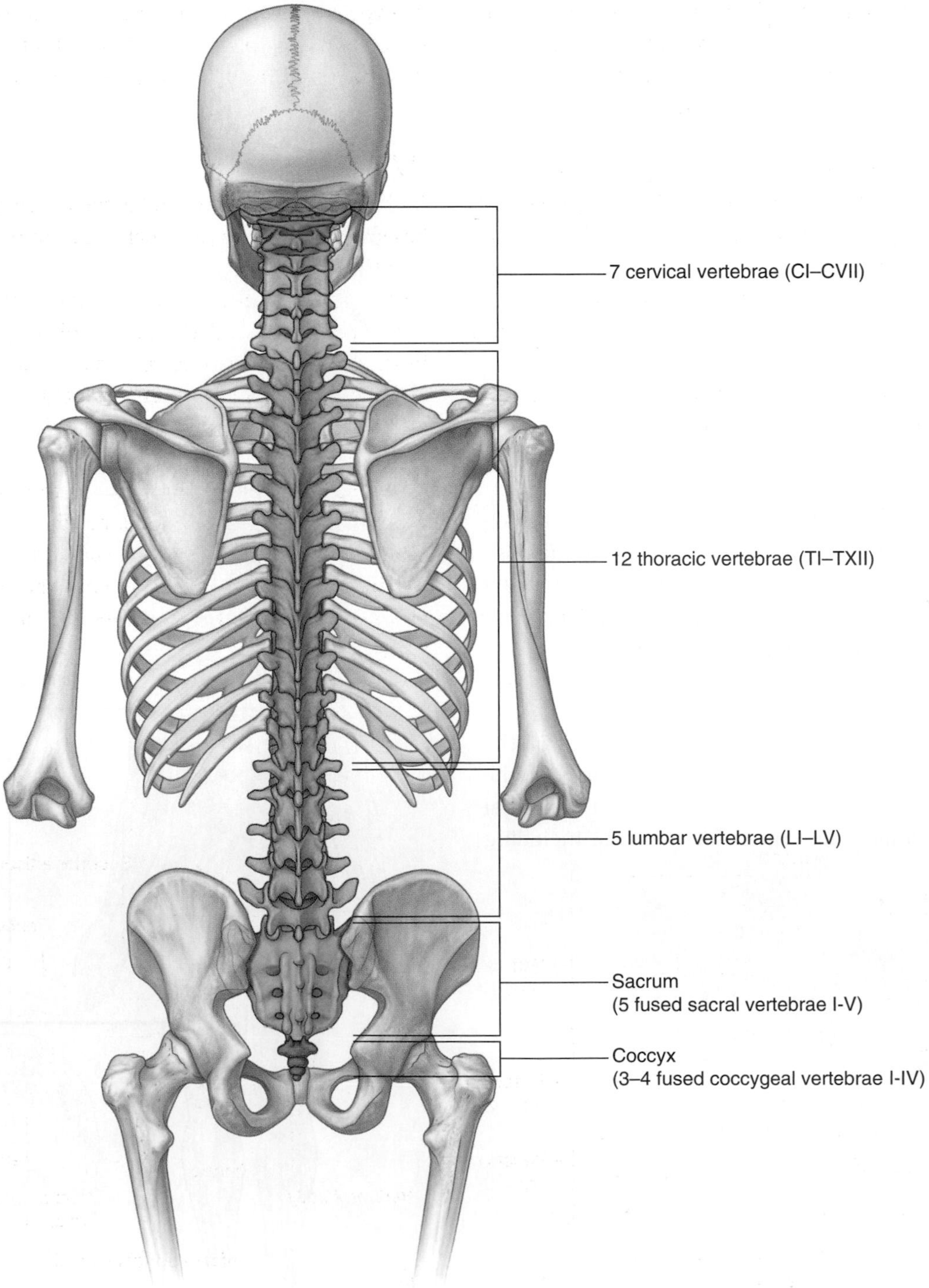

7 cervical vertebrae (CI–CVII)
12 thoracic vertebrae (TI–TXII)
5 lumbar vertebrae (LI–LV)
Sacrum
(5 fused sacral vertebrae I-V)
Coccyx
(3–4 fused coccygeal vertebrae I-IV)

**Fig. 2.5** Vertebrae.

### Typical vertebra

A typical vertebra consists of a vertebral body and a vertebral arch (Fig. 2.6).

The vertebral body is anterior and is the major weight-bearing component of the bone. It increases in size from vertebra CII to vertebra LV. Fibrocartilaginous intervertebral discs separate the vertebral bodies of adjacent vertebrae.

The vertebral arch is firmly anchored to the posterior surface of the vertebral body by two pedicles, which form the lateral pillars of the vertebral arch. The roof of the vertebral arch is formed by right and left laminae, which fuse at the midline.

The vertebral arches of the vertebrae are aligned to form the lateral and posterior walls of the vertebral canal, which extends from the first cervical vertebra (CI) to the last sacral vertebra (vertebra SV). This bony canal contains the spinal cord and its protective membranes, together with blood vessels, connective tissue, fat, and proximal parts of spinal nerves.

The vertebral arch of a typical vertebra has a number of characteristic projections, which serve as:

- attachments for muscles and ligaments,
- levers for the action of muscles, and
- sites of articulation with adjacent vertebrae.

A spinous process projects posteriorly and generally inferiorly from the roof of the vertebral arch.

On each side of the vertebral arch, a transverse process extends laterally from the region where a lamina meets a pedicle. From the same region, a superior articular process and an inferior articular process articulate with similar processes on adjacent vertebrae.

Each vertebra also contains rib elements. In the thorax, these costal elements are large and form ribs, which articulate with the vertebral bodies and transverse processes. In all other regions, these rib elements are small and are incorporated into the transverse processes. Occasionally, they develop into ribs in regions other than the thorax, usually in the lower cervical and upper lumbar regions.

## Muscles

Muscles in the back can be classified as extrinsic or intrinsic based on their embryological origin and type of innervation (Fig. 2.7).

The extrinsic muscles are involved with movements of the upper limbs and thoracic wall and, in general, are innervated by anterior rami of spinal nerves. The superficial group of these muscles is related to the upper limbs, while the intermediate layer of muscles is associated with the thoracic wall.

All of the intrinsic muscles of the back are deep in position and are innervated by the posterior rami of spinal nerves. They support and move the vertebral column and participate in moving the head. One group of intrinsic muscles also moves the ribs relative to the vertebral column.

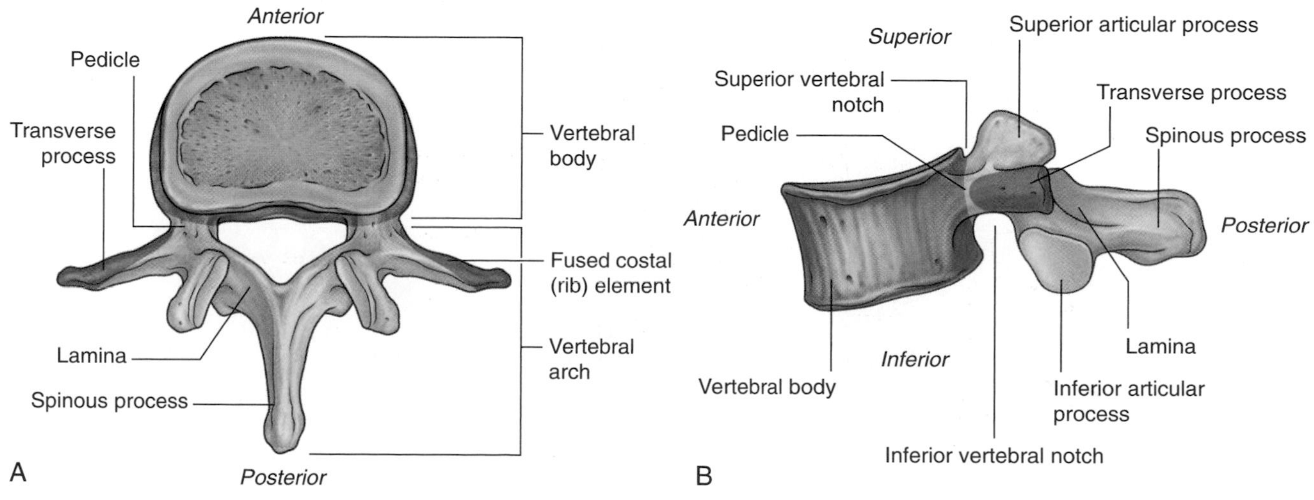


**Fig. 2.6** A typical vertebra. **A.** Superior view. **B.** Lateral view.

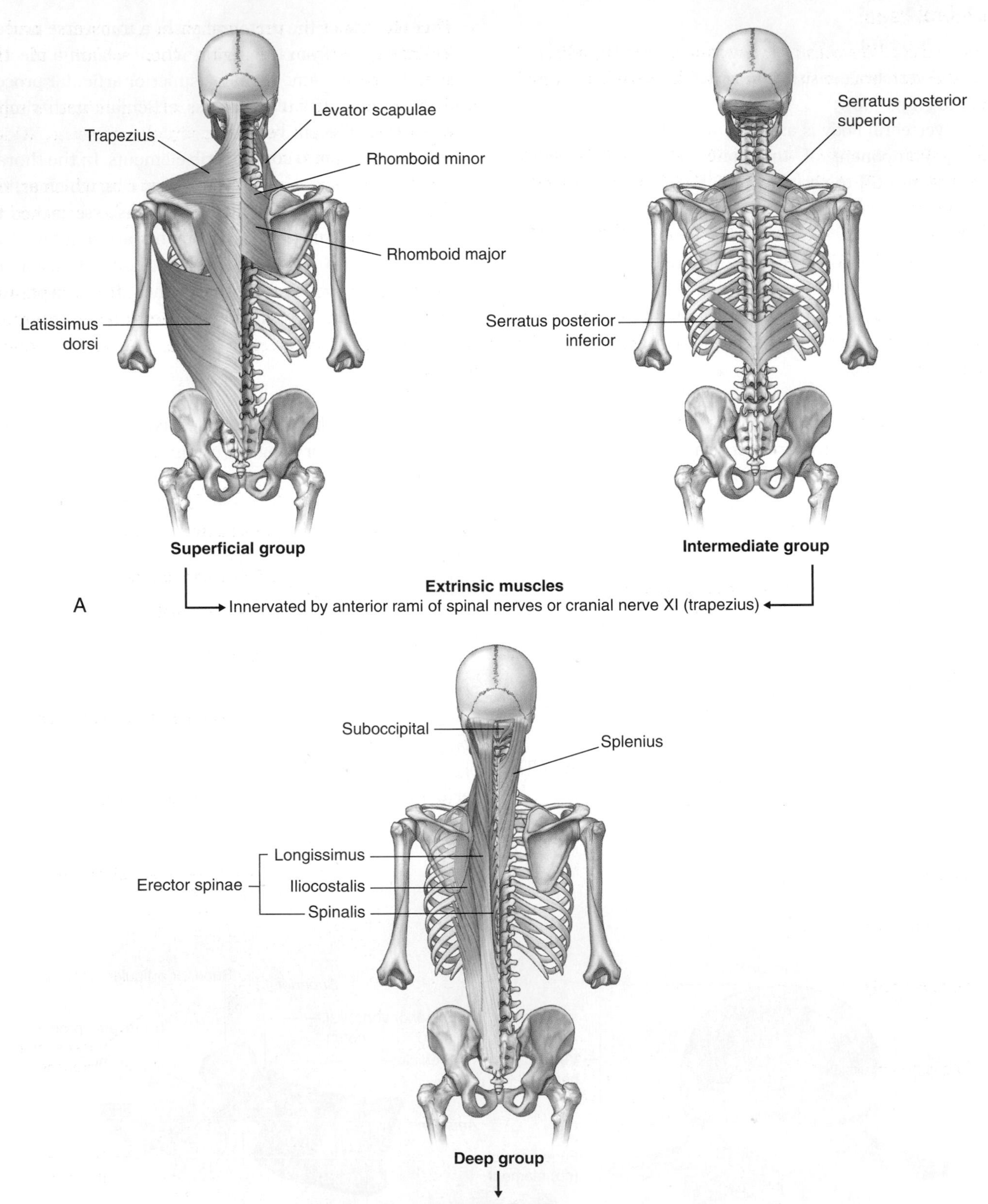


**Fig. 2.7** Back muscles. **A.** Extrinsic muscles. **B.** Intrinsic muscles.

### Vertebral canal

The spinal cord lies within a bony canal formed by adjacent vertebrae and soft tissue elements (the vertebral canal) (Fig. 2.8):

- The anterior wall is formed by the vertebral bodies of the vertebrae, intervertebral discs, and associated ligaments.
- The lateral walls and roof are formed by the vertebral arches and ligaments.

Within the vertebral canal, the spinal cord is surrounded by a series of three connective tissue membranes (the meninges):

- The pia mater is the innermost membrane and is intimately associated with the surface of the spinal cord.
- The second membrane, the arachnoid mater, is separated from the pia by the subarachnoid space, which contains cerebrospinal fluid.
- The thickest and most external of the membranes, the dura mater, lies directly against, but is not attached to, the arachnoid mater.

In the vertebral canal, the dura mater is separated from surrounding bone by an extradural (epidural) space containing loose connective tissue, fat, and a venous plexus.

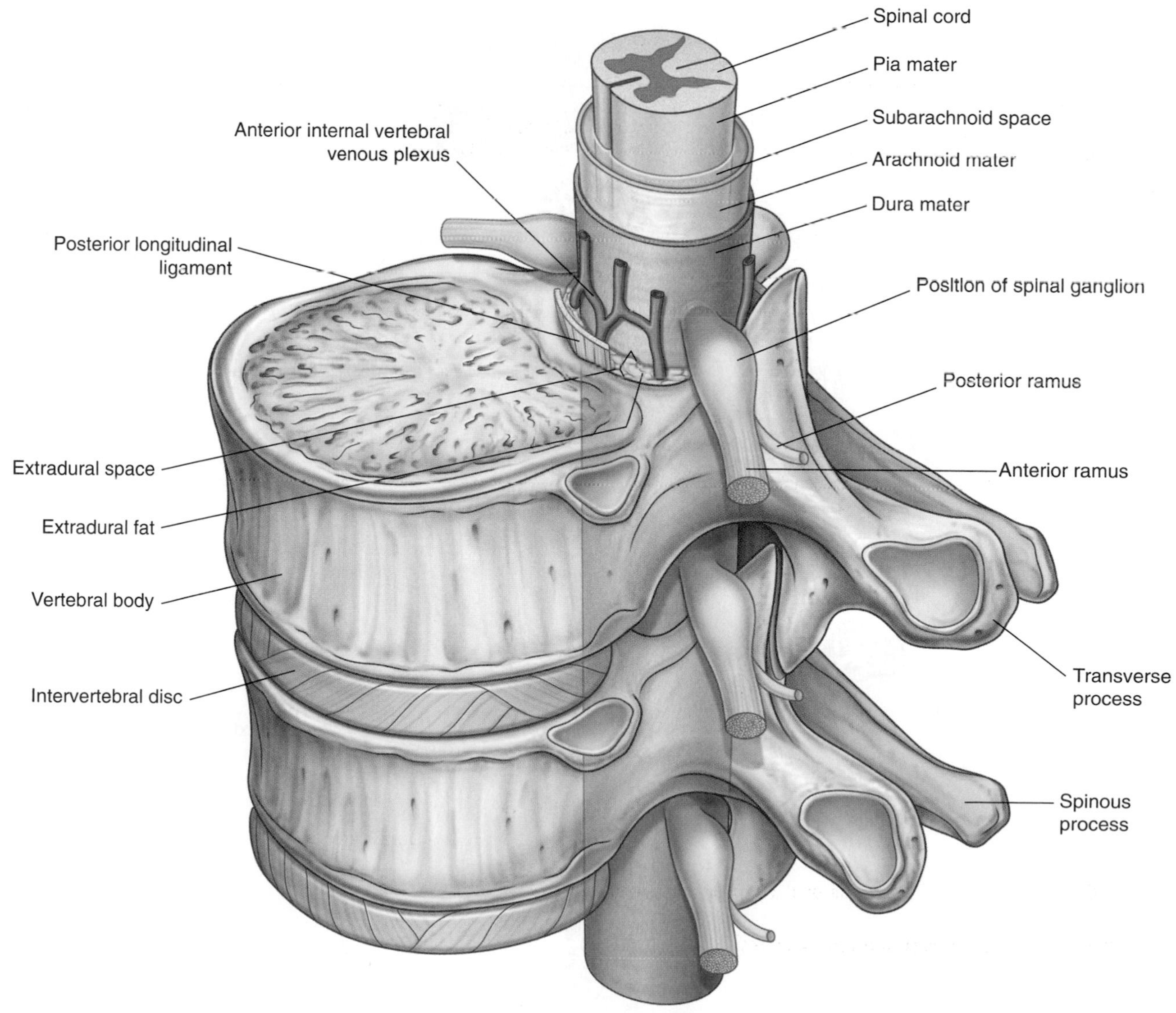


**Fig. 2.8** Vertebral canal.

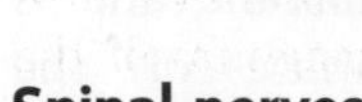

## Spinal nerves

The 31 pairs of spinal nerves are segmental in distribution and emerge from the vertebral canal between the pedicles of adjacent vertebrae. There are eight pairs of cervical nerves (C1 to C8), twelve thoracic (T1 to T12), five lumbar (L1 to L5), five sacral (S1 to S5), and one coccygeal (Co). Each nerve is attached to the spinal cord by a posterior root and an anterior root (Fig. 2.9).

After exiting the vertebral canal, each spinal nerve branches into:

- a posterior ramus—collectively, the small posterior rami innervate the back; and
- an anterior ramus—the much larger anterior rami innervate most other regions of the body except the head, which is innervated predominantly, but not exclusively, by cranial nerves.

The anterior rami form the major somatic plexuses (cervical, brachial, lumbar, and sacral) of the body. Major visceral components of the PNS (sympathetic trunk and prevertebral plexus) of the body are also associated mainly with the anterior rami of spinal nerves.



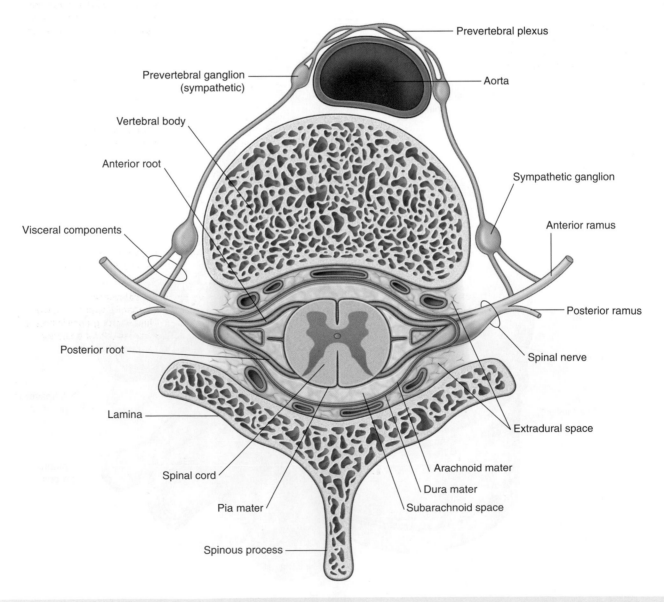

**Fig. 2.9** Spinal nerves (transverse section).

## RELATIONSHIP TO OTHER REGIONS

### Head

Cervical regions of the back constitute the skeletal and much of the muscular framework of the neck, which in turn supports and moves the head (Fig. 2.10).

The brain and cranial meninges are continuous with the spinal cord meninges at the foramen magnum of the skull. The paired vertebral arteries ascend, one on each side, through foramina in the transverse processes of cervical vertebrae and pass through the foramen magnum to participate, with the internal carotid arteries, in supplying blood to the brain.

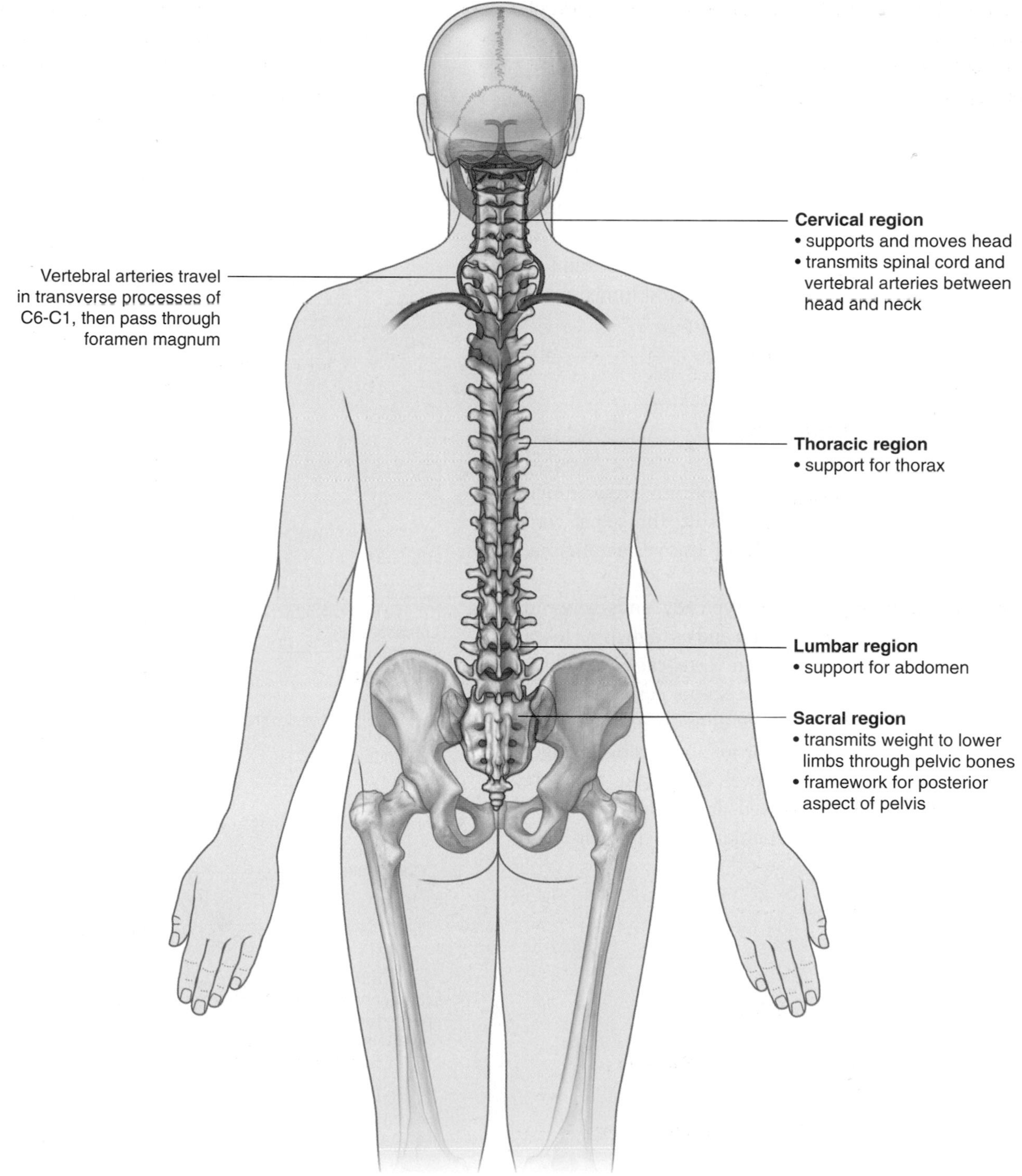
Cervical region
• supports and moves head
• transmits spinal cord and vertebral arteries between head and neck
Vertebral arteries travel in transverse processes of C6-C1, then pass through foramen magnum
Thoracic region
• support for thorax
Lumbar region
• support for abdomen
Sacral region
• transmits weight to lower limbs through pelvic bones
• framework for posterior aspect of pelvis

**Fig. 2.10** Relationships of the back to other regions.

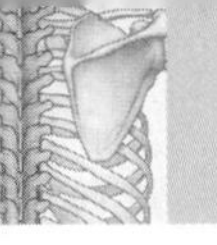



## Thorax, abdomen, and pelvis

The different regions of the vertebral column contribute to the skeletal framework of the thorax, abdomen, and pelvis (Fig. 2.10). In addition to providing support for each of these parts of the body, the vertebrae provide attachments for muscles and fascia, and articulation sites for other bones. The anterior rami of spinal nerves associated with the thorax, abdomen, and pelvis pass into these parts of the body from the back.

## Limbs

The bones of the back provide extensive attachments for muscles associated with anchoring and moving the upper limbs on the trunk. This is less true of the lower limbs, which are firmly anchored to the vertebral column through articulation of the pelvic bones with the sacrum. The upper and lower limbs are innervated by anterior rami of spinal nerves that emerge from cervical and lumbosacral levels, respectively, of the vertebral column.

# KEY FEATURES

## Long vertebral column and short spinal cord

During development, the vertebral column grows much faster than the spinal cord. As a result, the spinal cord does not extend the entire length of the vertebral canal (Fig. 2.11).

In the adult, the spinal cord typically ends between vertebrae LI and LII, although it can end as high as vertebra TXII and as low as the disc between vertebrae LII and LIII.

Spinal nerves originate from the spinal cord at increasingly oblique angles from vertebrae CI to Co, and the nerve roots pass in the vertebral canal for increasingly longer distances. Their spinal cord level of origin therefore becomes increasingly dissociated from their vertebral column level of exit. This is particularly evident for lumbar and sacral spinal nerves.

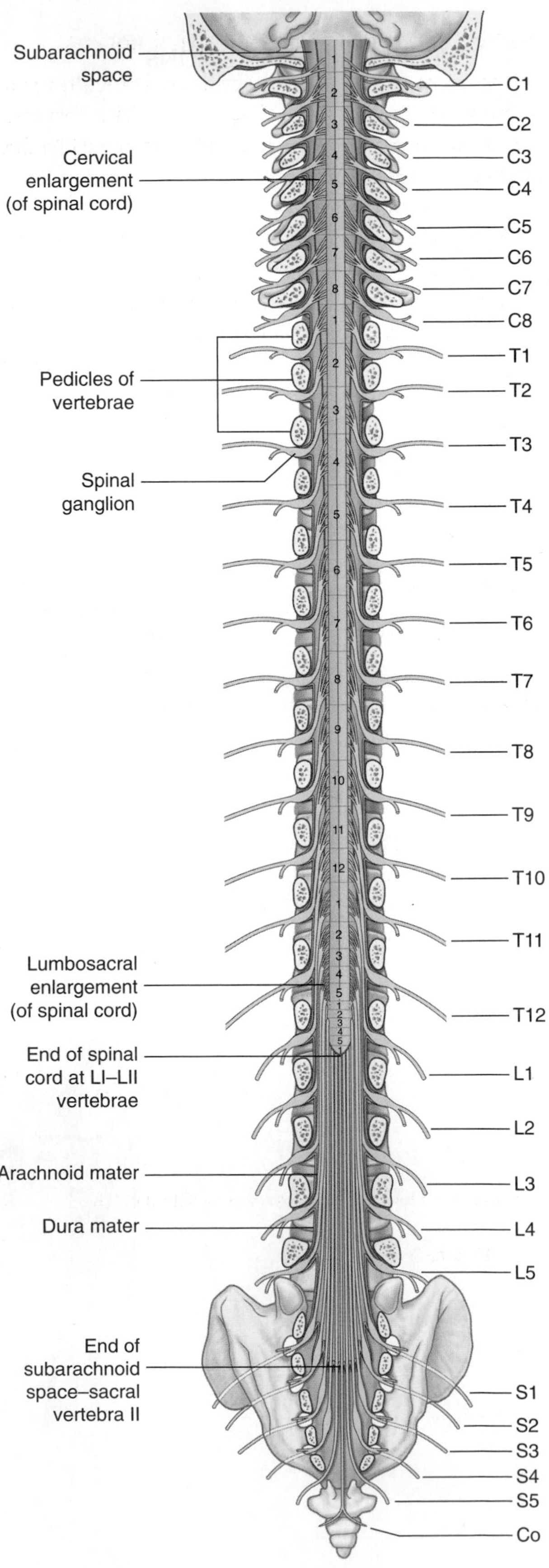


**Fig. 2.11** Vertebral canal, spinal cord, and spinal nerves.

## Intervertebral foramina and spinal nerves

Each spinal nerve exits the vertebral canal laterally through an intervertebral foramen (Fig. 2.12). The foramen is formed between adjacent vertebral arches and is closely related to intervertebral joints:

- The superior and inferior margins are formed by notches in adjacent pedicles.
- The posterior margin is formed by the articular processes of the vertebral arches and the associated joint.
- The anterior border is formed by the intervertebral disc between the vertebral bodies of the adjacent vertebrae.

Any pathology that occludes or reduces the size of an intervertebral foramen, such as bone loss, herniation of the intervertebral disc, or dislocation of the zygapophysial joint (the joint between the articular processes), can affect the function of the associated spinal nerve.

## Innervation of the back

Posterior branches of spinal nerves innervate the intrinsic muscles of the back and adjacent skin. The cutaneous distribution of these posterior rami extends into the gluteal region of the lower limb and the posterior aspect of the head. Parts of dermatomes innervated by the posterior rami of spinal nerves are shown in Fig. 2.13.

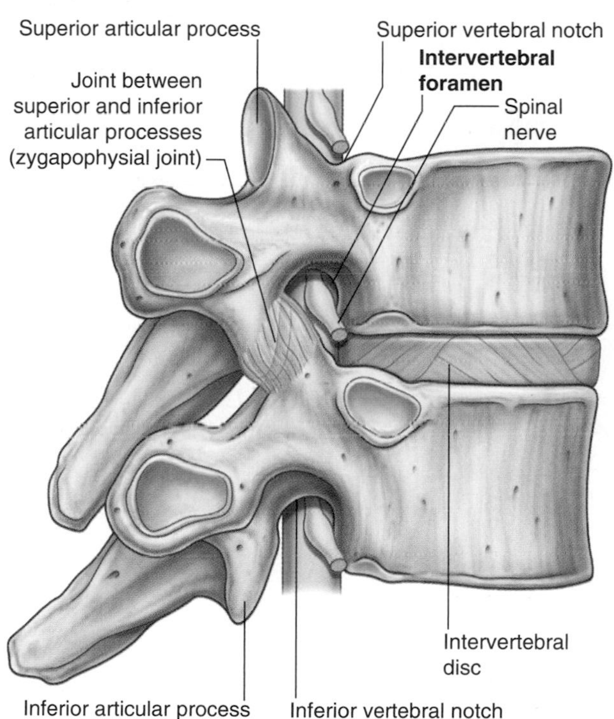


**Fig. 2.12** Intervertebral foramina.

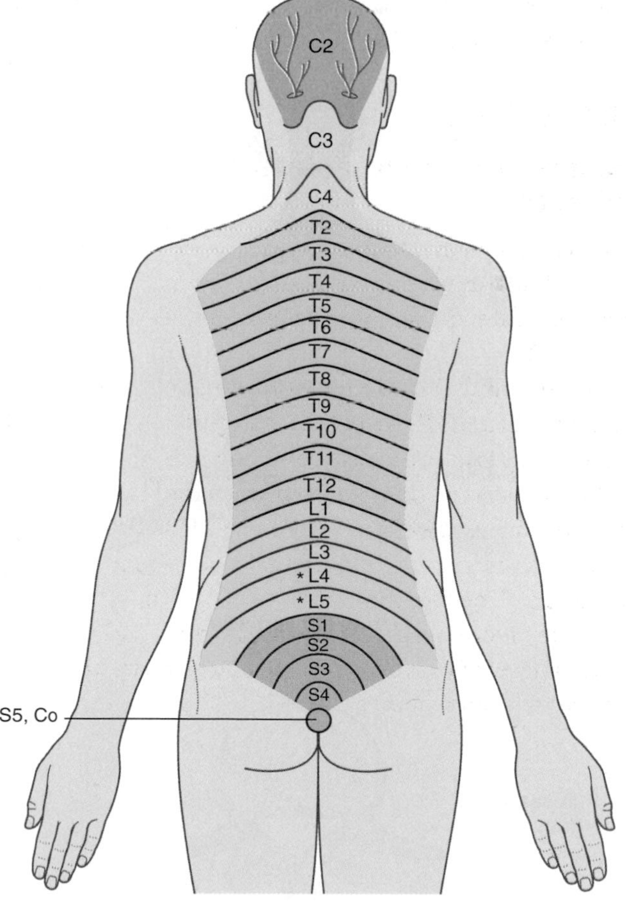


**Fig. 2.13** Dermatomes innervated by posterior rami of spinal nerves.

# Regional anatomy

## SKELETAL FRAMEWORK

Skeletal components of the back consist mainly of the vertebrae and associated intervertebral discs. The skull, scapulae, pelvic bones, and ribs also contribute to the bony framework of the back and provide sites for muscle attachment.

### Vertebrae

There are approximately 33 vertebrae, which are subdivided into five groups based on morphology and location (Fig. 2.14):

- The seven cervical vertebrae between the thorax and skull are characterized mainly by their small size and the presence of a foramen in each transverse process (Figs. 2.14 and 2.15).

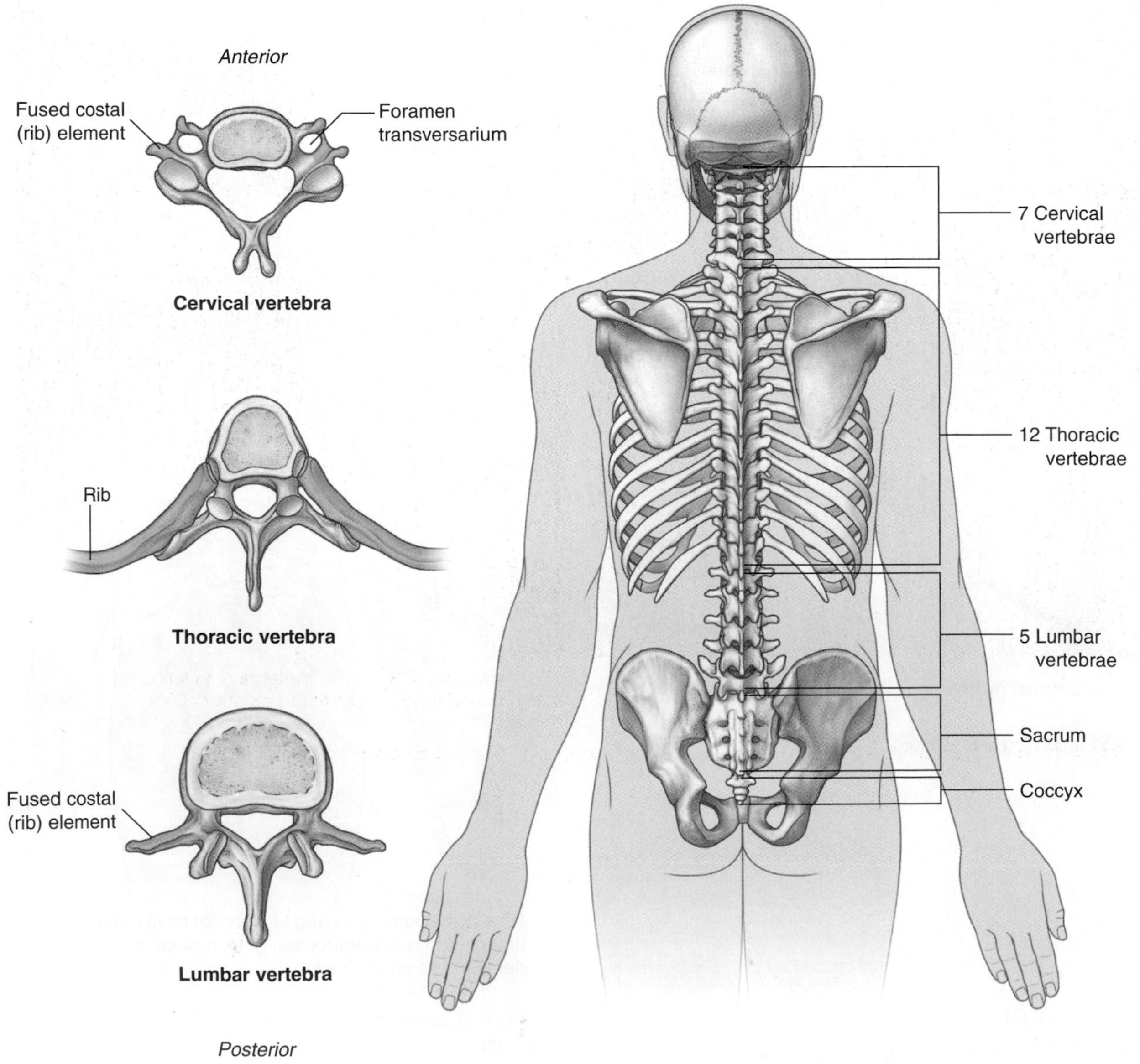

**Fig. 2.14** Vertebrae.

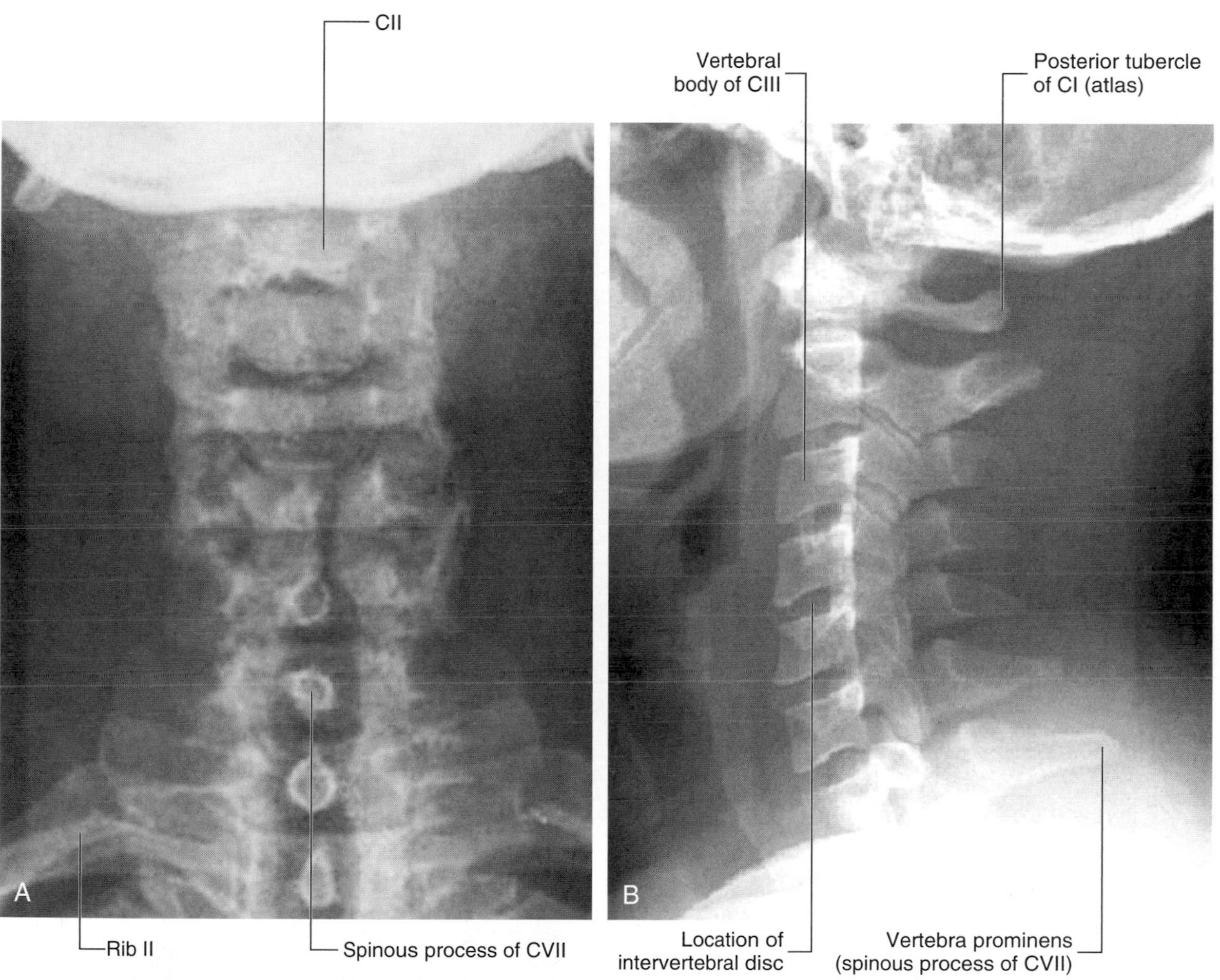


**Fig. 2.15** Radiograph of cervical region of vertebral column. **A.** Anteroposterior view. **B.** Lateral view.

- The 12 thoracic vertebrae are characterized by their articulated ribs (Figs. 2.14 and 2.16); although all vertebrae have rib elements, these elements are small and are incorporated into the transverse processes in regions other than the thorax; but in the thorax, the ribs are separate bones and articulate via synovial joints with the vertebral bodies and transverse processes of the associated vertebrae.
- Inferior to the thoracic vertebrae are five lumbar vertebrae, which form the skeletal support for the posterior abdominal wall and are characterized by their large size (Figs. 2.14 and 2.17).
- Next are five sacral vertebrae fused into one single bone called the sacrum, which articulates on each side with a pelvic bone and is a component of the pelvic wall.
- Inferior to the sacrum is a variable number, usually four, of coccygeal vertebrae, which fuse into a single small triangular bone called the coccyx.

In the embryo, the vertebrae are formed intersegmentally from cells called sclerotomes, which originate from adjacent somites (Fig. 2.18). Each vertebra is derived from the cranial parts of the two somites below, one on each side, and the caudal parts of the two somites above. The

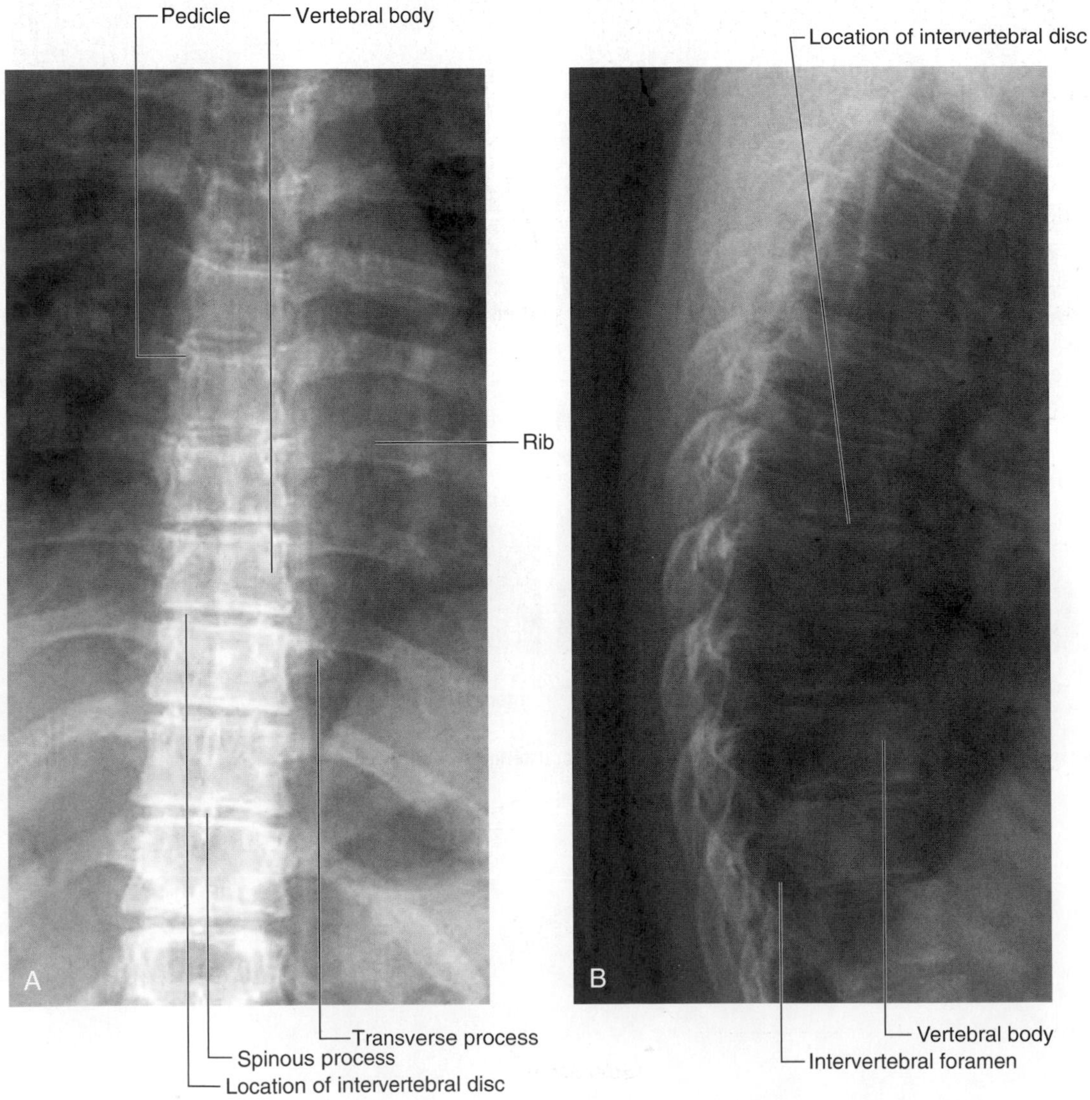

**Fig. 2.16** Radiograph of thoracic region of vertebral column. **A.** Anteroposterior view. **B.** Lateral view.

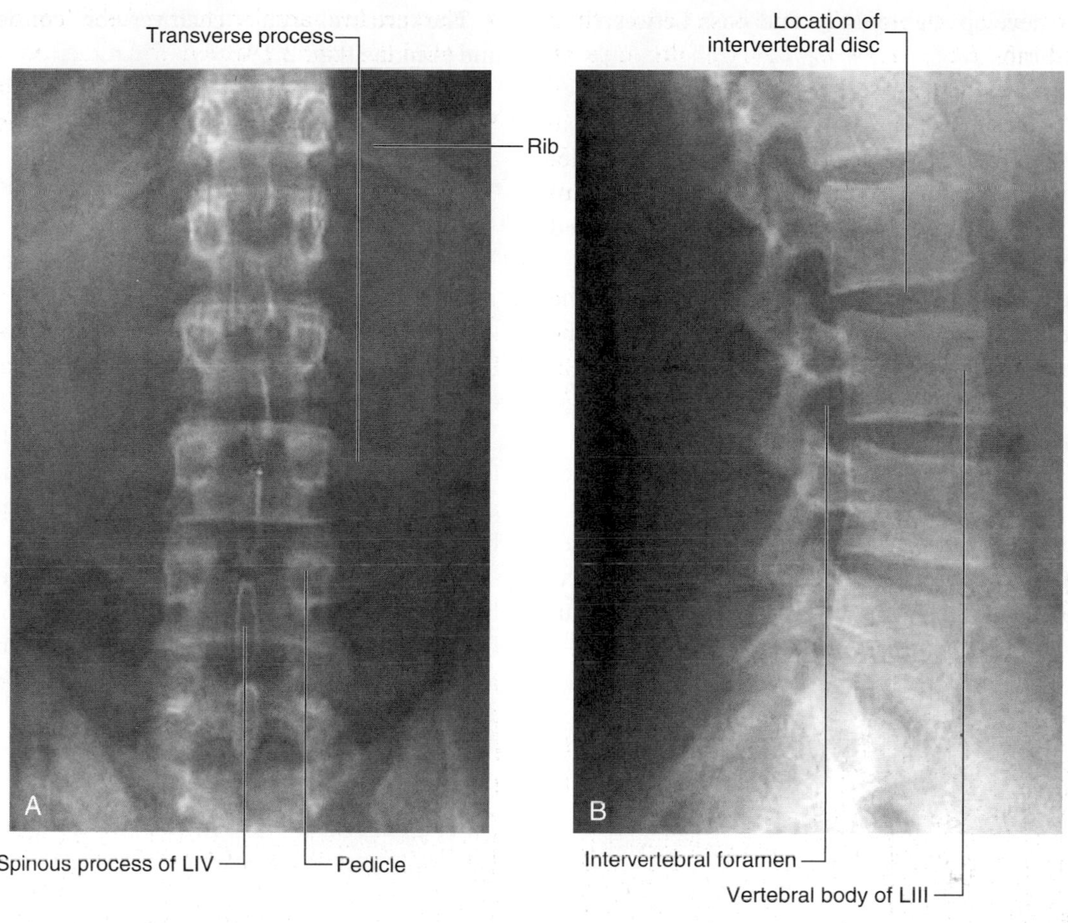


**Fig. 2.17** Radiograph of lumbar region of vertebral column. **A.** Anteroposterior view. **B.** Lateral view.

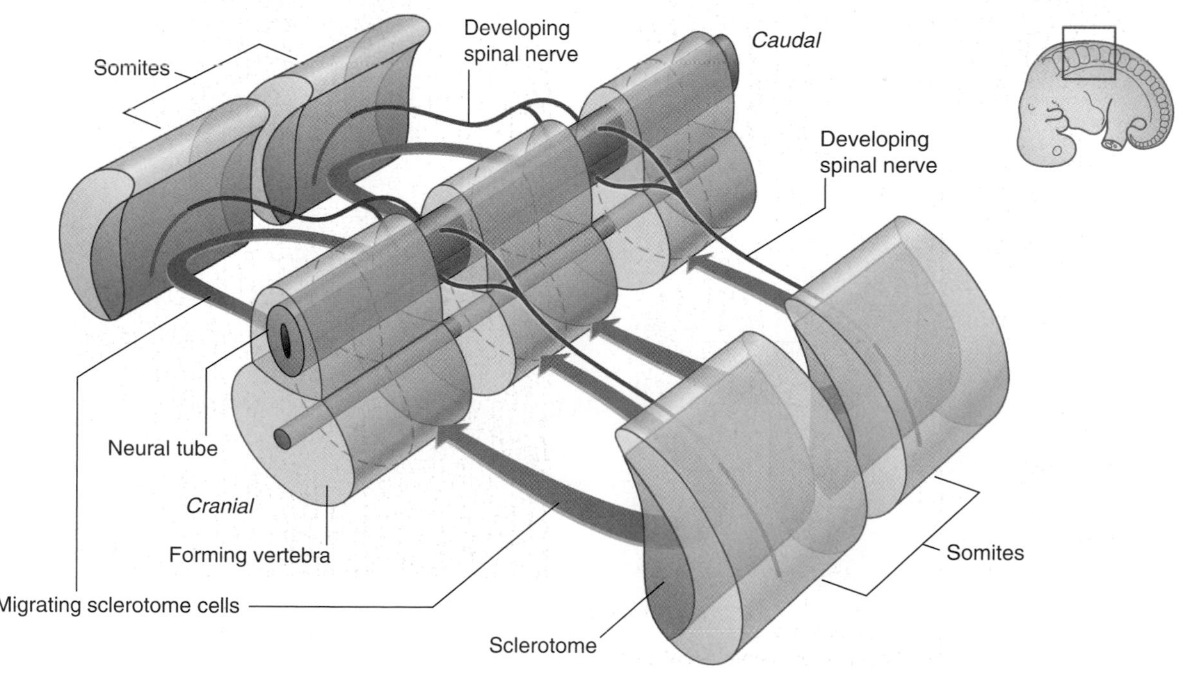


**Fig. 2.18** Development of the vertebrae.

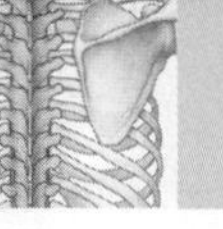


spinal nerves develop segmentally and pass between the forming vertebrae.

## Typical vertebra

A typical vertebra consists of a vertebral body and a posterior vertebral arch (Fig. 2.19). Extending from the vertebral arch are a number of processes for muscle attachment and articulation with adjacent bone.

The **vertebral body** is the weight-bearing part of the vertebra and is linked to adjacent vertebral bodies by intervertebral discs and ligaments. The size of vertebral bodies increases inferiorly as the amount of weight supported increases.

The **vertebral arch** forms the lateral and posterior parts of the vertebral foramen.

The vertebral foramina of all the vertebrae together form the **vertebral canal**, which contains and protects the spinal cord. Superiorly, the vertebral canal is continuous, through the foramen magnum of the skull, with the cranial cavity of the head.

The vertebral arch of each vertebra consists of pedicles and laminae (Fig. 2.19):

- The two **pedicles** are bony pillars that attach the vertebral arch to the vertebral body.
- The two **laminae** are flat sheets of bone that extend from each pedicle to meet in the midline and form the roof of the vertebral arch.

A **spinous process** projects posteriorly and inferiorly from the junction of the two laminae and is a site for muscle and ligament attachment.

A **transverse process** extends posterolaterally from the junction of the pedicle and lamina on each side and is a site for muscle and ligament attachment, and for articulation with ribs in the thoracic region.

Also projecting from the region where the pedicles join the laminae are **superior** and **inferior articular processes** (Fig. 2.19), which articulate with the inferior and superior articular processes, respectively, of adjacent vertebrae.

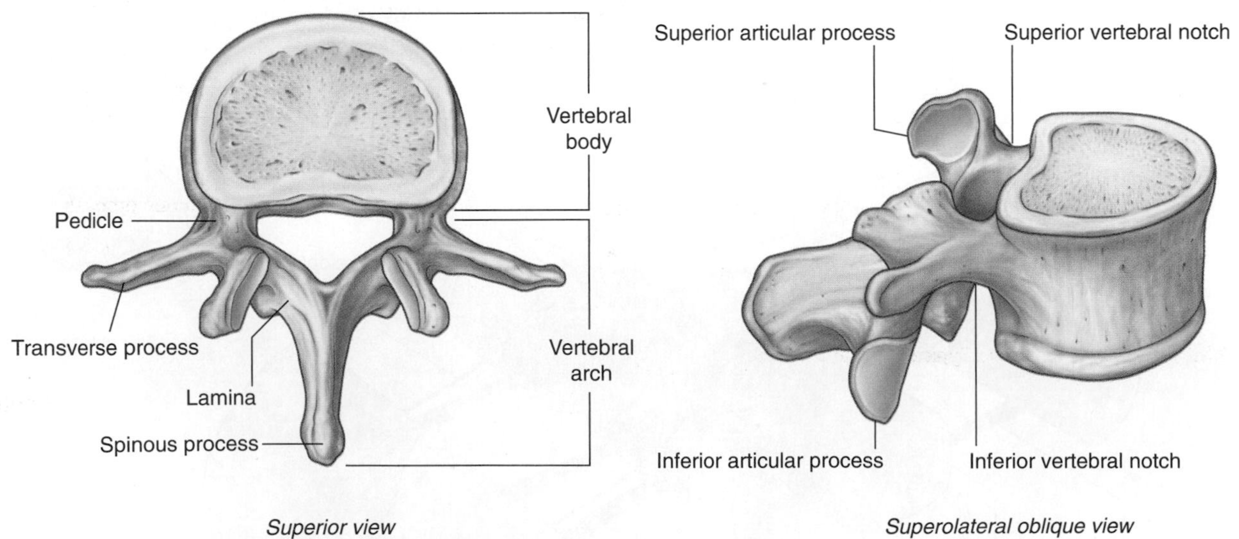


**Fig. 2.19** Typical vertebra.

Between the vertebral body and the origin of the articular processes, each pedicle is notched on its superior and inferior surfaces. These **superior** and **inferior vertebral notches** participate in forming intervertebral foramina.

## Cervical vertebrae

The seven cervical vertebrae are characterized by their small size and by the presence of a foramen in each transverse process. A typical cervical vertebra has the following features (Fig. 2.20A):

- The vertebral body is short in height and square shaped when viewed from above and has a concave superior surface and a convex inferior surface.
- Each transverse process is trough shaped and perforated by a round **foramen transversarium**.
- The spinous process is short and bifid.
- The vertebral foramen is triangular.

The first and second cervical vertebrae—the atlas and axis—are specialized to accommodate movement of the head.

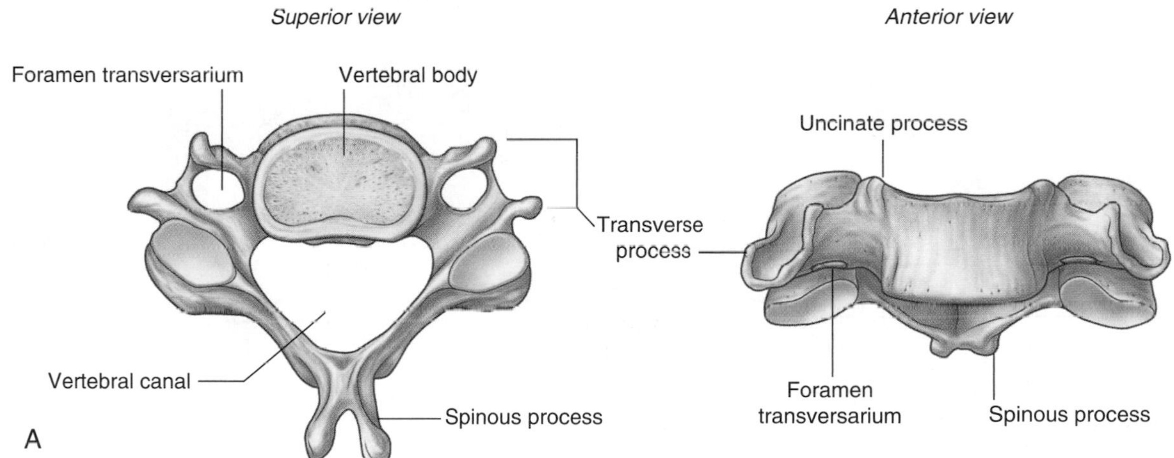


**Fig. 2.20** Regional vertebrae. **A.** Typical cervical vertebra.

*Continued*

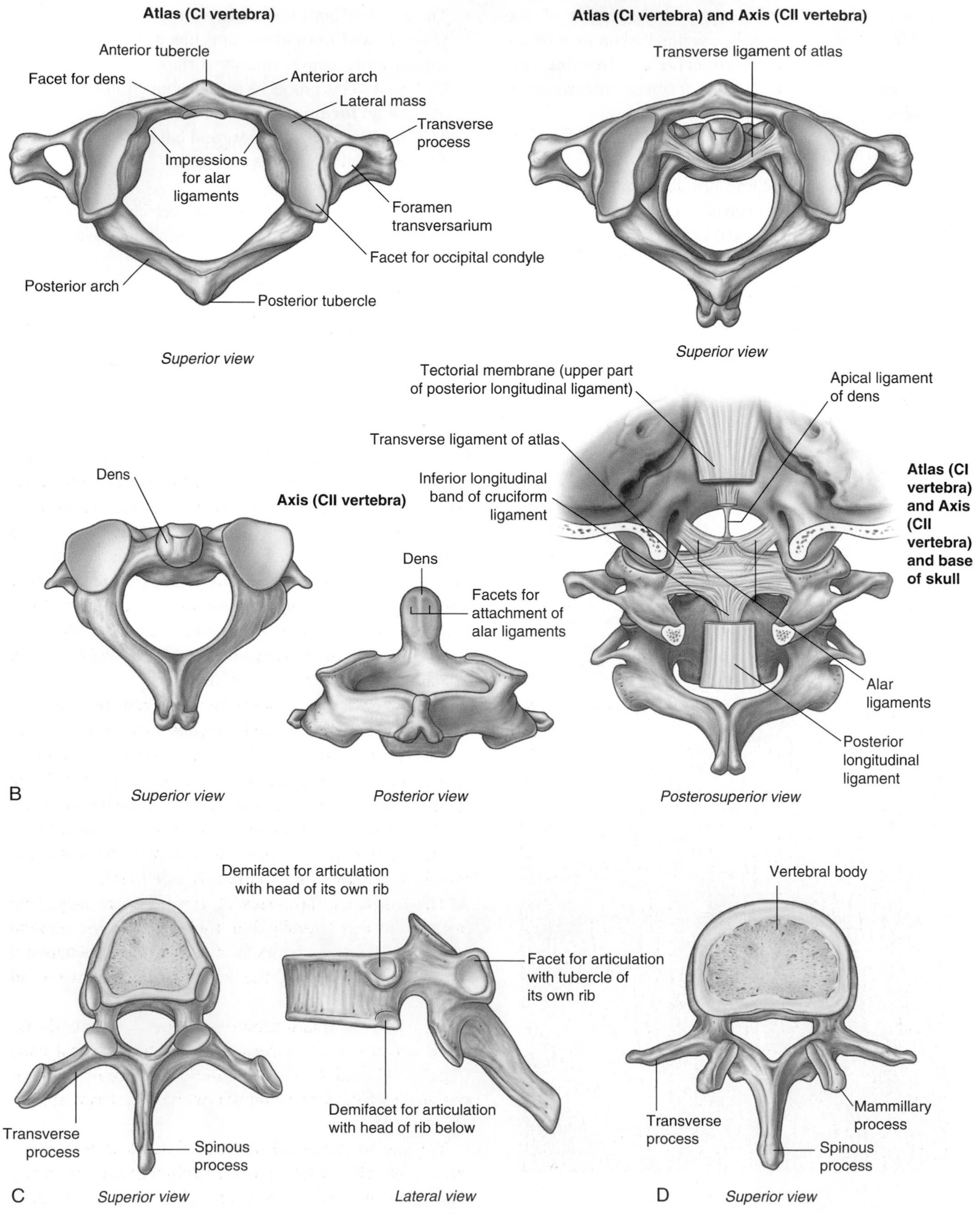

Atlas (CI vertebra)
Anterior tubercle
Facet for dens
Anterior arch
Lateral mass
Transverse process
Impressions for alar ligaments
Foramen transversarium
Facet for occipital condyle
Posterior arch
Posterior tubercle
Superior view
Atlas (CI vertebra) and Axis (CII vertebra)
Transverse ligament of atlas
Superior view
Tectorial membrane (upper part of posterior longitudinal ligament)
Apical ligament of dens
Transverse ligament of atlas
Inferior longitudinal band of cruciform ligament
Atlas (CI vertebra) and Axis (CII vertebra) and base of skull
Dens
Axis (CII vertebra)
Dens
Facets for attachment of alar ligaments
Alar ligaments
Posterior longitudinal ligament
B
Superior view
Posterior view
Posterosuperior view
Demifacet for articulation with head of its own rib
Facet for articulation with tubercle of its own rib
Demifacet for articulation with head of rib below
Transverse process
Spinous process
C
Superior view
Lateral view
Vertebral body
Mammillary process
Transverse process
Spinous process
D
Superior view

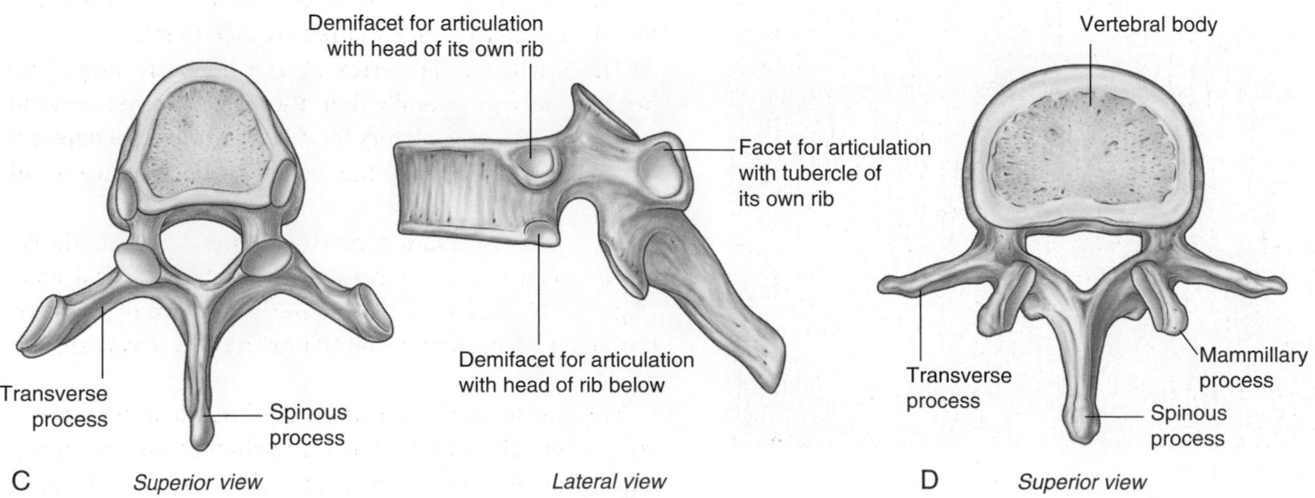

**Fig. 2.20, cont'd B.** Atlas and axis. **C.** Typical thoracic vertebra. **D.** Typical lumbar vertebra.

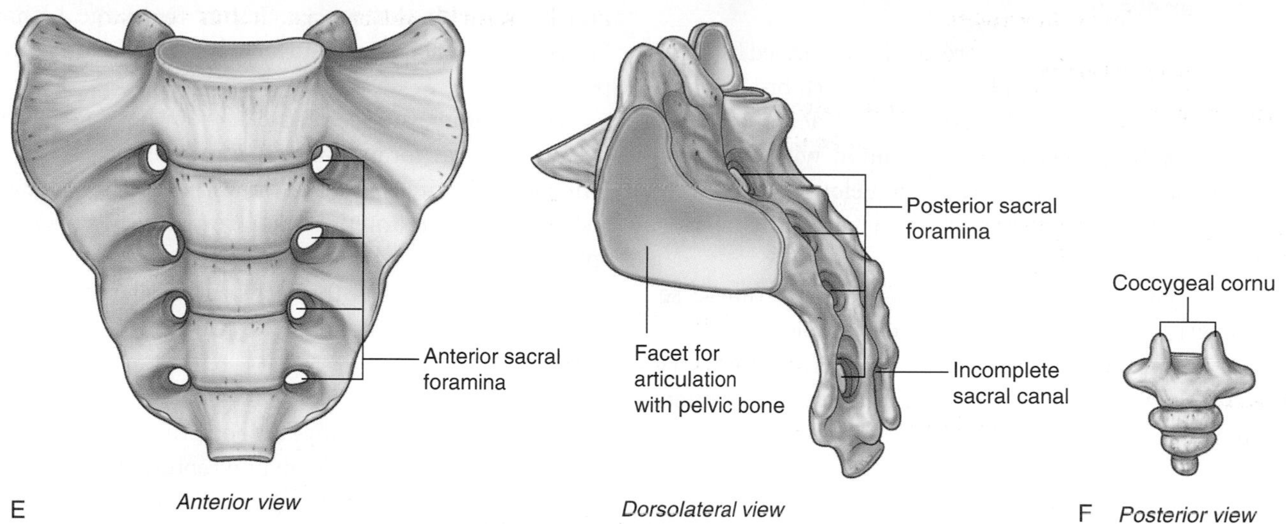


**Fig. 2.20, cont'd** E. Sacrum. F. Coccyx.

### Atlas and axis

Vertebra CI (the **atlas**) articulates with the head (Fig. 2.21). Its major distinguishing feature is that it lacks a vertebral body (Fig. 2.20B). In fact, the vertebral body of CI fuses onto the body of CII during development to become the dens of CII. As a result, there is no intervertebral disc between CI and CII. When viewed from above, the atlas is ring shaped and composed of two **lateral masses** interconnected by an **anterior arch** and a **posterior arch**.

Each lateral mass articulates above with an occipital **condyle** of the skull and below with the superior articular process of vertebra CII (the **axis**). The **superior articular surfaces** are bean shaped and concave, whereas the **inferior articular surfaces** are almost circular and flat.

The **atlanto-occipital joint** allows the head to nod up and down on the vertebral column.

The posterior surface of the anterior arch has an articular facet for the **dens**, which projects superiorly from the vertebral body of the axis. The dens is held in position by a strong **transverse ligament of atlas** posterior to it and spanning the distance between the oval attachment facets on the medial surfaces of the lateral masses of the atlas.

The dens acts as a pivot that allows the atlas and attached head to rotate on the axis, side to side.

The transverse processes of the atlas are large and protrude further laterally than those of the other cervical vertebrae and act as levers for muscle action, particularly for muscles that move the head at the **atlanto-axial joints**.

The axis is characterized by the large tooth-like dens, which extends superiorly from the vertebral body (Figs. 2.20B and 2.21). The anterior surface of the dens has an oval facet for articulation with the anterior arch of the atlas.

The two superolateral surfaces of the dens possess circular impressions that serve as attachment sites for strong alar ligaments, one on each side, which connect the dens to the medial surfaces of the occipital condyles. These **alar ligaments** check excessive rotation of the head and atlas relative to the axis.

Inferior articular facet on lateral mass of CI
Superior articular facet of CII
Dens

**Fig. 2.21** Radiograph showing CI (atlas) and CII (axis) vertebrae. Open mouth, anteroposterior (odontoid peg) view.

### Thoracic vertebrae

The twelve thoracic vertebrae are all characterized by their articulation with ribs. A typical thoracic vertebra has two partial facets (superior and inferior costal facets) on each side of the vertebral body for articulation with the head of its own rib and the head of the rib below (Fig. 2.20C). The superior costal facet is much larger than the inferior costal facet.

Each transverse process also has a facet (transverse costal facet) for articulation with the tubercle of its own rib. The vertebral body of the vertebra is somewhat heart shaped when viewed from above, and the vertebral foramen is circular.

### Lumbar vertebrae

The five lumbar vertebrae are distinguished from vertebrae in other regions by their large size (Fig. 2.20D). Also, they lack facets for articulation with ribs. The transverse processes are generally thin and long, with the exception of those on vertebra LV, which are massive and somewhat cone shaped for the attachment of **iliolumbar ligaments** to connect the transverse processes to the pelvic bones.

The vertebral body of a typical lumbar vertebra is cylindrical and the vertebral foramen is triangular in shape and larger than in the thoracic vertebrae.

### Sacrum

The sacrum is a single bone that represents the five fused sacral vertebrae (Fig. 2.20E). It is triangular in shape with the **apex** pointed inferiorly, and is curved so that it has a concave anterior surface and a correspondingly convex posterior surface. It articulates above with vertebra LV and below with the coccyx. It has two large L-shaped facets, one on each lateral surface, for articulation with the pelvic bones.

The posterior surface of the sacrum has four pairs of posterior sacral foramina, and the anterior surface has four pairs of anterior sacral foramina for the passage of the posterior and anterior rami, respectively, of S1 to S4 spinal nerves.

The posterior wall of the vertebral canal may be incomplete near the inferior end of the sacrum.

### Coccyx

The coccyx is a small triangular bone that articulates with the inferior end of the sacrum and represents three to four fused coccygeal vertebrae (Fig. 2.20F). It is characterized by its small size and by the absence of vertebral arches and therefore a vertebral canal.

## Intervertebral foramina

Intervertebral foramina are formed on each side between adjacent parts of vertebrae and associated intervertebral discs (Fig. 2.22). The foramina allow structures, such as spinal nerves and blood vessels, to pass in and out of the vertebral canal.

An intervertebral foramen is formed by the inferior vertebral notch on the pedicle of the vertebra above and the superior vertebral notch on the pedicle of the vertebra below. The foramen is bordered:

- posteriorly by the zygapophysial joint between the articular processes of the two vertebrae, and

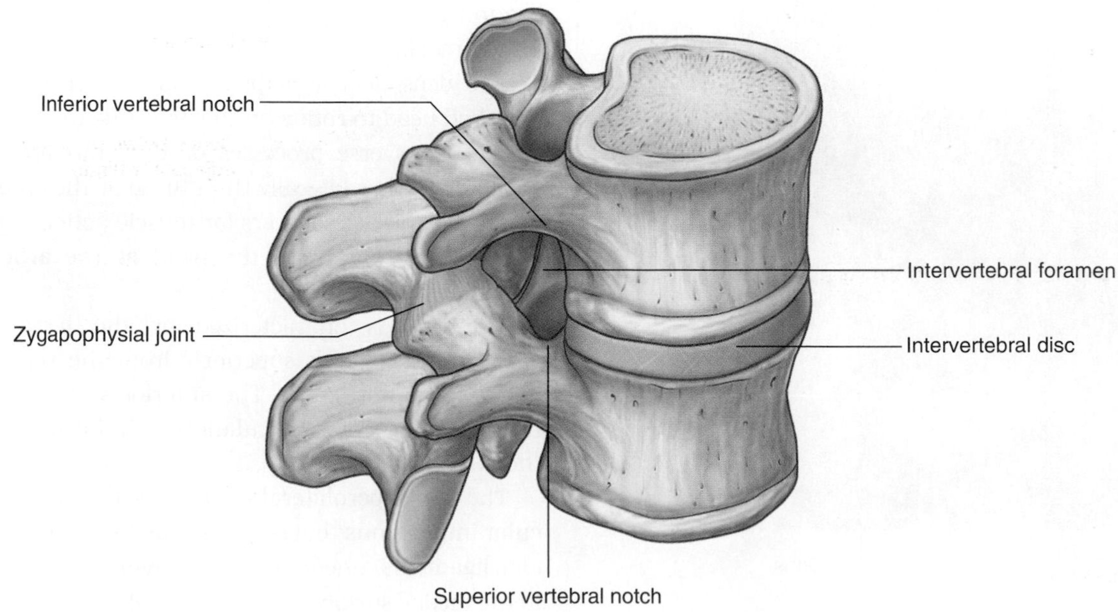


**Fig. 2.22** Intervertebral foramen.

- anteriorly by the intervertebral disc and adjacent vertebral bodies.

Each intervertebral foramen is a confined space surrounded by bone and ligament, and by joints. Pathology in any of these structures, and in the surrounding muscles, can affect structures within the foramen.

## Posterior spaces between vertebral arches

In most regions of the vertebral column, the laminae and spinous processes of adjacent vertebrae overlap to form a reasonably complete bony dorsal wall for the vertebral canal. However, in the lumbar region, large gaps exist between the posterior components of adjacent vertebral arches (Fig. 2.23). These gaps between adjacent laminae and spinous processes become increasingly wide from vertebra LI to vertebra LV. The spaces can be widened further by flexion of the vertebral column. These gaps allow relatively easy access to the vertebral canal for clinical procedures.

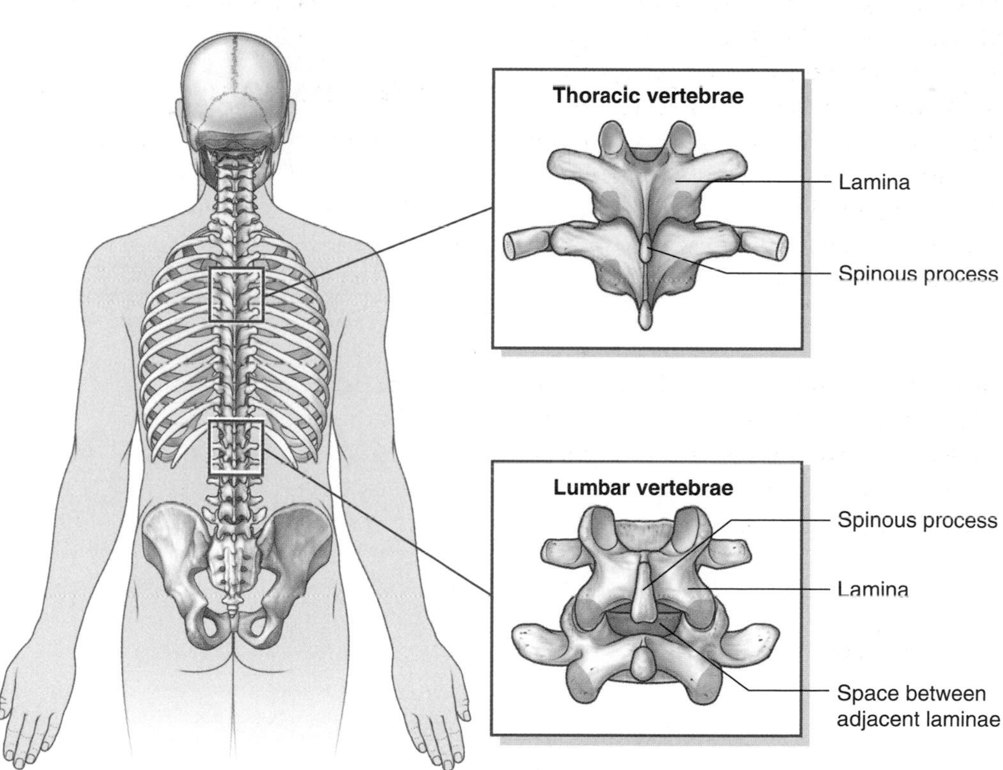


**Fig. 2.23** Spaces between adjacent vertebral arches in the lumbar region.

## In the clinic

### Spina bifida

Spina bifida is a disorder in which the two sides of vertebral arches, usually in lower vertebrae, fail to fuse during development, resulting in an "open" vertebral canal (Fig. 2.24). There are two types of spina bifida.

- The commonest type is spina bifida occulta, in which there is a defect in the vertebral arch of LV or SI. This defect occurs in as many as 10% of individuals and results in failure of the posterior arch to fuse in the midline. Clinically, the patient is asymptomatic, although physical examination may reveal a tuft of hair over the spinous processes.
- The more severe form of spina bifida involves complete failure of fusion of the posterior arch at the lumbosacral junction, with a large outpouching of the meninges. This may contain cerebrospinal fluid (a **meningocele**) or a portion of the spinal cord (a **myelomeningocele**). These abnormalities may result in a variety of neurological deficits, including problems with walking and bladder function.

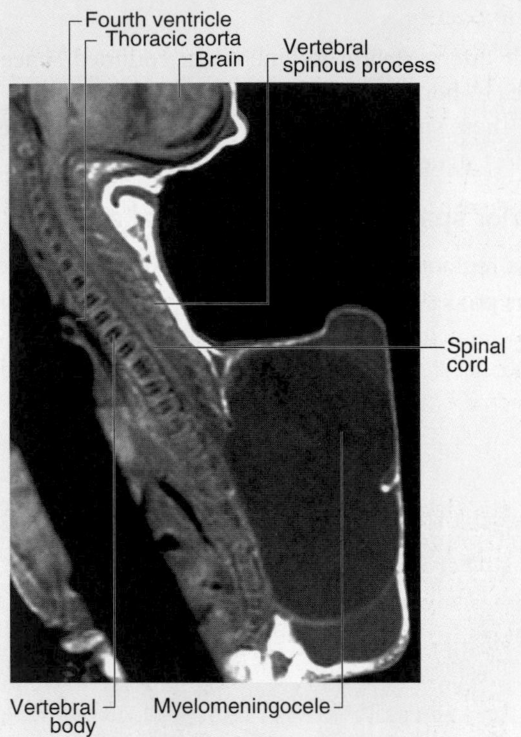


**Fig. 2.24** T1-weighted MR image in the sagittal plane demonstrating a lumbosacral myelomeningocele. There is an absence of laminae and spinous processes in the lumbosacral region.

## In the clinic

### Vertebroplasty

Vertebroplasty is a relatively new technique in which the body of a vertebra can be filled with bone cement (typically methyl methacrylate). The indications for the technique include vertebral body collapse and pain from the vertebral body, which may be secondary to tumor infiltration. The procedure is most commonly performed for osteoporotic wedge fractures, which are a considerable cause of morbidity and pain in older patients.

Osteoporotic wedge fractures (Fig. 2.25) typically occur in the thoracolumbar region, and the approach to performing vertebroplasty is novel and relatively straightforward. The procedure is performed under sedation or light general anesthetic. Using X-ray guidance the pedicle is identified on the anteroposterior image. A metal cannula is placed through the pedicle into the vertebral body. Liquid bone cement is injected via the cannula into the vertebral body (Fig. 2.26). The function of the bone cement is two-fold. First, it increases the strength of the vertebral body and prevents further loss of height. Furthermore, as the bone cement sets, there is a degree of heat generated that is believed to disrupt pain nerve endings. Kyphoplasty is a similar technique that aims to restore some or all of the lost vertebral body height from the wedge fracture by injecting liquid bone cement into the vertebral body.

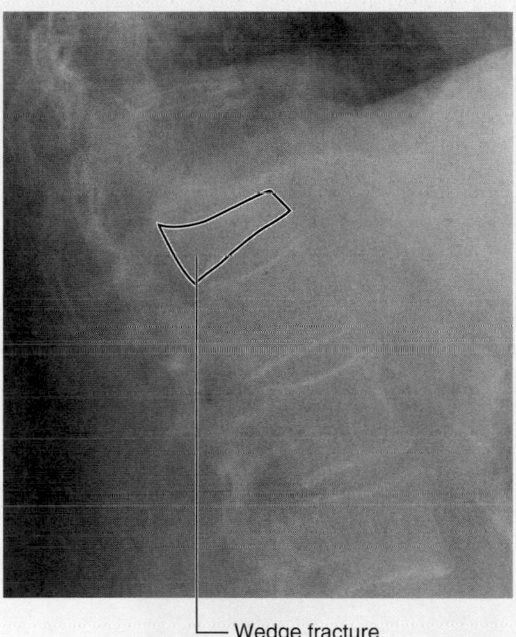


**Fig. 2.25** Radiograph of the lumbar region of the vertebral column demonstrating a wedge fracture of the L1 vertebra. This condition is typically seen in patients with osteoporosis.

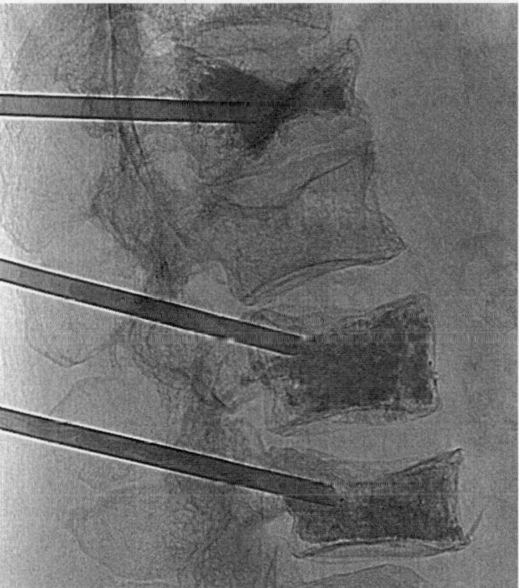

**Fig. 2.26** Radiograph of the lumbar region of the vertebral column demonstrating three intrapedicular needles, all of which have been placed into the middle of the vertebral bodies. The high-density material is radiopaque bone cement, which has been injected as a liquid that will harden.

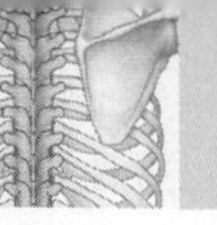



## In the clinic

### Scoliosis

Scoliosis is an abnormal lateral curvature of the vertebral column (Fig. 2.27).

A true scoliosis involves not only the curvature (right- or left-sided) but also a rotational element of one vertebra upon another.

The commonest types of scoliosis are those for which we have little understanding about how or why they occur and are termed idiopathic scoliosis. It is thought that there is some initial axial rotation of the vertebrae, which then alters the locations of the mechanical compressive and distractive forces applied through the vertebral growth plates, leading to changes in speed of bone growth and ultimately changes to spinal curvature. These are never present at birth and tend to occur in either the infantile, juvenile, or adolescent age groups. The vertebral bodies and posterior elements (pedicles and laminae) are normal in these patients.

When a scoliosis is present from birth (congenital scoliosis) it is usually associated with other developmental abnormalities. In these patients, there is a strong association with other abnormalities of the chest wall, genitourinary tract, and heart disease. This group of patients needs careful evaluation by many specialists.

A rare but important group of scoliosis is that in which the muscle is abnormal. Muscular dystrophy is the commonest example. The abnormal muscle does not retain the normal alignment of the vertebral column, and curvature develops as a result. A muscle biopsy is needed to make the diagnosis.

Other disorders that can produce scoliosis include bone tumors, spinal cord tumors, and localized disc protrusions.

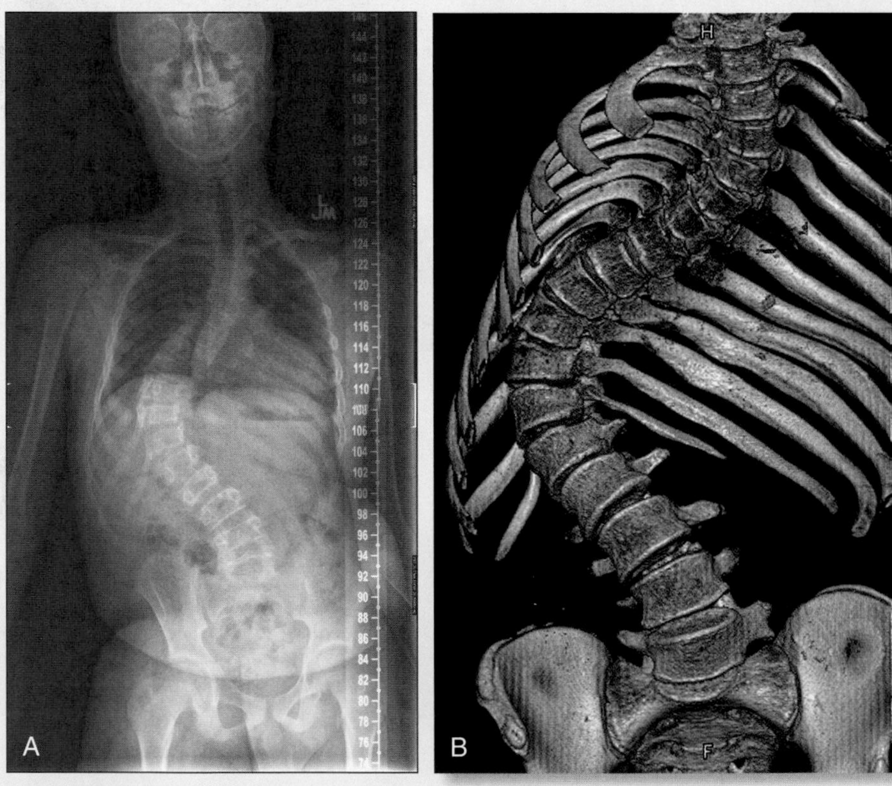


**Fig. 2.27** Severe scoliosis. **A.** Radiograph, anteroposterior view. **B.** Volume-rendered CT, anterior view.

## In the clinic

### Kyphosis

Kyphosis is abnormal curvature of the vertebral column in the thoracic region, producing a "hunchback" deformity. This condition occurs in certain disease states, the most dramatic of which is usually secondary to tuberculosis infection of a thoracic vertebral body, where the kyphosis becomes angulated at the site of the lesion. This produces the **gibbus deformity,** a deformity that was prevalent before the use of antituberculous medication (Fig. 2.28).

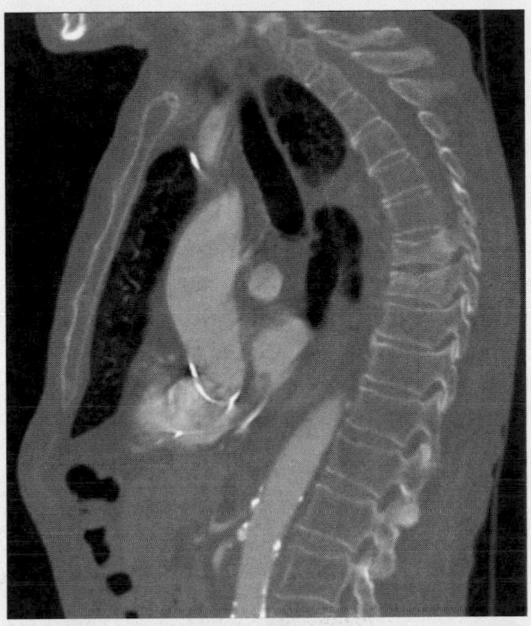

**Fig. 2.28** Sagittal CT showing kyphosis.

## In the clinic

### Lordosis

Lordosis is abnormal curvature of the vertebral column in the lumbar region, producing a swayback deformity.

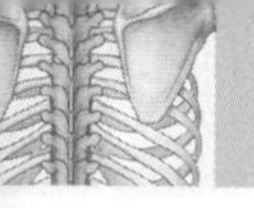



## In the clinic

### Variation in vertebral numbers

There are usually seven cervical vertebrae, although in certain diseases these may be fused. Fusion of cervical vertebrae (Fig. 2.29A) can be associated with other abnormalities, for example Klippel-Feil syndrome, in which there is fusion of vertebrae CI and CII or CV and CVI, and may be associated with a high-riding scapula (Sprengel's shoulder) and cardiac abnormalities.

Variations in the number of thoracic vertebrae also are well described.

One of the commonest abnormalities in the lumbar vertebrae is a partial fusion of vertebra LV with the sacrum (sacralization of the lumbar vertebra). Partial separation of vertebra SI from the sacrum (lumbarization of first sacral vertebra) may also occur (Fig. 2.29B). The LV vertebra can usually be identified by the iliolumbar ligament, which is a band of connective tissue that runs from the tip of the transverse process of LV to the iliac crest bilaterally (Fig. 2.29C).

A hemivertebra occurs when a vertebra develops only on one side (Fig. 2.29B).

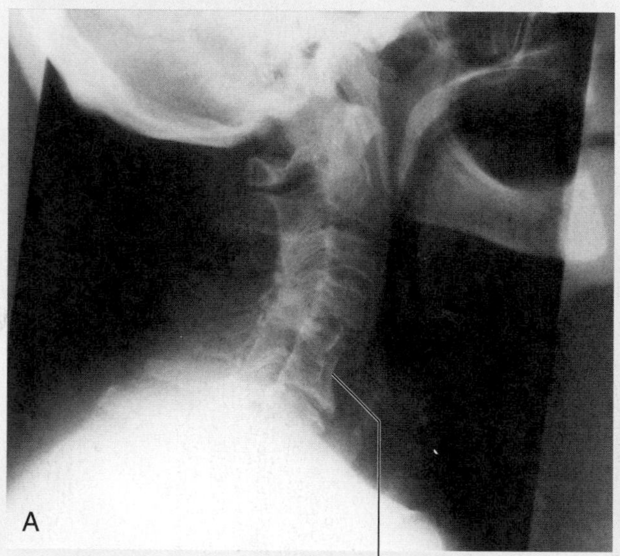



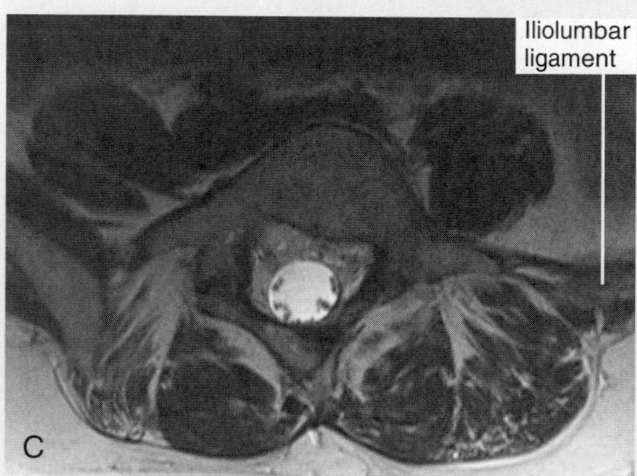



**Fig. 2.29** Variations in vertebral number. **A.** Fused vertebral bodies of cervical vertebrae. **B.** Hemivertebra. **C.** Axial slice MRI through the LV vertebra. The iliolumbar ligament runs from the tip of the LV vertebra transverse process to the iliac crest.

## In the clinic

### The vertebrae and cancer

The vertebrae are common sites for metastatic disease (secondary spread of cancer cells). When cancer cells grow within the vertebral bodies and the posterior elements, they interrupt normal bone cell turnover, leading to either bone destruction or formation and destroying the mechanical properties of the bone. A minor injury may therefore lead to vertebral collapse (Fig. 2.30A). Cancer cells have a much higher glucose metabolism compared with normal adjacent bone cells. These metastatic cancer cells can therefore be detected by administering radioisotope-labeled glucose to a patient and then tracing where the labeled glucose has been metabolized (Fig. 2.30B). Importantly, vertebrae that contain extensive metastatic disease may extrude fragments of tumor into the **vertebral canal**, compressing nerves and the spinal cord.

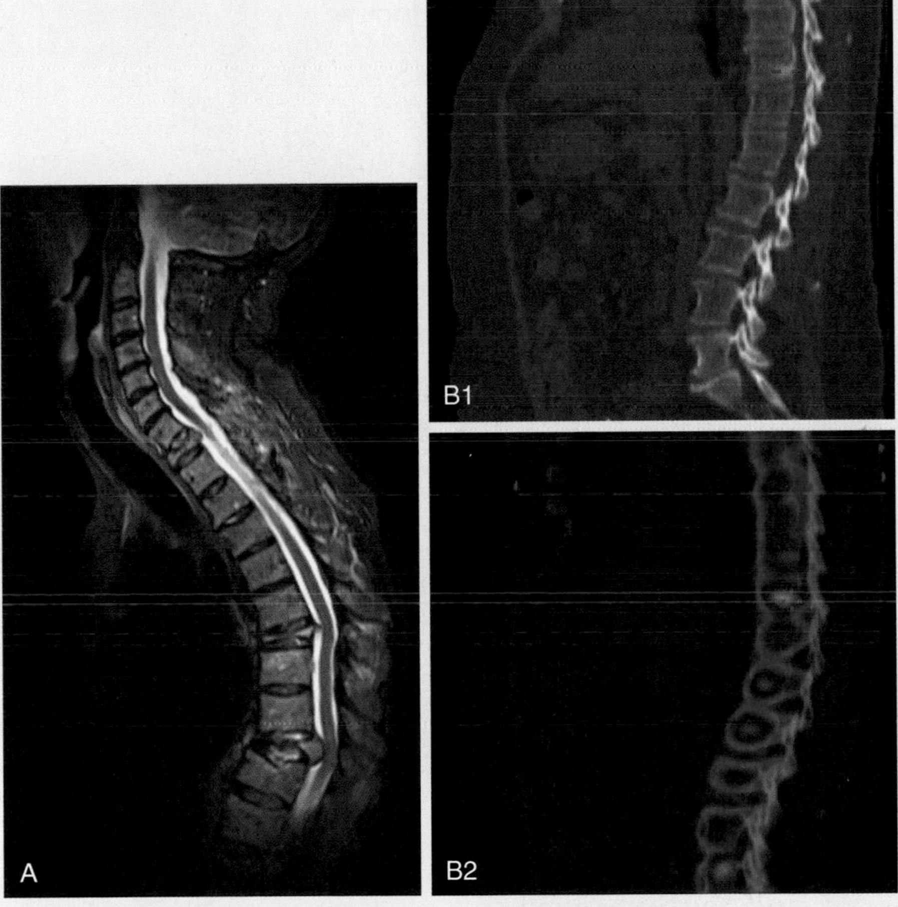


**Fig. 2.30 A.** MRI of a spine with multiple collapsed vertebrae due to diffuse metastatic myeloma infiltration. **B1, B2.** Positron emission tomography CT (PETCT) study detecting cancer cells in the spine that have high glucose metabolism.

## In the clinic

### Osteoporosis

Osteoporosis is a pathophysiologic condition in which bone quality is normal but the quantity of bone is deficient. It is a metabolic bone disorder that commonly occurs in women in their 50s and 60s and in men in their 70s.

Many factors influence the development of osteoporosis, including genetic predetermination, level of activity and nutritional status, and, in particular, estrogen levels in women.

Typical complications of osteoporosis include "crush" vertebral body fractures, distal fractures of the radius, and hip fractures.

With increasing age and poor-quality bone, patients are more susceptible to fracture. Healing tends to be impaired in these elderly patients, who consequently require long hospital stays and prolonged rehabilitation.

Patients likely to develop osteoporosis can be identified by dual-photon X-ray absorptiometry (DXA) scanning. Low-dose X-rays are passed through the bone, and by counting the number of photons detected and knowing the dose given, the number of X-rays absorbed by the bone can be calculated. The amount of X-ray absorption can be directly correlated with the bone mass, and this can be used to predict whether a patient is at risk for osteoporotic fractures.

## JOINTS

### Joints between vertebrae in the back

The two major types of joints between vertebrae are:

- symphyses between vertebral bodies (Fig. 2.31), and
- synovial joints between articular processes (Fig. 2.32).

A typical vertebra has a total of six joints with adjacent vertebrae: four synovial joints (two above and two below) and two symphyses (one above and one below). Each symphysis includes an intervertebral disc.

Although the movement between any two vertebrae is limited, the summation of movement among all vertebrae results in a large range of movement by the vertebral column.

Movements by the vertebral column include flexion, extension, lateral flexion, rotation, and circumduction.

Movements by vertebrae in a specific region (cervical, thoracic, and lumbar) are determined by the shape and orientation of joint surfaces on the articular processes and on the vertebral bodies.

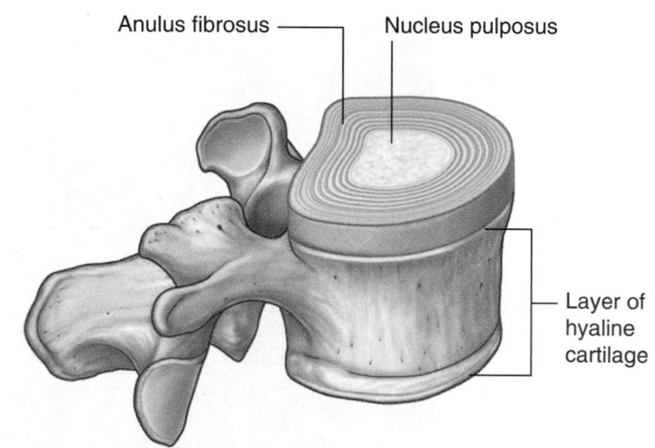


**Fig. 2.31** Intervertebral joints.

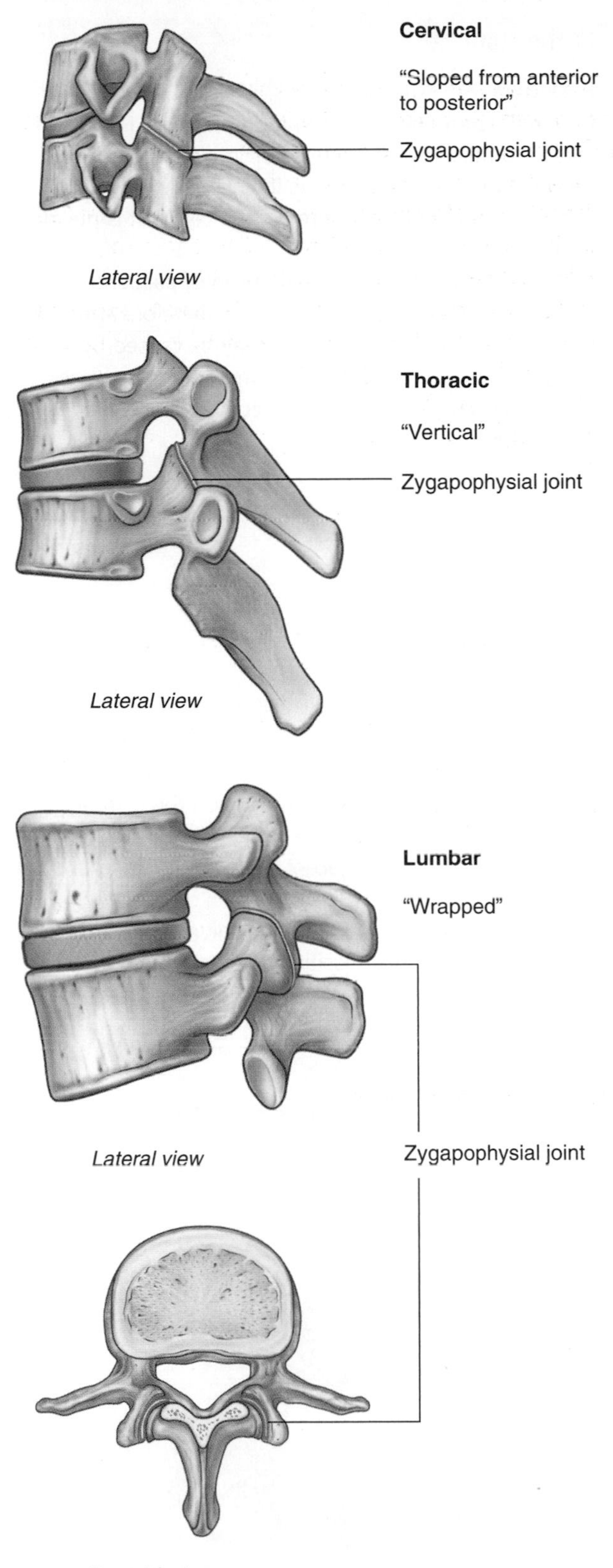


**Fig. 2.32** Zygapophysial joints.

### Symphyses between vertebral bodies (intervertebral discs)

The symphysis between adjacent vertebral bodies is formed by a layer of hyaline cartilage on each vertebral body and an intervertebral disc, which lies between the layers.

The **intervertebral disc** consists of an outer anulus fibrosus, which surrounds a central nucleus pulposus (Fig. 2.31).

- The **anulus fibrosus** consists of an outer ring of collagen surrounding a wider zone of fibrocartilage arranged in a lamellar configuration. This arrangement of fibers limits rotation between vertebrae.
- The **nucleus pulposus** fills the center of the intervertebral disc, is gelatinous, and absorbs compression forces between vertebrae.

Degenerative changes in the anulus fibrosus can lead to herniation of the nucleus pulposus. Posterolateral herniation can impinge on the roots of a spinal nerve in the intervertebral foramen.

### Joints between vertebral arches (zygapophysial joints)

The synovial joints between superior and inferior articular processes on adjacent vertebrae are the zygapophysial joints (Fig. 2.32). A thin articular capsule attached to the margins of the articular facets encloses each joint.

In cervical regions, the zygapophysial joints slope inferiorly from anterior to posterior and their shape facilitates flexion and extension. In thoracic regions, the joints are oriented vertically and their shape limits flexion and extension, but facilitates rotation. In lumbar regions, the joint surfaces are curved and adjacent processes interlock, thereby limiting range of movement, though flexion and extension are still major movements in the lumbar region.

### "Uncovertebral" joints

The lateral margins of the upper surfaces of typical cervical vertebrae are elevated into crests or lips termed uncinate processes. These may articulate with the body of the vertebra above to form small "uncovertebral" synovial joints (Fig. 2.33).

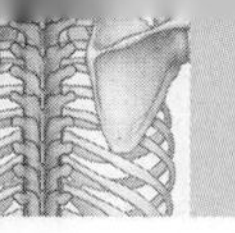

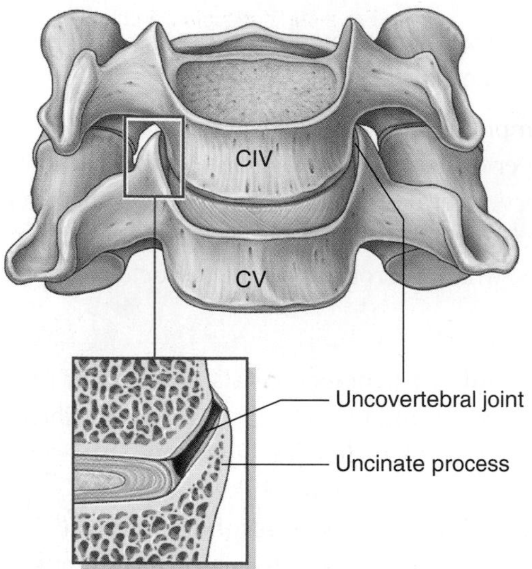


**Fig. 2.33** Uncovertebral joint.

## In the clinic

### Back pain

Back pain is an extremely common disorder. It can be related to mechanical problems or to disc protrusion impinging on a nerve. In cases involving discs, it may be necessary to operate and remove the disc that is pressing on the nerve.

Not infrequently, patients complain of pain and no immediate cause is found; the pain is therefore attributed to mechanical discomfort, which may be caused by degenerative disease. One of the treatments is to pass a needle into the facet joint and inject it with local anesthetic and corticosteroid.

## In the clinic

### Herniation of intervertebral discs

The discs between the vertebrae are made up of a central portion (the nucleus pulposus) and a complex series of fibrous rings (anulus fibrosus). A tear can occur within the anulus fibrosus through which the material of the nucleus pulposus can track. After a period of time, this material may track into the vertebral canal or into the intervertebral foramen to impinge on neural structures (Fig. 2.34). This is a common cause of back pain. A disc may protrude posteriorly to directly impinge on the cord or the roots of the lumbar nerves, depending on the level, or may protrude posterolaterally adjacent to the pedicle and impinge on the descending root.

In cervical regions of the vertebral column, cervical disc protrusions often become ossified and are termed disc osteophyte bars.

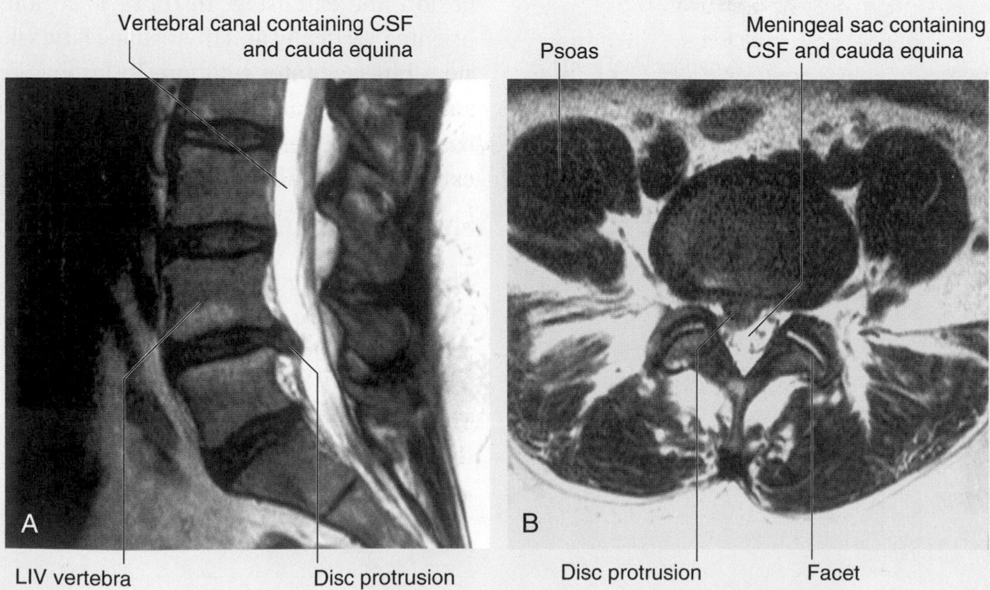


**Fig. 2.34** Disc protrusion. T2-weighted magnetic resonance images of the lumbar region of the vertebral column. **A.** Sagittal plane. **B.** Axial plane.

**In the clinic**

**Joint diseases**

Some diseases have a predilection for synovial joints rather than symphyses. A typical example is rheumatoid arthritis, which primarily affects synovial joints and synovial bursae, resulting in destruction of the joint and its lining. Symphyses are usually preserved.

## LIGAMENTS

Joints between vertebrae are reinforced and supported by numerous ligaments, which pass between vertebral bodies and interconnect components of the vertebral arches.

### Anterior and posterior longitudinal ligaments

The anterior and posterior longitudinal ligaments are on the anterior and posterior surfaces of the vertebral bodies and extend along most of the vertebral column (Fig. 2.35).

The **anterior longitudinal ligament** is attached superiorly to the base of the skull and extends inferiorly to attach to the anterior surface of the sacrum. Along its length it is attached to the vertebral bodies and intervertebral discs.

The **posterior longitudinal ligament** is on the posterior surfaces of the vertebral bodies and lines the anterior surface of the vertebral canal. Like the anterior longitudinal ligament, it is attached along its length to the vertebral bodies and intervertebral discs. The upper part of the posterior longitudinal ligament that connects CII to the intracranial aspect of the base of the skull is termed the **tectorial membrane** (see Fig. 2.20B).

### Ligamenta flava

The **ligamenta flava**, on each side, pass between the laminae of adjacent vertebrae (Fig. 2.36). These thin, broad ligaments consist predominantly of elastic tissue and form part of the posterior surface of the vertebral canal. Each ligamentum flavum runs between the posterior surface of the lamina on the vertebra below to the anterior surface of the lamina of the vertebra above. The ligamenta flava resist separation of the laminae in flexion and assist in extension back to the anatomical position.

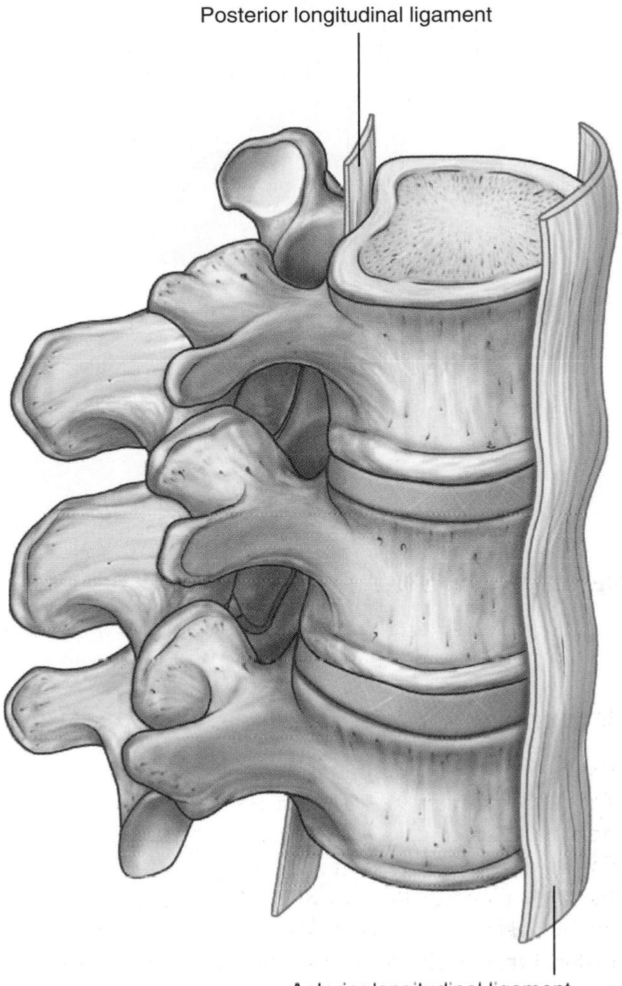


**Fig. 2.35** Anterior and posterior longitudinal ligaments of vertebral column.

### Supraspinous ligament and ligamentum nuchae

The supraspinous ligament connects and passes along the tips of the vertebral spinous processes from vertebra

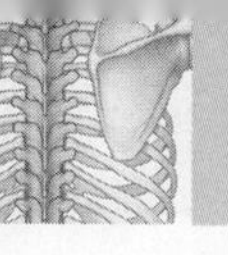



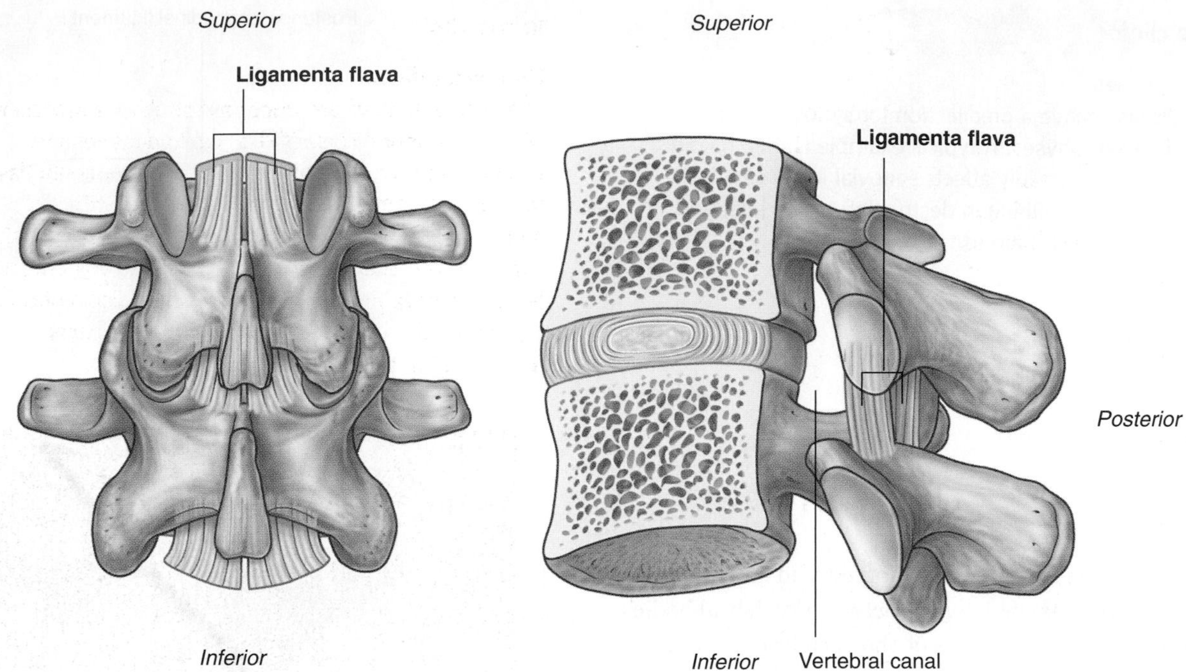


**Fig. 2.36** Ligamenta flava.

CVII to the sacrum (Fig. 2.37). From vertebra CVII to the skull, the ligament becomes structurally distinct from more caudal parts of the ligament and is called the ligamentum nuchae.

The **ligamentum nuchae** is a triangular, sheet-like structure in the median sagittal plane:

- The base of the triangle is attached to the skull, from the external occipital protuberance to the foramen magnum.
- The apex is attached to the tip of the spinous process of vertebra CVII.
- The deep side of the triangle is attached to the posterior tubercle of vertebra CI and the spinous processes of the other cervical vertebrae.

The ligamentum nuchae supports the head. It resists flexion and facilitates returning the head to the anatomical position. The broad lateral surfaces and the posterior edge of the ligament provide attachment for adjacent muscles.

## Interspinous ligaments

Interspinous ligaments pass between adjacent vertebral spinous processes (Fig. 2.38). They attach from the base to the apex of each spinous process and blend with the supraspinous ligament posteriorly and the ligamenta flava anteriorly on each side.

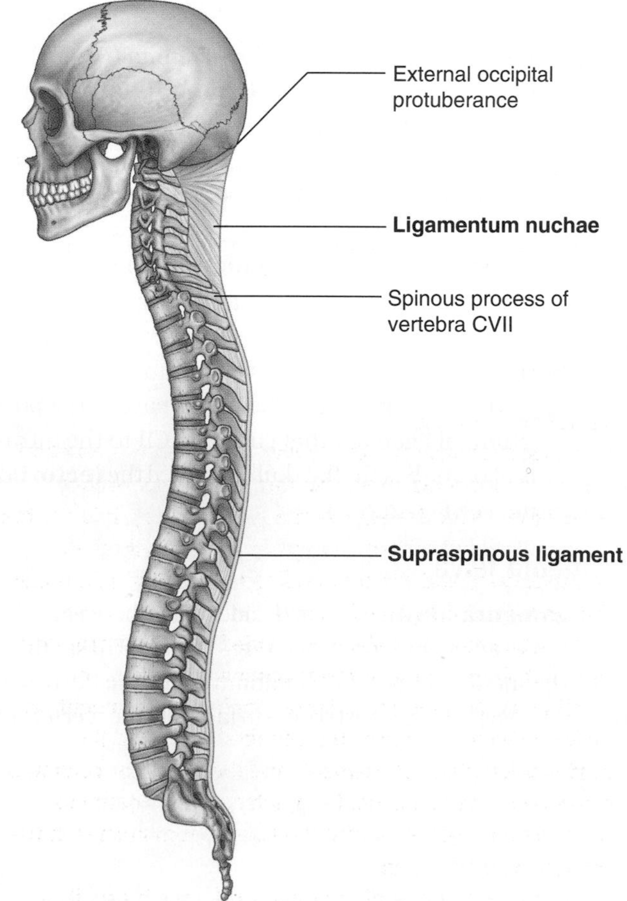


**Fig. 2.37** Supraspinous ligament and ligamentum nuchae.

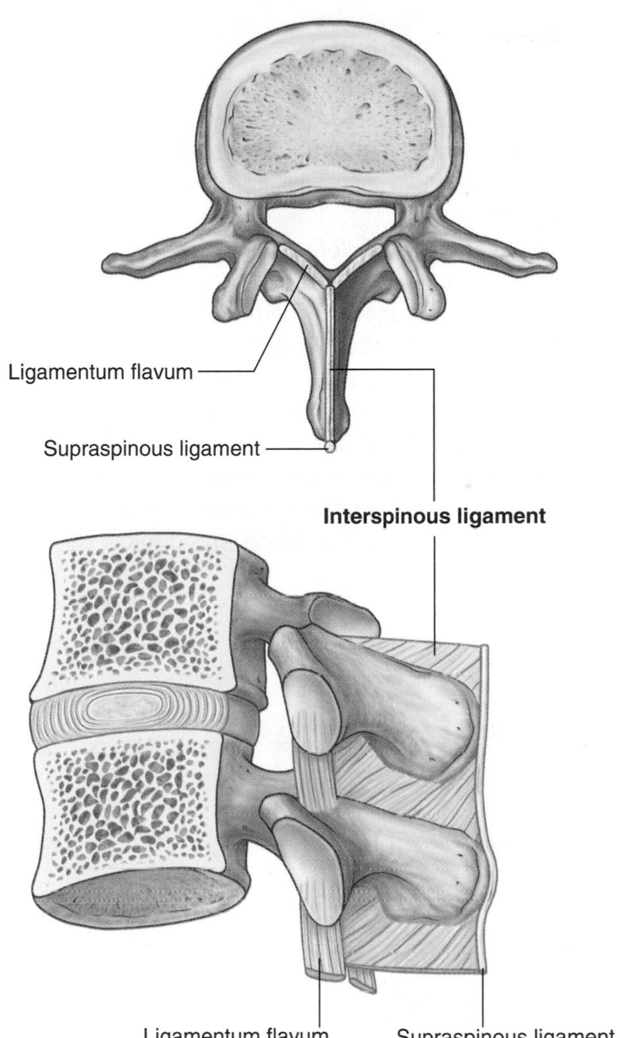


**Fig. 2.38** Interspinous ligaments.

## In the clinic

### Ligamenta flava

The ligamenta flava are important structures associated with the vertebral canal (Fig. 2.39). In degenerative conditions of the vertebral column, the ligamenta flava may hypertrophy. This is often associated with hypertrophy and arthritic change of the zygapophysial joints. In combination, zygapophysial joint hypertrophy, ligamenta flava hypertrophy, and a mild disc protrusion can reduce the dimensions of the vertebral canal, producing the syndrome of spinal stenosis.

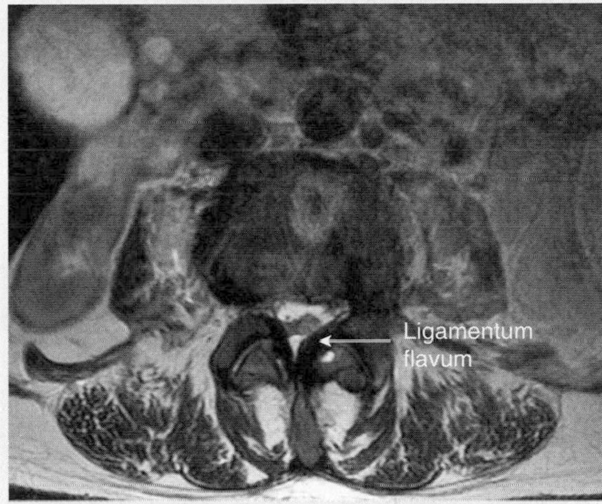


**Fig. 2.39** Axial slice MRI through the lumbar spine demonstrating bilateral hypertrophy of the ligamentum flavum.

## In the clinic

### Vertebral fractures

Vertebral fractures can occur anywhere along the vertebral column. In most instances, the fracture will heal under appropriate circumstances. At the time of injury, it is not the fracture itself but related damage to the contents of the vertebral canal and the surrounding tissues that determines the severity of the patient's condition.

Vertebral column stability is divided into three arbitrary clinical "columns": the **anterior column** consists of the vertebral bodies and the anterior longitudinal ligament; the **middle column** comprises the vertebral body and the posterior longitudinal ligament; and the **posterior column** is made up of the ligamenta flava, interspinous ligaments, supraspinous ligaments, and the ligamentum nuchae in the cervical vertebral column.

Destruction of one of the clinical columns is usually a stable injury requiring little more than rest and appropriate analgesia. Disruption of two columns is highly likely to be unstable and requires fixation and immobilization. A three-column spinal injury usually results in a significant neurological event and requires fixation to prevent further extension of the neurological defect and to create vertebral column stability.

At the craniocervical junction, a complex series of ligaments create stability. If the traumatic incident disrupts craniocervical stability, the chances of a significant spinal cord injury are extremely high. The consequences are quadriplegia. In addition, respiratory function may be compromised by paralysis of the phrenic nerve (which arises from spinal nerves C3 to C5), and severe hypotension (low blood pressure) may result from central disruption of the sympathetic part of the autonomic division of the nervous system.

*(continues)*

## In the clinic—cont'd

Mid and lower cervical vertebral column disruption may produce a range of complex neurological problems involving the upper and lower limbs, although below the level of C5, respiratory function is unlikely to be compromised.

Lumbar vertebral column injuries are rare. When they occur, they usually involve significant force. Knowing that a significant force is required to fracture a vertebra, one must assess the abdominal organs and the rest of the axial skeleton for further fractures and visceral rupture.

Vertebral injuries may also involve the soft tissues and supporting structures between the vertebrae. Typical examples of this are the unifacetal and bifacetal cervical vertebral dislocations that occur in hyperflexion injuries.

### *Pars interarticularis fractures*

The pars interarticularis is a clinical term to describe the specific region of a vertebra between the superior and inferior facet (zygapophysial) joints (Fig. 2.40A). This region is susceptible to trauma, especially in athletes.

If a fracture occurs around the pars interarticularis, the vertebral body may slip anteriorly and compress the vertebral canal.

The most common sites for pars interarticularis fractures are the LIV and LV levels (Fig. 2.40B,C). (Clinicians often refer to parts of the back in shorthand terms that are not strictly anatomical; for example, facet joints and apophyseal joints are terms used instead of zygapophysial joints, and spinal column is used instead of vertebral column.)

It is possible for a vertebra to slip anteriorly upon its inferior counterpart without a pars interarticularis fracture. Usually this is related to abnormal anatomy of the facet joints, facet joint degenerative change. This disorder is termed **spondylolisthesis.**

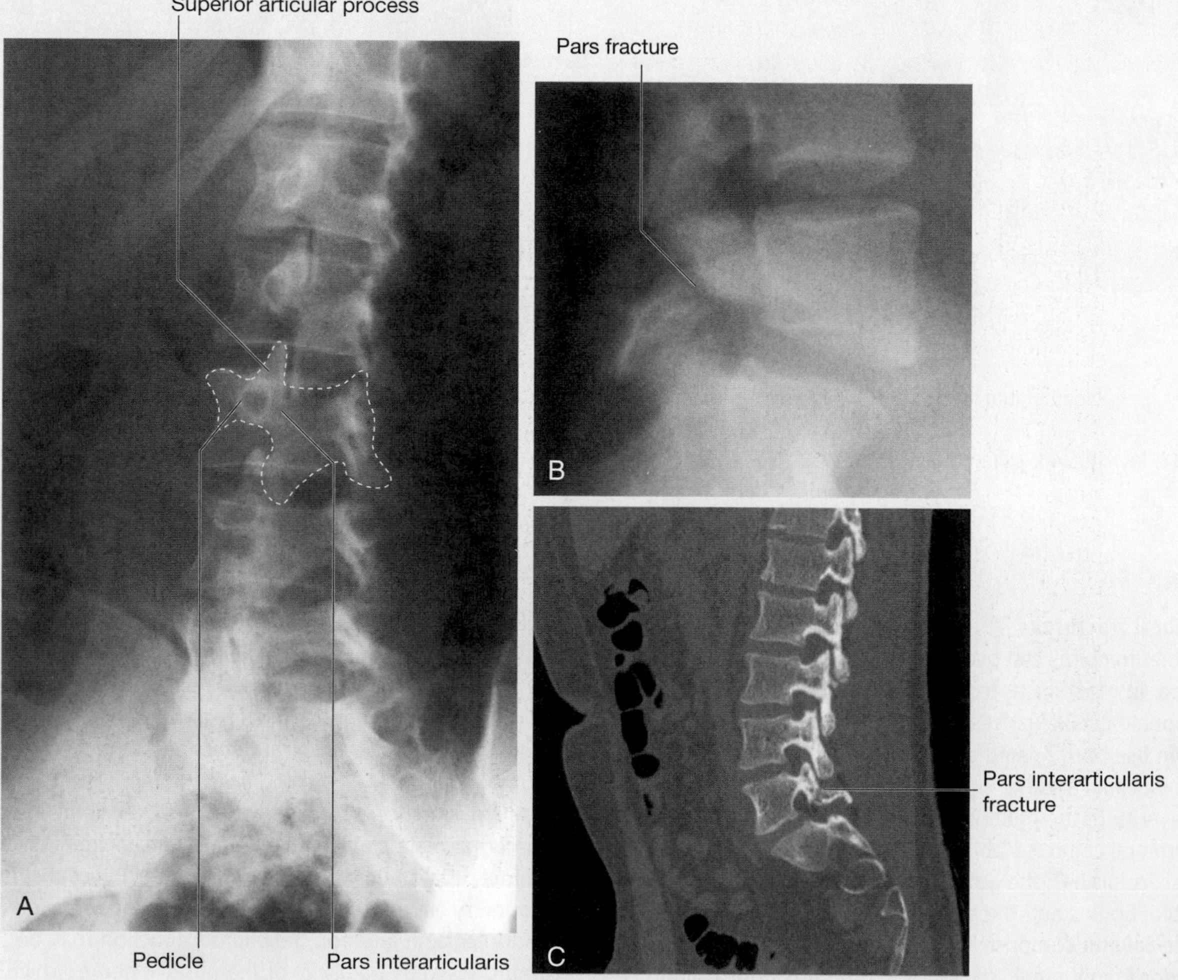


**Fig. 2.40** Radiograph of lumbar region of vertebral column, oblique view ("Scottie dog"). **A.** Normal radiograph of lumbar region of vertebral column, oblique view. In this view, the transverse process (nose), pedicle (eye), superior articular process (ear), inferior articular process (front leg), and pars interarticularis (neck) resemble a dog. A fracture of the pars interarticularis is visible as a break in the neck of the dog, or the appearance of a collar. **B.** Fracture of pars interarticularis. **C.** CT of lumbar spine shows fracture of the LV pars interarticularis.

## In the clinic

### Surgical procedures on the back

***Discectomy/laminectomy***

A prolapsed intervertebral disc may impinge upon the meningeal (thecal) sac, cord, and most commonly the nerve root, producing symptoms attributable to that level. In some instances the disc protrusion will undergo a degree of involution that may allow symptoms to resolve without intervention. In some instances pain, loss of function, and failure to resolve may require surgery to remove the disc protrusion.

It is of the utmost importance that the level of the disc protrusion is identified before surgery. This may require MRI scanning and on-table fluoroscopy to prevent operating on the wrong level. A midline approach to the right or to the left of the spinous processes will depend upon the most prominent site of the disc bulge. In some instances removal of the lamina will increase the potential space and may relieve symptoms. Some surgeons perform a small fenestration (windowing) within the ligamentum flavum. This provides access to the canal. The meningeal sac and its contents are gently retracted, exposing the nerve root and the offending disc. The disc is dissected free, removing its effect on the nerve root and the canal.

***Spinal Fusion***

Spinal fusion is performed when it is necessary to fuse one vertebra with the corresponding superior or inferior vertebra, and in some instances multilevel fusion may be necessary. Indications are varied, though they include stabilization after fracture, stabilization related to tumor infiltration, and stabilization when mechanical pain is produced either from the disc or from the posterior elements.

There are a number of surgical methods in which a fusion can be performed, through either a posterior approach and fusing the posterior elements, an anterior approach by removal of the disc and either disc replacement or anterior fusion, or in some instances a 360° fusion where the posterior elements and the vertebral bodies are fused (Fig. 2.41A,B).

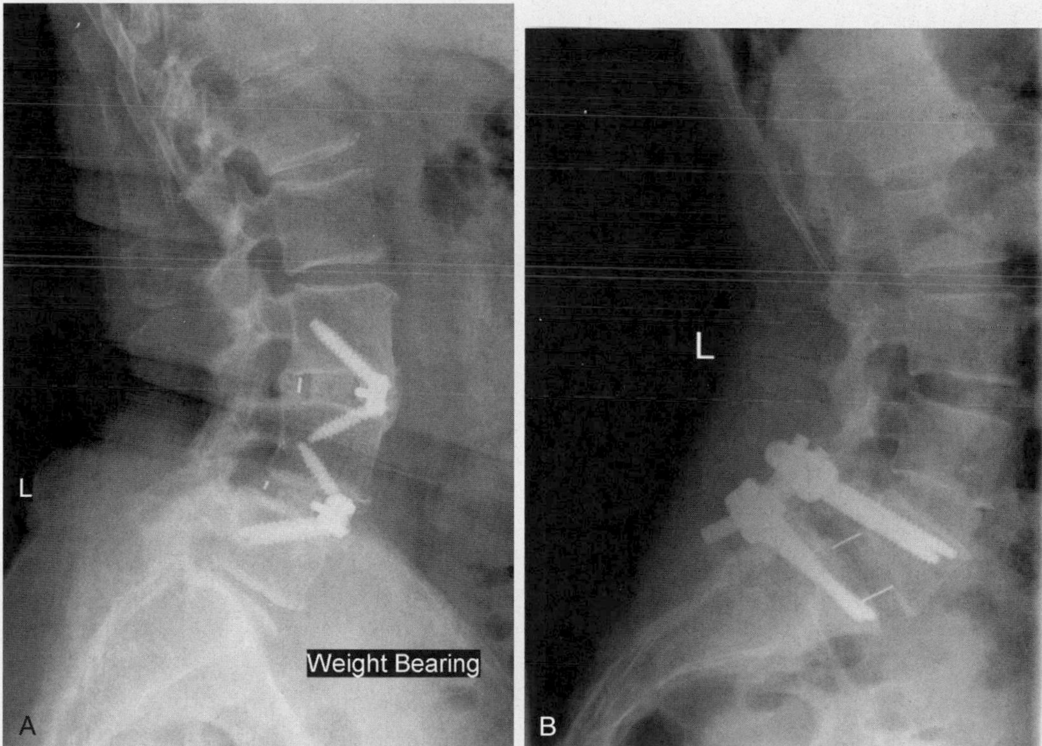


**Fig. 2.41 A.** Anterior lumbar interbody fusion (ALIF). **B.** Posterior lumbar interbody fusion (PLIF).

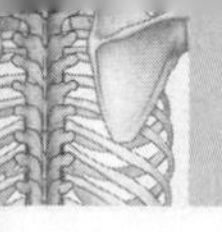



## BACK MUSCULATURE

Muscles of the back are organized into superficial, intermediate, and deep groups.

Muscles in the superficial and intermediate groups are extrinsic muscles because they originate embryologically from locations other than the back. They are innervated by anterior rami of spinal nerves:

- The superficial group consists of muscles related to and involved in movements of the upper limb.
- The intermediate group consists of muscles attached to the ribs and may serve a respiratory function.

Muscles of the deep group are intrinsic muscles because they develop in the back. They are innervated by posterior rami of spinal nerves and are directly related to movements of the vertebral column and head.

### Superficial group of back muscles

The muscles in the superficial group are immediately deep to the skin and superficial fascia (Figs. 2.42 to 2.45). They attach the superior part of the appendicular skeleton (clavicle, scapula, and humerus) to the axial skeleton (skull, ribs, and vertebral column). Because these muscles are primarily involved with movements of this part of the appendicular skeleton, they are sometimes referred to as the **appendicular group**.

Muscles in the superficial group include the trapezius, latissimus dorsi, rhomboid major, rhomboid minor, and levator scapulae. The rhomboid major, rhomboid minor, and levator scapulae muscles are located deep to the trapezius muscle in the superior part of the back.

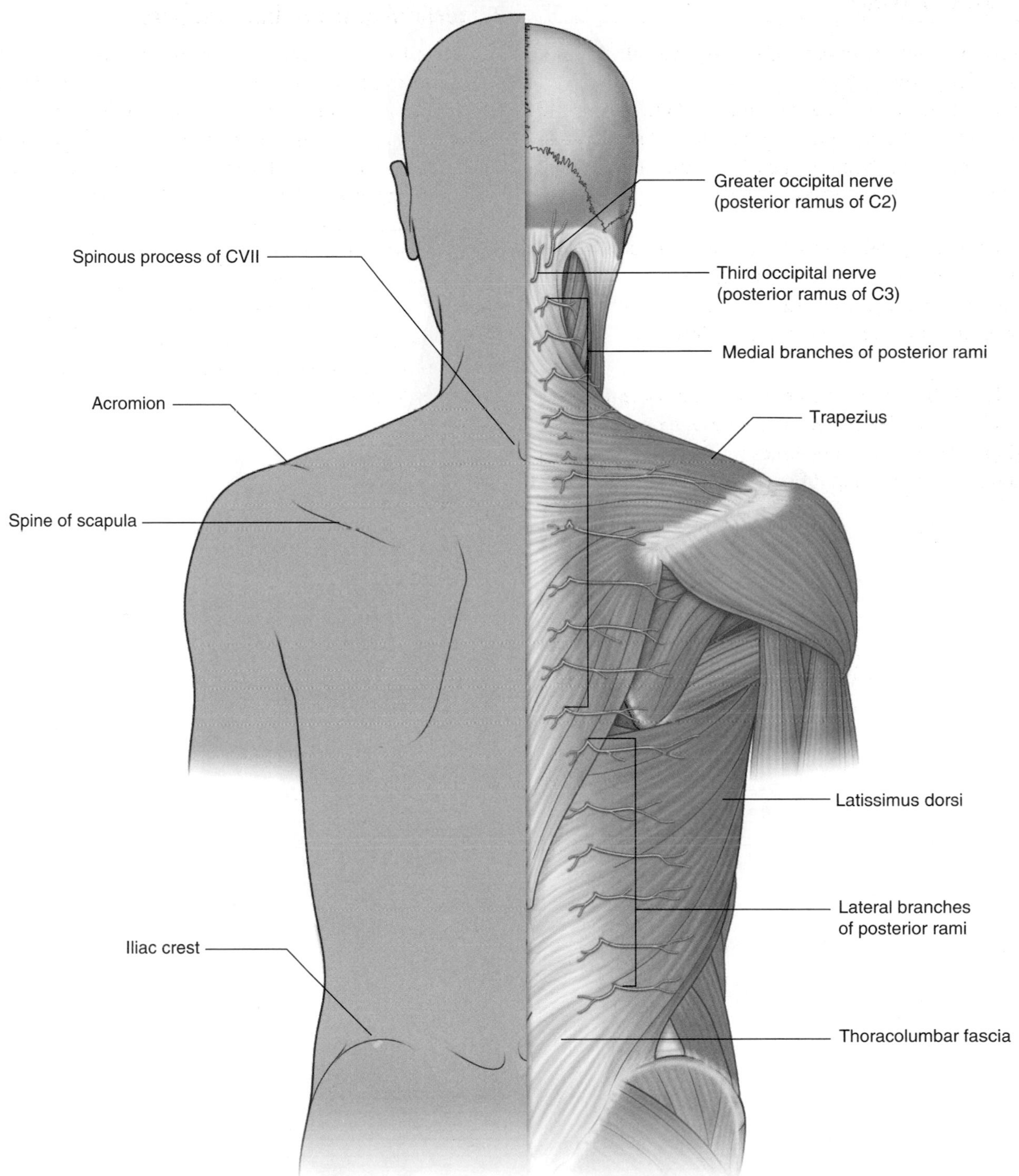


**Fig. 2.42** Superficial group of back muscles—trapezius and latissimus dorsi.

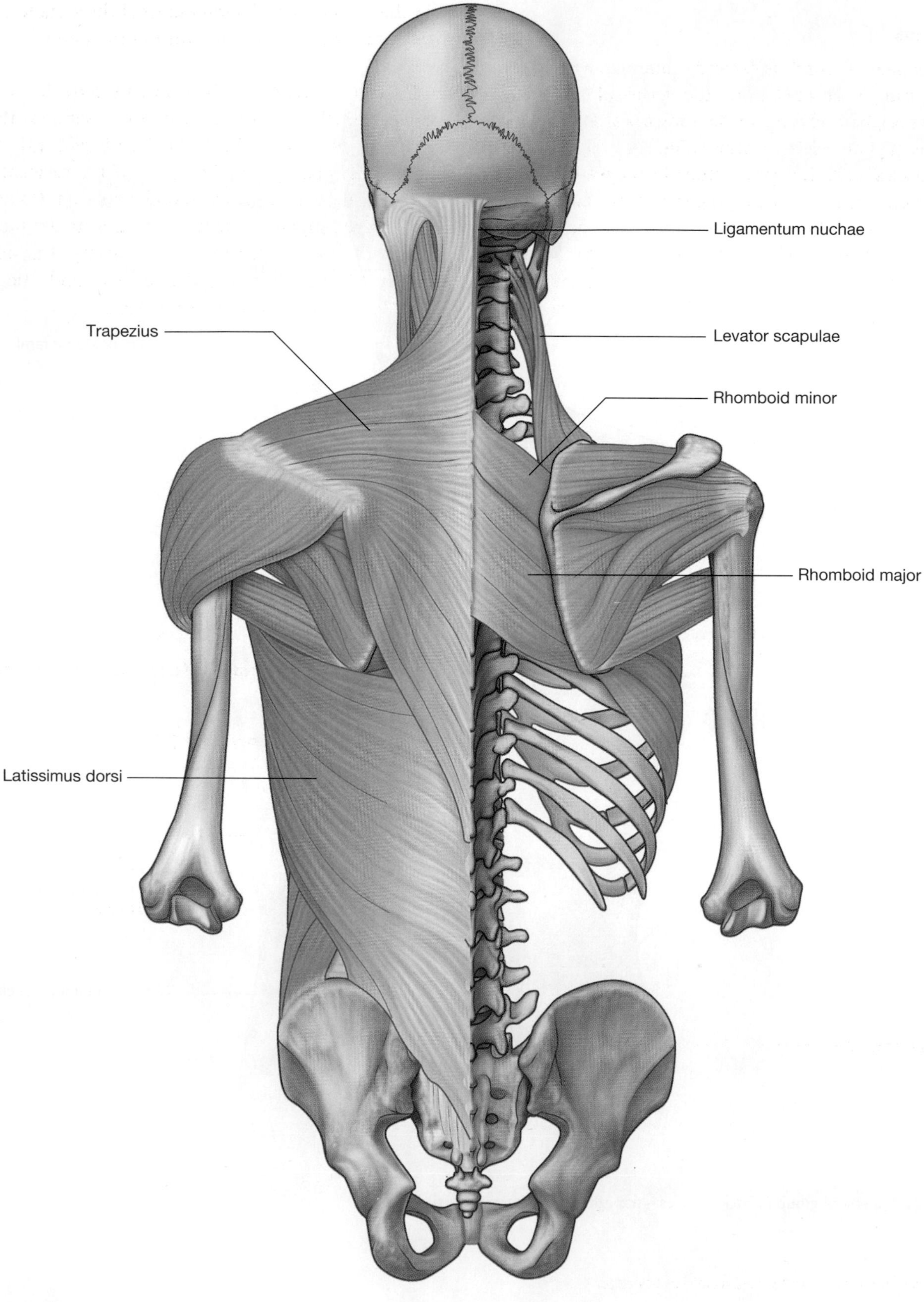

**Fig. 2.43** Superficial group of back muscles—trapezius and latissimus dorsi, with rhomboid major, rhomboid minor, and levator scapulae located deep to trapezius in the superior part of the back.

### Trapezius

Each **trapezius** muscle is flat and triangular, with the base of the triangle situated along the vertebral column (the muscle's origin) and the apex pointing toward the tip of the shoulder (the muscle's insertion) (Fig. 2.43 and Table 2.1). The muscles on both sides together form a trapezoid.

The superior fibers of the trapezius, from the skull and upper portion of the vertebral column, descend to attach to the lateral third of the clavicle and to the acromion of the scapula. Contraction of these fibers elevates the scapula. In addition, the superior and inferior fibers work together to rotate the lateral aspect of the scapula upward, which needs to occur when raising the upper limb above the head.

Motor innervation of the trapezius is by the accessory nerve [XI], which descends from the neck onto the deep surface of the muscle (Fig. 2.44). Proprioceptive fibers from the trapezius pass in the branches of the cervical plexus and enter the spinal cord at spinal cord levels C3 and C4.

The blood supply to the trapezius is from the superficial branch of the transverse cervical artery, the acromial branch of the suprascapular artery, and the dorsal branches of posterior intercostal arteries.

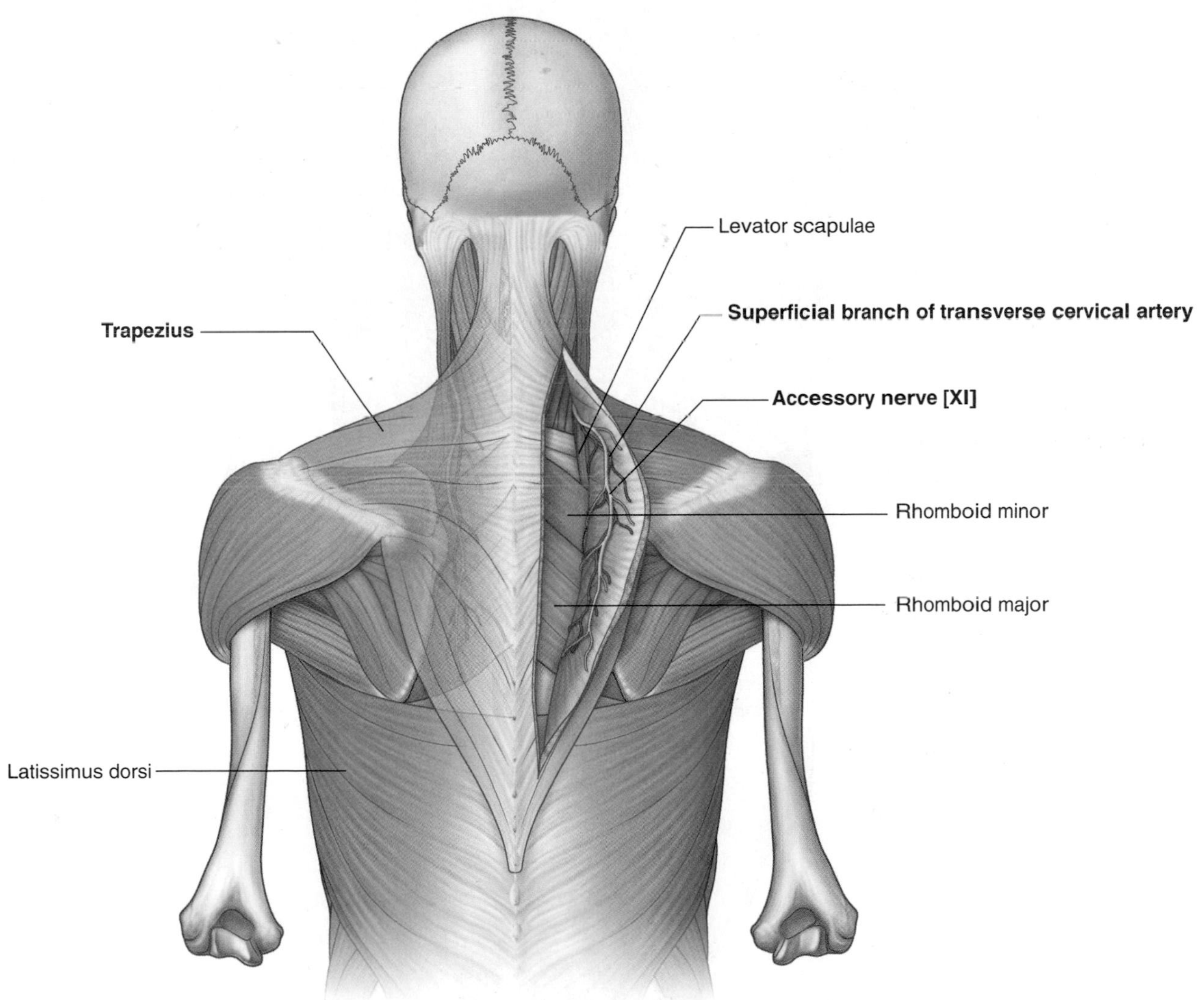


**Fig. 2.44** Innervation and blood supply of trapezius.

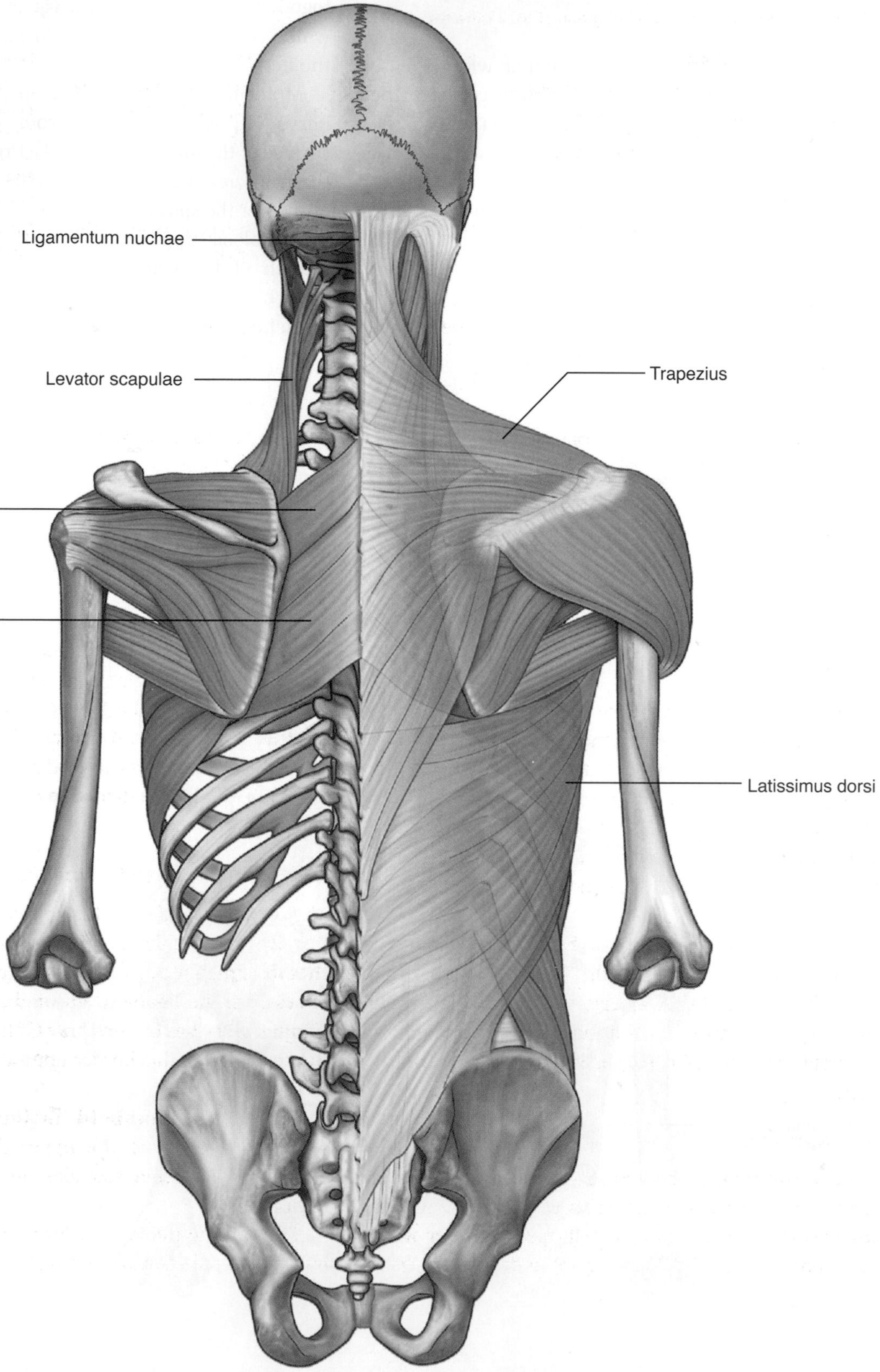

**Fig. 2.45** Rhomboid muscles and levator scapulae.

**Table 2.1** Superficial (appendicular) group of back muscles

| Muscle | Origin | Insertion | Innervation | Function |
|---|---|---|---|---|
| Trapezius | Superior nuchal line, external occipital protuberance, ligamentum nuchae, spinous processes of CVII to TXII | Lateral one third of clavicle, acromion, spine of scapula | Motor—accessory nerve [XI]; proprioception—C3 and C4 | Assists in rotating the scapula during abduction of humerus above horizontal; upper fibers elevate, middle fibers adduct, and lower fibers depress scapula |
| Latissimus dorsi | Spinous processes of TVII to LV and sacrum, iliac crest, ribs X to XII | Floor of intertubercular sulcus of humerus | Thoracodorsal nerve (C6 to C8) | Extends, adducts, and medially rotates humerus |
| Levator scapulae | Transverse processes of CI to CIV | Upper portion medial border of scapula | C3 to C4 and dorsal scapular nerve (C4, C5) | Elevates scapula |
| Rhomboid major | Spinous processes of TII to TV | Medial border of scapula between spine and inferior angle | Dorsal scapular nerve (C4, C5) | Retracts (adducts) and elevates scapula |
| Rhomboid minor | Lower portion of ligamentum nuchae, spinous processes of CVII and TI | Medial border of scapula at spine of scapula | Dorsal scapular nerve (C4, C5) | Retracts (adducts) and elevates scapula |

### Latissimus dorsi

**Latissimus dorsi** is a large, flat triangular muscle that begins in the lower portion of the back and tapers as it ascends to a narrow tendon that attaches to the humerus anteriorly (Figs. 2.42 to 2.45 and Table 2.1). As a result, movements associated with this muscle include extension, adduction, and medial rotation of the upper limb. The latissimus dorsi can also depress the shoulder, preventing its upward movement.

The thoracodorsal nerve of the brachial plexus innervates the latissimus dorsi muscle. Associated with this nerve is the thoracodorsal artery, which is the primary blood supply of the muscle. Additional small arteries come from dorsal branches of posterior intercostal and lumbar arteries.

### Levator scapulae

**Levator scapulae** is a slender muscle that descends from the transverse processes of the upper cervical vertebrae to the upper portion of the scapula on its medial border at the superior angle (Figs. 2.43 and 2.45 and Table 2.1). It elevates the scapula and may assist other muscles in rotating the lateral aspect of the scapula inferiorly.

The levator scapulae is innervated by branches from the anterior rami of spinal nerves C3 and C4 and the dorsal scapular nerve, and its arterial supply consists of branches primarily from the transverse and ascending cervical arteries.

### Rhomboid minor and rhomboid major

The two rhomboid muscles are inferior to levator scapulae (Fig. 2.45 and Table 2.1). **Rhomboid minor** is superior to rhomboid major, and is a small, cylindrical muscle that arises from the ligamentum nuchae of the neck and the spinous processes of vertebrae CVII and TI and attaches to the medial scapular border opposite the root of the spine of the scapula.

The larger **rhomboid major** originates from the spinous processes of the upper thoracic vertebrae and attaches to the medial scapular border inferior to rhomboid minor.

The two rhomboid muscles work together to retract or pull the scapula toward the vertebral column. With other muscles they may also rotate the lateral aspect of the scapula inferiorly.

The dorsal scapular nerve, a branch of the brachial plexus, innervates both rhomboid muscles (Fig. 2.46).

## Intermediate group of back muscles

The muscles in the intermediate group of back muscles consist of two thin muscular sheets in the superior and inferior regions of the back, immediately deep to the muscles in the superficial group (Fig. 2.47 and Table 2.2). Fibers from these two serratus posterior muscles (**serratus posterior superior** and **serratus posterior inferior**) pass obliquely outward from the vertebral column to attach to the ribs. This positioning suggests a respiratory function, and at times, these muscles have been referred to as the respiratory group.

Serratus posterior superior is deep to the rhomboid muscles, whereas serratus posterior inferior is deep to the latissimus dorsi. Both serratus posterior muscles are attached to the vertebral column and associated structures medially, and either descend (the fibers of the serratus posterior superior) or ascend (the fibers of the serratus posterior inferior) to attach to the ribs. These two muscles therefore elevate and depress the ribs.

The serratus posterior muscles are innervated by segmental branches of anterior rami of intercostal nerves. Their vascular supply is provided by a similar segmental pattern through the intercostal arteries.

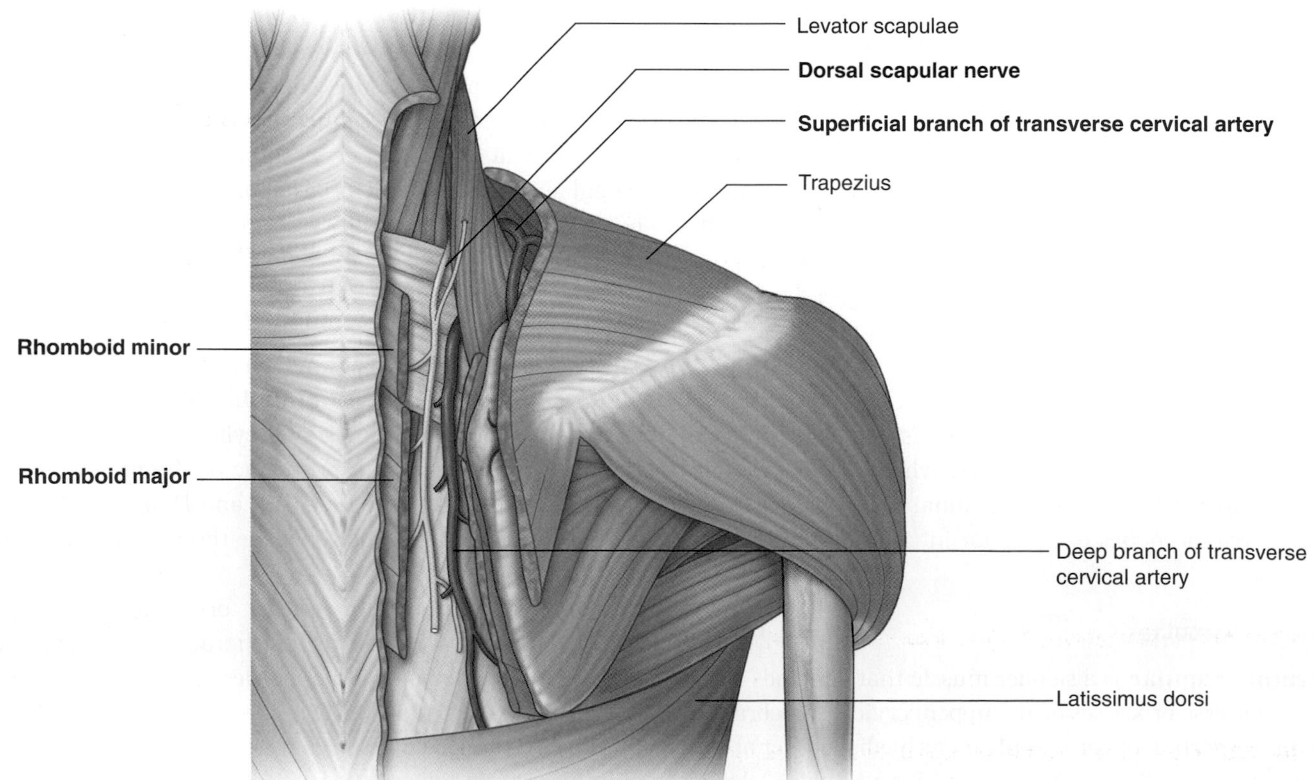


**Fig. 2.46** Innervation and blood supply of the rhomboid muscles.

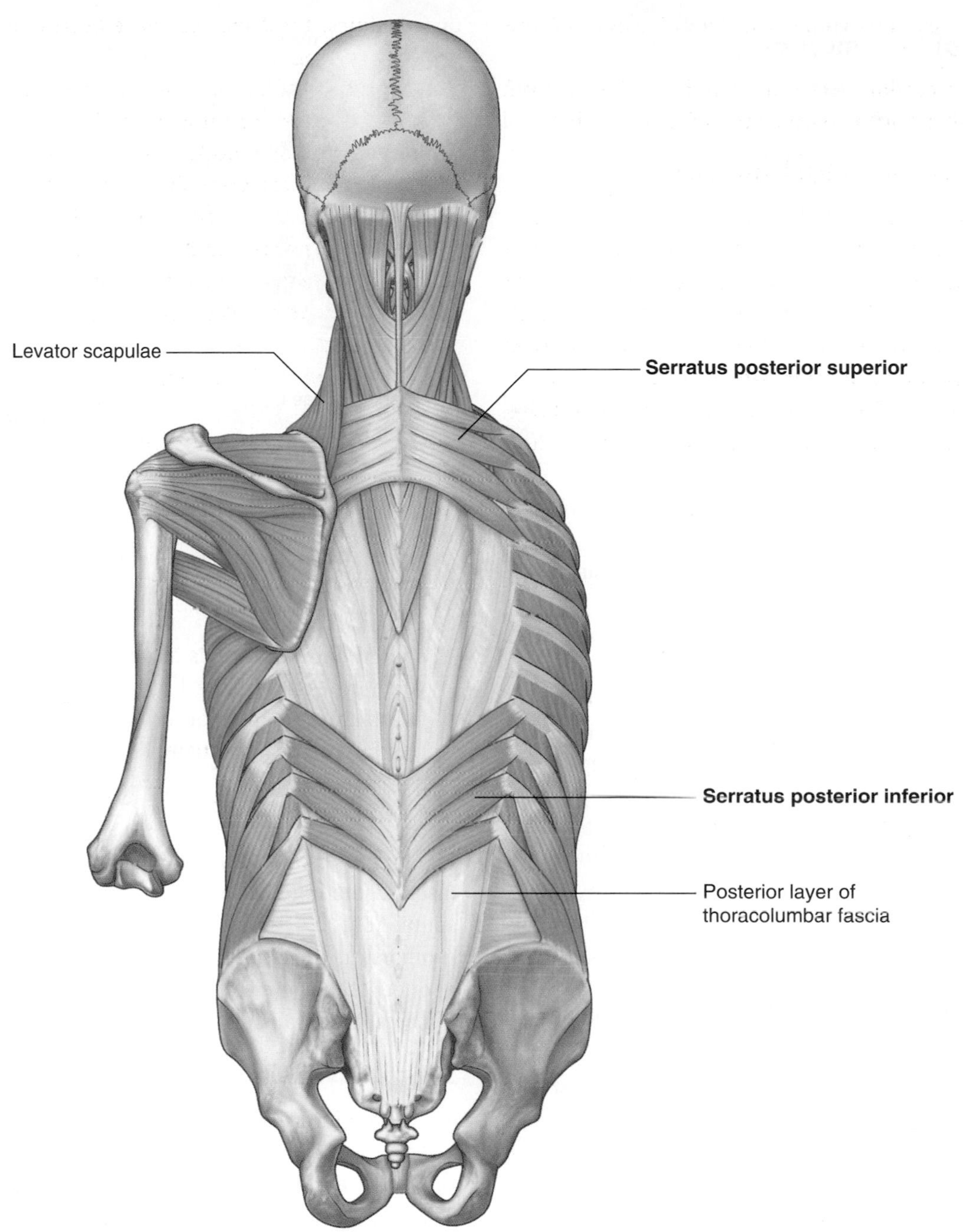


**Fig. 2.47** Intermediate group of back muscles—serratus posterior muscles.

**Table 2.2** Intermediate (respiratory) group of back muscles

| Muscle | Origin | Insertion | Innervation | Function |
|---|---|---|---|---|
| Serratus posterior superior | Lower portion of ligamentum nuchae, spinous processes of CVII to TIII, and supraspinous ligaments | Upper border of ribs II to V just lateral to their angles | Anterior rami of upper thoracic nerves (T2 to T5) | Elevates ribs II to V |
| Serratus posterior inferior | Spinous processes of TXI to LIII and supraspinous ligaments | Lower border of ribs IX to XII just lateral to their angles | Anterior rami of lower thoracic nerves (T9 to T12) | Depresses ribs IX to XII and may prevent lower ribs from being elevated when the diaphragm contracts |

## Deep group of back muscles

The deep or intrinsic muscles of the back extend from the pelvis to the skull and are innervated by segmental branches of the posterior rami of spinal nerves. They include:

- the extensors and rotators of the head and neck—the splenius capitis and cervicis (spinotransversales muscles),
- the extensors and rotators of the vertebral column—the erector spinae and transversospinales, and
- the short segmental muscles—the interspinales and intertransversarii.

The vascular supply to this deep group of muscles is through branches of the vertebral, deep cervical, occipital, transverse cervical, posterior intercostal, subcostal, lumbar, and lateral sacral arteries.

### Thoracolumbar fascia

The **thoracolumbar fascia** covers the deep muscles of the back and trunk (Fig. 2.48). This fascial layer is critical to the overall organization and integrity of the region:

- Superiorly, it passes anteriorly to the serratus posterior muscle and is continuous with deep fascia in the neck.
- In the thoracic region, it covers the deep muscles and separates them from the muscles in the superficial and intermediate groups.
- Medially, it attaches to the spinous processes of the thoracic vertebrae and, laterally, to the angles of the ribs.

The medial attachments of the latissimus dorsi and serratus posterior inferior muscles blend into the thoracolumbar fascia. In the lumbar region, the thoracolumbar fascia consists of three layers:

- The posterior layer is thick and is attached to the spinous processes of the lumbar vertebrae and sacral vertebrae and to the supraspinous ligament—from these attachments, it extends laterally to cover the erector spinae.
- The middle layer is attached medially to the tips of the transverse processes of the lumbar vertebrae and intertransverse ligaments—inferiorly, it is attached to the iliac crest and, superiorly, to the lower border of rib XII.
- The anterior layer covers the anterior surface of the quadratus lumborum muscle (a muscle of the posterior abdominal wall) and is attached medially to the transverse processes of the lumbar vertebrae—inferiorly, it is attached to the iliac crest and, superiorly, it forms the lateral arcuate ligament for attachment of the diaphragm.

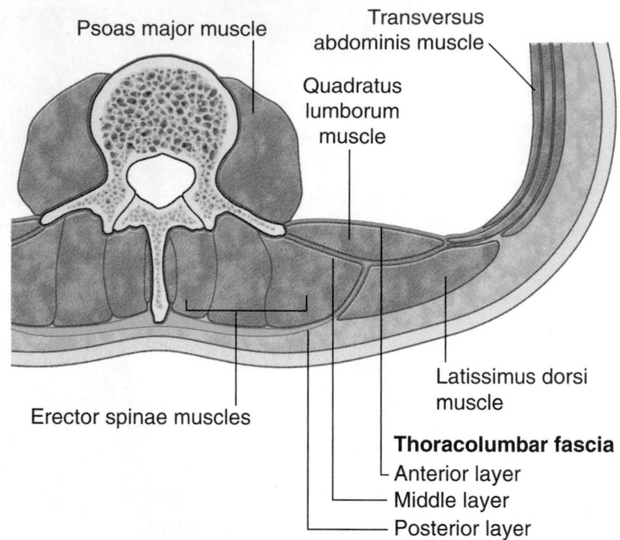


**Fig. 2.48** Thoracolumbar fascia and the deep back muscles (transverse section).

The posterior and middle layers of the thoracolumbar fascia come together at the lateral margin of the erector spinae (Fig. 2.48). At the lateral border of the quadratus lumborum, the anterior layer joins them and forms the aponeurotic origin for the transversus abdominis muscle of the abdominal wall.

### Spinotransversales muscles

The two spinotransversales muscles run from the spinous processes and ligamentum nuchae upward and laterally (Fig. 2.49 and Table 2.3):

- The splenius capitis is a broad muscle attached to the occipital bone and mastoid process of the temporal bone.
- The splenius cervicis is a narrow muscle attached to the transverse processes of the upper cervical vertebrae.

Together the spinotransversales muscles draw the head backward, extending the neck. Individually, each muscle rotates the head to one side—the same side as the contracting muscle.

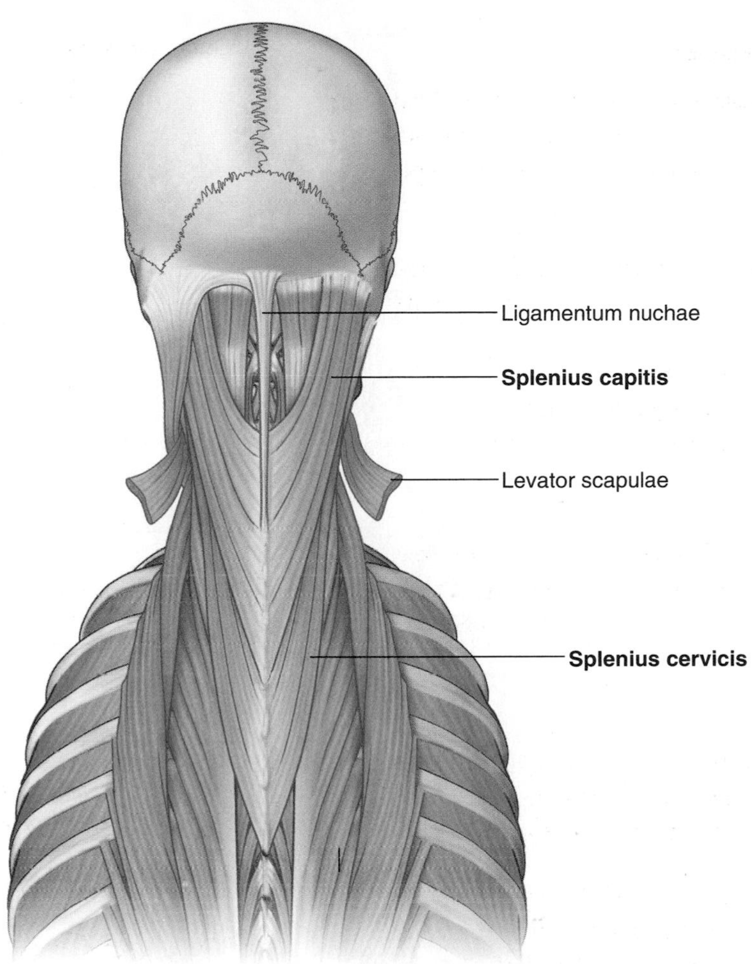


**Fig. 2.49** Deep group of back muscles—spinotransversales muscles (splenius capitis and splenius cervicis).

**Table 2.3** Spinotransversales muscles

| Muscle | Origin | Insertion | Innervation | Function |
|---|---|---|---|---|
| Splenius capitis | Lower half of ligamentum nuchae, spinous processes of CVII to TIV | Mastoid process, skull below lateral one third of superior nuchal line | Posterior rami of middle cervical nerves | Together—draw head backward, extending neck; individually—draw and rotate head to one side (turn face to same side) |
| Splenius cervicis | Spinous processes of TIII to TVI | Transverse processes of CI to CIII | Posterior rami of lower cervical nerves | Together—extend neck; individually—draw and rotate head to one side (turn face to same side) |

## Erector spinae muscles

The erector spinae is the largest group of intrinsic back muscles. The muscles lie posterolaterally to the vertebral column between the spinous processes medially and the angles of the ribs laterally. They are covered in the thoracic and lumbar regions by thoracolumbar fascia and the serratus posterior inferior, rhomboid, and splenius muscles. The mass arises from a broad, thick tendon attached to the sacrum, the spinous processes of the lumbar and lower thoracic vertebrae, and the iliac crest (Fig. 2.50 and Table 2.4). It divides in the upper lumbar region into three vertical columns of muscle, each of which is further subdivided regionally (lumborum, thoracis, cervicis, and capitis), depending on where the muscles attach superiorly.

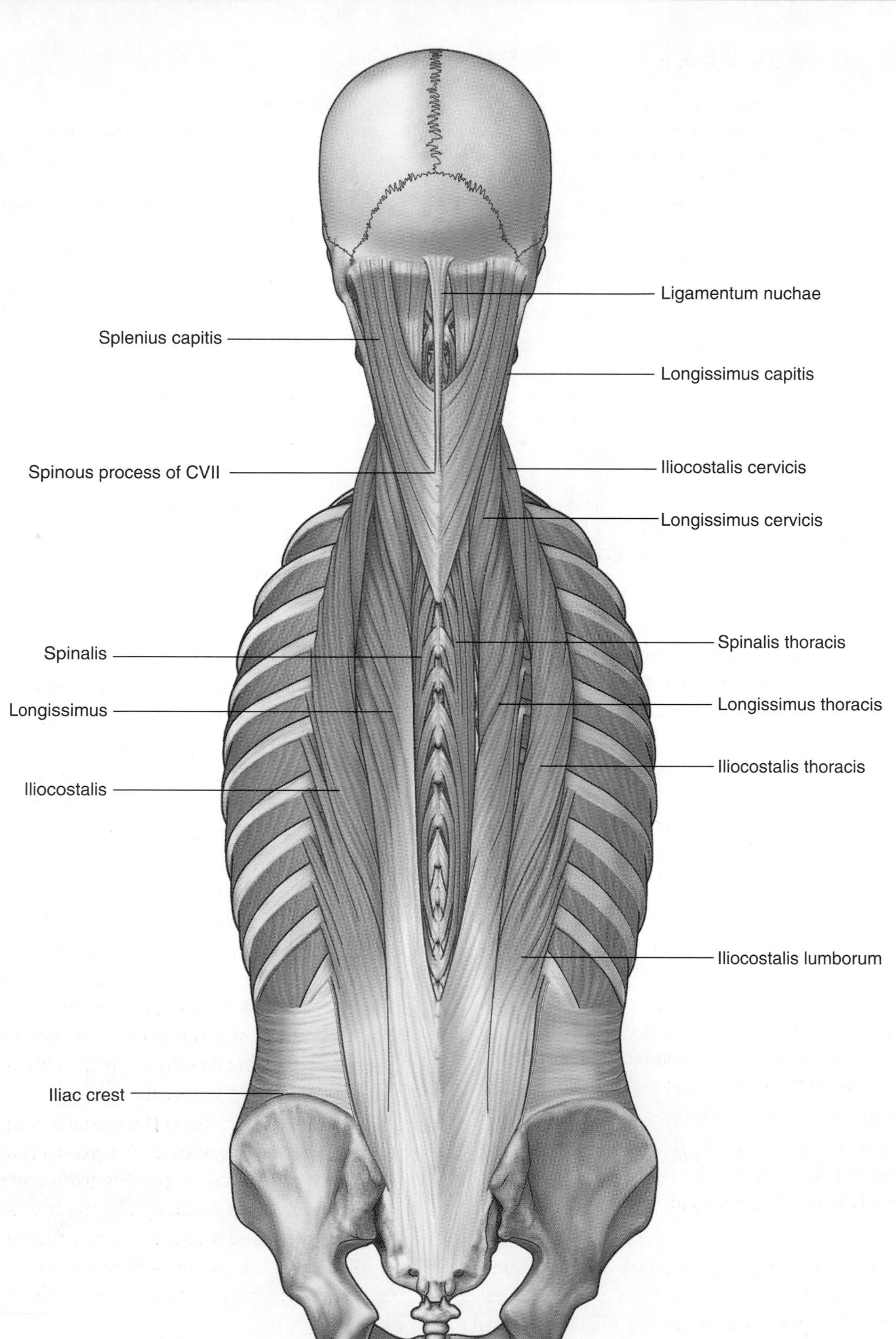


**Fig. 2.50** Deep group of back muscles—erector spinae muscles.

**Table 2.4** Erector spinae group of back muscles

| Muscle | Origin | Insertion |
|---|---|---|
| Iliocostalis lumborum | Sacrum, spinous processes of lumbar and lower two thoracic vertebrae and their supraspinous ligaments, and the iliac crest | Angles of the lower six or seven ribs |
| Iliocostalis thoracis | Angles of the lower six ribs | Angles of the upper six ribs and the transverse process of CVII |
| Iliocostalis cervicis | Angles of ribs III to VI | Transverse processes of CIV to CVI |
| Longissimus thoracis | Blends with iliocostalis in lumbar region and is attached to transverse processes of lumbar vertebrae | Transverse processes of all thoracic vertebrae and just lateral to the tubercles of the lower nine or ten ribs |
| Longissimus cervicis | Transverse processes of upper four or five thoracic vertebrae | Transverse processes of CII to CVI |
| Longissimus capitis | Transverse processes of upper four or five thoracic vertebrae and articular processes of lower three or four cervical vertebrae | Posterior margin of the mastoid process |
| Spinalis thoracis | Spinous processes of TX or TXI to LII | Spinous processes of TI to TVIII (varies) |
| Spinalis cervicis | Lower part of ligamentum nuchae and spinous process of CVII (sometimes TI to TII) | Spinous process of CII (axis) |
| Spinalis capitis | Usually blends with semispinalis capitis | With semispinalis capitis |

- The outer or most laterally placed column of the erector spinae muscles is the **iliocostalis**, which is associated with the costal elements and passes from the common tendon of origin to multiple insertions into the angles of the ribs and the transverse processes of the lower cervical vertebrae.
- The middle or intermediate column is the **longissimus**, which is the largest of the erector spinae subdivision extending from the common tendon of origin to the base of the skull. Throughout this vast expanse, the lateral positioning of the longissimus muscle is in the area of the transverse processes of the various vertebrae.
- The most medial muscle column is the **spinalis**, which is the smallest of the subdivisions and interconnects the spinous processes of adjacent vertebrae. The spinalis is most constant in the thoracic region and is generally absent in the cervical region. It is associated with a deeper muscle (the semispinalis capitis) as the erector spinae group approaches the skull.

The muscles in the erector spinae group are the primary extensors of the vertebral column and head. Acting bilaterally, they straighten the back, returning it to the upright position from a flexed position, and pull the head posteriorly. They also participate in controlling vertebral column flexion by contracting and relaxing in a coordinated fashion. Acting unilaterally, they bend the vertebral column laterally. In addition, unilateral contractions of muscles attached to the head turn the head to the actively contracting side.

### Transversospinales muscles

The transversospinales muscles run obliquely upward and medially from transverse processes to spinous processes, filling the groove between these two vertebral projections (Fig. 2.51 and Table 2.5). They are deep to the erector spinae and consist of three major subgroups—the semispinalis, multifidus, and rotatores muscles.

- The **semispinalis** muscles are the most superficial collection of muscle fibers in the transversospinales group. These muscles begin in the lower thoracic region and end by attaching to the skull, crossing between four and six vertebrae from their point of origin to point of attachment. Semispinalis muscles are found in the thoracic and cervical regions, and attach to the occipital bone at the base of the skull.
- Deep to the semispinalis is the second group of muscles, the **multifidus**. Muscles in this group span the length of the vertebral column, passing from a lateral point of origin upward and medially to attach to spinous processes and spanning between two and four vertebrae. The multifidus muscles are present throughout the length of the vertebral column but are best developed in the lumbar region.
- The small **rotatores** muscles are the deepest of the transversospinales group. They are present throughout the length of the vertebral column but are best developed in the thoracic region. Their fibers pass upward and medially from transverse processes to spinous processes crossing two vertebrae (long rotators) or attaching to an adjacent vertebra (short rotators).

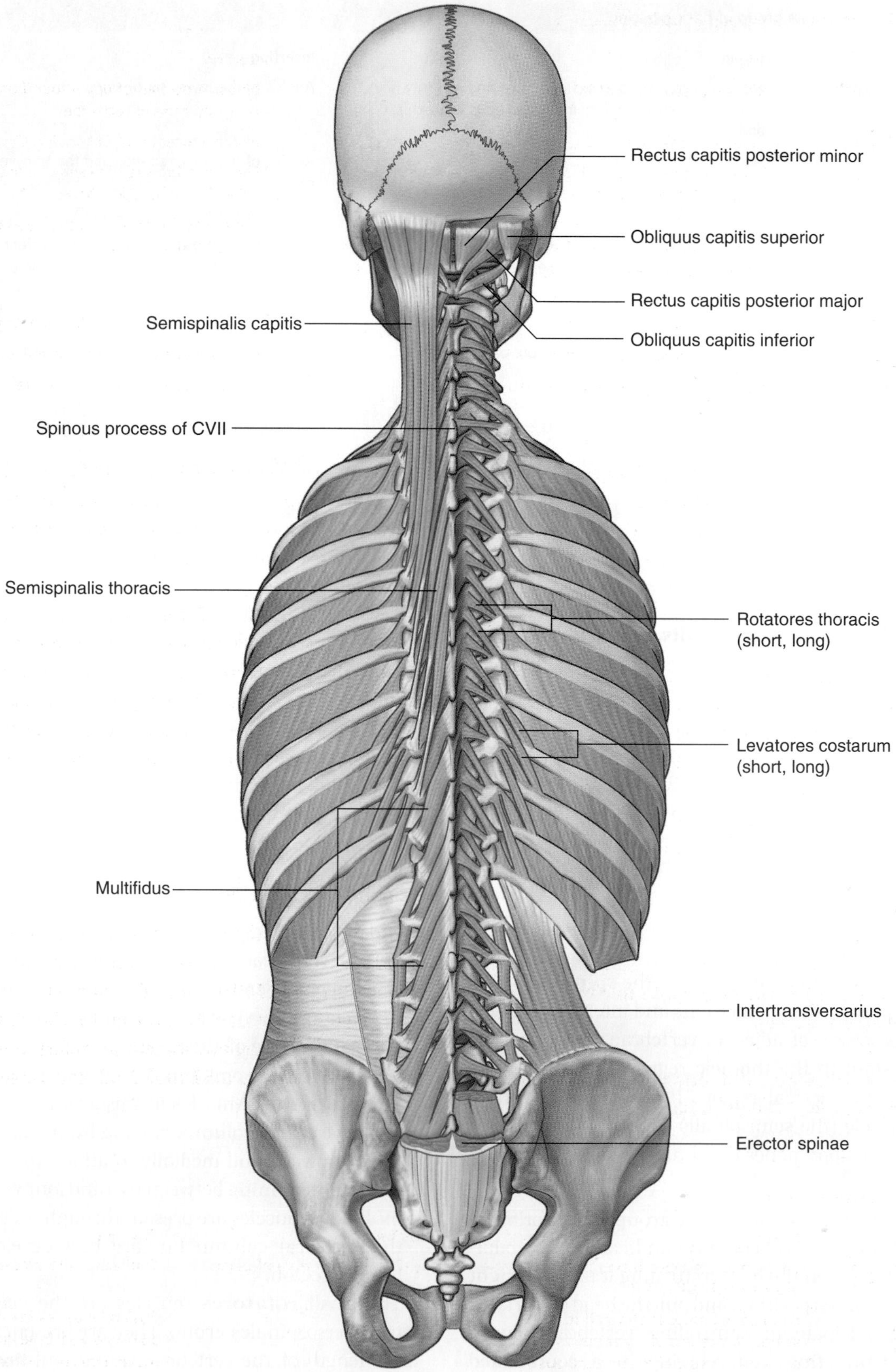

**Fig. 2.51** Deep group of back muscles—transversospinales and segmental muscles.

**Table 2.5** Transversospinales group of back muscles

| Muscle | Origin | Insertion |
|---|---|---|
| Semispinalis thoracis | Transverse processes of TVI to TX | Spinous processes of upper four thoracic and lower two cervical vertebrae |
| Semispinalis cervicis | Transverse processes of upper five or six thoracic vertebrae | Spinous processes of CII (axis) to CV |
| Semispinalis capitis | Transverse processes of TI to TVI (or TVII) and CVII and articular processes of CIV to CVI | Medial area between the superior and inferior nuchal lines of occipital bone |
| Multifidus | Sacrum, origin of erector spinae, posterior superior iliac spine, mammillary processes of lumbar vertebrae, transverse processes of thoracic vertebrae, and articular processes of lower four cervical vertebrae | Base of spinous processes of all vertebrae from LV to CII (axis) |
| Rotatores lumborum | Transverse processes of lumbar vertebrae | Spinous processes of lumbar vertebrae |
| Rotatores thoracis | Transverse processes of thoracic vertebrae | Spinous processes of thoracic vertebrae |
| Rotatores cervicis | Articular processes of cervical vertebrae | Spinous processes of cervical vertebrae |

When muscles in the transversospinales group contract bilaterally, they extend the vertebral column, an action similar to that of the erector spinae group. However, when muscles on only one side contract, they pull the spinous processes toward the transverse processes on that side, causing the trunk to turn or rotate in the opposite direction.

One muscle in the transversospinales group, the **semispinalis capitis**, has a unique action because it attaches to the skull. Contracting bilaterally, this muscle pulls the head posteriorly, whereas unilateral contraction pulls the head posteriorly and turns it, causing the chin to move superiorly and turn toward the side of the contracting muscle. These actions are similar to those of the upper erector spinae.

### Segmental muscles

The two groups of segmental muscles (Fig. 2.51 and Table 2.6) are deeply placed in the back and innervated by posterior rami of spinal nerves.

- The first group of segmental muscles are the **levatores costarum** muscles, which arise from the transverse processes of vertebrae CVII and TI to TXI. They have an oblique lateral and downward direction and insert into the rib below the vertebra of origin in the area of the tubercle. Contraction elevates the ribs.
- The second group of segmental muscles are the true segmental muscles of the back—the **interspinales**, which pass between adjacent spinous processes, and the **intertransversarii**, which pass between adjacent transverse processes. These postural muscles stabilize adjoining vertebrae during movements of the vertebral column to allow more effective action of the large muscle groups.

## Suboccipital muscles

A small group of deep muscles in the upper cervical region at the base of the occipital bone move the head. They connect vertebra CI (the atlas) to vertebra CII (the axis) and connect both vertebrae to the base of the skull. Because of their location they are sometimes referred to as suboccipital muscles (Figs. 2.51 and 2.52 and Table 2.7). They include, on each side:

**Table 2.6** Segmental back muscles

| Muscle | Origin | Insertion | Function |
|---|---|---|---|
| Levatores costarum | Short paired muscles arising from transverse processes of CVII to TXI | The rib below vertebra of origin near tubercle | Contraction elevates rib |
| Interspinales | Short paired muscles attached to the spinous processes of contiguous vertebrae, one on each side of the interspinous ligament | | Postural muscles that stabilize adjoining vertebrae during movements of vertebral column |
| Intertransversarii | Small muscles between the transverse processes of contiguous vertebrae | | Postural muscles that stabilize adjoining vertebrae during movements of vertebral column |

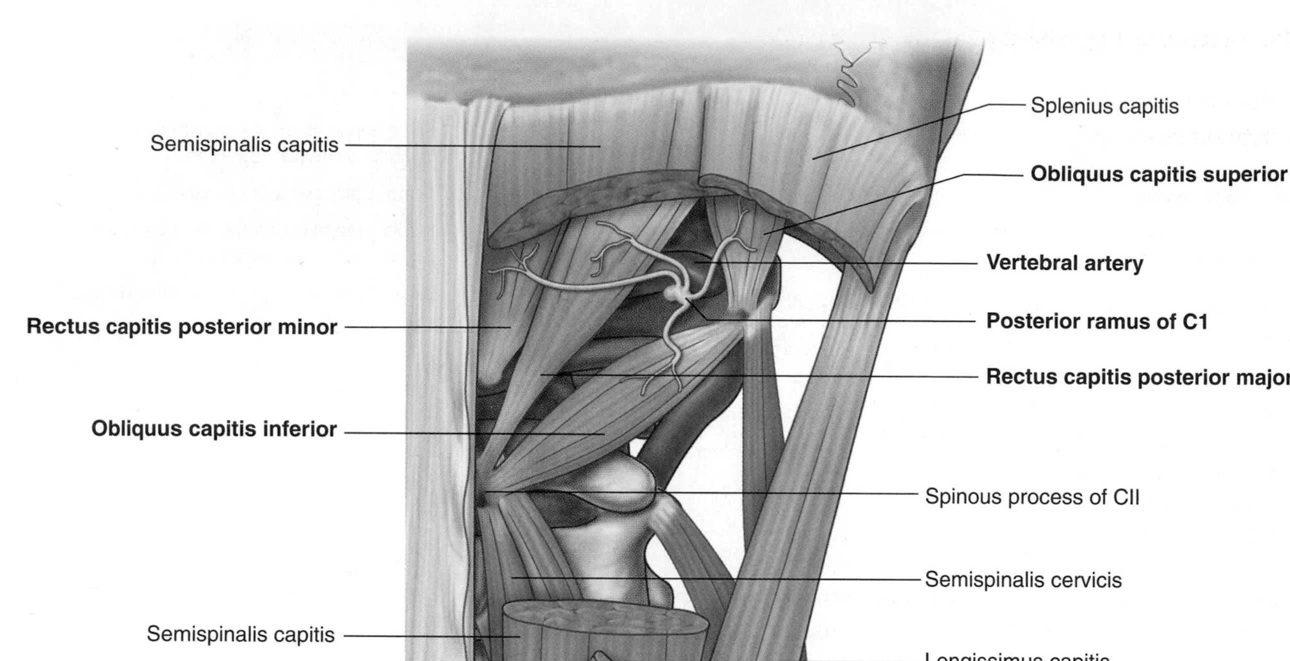


**Fig. 2.52** Deep group of back muscles—suboccipital muscles. This also shows the borders of the suboccipital triangle.

**Table 2.7** Suboccipital group of back muscles

| Muscle | Origin | Insertion | Innervation | Function |
|---|---|---|---|---|
| Rectus capitis posterior major | Spinous process of axis (CII) | Lateral portion of occipital bone below inferior nuchal line | Posterior ramus of C1 | Extension of head; rotation of face to same side as muscle |
| Rectus capitis posterior minor | Posterior tubercle of atlas (CI) | Medial portion of occipital bone below inferior nuchal line | Posterior ramus of C1 | Extension of head |
| Obliquus capitis superior | Transverse process of atlas (CI) | Occipital bone between superior and inferior nuchal lines | Posterior ramus of C1 | Extension of head and bends it to same side |
| Obliquus capitis inferior | Spinous process of axis (CII) | Transverse process of atlas (CI) | Posterior ramus of C1 | Rotation of face to same side |

- **rectus capitis posterior major,**
- **rectus capitis posterior minor,**
- **obliquus capitis inferior,** and
- **obliquus capitis superior.**

Contraction of the suboccipital muscles extends and rotates the head at the atlanto-occipital and atlanto-axial joints, respectively.

The suboccipital muscles are innervated by the posterior ramus of the first cervical nerve, which enters the area between the vertebral artery and the posterior arch of the atlas (Fig. 2.52). The vascular supply to the muscles in this area is from branches of the vertebral and occipital arteries.

The suboccipital muscles form the boundaries of the **suboccipital triangle**, an area that contains several important structures (Fig. 2.52):

- The rectus capitis posterior major muscle forms the medial border of the triangle.
- The obliquus capitis superior muscle forms the lateral border.
- The obliquus capitis inferior muscle forms the inferior border.

The contents of the suboccipital triangle include:

- posterior ramus of CI,
- vertebral artery, and
- veins

## In the clinic

**Nerve injuries affecting superficial back muscles**
Weakness in the trapezius, caused by an interruption of the accessory nerve [XI], may appear as drooping of the shoulder, inability to raise the arm above the head because of impaired rotation of the scapula, or weakness in attempting to raise the shoulder (i.e., shrug the shoulder against resistance).

A weakness in, or an inability to use, the latissimus dorsi, resulting from an injury to the thoracodorsal nerve, diminishes the capacity to pull the body upward while climbing or doing a pull-up.

An injury to the dorsal scapular nerve, which innervates the rhomboids, may result in a lateral shift in the position of the scapula on the affected side (i.e., the normal position of the scapula is lost because of the affected muscle's inability to prevent antagonistic muscles from pulling the scapula laterally).

## SPINAL CORD

The spinal cord extends from the foramen magnum to approximately the level of the disc between vertebrae LI and LII in adults, although it can end as high as vertebra TXII or as low as the disc between vertebrae LII and LIII (Fig. 2.53). In neonates, the spinal cord extends approximately to vertebra LIII but can reach as low as vertebra LIV. The distal end of the cord (the **conus medullaris**) is cone shaped. A fine filament of connective tissue (the pial part of the **filum terminale**) continues inferiorly from the apex of the **conus medullaris**.

The spinal cord is not uniform in diameter along its length. It has two major swellings or enlargements in regions associated with the origin of spinal nerves that innervate the upper and lower limbs. A **cervical enlargement** occurs in the region associated with the origins of spinal nerves C5 to T1, which innervate the upper limbs. A lumbosacral enlargement occurs in the region associated with the origins of spinal nerves L1 to S3, which innervate the lower limbs.

The external surface of the spinal cord is marked by a number of fissures and sulci (Fig. 2.54):

- The **anterior median fissure** extends the length of the anterior surface.

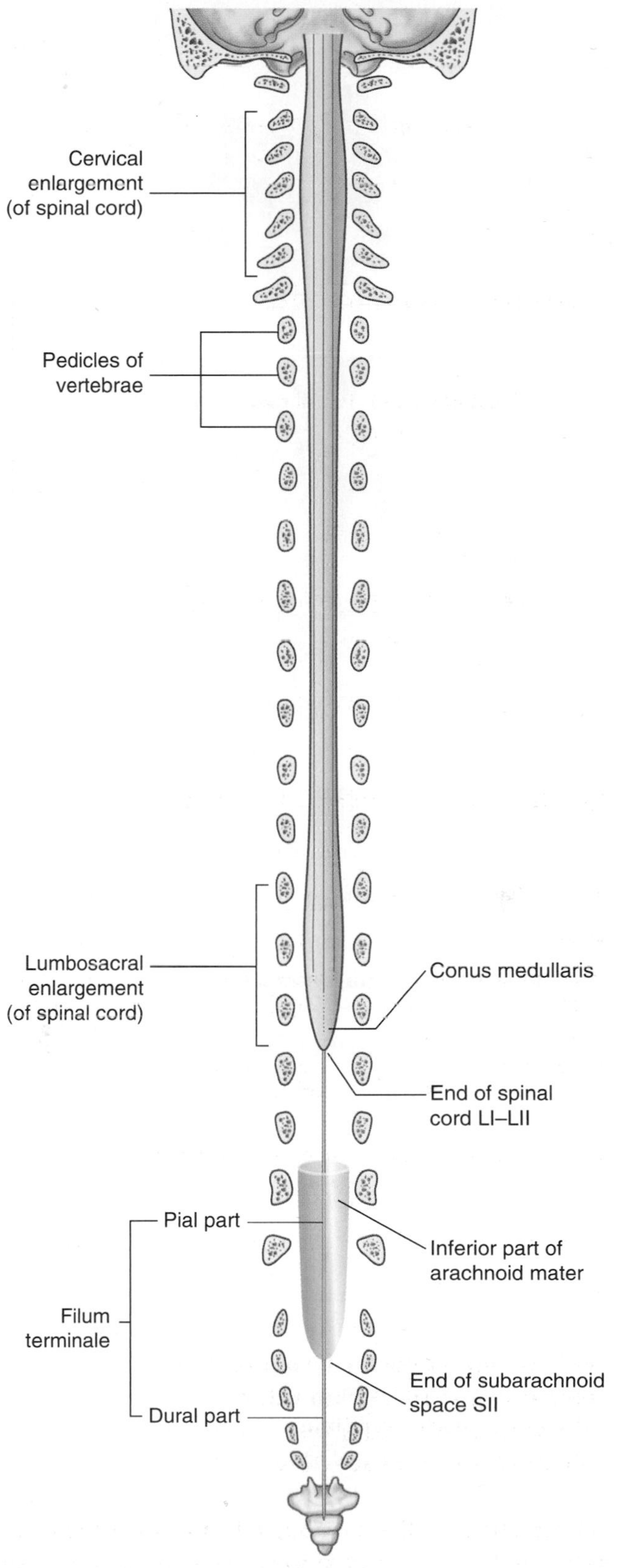


**Fig. 2.53** Spinal cord.

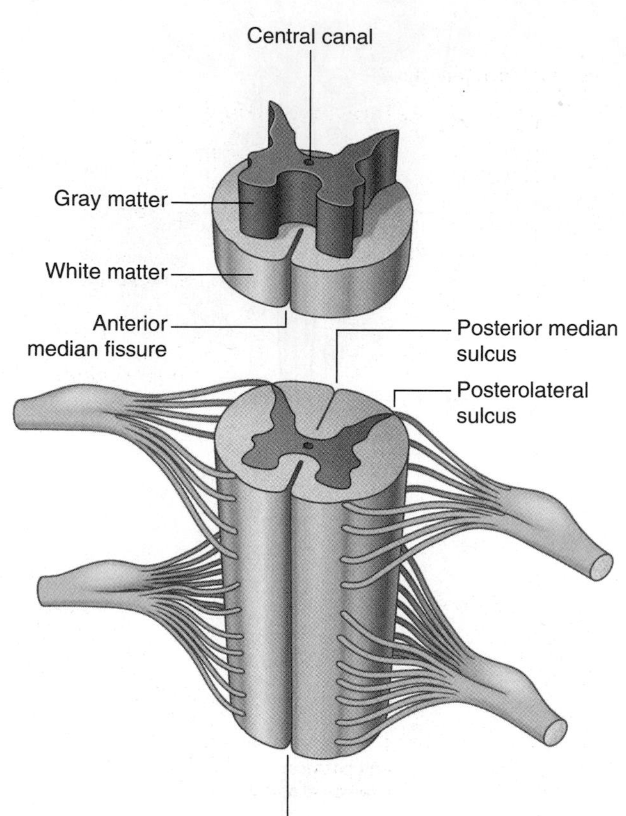


**Fig. 2.54** Features of the spinal cord.

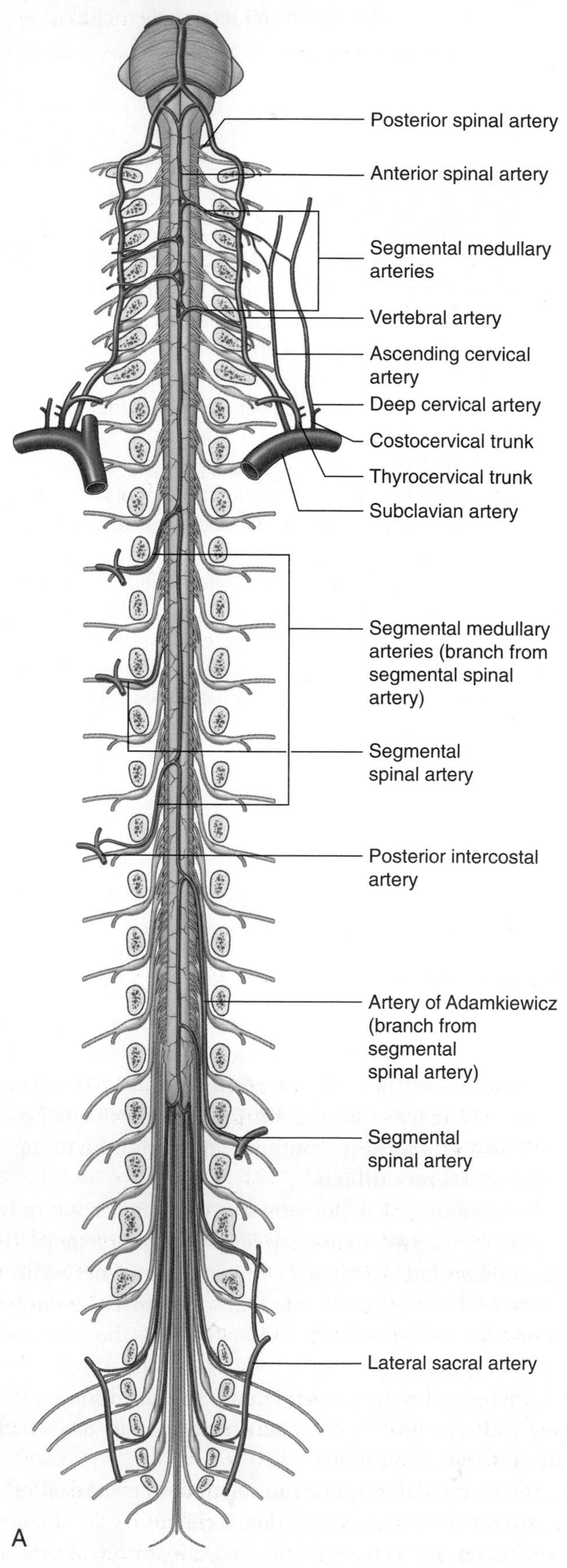


**Fig. 2.55** Arteries that supply the spinal cord. **A.** Anterior view of spinal cord (not all segmental spinal arteries are shown).

- The **posterior median sulcus** extends along the posterior surface.
- The **posterolateral sulcus** on each side of the posterior surface marks where the posterior rootlets of spinal nerves enter the cord.

Internally, the cord has a small central canal surrounded by gray and white matter:

- The gray matter is rich in nerve cell bodies, which form longitudinal columns along the cord, and in cross section these columns form a characteristic H-shaped appearance in the central regions of the cord.
- The white matter surrounds the gray matter and is rich in nerve cell processes, which form large bundles or tracts that ascend and descend in the cord to other spinal cord levels or carry information to and from the brain.

## Vasculature

### Arteries

The arterial supply to the spinal cord comes from two sources (Fig. 2.55). It consists of:

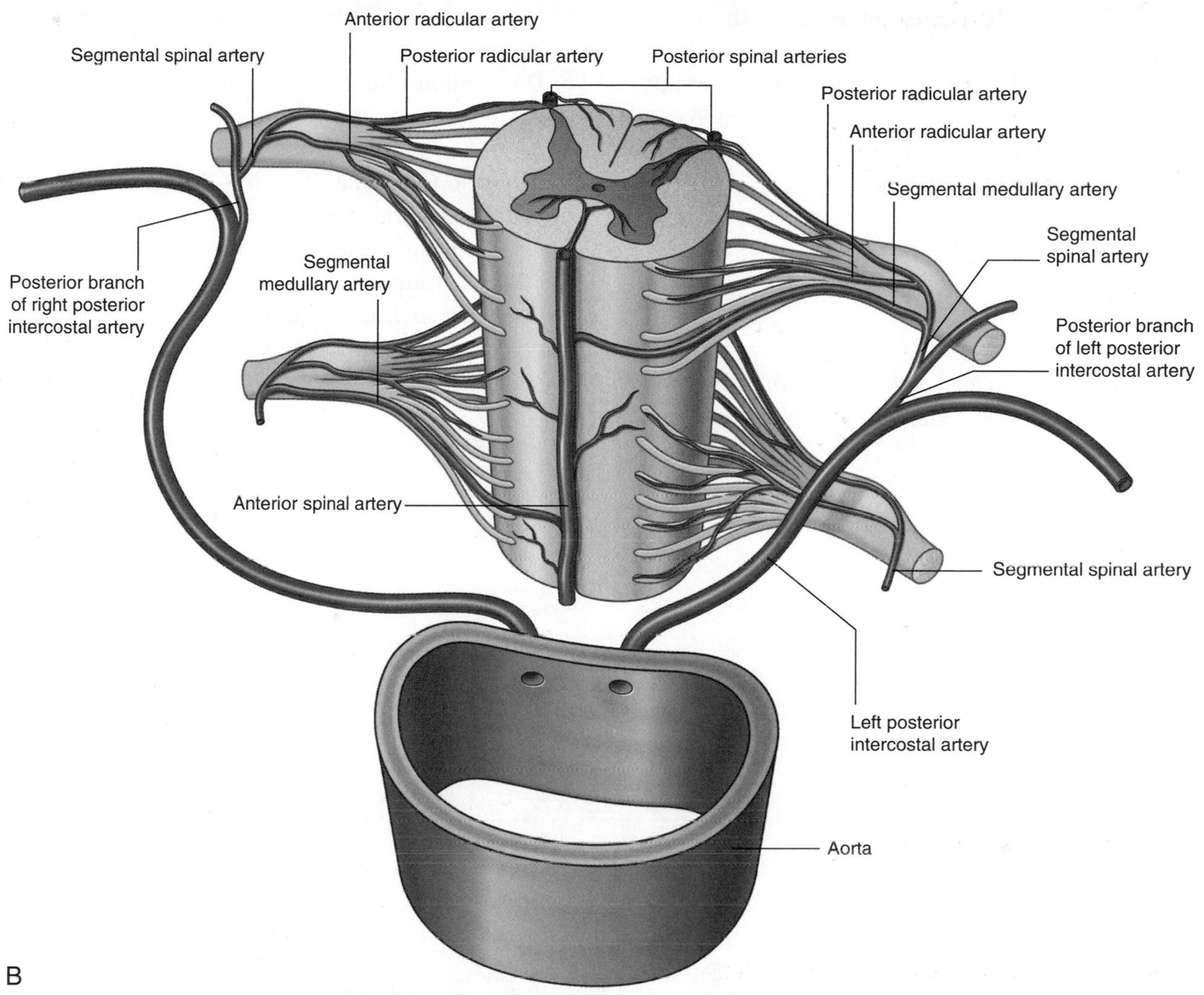


**Fig. 2.55, cont'd B.** Segmental supply of spinal cord.

- longitudinally oriented vessels, arising superior to the cervical portion of the cord, which descend on the surface of the cord; and
- feeder arteries that enter the vertebral canal through the intervertebral foramina at every level; these feeder vessels, or **segmental spinal arteries**, arise predominantly from the vertebral and deep cervical arteries in the neck, the posterior intercostal arteries in the thorax, and the lumbar arteries in the abdomen.

After entering an intervertebral foramen, the segmental spinal arteries give rise to **anterior** and **posterior radicular arteries** (Fig. 2.55). This occurs at every vertebral level. The radicular arteries follow, and supply, the anterior and posterior roots. At various vertebral levels, the **segmental spinal arteries** also give off **segmental medullary arteries** (Fig. 2.55). These vessels pass directly to the longitudinally oriented vessels, reinforcing these.

The longitudinal vessels consist of:

- a single **anterior spinal artery**, which originates within the cranial cavity as the union of two vessels that arise from the vertebral arteries—the resulting single anterior spinal artery passes inferiorly, approximately parallel to the anterior median fissure, along the surface of the spinal cord; and
- two **posterior spinal arteries**, which also originate in the cranial cavity, usually arising directly from a terminal branch of each vertebral artery (the posterior inferior cerebellar artery)—the right and left posterior spinal arteries descend along the spinal cord, each as two branches that bracket the posterolateral sulcus and the connection of posterior roots with the spinal cord.

The anterior and posterior spinal arteries are reinforced along their length by eight to ten segmental medullary arteries (Fig. 2.55). The largest of these is the **arteria radicularis magna** or the **artery of Adamkiewicz** (Fig. 2.55). This vessel arises in the lower thoracic or upper lumbar region, usually on the left side, and reinforces the arterial supply to the lower portion of the spinal cord, including the lumbar enlargement.

### Veins

Veins that drain the spinal cord form a number of longitudinal channels (Fig. 2.56):

- Two pairs of veins on each side bracket the connections of the posterior and anterior roots to the cord.
- One midline channel parallels the anterior median fissure.
- One midline channel passes along the posterior median sulcus.

These longitudinal channels drain into an extensive internal vertebral plexus in the extradural (epidural) space of the vertebral canal, which then drains into segmentally arranged vessels that connect with major systemic veins, such as the azygos system in the thorax. The internal vertebral plexus also communicates with intracranial veins.

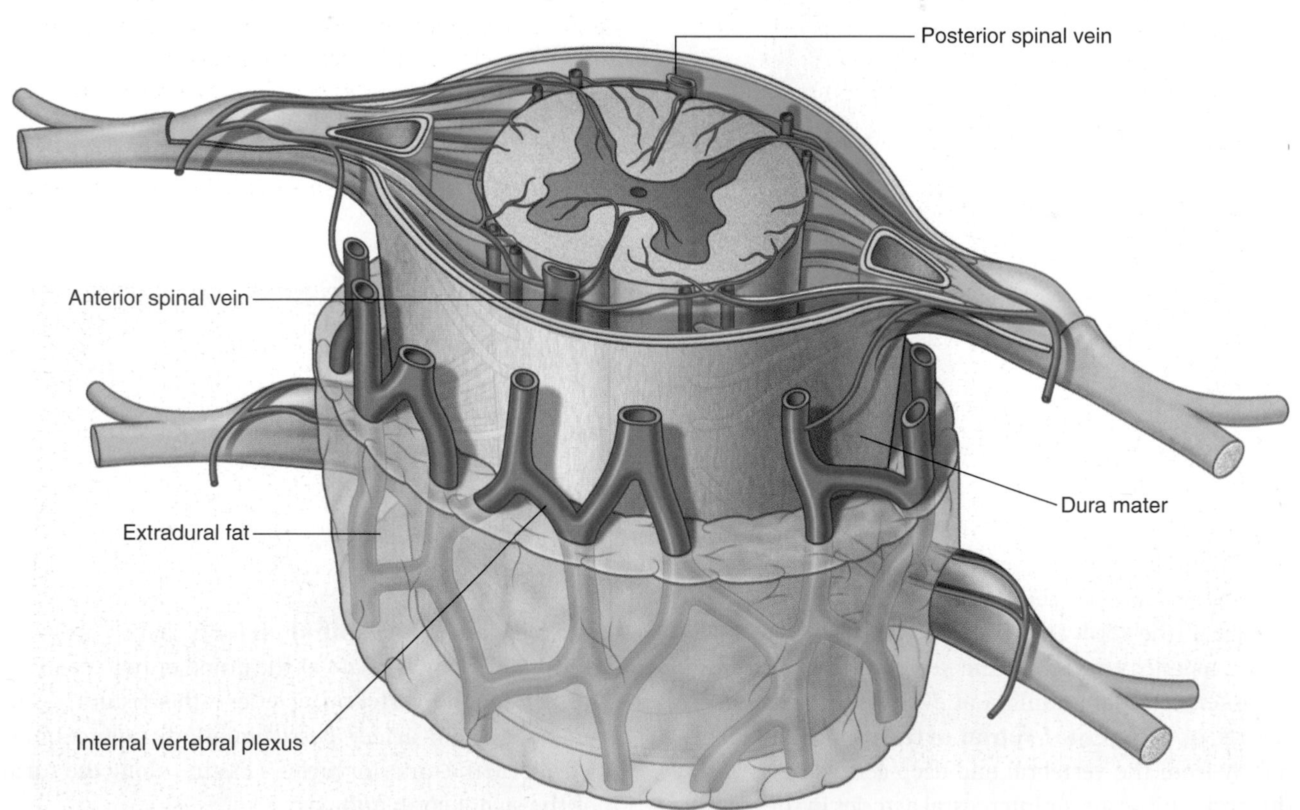


**Fig. 2.56** Veins that drain the spinal cord.

## In the clinic

### Discitis

The intervertebral discs are poorly vascularized; however, infection within the bloodstream can spread to the discs from the terminal branches of the spinal arteries within the vertebral body endplates, which lie immediately adjacent to the discs (Fig. 2.57). Common sources of infection include the lungs and urinary tract.

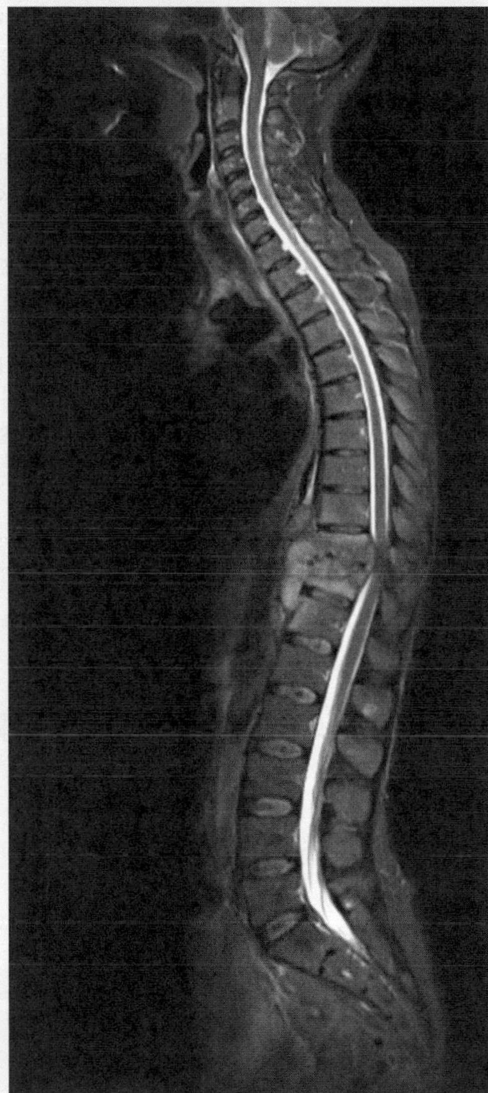

**Fig. 2.57** MRI of the spine. There is discitis of the T10-T11 intervertebral disc with destruction of the adjacent endplates. There is also a prevertebral abscess and an epidural abscess, which impinges the cord.

## In the clinic

### Fractures of the atlas and axis

Fractures of vertebra CI (the atlas) and vertebra CII (the axis) can potentially lead to the worst types of spinal cord injury including death and paralysis due to injury of the brainstem, which contains the cardiac and respiratory centers. The atlas is a closed ring with no vertebral body. Axial-loading injuries, such as hitting the head while diving into shallow water or hitting the head on the roof of a car in a motor vehicle accident, can cause a "burst" type of fracture, where the ring breaks at more than one site (Fig. 2.58). The British neurosurgeon, Geoffrey Jefferson, first described this fracture pattern in 1920, so these types of fractures are often called *Jefferson fractures.*

Fractures of the axis usually occur due to severe hyperextension and flexion, which can result in fracture of the tip of the dens, base of the dens, or through the body of the atlas. In judicial hangings, there is hyperextension and distraction injury causing fracture through the atlas pedicles and spondylolisthesis of C2 on C3. This type of fracture is often called a *hangman's fracture.*

In many cases of upper neck injuries, even in the absence of fractures to the atlas or axis, there may be injury to the atlanto-axial ligaments, which can render the neck unstable and pose severe risk to the brainstem and upper spinal cord.

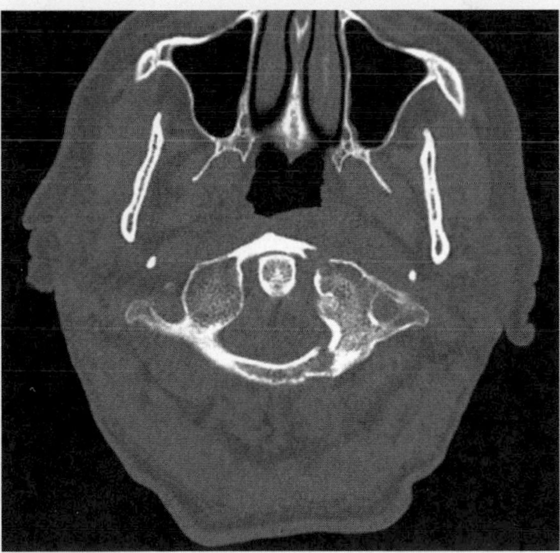

**Fig. 2.58** CT at the level of CI demonstrates two breaks in the closed ring of the atlas following an axial-loading injury.

## In the clinic

### Paraplegia and tetraplegia

An injury to the spinal cord in the cervical portion of the vertebral column can lead to varying degrees of impairment of sensory and motor function (paralysis) in all 4 limbs, termed *quadriplegia* or *tetraplegia*. An injury in upper levels of the cervical vertebral column can result in death because of loss of innervation to the diaphragm. An injury to the spinal cord below the level of TI can lead to varying degrees of impairment in motor and sensory function (paralysis) in the lower limbs, termed *paraplegia*.

## Meninges

### Spinal dura mater

The **spinal dura mater** is the outermost meningeal membrane and is separated from the bones forming the vertebral canal by an extradural space (Fig. 2.59). Superiorly, it is continuous with the inner meningeal layer of cranial dura mater at the foramen magnum of the skull. Inferiorly, the dural sac dramatically narrows at the level of the lower border of vertebra SII and forms an investing sheath for the pial part of the filum terminale of the spinal cord. This terminal cord-like extension of dura mater (the dural part of the filum terminale) attaches to the posterior surface of the vertebral bodies of the coccyx.

As spinal nerves and their roots pass laterally, they are surrounded by tubular sleeves of dura mater, which merge with and become part of the outer covering (epineurium) of the nerves.

### Arachnoid mater

The **arachnoid mater** is a thin delicate membrane against, but not adherent to, the deep surface of the dura mater (Fig. 2.59). It is separated from the pia mater by the subarachnoid space. The arachnoid mater ends at the level of vertebra SII (see Fig. 2.53).

### Subarachnoid space

The subarachnoid space between the arachnoid and pia mater contains CSF (Fig. 2.59). The subarachnoid space around the spinal cord is continuous at the foramen magnum with the subarachnoid space surrounding the brain. Inferiorly, the subarachnoid space terminates at approximately the level of the lower border of vertebra SII (see Fig. 2.53).

Delicate strands of tissue (**arachnoid trabeculae**) are continuous with the arachnoid mater on one side and the pia mater on the other; they span the subarachnoid space and interconnect the two adjacent membranes. Large blood vessels are suspended in the subarachnoid space by similar strands of material, which expand over the vessels to form a continuous external coat.

The subarachnoid space extends farther inferiorly than the spinal cord. The spinal cord ends at approximately the disc between vertebrae LI and LII, whereas the

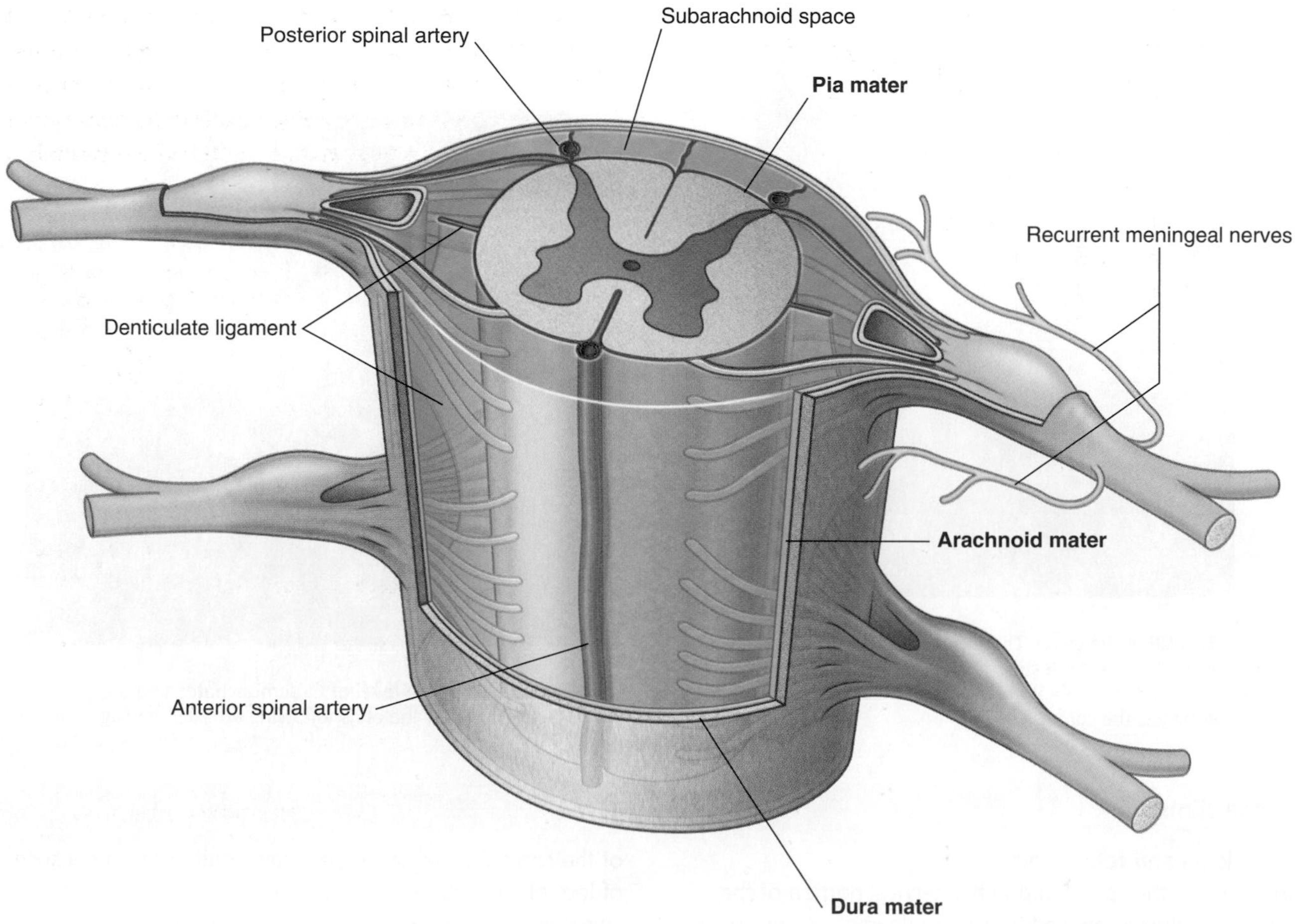

**Fig. 2.59** Meninges.

subarachnoid space extends to approximately the lower border of vertebra SII (see Fig. 2.53). The subarachnoid space is largest in the region inferior to the terminal end of the spinal cord, where it surrounds the cauda equina. As a consequence, CSF can be withdrawn from the subarachnoid space in the lower lumbar region without endangering the spinal cord.

### Pia mater

The spinal pia mater is a vascular membrane that firmly adheres to the surface of the spinal cord (Fig. 2.59). It extends into the anterior median fissure and reflects as sleeve-like coatings onto posterior and anterior rootlets and roots as they cross the subarachnoid space. As the roots exit the space, the sleeve-like coatings reflect onto the arachnoid mater.

On each side of the spinal cord, a longitudinally oriented sheet of pia mater (the **denticulate ligament**) extends laterally from the cord toward the arachnoid and dura mater (Fig. 2.59).

- Medially, each denticulate ligament is attached to the spinal cord in a plane that lies between the origins of the posterior and anterior rootlets.
- Laterally, each denticulate ligament forms a series of triangular extensions along its free border, with the apex of each extension being anchored through the arachnoid mater to the dura mater.

The lateral attachments of the denticulate ligaments generally occur between the exit points of adjacent posterior and anterior rootlets. The ligaments function to position the spinal cord in the center of the subarachnoid space.

## Arrangement of structures in the vertebral canal

The vertebral canal is bordered:

- anteriorly by the bodies of the vertebrae, intervertebral discs, and posterior longitudinal ligament (Fig. 2.60);
- laterally, on each side by the pedicles and intervertebral foramina; and
- posteriorly by the laminae and ligamenta flava, and in the median plane the roots of the interspinous ligaments and vertebral spinous processes.

Between the walls of the vertebral canal and the dural sac is an extradural space containing a vertebral plexus of veins embedded in fatty connective tissue.

The vertebral spinous processes can be palpated through the skin in the midline in thoracic and lumbar regions of the back. Between the skin and spinous processes is a layer of superficial fascia. In lumbar regions, the adjacent spinous processes and the associated laminae on either side of the midline do not overlap, resulting in gaps between adjacent vertebral arches.

When carrying out a lumbar puncture (spinal tap), the needle passes between adjacent vertebral spinous processes, through the supraspinous and interspinous ligaments, and enters the extradural space. The needle continues through the dura and arachnoid mater and enters the subarachnoid space, which contains CSF.

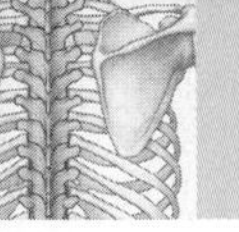



Crura of diaphragm
Posterior longitudinal ligament
Psoas
Dura
Internal vertebral plexus of veins in extradural space
Ligamenta flava
Interspinous ligament
Supraspinous ligament
Quadratus lumborum
Erector spinae muscles
Pedicle
Aorta
Cauda equina
Vein
Lumbar artery
Intervertebral foramen
Intervertebral disc
Vertebra
Skin
Lamina

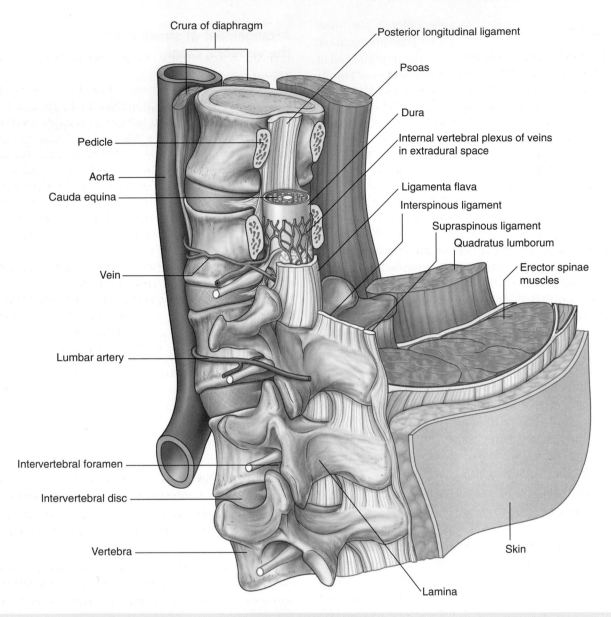

**Fig. 2.60** Arrangement of structures in the vertebral canal and the back (lumbar region).

## In the clinic

**Lumbar cerebrospinal fluid tap**

A lumbar tap (puncture) is carried out to obtain a sample of CSF for examination. In addition, passage of a needle or conduit into the subarachnoid space (CSF space) is used to inject antibiotics, chemotherapeutic agents, and anesthetics.

The lumbar region is an ideal site to access the subarachnoid space because the spinal cord terminates around the level of the disc between vertebrae LI and LII in the adult. The subarachnoid space extends to the region of the lower border of the SII vertebra. There is therefore a large CSF-filled space containing lumbar and sacral nerve roots but no spinal cord.

Depending on the clinician's preference, the patient is placed in the lateral or prone position. A needle is passed in the midline in between the spinous processes into the extradural space. Further advancement punctures the dura and arachnoid mater to enter the subarachnoid space. Most needles push the roots away from the tip without causing the patient any symptoms. Once the needle is in the subarachnoid space, fluid can be aspirated. In some situations, it is important to measure CSF pressure.

Local anesthetics can be injected into the extradural space or the subarachnoid space to anesthetize the sacral and lumbar nerve roots. Such anesthesia is useful for operations on the pelvis and the legs, which can then be carried out without the need for general anesthesia. When procedures are carried out, the patient must be in the erect position and not lying on his or her side or in the head-down position. If a patient lies on his or her side, the anesthesia is likely to be unilateral. If the patient is placed in the head-down position, the anesthetic can pass cranially and potentially depress respiration.

In some instances, anesthesiologists choose to carry out **extradural anesthesia.** A needle is placed through the skin, supraspinous ligament, interspinous ligament, and ligamenta flava into the areolar tissue and fat around the dura mater. Anesthetic agent is introduced and diffuses around the vertebral canal to anesthetize the exiting nerve roots and diffuse into the subarachnoid space.

## Spinal nerves

Each spinal nerve is connected to the spinal cord by posterior and anterior roots (Fig. 2.61):

- The **posterior root** contains the processes of sensory neurons carrying information to the CNS—the cell bodies of the sensory neurons, which are derived embryologically from neural crest cells, are clustered in a **spinal ganglion** at the distal end of the posterior root, usually in the intervertebral foramen.
- The **anterior root** contains motor nerve fibers, which carry signals away from the CNS—the cell bodies of the primary motor neurons are in anterior regions of the spinal cord.

Medially, the posterior and anterior roots divide into rootlets, which attach to the spinal cord.

A **spinal segment** is the area of the spinal cord that gives rise to the **posterior** and **anterior rootlets**, which will form a single pair of spinal nerves. Laterally, the posterior and anterior roots on each side join to form a spinal nerve.

Each spinal nerve divides, as it emerges from an intervertebral foramen, into two major branches: a small posterior ramus and a much larger anterior ramus (Fig. 2.61):

- The **posterior rami** innervate only intrinsic back muscles (the epaxial muscles) and an associated narrow strip of skin on the back.
- The **anterior rami** innervate most other skeletal muscles (the hypaxial muscles) of the body, including those of the limbs and trunk, and most remaining areas of the skin, except for certain regions of the head.

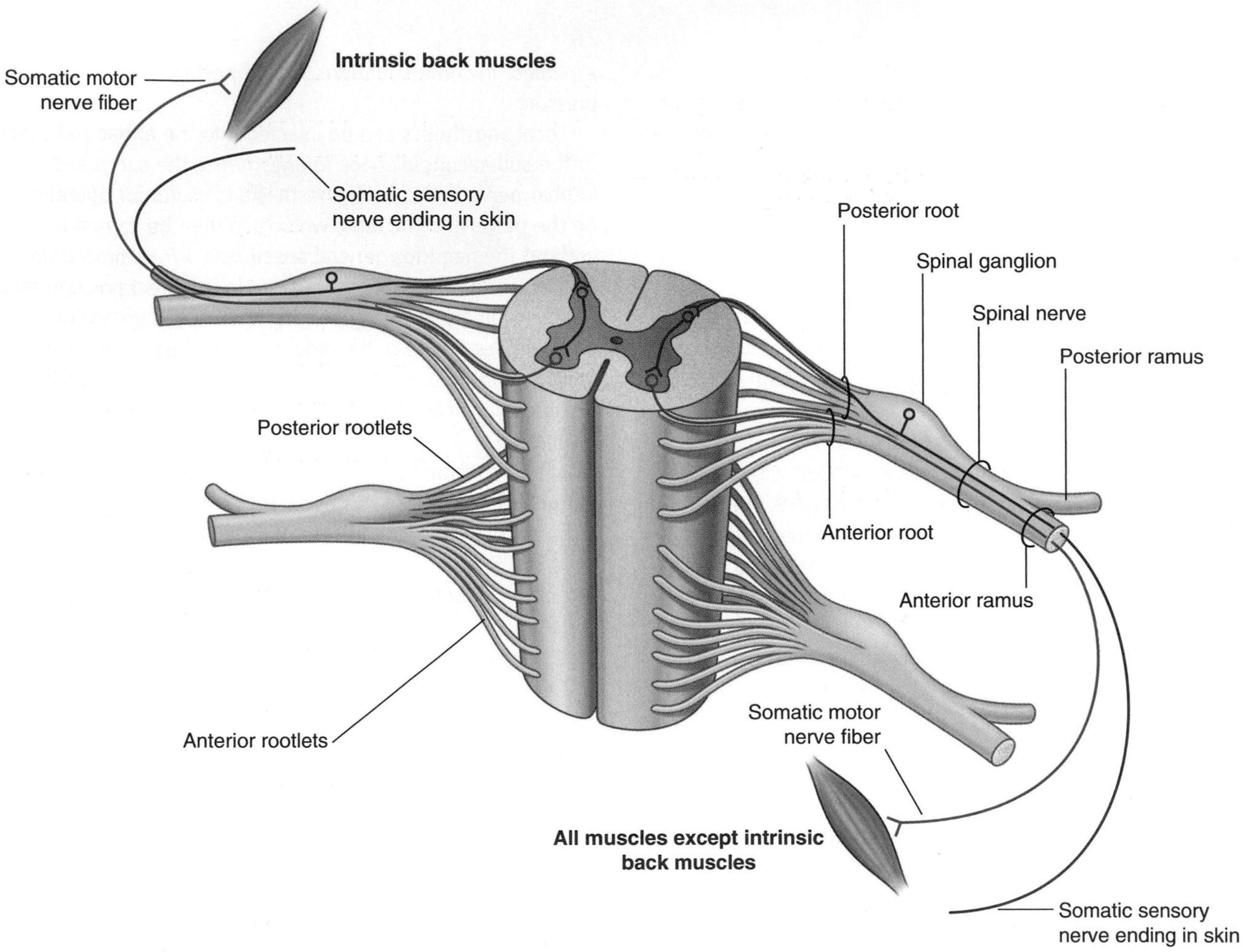

**Fig. 2.61** Basic organization of a spinal nerve.

Near the point of division into anterior and posterior rami, each spinal nerve gives rise to two to four small recurrent meningeal (sinuvertebral) nerves (see Fig. 2.59). These nerves reenter the intervertebral foramen to supply dura, ligaments, intervertebral discs, and blood vessels.

All major somatic plexuses (cervical, brachial, lumbar, and sacral) are formed by anterior rami.

Because the spinal cord is much shorter than the vertebral column, the roots of spinal nerves become longer and pass more obliquely from the cervical to coccygeal regions of the vertebral canal (Fig. 2.62).

In adults, the spinal cord terminates at a level approximately between vertebrae LI and LII, but this can range between vertebra TXII and the disc between vertebrae LII and LIII. Consequently, posterior and anterior roots forming spinal nerves emerging between vertebrae in the lower regions of the vertebral column are connected to the spinal cord at higher vertebral levels.

Below the end of the spinal cord, the posterior and anterior roots of lumbar, sacral, and coccygeal nerves pass inferiorly to reach their exit points from the vertebral canal. This terminal cluster of roots is the cauda equina.

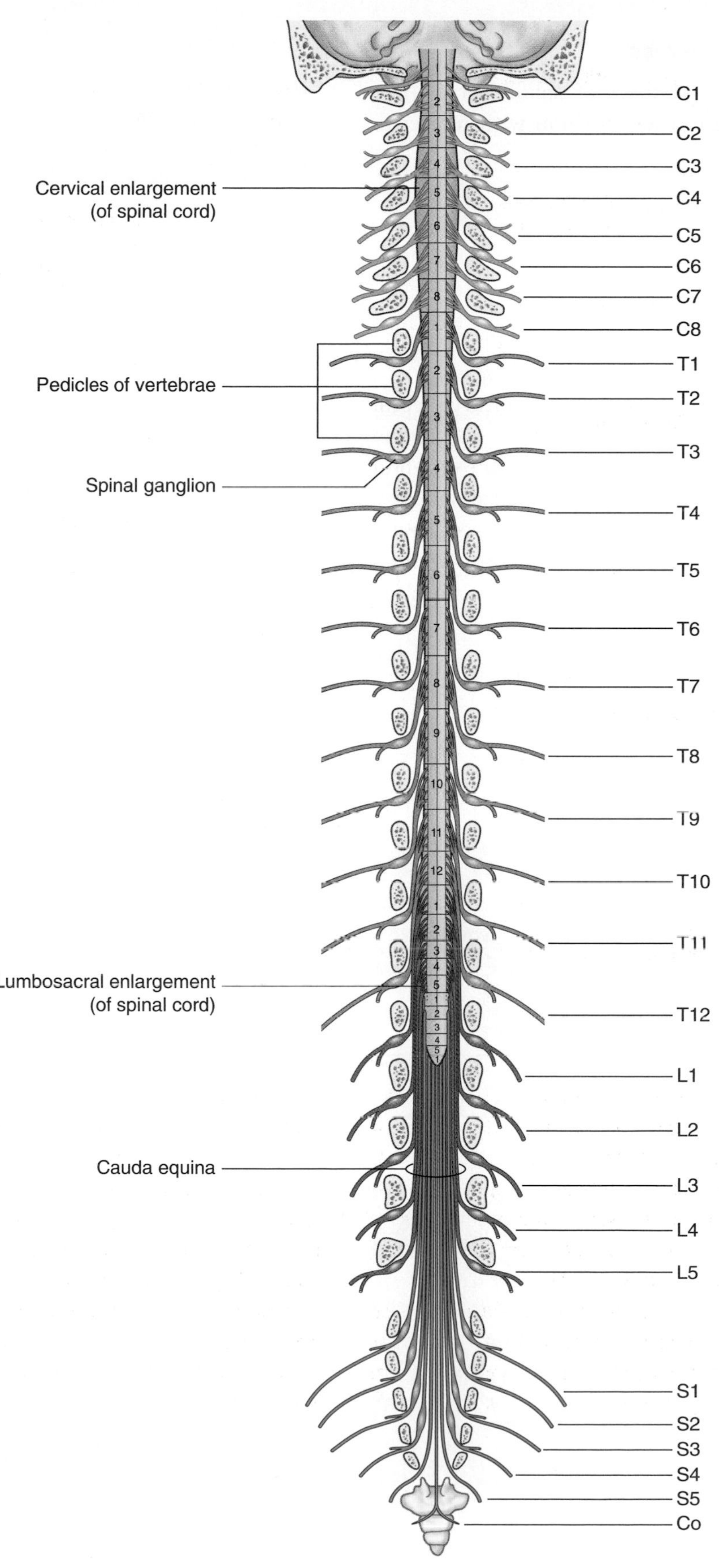


**Fig. 2.62** Course of spinal nerves in the vertebral canal.

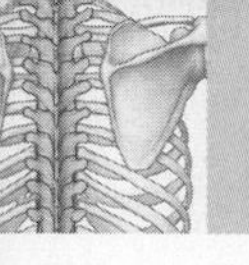



## Nomenclature of spinal nerves

There are approximately 31 pairs of spinal nerves (Fig. 2.62), named according to their position with respect to associated vertebrae:

- eight cervical nerves—C1 to C8,
- twelve thoracic nerves—T1 to T12,
- five lumbar nerves—L1 to L5,
- five sacral nerves—S1 to S5,
- one coccygeal nerve—Co.

The first cervical nerve (C1) emerges from the vertebral canal between the skull and vertebra CI (Fig. 2.63). Therefore cervical nerves C2 to C7 also emerge from the vertebral canal above their respective vertebrae. Because there are only seven cervical vertebrae, C8 emerges between vertebrae CVII and TI. As a consequence, all remaining spinal nerves, beginning with T1, emerge from the vertebral canal below their respective vertebrae.

### In the clinic

**Herpes zoster**

Herpes zoster is the virus that produces chickenpox in children. In some patients the virus remains dormant in the cells of the spinal ganglia. Under certain circumstances, the virus becomes activated and travels along the neuronal bundles to the areas supplied by that nerve (the dermatome). A rash ensues, which is characteristically exquisitely painful. Importantly, this typical dermatomal distribution is characteristic of this disorder.

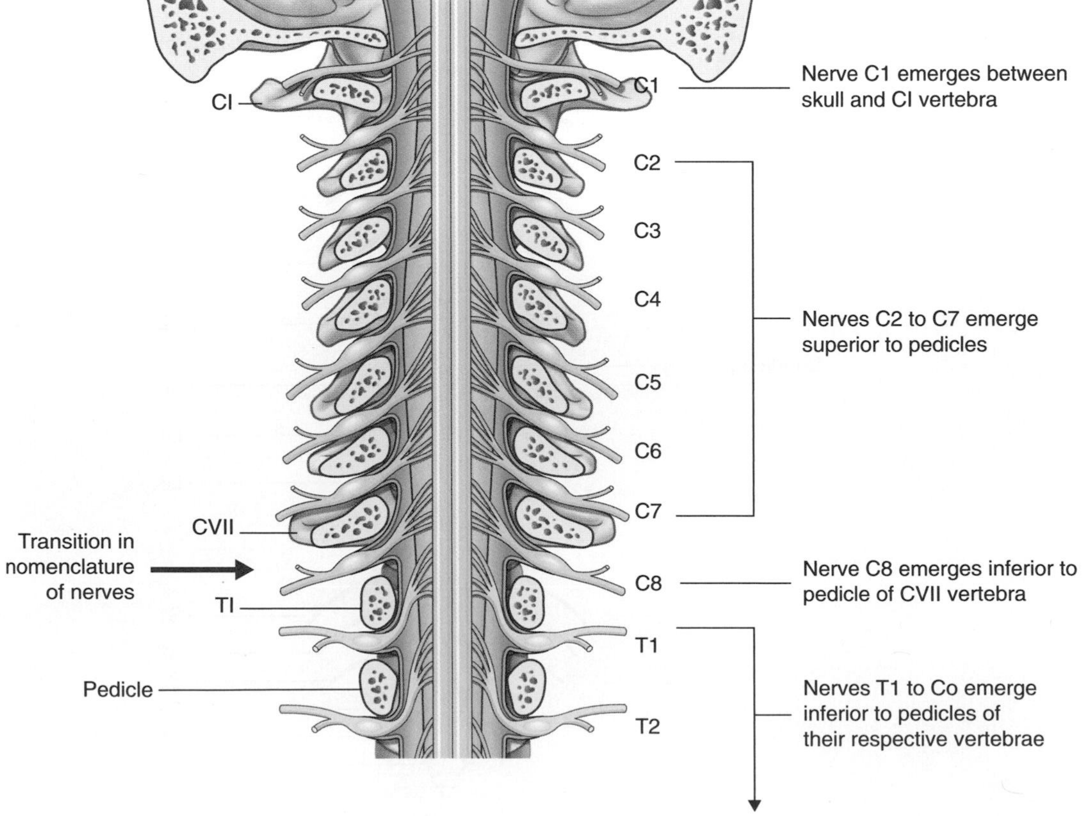


**Fig. 2.63** Nomenclature of the spinal nerves.

## In the clinic

**Back pain—alternative explanations**

Back pain is an extremely common condition affecting almost all individuals at some stage during their life. It is of key clinical importance to identify whether the back pain relates to the vertebral column and its attachments or relates to other structures.

The failure to consider other potential structures that may produce back pain can lead to significant mortality and morbidity. Pain may refer to the back from a number of organs situated in the retroperitoneum. Pancreatic pain in particular refers to the back and may be associated with pancreatic cancer and pancreatitis. Renal pain, which may be produced by stones in the renal collecting system or renal tumors, also typically refers to the back. More often than not this is usually unilateral; however, it can produce central posterior back pain. Enlarged lymph nodes in the pre- and para-aortic region may produce central posterior back pain and may be a sign of solid tumor malignancy, infection, or Hodgkin's lymphoma. An enlarging abdominal aorta (abdominal aortic aneurysm) may cause back pain as it enlarges without rupture. Therefore it is critical to think of this structure as a potential cause of back pain, because treatment will be lifesaving. Moreover, a ruptured abdominal aortic aneurysm may also cause acute back pain in the first instance.

In all patients back pain requires careful assessment not only of the vertebral column but also of the chest and abdomen in order not to miss other important anatomical structures that may produce signs and symptoms radiating to the back.

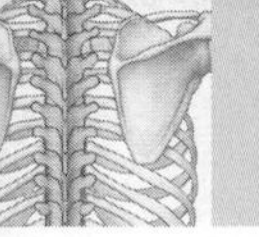

# Surface anatomy

## Back surface anatomy

Surface features of the back are used to locate muscle groups for testing peripheral nerves, to determine regions of the vertebral column, and to estimate the approximate position of the inferior end of the spinal cord. They are also used to locate organs that occur posteriorly in the thorax and abdomen.

### Absence of lateral curvatures

When viewed from behind, the normal vertebral column has no lateral curvatures. The vertical skin furrow between muscle masses on either side of the midline is straight (Fig. 2.64).

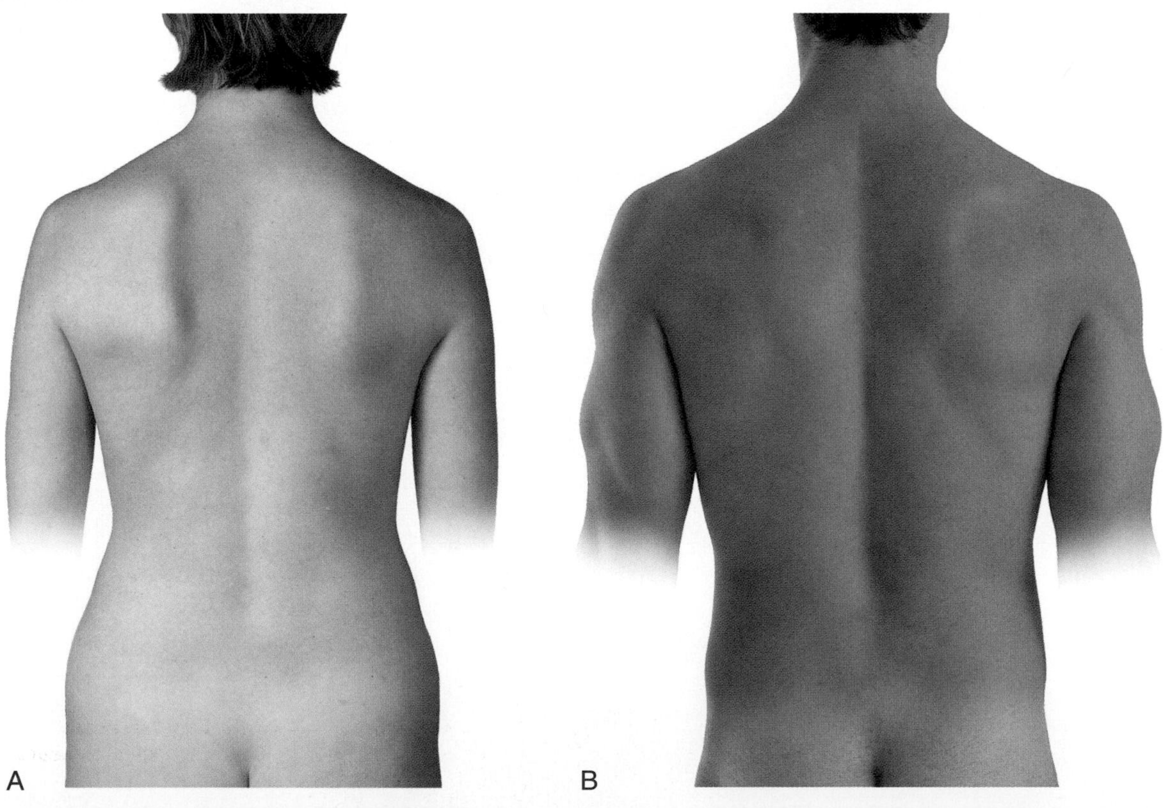


**Fig. 2.64** Normal appearance of the back. **A.** In women. **B.** In men.

## Primary and secondary curvatures in the sagittal plane

When viewed from the side, the normal vertebral column has primary curvatures in the thoracic and sacral/coccygeal regions and secondary curvatures in the cervical and lumbar regions (Fig. 2.65). The primary curvatures are concave anteriorly. The secondary curvatures are concave posteriorly.

## Useful nonvertebral skeletal landmarks

A number of readily palpable bony features provide useful landmarks for defining muscles and for locating structures associated with the vertebral column. Among these features are the external occipital protuberance, the scapula, and the iliac crest (Fig. 2.66).

The external occipital protuberance is palpable in the midline at the back of the head just superior to the hairline.

The spine, medial border, and inferior angle of the scapula are often visible and are easily palpable.

The iliac crest is palpable along its entire length, from the anterior superior iliac spine at the lower lateral margin of the anterior abdominal wall to the posterior superior iliac spine near the base of the back. The position of the posterior superior iliac spine is often visible as a "sacral dimple" just lateral to the midline.

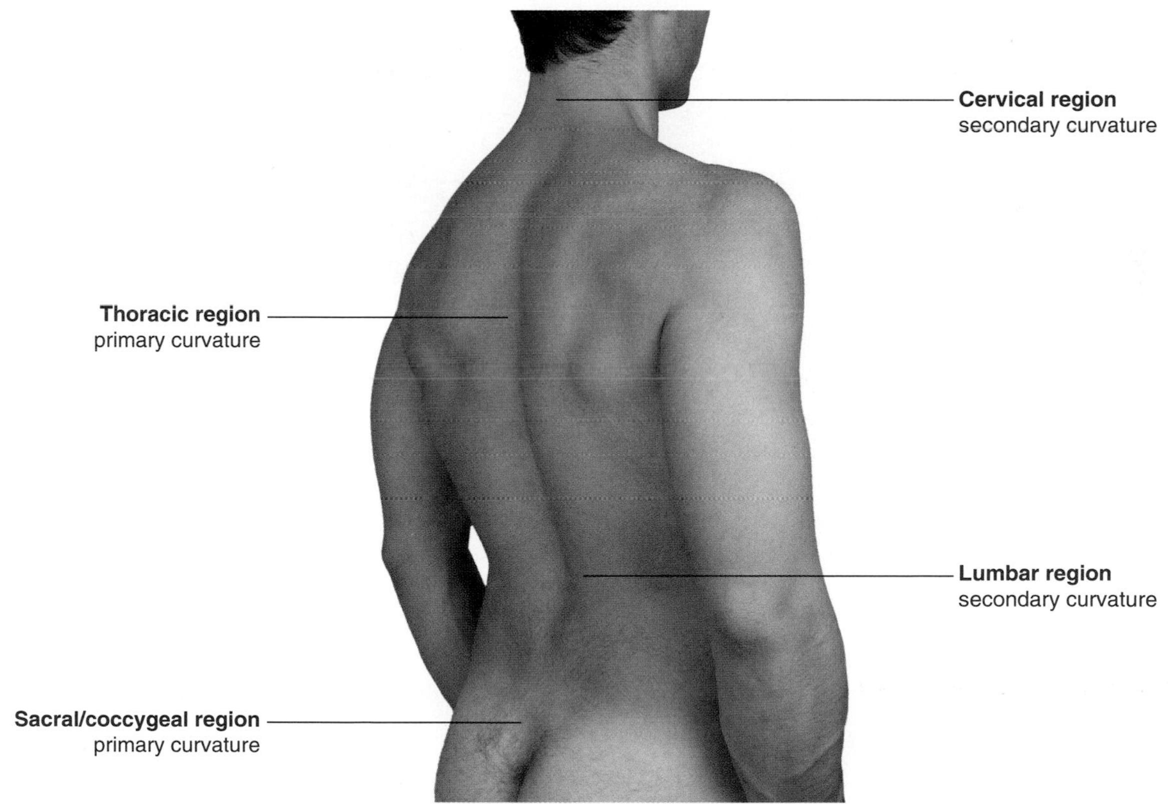


**Fig. 2.65** Normal curvatures of the vertebral column.

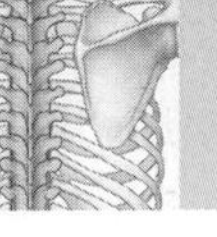



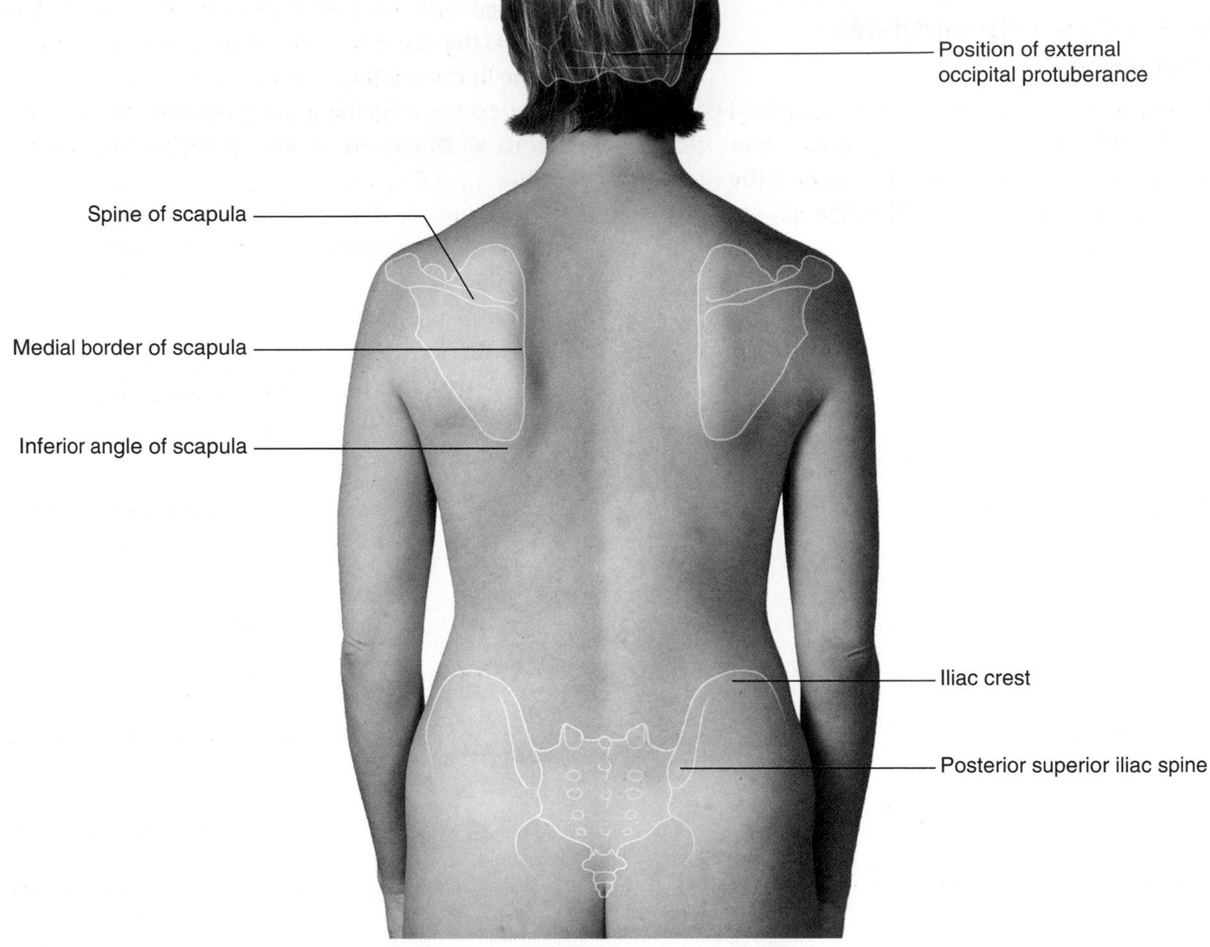


**Fig. 2.66** Back of a woman with major palpable bony landmarks indicated.

## How to identify specific vertebral spinous processes

Identification of vertebral spinous processes (Fig. 2.67A) can be used to differentiate between regions of the vertebral column and facilitate visualizing the position of deeper structures, such as the inferior ends of the spinal cord and subarachnoid space.

The spinous process of vertebra CII can be identified through deep palpation as the most superior bony protuberance in the midline inferior to the skull.

Most of the other spinous processes, except for that of vertebra CVII, are not readily palpable because they are obscured by soft tissue.

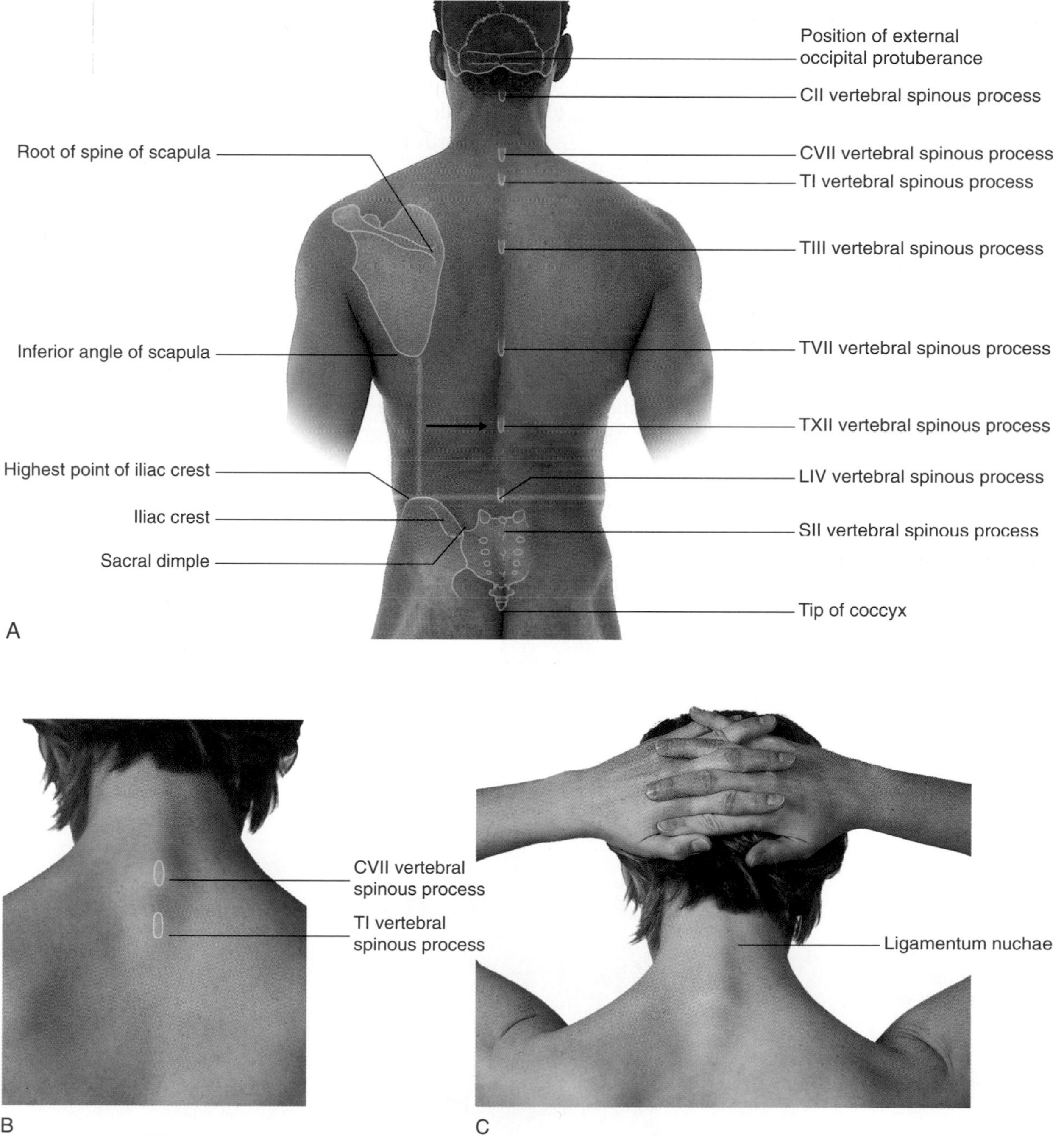


**Fig. 2.67** The back with the positions of vertebral spinous processes and associated structures indicated. **A.** In a man. **B.** In a woman with neck flexed. The prominent CVII and TI vertebral spinous processes are labeled. **C.** In a woman with neck flexed to accentuate the ligamentum nuchae.

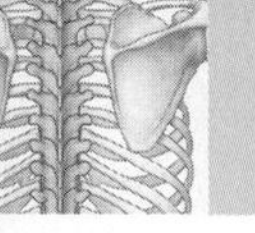



The spinous process of CVII is usually visible as a prominent eminence in the midline at the base of the neck (Fig. 2.67B), particularly when the neck is flexed.

Extending between CVII and the external occipital protuberance of the skull is the ligamentum nuchae, which is readily apparent as a longitudinal ridge when the neck is flexed (Fig. 2.67C).

Inferior to the spinous process of CVII is the spinous process of TI, which is also usually visible as a midline protuberance. Often it is more prominent than the spinous process of CVII (Fig. 2.67A,B).

The root of the spine of the scapula is at the same level as the spinous process of vertebra TIII, and the inferior angle of the scapula is level with the spinous process of vertebra TVII (Fig. 2.67A).

The spinous process of vertebra TXII is level with the midpoint of a vertical line between the inferior angle of the scapula and the iliac crest (Fig. 2.67A).

A horizontal line between the highest point of the iliac crest on each side crosses through the spinous process of vertebra LIV. The LIII and LV vertebral spinous processes can be palpated above and below the LIV spinous process, respectively (Fig. 2.67A).

The sacral dimples that mark the position of the posterior superior iliac spine are level with the SII vertebral spinous process (Fig. 2.67A).

The tip of the coccyx is palpable at the base of the vertebral column between the gluteal masses (Fig. 2.67A).

The tips of the vertebral spinous processes do not always lie in the same horizontal plane as their corresponding vertebral bodies. In thoracic regions, the spinous processes are long and sharply sloped downward so that their tips lie at the level of the vertebral body below. In other words, the tip of the TIII vertebral spinous process lies at vertebral level TIV.

In lumbar and sacral regions, the spinous processes are generally shorter and less sloped than in thoracic regions, and their palpable tips more closely reflect the position of their corresponding vertebral bodies. As a consequence, the palpable end of the spinous process of vertebra LIV lies at approximately the LIV vertebral level.

## Visualizing the inferior ends of the spinal cord and subarachnoid space

The spinal cord does not occupy the entire length of the vertebral canal. Normally in adults, it terminates at the level of the disc between vertebrae LI and LII; however, it may end as high as TXII or as low as the disc between vertebrae LII and LIII. The subarachnoid space ends at approximately the level of vertebra SII (Fig. 2.68A).

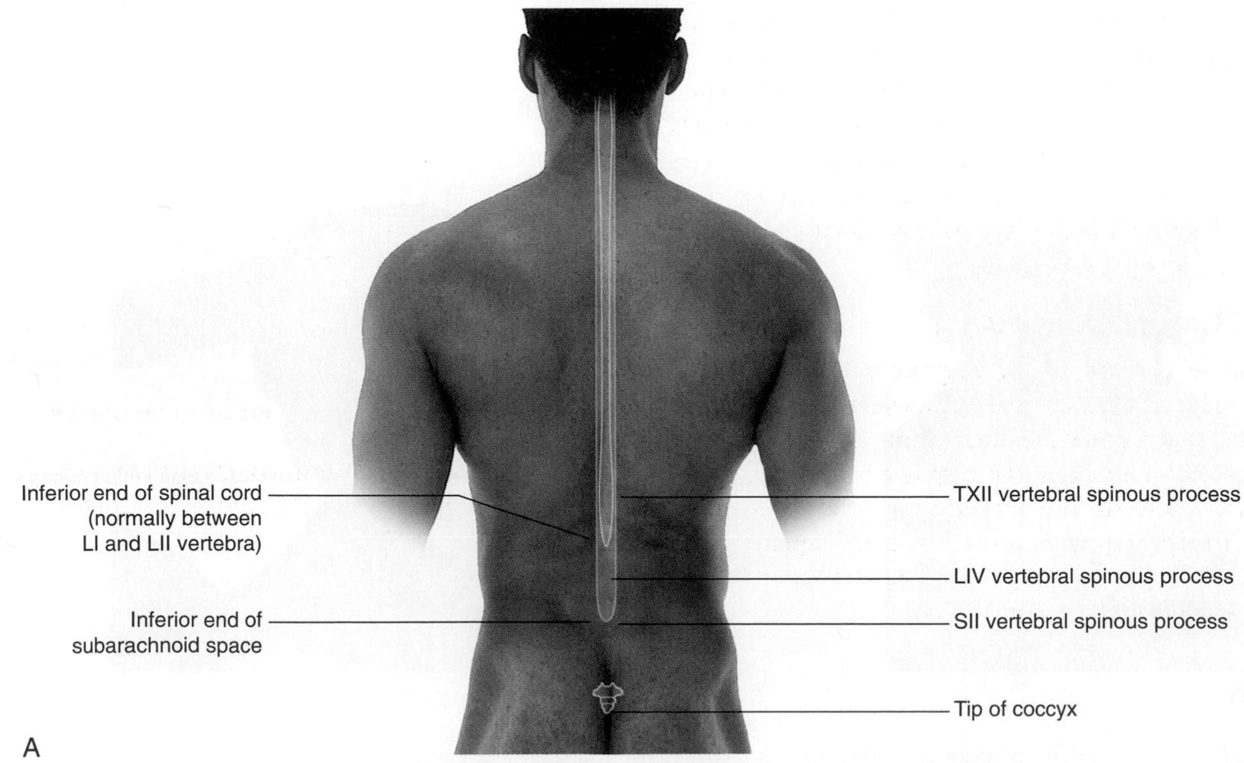


**Fig. 2.68** Back with the ends of the spinal cord and subarachnoid space indicated. **A.** In a man.

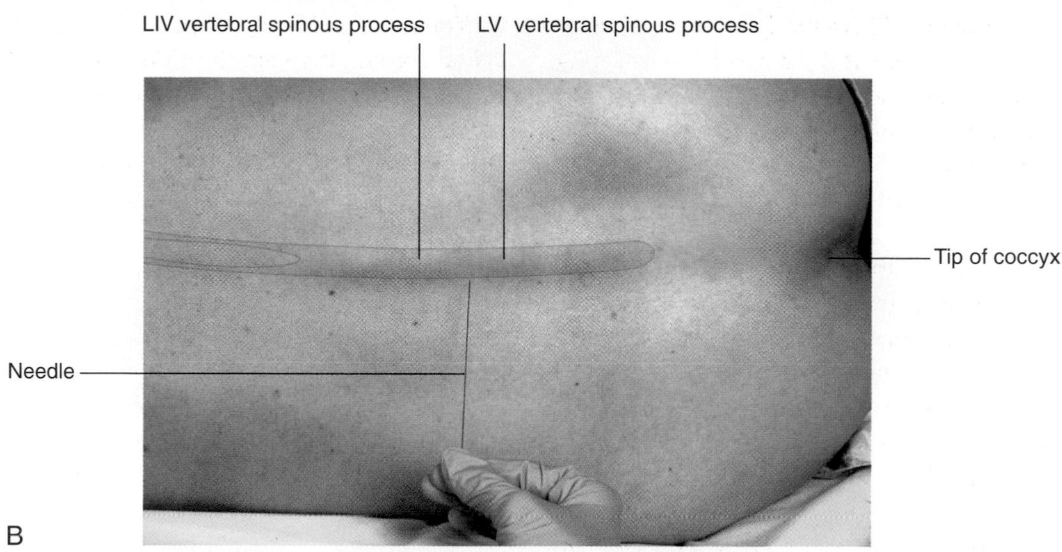


**Fig. 2.68, cont'd** Back with the ends of the spinal cord and subarachnoid space indicated. **B.** In a woman lying on her side in a fetal position, which accentuates the lumbar vertebral spinous processes and opens the spaces between adjacent vertebral arches. Cerebrospinal fluid can be withdrawn from the subarachnoid space in lower lumbar regions without endangering the spinal cord.

Because the subarachnoid space can be accessed in the lower lumbar region without endangering the spinal cord, it is important to be able to identify the position of the lumbar vertebral spinous processes. The LIV vertebral spinous process is level with a horizontal line between the highest points on the iliac crests. In the lumbar region, the palpable ends of the vertebral spinous processes lie opposite their corresponding vertebral bodies. The subarachnoid space can be accessed between vertebral levels LIII and LIV and between LIV and LV without endangering the spinal cord (Fig. 2.68B). The subarachnoid space ends at vertebral level SII, which is level with the sacral dimples marking the posterior superior iliac spines.

## Identifying major muscles

A number of intrinsic and extrinsic muscles of the back can readily be observed and palpated. The largest of these are the trapezius and latissimus dorsi muscles (Fig. 2.69A and 2.69B). Retracting the scapulae toward the midline can accentuate the rhomboid muscles (Fig. 2.69C), which lie deep to the trapezius muscle. The erector spinae muscles are visible as two longitudinal columns separated by a furrow in the midline (Fig. 2.69A).

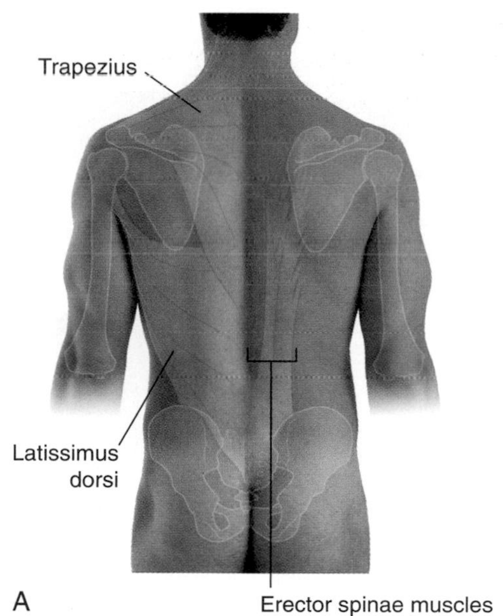


**Fig. 2.69** Back muscles. **A.** In a man with latissimus dorsi, trapezius, and erector spinae muscles outlined.

*Continued*

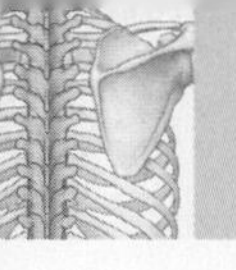

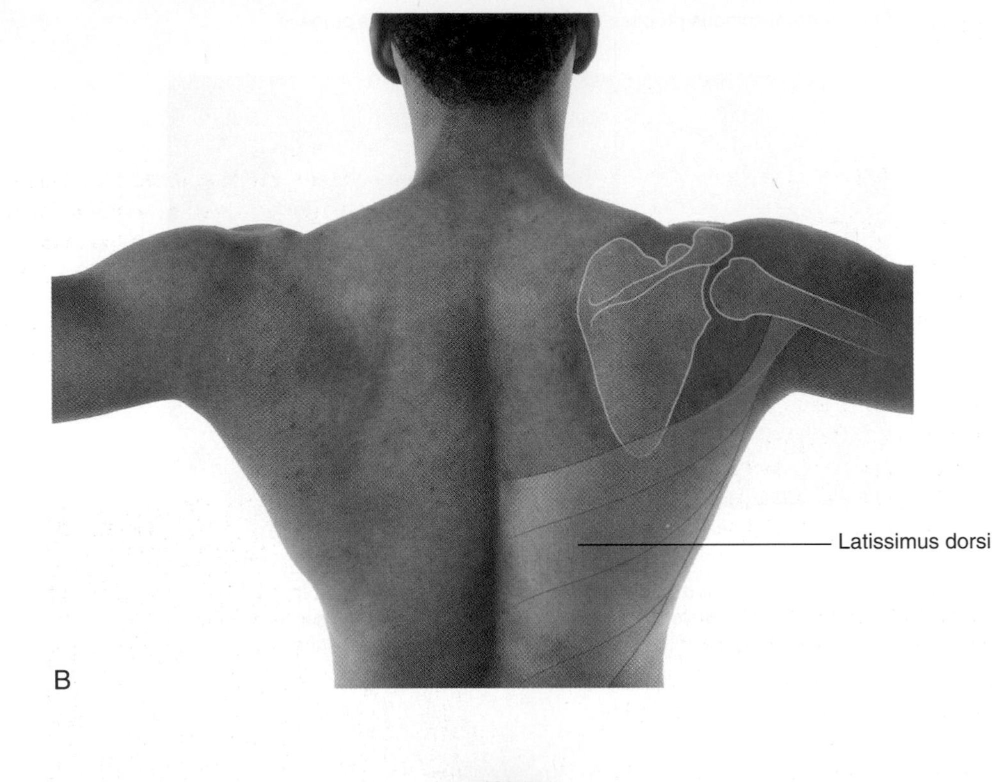


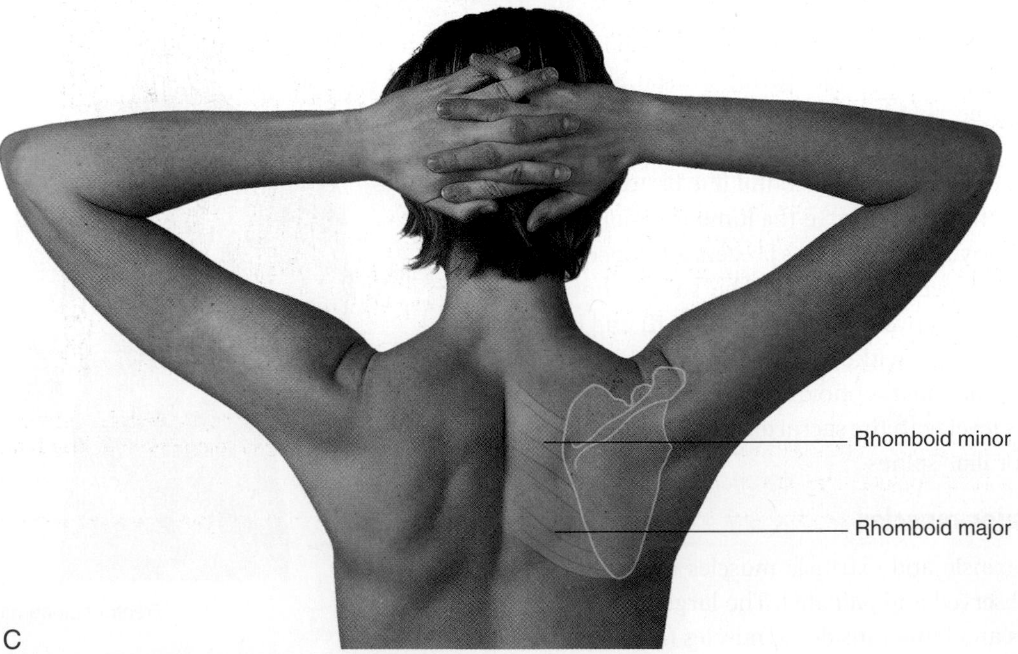


**Fig. 2.69, cont'd** Back muscles. **B.** In a man with arms abducted to accentuate the lateral margins of the latissimus dorsi muscles. **C.** In a woman with scapulae externally rotated and forcibly retracted to accentuate the rhomboid muscles.

# Clinical cases

## Case 1

CAUDA EQUINA SYNDROME

**A 50-year-old man was brought to the emergency department with severe lower back pain that had started several days ago. In the past 24 hours he has had two episodes of fecal incontinence and inability to pass urine and now reports numbness and weakness in both his legs.**

The attending physician performed a physical examination and found that the man had reduced strength during knee extension and when dorsiflexing his feet and toes. He also had reduced reflexes in his knees and ankles, numbness in the perineal (saddle) region, as well as reduced anal sphincter tone.

The patient's symptoms and physical examination findings raised serious concern for compression of multiple lumbar and sacral nerve roots in the spine, affecting both motor and sensory pathways. His reduced power in extending his knees and reduced knee reflexes was suggestive of compression of the L4 nerve roots. His reduced ability to dorsiflex his feet and toes was suggestive of compression of the L5 nerve roots. His reduced ankle reflexes was suggestive of compression of the S1 and S2 nerve roots, and his perineal numbness was suggestive of compression of the S3, S4, and S5 nerve roots.

A diagnosis of cauda equina syndrome was made, and the patient was transferred for an urgent MRI scan, which confirmed the presence of a severely herniating L2-3 disc compressing the cauda equina, giving rise to the cauda equina syndrome (Fig. 2.70). The patient underwent surgical decompression of the cauda equina and made a full recovery.

The collection of lumbar and sacral nerve roots beyond the conus medullaris has a horsetail-like appearance, from which it derives its name "cauda equina." Compression of the cauda equina may be caused by a herniating disc (as in this case), fracture fragments following traumatic injury, tumor, abscess, or severe degenerative stenosis of the central canal.

Cauda equina syndrome is classed as a surgical emergency to prevent permanent and irreversible damage to the compressed nerve roots.

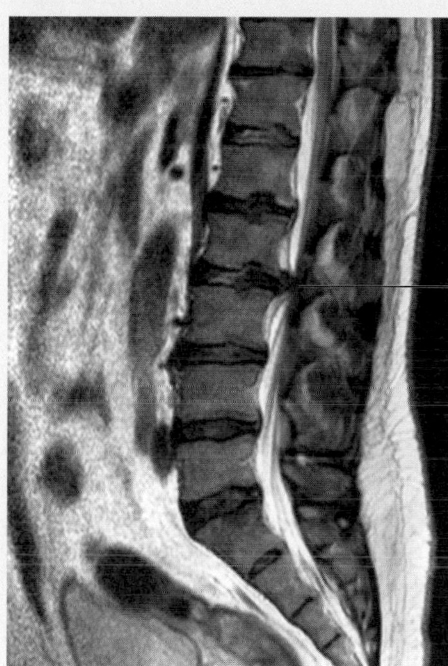


**Fig. 2.70** MRI of the lumbar spine reveals posterior herniation of the L2-3 disc resulting in compression of the cauda equina filaments.

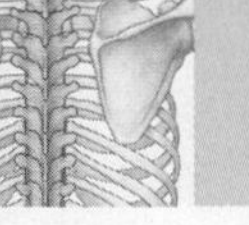



## Case 2

CERVICAL SPINAL CORD INJURY

**A 45-year-old man was involved in a serious car accident. On examination he had a severe injury to the cervical region of his vertebral column with damage to the spinal cord. In fact, his breathing became erratic and stopped.**

If the cervical spinal cord injury is above the level of C5, breathing is likely to stop. The phrenic nerve takes origin from C3, C4, and C5 and supplies the diaphragm. Breathing may not cease immediately if the lesion is just below C5, but does so as the cord becomes edematous and damage progresses superiorly. In addition, some respiratory and ventilatory exchange may occur by using neck muscles plus the sternocleidomastoid and trapezius muscles, which are innervated by the accessory nerve [XI].

The patient was unable to sense or move his upper and lower limbs.

The patient has paralysis of the upper and lower limbs and is therefore quadriplegic. If breathing is unaffected, the lesion is below the level of C5 or at the level of C5. The nerve supply to the upper limbs is via the brachial plexus, which begins at the C5 level. The site of the spinal cord injury is at or above the C5 level.

It is important to remember that although the cord has been transected in the cervical region, the cord below this level is intact. Reflex activity may therefore occur below the injury, but communication with the brain is lost.

# 3 Thorax

**ADDITIONAL LEARNING RESOURCES for Chapter 3, *Thorax*, on STUDENT CONSULT (www.studentconsult.com):**

- Self-Assessment—National Board style multiple-choice questions, Chapter 3
- Short Questions—these are questions requiring short responses, Chapter 3
- Interactive Surface Anatomy—interactive surface animations, Chapter 3
- Clinical Cases, Chapter 3
  Aortic dissection
  Broken pacemaker
  Cardiac tamponade
  Cervical rib
  Chest wound
  Coarctation of the aorta
  Cystic fibrosis with bronchiectasis
  Esophageal cancer
  Lung cancer
  Patent ductus arteriosus
  Pneumonia
  Sinus of Valsalva aneurysm
  Subclavian steal syndrome
  Venous access

**Free Online Anatomy and Embryology Self-Study Course**

- Anatomy modules 4 through 9
- Embryology modules 61 through 64

# Conceptual overview

## GENERAL DESCRIPTION

The **thorax** is an irregularly shaped cylinder with a narrow opening (superior thoracic aperture) superiorly and a relatively large opening (inferior thoracic aperture) inferiorly (Fig. 3.1). The superior thoracic aperture is open, allowing continuity with the neck; the inferior thoracic aperture is closed by the diaphragm.

The musculoskeletal wall of the thorax is flexible and consists of segmentally arranged vertebrae, ribs, and muscles and the sternum.

The **thoracic cavity** enclosed by the thoracic wall and the diaphragm is subdivided into three major compartments:

- a left and a right pleural cavity, each surrounding a lung, and
- the mediastinum.

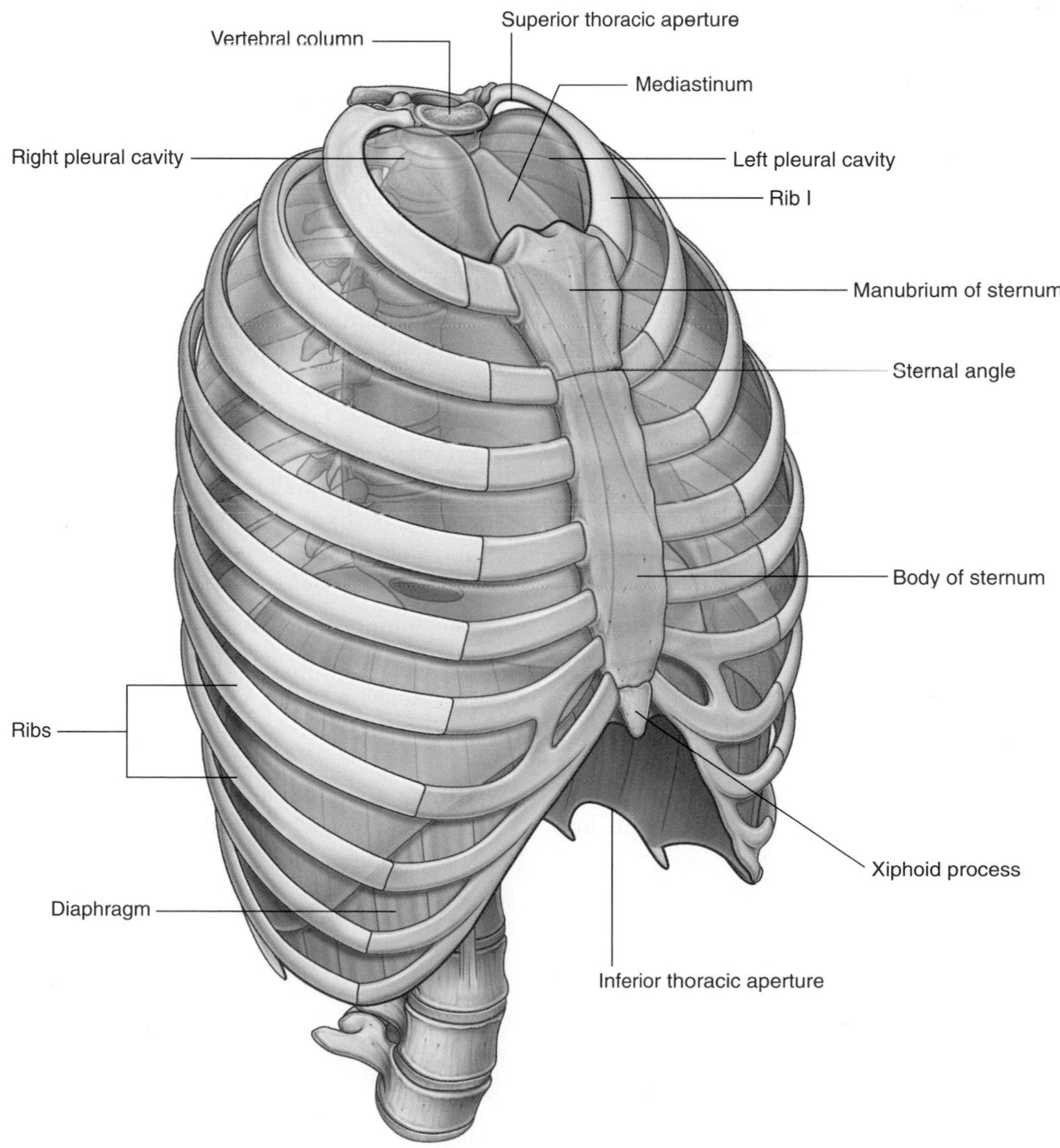


**Fig. 3.1** Thoracic wall and cavity.

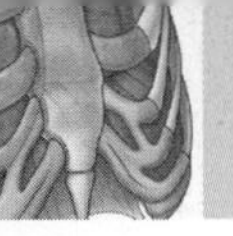



The mediastinum is a thick, flexible soft tissue partition oriented longitudinally in a median sagittal position. It contains the heart, esophagus, trachea, major nerves, and major systemic blood vessels.

The pleural cavities are completely separated from each other by the mediastinum. Therefore abnormal events in one pleural cavity do not necessarily affect the other cavity. This also means that the mediastinum can be entered surgically without opening the pleural cavities.

Another important feature of the pleural cavities is that they extend above the level of rib I. The apex of each lung actually extends into the root of the neck. As a consequence, abnormal events in the root of the neck can involve the adjacent pleura and lung, and events in the adjacent pleura and lung can involve the root of the neck.

## FUNCTIONS

### Breathing

One of the most important functions of the thorax is breathing. The thorax not only contains the lungs but also provides the machinery necessary—the diaphragm, thoracic wall, and ribs—for effectively moving air into and out of the lungs.

Up and down movements of the diaphragm and changes in the lateral and anterior dimensions of the thoracic wall, caused by movements of the ribs, alter the volume of the thoracic cavity and are key elements in breathing.

### Protection of vital organs

The thorax houses and protects the heart, lungs, and great vessels. Because of the upward domed shape of the diaphragm, the thoracic wall also offers protection to some important abdominal viscera.

Much of the liver lies under the right dome of the diaphragm, and the stomach and spleen lie under the left. The posterior aspects of the superior poles of the kidneys lie on the diaphragm and are anterior to rib XII, on the right, and to ribs XI and XII, on the left.

### Conduit

The mediastinum acts as a conduit for structures that pass completely through the thorax from one body region to another and for structures that connect organs in the thorax to other body regions.

The esophagus, vagus nerves, and thoracic duct pass through the mediastinum as they course between the abdomen and neck.

The phrenic nerves, which originate in the neck, also pass through the mediastinum to penetrate and supply the diaphragm.

Other structures such as the trachea, thoracic aorta, and superior vena cava course within the mediastinum en route to and from major visceral organs in the thorax.

## COMPONENT PARTS

### Thoracic wall

The **thoracic wall** consists of skeletal elements and muscles (Fig. 3.1):

- Posteriorly, it is made up of twelve thoracic vertebrae and their intervening intervertebral discs;
- Laterally, the wall is formed by **ribs** (twelve on each side) and three layers of flat muscles, which span the intercostal spaces between adjacent ribs, move the ribs, and provide support for the intercostal spaces;
- Anteriorly, the wall is made up of the **sternum**, which consists of the manubrium of sternum, body of sternum, and xiphoid process.

The manubrium of sternum, angled posteriorly on the body of sternum at the manubriosternal joint, forms the sternal angle, which is a major surface landmark used by clinicians in performing physical examinations of the thorax.

The anterior (distal) end of each rib is composed of costal cartilage, which contributes to the mobility and elasticity of the wall.

All ribs articulate with thoracic vertebrae posteriorly. Most ribs (from rib II to IX) have three articulations with the vertebral column. The head of each rib articulates with the body of its own vertebra and with the body of the vertebra above (Fig. 3.2). As these ribs curve posteriorly, each also articulates with the transverse process of its vertebra.

Anteriorly, the costal cartilages of ribs I to VII articulate with the sternum.

The costal cartilages of ribs VIII to X articulate with the inferior margins of the costal cartilages above them. Ribs XI and XII are called floating ribs because they do not articulate with other ribs, costal cartilages, or the sternum. Their costal cartilages are small, only covering their tips.

The skeletal framework of the thoracic wall provides extensive attachment sites for muscles of the neck, abdomen, back, and upper limbs.

A number of these muscles attach to ribs and function as accessory respiratory muscles; some of them also stabilize the position of the first and last ribs.

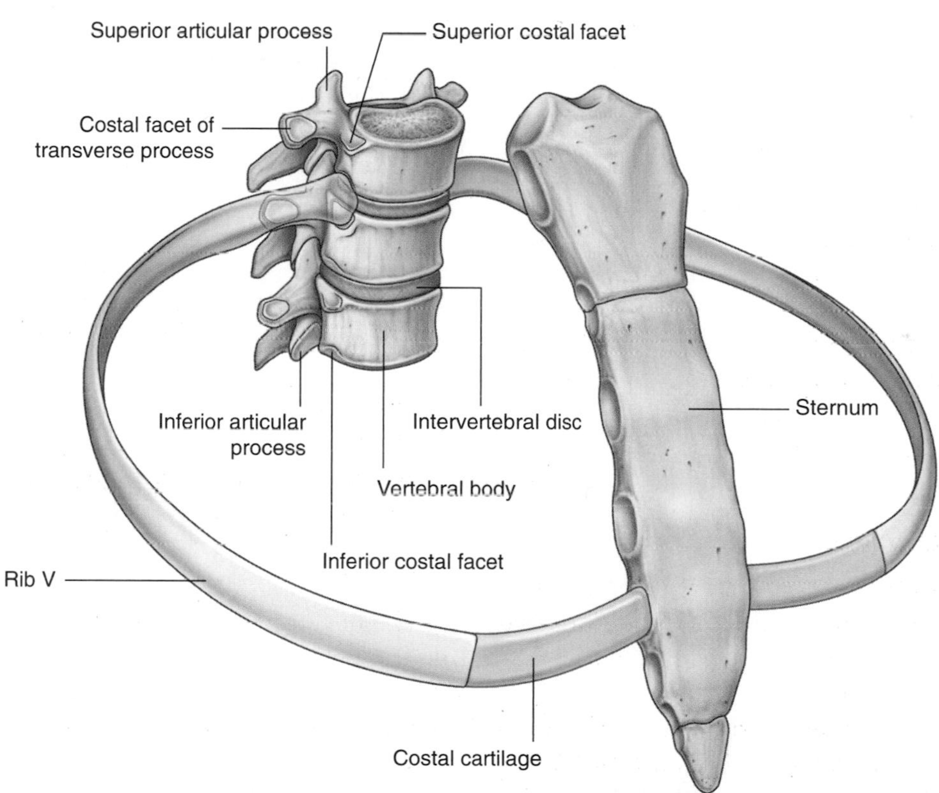


**Fig. 3.2** Joints between ribs and vertebrae.

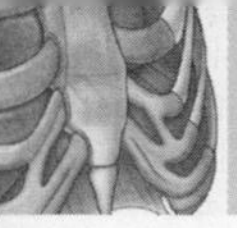



## Superior thoracic aperture

Completely surrounded by skeletal elements, the **superior thoracic aperture** consists of the body of vertebra TI posteriorly, the medial margin of rib I on each side, and the manubrium anteriorly.

The superior margin of the manubrium is in approximately the same horizontal plane as the intervertebral disc between vertebrae TII and TIII.

The first ribs slope inferiorly from their posterior articulation with vertebra TI to their anterior attachment to the manubrium. Consequently, the plane of the superior thoracic aperture is at an oblique angle, facing somewhat anteriorly.

At the superior thoracic aperture, the superior aspects of the pleural cavities, which surround the lungs, lie on either side of the entrance to the mediastinum (Fig. 3.3).

Structures that pass between the upper limb and thorax pass over rib I and the superior part of the pleural cavity as they enter and leave the mediastinum. Structures that pass between the neck and head and the thorax pass more vertically through the superior thoracic aperture.

## Inferior thoracic aperture

The **inferior thoracic aperture** is large and expandable. Bone, cartilage, and ligaments form its margin (Fig. 3.4A).

The inferior thoracic aperture is closed by the diaphragm, and structures passing between the abdomen and thorax pierce or pass posteriorly to the diaphragm.

Skeletal elements of the inferior thoracic aperture are:

- the body of vertebra TXII posteriorly,
- rib XII and the distal end of rib XI posterolaterally,
- the distal cartilaginous ends of ribs VII to X, which unite to form the costal margin anterolaterally, and
- the xiphoid process anteriorly.

The joint between the costal margin and sternum lies roughly in the same horizontal plane as the intervertebral disc between vertebrae TIX and TX. In other words, the posterior margin of the inferior thoracic aperture is inferior to the anterior margin.

When viewed anteriorly, the inferior thoracic aperture is tilted superiorly.

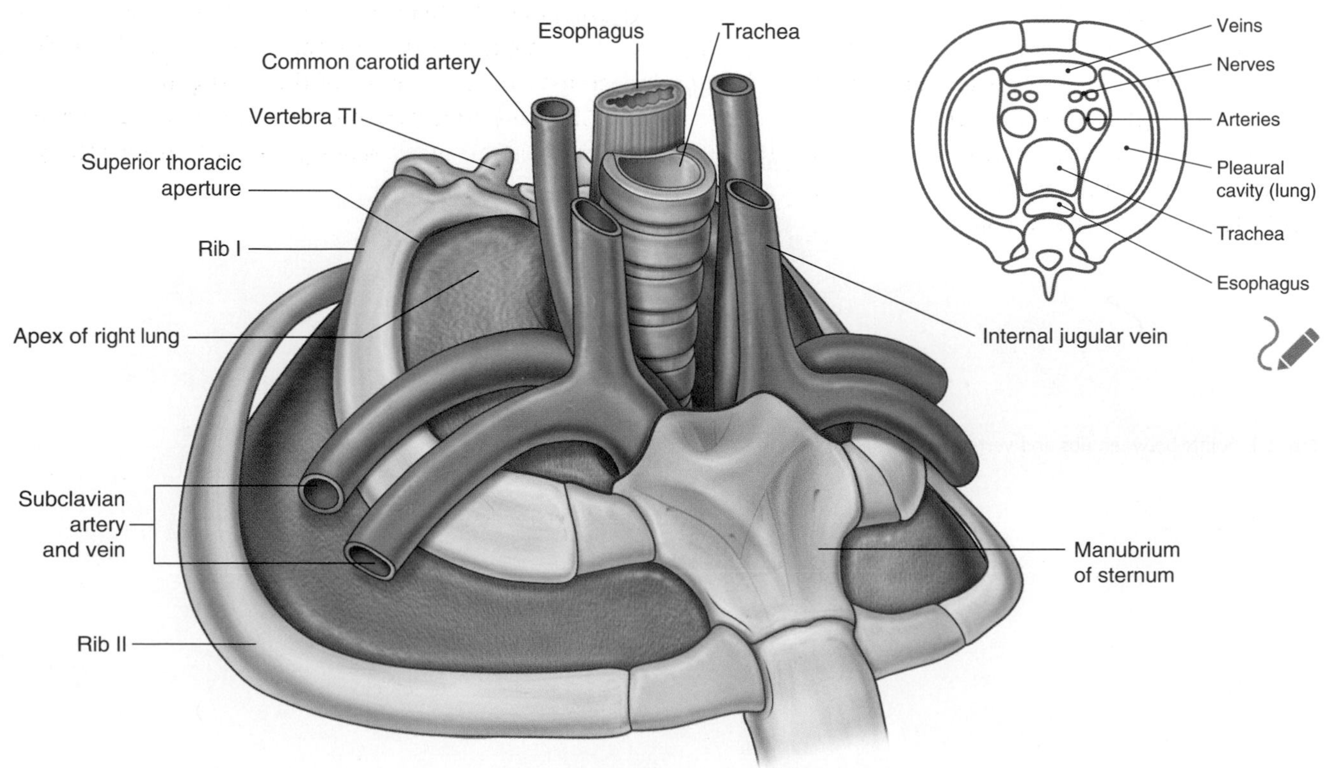




**Fig. 3.3** Superior thoracic aperture.

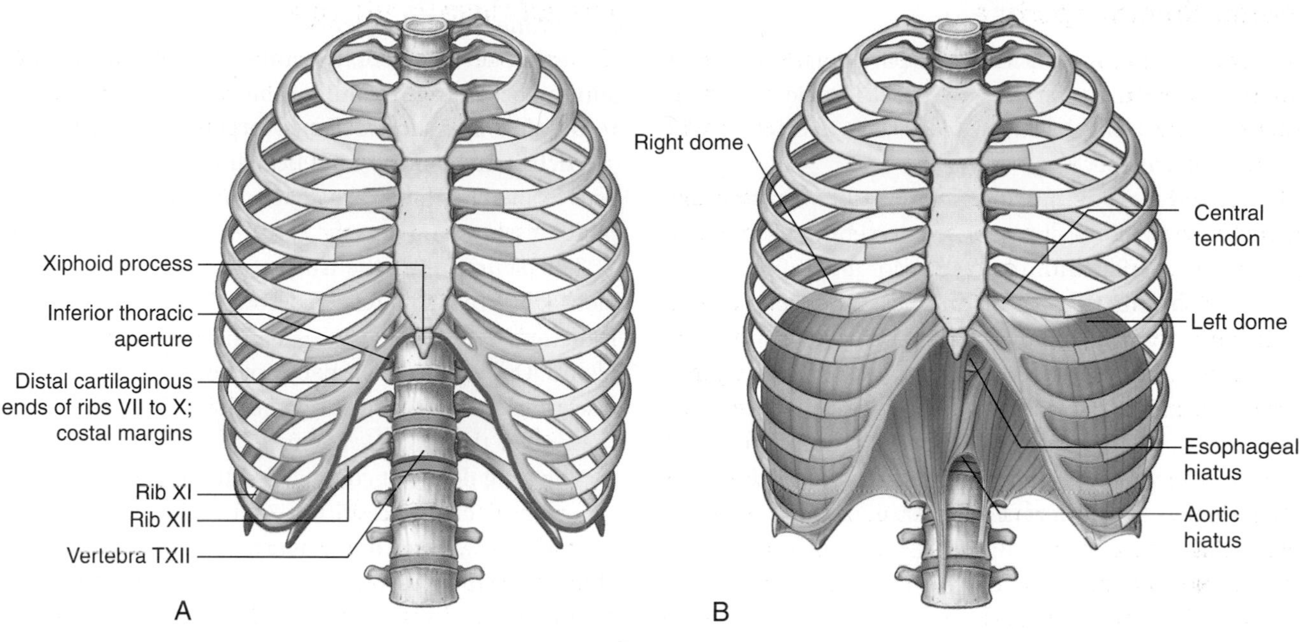


**Fig. 3.4 A.** Inferior thoracic aperture. **B.** Diaphragm.

## Diaphragm

The musculotendinous **diaphragm** seals the inferior thoracic aperture (Fig. 3.4B).

Generally, muscle fibers of the diaphragm arise radially, from the margins of the inferior thoracic aperture, and converge into a large central tendon.

Because of the oblique angle of the inferior thoracic aperture, the posterior attachment of the diaphragm is inferior to the anterior attachment.

The diaphragm is not flat; rather, it "balloons" superiorly, on both the right and left sides, to form domes. The right dome is higher than the left, reaching as far as rib V.

As the diaphragm contracts, the height of the domes decreases and the volume of the thorax increases.

The esophagus and inferior vena cava penetrate the diaphragm; the aorta passes posterior to the diaphragm.

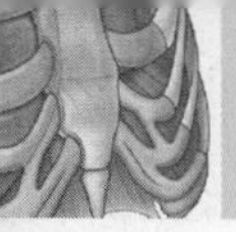



## Mediastinum

The **mediastinum** is a thick midline partition that extends from the sternum anteriorly to the thoracic vertebrae posteriorly, and from the superior thoracic aperture to the inferior thoracic aperture.

A horizontal plane passing through the sternal angle and the intervertebral disc between vertebrae TIV and TV separates the mediastinum into superior and inferior parts (Fig. 3.5). The inferior part is further subdivided by the pericardium, which encloses the pericardial cavity surrounding the heart. The pericardium and heart constitute the middle mediastinum.

The anterior mediastinum lies between the sternum and the pericardium; the posterior mediastinum lies between the pericardium and thoracic vertebrae.

## Pleural cavities

The two pleural cavities are situated on either side of the mediastinum (Fig. 3.6).

Each **pleural cavity** is completely lined by a mesothelial membrane called the pleura.

During development, the lungs grow out of the mediastinum, becoming surrounded by the pleural cavities. As a result, the outer surface of each organ is covered by pleura.

Each lung remains attached to the mediastinum by a root formed by the airway, pulmonary blood vessels, lymphatic tissues, and nerves.

The pleura lining the walls of the cavity is the parietal pleura, whereas that reflected from the mediastinum at the roots and onto the surfaces of the lungs is the visceral pleura. Only a potential space normally exists between the visceral pleura covering lung and the parietal pleura lining the wall of the thoracic cavity.

The lung does not completely fill the potential space of the pleural cavity, resulting in recesses, which do not contain lung and are important for accommodating changes in lung volume during breathing. The costodiaphragmatic recess, which is the largest and clinically most important recess, lies inferiorly between the thoracic wall and diaphragm.

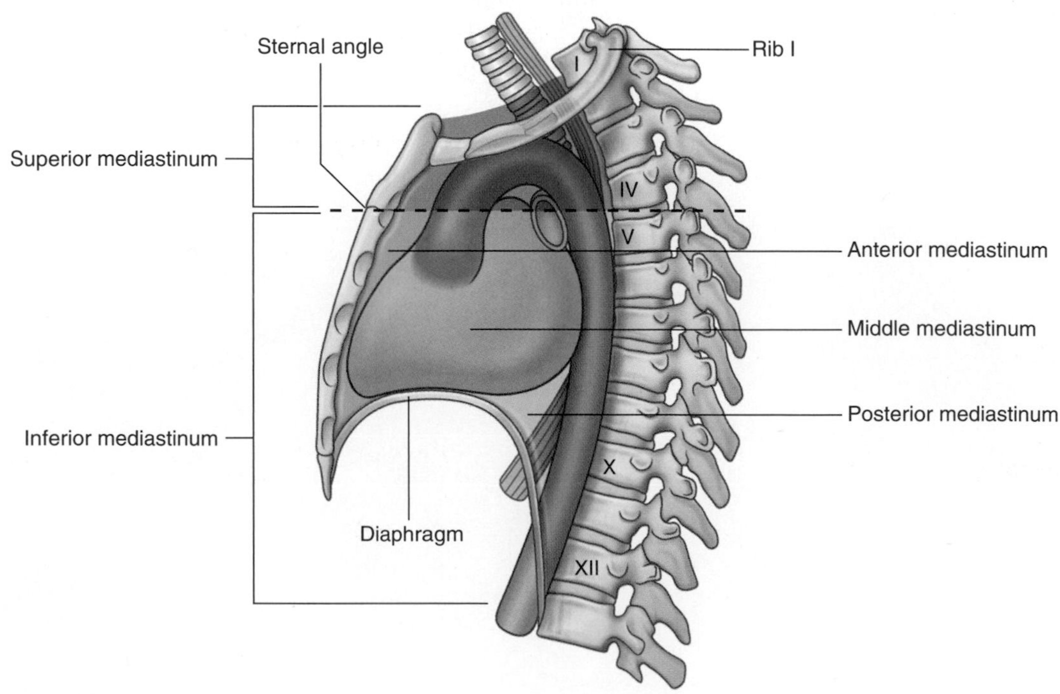


**Fig. 3.5** Subdivisions of the mediastinum.

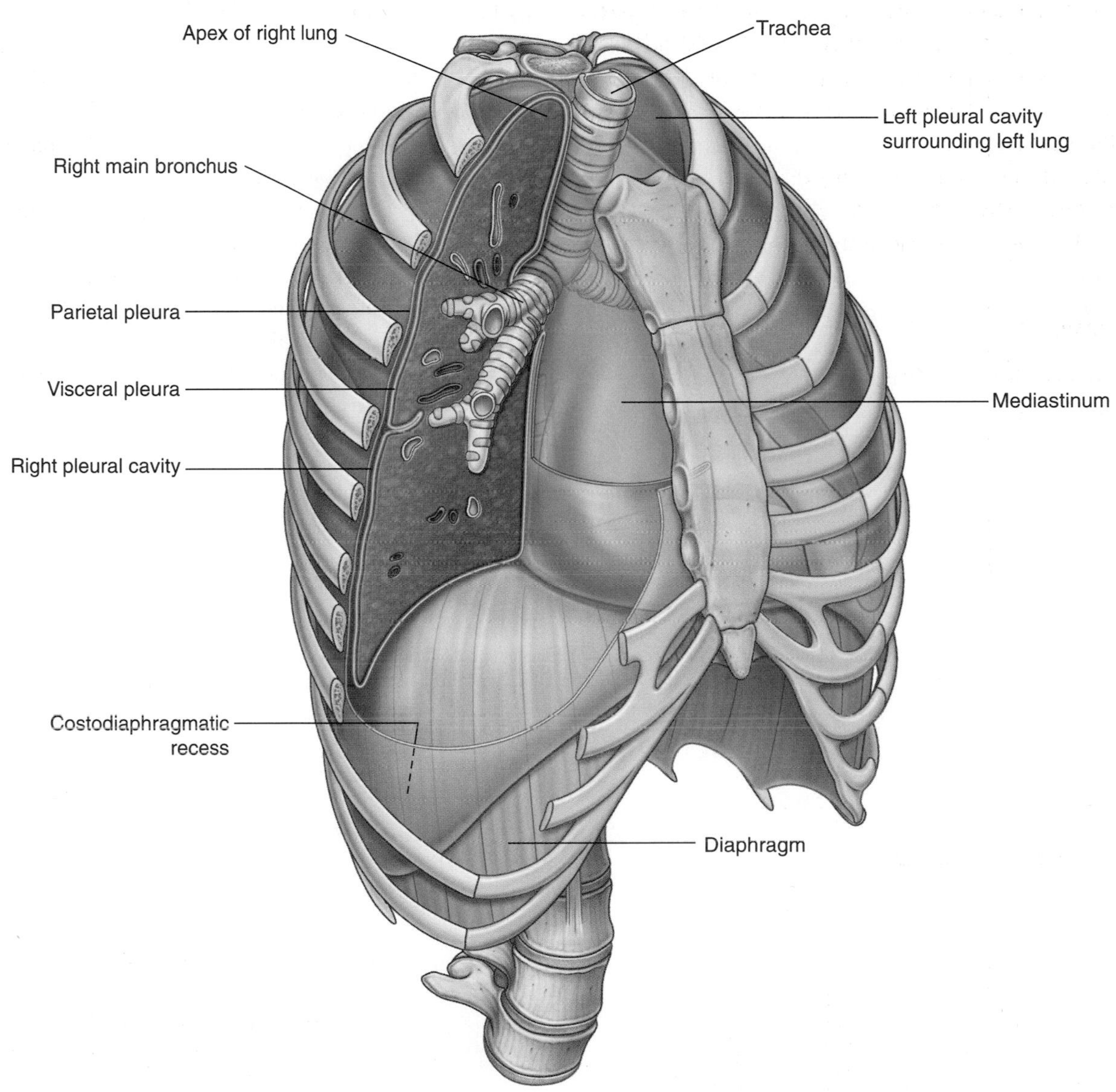


**Fig. 3.6** Pleural cavities.

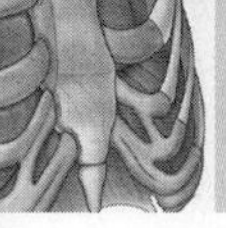


## RELATIONSHIP TO OTHER REGIONS

### Neck

The superior thoracic aperture opens directly into the root of the neck (Fig. 3.7).

The superior aspect of each pleural cavity extends approximately 2 to 3 cm above rib I and the costal cartilage into the neck. Between these pleural extensions, major visceral structures pass between the neck and superior mediastinum. In the midline, the trachea lies immediately anterior to the esophagus. Major blood vessels and nerves pass in and out of the thorax at the superior thoracic aperture anteriorly and laterally to these structures.

### Upper limb

An **axillary inlet**, or gateway to the upper limb, lies on each side of the superior thoracic aperture. These two axillary inlets and the superior thoracic aperture communicate superiorly with the root of the neck (Fig. 3.7).

Each axillary inlet is formed by:

- the superior margin of the scapula posteriorly,
- the clavicle anteriorly, and
- the lateral margin of rib I medially.

The apex of each triangular inlet is directed laterally and is formed by the medial margin of the coracoid process, which extends anteriorly from the superior margin of the scapula.

The base of the axillary inlet's triangular opening is the lateral margin of rib I.

Large blood vessels passing between the axillary inlet and superior thoracic aperture do so by passing over rib I.

Proximal parts of the brachial plexus also pass between the neck and upper limb by passing through the axillary inlet.

### Abdomen

The diaphragm separates the thorax from the abdomen. Structures that pass between the thorax and abdomen either penetrate the diaphragm or pass posteriorly to it (Fig. 3.8):

- The inferior vena cava pierces the **central tendon of the diaphragm** to enter the right side of the mediastinum near vertebral level TVIII.
- The esophagus penetrates the muscular part of the diaphragm to leave the mediastinum and enter the abdomen just to the left of the midline at vertebral level TX.

**Fig. 3.7** Superior thoracic aperture and axillary inlet.

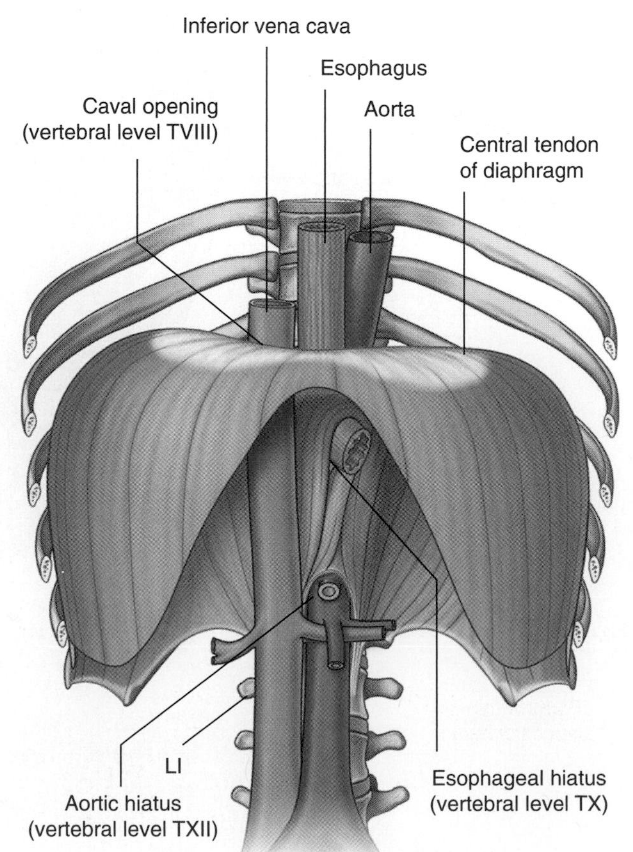


**Fig. 3.8** Major structures passing between abdomen and thorax.

- The aorta passes posteriorly to the diaphragm at the midline at vertebral level TXII.
- Numerous other structures that pass between the thorax and abdomen pass through or posterior to the diaphragm.

## Breast

The breasts, consisting of mammary glands, superficial fascia, and overlying skin, are in the **pectoral region** on each side of the anterior thoracic wall (Fig. 3.9).

Vessels, lymphatics, and nerves associated with the breast are as follows:

- Branches from the internal thoracic arteries and veins perforate the anterior chest wall on each side of the sternum to supply anterior aspects of the thoracic wall. Those branches associated mainly with the second to fourth intercostal spaces also supply the anteromedial parts of each breast.
- Lymphatic vessels from the medial part of the breast accompany the perforating arteries and drain into the parasternal nodes on the deep surface of the thoracic wall.
- Vessels and lymphatics associated with lateral parts of the breast emerge from or drain into the **axillary region** of the upper limb.
- Lateral and anterior branches of the fourth to sixth intercostal nerves carry general sensation from the skin of the breast.

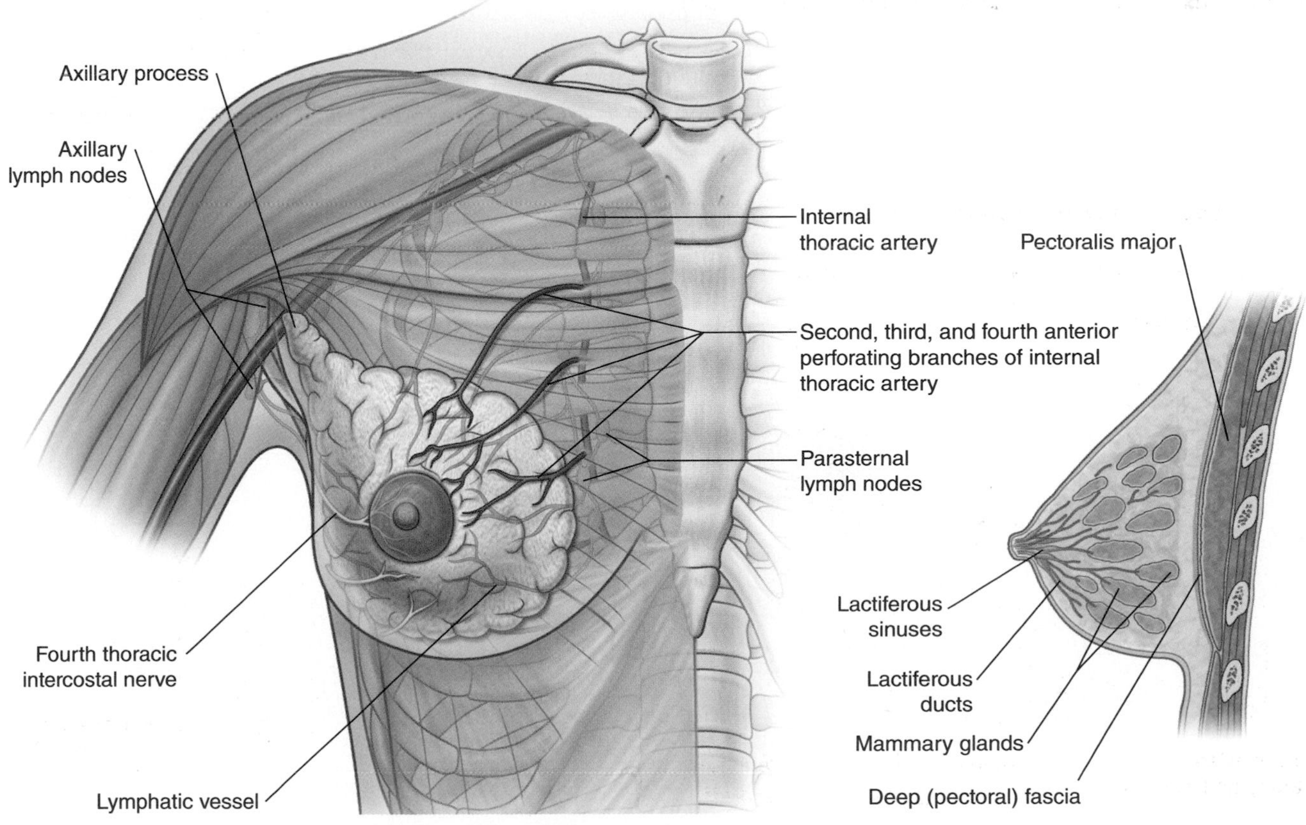


**Fig. 3.9** Right breast.

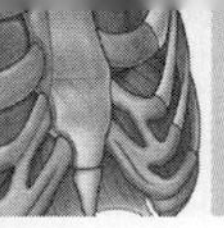



KEY FEATURES

## Vertebral level TIV/V

When working with patients, physicians use vertebral levels to determine the position of important anatomical structures within body regions.

The horizontal plane passing through the disc that separates thoracic vertebrae TIV and TV is one of the most significant planes in the body (Fig. 3.10) because it:

- passes through the sternal angle anteriorly, marking the position of the anterior articulation of the costal cartilage of rib II with the sternum. The sternal angle is used to find the position of rib II as a reference for counting ribs (because of the overlying clavicle, rib I is not palpable);
- separates the superior mediastinum from the inferior mediastinum and marks the position of the superior limit of the pericardium;
- marks where the arch of the aorta begins and ends;
- passes through the site where the superior vena cava penetrates the pericardium to enter the heart;
- is the level at which the trachea bifurcates into right and left main bronchi; and
- marks the superior limit of the pulmonary trunk.

## Venous shunts from left to right

The **right atrium** is the chamber of the heart that receives deoxygenated blood returning from the body. It lies on the right side of the midline, and the two major veins, the superior and inferior venae cavae, that drain into it are also located on the right side of the body. This means that, to get to the right side of the body, all blood coming from the left side has to cross the midline. This left-to-right shunting is carried out by a number of important and, in some cases, very large veins, several of which are in the thorax (Fig. 3.11).

In adults, the left brachiocephalic vein crosses the midline immediately posterior to the manubrium and delivers blood from the left side of the head and neck, the left upper limb, and part of the left thoracic wall into the superior vena cava.

The hemiazygos and accessory hemiazygos veins drain posterior and lateral parts of the left thoracic wall, pass immediately anterior to the bodies of thoracic vertebrae, and flow into the azygos vein on the right side, which ultimately connects with the superior vena cava.

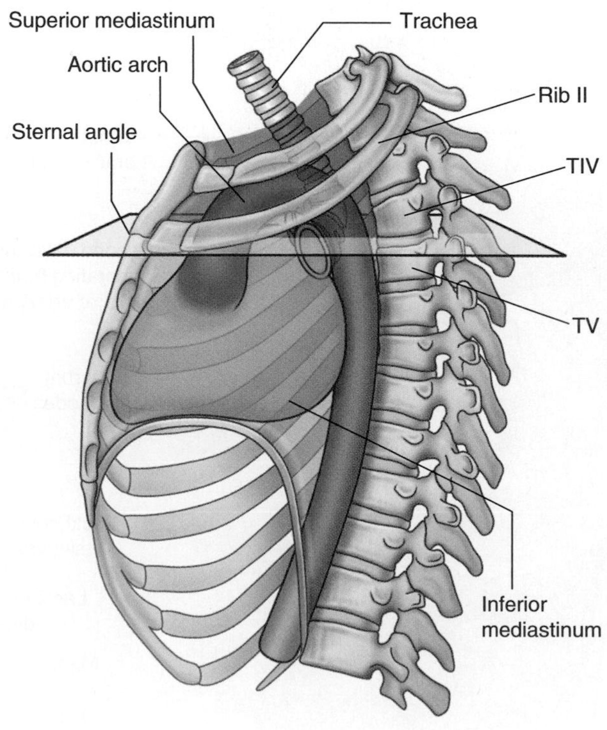


**Fig. 3.10** Vertebral level TIV/V.

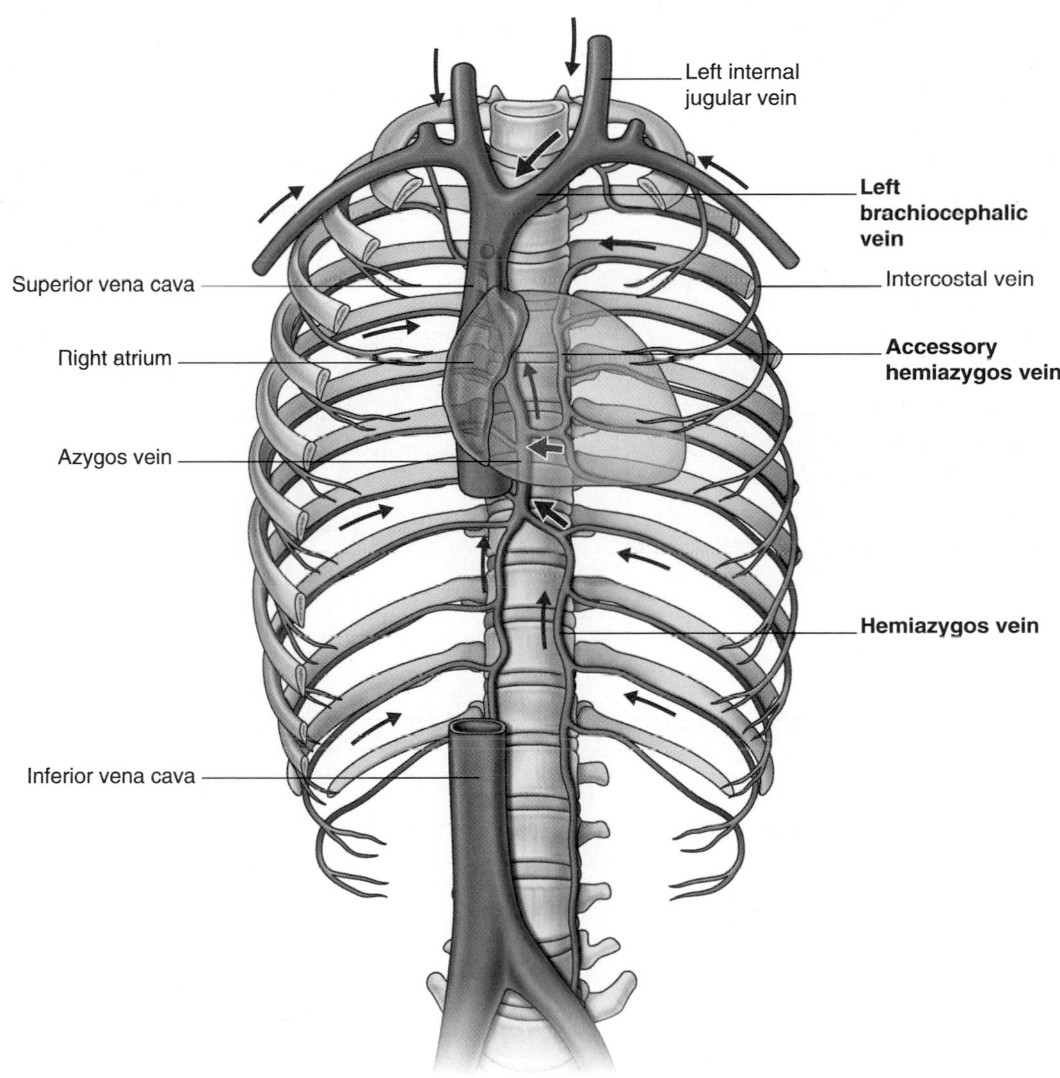


**Fig. 3.11** Left-to-right venous shunts.

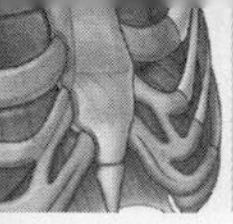



## Segmental neurovascular supply of thoracic wall

The arrangement of vessels and nerves that supply the thoracic wall reflects the segmental organization of the wall. Arteries to the wall arise from two sources:

- the thoracic aorta, which is in the posterior mediastinum, and
- a pair of vessels, the internal thoracic arteries, which run along the deep aspect of the anterior thoracic wall on either side of the sternum.

Posterior and anterior intercostal vessels branch segmentally from these arteries and pass laterally around the wall, mainly along the inferior margin of each rib (Fig. 3.12A). Running with these vessels are intercostal nerves (the anterior rami of thoracic spinal nerves), which innervate the wall, related parietal pleura, and associated



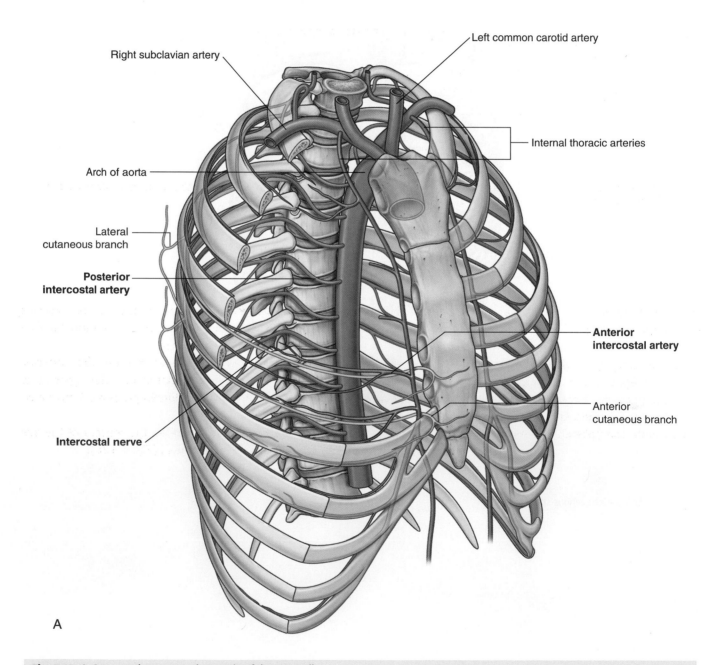

**Fig. 3.12 A.** Segmental neurovascular supply of thoracic wall.

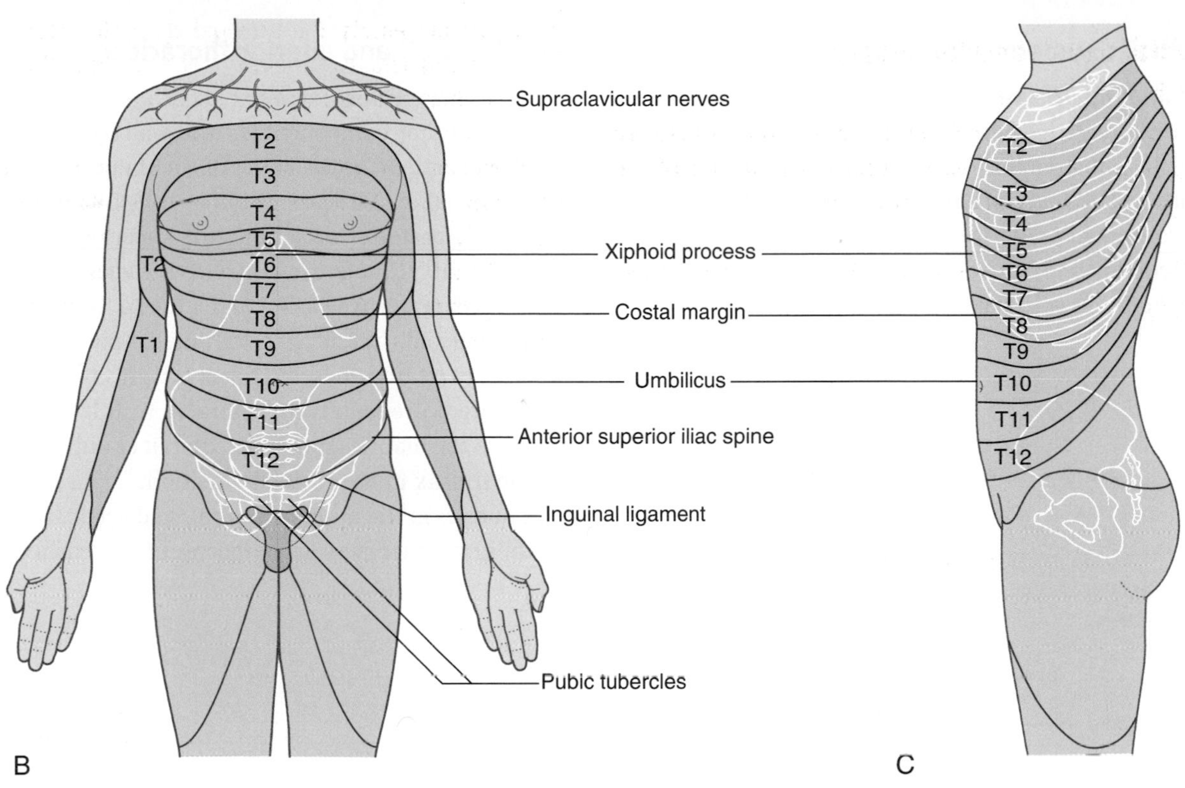


**Fig. 3.12, cont'd B.** Anterior view of thoracic dermatomes associated with thoracic spinal nerves. **C.** Lateral view of dermatomes associated with thoracic spinal nerves.

skin. The position of these nerves and vessels relative to the ribs must be considered when passing objects, such as chest tubes, through the thoracic wall.

Dermatomes of the thorax generally reflect the segmental organization of the thoracic spinal nerves (Fig. 3.12B). The exception occurs, anteriorly and superiorly, with the first thoracic dermatome, which is located mostly in the upper limb, and not on the trunk.

The anterosuperior region of the trunk receives branches from the anterior ramus of C4 via supraclavicular branches of the cervical plexus.

The highest thoracic dermatome on the anterior chest wall is T2, which also extends into the upper limb. In the midline, skin over the xiphoid process is innervated by T6.

Dermatomes of T7 to T12 follow the contour of the ribs onto the anterior abdominal wall (Fig. 3.12C).

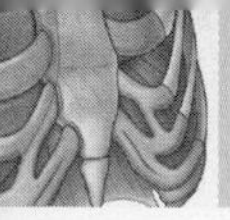



## Sympathetic system

All preganglionic nerve fibers of the sympathetic system are carried out of the spinal cord in spinal nerves T1 to L2 (Fig. 3.13). This means that sympathetic fibers found anywhere in the body ultimately emerge from the spinal cord as components of these spinal nerves. Preganglionic sympathetic fibers destined for the head are carried out of the spinal cord in spinal nerve T1.

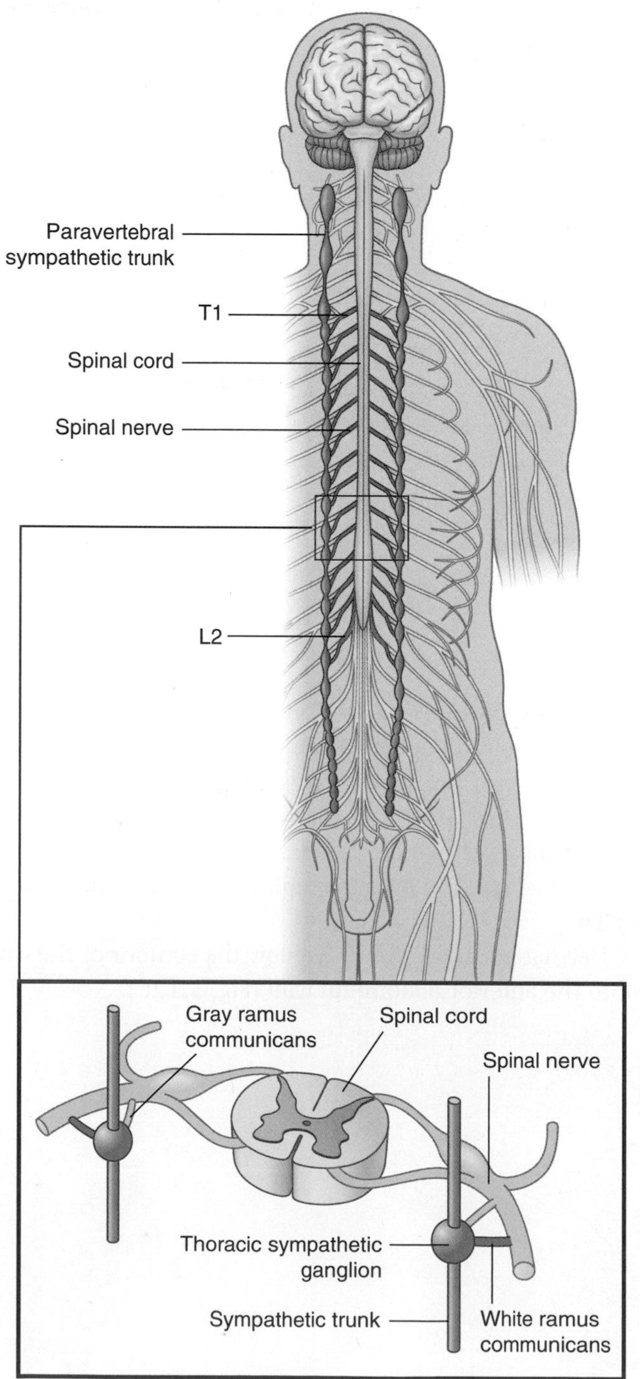


**Fig. 3.13** Sympathetic trunks.

## Flexible wall and inferior thoracic aperture

The thoracic wall is expandable because most ribs articulate with other components of the wall by true joints that allow movement, and because of the shape and orientation of the ribs (Fig. 3.14).

A rib's posterior attachment is superior to its anterior attachment. Therefore, when a rib is elevated, it moves the anterior thoracic wall forward relative to the posterior wall, which is fixed. In addition, the middle part of each rib is inferior to its two ends, so that when this region of the rib is elevated, it expands the thoracic wall laterally. Finally, because the diaphragm is muscular, it changes the volume of the thorax in the vertical direction.

Changes in the anterior, lateral, and vertical dimensions of the thoracic cavity are important for breathing.

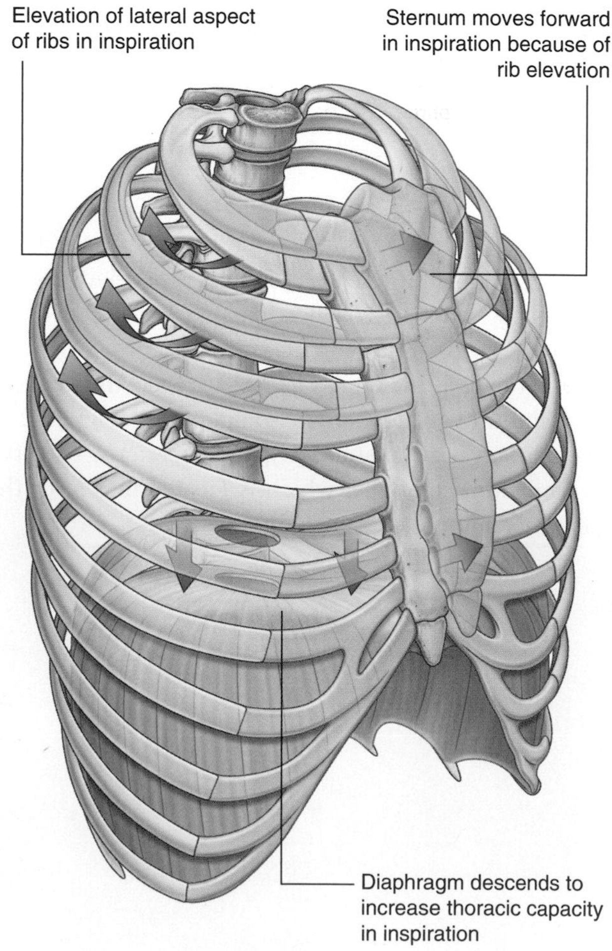


**Fig. 3.14** Flexible thoracic wall and inferior thoracic aperture.

## Innervation of the diaphragm

The diaphragm is innervated by two phrenic nerves that originate, one on each side, as branches of the cervical plexus in the neck (Fig. 3.15). They arise from the anterior rami of cervical nerves C3, C4, and C5, with the major contribution coming from C4.

The **phrenic nerves** pass vertically through the neck, the superior thoracic aperture, and the mediastinum to supply motor innervation to the entire diaphragm, including the crura (muscular extensions that attach the diaphragm to the upper lumbar vertebrae). In the mediastinum, the phrenic nerves pass anteriorly to the roots of the lungs.

The tissues that initially give rise to the diaphragm are in an anterior position on the embryological disc before the head fold develops, which explains the cervical origin of the nerves that innervate the diaphragm. In other words, the tissue that gives rise to the diaphragm originates superior to the ultimate location of the diaphragm.

Spinal cord injuries below the level of the origin of the phrenic nerve do not affect movement of the diaphragm.

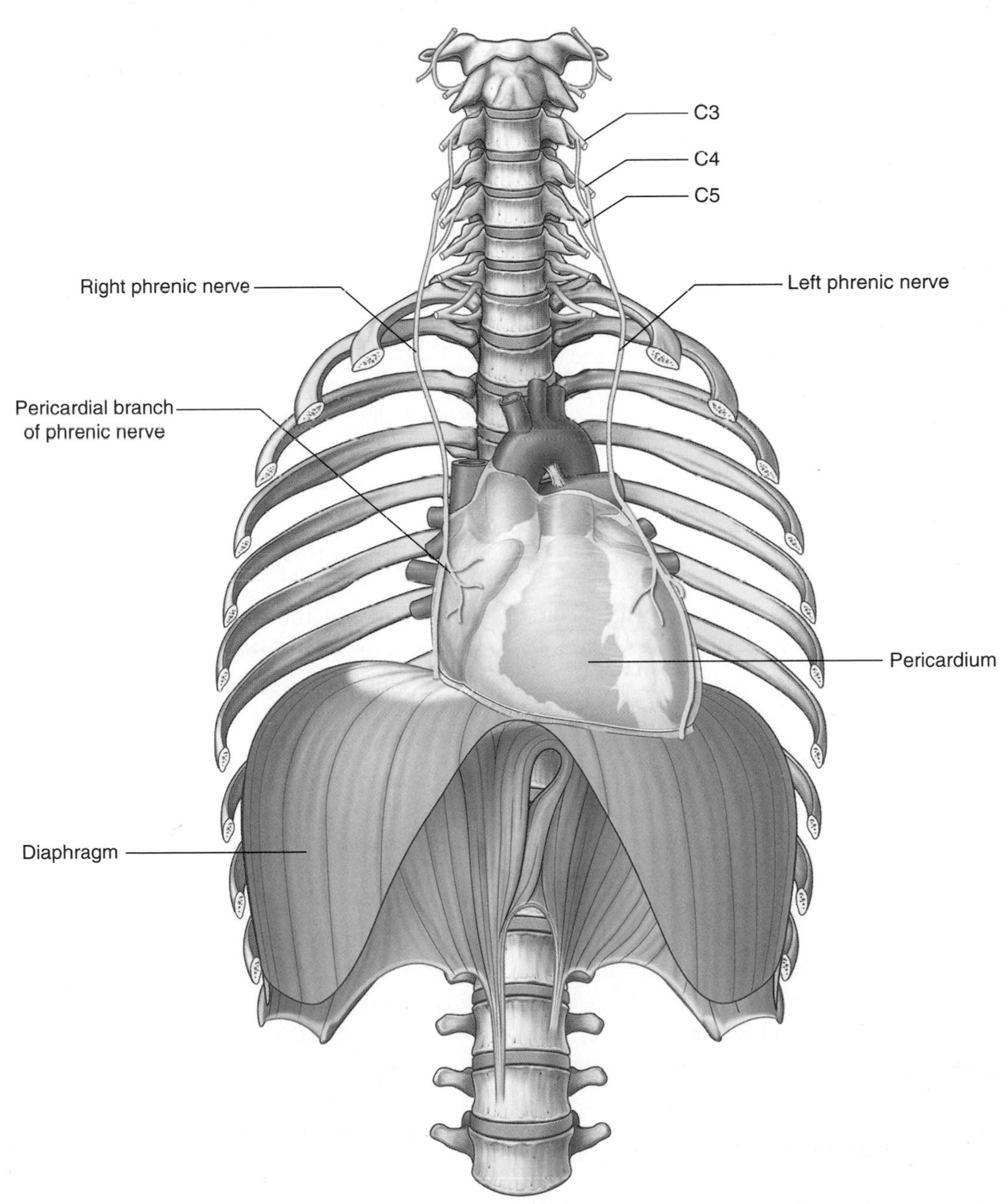


**Fig. 3.15** Innervation of the diaphragm.

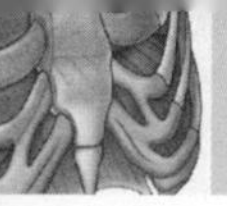



# Regional anatomy

The cylindrical thorax consists of:

- a wall,
- two pleural cavities,
- the lungs, and
- the mediastinum.

The thorax houses the heart and lungs, acts as a conduit for structures passing between the neck and the abdomen, and plays a principal role in breathing. In addition, the thoracic wall protects the heart and lungs and provides support for the upper limbs. Muscles anchored to the anterior thoracic wall provide some of this support, and together with their associated connective tissues, nerves, and vessels, and the overlying skin and superficial fascia, define the pectoral region.

## PECTORAL REGION

The pectoral region is external to the anterior thoracic wall and helps anchor the upper limb to the trunk. It consists of:

- a superficial compartment containing skin, superficial fascia, and breasts; and
- a deep compartment containing muscles and associated structures.

Nerves, vessels, and lymphatics in the superficial compartment emerge from the thoracic wall, the axilla, and the neck.

### Breast

The breasts consist of mammary glands and associated skin and connective tissues. The **mammary glands** are modified sweat glands in the superficial fascia anterior to the pectoral muscles and the anterior thoracic wall (Fig. 3.16).

The mammary glands consist of a series of ducts and associated secretory lobules. These converge to form 15 to 20 **lactiferous ducts**, which open independently onto the **nipple**. The nipple is surrounded by a circular pigmented area of skin termed the **areola**.

A well-developed, connective tissue stroma surrounds the ducts and lobules of the mammary gland. In certain regions, this condenses to form well-defined ligaments, the **suspensory ligaments of breast**, which are continuous with the dermis of the skin and support the breast.



Carcinoma of the breast creates tension on these ligaments, causing pitting of the skin.

In nonlactating women, the predominant component of the breasts is fat, while glandular tissue is more abundant in lactating women.

The breast lies on deep fascia related to the pectoralis major muscle and other surrounding muscles. A layer of loose connective tissue (the **retromammary space**) separates the breast from the deep fascia and provides some degree of movement over underlying structures.

The base, or attached surface, of each breast extends vertically from ribs II to VI, and transversely from the sternum to as far laterally as the midaxillary line.

#### Arterial supply

The breast is related to the thoracic wall and to structures associated with the upper limb; therefore, vascular supply and drainage can occur by multiple routes (Fig. 3.16):

- laterally, vessels from the axillary artery—superior thoracic, thoraco-acromial, lateral thoracic, and subscapular arteries;
- medially, branches from the internal thoracic artery;
- the second to fourth intercostal arteries via branches that perforate the thoracic wall and overlying muscle.

#### Venous drainage

Veins draining the breast parallel the arteries and ultimately drain into the axillary, internal thoracic, and intercostal veins.

#### Innervation

Innervation of the breast is via anterior and lateral cutaneous branches of the second to sixth intercostal nerves. The nipple is innervated by the fourth intercostal nerve.

#### Lymphatic drainage

Lymphatic drainage of the breast is as follows:

- Approximately 75% is via lymphatic vessels that drain laterally and superiorly into **axillary nodes** (Fig. 3.16).
- Most of the remaining drainage is into parasternal nodes deep to the anterior thoracic wall and associated with the internal thoracic artery.
- Some drainage may occur via lymphatic vessels that follow the lateral branches of posterior intercostal arteries and connect with intercostal nodes situated near the heads and necks of ribs.

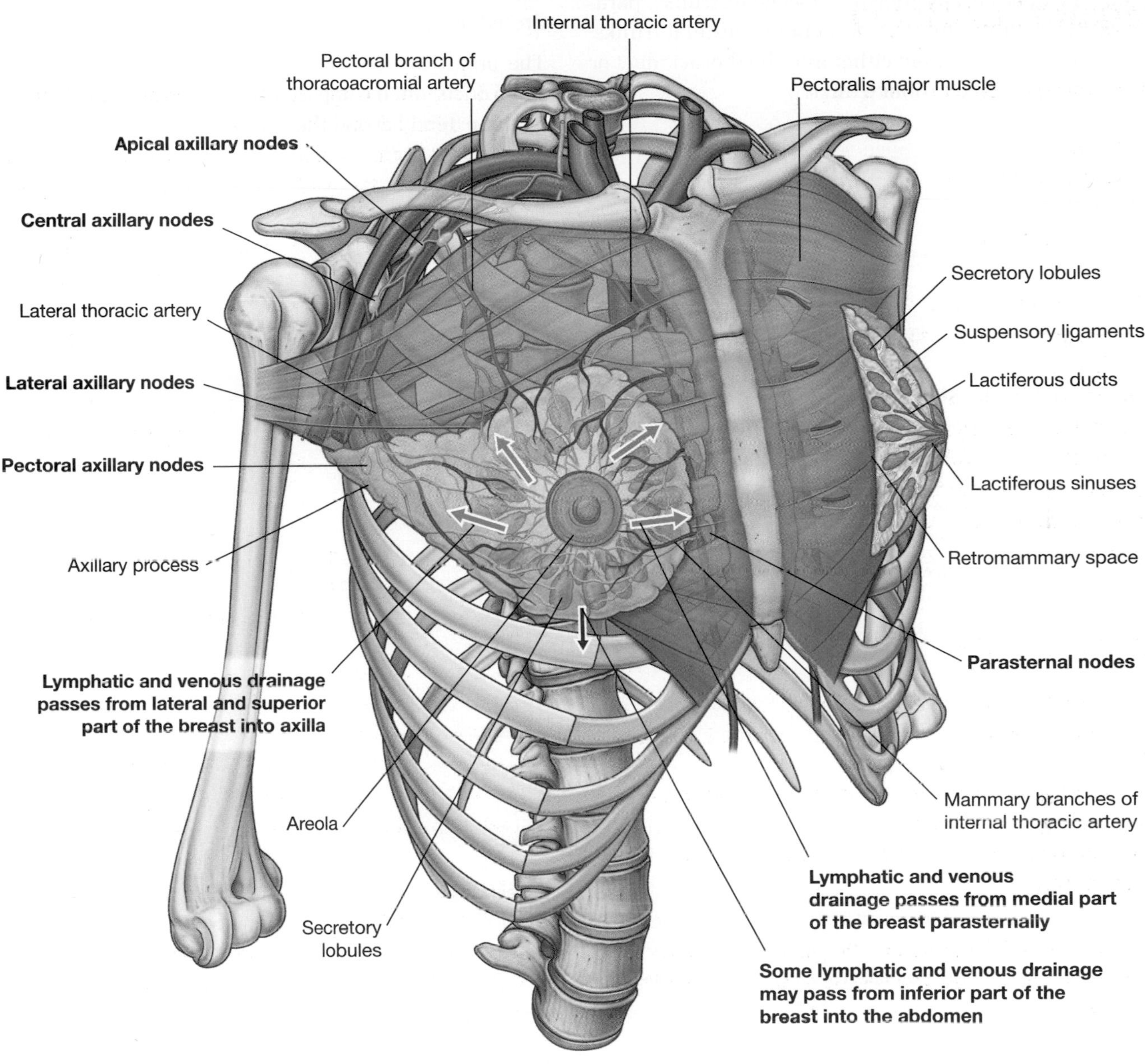

Internal thoracic artery
Pectoral branch of thoracoacromial artery
Pectoralis major muscle
Apical axillary nodes
Central axillary nodes
Secretory lobules
Lateral thoracic artery
Suspensory ligaments
Lateral axillary nodes
Lactiferous ducts
Pectoral axillary nodes
Lactiferous sinuses
Axillary process
Retromammary space
Parasternal nodes
Lymphatic and venous drainage passes from lateral and superior part of the breast into axilla
Mammary branches of internal thoracic artery
Areola
Lymphatic and venous drainage passes from medial part of the breast parasternally
Secretory lobules
Some lymphatic and venous drainage may pass from inferior part of the breast into the abdomen

**Fig. 3.16** Breasts.

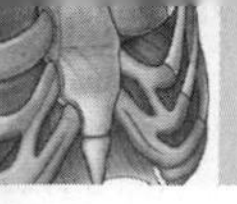



Axillary nodes drain into the subclavian trunks, parasternal nodes drain into the bronchomediastinal trunks, and intercostal nodes drain either into the thoracic duct or into the bronchomediastinal trunks.

### Breast in men

The breast in men is rudimentary and consists only of small ducts, often composed of cords of cells, that normally do not extend beyond the areola. Breast cancer can occur in men.

## In the clinic

### Axillary tail of breast

It is important for clinicians to remember when evaluating the breast for pathology that the upper lateral region of the breast can project around the lateral margin of the pectoralis major muscle and into the axilla. This axillary process (axillary tail) may perforate deep fascia and extend as far superiorly as the apex of the axilla.

## In the clinic

### Breast cancer

Breast cancer is one of the most common malignancies in women. It develops in the cells of the acini, lactiferous ducts, and lobules of the breast. Tumor growth and spread depends on the exact cellular site of origin of the cancer. These factors affect the response to surgery, chemotherapy, and radiotherapy. Breast tumors spread via the lymphatics and veins, or by direct invasion.

When a patient has a lump in the breast, a diagnosis of breast cancer is confirmed by a biopsy and histological evaluation. Once confirmed, the clinician must attempt to stage the tumor.

**Staging the tumor** means defining the:

- size of the primary tumor,
- exact site of the primary tumor,
- number and sites of lymph node spread, and
- organs to which the tumor may have spread.

Computed tomography (CT) scanning of the body may be carried out to look for any spread to the lungs (pulmonary metastases), liver (hepatic metastases), or bone (bony metastases).

Further imaging may include bone scanning using radioactive isotopes, which are avidly taken up by the tumor metastases in bone, and PET-CT, which can visualize active foci of the metastatic disease in the body.

Lymph drainage of the breast is complex. Lymph vessels pass to axillary, supraclavicular, and parasternal nodes and may even pass to abdominal lymph nodes, as well as to the opposite breast. Containment of nodal metastatic breast cancer is therefore potentially difficult because it can spread through many lymph node groups.

Subcutaneous lymphatic obstruction and tumor growth pull on connective tissue ligaments in the breast, resulting in the appearance of an orange peel texture (**peau d'orange**) on the surface of the breast. Further subcutaneous spread can induce a rare manifestation of breast cancer that produces a hard, woody texture to the skin (**cancer en cuirasse**).

A mastectomy (surgical removal of the breast) involves excision of breast tissue. Within the axilla the breast tissue must be removed from the medial axillary wall. Closely applied to the medial axillary wall is the long thoracic nerve. Damage to this nerve can result in paralysis of the serratus anterior muscle, producing a characteristic "winged" scapula. It is also possible to damage the nerve to the latissimus dorsi muscle, and this may affect extension, medial rotation, and adduction of the humerus.

## Muscles of the pectoral region

Each pectoral region contains the pectoralis major, pectoralis minor, and subclavius muscles (Fig. 3.17 and Table 3.1). All originate from the anterior thoracic wall and insert into bones of the upper limb.

### Pectoralis major

The **pectoralis major** muscle is the largest and most superficial of the pectoral region muscles. It directly underlies the breast and is separated from it by deep fascia and the loose connective tissue of the retromammary space.

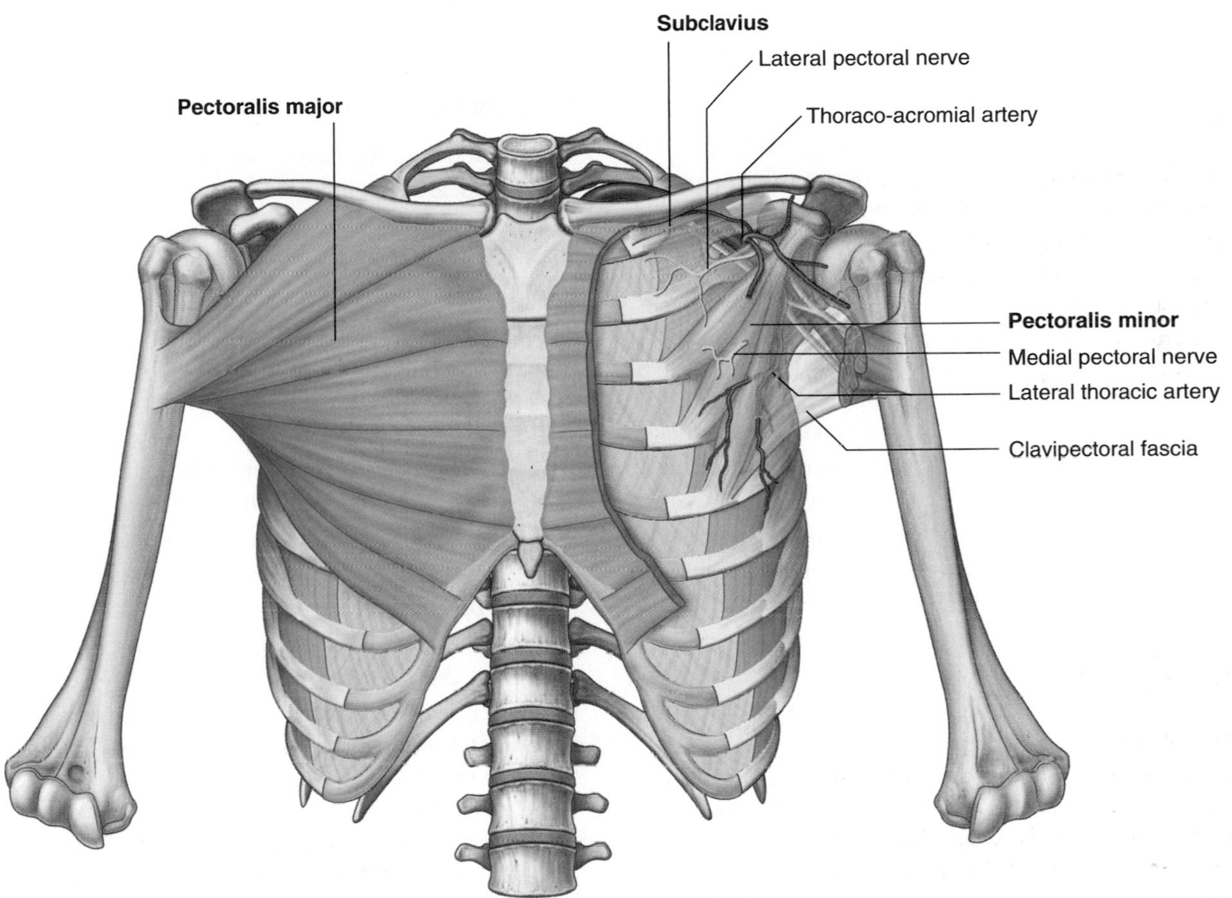


**Fig. 3.17** Muscles and fascia of the pectoral region.

**Table 3.1** Muscles of the pectoral region

| Muscle | Origin | Insertion | Innervation | Function |
|---|---|---|---|---|
| Pectoralis major | Medial half of clavicle and anterior surface of sternum, first seven costal cartilages, aponeurosis of external oblique | Lateral lip of intertubercular sulcus of humerus | Medial and lateral pectoral nerves | Adduction, medial rotation, and flexion of the humerus at the shoulder joint |
| Subclavius | Rib I at junction between rib and costal cartilage | Groove on inferior surface of middle third of clavicle | Nerve to subclavius | Pulls clavicle medially to stabilize sternoclavicular joint; depresses tip of shoulder |
| Pectoralis minor | Anterior surfaces of the third, fourth, and fifth ribs, and deep fascia overlying the related intercostal spaces | Coracoid process of scapula | Medial pectoral nerves | Depresses tip of shoulder; protracts scapula |

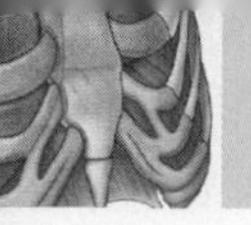



The pectoralis major has a broad origin that includes the anterior surfaces of the medial half of the clavicle, the sternum, and related costal cartilages. The muscle fibers converge to form a flat tendon, which inserts into the lateral lip of the intertubercular sulcus of the humerus.

The pectoralis major adducts, flexes, and medially rotates the arm.

### Subclavius and pectoralis minor muscles

The **subclavius** and **pectoralis minor muscles** underlie the pectoralis major:

- The subclavius is small and passes laterally from the anterior and medial part of rib I to the inferior surface of the clavicle.
- The pectoralis minor passes from the anterior surfaces of ribs III to V to the coracoid process of the scapula.

Both the subclavius and pectoralis minor pull the tip of the shoulder inferiorly.

A continuous layer of deep fascia, the **clavipectoral fascia**, encloses the subclavius and pectoralis minor and attaches to the clavicle above and to the floor of the axilla below.

The muscles of the pectoral region form the anterior wall of the axilla, a region between the upper limb and the neck through which all major structures pass. Nerves, vessels, and lymphatics that pass between the pectoral region and the axilla pass through the clavipectoral fascia between the subclavius and pectoralis minor or pass under the inferior margins of the pectoralis major and minor.

## THORACIC WALL

The thoracic wall is segmental in design and composed of skeletal elements and muscles. It extends between:

- the superior thoracic aperture, bordered by vertebra TI, rib I, and the manubrium of the sternum; and
- the inferior thoracic aperture, bordered by vertebra TXII, rib XII, the end of rib XI, the costal margin, and the xiphoid process of the sternum.

## Skeletal framework

The skeletal elements of the thoracic wall consist of the thoracic vertebrae, intervertebral discs, ribs, and sternum.

### Thoracic vertebrae

There are twelve **thoracic vertebrae**, each of which is characterized by articulations with ribs.

#### Typical thoracic vertebra

A typical thoracic vertebra has a heart-shaped **vertebral body**, with roughly equal dimensions in the transverse and anteroposterior directions, and a long spinous process (Fig. 3.18). The **vertebral foramen** is generally circular and the **laminae** are broad and overlap with those of the vertebra below. The **superior articular processes** are flat, with their articular surfaces facing almost directly posteriorly, while the **inferior articular processes** project from the laminae and their articular facets face anteriorly.

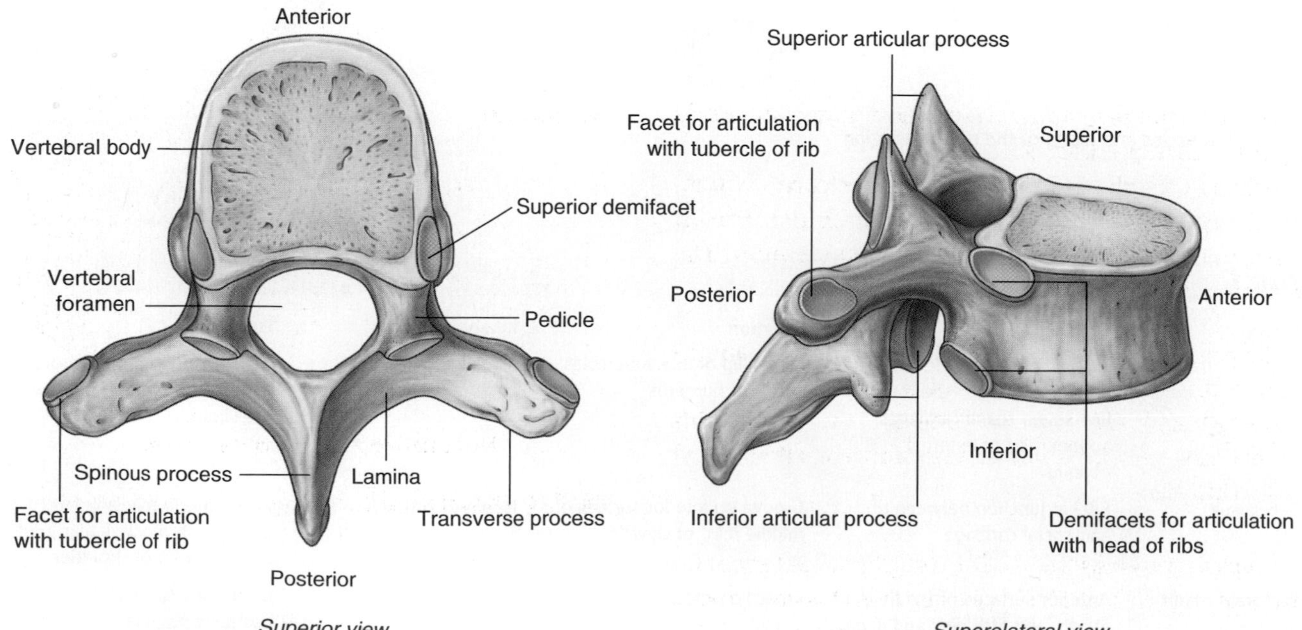


**Fig. 3.18** Typical thoracic vertebra.

The **transverse processes** are club shaped and project posterolaterally.

### Articulation with ribs

A typical thoracic vertebra has three sites on each side for articulation with ribs.

- Two demifacets (i.e., partial facets) are located on the superior and inferior aspects of the body for articulation with corresponding sites on the heads of adjacent ribs. The **superior costal facet** articulates with part of the head of its own rib, and the **inferior costal facet** articulates with part of the head of the rib below.
- An oval facet (**transverse costal facet**) at the end of the transverse process articulates with the tubercle of its own rib.

Not all vertebrae articulate with ribs in the same fashion (Fig. 3.19):

- The superior costal facets on the body of vertebra TI are complete and articulate with a single facet on the head of its own rib—in other words, the head of rib I does not articulate with vertebra CVII.
- Similarly, vertebra TX (and often TIX) articulates only with its own ribs and therefore lacks inferior demifacets on the body.
- Vertebrae TXI and TXII articulate only with the heads of their own ribs—they lack transverse costal facets and have only a single complete facet on each side of their bodies.

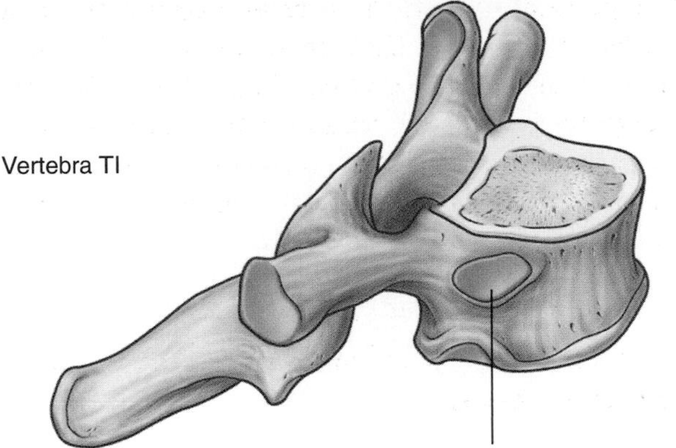


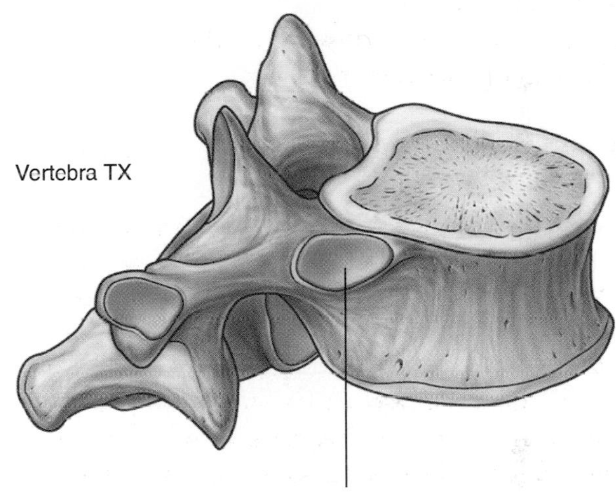


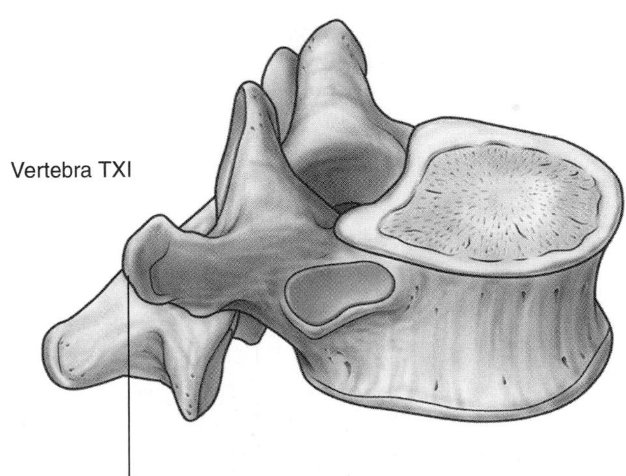


**Fig. 3.19** Atypical thoracic vertebrae.

## Ribs

There are twelve pairs of ribs, each terminating anteriorly in a costal cartilage (Fig. 3.20).

Although all ribs articulate with the vertebral column, only the costal cartilages of the upper seven ribs, known as **true ribs**, articulate directly with the sternum. The remaining five pairs of ribs are **false ribs**:

- The costal cartilages of ribs VIII to X articulate anteriorly with the costal cartilages of the ribs above.
- Ribs XI and XII have no anterior connection with other ribs or with the sternum and are often called **floating ribs**.

A typical rib consists of a curved shaft with anterior and posterior ends (Fig. 3.21). The anterior end is continuous with its costal cartilage. The posterior end articulates with the vertebral column and is characterized by a head, neck, and tubercle.

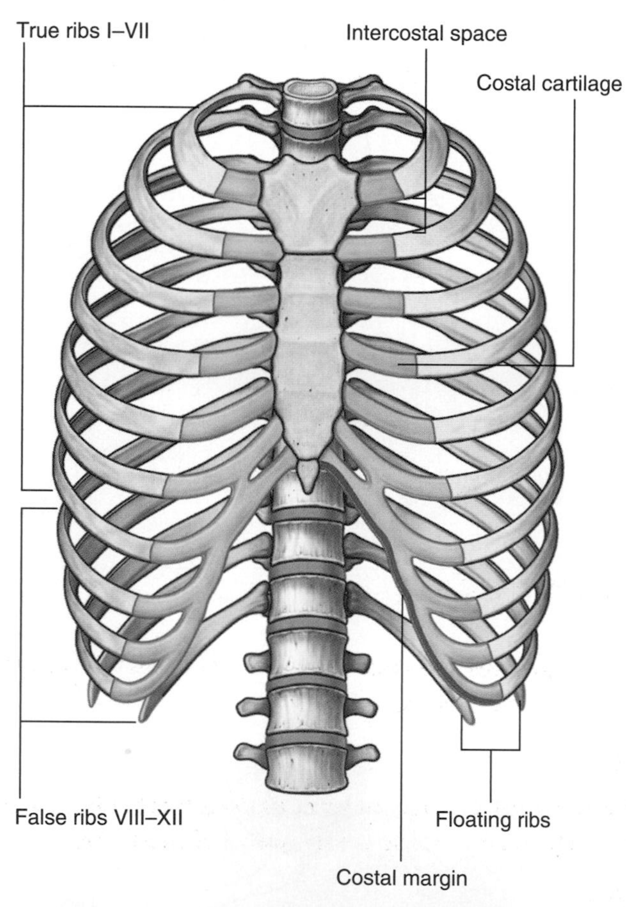


**Fig. 3.20** Ribs.

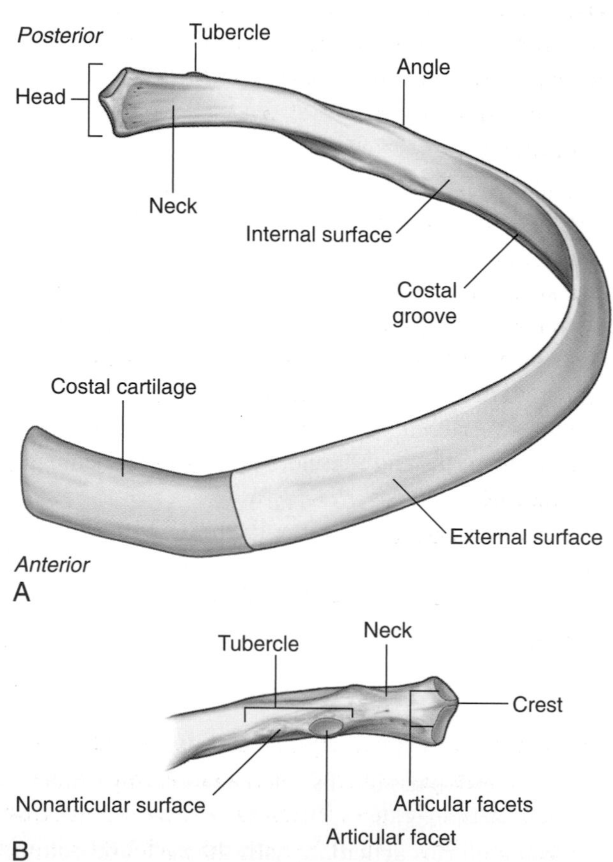


**Fig. 3.21** A typical rib. **A.** Anterior view. **B.** Posterior view of proximal end of rib.

The **head** is somewhat expanded and typically presents two articular surfaces separated by a **crest**. The smaller superior surface articulates with the inferior costal facet on the body of the vertebra above, whereas the larger inferior facet articulates with the superior costal facet of its own vertebra.

The **neck** is a short flat region of bone that separates the head from the tubercle.

The **tubercle** projects posteriorly from the junction of the neck with the shaft and consists of two regions, an articular part and a nonarticular part:

- The articular part is medial and has an oval facet for articulation with a corresponding facet on the transverse process of the associated vertebra.
- The raised nonarticular part is roughened by ligament attachments.

The shaft is generally thin and flat with internal and external surfaces.

The superior margin is smooth and rounded, whereas the inferior margin is sharp. The shaft bends forward just laterally to the tubercle at a site termed the **angle**. It also has a gentle twist around its longitudinal axis so that the external surface of the anterior part of the shaft faces somewhat superiorly relative to the posterior part. The inferior margin of the internal surface is marked by a distinct **costal groove**.

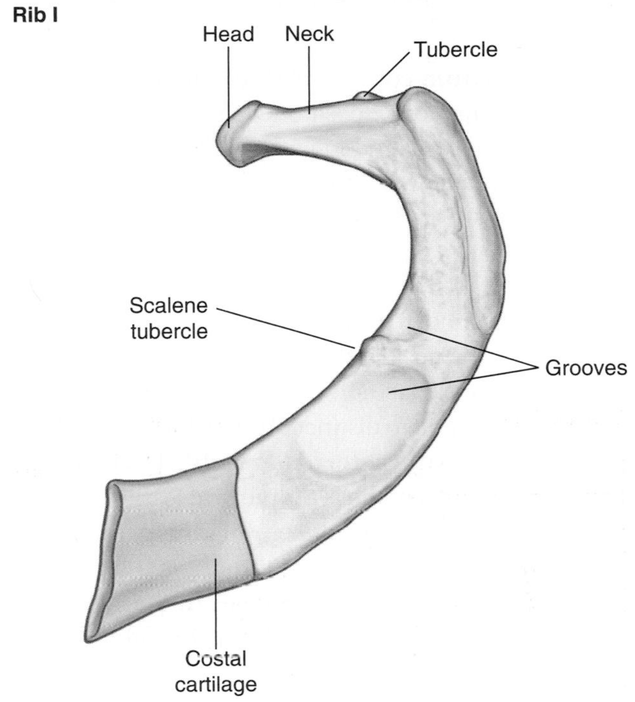


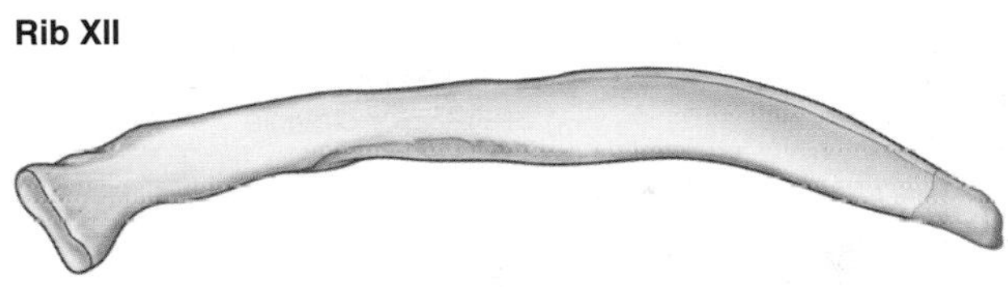


**Fig. 3.22** Atypical ribs.

### Distinct features of upper and lower ribs

The upper and lower ribs have distinct features (Fig. 3.22).

#### Rib I

**Rib I** is flat in the horizontal plane and has broad superior and inferior surfaces. From its articulation with vertebra TI, it slopes inferiorly to its attachment to the manubrium of the sternum. The head articulates only with the body of vertebra TI and therefore has only one articular surface. Like other ribs, the tubercle has a facet for articulation with the transverse process. The superior surface of the rib is characterized by a distinct tubercle, the **scalene tubercle**, which separates two smooth grooves that cross the rib approximately midway along the shaft. The anterior groove is caused by the subclavian vein, and the posterior groove is caused by the subclavian artery. Anterior and posterior to these grooves, the shaft is roughened by muscle and ligament attachments.

#### Rib II

**Rib II**, like rib I, is flat but twice as long. It articulates with the vertebral column in a way typical of most ribs.

#### Rib X

The head of **rib X** has a single facet for articulation with its own vertebra.

#### Ribs XI and XII

**Ribs XI** and **XII** articulate only with the bodies of their own vertebrae and have no tubercles or necks. Both ribs are short, have little curve, and are pointed anteriorly.

## Sternum

The adult **sternum** consists of three major elements: the broad and superiorly positioned manubrium of the sternum, the narrow and longitudinally oriented body of the sternum, and the small and inferiorly positioned xiphoid process (Fig. 3.23).

### Manubrium of the sternum

The **manubrium of the sternum** forms part of the bony framework of the neck and the thorax.

The superior surface of the manubrium is expanded laterally and bears a distinct and palpable notch, the **jugular notch** (**suprasternal notch**), in the midline. On either side of this notch is a large oval fossa for articulation with the clavicle. Immediately inferior to this fossa, on each lateral surface of the manubrium, is a facet for the attachment of the first costal cartilage. At the lower end of the lateral border is a demifacet for articulation with the upper half of the anterior end of the second costal cartilage.

### Body of the sternum

The **body of the sternum** is flat.

The anterior surface of the body of the sternum is often marked by transverse ridges that represent lines of fusion between the segmental elements called sternebrae, from which this part of the sternum arises embryologically.

The lateral margins of the body of the sternum have articular facets for costal cartilages. Superiorly, each lateral margin has a demifacet for articulation with the inferior aspect of the second costal cartilage. Inferior to this demifacet are four facets for articulation with the costal cartilages of ribs III to VI.

At the inferior end of the body of the sternum is a demifacet for articulation with the upper demifacet on the seventh costal cartilage. The inferior end of the body of the sternum is attached to the xiphoid process.

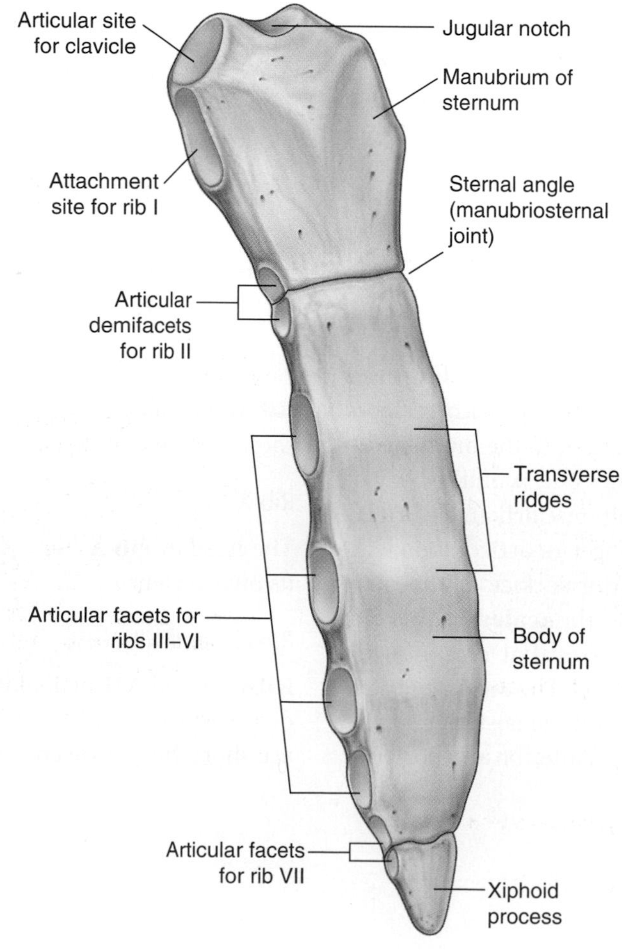


**Fig. 3.23** Sternum.

### Xiphoid process

The **xiphoid process** is the smallest part of the sternum. Its shape is variable: it may be wide, thin, pointed, bifid, curved, or perforated. It begins as a cartilaginous structure, which becomes ossified in the adult. On each side of its upper lateral margin is a demifacet for articulation with the inferior end of the seventh costal cartilage.

## Joints

### Costovertebral joints

A typical rib articulates with:

- the bodies of adjacent vertebrae, forming a joint with the head of the rib; and
- the transverse process of its related vertebra, forming a **costotransverse joint** (Fig. 3.24).

Together, the costovertebral joints and related ligaments allow the necks of the ribs either to rotate around their longitudinal axes, which occurs mainly in the upper ribs, or to ascend and descend relative to the vertebral column, which occurs mainly in the lower ribs. The combined movements of all of the ribs on the vertebral column are essential for altering the volume of the thoracic cavity during breathing.

#### Joint with head of rib

The two facets on the head of the rib articulate with the superior facet on the body of its own vertebra and with the inferior facet on the body of the vertebra above. This joint is divided into two synovial compartments by an intra-articular ligament, which attaches the crest to the adjacent intervertebral disc and separates the two articular surfaces on the head of the rib. The two synovial compartments and the intervening ligament are surrounded by a single joint

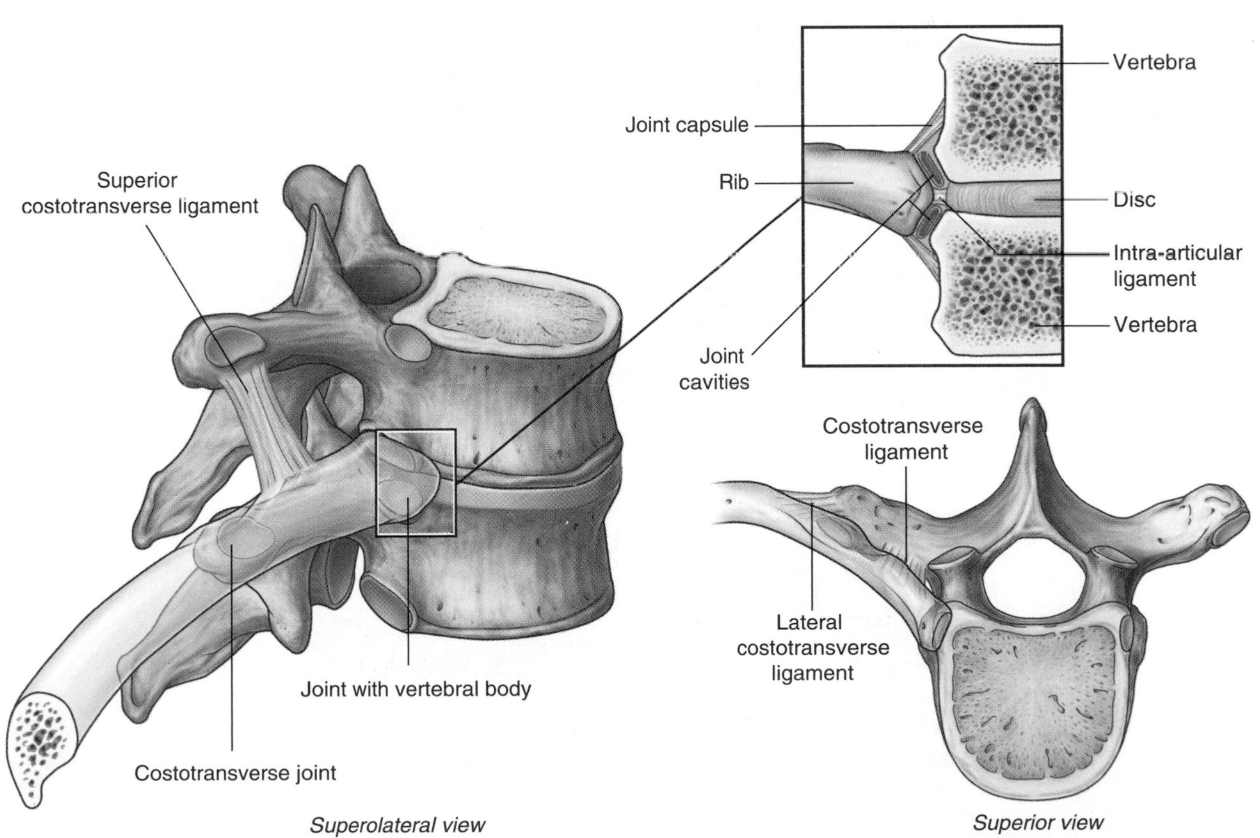


**Fig. 3.24** Costovertebral joints.

capsule attached to the outer margins of the combined articular surfaces of the head and vertebral column.

### Costotransverse joints

**Costotransverse joints** are synovial joints between the tubercle of a rib and the transverse process of the related vertebra (Fig. 3.24). The capsule surrounding each joint is thin. The joint is stabilized by two strong extracapsular ligaments that span the space between the transverse process and the rib on the medial and lateral sides of the joint:

- The **costotransverse ligament** is medial to the joint and attaches the neck of the rib to the transverse process.
- The **lateral costotransverse ligament** is lateral to the joint and attaches the tip of the transverse process to the roughened nonarticular part of the tubercle of the rib.

A third ligament, the **superior costotransverse ligament**, attaches the superior surface of the neck of the rib to the transverse process of the vertebra above.

Slight gliding movements occur at the costotransverse joints.

## Sternocostal joints

The sternocostal joints are joints between the upper seven costal cartilages and the sternum (Fig. 3.25).

The joint between rib I and the manubrium is not synovial and consists of a fibrocartilaginous connection

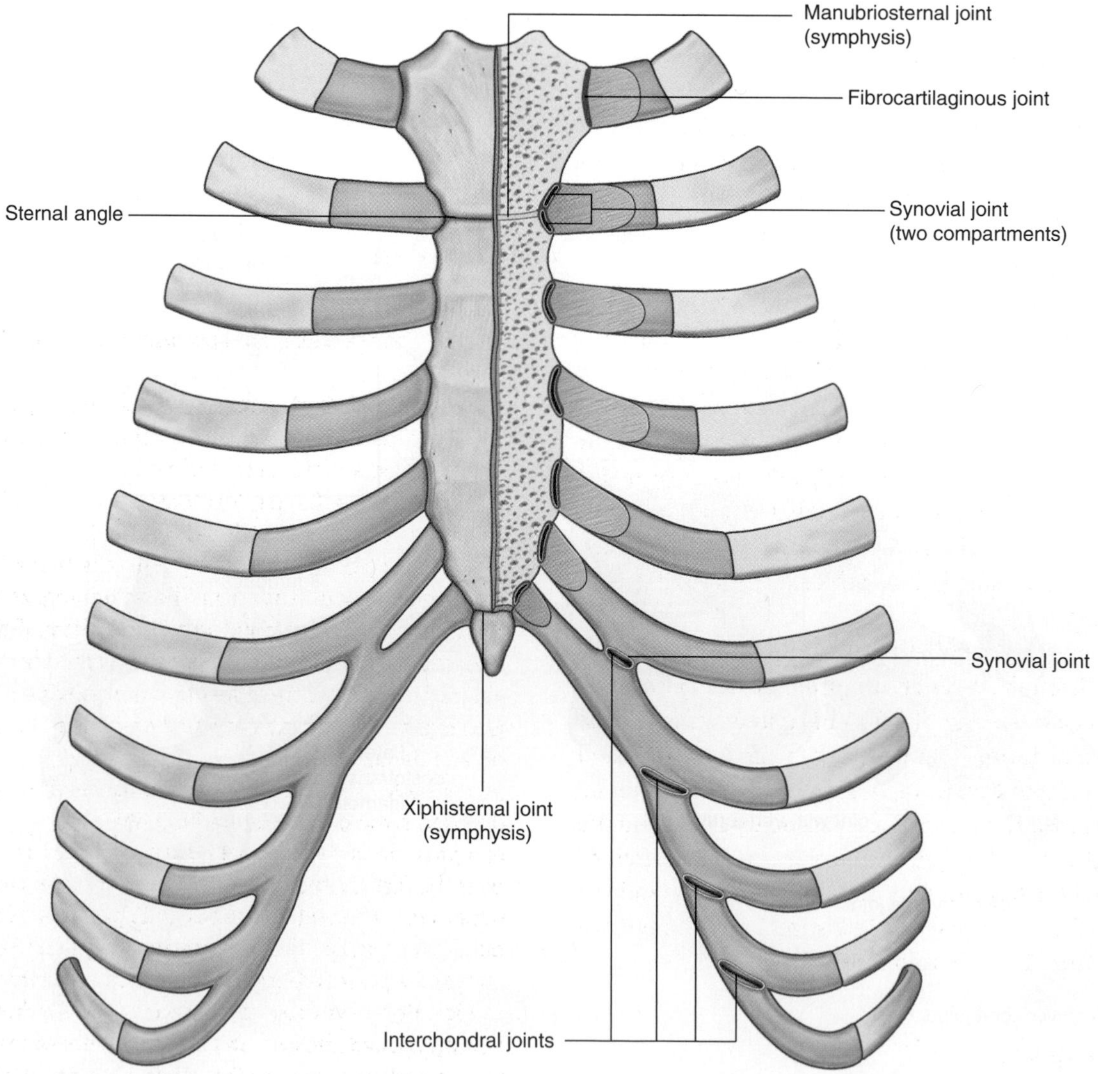

Fig. 3.25 Sternocostal joints.

between the manubrium and the costal cartilage. The second to seventh joints are synovial and have thin capsules reinforced by surrounding sternocostal ligaments.

The joint between the second costal cartilage and the sternum is divided into two compartments by an intra-articular ligament. This ligament attaches the second costal cartilage to the junction of the manubrium and the body of the sternum.

### Interchondral joints

Interchondral joints occur between the costal cartilages of adjacent ribs (Fig. 3.25), mainly between the costal cartilages of ribs VII to X, but may also involve the costal cartilages of ribs V and VI.

Interchondral joints provide indirect anchorage to the sternum and contribute to the formation of a smooth inferior costal margin. They are usually synovial, and the thin fibrous capsules are reinforced by interchondral ligaments.

### Manubriosternal and xiphisternal joints

The joints between the manubrium and the body of the sternum and between the body of the sternum and the xiphoid process are usually symphyses (Fig. 3.25). Only slight angular movements occur between the manubrium and the body of the sternum during respiration. The joint between the body of the sternum and the xiphoid process often becomes ossified with age.

A clinically useful feature of the manubriosternal joint is that it can be palpated easily. This is because the manubrium normally angles posteriorly on the body of the sternum, forming a raised feature referred to as the sternal angle. This elevation marks the site of articulation of rib II with the sternum. Rib I is not palpable, because it lies inferior to the clavicle and is embedded in tissues at the base of the neck. Therefore, rib II is used as a reference for counting ribs and can be felt immediately lateral to the sternal angle.

In addition, the sternal angle lies on a horizontal plane that passes through the intervertebral disc between vertebrae TIV and TV (see Fig. 3.10). This plane separates the superior mediastinum from the inferior mediastinum and marks the superior border of the pericardium. The plane also passes through the end of the ascending aorta and the beginning of the arch of the aorta, the end of the arch of the aorta and the beginning of the thoracic aorta, and the bifurcation of the trachea, and just superior to the pulmonary trunk (see Fig. 3.79 and 3.86).

## Intercostal spaces

**Intercostal spaces** lie between adjacent ribs and are filled by intercostal muscles (Fig. 3.26).

Intercostal nerves and associated major arteries and veins lie in the **costal groove** along the inferior margin of the superior rib and pass in the plane between the inner two layers of muscles.

In each space, the vein is the most superior structure and is therefore highest in the costal groove. The artery is inferior to the vein, and the nerve is inferior to the artery and often not protected by the groove. Therefore, the nerve is the structure most at risk when objects perforate the upper aspect of an intercostal space. Small collateral branches of the major intercostal nerves and vessels are often present superior to the inferior rib below.

Deep to the intercostal spaces and ribs, and separating these structures from the underlying pleura, is a layer of loose connective tissue, called **endothoracic fascia**, which contains variable amounts of fat.

Superficial to the spaces are deep fascia, superficial fascia, and skin. Muscles associated with the upper limbs and back overlie the spaces.

### In the clinic

#### Cervical ribs

Cervical ribs are present in approximately 1% of the population.

A cervical rib is an accessory rib articulating with vertebra CVII; the anterior end attaches to the superior border of the anterior aspect of rib I.

Plain radiographs may demonstrate cervical ribs as small horn-like structures (see Fig. 3.106).

It is often not appreciated by clinicians that a fibrous band commonly extends from the anterior tip of the small cervical ribs to rib I, producing a "cervical band" that is not visualized on radiography. In patients with cervical ribs and cervical bands, structures that normally pass over rib I (see Fig. 3.7) are elevated by, and pass over, the cervical rib and band.

Clinically, "thoracic outlet syndrome" is used to describe symptoms resulting from abnormal compression of the brachial plexus of nerves as it passes over the first rib and through the axillary inlet into the upper limb. The anterior ramus of T1 passes superiorly out of the superior thoracic aperture to join and become part of the brachial plexus. The cervical band from a cervical rib is one cause of thoracic outlet syndrome by putting upward stresses on the lower parts of the brachial plexus as they pass over the cervical band and related cervical rib.

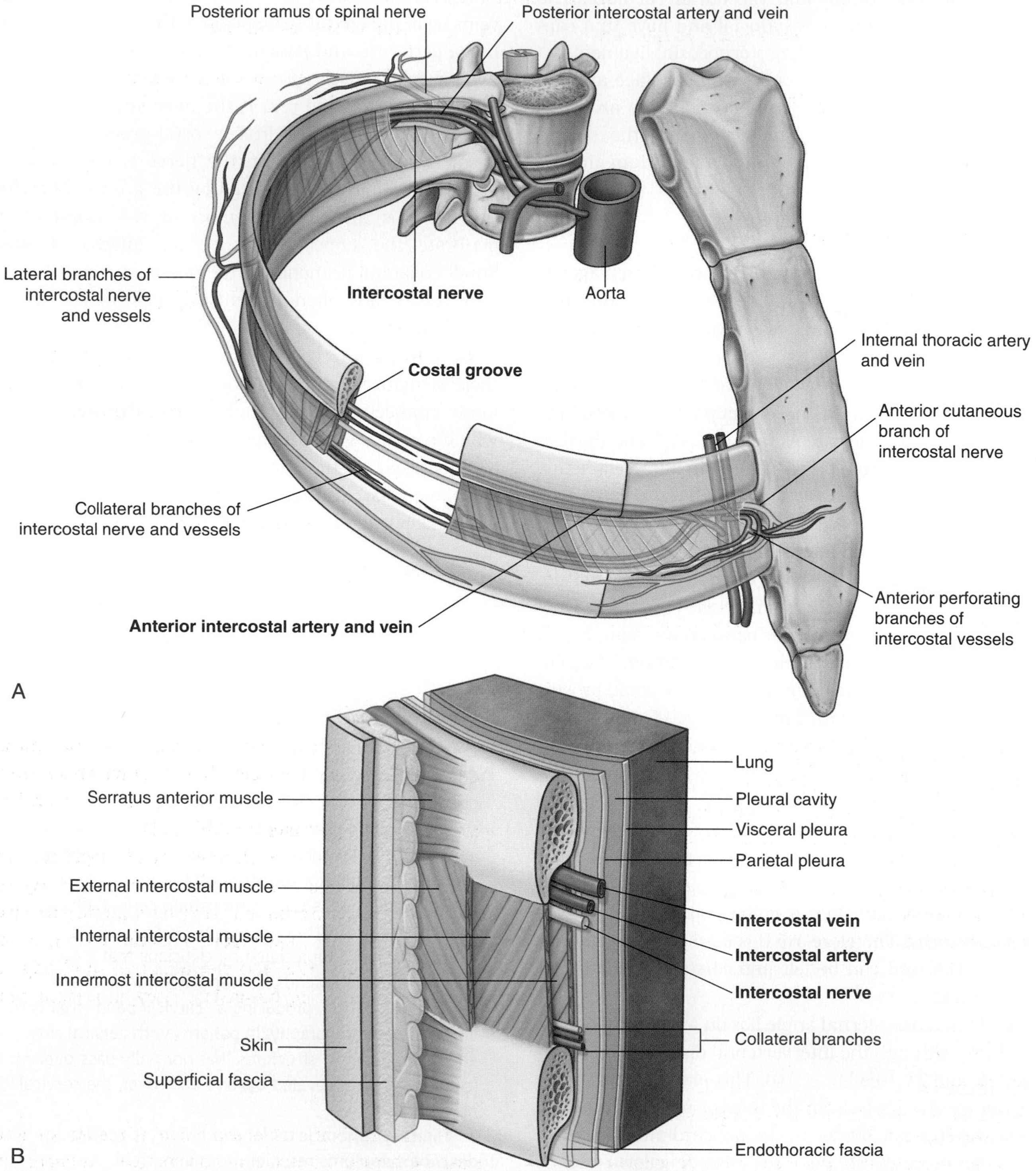

**Fig. 3.26** Intercostal space. **A.** Anterolateral view. **B.** Details of an intercostal space and relationships.

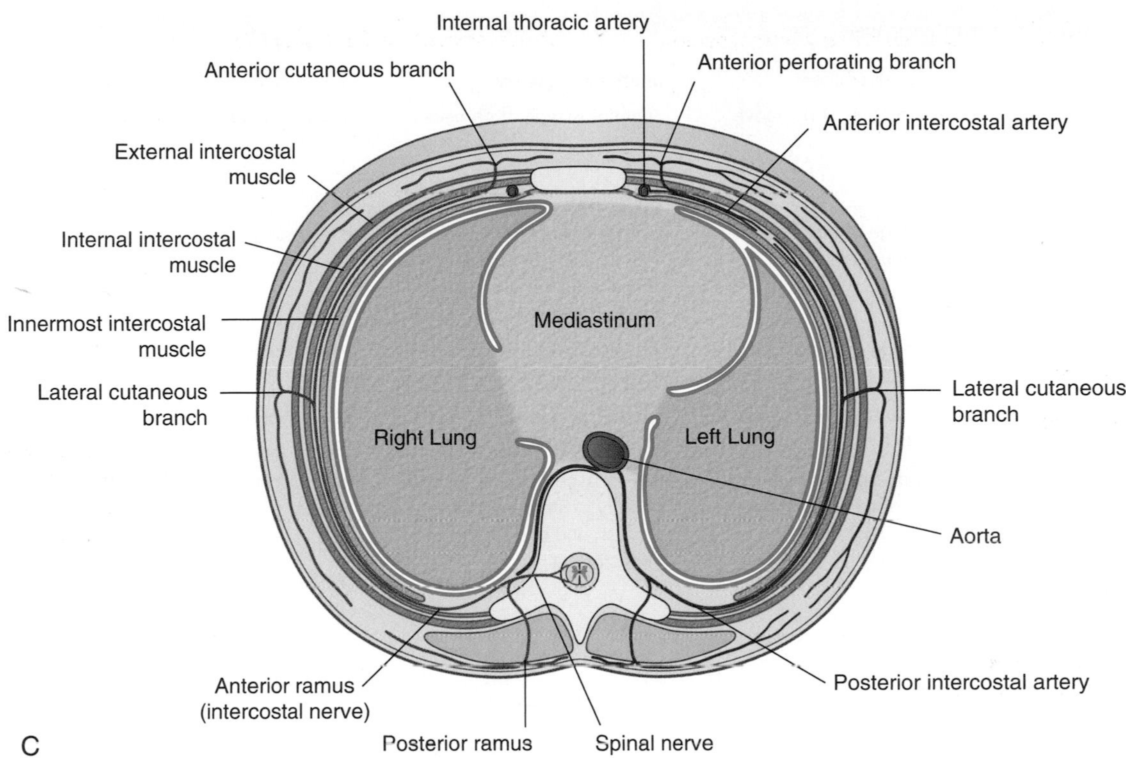


**Fig. 3.26, cont'd** Intercostal space. **C.** Transverse section.

## In the clinic

**Collection of sternal bone marrow**

The subcutaneous position of the sternum makes it possible to place a needle through the hard outer cortex into the internal (or medullary) cavity containing bone marrow. Once the needle is in this position, bone marrow can be aspirated. Evaluation of this material under the microscope helps clinicians diagnose certain blood diseases such as leukemia.

## In the clinic

**Rib fractures**

Single rib fractures are of little consequence, though extremely painful.

After severe trauma, ribs may be broken in two or more places. If enough ribs are broken, a loose segment of chest wall, a flail segment (**flail chest**), is produced. When the patient takes a deep inspiration, the flail segment moves in the opposite direction to the chest wall, preventing full lung expansion and creating a paradoxically moving segment. If a large enough segment of chest wall is affected, ventilation may be impaired and assisted ventilation may be required until the ribs have healed.

## Muscles

Muscles of the thoracic wall include those that fill and support the intercostal spaces, those that pass between the sternum and the ribs, and those that cross several ribs between costal attachments (Table 3.2).

The muscles of the thoracic wall, together with muscles between the vertebrae and ribs posteriorly (i.e., the **levatores costarum** and **serratus posterior superior** and **serratus posterior inferior** muscles) alter the position of the ribs and sternum and so change the thoracic volume during breathing. They also reinforce the thoracic wall.

### Intercostal muscles

The **intercostal muscles** are three flat muscles found in each intercostal space that pass between adjacent ribs (Fig. 3.27). Individual muscles in this group are named according to their positions:

- The external intercostal muscles are the most superficial.
- The internal intercostal muscles are sandwiched between the external and innermost muscles.
- The innermost intercostal muscles are the deepest of the three muscles.

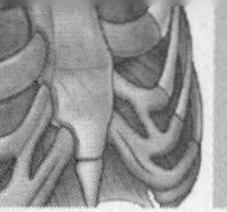



**Table 3.2** Muscles of the thoracic wall

| Muscle | Superior attachment | Inferior attachment | Innervation | Function |
|---|---|---|---|---|
| External intercostal | Inferior margin of rib above | Superior margin of rib below | Intercostal nerves; T1–T11 | Most active during inspiration; supports intercostal space; moves ribs superiorly |
| Internal intercostal | Lateral edge of costal groove of rib above | Superior margin of rib below deep to the attachment of the related external intercostal | Intercostal nerves; T1–T11 | Most active during expiration; supports intercostal space; moves ribs inferiorly |
| Innermost intercostal | Medial edge of costal groove of rib above | Internal aspect of superior margin of rib below | Intercostal nerves; T1–T11 | Acts with internal intercostal muscles |
| Subcostales | Internal surface (near angle) of lower ribs | Internal surface of second or third rib below | Related intercostal nerves | May depress ribs |
| Transversus thoracis | Inferior margins and internal surfaces of costal cartilages of second to sixth ribs | Inferior aspect of deep surface of body of sternum, xiphoid process, and costal cartilages of ribs IV–VII | Related intercostal nerves | Depresses costal cartilages |

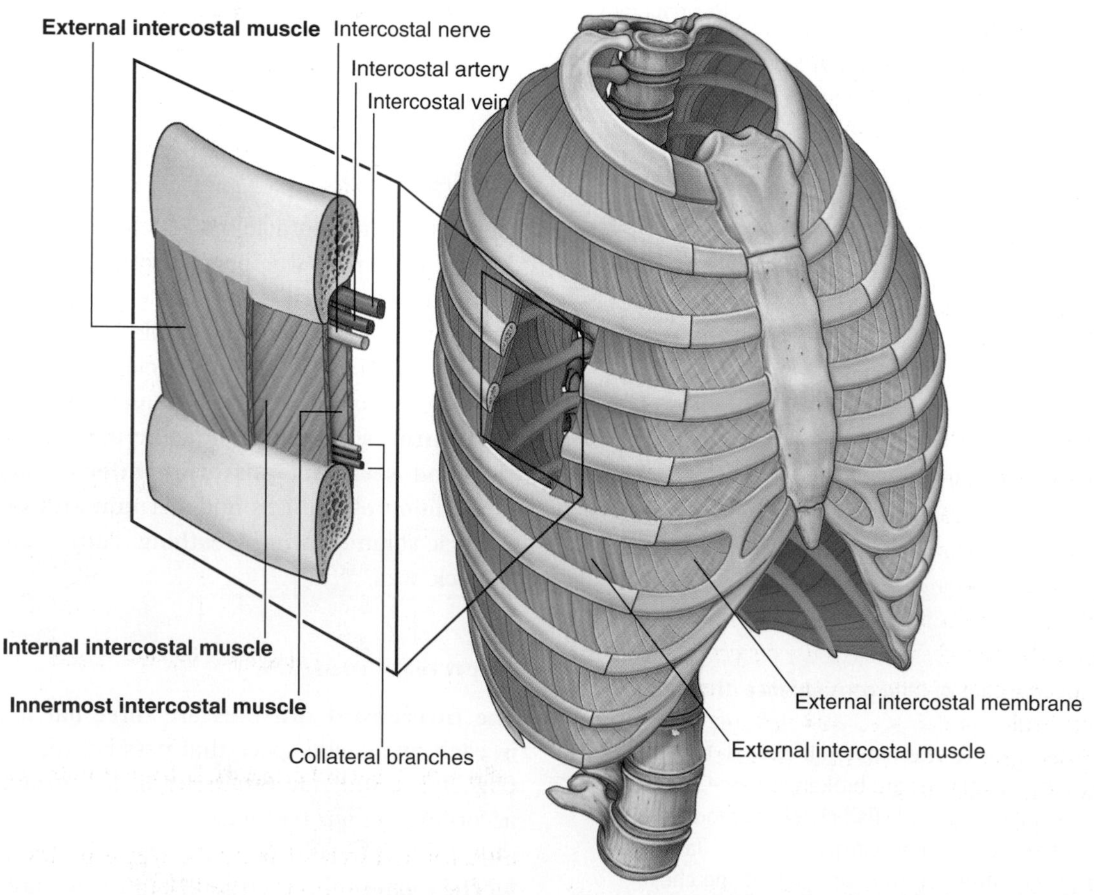


**Fig. 3.27** Intercostal muscles.

The intercostal muscles are innervated by the related intercostal nerves. As a group, the intercostal muscles provide structural support for the intercostal spaces during breathing. They can also move the ribs.

### External intercostal muscles

The eleven pairs of **external intercostal muscles** extend from the inferior margins (lateral edges of costal grooves) of the ribs above to the superior margins of the ribs below. When the thoracic wall is viewed from a lateral position, the muscle fibers pass obliquely anteroinferiorly (Fig. 3.27). The muscles extend around the thoracic wall from the regions of the tubercles of the ribs to the costal cartilages, where each layer continues as a thin connective tissue aponeurosis termed the **external intercostal membrane**. The external intercostal muscles are most active in inspiration.

### Internal intercostal muscles

The eleven pairs of **internal intercostal muscles** pass between the most inferior lateral edge of the costal grooves of the ribs above, to the superior margins of the ribs below. They extend from parasternal regions, where the muscles course between adjacent costal cartilages, to the angle of the ribs posteriorly (Fig. 3.27). This layer continues medially toward the vertebral column, in each intercostal space, as the **internal intercostal membrane**. The muscle fibers pass in the opposite direction to those of the external intercostal muscles. When the thoracic wall is viewed from a lateral position, the muscle fibers pass obliquely posteroinferiorly. The internal intercostal muscles are most active during expiration.

### Innermost intercostal muscles

The **innermost intercostal muscles** are the least distinct of the intercostal muscles, and the fibers have the same orientation as the internal intercostals (Fig. 3.27). These muscles are most evident in the lateral thoracic wall. They extend between the inner surfaces of adjacent ribs from the medial edge of the costal groove to the deep surface of the rib below. Importantly, the neurovascular bundles associated with the intercostal spaces pass around the thoracic wall in the costal grooves in a plane between the innermost and internal intercostal muscles.

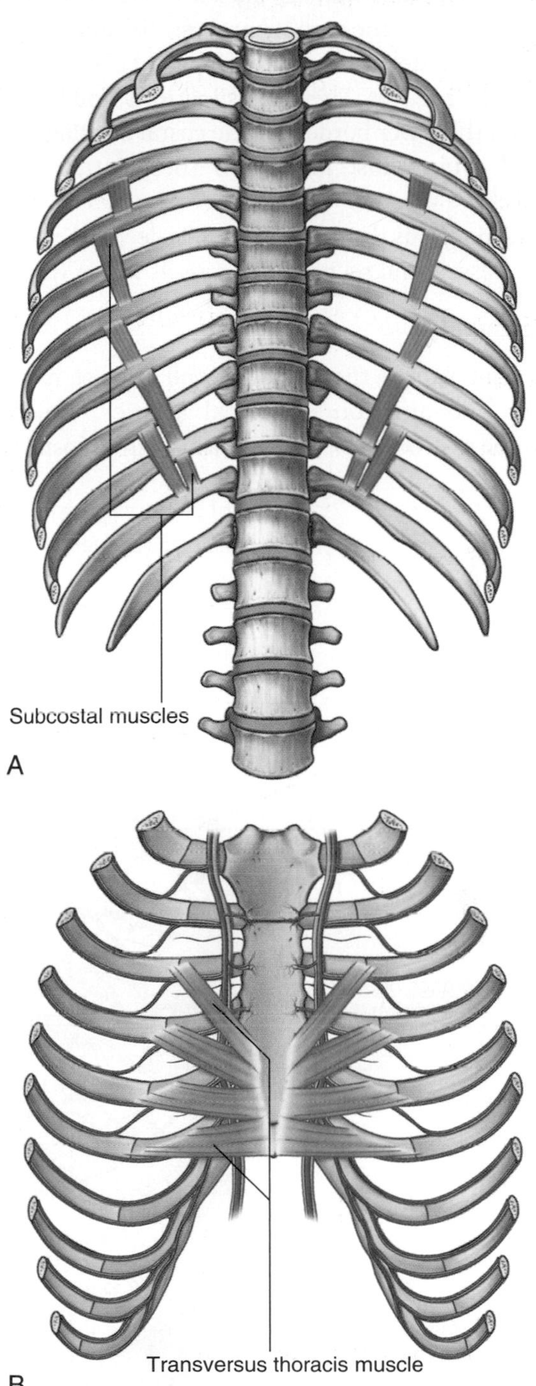


**Fig. 3.28 A.** Subcostal muscles. **B.** Transversus thoracis muscles.

### Subcostales

The **subcostales** are in the same plane as the innermost intercostals, span multiple ribs, and are more numerous in lower regions of the posterior thoracic wall (Fig. 3.28A). They extend from the internal surfaces of one rib to the internal surface of the second (next) or third rib below. Their fibers parallel the course of the internal intercostal muscles and extend from the angle of the ribs to more medial positions on the ribs below.

## Transversus thoracis muscles

The **transversus thoracis muscles** are found on the deep surface of the anterior thoracic wall (Fig. 3.28B) and in the same plane as the innermost intercostals.

The transversus thoracis muscles originate from the posterior aspect of the xiphoid process, the inferior part of the body of the sternum, and the adjacent costal cartilages of the lower true ribs. They pass superiorly and laterally to insert into the lower borders of the costal cartilages of ribs III to VI. They most likely pull these latter elements inferiorly.

The transversus thoracis muscles lie deep to the internal thoracic vessels and secure these vessels to the wall.

## Arterial supply

Vessels that supply the thoracic wall consist mainly of posterior and anterior intercostal arteries, which pass around the wall between adjacent ribs in intercostal spaces (Fig. 3.29). These arteries originate from the aorta and internal thoracic arteries, which in turn arise from the subclavian arteries in the root of the neck. Together, the intercostal arteries form a basket-like pattern of vascular supply around the thoracic wall.

### Posterior intercostal arteries

**Posterior intercostal arteries** originate from vessels associated with the posterior thoracic wall. The upper two posterior intercostal arteries on each side are derived from the **supreme intercostal artery**, which descends into the thorax as a branch of the costocervical trunk in the neck. The **costocervical trunk** is a posterior branch of the subclavian artery (Fig. 3.29).

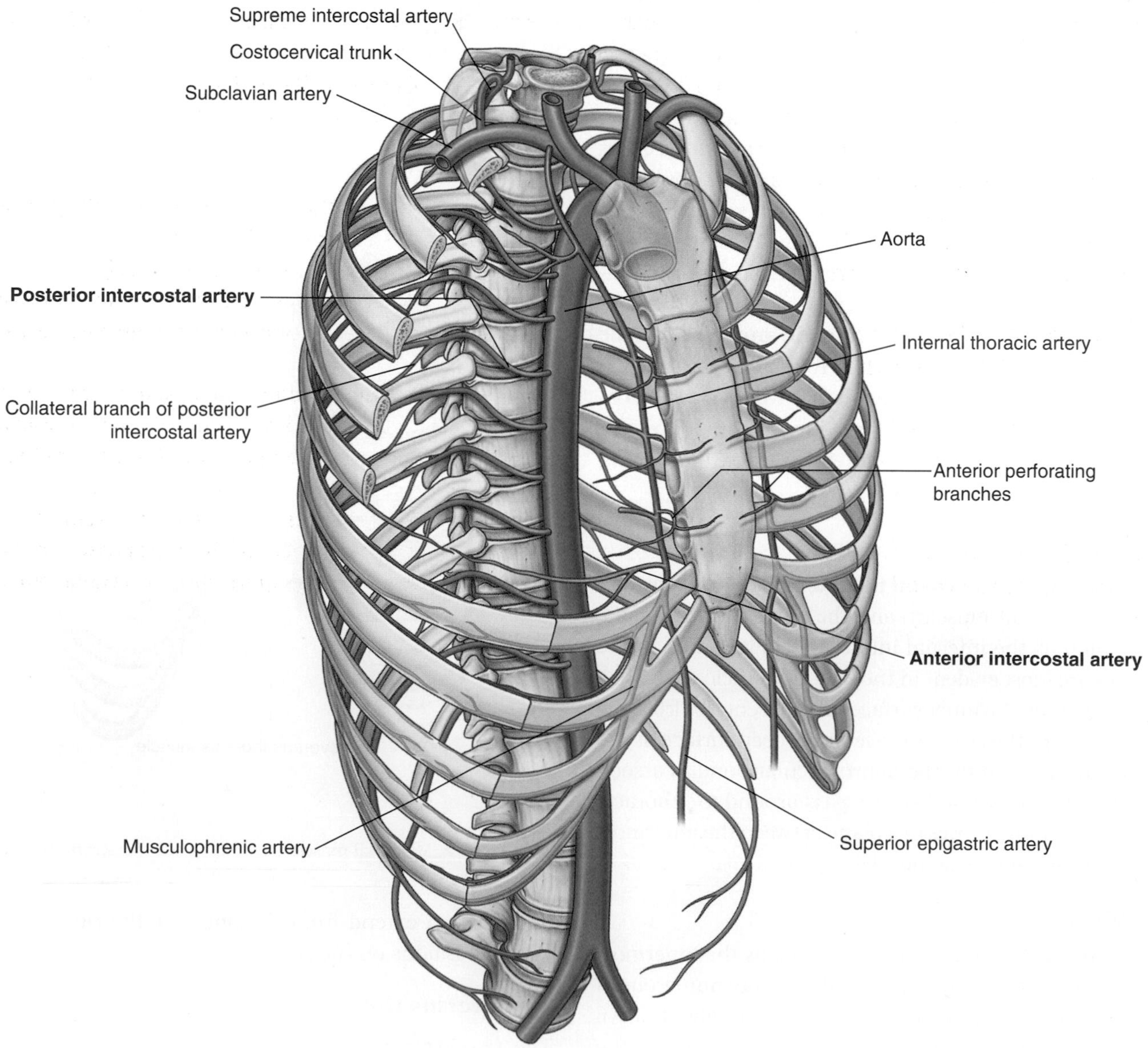

**Fig. 3.29** Arteries of the thoracic wall.

The remaining nine pairs of posterior intercostal arteries arise from the posterior surface of the thoracic aorta. Because the aorta is on the left side of the vertebral column, those posterior intercostal vessels passing to the right side of the thoracic wall cross the midline anterior to the bodies of the vertebrae and therefore are longer than the corresponding vessels on the left.

In addition to having numerous branches that supply various components of the wall, the posterior intercostal arteries have branches that accompany lateral cutaneous branches of the intercostal nerves to superficial regions.

### Anterior intercostal arteries

The **anterior intercostal arteries** originate directly or indirectly as lateral branches from the internal thoracic arteries (Fig. 3.29).

Each **internal thoracic artery** arises as a major branch of the subclavian artery in the neck. It passes anteriorly over the cervical dome of the pleura and descends vertically through the superior thoracic aperture and along the deep aspect of the anterior thoracic wall. On each side, the internal thoracic artery lies posterior to the costal cartilages of the upper six ribs and about 1 cm lateral to the sternum. At approximately the level of the sixth intercostal space, it divides into two terminal branches:

- the **superior epigastric artery**, which continues inferiorly into the anterior abdominal wall (Fig. 3.29); and
- the **musculophrenic artery**, which passes along the costal margin, goes through the diaphragm, and ends near the last intercostal space.

Anterior intercostal arteries that supply the upper six intercostal spaces arise as lateral branches from the internal thoracic artery, whereas those supplying the lower spaces arise from the musculophrenic artery.

In each intercostal space, the anterior intercostal arteries usually have two branches:

- One passes below the margin of the upper rib.
- The other passes above the margin of the lower rib and meets a collateral branch of the posterior intercostal artery.

The distributions of the anterior and posterior intercostal vessels overlap and can develop anastomotic connections. The anterior intercostal arteries are generally smaller than the posterior vessels.

In addition to anterior intercostal arteries and a number of other branches, the internal thoracic arteries give rise to perforating branches that pass directly forward between the costal cartilages to supply structures external to the thoracic wall. These vessels travel with the anterior cutaneous branches of the intercostal nerves.

### Venous drainage

Venous drainage from the thoracic wall generally parallels the pattern of arterial supply (Fig. 3.30).

Centrally, the intercostal veins ultimately drain into the azygos system of veins or into **internal thoracic veins**, which connect with the **brachiocephalic veins** in the neck.

Often the upper posterior intercostal veins on the left side come together and form the **left superior intercostal vein**, which empties into the left brachiocephalic vein.

Similarly, the upper posterior intercostal veins on the right side may come together and form the **right superior intercostal vein**, which empties into the **azygos vein**.

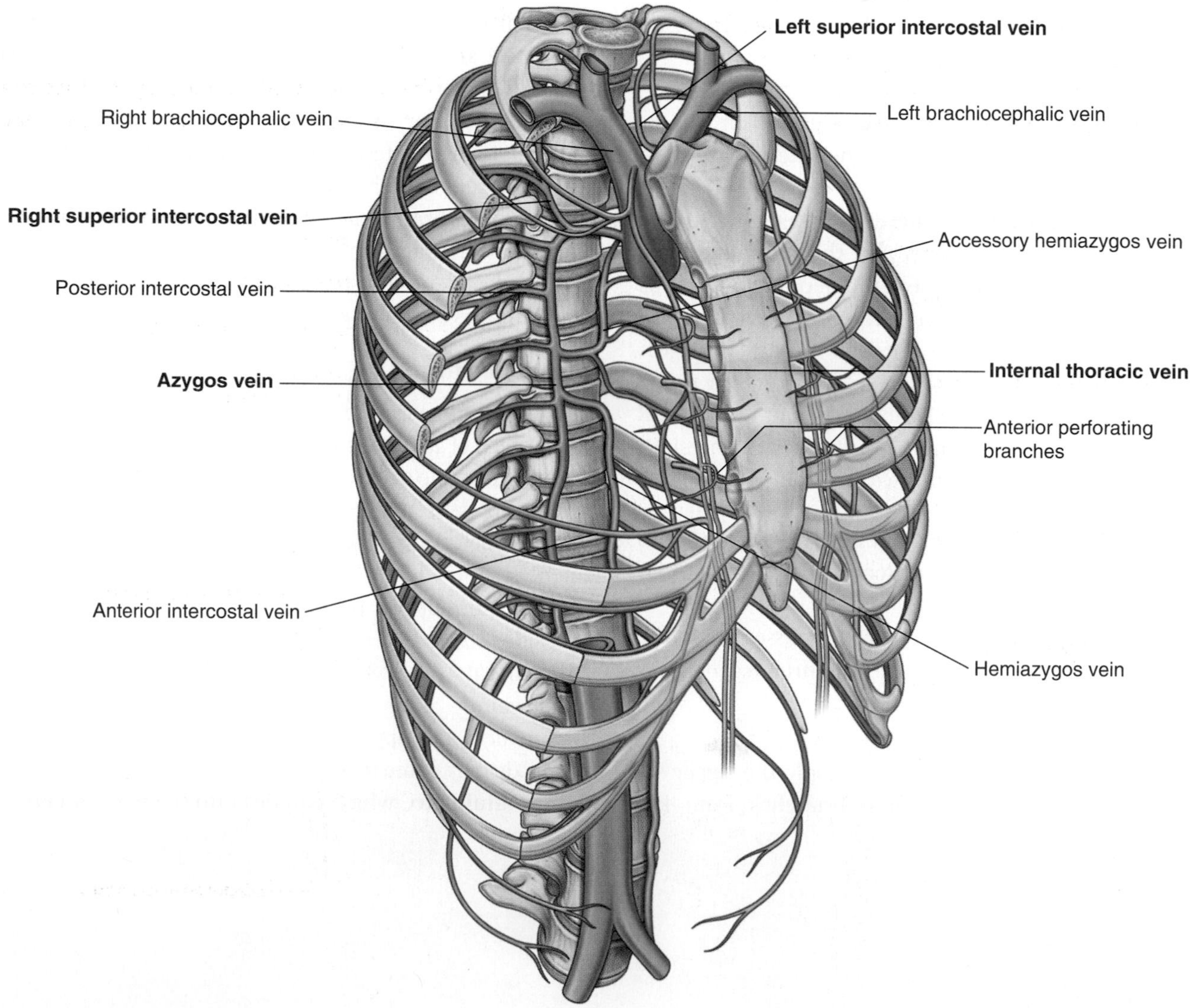

**Fig. 3.30** Veins of the thoracic wall.

### Lymphatic drainage

Lymphatic vessels of the thoracic wall drain mainly into lymph nodes associated with the internal thoracic arteries (**parasternal nodes**), with the heads and necks of ribs (**intercostal nodes**), and with the diaphragm (**diaphragmatic nodes**) (Fig. 3.31). Diaphragmatic nodes are posterior to the xiphoid and at sites where the phrenic nerves penetrate the diaphragm. They also occur in regions where the diaphragm is attached to the vertebral column.

Parasternal nodes drain into **bronchomediastinal trunks**. Intercostal nodes in the upper thorax also drain into bronchomediastinal trunks, whereas intercostal nodes in the lower thorax drain into the **thoracic duct**.

Nodes associated with the diaphragm interconnect with parasternal, prevertebral, and juxta-esophageal nodes, **brachiocephalic nodes** (anterior to the brachiocephalic veins in the superior mediastinum), and **lateral aortic/lumbar nodes** (in the abdomen).

Superficial regions of the thoracic wall drain mainly into **axillary lymph nodes** in the axilla or parasternal nodes.

### Innervation

#### Intercostal nerves

Innervation of the thoracic wall is mainly by the **intercostal nerves**, which are the anterior rami of spinal nerves

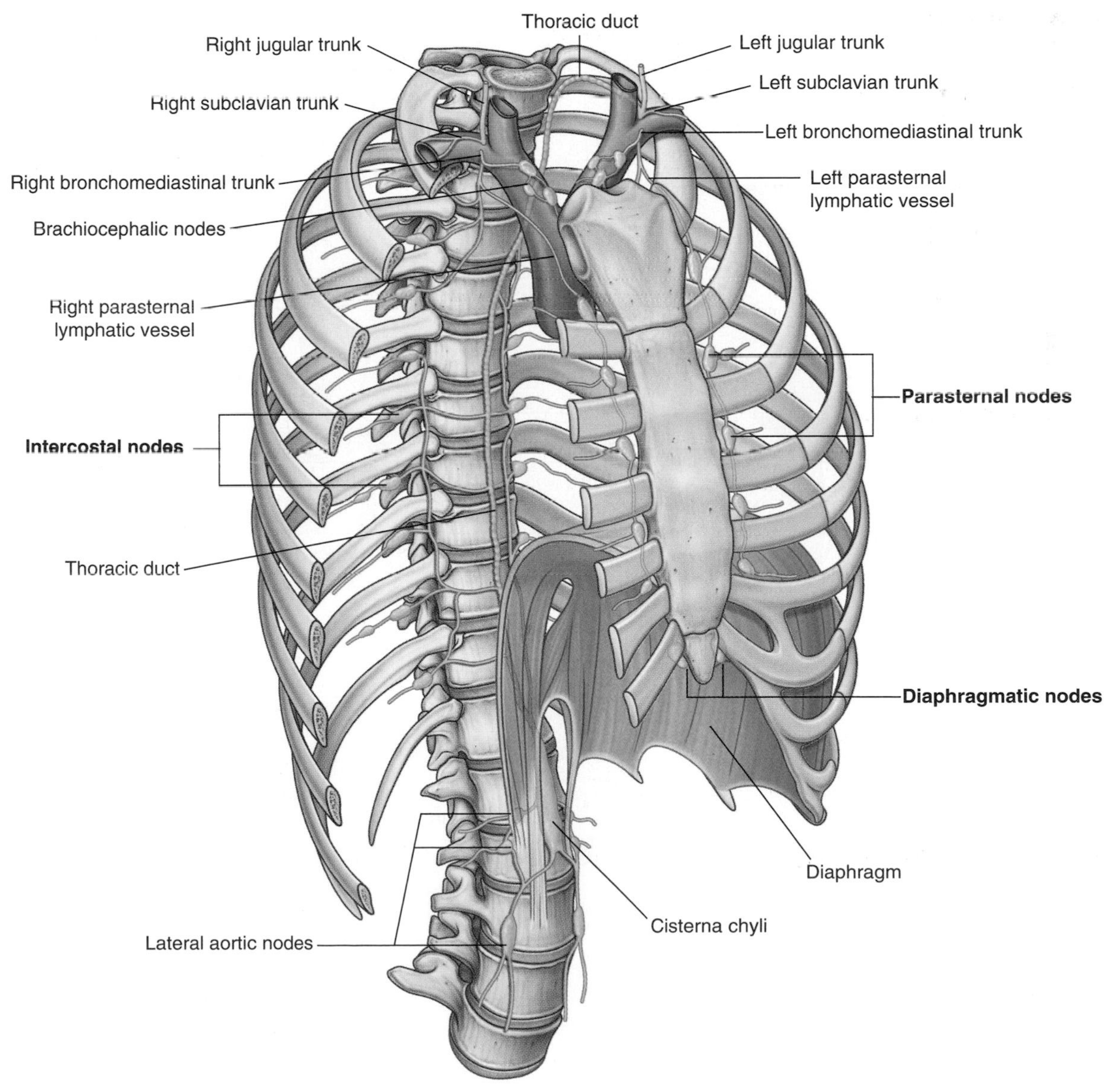


**Fig. 3.31** Major lymphatic vessels and nodes of the thoracic wall.

T1 to T11 and lie in the intercostal spaces between adjacent ribs. The anterior ramus of spinal nerve T12 (the **subcostal nerve**) is inferior to rib XII (Fig. 3.32).

A typical intercostal nerve passes laterally around the thoracic wall in an intercostal space. The largest of the branches is the **lateral cutaneous branch**, which pierces the lateral thoracic wall and divides into an anterior branch and a posterior branch that innervate the overlying skin.

The intercostal nerves end as **anterior cutaneous branches**, which emerge either parasternally, between adjacent costal cartilages, or laterally to the midline, on the anterior abdominal wall, to supply the skin.

In addition to these major branches, small collateral branches can be found in the intercostal space running along the superior border of the lower rib.

In the thorax, the intercostal nerves carry:

- somatic motor innervation to the muscles of the thoracic wall (intercostal, subcostal, and transversus thoracis muscles),
- somatic sensory innervation from the skin and parietal pleura, and
- postganglionic sympathetic fibers to the periphery.

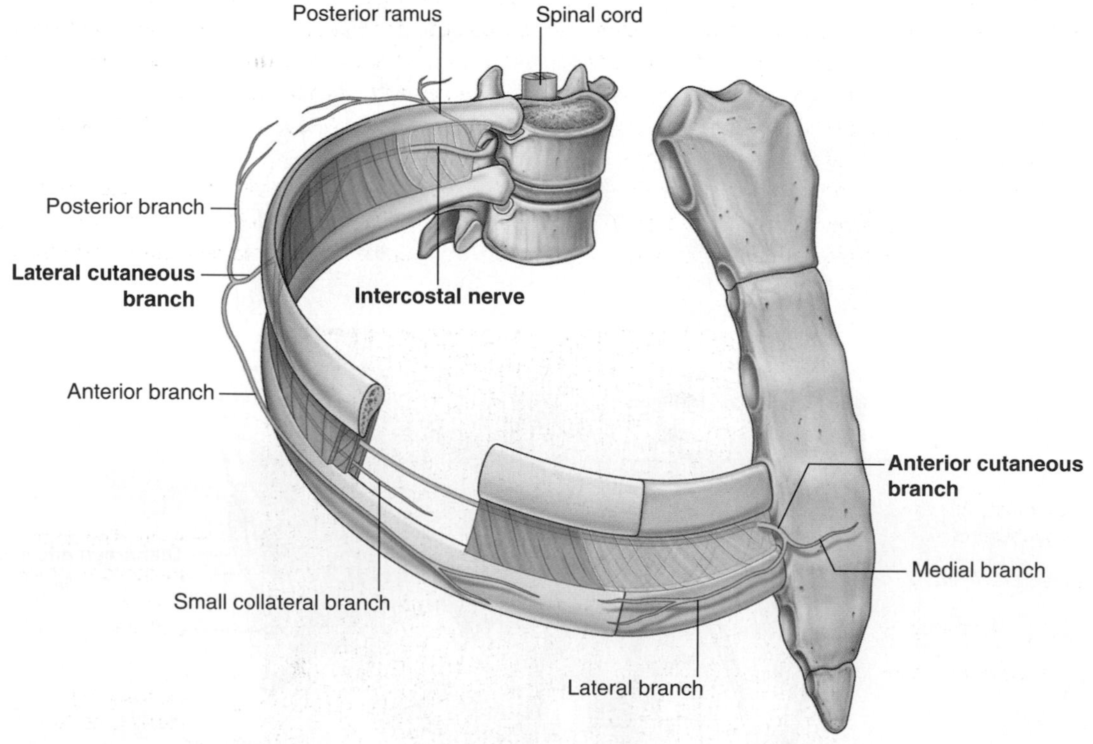


**Fig. 3.32** Intercostal nerves.

Sensory innervation of the skin overlying the upper thoracic wall is supplied by cutaneous branches (supraclavicular nerves), which descend from the cervical plexus in the neck.

In addition to innervating the thoracic wall, intercostal nerves innervate other regions:

- The anterior ramus of T1 contributes to the brachial plexus.
- The lateral cutaneous branch of the second intercostal nerve (the **intercostobrachial nerve**) contributes to cutaneous innervation of the medial surface of the upper arm.
- The lower intercostal nerves supply the muscles, skin, and peritoneum of the abdominal wall.

## In the clinic

### Surgical access to the chest

A surgical access is potentially more challenging in the chest given the rigid nature of the thoracic cage. Moreover, access is also dependent upon the organ that is operated upon and its relationships to subdiaphragmatic structures and structures in the neck.

The most common approaches are a median sternotomy and a lateral thoracotomy.

A median sternotomy involves making a vertical incision in the sternum from just below the sternal notch to the distal end of the xiphoid process. Care must be taken not to cause injury to the vessels, in particular to the brachiocephalic veins. Bleeding from the branches of the internal thoracic artery can occur and needs to be controlled. Opening the sternum causes traction on the upper ribs and may lead to rib fractures. Sometimes partial sternotomy is performed with the incision involving only the upper part of the sternum and ending at the level of manubriosternal junction or just below. A median sternotomy allows access to the heart, including coronary arteries and valves, pericardium, great vessels, anterior mediastinum, and thymus, as well as to the lower trachea. It can also be used for removal of retrosternal goiter or during esophagectomy. The incision can be extended laterally into the supraclavicular region, giving access to the subclavian and carotid arteries.

A lateral thoracotomy gives access to the ipsilateral hemithorax and its contents including the lung, mediastinum, esophagus, and heart (left lateral thoracotomy) (Fig. 3.33).

However, it involves division of muscles of the thoracic wall which leads to significant postoperative pain that needs to be well controlled to avoid restricted lung function. The

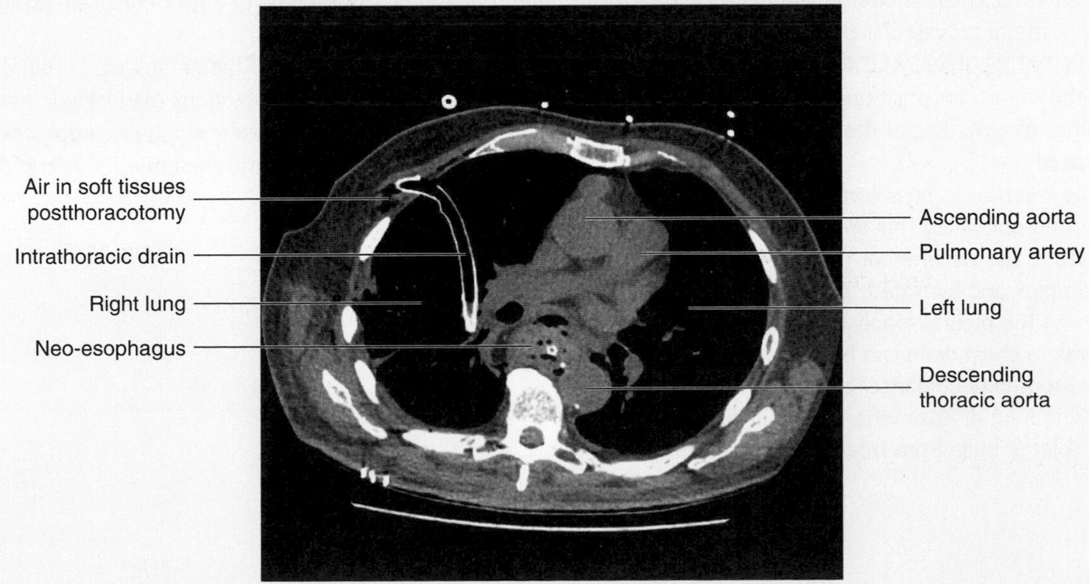


**Fig. 3.33** Right thoracotomy for esophageal cancer with intrathoracic large-bore drain. In this case, a neo-esophagus has been fashioned from the stomach.

*(continues)*

## In the clinic—cont'd

incision starts at the anterior axillary line and then passes below the tip of the scapula and is extended superiorly between the posterior midline and medial border of the scapula. The pleural cavity is entered through an intercostal space. In older patients and those with osteoporosis, a short segment of rib is often resected to minimize the risk of a rib fracture.

Minimally invasive thoracic surgery (video-assisted thoracic surgery [VATS]) involves making small (1-cm) incisions in the intercostal spaces, placing a small camera on a telescope, and manipulating other instruments through additional small incisions. A number of procedures can be performed in this manner, including lobectomy, lung biopsy, and esophagectomy.

## In the clinic

### Thoracostomy (chest) tube insertion

Insertion of a chest tube is a commonly performed procedure and is indicated to relieve air or fluid trapped in the thorax between the lung and the chest wall (pleural cavity). This procedure is done for pneumothorax, hemothorax, hemopneumothorax, malignant pleural effusion empyema, hydrothorax, and chylothorax, and also after thoracic surgery.

The position of the thoracostomy tube is usually between the anterior axillary and midaxillary anatomical lines from anterior to posterior and in either the fourth or fifth intercostal space. The position of the ribs in this region should be clearly marked. Anesthetic should be applied to the superior border of the rib and the inferior aspect of the intercostal space, including one rib and space above and one rib and space below. The neurovascular bundle runs in the neurovascular plane, which lies in the superior aspect of the intercostal space (just below the rib); hence, the reason for positioning the tube on the superior border of a rib (i.e., at the lowest position in the intercostal space).

Chest tube insertion is now commonly done with direct ultrasound guidance. This approach allows the physician both to assess whether the pleural effusion is simple or complex and loculated, and to select the safest site for entering the pleural space. In some cases of pneumothorax, a chest drain can be inserted under computed tomography-guidance, especially in patients with underlying lung disease where it is difficult to differentiate a large bulla from free air in the pleural space.

## In the clinic

### Intercostal nerve block

Local anesthesia of intercostal nerves produces excellent analgesia in patients with chest trauma and in those patients requiring anesthesia for a thoracotomy, mastectomy, or upper abdominal surgical procedures.

The intercostal nerves are situated inferior to the rib borders in the neurovascular bundle. Each neurovascular bundle is situated deep to the external and internal intercostal muscle groups.

The nerve block may be undertaken using a "blind" technique or under direct imaging guidance.

The patient is placed in the appropriate position to access the rib. Typically, under ultrasound guidance, a needle may be advanced into the region of the subcostal groove, followed by an injection with a local anesthetic. Depending on the type of anesthetic used, analgesia may be short- or long-acting.

Given the position of the neurovascular bundle and the subcostal groove, complications may include puncture of the parietal pleura and an ensuing pneumothorax. Bleeding may also occur if the artery or vein is damaged during the procedure.

## DIAPHRAGM

The **diaphragm** is a thin musculotendinous structure that fills the inferior thoracic aperture and separates the thoracic cavity from the abdominal cavity (Fig. 3.34 and see Chapter 4). It is attached peripherally to the:

- xiphoid process of the sternum,
- costal margin of the thoracic wall,
- ends of ribs XI and XII,
- ligaments that span across structures of the posterior abdominal wall, and
- vertebrae of the lumbar region.

From these peripheral attachments, muscle fibers converge to join the central tendon. The pericardium is attached to the middle part of the central tendon.

In the median sagittal plane, the diaphragm slopes inferiorly from its anterior attachment to the xiphoid, approximately at vertebral level TVIII/IX, to its posterior attachment to the **median arcuate ligament**, crossing anteriorly to the aorta at approximately vertebral level TXII.

Structures traveling between the thorax and abdomen pass through the diaphragm or between the diaphragm and its peripheral attachments:

- The inferior vena cava passes through the central tendon at approximately vertebral level TVIII.

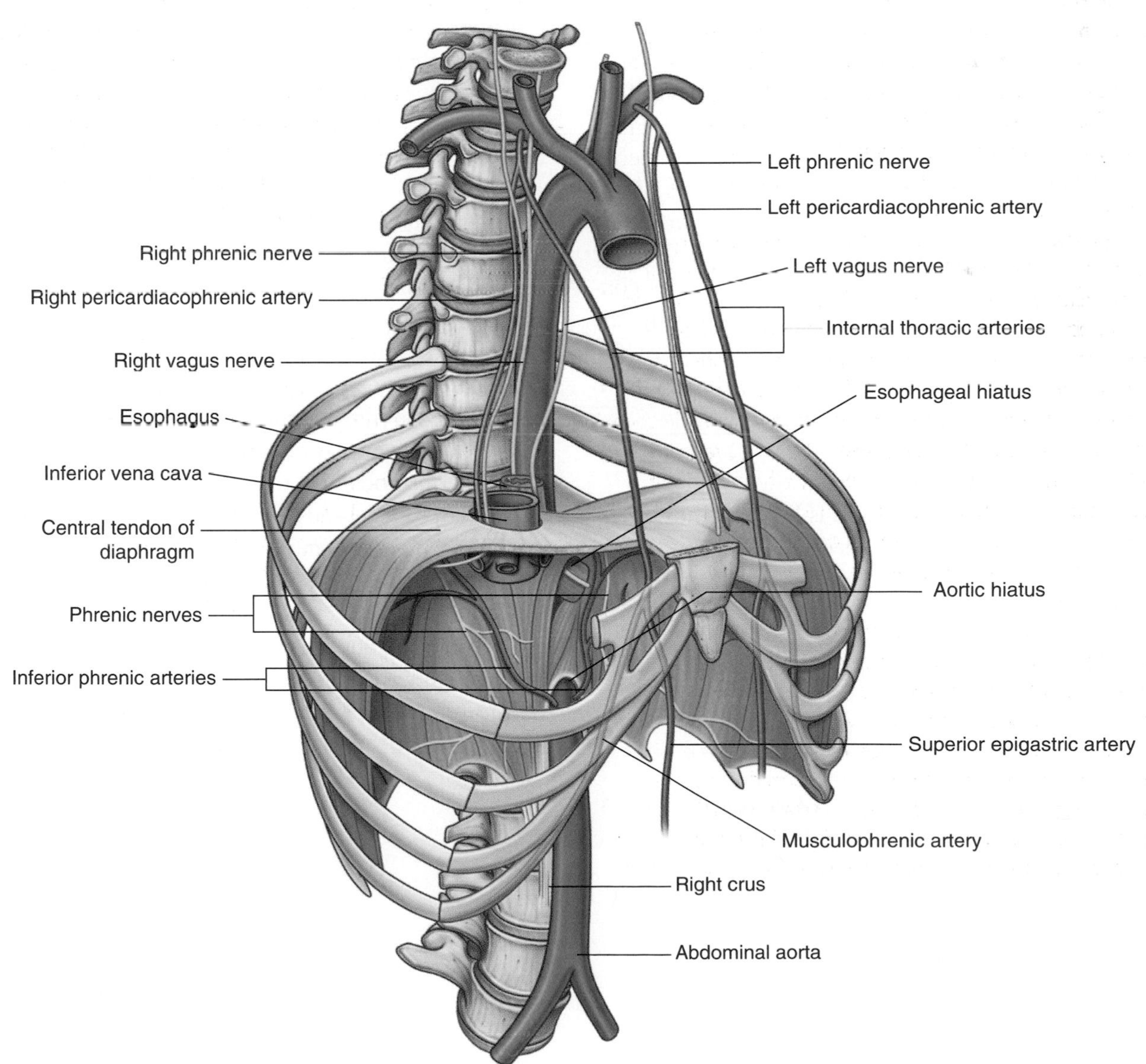


**Fig. 3.34** Diaphragm.

- The esophagus passes through the muscular part of the diaphragm, just to the left of midline, approximately at vertebral level TX.
- The vagus nerves pass through the diaphragm with the esophagus.
- The aorta passes behind the posterior attachment of the diaphragm at vertebral level TXII.
- The thoracic duct passes behind the diaphragm with the aorta.
- The azygos and hemiazygos veins may also pass through the aortic hiatus or through the crura of the diaphragm.

Other structures outside the posterior attachments of the diaphragm lateral to the aortic hiatus include the sympathetic trunks. The greater, lesser, and least splanchnic nerves penetrate the crura.

### Arterial supply

The arterial supply to the diaphragm is from vessels that arise superiorly and inferiorly to it (see Fig. 3.34). From above, pericardiacophrenic and musculophrenic arteries supply the diaphragm. These vessels are branches of the internal thoracic arteries. **Superior phrenic arteries**, which arise directly from lower parts of the thoracic aorta, and small branches from intercostal arteries contribute to the supply. The largest arteries supplying the diaphragm arise from below it. These arteries are the **inferior phrenic arteries**, which branch directly from the abdominal aorta.

### Venous drainage

Venous drainage of the diaphragm is by veins that generally parallel the arteries. The veins drain into:

- the brachiocephalic veins in the neck,
- the **azygos system of veins**, or
- abdominal veins (left suprarenal vein and inferior vena cava).

### Innervation

The diaphragm is innervated by the **phrenic nerves** (C3, C4, and C5), which penetrate the diaphragm and innervate it from its abdominal surface.

Contraction of the domes of the diaphragm flattens the diaphragm, thereby increasing thoracic volume. Movements of the diaphragm are essential for normal breathing.

## MOVEMENTS OF THE THORACIC WALL AND DIAPHRAGM DURING BREATHING

One of the principal functions of the thoracic wall and the diaphragm is to alter the volume of the thorax and thereby move air in and out of the lungs.

During breathing, the dimensions of the thorax change in the vertical, lateral, and anteroposterior directions. Elevation and depression of the diaphragm significantly alter the vertical dimensions of the thorax. Depression results when the muscle fibers of the diaphragm contract. Elevation occurs when the diaphragm relaxes.

Changes in the anteroposterior and lateral dimensions result from elevation and depression of the ribs (Fig. 3.35). The posterior ends of the ribs articulate with the vertebral column, whereas the anterior ends of most ribs articulate with the sternum or adjacent ribs.

Because the anterior ends of the ribs are inferior to the posterior ends, when the ribs are elevated, they move the sternum upward and forward. Also, the angle between the body of the sternum and the manubrium may become slightly less acute. When the ribs are depressed, the sternum moves downward and backward. This "pump handle" movement changes the dimensions of the thorax in the anteroposterior direction (Fig. 3.35A).

As well as the anterior ends of the ribs being lower than the posterior ends, the middles of the shafts tend to be lower than the two ends. When the shafts are elevated, the middles of the shafts move laterally. This "bucket handle" movement increases the lateral dimensions of the thorax (Fig. 3.35B).

Any muscles attaching to the ribs can potentially move one rib relative to another and therefore act as accessory respiratory muscles. Muscles in the neck and the abdomen can fix or alter the positions of upper and lower ribs.

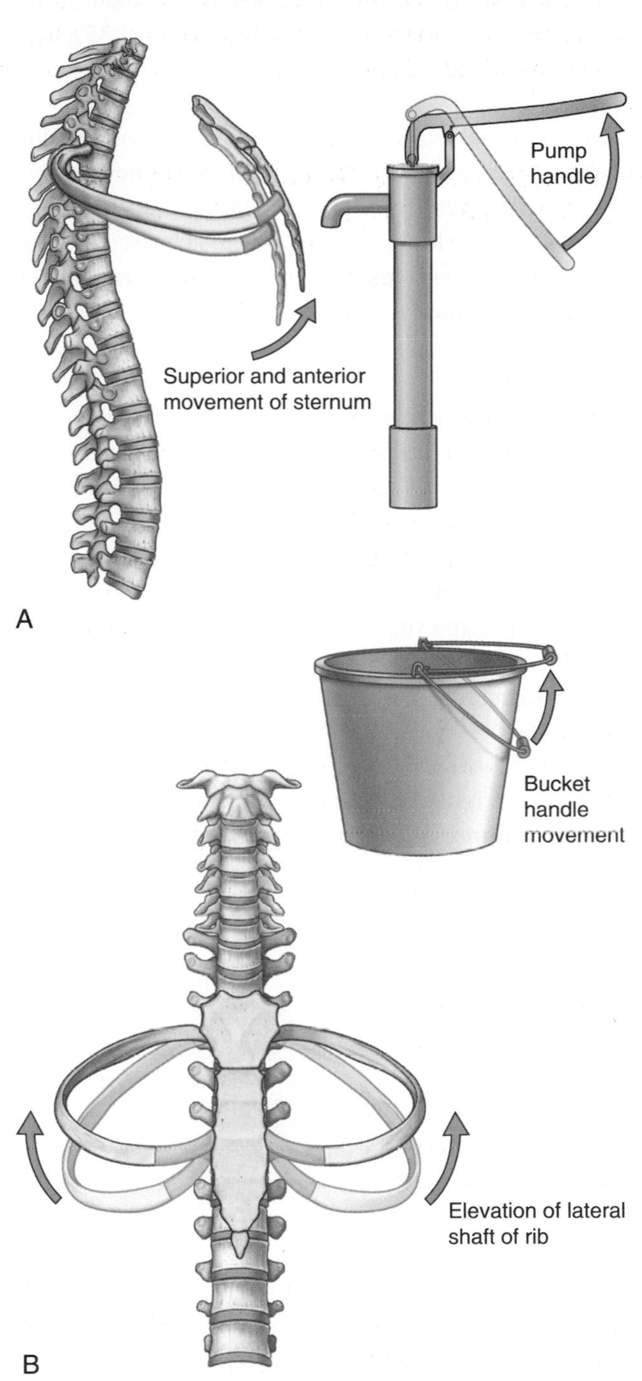


**Fig. 3.35** Movement of thoracic wall during breathing. **A.** Pump handle movement of ribs and sternum. **B.** Bucket handle movement of ribs.

## In the clinic

### Diaphragmatic paralysis

In cases of phrenic nerve palsy, diaphragmatic paralysis ensues, which is manifested by the elevation of the diaphragm muscle on the affected side (Fig. 3.36). The most important cause of the phrenic nerve palsy that should never be overlooked is malignant infiltration of the nerve by lung cancer. Other causes include postviral neuropathy (in particular, related to varicella zoster virus), trauma, iatrogenic injury during thoracic surgery, and degenerative changes in the cervical spine with compression of the C3–C5 nerve roots.

Most patients with unilateral diaphragmatic paralysis are asymptomatic and require no treatment. Some may report shortness of breath, particularly on exertion. Bilateral paralysis of the diaphragm is rare but can cause significant respiratory distress.

Surgical plication of the diaphragm can be performed in cases with respiratory compromise and is often done laparoscopically. The surgeon creates folds in the paralyzed diaphragm and sutures them in place, reducing the mobility of the diaphragmatic muscle. There is usually good improvement in lung function, exercise tolerance, and shortness of breath after the procedure.

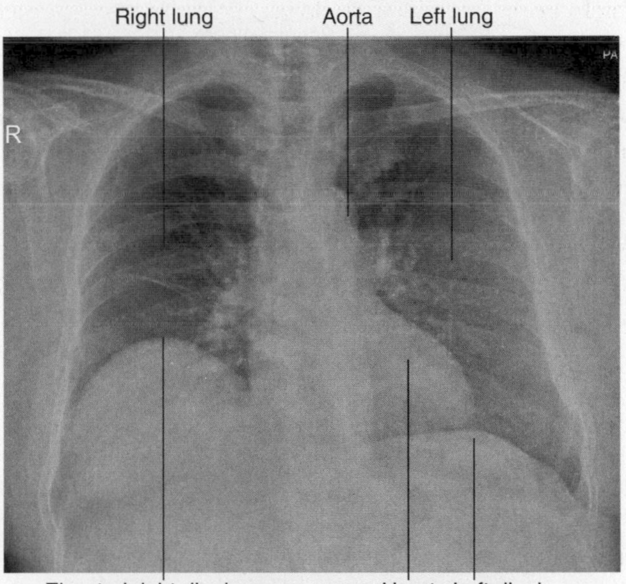


**Fig. 3.36** Chest radiograph showing an elevated right hemidiaphragm in a patient with right-sided diaphragmatic paralysis.

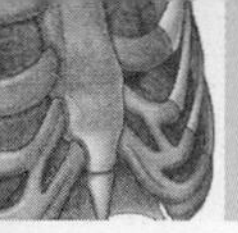

## PLEURAL CAVITIES

Two **pleural cavities**, one on either side of the mediastinum, surround the lungs (Fig. 3.37):

- Superiorly, they extend above rib I into the root of the neck.
- Inferiorly, they extend to a level just above the costal margin.
- The medial wall of each pleural cavity is the mediastinum.

### Pleura

Each pleural cavity is lined by a single layer of flat cells, mesothelium, and an associated layer of supporting connective tissue; together, they form the pleura.

The **pleura** is divided into two major types, based on location:

- Pleura associated with the walls of a pleural cavity is **parietal pleura** (Fig. 3.37).
- Pleura that reflects from the medial wall and onto the surface of the lung is **visceral pleura** (Fig. 3.37), which adheres to and covers the lung.

Each pleural cavity is the potential space enclosed between the visceral and parietal pleurae. They normally contain only a very thin layer of serous fluid. As a result, the surface of the lung, which is covered by visceral pleura, directly opposes and freely slides over the parietal pleura attached to the wall.

#### Parietal pleura

The names given to the parietal pleura correspond to the parts of the wall with which they are associated (Fig. 3.38):

- Pleura related to the ribs and intercostal spaces is termed the **costal part**.
- Pleura covering the diaphragm is the **diaphragmatic part**.
- Pleura covering the mediastinum is the **mediastinal part**.
- The dome-shaped layer of parietal pleura lining the cervical extension of the pleural cavity is **cervical pleura** (**dome of pleura** or **pleural cupola**).

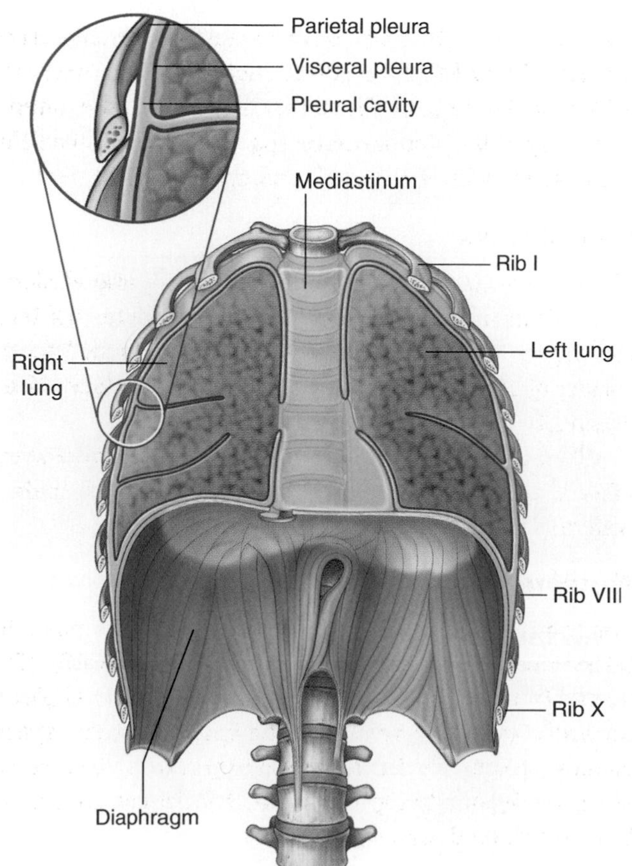


**Fig. 3.37** Pleural cavities.

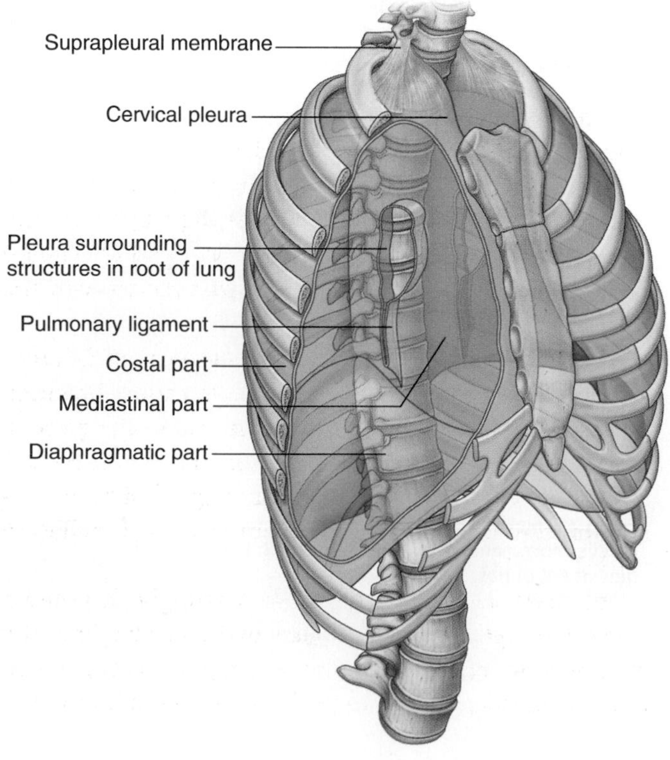


**Fig. 3.38** Parietal pleura.

Covering the superior surface of the cervical pleura is a distinct dome-like layer of fascia, the **suprapleural membrane** (Fig. 3.38). This connective tissue membrane is attached laterally to the medial margin of the first rib and behind to the transverse process of vertebra CVII. Superiorly, the membrane receives muscle fibers from some of the deep muscles in the neck (scalene muscles) that function to keep the membrane taut. The suprapleural membrane provides apical support for the pleural cavity in the root of the neck.

In the region of vertebrae TV to TVII, the mediastinal pleura reflects off the mediastinum as a tubular, sleeve-like covering for structures (i.e., airway, vessels, nerves, lymphatics) that pass between the lung and mediastinum. This sleeve-like covering and the structures it contains forms the **root of the lung**. The root joins the medial surface of the lung at an area referred to as the **hilum of the lung**. Here, the mediastinal pleura is continuous with the visceral pleura.

The parietal pleural is innervated by somatic afferent fibers. The costal pleura is innervated by branches from the intercostal nerves, and pain would be felt in relation to the thoracic wall. The diaphragmatic pleura and the mediastinal pleura are innervated mainly by the phrenic nerves (originating at spinal cord levels C3, C4, and C5). Pain from these areas would refer to the C3, C4, and C5 dermatomes (lateral neck and the supraclavicular region of the shoulder).

## Peripheral reflections

The peripheral reflections of parietal pleura mark the extent of the pleural cavities (Fig. 3.39).

Superiorly, the pleural cavity can project as much as 3 to 4 cm above the first costal cartilage but does not extend above the neck of rib I. This limitation is caused by the inferior slope of rib I to its articulation with the manubrium.

Anteriorly, the pleural cavities approach each other posterior to the upper part of the sternum. However, posterior to the lower part of the sternum, the parietal pleura does not come as close to the midline on the left side as it does on the right because the middle mediastinum, containing the pericardium and heart, bulges to the left.

Inferiorly, the costal pleura reflects onto the diaphragm above the costal margin. In the midclavicular line, the pleural cavity extends inferiorly to approximately rib VIII. In the midaxillary line, it extends to rib X. From this point, the inferior margin courses somewhat horizontally, crossing ribs XI and XII to reach vertebra TXII. From the midclavicular line to the vertebral column, the inferior boundary of the pleura can be approximated by a line that runs between rib VIII, rib X, and vertebra TXII.

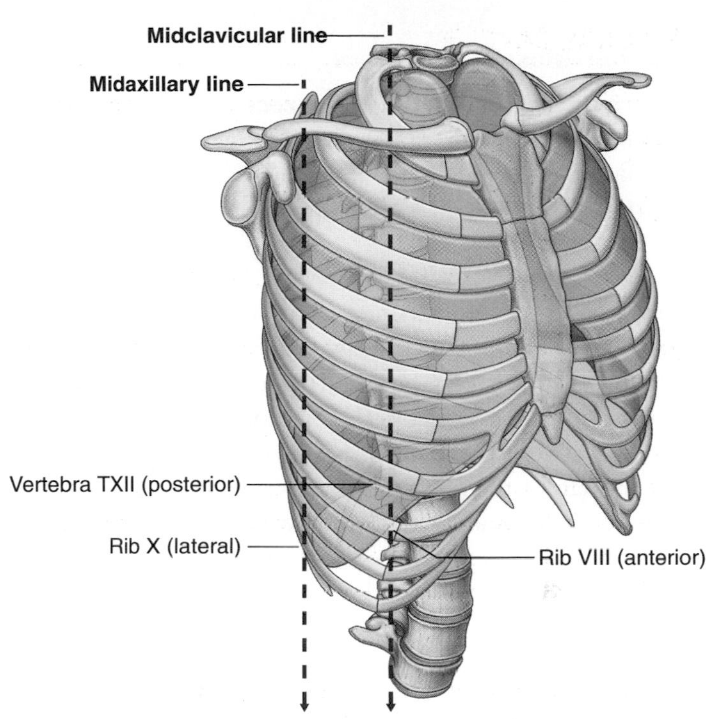


**Fig. 3.39** Pleural reflections.

### Visceral pleura

The visceral pleura is continuous with the parietal pleura at the hilum of each lung, where structures enter and leave the organ. The visceral pleura is firmly attached to the surface of the lung, including both opposed surfaces of the fissures that divide the lungs into lobes.

Although the visceral pleura is innervated by visceral afferent nerves that accompany bronchial vessels, pain is generally not elicited from this tissue.

### Pleural recesses

The lungs do not completely fill the anterior or posterior inferior regions of the pleural cavities (Fig. 3.40). This results in recesses in which two layers of parietal pleura become opposed. Expansion of the lungs into these spaces usually occurs only during forced inspiration; the recesses also provide potential spaces in which fluids can collect and from which fluids can be aspirated.

### Costomediastinal recesses

Anteriorly, a **costomediastinal recess** occurs on each side where costal pleura is opposed to mediastinal pleura. The largest is on the left side in the region overlying the heart (Fig. 3.40).

### Costodiaphragmatic recesses

The largest and clinically most important recesses are the **costodiaphragmatic recesses**, which occur in each pleural cavity between the costal pleura and diaphragmatic pleura (Fig. 3.40). The costodiaphragmatic recesses are the regions between the inferior margin of the lungs and inferior margin of the pleural cavities. They are deepest after forced expiration and shallowest after forced inspiration.

During quiet respiration, the inferior margin of the lung crosses rib VI in the midclavicular line and rib VIII in the midaxillary line, and then courses somewhat horizontally to reach the vertebral column at vertebral level TX. Thus, from the midclavicular line and around the thoracic wall to the vertebral column, the inferior margin of the lung can be approximated by a line running between rib VI, rib VIII, and vertebra TX. The inferior margin of the pleural cavity at the same points is rib VIII, rib X, and vertebra TXII. The costodiaphragmatic recess is the region between the two margins.

During expiration, the inferior margin of the lung rises and the costodiaphragmatic recess becomes larger.

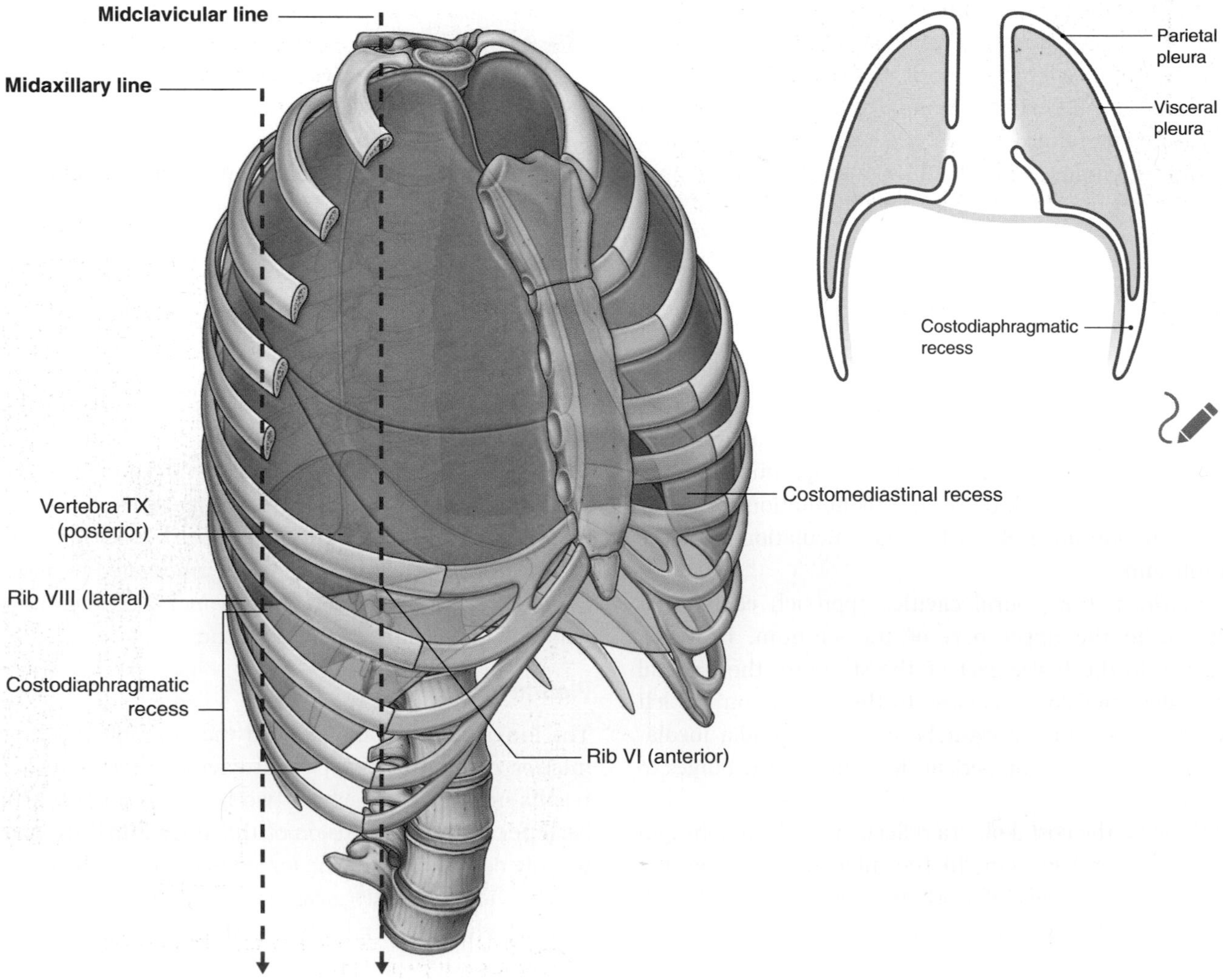

**Fig. 3.40** Parietal pleural reflections and recesses.

## In the clinic

### Pleural effusion

A pleural effusion occurs when excess fluid accumulates within the pleural space. As the fluid accumulates within the pleural space the underlying lung is compromised and may collapse as the volume of fluid increases. Once a pleural effusion has been diagnosed, fluid often will be aspirated to determine the cause, which can include infection, malignancy, cardiac failure, hepatic disease, and pulmonary embolism. A large pleural effusion needs to be drained to allow the collapsed part of the lung to reexpand and improve breathing (Fig. 3.41).

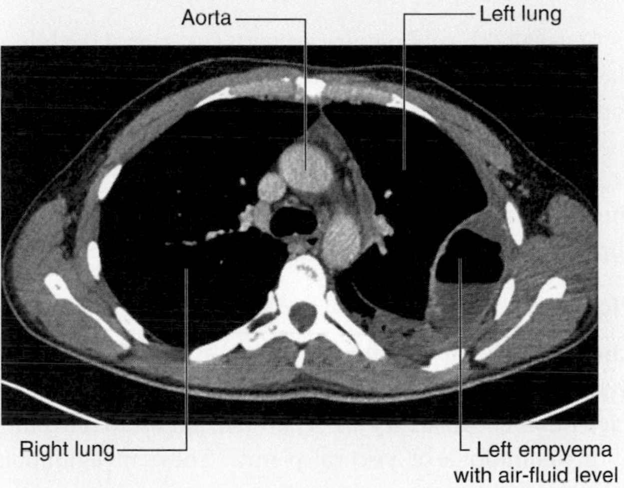


**Fig. 3.41** CT image of left pleural effusion.

## In the clinic

### Pneumothorax

A pneumothorax is a collection of gas or air within the pleural cavity (Fig. 3.42). When air enters the pleural cavity the tissue elasticity of the parenchyma causes the lung to collapse within the chest, impairing the lung function. Occasionally, the gas within the pleural cavity may accumulate to such an extent that the mediastinum is "pushed" to the opposite side, compromising the other lung. This is termed a tension pneumothorax and requires urgent treatment.

Most pneumothoraces are spontaneous (i.e., they occur in the absence of no known pathology and no known lung disease). In addition, pneumothoraces may occur as a result of trauma, inflammation, smoking, and other underlying pulmonary diseases. Certain pulmonary metastases, such as in patients with osteosarcoma, may cause spontaneous pneumothorax especially after chemotherapy. The occurrence of pneumothorax interferes with cancer treatment and increases mortality.

The symptoms of pneumothorax are often determined by the degree of air leak and the rate at which the accumulation of gas occurs and the ensuing lung collapses. They include pain, shortness of breath, and cardiorespiratory collapse, if severe.

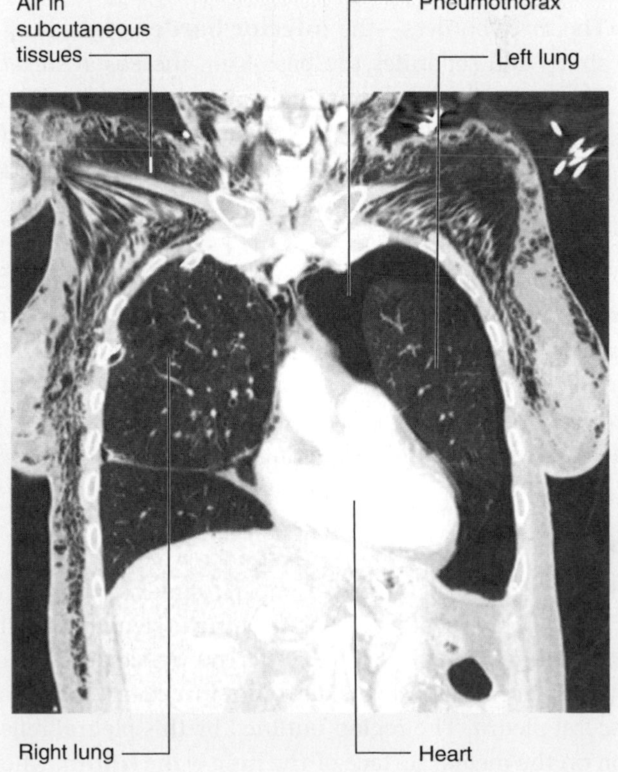


**Fig. 3.42** Pneumothorax in a patient with extensive subcutaneous emphysema.

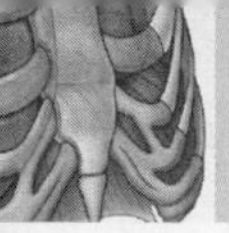



## Lungs

The two lungs are organs of respiration and lie on either side of the mediastinum surrounded by the right and left pleural cavities. Air enters and leaves the lungs via main bronchi, which are branches of the trachea.

The pulmonary arteries deliver deoxygenated blood to the lungs from the right ventricle of the heart. Oxygenated blood returns to the left atrium via the pulmonary veins.

The right lung is normally a little larger than the left lung because the middle mediastinum, containing the heart, bulges more to the left than to the right.

Each lung has a half-cone shape, with a base, apex, two surfaces, and three borders (Fig. 3.43).

- The **base** sits on the diaphragm.
- The **apex** projects above rib I and into the root of the neck.
- The two surfaces—the **costal surface** lies immediately adjacent to the ribs and intercostal spaces of the thoracic wall. The **mediastinal surface** lies against the mediastinum anteriorly and the vertebral column posteriorly and contains the comma-shaped hilum of the lung, through which structures enter and leave.
- The three borders—the **inferior border** of the lung is sharp and separates the base from the costal surface. The **anterior** and **posterior borders** separate the costal surface from the medial surface. Unlike the anterior and inferior borders, which are sharp, the posterior border is smooth and rounded.

The lungs lie directly adjacent to, and are indented by, structures contained in the overlying area. The heart and major vessels form bulges in the mediastinum that indent the medial surfaces of the lung; the ribs indent the costal surfaces. Pathology, such as tumors, or abnormalities in one structure can affect the related structure.

### Root and hilum

The **root** of each lung is a short tubular collection of structures that together attach the lung to structures in the mediastinum (Fig. 3.44). It is covered by a sleeve of mediastinal pleura that reflects onto the surface of the lung as visceral pleura. The region outlined by this pleural reflection on the medial surface of the lung is the **hilum**, where structures enter and leave.

A thin blade-like fold of pleura projects inferiorly from the root of the lung and extends from the hilum to the mediastinum. This structure is the **pulmonary ligament**. It may stabilize the position of the inferior lobe and may also accommodate the down-and-up translocation of structures in the root during breathing.

In the mediastinum, the vagus nerves pass immediately posterior to the roots of the lungs, while the phrenic nerves pass immediately anterior to them.

Within each root and located in the hilum are:

- a pulmonary artery,
- two pulmonary veins,
- a main bronchus,
- bronchial vessels,
- nerves, and
- lymphatics.

Generally, the pulmonary artery is superior at the hilum, the pulmonary veins are inferior, and the bronchi are somewhat posterior in position.

On the right side, the lobar bronchus to the superior lobe branches from the main bronchus in the root, unlike on the left where it branches within the lung itself, and is superior to the pulmonary artery.

### Right lung

The **right lung** has three lobes and two fissures (Fig. 3.45A). Normally, the lobes are freely movable against each other because they are separated, almost to the hilum, by invaginations of visceral pleura. These invaginations form the fissures:

- The **oblique fissure** separates the **inferior lobe** (**lower lobe**) from the **superior lobe** and the **middle lobe of the right lung**.
- The **horizontal fissure** separates the superior lobe (**upper lobe**) from the middle lobe.

The approximate position of the oblique fissure on a patient, in quiet respiration, can be marked by a curved line on the thoracic wall that begins roughly at the spinous process of the vertebra TIV level of the spine, crosses the fifth interspace laterally, and then follows the contour of rib VI anteriorly (see pp. 241–242).

The horizontal fissure follows the fourth intercostal space from the sternum until it meets the oblique fissure as it crosses rib V.

The orientations of the oblique and horizontal fissures determine where clinicians should listen for lung sounds from each lobe.

The largest surface of the superior lobe is in contact with the upper part of the anterolateral wall and the apex of this lobe projects into the root of the neck. The surface of the middle lobe lies mainly adjacent to the lower anterior

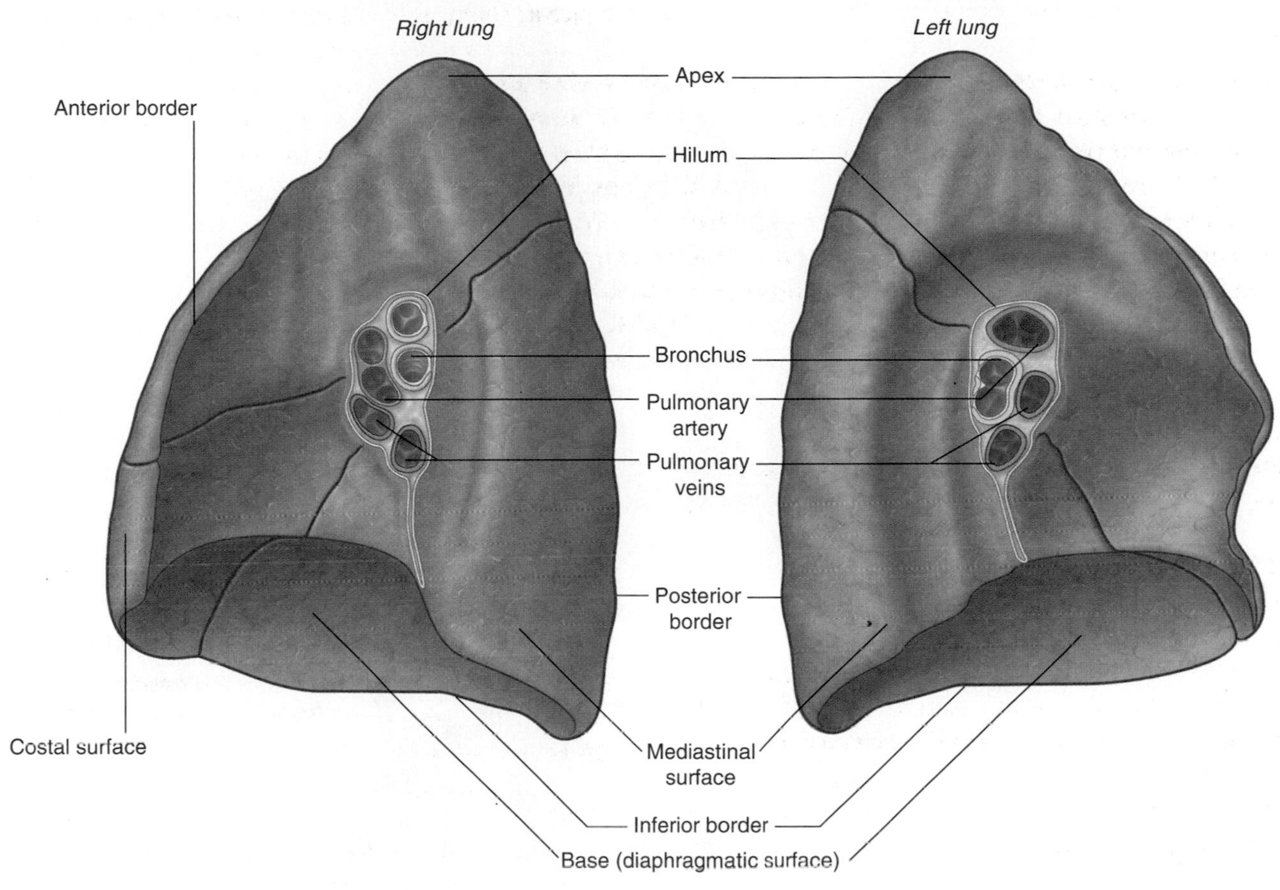


**Fig. 3.43** Lungs.

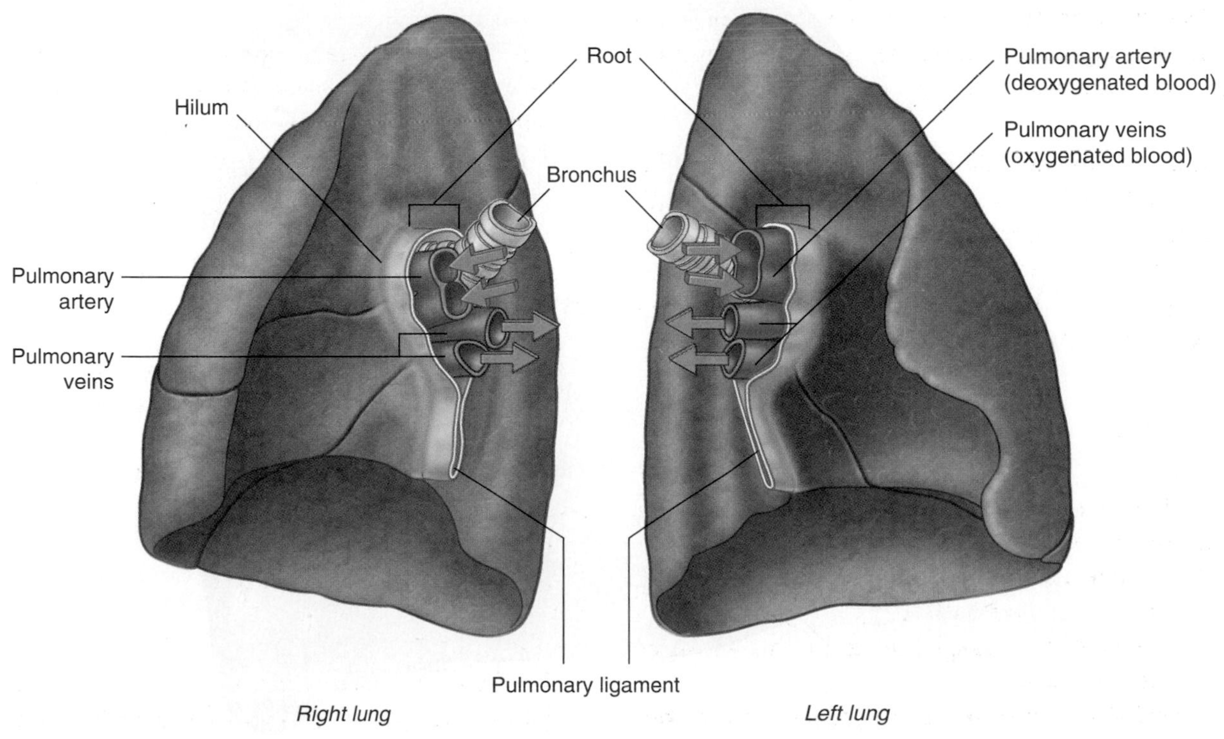


**Fig. 3.44** Roots and hila of the lungs.

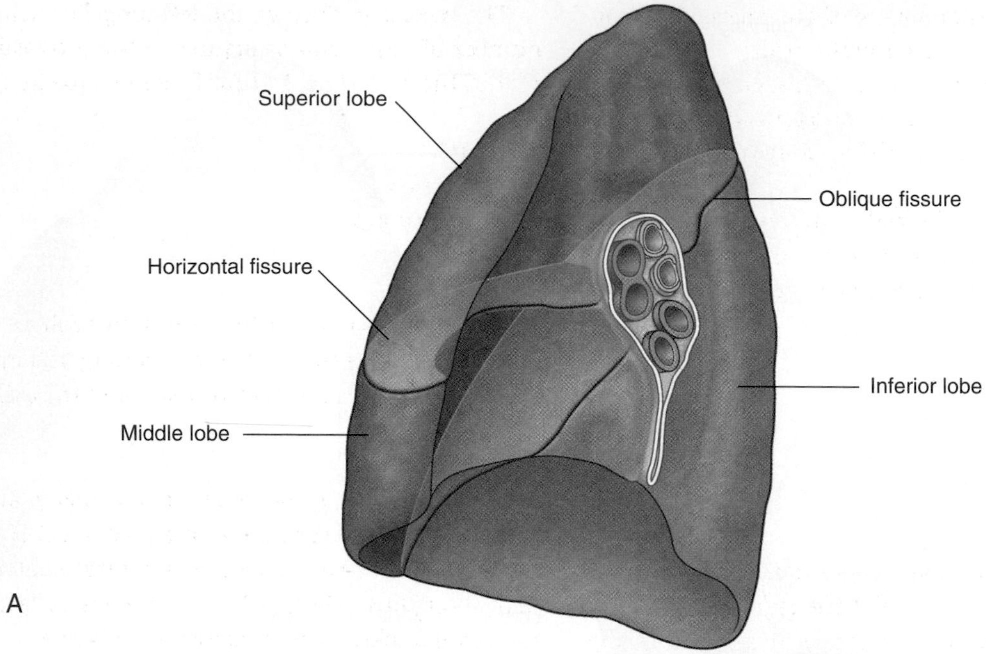

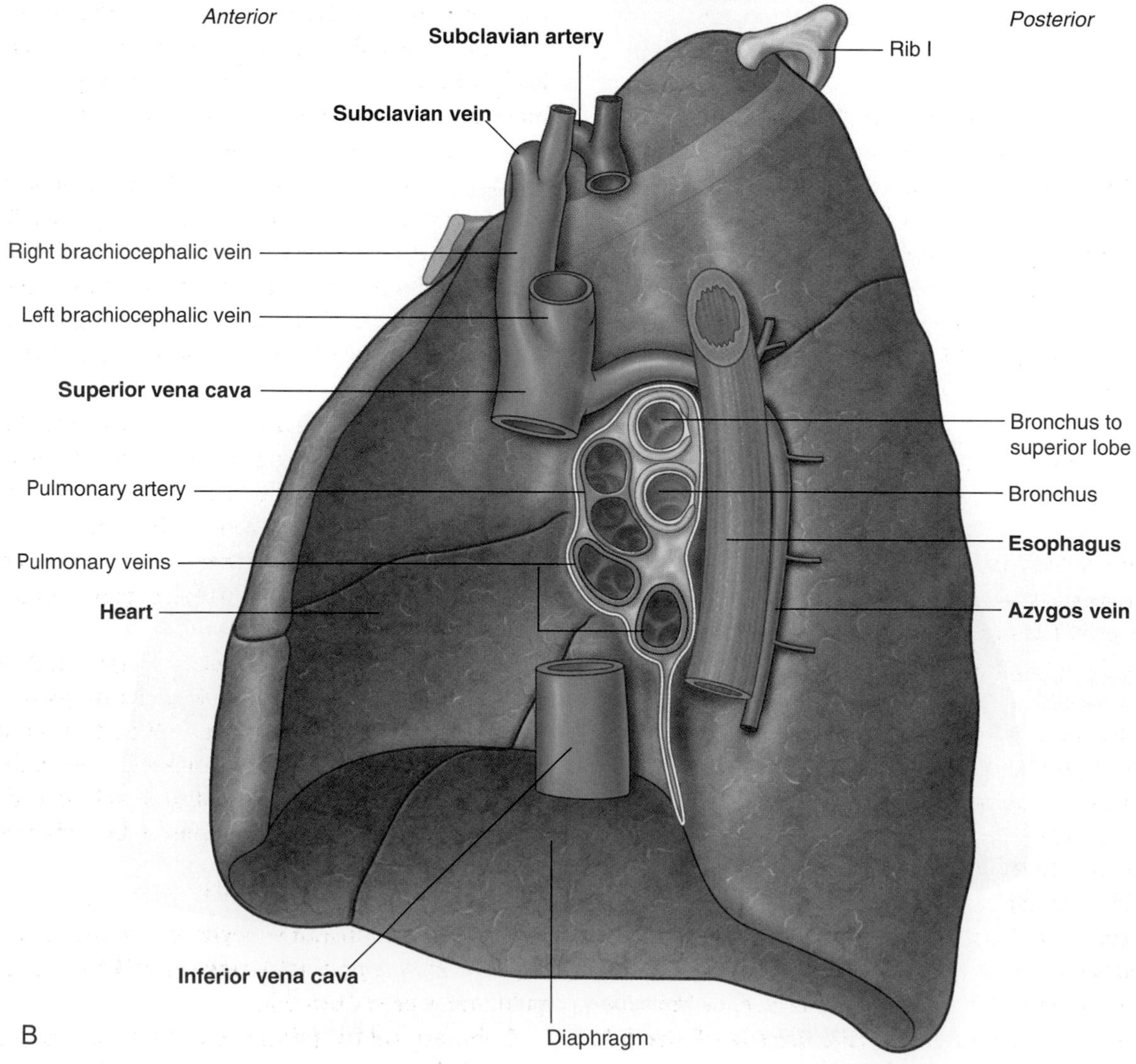

**Fig. 3.45 A.** Right lung. **B.** Major structures related to the right lung.

and lateral wall. The costal surface of the inferior lobe is in contact with the posterior and inferior walls.

When listening to lung sounds from each of the lobes, it is important to position the stethoscope on those areas of the thoracic wall related to the underlying positions of the lobes (see p. 243).

The medial surface of the right lung lies adjacent to a number of important structures in the mediastinum and the root of the neck (Fig. 3.45B). These include the:

- heart,
- inferior vena cava,
- superior vena cava,
- azygos vein, and
- esophagus.

The right subclavian artery and vein arch over and are related to the superior lobe of the right lung as they pass over the dome of the cervical pleura and into the axilla.

### Left lung

The **left lung** is smaller than the right lung and has two lobes separated by an oblique fissure (Fig. 3.46A). The **oblique fissure** of the left lung is slightly more oblique than the corresponding fissure of the right lung.

During quiet respiration, the approximate position of the left oblique fissure can be marked by a curved line on the thoracic wall that begins between the spinous processes of vertebrae TIII and TIV, crosses the fifth interspace laterally, and follows the contour of rib VI anteriorly (see pp. 241–242).

As with the right lung, the orientation of the oblique fissure determines where to listen for lung sounds from each lobe.

The largest surface of the superior lobe is in contact with the upper part of the anterolateral wall, and the apex of this lobe projects into the root of the neck. The costal surface of the inferior lobe is in contact with the posterior and inferior walls.

When listening to lung sounds from each of the lobes, the stethoscope should be placed on those areas of the thoracic wall related to the underlying positions of the lobes (see p. 243).

The inferior portion of the medial surface of the left lung, unlike the right lung, is notched because of the heart's projection into the left pleural cavity from the middle mediastinum.

From the anterior border of the lower part of the superior lobe a tongue-like extension (the **lingula of the left lung**) projects over the heart bulge.

The medial surface of the left lung lies adjacent to a number of important structures in the mediastinum and root of the neck (Fig. 3.46B). These include the:

- heart,
- aortic arch,
- thoracic aorta, and
- esophagus.

The left subclavian artery and vein arch over and are related to the superior lobe of the left lung as they pass over the dome of the cervical pleura and into the axilla.

### Bronchial tree

The **trachea** is a flexible tube that extends from vertebral level CVI in the lower neck to vertebral level TIV/V in the mediastinum where it bifurcates into a right and a left main bronchus (Fig. 3.47). The trachea is held open by C-shaped transverse cartilage rings embedded in its wall—the open part of the C facing posteriorly. The lowest tracheal ring has a hook-shaped structure, the carina, that projects backward in the midline between the origins of the two main bronchi. The posterior wall of the trachea is composed mainly of smooth muscle.

Each main bronchus enters the root of a lung and passes through the hilum into the lung itself. The right main bronchus is wider and takes a more vertical course through the root and hilum than the left main bronchus (Fig. 3.47A). Therefore, inhaled foreign bodies tend to lodge more frequently on the right side than on the left.

The main bronchus divides within the lung into **lobar bronchi** (secondary bronchi), each of which supplies a lobe. On the right side, the lobar bronchus to the superior lobe originates within the root of the lung.

The lobar bronchi further divide into **segmental bronchi** (tertiary bronchi), which supply bronchopulmonary segments (Fig. 3.47B).

Within each bronchopulmonary segment, the segmental bronchi give rise to multiple generations of divisions and, ultimately, to bronchioles, which further subdivide and supply the respiratory surfaces. The walls of the bronchi are held open by discontinuous elongated plates of cartilage, but these are not present in bronchioles.

### Bronchopulmonary segments

A **bronchopulmonary segment** is the area of lung supplied by a segmental bronchus and its accompanying pulmonary artery branch.

Tributaries of the pulmonary vein tend to pass intersegmentally between and around the margins of segments.

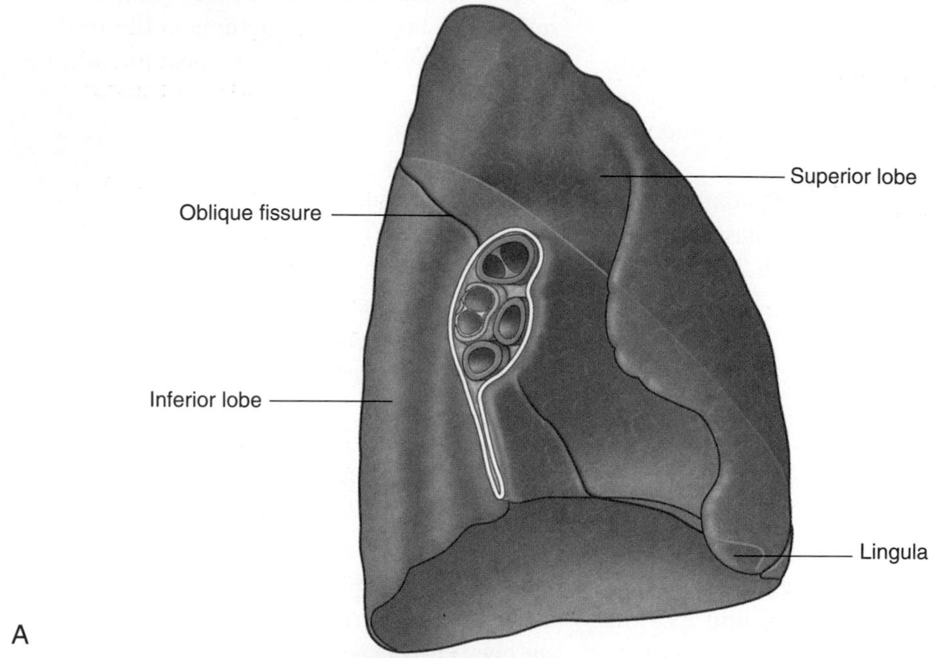

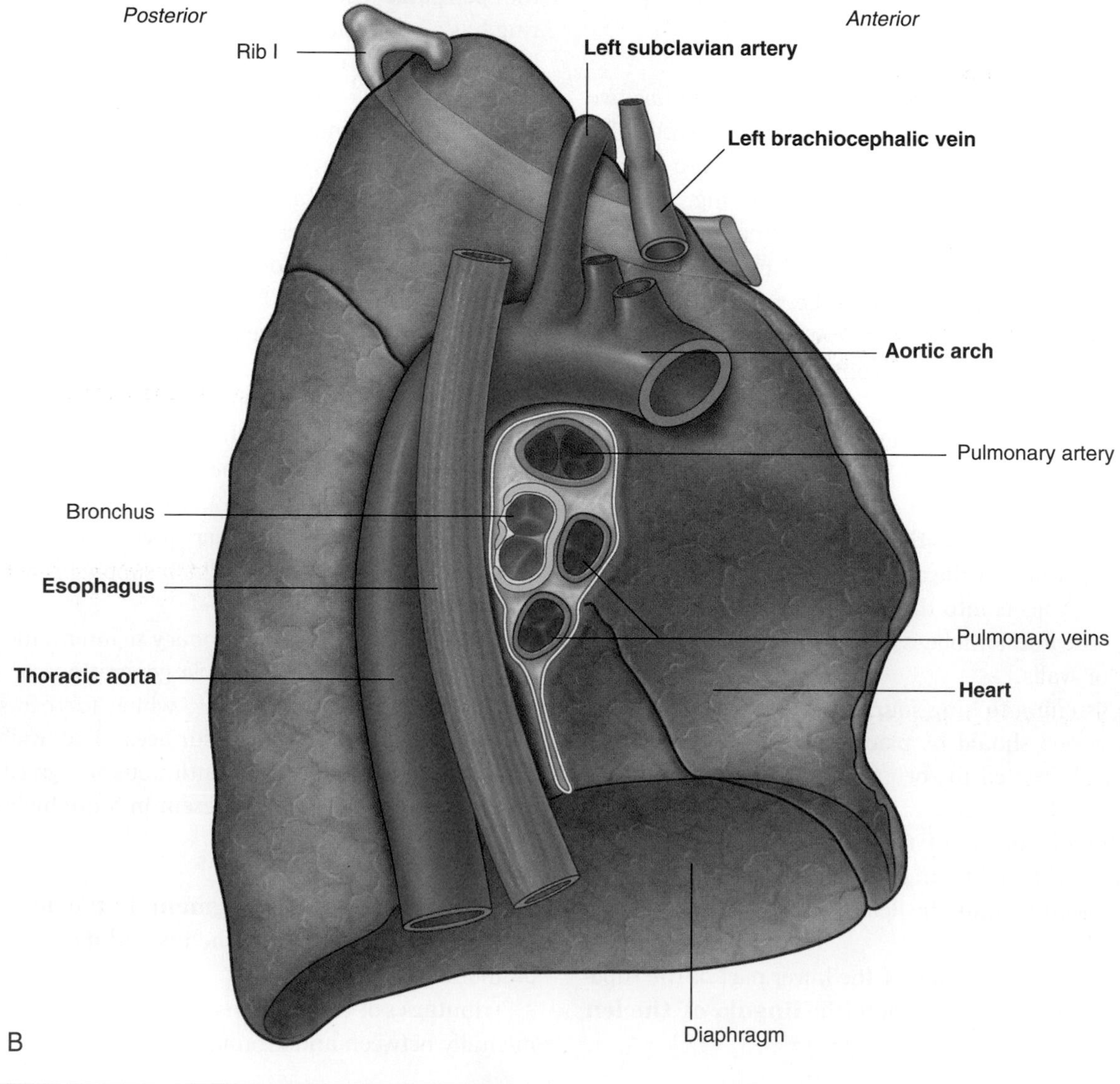

**Fig. 3.46** **A.** Left lung. **B.** Major structures related to the left lung.

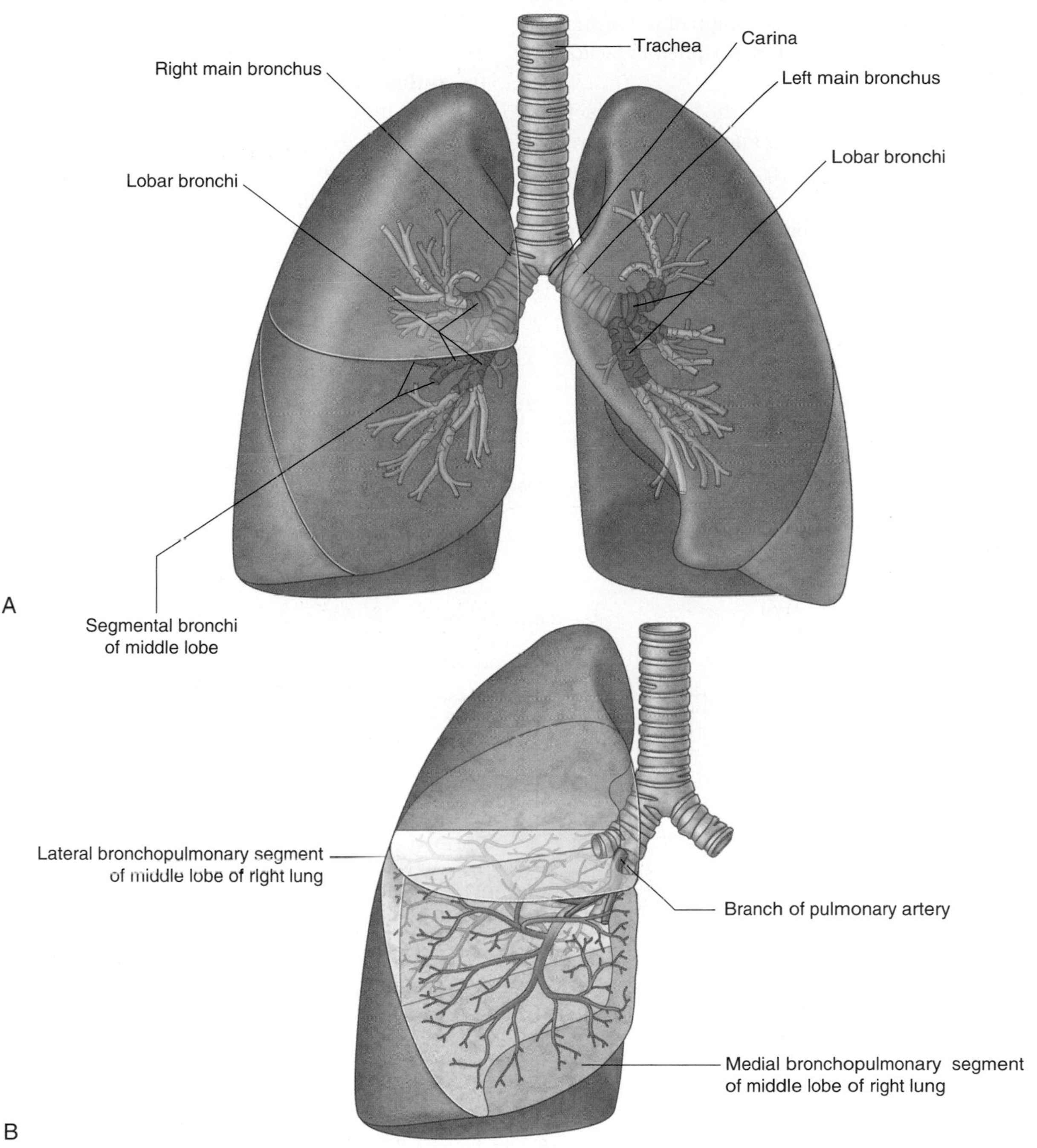


**Fig. 3.47 A.** Bronchial tree. **B.** Bronchopulmonary segments.

Each bronchopulmonary segment is shaped like an irregular cone, with the apex at the origin of the segmental bronchus and the base projected peripherally onto the surface of the lung.

A bronchopulmonary segment is the smallest functionally independent region of a lung and the smallest area of lung that can be isolated and removed without affecting adjacent regions.

There are ten bronchopulmonary segments in each lung (Fig. 3.48); some of them fuse in the left lung.

### Pulmonary arteries

The right and left pulmonary arteries originate from the **pulmonary trunk** and carry deoxygenated blood to the lungs from the right ventricle of the heart (Fig. 3.49).

The bifurcation of the pulmonary trunk occurs to the left of the midline just inferior to vertebral level TIV/V, and anteroinferiorly to the left of the bifurcation of the trachea.

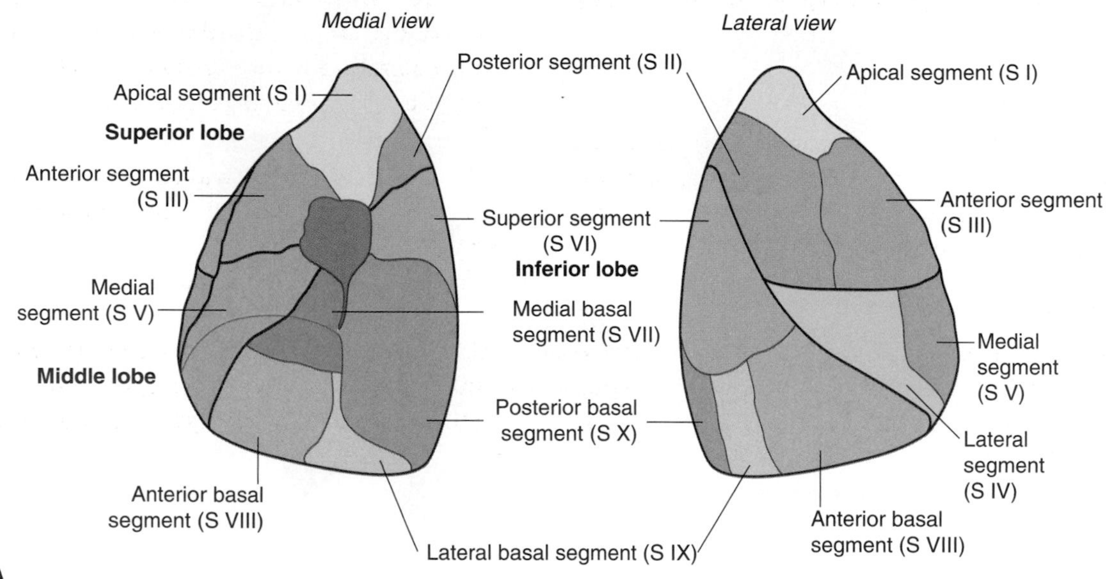


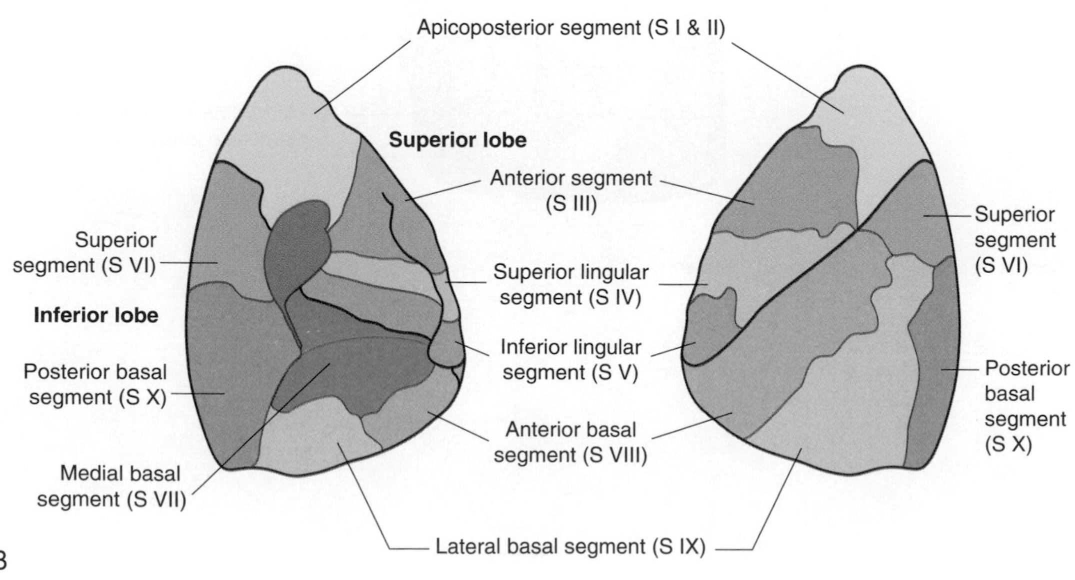


**Fig. 3.48** Bronchopulmonary segments. **A.** Right lung. **B.** Left lung. (Bronchopulmonary segments are numbered and named.)

### Right pulmonary artery

The **right pulmonary artery** is longer than the left and passes horizontally across the mediastinum (Fig. 3.49). It passes:

- anteriorly and slightly inferiorly to the tracheal bifurcation and anteriorly to the right main bronchus, and
- posteriorly to the ascending aorta, superior vena cava, and upper right pulmonary vein.

The right pulmonary artery enters the root of the lung and gives off a large branch to the superior lobe of the lung. The main vessel continues through the hilum of the lung, gives off a second (recurrent) branch to the superior lobe, and then divides to supply the middle and inferior lobes.

### Left pulmonary artery

The **left pulmonary artery** is shorter than the right and lies anterior to the descending aorta and posterior to the superior pulmonary vein (Fig. 3.49). It passes through the root and hilum and branches within the lung.

## Pulmonary veins

On each side a **superior pulmonary vein** and an **inferior pulmonary vein** carry oxygenated blood from the lungs back to the heart (Fig. 3.49). The veins begin at the hilum of the lung, pass through the root of the lung, and immediately drain into the left atrium.

## Bronchial arteries and veins

The bronchial arteries (Fig. 3.49) and veins constitute the "nutritive" vascular system of the pulmonary tissues (bronchial walls and glands, walls of large vessels, and visceral pleura). They interconnect within the lung with branches of the pulmonary arteries and veins.

The bronchial arteries originate from the thoracic aorta or one of its branches:

- A single **right bronchial artery** normally arises from the third posterior intercostal artery (but occasionally, it originates from the **upper left bronchial artery**).
- Two **left bronchial arteries** arise directly from the anterior surface of the thoracic aorta—the **superior left bronchial artery** arises at vertebral level TV, and the inferior one inferior to the left bronchus.

The bronchial arteries run on the posterior surfaces of the bronchi and ramify in the lungs to supply pulmonary tissues.

The **bronchial veins** drain into:

- either the pulmonary veins or the left atrium, and
- into the azygos vein on the right or into the superior intercostal vein or hemiazygos vein on the left.

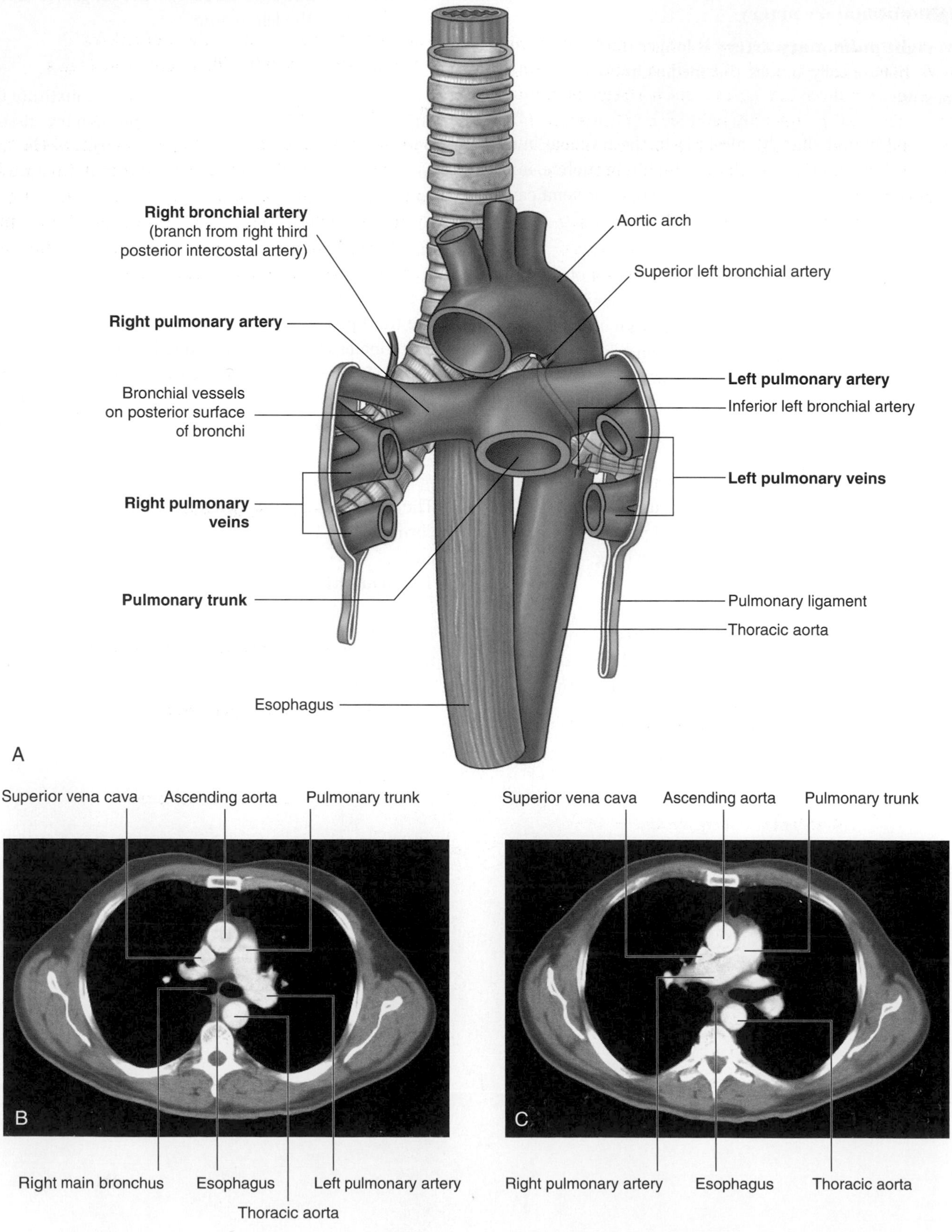

**Fig. 3.49** Pulmonary vessels. **A.** Diagram of an anterior view. **B.** Axial computed tomography image showing the left pulmonary artery branching from the pulmonary trunk. **C.** Axial computed tomography image (just inferior to the image in **B**) showing the right pulmonary artery branching from the pulmonary trunk.

### Innervation

Structures of the lung and the visceral pleura are supplied by visceral afferents and efferents distributed through the anterior pulmonary plexus and posterior pulmonary plexus (Fig. 3.50). These interconnected plexuses lie anteriorly and posteriorly to the tracheal bifurcation and main bronchi. The anterior plexus is much smaller than the posterior plexus.

Branches of these plexuses, which ultimately originate from the sympathetic trunks and vagus nerves, are distributed along branches of the airway and vessels.

Visceral efferents from:

- the vagus nerves constrict the bronchioles;
- the sympathetic system dilates the bronchioles.

### Lymphatic drainage

Superficial, or subpleural, and deep lymphatics of the lung drain into lymph nodes called **tracheobronchial nodes** around the roots of lobar and main bronchi and along the sides of the trachea (Fig. 3.51). As a group, these lymph nodes extend from within the lung, through the hilum and root, and into the posterior mediastinum.

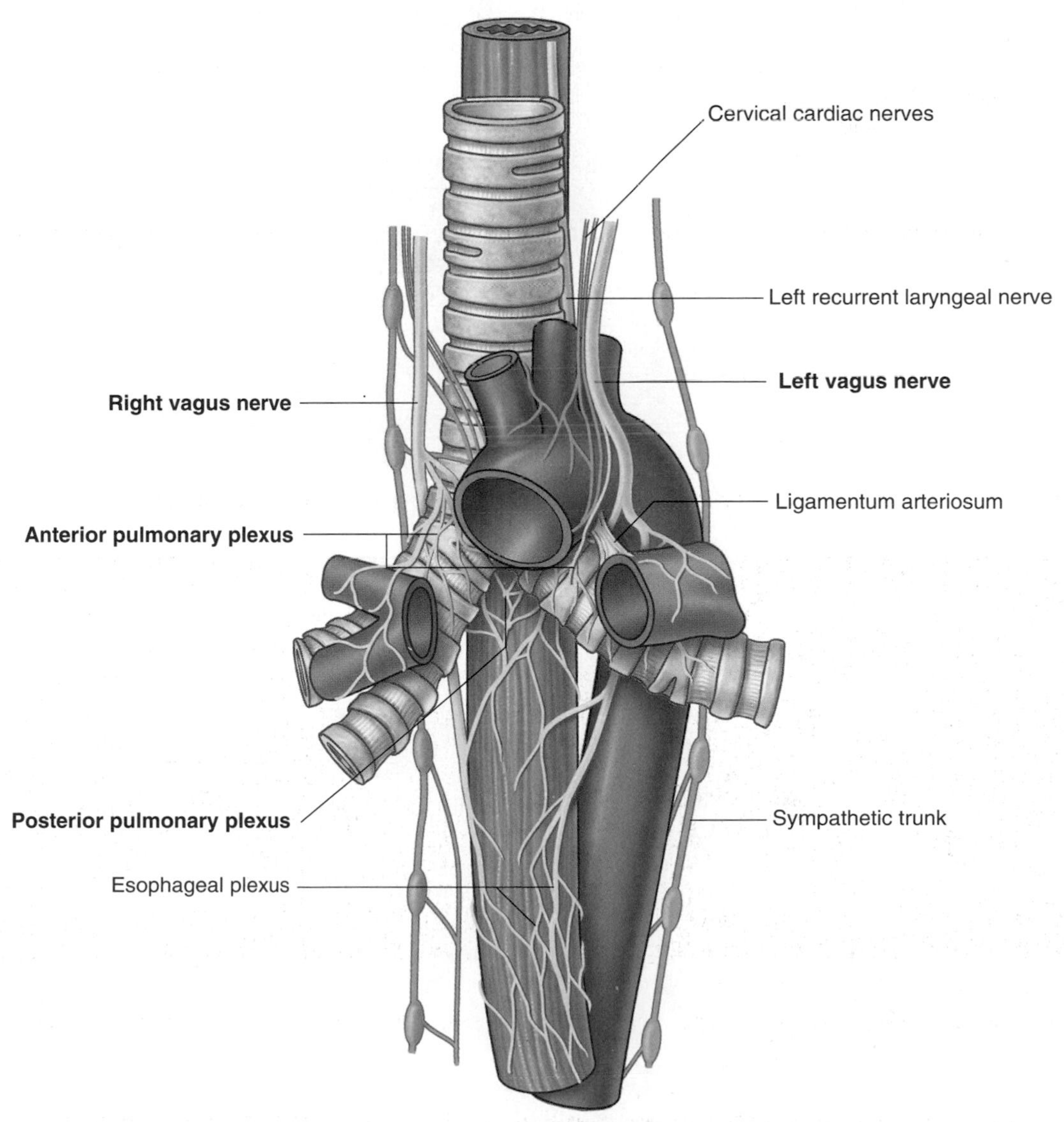


**Fig. 3.50** Pulmonary innervation.

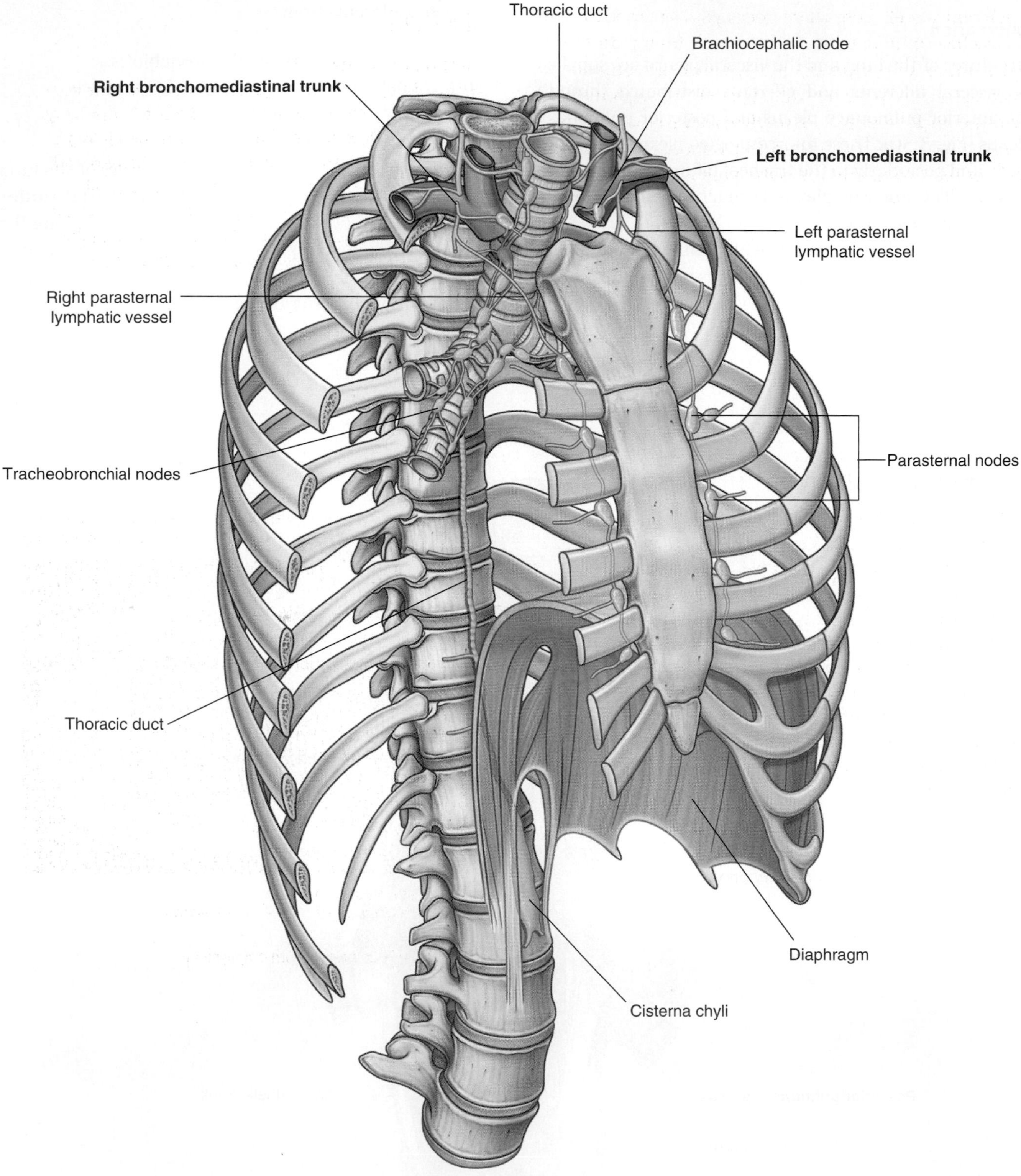

Thoracic duct
Brachiocephalic node
**Right bronchomediastinal trunk**
**Left bronchomediastinal trunk**
Left parasternal lymphatic vessel
Right parasternal lymphatic vessel
Parasternal nodes
Tracheobronchial nodes
Thoracic duct
Diaphragm
Cisterna chyli

**Fig. 3.51** Lymphatic drainage of lungs.

Efferent vessels from these nodes pass superiorly along the trachea to unite with similar vessels from parasternal nodes and brachiocephalic nodes, which are anterior to brachiocephalic veins in the superior mediastinum, to form the **right** and **left bronchomediastinal trunks**. These trunks drain directly into deep veins at the base of the neck, or may drain into the right lymphatic trunk or thoracic duct.

## In the clinic

### Imaging the lungs

Medical imaging of the lungs is important because they are one of the commonest sites for disease in the body. While the body is at rest, the lungs exchange up to 5 L of air per minute, and this may contain pathogens and other potentially harmful elements (e.g., allergens). Techniques to visualize the lung range from plain chest radiographs to high-resolution computed tomography (CT), which enables precise localization of a lesion within the lung.

## In the clinic

### High-resolution lung CT

High-resolution computed tomography (HRCT) is a diagnostic method for assessing the lungs but more specifically the interstitium of the lungs. The technique involves obtaining narrow cross-sectional slices of 1 to 2 mm. These scans enable the physician and radiologist to view the patterns of disease and their distribution. Diseases that may be easily demonstrated using this procedure include emphysema (Fig. 3.52), pneumoconiosis (coal worker's pneumoconiosis), and asbestosis. HRCT is also useful in regular follow-ups of patients with interstitial disease to monitor disease progression.

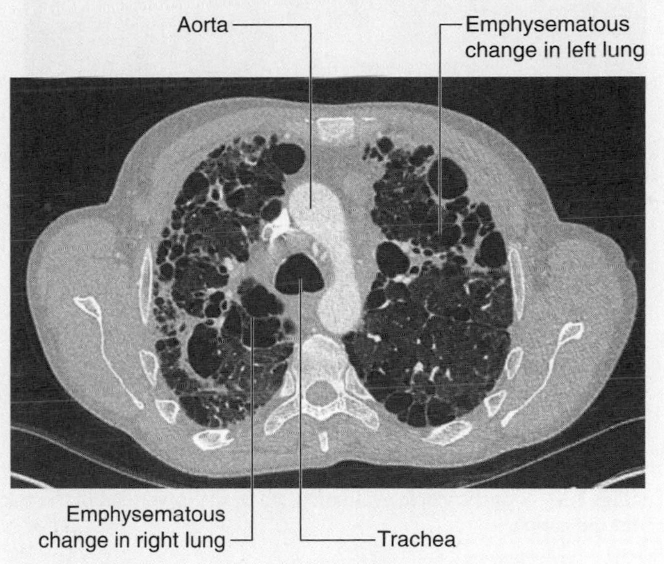


**Fig. 3.52** HRCT of patient with emphysema.

## In the clinic

### Bronchoscopy

Patients who have an endobronchial lesion (i.e., a lesion within a bronchus) may undergo bronchoscopic evaluation of the trachea and its main branches (Fig. 3.53). The bronchoscope is passed through the nose into the oropharynx and is then directed by a control system past the vocal cords into the trachea. The bronchi are inspected and, if necessary, small biopsies are obtained. Bronchoscopy can also be used in combination with ultrasound (a technique known as EBUS, endobronchial ultrasound). An ultrasound probe is inserted through a working channel of the bronchoscope to visualize the airway walls and adjacent structures. EBUS allows an accurate localization of the lesion and therefore provides a higher diagnostic yield. It can be used for sampling of mediastinal and hilar lymph nodes or to assist in transbronchial biopsy of pulmonary nodules.

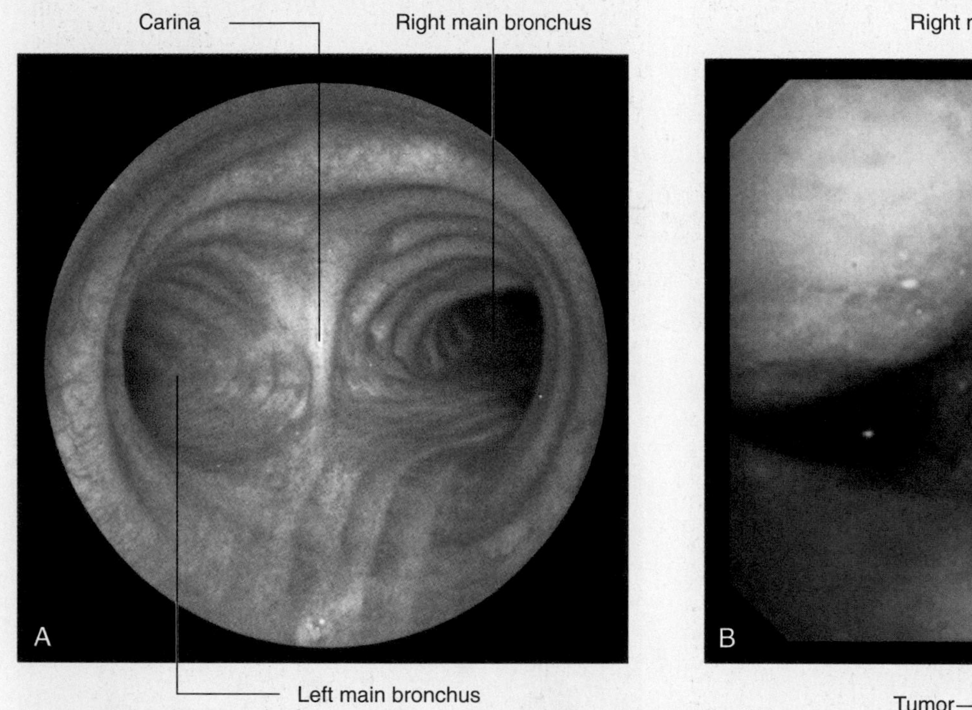

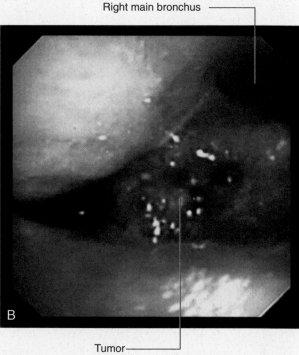

**Fig. 3.53** Bronchoscopic evaluation. **A.** Of the lower end of the trachea and its main branches. **B.** Of tracheal bifurcation showing a tumor at the carina.

## In the clinic

### Lung cancer

It is important to stage lung cancer because the treatment depends on its stage.

If a small malignant nodule is found within the lung, it can sometimes be excised and the prognosis is excellent. Unfortunately, many patients present with a tumor mass that has invaded structures in the mediastinum or the pleurae or has metastasized. The tumor may then be inoperable and is treated with radiotherapy and chemotherapy.

Spread of the tumor is by lymphatics to lymph nodes within the hila, mediastinum, and root of the neck.

A key factor affecting the prognosis and ability to cure the disease is the distant spread of metastases. Imaging methods to assess spread include plain radiography (Fig. 3.54A), computed tomography (CT; Fig. 3.54B,C), and magnetic resonance imaging (MRI). Increasingly, radionuclide studies using fluorodeoxyglucose positron emission tomography (FDG PET; Fig. 3.54D) are being used.

In FDG PET a gamma radiation emitter is attached to a glucose molecule. In areas of high metabolic activity (i.e., the tumor), excessive uptake occurs and is recorded by a gamma camera.

## In the clinic—cont'd

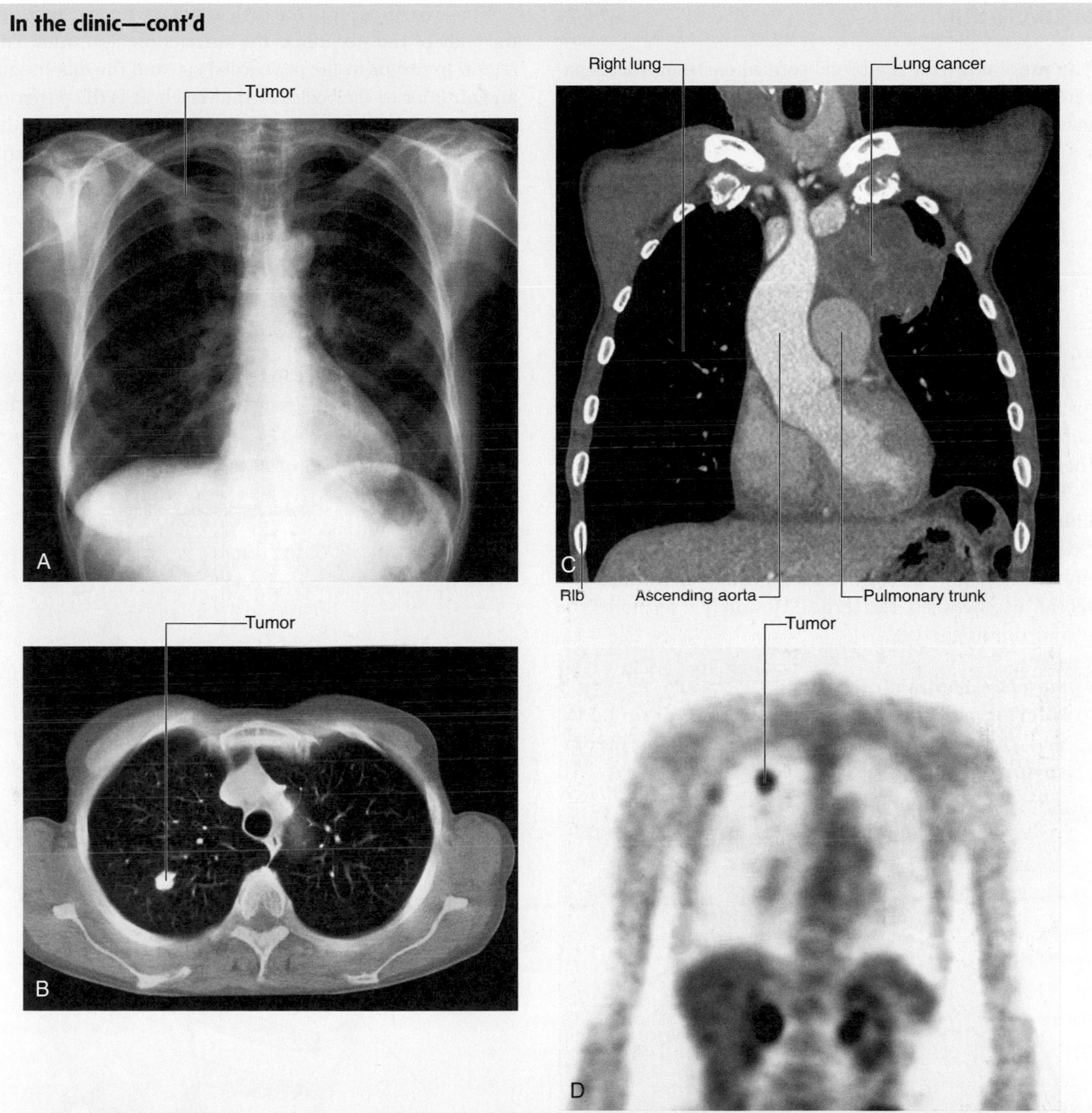


**Fig. 3.54** Imaging of the lungs. **A.** Standard posteroanterior view of the chest showing tumor in upper right lung. **B.** Axial CT image of lungs showing tumor in right lung. **C.** Coronal CT image of lungs showing tumor in left lung extending into mediastinum. **D.** Radionuclide study using FDG PET showing a tumor in the right lung.

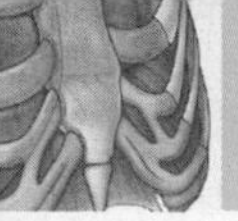



## MEDIASTINUM

The **mediastinum** is a broad central partition that separates the two laterally placed pleural cavities (Fig. 3.55). It extends:

- from the sternum to the bodies of the vertebrae, and
- from the superior thoracic aperture to the diaphragm (Fig. 3.56).

The mediastinum contains the thymus gland, the pericardial sac, the heart, the trachea, and the major arteries and veins.

Additionally, the mediastinum serves as a passageway for structures such as the esophagus, thoracic duct, and various components of the nervous system as they traverse the thorax on their way to the abdomen.

For organizational purposes, the mediastinum is subdivided into several smaller regions. A transverse plane extending from the sternal angle (the junction between the manubrium and the body of the sternum) to the intervertebral disc between vertebrae TIV and TV separates the mediastinum into the:

- **superior mediastinum**, and
- **inferior mediastinum**, which is further partitioned into the **anterior**, **middle**, and **posterior mediastinum** by the pericardial sac.

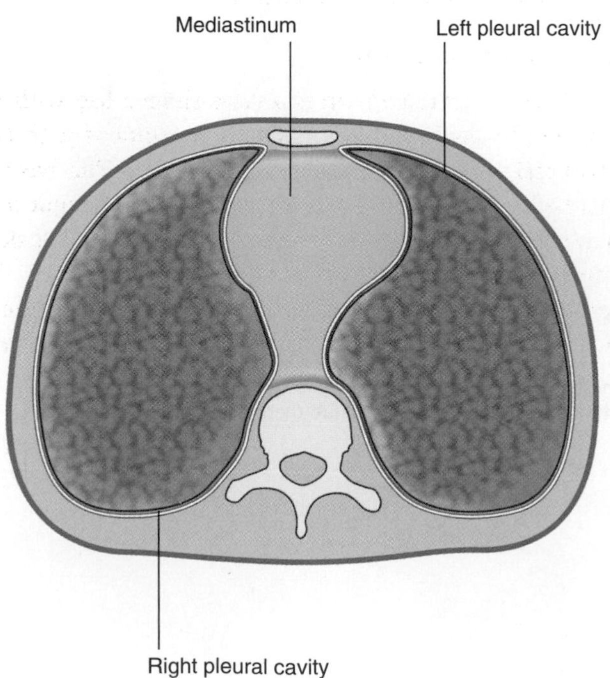


**Fig. 3.55** Cross-section of the thorax showing the position of the mediastinum.

The area anterior to the pericardial sac and posterior to the body of the sternum is the anterior mediastinum. The region posterior to the pericardial sac and the diaphragm and anterior to the bodies of the vertebrae is the posterior mediastinum. The area in the middle, which includes the pericardial sac and its contents, is the middle mediastinum (Fig. 3.57).

### Anterior mediastinum

The **anterior mediastinum** is posterior to the body of the sternum and anterior to the pericardial sac (see Fig. 3.57).

- Its superior boundary is a transverse plane passing from the sternal angle to the intervertebral disc between vertebra TIV and TV, separating it from the superior mediastinum.

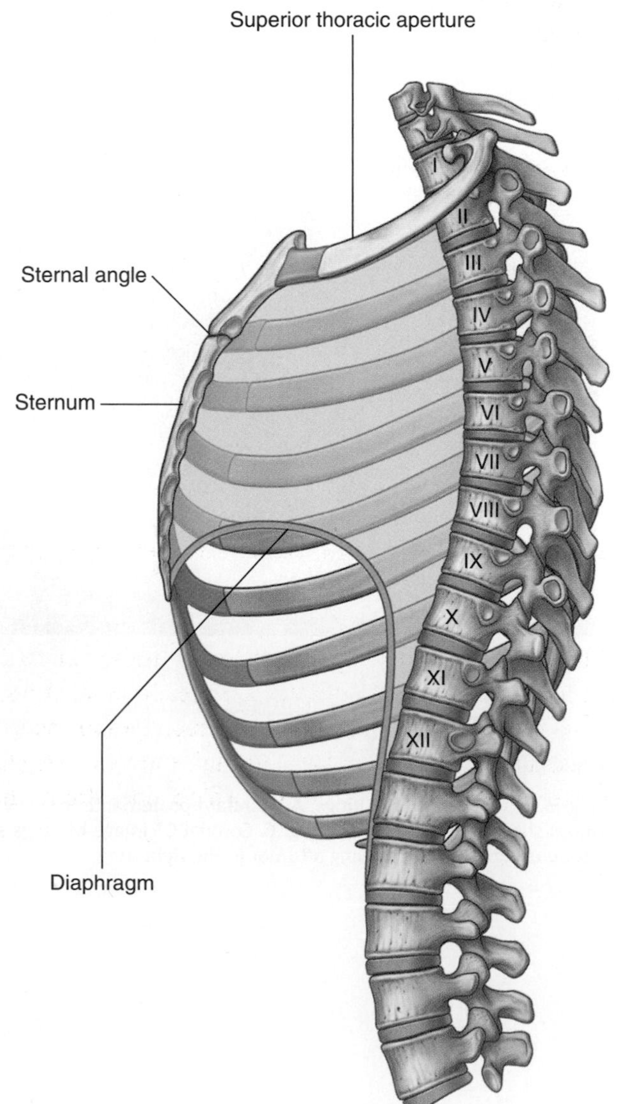


**Fig. 3.56** Lateral view of the mediastinum.

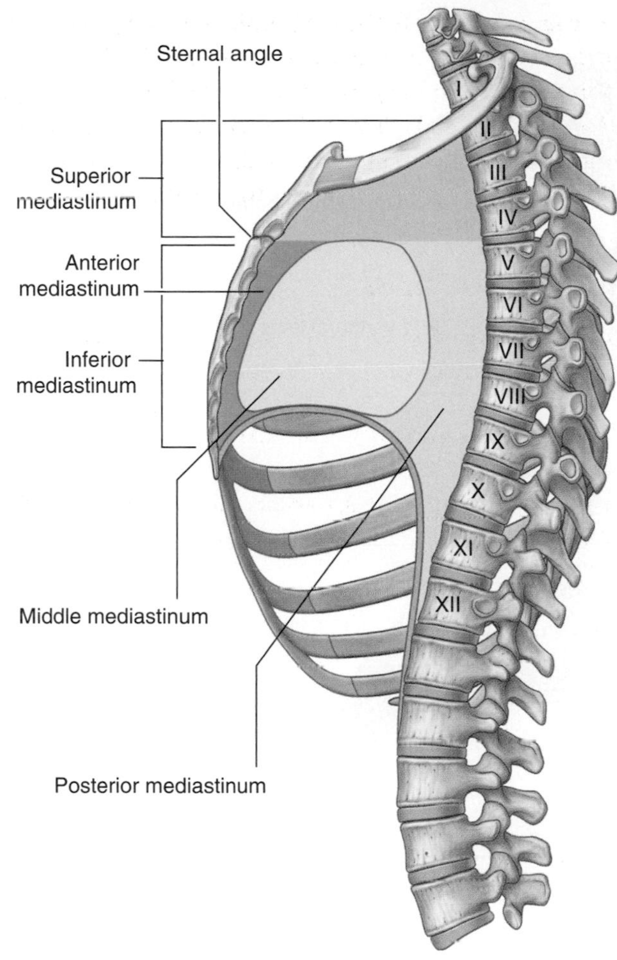


**Fig. 3.57** Subdivisions of the mediastinum.

- Its inferior boundary is the diaphragm.
- Laterally, it is bordered by the mediastinal part of parietal pleura on either side.

The major structure in the anterior mediastinum is an inferior extension of the thymus gland (Fig. 3.58). Also present are fat, connective tissue, lymph nodes, mediastinal branches of the internal thoracic vessels, and sternopericardial ligaments, which pass from the posterior surface of the body of the sternum to the fibrous pericardium.

## Middle mediastinum

The **middle mediastinum** is centrally located in the thoracic cavity. It contains the pericardium, heart, origins of the great vessels, various nerves, and smaller vessels.

### Pericardium

The **pericardium** is a fibroserous sac surrounding the heart and the roots of the great vessels. It consists of two components, the fibrous pericardium and the serous pericardium (Fig. 3.59).

The **fibrous pericardium** is a tough connective tissue outer layer that defines the boundaries of the middle mediastinum. The **serous pericardium** is thin and consists of two parts:

- The **parietal layer** of serous pericardium lines the inner surface of the fibrous pericardium.
- The **visceral layer** (**epicardium**) of serous pericardium adheres to the heart and forms its outer covering.

The parietal and visceral layers of serous pericardium are continuous at the roots of the great vessels. The narrow space created between the two layers of serous pericardium, containing a small amount of fluid, is the **pericardial cavity**. This potential space allows for the relatively uninhibited movement of the heart.

#### Fibrous pericardium

The **fibrous pericardium** is a cone-shaped bag with its base on the diaphragm and its apex continuous with the **adventitia** of the great vessels (Fig. 3.59). The base is attached to the **central tendon of the diaphragm** and to a small muscular area of the diaphragm on the left side. Anteriorly, it is attached to the posterior surface of the sternum by **sternopericardial ligaments**. These attachments help to retain the heart in its position in the thoracic cavity. The sac also limits cardiac distention.

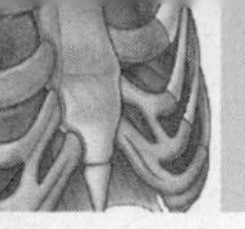

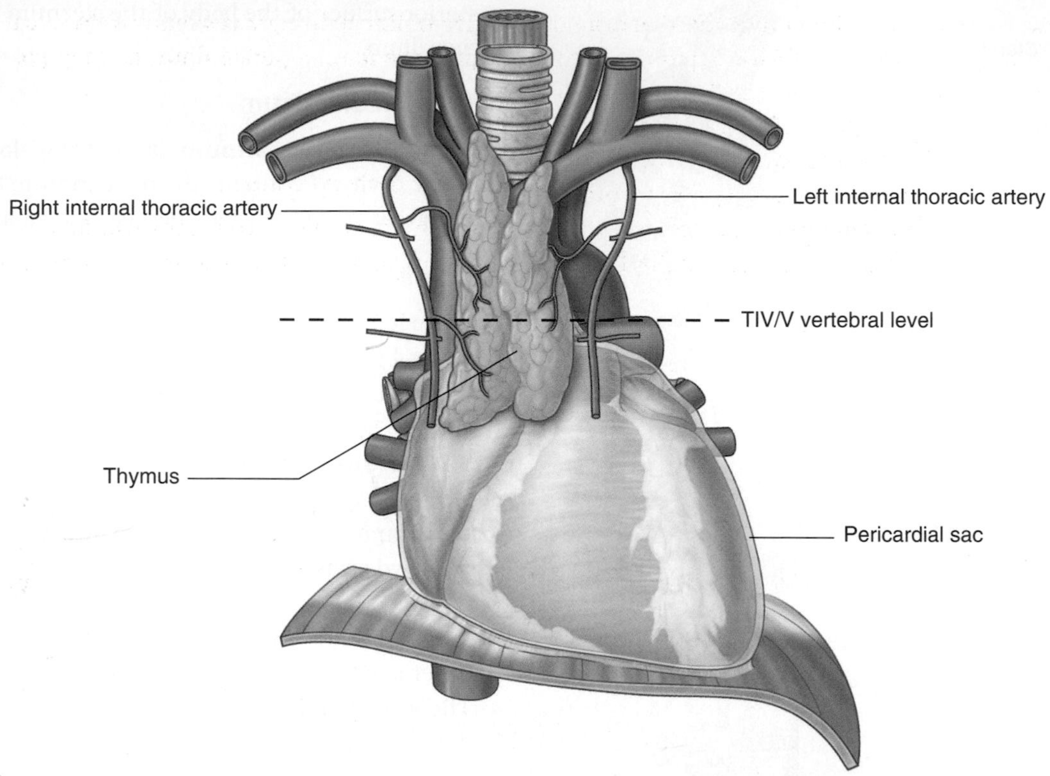


**Fig. 3.58** Thymus.

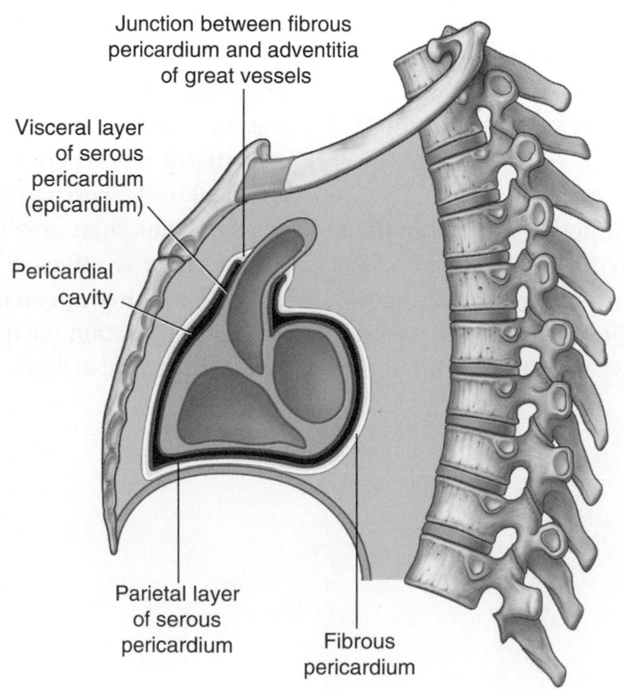


**Fig. 3.59** Sagittal section of the pericardium.

The phrenic nerves, which innervate the diaphragm and originate from spinal cord levels C3 to C5, pass through the fibrous pericardium and innervate the fibrous pericardium as they travel from their point of origin to their final destination (Fig. 3.60). Their location, within the fibrous pericardium, is directly related to the embryological origin of the diaphragm and the changes that occur during the formation of the pericardial cavity. Similarly, the **pericardiacophrenic vessels** are also located within and supply the fibrous pericardium as they pass through the thoracic cavity.

### Serous pericardium

The parietal layer of serous pericardium is continuous with the visceral layer of serous pericardium around the roots

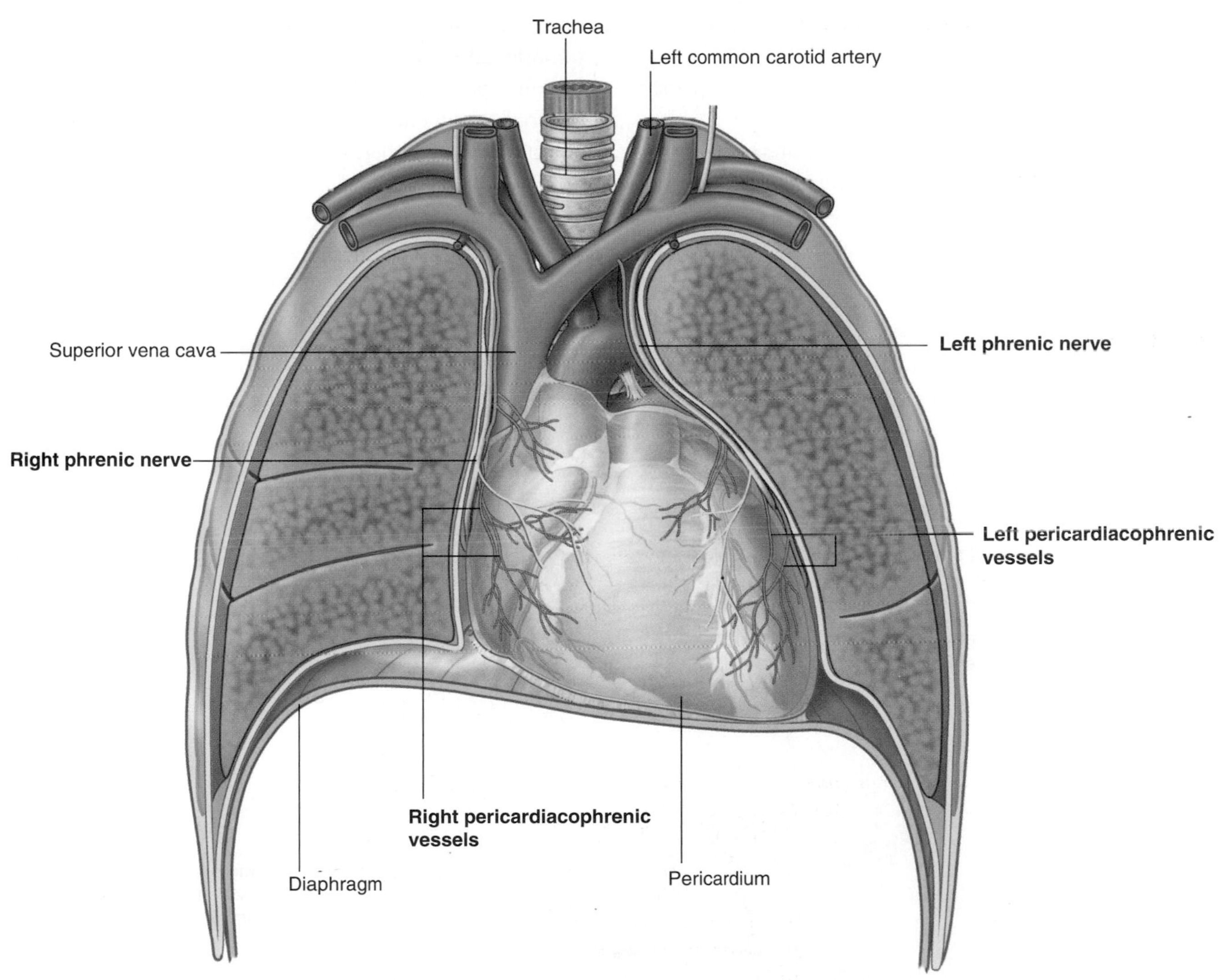


**Fig. 3.60** Phrenic nerves and pericardiacophrenic vessels.

of the great vessels. These reflections of serous pericardium (Fig. 3.61) occur in two locations:

- one superiorly, surrounding the arteries—the aorta and the pulmonary trunk;
- the second more posteriorly, surrounding the veins—the superior and inferior vena cava and the pulmonary veins.

The zone of reflection surrounding the veins is J-shaped, and the cul-de-sac formed within the J, posterior to the left atrium, is the **oblique pericardial sinus**.

A passage between the two sites of reflected serous pericardium is the **transverse pericardial sinus**. This sinus lies posterior to the ascending aorta and the pulmonary trunk, anterior to the superior vena cava, and superior to the left atrium.

When the pericardium is opened anteriorly during surgery, a finger placed in the transverse sinus separates arteries from veins. A hand placed under the apex of the heart and moved superiorly slips into the oblique sinus.

### Vessels and nerves

The pericardium is supplied by branches from the internal thoracic, pericardiacophrenic, musculophrenic, and inferior phrenic arteries, and the thoracic aorta.

Veins from the pericardium enter the azygos system of veins and the internal thoracic and superior phrenic veins.

Nerves supplying the pericardium arise from the vagus nerve [X], the sympathetic trunks, and the phrenic nerves.

It is important to note that the source of somatic sensation (pain) from the parietal pericardium is carried by somatic afferent fibers in the phrenic nerves. For this reason, "pain" related to a pericardial problem may be referred to the supraclavicular region of the shoulder or lateral neck area dermatomes for spinal cord segments C3, C4, and C5.

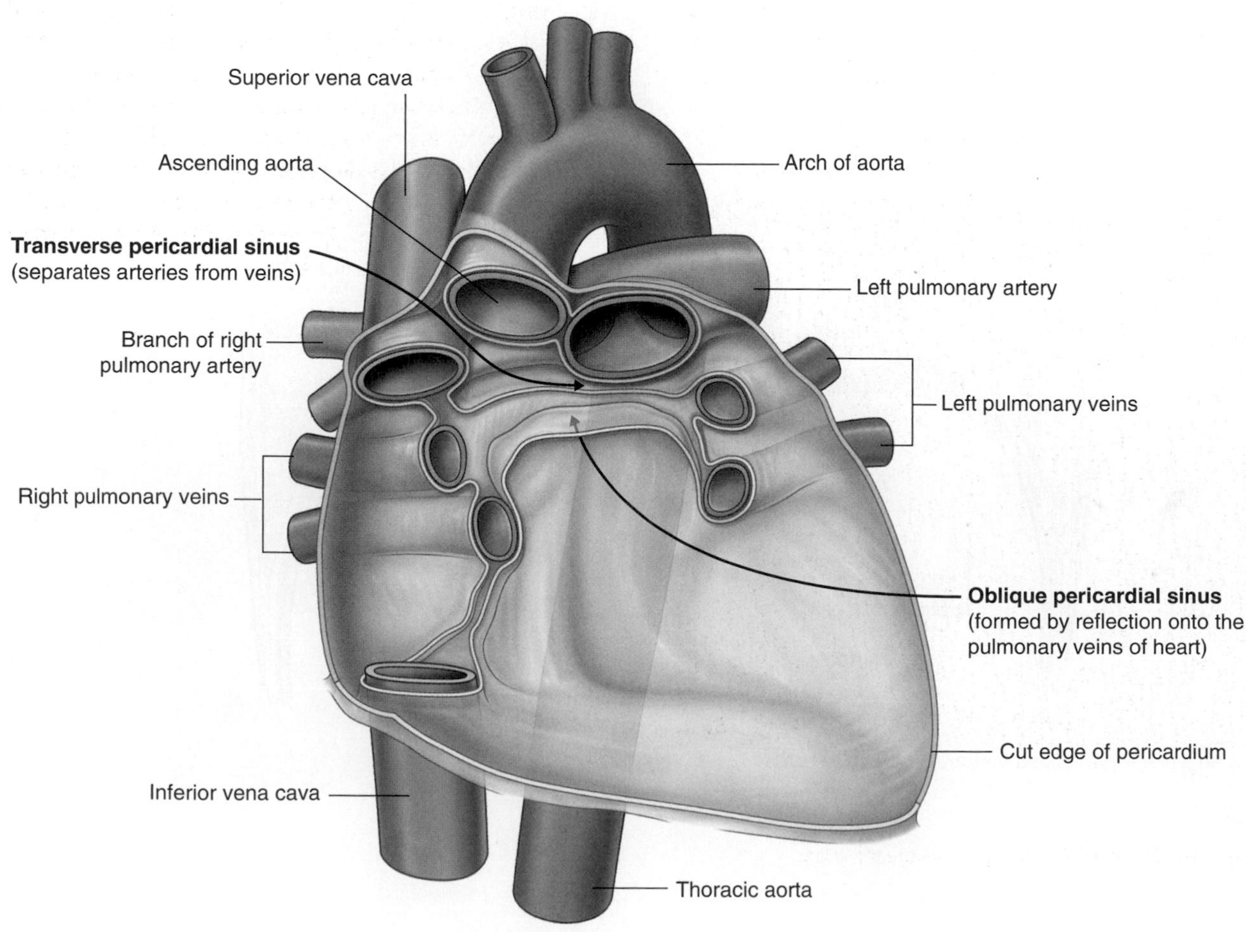


**Fig. 3.61** Posterior portion of pericardial sac showing reflections of serous pericardium.

## In the clinic

### Pericarditis

Pericarditis is an inflammatory condition of the pericardium. Common causes are viral and bacterial infections, systemic illnesses (e.g., chronic renal failure), and after myocardial infarction.

Pericarditis must be distinguished from myocardial infarction because the treatment and prognosis are quite different. As in patients with myocardial infarction, patients with pericarditis complain of continuous central chest pain that may radiate to one or both arms. Unlike myocardial infarction, however, the pain from pericarditis may be relieved by sitting forward. An electrocardiogram (ECG) is used to help differentiate between the two conditions. It usually shows diffuse ST elevation. Echocardiography can also be performed if there is clinical or radiographic suspicion of pericardial effusion.

## In the clinic

### Pericardial effusion

Normally, only a tiny amount of fluid is present between the visceral and parietal layers of the serous pericardium. In certain situations, this space can be filled with excess fluid (pericardial effusion) (Fig. 3.62).

Because the fibrous pericardium is a "relatively fixed" structure that cannot expand easily, a rapid accumulation of excess fluid within the pericardial sac compresses the heart (cardiac tamponade), resulting in biventricular failure. Removing the fluid with a needle inserted into the pericardial sac can relieve the symptoms.

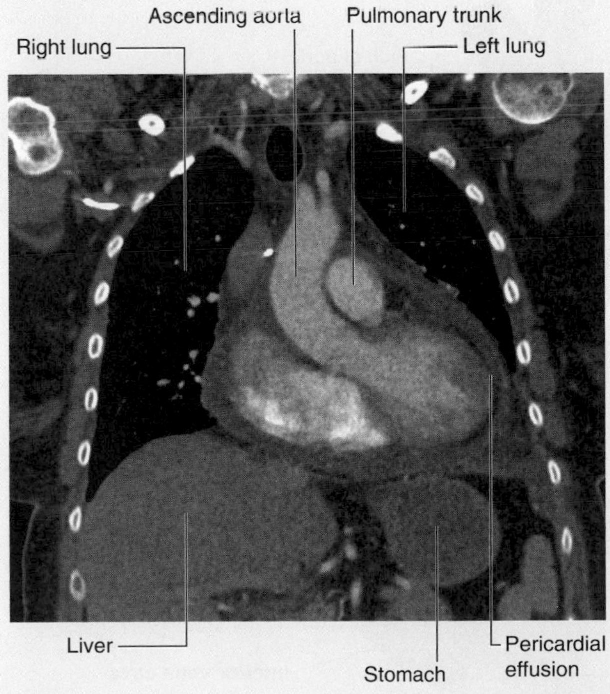


**Fig. 3.62** Coronal CT showing pericardial effusion.

## In the clinic

### Constrictive pericarditis

Abnormal thickening of the pericardial sac (constrictive pericarditis), which usually involves only the parietal pericardium, but can also less frequently involve the visceral layer, can compress the heart, impairing heart function and resulting in heart failure. It can present acutely but often results in a chronic condition when thickened pericardium with fibrin deposits causes pericardial inflammation, leading to chronic scarring and pericardial calcification. As a result, normal filling during the diastolic phase of the cardiac cycle is severely restricted. The diagnosis is made by inspecting the jugular venous pulse in the neck. In normal individuals, the jugular venous pulse drops on inspiration. In patients with constrictive pericarditis, the reverse happens and this is called Kussmaul's sign. Treatment often involves surgical opening of the pericardial sac.

## Heart

### Cardiac orientation

The general shape and orientation of the heart are that of a pyramid that has fallen over and is resting on one of its sides. Placed in the thoracic cavity, the apex of this pyramid projects forward, downward, and to the left, whereas the base is opposite the apex and faces in a posterior direction (Fig. 3.63). The sides of the pyramid consist of:

- a diaphragmatic (inferior) surface on which the pyramid rests,
- an anterior (sternocostal) surface oriented anteriorly,
- a right pulmonary surface, and
- a left pulmonary surface.

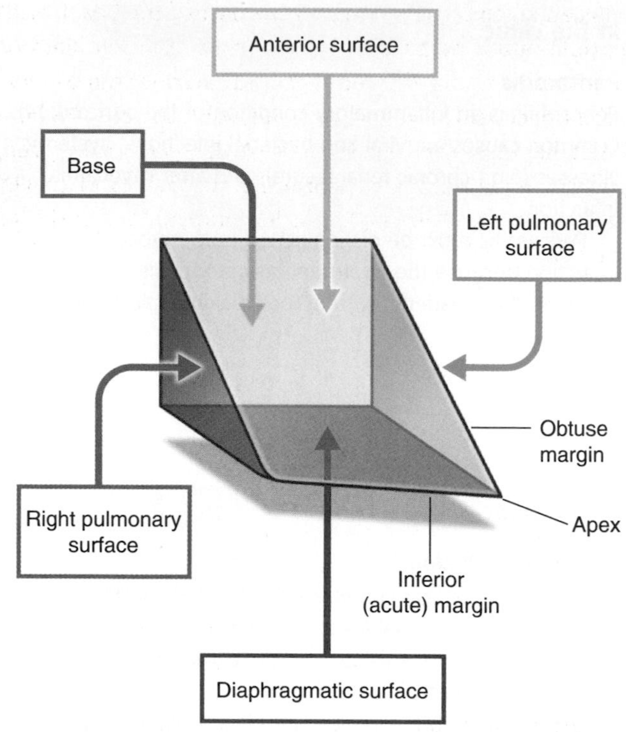


**Fig. 3.63** Schematic illustration of the heart showing orientation, surfaces, and margins.

### Base (posterior surface) and apex

The **base of the heart** is quadrilateral and directed posteriorly. It consists of:

- the left atrium,
- a small portion of the right atrium, and
- the proximal parts of the great veins (superior and inferior venae cavae and the pulmonary veins) (Fig. 3.64).

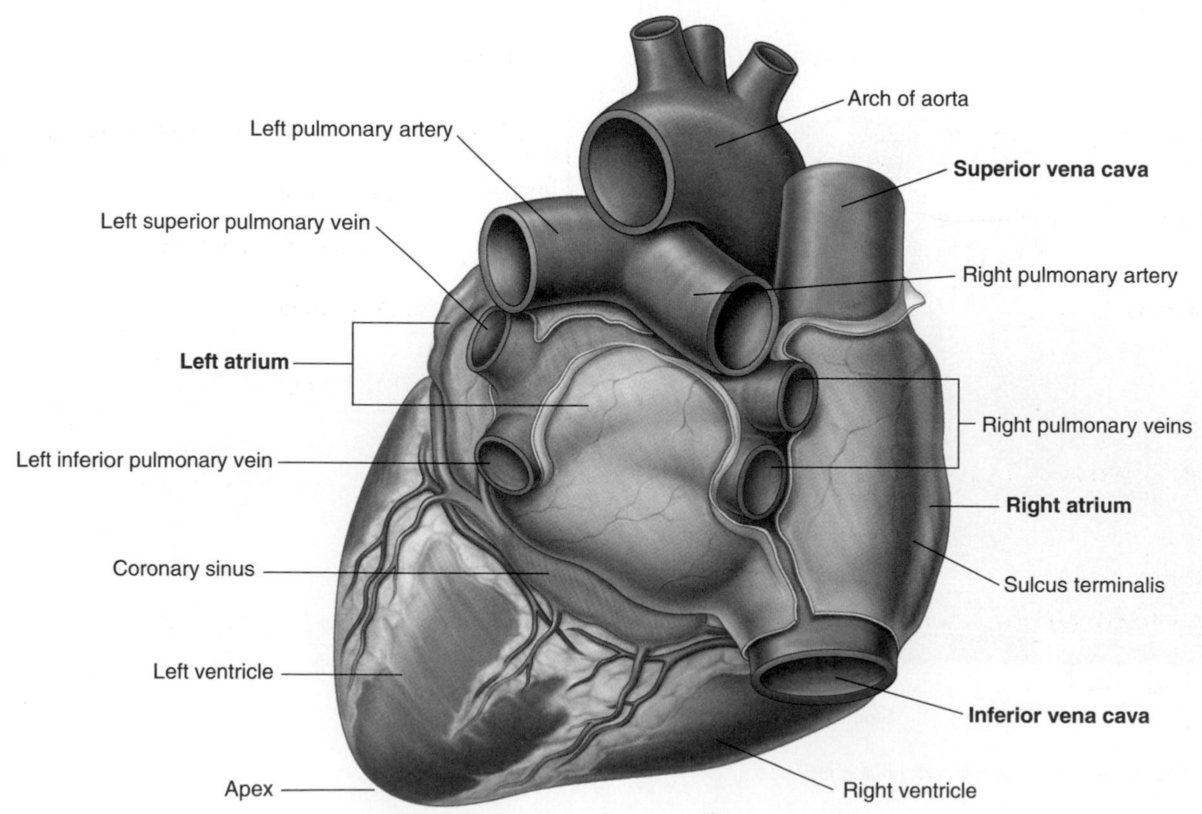


**Fig. 3.64** Base of the heart.

Because the great veins enter the base of the heart, with the pulmonary veins entering the right and left sides of the left atrium and the superior and inferior venae cavae at the upper and lower ends of the right atrium, the base of the heart is fixed posteriorly to the pericardial wall, opposite the bodies of vertebrae TV to TVIII (TVI to TIX when standing). The esophagus lies immediately posterior to the base.

From the base the heart projects forward, downward, and to the left, ending in the apex. The **apex of the heart** is formed by the inferolateral part of the left ventricle (Fig. 3.65) and is positioned deep to the left fifth intercostal space, 8 to 9 cm from the midsternal line.

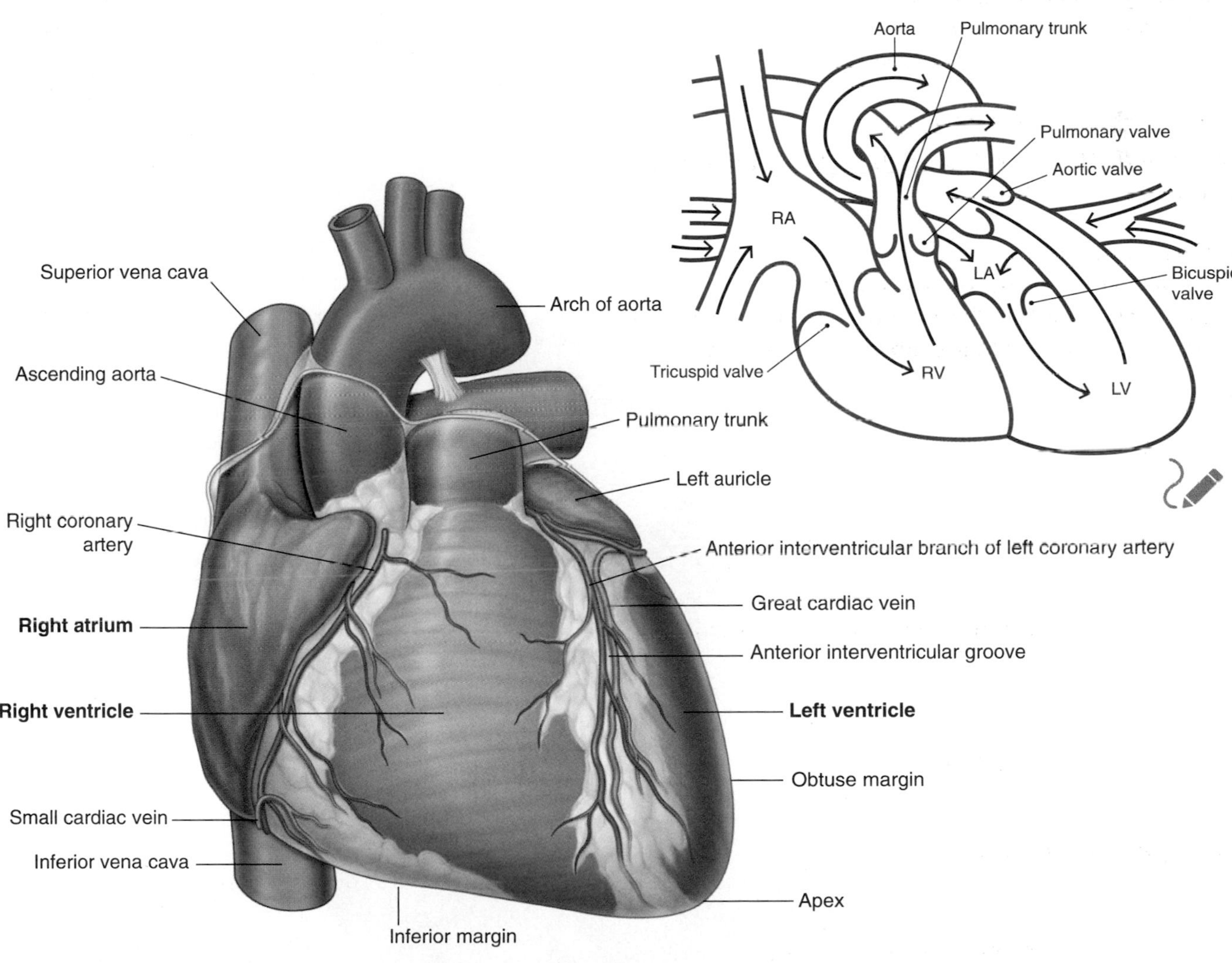


**Fig. 3.65** Anterior surface of the heart.

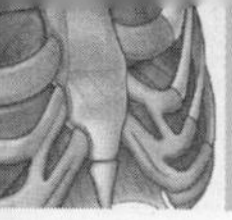



### Surfaces of the heart

The **anterior surface** faces anteriorly and consists mostly of the right ventricle, with some of the right atrium on the right and some of the left ventricle on the left (Fig. 3.65).

The heart in the anatomical position rests on the **diaphragmatic surface**, which consists of the left ventricle and a small portion of the right ventricle separated by the posterior interventricular groove (Fig. 3.66). This surface faces inferiorly, rests on the diaphragm, is separated from the base of the heart by the coronary sinus, and extends from the base to the apex of the heart.

The **left pulmonary surface** faces the left lung, is broad and convex, and consists of the left ventricle and a portion of the left atrium (Fig. 3.66).

The **right pulmonary surface** faces the right lung, is broad and convex, and consists of the right atrium (Fig. 3.66).

### Margins and borders

Some general descriptions of cardiac orientation refer to right, left, inferior (acute), and obtuse margins:

- The **right** and **left margins** are the same as the right and left pulmonary surfaces of the heart.
- The **inferior margin** is defined as the sharp edge between the anterior and diaphragmatic surfaces of the heart (Figs 3.63 and 3.65)—it is formed mostly by the right ventricle and a small portion of the left ventricle near the apex.

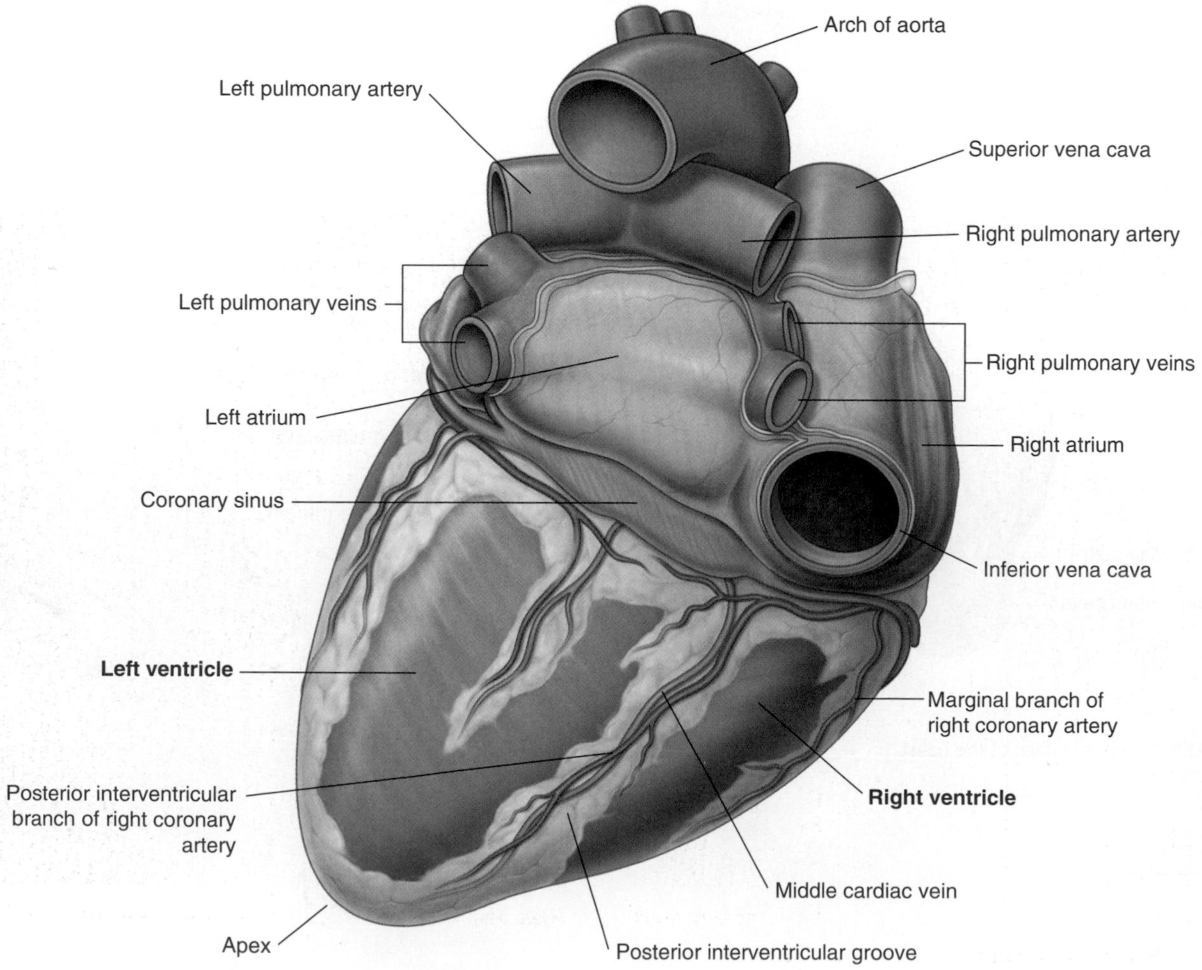

**Fig. 3.66** Diaphragmatic surface of the heart.

- The **obtuse margin** separates the anterior and left pulmonary surfaces (Fig. 3.63)—it is round and extends from the left auricle to the cardiac apex (Fig. 3.65), and is formed mostly by the left ventricle and superiorly by a small portion of the left auricle.

For radiological evaluations, a thorough understanding of the structures defining the cardiac borders is critical. The right border in a standard posteroanterior view consists of the superior vena cava, the right atrium, and the inferior vena cava (Fig. 3.67A). The left border in a similar view consists of the arch of the aorta, the pulmonary trunk, left auricle, and the left ventricle. The inferior border in this radiological study consists of the right ventricle and the left ventricle at the apex. In lateral views, the right ventricle is seen anteriorly, and the left atrium is visualized posteriorly (Fig. 3.67B).

### External sulci

Internal partitions divide the heart into four chambers (i.e., two atria and two ventricles) and produce surface or external grooves referred to as sulci.

- The **coronary sulcus** circles the heart, separating the atria from the ventricles (Fig. 3.68). As it circles the heart, it contains the right coronary artery, the small cardiac vein, the coronary sinus, and the circumflex branch of the left coronary artery.
- The **anterior** and **posterior interventricular sulci** separate the two ventricles—the anterior interventricular sulcus is on the anterior surface of the heart and contains the anterior interventricular artery and the great cardiac vein, and the posterior interventricular sulcus is on the diaphragmatic surface of the heart and contains the posterior interventricular artery and the middle cardiac vein.

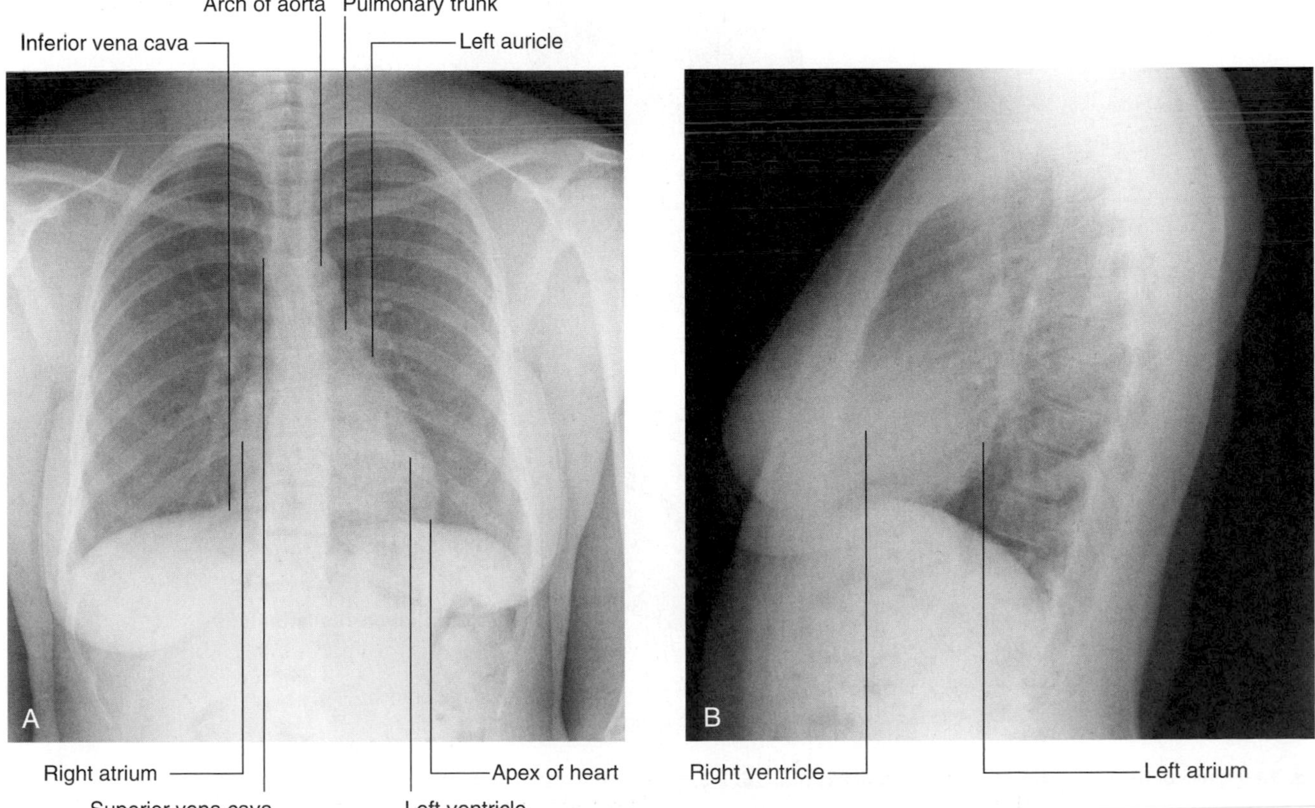


**Fig. 3.67** Chest radiographs. **A.** Standard posteroanterior view of the chest. **B.** Standard lateral view of the heart.

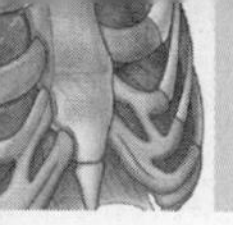



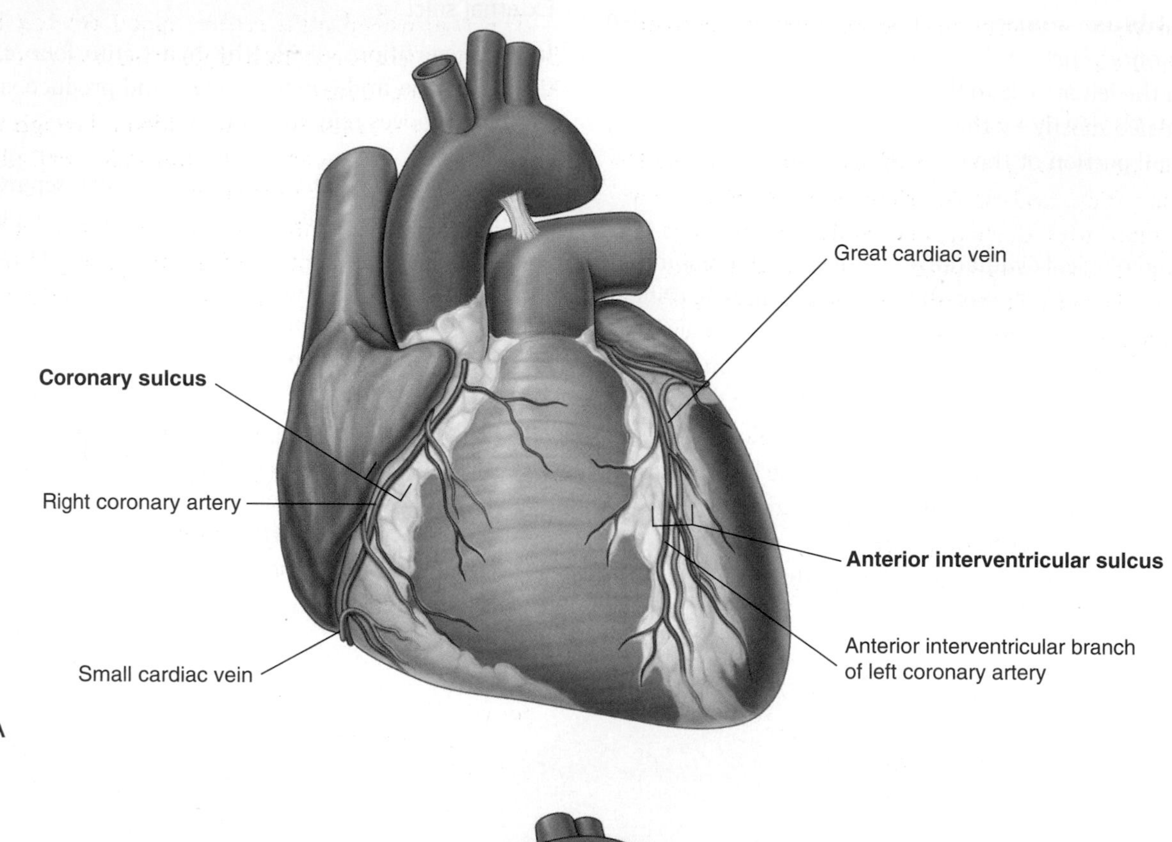


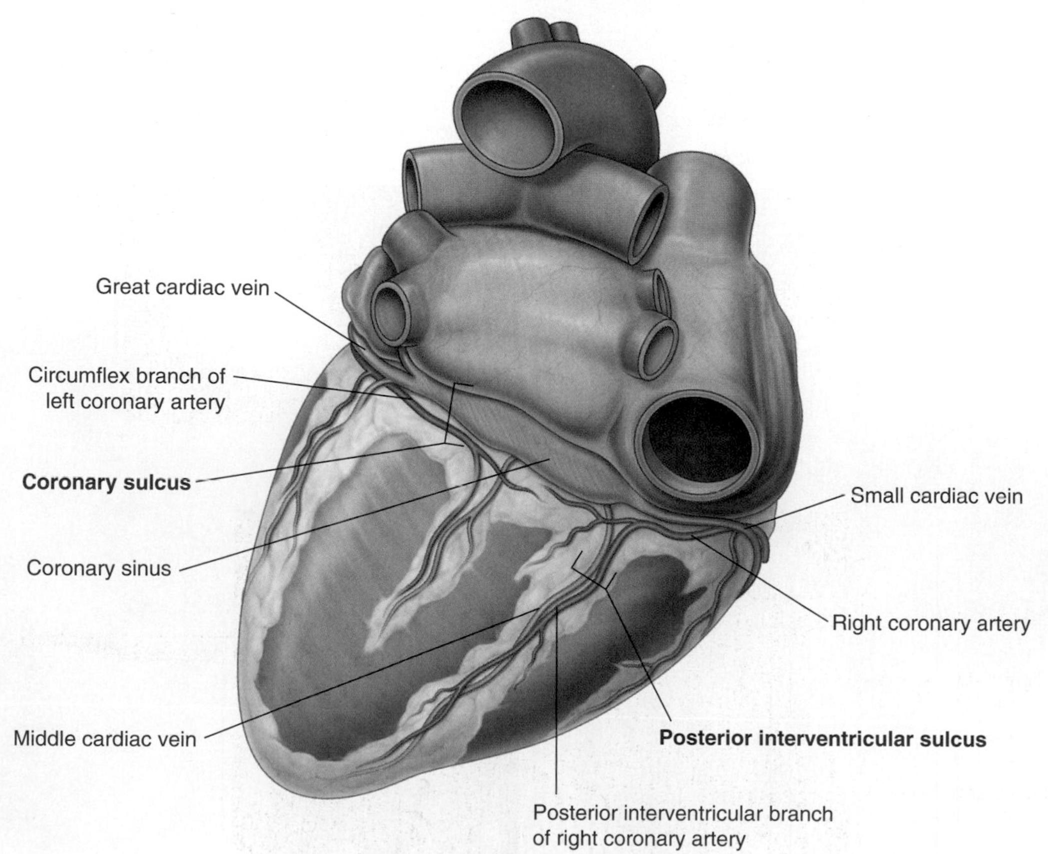


**Fig. 3.68** Sulci of the heart. **A.** Anterior surface of the heart. **B.** Diaphragmatic surface and base of the heart.

These sulci are continuous inferiorly, just to the right of the apex of the heart.

### Cardiac chambers

The heart functionally consists of two pumps separated by a partition (Fig. 3.69A). The right pump receives deoxygenated blood from the body and sends it to the lungs. The left pump receives oxygenated blood from the lungs and sends it to the body. Each pump consists of an atrium and a ventricle separated by a valve.

The thin-walled atria receive blood coming into the heart, whereas the relatively thick-walled ventricles pump blood out of the heart.

More force is required to pump blood through the body than through the lungs, so the muscular wall of the left ventricle is thicker than the right.

Interatrial, interventricular, and atrioventricular septa separate the four chambers of the heart (Fig. 3.69B). The internal anatomy of each chamber is critical to its function.

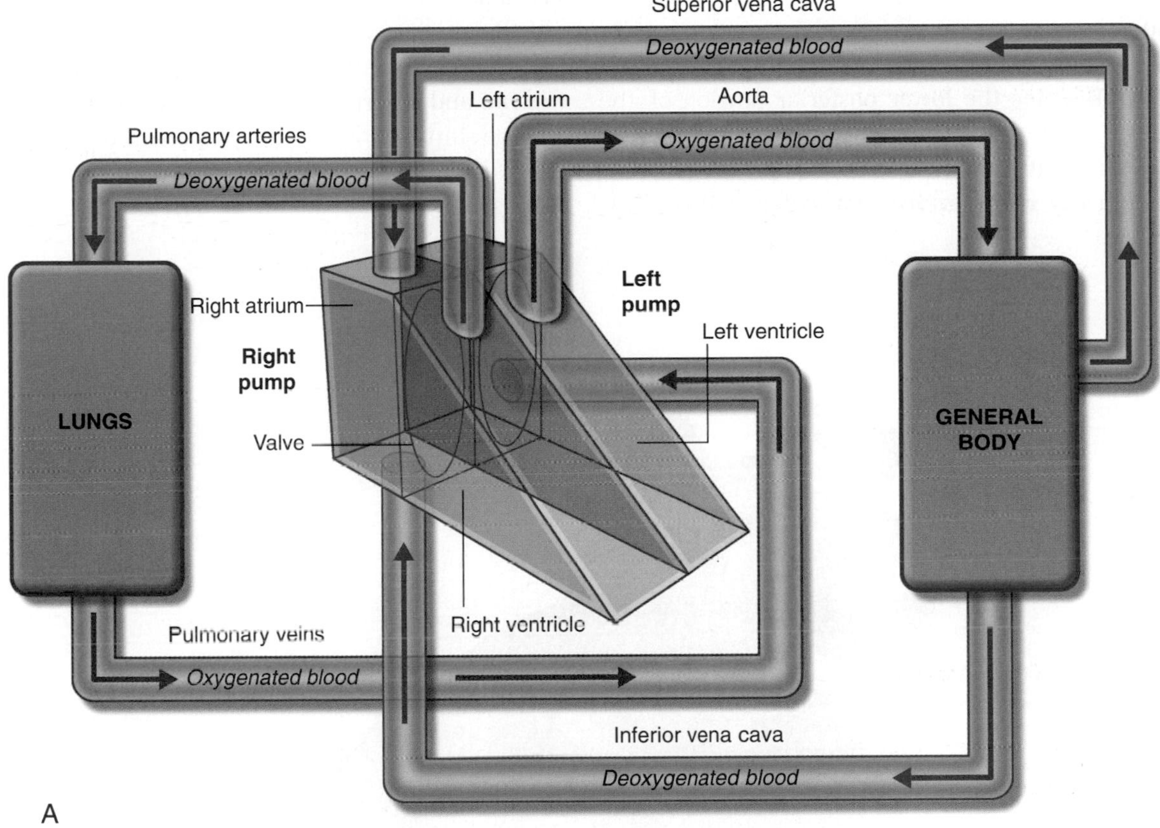


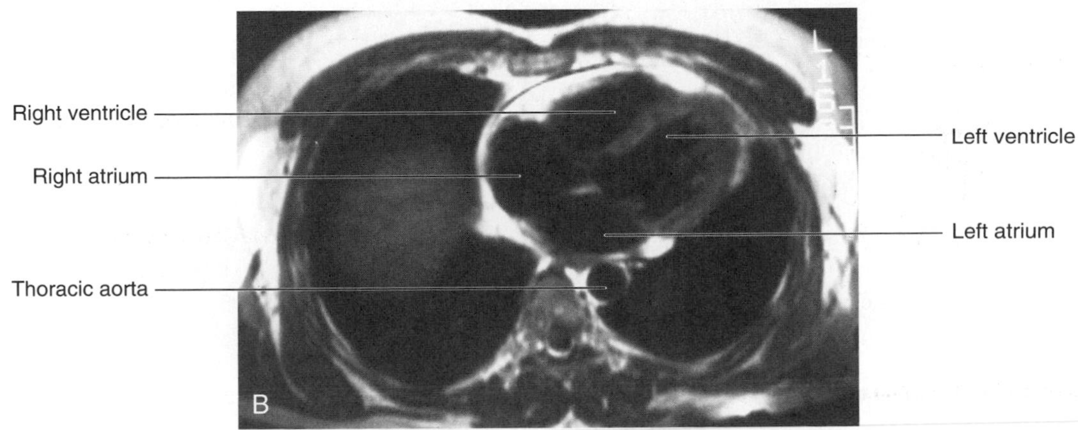


**Fig. 3.69 A.** The heart has two pumps. **B.** Magnetic resonance image of midthorax showing all four chambers and septa.

### Right atrium

In the anatomical position, the right border of the heart is formed by the **right atrium**. This chamber also contributes to the right portion of the heart's anterior surface.

Blood returning to the right atrium enters through one of three vessels. These are:

- the superior and inferior venae cavae, which together deliver blood to the heart from the body; and
- the coronary sinus, which returns blood from the walls of the heart itself.

The superior vena cava enters the upper posterior portion of the right atrium, and the inferior vena cava and coronary sinus enter the lower posterior portion of the right atrium.

From the right atrium, blood passes into the right ventricle through the **right atrioventricular orifice**. This opening faces forward and medially and is closed during ventricular contraction by the tricuspid valve.

The interior of the right atrium is divided into two continuous spaces. Externally, this separation is indicated by a shallow, vertical groove (the **sulcus terminalis cordis**), which extends from the right side of the opening of the superior vena cava to the right side of the opening of the inferior vena cava. Internally, this division is indicated by the **crista terminalis** (Fig. 3.70), which is a smooth, muscular ridge that begins on the roof of the atrium just in front of the opening of the superior vena cava and extends down the lateral wall to the anterior lip of the inferior vena cava.

The space posterior to the crista is the **sinus of venae cavae** and is derived embryologically from the right horn of the sinus venosus. This component of the right atrium has smooth, thin walls, and both venae cavae empty into this space.

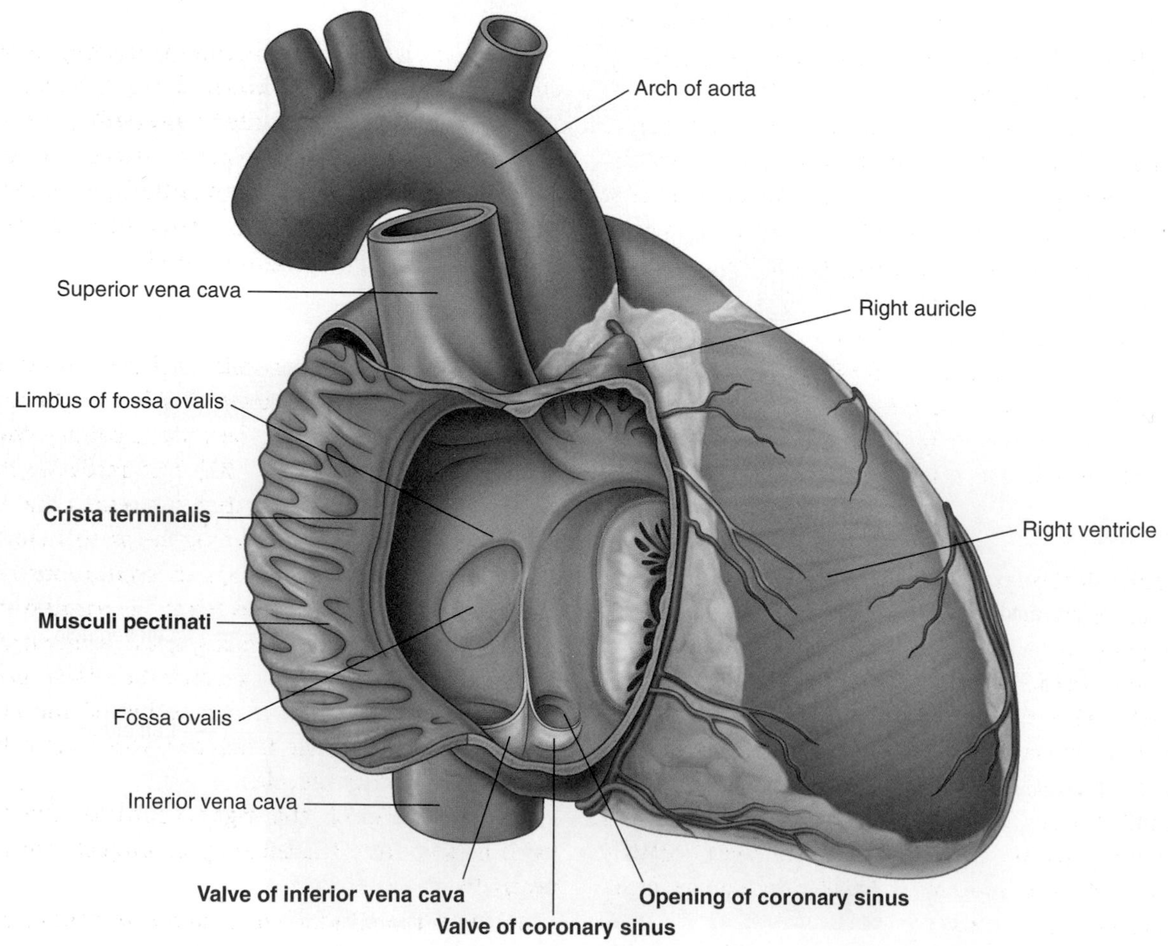


**Fig. 3.70** Internal view of right atrium.

The space anterior to the crista, including the **right auricle**, is sometimes referred to as the **atrium proper**. This terminology is based on its origin from the embryonic primitive atrium. Its walls are covered by ridges called the **musculi pectinati** (**pectinate muscles**), which fan out from the crista like the "teeth of a comb." These ridges are also found in the right auricle, which is an ear-like, conical, muscular pouch that externally overlaps the ascending aorta.

An additional structure in the right atrium is the **opening of the coronary sinus**, which receives blood from most of the cardiac veins and opens medially to the **opening of the inferior vena cava**. Associated with these openings are small folds of tissue derived from the valve of the embryonic sinus venosus (the **valve of the coronary sinus** and the **valve of inferior vena cava**, respectively). During development, the valve of the inferior vena cava helps direct incoming oxygenated blood through the foramen ovale and into the left atrium.

Separating the right atrium from the left atrium is the **interatrial septum**, which faces forward and to the right because the left atrium lies posteriorly and to the left of the right atrium. A depression is clearly visible in the septum just above the orifice of the inferior vena cava. This is the **fossa ovalis** (**oval fossa**), with its prominent margin, the **limbus fossa ovalis** (**border of the oval fossa**).

The fossa ovalis marks the location of the embryonic **foramen ovale**, which is an important part of fetal circulation. The foramen ovale allows oxygenated blood entering the right atrium through the inferior vena cava to pass directly to the left atrium and so bypass the lungs, which are nonfunctional before birth.

Finally, numerous small openings—the **openings of the smallest cardiac veins** (the **foramina of the venae cordis minimae**)—are scattered along the walls of the right atrium. These are small veins that drain the myocardium directly into the right atrium.

### Right ventricle

In the anatomical position, the right ventricle forms most of the anterior surface of the heart and a portion of the diaphragmatic surface. The right atrium is to the right of the right ventricle and the right ventricle is located in front of and to the left of the right atrioventricular orifice. Blood entering the right ventricle from the right atrium therefore moves in a horizontal and forward direction.

The outflow tract of the right ventricle, which leads to the pulmonary trunk, is the **conus arteriosus** (**infundibulum**). This area has smooth walls and derives from the embryonic bulbus cordis.

The walls of the inflow portion of the right ventricle have numerous muscular, irregular structures called **trabeculae carneae** (Fig. 3.71). Most of these are either attached to the ventricular walls throughout their length, forming ridges, or attached at both ends, forming bridges.

A few trabeculae carneae (**papillary muscles**) have only one end attached to the ventricular surface, while the other end serves as the point of attachment for tendon-like fibrous cords (the **chordae tendineae**), which connect to the free edges of the cusps of the tricuspid valve.

There are three papillary muscles in the right ventricle. Named relative to their point of origin on the ventricular surface, they are the anterior, posterior, and septal papillary muscles:

- The **anterior papillary muscle** is the largest and most constant papillary muscle, and arises from the anterior wall of the ventricle.
- The **posterior papillary muscle** may consist of one, two, or three structures, with some chordae tendineae arising directly from the ventricular wall.
- The **septal papillary muscle** is the most inconsistent papillary muscle, being either small or absent, with chordae tendineae emerging directly from the septal wall.

A single specialized trabeculum, the **septomarginal trabecula** (**moderator band**), forms a bridge between the lower portion of the **interventricular septum** and the base of the anterior papillary muscle. The septomarginal trabecula carries a portion of the cardiac conduction system, the right bundle of the atrioventricular bundle, to the anterior wall of the right ventricle.

### Tricuspid valve

The right atrioventricular orifice is closed during ventricular contraction by the **tricuspid valve** (**right atrioventricular valve**), so named because it usually consists of three cusps or leaflets (Fig. 3.71). The base of each cusp is secured to the fibrous ring that surrounds the atrioventricular orifice. This fibrous ring helps to maintain the shape of the opening. The cusps are continuous with each other near their bases at sites termed **commissures**.

The naming of the three cusps, the **anterior**, **septal**, and **posterior cusps**, is based on their relative position in the right ventricle. The free margins of the cusps are attached to the chordae tendineae, which arise from the tips of the papillary muscles.

During filling of the right ventricle, the tricuspid valve is open, and the three cusps project into the right ventricle.

Without the presence of a compensating mechanism, when the ventricular musculature contracts, the valve cusps would be forced upward with the flow of blood and

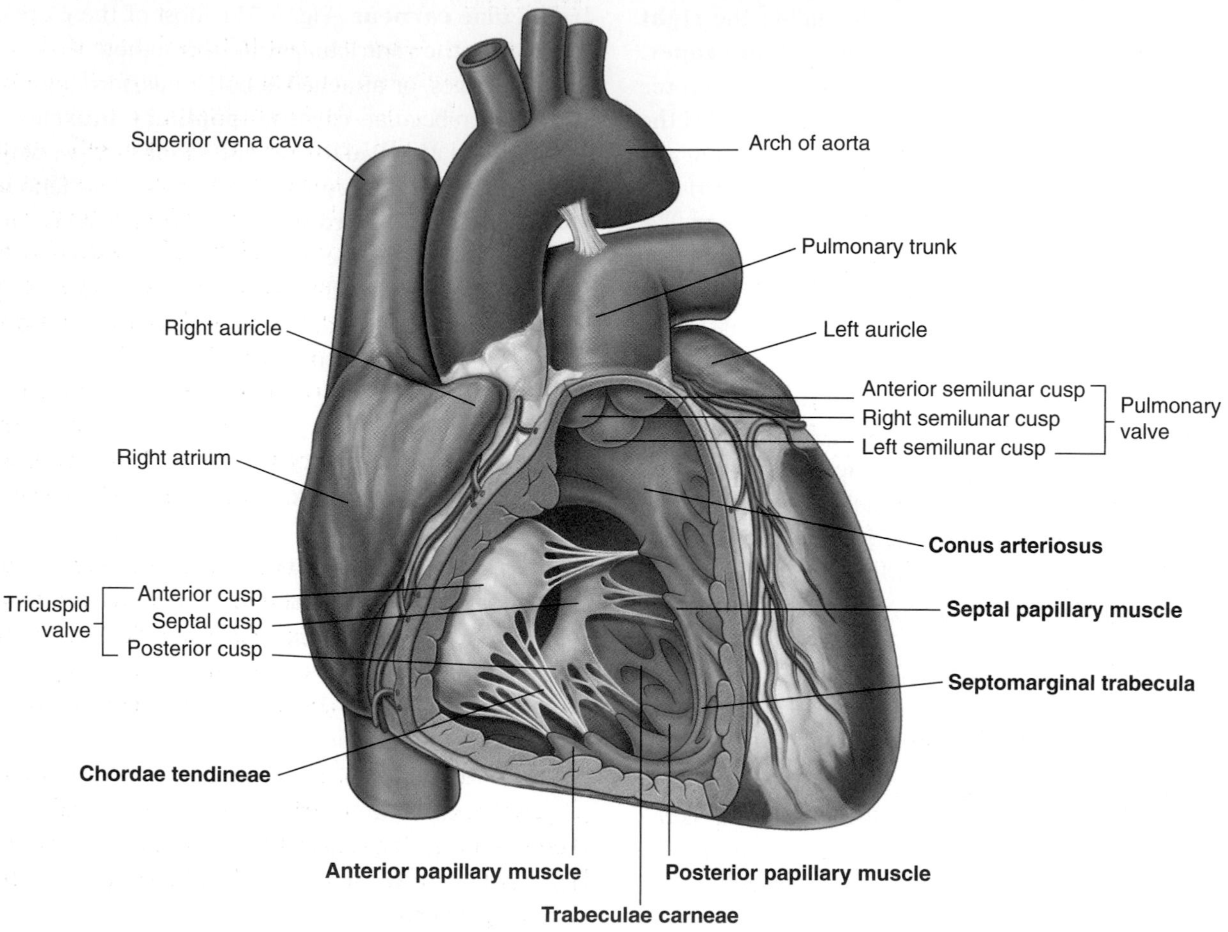

**Fig. 3.71** Internal view of the right ventricle.

blood would move back into the right atrium. However, contraction of the papillary muscles attached to the cusps by chordae tendineae prevents the cusps from being everted into the right atrium.

Simply put, the papillary muscles and associated chordae tendineae keep the valves closed during the dramatic changes in ventricular size that occur during contraction.

In addition, chordae tendineae from two papillary muscles attach to each cusp. This helps prevent separation of the cusps during ventricular contraction. Proper closing of the tricuspid valve causes blood to exit the right ventricle and move into the pulmonary trunk.

Necrosis of a papillary muscle following a myocardial infarction (heart attack) may result in prolapse of the related valve.

### Pulmonary valve

At the apex of the infundibulum, the outflow tract of the right ventricle, the opening into the pulmonary trunk is closed by the **pulmonary valve** (Fig. 3.71), which consists of three **semilunar cusps** with free edges projecting upward into the lumen of the pulmonary trunk. The free superior edge of each cusp has a middle, thickened portion, the **nodule of the semilunar cusp**, and a thin lateral portion, the **lunula of the semilunar cusp** (Fig. 3.72).

The cusps are named the **left**, **right**, and **anterior semilunar cusps**, relative to their fetal position before rotation of the outflow tracts from the ventricles is complete. Each cusp forms a pocket-like sinus (Fig. 3.72)—a dilation in the wall of the initial portion of the pulmonary trunk. After ventricular contraction, the recoil of blood fills these **pulmonary sinuses** and forces the cusps closed. This prevents blood in the pulmonary trunk from refilling the right ventricle.

### Left atrium

The **left atrium** forms most of the base or posterior surface of the heart.

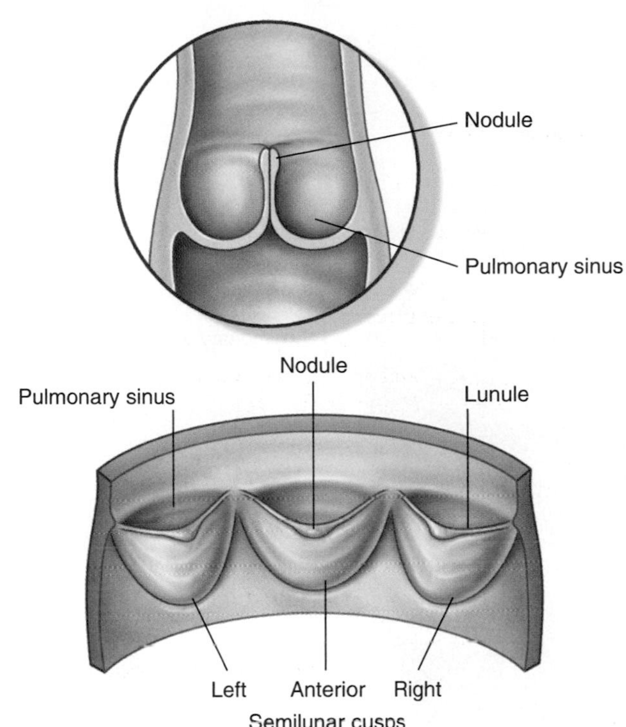


**Fig. 3.72** Posterior view of the pulmonary valve.

As with the right atrium, the left atrium is derived embryologically from two structures.

- The posterior half, or inflow portion, receives the four pulmonary veins (Fig. 3.73). It has smooth walls and derives from the proximal parts of the pulmonary veins that are incorporated into the left atrium during development.
- The anterior half is continuous with the left auricle. It contains musculi pectinati and derives from the embryonic primitive atrium. Unlike the crista terminalis in the right atrium, no distinct structure separates the two components of the left atrium.

The interatrial septum is part of the anterior wall of the left atrium. The thin area or depression in the septum is the valve of the foramen ovale and is opposite the floor of the fossa ovalis in the right atrium.

During development, the **valve of the foramen ovale** prevents blood from passing from the left atrium to the right atrium. This valve may not be completely fused in some adults, leaving a "probe patent" passage between the right atrium and the left atrium.

### Left ventricle

The left ventricle lies anterior to the left atrium. It contributes to the anterior, diaphragmatic, and left pulmonary surfaces of the heart, and forms the apex.

Blood enters the ventricle through the **left atrioventricular orifice** and flows in a forward direction to the apex. The chamber itself is conical, is longer than the right ventricle, and has the thickest layer of **myocardium**. The outflow tract (the **aortic vestibule**) is posterior to the infundibulum of the right ventricle, has smooth walls, and is derived from the embryonic bulbus cordis.

The **trabeculae carneae** in the left ventricle are fine and delicate in contrast to those in the right ventricle. The general appearance of the trabeculae with muscular ridges and bridges is similar to that of the right ventricle (Fig. 3.74).

Papillary muscles, together with chordae tendineae, are also observed and their structure is as described above for the right ventricle. Two papillary muscles, the **anterior** and **posterior papillary muscles**, are usually found in the left ventricle and are larger than those of the right ventricle.

In the anatomical position, the left ventricle is somewhat posterior to the right ventricle. The interventricular septum therefore forms the anterior wall and some of the wall on the right side of the left ventricle. The septum is described as having two parts:

- a **muscular part**, and
- a **membranous part**.

The muscular part is thick and forms the major part of the septum, whereas the membranous part is the thin, upper part of the septum. A third part of the septum may be considered an atrioventricular part because of its position above the septal cusp of the tricuspid valve. This superior location places this part of the septum between the left ventricle and right atrium.

### Mitral valve

The left atrioventricular orifice opens into the posterior right side of the superior part of the left ventricle. It is closed during ventricular contraction by the **mitral valve** (**left atrioventricular valve**), which is also referred to as the bicuspid valve because it has two cusps, the **anterior** and **posterior cusps** (Fig. 3.74). The bases of the cusps are secured to a fibrous ring surrounding the opening, and the cusps are continuous with each other at the commissures. The coordinated action of the papillary muscles and chordae tendineae is as described for the right ventricle.

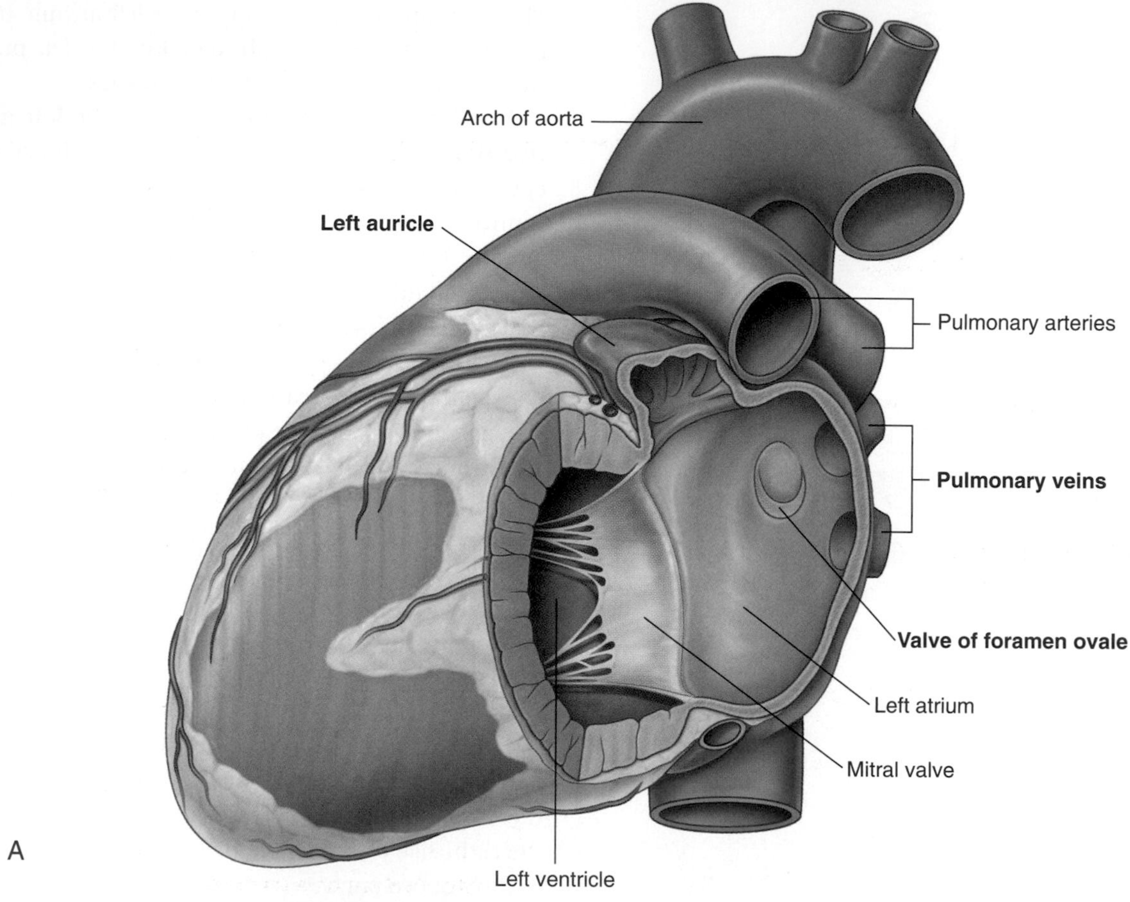


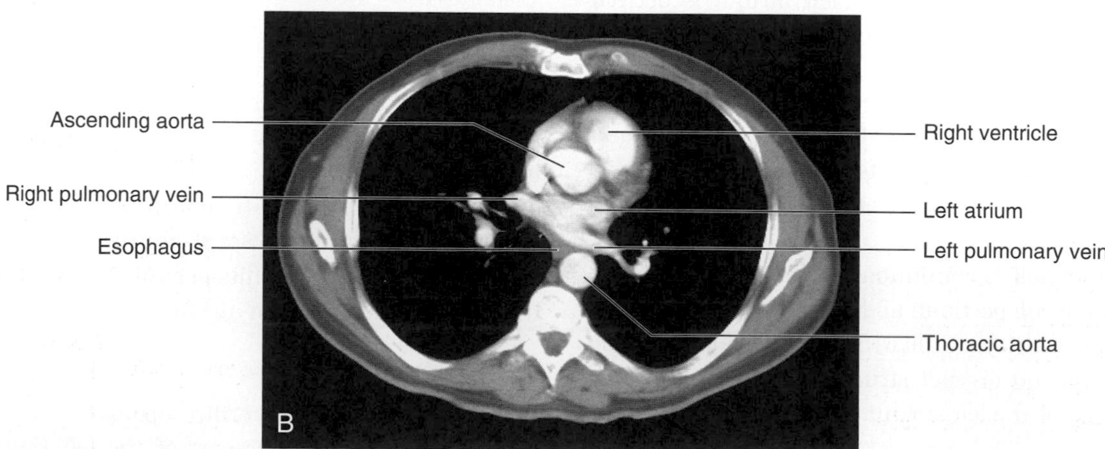


**Fig. 3.73** Left atrium. **A.** Internal view. **B.** Axial computed tomography image showing the pulmonary veins entering the left atrium.

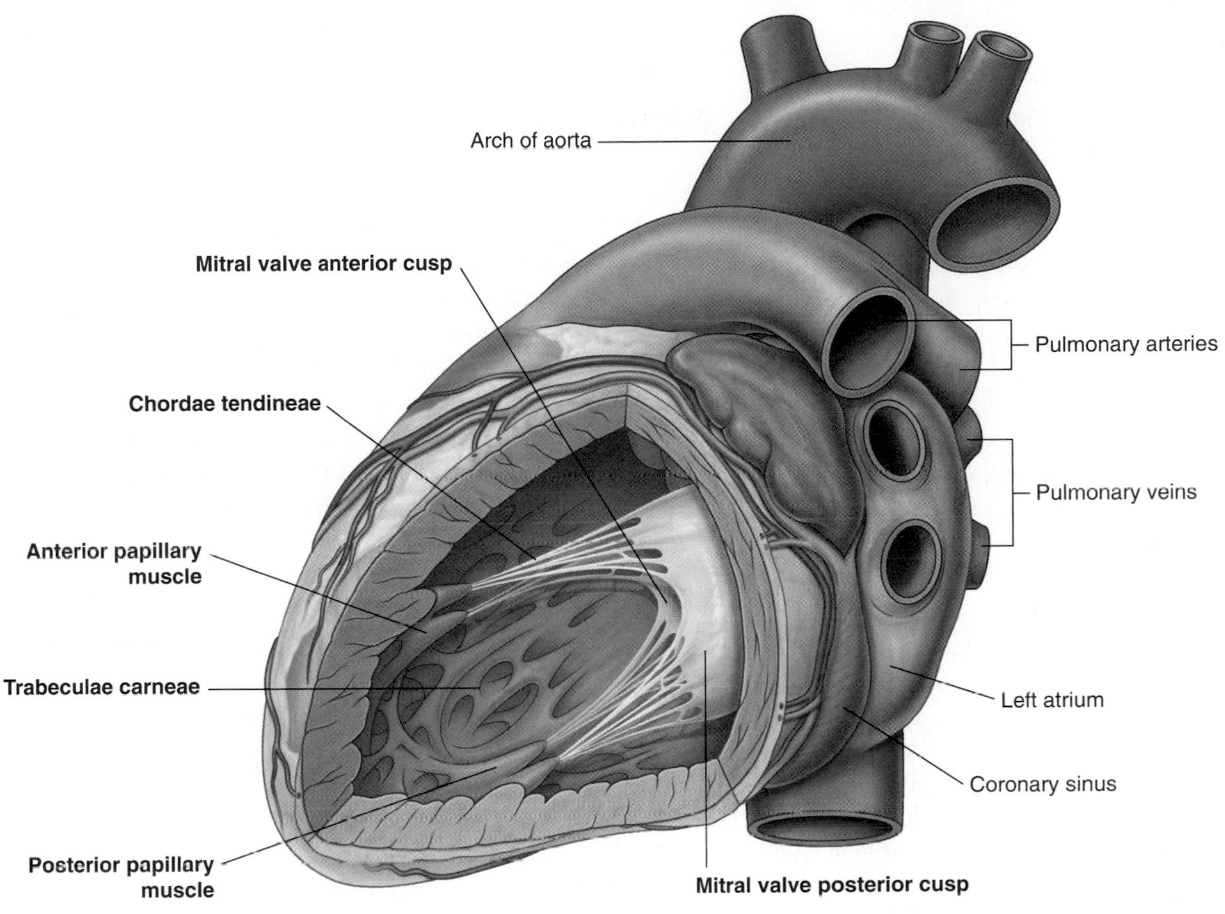


**Fig. 3.74** Internal view of the left ventricle.

## Aortic valve

The aortic vestibule, or outflow tract of the left ventricle, is continuous superiorly with the ascending aorta. The opening from the left ventricle into the aorta is closed by the **aortic valve**. This valve is similar in structure to the pulmonary valve. It consists of three **semilunar cusps** with the free edge of each projecting upward into the lumen of the ascending aorta (Fig. 3.75).

Between the semilunar cusps and the wall of the ascending aorta are pocket-like sinuses—the **right**, **left**, and **posterior aortic sinuses**. The right and left coronary arteries originate from the right and left aortic sinuses. Because of this, the posterior aortic sinus and cusp are sometimes referred to as the **noncoronary sinus and cusp**.

The functioning of the aortic valve is similar to that of the pulmonary valve with one important additional process: as blood recoils after ventricular contraction and fills the aortic sinuses, it is automatically forced into the coronary arteries because these vessels originate from the right and left aortic sinuses.

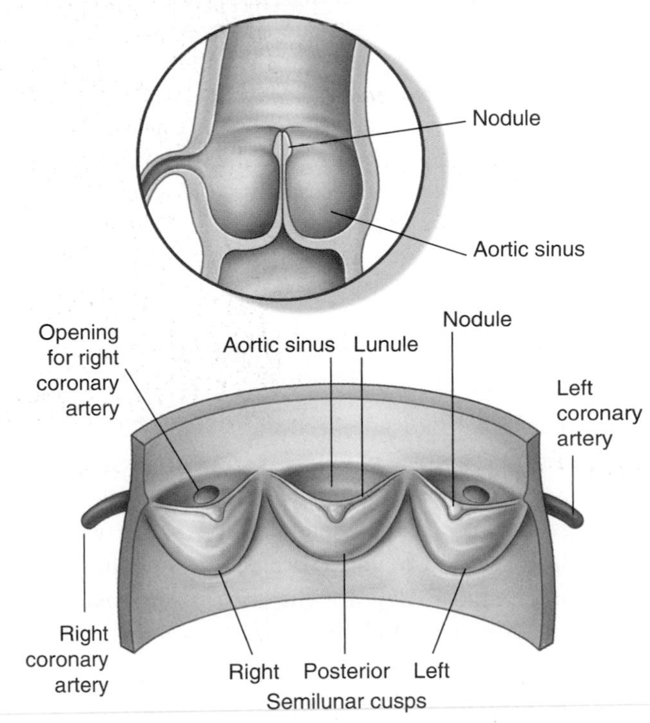


**Fig. 3.75** Anterior view of the aortic valve.

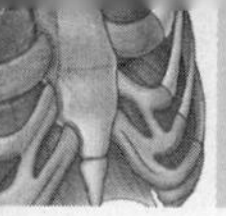



## In the clinic

### Valve disease

Valve problems consist of two basic types:

- incompetence (insufficiency), which results from poorly functioning valves; and
- stenosis, a narrowing of the orifice, caused by the valve's inability to open fully.

**Mitral valve disease** is usually a mixed pattern of stenosis and incompetence, one of which usually predominates. Both stenosis and incompetence lead to a poorly functioning valve and subsequent heart changes, which include:

- left ventricular hypertrophy (this is appreciably less marked in patients with mitral stenosis);
- increased pulmonary venous pressure;
- pulmonary edema; and
- enlargement (dilation) and hypertrophy of the left atrium.

Mitral valve stenosis can be congenital or acquired; in the latter, the most common cause is rheumatic fever. Stenosis usually occurs decades after an acute episode of rheumatic endocarditis.

**Aortic valve disease**, both aortic stenosis and aortic regurgitation (backflow), can produce marked heart failure. Aortic valve stenosis is the most common type of cardiac valve disease and results from atherosclerosis causing calcification of the valve leaflets. It can also be caused by postinflammatory or postrheumatic conditions. These may lead to aortic regurgitation such as infective endocarditis, degenerative valve disease, rheumatic fever, or trauma.

**Valve disease in the right side of the heart (affecting the tricuspid or pulmonary valve)** is most likely caused by infection. Intravenous drug use, alcoholism, indwelling catheters, and extensive burns predispose to infection of the valves, particularly the tricuspid valve. The resulting valve dysfunction produces abnormal pressure changes in the right atrium and right ventricle, and these can induce cardiac failure.

## Cardiac skeleton

The cardiac skeleton is a collection of dense, fibrous connective tissue in the form of four rings with interconnecting areas in a plane between the atria and the ventricles. The four rings of the cardiac skeleton surround the two atrioventricular orifices, the aortic orifice and opening of the pulmonary trunks. They are the **anulus fibrosus**. The interconnecting areas include:

- the **right fibrous trigone**, which is a thickened area of connective tissue between the aortic ring and right atrioventricular ring; and
- the **left fibrous trigone**, which is a thickened area of connective tissue between the aortic ring and the left atrioventricular ring (Fig. 3.76).

The cardiac skeleton helps maintain the integrity of the openings it surrounds and provides points of attachment for the cusps. It also separates the atrial musculature from the ventricular musculature. The atrial myocardium originates from the upper border of the rings, whereas the ventricular myocardium originates from the lower border of the rings.

The cardiac skeleton also serves as a dense connective tissue partition that electrically isolates the atria from the ventricles. The atrioventricular bundle, which passes through the anulus, is the single connection between these two groups of myocardium.

## Coronary vasculature

Two coronary arteries arise from the aortic sinuses in the initial portion of the ascending aorta and supply the muscle and other tissues of the heart. They circle the heart in the coronary sulcus, with marginal and interventricular branches, in the interventricular sulci, converging toward the apex of the heart (Fig. 3.77).

The returning venous blood passes through cardiac veins, most of which empty into the coronary sinus. This large venous structure is located in the coronary sulcus on the posterior surface of the heart between the left atrium and left ventricle. The coronary sinus empties into the right atrium between the opening of the inferior vena cava and the right atrioventricular orifice.

### Coronary arteries

***Right coronary artery.*** The right coronary artery originates from the right aortic sinus of the ascending aorta. It passes anteriorly and then descends vertically in the coronary sulcus, between the right atrium and right ventricle (Fig. 3.78A). On reaching the inferior margin of the heart, it turns posteriorly and continues in the sulcus onto the diaphragmatic surface and base of the heart. During this course, several branches arise from the main stem of the vessel:

- An early **atrial branch** passes in the groove between the right auricle and ascending aorta, and gives off the **sinu-atrial nodal branch** (Fig. 3.78A), which passes posteriorly around the superior vena cava to supply the sinu-atrial node.
- A **right marginal branch** is given off as the right coronary artery approaches the inferior (acute) margin of the heart (Fig. 3.78A,B) and continues along this border toward the apex of the heart.

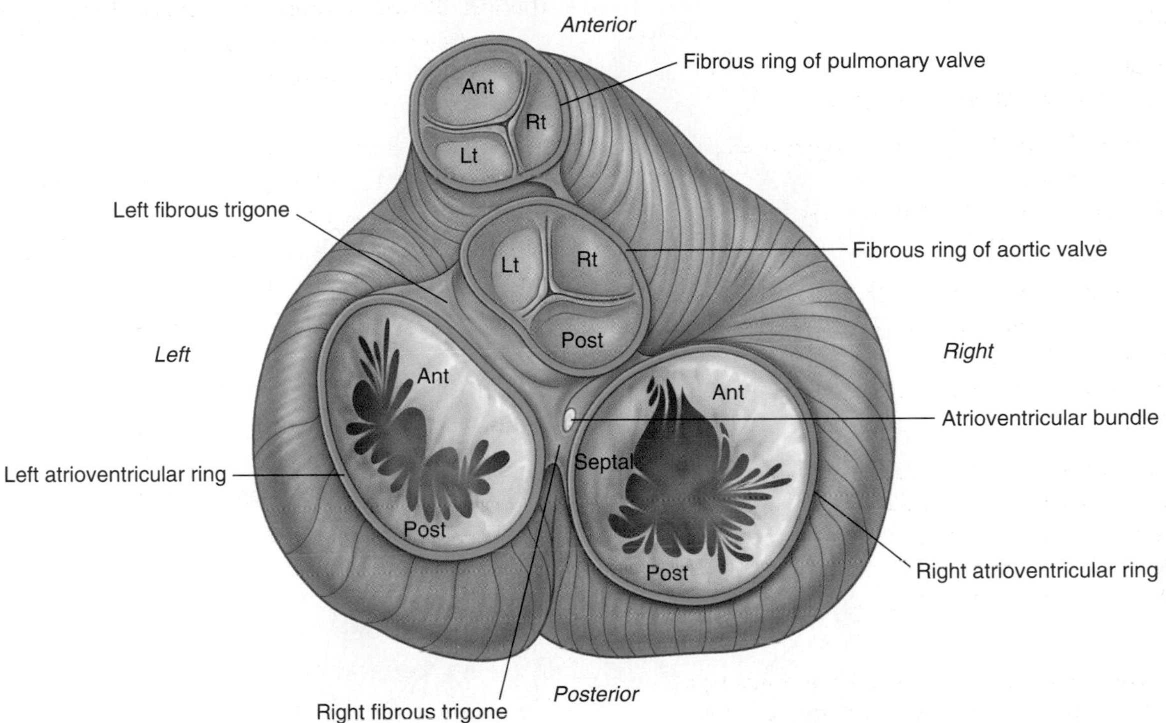


**Fig. 3.76** Cardiac skeleton (atria removed).

- As the right coronary artery continues on the base/diaphragmatic surface of the heart, it supplies a small branch to the atrioventricular node before giving off its final major branch, the **posterior interventricular branch** (Fig. 3.78A), which lies in the posterior interventricular sulcus.

The right coronary artery supplies the right atrium and right ventricle, the sinu-atrial and atrioventricular nodes, the interatrial septum, a portion of the left atrium, the posteroinferior one third of the interventricular septum, and a portion of the posterior part of the left ventricle.

***Left coronary artery.*** The left coronary artery originates from the left aortic sinus of the ascending aorta. It passes between the pulmonary trunk and the left auricle before entering the coronary sulcus. Emerging from behind the pulmonary trunk, the artery divides into its two terminal branches, the anterior interventricular and the circumflex (Fig. 3.78A).

- The **anterior interventricular branch** (**left anterior descending artery—LAD**) (Fig. 3.78A,C) continues around the left side of the pulmonary trunk and descends obliquely toward the apex of the heart in the anterior interventricular sulcus (Fig. 3.78A,C). During its course, one or two large **diagonal branches** may arise and descend diagonally across the anterior surface of the left ventricle.
- The **circumflex branch** (Fig. 3.78A,C) courses toward the left, in the coronary sulcus and onto the base/diaphragmatic surface of the heart, and usually ends before reaching the posterior interventricular sulcus. A large branch, the **left marginal artery** (Fig. 3.78A,C), usually arises from it and continues across the rounded obtuse margin of the heart.

The distribution pattern of the left coronary artery enables it to supply most of the left atrium and left ventricle, and most of the interventricular septum, including the atrioventricular bundle and its branches.

***Variations in the distribution patterns of coronary arteries.*** Several major variations in the basic distribution patterns of the coronary arteries occur.

- The distribution pattern described above for both right and left coronary arteries is the most common and consists of a right dominant coronary artery. This means that the posterior interventricular branch arises from the right coronary artery. The right coronary artery therefore supplies a large portion of the posterior wall of the left ventricle and the circumflex branch of the left coronary artery is relatively small.

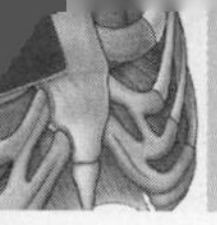

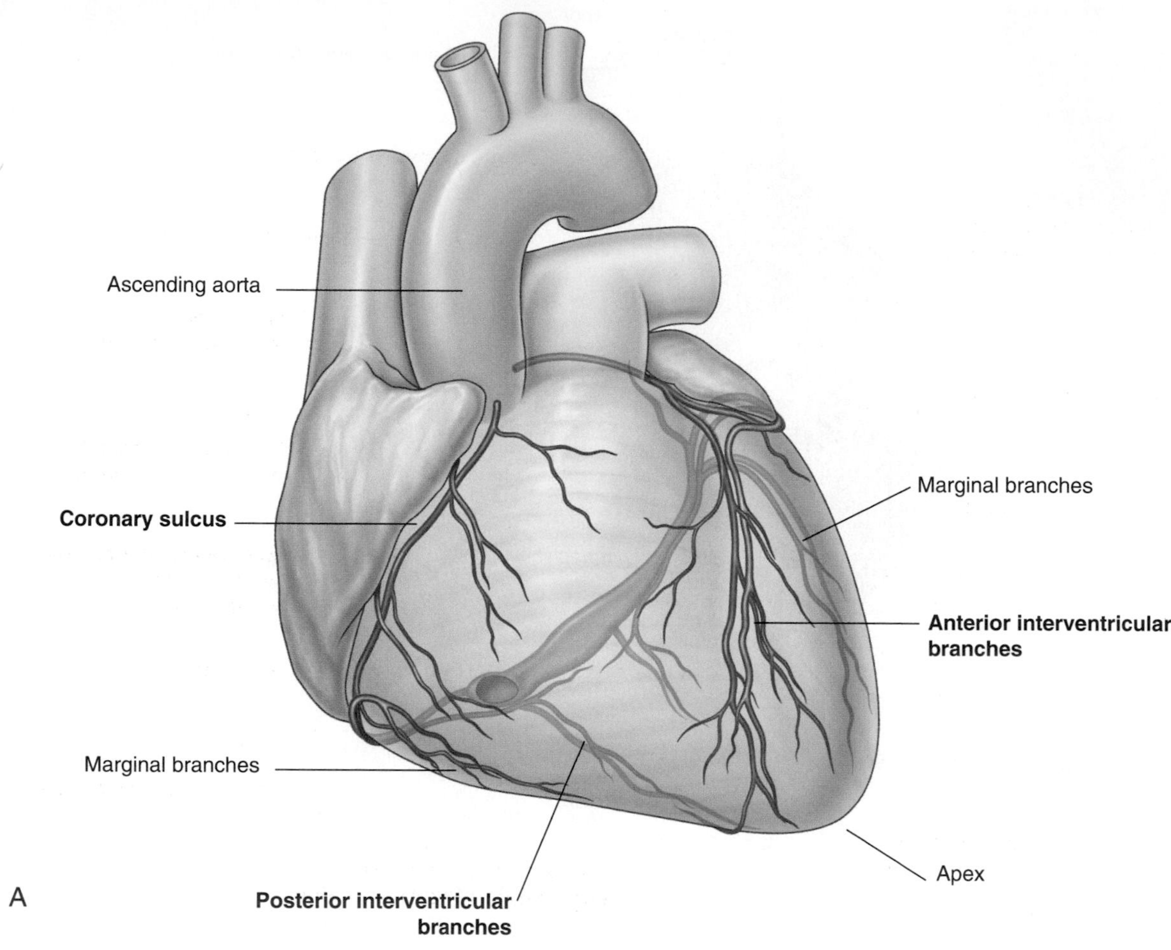


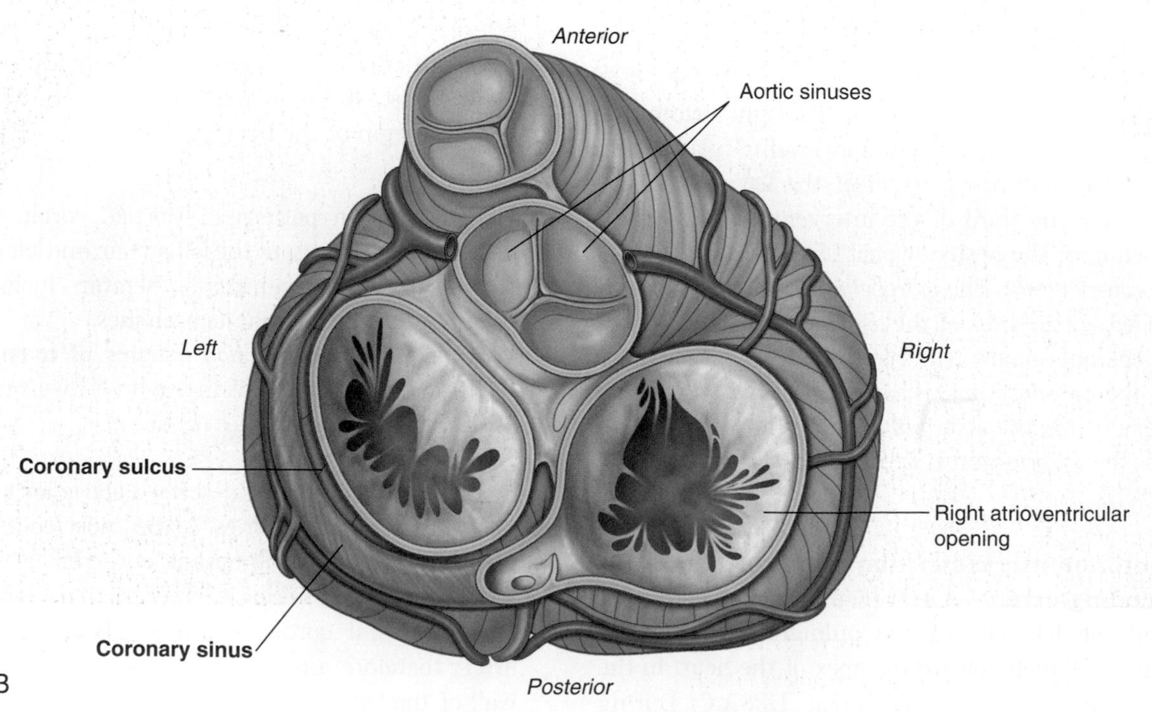




**Fig. 3.77** Cardiac vasculature. **A.** Anterior view. **B.** Superior view (atria removed).

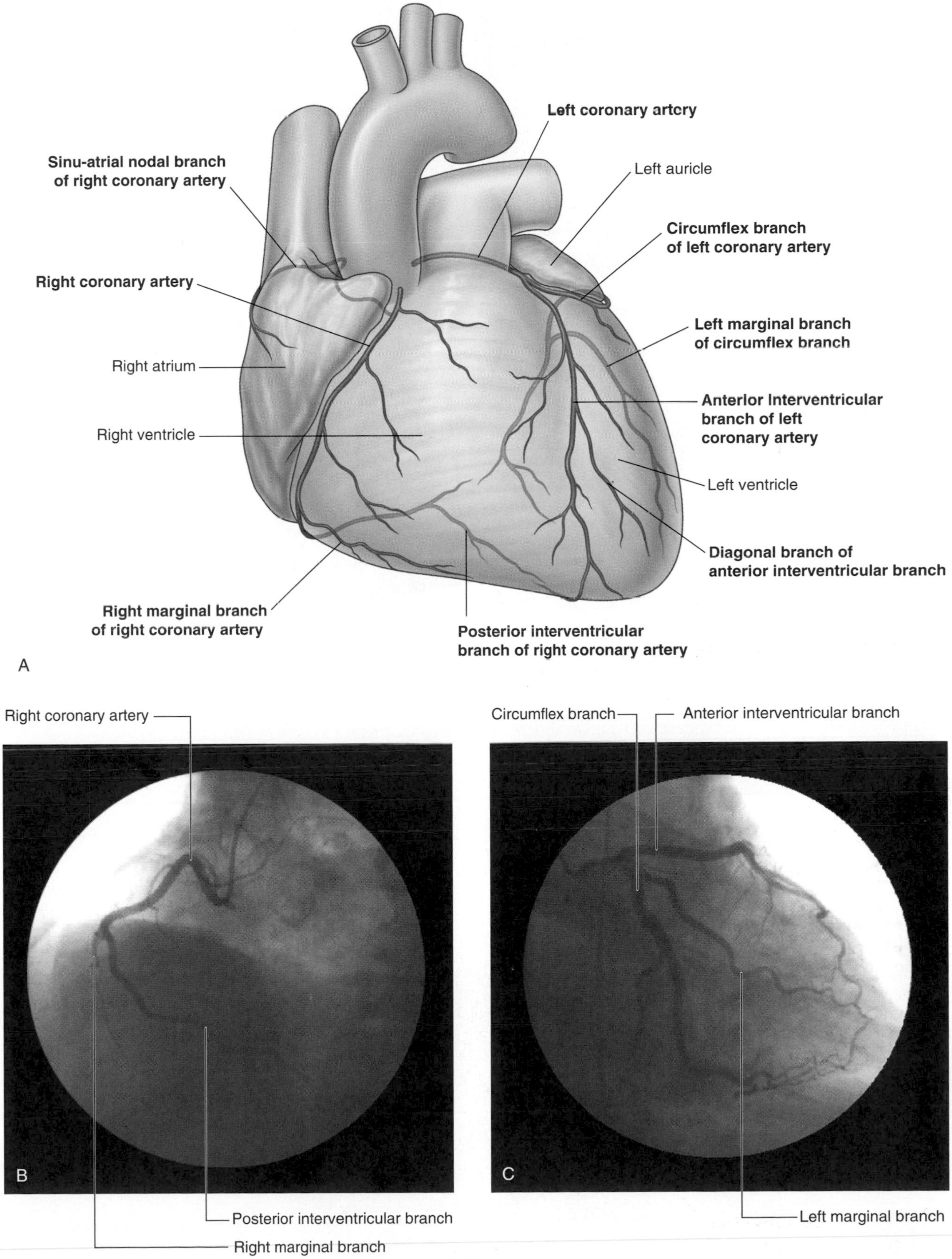

Left coronary artery
Sinu-atrial nodal branch of right coronary artery
Left auricle
Circumflex branch of left coronary artery
Right coronary artery
Left marginal branch of circumflex branch
Right atrium
Anterior interventricular branch of left coronary artery
Right ventricle
Left ventricle
Diagonal branch of anterior interventricular branch
Right marginal branch of right coronary artery
Posterior interventricular branch of right coronary artery
A

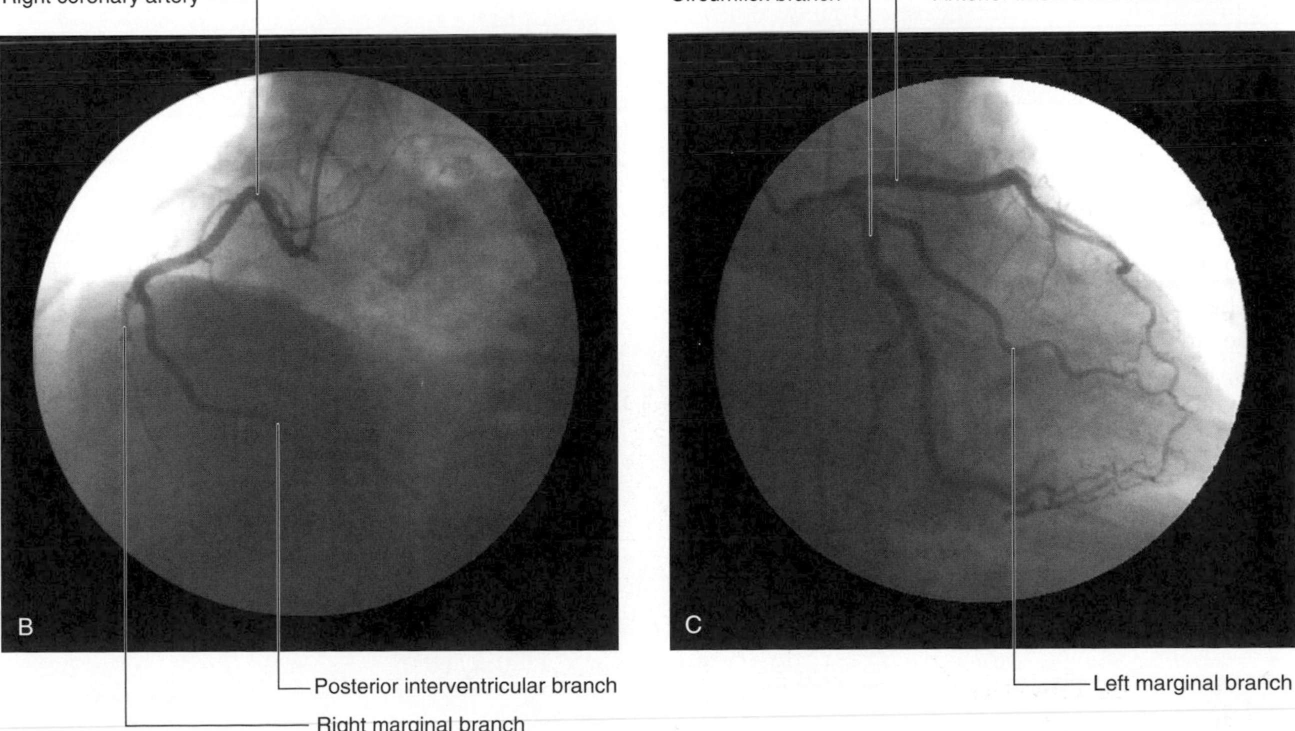


**Fig. 3.78 A.** Anterior view of coronary arterial system. Right dominant coronary artery. **B.** Left anterior oblique view of right coronary artery. **C.** Right anterior oblique view of left coronary artery.

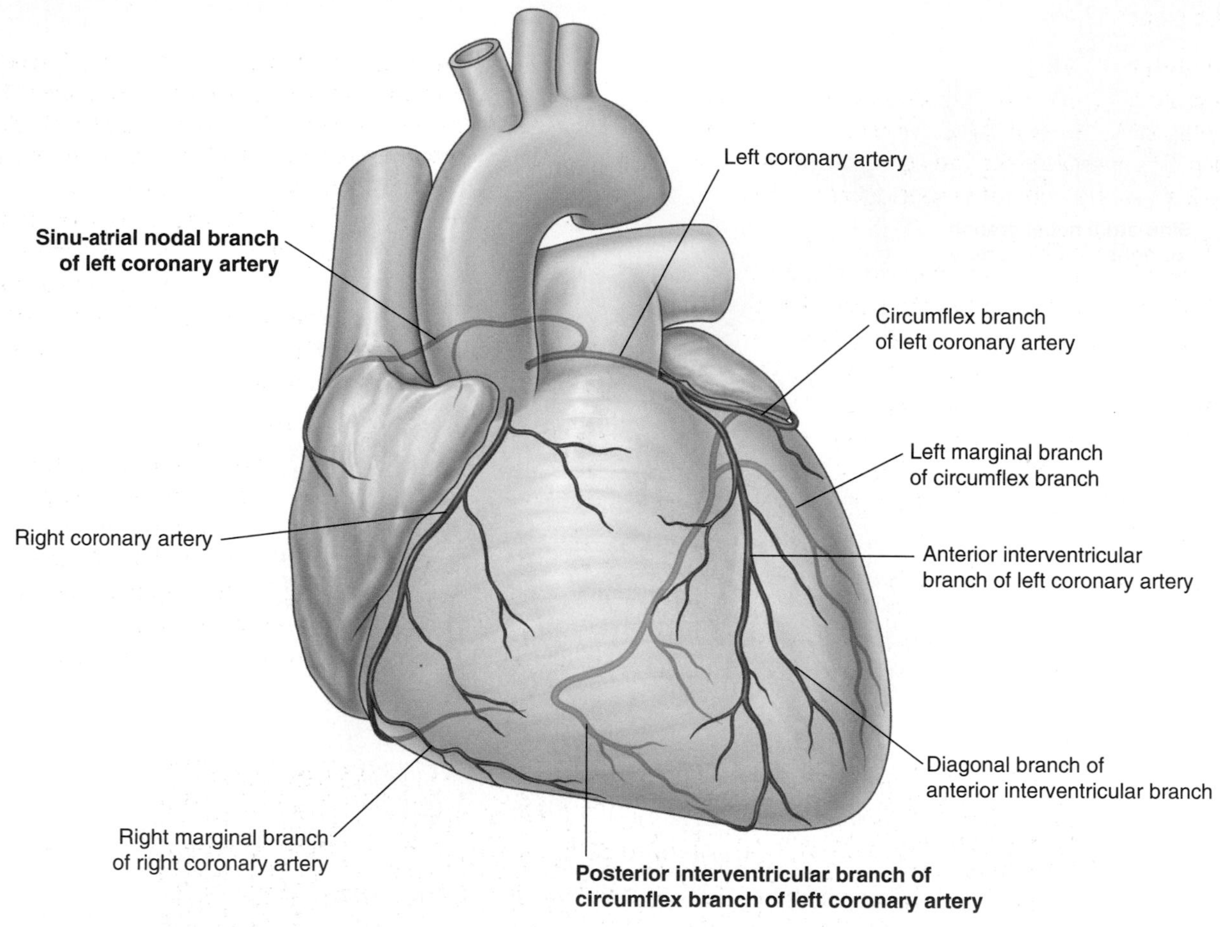


**Fig. 3.79** Left dominant coronary artery.

- In contrast, in hearts with a left dominant coronary artery, the posterior interventricular branch arises from an enlarged circumflex branch and supplies most of the posterior wall of the left ventricle (Fig. 3.79).
- Another point of variation relates to the arterial supply to the sinu-atrial and atrioventricular nodes. In most cases, these two structures are supplied by the right coronary artery. However, vessels from the circumflex branch of the left coronary artery occasionally supply these structures.

## In the clinic

### Clinical terminology for coronary arteries

In practice, physicians use alternative names for the coronary vessels. The short left coronary artery is referred to as the **left main stem vessel**. One of its primary branches, the anterior interventricular artery, is termed the **left anterior descending artery (LAD)**. Similarly, the terminal branch of the right coronary artery, the posterior interventricular artery, is termed the **posterior descending artery (PDA)**.

## In the clinic

### Heart attack

A heart attack occurs when the perfusion to the myocardium is insufficient to meet the metabolic needs of the tissue, leading to irreversible tissue damage. The most common cause is a total occlusion of a major coronary artery.

### *Coronary artery disease*

Occlusion of a major coronary artery, usually due to atherosclerosis, leads to inadequate oxygenation of an area of myocardium and cell death (Fig. 3.80). The severity of the problem will be related to the size and location of the artery involved, whether or not the blockage is complete, and whether there are collateral vessels to provide perfusion to the territory from other vessels. Depending on the severity, patients can develop pain (angina) or a myocardial infarction (MI).

### *Percutaneous coronary intervention*

This is a technique in which a long fine tube (a catheter) is inserted into the femoral artery in the thigh and passed through the external and common iliac arteries and into the abdominal aorta. It continues to be moved upward through the thoracic aorta to the origins of the coronary arteries. The coronaries may also be approached via the radial or brachial arteries. A fine wire is then passed into the coronary artery and is used to cross the stenosis. A fine balloon is then passed over the wire and may be inflated at the level of the obstruction, thus widening it; this is termed angioplasty. More commonly, this is augmented by placement of a fine wire mesh (a stent) inside the obstruction to hold it open. Other percutaneous interventions are suction extraction of a coronary thrombus and rotary ablation of a plaque.

### *Coronary artery bypass grafts*

If coronary artery disease is too extensive to be treated by percutaneous intervention, surgical coronary artery bypass grafting may be necessary. The great saphenous vein, in the lower limb, is harvested and used as a graft. It is divided into several pieces, each of which is used to bypass blocked sections of the coronary arteries. The internal thoracic and radial arteries can also be used.

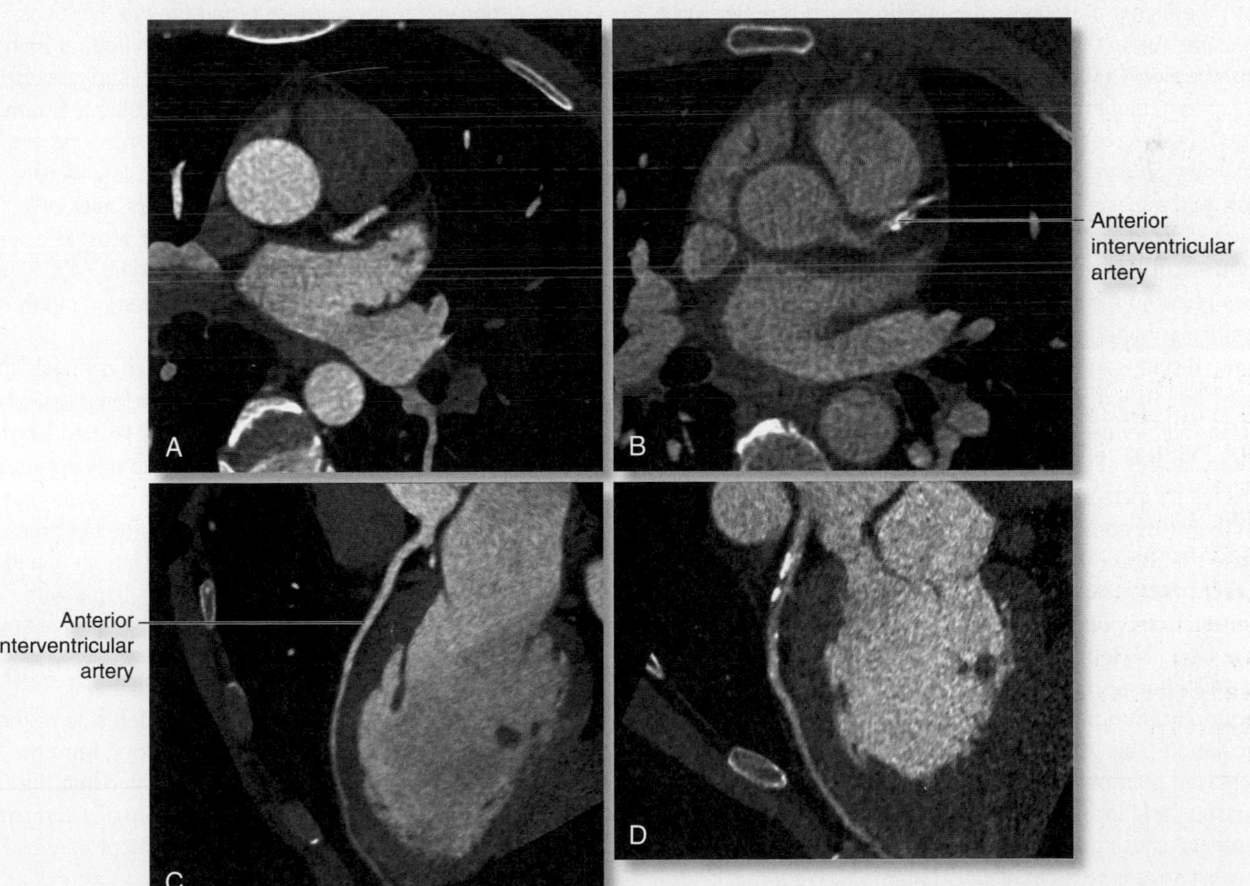


**Fig. 3.80 A** and **B.** Axial maximum intensity projection (MIP) CT image through the heart. **A.** Normal anterior interventricular (left anterior descending) artery. **B.** Stenotic (calcified) anterior interventricular (left anterior descending) artery. **C** and **D.** Vertical long axis multiplanar reformation (MRP) CT image through the heart. **C.** Normal anterior interventricular (left anterior descending) artery. **D.** Stenotic (calcified) anterior interventricular (left anterior descending) artery.

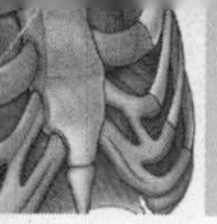



## In the clinic

### Classic symptoms of heart attack

The typical symptoms are chest heaviness or pressure, which can be severe, lasting more than 20 minutes, and often associated with sweating. The pain in the chest (which may be described as an "elephant sitting on my chest" or by using a clenched fist to describe the pain [Levine sign]) often radiates to the arms (left more common than the right), and can be associated with nausea. The severity of ischemia and infarction depends on the rate at which the occlusion or stenosis has occurred and whether or not collateral channels have had a chance to develop.

## In the clinic

### Are heart attack symptoms the same in men and women?

Although men and women can experience the typical symptoms of severe chest pain, cold sweats, and pain in the left arm, women are more likely than men to have subtler, less recognizable symptoms. These may include abdominal pain, achiness in the jaw or back, nausea, shortness of breath, or simply fatigue. The mechanism of this difference is not understood, but it is important to consider cardiac ischemia for a wide range of symptoms.

## In the clinic

### Common congenital heart defects

The most common abnormalities that occur during development are those produced by a defect in the atrial and ventricular septa.

A **defect in the interatrial septum** allows blood to pass from one side of the heart to the other from the chamber with the higher pressure to the chamber with the lower pressure; this is clinically referred to as a **shunt**. An **atrial septal defect (ASD)** allows oxygenated blood to flow from the left atrium (higher pressure) across the ASD into the right atrium (lower pressure), resulting in a left to right shunt and volume overload in the right-sided circulation. Many patients with ASD are asymptomatic, but in some cases the ASD may cause symptoms and needs to be closed surgically or by endovascular devices. Occasionally, increased blood flow into the right atrium over many years leads to right atrial and right ventricular hypertrophy and enlargement of the pulmonary trunk, resulting in pulmonary arterial hypertension. In such cases, the patients can present with shortness of breath, increasing tiredness, palpitations, fainting episodes and heart failure. In ASD, the left ventricle is not enlarged as it is not affected by increased returning blood volume.

The most common of all congenital heart defects are those that occur in the ventricular septum—**ventriculoseptal defect** (**VSD**). These lesions are most frequent in the membranous portion of the septum and they allow blood to flow from the left ventricle (higher pressure) to the right ventricle (lower pressure), leading to an abnormal communication between the systemic and pulmonary circulation. This leads to right ventricular hypertrophy, increased pulmonary blood flow, elevated arterial pulmonary pressure, and increased blood volume returning to the left ventricle, causing its dilation. Increased pulmonary pressure in most severe cases may cause pulmonary edema. If large enough and left untreated, VSDs can produce marked clinical problems that might require surgery. VSD may be an isolated abnormality or part of a syndromic constellation, such as the tetralogy of Fallot.

The **tetralogy of Fallot,** the most common cyanotic congenital heart disorder diagnosed soon after birth, classically consists of four abnormalities: pulmonary stenosis, VSD, overriding aorta (originating to a varying degree from the right ventricle), and right ventricular hypertrophy. The underdevelopment of the right ventricle and pulmonary stenosis reduce blood flow to the lungs, leading to reduced volume of oxygenated blood returning to the heart. The defect in the interventricular septum causes mixing of oxygenated and nonoxygenated blood. The mixed blood is then delivered by the aorta to the major organs, resulting in poor oxygenation and cyanosis. Infants can present with cyanosis at birth or develop episodes of cyanosis while feeding or crying (tet spells). Most affected infants require surgical intervention. The advent of cardiopulmonary bypass was crucial in delivering highly satisfactory surgical results.

Occasionally, the **ductus arteriosus**, which connects the left branch of the pulmonary artery to the inferior aspect of the aortic arch, fails to close at birth. This is termed a **patent** or **persistent ductus arteriosus** (**PDA**). When this occurs, the oxygenated blood in the aortic arch (higher pressure) passes into the left branch of the pulmonary artery (lower pressure) and produces pulmonary hypertension and left atrial and ventricular enlargement. The prognosis in patients with isolated PDA is extremely good, as most do not have any major sequelae after surgical closure.

All of these defects produce a left-to-right shunt, indicating that oxygenated blood from the left side of the heart is being mixed with deoxygenated blood from the right side of the heart before being recirculated into the pulmonary circulation. These shunts are normally compatible with life, but surgery or endovascular treatment may be necessary.

Rarely, a shunt is right-to-left. In isolation, this is fatal; however, this type of shunt is often associated with other anomalies, so some deoxygenated blood is returned to the lungs and the systemic circulation.

## In the clinic

### Cardiac auscultation

Auscultation of the heart reveals the normal audible cardiac cycle, which allows the clinician to assess heart rate, rhythm, and regularity. Furthermore, cardiac murmurs that have characteristic sounds within the phases of the cardiac cycle can be demonstrated (Fig. 3.81).

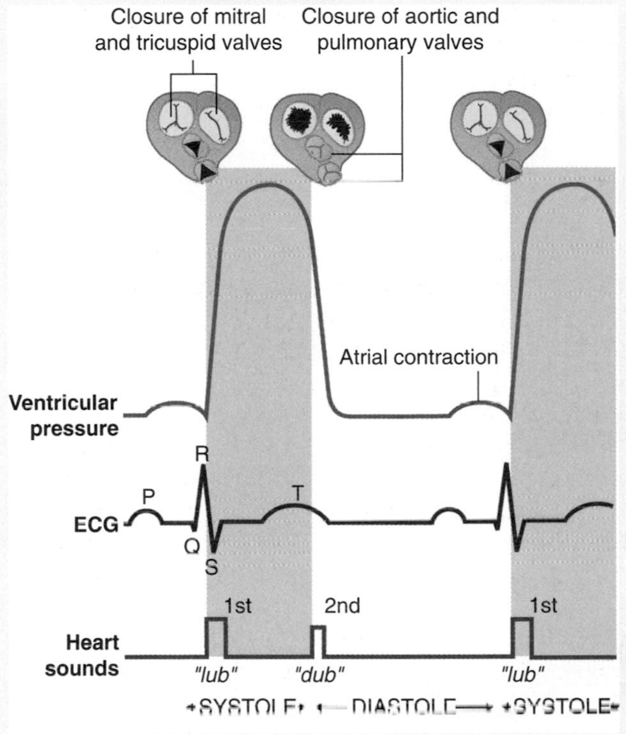


**Fig. 3.81** Heart sounds and how they relate to valve closure, the electrocardiogram (ECG), and ventricular pressure.

### Cardiac veins

The **coronary sinus** receives four major tributaries: the great, middle, small, and posterior cardiac veins.

***Great cardiac vein.*** The great cardiac vein begins at the apex of the heart (Fig. 3.82A). It ascends in the anterior interventricular sulcus, where it is related to the anterior interventricular artery and is often termed the anterior interventricular vein. Reaching the coronary sulcus, the great cardiac vein turns to the left and continues onto the base/diaphragmatic surface of the heart. At this point, it is associated with the circumflex branch of the left coronary artery. Continuing along its path in the coronary sulcus, the great cardiac vein gradually enlarges to form the coronary sinus, which enters the right atrium (Fig. 3.82B).

***Middle cardiac vein.*** The middle cardiac vein (posterior interventricular vein) begins near the apex of the heart and ascends in the posterior interventricular sulcus toward the coronary sinus (Fig. 3.82B). It is associated with the posterior interventricular branch of the right or left coronary artery throughout its course.

***Small cardiac vein.*** The small cardiac vein begins in the lower anterior section of the coronary sulcus between the right atrium and right ventricle (Fig. 3.82A). It continues in this groove onto the base/diaphragmatic surface of the heart where it enters the coronary sinus at its atrial end. It is a companion of the right coronary artery throughout its course and may receive the right marginal vein (Fig. 3.82A). This small vein accompanies the marginal branch of the right coronary artery along the acute margin of the heart. If the right marginal vein does not join the small cardiac vein, it enters the right atrium directly.

***Posterior cardiac vein.*** The posterior cardiac vein lies on the posterior surface of the left ventricle just to the left of the middle cardiac vein (Fig. 3.82B). It either enters the coronary sinus directly or joins the great cardiac vein.

***Other cardiac veins.*** Two additional groups of cardiac veins are also involved in the venous drainage of the heart.

- The **anterior veins of the right ventricle** (**anterior cardiac veins**) are small veins that arise on the anterior surface of the right ventricle (Fig. 3.82A). They cross the coronary sulcus and enter the anterior wall of the right atrium. They drain the anterior portion of the right ventricle. The right marginal vein may be part of this group if it does not enter the small cardiac vein.
- A group of smallest cardiac veins (**venae cordis minimae** or **veins of Thebesius**) have also been described. Draining directly into the cardiac chambers, they are numerous in the right atrium and right ventricle, are occasionally associated with the left atrium, and are rarely associated with the left ventricle.

### Coronary lymphatics

The lymphatic vessels of the heart follow the coronary arteries and drain mainly into:

- brachiocephalic nodes, anterior to the brachiocephalic veins; and
- tracheobronchial nodes, at the inferior end of the trachea.

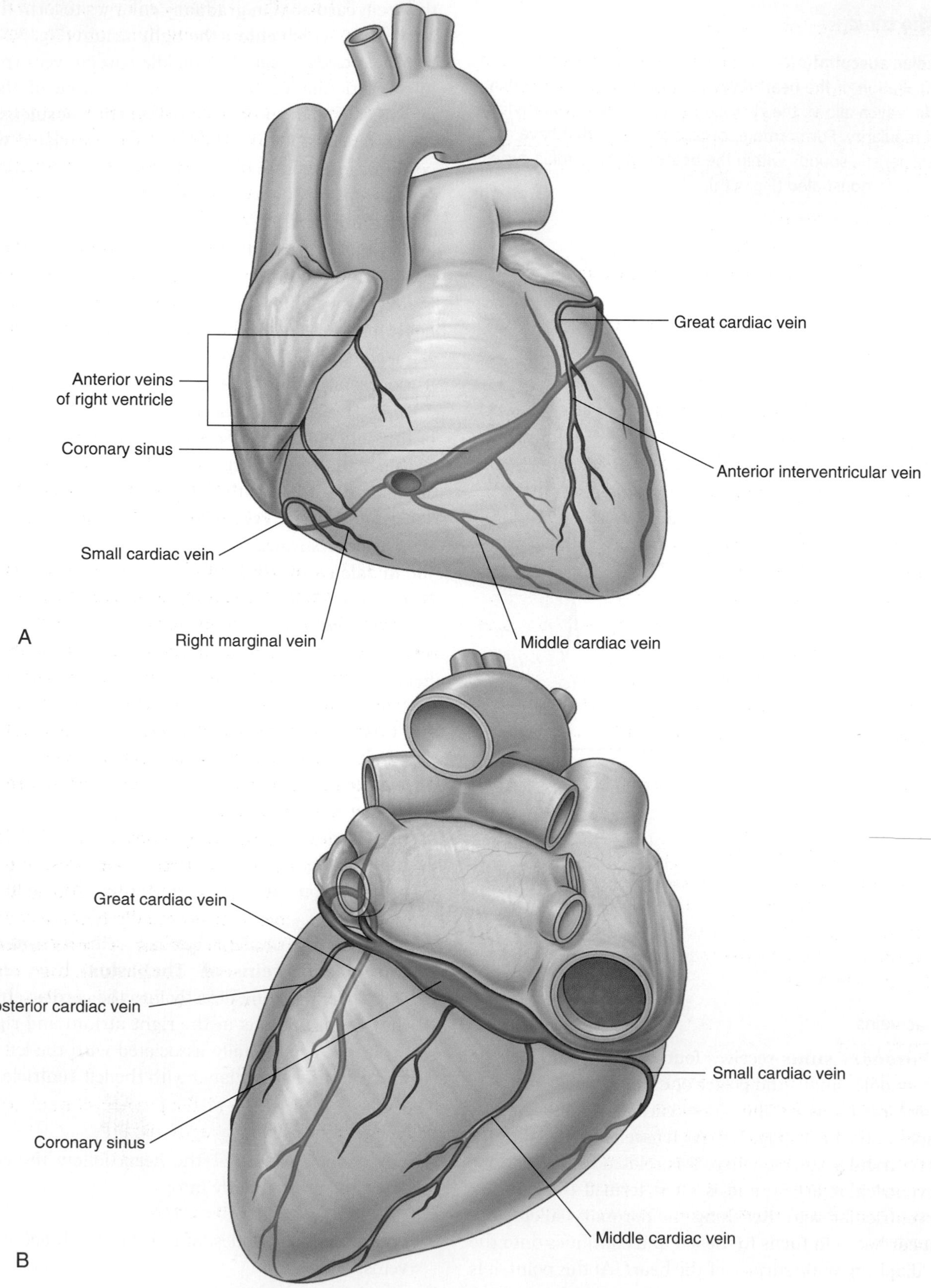

**Fig. 3.82** Major cardiac veins. **A.** Anterior view of major cardiac veins. **B.** Posteroinferior view of major cardiac veins.

### Cardiac conduction system

The musculature of the atria and ventricles is capable of contracting spontaneously. The cardiac conduction system initiates and coordinates contraction. The conduction system consists of nodes and networks of specialized cardiac muscle cells organized into four basic components:

- the sinu-atrial node,
- the atrioventricular node,
- the atrioventricular bundle with its right and left bundle branches, and
- the subendocardial plexus of conduction cells (the Purkinje fibers).

The unique distribution pattern of the cardiac conduction system establishes an important unidirectional pathway of excitation/contraction. Throughout its course, large branches of the conduction system are insulated from the surrounding myocardium by connective tissue. This tends to decrease inappropriate stimulation and contraction of cardiac muscle fibers.

The number of functional contacts between the conduction pathway and cardiac musculature greatly increases in the subendocardial network.

Thus, a unidirectional wave of excitation and contraction is established, which moves from the papillary muscles and apex of the ventricles to the arterial outflow tracts.

**In the clinic**

**Cardiac conduction system**

The cardiac conduction system can be affected by coronary artery disease. The normal rhythm may be disturbed if the blood supply to the coronary conduction system is disrupted. If a dysrhythmia affects the heart rate or the order in which the chambers contract, heart failure and death may ensue.

#### Sinu-atrial node

Impulses begin at the **sinu-atrial node**, the cardiac pacemaker. This collection of cells is located at the superior end of the crista terminalis at the junction of the superior vena cava and the right atrium (Fig. 3.83A). This is also the junction between the parts of the right atrium derived from the embryonic sinus venosus and the atrium proper.

The excitation signals generated by the sinu-atrial node spread across the atria, causing the muscle to contract.

#### Atrioventricular node

Concurrently, the wave of excitation in the atria stimulates the **atrioventricular node**, which is located near the opening of the coronary sinus, close to the attachment of the septal cusp of the tricuspid valve, and within the atrioventricular septum (Fig. 3.83A).

The atrioventricular node is a collection of specialized cells that forms the beginning of an elaborate system of conducting tissue, the atrioventricular bundle, which extends the excitatory impulse to all ventricular musculature.

#### Atrioventricular bundle

The **atrioventricular bundle** is a direct continuation of the atrioventricular node (Fig. 3.83A). It follows along the lower border of the membranous part of the interventricular septum before splitting into right and left bundles.

The **right bundle branch** continues on the right side of the interventricular septum toward the apex of the right ventricle. From the septum it enters the septomarginal trabecula to reach the base of the anterior papillary muscle. At this point, it divides and is continuous with the final component of the cardiac conduction system, the subendocardial plexus of ventricular conduction cells or Purkinje fibers. This network of specialized cells spreads throughout the ventricle to supply the ventricular musculature, including the papillary muscles.

The **left bundle branch** passes to the left side of the muscular interventricular septum and descends to the apex of the left ventricle (Fig. 3.83B). Along its course it gives off branches that eventually become continuous with the **subendocardial plexus of conduction cells** (**Purkinje fibers**). As with the right side, this network of specialized cells spreads the excitation impulses throughout the left ventricle.

### Cardiac innervation

The autonomic division of the peripheral nervous system is directly responsible for regulating:

- heart rate,
- force of each contraction, and
- cardiac output.

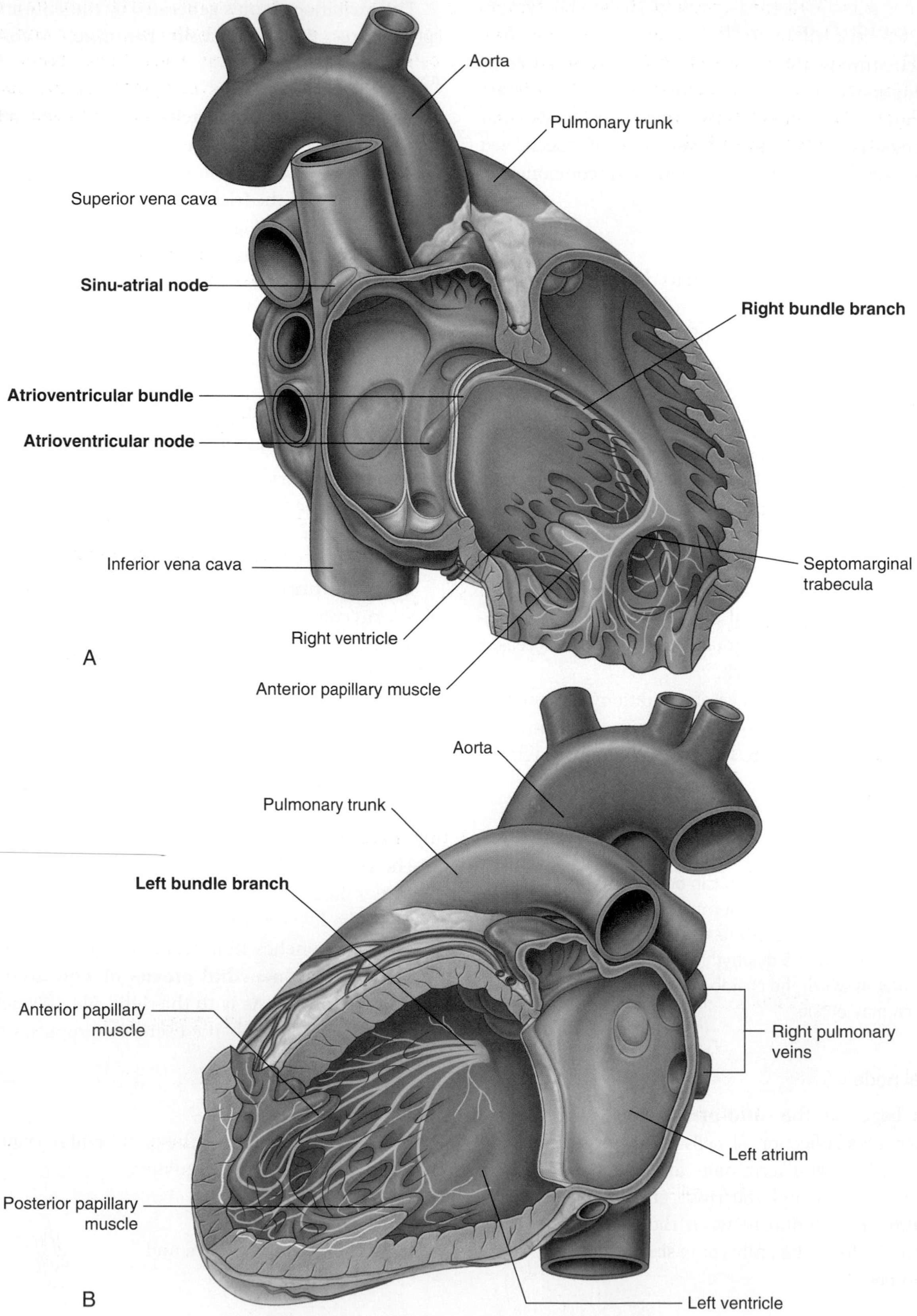

**Fig. 3.83** Conduction system of the heart. **A.** Right chambers. **B.** Left chambers.

Branches from both the parasympathetic and sympathetic systems contribute to the formation of the **cardiac plexus**. This plexus consists of a **superficial part**, inferior to the aortic arch and between it and the pulmonary trunk (Fig. 3.84A), and a **deep part**, between the aortic arch and the tracheal bifurcation (Fig. 3.84B).

From the cardiac plexus, small branches that are mixed nerves containing both sympathetic and parasympathetic fibers supply the heart. These branches affect nodal tissue and other components of the conduction system, coronary blood vessels, and atrial and ventricular musculature.

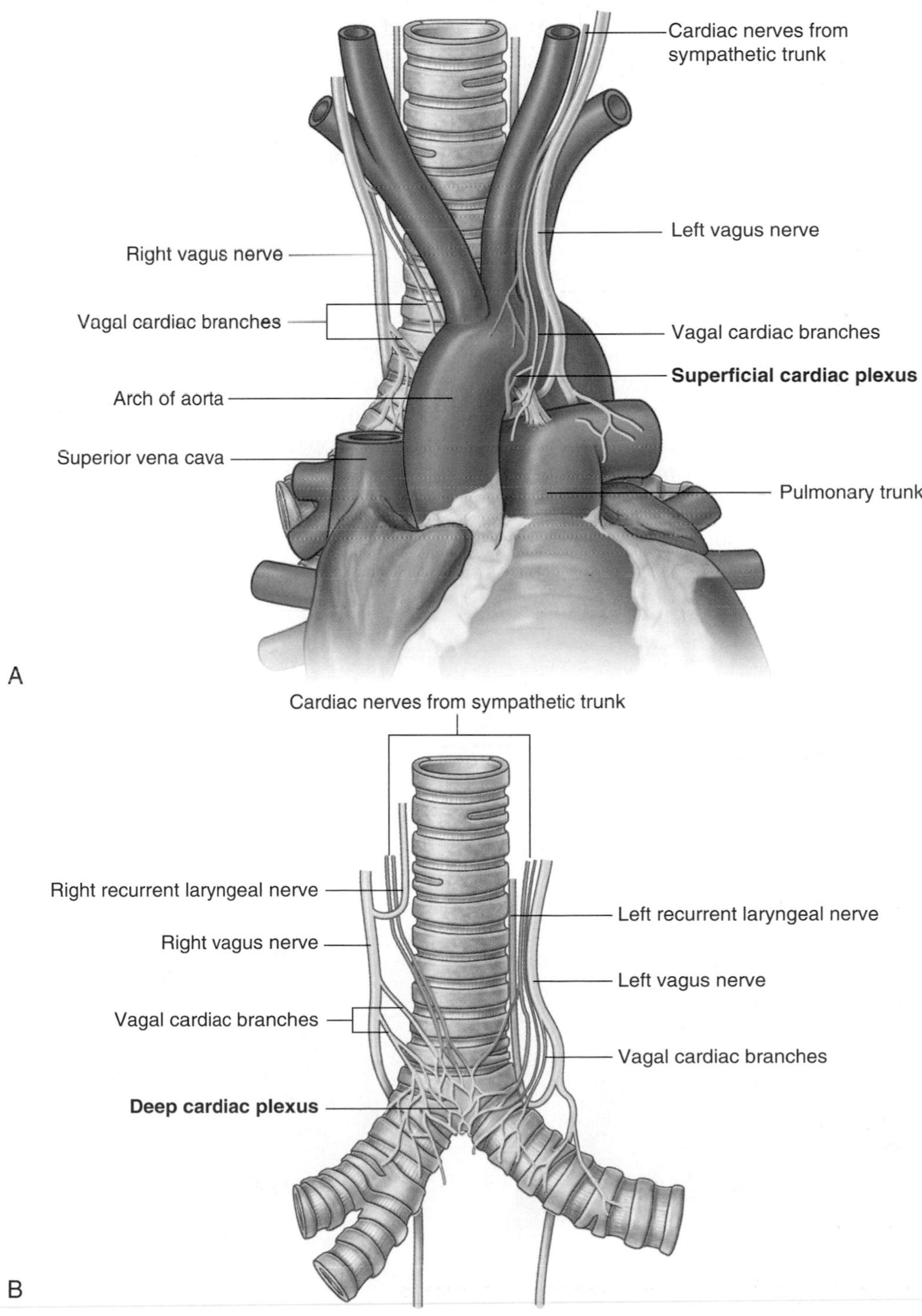


**Fig. 3.84** Cardiac plexus. **A.** Superficial. **B.** Deep.

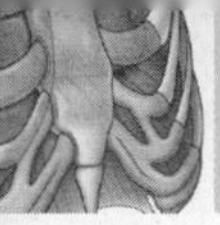



### Parasympathetic innervation

Stimulation of the parasympathetic system:

- decreases heart rate,
- reduces force of contraction, and
- constricts the coronary arteries.

The preganglionic parasympathetic fibers reach the heart as cardiac branches from the right and left vagus nerves. They enter the cardiac plexus and synapse in ganglia located either within the plexus or in the walls of the atria.

### Sympathetic innervation

Stimulation of the sympathetic system:

- increases heart rate, and
- increases the force of contraction.

Sympathetic fibers reach the cardiac plexus through the cardiac nerves from the sympathetic trunk. Preganglionic sympathetic fibers from the upper four or five segments of the thoracic spinal cord enter and move through the sympathetic trunk. They synapse in cervical and upper thoracic sympathetic ganglia, and postganglionic fibers proceed as bilateral branches from the sympathetic trunk to the cardiac plexus.

### Visceral afferents

Visceral afferents from the heart are also a component of the cardiac plexus. These fibers pass through the cardiac plexus and return to the central nervous system in the cardiac nerves from the sympathetic trunk and in the vagal cardiac branches.

The afferents associated with the vagal cardiac nerves return to the vagus nerve [X]. They sense alterations in blood pressure and blood chemistry and are therefore primarily concerned with cardiac reflexes.

The afferents associated with the cardiac nerves from the sympathetic trunks return to either the cervical or the thoracic portions of the sympathetic trunk. If they are in the cervical portion of the trunk, they normally descend to the thoracic region, where they reenter the upper four or five thoracic spinal cord segments, along with the afferents from the thoracic region of the sympathetic trunk. Visceral afferents associated with the sympathetic system conduct pain sensation from the heart, which is detected at the cellular level as tissue-damaging events (i.e., cardiac ischemia). This pain is often "referred" to cutaneous regions supplied by the same spinal cord levels (see "In the clinic: Referred pain," p. 46, and "Case 1," pp. 244–246).

## Pulmonary trunk

The **pulmonary trunk** is contained within the pericardial sac (Fig. 3.85), is covered by the visceral layer of serous

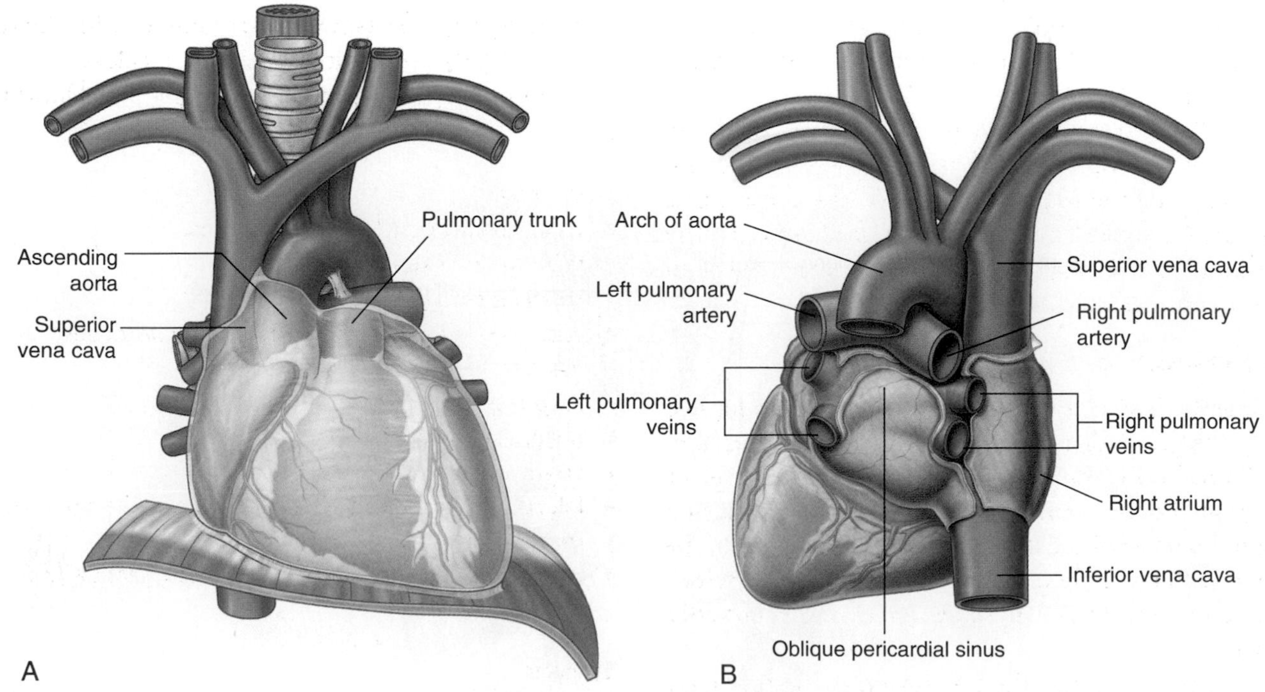


**Fig. 3.85** Major vessels within the middle mediastinum. **A.** Anterior view. **B.** Posterior view.

pericardium, and is associated with the ascending aorta in a common sheath. It arises from the conus arteriosus of the right ventricle at the opening of the pulmonary trunk slightly anterior to the aortic orifice and ascends, moving posteriorly and to the left, lying initially anterior and then to the left of the ascending aorta. At approximately the level of the intervertebral disc between vertebrae TV and TVI, opposite the left border of the sternum and posterior to the third left costal cartilage, the pulmonary trunk divides into:

- the right pulmonary artery, which passes to the right, posterior to the ascending aorta and the superior vena cava, to enter the right lung; and
- the left pulmonary artery, which passes inferiorly to the arch of the aorta and anteriorly to the descending aorta to enter the left lung.

### Ascending aorta

The **ascending aorta** is contained within the pericardial sac and is covered by a visceral layer of serous pericardium, which also surrounds the pulmonary trunk in a common sheath (Fig. 3.85A).

The origin of the ascending aorta is the aortic orifice at the base of the left ventricle, which is level with the lower edge of the third left costal cartilage, posterior to the left half of the sternum. Moving superiorly, slightly forward and to the right, the ascending aorta continues to the level of the second right costal cartilage. At this point, it enters the superior mediastinum and is then referred to as the **arch of the aorta**.

Immediately superior to the point where the ascending aorta arises from the left ventricle are three small outward bulges opposite the semilunar cusps of the aortic valve. These are the posterior, right, and left aortic sinuses. The right and left coronary arteries originate from the right and left aortic sinuses, respectively.

### Other vasculature

The inferior half of the **superior vena cava** is located within the pericardial sac (Fig. 3.85B). It passes through the fibrous pericardium at approximately the level of the second costal cartilage and enters the right atrium at the lower level of the third costal cartilage. The portion within the pericardial sac is covered with serous pericardium except for a small area on its posterior surface.

After passing through the diaphragm, at approximately the level of vertebra TVIII, the **inferior vena cava** enters the fibrous pericardium. A short portion of this vessel is within the pericardial sac before entering the right atrium. While within the pericardial sac, it is covered by serous pericardium except for a small portion of its posterior surface (Fig. 3.85B).

A very short segment of each of the pulmonary veins is also within the pericardial sac. These veins, usually two from each lung, pass through the fibrous pericardium and enter the superior region of the left atrium on its posterior surface. In the pericardial sac, all but a portion of the posterior surface of these veins is covered by serous pericardium. In addition, the **oblique pericardial sinus** is between the right and left pulmonary veins, within the pericardial sac (Fig. 3.85B).

## Superior mediastinum

The **superior mediastinum** is posterior to the manubrium of the sternum and anterior to the bodies of the first four thoracic vertebrae (see Fig. 3.57).

- Its superior boundary is an oblique plane passing from the jugular notch upward and posteriorly to the superior border of vertebra TI.
- Inferiorly, a transverse plane passing from the sternal angle to the intervertebral disc between vertebra TIV/V separates it from the inferior mediastinum.
- Laterally, it is bordered by the mediastinal part of the parietal pleura on either side.

The superior mediastinum is continuous with the neck above and with the inferior mediastinum below.

The major structures found in the superior mediastinum (Figs. 3.86 and 3.87) include the:

- thymus,
- right and left brachiocephalic veins,
- left superior intercostal vein,
- superior vena cava,
- arch of the aorta with its three large branches,
- trachea,
- esophagus,
- phrenic nerves,
- vagus nerves,
- left recurrent laryngeal branch of the left vagus nerve,
- thoracic duct, and
- other small nerves, blood vessels, and lymphatics.

### Thymus

The **thymus** is the most anterior component of the superior mediastinum, lying immediately posterior to the

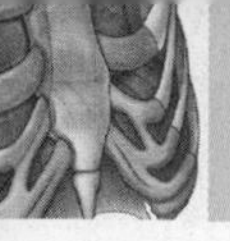

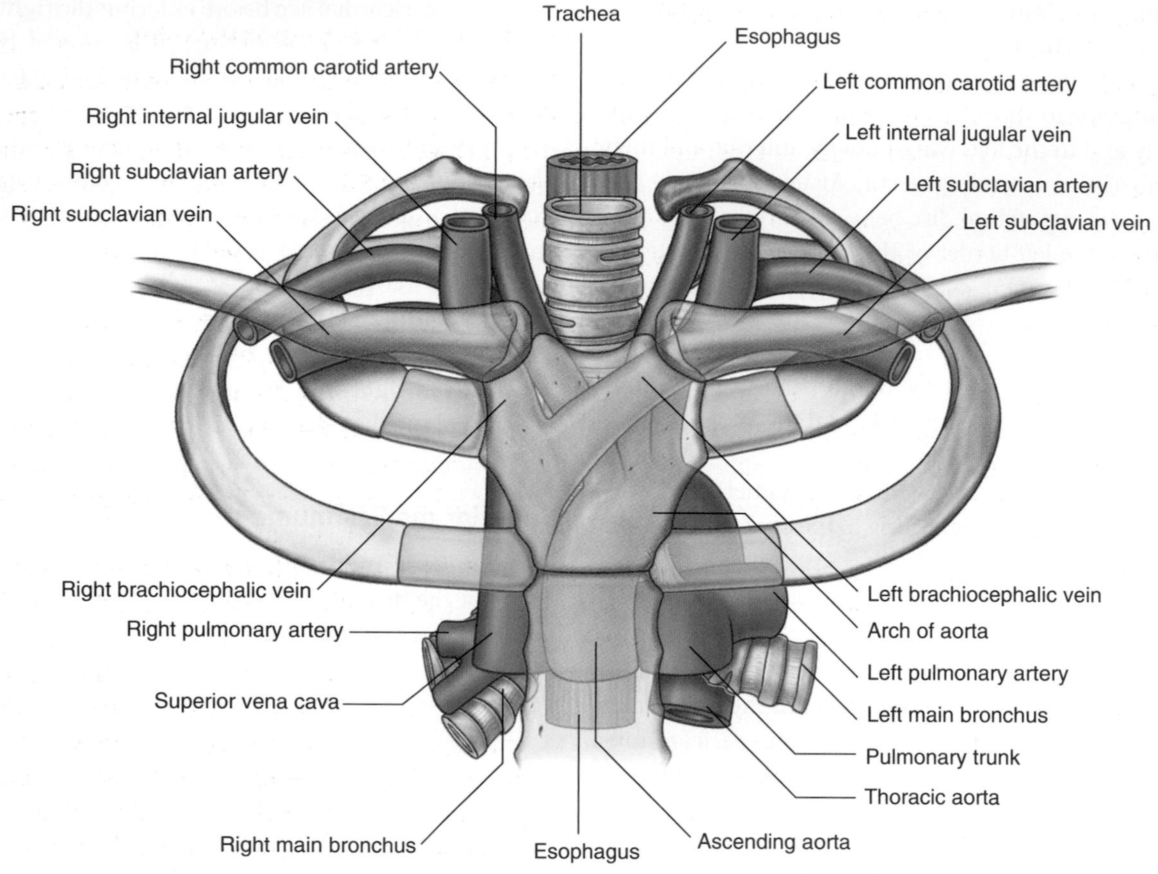


**Fig. 3.86** Structures in the superior mediastinum.

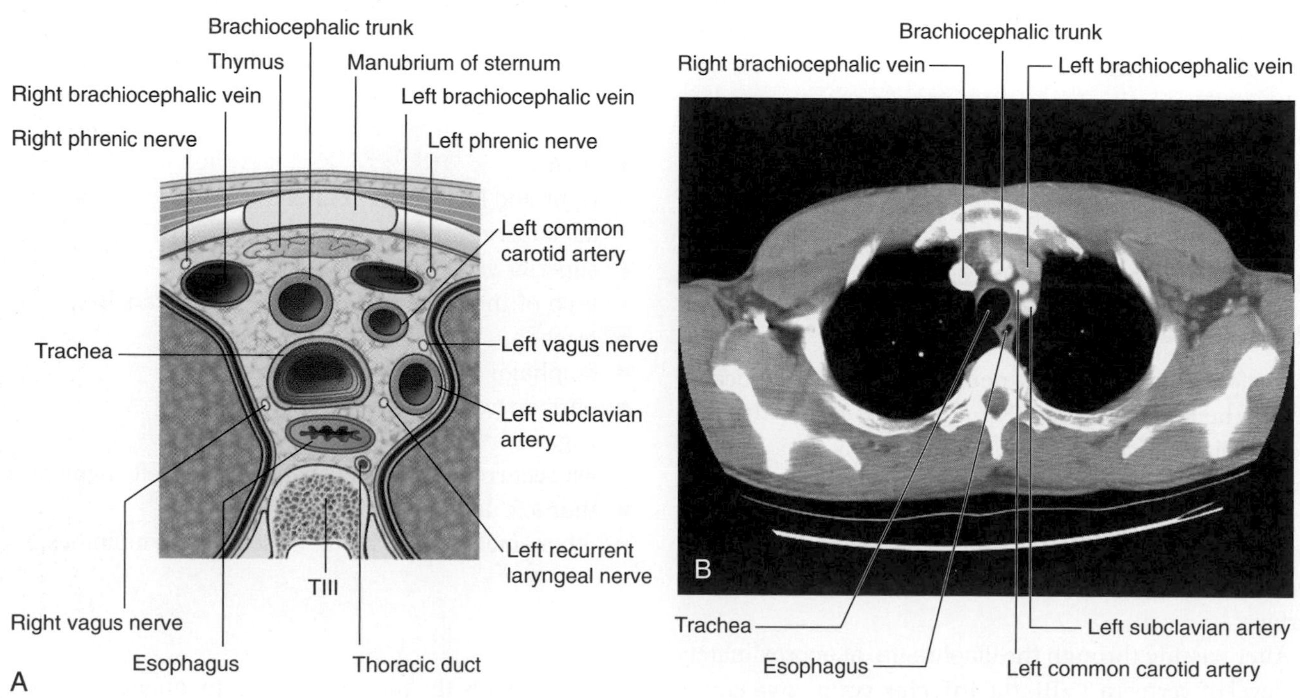


**Fig. 3.87** Cross section through the superior mediastinum at the level of vertebra TIII. **A.** Diagram. **B.** Axial computed tomography image.

manubrium of the sternum. It is an asymmetrical, bilobed structure (see Fig. 3.58).

The upper extent of the thymus can reach into the neck as high as the thyroid gland; a lower portion typically extends into the anterior mediastinum over the pericardial sac.

Involved in the early development of the immune system, the thymus is a large structure in the child, begins to atrophy after puberty, and shows considerable size variation in the adult. In the elderly adult, it is barely identifiable as an organ, consisting mostly of fatty tissue that is sometimes arranged as two lobulated fatty structures.

Arteries to the thymus consist of small branches originating from the internal thoracic arteries. Venous drainage is usually into the left brachiocephalic vein and possibly into the internal thoracic veins.

Lymphatic drainage returns to multiple groups of nodes at one or more of the following locations:

- along the internal thoracic arteries (parasternal);
- at the tracheal bifurcation (tracheobronchial); and
- in the root of the neck.

### In the clinic

**Ectopic parathyroid glands in the thymus**

The parathyroid glands develop from the third pharyngeal pouch, which also forms the thymus. The thymus is therefore a common site for ectopic parathyroid glands and, potentially, ectopic parathyroid hormone production.

### Right and left brachiocephalic veins

The left and right brachiocephalic veins are located immediately posterior to the thymus. They form on each side at the junction between the internal jugular and subclavian veins (see Fig. 3.86). The left brachiocephalic vein crosses the midline and joins with the right brachiocephalic vein to form the superior vena cava (Fig. 3.88).

- The **right brachiocephalic vein** begins posterior to the medial end of the right clavicle and passes vertically downward, forming the superior vena cava when it is joined by the left brachiocephalic vein. Venous tributaries include the vertebral, first posterior intercostal, and internal thoracic veins. The inferior thyroid and thymic veins may also drain into it.
- The **left brachiocephalic vein** begins posterior to the medial end of the left clavicle. It crosses to the right, moving in a slightly inferior direction, and joins with the right brachiocephalic vein to form the superior vena cava posterior to the lower edge of the right first costal cartilage close to the right sternal border. Venous tributaries include the vertebral, first posterior intercostal, left superior intercostal, inferior thyroid, and internal thoracic veins. It may also receive thymic and pericardial veins. The left brachiocephalic vein crosses the midline posterior to the manubrium in the adult. In infants and children the left brachiocephalic vein rises above the superior border of the manubrium and therefore is less protected.

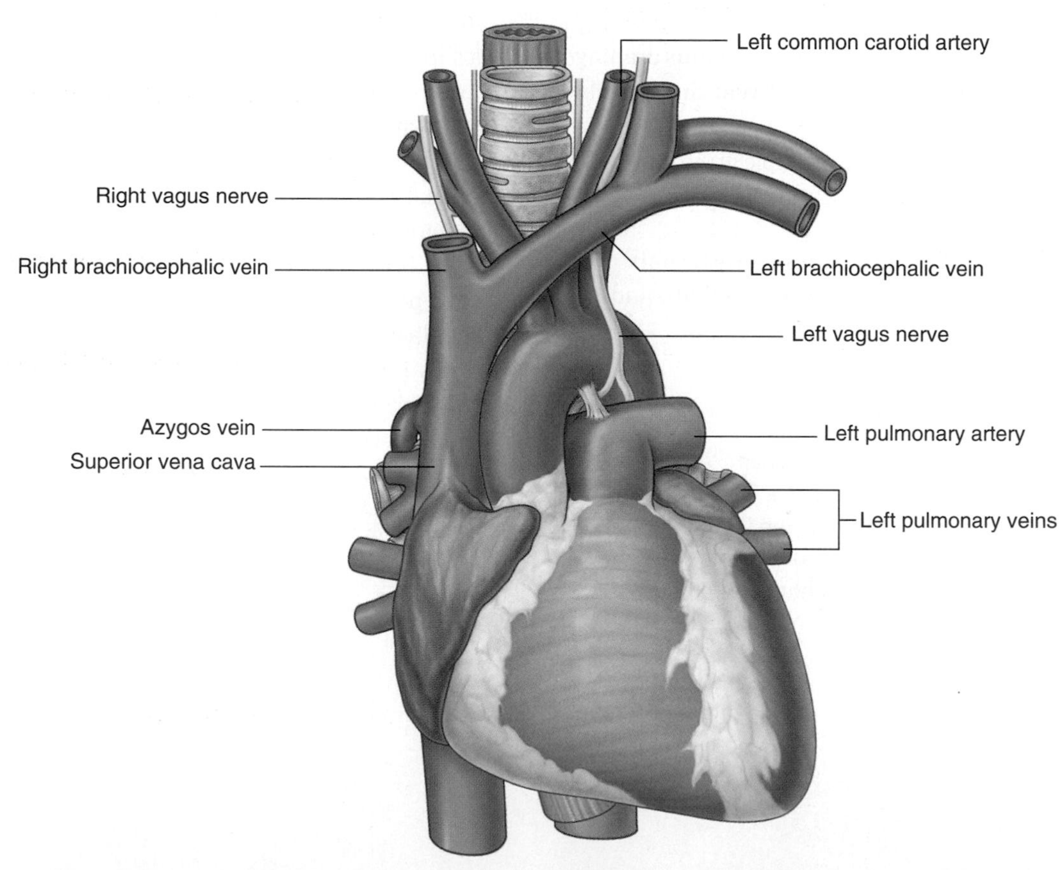


**Fig. 3.88** Superior mediastinum with thymus removed.

### Left superior intercostal vein

The **left superior intercostal vein** receives the second, third, and sometimes the fourth posterior intercostal veins, usually the left bronchial veins, and sometimes the left pericardiacophrenic vein. It passes over the left side of the aortic arch, lateral to the left vagus nerve and medial to the left phrenic nerve, before entering the left brachiocephalic vein (Fig. 3.89). Inferiorly, it may connect with the **accessory hemiazygos vein** (**superior hemiazygos vein**).

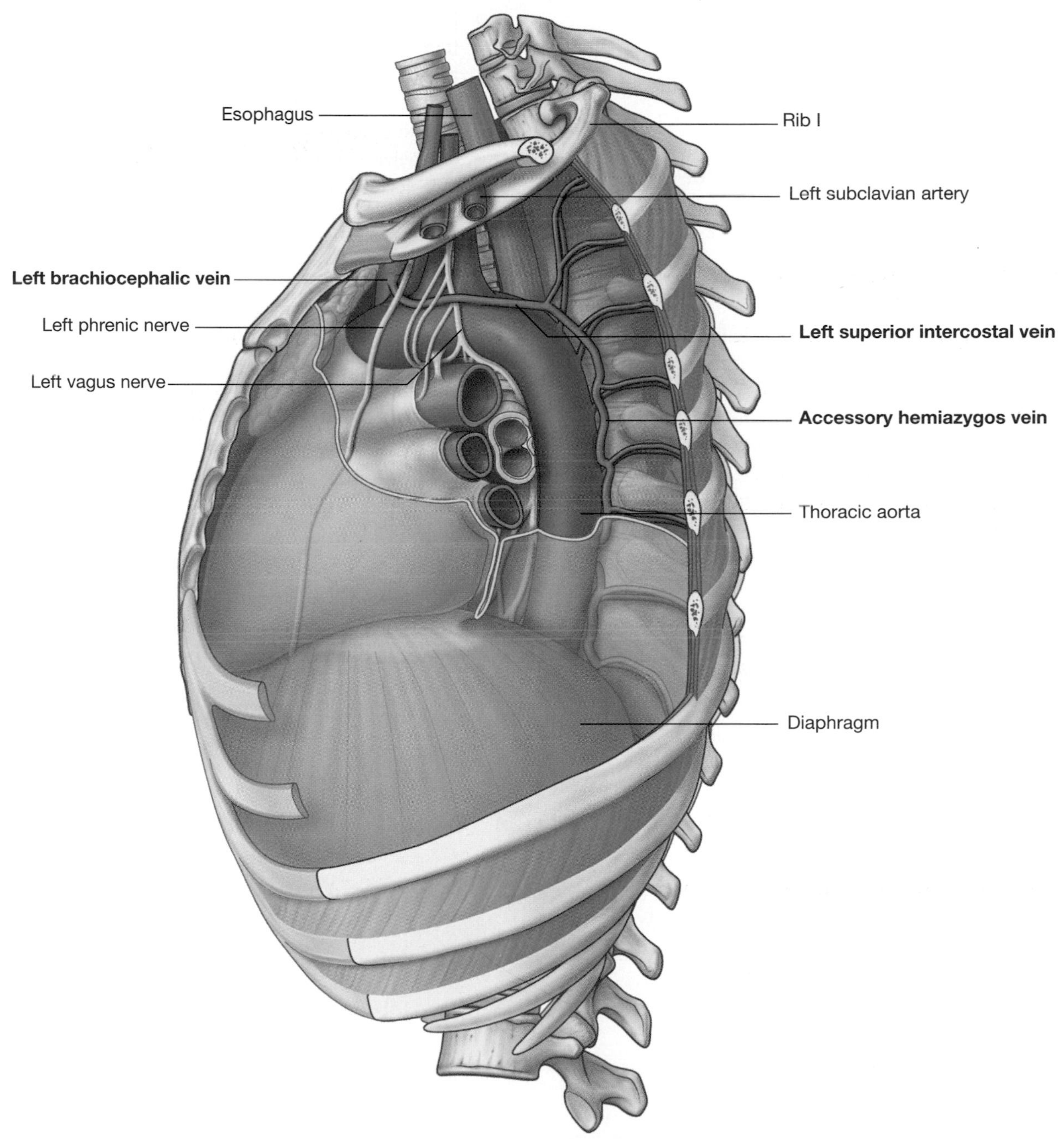


**Fig. 3.89** Left superior intercostal vein.

## Superior vena cava

The vertically oriented superior vena cava begins posterior to the lower edge of the right first costal cartilage, where the right and left brachiocephalic veins join, and terminates at the lower edge of the right third costal cartilage, where it joins the right atrium (see Fig. 3.86).

The lower half of the superior vena cava is within the pericardial sac and is therefore contained in the middle mediastinum.

The superior vena cava receives the azygos vein immediately before entering the pericardial sac and may also receive pericardial and mediastinal veins.

The superior vena cava can be easily visualized forming part of the right superolateral border of the mediastinum on a chest radiograph (see Fig. 3.67A).

### In the clinic

**Venous access for central and dialysis lines**

Large systemic veins are used to establish central venous access for administering large amounts of fluid, drugs, and blood. Most of these lines (small-bore tubes) are introduced through venous puncture into the axillary, subclavian, or internal jugular veins. The lines are then passed through the main veins of the superior mediastinum, with the tips of the lines usually residing in the distal portion of the superior vena cava or in the right atrium.

Similar devices, such as dialysis lines, are inserted into patients who have renal failure, so that a large volume of blood can be aspirated through one channel and reinfused through a second channel.

### In the clinic

**Using the superior vena cava to access the inferior vena cava**

Because the superior and inferior venae cavae are oriented along the same vertical axis, a guidewire, catheter, or line can be passed from the superior vena cava through the right atrium and into the inferior vena cava. This is a common route of access for such procedures as:

- transjugular liver biopsy,
- transjugular intrahepatic portosystemic shunts (TIPS), and
- insertion of an inferior vena cava filter to catch emboli dislodged from veins in the lower limb and pelvis (i.e., patients with deep vein thrombosis [DVT]).

## Arch of aorta and its branches

The thoracic portion of the aorta can be divided into **ascending aorta**, **arch of the aorta**, and **thoracic** (**descending**) **aorta**. Only the arch of the aorta is in the superior mediastinum. It begins when the ascending aorta emerges from the pericardial sac and courses upward, backward, and to the left as it passes through the superior mediastinum, ending on the left side at vertebral level TIV/V (see Fig. 3.86). Extending as high as the midlevel of the manubrium of the sternum, the arch is initially anterior and finally lateral to the trachea.

Three branches arise from the superior border of the arch of the aorta; at their origins, all three are crossed anteriorly by the left brachiocephalic vein.

### The first branch

Beginning on the right, the first branch of the arch of the aorta is the **brachiocephalic trunk** (Fig. 3.90). It is the largest of the three branches and, at its point of origin behind the manubrium of the sternum, is slightly anterior to the other two branches. It ascends slightly posteriorly and to the right. At the level of the upper edge of the right sternoclavicular joint, the brachiocephalic trunk divides into:

- the **right common carotid artery,** and
- the **right subclavian artery** (see Fig. 3.86).

The arteries mainly supply the right side of the head and neck and the right upper limb, respectively.

Occasionally, the brachiocephalic trunk has a small branch, the **thyroid ima artery**, which contributes to the vascular supply of the thyroid gland.

### The second branch

The second branch of the arch of the aorta is the **left common carotid artery** (Fig. 3.90). It arises from the arch immediately to the left and slightly posterior to the brachiocephalic trunk and ascends through the superior mediastinum along the left side of the trachea.

The left common carotid artery supplies the left side of the head and neck.

### The third branch

The third branch of the arch of the aorta is the **left subclavian artery** (Fig. 3.90). It arises from the arch of the aorta immediately to the left of, and slightly posterior to, the left common carotid artery and ascends through the superior mediastinum along the left side of the trachea.

The left subclavian artery is the major blood supply to the left upper limb.

#### Ligamentum arteriosum

The **ligamentum arteriosum** is also in the superior mediastinum and is important in embryonic circulation, when it is a patent vessel (the **ductus arteriosus**). It connects the pulmonary trunk with the arch of the aorta and allows blood to bypass the lungs during development (Fig. 3.90). The vessel closes soon after birth and forms the ligamentous connection observed in the adult.

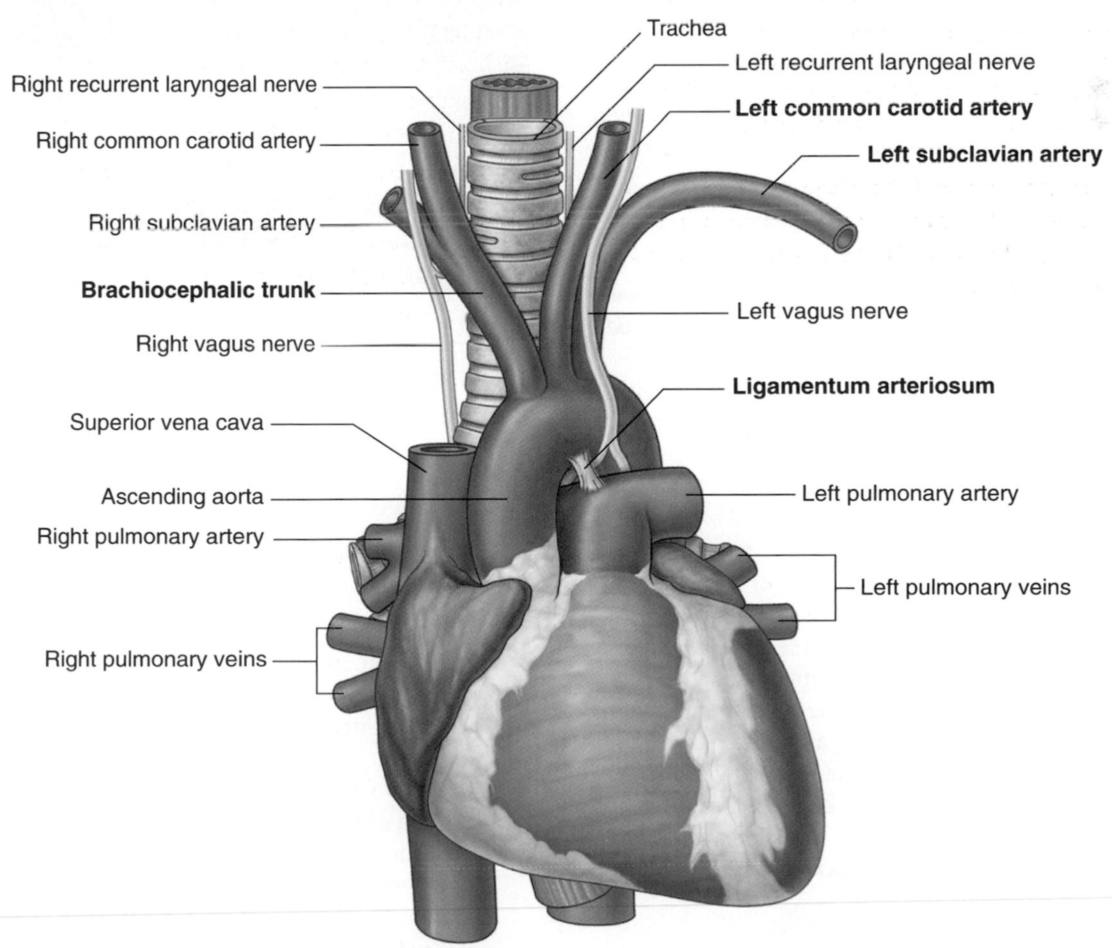


**Fig. 3.90** Superior mediastinum with thymus and venous channels removed.

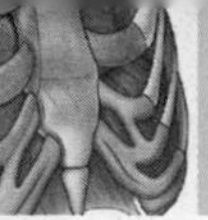



## In the clinic

### Coarctation of the aorta

Coarctation of the aorta is a congenital abnormality in which the aortic lumen is constricted just distal to the origin of the left subclavian artery. At this point, the aorta becomes significantly narrowed and the blood supply to the lower limbs and abdomen is diminished. Over time, collateral vessels develop around the chest wall and abdomen to supply the lower body. Dilated and tortuous intercostal vessels, which form a bypass to supply the descending thoracic aorta, may lead to erosions of the inferior margins of the ribs. This can be appreciated on chest radiographs as inferior rib notching and is usually seen in long standing cases. The coarctation also affects the heart, which has to pump the blood at higher pressure to maintain peripheral perfusion. This in turn may produce cardiac failure.

## In the clinic

### Thoracic aorta

Diffuse atherosclerosis of the thoracic aorta may occur in patients with vascular disease, but this rarely produces symptoms. There are, however, two clinical situations in which aortic pathology can produce life-threatening situations.

#### *Trauma*

The aorta has three fixed points of attachment:

- the aortic valve,
- the ligamentum arteriosum, and
- the point of passing behind the median arcuate ligament of the diaphragm to enter the abdomen.

The rest of the aorta is relatively free from attachment to other structures of the mediastinum. A serious deceleration injury (e.g., in a road traffic accident) is most likely to cause aortic trauma at these fixed points.

#### *Aortic dissection*

In certain conditions, such as in severe arteriovascular disease, the wall of the aorta can split longitudinally, creating a false channel, which may or may not rejoin into the true lumen distally (Fig. 3.91). This aortic dissection occurs between the intima and media anywhere along its length. If it occurs in the ascending aorta or arch of the aorta, blood flow in the coronary and cerebral arteries may be disrupted, resulting in myocardial infarction or stroke. In the abdomen the visceral vessels may be disrupted, producing ischemia to the gut or kidneys.

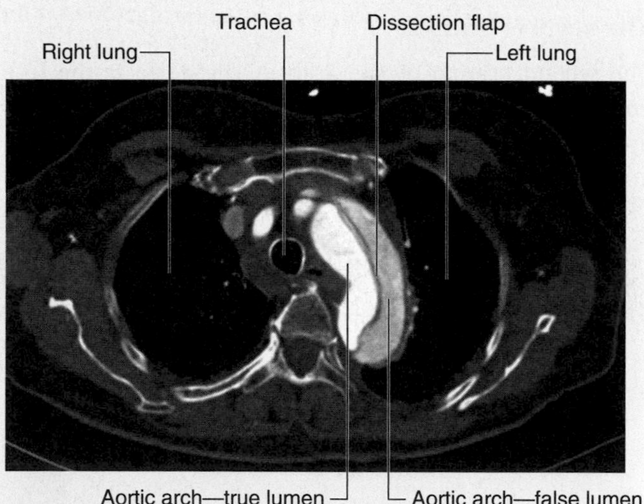


**Fig. 3.91** Axial CT showing aortic dissection.

## In the clinic

### Aortic arch and its anomalies

The normal aortic arch courses to the left of the trachea and passes over the left main bronchus. A right-sided aortic arch occurs when the vessel courses to the right of the trachea and passes over the right main bronchus. A right-sided arch of aorta is rare and may be asymptomatic. It can be associated with **dextrocardia** (right-sided heart) and, in some instances, with complete **situs inversus** (left-to-right inversion of the body's organs). It can also be associated with abnormal branching of the great vessels, particularly with an aberrant left subclavian artery.

## In the clinic

### Abnormal origin of great vessels

Great vessels occasionally have an abnormal origin, including:

- a common origin of the brachiocephalic trunk and the left common carotid artery,
- the left vertebral artery originating from the aortic arch, and
- the right subclavian artery originating from the distal portion of the aortic arch and passing behind the esophagus to supply the right arm—as a result, the great vessels form a vascular ring around the trachea and the esophagus, which can potentially produce difficulty swallowing. This configuration is one of the most common aortic arch abnormalities.

## Trachea and esophagus

The trachea is a midline structure that is palpable in the jugular notch as it enters the superior mediastinum. Posterior to it is the esophagus, which is immediately anterior to the vertebral column (Fig. 3.92, and see Figs. 3.86 and 3.87). Significant mobility exists in the vertical positioning of these structures as they pass through the superior mediastinum. Swallowing and breathing cause positional shifts, as may disease and the use of specialized instrumentation.

As the trachea and esophagus pass through the superior mediastinum, they are crossed laterally by the azygos vein on the right side and the arch of the aorta on the left side.

The trachea divides into the right and left main bronchi at, or just inferior to, the transverse plane between the sternal angle and vertebral level TIV/V (Fig. 3.93), whereas the esophagus continues into the posterior mediastinum.

## Nerves of the superior mediastinum

### Vagus nerves

The **vagus nerves** [X] pass through the superior and posterior divisions of the mediastinum on their way to the abdominal cavity. As they pass through the thorax, they provide parasympathetic innervation to the thoracic viscera and carry visceral afferents from the thoracic viscera.

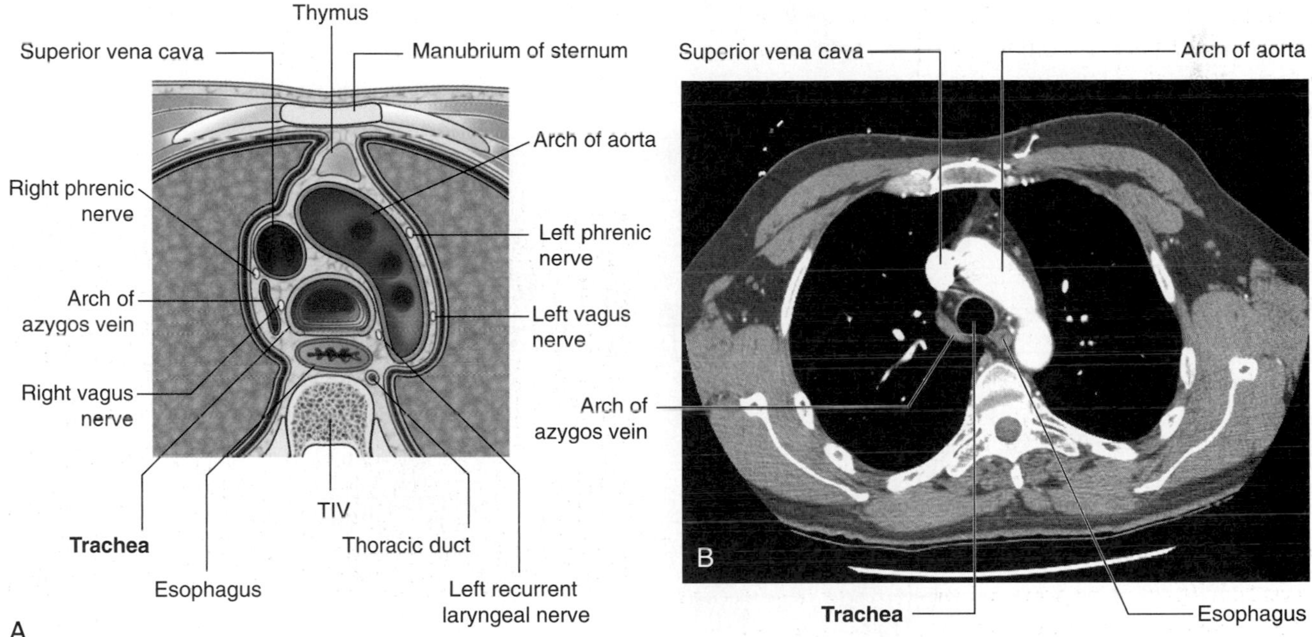


**Fig. 3.92** Cross section through the superior mediastinum at the level of vertebra TIV. **A.** Diagram. **B.** Axial computed tomography image.

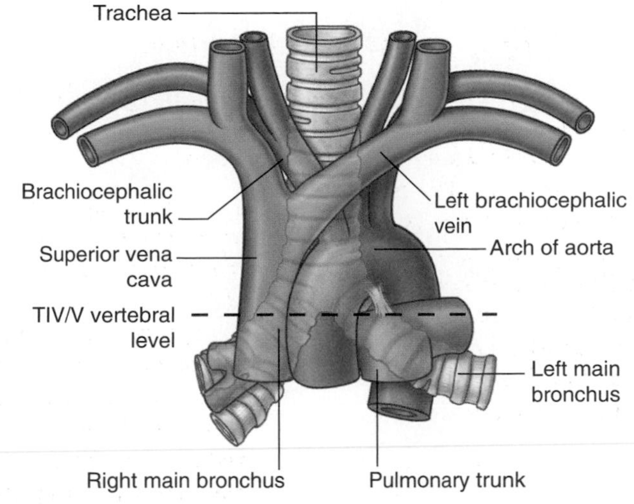


**Fig. 3.93** Trachea in the superior mediastinum.

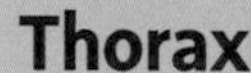


Visceral afferents in the vagus nerves relay information to the central nervous system about normal physiological processes and reflex activities. They do not transmit pain sensation.

### Right vagus nerve

The **right vagus nerve** enters the superior mediastinum and lies between the right brachiocephalic vein and the brachiocephalic trunk. It descends in a posterior direction toward the trachea (Fig. 3.94), crosses the lateral surface of the trachea, and passes posteriorly to the root of the right lung to reach the esophagus. Just before the esophagus, it is crossed by the arch of the azygos vein.

As the right vagus nerve passes through the superior mediastinum, it gives branches to the esophagus, cardiac plexus, and pulmonary plexus.

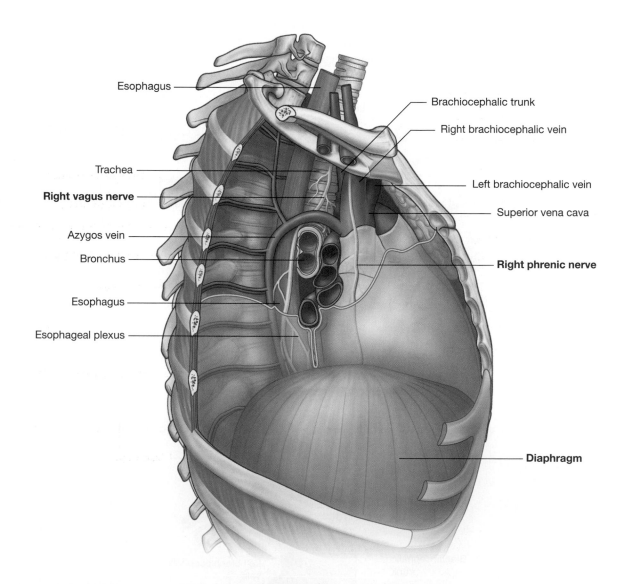

**Fig. 3.94** Right vagus nerve passing through the superior mediastinum.

### Left vagus nerve

The **left vagus nerve** enters the superior mediastinum posterior to the left brachiocephalic vein and between the left common carotid and left subclavian arteries (Fig. 3.95). As it passes into the superior mediastinum, it lies just deep to the mediastinal part of the parietal pleura and crosses the left side of the arch of the aorta. It continues to descend in a posterior direction and passes posterior to the root of the left lung to reach the esophagus in the posterior mediastinum.

As the left vagus nerve passes through the superior mediastinum, it gives branches to the esophagus, the cardiac plexus, and the pulmonary plexus.

The left vagus nerve also gives rise to the **left recurrent laryngeal nerve**, which arises from it at the inferior margin of the arch of the aorta just lateral to the ligamentum arteriosum. The left recurrent laryngeal nerve passes inferior to the arch of the aorta before ascending on its medial surface. Entering a groove between the trachea and esophagus, the left recurrent laryngeal nerve continues

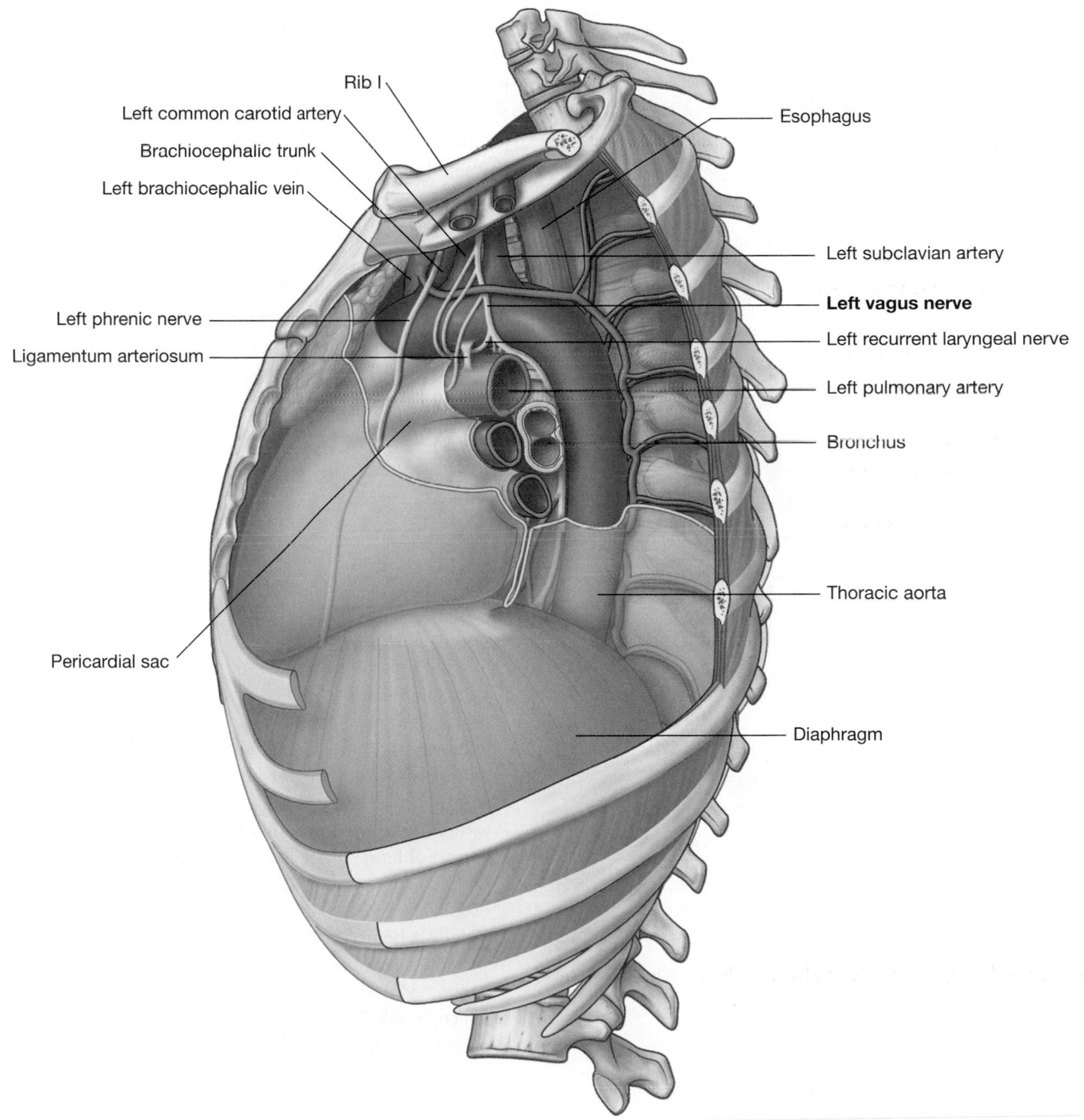


**Fig. 3.95** Left vagus nerve passing through the superior mediastinum.

superiorly to enter the neck and terminate in the larynx (Fig. 3.96).

### Phrenic nerves

The phrenic nerves arise in the cervical region mainly from the fourth, but also from the third and fifth, cervical spinal cord segments.

The phrenic nerves descend through the thorax to supply motor and sensory innervation to the diaphragm and its associated membranes. As they pass through the thorax, they provide innervation through somatic afferent fibers to the mediastinal pleura, fibrous pericardium, and parietal layer of serous pericardium.

#### Right phrenic nerve

The **right phrenic nerve** enters the superior mediastinum lateral to the right vagus nerve and lateral and slightly posterior to the beginning of the right brachiocephalic vein (see Fig. 3.94). It continues inferiorly along the right side of this vein and the right side of the superior vena cava.

On entering the middle mediastinum, the right phrenic nerve descends along the right side of the pericardial sac, within the fibrous pericardium, anterior to the root of the right lung. The pericardiacophrenic vessels accompany it through most of its course in the thorax (see Fig. 3.60). It leaves the thorax by passing through the diaphragm with the inferior vena cava.

#### Left phrenic nerve

The **left phrenic nerve** enters the superior mediastinum in a position similar to the path taken by the right phrenic nerve. It lies lateral to the left vagus nerve and lateral and slightly posterior to the beginning of the left brachiocephalic vein (see Fig. 3.89), and continues to descend across the left lateral surface of the arch of the aorta, passing superficially to the left vagus nerve and the left superior intercostal vein.

On entering the middle mediastinum, the left phrenic nerve follows the left side of the pericardial sac, within the fibrous pericardium, anterior to the root of the left lung, and is accompanied by the pericardiacophrenic vessels (see Fig. 3.60). It leaves the thorax by piercing the diaphragm near the apex of the heart.

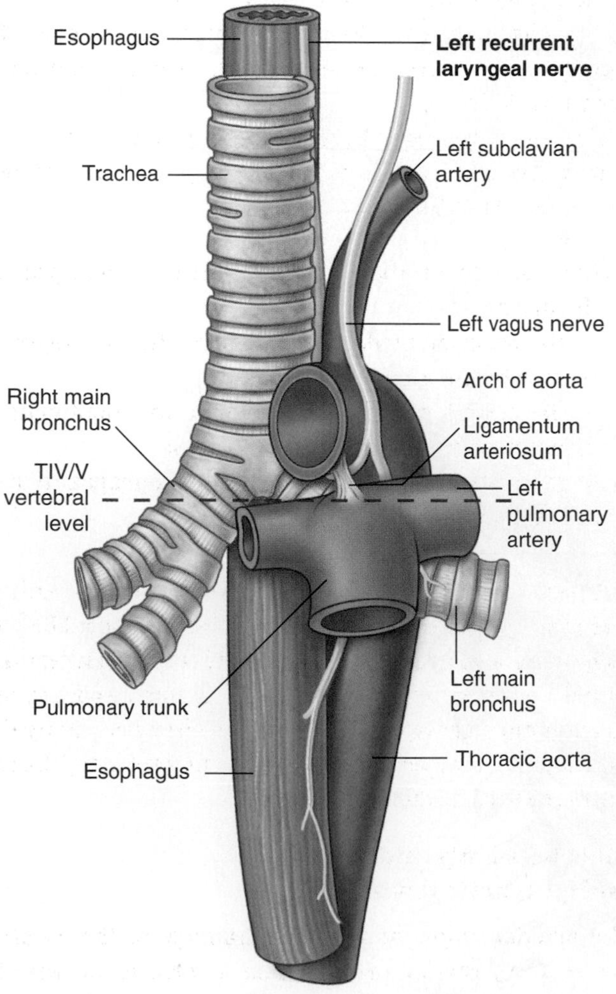

**Fig. 3.96** Left recurrent laryngeal nerve passing through the superior mediastinum.

## In the clinic

### The vagus nerves, recurrent laryngeal nerves, and hoarseness

The left recurrent laryngeal nerve is a branch of the left vagus nerve. It passes between the pulmonary artery and the aorta, a region known clinically as the **aortopulmonary window**, and may be compressed in any patient with a pathological mass in this region. This compression results in left vocal cord paralysis and hoarseness of the voice. Lymph node enlargement, often associated with the spread of lung cancer, is a common condition that may produce compression. Chest radiography is therefore usually carried out for all patients whose symptoms include a hoarse voice.

More superiorly, in the root of the neck, the right vagus nerve gives off the right recurrent laryngeal nerve, which "hooks" around the right subclavian artery as it passes over the cervical pleura. If a patient has a hoarse voice and a right vocal cord palsy is demonstrated at laryngoscopy, chest radiography with an apical lordotic view should be obtained to assess for cancer in the right lung apex (**Pancoast's tumor**).

### Thoracic duct in the superior mediastinum

The **thoracic duct**, which is the major lymphatic vessel in the body, passes through the posterior portion of the superior mediastinum (see Figs. 3.87 and 3.92). It:

- enters the superior mediastinum inferiorly, slightly to the left of the midline, having moved to this position just before leaving the posterior mediastinum opposite vertebral level TIV/V; and
- continues through the superior mediastinum, posterior to the arch of the aorta, and the initial portion of the left subclavian artery, between the esophagus and the left mediastinal part of the parietal pleura.

## Posterior mediastinum

The **posterior mediastinum** is posterior to the pericardial sac and diaphragm and anterior to the bodies of the mid and lower thoracic vertebrae (see Fig. 3.57).

- Its superior boundary is a transverse plane passing from the sternal angle to the intervertebral disc between vertebrae TIV and TV.
- Its inferior boundary is the diaphragm.
- Laterally, it is bordered by the mediastinal part of parietal pleura on either side.
- Superiorly, it is continuous with the superior mediastinum.

Major structures in the posterior mediastinum include the:

- esophagus and its associated nerve plexus,
- thoracic aorta and its branches,
- azygos system of veins,
- thoracic duct and associated lymph nodes,
- sympathetic trunks, and
- thoracic splanchnic nerves.

### Esophagus

The **esophagus** is a muscular tube passing between the pharynx in the neck and the stomach in the abdomen. It begins at the inferior border of the cricoid cartilage, opposite vertebra CVI, and ends at the cardiac opening of the stomach, opposite vertebra TXI.

The esophagus descends on the anterior aspect of the bodies of the vertebrae, generally in a midline position as it moves through the thorax (Fig. 3.97). As it approaches the diaphragm, it moves anteriorly and to the left, crossing from the right side of the thoracic aorta to eventually assume a position anterior to it. It then passes through the esophageal hiatus, an opening in the muscular part of the diaphragm, at vertebral level TX.

The esophagus has a slight anterior-to-posterior curvature that parallels the thoracic portion of the vertebral column, and is secured superiorly by its attachment to the pharynx and inferiorly by its attachment to the diaphragm.

#### Relationships to important structures in the posterior mediastinum

In the posterior mediastinum, the esophagus is related to a number of important structures. The right side is covered by the mediastinal part of the parietal pleura.

Posterior to the esophagus, the thoracic duct is on the right side inferiorly, but crosses to the left more superiorly. Also on the left side of the esophagus is the thoracic aorta.

Anterior to the esophagus, below the level of the tracheal bifurcation, are the right pulmonary artery and the left main bronchus. The esophagus then passes immediately posteriorly to the left atrium, separated from it only by pericardium. Inferior to the left atrium, the esophagus is related to the diaphragm.

Structures other than the thoracic duct posterior to the esophagus include portions of the hemiazygos veins, the right posterior intercostal vessels, and, near the diaphragm, the thoracic aorta.

The esophagus is a flexible, muscular tube that can be compressed or narrowed by surrounding structures at four locations (Fig. 3.98):

- the junction of the esophagus with the pharynx in the neck;
- in the superior mediastinum where the esophagus is crossed by the arch of the aorta;
- in the posterior mediastinum where the esophagus is compressed by the left main bronchus;
- in the posterior mediastinum at the esophageal hiatus in the diaphragm.

These constrictions have important clinical consequences. For example, a swallowed object is most likely to lodge at a constricted area. An ingested corrosive substance would move more slowly through a narrowed region, causing more damage at this site than elsewhere along the esophagus. Also, constrictions present problems during the passage of medical instruments.

#### Arterial supply and venous and lymphatic drainage

The arterial supply and venous drainage of the esophagus in the posterior mediastinum involve many vessels. Esophageal arteries arise from the thoracic aorta, bronchial arteries, and ascending branches of the left gastric artery in the abdomen.

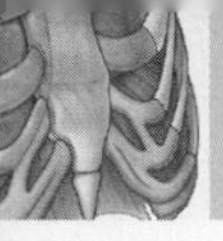





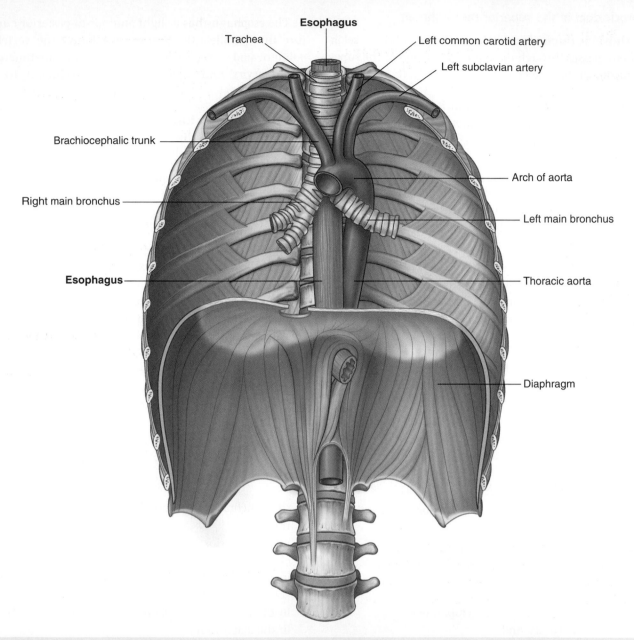

**Fig. 3.97** Esophagus.

Venous drainage involves small vessels returning to the azygos vein, hemiazygos vein, and esophageal branches to the left gastric vein in the abdomen.

Lymphatic drainage of the esophagus in the posterior mediastinum returns to posterior mediastinal and left gastric nodes.

### Innervation

Innervation of the esophagus, in general, is complex. Esophageal branches arise from the vagus nerves and sympathetic trunks.

Striated muscle fibers in the superior portion of the esophagus originate from the branchial arches and are innervated by branchial efferents from the vagus nerves.

Smooth muscle fibers are innervated by cranial components of the parasympathetic part of the autonomic division of the peripheral nervous system, visceral efferents from the vagus nerves. These are preganglionic fibers that synapse in the myenteric and submucosal plexuses of the enteric nervous system in the esophageal wall.

Sensory innervation of the esophagus involves visceral afferent fibers originating in the vagus nerves, sympathetic trunks, and splanchnic nerves.

The visceral afferents from the vagus nerves are involved in relaying information back to the central nervous system about normal physiological processes and reflex activities. They are not involved in the relay of pain recognition.

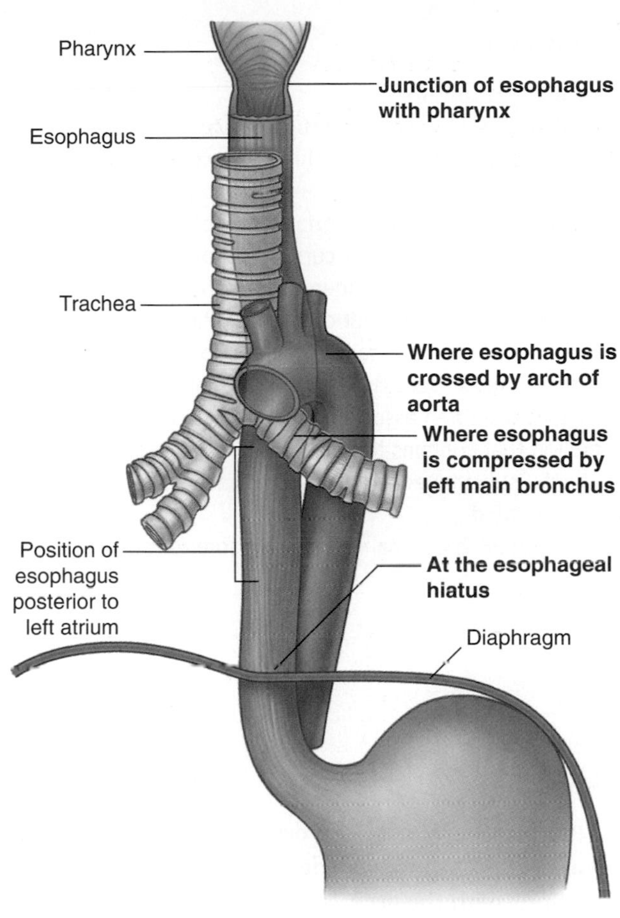


**Fig. 3.98** Sites of normal esophageal constrictions.

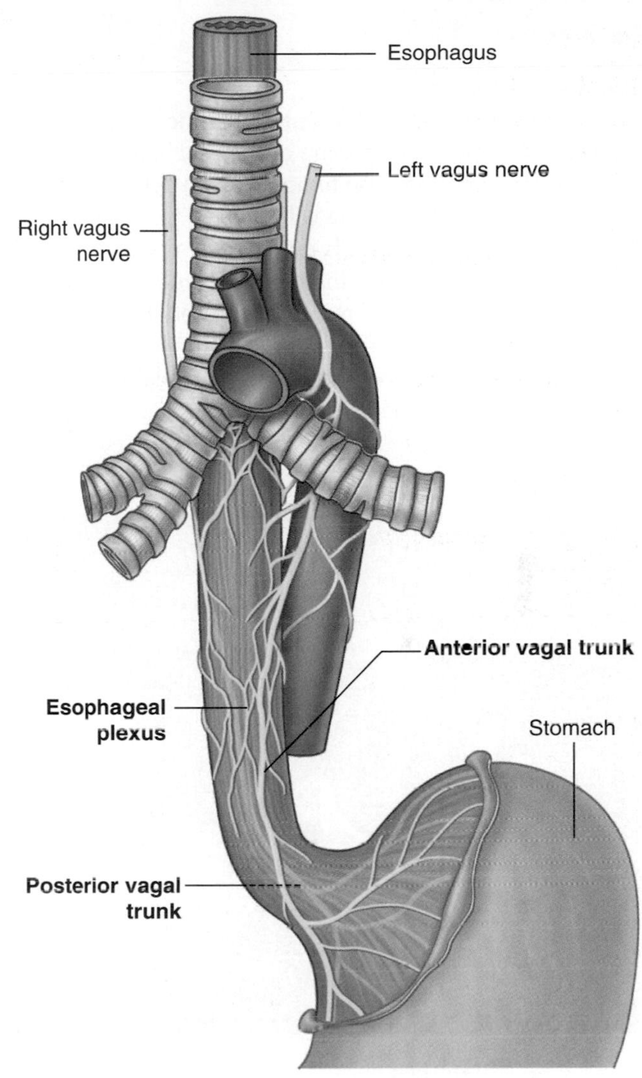


**Fig. 3.99** Esophageal plexus.

The visceral afferents that pass through the sympathetic trunks and the splanchnic nerves are the primary participants in detection of esophageal pain and transmission of this information to various levels of the central nervous system.

### Esophageal plexus

After passing posteriorly to the root of the lungs, the right and left vagus nerves approach the esophagus. As they reach the esophagus, each nerve divides into several branches that spread over this structure, forming the **esophageal plexus** (Fig. 3.99). There is some mixing of fibers from the two vagus nerves as the plexus continues inferiorly on the esophagus toward the diaphragm. Just above the diaphragm, fibers of the plexus converge to form two trunks:

- the **anterior vagal trunk** on the anterior surface of the esophagus, mainly from fibers originally in the left vagus nerve;
- the **posterior vagal trunk** on the posterior surface of the esophagus, mainly from fibers originally in the right vagus nerve.

The vagal trunks continue on the surface of the esophagus as it passes through the diaphragm into the abdomen.

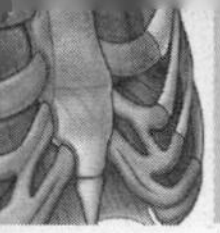

## In the clinic

### Esophageal cancer

When patients present with esophageal cancer, it is important to note which portion of the esophagus contains the tumor because tumor location determines the sites to which the disease will spread (Fig. 3.100).

Esophageal cancer spreads quickly to lymphatics, draining to lymph nodes in the neck and around the celiac artery. Endoscopy or barium swallow is used to assess the site. CT and MRI may be necessary to stage the disease.

Once the extent of the disease has been assessed, treatment can be planned.

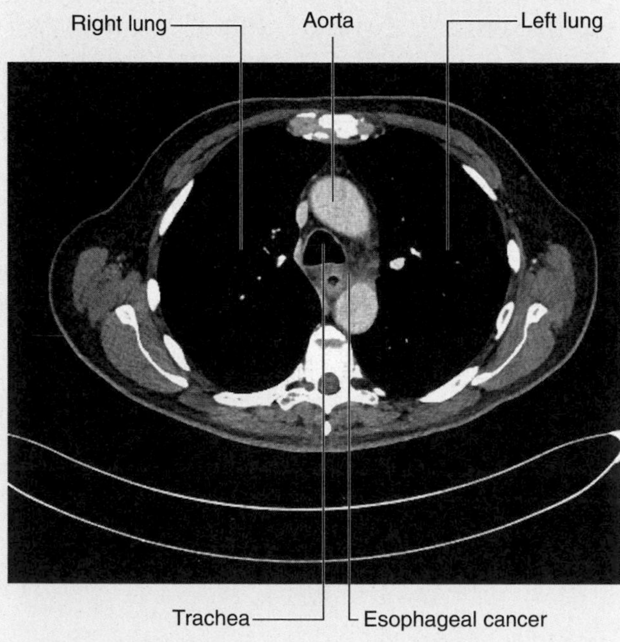


**Fig. 3.100** Axial CT showing esophageal cancer.

## In the clinic

### Esophageal rupture

The first case of esophageal rupture was described by Herman Boerhaave in 1724. This case was fatal, but early diagnosis has increased the survival rate up to 65%. If the disease is left untreated, mortality is 100%.

Typically, the rupture occurs in the lower third of the esophagus with a sudden rise in intraluminal esophageal pressure produced by vomiting secondary to an uncoordination and failure of the cricopharyngeus muscle to relax. Because the tears typically occur on the left, they are often associated with a large left pleural effusion that contains the gastric contents. In some patients, subcutaneous emphysema may be demonstrated.

Treatment is optimal with urgent surgical repair.

### Thoracic aorta

The thoracic portion of the descending aorta (**thoracic aorta**) begins at the lower edge of vertebra TIV, where it is continuous with the arch of the aorta. It ends anterior to the lower edge of vertebra TXII, where it passes through the aortic hiatus posterior to the diaphragm. Situated to the left of the vertebral column superiorly, it approaches the midline inferiorly, lying directly anterior to the lower thoracic vertebral bodies (Fig. 3.101). Throughout its course, it gives off a number of branches, which are summarized in Table 3.3.

**Table 3.3** Branches of the thoracic aorta

| Branches | Origin and course |
|---|---|
| Pericardial branches | A few small vessels to the posterior surface of the pericardial sac |
| Bronchial branches | Vary in number, size, and origin—usually, two left bronchial arteries from the thoracic aorta and one right bronchial artery from the third posterior intercostal artery or the superior left bronchial artery |
| Esophageal branches | Four or five vessels from the anterior aspect of the thoracic aorta, which form a continuous anastomotic chain—anastomotic connections include esophageal branches of the inferior thyroid artery superiorly, and esophageal branches of the left inferior phrenic and the left gastric arteries inferiorly |
| Mediastinal branches | Several small branches supplying lymph nodes, vessels, nerves, and areolar tissue in the posterior mediastinum |
| Posterior intercostal arteries | Usually, nine pairs of vessels branching from the posterior surface of the thoracic aorta—usually supply lower nine intercostal spaces (first two spaces are supplied by the supreme intercostal artery—a branch of the costocervical trunk) |
| Superior phrenic arteries | Small vessels from the lower part of the thoracic aorta supplying the posterior part of the superior surface of the diaphragm—they anastomose with the musculophrenic and pericardiacophrenic arteries |
| Subcostal artery | The lowest pair of branches from the thoracic aorta located inferior to rib XII |

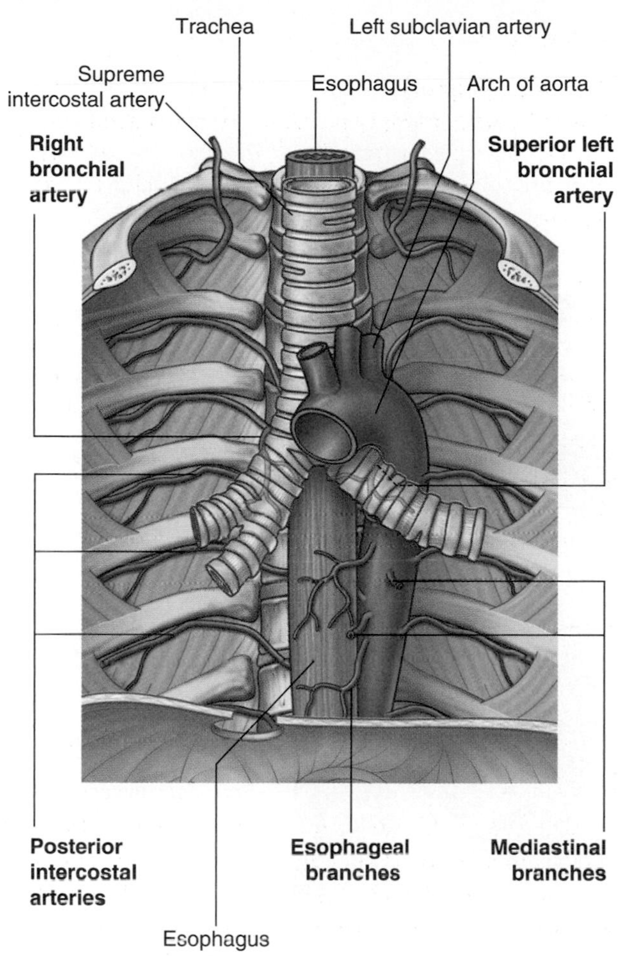


**Fig. 3.101** Thoracic aorta and branches.

## Azygos system of veins

The azygos system of veins consists of a series of longitudinal vessels on each side of the body that drain blood from the body wall and move it superiorly to empty into the superior vena cava. Blood from some of the thoracic viscera may also enter the system, and there are anastomotic connections with abdominal veins.

The longitudinal vessels may or may not be continuous and are connected to each other from side to side at various points throughout their course (Fig. 3.102).

The azygos system of veins serves as an important anastomotic pathway capable of returning venous blood from the lower part of the body to the heart if the inferior vena cava is blocked.

The major veins in the system are:

- the azygos vein, on the right; and
- the hemiazygos vein and the accessory hemiazygos vein, on the left.

There is significant variation in the origin, course, tributaries, anastomoses, and termination of these vessels.

### Azygos vein

The **azygos vein** arises opposite vertebra LI or LII at the junction between the **right ascending lumbar vein** and the **right subcostal vein** (Fig. 3.102). It may also arise as a direct branch of the inferior vena cava, which is joined by a common trunk from the junction of the right ascending lumbar vein and the right subcostal vein.

The azygos vein enters the thorax through the aortic hiatus of the diaphragm, or it enters through or posterior to the right crus of the diaphragm. It ascends through the posterior mediastinum, usually to the right of the thoracic duct. At approximately vertebral level TIV, it arches anteriorly, over the root of the right lung, to join the superior vena cava before the superior vena cava enters the pericardial sac.

Tributaries of the azygos vein include:

- the **right superior intercostal vein** (a single vessel formed by the junction of the second, third, and fourth intercostal veins),
- fifth to eleventh right posterior intercostal veins,
- the hemiazygos vein,
- the accessory hemiazygos vein,
- esophageal veins,
- mediastinal veins,
- pericardial veins, and
- right bronchial veins.

### Hemiazygos vein

The **hemiazygos vein** (**inferior hemiazygos vein**) usually arises at the junction between the **left ascending lumbar vein** and the **left subcostal vein** (Fig. 3.102). It may also arise from either of these veins alone and often has a connection to the left renal vein.

The hemiazygos vein usually enters the thorax through the left crus of the diaphragm, but may enter through the aortic hiatus. It ascends through the posterior mediastinum, on the left side, to approximately vertebral level TIX. At this point, it crosses the vertebral column, posterior to the thoracic aorta, esophagus, and thoracic duct, to enter the azygos vein.

Tributaries joining the hemiazygos vein include:

- the lowest four or five left posterior intercostal veins,
- esophageal veins, and
- mediastinal veins.

Right superior intercostal vein
Opening of azygos vein into superior vena cava
Azygos vein
Right subcostal vein
Right ascending lumbar vein
Left superior intercostal vein
Accessory hemiazygos vein
Posterior intercostal vein
Hemiazygos vein
Ascending lumbar vein
Inferior vena cava

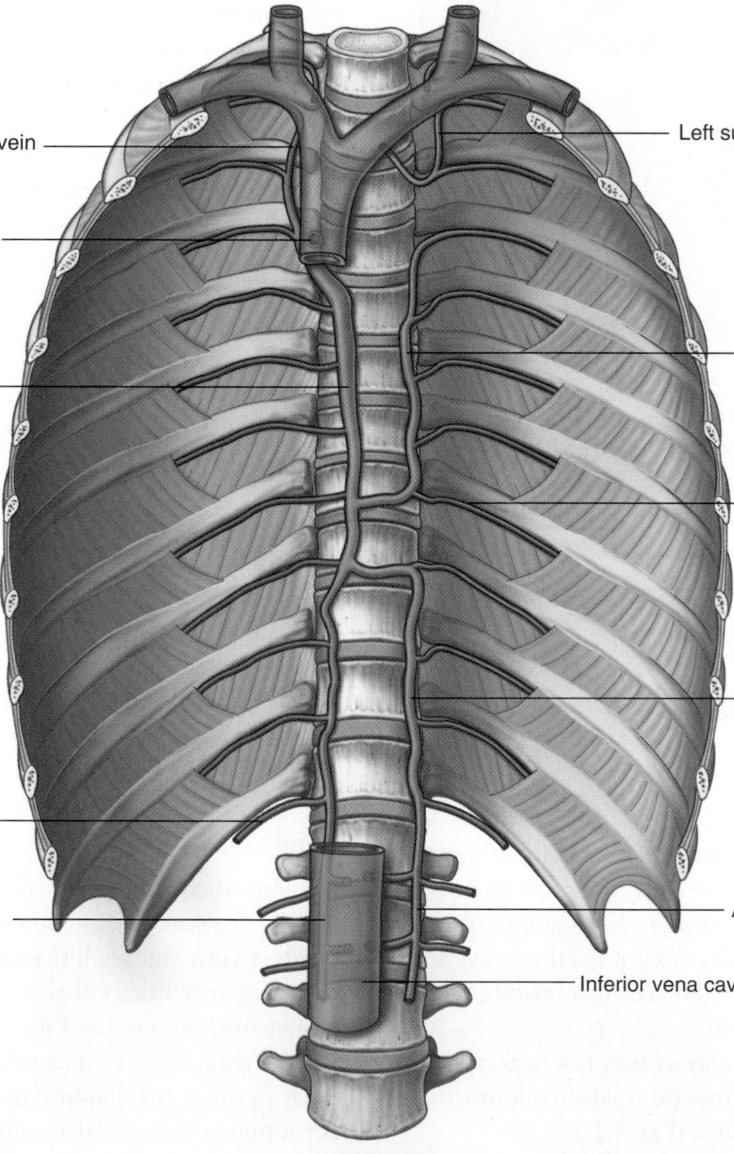

**Fig. 3.102** Azygos system of veins.

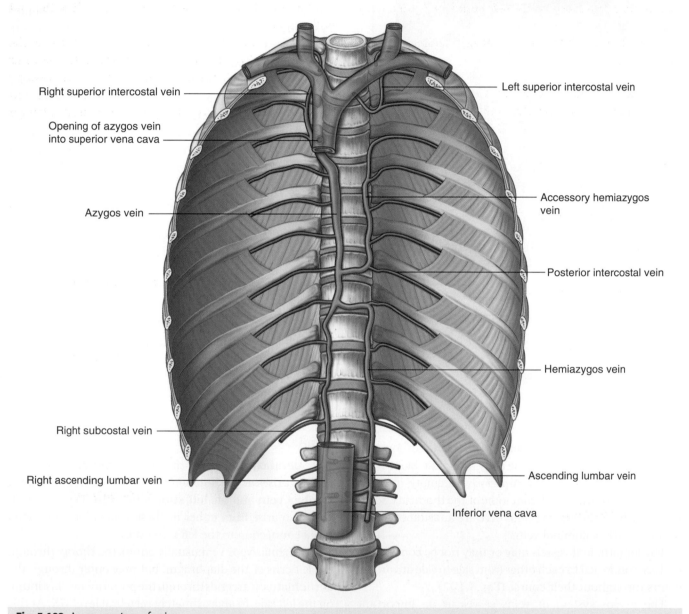

### Accessory hemiazygos vein

The **accessory hemiazygos vein** (**superior hemiazygos vein**) descends on the left side from the superior portion of the posterior mediastinum to approximately vertebral level TVIII (Fig. 3.102). At this point, it crosses the vertebral column to join the azygos vein, or ends in the hemiazygos vein, or has a connection to both veins. Usually, it also has a connection superiorly to the **left superior intercostal vein**.

Vessels that drain into the accessory hemiazygos vein include:

- the fourth to eighth left posterior intercostal veins, and
- sometimes, the left bronchial veins.

### Thoracic duct in the posterior mediastinum

The thoracic duct is the principal channel through which lymph from most of the body is returned to the venous system. It begins as a confluence of lymph trunks in the abdomen, sometimes forming a saccular dilation referred to as the **cisterna chyli** (**chyle cistern**), which drains the abdominal viscera and walls, pelvis, perineum, and lower limbs.

The thoracic duct extends from vertebra LII to the root of the neck.

Entering the thorax, posterior to the aorta, through the aortic hiatus of the diaphragm, the thoracic duct ascends through the posterior mediastinum to the right of midline between the thoracic aorta on the left and the azygos vein on the right (Fig. 3.103). It lies posterior to the diaphragm

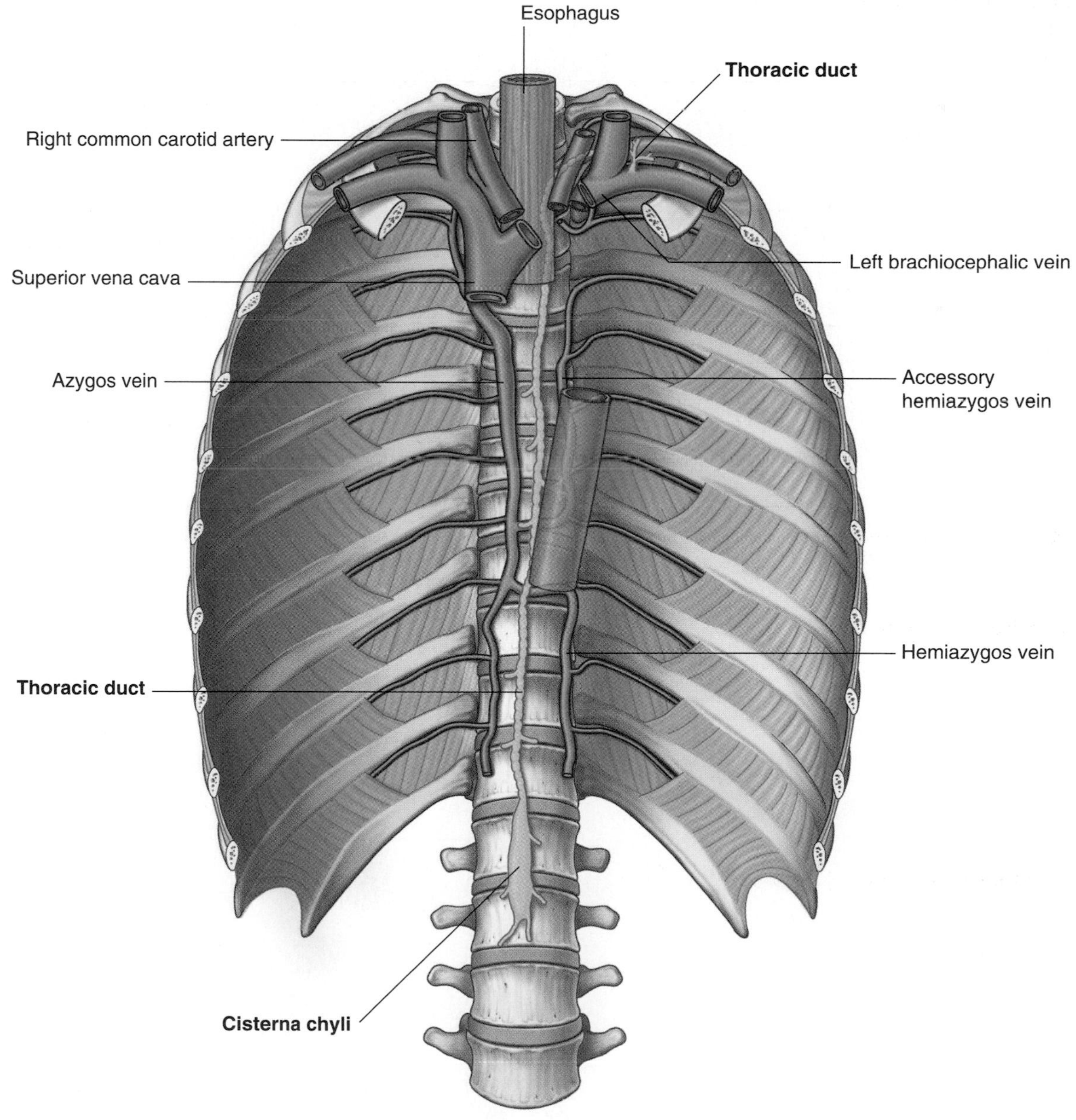


**Fig. 3.103** Thoracic duct.

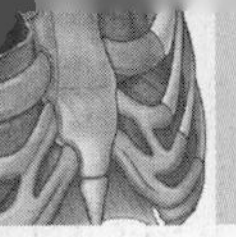



and the esophagus and anterior to the bodies of the vertebrae.

At vertebral level TV, the thoracic duct moves to the left of midline and enters the superior mediastinum. It continues through the superior mediastinum and into the neck.

After being joined, in most cases, by the **left jugular trunk**, which drains the left side of the head and neck, and the **left subclavian trunk**, which drains the left upper limb, the thoracic duct empties into the junction of the left subclavian and left internal jugular veins.

The thoracic duct usually receives the contents from:

- the confluence of lymph trunks in the abdomen,
- descending thoracic lymph trunks draining the lower six or seven intercostal spaces on both sides,
- upper intercostal lymph trunks draining the upper left five or six intercostal spaces,
- ducts from posterior mediastinal nodes, and
- ducts from posterior diaphragmatic nodes.

### Sympathetic trunks

The **sympathetic trunks** are an important component of the sympathetic part of the autonomic division of the peripheral nervous system and are usually considered a component of the posterior mediastinum as they pass through the thorax.

This portion of the sympathetic trunks consists of two parallel cords punctuated by 11 or 12 **ganglia** (Fig. 3.104). The ganglia are connected to adjacent thoracic spinal nerves by **white** and **gray rami communicantes** and are

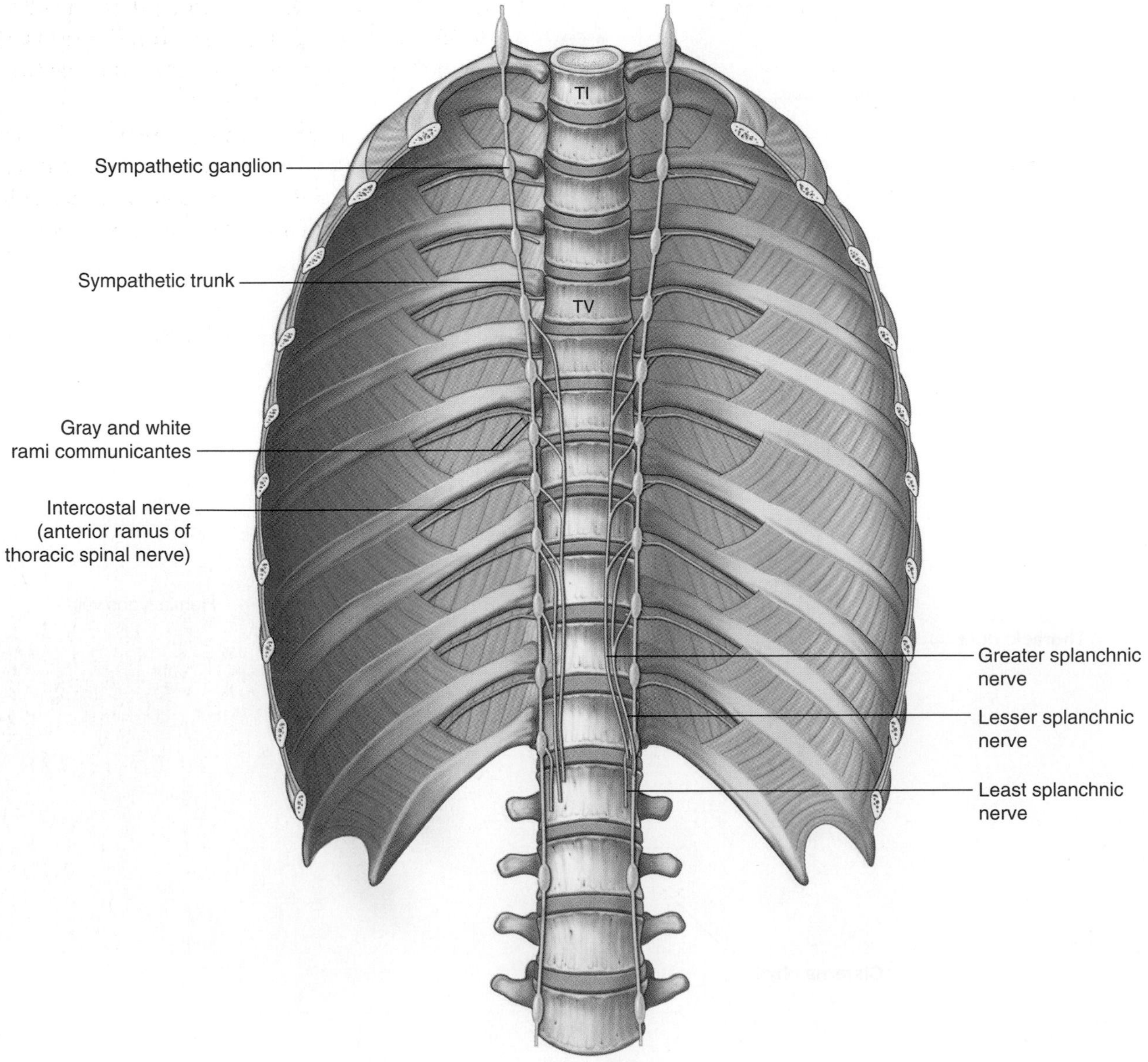

**Fig. 3.104** Thoracic portion of sympathetic trunks.

numbered according to the thoracic spinal nerve with which they are associated.

In the superior portion of the posterior mediastinum, the trunks are anterior to the neck of the ribs. Inferiorly, they become more medial in position until they lie on the lateral aspect of the vertebral bodies. The sympathetic trunks leave the thorax by passing posterior to the diaphragm under the medial arcuate ligament or through the crura of the diaphragm. Throughout their course the trunks are covered by parietal pleura.

### Branches from the ganglia

Two types of medial branches are given off by the ganglia:

- The first type includes branches from the upper five ganglia.
- The second type includes branches from the lower seven ganglia.

The first type, which includes branches from the upper five ganglia, consists mainly of postganglionic sympathetic fibers, which supply the various thoracic viscera. These branches are relatively small, and also contain visceral afferent fibers.

The second type, which includes branches from the lower seven ganglia, consists mainly of preganglionic sympathetic fibers, which supply the various abdominal and pelvic viscera. These branches are large, also carry visceral afferent fibers, and form the three thoracic splanchnic nerves referred to as the greater, lesser, and least splanchnic nerves (Fig. 3.104).

- The **greater splanchnic nerve** on each side usually arises from the fifth to ninth or tenth thoracic ganglia. It descends across the vertebral bodies moving in a medial direction, passes into the abdomen through the crus of the diaphragm, and ends in the celiac ganglion.
- The **lesser splanchnic nerve** usually arises from the ninth and tenth, or tenth and eleventh thoracic ganglia. It descends across the vertebral bodies moving in a medial direction, and passes into the abdomen through the crus of the diaphragm to end in the aorticorenal ganglion.
- The **least splanchnic nerve** (**lowest splanchnic nerve**) usually arises from the twelfth thoracic ganglion. It descends and passes into the abdomen through the crus of the diaphragm to end in the renal plexus.

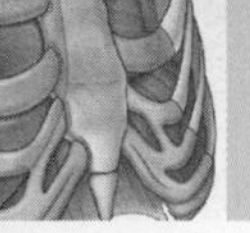



# Surface anatomy

## Thorax surface anatomy

The ability to visualize how anatomical structures in the thorax are related to surface features is fundamental to a physical examination. Landmarks on the body's surface can be used to locate deep structures and to assess function by auscultation and percussion.

## How to count ribs

Knowing how to count ribs is important because different ribs provide palpable landmarks for the positions of deeper structures. To determine the location of specific ribs, palpate the **jugular notch** at the superior extent of the manubrium of the sternum. Move down the sternum until a ridge is felt. This ridge is the **sternal angle**, which identifies the articulation between the manubrium of the sternum and the body of the sternum. The costal cartilage of rib II articulates with the sternum at this location. Identify rib II. Then continue counting the ribs, moving in a downward and lateral direction (Fig. 3.105).

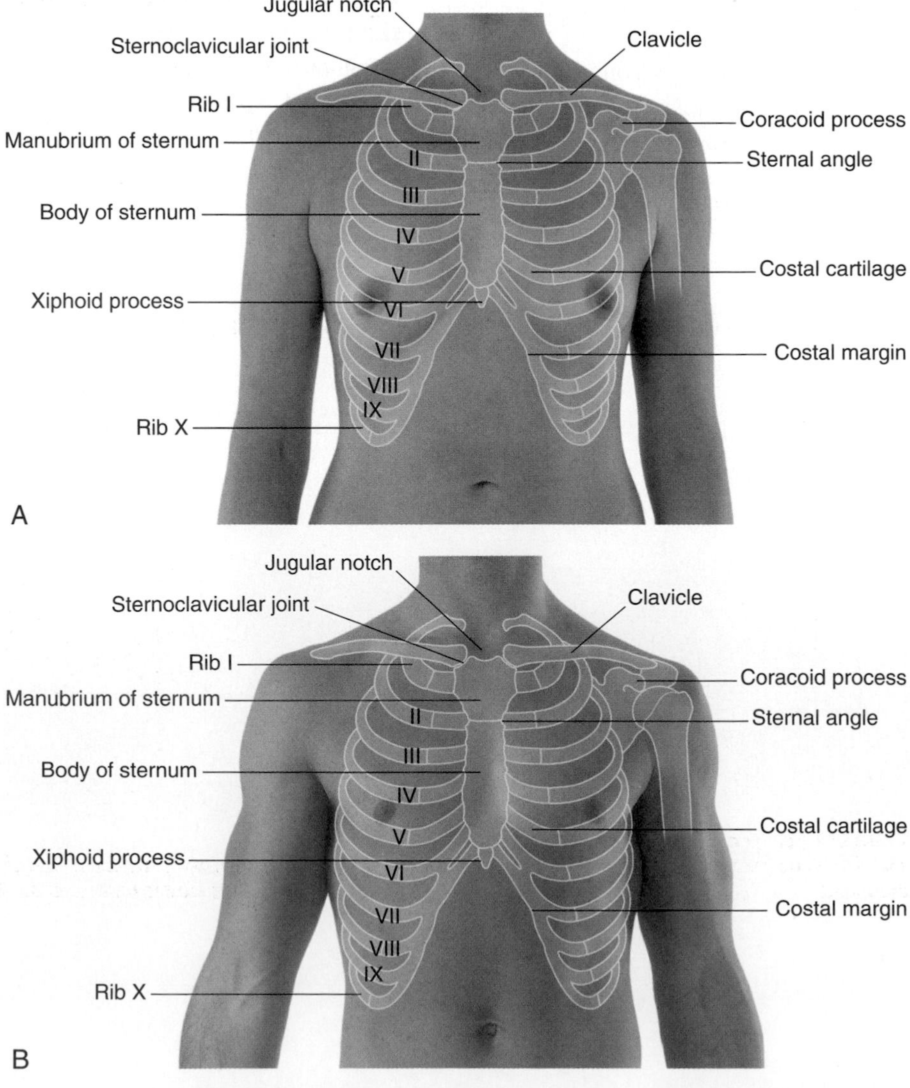


**Fig. 3.105** Anterior view of chest wall with the locations of skeletal structures shown. **A.** In women. The location of the nipple relative to a specific intercostal space varies depending on the size of the breasts, which may not be symmetrical. **B.** In men. Note the location of the nipple in the fourth intercostal space.

## Surface anatomy of the breast in women

Although breasts vary in size, they are normally positioned on the thoracic wall between ribs II and VI and overlie the pectoralis major muscles. Each mammary gland extends superolaterally around the lower margin of the pectoralis major muscle and enters the axilla (Fig. 3.106). This portion of the gland is the axillary tail or axillary process. The positions of the nipple and areola vary relative to the chest wall depending on breast size.

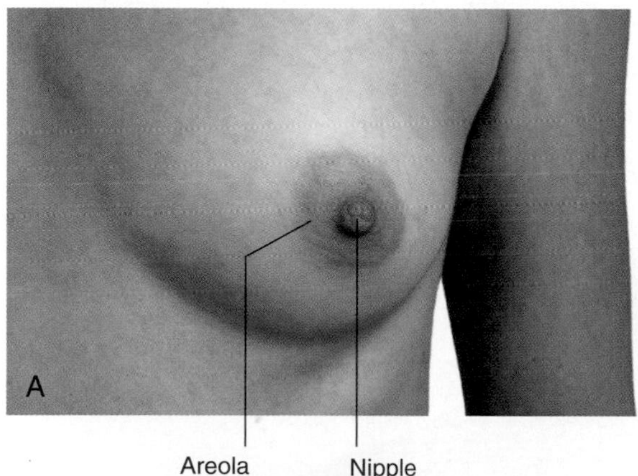


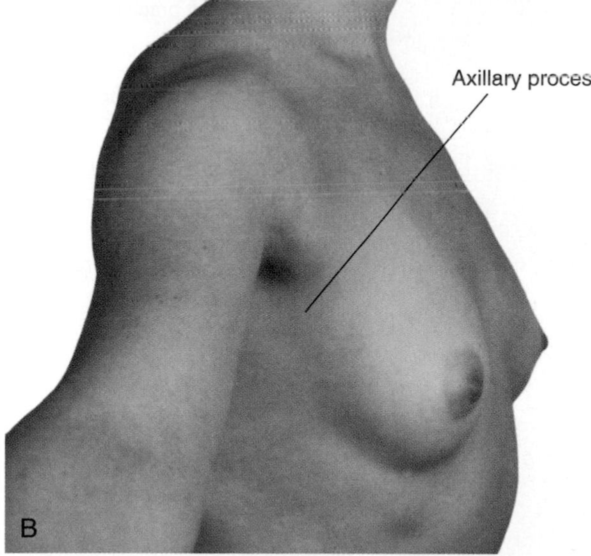


**Fig. 3.106 A.** Close-up view of nipple and surrounding areola of the breast. **B.** Lateral view of the chest wall of a woman showing the axillary process of the breast.

## Visualizing structures at the TIV/V vertebral level

The TIV/V vertebral level is a transverse plane that passes through the sternal angle on the anterior chest wall and the intervertebral disc between TIV and TV vertebrae posteriorly. This plane can easily be located, because the joint between the manubrium of the sternum and the body of the sternum forms a distinct bony protuberance that can be palpated. At the TIV/V level (Fig. 3.107):

- The costal cartilage of rib II articulates with the sternum.
- The superior mediastinum is separated from the inferior mediastinum.
- The ascending aorta ends and the arch of the aorta begins.
- The arch of the aorta ends and the thoracic aorta begins.
- The trachea bifurcates.

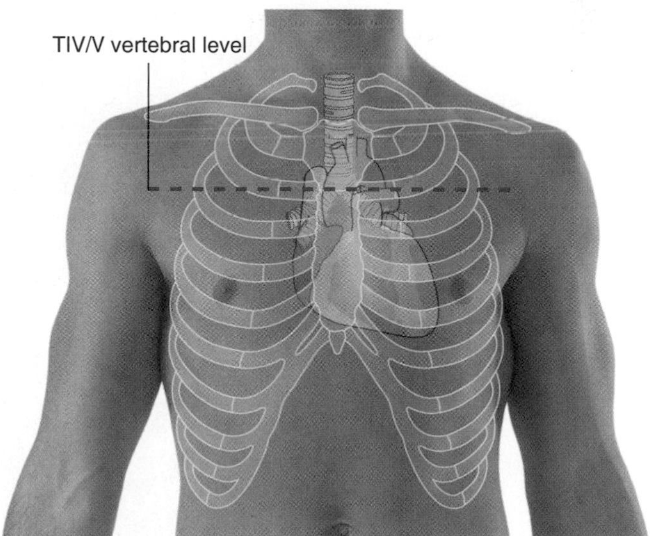


**Fig. 3.107** Anterior view of the chest wall of a man showing the locations of various structures related to the TIV/V level.

## Visualizing structures in the superior mediastinum

A number of structures in the superior mediastinum in adults can be visualized based on their positions relative to skeletal landmarks that can be palpated through the skin (Fig. 3.108).

- On each side, the internal jugular and subclavian veins join to form the brachiocephalic veins behind the sternal ends of the clavicles near the sternoclavicular joints.
- The left brachiocephalic vein crosses from left to right behind the manubrium of the sternum.
- The brachiocephalic veins unite to form the superior vena cava behind the lower border of the costal cartilage of the right first rib.
- The arch of the aorta begins and ends at the transverse plane between the sternal angle anteriorly and vertebral level TIV/V posteriorly. The arch may reach as high as the midlevel of the manubrium of the sternum.



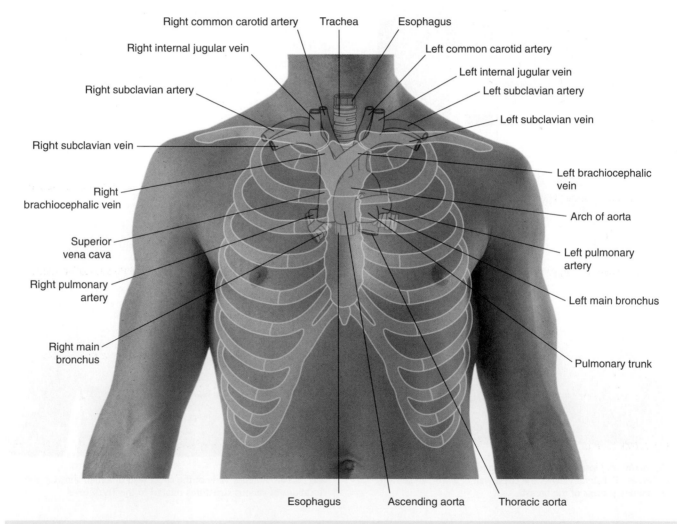

**Fig. 3.108** Anterior view of the chest wall of a man showing the locations of different structures in the superior mediastinum as they relate to the skeleton.

## Visualizing the margins of the heart

Surface landmarks can be palpated to visualize the outline of the heart (Fig. 3.109).

- The upper limit of the heart reaches as high as the third costal cartilage on the right side of the sternum and the second intercostal space on the left side of the sternum.
- The right margin of the heart extends from the right third costal cartilage to near the right sixth costal cartilage.
- The left margin of the heart descends laterally from the second intercostal space to the apex located near the midclavicular line in the fifth intercostal space.
- The lower margin of the heart extends from the sternal end of the right sixth costal cartilage to the apex in the fifth intercostal space near the midclavicular line.

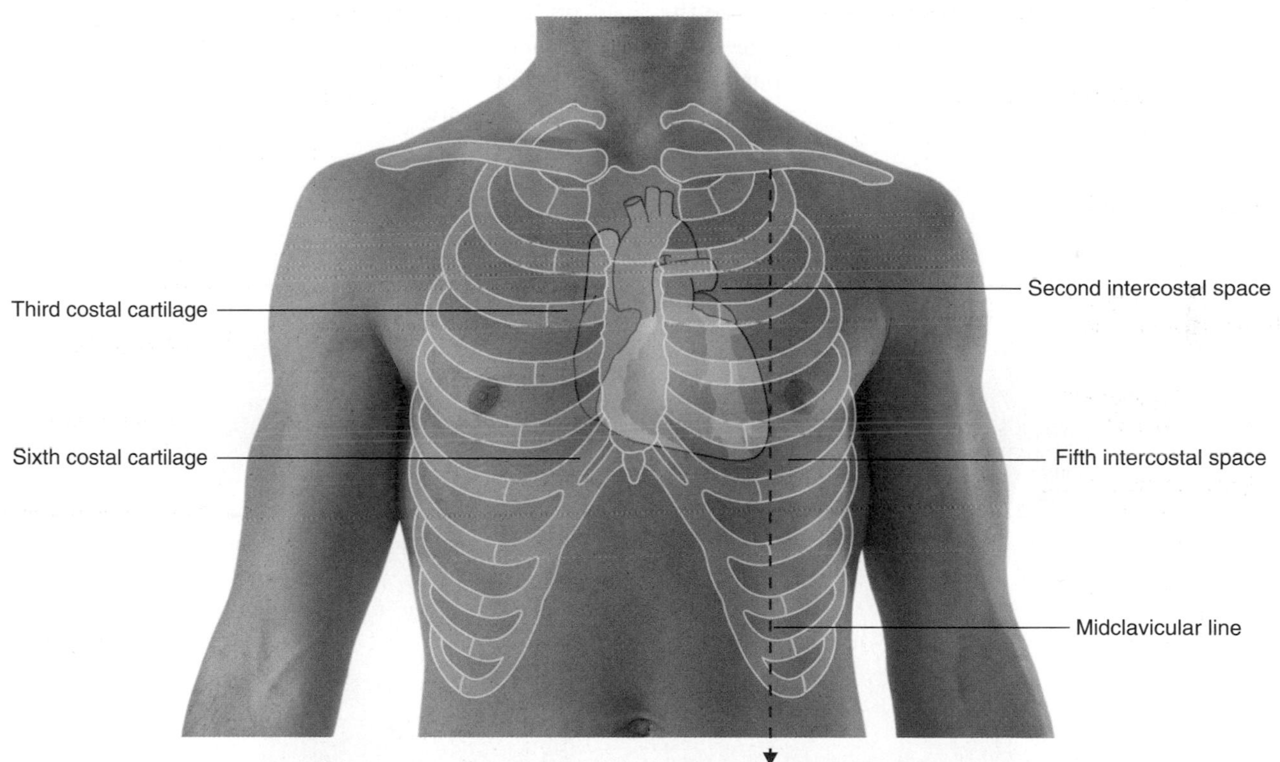


**Fig. 3.109** Anterior view of the chest wall of a man showing skeletal structures and the surface projection of the heart.

## Where to listen for heart sounds

To listen for valve sounds, position the stethoscope downstream from the flow of blood through the valves (Fig. 3.110).

- The tricuspid valve is heard just to the left of the lower part of the sternum near the fifth intercostal space.
- The mitral valve is heard over the apex of the heart in the left fifth intercostal space at the midclavicular line.
- The pulmonary valve is heard over the medial end of the left second intercostal space.
- The aortic valve is heard over the medial end of the right second intercostal space.

## Visualizing the pleural cavities and lungs, pleural recesses, and lung lobes and fissures

Palpable surface landmarks can be used to visualize the normal outlines of the pleural cavities and the lungs and to determine the positions of the pulmonary lobes and fissures.

Superiorly, the parietal pleura projects above the first costal cartilage. Anteriorly, the costal pleura approaches the midline posterior to the upper portion of the sternum. Posterior to the lower portion of the sternum, the left parietal pleura does not come as close to the midline as it does on the right side. This is because the heart bulges onto the left side (Fig. 3.111A).

Inferiorly, the pleura reflects onto the diaphragm above the costal margin and courses around the thoracic wall following an VIII, X, XII contour (i.e., rib VIII in the midclavicular line, rib X in the midaxillary line, and vertebra TXII posteriorly).

The lungs do not completely fill the area surrounded by the pleural cavities, particularly anteriorly and inferiorly.

- Costomediastinal recesses occur anteriorly, particularly on the left side in relationship to the heart bulge.
- Costodiaphragmatic recesses occur inferiorly between the lower lung margin and the lower margin of the pleural cavity.

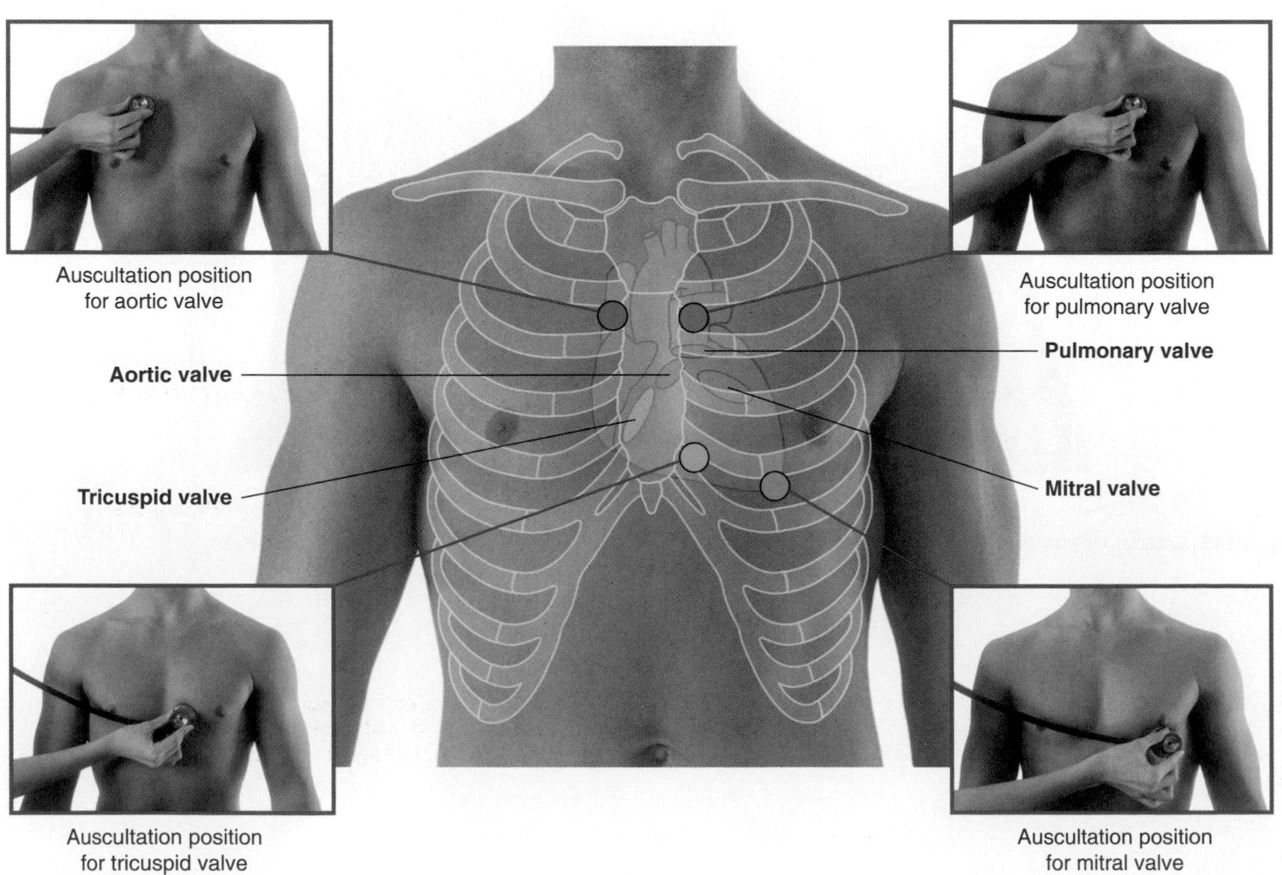


**Fig. 3.110** Anterior view of the chest wall of a man showing skeletal structures, heart, location of the heart valves, and auscultation points.

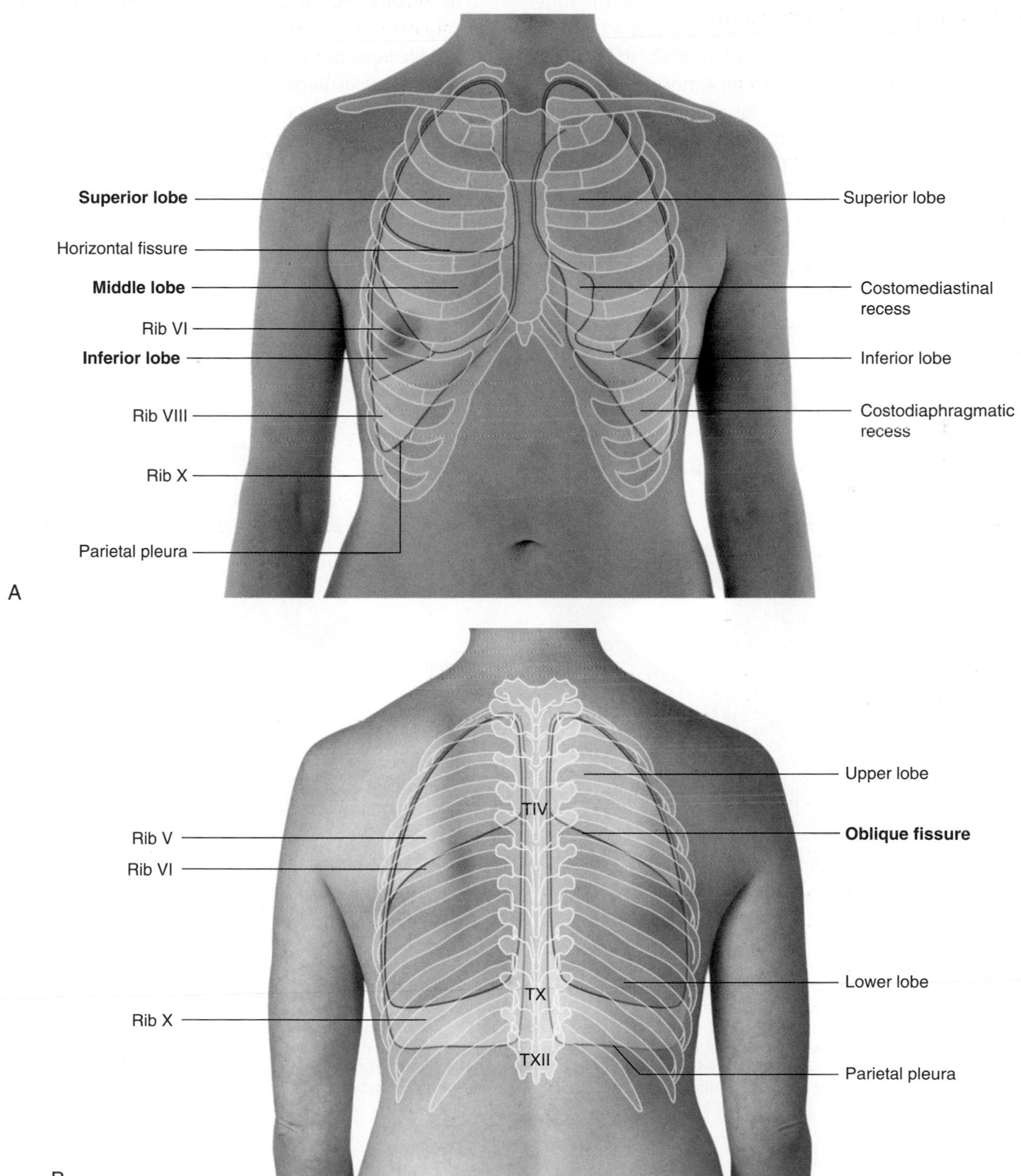


**Fig. 3.111** Views of the chest wall showing the surface projections of the lobes and the fissures of the lungs. **A.** Anterior view in a woman. On the right side, the superior, middle, and inferior lobes are illustrated. On the left side, the superior and inferior lobes are illustrated.
**B.** Posterior view in a woman. On both sides, the superior and inferior lobes are illustrated. The middle lobe on the right side is not visible in this view.

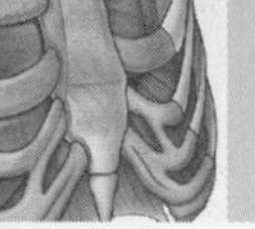


In quiet respiration, the inferior margin of the lungs travels around the thoracic wall following a VI, VIII, X contour (i.e., rib VI in the midclavicular line, rib VIII in the midaxillary line, and vertebra TX posteriorly).

In the posterior view, the oblique fissure on both sides is located in the midline near the spine of vertebra TIV (Figs. 3.111B and 3.112A). It moves laterally in a downward direction, crossing the fourth and fifth intercostal spaces and reaches rib VI laterally.

In the anterior view, the horizontal fissure on the right side follows the contour of rib IV and its costal cartilage and the oblique fissures on both sides follow the contour of rib VI and its costal cartilage (Fig. 3.112B).

## Where to listen for lung sounds

The stethoscope placements for listening for lung sounds are shown in Fig. 3.113.

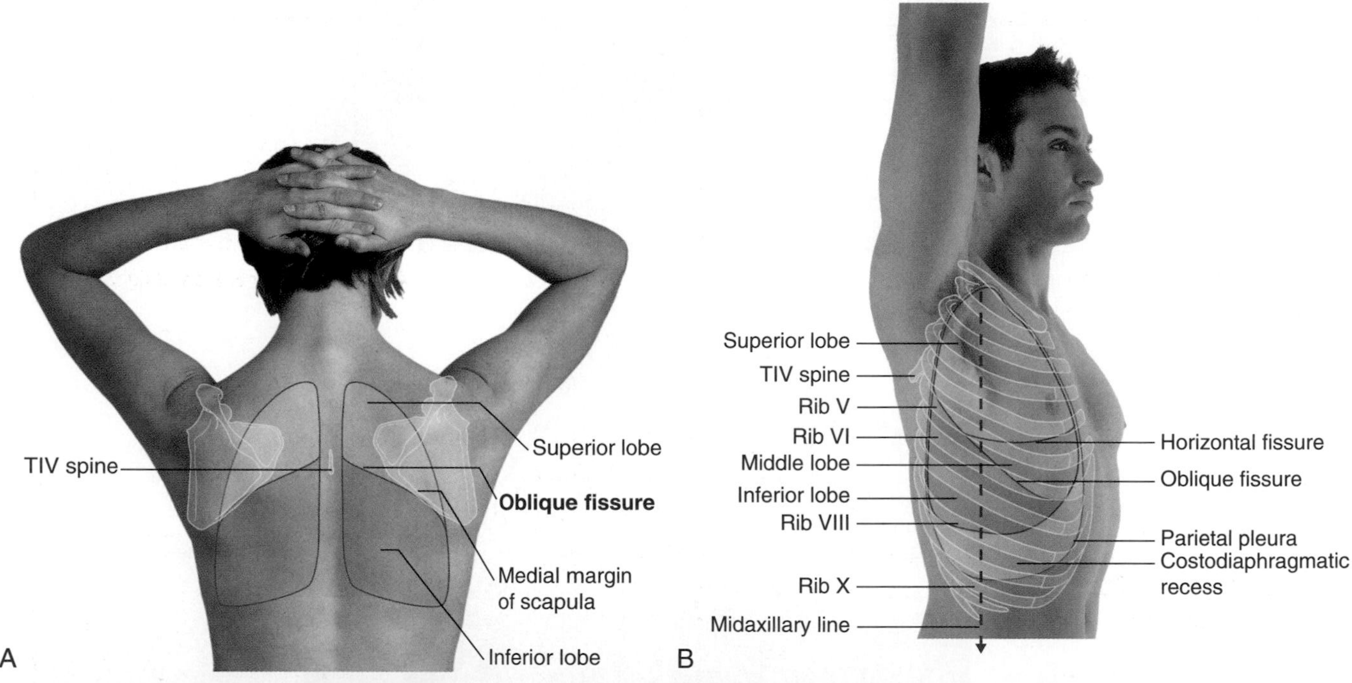


**Fig. 3.112** Views of the chest wall. **A.** Posterior view in a woman with arms abducted and hands positioned behind her head. On both sides, the superior and inferior lobes of the lungs are illustrated. When the scapula is rotated into this position, the medial border of the scapula parallels the position of the oblique fissure and can be used as a guide for determining the surface projection of the superior and inferior lobes of the lungs. **B.** Lateral view in a man with his right arm abducted. The superior, middle, and inferior lobes of the right lung are illustrated. The oblique fissure begins posteriorly at the level of the spine of vertebra TIV, passes inferiorly crossing rib IV, the fourth intercostal space, and rib V. It crosses the fifth intercostal space at the midaxillary line and continues anteriorly along the contour of rib VI. The horizontal fissure crosses rib V in the midaxillary space and continues anteriorly, crossing the fourth intercostal space and following the contour of rib IV and its costal cartilage to the sternum.

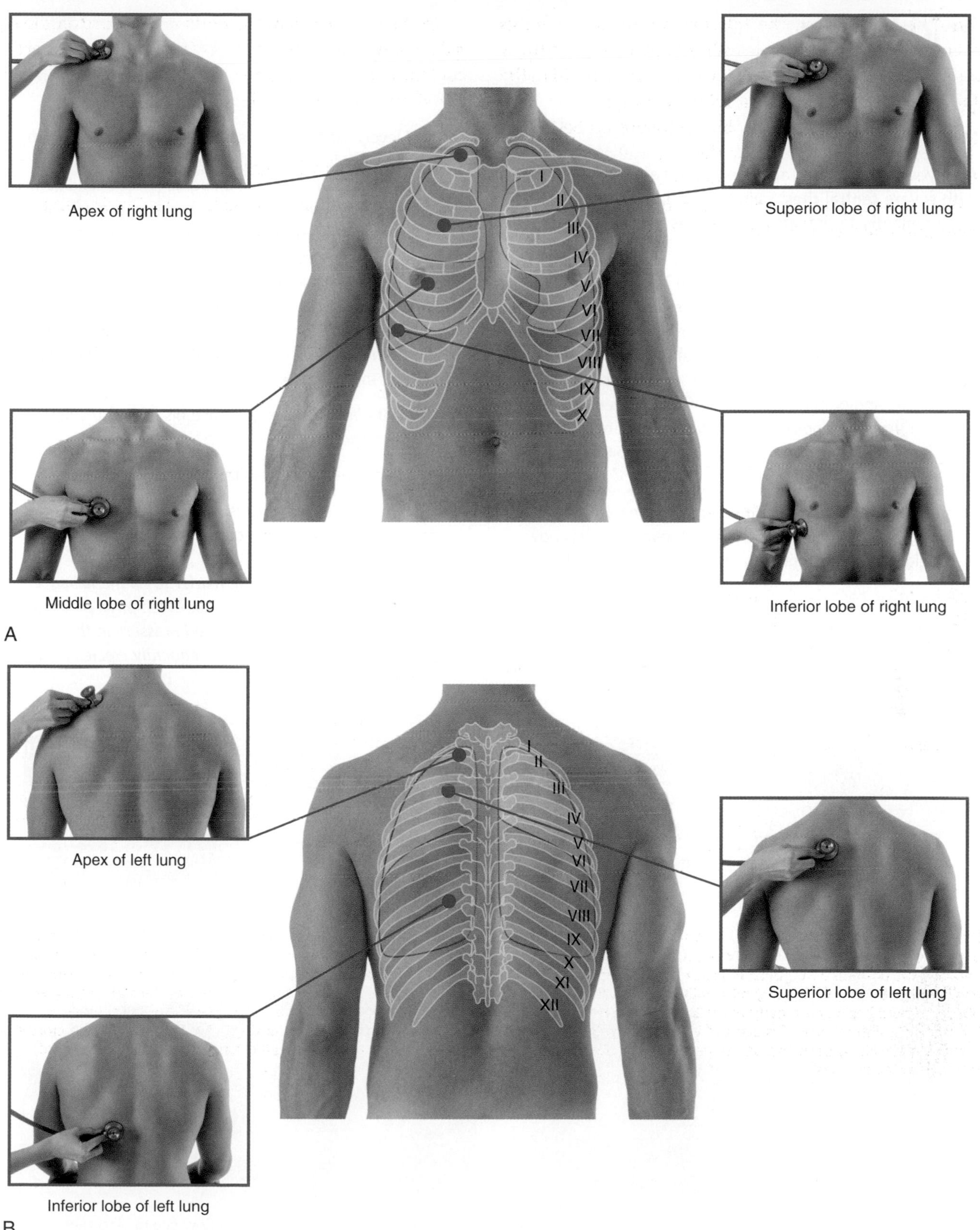


**Fig. 3.113** Views of the chest wall of a man with stethoscope placements for listening to the lobes of the lungs. **A.** Anterior views. **B.** Posterior views.

# Clinical cases

## Case 1

MYOCARDIAL INFARCTION

**A 65-year-old man was admitted to the emergency room with severe central chest pain that radiated to the neck and predominantly to the left arm. He was overweight and a known heavy smoker.**

On examination he appeared gray and sweaty. His blood pressure was 74/40 mm Hg (normal range 120/80 mm Hg). An electrocardiogram (ECG) was performed and demonstrated anterior myocardial infarction. An urgent echocardiograph demonstrated poor left ventricular function. The cardiac angiogram revealed an occluded vessel (Fig. 3.114A,B). Another approach to evaluating coronary arteries in patients is to perform maximum intensity projection (MIP) CT studies (Fig. 3.115A,B).

This patient underwent an emergency coronary artery bypass graft and made an excellent recovery. He has now lost weight, stopped smoking, and exercises regularly.

When cardiac cells die during a myocardial infarction, pain fibers (visceral afferents) are stimulated. These visceral sensory fibers follow the course of sympathetic fibers that innervate the heart and enter the spinal cord between the TI and TIV levels. At this level, somatic afferent nerves from spinal nerves T1 to T4 also enter the spinal cord via the posterior roots. Both types of afferents (visceral and somatic) synapse with interneurons, which then synapse with second neurons whose fibers pass across the cord and then ascend to the somatosensory areas of the brain that represent the T1 to T4 levels. The brain is unable to distinguish clearly between the visceral sensory distribution and the somatic sensory distribution and therefore the pain is interpreted as arising from the somatic regions rather than the visceral organ (i.e., the heart; Fig. 3.114C).

The patient was breathless because his left ventricular function was poor.

When the left ventricle fails, it produces two effects.

- *First, the contractile force is reduced. This reduces the pressure of the ejected blood and lowers the blood pressure.*
- *The left atrium has to work harder to fill the failing left ventricle. This extra work increases left atrial pressure, which is reflected in an increased pressure in the pulmonary veins, and this subsequently creates a higher*

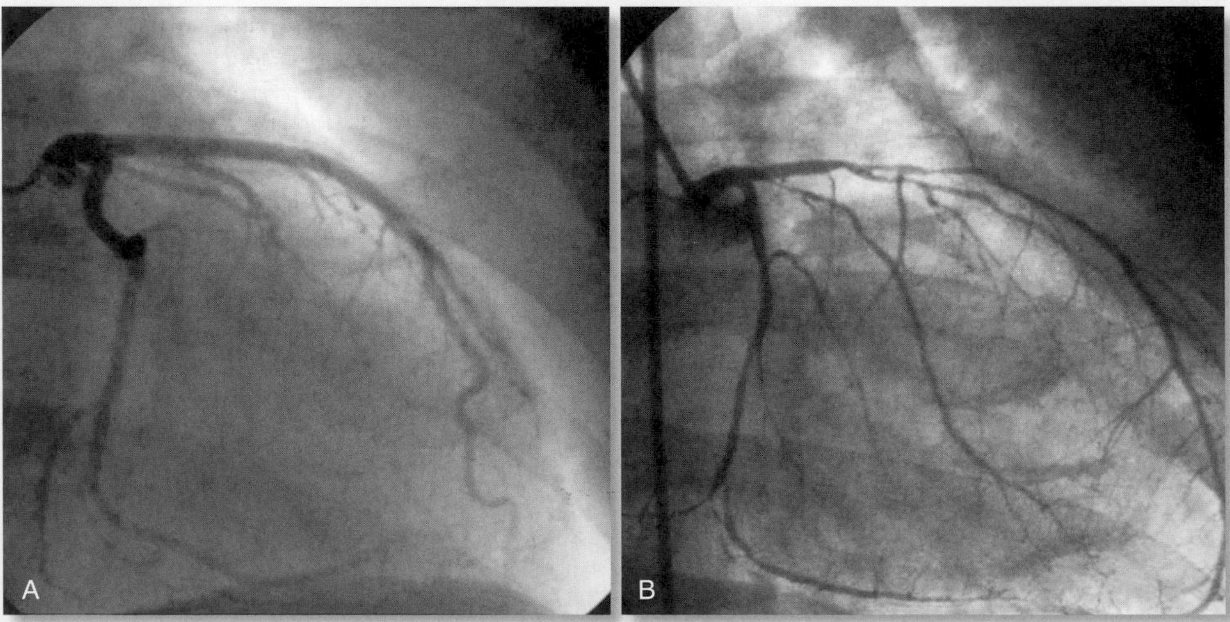


**Fig. 3.114 A.** Normal left coronary artery angiogram. **B.** Left coronary artery angiogram showing decreased flow due to blockages.

## Case 1—cont'd

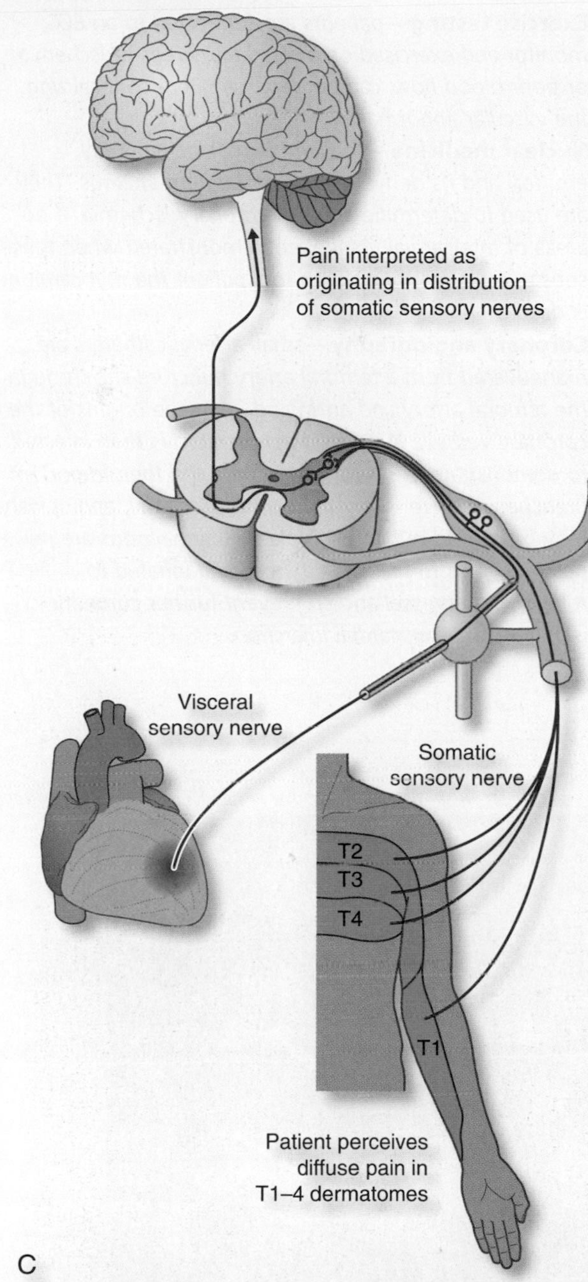


**Fig. 3.114, cont'd** **C.** Mechanism for perceiving heart pain in T1–4 dermatomes.

*pulmonary venular pressure. This rise in pressure will cause fluid to leak from the capillaries into the pulmonary interstitium and then into the alveoli. Such fluid is called pulmonary edema and it markedly restricts gas exchange. This results in shortness of breath.*

This man had a blocked left coronary artery, as shown in Fig. 3.114B.

It is important to know which coronary artery is blocked.

- *The left coronary artery supplies the majority of the left side of the heart. The left main stem vessel is approximately 2 cm long and divides into the circumflex artery, which lies between the atrium and the ventricle in the coronary sulcus, and the anterior interventricular artery, which is often referred to as the left anterior descending artery (LAD).*
- *When the right coronary artery is involved with arterial disease and occludes, associated disorders of cardiac rhythm often result because the sinu-atrial and the atrioventricular nodes derive their blood supplies predominantly from the right coronary artery.*

When this patient sought medical care, his myocardial function was assessed using ECG, echocardiography, and angiography.

During a patient's initial examination, the physician will usually assess myocardial function.

After obtaining a clinical history and carrying out a physical examination, a differential diagnosis for the cause of the malfunctioning heart is made. Objective assessment of myocardial and valve function is obtained in the following ways:

- **ECG/EKG (electrocardiography)**—*a series of electrical traces taken around the long and short axes of the heart that reveal heart rate and rhythm and conduction defects. In addition, it demonstrates the overall function of the right and left sides of the heart and points of dysfunction. Specific changes in the ECG relate to the areas of the heart that have been involved in a myocardial infarction. For example, a right coronary artery occlusion produces infarction in the area of myocardium it supplies, which*

*(continues)*

## Case 1—cont'd

*is predominantly the inferior aspect; the infarct is therefore called an inferior myocardial infarction. The ECG changes are demonstrated in the leads that visualize the inferior aspect of the myocardium (namely, leads II, III, and aVF).*

- **Chest radiography**—*reveals the size of the heart and chamber enlargement. Careful observation of the lungs will demonstrate excess fluid (pulmonary edema), which builds up when the left ventricle fails and can produce marked respiratory compromise and death unless promptly treated.*
- **Blood tests**—*the heart releases enzymes during myocardial infarction, namely lactate dehydrogenase (LDH), creatine kinase (CK), and aspartate transaminase (AST). These plasma enzymes are easily measured in the hospital laboratory and used to determine the diagnosis at an early stage. Further specific enzymes termed isoenzymes can also be determined (creatine kinase MB isoenzyme [CKMB]). Newer tests include an assessment for troponin (a specific component of the myocardium), which is released when cardiac cells die during myocardial infarction.*
- **Exercise testing**—*patients are connected to an ECG monitor and exercised on a treadmill. Areas of ischemia, or poor blood flow, can be demonstrated, so localizing the vascular abnormality.*
- **Nuclear medicine**—*thallium (a radioactive X-ray emitter) and its derivatives are potassium analogs. They are used to determine areas of coronary ischemia. If no areas of myocardial uptake are demonstrated when these substances are administered to a patient the myocardium is dead.*
- **Coronary angiography**—*small arterial catheters are maneuvered from a femoral artery puncture site through the femoral artery and aorta and up to the origins of the coronary vessels. X-ray contrast medium is then injected to demonstrate the coronary vessels and their important branches. If there is any narrowing (stenosis), angioplasty may be carried out. In angioplasty tiny balloons are passed across the narrowed areas and inflated to refashion the vessel and so prevent further coronary ischemia and myocardial infarction.*

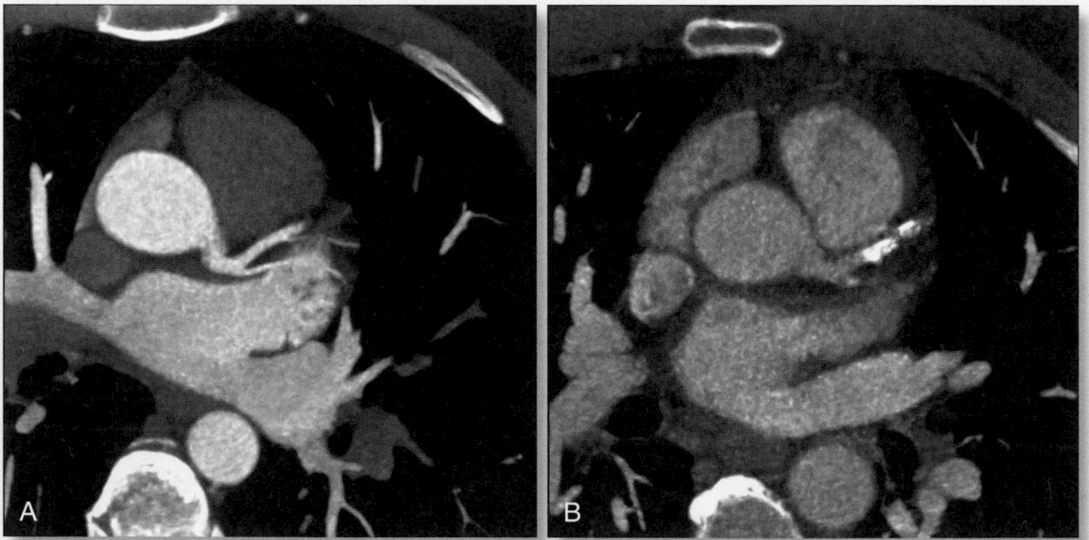


**Fig. 3.115** Axial maximum intensity projection (MIP) CT image through the heart. **A.** Normal anterior interventricular (left anterior descending) artery. **B.** Stenotic (calcified) anterior interventricular (left anterior descending) artery.

## Case 2

PULMONARY EMBOLISM

A 53-year-old man presented to the emergency department with a 5-hour history of sharp pleuritic chest pain and shortness of breath. The day before he was on a long haul flight, returning from his holidays. He was usually fit and well and was a keen mountain climber. He had no previous significant medical history.

On physical examination his lungs were clear, he was tachypneic at 24/min, and his saturation was reduced to 92% on room air. Pulmonary embolism was suspected and the patient was referred for a CT pulmonary angiogram. The study demonstrated clots within the right and left main pulmonary arteries. There was no pleural effusion, lung collapse, or consolidation.

He was immediately started on subcutaneous enoxaparin and converted to oral anticoagulation over the course of a couple of days. The whole treatment lasted 6 months as no other risk factors (except immobilization during a long haul flight) were identified. There were no permanent sequelae.

The embolic material usually originates in the peripheral deep veins of the lower limbs and less commonly in the pelvic, renal, or upper limb deep veins. The material gets detached from the main thrombus in the deep veins and travels into the pulmonary circulation, where it can lodge either in the pulmonary trunk and main pulmonary arteries, giving rise to central pulmonary embolism or in the lobar, segmental, or subsegmental branches, giving rise to peripheral embolism.

The gravity of symptoms is partly dependent on the thrombus load and on which part of the pulmonary arterial tree is affected. Large pulmonary embolisms can lead to severe hemodynamic and respiratory compromise and death (e.g., a saddle thrombus lodged in the pulmonary trunk and in both main pulmonary arteries).

Common risk factors include immobilization, surgery, trauma, malignancy, pregnancy, oral contraceptives, and hereditary factors.

# 4

# Abdomen

## ADDITIONAL LEARNING RESOURCES for Chapter 4, *Abdomen*, on STUDENT CONSULT (www.studentconsult.com):

- Self-Assessment—National Board style multiple-choice questions, Chapter 4
- Short Questions—these are questions requiring short responses, Chapter 4
- Interactive Surface Anatomy—interactive surface animations, Chapter 4
- Clinical Cases, Chapter 4
  - Aorto-iliac occlusive disease
  - Caval obstruction
  - Colon cancer
  - Complications of an abdominoperineal resection
  - Diverticular disease
  - Endoleak after endovascular repair of abdominal aortic aneurysm
  - Hodgkin's lymphoma
  - Inguinal hernia
  - Intraabdominal abscess
  - Intussusception
  - Liver biopsy in patients with suspected liver cirrhosis
  - Ureteric stone
  - Zollinger-Ellison syndrome

**Free Online Anatomy and Embryology Self-Study Course**

- Anatomy modules 10 through 17
- Embryology modules 65 and 66

# Conceptual overview

## GENERAL DESCRIPTION

The abdomen is a roughly cylindrical chamber extending from the inferior margin of the thorax to the superior margin of the pelvis and the lower limb (Fig. 4.1A).

The **inferior thoracic aperture** forms the superior opening to the abdomen and is closed by the diaphragm. Inferiorly, the deep abdominal wall is continuous with the pelvic wall at the **pelvic inlet**. Superficially, the inferior limit of the abdominal wall is the superior margin of the lower limb.

The chamber enclosed by the abdominal wall contains a single large **peritoneal cavity**, which freely communicates with the pelvic cavity.

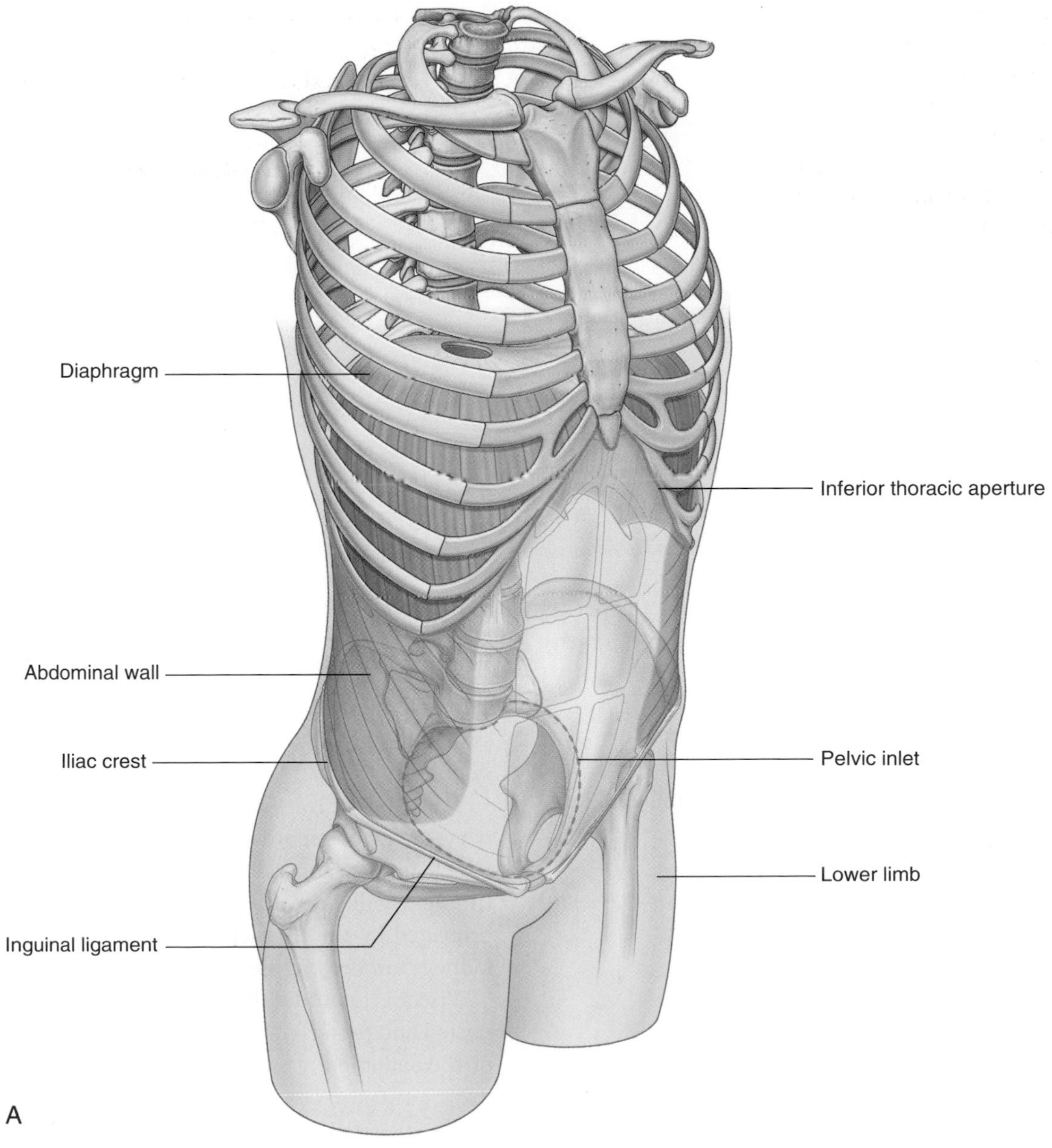


**Fig. 4.1** Abdomen. **A.** Boundaries.

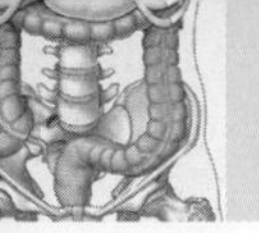

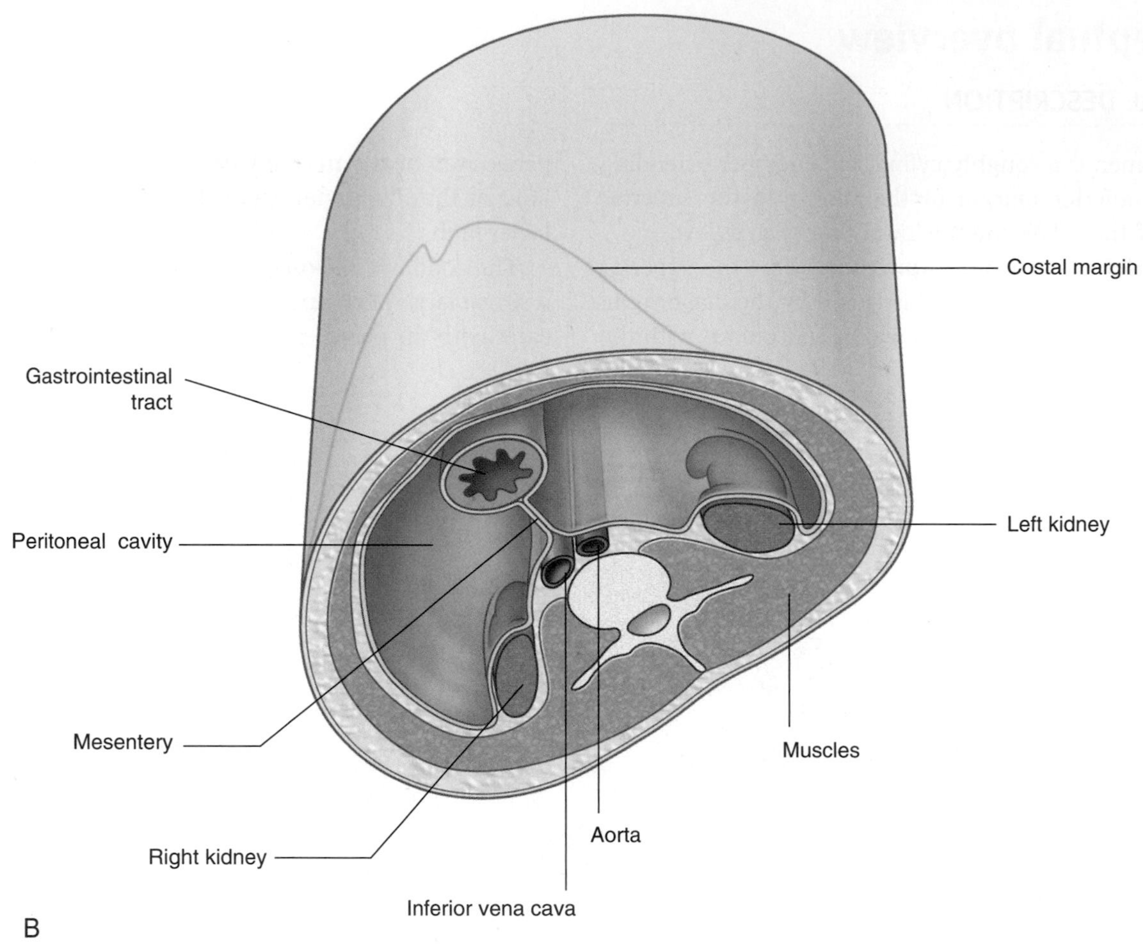


**Fig. 4.1, cont'd** **B.** Arrangement of abdominal contents. Inferior view.

Abdominal viscera are either suspended in the peritoneal cavity by mesenteries or positioned between the cavity and the musculoskeletal wall (Fig. 4.1B).

Abdominal viscera include:

- major elements of the gastrointestinal system—the caudal end of the esophagus, stomach, small and large intestines, liver, pancreas, and gallbladder;
- the spleen;
- components of the urinary system—kidneys and ureters;
- the suprarenal glands; and
- major neurovascular structures.

## FUNCTIONS

### Houses and protects major viscera

The abdomen houses major elements of the gastrointestinal system (Fig. 4.2), the spleen, and parts of the urinary system.

Much of the liver, gallbladder, stomach, and spleen and parts of the colon are under the domes of the diaphragm, which project superiorly above the costal margin of the thoracic wall, and as a result these abdominal viscera are protected by the thoracic wall. The superior poles of the kidneys are deep to the lower ribs.

Viscera not under the domes of the diaphragm are supported and protected predominantly by the muscular walls of the abdomen.

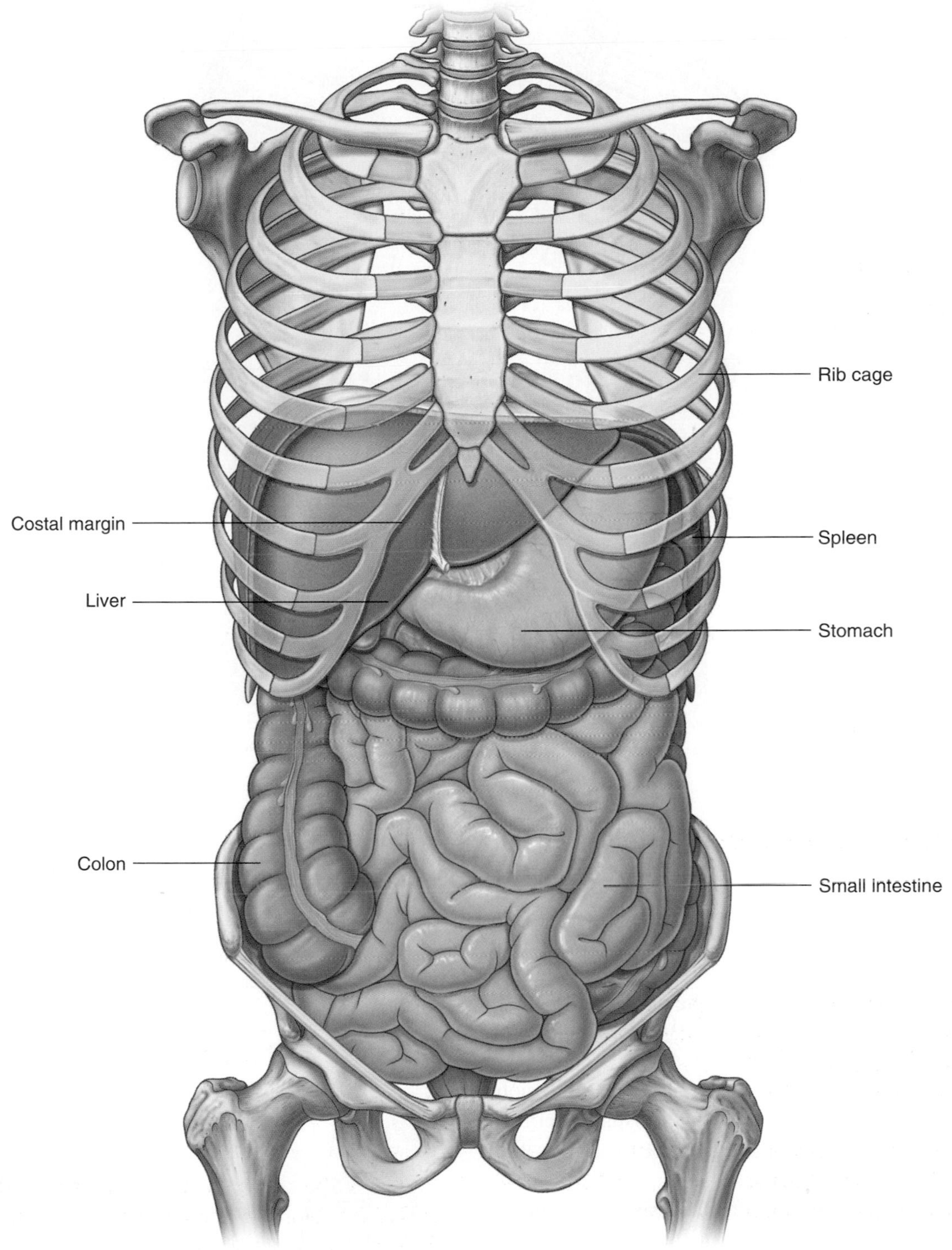


**Fig. 4.2** The abdomen contains and protects the abdominal viscera.

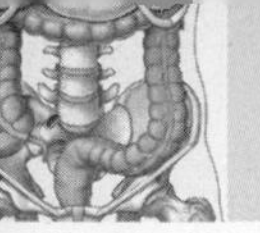


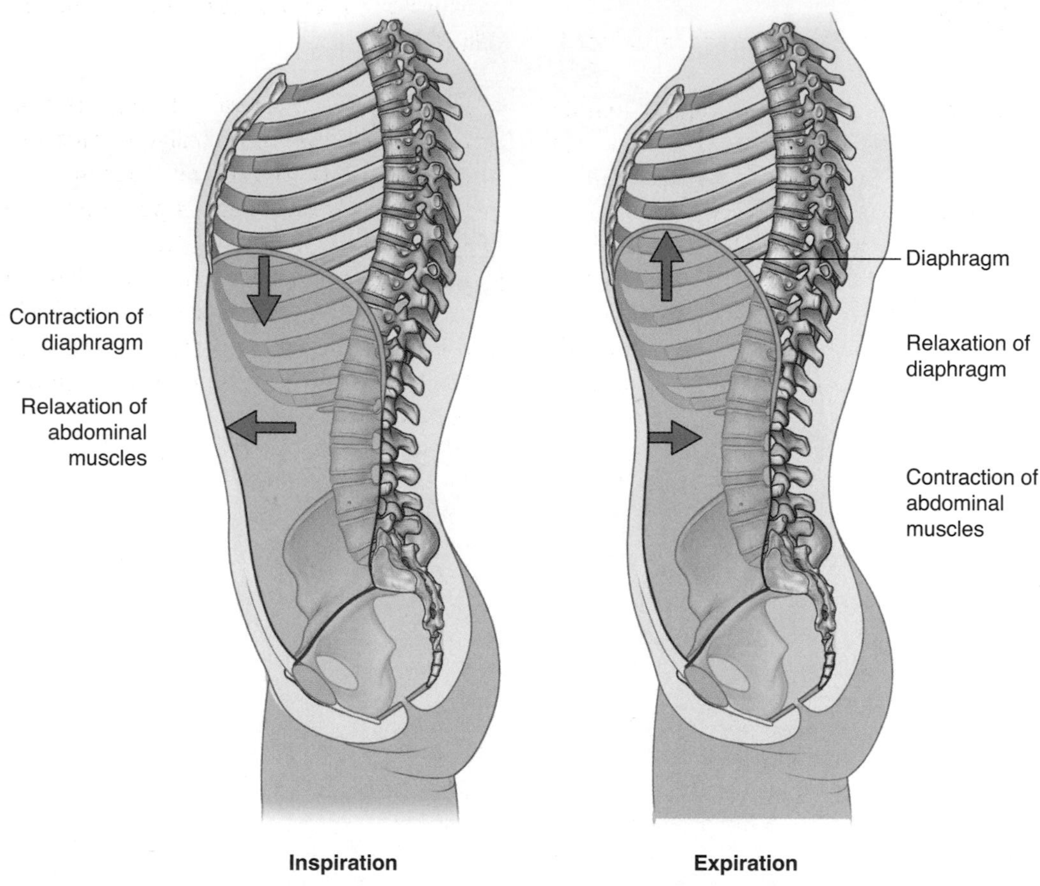


**Fig. 4.3** The abdomen assists in breathing.

## Breathing

One of the most important roles of the abdominal wall is to assist in breathing:

- It relaxes during inspiration to accommodate expansion of the thoracic cavity and the inferior displacement of abdominal viscera during contraction of the diaphragm (Fig. 4.3).
- During expiration, it contracts to assist in elevating the domes of the diaphragm, thus reducing thoracic volume.

Material can be expelled from the airway by forced expiration using the abdominal muscles, as in coughing or sneezing.

## Changes in intraabdominal pressure

Contraction of abdominal wall muscles can dramatically increase intraabdominal pressure when the diaphragm is in a fixed position (Fig. 4.4). Air is retained in the lungs by closing valves in the larynx in the neck. Increased intra-abdominal pressure assists in voiding the contents of the bladder and rectum and in giving birth.

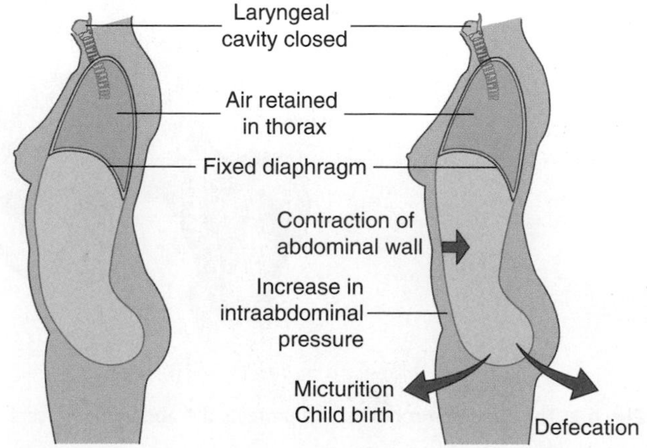


**Fig. 4.4** Increasing intraabdominal pressure to assist in micturition, defecation, and childbirth.

## COMPONENT PARTS

### Wall

The abdominal wall consists partly of bone but mainly of muscle (Fig. 4.5). The skeletal elements of the wall (Fig. 4.5A) are:

- the five lumbar vertebrae and their intervening intervertebral discs,
- the superior expanded parts of the pelvic bones, and
- bony components of the inferior thoracic wall, including the costal margin, rib XII, the end of rib XI, and the xiphoid process.

Muscles make up the rest of the abdominal wall (Fig. 4.5B):

- Lateral to the vertebral column, the quadratus lumborum, psoas major, and iliacus muscles reinforce the posterior aspect of the wall. The distal ends of the psoas major and iliacus muscles pass into the thigh and are major flexors of the hip joint.
- Lateral parts of the abdominal wall are predominantly formed by three layers of muscles, which are similar in orientation to the intercostal muscles of the thorax—transversus abdominis, internal oblique, and external oblique.

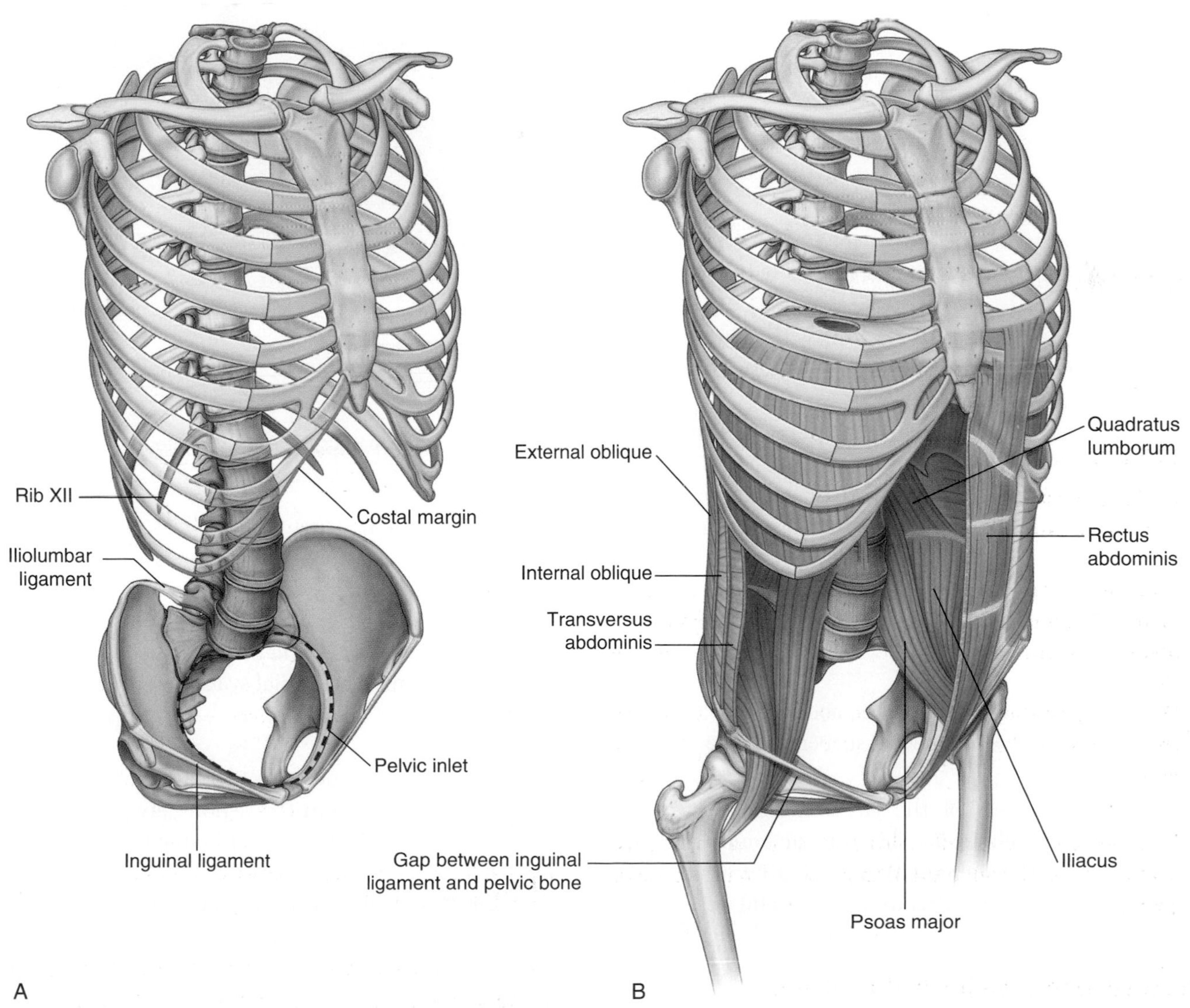


**Fig. 4.5** Abdominal wall. **A.** Skeletal elements. **B.** Muscles.

- Anteriorly, a segmented muscle (the rectus abdominis) on each side spans the distance between the inferior thoracic wall and the pelvis.

Structural continuity between posterior, lateral, and anterior parts of the abdominal wall is provided by thick fascia posteriorly and by flat tendinous sheets (aponeuroses) derived from muscles of the lateral wall. A fascial layer of varying thickness separates the abdominal wall from the peritoneum, which lines the abdominal cavity.

## Abdominal cavity

The general organization of the abdominal cavity is one in which a central gut tube (gastrointestinal system) is suspended from the posterior abdominal wall and partly from the anterior abdominal wall by thin sheets of tissue (**mesenteries**; Fig. 4.6):

- a ventral (anterior) mesentery for proximal regions of the gut tube;
- a dorsal (posterior) mesentery along the entire length of the system.

Different parts of these two mesenteries are named according to the organs they suspend or with which they are associated.

Major viscera, such as the kidneys, that are not suspended in the abdominal cavity by mesenteries are associated with the abdominal wall.

The abdominal cavity is lined by **peritoneum**, which consists of an epithelial-like single layer of cells (the **mesothelium**) together with a supportive layer of connective tissue. Peritoneum is similar to the pleura and serous pericardium in the thorax.

The peritoneum reflects off the abdominal wall to become a component of the mesenteries that suspend the viscera.

- **Parietal peritoneum** lines the abdominal wall.
- **Visceral peritoneum** covers suspended organs.

Normally, elements of the gastrointestinal tract and its derivatives completely fill the abdominal cavity, making the peritoneal cavity a potential space, and visceral peritoneum on organs and parietal peritoneum on the adjacent abdominal wall slide freely against one another.

Abdominal viscera are either intraperitoneal or retroperitoneal:

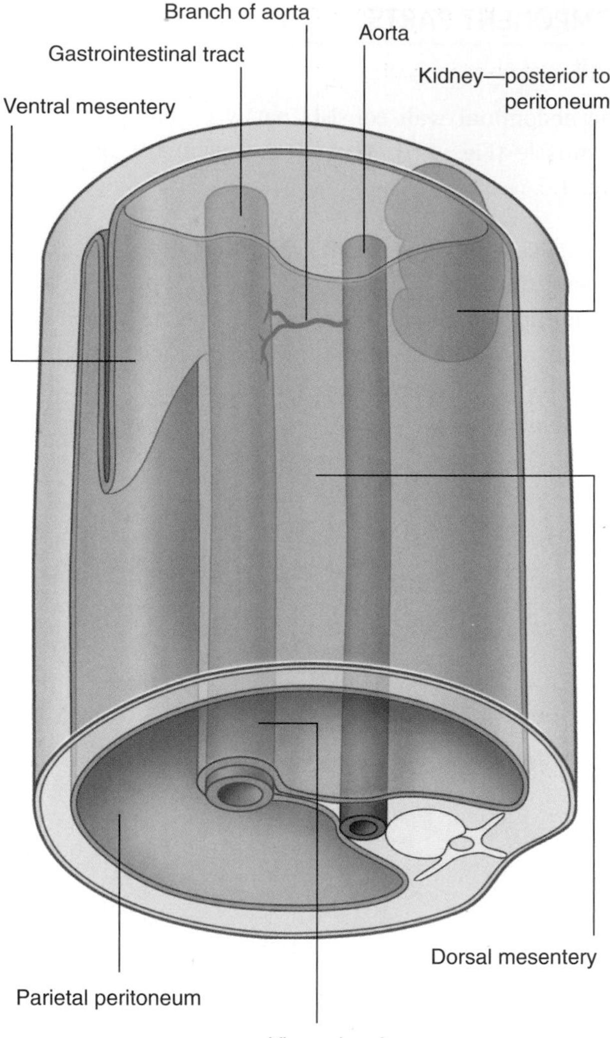


**Fig. 4.6** The gut tube is suspended by mesenteries.

- **Intraperitoneal** structures, such as elements of the gastrointestinal system, are suspended from the abdominal wall by mesenteries;
- Structures that are not suspended in the abdominal cavity by a mesentery and that lie between the parietal peritoneum and abdominal wall are **retroperitoneal** in position.

Retroperitoneal structures include the kidneys and ureters, which develop in the region between the peritoneum and the abdominal wall and remain in this position in the adult.

During development, some organs, such as parts of the small and large intestines, are suspended initially in the abdominal cavity by a mesentery, and later become retroperitoneal secondarily by fusing with the abdominal wall (Fig. 4.7).

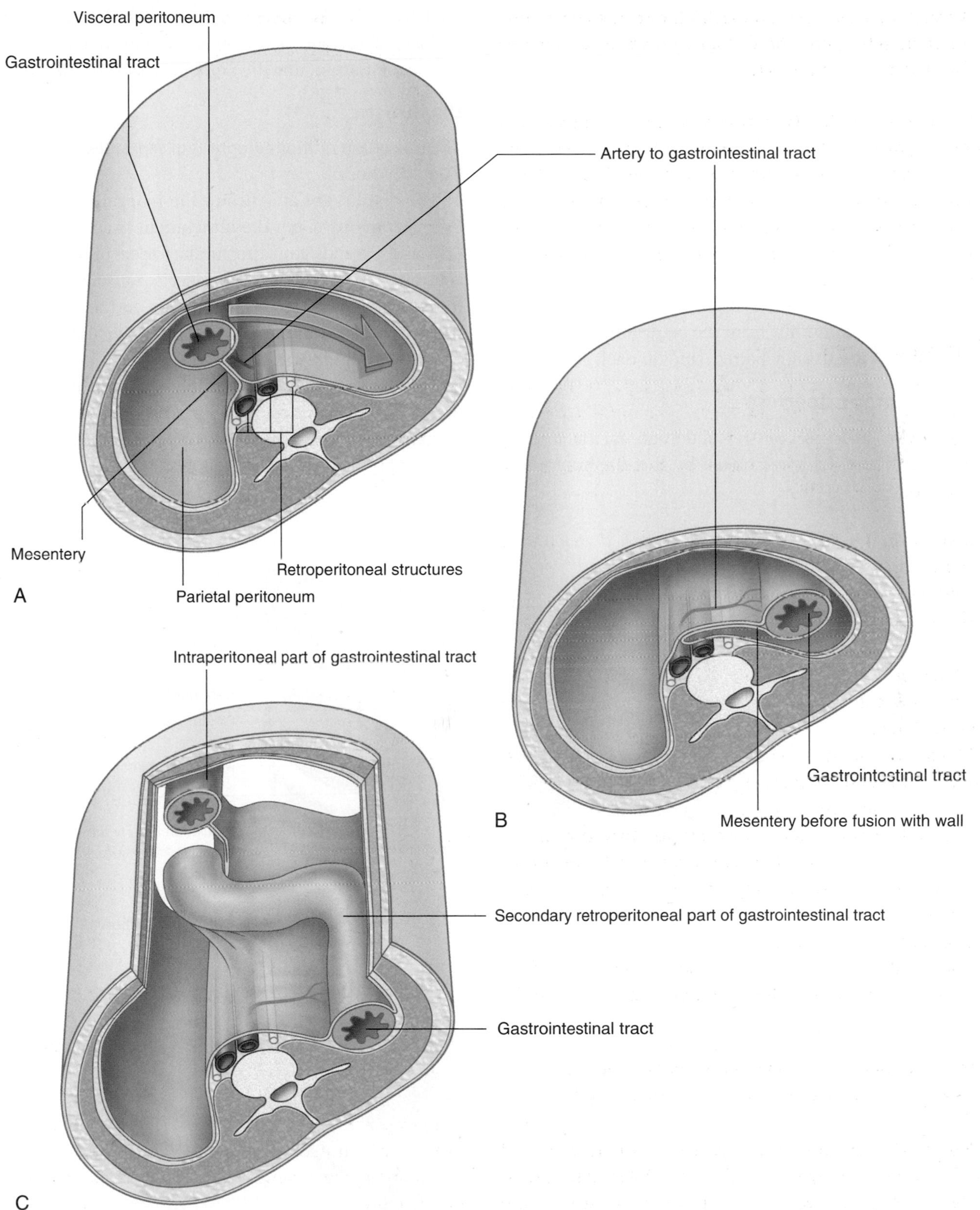


**Fig. 4.7** A series showing the progression (**A** to **C**) from an intraperitoneal organ to a secondarily retroperitoneal organ.

Large vessels, nerves, and lymphatics are associated with the posterior abdominal wall along the median axis of the body in the region where, during development, the peritoneum reflects off the wall as the dorsal mesentery, which supports the developing gut tube. As a consequence, branches of the neurovascular structures that pass to parts of the gastrointestinal system are unpaired, originate from the anterior aspects of their parent structures, and travel in mesenteries or pass retroperitoneally in areas where the mesenteries secondarily fuse to the wall.

Generally, vessels, nerves, and lymphatics to the abdominal wall and to organs that originate as retroperitoneal structures branch laterally from the central neurovascular structures and are usually paired, one on each side.

## Inferior thoracic aperture

The superior aperture of the abdomen is the inferior thoracic aperture, which is closed by the diaphragm (see pp. 126-127). The margin of the inferior thoracic aperture consists of vertebra TXII, rib XII, the distal end of rib XI, the costal margin, and the xiphoid process of the sternum.

## Diaphragm

The musculotendinous diaphragm separates the abdomen from the thorax.

The diaphragm attaches to the margin of the inferior thoracic aperture, but the attachment is complex posteriorly and extends into the lumbar area of the vertebral column (Fig. 4.8). On each side, a muscular extension (crus) firmly anchors the diaphragm to the anterolateral surface of the vertebral column as far down as vertebra LIII on the right and vertebra LII on the left.

Because the costal margin is not complete posteriorly, the diaphragm is anchored to arch-shaped (arcuate) ligaments, which span the distance between available bony points and the intervening soft tissues:

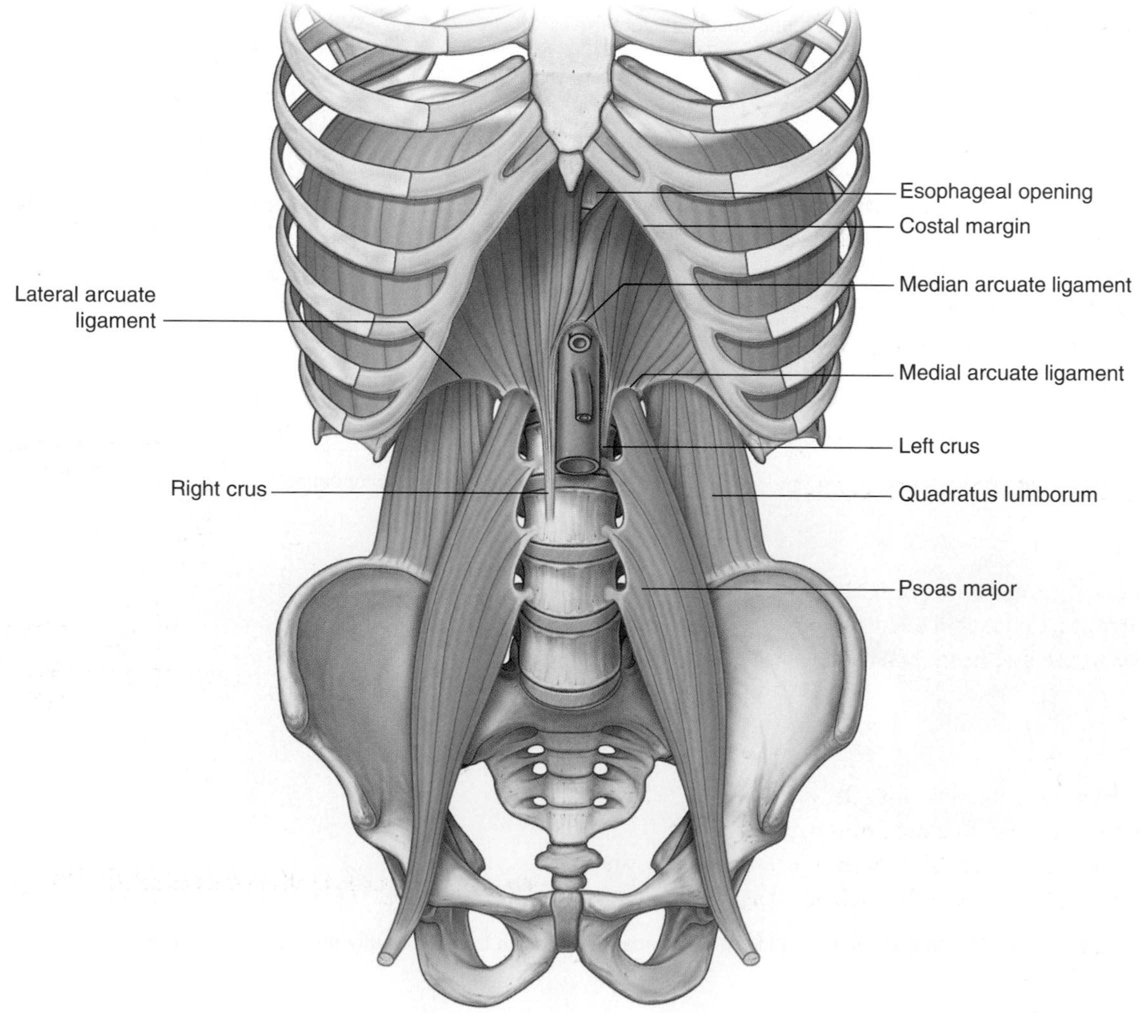


**Fig. 4.8** Inferior thoracic aperture and the diaphragm.

- **Medial** and **lateral arcuate ligaments** cross muscles of the posterior abdominal wall and attach to vertebrae, the transverse processes of vertebra LI and rib XII, respectively.
- A **median arcuate ligament** crosses the aorta and is continuous with the crus on each side.

The posterior attachment of the diaphragm extends much farther inferiorly than the anterior attachment. Consequently, the diaphragm is an important component of the posterior abdominal wall, to which a number of viscera are related.

## Pelvic inlet

The abdominal wall is continuous with the pelvic wall at the pelvic inlet, and the abdominal cavity is continuous with the pelvic cavity.

The circular margin of the pelvic inlet is formed entirely by bone:

- posteriorly by the sacrum,
- anteriorly by the pubic symphysis, and
- laterally, on each side, by a distinct bony rim on the pelvic bone (Fig. 4.9).

Because of the way in which the sacrum and attached pelvic bones are angled posteriorly on the vertebral column, the pelvic cavity is not oriented in the same vertical plane as the abdominal cavity. Instead, the pelvic cavity projects posteriorly, and the inlet opens anteriorly and somewhat superiorly (Fig. 4.10).

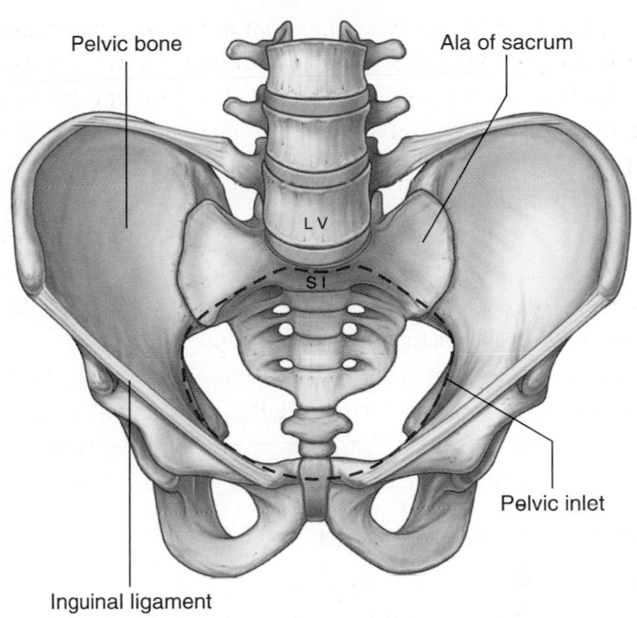


**Fig. 4.9** Pelvic inlet.

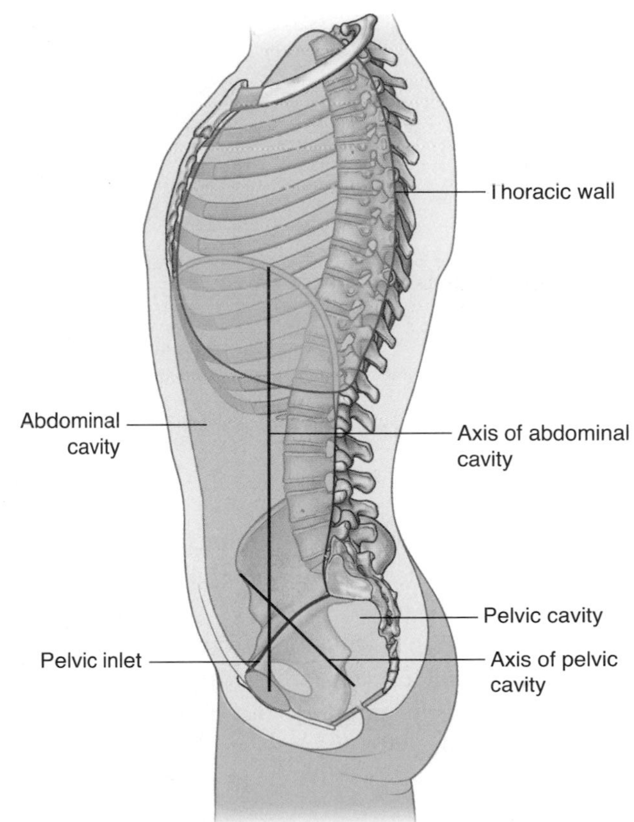


**Fig. 4.10** Orientation of abdominal and pelvic cavities.

## RELATIONSHIP TO OTHER REGIONS

### Thorax

The abdomen is separated from the thorax by the diaphragm. Structures pass between the two regions through or posterior to the diaphragm (see Fig. 4.8).

### Pelvis

The pelvic inlet opens directly into the abdomen and structures pass between the abdomen and pelvis through it.

The peritoneum lining the abdominal cavity is continuous with the peritoneum in the pelvis. Consequently, the

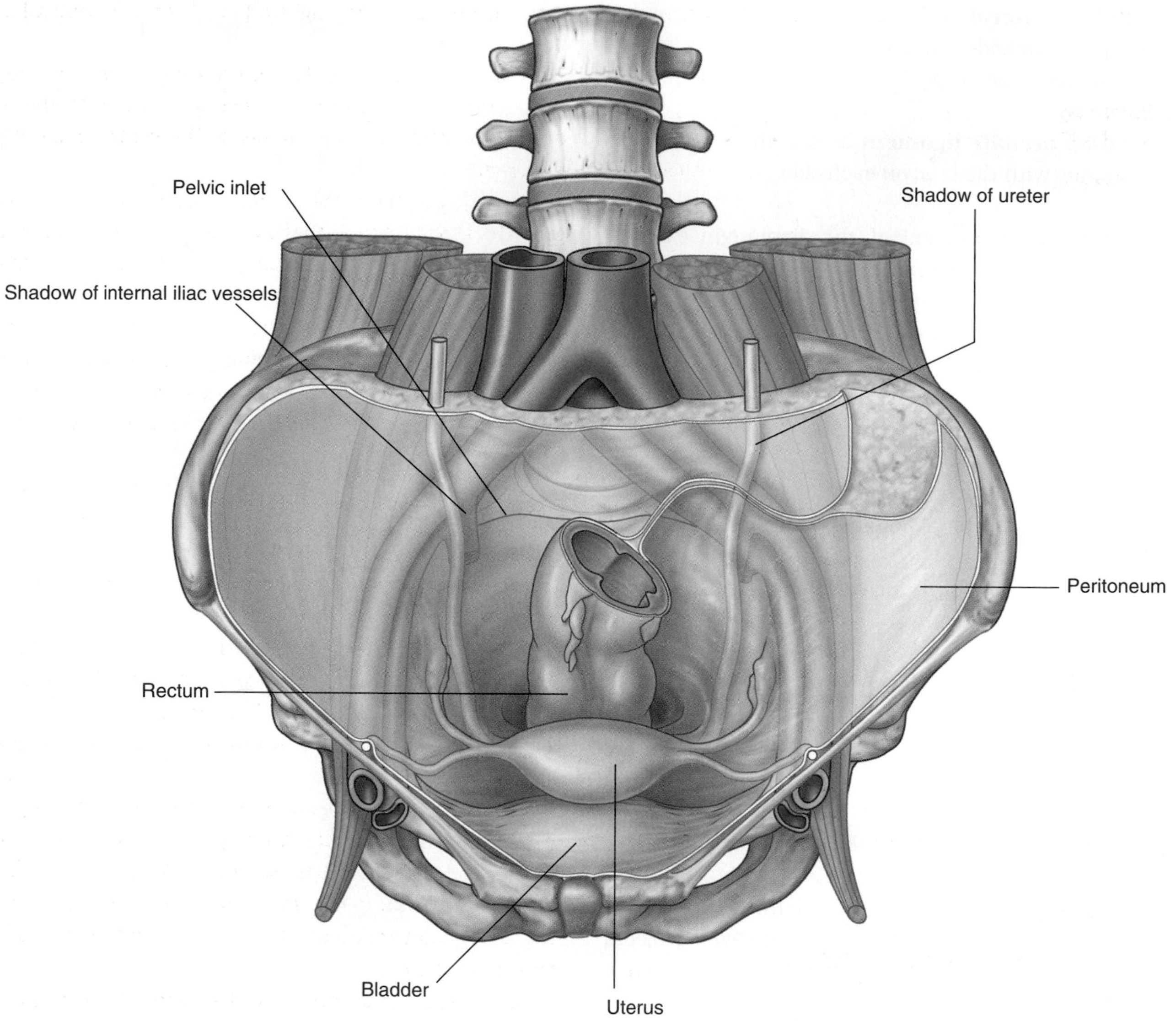

**Fig. 4.11** The abdominal cavity is continuous with the pelvic cavity.

abdominal cavity is entirely continuous with the pelvic cavity (Fig. 4.11). Infections in one region can therefore freely spread into the other.

The bladder expands superiorly from the pelvic cavity into the abdominal cavity and, during pregnancy, the uterus expands freely superiorly out of the pelvic cavity into the abdominal cavity.

## Lower limb

The abdomen communicates directly with the thigh through an aperture formed anteriorly between the inferior margin of the abdominal wall (marked by the inguinal ligament) and the pelvic bone (Fig. 4.12). Structures that pass through this aperture are:

- the major artery and vein of the lower limb;
- the femoral nerve, which innervates the quadriceps femoris muscle, which extends the knee;
- lymphatics; and
- the distal ends of psoas major and iliacus muscles, which flex the thigh at the hip joint.

As vessels pass inferior to the inguinal ligament, their names change—the external iliac artery and vein of the abdomen become the femoral artery and vein of the thigh.

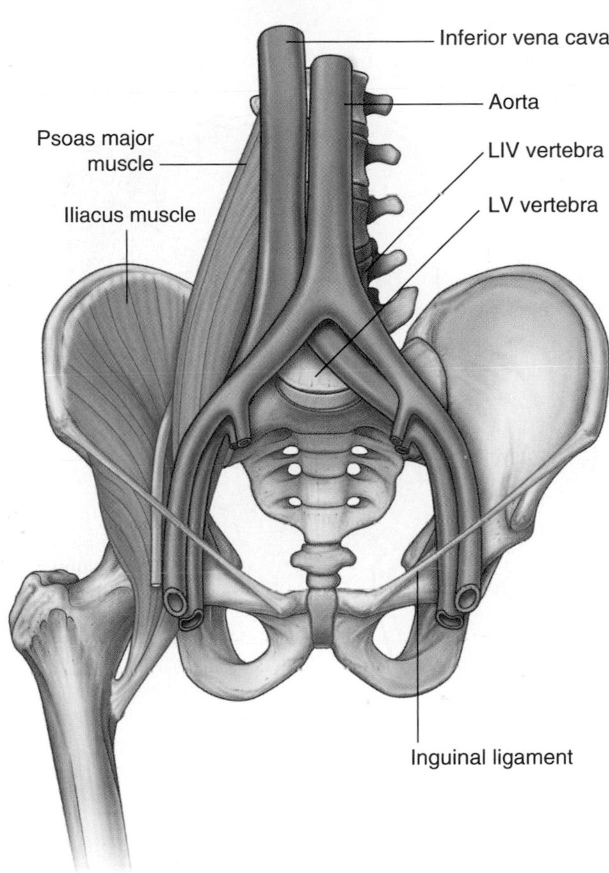


**Fig. 4.12** Structures passing between the abdomen and thigh.

## KEY FEATURES

### Arrangement of abdominal viscera in the adult

A basic knowledge of the development of the gastrointestinal tract is needed to understand the arrangement of viscera and mesenteries in the abdomen (Fig. 4.13).

The early gastrointestinal tract is oriented longitudinally in the body cavity and is suspended from surrounding walls by a large dorsal mesentery and a much smaller ventral mesentery.

Superiorly, the dorsal and ventral mesenteries are anchored to the diaphragm.

The primitive gut tube consists of the foregut, the midgut, and the hindgut. Massive longitudinal growth of the gut tube, rotation of selected parts of the tube, and secondary fusion of some viscera and their associated mesenteries to the body wall participate in generating the adult arrangement of abdominal organs.

#### Development of the foregut

In abdominal regions, the **foregut** gives rise to the distal end of the esophagus, the stomach, and the proximal part of the duodenum. The foregut is the only part of the gut tube suspended from the wall by both the ventral and dorsal mesenteries.

A diverticulum from the anterior aspect of the foregut grows into the ventral mesentery, giving rise to the liver and gallbladder, and, ultimately, to the ventral part of the pancreas.

The dorsal part of the pancreas develops from an outgrowth of the foregut into the dorsal mesentery. The spleen develops in the dorsal mesentery in the region between the body wall and presumptive stomach.

In the foregut, the developing stomach rotates clockwise and the associated dorsal mesentery, containing the spleen, moves to the left and greatly expands. During this process, part of the mesentery becomes associated with, and secondarily fuses with, the left side of the body wall.

At the same time, the duodenum, together with its dorsal mesentery and an appreciable part of the pancreas, swings to the right and fuses to the body wall.

Secondary fusion of the duodenum to the body wall, massive growth of the liver in the ventral mesentery, and fusion of the superior surface of the liver to the diaphragm restrict the opening to the space enclosed by the ballooned dorsal mesentery associated with the stomach. This restricted opening is the **omental foramen** (**epiploic foramen**).

The part of the abdominal cavity enclosed by the expanded dorsal mesentery, and posterior to the stomach, is the **omental bursa** (**lesser sac**). Access, through the omental foramen, to this space from the rest of the peritoneal cavity (**greater sac**) is inferior to the free edge of the ventral mesentery.

Part of the dorsal mesentery that initially forms part of the lesser sac greatly enlarges in an inferior direction, and the two opposing surfaces of the mesentery fuse to form an apron-like structure (the **greater omentum**). The greater omentum is suspended from the greater curvature of the stomach, lies over other viscera in the abdominal cavity, and is the first structure observed when the abdominal cavity is opened anteriorly.

#### Development of the midgut

The midgut develops into the distal part of the duodenum and the jejunum, ileum, ascending colon, and proximal two-thirds of the transverse colon. A small yolk sac projects anteriorly from the developing midgut into the umbilicus.

Rapid growth of the gastrointestinal system results in a loop of the midgut herniating out of the abdominal cavity and into the umbilical cord. As the body grows in size and the connection with the yolk sac is lost, the midgut returns to the abdominal cavity. While this process is occurring, the two limbs of the midgut loop rotate counterclockwise

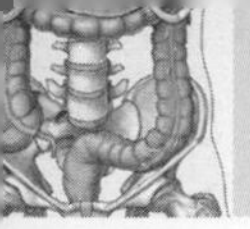

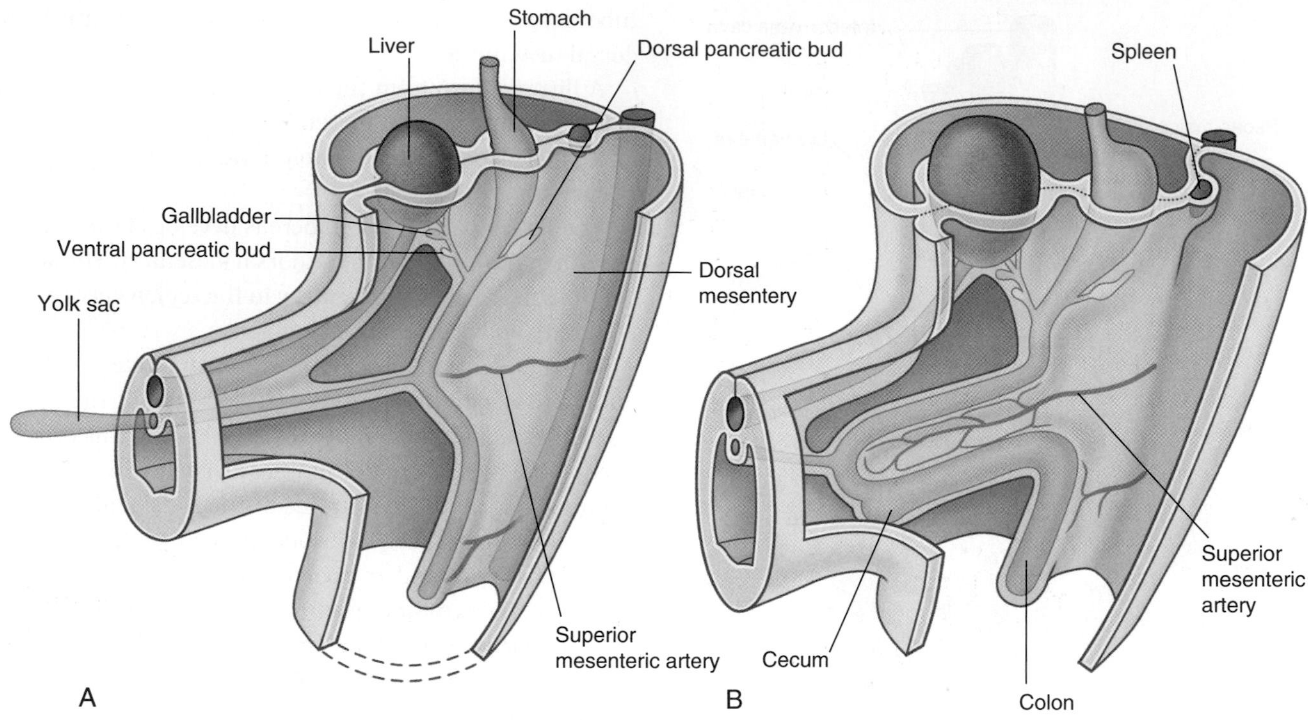


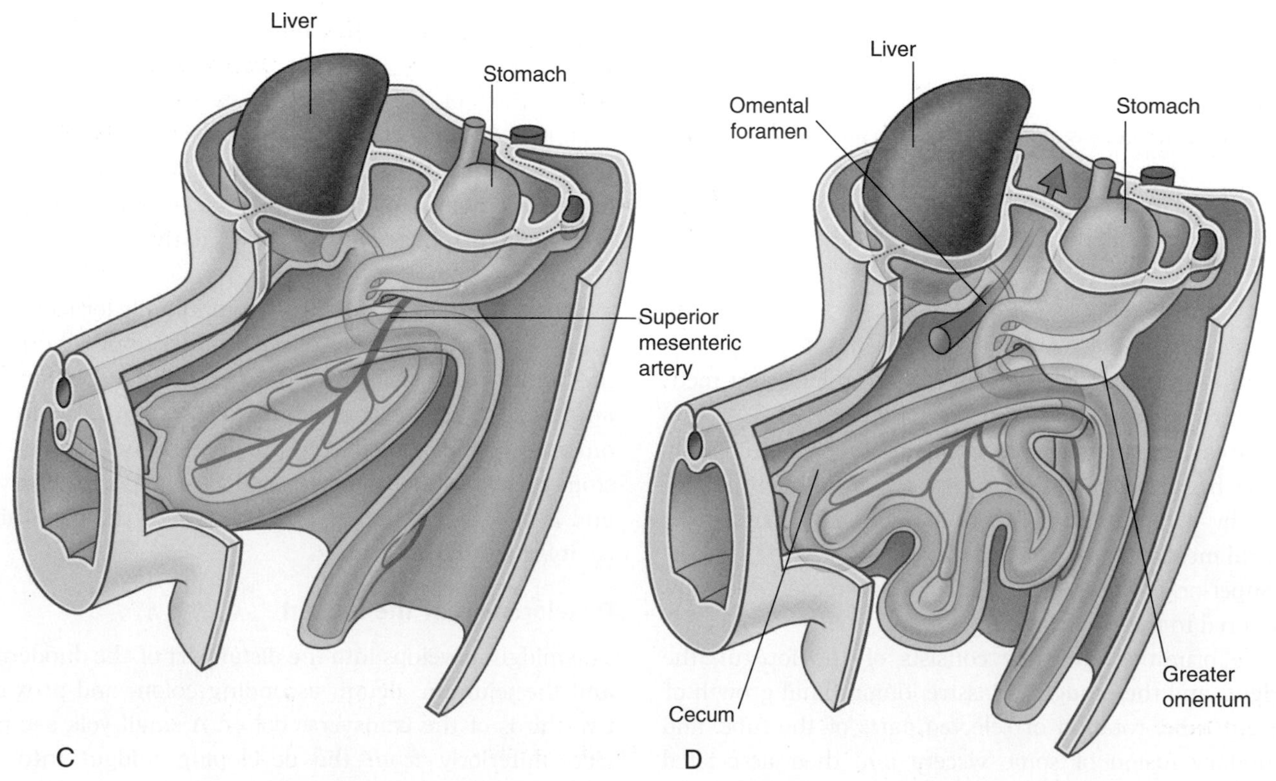


**Fig. 4.13** A series (**A** to **H**) showing the development of the gut and mesenteries.

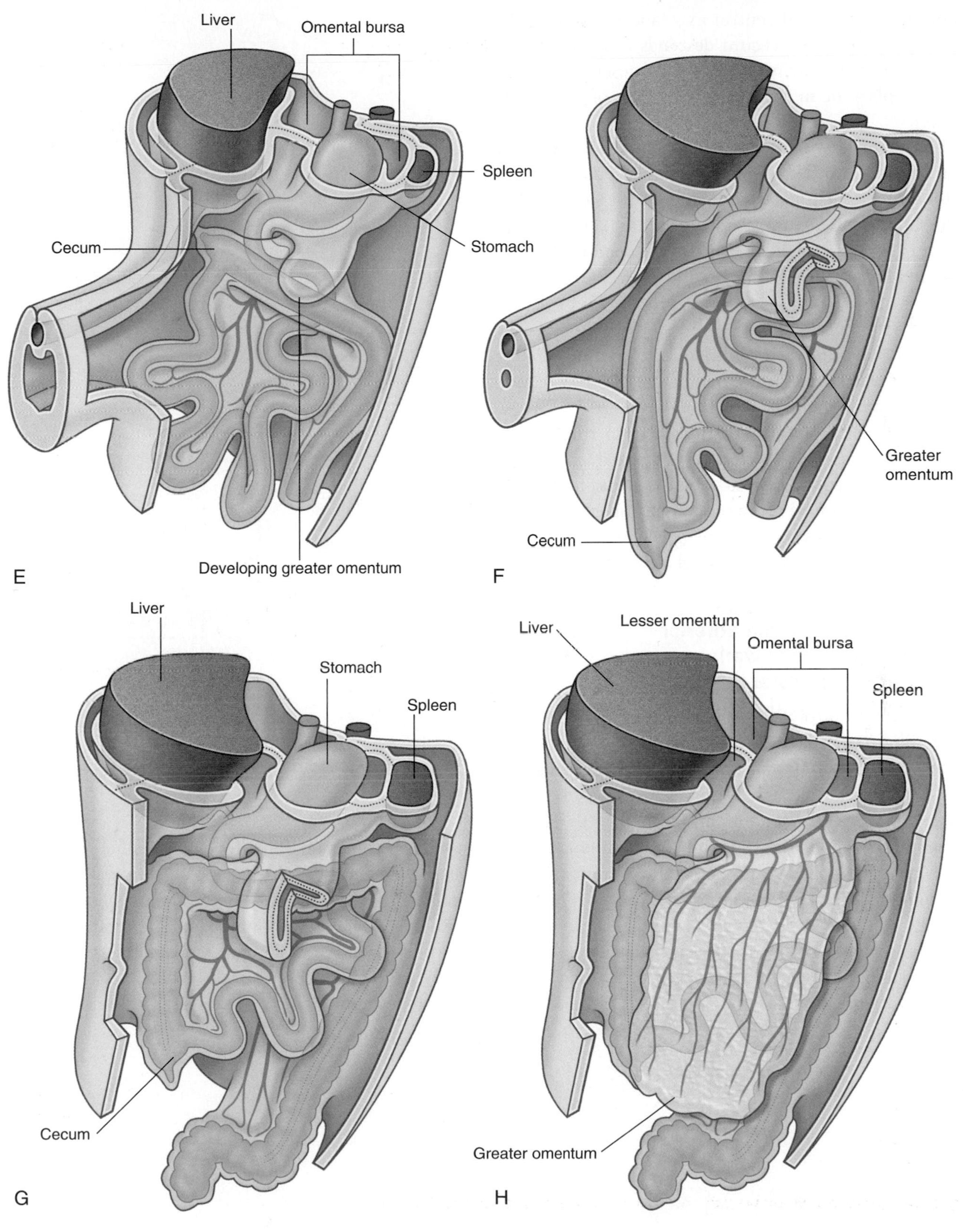
Liver
Omental bursa
Spleen
Cecum
Stomach
Developing greater omentum
E
Greater omentum
Cecum
F
Liver
Stomach
Spleen
Cecum
G
Liver
Lesser omentum
Omental bursa
Spleen
Greater omentum
H

**Fig. 4.13, cont'd**

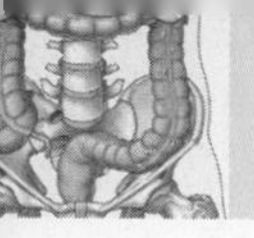



around their combined central axis, and the part of the loop that becomes the cecum descends into the inferior right aspect of the cavity. The superior mesenteric artery, which supplies the midgut, is at the center of the axis of rotation.

The cecum remains intraperitoneal, the ascending colon fuses with the body wall becoming secondarily retroperitoneal, and the transverse colon remains suspended by its dorsal mesentery (transverse mesocolon). The greater omentum hangs over the transverse colon and the mesocolon and usually fuses with these structures.

### Development of the hindgut

The distal one-third of the transverse colon, descending colon, sigmoid colon, and superior part of the rectum develop from the hindgut.

Proximal parts of the hindgut swing to the left and become the descending colon and sigmoid colon. The descending colon and its dorsal mesentery fuse to the body wall, while the sigmoid colon remains intraperitoneal. The sigmoid colon passes through the pelvic inlet and is continuous with the rectum at the level of vertebra SIII.

## Skin and muscles of the anterior and lateral abdominal wall and thoracic intercostal nerves

The anterior rami of thoracic spinal nerves T7 to T12 follow the inferior slope of the lateral parts of the ribs and cross the costal margin to enter the abdominal wall (Fig. 4.14). Intercostal nerves T7 to T11 supply skin and muscle of the abdominal wall, as does the subcostal nerve T12. In addition, T5 and T6 supply upper parts of the external oblique muscle of the abdominal wall; T6 also supplies cutaneous innervation to skin over the xiphoid.

Skin and muscle in the inguinal and suprapubic regions of the abdominal wall are innervated by L1 and not by thoracic nerves.

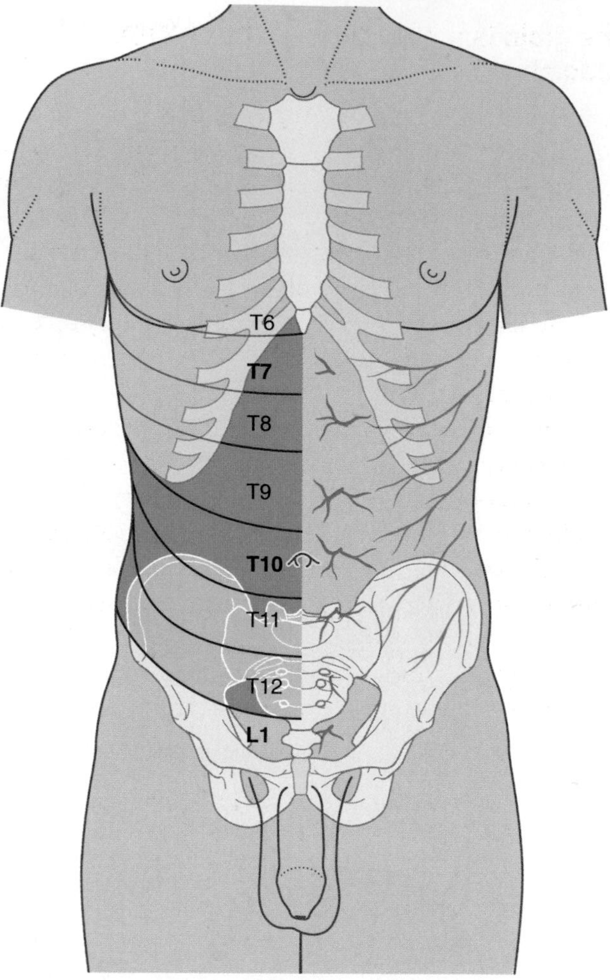


**Fig. 4.14** Innervation of the anterior abdominal wall.

Dermatomes of the anterior abdominal wall are indicated in Figure 4.14. In the midline, skin over the infrasternal angle is T6 and that around the umbilicus is T10. L1 innervates skin in the inguinal and suprapubic regions.

Muscles of the abdominal wall are innervated segmentally in patterns that generally reflect the patterns of the overlying dermatomes.

## The groin is a weak area in the anterior abdominal wall

During development, the gonads in both sexes descend from their sites of origin on the posterior abdominal wall into the pelvic cavity in women and the developing scrotum in men (Fig. 4.15).

Before descent, a cord of tissue (the **gubernaculum**) passes through the anterior abdominal wall and connects the inferior pole of each gonad with primordia of the scrotum in men and the labia majora in women (labioscrotal swellings).

A tubular extension (the **processus vaginalis**) of the peritoneal cavity and the accompanying muscular layers of the anterior abdominal wall project along the gubernaculum on each side into the labioscrotal swellings.

In men, the testis, together with its neurovascular structures and its efferent duct (the ductus deferens) descends into the scrotum along a path, initially defined by the gubernaculum, between the processus vaginalis and the accompanying coverings derived from the abdominal wall. All that remains of the gubernaculum is a connective tissue remnant that attaches the caudal pole of the testis to the scrotum.

The **inguinal canal** is the passage through the anterior abdominal wall created by the processus vaginalis. The **spermatic cord** is the tubular extension of the layers of the abdominal wall into the scrotum that contains all structures passing between the testis and the abdomen.

The distal sac-like terminal end of the spermatic cord on each side contains the testis, associated structures, and the now isolated part of the peritoneal cavity (the cavity of the tunica vaginalis).

In women, the gonads descend to a position just inside the pelvic cavity and never pass through the anterior abdominal wall. As a result, the only major structure passing through the inguinal canal is a derivative of the gubernaculum (the round ligament of the uterus).

In both men and women, the groin (inguinal region) is a weak area in the abdominal wall (Fig. 4.15) and is the site of inguinal hernias.

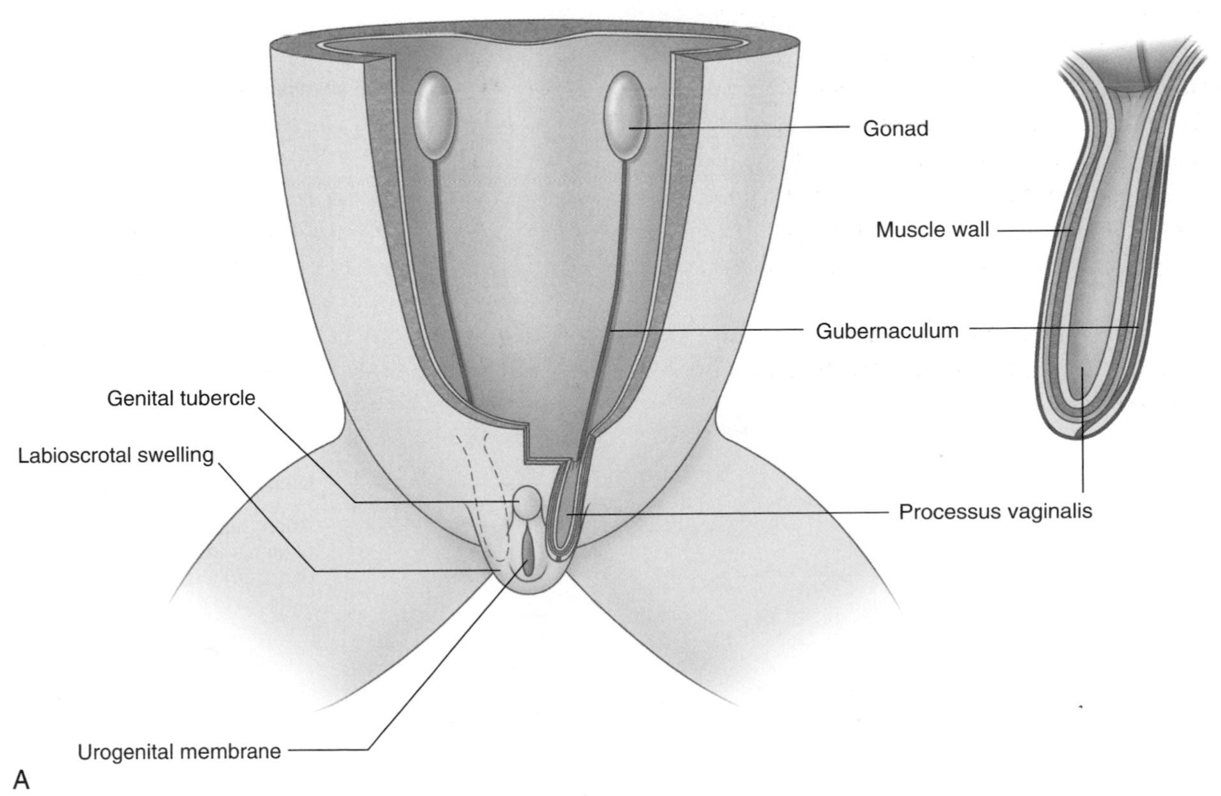


**Fig. 4.15** Inguinal region. **A.** Development.

Inferior vena cava
Aorta
Left testicular vein
Right testicular artery
Right testicular vein
Left testicular artery
Pelvic brim
Left ductus deferens
Deep inguinal ring
Inguinal canal
Superficial inguinal ring
Spermatic cord
Ductus deferens
Testicular artery and vein
Epididymis
Testis
Remnant of gubernaculum
Tunica vaginalis
B
Left renal artery
Inferior vena cava
Left renal vein
Left ovarian vein
Aorta
Left ovarian artery
Pelvic inlet
Uterine tube
Uterus
Round ligament of uterus (remnants of gubernaculum)
Superficial inguinal ring
C

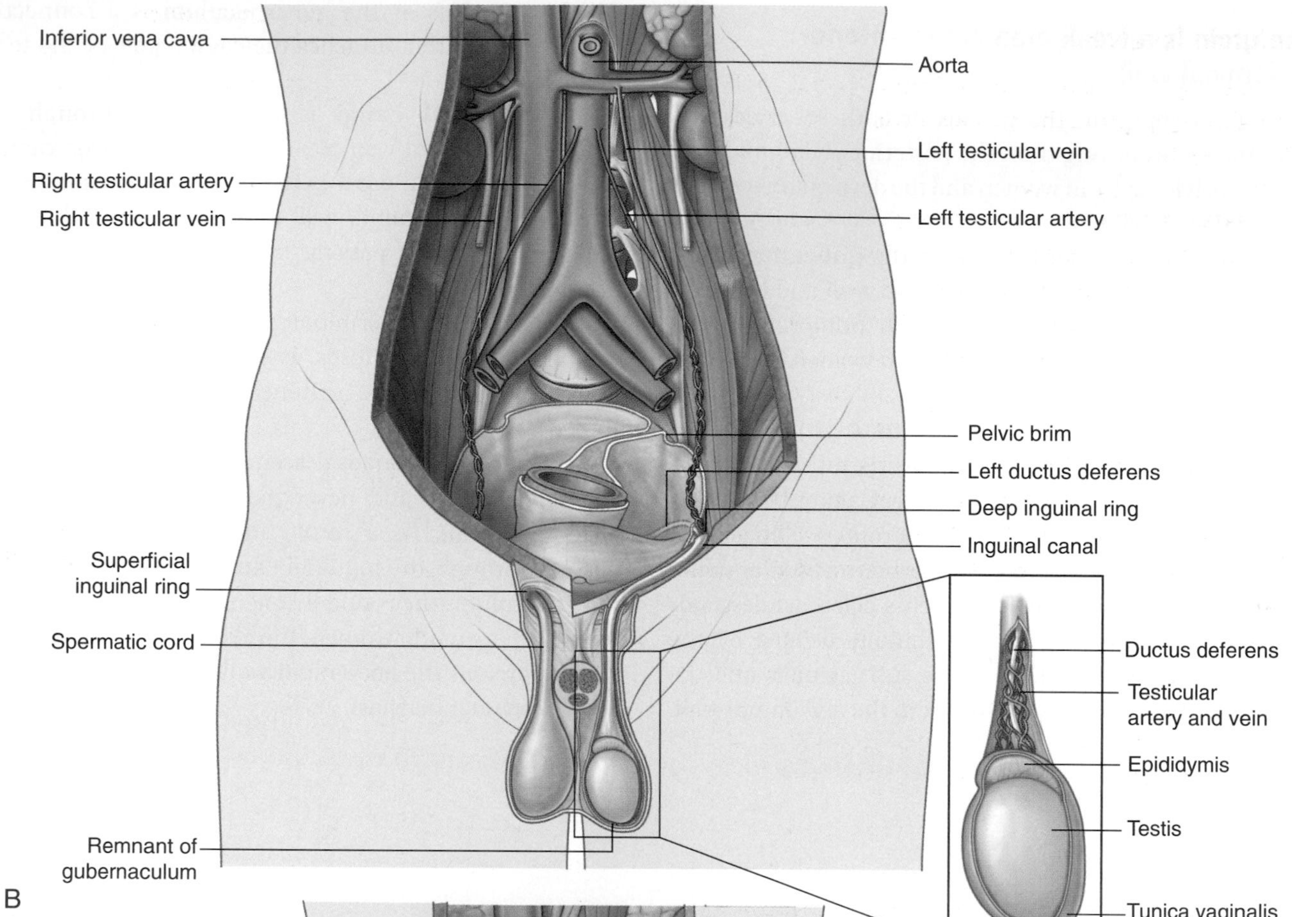

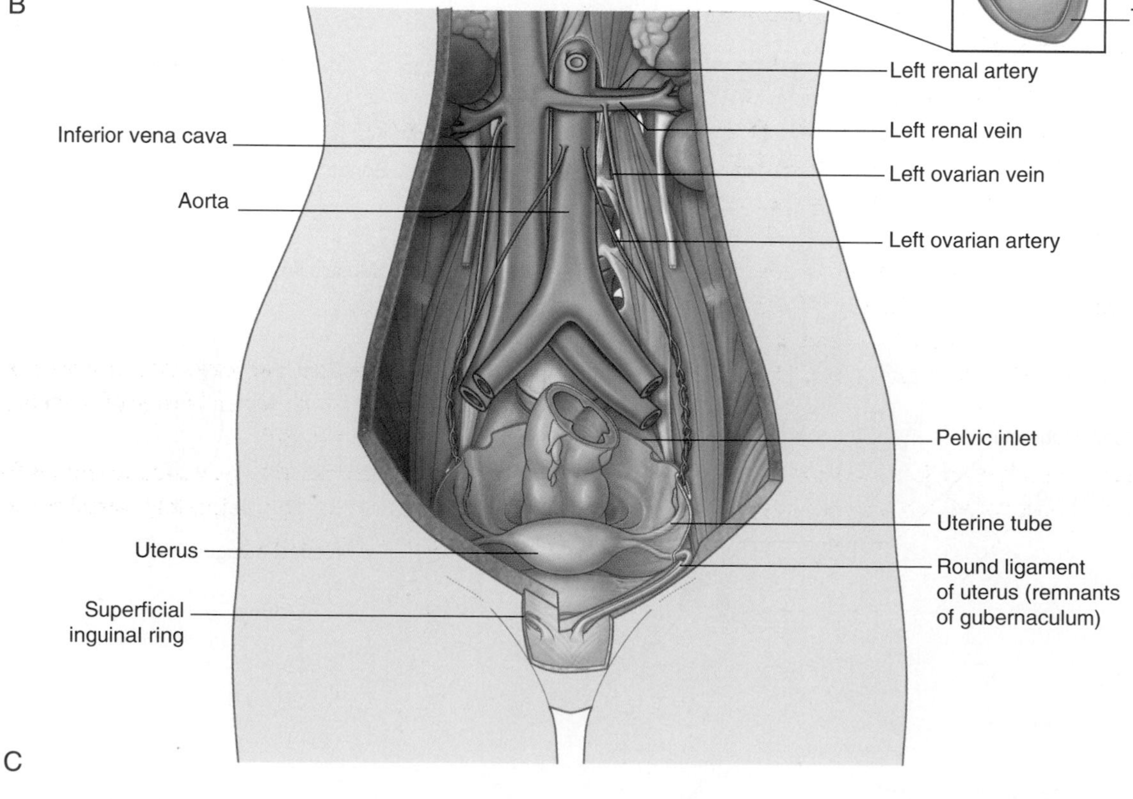

**Fig. 4.15, cont'd** **B.** In men. **C.** In women.

## Vertebral level LI

The transpyloric plane is a horizontal plane that transects the body through the lower aspect of vertebra LI (Fig. 4.16). It:

- is about midway between the jugular notch and the pubic symphysis, and crosses the costal margin on each side at roughly the ninth costal cartilage;
- crosses through the opening of the stomach into the duodenum (the pyloric orifice), which is just to the right of the body of LI; the duodenum then makes a characteristic C-shaped loop on the posterior abdominal wall and crosses the midline to open into the jejunum just to the left of the body of vertebra LII, whereas the head of the pancreas is enclosed by the loop of the duodenum, and the body of the pancreas extends across the midline to the left;
- crosses through the body of the pancreas; and
- approximates the position of the hila of the kidneys; though because the left kidney is slightly higher than the right, the transpyloric plane crosses through the inferior aspect of the left hilum and the superior part of the right hilum.

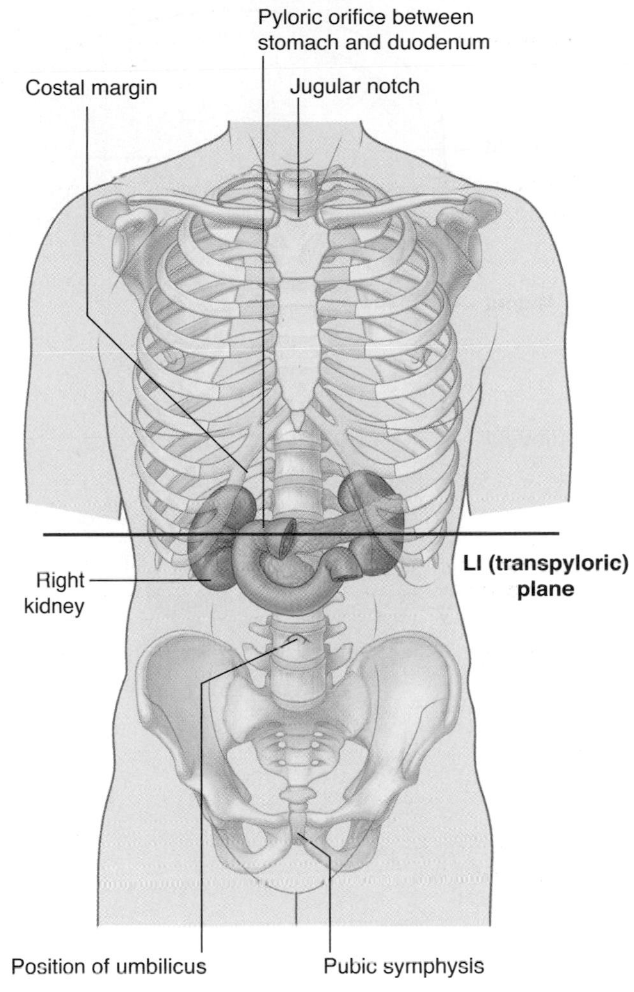


**Fig. 4.16** Vertebral level LI.

## The gastrointestinal system and its derivatives are supplied by three major arteries

Three large unpaired arteries branch from the anterior surface of the abdominal aorta to supply the abdominal part of the gastrointestinal tract and all of the structures (liver, pancreas, and gallbladder) to which this part of the gut gives rise during development (Fig. 4.17). These arteries pass through derivatives of the dorsal and ventral mesenteries to reach the target viscera. These vessels therefore also supply structures such as the spleen and lymph nodes that develop in the mesenteries. These three arteries are:

- the **celiac artery**, which branches from the abdominal aorta at the upper border of vertebra LI and supplies the foregut;
- the **superior mesenteric artery**, which arises from the abdominal aorta at the lower border of vertebra LI and supplies the midgut; and
- the **inferior mesenteric artery**, which branches from the abdominal aorta at approximately vertebral level LIII and supplies the hindgut.

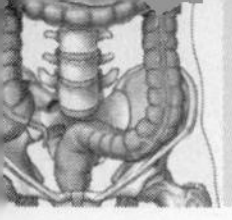



Foregut
Celiac trunk
Midgut
Superior mesenteric artery
Hindgut
Inferior mesenteric artery
A

Celiac trunk
Inferior vena cava
Superior mesenteric artery
Aorta
Inferior mesenteric artery
B

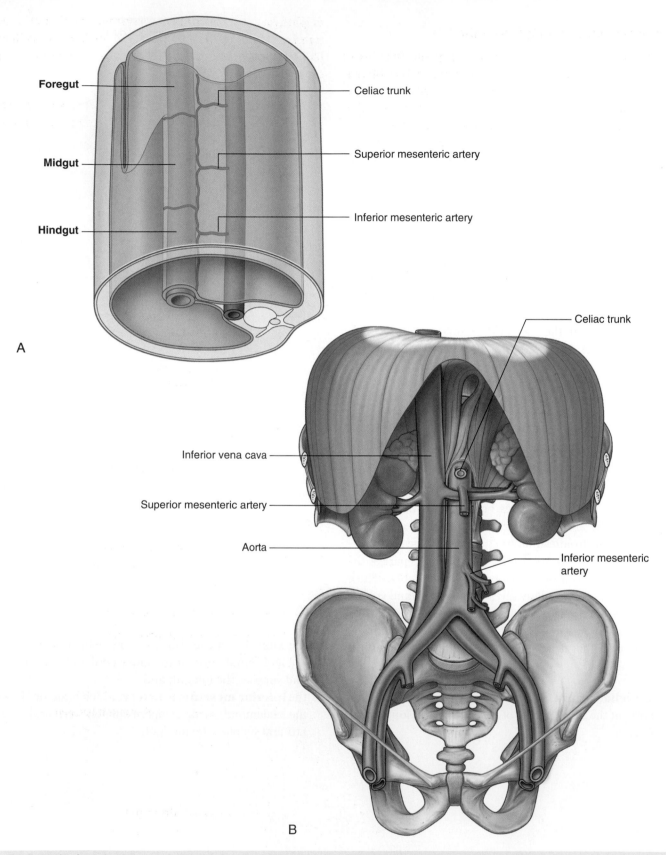

**Fig. 4.17** Blood supply of the gut. **A.** Relationship of vessels to the gut and mesenteries. **B.** Anterior view.

## Venous shunts from left to right

All blood returning to the heart from regions of the body other than the lungs flows into the right atrium of the heart. The inferior vena cava is the major systemic vein in the abdomen and drains this region together with the pelvis, perineum, and both lower limbs (Fig. 4.18).

The inferior vena cava lies to the right of the vertebral column and penetrates the central tendon of the diaphragm at approximately vertebral level TVIII. A number of large vessels cross the midline to deliver blood from the left side of the body to the inferior vena cava.

- One of the most significant is the left renal vein, which drains the kidney, suprarenal gland, and gonad on the same side.

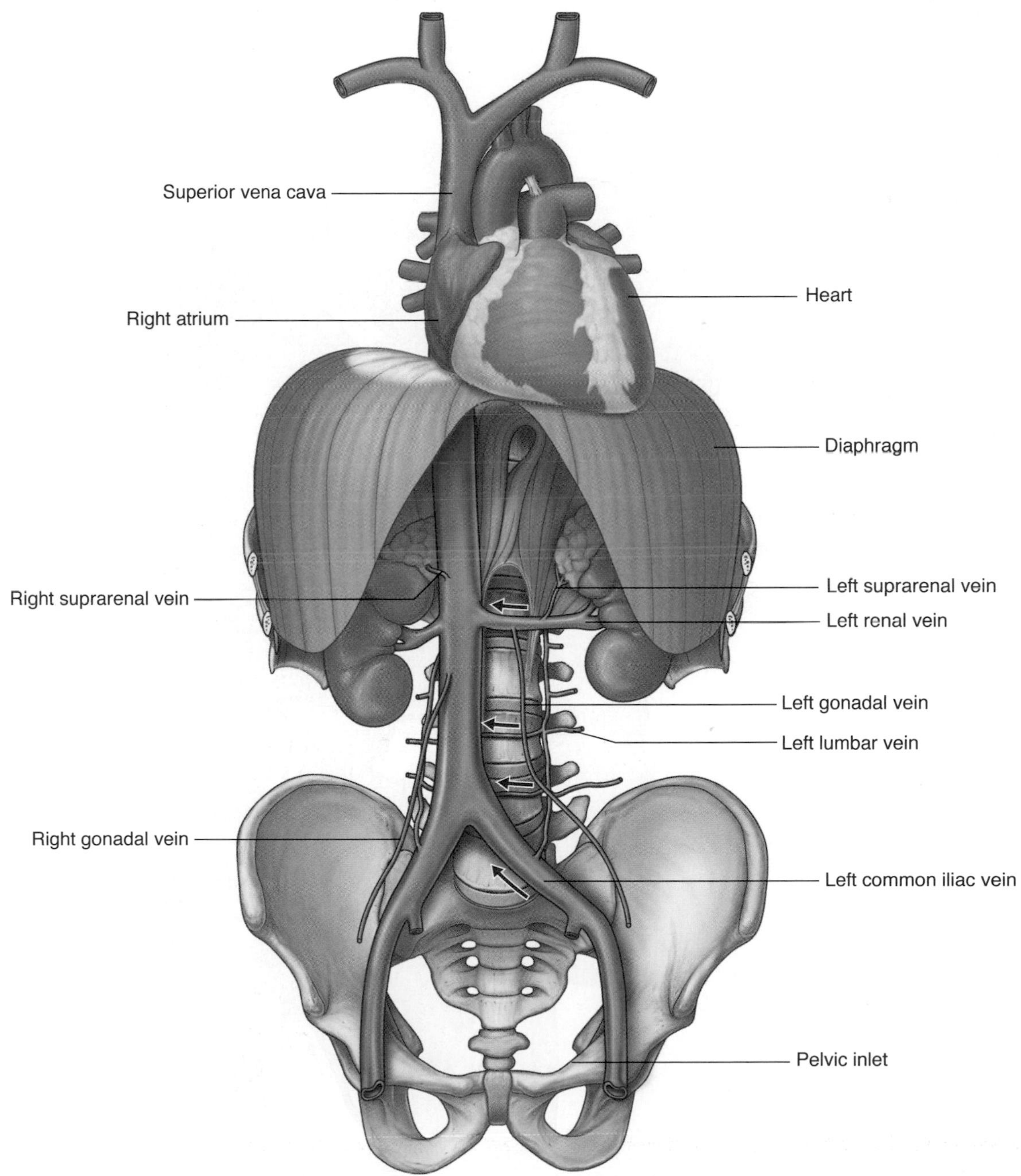


**Fig. 4.18** Left-to-right venous shunts.

- Another is the left common iliac vein, which crosses the midline at approximately vertebral level LV to join with its partner on the right to form the inferior vena cava. These veins drain the lower limbs, the pelvis, the perineum, and parts of the abdominal wall.
- Other vessels crossing the midline include the left lumbar veins, which drain the back and posterior abdominal wall on the left side.

### All venous drainage from the gastrointestinal system passes through the liver

Blood from abdominal parts of the gastrointestinal system and the spleen passes through a second vascular bed, in the liver, before ultimately returning to the heart (Fig. 4.19).

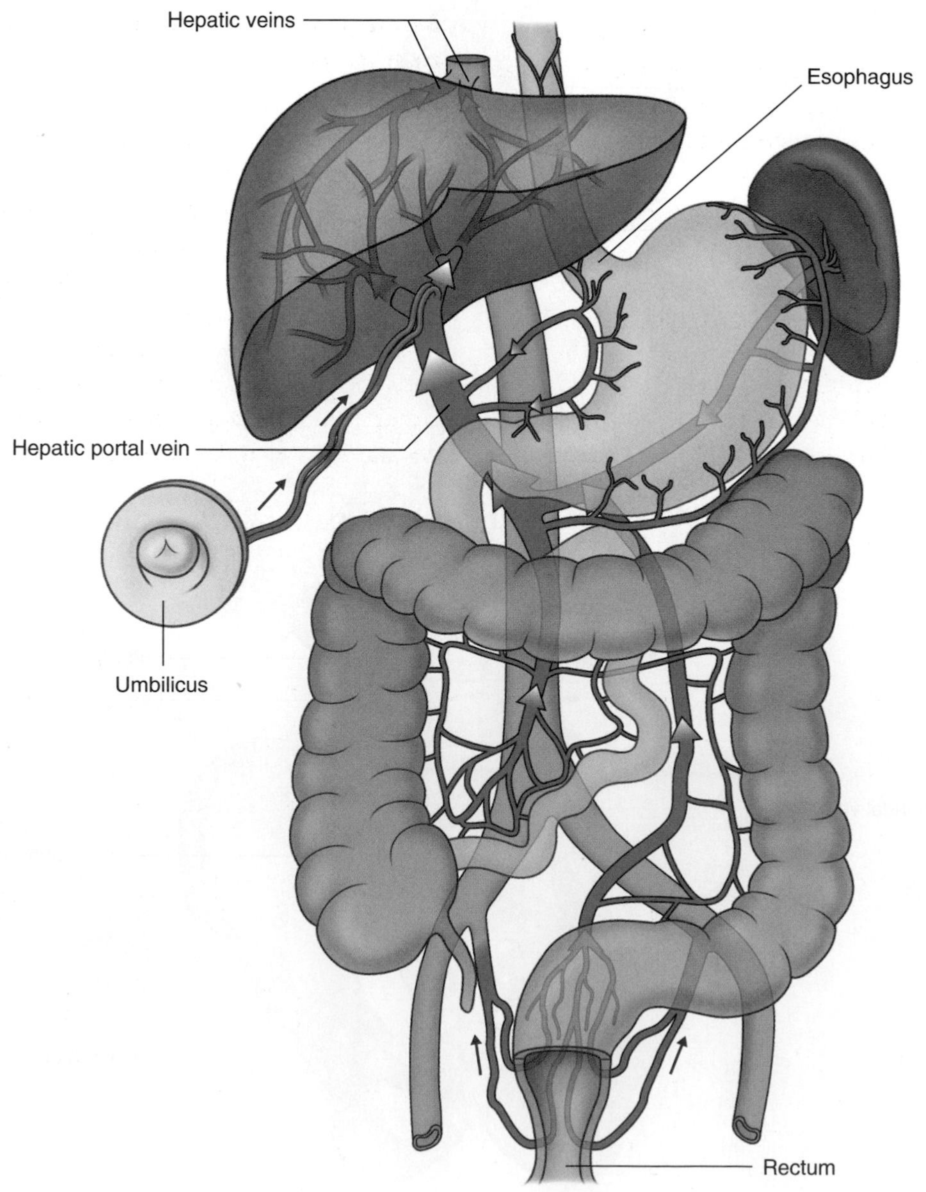


**Fig. 4.19** Hepatic portal system.

Venous blood from the digestive tract, pancreas, gallbladder, and spleen enters the inferior surface of the liver through the large **hepatic portal vein**. This vein then ramifies like an artery to distribute blood to small endothelial-lined hepatic sinusoids, which form the vascular exchange network of the liver.

After passing through the sinusoids, the blood collects in a number of short **hepatic veins**, which drain into the inferior vena cava just before the inferior vena cava penetrates the diaphragm and enters the right atrium of the heart.

Normally, vascular beds drained by the hepatic portal system interconnect, through small veins, with beds drained by systemic vessels, which ultimately connect directly with either the superior or inferior vena cava.

### Portacaval anastomoses

Among the clinically most important regions of overlap between the portal and caval systems are those at each end of the abdominal part of the gastrointestinal system:

- around the inferior end of the esophagus;
- around the inferior part of the rectum.

Small veins that accompany the degenerate umbilical vein (**round ligament of the liver**) establish another important portacaval anastomosis.

The round ligament of the liver connects the umbilicus of the anterior abdominal wall with the left branch of the portal vein as it enters the liver. The small veins that accompany this ligament form a connection between the portal system and para-umbilical regions of the abdominal wall, which drain into systemic veins.

Other regions where portal and caval systems interconnect include:

- where the liver is in direct contact with the diaphragm (the bare area of the liver);
- where the wall of the gastrointestinal tract is in direct contact with the posterior abdominal wall (retroperitoneal areas of the large and small intestine); and
- the posterior surface of the pancreas (much of the pancreas is secondarily retroperitoneal).

### Blockage of the hepatic portal vein or of vascular channels in the liver

Blockage of the hepatic portal vein or of vascular channels in the liver can affect the pattern of venous return from abdominal parts of the gastrointestinal system. Vessels that interconnect the portal and caval systems can become greatly enlarged and tortuous, allowing blood in tributaries of the portal system to bypass the liver, enter the caval system, and thereby return to the heart. Portal hypertension can result in esophageal and rectal varices and in caput medusae in which systemic vessels that radiate from para-umbilical veins enlarge and become visible on the abdominal wall.

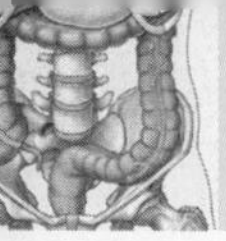



## Abdominal viscera are supplied by a large prevertebral plexus

Innervation of the abdominal viscera is derived from a large prevertebral plexus associated mainly with the anterior and lateral surfaces of the aorta (Fig. 4.20). Branches are distributed to target tissues along vessels that originate from the abdominal aorta.

The prevertebral plexus contains sympathetic, parasympathetic, and visceral sensory components:

- Sympathetic components originate from spinal cord levels T5 to L2.
- Parasympathetic components are from the vagus nerve [X] and spinal cord levels S2 to S4.
- Visceral sensory fibers generally parallel the motor pathways.

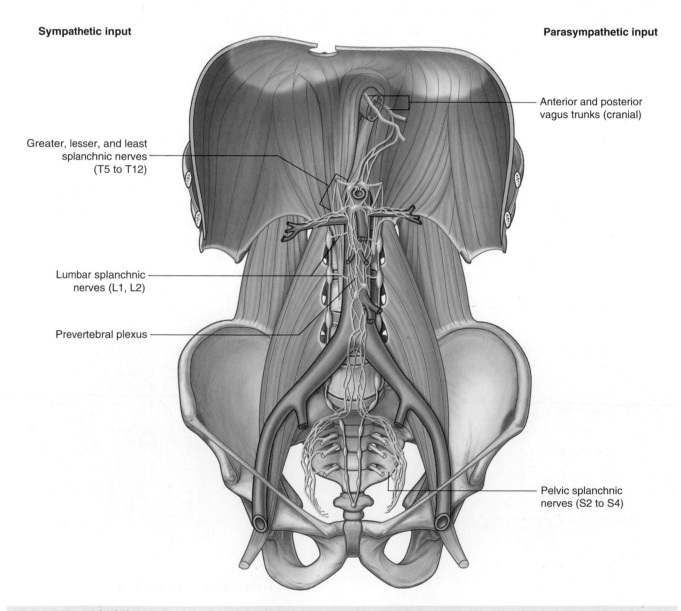



**Fig. 4.20** Prevertebral plexus.

# Regional anatomy

The abdomen is the part of the trunk inferior to the thorax (Fig. 4.21). Its musculomembranous walls surround a large cavity (the **abdominal cavity**), which is bounded superiorly by the diaphragm and inferiorly by the pelvic inlet.

The abdominal cavity may extend superiorly as high as the fourth intercostal space, and is continuous inferiorly with the pelvic cavity. It contains the **peritoneal cavity** and the abdominal viscera.

## SURFACE TOPOGRAPHY

Topographical divisions of the abdomen are used to describe the location of abdominal organs and the pain associated with abdominal problems. The two schemes most often used are:

- a four-quadrant pattern and
- a nine-region pattern.

### Four-quadrant pattern

A horizontal transumbilical plane passing through the umbilicus and the intervertebral disc between vertebrae LIII and LIV and intersecting with the vertical median plane divides the abdomen into four quadrants—the right upper, left upper, right lower, and left lower quadrants (Fig. 4.22).

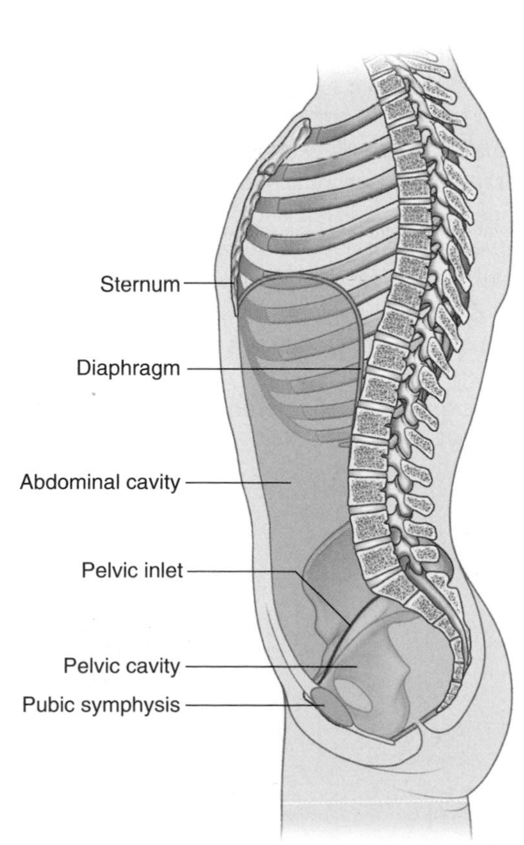


**Fig. 4.21** Boundaries of the abdominal cavity.

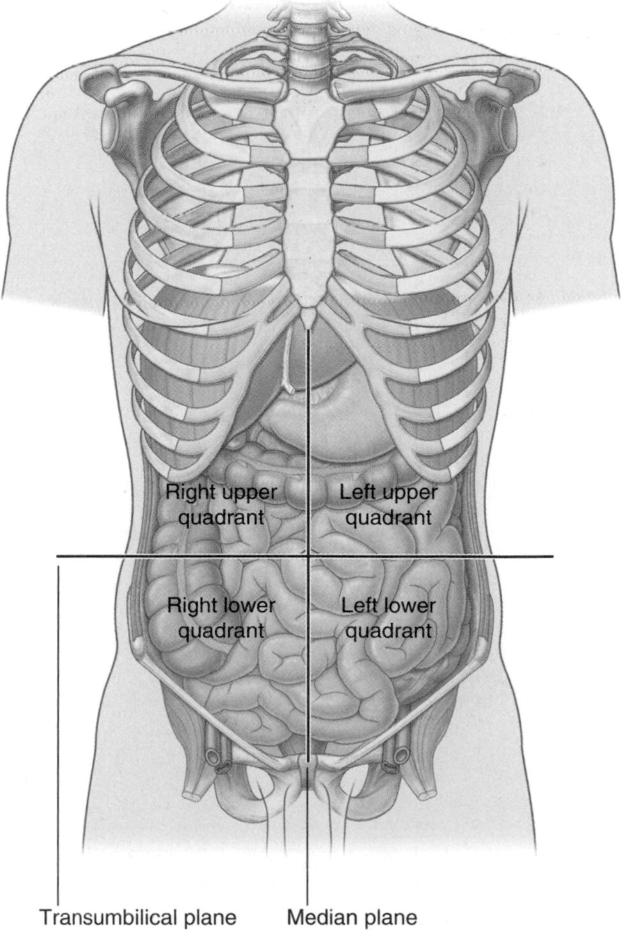


**Fig. 4.22** Four-quadrant topographical pattern.

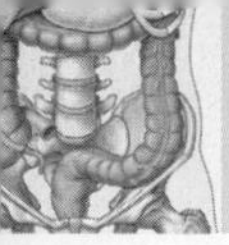



## Nine-region pattern

The nine-region pattern is based on two horizontal and two vertical planes (Fig. 4.23).

- The superior horizontal plane (the **subcostal plane**) is immediately inferior to the costal margins, which places it at the lower border of the costal cartilage of rib X and passing posteriorly through the body of vertebra LIII. (Note, however, that sometimes the **transpyloric plane,** halfway between the jugular notch and the symphysis pubis or halfway between the umbilicus and the inferior end of the body of the sternum, passing posteriorly through the lower border of vertebra LI and intersecting with the costal margin at the ends of the ninth costal cartilages, is used instead.)
- The inferior horizontal plane (the **intertubercular plane**) connects the tubercles of the iliac crests, which are palpable structures 5 cm posterior to the anterior superior iliac spines, and passes through the upper part of the body of vertebra LV.
- The vertical planes pass from the midpoint of the clavicles inferiorly to a point midway between the anterior superior iliac spine and pubic symphysis.

These four planes establish the topographical divisions in the nine-region organization. The following designations are used for each region: superiorly the right hypochondrium, the epigastric region, and the left hypochondrium; inferiorly the right groin (inguinal region), pubic region, and left groin (inguinal region); and in the middle the right flank (lateral region), the umbilical region, and the left flank (lateral region) (Fig. 4.23).

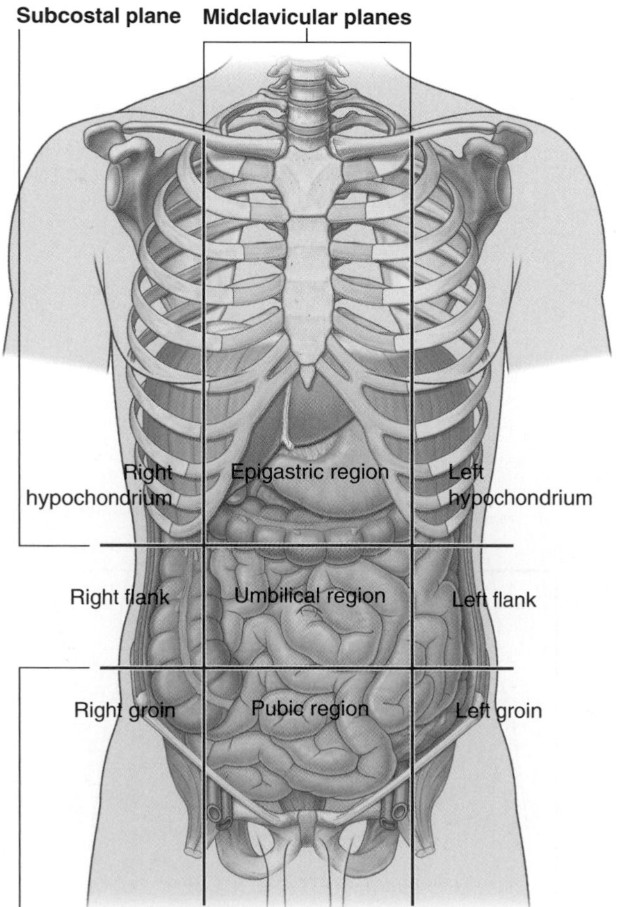


**Fig. 4.23** Nine-region organizational pattern.

### In the clinic

**Surgical incisions**

Access to the abdomen and its contents is usually obtained through incisions in the anterior abdominal wall. Traditionally, incisions have been placed at and around the region of surgical interest. The size of these incisions was usually large to allow good access and optimal visualization of the abdominal cavity. As anesthesia has developed and muscle-relaxing drugs have become widely used, the abdominal incisions have become smaller.

Currently, the most commonly used large abdominal incision is a central craniocaudad incision from the xiphoid process to the symphysis pubis, which provides wide access to the whole of the abdominal contents and allows an exploratory procedure to be performed (laparotomy).

## In the clinic

### Laparoscopic surgery

Laparoscopic surgery, also known as minimally invasive or keyhole surgery, is performed by operating through a series of small incisions no more than 1 to 2 cm in length. As the incisions are much smaller than those used in traditional abdominal surgery, patients experience less postoperative pain and have shorter recovery times. There is also a favorable cosmetic outcome with smaller scars. Several surgical procedures such as appendectomy, cholecystectomy, and hernia repair, as well as numerous orthopaedic, urological, and gynecological procedures, are now commonly performed laparoscopically.

During the operation, a camera known as a laparoscope is used to transmit live, magnified images of the surgical field to a monitor viewed by the surgeon. The camera is inserted into the abdominal cavity through a small incision, called a port-site, usually at the umbilicus. In order to create enough space to operate, the abdominal wall is elevated by inflating the cavity with gas, typically carbon dioxide. Other long, thin surgical instruments are then introduced through additional port-sites, which can be used by the surgeon to operate. The placement of these port-sites is carefully planned to allow optimal access to the surgical field.

Laparoscopic surgery has been further enhanced with the use of surgical robots. Using these systems the surgeon moves the surgical instruments indirectly by controlling robotic arms, which are inserted into the operating field through small incisions. Robot-assisted surgery is now routinely used worldwide and has helped overcome some of the limitations of laparoscopy by enhancing the surgeon's dexterity. The robotic system is precise, provides the surgeon with a 3D view of the surgical field, and allows improved degree of rotation and manipulation of the surgical instruments. Several procedures such as prostatectomy and cholecystectomy can now be performed with this method.

Laparoendoscopic single-site surgery, also known as single-port laparoscopy, is the most recent advance in laparoscopic surgery. This method uses a single incision, usually umbilical, to introduce a port with several operating channels and can be performed with or without robotic assistance. Benefits include less postoperative pain, a faster recovery time, and an even better cosmetic result than traditional laparoscopic surgery.

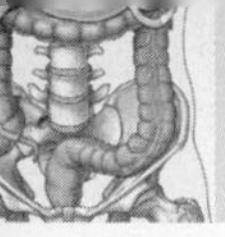



## ABDOMINAL WALL

The abdominal wall covers a large area. It is bounded superiorly by the xiphoid process and costal margins, posteriorly by the vertebral column, and inferiorly by the upper parts of the pelvic bones. Its layers consist of skin, superficial fascia (subcutaneous tissue), muscles and their associated deep fascias, extraperitoneal fascia, and parietal peritoneum (Fig. 4.24).

### Superficial fascia

The superficial fascia of the abdominal wall (subcutaneous tissue of abdomen) is a layer of fatty connective tissue. It is usually a single layer similar to, and continuous with, the superficial fascia throughout other regions of the body. However, in the lower region of the anterior part of the abdominal wall, below the umbilicus, it forms two layers: a superficial fatty layer and a deeper membranous layer.

#### Superficial layer

The superficial fatty layer of superficial fascia (**Camper's fascia**) contains fat and varies in thickness (Figs. 4.25 and 4.26). It is continuous over the inguinal ligament with the superficial fascia of the thigh and with a similar layer in the perineum.

In men, this superficial layer continues over the penis and, after losing its fat and fusing with the deeper layer of superficial fascia, continues into the scrotum where it forms a specialized fascial layer containing smooth muscle fibers (the **dartos fascia**). In women, this superficial layer retains some fat and is a component of the labia majora.

#### Deeper layer

The deeper membranous layer of superficial fascia (**Scarpa's fascia**) is thin and membranous, and contains little or no fat (Fig. 4.25). Inferiorly, it continues into the thigh, but just below the inguinal ligament, it fuses with the deep fascia of the thigh (the **fascia lata**; Fig. 4.26). In the midline, it is firmly attached to the linea alba and the symphysis pubis. It continues into the anterior part of the perineum where it is firmly attached to the ischiopubic rami and to the posterior margin of the perineal membrane. Here, it is referred to as the **superficial perineal fascia** (**Colles' fascia**).

In men, the deeper membranous layer of superficial fascia blends with the superficial layer as they both pass over the penis, forming the superficial fascia of the penis, before they continue into the scrotum where they form the dartos fascia (Fig. 4.25). Also in men, extensions of the deeper membranous layer of superficial fascia attached to the pubic symphysis pass inferiorly onto the dorsum and

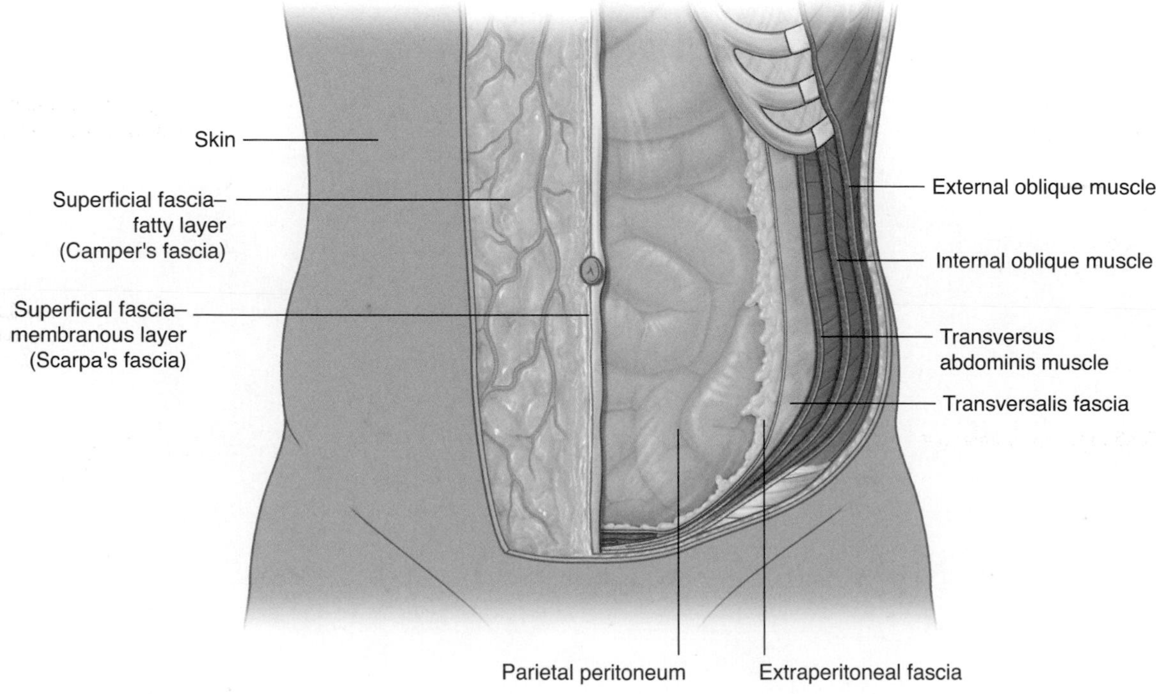


**Fig. 4.24** Layers of the abdominal wall.

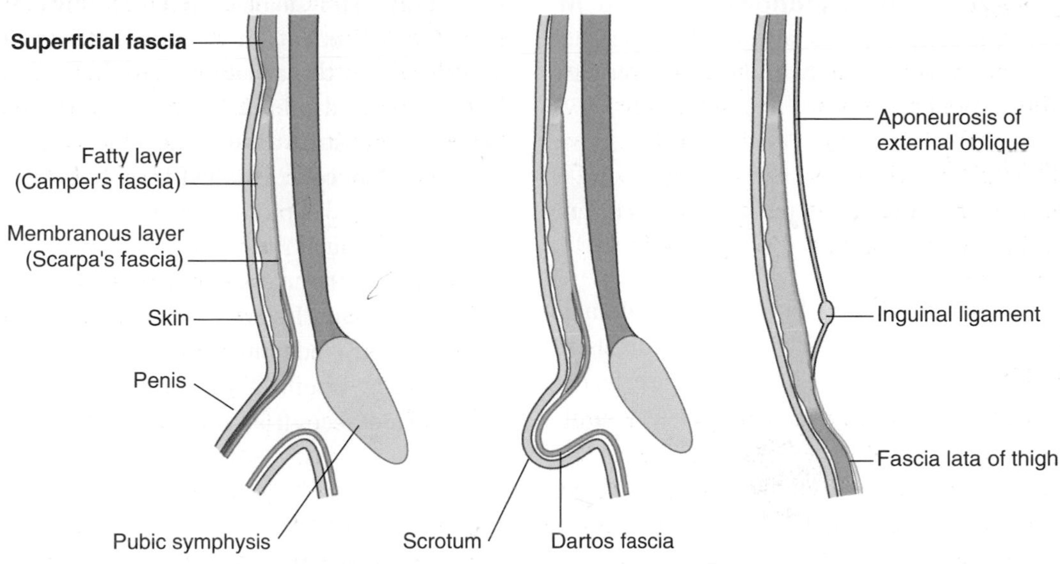


**Fig. 4.25** Superficial fascia.

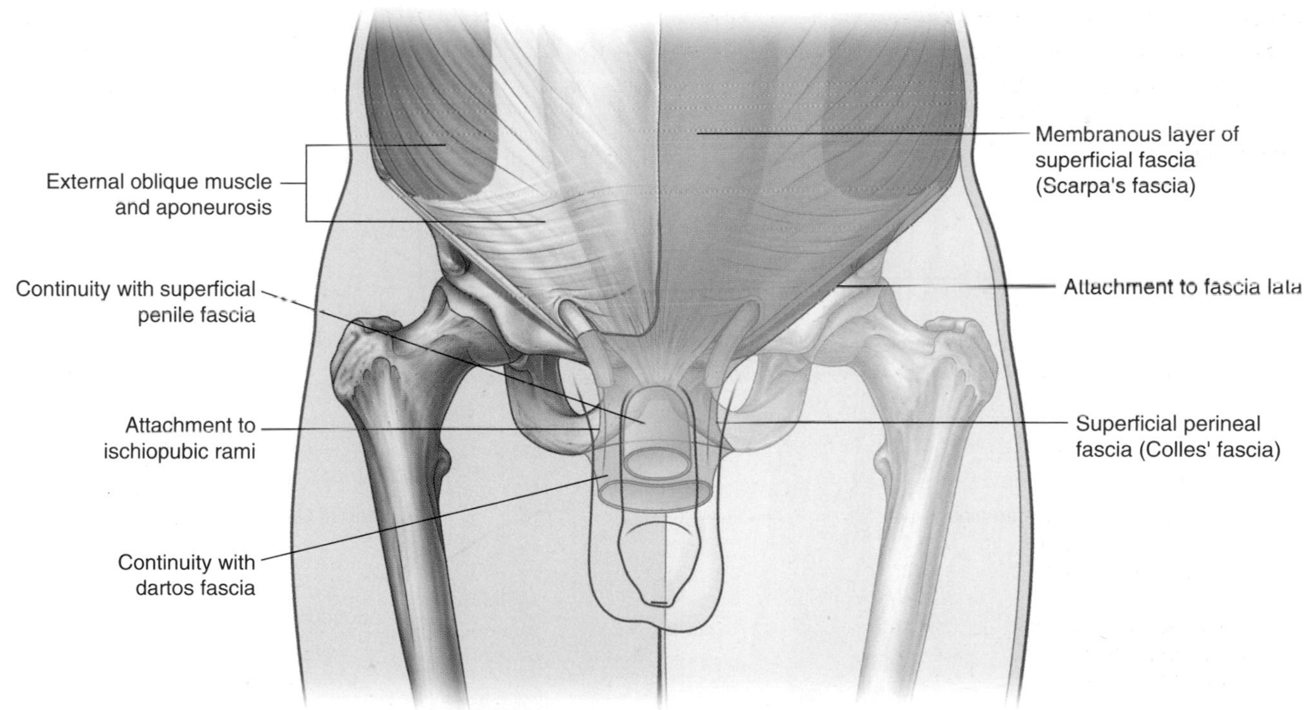


**Fig. 4.26** Continuity of membranous layer of superficial fascia into other areas.

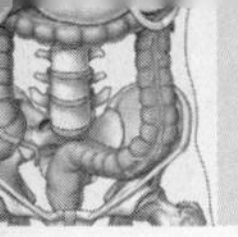



sides of the penis to form the **fundiform ligament of penis**. In women, the membranous layer of the superficial fascia continues into the labia majora and the anterior part of the perineum.

## Anterolateral muscles

There are five muscles in the anterolateral group of abdominal wall muscles:

- three flat muscles whose fibers begin posterolaterally, pass anteriorly, and are replaced by an aponeurosis as the muscle continues toward the midline—the external oblique, internal oblique, and transversus abdominis muscles;
- two vertical muscles, near the midline, which are enclosed within a tendinous sheath formed by the aponeuroses of the flat muscles—the rectus abdominis and pyramidalis muscles.

Each of these five muscles has specific actions, but together the muscles are critical for the maintenance of many normal physiological functions. By their positioning, they form a firm, but flexible, wall that keeps the abdominal viscera within the abdominal cavity, protects the viscera from injury, and helps maintain the position of the viscera in the erect posture against the action of gravity.

In addition, contraction of these muscles assists in both quiet and forced expiration by pushing the viscera upward (which helps push the relaxed diaphragm further into the thoracic cavity) and in coughing and vomiting.

All these muscles are also involved in any action that increases intraabdominal pressure, including parturition (childbirth), micturition (urination), and defecation (expulsion of feces from the rectum).

### Flat muscles

#### External oblique

The most superficial of the three flat muscles in the anterolateral group of abdominal wall muscles is the **external oblique**, which is immediately deep to the superficial fascia (Fig. 4.27, Table 4.1). Its laterally placed muscle fibers pass in an inferomedial direction, while its large

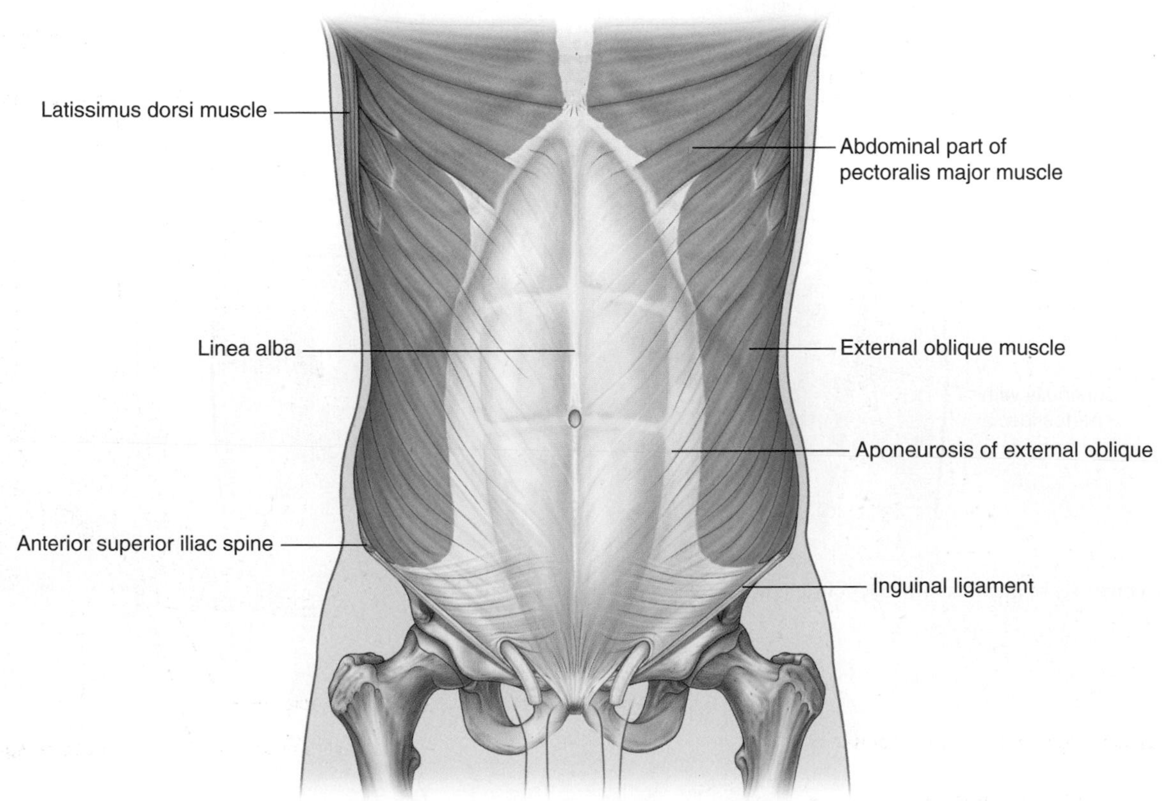


**Fig. 4.27** External oblique muscle and its aponeurosis.

aponeurotic component covers the anterior part of the abdominal wall to the midline. Approaching the midline, the aponeuroses are entwined, forming the linea alba, which extends from the xiphoid process to the pubic symphysis.

### Associated ligaments

The lower border of the external oblique aponeurosis forms the **inguinal ligament** on each side (Fig. 4.27). This thickened reinforced free edge of the external oblique aponeurosis passes between the anterior superior iliac spine laterally and the pubic tubercle medially (Fig. 4.28). It folds under itself forming a trough, which plays an important role in the formation of the inguinal canal.

Several other ligaments are also formed from extensions of the fibers at the medial end of the inguinal ligament:

- The **lacunar ligament** is a crescent-shaped extension of fibers at the medial end of the inguinal ligament that pass backward to attach to the **pecten pubis** on the superior ramus of the pubic bone (Figs. 4.28 and 4.29).
- Additional fibers extend from the lacunar ligament along the pecten pubis of the pelvic brim to form the **pectineal (Cooper's) ligament**.

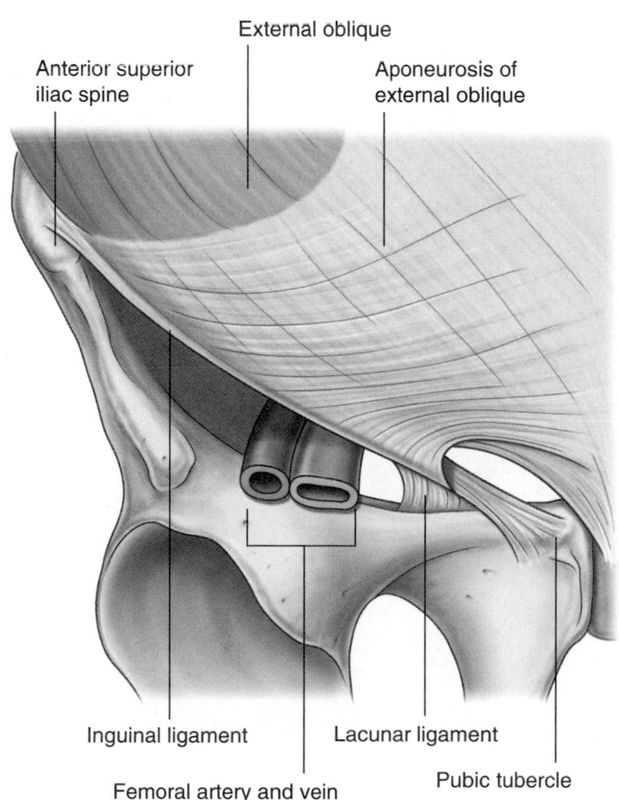


**Fig. 4.28** Ligaments formed from the external oblique aponeurosis.

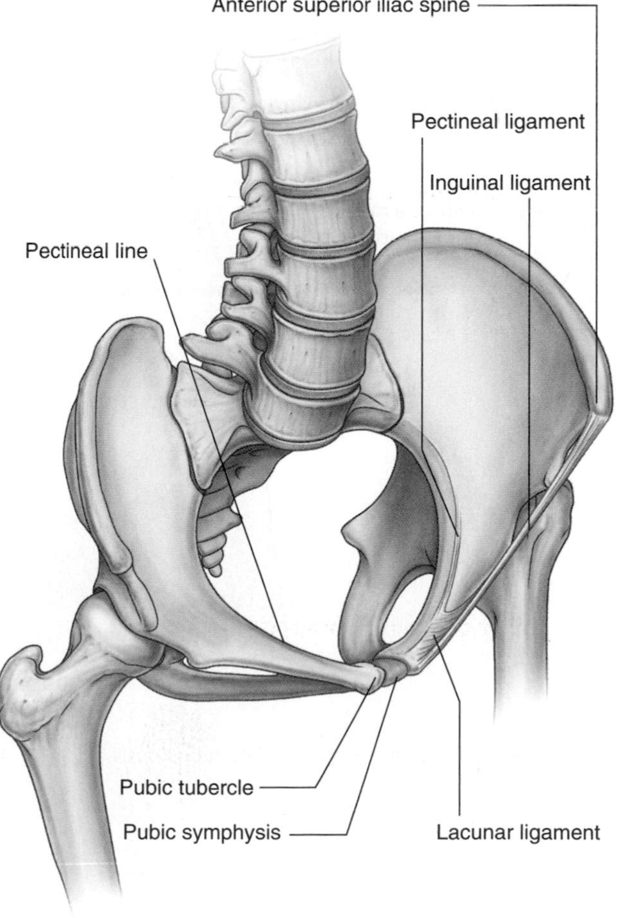


**Fig. 4.29** Ligaments of the inguinal region.

### Internal oblique

Deep to the external oblique muscle is the **internal oblique** muscle, which is the second of the three flat muscles (Fig. 4.30, Table 4.1). This muscle is smaller and thinner than the external oblique, with most of its muscle fibers passing in a superomedial direction. Its lateral muscular components end anteriorly as an aponeurosis that blends into the linea alba at the midline.

### Transversus abdominis

Deep to the internal oblique muscle is the **transversus abdominis** muscle (Fig. 4.31, Table 4.1), so named because of the direction of most of its muscle fibers. It ends in an anterior aponeurosis, which blends with the linea alba at the midline.

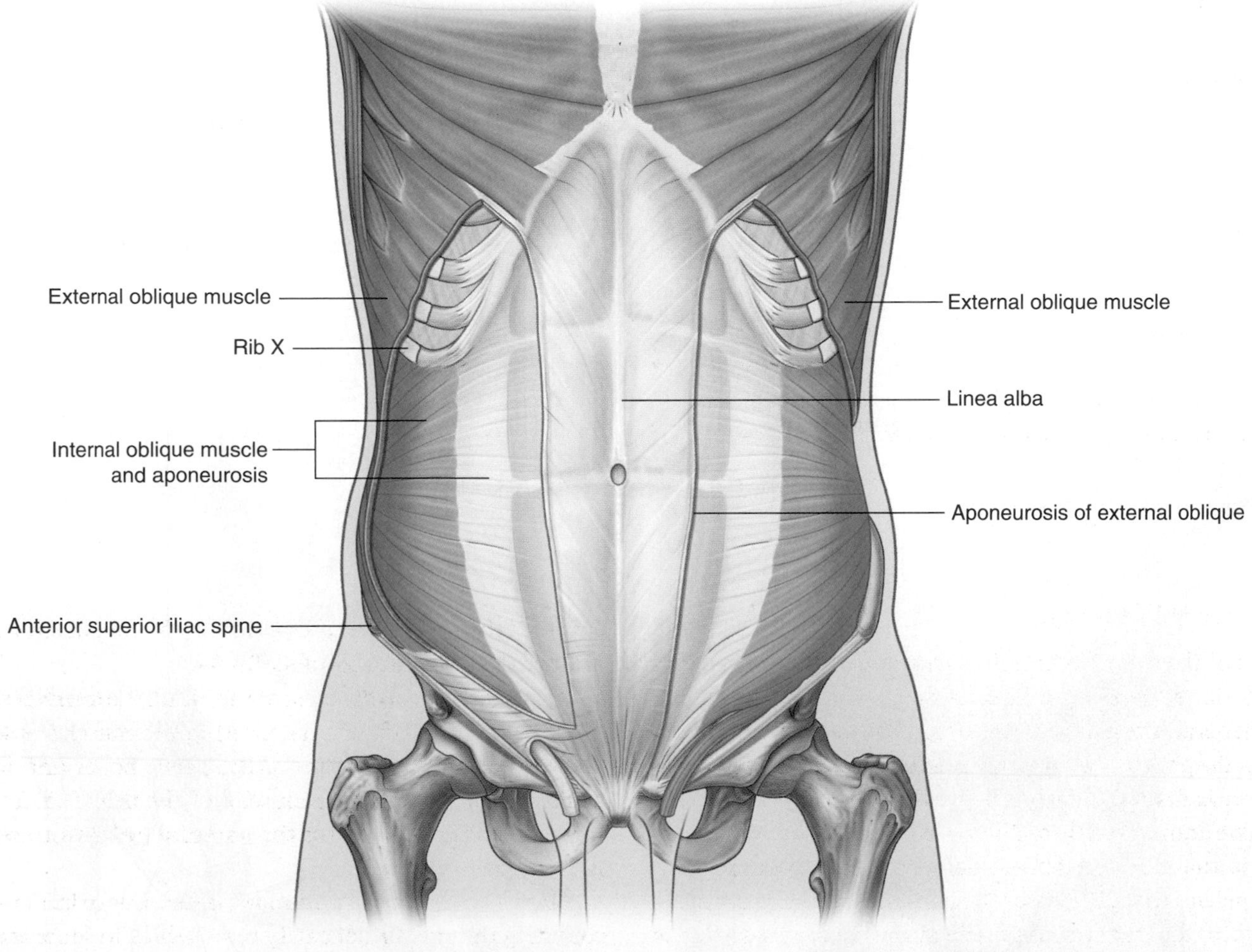

**Fig. 4.30** Internal oblique muscle and its aponeurosis.

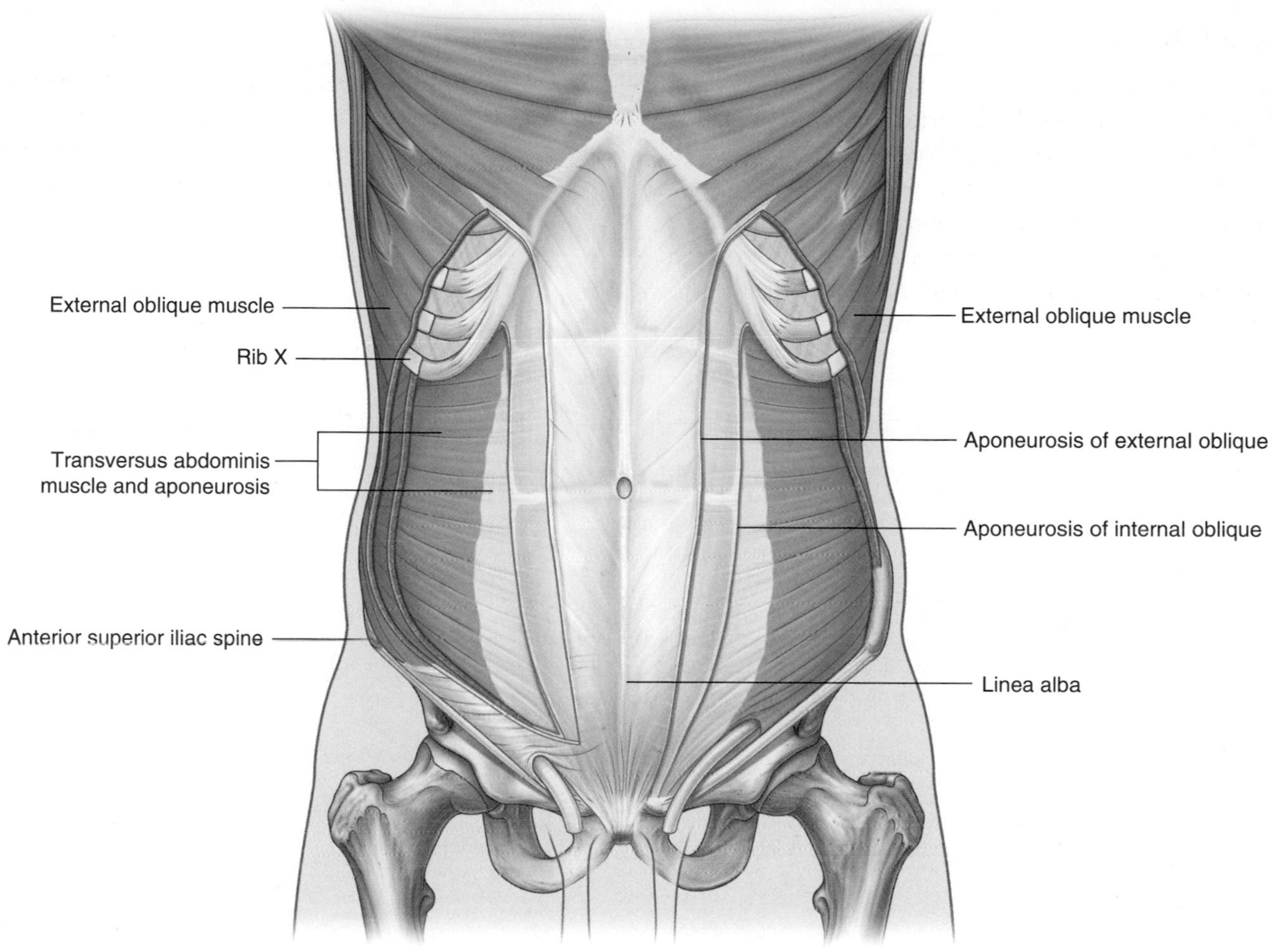


**Fig. 4.31** Transversus abdominis muscle and its aponeurosis.

## Transversalis fascia

Each of the three flat muscles is covered on its anterior and posterior surfaces by a layer of deep (or investing) fascia. In general, these layers are unremarkable except for the layer deep to the transversus abdominis muscle (the **transversalis fascia**), which is better developed.

The transversalis fascia is a continuous layer of deep fascia that lines the abdominal cavity and continues into the pelvic cavity. It crosses the midline anteriorly, associating with the transversalis fascia of the opposite side, and is continuous with the fascia on the inferior surface of the diaphragm. It is continuous posteriorly with the deep fascia covering the muscles of the posterior abdominal wall and attaches to the thoracolumbar fascia.

After attaching to the crest of the ilium, the transversalis fascia blends with the fascia covering the muscles associated with the upper regions of the pelvic bones and with similar fascia covering the muscles of the pelvic cavity. At this point, it is referred to as the **parietal pelvic** (or **endopelvic**) **fascia**.

There is therefore a continuous layer of deep fascia surrounding the abdominal cavity that is thick in some areas, thin in others, attached or free, and participates in the formation of specialized structures.

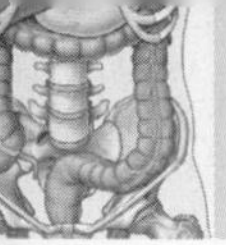



## Vertical muscles

The two vertical muscles in the anterolateral group of abdominal wall muscles are the large rectus abdominis and the small pyramidalis (Fig. 4.32, Table 4.1).

### Rectus abdominis

The **rectus abdominis** is a long, flat muscle and extends the length of the anterior abdominal wall. It is a paired muscle, separated in the midline by the linea alba, and it

**Table 4.1** Abdominal wall muscles

| Muscle | Origin | Insertion | Innervation | Function |
|---|---|---|---|---|
| External oblique | Muscular slips from the outer surfaces of the lower eight ribs (ribs V to XII) | Lateral lip of iliac crest; aponeurosis ending in midline raphe (linea alba) | Anterior rami of lower six thoracic spinal nerves (T7 to T12) | Compress abdominal contents; both muscles flex trunk; each muscle bends trunk to same side, turning anterior part of abdomen to opposite side |
| Internal oblique | Thoracolumbar fascia; iliac crest between origins of external and transversus; lateral two-thirds of inguinal ligament | Inferior border of the lower three or four ribs; aponeurosis ending in linea alba; pubic crest and pectineal line | Anterior rami of lower six thoracic spinal nerves (T7 to T12) and L1 | Compress abdominal contents; both muscles flex trunk; each muscle bends trunk and turns anterior part of abdomen to same side |
| Transversus abdominis | Thoracolumbar fascia; medial lip of iliac crest; lateral one-third of inguinal ligament; costal cartilages lower six ribs (ribs VII to XII) | Aponeurosis ending in linea alba; pubic crest and pectineal line | Anterior rami of lower six thoracic spinal nerves (T7 to T12) and L1 | Compress abdominal contents |
| Rectus abdominis | Pubic crest, pubic tubercle, and pubic symphysis | Costal cartilages of ribs V to VII; xiphoid process | Anterior rami of lower seven thoracic spinal nerves (T7 to T12) | Compress abdominal contents; flex vertebral column; tense abdominal wall |
| Pyramidalis | Front of pubis and pubic symphysis | Into linea alba | Anterior ramus of T12 | Tenses the linea alba |

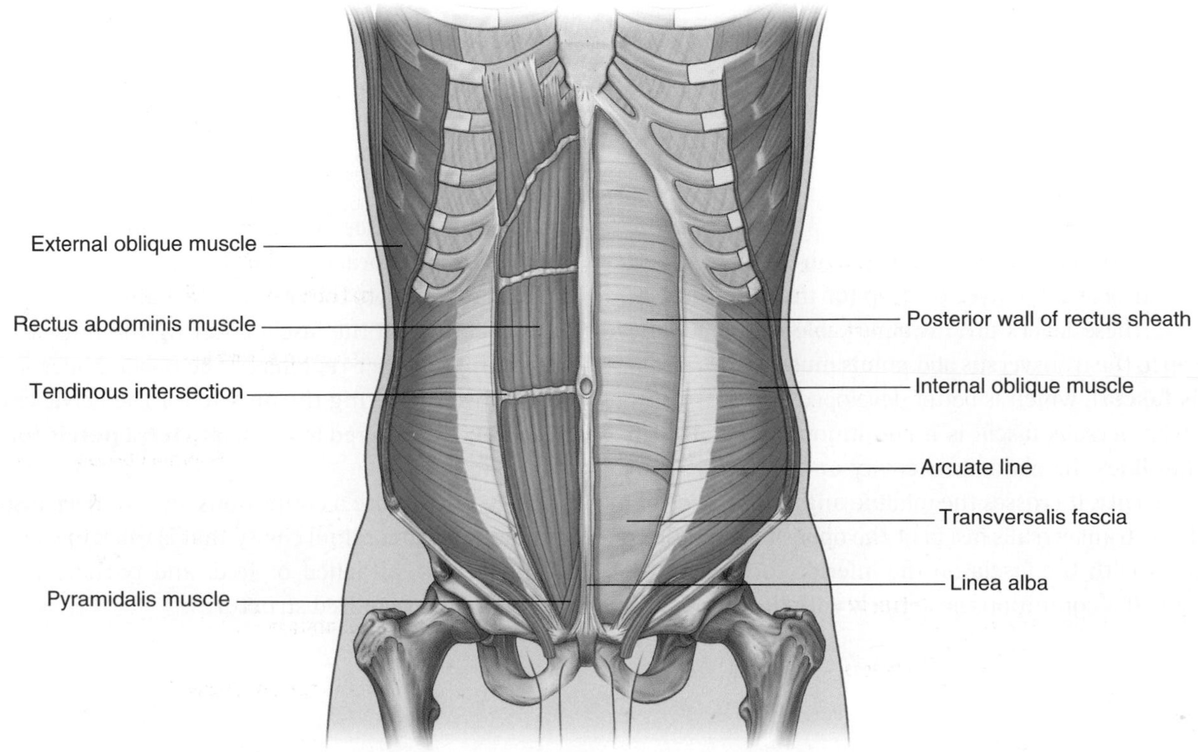


**Fig. 4.32** Rectus abdominis and pyramidalis muscles.

widens and thins as it ascends from the pubic symphysis to the costal margin. Along its course, it is intersected by three or four transverse fibrous bands or **tendinous intersections** (Fig. 4.32). These are easily visible on individuals with well-developed rectus abdominis muscles.

### Pyramidalis

The second vertical muscle is the **pyramidalis**. This small, triangular muscle, which may be absent, is anterior to the rectus abdominis and has its base on the pubis, and its apex is attached superiorly and medially to the linea alba (Fig. 4.32).

### Rectus sheath

The rectus abdominis and pyramidalis muscles are enclosed in an aponeurotic tendinous sheath (the **rectus sheath**) formed by a unique layering of the aponeuroses of the external and internal oblique, and transversus abdominis muscles (Fig. 4.33).

The rectus sheath completely encloses the upper three-quarters of the rectus abdominis and covers the anterior surface of the lower one-quarter of the muscle. As no sheath covers the posterior surface of the lower quarter of the rectus abdominis muscle, the muscle at this point is in direct contact with the transversalis fascia.

The formation of the rectus sheath surrounding the upper three-quarters of the rectus abdominis muscle has the following pattern:

- The anterior wall consists of the aponeurosis of the external oblique and half of the aponeurosis of the internal oblique, which splits at the lateral margin of the rectus abdominis.
- The posterior wall of the rectus sheath consists of the other half of the aponeurosis of the internal oblique and the aponeurosis of the transversus abdominis.

At a point midway between the umbilicus and the pubic symphysis, corresponding to the beginning of the lower one-quarter of the rectus abdominis muscle, all of the aponeuroses move anterior to the rectus muscle. There is no posterior wall of the rectus sheath and the anterior wall of the sheath consists of the aponeuroses of the external oblique, the internal oblique, and the transversus abdominis muscles. From this point inferiorly, the rectus abdominis muscle is in direct contact with the transversalis fascia. Marking this point of transition is an arch of fibers (the **arcuate line**; see Fig. 4.32).

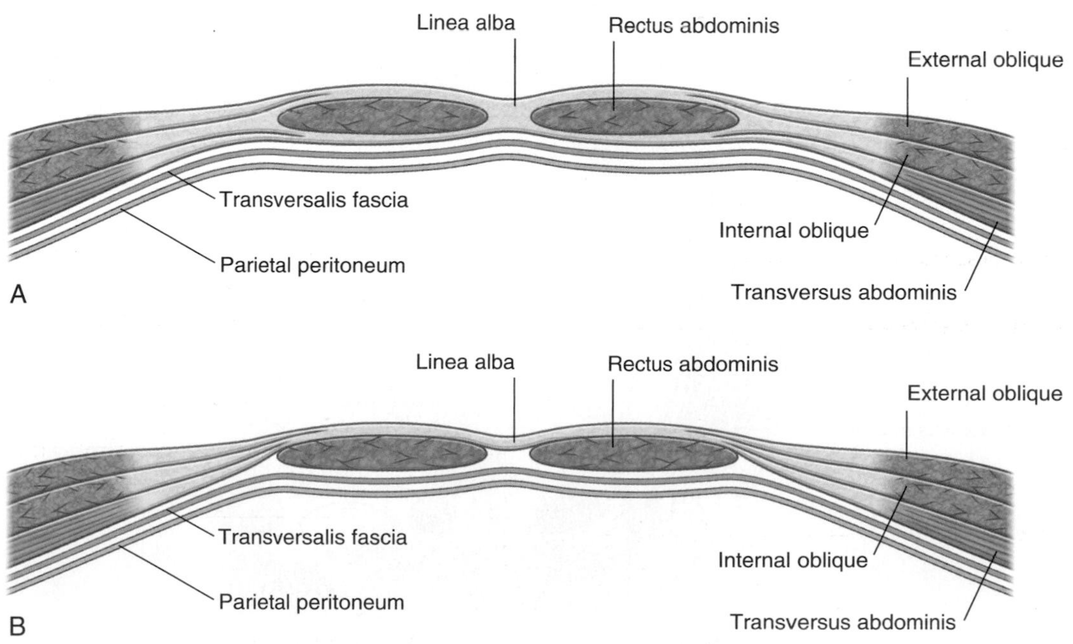


**Fig. 4.33** Organization of the rectus sheath. **A.** Transverse section through the upper three-quarters of the rectus sheath. **B.** Transverse section through the lower one-quarter of the rectus sheath.

## Extraperitoneal fascia

Deep to the transversalis fascia is a layer of connective tissue, the **extraperitoneal fascia**, which separates the transversalis fascia from the peritoneum (Fig. 4.34). Containing varying amounts of fat, this layer not only lines the abdominal cavity but is also continuous with a similar layer lining the pelvic cavity. It is abundant on the posterior abdominal wall, especially around the kidneys, continues over organs covered by peritoneal reflections, and, as the vasculature is located in this layer, extends into mesenteries with the blood vessels. Viscera in the extraperitoneal fascia are referred to as **retroperitoneal**.

In the description of specific surgical procedures, the terminology used to describe the extraperitoneal fascia is further modified. The fascia toward the anterior side of the body is described as preperitoneal (or, less commonly, properitoneal) and the fascia toward the posterior side of the body has been described as retroperitoneal (Fig. 4.35). Examples of the use of these terms would be the continuity of fat in the inguinal canal with the preperitoneal fat and a transabdominal preperitoneal laparoscopic repair of an inguinal hernia.

## Peritoneum

Deep to the extraperitoneal fascia is the peritoneum (see Figs. 4.6 and 4.7 on pp. 260-261). This thin serous membrane lines the walls of the abdominal cavity and, at various points, reflects onto the abdominal viscera, providing either a complete or a partial covering. The peritoneum lining the walls is the parietal peritoneum; the peritoneum covering the viscera is the visceral peritoneum.

The continuous lining of the abdominal walls by the parietal peritoneum forms a sac. This sac is closed in men but has two openings in women where the uterine tubes provide a passage to the outside. The closed sac in men and the semiclosed sac in women is called the peritoneal cavity.

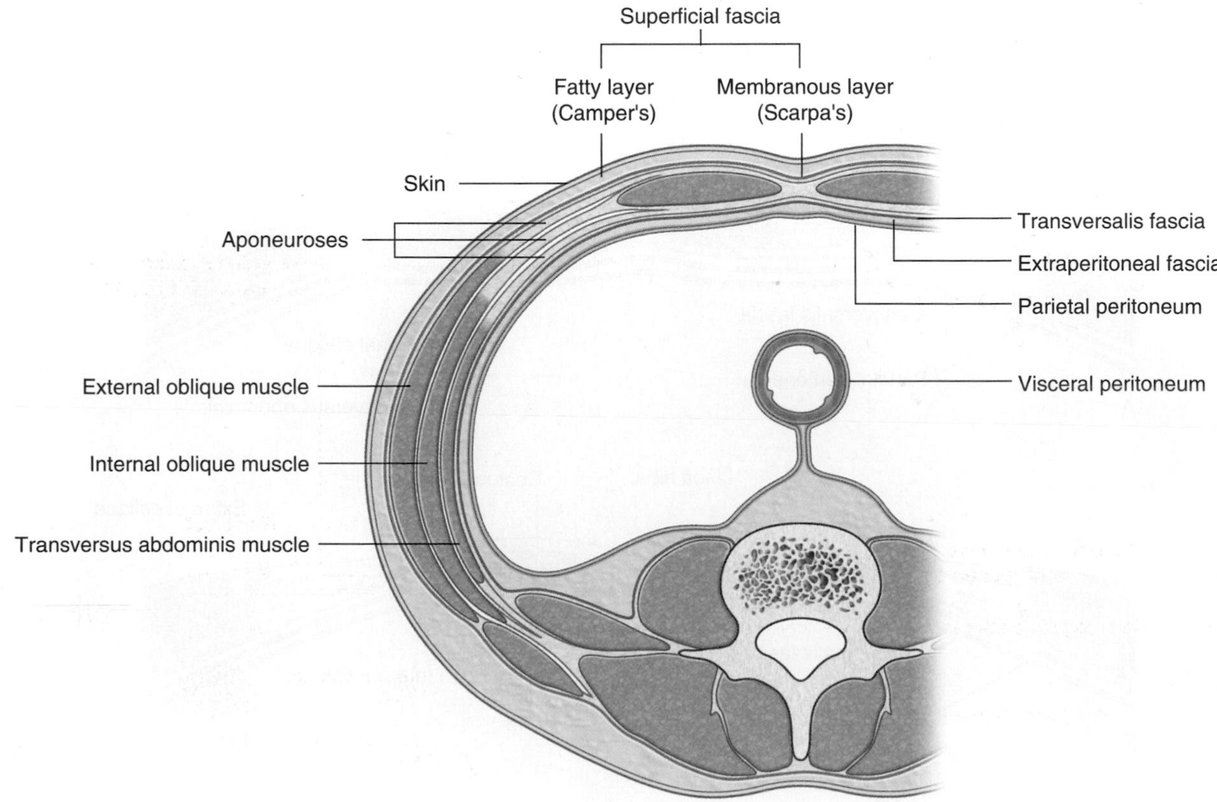


**Fig. 4.34** Transverse section showing the layers of the abdominal wall.

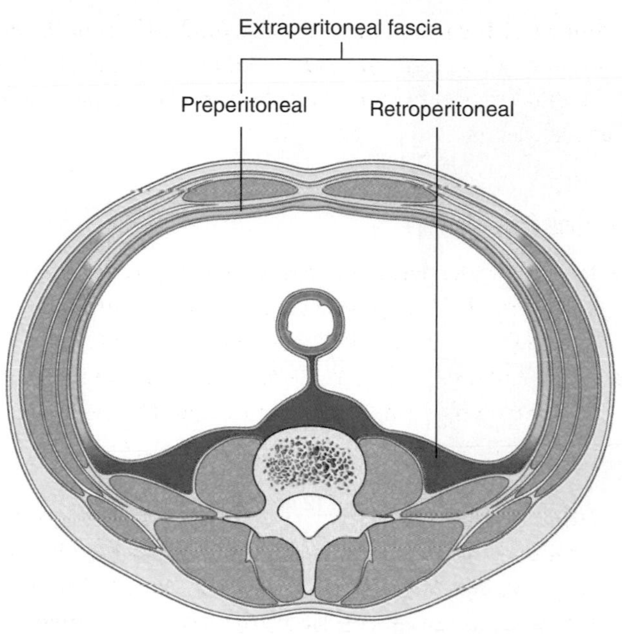


**Fig. 4.35** Subdivisions of the extraperitoneal fascia.

## Innervation

The skin, muscles, and parietal peritoneum of the anterolateral abdominal wall are supplied by T7 to T12 and L1 spinal nerves. The anterior rami of these spinal nerves pass around the body, from posterior to anterior, in an inferomedial direction (Fig. 4.36). As they proceed, they give off a lateral cutaneous branch and end as an anterior cutaneous branch.

The intercostal nerves (T7 to T11) leave their intercostal spaces, passing deep to the costal cartilages, and continue onto the anterolateral abdominal wall between the internal oblique and transversus abdominis muscles (Fig. 4.37). Reaching the lateral edge of the rectus sheath, they enter the rectus sheath and pass posterior to the lateral aspect of the rectus abdominis muscle. Approaching the midline, an anterior cutaneous branch passes through the rectus abdominis muscle and the anterior wall of the rectus sheath to supply the skin.

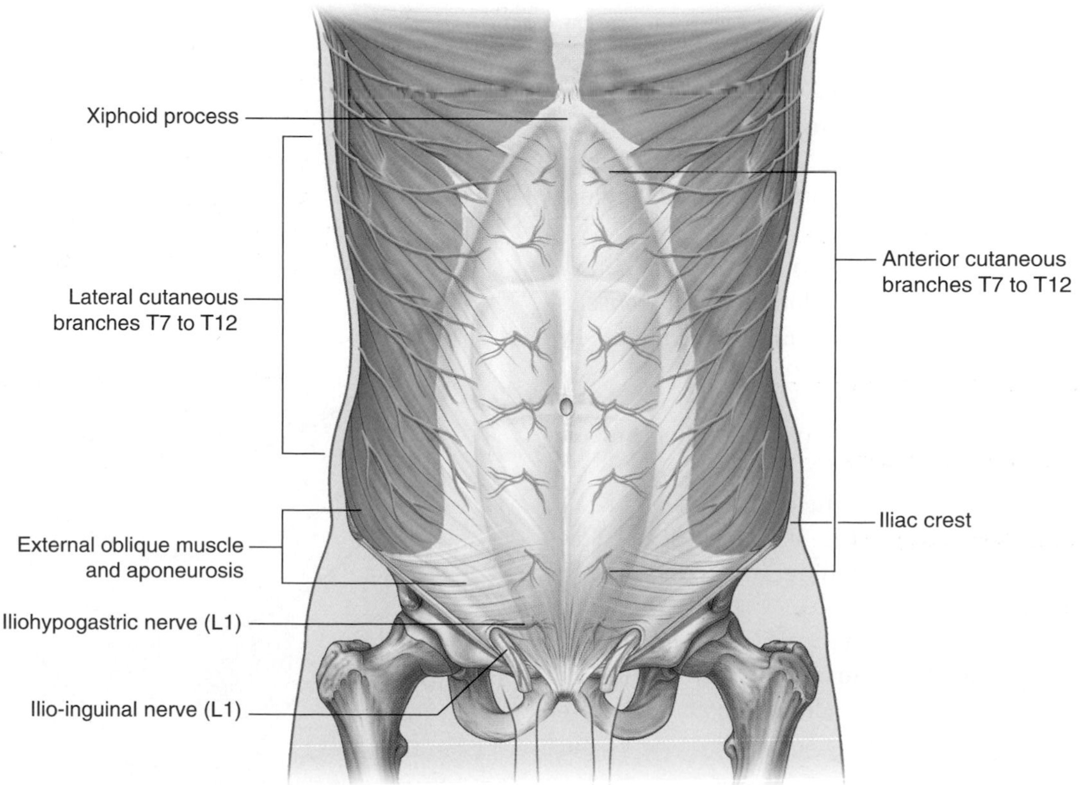


**Fig. 4.36** Innervation of the anterolateral abdominal wall.

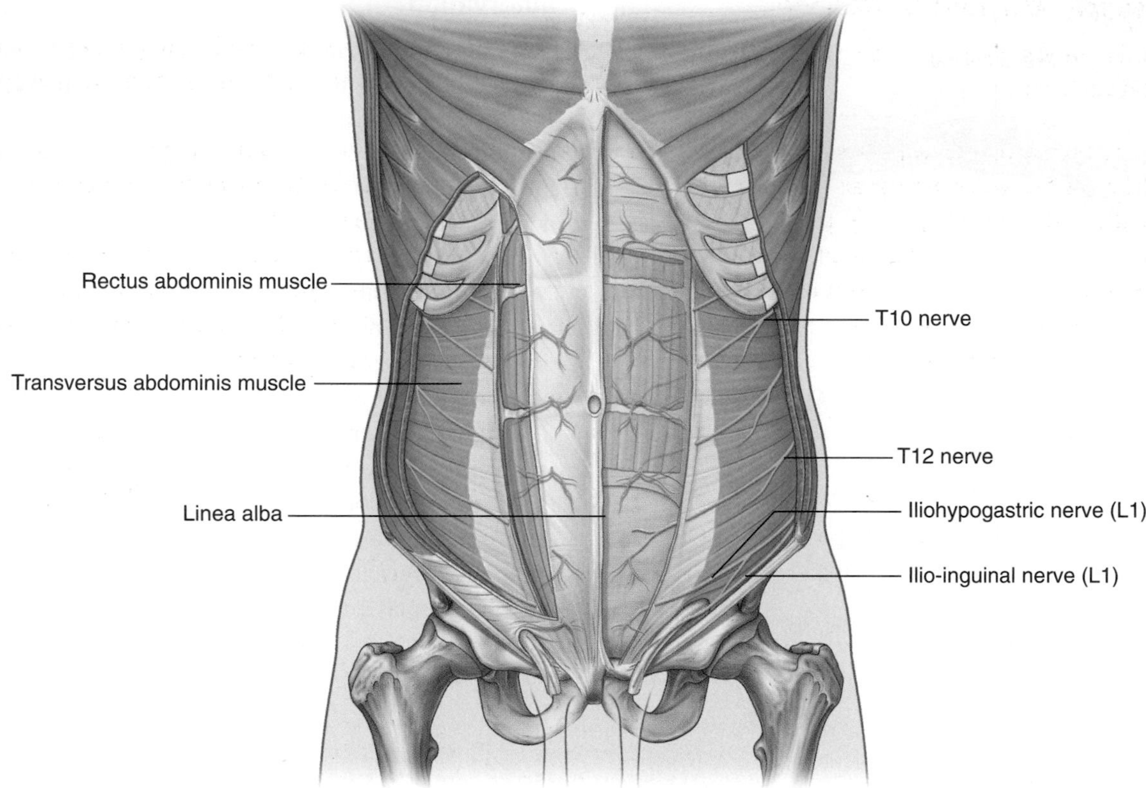


**Fig. 4.37** Path taken by the nerves innervating the anterolateral abdominal wall.

Spinal nerve T12 (the **subcostal nerve**) follows a similar course as the intercostals. Branches of L1 (the **iliohypogastric nerve** and **ilio-inguinal nerve**), which originate from the lumbar plexus, follow similar courses initially, but deviate from this pattern near their final destination.

Along their course, nerves T7 to T12 and L1 supply branches to the anterolateral abdominal wall muscles and the underlying parietal peritoneum. All terminate by supplying skin:

- Nerves T7 to T9 supply the skin from the xiphoid process to just above the umbilicus.
- T10 supplies the skin around the umbilicus.
- T11, T12, and L1 supply the skin from just below the umbilicus to, and including, the pubic region (Fig. 4.38).
- Additionally, the ilio-inguinal nerve (a branch of L1) supplies the anterior surface of the scrotum or labia majora, and sends a small cutaneous branch to the thigh.

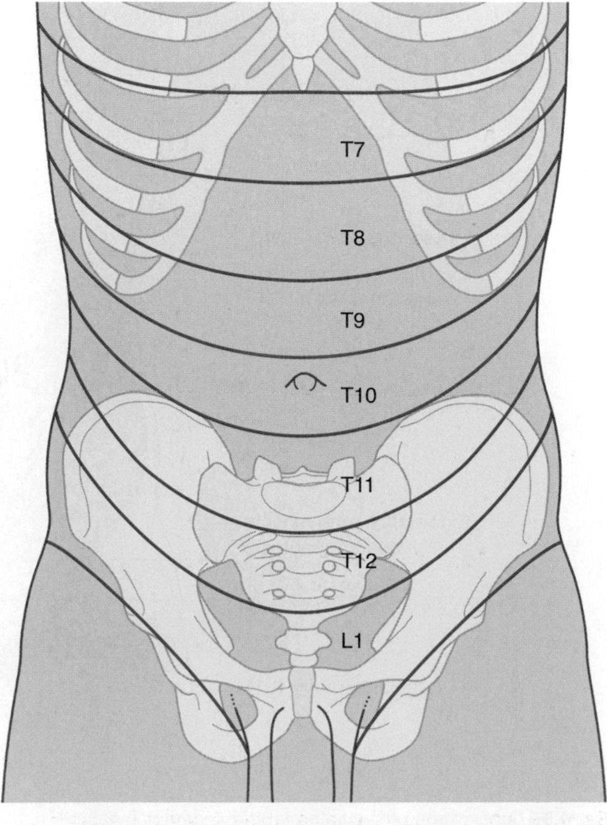


**Fig. 4.38** Dermatomes of the anterolateral abdominal wall.

## Arterial supply and venous drainage

Numerous blood vessels supply the anterolateral abdominal wall. Superficially:

- the superior part of the wall is supplied by branches from the **musculophrenic artery**, a terminal branch of the **internal thoracic artery**, and
- the inferior part of the wall is supplied by the medially placed **superficial epigastric artery** and the laterally placed **superficial circumflex iliac artery**, both branches of the **femoral artery** (Fig. 4.39).

At a deeper level:

- the superior part of the wall is supplied by the **superior epigastric artery**, a terminal branch of the internal thoracic artery;
- the lateral part of the wall is supplied by branches of the **tenth** and **eleventh intercostal arteries** and the **subcostal artery**; and
- the inferior part of the wall is supplied by the medially placed **inferior epigastric artery** and the laterally placed **deep circumflex iliac artery**, both branches of the **external iliac artery**.

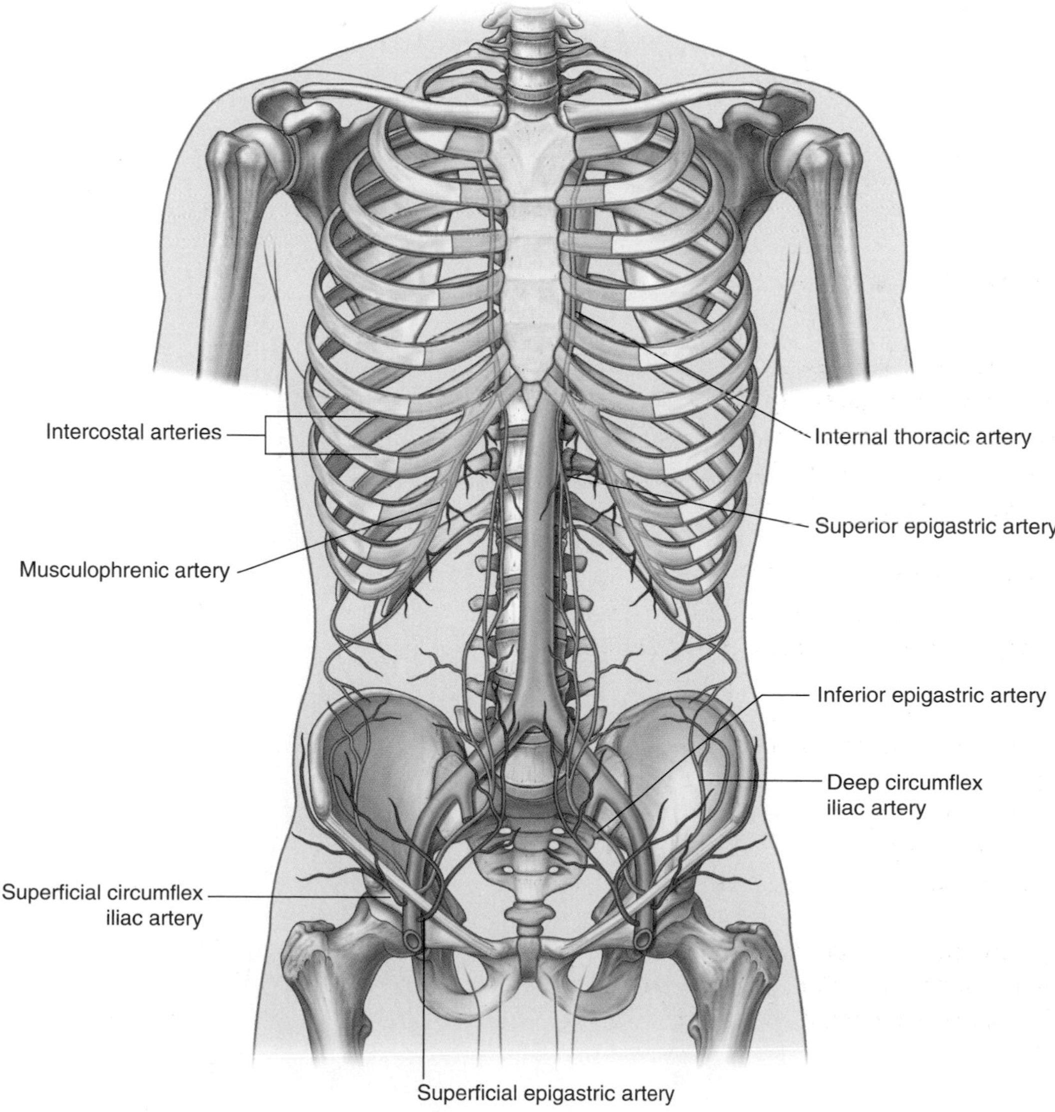


**Fig. 4.39** Arterial supply to the anterolateral abdominal wall.

The superior and inferior epigastric arteries both enter the rectus sheath. They are posterior to the rectus abdominis muscle throughout their course, and anastomose with each other (Fig. 4.40).

Veins of similar names follow the arteries and are responsible for venous drainage.

## Lymphatic drainage

Lymphatic drainage of the anterolateral abdominal wall follows the basic principles of lymphatic drainage:

- Superficial lymphatics above the umbilicus pass in a superior direction to the **axillary nodes**, while drainage below the umbilicus passes in an inferior direction to the **superficial inguinal nodes.**
- Deep lymphatic drainage follows the deep arteries back to **parasternal nodes** along the internal thoracic artery, lumbar nodes along the abdominal aorta, and external iliac nodes along the external iliac artery.

## GROIN

The groin (inguinal region) is the area of junction between the anterior abdominal wall and the thigh. In this area, the abdominal wall is weakened from changes that occur during development and a peritoneal sac or diverticulum, with or without abdominal contents, can therefore protrude through it, creating an inguinal hernia. This type of hernia can occur in both sexes, but it is most common in males.

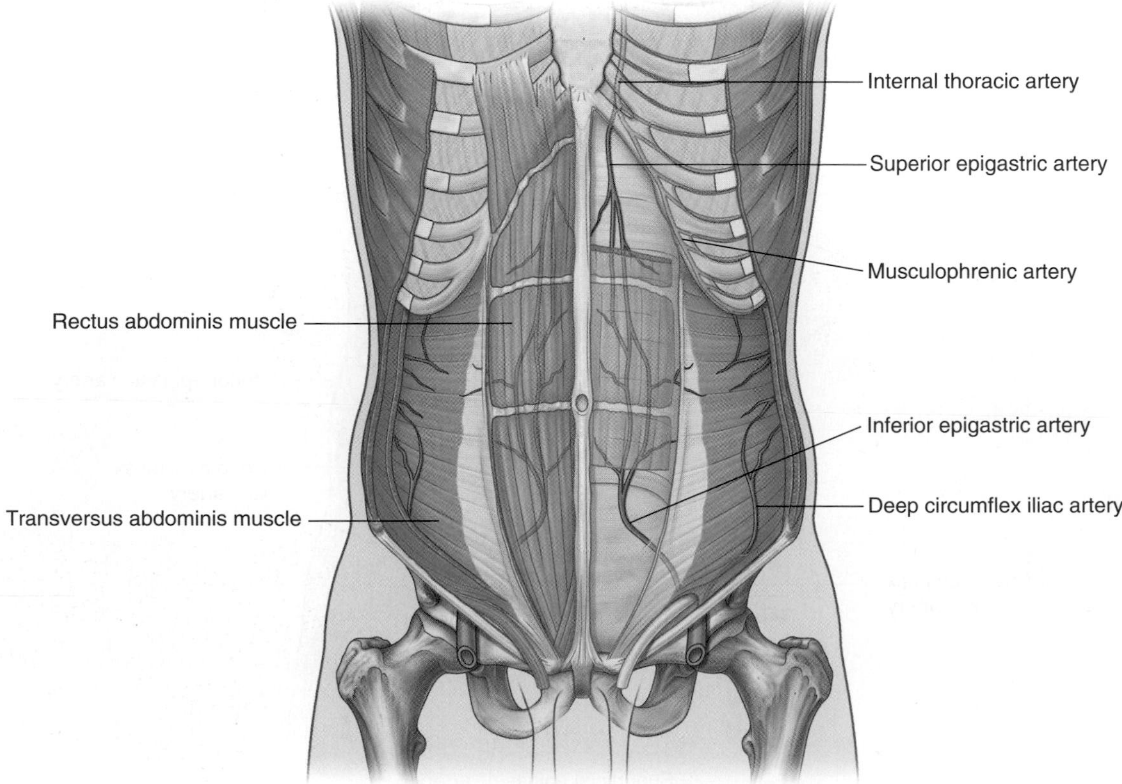


**Fig. 4.40** Superior and inferior epigastric arteries.

The inherent weakness in the anterior abdominal wall in the groin is caused by changes that occur during the development of the gonads. Before the descent of the testes and ovaries from their initial position high in the posterior abdominal wall, a peritoneal outpouching (the processus vaginalis) forms (Fig. 4.41), protruding through the various layers of the anterior abdominal wall and acquiring coverings from each:

- The transversalis fascia forms its deepest covering.
- The second covering is formed by the musculature of the internal oblique (a covering from the transversus abdominis muscle is not acquired because the processus vaginalis passes under the arching fibers of this abdominal wall muscle).
- Its most superficial covering is the aponeurosis of the external oblique.

As a result the processus vaginalis is transformed into a tubular structure with multiple coverings from the layers of the anterior abdominal wall. This forms the basic structure of the **inguinal canal**.

The final event in this development is the descent of the testes into the scrotum or of the ovaries into the pelvic cavity. This process depends on the development of the gubernaculum, which extends from the inferior border of the developing gonad to the labioscrotal swellings (Fig. 4.41).

The processus vaginalis is immediately anterior to the gubernaculum within the inguinal canal.

In men, as the testes descend, the testes and their accompanying vessels, ducts, and nerves pass through the inguinal canal and are therefore surrounded by the same fascial layers of the abdominal wall. Testicular descent completes the formation of the spermatic cord in men.

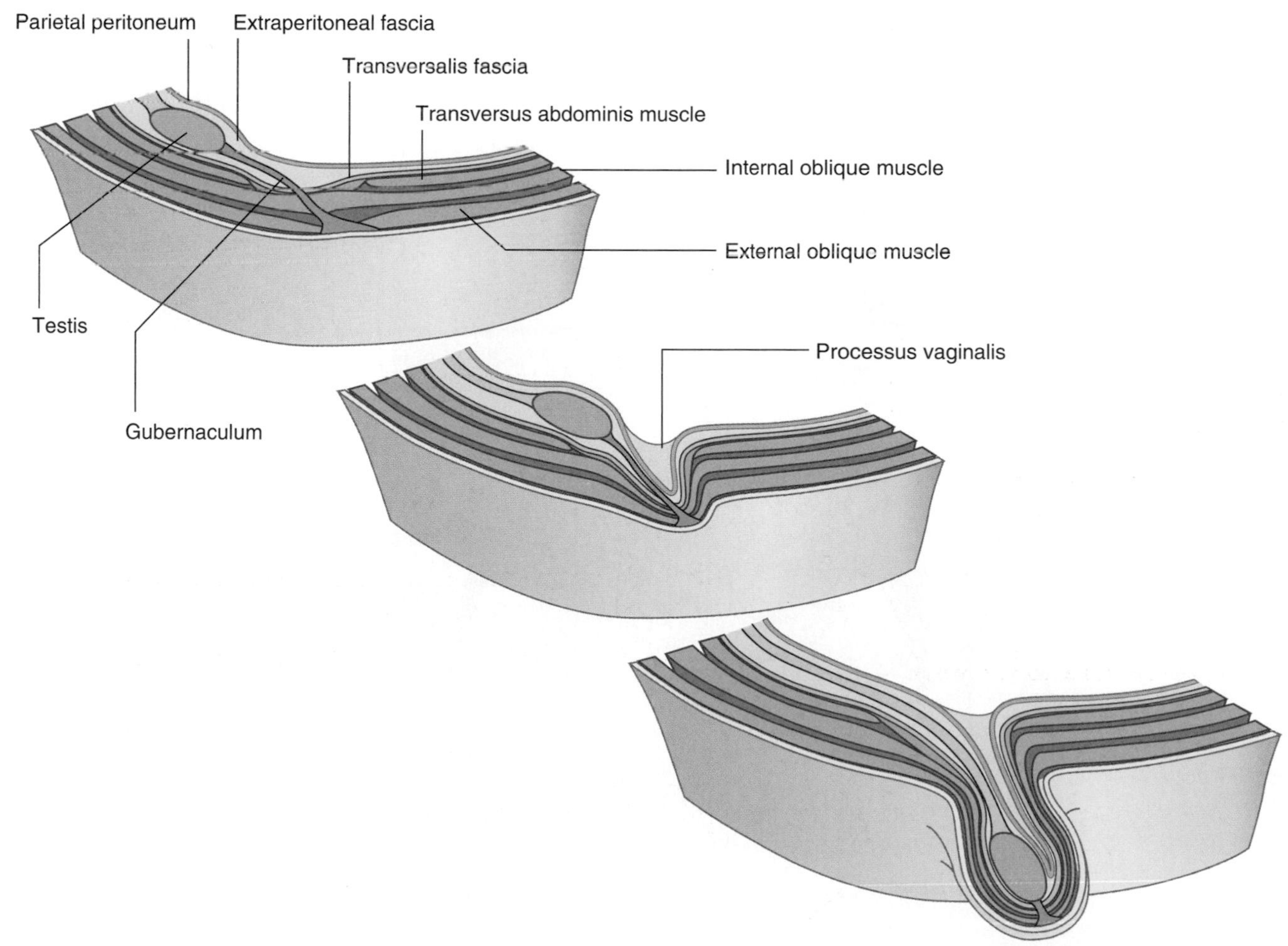


**Fig. 4.41** Descent of the testis from week 7 (postfertilization) to birth.

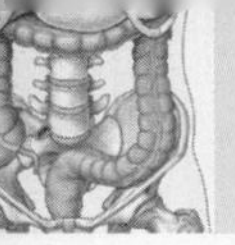

In women, the ovaries descend into the pelvic cavity and become associated with the developing uterus. Therefore, the only remaining structure passing through the inguinal canal is the round ligament of the uterus, which is a remnant of the gubernaculum.

The development sequence is concluded in both sexes when the processus vaginalis obliterates. If this does not occur or is incomplete, a potential weakness exists in the anterior abdominal wall and an inguinal hernia may develop. In males, only proximal regions of the processus vaginalis obliterate. The distal end expands to enclose most of the testis in the scrotum. In other words, the **cavity of the tunica vaginalis** in men forms as an extension of the developing peritoneal cavity that becomes separated off during development.

## Inguinal canal

The inguinal canal is a slit-like passage that extends in a downward and medial direction, just above and parallel to the lower half of the inguinal ligament. It begins at the deep inguinal ring and continues for approximately 4 cm, ending at the superficial inguinal ring (Fig. 4.42). The contents of the canal are the genital branch of the **genitofemoral nerve**, the **spermatic cord** in men, and the **round ligament of the uterus** in women. Additionally, in both sexes, the ilio-inguinal nerve passes through part of the canal, exiting through the superficial inguinal ring with the other contents.

### Deep inguinal ring

The deep (internal) inguinal ring is the beginning of the inguinal canal and is at a point midway between the anterior superior iliac spine and the pubic symphysis (Fig. 4.43). It is just above the inguinal ligament and immediately lateral to the inferior epigastric vessels. Although sometimes referred to as a defect or opening in the transversalis fascia, it is actually the beginning of the tubular evagination of transversalis fascia that forms one of the coverings (the **internal spermatic fascia**) of the spermatic cord in men or the round ligament of the uterus in women.

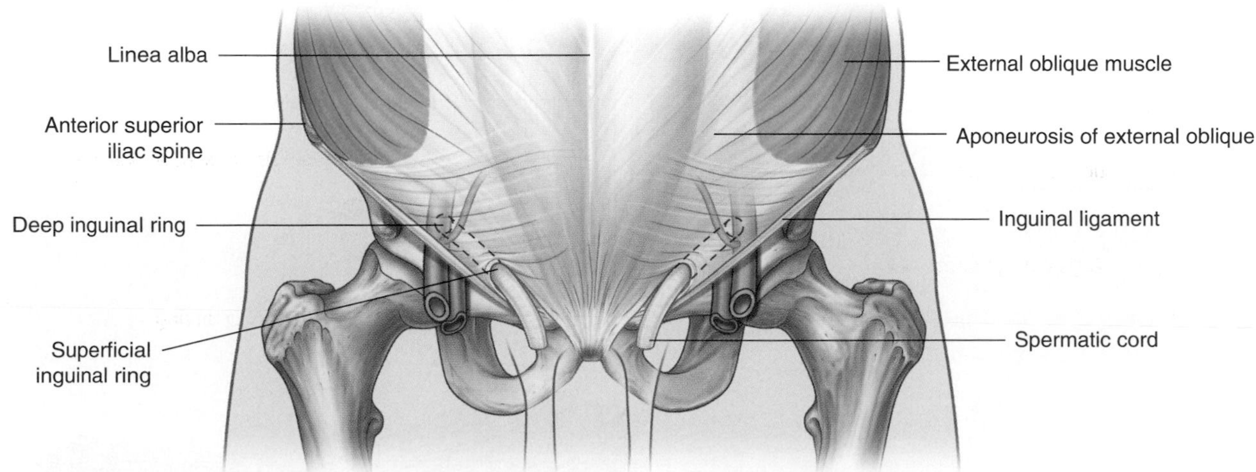


**Fig. 4.42** Inguinal canal.

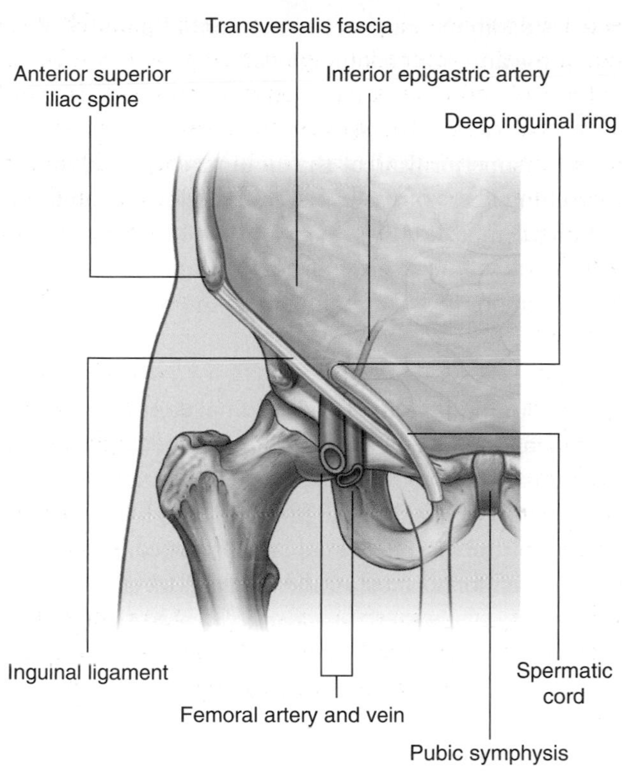


**Fig. 4.43** Deep inguinal ring and the transversalis fascia.

## Superficial inguinal ring

The superficial (external) inguinal ring is the end of the inguinal canal and is superior to the pubic tubercle (Fig. 4.44). It is a triangular opening in the aponeurosis of the external oblique, with its apex pointing superolaterally and its base formed by the pubic crest. The two remaining sides of the triangle (the **medial crus** and the **lateral crus**) are attached to the pubic symphysis and the pubic tubercle, respectively. At the apex of the triangle the two crura are held together by crossing (intercrural) fibers, which prevent further widening of the superficial ring.

As with the deep inguinal ring, the superficial inguinal ring is actually the beginning of the tubular evagination of the aponeurosis of the external oblique onto the structures traversing the inguinal canal and emerging from the superficial inguinal ring. This continuation of tissue over the spermatic cord is the **external spermatic fascia**.

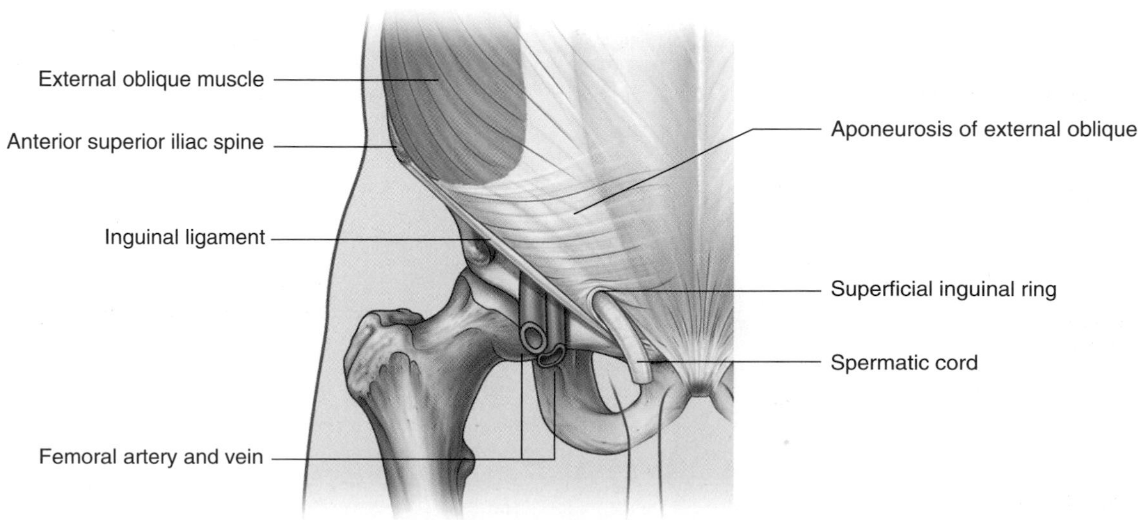


**Fig. 4.44** Superficial inguinal ring and the aponeurosis of the external oblique.

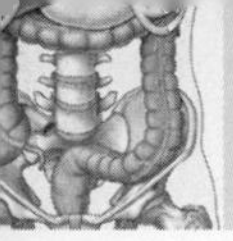



### Anterior wall

The anterior wall of the inguinal canal is formed along its entire length by the aponeurosis of the external oblique muscle (Fig. 4.44). It is also reinforced laterally by the lower fibers of the internal oblique that originate from the lateral two-thirds of the inguinal ligament (Fig. 4.45). This adds an additional covering over the deep inguinal ring, which is a potential point of weakness in the anterior abdominal wall. Furthermore, as the internal oblique muscle covers the deep inguinal ring, it also contributes a layer (the **cremasteric fascia** containing the **cremasteric muscle**) to the coverings of the structures traversing the inguinal canal.

### Posterior wall

The posterior wall of the inguinal canal is formed along its entire length by the transversalis fascia (see Fig. 4.43). It is reinforced along its medial one-third by the **conjoint tendon** (**inguinal falx**; Fig. 4.45). This tendon is the combined insertion of the transversus abdominis and internal oblique muscles into the pubic crest and pectineal line.

As with the internal oblique muscle's reinforcement of the area of the deep inguinal ring, the position of the conjoint tendon posterior to the superficial inguinal ring provides additional support to a potential point of weakness in the anterior abdominal wall.

### Roof

The roof (superior wall) of the inguinal canal is formed by the arching fibers of the transversus abdominis and internal oblique muscles (Figs. 4.45 and 4.46). They pass from their lateral points of origin from the inguinal ligament to their common medial attachment as the conjoint tendon.

### Floor

The floor (inferior wall) of the inguinal canal is formed by the medial one-half of the inguinal ligament. This rolled-under, free margin of the lowest part of the aponeurosis of the external oblique forms a gutter or trough on which the contents of the inguinal canal are positioned. The lacunar ligament reinforces most of the medial part of the gutter.

### Contents

The contents of the inguinal canal are:

- the spermatic cord in men, and
- the round ligament of the uterus and genital branch of the genitofemoral nerve in women.

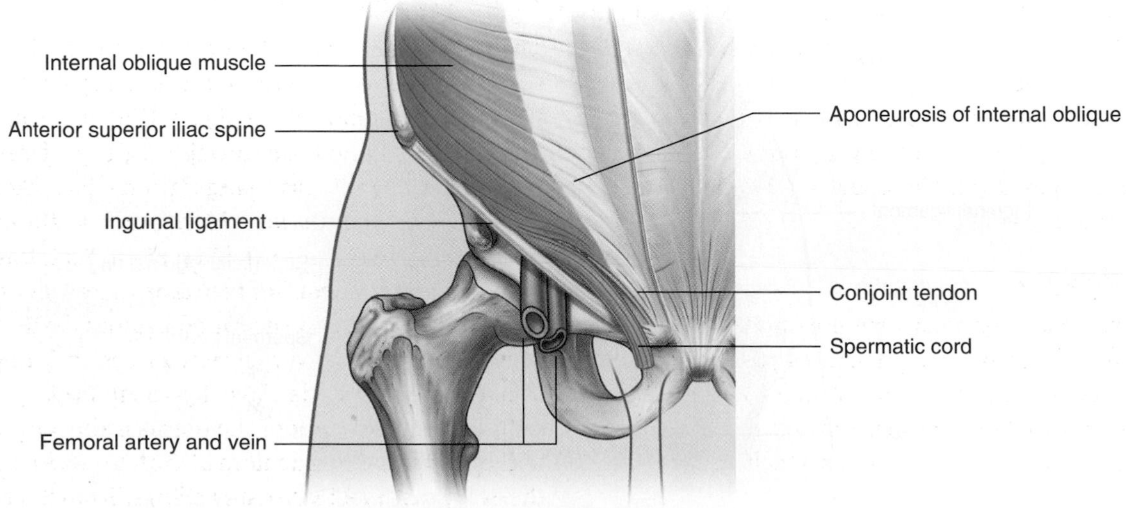


**Fig. 4.45** Internal oblique muscle and the inguinal canal.

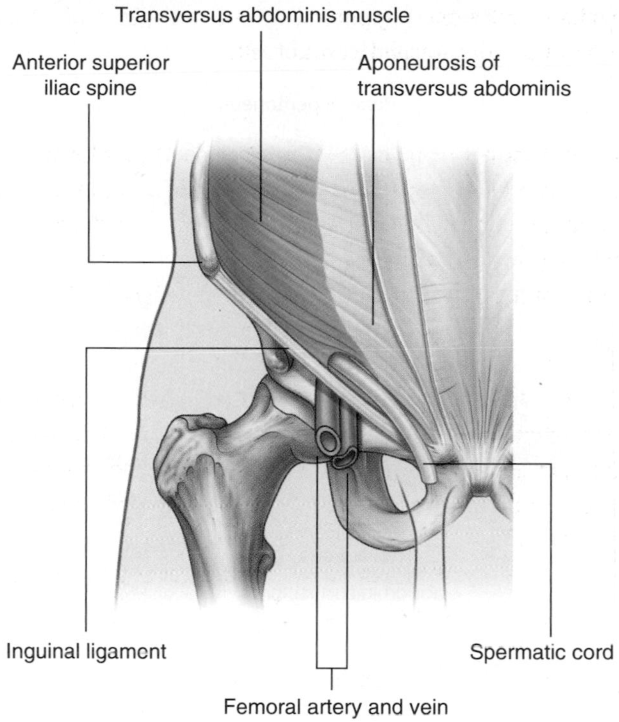


**Fig. 4.46** Transversus abdominis muscle and the inguinal canal.

These structures enter the inguinal canal through the deep inguinal ring and exit it through the superficial inguinal ring.

Additionally, the ilio-inguinal nerve (L1) passes through part of the inguinal canal. This nerve is a branch of the lumbar plexus, enters the abdominal wall posteriorly by piercing the internal surface of the transversus abdominis muscle, and continues through the layers of the anterior abdominal wall by piercing the internal oblique muscle. As it continues to pass inferomedially, it enters the inguinal canal. It continues down the canal to exit through the superficial inguinal ring.

### Spermatic cord

The spermatic cord begins to form proximally at the deep inguinal ring and consists of structures passing between the abdominopelvic cavities and the testis, and the three fascial coverings that enclose these structures (Fig. 4.47).

The structures in the spermatic cord include:

- the ductus deferens,
- the artery to the ductus deferens (from the inferior vesical artery),
- the testicular artery (from the abdominal aorta),
- the pampiniform plexus of veins (testicular veins),
- the cremasteric artery and vein (small vessels associated with the cremasteric fascia),
- the genital branch of the genitofemoral nerve (innervation to the cremasteric muscle),
- sympathetic and visceral afferent nerve fibers,
- lymphatics, and
- remnants of the processus vaginalis.

These structures enter the deep inguinal ring, proceed down the inguinal canal, and exit from the superficial inguinal ring, having acquired the three fascial coverings during their journey. This collection of structures and fascias continues into the scrotum where the structures connect with the testes and the fascias surround the testes.

Three fascias enclose the contents of the spermatic cord:

- The internal spermatic fascia, which is the deepest layer, arises from the transversalis fascia and is attached to the margins of the deep inguinal ring.
- The cremasteric fascia with the associated cremasteric muscle, which is the middle fascial layer, arises from the internal oblique muscle.
- The external spermatic fascia, which is the most superficial covering of the spermatic cord, arises from the aponeurosis of the external oblique muscle and is attached to the margins of the superficial inguinal ring (Fig. 4.47A).

### Round ligament of the uterus

The round ligament of the uterus is a cord-like structure that passes from the uterus to the deep inguinal ring where it enters the inguinal canal (Fig. 4.47B). It passes down the inguinal canal and exits through the superficial inguinal ring. At this point, it has changed from a cord-like structure to a few strands of tissue, which attach to the connective tissue associated with the labia majora. As it traverses the inguinal canal, it acquires the same coverings as the spermatic cord in men. As the round ligament exits the superficial inguinal ring, the coverings are indistinguishable from the tissue strands of the ligament itself.

The round ligament of the uterus is the long distal part of the original gubernaculum in the fetus that extends from the ovary to the labioscrotal swellings. From its attachment to the uterus, the round ligament of the uterus continues to the ovary as the ligament of the ovary that develops from the short proximal end of the gubernaculum.

Genital branch of genitofemoral nerve
Testicular artery and
pampiniform plexus of veins
Cremasteric vessels
Parietal peritoneum
Ductus deferens
Extraperitoneal fascia
Artery to ductus deferens
Transversalis
fascia
Inferior epigastric vessels
Ilioinguinal
nerve
External oblique
aponeurosis
Internal
oblique muscle
Transversus
abdominis muscle
Conjoint tendon
Deep inguinal ring
Ilioinguinal nerve
External oblique
aponeurosis
Superficial inguinal ring
External spermatic fascia
Cremasteric fascia
Internal spermatic fascia
Parietal layer of the tunica vaginalis
Cavity of the tunica vaginalis
Visceral layer of the tunica vaginalis
A

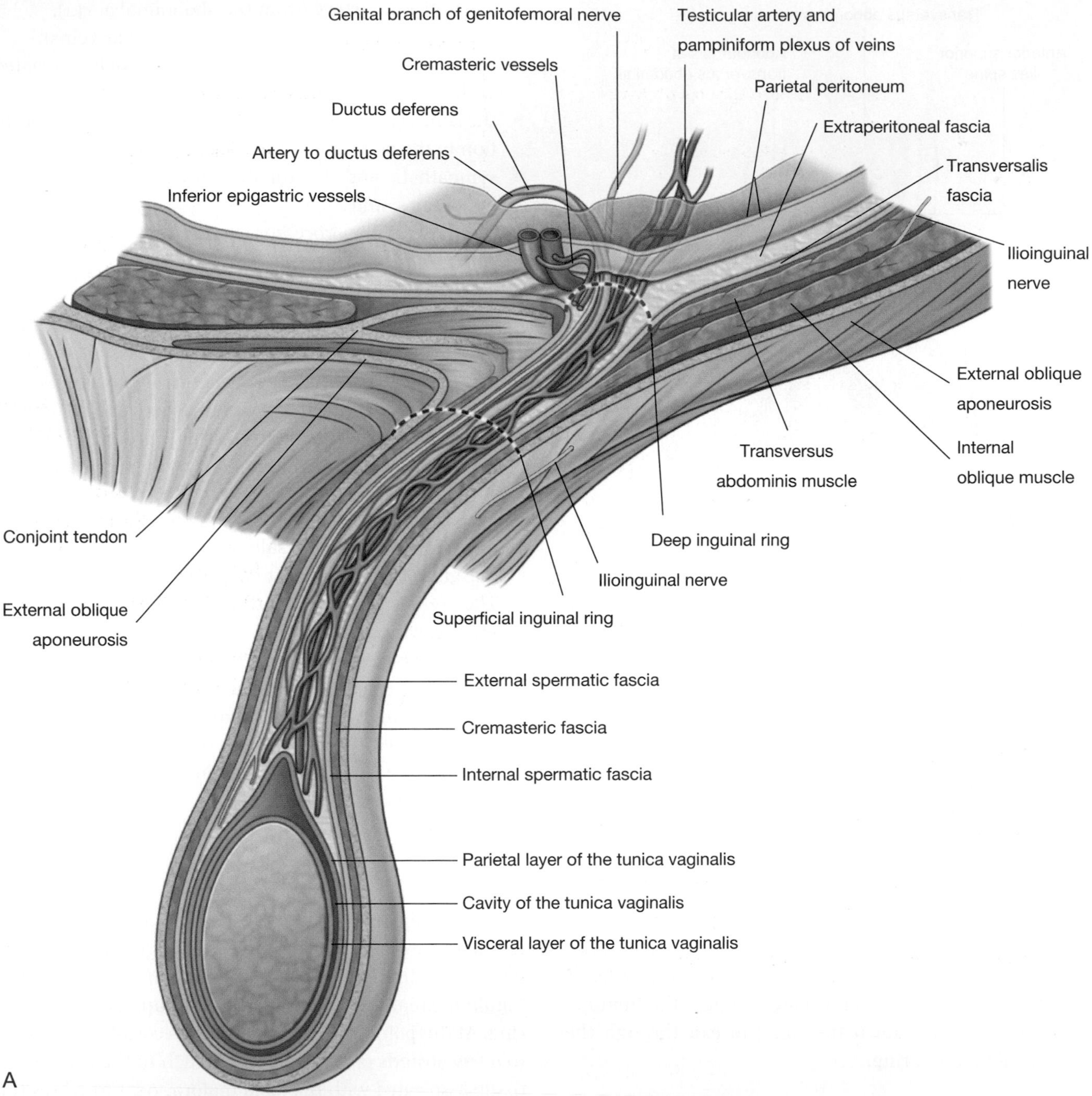

**Fig. 4.47** **A.** Spermatic cord (men).

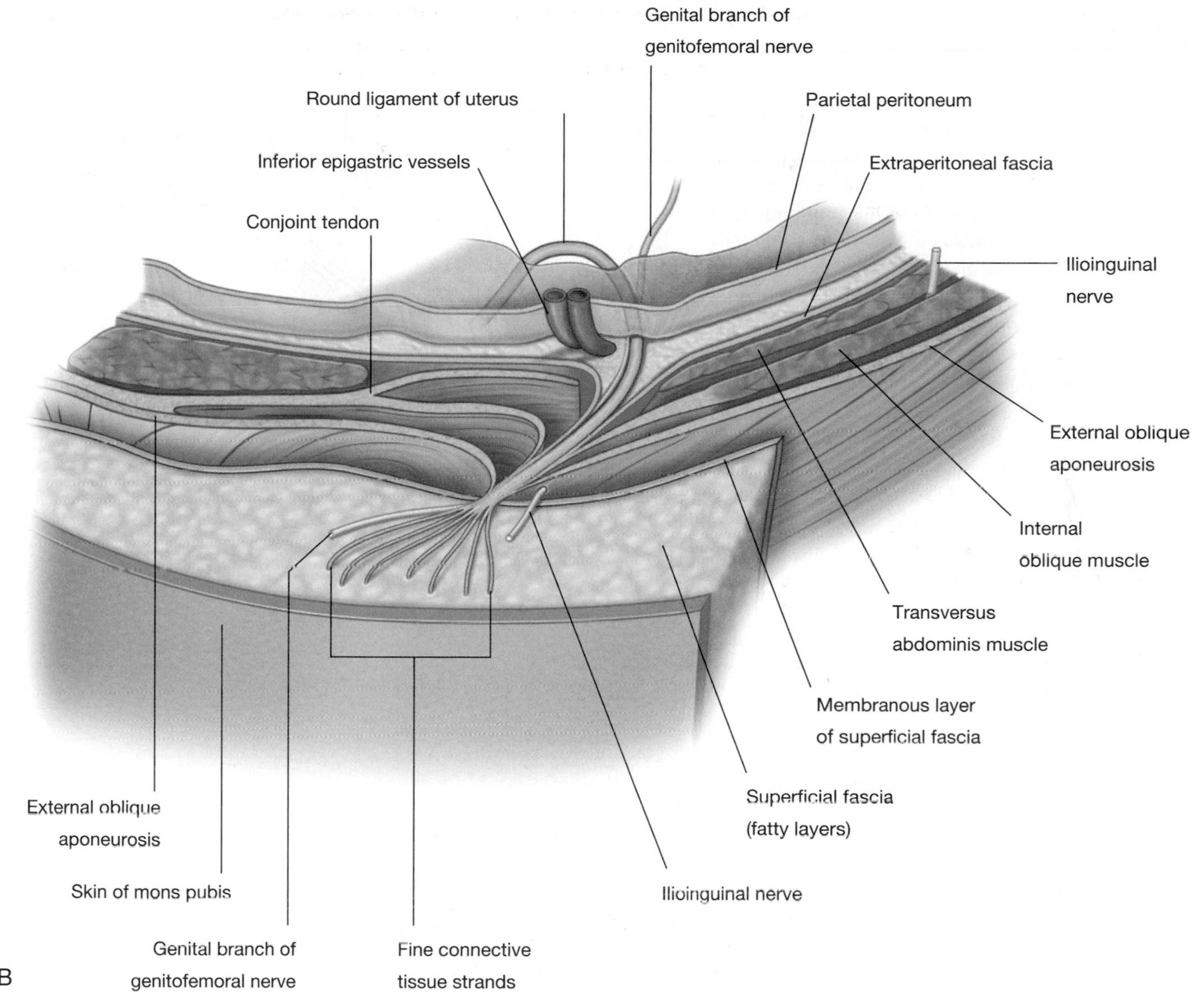
Genital branch of
genitofemoral nerve
Round ligament of uterus
Parietal peritoneum
Inferior epigastric vessels
Extraperitoneal fascia
Conjoint tendon
Ilioinguinal
nerve
External oblique
aponeurosis
Internal
oblique muscle
Transversus
abdominis muscle
Membranous layer
of superficial fascia
Superficial fascia
(fatty layers)
External oblique
aponeurosis
Skin of mons pubis
Ilioinguinal nerve
Genital branch of
genitofemoral nerve
Fine connective
tissue strands
B

**Fig. 4.47, cont'd. B.** Round ligament of uterus (women).

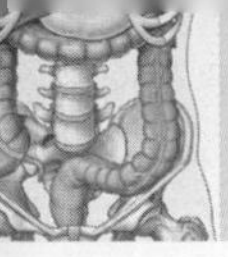

## In the clinic

### Cremasteric reflex

In men, the cremaster muscle and cremasteric fascia form the middle or second covering of the spermatic cord. This muscle and its associated fascia are supplied by the genital branch of the genitofemoral nerve (L1/L2). Contraction of this muscle and the resulting elevation of the testis can be stimulated by a reflex arc. Gently touching the skin at and around the anterior aspect of the superior part of the thigh stimulates the sensory fibers in the ilio-inguinal nerve. These sensory fibers enter the spinal cord at level L1. At this level, the sensory fibers stimulate the motor fibers carried in the genital branch of the genitofemoral nerve, which results in contraction of the cremaster muscle and elevation of the testis.

The cremasteric reflex is more active in children, tending to diminish with age. As with many reflexes, it may be absent in certain neurological disorders. Although it can be used for testing spinal cord function at level L1 in men, its clinical use is limited.

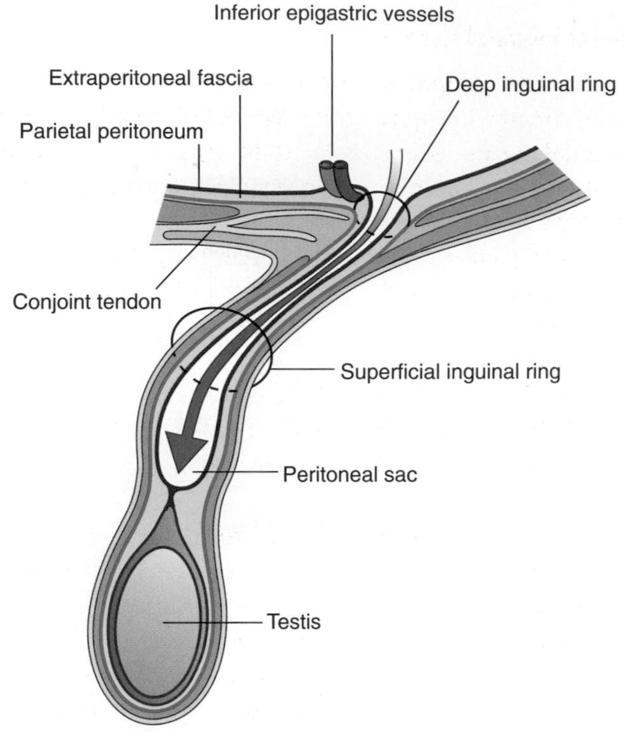


**Fig. 4.48** Indirect inguinal hernia.

## Inguinal hernias

An inguinal hernia is the protrusion or passage of a peritoneal sac, with or without abdominal contents, through a weakened part of the abdominal wall in the groin. It occurs because the peritoneal sac enters the inguinal canal either:

- indirectly, through the deep inguinal ring, or
- directly, through the posterior wall of the inguinal canal.

Inguinal hernias are therefore classified as either indirect or direct.

### Indirect inguinal hernias

The indirect inguinal hernia is the most common of the two types of inguinal hernia and is much more common in men than in women (Fig. 4.48). It occurs because some part, or all, of the embryonic processus vaginalis remains open or patent. It is therefore referred to as being congenital in origin.

The protruding peritoneal sac enters the inguinal canal by passing through the deep inguinal ring, just lateral to the inferior epigastric vessels. The extent of its excursion down the inguinal canal depends on the amount of processus vaginalis that remains patent. If the entire processus vaginalis remains patent, the peritoneal sac may traverse the length of the canal, exit the superficial inguinal ring, and continue into the scrotum in men or the labia majus in women. In this case, the protruding peritoneal sac acquires the same three coverings as those associated with the spermatic cord in men or the round ligament of the uterus in women.

### Direct inguinal hernias

A peritoneal sac that enters the medial end of the inguinal canal directly through a weakened posterior wall is a direct inguinal hernia (Fig. 4.49). It is usually described as acquired because it develops when abdominal musculature has been weakened, and is commonly seen in mature men. The bulging occurs medial to the inferior epigastric vessels in the inguinal triangle (Hesselbach's triangle), which is bounded:

- laterally by the inferior epigastric artery,
- medially by the rectus abdominis muscle, and
- inferiorly by the inguinal ligament (Fig. 4.50).

Internally, a thickening of the transversalis fascia (the iliopubic tract) follows the course of the inguinal ligament (Fig. 4.50).

A direct inguinal hernia does not traverse the entire length of the inguinal canal but may exit through the superficial inguinal ring. When this occurs, the peritoneal sac acquires a layer of external spermatic fascia and can extend, like an indirect hernia, into the scrotum.

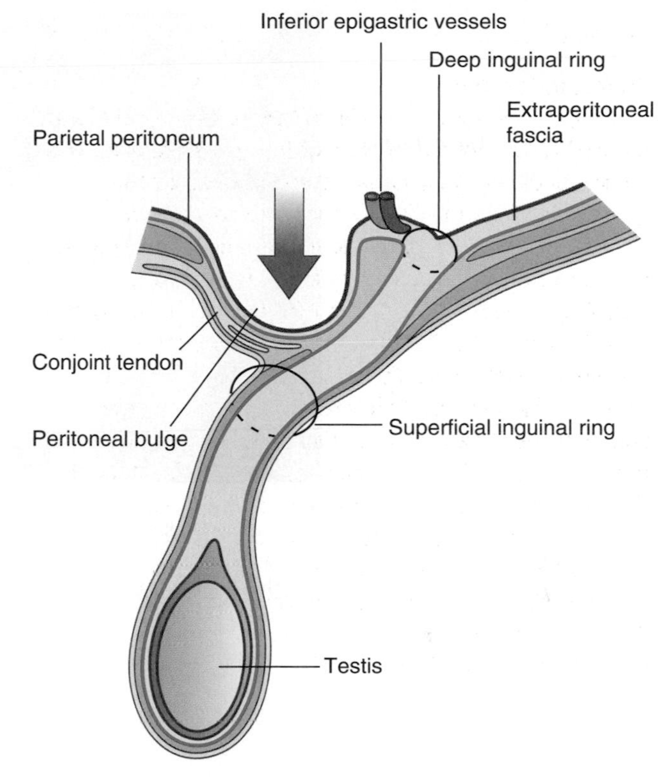


**Fig. 4.49** Direct inguinal hernia.

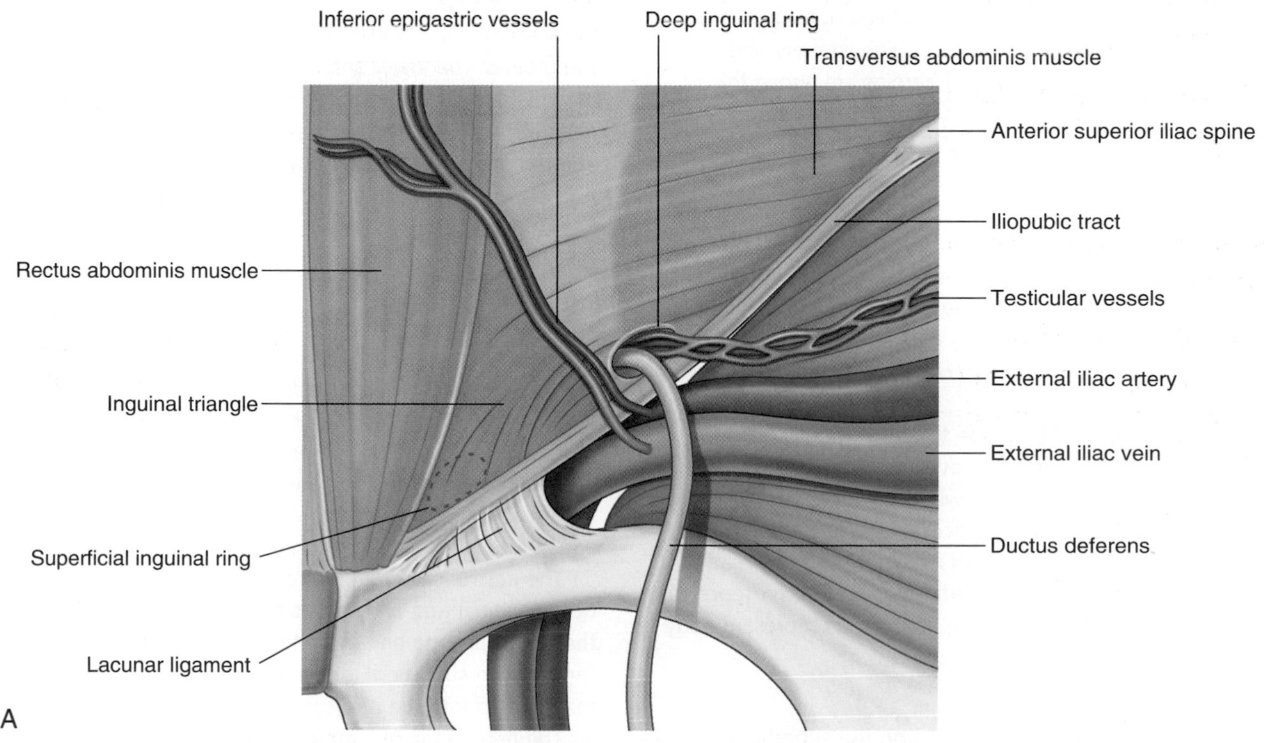


**Fig. 4.50** Right inguinal triangle. **A.** Internal view.

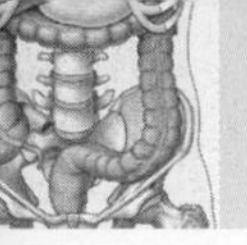

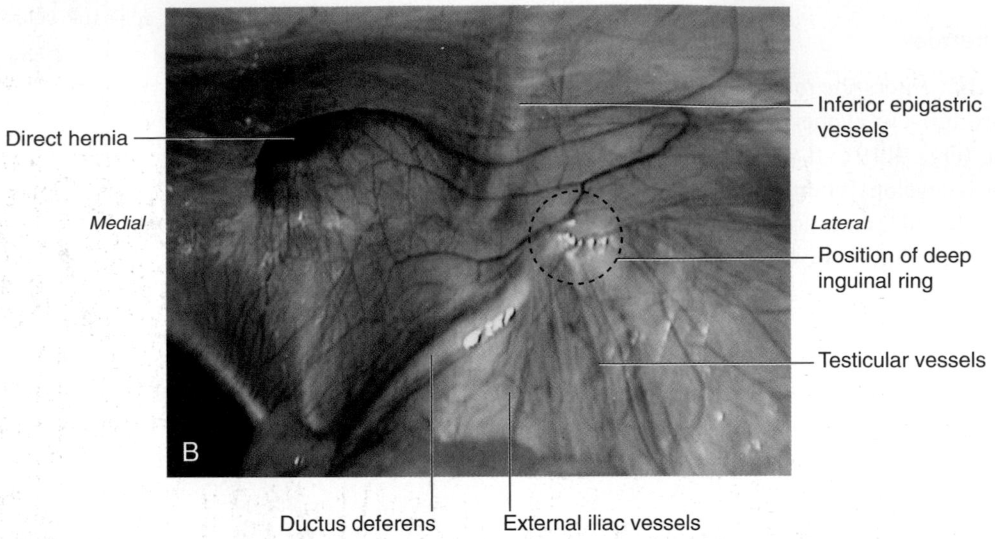


**Fig. 4.50, cont'd** **B.** Laparoscopic view showing the parietal peritoneum still covering the area.

## In the clinic

### Masses around the groin

Around the groin there is a complex confluence of anatomical structures. Careful examination and good anatomical knowledge allow determination of the correct anatomical structure from which the mass arises and therefore the diagnosis. The most common masses in the groin are hernias.

The key to groin examination is determining the position of the inguinal ligament. The inguinal ligament passes between the anterior superior iliac spine laterally and the pubic tubercle medially. Inguinal hernias are above the inguinal ligament and are usually more apparent on standing. A visual assessment of the lump is necessary, bearing in mind the anatomical landmarks of the inguinal ligament.

In men, it is wise to examine the scrotum to check for a lump. If an abnormal mass is present, an inability to feel its upper edge suggests that it may originate from the inguinal canal and might be a hernia. By placing the hand over the lump and asking the patient to cough, the lump bulges outward.

An attempt should be made to reduce the swelling by applying gentle, firm pressure over the lump. If the lump is reducible, the hand should be withdrawn and careful observation will reveal recurrence of the mass.

The position of an abnormal mass in the groin relative to the pubic tubercle is very important, as are the presence of increased temperature and pain, which may represent early signs of strangulation or infection.

As a general rule:

- An inguinal hernia appears through the superficial inguinal ring above the pubic tubercle and crest.
- A femoral hernia (see below) appears through the femoral canal below and lateral to the pubic tubercle.

A hernia is the protrusion of a viscus, in part or in whole, through a normal or abnormal opening. The viscus usually carries a covering of parietal peritoneum, which forms the lining of the hernial sac.

### *Inguinal hernias*

Hernias occur in a variety of regions. The commonest site is the groin of the lower anterior abdominal wall. In some patients, inguinal hernias are present from birth (congenital) and are caused by the persistence of the processus vaginalis and the passage of viscera through the inguinal canal. Acquired hernias occur in older patients and causes include raised intraabdominal pressure (e.g., from repeated coughing associated with lung disease), damage to nerves of the anterior abdominal wall (e.g., from surgical abdominal incisions), and weakening of the walls of the inguinal canal.

One of the potential problems with hernias is that bowel and fat may become stuck within the hernial sac. This can cause appreciable pain and bowel obstruction, necessitating urgent surgery. Another potential risk is **strangulation** of the hernia, in which the blood supply to the bowel is cut off at the neck of the hernial sac, rendering the bowel ischemic and susceptible to perforation (Fig. 4.51).

The hernial sac of an **indirect inguinal hernia** enters the deep inguinal ring and passes through the inguinal canal. If the hernia is large enough, the hernial sac may emerge through the superficial inguinal ring. In men, such a hernia may extend into the scrotum (Fig. 4.52).

The hernial sac of a **direct inguinal hernia** pushes forward through the posterior wall of the inguinal canal immediately

## In the clinic—cont'd

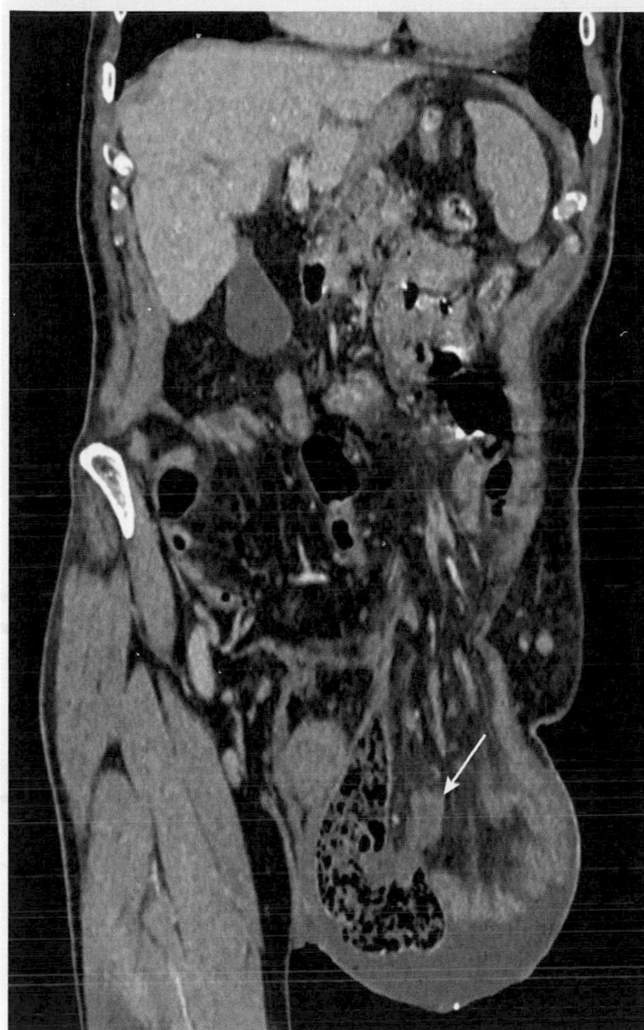

**Fig. 4.51** Coronal CT shows a large inguinal hernia containing loops of large and small bowel *(arrow)* on the left side of a male patient.

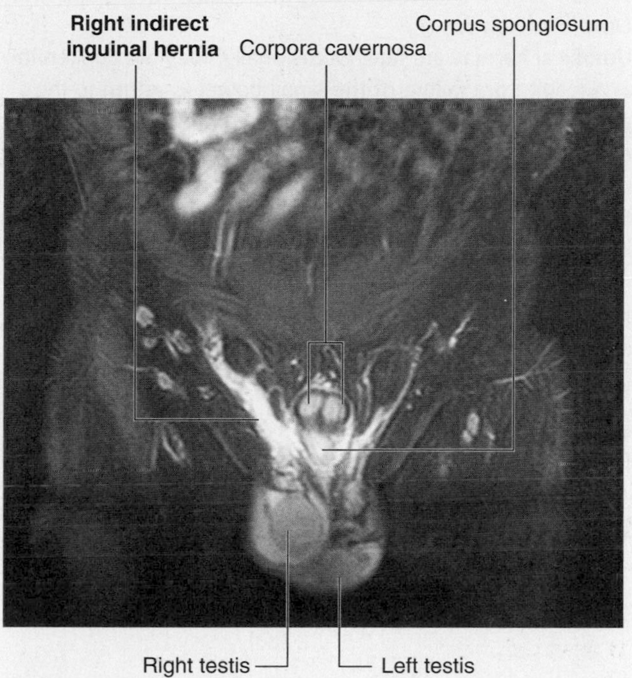


**Fig. 4.52** Right indirect inguinal hernia. T2, fat saturated, weighted magnetic resonance image in the coronal plane of a male groin.

posterior to the superficial inguinal ring. The hernia protrudes directly forward medial to the inferior epigastric vessels and through the superficial inguinal ring.

The differentiation between an indirect and a direct inguinal hernia is made during surgery when the inferior epigastric vessels are identified at the medial edge of the deep internal ring:

- An indirect hernial sac passes lateral to the inferior epigastric vessels.
- A direct hernia is medial to the inferior epigastric vessels.

Inguinal hernias occur more commonly in men than in women possibly because men have a much larger inguinal canal than women.

### *Femoral hernias*

A **femoral hernia** passes through the femoral canal and into the medial aspect of the anterior thigh. The femoral canal lies at the medial edge of the femoral sheath, which contains the femoral artery, femoral vein, and lymphatics. The neck of the femoral canal is extremely narrow and is prone to trapping bowel within the sac, so making this type of hernia irreducible and susceptible to bowel strangulation. Femoral hernias are usually acquired, are not congenital, and most commonly occur in middle-aged and elderly populations. In addition, because women generally have wider pelvises than men, they tend to occur more commonly in women.

### *Sportsmen's groin/sportsmen's hernia*

The groin can loosely be defined as the area where the leg meets the trunk near the midline. Here the abdominal muscles of the trunk blend in with the adductor muscles of the thigh, the medial end of the inguinal ligament attaches to the pubic tubercle, the pubic symphysis attaches the two pubic bones together, and the superficial (external) inguinal ring occurs. It also is in and around this region where there is considerable translation of force during most athletic and sporting activities. Pain in the groin or pubic region can be due to numerous causes, which include inflammatory changes at the pubic symphysis, insertional problems of the rectus abdominis/adductor longus, and hernias.

*(continues)*

## In the clinic—cont'd

### *Umbilical hernias*

**Umbilical hernias** are rare. Occasionally, they are congenital and result from failure of the small bowel to return to the abdominal cavity from the umbilical cord during development. After birth, umbilical hernias may result from incomplete closure of the umbilicus (navel). Overall, most of these hernias close in the first year of life, and surgical repair is not generally attempted until later.

**Para-umbilical hernias** may occur in adults at and around the umbilicus and often have small necks, so requiring surgical treatment.

### *Incisional hernias*

Incisional hernias occur through a defect in a scar of a previous abdominal operation. Usually, the necks of these hernias are wide and do not therefore strangulate the viscera they contain.

### *Other hernias*

A **spigelian hernia** passes upward through the arcuate line into the lateral border at the lower part of the posterior rectus sheath. It may appear as a tender mass on one side of the lower anterior abdominal wall.

Abdominopelvic cavity hernias can also develop in association with the pelvic walls, and sites include the obturator canal, the greater sciatic foramen and above and below the piriformis muscle.

# ABDOMINAL VISCERA

## Peritoneum

A thin membrane (the peritoneum) lines the walls of the abdominal cavity and covers much of the viscera. The parietal peritoneum lines the walls of the cavity and the visceral peritoneum covers the viscera. Between the parietal and visceral layers of peritoneum is a potential space (the peritoneal cavity). Abdominal viscera either are suspended in the peritoneal cavity by folds of peritoneum (**mesenteries**) or are outside the peritoneal cavity. Organs suspended in the cavity are referred to as intraperitoneal (Fig. 4.53); organs outside the peritoneal cavity, with only one surface or part of one surface covered by peritoneum, are retroperitoneal.

### Innervation of the peritoneum

The parietal peritoneum associated with the abdominal wall is innervated by somatic afferents carried in branches of the associated spinal nerves and is therefore sensitive to well-localized pain. The visceral peritoneum is innervated by visceral afferents that accompany autonomic nerves (sympathetic and parasympathetic) back to the central nervous system. Activation of these fibers can lead to referred and poorly localized sensations of discomfort, and to reflex visceral motor activity.

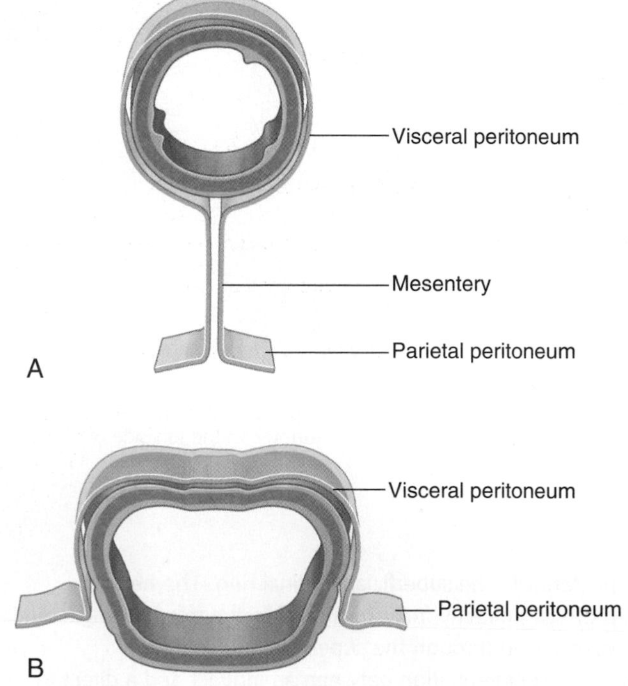


**Fig. 4.53 A.** Intraperitoneal. **B.** Retroperitoneal.

## Peritoneal cavity

The peritoneal cavity is subdivided into the greater sac and the omental bursa (lesser sac; Fig. 4.54).

- The greater sac accounts for most of the space in the peritoneal cavity, beginning superiorly at the diaphragm and continuing inferiorly into the pelvic cavity. It is entered once the parietal peritoneum has been penetrated.
- The omental bursa is a smaller subdivision of the peritoneal cavity posterior to the stomach and liver and is continuous with the greater sac through an opening, the omental (epiploic) foramen (Fig. 4.55).

Surrounding the omental (epiploic) foramen are numerous structures covered with peritoneum. They include the portal vein, hepatic artery proper, and bile duct anteriorly; the inferior vena cava posteriorly; the caudate lobe of the liver superiorly; and the first part of the duodenum inferiorly.

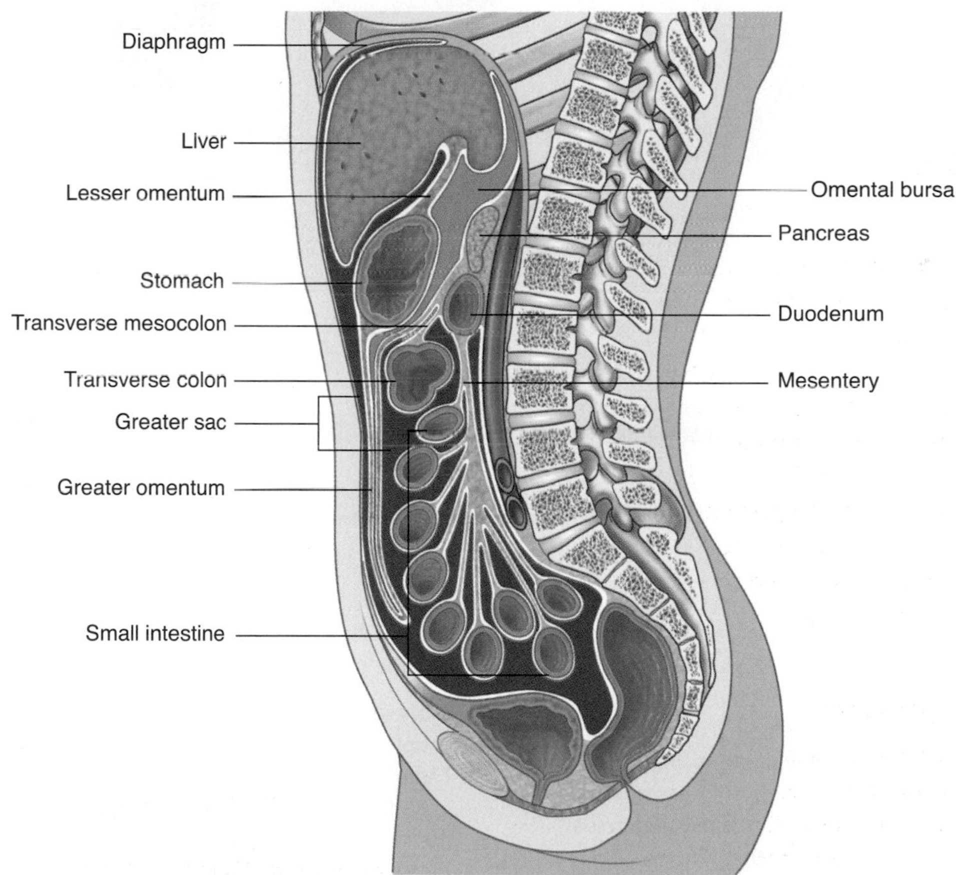


**Fig. 4.54** Greater and lesser sacs of the peritoneal cavity.

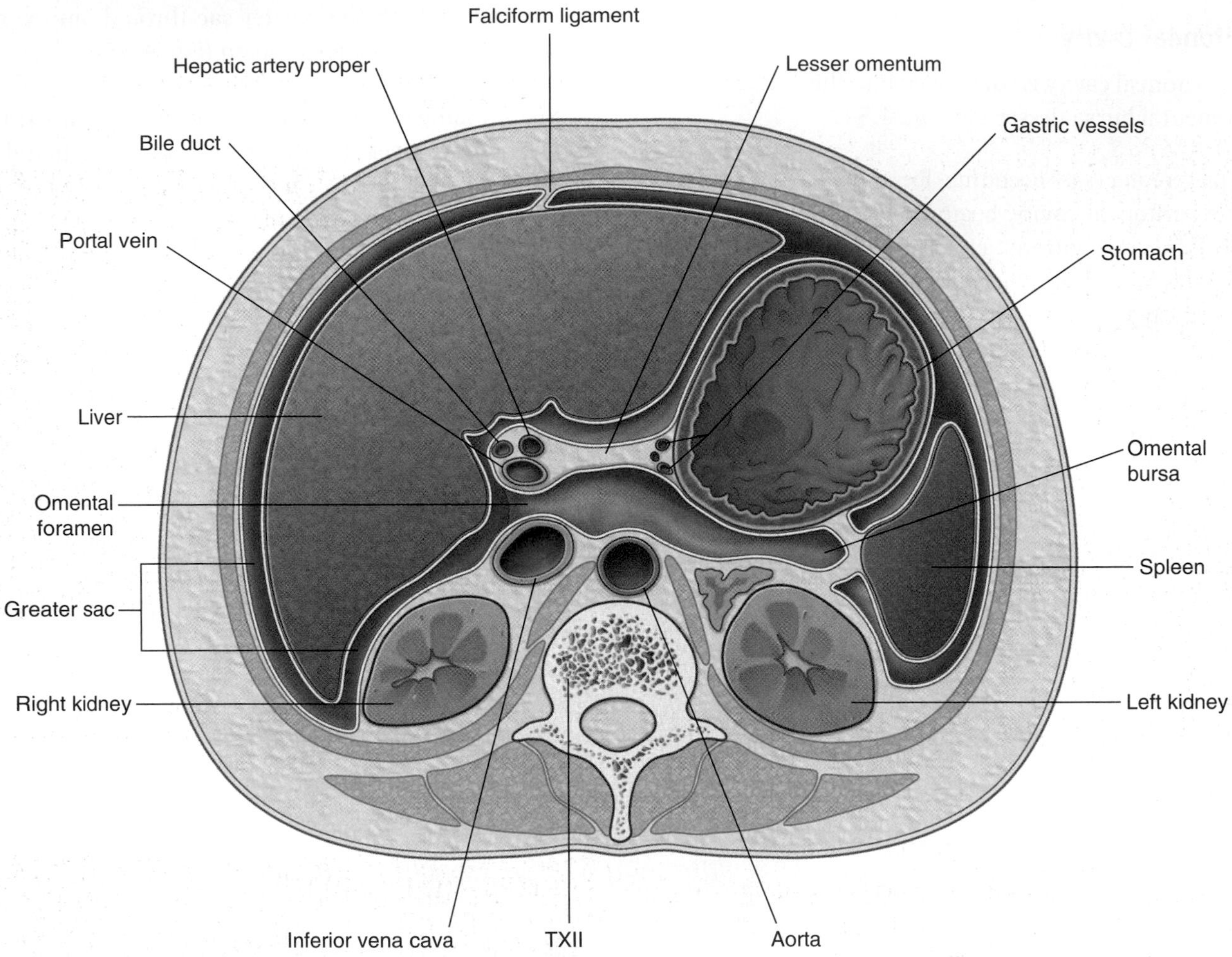

**Fig. 4.55** Transverse section illustrating the continuity between the greater and lesser sacs through the omental (epiploic) foramen.

## In the clinic

### Peritoneum

A small volume of peritoneal fluid within the peritoneal cavity lubricates movement of the viscera suspended in the abdominal cavity. It is not detectable on any available imaging such as ultrasound or computed tomography. In various pathological conditions (e.g., in liver cirrhosis, acute pancreatitis, or heart failure) the volume of peritoneal fluid can increase; this is known as ascites. In cases of high volume of free intraperitoneal fluid, marked abdominal distention can be observed (Fig. 4.56).

The peritoneal space has a large surface area, which facilitates the spread of disease through the peritoneal cavity and over the bowel and visceral surfaces. Conversely, this large surface area can be used for administering certain types of treatment and a number of procedures.

### *Ventriculoperitoneal shunts*

Patients with obstructive hydrocephalus (an excessive accumulation of cerebrospinal fluid within the cerebral ventricular system) require continuous drainage of this fluid. This is achieved by placing a fine-bore catheter through the skull into the cerebral ventricles and placing the extracranial part of the tube beneath the scalp and skin of the neck and chest wall, and then through the abdominal wall into the peritoneal cavity. Cerebrospinal fluid drains through the tube into the peritoneal cavity, where it is absorbed.

### *Dialysis and peritoneal dialysis*

People who develop renal failure require dialysis to live. There are two methods.

In the first method (**hemodialysis**), blood is taken from the circulation, dialyzed through a complex artificial membrane, and returned to the body. A high rate of blood flow is required to remove excess body fluid, exchange electrolytes, and remove noxious metabolites. To accomplish this, either an arteriovenous fistula is established surgically (by connecting an artery to a vein, usually in the upper limb, and requiring approximately six weeks to "mature") and is cannulated each time the patient returns for dialysis, or a large-bore cannula is placed into the right atrium, through which blood can be aspirated and returned.

In the second method **(peritoneal dialysis),** the peritoneum is used as the dialysis membrane. The large surface area of the peritoneal cavity is an ideal dialysis

## In the clinic—cont'd

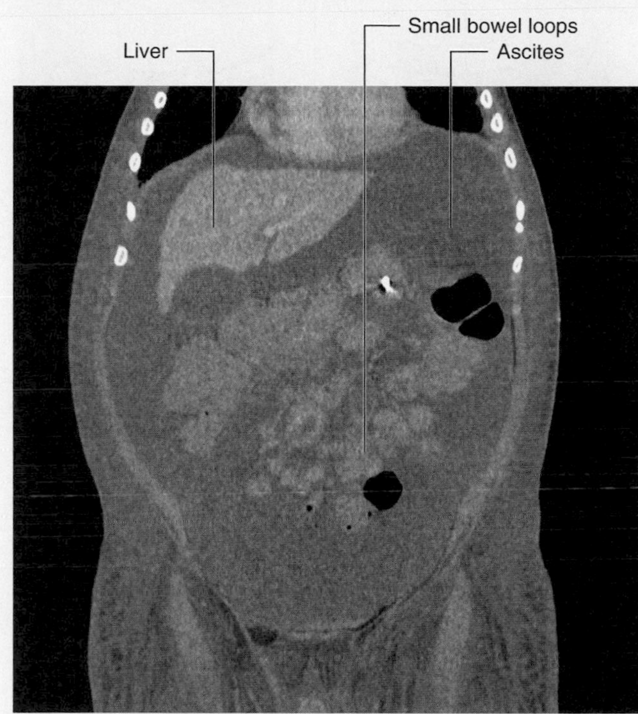


**Fig. 4.56** Coronal CT shows ascites fluid in abdominal cavity.

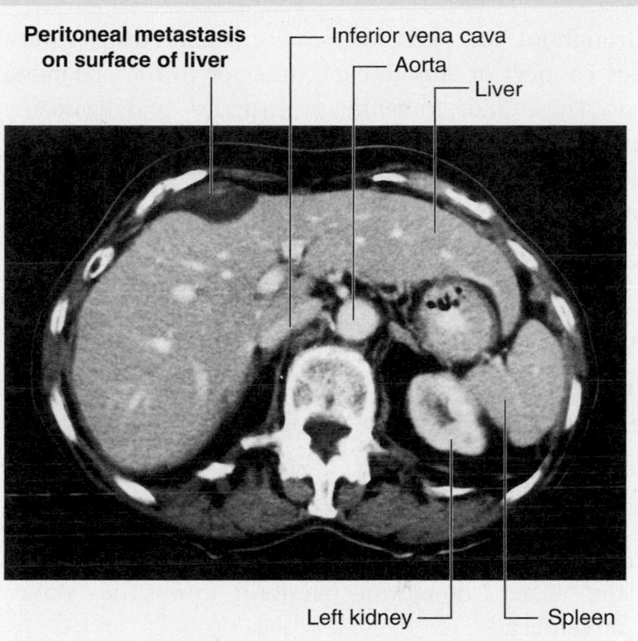


**Fig. 4.57** Peritoneal metastasis on the surface of the liver. Computed tomogram in the axial plane of the upper abdomen.

membrane for fluid and electrolyte exchange. To accomplish dialysis, a small tube is inserted through the abdominal wall and dialysis fluid is injected into the peritoneal cavity. Electrolytes and molecules are exchanged across the peritoneum between the fluid and blood. Once dialysis is completed, the fluid is drained.

### *Peritoneal spread of disease*

The large surface area of the peritoneal cavity allows infection and malignant disease to spread easily throughout the abdomen (Fig. 4.57). If malignant cells enter the peritoneal cavity by direct invasion (e.g., from colon or ovarian cancer), spread may be rapid. Similarly, a surgeon excising a malignant tumor and releasing malignant cells into the peritoneal cavity may cause an appreciable worsening of the patient's prognosis. Infection can also spread across the large surface area.

The peritoneal cavity can also act as a barrier to, and container of, disease. Intraabdominal infection therefore tends to remain below the diaphragm rather than spread into other body cavities.

### *Perforated bowel*

A perforated bowel (e.g., caused by a perforated duodenal ulcer) often leads to the release of gas into the peritoneal cavity. This peritoneal gas can be easily visualized on an erect chest radiograph—gas can be demonstrated in extremely small amounts beneath the diaphragm. A patient with severe abdominal pain and subdiaphragmatic gas needs a laparotomy (Fig. 4.58).

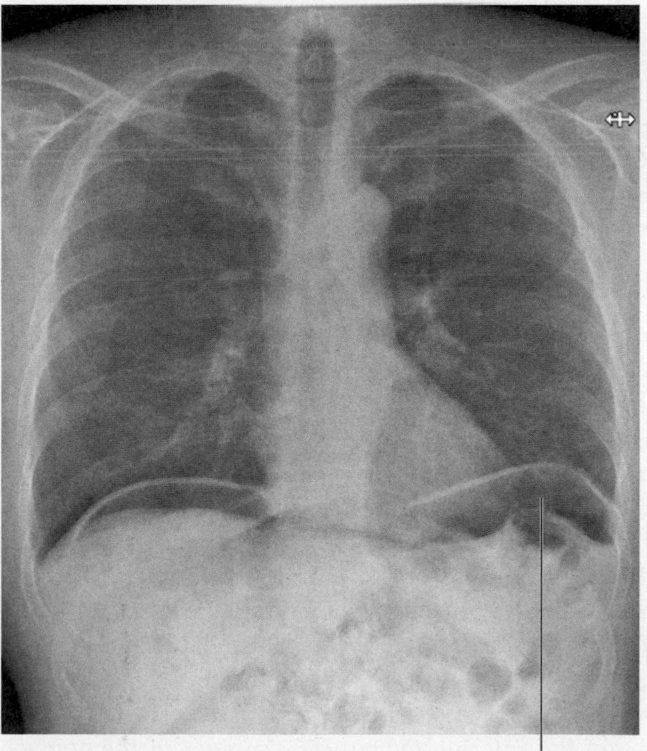


**Fig. 4.58** Radiograph of subdiaphragmatic gas.

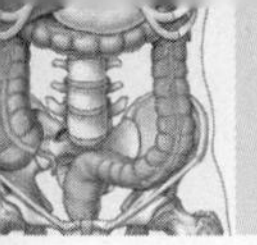



## Omenta, mesenteries, and ligaments

Throughout the peritoneal cavity numerous peritoneal folds connect organs to each other or to the abdominal wall. These folds (omenta, mesenteries, and ligaments) develop from the original dorsal and ventral mesenteries, which suspend the developing gastrointestinal tract in the embryonic coelomic cavity. Some contain vessels and nerves supplying the viscera, while others help maintain the proper positioning of the viscera.

### Omenta

The omenta consist of two layers of peritoneum, which pass from the stomach and the first part of the duodenum to other viscera. There are two:

- the greater omentum, derived from the dorsal mesentery, and
- the lesser omentum, derived from the ventral mesentery.

#### Greater omentum

The **greater omentum** is a large, apron-like, peritoneal fold that attaches to the greater curvature of the stomach and the first part of the duodenum (Fig. 4.59). It drapes inferiorly over the transverse colon and the coils of the jejunum and ileum (see Fig. 4.54). Turning posteriorly, it ascends to associate with, and become adherent to, the peritoneum on the superior surface of the transverse colon and the anterior layer of the transverse mesocolon before arriving at the posterior abdominal wall.

Usually a thin membrane, the greater omentum always contains an accumulation of fat, which may become substantial in some individuals. Additionally, there are two arteries and accompanying veins, the **right** and **left gastro-omental vessels**, between this double-layered peritoneal apron just inferior to the greater curvature of the stomach.

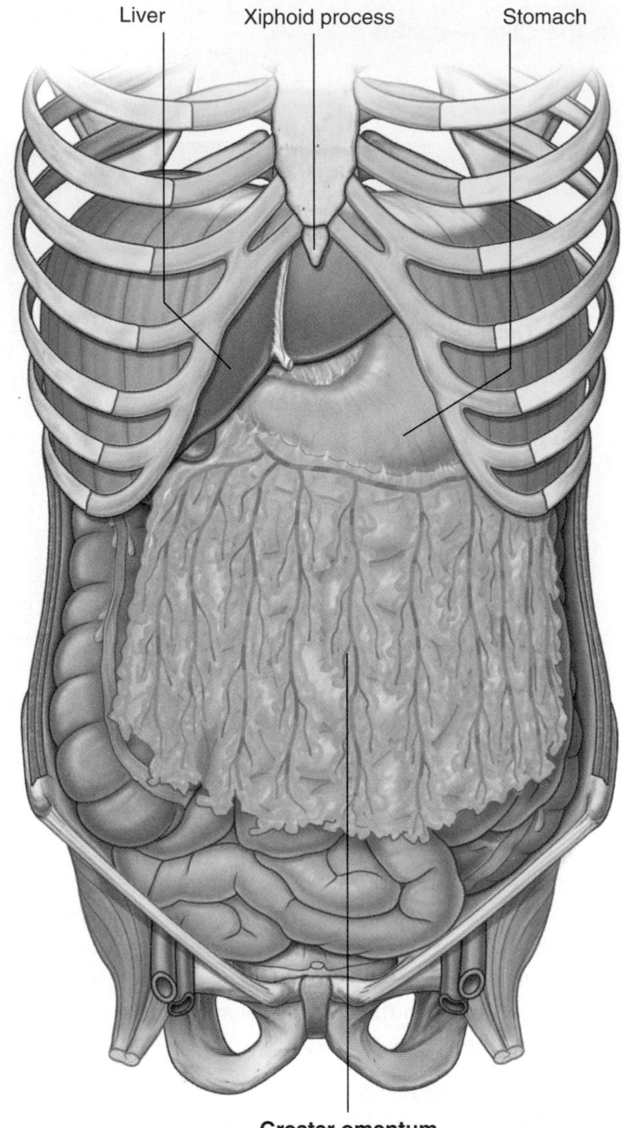


**Fig. 4.59** Greater omentum.

#### Lesser omentum

The other two-layered peritoneal omentum is the **lesser omentum** (Fig. 4.60). It extends from the lesser curvature of the stomach and the first part of the duodenum to the inferior surface of the liver (Figs. 4.54 and 4.60).

A thin membrane continuous with the peritoneal coverings of the anterior and posterior surfaces of the stomach and the first part of the duodenum, the lesser omentum is divided into:

- a medial hepatogastric ligament, which passes between the stomach and liver, and
- a lateral hepatoduodenal ligament, which passes between the duodenum and liver.

The hepatoduodenal ligament ends laterally as a free margin and serves as the anterior border of the omental foramen (Fig. 4.55). Enclosed in this free edge are the hepatic artery proper, the bile duct, and the portal vein. Additionally, the right and left gastric vessels are between the layers of the lesser omentum near the lesser curvature of the stomach.

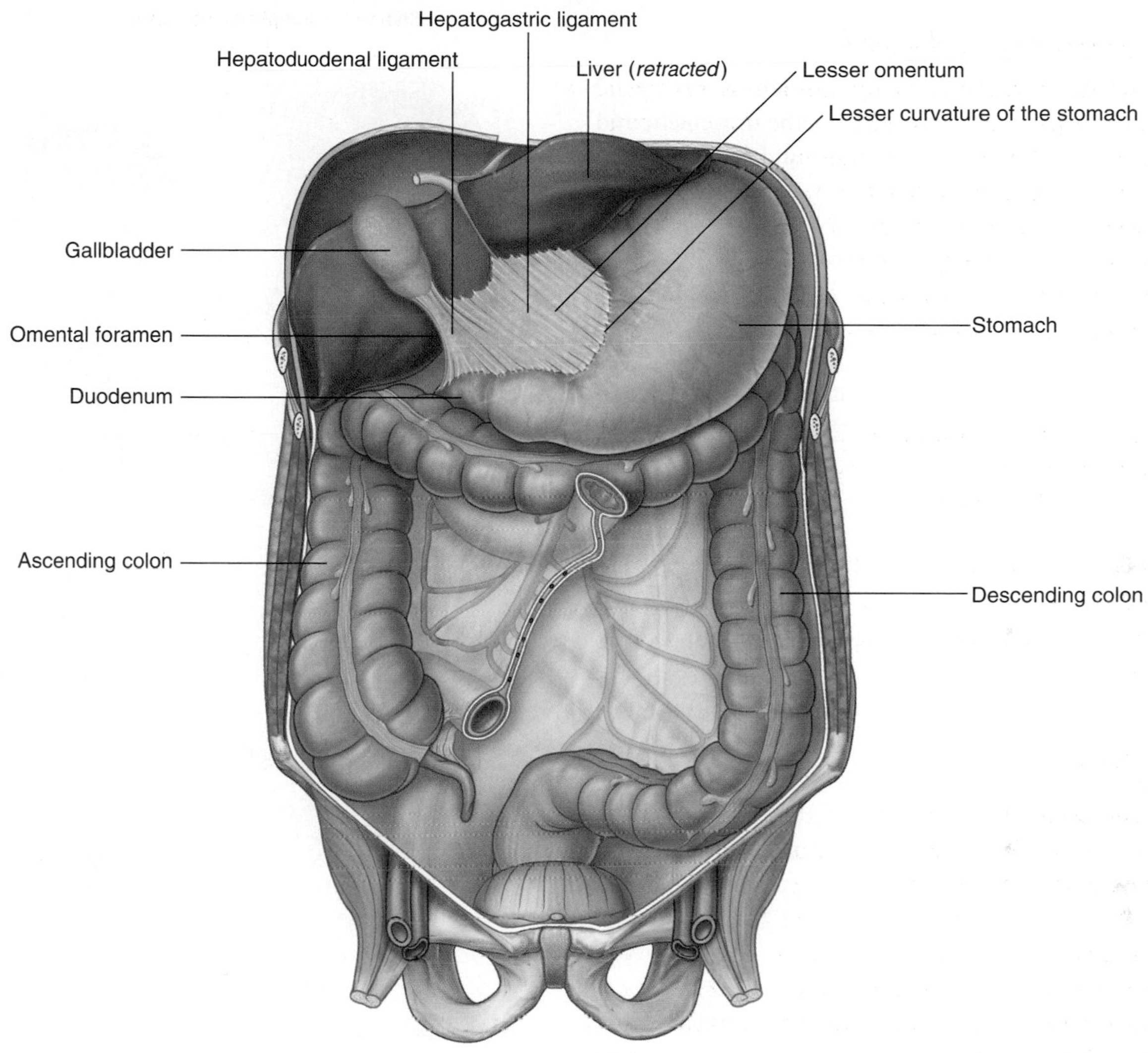


**Fig. 4.60** Lesser omentum.

## In the clinic

### The greater omentum

When a laparotomy is performed and the peritoneal cavity is opened, the first structure usually encountered is the greater omentum. This fatty double-layered vascular membrane hangs like an apron from the greater curvature of the stomach, drapes over the transverse colon, and lies freely suspended within the abdominal cavity. It is often referred to as the "policeman of the abdomen" because of its apparent ability to migrate to any inflamed area and wrap itself around the organ to wall off inflammation. When a part of bowel becomes inflamed, it ceases peristalsis. This aperistaltic area is referred to as a local paralytic ileus. The remaining noninflamed part of the bowel continues to move and "massages" the greater omentum to the region where there is no peristalsis. The localized inflammatory reaction spreads to the greater omentum, which then adheres to the diseased area of bowel.

The greater omentum is also an important site for metastatic tumor spread. Direct omental spread by a transcoelomic route is common for carcinoma of the ovary. As the metastases develop within the greater omentum, it becomes significantly thickened.

In computed tomography imaging and during laparotomy, the thickened omentum is referred to as an "omental cake."

## Mesenteries

Mesenteries are peritoneal folds that attach viscera to the posterior abdominal wall. They allow some movement and provide a conduit for vessels, nerves, and lymphatics to reach the viscera and include:

- the mesentery—associated with parts of the small intestine,
- the transverse mesocolon—associated with the transverse colon, and
- the sigmoid mesocolon—associated with the sigmoid colon.

All of these are derivatives of the dorsal mesentery.

### Mesentery

The **mesentery** is a large, fan-shaped, double-layered fold of peritoneum that connects the jejunum and ileum to the posterior abdominal wall (Fig. 4.61). Its superior attachment is at the duodenojejunal junction, just to the left of the upper lumbar part of the vertebral column. It passes obliquely downward and to the right, ending at the ileocecal junction near the upper border of the right sacro-iliac joint. In the fat between the two peritoneal layers of the mesentery are the arteries, veins, nerves, and lymphatics that supply the jejunum and ileum.

### Transverse mesocolon

The **transverse mesocolon** is a fold of peritoneum that connects the transverse colon to the posterior abdominal wall (Fig. 4.61). Its two layers of peritoneum leave the posterior abdominal wall across the anterior surface of the head and body of the pancreas and pass outward to surround the transverse colon. Between its layers are the arteries, veins, nerves, and lymphatics related to the transverse colon. The anterior layer of the transverse mesocolon is adherent to the posterior layer of the greater omentum.

### Sigmoid mesocolon

The **sigmoid mesocolon** is an inverted, V-shaped peritoneal fold that attaches the sigmoid colon to the abdominal wall (Fig. 4.61). The apex of the V is near the division of the left common iliac artery into its internal and external branches, with the left limb of the descending V along the medial border of the left psoas major muscle and the right limb descending into the pelvis to end at the level of vertebra SIII. The sigmoid and superior rectal vessels, along with the nerves and lymphatics associated with the sigmoid colon, pass through this peritoneal fold.

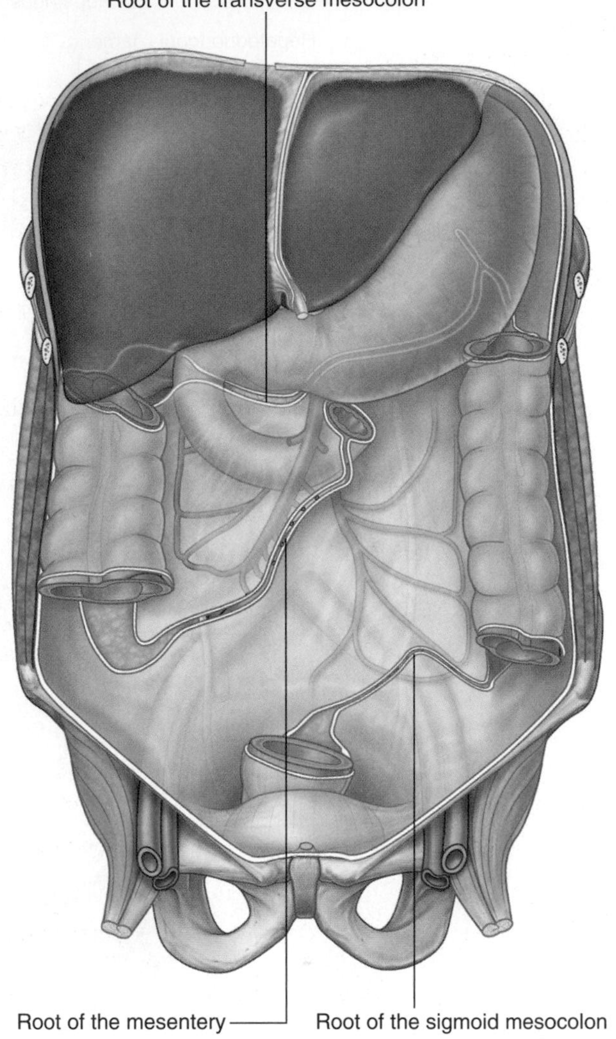


**Fig. 4.61** Peritoneal reflections, forming mesenteries, outlined on the posterior abdominal wall.

## Ligaments

Peritoneal ligaments consist of two layers of peritoneum that connect two organs to each other or attach an organ to the body wall, and may form part of an omentum. They are usually named after the structures being connected. For example, the splenorenal ligament connects the left kidney to the spleen and the gastrophrenic ligament connects the stomach to the diaphragm.

## Organs

### Abdominal esophagus

The abdominal esophagus represents the short distal part of the esophagus located in the abdominal cavity. Emerging through the right crus of the diaphragm, usually at the level of vertebra TX, it passes from the esophageal hiatus to the cardial orifice of the stomach just left of the midline (Fig. 4.62).

Associated with the esophagus, as it enters the abdominal cavity, are the anterior and posterior vagal trunks:

- The **anterior vagal trunk** consists of several smaller trunks whose fibers mostly come from the left vagus nerve; rotation of the gut during development moves these trunks to the anterior surface of the esophagus.
- Similarly, the **posterior vagal trunk** consists of a single trunk whose fibers mostly come from the right vagus nerve, and rotational changes during development move this trunk to the posterior surface of the esophagus.

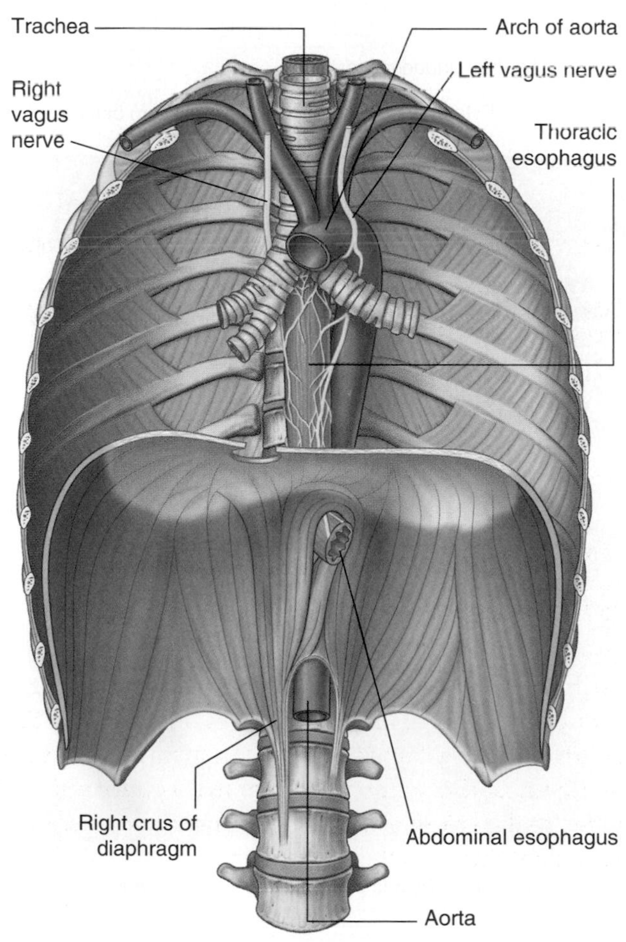


**Fig. 4.62** Abdominal esophagus.

The arterial supply to the abdominal esophagus (Fig. 4.63) includes:

- esophageal branches from the left gastric artery (from the celiac trunk), and
- esophageal branches from the left inferior phrenic artery (from the abdominal aorta).

### Stomach

The stomach is the most dilated part of the gastrointestinal tract and has a J-like shape (Figs. 4.64 and 4.65). Positioned between the abdominal esophagus and the small intestine, the stomach is in the epigastric, umbilical, and left hypochondrium regions of the abdomen.

The stomach is divided into four regions:

- the cardia, which surrounds the opening of the esophagus into the stomach;
- the fundus of the stomach, which is the area above the level of the cardial orifice;
- the body of the stomach, which is the largest region of the stomach; and
- the pyloric part, which is divided into the pyloric antrum and pyloric canal and is the distal end of the stomach.

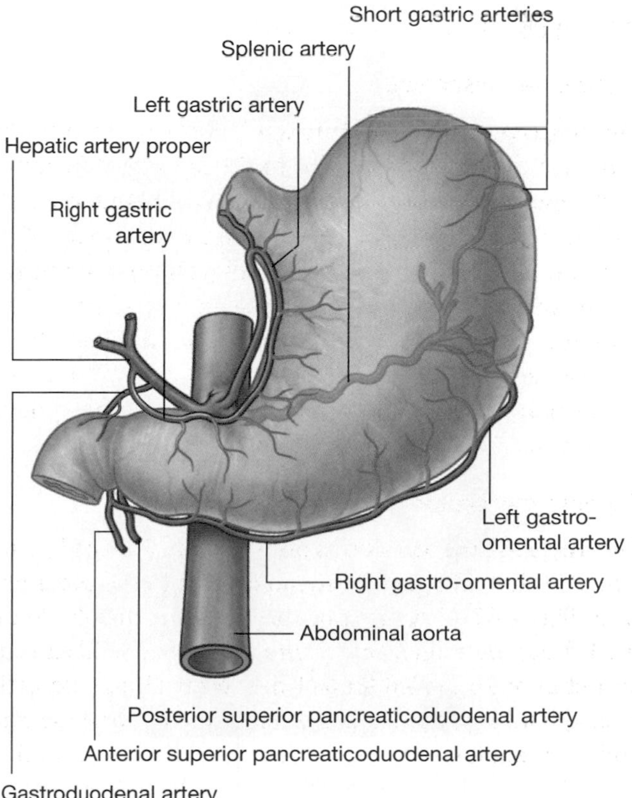


**Fig. 4.63** Arterial supply to the abdominal esophagus and stomach.

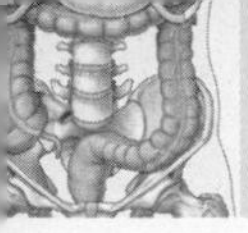

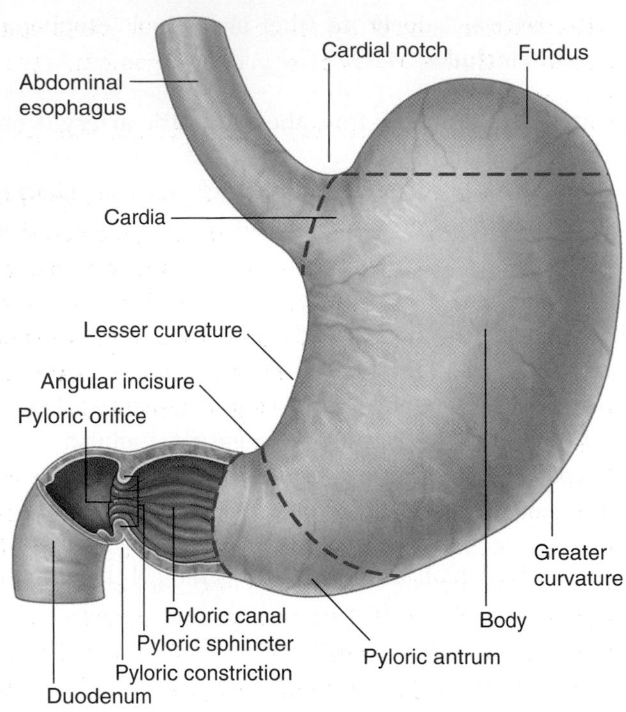


**Fig. 4.64** Stomach.

The most distal portion of the pyloric part of the stomach is the **pylorus** (Fig. 4.64). It is marked on the surface of the organ by the **pyloric constriction** and contains a thickened ring of gastric circular muscle, the **pyloric sphincter**, that surrounds the distal opening of the stomach, the **pyloric orifice** (Figs. 4.64 and 4.65B). The pyloric orifice is just to the right of midline in a plane that passes through the lower border of vertebra LI (the **transpyloric plane**).

Other features of the stomach include:

- the **greater curvature**, which is a point of attachment for the gastrosplenic ligament and the greater omentum;
- the **lesser curvature**, which is a point of attachment for the lesser omentum;
- the **cardial notch**, which is the superior angle created when the esophagus enters the stomach; and
- the **angular incisure**, which is a bend on the lesser curvature.

The arterial supply to the stomach (Fig. 4.63) includes:

- the left gastric artery from the celiac trunk,
- the right gastric artery, often from the hepatic artery proper,

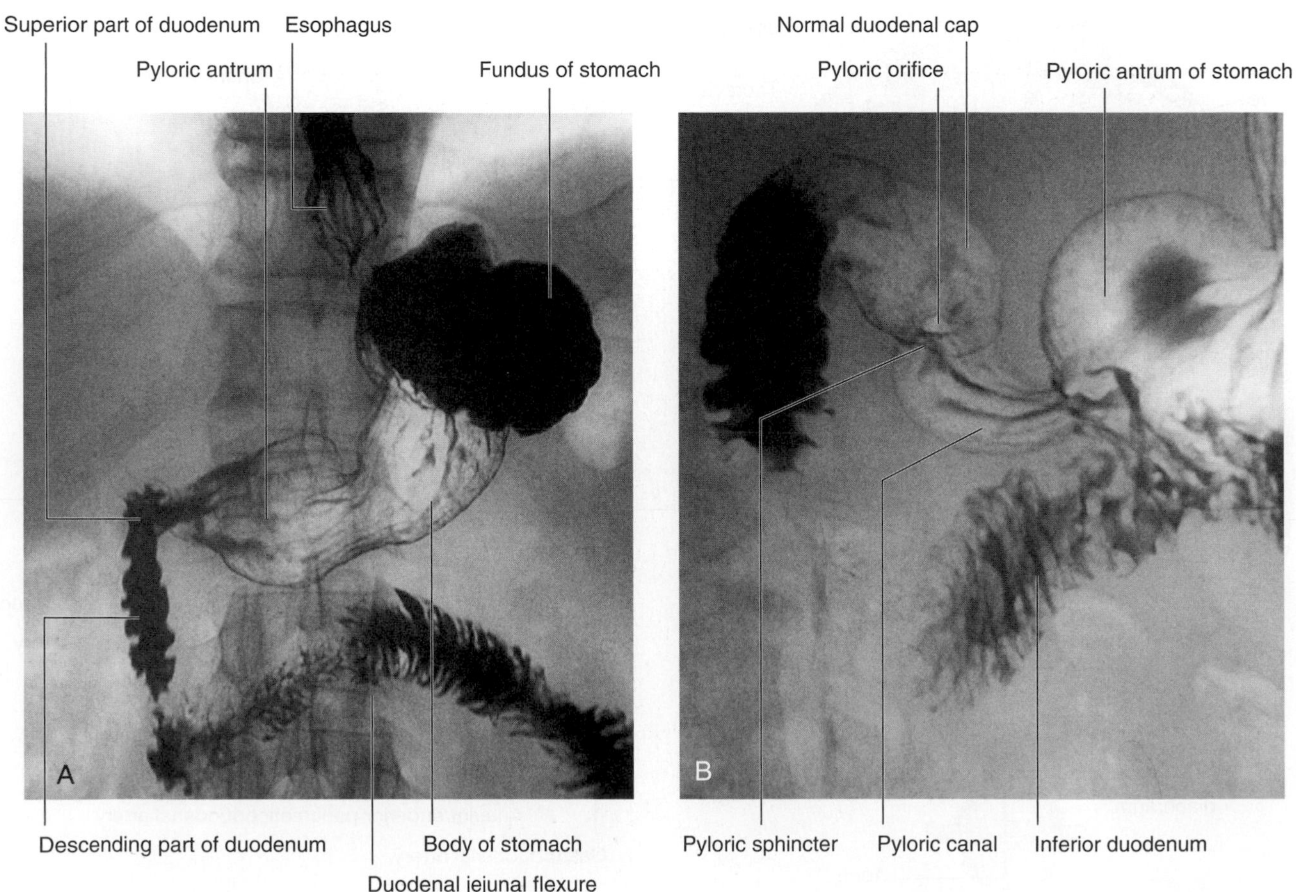


**Fig. 4.65** Radiograph, using barium, showing the stomach and duodenum. **A.** Double-contrast radiograph of the stomach. **B.** Double-contrast radiograph showing the duodenal cap.

- the right gastro-omental artery from the gastroduodenal artery,
- the left gastro-omental artery from the splenic artery, and
- the posterior gastric artery from the splenic artery (variant and not always present).

## Small intestine

The small intestine is the longest part of the gastrointestinal tract and extends from the pyloric orifice of the stomach to the ileocecal fold. This hollow tube, which is approximately 6 to 7 m long with a narrowing diameter from beginning to end, consists of the duodenum, the jejunum, and the ileum.

### Duodenum

The first part of the small intestine is the duodenum. This C-shaped structure, adjacent to the head of the pancreas, is 20 to 25 cm long and is above the level of the umbilicus; its lumen is the widest of the small intestine (Fig. 4.66). It is retroperitoneal except for its beginning, which is connected to the liver by the hepatoduodenal ligament, a part of the lesser omentum.

The duodenum is divided into four parts (Fig. 4.66).

- The **superior part** (first part) extends from the pyloric orifice of the stomach to the neck of the gallbladder, is just to the right of the body of vertebra LI, and passes anteriorly to the bile duct, gastroduodenal artery, portal vein, and inferior vena cava. Clinically, the beginning of this part of the duodenum is referred to as the ampulla or duodenal cap, and most duodenal ulcers occur in this part of the duodenum.
- The **descending part** (second part) of the duodenum is just to the right of midline and extends from the neck of the gallbladder to the lower border of vertebra LIII. Its anterior surface is crossed by the transverse colon, posterior to it is the right kidney, and medial to it is the head of the pancreas. This part of the duodenum contains the **major duodenal papilla**, which is the

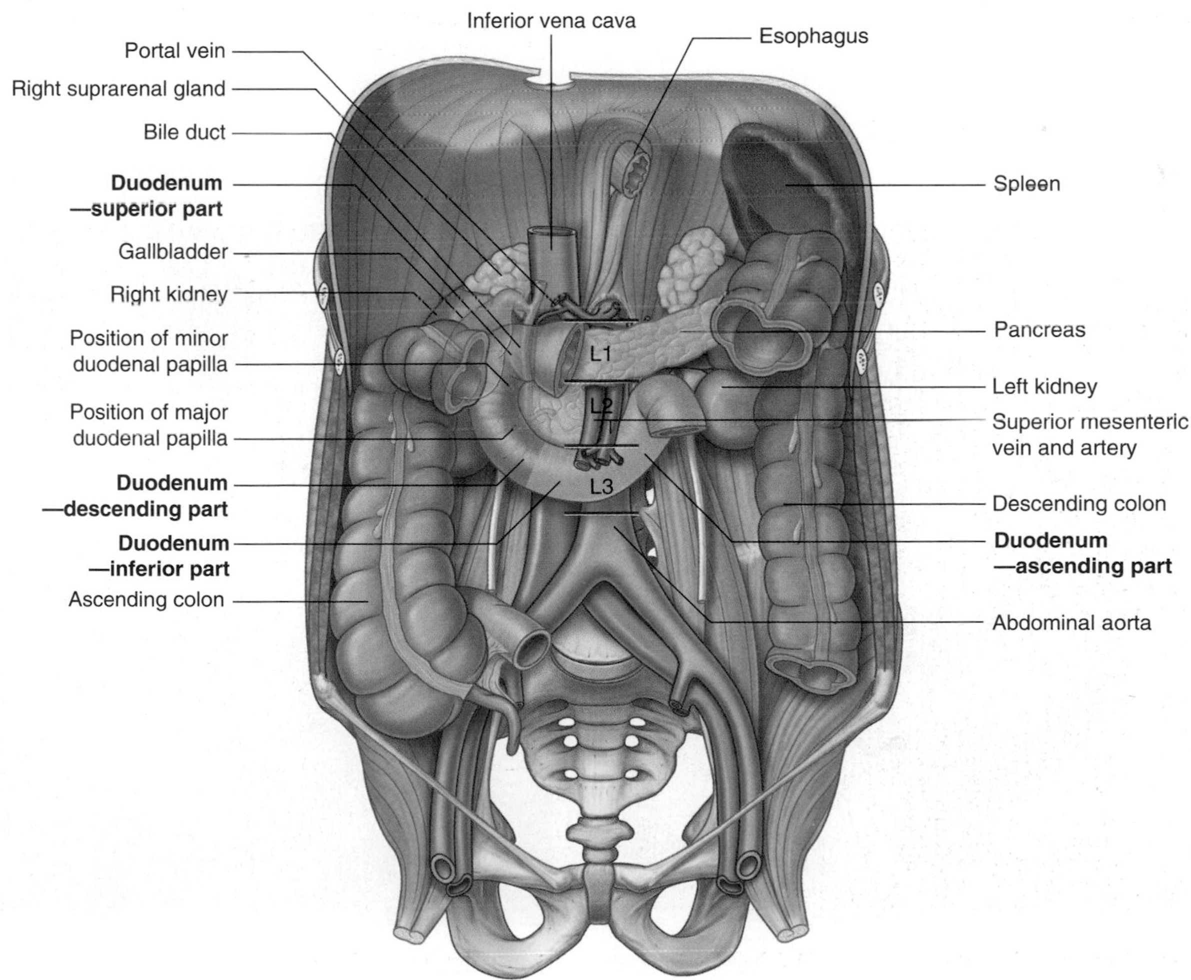


**Fig. 4.66** Duodenum.

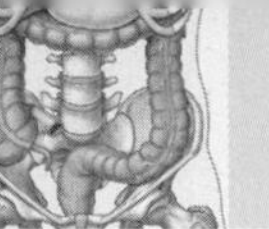

the **minor duodenal papilla**, which is the entrance for the accessory pancreatic duct. The junction of the foregut and the midgut occurs just below the major duodenal papilla.

common entrance for the bile and pancreatic ducts, and

- The **inferior part** (third part) of the duodenum is the longest section, crossing the inferior vena cava, the aorta, and the vertebral column (Figs. 4.65B and 4.66). It is crossed anteriorly by the superior mesenteric artery and vein.
- The **ascending part** (fourth part) of the duodenum passes upward on, or to the left of, the aorta to approximately the upper border of vertebra LII and terminates at the **duodenojejunal flexure**.

This duodenojejunal flexure is surrounded by a fold of peritoneum containing muscle fibers called the **suspensory muscle (ligament) of duodenum (ligament of Treitz).**

The arterial supply to the duodenum (Fig. 4.67) includes:

- branches from the gastroduodenal artery,
- the supraduodenal artery from the gastroduodenal artery,
- duodenal branches from the anterior superior pancreaticoduodenal artery (from the gastroduodenal artery),
- duodenal branches from the posterior superior pancreaticoduodenal artery (from the gastroduodenal artery),
- duodenal branches from the anterior inferior pancreaticoduodenal artery (from the inferior pancreaticoduodenal artery—a branch of the superior mesenteric artery),
- duodenal branches from the posterior inferior pancreaticoduodenal artery (from the inferior pancreaticoduodenal artery—a branch of the superior mesenteric artery), and
- the first jejunal branch from the superior mesenteric artery.

### Jejunum

The jejunum and ileum make up the last two sections of the small intestine (Fig. 4.68). The jejunum represents the proximal two-fifths. It is mostly in the left upper quadrant

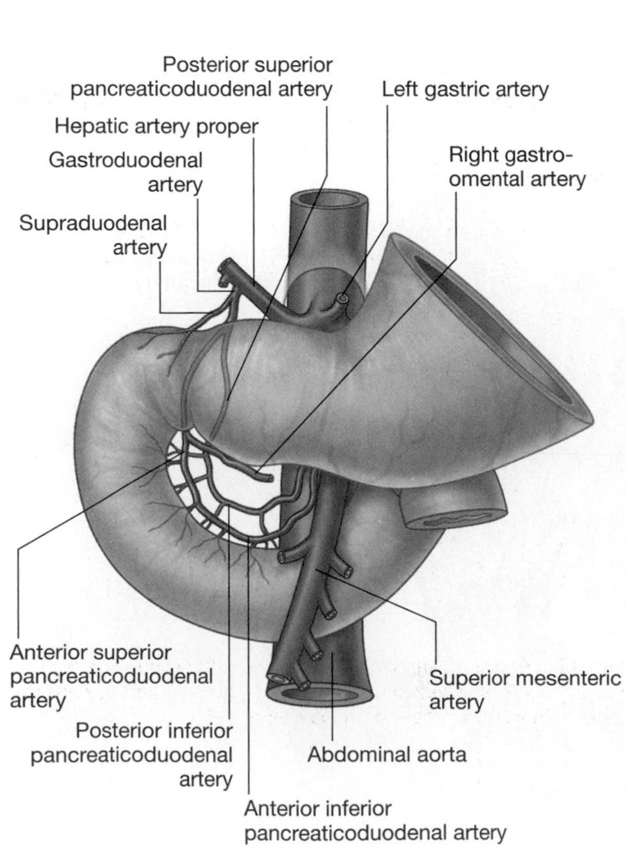


**Fig. 4.67** Arterial supply to the duodenum.

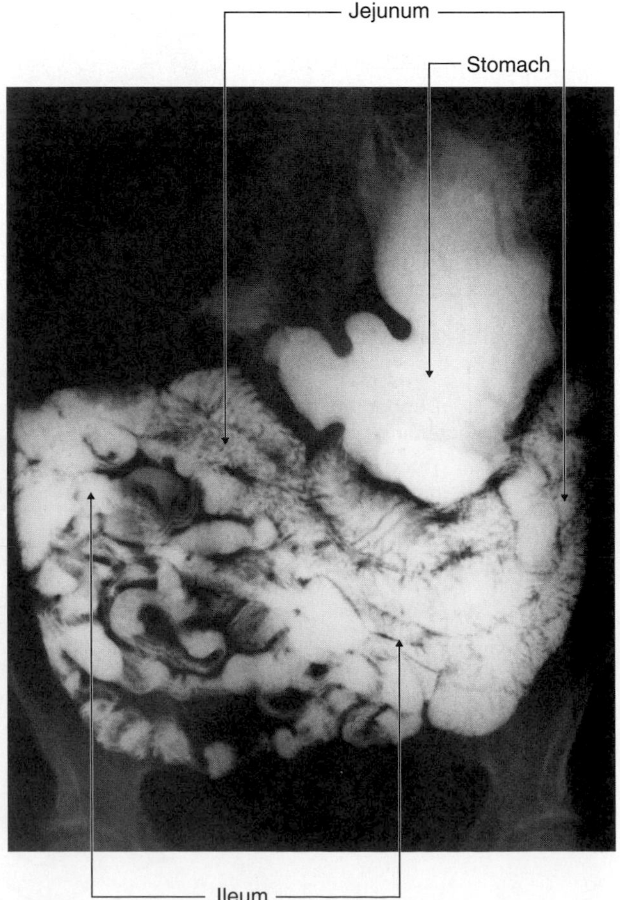


**Fig. 4.68** Radiograph, using barium, showing the jejunum and ileum.

of the abdomen and is larger in diameter and has a thicker wall than the ileum. Additionally, the inner mucosal lining of the jejunum is characterized by numerous prominent folds that circle the lumen (plicae circulares). The less prominent arterial arcades and longer vasa recta (straight arteries) compared to those of the ileum are a unique characteristic of the jejunum (Fig. 4.69).

The arterial supply to the jejunum includes jejunal arteries from the superior mesenteric artery.

### Ileum

The ileum makes up the distal three-fifths of the small intestine and is mostly in the right lower quadrant. Compared to the jejunum, the ileum has thinner walls, fewer and less prominent mucosal folds (plicae circulares), shorter vasa recta, more mesenteric fat, and more arterial arcades (Fig. 4.69).

The ileum opens into the large intestine, where the cecum and ascending colon join together. Two flaps projecting into the lumen of the large intestine (the **ileocecal fold**) surround the opening (Fig. 4.70). The flaps of the

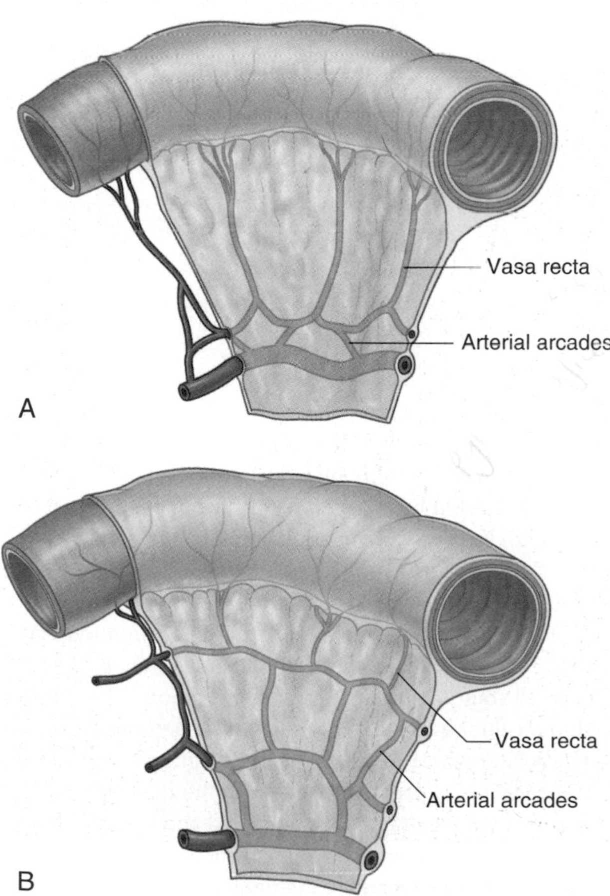


**Fig. 4.69** Differences in the arterial supply to the small intestine. **A.** Jejunum. **B.** Ileum.

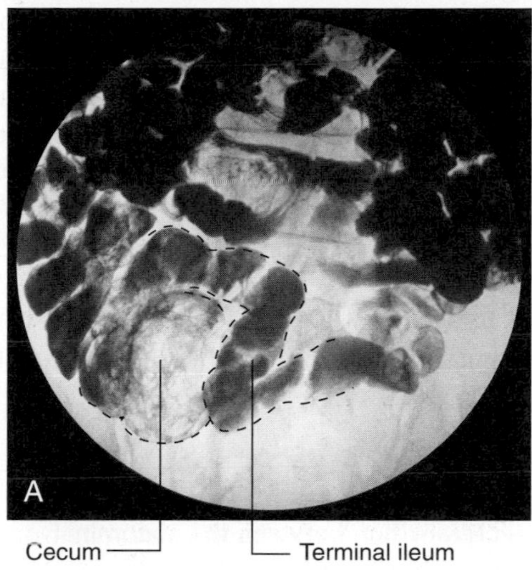


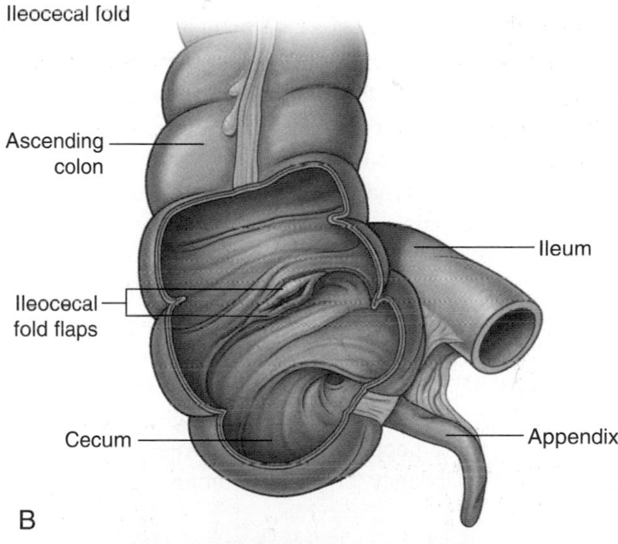


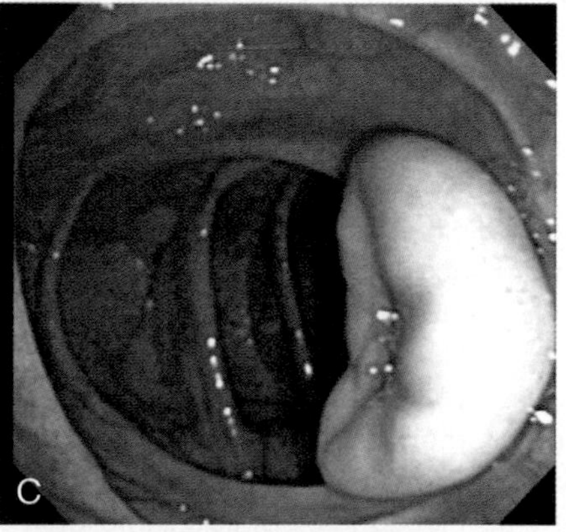


**Fig. 4.70** Ileocecal junction. **A.** Radiograph showing ileocecal junction. **B.** Illustration showing ileocecal junction and the ileocecal fold. **C.** Endoscopic image of the ileocecal fold.

ileocecal fold come together at their end, forming ridges. Musculature from the ileum continues into each flap, forming a sphincter. Possible functions of the ileocecal fold include preventing reflux from the cecum to the ileum, and regulating the passage of contents from the ileum to the cecum.

The arterial supply to the ileum (Fig. 4.71) includes:

- ileal arteries from the superior mesenteric artery, and
- an ileal branch from the ileocolic artery (from the superior mesenteric artery).

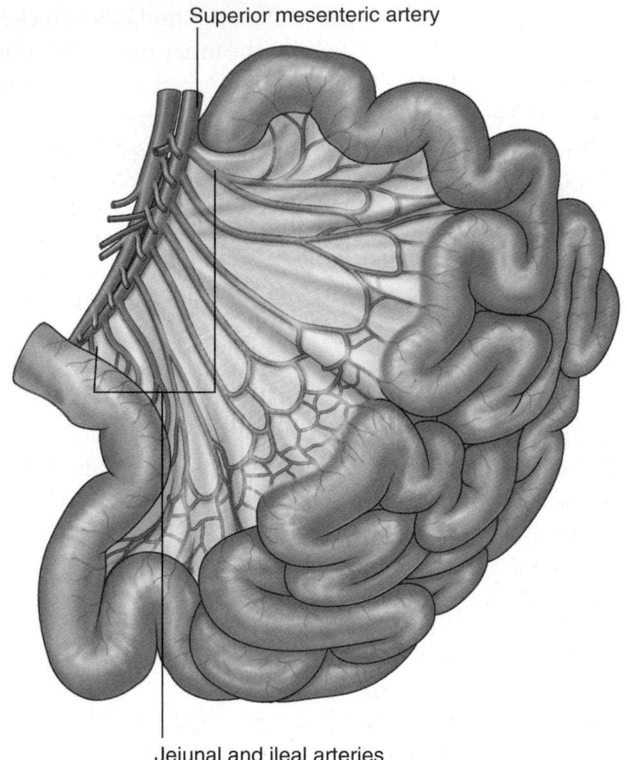


**Fig. 4.71** Arterial supply to the ileum.

## In the clinic

### Epithelial transition between the abdominal esophagus and stomach

The gastroesophageal junction is demarcated by a transition from one epithelial type (nonkeratinized stratified squamous epithelium) to another epithelial type (columnar epithelium). In some people, the histological junction does not lie at the anatomical gastroesophageal junction but occurs more proximally in the lower third of the esophagus. This may predispose these people to esophageal ulceration and is also associated with an increased risk of adenocarcinoma. In certain conditions, like gastroesophageal reflux, the stratified squamous epithelium in the esophagus can undergo metaplasia and the epithelium in the lower esophagus is replaced by columnar epithelium, a condition called Barrett's esophagus. The presence of Barrett's esophagus predisposes these people to the development of esophageal malignancy (adenocarcinoma).

## In the clinic

### Duodenal ulceration

Duodenal ulcers usually occur in the superior part of the duodenum and are much less common than they were 50 years ago. At first, there was no treatment and patients died from hemorrhage or peritonitis. As surgical techniques developed, patients with duodenal ulcers were subjected to extensive upper gastrointestinal surgery to prevent ulcer recurrence and for some patients the treatment was dangerous. As knowledge and understanding of the mechanisms for acid secretion in the stomach increased, drugs were developed to block acid stimulation and secretion indirectly (histamine $H_2$-receptor antagonists) and these have significantly reduced the morbidity and mortality rates of this disease. Pharmacological therapy can now directly inhibit the cells of the stomach that produce acid with, for example, proton pump inhibitors. Patients are also screened for the bacteria *Helicobacter pylori,* which when eradicated (by antibiotic treatment) significantly reduces duodenal ulcer formation.

Anatomically, duodenal ulcers tend to occur either anteriorly or posteriorly.

Posterior duodenal ulcers erode either directly onto the gastroduodenal artery or, more commonly, onto the posterior superior pancreaticoduodenal artery, which can produce torrential hemorrhage, which may be fatal in some patients. Treatment may involve extensive upper abdominal surgery with ligation of the vessels or by endovascular means whereby the radiologist may place a very fine catheter retrogradely from the femoral artery into the celiac artery. The common hepatic artery and the gastroduodenal artery are cannulated and the bleeding area may be blocked using small coils, which stem the flow of blood.

Anterior duodenal ulcers erode into the peritoneal cavity, causing peritonitis. This intense inflammatory reaction and the local ileus promote adhesion of the greater omentum, which attempts to seal off the perforation. The stomach and duodenum usually contain considerable amounts of gas, which enters the peritoneal cavity and can be observed on a chest radiograph of an erect patient as subdiaphragmatic gas. In most instances, treatment for the ulcer perforation is surgical.

## In the clinic

### Examination of the upper and lower gastrointestinal tract

It is often necessary to examine the esophagus, stomach, duodenum, proximal jejunum, and colon for disease. After taking an appropriate history and examining the patient, most physicians arrange a series of simple blood tests to look for bleeding, inflammation, and tumors. The next steps in the investigation assess the three components of any loop of bowel, namely, the lumen, the wall, and masses extrinsic to the bowel, which may compress or erode into it.

### Examination of the bowel lumen

Barium sulfate solutions may be swallowed by the patient and can be visualized using an X-ray fluoroscopy unit. The lumen can be examined for masses (e.g., polyps and tumors) and peristaltic waves can be assessed. Patients may also be given carbon dioxide–releasing granules to fill the stomach so that the barium thinly coats the mucosa, resulting in images displaying fine mucosal detail. These tests are relatively simple and can be used to image the esophagus, stomach, duodenum, and small bowel. For imaging the large bowel, a barium enema can be used to introduce barium sulfate into the colon. Colonoscopy and CT colonography are also used.

### Examination of the bowel wall and extrinsic masses

**Endoscopy** is a minimally invasive diagnostic medical procedure that can be used to assess the interior surfaces of an organ by inserting a tube into the body. The instrument is typically made of a flexible plastic material through which a light source and eyepiece are attached at one end. The images are then projected to a monitor. Some systems allow passage of small instruments through the main bore of the endoscope to obtain biopsies and to also undertake small procedures (e.g., the removal of polyps).

In gastrointestinal and abdominal medicine an endoscope is used to assess the esophagus, stomach, duodenum, and proximal small bowel (Figs. 4.72 to 4.75). The tube is swallowed by the patient under light sedation and is extremely well tolerated.

Assessment of the colon (colonoscopy) is performed by passage of the long flexible tube through the anus and into the rectum. The endoscope is then advanced into the colon to the cecum and sometimes to the terminal ileum. The patient undergoes bowel preparation before the examination to allow good visualization of the entire large bowel. Specially designed solutions are taken orally to help clear the bowel of fecal material. Air, water, and suction may be used during the examination to improve visualization. Biopsies, polyp removal, cauterization of bleeding, and stent placement can also be performed using additional instruments that can be passed through special openings in the colonoscope.

Cross sectional imaging using computed tomography or magnetic resonance is another way to assess the bowel lumen and wall. Magnetic resonance is particularly useful in assessment of the small bowel because it allows dynamic assessment of bowel distention and motility and provides good visualization of segmental or continuous bowel wall thickening and mural or mucosal ulcerations and also can demonstrate increased vascularity of the small bowel mesentery (Fig. 4.76). It is usually performed in patients with inflammatory bowel diseases, such as Crohn's disease.

### CT colonography

CT colonography (also called virtual colonoscopy or CT pneumocolon) is an alternative way to visualize and assess the colon for abnormal lesions such as polyps or strictures with the use spiral CT to produce high-resolution 3D views of the large bowel. It is less invasive than traditional colonoscopy, but to achieve good-quality images the patient needs to take bowel preparations to ensure bowel cleansing, and the colon needs to be insufflated with $CO_2$. If a tumor is present (Fig. 4.77), both CT and MRI are used to assess regional disease (MRI), abnormal lymph nodes (MRI, CT), and distant metastases (CT).

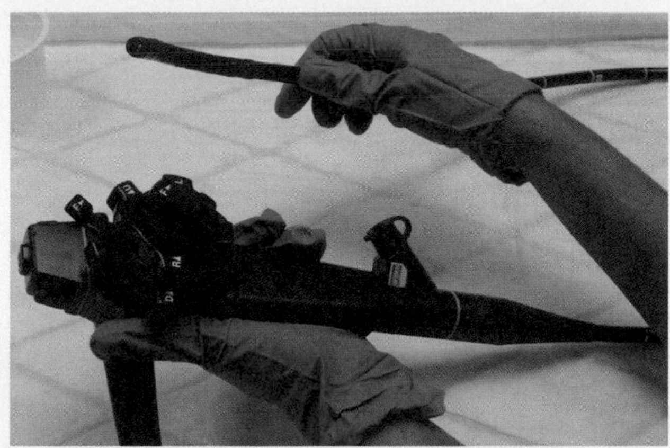

**Fig. 4.72** The endoscope is a flexible plastic tube that can be controlled from the proximal end. Through a side portal various devices can be inserted, which run through the endoscope and can be used to obtain biopsies and to perform minor endoluminal surgical procedures (e.g., excision of polyps).

*(continues)*

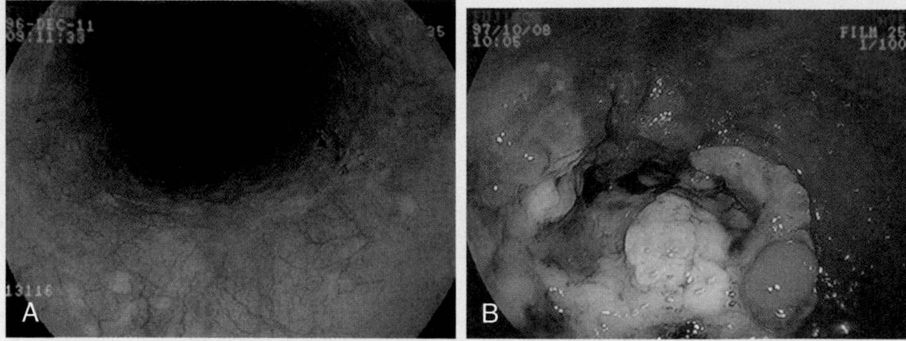


**Fig. 4.73** Endoscopic images of the gastroesophageal junction. **A.** Normal. **B.** Esophageal cancer at esophageal junction.

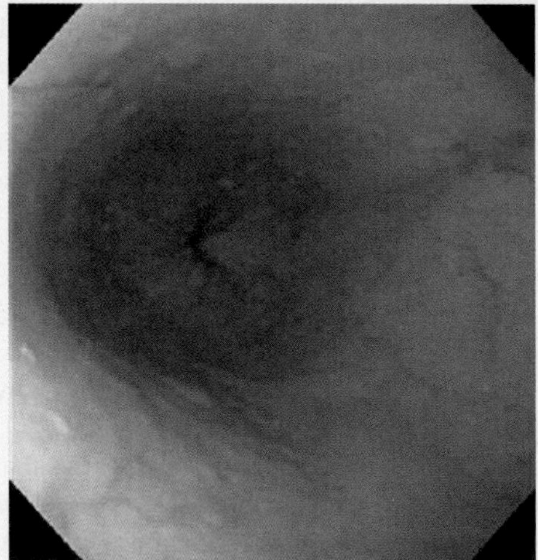

**Fig. 4.74** Endoscopic image of the pyloric antrum of the stomach looking toward the pylorus.

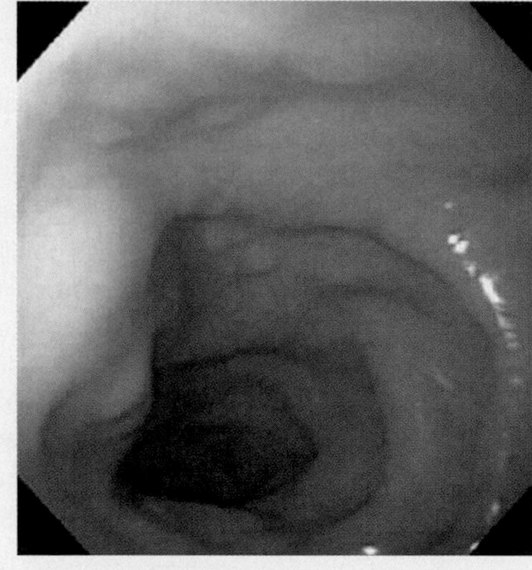

**Fig. 4.75** Endoscopic image showing normal appearance of the second part of the duodenum.

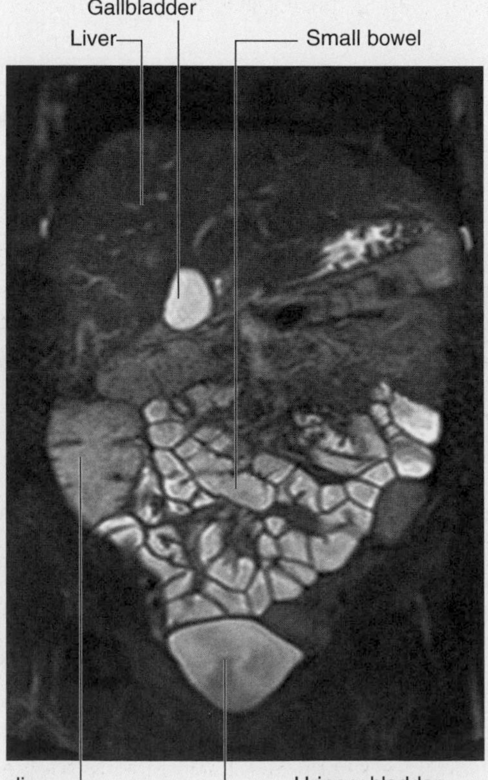


**Fig. 4.76** Small bowel visualization using MRI in coronal plane.

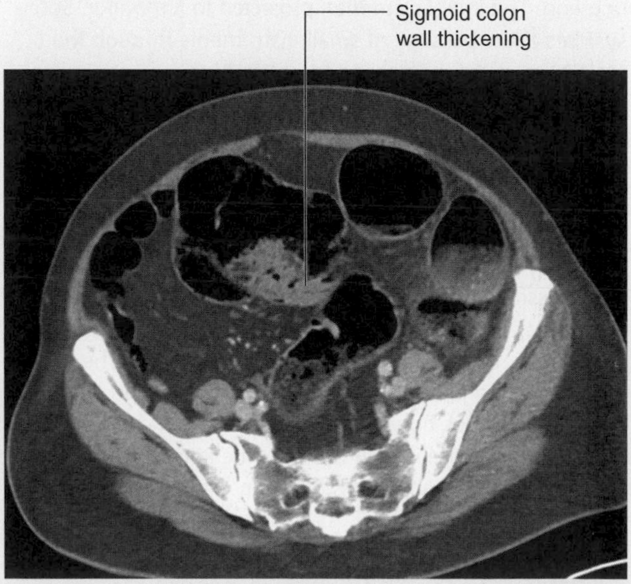

Sigmoid colon wall thickening

**Fig. 4.77** Axial CT shows sigmoid colon wall thickening caused by tumor.

## In the clinic

### Meckel's diverticulum

A Meckel's diverticulum (Fig. 4.78) is the remnant of the proximal part of the yolk stalk (vitelline duct) that extends into the umbilical cord in the embryo and lies on the antimesenteric border of the ileum. It appears as a blind-ended tubular outgrowth of bowel. Although it is an uncommon finding (occurring in approximately 2% of the population), it is always important to consider the diagnosis of Meckel's diverticulum because it does produce symptoms in a small number of patients. It may contain gastric mucosa and therefore lead to ulceration and hemorrhage. Other typical complications include intussusception, diverticulitis, and obstruction.

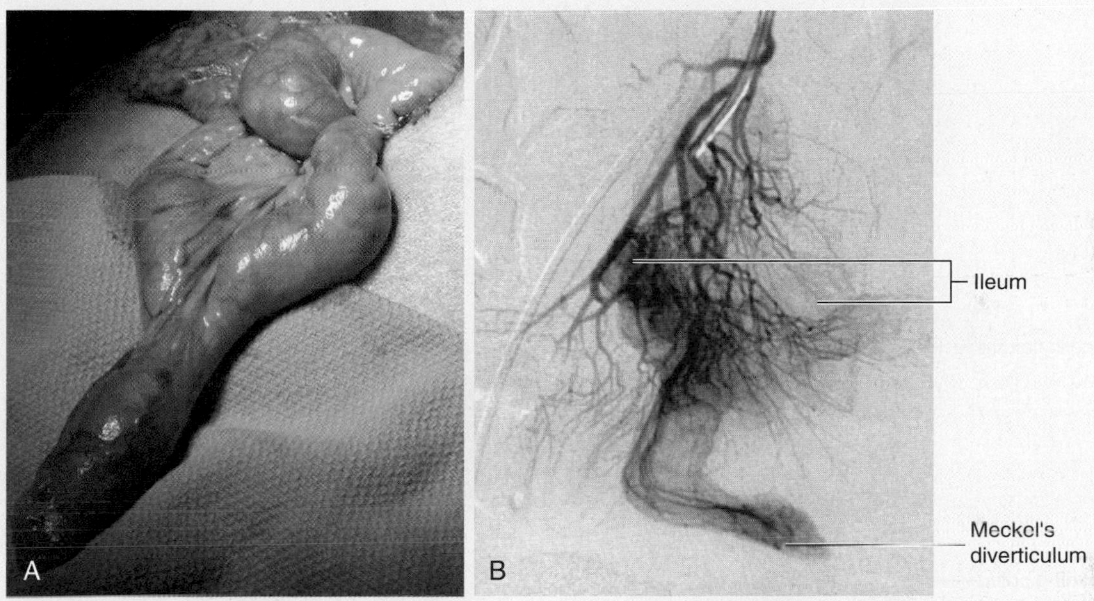


**Fig. 4.78** Vasculature associated with a Meckel's diverticulum. **A.** Surgical image of Meckel's diverticulum. **B.** Digital subtraction angiography.

## In the clinic

### Computed tomography (CT) scanning and magnetic resonance imaging (MRI)

These imaging techniques can provide important information about the wall of the bowel that may not be obtained from barium or endoscopic studies. Thickening of the wall may indicate inflammatory change or tumor and is always regarded with suspicion. If a tumor is demonstrated, the locoregional spread can be assessed, along with lymphadenopathy and metastatic spread.

### Advanced imaging methods

Endoscopic ultrasound (EUS) uses a small ultrasound device placed on the end of the endoscope to assess the upper gastrointestinal tract. It can produce extremely high-powered views of the mucosa and submucosa and therefore show whether a tumor is resectable. It also provides guidance to the clinician when taking a biopsy.

## In the clinic

### Carcinoma of the stomach

Carcinoma of the stomach is a common gastrointestinal malignancy. Chronic gastric inflammation (gastritis), pernicious anemia, and polyps predispose to the development of this aggressive cancer, which is usually not diagnosed until late in the course of the disease. Symptoms include vague epigastric pain, early fullness with eating, bleeding leading to chronic anemia, and obstruction.

The diagnosis may be made using barium and conventional radiology or endoscopy, which allows a biopsy to be obtained at the same time. Ultrasound scanning is used to check the liver for metastatic spread, and, if negative, computed tomography is carried out to assess for surgical resectability. If carcinoma of the stomach is diagnosed early, a curative surgical resection is possible. However, because most patients do not seek treatment until late in the disease, the overall 5-year survival rate is between 5% and 20%, with a mean survival time of between 5 and 8 months.

### Large intestine

The large intestine extends from the distal end of the ileum to the anus, a distance of approximately 1.5 m in adults. It absorbs fluids and salts from the gut contents, thus forming feces, and consists of the cecum, appendix, colon, rectum, and anal canal (Figs. 4.79 and 4.80).

Beginning in the right groin as the cecum, with its associated appendix, the large intestine continues upward as the **ascending colon** through the right flank and into the right hypochondrium (Fig. 4.81). Just below the liver, it bends to the left, forming the **right colic flexure** (**hepatic flexure**), and crosses the abdomen as the **transverse colon** to the left hypochondrium. At this position, just below the spleen, the large intestine bends downward, forming the **left colic flexure** (**splenic flexure**), and continues as the **descending colon** through the left flank and into the left groin.

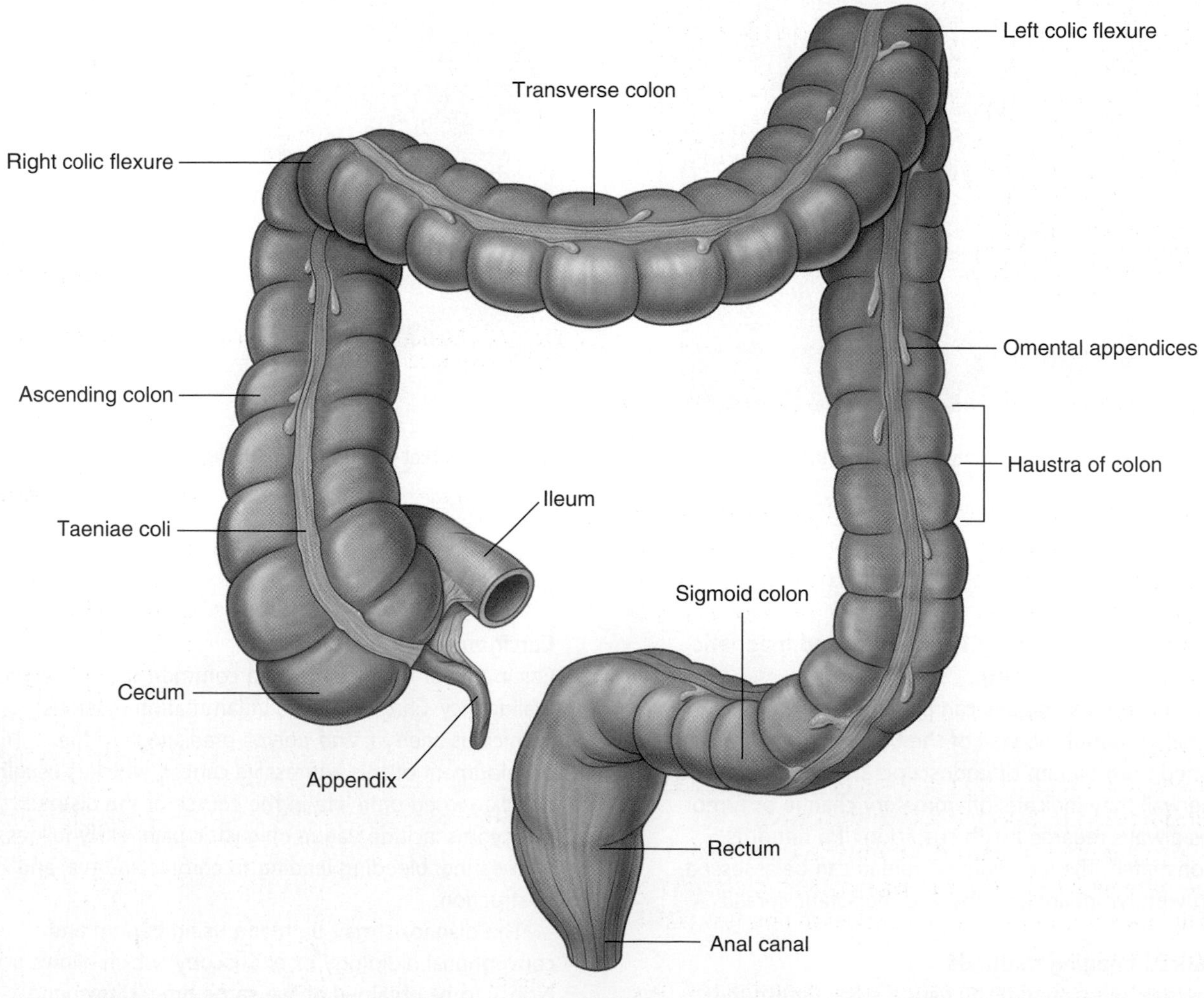

**Fig. 4.79** Large intestine.

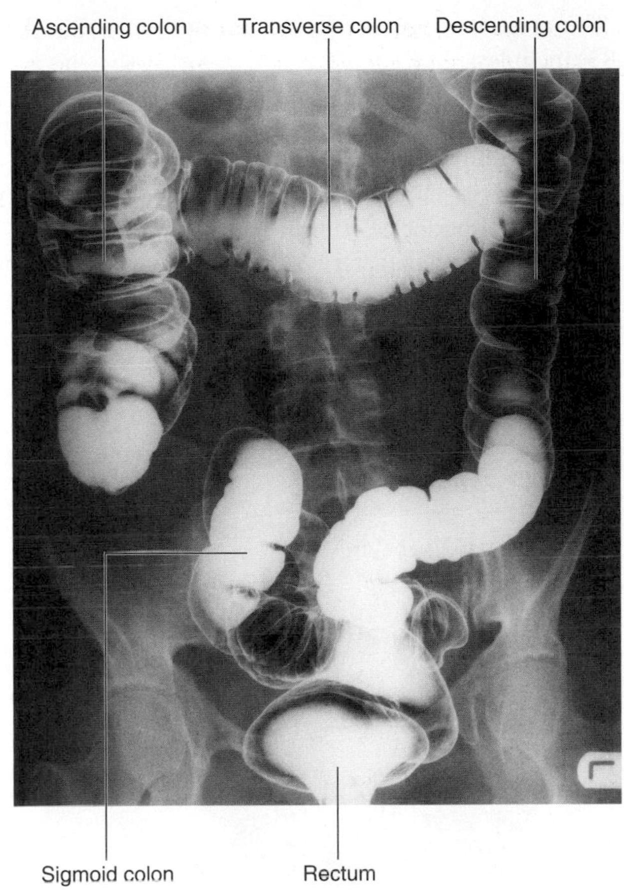


**Fig. 4.80** Radiograph, using barium, showing the large intestine.

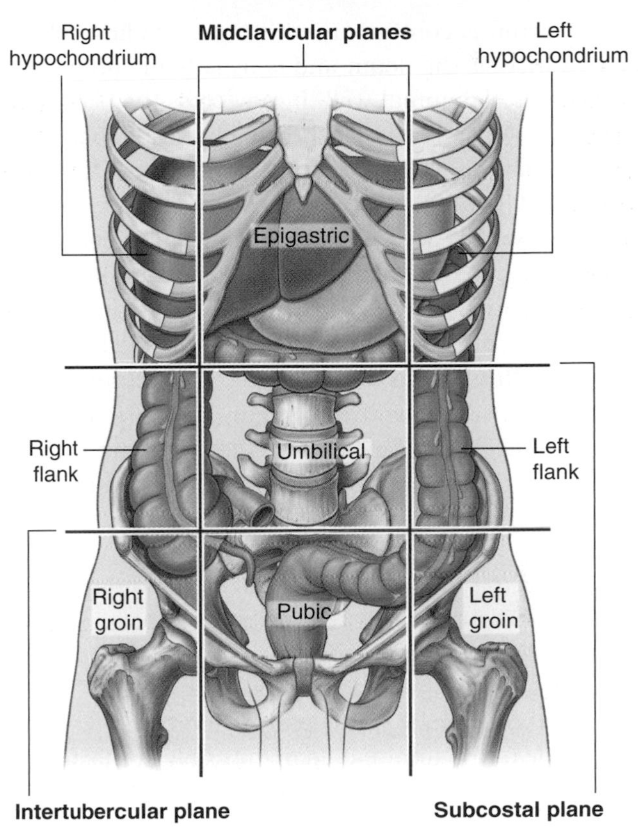


**Fig. 4.81** Position of the large intestine in the nine-region organizational pattern.

It enters the upper part of the pelvic cavity as the sigmoid colon, continues on the posterior wall of the pelvic cavity as the rectum, and terminates as the anal canal.

The general characteristics of most of the large intestine (Fig. 4.79) are:

- its large internal diameter compared to that of the small intestine;
- peritoneal-covered accumulations of fat (the **omental appendices**) are associated with the colon;
- the segregation of longitudinal muscle in its walls into three narrow bands (the **taeniae coli**), which are primarily observed in the cecum and colon and less visible in the rectum; and
- the sacculations of the colon (the **haustra of the colon**).

### Cecum and appendix

The **cecum** is the first part of the large intestine (Fig. 4.82). It is inferior to the ileocecal opening and in the right iliac fossa. It is generally considered to be an intraperitoneal structure because of its mobility, even though it normally is not suspended in the peritoneal cavity by a mesentery.

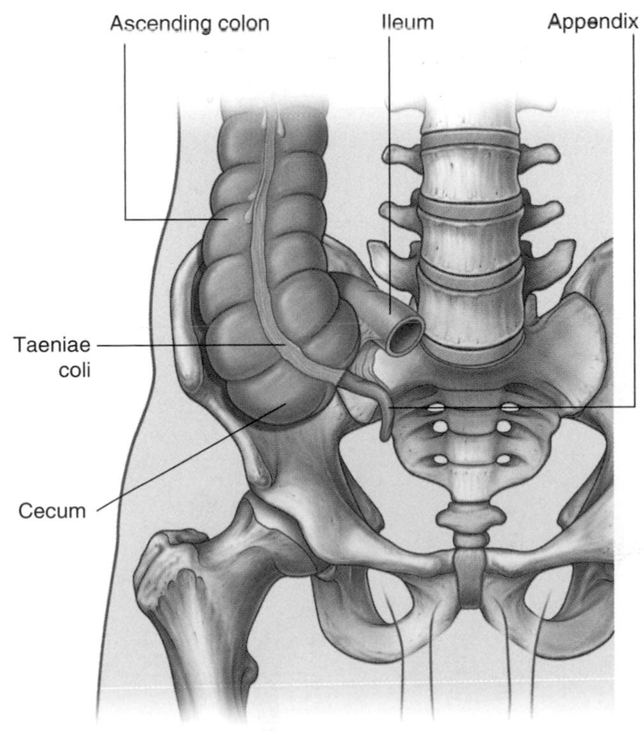


**Fig. 4.82** Cecum and appendix.

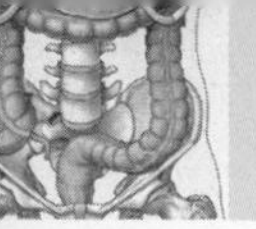



The cecum is continuous with the ascending colon at the entrance of the ileum and is usually in contact with the anterior abdominal wall. It may cross the pelvic brim to lie in the true pelvis. The appendix is attached to the posteromedial wall of the cecum, just inferior to the end of the ileum (Fig. 4.82).

The **appendix** is a narrow, hollow, blind-ended tube connected to the cecum. It has large aggregations of lymphoid tissue in its walls and is suspended from the terminal ileum by the **mesoappendix** (Fig. 4.83), which contains the **appendicular vessels**. Its point of attachment to the cecum is consistent with the highly visible free taeniae leading directly to the base of the appendix, but the location of the rest of the appendix varies considerably (Fig. 4.84). It may be:

- posterior to the cecum or the lower ascending colon, or both, in a retrocecal or retrocolic position;
- suspended over the pelvic brim in a pelvic or descending position;
- below the cecum in a subcecal location; or
- anterior to the terminal ileum, possibly contacting the body wall, in a pre-ileal position or posterior to the terminal ileum in a postileal position.

The surface projection of the base of the appendix is at the junction of the lateral and middle one-third of a line from the anterior superior iliac spine to the umbilicus (McBurney's point). People with appendicular problems may describe pain near this location.

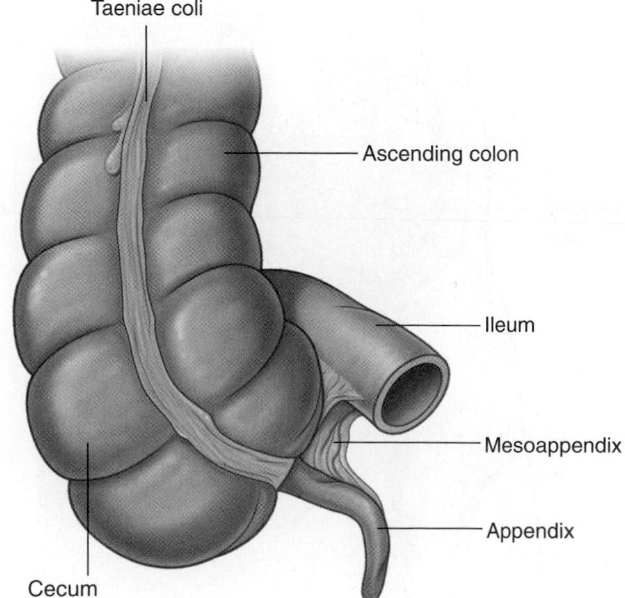


**Fig. 4.83** Mesoappendix and appendicular vessels.

The arterial supply to the cecum and appendix (Fig. 4.85) includes:

- the anterior cecal artery from the ileocolic artery (from the superior mesenteric artery),
- the posterior cecal artery from the ileocolic artery (from the superior mesenteric artery), and
- the appendicular artery from the ileocolic artery (from the superior mesenteric artery).

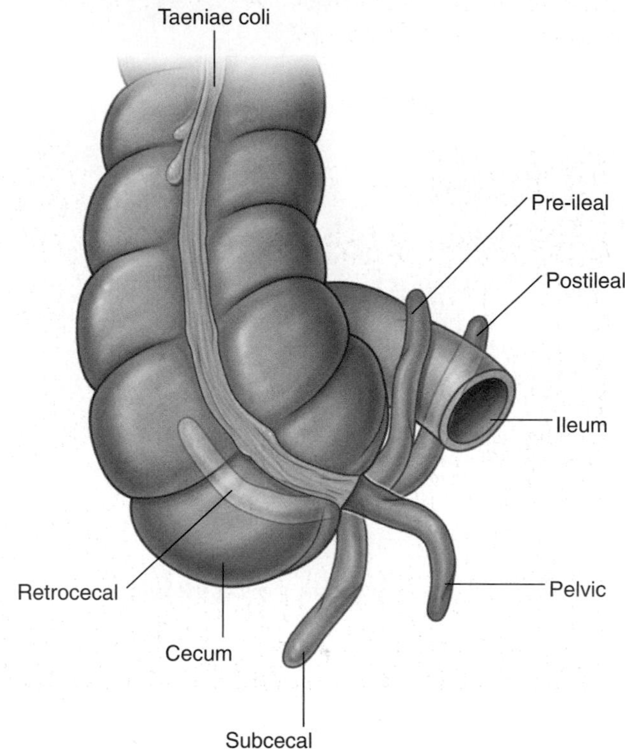


**Fig. 4.84** Positions of the appendix.

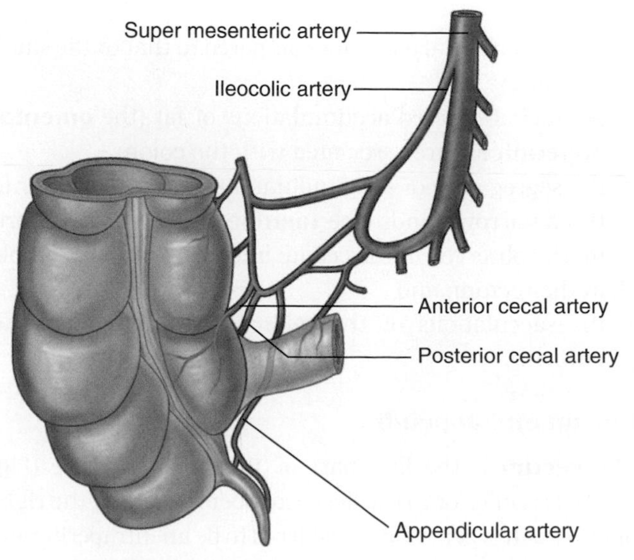


**Fig. 4.85** Arterial supply to the cecum and appendix.

## In the clinic

### Appendicitis

Acute appendicitis is an abdominal emergency. It usually occurs when the appendix is obstructed by either a fecalith or enlargement of the lymphoid nodules. Within the obstructed appendix, bacteria proliferate and invade the appendix wall, which becomes damaged by pressure necrosis. In some instances, this may resolve spontaneously; in other cases, inflammatory change (Figs. 4.86 and 4.87) continues and perforation ensues, which may lead to localized or generalized peritonitis.

Most patients with acute appendicitis have localized tenderness in the right groin. Initially, the pain begins as a central, periumbilical, colicky type of pain, which tends to come and go. After 6 to 10 hours, the pain tends to localize in the right iliac fossa and becomes constant. Patients may develop a fever, nausea, and vomiting. The etiology of the pain for appendicitis is described in Case 1 of Chapter 1 on p. 48.

The treatment for appendicitis is appendectomy.

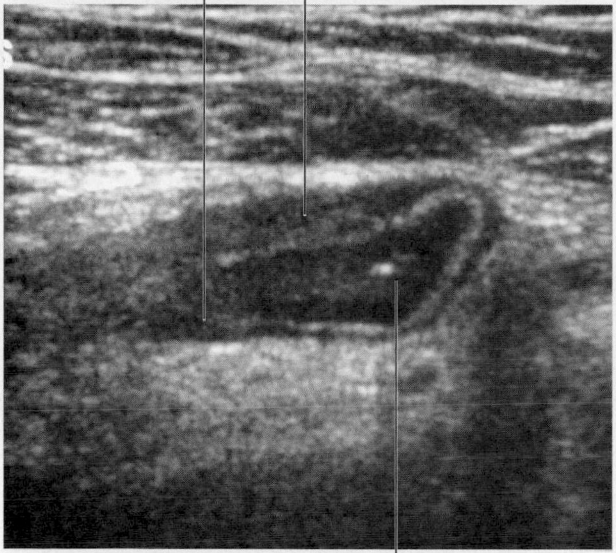

Thickened wall

Gas in lumen

**Fig. 4.86** Inflamed appendix. Ultrasound scan.

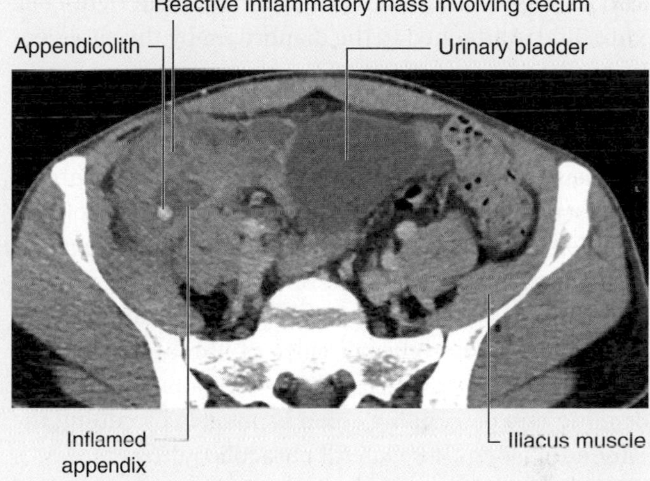


**Fig. 4.87** Axial CT shows inflamed appendix.

### Colon

The colon extends superiorly from the cecum and consists of the ascending, transverse, descending, and sigmoid colon (Fig. 4.88). Its ascending and descending segments are (secondarily) retroperitoneal and its transverse and sigmoid segments are intraperitoneal.

At the junction of the ascending and transverse colon is the right colic flexure, which is just inferior to the right lobe of the liver (Fig. 4.89). A similar, but more acute bend (the left colic flexure) occurs at the junction of the transverse and descending colon. This bend is just inferior to the spleen, is higher and more posterior than the right colic flexure, and is attached to the diaphragm by the phrenicocolic ligament.

Immediately lateral to the ascending and descending colon are the **right** and **left paracolic gutters** (Fig. 4.88). These depressions are formed between the lateral margins of the ascending and descending colon and the posterolateral abdominal wall and are gutters through which material can pass from one region of the peritoneal cavity to another. Because major vessels and lymphatics are on the medial or posteromedial sides of the ascending and descending colon, a relatively blood-free mobilization of the ascending and descending colon is possible by cutting the peritoneum along these lateral paracolic gutters.

The final segment of the colon (the sigmoid colon) begins above the pelvic inlet and extends to the level of vertebra SIII, where it is continuous with the rectum (Fig. 4.88). This S-shaped structure is quite mobile except at its beginning, where it continues from the descending colon, and at its end, where it continues as the rectum. Between these points, it is suspended by the sigmoid mesocolon.

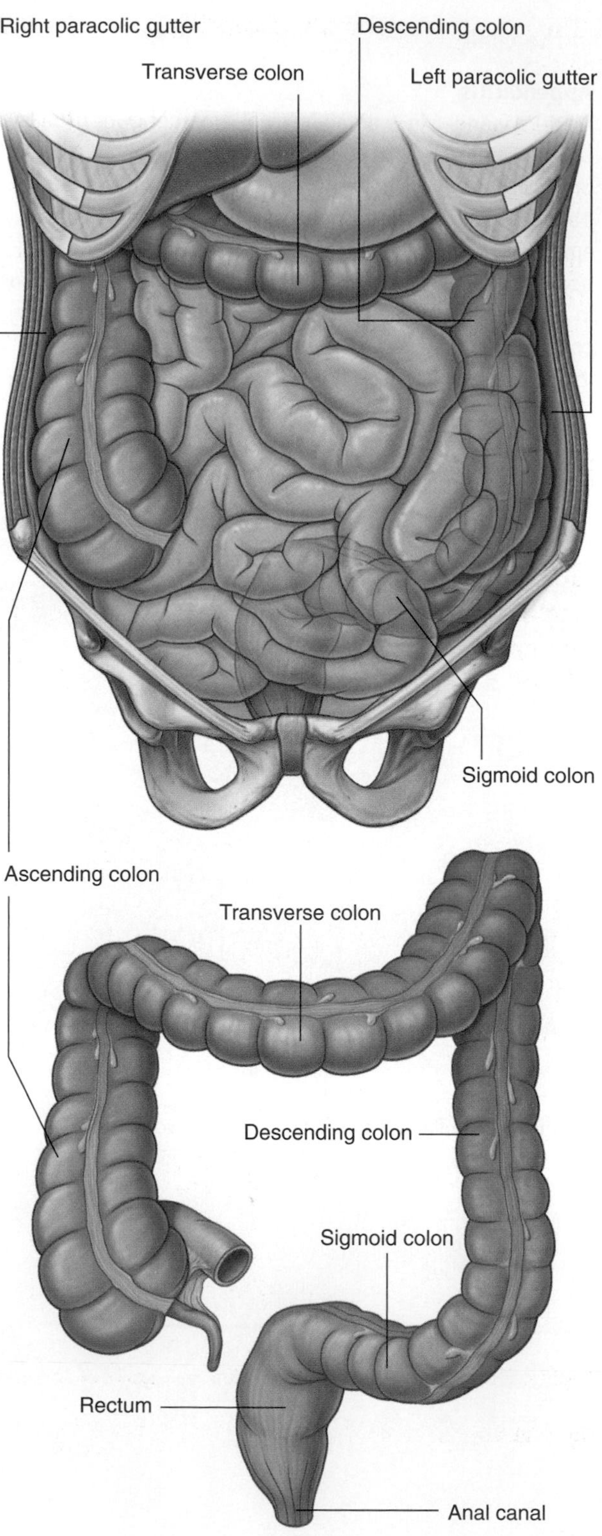


**Fig. 4.88** Colon.

The arterial supply to the ascending colon (Fig. 4.90) includes:

- the colic branch from the ileocolic artery (from the superior mesenteric artery),
- the anterior cecal artery from the ileocolic artery (from the superior mesenteric artery),
- the posterior cecal artery from the ileocolic artery (from the superior mesenteric artery), and
- the right colic artery from the superior mesenteric artery.

The arterial supply to the transverse colon (Fig. 4.90) includes:

- the right colic artery from the superior mesenteric artery,
- the middle colic artery from the superior mesenteric artery, and
- the left colic artery from the inferior mesenteric artery.

The arterial supply to the descending colon (Fig. 4.90) includes the left colic artery from the inferior mesenteric artery.

The arterial supply to the sigmoid colon (Fig. 4.90) includes sigmoidal arteries from the inferior mesenteric artery.

Anastomotic connections between arteries supplying the colon can result in a **marginal artery** that courses along the ascending, transverse, and descending parts of the large bowel (Fig. 4.90).

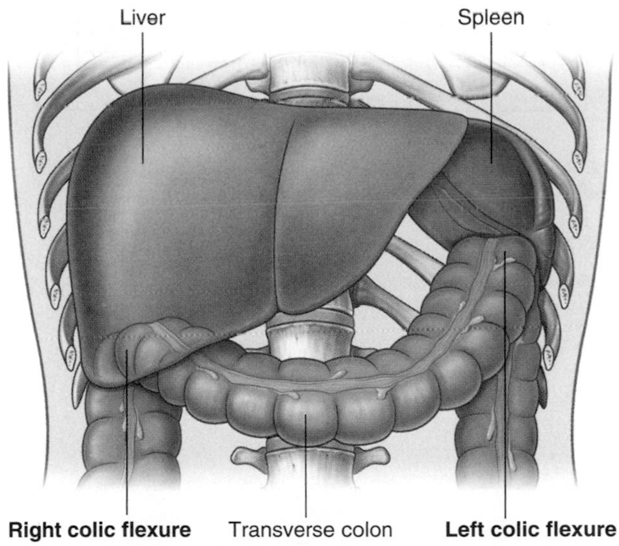


**Fig. 4.89** Right and left colic flexures.

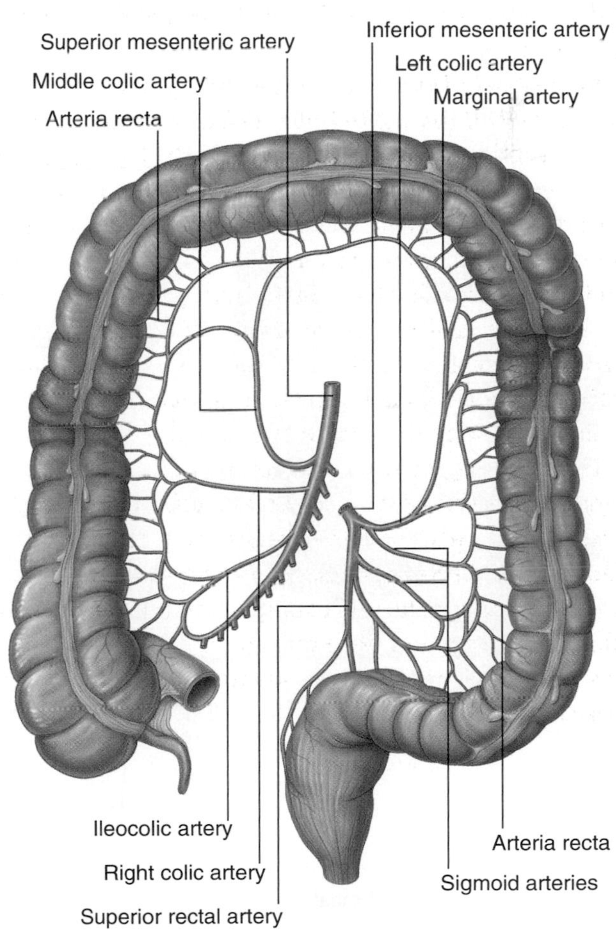


**Fig. 4.90** Arterial supply to the colon.

## Rectum and anal canal

Extending from the sigmoid colon is the rectum (Fig. 4.91). The rectosigmoid junction is usually described as being at the level of vertebra SIII or at the end of the sigmoid mesocolon because the rectum is a retroperitoneal structure.

The anal canal is the continuation of the large intestine inferior to the rectum.

The arterial supply to the rectum and anal canal (Fig. 4.92) includes:

- the superior rectal artery from the inferior mesenteric artery,
- the middle rectal artery from the internal iliac artery, and
- the inferior rectal artery from the internal pudendal artery (from the internal iliac artery).

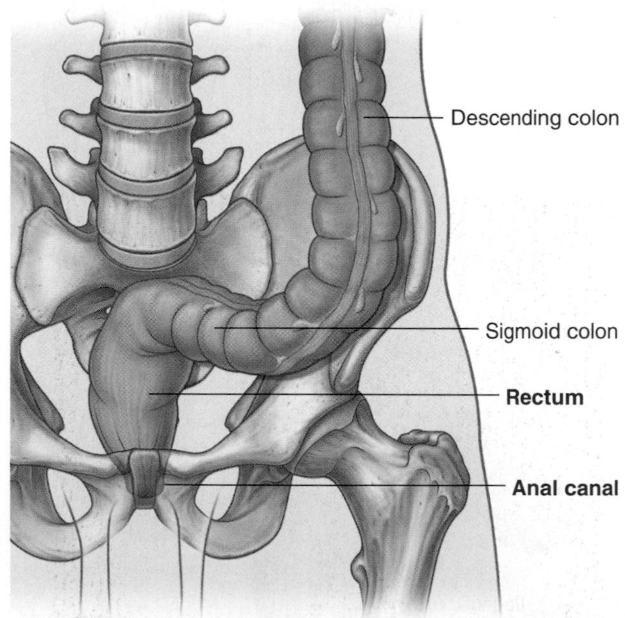


**Fig. 4.91** Rectum and anal canal.

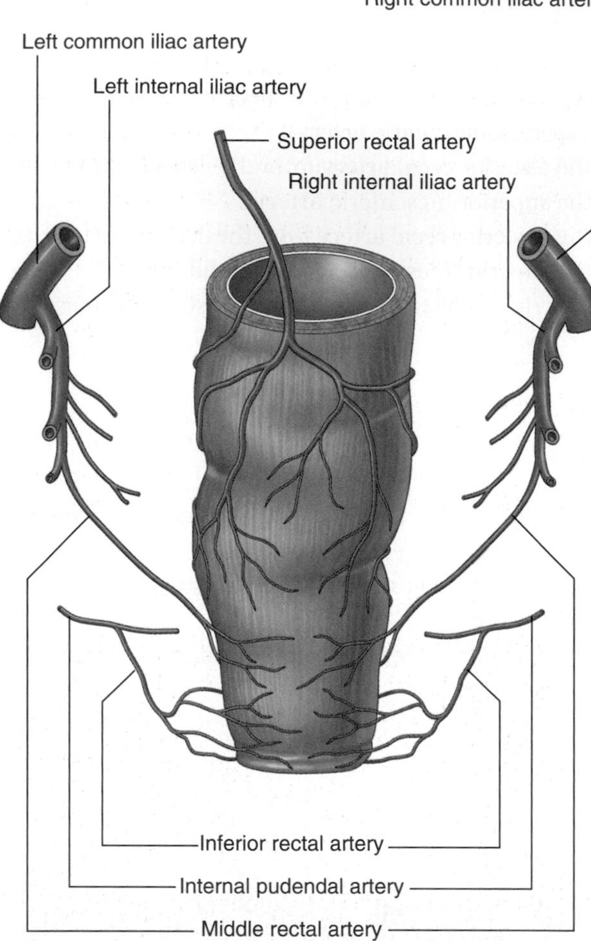


**Fig. 4.92** Arterial supply to the rectum and anal canal. Posterior view.

## In the clinic

**Congenital disorders of the gastrointestinal tract**

The normal positions of the abdominal viscera result from a complex series of rotations that the gut tube undergoes and from the growth of the abdominal cavity to accommodate changes in the size of the developing organs (see pp. 265-268). A number of developmental anomalies can occur during gut development, many of which appear in the neonate or infant, and some of which are surgical emergencies. Occasionally, such disorders are diagnosed only in adults.

***Malrotation and midgut volvulus***

Malrotation is incomplete rotation and fixation of the midgut after it has passed from the umbilical sac and returned to the abdominal coelom (Figs. 4.93 and 4.94). The proximal attachment of the small bowel mesentery begins at the **suspensory muscle of duodenum (ligament of Treitz)**, which determines the position of the duodenojejunal junction. The mesentery of the small bowel ends at the level of the ileocecal junction in the right lower quadrant. This long line of fixation of the mesentery prevents accidental twists of the gut.

If the duodenojejunal flexure or the cecum does not end up in its usual site, the origin of the small bowel mesentery shortens, which permits twisting of the small bowel around the axis of the superior mesenteric artery. Twisting of the bowel, in general, is termed **volvulus**. Volvulus of the small bowel may lead to a reduction of blood flow and infarction.

In some patients, the cecum ends up in the midabdomen. From the cecum and the right side of the colon a series of peritoneal folds (**Ladd's bands**) develop that extend to the right undersurface of the liver and compress the duodenum. A small bowel volvulus may then occur as well as duodenal obstruction. Emergency surgery may be necessary to divide the bands.

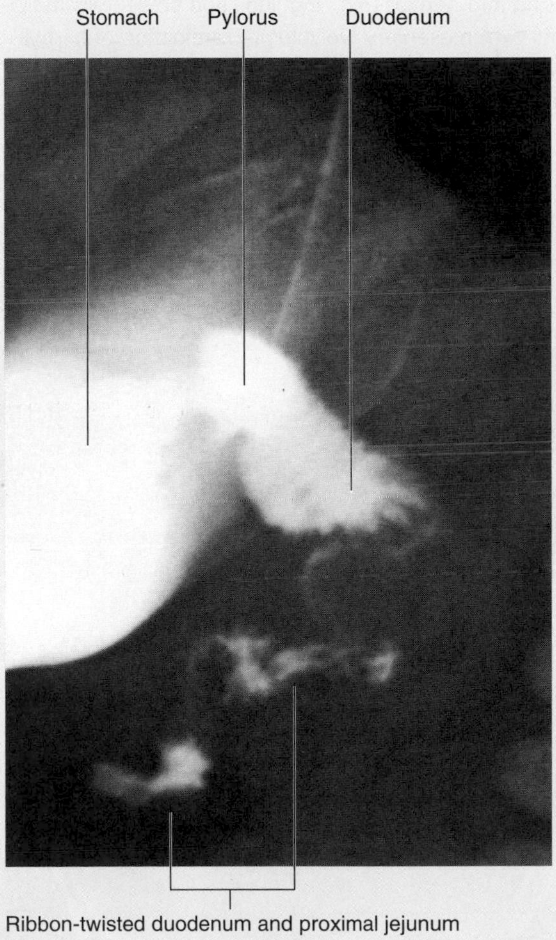

Stomach
Pylorus
Duodenum
Ribbon-twisted duodenum and proximal jejunum

**Fig. 4.93** Small bowel malrotation and volvulus. Radiograph of stomach, duodenum, and upper jejunum using barium.

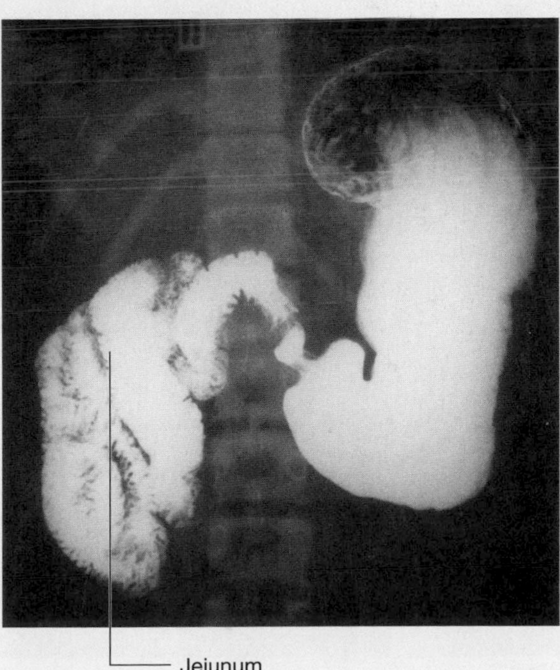


**Fig. 4.94** Small bowel malrotation. Radiograph of stomach, duodenum, and jejunum using barium.

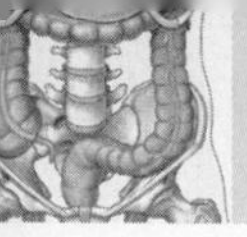



## In the clinic

### Bowel obstruction

A bowel obstruction can be either functional or due to a true obstruction. Mechanical obstruction is caused by an intraluminal, mural, or extrinsic mass which can be secondary to a foreign body, obstructing tumor in the wall, or extrinsic compression from an adhesion, or embryological band (Fig. 4.95).

A functional obstruction is usually due to an inability of the bowel to peristalse, which again has a number of causes, and most frequently is a postsurgical state due to excessive intraoperative bowel handling. Other causes may well include abnormality of electrolytes (e.g., sodium and potassium) rendering the bowel paralyzed until correction has occurred.

The signs and symptoms of obstruction depend on the level at which the obstruction has occurred. The primary symptom is central abdominal, intermittent, colicky pain as the peristaltic waves try to overcome the obstruction. Abdominal distention will occur if it is a low obstruction (distal), allowing more proximal loops of bowel to fill with fluid. A high obstruction (in the proximal small bowel) may not produce abdominal distention.

Vomiting and absolute constipation, including the inability to pass flatus, will ensue.

Early diagnosis is important because considerable fluid and electrolytes enter the bowel lumen and fail to be reabsorbed, which produces dehydration and electrolyte abnormalities. Furthermore, the bowel continues to distend, compromising the blood supply within the bowel wall, which may lead to ischemia and perforation. The symptoms and signs are variable and depend on the level of obstruction.

Small bowel obstruction is typically caused by adhesions following previous surgery, and history should always be sought for any operations or abdominal interventions (e.g., previous appendectomy). Other causes include bowel passing into hernias (e.g., inguinal) and bowel twisting on its own mesentery (volvulus). Examination of hernial orifices is mandatory in patients with bowel obstruction (Fig. 4.96).

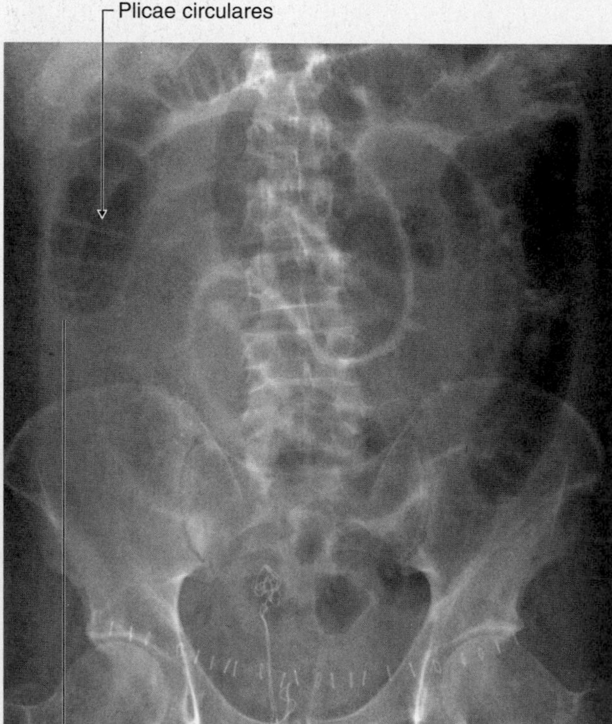


**Fig. 4.95** This radiograph of the abdomen, anteroposterior view, demonstrates a number of dilated loops of small bowel. Small bowel can be identified by the plicae circulares that pass from wall to wall as indicated. The large bowel is not dilated. The cause of the small bowel dilatation is an adhesion after pelvic surgery.

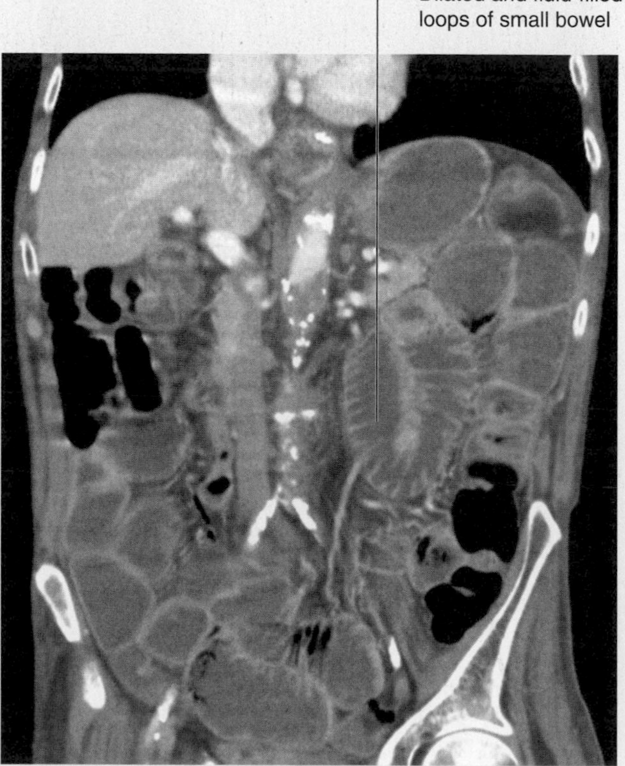


**Fig. 4.96** Coronal CT demonstrates dilated and fluid-filled loops of small bowel in patient with small bowel obstruction.

## In the clinic—cont'd

Large bowel obstruction is commonly caused by a tumor. Other potential causes include hernias and inflammatory diverticular disease of the sigmoid colon (Fig. 4.97).

The treatment is intravenous replacement of fluid and electrolytes, analgesia, and relief of obstruction. The passage of a nasogastric tube allows aspiration of fluid from the stomach. In many instances, small bowel obstruction, typically secondary to adhesions, will settle with nonoperative management. Large bowel obstruction may require an urgent operation to remove the obstructing lesion, or a temporary bypass procedure (e.g., defunctioning colostomy) (Fig. 4.98).

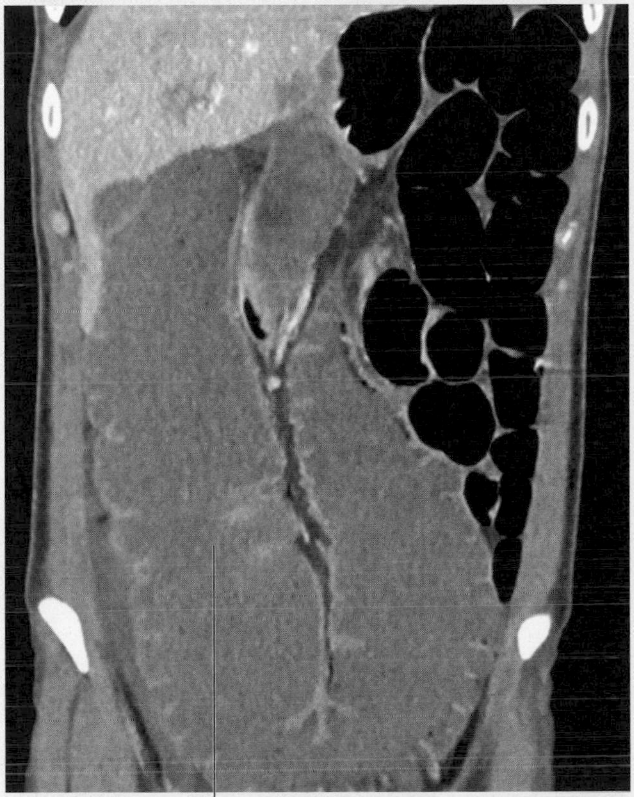

Fluid-filled and dilated ascending and transverse colon

**Fig. 4.97** Coronal CT of abdomen shows fluid-filled and dilated ascending and transverse colon in patient with large bowel obstruction.

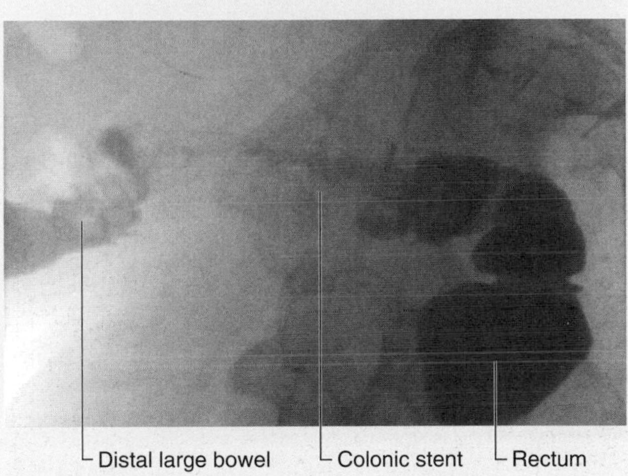


**Fig. 4.98** This oblique radiograph demonstrates contrast passing through a colonic stent that has been placed to relieve bowel obstruction prior to surgery.

## In the clinic

### Diverticular disease

Diverticular disease is the development of multiple colonic diverticula, predominantly throughout the sigmoid colon, though the whole colon may be affected (Fig. 4.99). The sigmoid colon has the smallest diameter of any portion of the colon and is therefore the site where intraluminal pressure is potentially the highest. Poor dietary fiber intake and obesity are also linked to diverticular disease.

The presence of multiple diverticula does not necessarily mean the patient requires any treatment. Moreover, many patients have no other symptoms or signs.

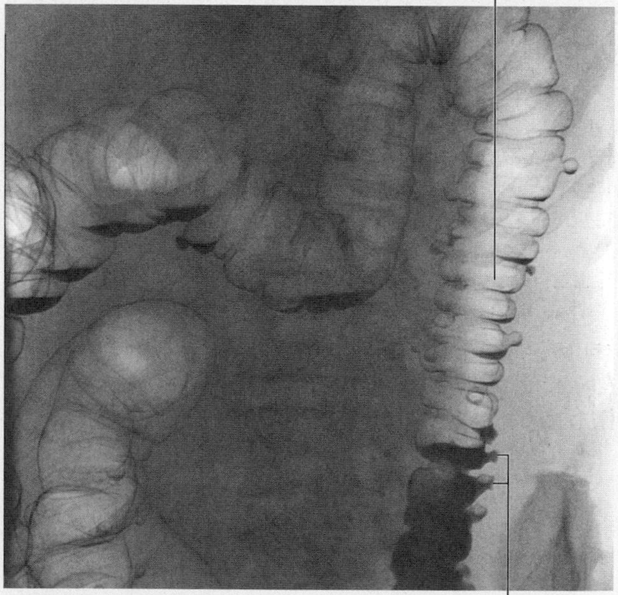

Descending colon

Diverticula

**Fig. 4.99** This double-contrast barium enema demonstrates numerous small outpouchings throughout the distal large bowel predominantly within the descending colon and the sigmoid colon. These small outpouchings are diverticula and in most instances remain quiescent.

Patients tend to develop symptoms and signs when the neck of the diverticulum becomes obstructed by feces and becomes infected. Inflammation may spread along the wall, causing abdominal pain. When the sigmoid colon becomes inflamed (diverticulitis), abdominal pain and fever ensue (Fig. 4.100).

Because of the anatomical position of the sigmoid colon there are a number of complications that may occur. The diverticula can perforate to form an abscess in the pelvis. The inflammation may produce an inflammatory mass, obstructing the left ureter. Inflammation may also spread to the bladder, producing a fistula between the sigmoid colon and the bladder. In these circumstances patients may develop a urinary tract infection and rarely have fecal material and gas passing per urethra.

The diagnosis is based upon clinical examination and often CT scanning. In the first instance, patients will be treated with antibiotic therapy; however, a surgical resection may be necessary if symptoms persist.

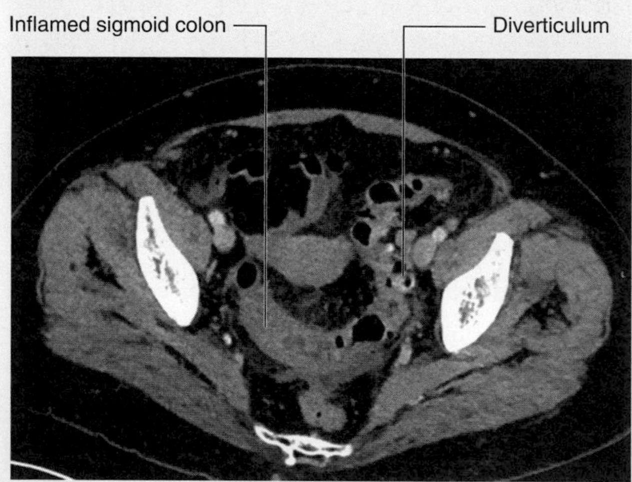


**Fig. 4.100** Axial CT of inflamed sigmoid colon in patient with diverticulitis.

## In the clinic

### Ostomies

It is occasionally necessary to surgically externalize bowel to the anterior abdominal wall. Externalization of bowel plays an important role in patient management. These extraanatomical bypass procedures use our anatomical knowledge and in many instances are life saving.

#### *Gastrostomy*

Gastrostomy is performed when the stomach is attached to the anterior abdominal wall and a tube is placed through the skin into the stomach. Typically this is performed to feed the patient when it is impossible to take food and fluid orally (e.g., complex head and neck cancer). The procedure can be performed either surgically or through a direct needlestick puncture under sedation in the anterior abdominal wall.

#### *Jejunostomy*

Similarly the jejunum is brought to the anterior abdominal wall and fixed. The jejunostomy is used as a site where a feeding tube is placed through the anterior abdominal wall into the proximal efferent small bowel.

#### *Ileostomy*

An ileostomy is performed when small bowel contents need to be diverted from the distal bowel. An ileostomy is often performed to protect a distal surgical anastomosis, such as in the colon to allow healing after surgery.

#### *Colostomy*

There are a number of instances when a colostomy may be necessary. In many circumstances it is performed to protect the distal large bowel after surgery. A further indication would include large bowel obstruction with imminent perforation wherein a colostomy allows decompression of the bowel and its contents. This is a safe and temporizing procedure performed when the patient is too unwell for extensive bowel surgery. It is relatively straightforward and carries reduced risk, preventing significant morbidity and mortality.

An end colostomy is necessary when the patient has undergone a surgical resection of the rectum and anus (typically for cancer).

#### *Ileal conduit*

An ileal conduit is an extraanatomical procedure and is performed after resection of the bladder for tumor. In this situation a short segment of small bowel is identified. The bowel is divided twice to produce a 20-cm segment of small bowel on its own mesentery. This isolated segment of bowel is used as a conduit. The remaining bowel is joined together. The proximal end is anastomosed to the ureters, and the distal end is anastomosed to the anterior abdominal wall. Hence, urine passes from the kidneys into the ureters and through the short segment of small bowel to the anterior abdominal wall.

When patients have either an ileostomy, colostomy, or ileal conduit it is necessary for them to fix a collecting bag onto the anterior abdominal wall. Contrary to one's initial thoughts these bags are tolerated extremely well by most patients and allow patients to live a nearly normal and healthy life.

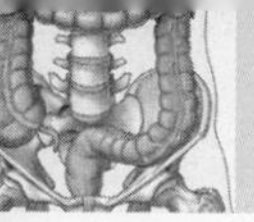



## Liver

The liver is the largest visceral organ in the body and is primarily in the right hypochondrium and epigastric region, extending into the left hypochondrium (or in the right upper quadrant, extending into the left upper quadrant) (Fig. 4.101).

Surfaces of the liver include:

- a **diaphragmatic surface** in the anterior, superior, and posterior directions; and
- a **visceral surface** in the inferior direction (Fig. 4.102).

### Diaphragmatic surface

The diaphragmatic surface of the liver, which is smooth and domed, lies against the inferior surface of the diaphragm (Fig. 4.103). Associated with it are the subphrenic and hepatorenal recesses (Fig. 4.102):

- The **subphrenic recess** separates the diaphragmatic surface of the liver from the diaphragm and is divided into right and left areas by the **falciform ligament**, a structure derived from the ventral mesentery in the embryo.
- The hepatorenal recess is a part of the peritoneal cavity on the right side between the liver and the right kidney and right suprarenal gland.

The subphrenic and hepatorenal recesses are continuous anteriorly.

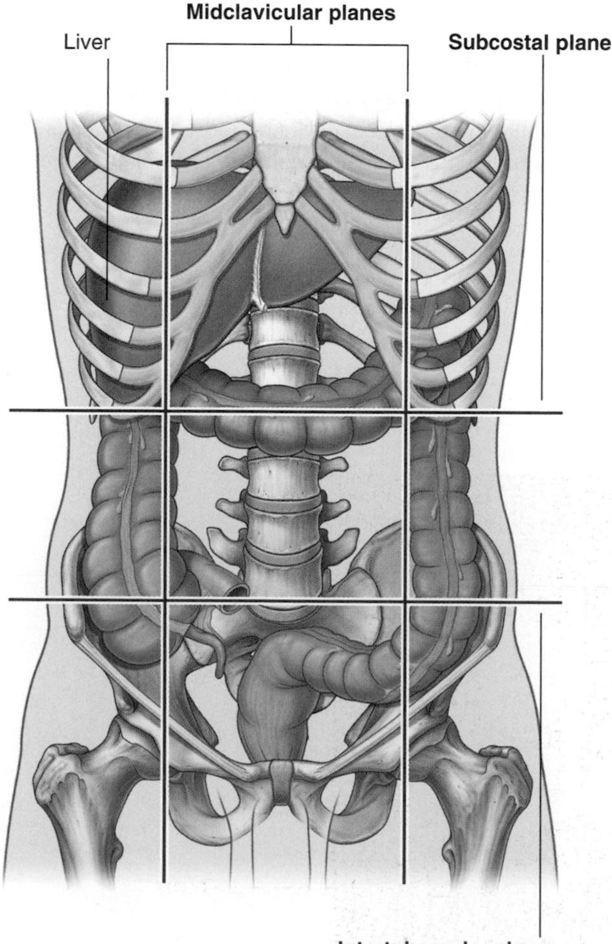


**Fig. 4.101** Position of the liver in the abdomen.

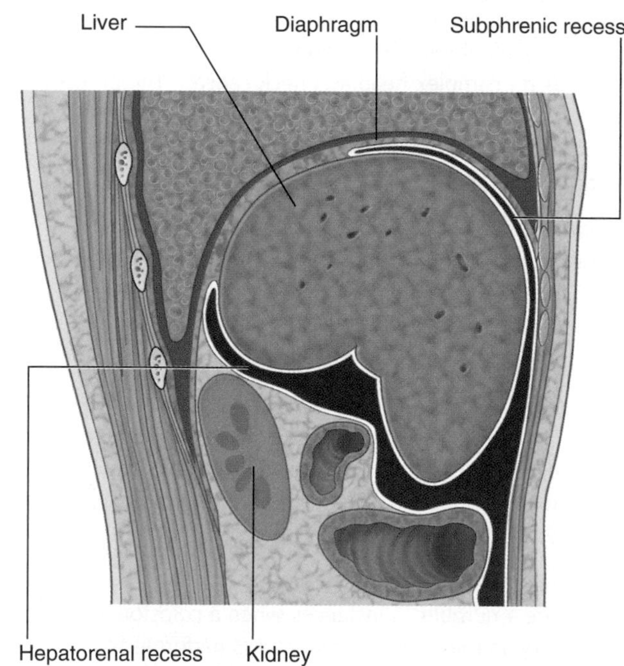


**Fig. 4.102** Surfaces of the liver and recesses associated with the liver.

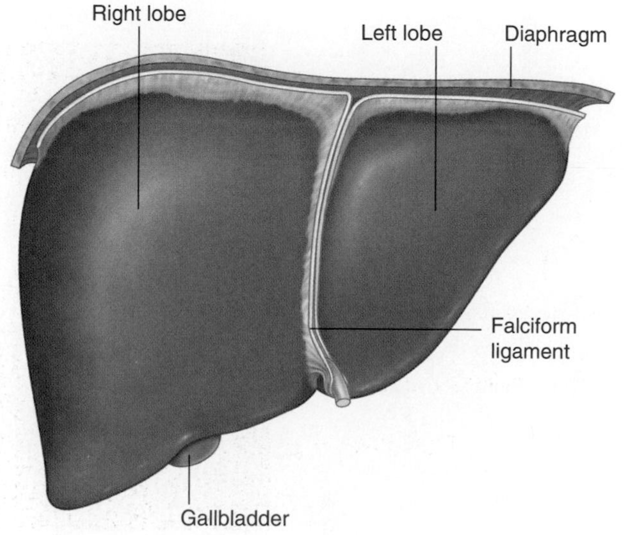


**Fig. 4.103** Diaphragmatic surface of the liver.

### Visceral surface

The visceral surface of the liver is covered with visceral peritoneum except in the **fossa for the gallbladder** and at the **porta hepatis** (gateway to the liver; Fig. 4.104), and structures related to it include the following (Fig. 4.105):

- esophagus,
- right anterior part of the stomach,
- superior part of the duodenum,
- lesser omentum,
- gallbladder,
- right colic flexure,
- right transverse colon,
- right kidney, and
- right suprarenal gland.

The **porta hepatis** serves as the point of entry into the liver for the hepatic arteries and the portal vein, and the exit point for the hepatic ducts (Fig. 4.104).

### Associated ligaments

The liver is attached to the anterior abdominal wall by the **falciform ligament** and, except for a small area of the

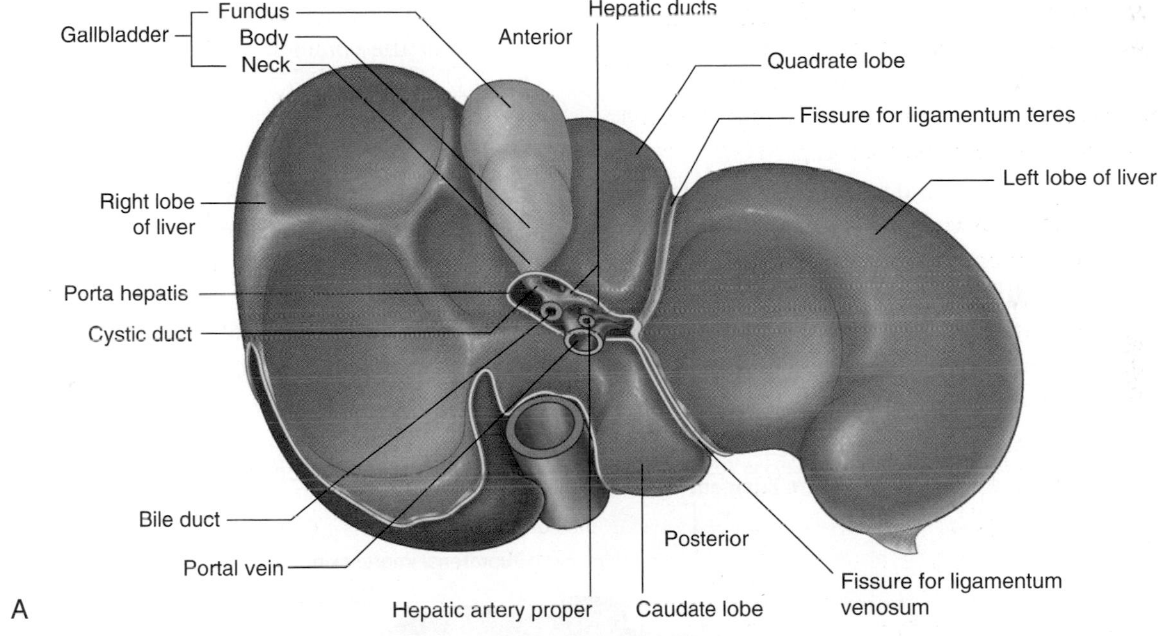


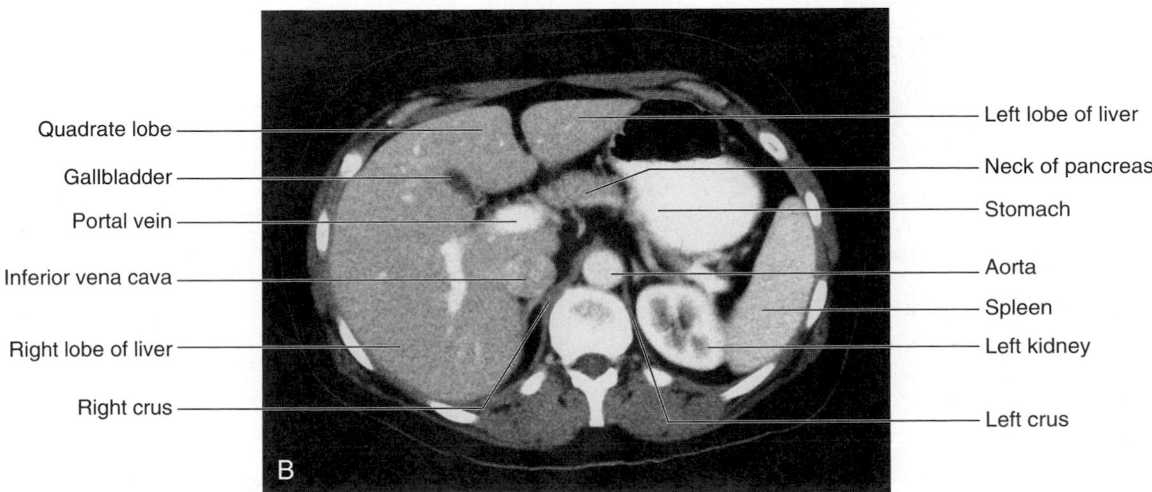


**Fig. 4.104** Visceral surface of the liver. **A.** Illustration. **B.** Abdominal computed tomogram, with contrast, in the axial plane.

liver against the diaphragm (the **bare area**), the liver is almost completely surrounded by visceral peritoneum (Fig. 4.105). Additional folds of peritoneum connect the liver to the stomach (**hepatogastric ligament**), the duodenum (**hepatoduodenal ligament**), and the diaphragm (**right** and **left triangular ligaments** and **anterior** and **posterior coronary ligaments**).

The bare area of the liver is a part of the liver on the diaphragmatic surface where there is no intervening peritoneum between the liver and the diaphragm (Fig. 4.105):

- The anterior boundary of the bare area is indicated by a reflection of peritoneum—the anterior coronary ligament.
- The posterior boundary of the bare area is indicated by a reflection of peritoneum—the posterior coronary ligament.
- Where the coronary ligaments come together laterally, they form the right and left triangular ligaments.

### Lobes

The liver is divided into right and left lobes by the falciform ligament anterosuperiorly and the fissure for the ligamentum venosum and ligamentum teres on the visceral surface. (Fig. 4.104). The **right lobe of the liver** is the largest lobe, whereas the **left lobe of the liver** is smaller. The quadrate and caudate lobes are described as arising from the right lobe of the liver but functionally are distinct.

- The **quadrate lobe** is visible on the anterior part of the visceral surface of the liver and is bounded on the left by the fissure for the ligamentum teres and on the right by the fossa for the gallbladder. Functionally, it is related to the left lobe of the liver.
- The **caudate lobe** is visible on the posterior part of the visceral surface of the liver. It is bounded on the left by the fissure for the ligamentum venosum and on the right by the groove for the inferior vena cava. Functionally, it is separate from the right and the left lobes of the liver.

The arterial supply to the liver includes:

- the right hepatic artery from the hepatic artery proper (a branch of the common hepatic artery from the celiac trunk), and
- the left hepatic artery from the hepatic artery proper (a branch of the common hepatic artery from the celiac trunk).

### Gallbladder

The **gallbladder** is a pear-shaped sac lying on the visceral surface of the right lobe of the liver in a fossa between the right and quadrate lobes (Fig. 4.104). It has:

- a rounded end (**fundus of the gallbladder**), which may project from the inferior border of the liver;

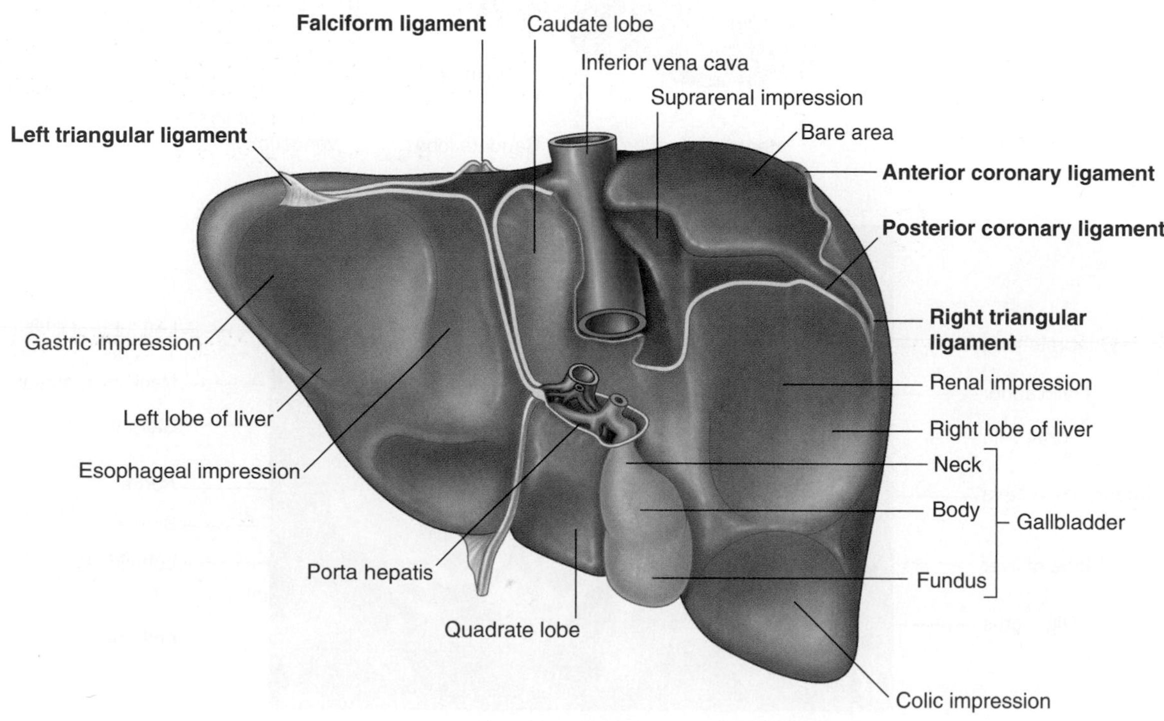


**Fig. 4.105** Posterior view of the bare area of the liver and associated ligaments.

- a major part in the fossa (**body of the gallbladder**), which may be against the transverse colon and the superior part of the duodenum; and
- a narrow part (**neck of the gallbladder**) with mucosal folds forming the spiral fold.

The arterial supply to the gallbladder (Fig. 4.106) is the cystic artery from the right hepatic artery (a branch of the hepatic artery proper).

The gallbladder receives, concentrates, and stores bile from the liver.

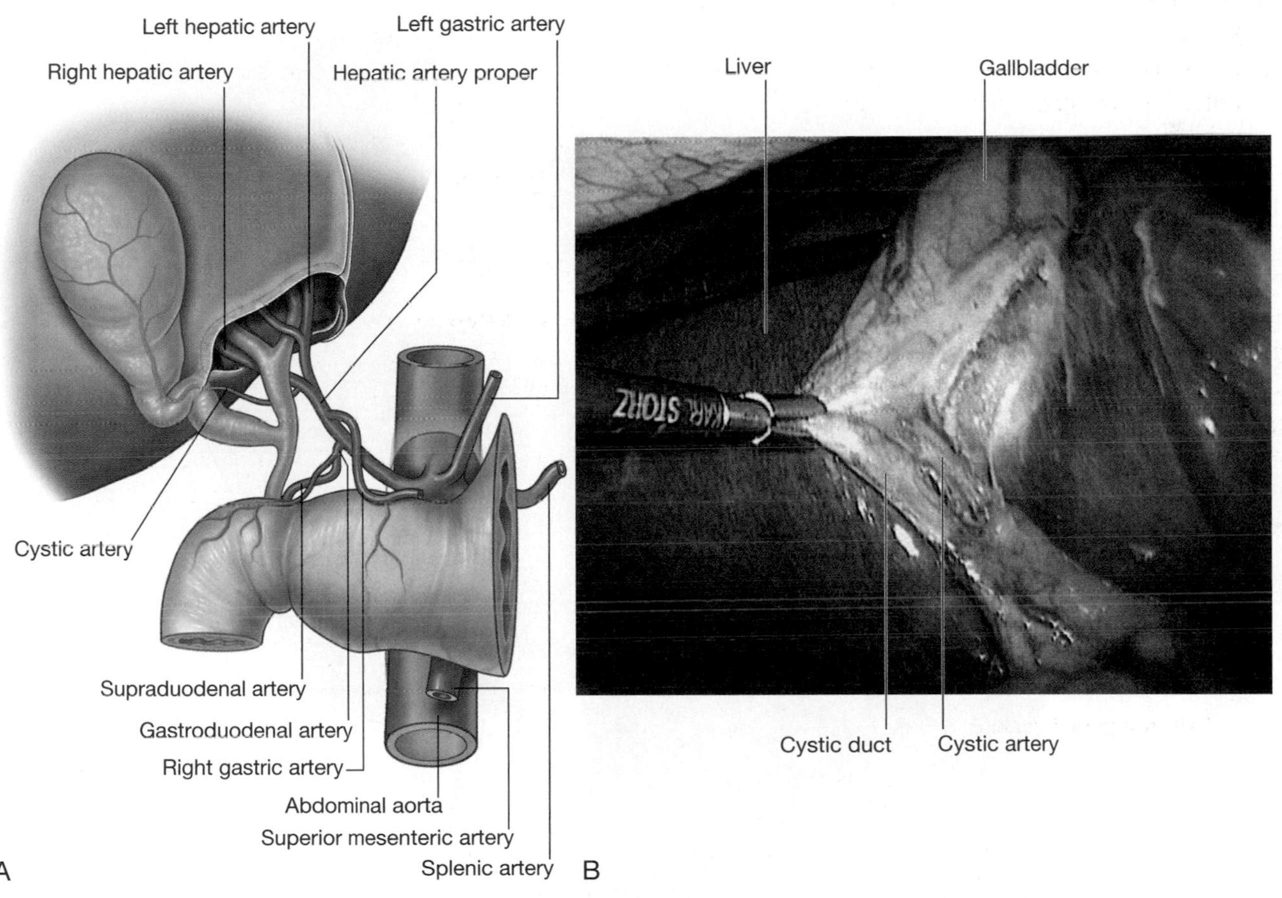


**Fig. 4.106** Arterial supply to the liver and gallbladder. **A.** Schematic. **B.** Laparoscopic surgical view of cystic duct and cystic artery.

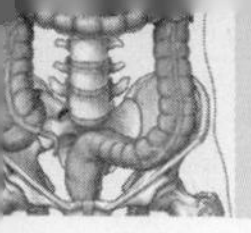

## Pancreas

The pancreas lies mostly posterior to the stomach (Figs. 4.107 and 4.108). It extends across the posterior abdominal wall from the duodenum, on the right, to the spleen, on the left.

The pancreas is (secondarily) retroperitoneal except for a small part of its tail and consists of a head, uncinate process, neck, body, and tail.

- The **head of the pancreas** lies within the C-shaped concavity of the duodenum.

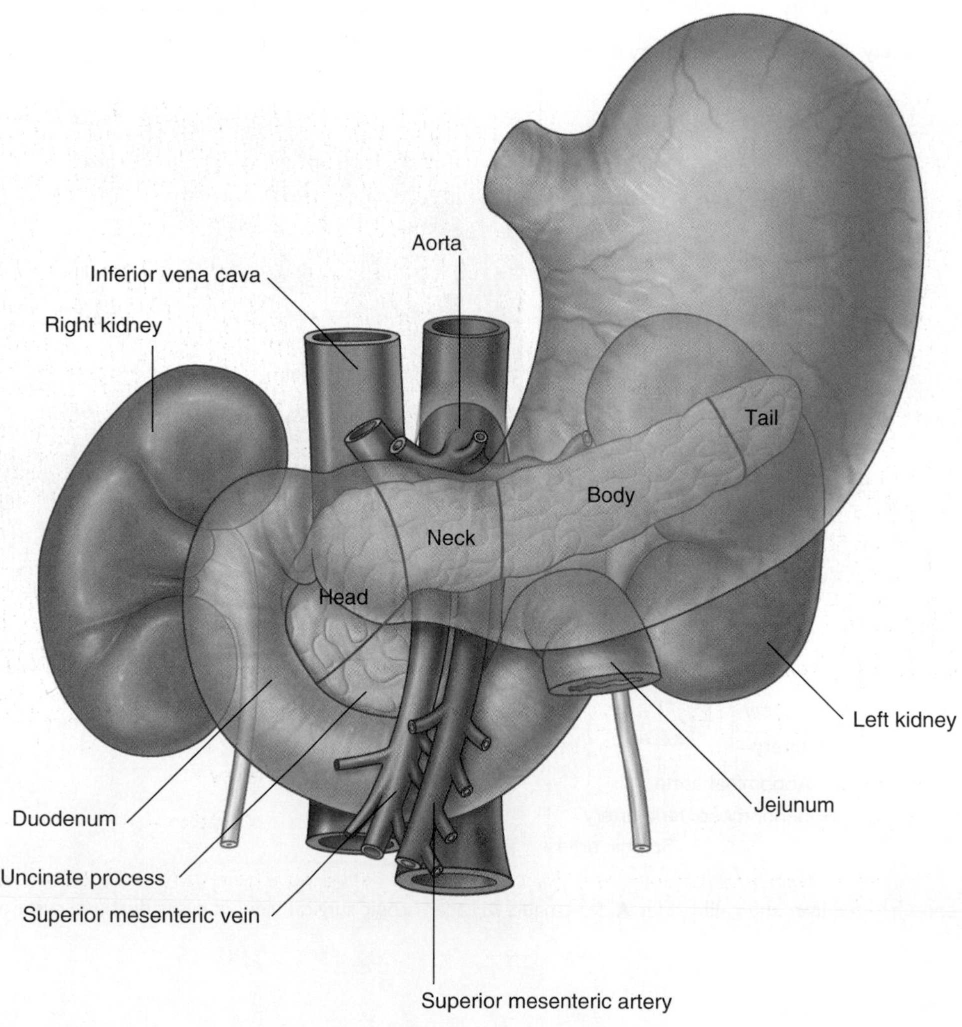



**Fig. 4.107** Pancreas.

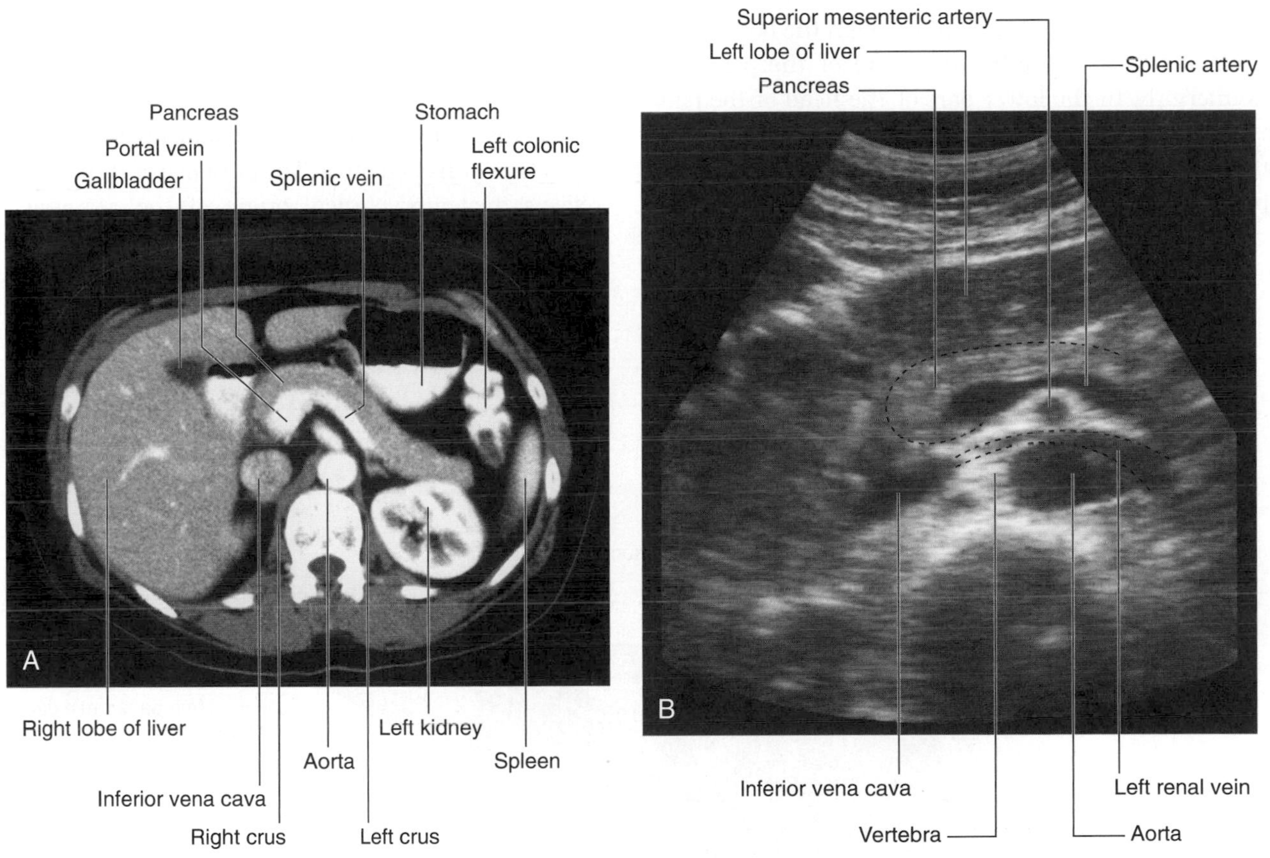


**Fig. 4.108** Abdominal images. **A.** Abdominal computed tomogram, with contrast, in the axial plane. **B.** Abdominal ultrasound scan.

- Projecting from the lower part of the head is the **uncinate process**, which passes posterior to the superior mesenteric vessels.
- The **neck of the pancreas** is anterior to the superior mesenteric vessels. Posterior to the neck of the pancreas, the superior mesenteric and splenic veins join to form the portal vein.
- The **body of the pancreas** is elongate and extends from the neck to the tail of the pancreas.
- The **tail of the pancreas** passes between layers of the splenorenal ligament.

The **pancreatic duct** begins in the tail of the pancreas (Fig. 4.109). It passes to the right through the body of the pancreas and, after entering the head of the pancreas, turns inferiorly. In the lower part of the head of the pancreas, the pancreatic duct joins the bile duct. The joining of these two structures forms the **hepatopancreatic ampulla** (ampulla of Vater), which enters the descending (second) part of the duodenum at the **major duodenal papilla**. Surrounding the ampulla is the **sphincter of ampulla** (sphincter of Oddi), which is a collection of smooth muscles.

The **accessory pancreatic duct** empties into the duodenum just above the major duodenal papilla at the **minor duodenal papilla** (Fig. 4.109). If the accessory duct is followed from the minor papilla into the head of the pancreas, a branch point is discovered:

- One branch continues to the left, through the head of the pancreas, and may connect with the pancreatic duct at the point where it turns inferiorly.
- A second branch descends into the lower part of the head of the pancreas, anterior to the pancreatic duct, and ends in the uncinate process.

The main and accessory pancreatic ducts usually communicate with each other. The presence of these two ducts reflects the embryological origin of the pancreas from dorsal and ventral buds from the foregut.

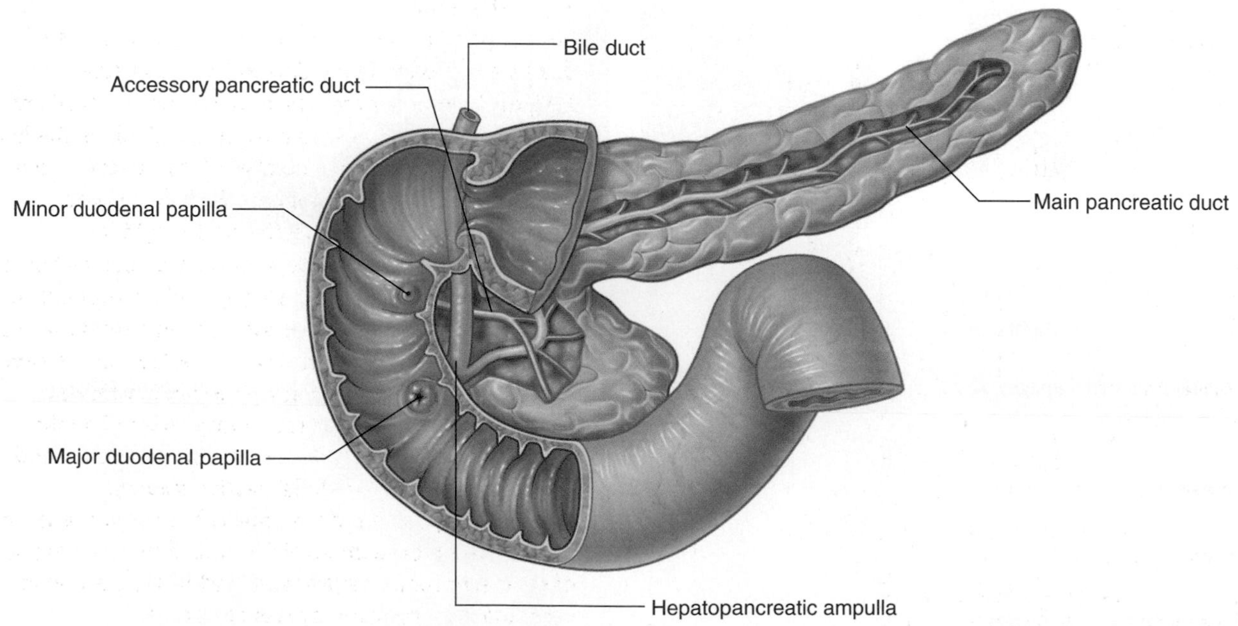


**Fig. 4.109** Pancreatic duct system.

The arterial supply to the pancreas (Fig. 4.110) includes the:

- gastroduodenal artery from the common hepatic artery (a branch of the celiac trunk),
- anterior superior pancreaticoduodenal artery from the gastroduodenal artery,
- posterior superior pancreaticoduodenal artery from the gastroduodenal artery,
- dorsal pancreatic artery from the inferior pancreatic artery (a branch of the splenic artery),
- great pancreatic artery from the inferior pancreatic artery (a branch of the splenic artery),
- anterior inferior pancreaticoduodenal artery from the inferior pancreaticoduodenal artery (a branch of the superior mesenteric artery), and
- posterior inferior pancreaticoduodenal artery from the inferior pancreaticoduodenal artery (a branch of the superior mesenteric artery).

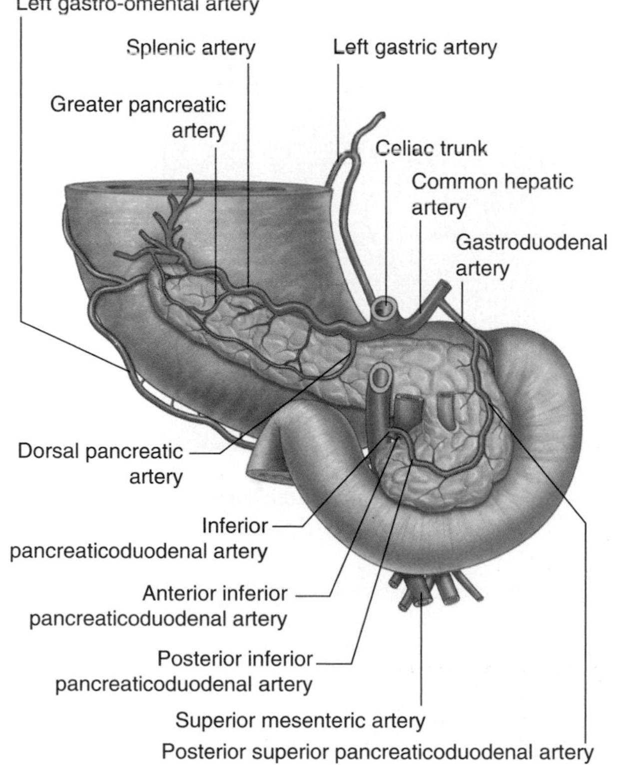


**Fig. 4.110** Arterial supply to the pancreas. Posterior view.

## In the clinic

### Annular pancreas

The pancreas develops from ventral and dorsal buds from the foregut. The dorsal bud forms most of the head, neck, and body of the pancreas. The ventral bud rotates around the bile duct to form part of the head and the uncinate process. If the ventral bud splits (becomes bifid), the two segments may encircle the duodenum. The duodenum is therefore constricted and may even undergo atresia, and be absent at birth because of developmental problems. After birth, the child may fail to thrive and may vomit due to poor gastric emptying.

Sometimes an annular pancreas is diagnosed in utero by ultrasound scanning. The obstruction of the duodenum may prevent the fetus from swallowing enough amniotic fluid, which may increase the overall volume of amniotic fluid in the amniotic sac surrounding the fetus (**polyhydramnios**).

## In the clinic

### Pancreatic cancer

Pancreatic cancer accounts for a significant number of deaths and is often referred to as the "silent killer." Malignant tumors of the pancreas may occur anywhere within the pancreas but are most frequent within the head and the neck. There are a number of nonspecific findings in patients with pancreatic cancer, including upper abdominal pain, loss of appetite, and weight loss. Depending on the exact site of the cancer, obstruction of the bile duct may occur, which can produce obstructive jaundice. Although surgery is indicated in patients where there is a possibility of cure, most detected cancers have typically spread locally, invading the portal vein and superior mesenteric vessels, and may extend into the porta hepatis. Lymph node spread also is common and these factors would preclude curative surgery.

Given the position of the pancreas, a surgical resection is a complex procedure involving resection of the region of pancreatic tumor usually with part of the duodenum, necessitating a complex bypass procedure.

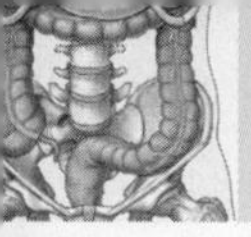

### Duct system for bile

The duct system for the passage of bile extends from the liver, connects with the gallbladder, and empties into the descending part of the duodenum (Fig. 4.111). The coalescence of ducts begins in the liver parenchyma and continues until the **right** and **left hepatic ducts** are formed. These drain the respective lobes of the liver.

The two hepatic ducts combine to form the **common hepatic duct**, which runs near the liver, with the hepatic artery proper and portal vein in the free margin of the lesser omentum.

As the common hepatic duct continues to descend, it is joined by the **cystic duct** from the gallbladder. This completes the formation of the **bile duct**. At this point, the bile duct lies to the right of the hepatic artery proper and usually to the right of, and anterior to, the portal vein in the free margin of the lesser omentum. The **omental foramen** is posterior to these structures at this point.

The bile duct continues to descend, passing posteriorly to the superior part of the duodenum before joining with the pancreatic duct to enter the descending part of the duodenum at the major duodenal papilla (Fig. 4.111).

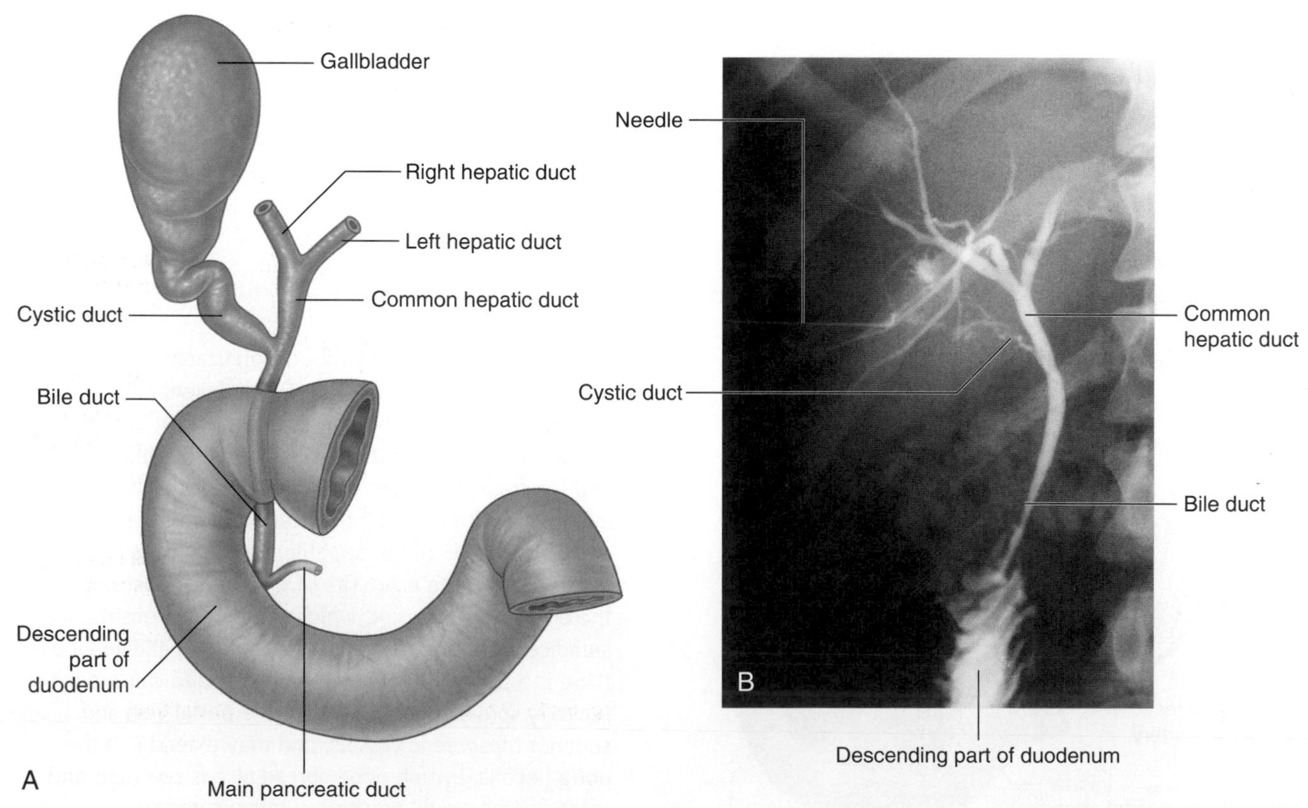


**Fig. 4.111** Bile drainage. **A.** Duct system for passage of bile. **B.** Percutaneous transhepatic cholangiogram demonstrating the bile duct system.

## Spleen

The spleen develops as part of the vascular system in the part of the dorsal mesentery that suspends the developing stomach from the body wall. In the adult, the spleen lies against the diaphragm, in the area of rib IX to rib X (Fig. 4.112). It is therefore in the left upper quadrant, or left hypochondrium, of the abdomen.

The spleen is connected to the:

- greater curvature of the stomach by the gastrosplenic ligament, which contains the short gastric and gastro-omental vessels; and
- left kidney by the splenorenal ligament (Fig. 4.113), which contains the splenic vessels.

Both these ligaments are parts of the greater omentum.

The spleen is surrounded by visceral peritoneum except in the area of the hilum on the medial surface of the spleen (Fig. 4.114). The **splenic hilum** is the entry point for the splenic vessels, and occasionally the tail of the pancreas reaches this area.

The arterial supply to the spleen (Fig. 4.115) is the splenic artery from the celiac trunk.

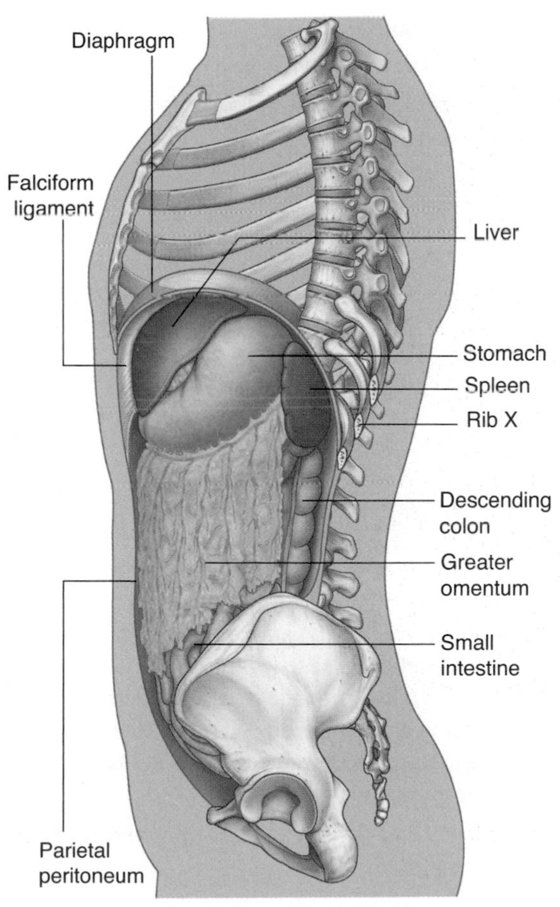


**Fig. 4.112** Spleen.

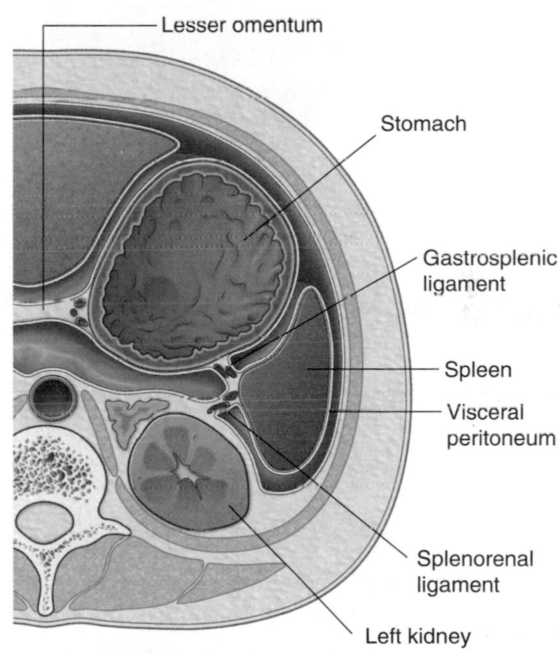


**Fig. 4.113** Splenic ligaments and related vasculature.

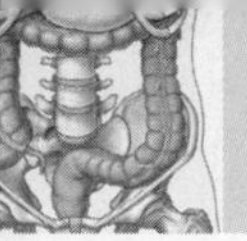

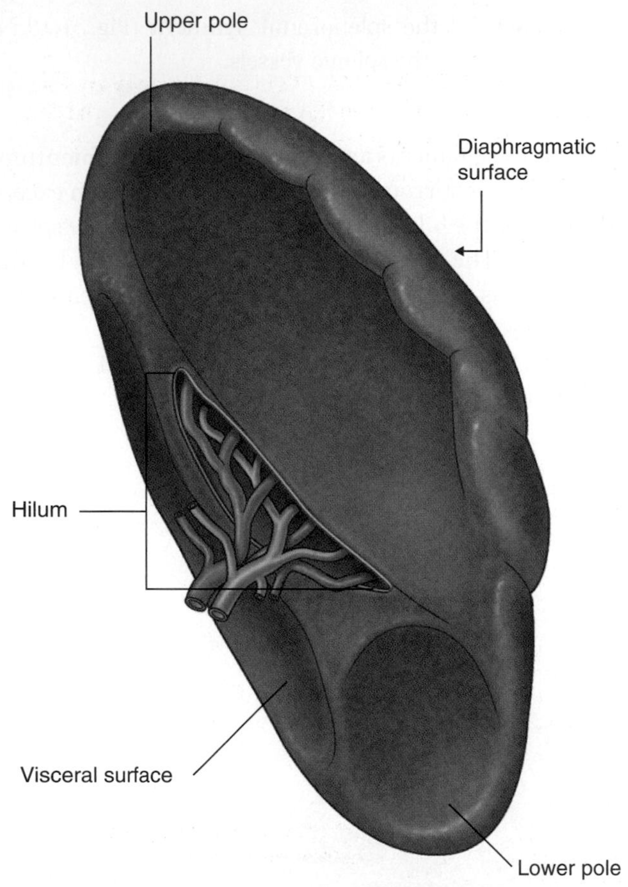


**Fig. 4.114** Surfaces and hilum of the spleen.

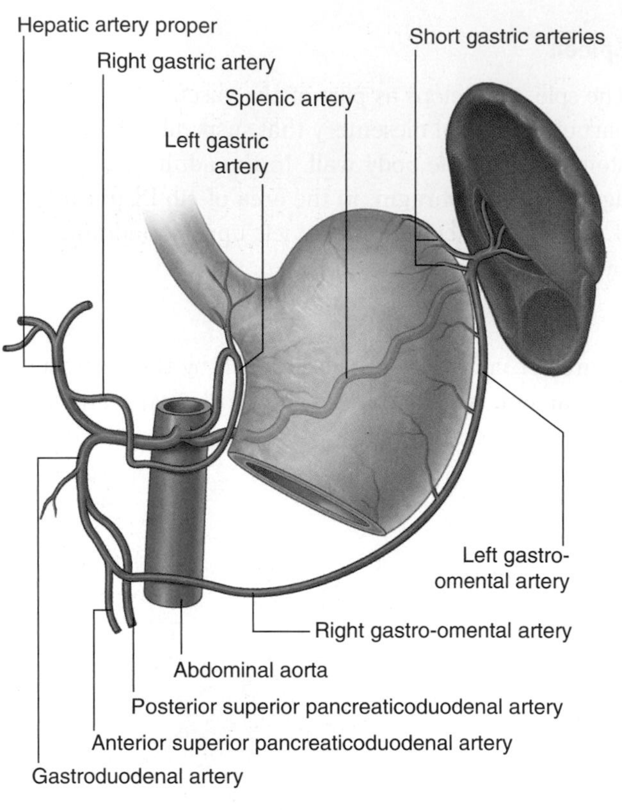


**Fig. 4.115** Arterial supply to the spleen.

## In the clinic

### Segmental anatomy of the liver

For many years the segmental anatomy of the liver was of little importance. However, since the development of liver resection surgery, the size, shape, and segmental anatomy of the liver have become clinically important, especially with regard to liver resection for metastatic disease. Indeed, with detailed knowledge of the segments, curative surgery can be performed in patients with tumor metastases.

The liver is divided by the **principal plane**, which divides the organ into halves of approximately equal size. This imaginary line is defined by a parasagittal line that passes through the gallbladder fossa to the inferior vena cava. It is in this plane that the middle hepatic vein is found. Importantly, the principal plane divides the left half of the liver from the right half. The lobes of the liver are unequal in size and bear only little relevance to operative anatomy.

## In the clinic—cont'd

The traditional eight-segment anatomy of the liver relates to the hepatic arterial, portal, and biliary drainage of these segments (Fig. 4.116).

The caudate lobe is defined as segment I, and the remaining segments are numbered in a clockwise fashion up to segment VIII. The features are extremely consistent between individuals.

From a surgical perspective, a right hepatectomy would involve division of the liver in the principal plane in which segments V, VI, VII, and VIII would be removed, leaving segments I, II, III, and IV.

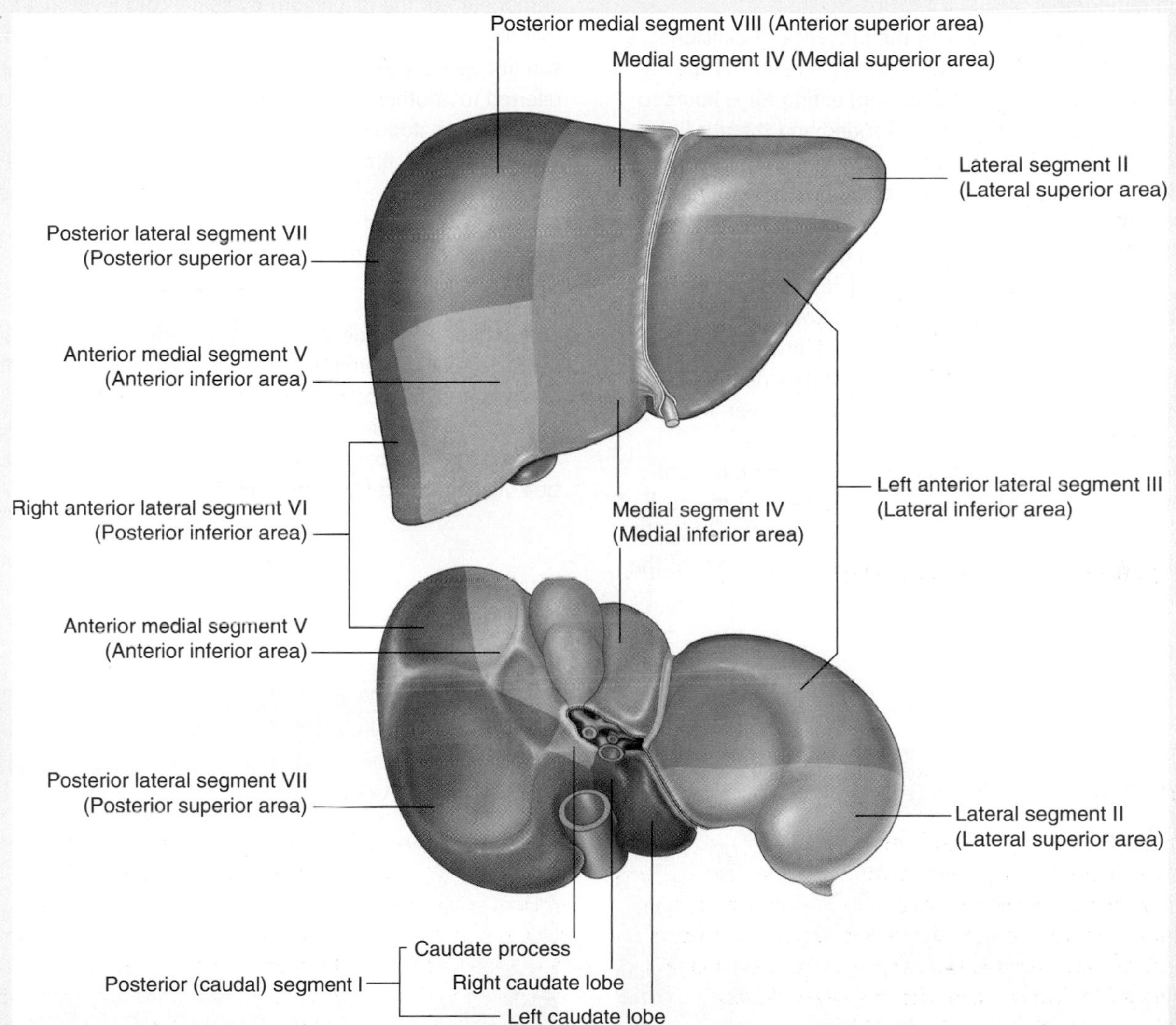


**Fig. 4.116** Division of the liver into segments based upon the distributions of the bile ducts and hepatic vessels (Couinaud's segments).

## In the clinic

### Gallstones

Gallstones are present in approximately 10% of people over the age of 40 and are more common in women. They consist of a variety of components but are predominantly a mixture of cholesterol and bile pigment. They may undergo calcification, which can be demonstrated on plain radiographs. Gallstones may be visualized incidentally as part of a routine abdominal ultrasound scan (Fig. 4.117) or on a plain radiograph.

The easiest way to confirm the presence of gallstones is by performing a fasting ultrasound examination of the gallbladder. The patient refrains from eating for 6 hours to ensure the gallbladder is well distended and there is little shadowing from overlying bowel gas. The examination may also identify bile duct dilation and the presence of cholecystitis. Magnetic resonance cholangiopancreatography (MRCP) is another way to image the gallbladder and biliary tree. MRCP uses fluid present in the bile ducts and in the pancreatic duct as a contrast agent to show stones as well as filling defects within the gallbladder and intrahepatic or extrahepatic bile ducts. It can demonstrate strictures in the biliary tree and can also be used to visualize liver and pancreatic anatomy (Fig. 4.118).

From time to time, gallstones impact in the region of **Hartmann's pouch**, which is a bulbous region of the neck of the gallbladder. When the gallstone lodges in this area, the gallbladder cannot empty normally and contractions of the gallbladder wall produce severe pain. If this persists, a **cholecystectomy** (removal of the gallbladder) may be necessary.

Sometimes the gallbladder may become inflamed (**cholecystitis**). If the inflammation involves the related parietal peritoneum of the diaphragm, pain may not only occur in the right upper quadrant of the abdomen but may also be referred to the shoulder on the right side. This referred pain is due to the innervation of the visceral peritoneum of the diaphragm by spinal cord levels (C3 to C5) that also innervate skin over the shoulder. In this case, one somatic sensory region of low sensory output (diaphragm) is referred to another somatic sensory region of high sensory output (dermatomes).

From time to time, small gallstones pass into the bile duct and are trapped in the region of the sphincter of the ampulla, which obstructs the flow of bile into the duodenum. This, in turn, produces jaundice.

### ERCP

Endoscopic retrograde cholangiopancreatography (ERCP) can be undertaken to remove obstructing gallstones within the biliary tree. This procedure combines endoluminal endoscopy with fluoroscopy to diagnose and treat problems in the biliary and pancreatic ducts. An endoscope with a side-viewing optical system is advanced through the

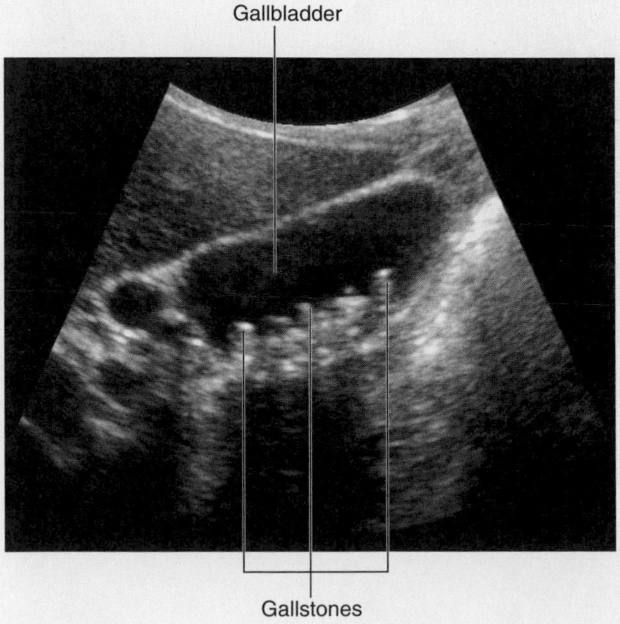

**Fig. 4.117** Gallbladder containing multiple stones. Ultrasound scan.

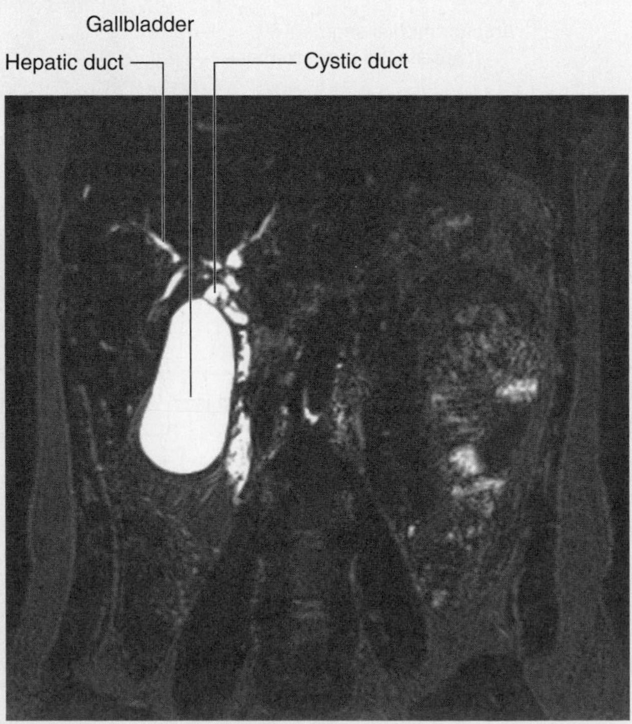


**Fig. 4.118** Magnetic resonance cholangiopancreatography (MRCP) in the coronal plane.

## In the clinic—cont'd

esophagus and stomach and placed in the second part of the duodenum where the major papilla (the ampulla of Vater) is identified. This is where the pancreatic duct converges with the common bile duct. The papilla is initially examined for possible abnormalities (stuck stone or malignant growth) and a biopsy may be taken if necessary. Then either the bile duct or pancreatic duct is cannulated and a small amount of radiopaque contrast medium is injected to visualize either the bile duct (cholangiogram) or pancreatic duct (pancreatogram) (Fig. 4.119). If a stone is present, it can be removed with a stone basket or an extraction balloon. Usually, a sphincterotomy is performed before stone removal to ease its passage through the distal bile duct.

In cases of biliary tree obstruction caused by benign or malignant strictures, a stent can be placed into the common bile duct or into one of the main hepatic ducts to allow opening of the narrowed segment. The patency of the newly inserted stent is confirmed by instillation of more contrast medium to demonstrate free flow of contrast through the stent.

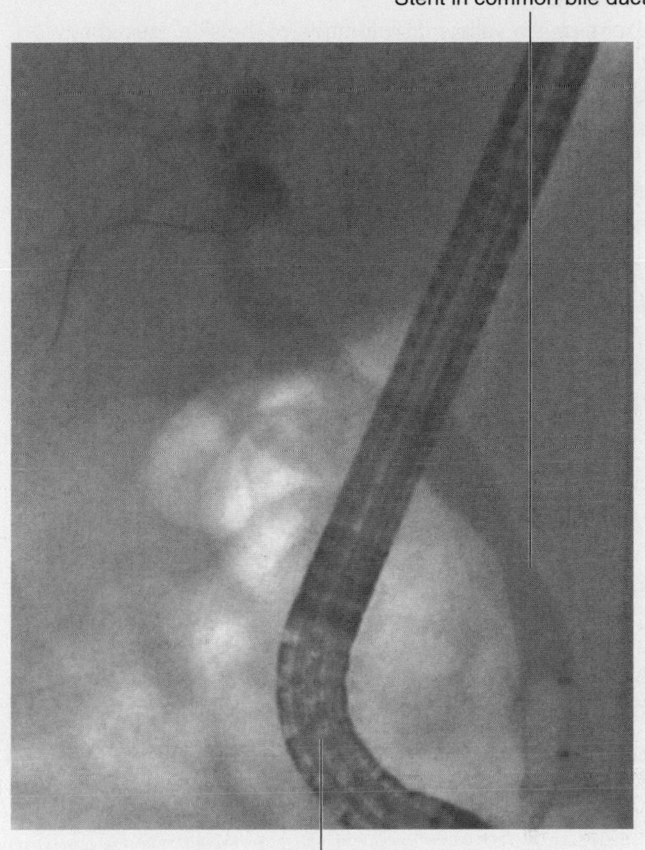


**Fig. 4.119** Endoscopic retrograde cholangiopancreatography (ERCP) of biliary system.

## In the clinic

### Jaundice

Jaundice is a yellow discoloration of the skin caused by excess bile pigment (bilirubin) within the plasma. The yellow color is best appreciated by looking at the normally white sclerae of the eyes, which turn yellow.

The extent of the elevation of the bile pigments and the duration for which they have been elevated account for the severity of jaundice.

***Simplified explanation to understanding the types of jaundice and their anatomical causes***

When red blood cells are destroyed by the reticuloendothelial system, the iron from the hemoglobin molecule is recycled, whereas the porphyrin ring (globin) compounds are broken down to form fat-soluble bilirubin. On reaching the liver via the bloodstream, the fat-soluble bilirubin is converted to a water-soluble form of bilirubin. This water-soluble bilirubin is secreted into the biliary tree and then in turn into the bowel, where it forms the dark color of the stool.

***Prehepatic jaundice***

This type of jaundice is usually produced by conditions where there is an excessive breakdown of red blood cells (e.g., in incompatible blood transfusion and hemolytic anemia).

***Hepatic jaundice***

The complex biochemical reactions for converting fat-soluble into water-soluble bilirubin may be affected by inflammatory change within the liver (e.g., from hepatitis or chronic liver disease, such as liver cirrhosis) and poisons (e.g., paracetamol overdose).

***Posthepatic jaundice***

Any obstruction of the biliary tree can produce jaundice, but the two most common causes are gallstones within the bile duct and an obstructing tumor at the head of the pancreas.

## In the clinic

### Spleen disorders

From a clinical point of view, there are two main categories of spleen disorders: rupture and enlargement.

***Splenic rupture***

This tends to occur when there is localized trauma to the left upper quadrant. It may be associated with left lower rib fractures. Because the spleen has such an extremely thin capsule, it is susceptible to injury even when there is no damage to surrounding structures, and because the spleen is highly vascular, when ruptured, it bleeds profusely into the peritoneal cavity. Splenic rupture should always be suspected with blunt abdominal injury. Current treatments preserve as much of the spleen as possible, but some patients require splenectomy.

***Splenic enlargement***

The spleen is an organ of the reticuloendothelial system involved in hematopoiesis and immunological surveillance. Diseases that affect the reticuloendothelial system (e.g., leukemia or lymphoma) may produce generalized lymphadenopathy and enlargement of the spleen (**splenomegaly**) (Fig. 4.120). The spleen often enlarges when performing its normal physiological functions, such as when clearing microorganisms and particulates from the circulation, producing increased antibodies in the course of sepsis, or removing deficient or destroyed erythrocytes (e.g., in thalassemia and spherocytosis). Splenomegaly may also be a result of increased venous pressure caused by congestive heart failure, splenic vein thrombosis, or portal hypertension. An enlarged spleen is prone to rupture.

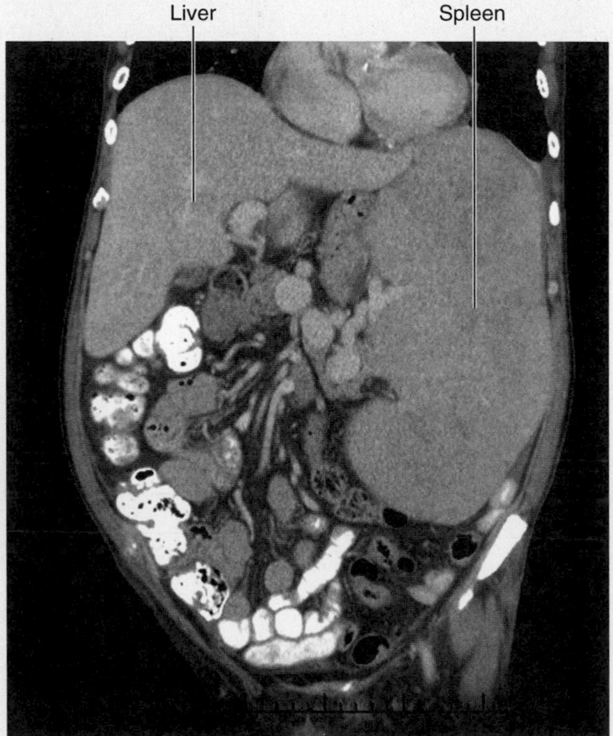


**Fig. 4.120** Coronal CT of the abdomen containing a massively enlarged spleen (splenomegaly).

## Arterial supply

The **abdominal aorta** begins at the aortic hiatus of the diaphragm, anterior to the lower border of vertebra TXII (Fig. 4.121). It descends through the abdomen, anterior to the vertebral bodies, and by the time it ends at the level of vertebra LIV it is slightly to the left of midline. The terminal branches of the abdominal aorta are the two **common iliac arteries**.

### Anterior branches of the abdominal aorta

The abdominal aorta has anterior, lateral, and posterior branches as it passes through the abdominal cavity. The three anterior branches supply the gastrointestinal viscera: the **celiac trunk** and the **superior mesenteric** and **inferior mesenteric arteries** (Fig. 4.121).

The primitive gut tube can be divided into foregut, midgut, and hindgut regions. The boundaries of these

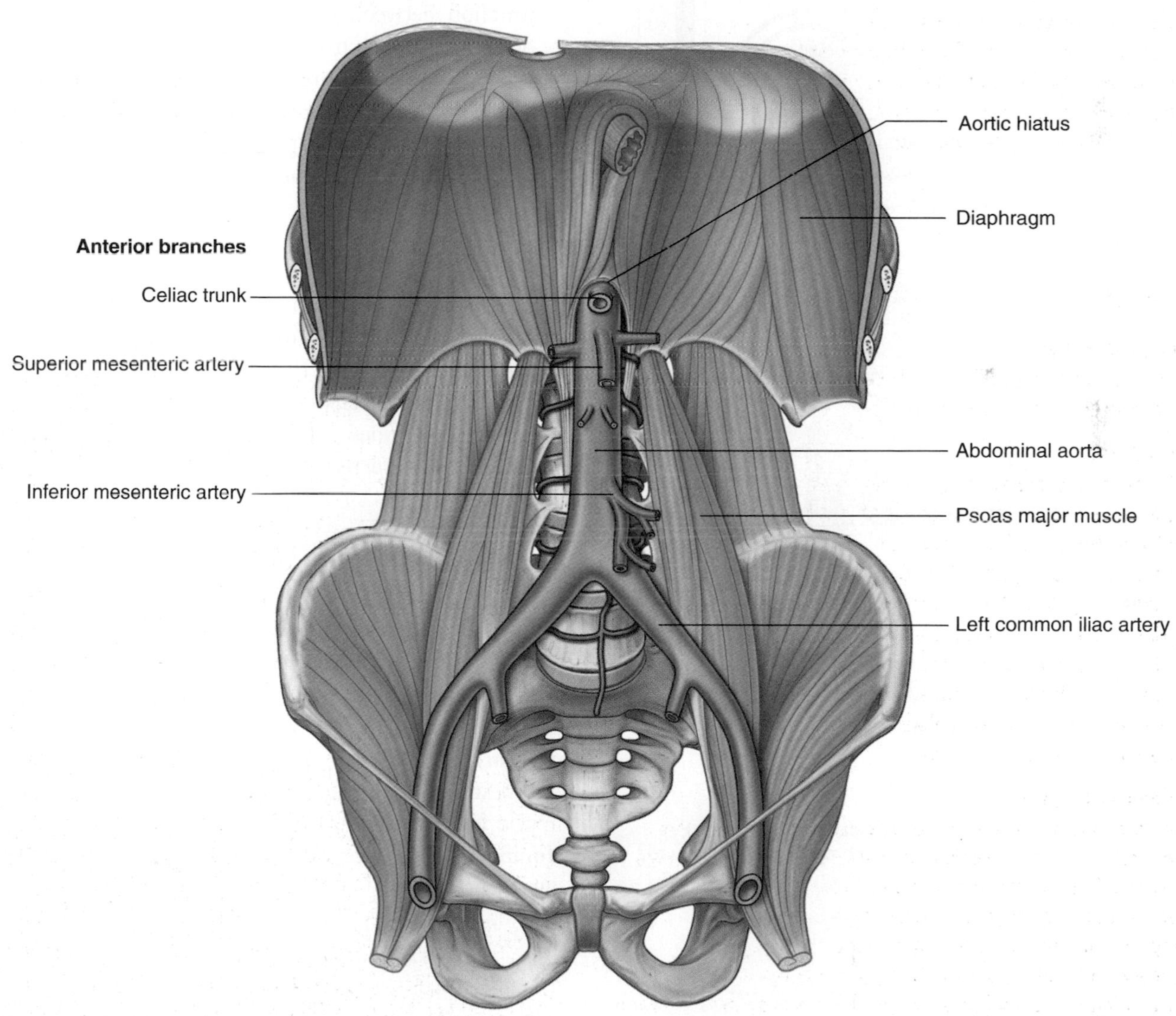


**Fig. 4.121** Anterior branches of the abdominal aorta.

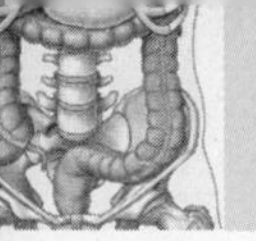

regions are directly related to the areas of distribution of the three anterior branches of the abdominal aorta (Fig. 4.122).

- The **foregut** begins with the abdominal esophagus and ends just inferior to the major duodenal papilla, midway along the descending part of the duodenum. It includes the abdominal esophagus, stomach, duodenum (superior to the major papilla), liver, pancreas, and gallbladder. The spleen also develops in relation to the foregut region. The foregut is supplied by the celiac trunk.
- The **midgut** begins just inferior to the major duodenal papilla, in the descending part of the duodenum, and ends at the junction between the proximal two-thirds and distal one-third of the transverse colon. It includes the duodenum (inferior to the major duodenal papilla), jejunum, ileum, cecum, appendix, ascending colon, and right two-thirds of the transverse colon. The midgut is supplied by the superior mesenteric artery (Fig. 4.122).
- The **hindgut** begins just before the left colic flexure (the junction between the proximal two-thirds and distal one-third of the transverse colon) and ends midway through the anal canal. It includes the left one-third of the transverse colon, descending colon, sigmoid colon, rectum, and upper part of the anal canal. The hindgut is supplied by the inferior mesenteric artery (Fig. 4.122).

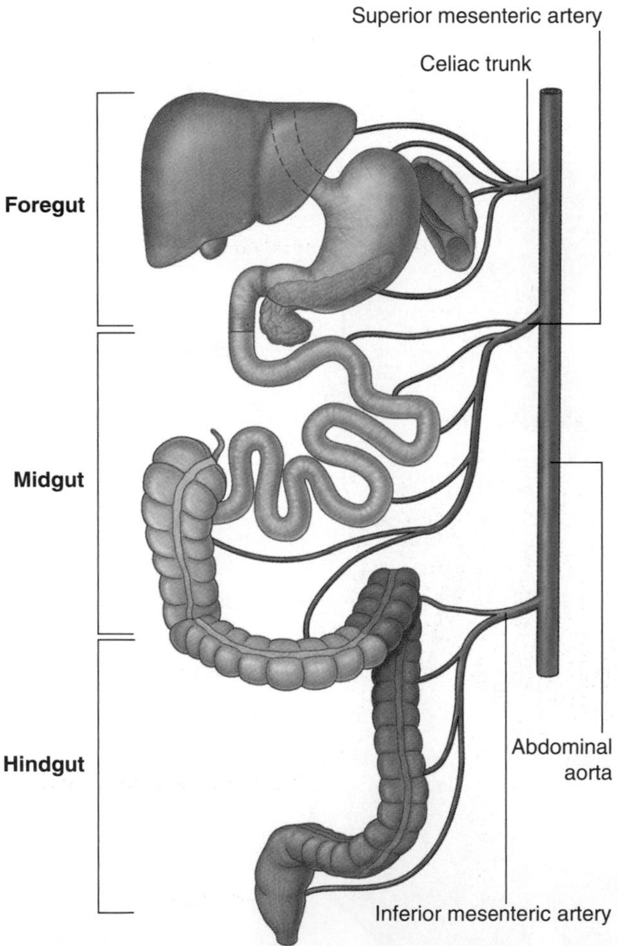


**Fig. 4.122** Divisions of the gastrointestinal tract into foregut, midgut, and hindgut, summarizing the primary arterial supply to each segment.

## Celiac trunk

The celiac trunk is the anterior branch of the abdominal aorta supplying the foregut. It arises from the abdominal aorta immediately below the aortic hiatus of the diaphragm (Fig. 4.123), anterior to the upper part of vertebra LI. It immediately divides into the left gastric, splenic, and common hepatic arteries.

### Left gastric artery

The **left gastric artery** is the smallest branch of the celiac trunk. It ascends to the cardioesophageal junction and sends **esophageal branches** upward to the abdominal part of the esophagus (Fig. 4.123). Some of these branches continue through the esophageal hiatus of the diaphragm and anastomose with esophageal branches from the thoracic aorta. The left gastric artery itself turns to the right and descends along the lesser curvature of the stomach in the lesser omentum. It supplies both surfaces of the stomach in this area and anastomoses with the right gastric artery.

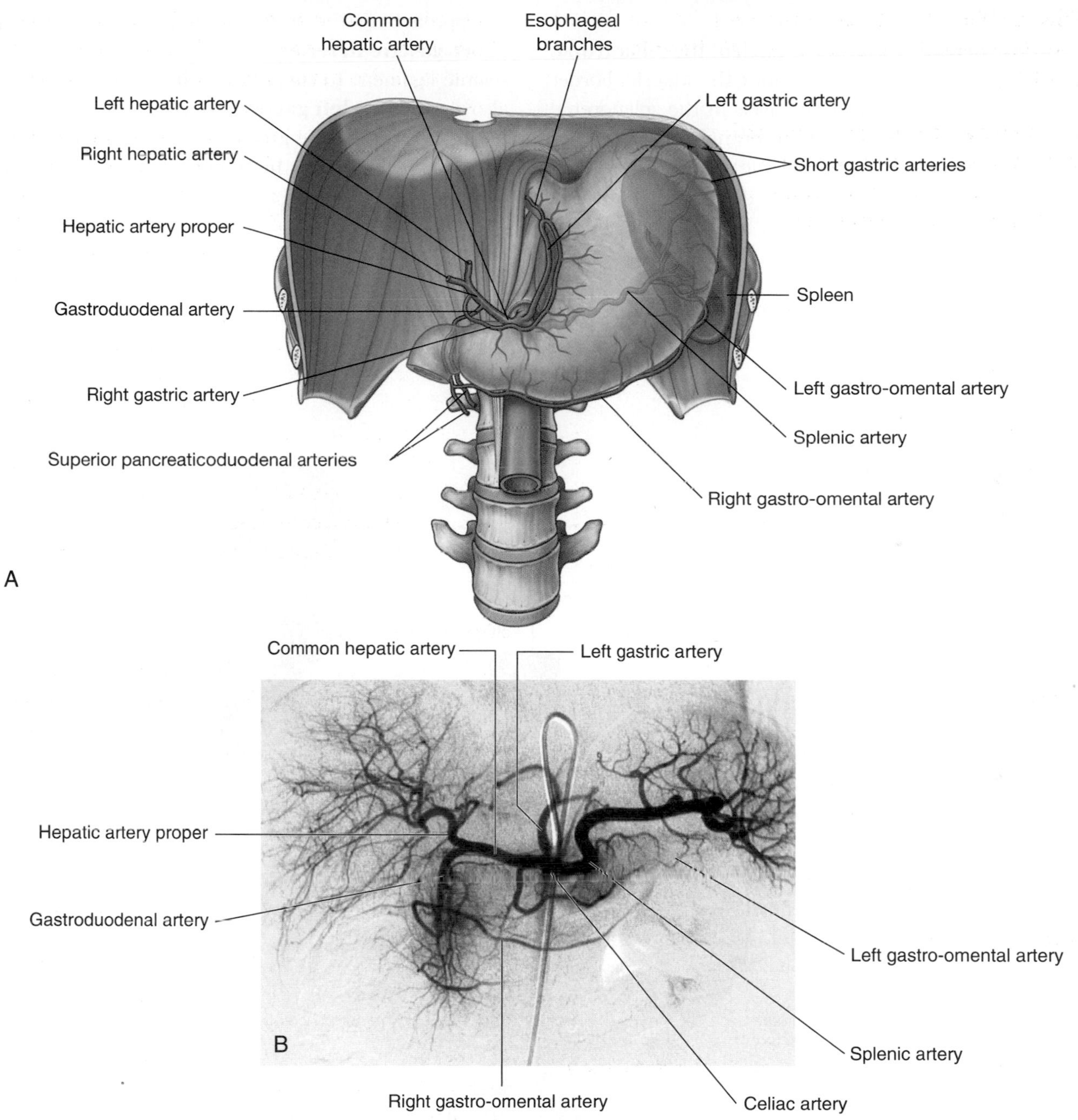


**Fig. 4.123** Celiac trunk. **A.** Distribution of the celiac trunk. **B.** Digital subtraction angiography of the celiac trunk and its branches.

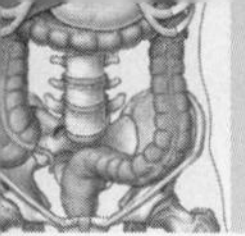



## Splenic artery

The **splenic artery**, the largest branch of the celiac trunk, takes a tortuous course to the left along the superior border of the pancreas (Fig. 4.123). It travels in the splenorenal ligament and divides into numerous branches, which enter the hilum of the spleen. As the splenic artery passes along the superior border of the pancreas, it gives off numerous small branches to supply the neck, body, and tail of the pancreas (Fig. 4.124).

Approaching the spleen, the splenic artery gives off **short gastric arteries**, which pass through the gastrosplenic ligament to supply the fundus of the stomach. It also gives off the **left gastro-omental artery**, which runs to the right along the greater curvature of the stomach, and anastomoses with the right gastro-omental artery.

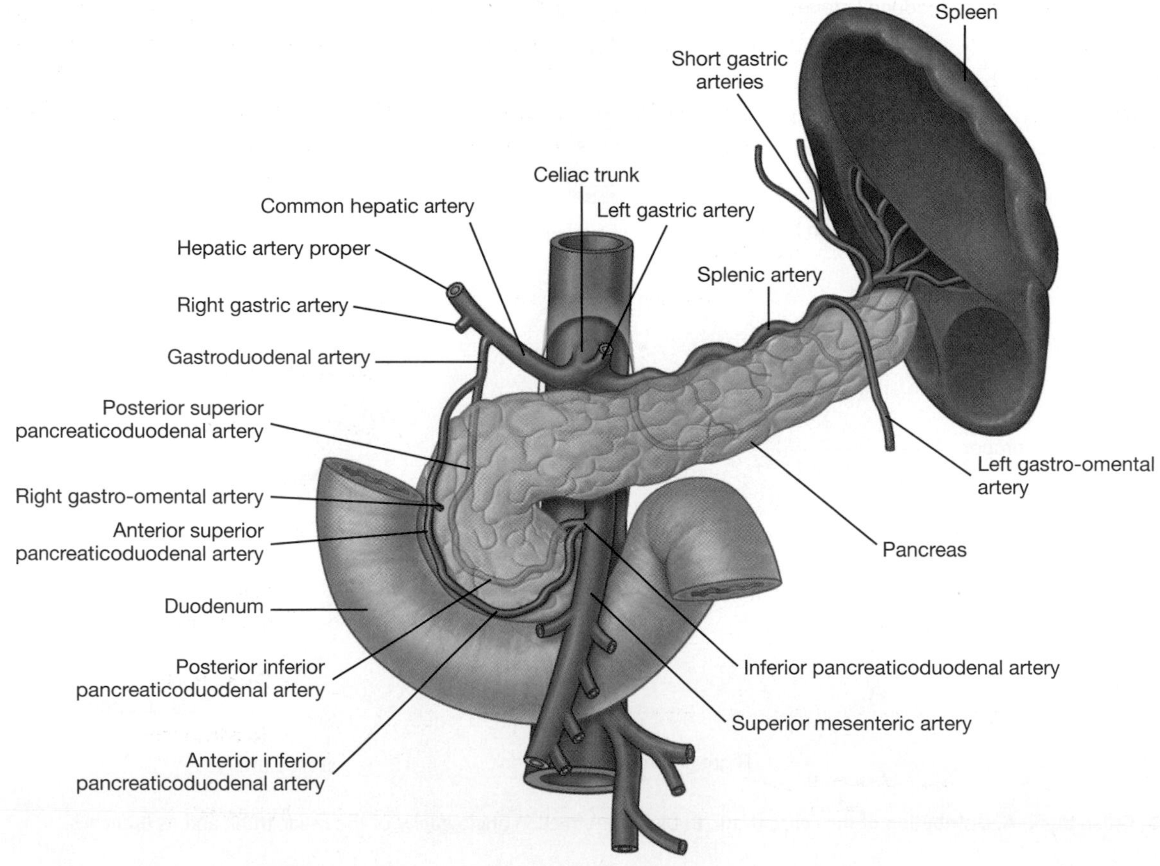


**Fig. 4.124** Arterial supply to the pancreas.

### Common hepatic artery

The **common hepatic artery** is a medium-sized branch of the celiac trunk that runs to the right and divides into its two terminal branches, the **hepatic artery proper** and the **gastroduodenal artery** (Figs. 4.123 and 4.124).

The hepatic artery proper ascends toward the liver in the free edge of the lesser omentum. It runs to the left of the bile duct and anterior to the portal vein, and divides into the **right** and **left hepatic arteries** near the porta hepatis (Fig. 4.125). As the right hepatic artery nears the liver, it gives off the cystic artery to the gallbladder.

The **right gastric artery** often originates from the hepatic artery proper but it can also arise from the common hepatic artery or from the left hepatic, gastroduodenal, or supraduodenal arteries. It courses to the left and ascends along the lesser curvature of the stomach in the lesser omentum, supplies adjacent areas of the stomach, and anastomoses with the left gastric artery.

The gastroduodenal artery may give off the **supraduodenal artery** and does give off the posterior superior pancreaticoduodenal artery near the upper border of the superior part of the duodenum. After these branch the gastroduodenal artery continues descending posterior to the superior part of the duodenum. Reaching the lower border of the superior part of the duodenum, the gastroduodenal artery divides into its terminal branches, the **right gastro-omental artery** and the **anterior superior pancreaticoduodenal artery** (Fig. 4.124).

The right gastro-omental artery passes to the left, along the greater curvature of the stomach, eventually anastomosing with the left gastro-omental artery from the splenic artery. The right gastro-omental artery sends branches to both surfaces of the stomach and additional branches descend into the greater omentum.

The anterior superior pancreaticoduodenal artery descends and, along with the posterior superior pancreaticoduodenal artery, supplies the head of the pancreas and the duodenum (Fig. 4.124). These vessels eventually anastomose with the anterior and posterior branches of the inferior pancreaticoduodenal artery.

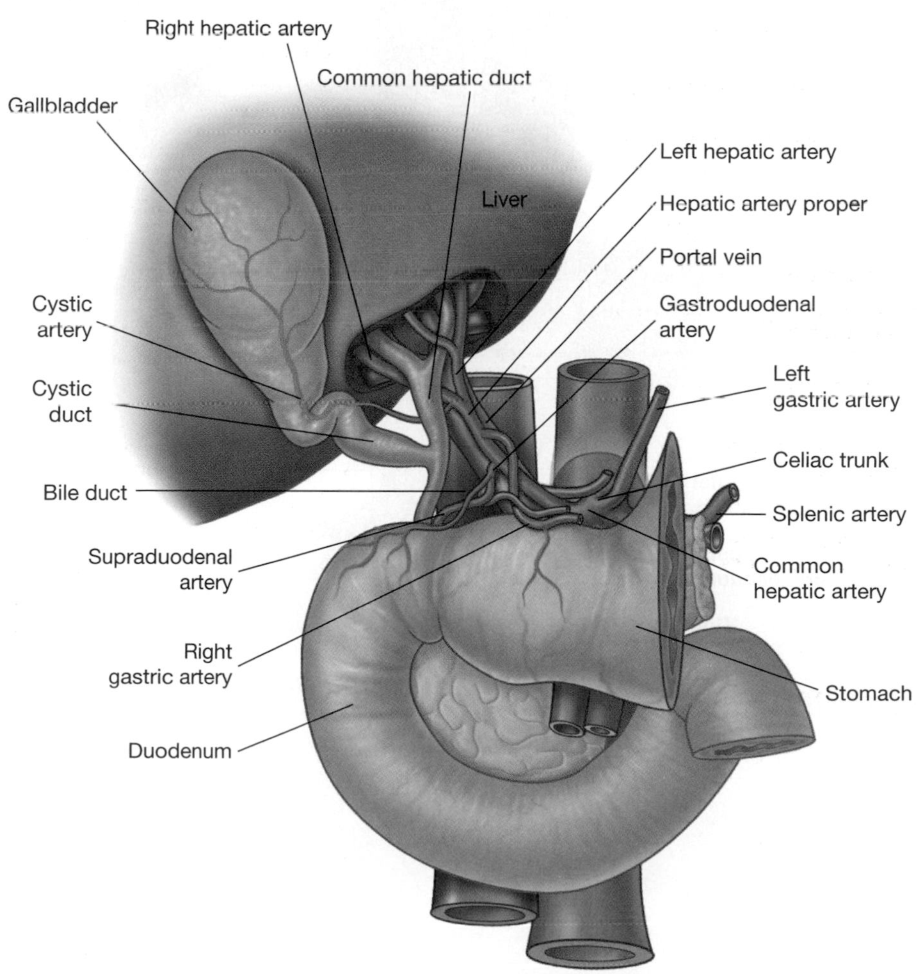


**Fig. 4.125** Distribution of the common hepatic artery.

### Superior mesenteric artery

The superior mesenteric artery is the anterior branch of the abdominal aorta supplying the midgut. It arises from the abdominal aorta immediately below the celiac artery (Fig. 4.126), anterior to the lower part of vertebra LI.

The superior mesenteric artery is crossed anteriorly by the splenic vein and the neck of the pancreas. Posterior to the artery are the left renal vein, the uncinate process of the pancreas, and the inferior part of the duodenum. After giving off its first branch (the **inferior pancreaticoduodenal artery**), the superior mesenteric artery gives off **jejunal** and **ileal arteries** on its left (Fig. 4.126). Branching from the right side of the main trunk of the superior mesenteric artery are three vessels—the **middle colic**, **right colic**, and **ileocolic arteries**—which supply the terminal ileum, cecum, ascending colon, and two-thirds of the transverse colon.

#### Inferior pancreaticoduodenal artery

The inferior pancreaticoduodenal artery is the first branch of the superior mesenteric artery. It divides immediately into anterior and posterior branches, which ascend on the corresponding sides of the head of the pancreas. Superiorly, these arteries anastomose with anterior and posterior superior pancreaticoduodenal arteries (see Figs. 4.125 and 4.126). This arterial network supplies the head and uncinate process of the pancreas and the duodenum.

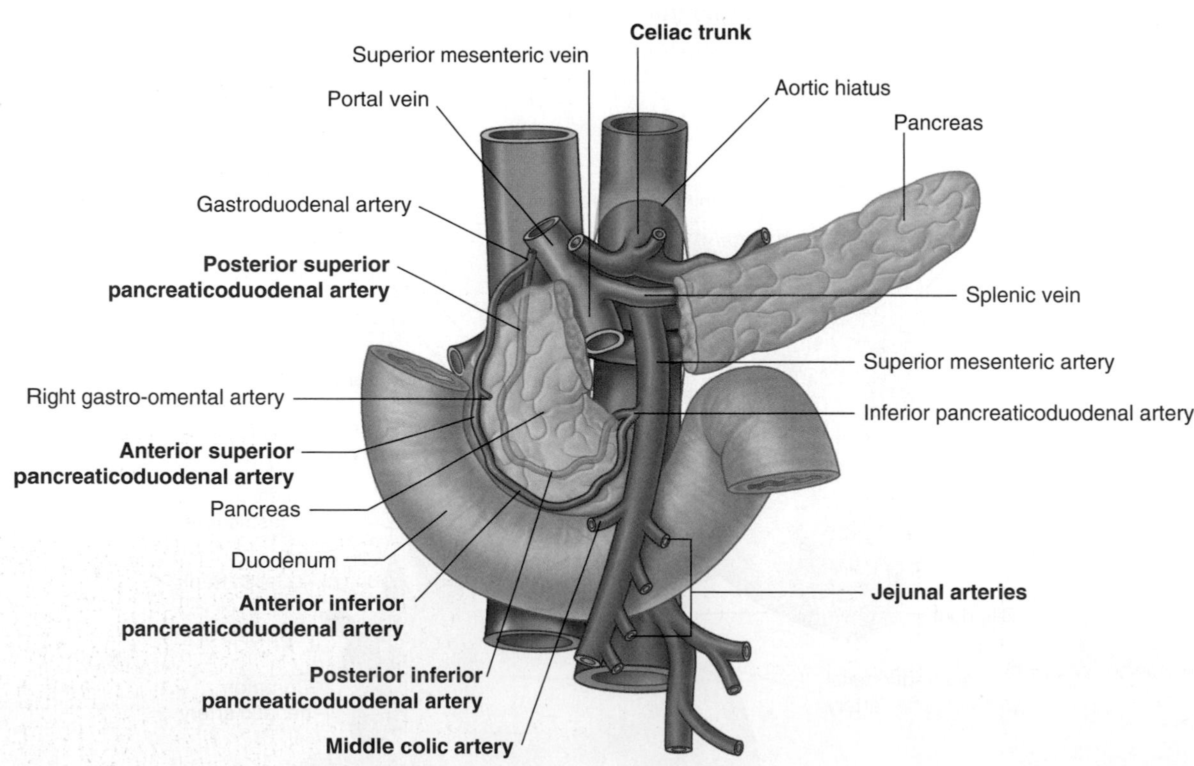


**Fig. 4.126** Initial branching and relationships of the superior mesenteric artery.

### Jejunal and ileal arteries

Distal to the inferior pancreaticoduodenal artery, the superior mesenteric artery gives off numerous branches. Arising on the left is a large number of jejunal and ileal arteries supplying the jejunum and most of the ileum (Fig. 4.127). These branches leave the main trunk of the artery, pass between two layers of the mesentery, and form anastomosing arches or arcades as they pass outward to supply the small intestine. The number of arterial arcades increases distally along the gut.

There may be single and then double arcades in the area of the jejunum, with a continued increase in the number

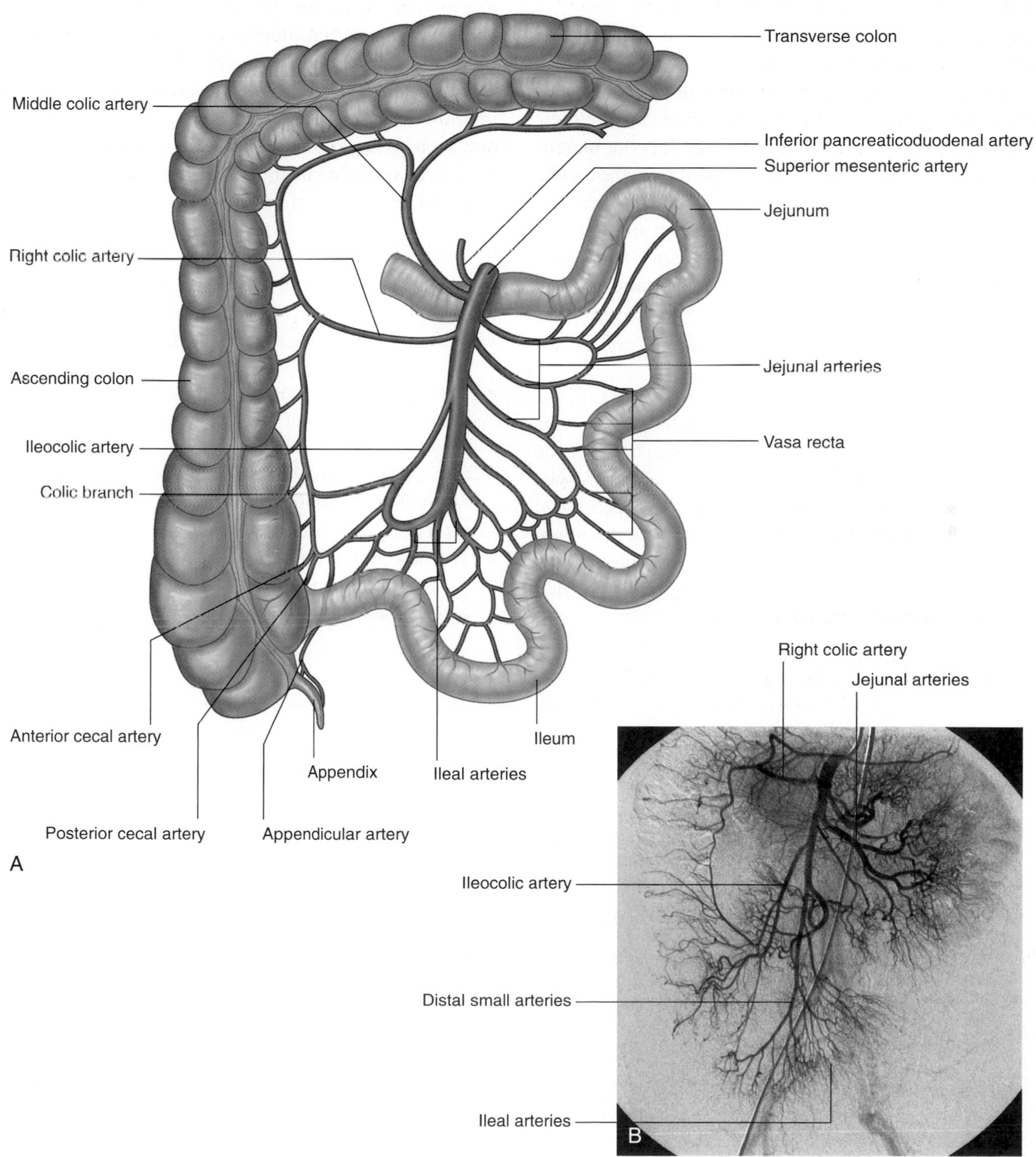


**Fig. 4.127** Superior mesenteric artery. **A.** Distribution of the superior mesenteric artery. **B.** Digital subtraction angiography of the superior mesenteric artery and its branches.

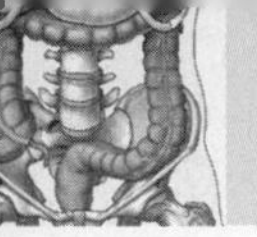



of arcades moving into and through the area of the ileum. Extending from the terminal arcade are vasa recta (straight arteries), which provide the final direct vascular supply to the walls of the small intestine. The **vasa recta** supplying the jejunum are usually long and close together, forming narrow windows visible in the mesentery. The vasa recta supplying the ileum are generally short and far apart, forming low broad windows.

#### Middle colic artery

The middle colic artery is the first of the three branches from the right side of the main trunk of the superior mesenteric artery (Fig. 4.127). Arising as the superior mesenteric artery emerges from beneath the pancreas, the middle colic artery enters the transverse mesocolon and divides into right and left branches. The right branch anastomoses with the right colic artery while the left branch anastomoses with the left colic artery, which is a branch of the inferior mesenteric artery.

#### Right colic artery

Continuing distally along the main trunk of the superior mesenteric artery, the right colic artery is the second of the three branches from the right side of the main trunk of the superior mesenteric artery (Fig. 4.126). It is an inconsistent branch, and passes to the right in a retroperitoneal position to supply the ascending colon. Nearing the colon, it divides into a descending branch, which anastomoses with the ileocolic artery, and an ascending branch, which anastomoses with the middle colic artery.

#### Ileocolic artery

The final branch arising from the right side of the superior mesenteric artery is the ileocolic artery (Fig. 4.127). This passes downward and to the right toward the right iliac fossa where it divides into superior and inferior branches:

- The superior branch passes upward along the ascending colon to anastomose with the right colic artery.
- The inferior branch continues toward the ileocolic junction, dividing into **colic, cecal, appendicular**, and **ileal branches** (Fig. 4.127).

The specific pattern of distribution and origin of these branches is variable:

- The colic branch crosses to the ascending colon and passes upward to supply the first part of the ascending colon.
- Anterior and posterior cecal branches, arising either as a common trunk or as separate branches, supply corresponding sides of the cecum.
- The appendicular branch enters the free margin of and supplies the mesoappendix and the appendix.
- The ileal branch passes to the left and ascends to supply the final part of the ileum before anastomosing with the superior mesenteric artery.

### Inferior mesenteric artery

The inferior mesenteric artery is the anterior branch of the abdominal aorta that supplies the hindgut. It is the smallest of the three anterior branches of the abdominal aorta and arises anterior to the body of vertebra LIII. Initially, the inferior mesenteric artery descends anteriorly to the aorta and then passes to the left as it continues inferiorly (Fig. 4.128). Its branches include the **left colic artery, several sigmoid arteries**, and the **superior rectal artery**.

#### Left colic artery

The left colic artery is the first branch of the inferior mesenteric artery (Fig. 4.128). It ascends retroperitoneally, dividing into ascending and descending branches:

- The ascending branch passes anteriorly to the left kidney, then enters the transverse mesocolon, and passes superiorly to supply the upper part of the descending colon and the distal part of the transverse colon; it anastomoses with branches of the middle colic artery.
- The descending branch passes inferiorly, supplying the lower part of the descending colon, and anastomoses with the first sigmoid artery.

#### Sigmoid arteries

The sigmoid arteries consist of two to four branches, which descend to the left, in the sigmoid mesocolon, to supply the lowest part of the descending colon and the sigmoid colon (Fig. 4.128). These branches anastomose superiorly with branches from the left colic artery and inferiorly with branches from the superior rectal artery.

#### Superior rectal artery

The terminal branch of the inferior mesenteric artery is the superior rectal artery (Fig. 4.128). This vessel descends into the pelvic cavity in the sigmoid mesocolon, crossing the left common iliac vessels. Opposite vertebra SIII, the superior rectal artery divides. The two terminal branches descend on each side of the rectum, dividing into smaller branches in the wall of the rectum. These smaller branches continue inferiorly to the level of the internal anal sphincter, anastomosing along the way with branches from the middle rectal arteries (from the internal iliac artery) and the inferior rectal arteries (from the internal pudendal artery).

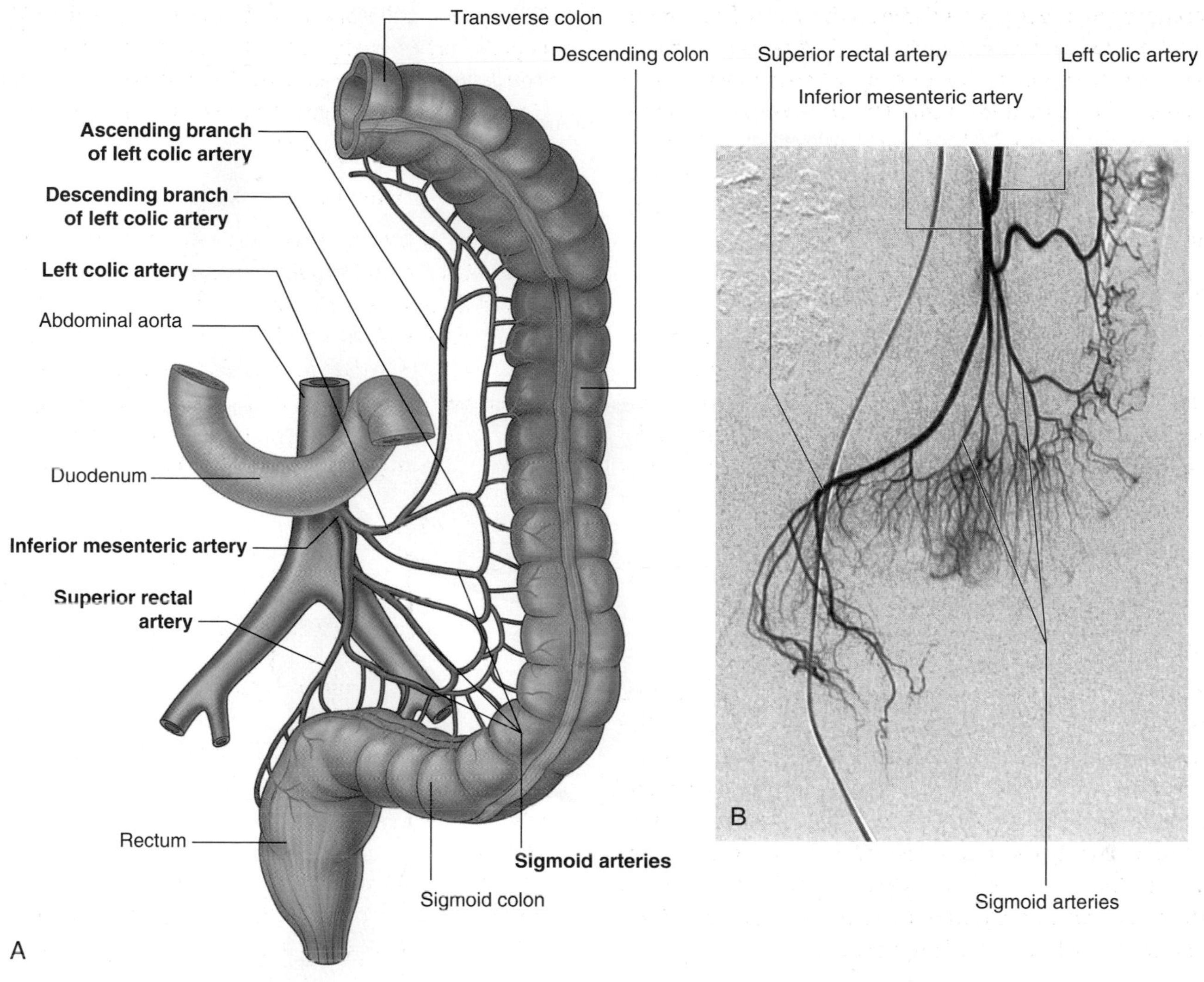


**Fig. 4.128** Inferior mesenteric artery. **A.** Distribution of the inferior mesenteric artery. **B.** Digital subtraction angiography of the inferior mesenteric artery and its branches.

## In the clinic

### Vascular supply to the gastrointestinal system

The abdominal parts of the gastrointestinal system are supplied mainly by the celiac trunk and the superior mesenteric and inferior mesenteric arteries (Fig. 4.129):

- The celiac trunk supplies the lower esophagus, stomach, superior part of the duodenum, and proximal half of the descending part of the duodenum.
- The superior mesenteric artery supplies the rest of the duodenum, the jejunum, the ileum, the ascending colon, and the proximal two-thirds of the transverse colon.
- The inferior mesenteric artery supplies the rest of the transverse colon, the descending colon, the sigmoid colon, and most of the rectum.

Along the descending part of the duodenum there is a potential watershed area between the celiac trunk blood supply and the superior mesenteric arterial blood supply. It is unusual for this area to become ischemic, whereas the watershed area between the superior mesenteric artery and the inferior mesenteric artery, at the splenic flexure, is extremely vulnerable to ischemia.

In certain disease states, the region of the splenic flexure of the colon can become ischemic. When this occurs, the mucosa sloughs off, rendering the patient susceptible to infection and perforation of the large bowel, which then requires urgent surgical attention.

*(continues)*

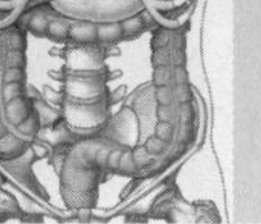



## In the clinic—cont'd



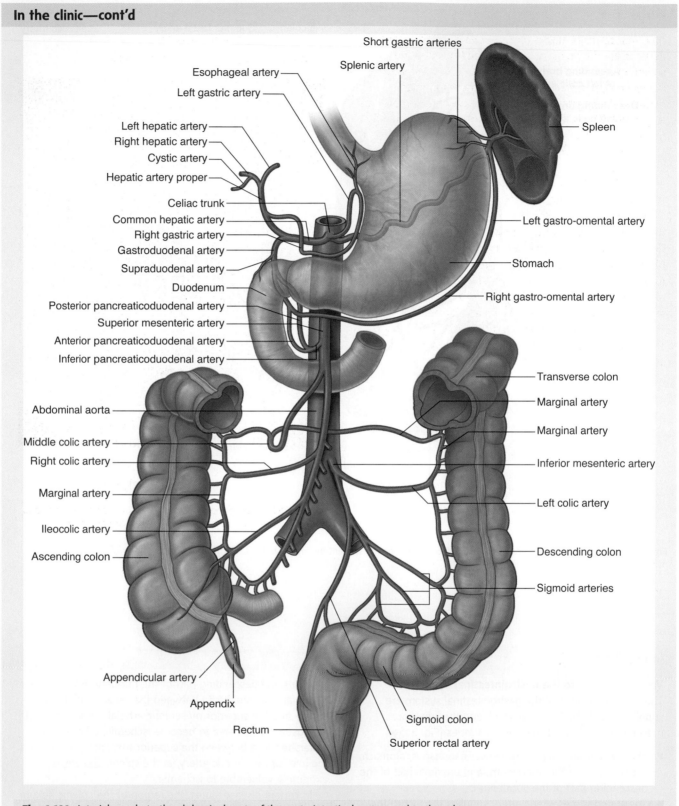

**Fig. 4.129** Arterial supply to the abdominal parts of the gastrointestinal system and to the spleen.

## In the clinic—cont'd

Arteriosclerosis may occur throughout the abdominal aorta and at the openings of the celiac trunk and the superior mesenteric and inferior mesenteric arteries. Not infrequently, the inferior mesenteric artery becomes occluded. Interestingly, many of these patients do not suffer any complications, because anastomoses between the right, middle, and left colic arteries gradually enlarge, forming a continuous **marginal artery**. The distal large bowel therefore becomes supplied by this enlarged marginal artery (marginal artery of Drummond), which replaces the blood supply of the inferior mesenteric artery (Fig. 4.130).

If the openings of the celiac trunk and superior mesenteric artery become narrowed, the blood supply to the gut is diminished. After a heavy meal, the oxygen demand of the bowel therefore outstrips the limited supply of blood through the stenosed vessels, resulting in severe pain and discomfort (**mesenteric angina**). Patients with this condition tend not to eat because of the pain and rapidly lose weight. The diagnosis is determined by aortic angiography, and the stenoses of the celiac trunk and superior mesenteric artery are best appreciated in the lateral view.

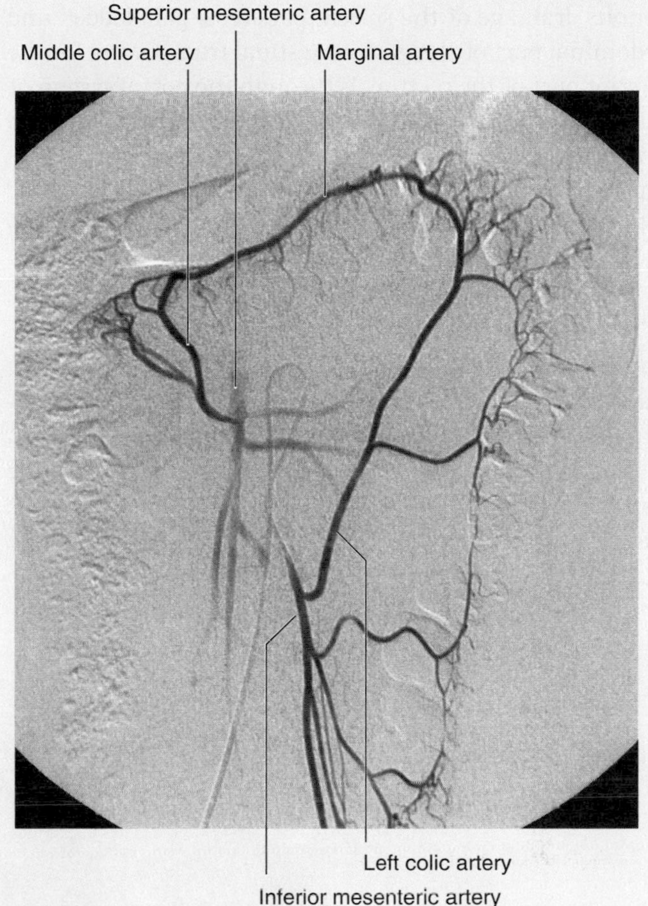


**Fig. 4.130** Enlarged marginal artery connecting the superior and inferior mesenteric arteries. Digital subtraction angiogram.

### Venous drainage

Venous drainage of the spleen, pancreas, gallbladder, and abdominal part of the gastrointestinal tract, except for the inferior part of the rectum, is through the portal system of veins, which deliver blood from these structures to the liver. Once blood passes through the hepatic sinusoids, it passes through progressively larger veins until it enters the hepatic veins, which return the venous blood to the inferior vena cava just inferior to the diaphragm.

#### Portal vein

The **portal vein** is the final common pathway for the transport of venous blood from the spleen, pancreas, gallbladder, and abdominal part of the gastrointestinal tract. It is formed by the union of the **splenic vein** and the **superior mesenteric vein** posterior to the neck of the pancreas at the level of vertebra LII (Fig. 4.131).

Ascending toward the liver, the portal vein passes posterior to the superior part of the duodenum and enters the right margin of the lesser omentum. As it passes through this part of the lesser omentum, it is anterior to the omental foramen and posterior to both the bile duct, which is slightly to its right, and the hepatic artery proper, which is slightly to its left (see Fig. 4.125, p. 347).

On approaching the liver, the portal vein divides into **right** and **left branches**, which enter the liver parenchyma. Tributaries to the portal vein include:

- **right** and **left gastric veins** draining the lesser curvature of the stomach and abdominal esophagus,
- **cystic veins** from the gallbladder, and
- the **para-umbilical veins**, which are associated with the obliterated umbilical vein and connect to veins on the anterior abdominal wall (Fig. 4.133 on p. 357).

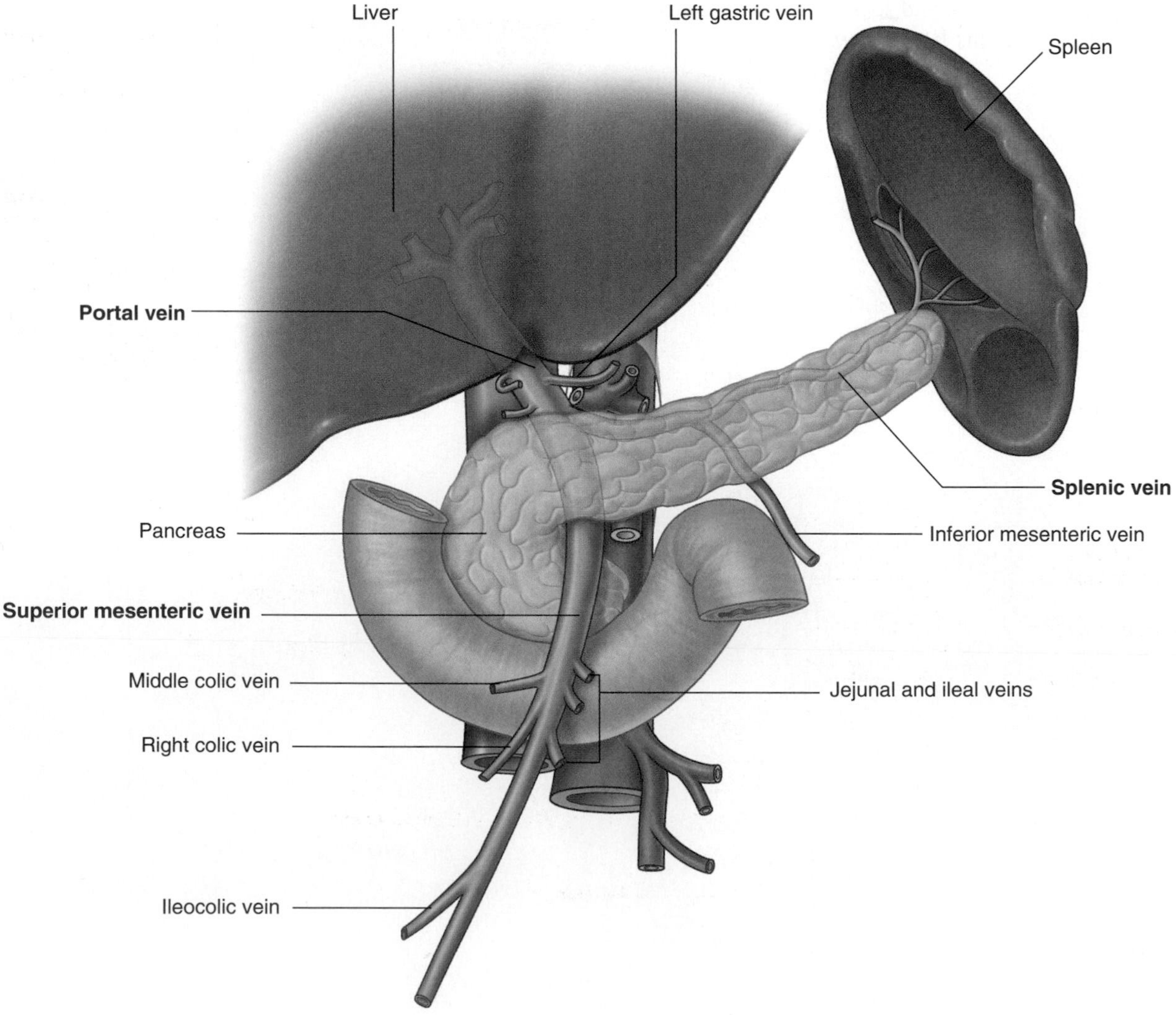

**Fig. 4.131** Portal vein.

### Splenic vein

The splenic vein forms from numerous smaller vessels leaving the hilum of the spleen (Fig. 4.132). It passes to the right, passing through the splenorenal ligament with the splenic artery and the tail of the pancreas. Continuing to the right, the large, straight splenic vein is in contact with the body of the pancreas as it crosses the posterior abdominal wall. Posterior to the neck of the pancreas, the splenic vein joins the superior mesenteric vein to form the portal vein.

Tributaries to the splenic vein include:

- **short gastric veins** from the fundus and left part of the greater curvature of the stomach,
- the **left gastro-omental vein** from the greater curvature of the stomach,
- **pancreatic veins** draining the body and tail of the pancreas, and
- usually the **inferior mesenteric vein**.

### Superior mesenteric vein

The superior mesenteric vein drains blood from the small intestine, cecum, ascending colon, and transverse colon (Fig. 4.132). It begins in the right iliac fossa as veins draining the terminal ileum, cecum, and appendix join, and ascends in the mesentery to the right of the superior mesenteric artery.

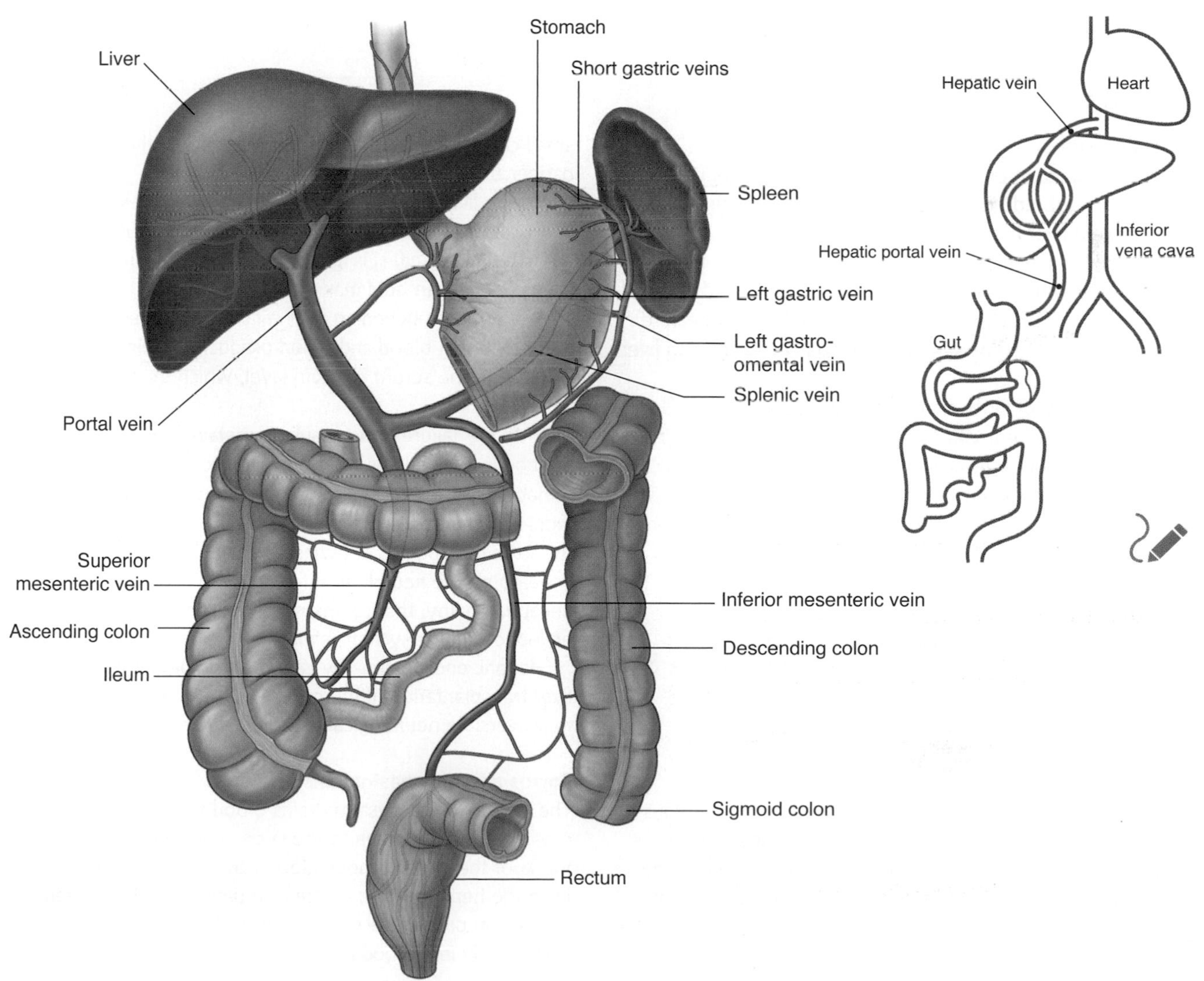


**Fig. 4.132** Venous drainage of the abdominal portion of the gastrointestinal tract.

Posterior to the neck of the pancreas, the superior mesenteric vein joins the splenic vein to form the portal vein.

As a corresponding vein accompanies each branch of the superior mesenteric artery, tributaries to the superior mesenteric vein include jejunal, ileal, ileocolic, right colic, and middle colic veins. Additional tributaries include:

- the **right gastro-omental vein**, draining the right part of the greater curvature of the stomach, and
- the **anterior** and **posterior inferior pancreaticoduodenal veins**, which pass alongside the arteries of the same name; the anterior superior pancreaticoduodenal vein usually empties into the right gastro-omental vein, and the posterior superior pancreaticoduodenal vein usually empties directly into the portal vein.

### Inferior mesenteric vein

The **inferior mesenteric vein** drains blood from the rectum, sigmoid colon, descending colon, and **splenic flexure** (Fig. 4.132). It begins as the **superior rectal vein** and ascends, receiving tributaries from the sigmoid veins and the **left colic vein**. All these veins accompany arteries of the same name. Continuing to ascend, the inferior mesenteric vein passes posterior to the body of the pancreas and usually joins the splenic vein. Occasionally, it ends at the junction of the splenic and superior mesenteric veins or joins the superior mesenteric vein.

## In the clinic

### Hepatic cirrhosis

Cirrhosis is a complex disorder of the liver, the diagnosis of which is confirmed histologically. When a diagnosis is suspected, a liver biopsy is necessary.

Cirrhosis is characterized by widespread hepatic fibrosis interspersed with areas of nodular regeneration and abnormal reconstruction of preexisting lobular architecture. The presence of cirrhosis implies previous or continuing liver cell damage.

The etiology of cirrhosis is complex and includes toxins (alcohol), viral inflammation, biliary obstruction, vascular outlet obstruction, nutritional (malnutrition) causes, and inherited anatomical and metabolic disorders.

As the cirrhosis progresses, the intrahepatic vasculature is distorted, which in turn leads to increased pressure in the portal vein and its draining tributaries (portal hypertension). Portal hypertension produces increased pressure in the splenic venules, leading to splenic enlargement. At the sites of portosystemic anastomosis (see below), large dilated veins (varices) develop. These veins are susceptible to bleeding and may produce marked blood loss, which in some instances can be fatal.

The liver is responsible for the production of numerous proteins, including those of the clotting cascade. Any disorder of the liver (including infection and cirrhosis) may decrease the production of these proteins and so prevent adequate blood clotting. Patients with severe cirrhosis of the liver have a significant risk of serious bleeding, even from small cuts; in addition, when varices rupture, there is a danger of rapid exsanguination.

As the liver progressively fails, the patient develops salt and water retention, which produces skin and subcutaneous edema. Fluid (ascites) is also retained in the peritoneal cavity, which can hold many liters.

The poorly functioning liver cells (hepatocytes) are unable to break down blood and blood products, leading to an increase in the serum bilirubin level, which manifests as jaundice.

With the failure of normal liver metabolism, toxic metabolic by-products do not convert to nontoxic metabolites. This buildup of noxious compounds is made worse by the numerous portosystemic shunts, which allow the toxic metabolites to bypass the liver. Patients may develop severe neurological features, called hepatic encephalopathy, that can manifest as acute confusion, epileptic fits, or psychotic state.

Hepatic encephalopathy is one of the urgent criteria for liver transplantation; if the condition is not reversed, it leads to irreversible neurological damage and death.

### Portosystemic anastomosis

The hepatic portal system drains blood from the visceral organs of the abdomen to the liver. In normal individuals, 100% of the portal venous blood flow can be recovered from the hepatic veins, whereas in patients with elevated portal vein pressure (e.g., from cirrhosis), there is significantly less blood flow to the liver. The rest of the

## In the clinic—cont'd

blood enters collateral channels, which drain into the systemic circulation at specific points (Fig. 4.133). The largest of these collaterals occur at:

- the gastroesophageal junction around the cardia of the stomach—where the left gastric vein and its tributaries form a portosystemic anastomosis with tributaries to the azygos system of veins of the caval system;
- the anus—the superior rectal vein of the portal system anastomoses with the middle and inferior rectal veins of the systemic venous system; and
- the anterior abdominal wall around the umbilicus—the para-umbilical veins anastomose with veins on the anterior abdominal wall.

When the pressure in the portal vein is elevated, venous enlargement (varices) tend to occur at and around the sites of portosystemic anastomoses and these enlarged veins are called:

- varices at the anorectal junction,
- esophageal varices at the gastroesophageal junction, and
- caput medusae at the umbilicus.

Esophageal varices are susceptible to trauma and, once damaged, may bleed profusely, requiring urgent surgical intervention.

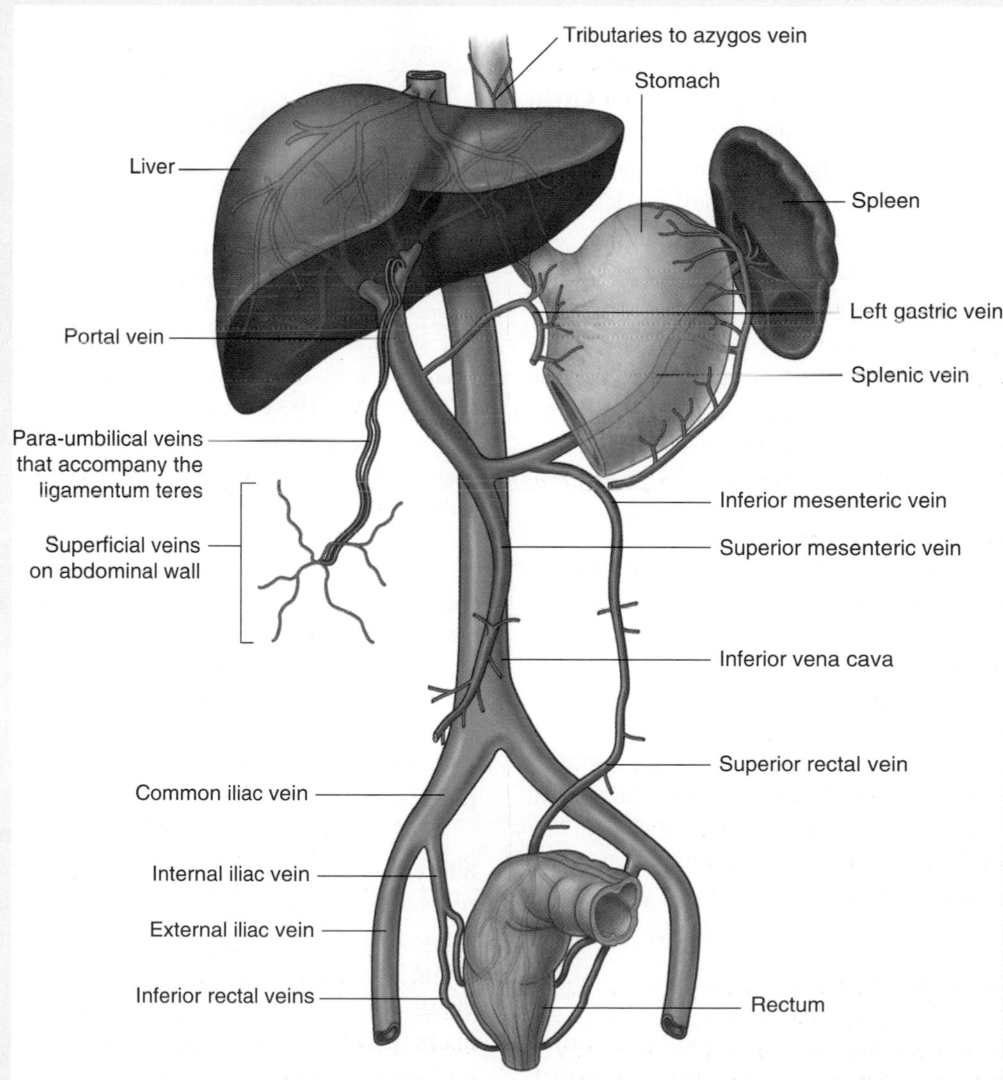


**Fig. 4.133** Portosystemic anastomoses.

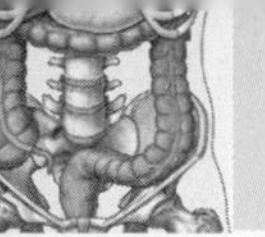



## Lymphatics

Lymphatic drainage of the abdominal part of the gastrointestinal tract, as low as the inferior part of the rectum, as well as the spleen, pancreas, gallbladder, and liver, is through vessels and nodes that eventually end in large collections of **pre-aortic lymph nodes** at the origins of the three anterior branches of the abdominal aorta, which supply these structures. These collections are therefore referred to as the **celiac, superior mesenteric**, and **inferior mesenteric** groups of pre-aortic lymph nodes. Lymph from viscera is supplied by three routes:

- The celiac trunk (i.e., structures that are part of the abdominal foregut) drains to pre-aortic nodes near the origin of the celiac trunk (Fig. 4.134)—these celiac nodes also receive lymph from the superior mesenteric and inferior mesenteric groups of pre-aortic nodes, and lymph from the celiac nodes enters the **cisterna chyli**.
- The superior mesenteric artery (i.e., structures that are part of the abdominal midgut) drains to pre-aortic nodes near the origin of the superior mesenteric artery (Fig. 4.134)—these superior mesenteric nodes also receive lymph from the inferior mesenteric groups of pre-aortic nodes, and lymph from the superior mesenteric nodes drains to the celiac nodes.
- The inferior mesenteric artery (i.e., structures that are part of the abdominal hindgut) drains to pre-aortic nodes near the origin of the inferior mesenteric artery (Fig. 4.134), and lymph from the inferior mesenteric nodes drains to the superior mesenteric nodes.

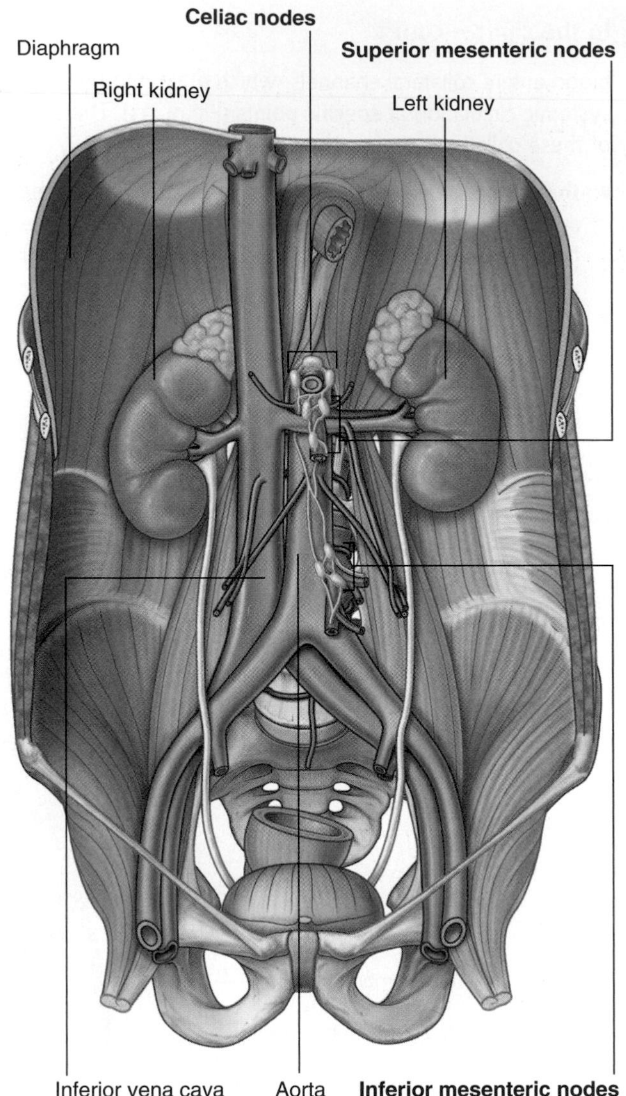


**Fig. 4.134** Lymphatic drainage of the abdominal portion of the gastrointestinal tract.

## Innervation

Abdominal viscera are innervated by both extrinsic and intrinsic components of the nervous system:

- Extrinsic innervation involves receiving motor impulses from, and sending sensory information to, the central nervous system.
- Intrinsic innervation involves the regulation of digestive tract activities by a generally self-sufficient network of sensory and motor neurons (the enteric nervous system).

Abdominal viscera receiving extrinsic innervation include the abdominal part of the gastrointestinal tract, the spleen, the pancreas, the gallbladder, and the liver. These viscera send sensory information back to the central nervous system through visceral afferent fibers and receive motor impulses from the central nervous system through visceral efferent fibers.

The visceral efferent fibers are part of the sympathetic and parasympathetic parts of the autonomic division of the peripheral nervous system.

Structural components serving as conduits for these afferent and efferent fibers include posterior and anterior roots of the spinal cord, respectively, spinal nerves, anterior rami, white and gray rami communicantes, the sympathetic trunks, splanchnic nerves carrying sympathetic fibers (thoracic, lumbar, and sacral), parasympathetic fibers (pelvic), the prevertebral plexus and related ganglia, and the vagus nerves [X].

The enteric nervous system consists of motor and sensory neurons in two interconnected plexuses in the walls of the gastrointestinal tract. These neurons control the coordinated contraction and relaxation of intestinal smooth muscle and regulate gastric secretion and blood flow.

## Sympathetic trunks

The sympathetic trunks are two parallel nerve cords extending on either side of the vertebral column from the base of the skull to the coccyx (Fig. 4.135). As they pass through the neck, they lie posterior to the carotid sheath. In the upper thorax, they are anterior to the necks of the ribs, while in the lower thorax they are on the lateral aspect of the vertebral bodies. In the abdomen, they are anterolateral to the lumbar vertebral bodies and, continuing into the pelvis, they are anterior to the sacrum. The two sympathetic trunks come together anterior to the coccyx to form the **ganglion impar**.

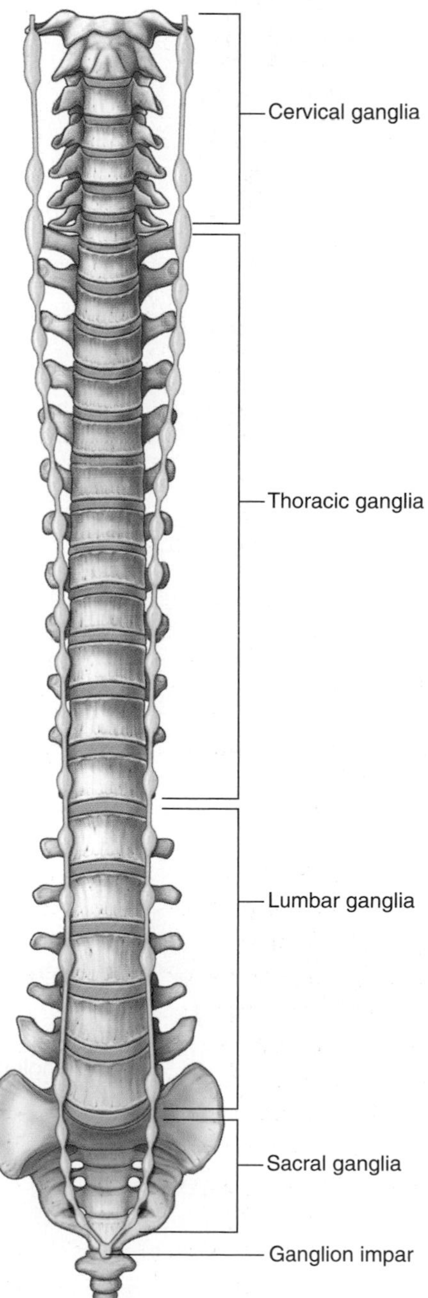


**Fig. 4.135** Sympathetic trunks.

Throughout the extent of the sympathetic trunks, small raised areas are visible. These collections of neuronal cell bodies outside the CNS are the paravertebral sympathetic ganglia. There are usually:

- three ganglia in the cervical region,
- eleven or twelve ganglia in the thoracic region,
- four ganglia in the lumbar region,
- four or five ganglia in the sacral region, and
- the ganglion impar anterior to the coccyx (Fig. 4.135).

The ganglia and trunks are connected to adjacent spinal nerves by gray rami communicantes throughout the length of the sympathetic trunk and by white rami communicantes in the thoracic and upper lumbar parts of the trunk (T1 to L2). Neuronal fibers found in the sympathetic trunks include **preganglionic** and **postganglionic sympathetic fibers** and **visceral afferent fibers**.

### Splanchnic nerves

The splanchnic nerves are important components in the innervation of the abdominal viscera. They pass from the sympathetic trunk or sympathetic ganglia associated with the trunk, to the prevertebral plexus and ganglia anterior to the abdominal aorta.

There are two different types of splanchnic nerves, depending on the type of visceral efferent fiber they are carrying:

- The thoracic, lumbar, and sacral splanchnic nerves carry preganglionic sympathetic fibers from the sympathetic trunk to ganglia in the prevertebral plexus, and also visceral afferent fibers.
- The pelvic splanchnic nerves carry preganglionic parasympathetic fibers from anterior rami of S2, S3, and S4 spinal nerves to an extension of the prevertebral plexus in the pelvis (the **inferior hypogastric plexus** or **pelvic plexus**).

### Thoracic splanchnic nerves

Three **thoracic splanchnic nerves** pass from sympathetic ganglia along the sympathetic trunk in the thorax to the prevertebral plexus and ganglia associated with the abdominal aorta in the abdomen (Fig. 4.136):

- The greater splanchnic nerve arises from the fifth to the ninth (or tenth) thoracic ganglia and travels to the

Cervical ganglia
Thoracic splanchnic nerves
Greater splanchnic nerve
Lesser splanchnic nerve
Least splanchnic nerve
Thoracic ganglia
Prevertebral plexus
Lumbar splanchnic nerves
Lumbar ganglia
Inferior hypogastric plexus
Sacral ganglia
Ganglion impar
Sacral splanchnic nerves
Pelvic splanchnic nerves

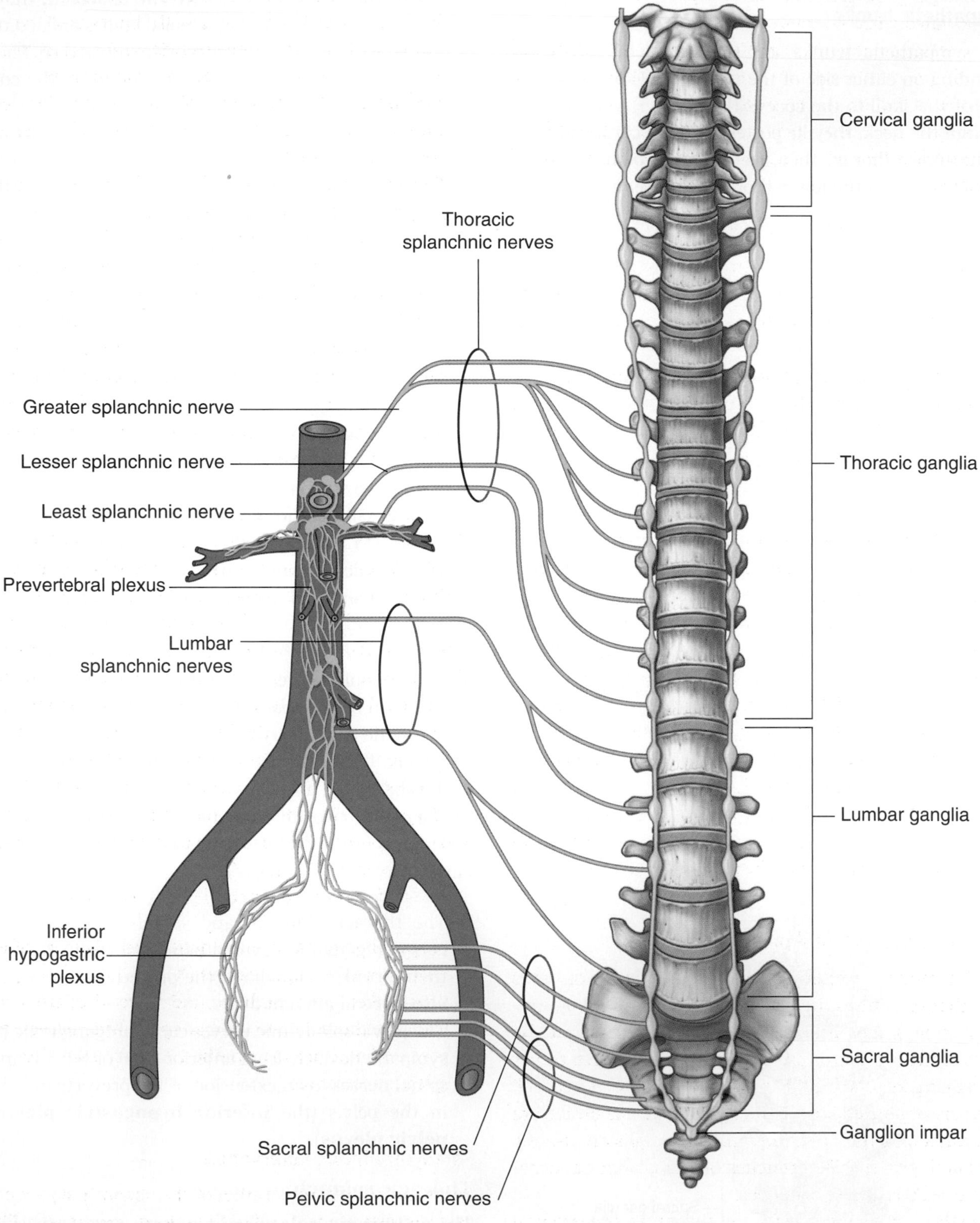

**Fig. 4.136** Splanchnic nerves.

celiac ganglion in the abdomen (a prevertebral ganglion associated with the celiac trunk).

- The lesser splanchnic nerve arises from the ninth and tenth (or tenth and eleventh) thoracic ganglia and travels to the aorticorenal ganglion.
- The least splanchnic nerve, when present, arises from the twelfth thoracic ganglion and travels to the renal plexus.

#### Lumbar and sacral splanchnic nerves

There are usually two to four **lumbar splanchnic nerves**, which pass from the lumbar part of the sympathetic trunk or associated ganglia and enter the prevertebral plexus (Fig. 4.136).

Similarly, the **sacral splanchnic nerves** pass from the sacral part of the sympathetic trunk or associated ganglia and enter the inferior hypogastric plexus, which is an extension of the prevertebral plexus into the pelvis.

#### Pelvic splanchnic nerves

The **pelvic splanchnic nerves** (**parasympathetic root**) are unique. They are the only splanchnic nerves that carry parasympathetic fibers. In other words, they do not originate from the sympathetic trunks. Rather, they originate directly from the anterior rami of S2 to S4. Preganglionic parasympathetic fibers originating in the sacral spinal cord pass from the S2 to S4 spinal nerves to the inferior hypogastric plexus (Fig. 4.136). Once in this plexus, some of these fibers pass upward, enter the abdominal prevertebral plexus, and distribute with the arteries supplying the hindgut. This provides the pathway for innervation of the distal one-third of the transverse colon, the descending colon, and the sigmoid colon by preganglionic parasympathetic fibers.

### Abdominal prevertebral plexus and ganglia

The abdominal prevertebral plexus is a collection of nerve fibers that surrounds the abdominal aorta and is continuous onto its major branches. Scattered throughout the length of the abdominal prevertebral plexus are cell bodies of postganglionic sympathetic fibers. Some of these cell bodies are organized into distinct ganglia, while others are more random in their distribution. The ganglia are usually associated with specific branches of the abdominal aorta and named after these branches.

The three major divisions of the abdominal prevertebral plexus and associated ganglia are the celiac, aortic, and superior hypogastric plexuses (Fig. 4.137).

- The celiac plexus is the large accumulation of nerve fibers and ganglia associated with the roots of the celiac trunk and superior mesenteric artery immediately below the aortic hiatus of the diaphragm. Ganglia associated with the celiac plexus include two celiac ganglia, a single superior mesenteric ganglion, and two aorticorenal ganglia.
- The aortic plexus consists of nerve fibers and associated ganglia on the anterior and lateral surfaces of the abdominal aorta extending from just below the origin of the superior mesenteric artery to the bifurcation of the aorta into the two common iliac arteries. The major ganglion in this plexus is the inferior mesenteric ganglion at the root of the inferior mesenteric artery.
- The superior hypogastric plexus contains numerous small ganglia and is the final part of the abdominal prevertebral plexus before the prevertebral plexus continues into the pelvic cavity.

Each of these major plexuses gives origin to a number of secondary plexuses, which may also contain small ganglia. These plexuses are usually named after the vessels with which they are associated. For example, the celiac plexus is usually described as giving origin to the superior mesenteric plexus and the renal plexus, as well as other plexuses that extend out along the various branches of the celiac trunk. Similarly, the aortic plexus has secondary plexuses consisting of the inferior mesenteric plexus, the spermatic plexus, and the external iliac plexus.

Inferiorly, the superior hypogastric plexus divides into the **hypogastric nerves**, which descend into the pelvis and contribute to the formation of the inferior hypogastric or pelvic plexus (Fig. 4.137).

The abdominal prevertebral plexus receives:

- preganglionic parasympathetic and visceral afferent fibers from the vagus nerves [X],
- preganglionic sympathetic and visceral afferent fibers from the thoracic and lumbar splanchnic nerves, and
- preganglionic parasympathetic fibers from the pelvic splanchnic nerves.

### Parasympathetic innervation

Parasympathetic innervation of the abdominal part of the gastrointestinal tract and of the spleen, pancreas, gallbladder, and liver is from two sources—the vagus nerves [X] and the pelvic splanchnic nerves.

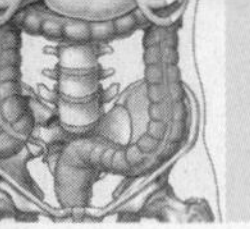

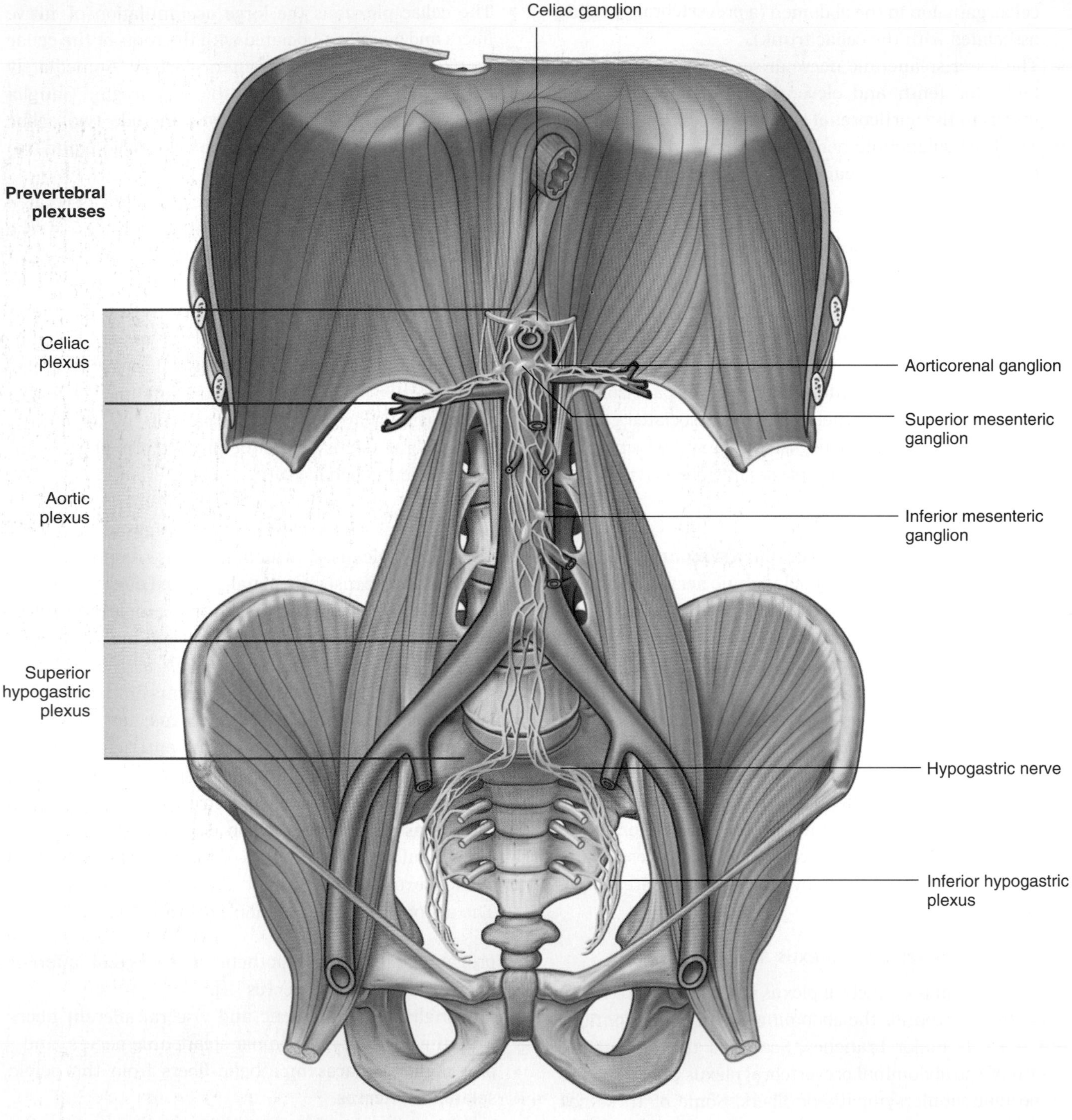

**Fig. 4.137** Abdominal prevertebral plexus and ganglia.

### Vagus nerves

The **vagus nerves** [X] enter the abdomen associated with the esophagus as the esophagus passes through the diaphragm (Fig. 4.138) and provide parasympathetic innervation to the foregut and midgut.

After entering the abdomen as the **anterior** and **posterior vagal trunks**, they send branches to the abdominal prevertebral plexus. These branches contain preganglionic parasympathetic fibers and visceral afferent fibers, which are distributed with the other components of the prevertebral plexus along the branches of the abdominal aorta.

### Pelvic splanchnic nerves

The **pelvic splanchnic nerves**, carrying preganglionic parasympathetic fibers from S2 to S4 spinal cord levels, enter the inferior hypogastric plexus in the pelvis. Some of these fibers move upward into the inferior mesenteric part of the prevertebral plexus in the abdomen (Fig. 4.138). Once there, these fibers are distributed with branches of

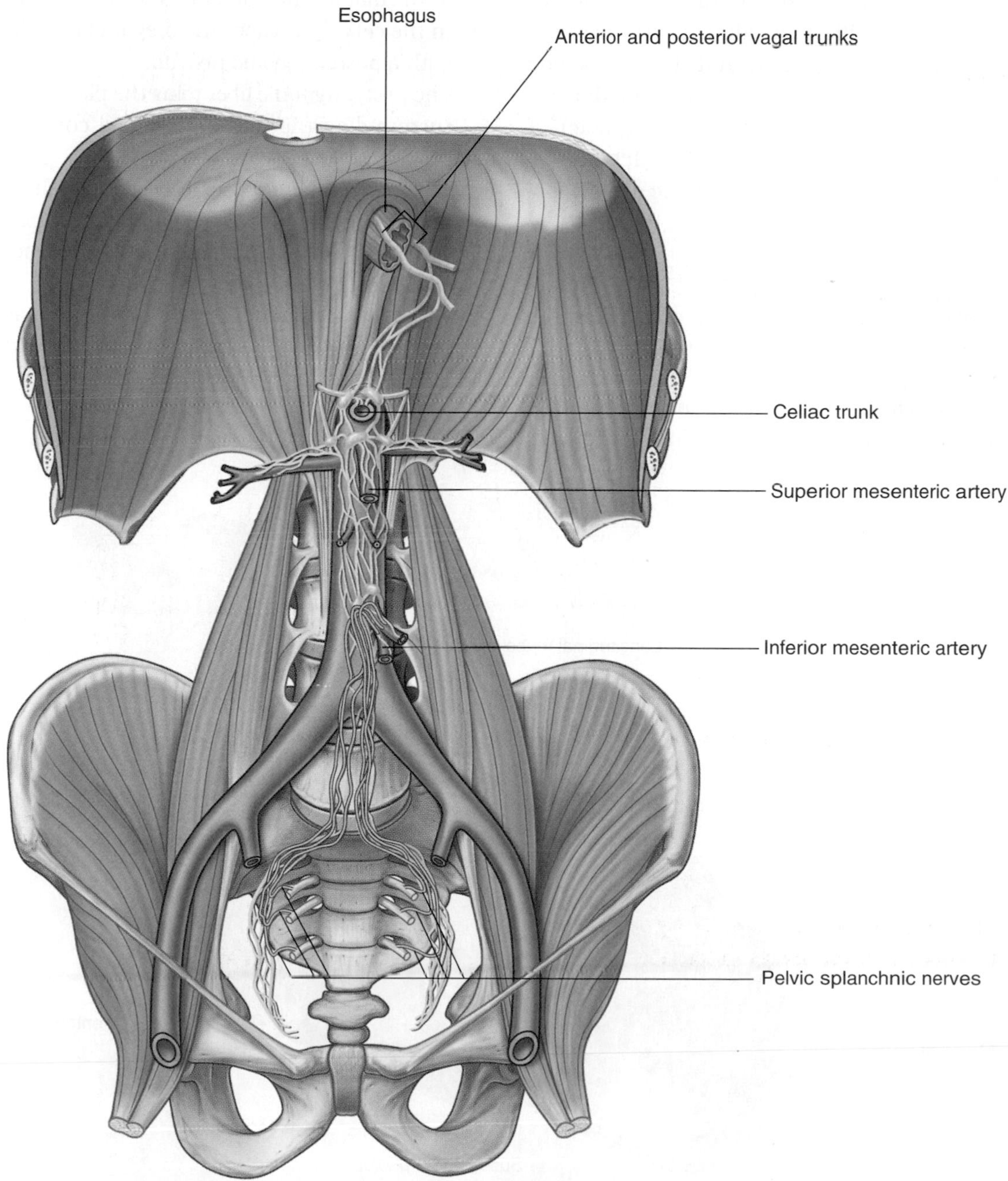


**Fig. 4.138** Parasympathetic innervation of the abdominal portion of the gastrointestinal tract.

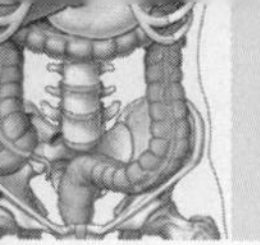



the inferior mesenteric artery and provide parasympathetic innervation to the hindgut.

### Enteric system

The enteric system is a division of the visceral part of the nervous system and is a local neuronal circuit in the wall of the gastrointestinal tract. It consists of motor and sensory neurons organized into two interconnected plexuses (the **myenteric** and **submucosal plexuses**) between the layers of the gastrointestinal wall, and the associated nerve fibers that pass between the plexuses and from the plexuses to the adjacent tissue (Fig. 4.139).

The enteric system regulates and coordinates numerous gastrointestinal tract activities, including gastric secretory activity, gastrointestinal blood flow, and the contraction and relaxation cycles of smooth muscle (**peristalsis**).

Although the enteric system is generally independent of the central nervous system, it does receive input from postganglionic sympathetic and preganglionic parasympathetic neurons that modifies its activities.

### Sympathetic innervation of the stomach

The pathway of sympathetic innervation of the stomach is as follows:

- A preganglionic sympathetic fiber originating at the T6 level of the spinal cord enters an anterior root to leave the spinal cord.
- At the level of the intervertebral foramen, the anterior root (which contains the preganglionic fiber) and a posterior root join to form a spinal nerve.
- Outside the vertebral column, the preganglionic fiber leaves the anterior ramus of the spinal nerve through the white ramus communicans.
- The white ramus communicans, containing the preganglionic fiber, connects to the sympathetic trunk.
- Entering the sympathetic trunk, the preganglionic fiber does not synapse but passes through the trunk and enters the greater splanchnic nerve.
- The greater splanchnic nerve passes through the crura of the diaphragm and enters the celiac ganglion.
- In the celiac ganglion, the preganglionic fiber synapses with a postganglionic neuron.
- The postganglionic fiber joins the plexus of nerve fibers surrounding the celiac trunk and continues along its branches.
- The postganglionic fiber travels through the plexus of nerves accompanying the branches of the celiac trunk supplying the stomach and eventually reaches its point of distribution.
- This input from the sympathetic system may modify the activities of the gastrointestinal tract controlled by the enteric nervous system.

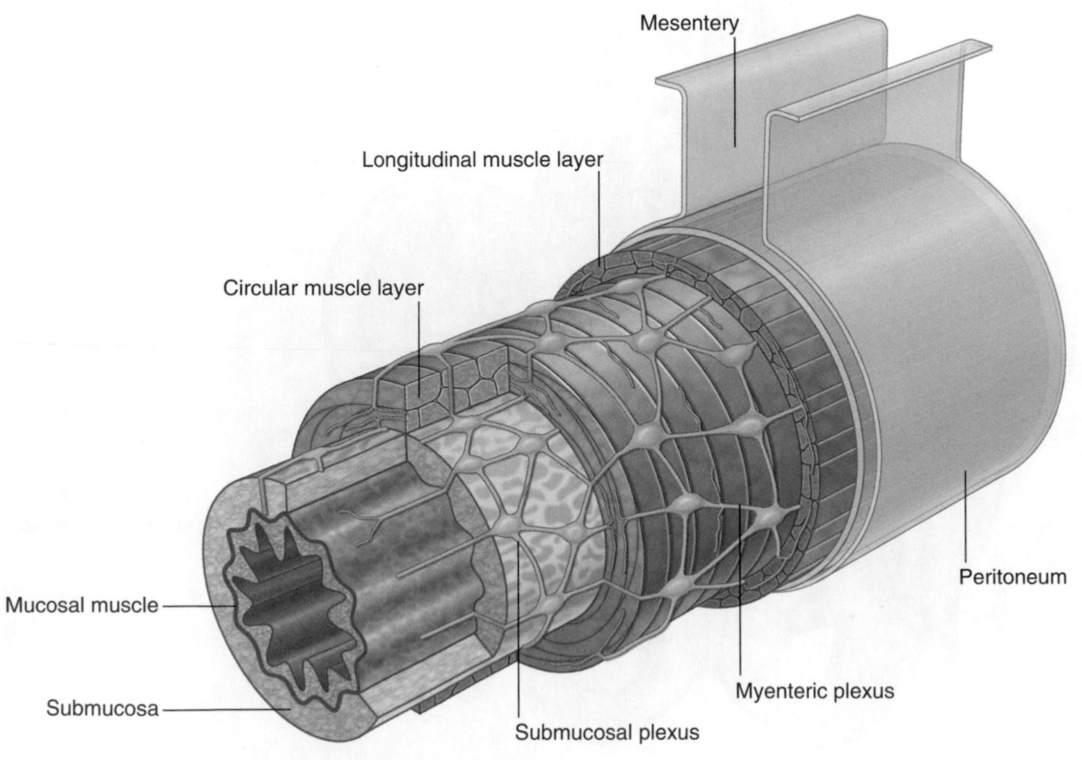


**Fig. 4.139** The enteric system.

## In the clinic

### Surgery for obesity

Surgery for obesity is also known as weight loss surgery and bariatric surgery. This type of surgery has become increasingly popular over the last few years for patients who are unable to achieve significant weight loss through appropriate diet modification and exercise programs. It is often regarded as a last resort. Importantly, we have to recognize the increasing medical impact that overweight patients pose. With obesity the patient is more likely to develop diabetes and cardiovascular problems and may suffer from increased general health disorders. All of these have a significant impact on health care budgeting and are regarded as serious conditions for the "health of a nation."

There are a number of surgical options to treat obesity. Surgery for patients who are morbidly obese can be categorized into two main groups: malabsorptive procedures and restrictive procedures.

#### *Malabsorptive procedures*

There are a variety of bypass procedures that produce a malabsorption state, preventing further weight gain and also producing weight loss. There are complications, which may include anemia, osteoporosis, and diarrhea (e.g., jejunoileal bypass).

#### *Predominantly restrictive procedures*

Restrictive procedures involve placing a band or stapling in or around the stomach to decrease the size of the organ. This reduction produces an earlier feeling of satiety and prevents the patient from overeating.

#### *Combination procedure*

Probably the most popular procedure currently in the United States is gastric bypass surgery. This procedure involves stapling the proximal stomach and joining a loop of small bowel to the small gastric remnant. The procedure is usually performed by fashioning a Roux-en-Y loop with alimentary and pancreaticobiliary limbs.

The other type of the procedure, sleeve gastrectomy, is increasing in popularity because it can be used in patients deemed to be at high risk for gastric bypass surgery. It involves reduction of the gastric lumen by removing a large portion of the stomach along the greater curvature.

Any overweight patient undergoing surgery faces significant risk and increased morbidity, with mortality rates from 1% to 5%.

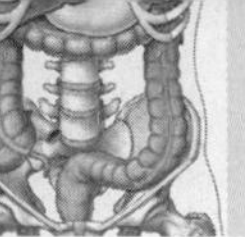


## POSTERIOR ABDOMINAL REGION

The posterior abdominal region is posterior to the abdominal part of the gastrointestinal tract, the spleen, and the pancreas (Fig. 4.140). This area, bounded by bones and muscles making up the posterior abdominal wall, contains numerous structures that not only are directly involved in the activities of the abdominal contents but also use this area as a conduit between body regions. Examples include the abdominal aorta and its associated nerve plexuses, the inferior vena cava, the sympathetic trunks, and lymphatics. There are also structures originating in this area that are critical to the normal function of other regions of the body (i.e., the lumbar plexus of nerves), and there are organs that associate with this area during development and remain in it in the adult (i.e., the kidneys and suprarenal glands).

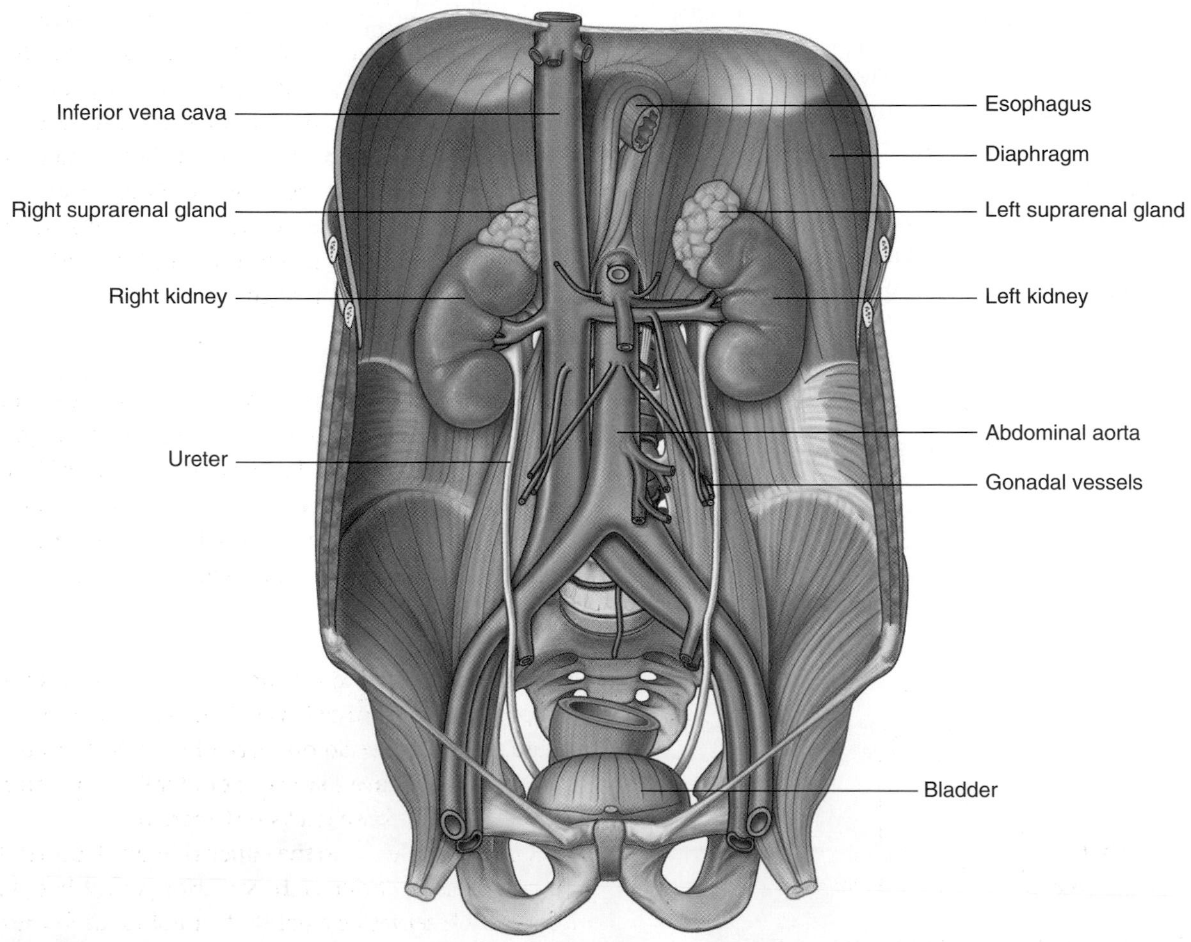


**Fig. 4.140** Posterior abdominal region.

## Posterior abdominal wall

### Bones

#### Lumbar vertebrae and the sacrum

Projecting into the midline of the posterior abdominal area are the bodies of the five lumbar vertebrae (Fig. 4.141). The prominence of these structures in this region is due to the secondary curvature (a forward convexity) of the lumbar part of the vertebral column.

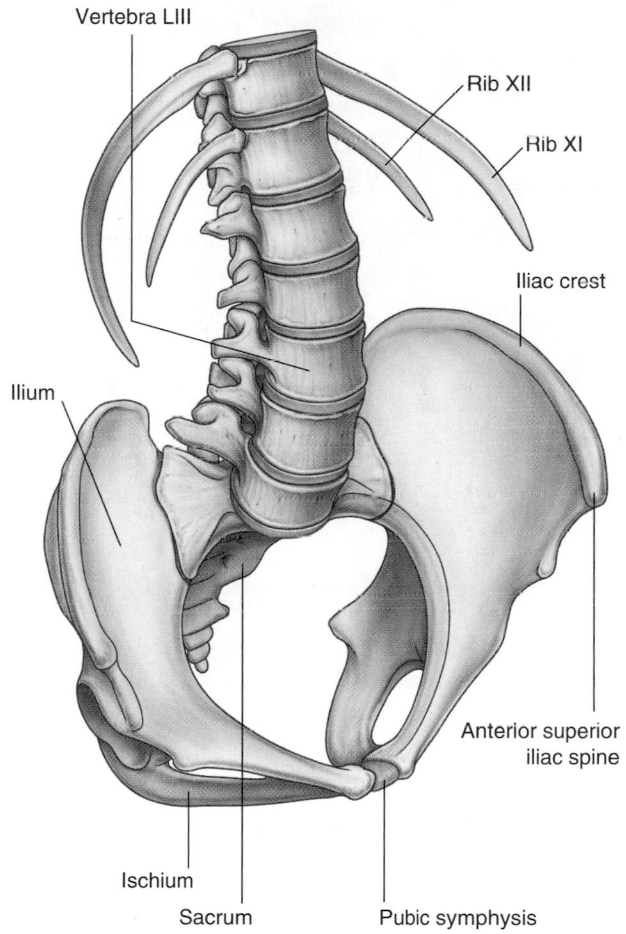


**Fig. 4.141** Osteology of the posterior abdominal wall.

The lumbar vertebrae can be distinguished from cervical and thoracic vertebrae because of their size. They are much larger than any other vertebrae in any other region. The vertebral bodies are massive and progressively increase in size from vertebra LI to LV. The pedicles are short and stocky, the transverse processes are long and slender, and the spinous processes are large and stubby. The articular processes are large and oriented medially and laterally, which promotes flexion and extension in this part of the vertebral column.

Between each lumbar vertebra is an intervertebral disc, which completes this part of the midline boundary of the posterior abdominal wall.

The midline boundary of the posterior abdominal wall, inferior to the lumbar vertebrae, consists of the upper margin of the sacrum (Fig. 4.141). The sacrum is formed by the fusion of the five sacral vertebrae into a single, wedge-shaped bony structure that is broad superiorly and narrows inferiorly. Its concave anterior surface and its convex posterior surface contain anterior and posterior sacral foramina for the anterior and posterior rami of spinal nerves to pass through.

#### Pelvic bones

The **ilia**, which are components of each pelvic bone, attach laterally to the sacrum at the sacro-iliac joints (Fig. 4.141). The upper part of each ilium expands outward into a thin wing-like area (the **iliac fossa**). The medial side of this region of each iliac bone, and the related muscles, are components of the posterior abdominal wall.

#### Ribs

Superiorly, ribs XI and XII complete the bony framework of the posterior abdominal wall (Fig. 4.141). These ribs are unique in that they do not articulate with the sternum or other ribs, they have a single articular facet on their heads, and they do not have necks or tubercles.

Rib XI is posterior to the superior part of the left kidney, and rib XII is posterior to the superior part of both kidneys. Also, rib XII serves as a point of attachment for numerous muscles and ligaments.

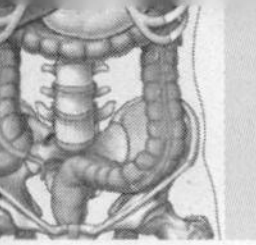



## Muscles

Muscles forming the medial, lateral, inferior, and superior boundaries of the posterior abdominal region fill in the bony framework of the posterior abdominal wall (Table 4.2). Medially are the psoas major and minor muscles, laterally is the quadratus lumborum muscle, inferiorly is the iliacus muscle, and superiorly is the diaphragm (Figs. 4.142 and 4.143).

### Psoas major and minor

Medially, the **psoas major** muscles cover the anterolateral surface of the bodies of the lumbar vertebrae, filling in the space between the vertebral bodies and the transverse processes (Fig. 4.142). Each of these muscles arises from the bodies of vertebra TXII and all five lumbar vertebrae, from the intervertebral discs between each vertebra, and from the transverse processes of the lumbar vertebrae. Passing inferiorly along the pelvic brim, each muscle continues into the anterior thigh, under the inguinal ligament, to attach to the lesser trochanter of the femur.

The psoas major muscle flexes the thigh at the hip joint when the trunk is stabilized and flexes the trunk against gravity when the body is supine. It is innervated by anterior rami of nerves L1 to L3.

Associated with the psoas major muscle is the **psoas minor** muscle, which is sometimes absent. Lying on the surface of the psoas major when present, this slender muscle arises from vertebrae TXII and LI and the intervening intervertebral disc; its long tendon inserts into the pectineal line of the pelvic brim and the iliopubic eminence.

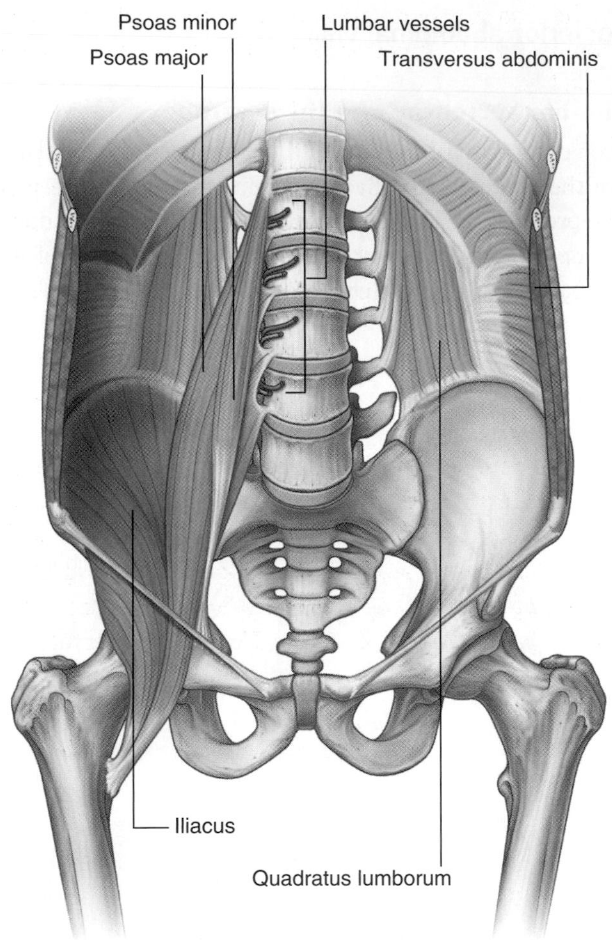


**Fig. 4.142** Muscles of the posterior abdominal wall.

**Table 4.2** Posterior abdominal wall muscles

| Muscle | Origin | Insertion | Innervation | Function |
|---|---|---|---|---|
| Psoas major | Lateral surface of bodies of TXII and LI to LV vertebrae, transverse processes of the lumbar vertebrae, and the intervertebral discs between TXII and LI to LV vertebrae | Lesser trochanter of the femur | Anterior rami of L1 to L3 | Flexion of thigh at hip joint |
| Psoas minor | Lateral surface of bodies of TXII and LI vertebrae and intervening intervertebral disc | Pectineal line of the pelvic brim and iliopubic eminence | Anterior rami of L1 | Weak flexion of lumbar vertebral column |
| Quadratus lumborum | Transverse process of LV vertebra, iliolumbar ligament, and iliac crest | Transverse processes of LI to LIV vertebrae and inferior border of rib XII | Anterior rami of T12 and L1 to L4 | Depress and stabilize rib XII and some lateral bending of trunk |
| Iliacus | Upper two-thirds of iliac fossa, anterior sacro-iliac and iliolumbar ligaments, and upper lateral surface of sacrum | Lesser trochanter of femur | Femoral nerve (L2 to L4) | Flexion of thigh at hip joint |

The psoas minor is a weak flexor of the lumbar vertebral column and is innervated by the anterior ramus of nerve L1.

### Quadratus lumborum

Laterally, the quadratus lumborum muscles fill the space between rib XII and the iliac crest on both sides of the vertebral column (Fig. 4.142). They are overlapped medially by the psoas major muscles; along their lateral borders are the transversus abdominis muscles.

Each quadratus lumborum muscle arises from the transverse process of vertebra LV, the iliolumbar ligament, and the adjoining part of the iliac crest. The muscle attaches superiorly to the transverse process of the first four lumbar vertebrae and the inferior border of rib XII.

The quadratus lumborum muscles depress and stabilize the twelfth ribs and contribute to lateral bending of the trunk. Acting together, the muscles may extend the lumbar part of the vertebral column. They are innervated by anterior rami of T12 and L1 to L4 spinal nerves.

### Iliacus

Inferiorly, an **iliacus** muscle fills the iliac fossa on each side (Fig. 4.142). From this expansive origin covering the iliac fossa, the muscle passes inferiorly, joins with the psoas major muscle, and attaches to the lesser trochanter of the femur. As they pass into the thigh, these combined muscles are referred to as the **iliopsoas** muscle.

Like the psoas major muscle, the iliacus flexes the thigh at the hip joint when the trunk is stabilized and flexes the trunk against gravity when the body is supine. It is innervated by branches of the femoral nerve.

### Diaphragm

Superiorly, the diaphragm forms the boundary of the posterior abdominal region. This musculotendinous sheet also separates the abdominal cavity from the thoracic cavity.

Structurally, the diaphragm consists of a central tendinous part into which the circumferentially arranged muscle fibers attach (Fig. 4.143). The diaphragm is

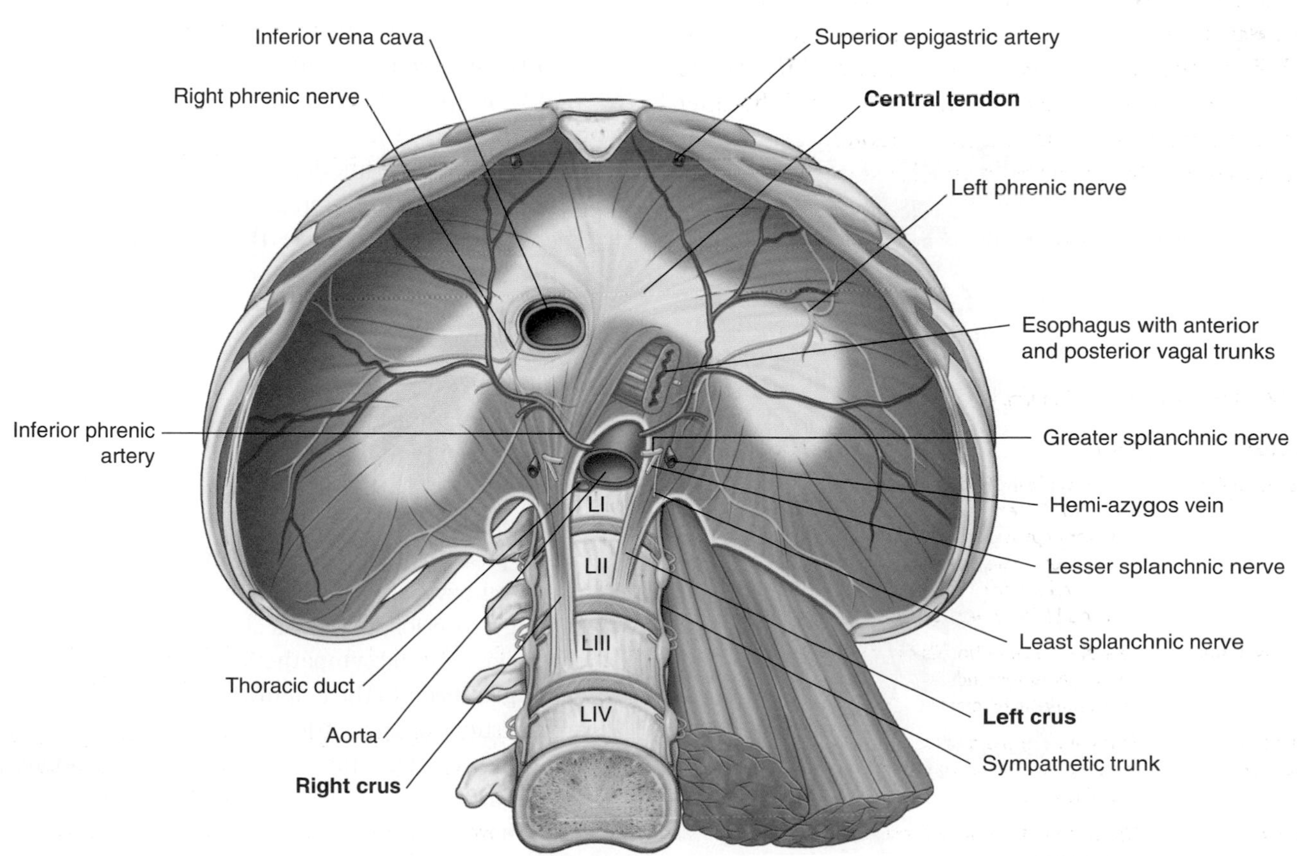


**Fig. 4.143** Diaphragm.

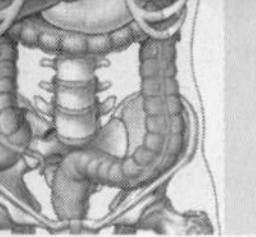



anchored to the lumbar vertebrae by musculotendinous crura, which blend with the anterior longitudinal ligament of the vertebral column:

- The right crus is the longest and broadest of the crura and is attached to the bodies of vertebrae LI to LIII and the intervening intervertebral discs (Fig. 4.144).
- Similarly, the left crus is attached to vertebrae LI and LII and the associated intervertebral disc.

The crura are connected across the midline by a tendinous arch (the **median arcuate ligament**), which passes anterior to the aorta (Fig. 4.144).

Lateral to the crura, a second tendinous arch is formed by the fascia covering the upper part of the psoas major muscle. This is the **medial arcuate ligament**, which is attached medially to the sides of vertebrae LI and LII and laterally to the transverse process of vertebra LI (Fig. 4.144).

A third tendinous arch, the **lateral arcuate ligament**, is formed by a thickening in the fascia that covers the quadratus lumborum. It is attached medially to the transverse process of vertebra LI and laterally to rib XII (Fig. 4.144).

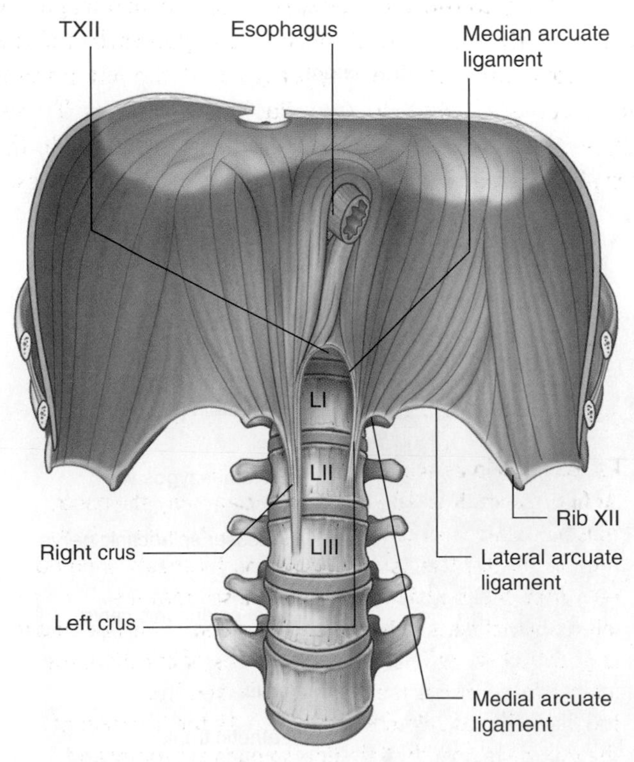


**Fig. 4.144** Crura of the diaphragm.

The medial and lateral arcuate ligaments serve as points of origin for some of the muscular components of the diaphragm.

### Structures passing through or around the diaphragm

Numerous structures pass through or around the diaphragm (Fig. 4.143):

- The aorta passes posterior to the diaphragm and anterior to the vertebral bodies at the lower level of vertebra TXII; it is between the two crura of the diaphragm and posterior to the median arcuate ligament, just to the left of midline.
- Accompanying the aorta through the aortic hiatus is the thoracic duct and, sometimes, the azygos vein.
- The esophagus passes through the musculature of the right crus of the diaphragm at the level of vertebra TX, just to the left of the aortic hiatus.
- Passing through the esophageal hiatus with the esophagus are the anterior and posterior vagal trunks, the esophageal branches of the left gastric artery and vein, and a few lymphatic vessels.
- The third large opening in the diaphragm is the caval opening, through which the inferior vena cava passes from the abdominal cavity to the thoracic cavity (Fig. 4.143) at approximately vertebra TVIII in the central tendinous part of the diaphragm.
- Accompanying the inferior vena cava through the caval opening is the right phrenic nerve.
- The left phrenic nerve passes through the muscular part of the diaphragm just anterior to the central tendon on the left side.

Additional structures pass through small openings either in or just outside the diaphragm as they pass from the thoracic cavity to the abdominal cavity (Fig. 4.143):

- The greater, lesser, and least (when present) splanchnic nerves pass through the crura, on either side.
- The hemi-azygos vein passes through the left crus.
- Passing posterior to the medial arcuate ligament, on either side, are the sympathetic trunks.
- Passing anterior to the diaphragm, just deep to the ribs, are the superior epigastric vessels.
- Other vessels and nerves (i.e., the musculophrenic vessels and intercostal nerves) also pass through the diaphragm at various points.

### Domes

The classic appearance of the right and left domes of the diaphragm is caused by the underlying abdominal contents pushing these lateral areas upward, and by the fibrous pericardium, which is attached centrally, causing a flattening of the diaphragm in this area (Fig. 4.145).

The domes are produced by:

- the liver on the right, with some contribution from the right kidney and the right suprarenal gland, and
- the fundus of the stomach and spleen on the left, with contributions from the left kidney and the left suprarenal gland.

Although the height of these domes varies during breathing, a reasonable estimate in normal expiration places the left dome at the fifth intercostal space and the right dome at rib V. This is important to remember when percussing the thorax.

During inspiration, the muscular part of the diaphragm contracts, causing the central tendon of the diaphragm to be drawn inferiorly. This results in some flattening of the domes, enlargement of the thoracic cavity, and a reduction in intrathoracic pressure. The physiological effect of these changes is that air enters the lungs and venous return to the heart is enhanced.

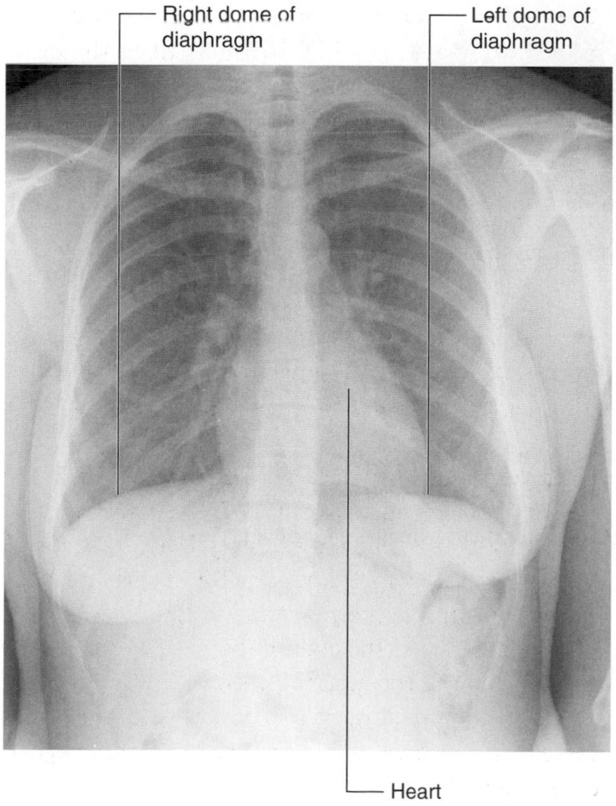

**Fig. 4.145** Right and left domes of the diaphragm. Chest radiograph.

### Blood supply

There is blood supply to the diaphragm on its superior and inferior surfaces:

- Superiorly, the musculophrenic and pericardiacophrenic arteries, both branches of the internal thoracic artery, and the superior phrenic artery, a branch of the thoracic aorta, supply the diaphragm.
- Inferiorly, the inferior phrenic arteries, branches of the abdominal aorta, supply the diaphragm (see Fig. 4.143).

Venous drainage is through companion veins to these arteries.

### Innervation

Innervation of the diaphragm is primarily by the **phrenic nerves**. These nerves, from the C3 to C5 spinal cord levels, provide all motor innervation to the diaphragm and sensory fibers to the central part. They pass through the thoracic cavity, between the mediastinal pleura and the pericardium, to the superior surface of the diaphragm. At this point, the right phrenic nerve accompanies the inferior vena cava through the diaphragm and the left phrenic nerve passes through the diaphragm by itself (see Fig. 4.143). Additional sensory fibers are supplied to the peripheral areas of the diaphragm by intercostal nerves.

### In the clinic

#### Psoas muscle abscess

At first glance, it is difficult to appreciate why the psoas muscle sheath is of greater importance than any other muscle sheath. The psoas muscle and its sheath arise not only from the lumbar vertebrae but also from the intervertebral discs between each vertebra. This disc origin is of critical importance. In certain types of infection, the intervertebral disc is preferentially affected (e.g., tuberculosis and salmonella discitis). As the infection of the disc develops, the infection spreads anteriorly and anterolaterally. In the anterolateral position, the infection passes into the psoas muscle sheath, and spreads within the muscle and sheath, and may appear below the inguinal ligament as a mass.

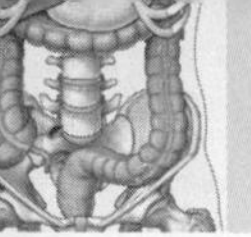



## In the clinic

### Diaphragmatic hernias

To understand why a hernia occurs through the diaphragm, it is necessary to consider the embryology of the diaphragm.

The diaphragm is formed from four structures—the septum transversum, the posterior esophageal mesentery, the pleuroperitoneal membrane, and the peripheral rim—which eventually fuse together, separating the abdominal cavity from the thoracic cavity. The septum transversum forms the central tendon, which develops from a mesodermal origin superior to the embryo's head and then moves to its more adult position during folding of the cephalic portion of the embryo.

Fusion of the various components of the diaphragm may fail, and hernias may occur through the failed points of fusion (Fig. 4.146). The commonest sites are:

- between the xiphoid process and the costal margins on the right (Morgagni's hernia), and
- through an opening on the left when the pleuroperitoneal membrane fails to close the pericardioperitoneal canal (Bochdalek's hernia).

Hernias may also occur through the central tendon and through a congenitally large esophageal hiatus.

Morgagni's and Bochdalek's hernias tend to appear at or around the time of birth or in early infancy. They allow abdominal bowel to enter the thoracic cavity, which may compress the lungs and reduce respiratory function. Most of these hernias require surgical closure of the diaphragmatic defect. However, large hernias can lead to pulmonary hypoplasia and the long-term outcome depends more on the degree of the hypoplasia rather than on the surgical repair itself.

Occasionally, small defects within the diaphragm fail to permit bowel through, but do allow free movement of fluid. Patients with ascites may develop pleural effusions, while patients with pleural effusions may develop ascites when these defects are present.

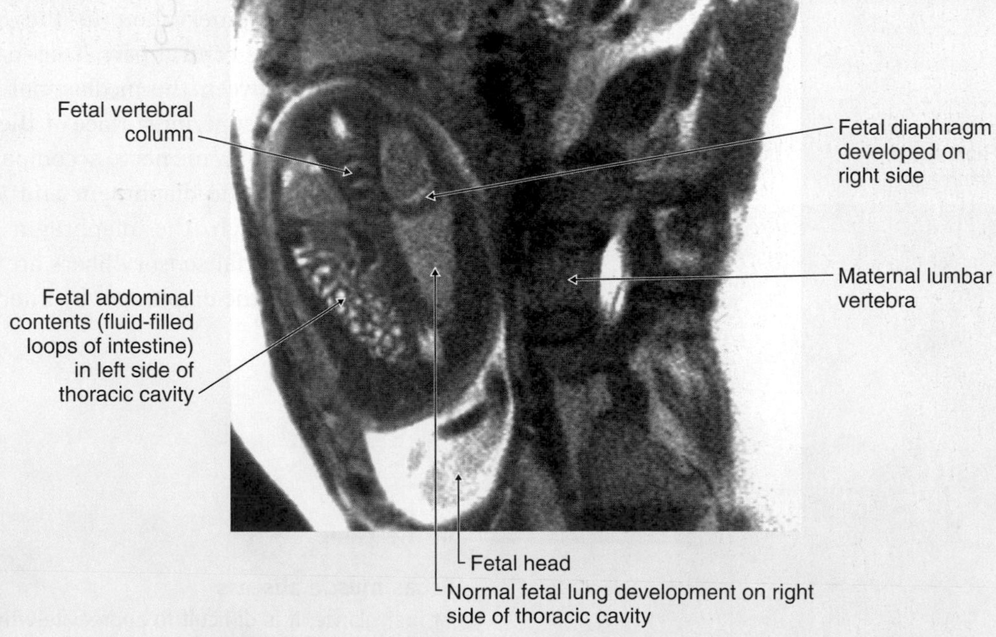


**Fig. 4.146** Fetal diaphragmatic hernia in utero. T2-weighted MR image. Fetus in coronal plane, mother in sagittal plane.

## In the clinic

### Hiatal hernia

At the level of the esophageal hiatus, the diaphragm may be lax, allowing the fundus of the stomach to herniate into the posterior mediastinum (Figs. 4.147 and 4.148). This typically causes symptoms of acid reflux. Ulceration may occur and may produce bleeding and anemia. The diagnosis is usually made by barium studies or endoscopy. Hiatal hernia is often asymptomatic and is frequently found incidentally on CT imaging performed for unrelated complaints. Treatment in the first instance is by medical management, although surgery may be necessary.

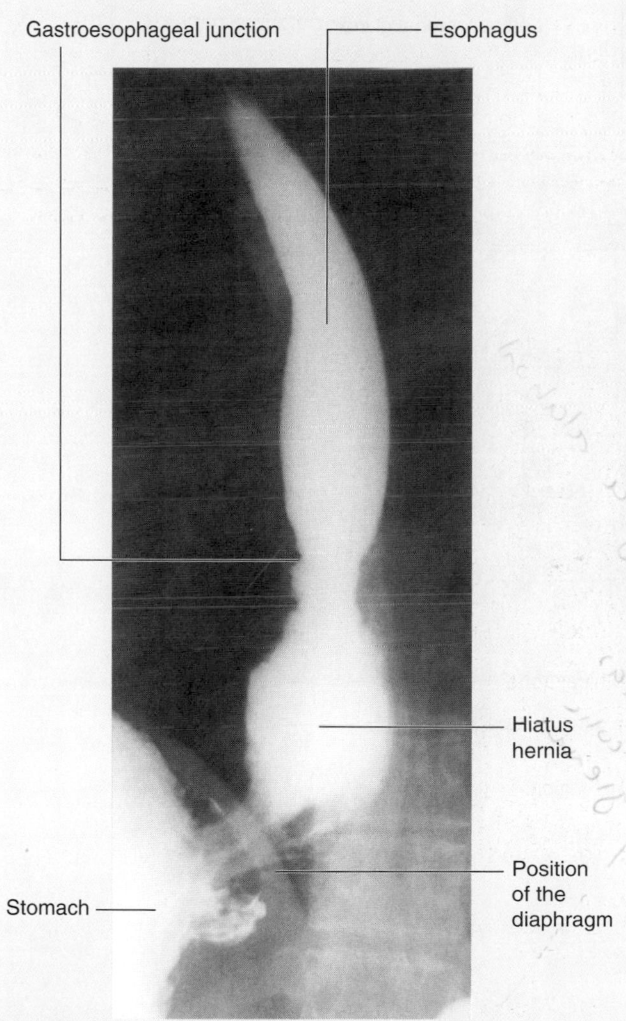


**Fig. 4.147** Lower esophagus and upper stomach showing a hiatal hernia. Radiograph using barium.

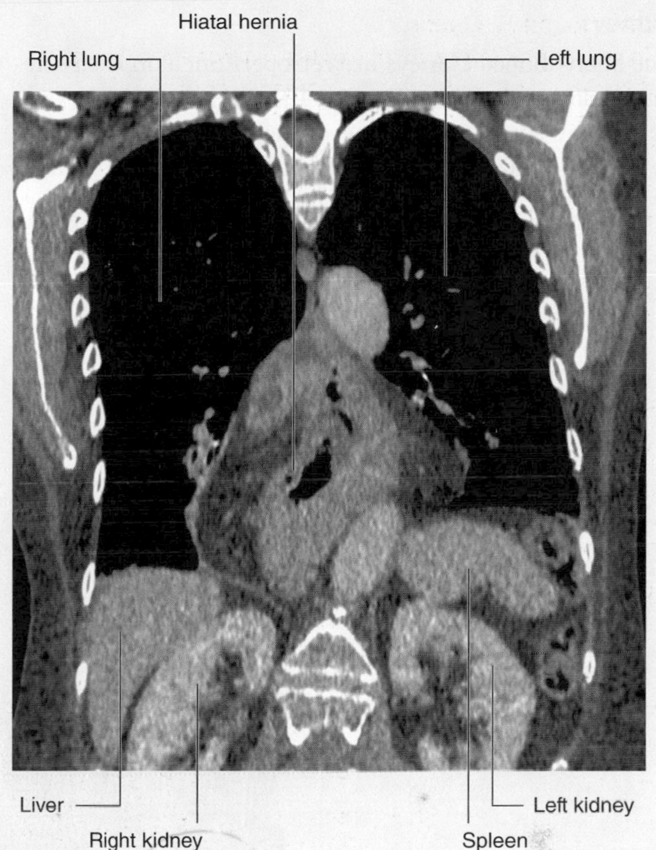


**Fig. 4.148** Coronal CT of hiatal hernia.

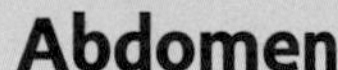

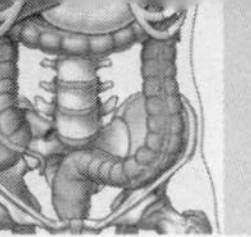

## Viscera

### Kidneys

The bean-shaped kidneys are retroperitoneal in the posterior abdominal region (Fig. 4.149). They lie in the extraperitoneal connective tissue immediately lateral to the vertebral column. In the supine position, the kidneys extend from approximately vertebra TXII superiorly to vertebra LIII inferiorly, with the right kidney somewhat lower than the left because of its relationship with the liver. Although they are similar in size and shape, the left kidney is a longer and more slender organ than the right kidney, and nearer to the midline.

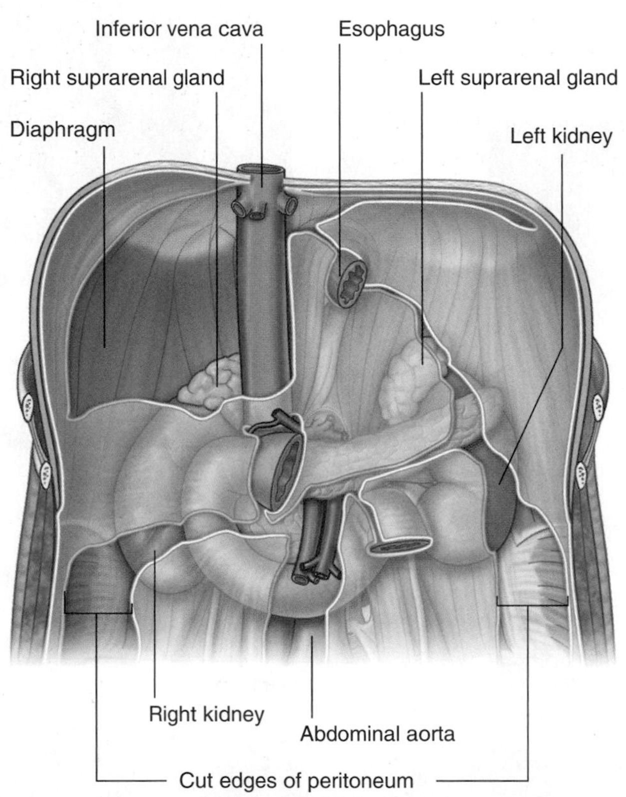


**Fig. 4.149** Retroperitoneal position of the kidneys in the posterior abdominal region.

### Relationships to other structures

The anterior surface of the right kidney is related to numerous structures, some of which are separated from the kidney by a layer of peritoneum and some of which are directly against the kidney (Fig. 4.150):

- A small part of the superior pole is covered by the right suprarenal gland.
- Moving inferiorly, a large part of the rest of the upper part of the anterior surface is against the liver and is separated from it by a layer of peritoneum.
- Medially, the descending part of the duodenum is retroperitoneal and contacts the kidney.
- The inferior pole of the kidney, on its lateral side, is directly associated with the right colic flexure and, on its medial side, is covered by a segment of the intraperitoneal small intestine.

The anterior surface of the left kidney is also related to numerous structures, some with an intervening layer of peritoneum and some directly against the kidney (Fig. 4.150):

- A small part of the superior pole, on its medial side, is covered by the left suprarenal gland.
- The rest of the superior pole is covered by the intraperitoneal stomach and spleen.
- Moving inferiorly, the retroperitoneal pancreas covers the middle part of the kidney.
- On its lateral side, the lower half of the kidney is covered by the left colic flexure and the beginning of the

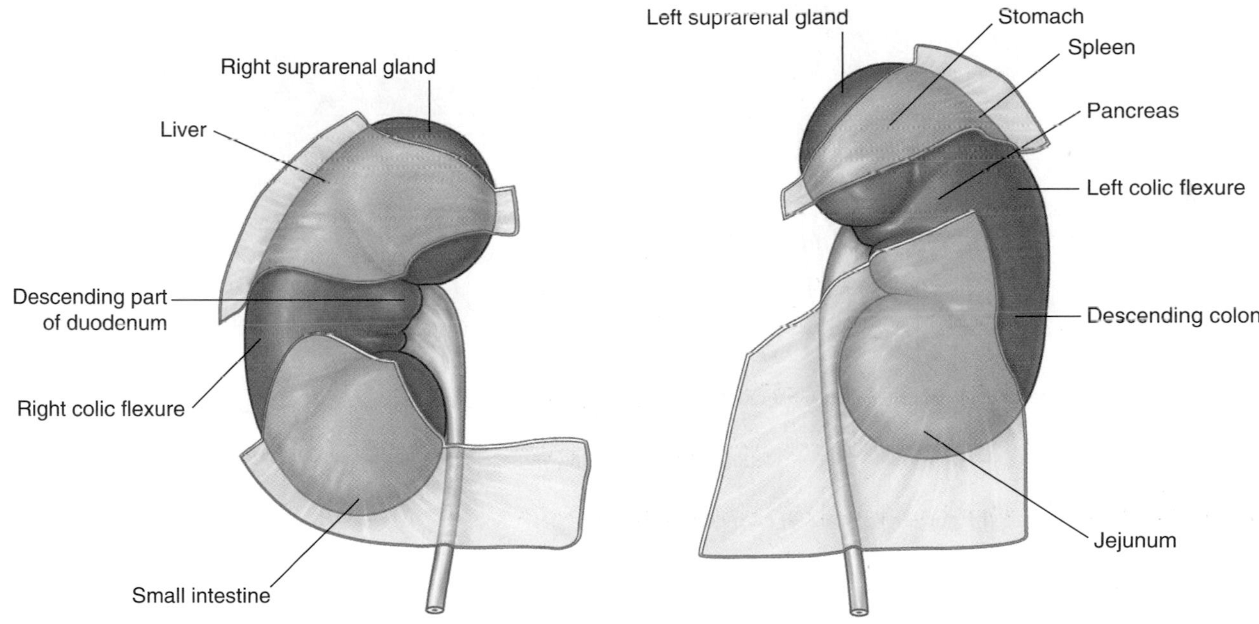


**Fig. 4.150** Structures related to the anterior surface of each kidney.

descending colon, and, on its medial side, by the parts of the intraperitoneal jejunum.

Posteriorly, the right and left kidneys are related to similar structures (Fig. 4.151). Superiorly is the diaphragm and inferior to this, moving in a medial to lateral direction, are the psoas major, quadratus lumborum, and transversus abdominis muscles.

The superior pole of the right kidney is anterior to rib XII, while the same region of the left kidney is anterior to ribs XI and XII. The pleural sacs and specifically the costodiaphragmatic recesses therefore extend posterior to the kidneys.

Also passing posterior to the kidneys are the subcostal vessels and nerves and the iliohypogastric and ilio-inguinal nerves.

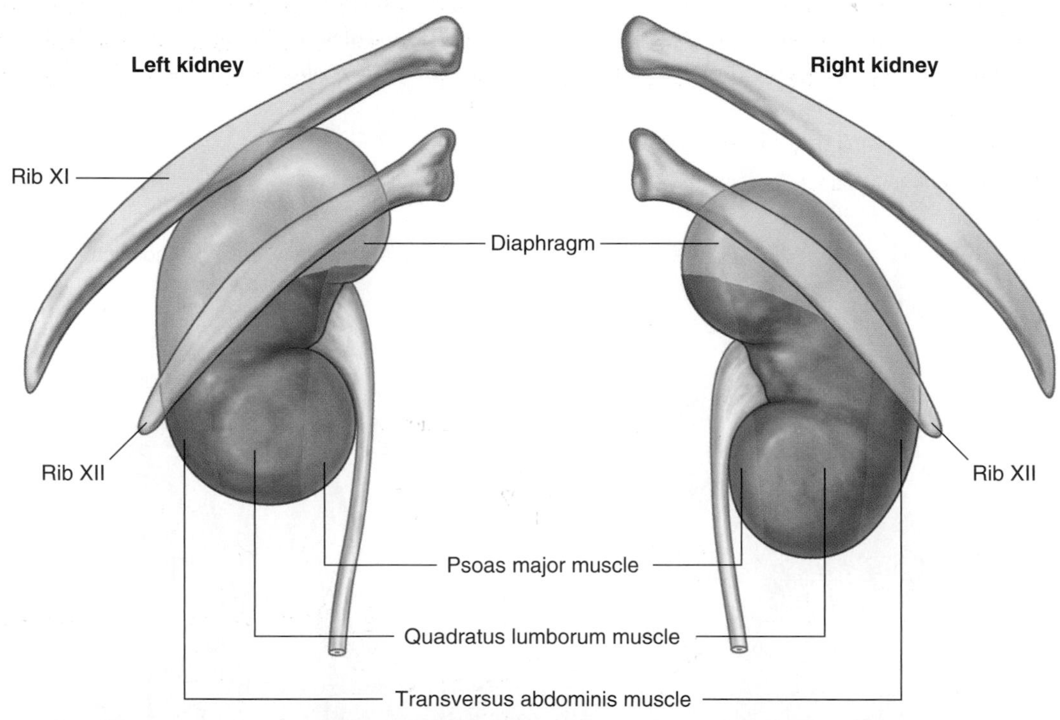


**Fig. 4.151** Structures related to the posterior surface of each kidney.

### Renal fat and fascia

The kidneys are enclosed in and associated with a unique arrangement of fascia and fat. Immediately outside the renal capsule, there is an accumulation of extraperitoneal fat—the **perinephric fat (perirenal fat)**, which completely surrounds the kidney (Fig. 4.152). Enclosing the perinephric fat is a membranous condensation of the extraperitoneal fascia (the **renal fascia**). The suprarenal glands are also enclosed in this fascial compartment, usually separated from the kidneys by a thin septum. The renal fascia must be incised in any surgical approach to this organ.

At the lateral margins of each kidney, the anterior and posterior layers of the renal fascia fuse (Fig. 4.152). This fused layer may connect with the transversalis fascia on the lateral abdominal wall.

Above each suprarenal gland, the anterior and posterior layers of the renal fascia fuse and blend with the fascia that covers the diaphragm.

Medially, the anterior layer of the renal fascia continues over the vessels in the hilum and fuses with the connective tissue associated with the abdominal aorta and the inferior vena cava (Fig. 4.152). In some cases, the anterior layer may cross the midline to the opposite side and blend with its companion layer.

The posterior layer of the renal fascia passes medially between the kidney and the fascia covering the quadratus lumborum muscle to fuse with the fascia covering the psoas major muscle.

Inferiorly, the anterior and posterior layers of the renal fascia enclose the ureters.

In addition to perinephric fat and the renal fascia, a final layer of **paranephric fat (pararenal fat)** completes the fat and fascias associated with the kidney (Fig. 4.152). This fat accumulates posterior and posterolateral to each kidney.

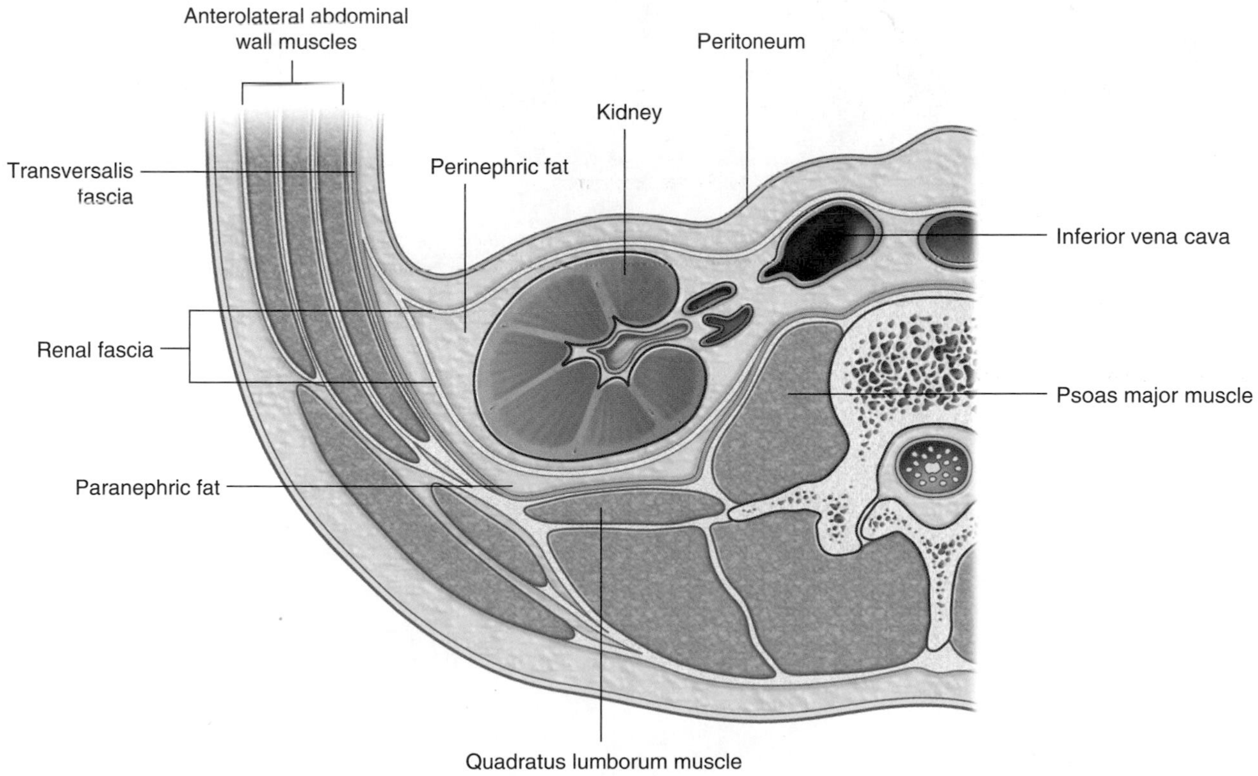


**Fig. 4.152** Organization of fat and fascia surrounding the kidney.

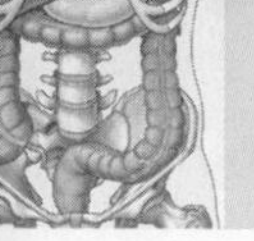

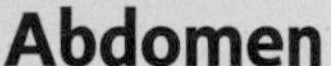

### Kidney structure

Each kidney has a smooth anterior and posterior surface covered by a fibrous capsule, which is easily removable except during disease.

On the medial margin of each kidney is the **hilum of the kidney**, which is a deep vertical slit through which renal vessels, lymphatics, and nerves enter and leave the substance of the kidney (Fig. 4.153). Internally, the hilum is continuous with the renal sinus. Perinephric fat continues into the hilum and sinus and surrounds all structures.

Each kidney consists of an outer **renal cortex** and an inner renal medulla. The renal cortex is a continuous band of pale tissue that completely surrounds the renal medulla. Extensions of the renal cortex (the **renal columns**) project into the inner aspect of the kidney, dividing the renal medulla into discontinuous aggregations of triangular-shaped tissue (the **renal pyramids**).

The bases of the renal pyramids are directed outward, toward the renal cortex, while the apex of each renal pyramid projects inward, toward the **renal sinus**. The apical projection (**renal papilla**) contains the openings of the papillary ducts draining the renal tubules and is surrounded by a **minor calyx**.

The minor calices receive urine from the papillary ducts and represent the proximal parts of the tube that will eventually form the ureter (Fig. 4.153). In the renal sinus, several minor calices unite to form a **major calyx**, and two or three major calices unite to form the **renal pelvis**, which is the funnel-shaped superior end of the ureters.

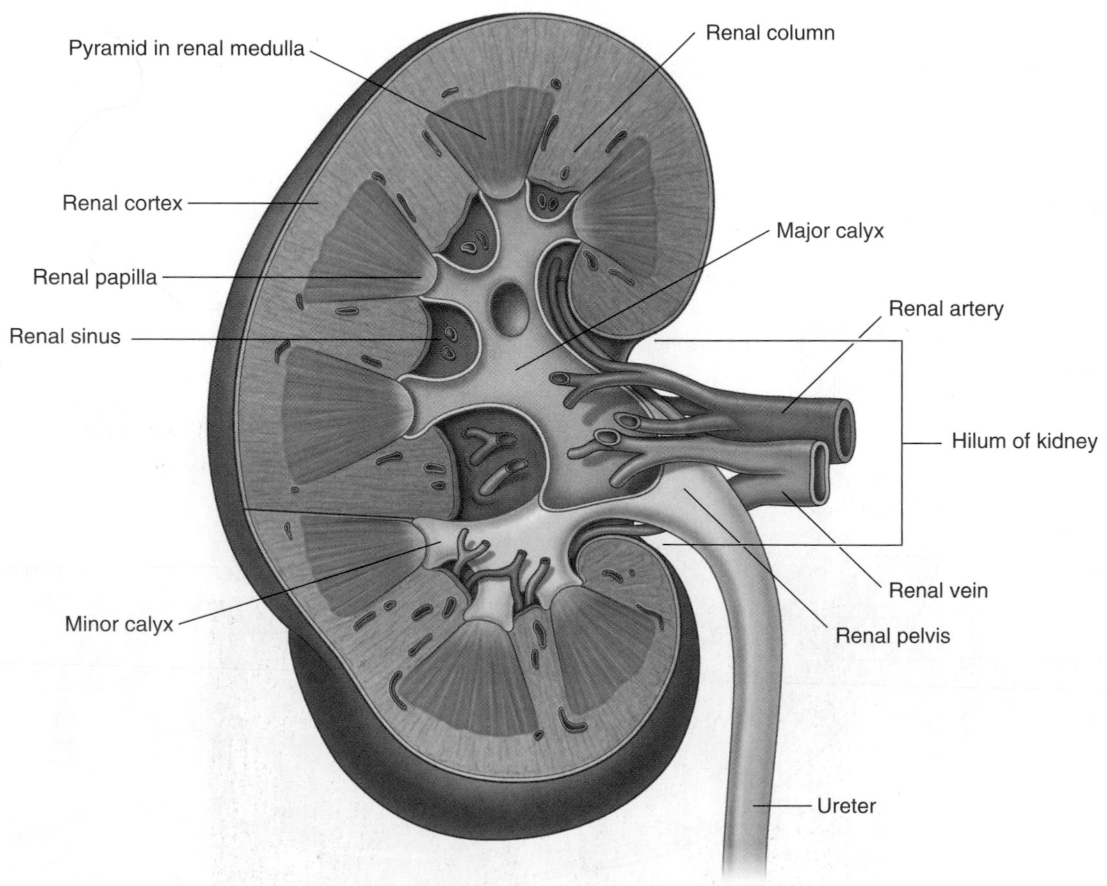


**Fig. 4.153** Internal structure of the kidney.

### Renal vasculature and lymphatics

A single large **renal artery**, a lateral branch of the abdominal aorta, supplies each kidney. These vessels usually arise just inferior to the origin of the superior mesenteric artery between vertebrae LI and LII (Fig. 4.154). The **left renal artery** usually arises a little higher than the right, and the **right renal artery** is longer and passes posterior to the inferior vena cava.

As each renal artery approaches the renal hilum, it divides into anterior and posterior branches, which supply the renal parenchyma. Accessory renal arteries are common. They originate from the lateral aspect of the abdominal aorta, either above or below the primary renal arteries, enter the hilum with the primary arteries or pass directly into the kidney at some other level, and are commonly called **extrahilar arteries**.

Multiple renal veins contribute to the formation of the **left** and **right renal veins**, both of which are anterior to the renal arteries (Fig. 4.154A). Importantly, the longer left renal vein crosses the midline anterior to the abdominal

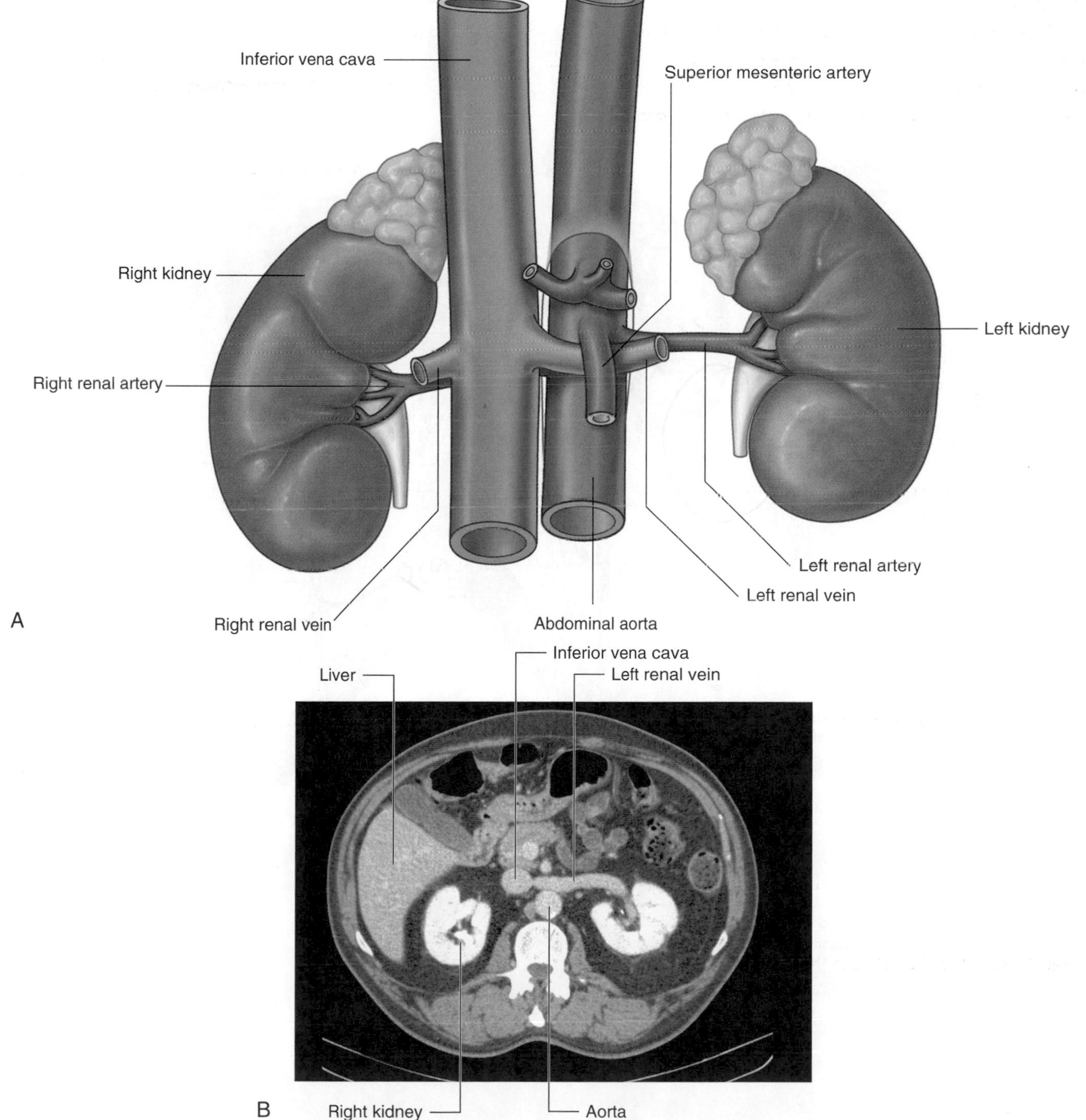


**Fig. 4.154 A.** Renal vasculature. **B.** CT image showing long left renal vein crossing the midline.

aorta and posterior to the superior mesenteric artery and can be compressed by an aneurysm in either of these two vessels (Fig. 4.154B).

The lymphatic drainage of each kidney is to the **lateral aortic (lumbar) nodes** around the origin of the renal artery.

### Ureters

The ureters are muscular tubes that transport urine from the kidneys to the bladder. They are continuous superiorly with the renal pelvis, which is a funnel-shaped structure in the renal sinus. The renal pelvis is formed from a condensation of two or three major calices, which in turn are formed by the condensation of several minor calices (see Fig. 4.153). The minor calices surround a renal papilla.

The renal pelvis narrows as it passes inferiorly through the hilum of the kidney and becomes continuous with the ureter at the **ureteropelvic junction** (Fig. 4.155). Inferior to this junction, the ureters descend retroperitoneally on the medial aspect of the psoas major muscle. At the pelvic brim, the ureters cross either the end of the common iliac artery or the beginning of the external iliac artery, enter the pelvic cavity, and continue their journey to the bladder.

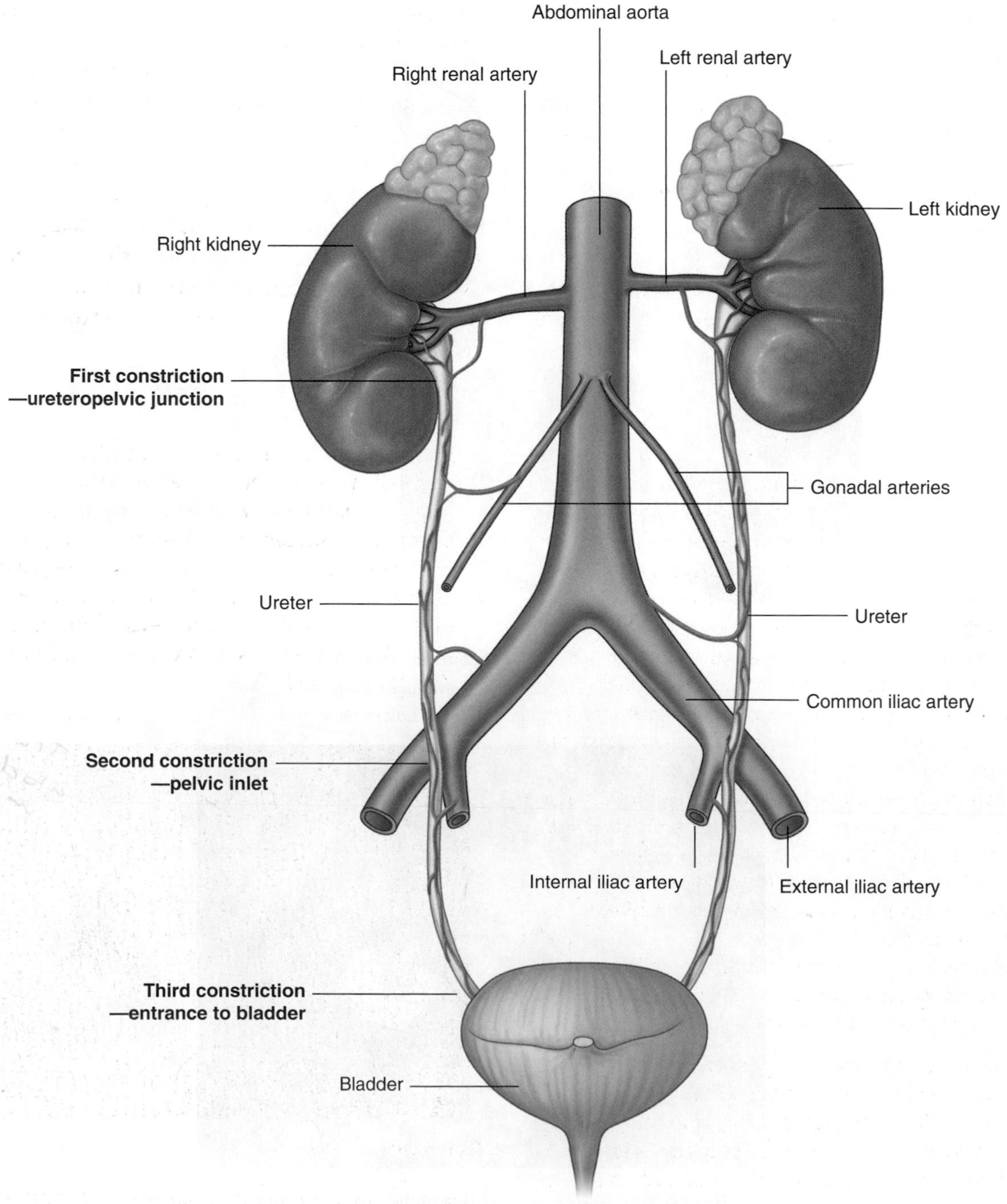

**Fig. 4.155** Ureters.

At three points along their course the ureters are constricted (Fig. 4.155):

- The first point is at the ureteropelvic junction.
- The second point is where the ureters cross the common iliac vessels at the pelvic brim.
- The third point is where the ureters enter the wall of the bladder.

Kidney stones can become lodged at these constrictions.

### Ureteric vasculature and lymphatics

The ureters receive arterial branches from adjacent vessels as they pass toward the bladder (Fig. 4.155):

- The renal arteries supply the upper end.
- The middle part may receive branches from the abdominal aorta, the testicular or ovarian arteries, and the common iliac arteries.
- In the pelvic cavity, the ureters are supplied by one or more arteries from branches of the internal iliac arteries.

In all cases, arteries reaching the ureters divide into ascending and descending branches, which form longitudinal anastomoses.

Lymphatic drainage of the ureters follows a pattern similar to that of the arterial supply. Lymph from:

- the upper part of each ureter drains to the lateral aortic (lumbar) nodes,
- the middle part of each ureter drains to lymph nodes associated with the common iliac vessels, and
- the inferior part of each ureter drains to lymph nodes associated with the external and internal iliac vessels.

### Ureteric innervation

Ureteric innervation is from the renal, aortic, superior hypogastric, and inferior hypogastric plexuses through nerves that follow the blood vessels.

Visceral efferent fibers come from both sympathetic and parasympathetic sources, whereas visceral afferent fibers return to T11 to L2 spinal cord levels. Ureteric pain, which is usually related to distention of the ureter, is therefore referred to cutaneous areas supplied by T11 to L2 spinal cord levels. These areas would most likely include the posterior and lateral abdominal wall below the ribs and above the iliac crest, the pubic region, the scrotum in males, the labia majora in females, and the proximal anterior aspect of the thigh.

## In the clinic

**Urinary tract stones**

Urinary tract stones (calculi) occur more frequently in men than in women, are most common in people aged between 20 and 60 years, and are usually associated with sedentary lifestyles. The stones are polycrystalline aggregates of calcium, phosphate, oxalate, urate, and other soluble salts within an organic matrix. The urine becomes saturated with these salts, and small variations in the pH cause the salts to precipitate.

Typically the patient has pain that radiates from the infrascapular region (loin) into the groin, and even into the scrotum or labia majora. Blood in the urine (**hematuria**) may also be noticed.

Infection must be excluded because certain species of bacteria are commonly associated with urinary tract stones.

The complications of urinary tract stones include infection, urinary obstruction, and renal failure. Stones may also develop within the bladder and produce marked irritation, causing pain and discomfort.

The diagnosis of urinary tract stones is based upon history and examination. Stones are often visible on abdominal radiographs. Special investigations include:

- ultrasound scanning, which may demonstrate the dilated renal pelvis and calices when the urinary system is obstructed. This is the preferred way of imaging in pregnant women or when clinical suspicion is low.
- low-dose CT of the urinary tract (CT KUB), which allows the detection of even smaller stones, shows the exact level of obstruction and, based on the size, density, and location of the stone, can help the urologist plan a procedure to remove the stone if necessary (extracorporeal shock wave lithotripsy versus ureteroscopy, percutaneous nephrolithotomy, or, extremely rare these days, open surgery) (Fig. 4.156).
- an intravenous urogram, which will demonstrate the obstruction, pinpoint the exact level of the stone is currently less often used because access to low-dose CT KUB has increased.

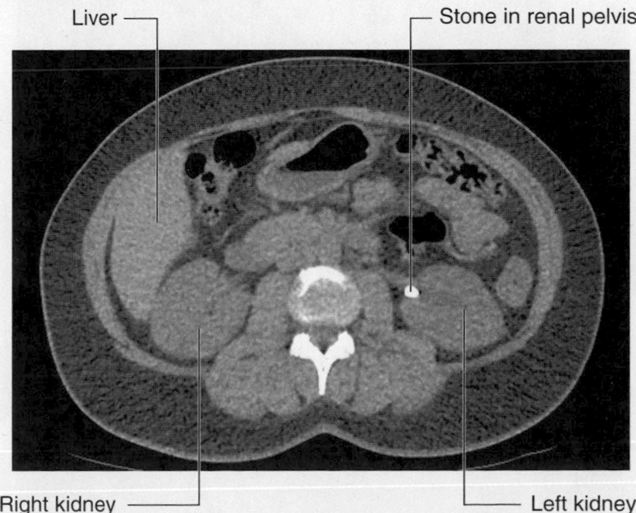


**Fig. 4.156** Low-dose axial CT of urinary tract (CT KUB) displays stone in left renal pelvis.

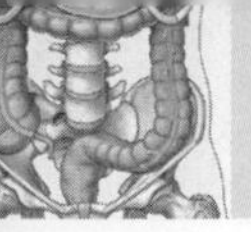



## In the clinic

### Urinary tract cancer

Most tumors that arise in the kidney are renal cell carcinomas. These tumors develop from the proximal tubular epithelium. Approximately 5% of tumors within the kidney are transitional cell tumors, which arise from the urothelium of the renal pelvis. Most patients typically have blood in the urine (hematuria), pain in the infrascapular region (loin), and a mass.

Renal cell tumors (Figs. 4.157 and 4.158) are unusual because not only do they grow outward from the kidney, invading the fat and fascia, but they also spread into the renal vein. This venous extension is rare for any other type of tumor, so, when seen, renal cell carcinoma should be suspected. In addition, the tumor may spread along the renal vein and into the inferior vena cava, and in rare cases can grow into the right atrium across the tricuspid valve and into the pulmonary artery.

Treatment for most renal cancers is surgical removal, even when metastatic spread is present, because some patients show regression of metastases.

Transitional cell carcinoma arises from the urothelium. The urothelium is present from the calices to the urethra and behaves as a "single unit." Therefore, when patients develop transitional carcinomas within the bladder, similar tumors may also be present within upper parts of the urinary tract. In patients with bladder cancer, the whole of the urinary tract must always be investigated to exclude the possibility of other tumors (Fig. 4.159). This is currently achieved by performing a dual-phase CT urogram that allows visualization of the renal parenchyma and the collecting system at the same time.

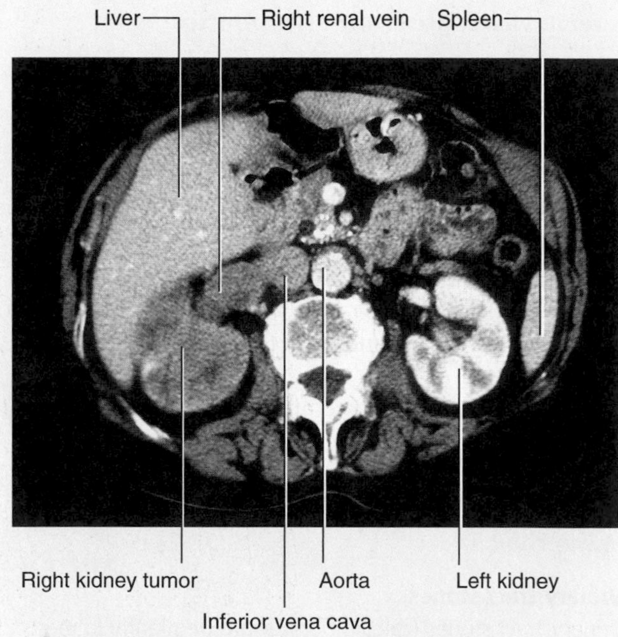


**Fig. 4.158** Tumor in the right kidney spreading into the right renal vein. Computed tomogram in the axial plane.

Duodenum
Left psoas major muscle
Right renal cancer
Aorta
Left kidney
Inferior vena cava

**Fig. 4.157** Tumor in the right kidney growing toward, and possibly invading, the duodenum. Computed tomogram in the axial plane.

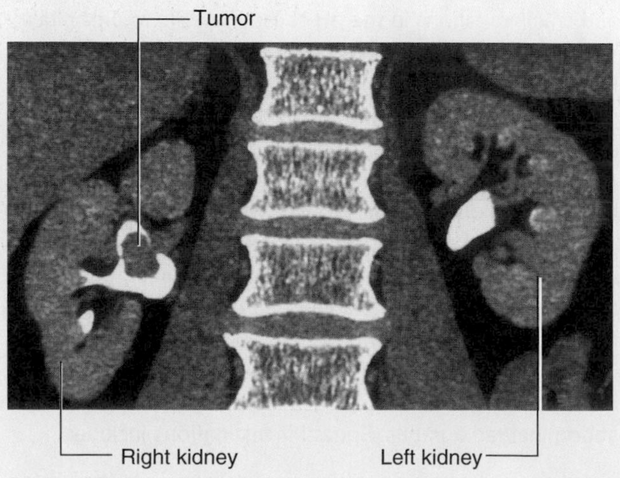


**Fig. 4.159** Transitional cell carcinoma in the pelvis of the right kidney. Coronal computed tomogram reconstruction.

## In the clinic

### Nephrostomy

A nephrostomy is a procedure where a tube is placed through the lateral or posterior abdominal wall into the renal cortex to lie within the renal pelvis. The function of this tube is to allow drainage of urine from the renal pelvis through the tube externally (Fig. 4.160).

The kidneys are situated on the posterior abdominal wall, and in thin healthy subjects may be only up to 2 to 3 cm from the skin. Access to the kidney is relatively straightforward, because the kidney can be easily visualized under ultrasound guidance. Using local anesthetic, a needle can be placed, under ultrasound direction, through the skin into the renal cortex and into the renal pelvis. A series of wires and tubes can be passed through the needle to position the drainage catheter.

The indications for such a procedure are many. In patients with distal ureteric obstruction the back pressure of urine within the ureters and the kidney significantly impairs the function of the kidney. This will produce renal failure and ultimately death. Furthermore, a dilated obstructed system is also susceptible to infection. In many cases, there is not only obstruction producing renal failure but also infected urine within the system.

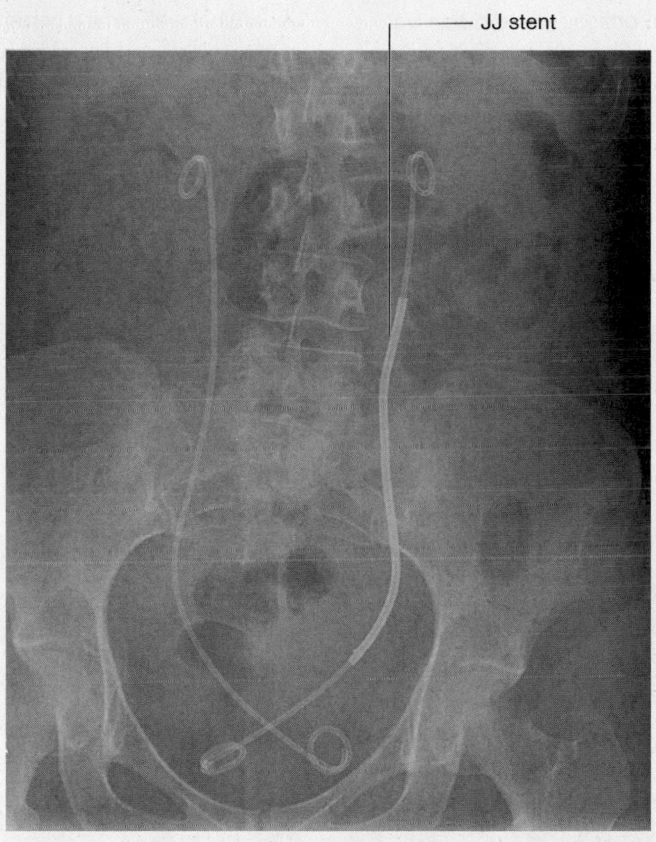


**Fig. 4.160** This radiograph demonstrates a double-J stent (anteroposterior view). The superior aspect of the double-J stent is situated within the renal pelvis. The stent passes through the ureter, describing the path of the ureter, and the tip of the double-J stent is projected over the bladder, which appears as a slightly dense area on the radiograph.

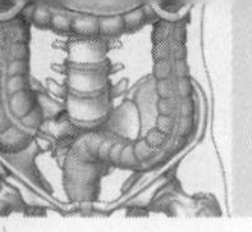

## In the clinic

### Kidney transplant

Renal transplantation is now a common procedure undertaken in patients with end-stage renal failure.

Transplant kidneys are obtained from either living or deceased donors. The living donors are carefully assessed, because harvesting a kidney from a normal healthy individual, even with modern-day medicine, carries a small risk.

Deceased kidney donors are brain dead or have suffered cardiac death. The donor kidney is harvested with a small cuff of aortic and venous tissue. The ureter is also harvested.

An ideal place to situate the transplant kidney is in the left or the right iliac fossa (Fig. 4.161). A curvilinear incision is made paralleling the iliac crest and pubic symphysis. The external oblique muscle, internal oblique muscle, transversus abdominis muscle, and transversalis fascia are divided. The surgeon identifies the parietal peritoneum but does not enter the peritoneal cavity. The parietal peritoneum is medially retracted to reveal the external iliac artery, external iliac vein, and bladder. In some instances the internal iliac artery of the recipient is mobilized and anastomosed directly as an end-to-end procedure onto the renal artery of the donor kidney. Similarly the internal iliac vein is anastomosed to the donor vein. In the presence of a small aortic cuff of tissue the donor artery is anastomosed to the recipient external iliac artery and similarly for the venous anastomosis. The ureter is easily tunneled obliquely through the bladder wall with a straightforward anastomosis.

The left and right iliac fossae are ideal locations for the transplant kidney because a new space can be created without compromise to other structures. The great advantage of this procedure is the proximity to the anterior abdominal wall, which permits easy ultrasound visualization of the kidney and Doppler vascular assessment. Furthermore, in this position biopsies are easily obtained. The extraperitoneal approach enables patients to make a swift recovery.

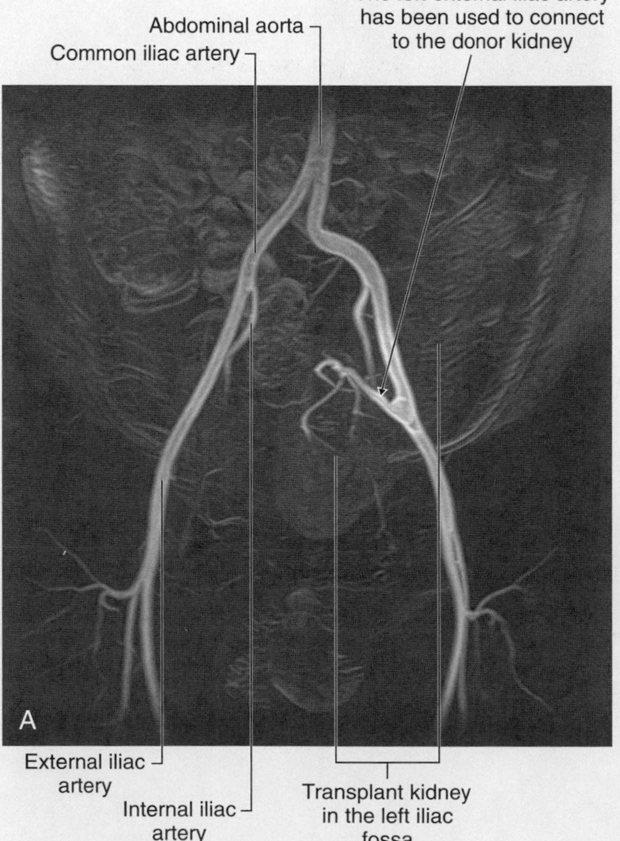


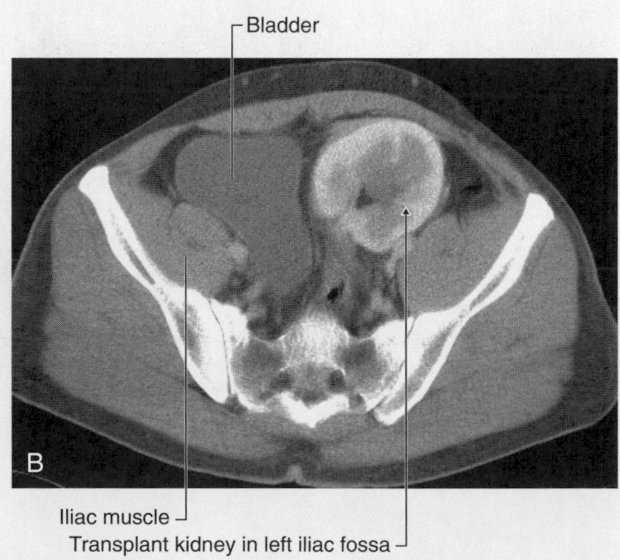


**Fig. 4.161** Kidney transplant. **A.** This image demonstrates an MR angiogram of the bifurcation of the aorta. Attaching to the left external iliac artery is the donor artery for a kidney that has been transplanted into the left iliac fossa. **B.** Abdominal computed tomogram, in the axial plane, showing the transplanted kidney in the left iliac fossa.

## In the clinic

### Investigation of the urinary tract

After an appropriate history and examination of the patient, including a digital rectal examination to assess the prostate in men, special investigations are required.

#### *Cystoscopy*

Cystoscopy is a technique that allows visualization of the urinary bladder and urethra using an optical system attached to a flexible or rigid tube (cystoscope). Images are displayed on a monitor, as done in other endoscopic studies. Biopsies, bladder stone removal, removal of foreign bodies from the bladder, and bleeding cauterization can be performed during cystoscopy. Cystoscopy is helpful in establishing the causes of macroscopic and microscopic hematuria, assessing bladder and urethral diverticula and fistulas, as well as serving as a tool to investigate patients with voiding problems.

#### *IVU (intravenous urogram)*

An IVU is one of the most important and commonly carried out radiological investigations (Fig. 4.162). The patient is injected with iodinated contrast medium. Most contrast media contain three iodine atoms spaced around a benzene ring. The relatively high atomic number of iodine compared to the atomic number of carbon, hydrogen, and oxygen attenuates the radiation beam. After intravenous injection, contrast media are excreted predominantly by glomerular filtration, although some are secreted by the renal tubules. This allows visualization of the collecting system as well as the ureters and bladder.

#### *Ultrasound*

Ultrasound can be used to assess kidney size and the size of the calices, which may be dilated when obstructed. Although the ureters are poorly visualized using ultrasound, the bladder can be easily seen when full. Ultrasound measurements of bladder volume can be obtained before and after micturition.

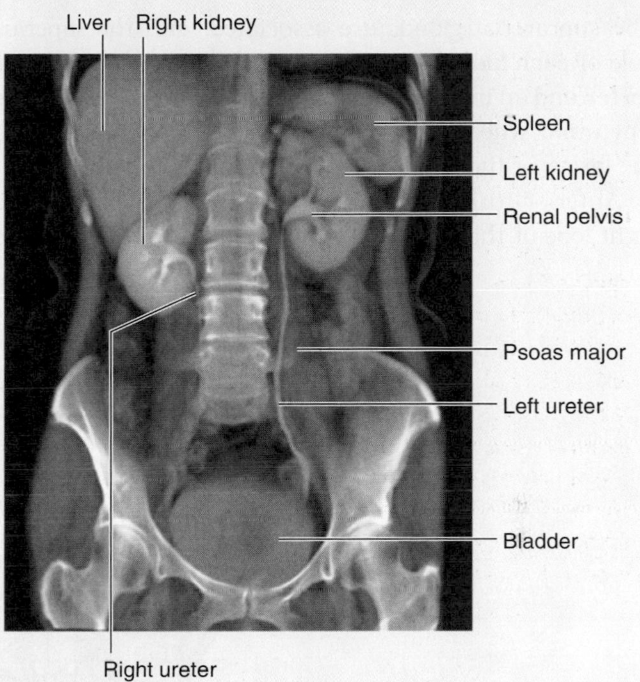


**Fig. 4.162** Coronal view of 3-D urogram using multidetector computed tomography.

#### *Nuclear medicine*

Nuclear medicine is an extremely useful tool for investigating the urinary tract because radioisotope compounds can be used to estimate renal cell mass and function and assess the parenchyma for renal scarring. These tests are often very useful in children when renal scarring and reflux disease is suspected.

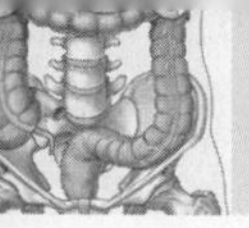
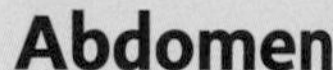


### Suprarenal glands

The suprarenal glands are associated with the superior pole of each kidney (Fig. 4.163). They consist of an outer cortex and an inner medulla. The right gland is shaped like a pyramid, whereas the left gland is semilunar in shape and the larger of the two.

Anterior to the right suprarenal gland is part of the right lobe of the liver and the inferior vena cava, whereas anterior to the left suprarenal gland is part of the stomach, pancreas, and, on occasion, the spleen. Parts of the diaphragm are posterior to both glands.

The suprarenal glands are surrounded by the perinephric fat and enclosed in the renal fascia, though a thin septum separates each gland from its associated kidney.

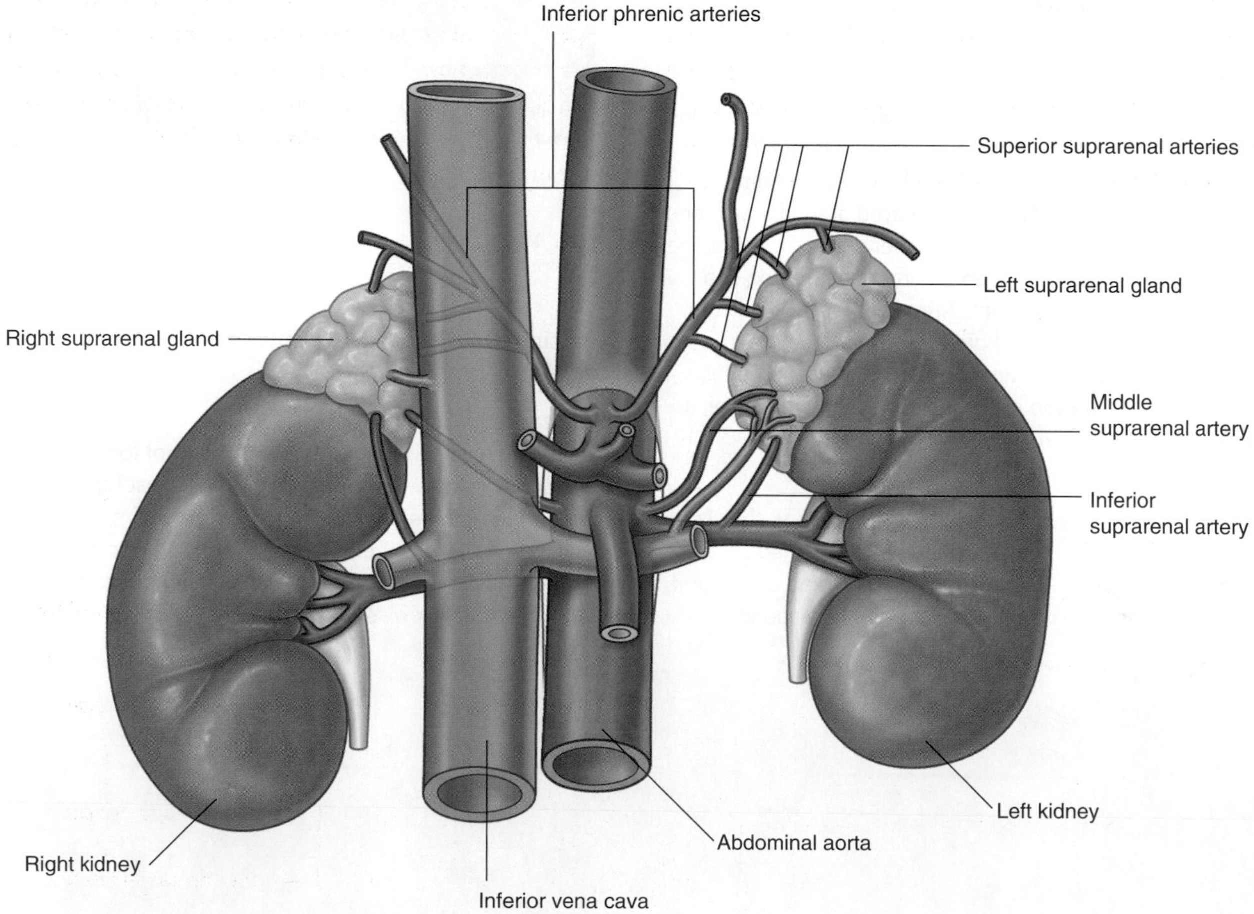

**Fig. 4.163** Arterial supply to the suprarenal glands.

### Suprarenal vasculature

The arterial supply to the suprarenal glands is extensive and arises from three primary sources (Fig. 4.163):

- As the bilateral inferior phrenic arteries pass upward from the abdominal aorta to the diaphragm, they give off multiple branches (superior suprarenal arteries) to the suprarenal glands.
- A middle branch (middle suprarenal artery) to the suprarenal glands usually arises directly from the abdominal aorta.
- Inferior branches (inferior suprarenal arteries) from the renal arteries pass upward to the suprarenal glands.

In contrast to this multiple arterial supply is the venous drainage, which usually consists of a single vein leaving the hilum of each gland. On the right side, the **right suprarenal vein** is short and almost immediately enters the inferior vena cava, while on the left side, the **left suprarenal vein** passes inferiorly to enter the left renal vein.

### Suprarenal innervation

The suprarenal gland is mainly innervated by preganglionic sympathetic fibers from spinal levels T8-L1 that pass through both the sympathetic trunk and the prevertebral plexus without synapsing. These preganglionic fibers directly innervate cells of the adrenal medulla.

## Vasculature

### Abdominal aorta

The abdominal aorta begins at the aortic hiatus of the diaphragm as a midline structure at approximately the lower level of vertebra TXII (Fig. 4.164). It passes downward on the anterior surface of the bodies of vertebrae LI to LIV, ending just to the left of midline at the lower level of vertebra LIV. At this point, it divides into the **right** and **left common iliac arteries**. This bifurcation can be visualized on the anterior abdominal wall as a point approximately 2.5 cm below the umbilicus or even with a line extending between the highest points of the iliac crest.

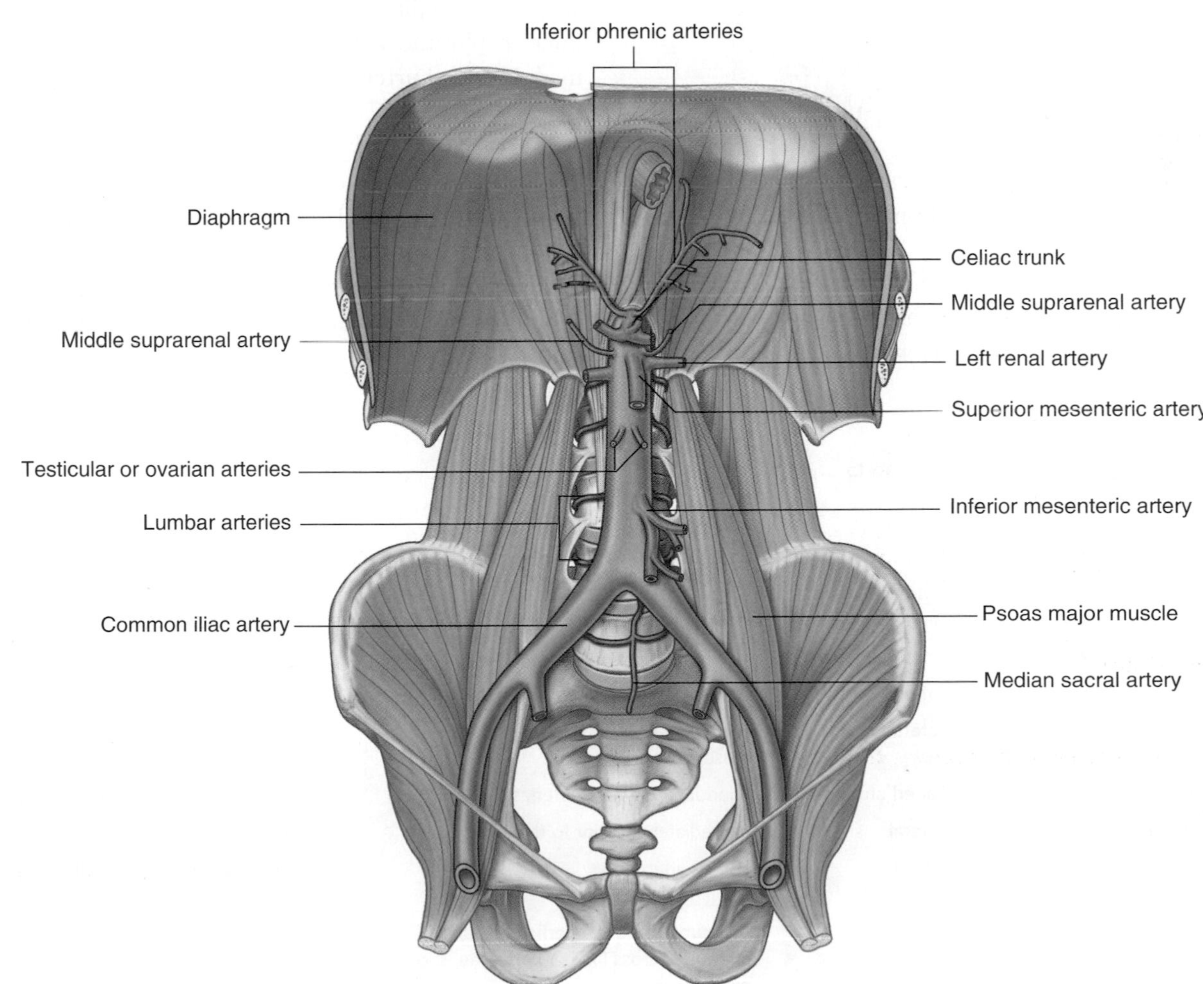


**Fig. 4.164** Abdominal aorta.

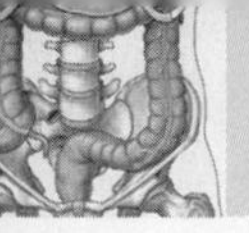



As the abdominal aorta passes through the posterior abdominal region, the prevertebral plexus of nerves and ganglia covers its anterior surface. It is also related to numerous other structures:

- Anterior to the abdominal aorta, as it descends, are the pancreas and splenic vein, the left renal vein, and the inferior part of the duodenum.
- Several left lumbar veins cross it posteriorly as they pass to the inferior vena cava.
- On its right side are the cisterna chyli, thoracic duct, azygos vein, right crus of the diaphragm, and the inferior vena cava.
- On its left side is the left crus of the diaphragm.

Branches of the abdominal aorta (Table 4.3) can be classified as:

- visceral branches supplying organs,
- posterior branches supplying the diaphragm or body wall, or
- terminal branches.

### Visceral branches

The visceral branches are either unpaired or paired vessels.

The three unpaired visceral branches that arise from the anterior surface of the abdominal aorta (Fig. 4.164) are:

- the celiac trunk, which supplies the abdominal foregut,
- the superior mesenteric artery, which supplies the abdominal midgut, and
- the inferior mesenteric artery, which supplies the abdominal hindgut.

The paired visceral branches of the abdominal aorta (Fig. 4.164) include:

- the middle suprarenal arteries—small, lateral branches of the abdominal aorta arising just above the renal arteries that are part of the multiple vascular supply to the suprarenal gland;
- the renal arteries—lateral branches of the abdominal aorta that arise just inferior to the origin of the superior mesenteric artery between vertebrae LI and LII, and supply the kidneys; and
- the testicular or ovarian arteries—anterior branches of the abdominal aorta that arise below the origin of the renal arteries, and pass downward and laterally on the anterior surface of the psoas major muscle.

### Posterior branches

The posterior branches of the abdominal aorta are vessels supplying the diaphragm or body wall. They consist of the inferior phrenic arteries, the lumbar arteries, and the median sacral artery (Fig. 4.164).

#### Inferior phrenic arteries

The **inferior phrenic arteries** arise immediately inferior to the aortic hiatus of the diaphragm either directly from the abdominal aorta, as a common trunk from the abdominal aorta, or from the base of the celiac trunk (Fig. 4.164). Whatever their origin, they pass upward, provide some arterial supply to the suprarenal gland, and continue onto the inferior surface of the diaphragm.

**Table 4.3** Branches of the abdominal aorta

| Artery | Branch | Origin | Parts supplied |
|---|---|---|---|
| Celiac trunk | Anterior | Immediately inferior to the aortic hiatus of the diaphragm | Abdominal foregut |
| Superior mesenteric artery | Anterior | Immediately inferior to the celiac trunk | Abdominal midgut |
| Inferior mesenteric artery | Anterior | Inferior to the renal arteries | Abdominal hindgut |
| Middle suprarenal arteries | Lateral | Immediately superior to the renal arteries | Suprarenal glands |
| Renal arteries | Lateral | Immediately inferior to the superior mesenteric artery | Kidneys |
| Testicular or ovarian arteries | Paired anterior | Inferior to the renal arteries | Testes in male and ovaries in female |
| Inferior phrenic arteries | Lateral | Immediately inferior to the aortic hiatus | Diaphragm |
| Lumbar arteries | Posterior | Usually four pairs | Posterior abdominal wall and spinal cord |
| Median sacral artery | Posterior | Just superior to the aortic bifurcation, passes inferiorly across lumbar vertebrae, sacrum, and coccyx | |
| Common iliac arteries | Terminal | Bifurcation usually occurs at the level of LIV vertebra | |

### Lumbar arteries

There are usually four pairs of **lumbar arteries** arising from the posterior surface of the abdominal aorta (Fig. 4.164). They run laterally and posteriorly over the bodies of the lumbar vertebrae, continue laterally, passing posterior to the sympathetic trunks and between the transverse processes of adjacent lumbar vertebrae, and reach the abdominal wall. From this point onward, they demonstrate a branching pattern similar to a posterior intercostal artery, which includes providing segmental branches that supply the spinal cord.

### Median sacral artery

The final posterior branch is the **median sacral artery** (Fig. 4.164). This vessel arises from the posterior surface of the abdominal aorta just superior to the bifurcation and passes in an inferior direction, first over the anterior surface of the lower lumbar vertebrae and then over the anterior surface of the sacrum and coccyx.

## In the clinic

**Abdominal aortic stent graft**

An abdominal aortic aneurysm is a dilation of the aorta and generally tends to occur in the infrarenal region (the region at or below the renal arteries). As the aorta expands, the risk of rupture increases, and it is now generally accepted that when an aneurysm reaches 5.5 cm or greater an operation will significantly benefit the patient.

With the aging population, the number of abdominal aortic aneurysms is increasing. Moreover, with the increasing use of imaging techniques, a number of abdominal aortic aneurysms are identified in asymptomatic patients.

For many years the standard treatment for repair was an open operative technique, which involved a large incision from the xiphoid process of the sternum to the symphysis pubis and dissection of the aneurysm. The aneurysm was excised and a tubular woven graft was sewn into place. Recovery may take a number of days, even weeks, and most patients would be placed in the intensive care unit after the operation.

Further developments and techniques have led to a new type of procedure being performed to treat abdominal aortic aneurysms—the endovascular graft (Fig. 4.165).

The technique involves surgically dissecting the femoral artery below the inguinal ligament. A small incision is made in the femoral artery and the preloaded compressed graft with metal support struts is passed on a large catheter into the abdominal aorta through the femoral artery. Using X-ray for guidance the graft is opened, lining the inside of the aorta. Limb attachments are made to the graft that extend into the common iliac vessels. This bifurcated tube device effectively excludes the abdominal aortic aneurysm.

This type of device is not suitable for all patients. Patients who receive this device do not need to go to the intensive care unit. Many patients leave the hospital within 24 to 48 hours. Importantly, this device can be used for patients who were deemed unfit for open surgical repair.

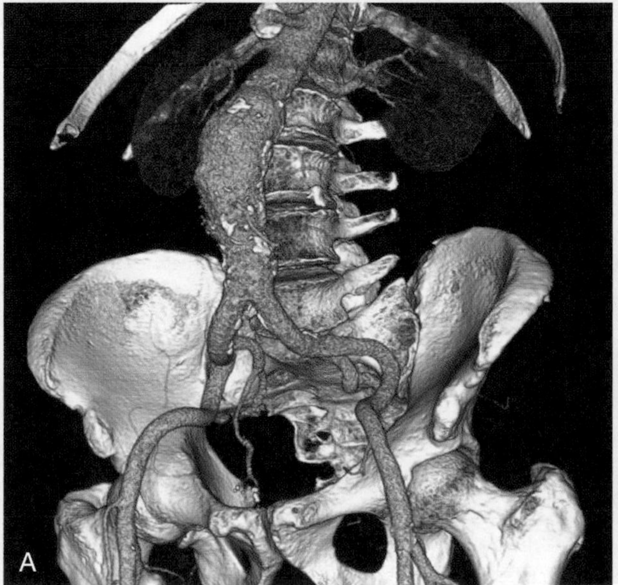


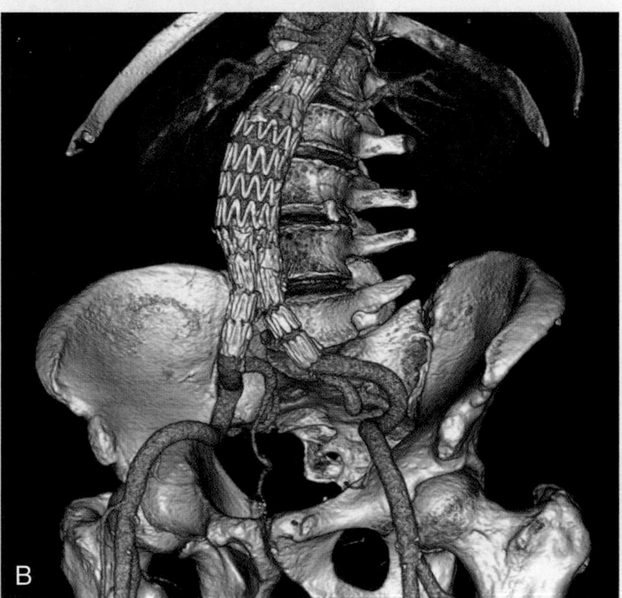


**Fig. 4.165** Volume-rendered reconstruction using multidetector computed tomography of patient with an infrarenal abdominal aortic aneurysm before (**A**) and after (**B**) endovascular aneurysm repair. Note the image only demonstrates the intraluminal contrast and not the entire vessel. White patches in the aorta represent intramural calcium.

### Inferior vena cava

The inferior vena cava returns blood from all structures below the diaphragm to the right atrium of the heart. It is formed when the two common iliac veins come together at the level of vertebra LV, just to the right of midline. It ascends through the posterior abdominal region anterior to the vertebral column immediately to the right of the abdominal aorta (Fig. 4.166), continues in a superior direction, and leaves the abdomen by piercing the central tendon of the diaphragm at the level of vertebra TVIII.

During its course, the anterior surface of the inferior vena cava is crossed by the right common iliac artery, the root of the mesentery, the right testicular or ovarian artery, the inferior part of the duodenum, the head of the pancreas, the superior part of the duodenum, the bile duct, the portal vein, and the liver, which overlaps and on occasion completely surrounds the vena cava (Fig. 4.166).

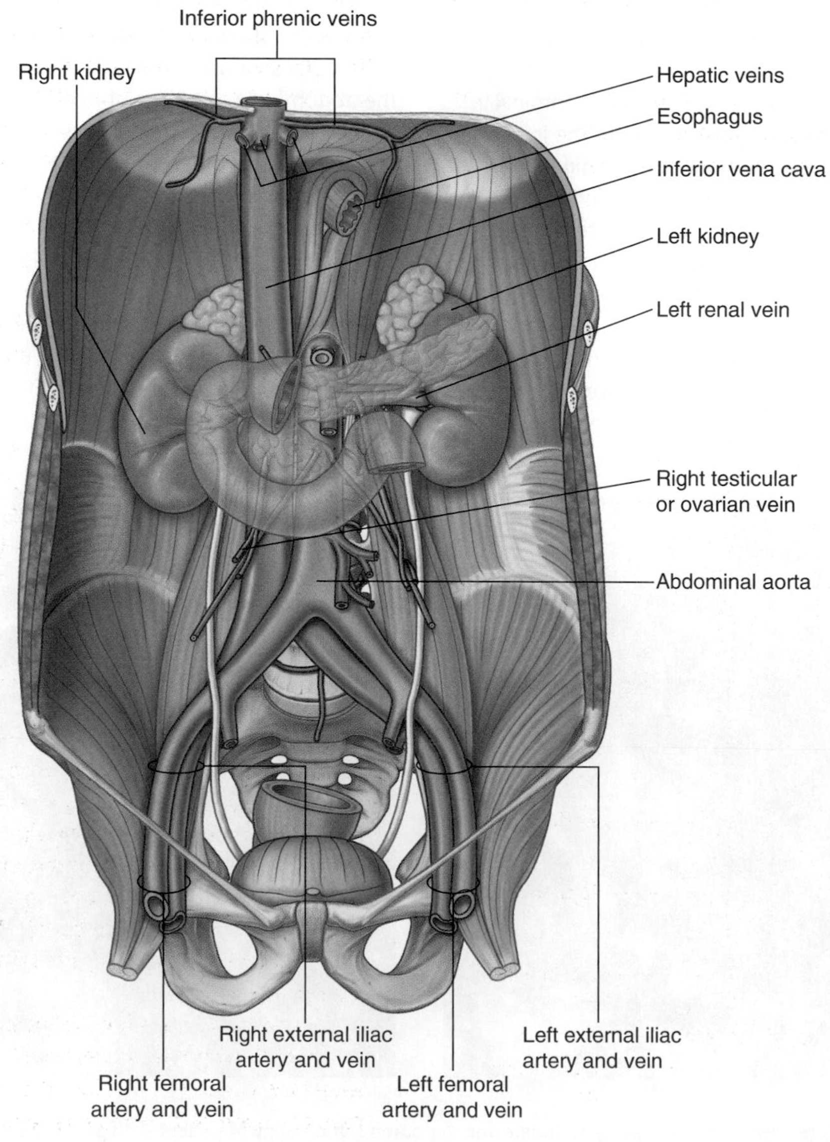


**Fig. 4.166** Inferior vena cava.

Tributaries to the inferior vena cava include the:

- common iliac veins,
- lumbar veins,
- right testicular or ovarian vein,
- renal veins,
- right suprarenal vein,
- inferior phrenic veins, and
- hepatic veins.

There are no tributaries from the abdominal part of the gastrointestinal tract, the spleen, the pancreas, or the gallbladder, because veins from these structures are components of the portal venous system, which first passes through the liver.

Of the venous tributaries mentioned above, the **lumbar veins** are unique in their connections and deserve special attention. Not all of the lumbar veins drain directly into the inferior vena cava (Fig. 4.167):

- The fifth lumbar vein generally drains into the iliolumbar vein, a tributary of the common iliac vein.
- The third and fourth lumbar veins usually drain into the inferior vena cava.
- The first and second lumbar veins may empty into the ascending lumbar veins.

The **ascending lumbar veins** are long, anastomosing venous channels that connect the common iliac, iliolumbar, and lumbar veins with the azygos and hemi-azygos veins of the thorax (Fig. 4.167).

If the inferior vena cava becomes blocked, the ascending lumbar veins become important collateral channels between the lower and upper parts of the body.

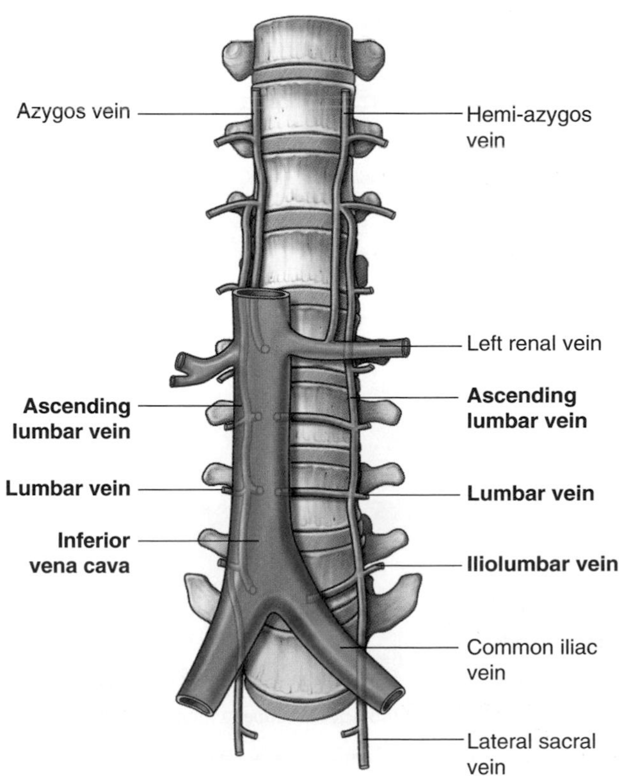


**Fig. 4.167** Lumbar veins.

## In the clinic

### Inferior vena cava filter

Deep vein thrombosis is a potentially fatal condition where a clot (thrombus) is formed in the deep venous system of the legs and the veins of the pelvis. Virchow described the reasons for thrombus formation as decreased blood flow, abnormality of the constituents of blood, and abnormalities of the vessel wall. Common predisposing factors include hospitalization and surgery, the oral contraceptive pill, smoking, and air travel. Other factors include clotting abnormalities (e.g., protein S and protein C deficiency).

The diagnosis of deep vein thrombosis may be difficult to establish, with symptoms including leg swelling and pain and discomfort in the calf. It may also be an incidental finding.

In practice, patients with suspected deep vein thrombosis undergo a D-dimer blood test, which measures levels of a fibrin degradation product. If this is positive there is a high association with deep vein thrombosis.

The consequences of deep vein thrombosis are twofold. Occasionally the clot may dislodge and pass into the venous system through the right side of the heart and into the main pulmonary arteries. If the clots are of significant size they obstruct blood flow to the lung and may produce instantaneous death. Secondary complications include destruction of the normal valvular system in the legs, which may lead to venous incompetency and chronic leg swelling with ulceration.

The treatment for deep vein thrombosis is prevention. In order to prevent deep vein thrombosis, patients are optimized by removing all potential risk factors. Subcutaneous heparin may be injected and the patient wears compression stockings to prevent venous stasis while in the hospital.

In certain situations it is not possible to optimize the patient with prophylactic treatment, and it may be necessary to insert a filter into the inferior vena cava that traps any large clots. It may be removed after the risk period has ended.

## Lymphatic system

Lymphatic drainage from most deep structures and regions of the body below the diaphragm converges mainly on collections of lymph nodes and vessels associated with the major blood vessels of the posterior abdominal region (Fig. 4.168). The lymph then predominantly drains into the thoracic duct. Major lymphatic channels that drain different regions of the body as a whole are summarized in Table 4.4 (also see Chapter 1, pp. 27–28, for discussion of lymphatics in general).

### Pre-aortic and lateral aortic or lumbar nodes (para-aortic nodes)

Approaching the aortic bifurcation, the collections of lymphatics associated with the two common iliac arteries and veins merge, and multiple groups of lymphatic vessels and nodes associated with the abdominal aorta and inferior

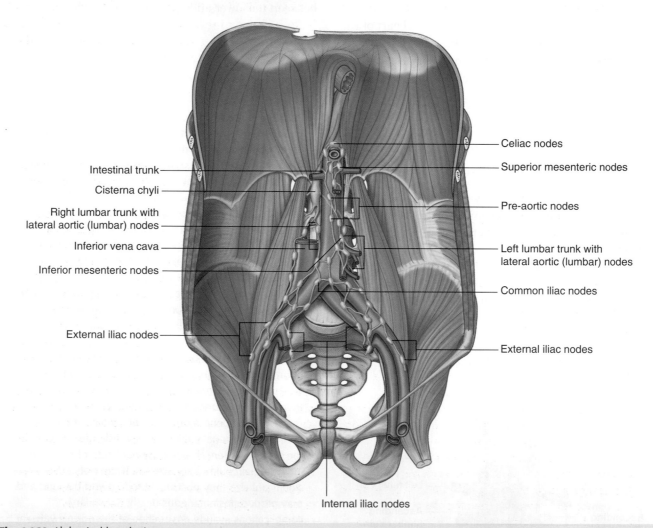

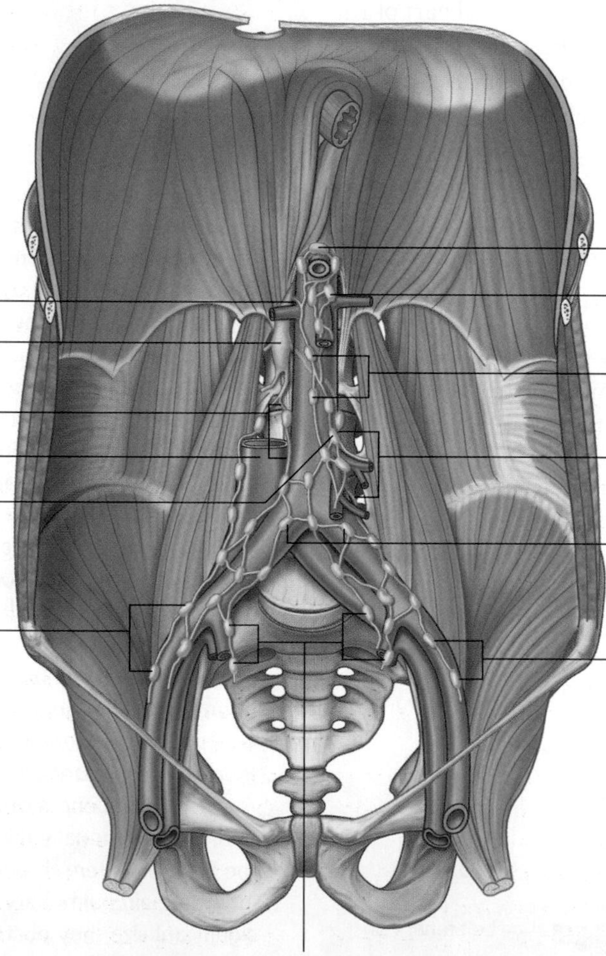

**Fig. 4.168** Abdominal lymphatics.

**Table 4.4** Lymphatic drainage

| Lymphatic vessel | Area drained |
|---|---|
| Right jugular trunk | Right side of head and neck |
| Left jugular trunk | Left side of head and neck |
| Right subclavian trunk | Right upper limb, superficial regions of thoracic and upper abdominal walls |
| Left subclavian trunk | Left upper limb, superficial regions of thoracic and upper abdominal walls |
| Right bronchomediastinal trunk | Right lung and bronchi, mediastinal structures, thoracic wall |
| Left bronchomediastinal trunk | Left lung and bronchi, mediastinal structures, thoracic wall |
| Thoracic duct | Lower limbs, abdominal walls and viscera, pelvic walls and viscera, thoracic wall |

vena cava pass superiorly. These collections may be subdivided into **pre-aortic nodes**, which are anterior to the abdominal aorta, and **right** and **left lateral aortic** or **lumbar nodes** (**para-aortic nodes**), which are positioned on either side of the abdominal aorta (Fig. 4.168).

As these collections of lymphatics pass through the posterior abdominal region, they continue to collect lymph from a variety of structures. The lateral aortic or lumbar lymph nodes (para-aortic nodes) receive lymphatics from the body wall, the kidneys, the suprarenal glands, and the testes or ovaries.

The pre-aortic nodes are organized around the three anterior branches of the abdominal aorta that supply the abdominal part of the gastrointestinal tract, as well as the spleen, pancreas, gallbladder, and liver. They are divided into celiac, superior mesenteric, and inferior mesenteric nodes, and receive lymph from the organs supplied by the similarly named arteries.

Finally, the lateral aortic or lumbar nodes form the right and left lumbar trunks, whereas the pre-aortic nodes form the intestinal trunk (Fig. 4.168). These trunks come together and form a confluence that, at times, appears as a saccular dilation (the cisterna chyli). This confluence of lymph trunks is posterior to the right side of the abdominal aorta and anterior to the bodies of vertebrae LI and LII. It marks the beginning of the thoracic duct.

## In the clinic

### Retroperitoneal lymph node surgery

From a clinical perspective, retroperitoneal lymph nodes are arranged in two groups. The pre-aortic lymph node group drains lymph from the embryological midline structures, such as the liver, bowel, and pancreas. The para-aortic lymph node group (the lateral aortic or lumbar nodes), on either side of the aorta, drain lymph from bilateral structures, such as the kidneys and adrenal glands. Organs embryologically derived from the posterior abdominal wall also drain lymph to these nodes. These organs include the ovaries and the testes (importantly, the testes do not drain lymph to the inguinal regions).

In general, lymphatic drainage follows standard predictable routes; however, in the presence of disease, alternate routes of lymphatic drainage will occur.

There are a number of causes for enlarged retroperitoneal lymph nodes. In the adult, massively enlarged lymph nodes are a feature of lymphoma, and smaller lymph node enlargement is observed in the presence of infection and metastatic malignant spread of disease (e.g., colon cancer).

The treatment for malignant lymph node disease is based upon a number of factors, including the site of the primary tumor (e.g., bowel) and its histological cell type. Normally, the primary tumor is surgically removed and the lymph node spread and metastatic organ spread (e.g., to the liver and the lungs) are often treated with chemotherapy and radiotherapy.

In certain instances it may be considered appropriate to resect the lymph nodes in the retroperitoneum (e.g., for testicular cancer).

The surgical approach to retroperitoneal lymph node resection involves a lateral paramedian incision in the midclavicular line. The three layers of the anterolateral abdominal wall (external oblique, internal oblique, and transversus abdominis) are opened and the transversalis fascia is divided. The next structure the surgeon sees is the parietal peritoneum. Instead of entering the parietal peritoneum, which is standard procedure for most intraabdominal operations, the surgeon gently pushes the parietal peritoneum toward the midline, which moves the intraabdominal contents and allows a clear view of the retroperitoneal structures. On the left, the para-aortic lymph node group is easily demonstrated, with a clear view of the abdominal aorta and kidney. On the right the inferior vena cava is demonstrated and has to be retracted to access the right para-aortic lymph node chain.

The procedure of retroperitoneal lymph node dissection is extremely well tolerated and lacks the problems of entering the peritoneal cavity (e.g., paralytic ileus). Unfortunately, a complication of a vertical incision in the midclavicular line is division of the segmental nerve supply to the rectus abdominis muscle. This produces muscle atrophy and asymmetrical proportions of the anterior abdominal wall.

## Nervous system in the posterior abdominal region

Several important components of the nervous system are in the posterior abdominal region. These include the sympathetic trunks and associated splanchnic nerves, the plexus of nerves and ganglia associated with the abdominal aorta, and the lumbar plexus of nerves.

### Sympathetic trunks and splanchnic nerves

The sympathetic trunks pass through the posterior abdominal region anterolateral to the lumbar vertebral bodies, before continuing across the sacral promontory and into the pelvic cavity (Fig. 4.169). Along their course, small raised areas are visible. These represent collections of neuronal cell bodies—primarily postganglionic neuronal cell bodies—which are located outside the central nervous system. They are sympathetic paravertebral ganglia. There are usually four ganglia along the sympathetic trunks in the posterior abdominal region.

Also associated with the sympathetic trunks in the posterior abdominal region are the lumbar splanchnic nerves (Fig. 4.169). These components of the nervous system pass from the sympathetic trunks to the plexus of nerves and ganglia associated with the abdominal aorta. Usually two to four lumbar splanchnic nerves carry preganglionic sympathetic fibers and visceral afferent fibers.

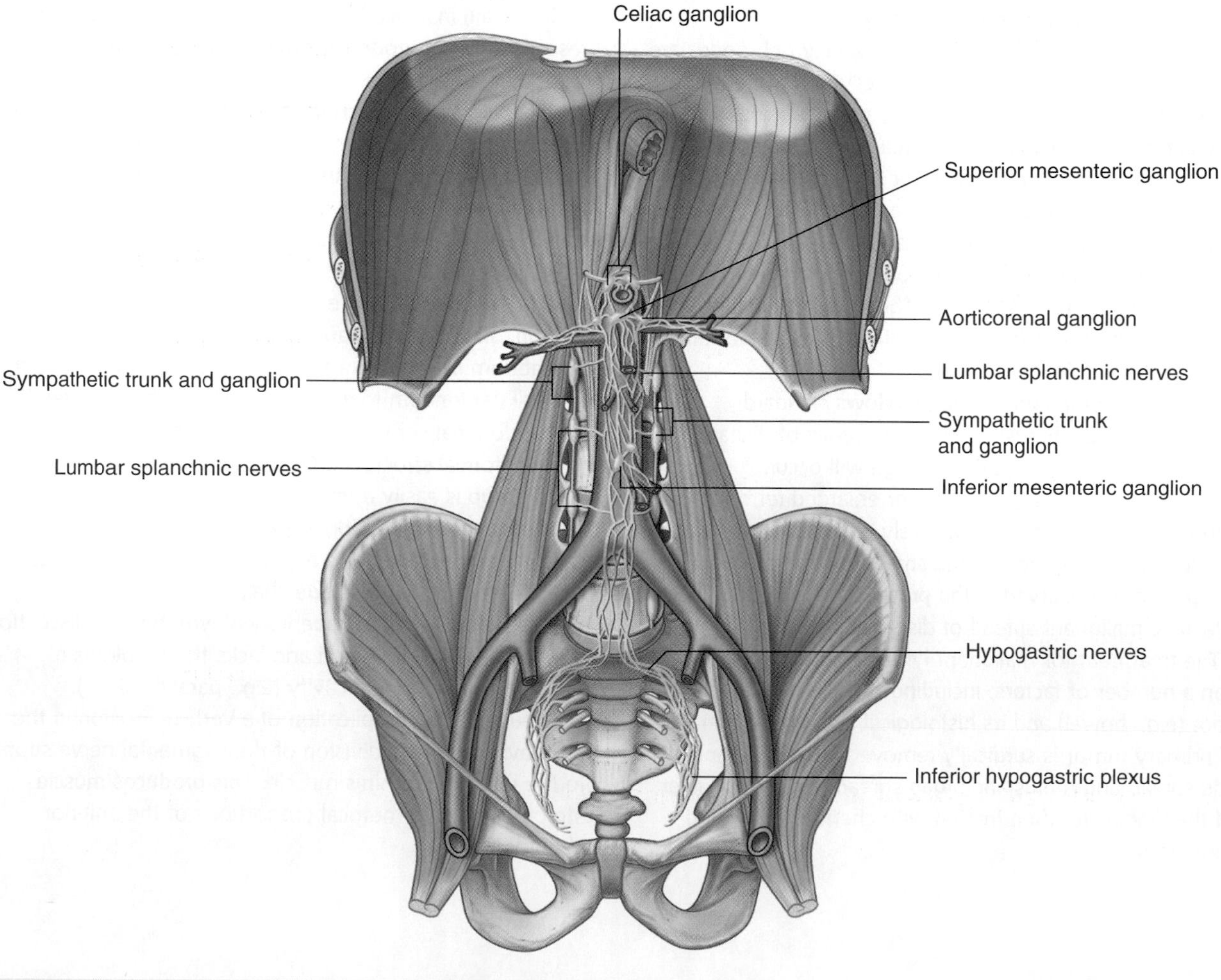

**Fig. 4.169** Sympathetic trunks passing through the posterior abdominal region.

### Abdominal prevertebral plexus and ganglia

The abdominal prevertebral plexus is a network of nerve fibers surrounding the abdominal aorta. It extends from the aortic hiatus of the diaphragm to the bifurcation of the aorta into the right and left common iliac arteries. Along its route, it is subdivided into smaller, named plexuses (Fig. 4.170):

- Beginning at the diaphragm and moving inferiorly, the initial accumulation of nerve fibers is referred to as the celiac plexus—this subdivision includes nerve fibers associated with the roots of the celiac trunk and superior mesenteric artery.
- Continuing inferiorly, the plexus of nerve fibers extending from just below the superior mesenteric artery to the aortic bifurcation is the abdominal aortic plexus (Fig. 4.170).
- At the bifurcation of the abdominal aorta, the abdominal prevertebral plexus continues inferiorly as the superior hypogastric plexus.

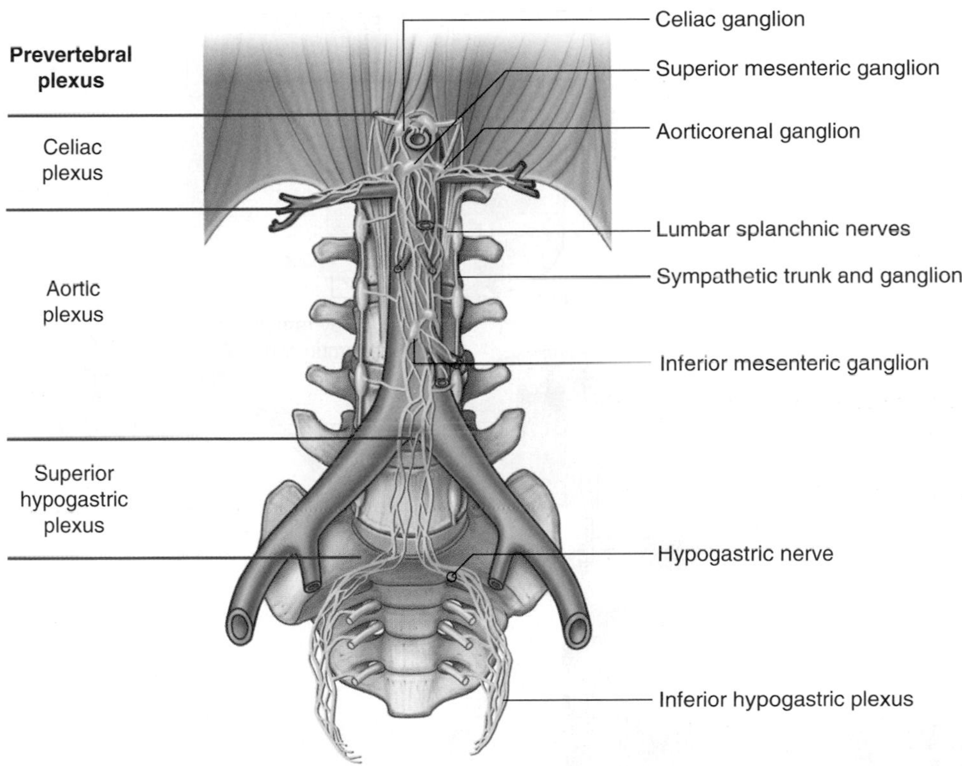


**Fig. 4.170** Prevertebral plexus and ganglia in the posterior abdominal region.

Throughout its length, the abdominal prevertebral plexus is a conduit for:

- preganglionic sympathetic and visceral afferent fibers from the thoracic and lumbar splanchnic nerves,
- preganglionic parasympathetic and visceral afferent fibers from the vagus nerves [X], and
- preganglionic parasympathetic fibers from the pelvic splanchnic nerves (Fig. 4.171).

Associated with the abdominal prevertebral plexus are clumps of nervous tissue (the **prevertebral ganglia**), which are collections of postganglionic sympathetic neuronal cell bodies in recognizable aggregations along the abdominal prevertebral plexus; they are usually named after the nearest branch of the abdominal aorta. They are therefore referred to as **celiac, superior mesenteric, aorticorenal,** and **inferior mesenteric ganglia** (Fig. 4.172). These structures, along with the abdominal prevertebral plexus, play a critical role in the innervation of the abdominal viscera.

Common sites for pain referred from the abdominal viscera and from the heart are given in Table 4.5.

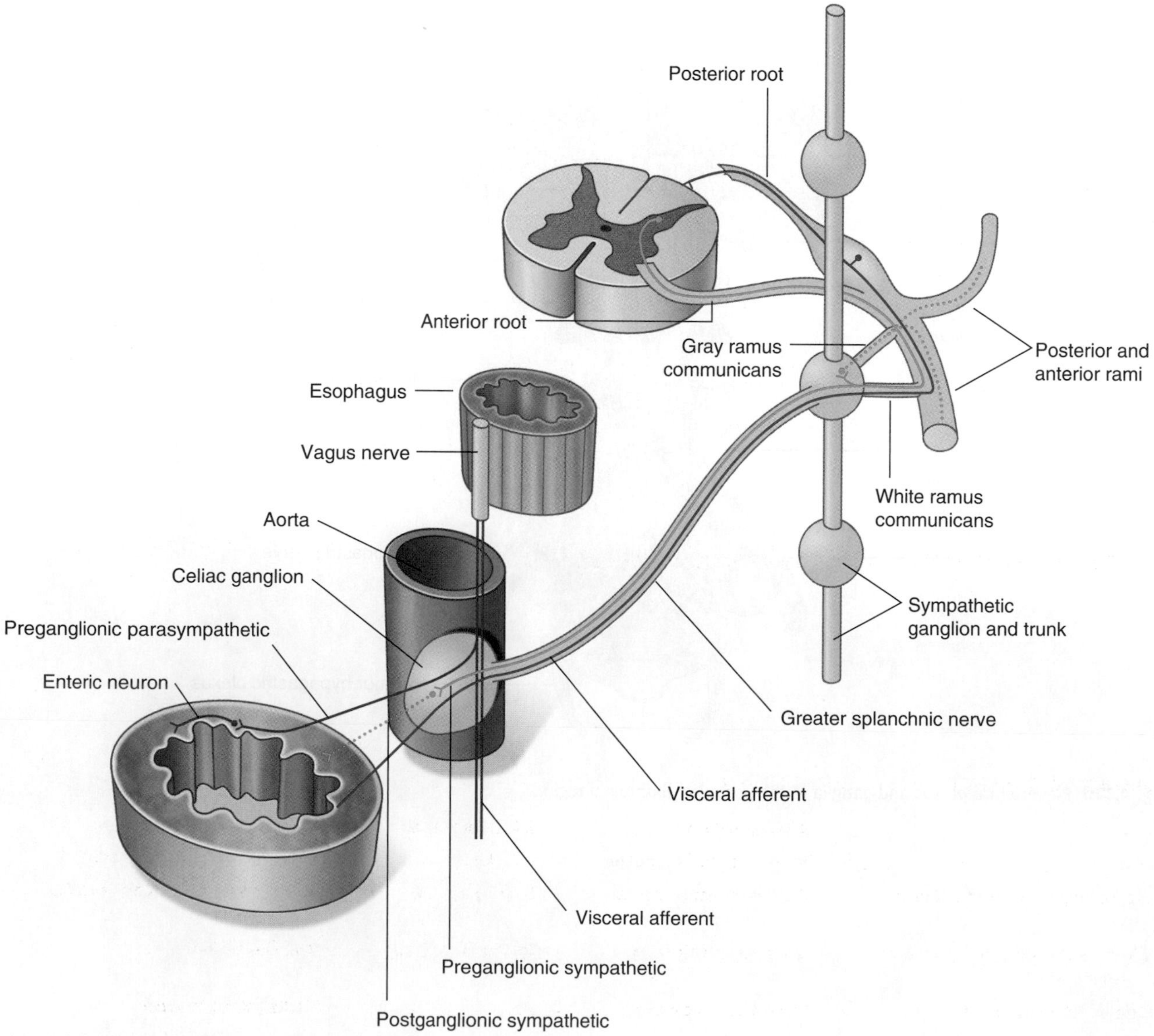

**Fig. 4.171** Nerve fibers passing through the abdominal prevertebral plexus and ganglia.

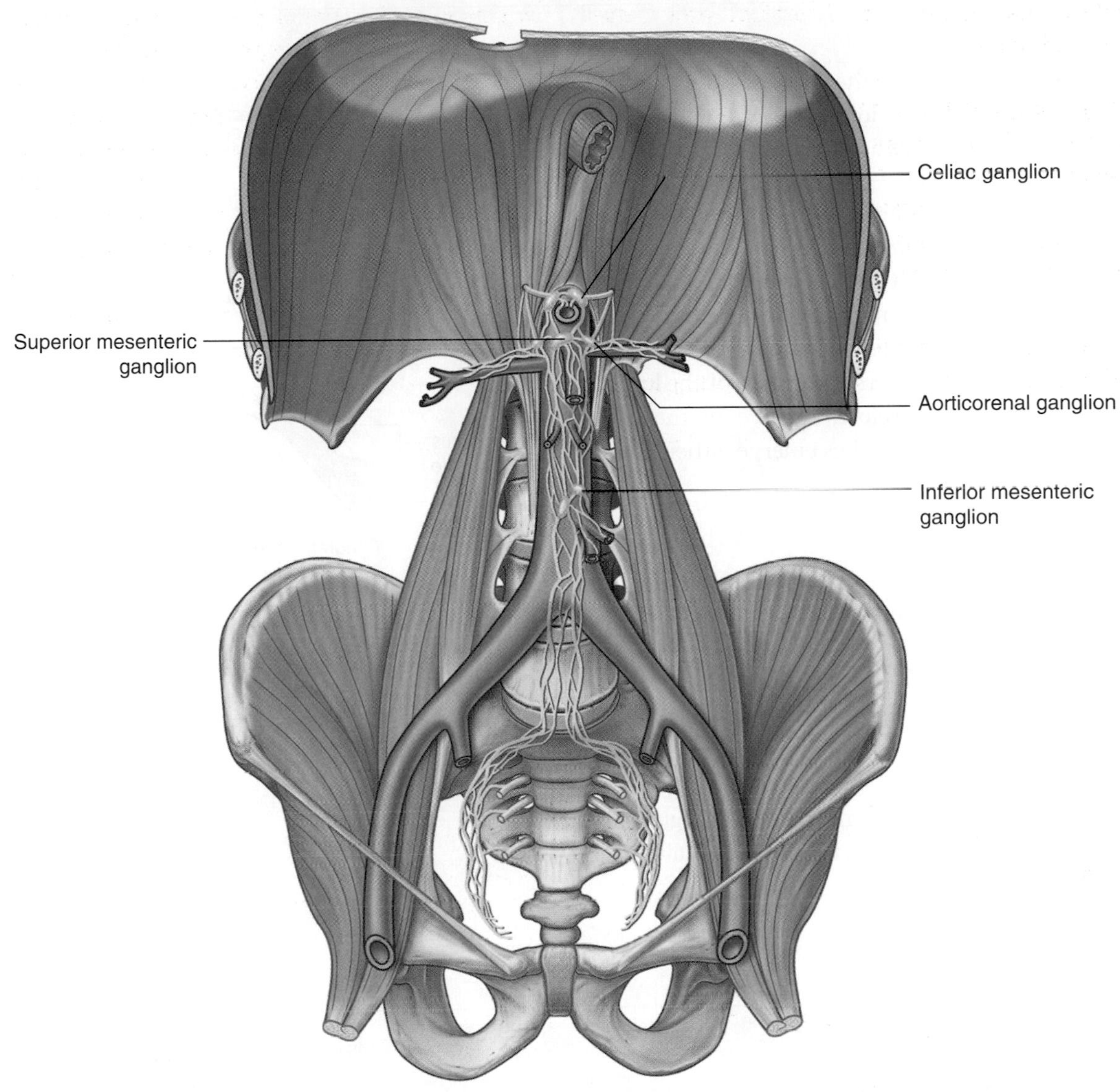


**Fig. 4.172** Prevertebral ganglia associated with the prevertebral plexus.

**Table 4.5** Referred pain pathways (visceral afferents)

| Organ | Afferent pathway | Spinal cord level | Referral area |
|---|---|---|---|
| Heart | Thoracic splanchnic nerves | T1 to T4 | Upper thorax and medial arm |
| Foregut (organs supplied by celiac trunk) | Greater splanchnic nerve | T5 to T9 (or T10) | Lower thorax and epigastric region |
| Midgut (organs supplied by superior mesenteric artery) | Lesser splanchnic nerve | T9, T10 (or T10, T11) | Umbilical region |
| Kidneys and upper ureter | Least splanchnic nerve | T12 | Flanks (lateral regions) |
| Hindgut (organs supplied by inferior mesenteric artery) and lower ureter | Lumbar splanchnic nerves | L1, L2 | Pubic region, lateral and anterior thighs, and groin |

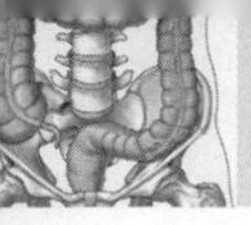



## Lumbar plexus

The lumbar plexus is formed by the anterior rami of nerves L1 to L3 and most of the anterior ramus of L4 (Fig. 4.173 and Table 4.6). It also receives a contribution from the T12 (subcostal) nerve.

Branches of the lumbar plexus include the iliohypogastric, ilio-inguinal, and genitofemoral nerves, the lateral cutaneous nerve of the thigh (lateral femoral cutaneous), and femoral and obturator nerves. The lumbar plexus forms in the substance of the psoas major muscle anterior to its attachment to the transverse processes of the lumbar vertebrae (Fig. 4.174). Therefore, relative to the psoas major muscle, the various branches emerge either:

- anterior—genitofemoral nerve,
- medial—obturator nerve, or
- lateral—iliohypogastric, ilio-inguinal, and femoral nerves and the lateral cutaneous nerve of the thigh.

### Iliohypogastric and ilio-inguinal nerves (L1)

The iliohypogastric and ilio-inguinal nerves arise as a single trunk from the anterior ramus of nerve L1 (Fig. 4.173). Either before or soon after emerging from the lateral border of the psoas major muscle, this single trunk divides into the iliohypogastric and the ilio-inguinal nerves.

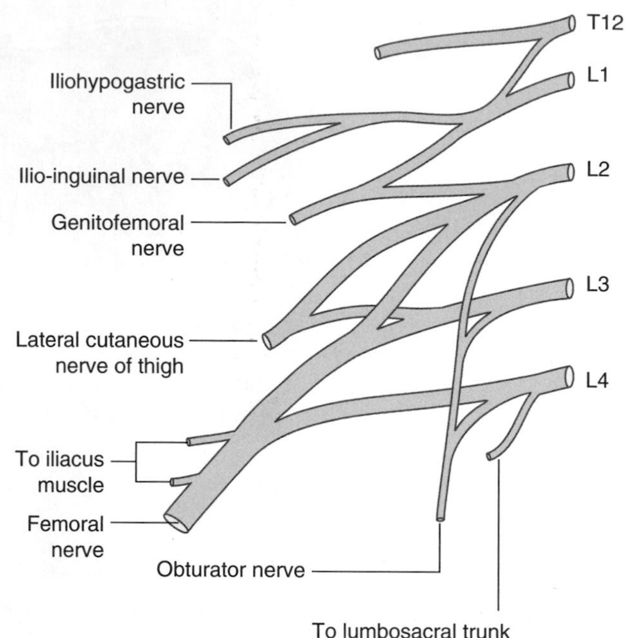


**Fig. 4.173** Lumbar plexus.

**Table 4.6** Branches of the lumbar plexus

| Branch | Origin | Spinal segments | Function: motor | Function: sensory |
|---|---|---|---|---|
| Iliohypogastric | Anterior ramus L1 | L1 | Internal oblique and transversus abdominis | Posterolateral gluteal skin and skin in pubic region |
| Ilio-inguinal | Anterior ramus L1 | L1 | Internal oblique and transversus abdominis | Skin in the upper medial thigh, and either the skin over the root of the penis and anterior scrotum or the mons pubis and labium majus |
| Genitofemoral | Anterior rami L1 and L2 | L1, L2 | Genital branch—male cremasteric muscle | Genital branch—skin of anterior scrotum or skin of mons pubis and labium majus; femoral branch—skin of upper anterior thigh |
| Lateral cutaneous nerve of thigh | Anterior rami L2 and L3 | L2, L3 | | Skin on anterior and lateral thigh to the knee |
| Obturator | Anterior rami L2 to L4 | L2 to L4 | Obturator externus, pectineus, and muscles in medial compartment of thigh | Skin on medial aspect of the thigh |
| Femoral | Anterior rami L2 to L4 | L2 to L4 | Iliacus, pectineus, and muscles in anterior compartment of thigh | Skin on anterior thigh and medial surface of leg |

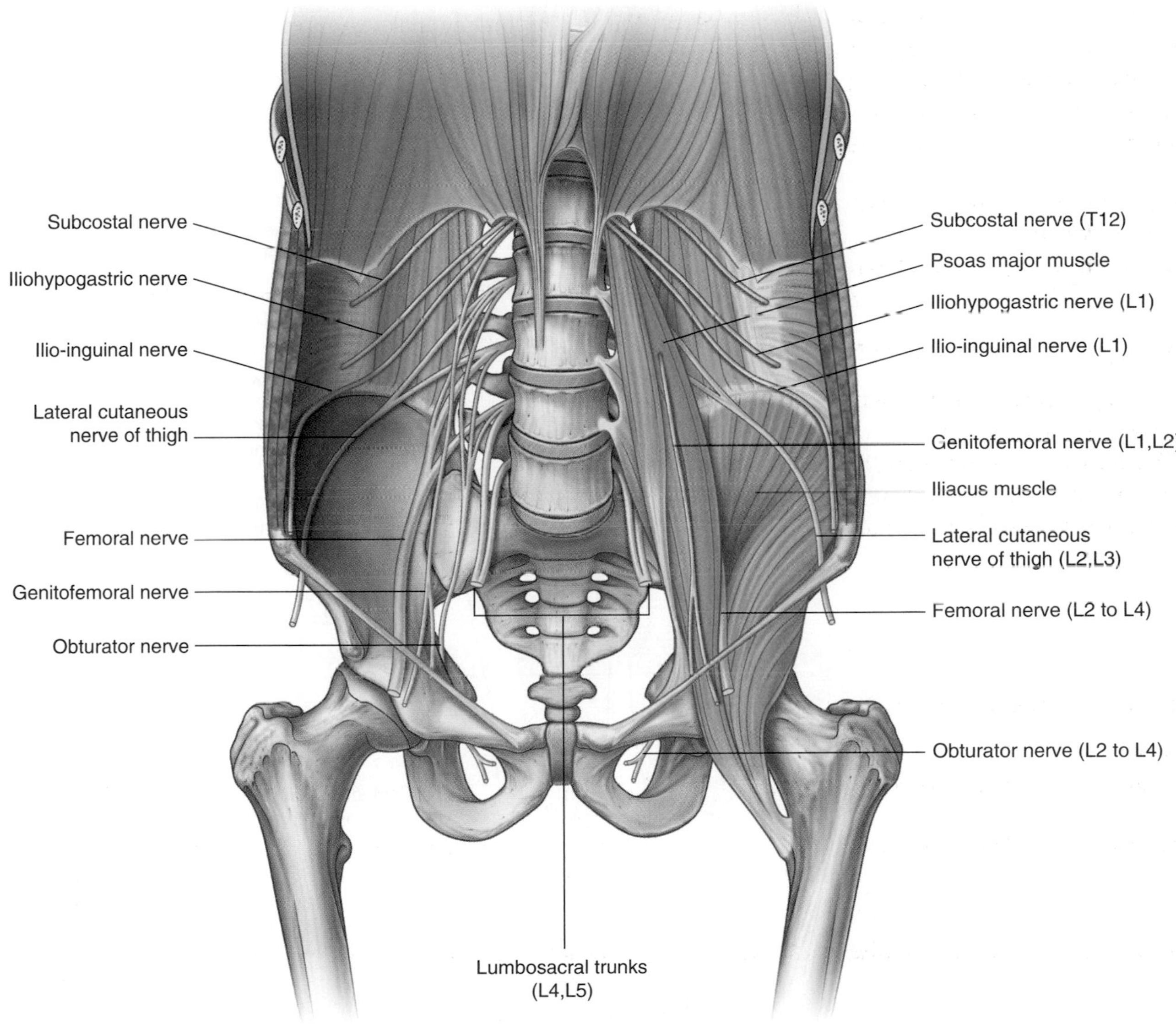


**Fig. 4.174** Lumbar plexus in the posterior abdominal region.

### Iliohypogastric nerve

The **iliohypogastric nerve** passes across the anterior surface of the quadratus lumborum muscle, posterior to the kidney. It pierces the transversus abdominis muscle and continues anteriorly around the body between the transversus abdominis and internal oblique muscles. Above the iliac crest, a **lateral cutaneous branch** pierces the internal and external oblique muscles to supply the posterolateral gluteal skin (Fig. 4.175).

The remaining part of the iliohypogastric nerve (the **anterior cutaneous branch**) continues in an anterior direction, piercing the internal oblique just medial to the anterior superior iliac spine as it continues in an obliquely downward and medial direction. Becoming cutaneous, just above the superficial inguinal ring, after piercing the aponeurosis of the external oblique, it distributes to the skin in the pubic region (Fig. 4.175). Throughout its course, it also supplies branches to the abdominal musculature.

### Ilio-inguinal nerve

The ilio-inguinal nerve is smaller than, and inferior to, the iliohypogastric nerve as it crosses the quadratus lumborum

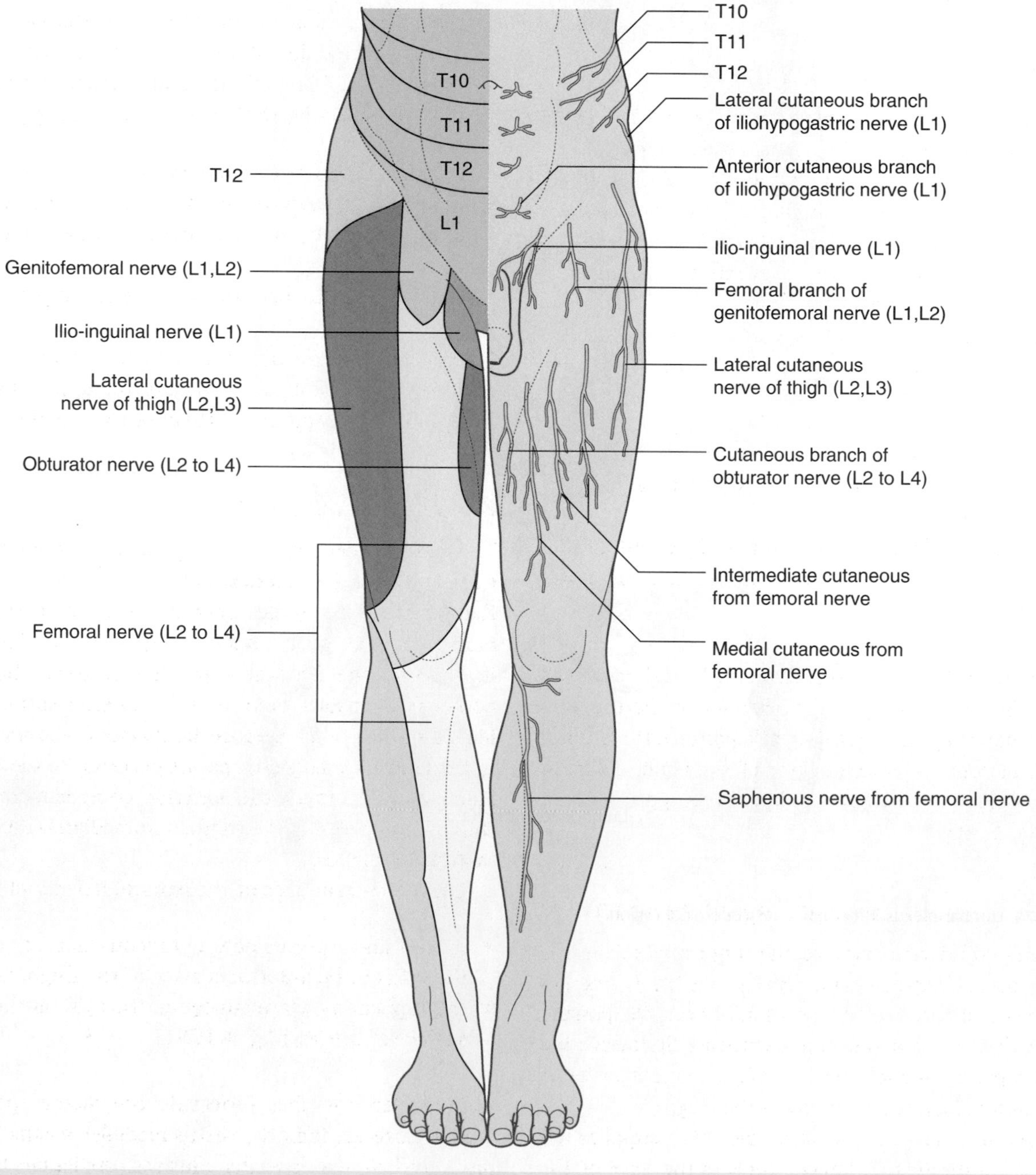

**Fig. 4.175** Cutaneous distribution of the nerves from the lumbar plexus.

muscle. Its course is more oblique than that of the iliohypogastric nerve, and it usually crosses part of the iliacus muscle on its way to the iliac crest. Near the anterior end of the iliac crest, it pierces the transversus abdominis muscle, and then pierces the internal oblique muscle and enters the inguinal canal.

The ilio-inguinal nerve emerges through the superficial inguinal ring, along with the spermatic cord, and provides cutaneous innervation to the upper medial thigh, the root of the penis, and the anterior surface of the scrotum in men, or the mons pubis and labium majus in women (Fig. 4.175). Throughout its course, it also supplies branches to the abdominal musculature.

### Genitofemoral nerve (L1 and L2)

The genitofemoral nerve arises from the anterior rami of nerves L1 and L2 (Fig. 4.173). It passes downward in the substance of the psoas major muscle until it emerges on the anterior surface of the psoas major. It then descends on the surface of the muscle, in a retroperitoneal position, passing posterior to the ureter. It eventually divides into genital and femoral branches.

The **genital branch** continues downward and enters the inguinal canal through the deep inguinal ring. It continues through the canal and:

- in men, innervates the cremasteric muscle and terminates on the skin in the upper anterior part of the scrotum, and
- in women, accompanies the round ligament of the uterus and terminates on the skin of the mons pubis and labium majus.

The **femoral branch** descends on the lateral side of the external iliac artery and passes posterior to the inguinal ligament, entering the femoral sheath lateral to the femoral artery. It pierces the anterior layer of the femoral sheath and the fascia lata to supply the skin of the upper anterior thigh (Fig. 4.175).

### Lateral cutaneous nerve of thigh (L2 and L3)

The lateral cutaneous nerve of the thigh arises from the anterior rami of nerves L2 and L3 (Fig. 4.173). It emerges from the lateral border of the psoas major muscle, passing obliquely downward across the iliacus muscle toward the anterior superior iliac spine (Fig. 4.175). It passes posterior to the inguinal ligament and enters the thigh.

The lateral cutaneous nerve of the thigh supplies the skin on the anterior and lateral thigh to the level of the knee (Fig. 4.175).

### Obturator nerve (L2 to L4)

The obturator nerve arises from the anterior rami of nerves L2 to L4 (Fig. 4.173). It descends in the psoas major muscle, emerging from its medial side near the pelvic brim (Fig. 4.174).

The obturator nerve continues posterior to the common iliac vessels, passes across the lateral wall of the pelvic cavity, and enters the obturator canal, through which the obturator nerve gains access to the medial compartment of the thigh.

In the area of the obturator canal, the obturator nerve divides into **anterior** and **posterior branches**. On entering the medial compartment of the thigh, the two branches are separated by the obturator externus and adductor brevis muscles. Throughout their course through the medial compartment, these two branches supply:

- articular branches to the hip joint,
- muscular branches to the obturator externus, pectineus, adductor longus, gracilis, adductor brevis, and adductor magnus muscles,
- cutaneous branches to the medial aspect of the thigh, and
- in association with the saphenous nerve, cutaneous branches to the medial aspect of the upper part of the leg and articular branches to the knee joint (Fig. 4.175).

### Femoral nerve (L2 to L4)

The femoral nerve arises from the anterior rami of nerves L2 to L4 (Fig. 4.173). It descends through the substance of the psoas major muscle, emerging from the lower lateral border of the psoas major (Fig. 4.174). Continuing its descent, the femoral nerve lies between the lateral border of the psoas major and the anterior surface of the iliacus muscle. It is deep to the iliacus fascia and lateral to the femoral artery as it passes posterior to the inguinal ligament and enters the anterior compartment of the thigh. Upon entering the thigh, it immediately divides into multiple branches.

Cutaneous branches of the femoral nerve include:

- medial and intermediate cutaneous nerves supplying the skin on the anterior surface of the thigh, and
- the saphenous nerve supplying the skin on the medial surface of the leg (Fig. 4.175).

Muscular branches innervate the iliacus, pectineus, sartorius, rectus femoris, vastus medialis, vastus intermedius, and vastus lateralis muscles. Articular branches supply the hip and knee joints.

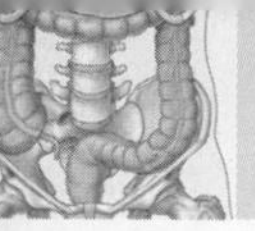



# Surface anatomy

## Abdomen surface anatomy

Visualization of the position of abdominal viscera is fundamental to a physical examination. Some of these viscera or their parts can be felt by palpating through the abdominal wall. Surface features can be used to establish the positions of deep structures.

### Defining the surface projection of the abdomen

Palpable landmarks can be used to delineate the extent of the abdomen on the surface of the body. These landmarks are:

- the costal margin above and
- the pubic tubercle, anterior superior iliac spine, and iliac crest below (Fig. 4.176).

The costal margin is readily palpable and separates the abdominal wall from the thoracic wall.

A line between the anterior superior iliac spine and the pubic tubercle marks the position of the inguinal ligament, which separates the anterior abdominal wall above from the thigh of the lower limb below.

The iliac crest separates the posterolateral abdominal wall from the gluteal region of the lower limb.

The upper part of the abdominal cavity projects above the costal margin to the diaphragm, and therefore abdominal viscera in this region of the abdomen are protected by the thoracic wall.

The level of the diaphragm varies during the breathing cycle. The dome of the diaphragm on the right can reach as high as the fourth costal cartilage during forced expiration.

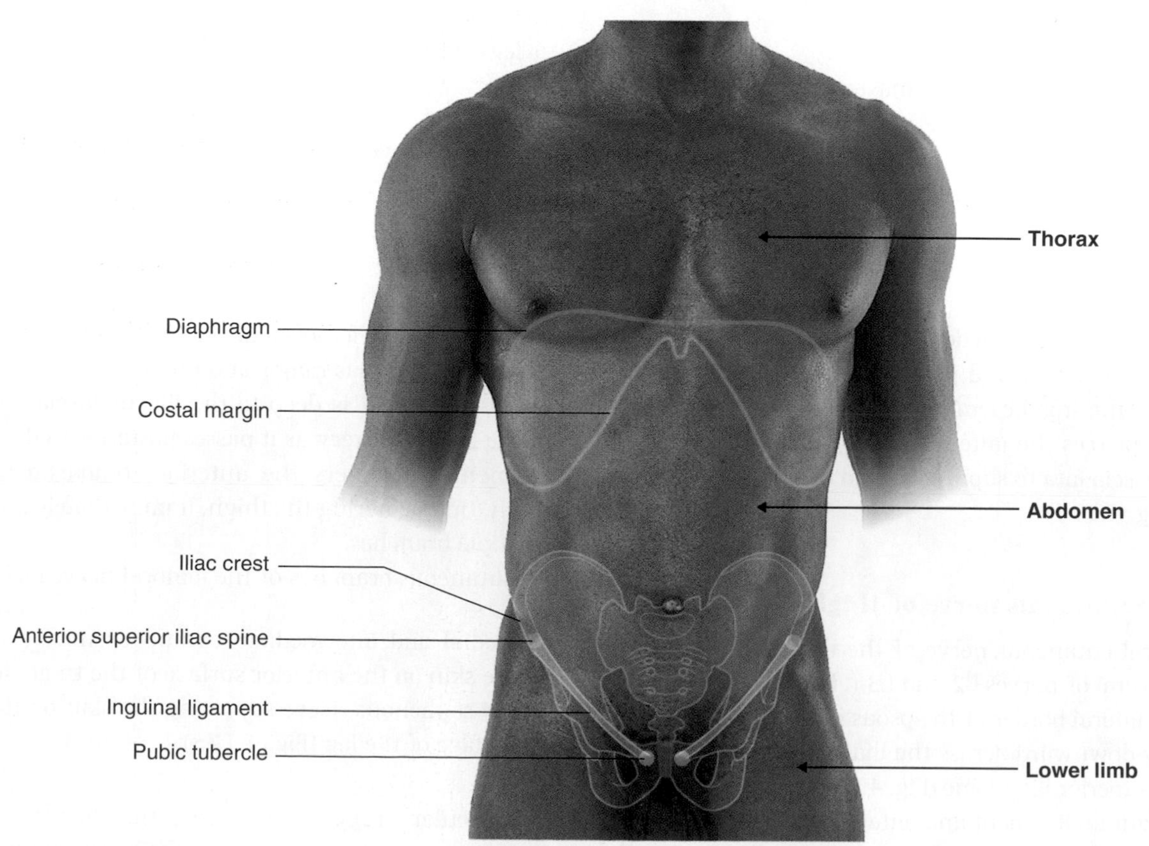


**Fig. 4.176** Interior view of the abdominal region of a man. Palpable bony landmarks, the inguinal ligament, and the position of the diaphragm are indicated.

## How to find the superficial inguinal ring

The superficial inguinal ring is an elongate triangular defect in the aponeurosis of the external oblique (Fig. 4.177). It lies in the lower medial aspect of the anterior abdominal wall and is the external opening of the inguinal canal. The inguinal canal and superficial ring are larger in men than in women:

- In men, structures that pass between the abdomen and the testis pass through the inguinal canal and superficial inguinal ring.

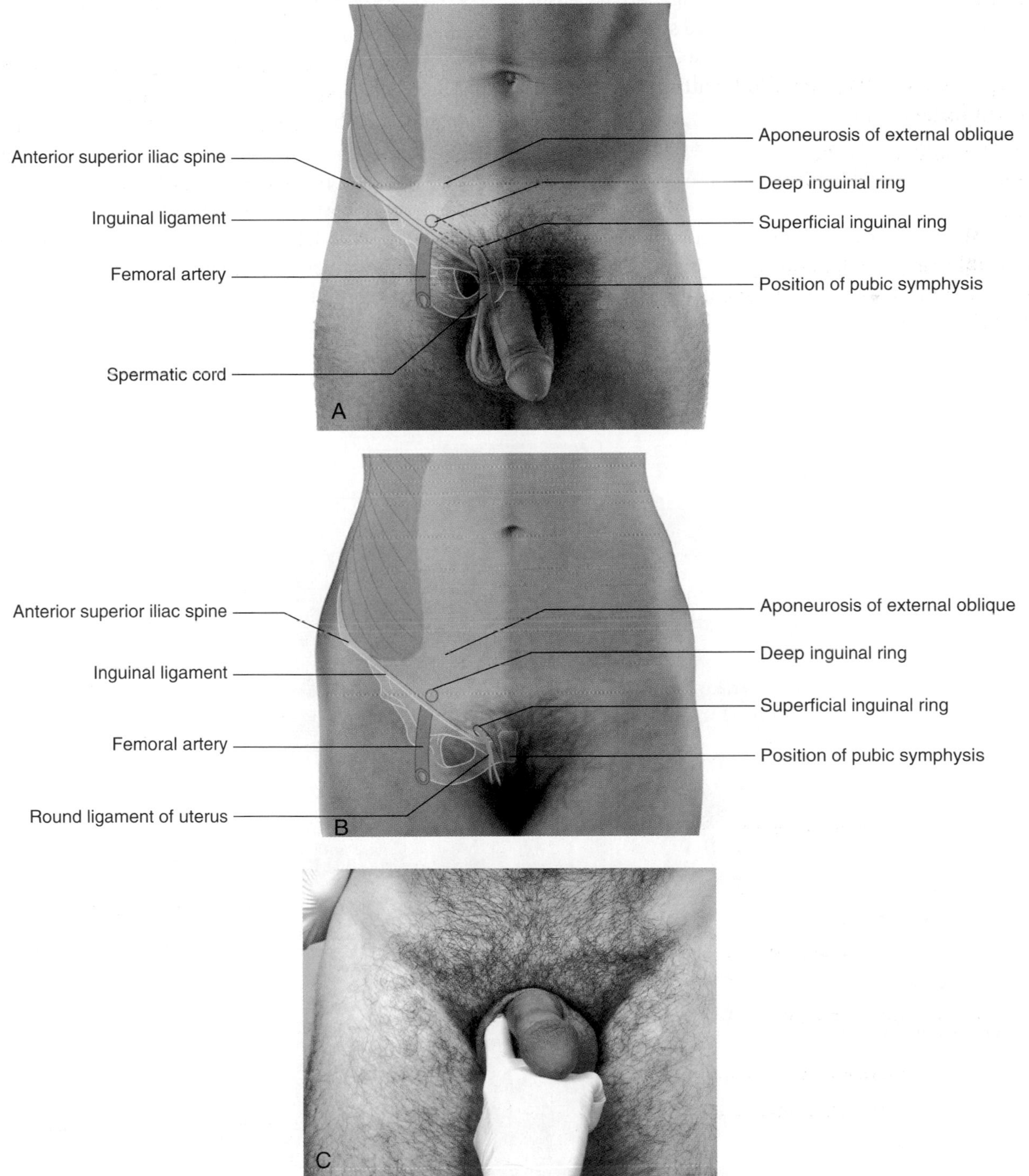


**Fig. 4.177** Groin. **A.** In a man. **B.** In a woman. **C.** Examination of the superficial inguinal ring and related regions of the inguinal canal in a man.

- In women, the round ligament of the uterus passes through the inguinal canal and superficial inguinal ring to merge with connective tissue of the labium majus.

The superficial inguinal ring is superior to the pubic crest and tubercle and to the medial end of the inguinal ligament:

- In men, the superficial inguinal ring can be easily located by following the spermatic cord superiorly to the lower abdominal wall—the external spermatic fascia of the spermatic cord is continuous with the margins of the superficial inguinal ring.
- In women, the pubic tubercle can be palpated and the ring is superior and lateral to it.

The deep inguinal ring, which is the internal opening to the inguinal canal, lies superior to the inguinal ligament, midway between the anterior superior iliac spine and pubic symphysis. The pulse of the femoral artery can be felt in the same position but below the inguinal ligament.

Because the superficial inguinal ring is the site where inguinal hernias appear, particularly in men, the ring and related parts of the inguinal canal are often evaluated during physical examination.

## How to determine lumbar vertebral levels

Lumbar vertebral levels are useful for visualizing the positions of viscera and major blood vessels. The approximate positions of the lumbar vertebrae can be established using palpable or visible landmarks (Fig. 4.178):

- A horizontal plane passes through the medial ends of the ninth costal cartilages and the body of the LI vertebra—this transpyloric plane cuts through the body midway between the suprasternal (jugular) notch and the pubic symphysis.

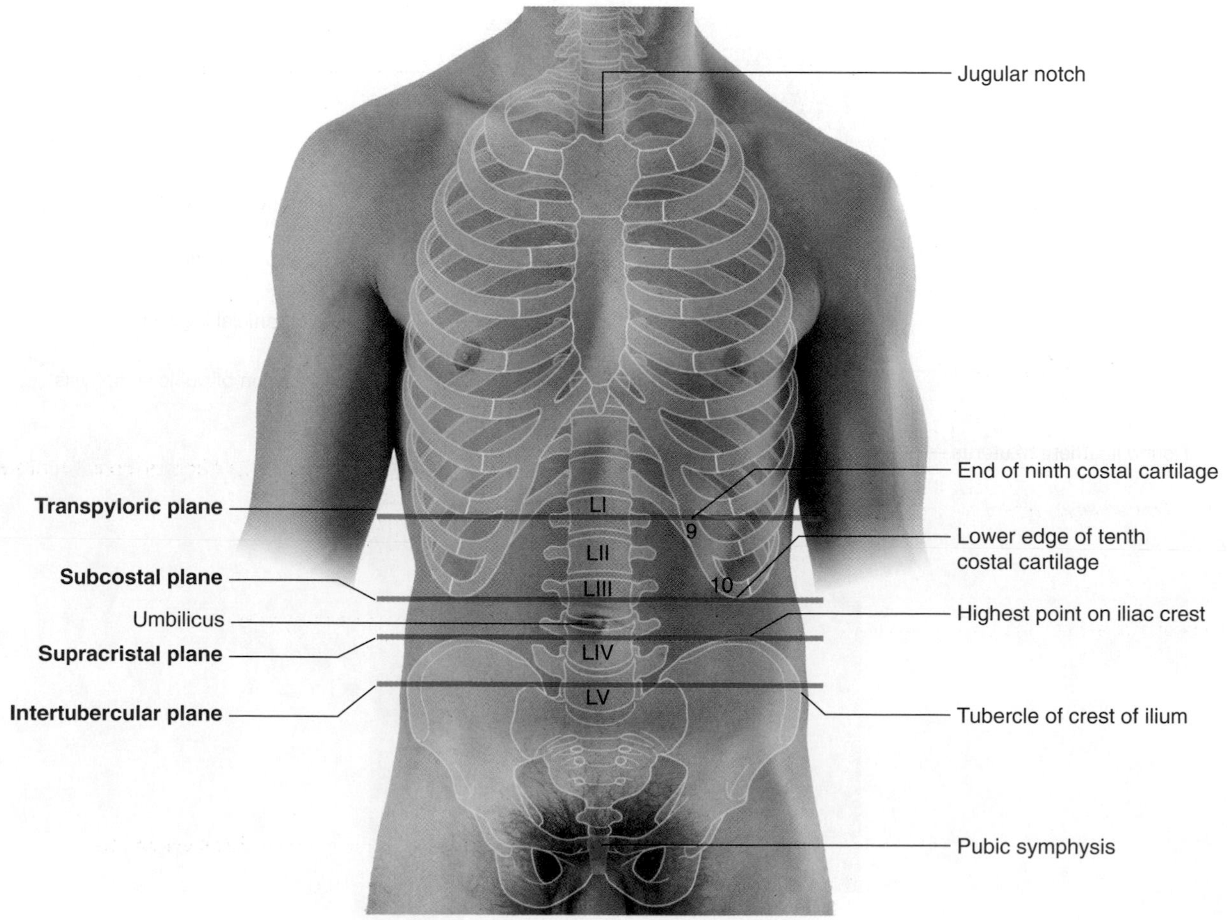


**Fig. 4.178** Landmarks used for establishing the positions of lumbar vertebrae are indicated. Anterior view of the abdominal region of a man.

- A horizontal plane passes through the lower edge of the costal margin (tenth costal cartilage) and the body of the LIII vertebra—the umbilicus is normally on a horizontal plane that passes through the disc between the LIII and LIV vertebrae.
- A horizontal plane (supracristal plane) through the highest point on the iliac crest passes through the spine and body of the LIV vertebra;
- A plane through the tubercles of the crest of the ilium passes through the body of the LV vertebra.

## Visualizing structures at the LI vertebral level

The LI vertebral level is marked by the transpyloric plane, which cuts transversely through the body midway between the jugular notch and pubic symphysis, and through the ends of the ninth costal cartilages (Fig. 4.179). At this level are:

- the beginning and upper limit of the end of the duodenum,
- the hila of the kidneys,
- the neck of the pancreas, and
- the origin of the superior mesenteric artery from the aorta.

The left and right colic flexures also are close to this level.

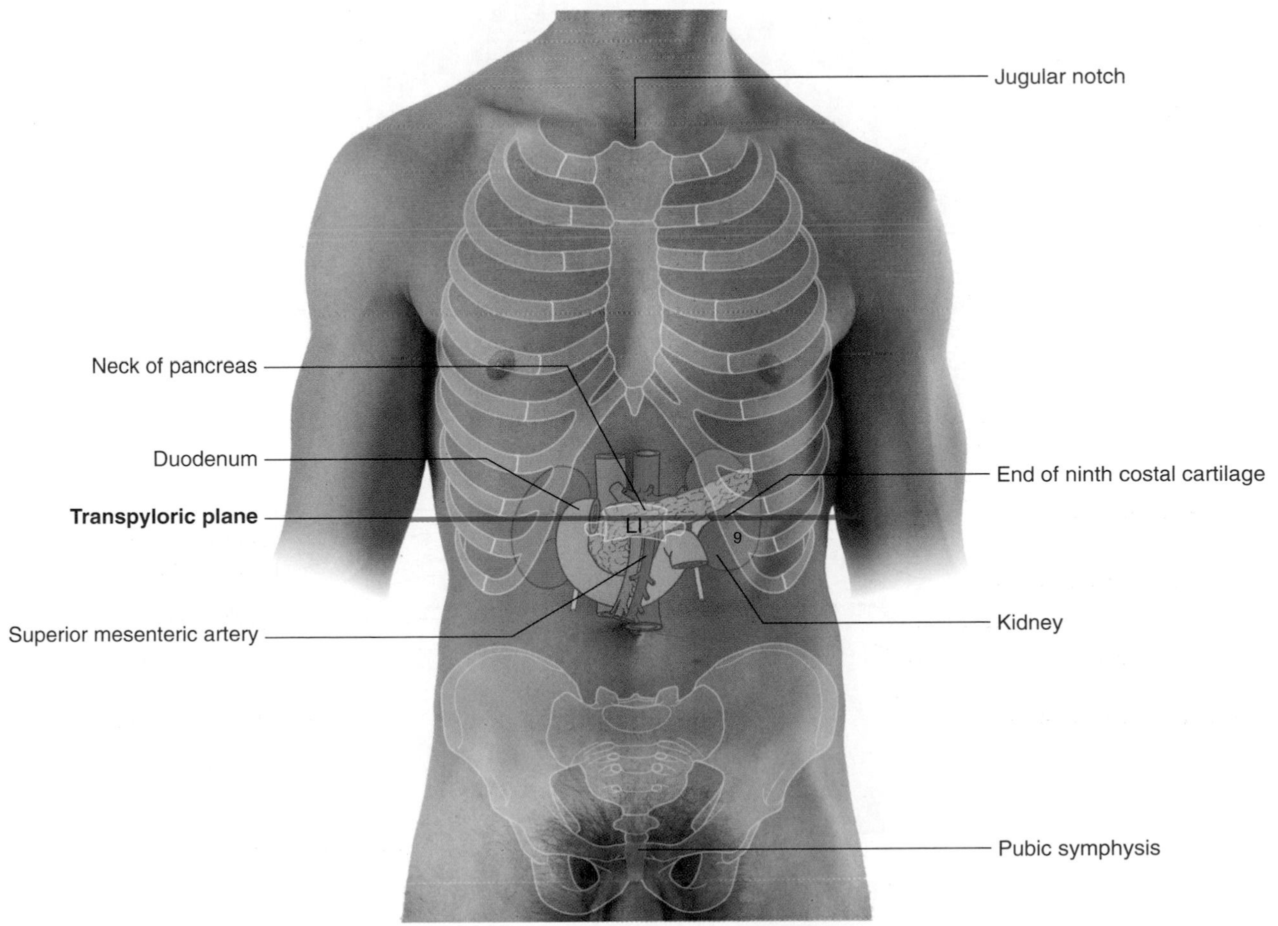


**Fig. 4.179** LI vertebral level and the important viscera associated with this level. Anterior view of the abdominal region of a man.

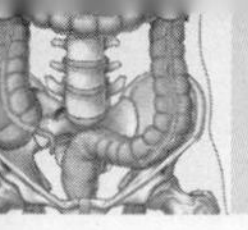



## Visualizing the position of major blood vessels

Each of the vertebral levels in the abdomen is related to the origin of major blood vessels (Fig. 4.180):

- The celiac trunk originates from the aorta at the upper border of the LI vertebra.
- The superior mesenteric artery originates at the lower border of the LI vertebra.
- The renal arteries originate at approximately the LII vertebra.
- The inferior mesenteric artery originates at the LIII vertebra.
- The aorta bifurcates into the right and left common iliac arteries at the level of the LIV vertebra.
- The left and right common iliac veins join to form the inferior vena cava at the LV vertebral level.

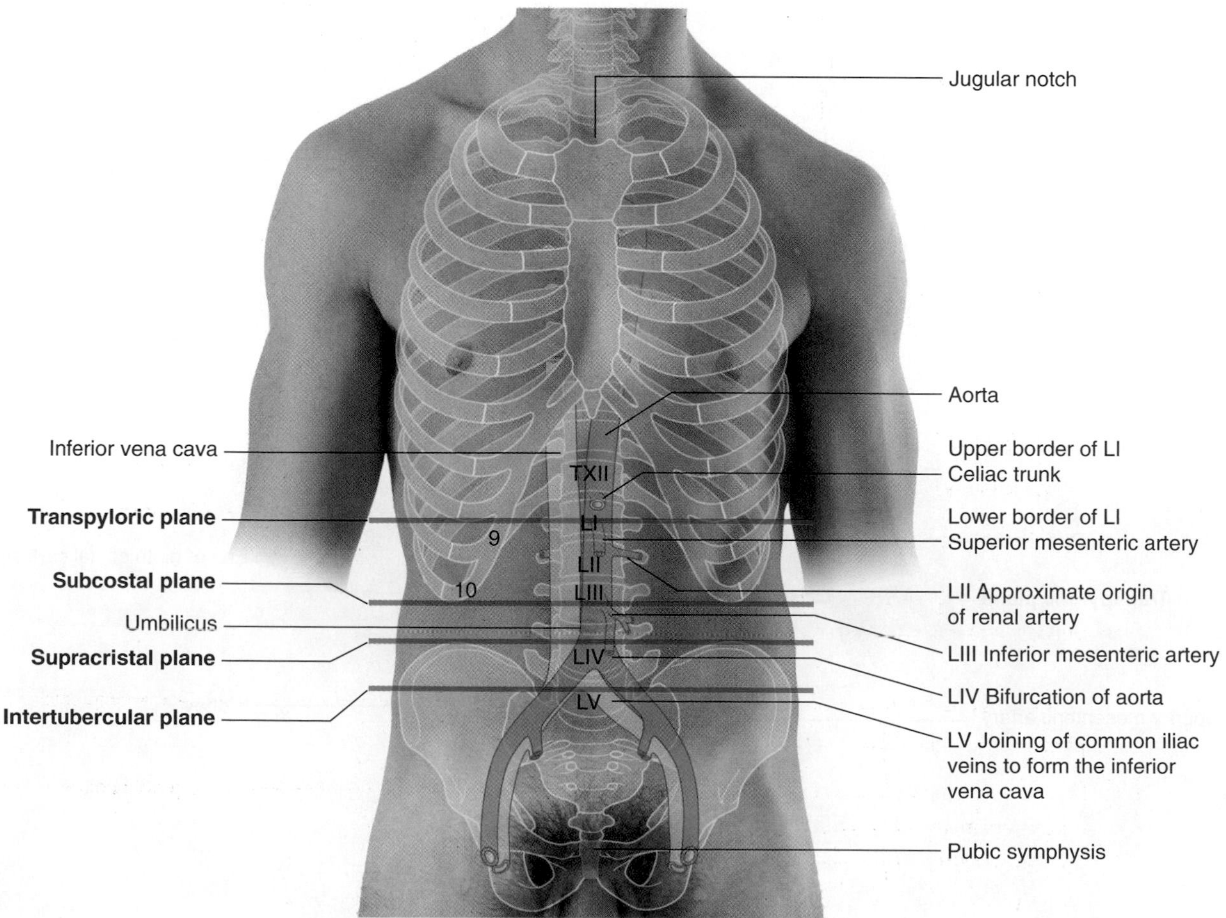


**Fig. 4.180** Major vessels projected onto the body's surface. Anterior view of the abdominal region of a man.

## Using abdominal quadrants to locate major viscera

The abdomen can be divided into quadrants by a vertical median plane and a horizontal transumbilical plane, which passes through the umbilicus (Fig. 4.181):

- The liver and gallbladder are in the right upper quadrant.
- The stomach and spleen are in the left upper quadrant.
- The cecum and appendix are in the right lower quadrant.
- The end of the descending colon and sigmoid colon are in the left lower quadrant.

Most of the liver is under the right dome of the diaphragm and is deep to the lower thoracic wall. The inferior margin of the liver can be palpated descending below the right costal margin when a patient is asked to inhale deeply. On deep inspiration, the edge of the liver can be felt "slipping" under the palpating fingers placed under the costal margin.

A common surface projection of the appendix is McBurney's point, which is one-third of the way up along a line from the right anterior superior iliac spine to the umbilicus.

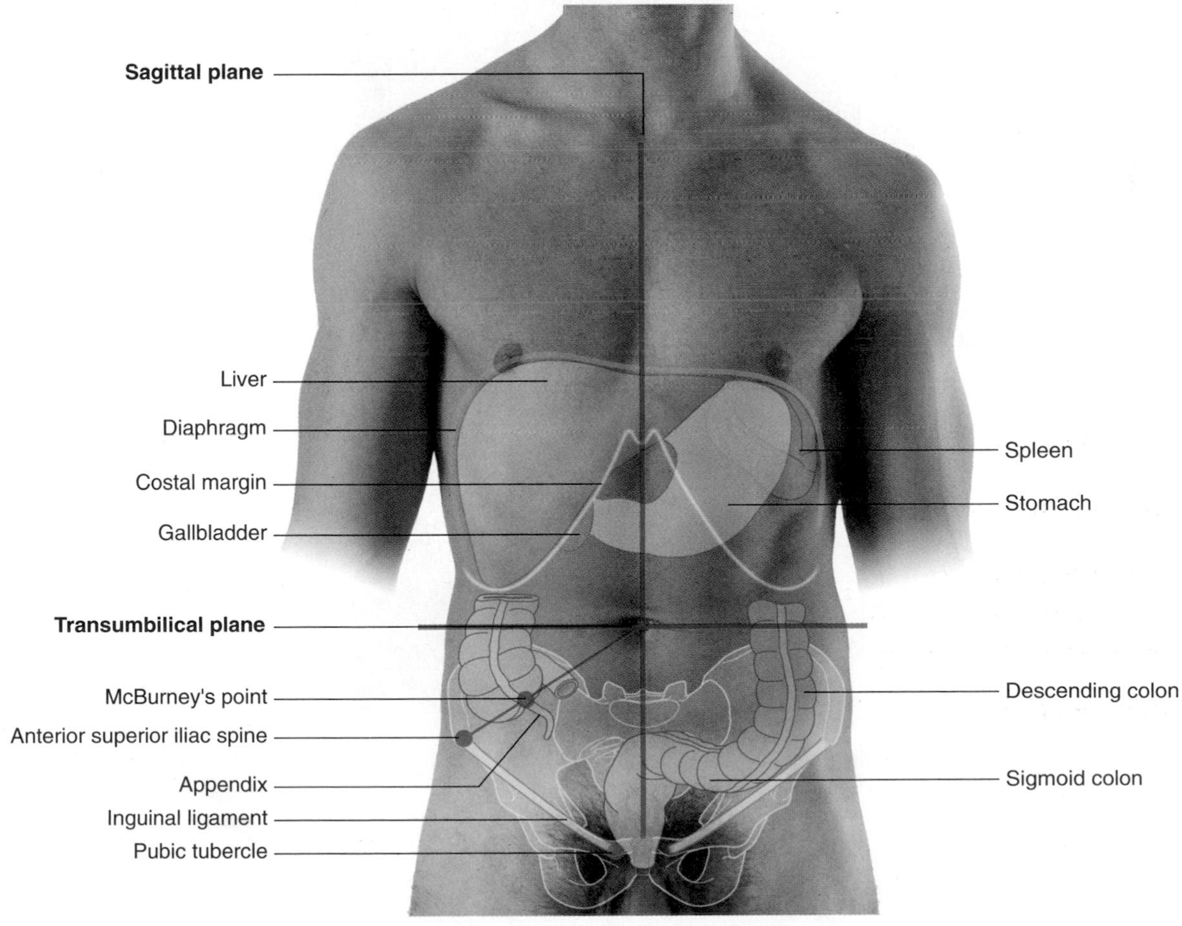


**Fig. 4.181** Abdominal quadrants and the positions of major viscera. Anterior view of a man.

## Defining surface regions to which pain from the gut is referred

The abdomen can be divided into nine regions by a midclavicular sagittal plane on each side and by the subcostal and intertubercular planes, which pass through the body transversely (Fig. 4.182). These planes separate the abdomen into:

- three central regions (epigastric, umbilical, pubic), and
- three regions on each side (hypochondrium, flank, groin).

Pain from the abdominal part of the foregut is referred to the epigastric region, pain from the midgut is referred to the umbilical region, and pain from the hindgut is referred to the pubic region.

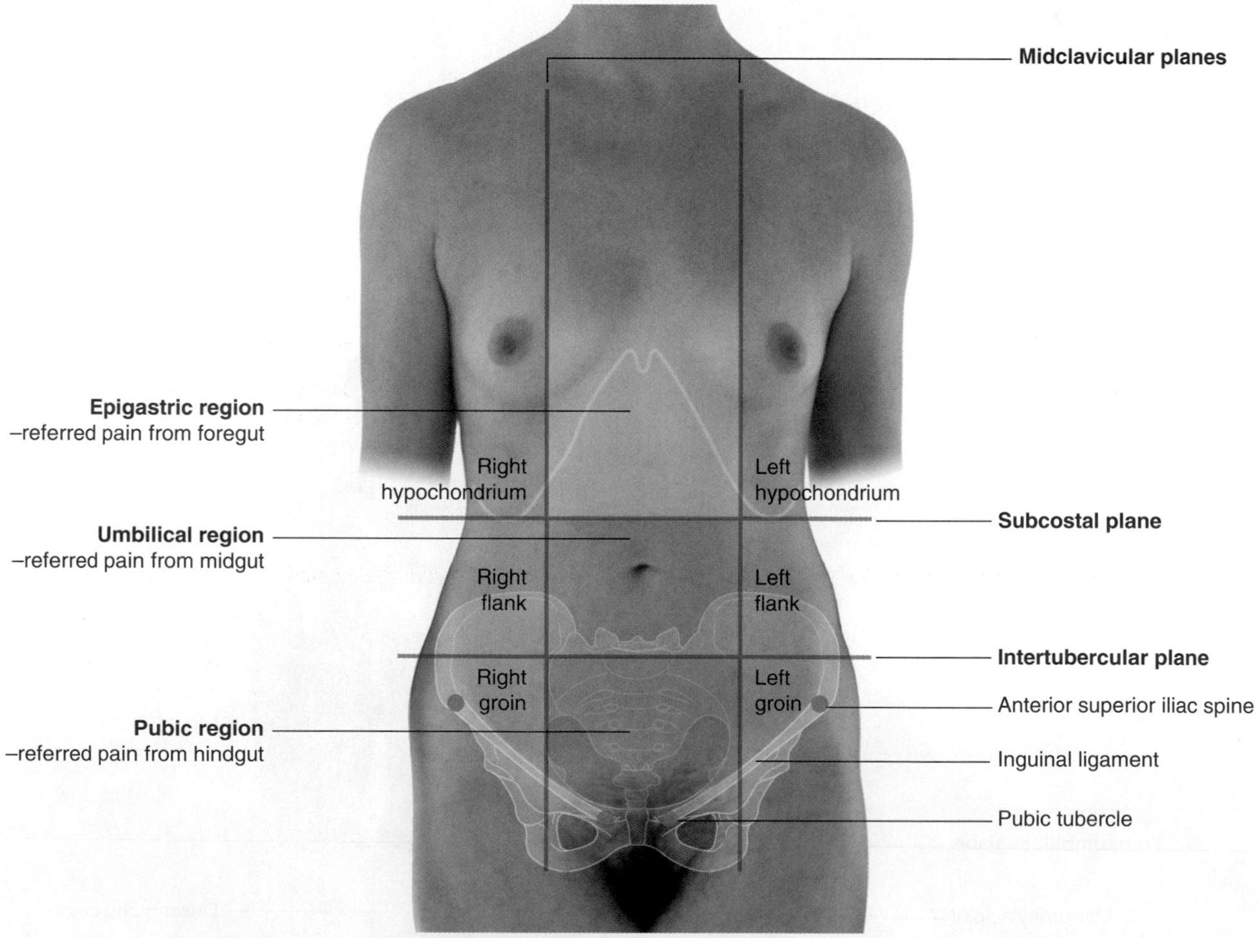

**Fig. 4.182** The nine regions of the abdomen. Anterior view of a woman.

## Where to find the kidneys

The kidneys project onto the back on either side of the midline and are related to the lower ribs (Fig. 4.183):

- The left kidney is a little higher than the right and reaches as high as rib XI.
- The superior pole of the right kidney reaches only as high as rib XII.

The lower poles of the kidneys occur around the level of the disc between the LIII and LIV vertebrae. The hila of the kidneys and the beginnings of the ureters are at approximately the LI vertebra.

The ureters descend vertically anterior to the tips of the transverse processes of the lower lumbar vertebrae and enter the pelvis.

## Where to find the spleen

The spleen projects onto the left side and back in the area of ribs IX to XI (Fig. 4.184). The spleen follows the contour of rib X and extends from the superior pole of the left kidney to just posterior to the midaxillary line.

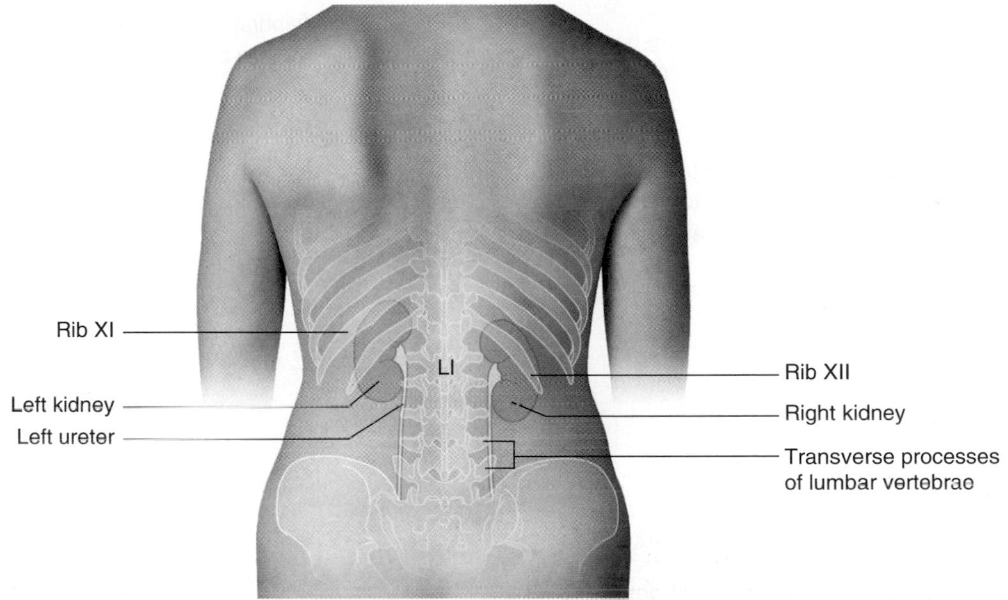


**Fig. 4.183** Surface projection of the kidneys and ureters. Posterior view of the abdominal region of a woman.

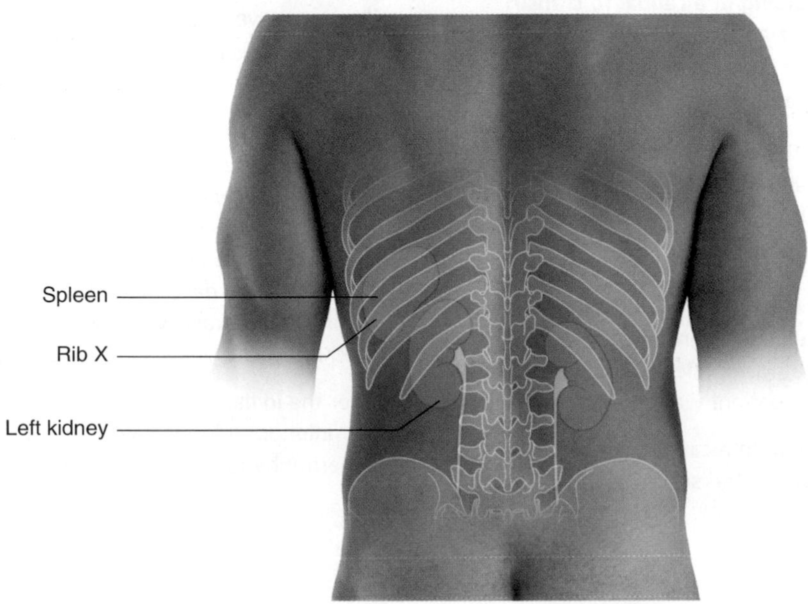


**Fig. 4.184** Surface projection of the spleen. Posterior view of a man.

# Clinical cases

## Case 1

TRAUMATIC RUPTURE OF THE DIAPHRAGM

**A 45-year-old man had mild epigastric pain, and a diagnosis of esophageal reflux was made. He was given appropriate medication, which worked well. However, at the time of the initial consultation, the family practitioner requested a chest radiograph, which demonstrated a prominent hump on the left side of the diaphragm and old rib fractures.**

The patient was recalled for further questioning.

He was extremely pleased with the treatment he had been given for his gastroesophageal reflux, but was concerned about being recalled for further history and examination. During the interview, he revealed that he had previously been involved in a motorcycle accident and had undergone a laparotomy for a "rupture." The patient did not recall what operation was performed, but was assured at the time that the operation was a great success.

The patient is likely to have undergone a splenectomy.

In any patient who has had severe blunt abdominal trauma (such as that caused by a motorcycle accident), lower left-sided rib fractures are an extremely important sign of appreciable trauma.

A review of the patient's old notes revealed that at the time of the injury the spleen was removed surgically, but it was not appreciated that there was a small rupture of the dome of the left hemidiaphragm. The patient gradually developed a hernia through which bowel could enter, producing the "hump" on the diaphragm seen on the chest radiograph.

Because this injury occurred many years ago and the patient has been asymptomatic, it is unlikely that the patient will come to any harm and was discharged.

## Case 2

CHRONIC THROMBOSIS OF THE INFERIOR VENA CAVA

**A medical student was asked to inspect the abdomen of two patients. On the first patient he noted irregular veins radiating from the umbilicus. On the second patient he noted irregular veins, coursing in a caudal to cranial direction, over the anterior abdominal wall from the groin to the chest. He was asked to explain his findings and determine the significance of these features.**

In the first patient the veins were draining radially away from the periumbilical region. In normal individuals, enlarged veins do not radiate from the umbilicus. In patients with portal hypertension the portal venous pressure is increased as a result of hepatic disease. Small collateral veins develop at and around the obliterated umbilical vein. These veins pass through the umbilicus and drain onto the anterior abdominal wall, forming a portosystemic anastomosis. The eventual diagnosis for this patient was cirrhosis of the liver.

The finding of veins draining in a caudocranial direction on the anterior abdominal wall in the second patient is not typical for veins on the anterior abdominal wall. When veins are so prominent, it usually implies that there is an obstruction to the normal route of venous drainage and an alternative route has been taken. Typically, blood from the lower limbs and the retroperitoneal organs drains into the inferior vena cava and from here to the right atrium of the heart. This patient had a chronic thrombosis of the inferior vena cava, preventing blood from returning to the heart by the "usual" route.

Blood from the lower limbs and the pelvis may drain via a series of collateral vessels, some of which include the superficial inferior epigastric veins, which run in the superficial fascia. These anastomose with the superior, superficial, and deep epigastric venous systems to drain into the internal thoracic veins, which in turn drain into the brachiocephalic veins and the superior vena cava.

After the initial inferior vena cava thrombosis, the veins of the anterior abdominal wall and other collateral pathways hypertrophy to accommodate the increase in blood flow.

## Case 3

CARCINOMA OF THE HEAD OF THE PANCREAS

**A 52-year-old woman visited her family physician with complaints of increasing lethargy and vomiting. The physician examined her and noted that compared to previous visits she had lost significant weight. She was also jaundiced, and on examination of the abdomen a well-defined 10-cm rounded mass was palpable below the liver edge in the right upper quadrant (Fig. 4.185).**

The clinical diagnosis was carcinoma of the head of the pancreas.

It is difficult to appreciate how such a precise diagnosis can be made clinically when only three clinical signs have been described.

The patient's obstruction was in the distal bile duct.

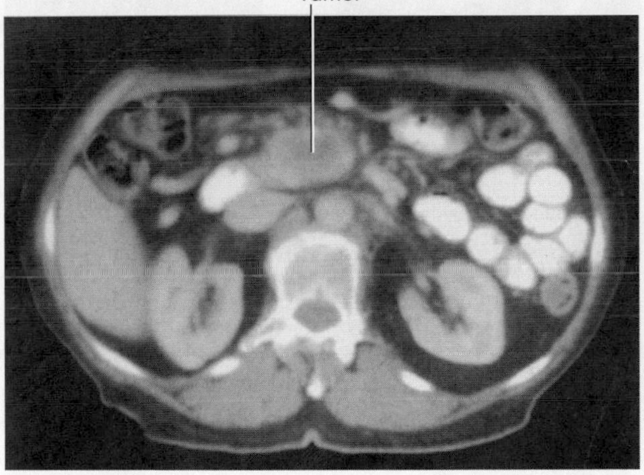

Tumor

**Fig. 4.185** Tumor in the head of the pancreas. Computed tomogram in the axial plane.

When a patient has jaundice, the causes are excessive breakdown of red blood cells (prehepatic), hepatic failure (hepatic jaundice), and posthepatic causes, which include obstruction along the length of the biliary tree.

The patient had a mass in her right upper quadrant that was palpable below the liver; this was the gallbladder.

In healthy individuals, the gallbladder is not palpable. An expanded gallbladder indicates obstruction either within the cystic duct or below the level of the cystic duct insertion (i.e., the bile duct).

The patient's vomiting was related to the position of the tumor.

It is not uncommon for vomiting and weight loss (cachexia) to occur in patients with a malignant disease. The head of the pancreas lies within the curve of the duodenum, primarily adjacent to the descending part of the duodenum. Any tumor mass in the region of the head of the pancreas is likely to expand and may encase and invade the duodenum. Unfortunately, in this patient's case, this happened, producing almost complete obstruction. Further discussion with the patient revealed that she was vomiting relatively undigested food soon after each meal.

A CT scan demonstrated further complications.

In the region of the head and neck of the pancreas are complex anatomical structures, which may be involved with a malignant process. The CT scan confirmed a mass in the region of the head of the pancreas, which invaded the descending part of the duodenum. The mass extended into the neck of the pancreas and had blocked the distal part of the bile duct and the pancreatic duct. Posteriorly the mass had directly invaded the portal venous confluence of the splenic and superior mesenteric veins, producing a series of gastric, splenic, and small bowel varices.

This patient underwent palliative chemotherapy, but died 7 months later.

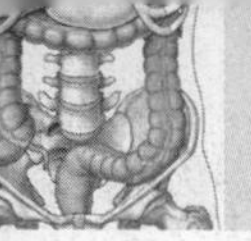



## Case 4

METASTATIC LESIONS IN THE LIVER

**A 44-year-old woman had been recently diagnosed with melanoma on the toe and underwent a series of investigations.**

Melanoma (properly called malignant melanoma) can be an aggressive form of skin cancer that spreads to lymph nodes and multiple other organs throughout the body. The malignant potential is dependent upon its cellular configuration and also the depth of its penetration through the skin.

The patient developed malignant melanoma in the foot, which spread to the lymph nodes of the groin. The inguinal lymph nodes were resected; however, it was noted on follow-up imaging that the patient had developed two metastatic lesions within the right lobe of the liver.

Surgeons and physicians considered the possibility of removing these lesions.

A CT scan was performed that demonstrated the lesions within segments V and VI of the liver (Fig. 4.186).

The segmental anatomy of the liver is important because it enables the surgical planning for resection.

The surgery was undertaken and involved identifying the portal vein and the confluence of the right and left hepatic ducts. The liver was divided in the imaginary principal plane of the middle hepatic vein. The main hepatic duct and biliary radicals were ligated and the right liver was successfully resected.

The segments remaining included the left lobe of the liver.

The patient underwent a surgical resection of segments V, VI, VII, and VIII. The remaining segments included IVa, IVb, I, II, and III. It is important to remember that the lobes of the liver do not correlate with the hepatic volume. The left lobe of the liver contains only segments II and III. The right lobe of the liver contains segments IV, V, VI, VII, and VIII. Hence, cross-sectional imaging is important when planning surgical segmental resection.

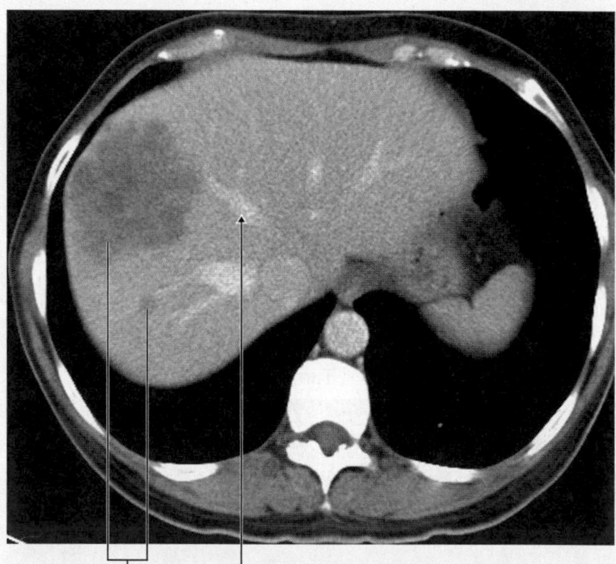


**Fig. 4.186** This postcontrast computed tomogram, in the axial plane, demonstrates two metastases situated within the right lobe of the liver. The left lobe of the liver is clear. The larger of the two metastases is situated to the right of the middle hepatic vein, which lies in the principal plane of the liver dividing the left and right sides of the liver.

5

# Pelvis and Perineum

**ADDITIONAL LEARNING RESOURCES for Chapter 5, Pelvis and Perineum, on STUDENT CONSULT (www.studentconsult.com):**

- Self-Assessment—National Board style multiple-choice questions, Chapter 5
- Short Questions—these are questions requiring short responses, Chapter 5
- Interactive Surface Anatomy—interactive surface animations, Chapter 5
- Clinical Cases, Chapter 5
  Ectopic pregnancy
  Iatrogenic ureteric injury
  Left common iliac artery obstruction
  Ovarian torsion
  Pelvic kidney
  Sciatic nerve compression
  Uterine fibroids
  Uterine tumor
  Varicocele

**Free Online Self-Study Course: Anatomy and Embryology**

- Anatomy modules 18 through 22
- Embryology modules 68 through 70

# Conceptual overview

## GENERAL DESCRIPTION

The pelvis and perineum are interrelated regions associated with the pelvic bones and terminal parts of the vertebral column. The pelvis is divided into two regions:

- The superior region related to upper parts of the pelvic bones and lower lumbar vertebrae is the **false pelvis** (**greater pelvis**) and is generally considered part of the abdominal cavity (Fig. 5.1).
- The **true pelvis** (**lesser pelvis**) is related to the inferior parts of the pelvic bones, sacrum, and coccyx, and has an inlet and an outlet.

The bowl-shaped **pelvic cavity** enclosed by the true pelvis consists of the pelvic inlet, walls, and floor. This cavity is continuous superiorly with the abdominal cavity and contains elements of the urinary, gastrointestinal, and reproductive systems.

The perineum (Fig. 5.1) is inferior to the floor of the pelvic cavity; its boundaries form the **pelvic outlet**. The perineum contains the external genitalia and external openings of the genitourinary and gastrointestinal systems.

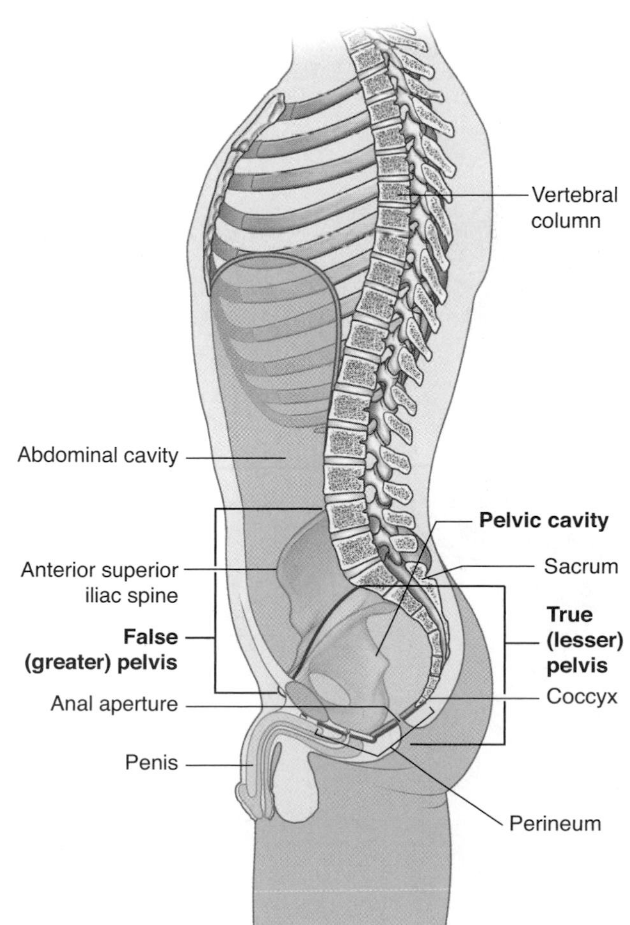


**Fig. 5.1** Pelvis and perineum.

## FUNCTIONS

### Contains and supports the bladder, rectum, anal canal, and reproductive tracts

Within the pelvic cavity, the bladder is positioned anteriorly and the rectum posteriorly in the midline.

As it fills, the bladder expands superiorly into the abdomen. It is supported by adjacent elements of the pelvic bone and by the pelvic floor. The urethra passes through the pelvic floor to the perineum, where, in women, it opens externally (Fig. 5.2A) and in men it enters the base of the penis (Fig. 5.2B).

Continuous with the sigmoid colon at the level of vertebra SIII, the rectum terminates at the anal canal, which penetrates the pelvic floor to open into the perineum. The anal canal is angled posteriorly on the rectum. This flexure is maintained by muscles of the pelvic floor and is relaxed during defecation. A skeletal muscle sphincter is associated with the anal canal and the urethra as each passes through the pelvic floor.

The pelvic cavity contains most of the reproductive tract in women and part of the reproductive tract in men.

- In women, the vagina penetrates the pelvic floor and connects with the uterus in the pelvic cavity. The uterus is positioned between the rectum and the bladder. A uterine (fallopian) tube extends laterally on each side toward the pelvic wall to open near the ovary.
- In men, the pelvic cavity contains the site of connection between the urinary and reproductive tracts. It also contains major glands associated with the reproductive system—the prostate and two seminal vesicles.

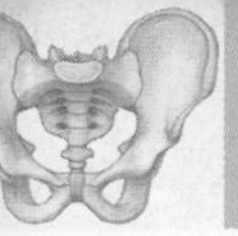



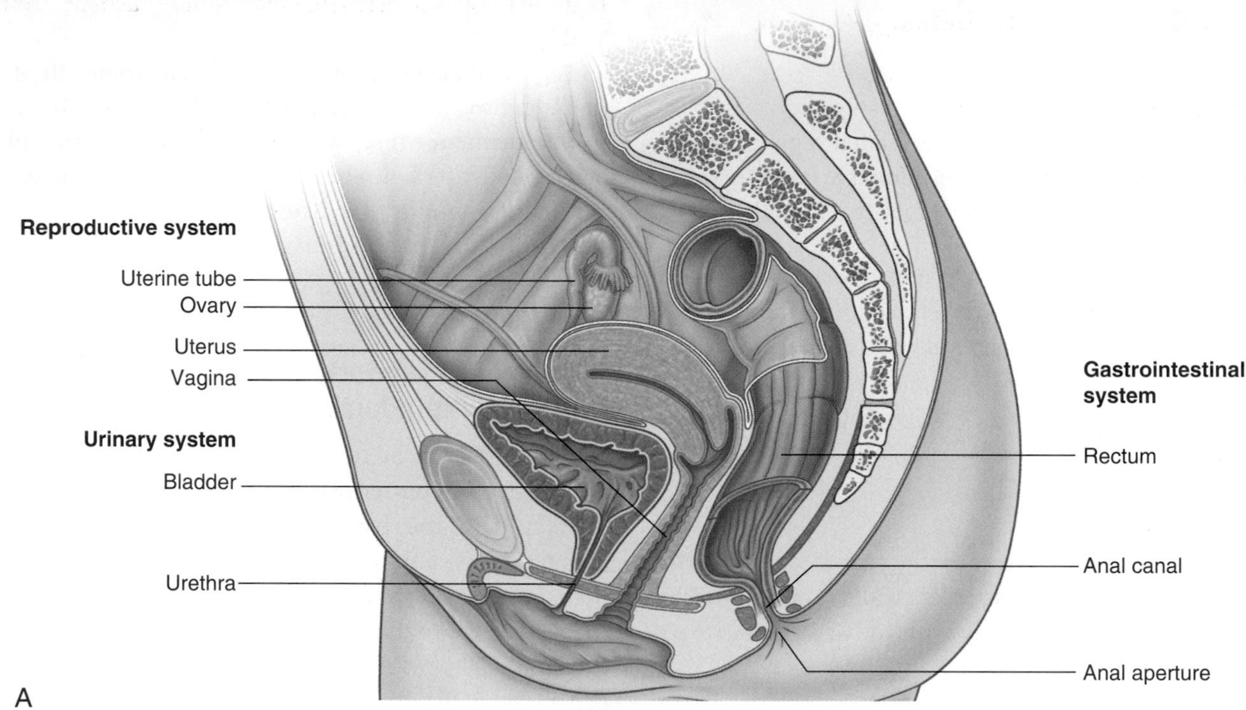


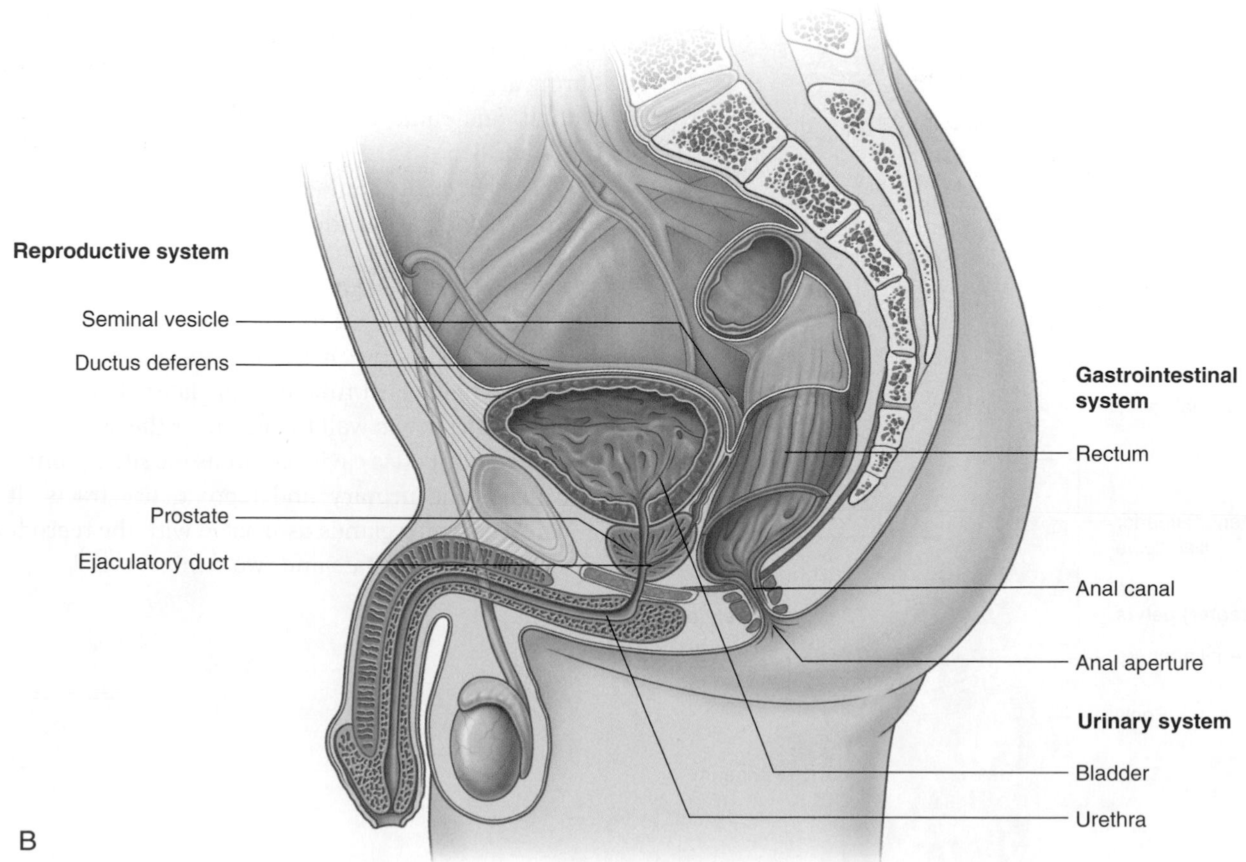


**Fig. 5.2** The pelvis and perineum contain and support terminal parts of the gastrointestinal, urinary, and reproductive systems. **A.** In women. **B.** In men.

## Anchors the roots of the external genitalia

In both genders, the roots of the external genitalia, the clitoris and the penis, are firmly anchored to:

- the bony margin of the anterior half of the pelvic outlet, and
- a thick, fibrous, perineal membrane, which fills the area (Fig. 5.3).

The roots of the external genitalia consist of erectile (vascular) tissues and associated skeletal muscles.

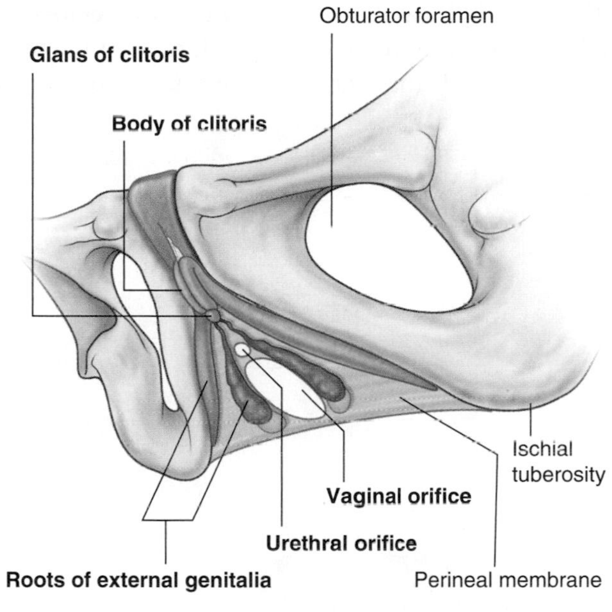


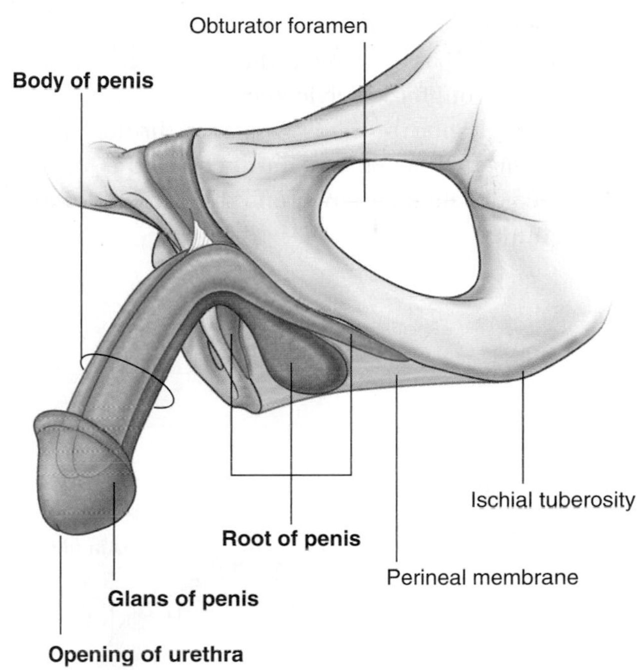


**Fig. 5.3** The perineum contains and anchors the roots of the external genitalia. **A.** In women. **B.** In men.

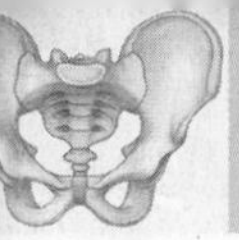



## COMPONENT PARTS

### Pelvic inlet

The pelvic inlet is somewhat heart shaped and completely ringed by bone (Fig. 5.4). Posteriorly, the inlet is bordered by the body of vertebra SI, which projects into the inlet as the sacral **promontory**. On each side of this vertebra, wing-like transverse processes called the **alae** (**wings**) contribute to the margin of the pelvic inlet. Laterally, a prominent rim on the pelvic bone continues the boundary of the inlet forward to the pubic symphysis, where the two pelvic bones are joined in the midline.

Structures pass between the pelvic cavity and the abdomen through the pelvic inlet.

During childbirth, the fetus passes through the pelvic inlet from the abdomen, into which the uterus has expanded during pregnancy, and then passes through the pelvic outlet.

### Pelvic walls

The walls of the true pelvis consist predominantly of bone, muscle, and ligaments, with the sacrum, coccyx, and inferior half of the pelvic bones forming much of them.

Two ligaments—the **sacrospinous** and the **sacrotuberous ligaments**—are important architectural elements of the walls because they link each pelvic bone to the sacrum and coccyx (Fig. 5.5A). These ligaments also convert two notches on the pelvic bones—the **greater** and **lesser sciatic notches**—into foramina on the lateral pelvic walls.

Completing the walls are the **obturator internus** and **piriformis** muscles (Fig. 5.5B), which arise in the pelvis and exit through the sciatic foramina to act on the hip joint.

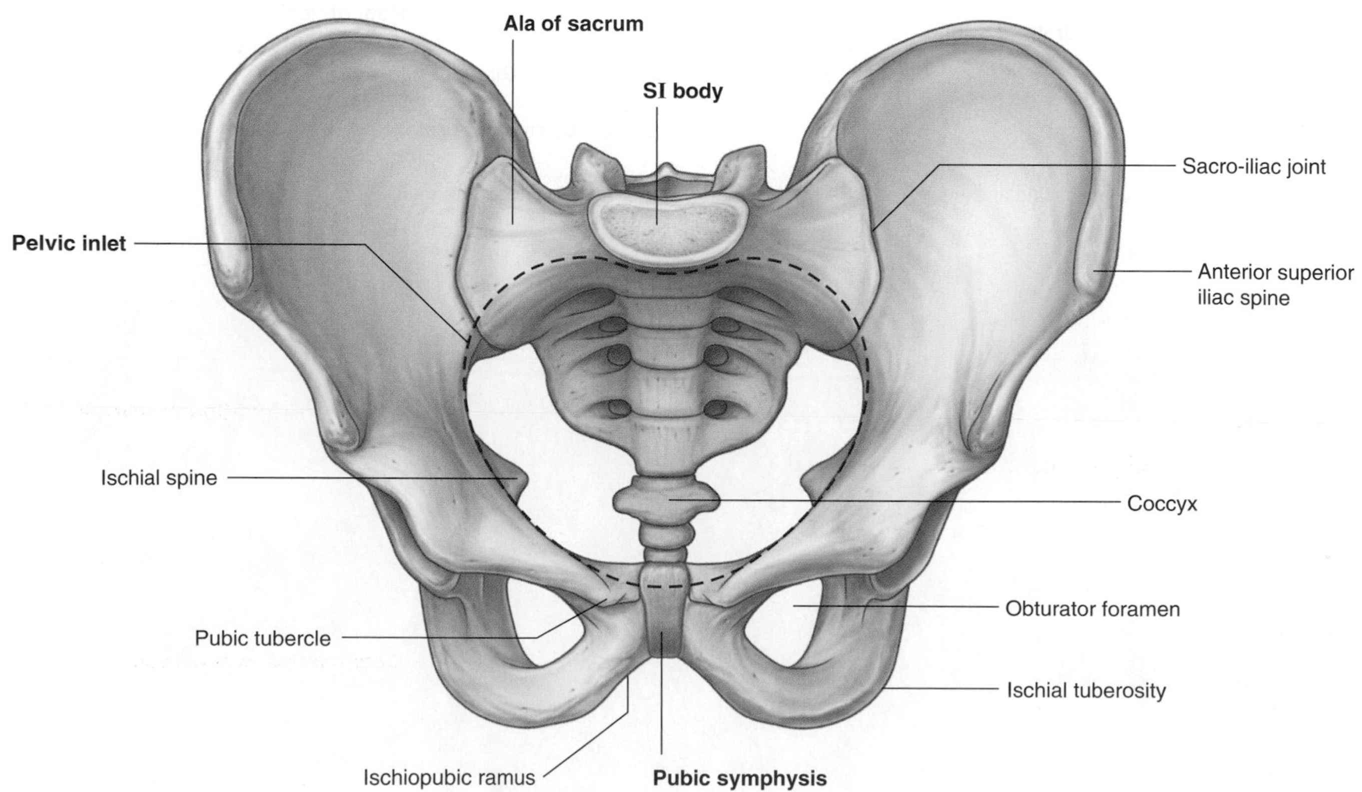


**Fig. 5.4** Pelvic inlet.

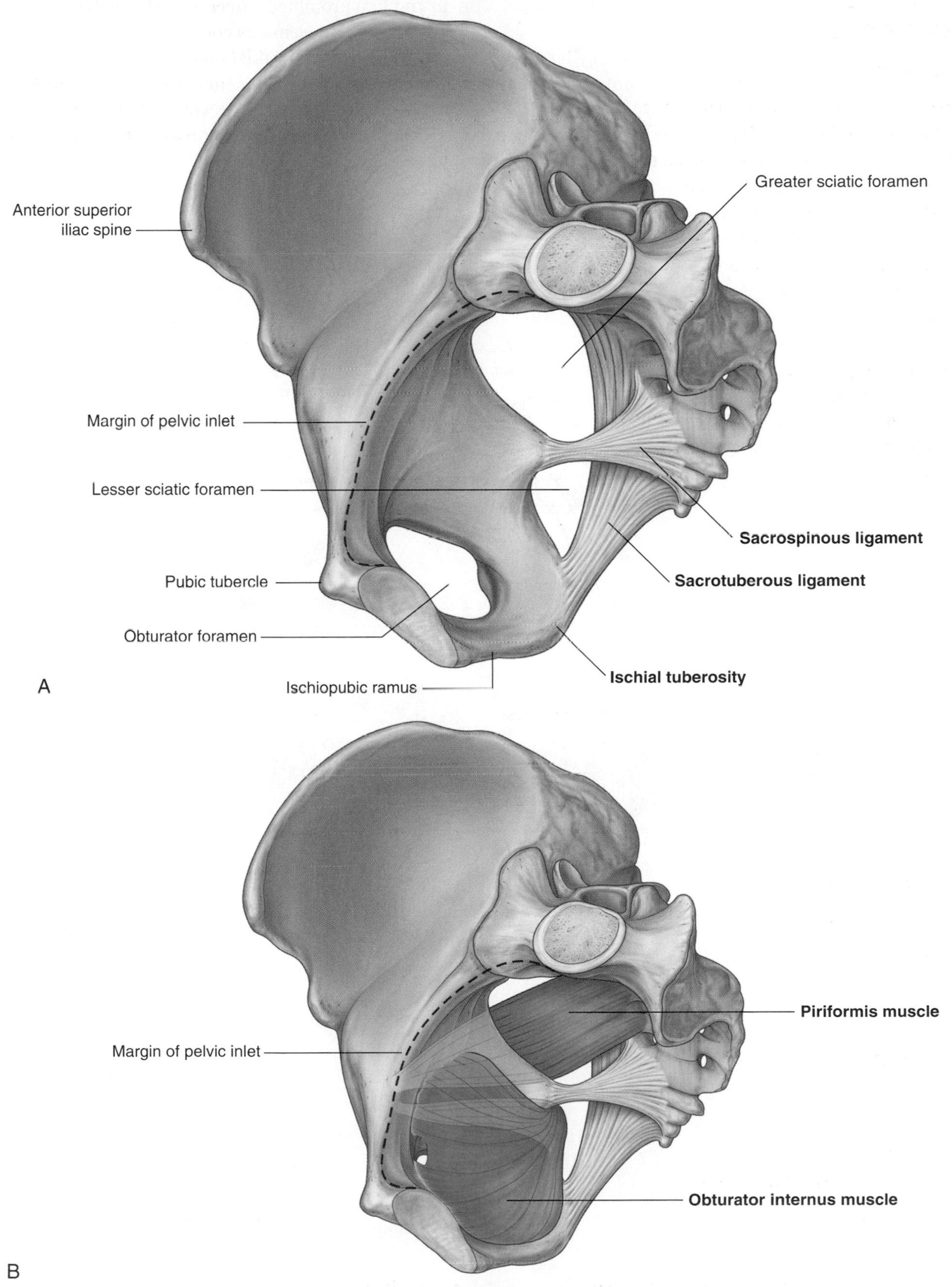


**Fig. 5.5** Pelvic walls. **A.** Bones and ligaments of the pelvic walls. **B.** Muscles of the pelvic walls.

## Pelvic outlet

The diamond-shaped pelvic outlet is formed by both bone and ligaments (Fig. 5.6). It is limited anteriorly in the midline by the pubic symphysis.

On each side, the inferior margin of the pelvic bone projects posteriorly and laterally from the pubic symphysis to end in a prominent tuberosity, the **ischial tuberosity**. Together, these elements construct the pubic arch, which forms the margin of the anterior half of the pelvic outlet. The sacrotuberous ligament continues this margin posteriorly from the ischial tuberosity to the coccyx and sacrum. The pubic symphysis, ischial tuberosities, and coccyx can all be palpated.

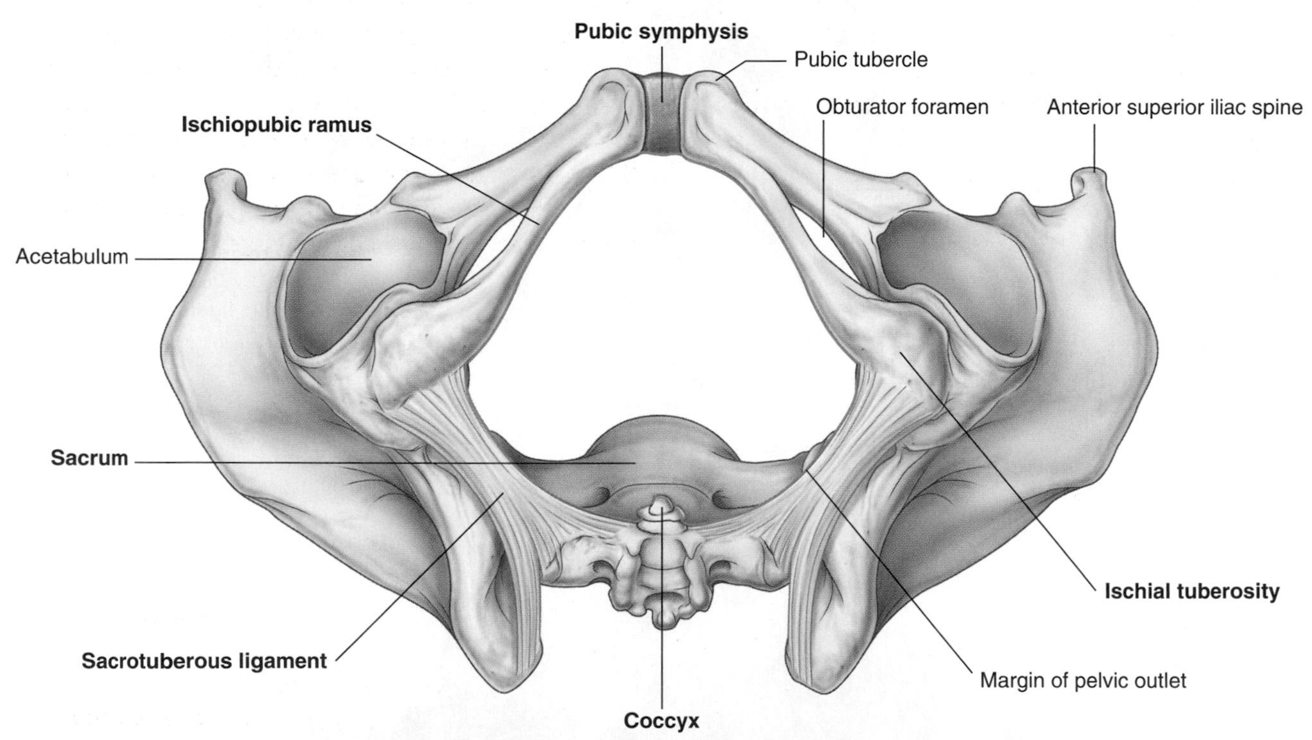


**Fig. 5.6** Pelvic outlet.

## Pelvic floor

The pelvic floor, which separates the pelvic cavity from the perineum, is formed by muscles and fascia (Fig. 5.7).

Two **levator ani** muscles attach peripherally to the pelvic walls and join each other at the midline by a connective tissue raphe. Together they are the largest components of the bowl- or funnel-shaped structure known as the **pelvic diaphragm**, which is completed posteriorly by the **coccygeus muscles**. These latter muscles overlie the sacrospinous ligaments and pass between the margins of the sacrum and the coccyx and a prominent spine on the pelvic bone, the **ischial spine**.

The pelvic diaphragm forms most of the pelvic floor and in its anterior regions contains a U-shaped defect, which is associated with elements of the urogenital system.

The anal canal passes from the pelvis to the perineum through a posterior circular orifice in the pelvic diaphragm.

The pelvic floor is supported anteriorly by:

- the perineal membrane, and
- muscles in the **deep perineal pouch**.

The **perineal membrane** is a thick, triangular fascial sheet that fills the space between the arms of the pubic arch, and has a free posterior border (Fig. 5.7). The deep perineal pouch is a narrow region superior to the perineal membrane.

The margins of the U-shaped defect in the pelvic diaphragm merge into the walls of the associated viscera and with muscles in the deep perineal pouch below.

The vagina and the urethra penetrate the pelvic floor to pass from the pelvic cavity to the perineum.

## Pelvic cavity

The pelvic cavity is lined by peritoneum continuous with the peritoneum of the abdominal cavity that drapes over the superior aspects of the pelvic viscera, but in most regions, does not reach the pelvic floor (Fig. 5.8A).

The pelvic viscera are located in the midline of the pelvic cavity. The bladder is anterior and the rectum is posterior. In women, the uterus lies between the bladder and rectum (Fig. 5.8B). Other structures, such as vessels and nerves, lie deep to the peritoneum in association with the pelvic walls and on either side of the pelvic viscera.

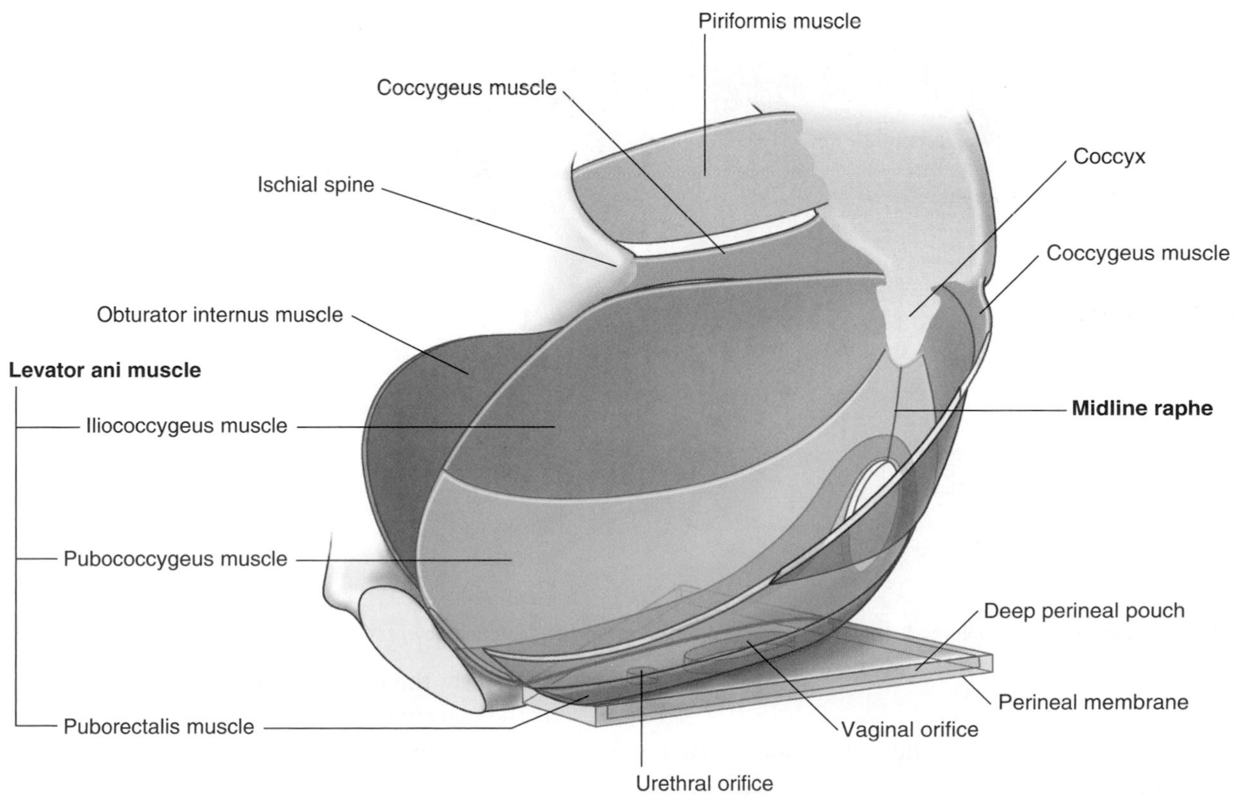


**Fig. 5.7** Pelvic floor.

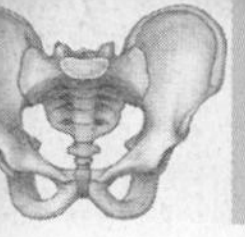

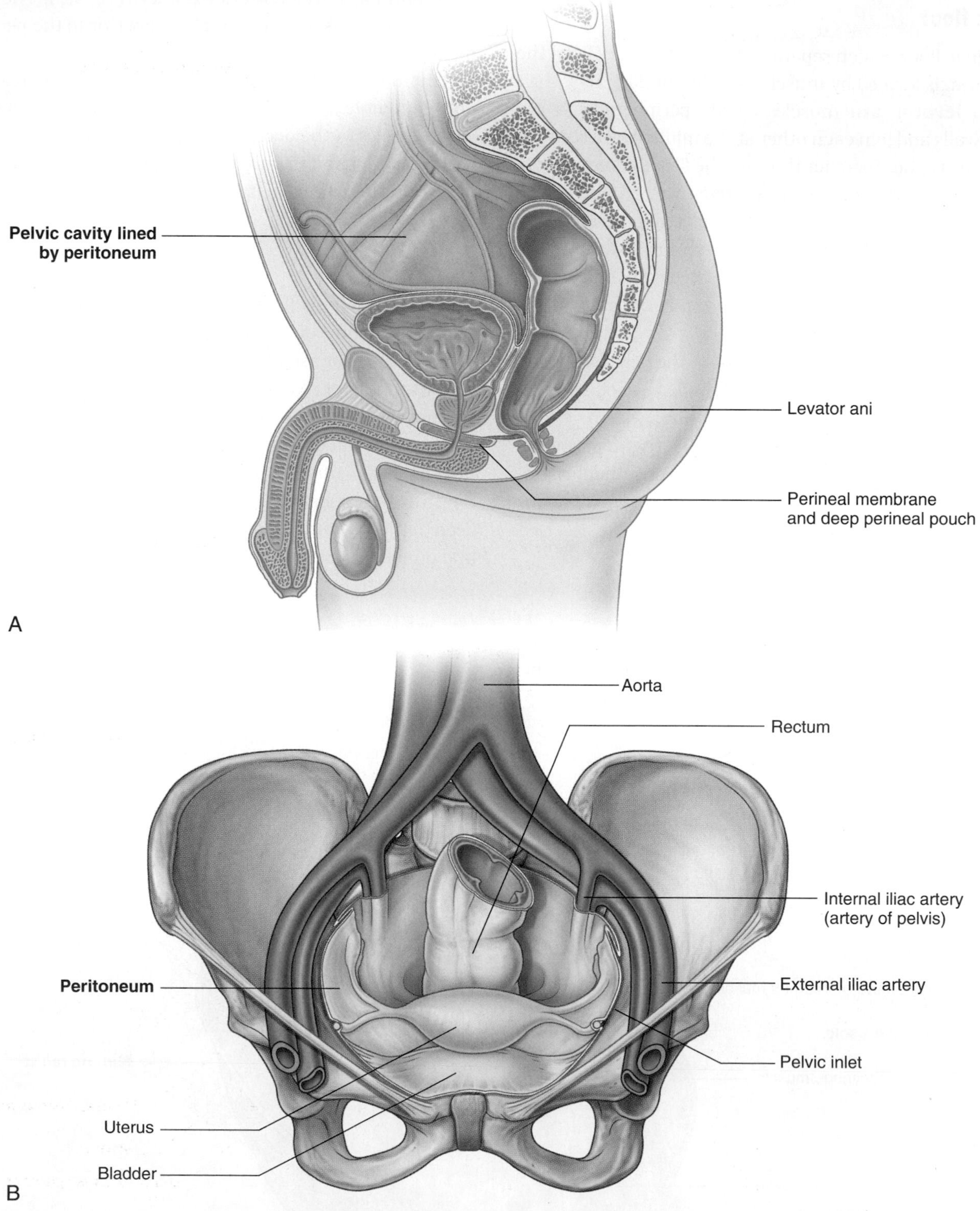

**Fig. 5.8** Pelvic cavity and peritoneum. **A.** In men (sagittal section). **B.** In women (anterior view).

## Perineum

The perineum lies inferior to the pelvic floor between the lower limbs (Fig. 5.9). Its margin is formed by the pelvic outlet. An imaginary line between the ischial tuberosities divides the perineum into two triangular regions.

- Anteriorly, the **urogenital triangle** contains the roots of the external genitalia and, in women, the openings of the urethra and the vagina (Fig. 5.9A). In men, the distal part of the urethra is enclosed by erectile tissues and opens at the end of the penis (Fig. 5.9B).
- Posteriorly, the **anal triangle** contains the anal aperture.

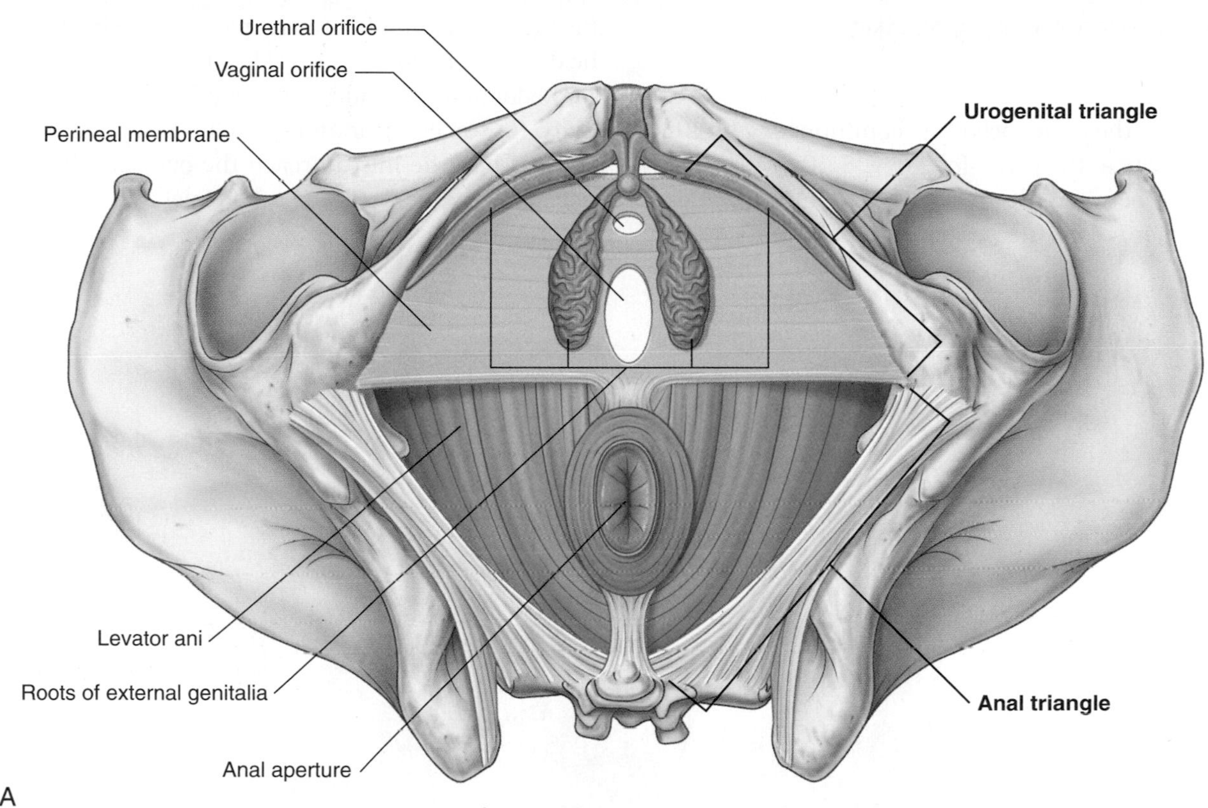
Urethral orifice
Vaginal orifice
Perineal membrane
Urogenital triangle
Levator ani
Roots of external genitalia
Anal triangle
Anal aperture
A

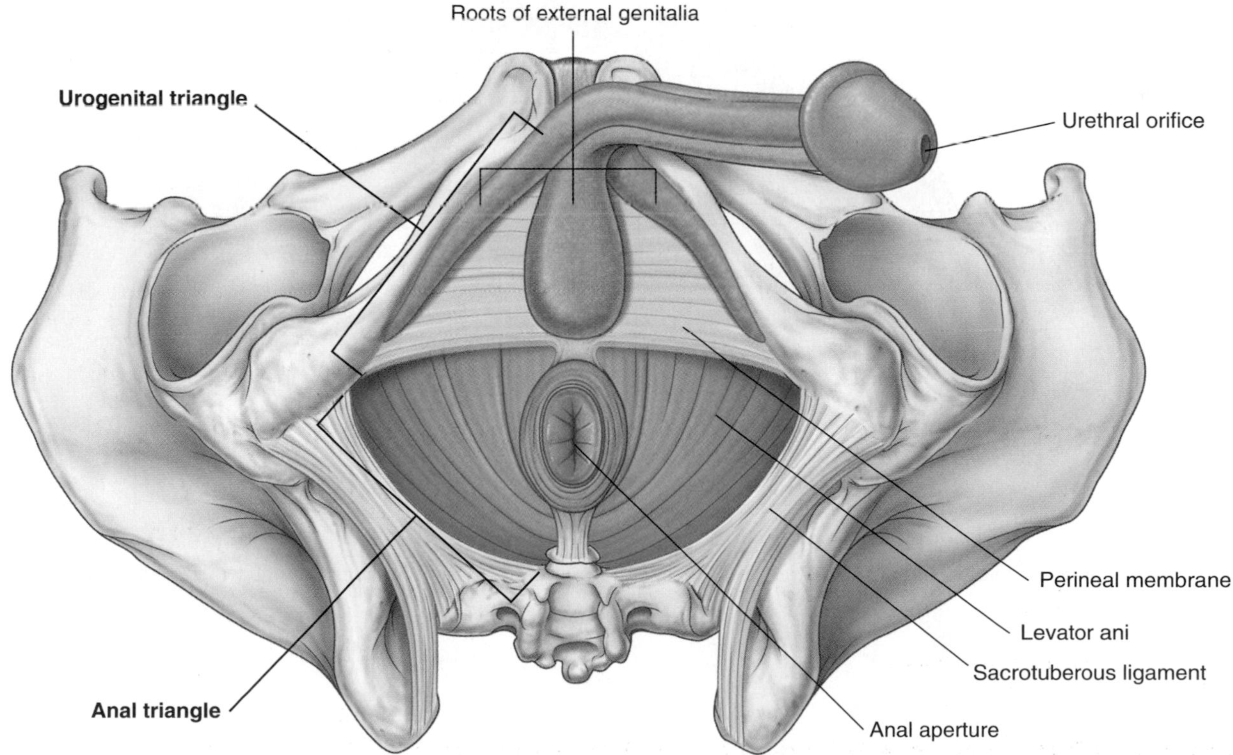
Roots of external genitalia
Urogenital triangle
Urethral orifice
Perineal membrane
Levator ani
Sacrotuberous ligament
Anal triangle
Anal aperture
B

**Fig. 5.9** Perineum. **A.** In women. **B.** In men.

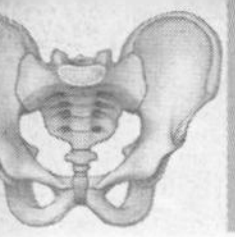

## RELATIONSHIP TO OTHER REGIONS

### Abdomen

The cavity of the true pelvis is continuous with the abdominal cavity at the pelvic inlet (Fig. 5.10A). All structures passing between the pelvic cavity and abdomen, including major vessels, nerves, and lymphatics, as well as the sigmoid colon and ureters, pass via the inlet. In men, the ductus deferens on each side passes through the anterior abdominal wall and over the inlet to enter the pelvic cavity. In women, ovarian vessels, nerves, and lymphatics pass through the inlet to reach the ovaries, which lie on each side just inferior to the pelvic inlet.

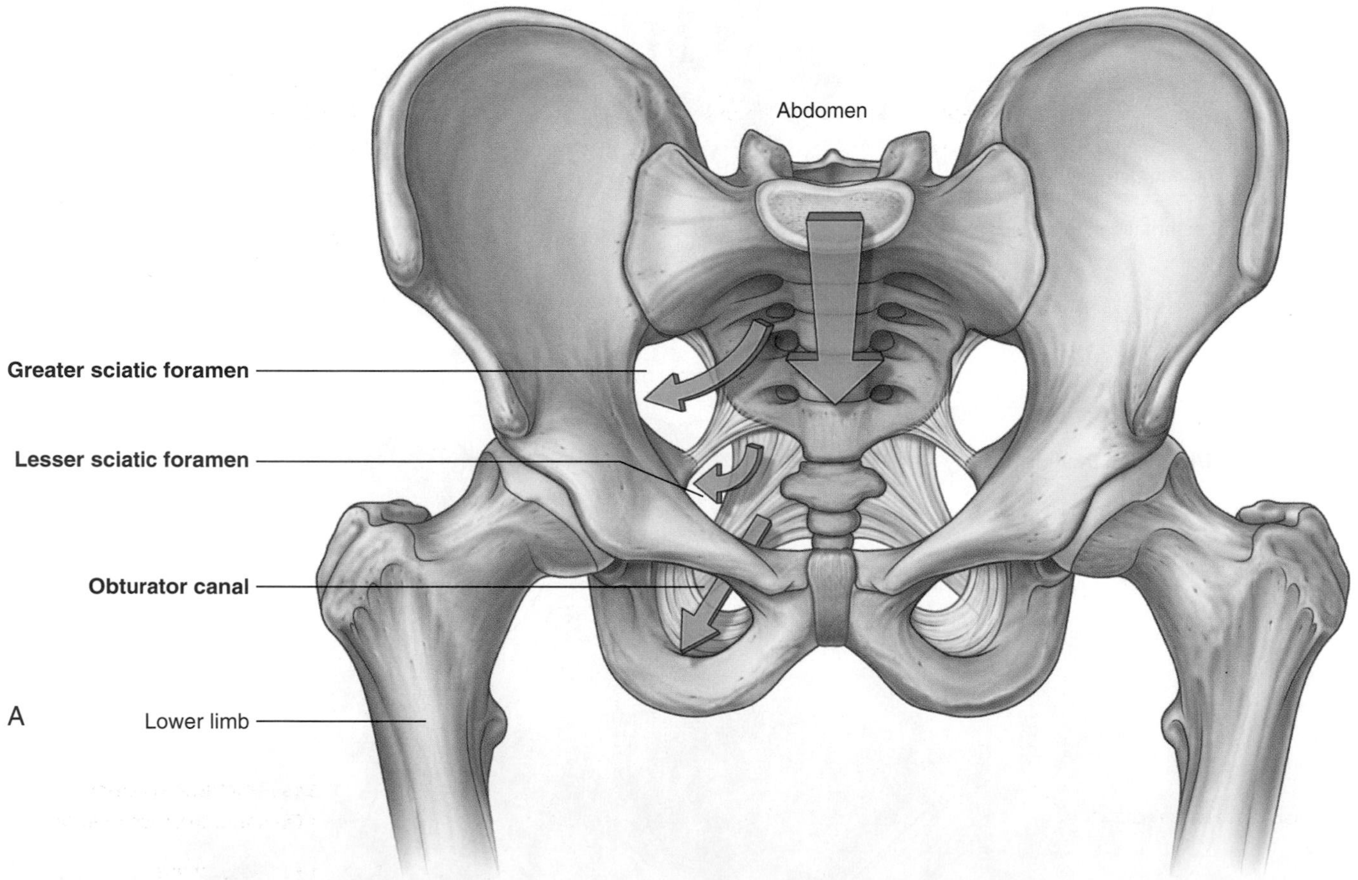

**Fig. 5.10** Areas of communication between the true pelvis and other regions. **A.** Between the true pelvis, abdomen, and lower limb.

## Lower limb

Three apertures in the pelvic wall communicate with the lower limb (Fig. 5.10A):

- the obturator canal,
- the greater sciatic foramen, and
- the lesser sciatic foramen.

The obturator canal forms a passageway between the pelvic cavity and the adductor region of the thigh, and is formed in the superior aspect of the obturator foramen, between bone, a connective tissue membrane, and muscles that fill the foramen.

The lesser sciatic foramen, which lies inferior to the pelvic floor, provides communication between the gluteal region and the perineum (Fig. 5.10B).

The pelvic cavity also communicates directly with the perineum through a small gap between the pubic symphysis and the perineal membrane (Fig. 5.10B).

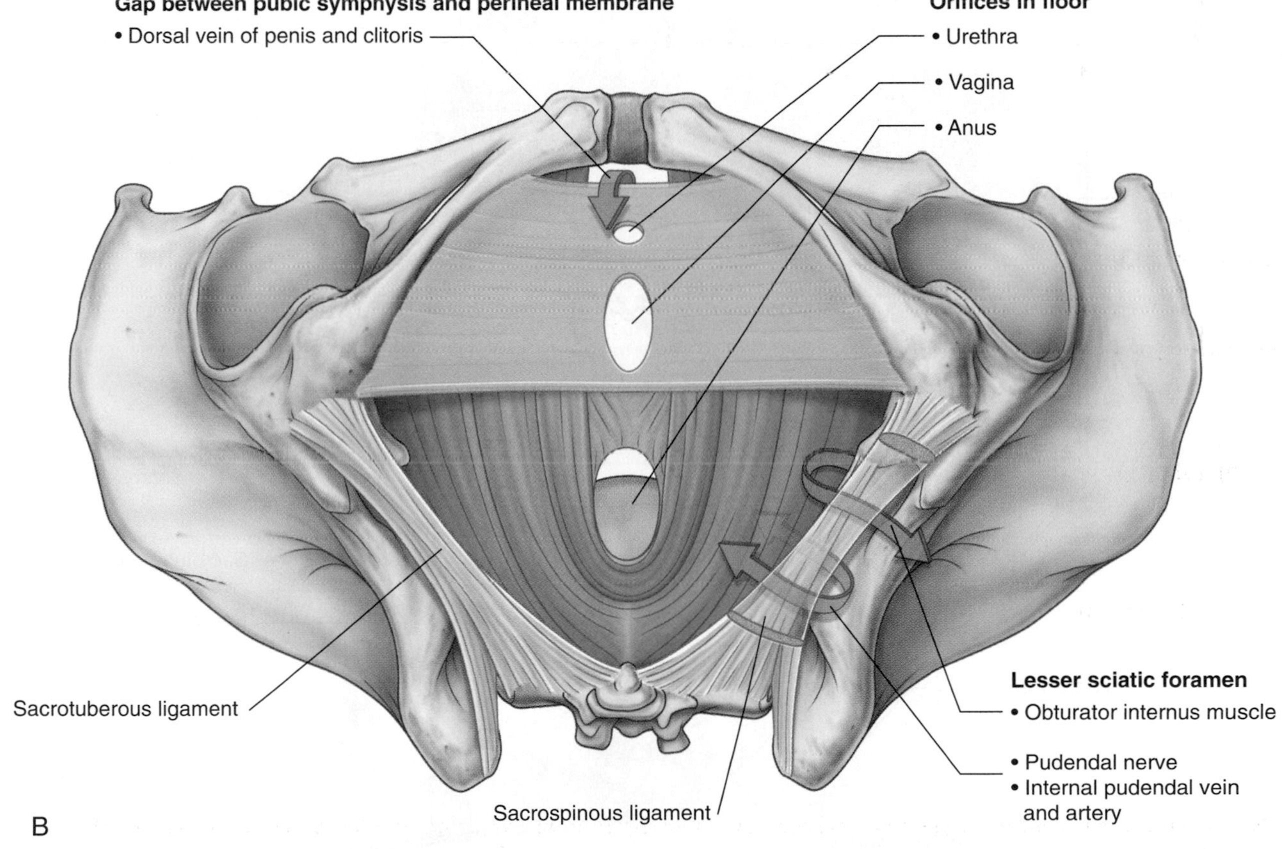


**Fig. 5.10, cont'd** **B.** Between the perineum and other regions.

## KEY FEATURES

### The pelvic cavity projects posteriorly

In the anatomical position, the anterior superior iliac spines and the superior edge of the pubic symphysis lie in the same vertical plane (Fig. 5.11). Consequently, the pelvic inlet is angled 50°–60° forward relative to the horizontal plane, and the pelvic cavity projects posteriorly from the abdominal cavity.

Meanwhile, the urogenital part of the pelvic outlet (the pubic arch) is oriented in a nearly horizontal plane, whereas the posterior part of the outlet is positioned more vertically. The urogenital triangle of the perineum therefore faces inferiorly, while the anal triangle faces more posteriorly.

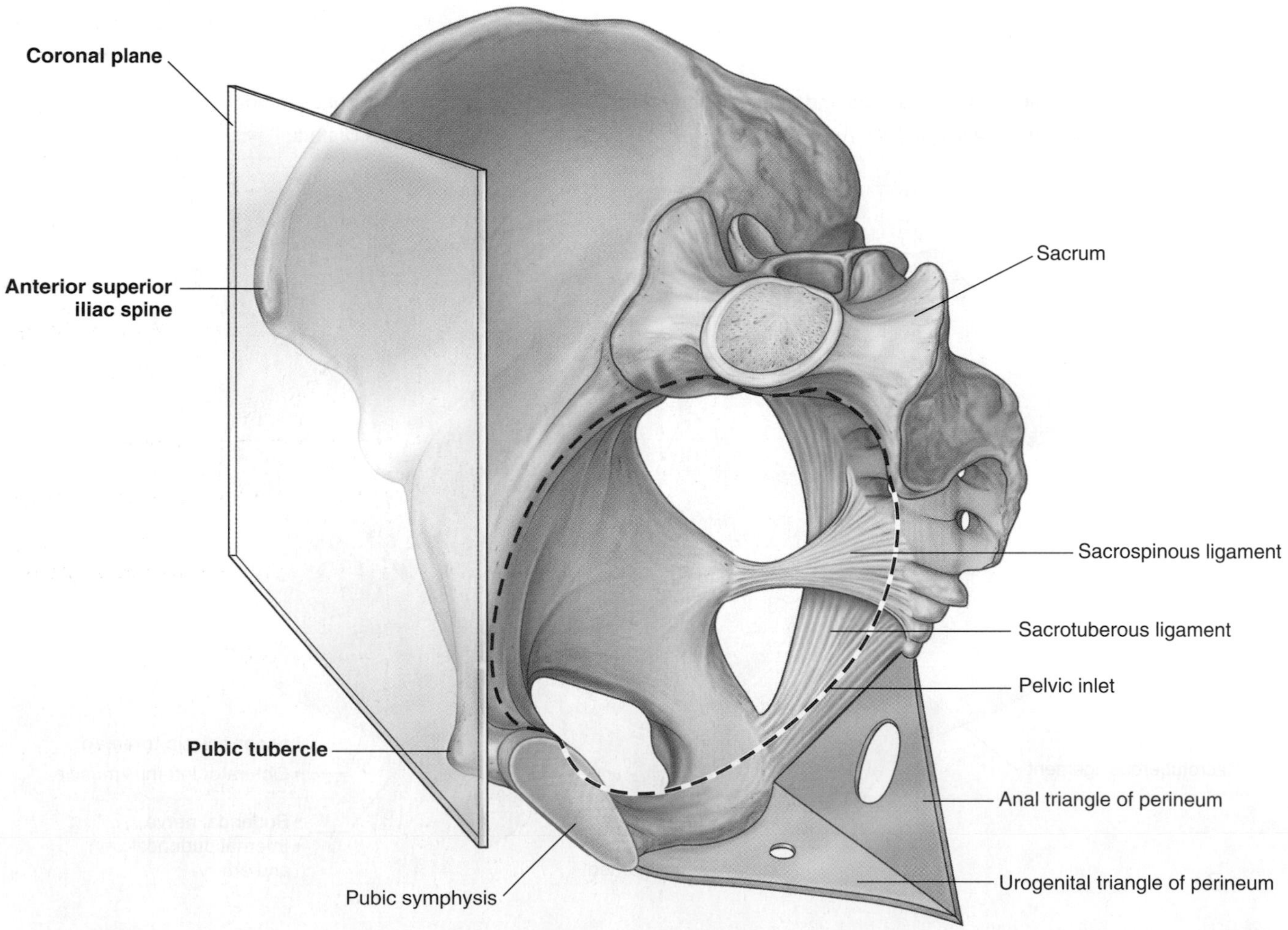

**Fig. 5.11** Orientation of the pelvis and pelvic cavity in the anatomical position.

## Important structures cross the ureters in the pelvic cavity

The ureters drain the kidneys, course down the posterior abdominal wall, and cross the pelvic inlet to enter the pelvic cavity. They continue inferiorly along the lateral pelvic wall and ultimately connect with the base of the bladder.

An important structure crosses the ureters in the pelvic cavity in both men and women—in women, the uterine artery crosses the ureter lateral to the cervix of the uterus (Fig. 5.12A), and in men, the ductus deferens crosses over the ureter just posterior to the bladder (Fig. 5.12B).

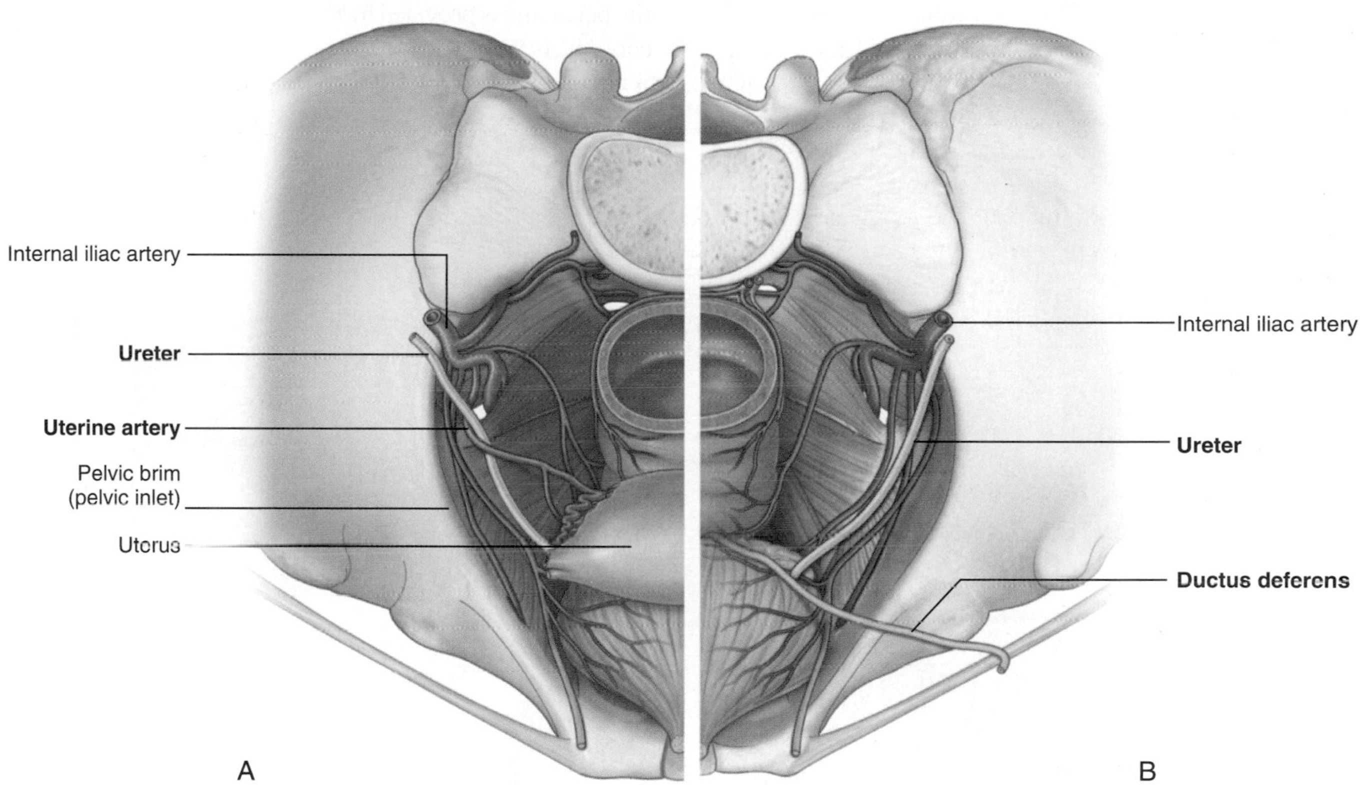


**Fig. 5.12** Structures that cross the ureters in the pelvic cavity. **A.** In women. **B.** In men.

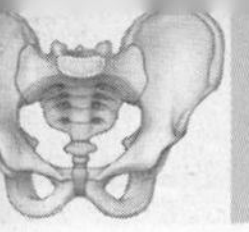

## The prostate in men and the uterus in women are anterior to the rectum

In men, the prostate gland is situated immediately anterior to the rectum, just above the pelvic floor (Fig. 5.13). It can be felt by digital palpation during a rectal examination.

In both sexes, the anal canal and the lower rectum also can be evaluated during a rectal examination by a clinician. In women, the cervix and lower part of the body of the uterus also are palpable. However, these structures can more easily be palpated with a bimanual examination where the index and middle fingers of a clinician's hand are placed in the vagina and the other hand is placed on the lower anterior abdominal wall. The organs are felt between the two hands. This bimanual technique can also be used to examine the ovaries and uterine tubes.

## The perineum is innervated by sacral spinal cord segments

Dermatomes of the perineum in both men and women are from spinal cord levels S3 to S5, except for the anterior regions, which tend to be innervated by spinal cord level L1 by nerves associated with the abdominal wall (Fig. 5.14). Dermatomes of L2 to S2 are predominantly in the lower limb.

Most of the skeletal muscles contained in the perineum and the pelvic floor, including the external anal sphincter and external urethral sphincter, are innervated by spinal cord levels S2 to S4.

Much of the somatic motor and sensory innervation of the perineum is provided by the pudendal nerve from spinal cord levels S2 to S4.

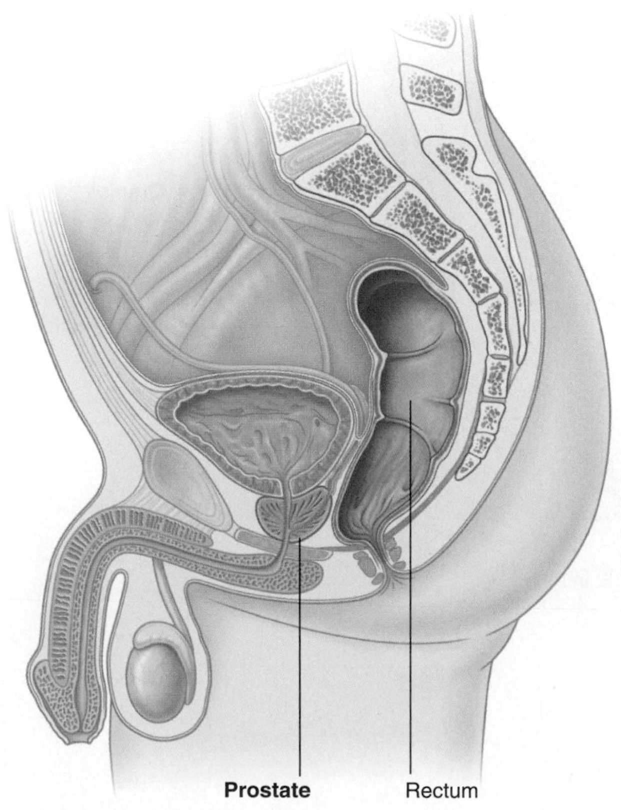


Fig. 5.13 Position of the prostate gland.

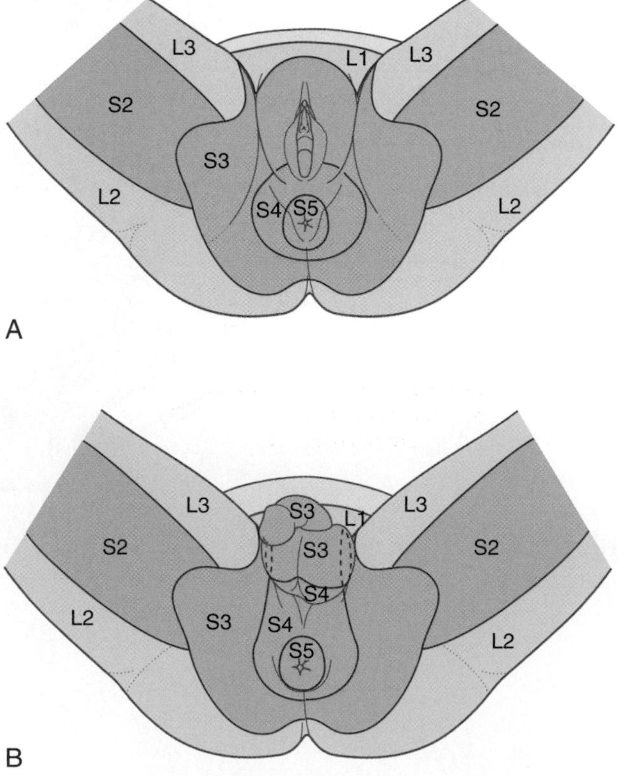


Fig. 5.14 Dermatomes of the perineum. A. In women. B. In men.

## Nerves are related to bone

The **pudendal nerve** is the major nerve of the perineum and is directly associated with the ischial spine of the pelvis (Fig. 5.15). On each side of the body, these spines and the attached sacrospinous ligaments separate the greater sciatic foramina from the lesser sciatic foramina on the lateral pelvic wall.

The pudendal nerve leaves the pelvic cavity through the greater sciatic foramen and then immediately enters the perineum inferiorly to the pelvic floor by passing around the ischial spine and through the lesser sciatic foramen (Fig. 5.15). The ischial spine can be palpated transvaginally in women and is the landmark that can be used for administering a pudendal nerve block.

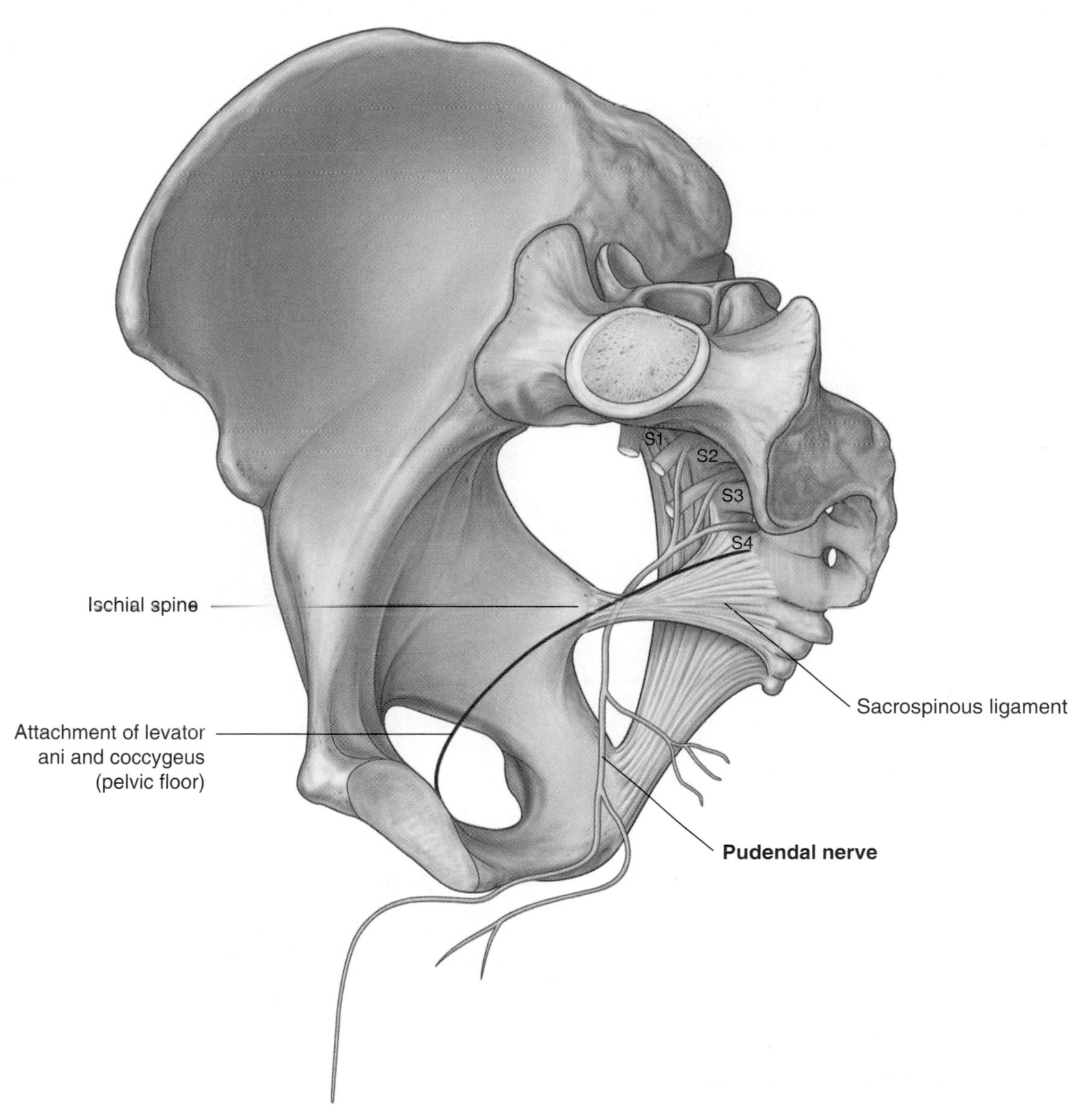


**Fig. 5.15** Pudendal nerve.

## Parasympathetic innervation from spinal cord levels S2 to S4 controls erection

The parasympathetic innervation from spinal cord levels S2 to S4 controls genital erection in both women and men (Fig. 5.16). On each side, preganglionic parasympathetic nerves leave the anterior rami of the sacral spinal nerves and enter the **inferior hypogastric plexus** (pelvic plexus) on the lateral pelvic wall.

The two inferior hypogastric plexuses are inferior extensions of the abdominal prevertebral plexus that forms on the posterior abdominal wall in association with the abdominal aorta. Nerves derived from these plexuses penetrate the pelvic floor to innervate the erectile tissues of the clitoris in women and the penis in men.

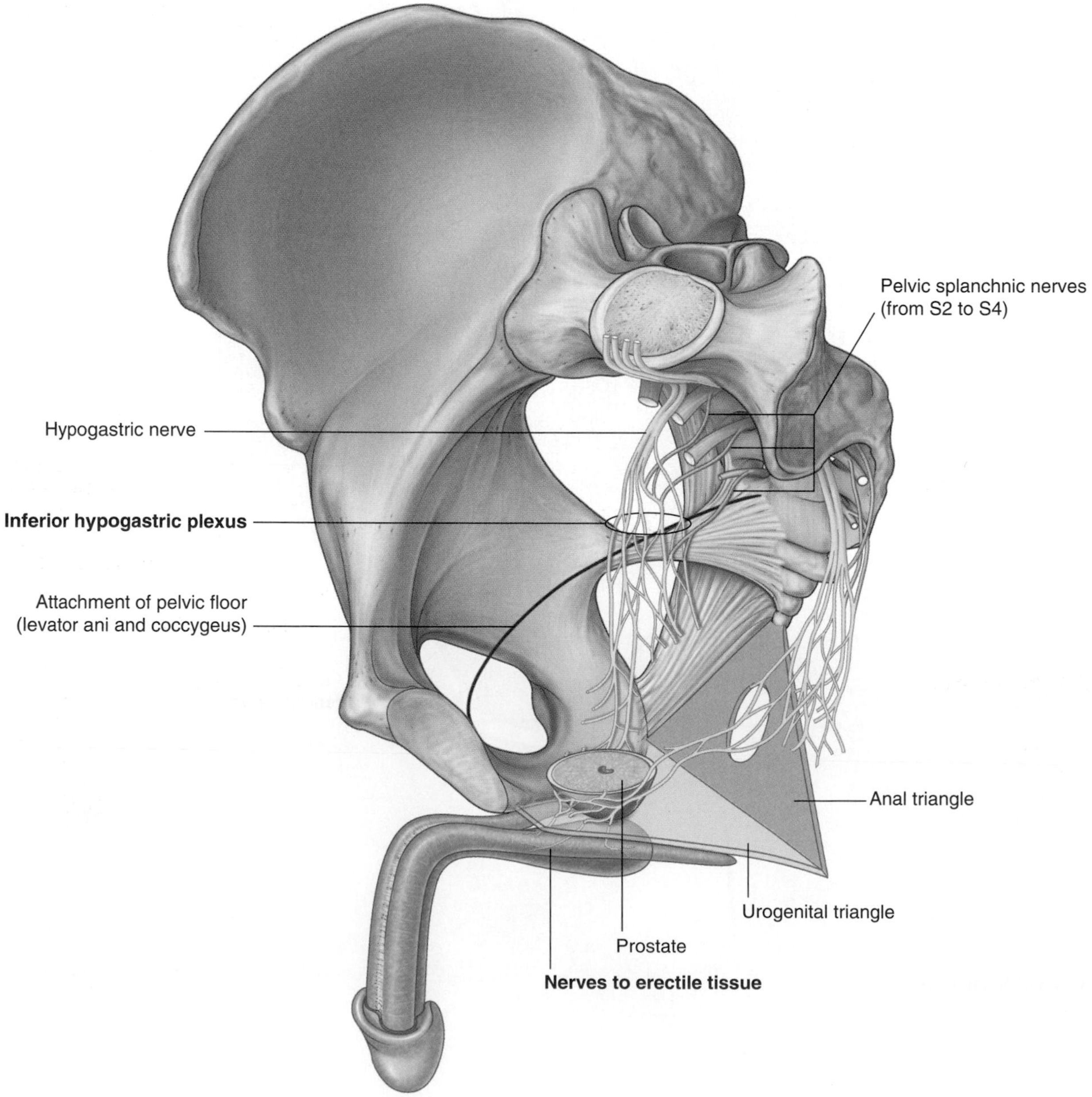

**Fig. 5.16** Pelvic splanchnic nerves from spinal levels S2 to S4 control erection.

## Muscles and fascia of the pelvic floor and perineum intersect at the perineal body

Structures of the pelvic floor intersect with structures in the perineum at the **perineal body** (Fig. 5.17). This poorly defined fibromuscular node lies at the center of the perineum, approximately midway between the two ischial tuberosities. Converging at the perineal body are:

- the levator ani muscles of the pelvic diaphragm, and
- muscles in the urogenital and anal triangles of the perineum, including the skeletal muscle sphincters associated with the urethra, vagina, and anus.

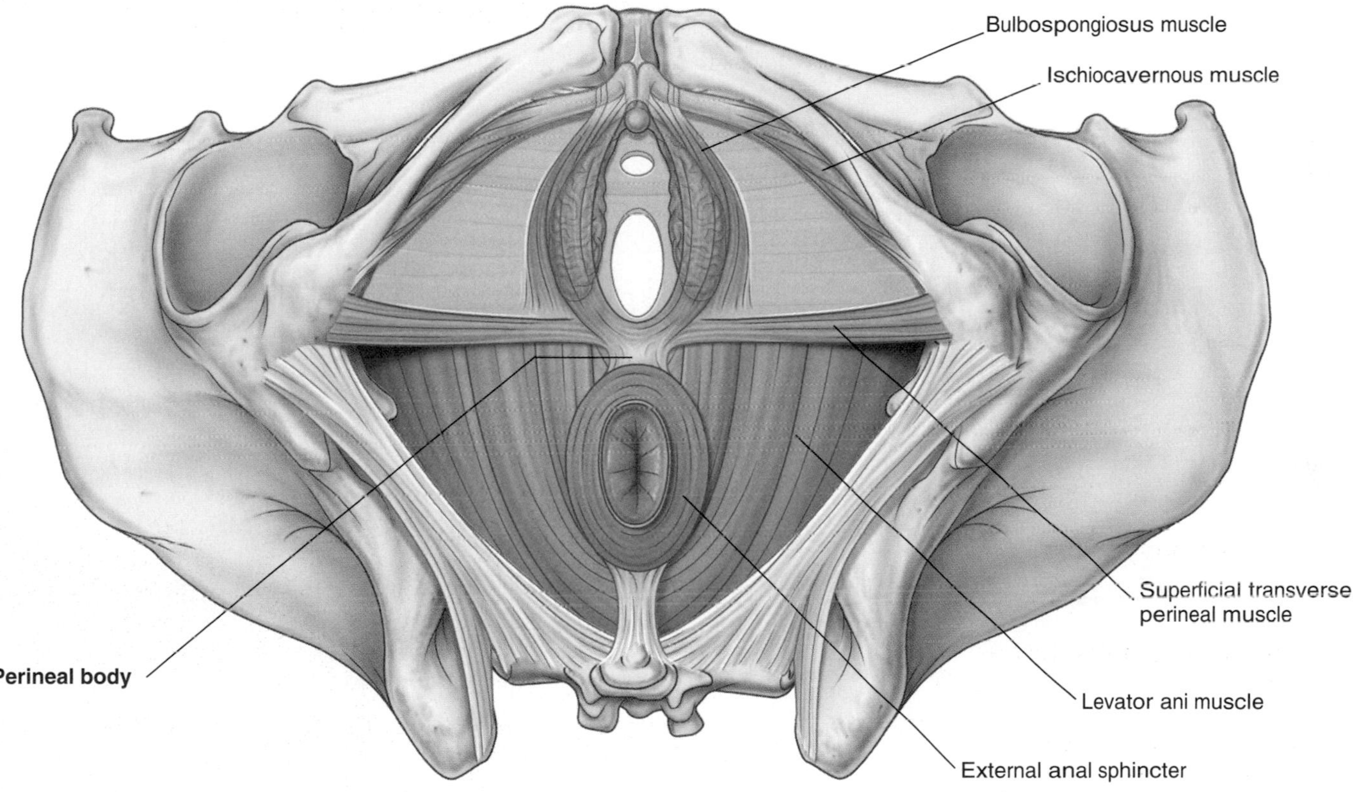


**Fig. 5.17** Perineal body.

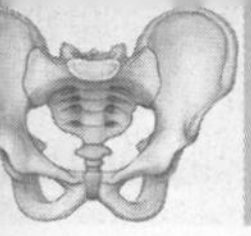


### The course of the urethra is different in men and women

In women, the urethra is short and passes inferiorly from the bladder through the pelvic floor and opens directly into the perineum (Fig. 5.18A).

In men the urethra passes through the prostate before coursing through the deep perineal pouch and perineal membrane and then becomes enclosed within the erectile tissues of the penis before opening at the end of the penis (Fig. 5.18B). The penile part of the male urethra has two angles:

- The more important of these is a fixed angle where the urethra bends anteriorly in the root of the penis after passing through the perineal membrane.
- Another angle occurs distally where the unattached part of the penis curves inferiorly—when the penis is erect, this second angle disappears.

It is important to consider the different courses of the urethra in men and women when catheterizing patients and when evaluating perineal injuries and pelvic pathology.

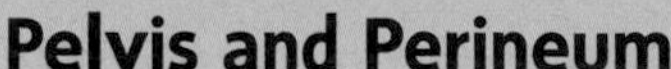

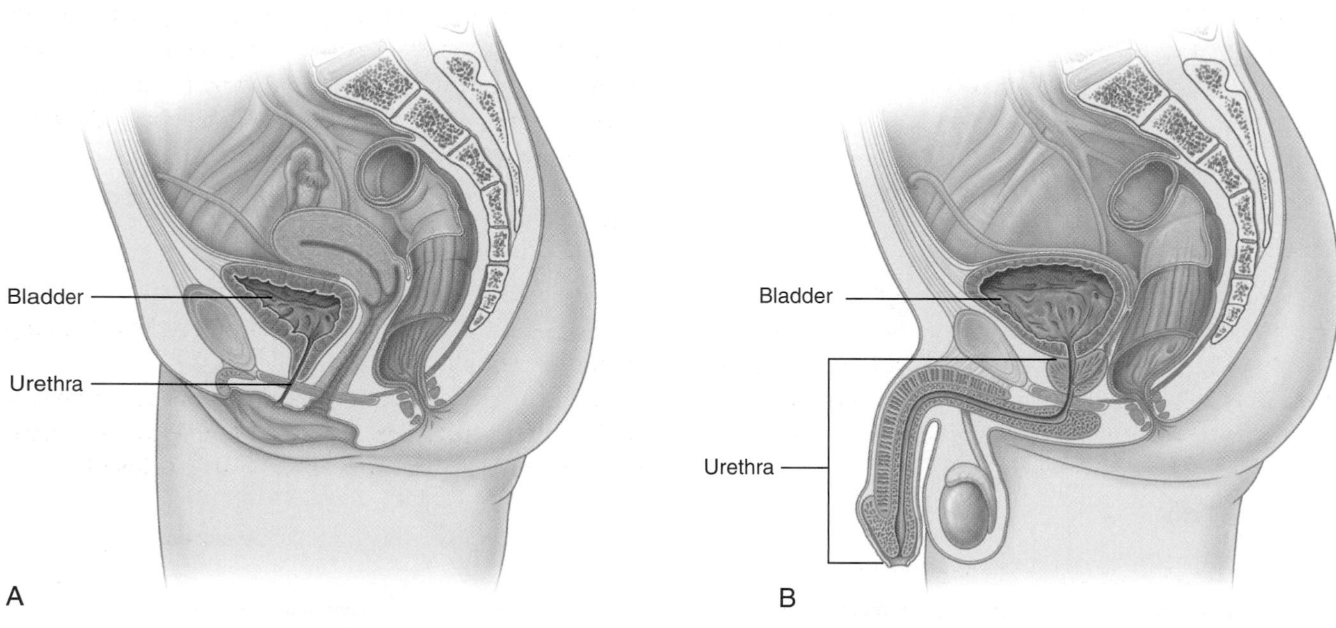


**Fig. 5.18** Course of the urethra. **A.** In women. **B.** In men.

# Regional anatomy

The pelvis is the region of the body surrounded by the pelvic bones and the inferior elements of the vertebral column. It is divided into two major regions: the superior region is the false (greater) pelvis and is part of the abdominal cavity; the inferior region is the true (lesser) pelvis, which encloses the pelvic cavity.

The bowl-shaped pelvic cavity is continuous above with the abdominal cavity. The rim of the pelvic cavity (the pelvic inlet) is completely encircled by bone. The pelvic floor is a fibromuscular structure separating the pelvic cavity above from the perineum below.

The perineum is inferior to the pelvic floor and its margin is formed by the pelvic outlet. The perineum contains:

- the terminal openings of the gastrointestinal and urinary systems,
- the external opening of the reproductive tract, and
- the roots of the external genitalia.

## PELVIS

### Bones

The bones of the pelvis consist of the right and left pelvic (hip) bones, the sacrum, and the coccyx. The sacrum articulates superiorly with vertebra LV at the lumbosacral joint. The pelvic bones articulate posteriorly with the sacrum at the sacro-iliac joints and with each other anteriorly at the pubic symphysis.

#### Pelvic bone

The pelvic bone is irregular in shape and has two major parts separated by an oblique line on the medial surface of the bone (Fig. 5.19A):

- The pelvic bone above this line represents the lateral wall of the false pelvis, which is part of the abdominal cavity.
- The pelvic bone below this line represents the lateral wall of the true pelvis, which contains the pelvic cavity.

The linea terminalis is the lower two-thirds of this line and contributes to the margin of the pelvic inlet.

The lateral surface of the pelvic bone has a large articular socket, the **acetabulum**, which, together with the head of the femur, forms the hip joint (Fig. 5.19B).

Inferior to the acetabulum is the large **obturator foramen**, most of which is closed by a flat connective tissue membrane, the **obturator membrane**. A small obturator canal remains open superiorly between the membrane and adjacent bone, providing a route of communication between the lower limb and the pelvic cavity.

The posterior margin of the bone is marked by two notches separated by the **ischial spine**:

- the **greater sciatic notch**, and
- the **lesser sciatic notch**.

The posterior margin terminates inferiorly as the large **ischial tuberosity**.

The irregular anterior margin of the pelvic bone is marked by the **anterior superior iliac spine**, the **anterior inferior iliac spine**, and the **pubic tubercle**.

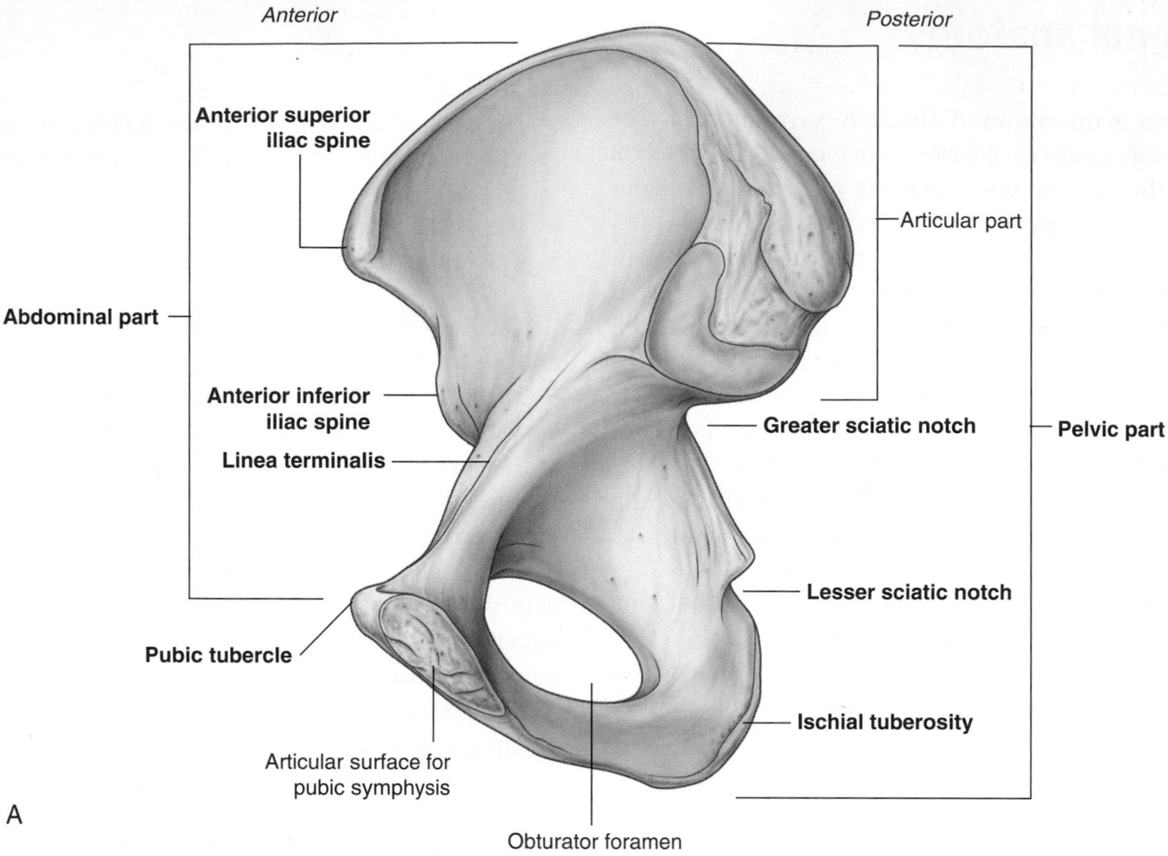

Anterior

Posterior

Anterior superior iliac spine

Articular part

Abdominal part

Anterior inferior iliac spine

Greater sciatic notch

Pelvic part

Linea terminalis

Lesser sciatic notch

Pubic tubercle

Ischial tuberosity

Articular surface for pubic symphysis

A

Obturator foramen

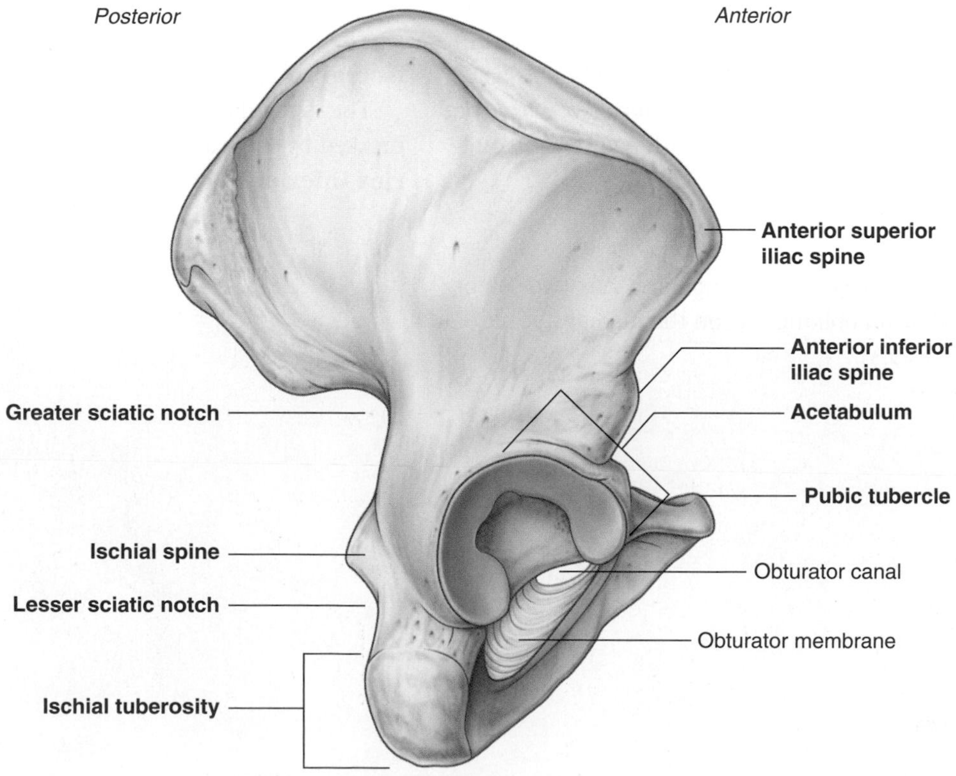

Posterior

Anterior

Anterior superior iliac spine

Anterior inferior iliac spine

Greater sciatic notch

Acetabulum

Pubic tubercle

Ischial spine

Obturator canal

Lesser sciatic notch

Obturator membrane

Ischial tuberosity

B

**Fig. 5.19** Right pelvic bone. **A.** Medial view. **B.** Lateral view.

## Components of the pelvic bone

Each pelvic bone is formed by three elements: the ilium, pubis, and ischium. At birth, these bones are connected by cartilage in the area of the acetabulum; later, at between 16 and 18 years of age, they fuse into a single bone (Fig. 5.20).

### Ilium

Of the three components of the pelvic bone, the **ilium** is the most superior in position.

The ilium is separated into upper and lower parts by a ridge on the medial surface (Fig. 5.21A).

- Posteriorly, the ridge is sharp and lies immediately superior to the surface of the bone that articulates with the sacrum. This sacral surface has a large L-shaped facet for articulating with the sacrum and an expanded, posterior roughened area for the attachment of the strong ligaments that support the sacro-iliac joint (Fig. 5.21).
- Anteriorly, the ridge separating the upper and lower parts of the ilium is rounded and termed the **arcuate line**.

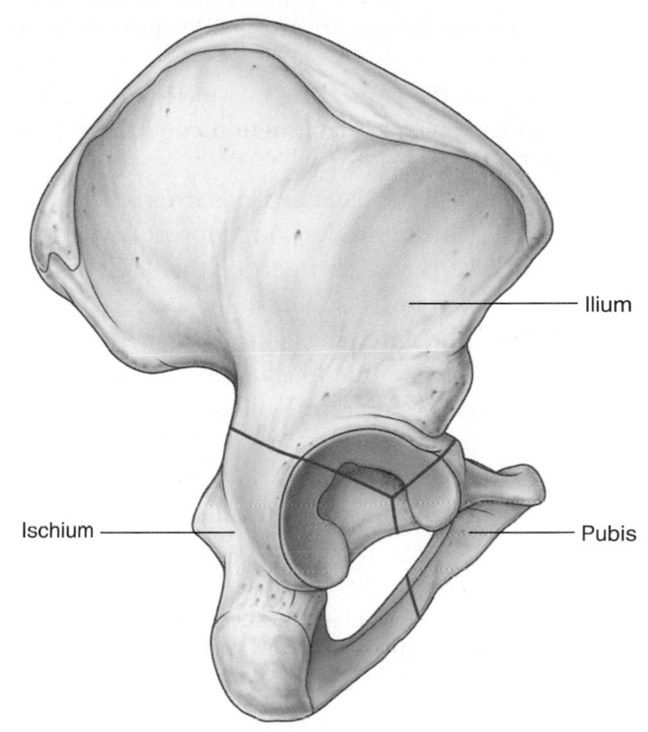


**Fig. 5.20** Ilium, ischium, and pubis.

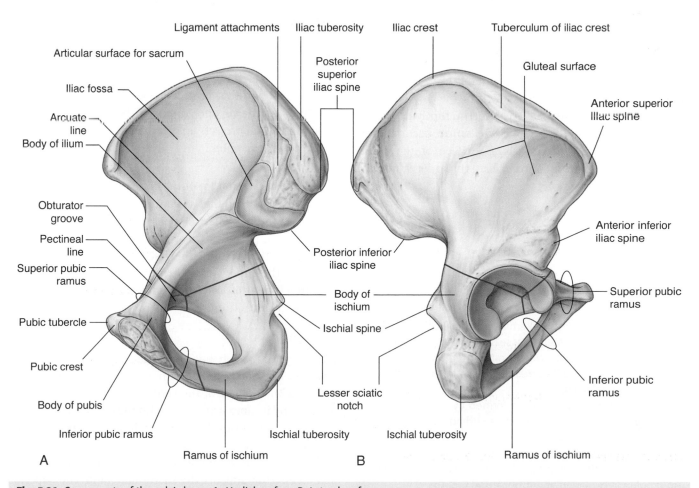

**Fig. 5.21** Components of the pelvic bone. **A.** Medial surface. **B.** Lateral surface.

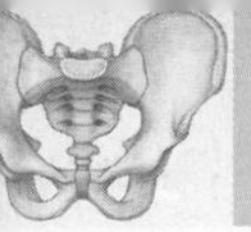



The arcuate line forms part of the linea terminalis and the pelvic brim.

The portion of the ilium lying inferiorly to the arcuate line is the pelvic part of the ilium and contributes to the wall of the lesser or true pelvis.

The upper part of the ilium expands to form a flat, fan-shaped "wing," which provides bony support for the lower abdomen, or false pelvis. This part of the ilium provides attachment for muscles functionally associated with the lower limb. The anteromedial surface of the wing is concave and forms the **iliac fossa**. The external (gluteal) surface of the wing is marked by lines and roughenings and is related to the gluteal region of the lower limb (Fig. 5.21B).

The entire superior margin of the ilium is thickened to form a prominent crest (the **iliac crest**), which is the site of attachment for muscles and fascia of the abdomen, back, and lower limb and terminates anteriorly as the **anterior superior iliac spine** and posteriorly as the **posterior superior iliac spine**.

A prominent tubercle, the **tuberculum of the iliac crest**, projects laterally near the anterior end of the crest; the posterior end of the crest thickens to form the **iliac tuberosity**.

Inferior to the anterior superior iliac spine of the crest, on the anterior margin of the ilium, is a rounded protuberance called the **anterior inferior iliac spine**. This structure serves as the point of attachment for the rectus femoris muscle of the anterior compartment of the thigh and the iliofemoral ligament associated with the hip joint. A less prominent **posterior inferior iliac spine** occurs along the posterior border of the sacral surface of the ilium, where the bone angles forward to form the superior margin of the greater sciatic notch.

### In the clinic

#### Bone marrow biopsy

In certain diseases (e.g., leukemia), a sample of bone marrow must be obtained to assess the stage and severity of the problem. The iliac crest is often used for such bone marrow biopsies. The iliac crest lies close to the surface and is easily palpated.

A bone marrow biopsy is performed by injecting anesthetic in the skin and passing a cutting needle through the cortical bone of the iliac crest. The bone marrow is aspirated and viewed under a microscope. Samples of cortical bone can also be obtained in this way to provide information about bone metabolism.

#### Pubis

The anterior and inferior part of the pelvic bone is the **pubis** (Fig. 5.21). It has a body and two arms (rami).

- The **body** is flattened dorsoventrally and articulates with the body of the pubic bone on the other side at the **pubic symphysis**. The body has a rounded pubic crest on its superior surface that ends laterally as the prominent **pubic tubercle**.
- The **superior pubic ramus** projects posterolaterally from the body and joins with the ilium and ischium at its base, which is positioned toward the acetabulum. The sharp superior margin of this triangular surface is termed the **pecten pubis** (**pectineal line**), which forms part of the linea terminalis of the pelvic bone and the pelvic inlet. Anteriorly, this line is continuous with the **pubic crest**, which also is part of the linea terminalis and pelvic inlet. The **superior pubic ramus** is marked on its inferior surface by the **obturator groove**, which forms the upper margin of the obturator canal.
- The inferior ramus projects laterally and inferiorly to join with the ramus of the ischium.

#### Ischium

The ischium is the posterior and inferior part of the pelvic bone (Fig. 5.21). It has:

- a large body that projects superiorly to join with the ilium and the superior ramus of the pubis, and
- a ramus that projects anteriorly to join with the inferior ramus of the pubis.

The posterior margin of the bone is marked by a prominent **ischial spine** that separates the lesser sciatic notch, below, from the greater sciatic notch, above.

The most prominent feature of the ischium is a large tuberosity (the **ischial tuberosity**) on the posteroinferior aspect of the bone. This tuberosity is an important site for the attachment of lower limb muscles and for supporting the body when sitting.

### Sacrum

The sacrum, which has the appearance of an inverted triangle, is formed by the fusion of the five sacral vertebrae (Fig. 5.22). The base of the sacrum articulates with vertebra LV, and its apex articulates with the coccyx. Each of the lateral surfaces of the bone bears a large L-shaped facet for articulation with the ilium of the pelvic bone. Posterior to the facet is a large roughened area for the attachment of ligaments that support the sacro-iliac joint. The superior surface of the sacrum is characterized by the superior

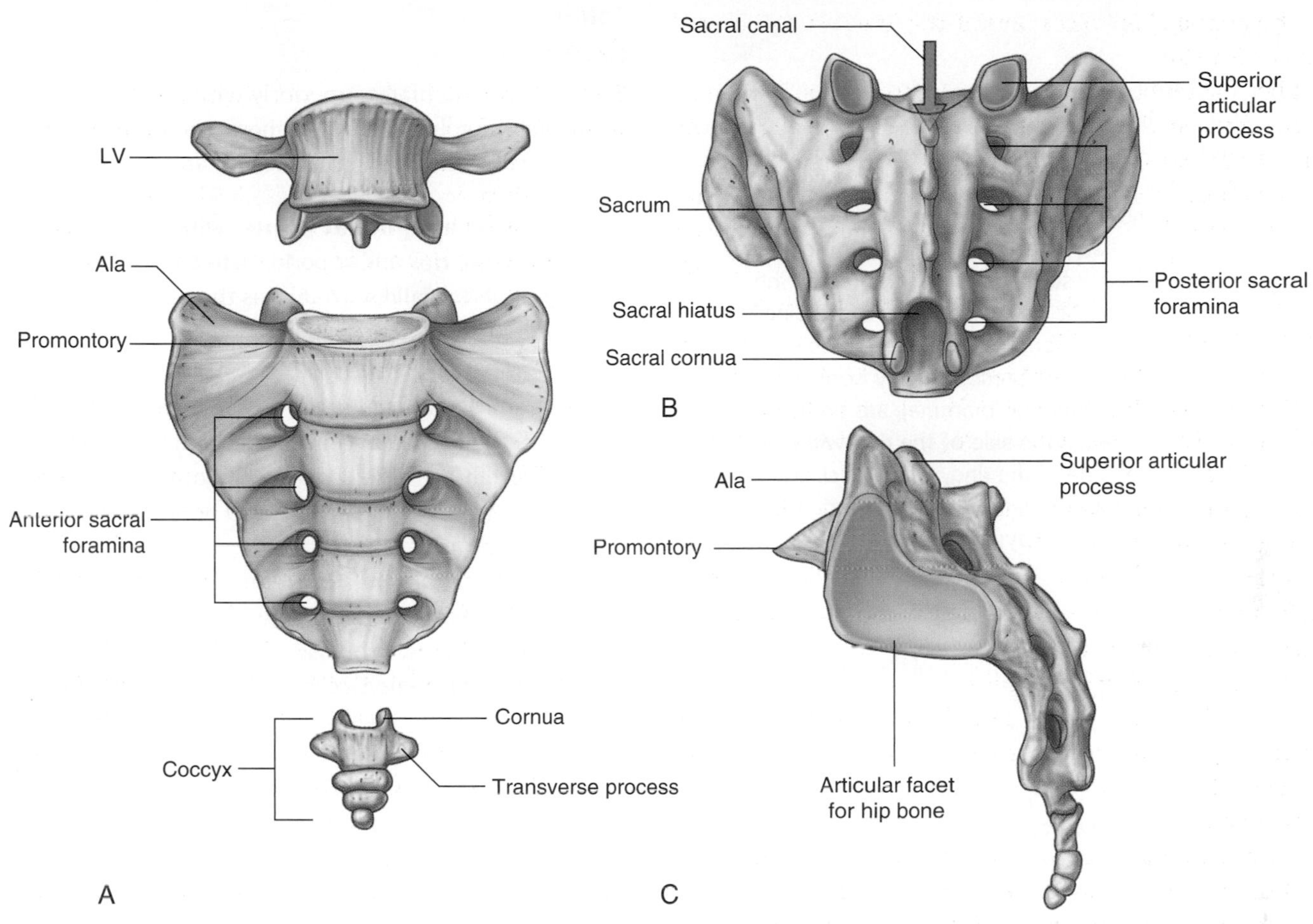


**Fig. 5.22** Sacrum and coccyx. **A.** Anterior view. **B.** Posterior view. **C.** Lateral view.

aspect of the body of vertebra SI and is flanked on each side by an expanded wing-like transverse process termed the **ala**. The anterior edge of the vertebral body projects forward as the **promontory**. The anterior surface of the sacrum is concave; the posterior surface is convex. Because the transverse processes of adjacent sacral vertebrae fuse lateral to the position of the intervertebral foramina and lateral to the bifurcation of spinal nerves into posterior and anterior rami, the posterior and anterior rami of spinal nerves S1 to S4 emerge from the sacrum through separate foramina. There are four pairs of **anterior sacral foramina** on the anterior surface of the sacrum for anterior rami, and four pairs of **posterior sacral foramina** on the posterior surface for the posterior rami. The **sacral canal** is a continuation of the vertebral canal that terminates as the **sacral hiatus**.

### Coccyx

The small terminal part of the vertebral column is the coccyx, which consists of four fused coccygeal vertebrae (Fig. 5.22) and, like the sacrum, has the shape of an inverted triangle. The base of the coccyx is directed superiorly. The superior surface bears a facet for articulation with the sacrum and two **horns**, or **cornua**, one on each side, that project upward to articulate or fuse with similar downward-projecting cornua from the sacrum. These processes are modified superior and inferior articular processes that are present on other vertebrae. Each lateral surface of the coccyx has a small rudimentary transverse process, extending from the first coccygeal vertebra. Vertebral arches are absent from coccygeal vertebrae; therefore no bony vertebral canal is present in the coccyx.

## In the clinic

### Pelvic fracture

The pelvis can be viewed as a series of anatomical rings. There are three bony rings and four fibro-osseous rings. The major bony pelvic ring consists of parts of the sacrum, ilium, and pubis, which forms the pelvic inlet. Two smaller subsidiary rings are the obturator foramina. The greater and lesser sciatic foramina formed by the greater and lesser sciatic notches and the sacrospinous and sacrotuberous ligaments form the four fibro-osseous rings. The rings, which are predominantly bony (i.e., the pelvic inlet and the obturator foramina), are brittle rings. It is not possible to break one side of the ring without breaking the other side of the ring, which in clinical terms means that if a fracture is demonstrated on one side, a second fracture should always be suspected.

Fractures of the pelvis may occur in isolation; however, they usually occur in trauma patients and warrant special mention.

Owing to the large bony surfaces of the pelvis, a fracture produces an area of bone that can bleed significantly. A large hematoma may be produced, which can compress organs such as the bladder and the ureters. This blood loss may occur rapidly, reducing the circulating blood volume and, unless this is replaced, the patient will become hypovolemic and shock will develop.

Pelvic fractures may also disrupt the contents of the pelvis, leading to urethral disruption, potential bowel rupture, and nerve damage.

## Joints

### Lumbosacral joints

The sacrum articulates superiorly with the lumbar part of the vertebral column. The lumbosacral joints are formed between vertebra LV and the sacrum and consist of:

- the two **zygapophysial joints**, which occur between adjacent inferior and superior articular processes, and
- an intervertebral disc that joins the bodies of vertebrae LV and SI (Fig. 5.23A).

These joints are similar to those between other vertebrae, with the exception that the sacrum is angled posteriorly on vertebra LV. As a result, the anterior part of the intervertebral disc between the two bones is thicker than the posterior part.

The lumbosacral joints are reinforced by strong iliolumbar and lumbosacral ligaments that extend from the expanded transverse processes of vertebra LV to the ilium and the sacrum, respectively (Fig. 5.23B).

### Sacro-iliac joints

The sacro-iliac joints transmit forces from the lower limbs to the vertebral column. They are synovial joints between the L-shaped articular facets on the lateral surfaces of the sacrum and similar facets on the iliac parts of the pelvic bones (Fig. 5.24A). The joint surfaces have an irregular contour and interlock to resist movement. The joints often

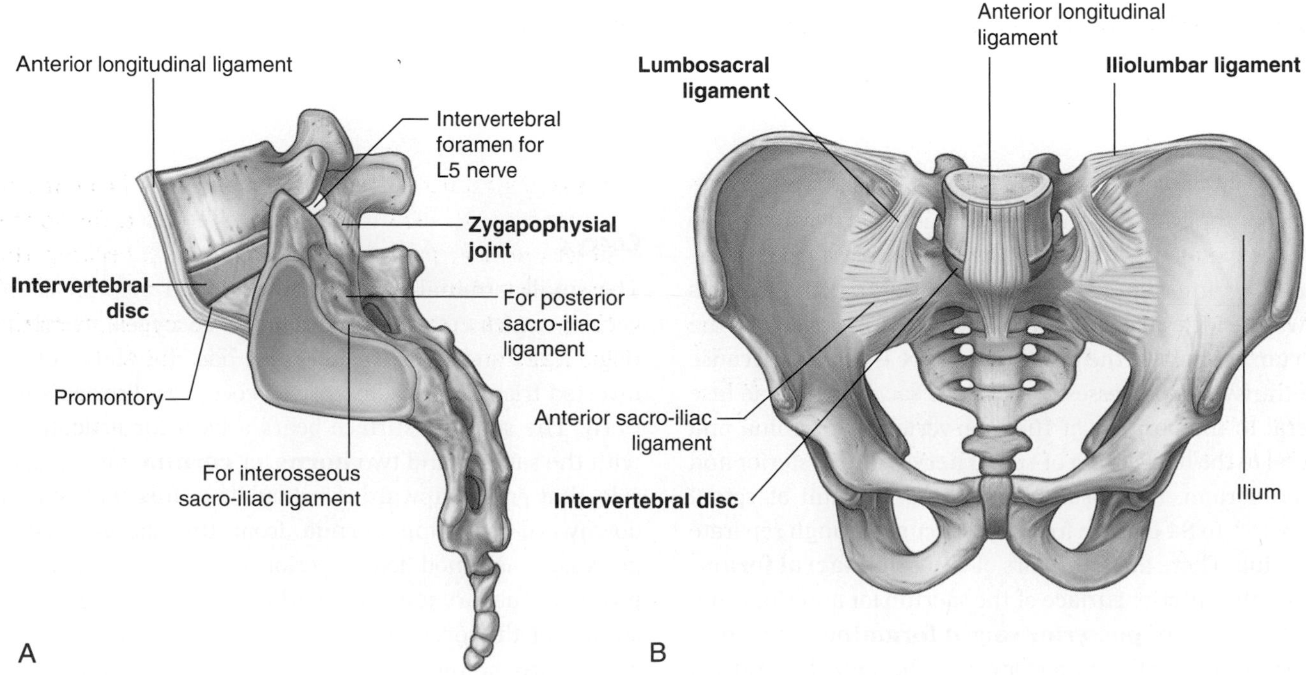


**Fig. 5.23** Lumbosacral joints and associated ligaments. **A.** Lateral view. **B.** Anterior view.

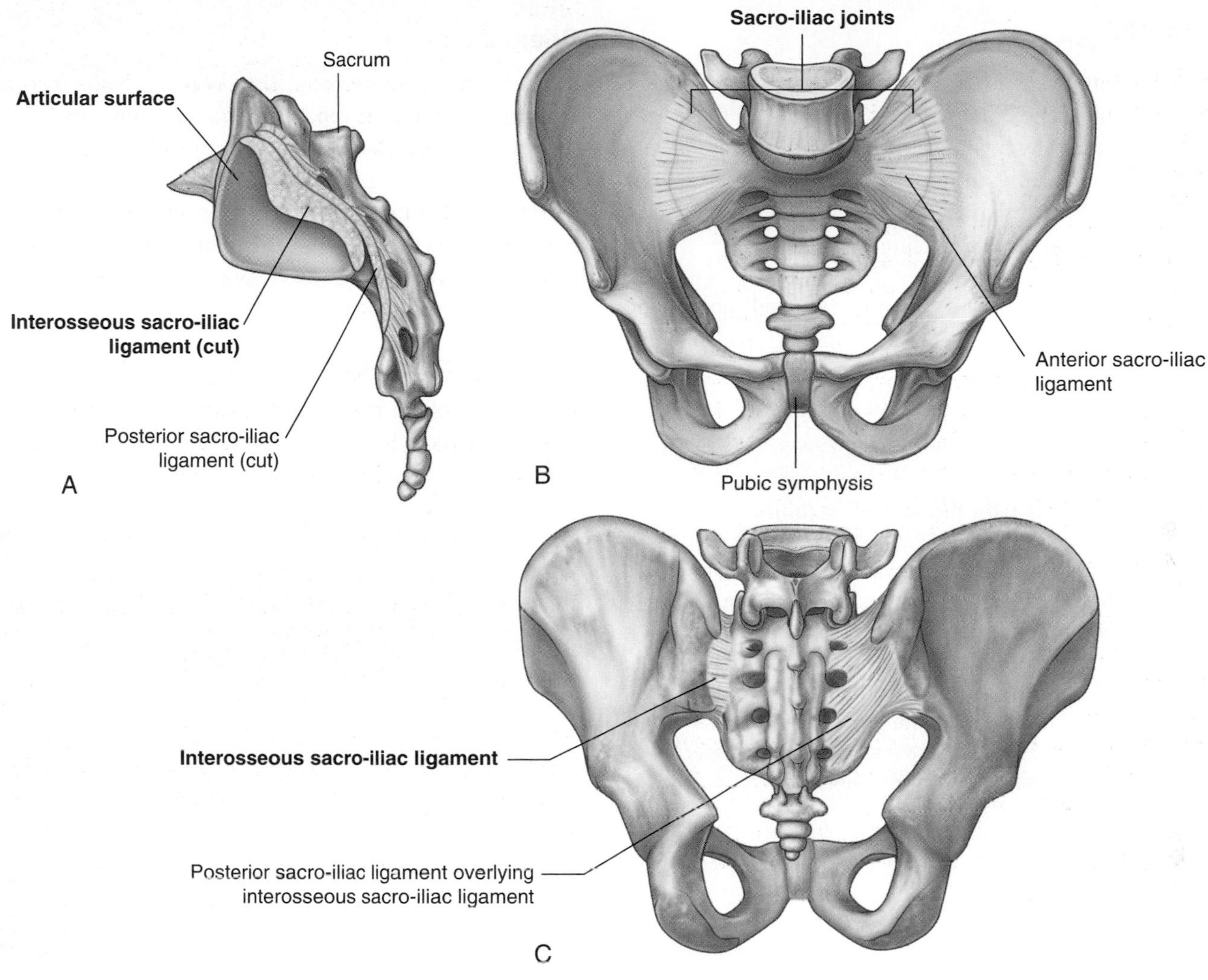


**Fig. 5.24** Sacro-iliac joints and associated ligaments. **A.** Lateral view. **B.** Anterior view. **C.** Posterior view.

become fibrous with age and may become completely ossified.

Each sacro-iliac joint is stabilized by three ligaments:

- the **anterior sacro-iliac ligament**, which is a thickening of the fibrous membrane of the joint capsule and runs anteriorly and inferiorly to the joint (Fig. 5.24B);
- the **interosseous sacro-iliac ligament**, which is the largest, strongest ligament of the three, and is positioned immediately posterosuperior to the joint and attaches to adjacent expansive roughened areas on the ilium and sacrum, thereby filling the gap between the two bones (Fig. 5.24A,C); and
- the **posterior sacro-iliac ligament**, which covers the interosseous sacro-iliac ligament (Fig. 5.24C).

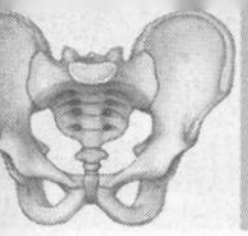



### Pubic symphysis joint

The pubic symphysis lies anteriorly between the adjacent surfaces of the pubic bones (Fig. 5.25). Each of the joint's surfaces is covered by hyaline cartilage and is linked across the midline to adjacent surfaces by fibrocartilage. The joint is surrounded by interwoven layers of collagen fibers and the two major ligaments associated with it are:

- the **superior pubic ligament**, located above the joint, and
- the **inferior pubic ligament**, located below it.

#### In the clinic

**Common problems with the sacro-iliac joints**

As with many weight-bearing joints, degenerative changes may occur with the sacro-iliac joints and cause pain and discomfort in the region. In addition, disorders associated with the major histocompatibility complex antigen HLA-B27, such as ankylosing spondylitis, psoriatic arthritis, inflammatory arthritis associated with inflammatory bowel disease, and reactive arthritis (the group referred to as seronegative spondyloarthropathies), can produce specific inflammatory changes within these joints.

## Orientation

In the anatomical position, the pelvis is oriented so that the front edge of the top of the pubic symphysis and the anterior superior iliac spines lie in the same vertical plane (Fig. 5.26). As a consequence, the pelvic inlet, which marks the entrance to the pelvic cavity, is tilted to face anteriorly, and the bodies of the pubic bones and the pubic arch are positioned in a nearly horizontal plane facing the ground.

## Differences between men and women

The pelvises of women and men differ in a number of ways, many of which have to do with the passing of a baby through a woman's pelvic cavity during childbirth.

- The pelvic inlet in women is circular (Fig. 5.27A) compared with the heart-shaped pelvic inlet (Fig. 5.27B) in men. The more circular shape is partly caused by the less distinct promontory and broader alae in women.
- The angle formed by the two arms of the pubic arch is larger in women (80°–85°) than it is in men (50°–60°).
- The ischial spines generally do not project as far medially into the pelvic cavity in women as they do in men.

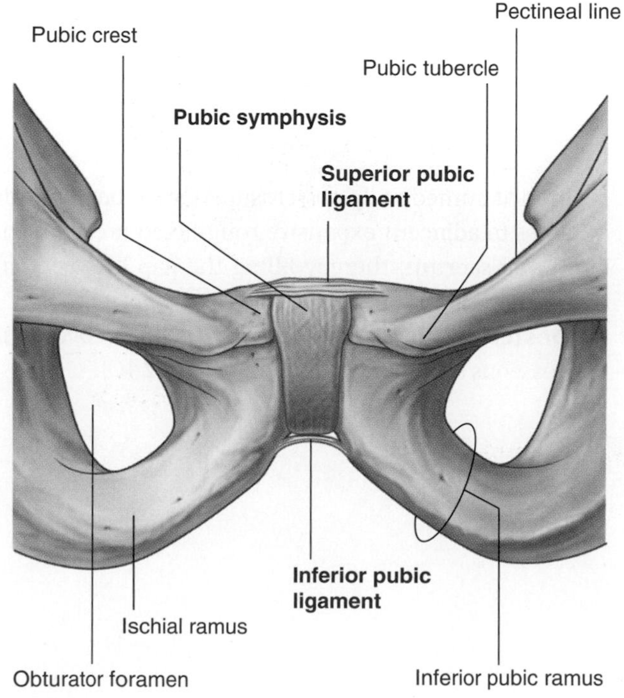


**Fig. 5.25** Pubic symphysis and associated ligaments.

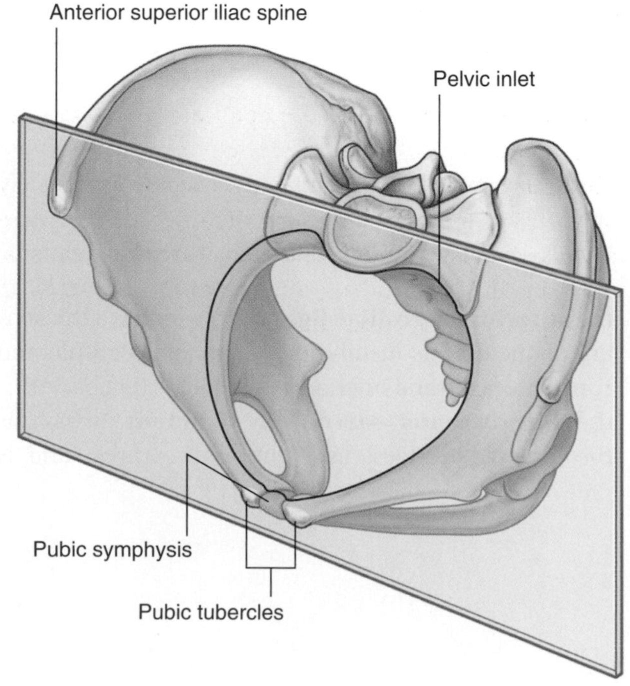


**Fig. 5.26** Orientation of the pelvis (anatomical position).

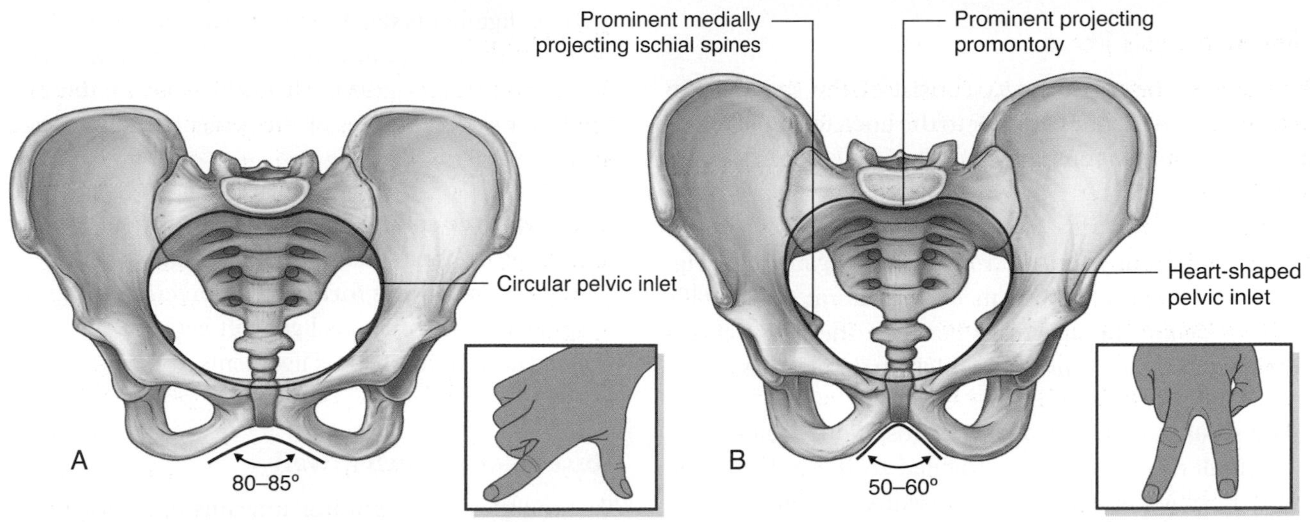


**Fig. 5.27** Structure of the bony pelvis. **A.** In women. **B.** In men. The angle formed by the pubic arch can be approximated by the angle between the thumb and index finger for women and the angle between the index finger and middle finger for men as shown in the insets.

## True pelvis

The true pelvis is cylindrical and has an inlet, a wall, and an outlet. The inlet is open, whereas the pelvic floor closes the outlet and separates the pelvic cavity, above, from the perineum, below.

### Pelvic inlet

The pelvic inlet is the circular opening between the abdominal cavity and the pelvic cavity through which structures traverse between the abdomen and pelvic cavity. It is completely surrounded by bones and joints (Fig. 5.28). The promontory of the sacrum protrudes into the inlet, forming its posterior margin in the midline. On either side of the promontory, the margin is formed by the alae of the sacrum. The margin of the pelvic inlet then crosses the sacro-iliac joint and continues along the linea terminalis (i.e., the arcuate line, the pecten pubis or pectineal line, and the pubic crest) to the pubic symphysis.

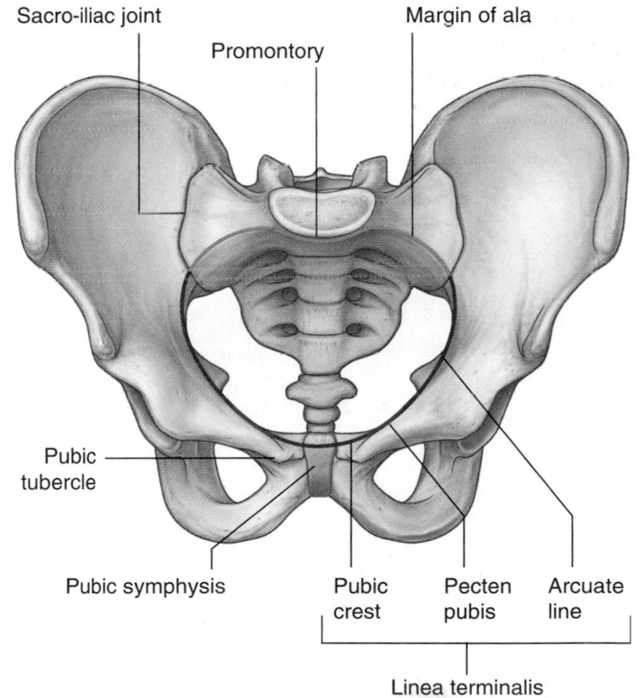


**Fig. 5.28** Pelvic inlet.

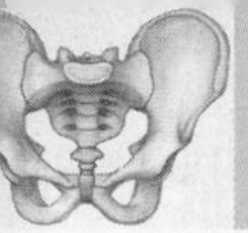



## Pelvic wall

The walls of the pelvic cavity consist of the sacrum, the coccyx, the pelvic bones inferior to the linea terminalis, two ligaments, and two muscles.

### Ligaments of the pelvic wall

The sacrospinous and sacrotuberous ligaments (Fig. 5.29A) are major components of the lateral pelvic walls that help define the apertures between the pelvic cavity and adjacent regions through which structures pass.

- The smaller of the two, the sacrospinous ligament, is triangular, with its apex attached to the ischial spine and its base attached to the related margins of the sacrum and the coccyx.
- The sacrotuberous ligament is also triangular and is superficial to the sacrospinous ligament. Its base has a broad attachment that extends from the posterior superior iliac spine of the pelvic bone, along the dorsal aspect and the lateral margin of the sacrum, and onto the dorsolateral surface of the coccyx. Laterally, the apex of the ligament is attached to the medial margin of the ischial tuberosity.

These ligaments stabilize the sacrum on the pelvic bones by resisting the upward tilting of the inferior aspect of the sacrum (Fig. 5.29B). They also convert the greater and lesser sciatic notches of the pelvic bone into foramina (Fig. 5.29A,B).

- The **greater sciatic foramen** lies superior to the sacrospinous ligament and the ischial spine.
- The **lesser sciatic foramen** lies inferior to the ischial spine and sacrospinous ligament between the sacrospinous and sacrotuberous ligaments.

### Muscles of the pelvic wall

Two muscles, the obturator internus and the piriformis, contribute to the lateral walls of the pelvic cavity. These muscles originate in the pelvic cavity but attach peripherally to the femur.

#### Obturator internus

The obturator internus is a flat, fan-shaped muscle that originates from the deep surface of the obturator membrane and from associated regions of the pelvic bone that surround the obturator foramen (Fig. 5.30 and Table 5.1).

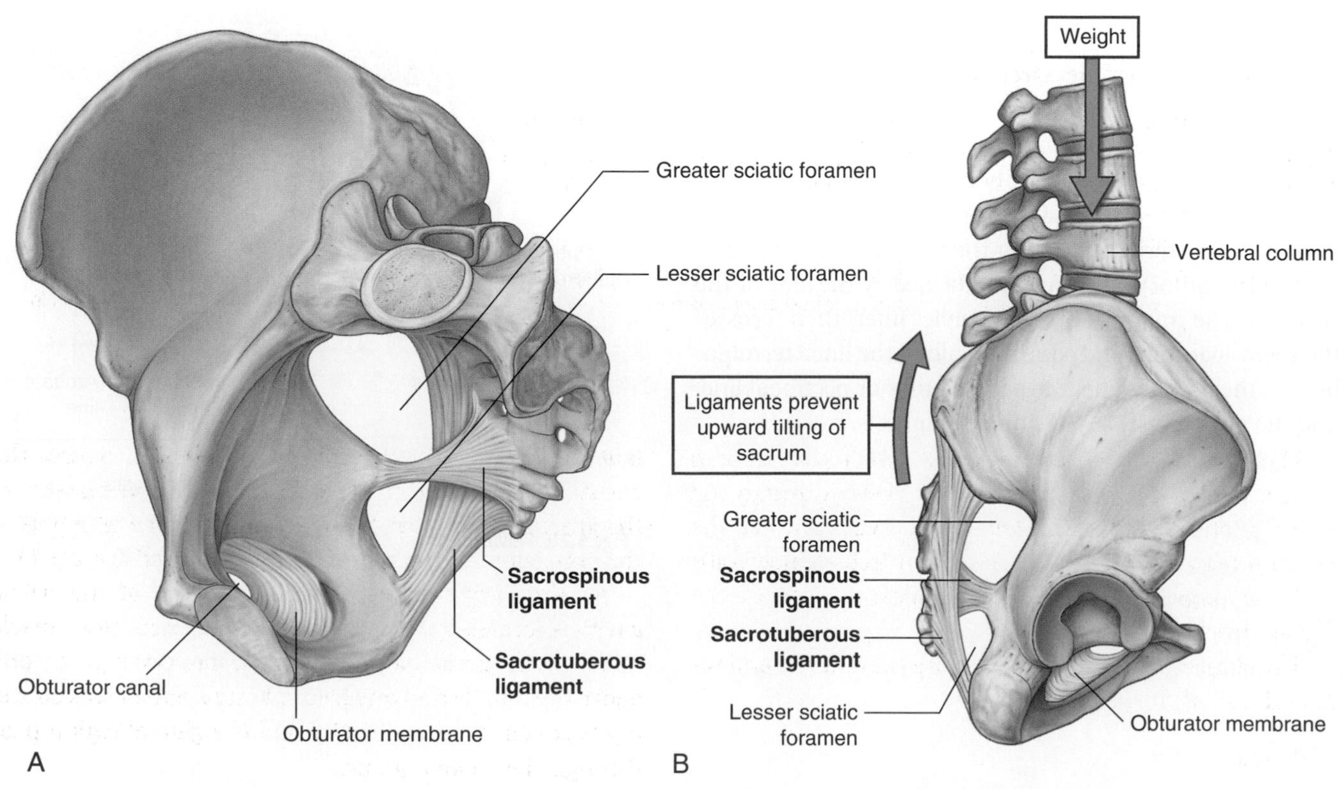


**Fig. 5.29** Sacrospinous and sacrotuberous ligaments. **A.** Medial view of right side of pelvis. **B.** Function of the ligaments.

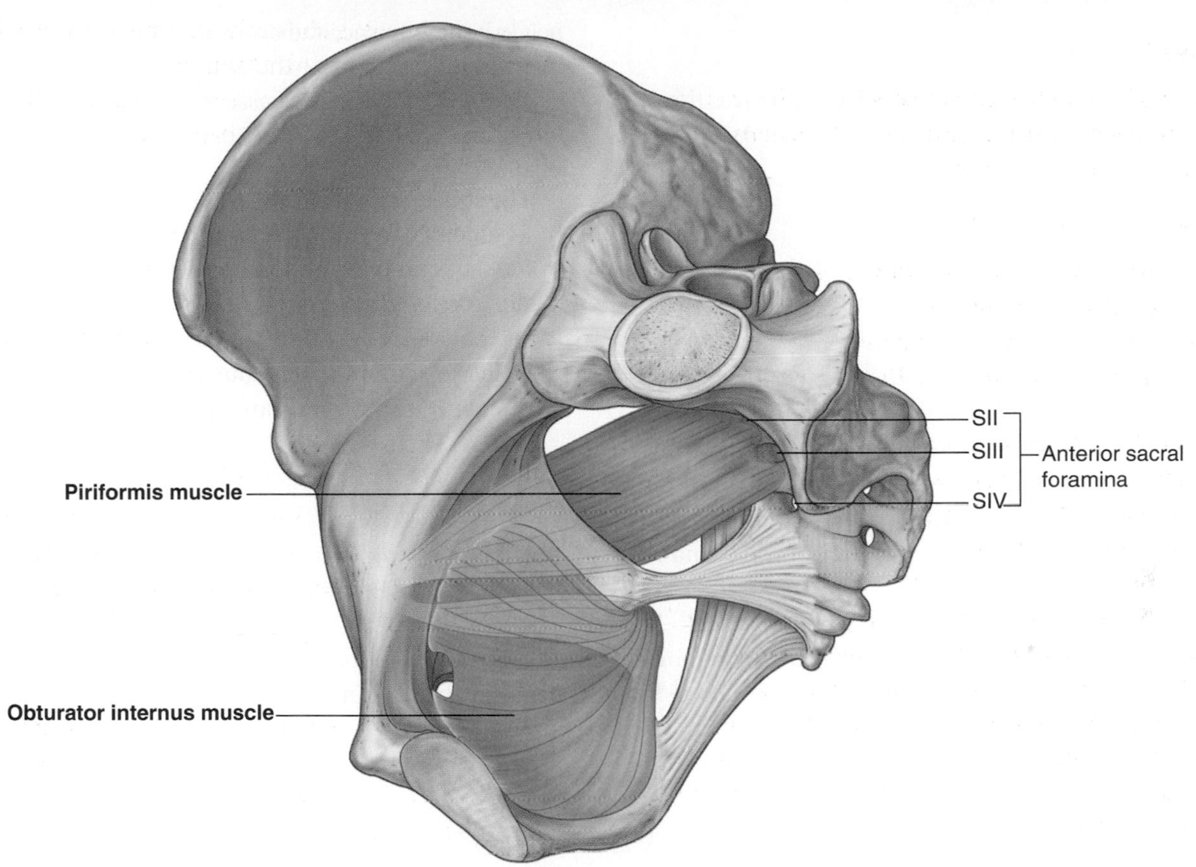


**Fig. 5.30** Obturator internus and piriformis muscles (medial view of right side of pelvis).

**Table 5.1** Muscles of the pelvic walls

| Muscle | Origin | Insertion | Innervation | Function |
|---|---|---|---|---|
| Obturator internus | Anterolateral wall of true pelvis (deep surface of obturator membrane and surrounding bone) | Medial surface of greater trochanter of femur | Nerve to obturator internus **L5, SI** | Lateral rotation of the extended hip joint; abduction of flexed hip |
| Piriformis | Anterior surface of sacrum between anterior sacral foramina | Medial side of superior border of greater trochanter of femur | Branches from **SI**, and **S2** | Lateral rotation of the extended hip joint; abduction of flexed hip |

The muscle fibers of the obturator internus converge to form a tendon that leaves the pelvic cavity through the lesser sciatic foramen, makes a 90° bend around the ischium between the ischial spine and ischial tuberosity, and then passes posterior to the hip joint to insert on the greater trochanter of the femur.

The obturator internus forms a large part of the anterolateral wall of the pelvic cavity.

## Piriformis

The piriformis is triangular and originates in the bridges of bone between the four anterior sacral foramina. It passes laterally through the greater sciatic foramen, crosses the posterosuperior aspect of the hip joint, and inserts on the greater trochanter of the femur above the insertion of the obturator internus muscle (Fig. 5.30 and Table 5.1).

A large part of the posterolateral wall of the pelvic cavity is formed by the piriformis. In addition, this muscle separates the greater sciatic foramen into two regions, one above the muscle and one below. Vessels and nerves coursing between the pelvic cavity and the gluteal region pass through these two regions.

### Apertures in the pelvic wall

Each lateral pelvic wall has three major apertures through which structures pass between the pelvic cavity and other regions:

- the obturator canal,
- the greater sciatic foramen, and
- the lesser sciatic foramen.

#### Obturator canal

At the top of the obturator foramen is the obturator canal, which is bordered by the obturator membrane, the associated obturator muscles, and the superior pubic ramus (Fig. 5.31). The obturator nerve and vessels pass from the pelvic cavity to the thigh through this canal.

#### Greater sciatic foramen

The greater sciatic foramen is a major route of communication between the pelvic cavity and the lower limb (Fig. 5.31). It is formed by the greater sciatic notch in the pelvic bone, the sacrotuberous and the sacrospinous ligaments, and the spine of the ischium.

The piriformis muscle passes through the greater sciatic foramen, dividing it into two parts.

- The superior gluteal nerves and vessels pass through the foramen above the piriformis.
- Passing through the foramen below the piriformis are the inferior gluteal nerves and vessels, the sciatic nerve, the pudendal nerve, the internal pudendal vessels, the posterior femoral cutaneous nerves, and the nerves to the obturator internus and quadratus femoris muscles.

#### Lesser sciatic foramen

The lesser sciatic foramen is formed by the lesser sciatic notch of the pelvic bone, the ischial spine, the sacrospinous ligament, and the sacrotuberous ligament (Fig. 5.31). The tendon of the obturator internus muscle passes through this foramen to enter the gluteal region of the lower limb.

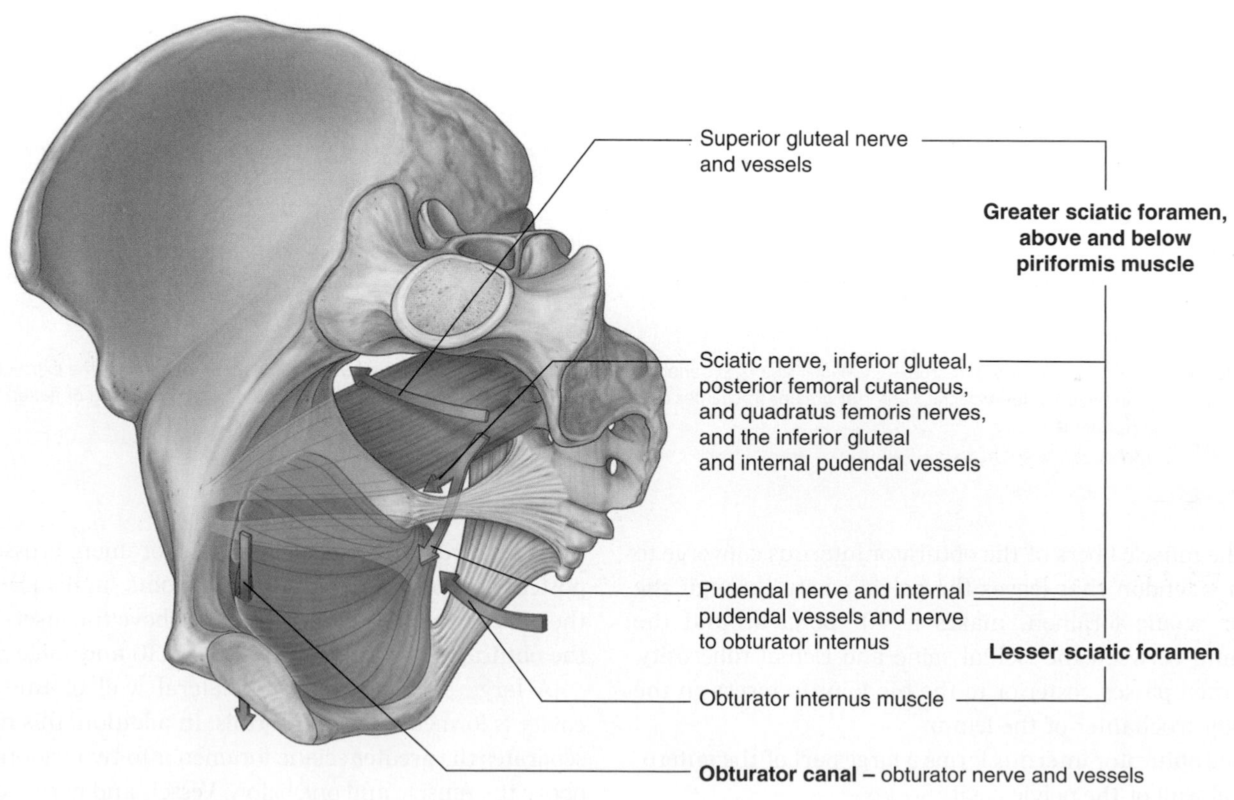


**Fig. 5.31** Apertures in the pelvic wall.

Because the lesser sciatic foramen is positioned below the attachment of the pelvic floor, it acts as a route of communication between the perineum and the gluteal region. The pudendal nerve and internal pudendal vessels pass between the pelvic cavity (above the pelvic floor) and the perineum (below the pelvic floor), by first passing out of the pelvic cavity through the greater sciatic foramen and then looping around the ischial spine and sacrospinous ligament to pass through the lesser sciatic foramen to enter the perineum. The nerve to obturator internus follows a similar course.

### Pelvic outlet

The pelvic outlet is diamond shaped, with the anterior part of the diamond defined predominantly by bone and the posterior part mainly by ligaments (Fig. 5.32). In the midline anteriorly, the boundary of the pelvic outlet is the pubic symphysis. Extending laterally and posteriorly, the boundary on each side is the inferior border of the body of the pubis, the inferior ramus of the pubis, the ramus of the ischium, and the ischial tuberosity. Together, the elements on both sides form the pubic arch.

From the ischial tuberosities, the boundaries continue posteriorly and medially along the sacrotuberous ligament on both sides to the coccyx.

Terminal parts of the urinary and gastrointestinal tracts and the vagina pass through the pelvic outlet.

The area enclosed by the boundaries of the pelvic outlet and below the pelvic floor is the **perineum**.

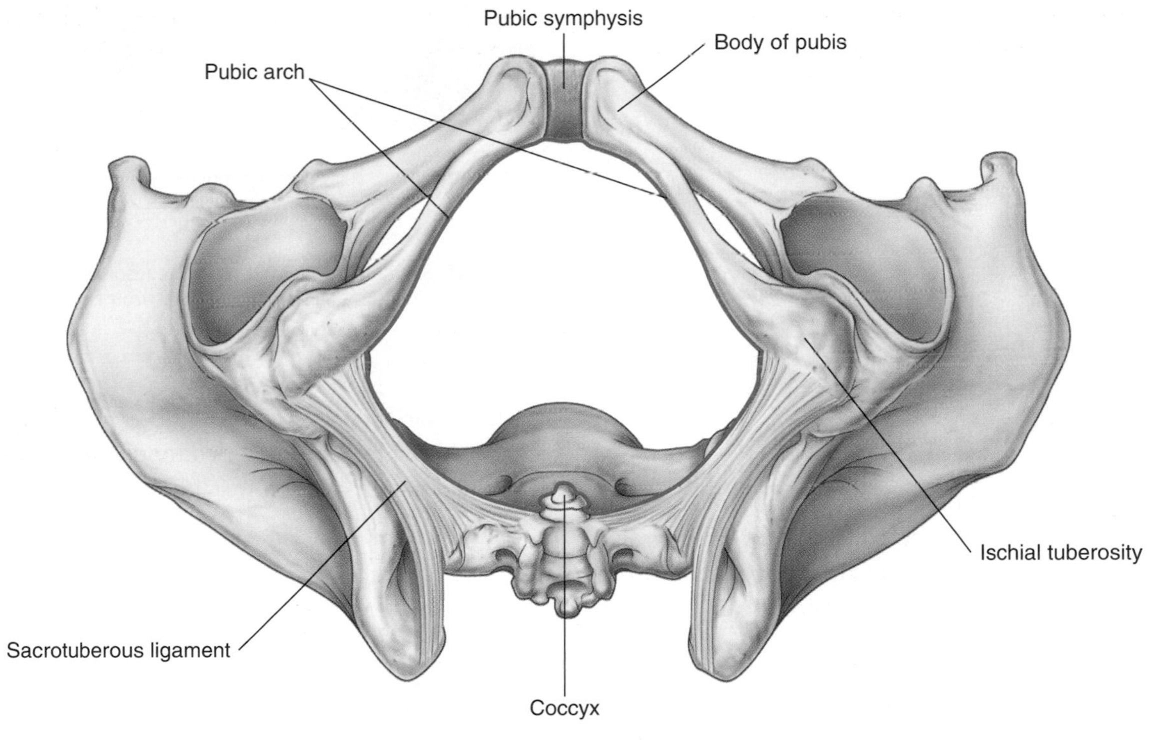


**Fig. 5.32** Pelvic outlet.

## In the clinic

**Pelvic measurements in obstetrics**

Transverse and sagittal measurements of a woman's pelvic inlet and outlet can help in predicting the likelihood of a successful vaginal delivery. These measurements include:

- the sagittal inlet (between the promontory and the top of the pubic symphysis),
- the maximum transverse diameter of the inlet,
- the bispinous outlet (the distance between ischial spines), and
- the sagittal outlet (the distance between the tip of the coccyx and the inferior margin of the pubic symphysis).

These measurements can be obtained using magnetic resonance imaging, which carries no radiation risk for the fetus or mother (Fig. 5.33).

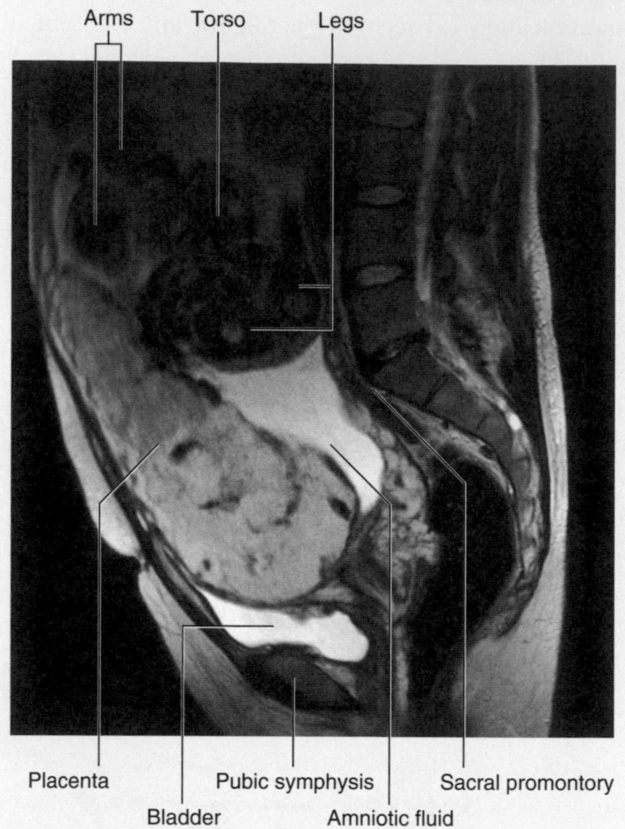


**Fig. 5.33** Sagittal T2-weighted magnetic resonance image of the lower abdomen and pelvis of a pregnant woman.

## Pelvic floor

The pelvic floor is formed by the pelvic diaphragm and, in the anterior midline, the perineal membrane and the muscles in the deep perineal pouch. The pelvic diaphragm is formed by the levator ani and the coccygeus muscles from both sides. The pelvic floor separates the pelvic cavity, above, from the perineum, below.

### The pelvic diaphragm

The pelvic diaphragm is the muscular part of the pelvic floor. Shaped like a bowl or funnel and attached superiorly to the pelvic walls, it consists of the levator ani and the coccygeus muscles (Fig. 5.34 and Table 5.2).

The pelvic diaphragm's circular line of attachment to the cylindrical pelvic wall passes, on each side, between the greater sciatic foramen and the lesser sciatic foramen. Thus:

- the greater sciatic foramen is situated above the level of the pelvic floor and is a route of communication between the pelvic cavity and the gluteal region of the lower limb; and
- the lesser sciatic foramen is situated below the pelvic floor, providing a route of communication between the gluteal region of the lower limb and the perineum.

#### Levator ani

The two levator ani muscles originate from each side of the pelvic wall, course medially and inferiorly, and join together in the midline. The attachment to the pelvic wall follows the circular contour of the wall and includes:

- the posterior aspect of the body of the pubic bone,
- a linear thickening called the **tendinous arch**, in the fascia covering the obturator internus muscle, and
- the spine of the ischium.

At the midline, the muscles blend together posterior to the vagina in women and around the anal aperture in both sexes. Posterior to the anal aperture, the muscles come together as a ligament or raphe called the **anococcygeal ligament** (**anococcygeal body**) and attaches to the coccyx. Anteriorly, the muscles are separated by a U-shaped

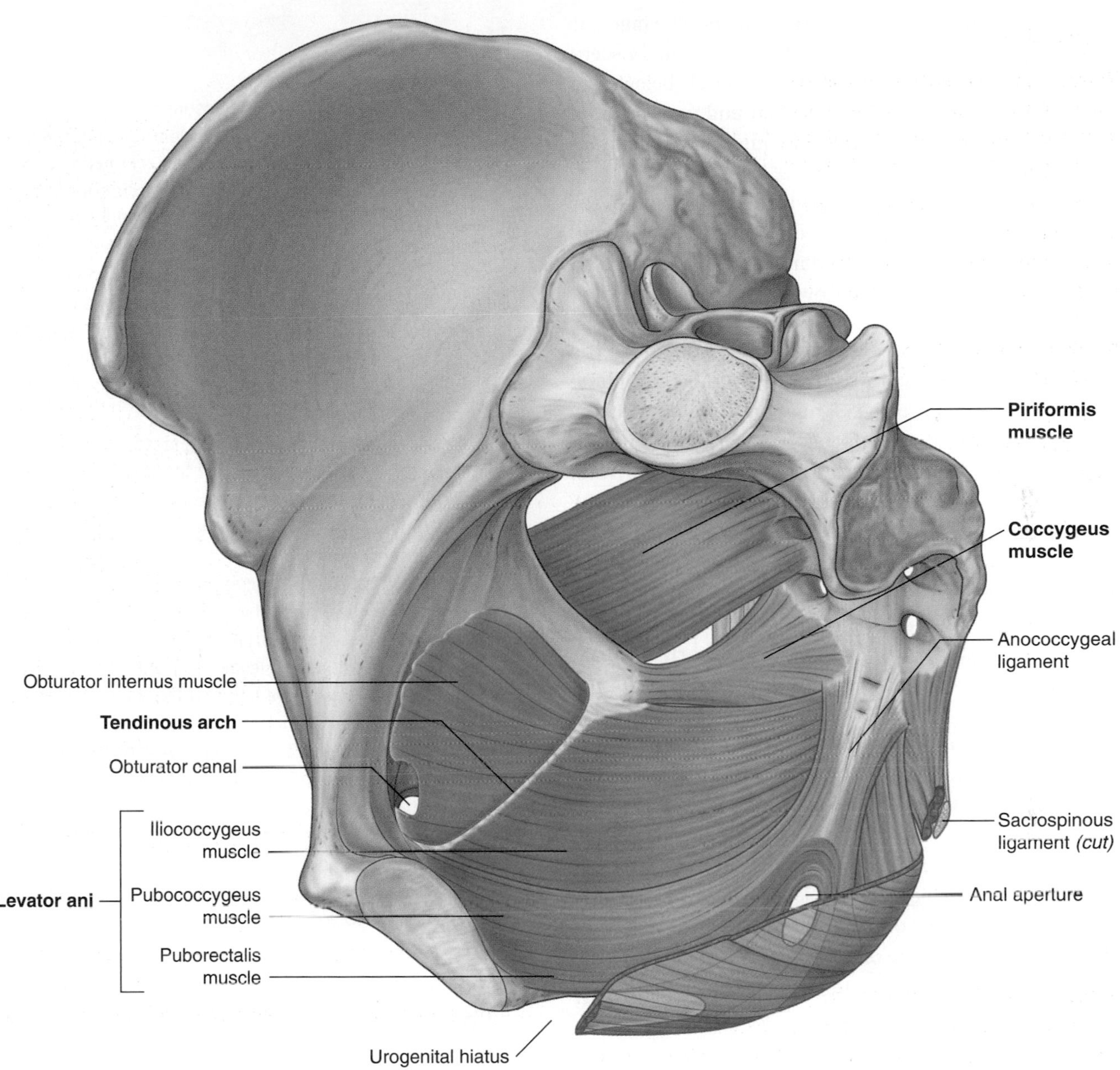


**Fig. 5.34** Pelvic diaphragm.

**Table 5.2** Muscles of the pelvic diaphragm

| Muscle | Origin | Insertion | Innervation | Function |
|---|---|---|---|---|
| Levator ani | In a line around the pelvic wall beginning on the posterior aspect of the pubic bone and extending across the obturator internus muscle as a tendinous arch (thickening of the obturator internus fascia) to the ischial spine | The anterior part is attached to the superior surface of the perineal membrane; the posterior part meets its partner on the other side at the perineal body, around the anal canal, and along the anococcygeal ligament | Branches direct from the anterior ramus of S4, and by the inferior rectal branch of the pudendal nerve (S2 to S4) | Contributes to the formation of the pelvic floor, which supports the pelvic viscera; maintains an angle between the rectum and anal canal; reinforces the external anal sphincter and, in women, functions as a vaginal sphincter |
| Coccygeus | Ischial spine and pelvic surface of the sacrospinous ligament | Lateral margin of coccyx and related border of sacrum | Branches from the anterior rami of S3 and S4 | Contributes to the formation of the pelvic floor, which supports the pelvic viscera; pulls coccyx forward after defecation |

defect or gap termed the **urogenital hiatus**. The margins of this hiatus merge with the walls of the associated viscera and with muscles in the deep perineal pouch below. The hiatus allows the urethra (in both men and women), and the vagina (in women), to pass through the pelvic diaphragm (Fig. 5.34).

The levator ani muscles are divided into at least three collections of muscle fibers, based on site of origin and relationship to viscera in the midline: the pubococcygeus, the puborectalis, and the iliococcygeus muscles.

- The **pubococcygeus** originates from the body of the pubis and courses posteriorly to attach along the midline as far back as the coccyx. This part of the muscle is further subdivided on the basis of association with structures in the midline into the **puboprostaticus (levator prostatae)**, the **pubovaginalis**, and the **puboanalis muscles**.
- A second major collection of muscle fibers, the **puborectalis** portion of the levator ani muscles, originates, in association with the pubococcygeus muscle, from the pubis and passes inferiorly on each side to form a sling around the terminal part of the gastrointestinal tract. This muscular sling maintains an angle or flexure, called the **perineal flexure**, at the anorectal junction. This angle functions as part of the mechanism that keeps the end of the gastrointestinal system closed.
- The final part of the levator ani muscle is the **iliococcygeus**. This part of the muscle originates from the fascia that covers the obturator internus muscle. It joins the same muscle on the other side in the midline to form a ligament or raphe that extends from the anal aperture to the coccyx.

The levator ani muscles help support the pelvic viscera and maintain closure of the rectum and vagina. They are innervated directly by branches from the anterior ramus of S4 and by branches of the pudendal nerve (S2 to S4).

### Coccygeus

The two coccygeus muscles, one on each side, are triangular and overlie the sacrospinous ligaments; together they complete the posterior part of the pelvic diaphragm (Fig. 5.34 and Table 5.2). They are attached, by their apices, to the tips of the ischial spines and, by their bases, to the lateral margins of the coccyx and adjacent margins of the sacrum.

The coccygeus muscles are innervated by branches from the anterior rami of S3 and S4 and participate in supporting the posterior aspect of the pelvic floor.

## In the clinic

### Defecation

At the beginning of defecation, closure of the larynx stabilizes the diaphragm and intraabdominal pressure is increased by contraction of abdominal wall muscles. As defecation proceeds, the puborectalis muscle surrounding the anorectal junction relaxes, which straightens the anorectal angle. Both the internal and the external anal sphincters also relax to allow feces to move through the anal canal. Normally, the puborectal sling maintains an angle of about 90° between the rectum and the anal canal and acts as a "pinch valve" to prevent defecation. When the puborectalis muscle relaxes, the anorectal angle increases to about 130° to 140°.

The fatty tissue of the ischio-anal fossa allows for changes in the position and size of the anal canal and anus during defecation. During evacuation, the anorectal junction moves down and back and the pelvic floor usually descends slightly.

During defecation, the circular muscles of the rectal wall undergo a wave of contraction to push feces toward the anus. As feces emerge from the anus, the longitudinal muscles of the rectum and levator ani bring the anal canal back up, the feces are expelled, and the anus and rectum return to their normal positions.

A magnetic resonance defecating proctogram is a fairly new imaging technique that allows assessment of different phases of defecation, including rectal function and behavior of the pelvic floor musculature during this process. It is useful in detecting pelvic organ abnormal descent/prolapse during dynamic scanning and potential formation of cystocele or rectocele (Fig. 5.35).

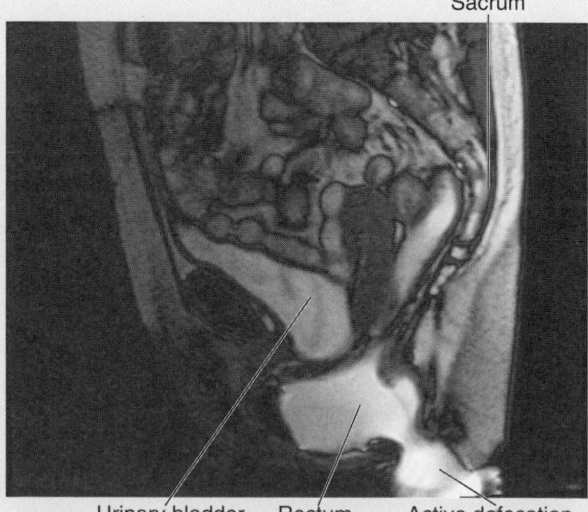


**Fig. 5.35** MRI defecating proctogram in sagittal plane showing active defecation.

### The perineal membrane and deep perineal pouch

The **perineal membrane** is a thick fascial, triangular structure attached to the bony framework of the pubic arch (Fig. 5.36A). It is oriented in the horizontal plane and has a free posterior margin. Anteriorly, there is a small gap (blue arrow in Fig. 5.36A) between the membrane and the **inferior pubic ligament** (a ligament associated with the pubic symphysis).

The perineal membrane is related above to a thin space called the **deep perineal pouch** (**deep perineal space**) (Fig. 5.36B), which contains a layer of skeletal muscle and various neurovascular elements.

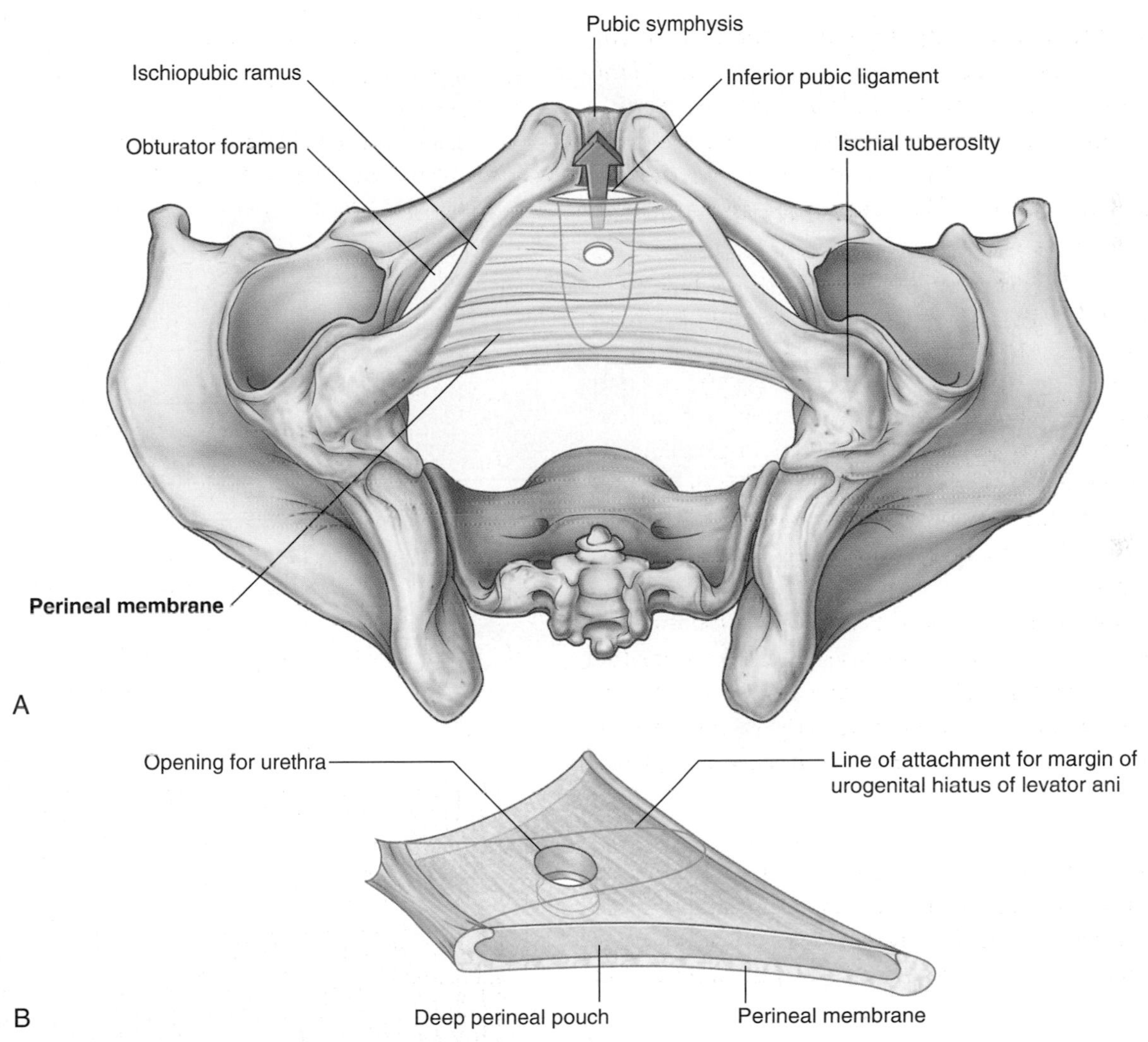


**Fig. 5.36** Perineal membrane and deep perineal pouch. **A.** Inferior view. **B.** Superolateral view.

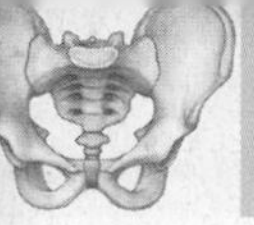





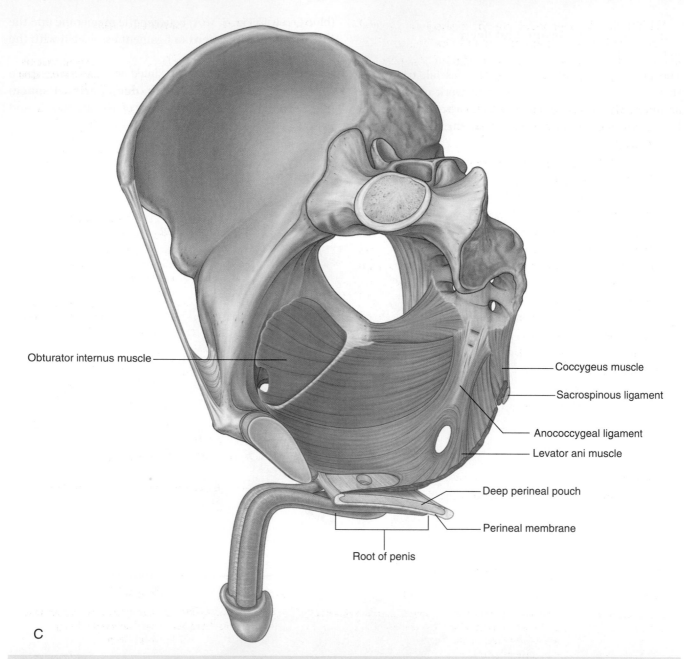

**Fig. 5.36, cont'd** Perineal membrane and deep perineal pouch. **C.** Medial view.

The deep perineal pouch is open above and is not separated from more superior structures by a distinct layer of fascia. The parts of the perineal membrane and structures in the deep perineal pouch, enclosed by the urogenital hiatus above, therefore contribute to the pelvic floor and support elements of the urogenital system in the pelvic cavity, even though the perineal membrane and deep perineal pouch are usually considered parts of the perineum.

The perineal membrane and adjacent pubic arch provide attachment for the roots of the external genitalia and the muscles associated with them (Fig. 5.36C).

The urethra penetrates vertically through a circular hiatus in the perineal membrane as it passes from the pelvic cavity, above, to the perineum, below. In women, the vagina also passes through a hiatus in the perineal membrane just posterior to the urethral hiatus.

Within the deep perineal pouch, a sheet of skeletal muscle functions as a sphincter, mainly for the urethra, and as a stabilizer of the posterior edge of the perineal membrane (Fig. 5.37 and Table 5.3).

- Anteriorly, a group of muscle fibers surround the urethra and collectively form the **external urethral sphincter**.
- Two additional groups of muscle fibers are associated with the urethra and vagina in women. One group forms the **sphincter urethrovaginalis**, which surrounds the urethra and vagina as a unit. The second group forms the **compressor urethrae**, on each side, which originate from the ischiopubic rami and meet anterior to the urethra. Together with the external urethral sphincter, the sphincter urethrovaginalis and compressor urethrae facilitate closing of the urethra.
- In both men and women, a **deep transverse perineal muscle** on each side parallels the free margin of the perineal membrane and joins with its partner at the midline. These muscles are thought to stabilize the position of the perineal body, which is a midline structure along the posterior edge of the perineal membrane.

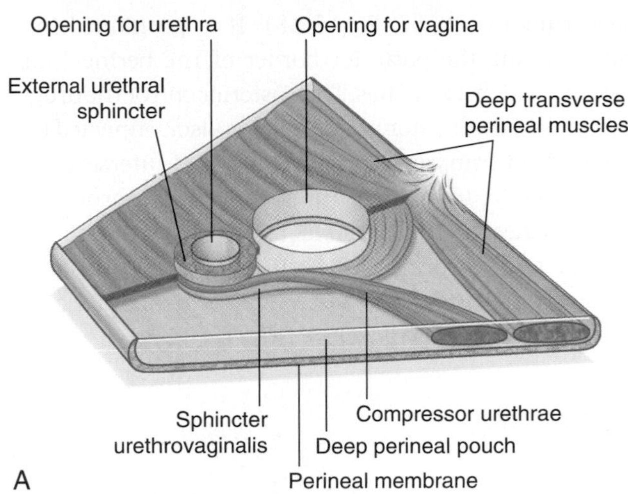


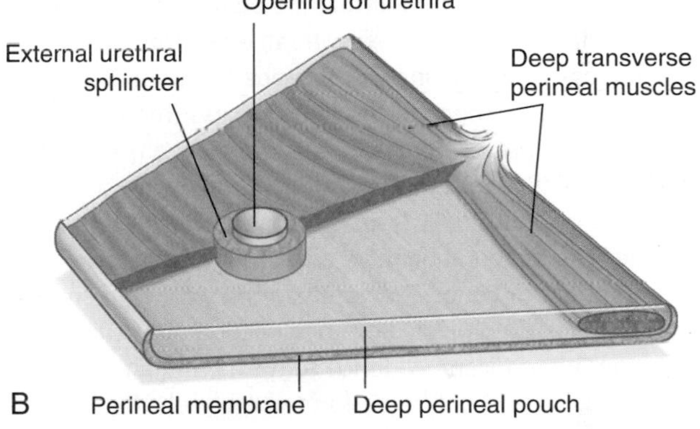


**Fig. 5.37** Muscles in the deep perineal pouch. **A.** In women. **B.** In men.

## Perineal body

The perineal body is an ill-defined but important connective tissue structure into which muscles of the pelvic floor and

**Table 5.3** Muscles within the deep perineal pouch

| Muscle | Origin | Insertion | Innervation | Function |
|---|---|---|---|---|
| External urethral sphincter | From the inferior ramus of the pubis on each side and adjacent walls of the deep perineal pouch | Surrounds membranous part of urethra | Perineal branches of the pudendal nerve (S2 to S4) | Compresses the membranous urethra; relaxes during micturition |
| Deep transverse perineal | Medial aspect of ischial ramus | Perineal body | Perineal branches of the pudendal nerve (S2 to S4) | Stabilizes the position of the perineal body |
| Compressor urethrae (in women only) | Ischiopubic ramus on each side | Blends with partner on other side anterior to the urethra | Perineal branches of the pudendal nerve (S2 to S4) | Functions as an accessory sphincter of the urethra |
| Sphincter urethrovaginalis (in women only) | Perineal body | Passes forward lateral to the vagina to blend with partner on other side anterior to the urethra | Perineal branches of the pudendal nerve (S2 to S4) | Functions as an accessory sphincter of the urethra (also may facilitate closing the vagina) |

the perineum attach (Fig. 5.38). It is positioned in the midline along the posterior border of the perineal membrane, to which it attaches. The posterior end of the urogenital hiatus in the levator ani muscles is also connected to it.

The deep transverse perineal muscles intersect at the perineal body; in women, the sphincter urethrovaginalis also attaches to the perineal body. Other muscles that connect to the perineal body include the external anal sphincter, the superficial transverse perineal muscles, and the bulbospongiosus muscles of the perineum.

### In the clinic

**Episiotomy**

During childbirth the perineal body may be stretched and torn. Traditionally it was felt that if a perineal tear is likely, the obstetrician may proceed with an episiotomy. This is a procedure in which an incision is made in the perineal body to allow the head of the fetus to pass through the vagina. There are two types of episiotomies: a median episiotomy cuts through the perineal body, while a mediolateral episiotomy is an incision 45° from the midline. The maternal benefits of this procedure have been thought to be less traumatic to the perineum and to result in decreased pelvic floor dysfunction after childbirth. However, more recent evidence suggests that an episiotomy should not be performed routinely. Review of data has failed to show a decrease in pelvic floor damage with routine use of episiotomies.

## Viscera

The pelvic viscera include parts of the gastrointestinal system, the urinary system, and the reproductive system.

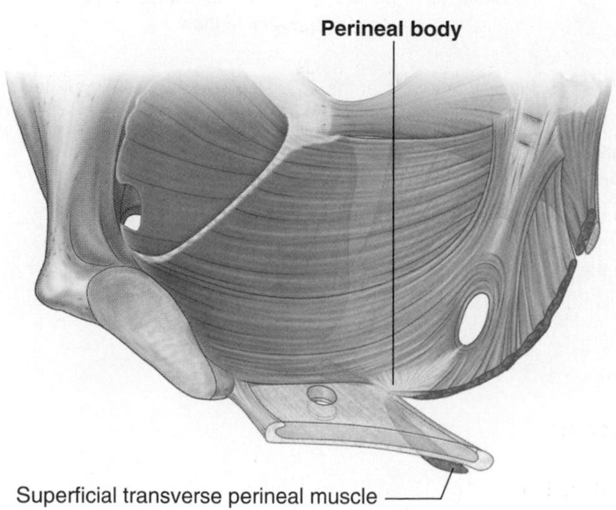


**Fig. 5.38** Perineal body.

The viscera are arranged in the midline, from front to back; the neurovascular supply is through branches that pass medially from vessels and nerves associated with the pelvic walls.

### Gastrointestinal system

Pelvic parts of the gastrointestinal system consist mainly of the rectum and the anal canal, although the terminal part of the sigmoid colon is also in the pelvic cavity (Fig. 5.39).

#### Rectum

The **rectum** is continuous:

- above, with the sigmoid colon at about the level of vertebra SIII, and
- below, with the anal canal as this structure penetrates the pelvic floor and passes through the perineum to end as the anus.

The rectum, the most posterior element of the pelvic viscera, is immediately anterior to and follows the concave contour of the sacrum.

The anorectal junction is pulled forward (perineal flexure) by the action of the puborectalis part of the levator ani muscle, so the anal canal moves in a posterior direction as it passes inferiorly through the pelvic floor.

In addition to conforming to the general curvature of the sacrum in the anteroposterior plane, the rectum has three lateral curvatures; the upper and lower curvatures to the right and the middle curvature to the left. The lower part of the rectum is expanded to form the **rectal ampulla**. Finally, unlike the colon, the rectum lacks distinct taeniae coli muscles, omental appendices, and sacculations (haustra of the colon).

#### Anal canal

The **anal canal** begins at the terminal end of the rectal ampulla where it narrows at the pelvic floor. It terminates as the anus after passing through the perineum. As it passes through the pelvic floor, the anal canal is surrounded along its entire length by the internal and external anal sphincters, which normally keep it closed.

The lining of the anal canal bears a number of characteristic structural features that reflect the approximate position of the anococcygeal membrane in the fetus (which closes the terminal end of the developing gastrointestinal system in the fetus) and the transition from gastrointestinal mucosa to skin in the adult (Fig. 5.39B).

- The upper part of the anal canal is lined by mucosa similar to that lining the rectum and is distinguished by a number of longitudinally oriented folds known as

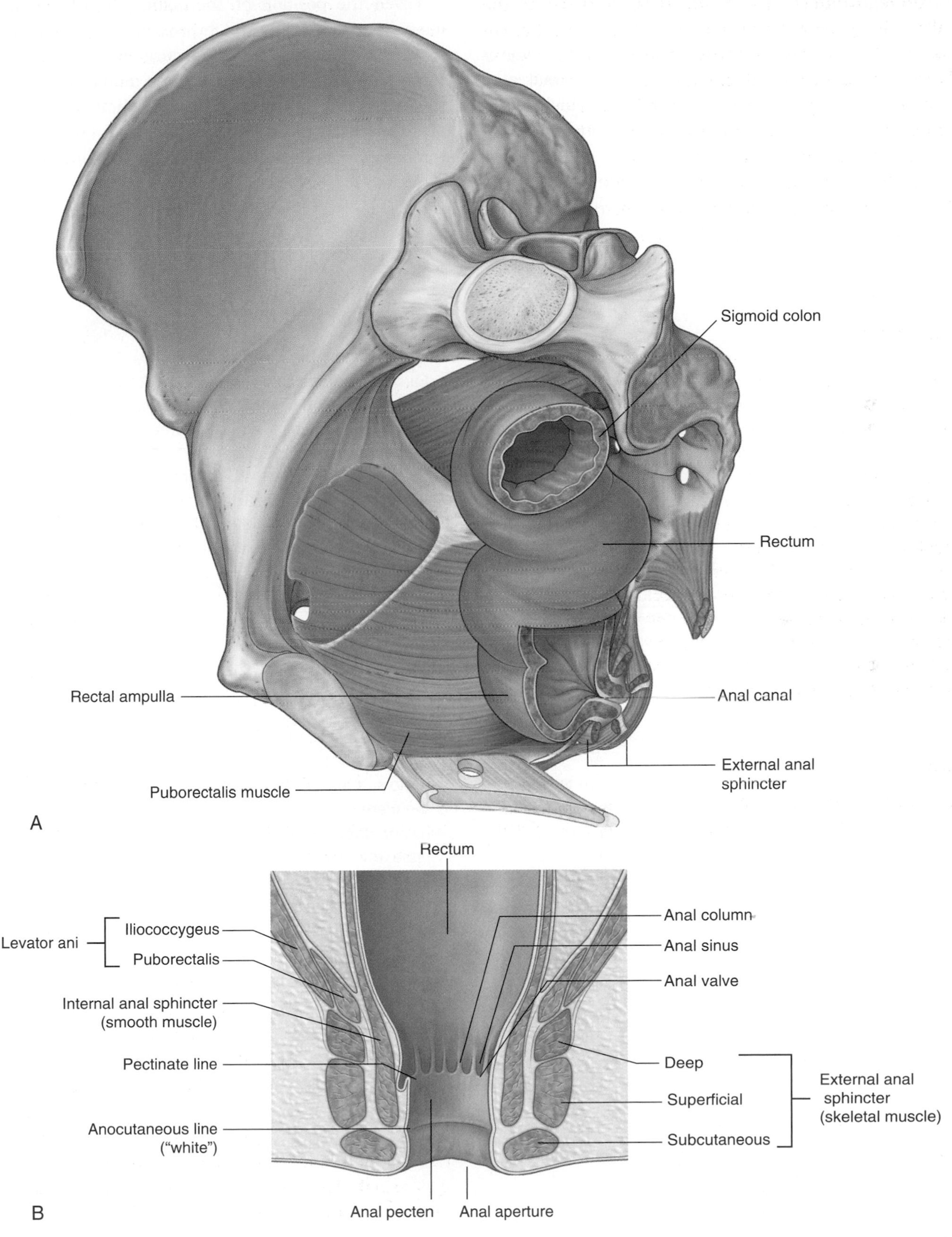


**Fig. 5.39** Rectum and anal canal. **A.** Left pelvic bone removed. **B.** Longitudinal section.

**anal columns**, which are united inferiorly by crescentic folds termed **anal valves**. Superior to each valve is a depression termed an **anal sinus**. The anal valves together form a circle around the anal canal at a location known as the **pectinate line**, which marks the approximate position of the anal membrane in the fetus.

- Inferior to the pectinate line is a transition zone known as the **anal pecten**, which is lined by nonkeratinized stratified squamous epithelium. The anal pecten ends inferiorly at the **anocutaneous line** ("white line"), or where the lining of the anal canal becomes true skin.

### In the clinic

**Digital rectal examination**

A digital rectal examination (DRE) is performed by placing the gloved and lubricated index finger into the rectum through the anus. The anal mucosa can be palpated for abnormal masses, and in women, the posterior wall of the vagina and the cervix can be palpated. In men, the prostate can be evaluated for any extraneous nodules or masses.

In many instances the digital rectal examination may be followed by proctoscopy or colonoscopy. An ultrasound probe may be placed into the rectum to assess the gynecological structures in females and the prostate in the male before performing a prostatic biopsy.

A digital rectal examination also allows detection of fresh or altered blood in the rectum in patients with acute gastrointestinal bleeding or chronic anemia.

Given the position of the colon and rectum in the abdominopelvic cavity and its proximity to other organs, it is extremely important to accurately stage colorectal tumors: a tumor in the pelvis, for example, could invade the uterus or bladder. Assessing whether spread has occurred may involve ultrasound scanning, computed tomography, and magnetic resonance imaging.

### In the clinic

**Carcinoma of the colon and rectum**

Carcinoma of the colon and rectum (colorectum) is a common and often lethal disease. Recent advances in surgery, radiotherapy, and chemotherapy have only slightly improved 5-year survival rates.

The biological behavior of tumors of the colon and rectum is relatively predictable. Most of the tumors develop from benign polyps, some of which undergo malignant change. The overall prognosis is related to:

- the degree of tumor penetration through the bowel wall,
- the presence or absence of lymphatic dissemination, and
- the presence or absence of systemic metastases.

Given the position of the colon and rectum in the abdominopelvic cavity and its proximity to other organs, it is extremely important to accurately stage colorectal tumors; a tumor in the pelvis, for example, could invade the uterus or bladder. Assessing whether or not spread has occurred usually involves computed tomography (assessment for distal metastases) and magnetic resonance imaging (local staging). Endoscopic ultrasound (EUS) is also used in some instances for local staging of rectal cancer.

## Urinary system

The pelvic parts of the urinary system consist of the terminal parts of the ureters, the bladder, and the proximal part of the urethra (Fig. 5.40).

### Ureters

The ureters enter the pelvic cavity from the abdomen by passing through the pelvic inlet. On each side, the ureter crosses the pelvic inlet and enters the pelvic cavity in the area anterior to the bifurcation of the common iliac artery. From this point, it continues along the pelvic wall and floor to join the base of the bladder.

In the pelvis, the ureter is crossed by:

- the ductus deferens in men, and
- the uterine artery in women.

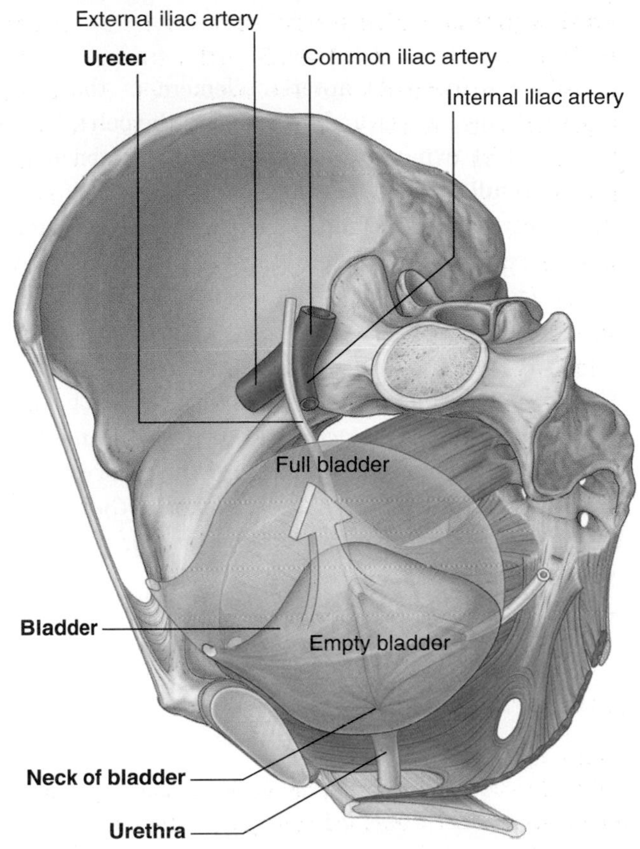


**Fig. 5.40** Pelvic parts of the urinary system.

#### In the clinic

**Iatrogenic injury of the ureters**

Ureters can be injured during various surgeries within the abdomen and pelvis as they lie close to the dissection planes. The most common surgeries that can result in ureteric injury are total abdominal hysterectomy and bilateral salpingo-oophorectomy (removal of the uterus, fallopian tubes and ovaries), laparoscopic vaginal hysterectomy, laparoscopic anterior resection of the rectum, and open left hemicolectomy. At increased risk of ureteric injury are patients with a bulky tumor (uterine, colonic, rectal) and those with a history of previous operations or pelvic irradiation, all of which make dissection of tissues more difficult. During surgery, the ureter can be crushed, cut open, devascularized, or avulsed. It can also be injured during cryoablation or electric cauterization to control intraoperative bleeding. Ureters can also undergo trauma during the course of ureteroscopy, a procedure where a small endoscope is introduced through the urethra and urinary bladder into one of the ureters to treat stones or tumors of the ureter (usually due to a tear or electrocauterization).

Ureteric injury leads to high morbidity due to infection and in most severe cases to renal impairment. The prognosis is improved when the diagnosis is made intraoperatively and the ureter is repaired immediately. Delayed diagnosis leads to urine leakage and contamination of the abdominal and pelvic cavity, development of sepsis, and in the case of injury near the vagina, a uretero-vaginal fistula can develop. When the diagnosis is made postoperatively, sometimes diversion of urine flow is required and percutaneous nephrostomy is performed.

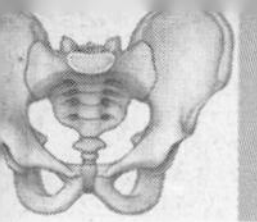



### Bladder

The bladder is the most anterior element of the pelvic viscera. Although it is entirely situated in the pelvic cavity when empty, it expands superiorly into the abdominal cavity when full (Fig. 5.40).

The empty bladder is shaped like a three-sided pyramid that has tipped over to lie on one of its margins (Fig. 5.41A). It has an apex, a base, a superior surface, and two inferolateral surfaces.

- The **apex** of the bladder is directed toward the top of the pubic symphysis; a structure known as the **median umbilical ligament** (a remnant of the embryological urachus that contributes to the formation of the bladder) continues from it superiorly up the anterior abdominal wall to the umbilicus.
- The **base** of the bladder is shaped like an inverted triangle and faces posteroinferiorly. The two ureters enter the bladder at each of the upper corners of the base, and the urethra drains inferiorly from the lower corner of the base. Inside, the mucosal lining on the base of the bladder is smooth and firmly attached to the underlying smooth muscle coat of the wall—unlike elsewhere in the bladder where the mucosa is folded and loosely attached to the wall. The smooth triangular area between the openings of the ureters and urethra on the inside of the bladder is known as the **trigone** (Fig. 5.41B).
- The **inferolateral surfaces** of the bladder are cradled between the levator ani muscles of the pelvic diaphragm and the adjacent obturator internus muscles above the attachment of the pelvic diaphragm. The superior surface is slightly domed when the bladder is empty; it balloons upward as the bladder fills.

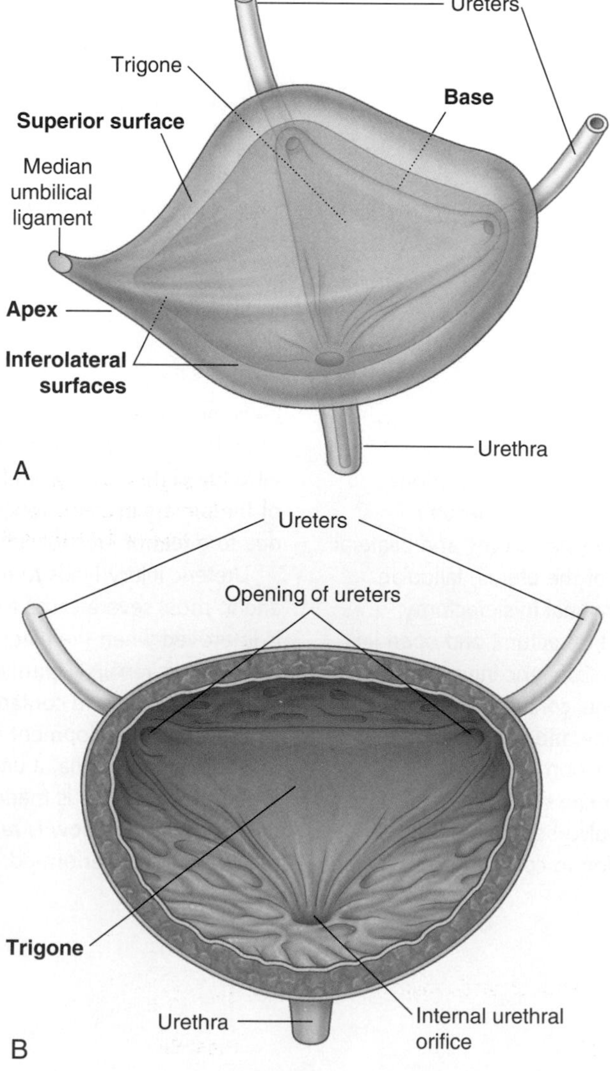


**Fig. 5.41** Bladder. **A.** Superolateral view. **B.** The trigone. Anterior view with the anterior part of the bladder cut away.

### Neck of bladder

The neck of the bladder surrounds the origin of the urethra at the point where the two inferolateral surfaces and the base intersect.

The neck is the most inferior part of the bladder and also the most "fixed" part. It is anchored into position by a pair of tough fibromuscular bands, which connect the neck and pelvic part of the urethra to the posteroinferior aspect of each pubic bone.

- In women, these fibromuscular bands are termed **pubovesical ligaments** (Fig. 5.42A). Together with the perineal membrane and associated muscles, the levator ani muscles, and the pubic bones, these ligaments help support the bladder.
- In men, the paired fibromuscular bands are known as **puboprostatic ligaments** because they blend with the fibrous capsule of the prostate, which surrounds the neck of the bladder and adjacent part of the urethra (Fig. 5.42B).

Although the bladder is considered to be pelvic in the adult, it has a higher position in children. At birth, the bladder is almost entirely abdominal; the urethra begins approximately at the upper margin of the pubic symphysis. With age, the bladder descends until after puberty when it assumes the adult position.

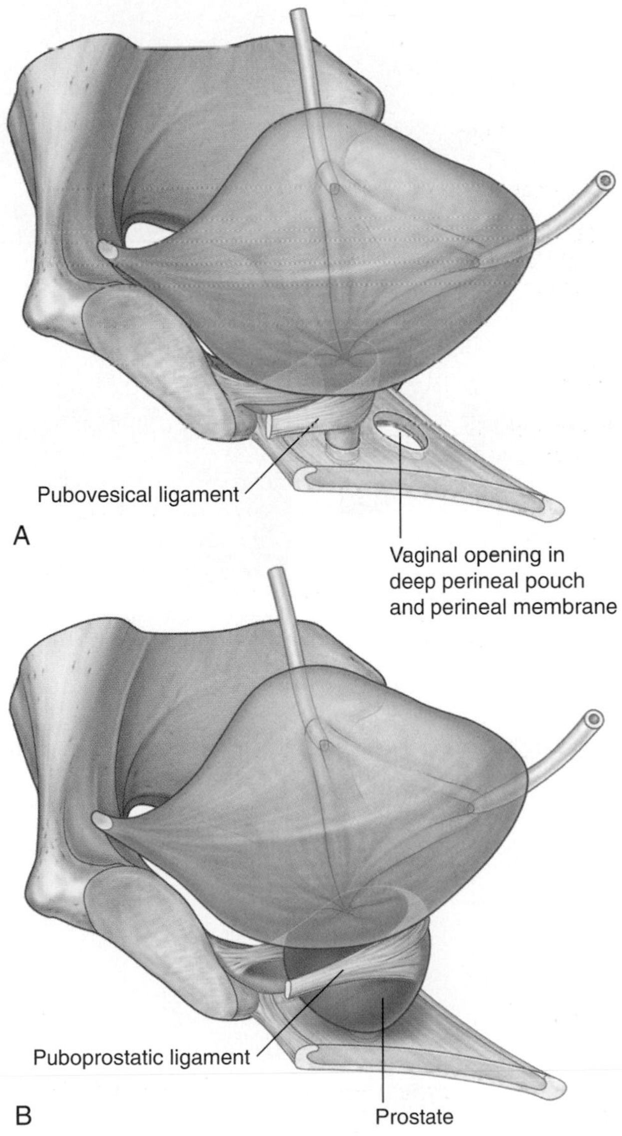


**Fig. 5.42** Ligaments that anchor the neck of the bladder and pelvic part of the urethra to the pelvic bones. **A.** In women. **B.** In men.

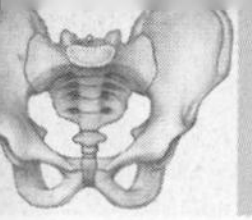

## In the clinic

### Bladder stones

In some patients, small calculi (stones) form in the kidneys. These may pass down the ureter, causing ureteric obstruction, and into the bladder (Fig. 5.43), where insoluble salts further precipitate on these small calculi to form larger calculi. Often, these patients develop (or may already have) problems with bladder emptying, which leaves residual urine in the bladder. This urine may become infected, which alters the pH of the urine, permitting further precipitation of insoluble salts.

If small enough, the stones may be removed via a transurethral route using specialized instruments. If the stones are too big, it may be necessary to make a suprapubic incision and enter the bladder retroperitoneally to remove them.

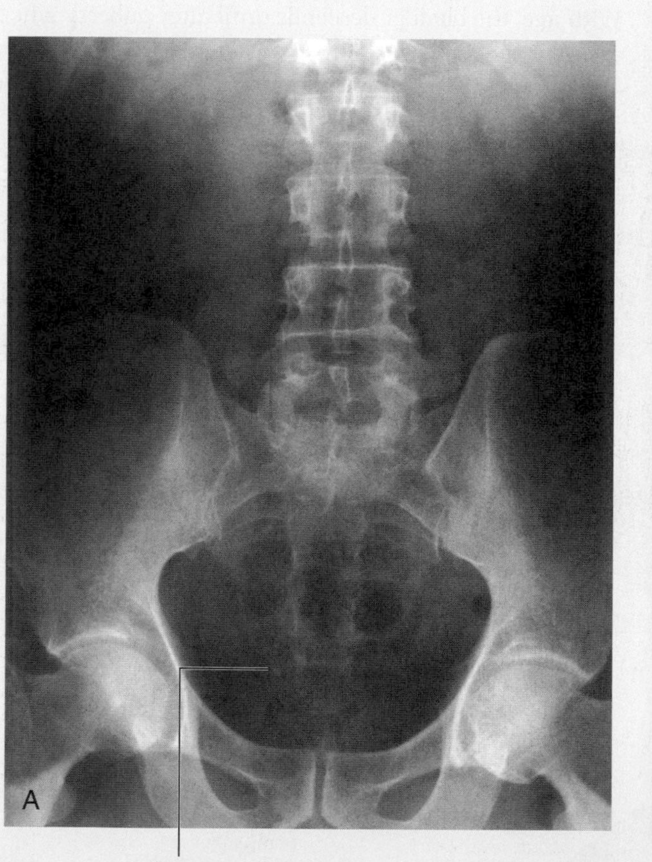


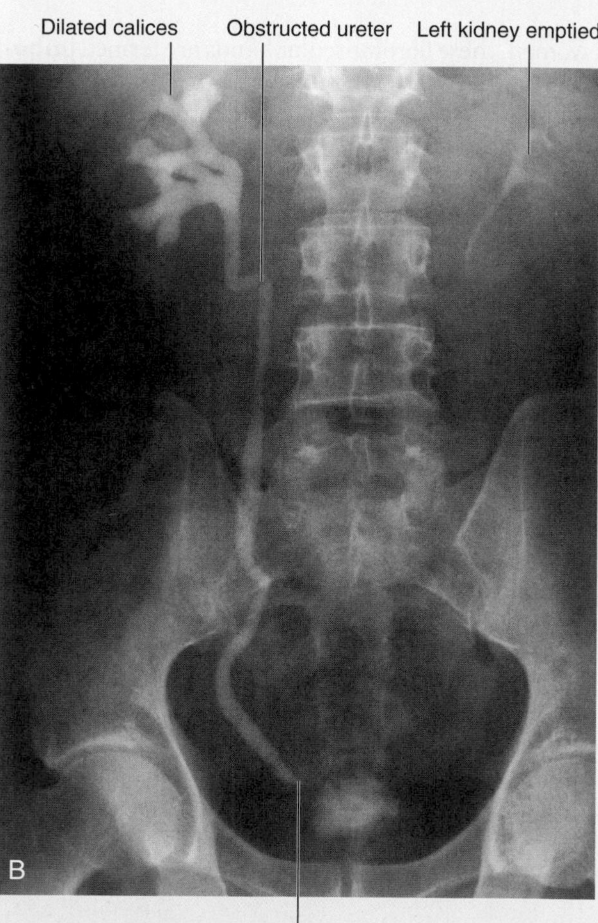


**Fig. 5.43** Intravenous urogram demonstrating a stone in the lower portion of the ureter. **A.** Control radiograph. **B.** Intravenous urogram, postmicturition.

## In the clinic

### Suprapubic catheterization

In certain instances it is necessary to catheterize the bladder through the anterior abdominal wall. For example, when the prostate is markedly enlarged and it is impossible to pass a urethral catheter, a suprapubic catheter may be placed.

The bladder is a retroperitoneal structure and when full lies adjacent to the anterior abdominal wall. Ultrasound visualization of the bladder may be useful in assessing the size of this structure and, importantly, differentiating this structure from other potential abdominal masses.

The procedure of suprapubic catheterization is straightforward and involves the passage of a small catheter on a needle in the midline approximately 2 cm above the pubic symphysis. The catheter passes easily into the bladder without compromise of other structures and permits free drainage.

## In the clinic

### Bladder cancer

Bladder cancer (Fig. 5.44) is the most common tumor of the urinary tract and is usually a disease of the sixth and seventh decades, although there is an increasing trend for younger patients to develop this disease.

Approximately one-third of bladder tumors are multifocal; fortunately, two-thirds are superficial tumors and amenable to local treatment.

Bladder tumors may spread through the bladder wall and invade local structures, including the rectum, uterus (in women), and lateral walls of the pelvic cavity. Prostatic involvement is not uncommon in male patients. The disease spreads via the internal iliac lymph nodes. Spread to distant metastatic sites rarely includes the lung.

Large bladder tumors may produce complications, including invasion and obstruction of the ureters. Ureteric obstruction can then obstruct the kidneys and induce kidney failure. Moreover, bladder tumors can invade other structures of the pelvic cavity.

Treatment for early-stage tumors includes local resection with preservation of the bladder. Diffuse tumors may be treated with local chemotherapy; more extensive tumors may require radical surgical removal of the bladder (cystectomy) and, in men, the prostate (prostatectomy). Bladder reconstruction (formation of so-called neobladder) is performed in patients after cystectomy using part of a bowel, most commonly the ileum.

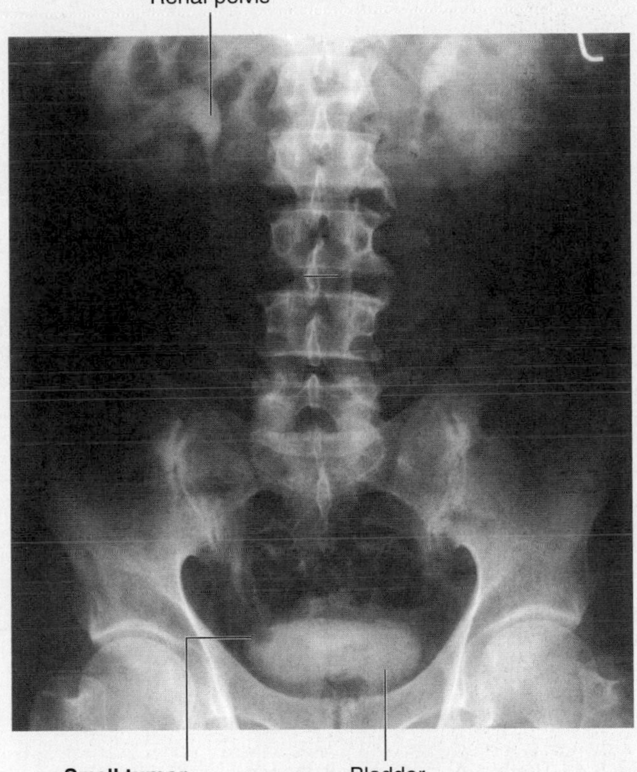


**Fig. 5.44** Intravenous urogram demonstrating a small tumor in the wall of the bladder.

## Urethra

The urethra begins at the base of the bladder and ends with an external opening in the perineum. The paths taken by the urethra differ significantly in women and men.

### In women

In women, the urethra is short, being about 4 cm long. It travels a slightly curved course as it passes inferiorly through the pelvic floor into the perineum, where it passes through the deep perineal pouch and perineal membrane before opening in the vestibule that lies between the labia minora (Fig. 5.45A).

The urethral opening is anterior to the vaginal opening in the vestibule. The inferior aspect of the urethra is bound to the anterior surface of the vagina. Two small para-urethral mucous glands (**Skene's glands**) are associated with the lower end of the urethra. Each drains via a duct that opens onto the lateral margin of the external urethral orifice.

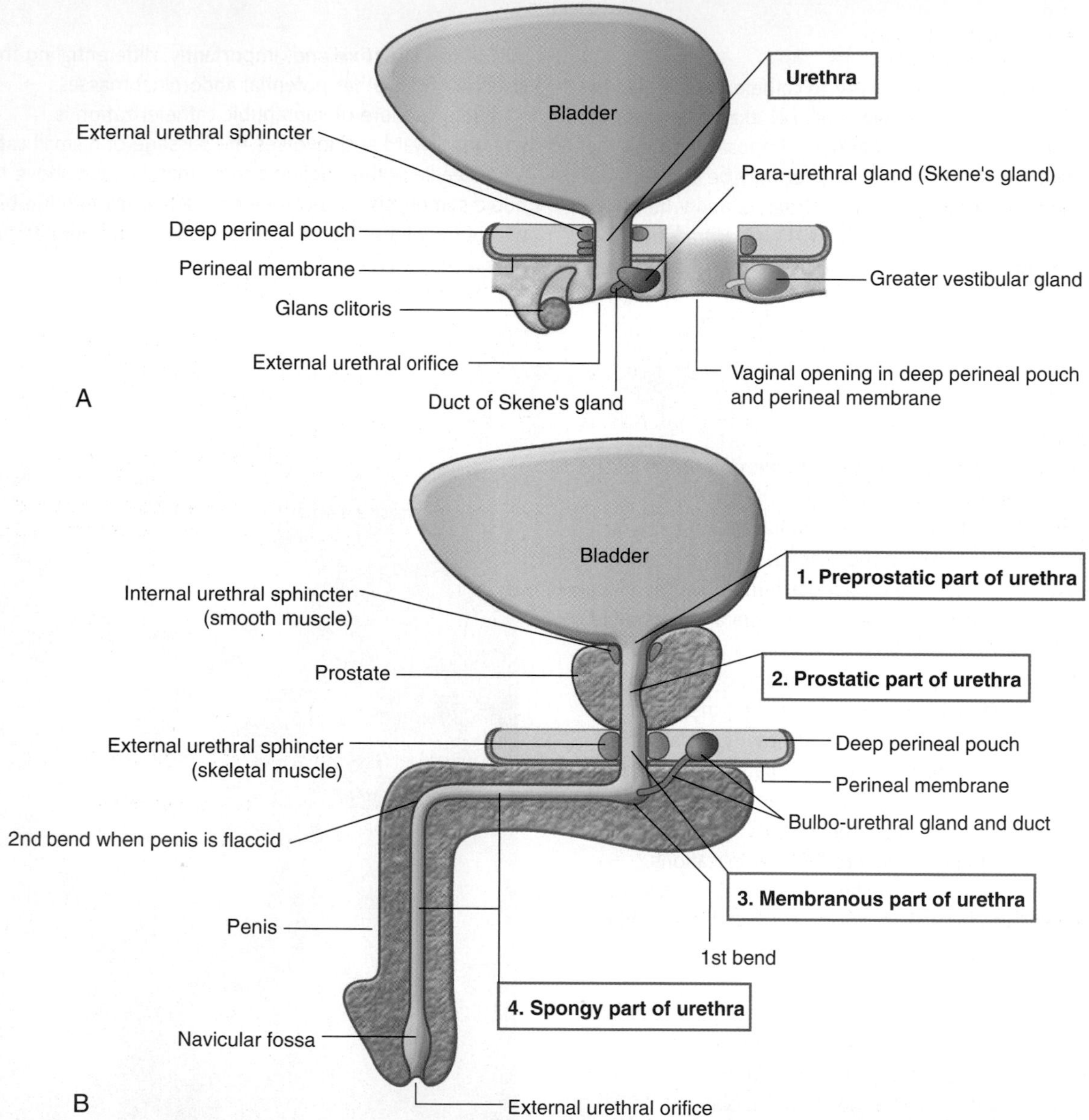

**Fig. 5.45** Urethra. **A.** In women. **B.** In men.

### In men

In men, the urethra is long, about 20 cm, and bends twice along its course (Fig. 5.45B). Beginning at the base of the bladder and passing inferiorly through the prostate, it passes through the deep perineal pouch and perineal membrane and immediately enters the root of the penis. As the urethra exits the deep perineal pouch, it bends forward to course anteriorly in the root of the penis. When the penis is flaccid, the urethra makes another bend, this time inferiorly, when passing from the root to the body of the penis. During erection, the bend between the root and body of the penis disappears.

The urethra in men is divided into preprostatic, prostatic, membranous, and spongy parts.

***Preprostatic part.*** The preprostatic part of the urethra is about 1 cm long, extends from the base of the bladder to the prostate, and is associated with a circular cuff of smooth muscle fibers (the **internal urethral sphincter**). Contraction of this sphincter prevents retrograde movement of semen into the bladder during ejaculation.

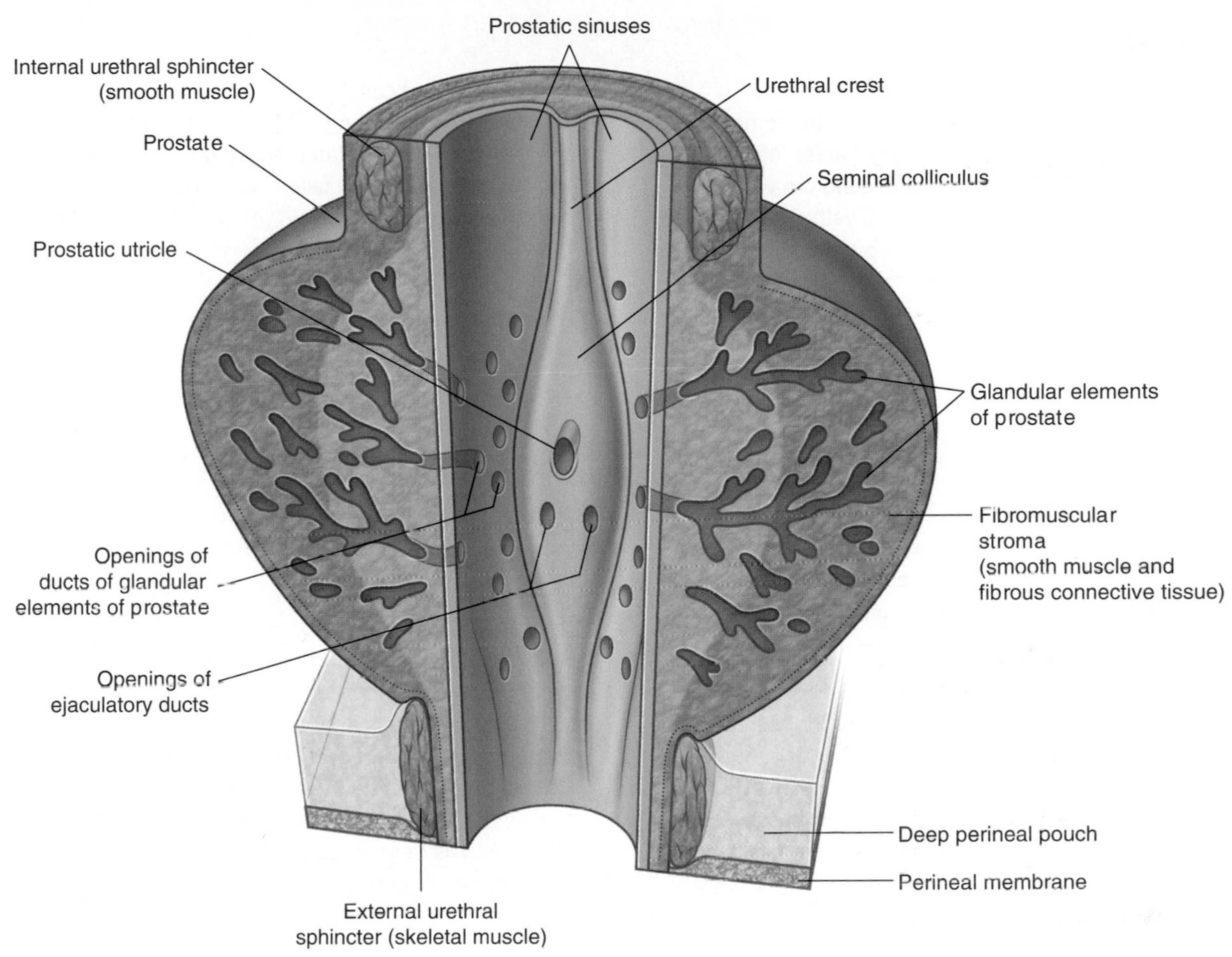


**Fig. 5.45, cont'd C.** Prostatic part of the urethra in men.

***Prostatic part.*** The prostatic part of the urethra (Fig. 5.45C) is 3 to 4 cm long and is surrounded by the prostate. In this region, the lumen of the urethra is marked by a longitudinal midline fold of mucosa (the **urethral crest**). The depression on each side of the crest is the **prostatic sinus**; the ducts of the prostate empty into these two sinuses.

Midway along its length, the urethral crest is enlarged to form a somewhat circular elevation (the seminal colliculus). In men, the seminal colliculus is used to determine the position of the prostate gland during transurethral transection of the prostate.

A small blind-ended pouch—the **prostatic utricle** (thought to be the homologue of the uterus in women)—opens onto the center of the seminal colliculus. On each side of the prostatic utricle is the opening of the ejaculatory duct of the male reproductive system. Therefore the connection between the urinary and reproductive tracts in men occurs in the prostatic part of the urethra.

***Membranous part.*** The membranous part of the urethra is narrow and passes through the deep perineal pouch (Fig. 5.45B). During its transit through this pouch, the urethra, in both men and women, is surrounded by skeletal muscle of the **external urethral sphincter**.

***Spongy urethra.*** The spongy urethra is surrounded by erectile tissue (the **corpus spongiosum**) of the penis. It is enlarged to form a bulb at the base of the penis and again at the end of the penis to form the **navicular fossa** (Fig. 5.45B). The two bulbo-urethral glands in the deep perineal pouch are part of the male reproductive system and open into the bulb of the spongy urethra. The external urethral orifice is the sagittal slit at the end of the penis.

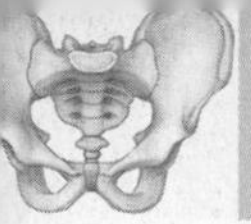



## In the clinic

### Bladder infection

The relatively short length of the urethra in women makes them more susceptible than men to bladder infection. The primary symptom of urinary tract infection in women is usually inflammation of the bladder (cystitis). The infection can be controlled in most instances by oral antibiotics and resolves without complication. In children under 1 year of age, infection from the bladder may spread via the ureters to the kidneys, where it can produce renal damage and ultimately lead to renal failure. Early diagnosis and treatment are necessary.

## In the clinic

### Urethral catheterization

Urethral catheterization is often performed to drain urine from a patient's bladder when the patient is unable to micturate. When inserting urinary catheters, it is important to appreciate the gender anatomy of the patient.

In men:

- The spongy urethra is surrounded by the erectile tissue of the bulb of the penis immediately inferior to the deep perineal pouch. The wall of this short segment of urethra is relatively thin and angles superiorly to pass through the deep perineal pouch; at this position the urethra is vulnerable to damage, notably during cystoscopy.
- The membranous part of the urethra runs superiorly as it passes through the deep perineal pouch.
- The prostatic part of the urethra takes a slight concave curve anteriorly as it passes through the prostate gland.

In women, it is much simpler to pass catheters and cystoscopes because the urethra is short and straight. Urine may therefore be readily drained from a distended bladder without significant concern for urethral rupture.

Occasionally, it is impossible to pass any form of instrumentation through the urethra to drain the bladder, usually because there is a urethral stricture or prostatic enlargement. In such cases, an ultrasound of the lower abdomen will demonstrate a full bladder (Fig. 5.46) behind the anterior abdominal wall. A suprapubic catheter may be inserted into the bladder with minimal trauma through a small incision under local anesthetic.

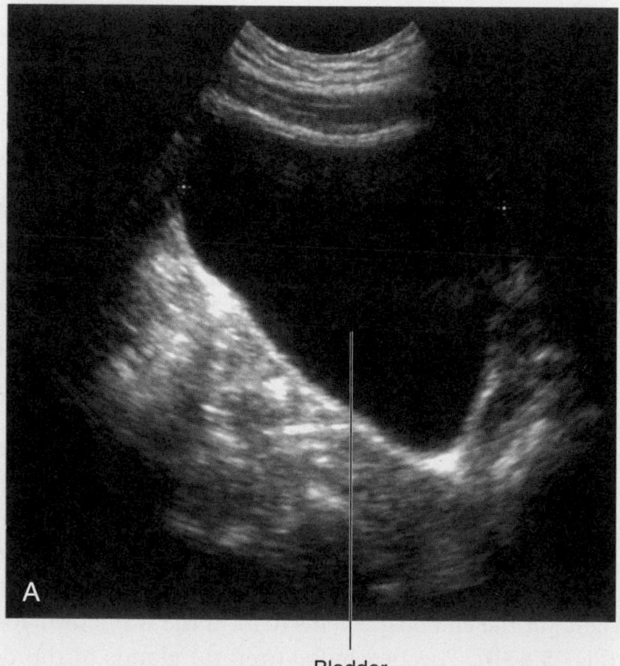


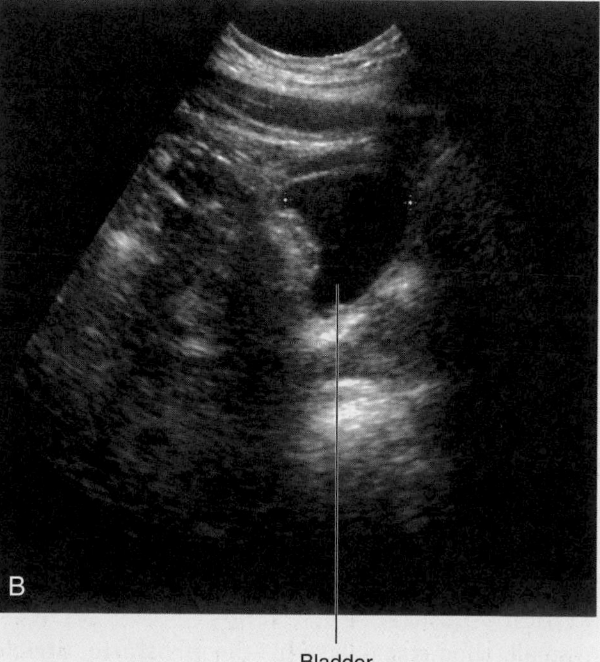


**Fig. 5.46** Ultrasound demonstrating the bladder. **A.** Full bladder. **B.** Postmicturition bladder.

## Reproductive system

### In men

The reproductive system in men has components in the abdomen, pelvis, and perineum (Fig. 5.47A). The major components are a testis, epididymis, ductus deferens, and ejaculatory duct on each side, and the urethra and penis in the midline. In addition, three types of accessory glands are associated with the system:

- a single prostate,
- a pair of seminal vesicles, and
- a pair of bulbo-urethral glands.

The design of the reproductive system in men is basically a series of ducts and tubules. The arrangement of parts and linkage to the urinary tract reflects its embryological development.

#### Testes

The **testes** originally develop high on the posterior abdominal wall and then descend, normally before birth, through the inguinal canal in the anterior abdominal wall and into the scrotum of the perineum. During descent, the testes carry their vessels, lymphatics, and nerves, as well as their principal drainage ducts, the **ductus deferens** (**vas deferens**) with them. The lymph drainage of the testes is therefore to the lateral aortic or lumbar nodes and pre-aortic nodes in the abdomen, and not to the inguinal or pelvic lymph nodes.

Each ellipsoid shaped testis is enclosed within the end of an elongated musculofascial pouch, which is continuous with the anterior abdominal wall and projects into the scrotum. The **spermatic cord** is the tube-shaped connection between the pouch in the scrotum and the abdominal wall.

The sides and anterior aspect of the testis are covered by a closed sac of peritoneum (the **tunica vaginalis**), which originally connected to the abdominal cavity. Normally after testicular descent, the connection closes, leaving a fibrous remnant.

Each testis (Fig. 5.47B) is composed of seminiferous tubules and interstitial tissue surrounded by a thick connective tissue capsule (the **tunica albuginea**). Spermatozoa are produced by the seminiferous tubules. The 400 to 600 highly coiled seminiferous tubules are modified at each end to become straight tubules, which connect to a collecting chamber (the **rete testis**) in a thick, vertically oriented linear wedge of connective tissue (the **mediastinum testis**), projecting from the capsule into the posterior aspect of the gonad. Approximately 12 to 20 **efferent ductules** originate from the upper end of the rete testis, penetrate the capsule, and connect with the epididymis.

### In the clinic

#### Testicular tumors

Tumors of the testis account for a small percentage of malignancies in men. However, they generally occur in younger patients (between 20 and 40 years of age). When diagnosed at an early stage, most of these tumors are curable by surgery and chemotherapy.

Early diagnosis of testicular tumors is extremely important. Abnormal lumps can be detected by palpation, and diagnosis can be made using ultrasound. Simple ultrasound scanning can reveal the extent of the local tumor, usually at an early stage.

Surgical removal of the malignant testis is often carried out using an inguinal approach. The testis is not usually removed through a scrotal incision, because it is possible to spread tumor cells into the subcutaneous tissues of the scrotum, which has a different lymphatic drainage than the testis.

### In the clinic

#### Ectopic testes

Interrupted descent of testis leads to an empty scrotal sac and abnormal location of the testis, which can lie anywhere along the usual route of descent. Most commonly the testis is present in the inguinal canal, where it can be palpated. This condition is usually diagnosed at birth or within the first year of life. A higher incidence of ectopic (undescended) testis occurs in premature births (30%) than in term births (3–5%). Normally, the ectopic testis can complete its descent within the first 3 months after a child is born; therefore watchful waiting is recommended for the first couple of months. A specialist referral is usually made at 6 months if the testis is still absent from the scrotal sac. It is crucial to make the diagnosis early so that an appropriate management plan can be initiated to avoid or reduce the risk of complications such as testicular malignancy, subfertility or infertility, testicular torsion, and inguinal hernia (due to patent processus vaginalis). If surgical correction is required, the ectopic testis is moved from the inguinal canal into the scrotum (orchiopexy). During mobilization of the testis, dissection of tissues must be performed carefully to avoid injuring the ilioinguinal nerve adjacent to the spermatic cord. At the time of orchiopexy, the patent processus vaginalis is closed and any inguinal hernia, if present, is repaired.

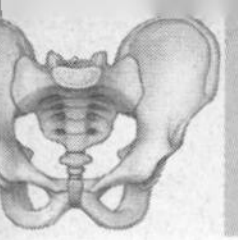



### Epididymis

The **epididymis** courses along the posterolateral side of the testis (Fig. 5.47B). It has two distinct components:

- the **efferent ductules**, which form an enlarged coiled mass that sits on the posterior superior pole of the testis and forms the **head of the epididymis**; and
- the **true epididymis**, which is a single, long coiled duct into which the efferent ductules all drain, and which continues inferiorly along the posterolateral margin of the testis as the **body of the epididymis** and enlarges to form the **tail of the epididymis** at the inferior pole of the testis.

During passage through the epididymis, spermatozoa acquire the ability to move and fertilize an egg. The epididymis also stores spermatozoa until ejaculation. The end of the epididymis is continuous with the ductus deferens.

### Ductus deferens

The ductus deferens is a long muscular duct that transports spermatozoa from the tail of the epididymis in the scrotum to the ejaculatory duct in the pelvic cavity (Fig. 5.47A). It ascends in the scrotum as a component of the spermatic cord and passes through the inguinal canal in the anterior abdominal wall.

After passing through the deep inguinal ring, the ductus deferens bends medially around the lateral side of the inferior epigastric artery and crosses the external iliac artery and the external iliac vein at the pelvic inlet to enter the pelvic cavity.

The duct descends medially on the pelvic wall, deep to the peritoneum, and crosses the ureter posterior to the bladder. It continues inferomedially along the base of the bladder, anterior to the rectum, almost to the midline, where it is joined by the duct of the seminal vesicle to form the ejaculatory duct.

Between the ureter and ejaculatory duct, the ductus deferens expands to form the ampulla of the ductus deferens. The ejaculatory duct penetrates through the prostate gland to connect with the prostatic urethra.

## In the clinic

### Vasectomy

The ductus deferens transports spermatozoa from the tail of the epididymis in the scrotum to the ejaculatory duct in the pelvic cavity. Because it has a thick smooth muscle wall, it can be easily palpated in the spermatic cord between the testes and the superficial inguinal ring. Also, because it can be accessed through skin and superficial fascia, it is amenable to surgical dissection and surgical division. When this is carried out bilaterally (vasectomy), the patient is rendered sterile—this is a useful method for male contraception.

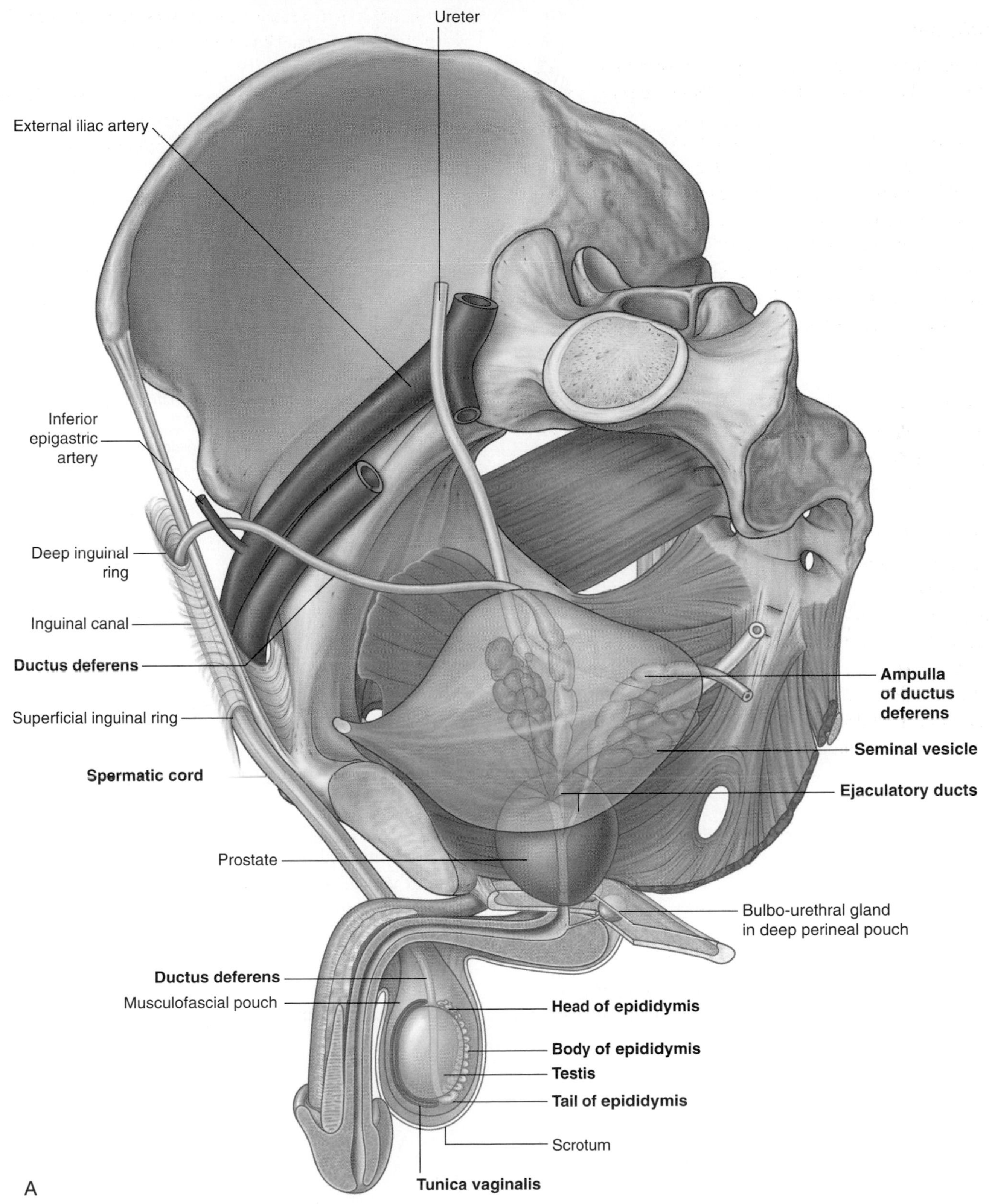


**Fig. 5.47** Reproductive system in men. **A.** Overview.

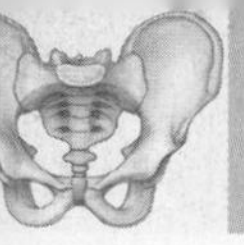

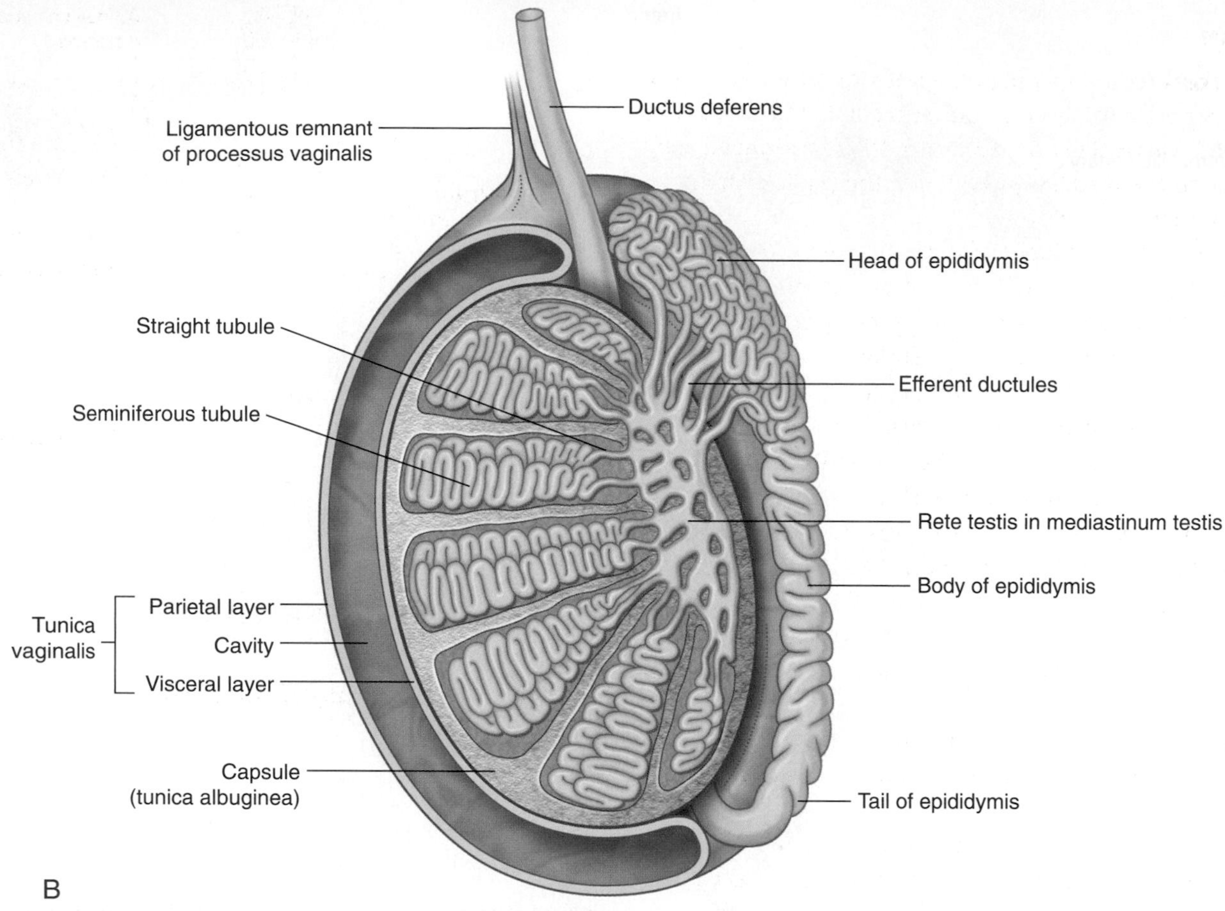


**Fig. 5.47, cont'd** **B.** Testis and surrounding structures.

## Seminal vesicle

Each **seminal vesicle** is an accessory gland of the male reproductive system that develops as a blind-ended tubular outgrowth from the ductus deferens (Fig. 5.47A). The tube is coiled with numerous pocket-like outgrowths and is encapsulated by connective tissue to form an elongate structure situated between the bladder and rectum. The seminal vesicle is immediately lateral to and follows the course of the ductus deferens at the base of the bladder.

The duct of the seminal vesicle joins the ductus deferens to form the **ejaculatory duct** (Fig. 5.48). Secretions from the seminal vesicle contribute significantly to the volume of the ejaculate (semen).

## Prostate

The **prostate** is an unpaired accessory structure of the male reproductive system that surrounds the urethra in the pelvic cavity (Figs. 5.47A and 5.48). It lies immediately inferior to the bladder, posterior to the pubic symphysis, and anterior to the rectum.

The prostate is shaped like an inverted rounded cone with a larger base, which is continuous above with the neck of the bladder, and a narrower apex, which rests below on the pelvic floor. The inferolateral surfaces of the prostate are in contact with the levator ani muscles that together cradle the prostate between them.

The prostate develops as 30 to 40 individual complex glands, which grow from the urethral epithelium into the surrounding wall of the urethra. Collectively, these glands enlarge the wall of the urethra into what is known as the prostate; however, the individual glands retain their own ducts, which empty independently into the prostatic sinuses on the posterior aspect of the urethral lumen (see Fig. 5.45C).

Secretions from the prostate, together with secretions from the seminal vesicles, contribute to the formation of semen during ejaculation.

The ejaculatory ducts pass almost vertically in an anteroinferior direction through the posterior aspect of the prostate to open into the prostatic urethra.

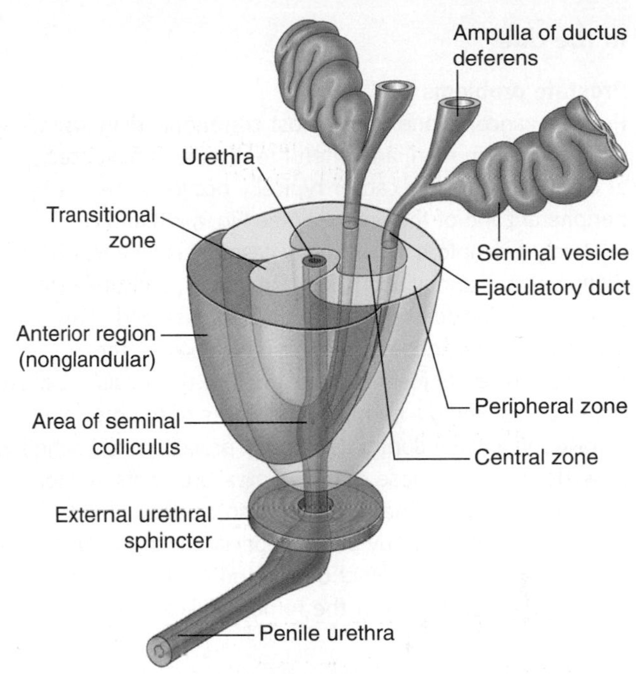


**Fig. 5.48** The prostate gland. Zonal anatomy.

## Bulbo-urethral glands

The **bulbo-urethral glands** (see Fig. 5.47A), one on each side, are small, pea-shaped mucous glands situated within the deep perineal pouch. They are lateral to the membranous part of the urethra. The duct from each gland passes inferomedially through the **perineal membrane**, to open into the bulb of the spongy urethra at the root of the penis.

Together with small glands positioned along the length of the spongy urethra, the bulbo-urethral glands contribute to lubrication of the urethra and the pre-ejaculatory emission from the penis.

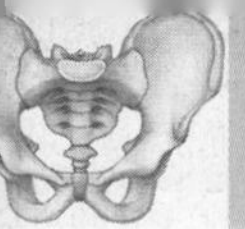

## In the clinic

### Prostate problems

Prostate cancer is one of the most commonly diagnosed malignancies in men, and often the disease is advanced at diagnosis. Prostate cancer typically occurs in the peripheral zone of the prostate (see Fig. 5.48) and is relatively asymptomatic. In many cases, it is diagnosed by a digital rectal examination (DRE) (Fig. 5.49A) and by blood tests, which include serum acid phosphatase and serum prostate-specific antigen (PSA). In rectal exams, the tumorous prostate feels "rock" hard. The diagnosis is usually made by obtaining a number of biopsies of the prostate. Ultrasound is used during the biopsy procedure to image the prostate for the purpose of taking measurements and for needle placement. Ultrasound can also be used to aid planning radiotherapy by placing special metal markers, called fiducials, under direct ultrasound guidance, through the rectal wall into or near the tumor. This allows maximization of the radiation dose to the tumor while protecting healthy tissue.

Benign prostatic hypertrophy is a disease of the prostate that occurs with increasing age in most men (Fig. 5.49B). It generally involves the more central regions of the prostate (see Fig. 5.48), which gradually enlarge. The prostate feels "bulky" on DRE. Owing to the more central hypertrophic change of the prostate, the urethra is compressed, and a urinary outflow obstruction develops in a number of patients. With time, the bladder may become hypertrophied in response to the urinary outflow obstruction. In some male patients, the obstruction becomes so severe that urine cannot be passed and transurethral or suprapubic catheterization is necessary. Despite being a benign disease, benign prostatic hypertrophy can therefore have a marked effect on the daily lives of many patients.

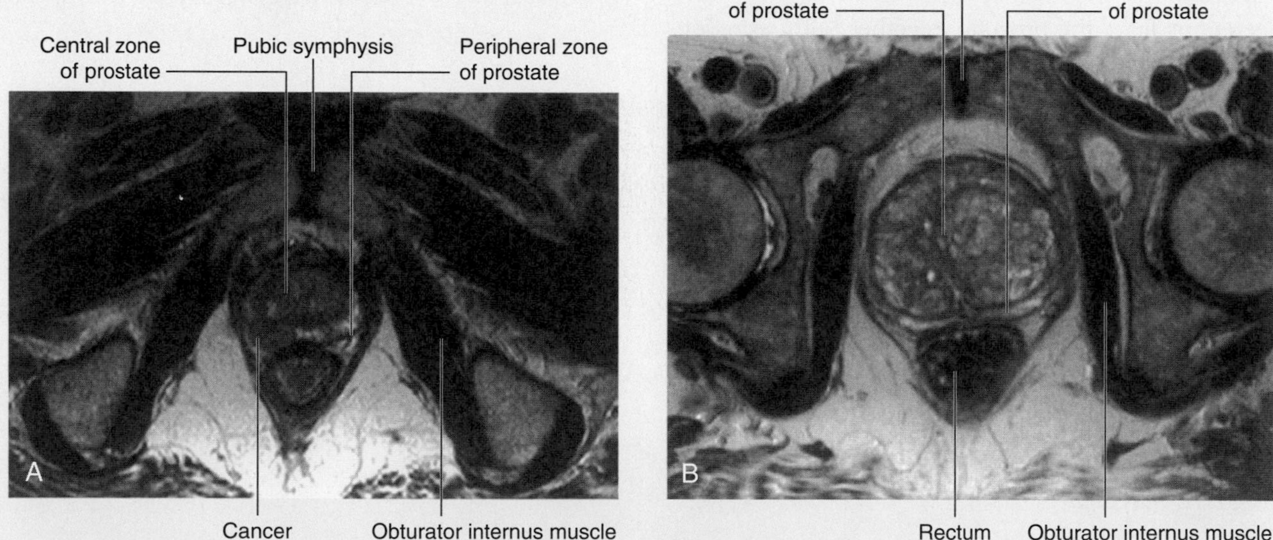


**Fig. 5.49** Axial T2-weighted magnetic resonance images of prostate problems. **A.** A small prostatic cancer in the peripheral zone of a normal-sized prostate. **B.** Benign prostatic hypertrophy.

### In women

The reproductive tract in women is contained mainly in the pelvic cavity and perineum, although during pregnancy, the uterus expands into the abdominal cavity. Major components of the system consist of:

- an ovary on each side, and
- a uterus, vagina, and clitoris in the midline (Fig. 5.50).

In addition, a pair of accessory glands (the **greater vestibular glands**) are associated with the tract.

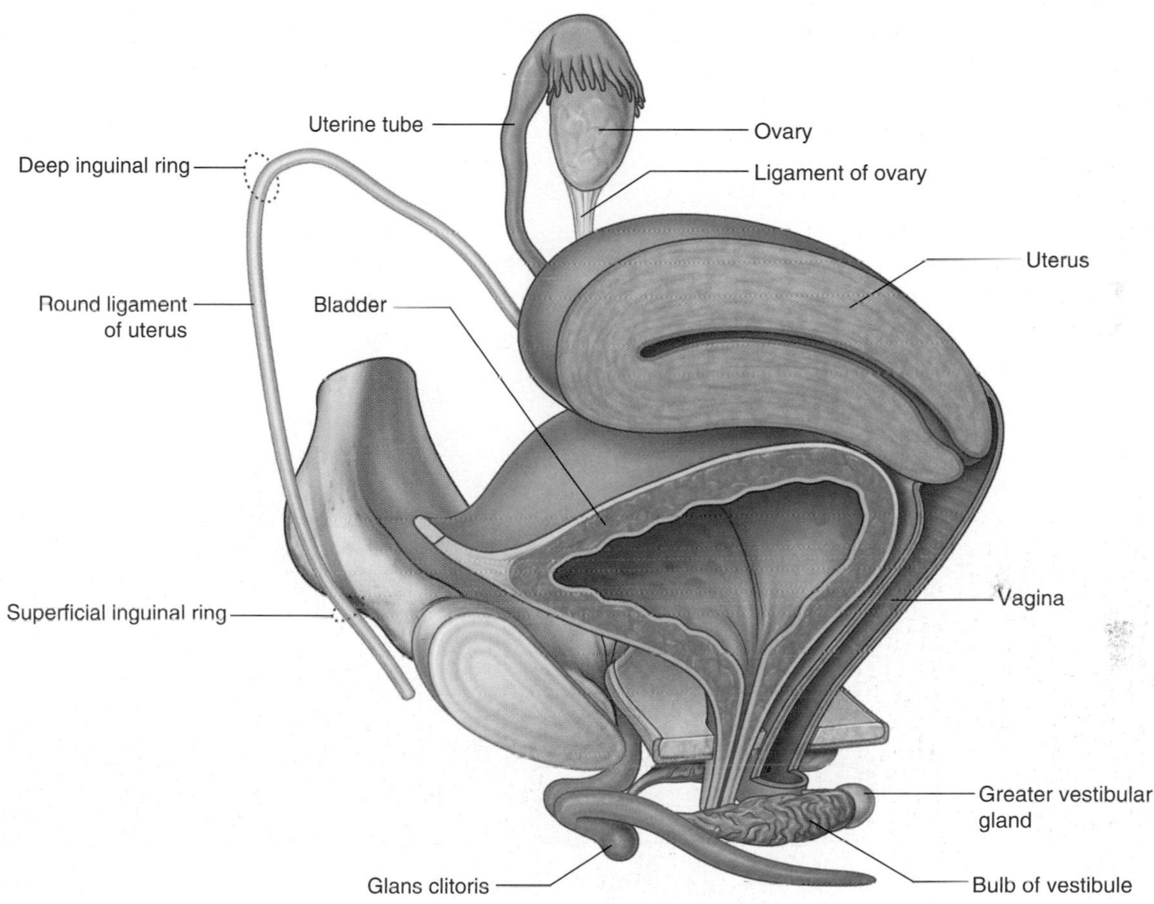


**Fig. 5.50** Reproductive system in women.

## Ovaries

Like the testes in men, the **ovaries** develop high on the posterior abdominal wall and then descend before birth, bringing with them their vessels, lymphatics, and nerves. Unlike the testes, the ovaries do not migrate through the inguinal canal into the perineum, but stop short and assume a position on the lateral wall of the pelvic cavity (Fig. 5.51).

The ovaries are the sites of egg production (oogenesis). Mature eggs are ovulated into the peritoneal cavity and normally directed into the adjacent openings of the uterine tubes by cilia on the ends of the uterine tubes.

The ovaries lie adjacent to the lateral pelvic wall just inferior to the pelvic inlet. Each of the two almond-shaped ovaries is about 3 cm long and is suspended by a mesentery (the **mesovarium**) that is a posterior extension of the broad ligament.

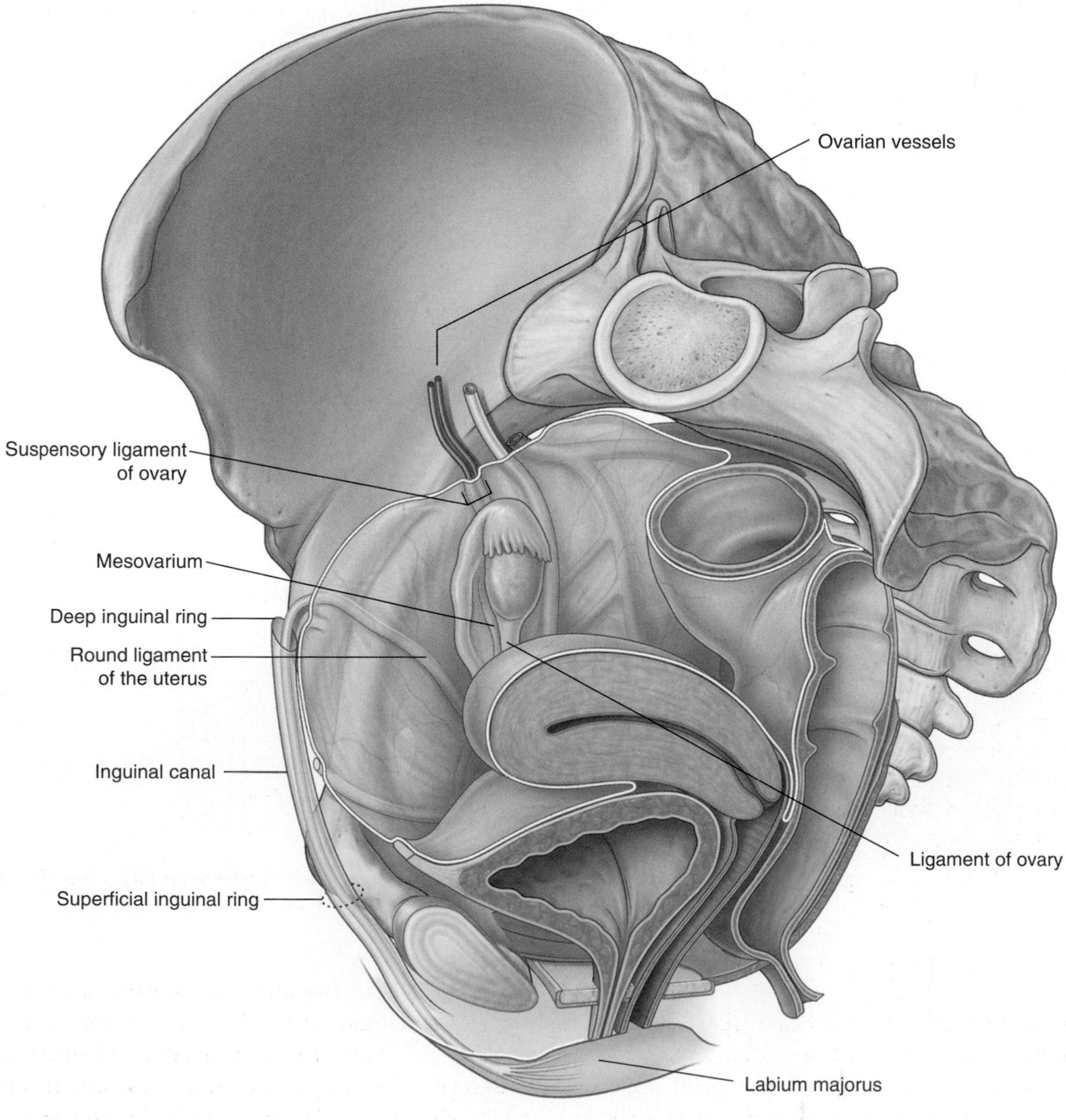

**Fig. 5.51** Ovaries and broad ligament.

## In the clinic

### Ovarian cancer

Ovarian cancer remains one of the major challenges in oncology. The ovaries contain numerous cell types, all of which can undergo malignant change and require different imaging and treatment protocols and ultimately have different prognoses.

Ovarian tumors most commonly originate from the ovarian surface (germinal) epithelium that is continuous at a sharp transition zone with the peritoneum of the mesovarium.

Many factors have been linked with the development of ovarian tumors, including a strong family history.

Ovarian cancer may occur at any age, but more typically it occurs in older women.

Cancer of the ovaries may spread via the blood and lymphatics, and frequently metastasizes directly into the peritoneal cavity. Such direct peritoneal cavity spread allows the passage of tumor cells along the paracolic gutters and over the liver from where this disease may disseminate easily. Unfortunately, many patients already have metastatic and diffuse disease (Fig. 5.52) at the time of diagnosis.

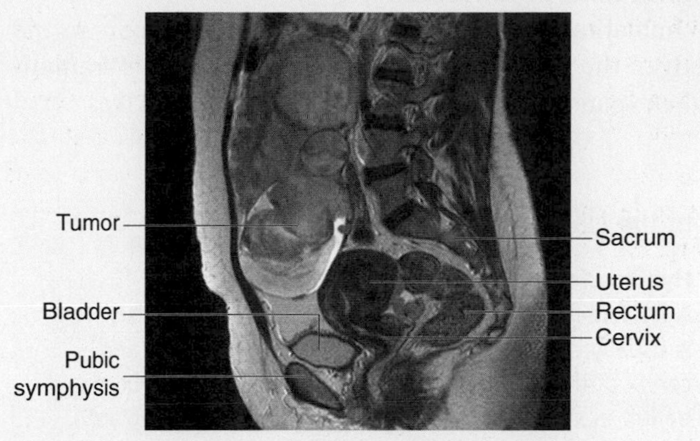


**Fig. 5.52** Sagittal magnetic resonance image demonstrating ovarian cancer.

## In the clinic

### Imaging the ovary

The ovaries can be visualized using ultrasound. If the patient drinks enough water, the bladder becomes enlarged and full. This fluid-filled cavity provides an excellent acoustic window, behind which the uterus and ovaries may be identified by transabdominal scanning with ultrasound. This technique also allows obstetricians and technicians to view a fetus and record its growth throughout pregnancy.

Some patients are not suitable for transabdominal scanning, in which case a probe may be passed into the vagina, permitting close visualization of the uterus, the contents of the recto-uterine pouch (pouch of Douglas), and the ovaries. The ovaries can also be visualized laparoscopically. Many countries have introduced screening programs for cervical cancer where women are regularly called for smear tests.

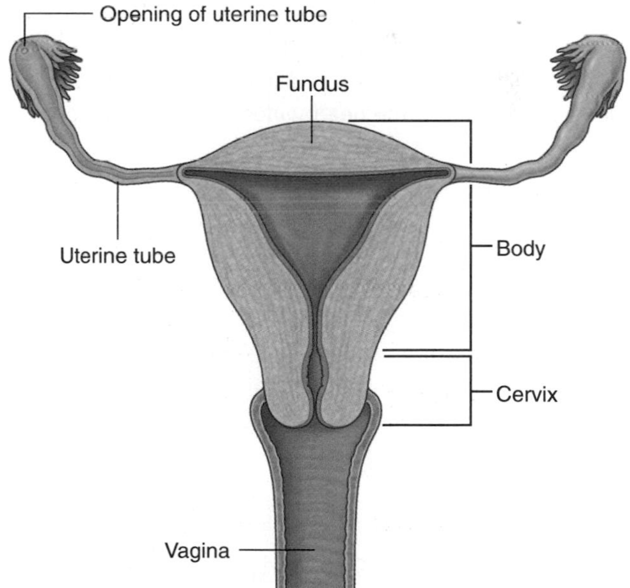


**Fig. 5.53** Uterus. Anterior view. The anterior halves of the uterus and vagina have been cut away.

### Uterus

The **uterus** is a thick-walled muscular organ in the midline between the bladder and rectum (see Fig. 5.51). It consists of a body and a cervix, and inferiorly it joins the vagina (Fig. 5.53). Superiorly, uterine tubes project laterally from the uterus and open into the peritoneal cavity immediately adjacent to the ovaries.

The body of the uterus is flattened anteroposteriorly and, above the level of origin of the uterine tubes (Fig. 5.53), has a rounded superior end (**fundus of the uterus**). The cavity of the body of the uterus is a narrow slit, when viewed laterally, and is shaped like an inverted triangle, when viewed anteriorly. Each of the superior corners of the cavity is continuous with the lumen of a

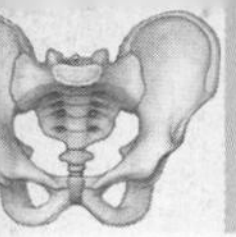



uterine tube; the inferior corner is continuous with the central canal of the cervix.

Implantation of the blastocyst normally occurs in the body of the uterus. During pregnancy, the uterus dramatically expands superiorly into the abdominal cavity.

## In the clinic

### Hysterectomy

A hysterectomy is the surgical removal of the uterus. This is usually complete excision of the body, fundus, and cervix of the uterus, though occasionally the cervix may be left in situ. In some instances the uterine (fallopian) tubes and ovaries are removed as well. This procedure is called a total abdominal hysterectomy and bilateral salpingo-oophorectomy.

Hysterectomy, oophorectomy, and salpingo-oophorectomy may be performed in patients who have reproductive malignancy, such as uterine, cervical, and ovarian cancers. Other indications include a strong family history of reproductive disorders, endometriosis, and excessive bleeding. Occasionally the uterus may need to be removed postpartum because of excessive postpartum bleeding.

A hysterectomy is performed through a transverse suprapubic incision (Pfannenstiel's incision). During the procedure tremendous care is taken to identify the distal ureters and to ligate the nearby uterine arteries without damage to the ureters.

### Uterine tubes

The **uterine tubes** extend from each side of the superior end of the body of the uterus to the lateral pelvic wall and are enclosed within the upper margins of the mesosalpinx portions of the broad ligaments (see p. 477). Because the ovaries are suspended from the posterior aspect of the broad ligaments, the uterine tubes pass superiorly over, and terminate laterally to, the ovaries.

Each uterine tube has an expanded trumpet-shaped end (the **infundibulum**), which curves around the superolateral pole of the related ovary (Fig. 5.54). The margin of the infundibulum is rimmed with small finger-like projections termed **fimbriae**. The lumen of the uterine tube opens into the peritoneal cavity at the narrowed end of the infundibulum. Medial to the infundibulum, the tube expands to form the **ampulla** and then narrows to form the **isthmus**, before joining with the body of the uterus.

The fimbriated infundibulum facilitates the collection of ovulated eggs from the ovary. Fertilization normally occurs in the ampulla.

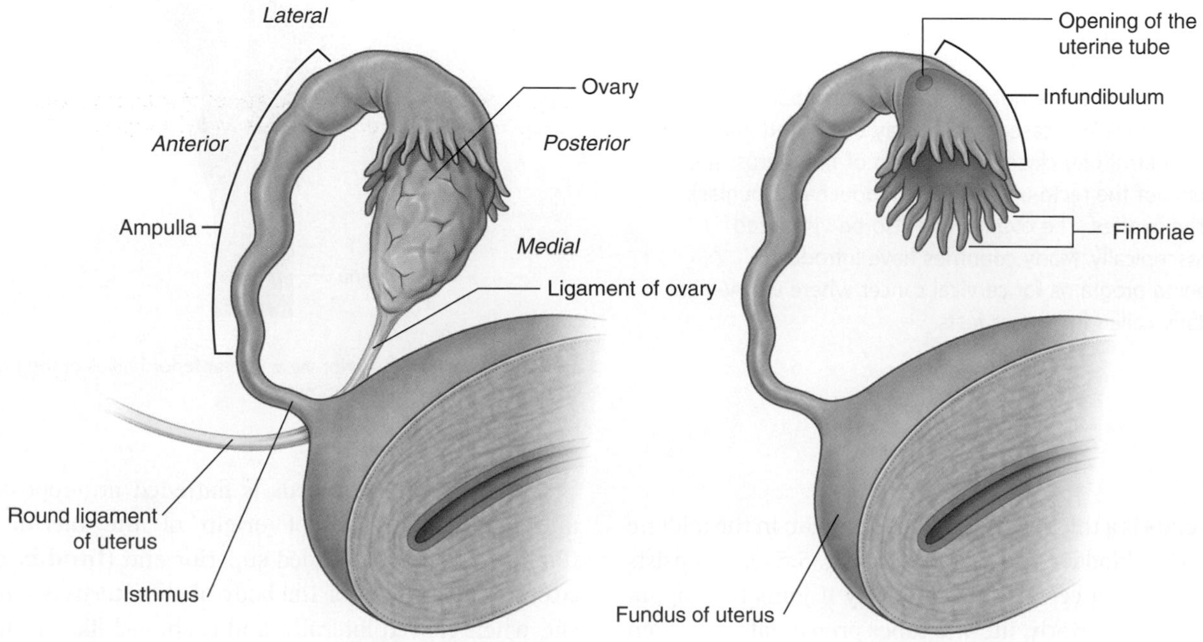


**Fig. 5.54** Uterine tubes.

## In the clinic

### Tubal ligation

After ovulation, the unfertilized egg is gathered by the fimbriae of the uterine tube. The egg passes into the uterine tube where it is normally fertilized in the ampulla. The zygote then begins development and passes into the uterine cavity where it implants in the uterine wall.

A simple and effective method of birth control is to surgically ligate (clip) the uterine tubes, preventing spermatozoa from reaching the ovum. This simple short procedure is performed under general anesthetic. A small laparoscope is passed into the peritoneal cavity and special equipment is used to identify the tubes.

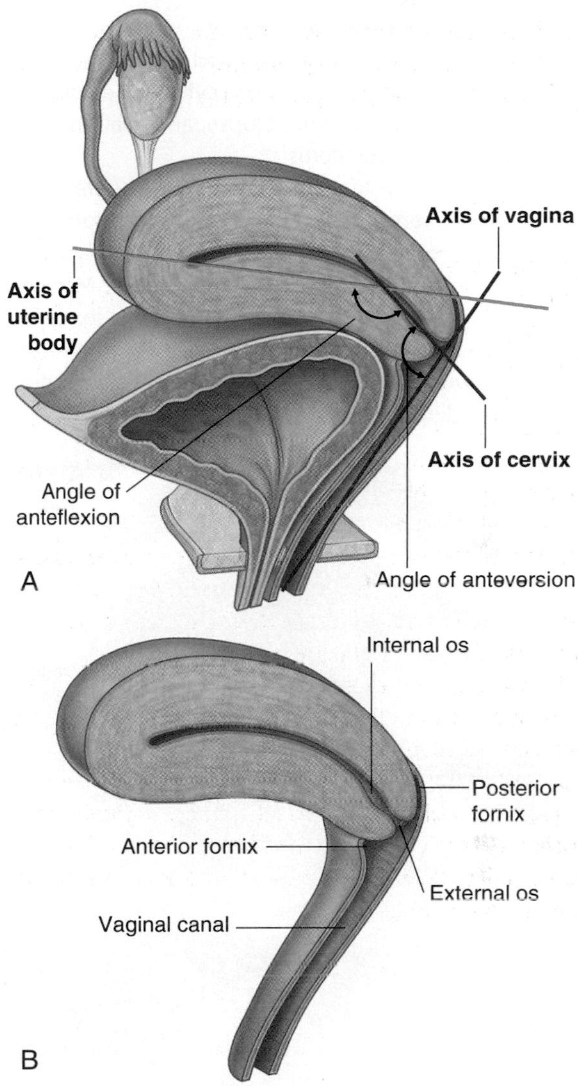


**Fig. 5.55** Uterus and vagina. **A.** Angles of anteflexion and anteversion. **B.** The cervix protrudes into the vagina.

### Cervix

The **cervix** forms the inferior part of the uterus and is shaped like a short, broad cylinder with a narrow central channel. The body of the uterus normally arches forward (anteflexed on the cervix) over the superior surface of the emptied bladder (Fig. 5.55A). In addition, the cervix is angled forward (anteverted) on the vagina so that the inferior end of the cervix projects into the upper anterior aspect of the vagina. Because the end of the cervix is dome shaped, it bulges into the vagina, and a gutter, or fornix, is formed around the margin of the cervix where it joins the vaginal wall (Fig. 5.55B). The tubular central canal of the cervix opens, below, as the **external os**, into the vaginal cavity and, above, as the **internal os**, into the uterine cavity.

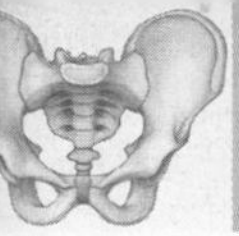

## In the clinic

### Carcinoma of the cervix and uterus

Carcinoma of the cervix (Fig. 5.56) and uterus is a common disease. Diagnosis is by inspection, cytology (examination of the cervical cells), imaging, biopsy, and dilation and curettage (D&C) of the uterus.

Carcinoma of the cervix and uterus may be treated by local resection, removal of the uterus (hysterectomy), and adjuvant chemotherapy. The tumor spreads via lymphatics to the internal and common iliac lymph nodes. Many countries have introduced screening programs for cervical cancer where women are regularly called for smear tests. The age of women included in the screening population varies depending on the country.

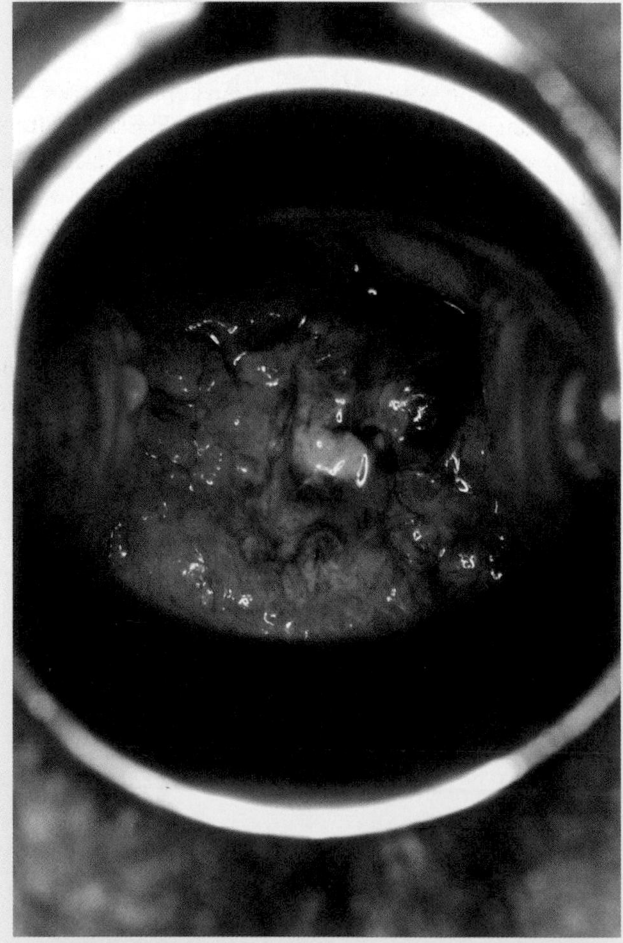

**Fig. 5.56** Picture taken through a speculum inserted into the vagina demonstrating cervical cancer. See Fig. 5.84E on p. 519 for a view of the normal cervix.

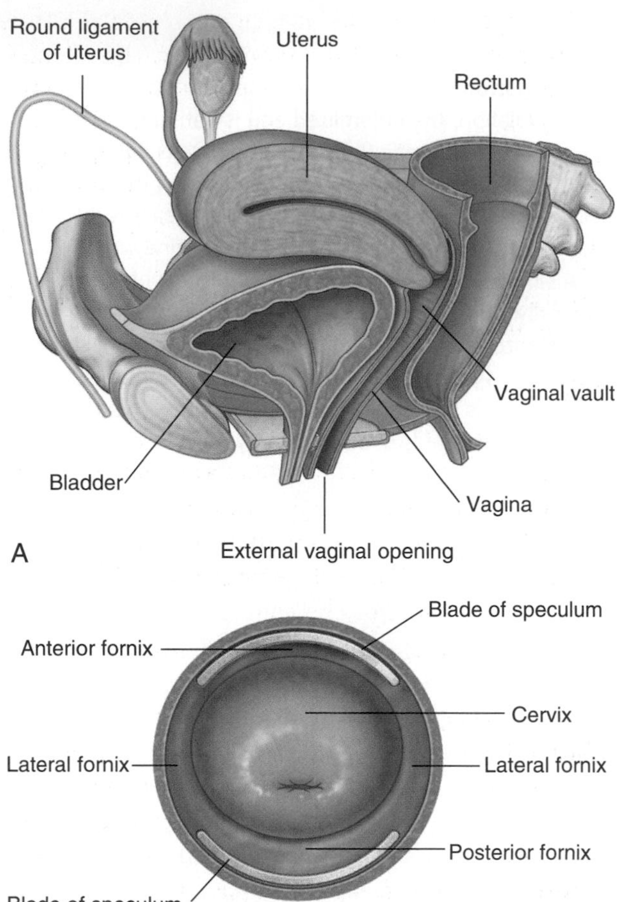


**Fig. 5.57** Vagina. **A.** Left half of pelvis cut away. **B.** Vaginal fornices and cervix as viewed through a speculum.

## Vagina

The **vagina** is the copulatory organ in women. It is a distensible fibromuscular tube that extends from the perineum through the pelvic floor and into the pelvic cavity (Fig. 5.57A). The internal end of the canal is enlarged to form a region called the **vaginal vault**.

The anterior wall of the vagina is related to the base of the bladder and to the urethra; in fact, the urethra is embedded in, or fused to, the anterior vaginal wall.

Posteriorly, the vagina is related principally to the rectum.

Inferiorly, the vagina opens into the vestibule of the perineum immediately posterior to the external opening of the urethra. From its external opening (the **introitus**), the vagina courses posterosuperiorly through the perineal membrane and into the pelvic cavity, where it is attached by its anterior wall to the circular margin of the cervix.

The **vaginal fornix** is the recess formed between the margin of the cervix and the vaginal wall. Based on position, the fornix is subdivided into a posterior fornix, an anterior fornix, and two lateral fornices (Fig. 5.57A and see Fig. 5.55).

The vaginal canal is normally collapsed so that the anterior wall is in contact with the posterior wall. By using a speculum to open the vaginal canal, a physician can see the domed inferior end of the cervix, the vaginal fornices, and the external os of the cervical canal in a patient (Fig. 5.57B).

During intercourse, semen is deposited in the vaginal vault. Spermatozoa make their way into the external os of the cervical canal, pass through the cervical canal into the uterine cavity, and then continue through the uterine cavity into the uterine tubes where fertilization normally occurs in the ampulla.

## Fascia

Fascia in the pelvic cavity lines the pelvic walls, surrounds the bases of the pelvic viscera, and forms sheaths around blood vessels and nerves that course medially from the pelvic walls to reach the viscera in the midline. This pelvic fascia is a continuation of the extraperitoneal connective tissue layer found in the abdomen.

### In women

In women, a **rectovaginal septum** separates the posterior surface of the vagina from the rectum (Fig. 5.58A). Condensations of fascia form ligaments that extend from the cervix to the anterior (**pubocervical ligament**), lateral (**transverse cervical** or **cardinal ligament**), and posterior (**uterosacral ligament**) pelvic walls (Fig. 5.58A). These ligaments, together with the perineal membrane, the levator ani muscles, and the perineal body, are thought to stabilize the uterus in the pelvic cavity. The most important of these ligaments are the transverse cervical or cardinal ligaments, which extend laterally from each side of the cervix and vaginal vault to the related pelvic wall.

#### In the clinic

**The recto-uterine pouch**

The recto-uterine pouch (**pouch of Douglas**) is an extremely important clinical region situated between the rectum and uterus. When the patient is in the supine position, the recto-uterine pouch is the lowest portion of the abdominopelvic cavity and is a site where infection and fluids typically collect. It is impossible to palpate this region transabdominally, but it can be examined by transvaginal and transrectal digital palpation. If an abscess is suspected, it may be drained through the vagina or the rectum without necessitating transabdominal surgery.

### In men

In men, a condensation of fascia around the anterior and lateral region of the prostate (**prostatic fascia**) contains and surrounds the prostatic plexus of veins and is continuous posteriorly with the **rectovesical septum**, which separates the posterior surface of the prostate and base of the bladder from the rectum (Fig. 5.58B).

## Peritoneum

The peritoneum of the pelvis is continuous at the pelvic inlet with the peritoneum of the abdomen. In the pelvis, the peritoneum drapes over the pelvic viscera in the midline, forming:

- pouches between adjacent viscera, and
- folds and ligaments between viscera and pelvic walls.

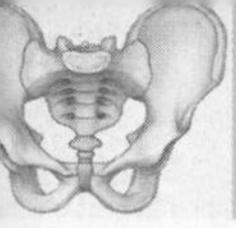

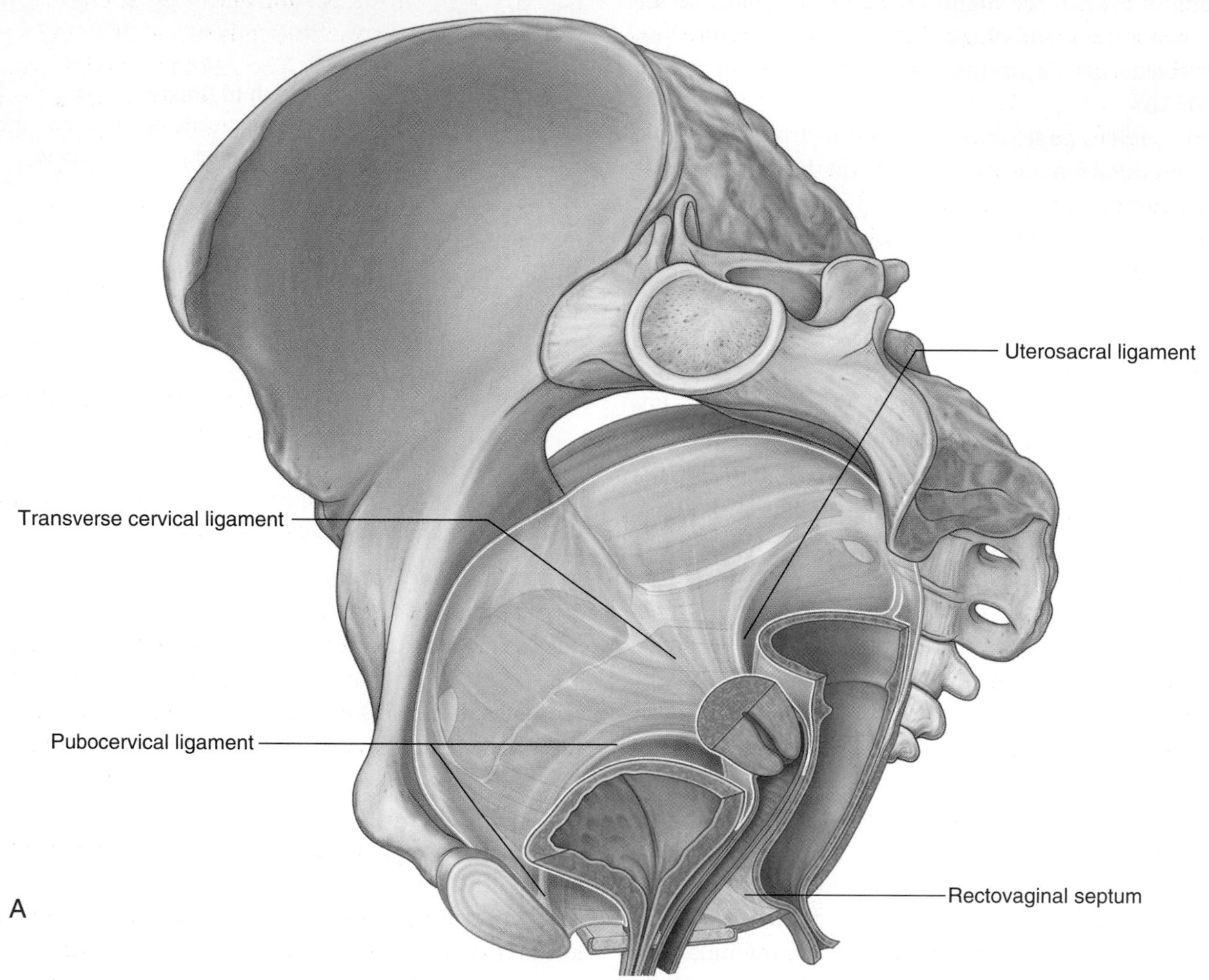

Uterosacral ligament
Transverse cervical ligament
Pubocervical ligament
A
Rectovaginal septum

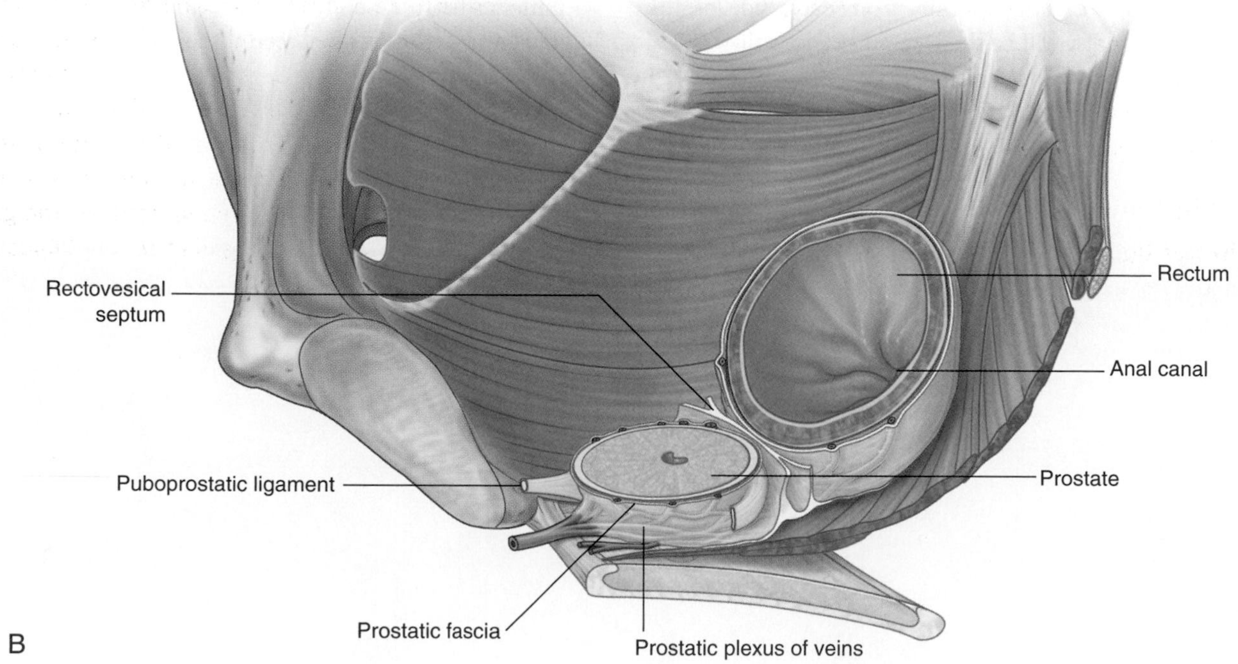

Rectovesical septum
Rectum
Anal canal
Puboprostatic ligament
Prostate
B
Prostatic fascia
Prostatic plexus of veins

**Fig. 5.58** Pelvic fascia. **A.** In women. **B.** In men.

Anteriorly, median and medial umbilical folds of peritoneum cover the embryological remnants of the urachus and umbilical arteries, respectively (Fig. 5.59). These folds ascend out of the pelvis and onto the anterior abdominal wall. Posteriorly, peritoneum drapes over the anterior and lateral aspects of the upper third of the rectum, but only the anterior surface of the middle third of the rectum is covered by peritoneum; the lower third of the rectum is not covered at all.

### In women

In women, the uterus lies between the bladder and rectum, and the uterine tubes extend from the superior aspect of the uterus to the lateral pelvic walls (Fig. 5.59A). As a consequence, a shallow **vesico-uterine pouch** occurs anteriorly, between the bladder and uterus, and a deep **recto-uterine pouch** (pouch of Douglas) occurs posteriorly, between the uterus and rectum. In addition, a large fold of peritoneum (the broad ligament), with a uterine tube enclosed in its superior margin and an ovary attached posteriorly, is located on each side of the uterus and extends to the lateral pelvic walls.

In the midline, the peritoneum descends over the posterior surface of the uterus and cervix and onto the vaginal wall adjacent to the posterior vaginal fornix. It then reflects onto the anterior and lateral walls of the rectum. The deep pouch of peritoneum formed between the anterior surface of the rectum and posterior surfaces of the uterus, cervix, and vagina is the recto-uterine pouch. A sharp sickle-shaped ridge of peritoneum (**recto-uterine fold**) occurs on each side near the base of the recto-uterine pouch. The **recto-uterine folds** overlie the **uterosacral ligaments**, which are condensations of pelvic fascia that extend from the cervix to the posterolateral pelvic walls.

#### Broad ligament

The **broad ligament** is a sheet-like fold of peritoneum, oriented in the coronal plane that runs from the lateral pelvic wall to the uterus, and encloses the uterine tube in its superior margin and suspends the ovary from its posterior aspect (Fig. 5.59A). The uterine arteries cross the ureters at the base of the broad ligaments, and the ligament of the ovary and round ligament of the uterus are enclosed within the parts of the broad ligament related to the ovary and uterus, respectively. The broad ligament has three parts:

- the mesometrium, the largest part of the broad ligament, which extends from the lateral pelvic walls to the body of the uterus;
- the mesosalpinx, the most superior part of the broad ligament, which suspends the uterine tube in the pelvic cavity; and
- the mesovarium, a posterior extension of the broad ligament, which attaches to the ovary.

The peritoneum of the mesovarium is continuous with the ovarian surface (germinal) epithelium (see Fig. 5.59A insert). The ovaries are positioned with their long axis in the vertical plane. The ovarian vessels, nerves, and lymphatics enter the superior pole of the ovary from a lateral position and are covered by another raised fold of peritoneum, which with the structures it contains forms the **suspensory ligament of the ovary** (**infundibulopelvic ligament**).

The inferior pole of the ovary is attached to a fibromuscular band of tissue (the **ligament of the ovary**), which courses medially in the margin of the mesovarium to the uterus and then continues anterolaterally as the **round ligament of the uterus** (Fig. 5.59A). The round ligament of the uterus passes over the pelvic inlet to reach the deep inguinal ring and then courses through the inguinal canal to end in connective tissue related to the labium majus in the perineum. Both the ligament of the ovary and the round ligament of the uterus are remnants of the gubernaculum, which attaches the gonad to the labioscrotal swellings in the embryo.

Ureter
Suspensory ligament of ovary
Recto-uterine fold
Broad ligament
Round ligament of uterus
Inferior epigastric artery
Lateral umbilical fold
Medial umbilical fold
Recto-uterine pouch
Ligament of ovary
Median umbilical fold
Vesico-uterine pouch
Sagittal section of broad ligament
Mesosalpinx
Ovarian surface (germinal) epithelium
Uterine tube
Ovary
Broad ligament
Mesovarium
Mesometrium
Ureter
Round ligament of uterus
Uterine artery
A

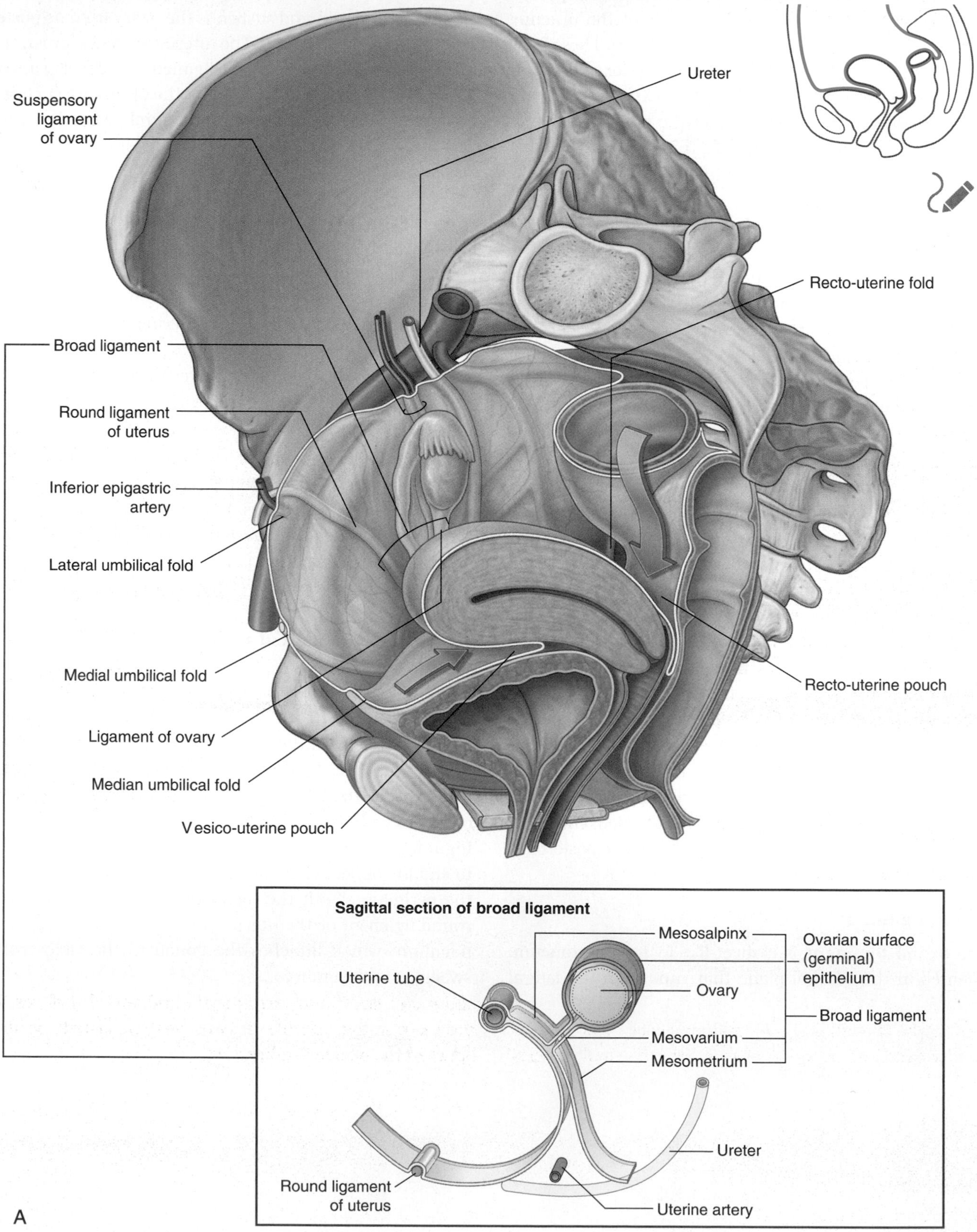

**Fig. 5.59** Peritoneum in the pelvis. **A.** In women.

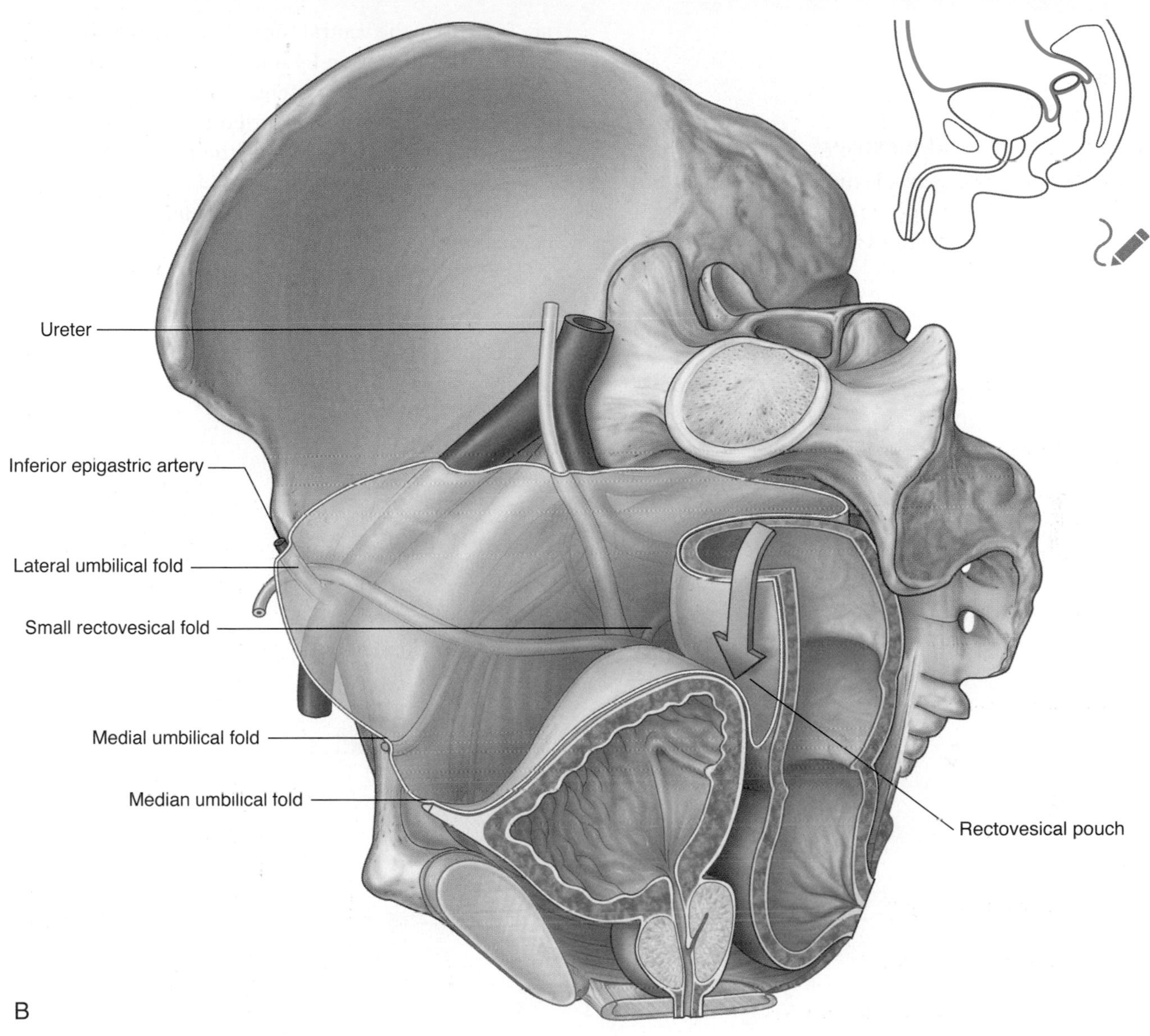


**Fig. 5.59, cont'd** **B.** In men.

## In men

In men, the visceral peritoneum drapes over the top of the bladder onto the superior poles of the seminal vesicles and then reflects onto the anterior and lateral surfaces of the rectum (Fig. 5.59B). A **rectovesical pouch** occurs between the bladder and rectum.

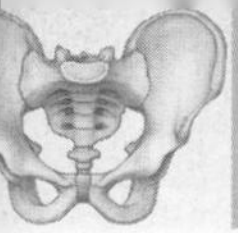



## Nerves

### Somatic plexuses

### Sacral and coccygeal plexuses

The sacral and coccygeal plexuses are situated on the posterolateral wall of the pelvic cavity and generally occur in the plane between the muscles and blood vessels. They are formed by the ventral rami of S1 to Co, with a significant contribution from L4 and L5, which enter the pelvis from the lumbar plexus (Fig. 5.60). Nerves from these mainly somatic plexuses contribute to the innervation of the lower limb and muscles of the pelvis and perineum. Cutaneous branches supply skin over the medial side of the foot, the posterior aspect of the lower limb, and most of the perineum.



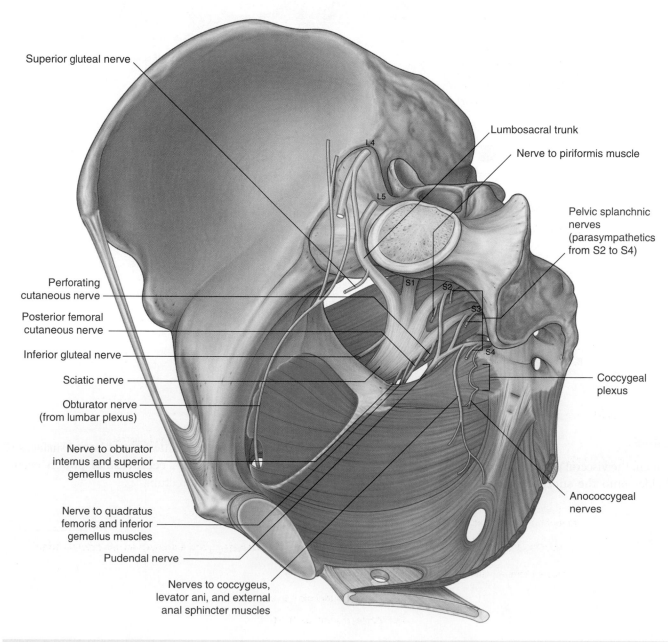

**Fig. 5.60** Sacral and coccygeal plexuses.

## Sacral plexus

The sacral plexus on each side is formed by the anterior rami of S1 to S4, and the lumbosacral trunk (L4 and L5) (Fig. 5.61). The plexus is formed in relation to the anterior surface of the piriformis muscle, which is part of the posterolateral pelvic wall. Sacral contributions to the plexus pass out of the anterior sacral foramina and course laterally and inferiorly on the pelvic wall. The lumbosacral trunk, consisting of part of the anterior ramus of L4 and all of the anterior ramus of L5, courses vertically into the pelvic cavity from the abdomen by passing immediately anterior to the sacro-iliac joint.

Gray rami communicantes from ganglia of the sympathetic trunk connect with each of the anterior rami and carry postganglionic sympathetic fibers destined for the

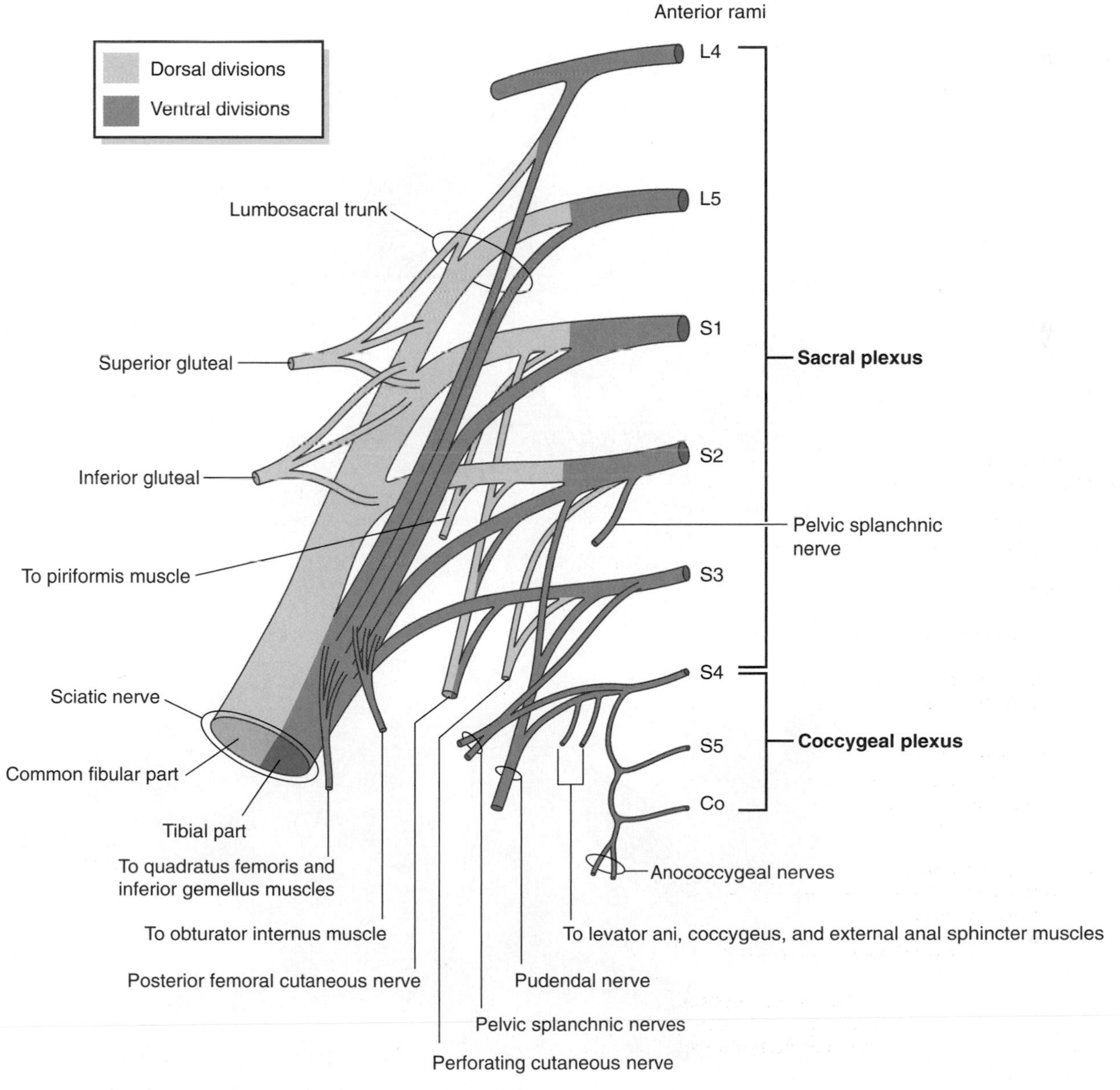


**Fig. 5.61** Components and branches of the sacral and coccygeal plexuses.

periphery to the somatic nerves (Fig. 5.62). In addition, special visceral nerves (**pelvic splanchnic nerves**) originating from S2 to S4 deliver preganglionic parasympathetic fibers to the pelvic part of the prevertebral plexus (Figs. 5.60 and 5.61).

Each anterior ramus has ventral and dorsal divisions that combine with similar divisions from other levels to form terminal nerves (Fig. 5.61). The anterior ramus of S4 has only a ventral division.

Branches of the sacral plexus include the sciatic nerve and gluteal nerves, which are major nerves of the lower limb, and the pudendal nerve, which is the nerve of the perineum (Table 5.4). Numerous smaller branches supply the pelvic wall, floor, and lower limb.

Most nerves originating from the sacral plexus leave the pelvic cavity by passing through the greater sciatic foramen inferior to the piriformis muscle, and enter the gluteal region of the lower limb. Other nerves leave the pelvic cavity using different routes; a few nerves do not leave the pelvic cavity and course directly into the muscles in the pelvic cavity. Finally, two nerves that leave the pelvic cavity through the greater sciatic foramen loop around the ischial spine and sacrospinous ligament and pass medially through the lesser sciatic foramen to supply structures in the perineum and lateral pelvic wall.

***Sciatic nerve.*** The **sciatic nerve** is the largest nerve of the body and carries contributions from L4 to S3 (Figs. 5.60 and 5.61). It:

- forms on the anterior surface of the piriformis muscle and leaves the pelvic cavity through the greater sciatic foramen inferior to the piriformis;

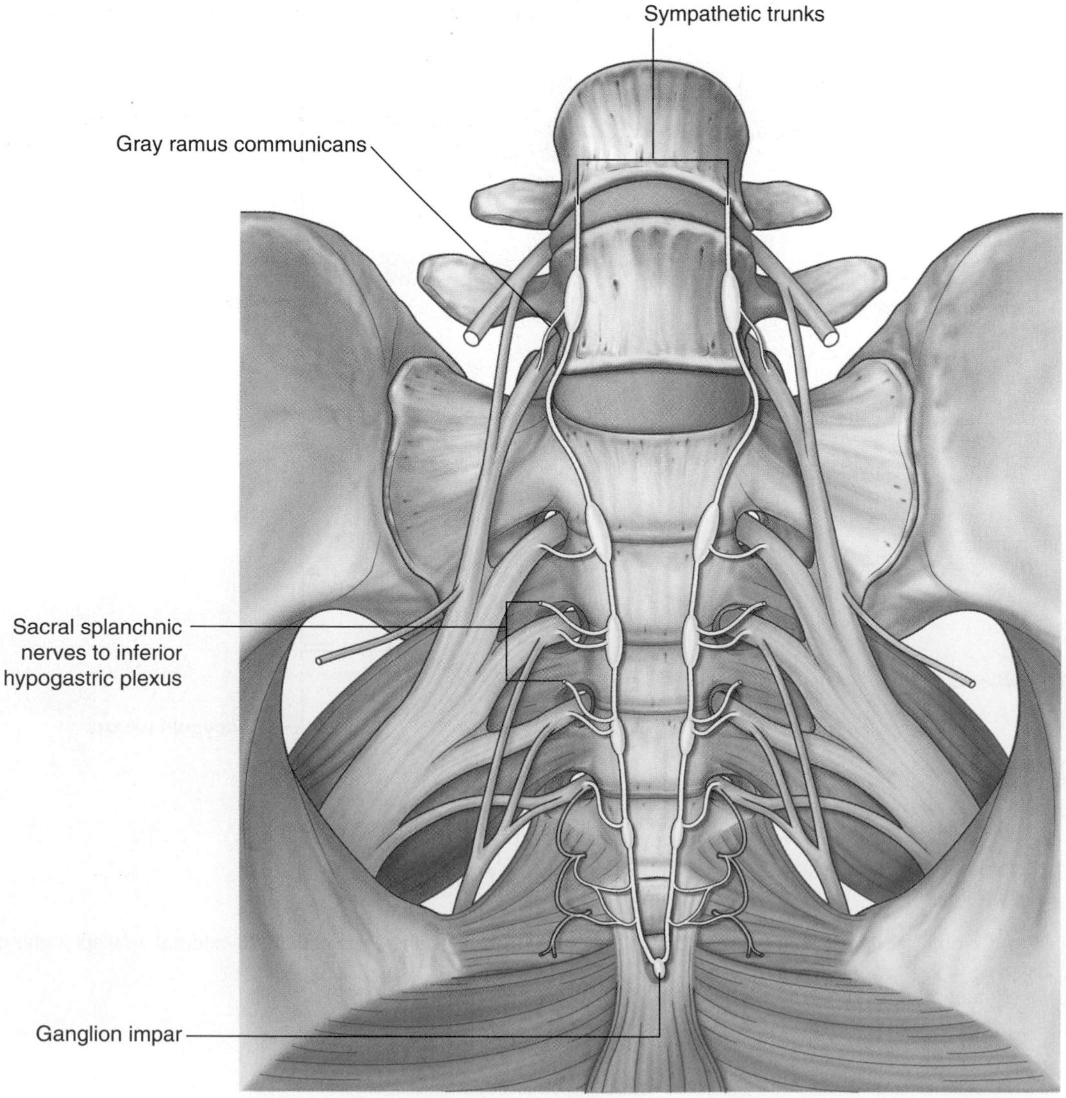

**Fig. 5.62** Sympathetic trunks in the pelvis.

**Table 5.4** Branches of the sacral and coccygeal plexuses (spinal segments in parentheses do not consistently participate)

| Branch | Spinal segments | Motor function |
|---|---|---|
| **SACRAL PLEXUS** | | |
| **Sciatic** | | |
| Tibial part | L4 to S3 | **Motor function**<br>All muscles in the posterior or hamstring compartment of the thigh (including the hamstring part of the adductor magnus) except for the short head of the biceps<br>All muscles in the posterior compartment of the leg<br>All muscles in the sole of the foot<br>**Sensory (cutaneous) function**<br>Skin on posterolateral and lateral surfaces of foot and sole of foot |
| Common fibular part | L4 to S2 | **Motor function**<br>Short head of biceps in the posterior compartment of the thigh<br>All muscles in the anterior and lateral compartments of the leg<br>Extensor digitorum brevis in the foot (also contributes to the supply of the first dorsal interosseous muscle)<br>**Sensory (cutaneous) function**<br>Skin on the anterolateral surface of the leg and dorsal surface of the foot |
| **Pudendal** | S2 to S4 | **Motor function**<br>Skeletal muscles in the perineum including the external urethral and anal sphincters and levator ani (overlaps in supply of the levator ani and external sphincter with branches directly from ventral division of S4)<br>**Sensory (cutaneous) function**<br>Most skin of the perineum. Penis and clitoris |
| **Superior gluteal** | L4 to S1 | **Motor function**<br>Gluteus medius, gluteus minimus, and tensor fasciae latae |
| **Inferior gluteal** | L5 to S2 | **Motor function**<br>Gluteus maximus |
| **Nerve to obturator internus and superior gemellus** | L5 to S2 | **Motor function**<br>Obturator internus and superior gemellus |

*(continues)*

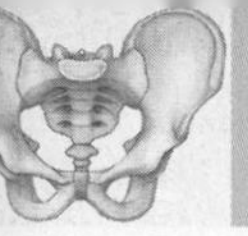



| **Table 5.4** Branches of the sacral and coccygeal plexuses (spinal segments in parentheses do not consistently participate)—cont'd | | |
|---|---|---|
| **Branch** | | |
| **Nerve to quadratus femoris and inferior gemellus** | L4 to S1 | **Motor function**<br>Quadratus femoris and inferior gemellus |
| **Posterior femoral cutaneous (posterior cutaneous nerve of thigh)** | S1, S3 | **Sensory (cutaneous) function**<br>Skin on the posterior aspect of the thigh |
| Perforating cutaneous | S2, S3 | **Sensory (cutaneous) function**<br>Skin over gluteal fold (overlaps with posterior femoral cutaneous) |
| **Nerve to piriformis** | S1, S2 | **Motor function**<br>Piriformis muscle |
| **Nerves to levator ani, coccygeus, and external anal sphincter** | S4 | **Motor function**<br>Levator ani, coccygeus, and external anal sphincter. (Overlaps with pudendal nerve)<br>**Sensory (cutaneous) function**<br>Small patch of skin between anus and coccyx |
| **Pelvic splanchnic nerves** | S2, S3 (4) | **Motor (visceral) function**<br>Visceral motor (preganglionic parasympathetic) to pelvic part of prevertebral plexus<br>Stimulate erection, modulate mobility in gastrointestinal system distal to the left colic flexure, inhibitory to internal urethral sphincter<br>**Sensory (visceral) function**<br>Visceral afferents (that follow the parasympathetics) from pelvic viscera and distal parts of colon. Pain from cervix and possibly from bladder and proximal urethra |
| **COCCYGEAL PLEXUS** | | |
| **Anococcygeal nerves** | S4 to Co | **Sensory (cutaneous) function**<br>Perianal skin |

- passes through the gluteal region into the thigh, where it divides into its two major branches, the common fibular nerve (common peroneal nerve) and the tibial nerve—dorsal divisions of L4, L5, S1, and S2 are carried in the common fibular part of the nerve and the ventral divisions of L4, L5, S1, S2, and S3 are carried in the tibial part;
- innervates muscles in the posterior compartment of the thigh and muscles in the leg and foot; and
- carries sensory fibers from the skin of the foot and lateral leg.

***Pudendal nerve.*** The **pudendal nerve** forms anteriorly to the lower part of the piriformis muscle from ventral divisions of S2 to S4 (Figs. 5.60 and 5.61). It:

- leaves the pelvic cavity through the greater sciatic foramen, inferior to the piriformis muscle, and enters the gluteal region;
- courses into the perineum by immediately passing around the sacrospinous ligament, where the ligament joins the ischial spine, and through the lesser sciatic foramen (this course takes the nerve out of the pelvic cavity, around the peripheral attachment of the pelvic floor, and into the perineum);
- is accompanied throughout its course by the internal pudendal vessels; and
- innervates skin and skeletal muscles of the perineum, including the external anal and external urethral sphincters.

## In the clinic

### Pudendal block

Pudendal block anesthesia is performed to relieve the pain associated with childbirth. Although the use of this procedure is less common since the widespread adoption of epidural anesthesia, it provides an excellent option for women who have a contraindication to neuraxial anesthesia (e.g., spinal anatomy, low platelets, too close to delivery). Pudendal blocks are also used for certain types of chronic pelvic pain and in some rectal or urological procedures. The injection is usually given where the pudendal nerve crosses the lateral aspect of the sacrospinous ligament near its attachment to the ischial spine. During childbirth, a finger inserted into the vagina can palpate the ischial spine. The needle is passed transcutaneously to the medial aspect of the ischial spine and around the sacrospinous ligament. Infiltration is performed and the perineum is anesthetized. Pudendal nerve blocks can also be performed with imaging guidance (using fluoroscopy, computed tomography, or ultrasound) to localize the nerve rather than relying purely on anatomical landmarks.

***Other branches of the sacral plexus.*** Other branches of the sacral plexus include:

- motor branches to muscles of the gluteal region, pelvic wall, and pelvic floor (superior and inferior gluteal nerves, nerve to obturator internus and superior gemellus, nerve to quadratus femoris and inferior gemellus, nerve to piriformis, nerves to levator ani); and
- sensory nerves to skin over the inferior gluteal region and posterior aspects of the thigh and upper leg (perforating cutaneous nerve and posterior cutaneous nerve of the thigh) (Figs. 5.60 and 5.61).

The **superior gluteal nerve**, formed by branches from the dorsal divisions of L4 to S1, leaves the pelvic cavity through the greater sciatic foramen superior to the piriformis muscle and supplies muscles in the gluteal region—**gluteus medius**, **gluteus minimus**, and **tensor fasciae latae** (**tensor of fascia lata**) **muscles**.

The **inferior gluteal nerve**, formed by branches from the dorsal divisions of L5 to S2, leaves the pelvic cavity through the greater sciatic foramen inferior to the piriformis muscle and supplies the **gluteus maximus**, the largest muscle in the gluteal region.

Both superior and inferior gluteal nerves are accompanied by corresponding arteries.

The **nerve to the obturator internus** and the associated **superior gemellus** muscle originates from the ventral divisions of L5 to S2 and leaves the pelvic cavity through the greater sciatic foramen inferior to the piriformis muscle. Like the pudendal nerve, it passes around the ischial spine and through the lesser sciatic foramen to enter the perineum and supply the obturator internus muscle from the medial side of the muscle, inferior to the attachment of the levator ani muscle.

The **nerve to the quadratus femoris** muscle and the **inferior gemellus** muscle, and the **posterior cutaneous nerve of the thigh** (**posterior femoral cutaneous nerve**) also leave the pelvic cavity through the greater sciatic foramen inferior to the piriformis muscle and course to muscles and skin, respectively, in the lower limb.

Unlike most of the other nerves originating from the sacral plexus, which leave the pelvic cavity through the greater sciatic foramen either above or below the piriformis muscle, the **perforating cutaneous nerve** leaves the pelvic cavity by penetrating directly through the sacrotuberous ligament and then courses to skin over the inferior aspect of the buttocks.

The **nerve to the piriformis** and a number of small nerves to the levator ani and coccygeus muscles originate from the sacral plexus and pass directly into their target muscles without leaving the pelvic cavity.

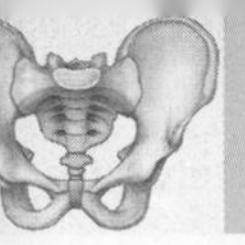



The **obturator nerve** (L2 to L4) is a branch of the lumbar plexus. It passes inferiorly along the posterior abdominal wall within the psoas muscle, emerges from the medial surface of the psoas, passes posteriorly to the common iliac artery and medially to the internal iliac artery at the pelvic inlet, and then courses along the lateral pelvic wall. It leaves the pelvic cavity by traveling through the obturator canal and supplies the adductor region of the thigh.

### Coccygeal plexus

The small coccygeal plexus has a minor contribution from S4 and is formed mainly by the anterior rami of S5 and Co, which originate inferiorly to the pelvic floor. They penetrate the coccygeus muscle to enter the pelvic cavity and join with the anterior ramus of S4 to form a single trunk, from which small **anococcygeal nerves** originate (Table 5.4). These nerves penetrate the muscle and the overlying sacrospinous and sacrotuberous ligaments and pass superficially to innervate skin in the anal triangle of the perineum.

## Visceral plexuses

### Paravertebral sympathetic chain

The paravertebral part of the visceral nervous system is represented in the pelvis by the inferior ends of the sympathetic trunks (Fig. 5.63A). Each trunk enters the pelvic cavity from the abdomen by passing over the ala of the

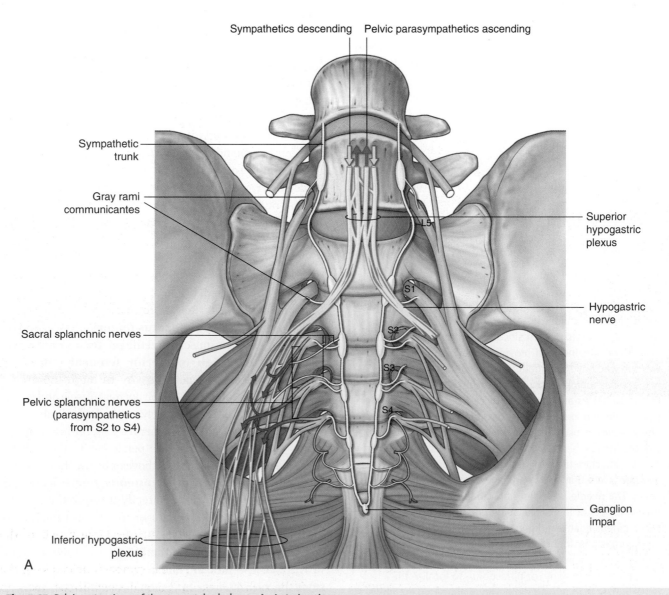



**Fig. 5.63** Pelvic extensions of the prevertebral plexus. **A.** Anterior view.

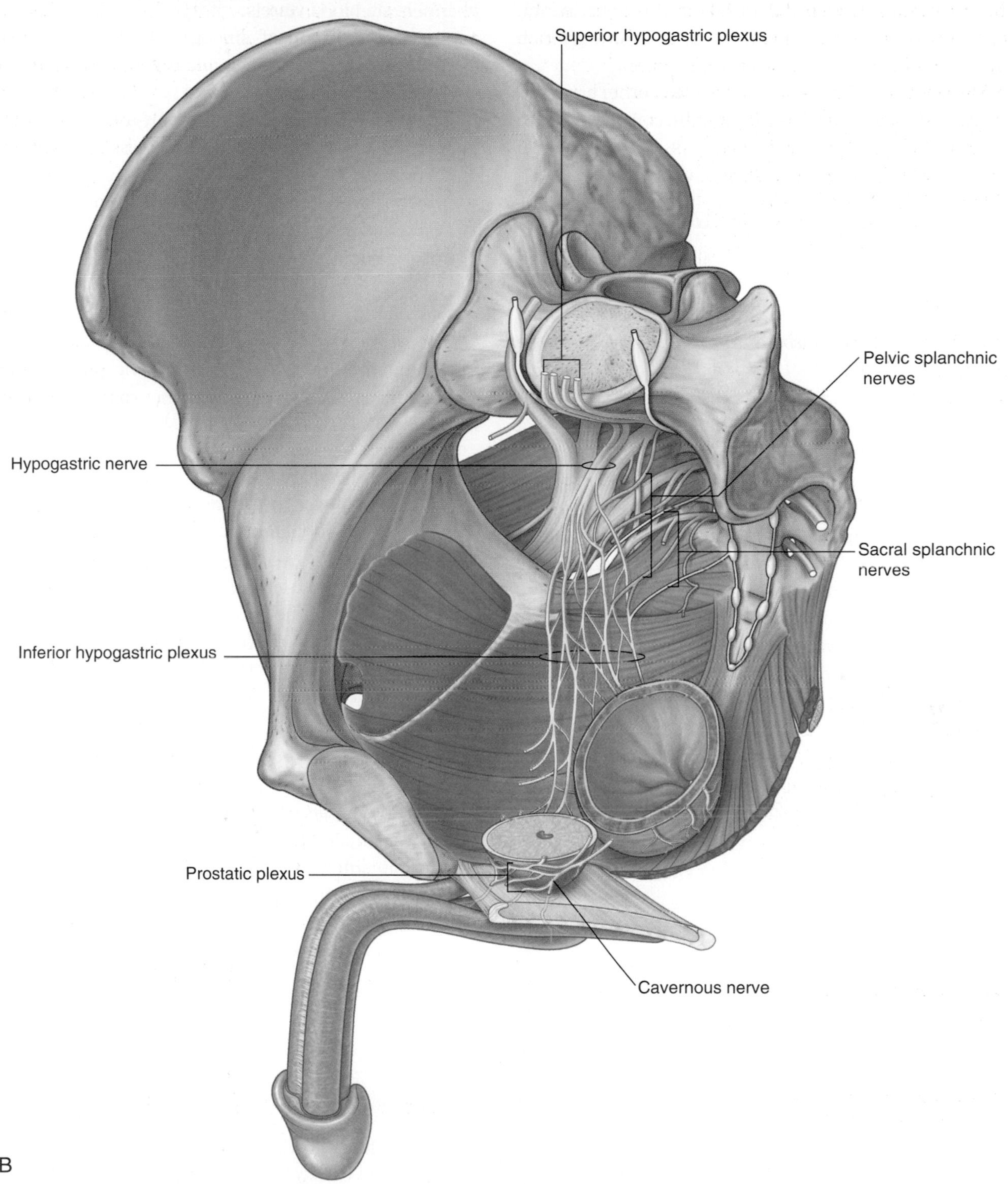


**Fig. 5.63, cont'd** **B.** Anteromedial view of right side of plexus.

sacrum medially to the lumbosacral trunks and posteriorly to the iliac vessels. The trunks course inferiorly along the anterior surface of the sacrum, where they are positioned medially to the anterior sacral foramina. Four ganglia occur along each trunk. Anteriorly to the coccyx, the two trunks join to form a single small terminal ganglion (the **ganglion impar**).

The principal function of the sympathetic trunks in the pelvis is to deliver postganglionic sympathetic fibers to the anterior rami of sacral nerves for distribution to the periphery, mainly to parts of the lower limb and perineum. This is accomplished by gray rami communicantes, which connect the trunks to the sacral anterior rami.

In addition to gray rami communicantes, other branches (the **sacral splanchnic nerves**) join and contribute to the pelvic part of the prevertebral plexus associated with innervating pelvic viscera (Fig. 5.63A).

### Pelvic extensions of the prevertebral plexus

The pelvic parts of the prevertebral plexus carry sympathetic, parasympathetic, and visceral afferent fibers (Fig. 5.63A). Pelvic parts of the plexus are associated with innervating pelvic viscera and erectile tissues of the perineum.

The prevertebral plexus enters the pelvis as two **hypogastric nerves**, one on each side, that cross the pelvic inlet medially to the internal iliac vessels (Fig. 5.63A). The hypogastric nerves are formed by the separation of the fibers in the **superior hypogastric plexus**, into right and left bundles. The superior hypogastric plexus is situated anterior to vertebra LV between the promontory of the sacrum and the bifurcation of the aorta.

When the hypogastric nerves are joined by pelvic splanchnic nerves carrying preganglionic parasympathetic fibers from S2 to S4, the **pelvic plexuses** (**inferior hypogastric plexuses**) are formed (Fig. 5.63). The inferior hypogastric plexuses, one on each side, course in an inferior direction around the pelvic walls, medially to major vessels and somatic nerves. They give origin to the following subsidiary plexuses, which innervate the pelvic viscera:

- the **rectal plexus**,
- the **uterovaginal plexus**,
- the **prostatic plexus**, and
- the **vesical plexus**.

Terminal branches of the inferior hypogastric plexuses penetrate and pass through the deep perineal pouch and innervate erectile tissues of the penis and the clitoris in the perineum (Fig. 5.63B). In men, these nerves, called **cavernous nerves**, are extensions of the prostatic plexus. The pattern of distribution of similar nerves in women is not entirely clear, but they are likely extensions of the uterovaginal plexus.

#### Sympathetic fibers

Sympathetic fibers enter the inferior hypogastric plexuses from the hypogastric nerves and from branches (sacral splanchnic nerves) of the upper sacral parts of the sympathetic trunks (Fig. 5.63A). Ultimately, these nerves are derived from preganglionic fibers that leave the spinal cord in the anterior roots, mainly of T10 to L2. These fibers:

- innervate blood vessels,
- cause contraction of smooth muscle in the internal urethral sphincter in men and the internal anal sphincters in both men and women,
- cause smooth muscle contraction associated with the reproductive tract and with the accessory glands of the reproductive system, and
- are important in moving secretions from the epididymis and associated glands into the urethra to form semen during ejaculation.

#### Parasympathetic fibers

Parasympathetic fibers enter the pelvic plexus in pelvic splanchnic nerves that originate from spinal cord levels S2 to S4 (Fig. 5.63A). They:

- are generally vasodilatory,
- stimulate bladder contraction,
- stimulate erection, and
- modulate activity of the enteric nervous system of the colon distal to the left colic flexure (in addition to pelvic viscera, some of the fibers from the pelvic plexus course superiorly in the prevertebral plexus, or as separate nerves, and pass into the inferior mesenteric plexus of the abdomen).

#### Visceral afferent fibers

Visceral afferent fibers follow the course of the sympathetic and parasympathetic fibers to the spinal cord. Afferent fibers that enter the cord in lower thoracic levels and lumbar levels with sympathetic fibers generally carry pain; however, pain fibers from the cervix and some pain fibers from the bladder and urethra may accompany parasympathetic nerves to sacral levels of the spinal cord.

### In the clinic

**Prostatectomy and impotence**

It may be necessary to perform radical surgery to cure cancer of the prostate. To do this, the prostate and its attachments around the base of the bladder, including the seminal vesicles, must be removed en masse. Parts of the inferior hypogastric plexus in this region give rise to nerves that innervate the erectile tissues of the penis. Impotence may occur if these nerves cannot be or are not preserved during removal of the prostate.

For the same reasons, women may experience sexual dysfunction if similar nerves are damaged during pelvic surgery, for example, during a total hysterectomy.

## In the clinic

### Robotic prostatectomy

This is a new and innovative way of performing radical prostatectomy in patients with prostate cancer. The patient is placed on an operating table near a so-called patient unit consisting of a high-resolution camera and three arms containing microsurgical instruments. The surgeon operates the robot from a computer console and views the surgical field on a monitor as magnified 3D images. The operator usually makes a number of incisions between 1 cm to 2 cm wide through which the camera and surgical instruments are inserted into the pelvis. The surgeon's hand movements are filtered and translated by the robot into very fine and precise movements of the microtools. This markedly increases the precision of prostate removal and reduces the risk of nerve damage and potential development of postsurgical erectile dysfunction.

## Blood vessels

### Arteries

The major artery of the pelvis and perineum is the internal iliac artery on each side (Fig. 5.64). In addition to providing a blood supply to most of the pelvic viscera, pelvic walls and floor, and structures in the perineum, including erectile tissues of the clitoris and the penis, this artery gives rise to branches that follow nerves into the gluteal region of the lower limb. Other vessels that originate in the abdomen and contribute to the supply of pelvic structures include the median sacral artery and, in women, the ovarian arteries.

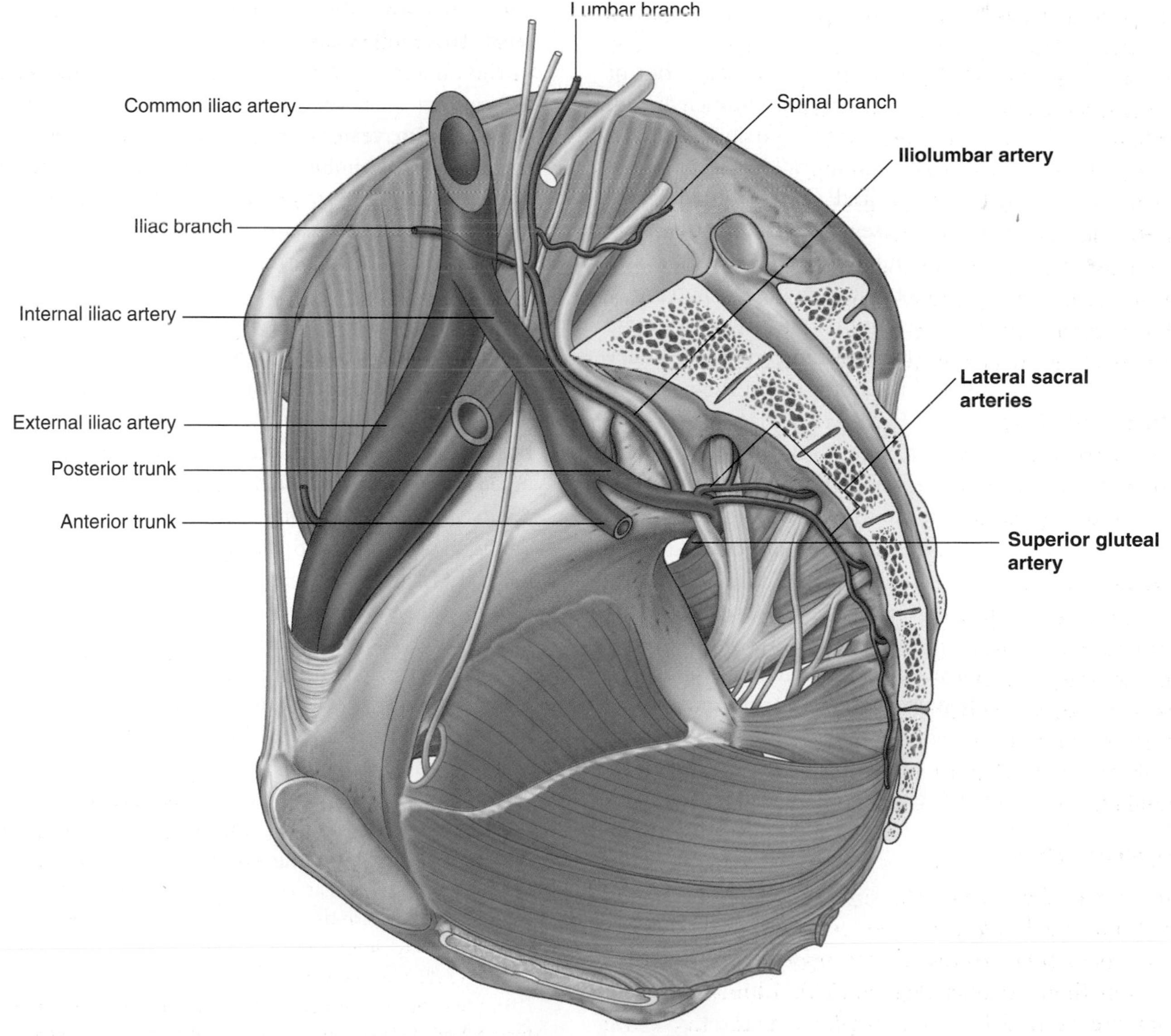


**Fig. 5.64** Branches of the posterior trunk of the internal iliac artery.

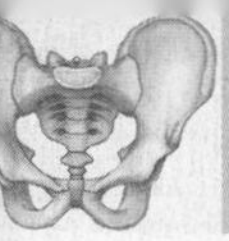



### Internal iliac artery

The internal iliac artery originates from the common iliac artery on each side, approximately at the level of the intervertebral disc between LV and SI, and lies anteromedial to the sacro-iliac joint (Fig. 5.64). The vessel courses inferiorly over the pelvic inlet and then divides into anterior and posterior trunks at the level of the superior border of the greater sciatic foramen. Branches from the posterior trunk contribute to the supply of the lower posterior abdominal wall, the posterior pelvic wall, and the gluteal region. Branches from the anterior trunk supply the pelvic viscera, the perineum, the gluteal region, the adductor region of the thigh, and, in the fetus, the placenta.

#### Posterior trunk

Branches of the posterior trunk of the internal iliac artery are the iliolumbar artery, the lateral sacral artery, and the superior gluteal artery (Fig. 5.64).

- The **iliolumbar artery** ascends laterally back out of the pelvic inlet and divides into a lumbar branch and an iliac branch. The lumbar branch contributes to the supply of the posterior abdominal wall, psoas and quadratus lumborum muscles, and cauda equina, via a small spinal branch that passes through the intervertebral foramen between LV and SI. The iliac branch passes laterally into the iliac fossa to supply muscle and bone.
- The **lateral sacral arteries**, usually two, originate from the posterior division of the internal iliac artery and course medially and inferiorly along the posterior pelvic wall. They give rise to branches that pass into the anterior sacral foramina to supply related bone and soft tissues, structures in the vertebral (sacral) canal, and skin and muscle posterior to the sacrum.
- The **superior gluteal artery** is the largest branch of the internal iliac artery and is the terminal continuation of the posterior trunk. It courses posteriorly, usually passing between the lumbosacral trunk and anterior ramus of S1, to leave the pelvic cavity through the greater sciatic foramen above the piriformis muscle and enter the gluteal region of the lower limb. This vessel makes a substantial contribution to the blood supply of muscles and skin in the gluteal region and also supplies branches to adjacent muscles and bones of the pelvic walls.

#### Anterior trunk

Branches of the anterior trunk of the internal iliac artery include the superior vesical artery, the umbilical artery, the inferior vesical artery, the middle rectal artery, the uterine artery, the vaginal artery, the obturator artery, the internal pudendal artery, and the inferior gluteal artery (Fig. 5.65).

- The first branch of the anterior trunk is the **umbilical artery**, which gives origin to the superior vesical artery and then travels forward just inferior to the margin of the pelvic inlet. Anteriorly, the vessel leaves the pelvic cavity and ascends on the internal aspect of the anterior abdominal wall to reach the umbilicus. In the fetus, the umbilical artery is large and carries blood from the fetus to the placenta. After birth, the vessel closes distally to the origin of the superior vesical artery and eventually becomes a solid fibrous cord. On the anterior abdominal wall, the cord raises a fold of peritoneum termed the **medial umbilical fold**. The fibrous remnant of the umbilical artery itself is the **medial umbilical ligament**.
- The **superior vesical artery** normally originates from the root of the umbilical artery and courses medially and inferiorly to supply the superior aspect of the bladder and distal parts of the ureter. In men, it also may give rise to an artery that supplies the ductus deferens.
- The **inferior vesical artery** occurs in men and supplies branches to the bladder, ureter, seminal vesicle, and prostate. The **vaginal artery** in women is the equivalent of the inferior vesical artery in men and, descending to the vagina, supplies branches to the vagina and to adjacent parts of the bladder and rectum. The vaginal artery and uterine artery may originate together as a common branch from the anterior trunk, or the vaginal artery may arise independently.
- The **middle rectal artery** courses medially to supply the rectum. The vessel anastomoses with the superior rectal artery, which originates from the inferior mesenteric artery in the abdomen, and the inferior rectal artery, which originates from the internal pudendal artery in the perineum.
- The **obturator artery** courses anteriorly along the pelvic wall and leaves the pelvic cavity via the obturator canal. Together with the obturator nerve, above, and

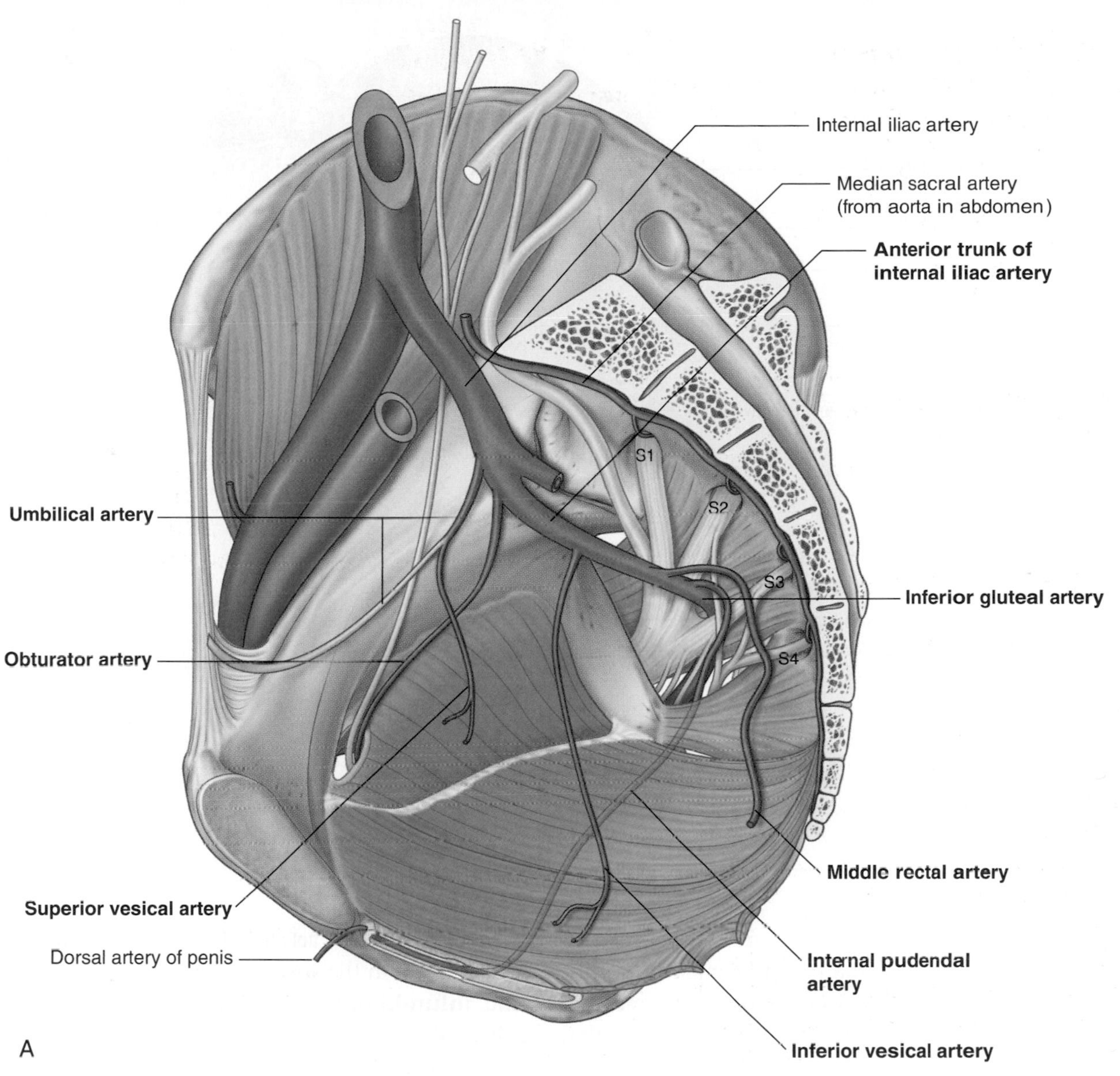


**Fig. 5.65** Branches of the anterior trunk of the internal iliac artery. **A.** Male.

obturator vein, below, it enters and supplies the adductor region of the thigh.

- The **internal pudendal artery** courses inferiorly from its origin in the anterior trunk and leaves the pelvic cavity through the greater sciatic foramen inferior to the piriformis muscle. In association with the pudendal nerve on its medial side, the vessel passes laterally to the ischial spine and then through the lesser sciatic foramen to enter the perineum. The internal pudendal artery is the main artery of the perineum. Among the structures it supplies are the erectile tissues of the clitoris and the penis.
- The **inferior gluteal artery** is a large terminal branch of the anterior trunk of the internal iliac artery. It passes between the anterior rami S1 and S2 or S2 and S3 of the sacral plexus and leaves the pelvic cavity through the greater sciatic foramen inferior to the piriformis muscle. It enters and contributes to the blood supply of the gluteal region and anastomoses with a network of vessels around the hip joint.

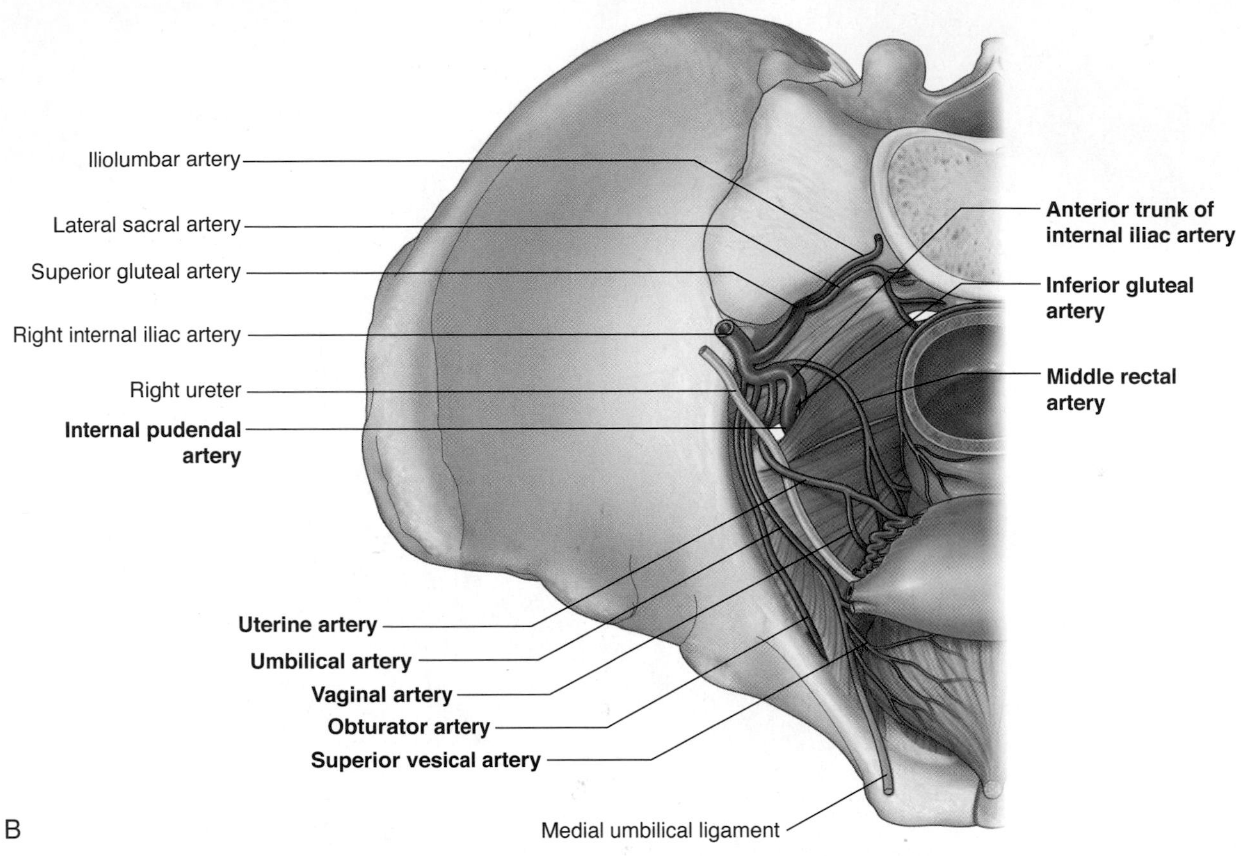


**Fig. 5.65, cont'd B.** Female.

- The **uterine artery** in women courses medially and anteriorly in the base of the broad ligament to reach the cervix (Figs. 5.65B and 5.66). Along its course, the vessel crosses the ureter and passes superiorly to the lateral vaginal fornix. Once the vessel reaches the cervix, it ascends along the lateral margin of the uterus to reach the uterine tube, where it curves laterally and anastomoses with the ovarian artery. The uterine artery is the major blood supply to the uterus and enlarges significantly during pregnancy. Through anastomoses with other arteries, the vessel contributes to the blood supply of the ovary and vagina as well.

### Ovarian arteries

In women, the gonadal (ovarian) vessels originate from the abdominal aorta and then descend to cross the pelvic inlet and supply the ovaries. They anastomose with terminal parts of the uterine arteries (Fig. 5.66). On each side, the vessels travel in the **suspensory ligament of the ovary** (the **infundibulopelvic ligament**) as they cross the pelvic inlet to the ovary. Branches pass through the mesovarium to reach the ovary and through the mesometrium of the broad ligament to anastomose with the uterine artery. The ovarian arteries enlarge significantly during pregnancy to augment the uterine blood supply.

### Median sacral artery

The median sacral artery (Figs. 5.65A and 5.66) originates from the posterior surface of the aorta just superior to the aortic bifurcation at vertebral level LIV in the abdomen. It descends in the midline, crosses the pelvic inlet, and then courses along the anterior surface of the sacrum and coccyx. It gives rise to the last pair of lumbar arteries and to branches that anastomose with the iliolumbar and lateral sacral arteries.

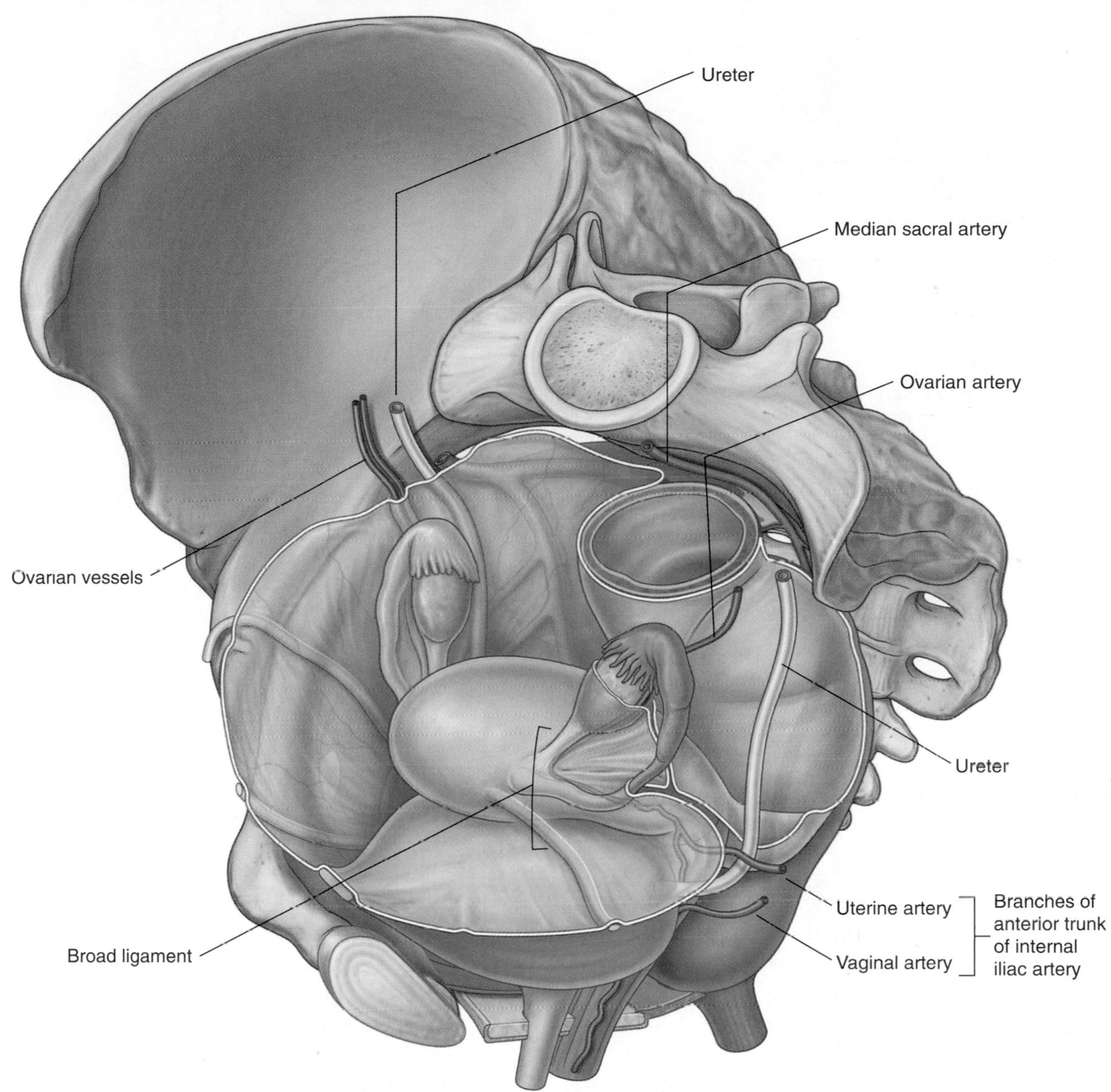


**Fig. 5.66** Uterine and vaginal arteries.

### Veins

Pelvic veins follow the course of all branches of the internal iliac artery except for the umbilical artery and the iliolumbar artery (Fig. 5.67A). On each side, the veins drain into internal iliac veins, which leave the pelvic cavity to join common iliac veins situated just superior and lateral to the pelvic inlet.

Within the pelvic cavity, extensive interconnected venous plexuses are associated with the surfaces of the viscera (bladder, rectum, prostate, uterus, and vagina). Together, these plexuses form the pelvic plexus of veins. The part of the venous plexus surrounding the rectum and anal canal drains via superior rectal veins (tributaries of inferior mesenteric veins) into the hepatic portal system, and via middle and inferior rectal veins into the caval system. This pelvic plexus is an important portacaval shunt when the hepatic portal system is blocked (Fig. 5.67B).

The inferior part of the rectal plexus around the anal canal has two parts, an internal and an external. The internal rectal plexus is in connective tissue between the internal anal sphincter and the epithelium lining the canal. This plexus connects superiorly with longitudinally arranged branches of the superior rectal vein that lie one

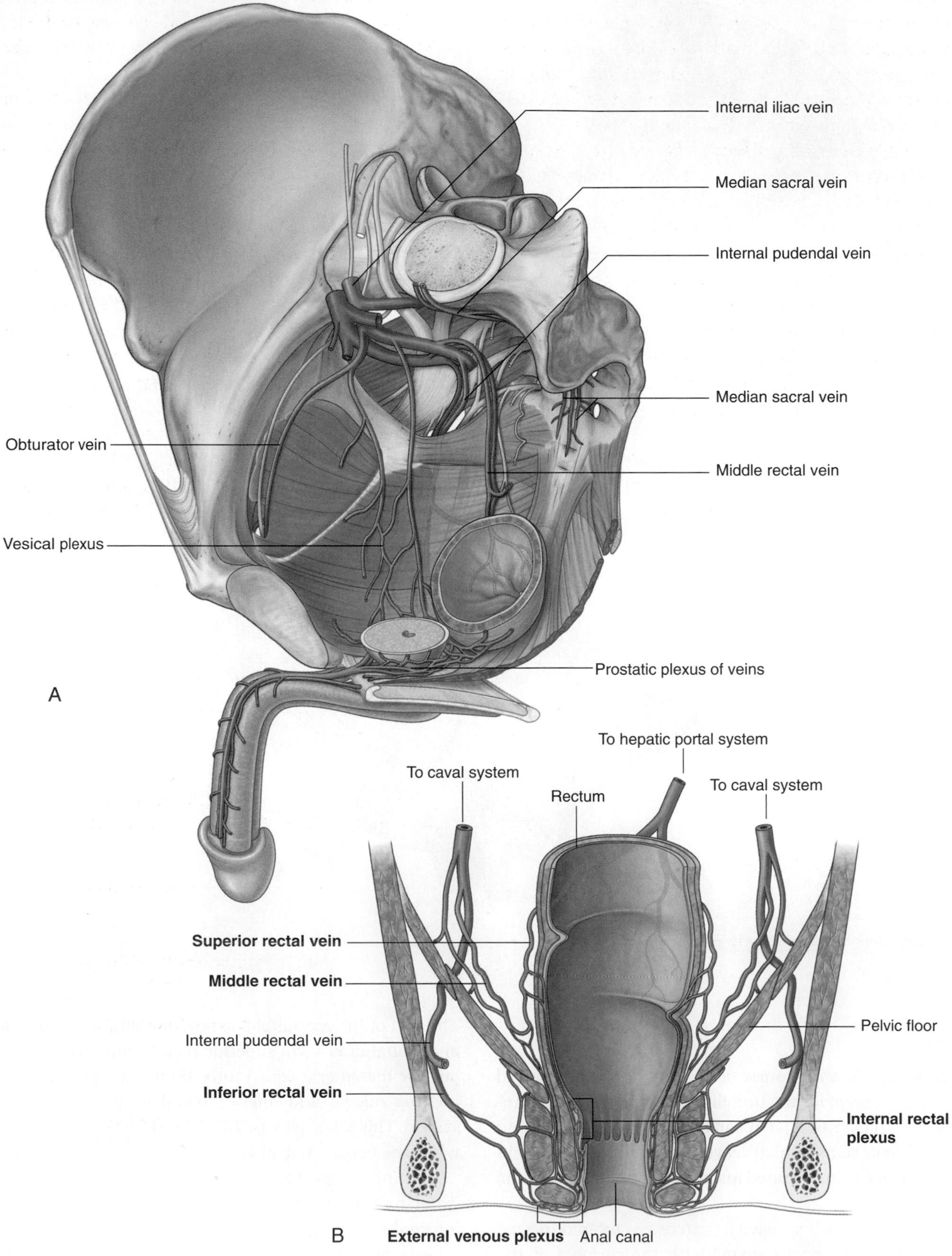

**Fig. 5.67** Pelvic veins. **A.** In a man with the left side of the pelvis and most of the viscera removed. **B.** Veins associated with the rectum and anal canal.

in each anal column. When enlarged, these branches form varices or internal hemorrhoids, which originate above the pectinate line and are covered by colonic mucosa. The external rectal plexus circles the external anal sphincter and is subcutaneous. Enlargement of vessels in the external rectal plexus results in external hemorrhoids.

The single **deep dorsal vein** that drains erectile tissues of the clitoris and the penis does not follow branches of the internal pudendal artery into the pelvic cavity. Instead, this vein passes directly into the pelvic cavity through a gap formed between the arcuate pubic ligament and the anterior margin of the perineal membrane. The vein joins the prostatic plexus of veins in men and the vesical (bladder) plexus of veins in women. (Superficial veins that drain the skin of the penis and corresponding regions of the clitoris drain into the external pudendal veins, which are tributaries of the great saphenous vein in the thigh.)

In addition to tributaries of the internal iliac vein, median sacral veins and ovarian veins parallel the courses of the median sacral artery and ovarian artery, respectively, and leave the pelvic cavity to join veins in the abdomen:

- The **median sacral veins** coalesce to form a single vein that joins either the left common iliac vein or the junction of the two common iliac veins to form the inferior vena cava.
- The **ovarian veins** follow the course of the corresponding arteries; on the left, they join the left renal vein and, on the right, they join the inferior vena cava in the abdomen.

## Lymphatics

Lymphatics from most pelvic viscera drain mainly into lymph nodes distributed along the internal iliac and external iliac arteries and their associated branches (Fig. 5.68), which drain into nodes associated with the common iliac arteries and then into the lateral aortic or lumbar nodes associated with the lateral surfaces of the abdominal aorta. In turn, these lateral aortic or lumbar nodes drain into the lumbar trunks, which continue to the origin of the thoracic duct at approximately vertebral level TXII.

Lymphatics from the ovaries and related parts of the uterus and uterine tubes leave the pelvic cavity superiorly

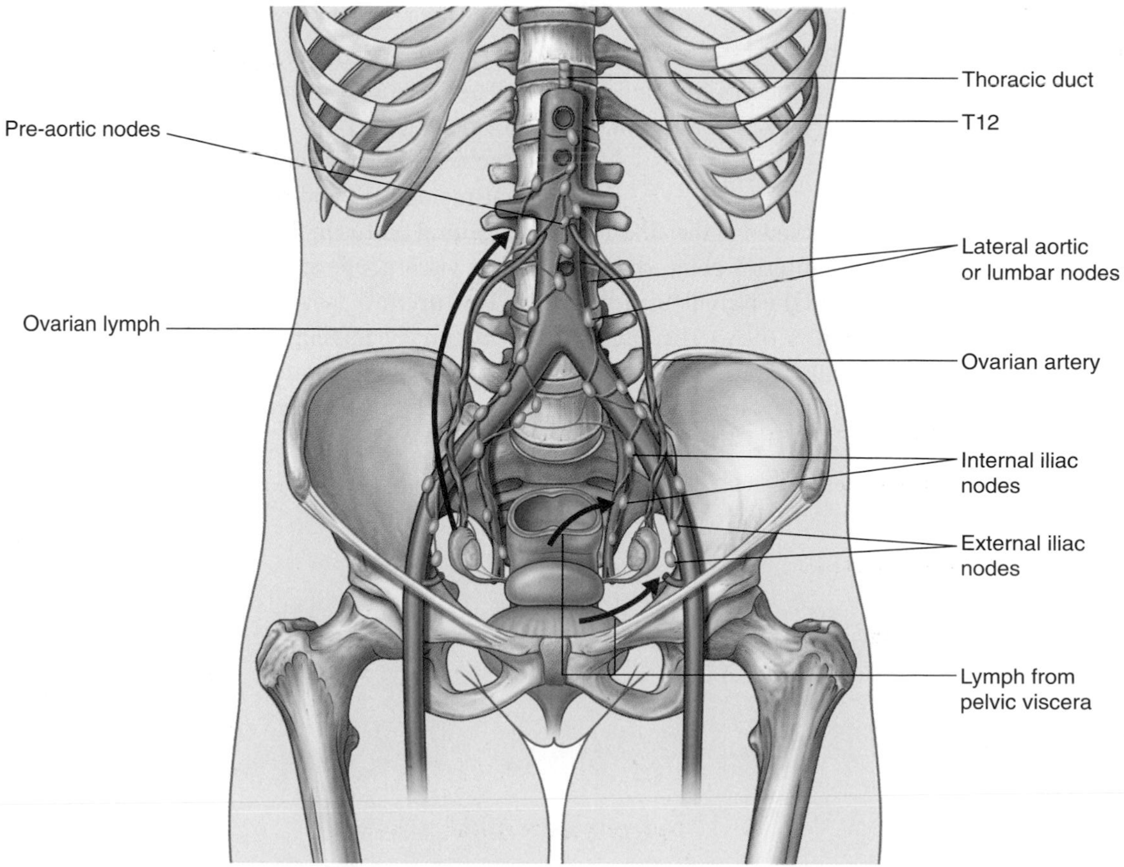


**Fig. 5.68** Pelvic lymphatics.

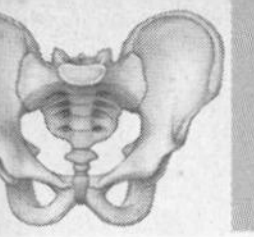


and drain, via vessels that accompany the ovarian arteries, directly into lateral aortic or lumbar nodes and, in some cases, into the pre-aortic nodes on the anterior surface of the aorta.

In addition to draining pelvic viscera, nodes along the internal iliac artery also receive drainage from the gluteal region of the lower limb and from deep areas of the perineum.

## PERINEUM

The perineum is a diamond-shaped region positioned inferiorly to the pelvic floor between the thighs. Its peripheral boundary is the pelvic outlet; its ceiling is the pelvic diaphragm (the levator ani and coccygeus muscles); and its narrow lateral walls are formed by the walls of the pelvic cavity below the attachment of the levator ani muscle (Fig. 5.69A).

The perineum is divided into an anterior urogenital triangle and a posterior anal triangle.

- The urogenital triangle is associated with the openings of the urinary systems and the reproductive systems and functions to anchor the external genitalia.
- The anal triangle contains the anus and the external anal sphincter.

The pudendal nerve (S2 to S4) and the internal pudendal artery are the major nerve and artery of the region.

### Borders and ceiling

The margin of the perineum is marked by the inferior border of the pubic symphysis at its anterior point, the tip of the coccyx at its posterior point, and the ischial tuberosities at each of the lateral points (Fig. 5.69A). The lateral margins are formed by the ischiopubic rami anteriorly and by the sacrotuberous ligaments posteriorly. The pubic symphysis, the ischial tuberosities, and the coccyx can be palpated on the patient.

The perineum is divided into two triangles by an imaginary line between the two ischial tuberosities (Fig. 5.69A). Anterior to the line is the urogenital triangle and posterior to the line is the anal triangle. Significantly, the two triangles are not in the same plane. In the anatomical position, the urogenital triangle is oriented in the horizontal plane, whereas the anal triangle is tilted upward at the transtubercular line so that it faces more posteriorly.

The roof of the perineum is formed mainly by the levator ani muscles that separate the pelvic cavity, above, from the perineum, below. These muscles, one on each side, form a cone- or funnel-shaped pelvic diaphragm, with the anal aperture at its inferior apex in the anal triangle.

Anteriorly, in the **urogenital triangle**, a U-shaped defect in the muscles, the **urogenital hiatus**, allows the passage of the urethra and vagina.

#### Perineal membrane and deep perineal pouch

The perineal membrane (see pp. 449–451) is a thick fibrous sheet that fills the urogenital triangle (Fig. 5.69B). It has a free posterior border, which is anchored in the midline to the perineal body and is attached laterally to the pubic arch. Immediately superior to the perineal membrane is a thin region termed the deep perineal pouch, containing a layer of skeletal muscle and neurovascular tissues. Among the skeletal muscles in the pouch (see p. 451, Fig. 5.37) is the external urethral sphincter.

The perineal membrane and deep perineal pouch provide support for the external genitalia, which are attached to its inferior surface. Also, the parts of the perineal membrane and deep perineal pouch inferior to the urogenital hiatus in the levator ani provide support for the pelvic viscera, above.

The urethra leaves the pelvic cavity and enters the perineum by passing through the deep perineal pouch and perineal membrane. In women, the vagina also passes through these structures posterior to the urethra.

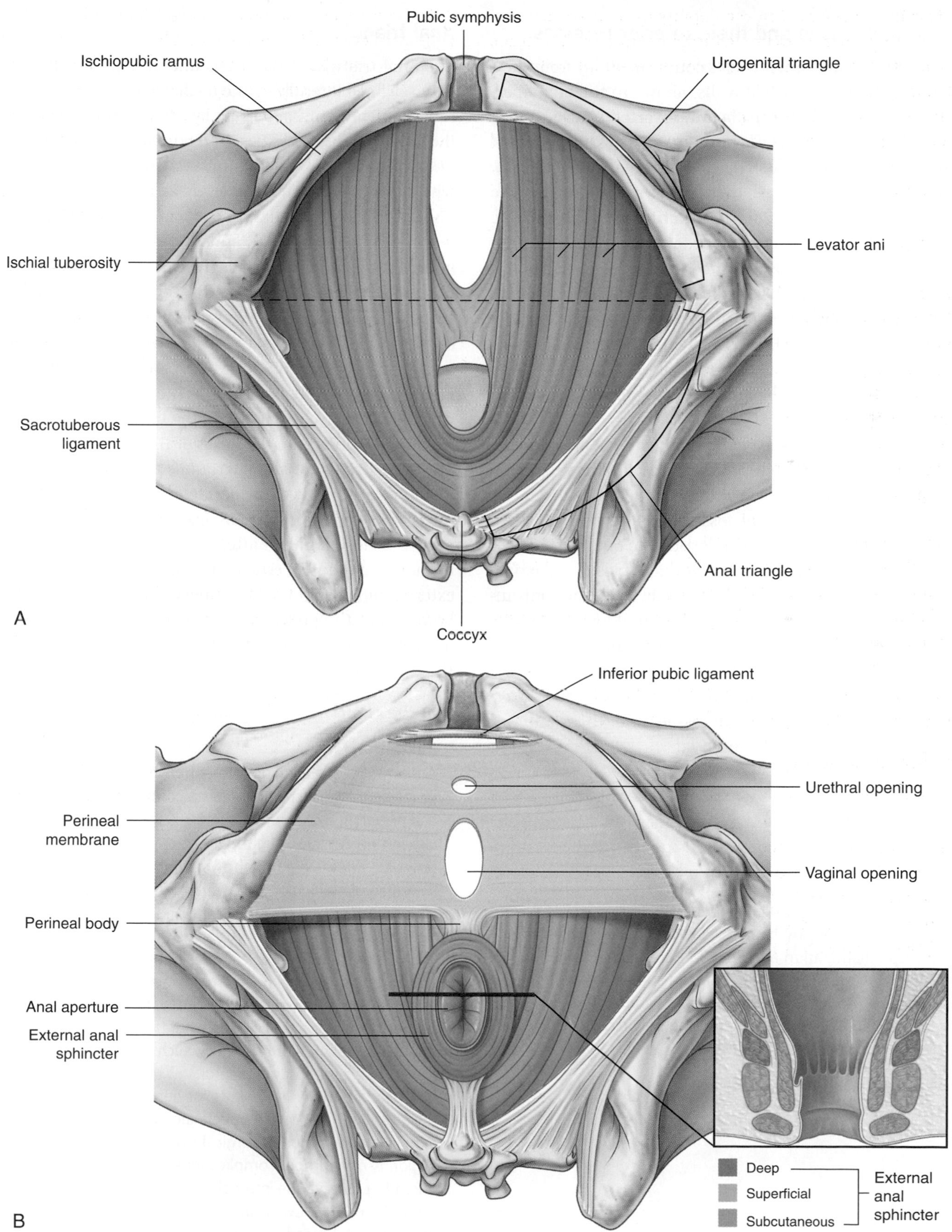


**Fig. 5.69** Borders and ceiling of the perineum. **A.** Boundaries of the perineum. **B.** Perineal membrane.

## Ischio-anal fossae and their anterior recesses

Because the levator ani muscles course medially from their origin on the lateral pelvic walls, above, to the anal aperture and urogenital hiatus, below, inverted wedge-shaped gutters occur between the levator ani muscles and adjacent pelvic walls as the two structures diverge inferiorly (Fig. 5.70). In the anal triangle, these gutters, one on each side of the anal aperture, are termed **ischio-anal fossae**. The lateral wall of each fossa is formed mainly by the ischium, obturator internus muscle, and sacrotuberous ligament. The medial wall is the levator ani muscle. The medial and lateral walls converge superiorly where the levator ani muscle attaches to the fascia overlying the obturator internus muscle. The ischio-anal fossae allow movement of the pelvic diaphragm and expansion of the anal canal during defecation.

The ischio-anal fossae of the anal triangle are continuous anteriorly with recesses that project into the urogenital triangle superior to the deep perineal pouch. These anterior recesses of the ischio-anal fossae are shaped like three-sided pyramids that have been tipped onto one of their sides (Fig. 5.70C). The apex of each pyramid is closed and points anteriorly toward the pubis. The base is open and continuous posteriorly with its related ischio-anal fossa. The inferior wall of each pyramid is the deep perineal pouch. The superomedial wall is the levator ani muscle, and the superolateral wall is formed mainly by the obturator internus muscle. The ischio-anal fossae and their anterior recesses are normally filled with fat.

### In the clinic

#### Abscesses in the ischio-anal fossae

The anal mucosa is particularly vulnerable to injury and may be easily torn by hard feces. Occasionally, patients develop inflammation and infection of the anal canal (sinuses or crypts). This infection can spread between the sphincters, producing intersphincteric fistulas. The infection can tract superiorly into the pelvic cavity or laterally into the ischio-anal fossae.

## Anal triangle

The anal triangle of the perineum faces posteroinferiorly and is defined laterally by the medial margins of the sacrotuberous ligaments, anteriorly by a horizontal line between the two ischial tuberosities, and posteriorly by the coccyx. The ceiling of the anal triangle is the pelvic diaphragm, which is formed by the levator ani and coccygeus muscles. The anal aperture occurs centrally in the anal triangle and is related on either side to an ischio-anal fossa. The major muscle in the anal triangle is the external anal sphincter.

The **external anal sphincter**, which surrounds the anal canal, is formed by skeletal muscle and consists of three parts—deep, superficial, and subcutaneous—arranged sequentially along the canal from superior to inferior (Fig. 5.69B, Table 5.5). The deep part is a thick ring-shaped muscle that circles the upper part of the anal canal and blends with the fibers of the levator ani muscle. The superficial part also surrounds the anal canal, but is anchored anteriorly to the perineal body and posteriorly to the coccyx and anococcygeal ligament. The subcutaneous part is a horizontally flattened disc of muscle that surrounds the anal aperture just beneath the skin. The external anal sphincter is innervated by inferior rectal branches of the pudendal nerve and by branches directly from the anterior ramus of S4.

### In the clinic

#### Hemorrhoids

A hemorrhoid is an engorgement of the venous plexus at or inside the anal sphincter. It is a common complaint and has prevalence of approximately 4% in the United States. Hemorrhoids have a slight genetic predisposition; however, straining during bowel movements, obesity, and sedentary lifestyle can also produce hemorrhoids. The symptoms include irritation, pain, and swelling. Hemorrhoids occurring at the anal verge (distal boundary of the anal canal) are typically called external hemorrhoids. Internal hemorrhoids occur inside the rectum and have a tendency to bleed. Prolapsed hemorrhoids are internal hemorrhoids that pass outside the anal canal and form lumps, which may undergo thrombosis and become painful.

There are many treatments for hemorrhoids, which include ligation above the pectinate (dentate) line using simple rubber bands or surgical excision. Surgery to this region is not without complications and care must be taken to preserve the internal anal sphincter.

In the back of every physician's mind is the concern that the rectal bleeding or symptoms may not be attributable to hemorrhoids. Therefore, excluding a tumor within the bowel is as important as treating the hemorrhoids.

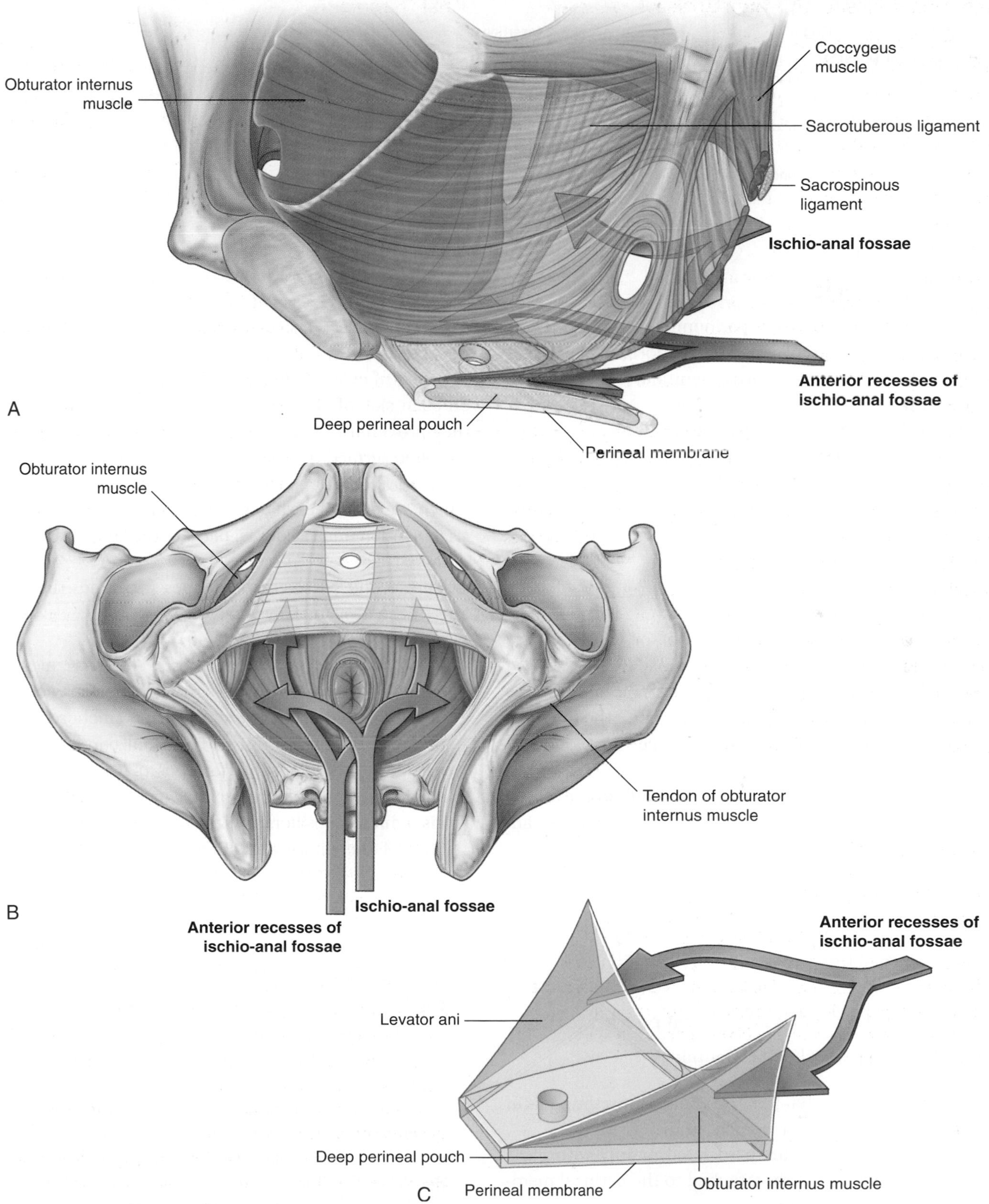


**Fig. 5.70** Ischio-anal fossae and their anterior recesses. **A.** Anterolateral view with left pelvic wall removed. **B.** Inferior view. **C.** Anterolateral view with pelvic walls and diaphragm removed.

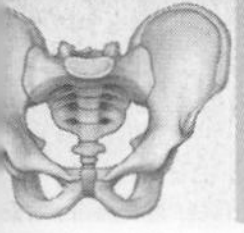



**Table 5.5** Muscles of the anal triangle

| Muscles | Origin | Insertion | Innervation | Function |
|---|---|---|---|---|
| EXTERNAL ANAL SPHINCTER | | | | Closes anal canal |
| Deep part | Surrounds superior aspect of anal canal | | Pudendal nerve (S2 and S3) and branches directly from S4 | |
| Superficial part | Surrounds lower part of anal canal | Anchored to perineal body and anococcygeal body | | |
| Subcutaneous part | Surrounds anal aperture | | | |

## Urogenital triangle

The urogenital triangle of the perineum is the anterior half of the perineum and is oriented in the horizontal plane. It contains the roots of the external genitalia (Fig. 5.71) and the openings of the urogenital system.

The urogenital triangle is defined:

- laterally by the ischiopubic rami,
- posteriorly by an imaginary line between the ischial tuberosities, and
- anteriorly by the inferior margin of the pubic symphysis.

As with the anal triangle, the roof or ceiling of the urogenital triangle is the levator ani muscle.

Unlike the anal triangle, the urogenital triangle contains a strong fibromuscular support platform, the perineal membrane and deep perineal pouch (see pp. 449–451), which is attached to the pubic arch.

Anterior extensions of the ischio-anal fossae occur between the deep perineal pouch and the levator ani muscle on each side.

Between the perineal membrane and the membranous layer of superficial fascia is the **superficial perineal pouch**. The principal structures in this pouch are the erectile tissues of the penis and clitoris and associated skeletal muscles.

### Structures in the superficial perineal pouch

The superficial perineal pouch contains:

- erectile structures that join together to form the penis in men and the clitoris in women, and
- skeletal muscles that are associated mainly with parts of the erectile structures attached to the perineal membrane and adjacent bone.

Each erectile structure consists of a central core of expandable vascular tissue and its surrounding connective tissue capsule.

#### Erectile tissues

Two sets of erectile structures join to form the penis and the clitoris.

A pair of cylindrically shaped **corpora cavernosa**, one on each side of the urogenital triangle, are anchored by their proximal ends to the pubic arch. These attached parts are often termed the **crura** (from the Latin for "legs") of the clitoris or the penis. The distal ends of the corpora, which are not attached to bone, form the body of the clitoris in women and the dorsal parts of the body of the penis in men.

The second set of erectile tissues surrounds the openings of the urogenital system.

- In women, a pair of erectile structures, termed the **bulbs of the vestibule**, are situated, one on each side, at the vaginal opening and are firmly anchored to the perineal membrane (Fig. 5.71A). Small bands of erectile tissues connect the anterior ends of these bulbs to a single, small, pea-shaped erectile mass, the **glans clitoris**, which is positioned in the midline at the end of the body of the clitoris and anterior to the opening of the urethra.
- In men, a single large erectile mass, the **corpus spongiosum**, is the structural equivalent to the bulbs of the vestibule, the glans clitoris, and the interconnecting bands of erectile tissues in women (Fig. 5.71B). The corpus spongiosum is anchored at its base to the perineal membrane. Its proximal end, which is not attached, forms the ventral part of the body of the penis and expands over the end of the body of the penis to form the glans penis. This pattern in men results from the absence of a vaginal opening and from the fusion of structures across the midline during embryological development. As the originally paired erectile structures fuse, they enclose the urethral opening and form an additional channel that ultimately becomes most of the penile part of the urethra. As a consequence of this fusion and growth in men, the urethra is enclosed by the corpus spongiosum and opens at the end of the

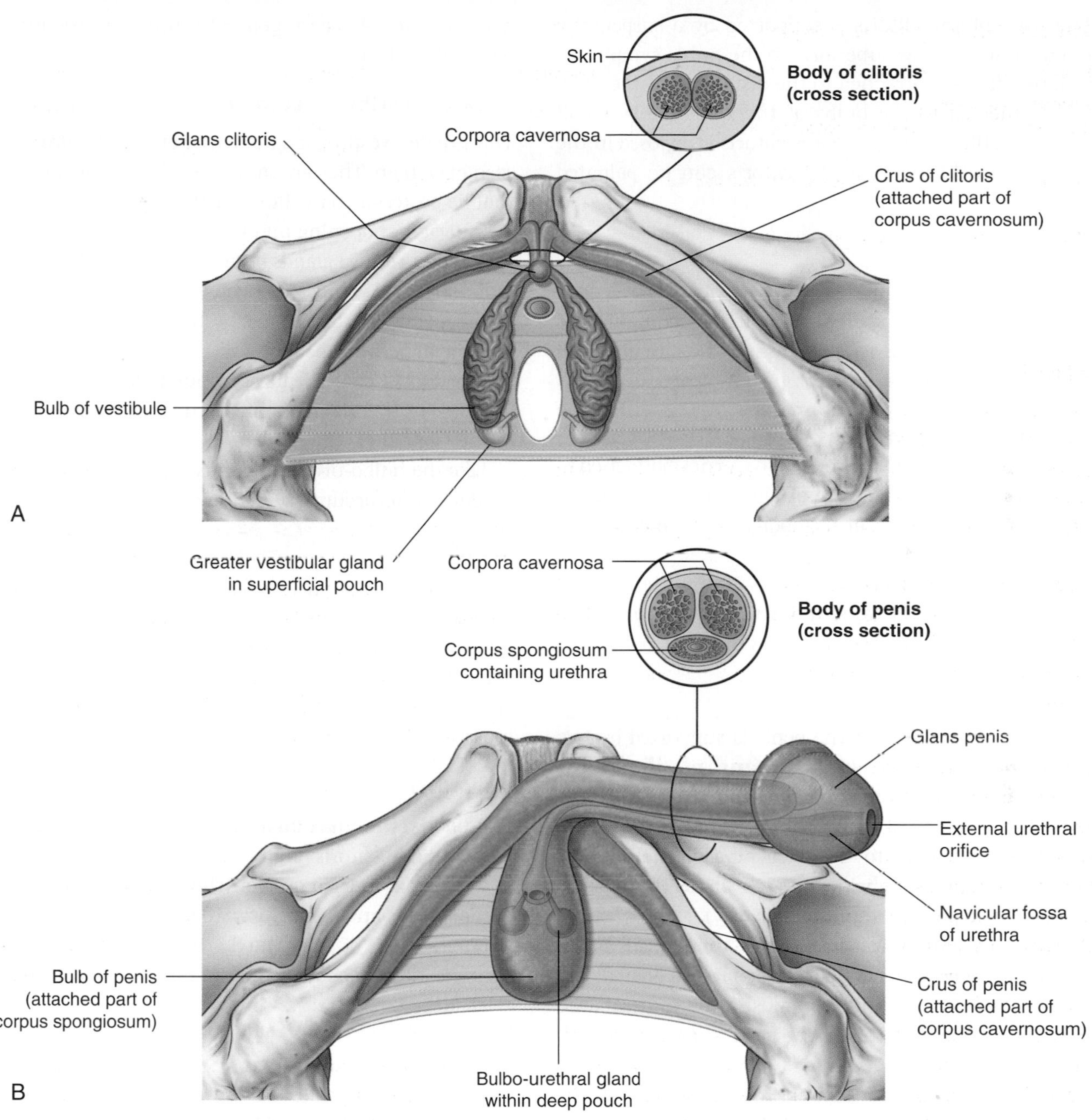


**Fig. 5.71** Erectile tissues of clitoris and penis. **A.** Clitoris. **B.** Penis.

penis. This is unlike the situation in women, where the urethra is not enclosed by erectile tissue of the clitoris and opens directly into the vestibule of the perineum.

## Clitoris

The clitoris is composed of two corpora cavernosa and the **glans clitoris** (Fig. 5.71A). As in the penis, it has an attached part (root) and a free part (body).

- Unlike the root of the penis, the **root of the clitoris** technically consists only of the two crura. (Although the bulbs of the vestibule are attached to the glans clitoris by thin bands of erectile tissue, they are not included in the attached part of the clitoris.)
- The **body of the clitoris**, which is formed only by the unattached parts of the two corpora cavernosa, angles posteriorly and is embedded in the connective tissues of the perineum.

The body of the clitoris is supported by a suspensory ligament that attaches superiorly to the pubic symphysis. The glans clitoris is attached to the distal end of the body and is connected to the bulbs of the vestibule by small bands of erectile tissue. The glans clitoris is exposed in the perineum and the body of the clitoris can be palpated through skin.

#### Penis

The penis is composed mainly of the two corpora cavernosa and the single corpus spongiosum, which contains the urethra (Fig. 5.71B.) As in the clitoris, it has an attached part (root) and a free part (body):

- The **root of the penis** consists of the two crura, which are proximal parts of the corpora cavernosa attached to the pubic arch, and the **bulb of the penis**, which is the proximal part of the corpus spongiosum anchored to the perineal membrane.
- The **body of the penis**, which is covered entirely by skin, is formed by the tethering of the two proximal free parts of the corpora cavernosa and the related free part of the corpus spongiosum.

The base of the body of the penis is supported by two ligaments: the **suspensory ligament of the penis** (attached superiorly to the pubic symphysis), and the more superficially positioned **fundiform ligament of the penis** (attached above to the linea alba of the anterior abdominal wall and split below into two bands that pass on each side of the penis and unite inferiorly).

Because the anatomical position of the penis is erect, the paired corpora are defined as dorsal in the body of the penis and the single corpus spongiosum as ventral, even though the positions are reversed in the nonerect (flaccid) penis.

The corpus spongiosum expands to form the head of the penis (**glans penis**) over the distal ends of the corpora cavernosa (Fig. 5.71B).

#### Erection

Erection of the penis and clitoris is a vascular event generated by parasympathetic fibers carried in pelvic splanchnic nerves from the anterior rami of S2 to S4, which enter the inferior hypogastric part of the prevertebral plexus and ultimately pass through the deep perineal pouch and perineal membrane to innervate the erectile tissues. Stimulation of these nerves causes specific arteries in the erectile tissues to relax. This allows blood to fill the tissues, causing the penis and clitoris to become erect.

Arteries supplying the penis and clitoris are branches of the internal pudendal artery; branches of the pudendal nerve (S2 to S4) carry general sensory nerves from the penis and clitoris.

### Greater vestibular glands

The greater vestibular glands (**Bartholin's glands**) are seen in women. They are small, pea-shaped mucous glands that lie posterior to the bulbs of the vestibule on each side of the vaginal opening and are the female homologues of the bulbo-urethral glands in men (Fig. 5.71). However, the bulbo-urethral glands are located within the deep perineal pouch, whereas the greater vestibular glands are in the superficial perineal pouch.

The duct of each greater vestibular gland opens into the vestibule of the perineum along the posterolateral margin of the vaginal opening.

Like the bulbo-urethral glands in men, the greater vestibular glands produce secretion during sexual arousal.

### Muscles

The superficial perineal pouch contains three pairs of muscles: the ischiocavernosus, bulbospongiosus, and superficial transverse perineal muscles (Fig. 5.72 and Table 5.6). Two of these three pairs of muscles are associated with the roots of the penis and clitoris; the other pair is associated with the perineal body.

#### Ischiocavernosus

The two **ischiocavernosus muscles** cover the crura of the penis and clitoris (Fig. 5.72). Each muscle is anchored to the medial margin of the ischial tuberosity and related ischial ramus and passes forward to attach to the sides and inferior surface of the related crus, and forces blood from the crus into the body of the erect penis and clitoris.

#### Bulbospongiosus

The two **bulbospongiosus muscles** are associated mainly with the bulbs of the vestibule in women and with the attached part of the corpus spongiosum in men (Fig. 5.72).

In women, each bulbospongiosus muscle is anchored posteriorly to the perineal body and courses anterolaterally over the inferior surface of the related greater vestibular gland and the bulb of the vestibule to attach to the surface of the bulb and to the perineal membrane (Fig. 5.72A). Other fibers course anterolaterally to blend with the fibers of the ischiocavernosus muscle, and still others travel anteriorly and arch over the body of the clitoris.

In men, the bulbospongiosus muscles are joined in the midline to a raphe on the inferior surface of the bulb of the penis. The raphe is anchored posteriorly to the perineal body. Muscle fibers course anterolaterally, on each side, from the raphe and perineal body to cover each side of the

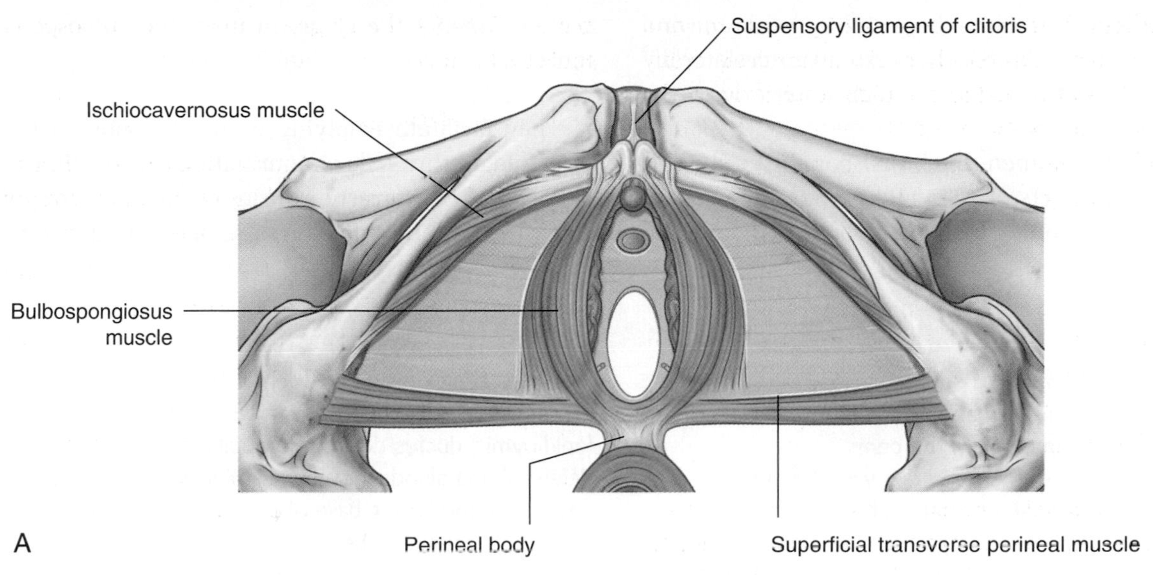


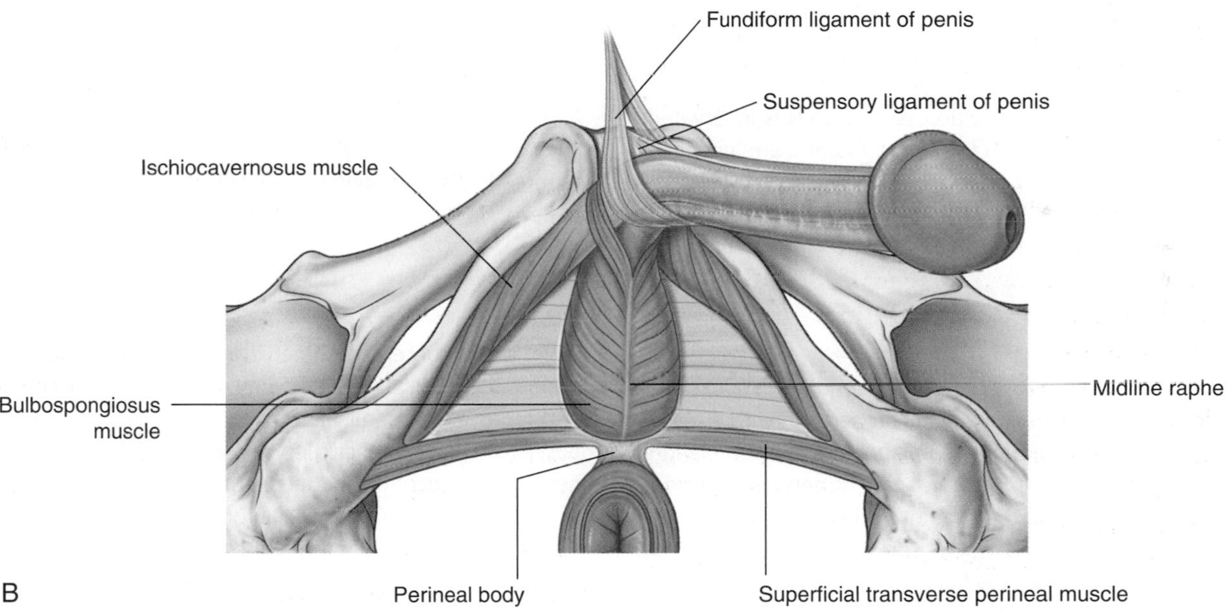


**Fig. 5.72** Muscles in the superficial perineal pouch. **A.** In women. **B.** In men.

**Table 5.6** Muscles of the superficial perineal pouch

| Muscles | Origin | Insertion | Innervation | Function |
|---|---|---|---|---|
| Ischiocavernosus | Ischial tuberosity and ramus | Crus of penis and clitoris | Pudendal nerve (S2 to S4) | Move blood from crura into the body of the erect penis and clitoris |
| Bulbospongiosus | In women: perineal body<br>In men: perineal body, midline raphe | In women: bulb of vestibule, perineal membrane, body of clitoris, and corpus cavernosum<br>In men: bulbospongiosus, perineal membrane, corpus cavernosum | Pudendal nerve (S2 to S4) | Move blood from attached parts of the clitoris and penis into the glans<br>In men: removal of residual urine from urethra after urination; pulsatile emission of semen during ejaculation |
| Superficial transverse perineal | Ischial tuberosity and ramus | Perineal body | Pudendal nerve (S2 to S4) | Stabilize the perineal body |

bulb of the penis and attach to the perineal membrane and connective tissue of the bulb. Others extend anterolaterally to associate with the crura and attach anteriorly to the ischiocavernosus muscles.

In both men and women, the bulbospongiosus muscles compress attached parts of the erect corpus spongiosum and bulbs of the vestibule and force blood into more distal regions, mainly the glans. In men, the bulbospongiosus muscles have two additional functions:

- They facilitate emptying of the bulbous part of the penile urethra following urination (micturition).
- Their reflex contraction during ejaculation is responsible for the pulsatile emission of semen from the penis.

## In the clinic

### Emission and ejaculation of semen

In men, emission is the formation of semen, and ejaculation is the expulsion of semen from the penis.

Although erection of the penis is a vascular event generated by parasympathetic nerves from spinal levels S2–S4, the formation of semen in the urethra is caused by the contraction of smooth muscle of the ducts and glands of the reproductive system that is innervated by the sympathetic part of the visceral nervous system. Ejaculation of semen from the penis is through the action of skeletal muscles innervated by somatic motor nerves.

Smooth muscle in the duct system of the male reproductive tract and in the accessory glands is innervated by sympathetic fibers from the lower thoracic and upper lumbar spinal levels (T12, L1,2). The fibers pass into the prevertebral plexus and are then distributed to target tissues. Semen is formed as luminal contents from the ducts (epididymis, ductus deferens, ampulla of the ductus deferens) and glands (prostate, seminal vesicles) are moved into the urethra at the base of the penis by the contraction of smooth muscle in the walls of the structures.

Pulsatile emission of semen from the penis is generated by the reflex contraction of the bulbospongiosus muscle that forces semen from the base of the penis and out of the external urethral meatus. Bulbospongiosus muscle is innervated by somatic motor fibers carried in the pudendal nerve (S2–S4). Contraction of the internal urethral sphincter and periurethral smooth muscle, innervated by the sympathetic part of the visceral nervous system, prevents retrograde ejaculation into the bladder.

## In the clinic

### Erectile dysfunction

Erectile dysfunction (ED) is a complex condition in which men are unable to initiate or maintain penile erection. When this affects erections during sleep, and with self-stimulation as well as with a partner, vascular and/or nerve impairment is present. This generalized type of ED increases with age and is recognized as a risk factor for coronary artery disease. It is frequently associated with cardiovascular disease, diabetes, and neurological conditions including Parkinson's disease, spinal cord injuries, multiple sclerosis, and as nerve damage from pelvic surgeries or radiation for pelvic malignancies. Low testosterone states can impair erections and consistently prevent sleep-induced erections. Medications including serotonin reuptake inhibitors (SSRIs), thiazides and anti-androgens can also underlie ED. When only partnered erections are problematic, psychological factors underlie the dysfunction—the normal erections from sleep confirming healthy vascular and neurological function. Most cases of ED are multifactorial in etiology, and all markedly lessen quality of life and a person's well-being and can lead to depression and low self-esteem as well as emotional and social isolation.

Delayed (or absent) ejaculation can result from nerve damage in conditions such as diabetes, Parkinson's disease, spinal cord injuries, multiple sclerosis, complications after major pelvic surgeries and pelvic irradiation. Ejaculation is absent after radical prostatectomy for prostate cancer (which also removes the seminal vesicles), but orgasm is still possible as the pudendal nerve is spared. SSRIs, neuroleptics, alcohol, and recreational drugs (marijuana, cocaine, and heroin) often delay orgasm and therefore ejaculation, as in health the two coincide (even though the nerves involved are different).

Because erectile tissues in the clitoris have similar innervation and blood supply to the penis, vulvar swelling is likely compromised by the same conditions that cause ED in men. However, it appears that this is a rare cause of female sexual dysfunction. Reduced clitoral swelling is rarely symptomatic. Note that phosphodiesterase type 5 (PDE5) inhibitors (sildenafil) do not improve female sexual dysfunction even in conditions such as diabetes. Research confirms that otherwise healthy women with complaints of low sexual arousal have a physiologically normal increase in genital congestion in response to visual sexual stimuli, even though they do not find the stimuli mentally sexually arousing. Loss of genital sexual sensitivity from somatic nerve damage from multiple sclerosis or diabetes can be highly symptomatic and preclude orgasm. Medications preventing orgasm in men can also affect women.

### Superficial transverse perineal muscles

The paired **superficial transverse perineal muscles** follow a course parallel to the posterior margin of the inferior surface of the perineal membrane (Fig. 5.72). These flat band-shaped muscles, which are attached to ischial tuberosities and rami, extend medially to the perineal body in the midline and stabilize the perineal body.

## Superficial features of the external genitalia

### In women

In women, the clitoris and vestibular apparatus, together with a number of skin and tissue folds, form the **vulva** (Fig. 5.73). On either side of the midline are two thin folds of skin termed the **labia minora**. The region enclosed between them, and into which the urethra and vagina

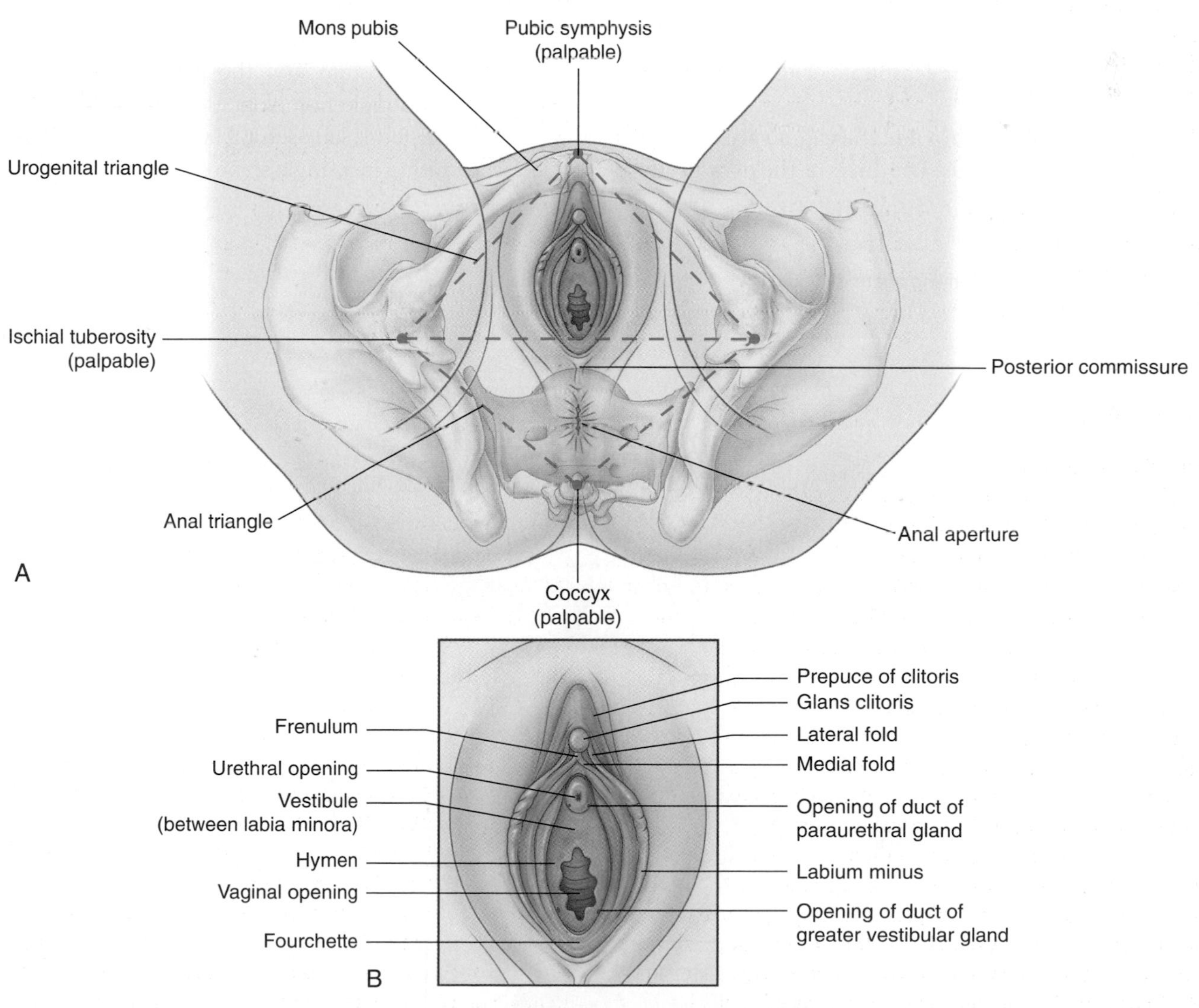


**Fig. 5.73** Superficial features of the perineum in women. **A.** Overview. **B.** Close-up of external genitalia.

open, is the **vestibule**. Anteriorly, the labia minora each bifurcate, forming a medial and a lateral fold. The medial folds unite to form the **frenulum of the clitoris**, that joins the glans clitoris. The lateral folds unite ventrally over the glans clitoris and the body of the clitoris to form the **prepuce of the clitoris** (hood). The body of the clitoris extends anteriorly from the glans clitoris and is palpable deep to the prepuce and related skin. Posterior to the vestibule, the labia minora unite, forming a small transverse fold, the **frenulum of the labia minora** (the **fourchette**).

Within the vestibule, the vaginal orifice is surrounded to varying degrees by a ring-like fold of membrane, the **hymen**, which may have a small central perforation or may completely close the vaginal opening. Following rupture of the hymen (resulting from first sexual intercourse or injury), irregular remnants of the hymen fringe the vaginal opening.

The orifices of the urethra and the vagina are associated with the openings of glands. The ducts of the para-urethral glands (**Skene's glands**) open into the vestibule, one on each side of the lateral margin of the urethra. The ducts of the greater vestibular glands (Bartholin's glands) open adjacent to the posterolateral margin of the vaginal opening in the crease between the vaginal orifice and remnants of the hymen.

Lateral to the labia minora are two broad folds, the **labia majora**, which unite anteriorly to form the mons pubis. The **mons pubis** overlies the inferior aspect of the pubic symphysis and is anterior to the vestibule and the clitoris. Posteriorly, the labia majora do not unite and are separated by a depression termed the **posterior commissure**, which overlies the position of the perineal body.

### In men

Superficial components of the genital organs in men consist of the scrotum and the penis (Fig. 5.74). The **scrotum** is the male homologue of the labia majora in women. In the fetus, labioscrotal swellings fuse across the midline, resulting in a single scrotum into which the testes

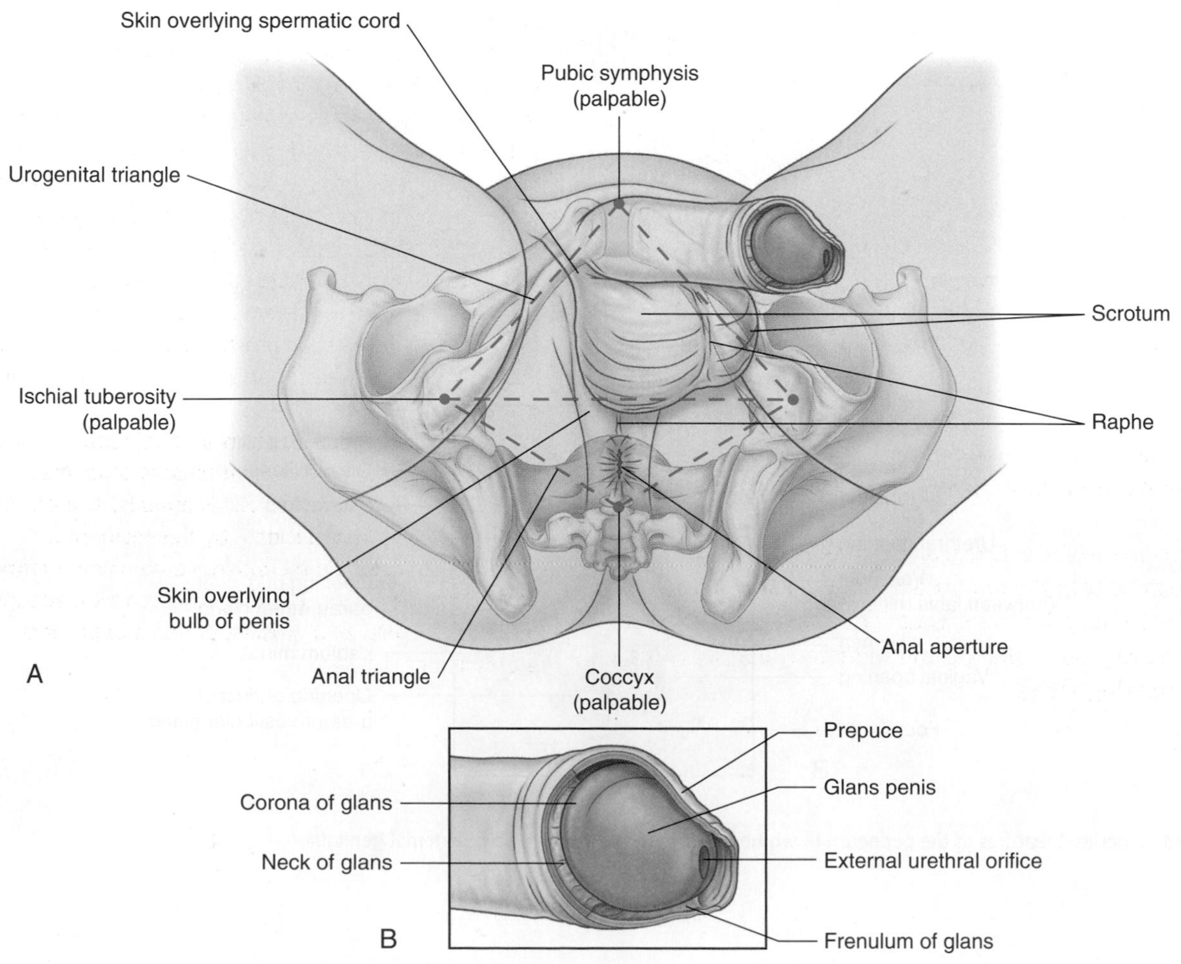


**Fig. 5.74** Superficial features of the perineum in men. **A.** Overview. **B.** Close-up of external genitalia.

and their associated musculofascial coverings, blood vessels, nerves, lymphatics, and drainage ducts descend from the abdomen. The remnant of the line of fusion between the labioscrotal swellings in the fetus is visible on the skin of the scrotum as a longitudinal midline **raphe** that extends from the anus, over the scrotal sac, and onto the inferior aspect of the body of the penis.

The **penis** consists of a root and body. The attached root of the penis is palpable posterior to the scrotum in the urogenital triangle of the perineum. The pendulous part of the penis (body of penis) is entirely covered by skin; the tip of the body is covered by the glans penis.

The external urethral orifice is a sagittal slit, normally positioned at the tip of the glans. The inferior margin of the urethral orifice is continuous with a midline **raphe of the penis**, which represents a line of fusion formed in the glans as the urethra develops in the fetus. The base of this raphe is continuous with the **frenulum** of the glans, which is a median fold of skin that attaches the glans to more loosely attached skin proximal to the glans. The base of the glans is expanded to form a raised circular margin (the **corona of the glans**); the two lateral ends of the corona join inferiorly at the midline raphe of the glans. The depression posterior to the corona is the neck of the glans. Normally, a fold of skin at the neck of the glans is continuous anteriorly with thin skin that tightly adheres to the glans and posteriorly with thicker skin loosely attached to the body. This fold, known as the prepuce, extends forward to cover the glans. The prepuce is removed during male circumcision, leaving the glans exposed.

### Superficial fascia of the urogenital triangle

The superficial fascia of the urogenital triangle is continuous with similar fascia on the anterior abdominal wall.

As with the superficial fascia of the abdominal wall, the perineal fascia has a membranous layer on its deep surface. This membranous layer (**Colles' fascia**), is attached:

- posteriorly to the perineal membrane and therefore does not extend into the anal triangle (Fig. 5.75), and
- to the ischiopubic rami that form the lateral borders of the urogenital triangle and therefore does not extend into the thigh (Fig. 5.75).

It defines the external limits of the superficial perineal pouch, lines the scrotum or labia, and extends around the body of the penis and clitoris.

Anteriorly, the membranous layer of fascia is continuous over the pubic symphysis and pubic bones with the membranous layer of fascia on the anterior abdominal wall. In the lower lateral abdominal wall, the membranous layer of abdominal fascia is attached to the deep fascia of the thigh just inferior to the inguinal ligament.

Because the membranous layer of fascia encloses the superficial perineal pouch and continues up the anterior abdominal wall, fluids or infectious materials that accumulate in the pouch can track out of the perineum and onto the lower abdominal wall. This material will not track into the anal triangle or the thigh because the fascia fuses with deep tissues at the borders of these regions.

#### In the clinic

**Urethral rupture**

Urethral rupture may occur at a series of well-defined anatomical points.

The commonest injury is a rupture of the proximal spongy urethra below the perineal membrane. The urethra is usually torn when structures of the perineum are caught between a hard object (e.g., a steel beam or crossbar of a bicycle) and the inferior pubic arch. Urine escapes through the rupture into the superficial perineal pouch and descends into the scrotum and up onto the anterior abdominal wall deep to the superficial fascia.

In association with severe pelvic fractures, urethral rupture may occur at the prostatomembranous junction above the deep perineal pouch. The urine will extravasate into the true pelvis.

The worst and most serious urethral rupture is related to serious pelvic injuries where there is complete disruption of the puboprostatic ligaments. The prostate is dislocated superiorly not only by the ligamentous disruption but also by the extensive hematoma formed within the true pelvis. The diagnosis can be made by palpating the elevated prostate during a digital rectal examination.

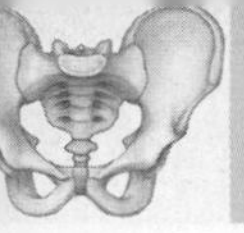

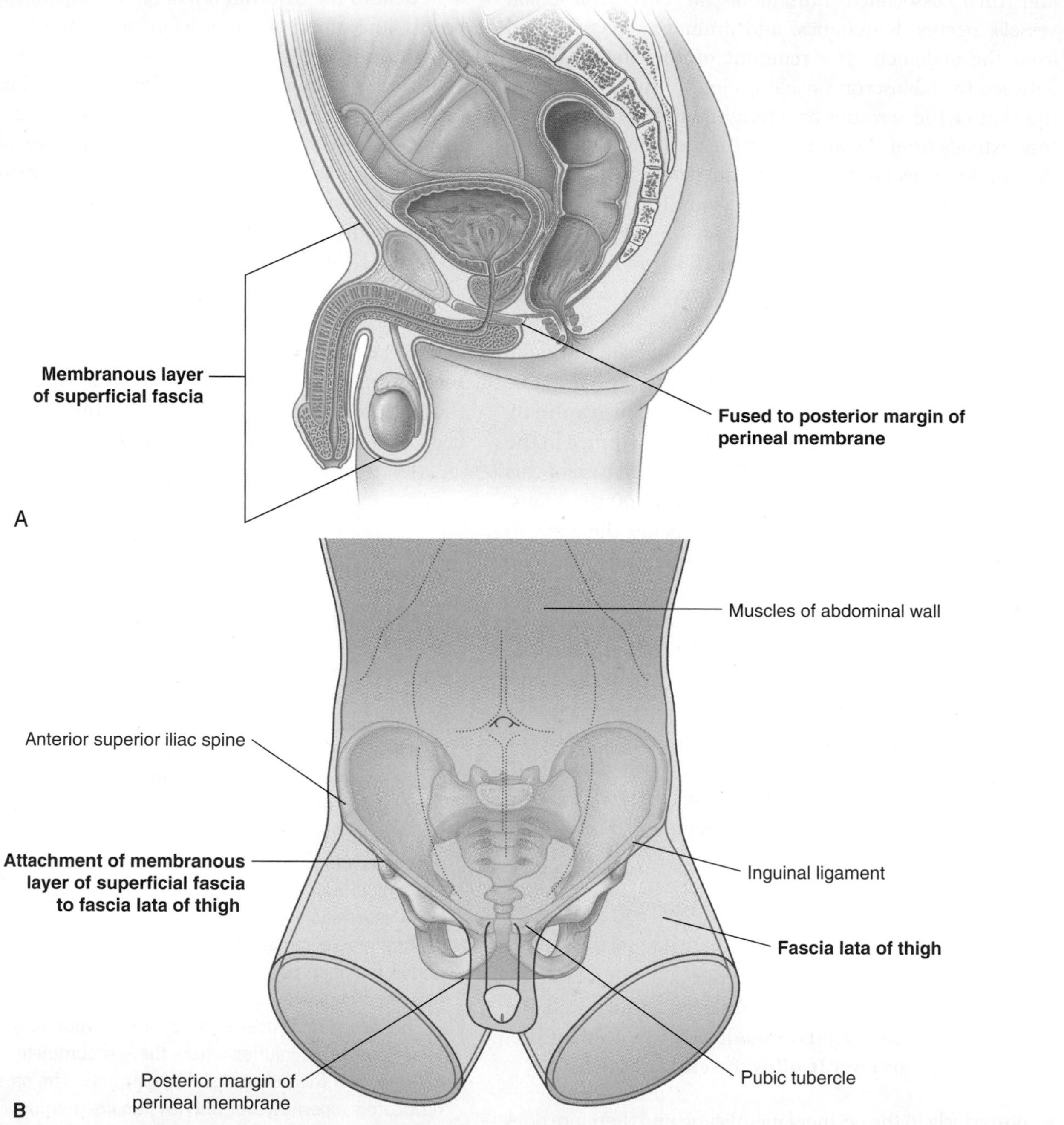

**Fig. 5.75** Superficial fascia. **A.** Lateral view. **B.** Anterior view.

## Somatic nerves

### Pudendal nerve

The major somatic nerve of the perineum is the pudendal nerve. This nerve originates from the sacral plexus and carries fibers from spinal cord levels S2 to S4. It leaves the pelvic cavity through the greater sciatic foramen inferior to the piriformis muscle, passes around the sacrospinous ligament, and then enters the anal triangle of the perineum by passing medially through the lesser sciatic foramen. As it enters and courses through the perineum, it travels along the lateral wall of the ischio-anal fossa in the **pudendal canal**, which is a tubular compartment formed in the fascia that covers the obturator internus muscle. This pudendal canal also contains the internal pudendal artery and accompanying veins.

The pudendal nerve (Fig. 5.76) has three major terminal branches—the inferior rectal and perineal nerves and the dorsal nerve of the penis or clitoris—which are

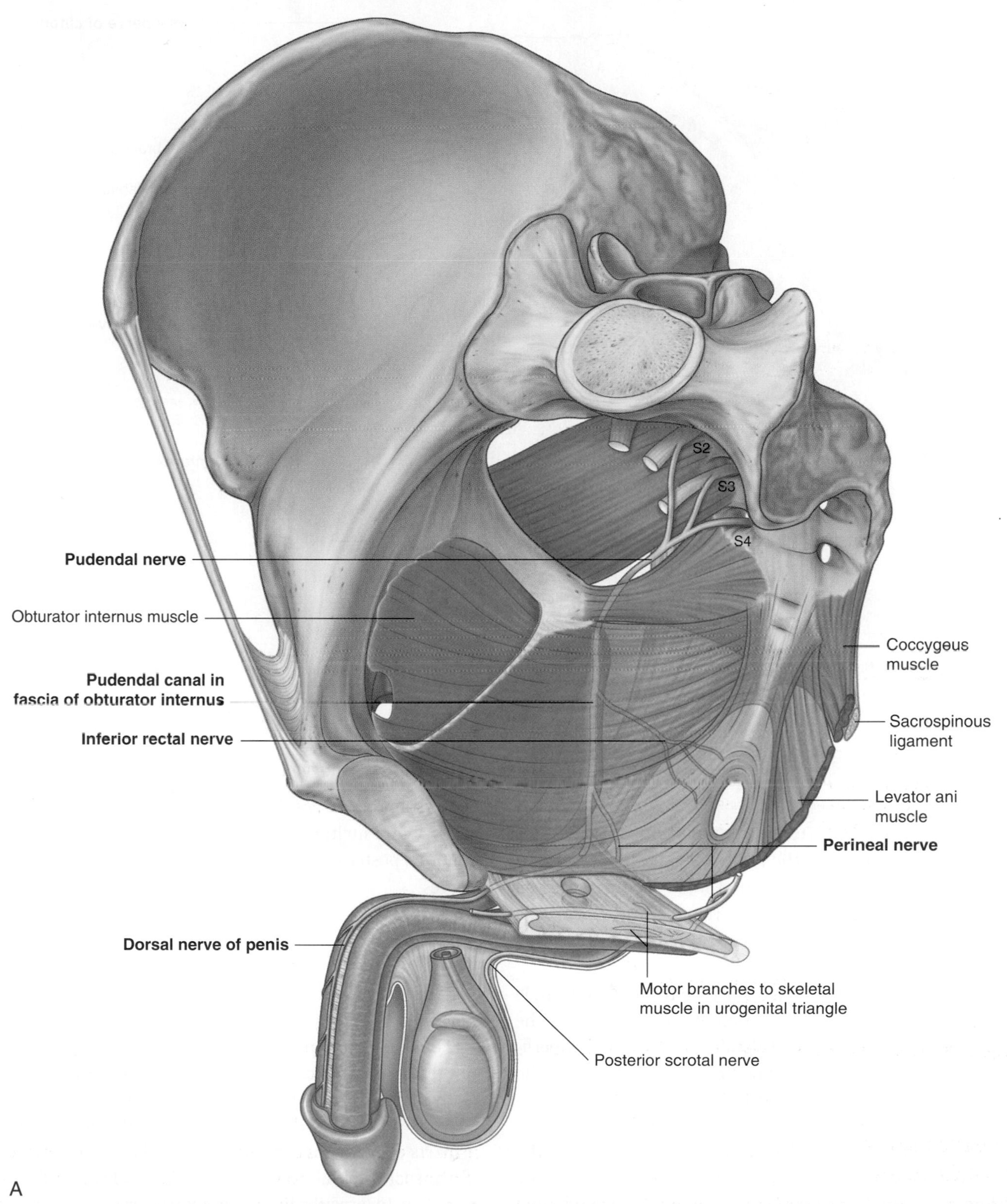
S2
S3
S4
Pudendal nerve
Obturator internus muscle
Pudendal canal in fascia of obturator internus
Inferior rectal nerve
Coccygeus muscle
Sacrospinous ligament
Levator ani muscle
Perineal nerve
Dorsal nerve of penis
Motor branches to skeletal muscle in urogenital triangle
Posterior scrotal nerve
A

**Fig. 5.76** Pudendal nerve. **A.** In men.

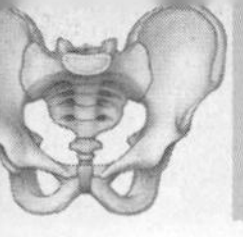

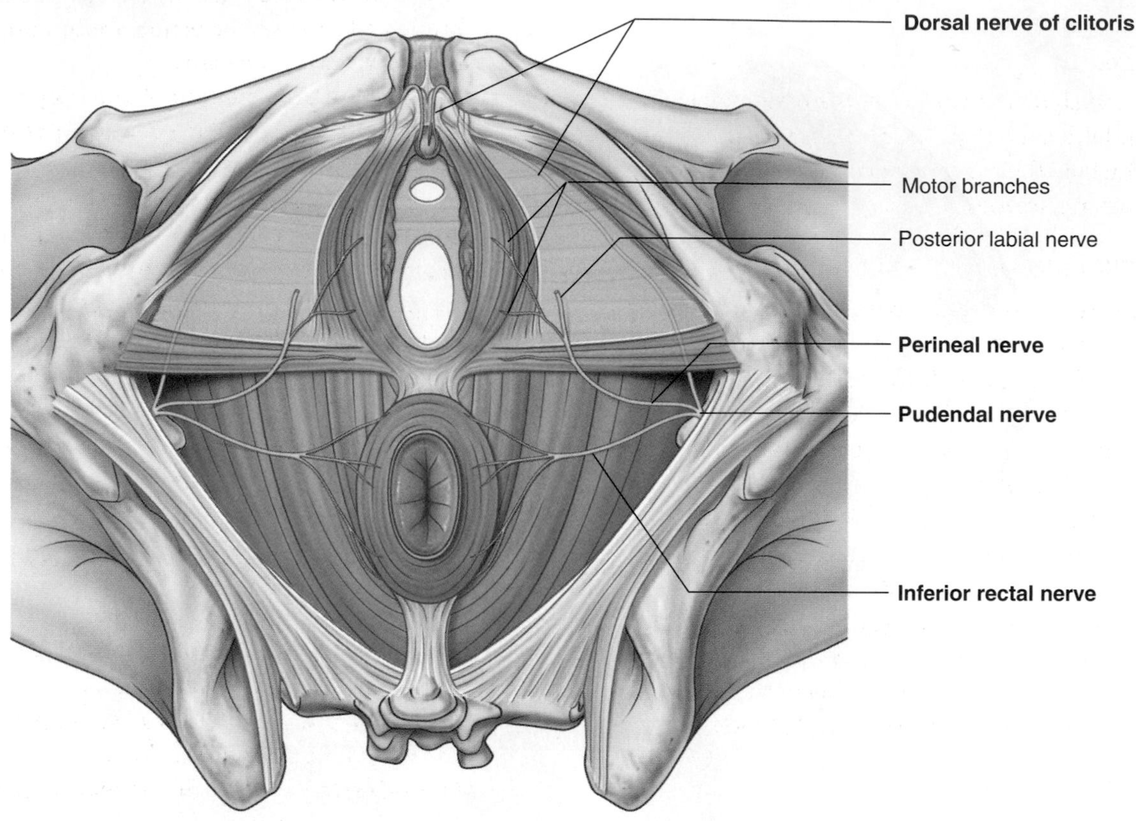


**Fig. 5.76, cont'd B.** In women.

accompanied by branches of the internal pudendal artery (Fig. 5.77).

- The **inferior rectal nerve** is often multiple, penetrates through the fascia of the pudendal canal, and courses medially across the ischio-anal fossa to innervate the external anal sphincter and related regions of the levator ani muscles. The nerve is also general sensory for the skin of the anal triangle.
- The **perineal nerve** passes into the urogenital triangle and gives rise to motor and cutaneous branches. The motor branches supply skeletal muscles in the superficial and deep perineal pouches. The largest of the sensory branches is the posterior scrotal nerve in men and the posterior labial nerve in women.
- The **dorsal nerve of the penis** and **clitoris** enters the deep perineal pouch (Fig. 5.76). It passes along the lateral margin of the pouch and then exits by passing inferiorly through the perineal membrane in a position just inferior to the pubic symphysis where it meets the body of the clitoris or the penis. It courses along the dorsal surface of the body to reach the glans. The dorsal nerve is sensory to the penis and clitoris, particularly to the glans.

### Other somatic nerves

Other somatic nerves that enter the perineum are mainly sensory and include branches of the ilio-inguinal, genitofemoral, posterior femoral cutaneous, and anococcygeal nerves.

## Visceral nerves

Visceral nerves enter the perineum by two routes:

- Those to the skin, which consist mainly of postganglionic sympathetics, are delivered into the region along the pudendal nerve. These fibers join the pudendal nerve from gray rami communicantes that connect pelvic parts of the sympathetic trunks to the anterior rami of the sacral spinal nerves (see p. 481 and Fig. 5.62).
- Those to erectile tissues enter the region mainly by passing through the deep perineal pouch from the inferior hypogastric plexus in the pelvic cavity (see p. 488 and Fig. 5.63B). The fibers that stimulate erection are parasympathetic fibers, which enter the inferior hypogastric plexus via pelvic splanchnic nerves from spinal cord levels of S2 to S4 (see Fig. 5.63A,B).

## Blood vessels

### Arteries

The most significant artery of the perineum is the internal pudendal artery (Fig. 5.77). Other arteries entering the area include the external pudendal, the testicular, and the cremasteric arteries.

#### Internal pudendal artery

The **internal pudendal artery** originates as a branch of the anterior trunk of the internal iliac artery in the pelvis (Fig. 5.77). Along with the pudendal nerve, it leaves the pelvis through the greater sciatic foramen inferior to the piriformis muscle. It passes around the ischial spine, where the artery lies lateral to the nerve, enters the perineum by coursing through the lesser sciatic foramen, and accompanies the pudendal nerve in the pudendal canal on the lateral wall of the ischio-anal fossa.

The branches of the internal pudendal artery are similar to those of the pudendal nerve in the perineum and include the inferior rectal and perineal arteries, and branches to the erectile tissues of the penis and clitoris (Fig. 5.77).

##### Inferior rectal arteries

One or more **inferior rectal arteries** originate from the internal pudendal artery in the anal triangle and cross the ischio-anal fossa medially to branch and supply muscle and related skin (Fig. 5.77). They anastomose with middle and superior rectal arteries from the internal iliac artery and the inferior mesenteric artery, respectively, to form a network of vessels that supply the rectum and anal canal.

##### Perineal artery

The **perineal artery** originates near the anterior end of the pudendal canal and gives off a transverse perineal branch, and a posterior scrotal or labial artery to surrounding tissues and skin (Fig. 5.77).

##### Terminal part of the internal pudendal artery

The terminal part of the internal pudendal artery accompanies the dorsal nerve of the penis or clitoris into the deep perineal pouch and supplies branches to the tissues in the deep perineal pouch and erectile tissues.

Branches that supply the erectile tissues in men include the artery to the bulb of the penis, the urethral artery, the deep artery of the penis, and the dorsal artery of the penis (Fig. 5.77).

- The **artery of the bulb of the penis** has a branch that supplies the bulbo-urethral gland and then penetrates the perineal membrane to supply the corpus spongiosum.
- A **urethral artery** also penetrates the perineal membrane and supplies the penile urethra and surrounding erectile tissue to the glans.
- Near the anterior margin of the deep perineal pouch, the internal pudendal artery bifurcates into two terminal branches. A **deep artery of the penis** penetrates the perineal membrane to enter the crus and supply the crus and corpus cavernosum of the body. The **dorsal artery of the penis** penetrates the anterior margin of the perineal membrane to meet the dorsal surface of the body of the penis. The vessel courses along the dorsal surface of the penis, medial to the dorsal nerve, and supplies the glans penis and superficial tissues of the penis; it also anastomoses with branches of the deep artery of the penis and the urethral artery.

Branches that supply the erectile tissues in women are similar to those in men.

- **Arteries of the bulb of the vestibule** supply the bulb of the vestibule and related vagina.
- **Deep arteries of the clitoris** supply the crura and corpus cavernosum of the body.
- **Dorsal arteries of the clitoris** supply surrounding tissues and the glans.

#### External pudendal arteries

The **external pudendal arteries** consist of a superficial vessel and a deep vessel, which originate from the femoral artery in the thigh. They course medially to enter the perineum anteriorly and supply related skin of the penis and scrotum or the clitoris and labia majora.

#### Testicular and cremasteric arteries

In men, the **testicular arteries** originate from the abdominal aorta and descend into the scrotum through the inguinal canal to supply the testes. Also, **cremasteric arteries**, which originate from the inferior epigastric branch of the external iliac artery, accompany the spermatic cord into the scrotum.

In women, small cremasteric arteries follow the round ligament of the uterus through the inguinal canal.

### Veins

Veins in the perineum generally accompany the arteries and join the **internal pudendal veins** that connect with the **internal iliac vein** in the pelvis (Fig. 5.78). The exception is the **deep dorsal vein of the penis or clitoris** that drains mainly the glans and the corpora cavernosa. The deep dorsal vein courses along the midline between the

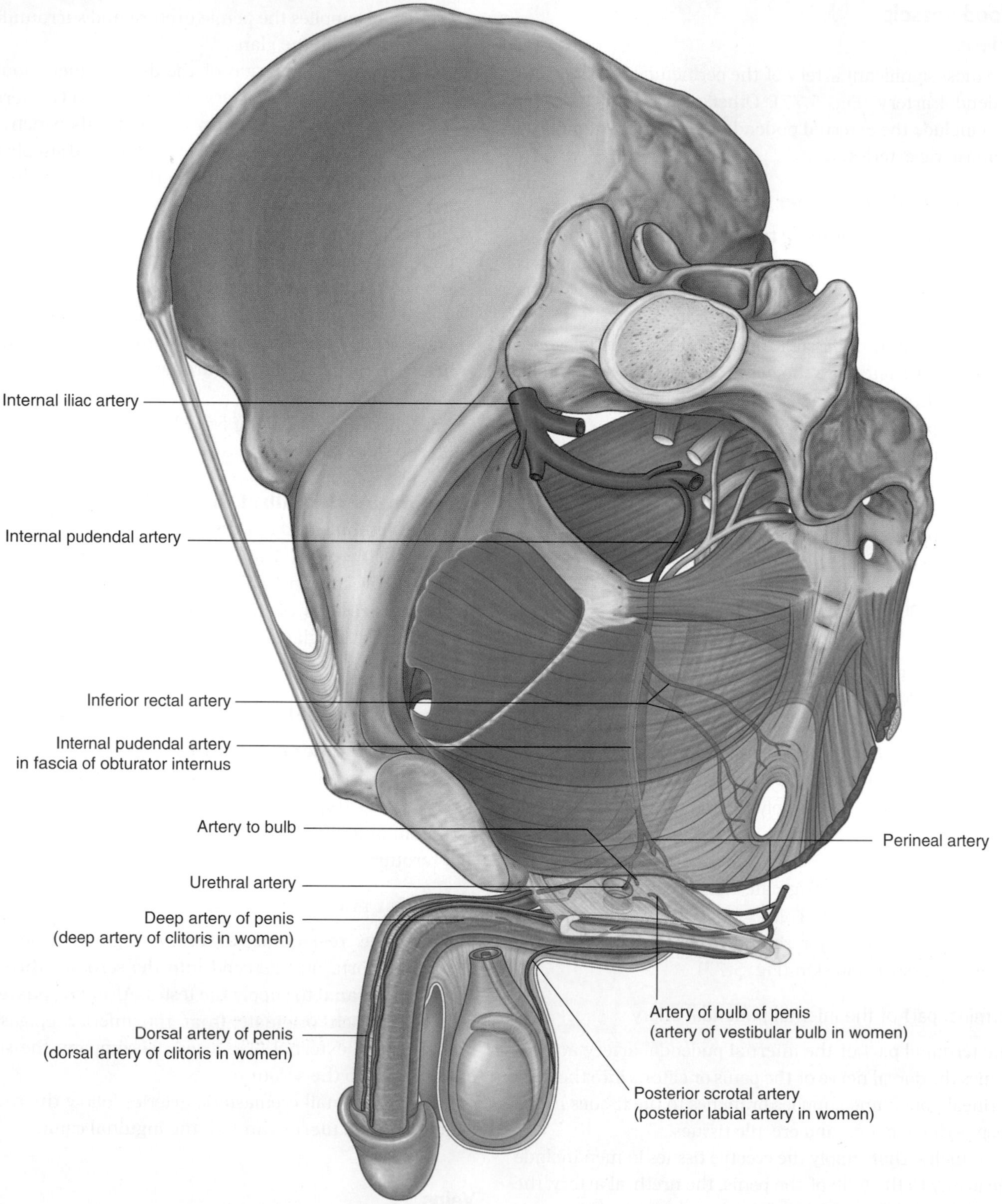

Internal iliac artery
Internal pudendal artery
Inferior rectal artery
Internal pudendal artery in fascia of obturator internus
Artery to bulb
Urethral artery
Deep artery of penis (deep artery of clitoris in women)
Dorsal artery of penis (dorsal artery of clitoris in women)
Perineal artery
Artery of bulb of penis (artery of vestibular bulb in women)
Posterior scrotal artery (posterior labial artery in women)

**Fig. 5.77** Arteries in the perineum.

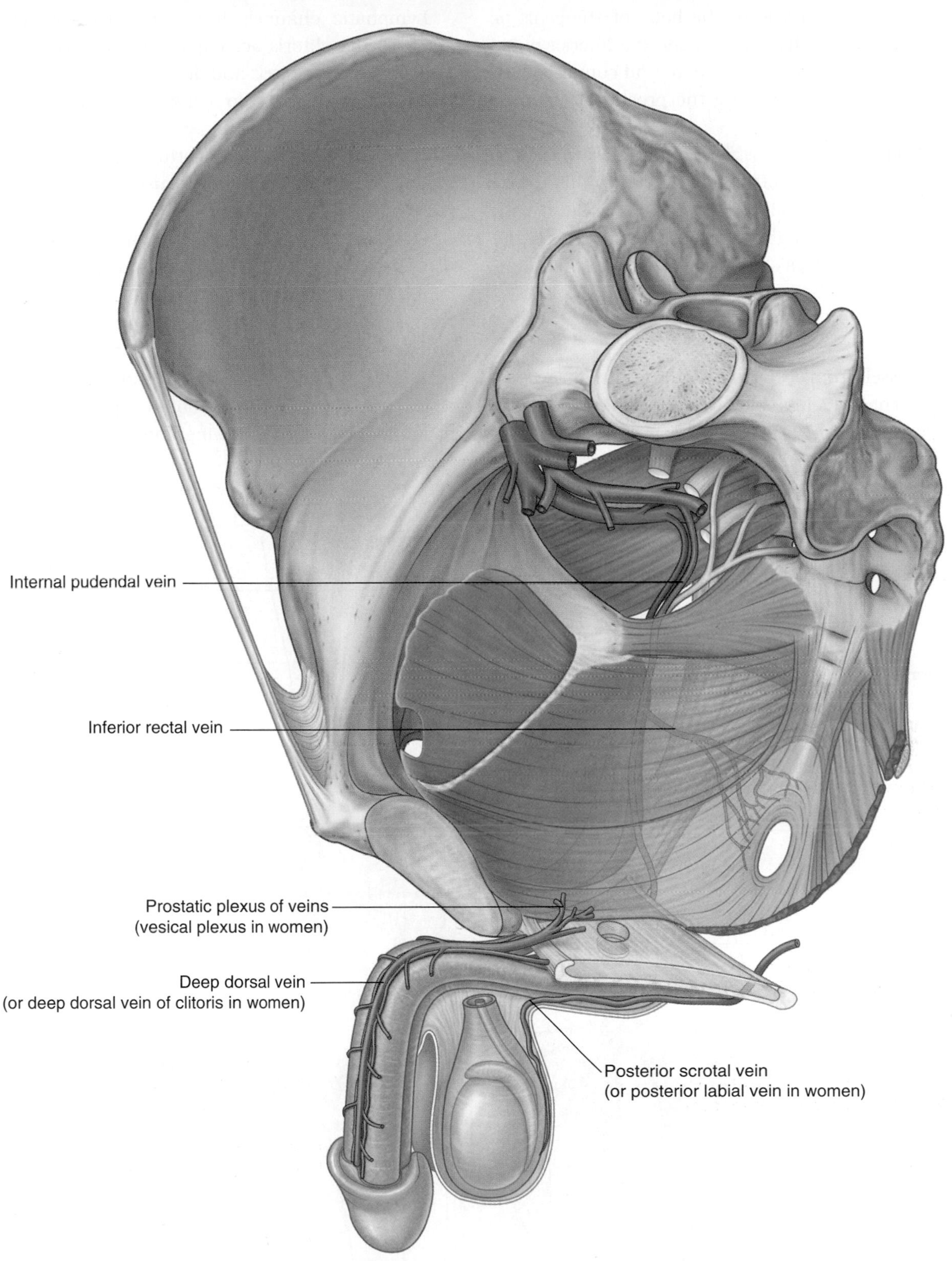
Internal pudendal vein
Inferior rectal vein
Prostatic plexus of veins
(vesical plexus in women)
Deep dorsal vein
(or deep dorsal vein of clitoris in women)
Posterior scrotal vein
(or posterior labial vein in women)

**Fig. 5.78** Perineal veins.

dorsal arteries on each side of the body of the penis or clitoris, passes though the gap between the inferior pubic ligament and the deep perineal pouch, and connects with the plexus of veins surrounding the prostate in men or bladder in women.

External pudendal veins, which drain anterior parts of the labia majora or the scrotum and overlap with the area of drainage of the internal pudendal veins, connect with the femoral vein in the thigh. Superficial dorsal veins of the penis or clitoris that drain skin are tributaries of the external pudendal veins.

## Lymphatics

Lymphatic vessels from deep parts of the perineum accompany the internal pudendal blood vessels and drain mainly into **internal iliac nodes** in the pelvis.

Lymphatic channels from superficial tissues of the penis or the clitoris accompany the superficial external pudendal blood vessels and drain mainly into **superficial inguinal nodes**, as do lymphatic channels from the scrotum or labia majora (Fig. 5.79). The glans penis, glans clitoris, labia minora, and terminal inferior end of the vagina drain into **deep inguinal nodes** and **external iliac nodes**.

Lymphatics from the testes drain via channels that ascend in the spermatic cord, pass through the inguinal canal, and course up the posterior abdominal wall to connect directly with **lateral aortic** or **lumbar nodes** and **pre-aortic nodes** around the aorta, at approximately vertebral levels LI and LII. Therefore disease from the testes tracks superiorly to nodes high in the posterior abdominal wall and not to inguinal or iliac nodes.

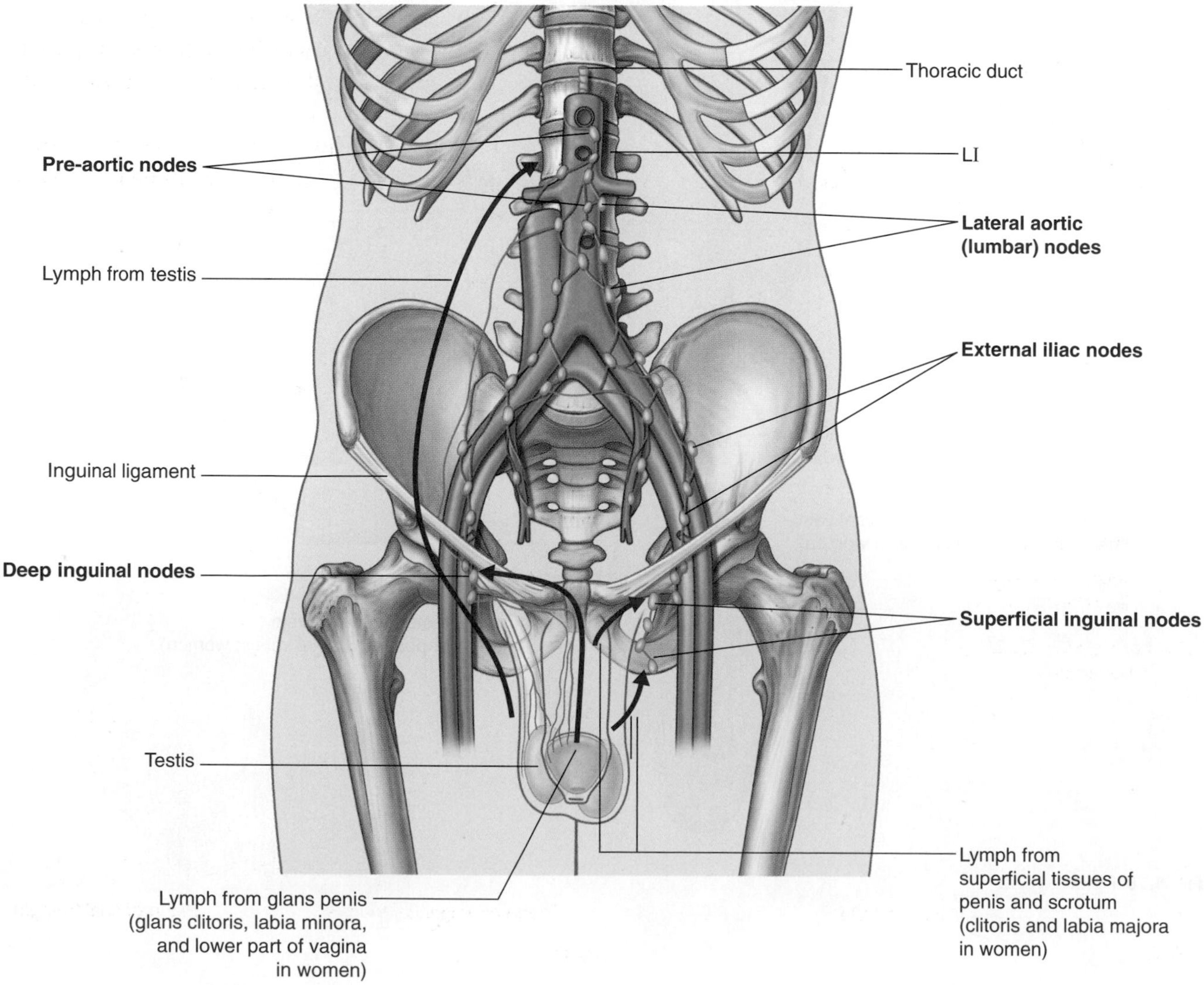

**Fig. 5.79** Lymphatic drainage of the perineum.

# Surface anatomy

## Surface anatomy of the pelvis and perineum

Palpable bony features of the pelvis are used as landmarks for:

- locating soft tissue structures,
- visualizing the orientation of the pelvic inlet, and
- defining the margins of the perineum.

The ability to recognize the normal appearance of structures in the perineum is an essential part of a physical examination.

In women, the cervix can be visualized directly by opening the vaginal canal using a speculum.

In men, the size and texture of the prostate in the pelvic cavity can be assessed by digital palpation through the anal aperture.

### Orientation of the pelvis and perineum in the anatomical position

In the anatomical position, the anterior superior iliac spines and the anterior superior edge of the pubic symphysis lie in the same vertical plane. The pelvic inlet faces anterosuperiorly. The urogenital triangle of the perineum is oriented in an almost horizontal plane and faces inferiorly, whereas the anal triangle is more vertical and faces posteriorly (Figs. 5.80 and 5.81).

### How to define the margins of the perineum

The pubic symphysis, ischial tuberosities, and tip of the sacrum are palpable on patients and can be used to define the boundaries of the perineum. This is best done with patients lying on their backs with their thighs flexed and abducted in the lithotomy position (Fig. 5.82).

- The ischial tuberosities are palpable on each side as large bony masses near the crease of skin (gluteal fold) between the thigh and gluteal region. They mark the lateral corners of the diamond-shaped perineum.
- The tip of the coccyx is palpable in the midline posterior to the anal aperture and marks the most posterior limit of the perineum.
- The anterior limit of the perineum is the pubic symphysis. In women, this is palpable in the midline deep to the

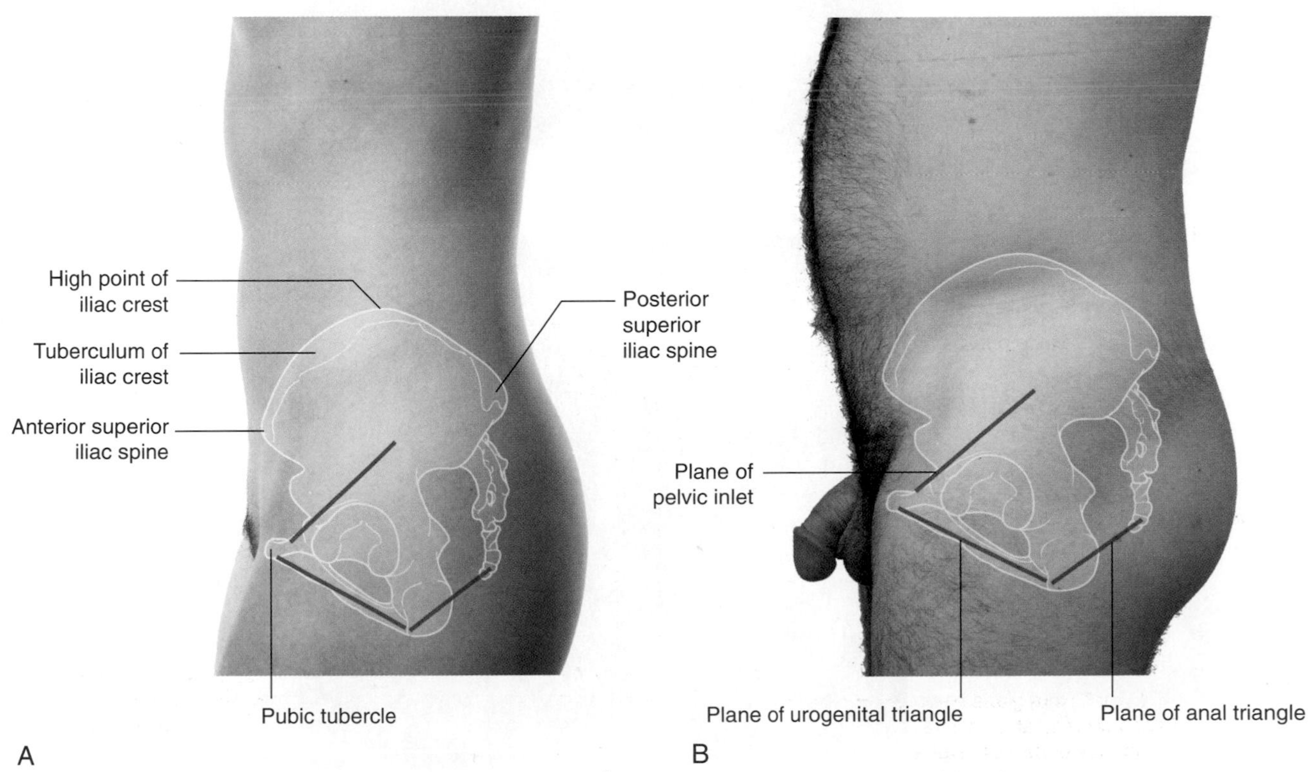


**Fig. 5.80** Lateral view of the pelvic area with the position of the skeletal features indicated. The orientation of the pelvic inlet, urogenital triangle, and anal triangle is also shown. **A.** In a woman. **B.** In a man.

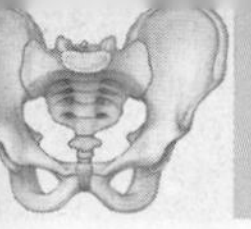



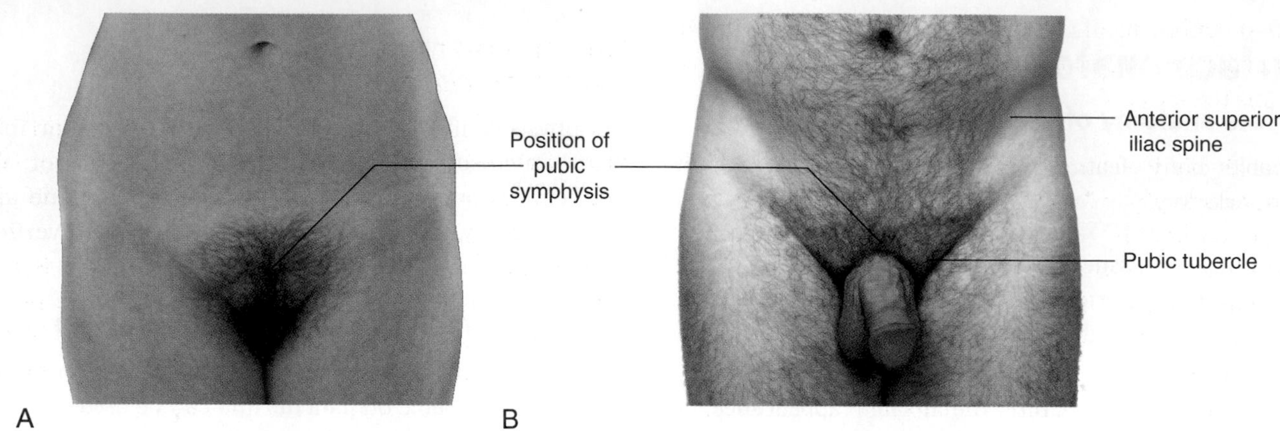


**Fig. 5.81** Anterior view of the pelvic area. **A.** In a woman showing the position of the pubic symphysis. **B.** In a man showing the position of the pubic tubercle, pubic symphysis, and anterior superior iliac spine.

Pubic symphysis
Urogenital triangle
Approximate position of
perineal body (central tendon)
Ischial tuberosity
Ischial tuberosity
Anal triangle
Coccyx
A

Mons pubis
Pubic symphysis
Urogenital triangle
Approximate position of
perineal body (central tendon)
Ischial tuberosity
Ischial tuberosity
Gluteal fold
Gluteal region
Anal triangle
Coccyx
B

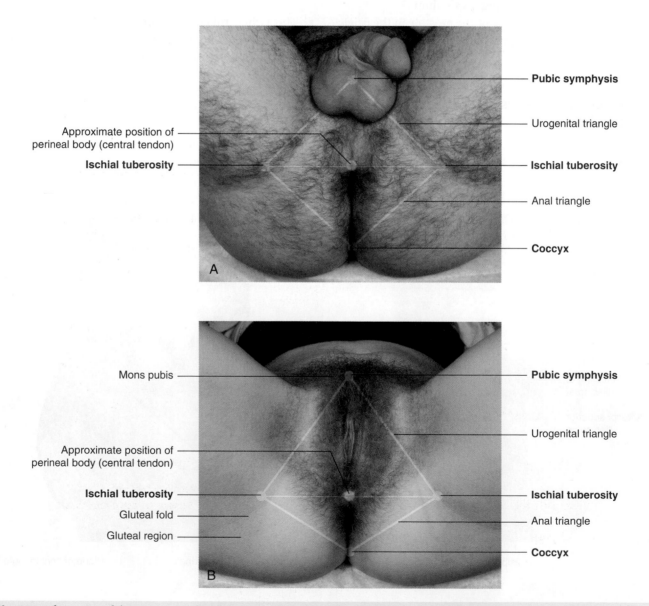

**Fig. 5.82** Inferior view of the perineum in the lithotomy position. Boundaries, subdivisions, and palpable landmarks are indicated. **A.** In a man. **B.** In a woman.

mons pubis. In men, the pubic symphysis is palpable immediately superior to where the body of the penis joins the lower abdominal wall.

Imaginary lines that join the ischial tuberosities with the pubic symphysis in front, and with the tip of the coccyx behind, outline the diamond-shaped perineum. An additional line between the ischial tuberosities divides the perineum into two triangles, the urogenital triangle anteriorly and anal triangle posteriorly. This line also approximates the position of the posterior margin of the perineal membrane. The midpoint of this line marks the location of the perineal body or central tendon of the perineum.

## Identification of structures in the anal triangle

The anal triangle is the posterior half of the perineum. The base of the triangle faces anteriorly and is an imaginary line joining the two ischial tuberosities. The apex of the triangle is the tip of the coccyx; the lateral margins can be approximated by lines joining the coccyx to the ischial tuberosities. In both women and men, the major feature of the anal triangle is the anal aperture in the center of the triangle. Fat fills the ischio-anal fossa on each side of the anal aperture (Fig. 5.83).

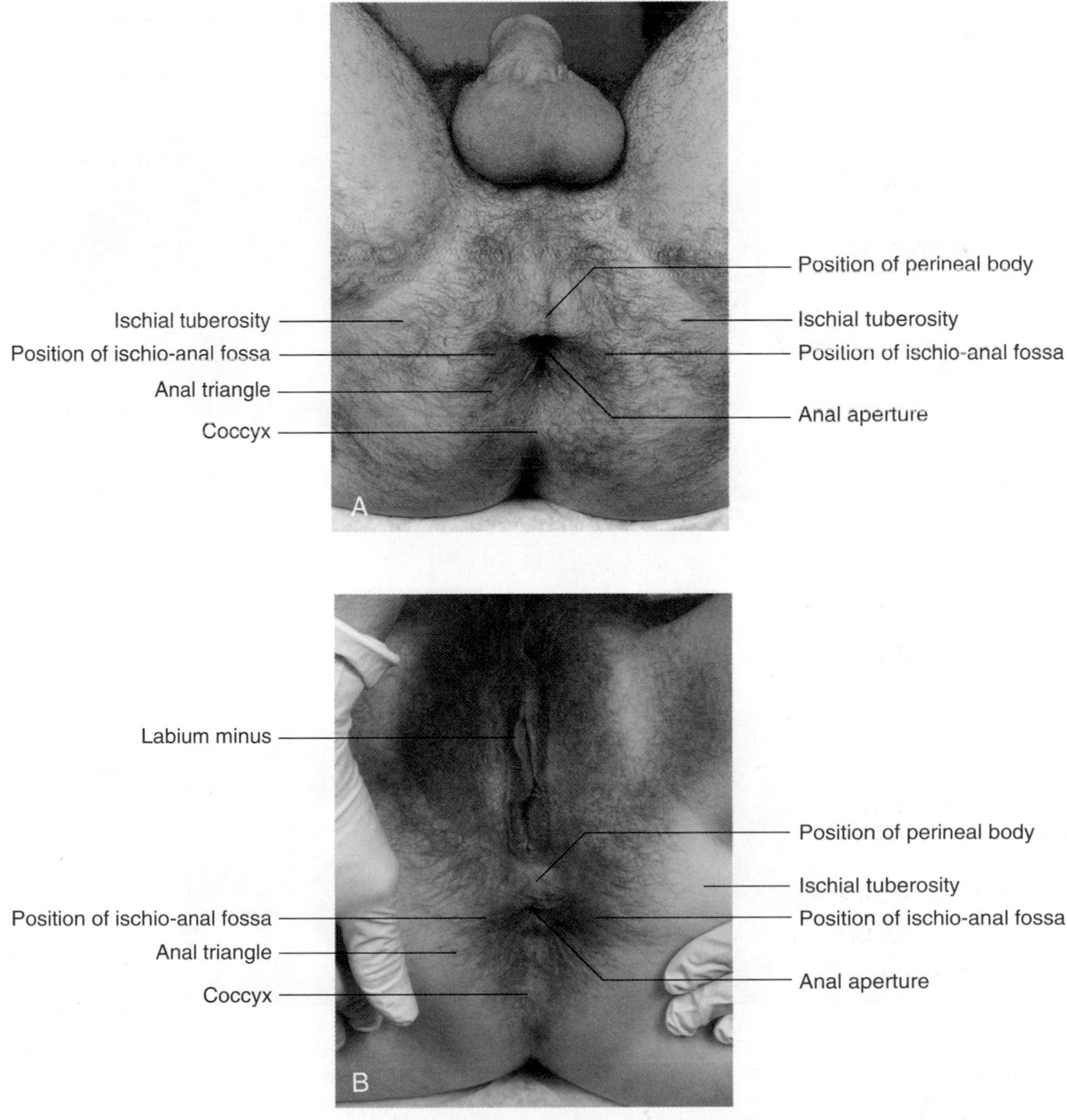


**Fig. 5.83** Anal triangle with the anal aperture and position of the ischio-anal fossae indicated. **A.** In a man. **B.** In a woman.

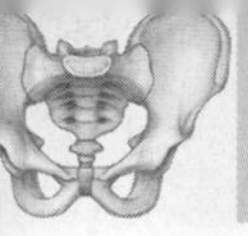



## Identification of structures in the urogenital triangle of women

The urogenital triangle is the anterior half of the perineum. The base of the triangle faces posteriorly and is an imaginary line joining the two ischial tuberosities. The apex of the triangle is the pubic symphysis. The lateral margins can be approximated by lines joining the pubic symphysis to the ischial tuberosities. These lines overlie the ischiopubic rami, which can be felt on deep palpation.

In women, the major contents of the urogenital triangle are the clitoris, the vestibule, and skin folds that together form the vulva (Fig. 5.84A,B).

Two thin skin folds, the labia minora, enclose between them a space termed the vestibule into which the vagina and the urethra open (Fig. 5.84C). Gentle lateral traction on the labia minora opens the vestibule and reveals a soft tissue mound on which the urethra opens. The para-urethral (Skene's) glands, one on each side, open into the skin crease between the urethra and the labia minora (Fig. 5.84D).

Posterior to the urethra is the vaginal opening. The vaginal opening (introitus) is ringed by remnants of the hymen that originally closes the vaginal orifice and is usually ruptured during the first sexual intercourse. The ducts of the greater vestibular (Bartholin's) glands, one on



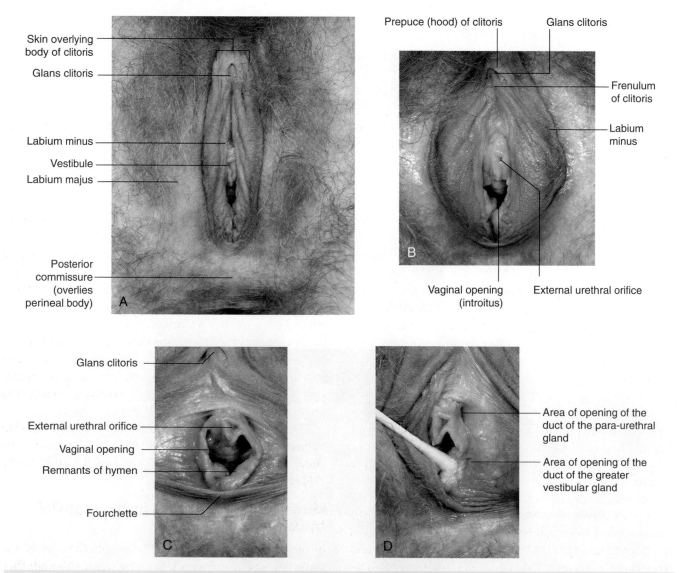

**Fig. 5.84** Structures in the urogenital triangle of a woman. **A.** Inferior view of the urogenital triangle of a woman with major features indicated. **B.** Inferior view of the vestibule. The labia minora have been pulled apart to open the vestibule. Also indicated are the glans clitoris, the clitoral hood, and the frenulum of the clitoris. **C.** Inferior view of the vestibule showing the urethral and vaginal orifices and the hymen. The labia minora have been pulled further apart than in Figure 5.84B. **D.** Inferior view of the vestibule with the left labium minus pulled to the side to show the regions of the vestibule into which the greater vestibular and para-urethral glands open.

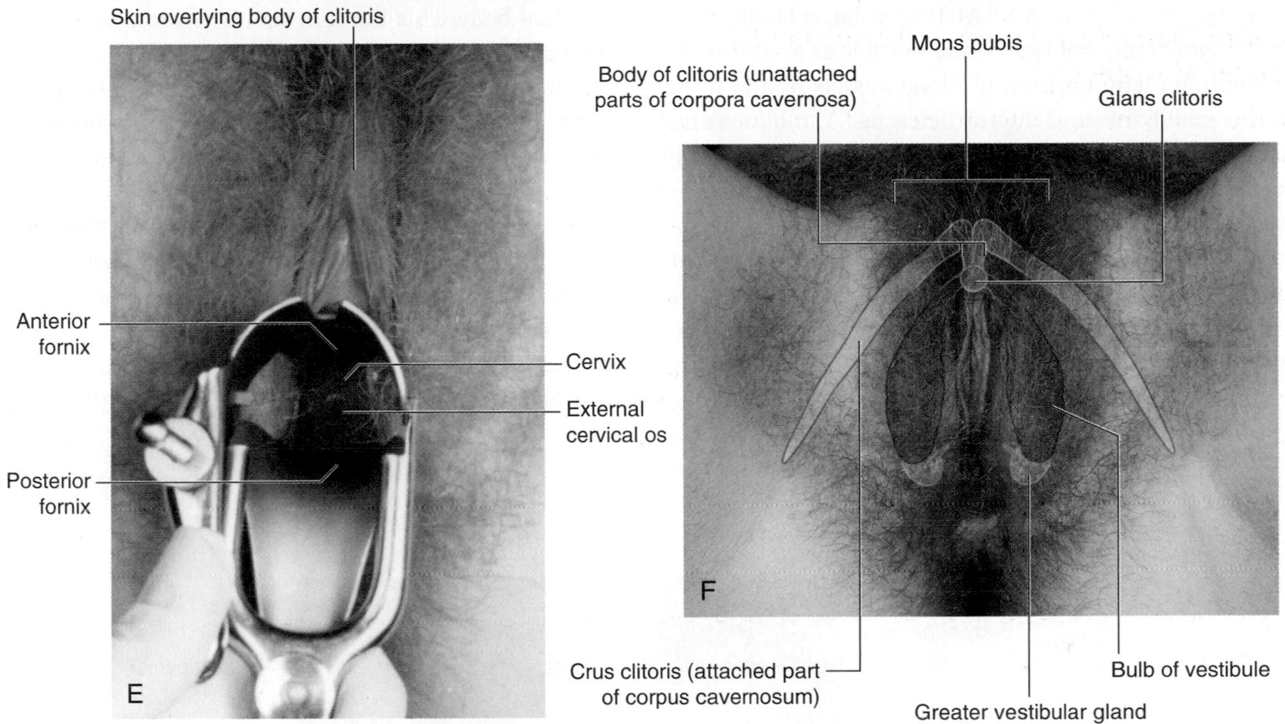


**Fig. 5.84, cont'd** **E.** View through the vaginal canal of the cervix. **F.** Inferior view of the urogenital triangle of a woman with the erectile tissues of the clitoris and vestibule and the greater vestibular glands indicated with overlays.

each side, open into the skin crease between the hymen and the adjacent labium minus (Fig. 5.84D).

The labia minora each bifurcate anteriorly into medial and lateral folds. The medial folds unite at the midline to form the frenulum of the clitoris. The larger lateral folds also unite across the midline to form the clitoral hood or prepuce that covers the glans clitoris and distal parts of the body of the clitoris. Posterior to the vaginal orifice, the labia minora join, forming a transverse skin fold (the fourchette).

The labia majora are broad folds positioned lateral to the labia minora. They come together in front to form the mons pubis, which overlies the inferior aspect of the pubic symphysis. The posterior ends of the labia majora are separated by a depression termed the posterior commissure, which overlies the position of the perineal body.

The cervix is visible when the vaginal canal is opened with a speculum (Fig. 5.84E). The external cervical os opens onto the surface of the dome-shaped cervix. A recess or gutter, termed the fornix, occurs between the cervix and the vaginal wall and is further subdivided, based on location, into anterior, posterior, and lateral fornices.

The roots of the clitoris occur deep to surface features of the perineum and are attached to the ischiopubic rami and the perineal membrane.

The bulbs of the vestibule (Fig. 5.84F), composed of erectile tissues, lie deep to the labia minora on either side of the vestibule. These erectile masses are continuous, via thin bands of erectile tissues, with the glans clitoris, which is visible under the clitoral hood. The greater vestibular glands occur posterior to the bulbs of the vestibule on either side of the vaginal orifice.

The crura of the clitoris are attached, one on each side, to the ischiopubic rami. Each crus is formed by the attached part of the corpus cavernosum. Anteriorly, these erectile corpora detach from bone, curve posteroinferiorly, and unite to form the body of the clitoris.

The body of the clitoris underlies the ridge of skin immediately anterior to the clitoral hood (prepuce). The glans clitoris is positioned at the end of the body of the clitoris.

## Identification of structures in the urogenital triangle of men

In men, the urogenital triangle contains the root of the penis. The testes and associated structures, although they migrate into the scrotum from the abdomen, are generally evaluated with the penis during a physical examination.

The scrotum in men is homologous to the labia majora in women. Each oval testis is readily palpable through the

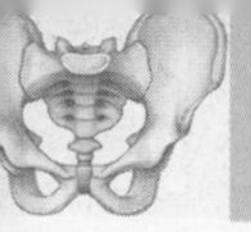



skin of the scrotum (Fig. 5.85A). Posterolateral to the testis is an elongated mass of tissue, often visible as a raised ridge that contains lymphatics and blood vessels of the testis, and the epididymis and ductus deferens. A midline raphe (Fig. 5.85B) is visible on the skin separating left and right sides of the scrotum. In some individuals, this raphe is prominent and extends from the anal aperture, over the scrotum and along the ventral surface of the body of the penis, to the frenulum of the glans.

The root of the penis is formed by the attached parts of the corpus spongiosum and the corpora cavernosa. The corpus spongiosum is attached to the perineal membrane and can be easily palpated as a large mass anterior to the perineal body. This mass, which is covered by the bulbospongiosus muscles, is the bulb of penis.

The corpus spongiosum detaches from the perineal membrane anteriorly, becomes the ventral part of the body of the penis (shaft of penis), and eventually terminates as the expanded glans penis (Fig. 5.85C,D).

The crura of the penis, one crus on each side, are the attached parts of the corpora cavernosa and are anchored to the ischiopubic rami (Fig. 5.85E). The corpora cavernosa are unattached anteriorly and become the paired erectile masses that form the dorsal part of the body of the penis. The glans penis caps the anterior ends of the corpora cavernosa.

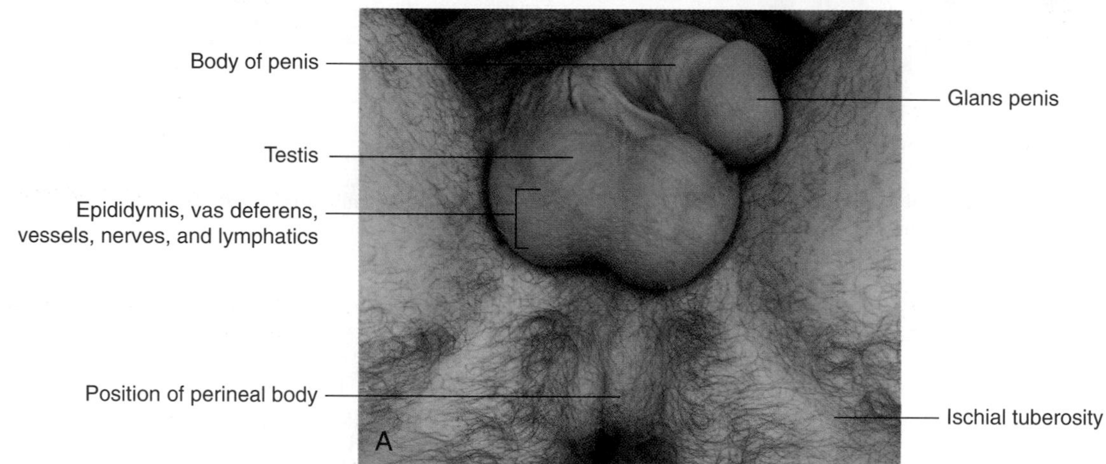


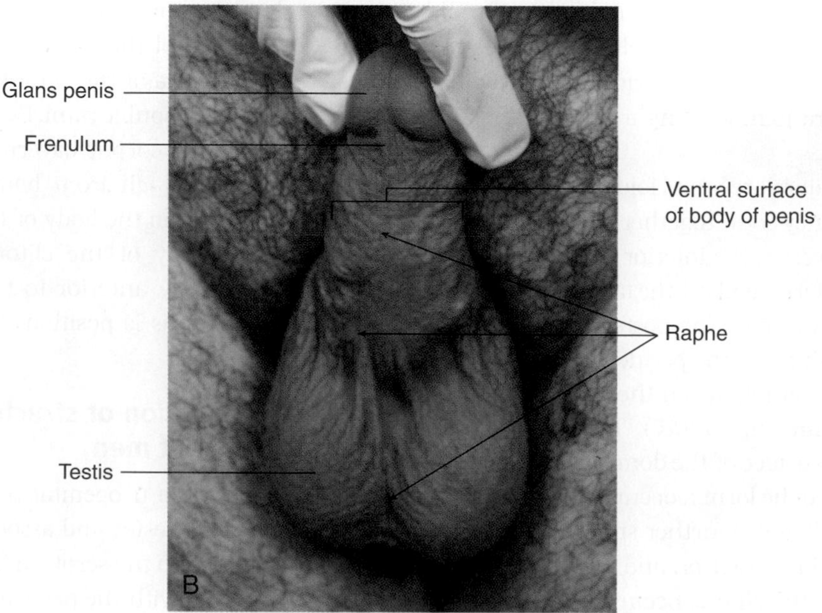


**Fig. 5.85** Structures in the urogenital triangle of a man. **A.** Inferior view. **B.** Ventral surface of the body of the penis.

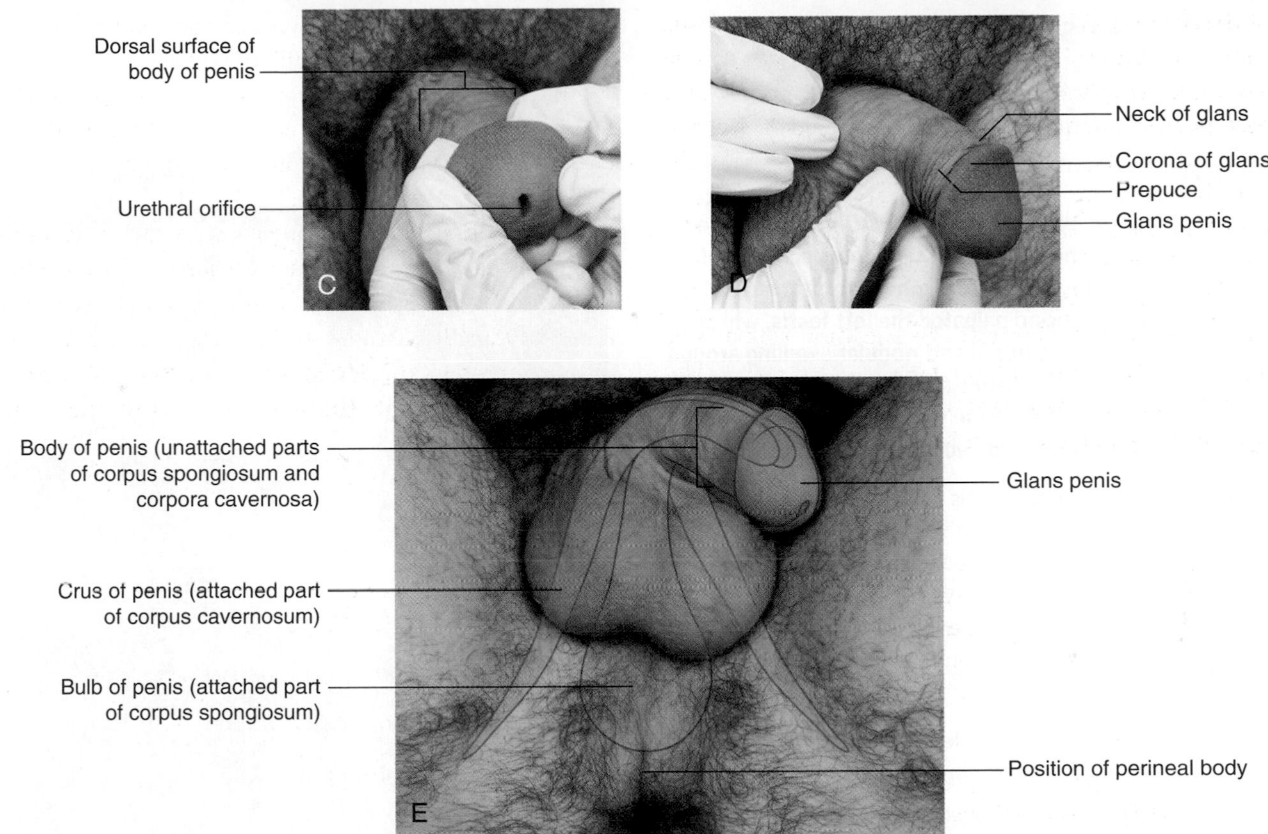


**Fig. 5.85, cont'd** **C.** Anterior view of the glans penis showing the urethral opening. **D.** Lateral view of the body of the penis and glans. **E.** Inferior view of the urogenital triangle of a man with the erectile tissues of the penis indicated with overlays.

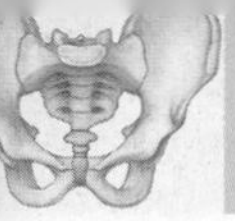



# Clinical cases

## Case 1

VARICOCELE

**A 25-year-old man visited his family physician because he had a "dragging feeling" in the left side of his scrotum. He was otherwise healthy and had no other symptoms. During examination, the physician palpated the left testis, which was normal, although he noted soft nodular swelling around the superior aspect of the testes and the epididymis. In his clinical notes, he described these findings as a "bag of worms" (Fig. 5.86). The bag of worms was a varicocele.**

The venous drainage of the testis is via the pampiniform plexus of veins that runs within the spermatic cord. A varicocele is a collection of dilated veins that arise from the pampiniform plexus. In many ways, they are similar to varicose veins that develop in the legs. Typically, the patient complains of a dragging feeling in the scrotum and around the testis, which is usually worse toward the end of the day.

The family physician recommended surgical treatment, with a recommendation for surgery through an inguinal incision.

A simple surgical technique divides the skin around the inguinal ligament. The aponeurosis of the external oblique muscle is divided in the anterior abdominal wall to display the spermatic cord. Careful inspection of the spermatic cord reveals the veins, which are surgically ligated.

Another option is to embolize the varicocele.

In this technique, a small catheter is placed via the right femoral vein. The catheter is advanced along the external iliac vein and the common iliac vein and into the inferior vena cava. The catheter is then positioned in the left renal vein, and a venogram is performed to demonstrate the origin of the left testicular vein. The catheter is advanced down the left testicular vein into the veins of the inguinal canal and the pampiniform plexus. Metal coils to occlude the vessels are injected, and the catheter is withdrawn.

The patient asked how blood would drain from the testis after the operation.

Although the major veins of the testis had been occluded, small collateral veins running within the scrotum and around the outer aspect of the spermatic cord permitted drainage without recurrence of the varicocele.

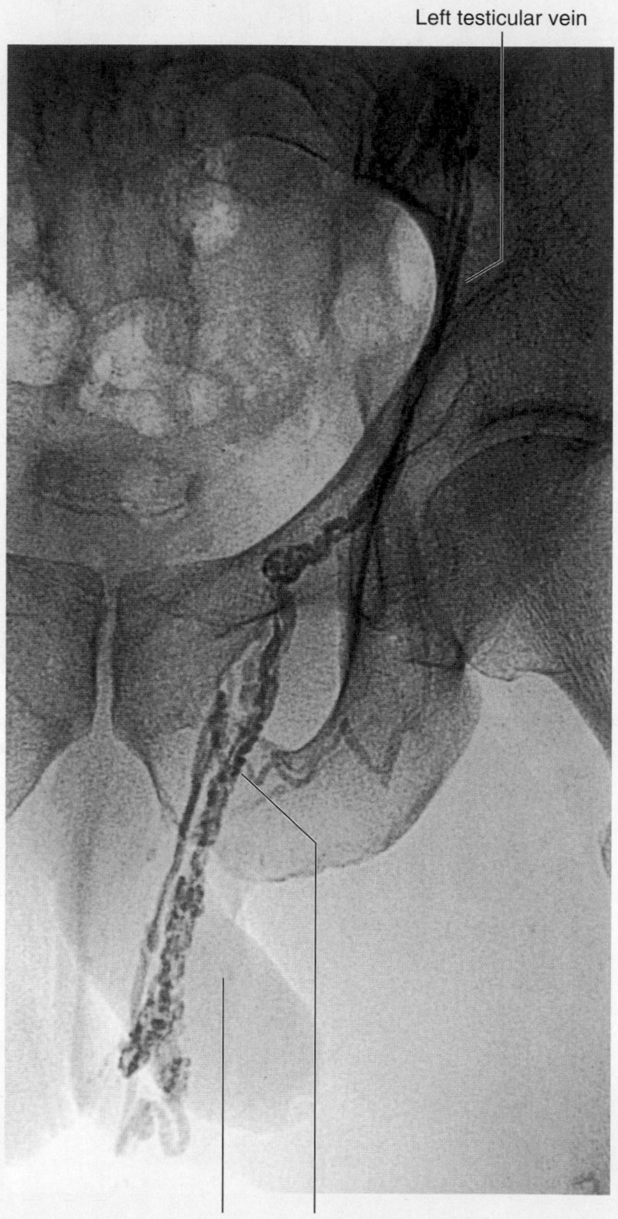


**Fig. 5.86** Left testicular venogram demonstrating the pampiniform plexus of veins.

## Case 2

PELVIC KIDNEY

**A young woman visited her family practitioner because she had mild upper abdominal pain. An ultrasound demonstrated gallstones within the gallbladder, which explained the patient's pain. However, when the technician assessed the pelvis, she noted a mass behind the bladder, which had sonographic findings similar to a kidney (Fig. 5.87).**

What did the sonographer do next? Having demonstrated this pelvic mass behind the bladder, the sonographer assessed both kidneys. The patient had a normal right kidney. However, the left kidney could not be found in its usual place. The technician diagnosed a pelvic kidney.

A pelvic kidney can be explained by the embryology. The kidneys develop from a complex series of structures that originate adjacent to the bladder within the fetal pelvis. As development proceeds and the functions of the various parts of the developing kidneys change, they attain a superior position in the upper abdomen adjacent to the abdominal aorta and inferior vena cava, on the posterior abdominal wall. A developmental arrest or complication may prevent the kidney from obtaining its usual position. Fortunately, it is unusual for patients to have any symptoms relating to a pelvic kidney.

This patient had no symptoms attributable to the pelvic kidney and she was discharged.

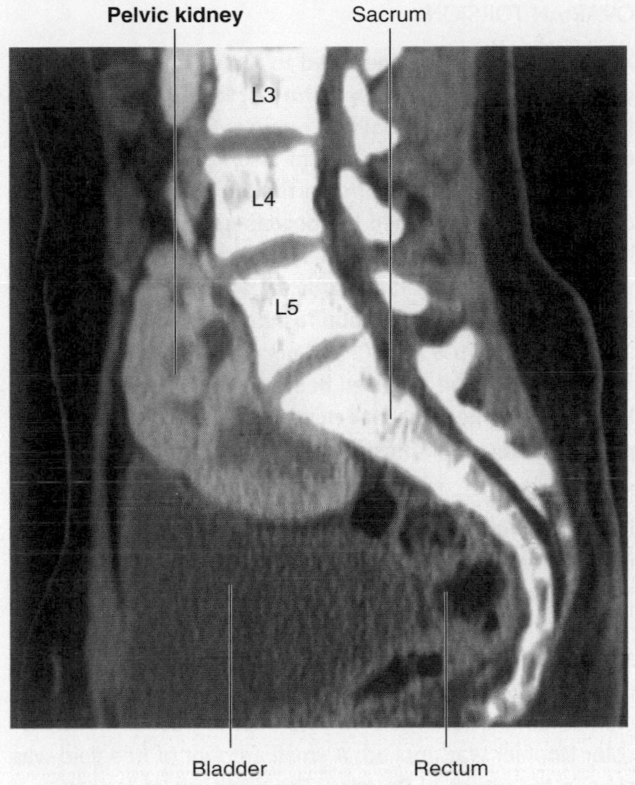


**Fig. 5.87** Sagittal computed tomogram demonstrating a pelvic kidney.

## Case 3

OVARIAN TORSION

**A 19-year-old woman presented to the emergency department with a 36-hour history of lower abdominal pain that was sharp and initially intermittent, later becoming constant and severe. The patient also reported feeling nauseated and vomited once in the ER. She did not have diarrhea and had opened her bowels normally 8 hours before admission. She had no symptoms of dysuria. She was afebrile, slightly tachycardic at 95/min, and had a normal blood pressure. Blood results showed mild leukocytosis of 11.6 x $10^9$/L and normal renal and liver function tests. She reported being sexually active with a long-term partner. She was never pregnant, and the urine pregnancy test on admission was negative.**

On physical examination there was tenderness in the right iliac fossa with guarding. On vaginal examination a tender mass in the right adnexal region was felt. The patient subsequently underwent a transvaginal ultrasound examination for evaluation of adnexal pathology. The scan showed a markedly enlarged right ovary measuring up to 8 cm in long axis with echogenic stroma and peripherally distributed follicles. There was no internal vascularity when color Doppler was applied. A small amount of free fluid was seen in the pouch of Douglas. The diagnosis of ovarian torsion was made.

Ovarian torsion is the twisting of an ovary on its suspensory ligament, which contains arterial, venous, and lymphatic vessels (forming so-called vascular pedicle), leading to a compromised blood supply. Initially, the venous and lymphatic circulation is compromised, resulting in ovarian edema and enlargement. The arterial flow is maintained longer due to thicker and less compressible arterial walls. Prolonged torsion leads to increased internal ovarian pressure that eventually results in arterial thrombosis, ischemia of the ovarian tissue, and necrosis. If the correct diagnosis and treatment are delayed, the patient may develop generalized sepsis.

The symptoms are nonspecific, making the diagnosis of ovarian torsion challenging. There is often no significant past medical history.

At surgery, the right ovary was hemorrhagic and necrotic with the pedicle twisted 360 degrees. The left ovary was normal in appearance. Right-sided salpingo-oophorectomy was performed, and histopathological examination confirmed completely necrotic ovary without any residual normal ovarian tissue. The patient made a quick recovery after surgical intervention.

Ovarian torsion is encountered in women of all ages, but those of reproductive age have much higher prevalence. Torsion of a normal ovary is uncommon and is seen more frequently in adolescent population, with elongated pelvic ligaments, fallopian tube spasm, or more mobile fallopian tubes or mesosalpinx cited as contributing factors.

# 6 Lower Limb

## ADDITIONAL LEARNING RESOURCES for Chapter 6, Lower Limb, on STUDENT CONSULT (www.studentconsult.com):

- Self-Assessment—National Board style multiple-choice questions, Chapter 6
- Short Questions—these are questions requiring short responses, Chapter 6
- Interactive Surface Anatomy—interactive surface animations, Chapter 6
- PT Case Studies, Chapter 6
  Plantar fasciitis
  Achilles tendinitis and tendinosis
  Eversion ankle sprain
  High ankle sprain
  Patellofemoral pain syndrome
  Anterior lateral shin splints
  *See more PT Case Studies online*
- Clinical Cases, Chapter 6
  Anterior talofibular ligament tear
  Deep vein thrombosis
  Femoral hernia
  Fracture of neck of femur
  Groin injury
  Iliopsoas tendinitis
  Iliotibial band syndrome
  Knee joint injury
  Nerve entrapment syndrome
  Osteomyelitis
  Pes cavus
  Popliteal artery aneurysm
  Ruptured calcaneal tendon
  Varicose veins

**FREE ONLINE Anatomy and Embryology Self-Study Course**
  Anatomy modules 26 through 31
  Embryology module 71

# Conceptual overview

## GENERAL INTRODUCTION

The lower limb is directly anchored to the axial skeleton by a sacroiliac joint and by strong ligaments, which link the pelvic bone to the sacrum. It is separated from the abdomen, back, and perineum by a continuous line (Fig. 6.1), which:

- joins the pubic tubercle with the anterior superior iliac spine (position of the inguinal ligament) and then continues along the iliac crest to the posterior superior iliac spine to separate the lower limb from the anterior and lateral abdominal walls;
- passes between the posterior superior iliac spine and along the dorsolateral surface of the sacrum to the coccyx to separate the lower limb from the muscles of the back; and
- joins the medial margin of the sacrotuberous ligament, the ischial tuberosity, the ischiopubic ramus, and the pubic symphysis to separate the lower limb from the perineum.

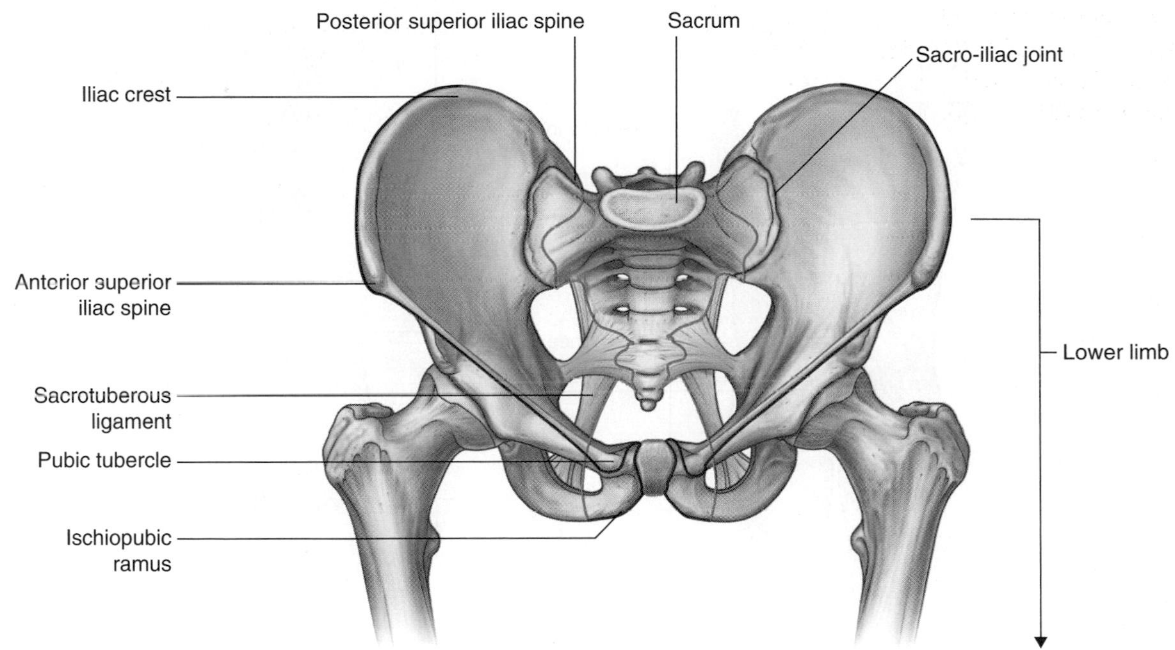


**Fig. 6.1** Upper margin of the lower limb.

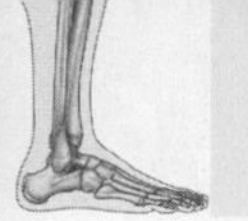
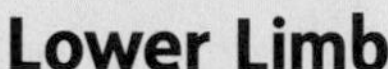



The lower limb is divided into the gluteal region, thigh, leg, and foot on the basis of major joints, component bones, and superficial landmarks (Fig. 6.2):

- The **gluteal region** is posterolateral and between the iliac crest and the fold of skin (gluteal fold) that defines the lower limit of the buttocks.
- Anteriorly, the **thigh** is between the inguinal ligament and the knee joint—the hip joint is just inferior to the middle third of the inguinal ligament, and the posterior thigh is between the gluteal fold and the knee.
- The **leg** is between the knee and ankle joint.
- The **foot** is distal to the ankle joint.



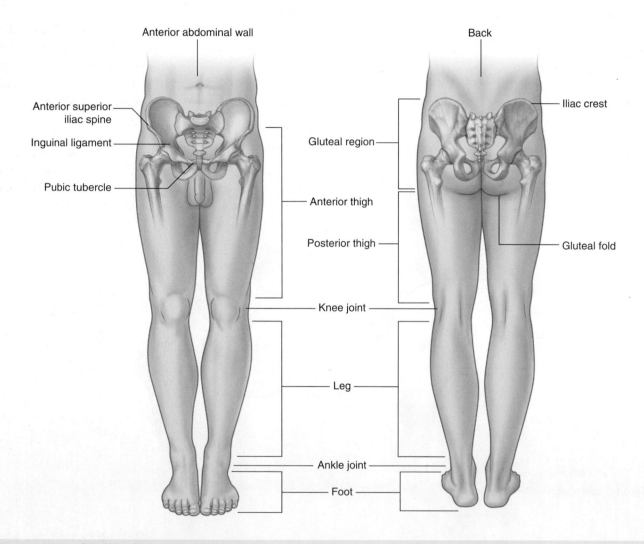

**Fig. 6.2** Regions of the lower limb.

The femoral triangle and popliteal fossa, as well as the posteromedial side of the ankle, are important areas of transition through which structures pass between regions (Fig. 6.3).

The **femoral triangle** is a pyramid-shaped depression formed by muscles in the proximal regions of the thigh and by the inguinal ligament, which forms the base of the triangle. The major blood supply and one of the nerves of the limb (femoral nerve) enter into the thigh from the abdomen by passing under the inguinal ligament and into the femoral triangle.

The **popliteal fossa** is posterior to the knee joint and is a diamond-shaped region formed by muscles of the thigh and leg. Major vessels and nerves pass between the thigh and leg through the popliteal fossa.

Most nerves, vessels, and flexor tendons that pass between the leg and foot pass through a series of canals (collectively termed the tarsal tunnel) on the posteromedial side of the ankle. The canals are formed by adjacent bones and a flexor retinaculum, which holds the tendons in position.

## FUNCTION

### Support the body weight

A major function of the lower limb is to support the weight of the body with minimal expenditure of energy. When standing erect, the center of gravity is anterior to the edge of the SII vertebra in the pelvis (Fig. 6.4). The vertical line through the center of gravity is slightly posterior to the hip joints, anterior to the knee and ankle joints, and directly over the almost circular support base formed by the feet on the ground and holds the knee and hip joints in extension.

The organization of ligaments at the hip and knee joints, together with the shape of the articular surfaces, particularly at the knee, facilitates "locking" of these joints into position when standing, thereby reducing the muscular energy required to maintain a standing position.

### Locomotion

A second major function of the lower limbs is to move the body through space. This involves the integration of movements at all joints in the lower limb to position the foot on the ground and to move the body over the foot.

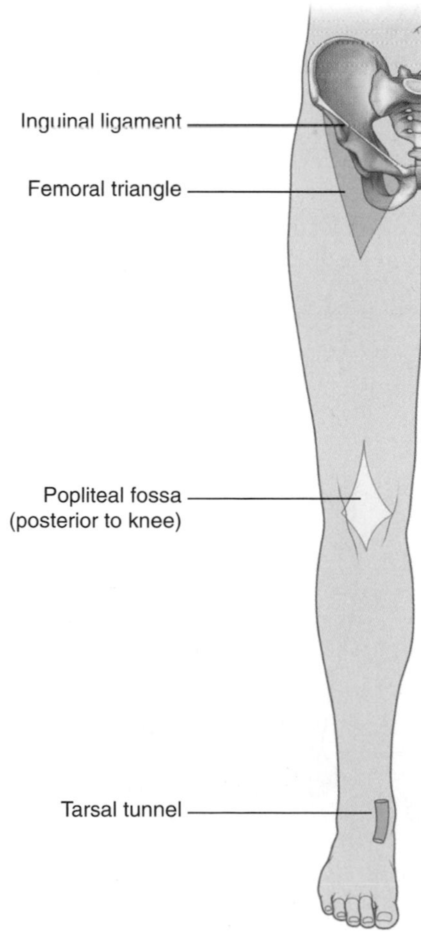


**Fig. 6.3** Areas of transition.

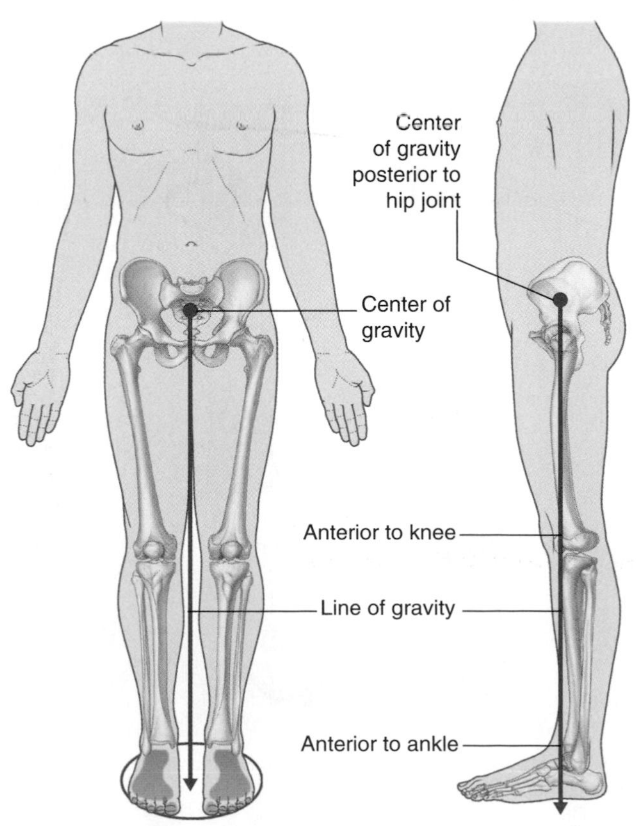


**Fig. 6.4** Center and line of gravity.

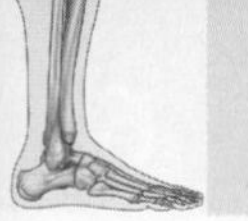



Movements at the hip joint are flexion, extension, abduction, adduction, medial and lateral rotation, and circumduction (Fig. 6.5).

The knee and ankle joints are primarily hinge joints. Movements at the knee are mainly flexion and extension (Fig. 6.6A). Movements at the ankle are dorsiflexion (movement of the dorsal side of the foot toward the leg) and plantarflexion (Fig. 6.6B).

During walking, many anatomical features of the lower limbs contribute to minimizing fluctuations in the body's center of gravity and thereby reduce the amount of energy needed to maintain locomotion and produce a smooth,

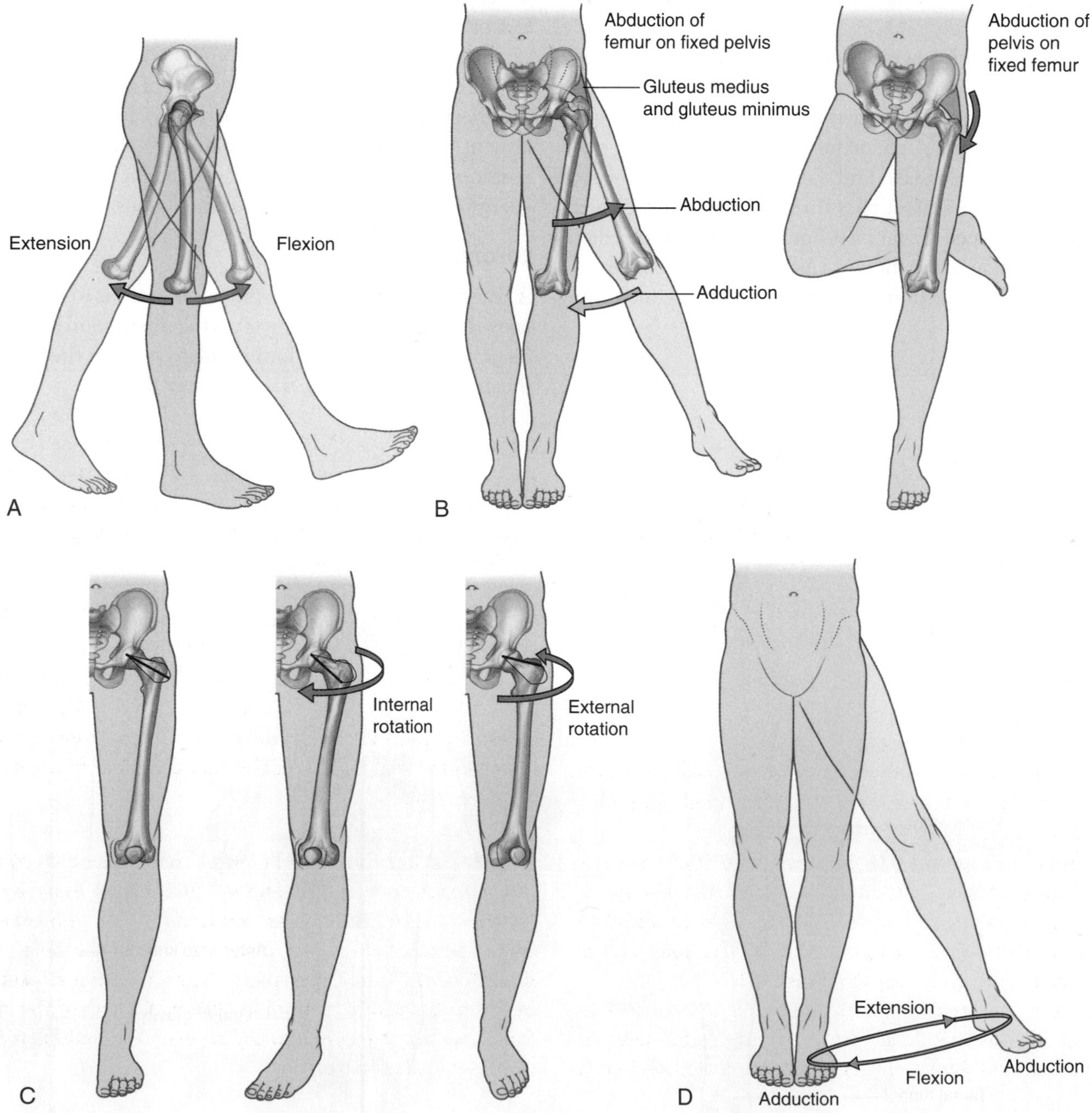

**Fig. 6.5** Movements of the hip joint. **A.** Flexion and extension. **B.** Abduction and adduction. **C.** External and internal rotation. **D.** Circumduction.

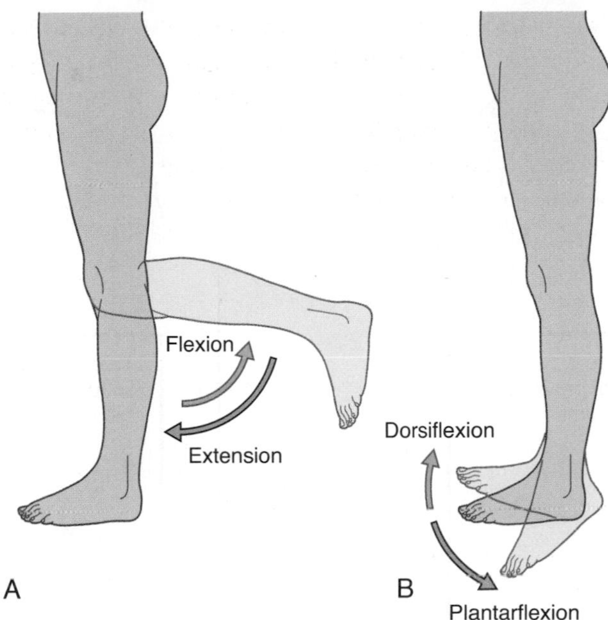


**Fig. 6.6** Movements of the knee and ankle. **A.** Knee flexion and extension. **B.** Ankle dorsiflexion and plantarflexion.

efficient gait (Fig. 6.7). They include pelvic tilt in the coronal plane, pelvic rotation in the transverse plane, movement of the knees toward the midline, flexion of the knees, and complex interactions between the hip, knee, and ankle. As a result, during walking, the body's center of gravity normally fluctuates only 5 cm in both vertical and lateral directions.

## COMPONENT PARTS

### Bones and joints

The bones of the gluteal region and the thigh are the pelvic bone and the femur (Fig. 6.8). The large ball and socket joint between these two bones is the hip joint.

The femur is the bone of the thigh. At its distal end, its major weight-bearing articulation is with the tibia, but it also articulates anteriorly with the patella (knee cap). The patella is the largest sesamoid bone in the body and is embedded in the quadriceps femoris tendon.

The joint between the femur and tibia is the principal articulation of the knee joint, but the joint between the patella and femur shares the same articular cavity. Although the main movements at the knee are flexion and extension, the knee joint also allows the femur to rotate on the tibia. This rotation contributes to "locking" of the knee when fully extended, particularly when standing.

The leg contains two bones:

- The tibia is medial in position, is larger than the laterally positioned fibula, and is the weight-bearing bone.
- The fibula does not take part in the knee joint and forms only the most lateral part of the ankle joint—proximally, it forms a small synovial joint (superior tibiofibular joint) with the inferolateral surface of the head of the tibia.

The tibia and fibula are linked along their lengths by an interosseous membrane, and at their distal ends by a fibrous inferior tibiofibular joint, and little movement occurs between them. The distal surfaces of the tibia and fibula together form a deep recess. The ankle joint is formed by this recess and part of one of the tarsal bones of the foot (talus), which projects into the recess. The ankle is most stable when dorsiflexed.

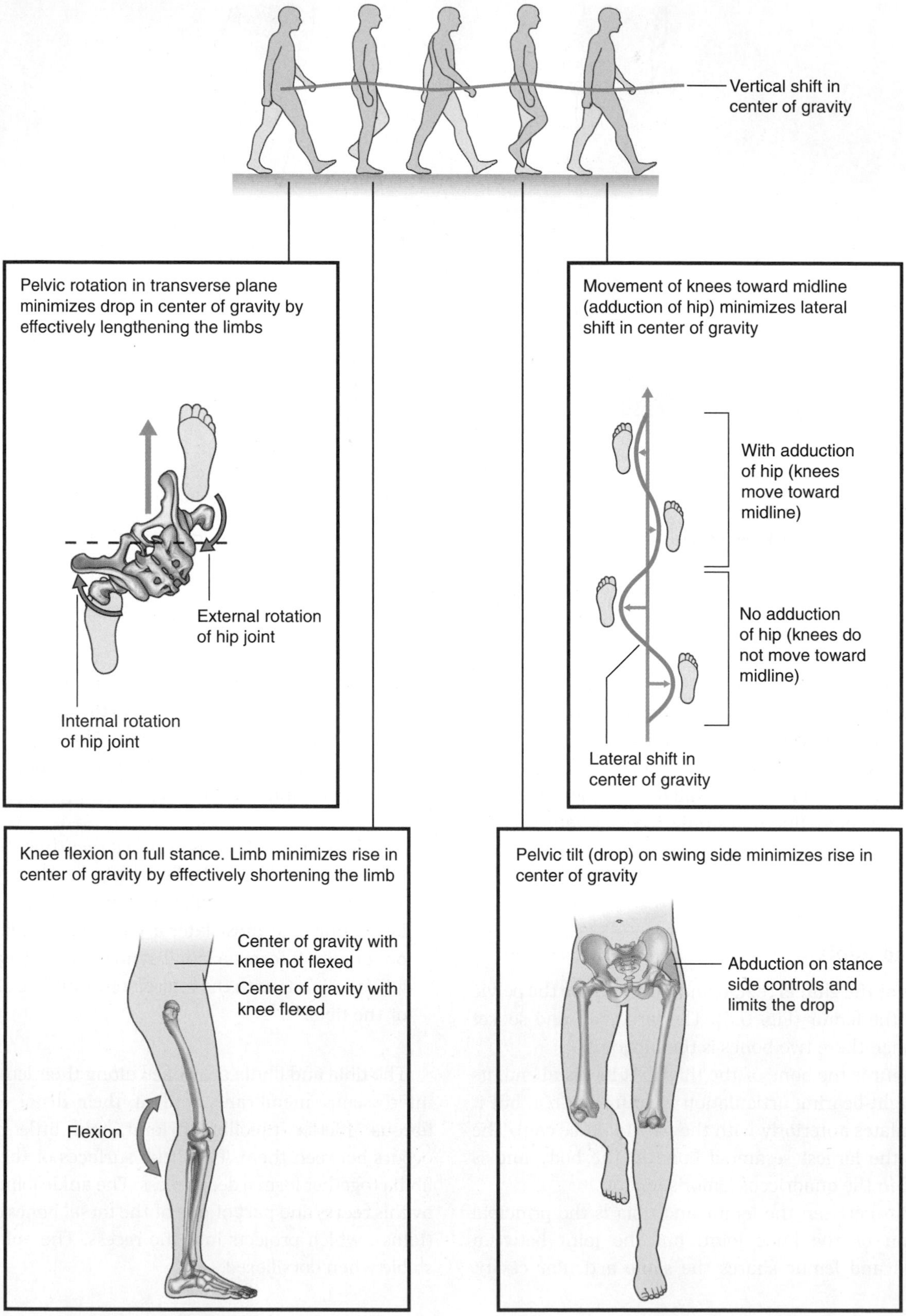

Vertical shift in center of gravity

Pelvic rotation in transverse plane minimizes drop in center of gravity by effectively lengthening the limbs

External rotation of hip joint

Internal rotation of hip joint

Movement of knees toward midline (adduction of hip) minimizes lateral shift in center of gravity

With adduction of hip (knees move toward midline)

No adduction of hip (knees do not move toward midline)

Lateral shift in center of gravity

Knee flexion on full stance. Limb minimizes rise in center of gravity by effectively shortening the limb

Center of gravity with knee not flexed

Center of gravity with knee flexed

Flexion

Pelvic tilt (drop) on swing side minimizes rise in center of gravity

Abduction on stance side controls and limits the drop

**Fig. 6.7** Some of the determinants of gait.

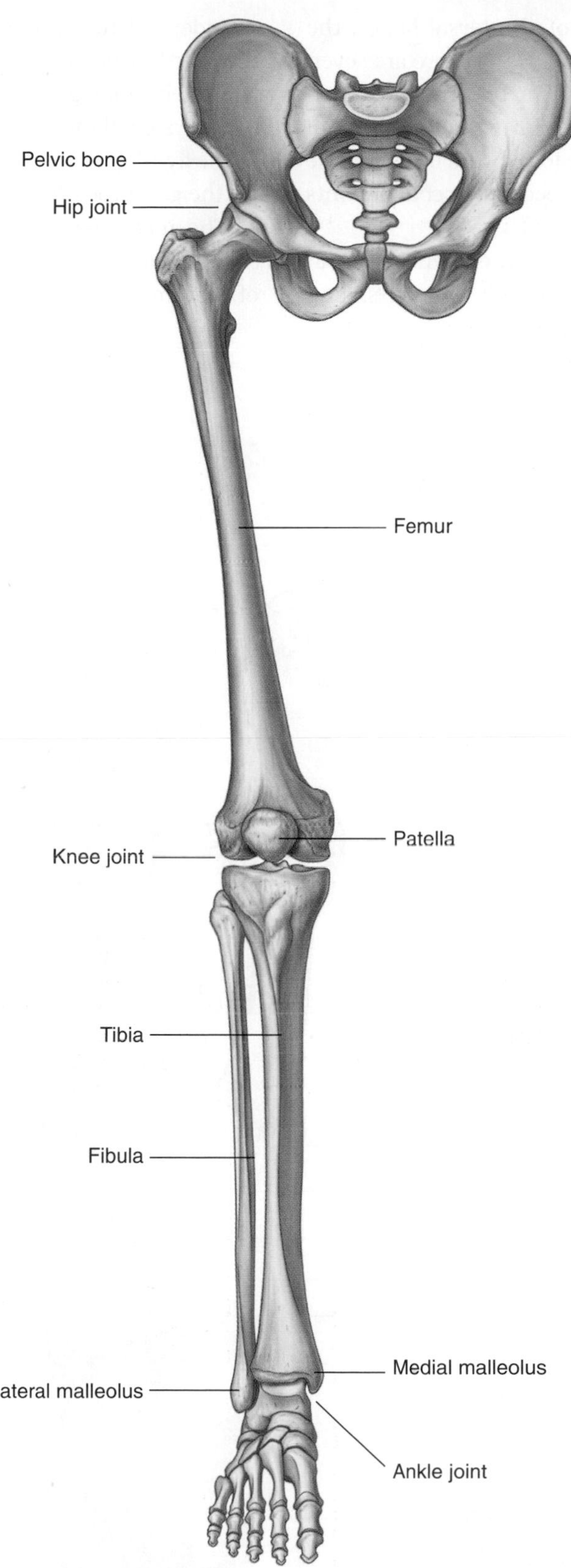


**Fig. 6.8** Bones and joints of the lower limb.

The bones of the foot consist of the tarsal bones, the metatarsals, and the phalanges (Fig. 6.9). There are seven tarsal bones, which are organized in two rows with an intermediate bone between the two rows on the medial side. Inversion and eversion of the foot, or turning the sole of the foot inward and outward, respectively, occur at joints between the tarsal bones.

The tarsal bones articulate with the metatarsals at tarsometatarsal joints, which allow only limited sliding movements.

Independent movements of the metatarsals are restricted by deep transverse metatarsal ligaments, which effectively link together the distal heads of the bones at the metatarsophalangeal joints. There is a metatarsal for each of the five digits, and each digit has three phalanges except for the great toe (digit I), which has only two.

The metatarsophalangeal joints allow flexion, extension, abduction, and adduction of the digits, but the range of movement is more restricted than in the hand.

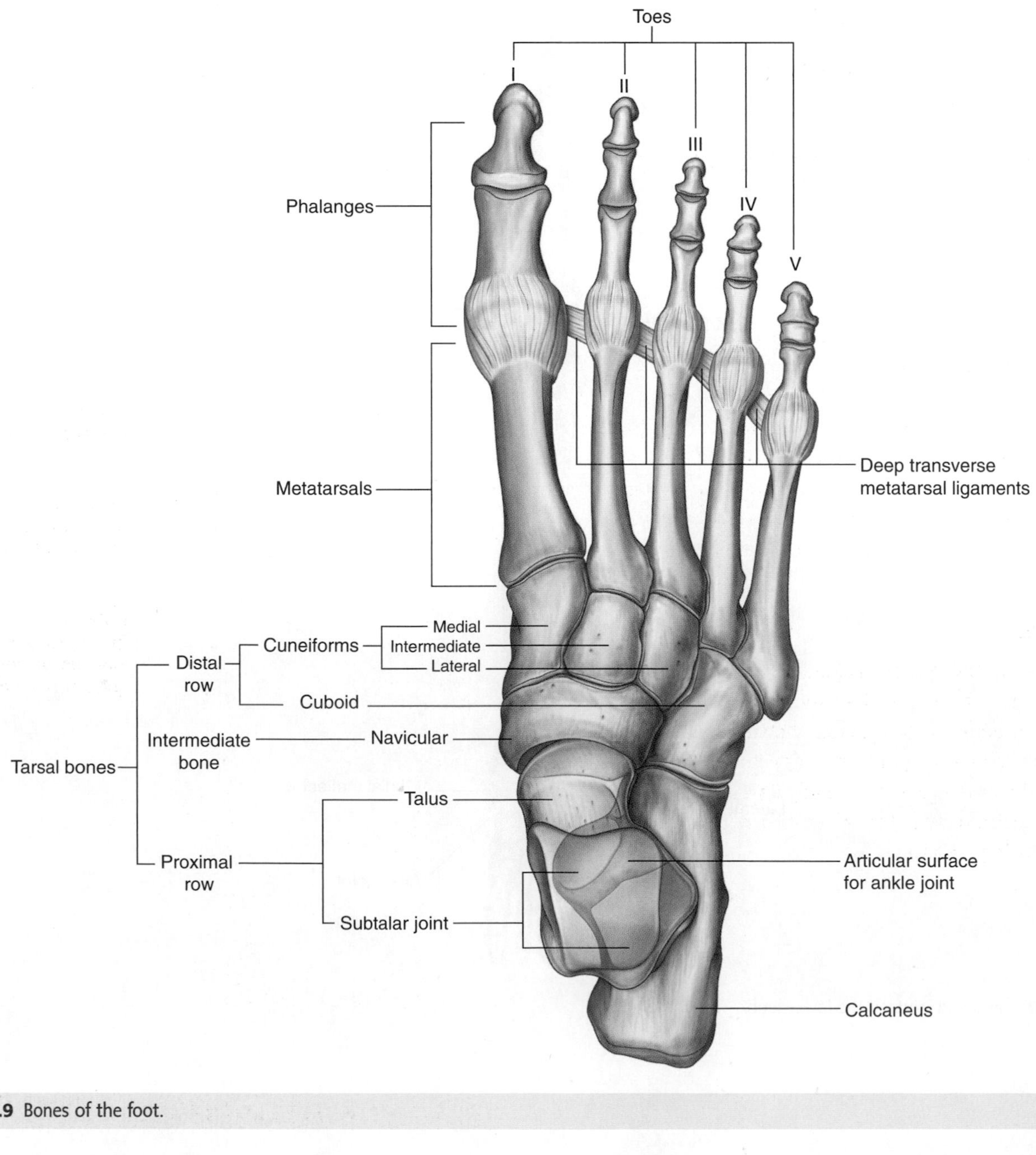

**Fig. 6.9** Bones of the foot.

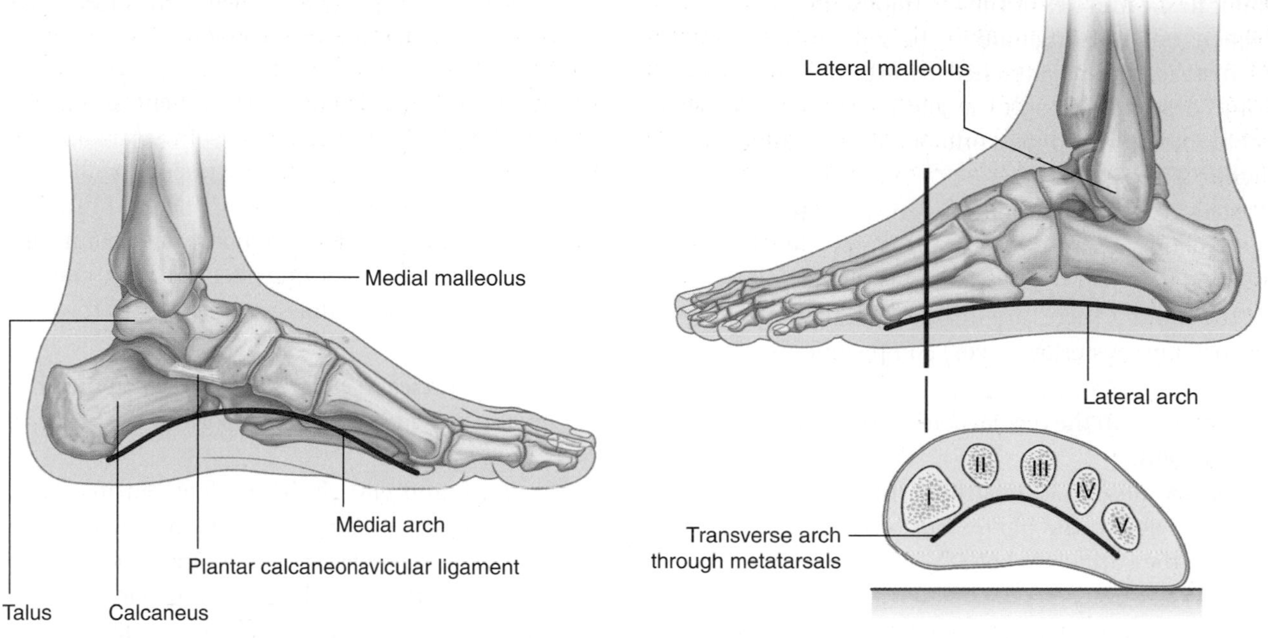


**Fig. 6.10** Longitudinal and transverse arches of the foot.

The interphalangeal joints are hinge joints and allow flexion and extension.

The bones of the foot are not organized in a single plane so that they lie flat on the ground. Rather, the metatarsals and tarsals form longitudinal and transverse arches (Fig. 6.10). The longitudinal arch is highest on the medial side of the foot. The arches are flexible in nature and are supported by muscles and ligaments. They absorb and transmit forces during walking and standing.

## Muscles

Muscles of the gluteal region consist predominantly of extensors, rotators, and abductors of the hip joint (Fig. 6.11). In addition to moving the thigh on a fixed pelvis, these muscles also control the movement of the pelvis relative to the limb bearing the body's weight (weight-bearing or stance limb) while the other limb swings forward (swing limb) during walking.

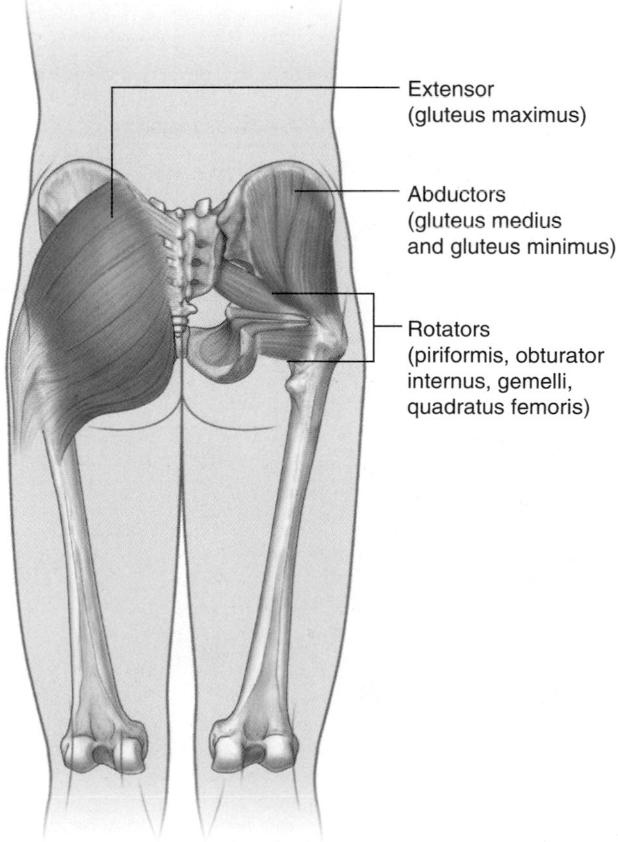


**Fig. 6.11** Muscles of the gluteal region.

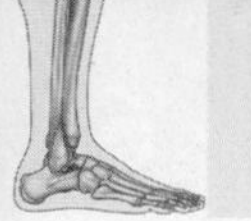

Major flexor muscles of the hip (iliopsoas—psoas major and iliacus) do not originate in the gluteal region or the thigh. Instead, they are attached to the posterior abdominal wall and descend through the gap between the inguinal ligament and pelvic bone to attach to the proximal end of the femur (Fig. 6.12).

Muscles in the thigh and leg are separated into three compartments by layers of fascia, bones, and ligaments (Fig. 6.13).

In the thigh, there are medial (adductor), anterior (extensor), and posterior (flexor) compartments:

- Most muscles in the medial compartment act mainly on the hip joint.
- The large muscles (hamstrings) in the posterior compartment act on the hip (extension) and knee (flexion) because they attach to both the pelvis and bones of the leg.
- Muscles in the anterior compartment (quadriceps femoris) predominantly extend the knee.

Muscles in the leg are divided into lateral (fibular), anterior, and posterior compartments:

- Muscles in the lateral compartment predominantly evert the foot.
- Muscles in the anterior compartment dorsiflex the foot and extend the digits.
- Muscles in the posterior compartment plantarflex the foot and flex the digits; one of the muscles can also flex the knee because it attaches superiorly to the femur.

Specific muscles in each of the three compartments in the leg also provide dynamic support for the arches of the foot.

Muscles found entirely in the foot (intrinsic muscles) modify the forces produced by tendons entering the toes from the leg and provide dynamic support for the longitudinal arches of the foot when walking, particularly when levering the body forward on the stance limb just before toe-off.

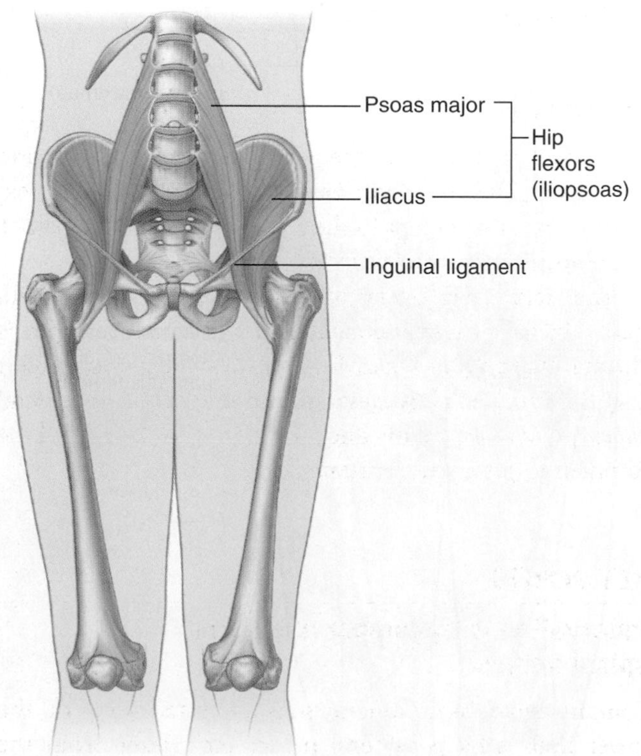


**Fig. 6.12** Major flexors of the hip.

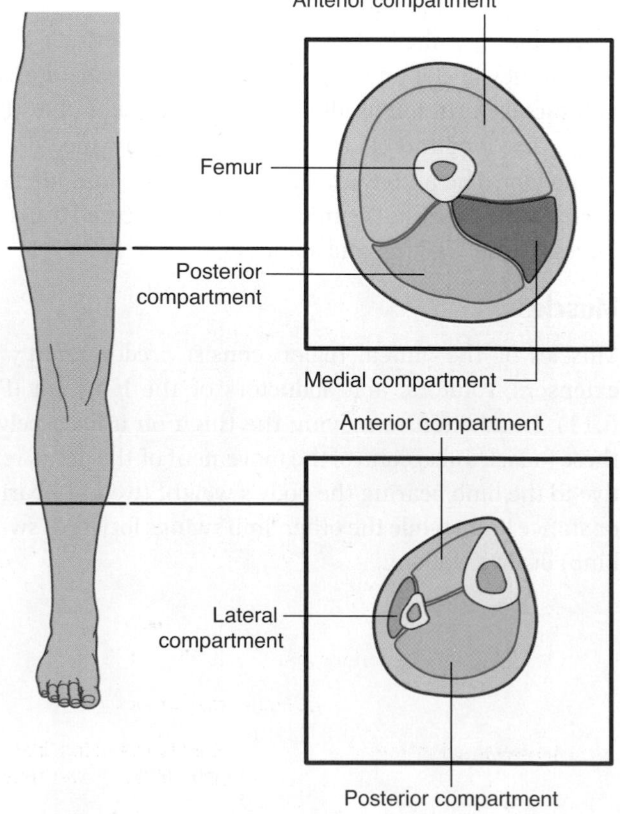


**Fig. 6.13** Muscle compartments in the thigh and leg.

## RELATIONSHIP TO OTHER REGIONS

Unlike in the upper limb where most structures pass between the neck and limb through a single axillary inlet, in the lower limb, there are four major entry and exit points between the lower limb and the abdomen, pelvis, and perineum (Fig. 6.14). These are:

- the gap between the inguinal ligament and pelvic bone,
- the greater sciatic foramen,
- the obturator canal (at the top of the obturator foramen), and
- the lesser sciatic foramen.

### Abdomen

The lower limb communicates directly with the abdomen through a gap between the pelvic bone and the inguinal ligament (Fig. 6.14). Structures passing though this gap include:

- muscles—psoas major, iliacus, and pectineus;
- nerves—femoral and femoral branch of the genitofemoral nerves, and the lateral cutaneous nerve of the thigh;
- vessels—femoral artery and vein; and
- lymphatics.

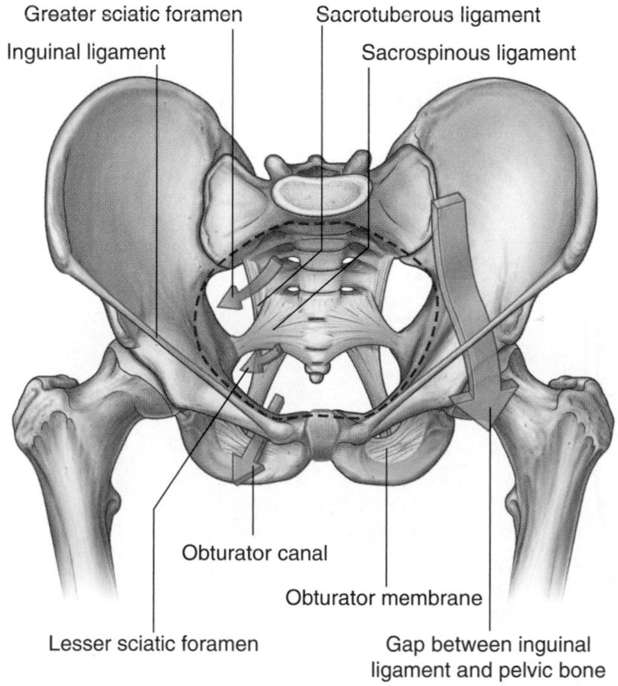

**Fig. 6.14** Apertures of communication between the lower limb and other regions.

This gap between the pelvic bone and the inguinal ligament is a weak area in the abdominal wall and often associated with abnormal protrusion of the abdominal cavity and contents into the thigh (femoral hernia). This type of hernia usually occurs where the lymphatic vessels pass through the gap (the femoral canal).

### Pelvis

Structures within the pelvis communicate with the lower limb through two major apertures (Fig. 6.14).

Posteriorly, structures communicate with the gluteal region through the greater sciatic foramen and include:

- a muscle—piriformis;
- nerves—sciatic, superior and inferior gluteal, and pudendal nerves; and
- vessels—superior and inferior gluteal arteries and veins, and the internal pudendal artery.

The sciatic nerve is the largest peripheral nerve of the body and is the major nerve of the lower limb.

Anteriorly, the obturator nerve and vessels pass between the pelvis and thigh through the obturator canal. This canal is formed between bone at the top of the obturator foramen and the obturator membrane, which closes most of the foramen during life.

### Perineum

Structures pass between the perineum and gluteal region through the lesser sciatic foramen (Fig. 6.14). The most important with respect to the lower limb is the tendon of the obturator internus muscle.

The nerve and artery of the perineum (the internal pudendal artery and pudendal nerve) pass out of the pelvis through the greater sciatic foramen into the gluteal region and then immediately pass around the ischial spine and sacrospinous ligament and through the lesser sciatic foramen to enter the perineum.

## KEY POINTS

### Innervation is by lumbar and sacral spinal nerves

Somatic motor and general sensory innervation of the lower limb is by peripheral nerves emanating from the lumbar and sacral plexuses on the posterior abdominal and pelvic walls. These plexuses are formed by the anterior rami of L1 to L3 and most of L4 (lumbar plexus) and L4 to S5 (sacral plexus).

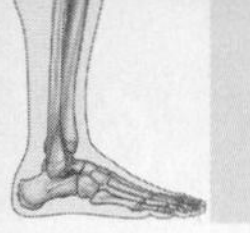

Nerves originating from the lumbar and sacral plexuses and entering the lower limb carry fibers from spinal cord levels L1 to S3 (Fig. 6.15). Nerves from lower sacral segments innervate the perineum. Terminal nerves exit the abdomen and pelvis through a number of apertures and foramina and enter the limb. As a consequence of this innervation, lumbar and upper sacral nerves are tested clinically by examining the lower limb. In addition, clinical signs (such as pain, pins-and-needles sensations, paresthesia, and fascicular muscle twitching) resulting from any disorder affecting these spinal nerves (e.g., herniated intervertebral disc in the lumbar region) appear in the lower limb.

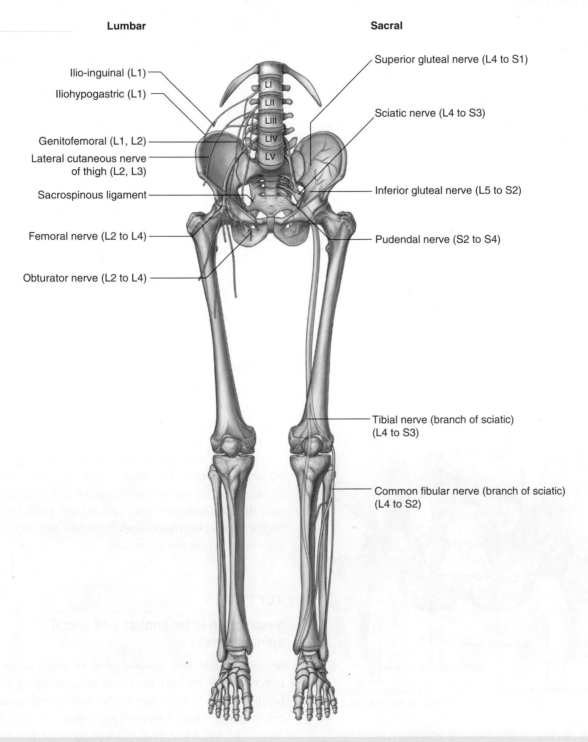



**Fig. 6.15** Innervation of the lower limb.

Dermatomes in the lower limb are shown in Fig. 6.16. Regions that can be tested for sensation and are reasonably autonomous (have minimal overlap) are:

- over the inguinal ligament—L1,
- lateral side of the thigh—L2,
- lower medial side of the thigh—L3,
- medial side of the great toe (digit I)—L4,
- medial side of digit II—L5,
- little toe (digit V)—S1,
- back of the thigh—S2, and
- skin over the gluteal fold—S3.

The dermatomes of S4 and S5 are tested in the perineum.

Selected joint movements are used to test myotomes (Fig. 6.17). For example:

- Flexion of the hip is controlled primarily by L1 and L2.
- Extension of the knee is controlled mainly by L3 and L4.
- Knee flexion is controlled mainly by L5 to S2.
- Plantarflexion of the foot is controlled predominantly by S1 and S2.
- Adduction of the digits is controlled by S2 and S3.

In an unconscious patient, both somatic sensory and somatic motor functions of spinal cord levels can be tested using tendon reflexes:

- A tap on the patellar ligament at the knee tests predominantly L3 and L4.
- A tendon tap on the calcaneal tendon posterior to the ankle (tendon of gastrocnemius and soleus) tests S1 and S2.

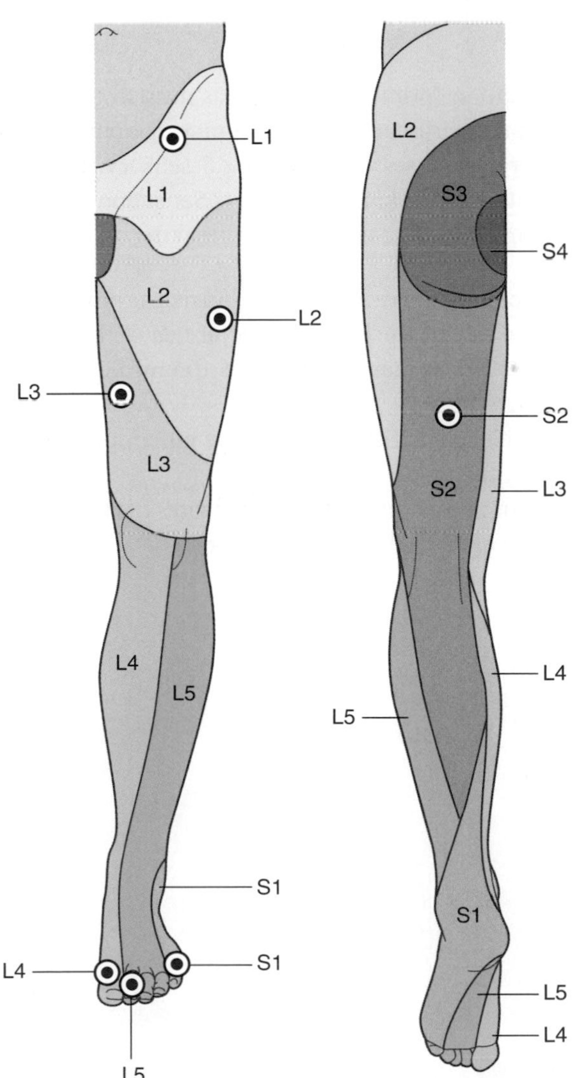


**Fig. 6.16** Dermatomes of the lower limb. Dots indicate autonomous zones (i.e., with minimal overlap).

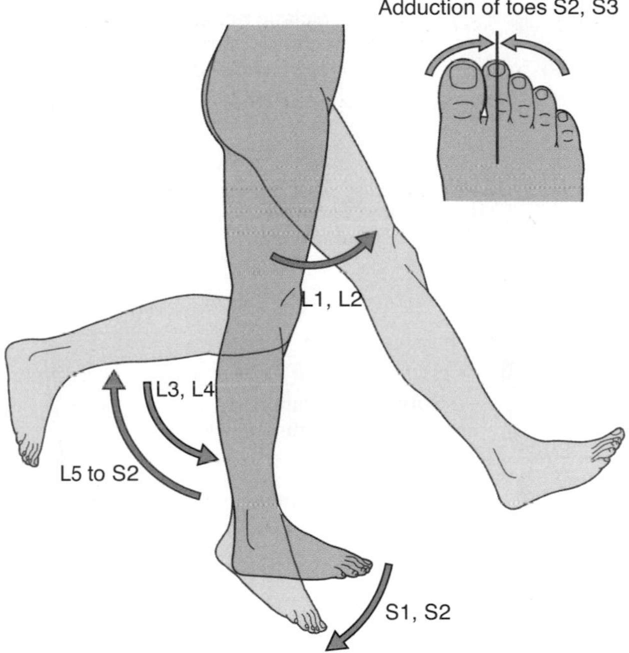


**Fig. 6.17** Movements generated by myotomes.

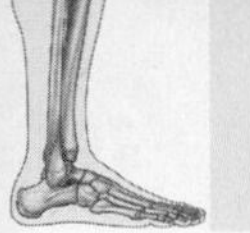

Each of the major muscle groups or compartments in the lower limb is innervated primarily by one or more of the major nerves that originate from the lumbar and sacral plexuses (Fig. 6.18):

- Large muscles in the gluteal region are innervated by the superior and inferior gluteal nerves.

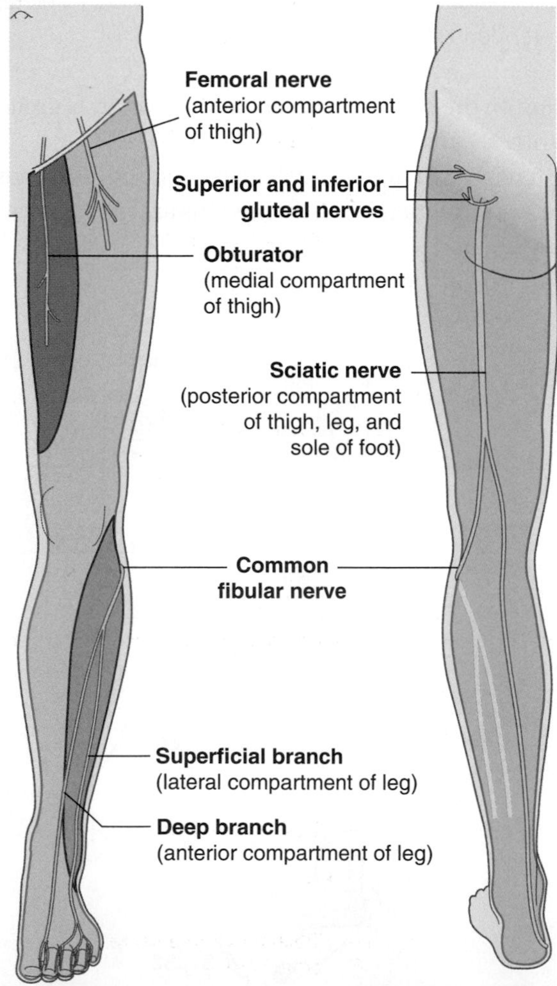


**Fig. 6.18** Major nerves of the lower limb (colors indicate regions of motor innervation).

- Most muscles in the anterior compartment of the thigh are innervated by the femoral nerve (except the tensor fasciae latae, which are innervated by the superior gluteal nerve).
- Most muscles in the medial compartment are innervated mainly by the obturator nerve (except the pectineus, which is innervated by the femoral nerve, and part of the adductor magnus, which is innervated by the tibial division of the sciatic nerve).
- Most muscles in the posterior compartment of the thigh and the leg and in the sole of the foot are innervated by the tibial part of the sciatic nerve (except the short head of the biceps femoris in the posterior thigh, which is innervated by the common fibular division of the sciatic nerve).
- The anterior and lateral compartments of the leg and muscles associated with the dorsal surface of the foot are innervated by the common fibular part of the sciatic nerve.

In addition to innervating major muscle groups, each of the major peripheral nerves originating from the lumbar and sacral plexuses carries general sensory information from patches of skin (Fig. 6.19). Sensation from these areas can be used to test for peripheral nerve lesions:

- The femoral nerve innervates skin on the anterior thigh, medial side of the leg, and medial side of the ankle.
- The obturator nerve innervates the medial side of the thigh.
- The tibial part of the sciatic nerve innervates the lateral side of the ankle and foot.
- The common fibular nerve innervates the lateral side of the leg and the dorsum of the foot.

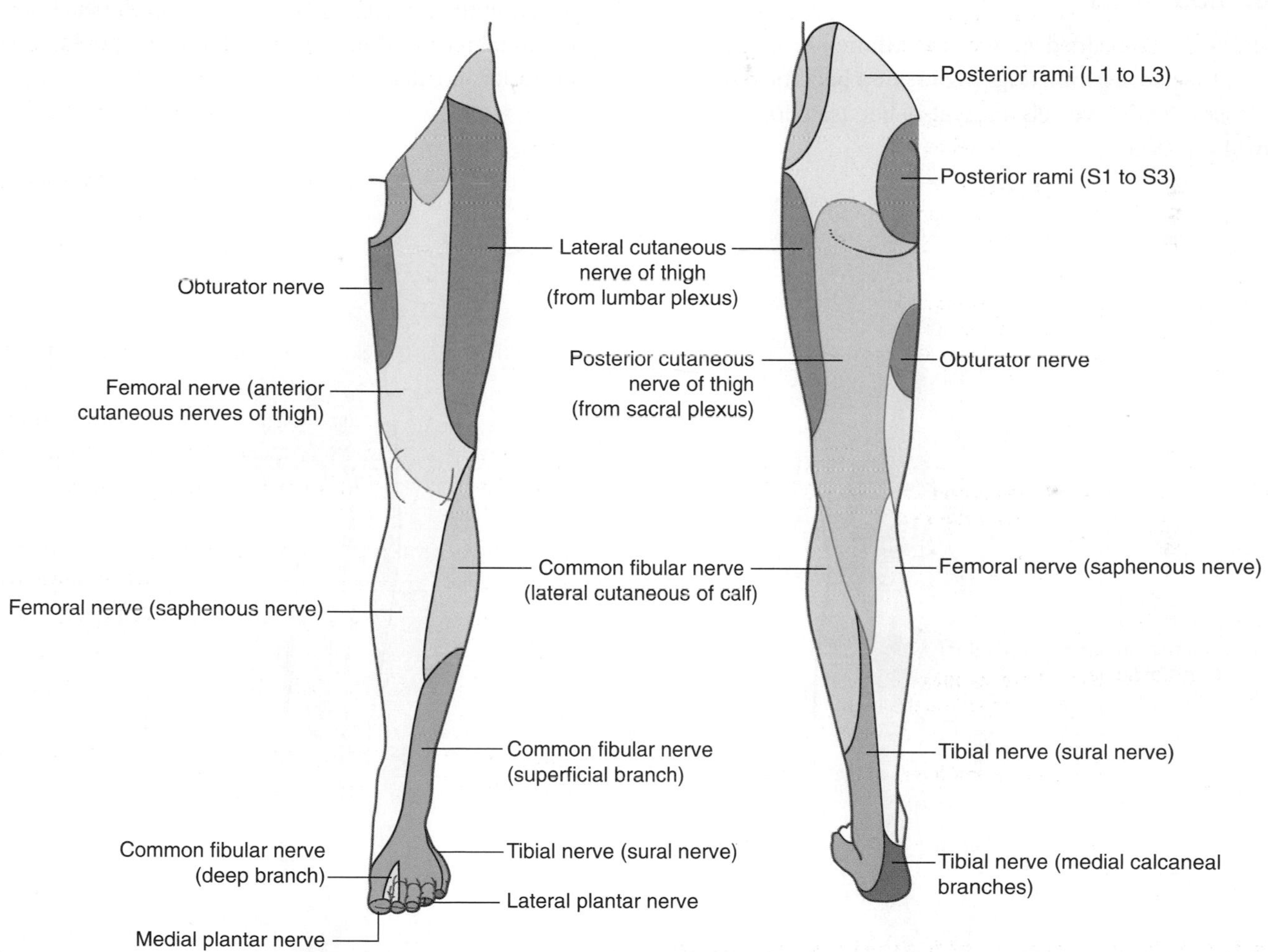
Obturator nerve
Femoral nerve (anterior cutaneous nerves of thigh)
Femoral nerve (saphenous nerve)
Common fibular nerve (deep branch)
Medial plantar nerve
Lateral cutaneous nerve of thigh (from lumbar plexus)
Posterior cutaneous nerve of thigh (from sacral plexus)
Common fibular nerve (lateral cutaneous of calf)
Common fibular nerve (superficial branch)
Tibial nerve (sural nerve)
Lateral plantar nerve
Posterior rami (L1 to L3)
Posterior rami (S1 to S3)
Obturator nerve
Femoral nerve (saphenous nerve)
Tibial nerve (sural nerve)
Tibial nerve (medial calcaneal branches)

**Fig. 6.19** Regions of skin innervated by peripheral nerves.

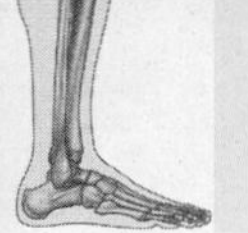



## Nerves related to bone

The common fibular branch of the sciatic nerve curves laterally around the neck of the fibula when passing from the popliteal fossa into the leg (Fig. 6.20). The nerve can be rolled against bone just distal to the attachment of biceps femoris to the head of the fibula. In this location, the nerve can be damaged by impact injuries, fractures to the bone, or leg casts that are placed too high.

## Superficial veins

Large veins embedded in the subcutaneous (superficial) fascia of the lower limb (Fig. 6.21) often become distended (varicose). These vessels can also be used for vascular transplantation.

The most important superficial veins are the great and small saphenous veins, which originate from the medial and lateral sides, respectively, of a dorsal venous arch in the foot.

- The great saphenous vein passes up the medial side of the leg, knee, and thigh to pass through an opening in deep fascia covering the femoral triangle and join with the femoral vein.
- The small saphenous vein passes behind the distal end of the fibula (lateral malleolus) and up the back of the leg to penetrate deep fascia and join the popliteal vein posterior to the knee.

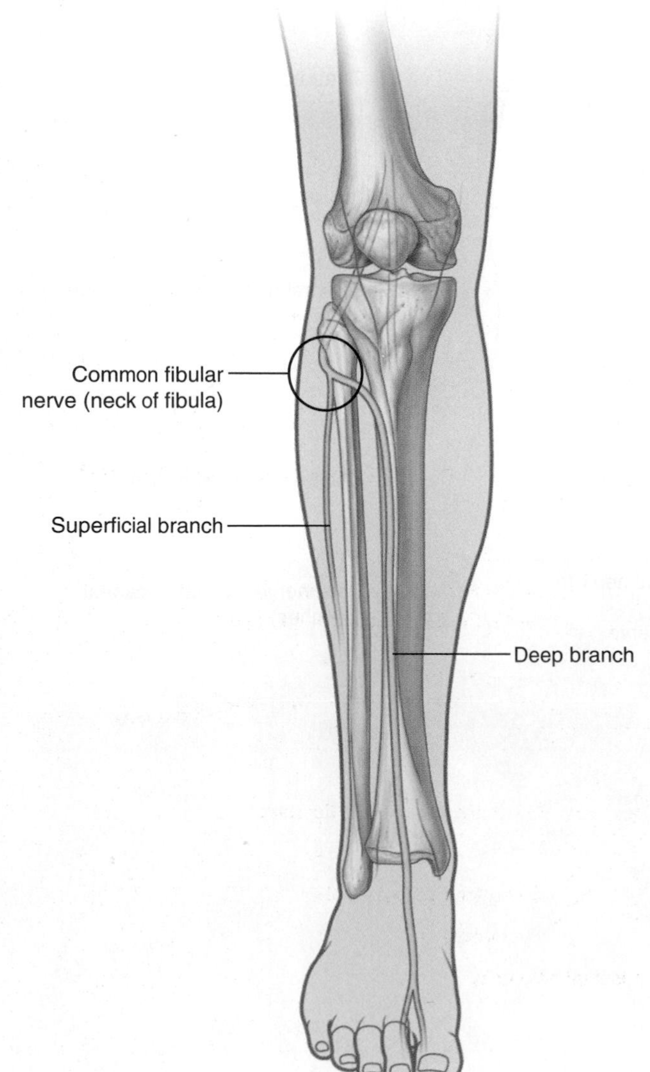


**Fig. 6.20** Nerves related to bone.

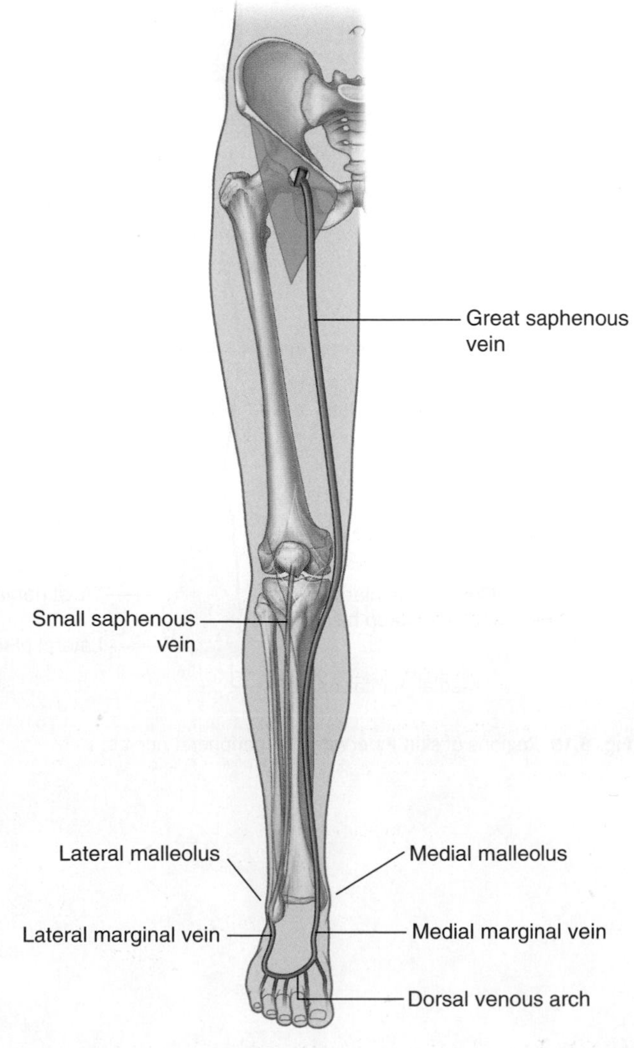


**Fig. 6.21** Superficial veins.

# Regional anatomy

## Bony pelvis

The external surfaces of the pelvic bones, sacrum, and coccyx are predominantly the regions of the pelvis associated with the lower limb, although some muscles do originate from the deep or internal surfaces of these bones and from the deep surfaces of the lumbar vertebrae, above (Fig. 6.22).

Each pelvic bone is formed by three bones (ilium, ischium, and pubis), which fuse during childhood. The **ilium** is superior and the **pubis** and **ischium** are anteroinferior and posteroinferior, respectively.

The ilium articulates with the sacrum. The pelvic bone is further anchored to the end of the vertebral column (sacrum and coccyx) by the sacrotuberous and sacrospinous ligaments, which attach to a tuberosity and spine on the ischium.

The outer surface of the ilium, and the adjacent surfaces of the sacrum, coccyx, and sacrotuberous ligament are associated with the gluteal region of the lower limb and provide extensive muscle attachment. The ischial tuberosity provides attachment for many of the muscles in the posterior compartment of the thigh, and the ischiopubic ramus and body of the pubis are associated mainly with muscles in the medial compartment of the thigh. The head of the femur articulates with the acetabulum on the lateral surface of the pelvic bone.

### Ilium

The upper fan-shaped part of the ilium is associated on its inner side with the abdomen and on its outer side with the lower limb. The top of this region is the **iliac crest**, which

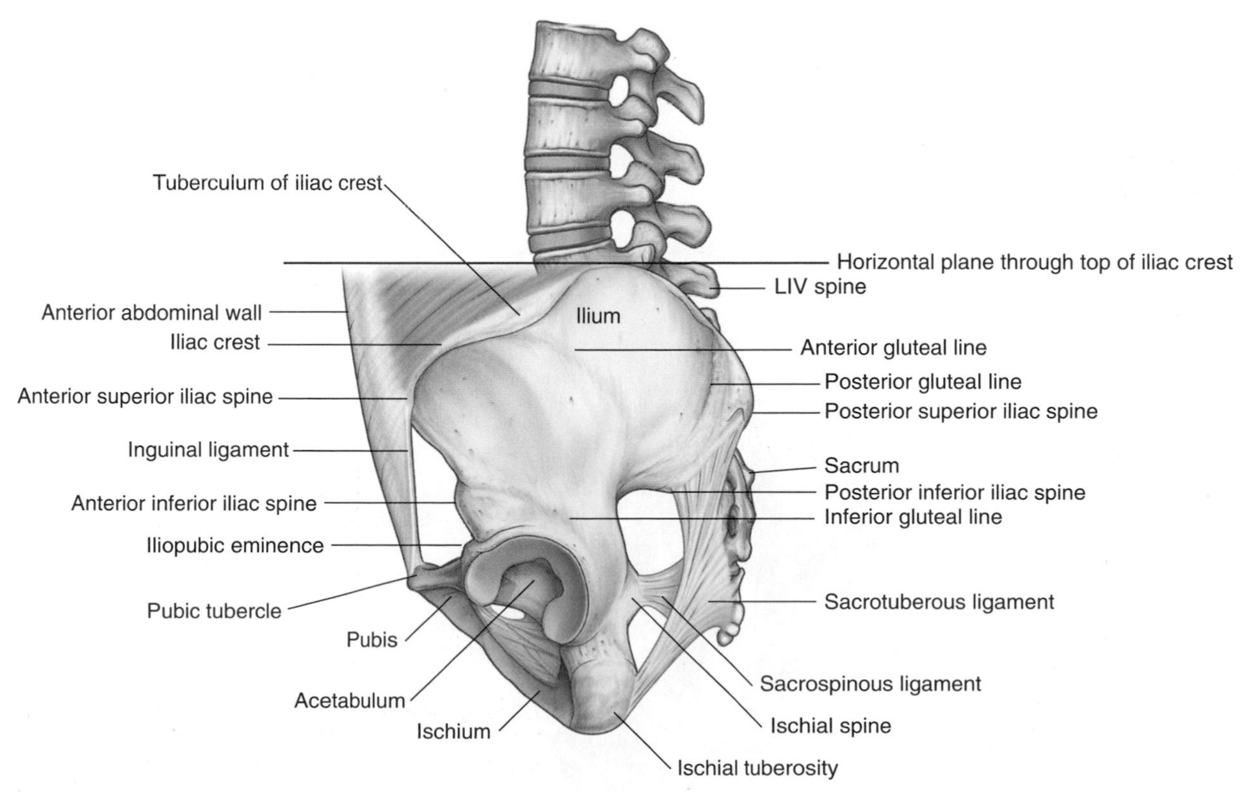


**Fig. 6.22** External surface of the bony pelvis. Lateral view.

ends anteriorly as the **anterior superior iliac spine** and posteriorly as the **posterior superior iliac spine**. A prominent lateral expansion of the crest just posterior to the anterior superior iliac spine is the **tuberculum of the iliac crest**.

The anterior inferior iliac spine is on the anterior margin of the ilium, and below this, where the ilium fuses with the pubis, is a raised area of bone (the **iliopubic eminence**).

The gluteal surface of the ilium faces posterolaterally and lies below the iliac crest. It is marked by three curved lines (inferior, anterior, and posterior gluteal lines), which divide the surface into four regions:

- The **inferior gluteal line** originates just superior to the anterior inferior iliac spine and curves inferiorly across the bone to end near the posterior margin of the acetabulum—the rectus femoris muscle attaches to the anterior inferior iliac spine and to a roughened patch of bone between the superior margin of the acetabulum and the inferior gluteal line.
- The **anterior gluteal line** originates from the lateral margin of the iliac crest between the anterior superior iliac spine and the tuberculum of the iliac crest, and arches inferiorly across the ilium to disappear just superior to the upper margin of the greater sciatic foramen—the gluteus minimus muscle originates from between the inferior and anterior gluteal lines.
- The **posterior gluteal line** descends almost vertically from the iliac crest to a position near the posterior inferior iliac spine—the gluteus medius muscle attaches to bone between the anterior and posterior gluteal lines, and the gluteus maximus muscle attaches posterior to the posterior gluteal line.

### Ischial tuberosity

The **ischial tuberosity** is posteroinferior to the acetabulum and is associated mainly with the hamstring muscles of the posterior thigh (Fig. 6.23). It is divided into upper and lower areas by a transverse line.

The upper area of the ischial tuberosity is oriented vertically and is further subdivided into two parts by an oblique line, which descends, from medial to lateral, across the surface:

- The more medial part of the upper area is for the attachment of the combined origin of the semitendinosus muscle and the long head of the biceps femoris muscle.
- The lateral part is for the attachment of the semimembranosus muscle.

The lower area of the ischial tuberosity is oriented horizontally and is divided into medial and lateral regions by a ridge of bone:

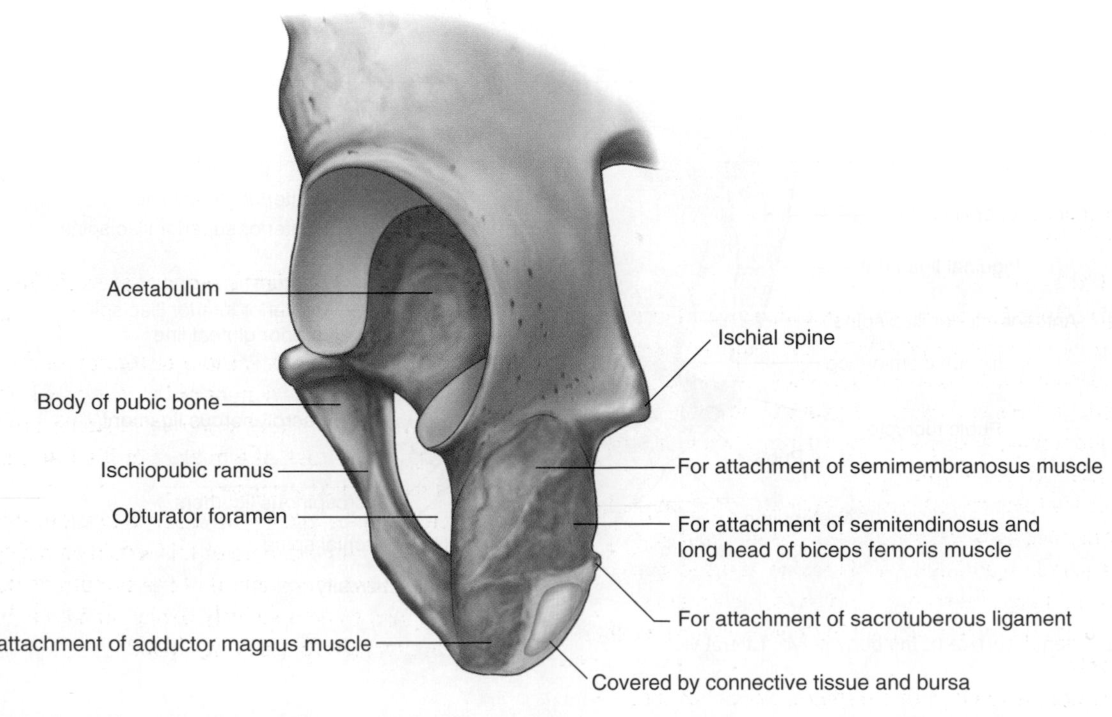


**Fig. 6.23** Ischial tuberosity. Posterolateral view.

- The lateral region provides attachment for part of the adductor magnus muscle.
- The medial part faces inferiorly and is covered by connective tissue and by a bursa.

When sitting, this medial part supports the body weight.

The sacrotuberous ligament is attached to a sharp ridge on the medial margin of the ischial tuberosity.

### Ischiopubic ramus and pubic bone

The external surfaces of the ischiopubic ramus anterior to the ischial tuberosity and the body of the pubis provide attachment for muscles of the medial compartment of the thigh (Fig. 6.23). These muscles include the adductor longus, adductor brevis, adductor magnus, pectineus, and gracilis.

### Acetabulum

The large cup-shaped **acetabulum** for articulation with the head of the femur is on the lateral surface of the pelvic bone in the region where the ilium, pubis, and ischium fuse (Fig. 6.24).

The margin of the acetabulum is marked inferiorly by a prominent notch (**acetabular notch**).

The wall of the acetabulum consists of nonarticular and articular parts:

- The nonarticular part is rough and forms a shallow circular depression (the **acetabular fossa**) in central and inferior parts of the acetabular floor the acetabular notch is continuous with the acetabular fossa.
- The articular surface is broad and surrounds the anterior, superior, and posterior margins of the acetabular fossa.

The smooth crescent-shaped articular surface (the **lunate surface**) is broadest superiorly where most of the body's weight is transmitted through the pelvis to the femur. The lunate surface is deficient inferiorly at the acetabular notch.

The acetabular fossa provides attachment for the ligament of the head of the femur, whereas blood vessels and nerves pass through the acetabular notch.

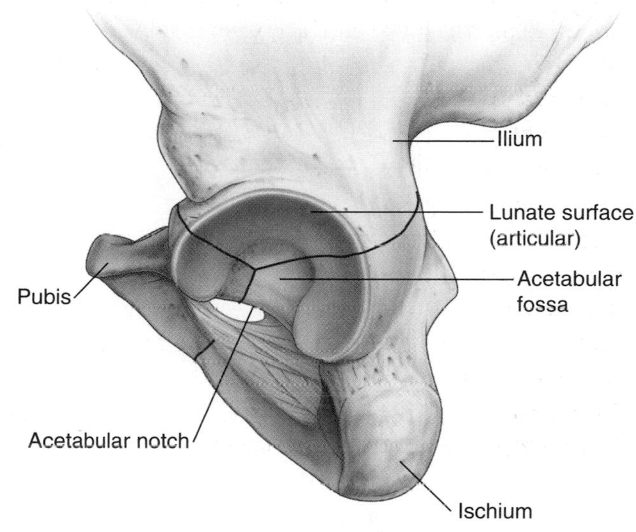


**Fig. 6.24** Acetabulum.

## In the clinic

### Pelvic fractures

The pelvic bones, sacrum, and associated joints form a bony ring surrounding the pelvic cavity. Soft tissue and visceral organ damage must be suspected when the pelvis is fractured. Patients with multiple injuries and evidence of chest, abdominal, and lower limb trauma should also be investigated for pelvic trauma.

Pelvic fractures can be associated with appreciable blood loss (concealed exsanguination) and blood transfusion is often required. In addition, this bleeding tends to form a significant pelvic hematoma, which can compress nerves, press on organs, and inhibit pelvic visceral function (Fig. 6.25).

There are many ways of classifying pelvic fractures, which enable the surgeon to determine the appropriate treatment and the patient's prognosis. Pelvic fractures are generally of four types.

- Type 1 injuries occur without disruption of the bony pelvic ring (e.g., a fracture of the iliac crest). These types of injuries are unlikely to represent significant trauma, though in the case of a fracture of the iliac crest, blood loss needs to be assessed.
- Type 2 injuries occur with a single break in the bony pelvic ring. An example of this would be a single fracture with diastasis (separation) of the symphysis pubis. Again, these injuries are relatively benign in nature, but it may be appropriate to assess for blood loss.

*(continues)*

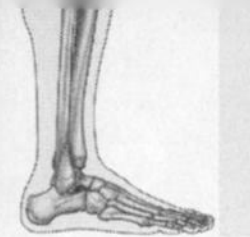



## In the clinic—cont'd

- Type 3 injuries occur with double breaks in the bony pelvic ring. These include bilateral fractures of the pubic rami, which may produce urethral damage.
- Type 4 injuries occur at and around the acetabulum.

Other types of pelvic ring injuries include fractures of the pubic rami and disruption of the sacro-iliac joint with or without dislocation. This may involve significant visceral pelvic trauma and hemorrhage.

Other general pelvic injuries include stress fractures and insufficiency fractures, as seen in athletes and elderly patients with osteoporosis, respectively.

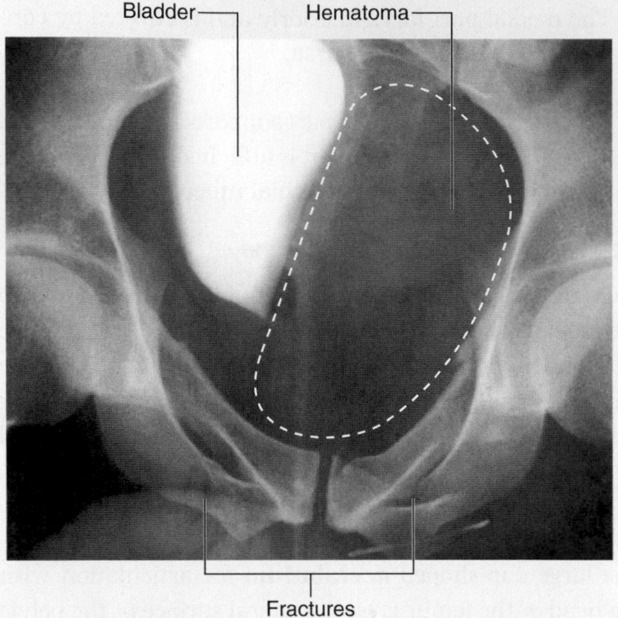


**Fig. 6.25** Multiple fractures of the pelvis. Radiograph with contrast in the bladder. A large accumulation of blood is deforming the bladder.

## Proximal femur

The femur is the bone of the thigh and the longest bone in the body. Its proximal end is characterized by a head and neck, and two large projections (the greater and lesser trochanters) on the upper part of the shaft (Fig. 6.26).

The **head** of the femur is spherical and articulates with the acetabulum of the pelvic bone. It is characterized by a nonarticular pit (**fovea**) on its medial surface for the attachment of the ligament of the head.

The **neck** of the femur is a cylindrical strut of bone that connects the head to the shaft of the femur. It projects superomedially from the shaft at an angle of approximately 125°, and projects slightly forward. The orientation of the neck relative to the shaft increases the range of movement of the hip joint.

The upper part of the **shaft** of the femur bears a greater and lesser trochanter, which are attachment sites for muscles that move the hip joint.

### Greater and lesser trochanters

The **greater trochanter** extends superiorly from the shaft of the femur just lateral to the region where the shaft joins the neck of the femur (Fig. 6.26). It continues posteriorly where its medial surface is deeply grooved to form the **trochanteric fossa**. The lateral wall of this fossa bears a distinct oval depression for attachment of the obturator externus muscle.

The greater trochanter has an elongate ridge on its anterolateral surface for attachment of the gluteus minimus and a similar ridge more posteriorly on its lateral surface for attachment of the gluteus medius. Between these two points, the greater trochanter is palpable.

On the medial side of the superior aspect of the greater trochanter and just above the trochanteric fossa is a small impression for attachment of the obturator internus and its associated gemelli muscles, and immediately above and behind this feature is an impression on the margin of the trochanter for attachment of the piriformis muscle.

The **lesser trochanter** is smaller than the greater trochanter and has a blunt conical shape. It projects posteromedially from the shaft of the femur just inferior to the junction with the neck (Fig. 6.26). It is the attachment site for the combined tendons of psoas major and iliacus muscles.

Extending between the two trochanters and separating the shaft from the neck of the femur are the intertrochanteric line and intertrochanteric crest.

#### Intertrochanteric line

The **intertrochanteric line** is a ridge of bone on the anterior surface of the upper margin of the shaft that

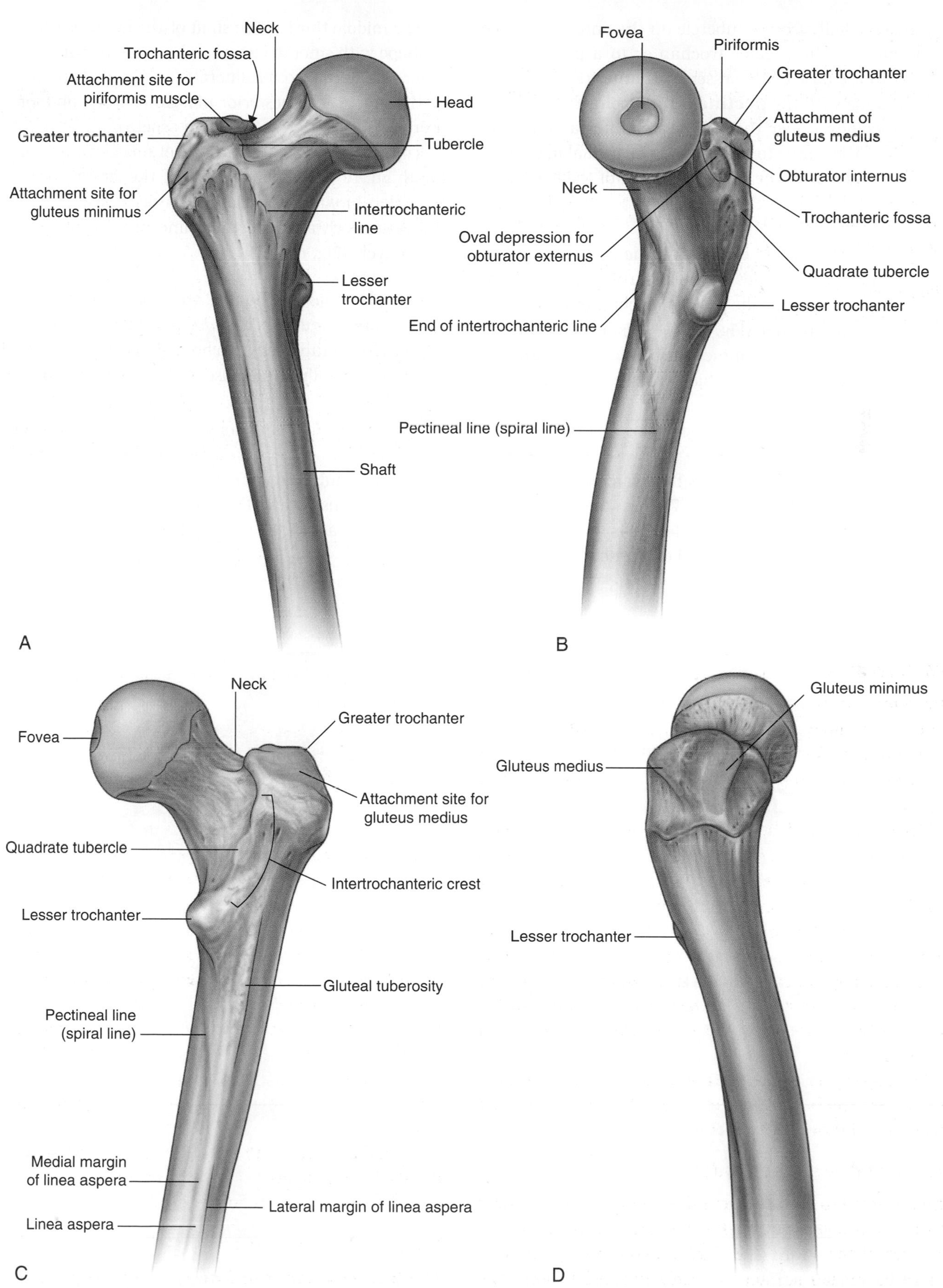


**Fig. 6.26** Proximal end of the femur (*right*). **A.** Anterior view. **B.** Medial view. **C.** Posterior view. **D.** Lateral view.

descends medially from a tubercle on the anterior surface of the base of the greater trochanter to a position just anterior to the base of the lesser trochanter (Fig. 6.26). It is continuous with the **pectineal line** (spiral line), which curves medially under the lesser trochanter and around the shaft of the femur to merge with the medial margin of the **linea aspera** on the posterior aspect of the femur.

### Intertrochanteric crest

The **intertrochanteric crest** is on the posterior surface of the femur and descends medially across the bone from the posterior margin of the greater trochanter to the base of the lesser trochanter (Fig. 6.26). It is a broad smooth ridge of bone with a prominent tubercle (the **quadrate tubercle**) on its upper half, which provides attachment for the quadratus femoris muscle.

### Shaft of the femur

The shaft of the femur descends from lateral to medial in the coronal plane at an angle of 7° from the vertical axis (Fig. 6.27). The distal end of the femur is therefore closer to the midline than the upper end of the shaft.

The middle third of the shaft of the femur is triangular in shape with smooth lateral and medial margins between anterior, lateral (posterolateral), and medial (posteromedial) surfaces. The posterior margin is broad and forms a prominent raised crest (the linea aspera).

The linea aspera is a major site of muscle attachment in the thigh. In the proximal third of the femur, the medial and lateral margins of the linea aspera diverge and continue superiorly as the pectineal line and gluteal tuberosity, respectively (Fig. 6.27):

- The pectineal line curves anteriorly under the lesser trochanter and joins the intertrochanteric line.
- The gluteal tuberosity is a broad linear roughening that curves laterally to the base of the greater trochanter.

The gluteus maximus muscle is attached to the gluteal tuberosity.

The triangular area enclosed by the pectineal line, the gluteal tuberosity, and the intertrochanteric crest is the posterior surface of the proximal end of the femur.

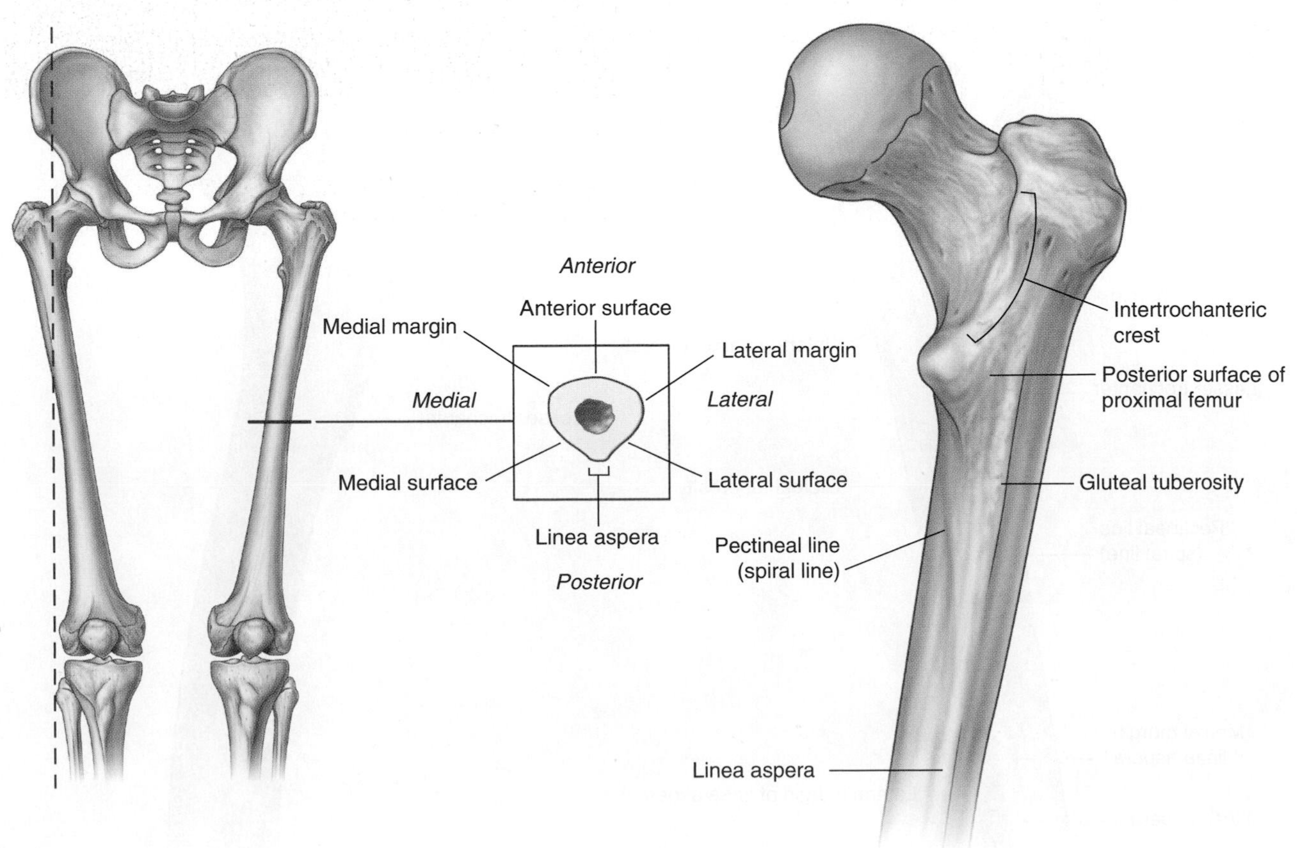


**Fig. 6.27** Shaft of the femur. On the right is a posterior view of proximal shaft of right femur.

## In the clinic

### Femoral neck fractures

Femoral neck fractures (Fig. 6.28) can interrupt the blood supply to the femoral head. The blood supply to the head and neck is primarily from an arterial ring formed by the branches of the medial and lateral circumflex femoral arteries around the base of the femoral neck. From here, vessels course along the neck, penetrate the capsule, and supply the femoral head. The blood supply to the femoral head and femoral neck is further enhanced by the artery of the ligamentum teres, a branch of the obturator artery, which is generally small and variable. Femoral neck fractures may disrupt associated vessels and lead to necrosis of the femoral head. Femoral neck fractures can be divided into three categories depending on the location of the fracture line: subcapital (fracture line passes across the femoral head-neck junction), transcervical (fracture line passes through the midportion of the femoral neck), and basicervical (fracture line passes across the base of the neck). Subcapital fractures have the highest risk of developing necrosis of the femoral head, and basicervical fractures have the lowest risk. Elderly patients with osteoporosis tend to have transverse subcapital fractures following low-energy trauma such as a fall from a standing height. Conversely, younger patients usually sustain more vertical fractures of the distal femoral neck (basicervical) after high-energy trauma such as a fall from a great height or due to axial load applied to an abducted knee, such as during a motor vehicle accident.

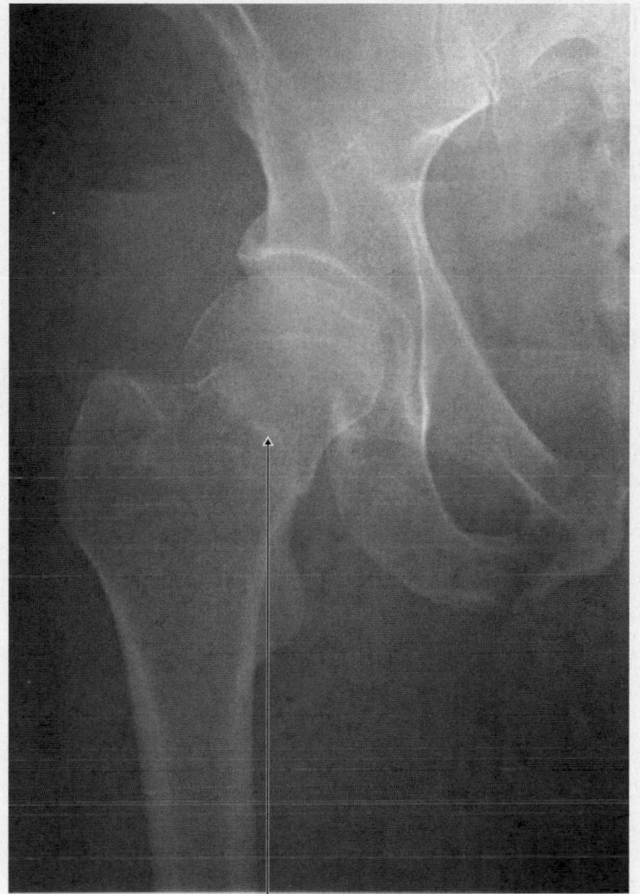


**Fig. 6.28** This radiograph of the pelvis, anteroposterior view, demonstrates a fracture of the neck of the femur.

## In the clinic

### Intertrochanteric fractures

In these fractures, the break usually runs from the greater trochanter through to the lesser trochanter and does not involve the femoral neck. Intertrochanteric fractures preserve the femoral neck blood supply and do not render the femoral head ischemic. They are most commonly seen in the elderly and result from low-energy impact (Fig. 6.29).

Sometimes isolated fractures of the greater or the lesser trochanter can occur. An isolated fracture of the lesser trochanter in adults is most commonly pathological and due to an underlying malignant deposit.

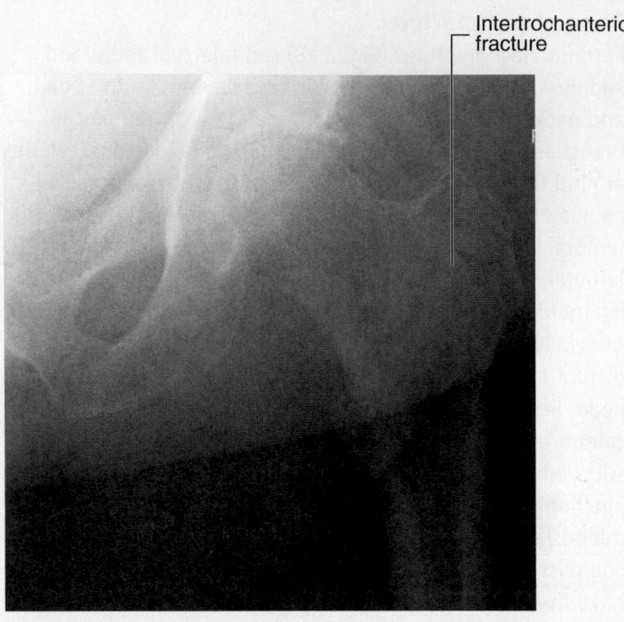


**Fig. 6.29** Anteroposterior radiograph showing an intertrochanteric fracture of proximal end of femur.

## In the clinic

### Femoral shaft fractures

An appreciable amount of energy is needed to fracture the femoral shaft. This type of injury is therefore accompanied by damage to the surrounding soft tissues, which include the muscle compartments and the structures they contain.

## Hip joint

The hip joint is a synovial articulation between the head of the femur and the acetabulum of the pelvic bone (Fig. 6.30A). The joint is a multiaxial ball and socket joint designed for stability and weight-bearing at the expense of mobility. Movements at the joint include flexion, extension, abduction, adduction, medial and lateral rotation, and circumduction.

When considering the effects of muscle action on the hip joint, the long neck of the femur and the angulation of the neck on the shaft of the femur must be borne in mind. For example, medial and lateral rotation of the femur involves muscles that move the greater trochanter forward and backward, respectively, relative to the acetabulum (Fig. 6.30B).

The articular surfaces of the hip joint are:

- the spherical head of the femur, and
- the lunate surface of the acetabulum of the pelvic bone.

The acetabulum almost entirely encompasses the hemispherical head of the femur and contributes substantially to joint stability. The nonarticular acetabular fossa contains loose connective tissue. The lunate surface is covered by hyaline cartilage and is broadest superiorly.

Except for the fovea, the head of the femur is also covered by hyaline cartilage.

The rim of the acetabulum is raised slightly by a fibrocartilaginous collar (the acetabular labrum). Inferiorly, the labrum bridges across the acetabular notch as the **transverse acetabular ligament** and converts the notch into a foramen (Fig. 6.31A).

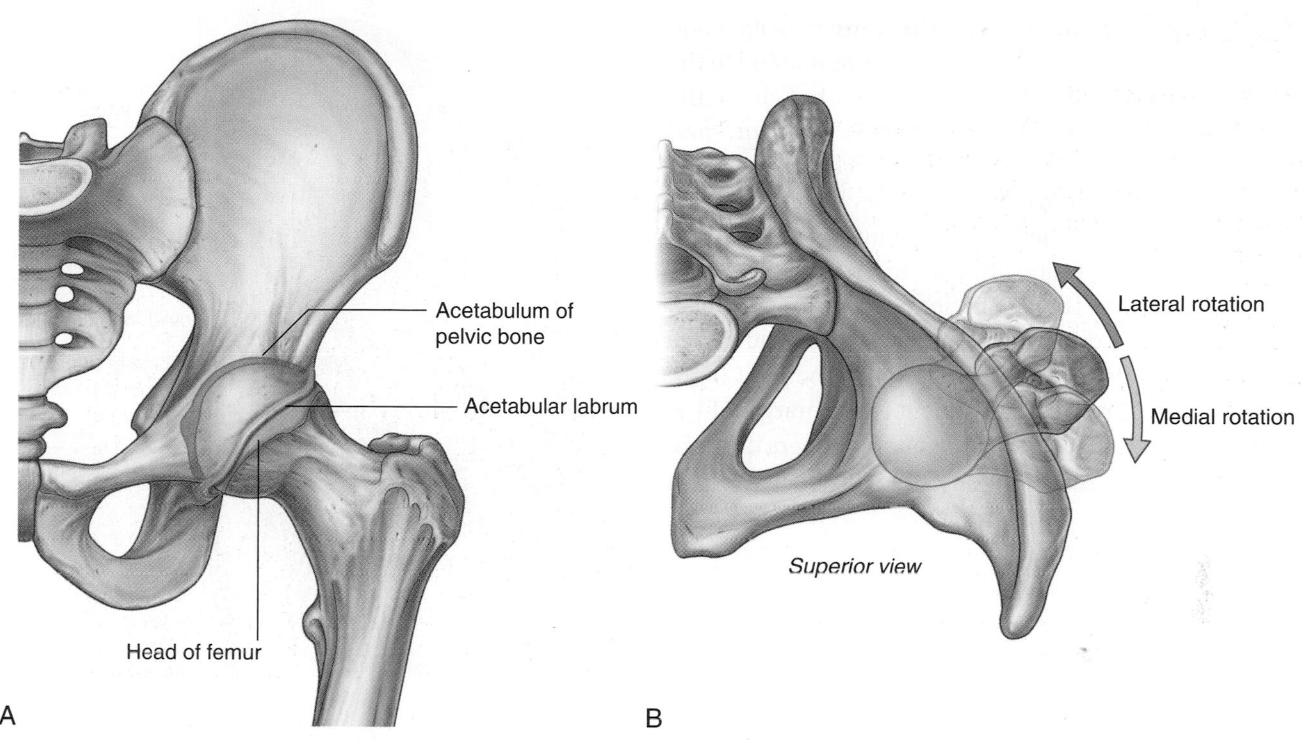


**Fig. 6.30** Hip joint. **A.** Articular surfaces. Anterior view. **B.** Movement of the neck of the femur during medial and lateral rotation. Superior view.

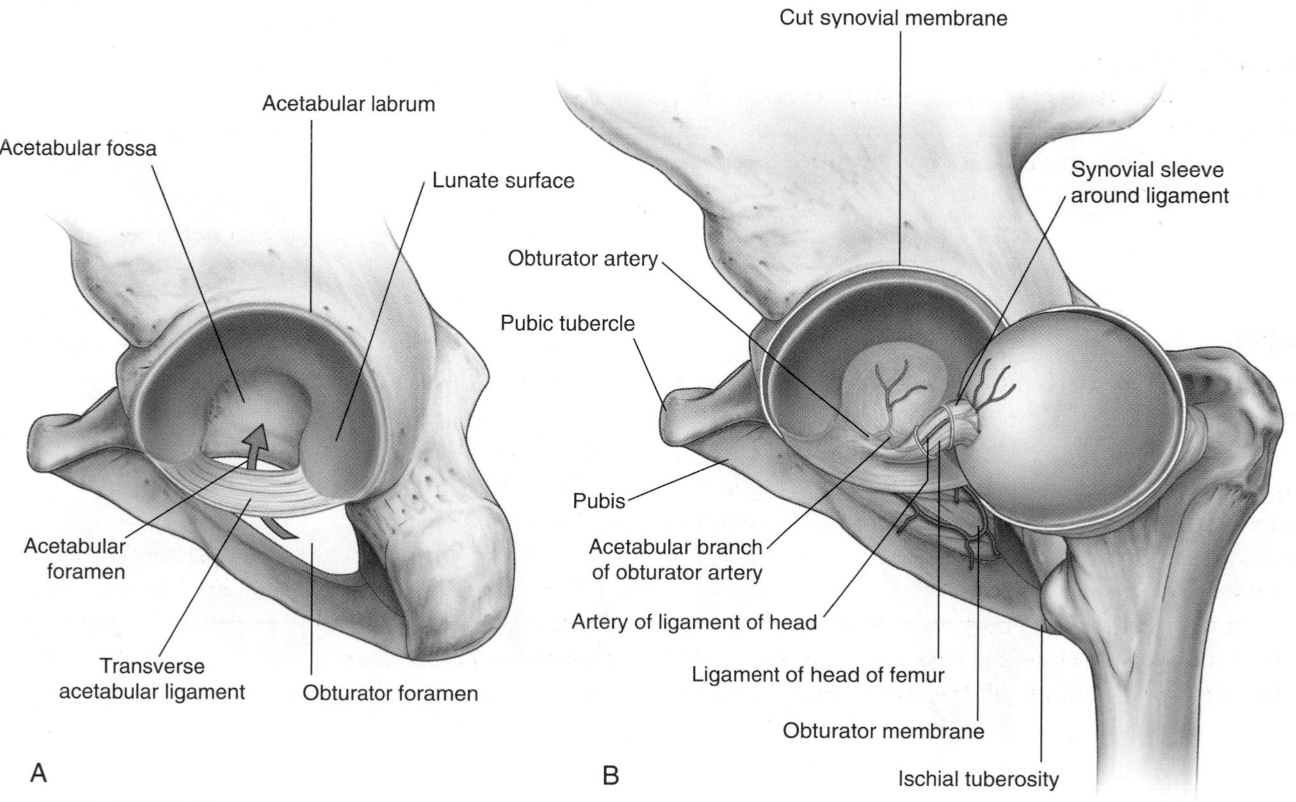


**Fig. 6.31** Hip joint. **A.** Transverse acetabular ligament. **B.** Ligament of the head of the femur. The head of the femur has been laterally rotated out of the acetabulum to show the ligament.

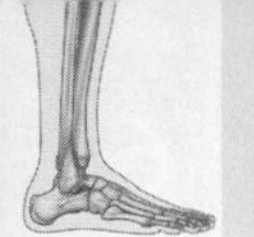

The **ligament of the head of the femur** is a flat band of delicate connective tissue that attaches at one end to the fovea on the head of the femur and at the other end to the acetabular fossa, transverse acetabular ligament, and margins of the acetabular notch (Fig. 6.31B). It carries a small branch of the obturator artery, which contributes to the blood supply of the head of the femur.

The synovial membrane attaches to the margins of the articular surfaces of the femur and acetabulum, forms a tubular covering around the ligament of the head of the femur, and lines the fibrous membrane of the joint (Figs. 6.31B and 6.32). From its attachment to the margin of the head of the femur, the synovial membrane covers the neck of the femur before reflecting onto the fibrous membrane (Fig. 6.32).

The fibrous membrane that encloses the hip joint is strong and generally thick. Medially, it is attached to the margin of the acetabulum, the transverse acetabular ligament, and the adjacent margin of the obturator foramen (Fig. 6.33A). Laterally, it is attached to the intertrochanteric line on the anterior aspect of the femur and to the neck of the femur just proximal to the intertrochanteric crest on the posterior surface.

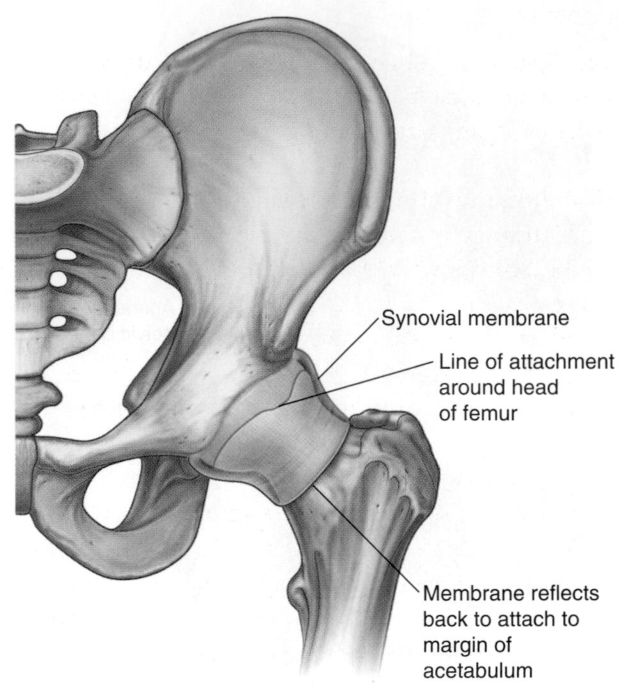


**Fig. 6.32** Synovial membrane of the hip joint.

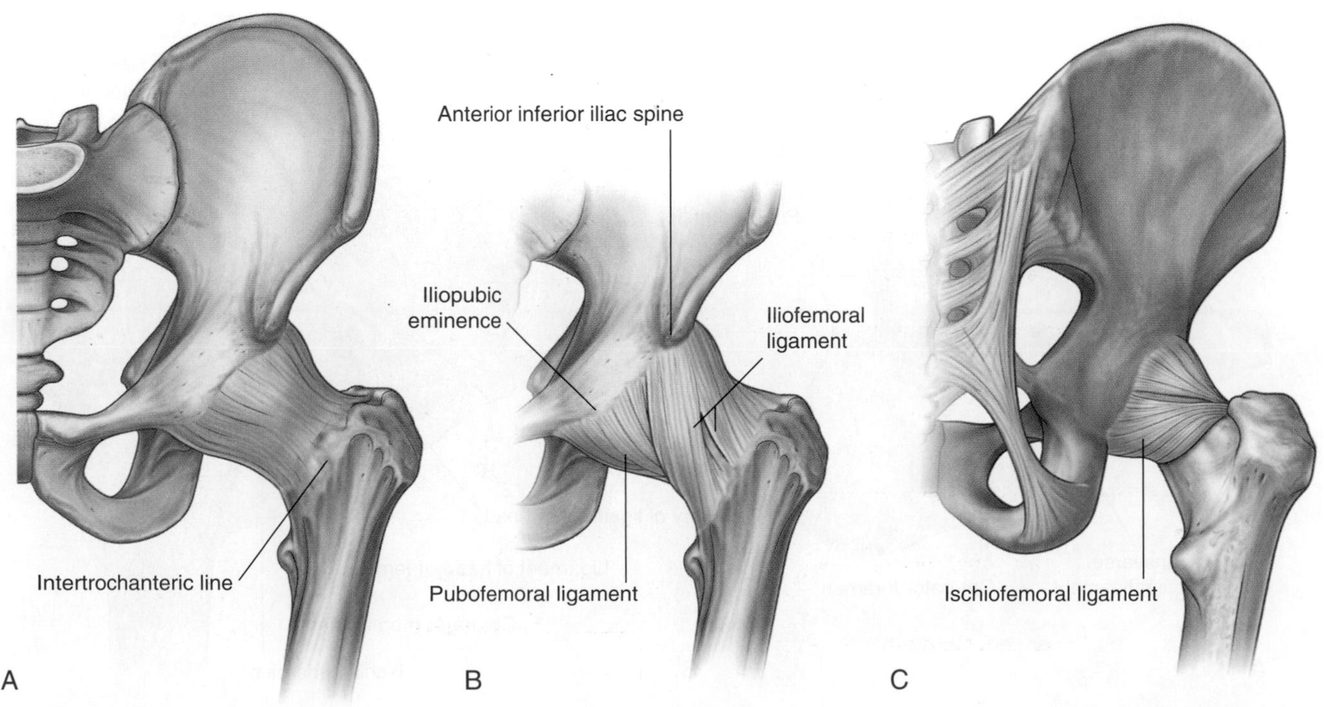


**Fig. 6.33** Fibrous membrane and ligaments of the hip joint. **A.** Fibrous membrane of the joint capsule. Anterior view. **B.** Iliofemoral and pubofemoral ligaments. Anterior view. **C.** Ischiofemoral ligament. Posterior view.

### Ligaments

Three ligaments reinforce the external surface of the fibrous membrane and stabilize the joint: the iliofemoral, pubofemoral, and ischiofemoral ligaments.

- The **iliofemoral ligament** is anterior to the hip joint and is triangular shaped (Fig. 6.33B). Its apex is attached to the ilium between the anterior inferior iliac spine and the margin of the acetabulum and its base is attached along the intertrochanteric line of the femur. Parts of the ligament attached above and below the intertrochanteric line are thicker than the part attached to the central part of the line. This results in the ligament having a Y appearance.
- The **pubofemoral ligament** is anteroinferior to the hip joint (Fig. 6.33B). It is also triangular in shape, with its base attached medially to the iliopubic eminence, adjacent bone, and obturator membrane. Laterally, it blends with the fibrous membrane and with the deep surface of the iliofemoral ligament.
- The **ischiofemoral ligament** reinforces the posterior aspect of the fibrous membrane (Fig. 6.33C). It is attached medially to the ischium, just posteroinferior to the acetabulum, and laterally to the greater trochanter deep to the iliofemoral ligament.

The fibers of all three ligaments are oriented in a spiral fashion around the hip joint so that they become taut when the joint is extended. This stabilizes the joint and reduces the amount of muscle energy required to maintain a standing position.

Vascular supply to the hip joint is predominantly through branches of the obturator artery, medial and lateral circumflex femoral arteries, superior and inferior gluteal arteries, and the first perforating branch of the deep artery of the thigh. The articular branches of these vessels form a network around the joint (Fig. 6.34).

The hip joint is innervated by articular branches from the femoral, obturator, and superior gluteal nerves, and the nerve to the quadratus femoris.

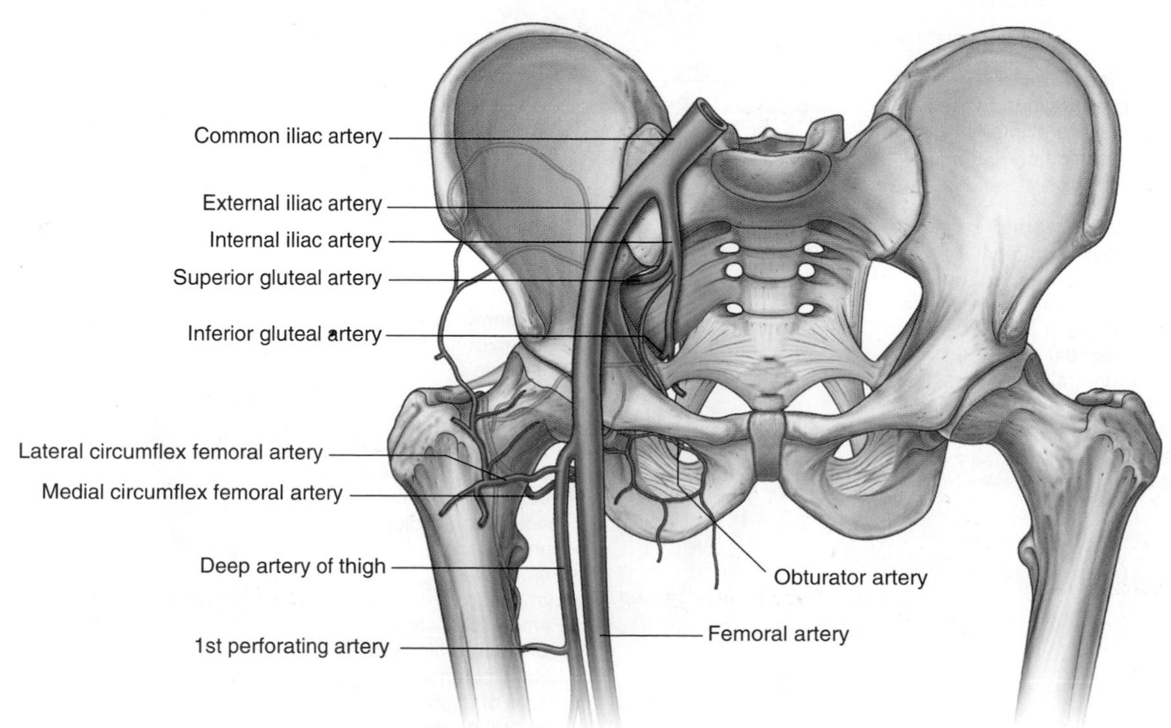


**Fig. 6.34** Blood supply of the hip joint.

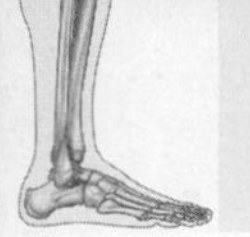



## Gateways to the lower limb

There are four major routes by which structures pass from the abdomen and pelvis into and out of the lower limb. These are the obturator canal, the greater sciatic foramen, the lesser sciatic foramen, and the gap between the inguinal ligament and the anterosuperior margin of the pelvis (Fig. 6.35).

### Obturator canal

The **obturator canal** is an almost vertically oriented passageway at the anterosuperior edge of the obturator foramen (Fig. 6.35). It is bordered:

- above by a groove (**obturator groove**) on the inferior surface of the superior ramus of the pubic bone, and
- below by the upper margin of the obturator membrane, which fills most of the obturator foramen, and by muscles (obturator internus and externus) attached to the inner and outer surfaces of the obturator membrane and surrounding bone.

The obturator canal connects the abdominopelvic region with the medial compartment of the thigh. The obturator nerve and vessels pass through the canal.

### Greater sciatic foramen

The **greater sciatic foramen** is formed on the posterolateral pelvic wall and is the major route for structures to pass between the pelvis and the gluteal region of the lower limb (Fig. 6.35). The margins of the foramen are formed by:

- the greater sciatic notch,
- parts of the upper borders of the sacrospinous and sacrotuberous ligaments, and
- the lateral border of the sacrum.

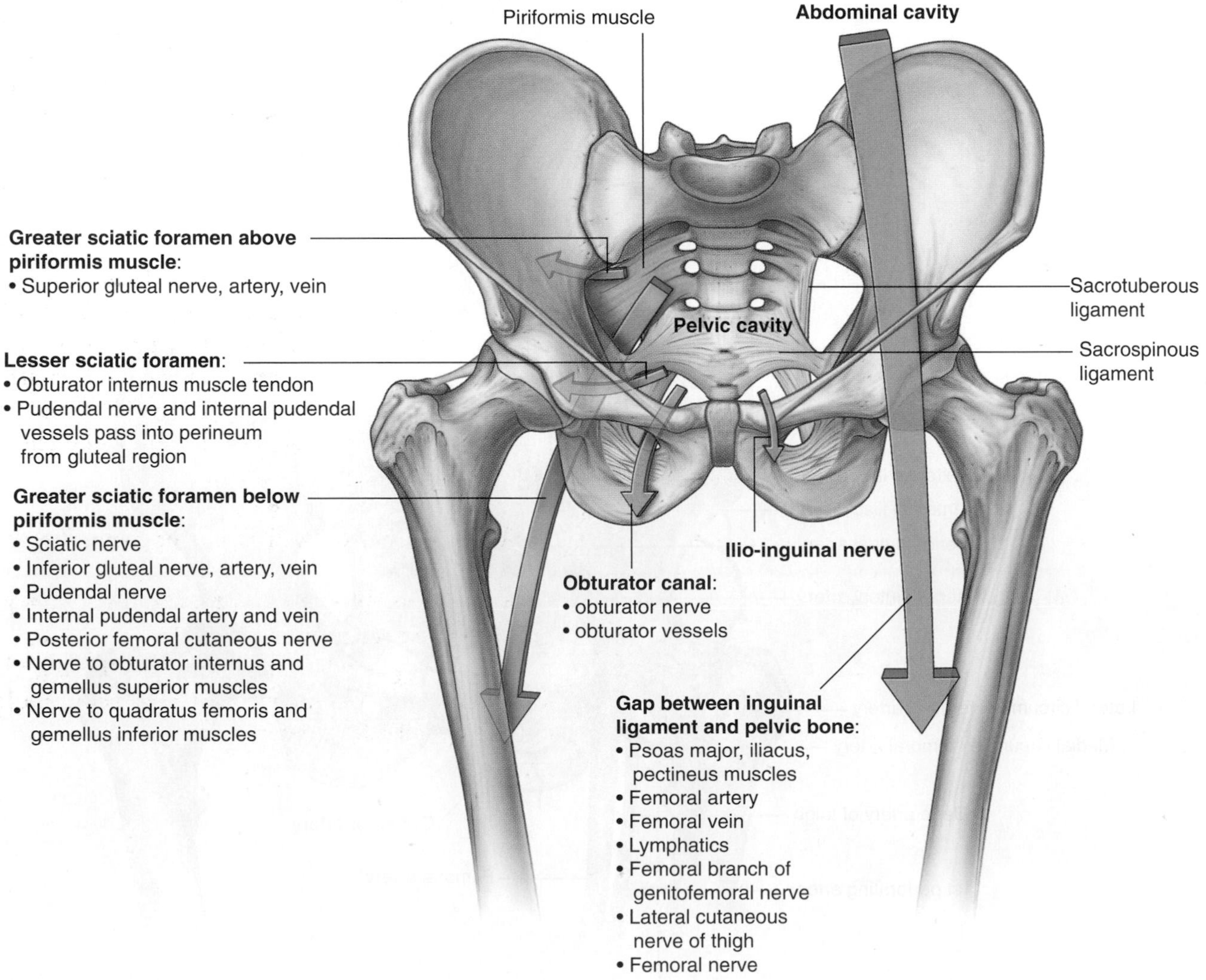

Piriformis muscle
Abdominal cavity
Greater sciatic foramen above piriformis muscle:
• Superior gluteal nerve, artery, vein
Sacrotuberous ligament
Pelvic cavity
Sacrospinous ligament
Lesser sciatic foramen:
• Obturator internus muscle tendon
• Pudendal nerve and internal pudendal vessels pass into perineum from gluteal region
Greater sciatic foramen below piriformis muscle:
• Sciatic nerve
• Inferior gluteal nerve, artery, vein
• Pudendal nerve
• Internal pudendal artery and vein
• Posterior femoral cutaneous nerve
• Nerve to obturator internus and gemellus superior muscles
• Nerve to quadratus femoris and gemellus inferior muscles
Ilio-inguinal nerve
Obturator canal:
• obturator nerve
• obturator vessels
Gap between inguinal ligament and pelvic bone:
• Psoas major, iliacus, pectineus muscles
• Femoral artery
• Femoral vein
• Lymphatics
• Femoral branch of genitofemoral nerve
• Lateral cutaneous nerve of thigh
• Femoral nerve

**Fig. 6.35** Gateways to the lower limb.

The piriformis muscle passes out of the pelvis into the gluteal region through the greater sciatic foramen and separates the foramen into two parts, a part above the muscle and a part below:

- The superior gluteal nerve and vessels pass through the greater sciatic foramen above the piriformis.
- The sciatic nerve, inferior gluteal nerve and vessels, pudendal nerve and internal pudendal vessels, posterior cutaneous nerve of the thigh, nerve to the obturator internus and gemellus superior, and nerve to the quadratus femoris and gemellus inferior pass through the greater sciatic foramen below the muscle.

### Lesser sciatic foramen

The **lesser sciatic foramen** is inferior to the greater sciatic foramen on the posterolateral pelvic wall (Fig. 6.35). It is also inferior to the lateral attachment of the pelvic floor (levator ani and coccygeus muscles) to the pelvic wall and therefore connects the gluteal region with the perineum:

- The tendon of the obturator internus passes from the lateral pelvic wall through the lesser sciatic foramen into the gluteal region to insert on the femur.
- The pudendal nerve and internal pudendal vessels, which first exit the pelvis by passing through the greater sciatic foramen below the piriformis muscle, enter the perineum below the pelvic floor by passing around the ischial spine and sacrospinous ligament and medially through the lesser sciatic foramen.

### Gap between the inguinal ligament and pelvic bone

The large crescent-shaped gap between the inguinal ligament above and the anterosuperior margin of the pelvic bone below is the major route of communication between the abdomen and the anteromedial aspect of the thigh (Fig. 6.35). The psoas major, iliacus, and pectineus muscles pass through this gap to insert onto the femur. The major blood vessels (femoral artery and vein) and lymphatics of the lower limb also pass through it, as does the femoral nerve, to enter the femoral triangle of the thigh.

## Nerves

Nerves that enter the lower limb from the abdomen and pelvis are terminal branches of the lumbosacral plexus on the posterior wall of the abdomen and the posterolateral walls of the pelvis (Fig. 6.36 and Table 6.1).

The **lumbar plexus** is formed by the anterior rami of spinal nerves L1 to L3 and part of L4 (see Chapter 4, pp. 398–401). The rest of the anterior ramus of L4 and the anterior ramus of L5 combine to form the **lumbosacral trunk**, which enters the pelvic cavity and joins with the anterior rami of S1 to S3 and part of S4 to form the **sacral plexus** (see Chapter 5, pp. 480–486).

Major nerves that originate from the lumbosacral plexus and leave the abdomen and pelvis to enter the lower limb include the femoral nerve, obturator nerve, sciatic nerve, superior gluteal nerve, and inferior gluteal nerve. Other nerves that also originate from the plexus and enter the lower limb to supply skin or muscle include the lateral cutaneous nerve of the thigh, nerve to the obturator internus, nerve to the quadratus femoris, posterior cutaneous nerve of the thigh, perforating cutaneous nerve, and branches of the ilio-inguinal and genitofemoral nerves.

### Femoral nerve

The **femoral nerve** carries contributions from the anterior rami of L2 to L4 and leaves the abdomen by passing through the gap between the inguinal ligament and superior margin of the pelvis to enter the femoral triangle on the anteromedial aspect of the thigh (Fig. 6.35 and Table 6.1). In the femoral triangle it is lateral to the femoral artery. The femoral nerve:

- innervates all muscles in the anterior compartment of the thigh,
- in the abdomen, gives rise to branches that innervate the iliacus and pectineus muscles, and
- innervates skin over the anterior aspect of the thigh, the anteromedial side of the knee, the medial side of the leg, and the medial side of the foot.

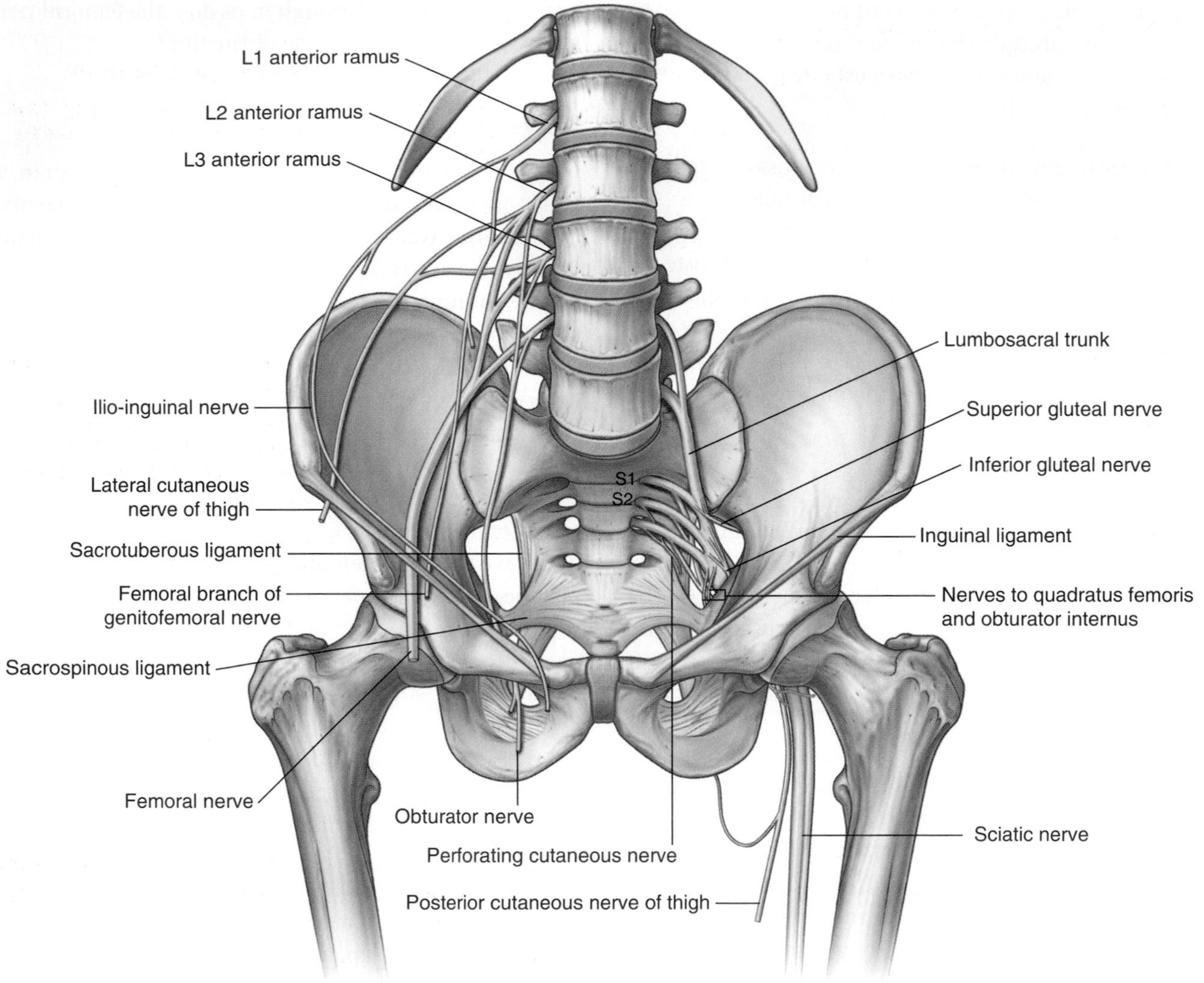

**Fig. 6.36** Branches of the lumbosacral plexus.

**Table 6.1** Branches of the lumbosacral plexus associated with the lower limb

| Branch | Spinal segments | Function: motor | Function: sensory (cutaneous) |
|---|---|---|---|
| Ilio-inguinal | L1 | No motor function in lower limb, but innervates muscles of the abdominal wall | Skin over anteromedial part of upper thigh and adjacent skin of perineum |
| Genitofemoral | L1, L2 | No motor function in lower limb, but genital branch innervates cremaster muscle in the wall of the spermatic cord in men | Femoral branch innervates skin on anterior central part of upper thigh; the genital branch innervates skin in anterior part of perineum (anterior scrotum in men, and mons pubis and anterior labia majora in women) |
| Femoral | L2 to L4 | All muscles in the anterior compartment of thigh; in the abdomen, also gives rise to branches that supply iliacus and pectineus | Skin over the anterior thigh, anteromedial knee, medial side of the leg, and the medial side of the foot |
| Obturator | L2 to L4 | All muscles in the medial compartment of thigh (except pectineus and the part of adductor magnus attached to the ischium); also innervates obturator externus | Skin over upper medial aspect of thigh |

**Table 6.1** Branches of the lumbosacral plexus associated with the lower limb—cont'd

| Branch | Spinal segments | Function: motor | Function: sensory (cutaneous) |
|---|---|---|---|
| Sciatic | L4 to S3 | All muscles in the posterior compartment of thigh and the part of adductor magnus attached to the ischium; all muscles in the leg and foot | Skin over lateral side of leg and foot, and over the sole and dorsal surface of foot |
| Superior gluteal | L4 to S1 | Muscles of the gluteal region (gluteus medius, gluteus minimus, tensor fasciae latae) | |
| Inferior gluteal | L5 to S2 | Muscle of the gluteal region (gluteus maximus) | |
| Lateral cutaneous nerve of thigh | L2, L3 | | Parietal peritoneum in iliac fossa; skin over anterolateral thigh |
| Posterior cutaneous nerve of thigh | S1 to S3 | | Skin over gluteal fold and upper medial aspect of thigh and adjacent perineum, posterior aspect of thigh and upper posterior leg |
| Nerve to quadratus femoris | L4 to S1 | Muscles of gluteal region (quadratus femoris and gemellus inferior) | |
| Nerve to obturator internus | L5 to S2 | Muscles of gluteal region (obturator internus and gemellus superior) | |
| Perforating cutaneous nerve | S2, S3 | | Skin over medial aspect of gluteal fold |

### Obturator nerve

The **obturator nerve**, like the femoral nerve, originates from L2 to L4. It descends along the posterior abdominal wall, passes through the pelvic cavity and enters the thigh by passing through the obturator canal (Fig. 6.36 and Table 6.1). The obturator nerve innervates:

- all muscles in the medial compartment of the thigh, except the part of the adductor magnus muscle that originates from the ischium and the pectineus muscle, which are innervated by the sciatic and the femoral nerves, respectively;
- the obturator externus muscle; and
- skin on the medial side of the upper thigh.

### Sciatic nerve

The **sciatic nerve** is the largest nerve of the body and carries contributions from L4 to S3. It leaves the pelvis through the greater sciatic foramen inferior to the piriformis muscle, enters and passes through the gluteal region (Fig. 6.36 and Table 6.1), and then enters the posterior compartment of the thigh where it divides into its two major branches:

- the common fibular nerve, and
- the tibial nerve.

Posterior divisions of L4 to S2 are carried in the common fibular part of the nerve and the anterior divisions of L4 to S3 are carried in the tibial part.

The sciatic nerve innervates:

- all muscles in the posterior compartment of the thigh,
- the part of the adductor magnus originating from the ischium,
- all muscles in the leg and foot, and
- skin on the lateral side of the leg and the lateral side and sole of the foot.

### Gluteal nerves

The gluteal nerves are major motor nerves of the gluteal region.

The **superior gluteal nerve** (Fig. 6.36 and Table 6.1) carries contributions from the anterior rami of L4 to S1, leaves the pelvis through the greater sciatic foramen above the piriformis muscle, and innervates:

- the gluteus medius and minimus muscles, and
- the tensor fasciae latae muscle.

The **inferior gluteal nerve** (Fig. 6.36 and Table 6.1) is formed by contributions from L5 to S2, leaves the pelvis through the greater sciatic foramen inferior to the

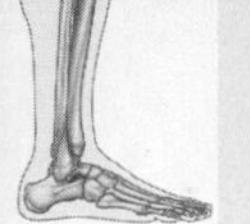



piriformis muscle, and enters the gluteal region to supply the gluteus maximus.

### Ilio-inguinal and genitofemoral nerves

Terminal sensory branches of the ilio-inguinal nerve (L1) and the genitofemoral nerve (L1, L2) descend into the upper thigh from the lumbar plexus.

The **ilio-inguinal nerve** originates from the superior part of the lumbar plexus, descends around the abdominal wall in the plane between the transversus abdominis and internal oblique muscles, and then passes through the inguinal canal to leave the abdominal wall through the superficial inguinal ring (Fig. 6.36 and Table 6.1). Its terminal branches innervate skin on the medial side of the upper thigh and adjacent parts of the perineum.

The **genitofemoral nerve** passes anteroinferiorly through the psoas major muscle on the posterior abdominal wall and descends on the anterior surface of the psoas major (Fig. 6.36 and Table 6.1). Its genital branch innervates anterior aspects of the perineum. Its femoral branch passes into the thigh by crossing under the inguinal ligament where it is lateral to the femoral artery. It passes superficially to innervate skin over the upper central part of the anterior thigh.

### Lateral cutaneous nerve of thigh

The **lateral cutaneous nerve of the thigh** originates from L2 and L3. It leaves the abdomen either by passing through the gap between the inguinal ligament and the pelvic bone just medial to the anterior superior iliac spine or by passing directly through the inguinal ligament (Fig. 6.36 and Table 6.1). It supplies skin on the lateral side of the thigh.

### Nerve to quadratus femoris and nerve to obturator internus

The **nerve to the quadratus femoris** (L4 to S1) and the **nerve to the obturator internus** (L5 to S2) are small motor nerves that originate from the sacral plexus. Both nerves pass through the greater sciatic foramen inferior to the piriformis muscle and enter the gluteal region (Fig. 6.36 and Table 6.1):

- The nerve to the obturator internus supplies the gemellus superior muscle in the gluteal region and then loops around the ischial spine and enters the perineum through the lesser sciatic foramen to penetrate the perineal surface of the obturator internus muscle.
- The nerve to the quadratus femoris supplies the gemellus inferior and quadratus femoris muscles.

### Posterior cutaneous nerve of thigh

The **posterior cutaneous nerve of the thigh** is formed by contributions from S1 to S3 and leaves the pelvic cavity through the greater sciatic foramen inferior to the piriformis muscle (Fig. 6.36 and Table 6.1). It passes vertically through the gluteal region deep to the gluteus maximus and enters the posterior thigh and innervates:

- a longitudinal band of skin over the posterior aspect of the thigh that continues into the upper leg, and
- skin over the gluteal fold, over the upper medial part of the thigh and in the adjacent regions of the perineum.

### Perforating cutaneous nerve

The **perforating cutaneous nerve** is a small sensory nerve formed by contributions from S2 and S3. It leaves the pelvic cavity by penetrating directly through the sacrotuberous ligament (Fig. 6.36 and Table 6.1) and passes inferiorly around the lower border of the gluteus maximus where it overlaps with the posterior cutaneous nerve of the thigh in innervating skin over the medial aspect of the gluteal fold.

## Arteries

### Femoral artery

The major artery supplying the lower limb is the **femoral artery** (Fig. 6.37), which is the continuation of the external iliac artery in the abdomen. The external iliac artery becomes the femoral artery as the vessel passes under the inguinal ligament to enter the femoral triangle in the anterior aspect of the thigh. Branches supply most of the thigh and all of the leg and foot.

### Superior and inferior gluteal arteries and the obturator artery

Other vessels supplying parts of the lower limb include the superior and inferior gluteal arteries and the obturator artery (Fig. 6.37).

The **superior and inferior gluteal arteries** originate in the pelvic cavity as branches of the internal iliac artery (see Chapter 5, pp. 489–492) and supply the gluteal region. The superior gluteal artery leaves the pelvis through the greater sciatic foramen above the piriformis muscle, and the inferior gluteal artery leaves through the same foramen but below the piriformis muscle.

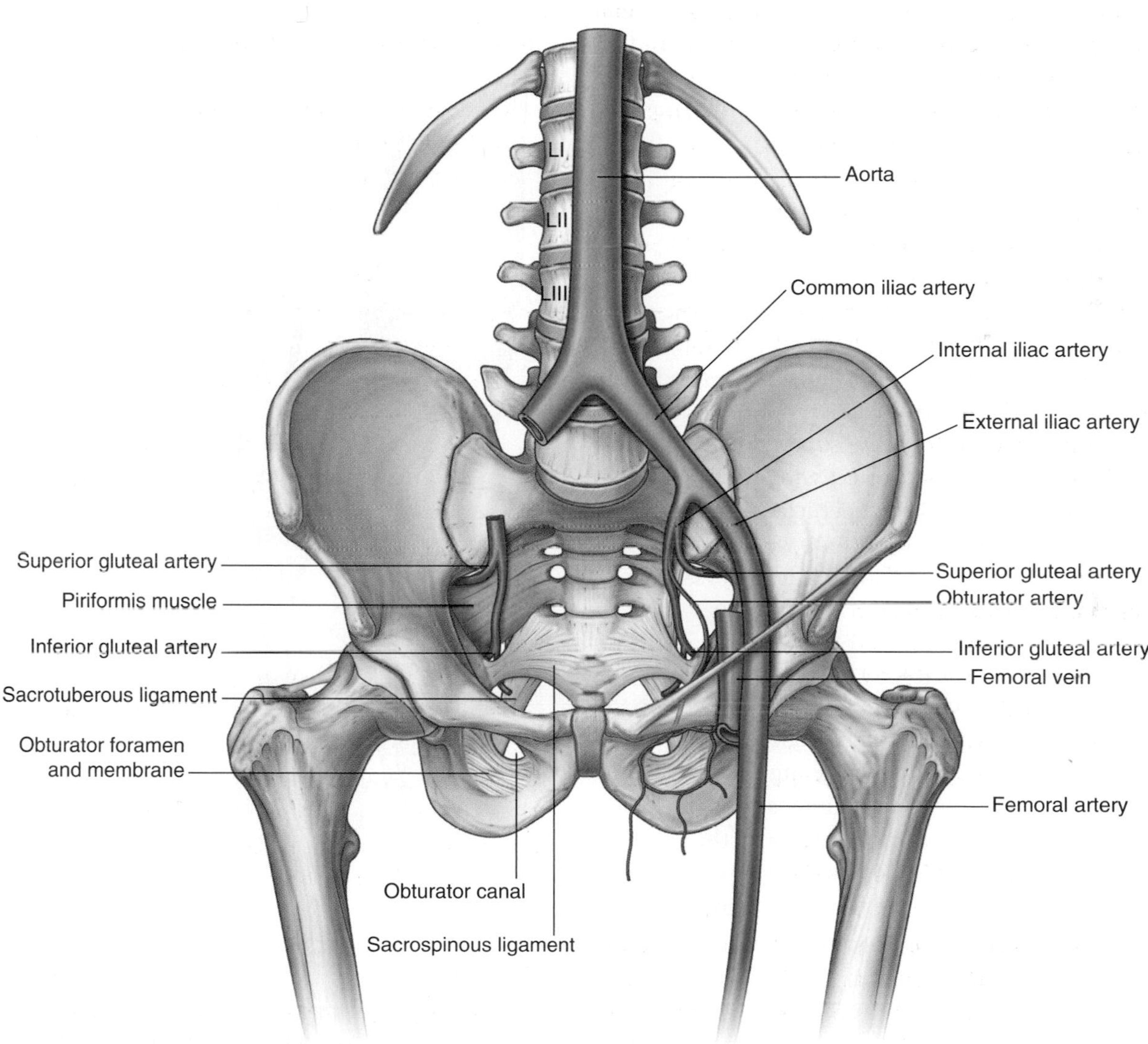
LI
LII
LIII
Aorta
Common iliac artery
Internal iliac artery
External iliac artery
Superior gluteal artery
Piriformis muscle
Inferior gluteal artery
Sacrotuberous ligament
Obturator foramen and membrane
Superior gluteal artery
Obturator artery
Inferior gluteal artery
Femoral vein
Femoral artery
Obturator canal
Sacrospinous ligament

**Fig. 6.37** Arteries of the lower limb.

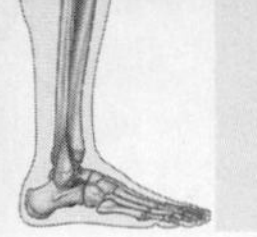

The **obturator artery** is also a branch of the internal iliac artery in the pelvic cavity (see Chapter 5, pp. 490–491) and passes through the obturator canal to enter and supply the medial compartment of the thigh.

Branches of the femoral, inferior gluteal, superior gluteal, and obturator arteries, together with branches from the internal pudendal artery of the perineum, interconnect to form an anastomotic network in the upper thigh and gluteal region. The presence of these anastomotic channels may provide collateral circulation when one of the vessels is interrupted.

## Veins

Veins draining the lower limb form superficial and deep groups.

The deep veins generally follow the arteries (femoral, superior gluteal, inferior gluteal, and obturator). The major deep vein draining the limb is the **femoral vein** (Fig. 6.38). It becomes the external iliac vein when it passes under the inguinal ligament to enter the abdomen.

The superficial veins are in the subcutaneous connective tissue and are interconnected with and ultimately drain into the deep veins. The superficial veins form two major channels—the great saphenous vein and the small saphenous vein. Both veins originate from a dorsal venous arch in the foot:

- The **great saphenous vein** originates from the medial side of the dorsal venous arch and then ascends up the medial side of the leg, knee, and thigh to connect with the femoral vein just inferior to the inguinal ligament.
- The **small saphenous vein** originates from the lateral side of the dorsal venous arch, ascends up the posterior surface of the leg, and then penetrates deep fascia to join the popliteal vein posterior to the knee; proximal to the knee, the popliteal vein becomes the femoral vein.

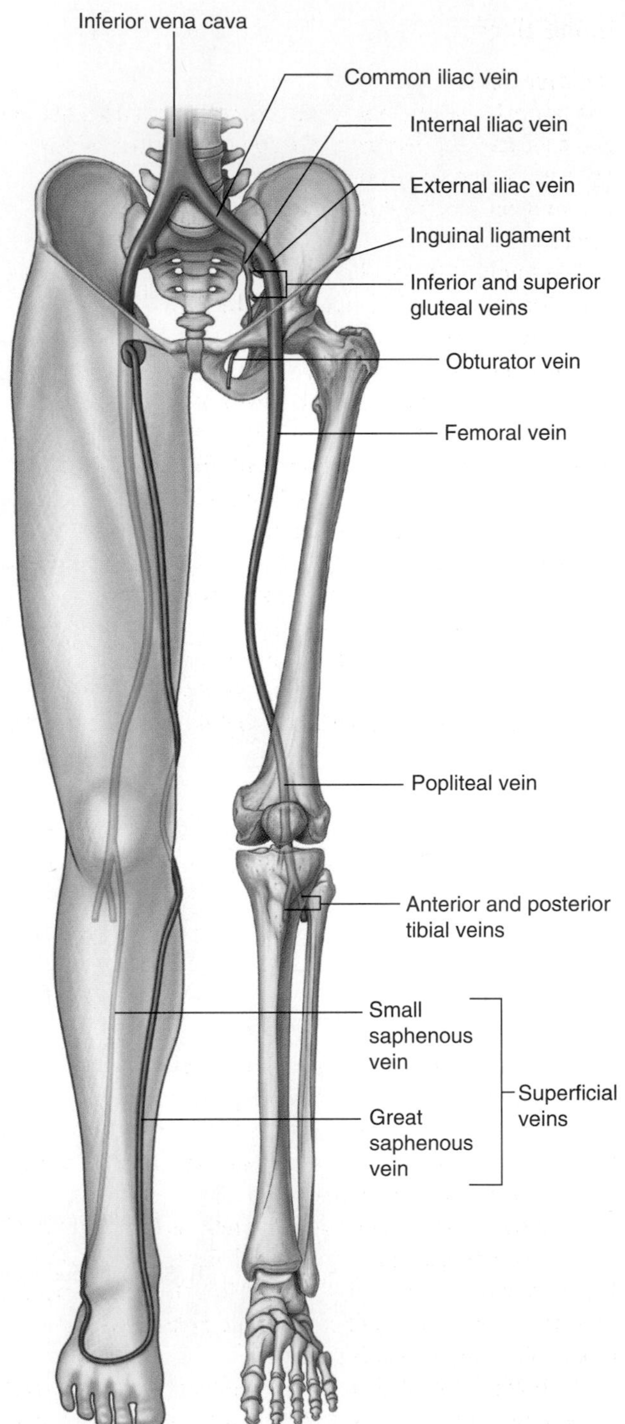


**Fig. 6.38** Veins of the lower limb.

## In the clinic

### Varicose veins

The normal flow of blood in the lower limbs is from the skin and subcutaneous tissues to the superficial veins, which drain via perforating veins to the deep veins, which in turn drain into the iliac veins and inferior vena cava.

The normal flow of blood in the venous system depends upon the presence of competent valves, which prevent reflux. Venous return is supplemented with contraction of the muscles in the lower limb, which pump the blood toward the heart. When venous valves become incompetent they tend to place extra pressure on more distal valves, which may also become incompetent. This condition produces dilated tortuous superficial veins (varicose veins) in the distribution of the great (long) and small (short) saphenous venous systems.

Varicose veins occur more commonly in women than in men, and symptoms are often aggravated by pregnancy. Some individuals have a genetic predisposition to developing varicose veins. Valves may also be destroyed when a deep vein thrombosis occurs if the clot incorporates the valve into its interstices; during the process of healing and recanalization the valve is destroyed, rendering it incompetent.

Typical sites for valvular incompetence include the junction between the great (long) saphenous vein and the femoral vein, perforating veins in the midthigh, and the junction between the small (short) saphenous vein and the popliteal vein.

Varicose veins may be unsightly, and soft tissue changes may occur with chronic venous incompetence. As the venous pressure rises, increased venular and capillary pressure damages the cells, and blood and blood products extrude into the soft tissue. This may produce a brown pigmentation in the skin, and venous eczema may develop. Furthermore, if the pressure remains high the skin may break down and ulcerate, and many weeks of hospitalization may be needed for this to heal.

Treatments for varicose veins include tying off the valve, "stripping" (removing) the great (long) and small (short) saphenous systems, and in some cases valvular reconstruction.

## In the clinic

### Deep vein thrombosis

Thrombosis may occur in the deep veins of the lower limb and within the pelvic veins. Its etiology was eloquently described by Virchow, who described the classic triad (venous stasis, injury to the vessel wall, and hypercoagulable states) that precipitates thrombosis.

In some patients a deep vein thrombosis (DVT) in the calf veins may propagate into the femoral veins. This clot may break off and pass through the heart to enter the pulmonary circulation, resulting in occlusion of the pulmonary artery, cardiopulmonary arrest, and death.

A significant number of patients undergoing surgery are likely to develop a DVT, so most surgical patients are given specific prophylactic treatment to prevent thrombosis. A typical DVT prophylactic regimen includes anticoagulant injections and graduated stockings (to prevent deep venous stasis and facilitate emptying of the deep veins).

Although physicians aim to prevent the formation of DVT, it is not always possible to detect it because there may be no clinical signs. Calf muscle tenderness, postoperative pyrexia, and limb swelling can be helpful clues. The diagnosis is predominantly made by duplex Doppler sonography or rarely by ascending venography.

If DVT is confirmed, intravenous and oral anticoagulation are started to prevent extension of the thrombus.

## Lymphatics

Most lymphatic vessels in the lower limb drain into superficial and deep inguinal nodes located in the fascia just inferior to the inguinal ligament (Fig. 6.39).

### Superficial inguinal nodes

The **superficial inguinal nodes**, approximately ten in number, are in the superficial fascia and parallel the course of the inguinal ligament in the upper thigh. Medially, they extend inferiorly along the terminal part of the great saphenous vein.

Superficial inguinal nodes receive lymph from the gluteal region, lower abdominal wall, perineum, and superficial regions of the lower limb. They drain, via vessels that accompany the femoral vessels, into **external iliac nodes** associated with the external iliac artery in the abdomen.

### Deep inguinal nodes

The **deep inguinal nodes**, up to three in number, are medial to the femoral vein (Fig. 6.39).

The deep inguinal nodes receive lymph from deep lymphatics associated with the femoral vessels and from the glans penis (or clitoris) in the perineum. They interconnect

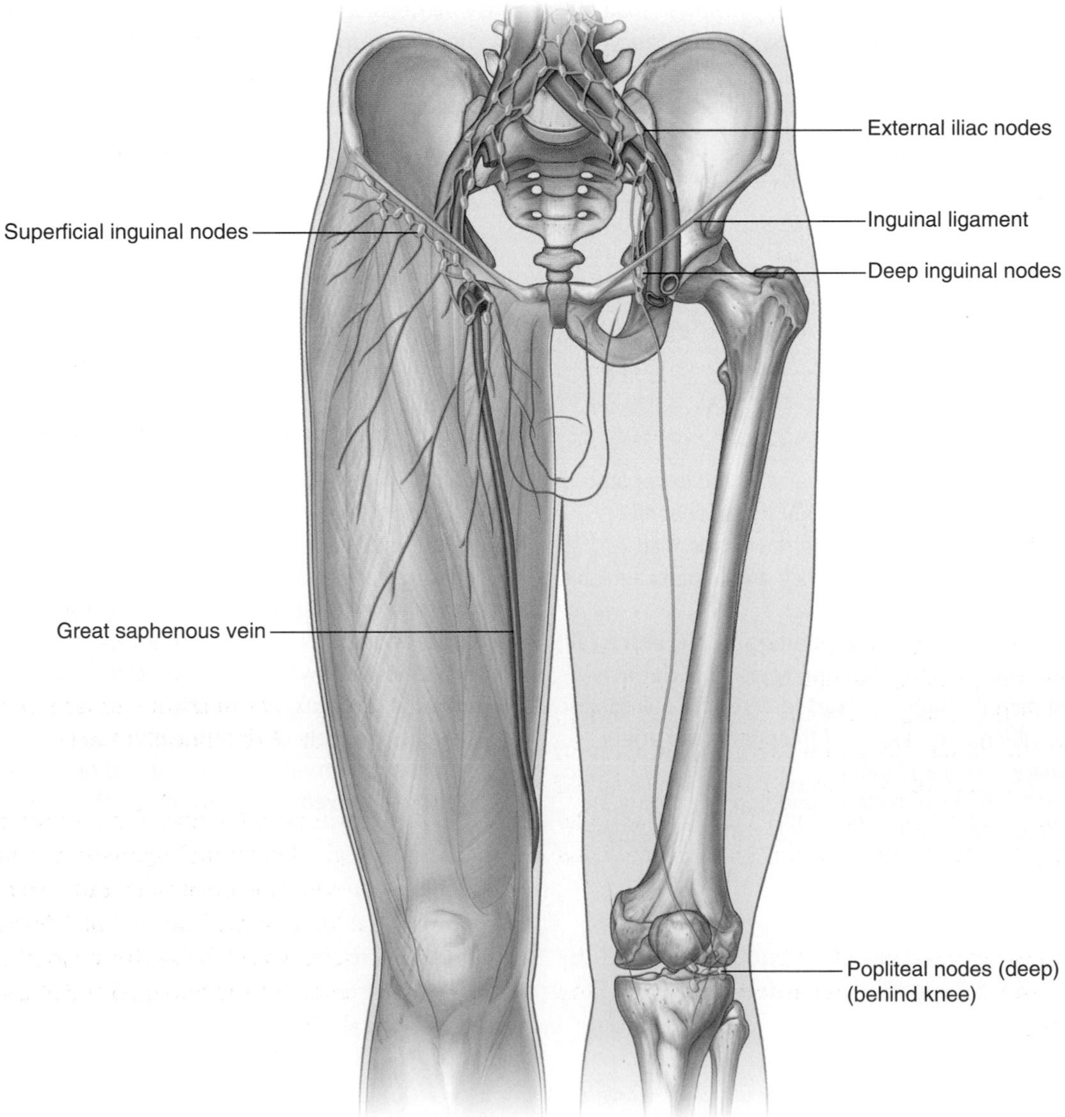

**Fig. 6.39** Lymphatic drainage of the lower limb.

with the superficial inguinal nodes and drain into the external iliac nodes via vessels that pass along the medial side of the femoral vein as it passes under the inguinal ligament. The space through which the lymphatic vessels pass under the inguinal ligament is the femoral canal.

### Popliteal nodes

In addition to the inguinal nodes, there is a small collection of deep nodes posterior to the knee close to the popliteal vessels (Fig. 6.39). These **popliteal nodes** receive lymph from superficial vessels, which accompany the small saphenous vein, and from deep areas of the leg and foot. They ultimately drain into the deep and superficial inguinal nodes.

## Deep fascia and the saphenous opening

### Fascia lata

The outer layer of deep fascia in the lower limb forms a thick "stocking-like" membrane, which covers the limb and lies beneath the superficial fascia (Fig. 6.40A). This deep fascia is particularly thick in the thigh and gluteal region and is termed the **fascia lata**.

The fascia lata is anchored superiorly to bone and soft tissues along a line of attachment that defines the upper margin of the lower limb. Beginning anteriorly and circling laterally around the limb, this line of attachment includes the inguinal ligament, iliac crest, sacrum, coccyx, sacrotuberous ligament, inferior ramus of the pubic bone, body of the pubic bone, and superior ramus of the pubic bone.

Inferiorly, the fascia lata is continuous with the deep fascia of the leg.

### Iliotibial tract

The fascia lata is thickened laterally into a longitudinal band (the **iliotibial tract**), which descends along the lateral margin of the limb from the tuberculum of the iliac crest to a bony attachment just below the knee (Fig. 6.40B).

The superior aspect of the fascia lata in the gluteal region splits anteriorly to enclose the tensor fasciae latae muscle and posteriorly to enclose the gluteus maximus muscle:

- The tensor fasciae latae muscle is partially enclosed by and inserts into the superior and anterior aspects of the iliotibial tract.

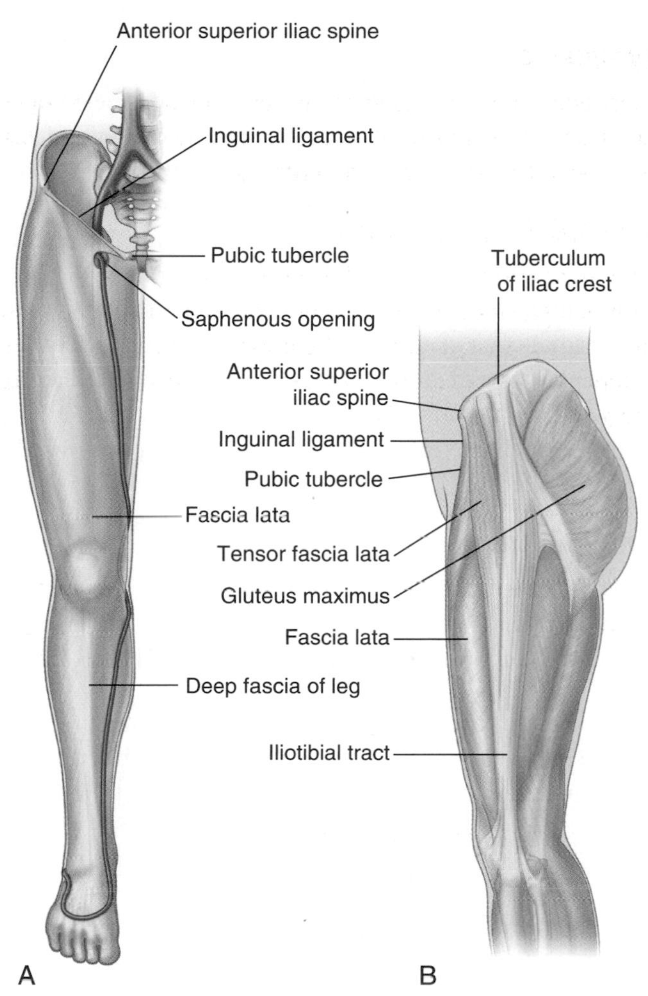


**Fig. 6.40** Fascia lata. **A.** Right limb. Anterior view. **B.** Lateral view.

- Most of the gluteus maximus muscle inserts into the posterior aspect of the iliotibial tract.

The tensor fasciae latae and gluteus maximus muscles, working through their attachments to the iliotibial tract, hold the leg in extension once other muscles have extended the leg at the knee joint. The iliotibial tract and its two associated muscles also stabilize the hip joint by preventing lateral displacement of the proximal end of the femur away from the acetabulum.

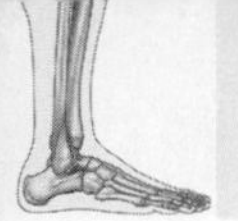



### Saphenous opening

The fascia lata has one prominent aperture on the anterior aspect of the thigh just inferior to the medial end of the inguinal ligament (the **saphenous opening**), which allows the great saphenous vein to pass from superficial fascia through the deep fascia to connect with the femoral vein (Fig. 6.41).

The margin of the saphenous opening is formed by the free medial edge of the fascia lata as it descends from the inguinal ligament and spirals around the lateral side of the great saphenous vein and medially under the femoral vein to attach to the pectineal line (pecten pubis) of the pelvic bone.

## Femoral triangle

The femoral triangle is a wedge-shaped depression formed by muscles in the upper thigh at the junction between the anterior abdominal wall and the lower limb (Fig. 6.42):

- The base of the triangle is the inguinal ligament.
- The medial border is the medial margin of the adductor longus muscle in the medial compartment of the thigh.
- The lateral margin is the medial margin of the sartorius muscle in the anterior compartment of the thigh.
- The floor of the triangle is formed medially by the pectineus and adductor longus muscles in the medial compartment of the thigh and laterally by the iliopsoas muscle descending from the abdomen.
- The apex of the femoral triangle points inferiorly and is continuous with a fascial canal (**adductor canal**), which descends medially down the thigh and posteriorly through an aperture in the lower end of one of the largest of the adductor muscles in the thigh (the adductor magnus muscle) to open into the popliteal fossa behind the knee.

The femoral nerve, artery, and vein and lymphatics pass between the abdomen and lower limb under the

**Fig. 6.41** Saphenous ring. Anterior view.

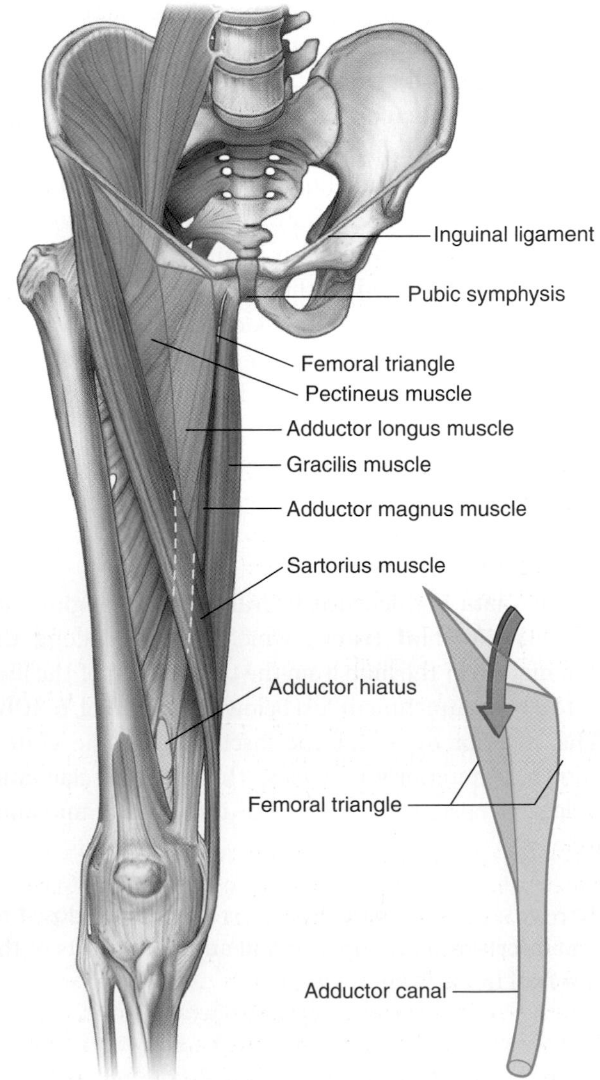


**Fig. 6.42** Boundaries of the femoral triangle.

inguinal ligament and in the femoral triangle (Fig. 6.43). The femoral artery and vein pass inferiorly through the adductor canal and become the popliteal vessels behind the knee where they meet and are distributed with branches of the sciatic nerve, which descends through the posterior thigh from the gluteal region.

From lateral to medial, major structures in the femoral triangle are the femoral nerve, the femoral artery, the femoral vein, and lymphatic vessels. The femoral artery can be palpated in the femoral triangle just inferior to the inguinal ligament and midway between the anterior superior iliac spine and the pubic symphysis.

### Femoral sheath

In the femoral triangle, the femoral artery and vein and the associated lymphatic vessels are surrounded by a funnel-shaped sleeve of fascia (the **femoral sheath**). The sheath is continuous superiorly with the transversalis fascia and iliac fascia of the abdomen and merges inferiorly with connective tissue associated with the vessels. Each of the three structures surrounded by the sheath is contained within a separate fascial compartment within the sheath. The most medial compartment (the femoral canal) contains the lymphatic vessels and is conical in shape. The opening of this canal superiorly is potentially a weak point in the lower abdomen and is the site for femoral hernias. The femoral nerve is lateral to and not contained within the femoral sheath.

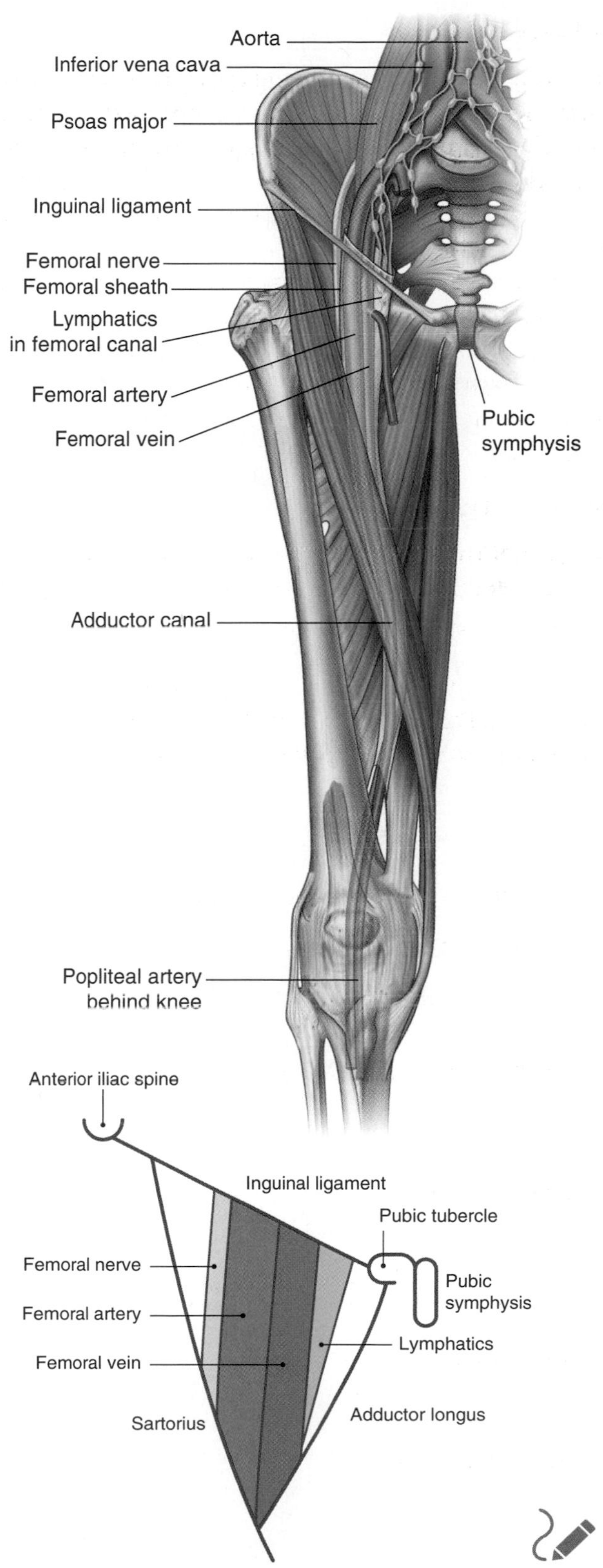


**Fig. 6.43** Contents of the femoral triangle.

#### In the clinic

**Vascular access to the lower limb**

Deep and inferior to the inguinal ligament are the femoral artery and femoral vein. The femoral artery is palpable as it passes over the femoral head and may be easily demonstrated using ultrasound. If arterial or venous access is needed rapidly, a physician can use the femoral approach to these vessels.

Many radiological procedures involve catheterization of the femoral artery or the femoral vein to obtain access to the contralateral lower limb, the ipsilateral lower limb, the vessels of the thorax and abdomen, and the cerebral vessels.

Cardiologists also use the femoral artery to place catheters in vessels around the arch of the aorta and into the coronary arteries to perform coronary angiography and angioplasty.

Access to the femoral vein permits catheters to be maneuvered into the renal veins, the gonadal veins, the right atrium, and the right side of the heart, including the pulmonary artery and distal vessels of the pulmonary tree. Access to the superior vena cava and the great veins of the neck is also possible.

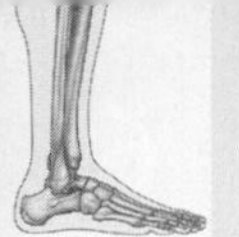

## GLUTEAL REGION

The gluteal region lies posterolateral to the bony pelvis and proximal end of the femur (Fig. 6.44). Muscles in the region mainly abduct, extend, and laterally rotate the femur relative to the pelvic bone.

The gluteal region communicates anteromedially with the pelvic cavity and perineum through the greater sciatic foramen and lesser sciatic foramen, respectively. Inferiorly, it is continuous with the posterior thigh.

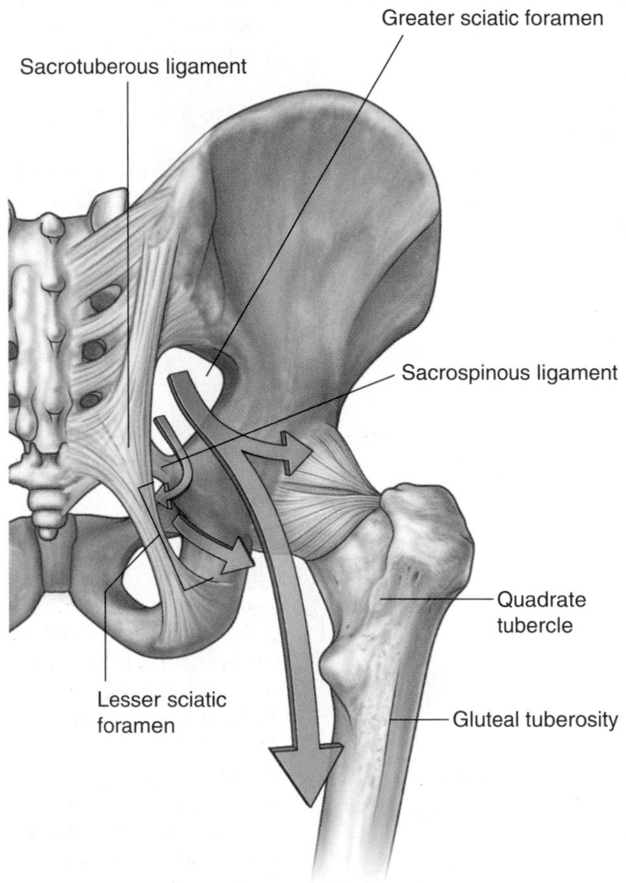


**Fig. 6.44** Gluteal region. Posterior view.

The sciatic nerve enters the lower limb from the pelvic cavity by passing through the greater sciatic foramen and descending through the gluteal region into the posterior thigh and then into the leg and foot.

The pudendal nerve and internal pudendal vessels pass between the pelvic cavity and perineum by passing first through the greater sciatic foramen to enter the gluteal region and then immediately passing through the lesser sciatic foramen to enter the perineum. The nerve to the obturator internus and gemellus superior follows a similar course. Other nerves and vessels that pass through the greater sciatic foramen from the pelvic cavity supply structures in the gluteal region itself.

### Muscles

Muscles of the gluteal region (Table 6.2) are composed mainly of two groups:

- a deep group of small muscles, which are mainly lateral rotators of the femur at the hip joint and include the piriformis, obturator internus, gemellus superior, gemellus inferior, and quadratus femoris;
- a more superficial group of larger muscles, which mainly abduct and extend the hip and include the gluteus minimus, gluteus medius, and gluteus maximus; an additional muscle in this group, the tensor fasciae latae, stabilizes the knee in extension by acting on a specialized longitudinal band of deep fascia (the iliotibial tract) that passes down the lateral side of the thigh to attach to the proximal end of the tibia in the leg.

Many of the important nerves in the gluteal region are in the plane between the superficial and deep groups of muscles.

**Table 6.2** Muscles of the gluteal region (spinal segments in bold are the major segments innervating the muscle)

| Muscle | Origin | Insertion | Innervation | Function |
|---|---|---|---|---|
| Piriformis | Anterior surface of sacrum between anterior sacral foramina | Medial side of superior border of greater trochanter of femur | Branches from S1 and **S2** | Laterally rotates the extended femur at hip joint; abducts flexed femur at hip joint |
| Obturator internus | Anterolateral wall of true pelvis; deep surface of obturator membrane and surrounding bone | Medial side of greater trochanter of femur | Nerve to obturator internus (L5, **S1**) | Laterally rotates the extended femur at hip joint; abducts flexed femur at hip joint |
| Gemellus superior | External surface of ischial spine | Along length of superior surface of the obturator internus tendon and into the medial side of greater trochanter of femur with obturator internus tendon | Nerve to obturator internus (L5, **S1**) | Laterally rotates the extended femur at hip joint; abducts flexed femur at hip joint |
| Gemellus inferior | Upper aspect of ischial tuberosity | Along length of inferior surface of the obturator internus tendon and into the medial side of greater trochanter of femur with obturator internus tendon | Nerve to quadratus femoris (**L5, S1**) | Laterally rotates the extended femur at hip joint; abducts flexed femur at hip joint |
| Quadratus femoris | Lateral aspect of the ischium just anterior to the ischial tuberosity | Quadrate tubercle on the intertrochanteric crest of the proximal femur | Nerve to quadratus femoris (**L5, S1**) | Laterally rotates femur at hip joint |
| Gluteus minimus | External surface of ilium between inferior and anterior gluteal lines | Linear facet on the anterolateral aspect of the greater trochanter | Superior gluteal nerve (**L4, L5**, S1) | Abducts femur at hip joint; holds pelvis secure over stance leg and prevents pelvic drop on the opposite swing side during walking; medially rotates thigh |
| Gluteus medius | External surface of ilium between anterior and posterior gluteal lines | Elongate facet on the lateral surface of the greater trochanter | Superior gluteal nerve (**L4, L5**, S1) | Abducts femur at hip joint; holds pelvis secure over stance leg and prevents pelvic drop on the opposite swing side during walking; medially rotates thigh |
| Gluteus maximus | Fascia covering gluteus medius, external surface of ilium behind posterior gluteal line, fascia of erector spinae, dorsal surface of lower sacrum, lateral margin of coccyx, external surface of sacrotuberous ligament | Posterior aspect of iliotibial tract of fascia lata and gluteal tuberosity of proximal femur | Inferior gluteal nerve (**L5, S1**, S2) | Powerful extensor of flexed femur at hip joint; lateral stabilizer of hip joint and knee joint; laterally rotates and abducts thigh |
| Tensor fasciae latae | Lateral aspect of crest of ilium between anterior superior iliac spine and tubercle of the crest | Iliotibial tract of fascia lata | Superior gluteal nerve (**L4, L5**, S1) | Stabilizes the knee in extension |

## Deep group

### Piriformis

The **piriformis** muscle is the most superior of the deep group of muscles (Fig. 6.45) and is a muscle of the pelvic wall and of the gluteal region (see Chapter 5, p. 443). It originates from between the anterior sacral foramina on the anterolateral surface of the sacrum and passes laterally and inferiorly through the greater sciatic foramen.

In the gluteal region, the piriformis passes posterior to the hip joint and attaches to a facet on the upper margin of the greater trochanter of the femur.

The piriformis externally rotates and abducts the femur at the hip joint and is innervated in the pelvic cavity by the nerve to the piriformis, which originates as branches from S1 and S2 of the sacral plexus (see Chapter 5, p. 485).

In addition to its action on the hip joint, the piriformis is an important landmark because it divides the greater sciatic foramen into two regions, one above and one below the piriformis. Vessels and nerves pass between the pelvis and gluteal region by passing through the greater sciatic foramen either above or below the piriformis.

### Obturator internus

The **obturator internus** muscle, like the piriformis muscle, is a muscle of the pelvic wall and of the gluteal region (Fig. 6.45). It is a flat fan-shaped muscle originating from the medial surface of the obturator membrane and adjacent bone of the obturator foramen (see Chapter 5, pp. 442–443). Because the pelvic floor attaches to a thickened band of fascia across the medial surface of the obturator internus, the obturator internus forms:

- the anterolateral wall of the pelvic cavity above the pelvic floor, and
- the lateral wall of the ischio-anal fossa in the perineum below the pelvic floor.

The muscle fibers of the obturator internus converge to form a tendon, which bends 90° around the ischium between the ischial spine and ischial tuberosity and passes through the lesser sciatic foramen to enter the gluteal region. The tendon then passes posteroinferiorly to the hip joint and attaches to the medial surface of the superior

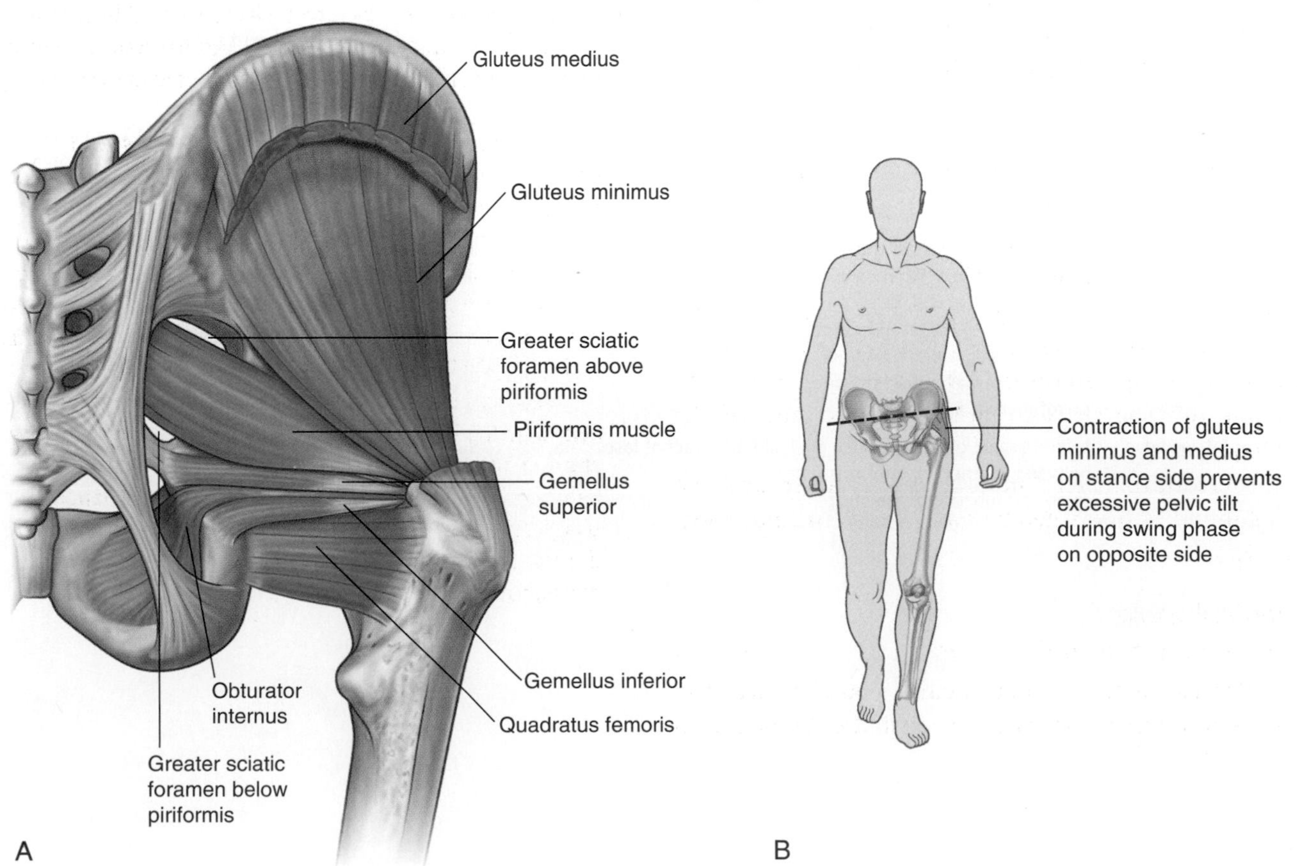


**Fig. 6.45** Deep muscles in the gluteal region. **A.** Posterior view. **B.** Function.

margin of the greater trochanter of the femur just inferior to the attachment of the piriformis muscle.

The obturator internus laterally rotates and abducts the femur at the hip joint and is innervated by the nerve to the obturator internus.

### Gemellus superior and inferior

The gemellus superior and inferior (*gemelli* is Latin for "twins") are a pair of triangular muscles associated with the upper and lower margins of the obturator internus tendon (Fig. 6.45):

- The base of the **gemellus superior** originates from the gluteal surface of the ischial spine.
- The base of the **gemellus inferior** originates from the upper gluteal and pelvic surfaces of the ischial tuberosity.

Fibers of the gemellus muscles attach along the length of the obturator internus tendon, and the apices of the two muscles insert with the tendon of the obturator internus on the greater trochanter of the femur.

The gemellus superior is innervated by the nerve to the obturator internus, and the gemellus inferior is innervated by the nerve to the quadratus femoris. The gemellus muscles act with the obturator internus muscle to laterally rotate and abduct the femur at the hip joint.

### Quadratus femoris

The **quadratus femoris** muscle is the most inferior of the deep group of muscles in the gluteal region (Fig. 6.45). It is a flat rectangular muscle below the obturator internus muscle and its associated gemellus muscles.

The quadratus femoris is attached at one end to a linear roughening on the lateral aspect of the ischium just anterior to the ischial tuberosity and at the other end to the quadrate tubercle on the intertrochanteric crest of the proximal femur.

The quadratus femoris laterally rotates the femur at the hip joint and is innervated by the nerve to the quadratus femoris.

## Superficial group

### Gluteus minimus and medius

The gluteus minimus and medius muscles are two muscles of the more superficial group in the gluteal region (Fig. 6.45).

The **gluteus minimus** is a fan-shaped muscle that originates from the external surface of the expanded upper part of the ilium, between the inferior gluteal line and the anterior gluteal line. The muscle fibers converge inferiorly and laterally to form a tendon, which inserts into a broad linear facet on the anterolateral aspect of the greater trochanter.

The **gluteus medius** overlies the gluteus minimus and is also fan shaped. It has a broad origin from the external surface of the ilium between the anterior gluteal line and posterior gluteal line and inserts on an elongate facet on the lateral surface of the greater trochanter.

The gluteus medius and minimus muscles abduct the lower limb at the hip joint and reduce pelvic drop over the opposite swing limb during walking by securing the position of the pelvis on the stance limb (Fig. 6.45B). Both muscles are innervated by the superior gluteal nerve.

### In the clinic

#### Trendelenburg's sign

Trendelenburg's sign occurs in people with weak or paralyzed abductor muscles (gluteus medius and gluteus minimus) of the hip. The sign is demonstrated by asking the patient to stand on one limb. When the patient stands on the affected limb, the pelvis severely drops over the swing limb.

Positive signs are typically found in patients with damage to the superior gluteal nerve. Damage to this nerve may occur with associated pelvic fractures, with space-occupying lesions within the pelvis extending into the greater sciatic foramen, and in some cases relating to hip surgery during which there has been disruption of and subsequent atrophy of the insertion of the gluteus medius and gluteus minimus tendons on the greater trochanter.

In patients with a positive Trendelenburg's sign, gait also is abnormal. Typically during the stance phase of the affected limb, the weakened abductor muscles allow the pelvis to tilt inferiorly over the swing limb. The patient compensates for the pelvic drop by lurching the trunk to the affected side to maintain the level of the pelvis throughout the gait cycle.

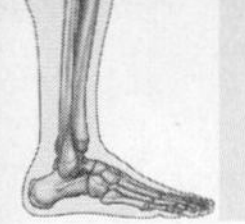


### Gluteus maximus

The gluteus maximus is the largest muscle in the gluteal region and overlies most of the other gluteal muscles (Fig. 6.46).

The gluteus maximus is quadrangular in shape and has a broad origin extending from a roughened area of the ilium behind the posterior gluteal line and along the dorsal surface of the lower sacrum and the lateral surface of the coccyx to the external surface of the sacrotuberous ligament. It is also attached to fascia overlying the gluteus medius muscle and, between the ilium and sacrum, to fascia covering the erector spinae muscle, and is often described as being enclosed within two layers of the fascia lata, which covers the thigh and gluteal region.

Laterally, the upper and superficial lower parts of the gluteus maximus insert into the posterior aspect of a tendinous thickening of the fascia lata (the iliotibial tract), which passes over the lateral surface of the greater trochanter and descends down the thigh and into the upper leg. Deep distal parts of the muscle attach to the elongate gluteal tuberosity of the proximal femur.

The gluteus maximus mainly extends the flexed thigh at the hip joint. Through its insertion into the iliotibial tract, it also stabilizes the knee and hip joints. It is innervated by the inferior gluteal nerve.

### Tensor fasciae latae

The tensor fasciae latae muscle is the most anterior of the superficial group of muscles in the gluteal region and overlies the gluteus minimus and the anterior part of the gluteus medius (Fig. 6.47).

The tensor fasciae latae originates from the outer margin of the iliac crest from the anterior superior iliac spine to approximately the tuberculum of the iliac crest. The muscle

**Fig. 6.46** Gluteus maximus muscle. Posterior view.

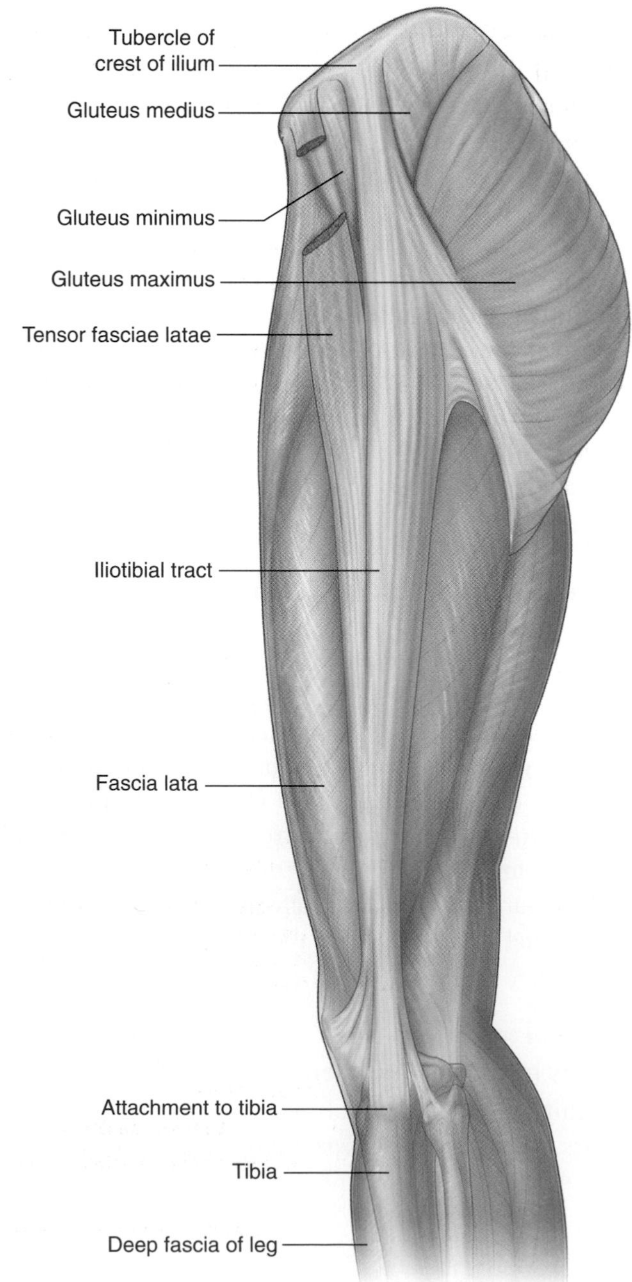


**Fig. 6.47** Tensor fasciae latae. Left gluteal region, lateral view.

fibers descend to insert into the anterior aspect of the iliotibial tract of deep fascia, which runs down the lateral side of the thigh and attaches to the upper tibia. Like the gluteus maximus muscle, the tensor fasciae latae is enclosed within a compartment of the fascia lata.

The tensor fasciae latae stabilizes the knee in extension and, working with the gluteus maximus muscle on the iliotibial tract lateral to the greater trochanter, stabilizes the hip joint by holding the head of the femur in the acetabulum (Fig. 6.47). It is innervated by the superior gluteal nerve.

## Nerves

Seven nerves enter the gluteal region from the pelvis through the greater sciatic foramen (Fig. 6.48): the superior gluteal nerve, sciatic nerve, nerve to the quadratus femoris, nerve to the obturator internus, posterior cutaneous nerve of the thigh, pudendal nerve, and inferior gluteal nerve.

An additional nerve, the perforating cutaneous nerve, enters the gluteal region by passing directly through the sacrotuberous ligament.

Some of these nerves, such as the sciatic and pudendal nerves, pass through the gluteal region en route to other areas. Nerves such as the superior and inferior gluteal nerves innervate structures in the gluteal region. Many of the nerves in the gluteal region are in the plane between the superficial and deep groups of muscles.

### Superior gluteal nerve

Of all the nerves that pass through the greater sciatic foramen, the superior gluteal nerve is the only one that passes above the piriformis muscle (Fig. 6.48). After entering the gluteal region, the nerve loops up over the inferior margin of the gluteus minimus and travels anteriorly and laterally in the plane between the gluteus minimus and medius muscles.

The superior gluteal nerve supplies branches to the gluteus minimus and medius muscles and terminates by innervating the tensor fasciae latae muscle.

### Sciatic nerve

The sciatic nerve enters the gluteal region through the greater sciatic foramen inferior to the piriformis muscle (Fig. 6.48). It descends in the plane between the superficial and deep group of gluteal region muscles, crossing the posterior surfaces of first the obturator internus and associated gemellus muscles and then the quadratus femoris

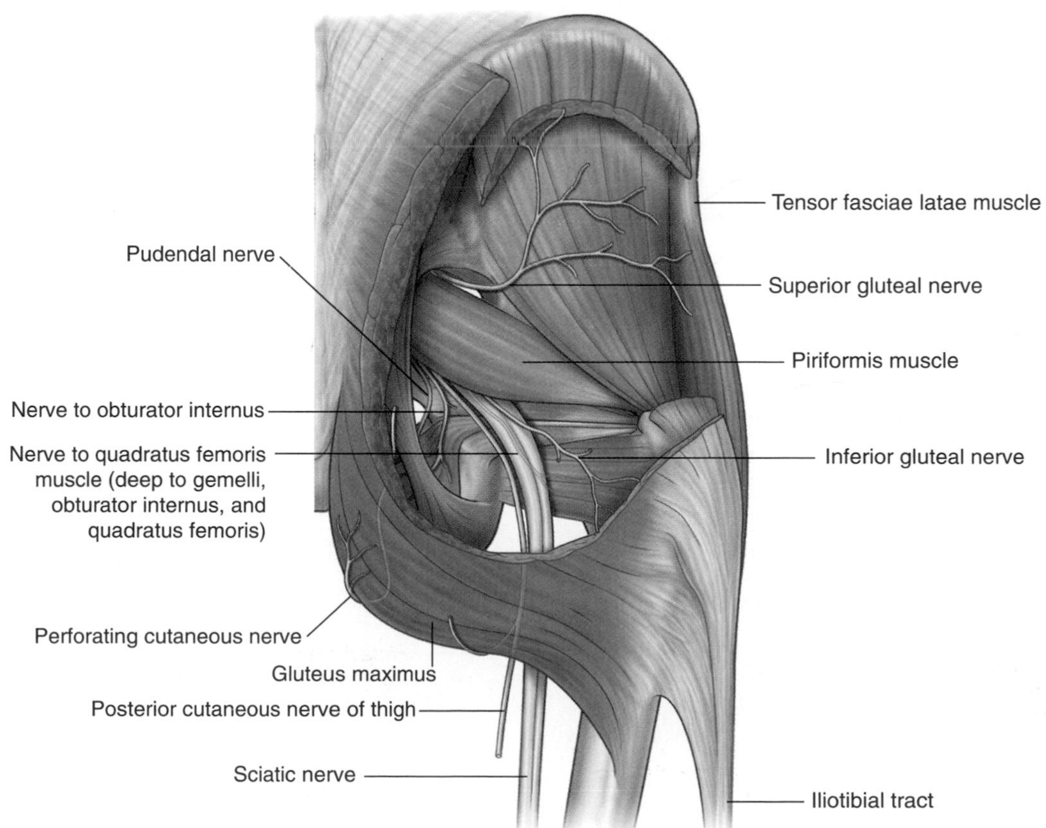


**Fig. 6.48** Nerves of the gluteal region. Posterior view.

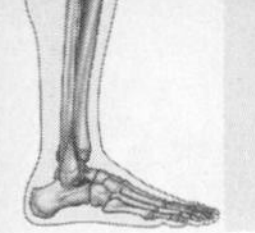



muscle. It lies just deep to the gluteus maximus at the midpoint between the ischial tuberosity and the greater trochanter. At the lower margin of the quadratus femoris muscle, the sciatic nerve enters the posterior thigh.

The sciatic nerve is the largest nerve in the body and innervates all muscles in the posterior compartment of the thigh that flex the knee and all muscles that work the ankle and foot. It also innervates a large area of skin in the lower limb.

### Nerve to quadratus femoris

The nerve to the quadratus femoris enters the gluteal region through the greater sciatic foramen inferior to the piriformis muscle and deep to the sciatic nerve (Fig. 6.48). Unlike other nerves in the gluteal region, the nerve to the quadratus femoris lies anterior to the plane of the deep muscles.

The nerve to the quadratus femoris descends along the ischium deep to the tendon of the obturator internus muscle and associated gemellus muscles to penetrate and innervate the quadratus femoris. It supplies a small branch to the gemellus inferior.

### Nerve to obturator internus

The nerve to the obturator internus enters the gluteal region through the greater sciatic foramen inferior to the piriformis muscle and between the posterior cutaneous nerve of the thigh and the pudendal nerve (Fig. 6.48). It supplies a small branch to the gemellus superior and then passes over the ischial spine and through the lesser sciatic foramen to innervate the obturator internus muscle from the medial surface of the muscle in the perineum.

### Posterior cutaneous nerve of the thigh

The posterior cutaneous nerve of the thigh enters the gluteal region through the greater sciatic foramen inferior to the piriformis muscle and immediately medial to the sciatic nerve (Fig. 6.48). It descends through the gluteal region just deep to the gluteus maximus and enters the posterior thigh.

The posterior cutaneous nerve of the thigh has a number of gluteal branches, which loop around the lower margin of the gluteus maximus muscle to innervate skin over the gluteal fold. A small perineal branch passes medially to contribute to the innervation of the skin of the scrotum or labia majora in the perineum. The main trunk of the posterior cutaneous nerve of the thigh passes inferiorly, giving rise to branches that innervate the skin on the posterior thigh and leg.

### Pudendal nerve

The pudendal nerve enters the gluteal region through the greater sciatic foramen inferior to the piriformis muscle and medial to the sciatic nerve (Fig. 6.48). It passes over the sacrospinous ligament and immediately passes through the lesser sciatic foramen to enter the perineum. The course of the pudendal nerve in the gluteal region is short and the nerve is often hidden by the overlying upper margin of the sacrotuberous ligament.

The pudendal nerve is the major somatic nerve of the perineum and has no branches in the gluteal region.

### Inferior gluteal nerve

The inferior gluteal nerve enters the gluteal region through the greater sciatic foramen inferior to the piriformis muscle and along the posterior surface of the sciatic nerve (Fig. 6.48). It penetrates and supplies the gluteus maximus muscle.

### Perforating cutaneous nerve

The perforating cutaneous nerve is the only nerve in the gluteal region that does not enter the area through the greater sciatic foramen. It is a small nerve that leaves the sacral plexus in the pelvic cavity by piercing the sacrotuberous ligament. It then loops around the lower border of the gluteus maximus to supply the skin over the medial aspect of the gluteus maximus (Fig. 6.48).

## In the clinic

### Intramuscular injections

From time to time it is necessary to administer drugs intramuscularly, that is, by direct injection into muscles. This procedure must be carried out without injuring neurovascular structures. A typical site for an intramuscular injection is the gluteal region. The sciatic nerve passes through this region and needs to be avoided. The safest place to inject is the upper outer quadrant of either gluteal region.

The gluteal region can be divided into quadrants by two imaginary lines positioned using palpable bony landmarks (Fig. 6.49). One line descends vertically from the highest point of the iliac crest. Another line is horizontal and passes through the first line midway between the highest point of the iliac crest and the horizontal plane through the ischial tuberosity.

It is important to remember that the gluteal region extends as far forward as the anterior superior iliac spine. The sciatic nerve curves through the upper lateral corner of the lower medial quadrant and descends along the medial margin of the lower lateral quadrant.

Occasionally, the sciatic nerve bifurcates into its tibial and common fibular branches in the pelvis, in which case the common fibular nerve passes into the gluteal region through, or even above, the piriformis muscle.

The superior gluteal nerve and vessels normally enter the gluteal region above the piriformis and pass superiorly and forward.

The anterior corner of the upper lateral quadrant is normally used for injections to avoid injuring any part of the sciatic nerve or other nerves and vessels in the gluteal region. A needle placed in this region enters the gluteus medius anterosuperior to the margin of the gluteus maximus.

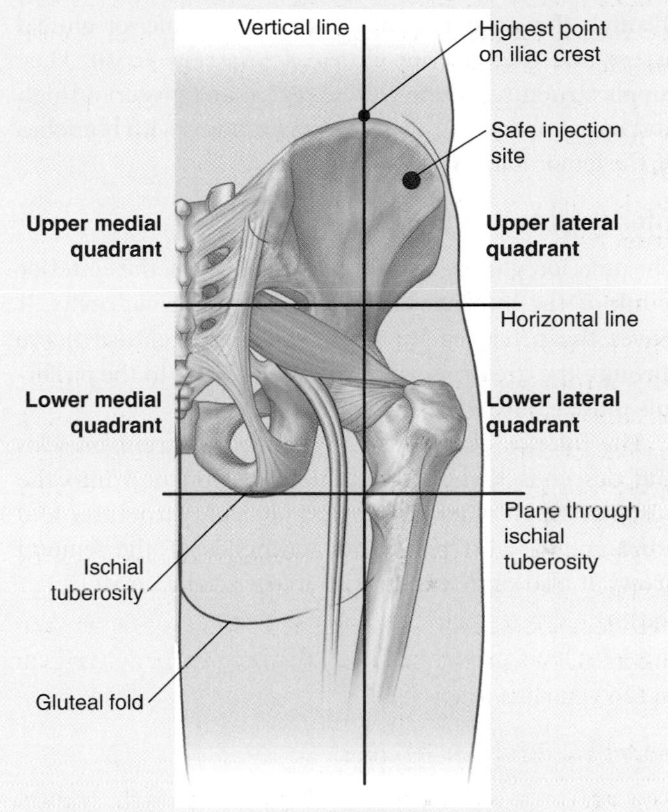


**Fig. 6.49** Site for intramuscular injections in the gluteal region.

## Arteries

Two arteries enter the gluteal region from the pelvic cavity through the greater sciatic foramen, the inferior gluteal artery and the superior gluteal artery (Fig. 6.50). They supply structures in the gluteal region and posterior thigh and have important collateral anastomoses with branches of the femoral artery.

### Inferior gluteal artery

The inferior gluteal artery originates from the anterior trunk of the internal iliac artery in the pelvic cavity. It leaves the pelvic cavity with the inferior gluteal nerve through the greater sciatic foramen inferior to the piriformis muscle (Fig. 6.50).

The inferior gluteal artery supplies adjacent muscles and descends through the gluteal region and into the posterior thigh where it supplies adjacent structures and anastomoses with perforating branches of the femoral artery. It also supplies a branch to the sciatic nerve.

### Superior gluteal artery

The superior gluteal artery originates from the posterior trunk of the internal iliac artery in the pelvic cavity. It leaves the pelvic cavity with the superior gluteal nerve through the greater sciatic foramen above the piriformis muscle (Fig. 6.50). In the gluteal region, it divides into a superficial branch and a deep branch:

- The superficial branch passes onto the deep surface of the gluteus maximus muscle.
- The deep branch passes between the gluteus medius and minimus muscles.

In addition to adjacent muscles, the superior gluteal artery contributes to the supply of the hip joint. Branches of the artery also anastomose with the lateral and medial femoral circumflex arteries from the deep femoral artery in the thigh, and with the inferior gluteal artery (Fig. 6.51).

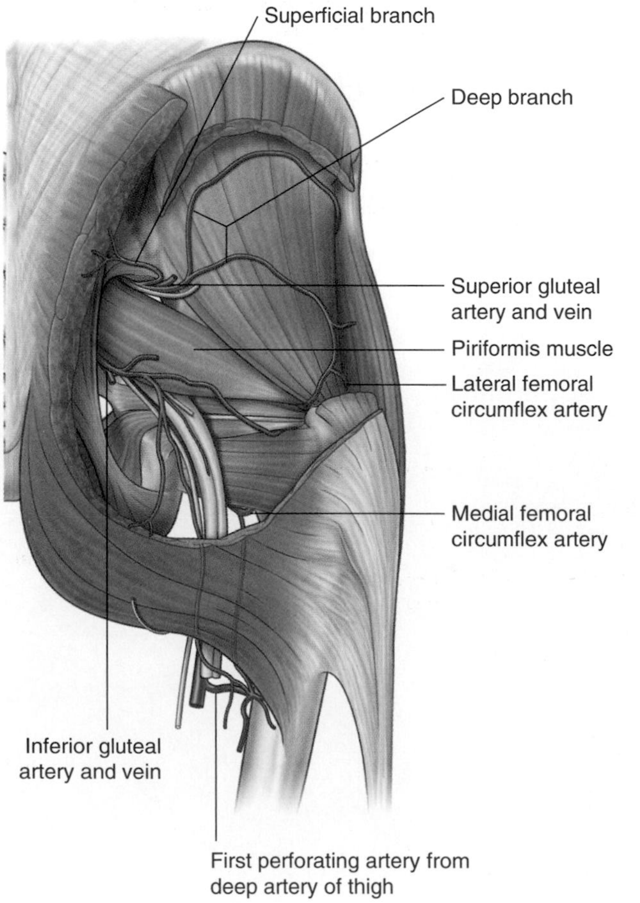

**Fig. 6.50** Arteries of the gluteal region.

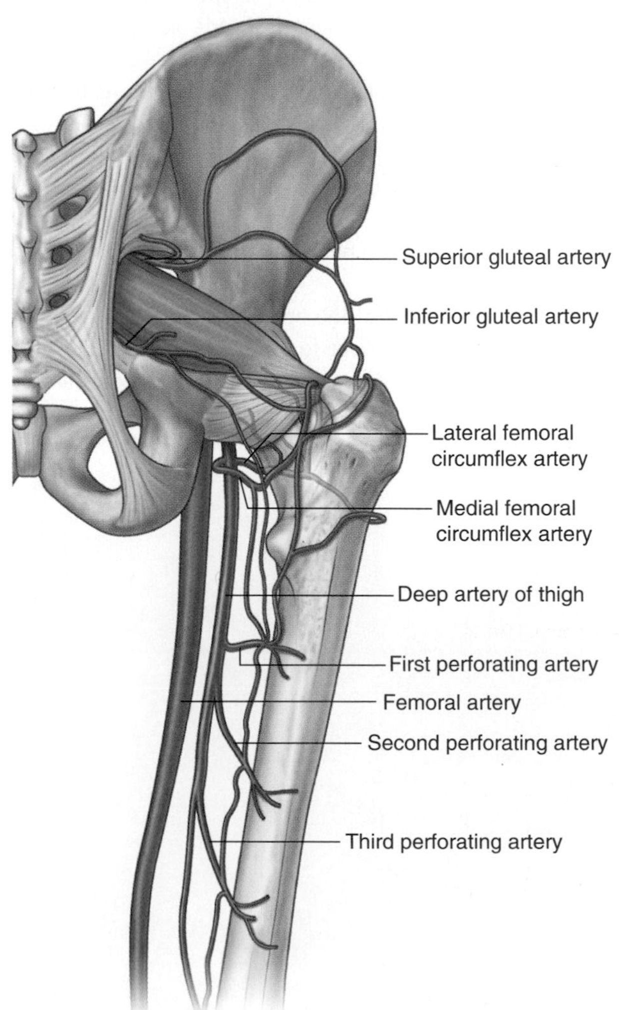


**Fig. 6.51** Anastomoses between gluteal arteries and vessels originating from the femoral artery in the thigh. Posterior view.

### Veins

Inferior and superior gluteal veins follow the inferior and superior gluteal arteries into the pelvis where they join the pelvic plexus of veins. Peripherally, the veins anastomose with superficial gluteal veins, which ultimately drain anteriorly into the femoral vein.

### Lymphatics

Deep lymphatic vessels of the gluteal region accompany the blood vessels into the pelvic cavity and connect with internal iliac nodes.

Superficial lymphatics drain into the superficial inguinal nodes on the anterior aspect of the thigh.

## THIGH

The thigh is the region of the lower limb that is approximately between the hip and knee joints (Fig. 6.52):

- Anteriorly, it is separated from the abdominal wall by the inguinal ligament.
- Posteriorly, it is separated from the gluteal region by the gluteal fold superficially, and by the inferior margins of the gluteus maximus and quadratus femoris on deeper planes.

Structures enter and leave the top of the thigh by three routes:

- Posteriorly, the thigh is continuous with the gluteal region and the major structure passing between the two regions is the sciatic nerve.
- Anteriorly, the thigh communicates with the abdominal cavity through the aperture between the inguinal ligament and pelvic bone, and major structures passing through this aperture are the iliopsoas and pectineus muscles; the femoral nerve, artery, and vein; and lymphatic vessels.

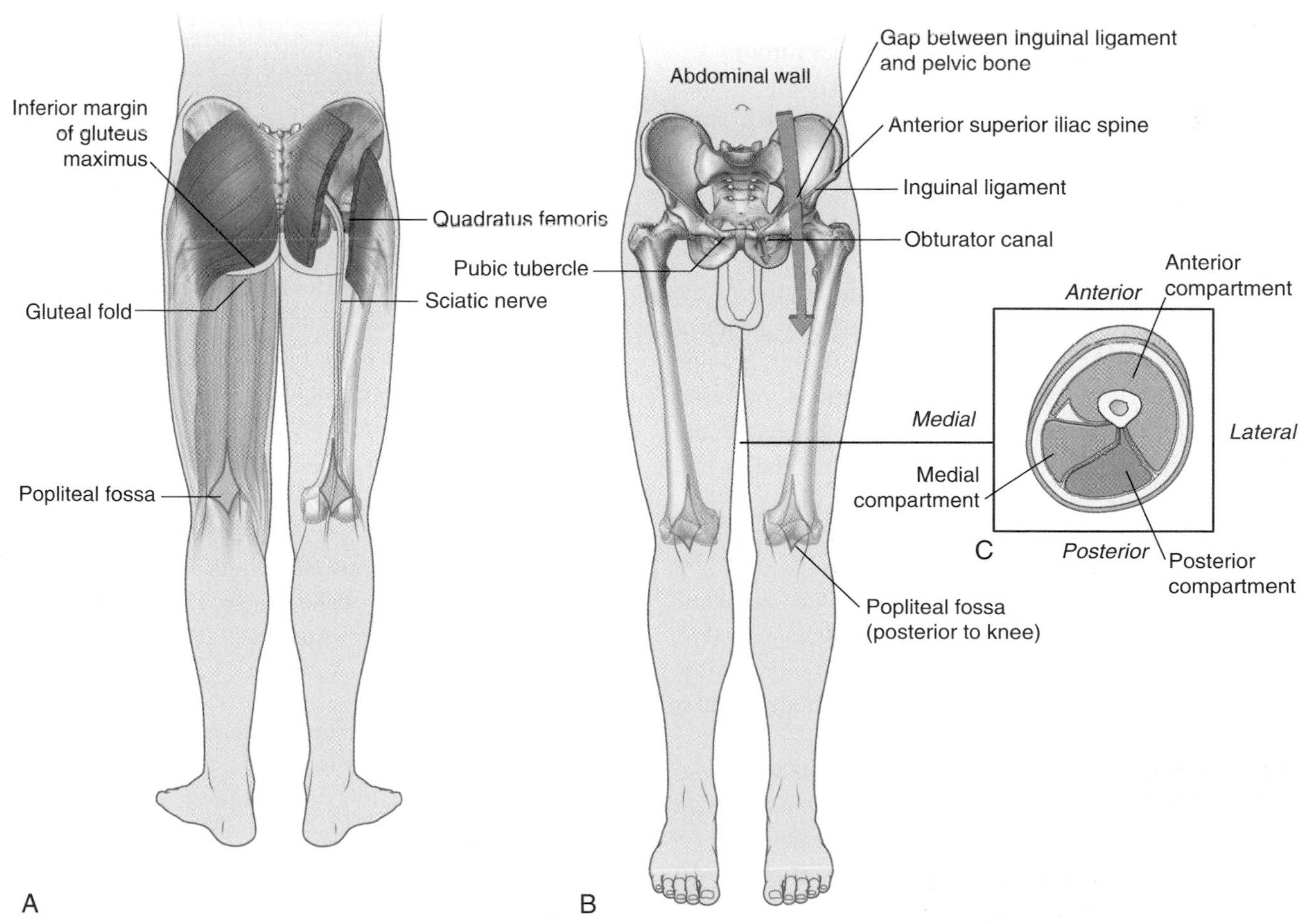


**Fig. 6.52** Thigh. **A.** Posterior view. **B.** Anterior view. **C.** Cross section through the midthigh.

- Medially, structures (including the obturator nerve and associated vessels) pass between the thigh and pelvic cavity through the obturator canal.

The thigh is divided into three compartments by intermuscular septa between the posterior aspect of the femur and the fascia lata (the thick layer of deep fascia that completely surrounds or invests the thigh; Fig. 6.52C):

- The **anterior compartment of the thigh** contains muscles that mainly extend the leg at the knee joint.
- The **posterior compartment of the thigh** contains muscles that mainly extend the thigh at the hip joint and flex the leg at the knee joint.
- The **medial compartment of the thigh** consists of muscles that mainly adduct the thigh at the hip joint.

The sciatic nerve innervates muscles in the posterior compartment of the thigh, the femoral nerve innervates muscles in the anterior compartment of the thigh, and the obturator nerve innervates most muscles in the medial compartment of the thigh.

The major artery, vein, and lymphatic channels enter the thigh anterior to the pelvic bone and pass through the femoral triangle inferior to the inguinal ligament. Vessels and nerves passing between the thigh and leg pass through the popliteal fossa posterior to the knee joint.

## Bones

The skeletal support for the thigh is the femur. Most of the large muscles in the thigh insert into the proximal ends of the two bones of the leg (tibia and fibula) and flex and extend the leg at the knee joint. The distal end of the femur provides origin for the gastrocnemius muscles, which are predominantly in the posterior compartment of the leg and plantarflex the foot.

### Shaft and distal end of femur

The shaft of the femur is bowed forward and has an oblique course from the neck of the femur to the distal end (Fig. 6.53). As a consequence of this oblique orientation, the knee is close to the midline under the body's center of gravity.

The middle part of the shaft of the femur is triangular in cross section (Fig. 6.53D). In the middle part of the shaft, the femur has smooth medial (posteromedial), lateral (posterolateral), and anterior surfaces and medial, lateral, and posterior borders. The medial and lateral borders are rounded, whereas the posterior border forms a broad roughened crest—the **linea aspera**.

In proximal and distal regions of the femur, the linea aspera widens to form an additional posterior surface. At the distal end of the femur, this posterior surface forms the floor of the popliteal fossa, and its margins form the **medial** and **lateral supracondylar lines**. The medial supracondylar line terminates at a prominent tubercle (the **adductor tubercle**) on the superior aspect of the **medial condyle** of the distal end. Just lateral to the lower end of the medial supracondylar line is an elongate roughened area of bone for the proximal attachment of the medial head of the gastrocnemius muscle (Fig. 6.52).

The distal end of the femur is characterized by two large condyles, which articulate with the proximal head of the tibia. The condyles are separated posteriorly by an **intercondylar fossa** and are joined anteriorly where they articulate with the patella.

The surfaces of the condyles that articulate with the tibia are rounded posteriorly and become flatter inferiorly. On each condyle, a shallow oblique groove separates the surface that articulates with the tibia from the more anterior surface that articulates with the patella. The surfaces of the medial and lateral condyles that articulate with the patella form a V-shaped trench, which faces anteriorly. The lateral surface of the trench is larger and steeper than the medial surface.

The walls of the intercondylar fossa bear two facets for the superior attachment of the cruciate ligaments, which stabilize the knee joint (Fig. 6.53):

- The wall formed by the lateral surface of the medial condyle has a large oval facet, which covers most of the inferior half of the wall, for attachment of the proximal end of the **posterior cruciate ligament**.
- The wall formed by the medial surface of the lateral condyle has a posterosuperior smaller oval facet for attachment of the proximal end of the **anterior cruciate ligament**.

Epicondyles, for the attachment of collateral ligaments of the knee joint, are bony elevations on the nonarticular outer surfaces of the condyles (Fig. 6.53). Two facets separated by a groove are just posterior to the **lateral epicondyle**:

- The upper facet is for attachment of the lateral head of the gastrocnemius muscle.
- The inferior facet is for attachment of the popliteus muscle.

The tendon of the popliteus muscle lies in the groove separating the two facets.

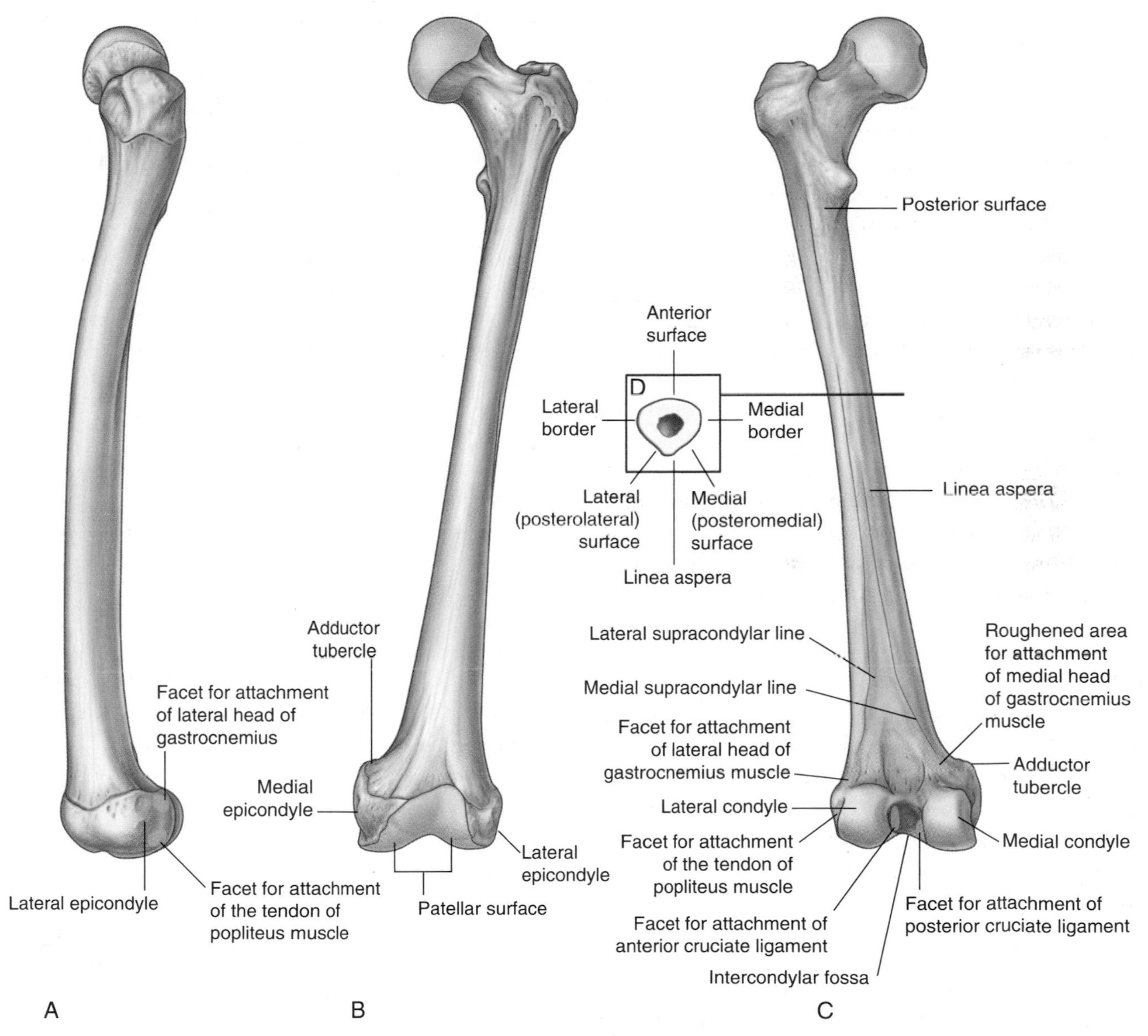


**Fig. 6.53** Shaft and distal end of femur. **A.** Lateral view. **B.** Anterior view. **C.** Posterior view. **D.** Cross section through shaft of femur.

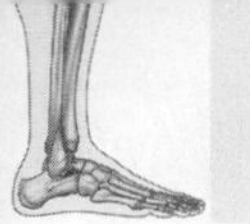


The **medial epicondyle** is a rounded eminence on the medial surface of the medial condyle. Just posterosuperior to the medial epicondyle is the adductor tubercle.

### Patella

The patella (knee cap) is the largest sesamoid bone (a bone formed within the tendon of a muscle) in the body and is formed within the tendon of the quadriceps femoris muscle as it crosses anterior to the knee joint to insert on the tibia.

The patella is triangular:

- Its apex is pointed inferiorly for attachment to the patellar ligament, which connects the patella to the tibia (Fig. 6.54).
- Its base is broad and thick for the attachment of the quadriceps tendon from above.
- Its posterior surface articulates with the femur and has medial and lateral facets, which slope away from a raised smooth ridge—the lateral facet is larger than the medial facet for articulation with the larger corresponding surface on the lateral condyle of the femur.

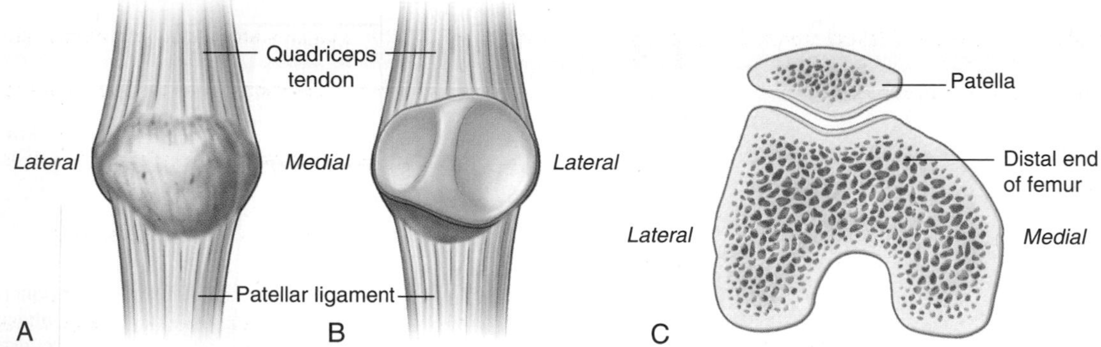


**Fig. 6.54** Patella. **A.** Anterior view. **B.** Posterior view. **C.** Superior view.

### Proximal end of tibia

The tibia is the medial and larger of the two bones in the leg, and is the only one that articulates with the femur at the knee joint.

The proximal end of the tibia is expanded in the transverse plane for weight-bearing and consists of a **medial condyle** and a **lateral condyle**, which are both flattened in the horizontal plane and overhang the shaft (Fig. 6.55).

The superior surfaces of the medial and lateral condyles are articular and separated by an intercondylar region, which contains sites of attachment for strong ligaments (cruciate ligaments) and interarticular cartilages (menisci) of the knee joint.

The articular surfaces of the medial and lateral condyles and the intercondylar region together form a "tibial plateau," which articulates with and is anchored to the distal end of the femur. Inferior to the condyles on the proximal part of the shaft is a large **tibial tuberosity** and roughenings for muscle and ligament attachments.

#### Tibial condyles and intercondylar areas

The tibial condyles are thick horizontal discs of bone attached to the top of the tibial shaft (Fig. 6.55).

The medial condyle is larger than the lateral condyle and is better supported over the shaft of the tibia. Its superior surface is oval for articulation with the medial condyle

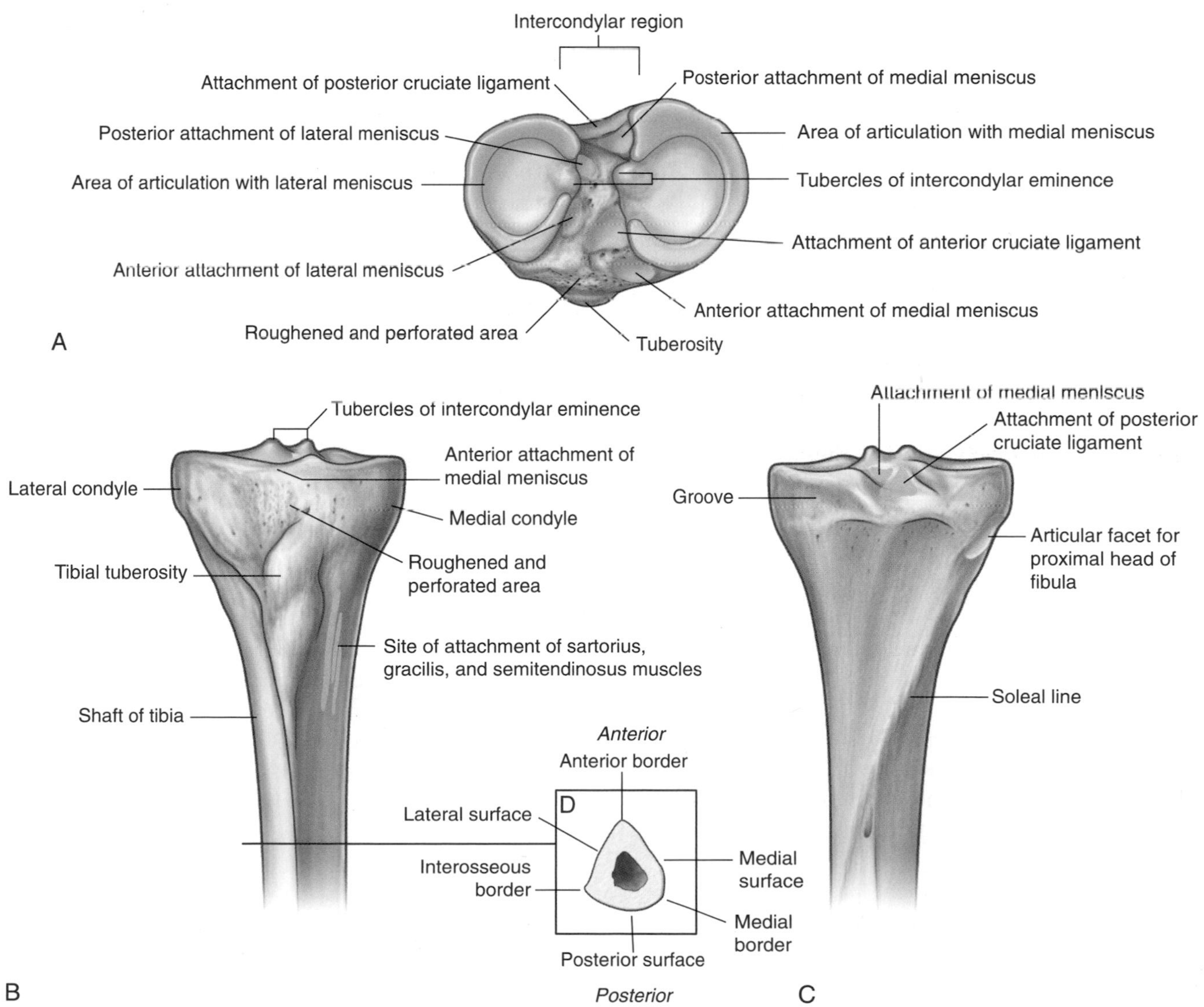


**Fig. 6.55** Proximal end of the tibia. **A.** Superior view, tibial plateau. **B.** Anterior view. **C.** Posterior view. **D.** Cross section through the shaft of tibia.

of the femur. The articular surface extends laterally onto the side of the raised **medial intercondylar tubercle**.

The superior surface of the lateral condyle is circular and articulates above with the lateral condyle of the femur. The medial edge of this surface extends onto the side of the **lateral intercondylar tubercle**.

The superior articular surfaces of both the lateral and medial condyles are concave, particularly centrally. The outer margins of the surfaces are flatter and are the regions in contact with the interarticular discs (menisci) of fibrocartilage in the knee joint.

The nonarticular posterior surface of the medial condyle bears a distinct horizontal groove for part of the attachment of the semimembranosus muscle, and the undersurface of the lateral condyle bears a distinct circular facet for articulation with the proximal head of the fibula.

The intercondylar region of the tibial plateau lies between the articular surfaces of the medial and lateral condyles (Fig. 6.55). It is narrow centrally where it is raised to form the **intercondylar eminence**, the sides of which are elevated further to form medial and lateral intercondylar tubercles.

The intercondylar region bears six distinct facets for the attachment of menisci and cruciate ligaments. The anterior intercondylar area widens anteriorly and bears three facets:

- The most anterior facet is for attachment of the anterior end (horn) of the medial meniscus.
- Immediately posterior to the most anterior facet is a facet for the attachment of the anterior cruciate ligament.
- A small facet for the attachment of the anterior end (horn) of the lateral meniscus is just lateral to the site of attachment of the anterior cruciate ligament.

The posterior intercondylar area also bears three attachment facets:

- The most anterior is for attachment of the posterior horn of the lateral meniscus.
- Posteromedial to the most anterior facet is the site of attachment for the posterior horn of the medial meniscus.
- Behind the site of attachment for the posterior horn of the medial meniscus is a large facet for the attachment of the posterior cruciate ligament.

In addition to these six sites of attachment for menisci and cruciate ligaments, a large anterolateral region of the anterior intercondylar area is roughened and perforated by numerous small nutrient foramina for blood vessels. This region is continuous with a similar surface on the front of the tibia above the tuberosity and lies against infrapatellar connective tissue.

### Tibial tuberosity

The **tibial tuberosity** is a palpable inverted triangular area on the anterior aspect of the tibia below the site of junction between the two condyles (Fig. 6.55). It is the site of attachment for the **patellar ligament**, which is a continuation of the quadriceps femoris tendon below the patella.

### Shaft of tibia

The shaft of the tibia is triangular in cross section and has three surfaces (posterior, medial, and lateral) and three borders (anterior, interosseous, and medial) (Fig. 6.55D):

- The **anterior border** is sharp and descends from the tibial tuberosity where it is continuous superiorly with a ridge that passes along the lateral margin of the tuberosity and onto the lateral condyle.
- The **interosseous border** is a subtle vertical ridge that descends along the lateral aspect of the tibia from the region of bone anterior and inferior to the articular facet for the head of the fibula.
- The medial border is indistinct superiorly where it begins at the anterior end of the groove on the posterior surface of the medial tibial condyle, but is sharp in midshaft.

The large **medial surface** of the shaft of the tibia, between the anterior and medial borders, is smooth and subcutaneous, and is palpable along almost its entire extent. Medial and somewhat inferior to the tibial tuberosity, this medial surface bears a subtle, slightly roughened elongate elevation. This elevation is the site of the combined attachment of three muscles (sartorius, gracilis, and semitendinosus), which descend from the thigh.

The **posterior surface** of the shaft of the tibia, between the interosseous and medial borders, is widest superiorly where it is crossed by a roughened oblique line (the **soleal line**).

The **lateral surface**, between the anterior and interosseous borders, is smooth and unremarkable.

### Proximal end of fibula

The fibula is the lateral bone of the leg and does not take part in formation of the knee joint or in weight-bearing. It is much smaller than the tibia and has a small proximal head, a narrow neck, and a delicate shaft, which ends as the lateral malleolus at the ankle.

The **head** of the fibula is a globe-shaped expansion at the proximal end of the fibula (Fig. 6.56). A circular facet on the superomedial surface is for articulation above with a similar facet on the inferior aspect of the lateral condyle of the tibia. Just posterolateral to this facet, the bone projects superiorly as a blunt apex (styloid process).

The lateral surface of the head of the fibula bears a large impression for the attachment of the biceps femoris muscle. A depression near the upper margin of this impression is for attachment of the fibular collateral ligament of the knee joint.

The **neck** of the fibula separates the expanded head from the **shaft**. The common fibular nerve lies against the posterolateral aspect of the neck.

Like the tibia, the shaft of the fibula has three borders (anterior, posterior, and interosseous) and three surfaces (lateral, posterior, and medial), which lie between the borders (Fig. 6.56):

- The **anterior border** is sharp midshaft and begins superiorly from the anterior aspect of the head.
- The **posterior border** is rounded and descends from the region of the styloid process of the head.
- The **interosseous border** is medial in position.

The three surfaces of the fibula are associated with the three muscular compartments (lateral, posterior, and anterior) of the leg.

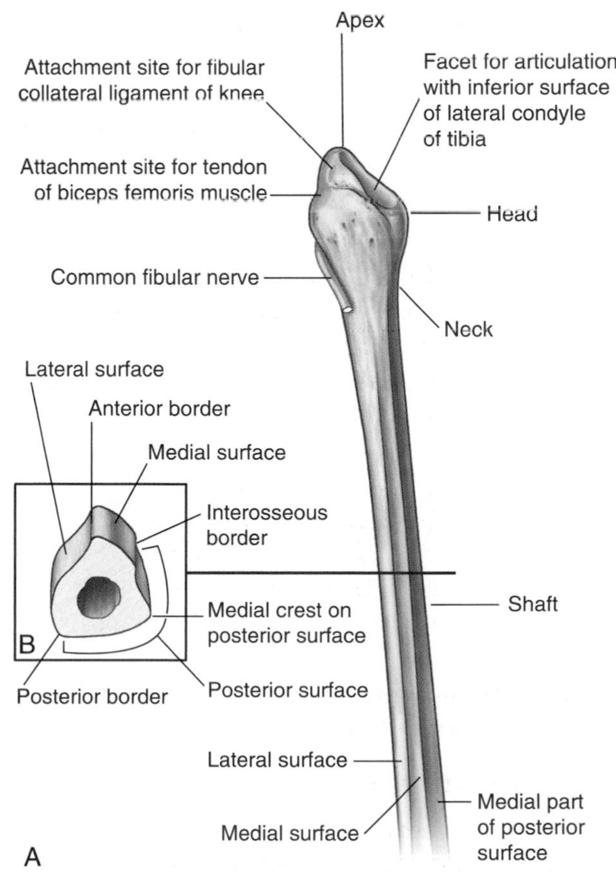


**Fig. 6.56** Proximal end of the fibula. **A.** Anterior view. **B.** Cross section through the shaft of fibula.

## Muscles

Muscles of the thigh are arranged in three compartments separated by intermuscular septa (Fig. 6.57).

The **anterior compartment of the thigh** contains the sartorius and the four large quadriceps femoris muscles (rectus femoris, vastus lateralis, vastus medialis, and vastus intermedius). All are innervated by the femoral nerve. In addition, the terminal ends of the psoas major and iliacus muscles pass into the upper part of the anterior compartment from sites of origin on the posterior abdominal wall. These muscles are innervated by branches directly from the anterior rami of L1 to L3 (psoas major) or from the femoral nerve (iliacus) as it passes down the abdominal wall.

The **medial compartment of the thigh** contains six muscles (gracilis, pectineus, adductor longus, adductor brevis, adductor magnus, and obturator externus). All except the pectineus, which is innervated by the femoral nerve, and part of the adductor magnus, which is innervated by the sciatic nerve, are innervated by the obturator nerve.

The **posterior compartment of the thigh** contains three large muscles termed the "hamstrings." All are innervated by the sciatic nerve.

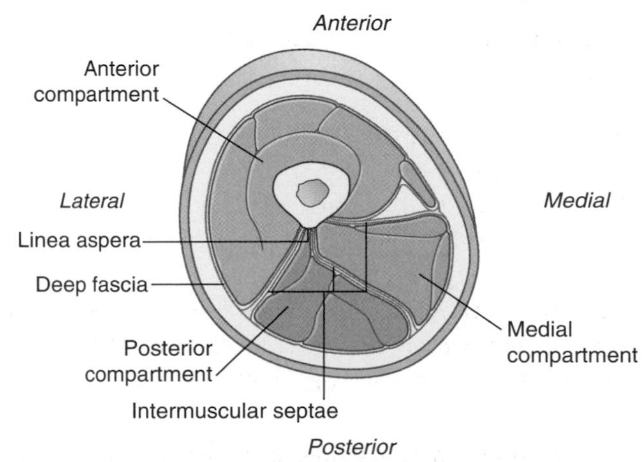


**Fig. 6.57** Transverse section through the midthigh.

## In the clinic

### Compartment syndrome

Compartment syndrome occurs when there is swelling within a fascial enclosed muscle compartment in the limbs. Typical causes include limb trauma, intracompartment hemorrhage, and limb compression. As pressure within the compartment elevates, capillary blood flow and tissue perfusion is compromised, which can ultimately lead to neuromuscular damage if not treated.

### Anterior compartment

Muscles in the anterior compartment (Table 6.3) act on the hip and knee joints:

- the psoas major and iliacus act on the hip joint,
- the sartorius and rectus femoris act on both the hip and knee joints, and
- the vastus muscles act on the knee joint.

**Table 6.3** Muscles of the anterior compartment of thigh (spinal segments in bold are the major segments innervating the muscle)

| Muscle | Origin | Insertion | Innervation | Function |
|---|---|---|---|---|
| Psoas major | Posterior abdominal wall (lumbar transverse processes, intervertebral discs, and adjacent bodies from TXII to LV and tendinous arches between these points) | Lesser trochanter of femur | Anterior rami (**L1, L2**, L3) | Flexes the thigh at the hip joint |
| Iliacus | Posterior abdominal wall (iliac fossa) | Lesser trochanter of femur | Femoral nerve (**L2**, L3) | Flexes the thigh at the hip joint |
| Vastus medialis | Femur—medial part of intertrochanteric line, pectineal line, medial lip of the linea aspera, medial supracondylar line | Quadriceps femoris tendon and medial border of patella | Femoral nerve (L2, **L3, L4**) | Extends the leg at the knee joint |
| Vastus intermedius | Femur—upper two-thirds of anterior and lateral surfaces | Quadriceps femoris tendon, lateral margin of patella, and lateral condyle of tibia | Femoral nerve (L2, **L3, L4**) | Extends the leg at the knee joint |
| Vastus lateralis | Femur—lateral part of intertrochanteric line, margin of greater trochanter, lateral margin of gluteal tuberosity, lateral lip of the linea aspera | Quadriceps femoris tendon and lateral margin of patella | Femoral nerve (L2, **L3, L4**) | Extends the leg at the knee joint |
| Rectus femoris | Straight head originates from the anterior inferior iliac spine; reflected head originates from the ilium just superior to the acetabulum | Quadriceps femoris tendon | Femoral nerve (L2, **L3, L4**) | Flexes the thigh at the hip joint and extends the leg at the knee joint |
| Sartorius | Anterior superior iliac spine | Medial surface of tibia just inferomedial to tibial tuberosity | Femoral nerve (**L2, L3**) | Flexes the thigh at the hip joint and flexes the leg at the knee joint |

### Iliopsoas—psoas major and iliacus

The **psoas major** and **iliacus** muscles originate on the posterior abdominal wall and descend into the upper part of the anterior compartment of the thigh through the lateral half of the gap between the inguinal ligament and the pelvic bone (Fig. 6.58).

Although the iliacus and psoas major originate as separate muscles in the abdomen, both insert by a common tendon onto the lesser trochanter of the femur and together are usually referred to as the **iliopsoas** muscle.

The iliopsoas is a powerful flexor of the thigh at the hip joint and can also contribute to lateral rotation of the thigh. The psoas major is innervated by branches from the anterior rami of L1 to L3 and the iliacus is innervated by branches from the femoral nerve in the abdomen.

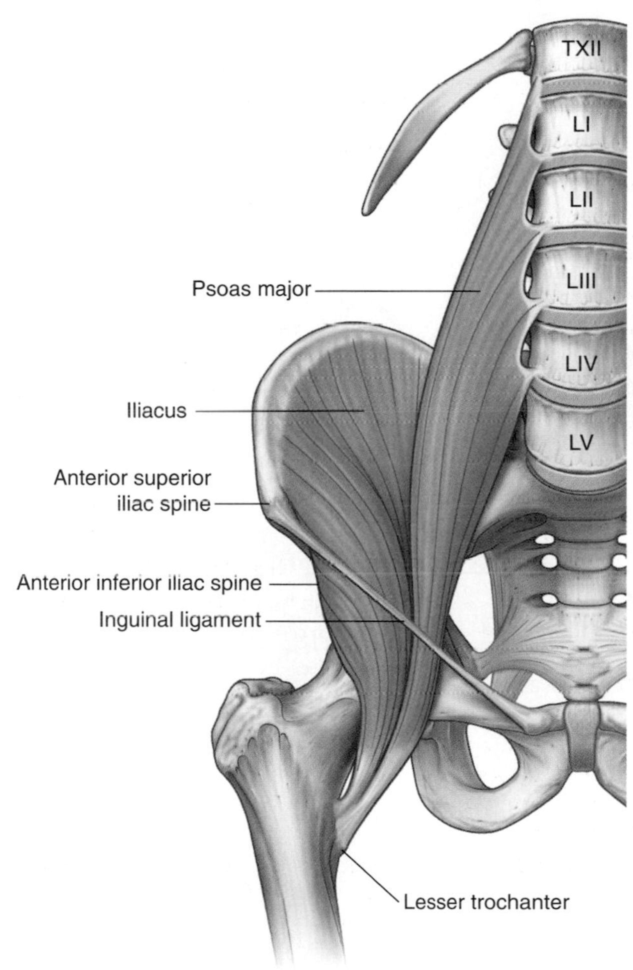


**Fig. 6.58** Psoas major and iliacus muscles.

### Quadriceps femoris—vastus medialis, intermedius, and lateralis and rectus femoris

The large **quadriceps femoris** muscle consists of three vastus muscles (vastus medialis, vastus intermedius, and vastus lateralis) and the rectus femoris muscle (Fig. 6.59).

The quadriceps femoris muscle mainly extends the leg at the knee joint, but the rectus femoris component also assists flexion of the thigh at the hip joint. Because the vastus muscles insert into the margins of the patella as well as into the quadriceps femoris tendon, they stabilize the position of the patella during knee joint movement.

The quadriceps femoris is innervated by the femoral nerve with contributions mainly from spinal segments L3 and L4. A tap with a tendon hammer on the patellar ligament therefore tests reflex activity mainly at spinal cord levels L3 and L4.

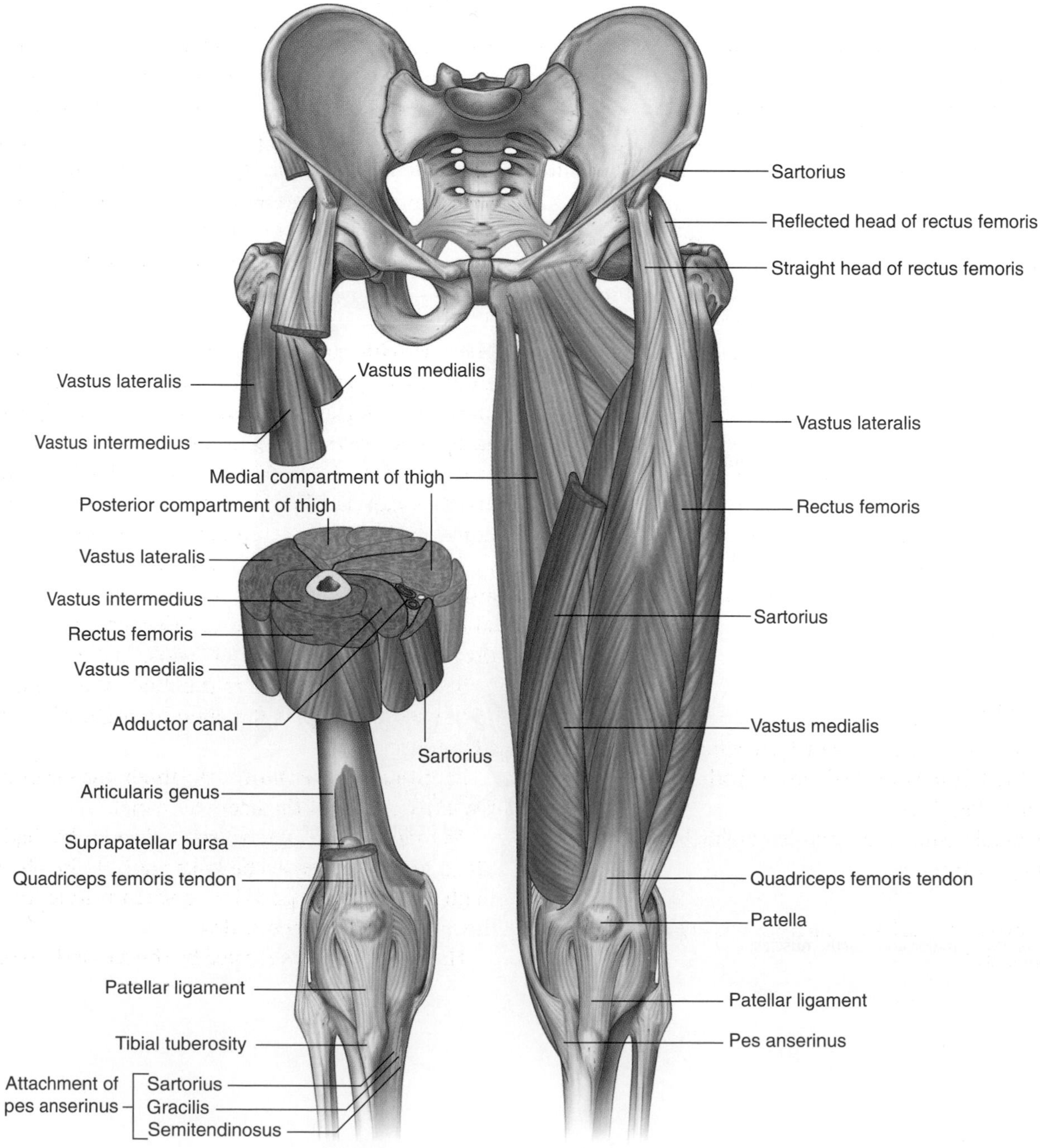

**Fig. 6.59** Muscles of the anterior compartment of thigh.

### Vastus muscles

The vastus muscles originate from the femur, whereas the rectus femoris muscle originates from the pelvic bone. All attach first to the patella by the quadriceps femoris tendon and then to the tibia by the **patellar ligament**.

The **vastus medialis** originates from a continuous line of attachment on the femur, which begins anteromedially on the intertrochanteric line and continues posteroinferiorly along the pectineal line and then descends along the medial lip of the linea aspera and onto the medial supracondylar line. The fibers converge onto the medial aspect of the quadriceps femoris tendon and the medial border of the patella (Fig. 6.59).

The **vastus intermedius** originates mainly from the upper two-thirds of the anterior and lateral surfaces of the femur and the adjacent intermuscular septum (Fig. 6.59). It merges into the deep aspect of the quadriceps femoris tendon and also attaches to the lateral margin of the patella and lateral condyle of the tibia.

A tiny muscle (**articularis genus**) originates from the femur just inferior to the origin of the vastus intermedius and inserts into the suprapatellar bursa associated with the knee joint (Fig. 6.59). This articular muscle, which is often part of the vastus intermedius muscle, pulls the bursa away from the knee joint during extension.

The **vastus lateralis** is the largest of the vastus muscles (Fig. 6.59). It originates from a continuous line of attachment, which begins anterolaterally from the superior part of the intertrochanteric line of the femur and then circles laterally around the bone to attach to the lateral margin of the gluteal tuberosity and continues down the upper part of the lateral lip of the linea aspera. Muscle fibers converge mainly onto the quadriceps femoris tendon and the lateral margin of the patella.

### Rectus femoris

Unlike the vastus muscles, which cross only the knee joint, the **rectus femoris** muscle crosses both the hip and the knee joints (Fig. 6.59).

The rectus femoris has two tendinous heads of origin from the pelvic bone:

- one from the anterior inferior iliac spine (**straight head**), and
- the other from a roughened area of the ilium immediately superior to the acetabulum (**reflected head**) (Fig. 6.59).

The two heads of the rectus femoris unite to form an elongate muscle belly, which lies anterior to the vastus intermedius muscle and between the vastus lateralis and vastus medialis muscles, to which it is attached on either side. At the distal end, the rectus femoris muscle converges on the quadriceps femoris tendon and inserts on the base of the patella.

### Patellar ligament

The patellar ligament is functionally the continuation of the quadriceps femoris tendon below the patella and is attached above to the apex and margins of the patella and below to the tibial tuberosity (Fig. 6.59). The more superficial fibers of the quadriceps femoris tendon and the patellar ligament are continuous over the anterior surface of the patella, and lateral and medial fibers are continuous with the ligament beside the margins of the patella.

## Sartorius

The **sartorius** muscle is the most superficial muscle in the anterior compartment of the thigh and is a long strap-like muscle that descends obliquely through the thigh from the anterior superior iliac spine to the medial surface of the proximal shaft of the tibia (Fig. 6.59). Its flat aponeurotic insertion into the tibia is immediately anterior to the insertion of the gracilis and semitendinosus muscles.

The sartorius, gracilis, and semitendinosus muscles attach to the tibia in a three-pronged pattern on the tibia, so their combined tendons of insertion are often termed the **pes anserinus** (Latin for "goose foot").

In the upper one-third of the thigh, the medial margin of the sartorius forms the lateral margin of the femoral triangle.

In the middle one-third of the thigh, the sartorius forms the anterior wall of the adductor canal.

The sartorius muscle assists in flexing the thigh at the hip joint and the leg at the knee joint. It also abducts the thigh and rotates it laterally, as when resting the foot on the opposite knee when sitting.

The sartorius is innervated by the femoral nerve.

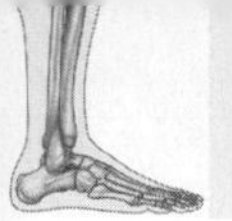


### Medial compartment

There are six muscles in the medial compartment of the thigh (Table 6.4): gracilis, pectineus, adductor longus, adductor brevis, adductor magnus, and obturator externus (Fig. 6.60). Collectively, all these muscles except the obturator externus mainly adduct the thigh at the hip joint; the adductor muscles may also medially rotate the thigh. Obturator externus is a lateral rotator of the thigh at the hip joint.

#### Gracilis

The **gracilis** is the most superficial of the muscles in the medial compartment of thigh and descends almost vertically down the medial side of the thigh (Fig. 6.60). It is attached above to the outer surface of the ischiopubic ramus of the pelvic bone and below to the medial surface of the proximal shaft of the tibia, where it lies sandwiched between the tendon of sartorius in front and the tendon of the semitendinosus behind.

**Table 6.4** Muscles of the medial compartment of thigh (spinal segments in bold are the major segments innervating the muscle)

| Muscle | Origin | Insertion | Innervation | Function |
|---|---|---|---|---|
| Gracilis | A line on the external surfaces of the body of the pubis, the inferior pubic ramus, and the ramus of the ischium | Medial surface of proximal shaft of tibia | Obturator nerve (**L2**, L3) | Adducts thigh at hip joint and flexes leg at knee joint |
| Pectineus | Pectineal line (pecten pubis) and adjacent bone of pelvis | Oblique line extending from base of lesser trochanter to linea aspera on posterior surface of proximal femur | Femoral nerve (**L2**, L3) | Adducts and flexes thigh at hip joint |
| Adductor longus | External surface of body of pubis (triangular depression inferior to pubic crest and lateral to pubic symphysis) | Linea aspera on middle one-third of shaft of femur | Obturator nerve (anterior division) (**L2**, **L3**, L4) | Adducts and medially rotates thigh at hip joint |
| Adductor brevis | External surface of body of pubis and inferior pubic ramus | Posterior surface of proximal femur and upper one-third of linea aspera | Obturator nerve (**L2**, **L3**) | Adducts and medially rotates thigh at hip joint |
| Adductor magnus | Adductor part—ischiopubic ramus | Posterior surface of proximal femur, linea aspera, medial supracondylar line | Obturator nerve (**L2**, **L3**, L4) | Adducts and medially rotates thigh at hip joint |
| | Hamstring part—ischial tuberosity | Adductor tubercle and supracondylar line | Sciatic nerve (tibial division) (**L2**, **L3**, L4) | |
| Obturator externus | External surface of obturator membrane and adjacent bone | Trochanteric fossa | Obturator nerve (posterior division) (L3, **L4**) | Laterally rotates thigh at hip joint |

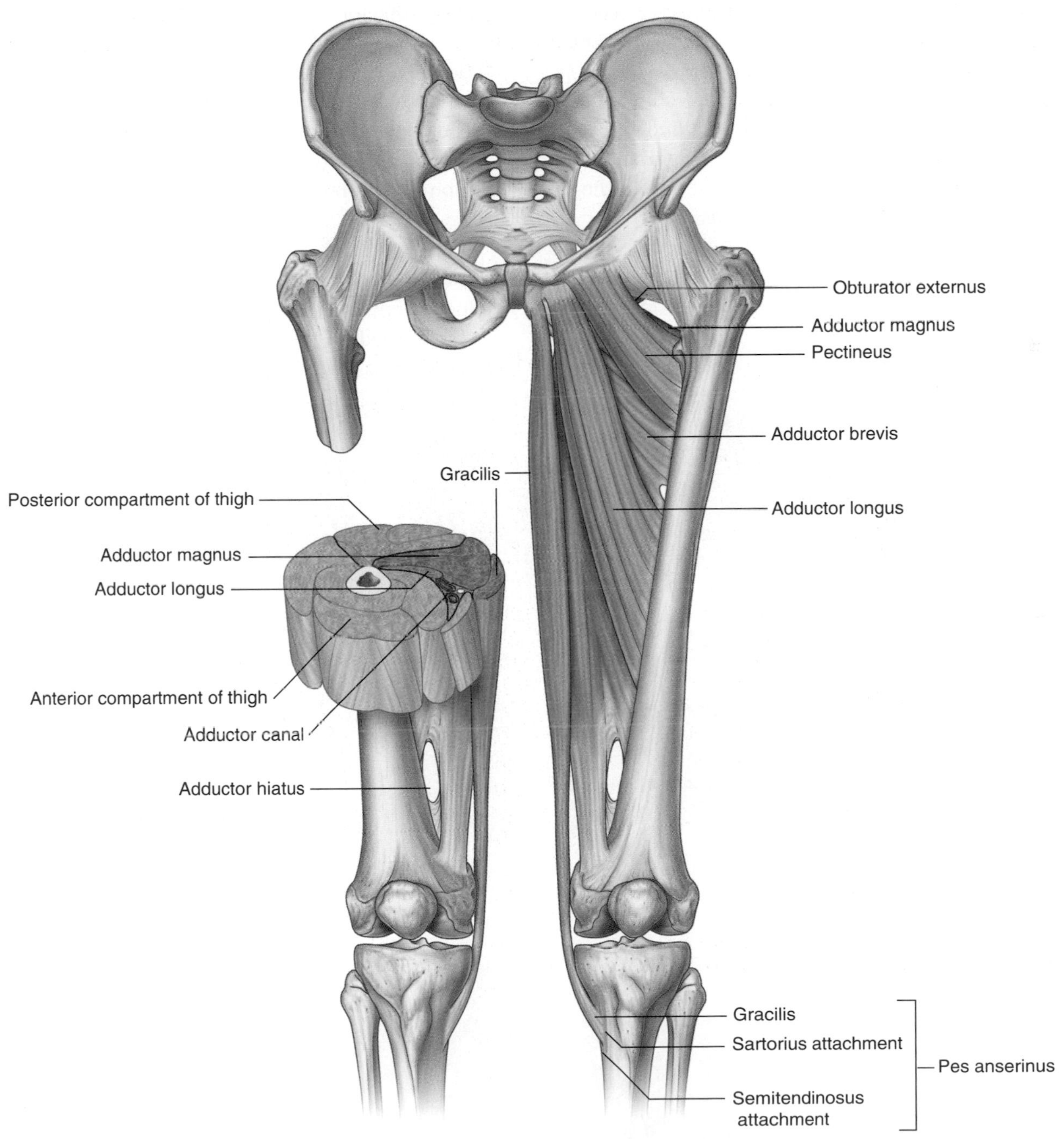


**Fig. 6.60** Muscles of the medial compartment of thigh. Anterior view.

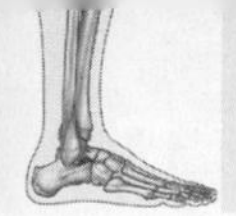



### Pectineus

The **pectineus** is a flat quadrangular muscle (Fig. 6.61). It is attached above to the pectineal line of the pelvic bone and adjacent bone, and descends laterally to attach to an oblique line extending from the base of the lesser trochanter to the linea aspera on the posterior surface of the proximal femur.

From its origin on the pelvic bone, the pectineus passes into the thigh below the inguinal ligament and forms part of the floor of the medial half of the femoral triangle.

The pectineus adducts and flexes the thigh at the hip joint and is innervated by the femoral nerve.

### Adductor longus

The **adductor longus** is a flat fan-shaped muscle that originates from a small rough triangular area on the external surface of the body of the pubis just inferior to the pubic crest and lateral to the pubic symphysis (Fig. 6.61). It expands as it descends posterolaterally to insert via an aponeurosis into the middle third of the linea aspera.

The adductor longus contributes to the floor of the femoral triangle, and its medial margin forms the medial border of the femoral triangle. The muscle also forms the proximal posterior wall of the adductor canal.

The adductor longus adducts and medially rotates the thigh at the hip joint and is innervated by the anterior division of the obturator nerve.

### Adductor brevis

The **adductor brevis** lies posterior to the pectineus and adductor longus. It is a triangular muscle attached at its apex to the body of the pubis and inferior pubic ramus just

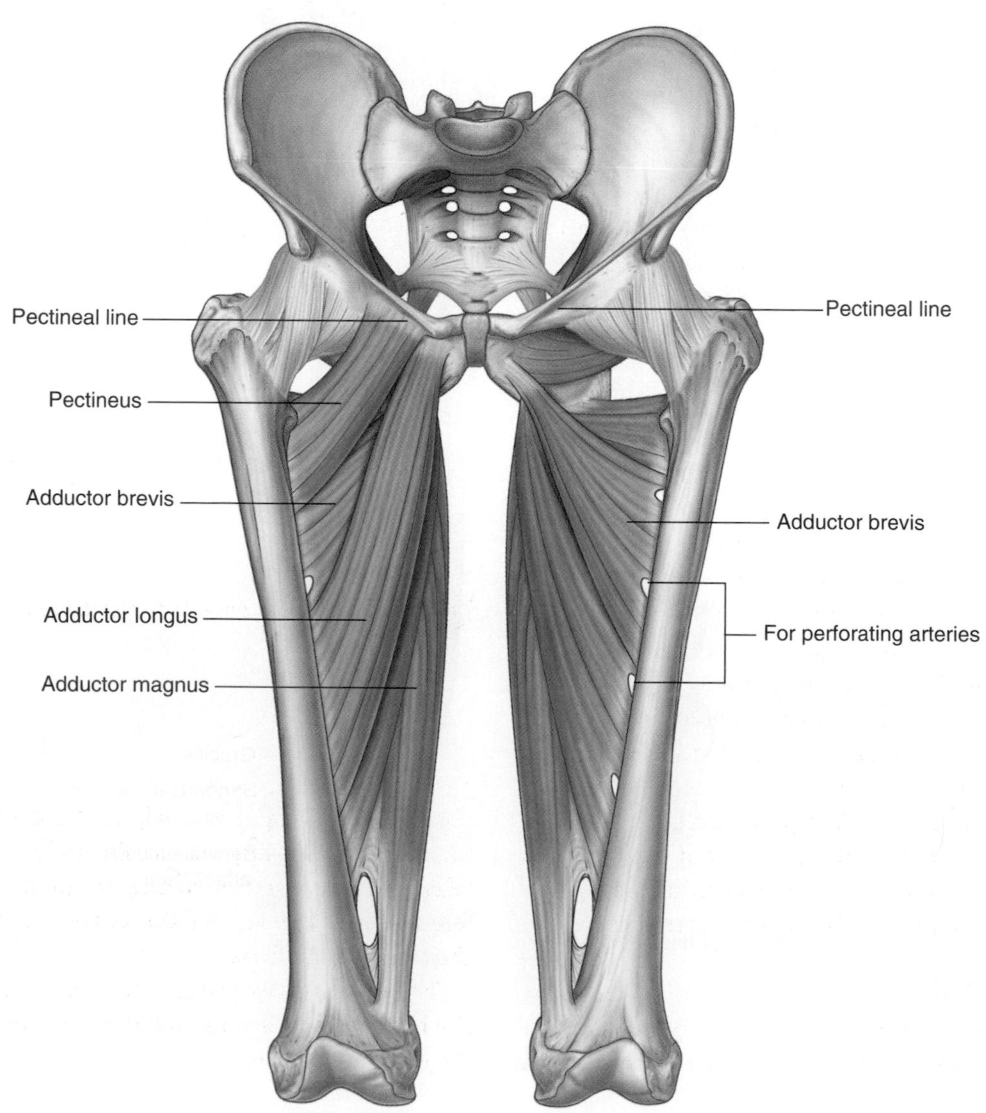


**Fig. 6.61** Pectineus, adductor longus, and adductor brevis muscles. Anterior view.

superior to the origin of the gracilis muscle (Fig. 6.61). The muscle is attached by its expanded base via an aponeurosis to a vertical line extending from lateral to the insertion of the pectineus into the upper aspect of the linea aspera lateral to the attachment of the adductor longus.

The adductor brevis adducts and medially rotates the thigh at the hip joint and is innervated by the obturator nerve.

### Adductor magnus

The **adductor magnus** is the largest and deepest of the muscles in the medial compartment of the thigh (Fig. 6.62). The muscle forms the distal posterior wall of the adductor canal. Like the adductor longus and brevis muscles, the adductor magnus is a triangular or fan-shaped muscle anchored by its apex to the pelvis and attached by its expanded base to the femur.

On the pelvis, the adductor magnus is attached along a line that extends from the inferior pubic ramus, above the attachments of the adductor longus and brevis muscles, and along the ramus of the ischium to the ischial tuberosity. The part of the muscle that originates from the ischiopubic ramus expands laterally and inferiorly to insert on the femur along a vertical line of attachment that extends from just inferior to the quadrate tubercle and medial to the gluteal tuberosity, along the linea aspera and onto the medial supracondylar line. This lateral part of the muscle is often termed the "adductor part" of the adductor magnus.

The medial part of the adductor magnus, often called the "hamstring part," originates from the ischial tuberosity of the pelvic bone and descends almost vertically along the thigh to insert via a rounded tendon into the adductor tubercle on the medial condyle of the distal head of the femur. It also inserts via an aponeurosis up onto the medial supracondylar line. A large circular gap inferiorly between the hamstring and adductor parts of the muscle is the **adductor hiatus** (Fig. 6.62), which allows the femoral artery and associated veins to pass between the adductor canal on the anteromedial aspect of the thigh and the popliteal fossa posterior to the knee.

The adductor magnus adducts and medially rotates the thigh at the hip joint. The adductor part of the muscle is innervated by the obturator nerve and the hamstring part is innervated by the tibial division of the sciatic nerve.

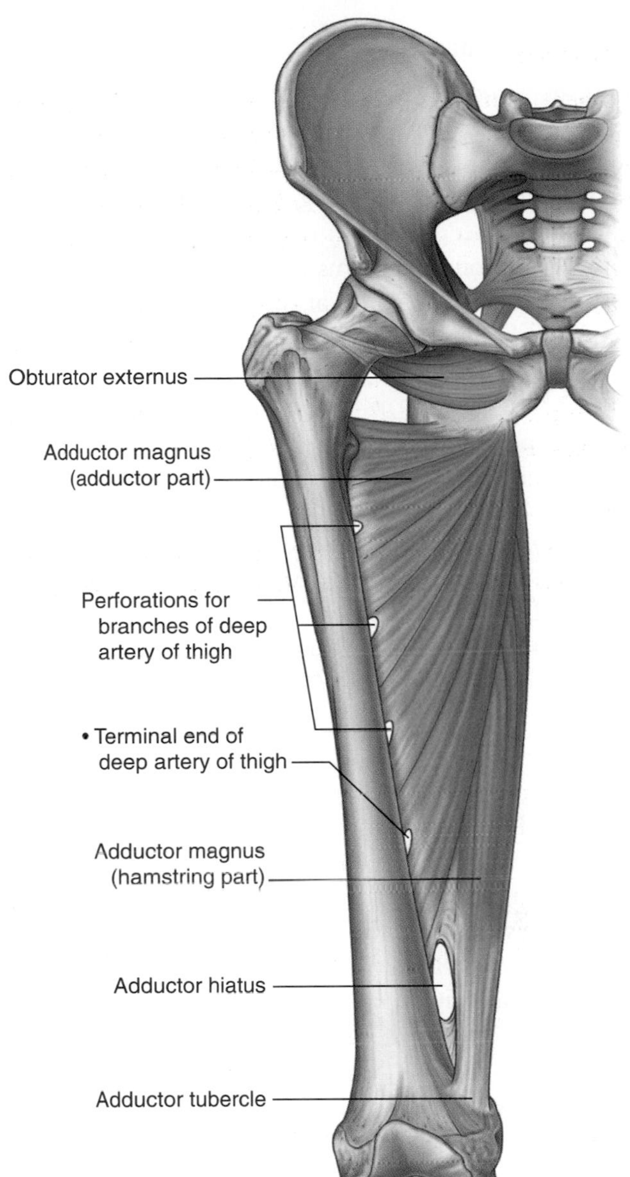


**Fig. 6.62** Adductor magnus and obturator externus muscles. Anterior view.

### Obturator externus

The **obturator externus** is a flat fan-shaped muscle. Its expansive body is attached to the external aspect of the obturator membrane and adjacent bone (Fig. 6.62). The muscle fibers converge posterolaterally to form a tendon, which passes posterior to the hip joint and neck of the femur to insert on an oval depression on the lateral wall of the trochanteric fossa.

The obturator externus externally rotates the thigh at the hip joint and is innervated by the posterior branch of the obturator nerve.

### Posterior compartment

There are three long muscles in the posterior compartment of the thigh: biceps femoris, semitendinosus, and semimembranosus (Table 6.5)—and they are collectively known as the hamstrings (Fig. 6.63). All except the short head of the biceps femoris cross both the hip and knee joints. As a group, the hamstrings flex the leg at the knee joint and extend the thigh at the hip joint. They are also rotators at both joints.

#### Biceps femoris

The **biceps femoris** muscle is lateral in the posterior compartment of the thigh and has two heads (Fig. 6.63):

- The **long head** originates with the semitendinosus muscle from the inferomedial part of the upper area of the ischial tuberosity.
- The **short head** arises from the lateral lip of the linea aspera on the shaft of the femur.

The muscle belly of the long head crosses the posterior thigh obliquely from medial to lateral and is joined by the short head distally. Together, fibers from the two heads form a tendon, which is palpable on the lateral side of the distal thigh. The main part of the tendon inserts into the lateral surface of the head of the fibula. Extensions from the tendon blend with the fibular collateral ligament and with ligaments associated with the lateral side of the knee joint.

The biceps femoris flexes the leg at the knee joint. The long head also extends and laterally rotates the hip. When the knee is partly flexed, the biceps femoris can laterally rotate the leg at the knee joint.

The long head is innervated by the tibial division of the sciatic nerve and the short head is innervated by the common fibular division of the sciatic nerve.

#### Semitendinosus

The **semitendinosus** muscle is medial to the biceps femoris muscle in the posterior compartment of the thigh (Fig. 6.63). It originates with the long head of the biceps femoris muscle from the inferomedial part of the upper area of the ischial tuberosity. The spindle-shaped muscle belly ends in the lower half of the thigh and forms a long cord-like tendon, which lies on the semimembranosus muscle and descends to the knee. The tendon curves around the medial condyle of the tibia and inserts into the medial surface of the tibia just posterior to the tendons of the gracilis and sartorius muscles as part of the pes anserinus.

The semitendinosus flexes the leg at the knee joint and extends the thigh at the hip joint. Working with the semimembranosus, it also medially rotates the thigh at the hip joint and medially rotates the leg at the knee joint.

The semitendinosus muscle is innervated by the tibial division of the sciatic nerve.

#### Semimembranosus

The **semimembranosus** muscle lies deep to the semitendinosus muscle in the posterior compartment of the thigh (Fig. 6.63). It is attached above to the superolateral impression on the ischial tuberosity and below mainly to the

**Table 6.5** Muscles of the posterior compartment of thigh (spinal segments in bold are the major segments innervating the muscle)

| Muscle | Origin | Insertion | Innervation | Function |
|---|---|---|---|---|
| Biceps femoris | Long head—inferomedial part of the upper area of the ischial tuberosity; short head—lateral lip of linea aspera | Head of fibula | Sciatic nerve (L5, **S1**, S2) | Flexes leg at knee joint; extends and laterally rotates thigh at hip joint and laterally rotates leg at knee joint |
| Semitendinosus | Inferomedial part of the upper area of the ischial tuberosity | Medial surface of proximal tibia | Sciatic nerve (L5, **S1**, S2) | Flexes leg at knee joint and extends thigh at hip joint; medially rotates thigh at hip joint and leg at knee joint |
| Semimembranosus | Superolateral impression on the ischial tuberosity | Groove and adjacent bone on medial and posterior surface of medial tibial condyle | Sciatic nerve (L5, **S1**, S2) | Flexes leg at knee joint and extends thigh at hip joint; medially rotates thigh at hip joint and leg at knee joint |

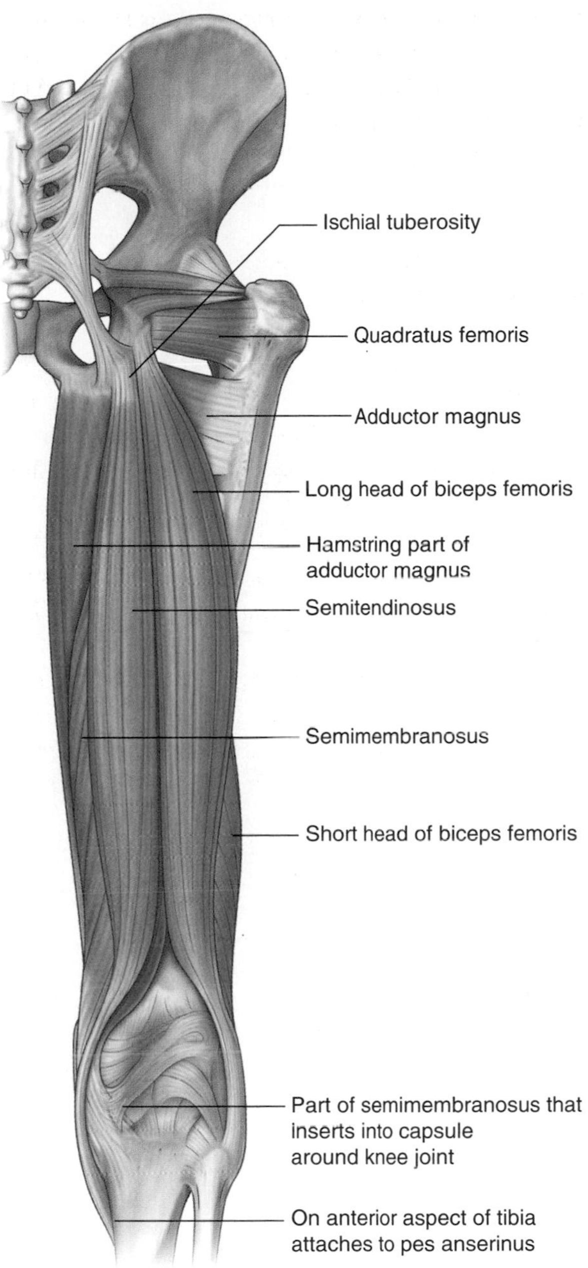


**Fig. 6.63** Muscles of the posterior compartment of thigh. Posterior view.

groove and adjacent bone on the medial and posterior surfaces of the medial tibial condyle. Expansions from the tendon also insert into and contribute to the formation of ligaments and fascia around the knee joint.

The semimembranosus flexes the leg at the knee joint and extends the thigh at the hip joint. Working with the semitendinosus muscle, it medially rotates the thigh at the hip joint and the leg at the knee joint.

The semimembranosus muscle is innervated by the tibial division of the sciatic nerve.

## In the clinic

### Muscle injuries to the lower limb

Muscle injuries may occur as a result of direct trauma or as part of an overuse syndrome.

Muscle injuries may occur as a minor muscle tear, which may be demonstrated as a focal area of fluid within the muscle. With increasingly severe injuries, more muscle fibers are torn and this may eventually result in a complete muscle tear. The usual muscles in the thigh that tear are the hamstring muscles. Tears in the muscles below the knee typically occur within the soleus muscle, though other muscles may be affected.

### Hamstring muscle injury

Injury to the hamstring muscles is a common source of pain in athletes, particularly in those competing in sports requiring a high degree of power and speed (such as sprinting, track and field, football) where the hamstring muscles are very susceptible to injury from excessive stretching.

The injury can range from a mild muscle strain to a complete tear of a muscle or a tendon. It usually occurs during sudden accelerations and decelerations or rapid change in direction. In adults, the most commonly injured is the muscle-tendon junction, which is a wide transition zone between the muscle and the tendon. An avulsion of the ischial tuberosity with proximal hamstring origin attachment is common in the adolescent population, particularly during sudden hip flexion because the ischial apophysis is the weakest element of the proximal hamstring unit in this age group (Fig. 6.64). Both ultrasound and MRI can be used to assess the hamstring injury with the MRI providing not only the information about the extent of the injury but also give some indication about the prognosis (future risk of re-tear, loss of function, etc).

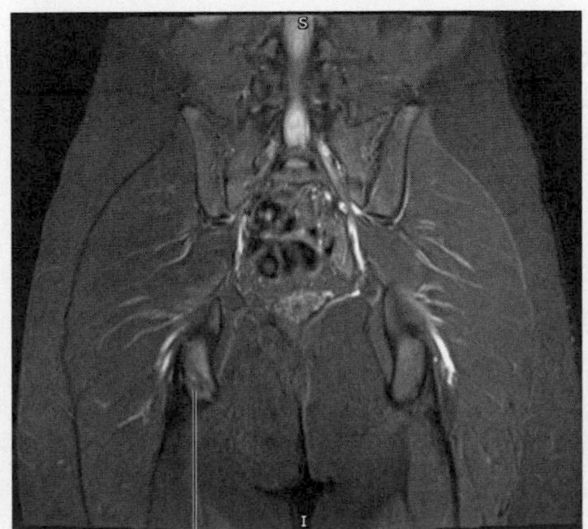


**Fig. 6.64** Coronal MRI of the posterior pelvis and thigh showing a hamstring avulsion injury.

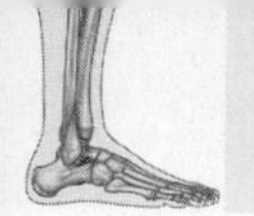

## Arteries

Three arteries enter the thigh: the femoral artery, the obturator artery, and the inferior gluteal artery. Of these, the femoral artery is the largest and supplies most of the lower limb. The three arteries contribute to an anastomotic network of vessels around the hip joint.

### Femoral artery

The femoral artery is the continuation of the external iliac artery and begins as the external iliac artery passes under the inguinal ligament to enter the femoral triangle on the anterior aspect of the upper thigh (Fig. 6.65). The femoral artery is palpable in the femoral triangle just inferior to the inguinal ligament midway between the anterior superior iliac spine and the pubic symphysis.

The femoral artery passes vertically through the femoral triangle and then continues down the thigh in the adductor canal. It leaves the canal by passing through the adductor hiatus in the adductor magnus muscle and becomes the popliteal artery behind the knee.

A cluster of four small branches—**superficial epigastric artery**, **superficial circumflex iliac artery**, **superficial external pudendal artery**, and **deep external pudendal artery**—originate from the femoral artery in the femoral triangle and supply cutaneous regions of the upper thigh, lower abdomen, and perineum.

### Deep artery of thigh

The largest branch of the femoral artery in the thigh is the **deep artery of the thigh** (profunda femoris artery), which originates from the lateral side of the femoral artery in the femoral triangle and is the major source of blood supply to the thigh (Fig. 6.65). The deep artery of the thigh immediately passes:

- posteriorly between the pectineus and adductor longus muscles and then between the adductor longus and adductor brevis muscles, and
- then travels inferiorly between the adductor longus and adductor magnus, eventually penetrating through the adductor magnus to connect with branches of the popliteal artery behind the knee.

The deep artery of the thigh has lateral and medial circumflex femoral branches and three perforating branches.

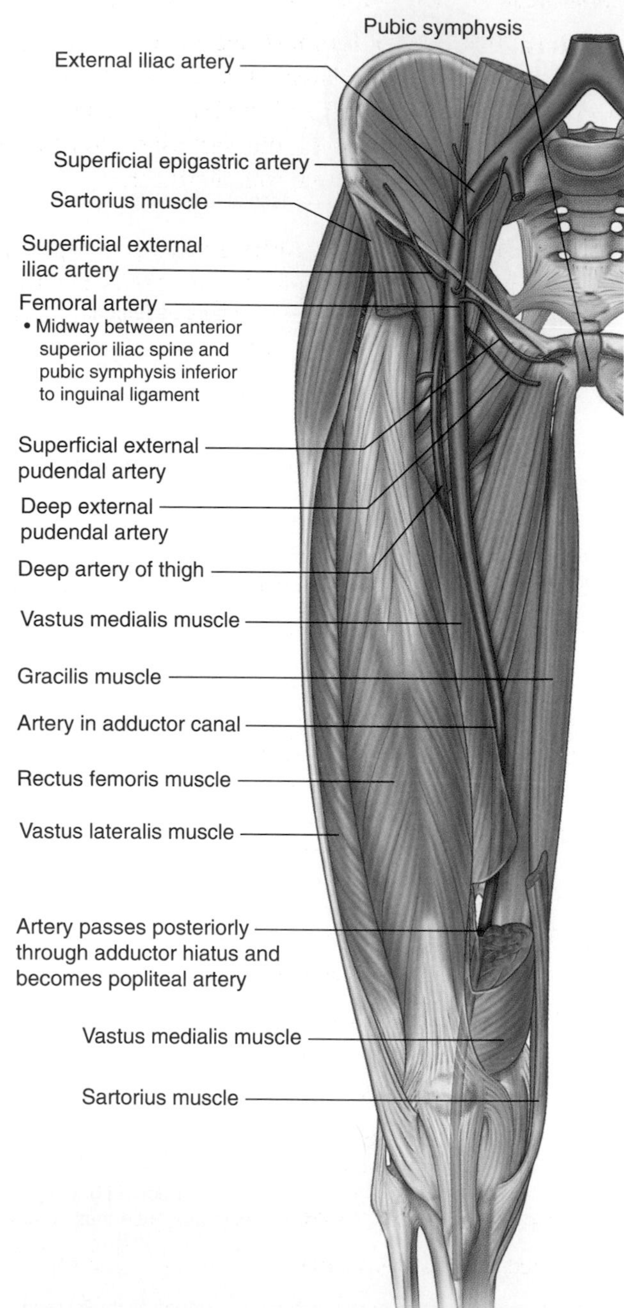


**Fig. 6.65** Femoral artery.

### Lateral circumflex femoral artery

The **lateral circumflex femoral artery** normally originates proximally from the lateral side of the deep artery of the thigh, but may arise directly from the femoral artery (Fig. 6.66). It passes deep to the sartorius and rectus femoris and divides into three terminal branches:

- One vessel (**ascending branch**) ascends laterally deep to the tensor fasciae latae muscle and connects with a branch of the medial circumflex femoral artery to form a channel, which circles the neck of the femur and supplies the neck and head of the femur.
- One vessel (**descending branch**) descends deep to the rectus femoris, penetrates the vastus lateralis muscle, and connects with a branch of the popliteal artery near the knee.
- One vessel (**transverse branch**) passes laterally to pierce the vastus lateralis and then circles around the proximal shaft of the femur to anastomose with branches from the medial femoral circumflex artery, the inferior gluteal artery, and the first perforating artery to form the cruciate anastomosis around the hip.

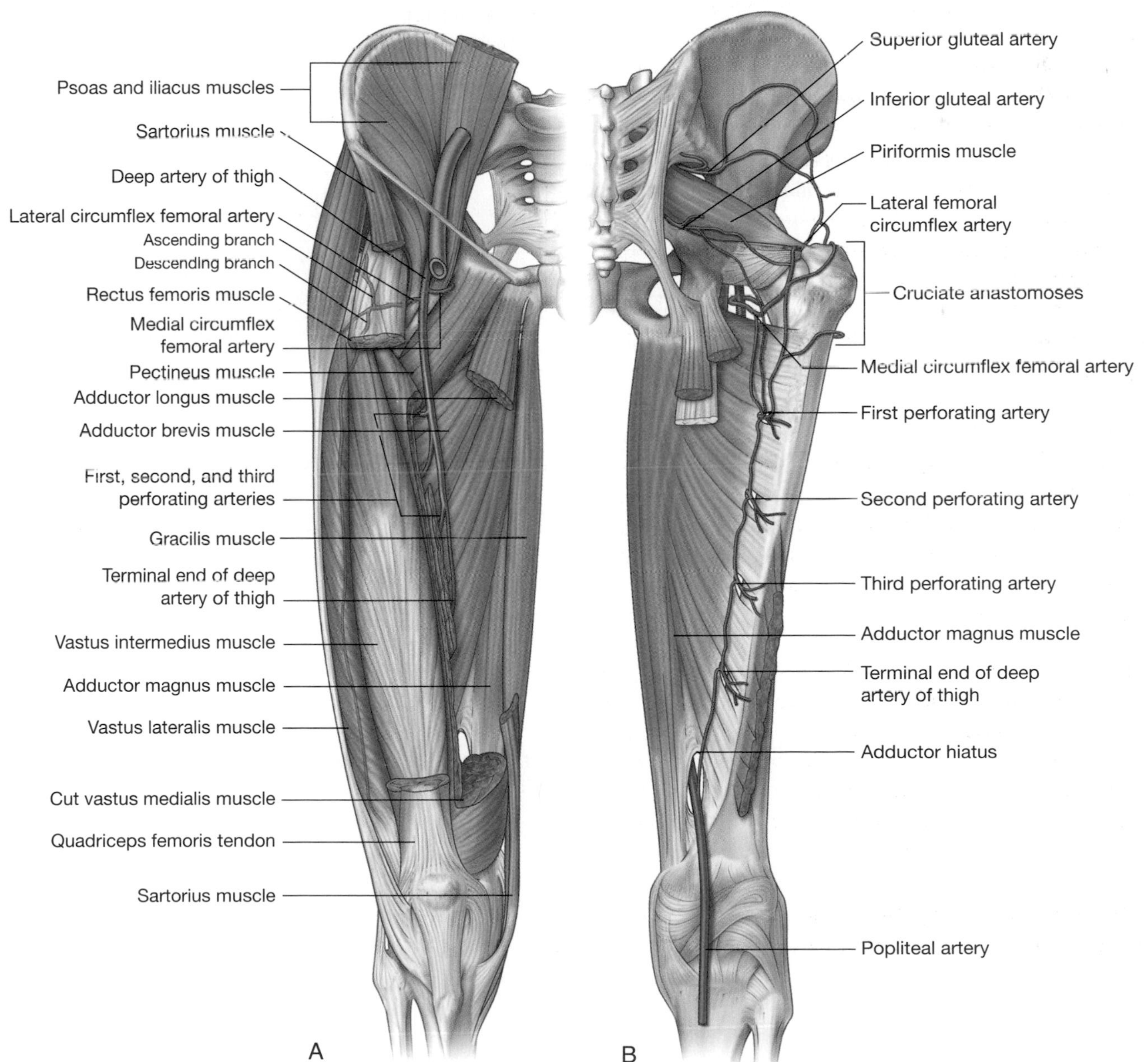


**Fig. 6.66** Deep artery of thigh. **A.** Anterior view. **B.** Posterior view.

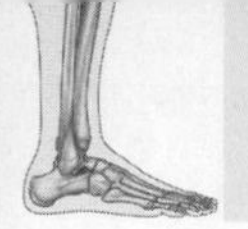



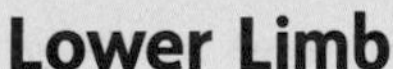

### Medial circumflex femoral artery

The **medial circumflex femoral artery** normally originates proximally from the posteromedial aspect of the deep artery of the thigh, but may originate from the femoral artery (Fig. 6.66). It passes medially around the shaft of the femur, first between the pectineus and iliopsoas and then between the obturator externus and adductor brevis muscles. Near the margin of the adductor brevis the vessel gives off a small branch, which enters the hip joint through the acetabular notch and anastomoses with the acetabular branch of the obturator artery.

The main trunk of the medial circumflex femoral artery passes over the superior margin of the adductor magnus and divides into two major branches deep to the quadratus femoris muscle:

- One branch ascends to the trochanteric fossa and connects with branches of the gluteal and lateral circumflex femoral arteries.
- The other branch passes laterally to participate with branches from the lateral circumflex femoral artery, the inferior gluteal artery, and the first perforating artery in forming an anastomotic network of vessels around the hip.

### Perforating arteries

The three **perforating arteries** branch from the deep artery of the thigh (Fig. 6.66) as it descends anterior to the adductor brevis muscle—the first originates above the muscle, the second originates anterior to the muscle, and the third originates below the muscle. All three penetrate through the adductor magnus near its attachment to the linea aspera to enter and supply the posterior compartment of the thigh. Here, the vessels have ascending and descending branches, which interconnect to form a longitudinal channel, which participates above in forming an anastomotic network of vessels around the hip and inferiorly anastomoses with branches of the popliteal artery behind the knee.

## Obturator artery

The **obturator artery** originates as a branch of the internal iliac artery in the pelvic cavity and enters the medial compartment of the thigh through the obturator canal (Fig. 6.67). As it passes through the canal, it bifurcates into an **anterior branch** and a **posterior branch**, which together form a channel that circles the margin of the obturator membrane and lies within the attachment of the obturator externus muscle.

Vessels arising from the anterior and posterior branches supply adjacent muscles and anastomose with the inferior gluteal and medial circumflex femoral arteries. In addition, an acetabular vessel originates from the posterior branch, enters the hip joint through the acetabular notch, and contributes to the supply of the head of the femur.

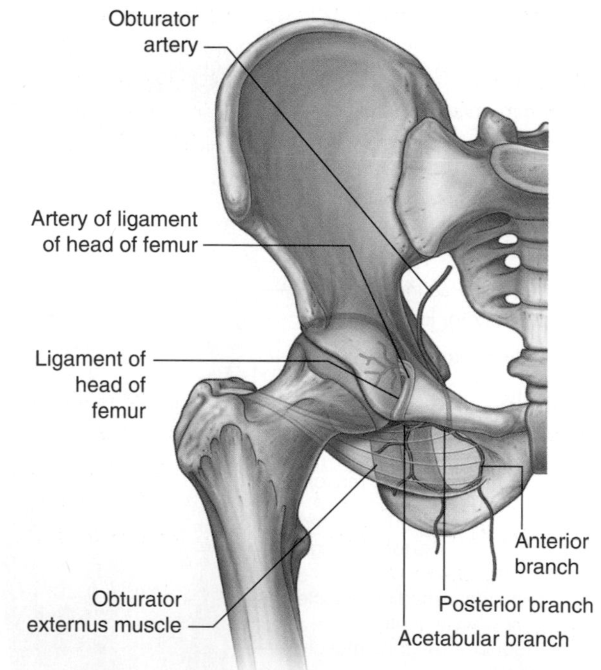


**Fig. 6.67** Obturator artery.

## In the clinic

**Peripheral vascular disease**

Peripheral vascular disease is often characterized by reduced blood flow to the legs. This disorder may be caused by stenoses (narrowing) and/or occlusions (blockages) in the lower aorta and the iliac, femoral, tibial, and fibular vessels. Patients typically have chronic leg ischemia and "acute on chronic" leg ischemia.

***Chronic leg ischemia***

Chronic leg ischemia is a disorder in which vessels have undergone atheromatous change, and often there is significant luminal narrowing (usually over 50%). Most patients with peripheral arterial disease have widespread arterial disease (including cardiovascular and cerebrovascular disease), which may be clinically asymptomatic. Some of these patients develop such severe ischemia that the viability of the limb is threatened (**critical limb ischemia**).

The commonest symptom of chronic leg ischemia is **intermittent claudication.** Patients typically have a history of pain that develops in the calf muscles (usually associated with occlusions or narrowing in the femoral artery) or the buttocks (usually associated with occlusion or narrowing in the aorto-iliac segments). The pain experienced in these muscles is often cramplike and occurs with walking. The patient rests and is able to continue walking up to the same distance until the pain recurs and stops walking as before.

***Acute on chronic ischemia***

In some patients with chronic limb ischemia, an acute event blocks the vessels or reduces the blood supply to such a degree that the viability of the limb is threatened.

Occasionally a leg may become acutely ischemic with no evidence of underlying atheromatous disease. In these instances a blood clot is likely to have embolized from the heart. Patients with mitral valve disease and atrial fibrillation are prone to embolic disease.

***Critical limb ischemia***

Critical limb ischemia occurs when the blood supply to the limb is so poor that the viability of the limb is severely threatened, and in this case many patients develop gangrene, ulceration, and severe rest pain in the foot. These patients require urgent treatment, which may be in the form of surgical reconstruction, radiological angioplasty, or even amputation.

## Veins

Veins in the thigh consist of superficial and deep veins. Deep veins generally follow the arteries and have similar names. Superficial veins are in the superficial fascia, interconnect with deep veins, and do not generally accompany arteries. The largest of the superficial veins in the thigh is the great saphenous vein.

### Great saphenous vein

The great saphenous vein originates from a venous arch on the dorsal aspect of the foot and ascends along the medial side of the lower limb to the proximal thigh (see p. 560). Here it passes through the saphenous ring in deep fascia covering the anterior thigh to connect with the femoral vein in the femoral triangle (see p. 566).

## Nerves

There are three major nerves in the thigh, each associated with one of the three compartments. The femoral nerve is associated with the anterior compartment of the thigh, the obturator nerve is associated with the medial compartment of the thigh, and the sciatic nerve is associated with the posterior compartment of the thigh.

### Femoral nerve

The femoral nerve originates from the lumbar plexus (spinal cord segments L2–L4) on the posterior abdominal wall and enters the femoral triangle of the thigh by passing under the inguinal ligament (Fig. 6.68). In the femoral triangle the femoral nerve lies on the lateral side of the femoral artery and is outside the femoral sheath, which surrounds the vessels.

Before entering the thigh, the femoral nerve supplies branches to the iliacus and pectineus muscles.

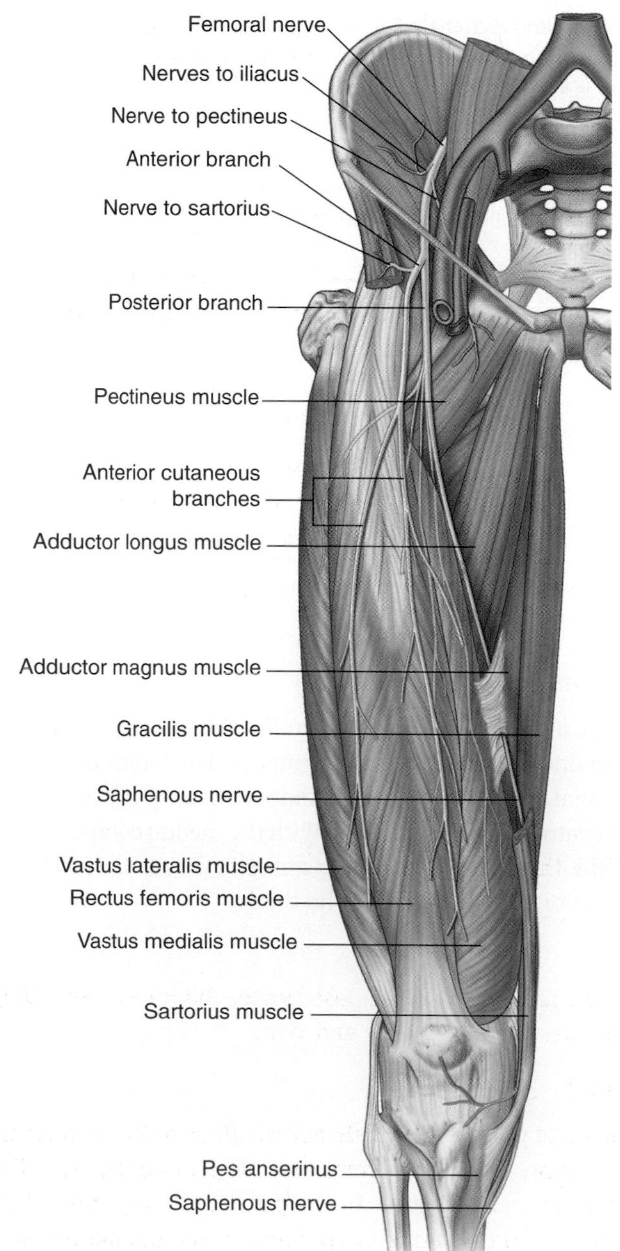


**Fig. 6.68** Femoral nerve.

Immediately after passing under the inguinal ligament, the femoral nerve divides into anterior and posterior branches, which supply muscles of the anterior compartment of the thigh and skin on the anterior and medial aspects of the thigh and on the medial sides of the leg and foot.

Branches of the femoral nerve (Fig. 6.68) include:

- anterior cutaneous branches, which penetrate deep fascia to supply skin on the front of the thigh and knee;
- numerous motor nerves, which supply the quadriceps femoris muscles (rectus femoris, vastus lateralis, vastus intermedius, and vastus medialis muscles) and the sartorius muscle; and
- one long cutaneous nerve, the saphenous nerve, which supplies skin as far distally as the medial side of the foot.

The **saphenous nerve** accompanies the femoral artery through the adductor canal, but does not pass through the adductor hiatus with the femoral artery. Rather, the saphenous nerve penetrates directly through connective tissues near the end of the canal to appear between the sartorius and gracilis muscles on the medial side of the knee. Here the saphenous nerve penetrates deep fascia and continues down the medial side of the leg to the foot, and supplies skin on the medial side of the knee, leg, and foot.

### Obturator nerve

The obturator nerve is a branch of the lumbar plexus (spinal cord segments L2–L4) on the posterior abdominal wall. It descends in the psoas muscle, and then passes out of the medial margin of the psoas muscle to enter the pelvis (Fig. 6.69). The obturator nerve continues along the lateral pelvic wall and then enters the medial compartment of the thigh by passing through the obturator canal. It supplies most of the adductor muscles and skin on the medial aspect of the thigh. As the obturator nerve enters the thigh, it divides into two branches, an anterior branch and a posterior branch, which are separated by the adductor brevis muscle:

- The **posterior branch** descends behind the adductor brevis muscle and on the anterior surface of the adductor magnus muscle, and supplies the obturator externus and adductor brevis muscles and the part of the adductor magnus that attaches to the linea aspera.
- The **anterior branch** descends on the anterior surface of the adductor brevis muscle and is behind the pectineus and adductor longus muscles—it supplies branches to the adductor longus, gracilis, and adductor brevis muscles, and often contributes to the supply of the pectineus muscle, and cutaneous branches innervate the skin on the medial side of the thigh.

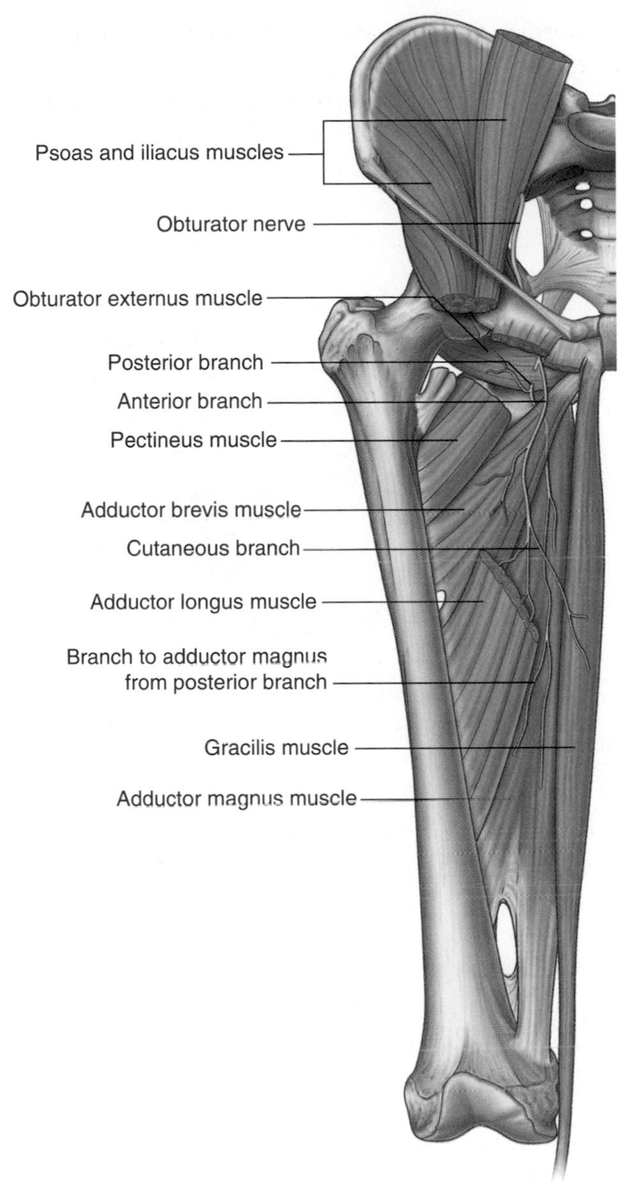


**Fig. 6.69** Obturator nerve.

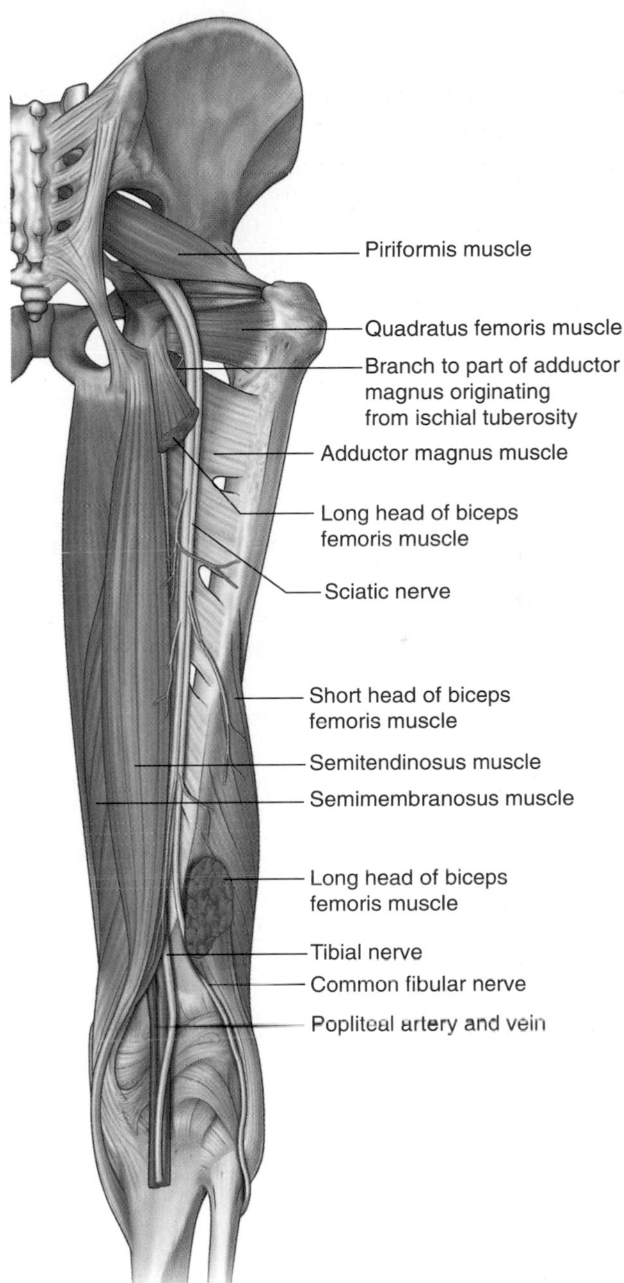


**Fig. 6.70** Sciatic nerve.

## Sciatic nerve

The sciatic nerve is a branch of the lumbosacral plexus (spinal cord segments L4–S3) and descends into the posterior compartment of the thigh from the gluteal region (Fig. 6.70). It innervates all muscles in the posterior compartment of the thigh and then its branches continue into the leg and foot.

In the posterior compartment of the thigh, the sciatic nerve lies on the adductor magnus muscle and is crossed by the long head of the biceps femoris muscle.

Proximal to the knee, and sometimes within the pelvis, the sciatic nerve divides into its two terminal branches: the **tibial nerve** and the **common fibular nerve**. These nerves travel vertically down the thigh and enter the popliteal fossa posterior to the knee. Here, they meet the popliteal artery and vein.

### Tibial nerve

The tibial part of the sciatic nerve, either before or after its separation from the common fibular nerve, supplies branches to all muscles in the posterior compartment of the thigh (long head of biceps femoris, semimembranosus, semitendinosus) except the short head of the biceps femoris, which is innervated by the common fibular part (Fig. 6.70).

The tibial nerve descends through the popliteal fossa, enters the posterior compartment of the leg, and continues into the sole of the foot.

The tibial nerve innervates:

- all muscles in the posterior compartment of the leg,
- all intrinsic muscles in the sole of the foot including the first two dorsal interossei muscles, which also may receive innervation from the deep fibular nerve, and
- skin on the posterolateral side of the lower half of the leg and lateral side of the ankle, foot, and little toe, and skin on the sole of the foot and toes.

### Common fibular nerve

The common fibular part of the sciatic nerve innervates the short head of the biceps femoris in the posterior compartment of the thigh and then continues into the lateral and anterior compartments of the leg and onto the foot (Fig. 6.70).

The common fibular nerve innervates:

- all muscles in the anterior and lateral compartments of the leg,
- one muscle (extensor digitorum brevis) on the dorsal aspect of the foot,
- the first two dorsal interossei muscles in the sole of the foot, and
- skin over the lateral aspect of the leg, and ankle, and over the dorsal aspect of the foot and toes.

## Knee joint

The knee joint is the largest synovial joint in the body. It consists of:

- the articulation between the femur and tibia, which is weight-bearing, and
- the articulation between the patella and the femur, which allows the pull of the quadriceps femoris muscle to be directed anteriorly over the knee to the tibia without tendon wear (Fig. 6.71).

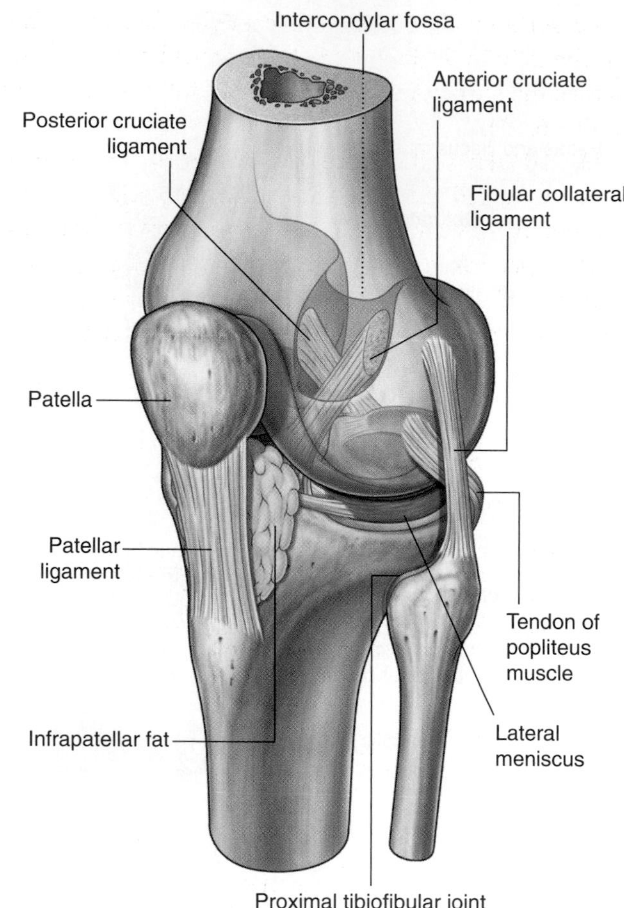


**Fig. 6.71** Knee joint. Joint capsule is not shown.

Two fibrocartilaginous menisci, one on each side, between the femoral condyles and tibia accommodate changes in the shape of the articular surfaces during joint movements.

The detailed movements of the knee joint are complex, but basically the joint is a hinge joint that allows mainly flexion and extension. Like all hinge joints, the knee joint is reinforced by collateral ligaments, one on each side of the joint. In addition, two very strong ligaments (the cruciate ligaments) interconnect the adjacent ends of the femur and tibia and maintain their opposed positions during movement.

Because the knee joint is involved in weight-bearing, it has an efficient "locking" mechanism to reduce the amount of muscle energy required to keep the joint extended when standing.

### Articular surfaces

The articular surfaces of the bones that contribute to the knee joint are covered by hyaline cartilage. The major surfaces involved include:

- the two femoral condyles, and
- the adjacent surfaces of the superior aspect of the tibial condyles.

The surfaces of the femoral condyles that articulate with the tibia in flexion of the knee are curved or round, whereas the surfaces that articulate in full extension are flat (Fig. 6.72).

The articular surfaces between the femur and patella are the V-shaped trench on the anterior surface of the distal end of the femur where the two condyles join and the adjacent surfaces on the posterior aspect of the patella. The joint surfaces are all enclosed within a single articular cavity, as are the intraarticular menisci between the femoral and tibial condyles.

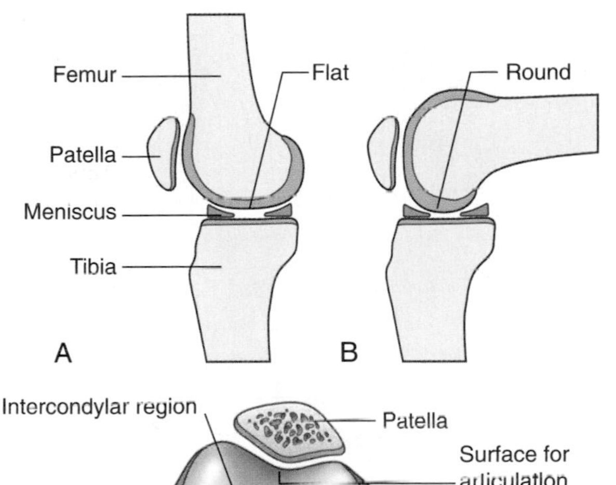


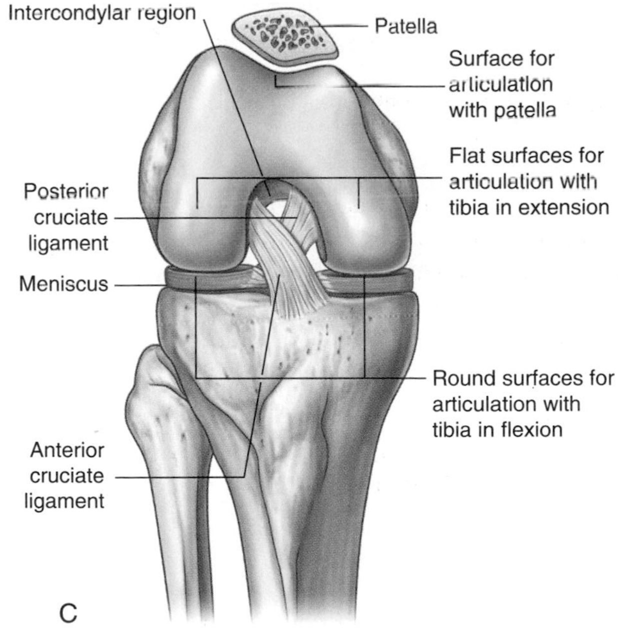


**Fig. 6.72** Articular surfaces of the knee joint. **A.** Extended. **B.** Flexed. **C.** Anterior view (flexed).

## Menisci

There are two menisci, which are fibrocartilaginous C-shaped cartilages, in the knee joint, one medial (**medial meniscus**) and the other lateral (**lateral meniscus**) (Fig. 6.73). Both are attached at each end to facets in the intercondylar region of the tibial plateau.

The medial meniscus is attached around its margin to the capsule of the joint and to the tibial collateral ligament, whereas the lateral meniscus is unattached to the capsule. Therefore, the lateral meniscus is more mobile than the medial meniscus.

The menisci are interconnected anteriorly by a transverse ligament of the knee. The lateral meniscus is also connected to the tendon of the popliteus muscle, which passes superolaterally between this meniscus and the capsule to insert on the femur.

The menisci improve congruency between the femoral and tibial condyles during joint movements where the surfaces of the femoral condyles articulating with the tibial plateau change from small curved surfaces in flexion to large flat surfaces in extension.

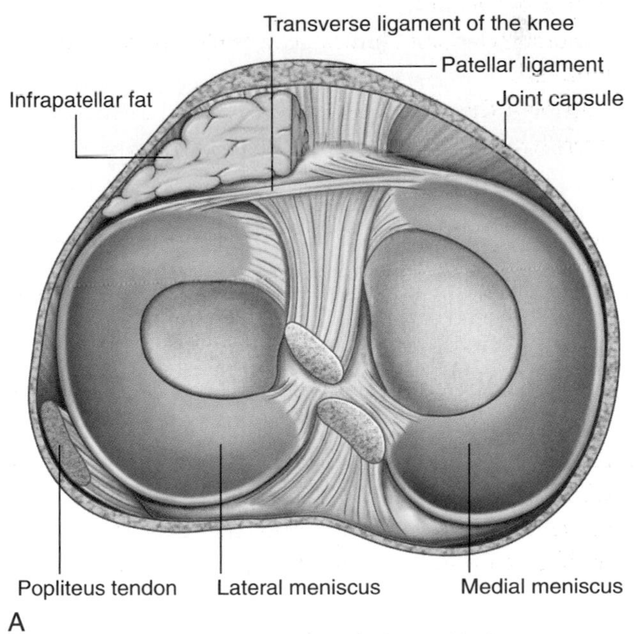


**Fig. 6.73** Menisci of the knee joint. **A.** Superior view.

*Continues*

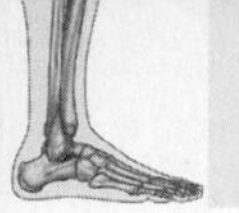

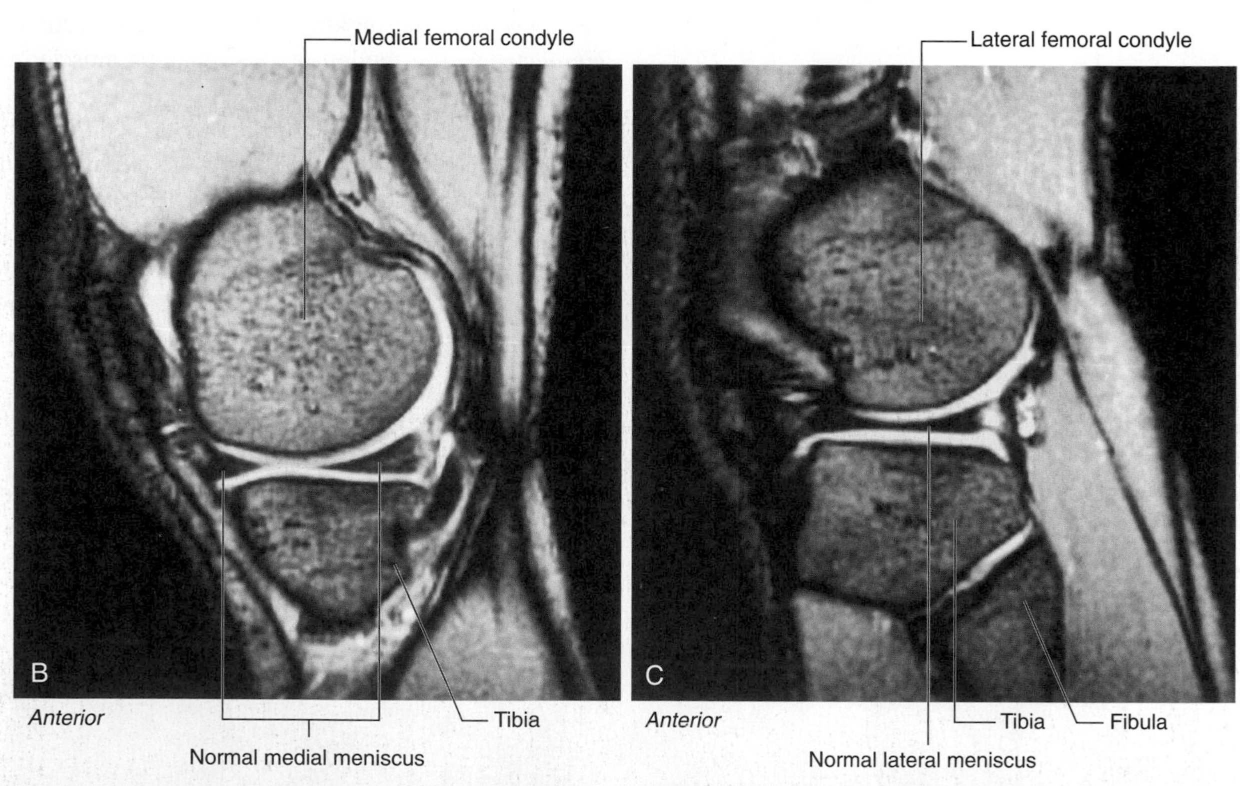


**Fig. 6.73, cont'd** Menisci of the knee joint. **B.** Normal knee joint showing the medial meniscus. T2-weighted magnetic resonance image in the sagittal plane. **C.** Normal knee joint showing the lateral meniscus. T2-weighted magnetic resonance image in the sagittal plane.

## In the clinic

### Meniscal injuries

Menisci can get torn during forceful rotation or twisting of the knee, but significant trauma is not always necessary for a tear to occur. There are various patterns of meniscal tearing depending on the cleavage plane such as vertical tears (perpendicular to the tibial plateau), horizontal tears (parallel to the long axis of the meniscus and perpendicular to the tibial plateau), or bucket handle tears (longitudinal tear where the torn portion of the meniscus forms a handle shaped fragment which gets displaced into the intercondylar notch).

The patient usually complains of pain localized to the medial or lateral side of the knee, knee locking or clicking, sensation of knee giving way, and swelling, which can be intermittent and usually delayed.

MRI is the modality of choice to assess meniscal tears and detect other associated injuries, such as ligamentous tears and articular cartilage damage (Fig. 6.74A). Arthroscopy is usually performed to repair a tear, debride the damaged meniscal material, or rarely remove the entire torn meniscus (Fig. 6.74B).

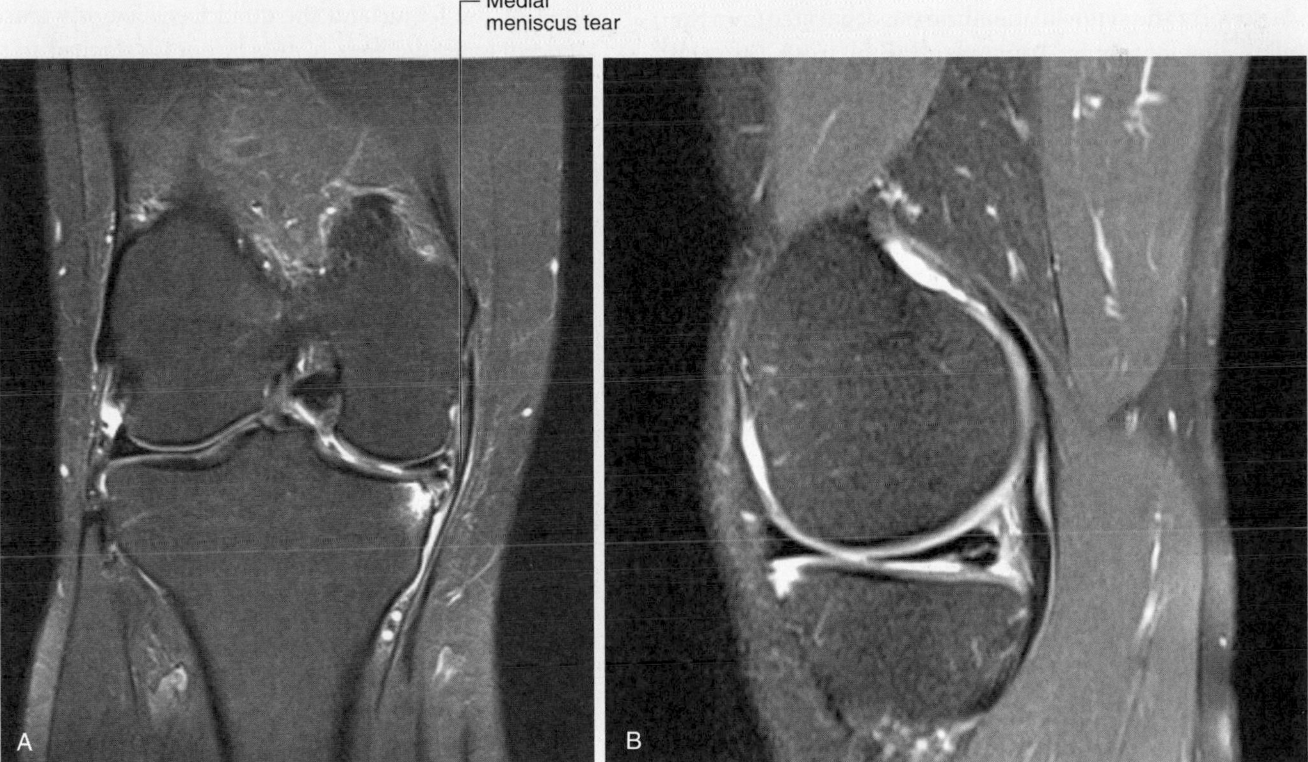


**Fig. 6.74** Meniscal injury and repair. **A.** Sagittal MRI of a knee joint showing tear of the medial meniscus. **B.** Coronal MRI of a knee showing a truncated lateral meniscus after partial meniscectomy to treat a tear.

### Synovial membrane

The synovial membrane of the knee joint attaches to the margins of the articular surfaces and to the superior and inferior outer margins of the menisci (Fig. 6.75A). The two cruciate ligaments, which attach in the intercondylar region of the tibia below and the intercondylar fossa of the femur above, are outside the articular cavity, but enclosed within the fibrous membrane of the knee joint.

Posteriorly, the synovial membrane reflects off the fibrous membrane of the joint capsule on either side of the posterior cruciate ligament and loops forward around both ligaments thereby excluding them from the articular cavity.

Anteriorly, the synovial membrane is separated from the patellar ligament by an **infrapatellar fat pad**. On each side of the pad, the synovial membrane forms a fringed margin (an **alar fold**), which projects into the articular cavity. In addition, the synovial membrane covering the lower part of the infrapatellar fat pad is raised into a sharp midline fold directed posteriorly (the **infrapatellar synovial fold**), which attaches to the margin of the intercondylar fossa of the femur.

The synovial membrane of the knee joint forms pouches in two locations to provide low-friction surfaces for the movement of tendons associated with the joint:

- The smallest of these expansions is the **subpopliteal recess** (Fig. 6.75A), which extends posterolaterally from the articular cavity and lies between the lateral meniscus and the tendon of the popliteus muscle, which passes through the joint capsule.
- The second expansion is the **suprapatellar bursa** (Fig. 6.75B), a large bursa that is a continuation of the articular cavity superiorly between the distal end of the shaft of the femur and the quadriceps femoris muscle and tendon—the apex of this bursa is attached to the small articularis genus muscle, which pulls the bursa away from the joint during extension of the knee.

Other bursae associated with the knee but not normally communicating with the articular cavity include the

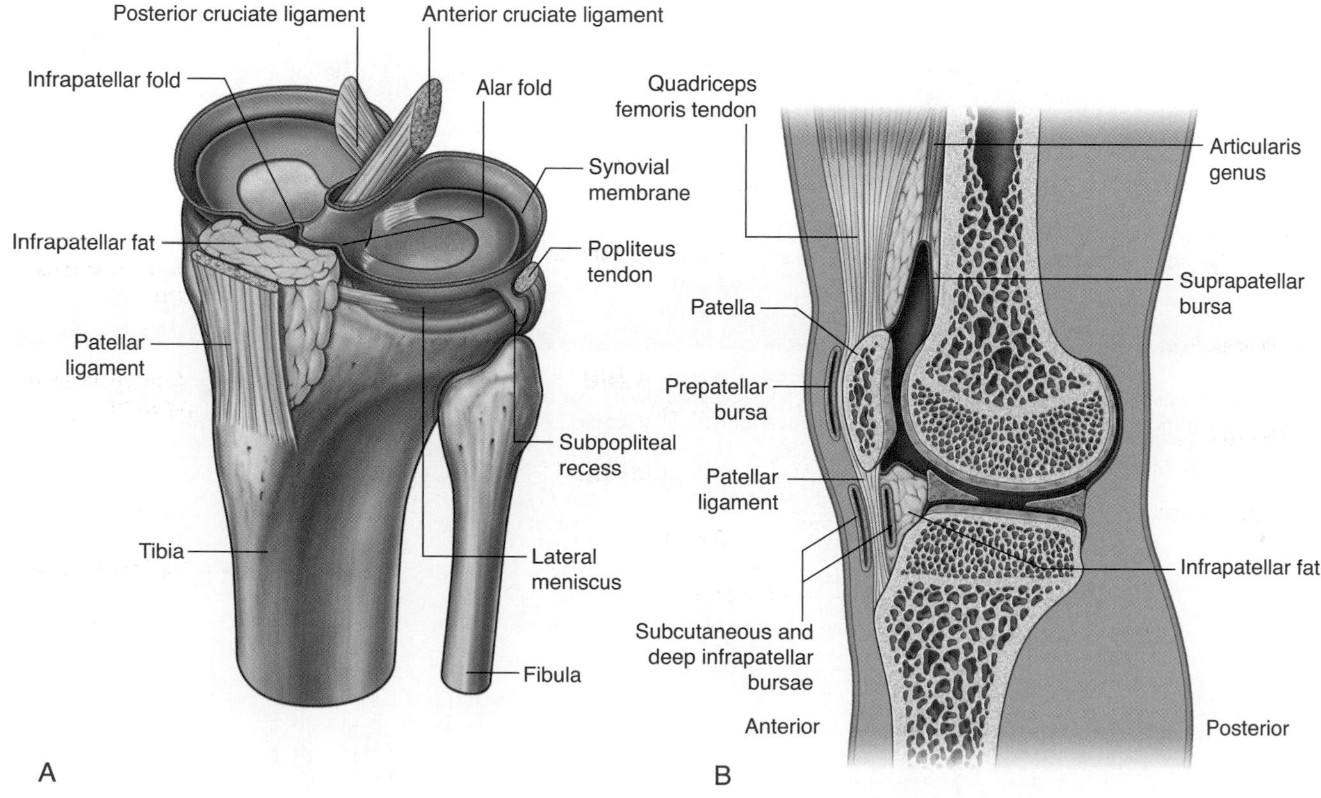


**Fig. 6.75** Synovial membrane of the knee joint and associated bursae. **A.** Superolateral view; patella and femur not shown. **B.** Paramedial sagittal section through the knee.

subcutaneous prepatellar bursa, deep and subcutaneous infrapatellar bursae, and numerous other bursae associated with tendons and ligaments around the joint (Fig. 6.75B).

The prepatellar bursa is subcutaneous and anterior to the patella. The deep and subcutaneous infrapatellar bursae are on the deep and subcutaneous sides of the patellar ligament, respectively.

### Fibrous membrane

The fibrous membrane of the knee joint is extensive and is partly formed and reinforced by extensions from tendons of the surrounding muscles (Fig. 6.76). In general, the fibrous membrane encloses the articular cavity and the intercondylar region:

- On the medial side of the knee joint, the fibrous membrane blends with the tibial collateral ligament and is attached on its internal surface to the medial meniscus.
- Laterally, the external surface of the fibrous membrane is separated by a space from the fibular collateral ligament and the internal surface of the fibrous membrane is not attached to the lateral meniscus.
- Anteriorly, the fibrous membrane is attached to the margins of the patella where it is reinforced with tendinous expansions from the vastus lateralis and vastus medialis muscles, which also merge above with the quadriceps femoris tendon and below with the patellar ligament.

The fibrous membrane is reinforced anterolaterally by a fibrous extension from the iliotibial tract and posteromedially by an extension from the tendon of the semimembranosus (the **oblique popliteal ligament**), which reflects superiorly across the back of the fibrous membrane from medial to lateral.

The upper end of the popliteus muscle passes through an aperture in the posterolateral aspect of the fibrous membrane of the knee and is enclosed by the fibrous membrane as its tendon travels around the joint to insert on the lateral aspect of the lateral femoral condyle.

### Ligaments

The major ligaments associated with the knee joint are the patellar ligament, the tibial (medial) and fibular (lateral) collateral ligaments, and the anterior and posterior cruciate ligaments.

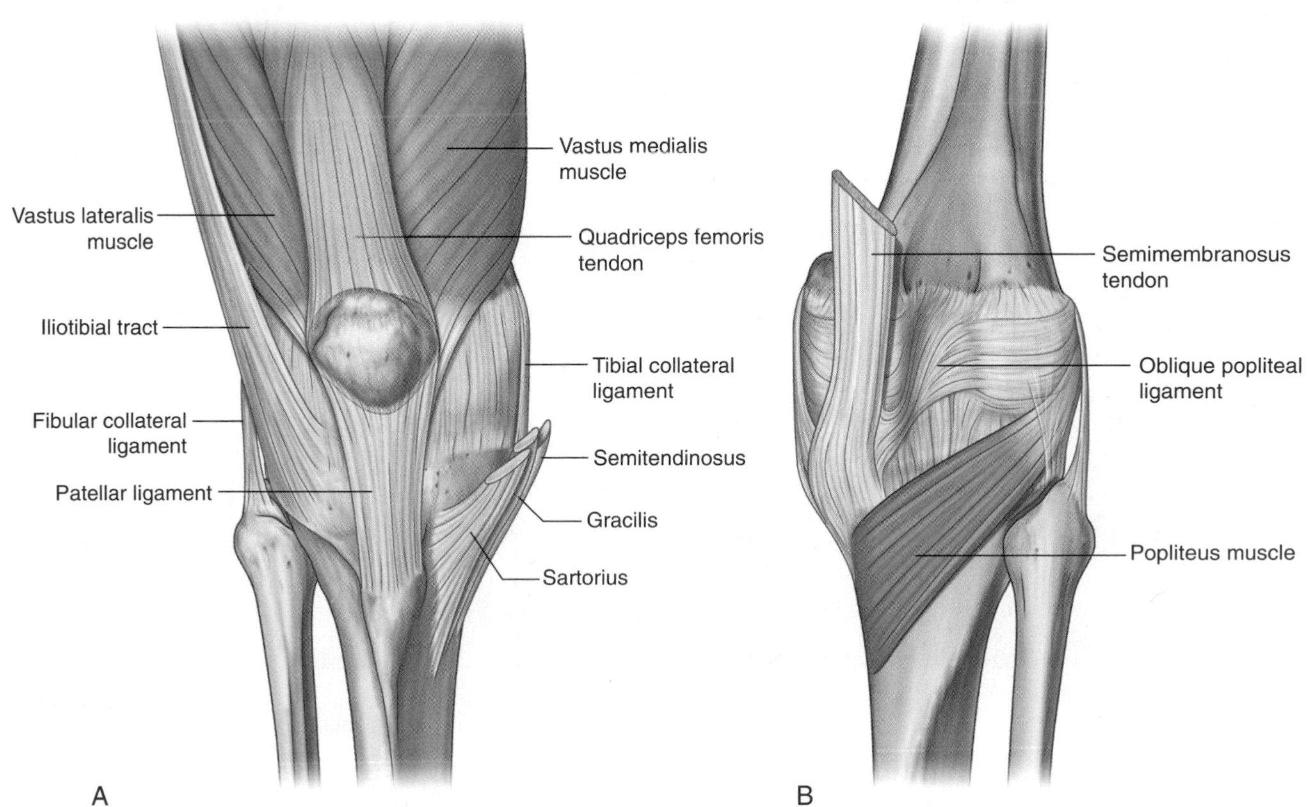


**Fig. 6.76** Fibrous membrane of the knee joint capsule. **A.** Anterior view. **B.** Posterior view.

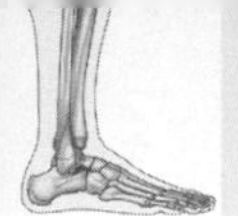



### Patellar ligament

The **patellar ligament** is basically the continuation of the quadriceps femoris tendon inferior to the patella (Fig. 6.76). It is attached above to the margins and apex of the patella and below to the tibial tuberosity.

### Collateral ligaments

The collateral ligaments, one on each side of the joint, stabilize the hinge-like motion of the knee (Fig. 6.77).

The cord-like **fibular collateral ligament** is attached superiorly to the lateral femoral epicondyle just above the groove for the popliteus tendon. Inferiorly, it is attached to a depression on the lateral surface of the fibular head. It is separated from the fibrous membrane by a bursa.

The broad and flat **tibial collateral ligament** is attached by much of its deep surface to the underlying fibrous membrane. It is anchored superiorly to the medial femoral epicondyle just inferior to the adductor tubercle and descends anteriorly to attach to the medial margin and

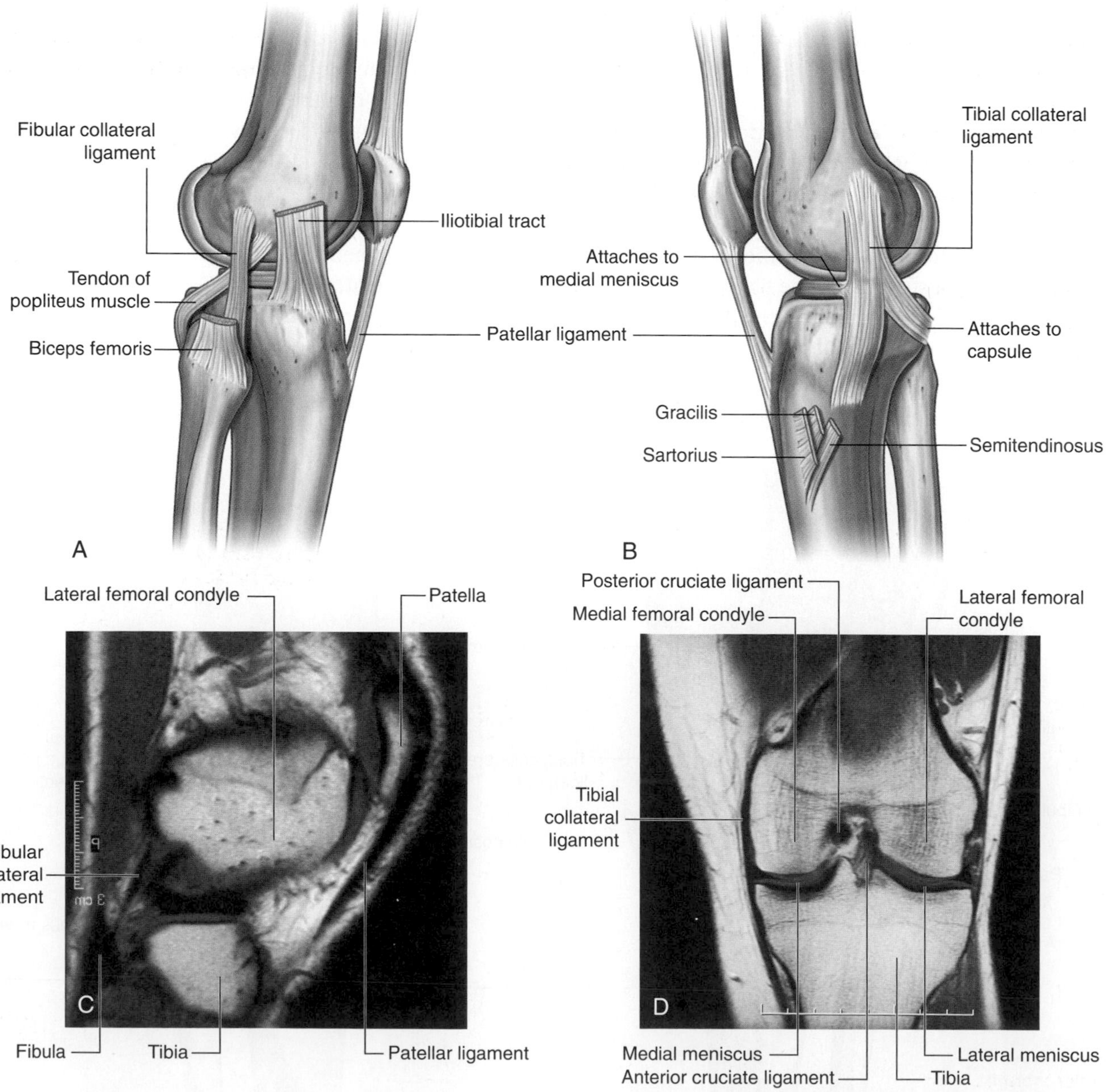

**Fig. 6.77** Collateral ligaments of the knee joint. **A.** Lateral view. **B.** Medial view. **C.** Normal knee joint showing the patellar ligament and the fibular collateral ligament. T1-weighted magnetic resonance image in the sagittal plane. **D.** Normal knee joint showing the tibial collateral ligament, the medial and lateral menisci, and the anterior and posterior cruciate ligaments. T1-weighted magnetic resonance image in the coronal plane.

medial surface of the tibia above and behind the attachment of the sartorius, gracilis, and semitendinosus tendons.

### Cruciate ligaments

The two cruciate ligaments are in the intercondylar region of the knee and interconnect the femur and tibia (Figs. 6.77D and 6.78). They are termed "cruciate" (Latin for "shaped like a cross") because they cross each other in the sagittal plane between their femoral and tibial attachments:

- The **anterior cruciate ligament** attaches to a facet on the anterior part of the intercondylar area of the tibia and ascends posteriorly to attach to a facet at the back of the lateral wall of the intercondylar fossa of the femur.
- The **posterior cruciate ligament** attaches to the posterior aspect of the intercondylar area of the tibia and ascends anteriorly to attach to the medial wall of the intercondylar fossa of the femur.

The anterior cruciate ligament crosses lateral to the posterior cruciate ligament as they pass through the intercondylar region.

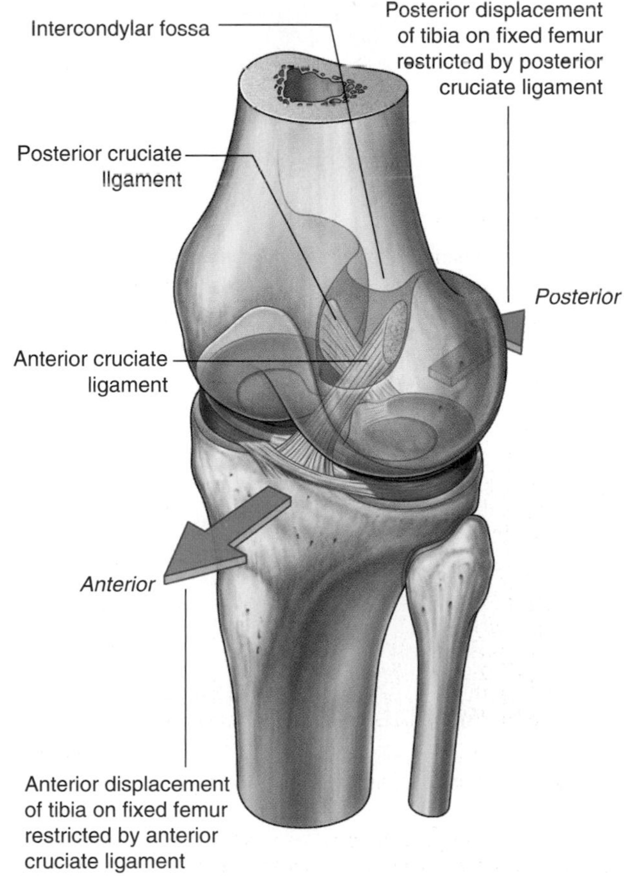


**Fig. 6.78** Cruciate ligaments of the knee joint. Superolateral view.

The anterior cruciate ligament prevents anterior displacement of the tibia relative to the femur and the posterior cruciate ligament restricts posterior displacement (Fig. 6.78).

### Locking mechanism

When standing, the knee joint is locked into position, thereby reducing the amount of muscle work needed to maintain the standing position (Fig. 6.79).

One component of the locking mechanism is a change in the shape and size of the femoral surfaces that articulate with the tibia:

- In flexion, the surfaces are the curved and rounded areas on the posterior aspects of the femoral condyles.

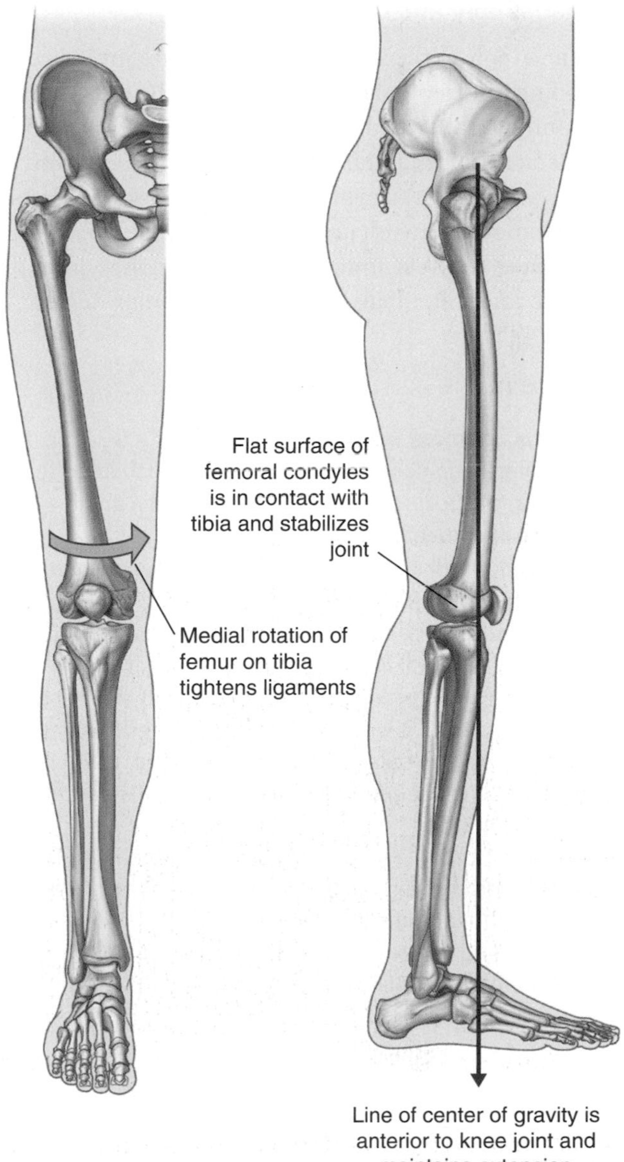


**Fig. 6.79** Knee "locking" mechanism.

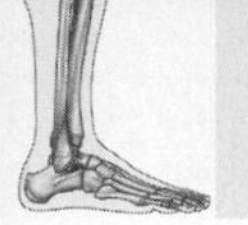

- As the knee is extended, the surfaces move to the broad and flat areas on the inferior aspects of the femoral condyles.

Consequently the joint surfaces become larger and more stable in extension.

Another component of the locking mechanism is medial rotation of the femur on the tibia during extension. Medial rotation and full extension tightens all the associated ligaments.

Another feature that keeps the knee extended when standing is that the body's center of gravity is positioned along a vertical line that passes anterior to the knee joint.

The popliteus muscle unlocks the knee by initiating lateral rotation of the femur on the tibia.

### Vascular supply and innervation

Vascular supply to the knee joint is predominantly through descending and genicular branches from the femoral, popliteal, and lateral circumflex femoral arteries in the thigh and the circumflex fibular artery and recurrent branches from the anterior tibial artery in the leg. These vessels form an anastomotic network around the joint (Fig. 6.80).

The knee joint is innervated by branches from the obturator, femoral, tibial, and common fibular nerves.

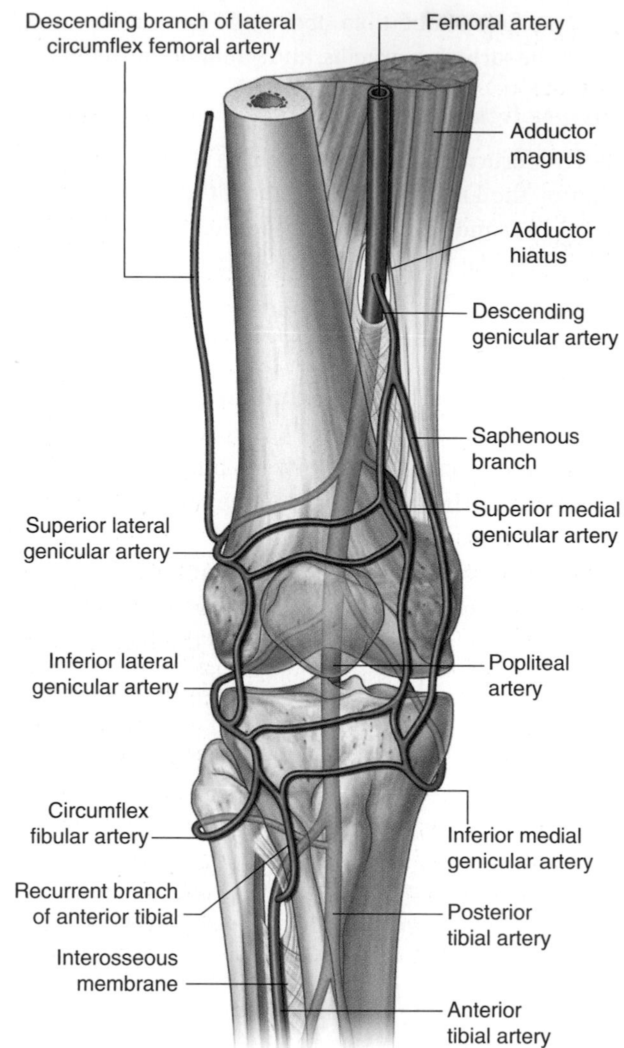


**Fig. 6.80** Anastomoses of arteries around the knee. Anterior view.

## In the clinic

### Collateral ligament injuries

The collateral ligaments are responsible for stabilizing the knee joint, controlling its sideway movements, and protecting the knee from excessive motion.

Injury to the fibular collateral ligament occurs when excessive outward force is applied to the medial side of the knee (varus force), and is less common than an injury to the tibial collateral ligament that is damaged when excessive force is applied inward to the lateral side of the joint (valgus force). Injuries to the tibial collateral ligament can be part of a so called "unhappy triad" that also involves tears of the medial meniscus and the anterior cruciate ligament.

The spectrum of injuries to collateral ligaments of the knee range from minor sprains where the ligaments are slightly stretched, but still able to stabilize the knee joint, to full thickness tears where all fibers are torn and the ligaments lose their stabilizing function.

## In the clinic

### Cruciate ligament injuries

The anterior cruciate ligament (ACL) is most frequently injured during non-contact activities when there is a sudden change in the direction of movement (cutting or pivoting) (Fig. 6.81). Contact sports may also result in ACL injury due to sudden twisting, hyperextension, and valgus force related to direct collision. The injury usually affects the mid-portion of the ligament and manifests itself as a complete or partial discontinuity of the fibers or abnormal orientation and contour of the ligament. With an acute ACL tear, a sudden click or pop can be heard and the knee becomes rapidly swollen. Several tests are used to clinically assess the injury, and the diagnosis is usually confirmed by MRI. A full thickness ACL tear causes instability of the knee joint. The treatment depends on the desired level of activity of the patient. In those with high activity levels, surgical reconstruction of the ligament is required. Those with low activity levels may opt for knee bracing and physiotherapy; however, in the long term the internal damage to the knee leads to the development of early osteoarthritis.

A tear to the posterior cruciate ligament (PCL) requires significant force, so it rarely occurs in isolation. It usually occurs during hyperextension of the knee or as a result of a direct blow to a bent knee such as when striking the knee against the dashboard in a motor vehicle accident. Typically, the injury presents as posterior displacement of the tibia on physical examination (the so called tibial sag sign). Patients complain of knee pain and swelling, inability to bear weight, and instability. The diagnosis is confirmed on MRI. The management, as in ACL injury, depends on the degree of the injury (sprain, partial thickness, full thickness) and the level of desired activity.

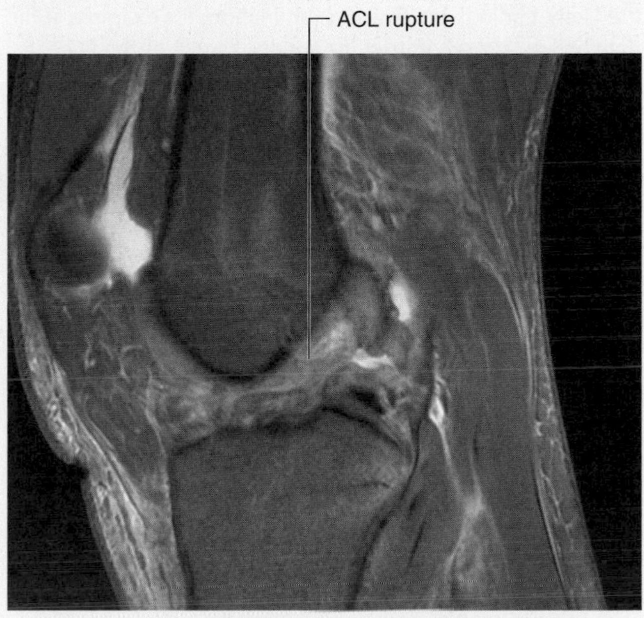


**Fig. 6.81** Sagittal MRI of knee joint showing rupture of the anterior cruciate ligament.

## In the clinic

### Degenerative joint disease/osteoarthritis

Degenerative joint disease occurs throughout many joints within the body. Articular degeneration may result from an abnormal force across the joint with a normal cartilage or a normal force with abnormal cartilage.

Typically degenerative joint disease occurs in synovial joints and the process is called osteoarthritis. In the joints where osteoarthritis occurs the cartilage and bony tissues are usually involved, with limited change within the synovial membrane. The typical findings include reduction in the joint space, eburnation (joint sclerosis), osteophytosis (small bony outgrowths), and bony cyst formation. As the disease progresses the joint may become malaligned, its movement may become severely limited, and there may be significant pain.

The commonest sites for osteoarthritis include the small joints of the hands and wrist, and in the lower limb, the hip and knee are typically affected, though the tarsometatarsal and metatarsophalangeal articulations may undergo similar changes.

The etiology of degenerative joint disease is unclear, but there are some associations, including genetic predisposition, increasing age (males tend to be affected younger than females), overuse or underuse of joints, and nutritional and metabolic abnormalities. Further factors include joint trauma and pre-existing articular disease or deformity.

The histological findings of osteoarthritis consist of degenerative changes within the cartilage and the subchondral bone. Further articular damage worsens these changes, which promote further abnormal stresses upon the joint. As the disease progresses the typical finding is pain, which is usually worse on rising from bed and at the end of a day's activity. Commonly it is aggravated by the extremes of movement or unaccustomed exertion. Stiffness and limitation of movement may ensue.

Treatment in the first instance includes alteration of lifestyle to prevent pain and simple analgesia. As symptoms progress a joint replacement may be necessary, but although joint replacement appears to be the panacea for degenerative joint disease, it is not without risks and complications, which include infection and failure in the short and long term.

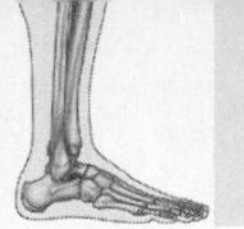



## In the clinic

### Examination of the knee joint

It is important to establish the nature of the patient's complaint before any examination. The history should include information about the complaint, the signs and symptoms, and the patient's lifestyle (level of activity). This history may give a significant clue to the type of injury and the likely findings on clinical examination, for example, if the patient was kicked around the medial aspect of the knee, a valgus deformity injury to the tibial collateral ligament might be suspected.

The examination should include assessment in the erect position, while walking, and on the couch. The affected side must be compared with the unaffected side.

There are many tests and techniques for examining the knee joint, including the following.

#### *Tests for anterior instability*

- Lachman's test—the patient lies on the couch. The examiner places one hand around the distal femur and the other around the proximal tibia and then elevates the knee, producing 20° of flexion. The patient's heel rests on the couch. The examiner's thumb must be on the tibial tuberosity. The hand on the tibia applies a brisk anteriorly directed force. If the movement of the tibia on the femur comes to a sudden stop, it is a firm endpoint. If it does not come to a sudden stop, the endpoint is described as soft and is associated with a tear of the anterior cruciate ligament.
- Anterior drawer test—a positive anterior drawer test is when the proximal head of a patient's tibia can be pulled anteriorly on the femur. The patient lies supine on the couch. The knee is flexed to 90° and the heel and sole of the foot are placed on the couch. The examiner sits gently on the patient's foot, which has been placed in a neutral position. The index fingers are used to check that the hamstrings are relaxed while the other fingers encircle the upper end of the tibia and pull the tibia. If the tibia moves forward, the anterior cruciate ligament is torn. Other peripheral structures, such as the medial meniscus or meniscotibial ligaments, must also be damaged to elicit this sign.
- Pivot shift test—there are many variations of this test. The patient's foot is wedged between the examiner's body and elbow. The examiner places one hand flat under the tibia pushing it forward with the knee in extension. The other hand is placed against the patient's thigh pushing it the other way. The lower limb is taken into slight abduction by the examiner's elbow with the examiner's body acting as a fulcrum to produce the valgus. The examiner maintains the anterior tibial translation and the valgus and initiates flexion of the patient's knee. At about 20°–30° the pivot shift will occur as the lateral tibial plateau reduces. This test demonstrates damage to the posterolateral corner of the knee joint and the anterior cruciate ligament.

#### *Tests for posterior instability*

- Posterior drawer test—a positive posterior drawer test occurs when the proximal head of a patient's tibia can be pushed posteriorly on the femur. The patient is placed in a supine position and the knee is flexed to approximately 90° with the foot in the neutral position. The examiner sits gently on the patient's foot placing both thumbs on the tibial tuberosity and pushing the tibia backward. If the tibial plateau moves, the posterior cruciate ligament is torn.

#### *Assessment of other structures of the knee*

- Assessment of the tibial collateral ligament can be performed by placing a valgus stress on the knee.
- Assessment of lateral and posterolateral knee structures requires more complex clinical testing.

The knee will also be assessed for:

- joint line tenderness,
- patellofemoral movement and instability,
- presence of an effusion,
- muscle injury, and
- popliteal fossa masses.

#### *Further investigations*

After the clinical examination has been carried out, further investigations usually include **plain radiography** and possibly **magnetic resonance imaging**, which allows the radiologist to assess the menisci, cruciate ligaments, collateral ligaments, bony and cartilaginous surfaces, and soft tissues.

**Arthroscopy** may be carried out and damage to any internal structures repaired or trimmed. An arthroscope is a small camera that is placed into the knee joint through the anterolateral or anteromedial aspect of the knee joint. The joint is filled with a saline solution and the telescope is manipulated around the knee joint to assess the cruciate ligaments, menisci, and cartilaginous surfaces.

## In the clinic

**Anterolateral ligament of the knee**

A ligament associated at its origin with the fibular collateral ligament of the knee has been described. This ligament (anterolateral ligament of the knee) courses from the lateral femoral epicondyle to the anterolateral region of the proximal end of the tibia and may control internal rotation of the tibia. (*J Anat* 2013;223:321–328)

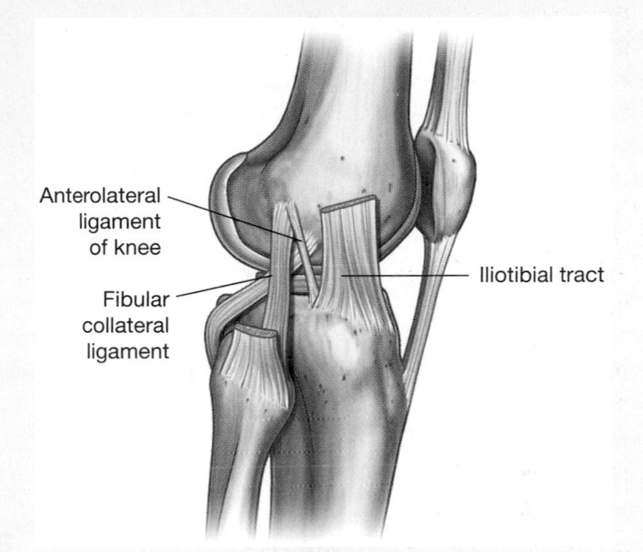


## Tibiofibular joint

The small proximal tibiofibular joint is synovial in type and allows very little movement (Fig. 6.82). The opposing joint surfaces, on the undersurface of the lateral condyle of the tibia and on the superomedial surface of the head of the fibula, are flat and circular. The capsule is reinforced by anterior and posterior ligaments.

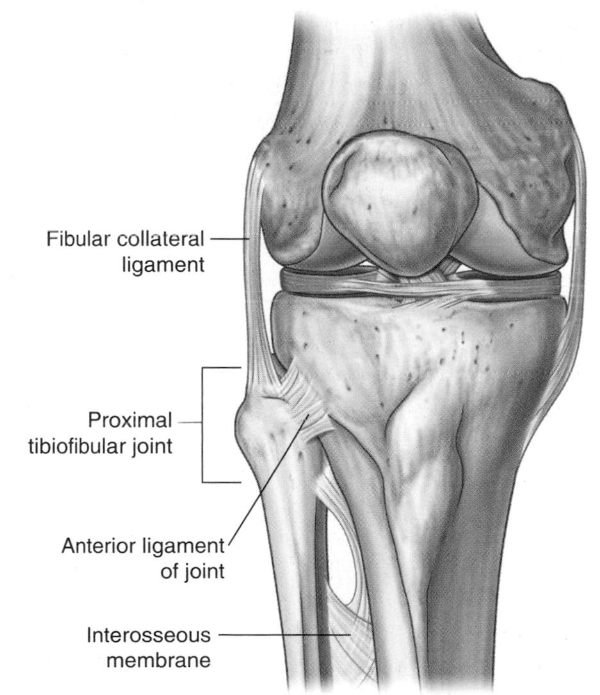


**Fig. 6.82** Tibiofibular joint.

## Popliteal fossa

The **popliteal fossa** is an important area of transition between the thigh and leg and is the major route by which structures pass from one region to the other.

The popliteal fossa is a diamond-shaped space behind the knee joint formed between muscles in the posterior compartments of the thigh and leg (Fig. 6.83A):

- The margins of the upper part of the diamond are formed medially by the distal ends of the semitendinosus and semimembranosus muscles and laterally by the distal end of the biceps femoris muscle.
- The margins of the smaller lower part of the space are formed medially by the medial head of the gastrocnemius muscle and laterally by the plantaris muscle and the lateral head of the gastrocnemius muscle.
- The floor of the fossa is formed by the capsule of the knee joint and adjacent surfaces of the femur and tibia, and, more inferiorly, by the popliteus muscle.
- The roof is formed by deep fascia, which is continuous above with the fascia lata of the thigh and below with deep fascia of the leg.

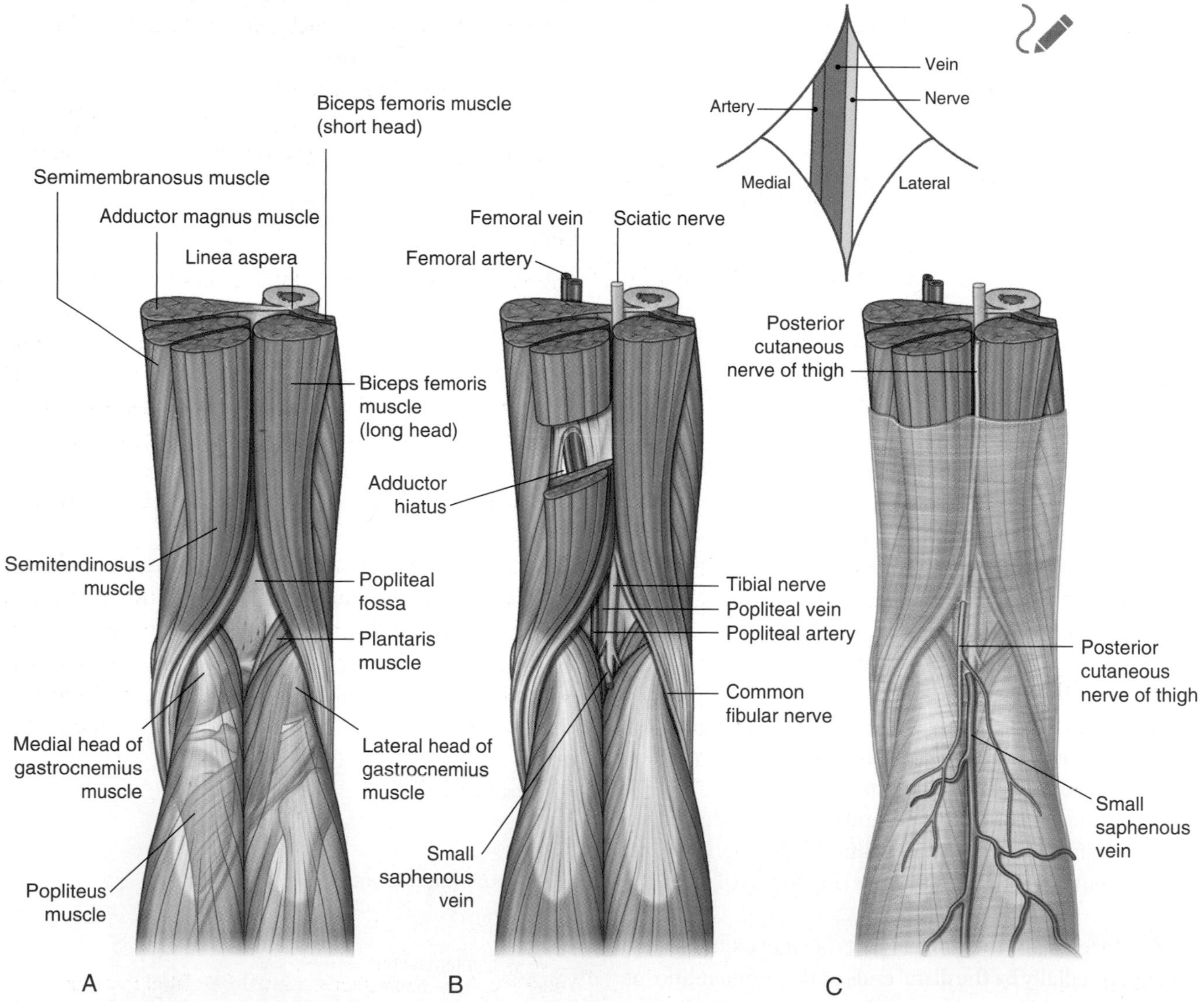

**Fig. 6.83** Popliteal fossa. **A.** Boundaries. **B.** Nerves and vessels. **C.** Superficial structures.

### Contents

The major contents of the popliteal fossa are the popliteal artery, the popliteal vein, and the tibial and common fibular nerves (Fig. 6.83B).

#### Tibial and common fibular nerves

The tibial and common fibular nerves originate proximal to the popliteal fossa as the two major branches of the sciatic nerve. They are the most superficial of the neurovascular structures in the popliteal fossa and enter the region directly from above under the margin of the biceps femoris muscle:

- The tibial nerve descends vertically through the popliteal fossa and exits deep to the margin of the plantaris muscle to enter the posterior compartment of the leg.
- The common fibular nerve exits by following the biceps femoris tendon over the lower lateral margin of the popliteal fossa, and continues to the lateral side of the leg where it swings around the neck of the fibula and enters the lateral compartment of the leg.

#### Popliteal artery and vein

The popliteal artery is the continuation of the femoral artery in the anterior compartment of the thigh, and begins as the femoral artery passes posteriorly through the adductor hiatus in the adductor magnus muscle.

The popliteal artery appears in the popliteal fossa on the upper medial side under the margin of the semimembranosus muscle. It descends obliquely through the fossa with the tibial nerve and enters the posterior compartment of the leg where it ends just lateral to the midline of the leg by dividing into the anterior and posterior tibial arteries.

The popliteal artery is the deepest of the neurovascular structures in the popliteal fossa and is therefore difficult to palpate; however, a pulse can usually be detected by deep palpation near the midline.

In the popliteal fossa, the popliteal artery gives rise to branches, which supply adjacent muscles, and to a series of geniculate arteries, which contribute to vascular anastomoses around the knee.

The popliteal vein is superficial to and travels with the popliteal artery. It exits the popliteal fossa superiorly to become the femoral vein by passing through the adductor hiatus.

### Roof of popliteal fossa

The roof of the popliteal fossa is covered by superficial fascia and skin (Fig. 6.83C). The most important structure in the superficial fascia is the small saphenous vein. This vessel ascends vertically in the superficial fascia on the back of the leg from the lateral side of the dorsal venous arch in the foot. It ascends to the back of the knee where it penetrates deep fascia, which forms the roof of the popliteal fossa, and joins with the popliteal vein.

One other structure that passes through the roof of the fossa is the posterior cutaneous nerve of the thigh, which descends through the thigh superficial to the hamstring muscles, passes through the roof of the popliteal fossa, and then continues inferiorly with the small saphenous vein to innervate skin on the upper half of the back of the leg.

### In the clinic

**Popliteal artery aneurysm**

The popliteal artery can become abnormally dilated, forming an aneurysm. The artery is considered aneurysmal when its diameter exceeds 7 mm. Although popliteal artery aneurysms can occur in isolation, they are most commonly associated with aneurysms in other large vessels such as the femoral artery or the thoracic or abdominal aorta. Therefore, once a popliteal aneurysm has been detected, the entire arterial tree needs to be investigated for the presence of coexisting aneurysms elsewhere in the body.

Popliteal artery aneurysms tend to undergo thrombosis and are less likely to rupture than other aneurysms. Therefore the complications are mainly related to distal embolization of the arterial tree and lower limb ischemia, which in the most severe cases can lead to leg amputation.

Ultrasound with duplex Doppler is the most helpful way of diagnosing a popliteal artery aneurysm because it can demonstrate abnormal dilation of the artery, confirm or rule out thrombus within the aneurysm, and help distinguish it from other masses of the popliteal fossa such as a synovial cyst (Baker's cyst). Popliteal artery aneurysms are usually repaired surgically in view of high risk of thromboembolic complications.

## LEG

The leg is that part of the lower limb between the knee joint and ankle joint (Fig. 6.84):

- Proximally, most major structures pass between the thigh and leg through or in relation to the popliteal fossa behind the knee.
- Distally, structures pass between the leg and foot mainly through the tarsal tunnel on the posteromedial side of the ankle, the exceptions being the anterior tibial artery and the ends of the deep and superficial fibular nerves, which enter the foot anterior to the ankle.

The bony framework of the leg consists of two bones, the tibia and fibula, arranged in parallel.

The **fibula** is much smaller than the tibia and is on the lateral side of the leg. It articulates superiorly with the inferior aspect of the lateral condyle of the proximal tibia, but does not take part in formation of the knee joint. The distal end of the fibula is firmly anchored to the tibia by a fibrous joint and forms the lateral malleolus of the ankle joint.

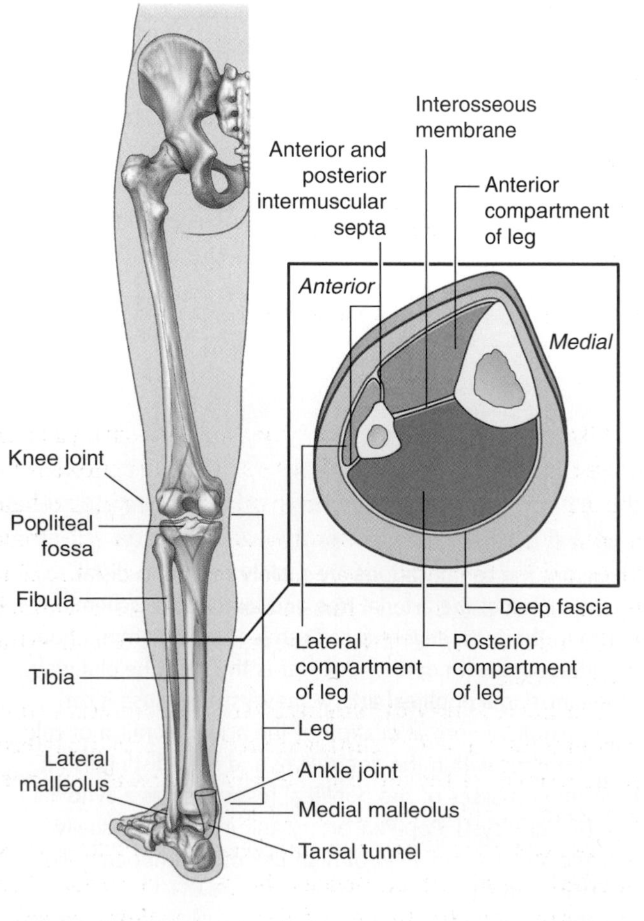

**Fig. 6.84** Posterior view of leg; cross section through the left leg (*inset*).

The **tibia** is the weight-bearing bone of the leg and is therefore much larger than the fibula. Above, it takes part in the formation of the knee joint and below it forms the medial malleolus and most of the bony surface for articulation of the leg with the foot at the ankle joint.

The leg is divided into anterior (extensor), posterior (flexor), and lateral (fibular) compartments by:

- an interosseous membrane, which links adjacent borders of the tibia and fibula along most of their length;
- two intermuscular septa, which pass between the fibula and deep fascia surrounding the limb; and
- direct attachment of the deep fascia to the periosteum of the anterior and medial borders of the tibia (Fig. 6.84).

Muscles in the anterior compartment of the leg dorsiflex the ankle, extend the toes, and invert the foot. Muscles in the posterior compartment plantarflex the ankle, flex the toes, and invert the foot. Muscles in the lateral compartment evert the foot. Major nerves and vessels supply or pass through each compartment.

### Bones

#### Shaft and distal end of tibia

The shaft of the tibia is triangular in cross section and has anterior, interosseous, and medial borders and medial, lateral, and posterior surfaces (Fig. 6.85):

- The anterior and medial borders and the entire medial surface are subcutaneous and easily palpable.
- The interosseous border of the tibia is connected, by the interosseous membrane, along its length to the interosseous border of the fibula.
- The posterior surface is marked by an oblique line (the soleal line).

The soleal line descends across the bone from the lateral side to the medial side where it merges with the medial border. In addition, a vertical line descends down the upper part of the posterior surface from the midpoint of the soleal line. It disappears in the lower one-third of the tibia.

The shaft of the tibia expands at both the upper and lower ends to support the body's weight at the knee and ankle joints.

The distal end of the tibia is shaped like a rectangular box with a bony protuberance on the medial side (the **medial malleolus**; Fig. 6.81). The upper part of the box is continuous with the shaft of the tibia while the lower

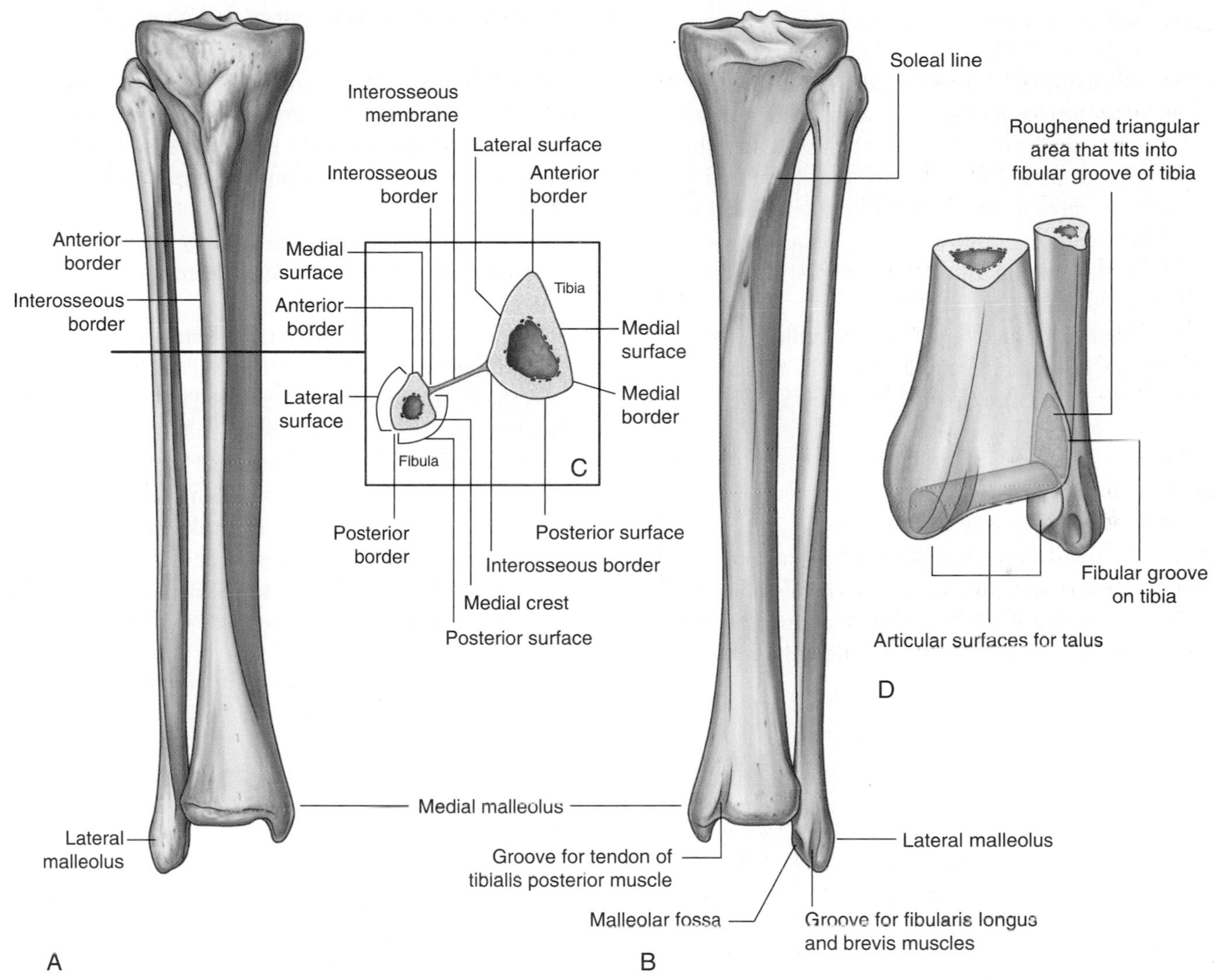


**Fig. 6.85** Tibia and fibula. **A.** Anterior view. **B.** Posterior view. **C.** Cross section through shafts. **D.** Posteromedial view of distal ends.

surface and the medial malleolus articulate with one of the tarsal bones (talus) to form a large part of the ankle joint.

The posterior surface of the box-like distal end of the tibia is marked by a vertical groove, which continues inferiorly and medially onto the posterior surface of the medial malleolus. The groove is for the tendon of the tibialis posterior muscle.

The lateral surface of the distal end of the tibia is occupied by a deep triangular notch (the **fibular notch**), to which the distal head of the fibula is anchored by a thickened part of the interosseous membrane.

### Shaft and distal end of fibula

The fibula is not involved in weight-bearing. The fibular shaft is therefore much narrower than the shaft of the tibia. Also, and except for the ends, the fibula is enclosed by muscles.

Like the tibia, the shaft of the fibula is triangular in cross section and has three borders and three surfaces for the attachment of muscles, intermuscular septa, and ligaments (Fig. 6.85). The interosseous border of the fibula faces and is attached to the interosseous border of the tibia by the interosseous membrane. Intermuscular septa attach to the anterior and posterior borders. Muscles attach to the three surfaces.

The narrow **medial surface** faces the anterior compartment of the leg, the **lateral surface** faces the lateral compartment of the leg, and the **posterior surface** faces the posterior compartment of the leg.

The posterior surface is marked by a vertical crest (**medial crest**), which divides the posterior surface into two parts each attached to a different deep flexor muscle.

The distal end of the fibula expands to form the spade-shaped **lateral malleolus** (Fig. 6.85).

The medial surface of the lateral malleolus bears a facet for articulation with the lateral surface of the talus, thereby forming the lateral part of the ankle joint. Just superior to this articular facet is a triangular area, which fits into the fibular notch on the distal end of the tibia. Here the tibia and fibula are joined together by the distal end of the interosseous membrane. Posteroinferior to the facet for articulation with the talus is a pit or fossa (the **malleolar fossa**) for the attachment of the posterior talofibular ligament associated with the ankle joint.

The posterior surface of the lateral malleolus is marked by a shallow groove for the tendons of the fibularis longus and fibularis brevis muscles.

## Joints

### Interosseous membrane of leg

The interosseous membrane of the leg is a tough fibrous sheet of connective tissue that spans the distance between facing interosseous borders of the tibial and fibular shafts (Fig. 6.86). The collagen fibers descend obliquely from the interosseous border of the tibia to the interosseous border of the fibula, except superiorly where there is a ligamentous band, which ascends from the tibia to fibula.

There are two apertures in the interosseous membrane, one at the top and the other at the bottom, for vessels to pass between the anterior and posterior compartments of the leg.

The interosseous membrane not only links the tibia and fibula together, but also provides an increased surface area for muscle attachment.

The distal ends of the fibula and tibia are held together by the inferior aspect of the interosseous membrane, which spans the narrow space between the fibular notch on the lateral surface of the distal end of the tibia and the corresponding surface on the distal end of the fibula. This expanded end of the interosseous membrane is reinforced by **anterior** and **posterior tibiofibular ligaments**. This firm linking together of the distal ends of the tibia and fibula is essential to produce the skeletal framework for articulation with the foot at the ankle joint.

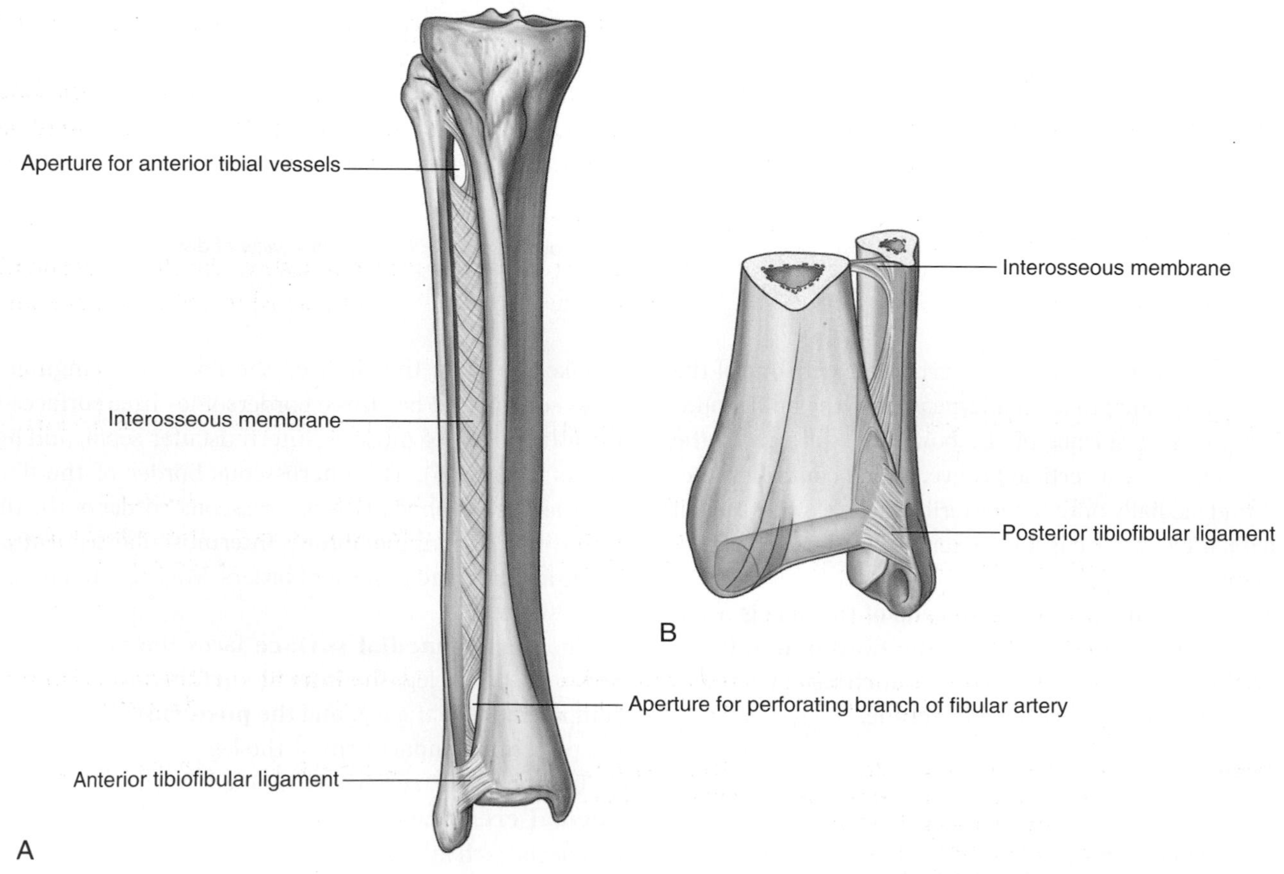


**Fig. 6.86** Interosseous membrane. **A.** Anterior view. **B.** Posteromedial view.

## Posterior compartment of leg

### Muscles

Muscles in the posterior (flexor) compartment of the leg are organized into two groups, superficial and deep, separated by a layer of deep fascia. Generally, the muscles mainly plantarflex and invert the foot and flex the toes. All are innervated by the tibial nerve.

#### Superficial group

The superficial group of muscles in the posterior compartment of the leg comprises three muscles—the gastrocnemius, plantaris, and soleus (Table 6.6)—all of which insert onto the heel (calcaneus) of the foot and plantarflex the foot at the ankle joint (Fig. 6.87). As a unit, these muscles are large and powerful because they propel the body forward off the planted foot during walking and can elevate the body upward onto the toes when standing. Two of the muscles (gastrocnemius and plantaris) originate on the distal end of the femur and can also flex the knee.

##### Gastrocnemius

The **gastrocnemius** muscle is the most superficial of the muscles in the posterior compartment and is one of the largest muscles in the leg (Fig. 6.87). It originates from two heads, one lateral and one medial:

- The **medial head** is attached to an elongate roughening on the posterior aspect of the distal femur just behind the adductor tubercle and above the articular surface of the medial condyle.
- The **lateral head** originates from a distinct facet on the upper lateral surface of the lateral femoral condyle where it joins the lateral supracondylar line.

At the knee, the facing margins of the two heads of the gastrocnemius form the lateral and medial borders of the lower end of the popliteal fossa.

In the upper leg, the heads of the gastrocnemius combine to form a single elongate muscle belly, which forms much of the soft tissue bulge identified as the **calf**.

In the lower leg, the muscle fibers of the gastrocnemius converge with those of the deeper soleus muscle to form the **calcaneal tendon**, which attaches to the calcaneus (heel) of the foot.

The gastrocnemius plantarflexes the foot at the ankle joint and can also flex the leg at the knee joint. It is innervated by the tibial nerve.

##### Plantaris

The **plantaris** has a small muscle belly proximally and a long thin tendon, which descends through the leg and joins the calcaneal tendon (Fig. 6.87). The muscle takes origin superiorly from the lower part of the lateral supracondylar ridge of the femur and from the oblique popliteal ligament associated with the knee joint.

The short spindle-shaped muscle body of the plantaris descends medially, deep to the lateral head of the gastrocnemius, and forms a thin tendon, which passes between the gastrocnemius and soleus muscles and eventually fuses with the medial side of the calcaneal tendon near its attachment to the calcaneus.

The plantaris contributes to plantarflexion of the foot at the ankle joint and flexion of the leg at the knee joint, and is innervated by the tibial nerve.

##### Soleus

The **soleus** is a large flat muscle under the gastrocnemius muscle (Fig. 6.87). It is attached to the proximal ends of

**Table 6.6** Superficial group of muscles in the posterior compartment of leg (spinal segments in bold are the major segments innervating the muscle)

| Muscle | Origin | Insertion | Innervation | Function |
|---|---|---|---|---|
| Gastrocnemius | Medial head—posterior surface of distal femur just superior to medial condyle; lateral head—upper posterolateral surface of lateral femoral condyle | Via calcaneal tendon, to posterior surface of calcaneus | Tibial nerve (**S1, S2**) | Plantarflexes foot and flexes knee |
| Plantaris | Inferior part of lateral supracondylar line of femur and oblique popliteal ligament of knee | Via calcaneal tendon, to posterior surface of calcaneus | Tibial nerve (**S1, S2**) | Plantarflexes foot and flexes knee |
| Soleus | Soleal line and medial border of tibia; posterior aspect of fibular head and adjacent surfaces of neck and proximal shaft; tendinous arch between tibial and fibular attachments | Via calcaneal tendon, to posterior surface of calcaneus | Tibial nerve (**S1, S2**) | Plantarflexes the foot |

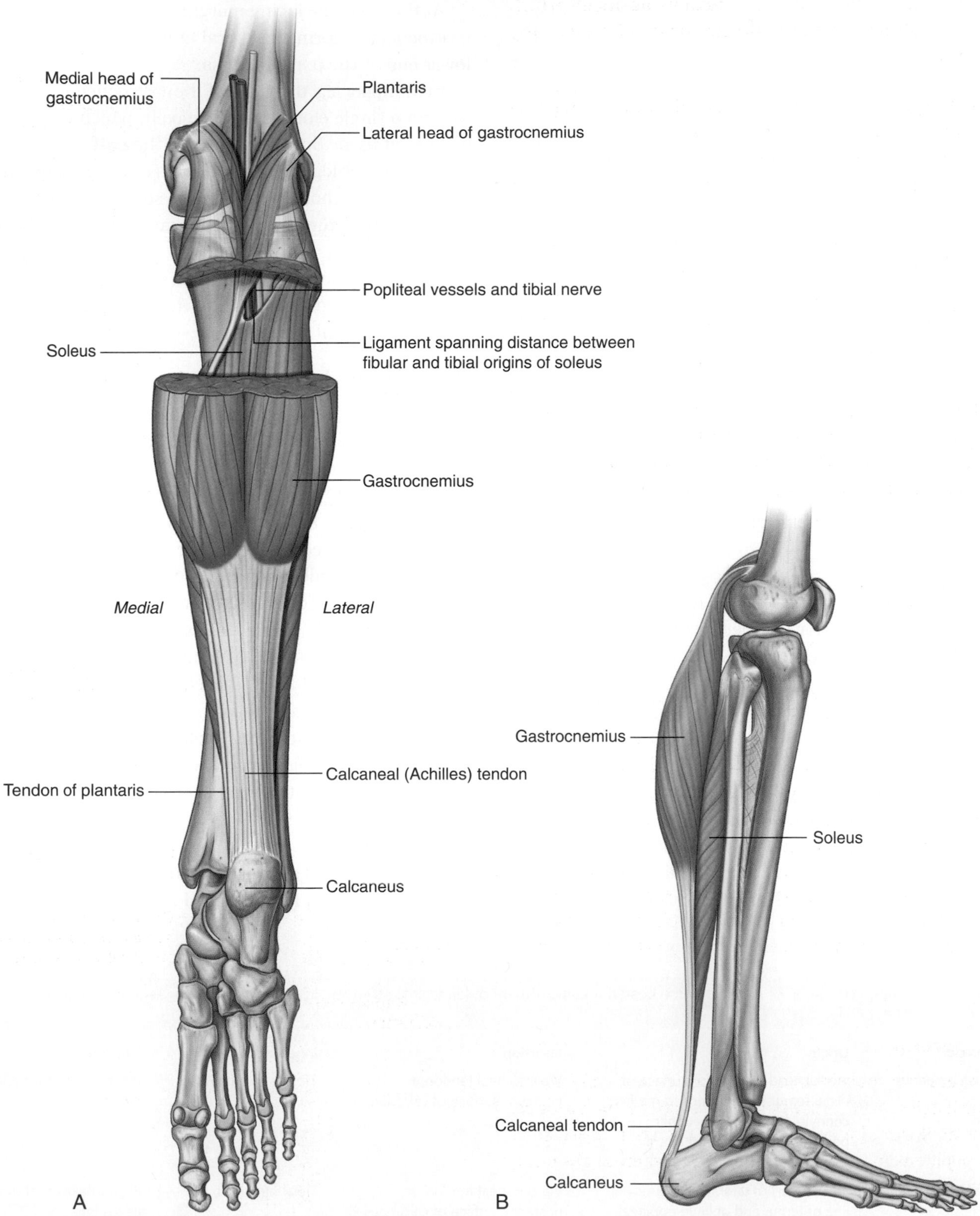

Medial head of gastrocnemius
Plantaris
Lateral head of gastrocnemius
Popliteal vessels and tibial nerve
Soleus
Ligament spanning distance between fibular and tibial origins of soleus
Gastrocnemius
*Medial*
*Lateral*
Tendon of plantaris
Calcaneal (Achilles) tendon
Calcaneus
A

Gastrocnemius
Soleus
Calcaneal tendon
Calcaneus
B

**Fig. 6.87** Superficial group of muscles in the posterior compartment of leg. **A.** Posterior view. **B.** Lateral view.

the fibula and tibia, and to a tendinous ligament, which spans the distance between the two heads of attachment to the fibula and tibia:

- On the proximal end of the fibula, the soleus originates from the posterior aspect of the head and adjacent surface of the neck and upper shaft of the fibula.
- On the tibia, the soleus originates from the soleal line and adjacent medial border.
- The ligament, which spans the distance between the attachments to the tibia and fibula, arches over the popliteal vessels and tibial nerve as they pass from the popliteal fossa into the deep region of the posterior compartment of the leg.

In the lower leg, the soleus muscle narrows to join the calcaneal tendon that attaches to the calcaneus.

The soleus muscle, together with the gastrocnemius and plantaris, plantarflexes the foot at the ankle joint. It is innervated by the tibial nerve.

## In the clinic

**Calcaneal (Achilles) tendon rupture**

Rupture of the calcaneal tendon is often related to sudden or direct trauma. This type of injury frequently occurs in a normal healthy tendon. In addition, there are certain conditions that may predispose the tendon to rupture. Among these conditions are tendinopathy (due to overuse, or to age-related degenerative changes) and previous calcaneal tendon interventions such as injections of pharmaceuticals and the use of certain antibiotics (quinolone group). The diagnosis of calcaneal tendon rupture is relatively straightforward. The patient typically complains of "being kicked" or "shot" behind the ankle, and clinical examination often reveals a gap in the tendon.

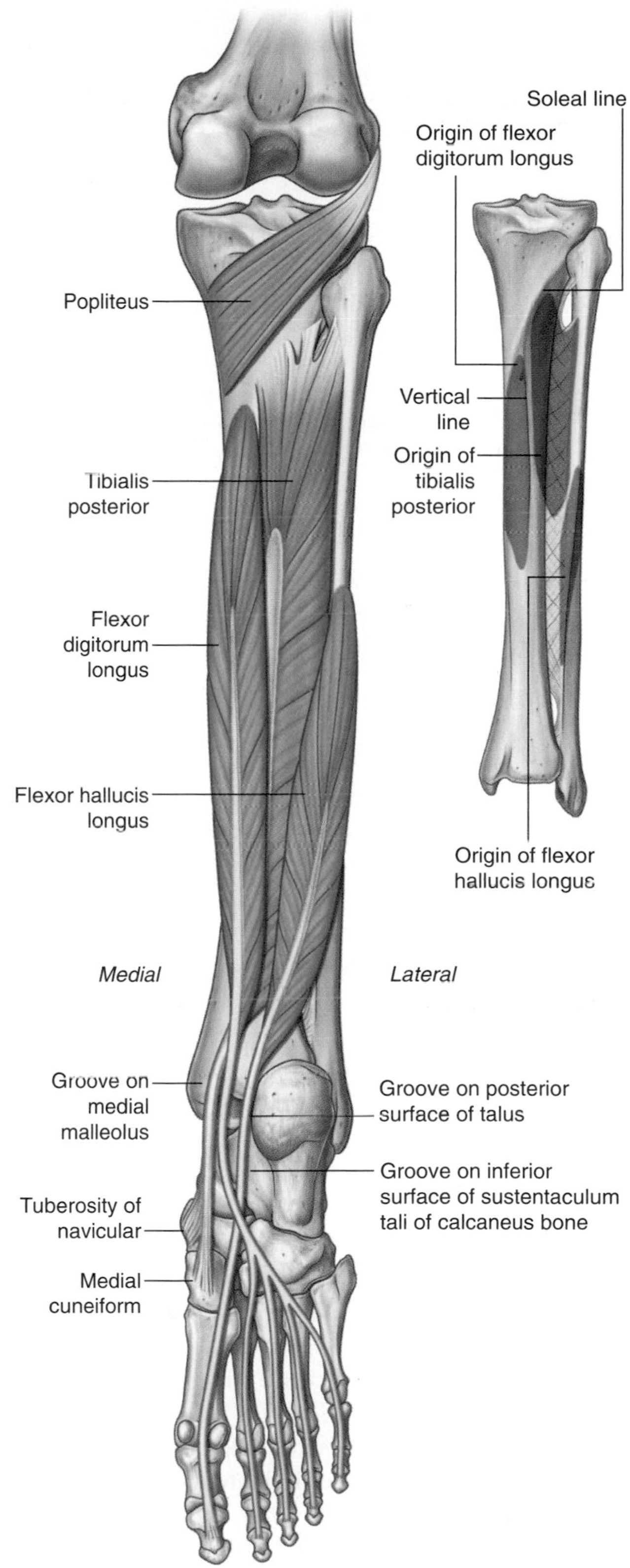


**Fig. 6.88** Deep group of muscles in the posterior compartment of leg.

### Deep group

There are four muscles in the deep posterior compartment of the leg (Fig. 6.88)—the popliteus, flexor hallucis longus, flexor digitorum longus, and tibialis posterior (Table 6.7). The popliteus muscle acts on the knee, whereas the other three muscles act mainly on the foot.

#### Popliteus

The **popliteus** is the smallest and most superior of the deep muscles in the posterior compartment of the leg. It unlocks the extended knee at the initiation of flexion and stabilizes the knee by resisting lateral (external) rotation of the tibia on the femur. It is flat and triangular in shape,

**Table 6.7** Deep group of muscles in the posterior compartment of leg (spinal segments in bold are the major segments innervating the muscle)

| Muscle | Origin | Insertion | Innervation | Function |
|---|---|---|---|---|
| Popliteus | Lateral femoral condyle | Posterior surface of proximal tibia | Tibial nerve (L4 to S1) | Stabilizes knee joint (resists lateral rotation of tibia on femur)<br>Unlocks knee joint (laterally rotates femur on fixed tibia) |
| Flexor hallucis longus | Posterior surface of fibula and adjacent interosseous membrane | Plantar surface of distal phalanx of great toe | Tibial nerve (**S2**, S3) | Flexes great toe |
| Flexor digitorum longus | Medial side of posterior surface of the tibia | Plantar surfaces of bases of distal phalanges of the lateral four toes | Tibial nerve (**S2**, S3) | Flexes lateral four toes |
| Tibialis posterior | Posterior surfaces of interosseous membrane and adjacent regions of tibia and fibula | Mainly to tuberosity of navicular and adjacent region of medial cuneiform | Tibial nerve (L4, L5) | Inversion and plantarflexion of foot; support of medial arch of foot during walking |

forms part of the floor of the popliteal fossa (Fig. 6.88), and is inserted into a broad triangular region above the soleal line on the posterior surface of the tibia.

The popliteus muscle ascends laterally across the lower aspect of the knee and originates from a tendon, which penetrates the fibrous membrane of the joint capsule of the knee. The tendon ascends laterally around the joint where it passes between the lateral meniscus and the fibrous membrane and then into a groove on the inferolateral aspect of the lateral femoral condyle. The tendon attaches to and originates from a depression at the anterior end of the groove.

When initiating gait from a standing position, contraction of the popliteus laterally rotates the femur on the fixed tibia, unlocking the knee joint. The popliteus muscle is innervated by the tibial nerve.

### Flexor hallucis longus

The flexor hallucis longus muscle originates on the lateral side of the posterior compartment of the leg and inserts into the plantar surface of the great toe on the medial side of the foot (Fig. 6.88). It arises mainly from the lower two-thirds of the posterior surface of the fibula and adjacent interosseous membrane.

The muscle fibers of the flexor hallucis longus converge inferiorly to form a large cord-like tendon, which passes behind the distal head of the tibia and then slips into a distinct groove on the posterior surface of the adjacent tarsal bone (talus) of the foot. The tendon curves anteriorly first under the talus and then under a shelf of bone (the sustentaculum tali), which projects medially from the calcaneus, and then continues anteriorly through the sole of the foot to insert on the inferior surface of the base of the distal phalanx of the great toe.

The flexor hallucis longus flexes the great toe. It is particularly active during the toe-off phase of walking when the body is propelled forward off the stance leg and the great toe is the last part of the foot to leave the ground. It can also contribute to plantarflexion of the foot at the ankle joint and is innervated by the tibial nerve.

### Flexor digitorum longus

The flexor digitorum longus muscle originates on the medial side of the posterior compartment of the leg and inserts into the lateral four digits of the foot (Fig. 6.88). It arises mainly from the medial side of the posterior surface of the tibia inferior to the soleal line.

The flexor digitorum longus descends in the leg and forms a tendon, which crosses posterior to the tendon of the tibialis posterior muscle near the ankle joint. The tendon continues inferiorly in a shallow groove behind the medial malleolus and then swings forward to enter the sole of the foot. It crosses inferior to the tendon of the flexor hallucis longus muscle to reach the medial side of the foot and then divides into four tendons, which insert on the plantar surfaces of the bases of the distal phalanges of digits II to V.

The flexor digitorum longus flexes the lateral four toes. It is involved with gripping the ground during walking and propelling the body forward off the toes at the end of the stance phase of gait. It is innervated by the tibial nerve.

### Tibialis posterior

The tibialis posterior muscle originates from the interosseous membrane and the adjacent posterior surfaces of the tibia and fibula (Fig. 6.88). It lies between and is overlapped by the flexor digitorum longus and the flexor hallucis longus muscles.

Near the ankle, the tendon of the tibialis posterior is crossed superficially by the tendon of the flexor digitorum longus muscle and lies medial to this tendon in the groove on the posterior surface of the medial malleolus. The tendon curves forward under the medial malleolus and enters the medial side of the foot. It wraps around the medial margin of the foot to attach to the plantar surfaces of the medial tarsal bones, mainly to the tuberosity of the navicular and to the adjacent region of the medial cuneiform.

The tibialis posterior inverts and plantarflexes the foot, and supports the medial arch of the foot during walking. It is innervated by the tibial nerve.

## In the clinic

**Neurological examination of the legs**

Some of the commonest conditions that affect the legs are peripheral neuropathy (particularly associated with diabetes mellitus), lumbar nerve root lesions (associated with pathology of the intervertebral discs), fibular nerve palsy, and spastic paraparesis.

- Look for muscle wasting—loss of muscle mass may indicate loss of or reduced innervation.
- Test the power in muscle groups—hip flexion (L1, L2—iliopsoas—straight leg raise); knee flexion (L5 to S2—hamstrings—the patient tries to bend the knee while the examiner applies force to the leg to hold the knee in extension); knee extension (L3, L4—quadriceps femoris—the patient attempts to keep the leg straight while the examiner applies a force to the leg to flex the knee joint); ankle plantarflexion (S1, S2—the patient pushes the foot down while the examiner applies a force to the plantar surface of the foot to dorsiflex the ankle joint); ankle dorsiflexion (L4, L5—the patient pulls the foot upward while the examiner applies a force to the dorsal aspect of the foot to plantarflex the ankle joint).
- Examine knee and ankle reflexes—a tap with a tendon hammer on the patellar ligament (tendon) tests reflexes at the L3–L4 spinal levels, and tapping the calcaneal tendon tests reflexes at the S1–S2 spinal levels.
- Assess status of general sensory input to lumbar and upper sacral spinal cord levels—test light touch, pin prick, and vibration sense at dermatomes in the lower limb.

## Arteries

### Popliteal artery

The **popliteal artery** is the major blood supply to the leg and foot and enters the posterior compartment of the leg from the popliteal fossa behind the knee (Fig. 6.89).

The popliteal artery passes into the posterior compartment of the leg between the gastrocnemius and popliteus muscles. As it continues inferiorly it passes under the tendinous arch formed between the fibular and tibial heads of the soleus muscle and enters the deep region of the posterior compartment of the leg where it immediately divides into an anterior tibial artery and a posterior tibial artery.

Two large sural arteries, one on each side, branch from the popliteal artery to supply the gastrocnemius, soleus, and plantaris muscles (Fig. 6.89). In addition, the popliteal artery gives rise to branches that contribute to a collateral network of vessels around the knee joint (see Fig. 6.80).

#### Anterior tibial artery

The **anterior tibial artery** passes forward through the aperture in the upper part of the interosseous membrane and enters and supplies the anterior compartment of the leg. It continues inferiorly onto the dorsal aspect of the foot.

#### Posterior tibial artery

The **posterior tibial artery** supplies the posterior and lateral compartments of the leg and continues into the sole of the foot (Fig. 6.89).

The posterior tibial artery descends through the deep region of the posterior compartment of the leg on the superficial surfaces of the tibialis posterior and flexor digitorum longus muscles. It passes through the tarsal tunnel behind the medial malleolus and into the sole of the foot.

In the leg, the posterior tibial artery supplies adjacent muscles and bone and has two major branches, the circumflex fibular artery and the fibular artery:

- The **circumflex fibular artery** passes laterally through the soleus muscle and around the neck of the fibula to connect with the anastomotic network of vessels surrounding the knee (Fig. 6.89; see also Fig. 6.80).
- The **fibular artery** parallels the course of the tibial artery, but descends along the lateral side of the posterior compartment adjacent to the medial crest on the posterior surface of the fibula, which separates the attachments of the tibialis posterior and flexor hallucis longus muscles.

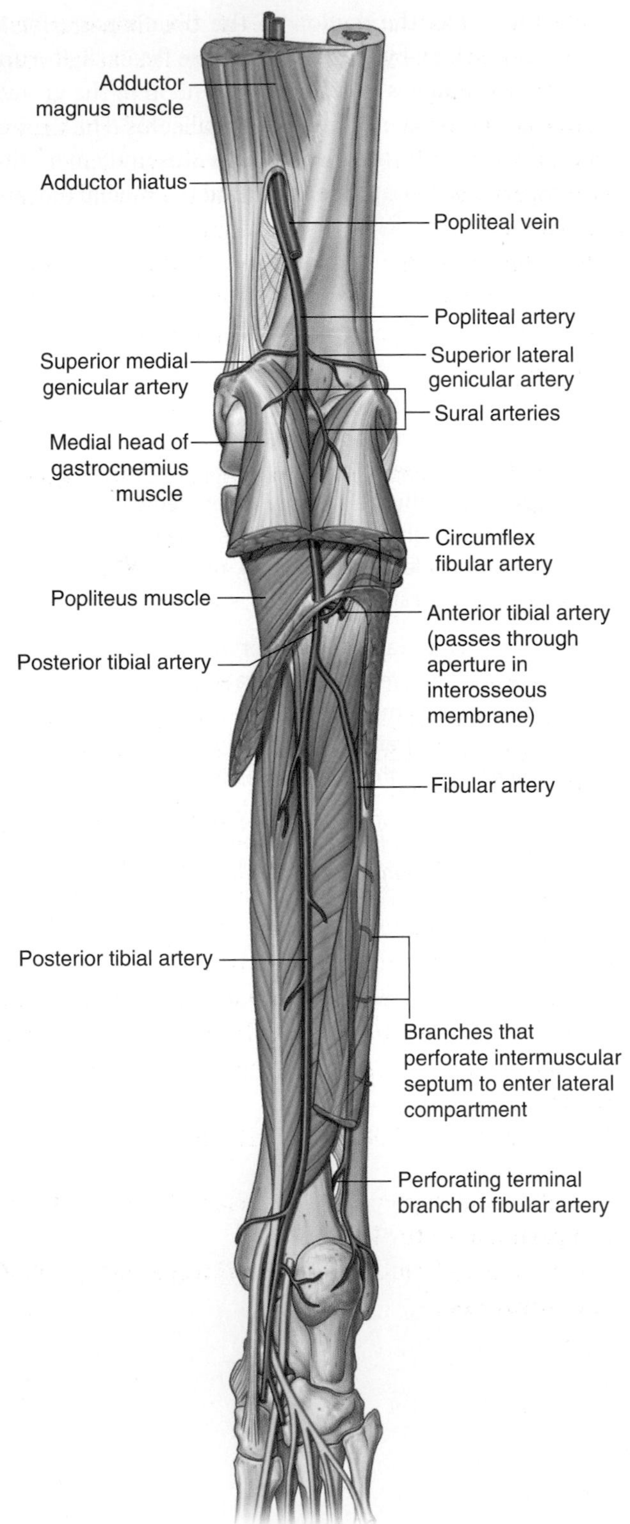


**Fig. 6.89** Arteries in the posterior compartment of leg.

The fibular artery supplies adjacent muscles and bone in the posterior compartment of the leg and also has branches that pass laterally through the intermuscular septum to supply the fibularis muscles in the lateral compartment of the leg.

A **perforating branch** that originates from the fibular artery distally in the leg passes anteriorly through the inferior aperture in the interosseous membrane to anastomose with a branch of the anterior tibial artery.

The fibular artery passes behind the attachment between the distal ends of the tibia and fibula and terminates in a network of vessels over the lateral surface of the calcaneus.

## Veins

Deep veins in the posterior compartment generally follow the arteries.

## Nerves

### Tibial nerve

The nerve associated with the posterior compartment of the leg is the tibial nerve (Fig. 6.90), a major branch of the sciatic nerve that descends into the posterior compartment from the popliteal fossa.

The tibial nerve passes under the tendinous arch formed between the fibular and tibial heads of the soleus muscle and passes vertically through the deep region of the posterior compartment of the leg on the surface of the tibialis posterior muscle with the posterior tibial vessels.

The tibial nerve leaves the posterior compartment of the leg at the ankle by passing through the tarsal tunnel behind the medial malleolus. It enters the foot to supply most intrinsic muscles and skin.

In the leg, the tibial nerve gives rise to:

- branches that supply all the muscles in the posterior compartment of the leg, and
- two cutaneous branches, the **sural nerve** and **medial calcaneal nerve**.

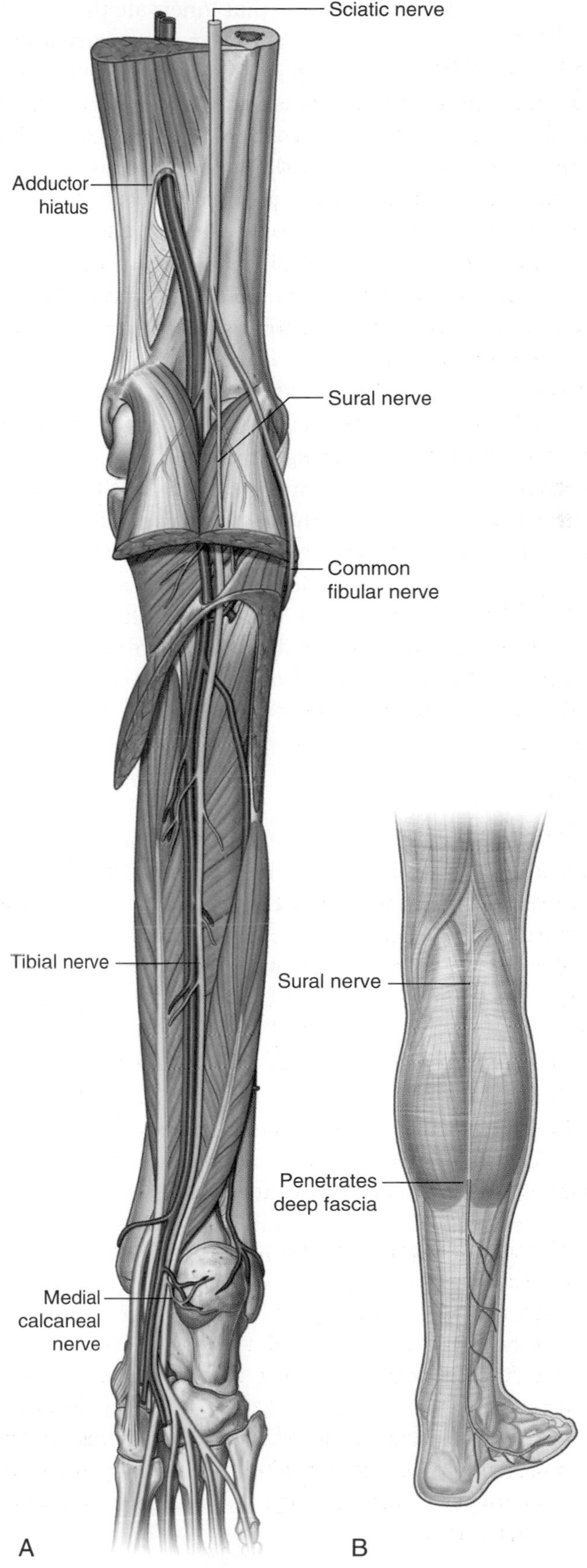


**Fig. 6.90** Tibial nerve. **A.** Posterior view. **B.** Sural nerve.

Branches of the tibial nerve that innervate the superficial group of muscles of the posterior compartment and popliteus muscle of the deep group originate high in the leg between the two heads of the gastrocnemius muscle in the distal region of the popliteal fossa (Fig. 6.91). Branches innervate the gastrocnemius, plantaris, and soleus muscles, and pass more deeply into the popliteus muscle.

Branches to the deep muscles of the posterior compartment originate from the tibial nerve deep to the soleus muscle in the upper half of the leg and innervate the tibialis posterior, flexor hallucis longus, and flexor digitorum longus muscles.

### Sural nerve

The sural nerve originates high in the leg between the two heads of the gastrocnemius muscle (Fig. 6.90). It descends superficial to the belly of the gastrocnemius muscle and penetrates through the deep fascia approximately in the middle of the leg where it is joined by a sural communicating branch from the common fibular nerve. It passes down the leg, around the lateral malleolus, and into the foot.

The sural nerve supplies skin on the lower posterolateral surface of the leg and the lateral side of the foot and little toe.

### Medial calcaneal nerve

The medial calcaneal nerve is often multiple and originates from the tibial nerve low in the leg near the ankle and descends onto the medial side of the heel.

The medial calcaneal nerve innervates skin on the medial surface and sole of the heel (Fig. 6.90).

## Lateral compartment of leg

### Muscles

There are two muscles in the lateral compartment of the leg—the fibularis longus and fibularis brevis (Fig. 6.91 and Table 6.8). Both evert the foot (turn the sole outward) and are innervated by the superficial fibular nerve, which is a branch of the common fibular nerve.

#### Fibularis longus

The **fibularis longus** muscle arises in the lateral compartment of the leg, but its tendon crosses under the foot to attach to bones on the medial side (Fig. 6.91). It originates from both the upper lateral surface of the fibula and from

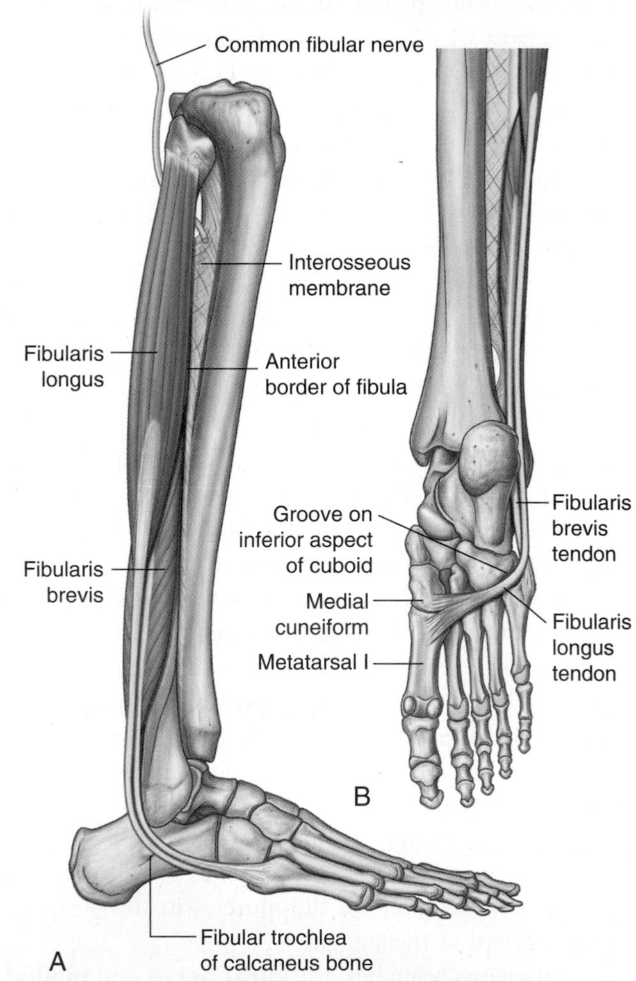


**Fig. 6.91** Muscles in the lateral compartment of leg. **A.** Lateral view. **B.** Inferior view of the right foot, with the foot plantarflexed at the ankle.

**Table 6.8** Muscles of the lateral compartment of leg (spinal segments in bold are the major segments innervating the muscle)

| Muscle | Origin | Insertion | Innervation | Function |
|---|---|---|---|---|
| Fibularis longus | Upper lateral surface of fibula, head of fibula, and occasionally the lateral tibial condyle | Undersurface of lateral sides of distal end of medial cuneiform and base of metatarsal I | Superficial fibular nerve (**L5, S1**, S2) | Eversion and plantarflexion of foot; supports arches of foot |
| Fibularis brevis | Lower two-thirds of lateral surface of shaft of fibula | Lateral tubercle at base of metatarsal V | Superficial fibular nerve (**L5, S1**, S2) | Eversion of foot |

the anterior aspect of the fibular head and occasionally up onto the adjacent region of the lateral tibial condyle.

The common fibular nerve passes anteriorly around the fibular neck between the attachments of the fibularis longus to the fibular head and shaft.

Distally, the fibularis longus descends in the leg to form a tendon, which, in order:

- passes posterior to the lateral malleolus in a shallow bony groove,
- swings forward to enter the lateral side of the foot,
- descends obliquely down the lateral side of the foot where it curves forward under a bony tubercle (fibular trochlea) of the calcaneus,
- enters a deep groove on the inferior surface of one of the other tarsal bones (the cuboid), and
- swings under the foot to cross the sole and attach to the inferior surfaces of bones on the medial side of the foot (lateral sides of the base of metatarsal I and the distal end of the medial cuneiform).

The fibularis longus everts and plantarflexes the foot. In addition, the fibularis longus, tibialis anterior, and tibialis posterior muscles, which all insert on the undersurfaces of bones on the medial side of the foot, together act as a stirrup to support the arches of the foot. The fibularis longus supports mainly the lateral and transverse arches.

The fibularis longus is innervated by the superficial fibular nerve.

### Fibularis brevis

The fibularis brevis muscle is deep to the fibularis longus muscle in the leg and originates from the lower two-thirds of the lateral surface of the shaft of the fibula (Fig. 6.91).

The tendon of the fibularis brevis passes behind the lateral malleolus with the tendon of the fibularis longus muscle and then curves forward across the lateral surface of the calcaneus to attach to a tubercle on the lateral surface of the base of metatarsal V (the metatarsal associated with the little toe).

The fibularis brevis assists in eversion of the foot and is innervated by the superficial fibular nerve.

### Arteries

No major artery passes vertically through the lateral compartment of the leg. It is supplied by branches (mainly from the fibular artery in the posterior compartment of the leg) that penetrate into the lateral compartment (Fig. 6.92).

### Veins

Deep veins generally follow the arteries.

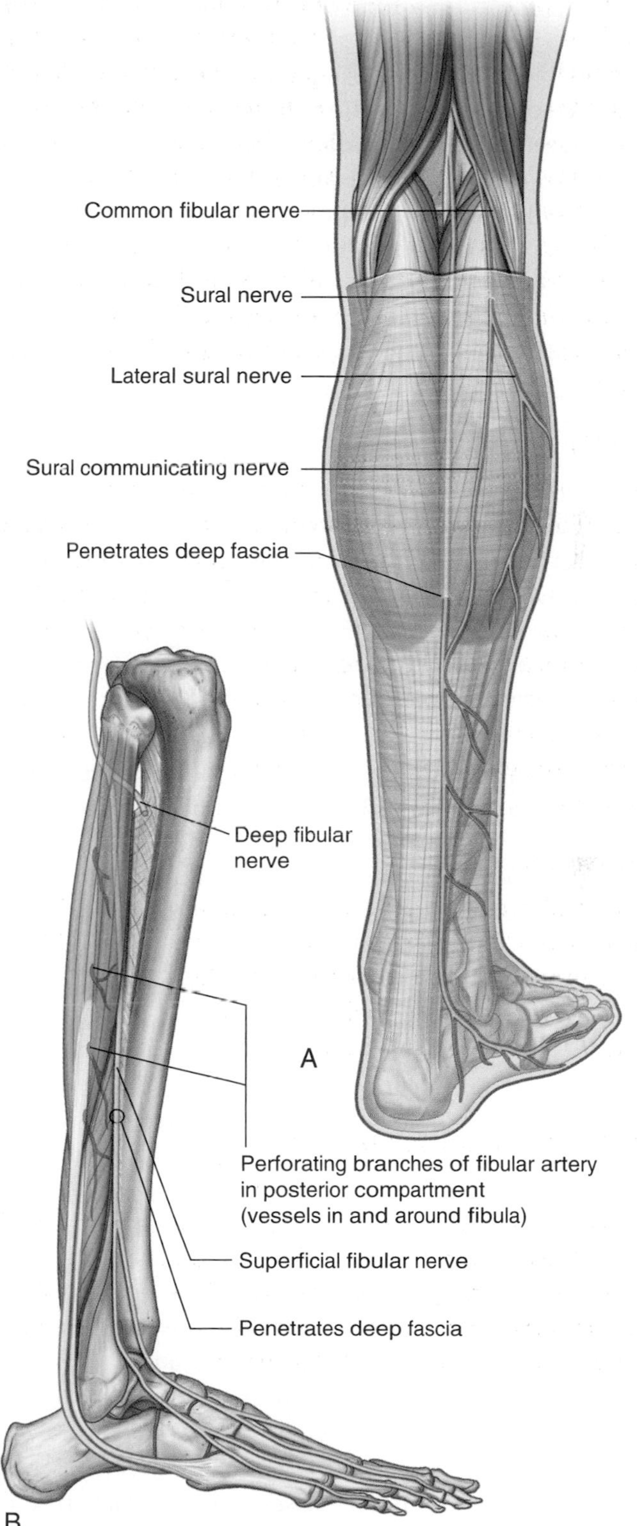


**Fig. 6.92** Common fibular nerve, and nerves and arteries of the lateral compartment of leg. **A.** Posterior view, right leg. **B.** Lateral view, right leg.

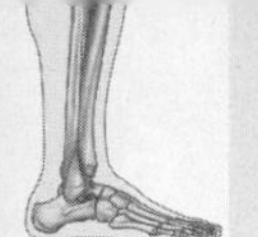



## Nerves

### Superficial fibular nerve

The nerve associated with the lateral compartment of the leg is the **superficial fibular nerve**. This nerve originates as one of the two major branches of the common fibular nerve, which enters the lateral compartment of the leg from the popliteal fossa (Fig. 6.92B).

The common fibular nerve originates from the sciatic nerve in the posterior compartment of the thigh or in the popliteal fossa (Fig. 6.92A), and follows the medial margin of the biceps femoris tendon over the lateral head of the gastrocnemius muscle and toward the fibula. Here it gives origin to two cutaneous branches, which descend in the leg:

- the **sural communicating nerve**, which joins the sural branch of the tibial nerve and contributes to innervation of skin over the lower posterolateral side of the leg; and
- the **lateral sural cutaneous nerve**, which innervates skin over the upper lateral leg.

The common fibular nerve continues around the neck of the fibula and enters the lateral compartment by passing between the attachments of the fibularis longus muscle to the head and shaft of the fibula. Here the common fibular nerve divides into its two terminal branches:

- the superficial fibular nerve, and
- the deep fibular nerve.

The superficial fibular nerve descends in the lateral compartment deep to the fibularis longus and innervates the fibularis longus and fibularis brevis (Fig. 6.91B). It then penetrates deep fascia in the lower leg and enters the foot where it divides into medial and lateral branches, which supply dorsal areas of the foot and toes except for:

- the web space between the great and second toes, which is supplied by the deep fibular nerve; and
- the lateral side of the little toe, which is supplied by the sural branch of the tibial nerve.

The deep fibular nerve passes anteromedially through the intermuscular septum into the anterior compartment of the leg, which it supplies.

## Anterior compartment of leg

### Muscles

There are four muscles in the anterior compartment of the leg—the tibialis anterior, extensor hallucis longus, extensor digitorum longus, and fibularis tertius (Fig. 6.93 and Table 6.9). Collectively they dorsiflex the foot at the ankle joint, extend the toes, and invert the foot. All are innervated by the deep fibular nerve, which is a branch of the common fibular nerve.



### Tibialis anterior

The **tibialis anterior** muscle is the most anterior and medial of the muscles in the anterior compartment of the

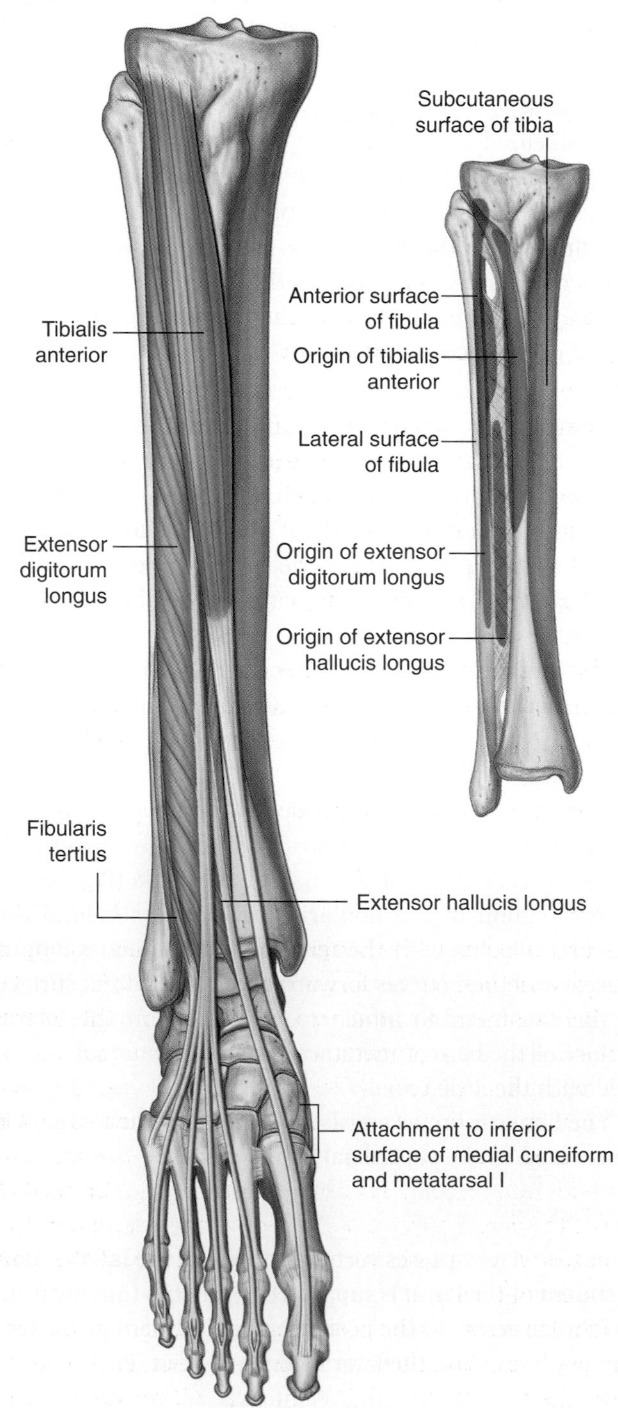


**Fig. 6.93** Muscles of the anterior compartment of leg.

**Table 6.9** Muscles of the anterior compartment of leg (spinal segments in bold are the major segments innervating the muscle)

| Muscle | Origin | Insertion | Innervation | Function |
|---|---|---|---|---|
| Tibialis anterior | Lateral surface of tibia and adjacent interosseous membrane | Medial and inferior surfaces of medial cuneiform and adjacent surfaces on base of metatarsal I | Deep fibular nerve (**L4**, L5) | Dorsiflexion of foot at ankle joint; inversion of foot; dynamic support of medial arch of foot |
| Extensor hallucis longus | Middle one-half of medial surface of fibula and adjacent surface of interosseous membrane | Dorsal surface of base of distal phalanx of great toe | Deep fibular nerve (**L5, S1**) | Extension of great toe and dorsiflexion of foot |
| Extensor digitorum longus | Proximal one-half of medial surface of fibula and related surface of lateral tibial condyle | Via dorsal digital expansions into bases of distal and middle phalanges of lateral four toes | Deep fibular nerve (**L5, S1**) | Extension of lateral four toes and dorsiflexion of foot |
| Fibularis tertius | Distal part of medial surface of fibula | Dorsomedial surface of base of metatarsal V | Deep fibular nerve (**L5, S1**) | Dorsiflexion and eversion of foot |

leg (Fig. 6.93). It originates mainly from the upper two-thirds of the lateral surface of the shaft of the tibia and adjacent surface of the interosseous membrane. It also originates from deep fascia.

The muscle fibers of the tibialis anterior converge in the lower one-third of the leg to form a tendon, which descends into the medial side of the foot, where it attaches to the medial and inferior surfaces of one of the tarsal bones (medial cuneiform) and adjacent parts of metatarsal I associated with the great toe.

The tibialis anterior dorsiflexes the foot at the ankle joint and inverts the foot at the intertarsal joints. During walking, it provides dynamic support for the medial arch of the foot.

The tibialis anterior is innervated by the deep fibular nerve.

### Extensor hallucis longus

The **extensor hallucis longus** muscle lies next to and is partly overlapped by the tibialis anterior muscle (Fig. 6.93). It originates from the middle one-half of the medial surface of the fibula and adjacent interosseous membrane.

The tendon of the extensor hallucis longus appears between the tendons of the tibialis anterior and extensor digitorum longus in the lower one-half of the leg and descends into the foot. It continues anteriorly on the medial side of the dorsal surface of the foot to near the end of the great toe where it inserts on the upper surface of the base of the distal phalanx.

The extensor hallucis longus extends the great toe. Because it crosses anterior to the ankle joint, it also dorsiflexes the foot at the ankle joint. Like all muscles in the anterior compartment of the leg, the extensor hallucis longus muscle is innervated by the deep fibular nerve.

### Extensor digitorum longus

The **extensor digitorum longus** muscle is the most posterior and lateral of the muscles in the anterior compartment of the leg (Fig. 6.93). It originates mainly from the upper one-half of the medial surface of the fibula lateral to and above the origin of the extensor hallucis longus muscle, and extends superiorly onto the lateral condyle of the tibia. Like the tibialis anterior muscle, it also originates from deep fascia.

The extensor digitorum longus muscle descends to form a tendon, which continues into the dorsal aspect of the foot, where it divides into four tendons, which insert, via dorsal digital expansions, into the dorsal surfaces of the bases of the middle and distal phalanges of the lateral four toes.

The extensor digitorum longus extends the toes and dorsiflexes the foot at the ankle joint, and is innervated by the deep fibular nerve.

### Fibularis tertius

The **fibularis tertius** muscle is normally considered part of the extensor digitorum longus (Fig. 6.93). The fibularis tertius originates from the medial surface of the fibula immediately below the origin of the extensor digitorum longus muscle and the two muscles are normally connected.

The tendon of the fibularis tertius descends into the foot with the tendon of the extensor digitorum longus. On the dorsal aspect of the foot, it deviates laterally to insert into the dorsomedial surface of the base of metatarsal V (the metatarsal associated with the little toe).

The fibularis tertius assists in dorsiflexion and possibly eversion of the foot, and is innervated by the deep fibular nerve.

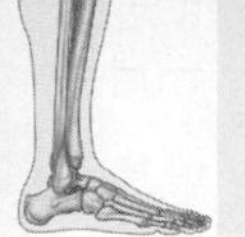



## Arteries

### Anterior tibial artery

The artery associated with the anterior compartment of the leg is the **anterior tibial artery**, which originates from the popliteal artery in the posterior compartment of the leg and passes forward into the anterior compartment of the leg through an aperture in the interosseous membrane.

The anterior tibial artery descends through the anterior compartment on the interosseous membrane (Fig. 6.94). In the distal leg, it lies between the tendons of the tibialis anterior and extensor hallucis longus muscles. It leaves the leg by passing anterior to the distal end of the tibia and ankle joint and continues onto the dorsal aspect of the foot as the dorsalis pedis artery.

In the proximal leg, the anterior tibial artery has a recurrent branch, which connects with the anastomotic network of vessels around the knee joint.

Along its course, the anterior tibial artery supplies numerous branches to adjacent muscles and is joined by the perforating branch of the fibular artery, which passes forward through the lower aspect of the interosseous membrane from the posterior compartment of the leg.

Distally, the anterior tibial artery gives rise to an **anterior medial malleolar artery** and an **anterior lateral malleolar artery**, which pass posteriorly around the distal ends of the tibia and fibula, respectively, and connect with vessels from the posterior tibial and fibular arteries to form an anastomotic network around the ankle.

## Veins

Deep veins follow the arteries and have similar names.

## Nerves

### Deep fibular nerve

The nerve associated with the anterior compartment of the leg is the **deep fibular nerve** (Fig. 6.94). This nerve originates in the lateral compartment of the leg as one of the two divisions of the common fibular nerve.

The deep fibular nerve passes anteromedially through the intermuscular septum that separates the lateral from the anterior compartments of the leg and then passes deep to the extensor digitorum longus. It reaches the anterior interosseous membrane where it meets and descends with the anterior tibial artery.

The deep fibular nerve:

- innervates all muscles in the anterior compartment;
- then continues into the dorsal aspect of the foot where it innervates the extensor digitorum brevis, contributes to the innervation of the first two dorsal interossei muscles, and supplies the skin between the great and second toes.

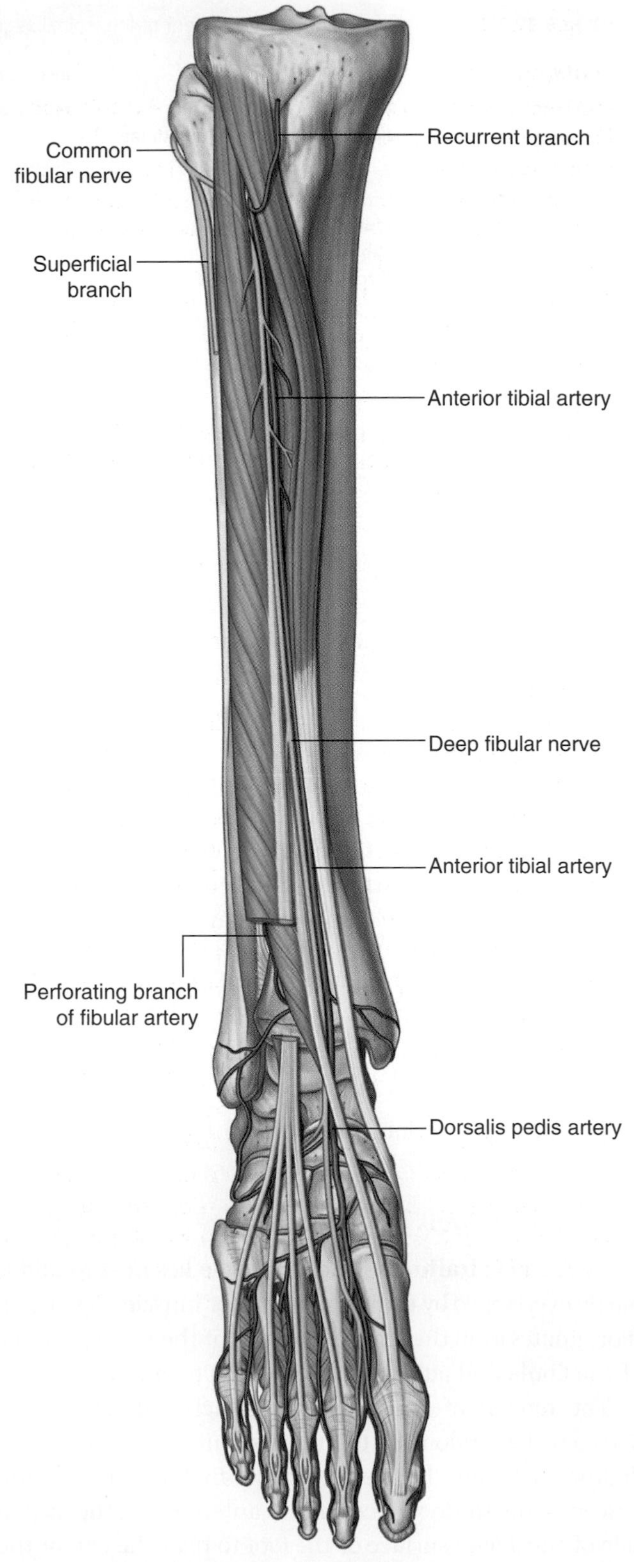


**Fig. 6.94** Anterior tibial artery and deep fibular nerve.

## In the clinic

### Footdrop

Footdrop is an inability to dorsiflex the foot. Patients with footdrop have a characteristic "steppage" gait. As the patient walks, the knee of the affected limb is elevated to an abnormal height during the swing phase to prevent the foot from dragging. At the end of the swing phase, the foot "slaps" the ground. Also, the unaffected limb often acquires a characteristic tiptoe pattern of gait during the stance phase. A typical cause of footdrop is damage to the common fibular nerve, which may occur with fractures of the fibular neck. Other causes include disc protrusion compressing the L5 nerve root, disorders of the sciatic nerve and the lumbosacral plexus, and pathologies of the spinal cord and brain.

## In the clinic

### Common fibular nerve injury

The common fibular nerve is susceptible to injury as it passes around the lateral aspect of the neck of the fibula. It can be injured as a result of a direct trauma (blow or laceration), secondary to knee injury (knee dislocation), or as a consequence of a proximal fibular fracture. Sometimes damage to the nerve can be iatrogenic, that is, damaged during arthroscopy or knee surgery.

Symptoms of common fibular nerve injury are often observed in bed-bound patients, particularly in those with decreased levels of consciousness, due to prolonged external pressure to the knee leading to nerve compression and neuropathy. Similarly, application of a tight cast or a brace to the leg can compress the nerve, producing symptoms of fibular muscle palsy.

Apart from a foot drop, other symptoms of common fibular nerve injury include loss of sensation over the lateral aspect of the leg and dorsum of the foot, and wasting of fibular and anterior tibial muscles.

## FOOT

The foot is the region of the lower limb distal to the ankle joint. It is subdivided into the ankle, the metatarsus, and the digits.

There are five digits consisting of the medially positioned great toe (digit I) and four more laterally placed digits, ending laterally with the little toe (digit V) (Fig. 6.95).

The foot has a superior surface (**dorsum of foot**) and an inferior surface (**sole**; Fig. 6.95).

Abduction and adduction of the toes are defined with respect to the long axis of the second digit. Unlike in the hand, where the thumb is oriented 90° to the other fingers, the great toe is oriented in the same position as the other toes. The foot is the body's point of contact with the ground and provides a stable platform for upright stance. It also levers the body forward during walking.

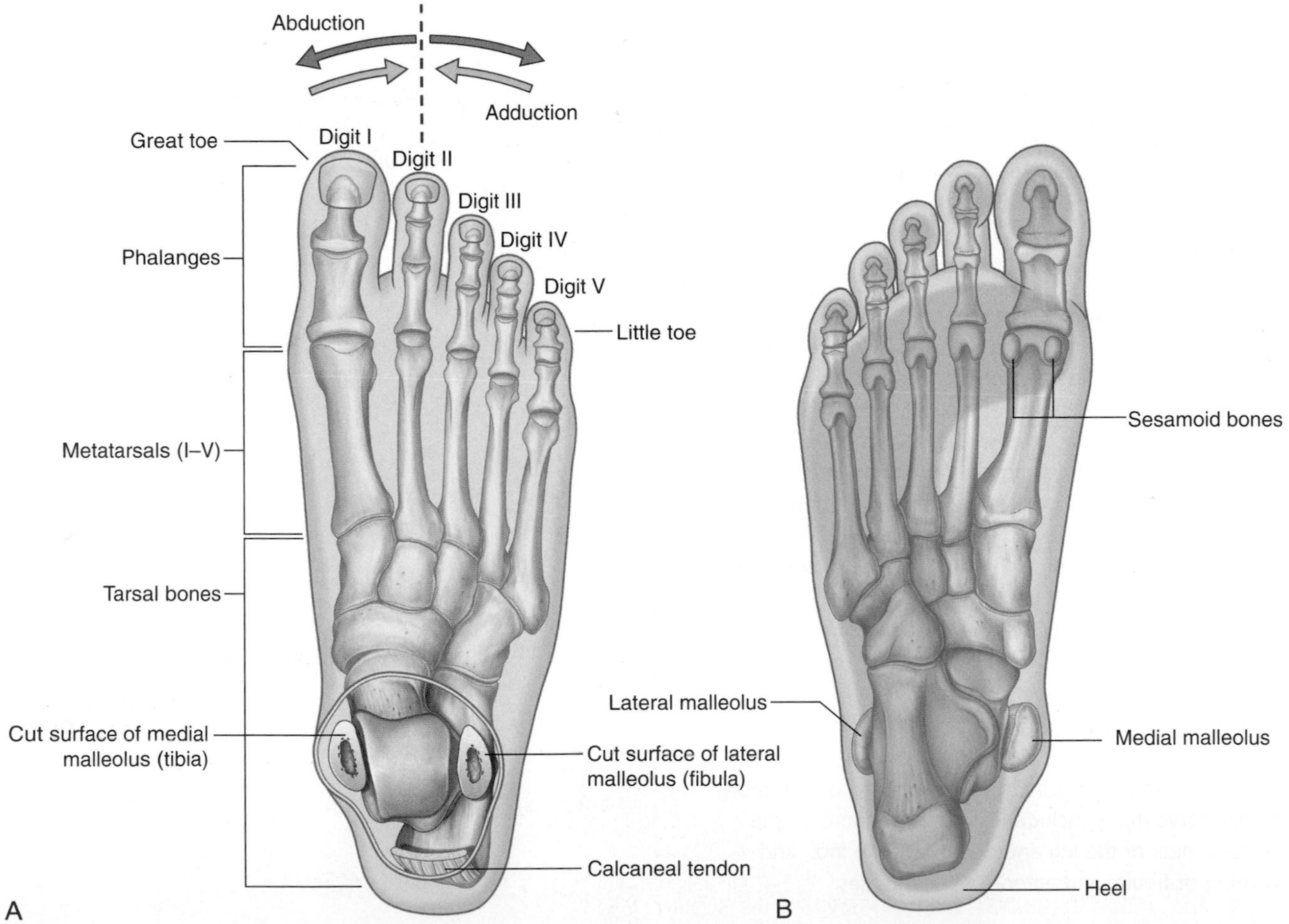

**Fig. 6.95** Foot. **A.** Dorsal aspect, right foot. **B.** Plantar aspect, right foot, showing the surface in contact with the ground when standing.

## Bones

There are three groups of bones in the foot (Fig. 6.96):

- the seven **tarsal bones**, which form the skeletal framework for the ankle;
- **metatarsals** (**I to V**), which are the bones of the metatarsus; and
- the **phalanges**, which are the bones of the toes—each toe has three phalanges, except for the great toe, which has two.

### Tarsal bones

The tarsal bones are arranged in a proximal group and a distal group with an intermediate bone between the two groups on the medial side of the foot (Fig. 6.96A).

#### Proximal group

The proximal group consists of two large bones, the talus (Latin for "ankle") and the calcaneus (Latin for "heel"):

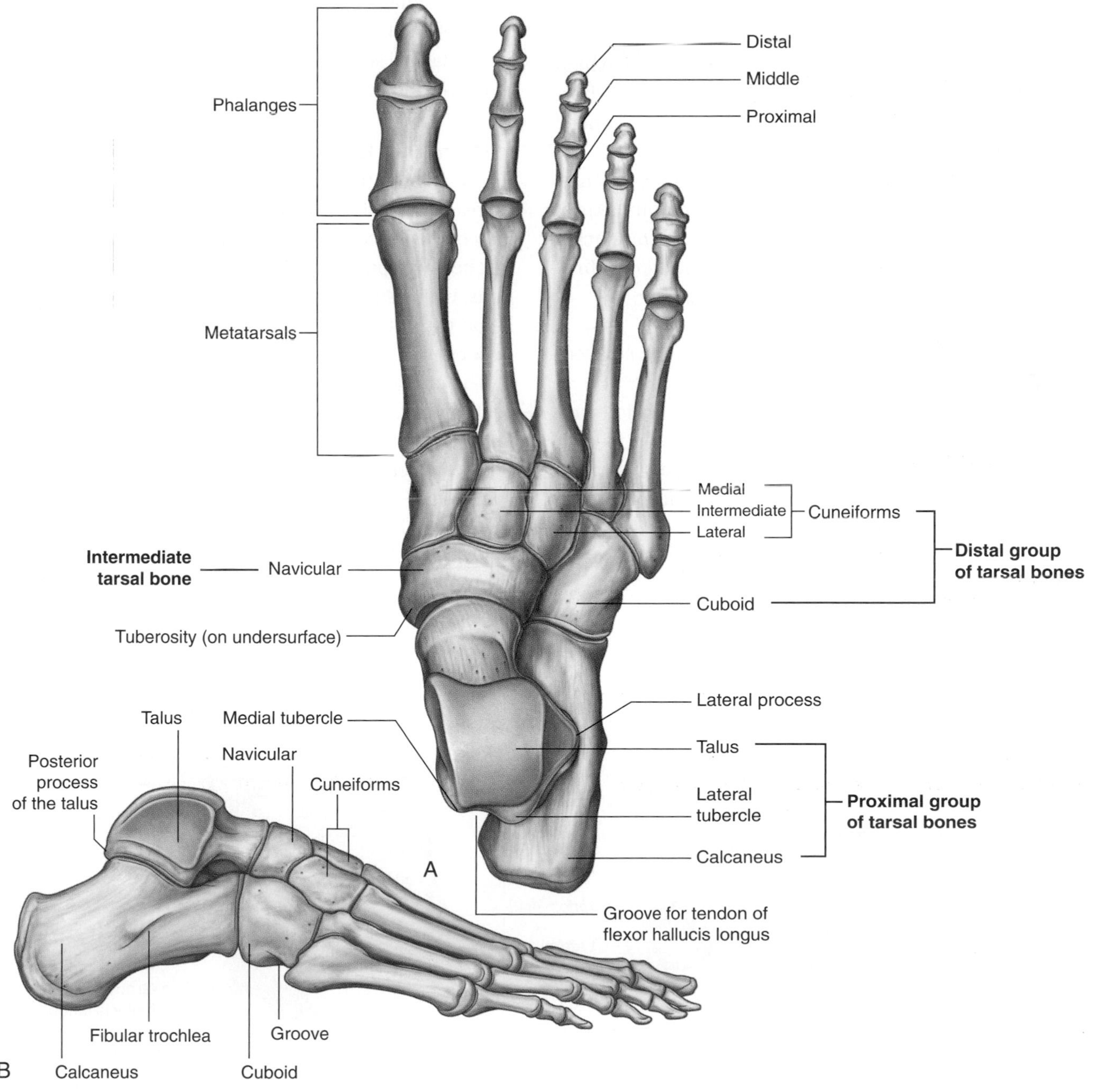


**Fig. 6.96** Bones of the foot. **A.** Dorsal view, right foot. **B.** Lateral view, right foot.

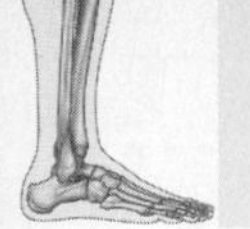



- The **talus** is the most superior bone of the foot and sits on top of and is supported by the calcaneus (Fig. 6.96B)—it articulates above with the tibia and fibula to form the ankle joint and also projects forward to articulate with the intermediate tarsal bone (navicular) on the medial side of the foot.
- The **calcaneus** is the largest of the tarsal bones—posteriorly it forms the bony framework of the heel and anteriorly it projects forward to articulate with one of the distal group of tarsal bones (cuboid) on the lateral side of the foot.

### Talus

The talus, when viewed from the medial or lateral sides, is snail-shaped (Fig. 6.97A,B). It has a rounded **head**, which is projected forward and medially at the end of a short broad **neck**, which is connected posteriorly to an expanded body.

Anteriorly, the head of the talus is domed for articulation with a corresponding circular depression on the posterior surface of the navicular bone. Inferiorly, this domed articular surface is continuous with an additional three articular facets separated by smooth ridges:

- The anterior and middle facets articulate with adjacent surfaces on the calcaneus bone.
- The other facet, medial to the facets for articulation with the calcaneus, articulates with a ligament—the plantar calcaneonavicular ligament (spring ligament)—which connects the calcaneus to the navicular under the head of the talus.

The neck of the talus is marked by a deep groove (the **sulcus tali**), which passes obliquely forward across the inferior surface from medial to lateral, and expands dramatically on the lateral side. Posterior to the sulcus tali is a large facet (posterior calcaneal surface) for articulation with the calcaneus.

The superior aspect of the body of the talus is elevated to fit into the socket formed by the distal ends of the tibia and fibula to form the ankle joint:

- The upper (trochlear) surface of this elevated region articulates with the inferior end of the tibia.
- The medial surface articulates with the medial malleolus of the tibia.
- The lateral surface articulates with the lateral malleolus of the fibula.

Because the lateral malleolus is larger and projects more inferiorly than the medial malleolus at the ankle joint, the corresponding lateral articular surface on the talus is larger and projects more inferiorly than the medial surface.

The lower part of the lateral surface of the body of the talus, which supports the lower part of the facet for articulation with the fibula, forms a bony projection (the **lateral process**).

The inferior surface of the body of the talus has a large oval concave facet (the **posterior calcaneal articular facet**) for articulation with the calcaneus.

The posterior aspect of the body of the talus consists of a backward and medially facing projection (the **posterior process**). The posterior process is marked on its surface by

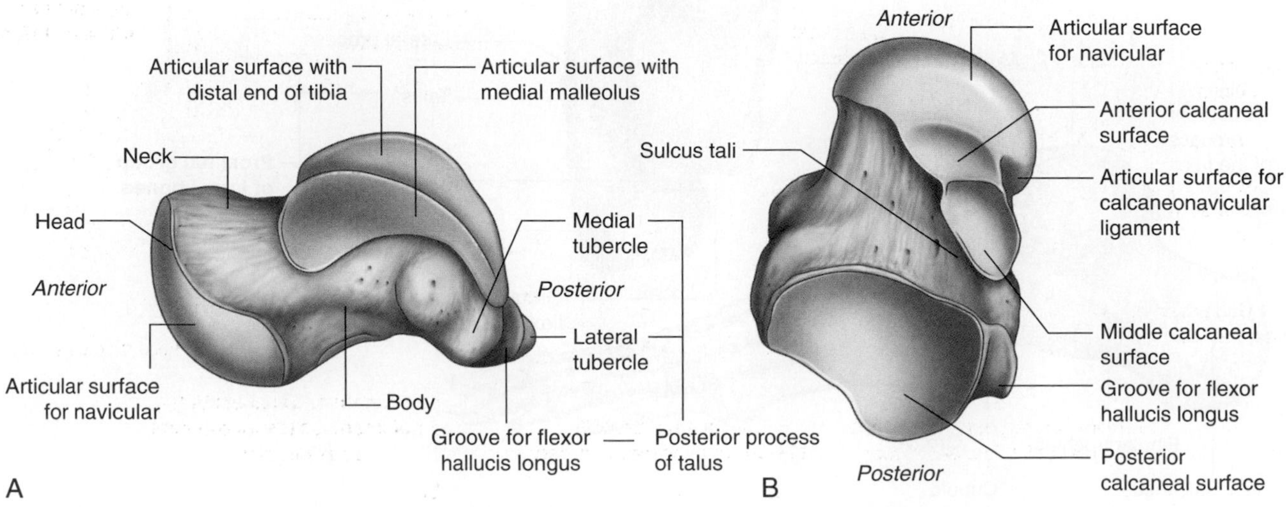


**Fig. 6.97** Talus. **A.** Medial view. **B.** Inferior view.

a lateral tubercle and a medial tubercle, which bracket between them the **groove for the tendon of the flexor hallucis longus** as it passes from the leg into the foot.

### Calcaneus

The calcaneus sits under and supports the talus. It is an elongate, irregular, box-shaped bone with its long axis generally oriented along the midline of the foot, but deviating lateral to the midline anteriorly (Fig. 6.98).

The calcaneus projects behind the ankle joint to form the skeletal framework of the heel. The posterior surface of this heel region is circular and divided into upper, middle, and lower parts. The calcaneal tendon (Achilles tendon) attaches to the middle part:

- The upper part is separated from the calcaneal tendon by a bursa.
- The lower part curves forward, is covered by subcutaneous tissue, is the weight-bearing region of the heel, and is continuous onto the plantar surface of the bone as the **calcaneal tuberosity**.

The calcaneal tuberosity projects forward on the plantar surface as a large medial process and a small lateral process separated from each other by a V-shaped notch (Fig. 6.98B). At the anterior end of the plantar surface is a tubercle (the **calcaneal tubercle**) for the posterior attachment of the short plantar ligament of the sole of the foot.

The lateral surface of the calcaneus has a smooth contour except for two slightly raised regions (Fig. 6.98C). One of these raised areas—the **fibular trochlea** (peroneal tubercle)—is anterior to the middle of the surface and often has two shallow grooves, which pass, one above the other, obliquely across its surface. The tendons of the fibularis brevis and longus muscles are bound to the trochlea as they pass over the lateral side of the calcaneus.

Superior and posterior to the fibular trochlea is a second raised area or tubercle for attachment of the calcaneofibular part of the lateral collateral ligament of the ankle joint.

The medial surface of the calcaneus is concave and has one prominent feature associated with its upper margin (the **sustentaculum tali**; Fig. 6.98A), which is a shelf of bone projecting medially and supporting the more posterior part of the head of the talus.

The underside of the sustentaculum tali has a distinct groove running from posterior to anterior and along which the tendon of the flexor hallucis longus muscle travels into the sole of the foot.

The superior surface of the sustentaculum tali has a facet (**middle talar articular surface**) for articulation with the corresponding middle facet on the head of the talus.

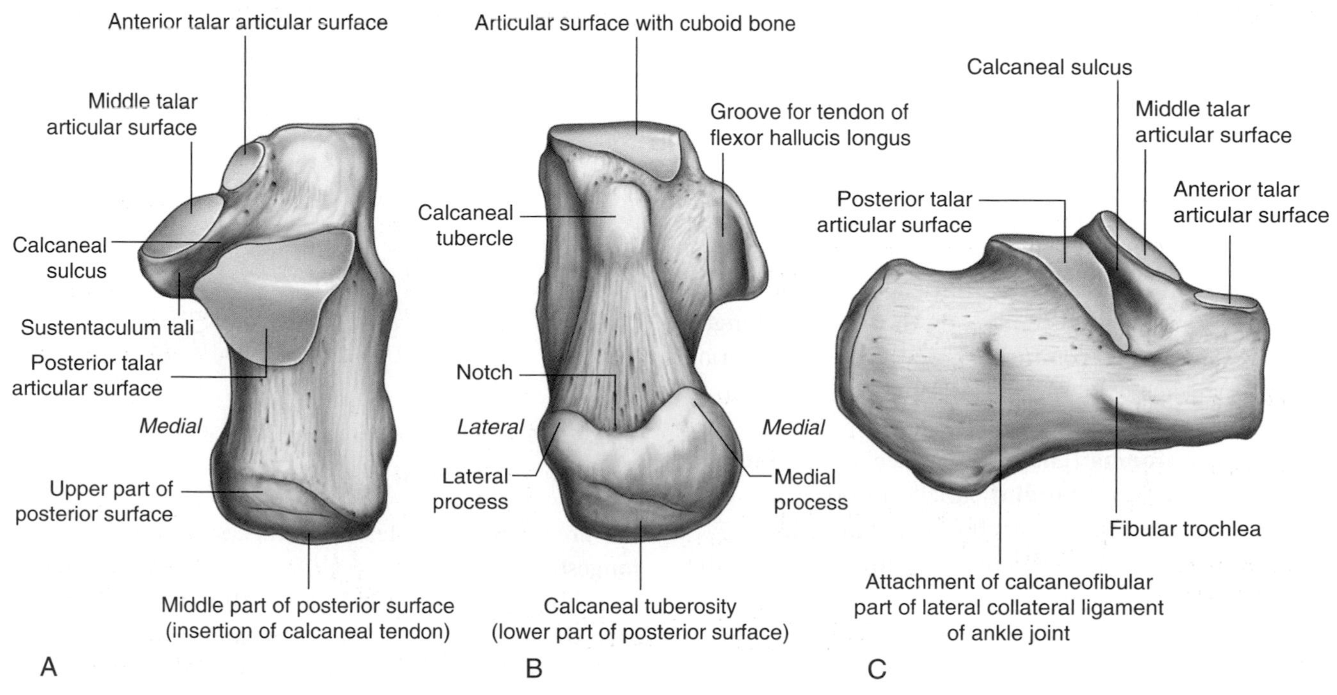


**Fig. 6.98** Calcaneus. **A.** Superior view. **B.** Inferior view. **C.** Lateral view.

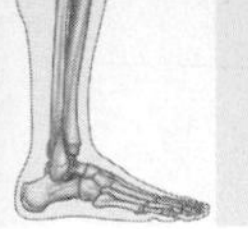



**Anterior** and **posterior talar articular surfaces** are on the superior surface of the calcaneus itself (Fig. 6.98A):

- The anterior talar articular surface is small and articulates with the corresponding anterior facet on the head of the talus.
- The posterior talar articular surface is large and is approximately near the middle of the superior surface of the calcaneus.

Between the posterior talar articular surface, which articulates with the body of the talus, and the other two articular surfaces, which articulate with the head of the talus, is a deep groove (the **calcaneal sulcus**; Fig. 6.98A,C).

The calcaneal sulcus on the superior surface of the calcaneus and the sulcus tali on the inferior surface of the talus together form the **tarsal sinus**, which is a large gap between the anterior ends of the calcaneus and talus that is visible when the skeleton of the foot is viewed from its lateral aspect (Fig. 6.99).

### Intermediate tarsal bone

The intermediate tarsal bone on the medial side of the foot is the **navicular** (boat shaped) (Fig. 6.96). This bone articulates behind with the talus and articulates in front and on the lateral side with the distal group of tarsal bones.

One distinctive feature of the navicular is a prominent rounded tuberosity for the attachment of the tibialis posterior tendon, which projects inferiorly on the medial side of the plantar surface of the bone.

### Distal group

From lateral to medial, the distal group of tarsal bones consists of (Fig. 6.96):

- The **cuboid** (Greek for "cube"), which articulates posteriorly with the calcaneus, medially with the lateral cuneiform, and anteriorly with the bases of the lateral two metatarsals—the tendon of the fibularis longus muscle lies in a prominent groove on the anterior plantar surface, which passes obliquely forward across the bone from lateral to medial.
- Three **cuneiforms** (Latin for "wedge")—the **lateral**, **intermediate**, and **medial** cuneiform bones, in addition to articulating with each other, articulate posteriorly with the navicular bone and anteriorly with the bases of the medial three metatarsals.

### Metatarsals

There are five metatarsals in the foot, numbered I to V from medial to lateral (Fig. 6.100). Metatarsal I, associated with the great toe, is shortest and thickest. The second is the longest.

Each metatarsal has a **head** at the distal end, an elongate **shaft** in the middle, and a proximal **base**.

The head of each metatarsal articulates with the proximal phalanx of a toe and the base articulates with one or more of the distal group of tarsal bones. The plantar

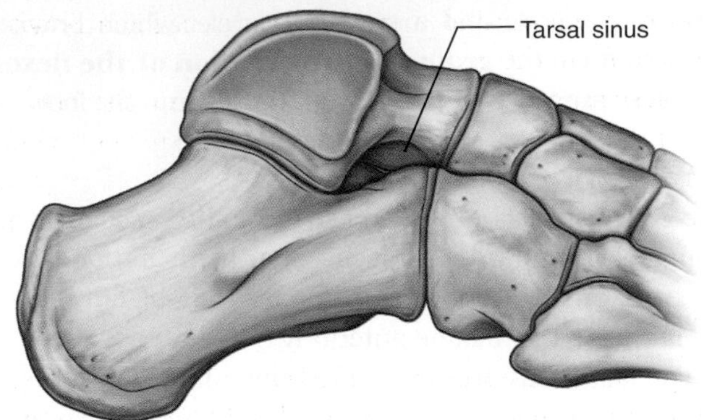


**Fig. 6.99** Tarsal sinus. Lateral view, right foot.

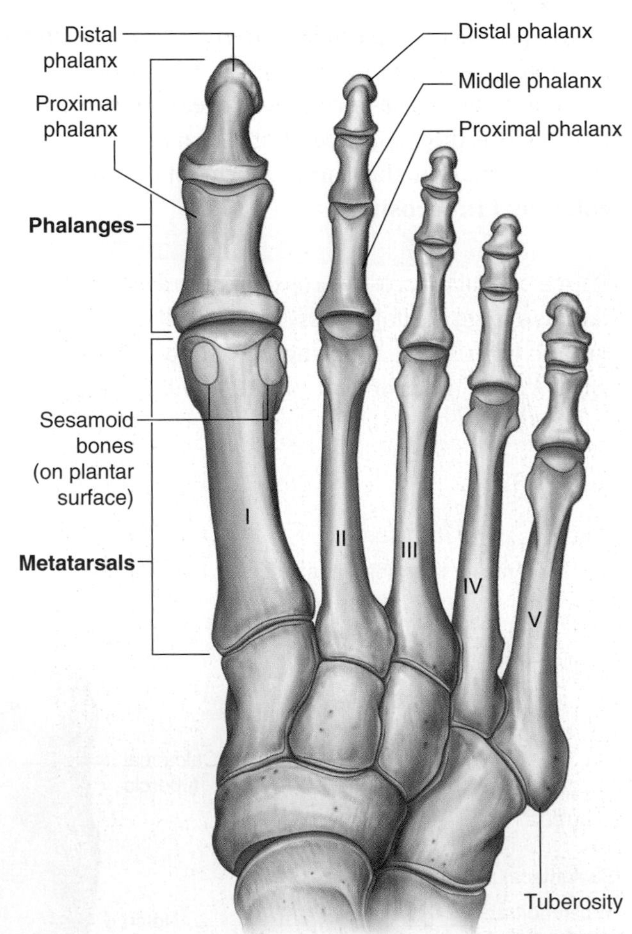


**Fig. 6.100** Metatarsals and phalanges. Dorsal view, right foot.

surface of the head of metatarsal I also articulates with two sesamoid bones.

The sides of the bases of metatarsals II to V also articulate with each other. The lateral side of the base of metatarsal V has a prominent **tuberosity**, which projects posteriorly and is the attachment site for the tendon of the fibularis brevis muscle.

### Phalanges

The phalanges are the bones of the toes (Fig. 6.100). Each toe has three phalanges (**proximal**, **middle**, and **distal**), except for the great toe, which has only two (proximal and distal).

Each phalanx consists of a **base**, a **shaft**, and a distal **head**:

- The base of each proximal phalanx articulates with the head of the related metatarsal.
- The head of each distal phalanx is nonarticular and flattened into a crescent-shaped plantar tuberosity under the plantar pad at the end of the digit.

In each toe, the total length of the phalanges combined is much shorter than the length of the associated metatarsal.

## Joints

### Ankle joint

The ankle joint is synovial in type and involves the talus of the foot and the tibia and fibula of the leg (Fig. 6.102).

The ankle joint mainly allows hinge-like dorsiflexion and plantarflexion of the foot on the leg.

The distal end of the fibula is firmly anchored to the larger distal end of the tibia by strong ligaments. Together, the fibula and tibia create a deep bracket-shaped socket for the upper expanded part of the body of the talus:

- The roof of the socket is formed by the inferior surface of the distal end of the tibia.
- The medial side of the socket is formed by the medial malleolus of the tibia.
- The longer lateral side of the socket is formed by the lateral malleolus of the fibula.

The articular surfaces are covered by hyaline cartilage.

The articular part of the talus is shaped like a short half-cylinder tipped onto its flat side with one end facing lateral and the other end facing medial. The curved upper surface of the half-cylinder and the two ends are covered by hyaline cartilage and fit into the bracket-shaped socket formed by the distal ends of the tibia and fibula.

When viewed from above, the articular surface of the talus is much wider anteriorly than it is posteriorly. As a result, the bone fits tighter into its socket when the foot is dorsiflexed and the wider surface of the talus moves into the ankle joint than when the foot is plantarflexed and the narrower part of the talus is in the joint. The joint is therefore most stable when the foot is dorsiflexed.

The articular cavity is enclosed by a synovial membrane, which attaches around the margins of the articular surfaces, and by a fibrous membrane, which covers the synovial membrane and is also attached to the adjacent bones.

### In the clinic

#### Talar beak

Occasionally there is a superior projection of the distal aspect of the talus, which is beak-shaped (Fig. 6.101). It is often associated with the presence of a bony or fibrous joint between the talus and calcaneus.

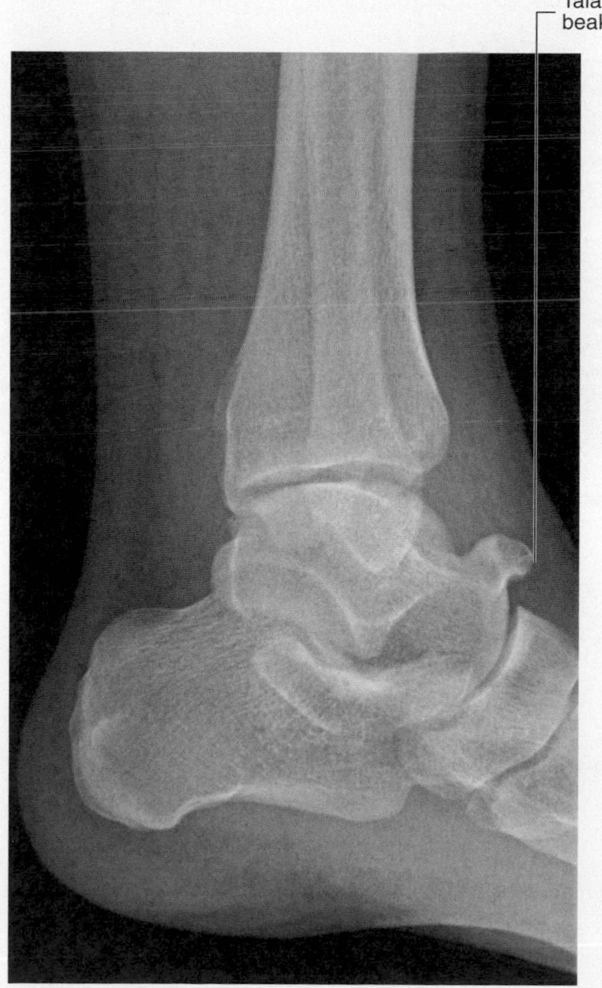


**Fig. 6.101** Radiograph of ankle showing talar beak.

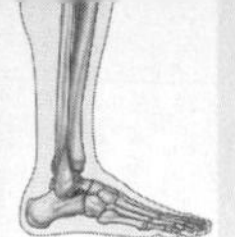



## In the clinic

### Fracture of the talus

The talus is an unusual bone because it ossifies from a single primary ossification center, which initially appears in the neck. The posterior aspect of the talus appears to ossify last, normally after puberty. In up to 50% of people there is a small accessory ossicle (the os trigonum) posterior to the lateral tubercle of the posterior process. Articular cartilage covers approximately 60% of the talar surface and there are no direct tendon or muscle attachments to the bone.

One of the problems with fractures of the talus is that the blood supply to the bone is vulnerable to damage. The main blood supply to the bone enters the talus through the tarsal sinus from a branch of the posterior tibial artery. This vessel supplies most of the neck and the body of the talus. Branches of the dorsalis pedis artery enter the superior aspect of the talar neck and supply the dorsal portion of the head and neck, and branches from the fibular artery supply a small portion of the lateral talus.

Fractures of the neck of the talus often interrupt the blood supply to the talus, so making the body and posterior aspect of the talus susceptible to osteonecrosis, which may in turn lead to premature osteoarthritis and require extensive surgery.

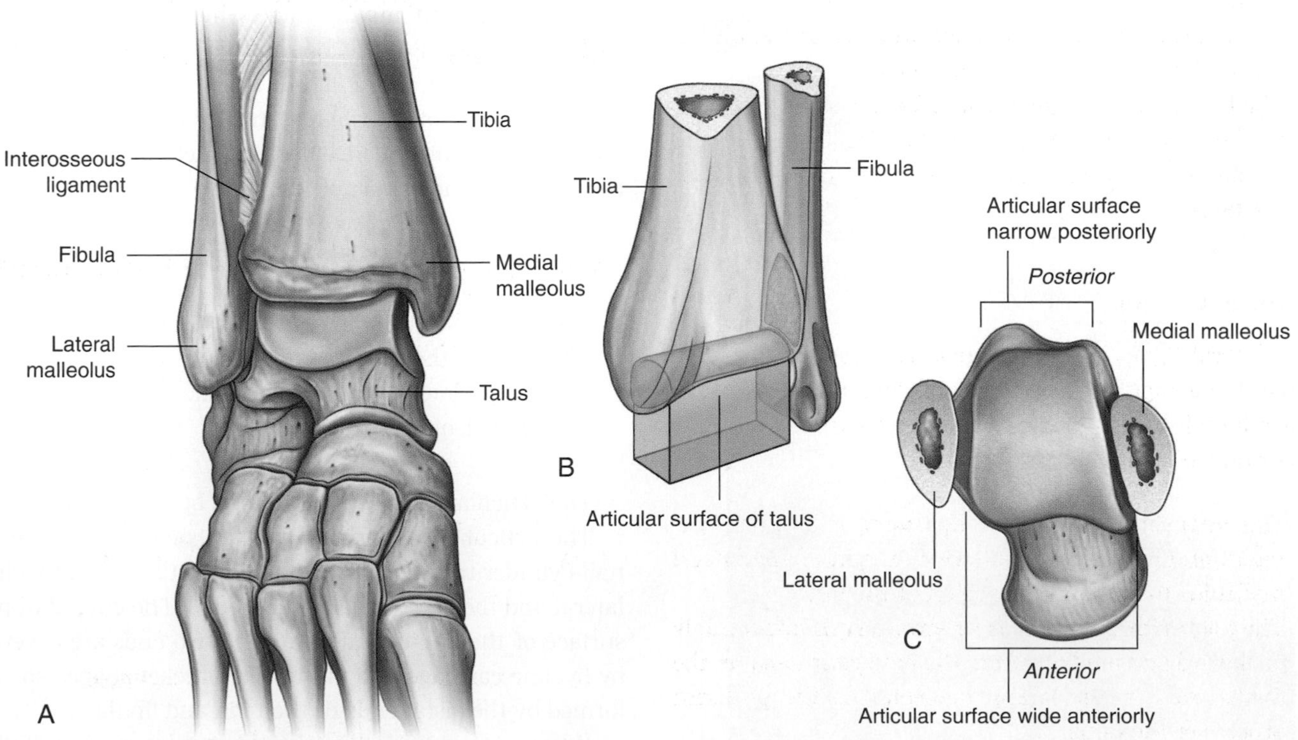


**Fig. 6.102** Ankle joint. **A.** Anterior view with right foot plantarflexed. **B.** Schematic of joint, posterior view. **C.** Superior view of the talus to show the shape of the articular surface.

The ankle joint is stabilized by **medial** (deltoid) and **lateral ligaments**.

### Medial ligament (deltoid ligament)

The medial (deltoid) ligament is large, strong (Fig. 6.103), and triangular in shape. Its apex is attached above to the medial malleolus and its broad base is attached below to a line that extends from the tuberosity of the navicular bone in front to the medial tubercle of the talus behind.

The medial ligament is subdivided into four parts based on the inferior points of attachment:

- The part that attaches in front to the tuberosity of the navicular and the associated margin of the plantar calcaneonavicular ligament (spring ligament), which connects the navicular bone to the sustentaculum tali of the calcaneus bone behind, is the **tibionavicular part** of the medial ligament.
- The **tibiocalcaneal part**, which is more central, attaches to the sustentaculum tali of the calcaneus bone.
- The **posterior tibiotalar part** attaches to the medial side and medial tubercle of the talus.
- The fourth part (the **anterior tibiotalar part**) is deep to the tibionavicular and tibiocalcaneal parts of the medial ligament and attaches to the medial surface of the talus.

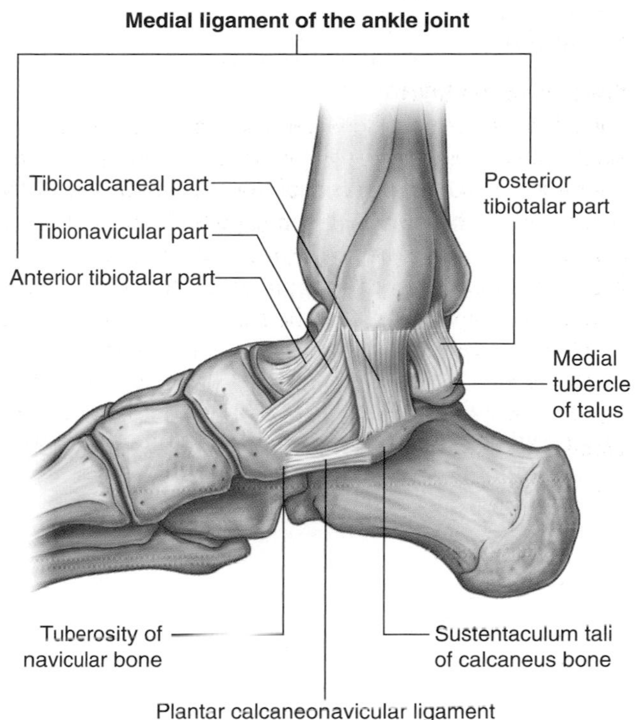


**Fig. 6.103** Medial ligament of the ankle joint, right foot.

### Lateral ligament

The lateral ligament of the ankle is composed of three separate ligaments, the anterior talofibular ligament, the posterior talofibular ligament, and the calcaneofibular ligament (Fig. 6.104):

- The **anterior talofibular ligament** is a short ligament, and attaches the anterior margin of the lateral malleolus to the adjacent region of the talus.
- The **posterior talofibular ligament** runs horizontally backward and medially from the malleolar fossa on the medial side of the lateral malleolus to the posterior process of the talus.
- The **calcaneofibular ligament** is attached above to the malleolar fossa on the posteromedial side of the lateral malleolus and passes posteroinferiorly to attach below to a tubercle on the lateral surface of the calcaneus.

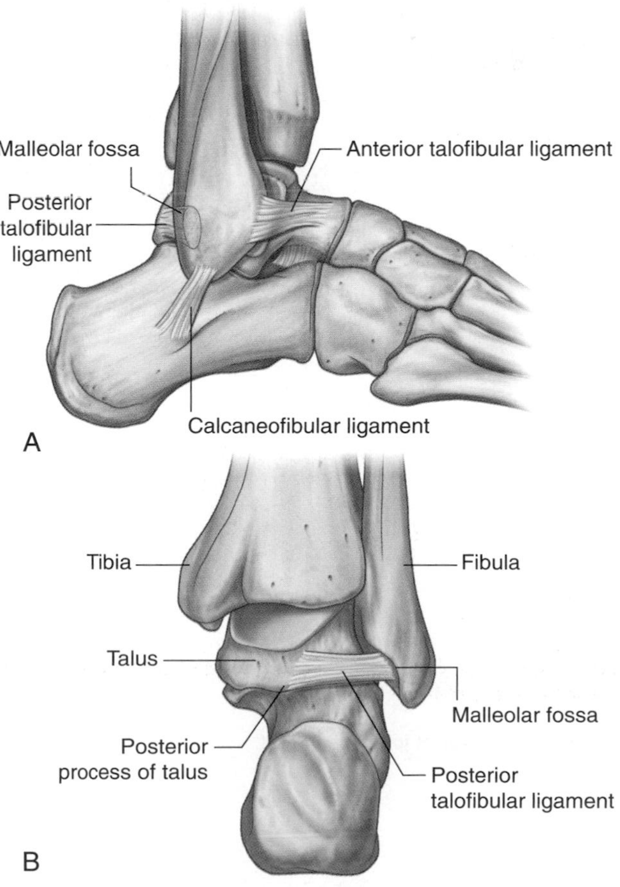


**Fig. 6.104** Lateral ligament of the ankle joint. **A.** Lateral view, right foot. **B.** Posterior view, right foot.

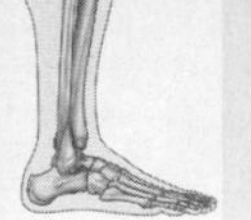
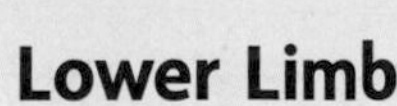


### Intertarsal joints

The numerous synovial joints between the individual tarsal bones mainly invert, evert, supinate, and pronate the foot:

- Inversion and eversion is turning the whole sole of the foot inward and outward, respectively.
- Pronation is rotating the front of the foot laterally relative to the back of the foot, and supination is the reverse movement.

Pronation and supination allow the foot to maintain normal contact with the ground when in different stances or when standing on irregular surfaces.

The major joints at which movements occur include the subtalar, talocalcaneonavicular, and calcaneocuboid joints (Fig. 6.105). The talocalcaneonavicular and calcaneocuboid joints together form what is often referred to as the **transverse tarsal joint**.

Intertarsal joints between the cuneiforms and between the cuneiforms and the navicular allow only limited movement.

The joint between the cuboid and navicular is normally fibrous.

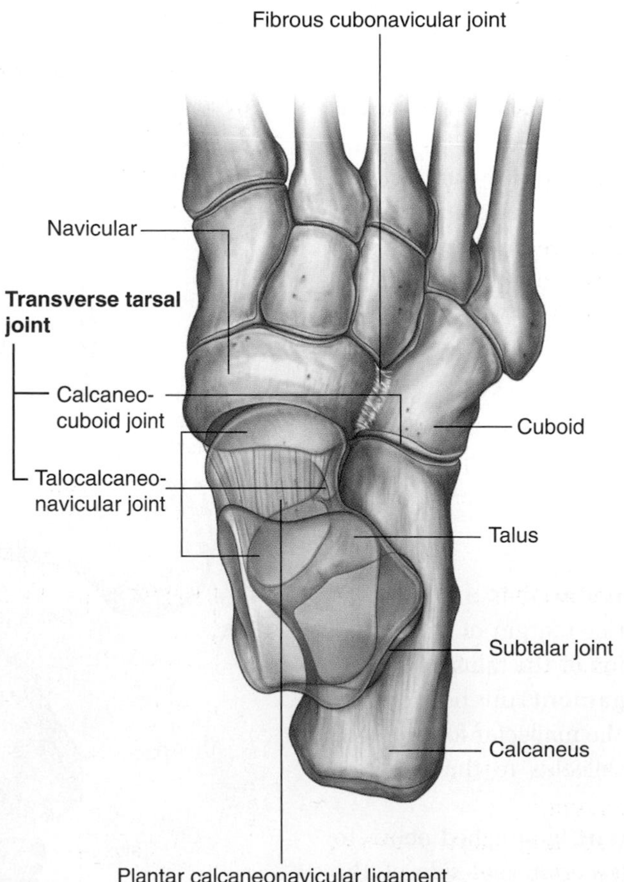


**Fig. 6.105** Intertarsal joints, right foot.

## In the clinic

**Ankle fractures**

An appreciation of ankle anatomy is essential to understand the wide variety of fractures that may occur at and around the ankle joint.

The ankle joint and related structures can be regarded as a fibro-osseous ring oriented in the coronal plane.

- The upper part of the ring is formed by the joint between the distal ends of the fibula and tibia and by the ankle joint itself.
- The sides of the ring are formed by the ligaments that connect the medial malleolus and lateral malleolus to the adjacent tarsal bones.
- The bottom of the ring is not part of the ankle joint, but consists of the subtalar joint and the associated ligaments.

Visualizing the ankle joint and surrounding structures as a fibro-osseous ring allows the physician to predict the type of damage likely to result from a particular type of injury. For example, an inversion injury may fracture the medial malleolus and tear ligaments anchoring the lateral malleolus to the tarsal bones.

The ring may be disrupted not only by damage to the bones (which produces fractures), but also by damage to the ligaments. Unlike bone fractures, damage to ligaments is unlikely to be appreciated on plain radiographs. When a fracture is noted on a plain radiograph, the physician must always be aware that there may also be appreciable ligamentous disruption.

***Ottawa Ankle Rules***

The Ottawa ankle rules were developed to assist clinicians in deciding whether patients with acute ankle injuries require investigation with radiographs in order to avoid unnecessary studies. Named after the hospital where they were developed, the rules are highly sensitive and have reduced the utilization of unwarranted ankle radiographs since their implementation.

An ankle x-ray series is required if there is ankle pain and any of the following:

- Bone tenderness along the distal 6 cm of the posterior tibia or tip of the medial malleolus
- Bone tenderness along the distal 6 cm of the posterior fibula or tip of the lateral malleolus
- Inability to bear weight for four steps both immediately after the injury and in the emergency department

A foot x-ray series is required if there is midfoot pain and any of the following:

- Bone tenderness at the base of the fifth metatarsal bone
- Bone tenderness at the navicular bone
- Inability to bear weight for four steps both immediately after the injury and in the emergency department

### Subtalar joint

The **subtalar joint** is between:

- the large posterior calcaneal facet on the inferior surface of the talus, and
- the corresponding posterior talar facet on the superior surface of the calcaneus.

The articular cavity is enclosed by synovial membrane, which is covered by a fibrous membrane.

The subtalar joint allows gliding and rotation, which are involved in inversion and eversion of the foot. **Lateral**, **medial**, **posterior**, and **interosseous talocalcaneal ligaments** stabilize the joint. The interosseous talocalcaneal ligament lies in the tarsal sinus (Fig. 6.106).

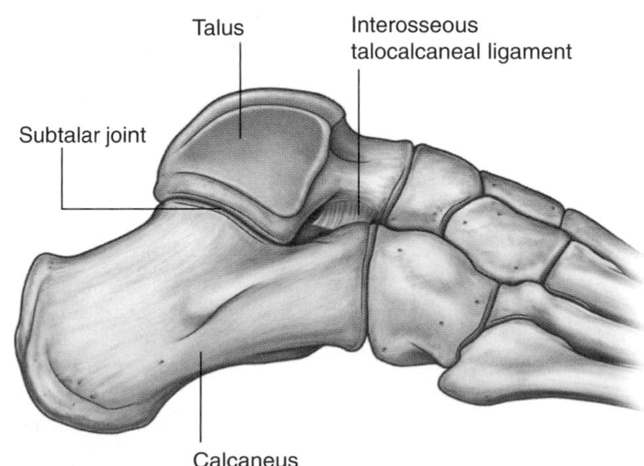


**Fig. 6.106** Interosseous talocalcaneal ligament. Lateral view, right foot.

### Talocalcaneonavicular joint

The **talocalcaneonavicular joint** is a complex joint in which the head of the talus articulates with the calcaneus and plantar calcaneonavicular ligament (spring ligament) below and the navicular in front (Fig. 6.107A).

The talocalcaneonavicular joint allows gliding and rotation movements, which together with similar movements of the subtalar joint are involved with inversion and eversion of the foot. It also participates in pronation and supination.

The parts of the talocalcaneonavicular joint between the talus and calcaneus are:

- the anterior and middle calcaneal facets on the inferior surface of the talar head, and
- the corresponding anterior and middle talar facets on the superior surface and sustentaculum tali, respectively, of the calcaneus (Fig. 6.107B).

The part of the joint between the talus and the plantar calcaneonavicular ligament (spring ligament) is between the ligament and the medial facet on the inferior surface of the talar head.

The joint between the navicular and talus is the largest part of the talocalcaneonavicular joint and is between the ovoid anterior end of the talar head and the corresponding concave posterior surface of the navicular.

#### Ligaments

The capsule of the talocalcaneonavicular joint, which is a synovial joint, is reinforced:

- posteriorly by the interosseous talocalcaneal ligament,
- superiorly by the **talonavicular ligament**, which passes between the neck of the talus and adjacent regions of the navicular, and
- inferiorly by the plantar calcaneonavicular ligament (spring ligament) (Fig. 6.107C,D).

The lateral part of the talocalcaneonavicular joint is reinforced by the calcaneonavicular part of the **bifurcate ligament**, which is a Y-shaped ligament superior to the joint. The base of the bifurcate ligament is attached to the anterior aspect of the superior surface of the calcaneus and its arms are attached to:

- the dorsomedial surface of the cuboid (**calcaneocuboid ligament**), and
- the dorsolateral part of the navicular (**calcaneonavicular ligament**).

The **plantar calcaneonavicular ligament** (spring ligament) is a broad thick ligament that spans the space between the sustentaculum tali behind and the navicular bone in front (Fig. 6.107B,C). It supports the head of the

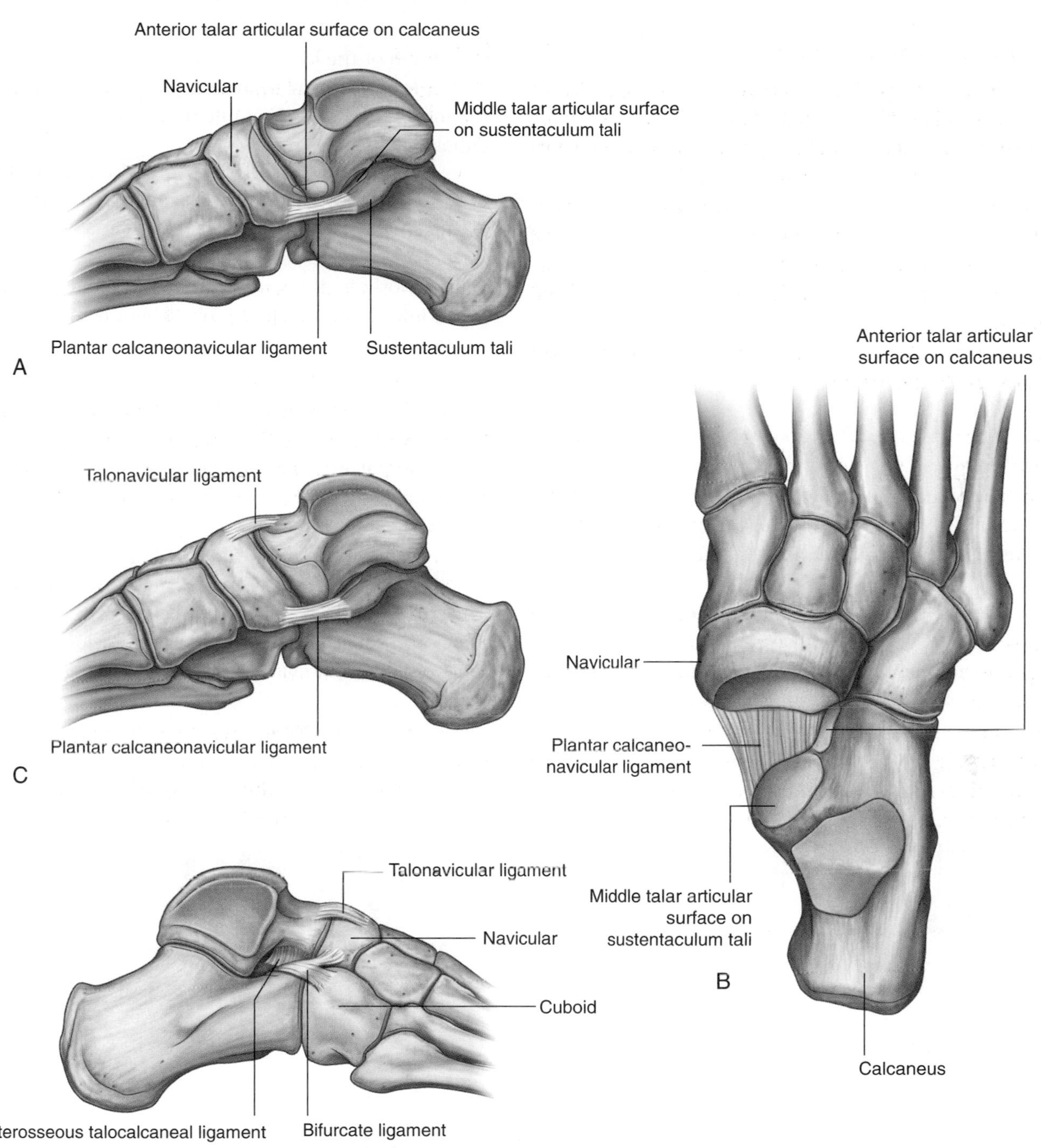


**Fig. 6.107** Talocalcaneonavicular joint. **A.** Medial view, right foot. **B.** Superior view, right foot, talus removed. **C.** Ligaments, medial view, right foot. **D.** Ligaments, lateral view, right foot.

talus, takes part in the talocalcaneonavicular joint, and resists depression of the medial arch of the foot.

### Calcaneocuboid joint

The **calcaneocuboid joint** is a synovial joint between:

- the facet on the anterior surface of the calcaneus, and
- the corresponding facet on the posterior surface of the cuboid.

The calcaneocuboid joint allows sliding and rotating movements involved with inversion and eversion of the foot, and also contributes to pronation and supination of the forefoot on the hindfoot.

#### Ligaments

The calcaneocuboid joint is reinforced by the bifurcate ligament (see above) and by the long plantar ligament and the plantar calcaneocuboid ligament (short plantar ligament).

The **plantar calcaneocuboid ligament** (short plantar ligament) is short, wide, and very strong, and connects the calcaneal tubercle to the inferior surface of the cuboid (Fig. 6.108A). It not only supports the calcaneocuboid joint, but also assists the long plantar ligament in resisting depression of the lateral arch of the foot.

The **long plantar ligament** is the longest ligament in the sole of the foot and lies inferior to the plantar calcaneocuboid ligament (Fig. 6.108B):

- Posteriorly, it attaches to the inferior surface of the calcaneus between the tuberosity and the calcaneal tubercle.
- Anteriorly, it attaches to a broad ridge and a tubercle on the inferior surface of the cuboid bone behind the groove for the fibularis longus tendon.

More superficial fibers of the long plantar ligament extend to the bases of the metatarsal bones.

The long plantar ligament supports the calcaneocuboid joint and is the strongest ligament, resisting depression of the lateral arch of the foot.

### Tarsometatarsal joints

The **tarsometatarsal joints** between the metatarsal bones and adjacent tarsal bones are plane joints and allow limited sliding movements (Fig. 6.109).

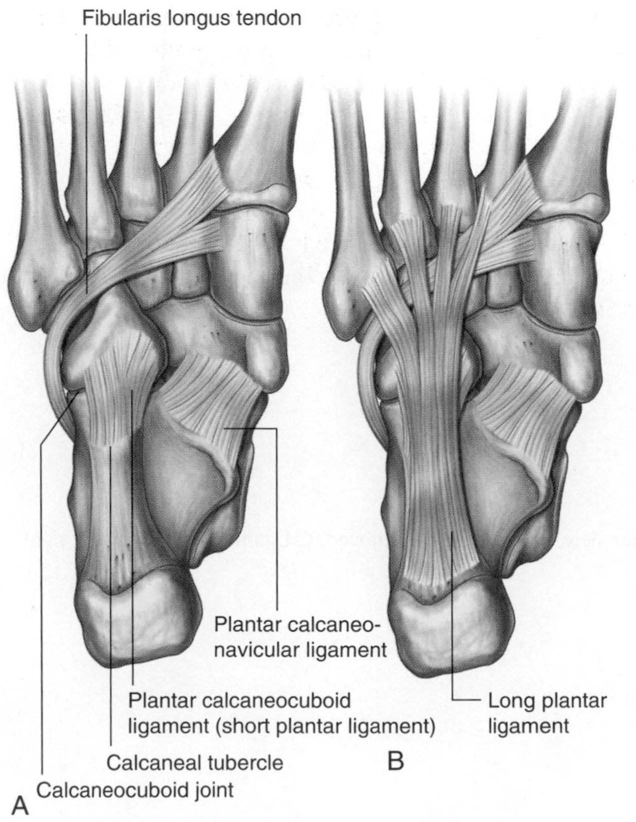

**Fig. 6.108** Plantar ligaments, right foot. **A.** Plantar calcaneocuboid ligament (short plantar ligament). **B.** Long plantar ligament.

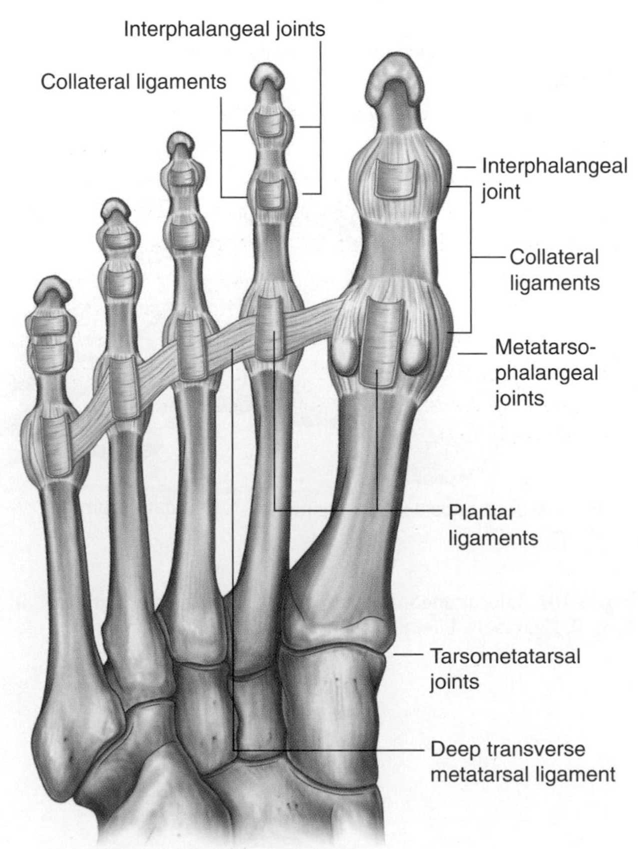


**Fig. 6.109** Tarsometatarsal, metatarsophalangeal, and interphalangeal joints, and the deep transverse metatarsal ligaments, right foot.

The range of movement of the tarsometatarsal joint between the metatarsal of the great toe and the medial cuneiform is greater than that of the other tarsometatarsal joints and allows flexion, extension, and rotation. The tarsometatarsal joints, with the transverse tarsal joint, take part in pronation and supination of the foot.

### Metatarsophalangeal joints

The metatarsophalangeal joints are ellipsoid synovial joints between the sphere-shaped heads of the metatarsals and the corresponding bases of the proximal phalanges of the digits.

The metatarsophalangeal joints allow extension and flexion, and limited abduction, adduction, rotation, and circumduction.

The joint capsules are reinforced by medial and lateral **collateral ligaments**, and by **plantar ligaments**, which have grooves on their plantar surfaces for the long tendons of the digits (Fig. 6.109).

### Deep transverse metatarsal ligaments

Four **deep transverse metatarsal ligaments** link the heads of the metatarsals together and enable the metatarsals to act as a single unified structure (Fig. 6.109). The ligaments blend with the plantar ligaments of the adjacent metatarsophalangeal joints.

The metatarsal of the great toe is oriented in the same plane as the metatarsals of the other toes and is linked to the metatarsal of the second toe by a deep transverse metatarsal ligament. In addition, the joint between the metatarsal of the great toe and medial cuneiform has a limited range of motion. The great toe therefore has a very restricted independent function—unlike the thumb in the hand, where the metacarpal is oriented 90° to the metacarpals of the fingers, there is no deep transverse metacarpal ligament between the metacarpals of the thumb and index finger, and the joint between the metacarpal and carpal bone allows a wide range of motion.

### In the clinic

#### Bunions

A bunion occurs on the medial aspect of the first metatarsophalangeal joint. This is an extremely important area of the foot because it is crossed by tendons and ligaments, which transmit and distribute the body's weight during movement. It is postulated that abnormal stresses in this region of the joint may produce the bunion deformity.

Clinically, a bunion is a significant protuberance of bone that may include soft tissue around the medial aspect of the first metatarsophalangeal joint. As it progresses, the toe appears to move toward the smaller toes, producing crowding of the digits.

This deformity tends to occur among people who wear high-heeled or pointed shoes, but osteoporosis and a hereditary predisposition are also risk factors.

Typically the patient's symptoms are pain, swelling, and inflammation. The bunion tends to enlarge and may cause problems in obtaining appropriate footwear.

Initial treatment is by adding padding to shoes, changing the type of footwear used, and taking anti-inflammatory drugs. Some patients may need surgery to correct the deformity and realign the toe.

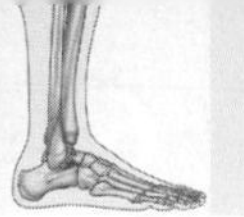

### Interphalangeal joints

The interphalangeal joints are hinge joints that allow mainly flexion and extension. They are reinforced by medial and lateral **collateral ligaments** and by **plantar ligaments** (Fig. 6.109).

## Tarsal tunnel, retinacula, and arrangement of major structures at the ankle

The tarsal tunnel is formed on the posteromedial side of the ankle by:

- a depression formed by the medial malleolus of the tibia, the medial and posterior surfaces of the talus, the medial surface of the calcaneus, and the inferior surface of the sustentaculum tali of the calcaneus; and
- an overlying flexor retinaculum (Fig. 6.110).

### Flexor retinaculum

The flexor retinaculum is a strap-like layer of connective tissue that spans the bony depression formed by the medial malleolus, the medial and posterior surfaces of the talus, the medial surface of the calcaneus, and the inferior surface of the sustentaculum tali (Fig. 6.110). It attaches above to the medial malleolus and below and behind to the inferomedial margin of the calcaneus.

The retinaculum is continuous above with the deep fascia of the leg and below with the deep fascia (plantar aponeurosis) of the foot.

Septa from the flexor retinaculum convert grooves on the bones into tubular connective tissue channels for the tendons of the flexor muscles as they pass into the sole of the foot from the posterior compartment of the leg (Fig. 6.110). Free movement of the tendons in the channels is facilitated by synovial sheaths, which surround the tendons.

Two compartments on the posterior surface of the medial malleolus are for the tendons of the tibialis posterior and flexor digitorum longus muscles. The tendon of the tibialis posterior is medial to the tendon of the flexor digitorum longus.

Immediately lateral to the tendons of the tibialis posterior and flexor digitorum longus, the posterior tibial artery with its associated veins and the tibial nerve pass through the tarsal tunnel into the sole of the foot. The pulse of the posterior tibial artery can be felt through the flexor retinaculum midway between the medial malleolus and the calcaneus.

Lateral to the tibial nerve is the compartment on the posterior surface of the talus and the undersurface of the sustentaculum tali for the tendon of the flexor hallucis longus muscle.

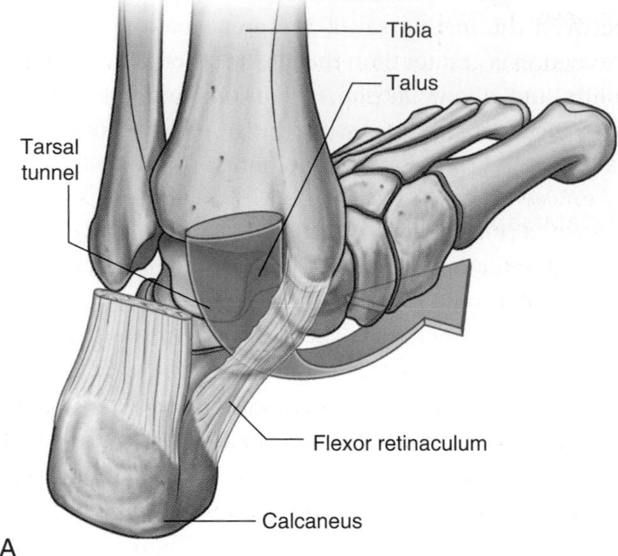


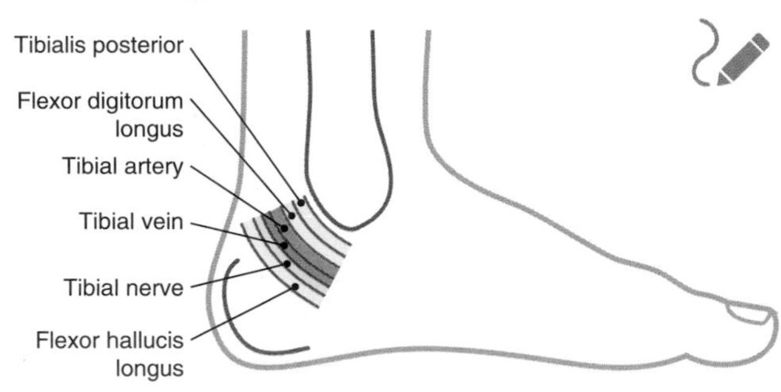


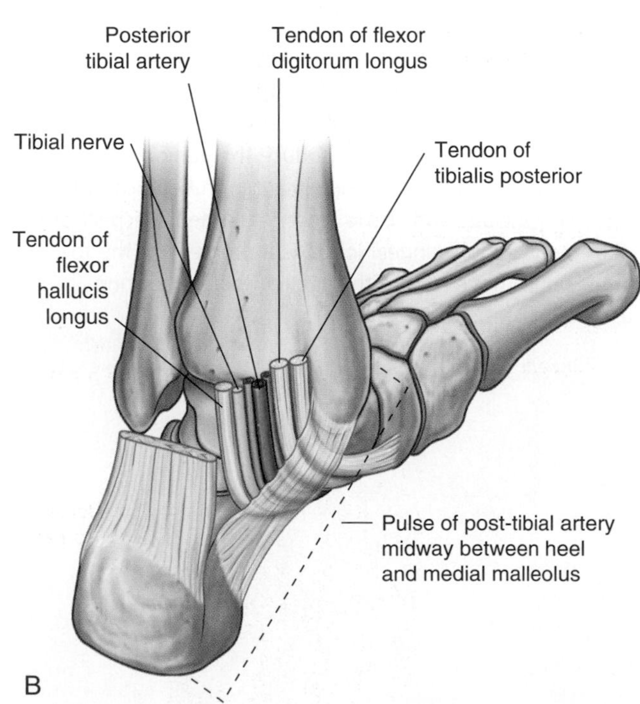


**Fig. 6.110** Tarsal tunnel and flexor retinaculum. Posteromedial view, right foot. **A.** Bones. **B.** Tarsal tunnel and flexor retinaculum.

### Extensor retinacula

Two extensor retinacula strap the tendons of the extensor muscles to the ankle region and prevent tendon bowing during extension of the foot and toes (Fig. 6.111):

- A **superior extensor retinaculum** is a thickening of deep fascia in the distal leg just superior to the ankle joint and attached to the anterior borders of the fibula and tibia.
- An **inferior retinaculum** is Y-shaped, attached by its base to the lateral side of the upper surface of the calcaneus, and crosses medially over the foot to attach by one of its arms to the medial malleolus, whereas the other arm wraps medially around the foot and attaches to the medial side of the plantar aponeurosis.

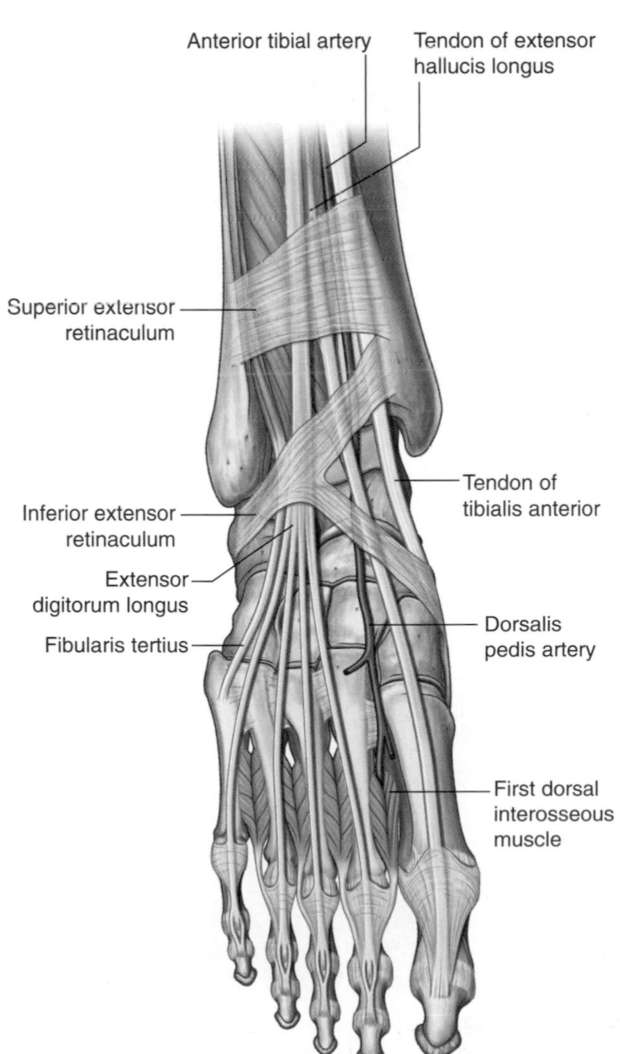


**Fig. 6.111** Extensor retinacula, right foot.

The tendons of the extensor digitorum longus and fibularis tertius pass through a compartment on the lateral side of the proximal foot. Medial to these tendons, the dorsalis pedis artery (terminal branch of the anterior tibial artery), the tendon of the extensor hallucis longus muscle, and finally the tendon of the tibialis anterior muscle pass under the extensor retinacula.

### Fibular retinacula

Fibular (peroneal) retinacula bind the tendons of the fibularis longus and fibularis brevis muscles to the lateral side of the foot (Fig. 6.112):

- **A superior fibular retinaculum** extends between the lateral malleolus and the calcaneus.
- **An inferior fibular retinaculum** attaches to the lateral surface of the calcaneus around the fibular trochlea and blends above with the fibers of the inferior extensor retinaculum.

At the fibular trochlea, a septum separates the compartment for the tendon of the fibularis brevis muscle above from that for the fibularis longus below.

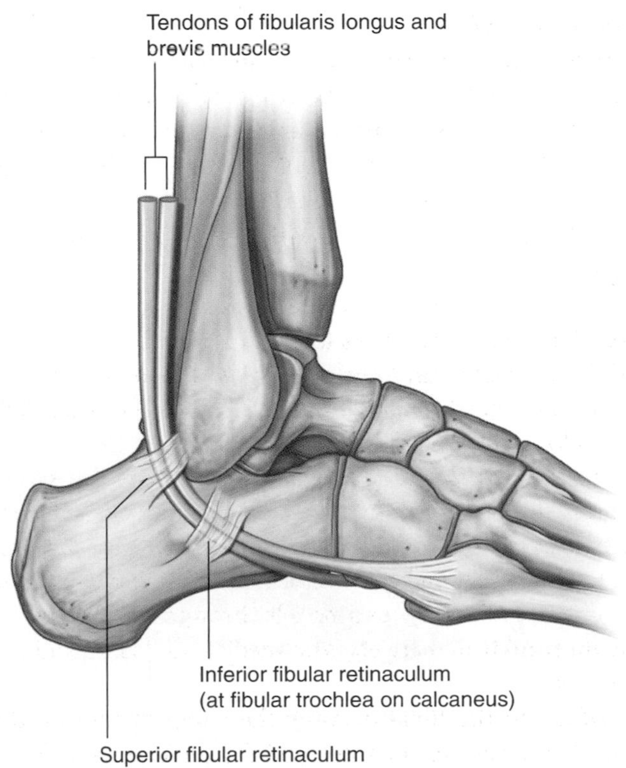


**Fig. 6.112** Fibular retinacula. Lateral view, right foot.

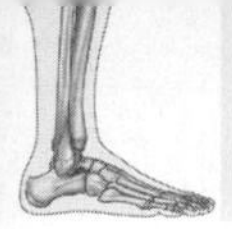



## Arches of the foot

The bones of the foot do not lie in a horizontal plane. Instead, they form longitudinal and transverse arches relative to the ground (Fig. 6.113), which absorb and distribute downward forces from the body during standing and moving on different surfaces.

### Longitudinal arch

The longitudinal arch of the foot is formed between the posterior end of the calcaneus and the heads of the metatarsals (Fig. 6.113A). It is highest on the medial side, where it forms the medial part of the longitudinal arch, and lowest on the lateral side, where it forms the lateral part.

### Transverse arch

The transverse arch of the foot is highest in a coronal plane that cuts through the head of the talus and disappears near the heads of the metatarsals, where these bones are held together by the deep transverse metatarsal ligaments (Fig. 6.113B).

### Ligament and muscle support

Ligaments and muscles support the arches of the foot (Fig. 6.114):

- Ligaments that support the arches include the plantar calcaneonavicular (spring ligament), plantar calcaneocuboid (short plantar ligament), and long plantar ligaments, and the plantar aponeurosis.

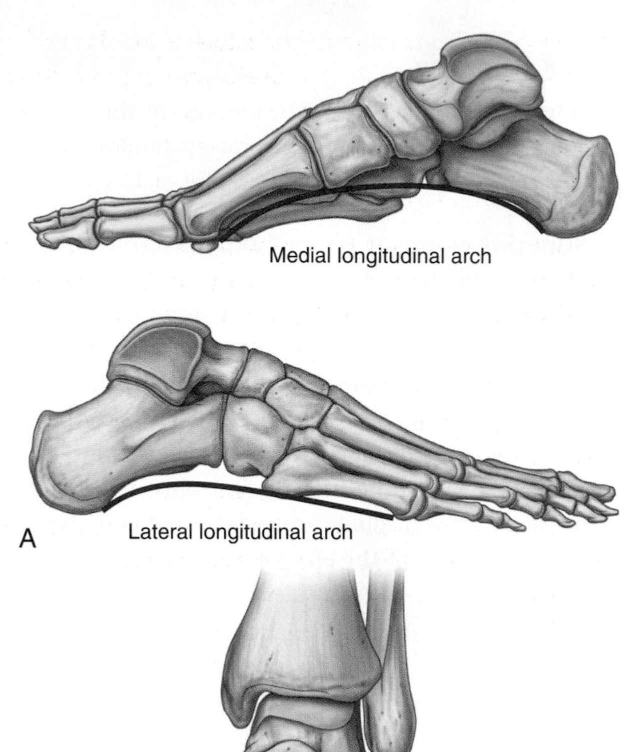


**Fig. 6.113** Arches of the foot. **A.** Longitudinal arches, right foot. **B.** Transverse arch, left foot.

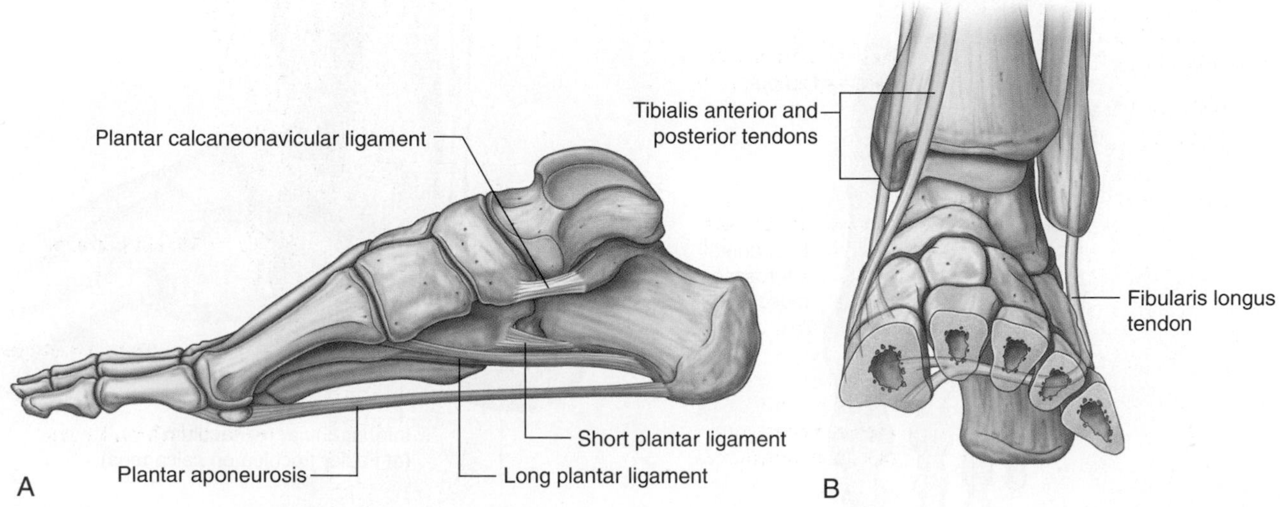


**Fig. 6.114** Support for arches of the foot. **A.** Ligaments. Medial view, right foot. **B.** Cross section through the foot to show tendons of muscles supporting the arches, left foot.

- Muscles that provide dynamic support for the arches during walking include the tibialis anterior and posterior and the fibularis longus.

## Plantar aponeurosis

The plantar aponeurosis is a thickening of deep fascia in the sole of the foot (Fig. 6.115). It is firmly anchored to the medial process of the calcaneal tuberosity and extends forward as a thick band of longitudinally arranged connective tissue fibers. The fibers diverge as they pass anteriorly and form digital bands, which enter the toes and connect with bones, ligaments, and dermis of the skin.

Distal to the metatarsophalangeal joints, the digital bands of the plantar aponeurosis are interconnected by transverse fibers, which form superficial transverse metatarsal ligaments.

The plantar aponeurosis supports the longitudinal arch of the foot and protects deeper structures in the sole.

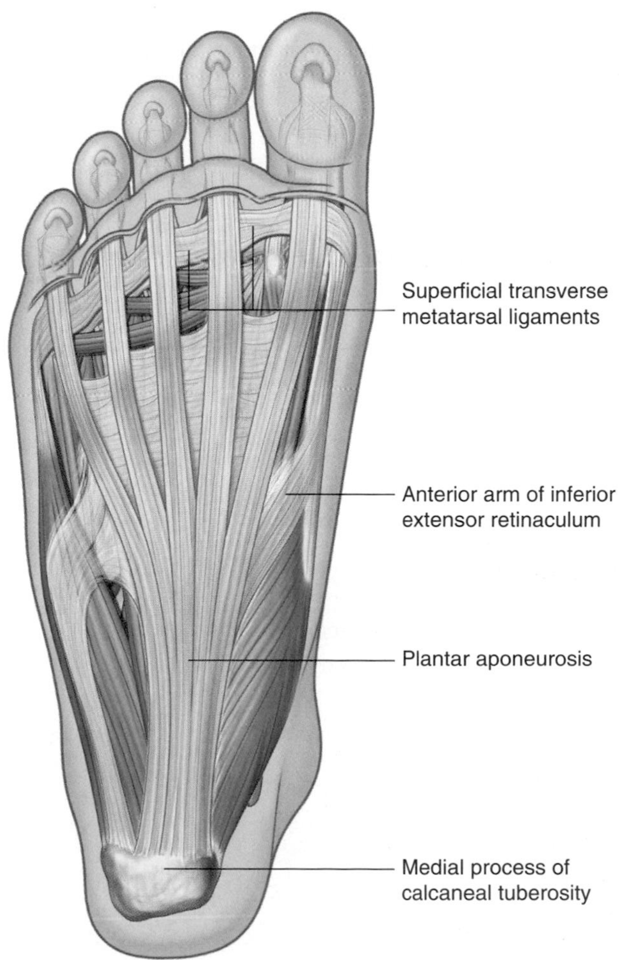


**Fig. 6.115** Plantar aponeurosis, right foot.

## Fibrous sheaths of toes

The tendons of the flexor digitorum longus, flexor digitorum brevis, and flexor hallucis longus muscles enter fibrous digital sheaths or tunnels on the plantar aspect of the digits (Fig. 6.116). These fibrous sheaths begin anterior to the metatarsophalangeal joints and extend to the distal phalanges. They are formed by fibrous arches and cruciate (cross-shaped) ligaments attached posteriorly to the margins of the phalanges and to the plantar ligaments associated with the metatarsophalangeal and interphalangeal joints.

These fibrous tunnels hold the tendons to the bony plane and prevent tendon bowing when the toes are flexed.

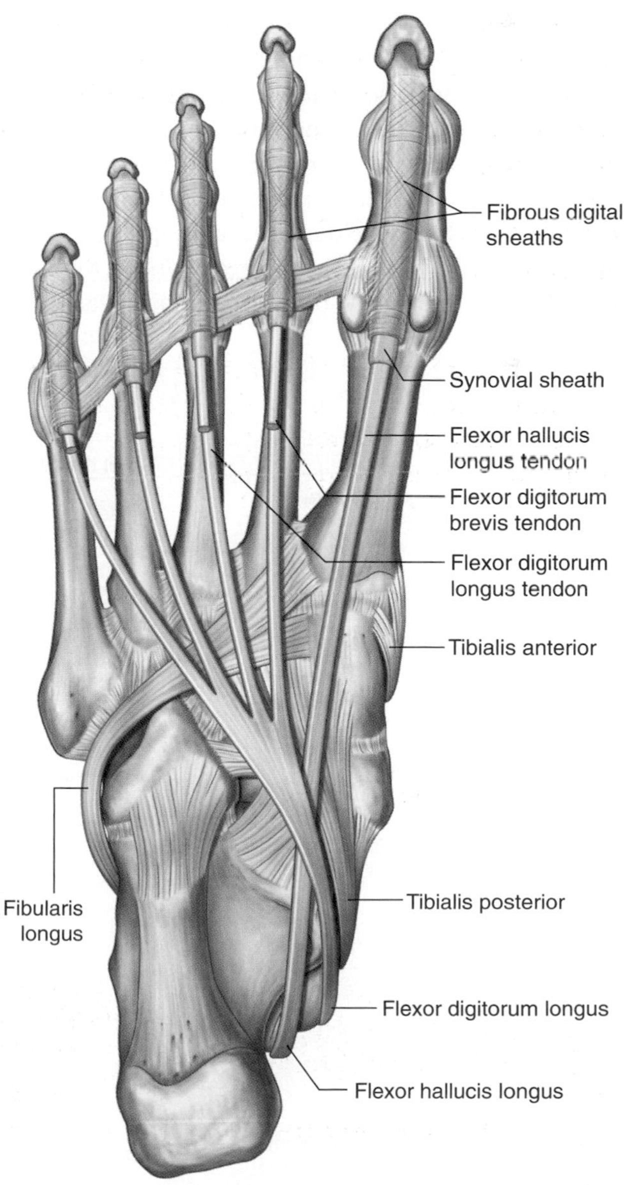


**Fig. 6.116** Fibrous digital sheaths, right foot.

Within each tunnel, the tendons are surrounded by a synovial sheath.

## Extensor hoods

The tendons of the extensor digitorum longus, extensor digitorum brevis, and extensor hallucis longus pass into the dorsal aspect of the digits and expand over the proximal phalanges to form complex dorsal digital expansions ("extensor hoods") (Fig. 6.117).

Each extensor hood is triangular in shape with the apex attached to the distal phalanx, the central region attached to the middle (toes II to V) or proximal (toe I) phalanx, and each corner of the base wrapped around the sides of the metatarsophalangeal joint. The corners of the hoods attach mainly to the deep transverse metatarsal ligaments.

Many of the intrinsic muscles of the foot insert into the free margin of the hood on each side. The attachment of these muscles into the extensor hoods allows the forces from these muscles to be distributed over the toes to cause flexion of the metatarsophalangeal joints while at the same time extending the interphalangeal joints (Fig. 6.117). The function of these movements in the foot is uncertain, but they may prevent overextension of the metatarsophalangeal joints and flexion of the interphalangeal joints when the heel is elevated off the ground and the toes grip the ground during walking.

## Intrinsic muscles

Intrinsic muscles of the foot originate and insert in the foot:

- the extensor digitorum brevis and extensor hallucis brevis on the dorsal aspect of the foot;
- all other intrinsic muscles—the dorsal and plantar interossei, flexor digiti minimi brevis, flexor hallucis brevis, flexor digitorum brevis, quadratus plantae (flexor accessorius), abductor digiti minimi, abductor hallucis, and lumbricals—are on the plantar side of the foot in the sole where they are organized into four layers.

Intrinsic muscles mainly modify the actions of the long tendons and generate fine movements of the toes.

All intrinsic muscles of the foot are innervated by the medial and lateral plantar branches of the tibial nerve except for the extensor digitorum brevis, which is innervated by the deep fibular nerve. The first two dorsal interossei also may receive part of their innervation from the deep fibular nerve.

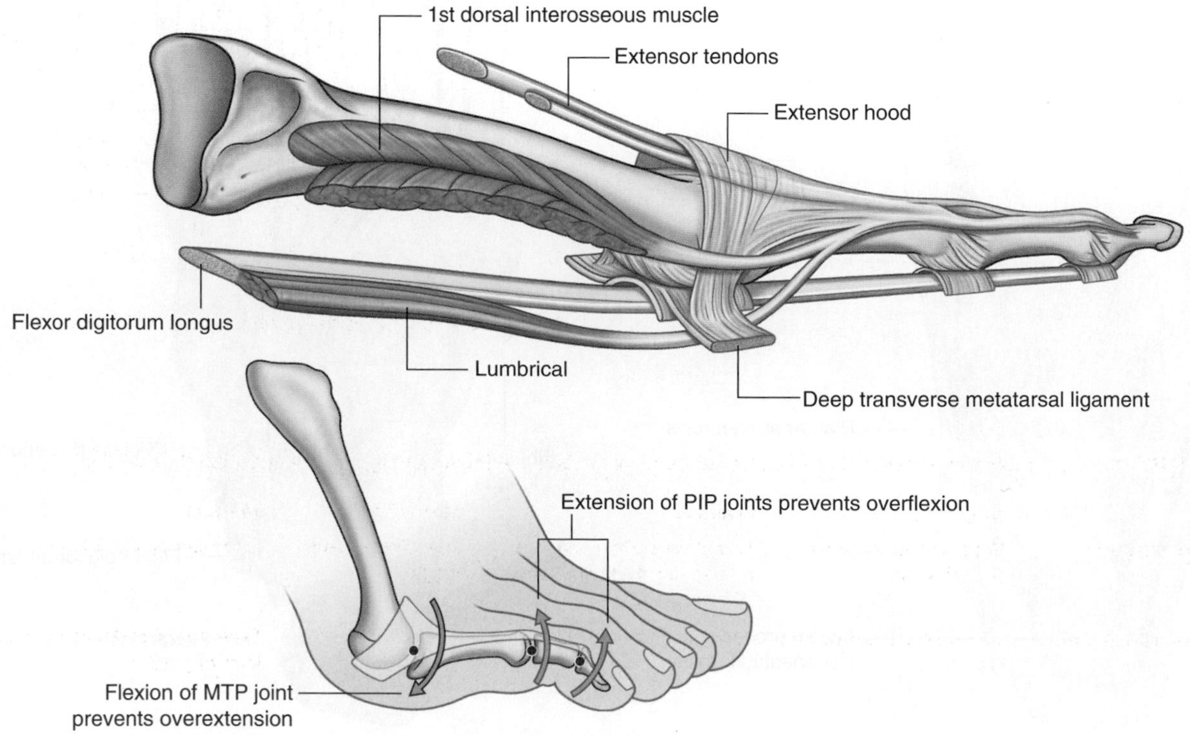


**Fig. 6.117** Extensor hoods.

### On the dorsal aspect

#### Extensor digitorum brevis and extensor hallucis brevis

The **extensor digitorum brevis** is attached to a roughened area on the superolateral surface of the calcaneus lateral to the tarsal sinus (Fig. 6.118 and Table 6.10).

The flat muscle belly passes anteromedially over the foot, deep to the tendons of the extensor digitorum longus, and forms three tendons, which enter digits II, III, and IV. The tendons join the lateral sides of the tendons of the extensor digitorum longus. The extensor digitorum brevis extends the middle three toes through attachments to the long extensor tendons and extensor hoods. It is innervated by the deep fibular nerve.

The **extensor hallucis brevis** originates in conjunction with the extensor digitorum brevis. Its tendon attaches to the base of the proximal phalanx of the great toes. The muscle extends the metatarsophalangeal joint of the great toe and is innervated by the deep fibular nerve.

### In the sole

The muscles in the sole of the foot are organized into four layers. From superficial to deep, or plantar to dorsal, these layers are the first, second, third, and fourth layers.

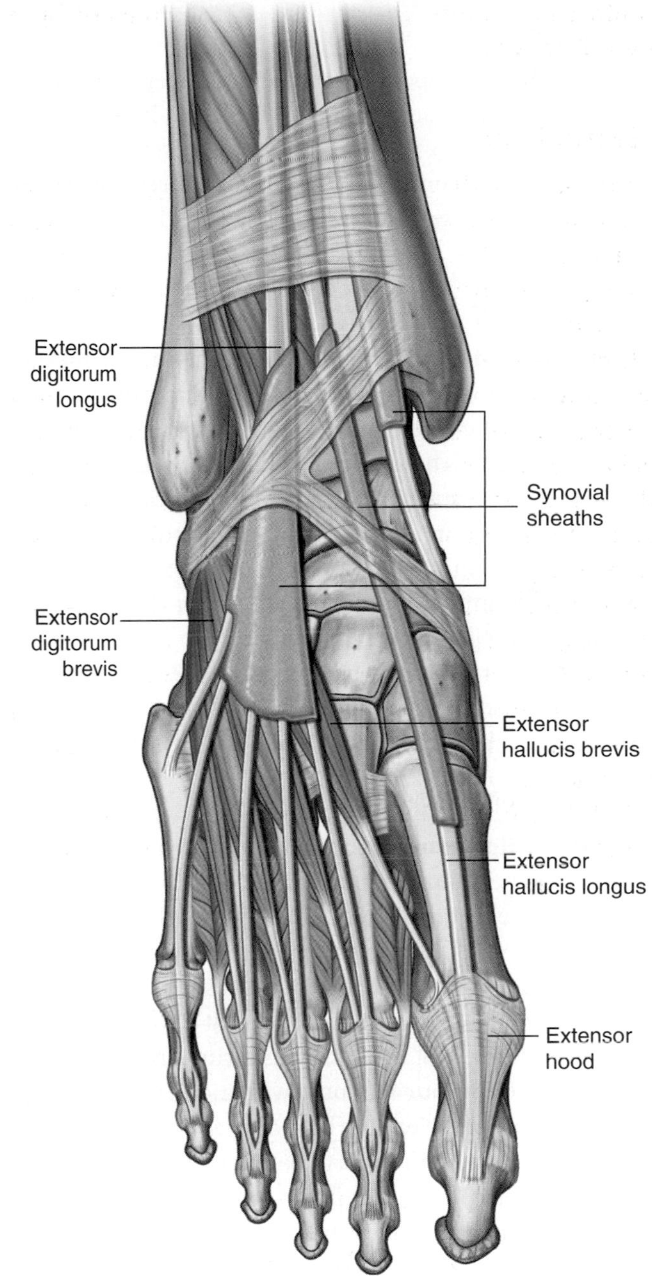


**Fig. 6.118** Extensor digitorum brevis muscle, right foot.

**Table 6.10** Muscles of the dorsal aspect of the foot (spinal segments in bold are the major segments innervating the muscle)

| Muscle | Origin | Insertion | Innervation | Function |
|---|---|---|---|---|
| Extensor digitorum brevis | Superolateral surface of the calcaneus | Lateral sides of the tendons of extensor digitorum longus of toes II to IV | Deep fibular nerve (**S1, S2**) | Extension of toes II to IV |
| Extensor hallucis brevis | Superolateral surface of calcaneus | Base of proximal phalanx of great toe | Deep fibular nerve (**S1, S2**) | Extension of metatarsophalangeal joint of great toe |

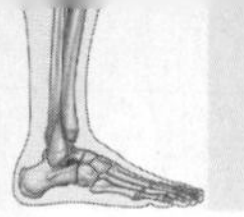

### First layer

There are three components in the first layer of muscles, which is the most superficial of the four layers and is immediately deep to the plantar aponeurosis (Fig. 6.119 and Table 6.11). From medial to lateral, these muscles are the abductor hallucis, flexor digitorum brevis, and abductor digiti minimi.

#### Abductor hallucis

The **abductor hallucis** muscle forms the medial margin of the foot and contributes to a soft tissue bulge on the medial side of the sole (Fig. 6.119). It originates from the medial process of the calcaneal tuberosity and adjacent margins of the flexor retinaculum and plantar aponeurosis. It forms a tendon that inserts on the medial side of the base of the proximal phalanx of the great toe and on the medial sesamoid bone associated with the tendon of the flexor hallucis brevis muscle.

The abductor hallucis abducts and flexes the great toe at the metatarsophalangeal joint and is innervated by the medial plantar branch of the tibial nerve.

#### Flexor digitorum brevis

The **flexor digitorum brevis** muscle lies immediately superior to the plantar aponeurosis and inferior to the tendons of the flexor digitorum longus in the sole of the foot (Fig. 6.119). The flat spindle-shaped muscle belly originates as a tendon from the medial process of the calcaneal tuberosity and from the adjacent plantar aponeurosis.

The muscle fibers of the flexor digitorum brevis converge anteriorly to form four tendons, which each enter one of the lateral four toes. Near the base of the proximal phalanx of the toe, each tendon splits to pass dorsally around each side of the tendon of the flexor digitorum longus and attach to the margins of the middle phalanx.

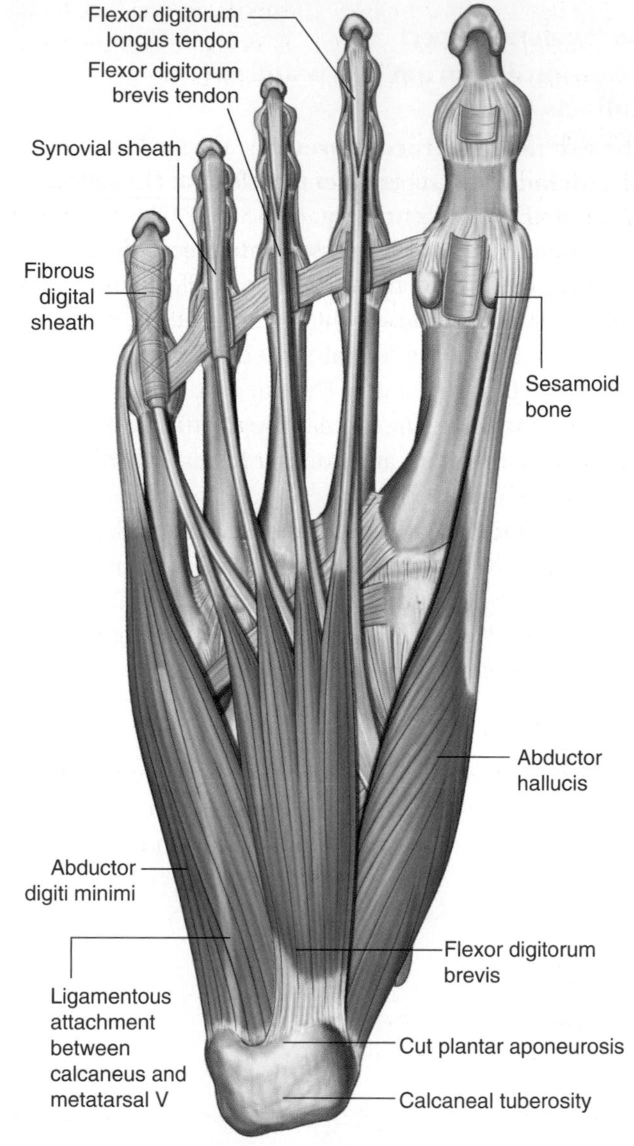


**Fig. 6.119** First layer of muscles in the sole of the right foot.

**Table 6.11** First layer of muscles in the sole of the foot (spinal segments in bold are the major segments innervating the muscle)

| Muscle | Origin | Insertion | Innervation | Function |
|---|---|---|---|---|
| Abductor hallucis | Medial process of calcaneal tuberosity | Medial side of base of proximal phalanx of great toe | Medial plantar nerve from the tibial nerve (**S1, S2, S3**) | Abducts and flexes great toe at metatarsophalangeal joint |
| Flexor digitorum brevis | Medial process of calcaneal tuberosity and plantar aponeurosis | Sides of plantar surface of middle phalanges of lateral four toes | Medial plantar nerve from the tibial nerve (**S1, S2, S3**) | Flexes lateral four toes at proximal interphalangeal joint |
| Abductor digiti minimi | Lateral and medial processes of calcaneal tuberosity, and band of connective tissue connecting calcaneus with base of metatarsal V | Lateral side of base of proximal phalanx of little toe | Lateral plantar nerve from the tibial nerve (**S1, S2, S3**) | Abducts little toe at the metatarsophalangeal joint |

The flexor digitorum brevis flexes the lateral four toes at the proximal interphalangeal joints and is innervated by the medial plantar branch of the tibial nerve.

### Abductor digiti minimi

The **abductor digiti minimi** muscle is on the lateral side of the foot and contributes to the large lateral plantar eminence on the sole (Fig. 6.119). It has a broad base of origin, mainly from the lateral and medial processes of the calcaneal tuberosity and from a fibrous band of connective tissue, which connects the calcaneus with the base of metatarsal V.

The abductor digiti minimi forms a tendon, which travels in a shallow groove on the plantar surface of the base of metatarsal V and continues forward to attach to the lateral side of the base of the proximal phalanx of the little toe.

The abductor digiti minimi abducts the little toe at the metatarsophalangeal joint and is innervated by the lateral plantar branch of the tibial nerve.

## Second layer

The second muscle layer in the sole of the foot is associated with the tendons of the flexor digitorum longus muscle, which pass through this layer, and consists of the quadratus plantae and four lumbrical muscles (Fig. 6.120 and Table 6.12).

### Quadratus plantae

The **quadratus plantae** muscle is a flat quadrangular muscle with two heads of origin (Fig. 6.120):

- One of the heads originates from the medial surface of the calcaneus inferior to the sustentaculum tali.
- The other head originates from the inferior surface of the calcaneus anterior to the lateral process of the calcaneal tuberosity and the attachment of the long plantar ligament.

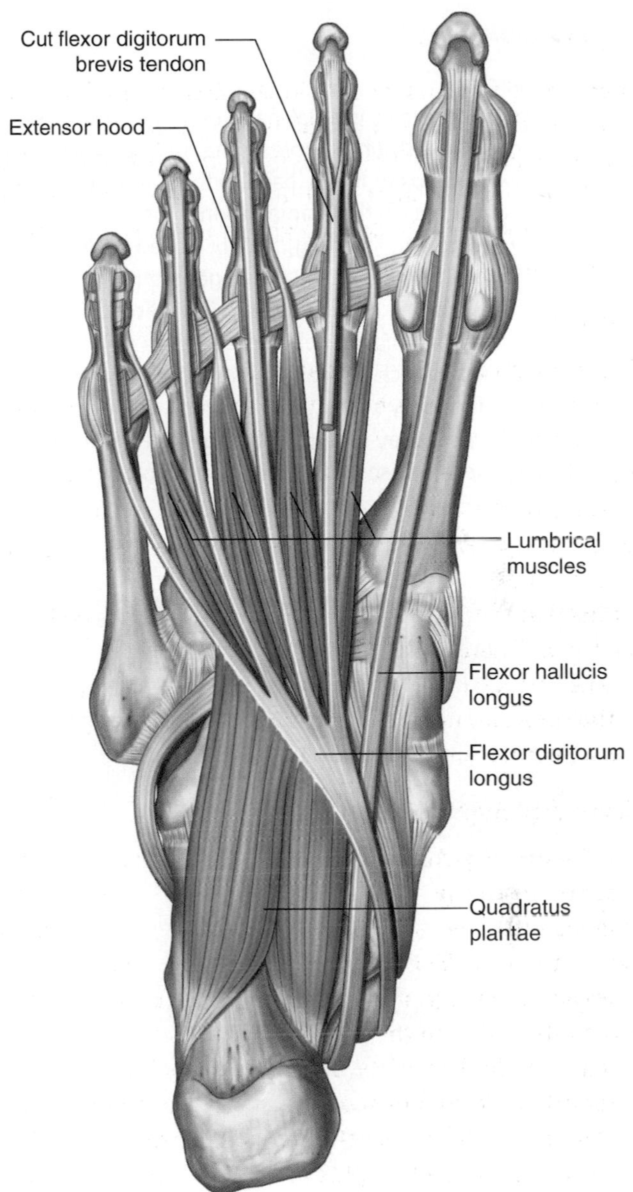


**Fig. 6.120** Second layer of muscles in the sole of the right foot.

**Table 6.12** Second layer of muscles in the sole of the foot (spinal segments in bold are the major segments innervating the muscle)

| Muscle | Origin | Insertion | Innervation | Function |
|---|---|---|---|---|
| Quadratus plantae | Medial surface of calcaneus and lateral process of calcaneal tuberosity | Lateral side of tendon of flexor digitorum longus in proximal sole of the foot | Lateral plantar nerve from tibial nerve (**S1, S2, S3**) | Assists flexor digitorum longus tendon in flexing toes II to V |
| Lumbricals | First lumbrical—medial side of tendon of flexor digitorum longus associated with toe II; second, third, and fourth lumbricals—adjacent surfaces of adjacent tendons of flexor digitorum longus | Medial free margins of extensor hoods of toes II to V | First lumbrical—medial plantar nerve from the tibial nerve; second, third, and fourth lumbricals—lateral plantar nerve from the tibial nerve (**S2, S3**) | Flexion of metatarsophalangeal joint and extension of interphalangeal joints |

## In the clinic

### Plantar fasciitis

The plantar aponeurosis is a flat band of connective tissue that supports the arch of the sole of the foot. It runs from the calcaneal tuberosity to the base of the toes. Overuse and increased strain on the plantar aponeurosis, such as excessive running and standing, and increased body weight, can lead to micro-tears and degeneration within the aponeurosis at the heel with disorganization of the collagen fibers. Patients typically present with mild to severe heel pain, which appears thickened on imaging (Fig. 6.121). It is usually successfully treated with intense physiotherapy, but may require image-guided injection therapies. In severe cases, the diseased section of the fascia needs to be surgically removed.

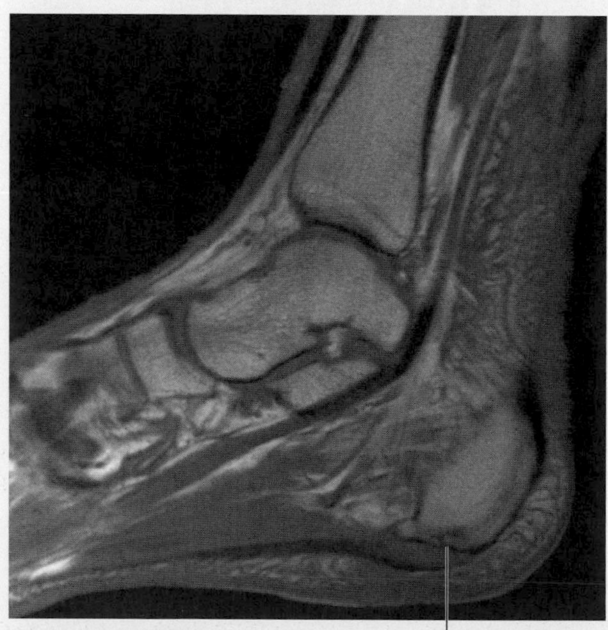


**Fig. 6.121** MRI of posterior foot in a patient with plantar fasciitis shows thickening of the plantar aponeurosis at the calcaneal insertion and a calcaneal spur.

The quadratus plantae muscle inserts into the lateral side of the tendon of the flexor digitorum longus in the proximal half of the sole of the foot near where the tendon divides.

The quadratus plantae assists the flexor digitorum longus tendon in flexing the toes and may also adjust the "line of pull" of this tendon as it enters the sole of the foot from the medial side. The muscle is innervated by the lateral plantar nerve.

#### Lumbricals

The lumbrical muscles are four worm-like muscles that originate from the tendons of the flexor digitorum longus and pass dorsally to insert into the free medial margins of the extensor hoods of the four lateral toes (Fig. 6.120).

The first lumbrical originates from the medial side of the tendon of the flexor digitorum longus that is associated with the second toe. The remaining three muscles are bipennate and originate from the sides of adjacent tendons.

The lumbrical muscles act through the extensor hoods to resist excessive extension of the metatarsophalangeal joints and flexion of the interphalangeal joints when the heel leaves the ground during walking.

The first lumbrical is innervated by the medial plantar nerve, while the other three are innervated by the lateral plantar nerve.

### Third layer

There are three muscles in the third layer in the sole of the foot (Fig. 6.122 and Table 6.13):

- Two (the flexor hallucis brevis and adductor hallucis) are associated with the great toe.
- The third (the flexor digiti minimi brevis) is associated with the little toe.

#### Flexor hallucis brevis

The **flexor hallucis brevis** muscle has two tendinous heads of origin (Fig. 6.122):

- The **lateral head** originates from the plantar surfaces of the cuboid, behind the groove for the fibularis longus, and adjacent surface of the lateral cuneiform.
- The **medial head** originates from the tendon of the tibialis posterior muscle as it passes into the sole of the foot.

The medial and lateral heads unite and give rise to a muscle belly, which itself is separated into medial and lateral parts adjacent to the plantar surface of metatarsal I. Each part of the muscle gives rise to a tendon that inserts on either the lateral or medial side of the base of the proximal phalanx of the great toe.

A sesamoid bone occurs in each tendon of the flexor hallucis brevis as it crosses the plantar surface of the head of metatarsal I. The tendon of the flexor hallucis longus passes between the sesamoid bones.

The flexor hallucis brevis flexes the metatarsophalangeal joint of the great toe and is innervated by the medial plantar nerve.

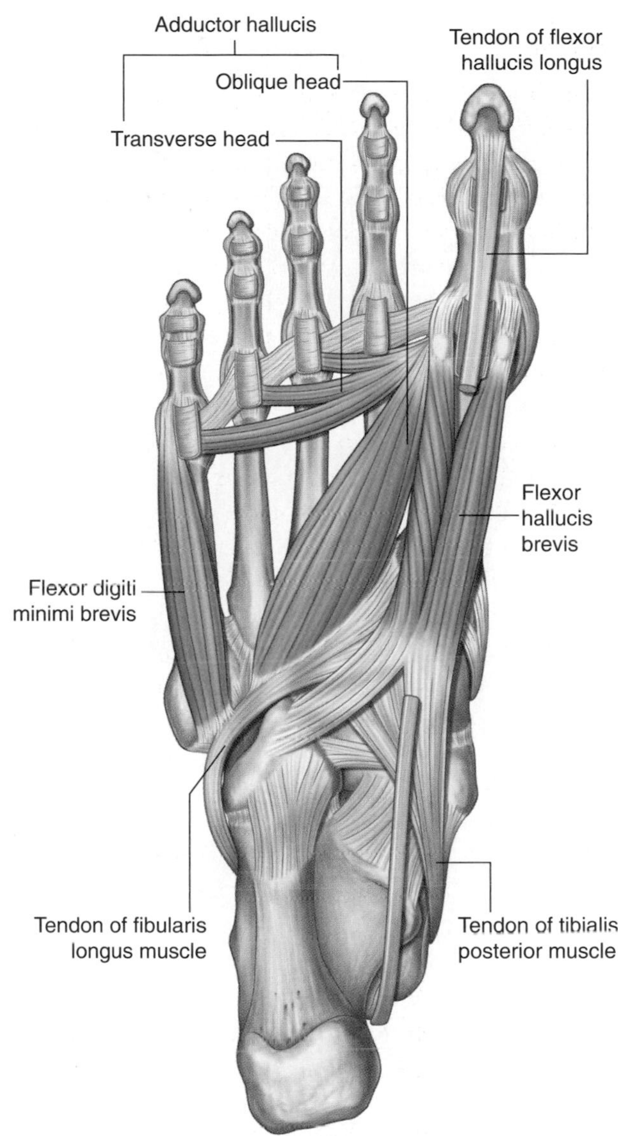


**Fig. 6.122** Third layer of muscles in the sole of the right foot.

### Adductor hallucis

The **adductor hallucis** muscle originates by two muscular heads, transverse and oblique, which join near their ends to insert into the lateral side of the base of the proximal phalanx of the great toe (Fig. 6.122):

- The **transverse head** originates from the plantar ligaments associated with the metatarsophalangeal joints of the lateral three toes and from the associated deep transverse metatarsal ligaments—the muscle crosses the sole of the foot transversely from lateral to medial and joins the oblique head near the base of the great toe.
- The **oblique head** is larger than the transverse head and originates from the plantar surfaces of the bases of metatarsals II to IV and from the sheath covering the fibularis longus muscle—this head passes anterolaterally through the sole of the foot and joins the transverse head.

The tendon of insertion of the adductor hallucis attaches to the lateral sesamoid bone associated with the tendon of the flexor hallucis brevis muscle in addition to attaching to the proximal phalanx.

The adductor hallucis adducts the great toe at the metatarsophalangeal joint and is innervated by the lateral plantar nerve.

### Flexor digiti minimi brevis

The **flexor digiti minimi brevis** muscle originates from the plantar surface of the base of metatarsal V and adjacent sheath of the fibularis longus tendon (Fig. 6.122). It inserts on the lateral side of the base of the proximal phalanx of the little toe.

The flexor digiti minimi brevis flexes the little toe at the metatarsophalangeal joint and is innervated by the lateral plantar nerve.

**Table 6.13** Third layer of muscles in the sole of the foot (spinal segments in bold are the major segments innervating the muscle)

| Muscle | Origin | Insertion | Innervation | Function |
|---|---|---|---|---|
| Flexor hallucis brevis | Plantar surface of cuboid and lateral cuneiform; tendon of tibialis posterior | Lateral and medial sides of base of proximal phalanx of the great toe | Medial plantar nerve from tibial nerve (**S1**, S2) | Flexes metatarsophalangeal joint of the great toe |
| Adductor hallucis | Transverse head—ligaments associated with metatarsophalangeal joints of lateral three toes; oblique head—bases of metatarsals II to IV and from sheath covering fibularis longus | Lateral side of base of proximal phalanx of great toe | Lateral plantar nerve from tibial nerve (**S2, S3**) | Adducts great toe at metatarsophalangeal joint |
| Flexor digiti minimi brevis | Base of metatarsal V and related sheath of fibularis longus tendon | Lateral side of base of proximal phalanx of little toe | Lateral plantar nerve from tibial nerve (**S2, S3**) | Flexes little toe at metatarsophalangeal joint |

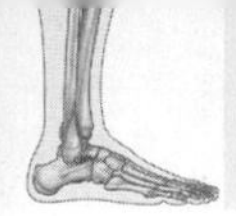

## Fourth layer

There are two muscle groups in the deepest muscle layer in the sole of the foot, the dorsal and plantar interossei (Fig. 6.123 and Table 6.14).

### Dorsal interossei

The four **dorsal interossei** are the most superior muscles in the sole of the foot and abduct the second to fourth toes relative to the long axis through the second toe (Fig. 6.123). All four muscles are bipennate and originate from the sides of adjacent metatarsals.

The tendons of the dorsal interossei insert into the free margin of the extensor hoods and base of the proximal phalanges of the toes.

The second toe can be abducted to either side of its long axis, so it has two dorsal interossei associated with it, one on each side. The third and fourth toes have a dorsal interosseous muscle on their lateral sides only. The great and little toes have their own abductors (the abductor hallucis and abductor digiti minimi) in the first layer of muscles in the sole of the foot.

In addition to abduction, the dorsal interossei act through the extensor hoods to resist extension of the metatarsophalangeal joints and flexion of the interphalangeal joints.

The dorsal interossei are innervated by the lateral plantar nerve. The first and second dorsal interossei also receive branches on their superior surfaces from the deep fibular nerve.

### Plantar interossei

The three plantar interossei adduct the third, fourth, and little toes toward the long axis through the second toe (Fig. 6.123).

Each plantar interosseous muscle originates from the medial side of its associated metatarsal and inserts into

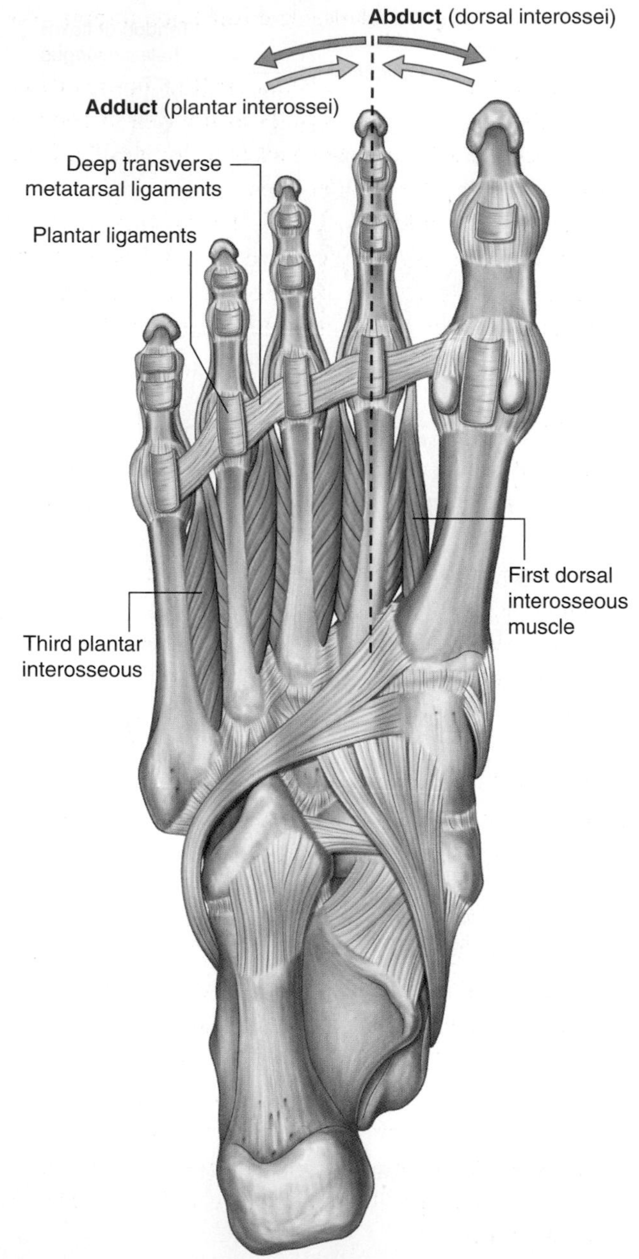


**Fig. 6.123** Fourth layer of muscles in the sole of the right foot.

**Table 6.14** Fourth layer of muscles in the sole of the foot (spinal segments in bold are the major segments innervating the muscle)

| Muscle | Origin | Insertion | Innervation | Function |
|---|---|---|---|---|
| Dorsal interossei | Sides of adjacent metatarsals | Extensor hoods and bases of proximal phalanges of toes II to IV | Lateral plantar nerve from tibial nerve; first and second dorsal interossei also innervated by deep fibular nerve (**S2, S3**) | Abduction of toes II to IV at metatarsophalangeal joints; resist extension of metatarsophalangeal joints and flexion of interphalangeal joints |
| Plantar interossei | Medial sides of metatarsals of toes III to V | Extensor hoods and bases of proximal phalanges of toes III to V | Lateral plantar nerve from tibial nerve (**S2, S3**) | Adduction of toes III to V at metatarsophalangeal joints; resist extension of the metatarsophalangeal joints and flexion of the interphalangeal joints |

the medial free margin of the extensor hood and base of the proximal phalanx.

The great toe has its own adductor (the adductor hallucis) in the third layer of muscles in the sole of the foot and the second toe is adducted back to its longitudinal axis by using one of its dorsal interossei.

In addition to adduction, the plantar interossei act through the extensor hoods to resist extension of the metatarsophalangeal joints and flexion of the interphalangeal joints. All are innervated by the lateral plantar nerve.

## Arteries

Blood supply to the foot is by branches of the posterior tibial and dorsalis pedis (dorsal artery of the foot) arteries.

The posterior tibial artery enters the sole and bifurcates into lateral and medial plantar arteries. The lateral plantar artery joins with the terminal end of the dorsalis pedis artery (the deep plantar artery) to form the deep plantar arch. Branches from this arch supply the toes.

The dorsalis pedis artery is the continuation of the anterior tibial artery, passes onto the dorsal aspect of the foot and then inferiorly, as the deep plantar artery, between metatarsals I and II to enter the sole of the foot.

### Posterior tibial artery and plantar arch

The posterior tibial artery enters the foot through the tarsal tunnel on the medial side of the ankle and posterior to the medial malleolus. Midway between the medial malleolus and the heel, the pulse of the posterior tibial artery is palpable because here the artery is covered only by a thin layer of retinaculum, by superficial connective tissue, and by skin. Near this location, the posterior tibial artery bifurcates into a small medial plantar artery and a much larger lateral plantar artery.

#### Lateral plantar artery

The **lateral plantar artery** passes anterolaterally into the sole of the foot, first deep to the proximal end of the abductor hallucis muscle and then between the quadratus plantae and flexor digitorum brevis muscles (Fig. 6.124). It reaches the base of metatarsal V where it lies in the groove between the flexor digitorum brevis and abductor digiti minimi muscles. From here, the lateral plantar artery curves medially to form the **deep plantar arch**, which crosses the deep plane of the sole on the metatarsal bases and the interossei muscles.

Between the bases of metatarsals I and II, the deep plantar arch joins with the terminal branch (deep plantar artery) of the dorsalis pedis artery, which enters the sole from the dorsal side of the foot.

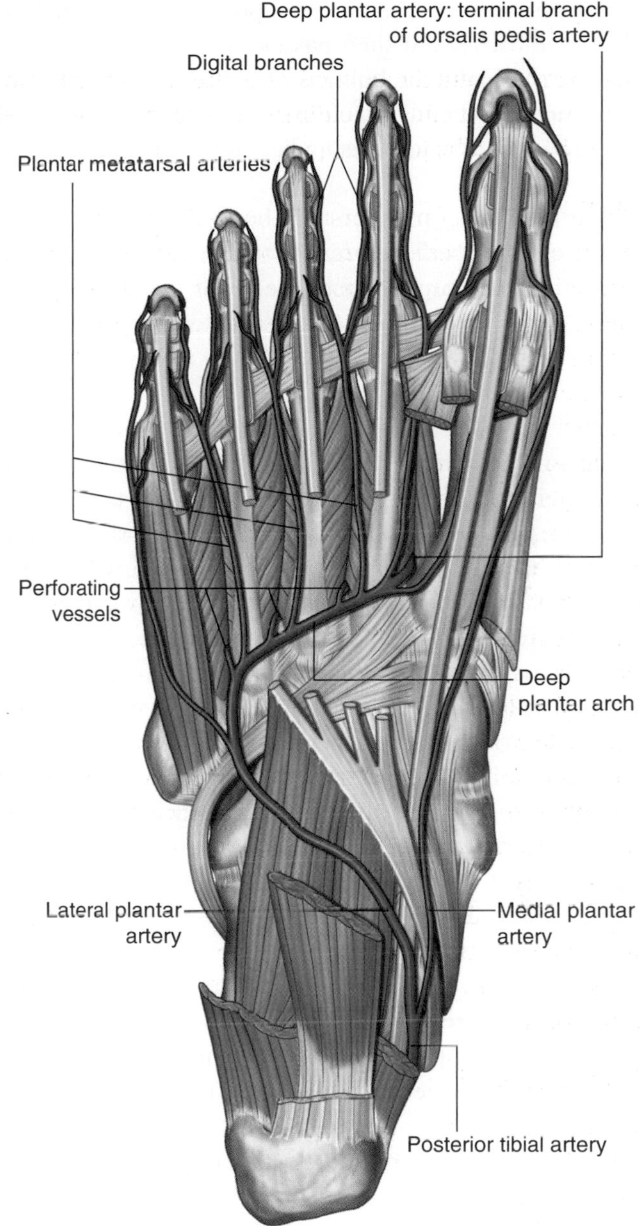


**Fig. 6.124** Arteries in the sole of the right foot.

Major branches of the deep plantar arch include:

- a digital branch to the lateral side of the little toe;
- four plantar metatarsal arteries, which supply digital branches to adjacent sides of toes I to V and the medial side of the great toe; and
- three perforating arteries, which pass between the bases of metatarsals II to V to anastomose with vessels on the dorsal aspect of the foot.

#### Medial plantar artery

The **medial plantar artery** passes into the sole of the foot by passing deep to the proximal end of the abductor hallucis muscle (Fig. 6.124). It supplies a deep branch to adjacent muscles and then passes forward in the groove between the abductor hallucis and the flexor digitorum brevis muscles. It ends by joining the digital branch of the deep plantar arch, which supplies the medial side of the great toe.

Near the base of metatarsal I, the medial plantar artery gives rise to a superficial branch, which divides into three vessels that pass superficial to the flexor digitorum brevis muscle to join the plantar metatarsal arteries from the deep plantar arch.

### Dorsalis pedis artery

The dorsalis pedis artery is the continuation of the anterior tibial artery and begins as the anterior tibial artery crosses the ankle joint (Fig. 6.125). It passes anteriorly over the dorsal aspect of the talus, navicular, and intermediate cuneiform bones, and then passes inferiorly, as the deep plantar artery, between the two heads of the first dorsal interosseous muscle to join the deep plantar arch in the sole of the foot. The pulse of the dorsalis pedis artery on the dorsal surface of the foot can be felt by gently palpating the vessel against the underlying tarsal bones between the tendons of the extensor hallucis longus and the extensor digitorum longus to the second toe.

Branches of the dorsalis pedis artery include lateral and medial tarsal branches, an arcuate artery, and a first dorsal metatarsal artery:

- The **tarsal arteries** pass medially and laterally over the tarsal bones, supplying adjacent structures and anastomosing with a network of vessels formed around the ankle.
- The **arcuate artery** passes laterally over the dorsal aspect of the metatarsals near their bases and gives rise to three **dorsal metatarsal arteries**, which supply **dorsal digital arteries** to adjacent sides of digits II to V, and to a dorsal digital artery that supplies the lateral side of digit V.
- The **first dorsal metatarsal artery** (the last branch of the dorsalis pedis artery before the dorsalis pedis artery continues as the deep plantar artery into the sole of the foot) supplies dorsal digital branches to adjacent sides of the great and second toes.

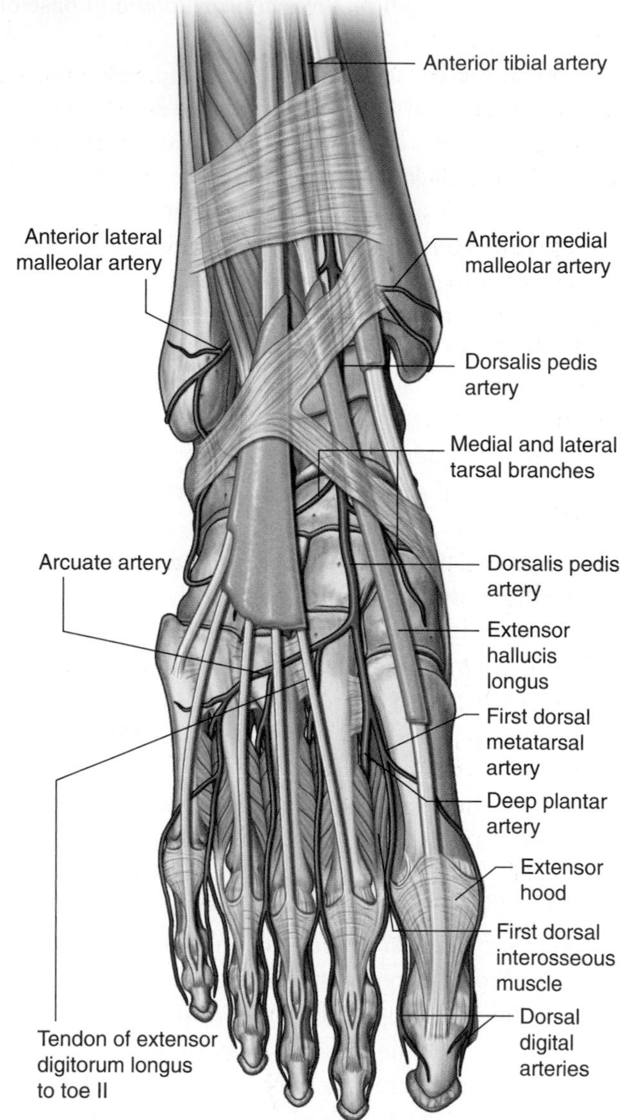


**Fig. 6.125** Dorsalis pedis artery right foot.

The dorsal metatarsal arteries connect with perforating branches from the deep plantar arch and similar branches from the plantar metatarsal arteries.

## Veins

There are interconnected networks of deep and superficial veins in the foot. The deep veins follow the arteries. Superficial veins drain into a dorsal venous arch on the dorsal surface of the foot over the metatarsals (Fig. 6.126):

- The **great saphenous vein** originates from the medial side of the arch and passes anterior to the medial malleolus and onto the medial side of the leg.
- The **small saphenous vein** originates from the lateral side of the arch and passes posterior to the lateral malleolus and onto the back of the leg.

## Nerves

The foot is supplied by the tibial, deep fibular, superficial fibular, sural, and saphenous nerves:

- All five nerves contribute to cutaneous or general sensory innervation.
- The tibial nerve innervates all intrinsic muscles of the foot except for the extensor digitorum brevis, which is innervated by the deep fibular nerve.
- The deep fibular nerve often also contributes to the innervation of the first and second dorsal interossei.

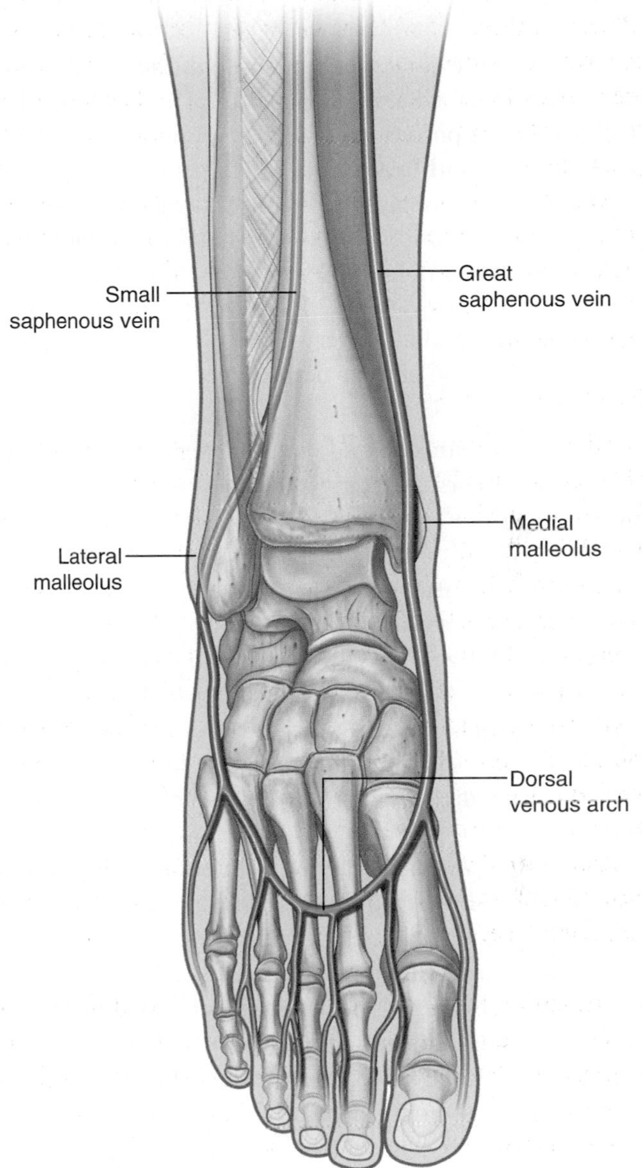


**Fig. 6.126** Superficial veins of the right foot.

## Tibial nerve

The **tibial nerve** enters the foot through the tarsal tunnel posterior to the medial malleolus. In the tunnel, the nerve is lateral to the posterior tibial artery, and gives origin to **medial calcaneal branches**, which penetrate the flexor retinaculum to supply the heel. Midway between the medial malleolus and the heel, the tibial nerve bifurcates with the posterior tibial artery into:

- a large medial plantar nerve, and
- a smaller lateral plantar nerve (Fig. 6.127).

The medial and lateral plantar nerves lie together between their corresponding arteries.

### Medial plantar nerve

The **medial plantar nerve** is the major sensory nerve in the sole of the foot (Fig. 6.127). It innervates skin on most of the anterior two-thirds of the sole and adjacent surfaces of the medial three and one-half toes, which includes the great toe. In addition to this large area of plantar skin, the nerve also innervates four intrinsic muscles—the abductor hallucis, flexor digitorum brevis, flexor hallucis brevis, and first lumbrical.

The medial plantar nerve passes into the sole of the foot deep to the abductor hallucis muscle and forward in the groove between the abductor hallucis and flexor digitorum brevis, supplying branches to both these muscles.

The medial plantar nerve supplies a digital branch (**proper plantar digital nerve**) to the medial side of the great toe and then divides into three nerves (**common plantar digital nerves**) on the plantar surface of the flexor digitorum brevis, which continue forward to supply proper plantar digital branches to adjacent surfaces of toes I to IV. The nerve to the first lumbrical originates from the first common plantar digital nerve.

### Lateral plantar nerve

The **lateral plantar nerve** is an important motor nerve in the foot because it innervates all intrinsic muscles in the sole, except for the muscles supplied by the medial plantar nerve (the abductor hallucis, flexor digitorum brevis, flexor hallucis brevis, and first lumbrical) (Fig. 6.127). It also innervates a strip of skin on the lateral side of the anterior two-thirds of the sole and the adjacent plantar surfaces of the lateral one and one-half digits.

The lateral plantar nerve enters the sole of the foot by passing deep to the proximal attachment of the abductor hallucis muscle. It continues laterally and anteriorly across the sole between the flexor digitorum brevis and quadratus plantae muscles, supplying branches to both these muscles,

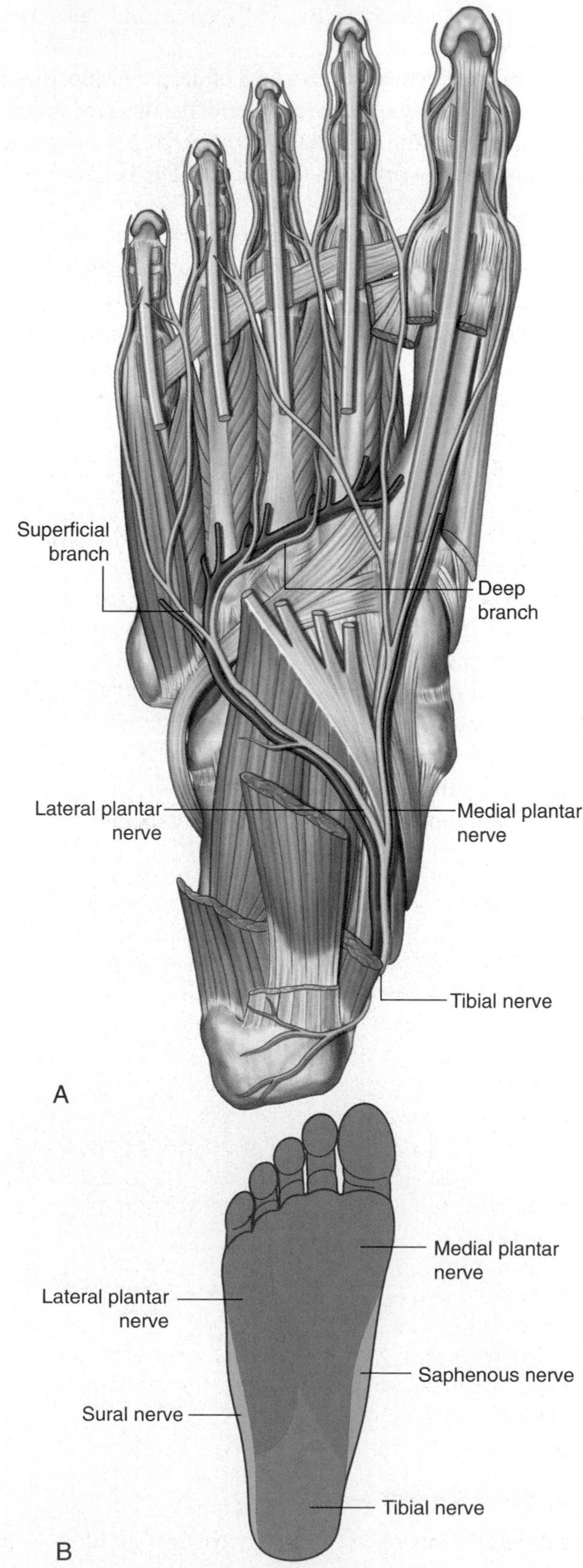


**Fig. 6.127** Lateral and medial plantar nerves. **A.** Sole of the right foot. **B.** Cutaneous distribution of right foot.

and then divides near the head of metatarsal V into deep and superficial branches.

The **superficial branch** of the lateral plantar nerve gives rise to a **proper plantar digital nerve**, which supplies skin on the lateral side of the little toe, and to a **common plantar digital nerve**, which divides to supply proper plantar digital nerves to skin on the adjacent sides of toes IV and V.

The proper plantar digital nerve to the lateral side of the little toe also innervates the flexor digiti minimi brevis and the dorsal and plantar interossei muscles between metatarsals IV and V.

The **deep branch** of the lateral plantar nerve is motor and accompanies the lateral plantar artery deep to the long flexor tendons and the adductor hallucis muscle. It supplies branches to the second to fourth lumbrical muscles, the adductor hallucis muscle, and all interossei except those between metatarsals IV and V, which are innervated by the superficial branch.

## In the clinic

### Morton's neuroma

A Morton's neuroma is an enlarged common plantar nerve, usually in the third interspace between the third and fourth toes. In this region of the foot the lateral plantar nerve often unites with the medial plantar nerve. As the two nerves join, the resulting nerve is typically larger in diameter than those of the other toes. Also, it is in a relatively subcutaneous position, just above the fat pad of the foot close to the artery and the vein. Above the nerve is the deep transverse metatarsal ligament, which is a broad strong structure holding the metatarsals together. Typically, as the patient enters the "push-off" phase of walking the interdigital nerve is sandwiched between the ground and the deep transverse metatarsal ligament. The forces tend to compress the common plantar nerve, which can be irritated, in which case there is usually some associated inflammatory change and thickening.

Typically, patients experience pain in the third interspace, which may be sharp or dull and is usually worsened by wearing shoes and walking.

Treatment may include injection of anti-inflammatory drugs, or it may be necessary to surgically remove the lesion.

## Deep fibular nerve

The **deep fibular nerve** innervates the extensor digitorum brevis, contributes to the innervation of the first two dorsal interossei muscles, and supplies general sensory branches to the skin on the adjacent dorsal sides of the first and second toes and to the web space between them (Fig. 6.128).

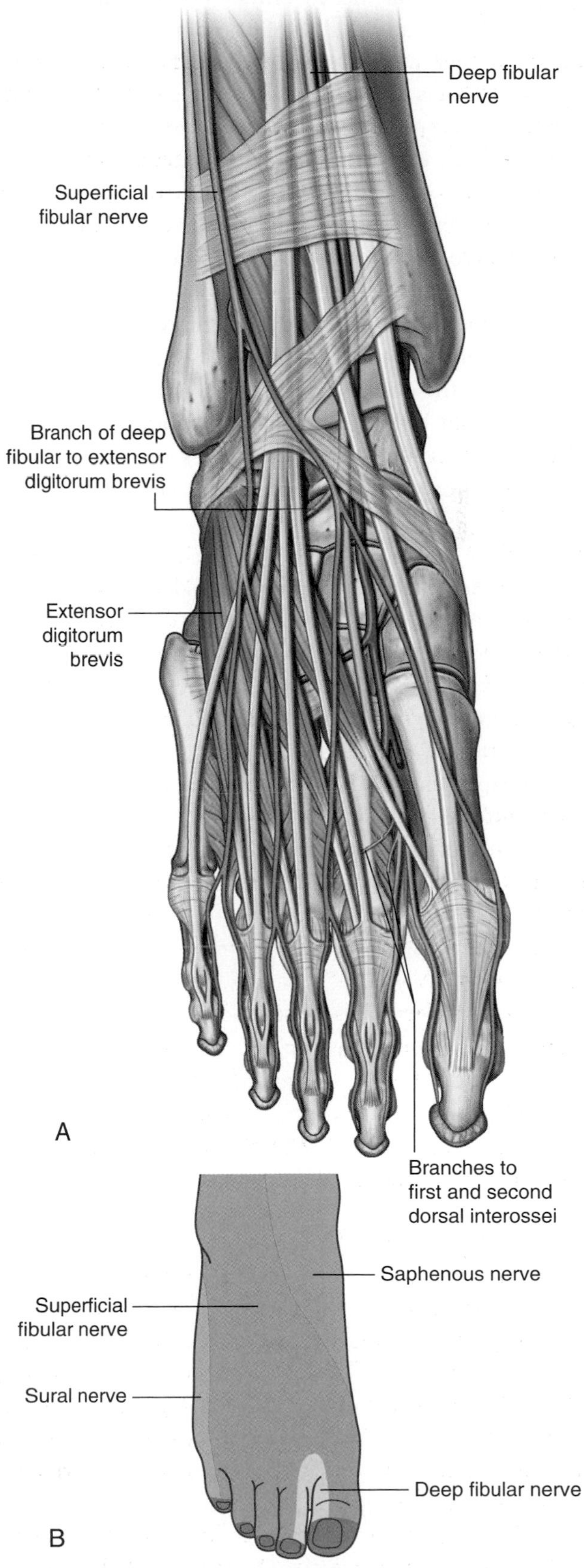


**Fig. 6.128 A.** Terminal branches of superficial and deep fibular nerves in the right foot. **B.** Cutaneous distribution right foot.

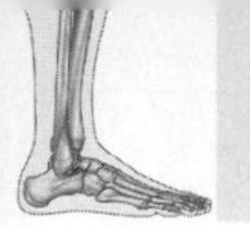



The deep fibular nerve enters the dorsal aspect of the foot on the lateral side of the dorsalis pedis artery, and is parallel with and lateral to the tendon of the extensor hallucis longus muscle. Just distal to the ankle joint, the nerve gives origin to a lateral branch, which innervates the extensor digitorum brevis from its deep surface.

The deep fibular nerve continues forward on the dorsal surface of the foot, penetrates deep fascia between metatarsals I and II near the metatarsophalangeal joints, and then divides into two **dorsal digital nerves**, which supply skin over adjacent surfaces of toes I and II down to the beginning of the nail beds.

Small motor branches, which contribute to the supply of the first two dorsal interossei muscles, originate from the deep fibular nerve before it penetrates deep fascia.

### Superficial fibular nerve

The **superficial fibular nerve** is sensory to most skin on the dorsal aspect of the foot and toes except for skin on adjacent sides of toes I and II (which is innervated by the deep fibular nerve) and skin on the lateral side of the foot and little toe (which is innervated by the sural nerve; Fig. 6.128).

The superficial fibular nerve penetrates deep fascia on the anterolateral side of the lower leg and enters the dorsal aspect of the foot in superficial fascia. It gives rise to cutaneous branches and **dorsal digital nerves** along its course.

### Sural nerve

The sural nerve is a cutaneous branch of the tibial nerve that originates high in the leg. It enters the foot in superficial fascia posterior to the lateral malleolus close to the short saphenous vein. Terminal branches innervate skin on the lateral side of the foot and dorsolateral surface of the little toe (Fig. 6.128B).

### Saphenous nerve

The saphenous nerve is a cutaneous branch of the femoral nerve that originates in the thigh. Terminal branches enter the foot in superficial fascia on the medial side of the ankle and supply skin on the medial side of the proximal foot (Fig. 6.128B).

## In the clinic

**Clubfoot**

Clubfoot is a congenital deformity in which babies are born with one or both feet pointing inward and downward. It is treated with gentle manipulation of the affected foot and with plaster casts to straighten the foot, which is usually followed by a minor surgical procedure where the calcaneal tendon is cut to release the foot into a better position.

# Surface anatomy

## Lower limb surface anatomy

Tendons, muscles, and bony landmarks in the lower limb are used to locate major arteries, veins, and nerves.

Because vessels are large, they can be used as entry points to the vascular system. In addition, vessels in the lower limb are farthest from the heart and the most inferior in the body. Therefore, the nature of peripheral pulses in the lower limb can give important information about the status of the circulatory system in general.

Sensation and muscle action in the lower limb are tested to assess lumbar and sacral regions of the spinal cord.

## Avoiding the sciatic nerve

The sciatic nerve innervates muscles in the posterior compartment of the thigh, muscles in the leg and foot, and an appreciable area of skin. It enters the lower limb in the gluteal region (Fig. 6.129) and passes inferiorly midway between two major palpable bony landmarks, the greater trochanter and the ischial tuberosity. The greater trochanter can be easily felt as a hard bony protuberance about one hand's width inferior to the midpoint of the iliac crest. The ischial tuberosity is palpable just above the gluteal fold.

The gluteal region can be divided into quadrants by two lines positioned using palpable bony landmarks.

- One line descends vertically from the highest point of the iliac crest.
- The other line passes horizontally through the first line midway between the highest point of the iliac crest and the horizontal plane through the ischial tuberosity.

The sciatic nerve curves through the upper lateral corner of the lower medial quadrant and descends along the lateral margin of the lower medial quadrant. Injections can be carried out in the anterior corner of the upper lateral quadrant to avoid injury to the sciatic nerve and major vessels in the region (Fig. 6.129B).

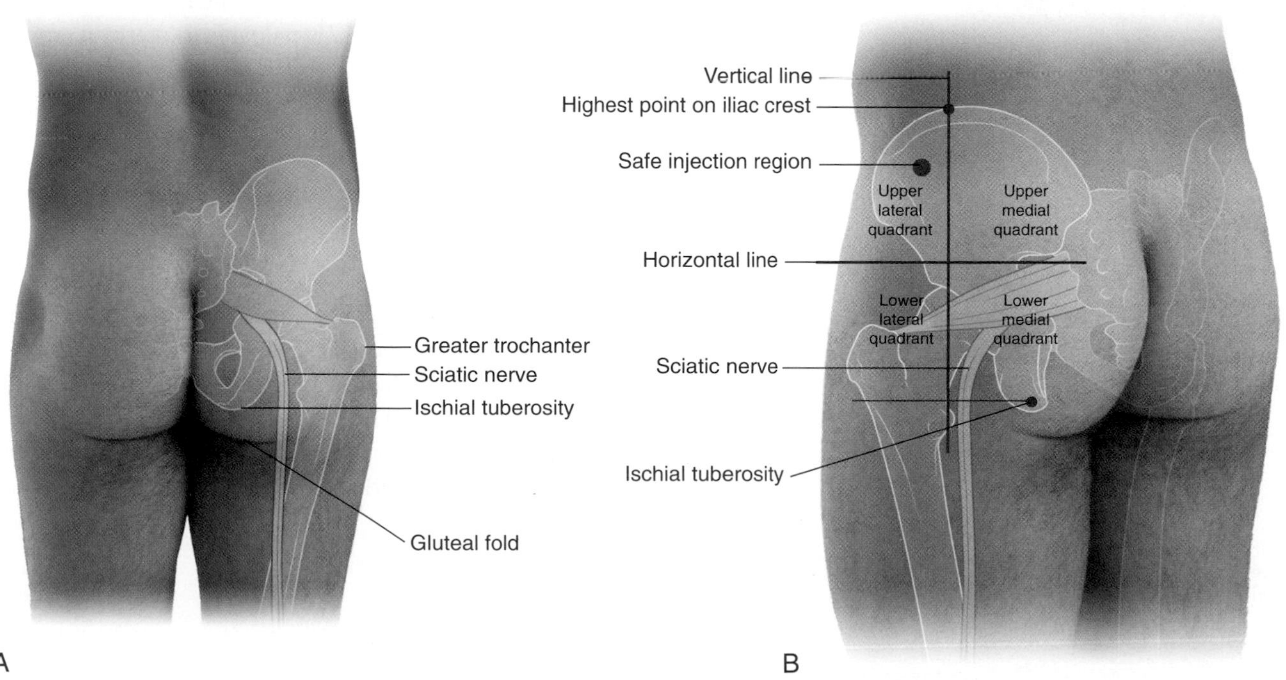


**Fig. 6.129** Avoiding the sciatic nerve. **A.** Posterior view of the gluteal region of a man with the position of the sciatic nerve indicated. **B.** Posterolateral view of the left gluteal region with gluteal quadrants and the position of the sciatic nerve indicated.

## Finding the femoral artery in the femoral triangle

The femoral artery passes into the femoral triangle (Fig. 6.130) of the lower limb from the abdomen.

The femoral triangle is the depression formed in the anterior thigh between the medial margin of the adductor longus muscle, the medial margin of the sartorius muscle, and the inguinal ligament.

The tendon of the adductor longus muscle can be palpated as a cord-like structure that attaches to bone immediately inferior to the pubic tubercle.

The sartorius muscle originates from the anterior superior iliac spine and crosses anteriorly over the thigh to attach to the medial aspect of the tibia below the knee joint.

The inguinal ligament attaches to the anterior superior iliac spine laterally and the pubic tubercle medially.

The femoral artery descends into the thigh from the abdomen by passing under the inguinal ligament and into the femoral triangle. In the femoral triangle, its pulse is easily felt just inferior to the inguinal ligament midway between the pubic symphysis and the anterior superior iliac spine. Medial to the artery is the femoral vein and medial to the vein is the femoral canal, which contains lymphatics and lies immediately lateral to the pubic tubercle. The femoral nerve lies lateral to the femoral artery.

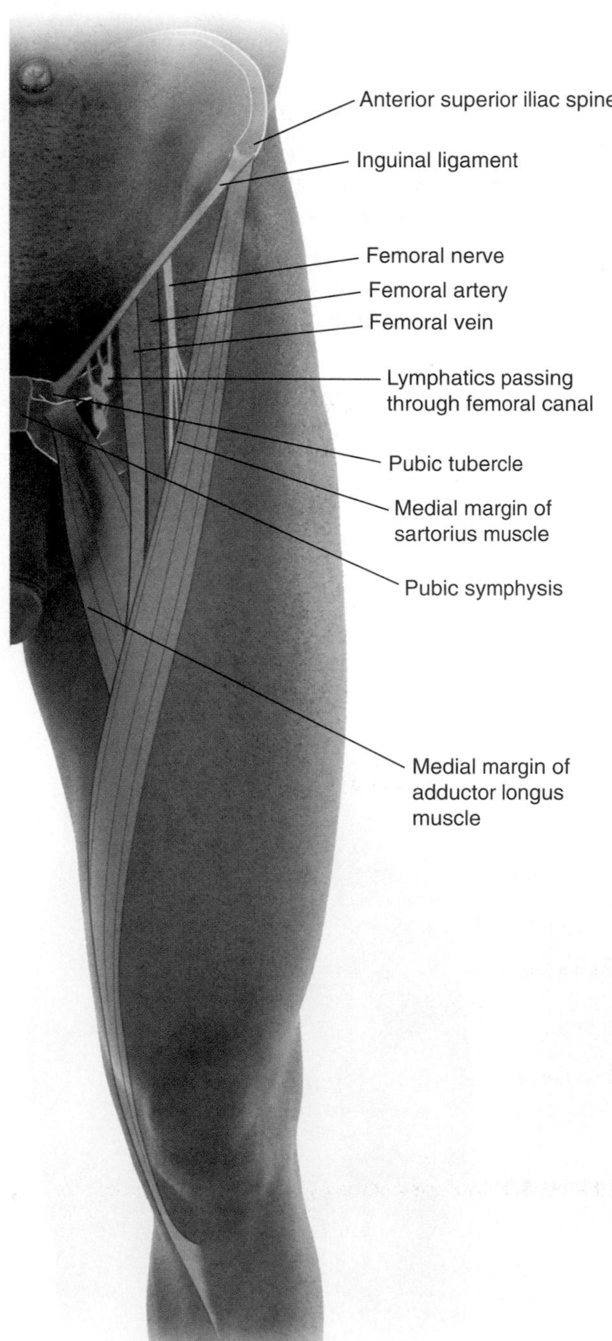


**Fig. 6.130** Position of the femoral artery in the femoral triangle. Anterior thigh.

## Identifying structures around the knee

The patella is a prominent palpable feature at the knee. The quadriceps femoris tendon attaches superiorly to it and the patellar ligament connects the inferior surface of the patella to the tibial tuberosity (Fig. 6.131). The patellar ligament and the tibial tuberosity are easily palpable. A tap on the patellar ligament (tendon) tests reflex activity mainly at spinal cord levels L3 and L4.

The head of the fibula is palpable as a protuberance on the lateral surface of the knee just inferior to the lateral condyle of the tibia. It can also be located by following the tendon of the biceps femoris inferiorly.

The common fibular nerve passes around the lateral surface of the neck of the fibula just inferior to the head and can often be felt as a cord-like structure in this position.

Another structure that can usually be located on the lateral side of the knee is the iliotibial tract. This flat tendinous structure, which attaches to the lateral tibial condyle, is most prominent when the knee is fully extended. In this position, the anterior edge of the tract raises a sharp vertical fold of skin posterior to the lateral edge of the patella.

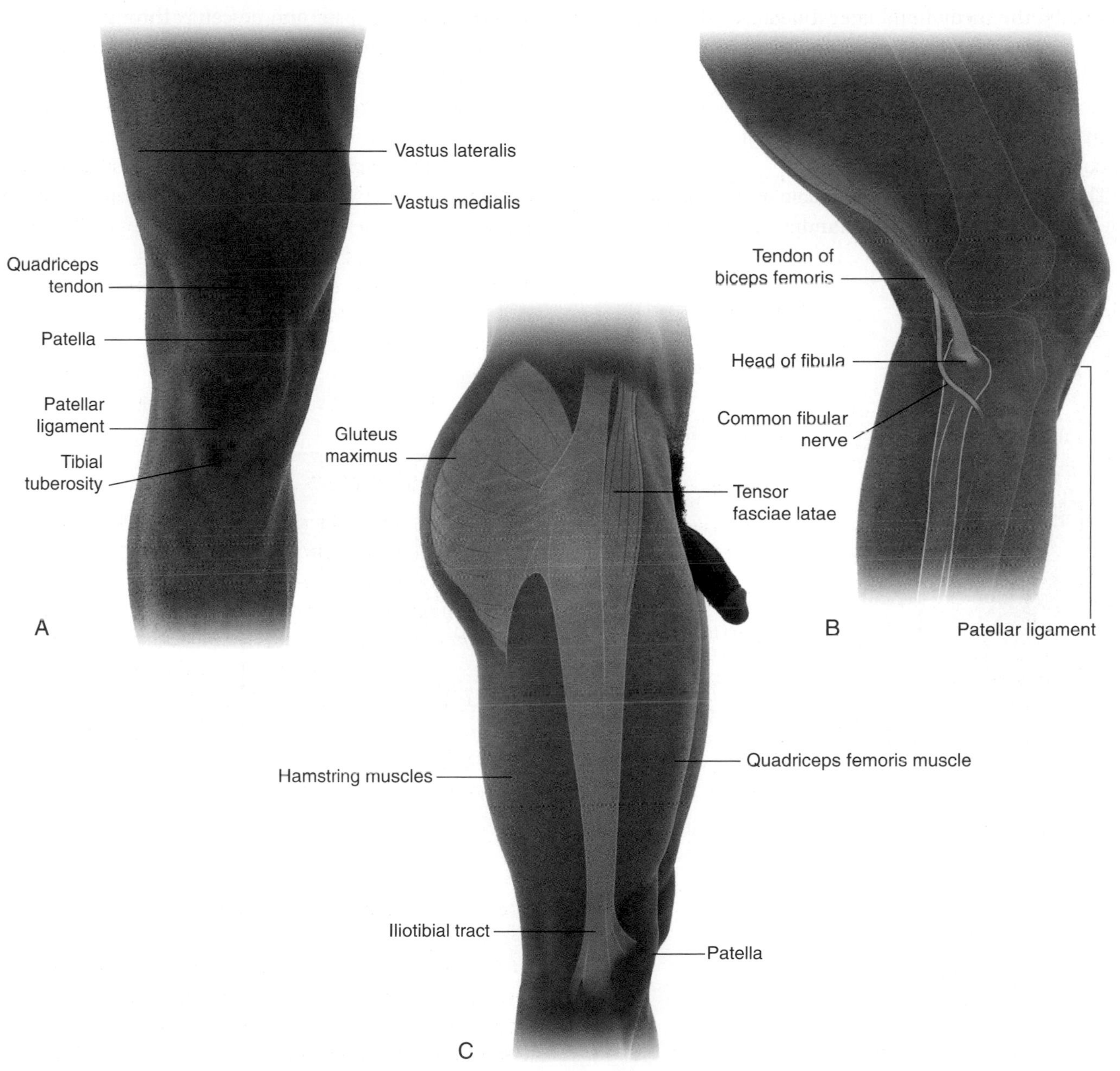


**Fig. 6.131** Identifying structures around the knee. **A.** Anterior view of the right knee. **B.** Lateral view of the partially flexed right knee. **C.** Lateral view of the extended right knee, thigh, and gluteal region.

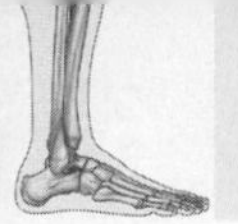



## Visualizing the contents of the popliteal fossa

The popliteal fossa is a diamond-shaped depression formed between the hamstrings and gastrocnemius muscle posterior to the knee. The inferior margins of the diamond are formed by the medial and lateral heads of the gastrocnemius muscle. The superior margins are formed laterally by the biceps femoris muscle and medially by the semimembranosus and semitendinosus muscles. The tendons of the biceps femoris muscle and the semitendinosus muscle are palpable and often visible.

The head of the fibula is palpable on the lateral side of the knee and can be used as a landmark for identifying the biceps femoris tendon and the common fibular nerve, which curves laterally out of the popliteal fossa and crosses the neck of the fibula just inferior to the head.

The popliteal fossa contains the popliteal artery, the popliteal vein, the tibial nerve, and the common fibular nerve (Fig. 6.132). The popliteal artery is the deepest of the structures in the fossa and descends through the region from the upper medial side. As a consequence of its position, the popliteal artery pulse is difficult to find, but usually can be detected on deep palpation just medial to the midline of the fossa.

The small saphenous vein penetrates deep fascia in the upper part of the posterior leg and joins the popliteal vein.

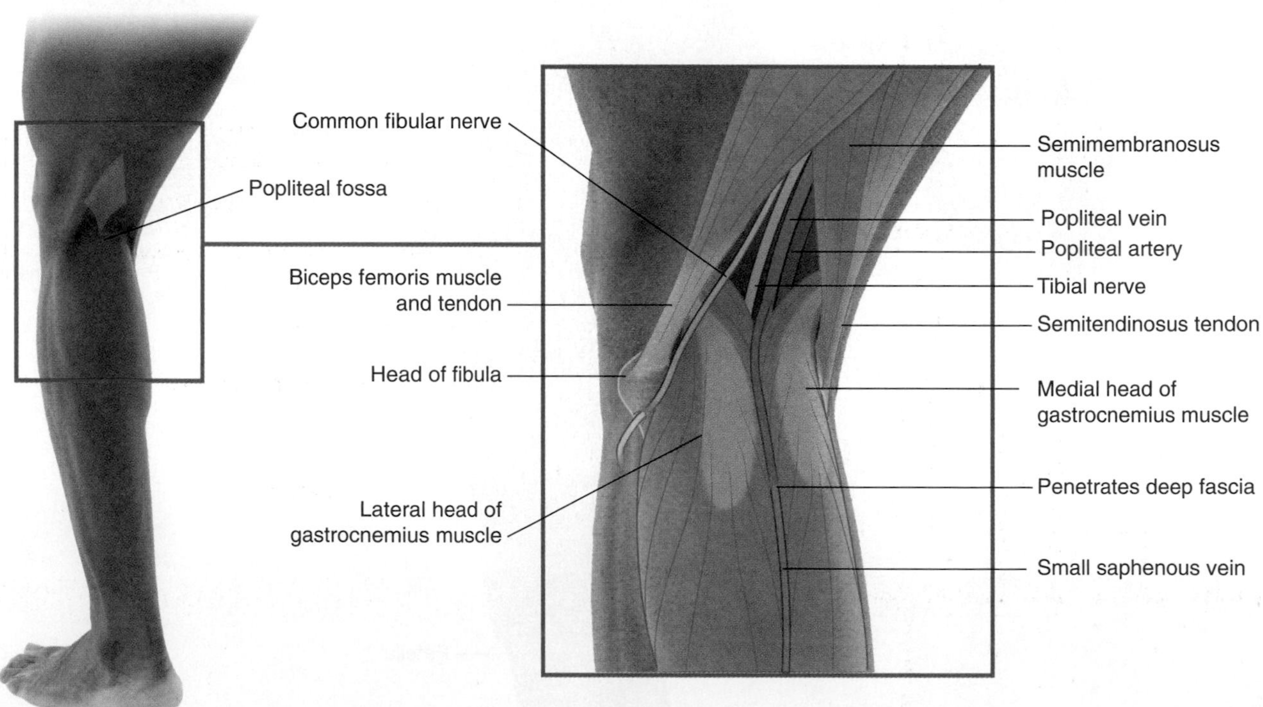


**Fig. 6.132** Visualizing the contents of the popliteal fossa. Posterior view of the left knee.

## Finding the tarsal tunnel—the gateway to the foot

The tarsal tunnel (Fig. 6.133) is formed on the medial side of the foot in the groove between the medial malleolus and the heel (calcaneal tuberosity) and by the overlying flexor retinaculum.

The posterior tibial artery and tibial nerve enter the foot through the tarsal tunnel. The tendons of the tibialis posterior, flexor digitorum longus, and flexor hallucis longus also pass through the tarsal tunnel in compartments formed by septa of the flexor retinaculum.

The order of structures passing through the tunnel from the anteromedial to posterolateral are the tendon of the tibialis posterior, the tendon of the flexor digitorum longus, the posterior tibial artery and associated veins, the tibial nerve, and the tendon of the flexor hallucis longus ("Tom, Dick, and a very nervous Harry").

The tibial artery is palpable just posteroinferior to the medial malleolus on the anterior face of the visible groove between the heel and medial malleolus.

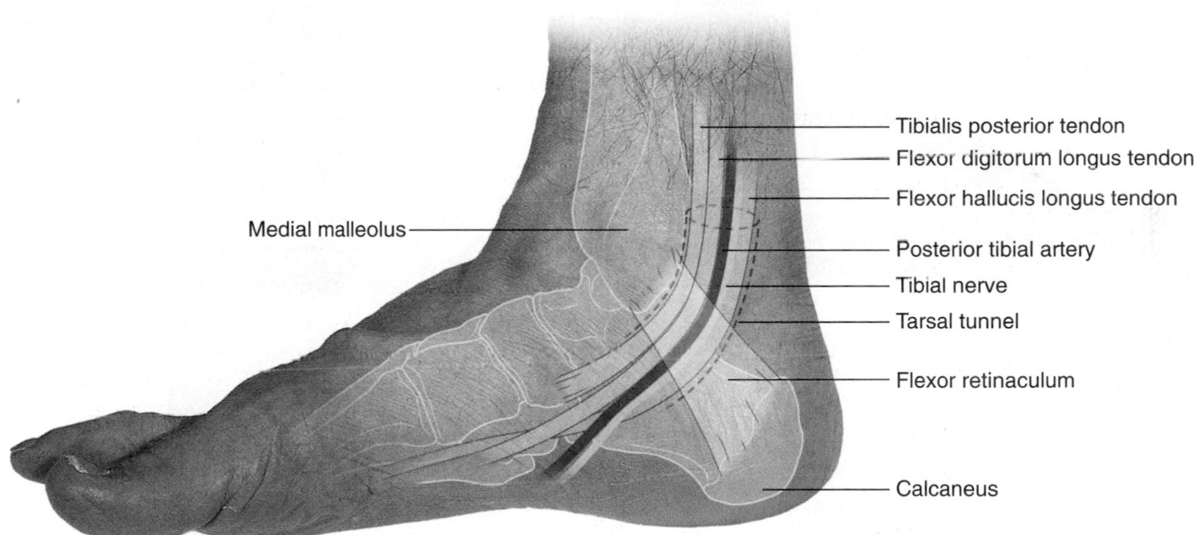


**Fig. 6.133** Finding the tarsal tunnel—the gateway to the foot.

## Identifying tendons around the ankle and in the foot

Numerous tendons can be identified around the ankle and in the foot (Fig. 6.134) and can be used as useful landmarks for locating vessels or testing spinal reflexes.

The tibialis anterior tendon is visible on the medial side of the ankle anterior to the medial malleolus.

The calcaneal tendon is the largest tendon entering the foot and is prominent on the posterior aspect of the foot as it descends from the leg to the heel. A tap with a tendon hammer on this tendon tests reflex activity of spinal cord levels S1 and S2.

When the foot is everted, the tendons of the fibularis longus and fibularis brevis raise a linear fold of skin, which descends from the lower leg to the posterior edge of the lateral malleolus.

The tendon of the fibularis brevis is often evident on the lateral surface of the foot descending obliquely to the base of metatarsal V. The tendons of the fibularis tertius,

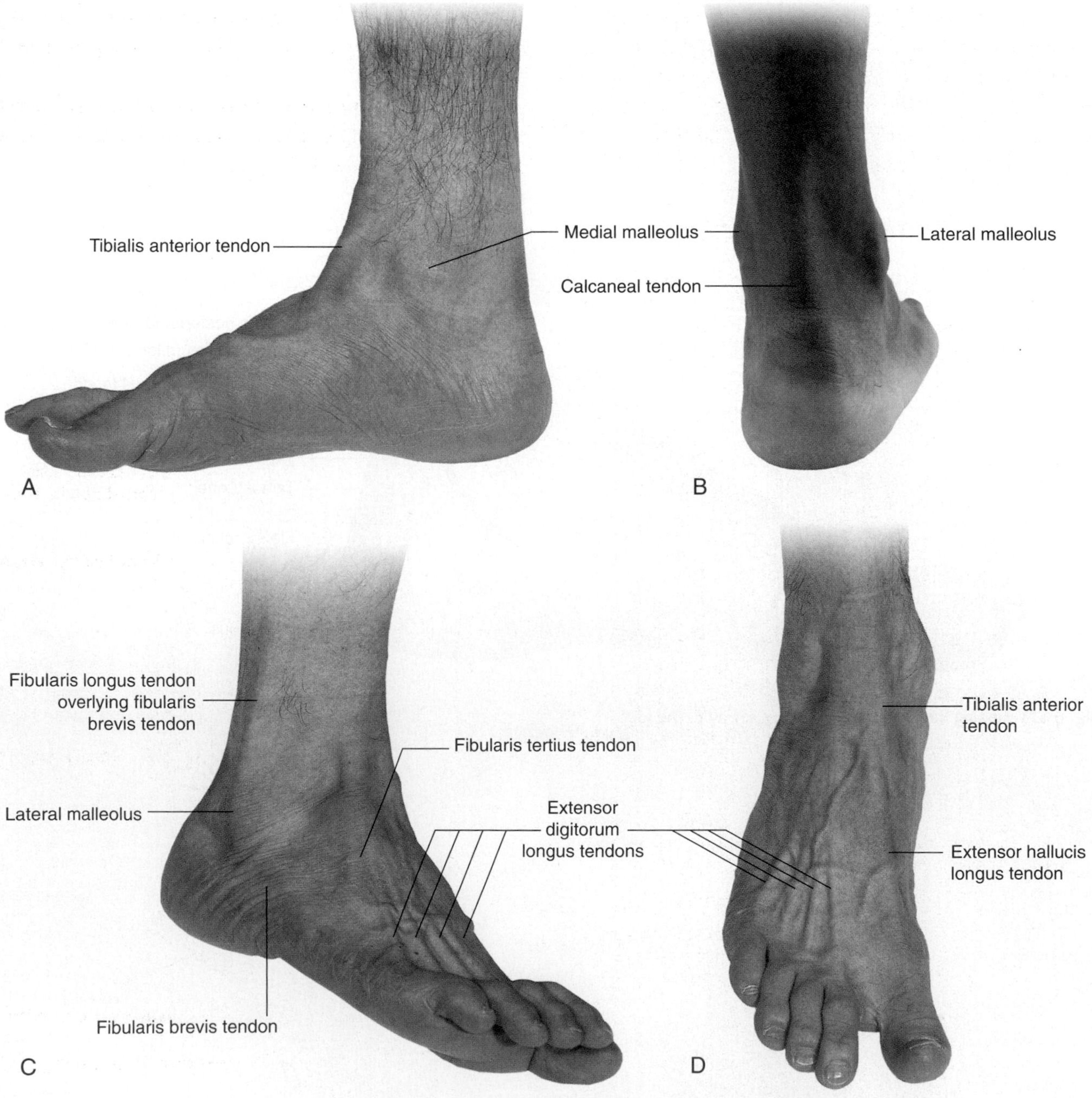

**Fig. 6.134** Identifying tendons around the ankle and in the foot. **A.** Medial side of the right foot. **B.** Posterior aspect of the right foot. **C.** Lateral side of the right foot. **D.** Dorsal aspect of the right foot.

extensor digitorum longus, and extensor hallucis longus are visible on the dorsal aspect of the foot from lateral to medial.

## Finding the dorsalis pedis artery

The nature of the dorsalis pedis pulse (Fig. 6.135) is important for assessing peripheral circulation because the dorsalis pedis artery is the farthest palpable vessel from the heart. Also, it is the lowest palpable artery in the body when a person is standing.

The dorsalis pedis artery passes onto the dorsal aspect of the foot and anteriorly over the tarsal bones where it lies between and is parallel to the tendon of the extensor hallucis longus and the tendon of the extensor digitorum longus to the second toe. It is palpable in this position. The terminal branch of the dorsalis pedis artery passes into the plantar surface of the foot between the two heads of the first dorsal interosseous muscle.

## Approximating the position of the plantar arterial arch

The blood supply of the foot is provided by branches of the posterior tibial and dorsalis pedis arteries.

The posterior tibial artery enters the plantar surface of the foot through the tarsal tunnel and divides into a lateral and a medial plantar artery.

The lateral plantar artery curves laterally across the posterior half of the sole and then curves medially as the plantar arch (Fig. 6.136) through the anterior sole. Between the bases of metatarsals I and II, the plantar arch joins the terminal branch (deep plantar artery) of the dorsalis pedis artery. Most of the foot is supplied by the plantar arch.

The medial plantar artery passes anteriorly through the sole, connects with branches of the plantar arch, and supplies the medial side of the great toe.

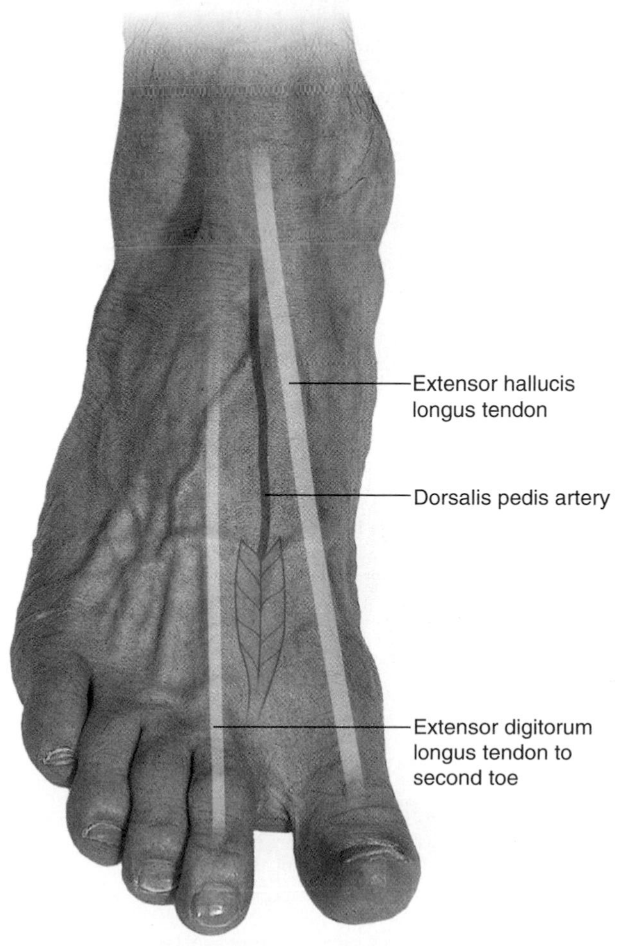


**Fig. 6.135** Finding the dorsalis pedis artery.

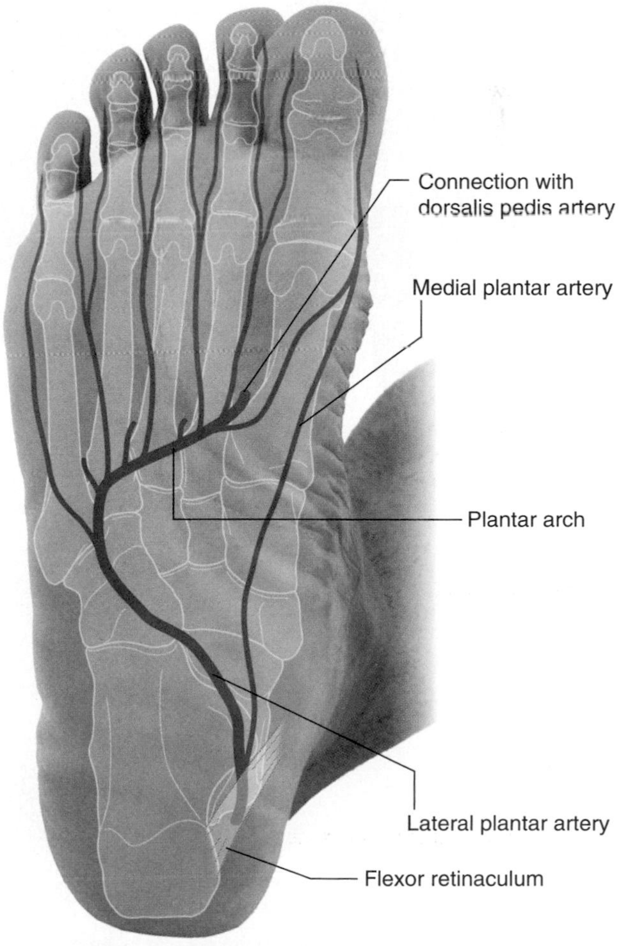


**Fig. 6.136** Position of the plantar arch.

## Major superficial veins

Superficial veins in the lower limb often become enlarged. Also, because the veins are long, they can be removed and used elsewhere in the body as vascular grafts.

Superficial veins (Fig. 6.137) in the lower limb begin as a dorsal venous arch in the foot. The medial side of the arch curves superiorly anterior to the medial malleolus and passes up the leg and thigh as the great saphenous vein. This vein passes through an aperture in the fascia lata (saphenous ring) to join with the femoral vein in the femoral triangle.

The lateral side of the dorsal venous arch in the foot passes posterior to the lateral malleolus and up the posterior surface of the leg as the small saphenous vein. This vessel passes through the deep fascia in the upper one-third of the leg and connects with the popliteal vein in the popliteal fossa posterior to the knee.

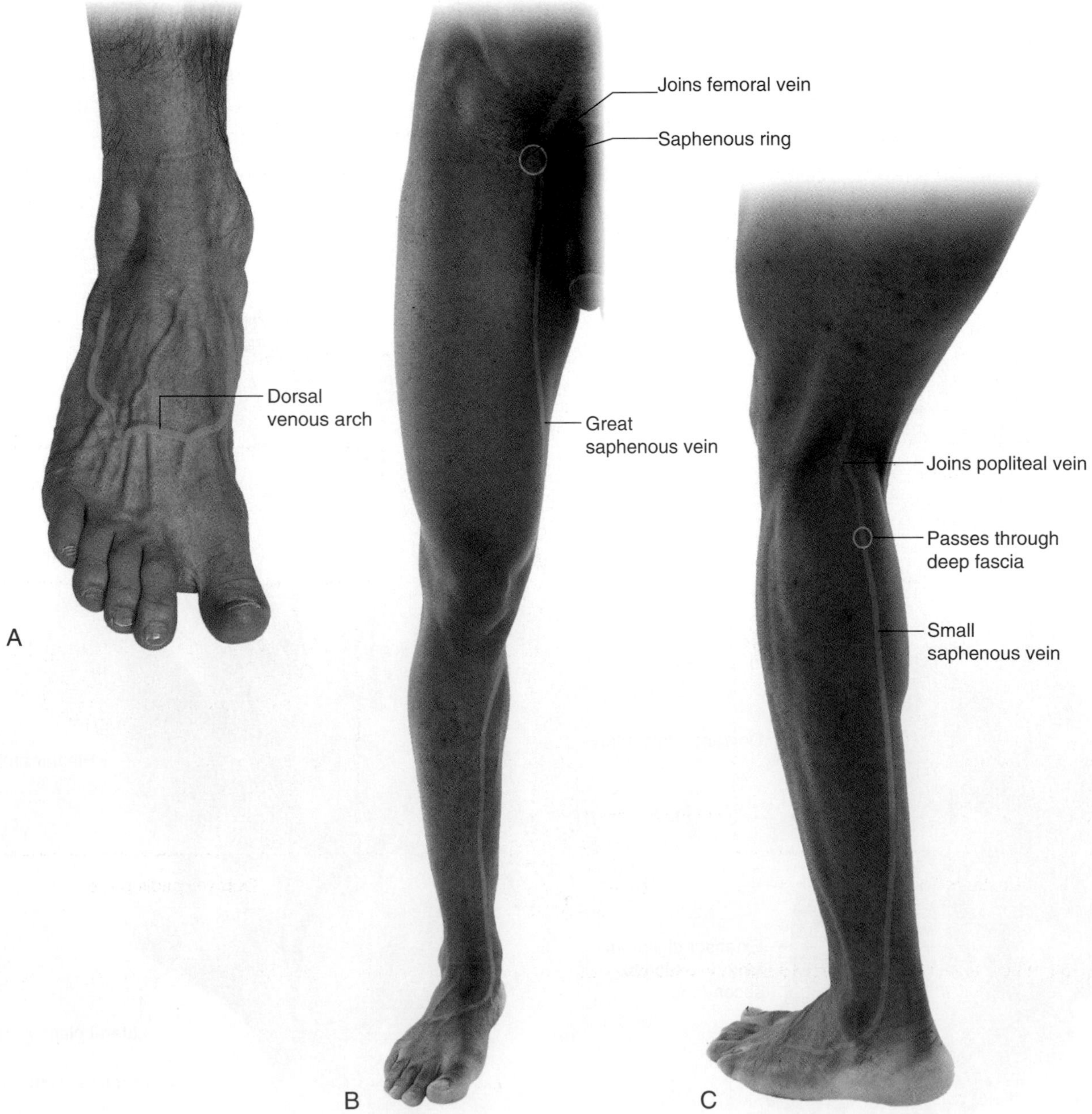

**Fig. 6.137** Major superficial veins. **A.** Dorsal aspect of the right foot. **B.** Anterior view of right lower limb. **C.** Posterior aspect of the left thigh, leg, and foot.

## Pulse points

Peripheral pulses can be felt at four locations in the lower limb (Fig. 6.138):

- **femoral pulse** in the femoral triangle—femoral artery inferior to the inguinal ligament and midway between the anterior superior iliac spine and the pubic symphysis;
- **popliteal pulse** in the popliteal fossa—popliteal artery deep in the popliteal fossa near the midline;
- **posterior tibial pulse** in the tarsal tunnel—posterior tibial artery posteroinferior to the medial malleolus in the groove between the medial malleolus and the heel (calcaneal tuberosity);
- **dorsalis pedis pulse** on the dorsal aspect of the foot—dorsalis pedis artery as it passes distally over the tarsal bones between the tendon of the extensor hallucis longus and the tendon of the extensor digitorum longus to the second toe.

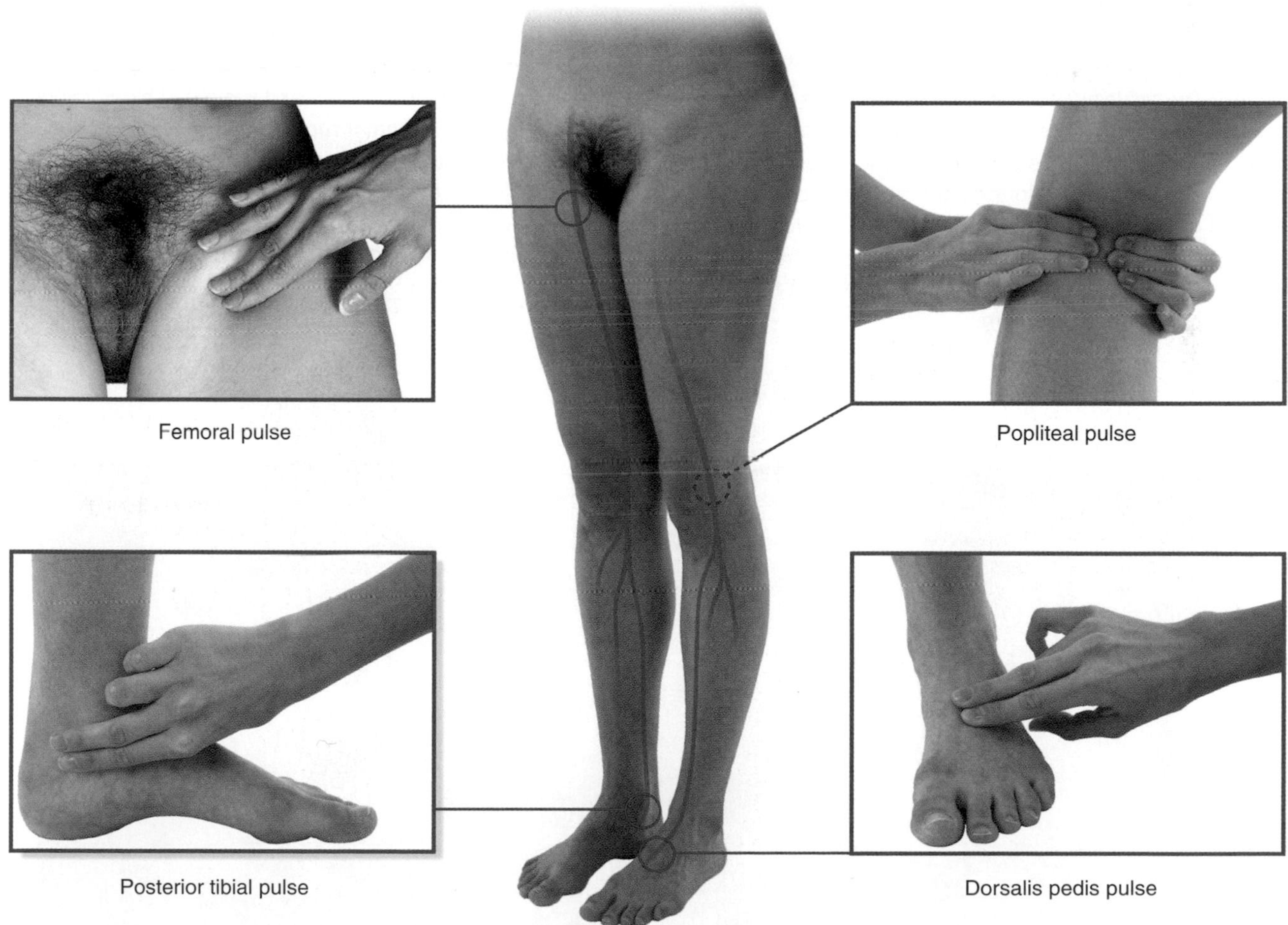


**Fig. 6.138** Where to feel peripheral arterial pulses in the lower limb.

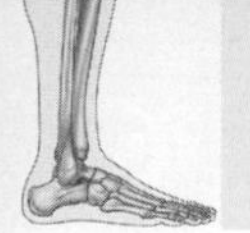



# Clinical cases

## Case 1

KNEE JOINT INJURY

**A young man was enjoying a long weekend skiing at a European ski resort. While racing a friend he caught an inner edge of his right ski. He lost his balance and fell. During his tumble he heard an audible "click." After recovering from his spill, he developed tremendous pain in his right knee. He was unable to carry on skiing for that day, and by the time he returned to his chalet, his knee was significantly swollen. He went immediately to see an orthopedic surgeon.**

The orthopedic surgeon carefully reviewed the mechanism of injury.

The man was skiing down the slope with both skis in parallel. The ankles were held rigid in the boots and the knees were slightly flexed. A momentary loss of concentration led to the skier catching the inner edge of his right ski. This effect was to force the boot and calf into external rotation. Furthermore, the knee was forced into a valgus position (bowed laterally away from the midline) and the skier tumbled. Both skis were detached from the boots as the bindings released them.

A series of structures within the knee joint were damaged sequentially.

As the knee went into external rotation and valgus, the anterior cruciate ligament became taut, acting as a fulcrum. The tibial collateral ligament was stressed and the lateral compartment of the knee compressed. As the force increased, the tibial collateral ligament was torn (Fig. 6.139A,B), as was the medial meniscus (Fig. 6.140C). Finally, the anterior cruciate ligament, which was taut, gave way (Fig. 6.140A,B).

The joint became swollen some hours afterward.

Disruption of the anterior cruciate ligament characteristically produces marked joint swelling. The ligament is extrasynovial and intracapsular and has a rich blood supply. As the ligament was torn it ruptured into the joint. Blood from the tear irritates the synovial membrane and also enters the joint. These factors produce gradual swelling of the joint over the ensuing hours with significant fluid accumulation in the joint cavity.

The patient had a surgical reconstruction of the anterior cruciate ligament.

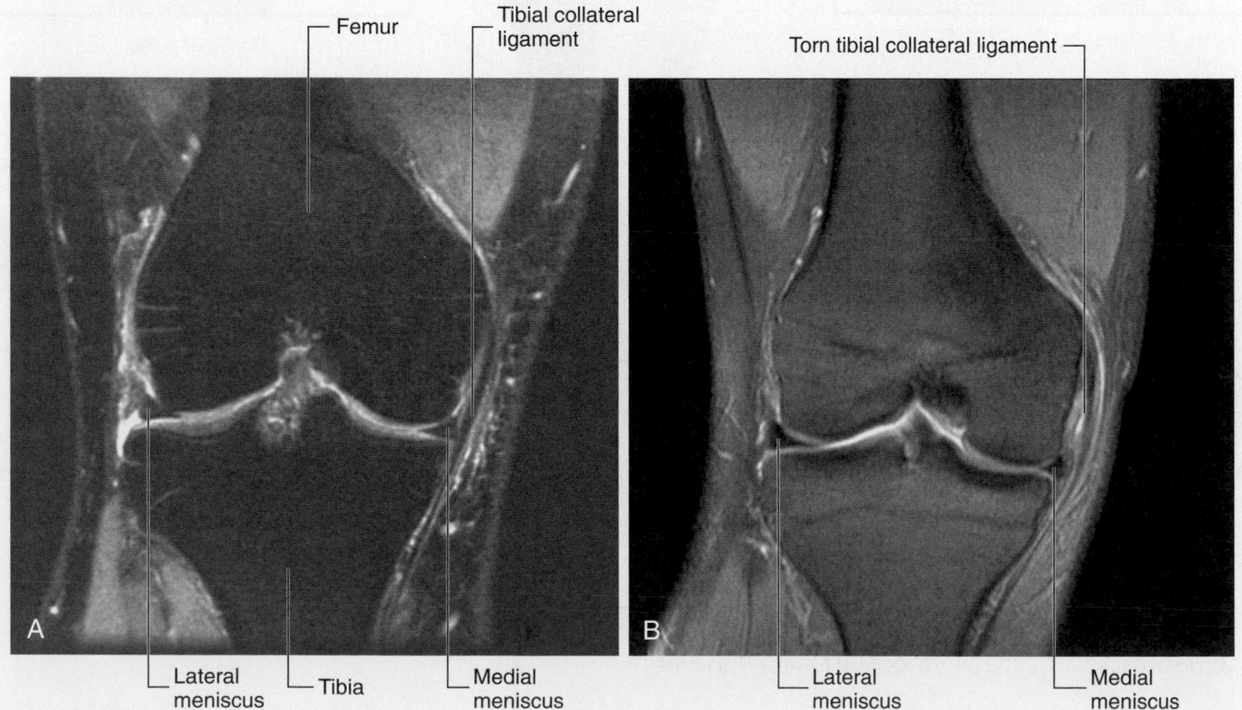


**Fig. 6.139 A.** Normal knee joint showing the tibial collateral ligament and the medial and lateral menisci. Proton density (PD)-weighted magnetic resonance image in the coronal plane. **B.** Knee joint showing a torn tibial collateral ligament. PD-weighted magnetic resonance image in the coronal plane.

## Case 1—cont'd

It is difficult to find a man-made substance that can act in the same way as the anterior cruciate ligament and demonstrate the same physical properties. Surgeons have devised ingenious ways of reconstructing the anterior cruciate ligament. Two of the commonest methods use the patellar ligament (tendon) and hamstrings to reconstruct the ligament.

The patient had further surgical procedures.

The tibial collateral ligament was explored and resutured. Using arthroscopic techniques, the tear in the medial meniscus was débrided to prevent further complications.

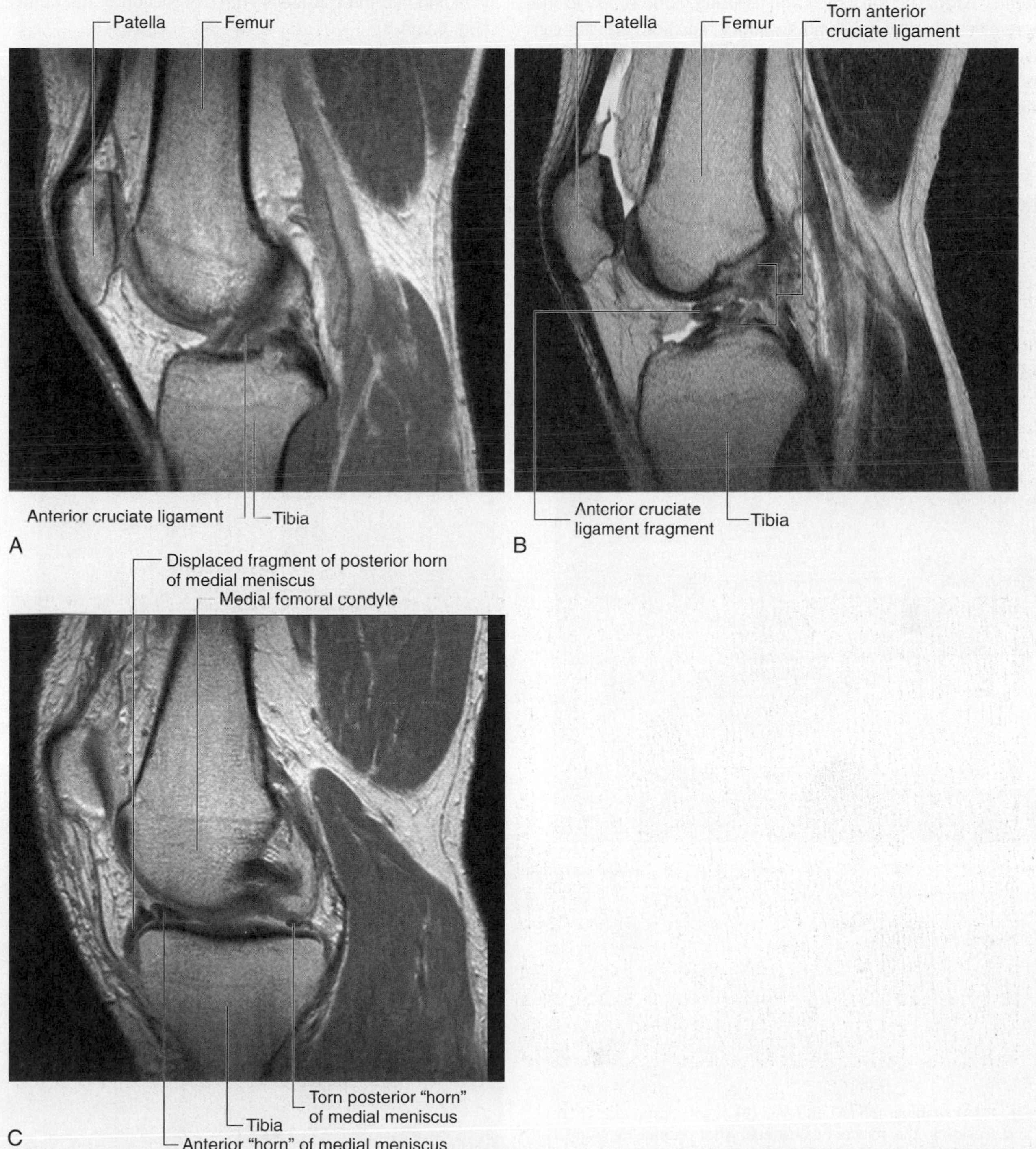


**Fig. 6.140** **A.** Knee joint showing an intact anterior cruciate ligament. T2-weighted magnetic resonance image in the sagittal plane. **B.** Knee joint showing a torn anterior cruciate ligament. T2-weighted magnetic resonance image in the sagittal plane. **C.** Knee joint showing a torn medial meniscus (the broken off portion of the posterior horn has moved into the anterior aspect of the joint giving the impression of a 'double meniscus' in this location). Proton density-weighted magnetic resonance image in the sagittal plane.

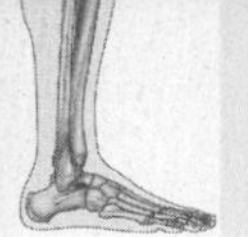



## Case 2

OSTEOMYELITIS

**A 45-year-old man with diabetes mellitus visited his nurse because he had an ulcer on his foot that was not healing despite daily dressings.**

Diabetes can lead to vascular disease of large and medium arteries, narrowing the lumen and reducing blood supply to the extremities, thereby impairing healing. In addition, diabetes can also affect blood supply to nerves, which leads to peripheral neuropathy. Peripheral neuropathy results in reduced sensation, and therefore minor injuries can often go unnoticed.

This patient has developed an ulcer on his heel, which is a pressure point and likely to be under repeated strain. The nurse examined the ulcer and found that the ulcer was looking infected with pus at the base of the ulcer and asked for a specialist orthopedic opinion, who requested an x-ray and an MRI. The MRI and x-ray both demonstrated infection invading into the calcaneus with destruction of the bone (Fig. 6.141A,B).

The patient required surgical washout with removal of the dead and infected bone (debridement) and was given long-term antibiotic treatment (Fig. 6.141C).

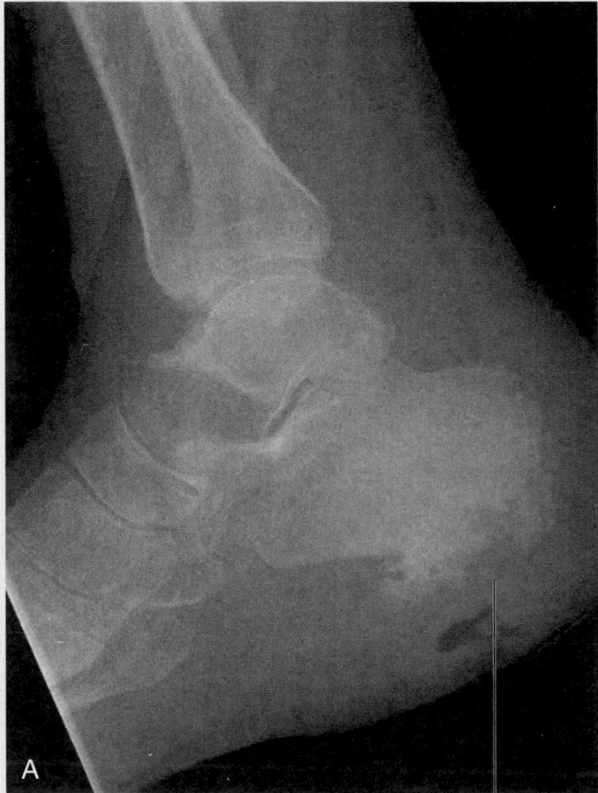


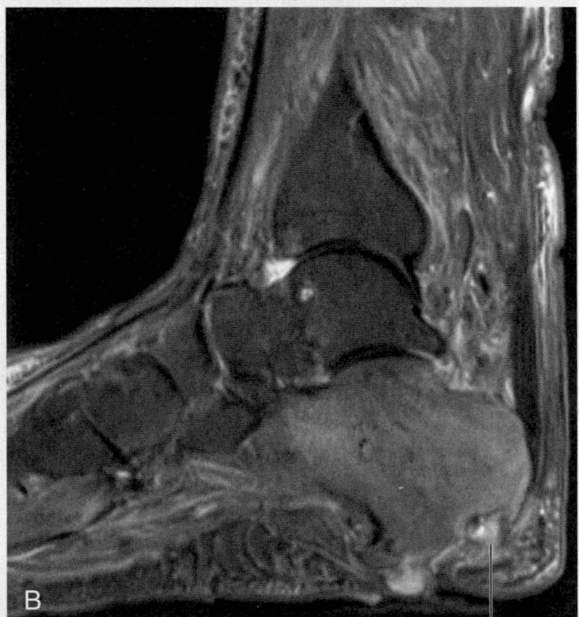


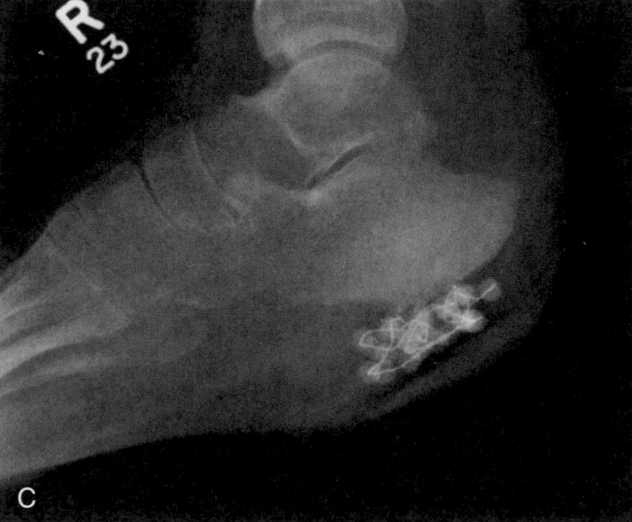


**Fig. 6.141** Radiograph **(A)** and MRI **(B)** of soft tissue ulceration and erosion in the adjacent calcaneus. After debridement and placement of antibiotics beads in the wound there is progressive healing **(C)**.

# 7

# Upper Limb

## ADDITIONAL LEARNING RESOURCES for Chapter 7, *Upper Limb*, on STUDENT CONSULT (www.studentconsult.com):

- Self-Assessment—National Board style multiple-choice questions, Chapter 7
- Short Questions—these are questions requiring short responses, Chapter 7
- Interactive Surface Anatomy—interactive surface animations, Chapter 7
- PT Case Studies, Chapter 7
  - Upper quarter screening examination
  - Thoracic outlet syndrome
  - Rotator cuff tendinitis
  - Adhesive capsulitis
  - Subacromial impingement
  - *See more PT Case Studies online*
- Clinical Cases, Chapter 7
  - Clavicular fracture
  - Complication of a fractured first rib
  - Extensor tendon injury of the hand
  - Hand infection
  - High median nerve palsy
  - How to examine the hand
  - Immobilizing the extensor digitorum muscle
  - Median nerve compression
  - Pancoast tumor
  - Shoulder joint problem
  - Shoulder problem after falling on an outstretched hand
  - Torn supraspinatus tendon
  - Winged scapula

**FREE Online Anatomy and Embryology Self-Study Course**

- Anatomy modules 32 through 38
- Embryology module 71

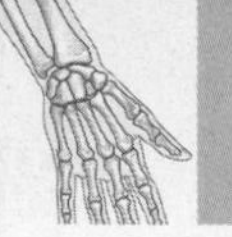

# Upper Limb

# Conceptual overview

## GENERAL DESCRIPTION

The upper limb is associated with the lateral aspect of the lower portion of the neck and with the thoracic wall. It is suspended from the trunk by muscles and a small skeletal articulation between the clavicle and the sternum—the sternoclavicular joint. Based on the position of its major joints and component bones, the upper limb is divided into shoulder, arm, forearm, and hand (Fig. 7.1A).

The shoulder is the area of upper limb attachment to the trunk (Fig. 7.1B).

The arm is the part of the upper limb between the shoulder and the elbow joint; the forearm is between the elbow joint and the wrist joint; and the hand is distal to the wrist joint.

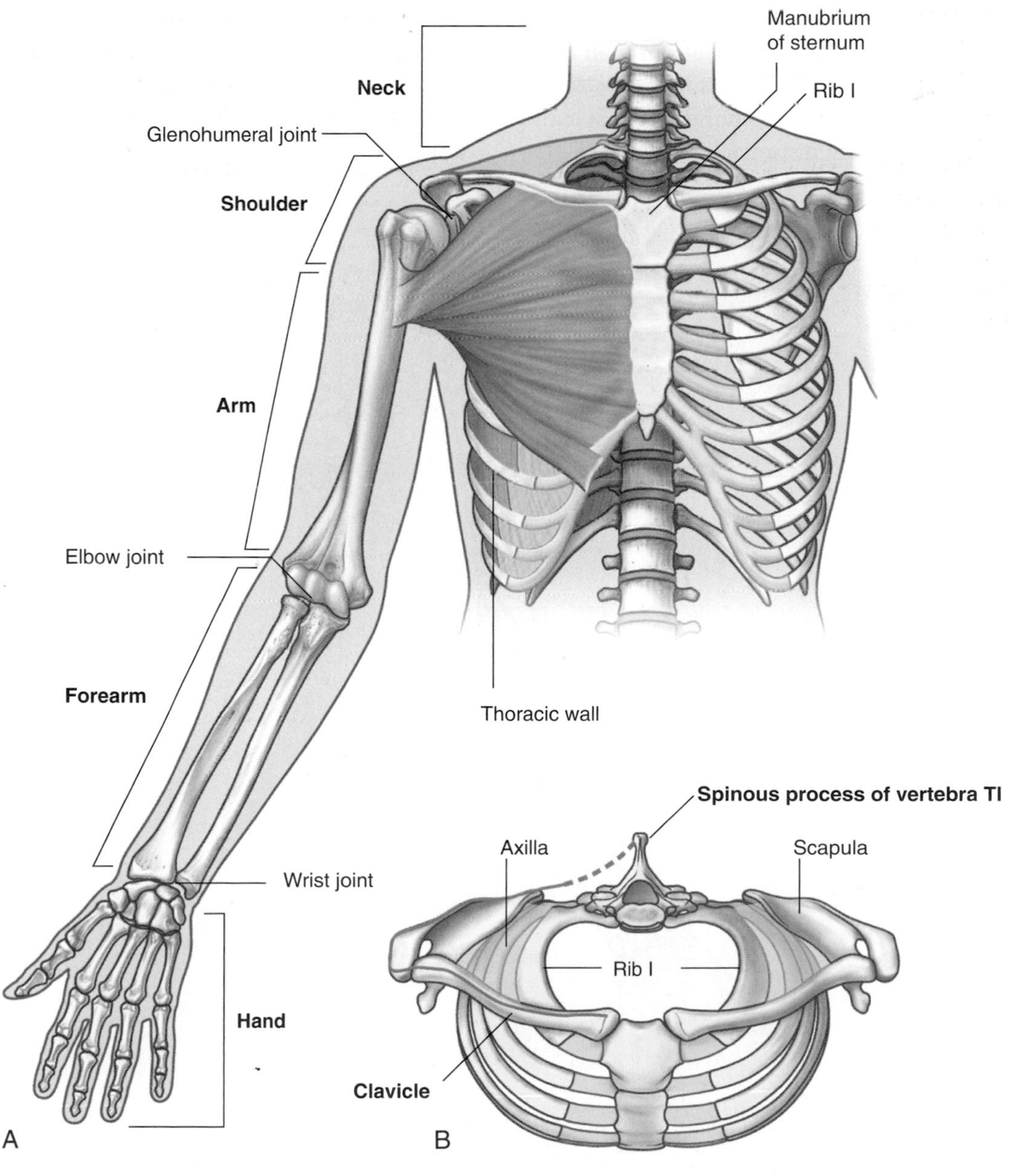


**Fig. 7.1** Upper limb. **A.** Anterior view of the upper limb. **B.** Superior view of the shoulder.

The axilla, cubital fossa, and carpal tunnel are significant areas of transition between the different parts of the limb (Fig. 7.2). Important structures pass through, or are related to, each of these areas.

The axilla is an irregularly shaped pyramidal area formed by muscles and bones of the shoulder and the lateral surface of the thoracic wall. The apex or inlet opens directly into the lower portion of the neck. The skin of the armpit forms the floor. All major structures that pass between the neck and arm pass through the axilla.

The cubital fossa is a triangularly shaped depression formed by muscles anterior to the elbow joint. The major artery, the brachial artery, passing from the arm to the forearm passes through this fossa, as does one of the major nerves of the upper limb, the median nerve.

The carpal tunnel is the gateway to the palm of the hand. Its posterior, lateral, and medial walls form an arch, which is made up of small carpal bones in the proximal region of the hand. A thick band of connective tissue, the flexor retinaculum, spans the distance between each side of the arch and forms the anterior wall of the tunnel. The median nerve and all the long flexor tendons passing from the forearm to the digits of the hand pass through the carpal tunnel.

## FUNCTIONS

### Positioning the hand

Unlike the lower limb, which is used for support, stability, and locomotion, the upper limb is highly mobile for positioning the hand in space.

The shoulder is suspended from the trunk predominantly by muscles and can therefore be moved relative to the body. Sliding (protraction and retraction) and rotating the scapula on the thoracic wall changes the position of the **glenohumeral joint** (**shoulder joint**) and extends the reach of the hand (Fig. 7.3). The glenohumeral joint allows the arm to move around three axes with a wide range of motion. Movements of the arm at this joint are

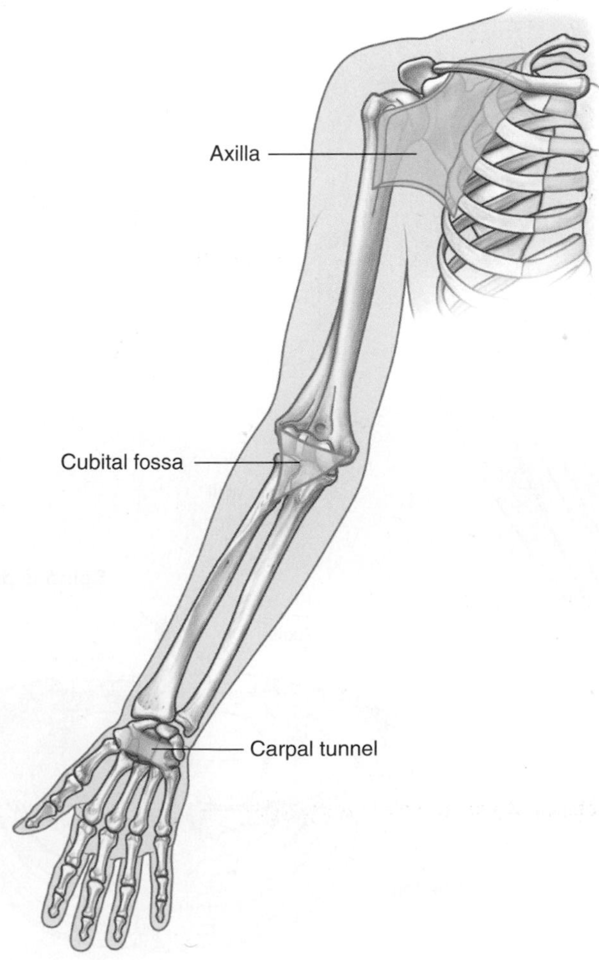


**Fig. 7.2** Areas of transition in the upper limb.

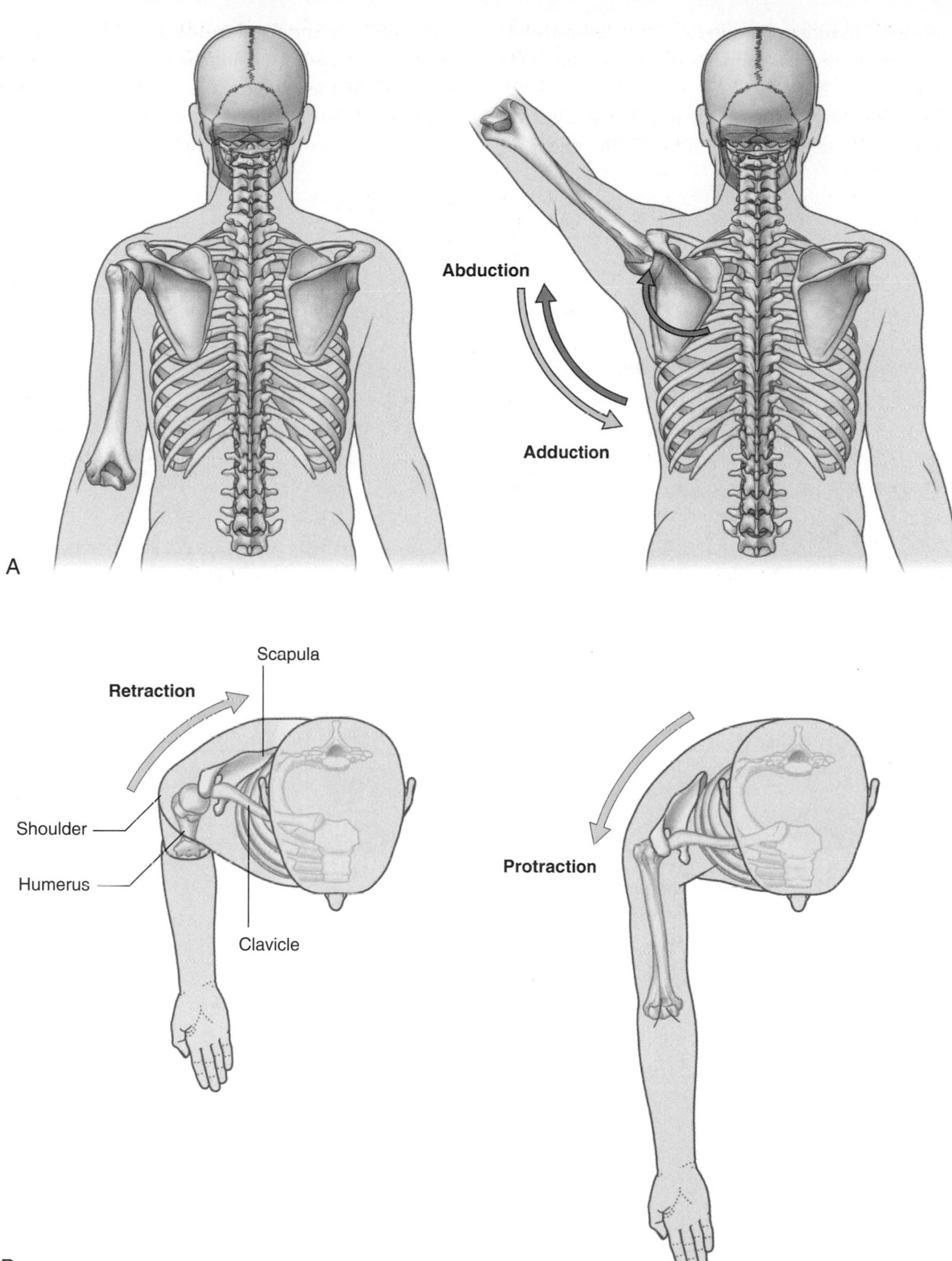


**Fig. 7.3** Movements of the scapula. **A.** Rotation. **B.** Protraction and retraction.

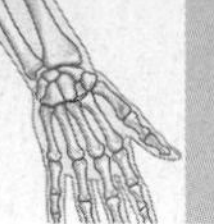



flexion, extension, abduction, adduction, medial rotation (internal rotation), lateral rotation (external rotation), and circumduction (Fig. 7.4).

The major movements at the **elbow joint** are flexion and extension of the forearm (Fig. 7.5A). At the other end of the forearm, the distal end of the lateral bone, the radius, can be flipped over the adjacent head of the medial bone, the ulna. Because the hand is articulated with the radius, it can be efficiently moved from a palm-anterior position to a palm-posterior position simply by crossing the distal end

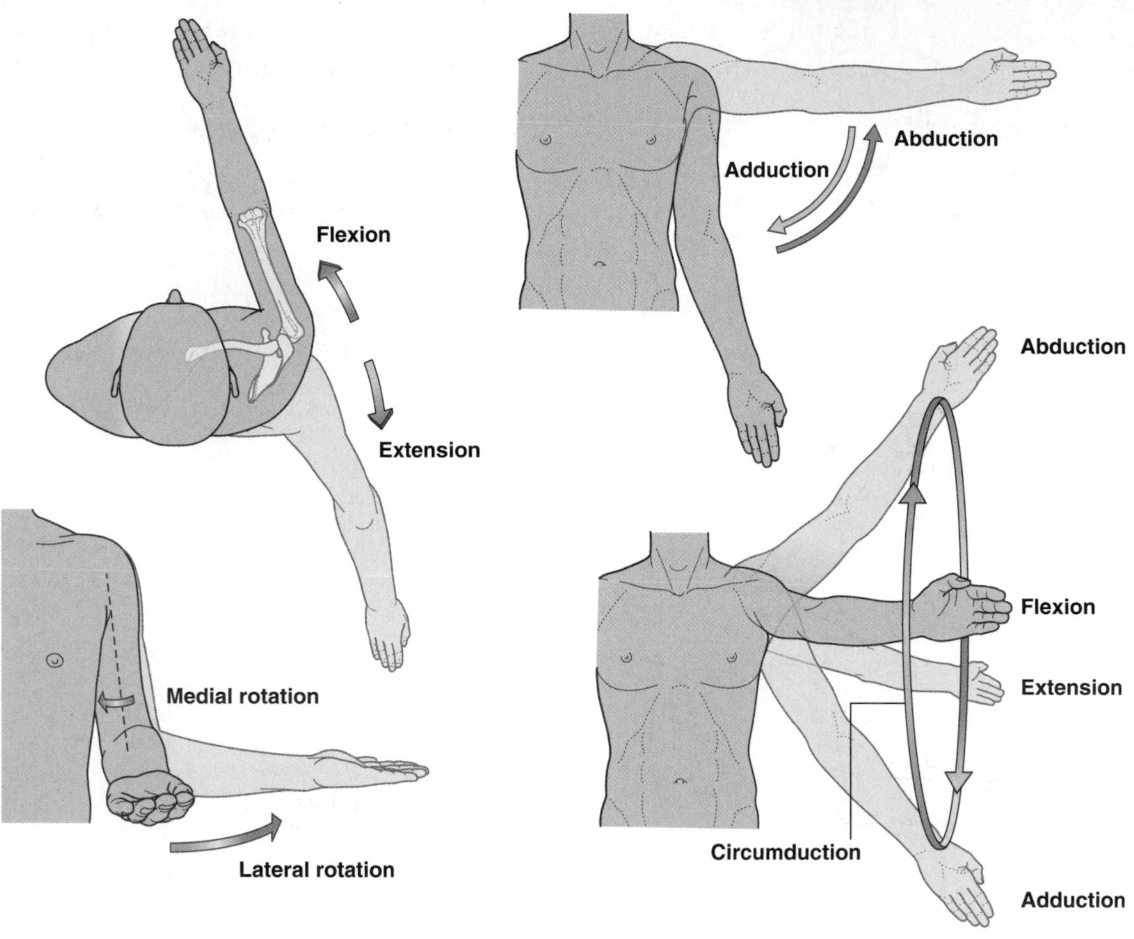


**Fig. 7.4** Movements of the arm at the glenohumeral joint.

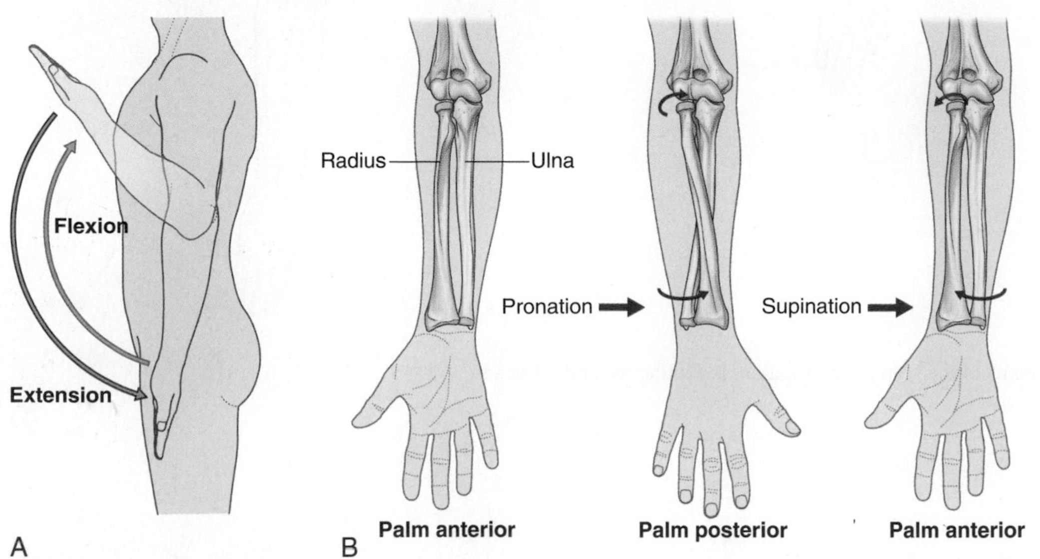


**Fig. 7.5** Movements of the forearm. **A.** Flexion and extension at the elbow joint. **B.** Pronation and supination.

of the radius over the ulna (Fig. 7.5B). This movement, termed pronation, occurs solely in the forearm. Supination returns the hand to the anatomical position.

At the **wrist joint**, the hand can be abducted, adducted, flexed, extended, and circumducted (Fig. 7.6). These movements, combined with those of the shoulder, arm, and forearm, enable the hand to be placed in a wide range of positions relative to the body.

## The hand as a mechanical tool

One of the major functions of the hand is to grip and manipulate objects. Gripping objects generally involves flexing the fingers against the thumb. Depending on the type of grip, muscles in the hand act to:

- modify the actions of long tendons that emerge from the forearm and insert into the digits of the hand, and
- produce combinations of joint movements within each digit that cannot be generated by the long flexor and extensor tendons alone coming from the forearm.

## The hand as a sensory tool

The hand is used to discriminate between objects on the basis of touch. The pads on the palmar aspect of the fingers

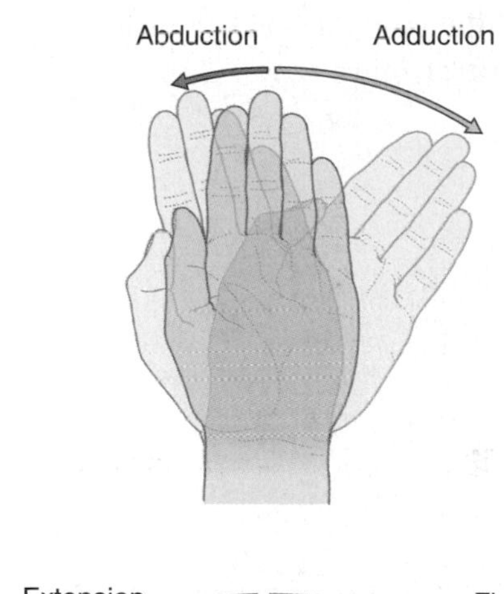


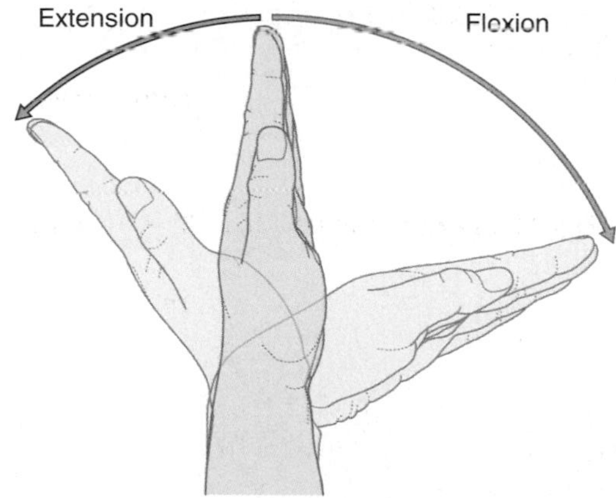


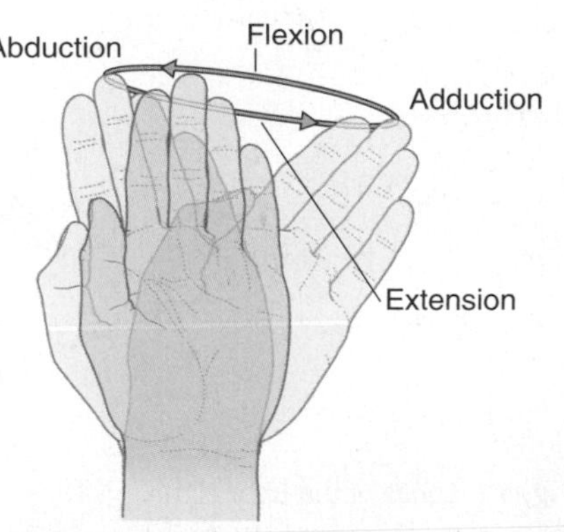


**Fig. 7.6** Movements of the hand at the wrist joint.

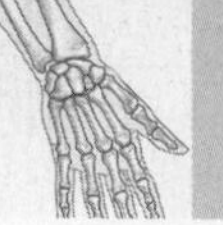



contain a high density of somatic sensory receptors. Also, the sensory cortex of the brain devoted to interpreting information from the hand, particularly from the thumb, is disproportionately large relative to that for many other regions of skin.

## COMPONENT PARTS

### Bones and joints

The bones of the shoulder consist of the scapula, clavicle, and proximal end of the humerus (Fig. 7.7).

The clavicle articulates medially with the manubrium of the sternum and laterally with the acromion of the scapula, which arches over the joint between the glenoid cavity of the scapula and the head of the humerus (the glenohumeral joint).

The humerus is the bone of the arm (Fig. 7.7). The distal end of the humerus articulates with the bones of the forearm at the elbow joint, which is a hinge joint that allows flexion and extension of the forearm.

The forearm contains two bones:

- The lateral bone is the radius.
- The medial bone is the ulna (Fig. 7.7).

At the elbow joint, the proximal ends of the radius and ulna articulate with each other as well as with the humerus.

In addition to flexing and extending the forearm, the elbow joint allows the radius to spin on the humerus while sliding against the head of the ulna during pronation and supination of the hand.

The distal portions of the radius and the ulna also articulate with each other. This joint allows the end of the

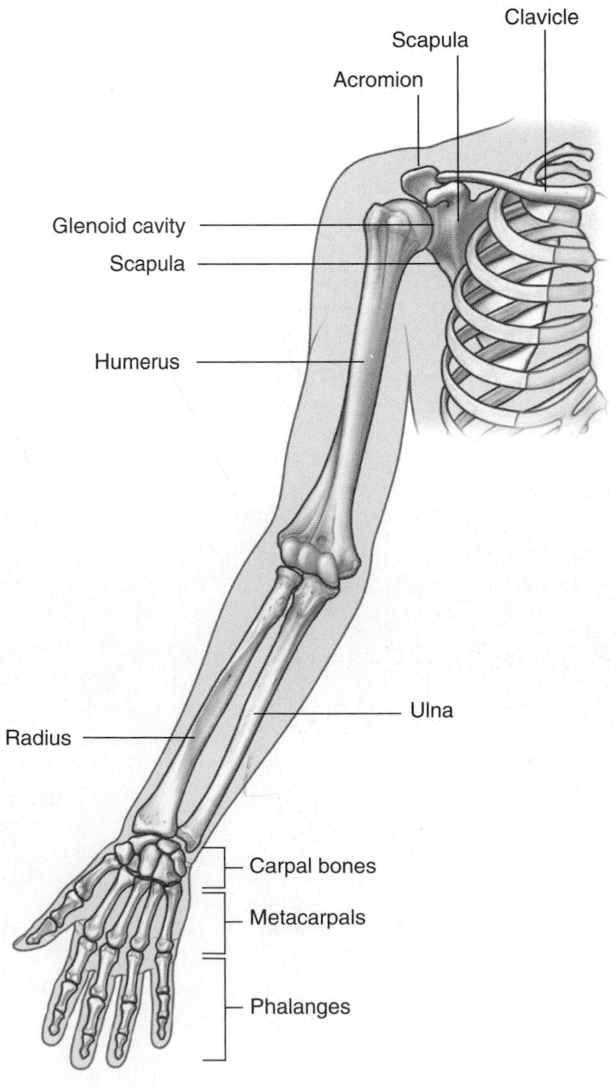


**Fig. 7.7** Bones of the upper limb.

radius to flip from the lateral side to the medial side of the ulna during pronation of the hand.

The wrist joint is formed between the radius and carpal bones of the hand and between an articular disc, distal to the ulna, and carpal bones.

The bones of the hand consist of the carpal bones, the metacarpals, and the phalanges (Fig. 7.7).

The five digits in the hand are the thumb and the index, middle, ring, and little fingers.

Joints between the eight small carpal bones allow only limited amounts of movement; as a result, the bones work together as a unit.

The five metacarpals, one for each digit, are the primary skeletal foundation of the palm (Fig. 7.7).

The joint between the metacarpal of the thumb (metacarpal I) and one of the carpal bones allows greater mobility than the limited sliding movement that occurs at the carpometacarpal joints of the fingers.

Distally, the heads of metacarpals II to V (i.e., except that of the thumb) are interconnected by strong ligaments. Lack of this ligamentous connection between the metacarpal bones of the thumb and index finger together with the biaxial **saddle joint** between the metacarpal bone of the thumb and the carpus provide the thumb with greater freedom of movement than the other digits of the hand.

The bones of the digits are the phalanges (Fig. 7.7). The thumb has two phalanges, while each of the other digits has three.

The metacarpophalangeal joints are biaxial **condylar joints** (**ellipsoid joints**) that allow abduction, adduction, flexion, extension, and circumduction (Fig. 7.8). Abduction and adduction of the fingers is defined in reference to an axis passing through the center of the middle finger in the anatomical position. The middle finger can therefore abduct both medially and laterally and adduct back to the central axis from either side. The interphalangeal joints are primarily **hinge joints** that allow only flexion and extension.

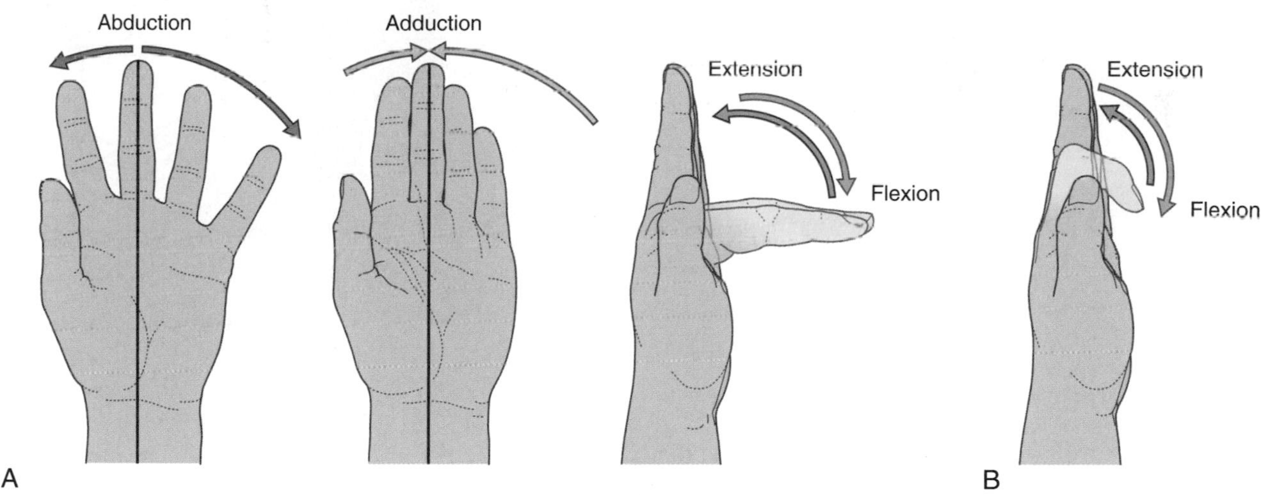


**Fig. 7.8** Movements of the metacarpophalangeal (**A**) and interphalangeal (**B**) joints.

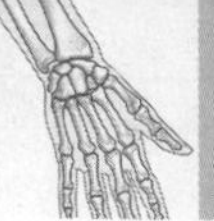



## Muscles

Some muscles of the shoulder, such as the trapezius, levator scapulae, and rhomboids, connect the scapula and clavicle to the trunk. Other muscles connect the clavicle, scapula, and body wall to the proximal end of the humerus. These include the pectoralis major, pectoralis minor, latissimus dorsi, teres major, and deltoid (Fig. 7.9A,B). The most important of these muscles are the four rotator cuff muscles—the subscapularis, supraspinatus, infraspinatus, and teres minor muscles—which connect the scapula to the humerus and provide support for the glenohumeral joint (Fig. 7.9C).



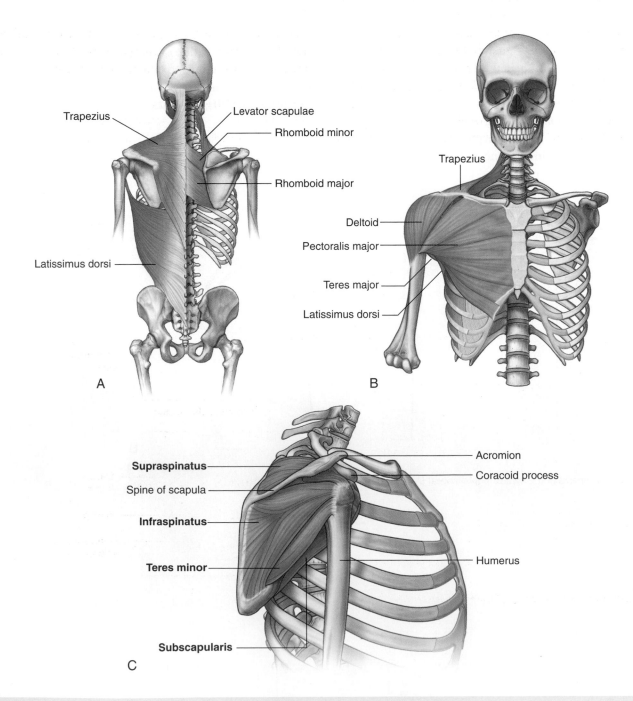

**Fig. 7.9** Muscles of the shoulder. **A.** Posterior shoulder. **B.** Anterior shoulder. **C.** Rotator cuff muscles.

Muscles in the arm and forearm are separated into anterior (flexor) and posterior (extensor) compartments by layers of fascia, bones, and ligaments (Fig. 7.10).

The anterior compartment of the arm lies anteriorly in position and is separated from muscles of the posterior compartment by the humerus and by medial and lateral intermuscular septa. These intermuscular septa are continuous with the deep fascia enclosing the arm and attach to the sides of the humerus.

In the forearm, the anterior and posterior compartments are separated by a lateral intermuscular septum, the radius, the ulna, and an interosseous membrane, which joins adjacent sides of the radius and ulna (Fig. 7.10).

Muscles in the arm act mainly to move the forearm at the elbow joint, while those in the forearm function predominantly to move the hand at the wrist joint and the fingers and thumb.

Muscles found entirely in the hand, the intrinsic muscles, generate delicate movements of the digits of the hand and modify the forces produced by tendons coming into the fingers and thumb from the forearm. Included among the intrinsic muscles of the hand are three small thenar muscles, which form a soft tissue mound, called the **thenar eminence**, over the palmar aspect of metacarpal I. The thenar muscles allow the thumb to move freely relative to the other fingers.

## RELATIONSHIP TO OTHER REGIONS

### Neck

The upper limb is directly related to the neck. Lying on each side of the **superior thoracic aperture** at the base of the neck is an **axillary inlet**, which is formed by:

- the lateral margin of rib I,
- the posterior surface of the clavicle,
- the superior margin of the scapula, and

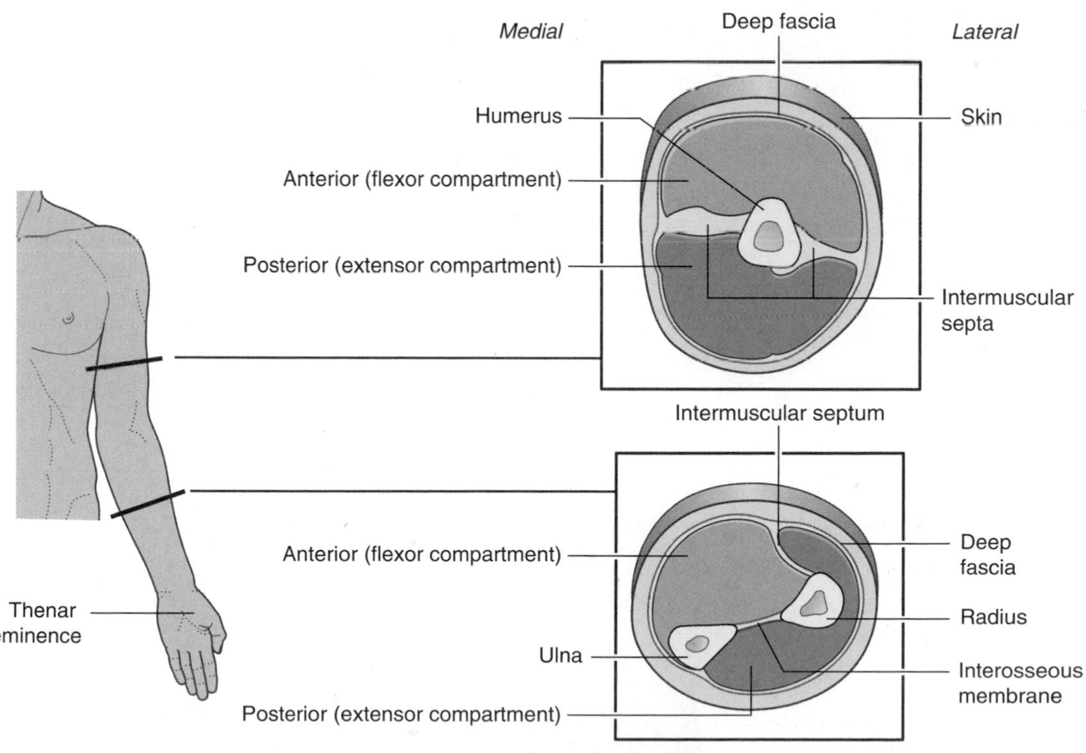


**Fig. 7.10** Muscle components in the arm and forearm.

- the medial surface of the coracoid process of the scapula (Fig. 7.11).

The major artery and vein of the upper limb pass between the thorax and the limb by passing over rib I and through the axillary inlet. Nerves, predominantly derived from the cervical portion of the spinal cord, also pass through the axillary inlet and the axilla to supply the upper limb.

## Back and thoracic wall

Muscles that attach the bones of the shoulder to the trunk are associated with the back and the thoracic wall and

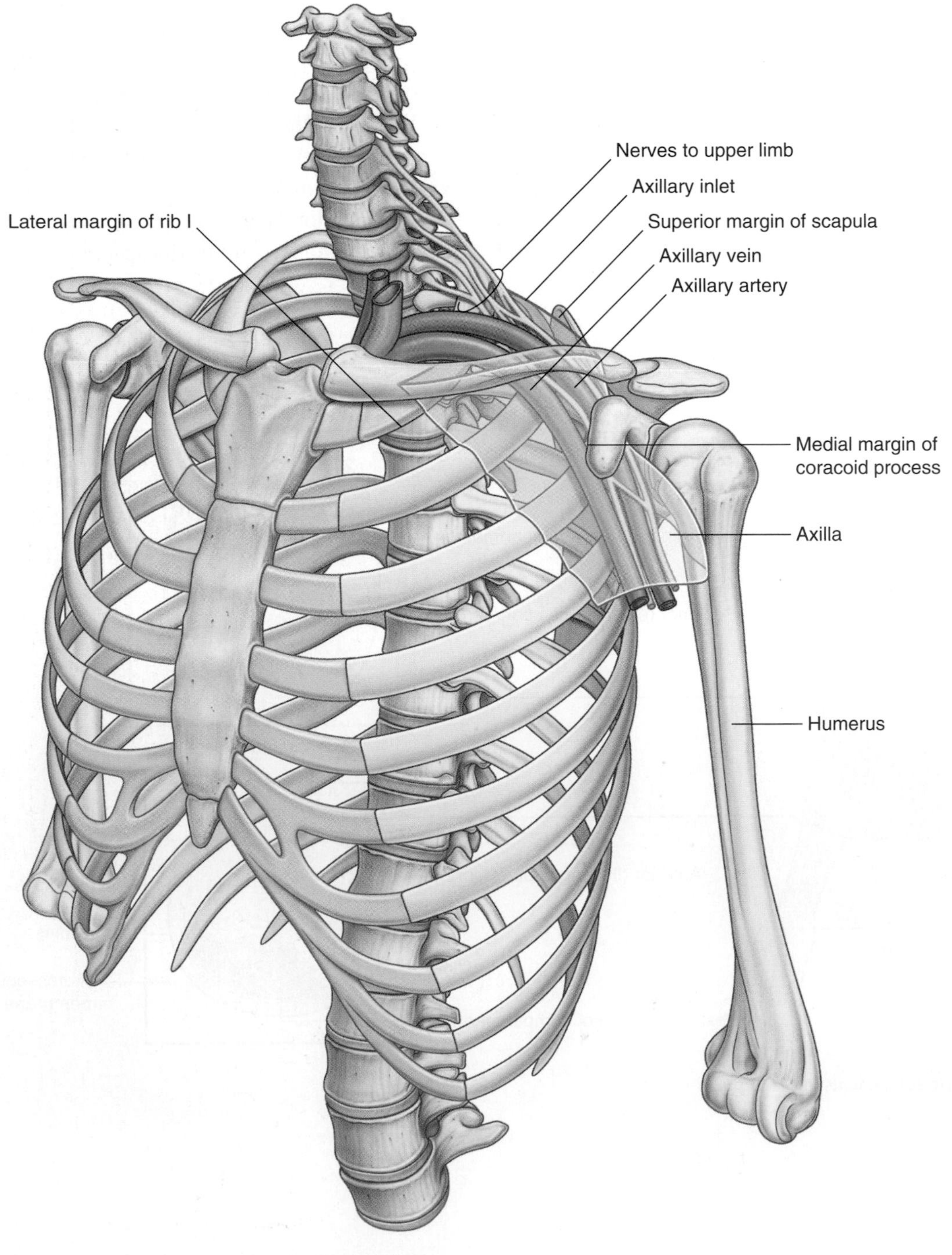

**Fig. 7.11** Relationship of the upper limb to the neck.

include the trapezius, levator scapulae, rhomboid major, rhomboid minor, and latissimus dorsi (Fig. 7.12).

The breast on the anterior thoracic wall has a number of significant relationships with the axilla and upper limb. It overlies the pectoralis major muscle, which forms most of the anterior wall of the axilla and attaches the humerus to the chest wall (Fig. 7.13). Often, part of the breast known as the axillary process extends around the lateral margin of the pectoralis major into the axilla.

Lymphatic drainage from lateral and superior parts of the breast is predominantly into lymph nodes in the axilla. Several arteries and veins that supply or drain the gland also originate from, or drain into, major axillary vessels.

## KEY POINTS

### Innervation by cervical and upper thoracic nerves

Innervation of the upper limb is by the brachial plexus, which is formed by the anterior rami of cervical spinal

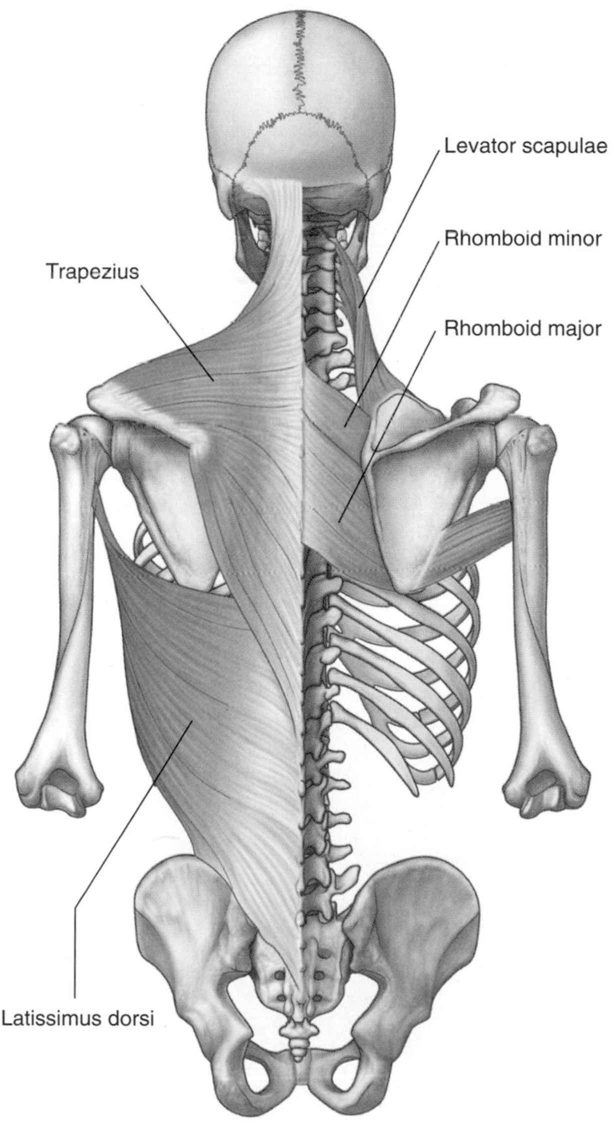


**Fig. 7.12** Muscles of the back and thoracic wall.

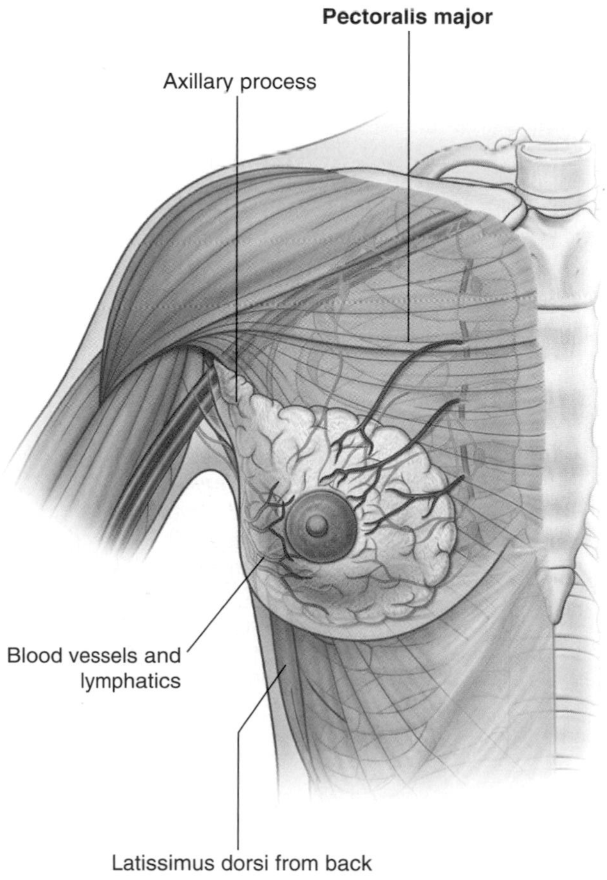


**Fig. 7.13** Breast.

nerves C5 to C8, and T1 (Fig. 7.14). This plexus is initially formed in the neck and then continues through the axillary inlet into the axilla. Major nerves that ultimately innervate the arm, forearm, and hand originate from the brachial plexus in the axilla.

As a consequence of this innervation pattern, clinical testing of lower cervical and T1 nerves is carried out by examining dermatomes, myotomes, and tendon reflexes in the upper limb. Another consequence is that the clinical signs of problems related to lower cervical nerves—pain; pins-and-needles sensations, or paresthesia; and muscle twitching—appear in the upper limb.

Dermatomes of the upper limb (Fig. 7.15A) are often tested for sensation. Areas where overlap of dermatomes is minimal include the:

- upper lateral region of the arm for spinal cord level C5,
- palmar pad of the thumb for spinal cord level C6,
- pad of the index finger for spinal cord level C7,
- pad of the little finger for spinal cord level C8, and
- skin on the medial aspect of the elbow for spinal cord level T1.

Selected joint movements are used to test myotomes (Fig. 7.15B):

- Abduction of the arm at the glenohumeral joint is controlled predominantly by C5.
- Flexion of the forearm at the elbow joint is controlled primarily by C6.
- Extension of the forearm at the elbow joint is controlled mainly by C7.
- Flexion of the fingers is controlled mainly by C8.
- Abduction and adduction of the index, middle, and ring fingers is controlled predominantly by T1.

In an unconscious patient, both somatic sensory and motor functions of spinal cord levels can be tested using tendon reflexes:

- A tap on the tendon of the biceps in the cubital fossa tests mainly for spinal cord level C6.
- A tap on the tendon of the triceps posterior to the elbow tests mainly for C7.

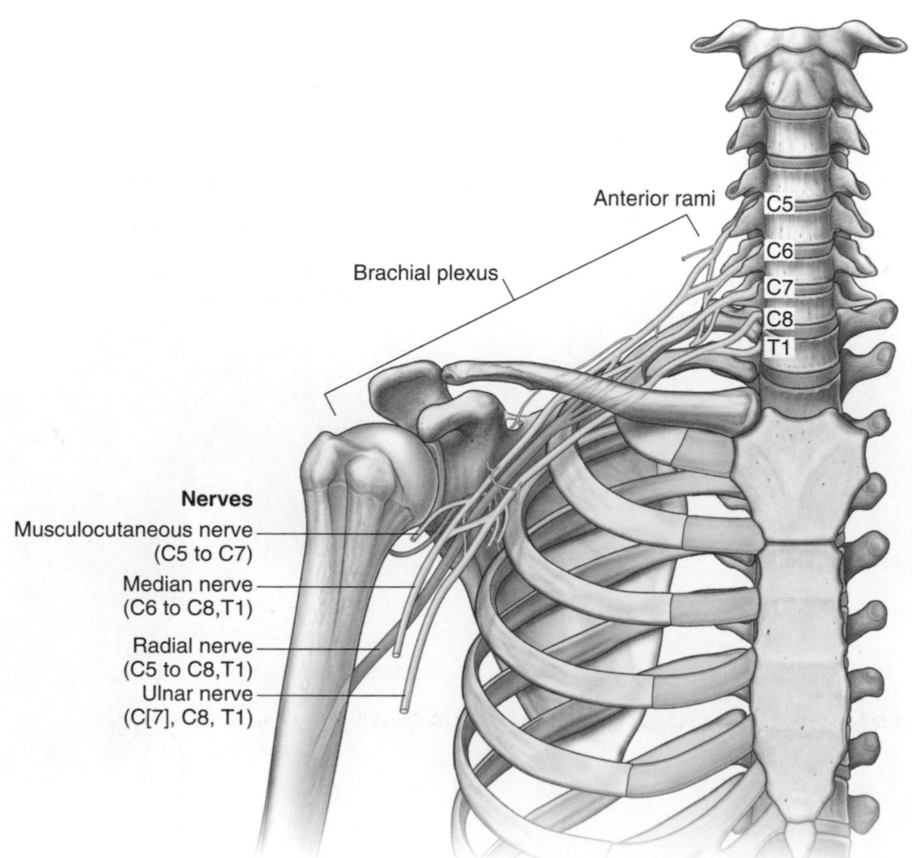


**Fig. 7.14** Innervation of the upper limb.

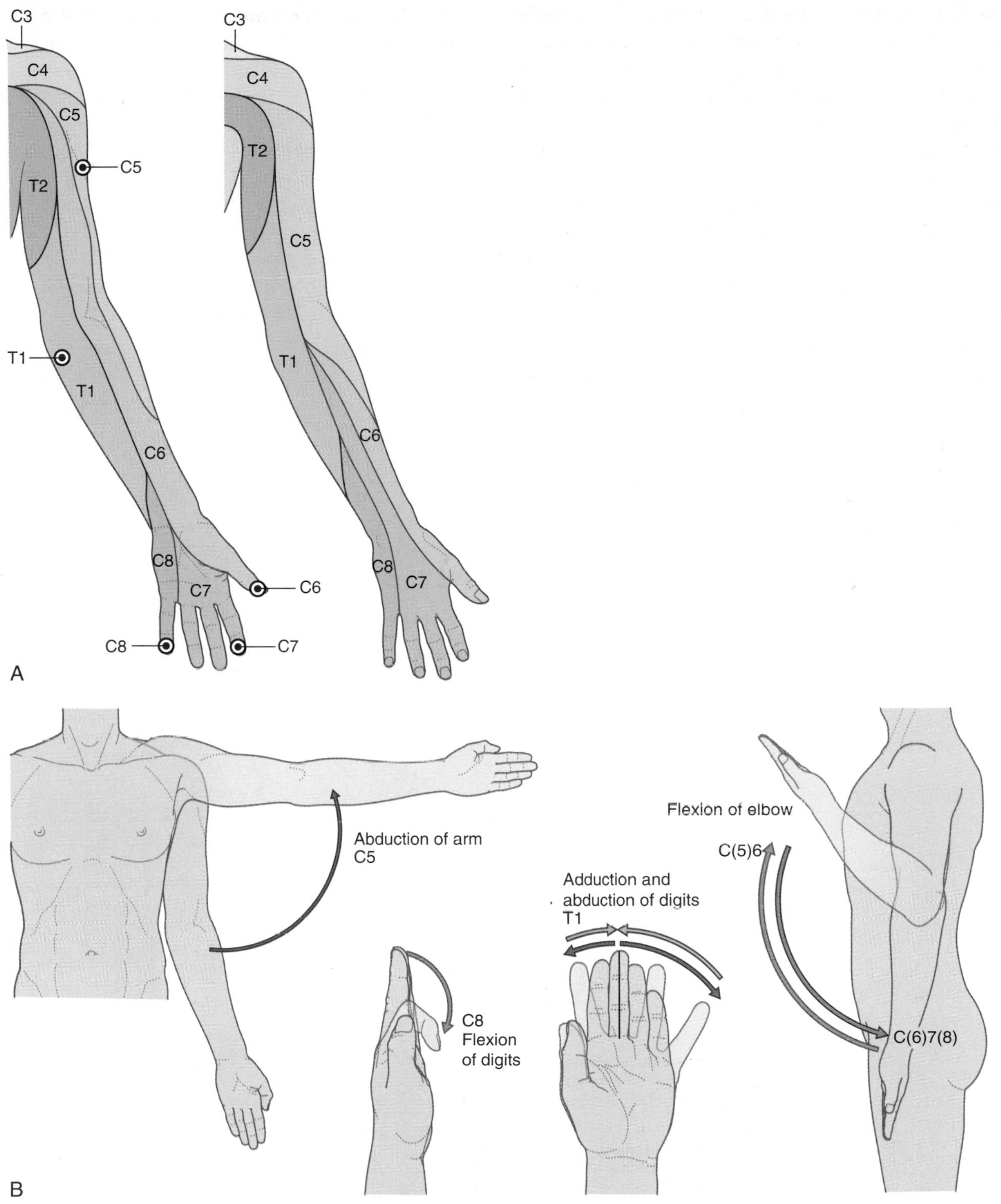


**Fig. 7.15** Dermatomes and myotomes in the upper limb. **A.** Dermatomes. **B.** Movements produced by myotomes.

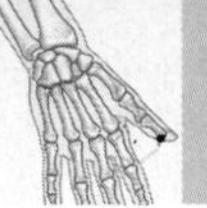



The major spinal cord level associated with innervation of the diaphragm, C4, is immediately above the spinal cord levels associated with the upper limb.

Evaluation of dermatomes and myotomes in the upper limb can provide important information about potential breathing problems that might develop as complications of damage to the spinal cord in regions just below the C4 spinal level.

Each of the major muscle compartments in the arm and forearm and each of the intrinsic muscles of the hand is innervated predominantly by one of the major nerves that originate from the brachial plexus in the axilla (Fig. 7.16A):

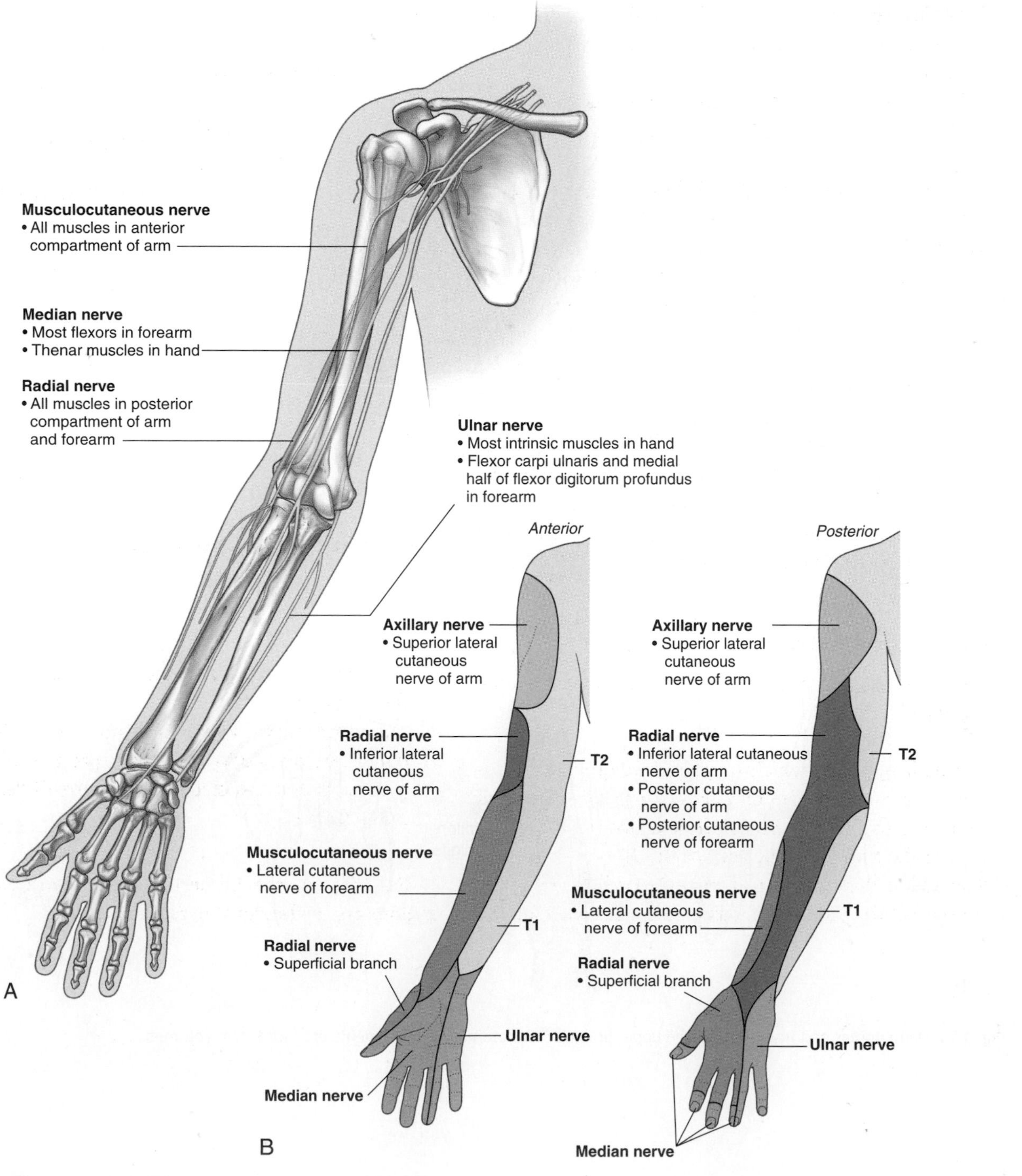

**Fig. 7.16** Nerves of upper limb. **A.** Major nerves in the arm and forearm. **B.** Anterior and posterior areas of skin innervated by major peripheral nerves in the arm and forearm.

- All muscles in the anterior compartment of the arm are innervated by the musculocutaneous nerve.
- The median nerve innervates the muscles in the anterior compartment of the forearm, with two exceptions—one flexor of the wrist (the flexor carpi ulnaris muscle) and part of one flexor of the fingers (the medial half of the flexor digitorum profundus muscle) are innervated by the ulnar nerve.
- Most intrinsic muscles of the hand are innervated by the ulnar nerve, except for the thenar muscles and two lateral lumbrical muscles, which are innervated by the median nerve.
- All muscles in the posterior compartments of the arm and forearm are innervated by the radial nerve.

In addition to innervating major muscle groups, each of the major peripheral nerves originating from the brachial plexus carries somatic sensory information from patches of skin quite different from dermatomes (Fig. 7.16B). Sensation in these areas can be used to test for peripheral nerve lesions:

- The musculocutaneous nerve innervates skin on the anterolateral side of the forearm.
- The median nerve innervates the palmar surface of the lateral three and one-half digits, and the ulnar nerve innervates the medial one and one-half digits.
- The radial nerve supplies skin on the posterior surface of the forearm and the dorsolateral surface of the hand.

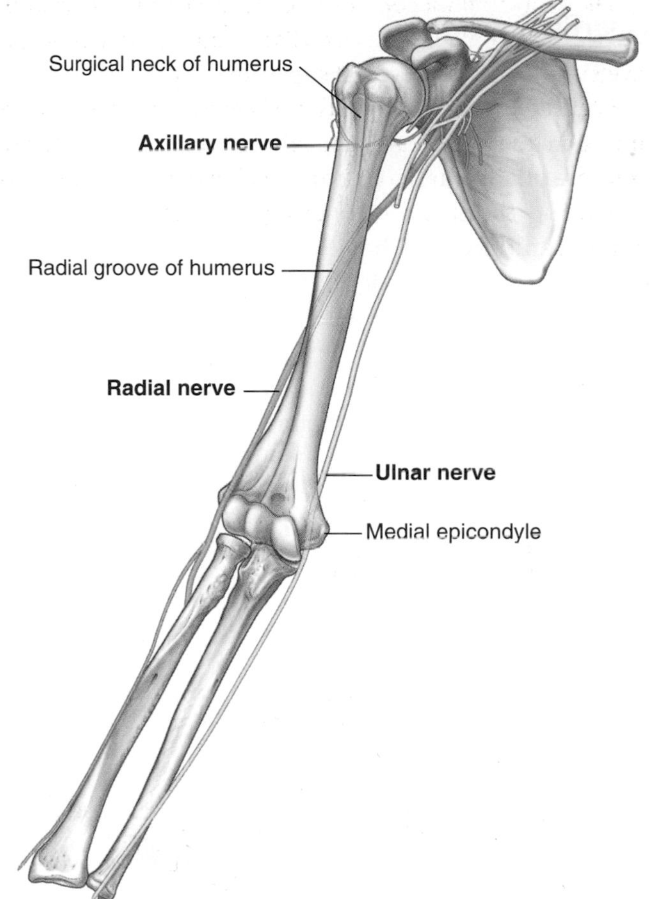


**Fig. 7.17** Nerves related to the humerus.

## Nerves related to bone

Three important nerves are directly related to parts of the humerus (Fig. 7.17):

- The axillary nerve, which supplies the deltoid muscle, a major abductor of the humerus at the glenohumeral joint, passes around the posterior aspect of the upper part of the humerus (the surgical neck).
- The radial nerve, which supplies all of the extensor muscles of the upper limb, passes diagonally around the posterior surface of the middle of the humerus in the radial groove.
- The ulnar nerve, which is ultimately destined for the hand, passes posteriorly to a bony protrusion, the medial epicondyle, on the medial side of the distal end of the humerus.

Fractures of the humerus in any one of these three regions can endanger the related nerve.

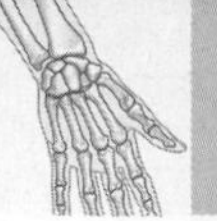

## Superficial veins

Large veins embedded in the superficial fascia of the upper limb are often used to access a patient's vascular system and to withdraw blood. The most significant of these veins are the cephalic, basilic, and median cubital veins (Fig. 7.18).

The **cephalic** and **basilic veins** originate from the **dorsal venous network** on the back of the hand.

The cephalic vein originates over the anatomical snuffbox at the base of the thumb, passes laterally around the distal forearm to reach the anterolateral surface of the limb, and then continues proximally. It crosses the elbow, then passes up the arm into a triangular depression—the **clavipectoral triangle** (**deltopectoral triangle**)—between the pectoralis major muscle, deltoid muscle, and clavicle. In this depression, the vein passes into the axilla by penetrating deep fascia just inferior to the clavicle.

The basilic vein originates from the medial side of the dorsal venous network of the hand and passes proximally up the posteromedial surface of the forearm. It passes onto the anterior surface of the limb just inferior to the elbow and then continues proximally to penetrate deep fascia about midway up the arm.

At the elbow, the cephalic and basilic veins are connected by the **median cubital vein**, which crosses the roof of the cubital fossa.

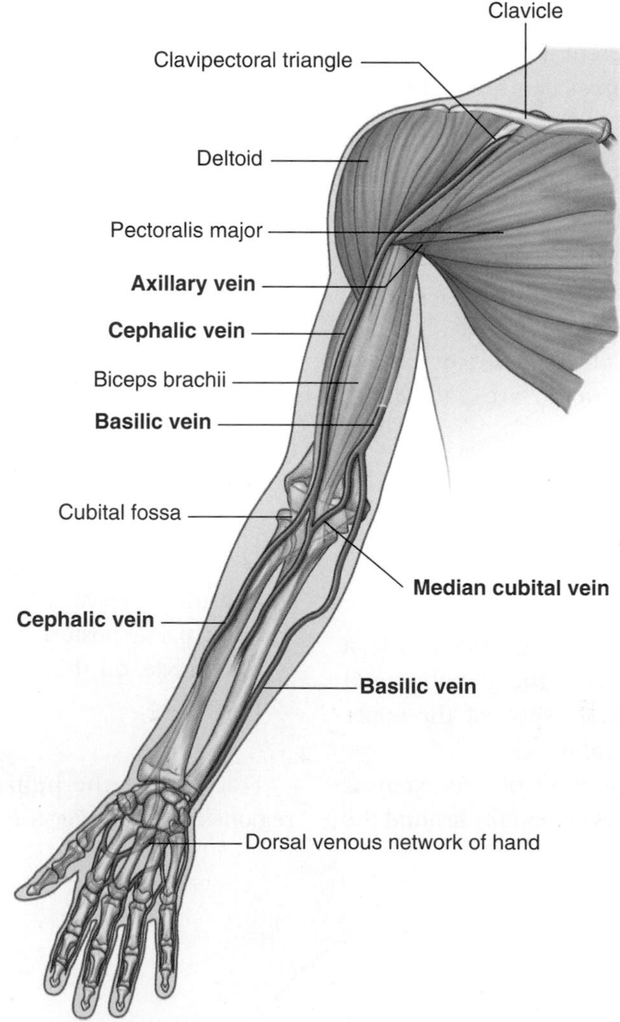


**Fig. 7.18** Veins in the superficial fascia of upper limb. The area of the cubital fossa is shown in yellow.

## Orientation of the thumb

The thumb is positioned at right angles to the orientation of the index, middle, ring, and little fingers (Fig. 7.19). As a result, movements of the thumb occur at right angles to those of the other digits. For example, flexion brings the thumb across the palm, whereas abduction moves it away from the fingers at right angles to the palm.

Importantly, with the thumb positioned at right angles to the palm, only a slight rotation of metacarpal I on the wrist brings the pad of the thumb into a position directly facing the pads of the other fingers. This opposition of the thumb is essential for normal hand function.

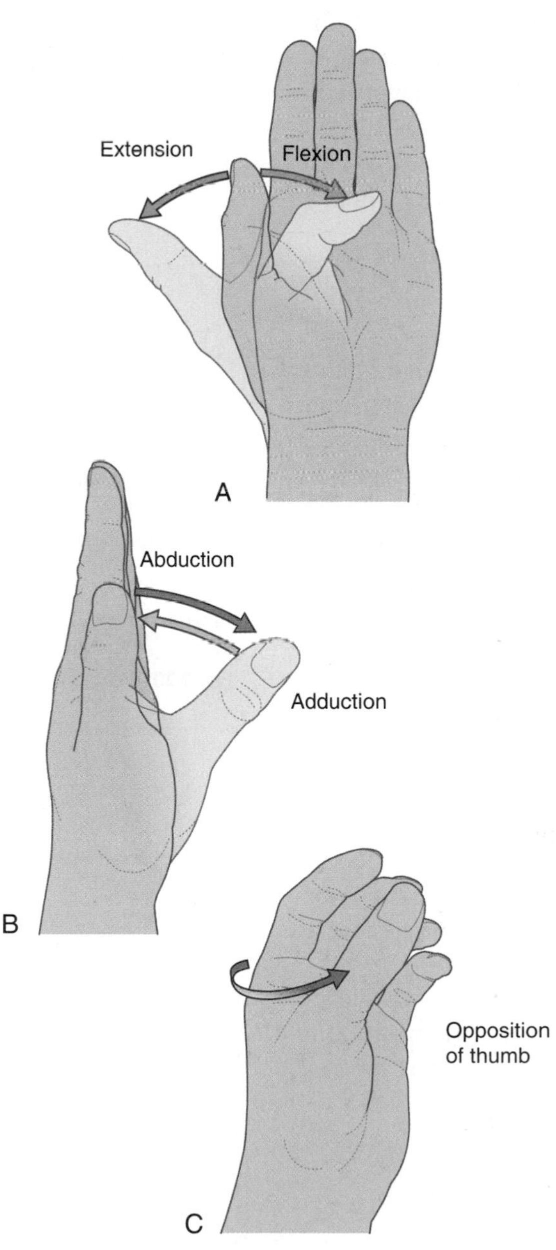


**Fig. 7.19 A** to **C.** Movements of the thumb.

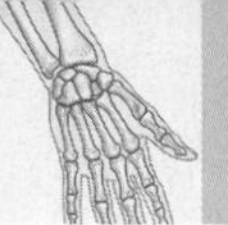



# Regional anatomy

## SHOULDER

The shoulder is the region of upper limb attachment to the trunk.

The bone framework of the shoulder consists of:

- the clavicle and scapula, which form the **pectoral girdle** (**shoulder girdle**), and
- the proximal end of the humerus.

The superficial muscles of the shoulder consist of the trapezius and deltoid muscles, which together form the smooth muscular contour over the lateral part of the shoulder. These muscles connect the scapula and clavicle to the trunk and to the arm, respectively.

### Bones

#### Clavicle

The clavicle is the only bony attachment between the trunk and the upper limb. It is palpable along its entire length and has a gentle S-shaped contour, with the forward-facing convex part medial and the forward-facing concave part lateral. The acromial (lateral) end of the clavicle is flat, whereas the sternal (medial) end is more robust and somewhat quadrangular in shape (Fig. 7.20).

The acromial end of the clavicle has a small oval facet on its surface for articulation with a similar facet on the medial surface of the acromion of the scapula.

The sternal end has a much larger facet for articulation mainly with the manubrium of the sternum, and to a lesser extent, with the first costal cartilage.

The inferior surface of the lateral third of the clavicle possesses a distinct tuberosity consisting of a tubercle (the **conoid tubercle**) and lateral roughening (the **trapezoid line**), for attachment of the important coracoclavicular ligament.

In addition, the surfaces and margins of the clavicle are roughened by the attachment of muscles that connect the clavicle to the thorax, neck, and upper limb. The superior surface is smoother than the inferior surface.

#### Scapula

The scapula is a large, flat triangular bone with:

- three angles (lateral, superior, and inferior),
- three borders (superior, lateral, and medial),
- two surfaces (costal and posterior), and
- three processes (acromion, spine, and coracoid process) (Fig. 7.21).

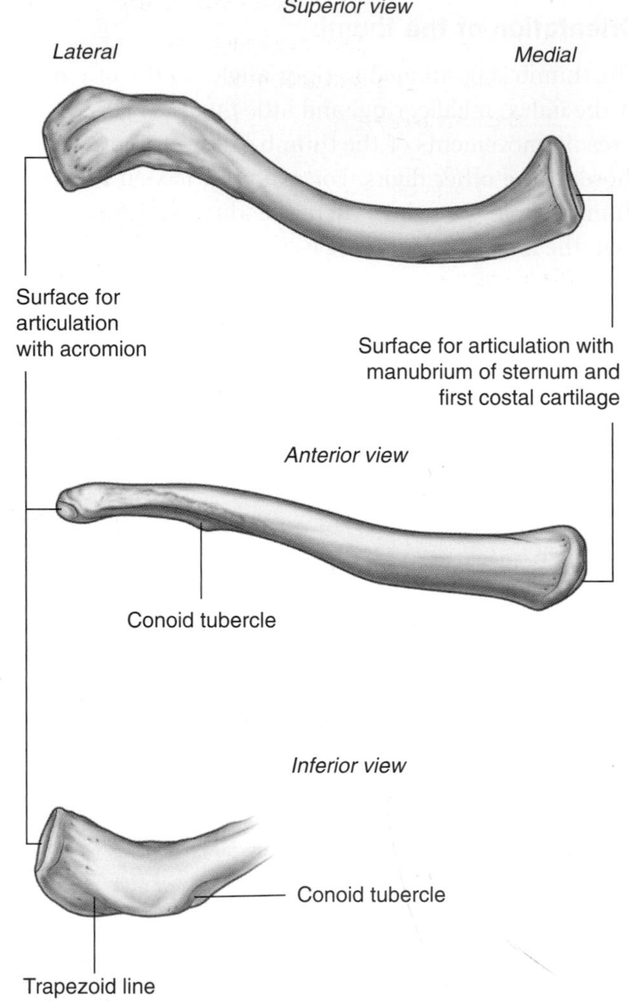


**Fig. 7.20** Right clavicle.

The **lateral angle** of the scapula is marked by a shallow, somewhat comma-shaped **glenoid cavity**, which articulates with the head of the humerus to form the glenohumeral joint (Fig. 7.21B,C).

A large triangular-shaped roughening (the **infraglenoid tubercle**) inferior to the glenoid cavity is the site of attachment for the long head of the triceps brachii muscle.

A less distinct **supraglenoid tubercle** is located superior to the glenoid cavity and is the site of attachment for the long head of the biceps brachii muscle.

A prominent **spine** subdivides the **posterior surface** of the scapula into a small, superior **supraspinous**

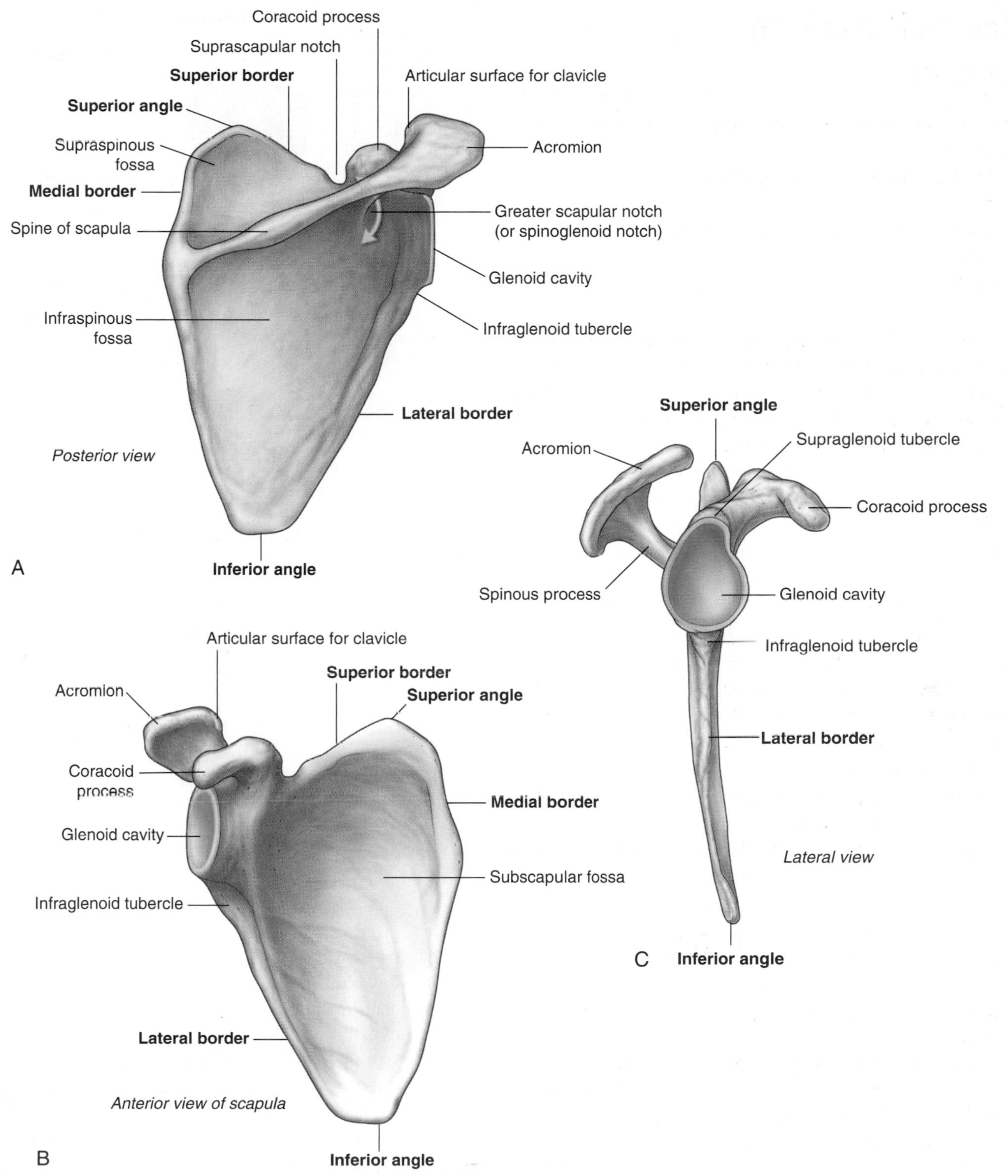


**Fig. 7.21** Scapula. **A.** Posterior view of right scapula. **B.** Anterior view of costal surface. **C.** Lateral view.

**fossa** and a much larger, inferior **infraspinous fossa** (Fig. 7.21A).

The **acromion**, which is an anterolateral projection of the spine, arches over the glenohumeral joint and articulates, via a small oval facet on its distal end, with the clavicle.

The region between the lateral angle of the scapula and the attachment of the spine to the posterior surface of the scapula is the **greater scapular notch** (**spinoglenoid notch**).

Unlike the posterior surface, the **costal surface** of the scapula is unremarkable, being characterized by a shallow concave **subscapular fossa** over much of its extent (Fig. 7.21B). The costal surface and margins provide for muscle attachment, and the costal surface, together with its related muscle (**subscapularis**), moves freely over the underlying thoracic wall.

The lateral border of the scapula is strong and thick for muscle attachment, whereas the medial border and much of the superior border is thin and sharp.

The superior border is marked on its lateral end by:

- the **coracoid process**, a hook-like structure that projects anterolaterally and is positioned directly inferior to the lateral part of the clavicle; and
- the small but distinct **suprascapular notch**, which lies immediately medial to the root of the coracoid process.

The spine and acromion can be readily palpated on a patient, as can the tip of the coracoid process, the inferior angle, and much of the medial border of the scapula.

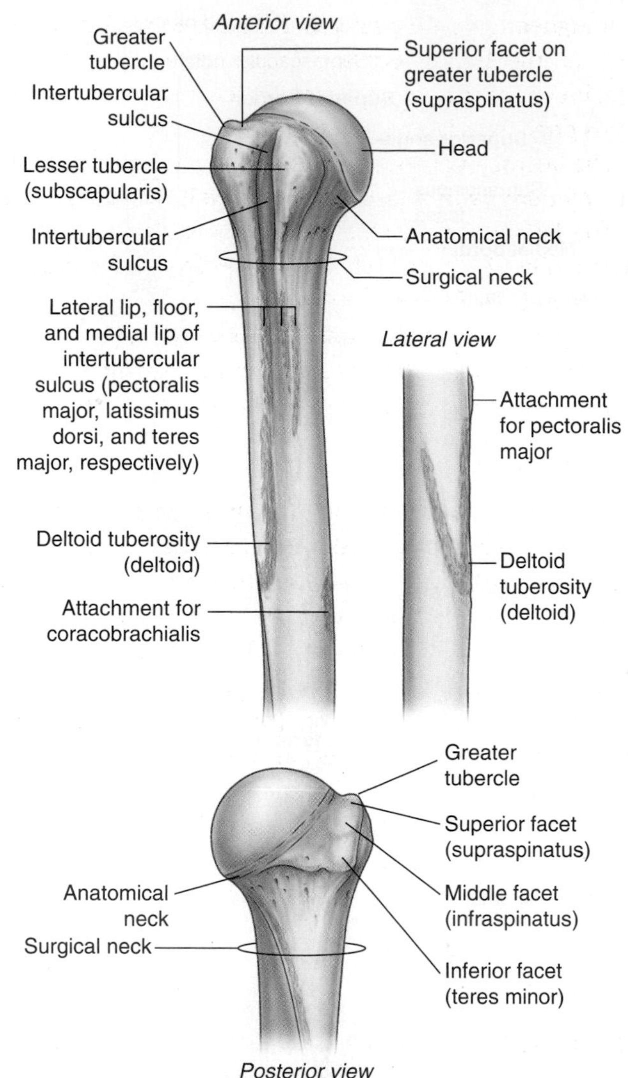


**Fig. 7.22** Proximal end of right humerus.

## Proximal humerus

The proximal end of the humerus consists of the head, the anatomical neck, the greater and lesser tubercles, the surgical neck, and the superior half of the shaft of the humerus (Fig. 7.22).

The **head** is half-spherical in shape and projects medially and somewhat superiorly to articulate with the much smaller glenoid cavity of the scapula.

The **anatomical neck** is very short and is formed by a narrow constriction immediately distal to the head. It lies between the head and the greater and lesser tubercles laterally, and between the head and the shaft more medially.

### Greater and lesser tubercles

The **greater** and **lesser tubercles** are prominent landmarks on the proximal end of the humerus and serve as attachment sites for the four rotator cuff muscles of the glenohumeral joint.

The greater tubercle is lateral in position. Its superior surface and posterior surface are marked by three large smooth facets for muscle tendon attachments:

- The superior facet is for attachment of the supraspinatus muscle.
- The middle facet is for attachment of the infraspinatus.
- The inferior facet is for attachment of the teres minor.

The lesser tubercle is anterior in position and its surface is marked by a large smooth impression for attachment of the subscapularis muscle.

A deep **intertubercular sulcus** (**bicipital groove**) separates the lesser and greater tubercles and continues inferiorly onto the proximal shaft of the humerus (Fig. 7.22). The tendon of the long head of the biceps brachii passes through this sulcus.

Roughenings on the lateral and medial lips and on the floor of the intertubercular sulcus mark sites for the attachment of the pectoralis major, teres major, and latissimus dorsi muscles, respectively.

The lateral lip of the intertubercular sulcus is continuous inferiorly with a large V-shaped **deltoid tuberosity** on the lateral surface of the humerus midway along its length (Fig. 7.22), which is where the deltoid muscle inserts onto the humerus.

In approximately the same position, but on the medial surface of the bone, there is a thin vertical roughening for attachment of the coracobrachialis muscle.

### Surgical neck

One of the most important features of the proximal end of the humerus is the **surgical neck** (Fig. 7.22). This region is oriented in the horizontal plane between the expanded proximal part of the humerus (head, anatomical neck, and tubercles) and the narrower shaft. The axillary nerve and the posterior circumflex humeral artery, which pass into the deltoid region from the axilla, do so immediately posterior to the surgical neck. Because the surgical neck is weaker than more proximal regions of the bone, it is one of the sites where the humerus commonly fractures. The associated nerve (axillary) and artery (posterior circumflex humeral) can be damaged by fractures in this region.

#### In the clinic

**Fracture of the proximal humerus**

It is extremely rare for fractures to occur across the anatomical neck of the humerus because the obliquity of such a fracture would have to traverse the thickest region of bone. Typically fractures occur around the surgical neck of the humerus. Although the axillary nerve and posterior circumflex humeral artery may be damaged with this type of fracture, this rarely happens. It is important that the axillary nerve is tested before relocation to be sure that the injury has not damaged the nerve and that the treatment itself does not cause a neurological deficit.

## Joints

The three joints in the shoulder complex are the sternoclavicular, acromioclavicular, and glenohumeral joints.

The sternoclavicular joint and the acromioclavicular joint link the two bones of the pectoral girdle to each other and to the trunk. The combined movements at these two joints enable the scapula to be positioned over a wide range on the thoracic wall, substantially increasing "reach" by the upper limb.

The glenohumeral joint (shoulder joint) is the articulation between the humerus of the arm and the scapula.

### Sternoclavicular joint

The sternoclavicular joint occurs between the proximal end of the clavicle and the **clavicular notch** of the **manubrium of the sternum** together with a small part of the first costal cartilage (Fig. 7.23). It is synovial and saddle shaped. The articular cavity is completely separated into two compartments by an articular disc. The sternoclavicular joint allows movement of the clavicle, predominantly in the anteroposterior and vertical planes, although some rotation also occurs.

The sternoclavicular joint is surrounded by a joint capsule and is reinforced by four ligaments:

- The **anterior** and **posterior sternoclavicular ligaments** are anterior and posterior, respectively, to the joint.

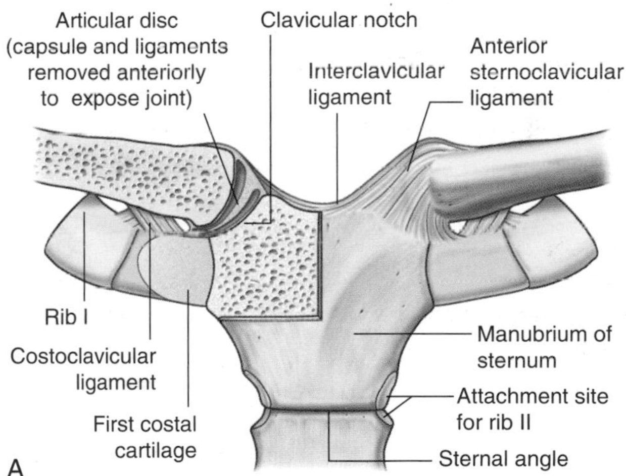


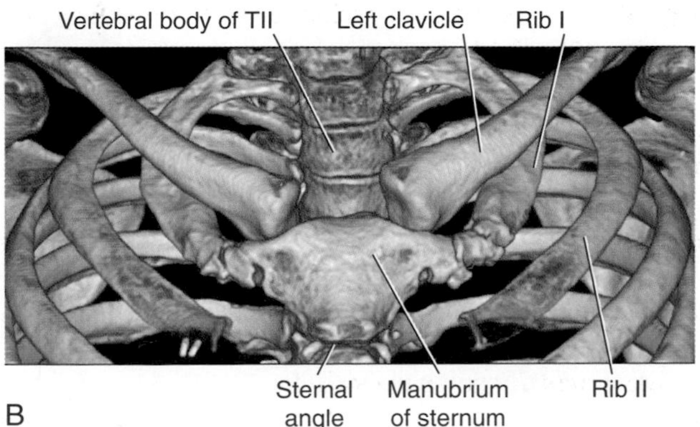


**Fig. 7.23** Sternoclavicular joint. **A.** Bones and ligaments. **B.** Volume-rendered reconstruction using multidetector computed tomography.

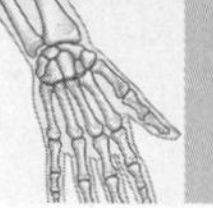

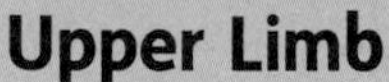

- An **interclavicular ligament** links the ends of the two clavicles to each other and to the superior surface of the manubrium of the sternum.
- The **costoclavicular ligament** is positioned laterally to the joint and links the proximal end of the clavicle to the first rib and related costal cartilage.

### Acromioclavicular joint

The acromioclavicular joint is a small synovial joint between an oval facet on the medial surface of the acromion and a similar facet on the acromial end of the clavicle (Fig. 7.24, also see Fig. 7.31). It allows movement in the anteroposterior and vertical planes together with some axial rotation.

The acromioclavicular joint is surrounded by a joint capsule and is reinforced by:

- a small **acromioclavicular ligament** superior to the joint and passing between adjacent regions of the clavicle and acromion, and
- a much larger **coracoclavicular ligament**, which is not directly related to the joint, but is an important strong accessory ligament, providing much of the weight-bearing support for the upper limb on the clavicle and maintaining the position of the clavicle on the acromion—it spans the distance between the coracoid process of the scapula and the inferior surface of the acromial end of the clavicle and comprises an anterior **trapezoid ligament** (which attaches to the trapezoid line on the clavicle) and a posterior **conoid ligament** (which attaches to the related conoid tubercle).

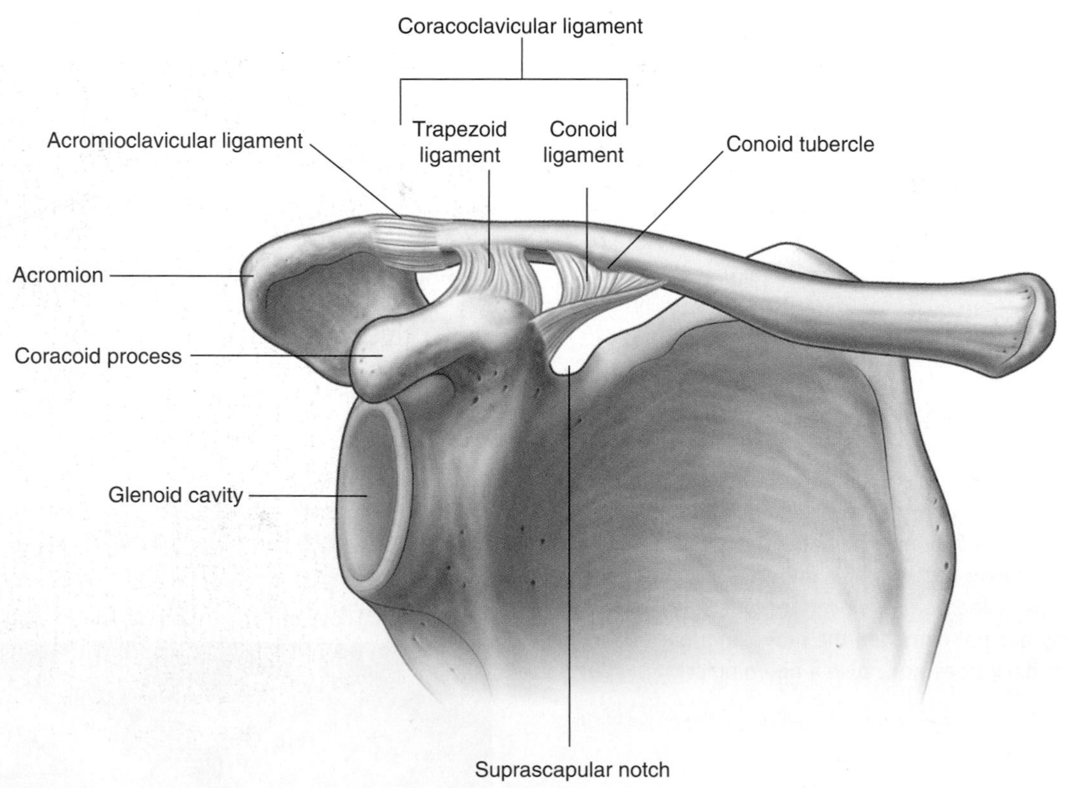


**Fig. 7.24** Right acromioclavicular joint.

## Glenohumeral joint

The **glenohumeral joint** is a synovial ball and socket articulation between the head of the humerus and the glenoid cavity of the scapula (Fig. 7.25). It is multiaxial with a wide range of movements provided at the cost of skeletal stability. Joint stability is provided, instead, by the rotator cuff muscles, the long head of the biceps brachii muscle, related bony processes, and extracapsular ligaments. Movements at the joint include flexion, extension, abduction, adduction, medial rotation, lateral rotation, and circumduction.

The articular surfaces of the glenohumeral joint are the large spherical head of the humerus and the small glenoid cavity of the scapula (Fig. 7.25). Each of the surfaces is covered by hyaline cartilage.

The glenoid cavity is deepened and expanded peripherally by a fibrocartilaginous collar (the **glenoid labrum**), which attaches to the margin of the fossa. Superiorly, this labrum is continuous with the tendon of the long head of the biceps brachii muscle, which attaches to the supraglenoid tubercle and passes through the articular cavity superior to the head of the humerus.

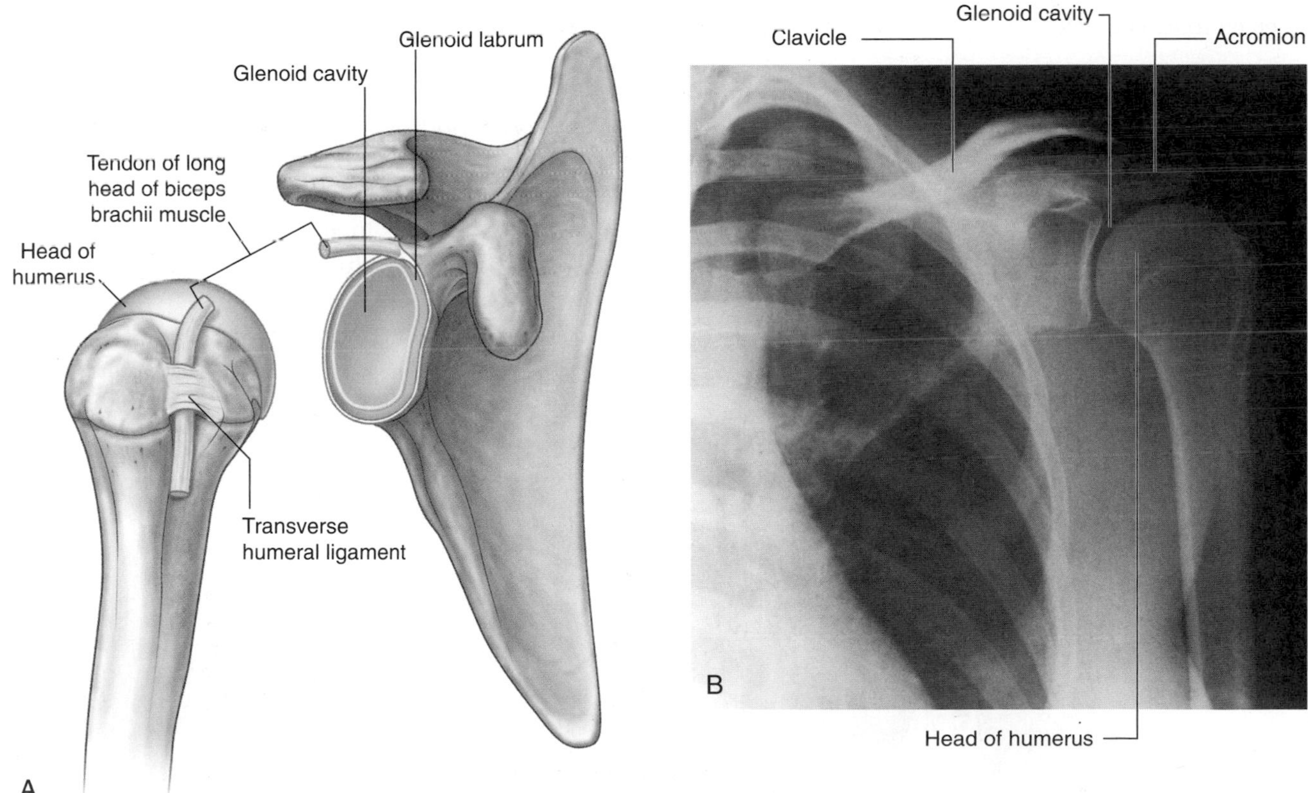


**Fig. 7.25** Glenohumeral joint. **A.** Articular surfaces of right glenohumeral joint. **B.** Radiograph of a normal glenohumeral joint.

The synovial membrane attaches to the margins of the articular surfaces and lines the fibrous membrane of the joint capsule (Fig. 7.26). The synovial membrane is loose inferiorly. This redundant region of synovial membrane and related fibrous membrane accommodates abduction of the arm.

The synovial membrane protrudes through apertures in the fibrous membrane to form bursae, which lie between the tendons of surrounding muscles and the fibrous membrane. The most consistent of these is the **subtendinous bursa of the subscapularis**, which lies between the subscapularis muscle and the fibrous membrane. The synovial membrane also folds around the tendon of the long head of the biceps brachii muscle in the joint and extends along the tendon as it passes into the intertubercular sulcus. All these synovial structures reduce friction between the tendons and adjacent joint capsule and bone.

In addition to bursae that communicate with the articular cavity through apertures in the fibrous membrane, other bursae are associated with the joint but are not connected to it. These occur:

- between the acromion (or deltoid muscle) and supraspinatus muscle (or joint capsule) (the **subacromial** or **subdeltoid bursa**),
- between the acromion and skin,
- between the coracoid process and the joint capsule, and
- in relationship to tendons of muscles around the joint (coracobrachialis, teres major, long head of triceps brachii, and latissimus dorsi muscles).

The fibrous membrane of the joint capsule attaches to the margin of the glenoid cavity, outside the attachment of the glenoid labrum and the long head of the biceps brachii muscle, and to the anatomical neck of the humerus (Fig. 7.27).

On the humerus, the medial attachment occurs more inferiorly than the neck and extends onto the shaft. In this region, the fibrous membrane is also loose or folded in the anatomical position. This redundant area of the fibrous membrane accommodates abduction of the arm.

Openings in the fibrous membrane provide continuity of the articular cavity with bursae that occur between the joint capsule and surrounding muscles and around the tendon of the long head of the biceps brachii muscle in the intertubercular sulcus.

The fibrous membrane of the joint capsule is thickened:

- anterosuperiorly in three locations to form **superior**, **middle**, and **inferior glenohumeral ligaments**, which pass from the superomedial margin of the glenoid cavity to the lesser tubercle and inferiorly related anatomical neck of the humerus (Fig. 7.27);

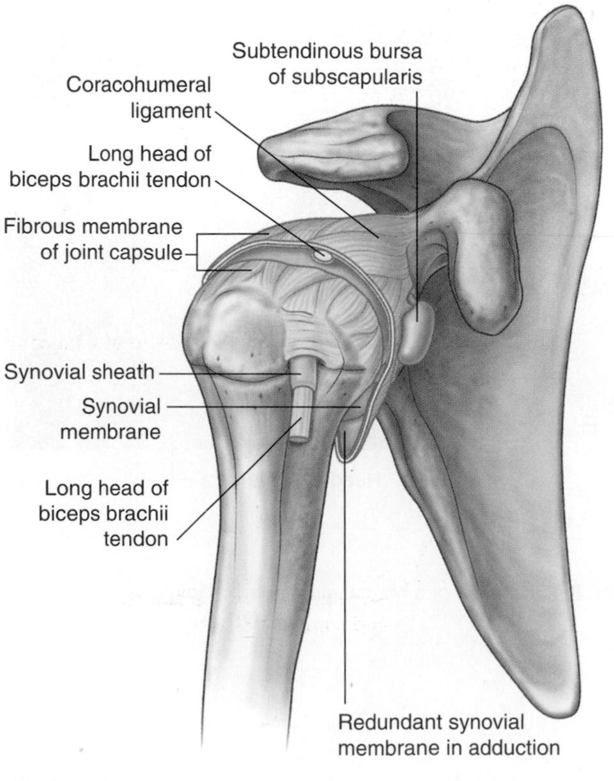

**Fig. 7.26** Synovial membrane and joint capsule of right glenohumeral joint.

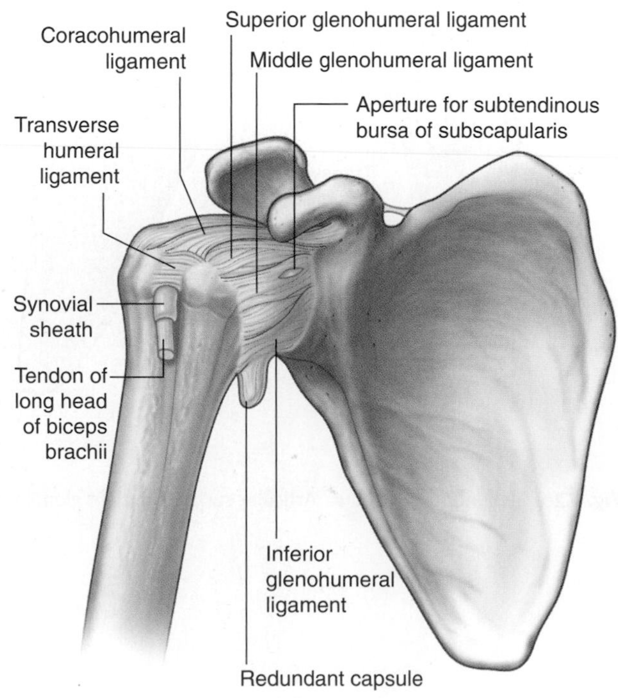


**Fig. 7.27** Capsule of right glenohumeral joint.

- superiorly between the base of the coracoid process and the greater tubercle of the humerus (the **coracohumeral ligament**); and
- between the greater and lesser tubercles of the humerus (**transverse humeral ligament**)—this holds the tendon of the long head of the biceps brachii muscle in the intertubercular sulcus (Fig. 7.27).

Joint stability is provided by surrounding muscle tendons and a skeletal arch formed superiorly by the coracoid process and acromion and the coraco-acromial ligament (Fig. 7.28).

Tendons of the rotator cuff muscles (the supraspinatus, infraspinatus, teres minor, and subscapularis muscles) blend with the joint capsule and form a musculotendinous

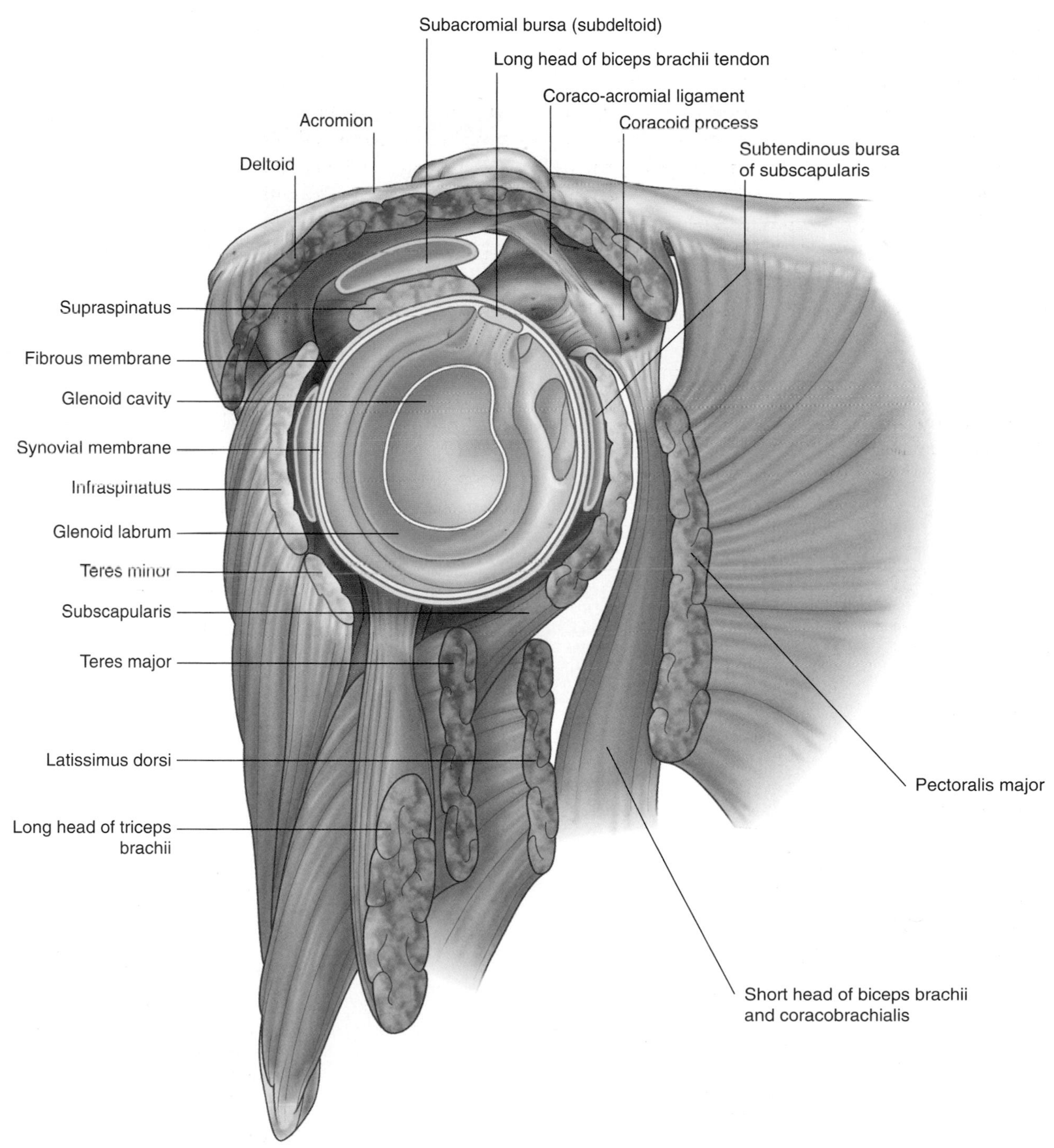


**Fig. 7.28** Lateral view of right glenohumeral joint and surrounding muscles with proximal end of humerus removed.

collar that surrounds the posterior, superior, and anterior aspects of the glenohumeral joint (Figs. 7.28 and 7.29). This cuff of muscles stabilizes and holds the head of the humerus in the glenoid cavity of the scapula without compromising the arm's flexibility and range of motion. The tendon of the long head of the biceps brachii muscle passes superiorly through the joint and restricts upward movement of the humeral head on the glenoid cavity.

Vascular supply to the glenohumeral joint is predominantly through branches of the anterior and posterior circumflex humeral and suprascapular arteries.

The glenohumeral joint is innervated by branches from the posterior cord of the brachial plexus, and from the suprascapular, axillary, and lateral pectoral nerves.

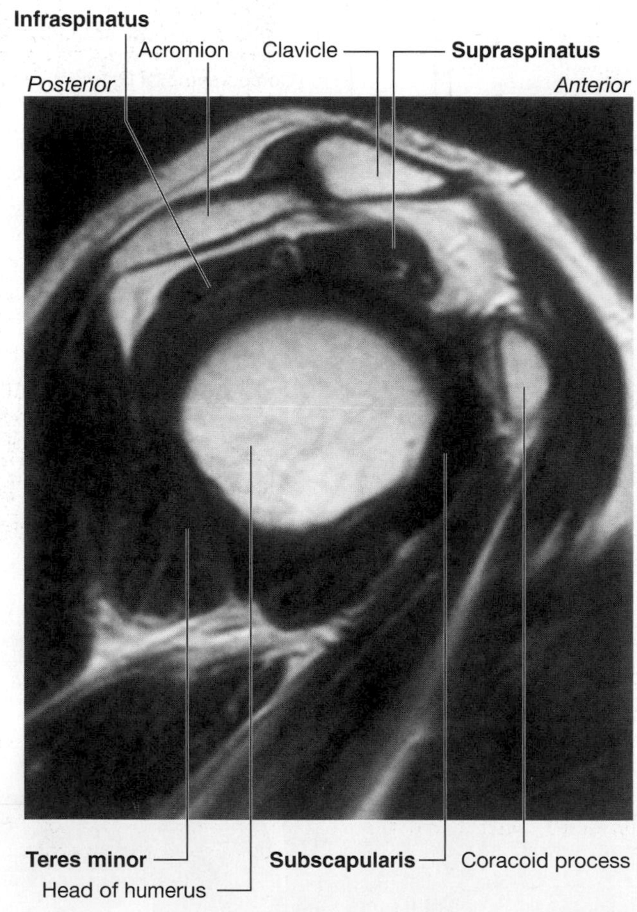


**Fig. 7.29** Magnetic resonance image (T1-weighted) of a normal glenohumeral joint in the sagittal plane.

## In the clinic

### Fractures of the clavicle and dislocations of the acromioclavicular and sternoclavicular joints

The clavicle provides the only bony connection between the upper limb and trunk. Given its relative size and the potential forces that it transmits from the upper limb to the trunk, it is not surprising that it is often fractured. The typical site of fracture is the middle third (Fig. 7.30). The medial and lateral thirds are rarely fractured.

The acromial end of the clavicle tends to dislocate at the acromioclavicular joint with trauma (Fig. 7.31). The outer third of the clavicle is joined to the scapula by the conoid and trapezoid ligaments of the coracoclavicular ligament.

A minor injury tends to tear the fibrous joint capsule and ligaments of the acromioclavicular joint, resulting in acromioclavicular separation on a plain radiograph. More severe trauma will disrupt the conoid and trapezoid ligaments of the coracoclavicular ligament, which results in elevation and upward subluxation of the clavicle.

Fracture of clavicle

Acromioclavicular joint

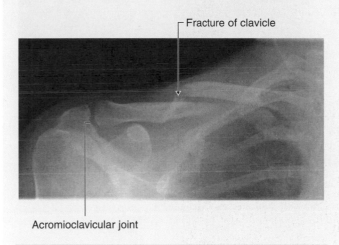

**Fig. 7.30** There is an oblique fracture of the middle third of the right clavicle.

The typical injury at the medial end of the clavicle is an anterior or posterior dislocation of the sternoclavicular joint. Importantly, a posterior dislocation of the clavicle may impinge on the great vessels in the root of the neck and compress or disrupt them.

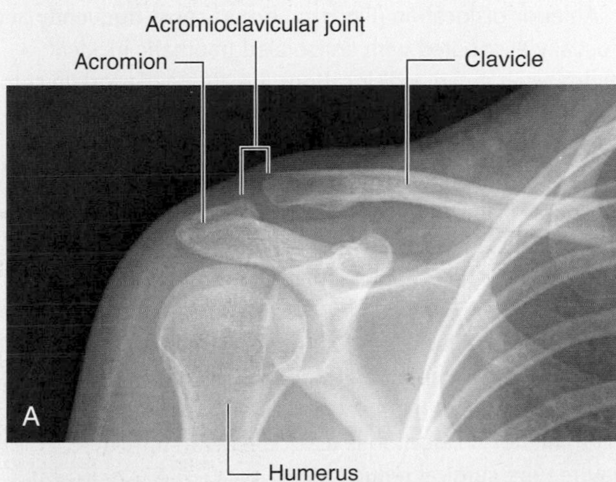


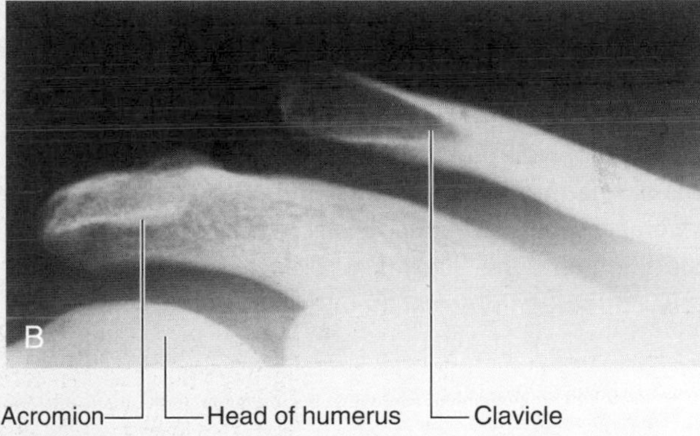


**Fig. 7.31** Radiographs of acromioclavicular joints. **A.** Normal right acromioclavicular joint. **B.** Dislocated right acromioclavicular joint (shoulder separation).

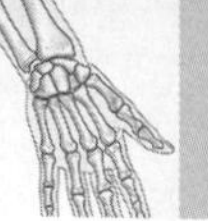



## In the clinic

### Dislocations of the glenohumeral joint

The glenohumeral joint is extremely mobile, providing a wide range of movement at the expense of stability. The relatively small bony glenoid cavity, supplemented by the less robust fibrocartilaginous glenoid labrum and the ligamentous support, make it susceptible to dislocation.

Anterior dislocation (Fig. 7.32) occurs most frequently and is usually associated with an isolated traumatic incident (clinically, all anterior dislocations are anteroinferior). In some cases, the anteroinferior glenoid labrum is torn with or without a small bony fragment. Once the joint capsule and cartilage are disrupted, the joint is susceptible to further (recurrent) dislocations. When an anteroinferior dislocation occurs, the axillary nerve may be injured by direct compression of the humeral head on the nerve inferiorly as it passes through the quadrangular space. Furthermore, the "lengthening" effect of the humerus may stretch the radial nerve, which is tightly bound within the radial groove, and produce a radial nerve paralysis. Occasionally, an anteroinferior dislocation is associated with a fracture, which may require surgical reduction.

Posterior dislocation is extremely rare; when seen, the clinician should focus on its cause, the most common being extremely vigorous muscle contractions, which may be associated with an epileptic seizure caused by electrocution. Treatment of recurrent instability can be challenging. The aims of treatment are to maintain function and range of movement while preventing instability (subluxation, dislocation, and the "feeling" of dislocation). This can be achieved through physical therapy and shoulder "re-education." If this fails, capsular tightening and stabilization of the labrum can be achieved arthroscopically. If the problem persists, the coracoid process can be divided at the base, maintaining continuity of the muscular attachments. The process is transferred and a screw fixed to the anterior inferior border of the glenoid to form a buttress to prevent future dislocations.

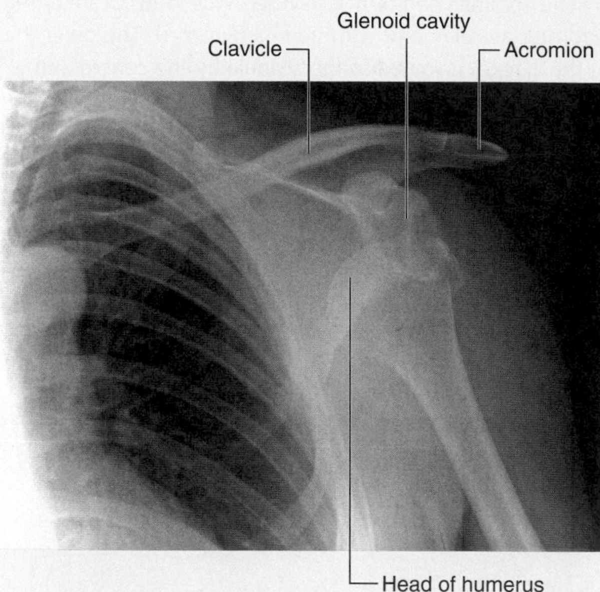


**Fig. 7.32** Radiograph showing an anteroinferior dislocation of the shoulder joint.

## In the clinic

### Rotator cuff disorders

The two main disorders of the rotator cuff are impingement and tendinopathy. The muscle most commonly involved is supraspinatus as it passes beneath the acromion and the acromioclavicular ligament. This space, beneath which the supraspinatus tendon passes, is of fixed dimensions. Swelling of the supraspinatus muscle, excessive fluid within the subacromial/subdeltoid bursa, or subacromial bony spurs may produce significant impingement when the arm is abducted.

The blood supply to the supraspinatus tendon is relatively poor. Repeated trauma, in certain circumstances, makes the tendon susceptible to degenerative change, which may result in calcium deposition, producing extreme pain. The calcium deposits can be extracted through a needle under image guidance and often have the consistency of toothpaste.

When the supraspinatus tendon has undergone significant degenerative change, it is more susceptible to trauma, and partial- or full-thickness tears may develop (Fig. 7.33). These tears are most common in older patients and may result in considerable difficulty in carrying out normal activities of daily living such as combing hair. However, complete tears may be entirely asymptomatic.

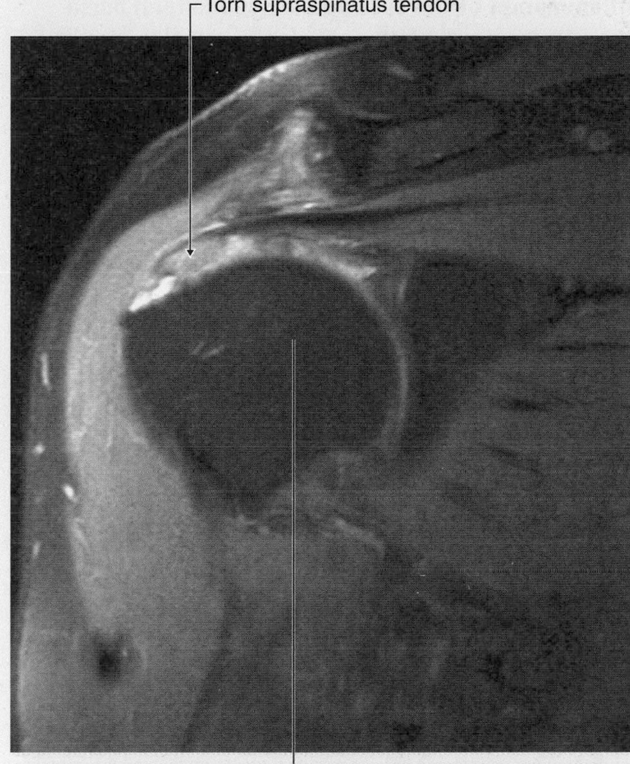


**Fig. 7.33** Magnetic resonance image of a full-thickness tear of the supraspinatus tendon as it inserts onto the greater tubercle of the humerus.

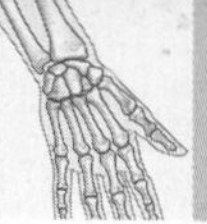



## In the clinic

**Inflammation of the subacromial (subdeltoid) bursa**
Between the supraspinatus and deltoid muscles laterally and the acromion medially, there is a bursa referred to clinically as the subacromial or subdeltoid bursa. In patients who have injured the shoulder or who have supraspinatus tendinopathy, this bursa may become inflamed, making movements of the glenohumeral joint painful. These inflammatory changes may be treated by injection of a corticosteroid and local anesthetic agent (Fig. 7.34).

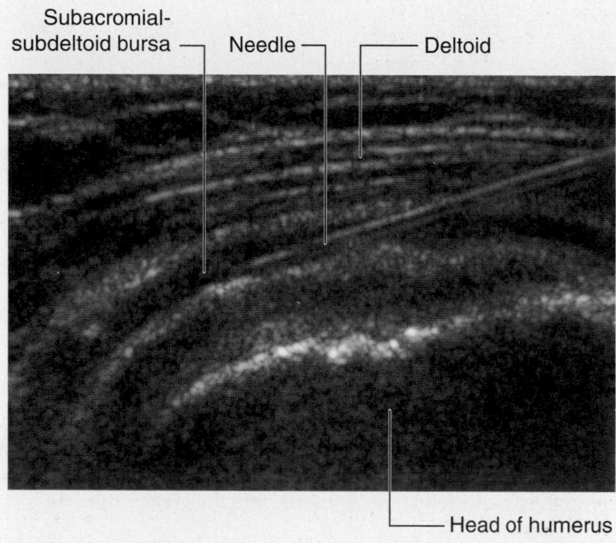


**Fig. 7.34** Ultrasound of shoulder showing needle placement into the subdeltoid/subacromial bursa.

## Muscles

The two most superficial muscles of the shoulder are the trapezius and deltoid muscles (Fig. 7.35 and Table 7.1). Together, they provide the characteristic contour of the shoulder:

- The trapezius attaches the scapula and clavicle to the trunk.
- The deltoid attaches the scapula and clavicle to the humerus.

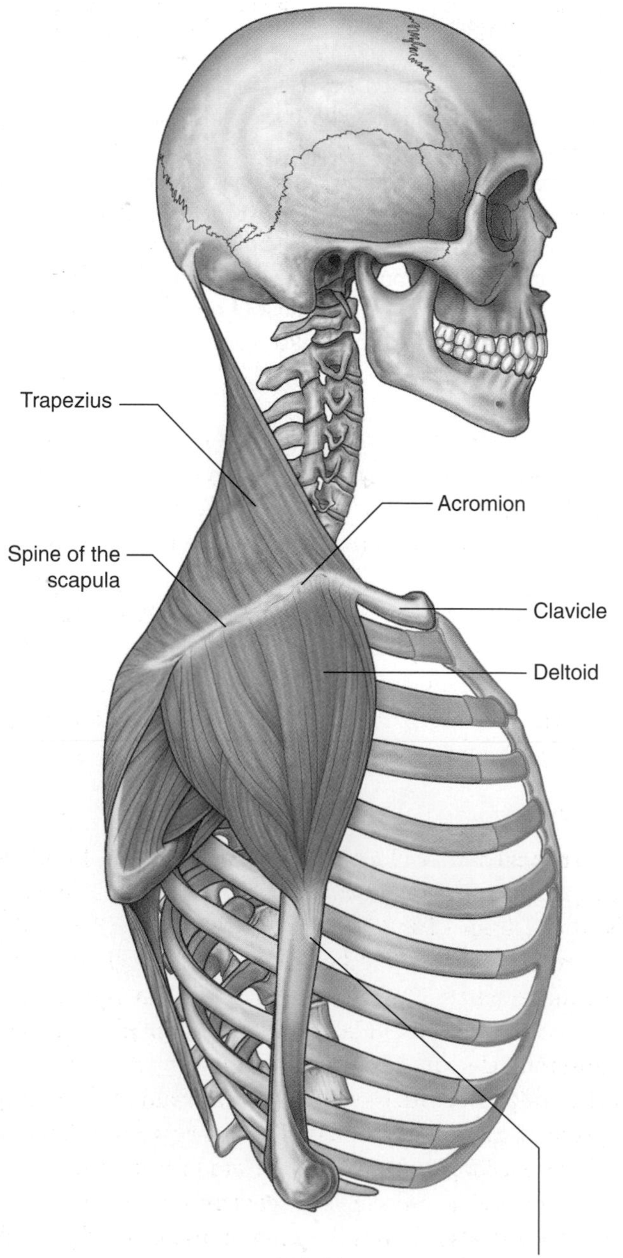


**Fig. 7.35** Lateral view of trapezius and deltoid muscles.

**Table 7.1** Muscles of the shoulder (spinal segments in bold are the major segments innervating the muscle)

| Muscle | Origin | Insertion | Innervation | Function |
|---|---|---|---|---|
| Trapezius | Superior nuchal line, external occipital protuberance, medial margin of the ligamentum nuchae, spinous processes of CVII to TXII and the related supraspinous ligaments | Superior edge of the crest of the spine of the scapula, acromion, posterior border of lateral one-third of clavicle | Motor spinal part of accessory nerve (XI). Sensory (proprioception) anterior rami of C3 and C4 | Powerful elevator of the scapula; rotates the scapula during abduction of humerus above horizontal; middle fibers retract scapula; lower fibers depress scapula |
| Deltoid | Inferior edge of the crest of the spine of the scapula, lateral margin of the acromion, anterior border of lateral one-third of clavicle | Deltoid tuberosity of humerus | Axillary nerve (**C5**, C6) | Major abductor of arm; clavicular fibers assist in flexing the arm; posterior fibers assist in extending the arm |
| Levator scapulae | Transverse processes of CI and CII vertebrae and posterior tubercles of transverse processes of CIII and CIV vertebrae | Posterior surface of medial border of scapula from superior angle to root of spine of the scapula | Branches directly from anterior rami of **C3** and **C4** spinal nerves and by branches (**C5**) from the dorsal scapular nerve | Elevates the scapula |
| Rhomboid minor | Lower end of ligamentum nuchae and spinous processes of CVII and TI vertebrae | Posterior surface of medial border of scapula at the root of the spine of the scapula | Dorsal scapular nerve (**C4, C5**) | Elevates and retracts the scapula |
| Rhomboid major | Spinous processes of TII–TV vertebrae and intervening supraspinous ligaments | Posterior surface of medial border of scapula from the root of the spine of the scapula to the inferior angle | Dorsal scapular nerve (**C4, C5**) | Elevates and retracts the scapula |

Both the trapezius and deltoid are attached to opposing surfaces and margins of the spine of the scapula, acromion, and clavicle. The scapula, acromion, and clavicle can be palpated between the attachments of the trapezius and deltoid.

Deep to the trapezius the scapula is attached to the vertebral column by three muscles—the levator scapulae, rhomboid minor, and rhomboid major. These three muscles work with the trapezius (and with muscles found anteriorly) to position the scapula on the trunk.

### Trapezius

The **trapezius** muscle has an extensive origin from the axial skeleton, which includes sites on the skull and the vertebrae, from CI to TXII (Fig. 7.36). From CI to CVII, the muscle attaches to the vertebrae through the ligamentum nuchae. The muscle inserts onto the skeletal framework of the shoulder along the inner margins of a continuous U-shaped line of attachment oriented in the horizontal plane, with the bottom of the U directed laterally. Together, the left and right trapezius muscles form a diamond or trapezoid shape, from which the name is derived.

The trapezius muscle is a powerful elevator of the shoulder and also rotates the scapula to extend the reach superiorly.

Innervation of the trapezius muscle is by the accessory nerve [XI] and the anterior rami of cervical nerves C3 and C4 (Fig. 7.36). These nerves pass vertically along the deep surface of the muscle. The accessory nerve can be evaluated by testing the function of the trapezius muscle. This is most easily done by asking patients to shrug their shoulders against resistance.

### Deltoid

The **deltoid** muscle is large and triangular in shape, with its base attached to the scapula and clavicle and its apex attached to the humerus (Fig. 7.36). It originates along a continuous U-shaped line of attachment to the clavicle and the scapula, mirroring the adjacent insertion sites of the trapezius muscle. It inserts into the deltoid tuberosity on the lateral surface of the shaft of the humerus.

The major function of the deltoid muscle is abduction of the arm.

The deltoid muscle is innervated by the axillary nerve, which is a branch of the posterior cord of the brachial plexus. The axillary nerve and associated blood vessels (the posterior circumflex humeral artery and vein) enter the deltoid by passing posteriorly around the surgical neck of the humerus.

### Levator scapulae

The levator scapulae originates from the transverse processes of CI to CIV vertebrae (Fig. 7.36). It descends laterally

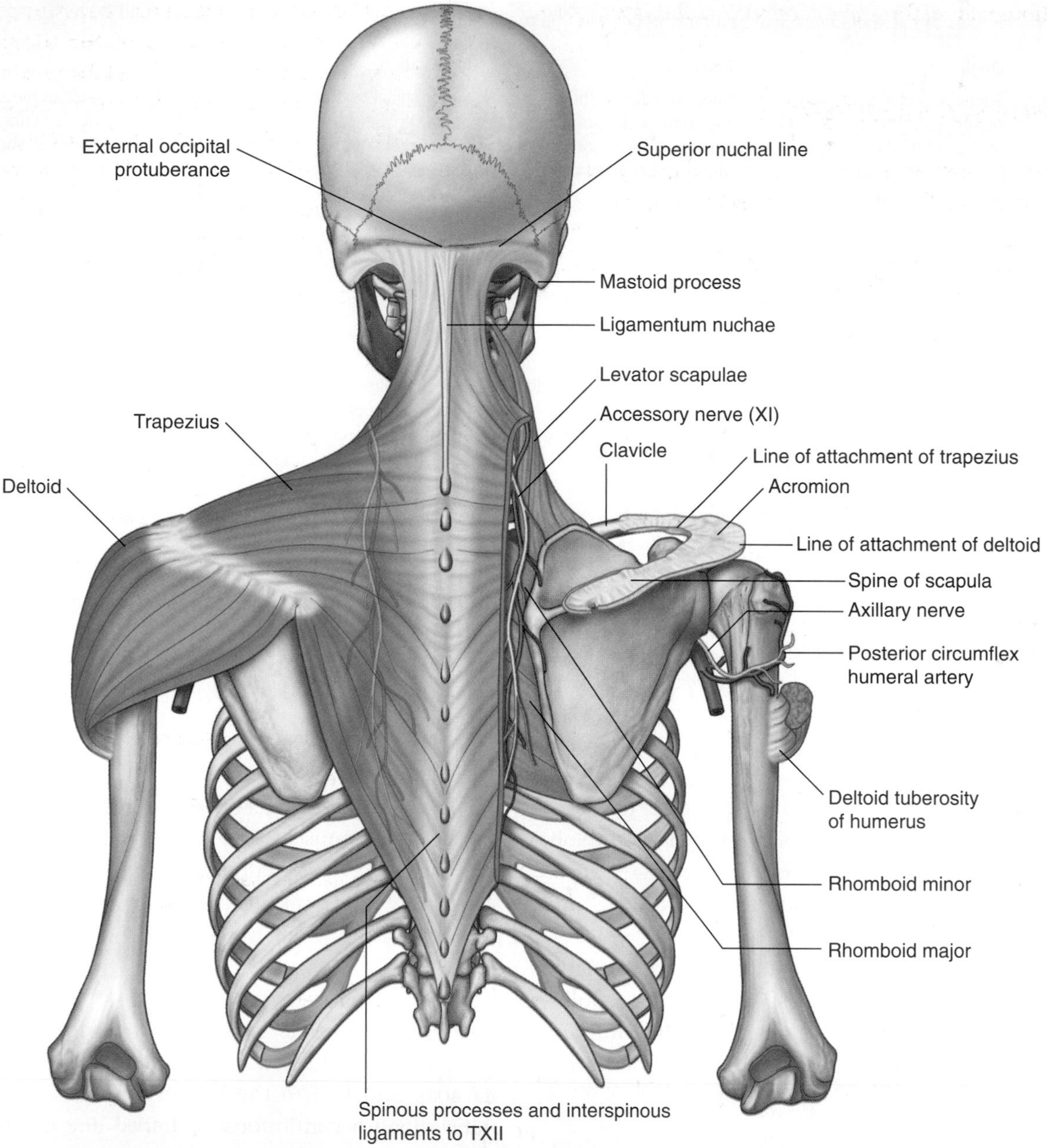

**Fig. 7.36** Attachments and neurovascular supply of the trapezius and deltoid muscles.

to attach to the posterior surface of the medial border of the scapula from the superior angle to the smooth triangular area of bone at the root of the spine.

The levator scapulae muscle is innervated by the dorsal scapular nerve and directly from C3 and C4 spinal nerves.

The levator scapulae elevates the scapula.

### Rhomboid minor and major

The rhomboid minor and major muscles attach medially to the vertebral column and descend laterally to attach to the medial border of the scapula inferior to the levator scapulae muscle (Fig. 7.36).

The rhomboid minor originates from the lower end of the ligamentum nuchae and the spines of CVII and TI vertebrae. It inserts laterally into the smooth triangular area of bone at the root of the spine of the scapula on the posterior surface.

The rhomboid major originates from the spines of vertebrae TII to TV and from the intervening supraspinous ligaments. It descends laterally to insert along the posterior surface of the medial border of the scapula from the insertion of the rhomboid minor to the inferior angle.

The rhomboid muscles are innervated by the dorsal scapular nerve, which is a branch of the brachial plexus.

The rhomboid minor and major retract and elevate the scapula.

## POSTERIOR SCAPULAR REGION

The posterior scapular region occupies the posterior aspect of the scapula and is located deep to the trapezius and deltoid muscles (Fig. 7.37 and Table 7.2). It contains four muscles, which pass between the scapula and proximal end of the humerus: the supraspinatus, infraspinatus, teres minor, and teres major muscles.

The posterior scapular region also contains part of one additional muscle, the long head of the triceps brachii, which passes between the scapula and the proximal end of the forearm. This muscle, along with other muscles of the region and the humerus, participates in forming a number of spaces through which nerves and vessels enter and leave the region.

The supraspinatus, infraspinatus, and teres minor muscles are components of the rotator cuff, which stabilizes the glenohumeral joint.

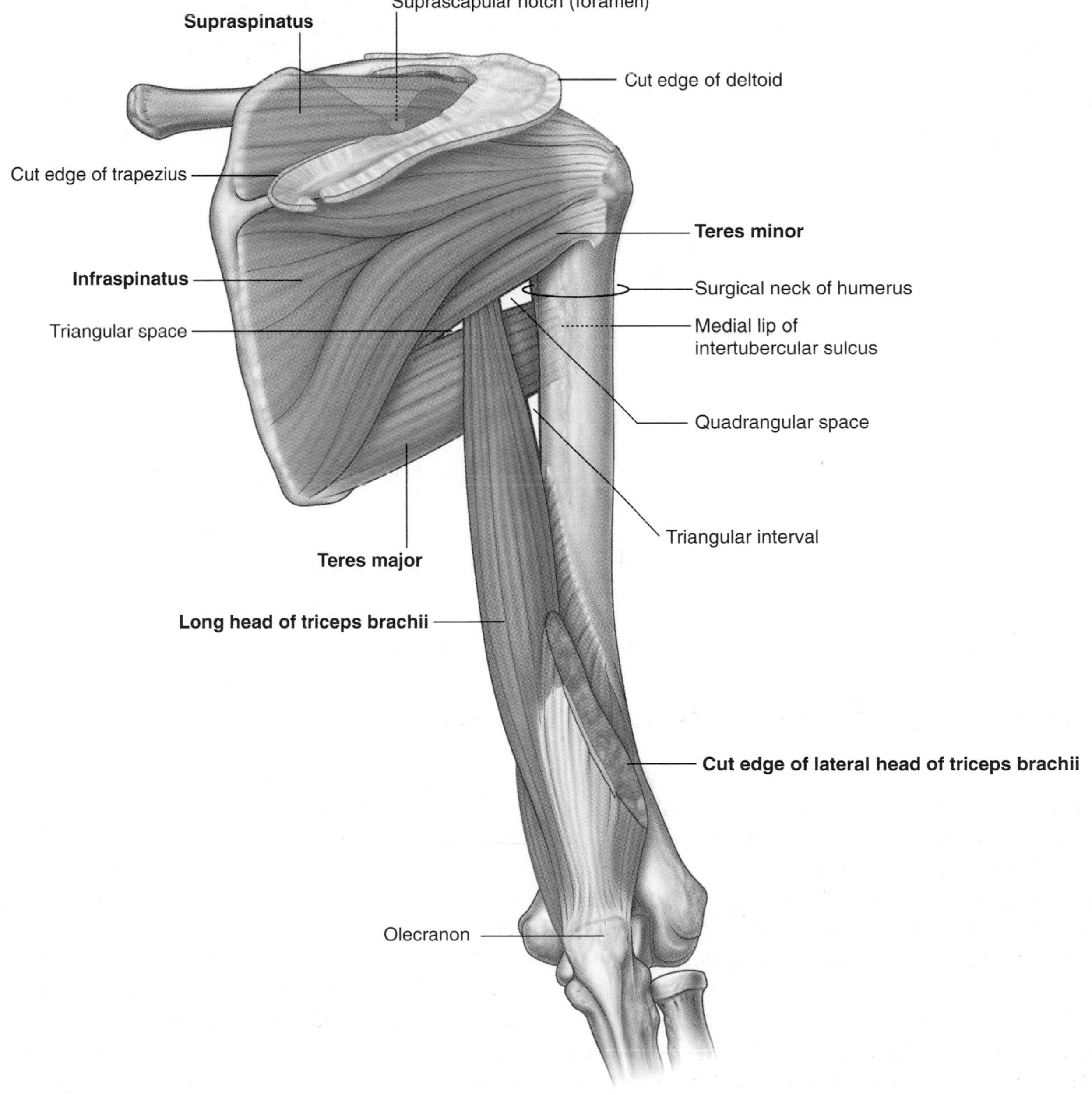


**Fig. 7.37** Right posterior scapular region.

**Table 7.2** Muscles of the posterior scapular region (spinal segments in bold are the major segments innervating the muscle)

| Muscle | Origin | Insertion | Innervation | Function |
|---|---|---|---|---|
| Supraspinatus | Medial two-thirds of the supraspinous fossa of the scapula and the deep fascia that covers the muscle | Most superior facet on the greater tubercle of the humerus | Suprascapular nerve (**C5**, C6) | **Rotator cuff muscle**; participates in abduction of the glenohumeral joint; stabilization of glenohumeral joint |
| Infraspinatus | Medial two-thirds of the infraspinous fossa of the scapula and the deep fascia that covers the muscle | Middle facet on posterior surface of the greater tubercle of the humerus | Suprascapular nerve (**C5**, C6) | **Rotator cuff muscle**; lateral rotation of arm at the glenohumeral joint; stabilization of glenohumeral joint |
| Teres minor | Upper two-thirds of a flattened strip of bone on the posterior surface of the scapula immediately adjacent to the lateral border of the scapula | Inferior facet on the posterior surface of the greater tubercle of the humerus | Axillary nerve (**C5**, C6) | **Rotator cuff muscle**; lateral rotation of arm at the glenohumeral joint; stabilization of glenohumeral joint |
| Teres major | Elongate oval area on the posterior surface of the inferior angle of the scapula | Medial lip of the intertubercular sulcus on the anterior surface of the humerus | Inferior subscapular nerve (**C5**, **C6**, **C7**) | Medial rotation and extension of the arm at the glenohumeral joint; stabilization of glenohumeral joint |
| Long head of triceps brachii | Infraglenoid tubercle on scapula | Common tendon of insertion with medial and lateral heads on the olecranon process of ulna | Radial nerve (C6, **C7**, C8) | Extension of the forearm at the elbow joint; accessory adductor and extensor of the arm at the glenohumeral joint |

## Muscles

### Supraspinatus and infraspinatus

The **supraspinatus** and **infraspinatus** muscles originate from two large fossae, one above and one below the spine, on the posterior surface of the scapula (Fig. 7.37). They form tendons that insert on the greater tubercle of the humerus.

- The tendon of the supraspinatus passes under the acromion, where it is separated from the bone by a subacromial bursa, passes over the glenohumeral joint, and inserts on the superior facet of the greater tubercle.
- The tendon of the infraspinatus passes posteriorly to the glenohumeral joint and inserts on the middle facet of the greater tubercle.

The supraspinatus participates in abduction of the arm. The infraspinatus laterally rotates the humerus.

### Teres minor and teres major

The **teres minor** muscle is a cord-like muscle that originates from a flattened area of the scapula immediately adjacent to its lateral border below the infraglenoid tubercle (Fig. 7.37). Its tendon inserts on the inferior facet of the greater tubercle of the humerus. The teres minor laterally rotates the humerus and is a component of the rotator cuff.

The **teres major** muscle originates from a large oval region on the posterior surface of the inferior angle of the scapula (Fig. 7.37). This broad cord-like muscle passes superiorly and laterally and ends as a flat tendon that attaches to the medial lip of the intertubercular sulcus on the anterior surface of the humerus. The teres major medially rotates and extends the humerus.

### Long head of triceps brachii

The **long head of the triceps brachii** muscle originates from the infraglenoid tubercle and passes somewhat vertically down the arm to insert, with the medial and lateral heads of this muscle, on the olecranon of the ulna (Fig. 7.37).

The triceps brachii is the primary extensor of the forearm at the elbow joint. Because the long head crosses the glenohumeral joint, it can also extend and adduct the humerus.

The importance of the triceps brachii in the posterior scapular region is that its vertical course between the teres minor and teres major, together with these muscles and the humerus, forms spaces through which nerves and vessels pass between regions.

## Gateways to the posterior scapular region

### Suprascapular foramen

The suprascapular foramen is the route through which structures pass between the base of the neck and the posterior scapular region (Fig. 7.37). It is formed by the suprascapular notch of the scapula and the superior transverse scapular (suprascapular) ligament, which converts the notch into a foramen.

The suprascapular nerve passes through the suprascapular foramen; the suprascapular artery and the suprascapular vein follow the same course as the nerve, but normally pass immediately superior to the superior transverse scapular ligament and not through the foramen (Fig. 7.38).

## Quadrangular space (from posterior)

The quadrangular space provides a passageway for nerves and vessels passing between more anterior regions (the axilla) and the posterior scapular region (Fig. 7.37). In the posterior scapular region, its boundaries are formed by:

- the inferior margin of the teres minor,
- the surgical neck of the humerus,
- the superior margin of the teres major, and
- the lateral margin of the long head of the triceps brachii.

The axillary nerve and the posterior circumflex humeral artery and vein pass through this space (Fig. 7.38).

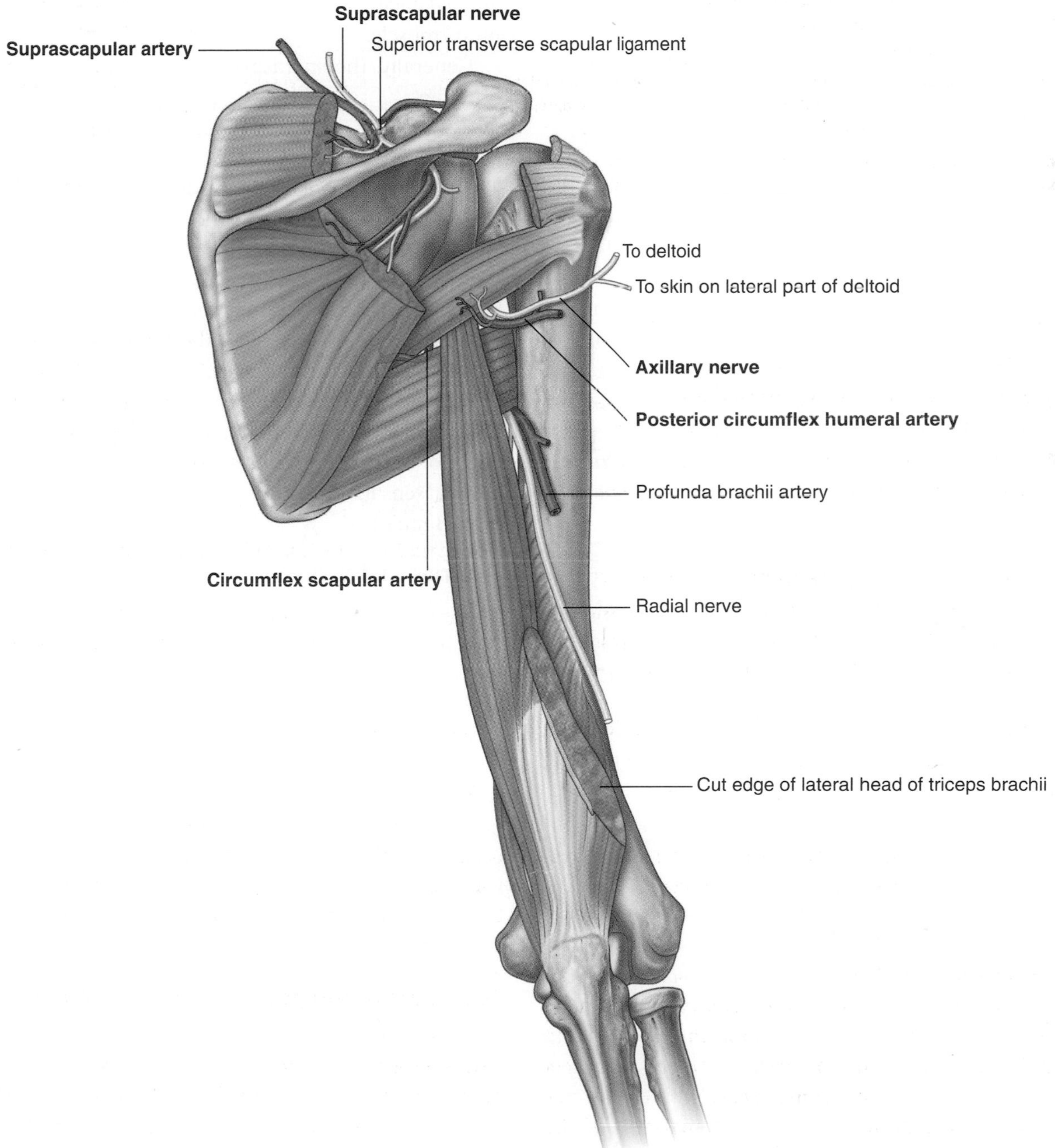


**Fig. 7.38** Arteries and nerves associated with gateways in the posterior scapular region.

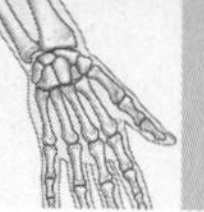



### Triangular space

The triangular space is an area of communication between the axilla and the posterior scapular region (Fig. 7.37). When viewed from the posterior scapular region, the triangular space is formed by:

- the medial margin of the long head of the triceps brachii,
- the superior margin of the teres major, and
- the inferior margin of the teres minor.

The circumflex scapular artery and vein pass through this gap (Fig. 7.38).

### Triangular interval

The triangular interval is formed by:

- the lateral margin of the long head of the triceps brachii,
- the shaft of the humerus, and
- the inferior margin of the teres major (Fig. 7.37).

Because this space is below the inferior margin of the teres major, which defines the inferior boundary of the axilla, the triangular interval serves as a passageway between the anterior and posterior compartments of the arm and between the posterior compartment of the arm and the axilla. The radial nerve, the **profunda brachii artery** (**deep artery of arm**), and associated veins pass through it (Fig. 7.38).

## Nerves

The two major nerves of the posterior scapular region are the suprascapular and axillary nerves, both of which originate from the brachial plexus in the axilla (Fig. 7.38).

### Suprascapular nerve

The **suprascapular nerve** originates in the base of the neck from the superior trunk of the brachial plexus. It passes posterolaterally from its origin, through the suprascapular foramen to reach the posterior scapular region, where it lies in the plane between bone and muscle (Fig. 7.38).

It innervates the supraspinatus muscle and then passes through the greater scapular (spinoglenoid) notch, between the root of the spine of the scapula and the glenoid cavity, to terminate in and innervate the infraspinatus muscle.

Generally, the suprascapular nerve has no cutaneous branches.

### Axillary nerve

The **axillary nerve** originates from the posterior cord of the brachial plexus. It exits the axilla by passing through the quadrangular space in the posterior wall of the axilla, and enters the posterior scapular region (Fig. 7.38). Together with the posterior circumflex humeral artery and vein, it is directly related to the posterior surface of the surgical neck of the humerus.

The axillary nerve innervates the deltoid and teres minor muscles. In addition, it has a cutaneous branch, the superior lateral cutaneous nerve of the arm, which carries general sensation from the skin over the inferior part of the deltoid muscle.

## Arteries and veins

Three major arteries are found in the posterior scapular region: the suprascapular, posterior circumflex humeral, and circumflex scapular arteries. These arteries contribute to an interconnected vascular network around the scapula (Fig. 7.39).

## In the clinic

### Quadrangular space syndrome

Hypertrophy of the quadrangular space muscles or fibrosis of the muscle edges may impinge on the axillary nerve. Uncommonly, this produces weakness of the deltoid muscle. Typically it produces atrophy of the teres minor muscle, which may affect the control that the rotator cuff muscles exert on shoulder movement.

### Suprascapular artery

The **suprascapular artery** originates in the base of the neck as a branch of the thyrocervical trunk, which, in turn, is a major branch of the subclavian artery (Figs. 7.38 and 7.39). The vessel may also originate directly from the third part of the subclavian artery.

The suprascapular artery normally enters the posterior scapular region superior to the suprascapular foramen, whereas the nerve passes through the foramen. In the posterior scapular region, the vessel runs with the suprascapular nerve.

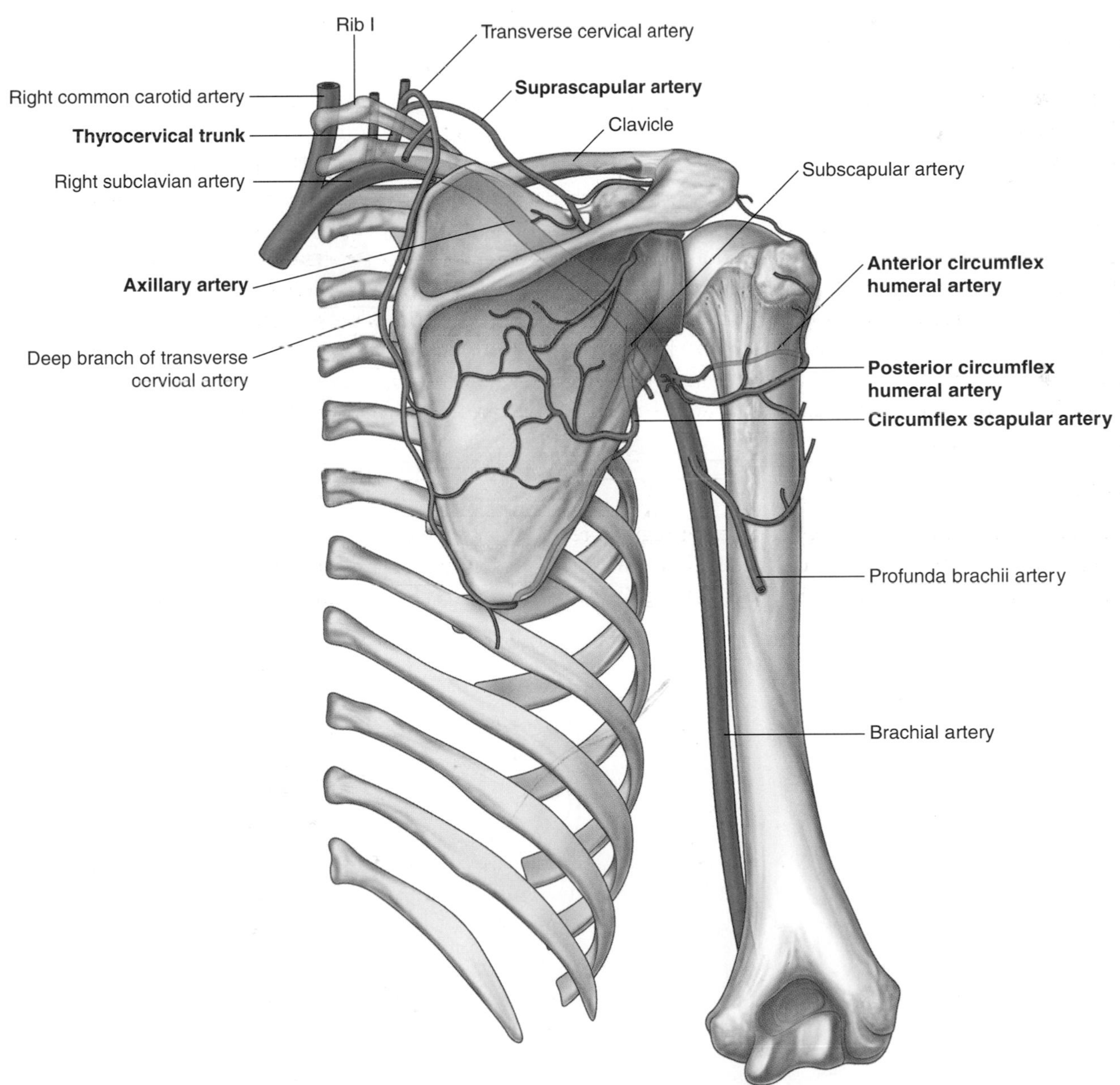


**Fig. 7.39** Arterial anastomoses around the shoulder.

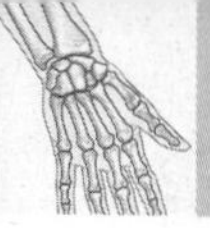



In addition to supplying the supraspinatus and infraspinatus muscles, the suprascapular artery contributes branches to numerous structures along its course.

### Posterior circumflex humeral artery

The **posterior circumflex humeral artery** originates from the third part of the axillary artery in the axilla (Figs. 7.38 and 7.39).

The posterior circumflex humeral artery and axillary nerve leave the axilla through the quadrangular space in the posterior wall and enter the posterior scapular region. The vessel supplies the related muscles and the glenohumeral joint.

### Circumflex scapular artery

The **circumflex scapular artery** is a branch of the subscapular artery that also originates from the third part of the axillary artery in the axilla (Figs. 7.38 and 7.39). The circumflex scapular artery leaves the axilla through the triangular space and enters the posterior scapular region, passes through the origin of the teres minor muscle, and forms anastomotic connections with other arteries in the region.

### Veins

Veins in the posterior scapular region generally follow the arteries and connect with vessels in the neck, back, arm, and axilla.

## AXILLA

The axilla is the gateway to the upper limb, providing an area of transition between the neck and the arm (Fig. 7.40A). Formed by the clavicle, the scapula, the upper

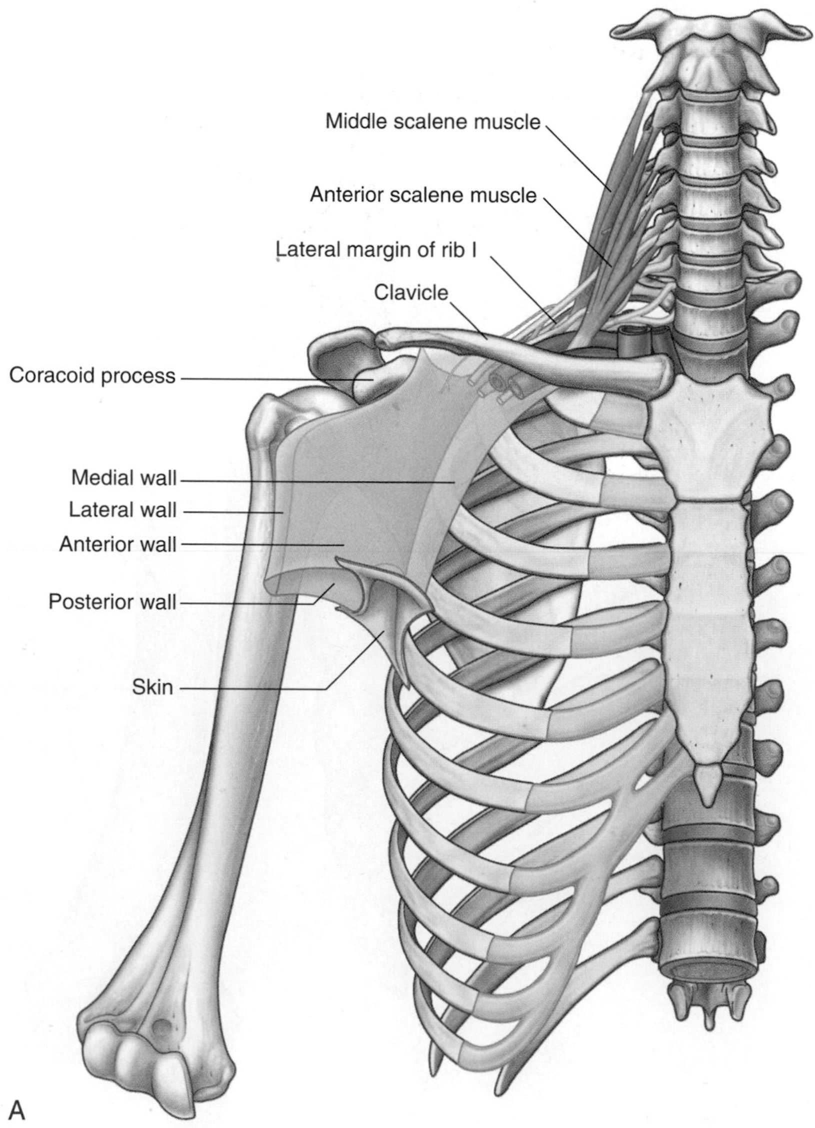


**Fig. 7.40** Axilla. **A.** Walls and transition between neck and arm.

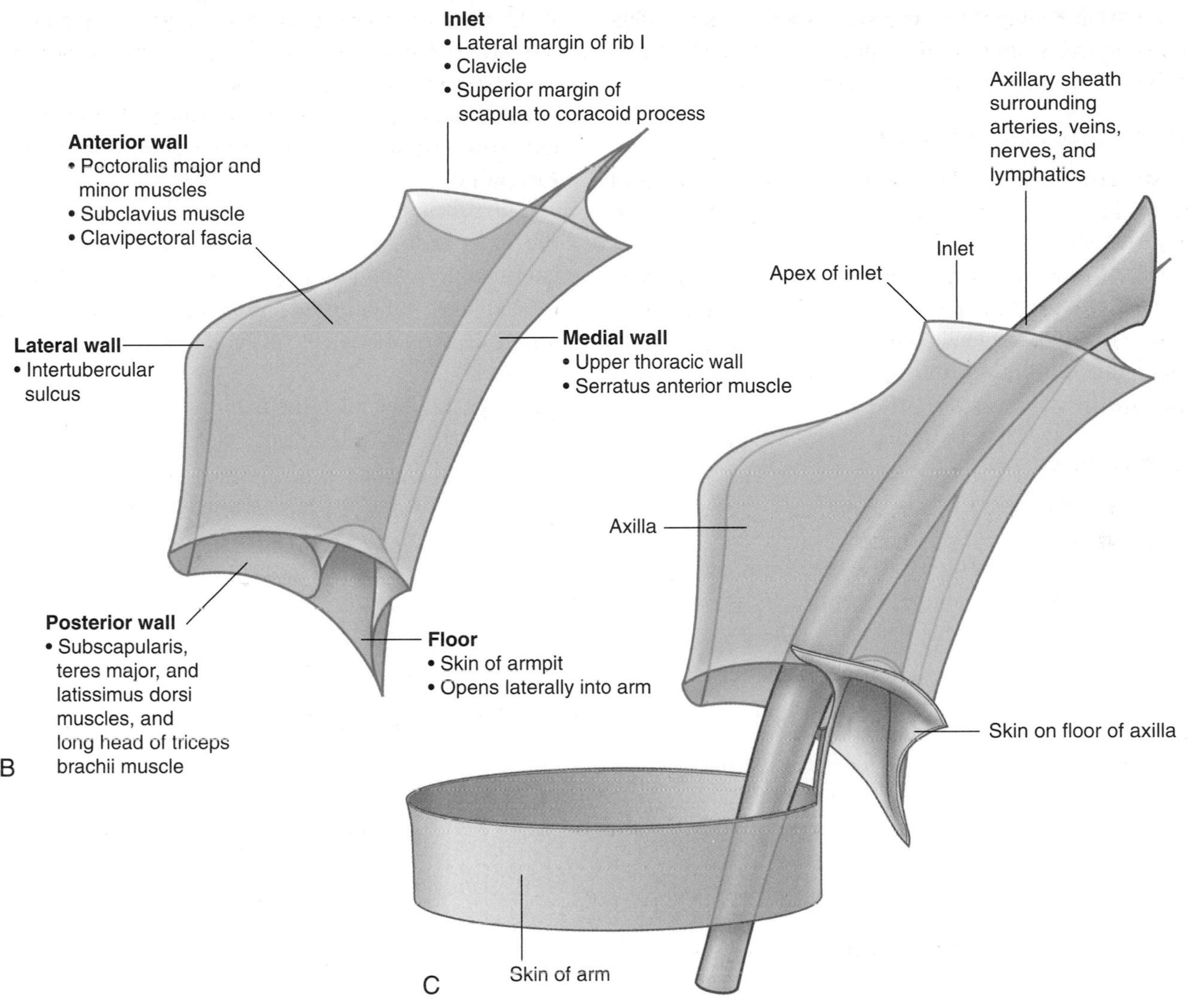
Inlet
• Lateral margin of rib I
• Clavicle
• Superior margin of scapula to coracoid process
Anterior wall
• Pectoralis major and minor muscles
• Subclavius muscle
• Clavipectoral fascia
Lateral wall
• Intertubercular sulcus
Medial wall
• Upper thoracic wall
• Serratus anterior muscle
Posterior wall
• Subscapularis, teres major, and latissimus dorsi muscles, and long head of triceps brachii muscle
Floor
• Skin of armpit
• Opens laterally into arm
B
Axillary sheath surrounding arteries, veins, nerves, and lymphatics
Inlet
Apex of inlet
Axilla
Skin on floor of axilla
Skin of arm
C

**Fig. 7.40, cont'd** Axilla. **B.** Boundaries. **C.** Continuity with the arm.

thoracic wall, the humerus, and related muscles, the axilla is an irregularly shaped pyramidal space with:

- four sides,
- an inlet, and
- a floor (base) (Fig. 7.40A,B).

The axillary inlet is continuous superiorly with the neck, and the lateral part of the floor opens into the arm.

All major structures passing into and out of the upper limb pass through the axilla (Fig. 7.40C). Apertures formed between muscles in the anterior and posterior walls enable structures to pass between the axilla and immediately adjacent regions (the posterior scapular, pectoral, and deltoid regions).

## Axillary inlet

The axillary inlet is oriented in the horizontal plane and is somewhat triangular in shape, with its apex directed laterally (Fig. 7.40A,B). The margins of the inlet are completely formed by bone:

- The medial margin is the lateral border of rib I.
- The anterior margin is the posterior surface of the clavicle.
- The posterior margin is the superior border of the scapula up to the coracoid process.

The apex of the triangularly shaped axillary inlet is lateral in position and is formed by the medial aspect of the coracoid process.

Major vessels and nerves pass between the neck and the axilla by crossing over the lateral border of rib I and through the axillary inlet (Fig. 7.40A).

The subclavian artery, the major blood vessel supplying the upper limb, becomes the axillary artery as it crosses the lateral margin of rib I and enters the axilla. Similarly, the axillary vein becomes the subclavian vein as it passes over the lateral margin of rib I and leaves the axilla to enter the neck.

At the axillary inlet, the axillary vein is anterior to the axillary artery, which, in turn, is anterior to the trunks of the brachial plexus.

The inferior trunk (lower trunk) of the brachial plexus lies directly on rib I in the neck, as does the subclavian artery and vein. As they pass over rib I, the vein and artery are separated by the insertion of the anterior scalene muscle (Fig. 7.40A).

## Anterior wall

The anterior wall of the axilla is formed by the lateral part of the pectoralis major muscle, the underlying pectoralis minor and subclavius muscles, and the clavipectoral fascia (Table 7.3).

**Table 7.3** Muscles of the anterior wall of the axilla (spinal segments in bold are the major segments innervating the muscle)

| Muscle | Origin | Insertion | Innervation | Function |
|---|---|---|---|---|
| Pectoralis major | Clavicular head—anterior surface of medial half of clavicle; sternocostal head—anterior surface of sternum; first seven costal cartilages; sternal end of sixth rib; aponeurosis of external oblique | Lateral lip of intertubercular sulcus of humerus | Medial and lateral pectoral nerves; clavicular head (**C5**, C6); sternocostal head (C6, **C7**, C8, T1) | Flexion, adduction, and medial rotation of arm at glenohumeral joint; clavicular head—flexion of extended arm; sternocostal head—extension of flexed arm |
| Subclavius | First rib at junction between rib and costal cartilage | Groove on inferior surface of middle one-third of clavicle | Nerve to subclavius (**C5, C6**) | Pulls tip of shoulder down; pulls clavicle medially to stabilize sternoclavicular joint |
| Pectoralis minor | Anterior surfaces and superior borders of ribs III to V; and from deep fascia overlying the related intercostal spaces | Coracoid process of scapula (medial border and upper surface) | Medial pectoral nerve (C5, C6, **C7**, **C8**, T1) | Pulls tip of shoulder down; protracts scapula |

### Pectoralis major

The **pectoralis major** muscle is the largest and most superficial muscle of the anterior wall (Fig. 7.41). Its inferior margin underlies the anterior axillary fold, which marks the anteroinferior border of the axilla. The muscle has two heads:

- The clavicular head originates from the medial half of the clavicle.
- The sternocostal head originates from the medial part of the anterior thoracic wall—often, fibers from this head continue inferiorly and medially to attach to the anterior abdominal wall, forming an additional abdominal part of the muscle.

The muscle inserts into the lateral lip of the intertubercular sulcus of the humerus. The parts of the muscle that have a superior origin on the trunk insert lower and more anteriorly on the lateral lip of the intertubercular sulcus than the parts of the muscle that originate inferiorly.

Acting together, the two heads of the pectoralis major flex, adduct, and medially rotate the arm at the glenohumeral joint. The clavicular head flexes the arm from an extended position, whereas the sternocostal head extends the arm from a flexed position, particularly against resistance.

The pectoralis major is innervated by the lateral and medial pectoral nerves, which originate from the brachial plexus in the axilla.

### Subclavius

The **subclavius** muscle is a small muscle that lies deep to the pectoralis major muscle and passes between the clavicle and rib I (Fig. 7.42). It originates medially, as a tendon, from rib I at the junction between the rib and its costal cartilage. It passes laterally and superiorly to insert via a muscular attachment into an elongate shallow groove on the inferior surface of the middle third of the clavicle.

The function of the subclavius is not entirely clear, but it may act to pull the shoulder down by depressing the

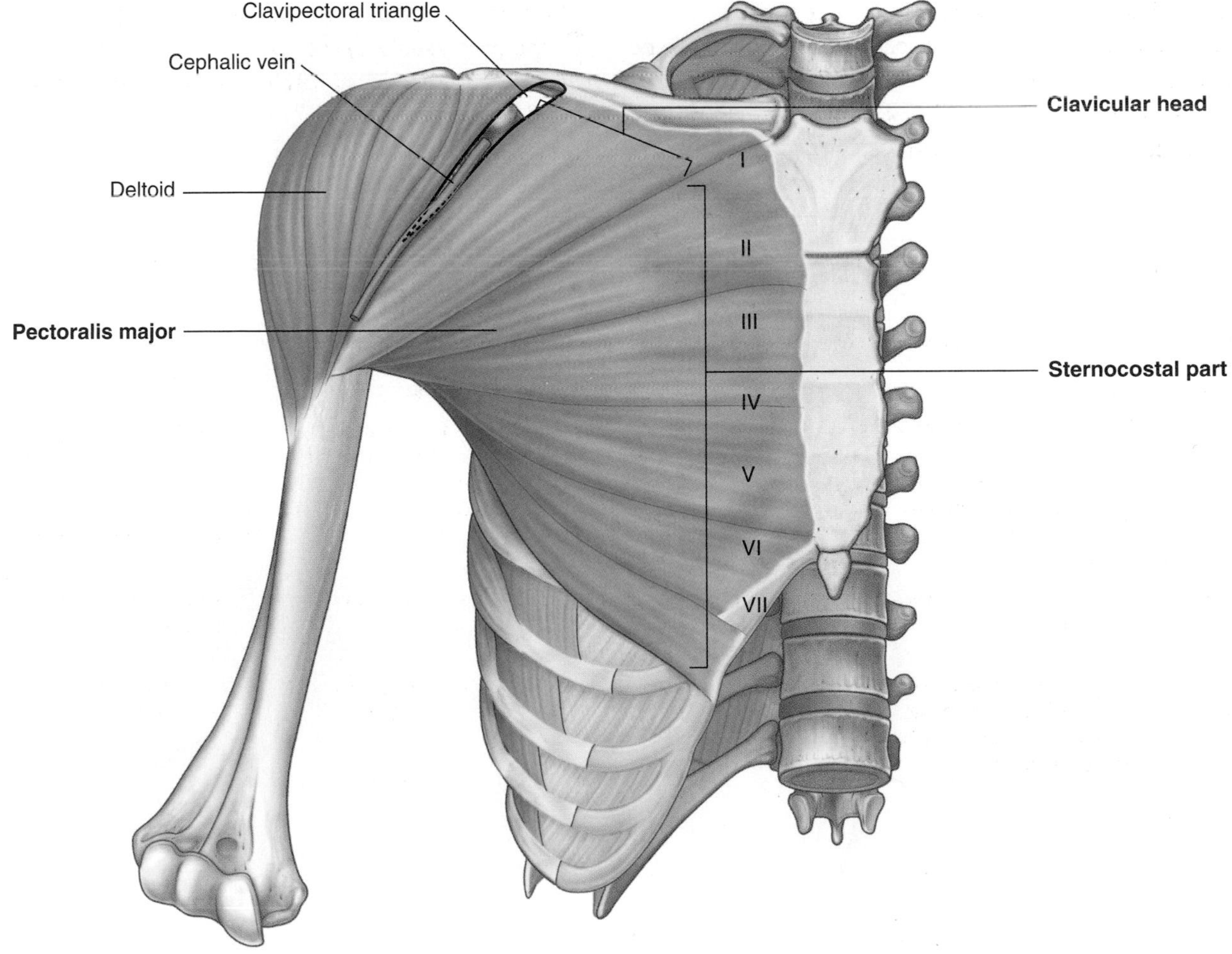


**Fig. 7.41** Pectoralis major muscle.

clavicle and may also stabilize the sternoclavicular joint by pulling the clavicle medially.

The subclavius muscle is innervated by a small branch from the superior trunk of the brachial plexus.

### Pectoralis minor

The **pectoralis minor** muscle is a small triangular-shaped muscle that lies deep to the pectoralis major muscle and passes from the thoracic wall to the coracoid process of the scapula (Fig. 7.42). It originates as three muscular slips from the anterior surfaces and upper margins of ribs III to V and from the fascia overlying muscles of the related intercostal spaces. The muscle fibers pass superiorly and laterally to insert into the medial and upper aspects of the coracoid process.

The pectoralis minor muscle protracts the scapula (by pulling the scapula anteriorly on the thoracic wall) and depresses the lateral angle of the scapula.

The pectoralis minor is innervated by the medial pectoral nerve, which originates from the brachial plexus in the axilla.

### Clavipectoral fascia

The clavipectoral fascia is a thick sheet of connective tissue that connects the clavicle to the floor of the axilla (Fig. 7.42). It encloses the subclavius and pectoralis minor muscles and spans the gap between them.

Structures travel between the axilla and the anterior wall of the axilla by passing through the clavipectoral

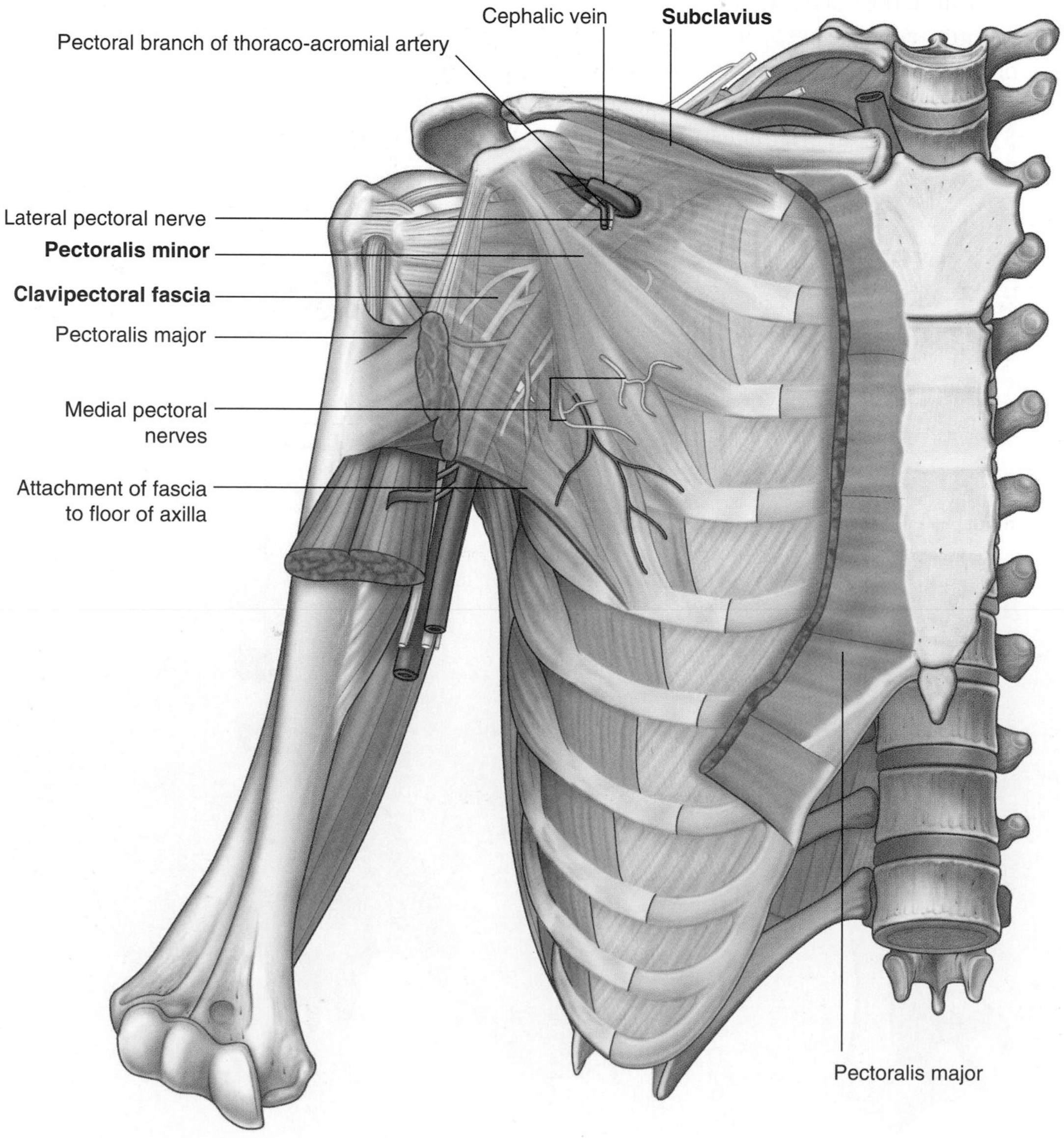

**Fig. 7.42** Pectoralis minor and subclavius muscles and clavipectoral fascia.

fascia either between the pectoralis minor and subclavius muscles or inferior to the pectoralis minor muscle.

Important structures that pass between the subclavius and pectoralis minor muscles include the cephalic vein, the thoraco-acromial artery, and the lateral pectoral nerve.

The lateral thoracic artery leaves the axilla by passing through the fascia inferior to the pectoralis minor muscle.

The medial pectoral nerve leaves the axilla by penetrating directly through the pectoralis minor muscle to supply this muscle and to reach the pectoralis major muscle. Occasionally, branches of the medial pectoral nerve pass around the lower margin of the pectoralis minor to reach and innervate the overlying pectoralis major muscle.

## Medial wall

The medial wall of the axilla consists of the upper thoracic wall (the ribs and related intercostal tissues) and the serratus anterior muscle (Fig. 7.43 and Table 7.4, and see Fig. 7.40).

### Serratus anterior

The **serratus anterior** muscle originates as a number of muscular slips from the lateral surfaces of ribs I to IX and the intervening deep fascia overlying the related intercostal spaces (Fig. 7.43). The muscle forms a flattened sheet, which passes posteriorly around the thoracic wall to insert

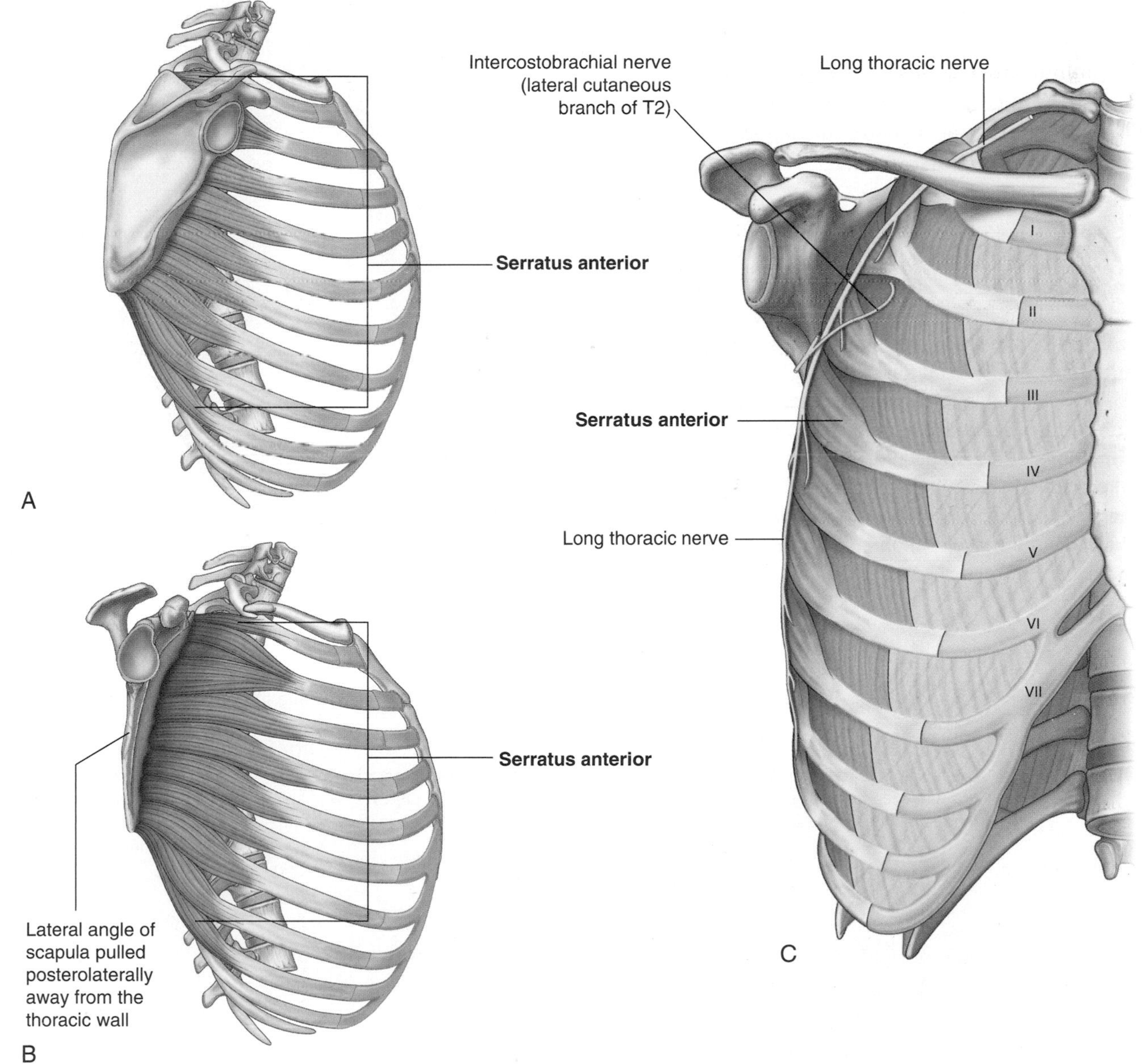


**Fig. 7.43** Medial wall of the axilla. **A.** Lateral view. **B.** Lateral view with lateral angle of scapula retracted posteriorly. **C.** Anterior view.

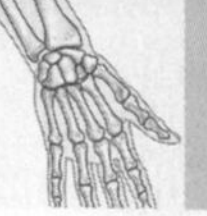

primarily on the costal surface of the medial border of the scapula.

The serratus anterior pulls the scapula forward over the thoracic wall and facilitates scapular rotation. It also keeps the costal surface of the scapula closely opposed to the thoracic wall.

The serratus anterior is innervated by the long thoracic nerve, which is derived from the roots of the brachial plexus, passes through the axilla along the medial wall, and passes vertically down the serratus anterior muscle on its external surface, just deep to skin and superficial fascia.

### Intercostobrachial nerve

The only major structure that passes directly through the medial wall and into the axilla is the intercostobrachial nerve (Fig. 7.43). This nerve is the lateral cutaneous branch of the second intercostal nerve (anterior ramus of T2). It communicates with a branch of the brachial plexus (the medial cutaneous nerve of the arm) in the axilla and supplies skin on the upper posteromedial side of the arm, which is part of the T2 dermatome.

## In the clinic

**"Winging" of the scapula**

Because the long thoracic nerve passes down the lateral thoracic wall on the external surface of the serratus anterior muscle, just deep to skin and subcutaneous fascia, it is vulnerable to damage. Loss of function of this muscle causes the medial border, and particularly the inferior angle, of the scapula to elevate away from the thoracic wall, resulting in characteristic "winging" of the scapula, on pushing forward with the arm. Furthermore, normal elevation at the arm is no longer possible.

## Lateral wall

The lateral wall of the axilla is narrow and formed entirely by the intertubercular sulcus of the humerus (Fig. 7.44). The pectoralis major muscle of the anterior wall attaches to the lateral lip of the intertubercular sulcus. The latissimus dorsi and teres major muscles of the posterior wall

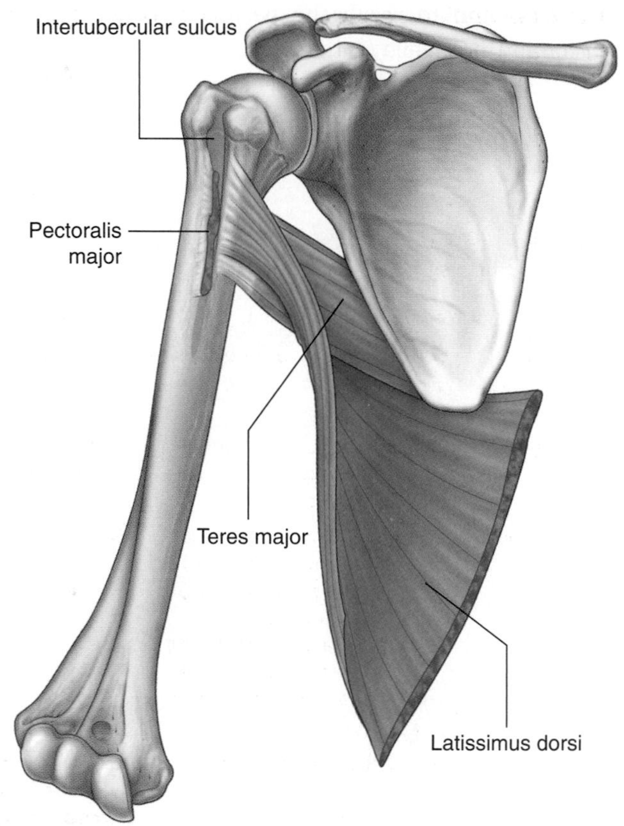


**Fig. 7.44** Lateral wall of the axilla.

**Table 7.4** Muscle of the medial wall of the axilla (spinal segment in bold is the major segment innervating the muscle)

| Muscle | Origin | Insertion | Innervation | Function |
|---|---|---|---|---|
| Serratus anterior | Lateral surfaces of upper 8–9 ribs and deep fascia overlying the related intercostal spaces | Costal surface of medial border of scapula | Long thoracic nerve (**C5**, C6, C7) | Protraction and rotation of the scapula; keeps medial border and inferior angle of scapula opposed to thoracic wall |

attach to the floor and medial lip of the intertubercular sulcus, respectively (Table 7.5).

## Posterior wall

The posterior wall of the axilla is complex (Fig. 7.45 and see Fig. 7.50). Its bone framework is formed by the costal surface of the scapula. Muscles of the wall are:

- the subscapularis muscle (associated with the costal surface of the scapula),
- the distal parts of the latissimus dorsi and teres major muscles (which pass into the wall from the back and posterior scapular region), and
- the proximal part of the long head of the triceps brachii muscle (which passes vertically down the wall and into the arm).

Gaps between the muscles of the posterior wall form apertures through which structures pass between the axilla, posterior scapular region, and posterior compartment of the arm.

**Table 7.5** Muscles of the lateral and posterior wall of the axilla (spinal segments in bold are the major segments innervating the muscle; spinal segments in parentheses do not consistently innervate the muscle)

| Muscle | Origin | Insertion | Innervation | Function |
|---|---|---|---|---|
| Subscapularis | Medial two-thirds of subscapular fossa | Lesser tubercle of humerus | Upper and lower subscapular nerves (C5, **C6**, (C7)) | **Rotator cuff muscle**; medial rotation of the arm at the glenohumeral joint |
| Teres major | Elongate oval area on the posterior surface of the inferior angle of the scapula | Medial lip of the intertubercular sulcus on the anterior surface of the humerus | Lower subscapular nerve (**C5, C6, C7**) | Medial rotation and extension of the arm at the glenohumeral joint |
| Latissimus dorsi | Spinous processes of lower six thoracic vertebrae and related interspinous ligaments; via the thoracolumbar fascia to the spinous processes of the lumbar vertebrae, related interspinous ligaments, and iliac crest; lower 3–4 ribs | Floor of intertubercular sulcus | Thoracodorsal nerve (C6, **C7**, C8) | Adduction, medial rotation, and extension of the arm at the glenohumeral joint |
| Long head of triceps brachii | Infraglenoid tubercle on scapula | Common tendon of insertion with medial and lateral heads on the olecranon process of ulna | Radial nerve (C6, **C7**, C8) | Extension of the forearm at the elbow joint; accessory adductor and extensor of the arm at the glenohumeral joint |

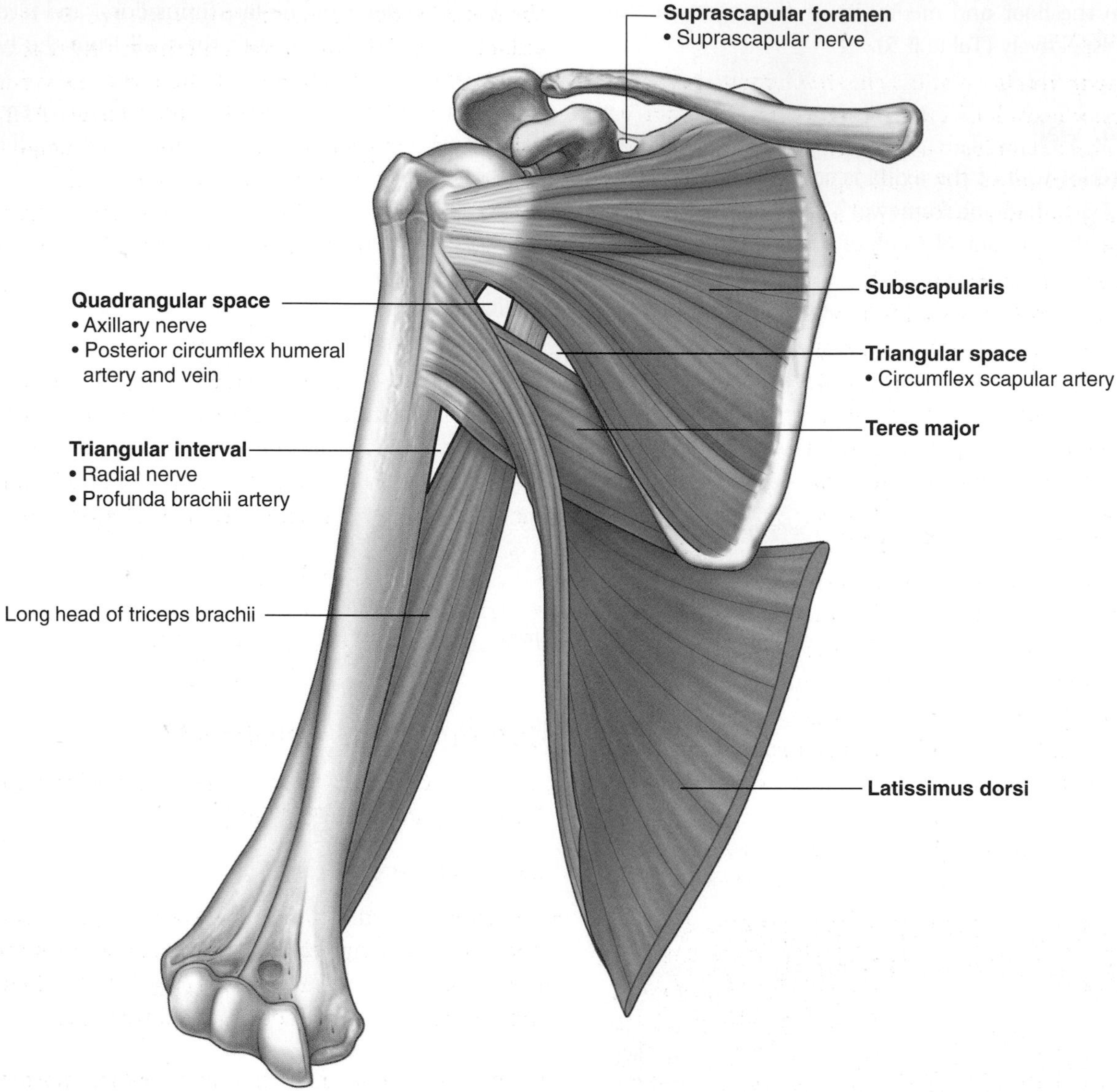

**Fig. 7.45** Posterior wall of the axilla.

### Subscapularis

The **subscapularis** muscle forms the largest component of the posterior wall of the axilla. It originates from, and fills, the subscapular fossa and inserts on the lesser tubercle of the humerus (Figs. 7.45 and 7.46). The tendon crosses immediately anterior to the joint capsule of the glenohumeral joint.

Together with three muscles of the posterior scapular region (the supraspinatus, infraspinatus, and teres minor muscles), the subscapularis is a member of the rotator cuff muscle group, which stabilizes the glenohumeral joint.

The subscapularis is innervated by branches of the brachial plexus (the **superior** and **inferior subscapular nerves**), which originate in the axilla.

### Teres major and latissimus dorsi

The inferolateral aspect of the posterior wall of the axilla is formed by the terminal part of the **teres major** muscle and the tendon of the **latissimus dorsi** muscle (Fig. 7.45). These two structures lie under the posterior axillary fold, which marks the posteroinferior border of the axilla.

The flat tendon of the latissimus dorsi muscle curves around the inferior margin of the teres major muscle on the posterior wall to insert into the floor of the intertubercular sulcus of the humerus, anterior to and slightly above the most distal attachment of the teres major muscle to the medial lip of the intertubercular sulcus. As a consequence, the inferior margin of the teres major muscle defines the inferior limit of the axilla laterally.

The axillary artery becomes the brachial artery of the arm as it crosses the inferior margin of the teres major muscle.

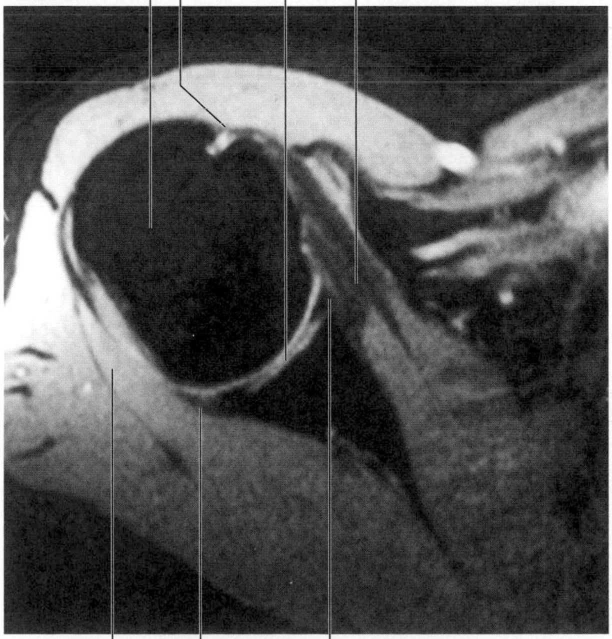

**Fig. 7.46** Magnetic resonance image of the glenohumeral joint in the transverse or horizontal plane.

### Long head of the triceps brachii

The **long head of the triceps brachii** muscle passes vertically through the posterior wall of the axilla, and, together with surrounding muscles and adjacent bones, results in the formation of three apertures through which major structures pass through the posterior wall:

- the quadrangular space,
- the triangular space, and
- the triangular interval (Fig. 7.45).

## Gateways in the posterior wall

(See also "Gateways to the posterior scapular region," pp. 706–710, and Figs. 7.37 and 7.38.)

### Quadrangular space

The quadrangular space provides a passageway for nerves and vessels passing between the axilla and the more posterior scapular and deltoid regions (Fig. 7.45). When viewed from anteriorly, its boundaries are formed by:

- the inferior margin of the subscapularis muscle,
- the surgical neck of the humerus,
- the superior margin of the teres major muscle, and
- the lateral margin of the long head of the triceps brachii muscle.

Passing through the quadrangular space are the axillary nerve and the posterior circumflex humeral artery and vein.

### Triangular space

The **triangular space** is an area of communication between the axilla and the posterior scapular region (Fig. 7.45). When viewed from anteriorly, it is formed by:

- the medial margin of the long head of the triceps brachii muscle,
- the superior margin of the teres major muscle, and
- the inferior margin of the subscapularis muscle.

The circumflex scapular artery and vein pass into this space.

### Triangular interval

This triangular interval is formed by:

- the lateral margin of the long head of the triceps brachii muscle,
- the shaft of the humerus, and
- the inferior margin of the teres major muscle (Fig. 7.45).

The radial nerve passes out of the axilla traveling through this interval to reach the posterior compartment of the arm.

## Floor

The floor of the axilla is formed by fascia and a dome of skin that spans the distance between the inferior margins of the walls (Fig. 7.47 and see Fig. 7.40B). It is supported by the clavipectoral fascia. On a patient, the anterior axillary fold is more superior in position than is the posterior axillary fold.

Inferiorly, structures pass into and out of the axilla immediately lateral to the floor where the anterior and posterior walls of the axilla converge and where the axilla is continuous with the anterior compartment of the arm.

## Contents of the axilla

Passing through the axilla are the major vessels, nerves, and lymphatics of the upper limb. The space also contains the proximal parts of two muscles of the arm, the axillary process of the breast, and collections of lymph nodes, which drain the upper limb, chest wall, and breast.

The proximal parts of the biceps brachii and coracobrachialis muscles pass through the axilla (Table 7.6).

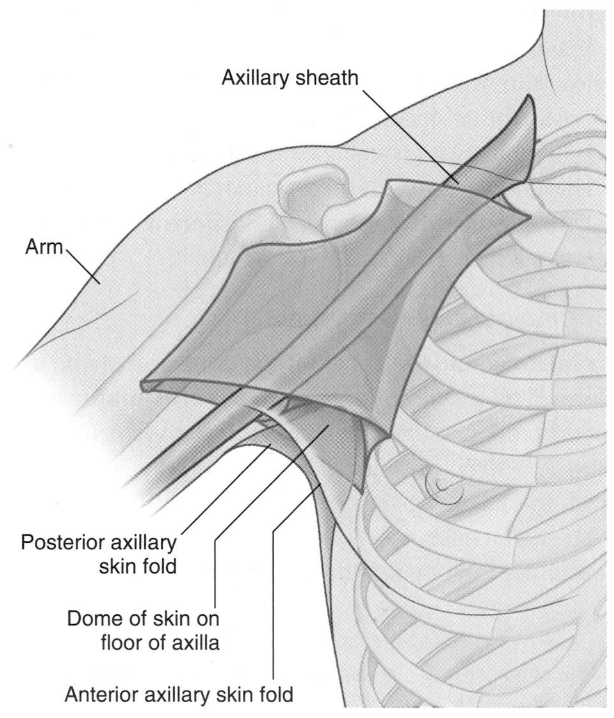


**Fig. 7.47** Floor of the axilla.

**Table 7.6** Muscles having parts that pass through the axilla (spinal segments in bold are the major segments innervating the muscle)

| Muscle | Origin | Insertion | Innervation | Function |
|---|---|---|---|---|
| Biceps brachii | Long head—supraglenoid tubercle of scapula; short head—apex of coracoid process | Tuberosity of radius | Musculocutaneous nerve (**C5, C6**) | Powerful flexor of the forearm at the elbow joint and supinator of the forearm; accessory flexor of the arm at the glenohumeral joint |
| Coracobrachialis | Apex of coracoid process | Linear roughening on midshaft of humerus on medial side | Musculocutaneous nerve (**C5, C6, C7**) | Flexor of the arm at the glenohumeral joint; adducts arm |

### Biceps brachii

The **biceps brachii** muscle originates as two heads (Fig. 7.48):

- The short head originates from the apex of the coracoid process of the scapula and passes vertically through the axilla and into the arm where it joins the long head.
- The long head originates as a tendon from the supraglenoid tubercle of the scapula, passes over the head of the humerus deep to the joint capsule of the glenohumeral joint, and enters the intertubercular sulcus where it is held in position by a ligament, the transverse humeral ligament, which spans the distance between the greater and lesser tubercles; the tendon passes through the axilla in the intertubercular sulcus and forms a muscle belly in the proximal part of the arm.

The long and short heads of the muscle join in distal regions of the arm and primarily insert as a single tendon into the radial tuberosity in the forearm.

The biceps brachii muscle is primarily a powerful flexor of the forearm at the elbow joint and a powerful supinator in the forearm. Because both heads originate from the scapula, the muscle also acts as an accessory flexor of the arm at the glenohumeral joint. In addition, the long head prevents superior movement of the humerus on the glenoid cavity.

The biceps brachii muscle is innervated by the musculocutaneous nerve.

### Coracobrachialis

The **coracobrachialis** muscle, together with the short head of the biceps brachii muscle, originates from the apex of the coracoid process (Fig. 7.48). It passes vertically through the axilla to insert on a small linear roughening on the medial aspect of the humerus, approximately midshaft.

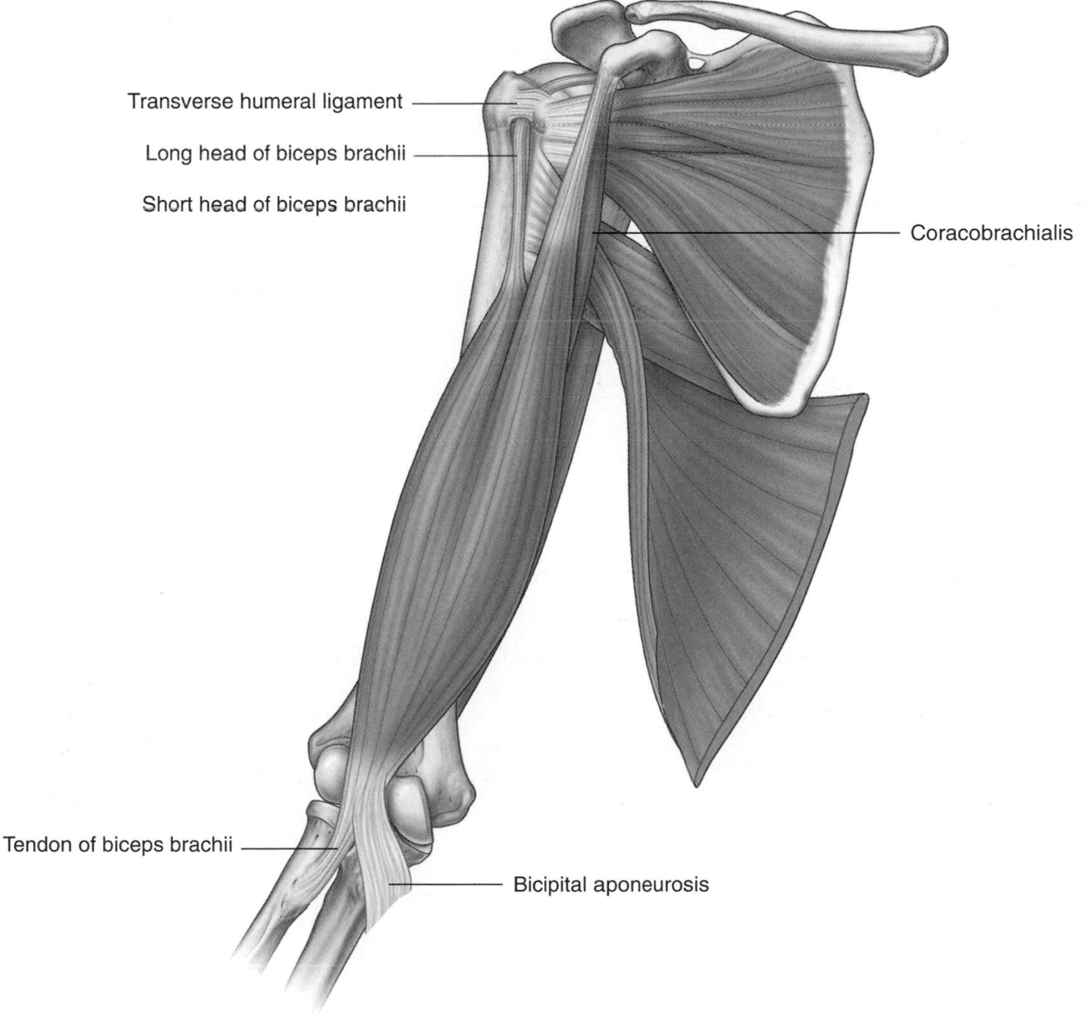


**Fig. 7.48** Contents of the axilla: muscles.

The coracobrachialis muscle flexes the arm at the glenohumeral joint.

In the axilla, the medial surface of the coracobrachialis muscle is pierced by the musculocutaneous nerve, which innervates and then passes through the muscle to enter the arm.

### Axillary artery

The axillary artery supplies the walls of the axilla and related regions, and continues as the major blood supply to the more distal parts of the upper limb (Fig. 7.49).

The subclavian artery in the neck becomes the axillary artery at the lateral margin of rib I and passes through the axilla, becoming the brachial artery at the inferior margin of the teres major muscle.

The axillary artery is separated into three parts by the pectoralis minor muscle, which crosses anteriorly to the vessel (Fig. 7.49):

- The first part is proximal to the pectoralis minor.
- The second part is posterior to the pectoralis minor.
- The third part is distal to the pectoralis minor.

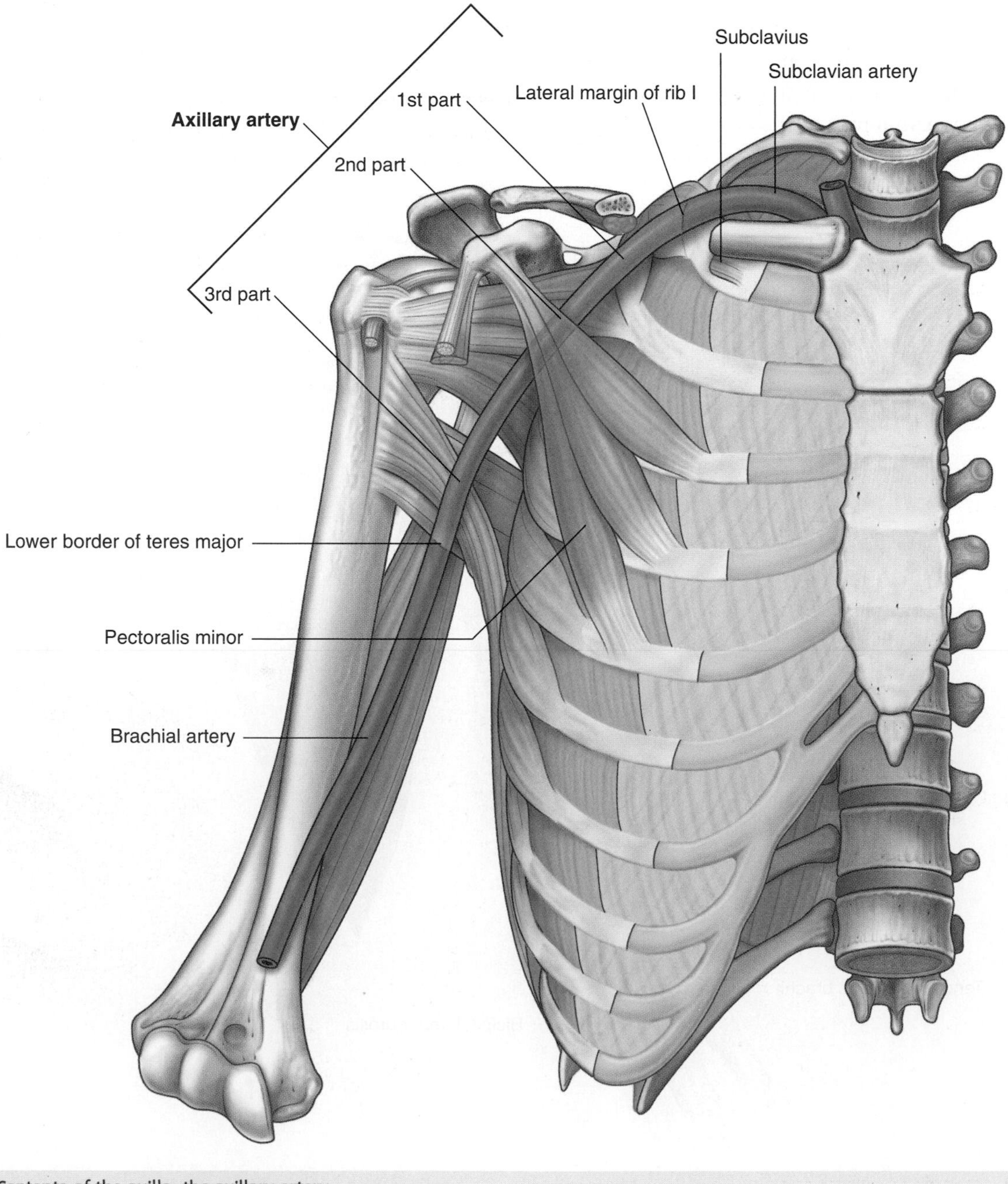

**Fig. 7.49** Contents of the axilla: the axillary artery.

Generally, six branches arise from the axillary artery:

- One branch, the **superior thoracic artery**, originates from the first part.
- Two branches, the **thoraco-acromial artery** and the **lateral thoracic artery**, originate from the second part.
- Three branches, the **subscapular artery**, the **anterior circumflex humeral artery**, and the **posterior circumflex humeral artery**, originate from the third part (Fig. 7.50).

### Superior thoracic artery

The superior thoracic artery is small and originates from the anterior surface of the first part of the axillary artery (Fig. 7.50). It supplies upper regions of the medial and anterior axillary walls.

### Thoraco-acromial artery

The thoraco-acromial artery is short and originates from the anterior surface of the second part of the axillary

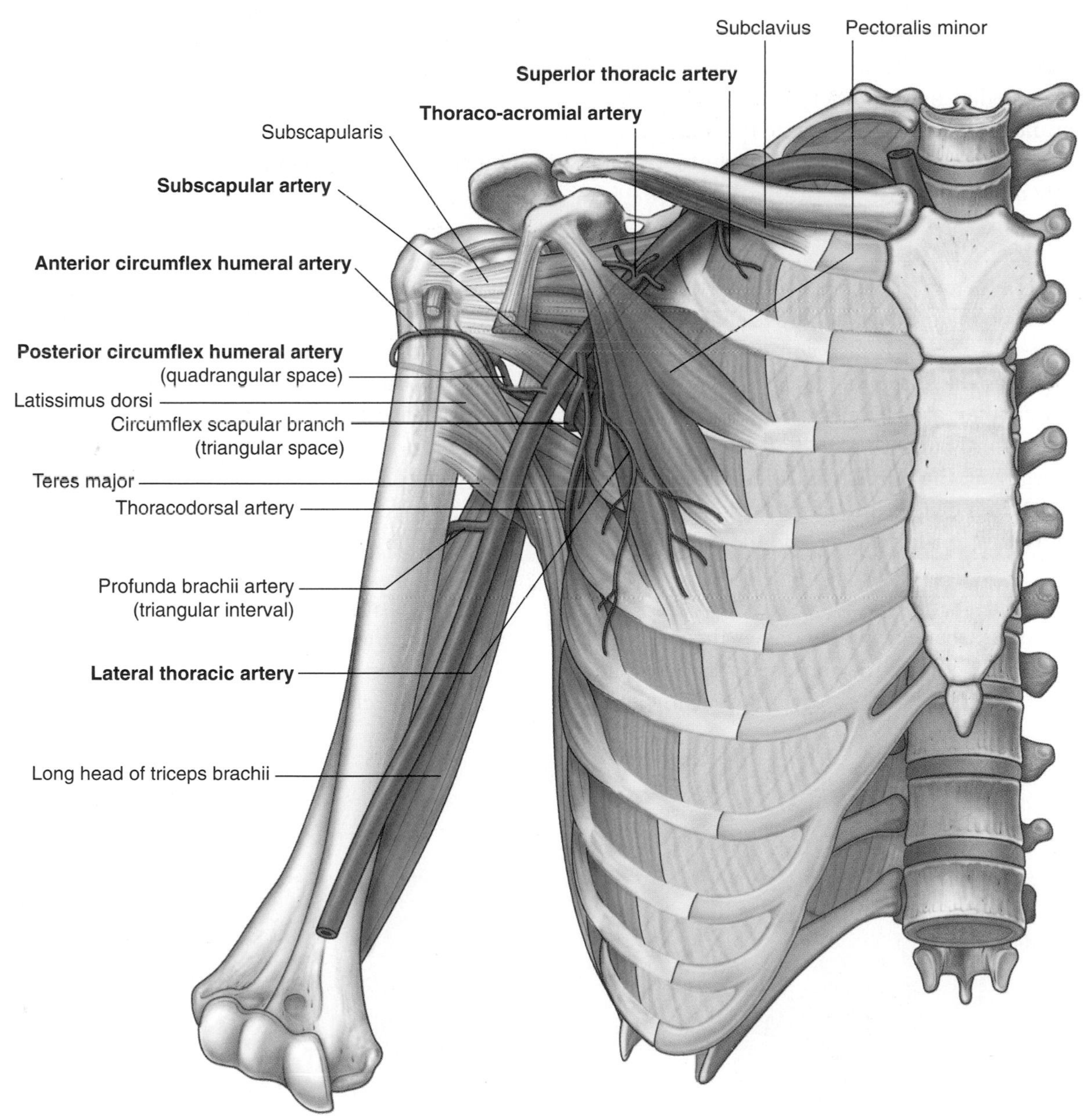


**Fig. 7.50** Branches of the axillary artery.

artery just posterior to the medial (superior) margin of the pectoralis minor muscle (Fig. 7.50). It curves around the superior margin of the muscle, penetrates the clavipectoral fascia, and immediately divides into four branches—the pectoral, deltoid, clavicular, and acromial branches, which supply the anterior axillary wall and related regions.

Additionally, the pectoral branch contributes vascular supply to the breast, and the deltoid branch passes into the clavipectoral triangle where it accompanies the cephalic vein and supplies adjacent structures (see Fig. 7.41).

### Lateral thoracic artery

The lateral thoracic artery arises from the anterior surface of the second part of the axillary artery posterior to the lateral (inferior) margin of the pectoralis minor (Fig. 7.50). It follows the margin of the muscle to the thoracic wall and supplies the medial and anterior walls of the axilla. In women, branches emerge from around the inferior margin of the pectoralis major muscle and contribute to the vascular supply of the breast.

### Subscapular artery

The subscapular artery is the largest branch of the axillary artery and is the major blood supply to the posterior wall of the axilla (Fig. 7.50). It also contributes to the blood supply of the posterior scapular region.

The subscapular artery originates from the posterior surface of the third part of the axillary artery, follows the inferior margin of the subscapularis muscle for a short distance, and then divides into its two terminal branches, the **circumflex scapular artery** and the **thoracodorsal artery**.

- The circumflex scapular artery passes through the triangular space between the subscapularis, teres major, and long head of the triceps muscle. Posteriorly, it passes inferior to, or pierces, the origin of the teres minor muscle to enter the infraspinous fossa. It anastomoses with the suprascapular artery and the **deep branch (dorsal scapular artery)** of the transverse cervical artery, thereby contributing to an anastomotic network of vessels around the scapula.
- The thoracodorsal artery approximately follows the lateral border of the scapula to the inferior angle. It contributes to the vascular supply of the posterior and medial walls of the axilla.

### Anterior circumflex humeral artery

The **anterior circumflex humeral artery** is small compared to the posterior circumflex humeral artery, and originates from the lateral side of the third part of the axillary artery (Fig. 7.50). It passes anterior to the surgical neck of the humerus and anastomoses with the posterior circumflex humeral artery.

This anterior circumflex humeral artery supplies branches to surrounding tissues, which include the glenohumeral joint and the head of the humerus.

### Posterior circumflex humeral artery

The **posterior circumflex humeral artery** originates from the lateral surface of the third part of the axillary artery immediately posterior to the origin of the anterior circumflex humeral artery (Fig. 7.50). With the axillary nerve, it leaves the axilla by passing through the quadrangular space between the teres major, teres minor, and long head of the triceps brachii muscle and the surgical neck of the humerus.

The posterior circumflex humeral artery curves around the surgical neck of the humerus and supplies the surrounding muscles and the glenohumeral joint. It anastomoses with the anterior circumflex humeral artery and with branches from the profunda brachii, suprascapular, and thoraco-acromial arteries.

### Axillary vein

The axillary vein begins at the lower margin of the teres major muscle and is the continuation of the basilic vein (Fig. 7.51), which is a superficial vein that drains the posteromedial surface of the hand and forearm and penetrates the deep fascia in the middle of the arm.

The axillary vein passes through the axilla medial and anterior to the axillary artery and becomes the subclavian vein as the vessel crosses the lateral border of rib I at the axillary inlet. Tributaries of the axillary vein generally follow the branches of the axillary artery. Other tributaries include brachial veins that follow the brachial artery and the cephalic vein.

The cephalic vein is a superficial vein that drains the lateral and posterior parts of the hand, the forearm, and the arm. In the area of the shoulder, it passes into an inverted triangular cleft (the clavipectoral triangle) between the deltoid muscle, pectoralis major muscle, and clavicle. In the superior part of the clavipectoral triangle, the cephalic vein passes deep to the clavicular head of the pectoralis major muscle and pierces the clavipectoral fascia to join the axillary vein. Many patients who are critically ill have lost blood or fluid, which requires replacement. Access to a peripheral vein is necessary to replace the fluid. The typical sites for venous access are the cephalic vein in the hand or veins that lie within the superficial tissues of the cubital fossa.

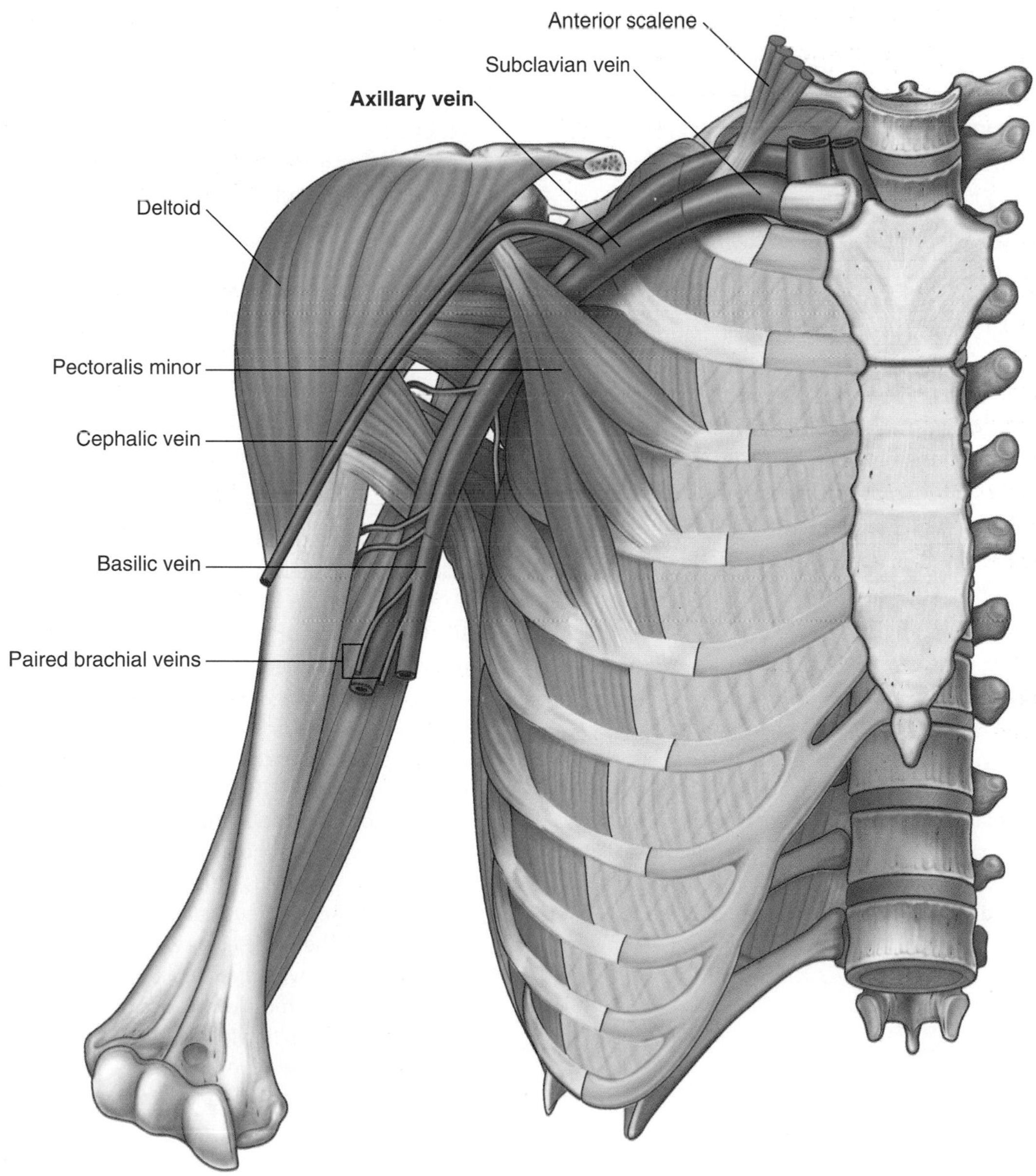


**Fig. 7.51** Axillary vein.

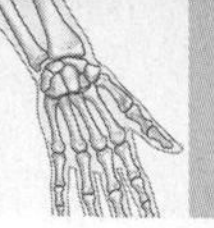



## In the clinic

### Imaging the blood supply to the upper limb

When there is clinical evidence of vascular compromise to the upper limb, or vessels are needed to form an arteriovenous fistula (which is necessary for renal dialysis), imaging is required to assess the vessels.

Ultrasound is a useful tool for carrying out a noninvasive assessment of the vessels of the upper limb from the third part of the subclavian artery to the deep and superficial palmar arteries. Blood flow can be quantified and anatomical variants can be noted.

Angiography is carried out in certain cases. The femoral artery is punctured below the inguinal ligament and a long catheter is placed through the iliac arteries and around the arch of the aorta to enter either the left subclavian artery or the brachiocephalic trunk and then the right subclavian artery. Radiopaque contrast agents are injected into the vessel and radiographs are obtained as the contrast agents pass first through the arteries, then the capillaries, and finally the veins.

## In the clinic

### Trauma to the arteries of the upper limb

The arterial supply to the upper limb is particularly susceptible to trauma in places where it is relatively fixed or in a subcutaneous position.

#### *Fracture of rib I*

As the subclavian artery passes out of the neck and into the axilla, it is fixed in position by the surrounding muscles to the superior surface of rib I. A rapid deceleration injury involving upper thoracic trauma may cause a first rib fracture, which may significantly compromise the distal part of the subclavian artery or the first part of the axillary artery. Fortunately, there are anastomotic connections between branches of the subclavian artery and the axillary artery, which form a network around the scapula and proximal end of the humerus; therefore, even with complete vessel transection, the arm is rarely rendered completely ischemic (ischemia is poor blood supply to an organ or a limb).

#### *Anterior dislocation of the humeral head*

Anterior dislocation of the humeral head may compress the axillary artery, resulting in vessel occlusion. This is unlikely to render the upper limb completely ischemic, but it may be necessary to surgically reconstruct the axillary artery to obtain pain-free function. Importantly, the axillary artery is intimately related to the brachial plexus, which may be damaged at the time of anterior dislocation.

## In the clinic

### Subclavian/axillary venous access

There are a number of routes through which central venous access may be obtained. The "subclavian route" and the jugular routes are commonly used by clinicians. The subclavian route is a misnomer that remains the preferred term in clinical practice. In fact, most clinicians enter the first part of the axillary vein.

There are a number of patients that undergo catheterization of the subclavian vein/axillary vein. Entering the subclavian vein/axillary vein is a relatively straightforward technique. The clavicle is identified and a sharp needle is placed in the infraclavicular region, aiming superomedially. When venous blood is aspirated, access has been obtained. This route is popular for long-term venous access, such as Hickman lines, and for shorter-term access where multiple-lumen catheters are inserted (e.g., intensive care unit).

The subclavian vein/axillary vein is also the preferred site for insertion of pacemaker wires. There is, however, a preferred point of entry into the vein to prevent complications. The vein should be punctured in the midclavicular line or lateral to this line. The reason for this puncture site is the course of the vein and its relationship to other structures. The vein passes anterior to the artery, superior to the first rib, and inferior to the clavicle as it courses toward the thoracic inlet. Beneath the clavicle is situated the subclavius muscle. Should the puncture of the vein enter where the subclavius muscle is related to the axillary vein, the catheter or wire may become kinked at this point. Moreover, the constant contraction and relaxation of this muscle will induce fatigue in the line and wire, which may ultimately lead to fracture. A fractured pacemaker wire or a rupture in a chemotherapy catheter can have severe consequences for the patient.

## Brachial plexus

The brachial plexus is a somatic nerve plexus formed by the **anterior rami** of C5 to C8, and most of the anterior ramus of T1 (Fig. 7.52). The plexus originates in the neck, passes laterally and inferiorly over rib I, and enters the axilla.

The parts of the brachial plexus, from medial to lateral, are roots, trunks, divisions, and cords. All major nerves that innervate the upper limb originate from the brachial plexus, mostly from the cords. Proximal parts of the brachial plexus are posterior to the subclavian artery in the neck, while more distal regions of the plexus surround the axillary artery.

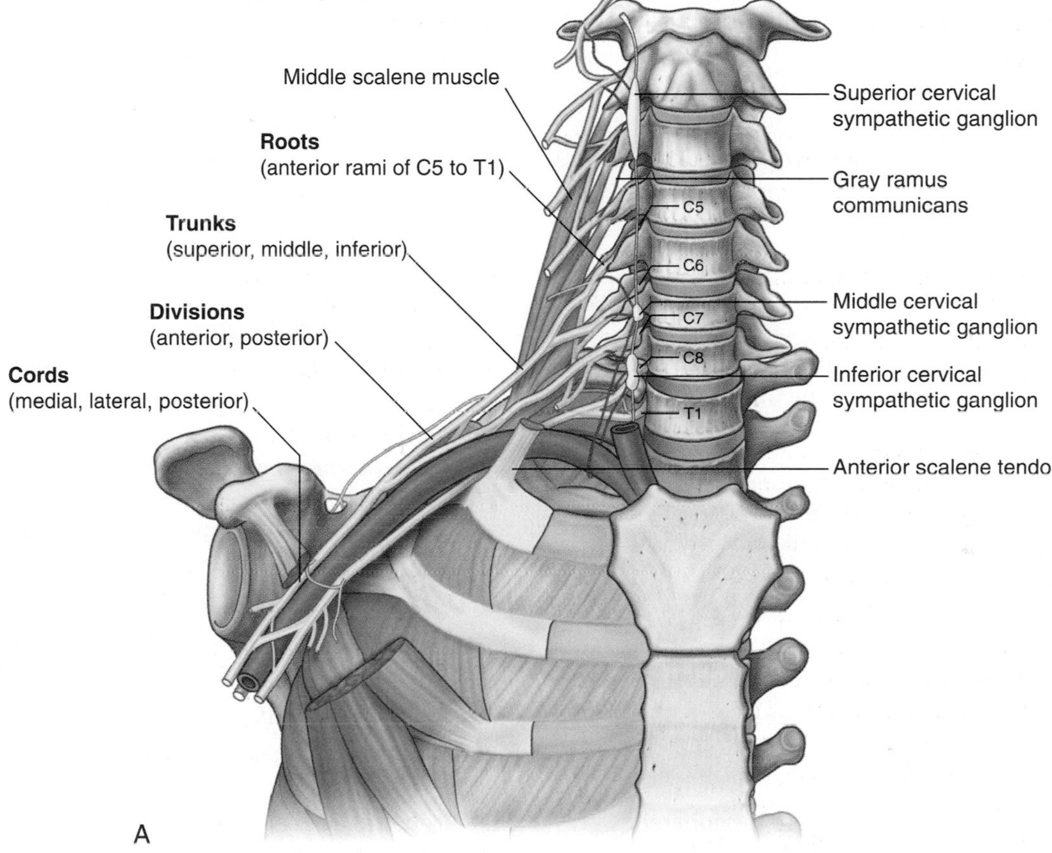


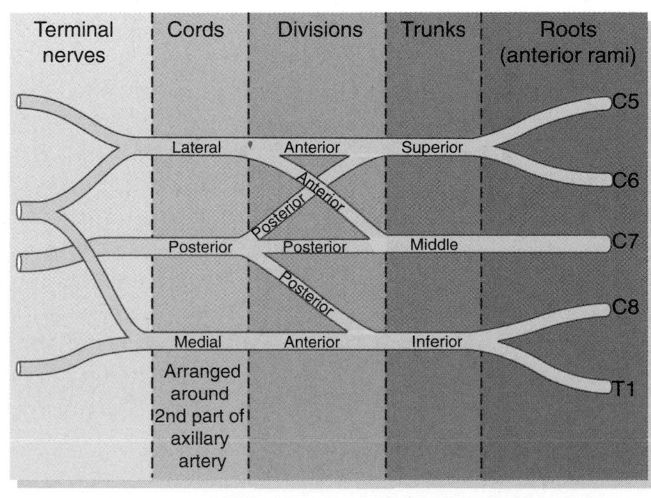


**Fig. 7.52** Brachial plexus. **A.** Major components in the neck and axilla. **B.** Schematic showing parts of the brachial plexus.

### Roots

The roots of the brachial plexus are the anterior rami of C5 to C8, and most of T1. Close to their origin, the roots receive **gray rami communicantes** from the sympathetic trunk (Fig. 7.52). These carry postganglionic sympathetic fibers onto the roots for distribution to the periphery. The roots and trunks enter the **posterior triangle** of the neck by passing between the anterior scalene and middle scalene muscles and lie superior and posterior to the subclavian artery.

### Trunks

The three trunks of the brachial plexus originate from the roots, pass laterally over rib I, and enter the axilla (Fig. 7.52):

- The superior trunk is formed by the union of C5 and C6 roots.
- The middle trunk is a continuation of the C7 root.
- The inferior trunk is formed by the union of the C8 and T1 roots.

The inferior trunk lies on rib I posterior to the subclavian artery; the middle and superior trunks are more superior in position.

### Divisions

Each of the three trunks of the brachial plexus divides into an **anterior** and a **posterior division** (Fig. 7.52):

- The three anterior divisions form parts of the brachial plexus that ultimately give rise to peripheral nerves associated with the anterior compartments of the arm and forearm.
- The three posterior divisions combine to form parts of the brachial plexus that give rise to nerves associated with the posterior compartments.

No peripheral nerves originate directly from the divisions of the brachial plexus.

### Cords

The three cords of the brachial plexus originate from the divisions and are related to the second part of the axillary artery (Fig. 7.52):

- The **lateral cord** results from the union of the anterior divisions of the upper and middle trunks and therefore has contributions from C5 to C7—it is positioned lateral to the second part of the axillary artery.
- The **medial cord** is medial to the second part of the axillary artery and is the continuation of the anterior division of the inferior trunk—it contains contributions from C8 and T1.
- The **posterior cord** occurs posterior to the second part of the axillary artery and originates as the union of all three posterior divisions—it contains contributions from all roots of the brachial plexus (C5 to T1).

Most of the major peripheral nerves of the upper limb originate from the cords of the brachial plexus. Generally, nerves associated with the anterior compartments of the upper limb arise from the medial and lateral cords and nerves associated with the posterior compartments originate from the posterior cord.

### Branches (Table 7.7)

#### Branches of the roots

In addition to small segmental branches from C5 to C8 to muscles of the neck and a contribution of C5 to the phrenic nerve, the roots of the brachial plexus give rise to the dorsal scapular and long thoracic nerves (Fig. 7.53).

The **dorsal scapular nerve**:

- originates from the C5 root of the brachial plexus,
- passes posteriorly, often piercing the middle scalene muscle in the neck, to reach and travel along the medial border of the scapula (Fig. 7.54), and
- innervates the rhomboid major and minor muscles from their deep surfaces.

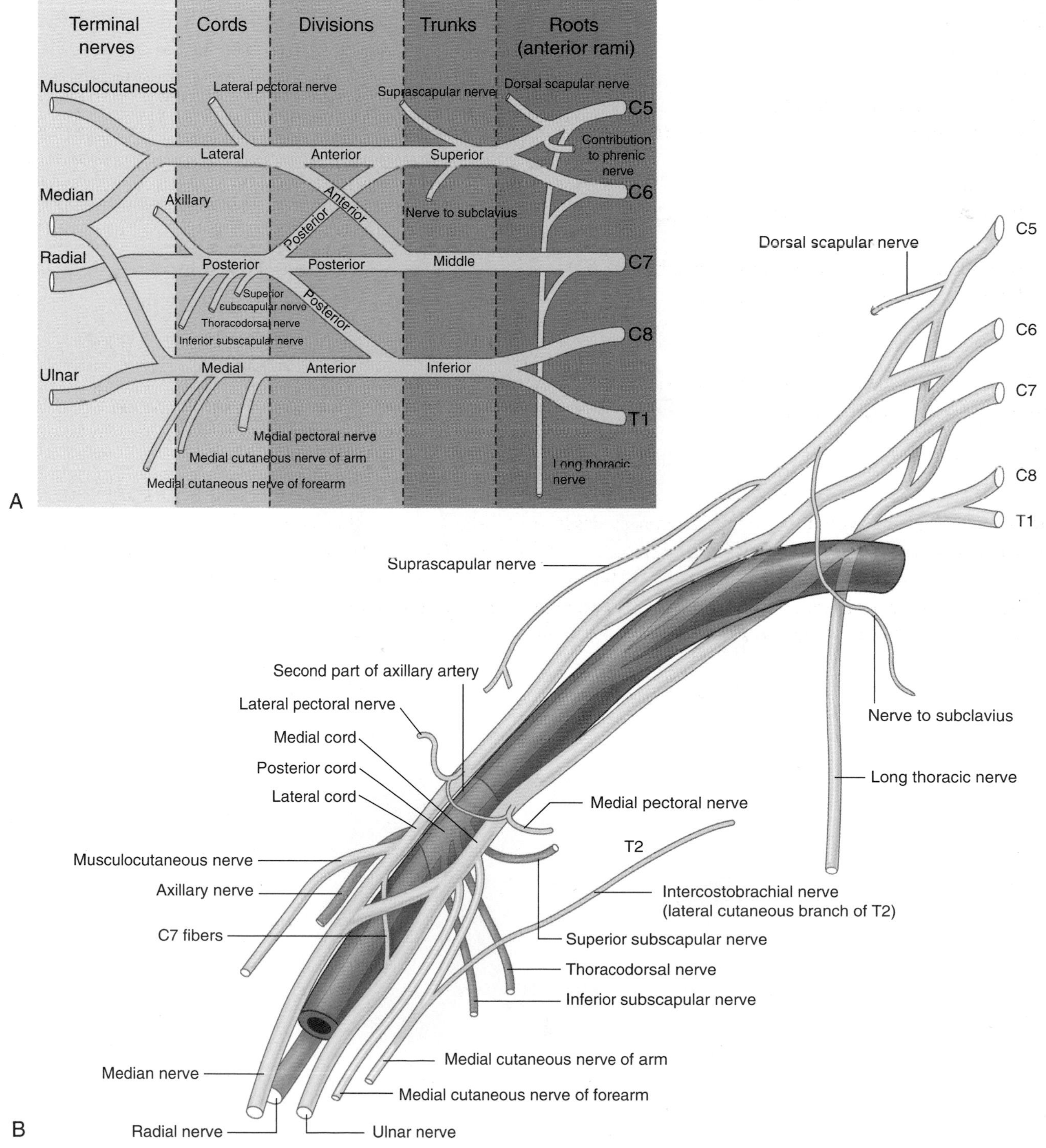


**Fig. 7.53** Brachial plexus. **A.** Schematic showing branches of the brachial plexus. **B.** Relationships to the axillary artery.

The **long thoracic nerve**:

- originates from the anterior rami of C5 to C7,
- passes vertically down the neck, through the axillary inlet, and down the medial wall of the axilla to supply the serratus anterior muscle (Fig. 7.54), and
- lies on the superficial aspect of the serratus anterior muscle.

### Branches of the trunks

The only branches from the trunks of the brachial plexus are two nerves that originate from the superior trunk (upper trunk): the suprascapular nerve and the nerve to the subclavius muscle (Fig. 7.53).

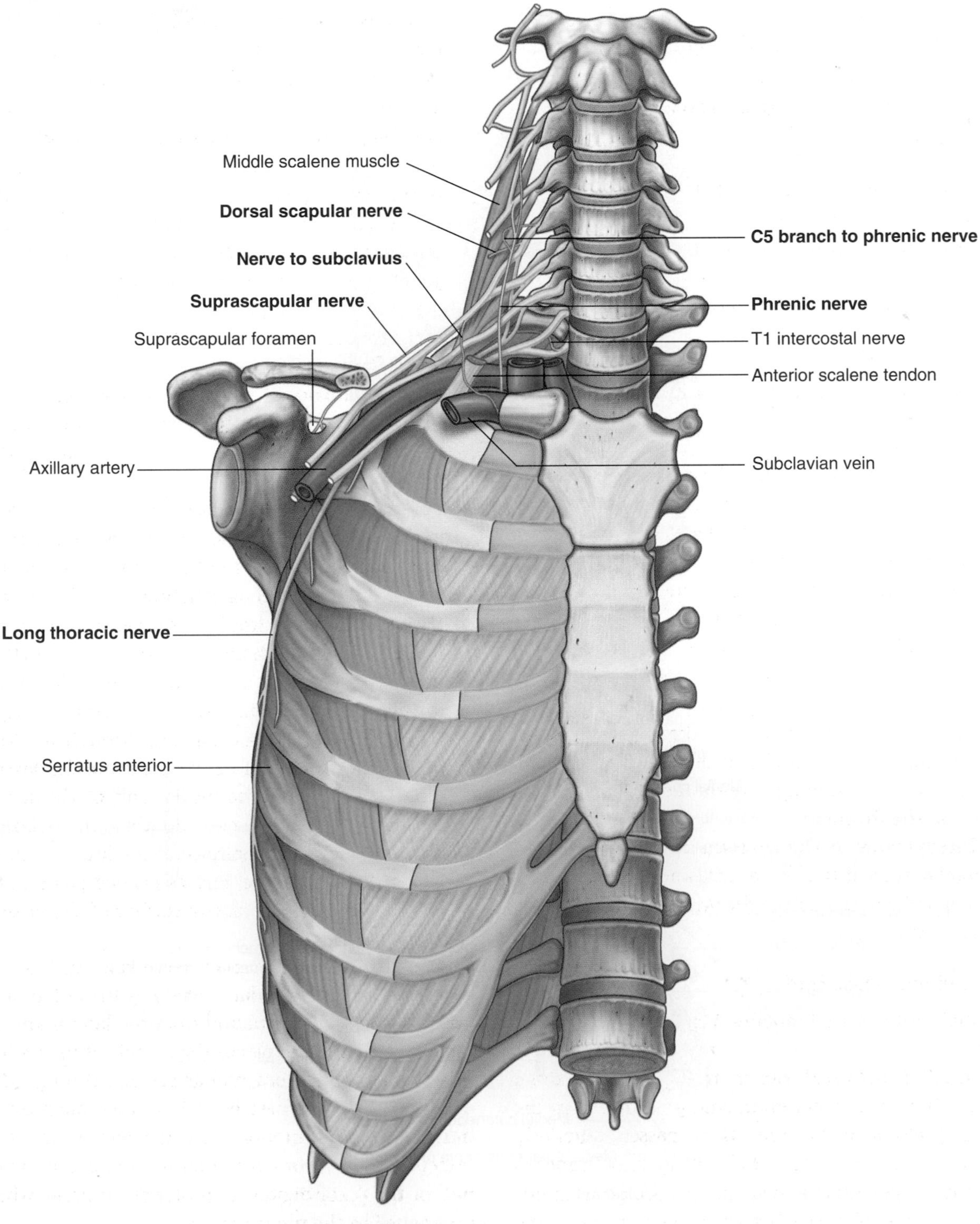

**Fig. 7.54** Branches of the roots and trunks of the brachial plexus.

The **suprascapular nerve** (C5 and C6):

- originates from the superior trunk of the brachial plexus,
- passes laterally through the posterior triangle of the neck (Fig. 7.54) and through the suprascapular foramen to enter the posterior scapular region,
- innervates the supraspinatus and infraspinatus muscles, and
- is accompanied in the lateral parts of the neck and in the posterior scapular region by the suprascapular artery.

The **nerve to the subclavius muscle** (C5 and C6) is a small nerve that:

- originates from the superior trunk of the brachial plexus,
- passes anteroinferiorly over the subclavian artery and vein, and
- innervates the subclavius muscle.

### Branches of the lateral cord

Three nerves originate entirely or partly from the lateral cord (Fig. 7.53).

- The **lateral pectoral nerve** is the most proximal of the branches from the lateral cord. It passes anteriorly, together with the thoraco-acromial artery, to penetrate the clavipectoral fascia that spans the gap between the subclavius and pectoralis minor muscles (Fig. 7.55), and innervates the pectoralis major muscle.
- The **musculocutaneous nerve** is a large terminal branch of the lateral cord. It passes laterally to penetrate the coracobrachialis muscle and pass between the biceps brachii and brachialis muscles in the arm, and innervates all three flexor muscles in the anterior compartment of the arm, terminating as the **lateral cutaneous nerve of the forearm**.
- The **lateral root of the median nerve** is the largest terminal branch of the lateral cord and passes medially to join a similar branch from the medial cord to form the median nerve (Fig. 7.55).

### Branches of the medial cord

The medial cord has five branches (Fig. 7.55).

- The **medial pectoral nerve** is the most proximal branch. It receives a communicating branch from the lateral pectoral nerve and then passes anteriorly between the axillary artery and axillary vein. Branches of the nerve penetrate and supply the pectoralis minor muscle. Some of these branches pass through the muscle to reach and supply the pectoralis major muscle. Other branches occasionally pass around the inferior or lateral margin of the pectoralis minor muscle to reach the pectoralis major muscle.
- The **medial cutaneous nerve of the arm** (**medial brachial cutaneous nerve**) passes through the axilla and into the arm where it penetrates deep fascia and supplies skin over the medial side of the distal third of the arm. In the axilla, the nerve communicates with the **intercostobrachial nerve** of T2. Fibers of the medial cutaneous nerve of the arm innervate the upper part of the medial surface of the arm and floor of the axilla.
- The **medial cutaneous nerve of the forearm (medial antebrachial cutaneous nerve)** originates just distal to the origin of the medial cutaneous nerve of the arm. It passes out of the axilla and into the arm where it gives off a branch to the skin over the biceps brachii muscle, and then continues down the arm to penetrate the deep fascia with the basilic vein, continuing inferiorly to supply the skin over the anterior surface of the forearm. It innervates skin over the medial surface of the forearm down to the wrist.
- The **medial root of the median nerve** passes laterally to join with a similar root from the lateral cord to form the median nerve anterior to the third part of the axillary artery.
- The **ulnar nerve** is a large terminal branch of the medial cord (Fig. 7.55). However, near its origin, it often receives a communicating branch from the lateral root of the median nerve originating from the lateral cord and carrying fibers from C7 (see Fig. 5.73B). The ulnar nerve passes through the arm and forearm into the hand where it innervates all intrinsic muscles of the hand (except for the three thenar muscles and the two lateral lumbrical muscles). On passing through the forearm, branches of the ulnar nerve innervate the flexor carpi ulnaris muscle and the medial half of the flexor digitorum profundus muscle. The ulnar nerve innervates skin over the palmar surface of the little finger, medial half of the ring finger, and associated palm and wrist, and the skin over the dorsal surface of the medial part of the hand.

***Median nerve.*** The median nerve is formed anterior to the third part of the axillary artery by the union of lateral and medial roots originating from the lateral and medial cords of the brachial plexus (Fig. 7.55). It passes into the arm anterior to the brachial artery and through the arm into the forearm, where branches innervate most of the muscles in the anterior compartment of the forearm (except for the flexor carpi ulnaris muscle and the medial half of the flexor digitorum profundus muscle, which are innervated by the ulnar nerve).

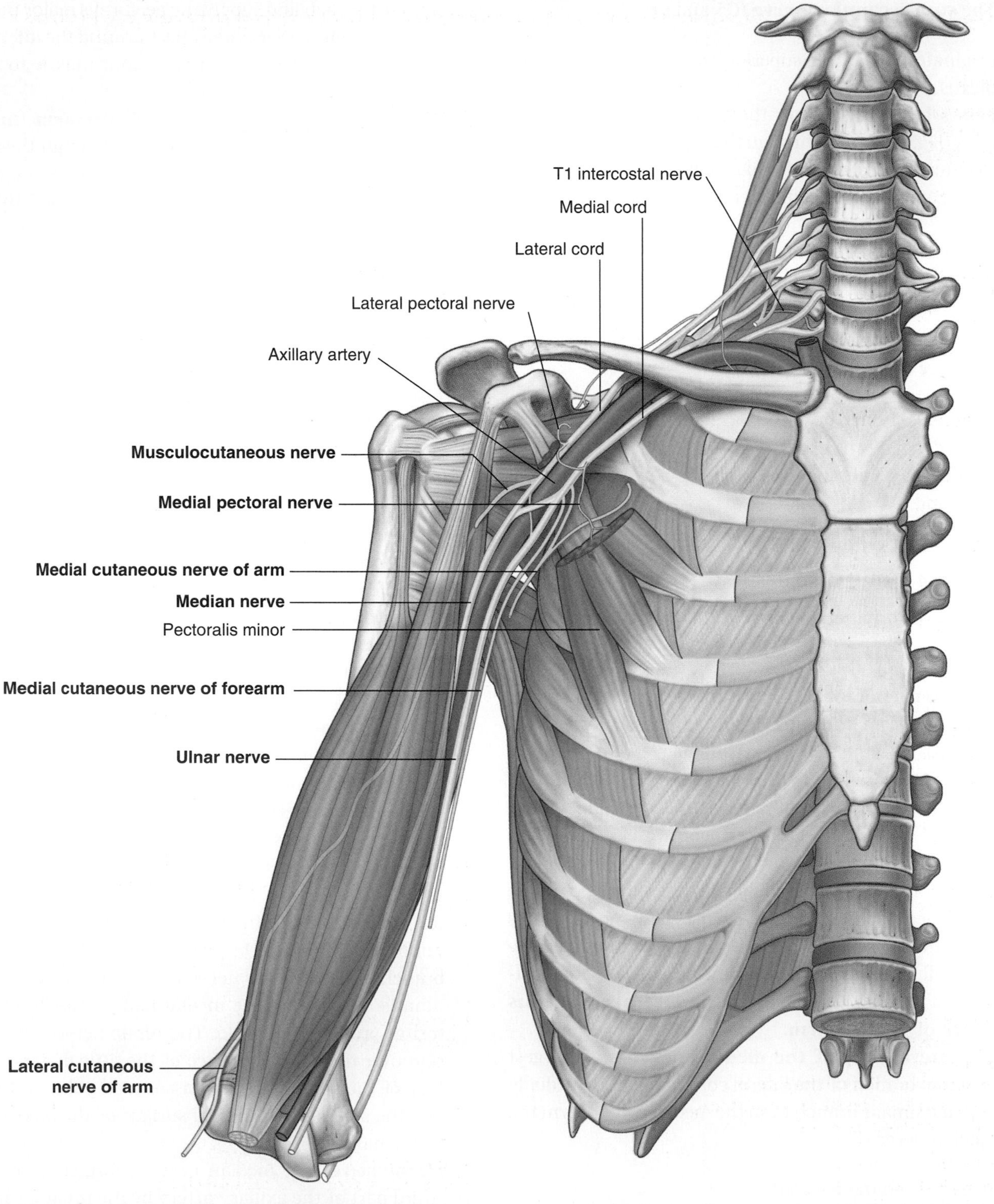

T1 intercostal nerve
Medial cord
Lateral cord
Lateral pectoral nerve
Axillary artery
**Musculocutaneous nerve**
**Medial pectoral nerve**
**Medial cutaneous nerve of arm**
**Median nerve**
Pectoralis minor
**Medial cutaneous nerve of forearm**
**Ulnar nerve**
**Lateral cutaneous nerve of arm**

**Fig. 7.55** Branches of the lateral and medial cords of the brachial plexus.

**Table 7.7** Branches of brachial plexus (parentheses indicate that a spinal segment is a minor component of the nerve or is inconsistently present in the nerve)

| **Branch** | |
|---|---|
| Dorsal scapular<br>Origin: C5 root<br>Spinal segment: C5 | Function: motor<br>Rhomboid major, rhomboid minor |
| Long thoracic<br>Origin: C5 to C7 roots<br>Spinal segments: C5 to C7 | Function: motor<br>Serratus anterior |
| Suprascapular<br>Origin: Superior trunk<br>Spinal segments: C5, C6 | Function: motor<br>Supraspinatus, infraspinatus |
| Nerve to subclavius<br>Origin: Superior trunk<br>Spinal segments: C5, C6 | Function: motor<br>Subclavius |
| Lateral pectoral<br>Origin: Lateral cord<br>Spinal segments: C5 to C7 | Function: motor<br>Pectoralis major |
| Musculocutaneous<br>Origin: Lateral cord<br>Spinal segments: C5 to C7 | Function: motor<br>All muscles in the anterior compartment of the arm<br>Function: sensory<br>Skin on lateral side of forearm |
| Medial pectoral<br>Origin: Medial cord<br>Spinal segments: C8, T1<br>(also receives contributions from spinal segments C5 to C7 through a communication with the lateral pectoral nerve) | Function: motor<br>Pectoralis major, pectoralis minor |
| Medial cutaneous of arm<br>Origin: Medial cord<br>Spinal segments: C8, T1 | Function: sensory<br>Skin on medial side of distal one-third of arm |

*Continued*

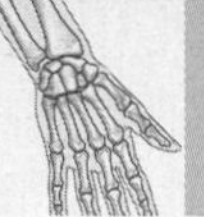

**Table 7.7** Branches of brachial plexus (parentheses indicate that a spinal segment is a minor component of the nerve or is inconsistently present in the nerve)—cont'd

| Branch | |
|---|---|
| Medial cutaneous of forearm<br>Origin: Medial cord<br>Spinal segments: C8, T1 | Function: sensory<br>Skin on medial side of forearm |
| Median<br>Origin: Medial and lateral cords<br>Spinal segments: (C5), C6 to T1 | Function: motor<br>All muscles in the anterior compartment of the forearm (except flexor carpi ulnaris and medial half of flexor digitorum profundus), three thenar muscles of the thumb and two lateral lumbrical muscles<br>Function: sensory<br>Skin over the palmar surface of the lateral three and one-half digits and over the lateral side of the palm and middle of the wrist |
| Ulnar<br>Origin: Medial cord<br>Spinal segments: (C7), C8, T1 | Function: motor<br>All intrinsic muscles of the hand (except three thenar muscles and two lateral lumbricals); also flexor carpi ulnaris and the medial half of flexor digitorum profundus in the forearm<br>Function: sensory<br>Skin over the palmar surface of the medial one and one-half digits and associated palm and wrist, and skin over the dorsal surface of the medial one and one-half digits |
| Superior subscapular<br>Origin: Posterior cord<br>Spinal segments: C5, C6 | Function: motor<br>Subscapularis |
| Thoracodorsal<br>Origin: Posterior cord<br>Spinal segments: C6 to C8 | Function: motor<br>Latissimus dorsi |
| Inferior subscapular<br>Origin: Posterior cord<br>Spinal segments: C5, C6 | Function: motor<br>Subscapularis, teres major |
| Axillary<br>Origin: Posterior cord<br>Spinal segments: C5, C6 | Function: motor<br>Deltoid, teres minor<br>Function: sensory<br>Skin over upper lateral part of arm |
| Radial<br>Origin: Posterior cord<br>Spinal segments: C5 to C8, (T1) | Function: motor<br>All muscles in the posterior compartments of arm and forearm<br>Function: sensory<br>Skin on the posterior aspects of the arm and forearm, the lower lateral surface of the arm, and the dorsal lateral surface of the hand |

The median nerve continues into the hand to innervate:

- the three thenar muscles associated with the thumb,
- the two lateral lumbrical muscles associated with movement of the index and middle fingers, and
- the skin over the palmar surface of the lateral three and one-half digits and over the lateral side of the palm and middle of the wrist.

The musculocutaneous nerve, the lateral root of the median nerve, the median nerve, the medial root of the median nerve, and the ulnar nerve form an M over the third part of the axillary artery (Fig. 7.55). This feature, together with penetration of the coracobrachialis muscle by the musculocutaneous nerve, can be used to identify components of the brachial plexus in the axilla.

### Branches of the posterior cord

Five nerves originate from the posterior cord of the brachial plexus:

- the superior subscapular nerve,
- the thoracodorsal nerve,
- the inferior subscapular nerve,
- the axillary nerve, and
- the radial nerve (Fig. 7.53).

All these nerves except the radial nerve innervate muscles associated with the shoulder region or the posterior wall of the axilla; the radial nerve passes into the arm and forearm.

The superior subscapular, thoracodorsal, and inferior subscapular nerves originate sequentially from the posterior cord and pass directly into muscles associated with the posterior axillary wall (Fig. 7.56). The **superior subscapular nerve** is short and passes into and supplies the subscapularis muscle. The **thoracodorsal nerve** is the longest of these three nerves and passes vertically along the posterior axillary wall. It penetrates and innervates the latissimus dorsi muscle. The **inferior subscapular nerve** also passes inferiorly along the posterior axillary wall and innervates the subscapularis and teres major muscles.

The **axillary nerve** originates from the posterior cord and passes inferiorly and laterally along the posterior wall to exit the axilla through the quadrangular space (Fig. 7.56). It passes posteriorly around the surgical neck of the humerus and innervates both the deltoid and teres minor muscles. A **superior lateral cutaneous nerve of the arm** originates from the axillary nerve after passing through the quadrangular space and loops around the posterior margin of the deltoid muscle to innervate skin in that region. The axillary nerve is accompanied by the posterior circumflex humeral artery.

The **radial nerve** is the largest terminal branch of the posterior cord (Fig. 7.56). It passes out of the axilla and into the posterior compartment of the arm by passing through the triangular interval between the inferior border of the teres major muscle, the long head of the triceps brachii muscle, and the shaft of the humerus. It is accompanied through the triangular interval by the profunda brachii artery, which originates from the brachial artery in the anterior compartment of the arm. The radial nerve and its branches innervate:

- all muscles in the posterior compartments of the arm and forearm, and
- the skin on the posterior aspect of the arm and forearm, the lower lateral surface of the arm, and the dorsal lateral surface of the hand.

The **posterior cutaneous nerve of the arm** (**posterior brachial cutaneous nerve**) originates from the radial nerve in the axilla and innervates skin on the posterior surface of the arm.

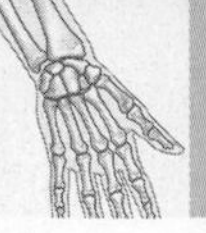





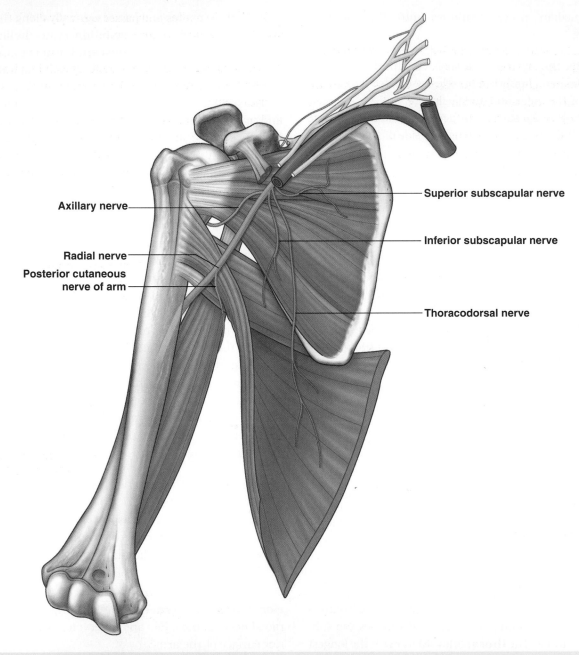

**Fig. 7.56** Branches of the posterior cord of the brachial plexus.

## In the clinic

### Injuries to the brachial plexus

The brachial plexus is an extremely complex structure. When damaged, it requires meticulous clinical history taking and examination. Assessment of the individual nerve functions can be obtained by nerve conduction studies and electromyography, which assess the latency of muscle contraction when the nerve is artificially stimulated.

Brachial plexus injuries are usually the result of blunt trauma producing nerve avulsions and disruption. These injuries are usually devastating for the function of the upper limb and require many months of dedicated rehabilitation for even a small amount of function to return.

Spinal cord injuries in the cervical region and direct pulling injuries tend to affect the roots of the brachial plexus. Severe trauma to the first rib usually affects the trunks. The divisions and cords of the brachial plexus can be injured by dislocation of the glenohumeral joint.

## Lymphatics

All lymphatics from the upper limb drain into lymph nodes in the axilla (Fig. 7.57).

In addition, axillary nodes receive drainage from an extensive area on the adjacent trunk, which includes regions of the upper back and shoulder, the lower neck, the chest, and the upper anterolateral abdominal wall. Axillary nodes also receive drainage from approximately 75% of the mammary gland.

The 20–30 axillary nodes are generally divided into five groups on the basis of location.

- **Humeral (lateral) nodes** posteromedial to the axillary vein receive most of the lymphatic drainage from the upper limb.
- **Pectoral (anterior) nodes** occur along the inferior margin of the pectoralis minor muscle along the course of the lateral thoracic vessels and receive drainage from the abdominal wall, the chest, and the mammary gland.

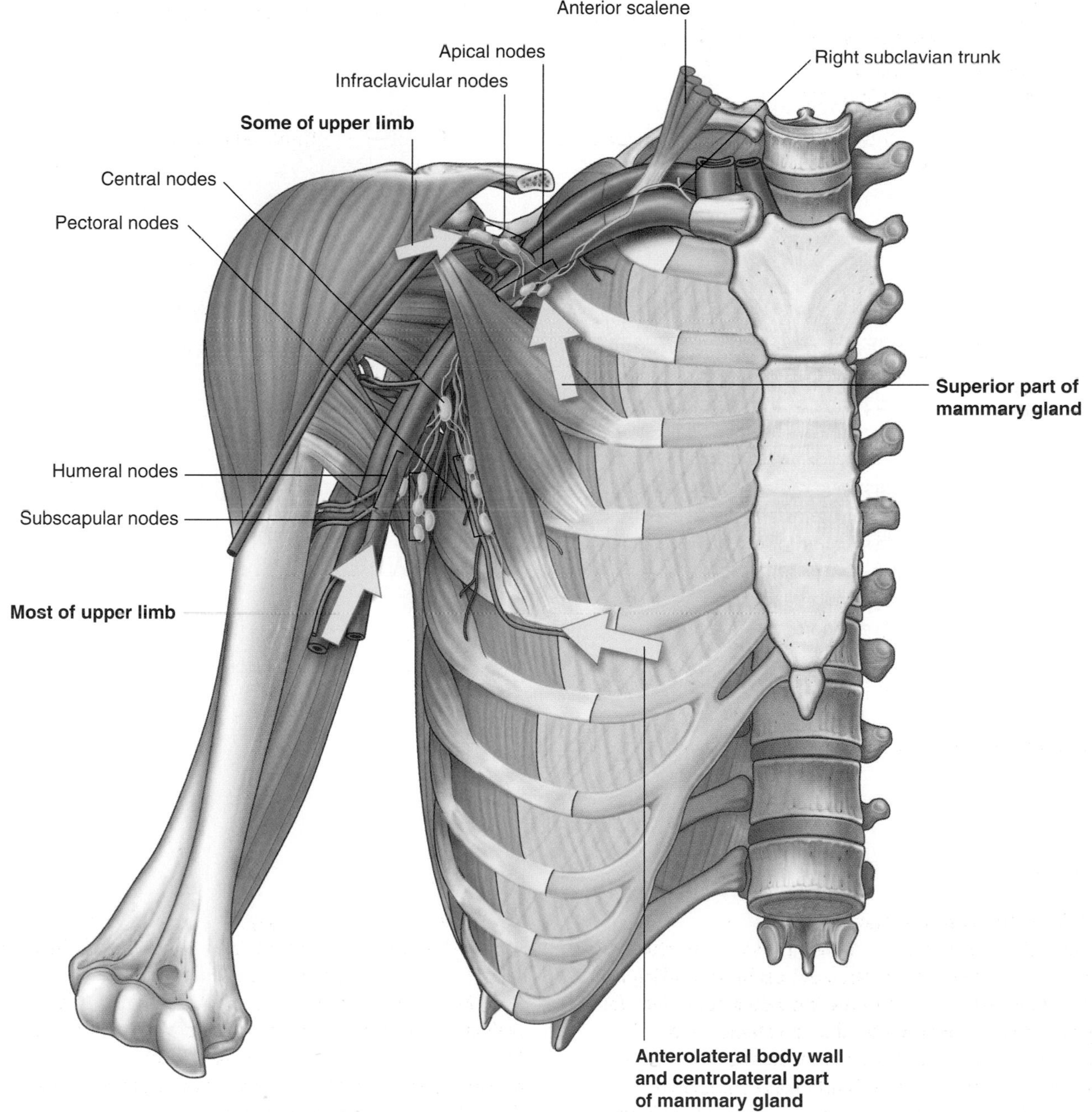


**Fig. 7.57** Lymph nodes and vessels in the axilla.

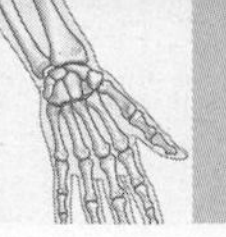

- **Subscapular (posterior) nodes** on the posterior axillary wall in association with the subscapular vessels drain the posterior axillary wall and receive lymphatics from the back, the shoulder, and the neck.
- **Central nodes** are embedded in axillary fat and receive tributaries from humeral, subscapular, and pectoral groups of nodes.
- **Apical nodes** are the most superior group of nodes in the axilla and drain all other groups of nodes in the region. In addition, they receive lymphatic vessels that accompany the cephalic vein as well as vessels that drain the superior region of the mammary gland.

Efferent vessels from the apical group converge to form the subclavian trunk, which usually joins the venous system at the junction between the right subclavian vein and the right internal jugular vein in the neck. On the left, the subclavian trunk usually joins the thoracic duct in the base of the neck.

## In the clinic

### Breast cancer

Lymphatic drainage from the lateral part of the breast passes through nodes in the axilla. Significant disruption to the normal lymphatic drainage of the upper limb may occur if a mastectomy or a surgical axillary nodal clearance has been carried out for breast cancer. Furthermore, some patients have radiotherapy to the axilla to prevent the spread of metastatic disease, but a side effect of this is the destruction of the tiny lymphatics as well as the cancer cells.

If the lymphatic drainage of the upper limb is damaged, the arm may swell and pitting edema (lymphedema) may develop.

### Axillary process of the mammary gland

Although the mammary gland is in superficial fascia overlying the thoracic wall, its superolateral region extends along the inferior margin of the pectoralis major muscle toward the axilla. In some cases, this may pass around the margin of the muscle to penetrate deep fascia and enter the axilla (Fig. 7.58). This axillary process rarely reaches as high as the apex of the axilla.

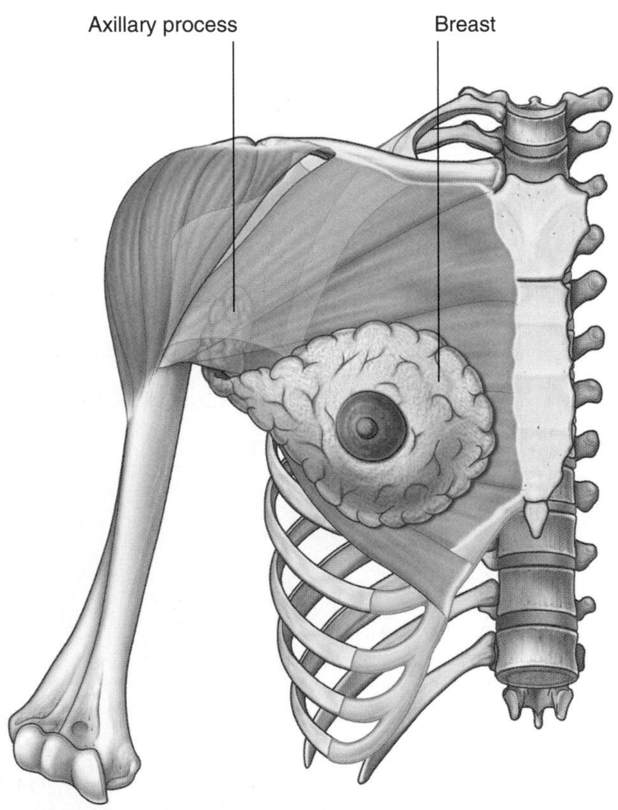


**Fig. 7.58** Axillary process of the breast.

## ARM

The arm is the region of the upper limb between the shoulder and the elbow (Fig. 7.59). The superior aspect of the arm communicates medially with the axilla. Inferiorly, a number of important structures pass between the arm and the forearm through the cubital fossa, which is positioned anterior to the elbow joint.

The arm is divided into two compartments by medial and lateral intermuscular septa, which pass from each side of the humerus to the outer sleeve of deep fascia that surrounds the limb (Fig. 7.59).

The anterior compartment of the arm contains muscles that predominantly flex the elbow joint; the posterior compartment contains muscles that extend the joint. Major nerves and vessels supply and pass through each compartment.

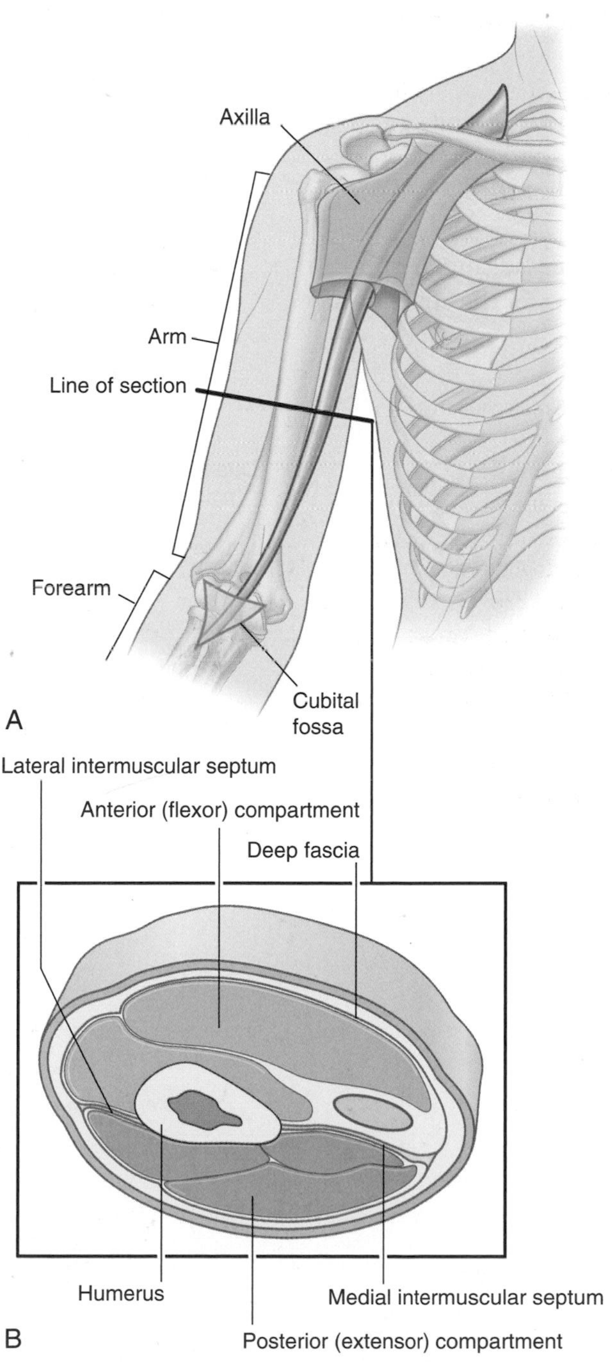


**Fig. 7.59** Arm. **A.** Proximal and distal relationships. **B.** Transverse section through the middle of the arm.

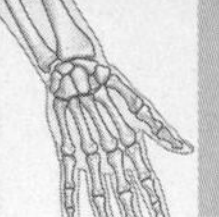



## Bones

The skeletal support for the arm is the humerus (Fig. 7.60). Most of the large muscles of the arm insert into the proximal ends of the two bones of the forearm, the radius and the ulna, and flex and extend the forearm at the elbow joint. In addition, the muscles predominantly situated in the forearm that move the hand originate at the distal end of the humerus.

### Shaft and distal end of the humerus

In cross section, the shaft of the humerus is somewhat triangular with:

- **anterior, lateral**, and **medial borders**, and
- **anterolateral, anteromedial**, and **posterior surfaces** (Fig. 7.60).

The posterior surface of the humerus is marked on its superior aspect by a linear roughening for the attachment of the lateral head of the triceps brachii muscle, beginning just inferior to the surgical neck and passing diagonally across the bone to the **deltoid tuberosity**.

The middle part of the posterior surface and adjacent part of the anterolateral surface are marked by the shallow **radial groove**, which passes diagonally down the bone and parallel to the sloping posterior margin of the deltoid tuberosity. The radial nerve and the profunda brachii artery lie in this groove.

Approximately in the middle of the shaft, the medial border is marked by thin elongate roughening for the attachment of the coracobrachialis muscle.

Intermuscular septa, which separate the anterior compartment from the posterior compartment, attach to the medial and lateral borders (Fig. 7.61).

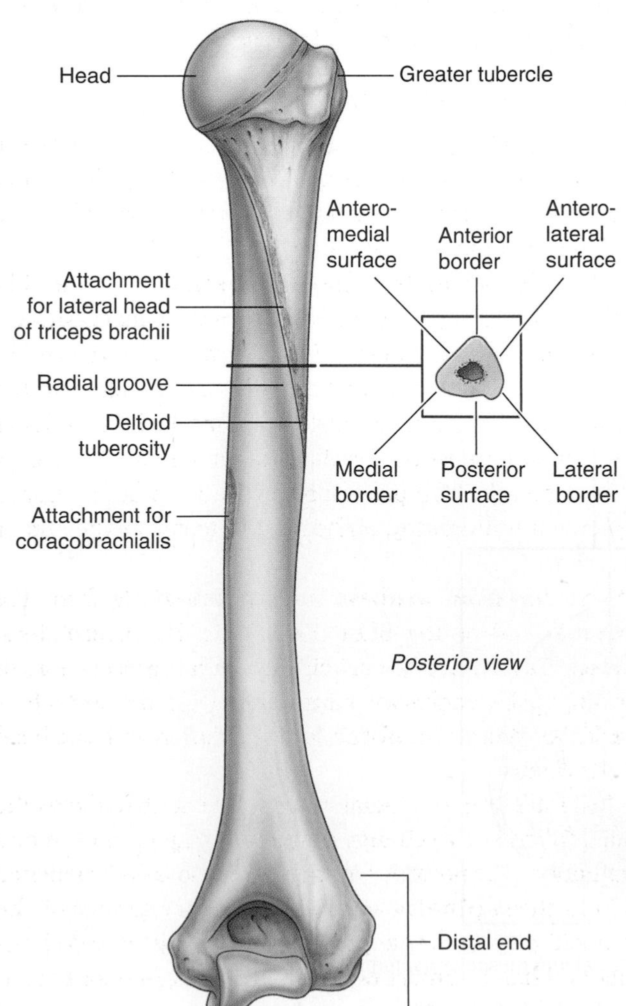


**Fig. 7.60** Humerus. Posterior view.

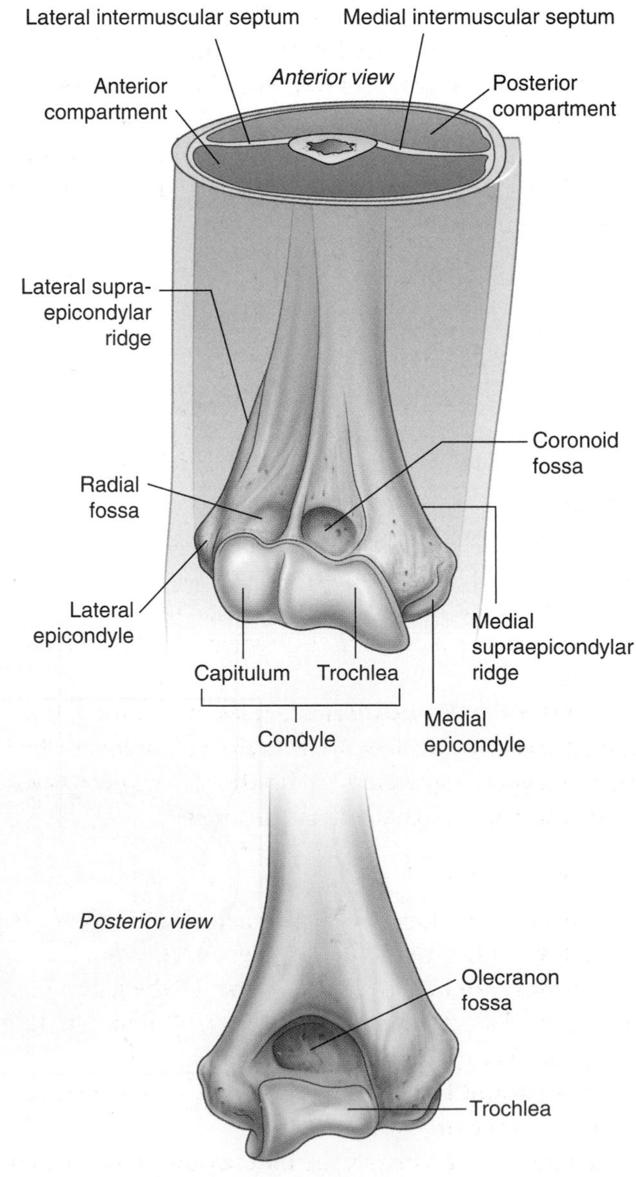


**Fig. 7.61** Distal end of the humerus.

Distally, the bone becomes flattened, and these borders expand as the **lateral supraepicondylar ridge** (**lateral supracondylar ridge**) and the **medial supraepicondylar ridge** (**medial supracondylar ridge**). The lateral supraepicondylar ridge is more pronounced than the medial ridge and is roughened for the attachment of muscles found in the posterior compartment of the forearm.

The distal end of the humerus, which is flattened in the anteroposterior plane, bears a condyle, two epicondyles, and three fossae, as follows (Fig. 7.61).

### The condyle

The two articular parts of the condyle, the **capitulum** and the **trochlea**, articulate with the two bones of the forearm.

The **capitulum** articulates with the radius of the forearm. Lateral in position and hemispherical in shape, it projects anteriorly and somewhat inferiorly and is not visible when the humerus is viewed from the posterior aspect.

The **trochlea** articulates with the ulna of the forearm. It is pulley shaped and lies medial to the capitulum. Its medial edge is more pronounced than its lateral edge and, unlike the capitulum, it extends onto the posterior surface of the bone.

### The two epicondyles

The two epicondyles lie adjacent, and somewhat superior, to the trochlea and capitulum (Fig. 7.61).

The **medial epicondyle**, a large bony protuberance, is the major palpable landmark on the medial side of the elbow, and projects medially from the distal end of the humerus. On its surface, it bears a large oval impression for the attachment of muscles in the anterior compartment of the forearm. The ulnar nerve passes from the arm into the forearm around the posterior surface of the medial epicondyle and can be palpated against the bone in this location.

The **lateral epicondyle** is much less pronounced than the medial epicondyle. It is lateral to the capitulum and has a large irregular impression for the attachment of muscles in the posterior compartment of the forearm.

### The three fossae

Three fossae occur superior to the trochlea and capitulum on the distal end of the humerus (Fig. 7.61).

The **radial fossa** is the least distinct of the fossae and occurs immediately superior to the capitulum on the anterior surface of the humerus.

The **coronoid fossa** is adjacent to the radial fossa and is superior to the trochlea.

The largest of the fossae, the **olecranon fossa**, occurs immediately superior to the trochlea on the posterior surface of the distal end of the humerus.

These three fossae accommodate projections from the bones in the forearm during movements of the elbow joint.

## Proximal end of the radius

The proximal end of the radius consists of a head, a neck, and the radial tuberosity (Fig. 7.62A,B).

The **head** of the radius is a thick disc-shaped structure oriented in the horizontal plane. The circular superior surface is concave for articulation with the capitulum of the humerus. The thick margin of the disc is broad medially where it articulates with the radial notch on the proximal end of the ulna.

The **neck** of the radius is a short and narrow cylinder of bone between the expanded head and the radial tuberosity on the shaft.

The **radial tuberosity** is a large blunt projection on the medial surface of the radius immediately inferior to the neck. Much of its surface is roughened for the attachment of the biceps brachii tendon. The oblique line of the radius continues diagonally across the shaft of the bone from the inferior margin of the radial tuberosity.

## Proximal end of the ulna

The proximal end of the ulna is much larger than the proximal end of the radius and consists of the olecranon, the coronoid process, the trochlear notch, the radial notch, and the tuberosity of the ulna (Fig. 7.63A,B).

The **olecranon** is a large projection of bone that extends proximally from the ulna. Its anterolateral surface is articular and contributes to the formation of the trochlear notch, which articulates with the trochlea of the humerus. The superior surface is marked by a large roughened impression for the attachment of the triceps brachii muscle. The posterior surface is smooth, shaped somewhat triangularly, and can be palpated as the "tip of the elbow."

The **coronoid process** projects anteriorly from the proximal end of the ulna (Fig. 7.63). Its superolateral surface is articular and participates, with the olecranon, in forming the **trochlear notch**. The lateral surface is marked by the **radial notch** for articulation with the head of the radius.

Just inferior to the radial notch is a fossa that allows the radial tuberosity to change position during pronation and supination. The posterior margin of this fossa is broadened to form the **supinator crest**. The anterior surface of the coronoid process is triangular, with the apex directed distally, and has a number of roughenings for muscle attachment. The largest of these roughenings, the **tuberosity of the ulna**, is at the apex of the anterior surface and is the attachment site for the brachialis muscle.

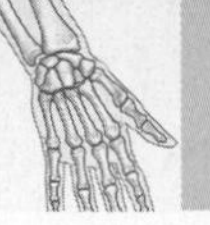

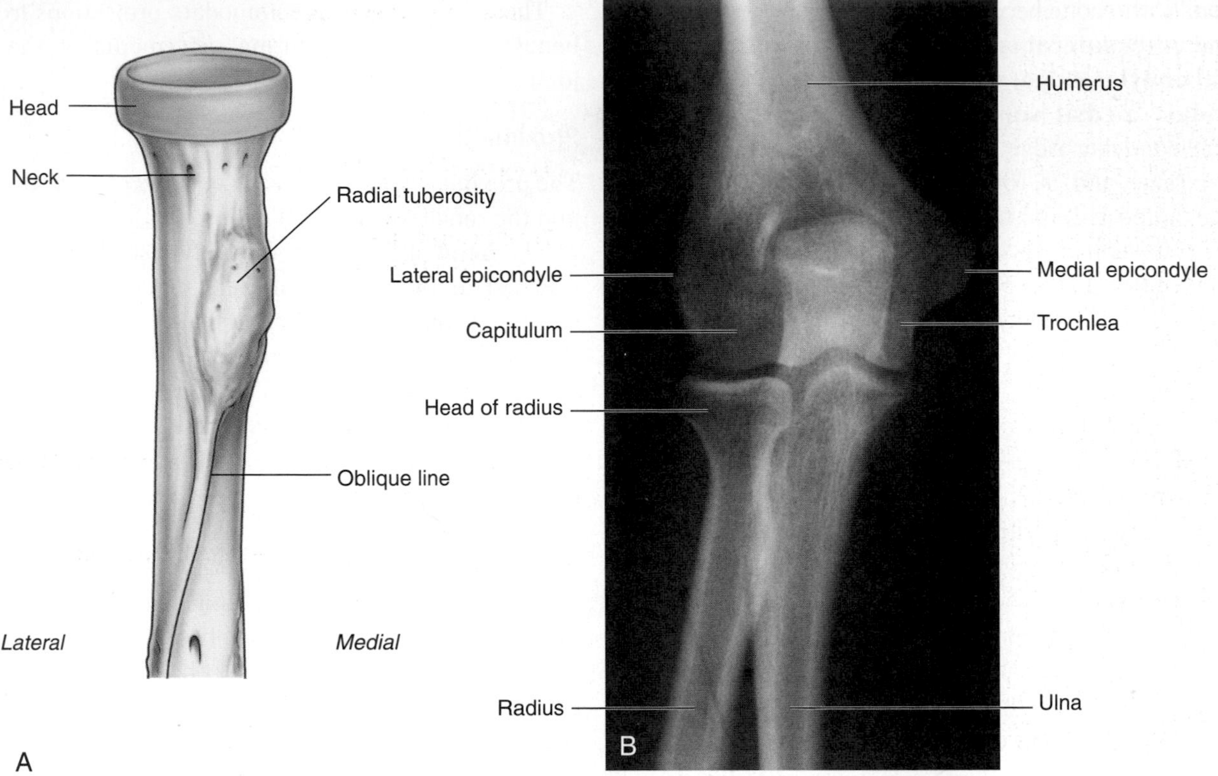


**Fig. 7.62** **A.** Anterior view of the proximal end of the radius. **B.** Radiograph of the elbow joint (anteroposterior view).

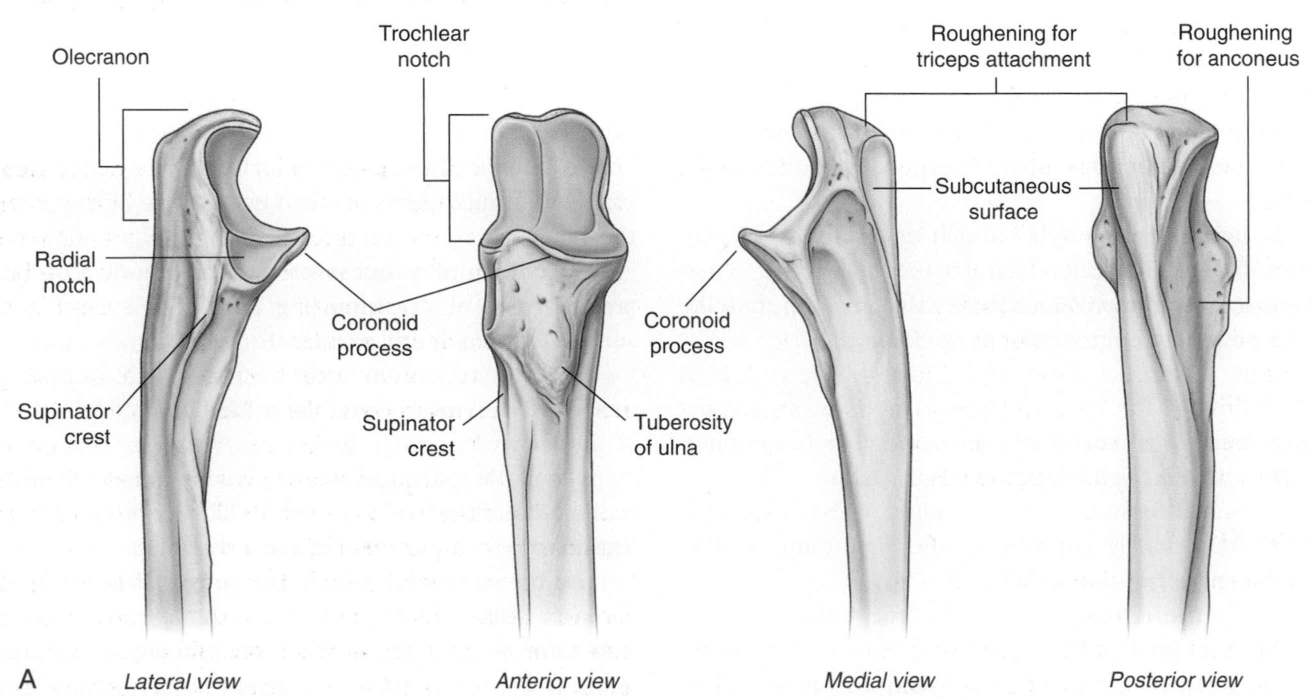


**Fig. 7.63** **A.** Lateral, anterior, medial, and posterior views of the proximal end of the ulna.

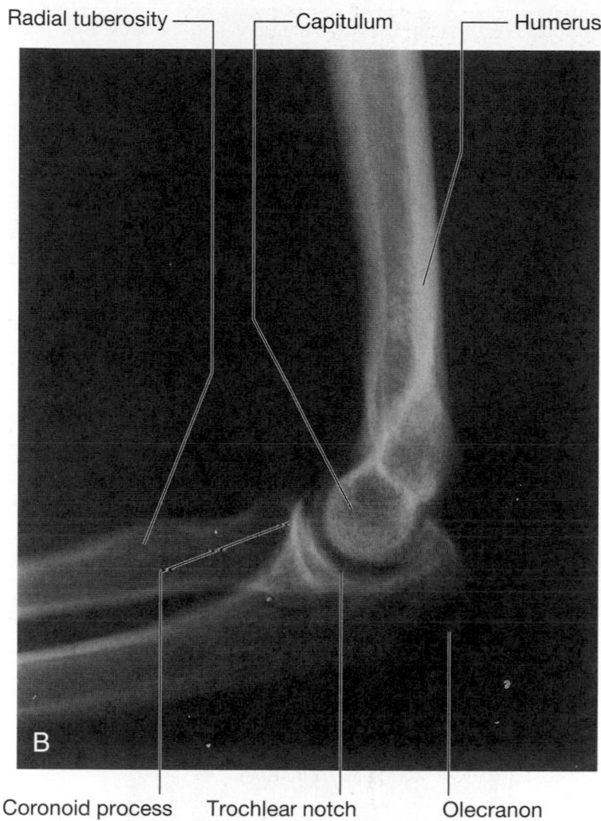


**Fig. 7.63, cont'd B.** Radiograph of the elbow joint (lateral view).

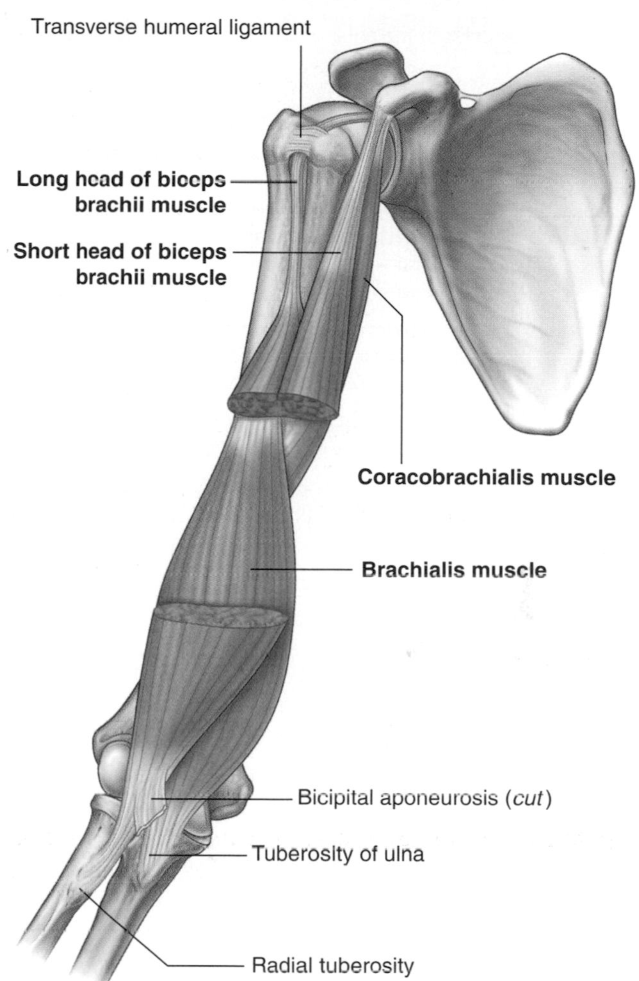


**Fig. 7.64** Coracobrachialis, biceps brachii, and brachialis muscles.

## Muscles

The anterior compartment of the arm contains three muscles—the coracobrachialis, brachialis, and biceps brachii muscles—which are innervated predominantly by the musculocutaneous nerve.

The posterior compartment contains one muscle—the triceps brachii muscle—which is innervated by the radial nerve.

### Coracobrachialis

The **coracobrachialis muscle** extends from the tip of the coracoid process of the scapula to the medial side of the midshaft of the humerus (Fig. 7.64 and Table 7.8). It passes through the axilla and is penetrated and innervated by the musculocutaneous nerve.

The coracobrachialis muscle flexes the arm.

### Biceps brachii

The **biceps brachii** muscle has two heads:

- The short head of the muscle originates from the coracoid process in conjunction with the coracobrachialis.
- The long head originates as a tendon from the supraglenoid tubercle of the scapula (Fig. 7.64 and Table 7.8).

The tendon of the long head passes through the glenohumeral joint superior to the head of the humerus and then passes through the intertubercular sulcus and enters the arm. In the arm, the tendon joins with its muscle belly and, together with the muscle belly of the short head, overlies the brachialis muscle.

The long and short heads converge to form a single tendon, which inserts onto the radial tuberosity.

As the tendon enters the forearm, a flat sheet of connective tissue (the **bicipital aponeurosis**) fans out from the medial side of the tendon to blend with deep fascia covering the anterior compartment of the forearm.

The biceps brachii muscle is a powerful flexor of the forearm at the elbow joint; it is also the most powerful supinator of the forearm when the elbow joint is flexed. Because the two heads of the biceps brachii muscle cross the glenohumeral joint, the muscle can also flex the glenohumeral joint.

**Table 7.8** Muscles of the anterior compartment of the arm (spinal segments in bold are the major segments innervating the muscle)

| Muscle | Origin | Insertion | Innervation | Function |
|---|---|---|---|---|
| Coracobrachialis | Apex of coracoid process | Linear roughening on midshaft of humerus on medial side | Musculocutaneous nerve (**C5, C6, C7**) | Flexor of the arm at the glenohumeral joint |
| Biceps brachii | Long head—supraglenoid tubercle of scapula; short head—apex of coracoid process | Radial tuberosity | Musculocutaneous nerve (**C5, C6**) | Powerful flexor of the forearm at the elbow joint and supinator of the forearm; accessory flexor of the arm at the glenohumeral joint |
| Brachialis | Anterior aspect of humerus (medial and lateral surfaces) and adjacent intermuscular septae | Tuberosity of the ulna | Musculocutaneous nerve (C5, **C6**); small contribution by the radial nerve (C7) to lateral part of muscle | Powerful flexor of the forearm at the elbow joint |

The biceps brachii muscle is innervated by the musculocutaneous nerve. A tap on the tendon of the biceps brachii at the elbow is used to test predominantly spinal cord segment C6.

## In the clinic

### Rupture of biceps tendon

It is relatively unusual for muscles and their tendons to rupture in the upper limb; however, the tendon that most commonly ruptures is the tendon of the long head of the biceps brachii muscle. In isolation, this has relatively little effect on the upper limb, but it does produce a characteristic deformity—on flexing the elbow, there is an extremely prominent bulge of the muscle belly as its unrestrained fibers contract—the "Popeye" sign.

Distal biceps tendon rupture also occurs. It is important to determine the site of the rupture, whether it's at the musculotendinous junction, midtendon, or at the insertion because this will determine the surgical approach for repair.

### Brachialis

The **brachialis muscle** originates from the distal half of the anterior aspect of the humerus and from adjacent parts of the intermuscular septa, particularly on the medial side (Fig. 7.64 and Table 7.8). It lies beneath the biceps brachii muscle, is flattened dorsoventrally, and converges to form a tendon, which attaches to the tuberosity of the ulna.

The brachialis muscle flexes the forearm at the elbow joint.

Innervation of the brachialis muscle is predominantly by the musculocutaneous nerve. A small component of the lateral part is innervated by the radial nerve.

### Posterior compartment

The only muscle of the posterior compartment of the arm is the **triceps brachii muscle** (Fig. 7.65 and Table 7.9). The triceps brachii muscle has three heads:

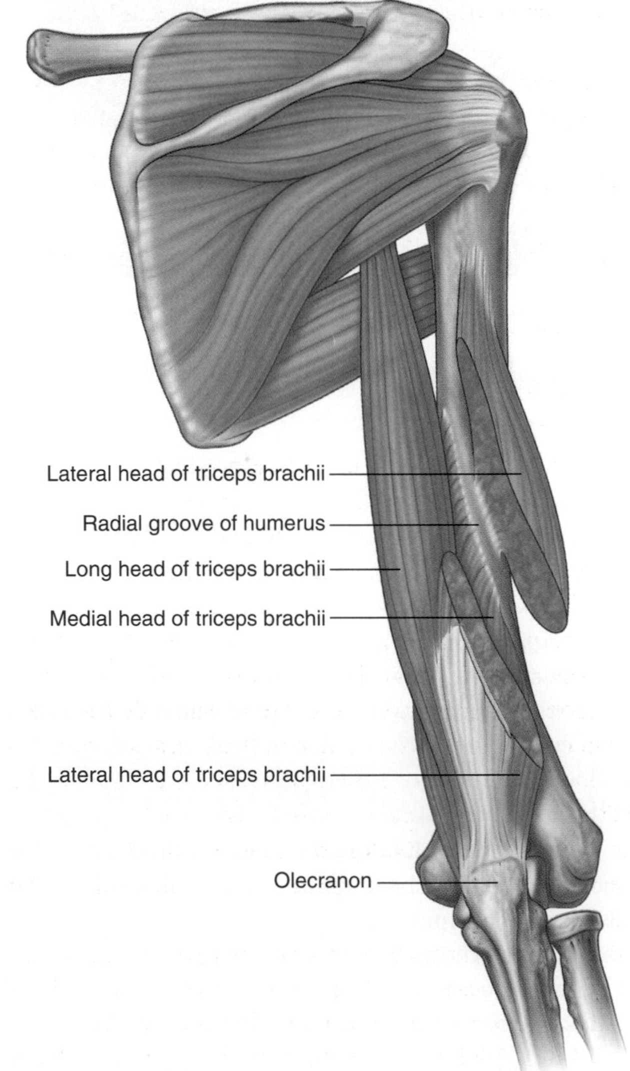


**Fig. 7.65** Triceps muscle.

Table 7.9 Muscle of the posterior compartment of the arm (spinal segment indicated in bold is the major segment innervating the muscle)

| Muscle | Origin | Insertion | Innervation | Function |
|---|---|---|---|---|
| Triceps brachii | Long head—infraglenoid tubercle of scapula; medial head—posterior surface of humerus; lateral head—posterior surface of humerus | Olecranon | Radial nerve (C6, **C7**, C8) | Extension of the forearm at the elbow joint; long head can also extend and adduct the arm at the shoulder joint |

- The long head originates from the infraglenoid tubercle of the scapula.
- The medial head originates from the extensive area on the posterior surface of the shaft of the humerus inferior to the radial groove.
- The lateral head originates from a linear roughening superior to the radial groove of the humerus.

The three heads converge to form a large tendon, which inserts on the superior surface of the olecranon of the ulna.

The triceps brachii muscle extends the forearm at the elbow joint.

Innervation of the triceps brachii is by branches of the radial nerve. A tap on the tendon of the triceps brachii tests predominantly spinal cord segment C7.

## Arteries and veins

### Brachial artery

The major artery of the arm, the **brachial artery**, is found in the anterior compartment (Fig. 7.66A). Beginning as a continuation of the axillary artery at the lower border of the teres major muscle, it terminates just distal to the elbow joint where it divides into the radial and ulnar arteries.

In the proximal arm, the brachial artery lies on the medial side. In the distal arm, it moves laterally to assume a position midway between the lateral epicondyle and the medial epicondyle of the humerus. It crosses anteriorly to the elbow joint where it lies immediately medial to the tendon of the biceps brachii muscle. The brachial artery is palpable along its length. In proximal regions, the brachial artery can be compressed against the medial side of the humerus.

Branches of the brachial artery in the arm include those to adjacent muscles and two ulnar collateral vessels, which contribute to a network of arteries around the elbow joint (Fig. 7.66B). Additional branches are the profunda brachii artery and nutrient arteries to the humerus, which pass through a foramen in the anteromedial surface of the humeral shaft.

### Profunda brachii artery

The **profunda brachii artery,** the largest branch of the brachial artery, passes into and supplies the posterior compartment of the arm (Fig. 7.66A,B). It enters the posterior compartment with the radial nerve and together they pass through the triangular interval, which is formed by the shaft of the humerus, the inferior margin of the teres major muscle, and the lateral margin of the long head of the triceps muscle. They then pass along the radial groove on the posterior surface of the humerus deep to the lateral head of the triceps brachii muscle.

Branches of the profunda brachii artery supply adjacent muscles and anastomose with the posterior circumflex humeral artery. The artery terminates as two collateral vessels, which contribute to an anastomotic network of arteries around the elbow joint (Fig. 7.66B).

## In the clinic

### Blood pressure measurement

Blood pressure measurement is an extremely important physiological parameter. High blood pressure (hypertension) requires treatment to prevent long-term complications such as stroke. Low blood pressure may be caused by extreme blood loss, widespread infection, or poor cardiac output (e.g., after myocardial infarction). Accurate measurement of blood pressure is essential.

Most clinicians use a sphygmomanometer and a stethoscope. The sphygmomanometer is a device that inflates a cuff around the midportion of the arm to compress the brachial artery against the humerus. The cuff is inflated so it exceeds the systolic blood pressure (greater than 120 mm Hg). The clinician places a stethoscope over the brachial artery in the cubital fossa and listens (auscultates) for the pulse. As the pressure in the arm cuff of the sphygmomanometer is reduced just below the level of the systolic blood pressure, the pulse becomes audible as a regular thumping sound. As the pressure in the sphygmomanometer continues to drop, the regular thumping sound becomes clearer. When the pressure in the sphygmomanometer is less than that of the diastolic blood pressure, the audible thumping sound becomes inaudible. Using the simple scale on the sphygmomanometer, the patient's blood pressure can be determined. The normal range is 90–120/60–80 mm Hg (systolic blood pressure/diastolic blood pressure).

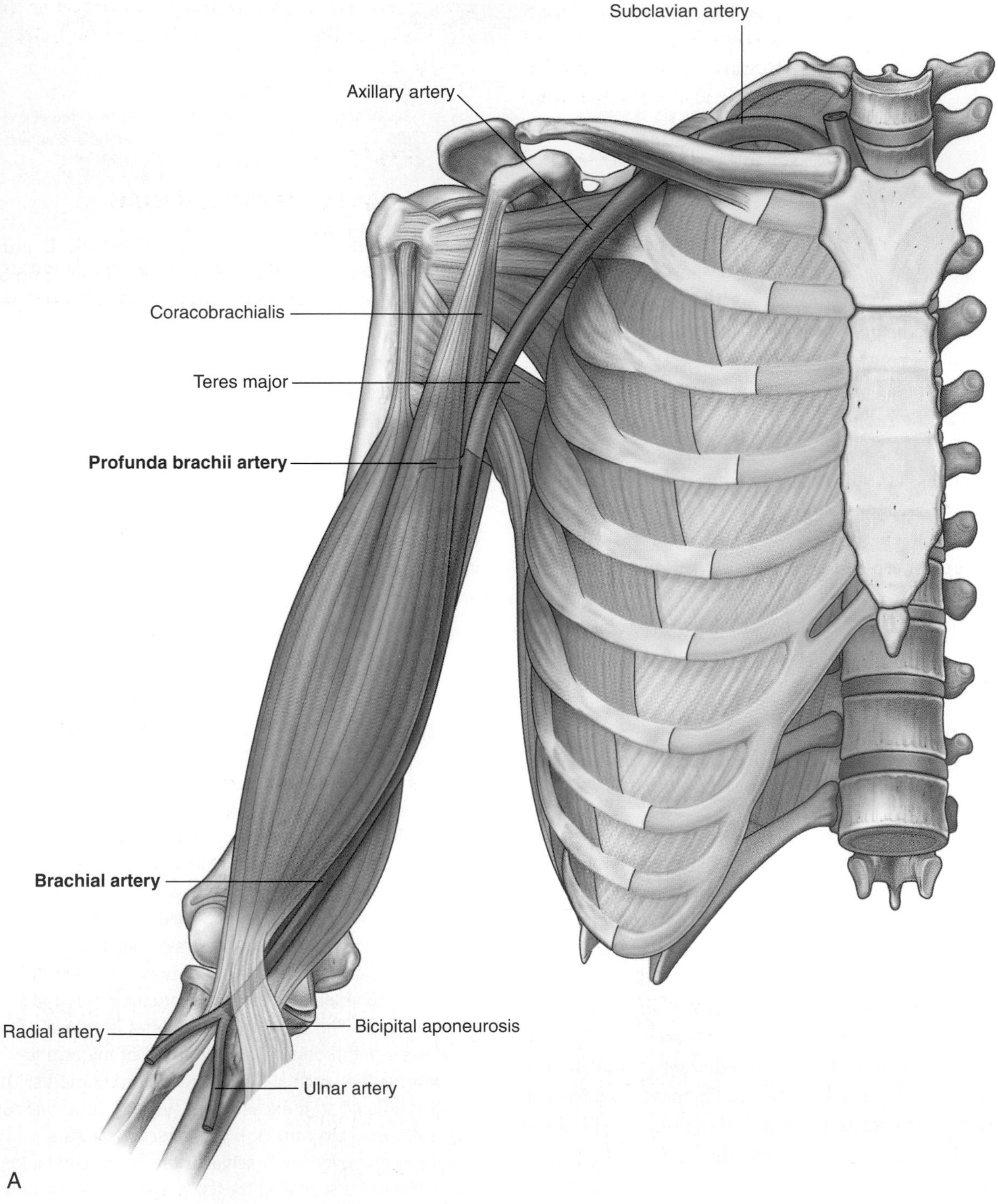

**Fig. 7.66** Brachial artery. **A.** In context.

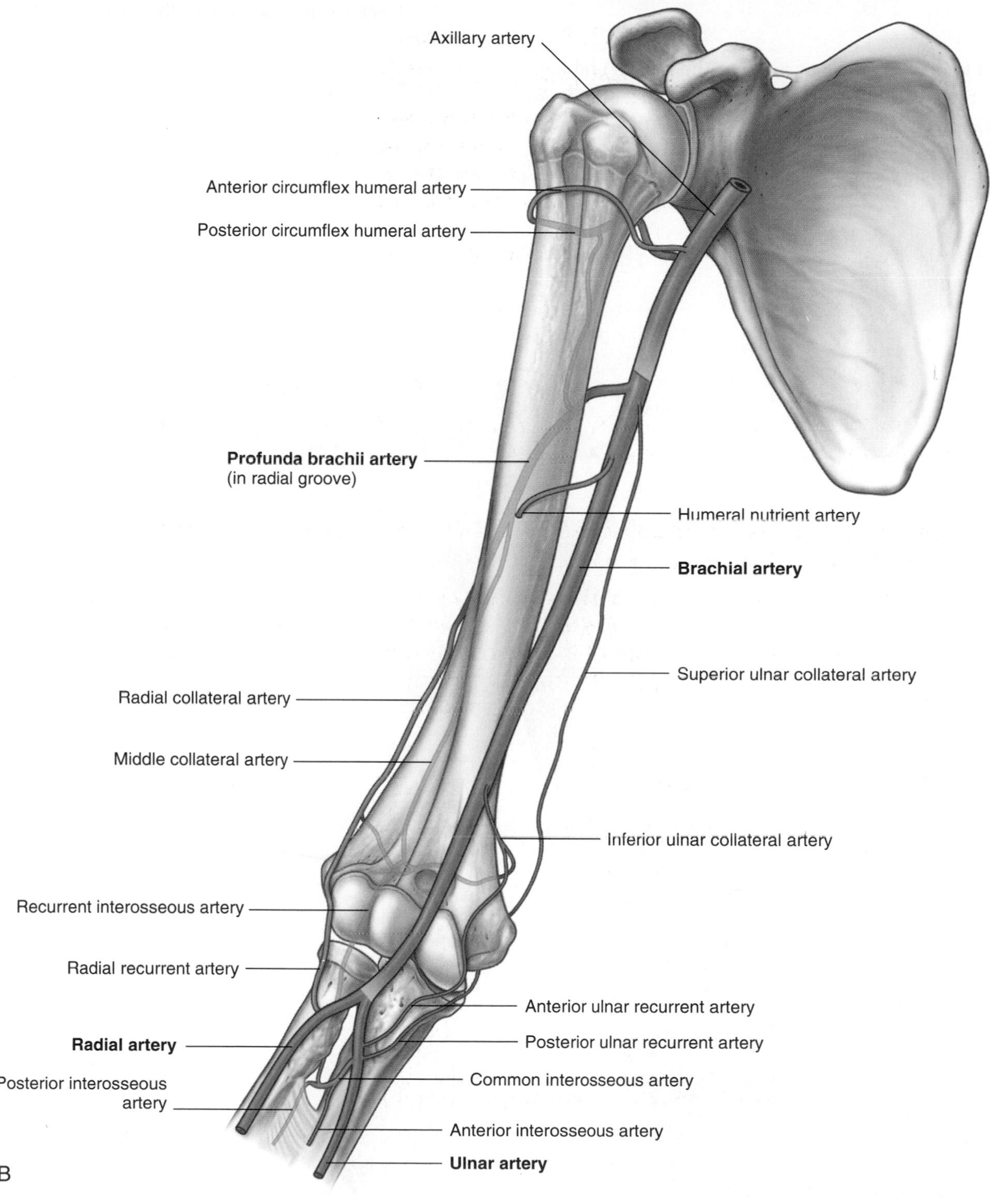

Axillary artery
Anterior circumflex humeral artery
Posterior circumflex humeral artery
Profunda brachii artery
(in radial groove)
Humeral nutrient artery
Brachial artery
Superior ulnar collateral artery
Radial collateral artery
Middle collateral artery
Inferior ulnar collateral artery
Recurrent interosseous artery
Radial recurrent artery
Anterior ulnar recurrent artery
Radial artery
Posterior ulnar recurrent artery
Common interosseous artery
Posterior interosseous artery
Anterior interosseous artery
Ulnar artery
B

**Fig. 7.66, cont'd** Brachial artery. **B.** Branches.

### Veins

**Paired brachial veins** pass along the medial and lateral sides of the brachial artery, receiving tributaries that accompany branches of the artery (Fig. 7.67).

In addition to these deep veins, two large subcutaneous veins, the basilic vein and the cephalic vein, are located in the arm.

The basilic vein passes vertically in the distal half of the arm, penetrates deep fascia to assume a position medial to the brachial artery, and then becomes the axillary vein at the lower border of the teres major muscle. The brachial veins join the basilic, or axillary, vein.

The cephalic vein passes superiorly on the anterolateral aspect of the arm and through the anterior wall of the axilla to reach the axillary vein.

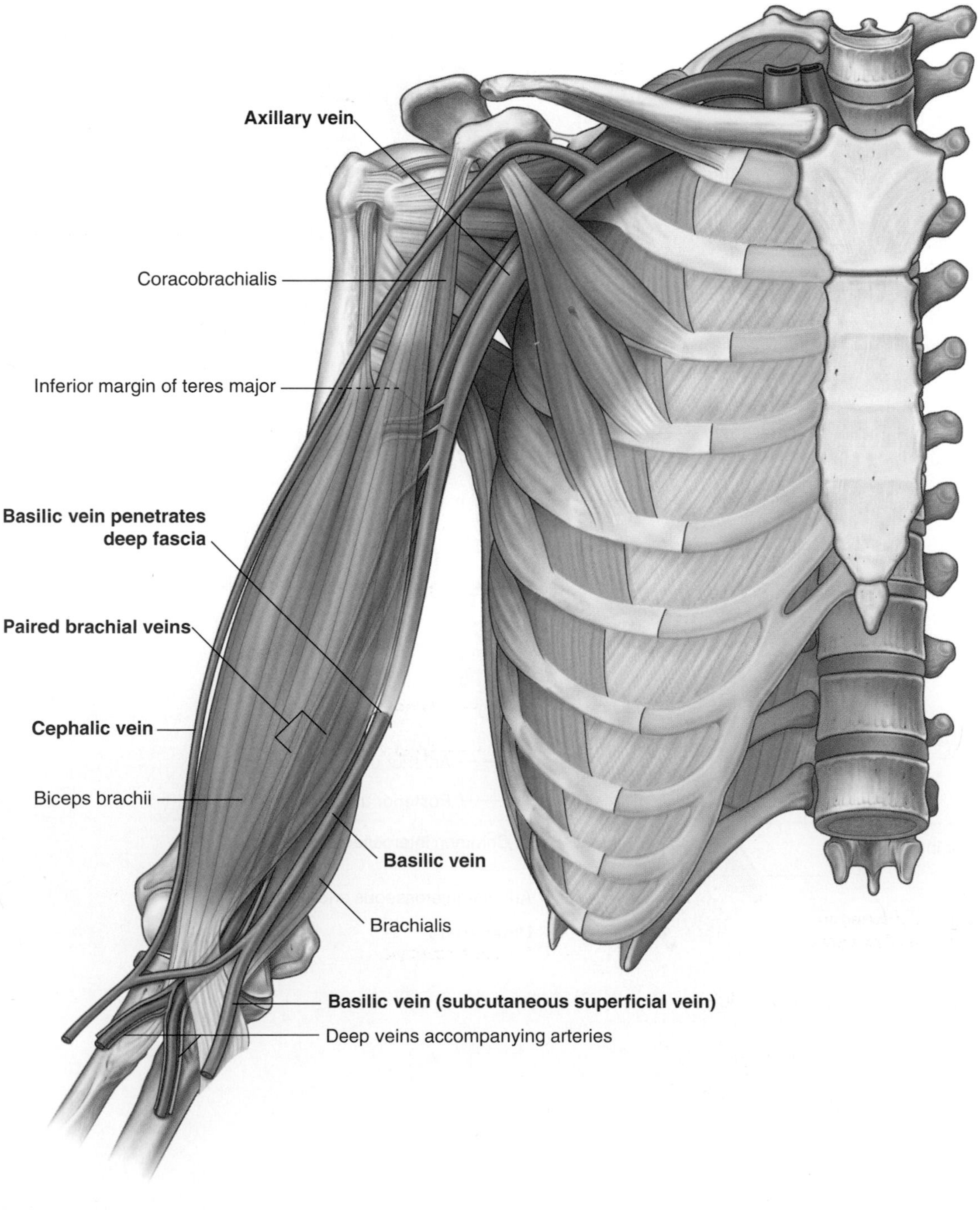

**Fig. 7.67** Veins of the arm.

## Nerves

### Musculocutaneous nerve

The musculocutaneous nerve leaves the axilla and enters the arm by passing through the coracobrachialis muscle (Fig. 7.68). It passes diagonally down the arm in the plane between the biceps brachii and brachialis muscles. After giving rise to motor branches in the arm, it emerges laterally to the tendon of the biceps brachii muscle at the elbow, penetrates deep fascia, and continues as the **lateral cutaneous nerve of the forearm**.

The musculocutaneous nerve provides:

- motor innervation to all muscles in the anterior compartment of the arm, and
- sensory innervation to skin on the lateral surface of the forearm.

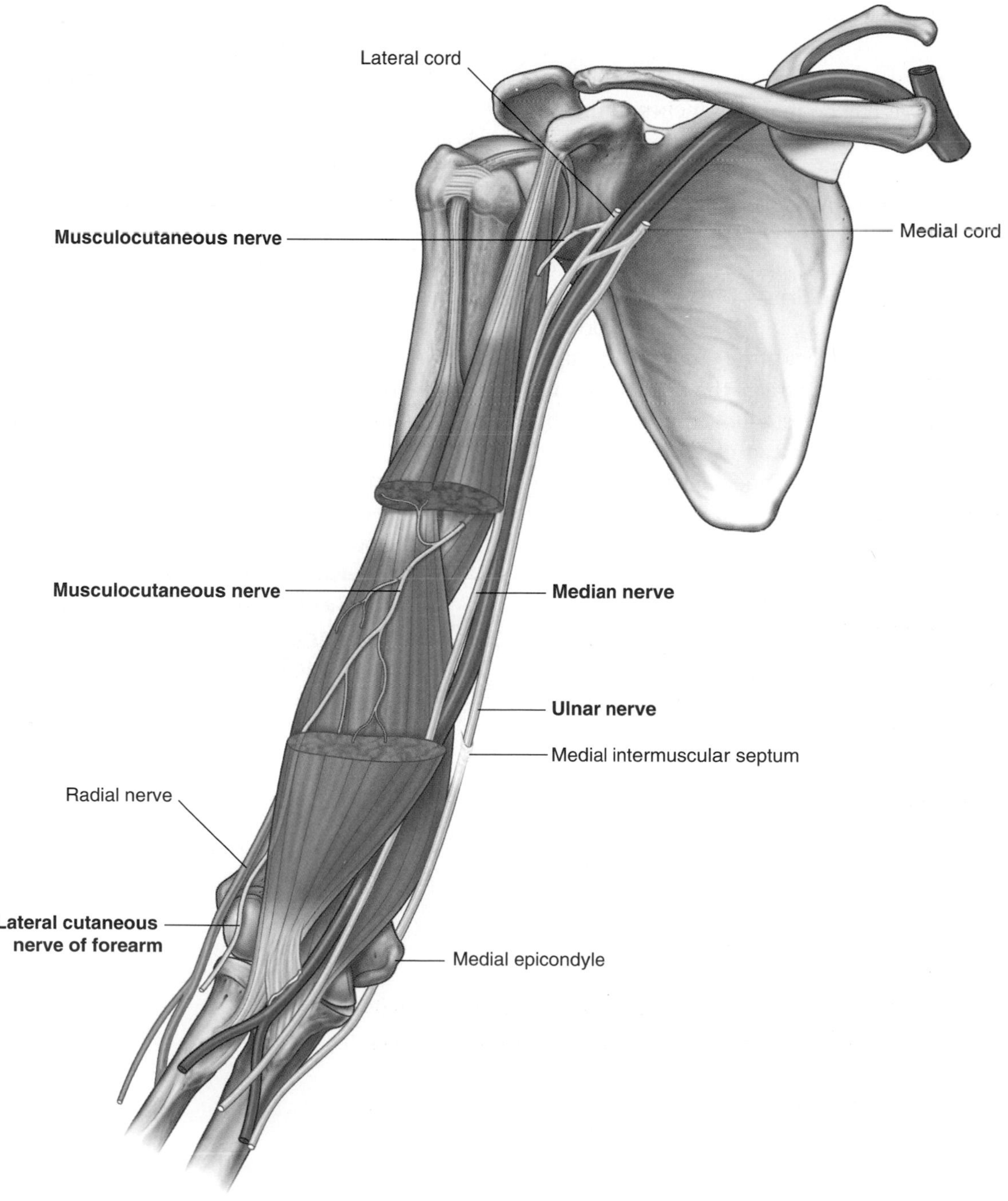


**Fig. 7.68** Musculocutaneous, median, and ulnar nerves in the arm.

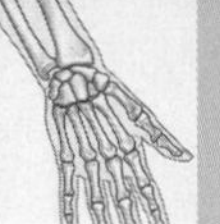



### Median nerve

The median nerve enters the arm from the axilla at the inferior margin of the teres major muscle (Fig. 7.68). It passes vertically down the medial side of the arm in the anterior compartment and is related to the brachial artery throughout its course:

- In proximal regions, the median nerve is immediately lateral to the brachial artery.
- In more distal regions, the median nerve crosses to the medial side of the brachial artery and lies anterior to the elbow joint.

The median nerve has no major branches in the arm, but a branch to one of the muscles of the forearm, the pronator teres muscle, may originate from the nerve immediately proximal to the elbow joint.

### Ulnar nerve

The ulnar nerve enters the arm with the median nerve and axillary artery (Fig. 7.68). It passes through proximal regions medial to the axillary artery. In the middle of the arm, the ulnar nerve penetrates the medial intermuscular septum and enters the posterior compartment where it lies anterior to the medial head of the triceps brachii muscle. It passes posterior to the medial epicondyle of the humerus and then into the anterior compartment of the forearm.

The ulnar nerve has no major branches in the arm.

### Radial nerve

The radial nerve originates from the posterior cord of the brachial plexus and enters the arm by crossing the inferior margin of the teres major muscle (Fig. 7.69). As it enters the arm, it lies posterior to the brachial artery. Accompanied by the profunda brachii artery, the radial nerve enters the posterior compartment of the arm by passing through the triangular interval.

As the radial nerve passes diagonally, from medial to lateral, through the posterior compartment, it lies in the radial groove directly on bone. On the lateral side of the arm, it passes anteriorly through the lateral intermuscular septum and enters the anterior compartment where it lies between the brachialis muscle and a muscle of the posterior compartment of the forearm—the brachioradialis muscle, which attaches to the lateral supraepicondylar ridge of the humerus. The radial nerve enters the forearm anterior to the lateral epicondyle of the humerus, just deep to the brachioradialis muscle.

In the arm, the radial nerve has muscular and cutaneous branches (Fig. 7.69).

- Muscular branches include those to the triceps brachii, brachioradialis, and extensor carpi radialis longus muscles. In addition, the radial nerve contributes to the innervation of the lateral part of the brachialis muscle. One of the branches to the medial head of the triceps brachii muscle arises before the radial nerve's entrance into the posterior compartment and passes vertically down the arm in association with the ulnar nerve.
- Cutaneous branches of the radial nerve that originate in the posterior compartment of the arm are the **inferior lateral cutaneous nerve of the arm** and the **posterior cutaneous nerve of the forearm**, both of which penetrate through the lateral head of the triceps brachii muscle and the overlying deep fascia to become subcutaneous.

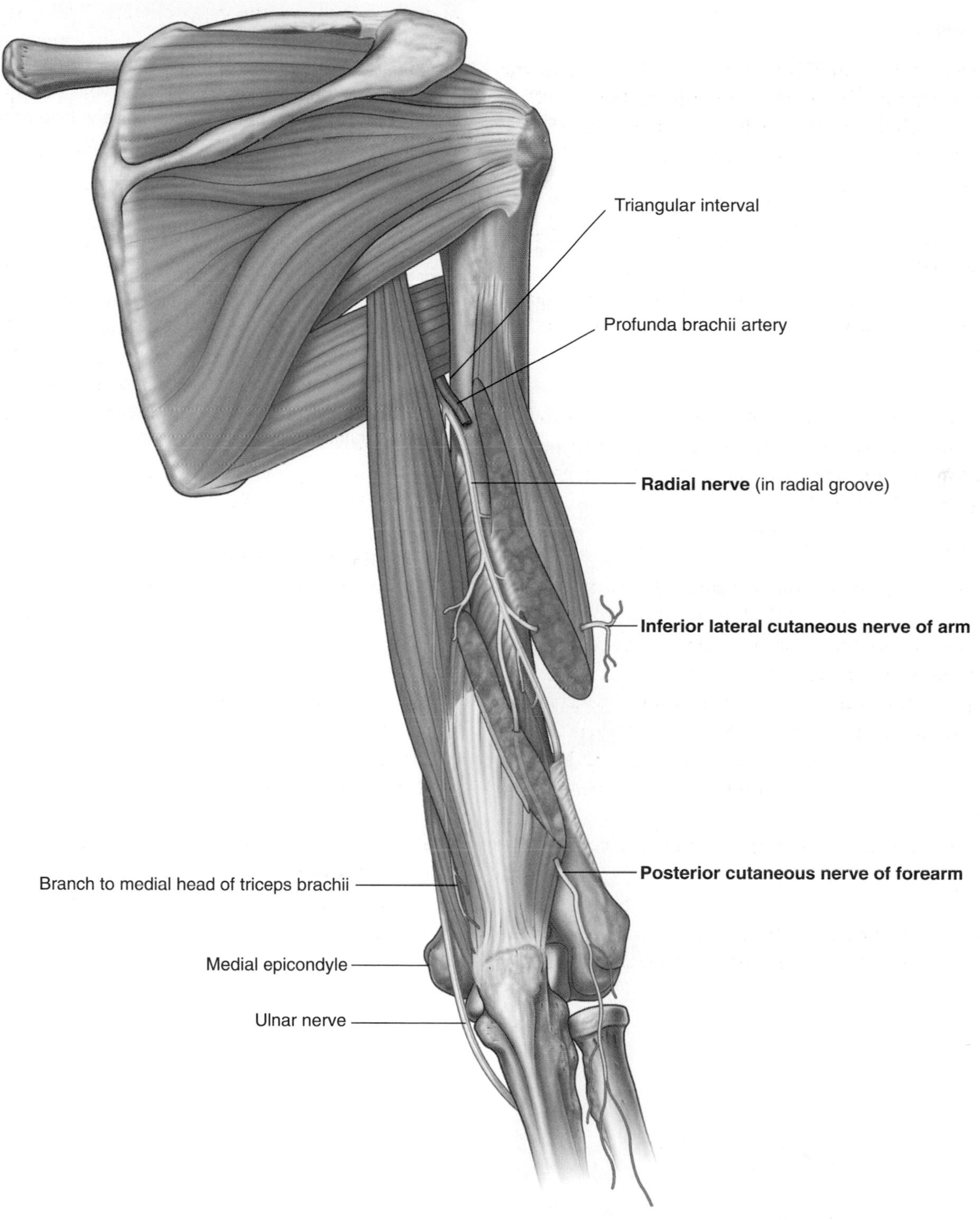


**Fig. 7.69** Radial nerve in the arm.

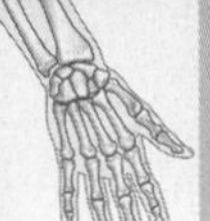



## In the clinic

### Radial nerve injury in the arm

The radial nerve is tightly bound with the profunda brachii artery between the medial and lateral heads of the triceps brachii muscle in the radial groove. If the humerus is fractured, the radial nerve may become stretched or transected in this region, leading to permanent damage and loss of function. This injury is typical (Fig. 7.70) and the nerve should always be tested when a fracture of the midshaft of the humerus is suspected. The patient's symptoms usually include wrist drop (due to denervation of the extensor muscles) and sensory changes over the dorsum of the hand.

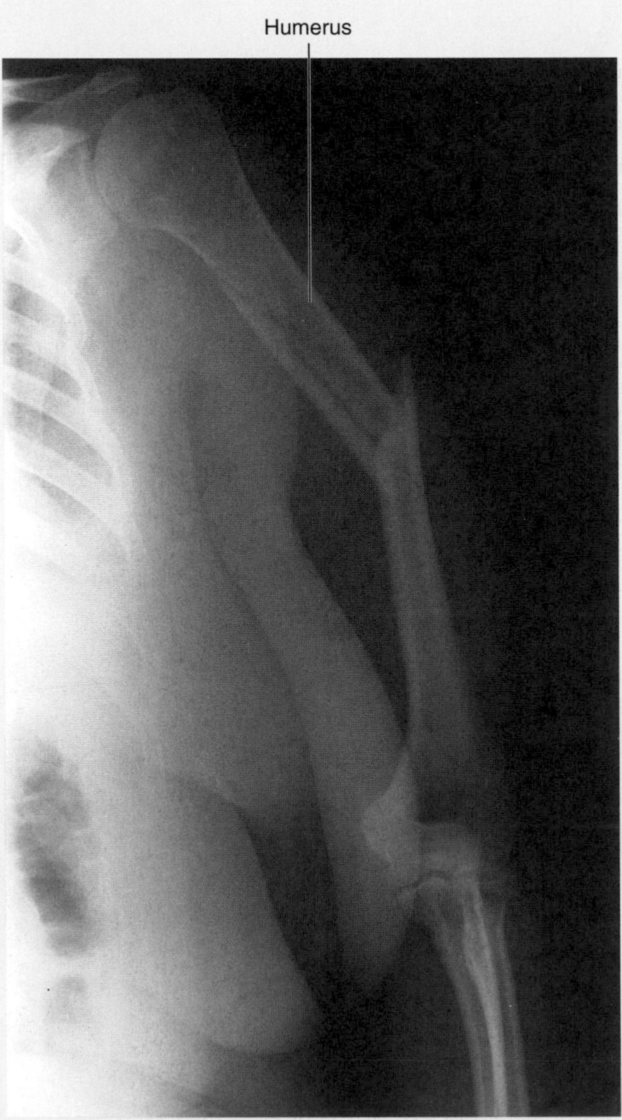


**Fig. 7.70** Radiograph of the humerus demonstrating a midshaft fracture, which may disrupt the radial nerve.

## In the clinic

### Median nerve injury in the arm

In the arm and forearm the median nerve is usually not injured by trauma because of its relatively deep position. The commonest neurological problem associated with the median nerve is compression beneath the flexor retinaculum at the wrist (carpal tunnel syndrome).

On very rare occasions, a fibrous band may arise from the anterior aspect of the humerus beneath which the median nerve passes. This is an embryological remnant of the coracobrachialis muscle and is sometimes called the ligament of Struthers; occasionally, it may calcify. This band can compress the median nerve, resulting in weakness of the flexor muscles in the forearm and the thenar muscles. Nerve conduction studies will demonstrate the site of nerve compression.

## ELBOW JOINT

The elbow joint is a complex joint involving three separate articulations, which share a common synovial cavity (Fig. 7.71).

- The joints between the trochlear notch of the ulna and the trochlea of the humerus and between the head of the radius and the capitulum of the humerus are primarily involved with hinge-like flexion and extension of the forearm on the arm and, together, are the principal articulations of the elbow joint.
- The joint between the head of the radius and the radial notch of the ulna, the proximal radio-ulnar joint, is involved with pronation and supination of the forearm.

The articular surfaces of the bones are covered with hyaline cartilage.

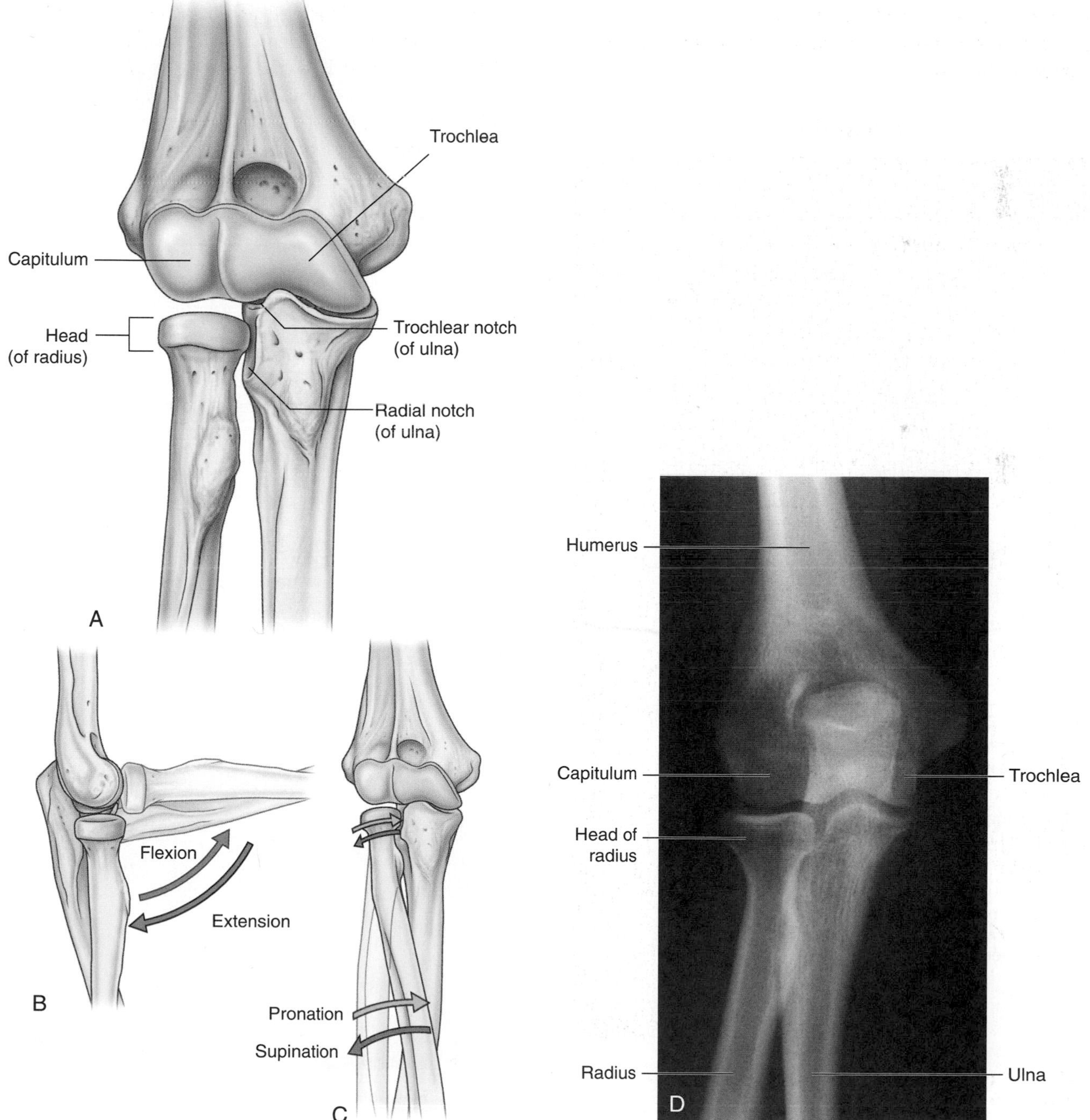


**Fig. 7.71** Components and movements of the elbow joint. **A.** Bones and joint surfaces. **B.** Flexion and extension. **C.** Pronation and supination. **D.** Radiograph of a normal elbow joint (anteroposterior view).

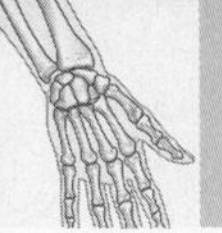



The synovial membrane originates from the edges of the articular cartilage and lines the radial fossa, the coronoid fossa, the olecranon fossa, the deep surface of the joint capsule, and the medial surface of the trochlea (Fig. 7.72).

The synovial membrane is separated from the fibrous membrane of the joint capsule by pads of fat in regions overlying the coronoid fossa, the olecranon fossa, and the radial fossa. These fat pads accommodate the related bony processes during extension and flexion of the elbow. Attachments of the brachialis and triceps brachii muscles to the joint capsule overlying these regions pull the attached fat pads out of the way when the adjacent bony processes are moved into the fossae.

The fibrous membrane of the joint capsule overlies the synovial membrane, encloses the joint, and attaches to the medial epicondyle and the margins of the olecranon, coronoid, and radial fossae of the humerus (Fig. 7.73). It also attaches to the coronoid process and olecranon of the ulna. On the lateral side, the free inferior margin of the joint capsule passes around the neck of the radius from an anterior attachment to the coronoid process of the ulna to a posterior attachment to the base of the olecranon.

The fibrous membrane of the joint capsule is thickened medially and laterally to form collateral ligaments, which

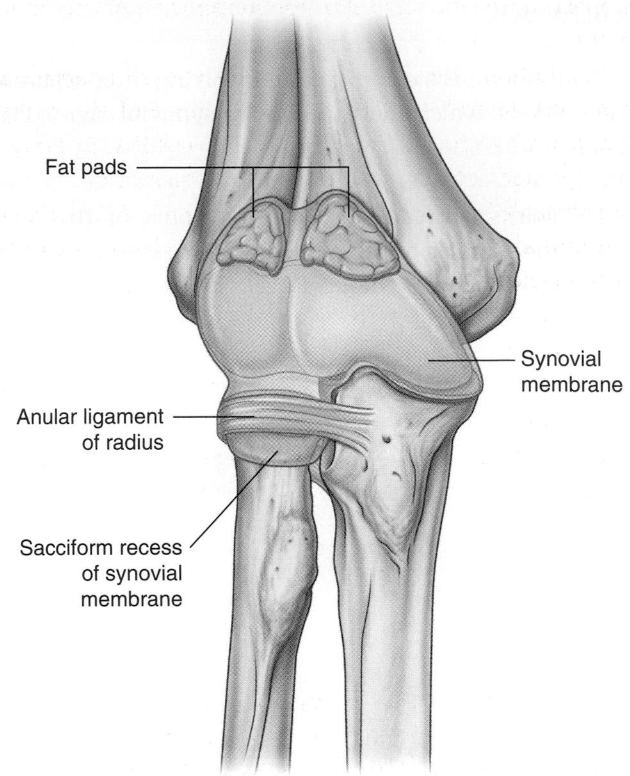


**Fig. 7.72** Synovial membrane of elbow joint (anterior view).

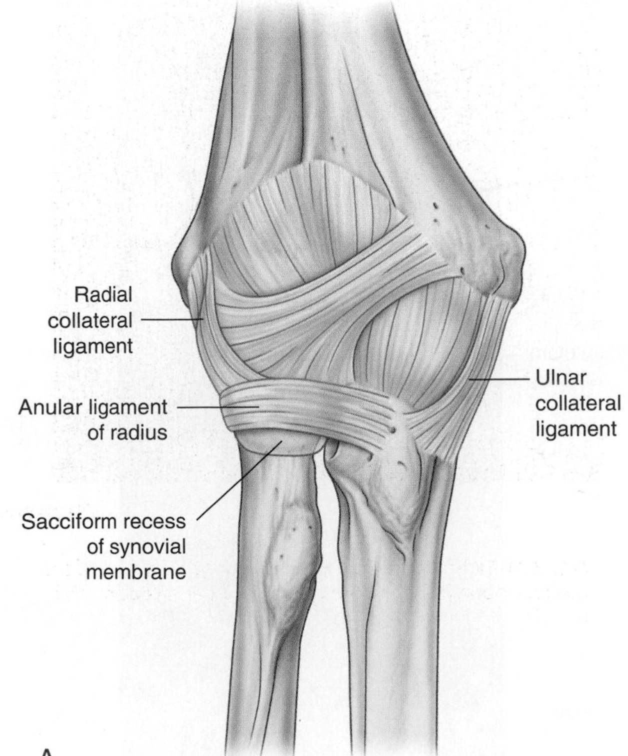


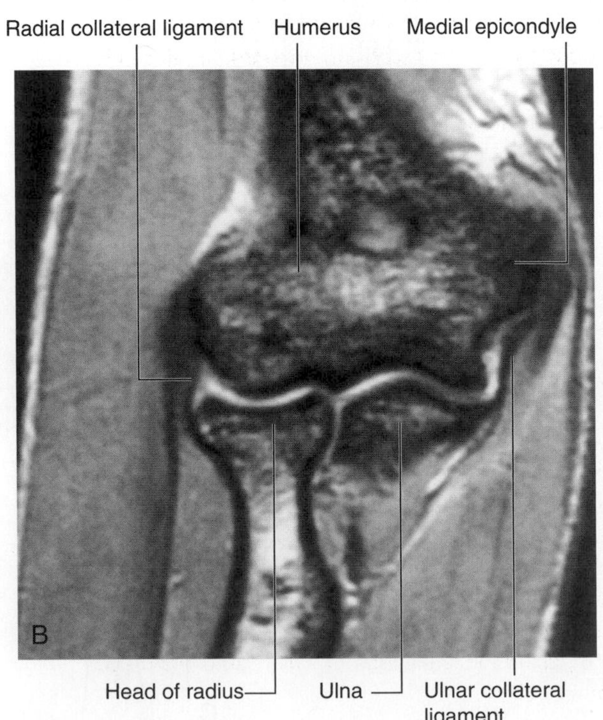


**Fig. 7.73** Elbow joint. **A.** Joint capsule and ligaments of the right elbow joint. **B.** Magnetic resonance image of the elbow joint in the coronal plane.

support the flexion and extension movements of the elbow joint (Fig. 7.73).

In addition, the external surface of the joint capsule is reinforced laterally where it cuffs the head of the radius with a strong **anular ligament of the radius**. Although this ligament blends with the fibrous membrane of the joint capsule in most regions, they are separate posteriorly. The anular ligament of the radius also blends with the **radial collateral ligament**.

The anular ligament of the radius and related joint capsule allow the radial head to slide against the radial notch of the ulna and pivot on the capitulum during pronation and supination of the forearm.

The deep surface of the fibrous membrane of the joint capsule and the related anular ligament of the radius that articulate with the sides of the radial head are lined by cartilage. A pocket of synovial membrane (sacciform recess) protrudes from the inferior free margin of the joint capsule and facilitates rotation of the radial head during pronation and supination.

Vascular supply to the elbow joint is through an anastomotic network of vessels derived from collateral and recurrent branches of the brachial, profunda brachii, radial, and ulnar arteries.

The elbow joint is innervated predominantly by branches of the radial and musculocutaneous nerves, but there may be some innervation by branches of the ulnar and median nerves.

## In the clinic

### Supracondylar fracture of the humerus

Elbow injuries in children may result in a transverse fracture of the distal end of the humerus, above the level of the epicondyles. This fracture is termed a supracondylar fracture. The distal fragment and its soft tissues are pulled posteriorly by the triceps muscle. This posterior displacement effectively "bowstrings" the brachial artery over the irregular proximal fracture fragment. In children, this is a relatively devastating injury: the muscles of the anterior compartment of the forearm are rendered ischemic and form severe contractions, significantly reducing the function of the anterior compartment and flexor muscles (Volkmann's ischemic contracture).

## In the clinic

### Pulled elbow

Pulled elbow is a disorder that typically occurs in children under 5 years of age. It is commonly caused by a sharp pull of the child's hand, usually when the child is pulled up a curb. The not-yet-developed head of the radius and the laxity of the anular ligament of the radius allow the head to sublux from this cuff of tissue. Pulled elbow is extremely painful, but can be treated easily by simple supination and compression of the elbow joint by the clinician. When the radial head is relocated the pain subsides immediately and the child can continue with normal activity.

## In the clinic

### Fracture of the olecranon

Fractures of the olecranon can result from a direct blow to the olecranon or from a fall onto an outstretched hand (Fig. 7.74). The triceps inserts into the olecranon and injuries can cause avulsion of the muscle.

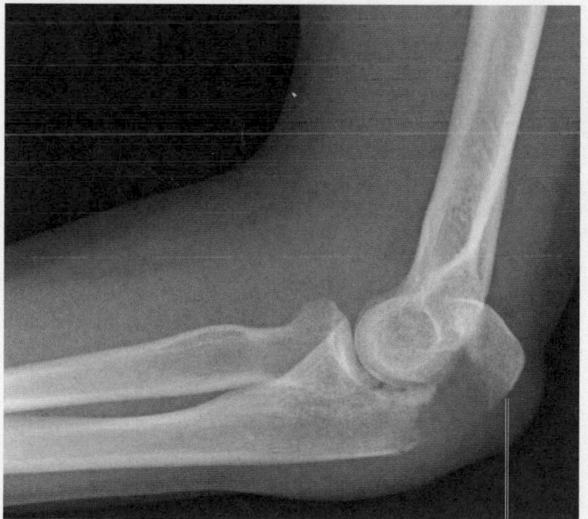


**Fig. 7.74** Radiograph of an elbow showing a fracture of the olecranon and involving the insertion of the triceps brachii muscle.

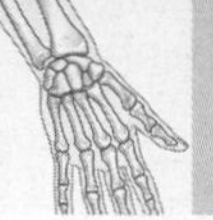



## In the clinic

### Developmental changes in the elbow joint

The elbow joint can be injured in many ways; the types of injuries are age dependent. When a fracture or soft tissue trauma is suspected, a plain lateral and an anteroposterior radiograph are obtained. In an adult it is usually not difficult to interpret the radiograph, but in children additional factors require interpretation.

As the elbow develops in children, numerous secondary ossification centers appear before and around puberty. It is easy to mistakenly interpret these as fractures. In addition, it is also possible for the epiphyses and apophyses to be "pulled off" or disrupted. Therefore, when interpreting a child's radiograph of the elbow, the physician must know the child's age (Fig. 7.75). Fusion occurs at around the time of puberty. An understanding of the normal epiphyses and apophyses and their normal relationship to the bones will secure a correct diagnosis. The approximate ages of appearance of the secondary ossification centers around the elbow joint are:

- capitulum—1 year,
- head (of radius)—5 years,
- medial epicondyle—5 years,
- trochlea—11 years,
- olecranon—12 years, and
- lateral epicondyle—13 years.

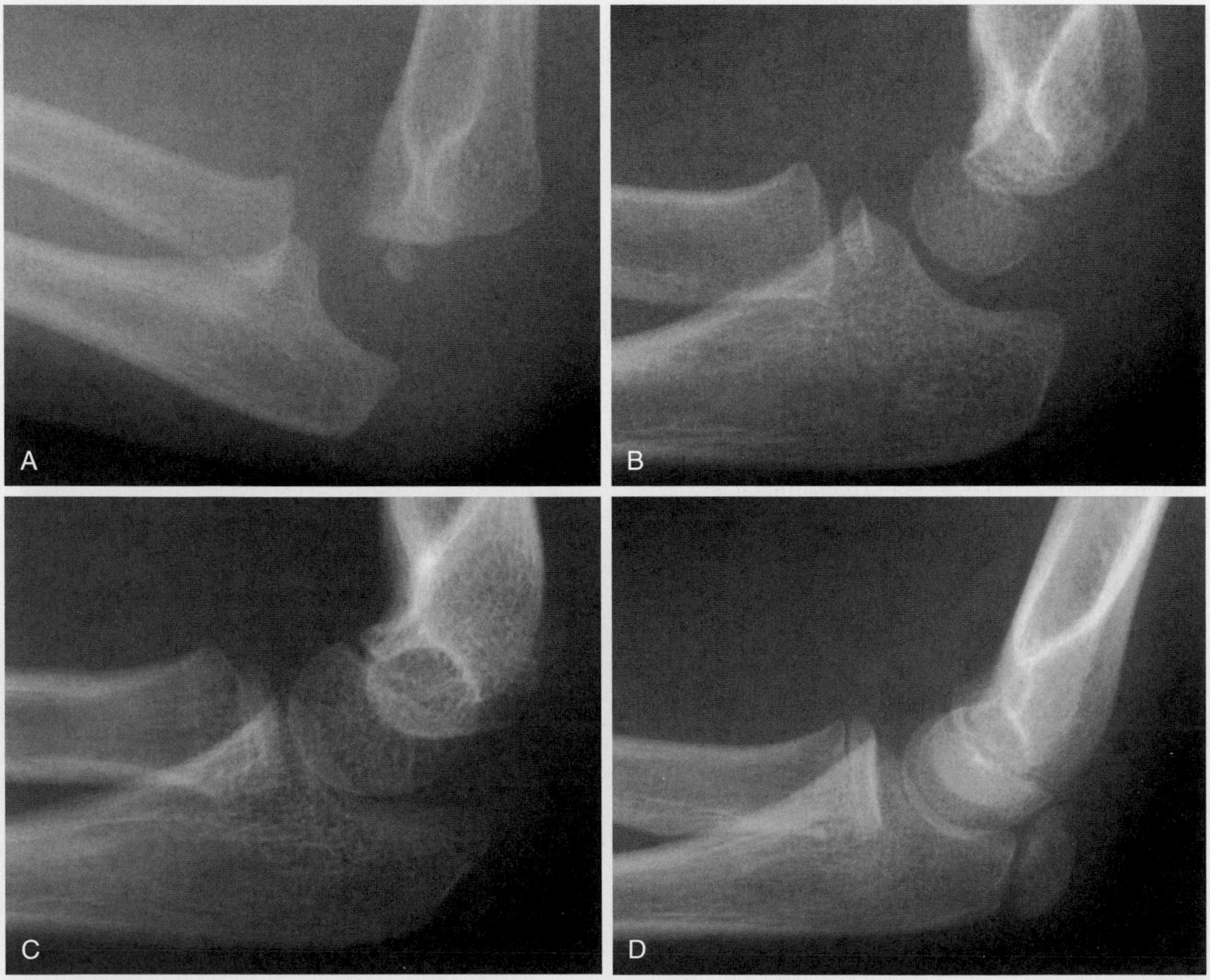

**Fig. 7.75** Radiographs of elbow joint development. **A.** At age 2 years. **B.** At age 5 years. **C.** At age 5–6 years. **D.** At age 12 years.

### In the clinic

**Fracture of the head of the radius**

A fracture of the head of the radius is a common injury and can cause appreciable morbidity. It is one of the typical injuries that occur with a fall on the outstretched hand. On falling, the force is transmitted to the radial head, which fractures. These fractures typically result in loss of full extension, and potential surgical reconstruction may require long periods of physiotherapy to obtain a full range of movement at the elbow joint.

A lateral radiograph of a fracture of the head of the radius typically demonstrates the secondary phenomenon of this injury. When the bone is fractured, fluid fills the synovial cavity, elevating the small pad of fat within the coronoid and olecranon fossae. These fat pads appear as areas of lucency on the lateral radiograph—the "fat pad" sign. This radiological finding is useful because fracture of the head of the radius is not always clearly visible. If there is an appropriate clinical history, tenderness around the head of the radius, and a positive fat pad sign, a fracture can be inferred clinically even if no fracture can be identified on the radiograph, and appropriate treatment can be instituted.

### In the clinic

**"Tennis" and "golfer's" elbow (epicondylitis)**

It is not uncommon for people who are involved in sports such as golf and tennis to develop an overuse strain of the origins of the flexor and extensor muscles of the forearm. The pain is typically around the epicondyles and usually resolves after rest and physical therapy. It may also be treated with injection of the patient's own plasma, rich in platelets, into the tendon to promote tendon healing and repair. If pain and inflammation persist, surgical division of the extensor or flexor origin from the bone may be necessary. Typically, in tennis players this pain occurs on the lateral epicondyle and common extensor origin (tennis elbow), whereas in golfers it occurs on the medial epicondyle and common flexor origin.

### In the clinic

**Elbow arthritis**

Osteoarthritis is extremely common and is usually most severe in the dominant limb. From time to time an arthritic elbow may undergo such degenerative change that small bone fragments appear in the articular cavity. Given the relatively small joint space, these fragments can result in an appreciable reduction in flexion and extension, and typically lodge within the olecranon and coronoid fossae.

## In the clinic

### Ulnar nerve injury at the elbow

Posterior to the medial epicondyle of the humerus the ulnar nerve is bound in a fibro-osseous tunnel (the cubital tunnel) by a retinaculum. Older patients may develop degenerative changes within this tunnel, which compresses the ulnar nerve when flexed. The repeated action of flexion and extension of the elbow may cause local nerve damage, resulting in impaired function of the ulnar nerve. Accessory muscles and localized neuritis in this region secondary to direct trauma may also produce ulnar nerve damage (Fig. 7.76).

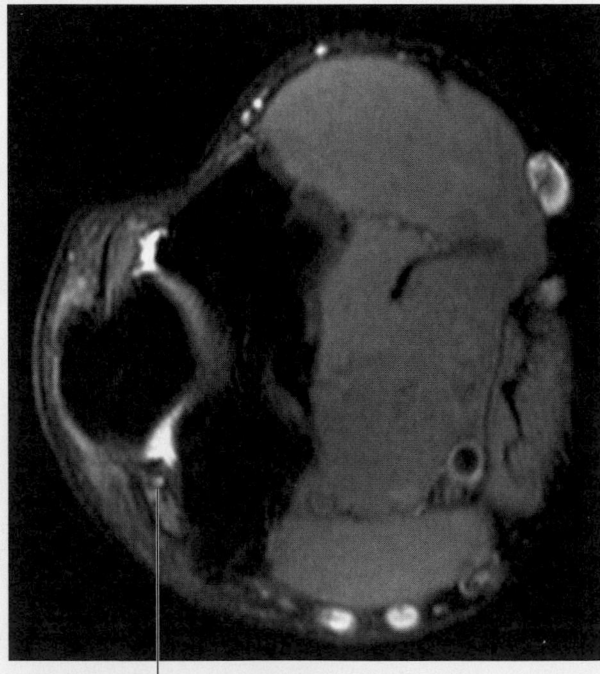


**Fig. 7.76** MRI of right elbow showing swelling of the ulnar nerve in the cubital tunnel posterior to the medial epicondyle, consistent with nerve compression.

## CUBITAL FOSSA

The cubital fossa is an important area of transition between the arm and the forearm. It is located anterior to the elbow joint and is a triangular depression formed between two forearm muscles:

- the brachioradialis muscle originating from the lateral supra-epicondylar ridge of the humerus, and
- the pronator teres muscle originating from the medial epicondyle of the humerus (Fig. 7.77A).

The base of the triangle is an imaginary horizontal line between the medial and lateral epicondyles. The bed or floor of the fossa is formed mainly by the brachialis muscle.

The major contents of the cubital fossa, from lateral to medial, are:

- the tendon of the biceps brachii muscle,
- the brachial artery, and
- the median nerve (Fig. 7.77B).

The brachial artery normally bifurcates into the radial and ulnar arteries in the apex of the fossa (Fig. 7.77B), although this bifurcation may occur much higher in the arm, even in the axilla. When taking a blood pressure reading from a patient, the clinician places the stethoscope over the brachial artery in the cubital fossa.

The median nerve lies immediately medial to the brachial artery and leaves the fossa by passing between the ulnar and humeral heads of the pronator teres muscle (Fig. 7.77C).

The brachial artery and the median nerve are covered and protected anteriorly in the distal part of the cubital fossa by the bicipital aponeurosis (Fig. 7.77B). This flat connective tissue membrane passes between the medial side of the tendon of the biceps brachii muscle and deep

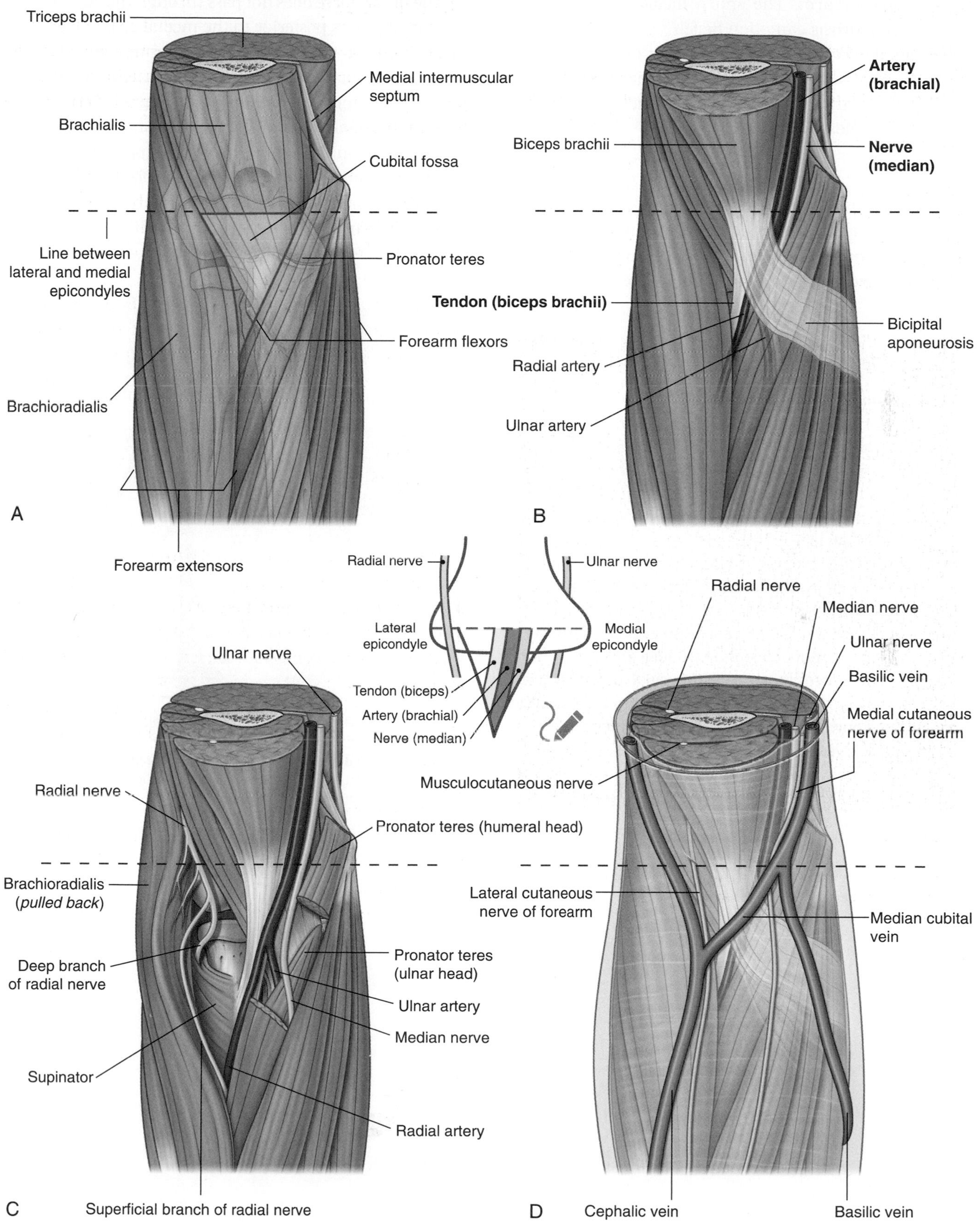


**Fig. 7.77** Cubital fossa. **A.** Margins. **B.** Contents. **C.** Position of the radial nerve. **D.** Superficial structures.

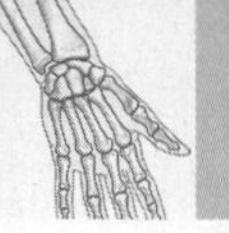


fascia of the forearm. The sharp medial margin of the bicipital aponeurosis can often be felt.

The radial nerve lies just under the lip of the brachioradialis muscle, which forms the lateral margin of the fossa (Fig. 7.77C). In this position, the radial nerve divides into superficial and deep branches:

- The superficial branch continues into the forearm just deep to the brachioradialis muscle.
- The deep branch passes between the two heads of the supinator muscle (see pp. 778–780 and Fig. 7.92) to access the posterior compartment of the forearm.

The ulnar nerve does not pass through the cubital fossa. Instead, it passes posterior to the medial epicondyle.

The roof of the cubital fossa is formed by superficial fascia and skin. The most important structure within the roof is the median cubital vein (Fig. 7.77D), which passes diagonally across the roof and connects the cephalic vein on the lateral side of the upper limb with the basilic vein on the medial side. The bicipital aponeurosis separates the median cubital vein from the brachial artery and median nerve. Other structures within the roof are cutaneous nerves—the medial cutaneous and lateral cutaneous nerves of the forearm.

## In the clinic

### Construction of a dialysis fistula

Many patients throughout the world require renal dialysis for kidney failure. The patient's blood is filtered and cleaned by the dialysis machine. Blood therefore has to be taken from patients into the filtering device and then returned to them. This process of dialysis occurs over many hours and requires considerable flow rates of 250–500 mL per minute. To enable such large volumes of blood to be removed from and returned to the body, the blood is taken from vessels that have a high flow. As no veins in the peripheral limbs have such high flow, a surgical procedure is necessary to create such a system. In most patients, the radial artery is anastomosed (joined) to the cephalic vein (Fig. 7.78) at the wrist, or the brachial artery is anastomosed to the cephalic vein at the elbow. Some surgeons place an arterial graft between these vessels.

After six weeks, the veins increase in size in response to their arterial blood flow and are amenable to direct cannulation or dialysis.

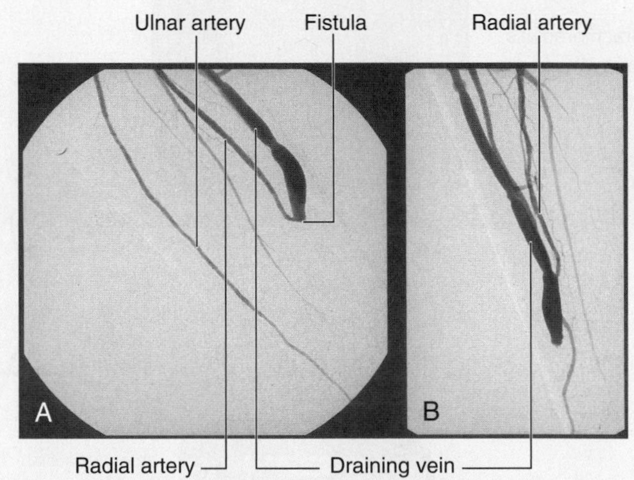


**Fig. 7.78** Digital subtraction angiograms of forearm demonstrating a surgically created radiocephalic fistula. **A.** Anteroposterior view. **B.** Lateral view.

## FOREARM

The forearm is the part of the upper limb that extends between the elbow joint and the wrist joint. Proximally, most major structures pass between the arm and forearm through, or in relation to, the cubital fossa, which is anterior to the elbow joint (Fig. 7.79). The exception is the ulnar nerve, which passes posterior to the medial epicondyle of the humerus.

Distally, structures pass between the forearm and the hand through, or anterior to, the carpal tunnel (Fig. 7.79). The major exception is the radial artery, which passes dorsally around the wrist to enter the hand posteriorly.

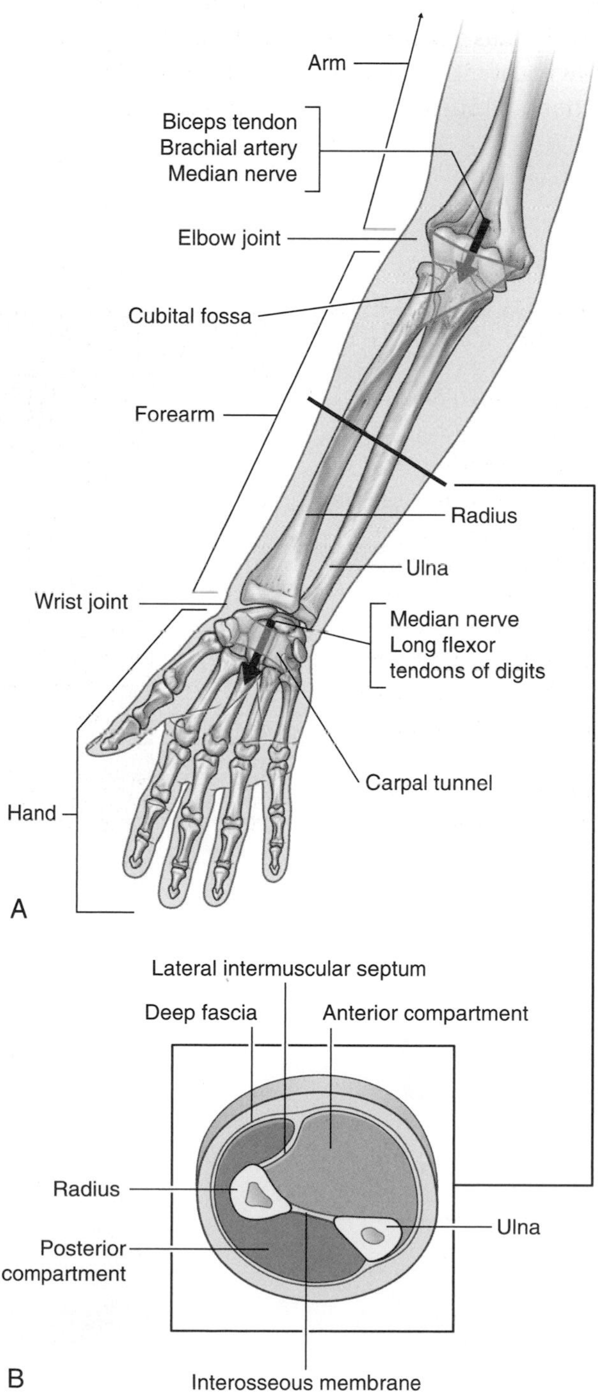


**Fig. 7.79** Forearm. **A.** Proximal and distal relationships of the forearm. **B.** Transverse section through the middle of the forearm.

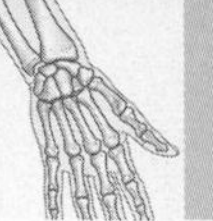



The bone framework of the forearm consists of two parallel bones, the radius and the ulna (Figs. 7.79 and 7.80B). The radius is lateral in position and is small proximally, where it articulates with the humerus, and large distally, where it forms the wrist joint with the carpal bones of the hand.

The ulna is medial in the forearm, and its proximal and distal dimensions are the reverse of those for the radius: the ulna is large proximally and small distally. Proximal and distal joints between the radius and the ulna allow the distal end of the radius to swing over the adjacent end of the ulna, resulting in pronation and supination of the hand.

As in the arm, the forearm is divided into anterior and posterior compartments (Fig. 7.79). In the forearm, these compartments are separated by:

- a lateral intermuscular septum, which passes from the anterior border of the radius to deep fascia surrounding the limb;
- an interosseous membrane, which links adjacent borders of the radius and ulna along most of their length; and
- the attachment of deep fascia along the posterior border of the ulna.

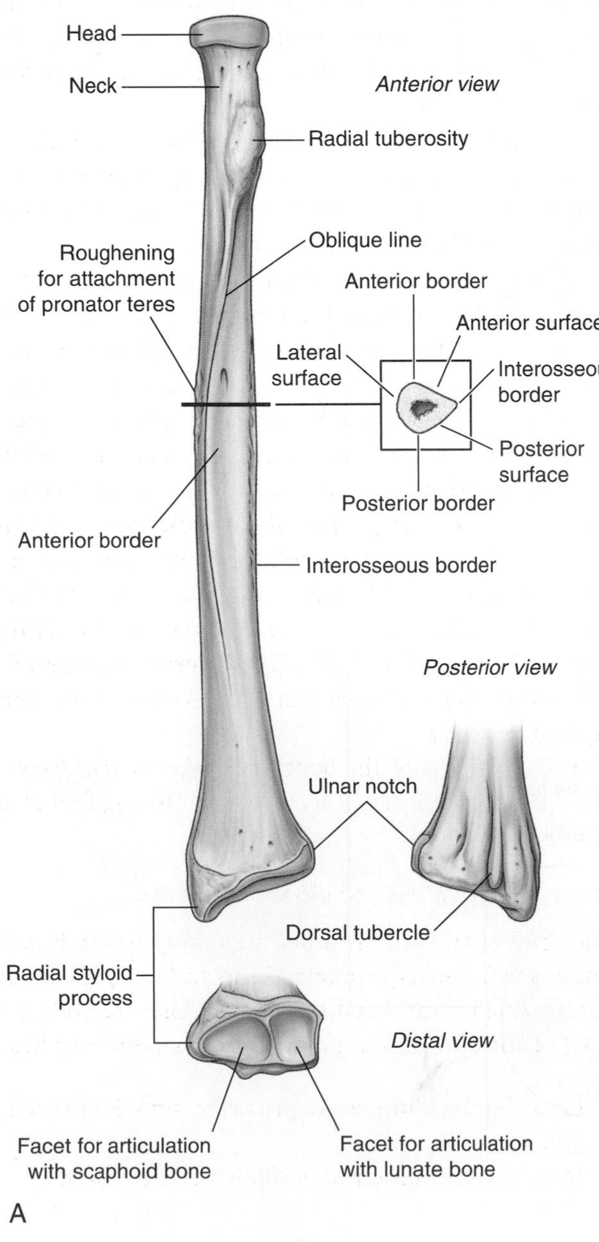


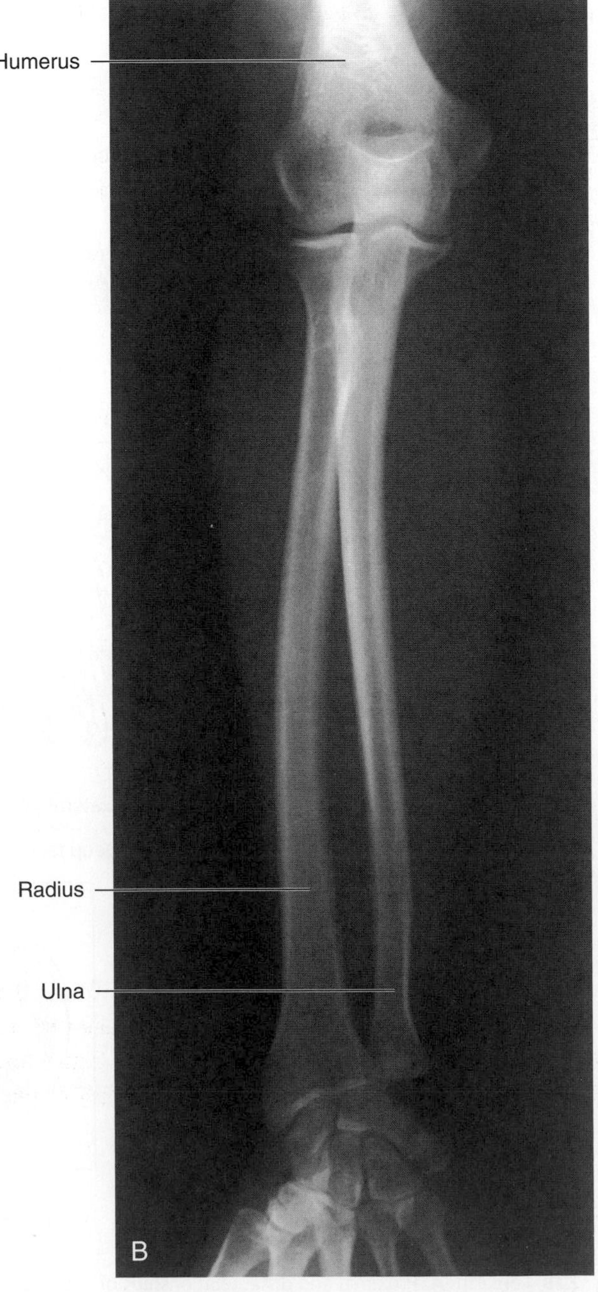


**Fig. 7.80** Radius. **A.** Shaft and distal end of the right radius. **B.** Radiograph of the forearm (anteroposterior view).

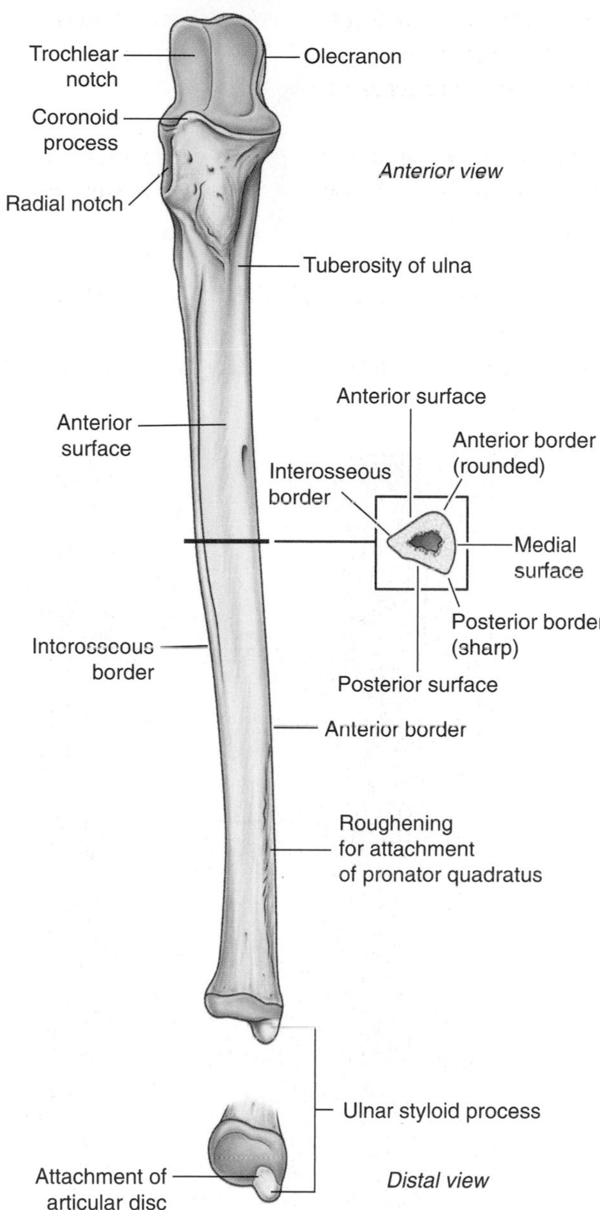


**Fig. 7.81** Shaft and distal end of right ulna.

Muscles in the anterior compartment of the forearm flex the wrist and digits and pronate the hand. Muscles in the posterior compartment extend the wrist and digits and supinate the hand. Major nerves and vessels supply or pass through each compartment.

## Bones

### Shaft and distal end of radius

The shaft of the radius is narrow proximally, where it is continuous with the radial tuberosity and neck, and much broader distally, where it expands to form the distal end (Fig. 7.80).

Throughout most of its length, the shaft of the radius is triangular in cross section, with:

- three borders (anterior, posterior, and interosseous), and
- three surfaces (anterior, posterior, and lateral).

The **anterior border** begins on the medial side of the bone as a continuation of the radial tuberosity. In the superior third of the bone, it crosses the shaft diagonally, from medial to lateral, as the oblique line of the radius. The **posterior border** is distinct only in the middle third of the bone. The **interosseous border** is sharp and is the attachment site for the interosseous membrane, which links the radius to the ulna.

The anterior and posterior surfaces of the radius are generally smooth, whereas an oval roughening for the attachment of the pronator teres marks approximately the middle of the lateral surface of the radius.

Viewed anteriorly, the distal end of the radius is broad and somewhat flattened anteroposteriorly (Fig. 7.80). Consequently, the radius has expansive anterior and posterior surfaces and narrow medial and lateral surfaces. Its anterior surface is smooth and unremarkable, except for the prominent sharp ridge that forms its lateral margin.

The **posterior surface** of the radius is characterized by the presence of a large **dorsal tubercle**, which acts as a pulley for the tendon of one of the extensor muscles of the thumb (extensor pollicis longus). The medial surface is marked by a prominent facet for articulation with the distal end of the ulna (Fig. 7.80). The **lateral surface** of the radius is diamond shaped and extends distally as a **radial styloid process**.

The distal end of the bone is marked by two facets for articulation with two carpal bones (the scaphoid and lunate).

### Shaft and distal end of ulna

The shaft of the ulna is broad superiorly where it is continuous with the large proximal end and narrow distally to form a small distal head (Fig. 7.81). Like the radius, the shaft of the ulna is triangular in cross section and has:

- three borders (anterior, posterior, and interosseous), and
- three surfaces (anterior, posterior, and medial).

The **anterior border** is smooth and rounded. The **posterior border** is sharp and palpable along its entire length. The **interosseous border** is also sharp and is the

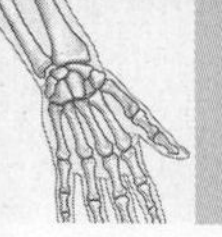



attachment site for the interosseous membrane, which joins the ulna to the radius.

The **anterior surface** of the ulna is smooth, except distally where there is a prominent linear roughening for the attachment of the pronator quadratus muscle. The **medial surface** is smooth and unremarkable. The **posterior surface** is marked by lines, which separate different regions of muscle attachments to bone.

The distal end of the ulna is small and characterized by a rounded head and the **ulnar styloid process** (Fig. 7.81). The anterolateral and distal part of the head is covered by articular cartilage. The ulnar styloid process originates from the posteromedial aspect of the ulna and projects distally.

### In the clinic

#### Fractures of the radius and ulna

The radius and ulna are attached to the humerus proximally and the carpal bones distally by a complex series of ligaments. Although the bones are separate, they behave as one. When a severe injury occurs to the forearm it usually involves both bones, resulting in either fracture of both bones or more commonly a fracture of one bone and a dislocation of the other. Commonly, the mechanism of injury and the age of the patient determine which of these are likely to occur.

There are three classic injuries to the radius and ulna:

- Monteggia's fracture is a fracture of the proximal third of the ulna and an anterior dislocation of the head of the radius at the elbow.
- Galeazzi's fracture is a fracture of the distal third of the radius associated with subluxation (partial dislocation) of the head of the ulna at the wrist joint.
- Colles' fracture is a fracture, and posterior displacement, of the distal end of the radius.

Whenever a fracture of the radius or ulna is demonstrated radiographically, further images of the elbow and wrist should be obtained to exclude dislocations.

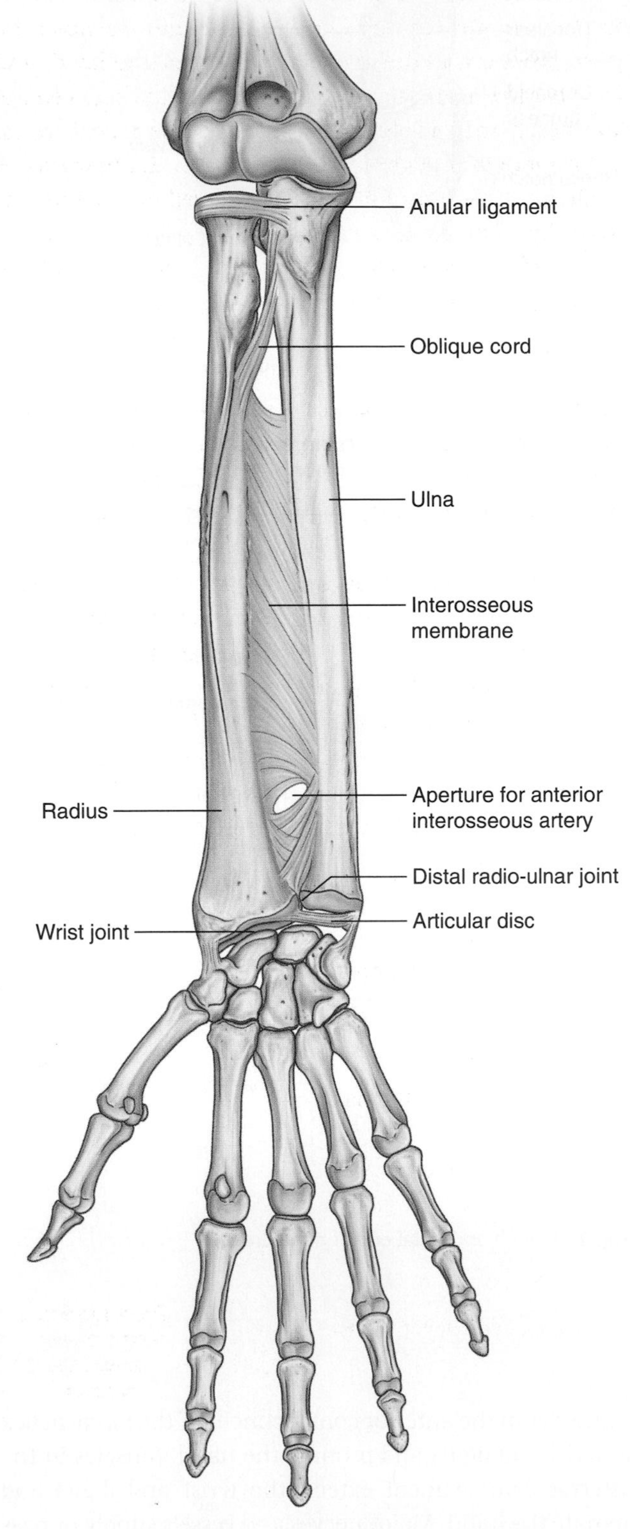


**Fig. 7.82** Distal radio-ulnar joint and the interosseous membrane.

## Joints

### Distal radio-ulnar joint

The distal radio-ulnar joint occurs between the articular surface of the head of the ulna, with the ulnar notch on the end of the radius, and with a fibrous articular disc, which separates the radio-ulnar joint from the wrist joint (Fig. 7.82).

The triangular-shaped articular disc is attached by its apex to a roughened depression on the ulna between the styloid process and the articular surface of the head, and by its base to the angular margin of the radius between the ulnar notch and the articular surface for the carpal bones.

The synovial membrane is attached to the margins of the distal radio-ulnar joint and is covered on its external surface by a fibrous joint capsule.

The distal radio-ulnar joint allows the distal end of the radius to move anteromedially over the ulna.

### Interosseous membrane

The interosseous membrane is a thin fibrous sheet that connects the medial and lateral borders of the radius and ulna, respectively (Fig. 7.82). Collagen fibers within the sheet pass predominantly inferiorly from the radius to the ulna.

The interosseous membrane has a free upper margin, which is situated just inferior to the radial tuberosity, and a small circular aperture in its distal third. Vessels pass between the anterior and posterior compartments superior to the upper margin and through the inferior aperture.

The interosseous membrane connects the radius and ulna without restricting pronation and supination and provides attachment for muscles in the anterior and posterior compartments. The orientation of fibers in the membrane is also consistent with its role in transferring forces from the radius to the ulna and ultimately, therefore, from the hand to the humerus.

### Pronation and supination

Pronation and supination of the hand occur entirely in the forearm and involve rotation of the radius at the elbow and movement of the distal end of the radius over the ulna (Fig. 7.83).

At the elbow, the superior articular surface of the radial head spins on the capitulum while, at the same time, the articular surface on the side of the head slides against the radial notch of the ulna and adjacent areas of the joint capsule and anular ligament of the radius. At the distal radio-ulnar joint, the ulnar notch of the radius slides anteriorly over the convex surface of the head of the ulna. During these movements, the bones are held together by:

- the anular ligament of the radius at the proximal radio-ulnar joint,
- the interosseous membrane along the lengths of the radius and ulna, and
- the articular disc at the distal radio-ulnar joint (Fig. 7.83).

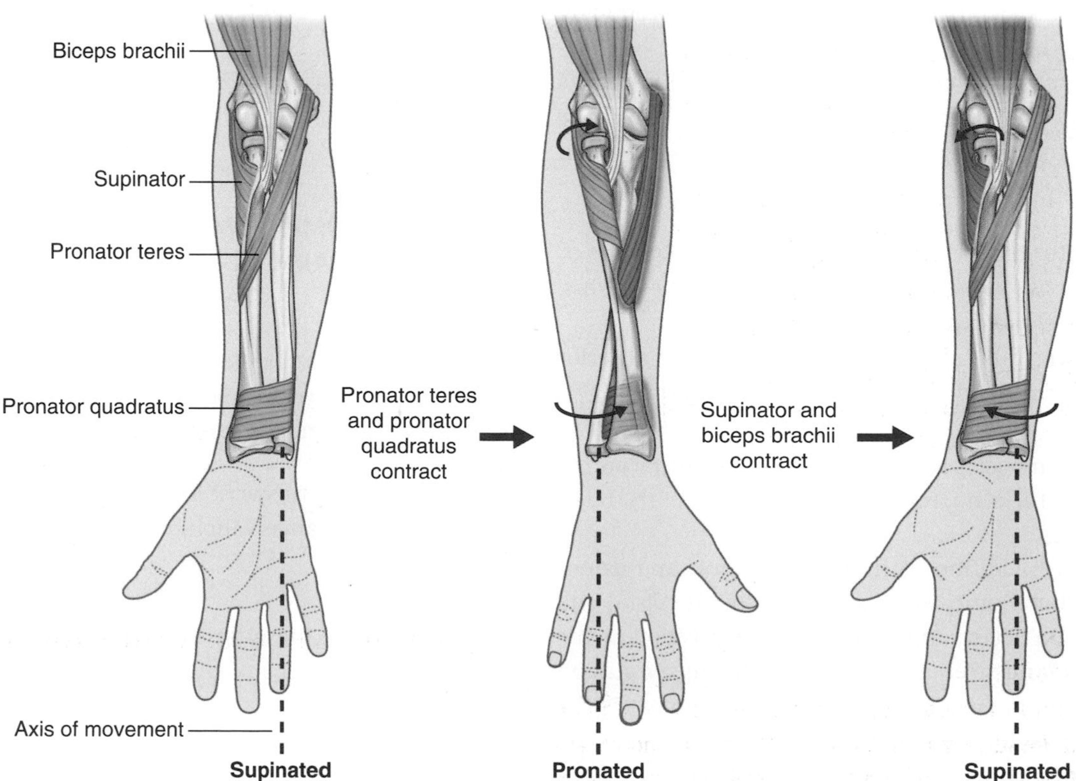


**Fig. 7.83** Pronation and supination.

Because the hand articulates predominantly with the radius, the translocation of the distal end of the radius medially over the ulna moves the hand from the palm-anterior (supinated) position to the palm-posterior (pronated) position.

Two muscles supinate and two muscles pronate the hand (Fig. 7.83).

### Muscles involved in pronation and supination

***Biceps brachii.*** The biceps brachii muscle, the largest of the four muscles that supinate and pronate the hand, is a powerful supinator as well as a flexor of the elbow joint. It is most effective as a supinator when the forearm is flexed.

***Supinator.*** The second of the muscles involved with supination is the **supinator** muscle. Located in the posterior compartment of the forearm, it has a broad origin, from the supinator crest of the ulna and the lateral epicondyle of the humerus and from ligaments associated with the elbow joint.

The supinator muscle curves around the posterior surface and the lateral surface of the upper third of the radius to attach to the shaft of the radius superior to the oblique line.

The tendon of the biceps brachii muscle and the supinator muscle both become wrapped around the proximal end of the radius when the hand is pronated (Fig. 7.83). When they contract, they unwrap from the bone, producing supination of the hand.

***Pronator teres and pronator quadratus.*** Pronation results from the action of the **pronator teres** and **pronator quadratus** muscles (Fig. 7.83). Both these muscles are in the anterior compartment of the forearm:

- The pronator teres runs from the medial epicondyle of the humerus to the lateral surface of the radius, approximately midway along the shaft.
- The pronator quadratus extends between the anterior surfaces of the distal ends of the radius and ulna.

When these muscles contract, they pull the distal end of the radius over the ulna, resulting in pronation of the hand (Fig. 7.83).

***Anconeus.*** In addition to hinge-like flexion and extension at the elbow joint, some abduction of the distal end of the ulna also occurs and maintains the position of the palm of the hand over a central axis during pronation (Fig. 7.84). The muscle involved in this movement is the **anconeus muscle,** which is a triangular muscle in the posterior compartment of the forearm that runs from the lateral epicondyle to the lateral surface of the proximal end of the ulna.

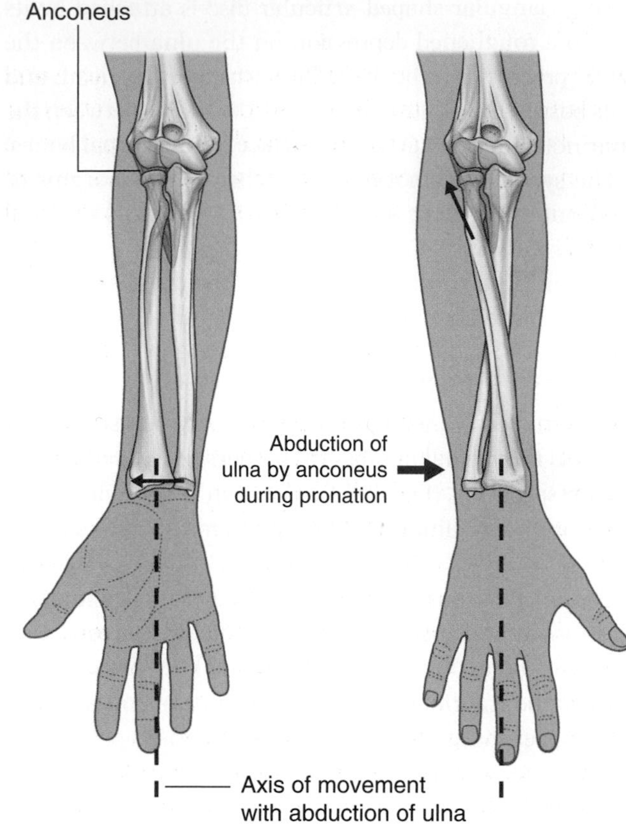


**Fig. 7.84** Abduction of the distal end of the ulna by the anconeus during pronation.

## ANTERIOR COMPARTMENT OF THE FOREARM

### Muscles

Muscles in the anterior (flexor) compartment of the forearm occur in three layers: superficial, intermediate, and deep. Generally, these muscles are associated with:

- movements of the wrist joint,
- flexion of the fingers including the thumb, and
- pronation.

All muscles in the anterior compartment of the forearm are innervated by the median nerve, except for the flexor carpi ulnaris muscle and the medial half of the flexor digitorum profundus muscle, which are innervated by the ulnar nerve.

### Superficial layer

All four muscles in the superficial layer—the flexor carpi ulnaris, palmaris longus, flexor carpi radialis, and pronator teres—have a common origin from the medial epicondyle of the humerus, and, except for the pronator teres, extend distally from the forearm into the hand (Fig. 7.85 and Table 7.10).

#### Flexor carpi ulnaris

The **flexor carpi ulnaris** muscle is the most medial of the muscles in the superficial layer of flexors, having a long linear origin from the olecranon and posterior border of the ulna, in addition to an origin from the medial epicondyle of the humerus (Fig. 7.85A,B).

The ulnar nerve enters the anterior compartment of the forearm by passing through the triangular gap between the humeral and ulnar heads of the flexor carpi ulnaris (Fig. 7.85B). The muscle fibers converge on a tendon that passes distally and attaches to the pisiform bone of the wrist. From this point, force is transferred to the hamate bone of the wrist and to the base of metacarpal V by the **pisohamate** and **pisometacarpal ligaments**.

The flexor carpi ulnaris muscle is a powerful flexor and adductor of the wrist and is innervated by the ulnar nerve (Table 7.10).

#### Palmaris longus

The **palmaris longus** muscle, which is absent in about 15% of the population, lies between the flexor carpi ulnaris and the flexor carpi radialis muscles (Fig. 7.85A). It is a spindle-shaped muscle with a long tendon, which passes into the hand and attaches to the flexor retinaculum and to a thick layer of deep fascia, the palmar aponeurosis, which underlies and is attached to the skin of the palm and fingers.

In addition to its role as an accessory flexor of the wrist joint, the palmaris longus muscle also opposes shearing forces on the skin of the palm during gripping (Table 7.10).

#### Flexor carpi radialis

The **flexor carpi radialis** muscle is lateral to the palmaris longus and has a large and prominent tendon in the distal half of the forearm (Fig. 7.85A and Table 7.10). Unlike the tendon of the flexor carpi ulnaris, which forms the medial margin of the distal forearm, the tendon of the flexor carpi radialis muscle is positioned just lateral to the midline. In this position, the tendon can be easily palpated, making it an important landmark for finding the pulse in the radial artery, which lies immediately lateral to it.

The tendon of the flexor carpi radialis passes through a compartment formed by bone and fascia on the lateral side of the anterior surface of the wrist and attaches to the anterior surfaces of the bases of metacarpals II and III.

The flexor carpi radialis is a powerful flexor of the wrist and can also abduct the wrist.

#### Pronator teres

The **pronator teres** muscle originates from the medial epicondyle and supraepicondylar ridge of the humerus and from a small linear region on the medial edge of the coronoid process of the ulna (Fig. 7.85A). The median nerve often exits the cubital fossa by passing between the humeral and ulnar heads of this muscle. The pronator teres crosses the forearm and attaches to an oval roughened area on the lateral surface of the radius approximately midway along the bone.

**Table 7.10** Superficial layer of muscles in the anterior compartment of the forearm (spinal segments indicated in bold are the major segments innervating the muscle)

| Muscle | Origin | Insertion | Innervation | Function |
|---|---|---|---|---|
| Flexor carpi ulnaris | Humeral head—medial epicondyle of humerus; ulnar head—olecranon and posterior border of ulna | Pisiform bone, and then via pisohamate and pisometacarpal ligaments into the hamate and base of metacarpal V | Ulnar nerve (C7, **C8**, T1) | Flexes and adducts the wrist joint |
| Palmaris longus | Medial epicondyle of humerus | Palmar aponeurosis of hand | Median nerve (**C7, C8**) | Flexes wrist joint; because the palmar aponeurosis anchors skin of the hand, contraction of the muscle resists shearing forces when gripping |
| Flexor carpi radialis | Medial epicondyle of humerus | Base of metacarpals II and III | Median nerve (**C6, C7**) | Flexes and abducts the wrist |
| Pronator teres | Humeral head—medial epicondyle and adjacent supra-epicondylar ridge; ulnar head—medial side of coronoid process | Roughening on lateral surface, midshaft, of radius | Median nerve (**C6, C7**) | Pronation |

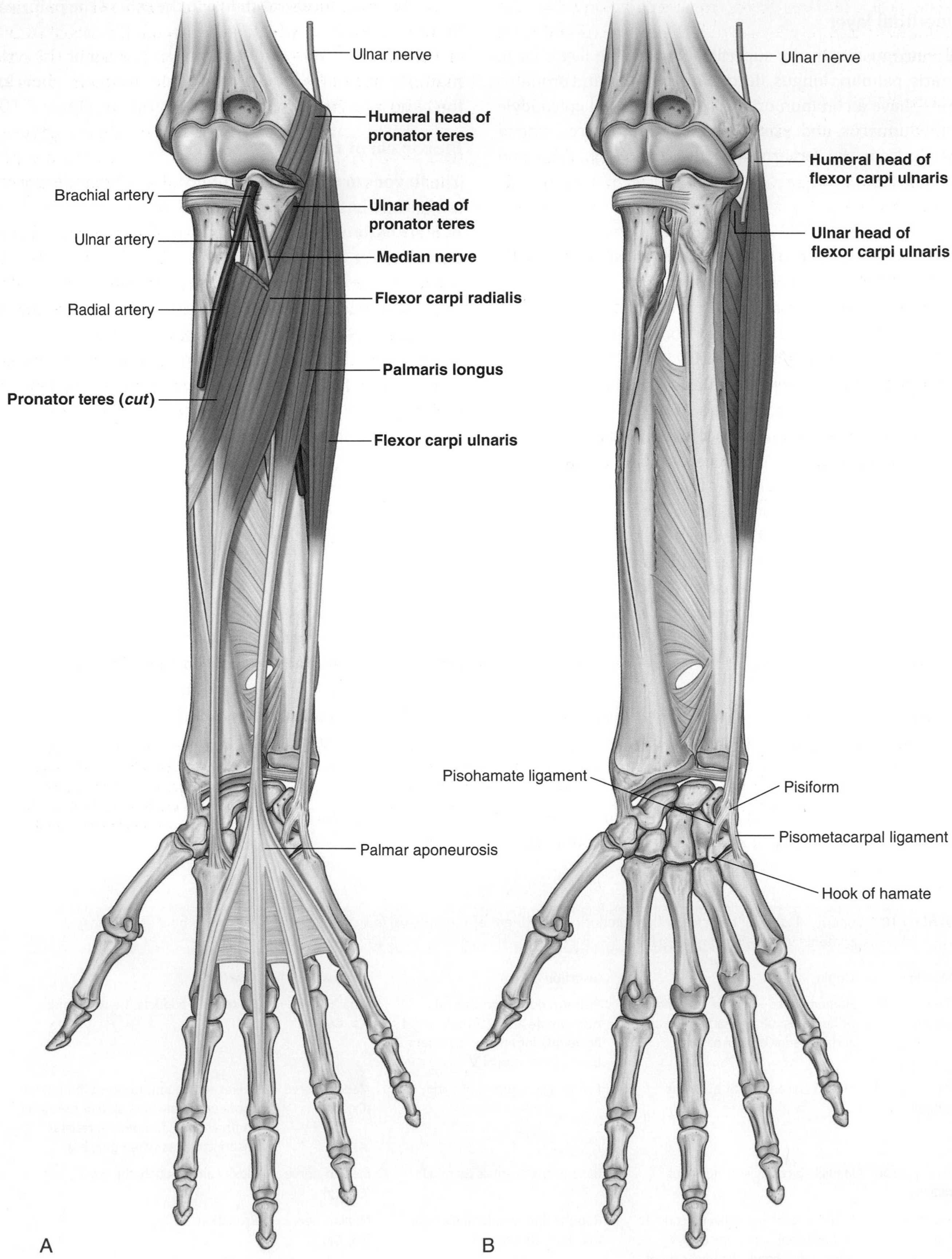

**Fig. 7.85** Superficial layer of forearm muscles. **A.** Superficial muscles (flexor retinaculum not shown). **B.** Flexor carpi ulnaris muscle.

The pronator teres forms the medial border of the cubital fossa and rotates the radius over the ulna during pronation (Table 7.10).

### Intermediate layer

#### Flexor digitorum superficialis

The muscle in the intermediate layer of the anterior compartment of the forearm is the **flexor digitorum superficialis** muscle (Fig. 7.86). This large muscle has two heads:

- the humero-ulnar head, which originates mainly from the medial epicondyle of the humerus and from the adjacent medial edge of the coronoid process of the ulna; and
- the radial head, which originates from the anterior oblique line of the radius.

The median nerve and ulnar artery pass deep to the flexor digitorum superficialis between the two heads.

In the distal forearm, the flexor digitorum superficialis forms four tendons, which pass through the carpal tunnel of the wrist and into the four fingers. The tendons for the ring and middle fingers are superficial to the tendons for the index and little fingers.

In the forearm, carpal tunnel, and proximal regions of the four fingers, the tendons of the flexor digitorum superficialis are anterior to the tendons of the flexor digitorum profundus muscle.

Near the base of the proximal phalanx of each finger, the tendon of the flexor digitorum superficialis splits into two parts to pass posteriorly around each side of the tendon of the flexor digitorum profundus and ultimately attach to the margins of the middle phalanx (Fig. 7.86).

The flexor digitorum superficialis flexes the metacarpophalangeal joint and proximal interphalangeal joint of each finger; it also flexes the wrist joint (Table 7.11).

**Table 7.11** Intermediate layer of muscles in the anterior compartment of the forearm (spinal segment indicated in bold is the major segment innervating the muscle)

| Muscle | Origin | Insertion | Innervation | Function |
|---|---|---|---|---|
| Flexor digitorum superficialis | Humero-ulnar head—medial epicondyle of humerus and adjacent margin of coronoid process; radial head—oblique line of radius | Four tendons, which attach to the palmar surfaces of the middle phalanges of the index, middle, ring, and little fingers | Median nerve (**C8**, T1) | Flexes proximal interphalangeal joints of the index, middle, ring, and little fingers; can also flex metacarpophalangeal joints of the same fingers and the wrist joint |

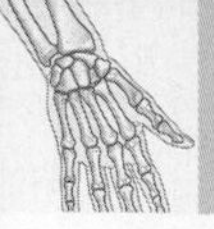

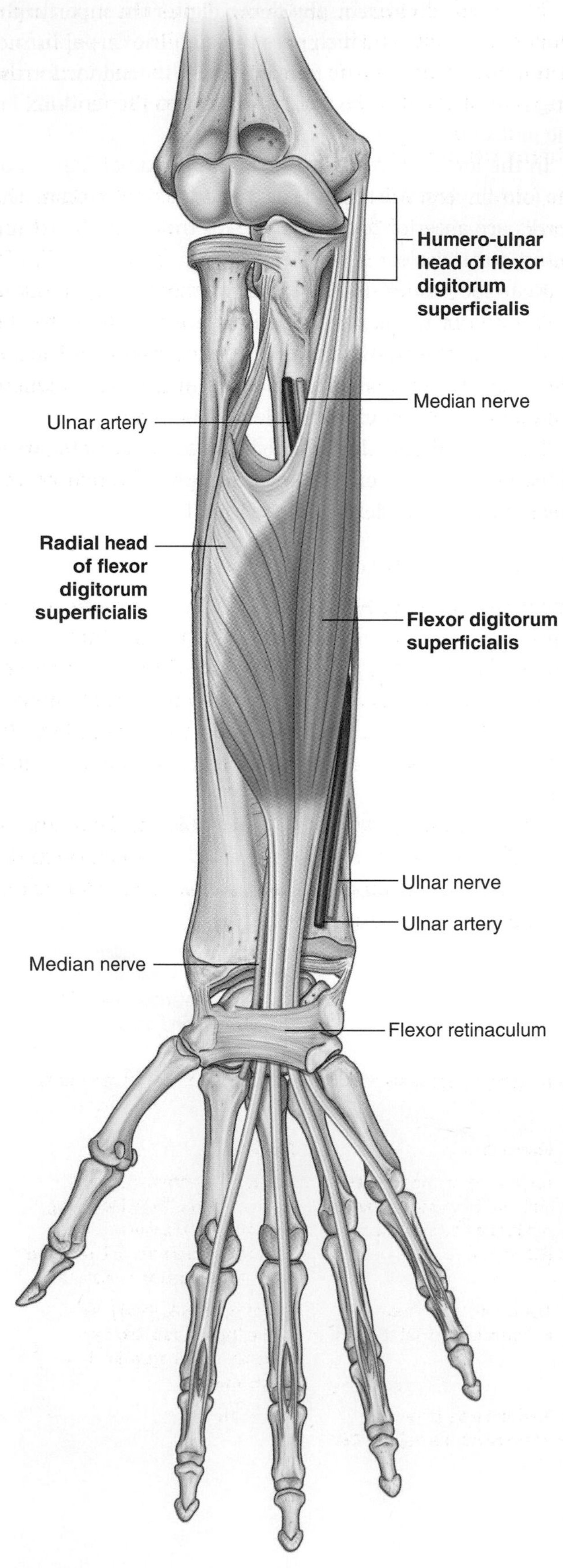


**Fig. 7.86** Intermediate layer of forearm muscles.

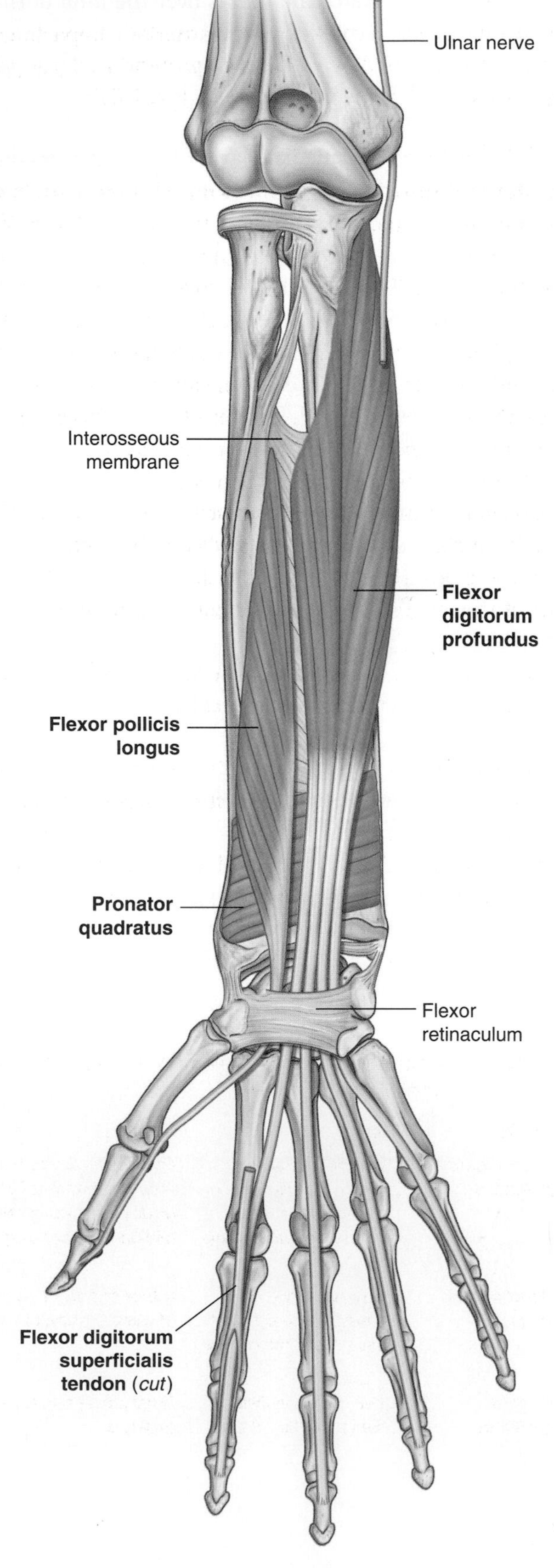


**Fig. 7.87** Deep layer of forearm muscles.

## Deep layer

There are three deep muscles in the anterior compartment of the forearm: the flexor digitorum profundus, flexor pollicis longus, and pronator quadratus (Fig. 7.87).

### Flexor digitorum profundus

The **flexor digitorum profundus** muscle originates from the anterior and medial surfaces of the ulna and from the adjacent half of the anterior surface of the interosseous membrane (Fig. 7.87). It gives rise to four tendons, which pass through the carpal tunnel into the four medial fingers. Throughout most of their course, the tendons are deep to the tendons of the flexor digitorum superficialis muscle.

Opposite the proximal phalanx of each finger, each tendon of the flexor digitorum profundus passes through a split formed in the overlying tendon of the flexor digitorum superficialis muscle and passes distally to insert into the anterior surface of the base of the distal phalanx.

In the palm, the lumbrical muscles originate from the sides of the tendons of the flexor digitorum profundus (see Fig. 7.108).

Innervation of the medial and lateral halves of the flexor digitorum profundus varies as follows:

- The lateral half (associated with the index and middle fingers) is innervated by the anterior interosseous nerve (branch of the median nerve).
- The medial half (the part associated with the ring and little fingers) is innervated by the ulnar nerve.

The flexor digitorum profundus flexes the metacarpophalangeal joints and the proximal and distal interphalangeal joints of the four fingers. Because the tendons cross the wrist, it can flex the wrist joint as well (Table 7.12).

### Flexor pollicis longus

The **flexor pollicis longus** muscle originates from the anterior surface of the radius and the adjacent half of the anterior surface of the interosseous membrane (Fig. 7.87). It is a powerful muscle and forms a single large tendon, which passes through the carpal tunnel, lateral to the tendons of the flexor digitorum superficialis and flexor digitorum profundus muscles, and into the thumb where it attaches to the base of the distal phalanx.

The flexor pollicis longus flexes the thumb and is innervated by the anterior interosseous nerve (branch of the median nerve) (Table 7.12).

### Pronator quadratus

The **pronator quadratus** muscle is a flat square-shaped muscle in the distal forearm (Fig. 7.87). It originates from a linear ridge on the anterior surface of the lower end of the ulna and passes laterally to insert onto the flat anterior surface of the radius. It lies deep to, and is crossed by, the tendons of the flexor digitorum profundus and flexor pollicis longus muscles.

The pronator quadratus muscle pulls the distal end of the radius anteriorly over the ulna during pronation and is innervated by the anterior interosseous nerve (branch of the median nerve) (Table 7.12).

**Table 7.12** Deep layer of muscles in the anterior compartment of the forearm (spinal segments indicated in bold are the major segments innervating the muscle)

| Muscle | Origin | Insertion | Innervation | Function |
|---|---|---|---|---|
| Flexor digitorum profundus | Anterior and medial surfaces of ulna and anterior medial half of interosseous membrane | Four tendons, which attach to the palmar surfaces of the distal phalanges of the index, middle, ring, and little fingers | Lateral half by median nerve (anterior interosseous nerve); medial half by ulnar nerve (**C8**, T1) | Flexes distal interphalangeal joints of the index, middle, ring, and little fingers; can also flex metacarpophalangeal joints of the same fingers and the wrist joint |
| Flexor pollicis longus | Anterior surface of radius and radial half of interosseous membrane | Palmar surface of base of distal phalanx of thumb | Median nerve (anterior interosseous nerve) (C7, **C8**) | Flexes interphalangeal joint of the thumb; can also flex metacarpophalangeal joint of the thumb |
| Pronator quadratus | Linear ridge on distal anterior surface of ulna | Distal anterior surface of radius | Median nerve (anterior interosseous nerve) (C7, **C8**) | Pronation |

## Arteries and veins

The largest arteries in the forearm are in the anterior compartment, pass distally to supply the hand, and give rise to vessels that supply the posterior compartment (Fig. 7.88).

The brachial artery enters the forearm from the arm by passing through the cubital fossa. At the apex of the cubital fossa, it divides into its two major branches, the radial and ulnar arteries.

### Radial artery

The radial artery originates from the brachial artery at approximately the neck of the radius and passes along the lateral aspect of the forearm (Fig. 7.88). It is:

- just deep to the brachioradialis muscle in the proximal half of the forearm,
- related on its lateral side to the superficial branch of the radial nerve in the middle third of the forearm, and
- medial to the tendon of the brachioradialis muscle and covered only by deep fascia, superficial fascia, and skin in the distal forearm.

In the distal forearm, the radial artery lies immediately lateral to the large tendon of the flexor carpi radialis muscle and directly anterior to the pronator quadratus muscle and the distal end of the radius (Fig. 7.88). In the distal forearm, the radial artery can be located using the flexor carpi radialis muscle as a landmark. The radial pulse can be felt by gently palpating the radial artery against the underlying muscle and bone.

The radial artery leaves the forearm, passes around the lateral side of the wrist, and penetrates the posterolateral aspect of the hand between the bases of metacarpals I and II (Fig. 7.88). Branches of the radial artery in the hand often provide the major blood supply to the thumb and lateral side of the index finger.

Branches of the radial artery originating in the forearm include:

- a **radial recurrent artery**, which contributes to an anastomotic network around the elbow joint and to

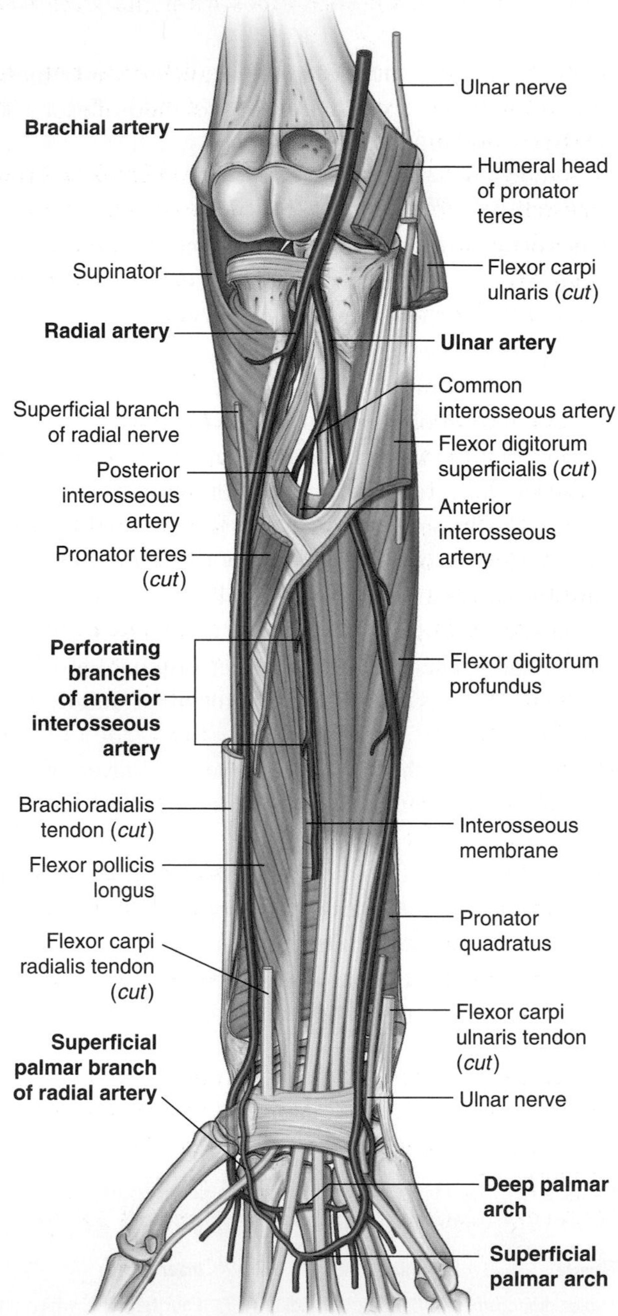


**Fig. 7.88** Arteries of the anterior compartment of the forearm.

numerous vessels that supply muscles on the lateral side of the forearm (see Fig. 7.66B);

- a small **palmar carpal branch,** which contributes to an anastomotic network of vessels that supply the carpal bones and joints;
- a somewhat larger branch, the **superficial palmar branch,** which enters the hand by passing through, or superficial to, the thenar muscles at the base of the thumb (Fig. 7.88) and anastomoses with the superficial palmar arch formed by the ulnar artery.

### Ulnar artery

The ulnar artery is larger than the radial artery and passes down the medial side of the forearm (Fig. 7.88). It leaves the cubital fossa by passing deep to the pronator teres muscle, and then passes through the forearm in the fascial plane between the flexor carpi ulnaris and flexor digitorum profundus muscles.

In the distal forearm, the ulnar artery often remains tucked under the anterolateral lip of the flexor carpi ulnaris tendon, and is therefore not easily palpable.

In distal regions of the forearm, the ulnar nerve is immediately medial to the ulnar artery.

The ulnar artery leaves the forearm, enters the hand by passing lateral to the pisiform bone and superficial to the flexor retinaculum of the wrist, and arches over the palm (Fig. 7.88). It is often the major blood supply to the medial three and one-half digits.

Branches of the ulnar artery that arise in the forearm include:

- the **ulnar recurrent artery** with **anterior** and **posterior branches**, which contribute to an anastomotic network of vessels around the elbow joint (see Fig. 7.66B);
- numerous muscular arteries, which supply surrounding muscles;
- the **common interosseous artery**, which divides into anterior and posterior interosseous arteries (Fig. 7.88); and
- two small carpal arteries (**dorsal carpal branch** and **palmar carpal branch**), which supply the wrist.

The **posterior interosseous artery** passes dorsally over the proximal margin of the interosseous membrane into the posterior compartment of the forearm.

The **anterior interosseous artery** passes distally along the anterior aspect of the interosseous membrane and supplies muscles of the deep compartment of the forearm and the radius and ulna. It has numerous branches, which perforate the interosseous membrane to supply deep muscles of the posterior compartment; it also has a small branch, which contributes to the vascular network around the carpal bones and joints. Perforating the interosseous membrane in the distal forearm, the anterior interosseous artery terminates by joining the posterior interosseous artery.

### Veins

Deep veins of the anterior compartment generally accompany the arteries and ultimately drain into brachial veins associated with the brachial artery in the cubital fossa.

#### In the clinic

**Transection of the radial or ulnar artery**

Adult patients may transect the radial or ulnar artery because these vessels are relatively subcutaneous. A typical method of injury is when the hand is forced through a plate glass window. Fortunately, the dual supply to the hand usually enables the surgeon to tie off either the ulnar or the radial artery, without significant consequence.

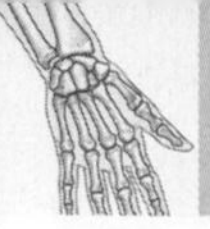

## Nerves

Nerves in the anterior compartment of the forearm are the median and ulnar nerves and the superficial branch of the radial nerve (Fig. 7.89).

### Median nerve

The median nerve innervates the muscles in the anterior compartment of the forearm except for the flexor carpi ulnaris and the medial part of the flexor digitorum profundus (ring and little fingers). It leaves the cubital fossa by passing between the two heads of the pronator teres muscle and passing between the humero-ulnar and radial heads of the flexor digitorum superficialis muscle (Fig. 7.89).

The median nerve continues a straight linear course distally down the forearm in the fascia on the deep surface of the flexor digitorum superficialis muscle. Just proximal to the wrist, it moves around the lateral side of the muscle and becomes more superficial in position, lying between the tendons of the palmaris longus and flexor carpi radialis muscles. It leaves the forearm and enters the palm of the hand by passing through the carpal tunnel deep to the flexor retinaculum.

Most branches to the muscles in the superficial and intermediate layers of the forearm originate medially from the nerve just distal to the elbow joint.

- The largest branch of the median nerve in the forearm is the **anterior interosseous nerve,** which originates between the two heads of the pronator teres, passes distally down the forearm with the anterior interosseous artery, innervates the muscles in the deep layer (the flexor pollicis longus, the lateral half of the flexor digitorum profundus, and the pronator quadratus) and terminates as articular branches to joints of the distal forearm and wrist.
- A small **palmar branch** originates from the median nerve in the distal forearm immediately proximal to the flexor retinaculum (Fig. 7.89), passes superficially into the hand, and innervates the skin over the base and central palm. This palmar branch is spared in carpal tunnel syndrome because it passes into the hand superficial to the flexor retinaculum of the wrist.

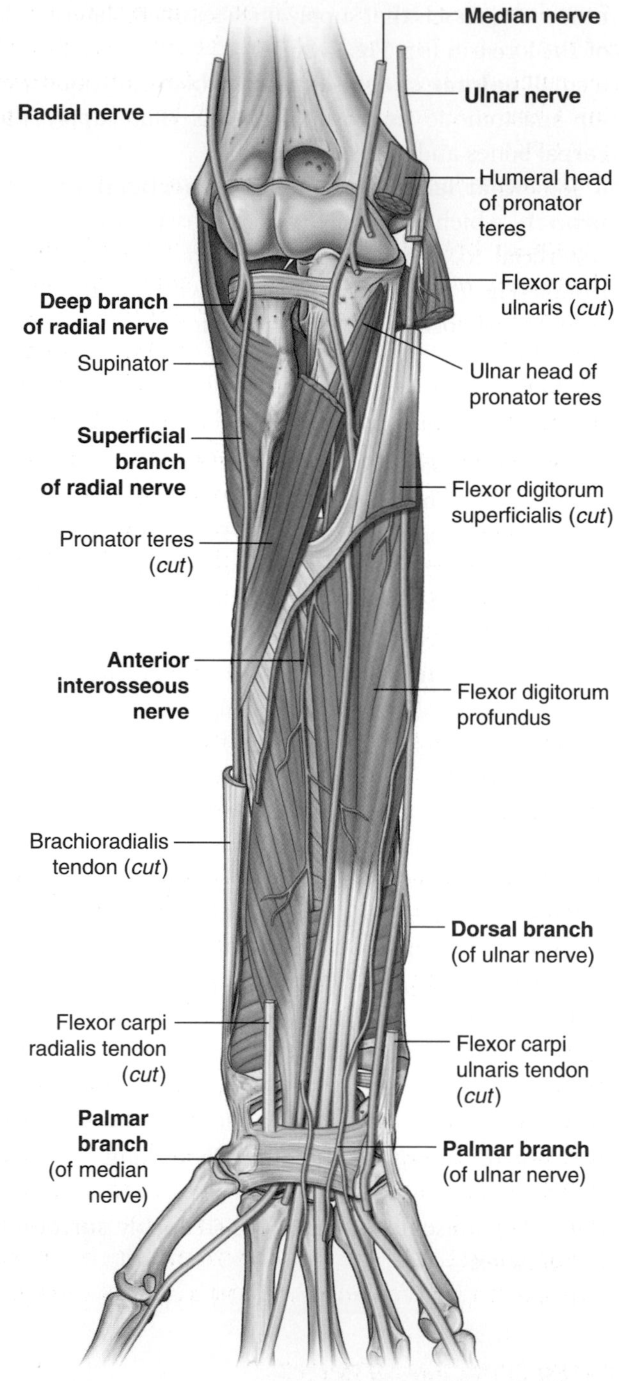


**Fig. 7.89** Nerves of anterior forearm.

### Ulnar nerve

The ulnar nerve passes through the forearm and into the hand, where most of its major branches occur. In the forearm, the ulnar nerve innervates only the flexor carpi ulnaris muscle and the medial part (ring and little fingers) of the flexor digitorum profundus muscle (Fig. 7.89).

The ulnar nerve enters the anterior compartment of the forearm by passing posteriorly around the medial epicondyle of the humerus and between the humeral and ulnar heads of the flexor carpi ulnaris muscle. After passing down the medial side of the forearm in the plane between the flexor carpi ulnaris and the flexor digitorum profundus muscles, it lies under the lateral lip of the tendon of the flexor carpi ulnaris proximal to the wrist.

The ulnar artery is lateral to the ulnar nerve in the distal two-thirds of the forearm, and both the ulnar artery and nerve enter the hand by passing superficial to the flexor retinaculum and immediately lateral to the pisiform bone (Fig. 7.89).

In the forearm the ulnar nerve gives rise to:

- **muscular branches** to the flexor carpi ulnaris and to the medial half of the flexor digitorum profundus that arise soon after the ulnar nerve enters the forearm; and
- two small cutaneous branches—the **palmar branch** originates in the middle of the forearm and passes into the hand to supply skin on the medial side of the palm; the larger **dorsal branch** originates from the ulnar nerve in the distal forearm and passes posteriorly deep to the tendon of the flexor carpi ulnaris and innervates skin on the posteromedial side of the back of the hand and most skin on the posterior surfaces of the medial one and one-half digits.

### Radial nerve

The radial nerve bifurcates into deep and superficial branches under the margin of the brachioradialis muscle in the lateral border of the cubital fossa (Fig. 7.89).

- The **deep branch** is predominantly motor and passes between the superficial and deep layers of the supinator muscle to access and supply muscles in the posterior compartment of the forearm.
- The **superficial branch** of the radial nerve is sensory. It passes down the anterolateral aspect of the forearm deep to the brachioradialis muscle and in association with the radial artery. Approximately two-thirds of the way down the forearm, the superficial branch of the radial nerve passes laterally and posteriorly around the radial side of the forearm deep to the tendon of the brachioradialis. The nerve continues into the hand where it innervates skin on the posterolateral surface.

# POSTERIOR COMPARTMENT OF THE FOREARM

## Muscles

Muscles in the posterior compartment of the forearm occur in two layers: a superficial and a deep layer. The muscles are associated with:

- movement of the wrist joint,
- extension of the fingers and thumb, and
- supination.

All muscles in the posterior compartment of the forearm are innervated by the radial nerve.

### Superficial layer

The seven muscles in the superficial layer are the brachioradialis, extensor carpi radialis longus, extensor carpi radialis brevis, extensor digitorum, extensor digiti minimi, extensor carpi ulnaris, and anconeus (Fig. 7.90). All have a common origin from the supraepicondylar ridge and lateral epicondyle of the humerus and, except for the brachioradialis and anconeus, extend as tendons into the hand.

#### Brachioradialis

The **brachioradialis** muscle originates from the proximal part of the supraepicondylar ridge of the humerus and passes through the forearm to insert on the lateral side of the distal end of the radius just proximal to the radial styloid process (Fig. 7.90).

In the anatomical position, the brachioradialis is part of the muscle mass overlying the anterolateral surface of the forearm and forms the lateral boundary of the cubital fossa.

Because the brachioradialis is anterior to the elbow joint, it acts as an accessory flexor of this joint even though it is in the posterior compartment of the forearm. Its action is most efficient when the forearm is midpronated and it forms a prominent bulge as it acts against resistance.

The radial nerve emerges from the posterior compartment of the arm just deep to the brachioradialis in the distal arm and innervates the brachioradialis. Lateral to the cubital fossa, the brachioradialis lies over the radial nerve and its bifurcation into deep and superficial branches. In more distal regions, the brachioradialis lies over the

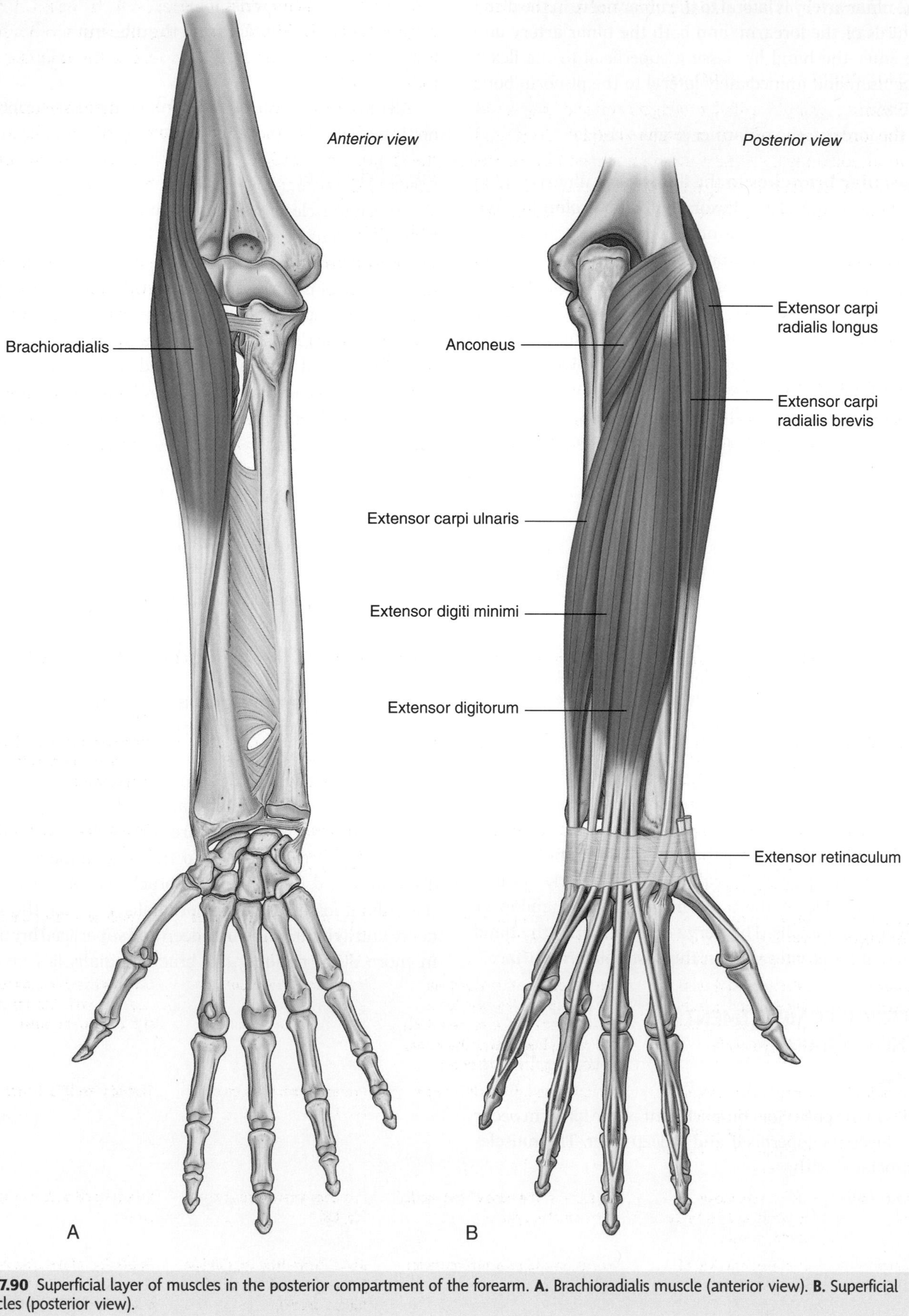

**Fig. 7.90** Superficial layer of muscles in the posterior compartment of the forearm. **A.** Brachioradialis muscle (anterior view). **B.** Superficial muscles (posterior view).

superficial branch of the radial nerve and radial artery (Table 7.13).

### Extensor carpi radialis longus

The **extensor carpi radialis longus** muscle originates from the distal part of the supraepicondylar ridge and the lateral epicondyle of the humerus; its tendon inserts on the dorsal surface of the base of metacarpal II (Fig. 7.90). In proximal regions, it is deep to the brachioradialis muscle.

The extensor carpi radialis longus muscle extends and abducts the wrist, and is innervated by the radial nerve before the nerve divides into superficial and deep branches (Table 7.13).

### Extensor carpi radialis brevis

The **extensor carpi radialis brevis** muscle originates from the lateral epicondyle of the humerus, and the tendon inserts onto adjacent dorsal surfaces of the bases of metacarpals II and III (Fig. 7.90). Along much of its course, the extensor carpi radialis brevis lies deep to the extensor carpi radialis longus.

The extensor carpi radialis brevis muscle extends and abducts the wrist, and is innervated by the deep branch of the radial nerve before the nerve passes between the two heads of the supinator muscle (Table 7.13).

### Extensor digitorum

The **extensor digitorum** muscle is the major extensor of the four fingers (index, middle, ring, and little fingers). It originates from the lateral epicondyle of the humerus and forms four tendons, each of which passes into a finger (Fig. 7.90).

On the dorsal surface of the hand, adjacent tendons of the extensor digitorum are interconnected. In the fingers, each tendon inserts, via a triangular-shaped connective

**Table 7.13** Superficial layer of muscles in the posterior compartment of the forearm (spinal segments indicated in bold are the major segments innervating the muscle)

| Muscle | Origin | Insertion | Innervation | Function |
|---|---|---|---|---|
| Brachioradialis | Proximal part of lateral supraepicondylar ridge of humerus and adjacent intermuscular septum | Lateral surface of distal end of radius | Radial nerve (C5, **C6**) before division into superficial and deep branches | Accessory flexor of elbow joint when forearm is midpronated |
| Extensor carpi radialis longus | Distal part of lateral supraepicondylar ridge of humerus and adjacent intermuscular septum | Dorsal surface of base of metacarpal II | Radial nerve (**C6**, C7) before division into superficial and deep branches | Extends and abducts the wrist |
| Extensor carpi radialis brevis | Lateral epicondyle of humerus and adjacent intermuscular septum | Dorsal surface of base of metacarpals II and III | Deep branch of radial nerve (**C7**, C8) before penetrating supinator muscle | Extends and abducts the wrist |
| Extensor digitorum | Lateral epicondyle of humerus and adjacent intermuscular septum and deep fascia | Four tendons, which insert via extensor hoods into the dorsal aspects of the bases of the middle and distal phalanges of the index, middle, ring, and little fingers | Posterior interosseous nerve (**C7**, C8) | Extends the index, middle, ring, and little fingers; can also extend the wrist |
| Extensor digiti minimi | Lateral epicondyle of humerus and adjacent intermuscular septum together with extensor digitorum | Extensor hood of the little finger | Posterior interosseous nerve (**C7**, C8) | Extends the little finger |
| Extensor carpi ulnaris | Lateral epicondyle of humerus and posterior border of ulna | Tubercle on the base of the medial side of metacarpal V | Posterior interosseous nerve (**C7**, C8) | Extends and adducts the wrist |
| Anconeus | Lateral epicondyle of humerus | Olecranon and proximal posterior surface of ulna | Radial nerve (**C6, C7, C8**) (via branch to medial head of triceps brachii) | Abduction of the ulna in pronation; accessory extensor of the elbow joint |

tissue aponeurosis (the extensor hood), into the base of the dorsal surfaces of the middle and distal phalanges.

The extensor digitorum muscle is innervated by the posterior interosseous nerve, which is the continuation of the deep branch of the radial nerve after it emerges from the supinator muscle (Table 7.13).

### Extensor digiti minimi

The **extensor digiti minimi** muscle is an accessory extensor of the little finger and is medial to the extensor digitorum in the forearm (Fig. 7.90). It originates from the lateral epicondyle of the humerus and inserts, together with the tendon of the extensor digitorum, into the extensor hood of the little finger.

The extensor digiti minimi is innervated by the posterior interosseous nerve (Table 7.13).

### Extensor carpi ulnaris

The **extensor carpi ulnaris** muscle is medial to the extensor digiti minimi (Fig. 7.90). It originates from the lateral epicondyle, and its tendon inserts into the medial side of the base of metacarpal V.

The extensor carpi ulnaris extends and adducts the wrist, and is innervated by the posterior interosseous nerve (Table 7.13).

### Anconeus

The **anconeus muscle** is the most medial of the superficial extensors and has a triangular shape. It originates from the lateral epicondyle of the humerus and has a broad insertion into the posterolateral surface of the olecranon and related posterior surface of the ulna (see Fig. 7.84).

The anconeus abducts the ulna during pronation to maintain the center of the palm over the same point when the hand is flipped. It is also considered to be an accessory extensor of the elbow joint.

The anconeus is innervated by the branch of the radial nerve that innervates the medial head of the triceps brachii muscle (Table 7.13).

## Deep layer

The deep layer of the posterior compartment of the forearm consists of five muscles: supinator, abductor pollicis longus, extensor pollicis brevis, extensor pollicis longus, and extensor indicis (Fig. 7.91).

Except for the supinator muscle, all these deep layer muscles originate from the posterior surfaces of the radius, ulna, and interosseous membrane and pass into the thumb and fingers.

- Three of these muscles—the abductor pollicis longus, extensor pollicis brevis, and extensor pollicis longus—emerge from between the extensor digitorum and the extensor carpi radialis brevis tendons of the superficial layer and pass into the thumb.
- Two of the three "outcropping" muscles (the abductor pollicis longus and extensor pollicis brevis) form a distinct muscular bulge in the distal posterolateral surface of the forearm.

All muscles of the deep layer are innervated by the posterior interosseous nerve, the continuation of the deep branch of the radial nerve.

### Supinator

The **supinator** muscle has two layers, which insert together on the proximal aspect of the radius (Fig. 7.91):

- The more superficial (humeral) layer originates mainly from the lateral epicondyle of the humerus and the related anular ligament and the radial collateral ligament of the elbow joint.
- The deep (ulnar) layer originates mainly from the supinator crest on the posterolateral surface of the ulna.

From their sites of origin, the two layers wrap around the posterior and lateral aspect of the head, neck, and proximal shaft of the radius to insert on the lateral surface of the radius superior to the anterior oblique line and to the insertion of the pronator teres muscle.

The supinator muscle supinates the forearm and hand.

The deep branch of the radial nerve innervates the supinator muscle and passes to the posterior compartment

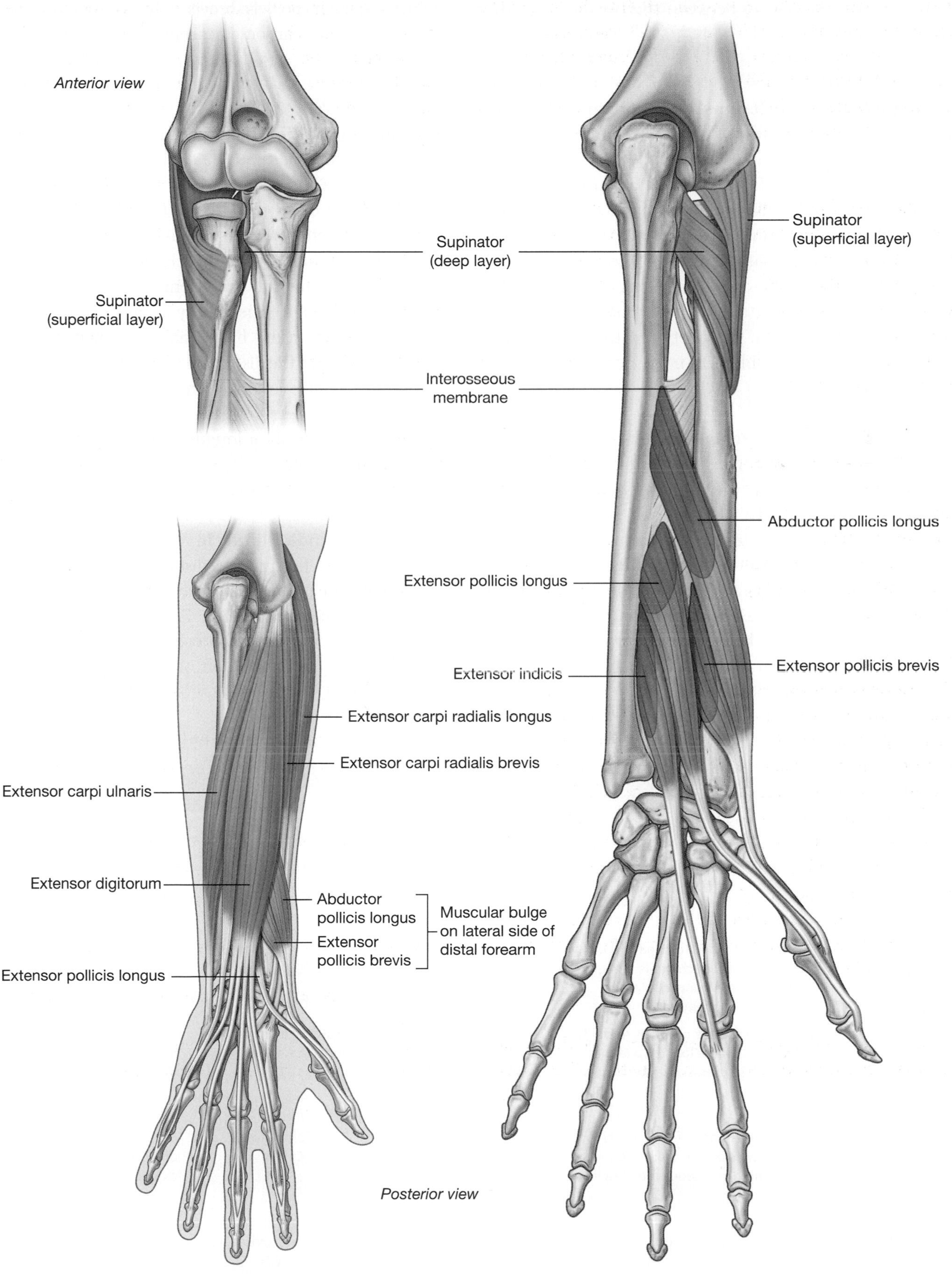




**Fig. 7.91** Deep layer of muscles in the posterior compartment of the forearm.

of the forearm by passing between the two heads of this muscle (Table 7.14).

### Abductor pollicis longus

The **abductor pollicis longus** muscle originates from the proximal posterior surfaces of the radius and the ulna and from the related interosseous membrane (Fig. 7.91). In the distal forearm, it emerges between the extensor digitorum and extensor carpi radialis brevis muscles to form a tendon that passes into the thumb and inserts on the lateral side of the base of metacarpal I. The tendon contributes to the lateral border of the anatomical snuffbox at the wrist.

The major function of the abductor pollicis longus is to abduct the thumb at the joint between the metacarpal I and trapezium bones (Table 7.14).

### Extensor pollicis brevis

The **extensor pollicis brevis** muscle arises distal to the origin of the abductor pollicis longus from the posterior surface of the radius and interosseous membrane (Fig. 7.91). Together with the abductor pollicis longus, it emerges between the extensor digitorum and extensor carpi radialis brevis muscles to form a bulge on the posterolateral surface of the distal forearm. The tendon of the extensor pollicis brevis passes into the thumb and inserts on the dorsal surface of the base of the proximal phalanx. At the wrist, the tendon contributes to the lateral border of the anatomical snuffbox.

The extensor pollicis brevis extends the metacarpophalangeal and carpometacarpal joints of the thumb (Table 7.14).

### Extensor pollicis longus

The **extensor pollicis longus** muscle originates from the posterior surface of the ulna and adjacent interosseous membrane and inserts via a long tendon into the dorsal surface of the distal phalanx of the thumb (Fig. 7.91). Like the abductor pollicis longus and extensor pollicis brevis, the tendon of this muscle emerges between the extensor digitorum and the extensor carpi radialis brevis muscles. However, it is held away from the other two deep muscles of the thumb by passing medially around the dorsal tubercle on the distal end of the radius. The tendon forms the medial margin of the anatomical snuffbox at the wrist.

The extensor pollicis longus extends all joints of the thumb (Table 7.14).

### Extensor indicis

The **extensor indicis muscle** is an accessory extensor of the index finger. It originates distal to the extensor pollicis longus from the posterior surface of the ulna and adjacent interosseous membrane (Fig. 7.91). The tendon passes into the hand and inserts into the extensor hood of the index finger with the tendon of the extensor digitorum (Table 7.14).

**Table 7.14** Deep layer of muscles in the posterior compartment of the forearm (spinal segments indicated in bold are the major segments innervating the muscle)

| Muscle | Origin | Insertion | Innervation | Function |
|---|---|---|---|---|
| Supinator | Superficial layer—lateral epicondyle of humerus, radial collateral and anular ligaments; deep layer—supinator crest of the ulna | Lateral surface of radius superior to the anterior oblique line | Posterior interosseous nerve (**C6**, C7) | Supination |
| Abductor pollicis longus | Posterior surfaces of ulna and radius (distal to the attachments of supinator and anconeus), and intervening interosseous membrane | Lateral side of base of metacarpal I | Posterior interosseous nerve (**C7**, C8) | Abducts carpometacarpal joint of thumb; accessory extensor of the thumb |
| Extensor pollicis brevis | Posterior surface of radius (distal to abductor pollicis longus) and the adjacent interosseous membrane | Dorsal surface of base of proximal phalanx of the thumb | Posterior interosseous nerve (**C7**, C8) | Extends metacarpophalangeal joint of the thumb; can also extend the carpometacarpal joint of the thumb |
| Extensor pollicis longus | Posterior surface of ulna (distal to the abductor pollicis longus) and the adjacent interosseous membrane | Dorsal surface of base of distal phalanx of thumb | Posterior interosseous nerve (**C7**, C8) | Extends interphalangeal joint of the thumb; can also extend carpometacarpal and metacarpophalangeal joints of the thumb |
| Extensor indicis | Posterior surface of ulna (distal to extensor pollicis longus) and adjacent interosseous membrane | Extensor hood of index finger | Posterior interosseous nerve (**C7**, C8) | Extends index finger |

## Arteries and veins

The blood supply to the posterior compartment of the forearm occurs predominantly through branches of the radial, posterior interosseous, and anterior interosseous arteries (Fig. 7.92).

### Posterior interosseous artery

The posterior interosseous artery originates in the anterior compartment from the common interosseous branch of the ulnar artery and passes posteriorly over the proximal margin of the interosseous membrane and into the posterior compartment of the forearm. It contributes a branch, the **recurrent interosseous artery** (see Fig. 7.66B), to the vascular network around the elbow joint and then passes between the supinator and abductor pollicis longus muscles to supply the superficial extensors. After receiving the terminal end of the anterior interosseous artery, the posterior interosseous artery terminates by joining the dorsal carpal arch of the wrist.

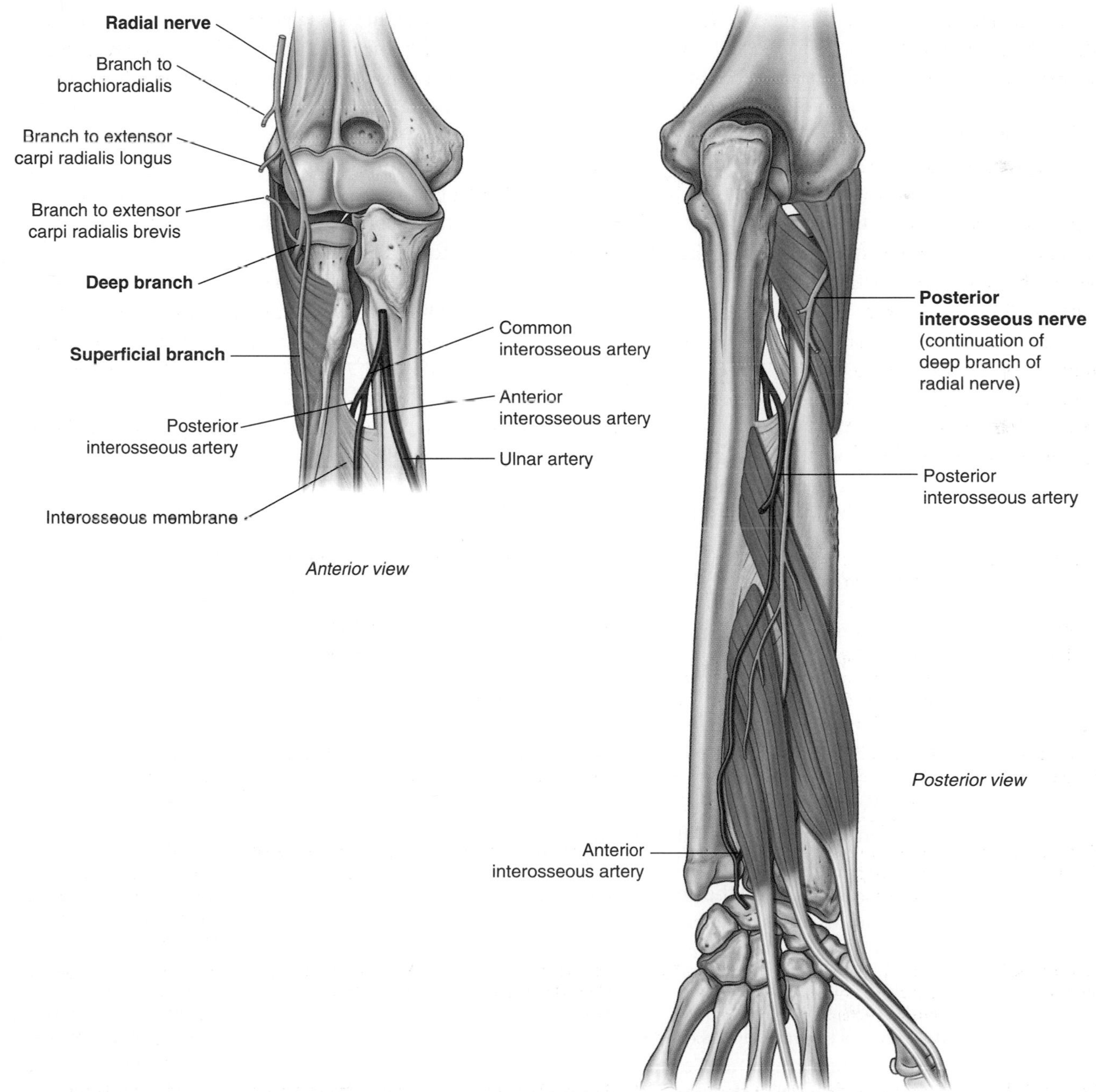


**Fig. 7.92** Posterior interosseous artery and radial nerve in posterior compartment of forearm.

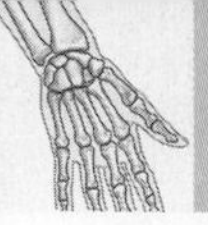



### Anterior interosseous artery

The anterior interosseous artery, also a branch of the common interosseous branch of the ulnar artery, is situated in the anterior compartment of the forearm on the interosseous membrane. It has numerous perforating branches, which pass directly through the interosseous membrane to supply deep muscles of the posterior compartment. The terminal end of the anterior interosseous artery passes posteriorly through an aperture in the interosseous membrane in distal regions of the forearm to join the posterior interosseous artery.

### Radial artery

The radial artery has muscular branches, which contribute to the supply of the extensor muscles on the radial side of the forearm.

### Veins

Deep veins of the posterior compartment generally accompany the arteries. They ultimately drain into brachial veins associated with the brachial artery in the cubital fossa.

## Nerves

### Radial nerve

The nerve of the posterior compartment of the forearm is the radial nerve (Fig. 7.92). Most of the muscles are innervated by the deep branch, which originates from the radial nerve in the lateral wall of the cubital fossa deep to the brachioradialis muscle and becomes the **posterior interosseous nerve** after emerging from between the superficial and deep layers of the supinator muscle in the posterior compartment of the forearm.

In the lateral wall of the cubital fossa, and before dividing into **superficial** and **deep branches**, the radial nerve innervates the brachioradialis and extensor carpi radialis longus muscles.

The deep branch innervates the extensor carpi radialis brevis, then passes between the two layers of the supinator muscle and follows the plane of separation between the two layers dorsally and laterally around the proximal shaft of the radius to the posterior aspect of the forearm. It supplies the supinator muscle and then emerges, as the posterior interosseous nerve, from the muscle to lie between the superficial and deep layers of muscles.

The posterior interosseous nerve supplies the remaining muscles in the posterior compartment and terminates as articular branches, which pass deep to the extensor pollicis longus muscle to reach the wrist.

## HAND

The hand (Fig. 7.93) is the region of the upper limb distal to the wrist joint. It is subdivided into three parts:

- the wrist (carpus),
- the metacarpus, and
- the digits (five fingers including the thumb).

The five digits consist of the laterally positioned thumb and, medial to the thumb, the four fingers—the index, middle, ring, and little fingers.

In the normal resting position, the fingers form a flexed arcade, with the little finger flexed most and the index finger flexed least. In the anatomical position, the fingers are extended.

The hand has an anterior surface (**palm**) and a dorsal surface (**dorsum of hand**).

Abduction and adduction of the fingers are defined with respect to the long axis of the middle finger (Fig. 7.93). In the anatomical position, the long axis of the thumb is

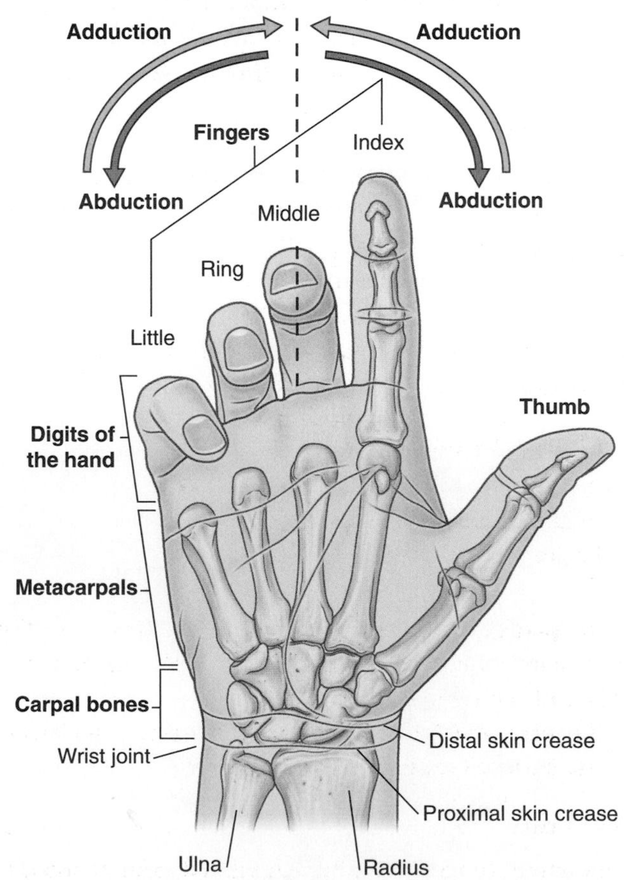


**Fig. 7.93** Right hand. The fingers are shown in a normal resting arcade in which they are flexed. In the anatomical position, the digits are straight and adducted.

rotated 90° to the rest of the digits so that the pad of the thumb points medially; consequently, movements of the thumb are defined at right angles to the movements of the other digits of the hand.

The hand is a mechanical and sensory tool. Many of the features of the upper limb are designed to facilitate positioning the hand in space.

## Bones

There are three groups of bones in the hand:

- The eight **carpal bones** are the bones of the wrist.
- The five **metacarpals** (**I to V**) are the bones of the metacarpus.
- The **phalanges** are the bones of the digits—the thumb has only two; the rest of the digits have three (Fig. 7.94).

The carpal bones and metacarpals of the index, middle, ring, and little fingers (metacarpals II to V) tend to function as a unit and form much of the bony framework of the palm. The metacarpal of the thumb functions independently and has increased flexibility at the carpometacarpal joint to provide opposition of the thumb to the fingers.

### Carpal bones

The small carpal bones of the wrist are arranged in two rows, a proximal and a distal row, each consisting of four bones (Fig. 7.94).

#### Proximal row

From lateral to medial and when viewed from anteriorly, the proximal row of bones consists of:

- the boat-shaped **scaphoid**,
- the **lunate**, which has a crescent shape,
- the three-sided **triquetrum** bone, and
- the pea-shaped **pisiform** (Fig. 7.94).

The **pisiform** is a sesamoid bone in the tendon of the flexor carpi ulnaris and articulates with the anterior surface of the **triquetrum**.

The **scaphoid** has a prominent **tubercle** on its lateral palmar surface that is directed anteriorly.

#### Distal row

From lateral to medial and when viewed from anteriorly, the distal row of carpal bones consists of:

- the irregular four-sided **trapezium** bone,
- the four-sided **trapezoid**,
- the **capitate**, which has a head, and
- the **hamate**, which has a hook (Fig. 7.94).

The **trapezium** articulates with the metacarpal bone of the thumb and has a distinct **tubercle** on its palmar surface that projects anteriorly.

The largest of the carpal bones, the **capitate**, articulates with the base of metacarpal III.

The **hamate**, which is positioned just lateral and distal to the pisiform, has a prominent hook (**hook of hamate**) on its palmar surface that projects anteriorly.

#### Articular surfaces

The carpal bones have numerous articular surfaces (Fig. 7.94). All of them articulate with each other, and the carpal bones in the distal row articulate with the metacarpals of the digits. With the exception of the metacarpal of the thumb, all movements of the metacarpal bones on the carpal bones are limited.

The expansive proximal surfaces of the scaphoid and lunate articulate with the radius to form the wrist joint.

#### Carpal arch

The carpal bones do not lie in a flat plane; rather, they form an arch, whose base is directed anteriorly (Fig. 7.94). The lateral side of this base is formed by the tubercles of the scaphoid and trapezium. The medial side is formed by the pisiform and the hook of the hamate.

The flexor retinaculum attaches to, and spans the distance between, the medial and lateral sides of the base to form the anterior wall of the so-called carpal tunnel. The sides and roof of the carpal tunnel are formed by the arch of the carpal bones.

### Metacarpals

Each of the five metacarpals is related to one digit:

- Metacarpal I is related to the thumb.
- Metacarpals II to V are related to the index, middle, ring, and little fingers, respectively (Fig. 7.94).

Each metacarpal consists of a **base**, a **shaft** (**body**), and distally, a **head**.

All of the bases of the metacarpals articulate with the carpal bones; in addition, the bases of the metacarpal bones of the fingers articulate with each other.

All of the heads of the metacarpals articulate with the proximal phalanges of the digits. The heads form the knuckles on the dorsal surface of the hand when the fingers are flexed.

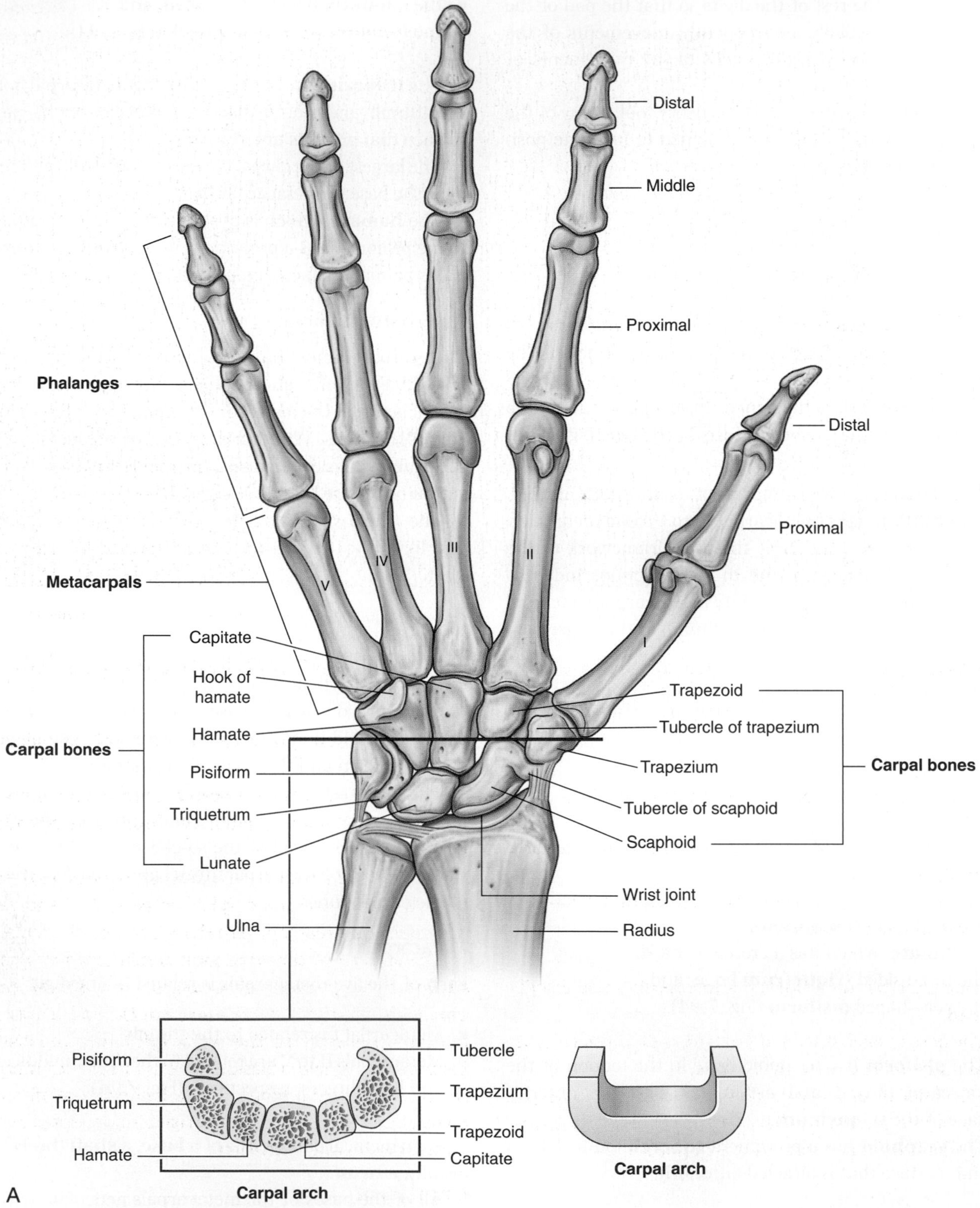

**Fig. 7.94** Right hand and wrist joint. **A.** Bones.

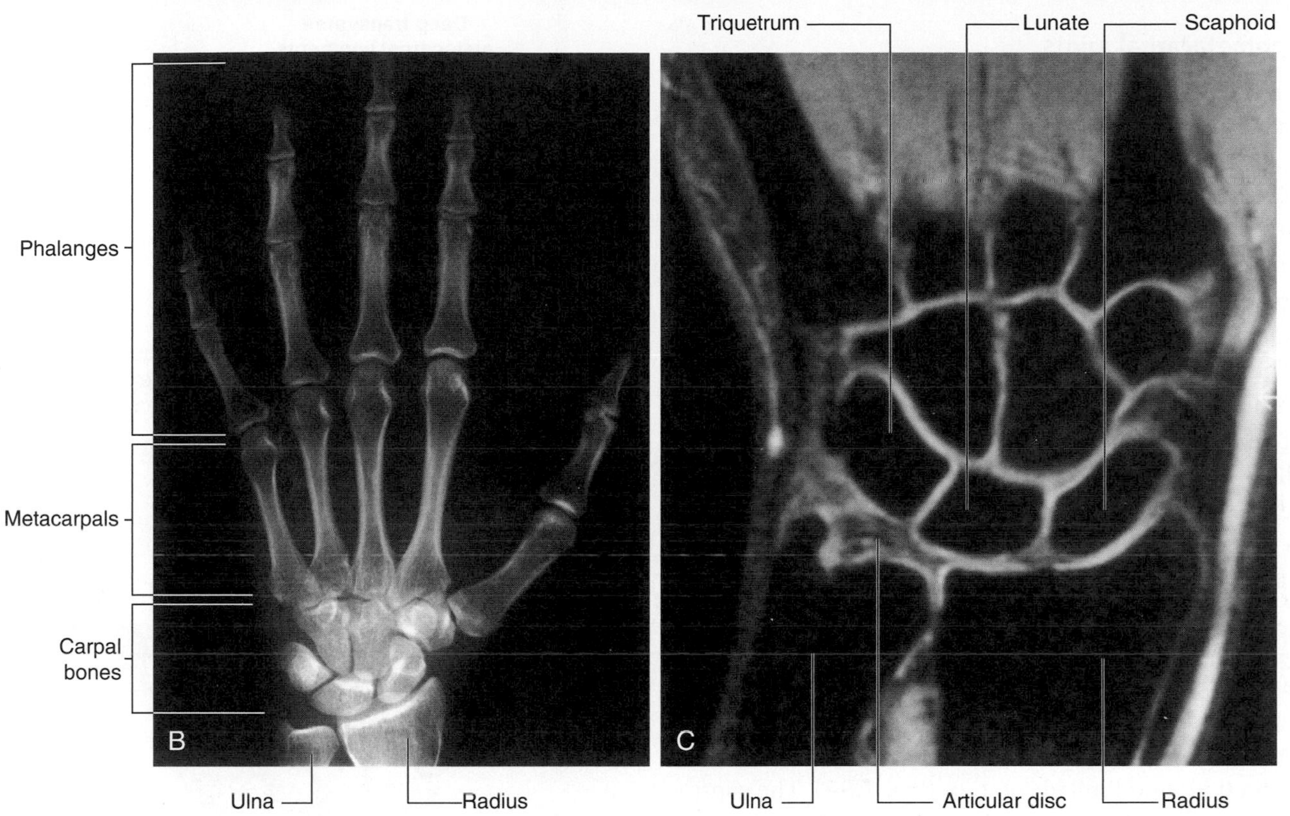


**Fig. 7.94, cont'd** Right hand and wrist joint. **B.** Radiograph of a normal hand and wrist joint (anteroposterior view). **C.** Magnetic resonance image of a normal wrist joint in the coronal plane.

### Phalanges

The phalanges are the bones of the digits (Fig. 7.94):

- The thumb has two—a **proximal** and a **distal phalanx**.
- The rest of the digits have three—a **proximal**, a **middle**, and a **distal phalanx**.

Each phalanx has a **base**, a **shaft** (**body**), and distally, a **head**.

The base of each proximal phalanx articulates with the head of the related metacarpal bone.

The head of each distal phalanx is nonarticular and flattened into a crescent-shaped palmar tuberosity, which lies under the palmar pad at the end of the digit.

## Joints

### Wrist joint

The wrist joint is a synovial joint between the distal end of the radius and the articular disc overlying the distal end of the ulna, and the scaphoid, lunate, and triquetrum (Fig. 7.94). Together, the articular surfaces of the carpals form an oval shape with a convex contour, which articulates with the corresponding concave surface of the radius and articular disc.

The wrist joint allows movement around two axes. The hand can be abducted, adducted, flexed, and extended at the wrist joint.

Because the radial styloid process extends further distally than does the ulnar styloid process, the hand can be adducted to a greater degree than it can be abducted.

The capsule of the wrist joint is reinforced by **palmar radiocarpal**, **palmar ulnocarpal**, and **dorsal radiocarpal ligaments**. In addition, **radial** and **ulnar collateral ligaments of the wrist joint** span the distance between the styloid processes of the radius and ulna and the adjacent carpal bones. These ligaments reinforce the medial and lateral sides of the wrist joint and support them during flexion and extension.

### Carpal joints

The synovial joints between the carpal bones share a common articular cavity. The joint capsule of the joints is reinforced by numerous ligaments.

Although movement at the **carpal joints** (**intercarpal joints**) is limited, the joints do contribute to the positioning of the hand in abduction, adduction, flexion, and, particularly, extension.

## Carpometacarpal joints

There are five carpometacarpal joints between the metacarpals and the related distal row of carpal bones (Fig. 7.94).

The saddle joint, between metacarpal I and the trapezium, imparts a wide range of mobility to the thumb that is not a feature of the rest of the digits. Movements at this carpometacarpal joint are flexion, extension, abduction, adduction, rotation, and circumduction.

The carpometacarpal joints between metacarpals II to V and the carpal bones are much less mobile than the carpometacarpal joint of the thumb, allowing only limited gliding movements. Movement of the joints increases medially, so metacarpal V slides to the greatest degree. This can be best observed on the dorsal surface of the hand as it makes a fist.

## Metacarpophalangeal joints

The joints between the distal heads of the metacarpals and the proximal phalanges of the digits are condylar joints, which allow flexion, extension, abduction, adduction, circumduction, and limited rotation (Fig. 7.94). The capsule of each joint is reinforced by the **palmar ligament** and by medial and lateral **collateral ligaments**.

### Deep transverse metacarpal ligaments

The three deep **transverse metacarpal ligaments** (Fig. 7.95) are thick bands of connective tissue connecting the palmar ligaments of the metacarpophalangeal joints of the fingers to each other. They are important because, by linking the heads of the metacarpal bones together, they restrict the movement of these bones relative to each other. As a result, they help form a unified skeletal framework for the palm of the hand.

Significantly, a deep transverse metacarpal ligament does not occur between the palmar ligament of the metacarpophalangeal joint of the thumb and the palmar ligament of the index finger. The absence of this ligament, and the presence of a saddle joint between metacarpal I and the trapezium, are responsible for the increased mobility of the thumb relative to the rest of the digits of the hand.

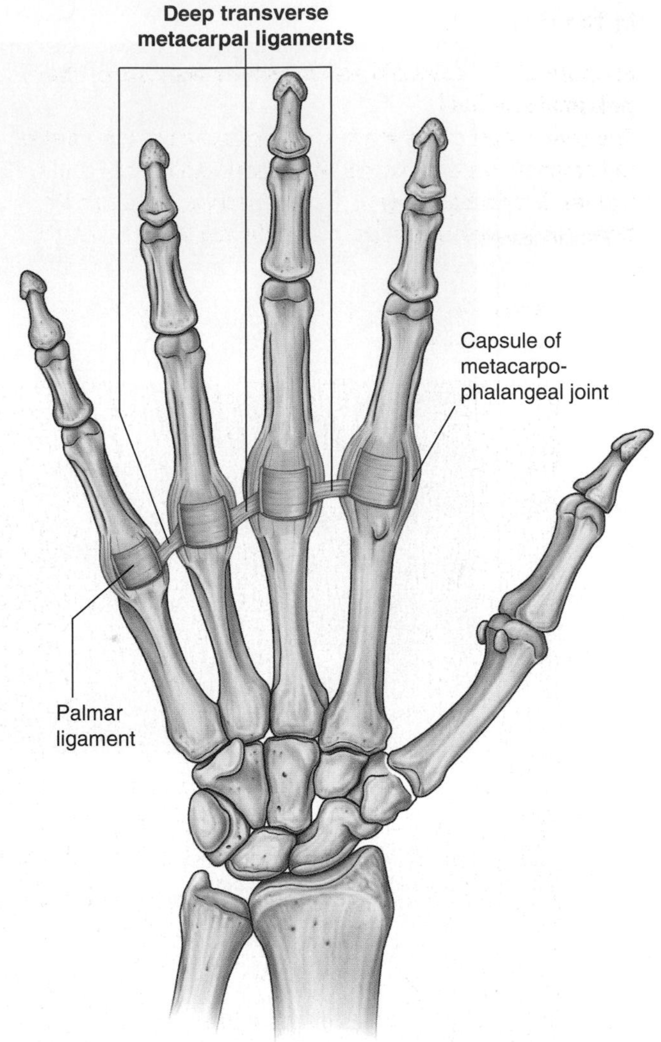


**Fig. 7.95** Deep transverse metacarpal ligaments, right hand.

## Interphalangeal joints of hand

The **interphalangeal joints of the hand** are hinge joints that allow mainly flexion and extension. They are reinforced by medial and lateral **collateral ligaments** and **palmar ligaments**.

## In the clinic

**Fracture of the scaphoid and avascular necrosis of the proximal scaphoid**

The commonest carpal injury is a fracture across the waist of the scaphoid bone (Fig. 7.96). It is uncommon to see other injuries. In approximately 10% of individuals, the scaphoid bone has a sole blood supply from the radial artery, which enters through the distal portion of the bone to supply the proximal portion. When a fracture occurs across the waist of the scaphoid, the proximal portion therefore undergoes avascular necrosis. It is impossible to predict which patients have this blood supply.

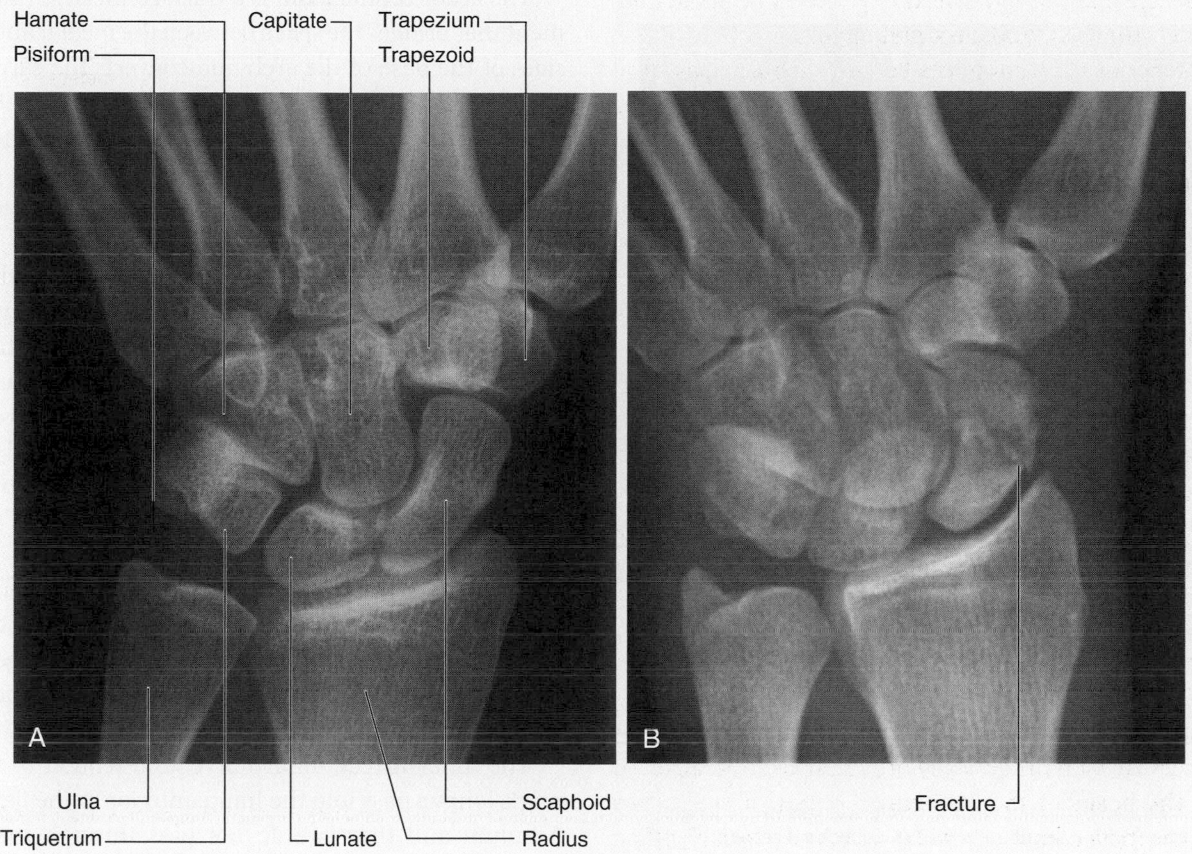


**Fig. 7.96** Wrist radiographs (posteroanterior view). **A.** Normal. **B.** Scaphoid fracture.

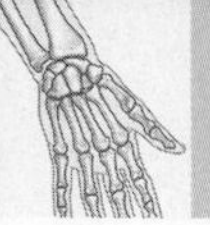



**In the clinic**

**Kienbock's disease**

Interruption of the blood supply to the lunate can lead to avascular necrosis of the lunate, known as Kienbock's disease (Fig. 7.97). This can cause pain and stiffness and arthritis in the longer term.

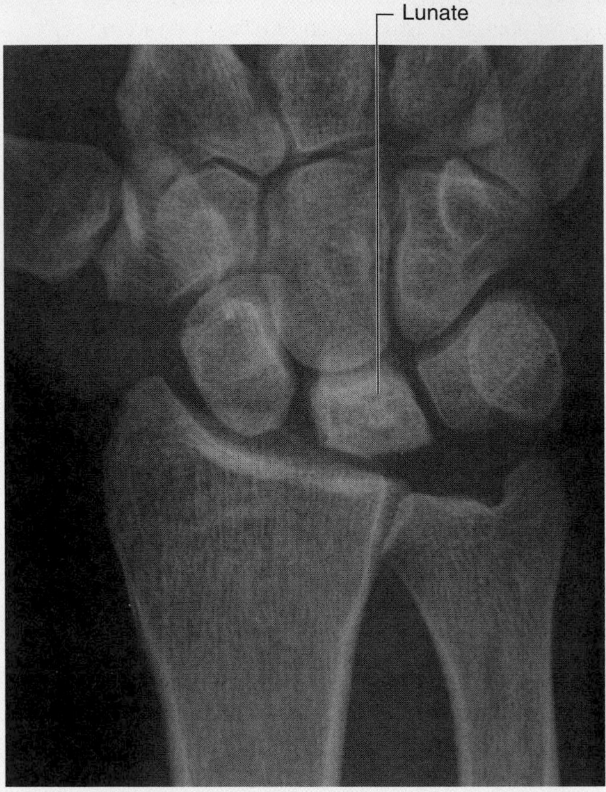


**Fig. 7.97** Radiograph of wrist showing sclerosis in the lunate consistent with avascular necrosis (Kienbock's disease).

## Carpal tunnel and structures at the wrist

The carpal tunnel is formed anteriorly at the wrist by a deep arch formed by the carpal bones and the flexor retinaculum (see Fig. 7.94).

The base of the carpal arch is formed medially by the pisiform and the hook of the hamate and laterally by the tubercles of the scaphoid and trapezium.

The flexor retinaculum is a thick connective tissue ligament that bridges the space between the medial and lateral sides of the base of the arch and converts the carpal arch into the carpal tunnel.

The four tendons of the flexor digitorum profundus, the four tendons of the flexor digitorum superficialis, and the tendon of the flexor pollicis longus pass through the carpal tunnel, as does the median nerve (Fig. 7.98).

The flexor retinaculum holds the tendons to the bony plane at the wrist and prevents them from "bowing."

Free movement of the tendons in the carpal tunnel is facilitated by synovial sheaths, which surround the tendons. All the tendons of the flexor digitorum profundus and flexor digitorum superficialis are surrounded by a single synovial sheath; a separate sheath surrounds the tendon of the flexor pollicis longus. The median nerve is anterior to the tendons in the carpal tunnel.

The tendon of the flexor carpi radialis is surrounded by a synovial sheath and passes through a tubular compartment formed by the attachment of the lateral aspect of the flexor retinaculum to the margins of a groove on the medial side of the tubercle of the trapezium.

The ulnar artery, ulnar nerve, and tendon of the palmaris longus pass into the hand anterior to the flexor retinaculum and therefore do not pass through the carpal tunnel (Fig. 7.98). The tendon of the palmaris longus is not surrounded by a synovial sheath.

The radial artery passes dorsally around the lateral side of the wrist and lies adjacent to the external surface of the scaphoid.

The extensor tendons pass into the hand on the medial, lateral, and posterior surfaces of the wrist in six compartments defined by an extensor retinaculum and lined by synovial sheaths (Fig. 7.98):

- The tendons of the extensor digitorum and extensor indicis share a compartment and synovial sheath on the posterior surface of the wrist.
- The tendons of the extensor carpi ulnaris and extensor digiti minimi have separate compartments and sheaths on the medial side of the wrist.
- The tendons of the abductor pollicis longus and extensor pollicis brevis muscles, the extensor carpi radialis longus and extensor carpi radialis brevis muscles, and the extensor pollicis longus muscle pass through three compartments on the lateral surface of the wrist.

## In the clinic

**Median artery**

A large median artery is an anatomical variant found in some individuals, where a persistent artery runs alongside the median nerve in one or both forearms and through the carpal tunnel. Individuals are at risk from heavy bleeding from deep cuts to the wrist.

## In the clinic

**Carpal tunnel syndrome**

Carpal tunnel syndrome is an entrapment syndrome caused by pressure on the median nerve within the carpal tunnel. The etiology of this condition is often obscure, though in some instances the nerve injury may be a direct effect of increased pressure on the median nerve caused by overuse, swelling of the tendons and tendon sheaths (e.g., rheumatoid arthritis), and cysts arising from the carpal joints. Increased pressure in the carpal tunnel is thought to cause venous congestion that produces nerve edema and anoxic damage to the capillary endothelium of the median nerve itself.

Patients typically report pain and pins-and-needles sensations in the distribution of the median nerve. Weakness and loss of muscle bulk of the thenar muscles may also occur. Gently tapping over the median nerve (in the region of the flexor retinaculum) readily produces these symptoms (Tinel's sign).

Initial treatment is aimed at reducing the inflammation and removing any repetitive insults that produce the symptoms. If this does not lead to improvement, nerve conduction studies will be necessary to confirm nerve entrapment, which may require surgical decompression of the flexor retinaculum.

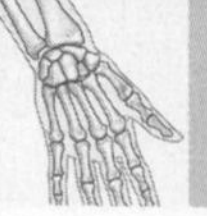

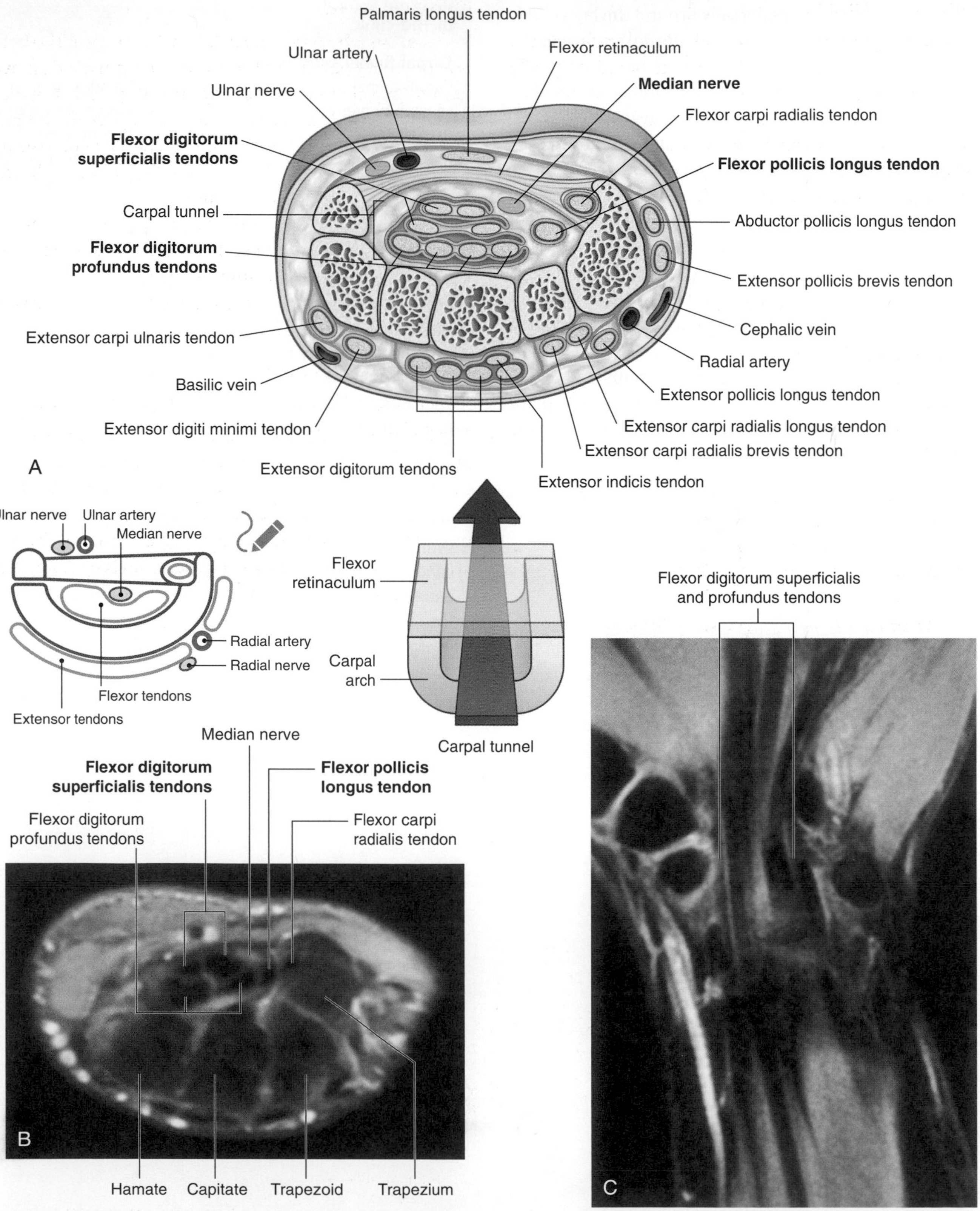

**Fig. 7.98** Carpal tunnel. **A.** Structure and relations. **B.** Magnetic resonance image of a normal wrist in the axial plane. **C.** Magnetic resonance image of a normal wrist in the coronal plane.

## Palmar aponeurosis

The **palmar aponeurosis** is a triangular condensation of deep fascia that covers the palm and is anchored to the skin in distal regions (Fig. 7.99).

The apex of the triangle is continuous with the palmaris longus tendon, when present; otherwise, it is anchored to the flexor retinaculum. From this point, fibers radiate to extensions at the bases of the digits that project into each of the index, middle, ring, and little fingers and, to a lesser extent, the thumb.

Transverse fibers interconnect the more longitudinally arranged bundles that continue into the digits.

Vessels, nerves, and long flexor tendons lie deep to the palmar aponeurosis in the palm.

### In the clinic

**Dupuytren's contracture**

The palmar fascia can become abnormally thickened in certain individuals, causing the fingers to progressively develop a fixed flexion position. This results in loss of dexterity and function, and in severe cases requires surgical removal of the abnormal tissue.

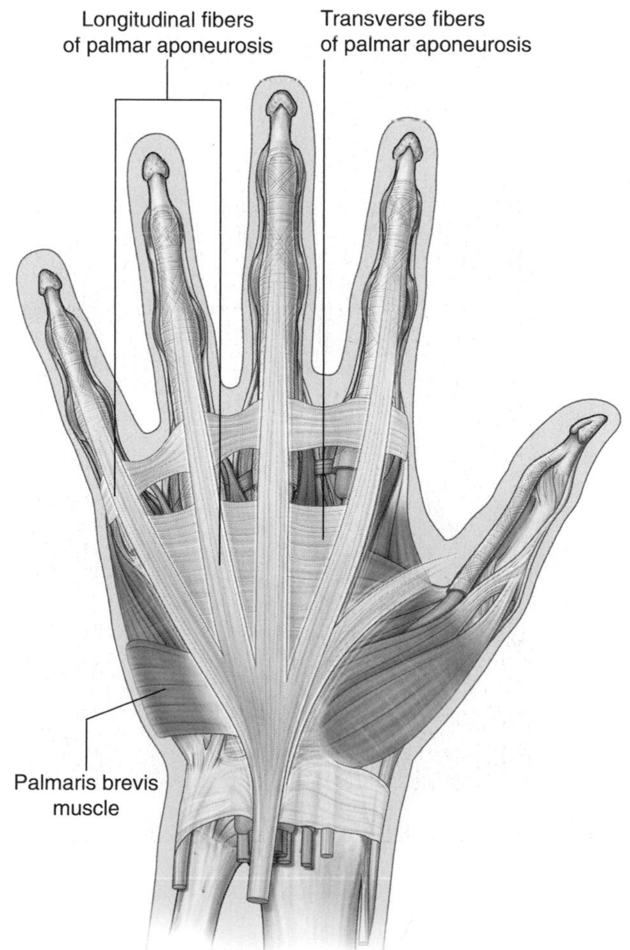


**Fig. 7.99** Palmar aponeurosis, right hand.

## Palmaris brevis

The **palmaris brevis**, a small intrinsic muscle of the hand, is a quadrangular-shaped subcutaneous muscle that overlies the hypothenar muscles, ulnar artery, and superficial branch of the ulnar nerve at the medial side of the palm (Fig. 7.99). It originates from the palmar aponeurosis and flexor retinaculum and inserts into the dermis of the skin on the medial margin of the hand.

The palmaris brevis deepens the cup of the palm by pulling on skin over the hypothenar eminence and forming a distinct ridge. This may improve grip.

The palmaris brevis is innervated by the superficial branch of the ulnar nerve.

## Anatomical snuffbox

The "anatomical snuffbox" is a term given to the triangular depression formed on the posterolateral side of the wrist and metacarpal I by the extensor tendons passing into the thumb (Fig. 7.100). Historically, ground tobacco (snuff) was placed in this depression before being inhaled into the nose. The base of the triangle is at the wrist and the apex is directed into the thumb. The impression is most apparent when the thumb is extended:

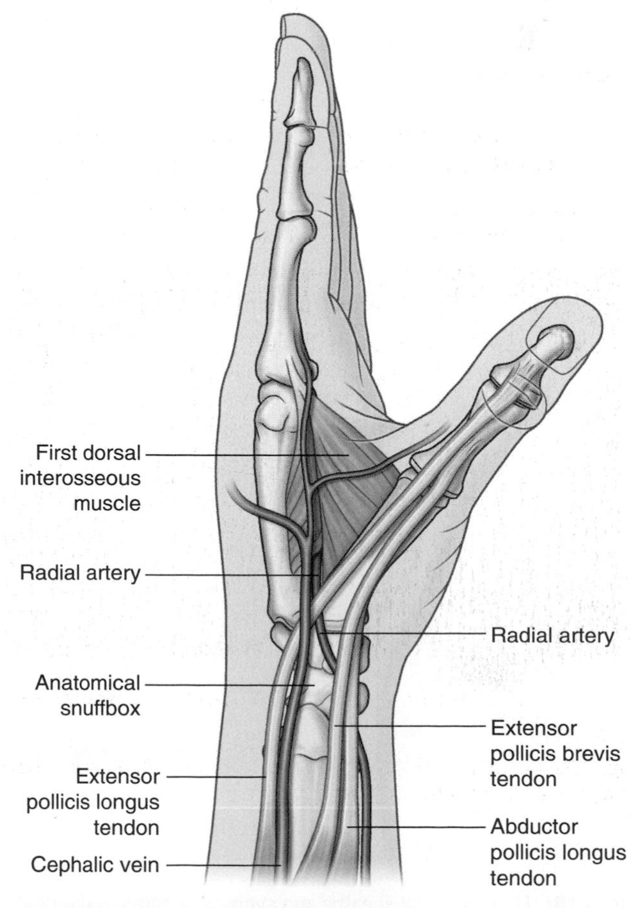


**Fig. 7.100** Anatomical snuffbox, left hand.

- The lateral border is formed by the tendons of the abductor pollicis longus and extensor pollicis brevis.
- The medial border is formed by the tendon of the extensor pollicis longus.
- The floor of the impression is formed by the scaphoid and trapezium, and the distal ends of the tendons of the extensor carpi radialis longus and extensor carpi radialis brevis.

The radial artery passes obliquely through the anatomical snuffbox, deep to the extensor tendons of the thumb and lies adjacent to the scaphoid and trapezium.

Terminal parts of the superficial branch of the radial nerve pass subcutaneously over the snuffbox as does the origin of the cephalic vein from the dorsal venous arch of the hand.

### In the clinic

**Snuffbox**

The anatomical snuffbox is an important clinical region. When the hand is in ulnar deviation, the scaphoid becomes palpable within the snuffbox. This position enables the physician to palpate the bone to assess for a fracture. The pulse of the radial artery can also be felt in the snuffbox.

## Fibrous digital sheaths

After exiting the carpal tunnel, the tendons of the flexor digitorum superficialis and profundus muscles cross the palm and enter fibrous sheaths on the palmar aspect of the digits (Fig. 7.101). These fibrous sheaths:

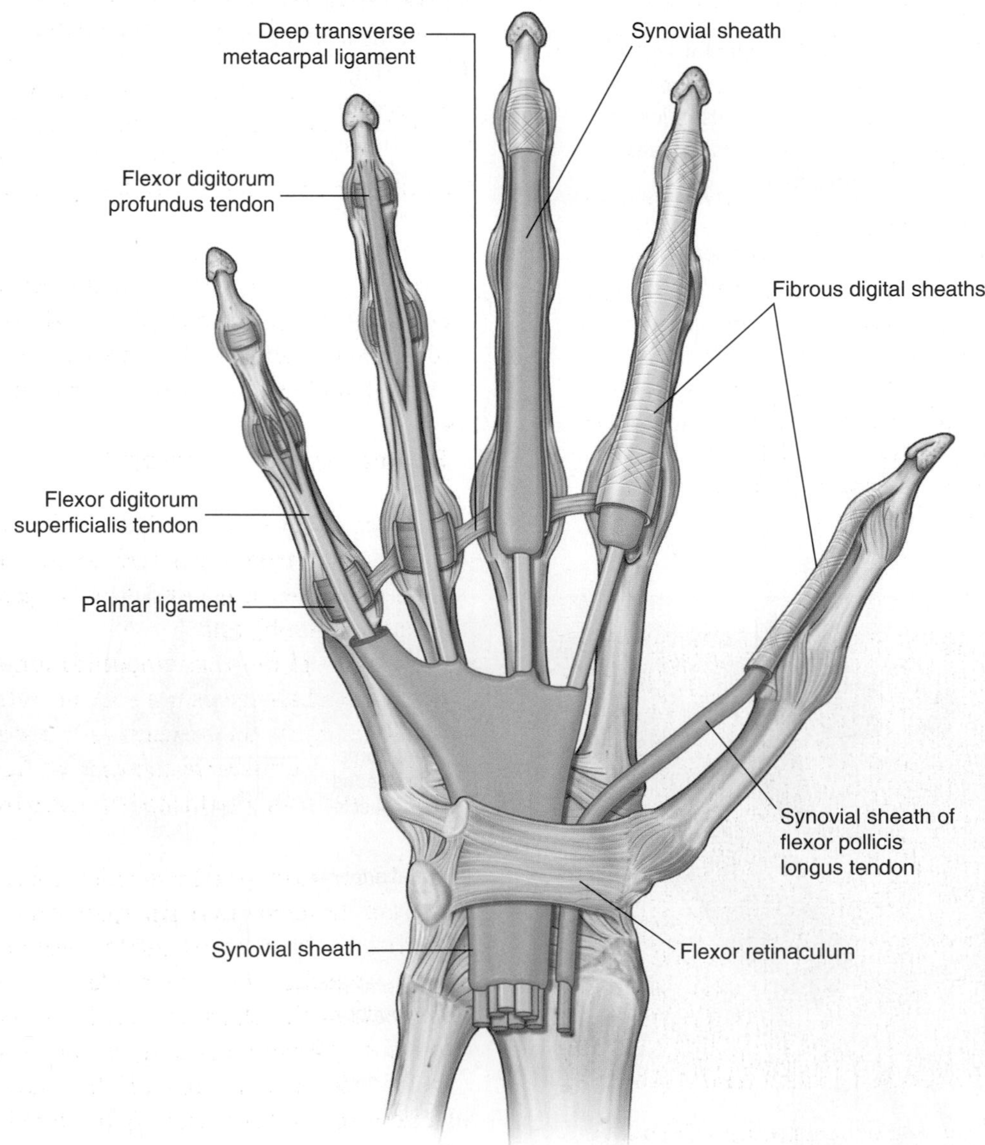


**Fig. 7.101** Fibrous digital sheaths and synovial sheaths of the right hand.

- begin proximally, anterior to the metacarpophalangeal joints, and extend to the distal phalanges;
- are formed by fibrous arches and cruciate (cross-shaped) ligaments, which are attached posteriorly to the margins of the phalanges and to the palmar ligaments associated with the metacarpophalangeal and interphalangeal joints; and
- hold the tendons to the bony plane and prevent the tendons from bowing when the digits are flexed.

Within each tunnel, the tendons are surrounded by a synovial sheath. The synovial sheaths of the thumb and little finger are continuous with the sheaths associated with the tendons in the carpal tunnel (Fig. 7.101).

## Extensor hoods

The tendons of the extensor digitorum and extensor pollicis longus muscles pass onto the dorsal aspect of the digits and expand over the proximal phalanges to form complex "**extensor hoods**" or "**dorsal digital expansions**" (Fig. 7.103A). The tendons of the extensor digiti minimi, extensor indicis, and extensor pollicis brevis muscles join these hoods.

Each extensor hood is triangular, with:

- the apex attached to the distal phalanx,
- the central region attached to the middle phalanx (index, middle, ring, and little fingers) or proximal phalanx (thumb), and
- each corner of the base wrapped around the sides of the metacarpophalangeal joint—in the index, middle, ring, and little fingers, the corners of the hoods attach mainly to the deep transverse metacarpal ligaments; in the thumb, the hood is attached on each side to muscles.

In addition to other attachments, many of the intrinsic muscles of the hand insert into the free margin of the hood on each side. By inserting into the extensor hood, these intrinsic muscles are responsible for complex delicate movements of the digits that could not be accomplished with the long flexor and extensor tendons alone.

In the index, middle, ring, and little fingers, the lumbrical, interossei, and abductor digiti minimi muscles attach to the extensor hoods. In the thumb, the adductor pollicis and abductor pollicis brevis muscles insert into and anchor the extensor hood.

### In the clinic

**De Quervain's syndrome**

De Quervain's syndrome is an inflammatory disorder that occurs within the first dorsal extensor compartment and involves the extensor pollicis brevis tendon and abductor pollicis longus tendon and their common tendon sheath (Fig. 7.102). Patients typically present with significant wrist pain preventing appropriate flexion/extension and abduction of the thumb. The cause of this disorder is often overuse. For example, the syndrome is common in young mothers who are constantly lifting young children. Other causes include inflammatory disorders such as rheumatoid arthritis.

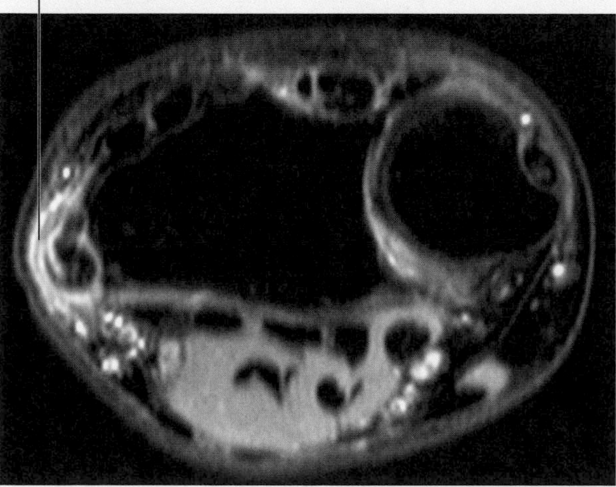


**Fig. 7.102** MRI of the wrist showing fluid and inflammation associated with the first extensor compartment, consistent with De Quervain's tenosynovitis.

### In the clinic

**Tenosynovitis**

Tenosynovitis is inflammation of a tendon and its sheath. The condition may be caused by overuse; however, it can also be associated with other disorders such as rheumatoid arthritis and connective tissue pathologies. If the inflammation becomes severe and ensuing fibrosis occurs, the tendon will not run smoothly within the tendon sheath, and typically within the fingers the tendon may stick or require excess force to fully extend and flex, producing a "triggering" phenomenon.

### In the clinic

**Trigger finger**

Trigger finger is a common disorder of late childhood and adulthood and is typically characterized by catching or snapping and occasionally locking of the flexor tendon(s) in the hand. Trigger finger can be associated with significant dysfunction and pain. The triggering is usually related to fibrosis and tightening of the flexor tendon sheath at the level of the metacarpophalangeal joint.

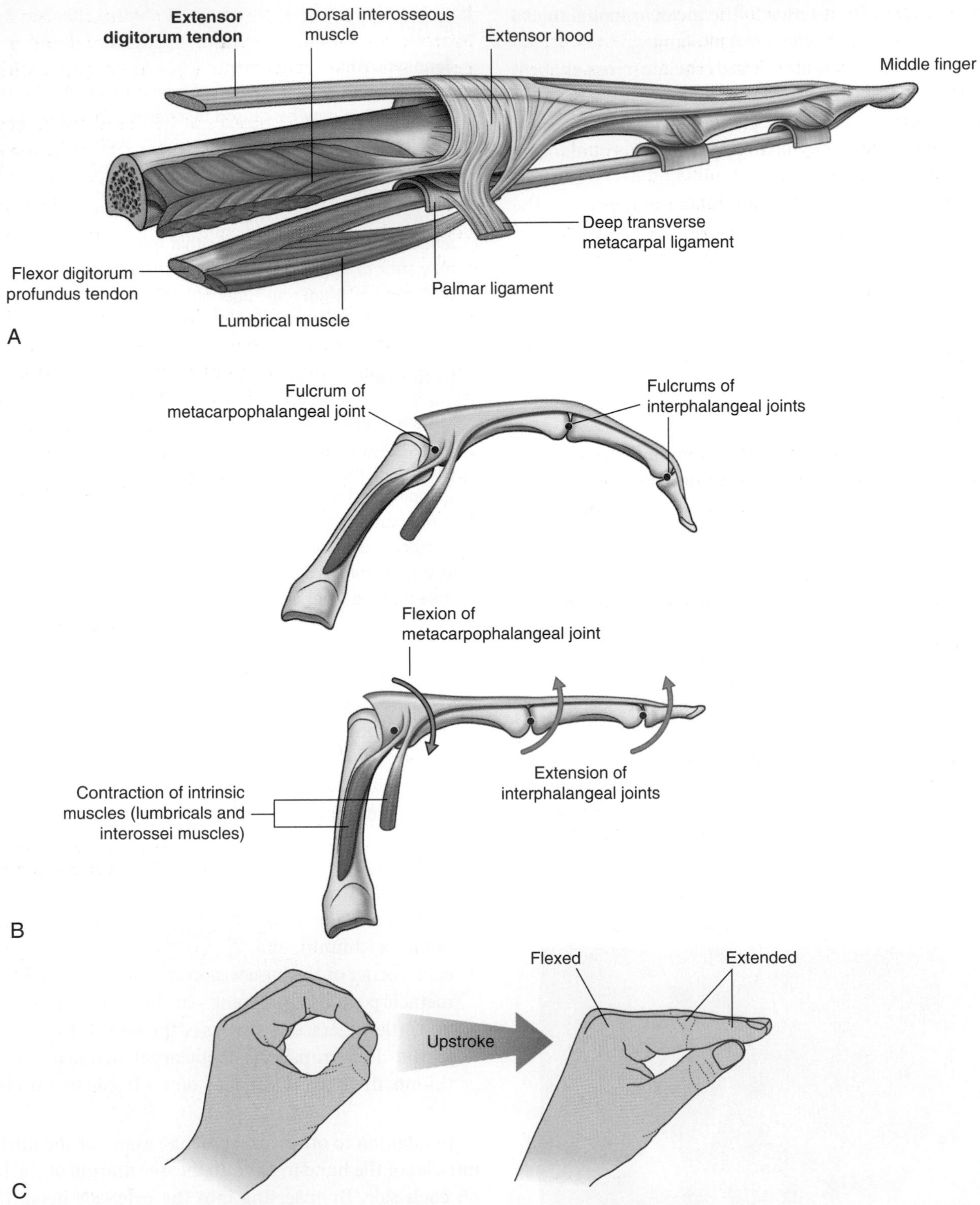

**Fig. 7.103** Extensor hood. **A** and **B.** Middle finger, left hand. **C.** Function of extensor hoods and intrinsic muscles.

Because force from the small intrinsic muscles of the hand is applied to the extensor hood distal to the fulcrum of the metacarpophalangeal joints, the muscles flex these joints (Fig. 7.103B). Simultaneously, the force is transferred dorsally through the hood to extend the interphalangeal joints. This ability to flex the metacarpophalangeal joints, while at the same time extending the interphalangeal joints, is entirely due to the intrinsic muscles of the hand working through the extensor hoods. This type of precision movement is used in the upstroke when writing a *t* (Fig. 7.103C).

## Muscles

The intrinsic muscles of the hand are the palmaris brevis (described on p. 791; see Fig. 7.99), interossei, adductor pollicis, thenar, hypothenar, and lumbrical muscles (Figs. 7.104 to 7.108). Unlike the extrinsic muscles that originate in the forearm, insert in the hand, and function in forcefully gripping ("power grip") with the hand, the intrinsic muscles occur entirely in the hand and mainly execute precision movements ("precision grip") with the fingers and thumb.

All of the intrinsic muscles of the hand are innervated by the deep branch of the ulnar nerve except for the three thenar and two lateral lumbrical muscles, which are innervated by the median nerve. The intrinsic muscles are predominantly innervated by spinal cord segment T1 with a contribution from C8.

The interossei are muscles between and attached to the metacarpals (Figs. 7.104 and 7.105). They insert into the proximal phalanx of each digit and into the extensor hood and are divided into two groups, the dorsal interossei and the palmar interossei. All of the interossei are innervated by the deep branch of the ulnar nerve. Collectively, the interossei abduct and adduct the digits and contribute to the complex flexion and extension movements generated by the extensor hoods.

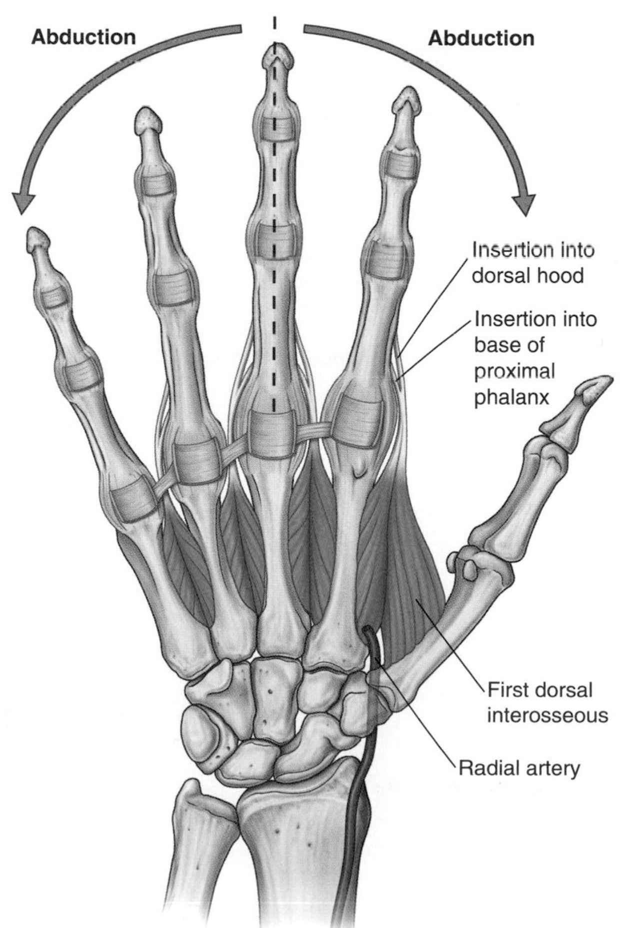


**Fig. 7.104** Dorsal interossei (palmar view), right hand.

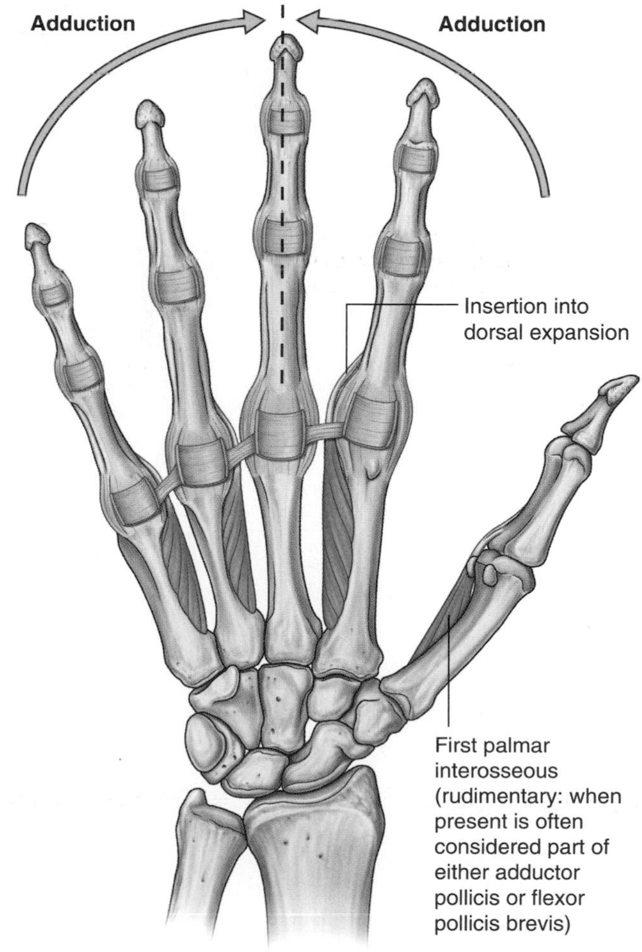


**Fig. 7.105** Palmar interossei (palmar view), right hand.

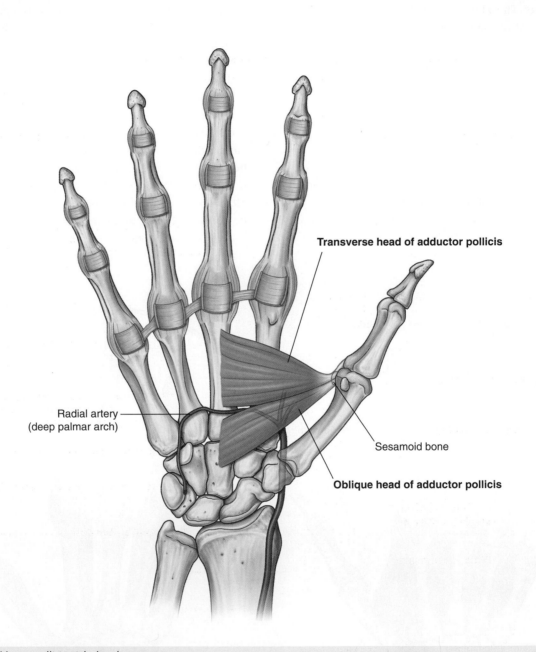

Transverse head of adductor pollicis

Radial artery
(deep palmar arch)

Sesamoid bone

Oblique head of adductor pollicis

**Fig. 7.106** Adductor pollicis, right hand.

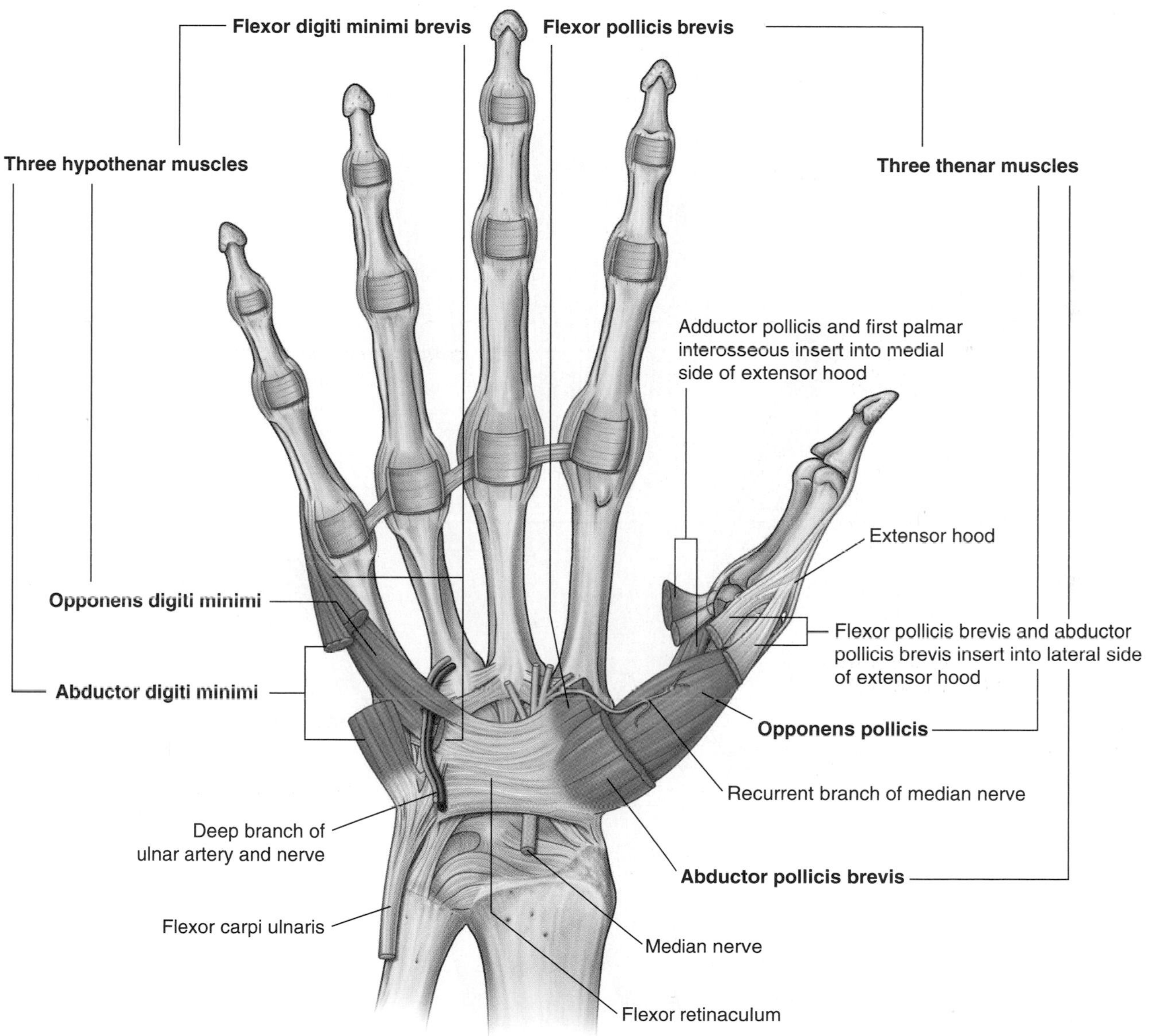
Flexor digiti minimi brevis
Flexor pollicis brevis
Three hypothenar muscles
Three thenar muscles
Adductor pollicis and first palmar interosseous insert into medial side of extensor hood
Extensor hood
Opponens digiti minimi
Flexor pollicis brevis and abductor pollicis brevis insert into lateral side of extensor hood
Abductor digiti minimi
Opponens pollicis
Recurrent branch of median nerve
Deep branch of ulnar artery and nerve
Abductor pollicis brevis
Flexor carpi ulnaris
Median nerve
Flexor retinaculum

**Fig. 7.107** Thenar and hypothenar muscles, right hand.

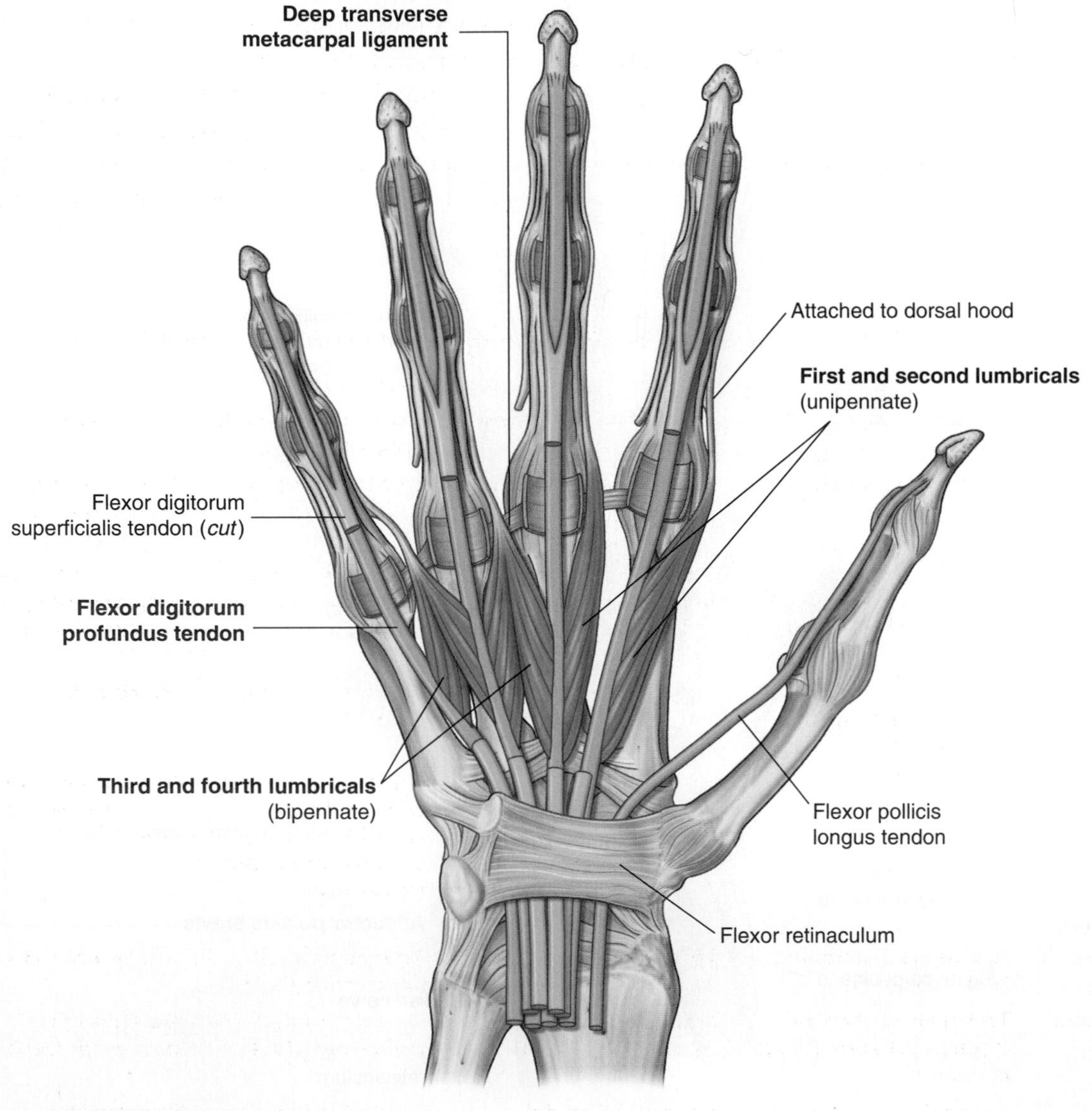

**Fig. 7.108** Lumbrical muscles, right hand.

### Dorsal interossei

**Dorsal interossei** are the most dorsally situated of all of the intrinsic muscles and can be palpated through the skin on the dorsal aspect of the hand (Fig. 7.104). There are four bipennate dorsal interosseous muscles between, and attached to, the shafts of adjacent metacarpal bones (Fig. 7.104). Each muscle inserts both into the base of the proximal phalanx and into the extensor hood of its related digit.

The tendons of the dorsal interossei pass dorsal to the deep transverse metacarpal ligaments:

- The first dorsal interosseous muscle is the largest and inserts into the lateral side of the index finger.
- The second and third dorsal interossei insert into the lateral and medial sides, respectively, of the middle finger.
- The fourth dorsal interosseous muscle inserts into the medial side of the ring finger.

In addition to generating flexion and extension movements of the fingers through their attachments to the extensor hoods, the dorsal interossei are the major abductors of the index, middle, and ring fingers, at the metacarpophalangeal joints (Table 7.15).

The middle finger can abduct medially and laterally with respect to the long axis of the middle finger and consequently has a dorsal interosseous muscle on each side. The thumb and little finger have their own abductors

**Table 7.15** Intrinsic muscles of the hand (spinal segments indicated in bold are the major segments innervating the muscle)

| Muscle | Origin | Insertion | Innervation | Function |
|---|---|---|---|---|
| Palmaris brevis | Palmar aponeurosis and flexor retinaculum | Dermis of skin on the medial margin of the hand | Superficial branch of the ulnar nerve (C8, **T1**) | Improves grip |
| Dorsal interossei (four muscles) | Adjacent sides of metacarpals | Extensor hood and base of proximal phalanges of index, middle, and ring fingers | Deep branch of ulnar nerve (C8, **T1**) | Abduction of index, middle, and ring fingers at the metacarpophalangeal joints |
| Palmar interossei (three or four muscles) | Sides of metacarpals | Extensor hoods of the thumb, index, ring, and little fingers and the proximal phalanx of thumb | Deep branch of ulnar nerve (C8, **T1**) | Adduction of the thumb, index, ring, and little fingers at the metacarpophalangeal joints |
| Adductor pollicis | Transverse head—metacarpal III; oblique head—capitate and bases of metacarpals II and III | Base of proximal phalanx and extensor hood of thumb | Deep branch of ulnar nerve (C8, **T1**) | Adducts thumb |
| Lumbricals (four muscles) | Tendons of flexor digitorum profundus | Extensor hoods of index, ring, middle, and little fingers | Medial two by the deep branch of the ulnar nerve; lateral two by digital branches of the median nerve | Flex metacarpophalangeal joints while extending interphalangeal joints |
| **THENAR MUSCLES** | | | | |
| Opponens pollicis | Tubercle of trapezium and flexor retinaculum | Lateral margin and adjacent palmar surface of metacarpal I | Recurrent branch of median nerve (C8, **T1**) | Medially rotates thumb |
| Abductor pollicis brevis | Tubercles of scaphoid and trapezium and adjacent flexor retinaculum | Proximal phalanx and extensor hood of thumb | Recurrent branch of median nerve (C8, **T1**) | Abducts thumb at metacarpophalangeal joint |
| Flexor pollicis brevis | Tubercle of the trapezium and flexor retinaculum | Proximal phalanx of the thumb | Recurrent branch of median nerve (C8, **T1**) | Flexes thumb at metacarpophalangeal joint |
| **HYPOTHENAR MUSCLES** | | | | |
| Opponens digiti minimi | Hook of hamate and flexor retinaculum | Medial aspect of metacarpal V | Deep branch of ulnar nerve (C8, **T1**) | Laterally rotates metacarpal V |
| Abductor digiti minimi | Pisiform, the pisohamate ligament, and tendon of flexor carpi ulnaris | Proximal phalanx of little finger | Deep branch of ulnar nerve (C8, **T1**) | Abducts little finger at metacarpophalangeal joint |
| Flexor digiti minimi brevis | Hook of the hamate and flexor retinaculum | Proximal phalanx of little finger | Deep branch of ulnar nerve (C8, **T1**) | Flexes little finger at metacarpophalangeal joint |

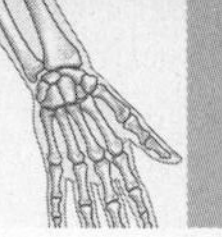



in the thenar and hypothenar muscle groups, respectively, and therefore do not have dorsal interossei.

The radial artery passes between the two heads of the first dorsal interosseous muscle as it passes from the anatomical snuffbox on the posterolateral side of the wrist into the deep aspect of the palm.

### Palmar interossei

The three (or four) **palmar interossei** are anterior to the dorsal interossei, and are unipennate muscles originating from the metacarpals of the digits with which each is associated (Fig. 7.105).

The first palmar interosseous muscle is rudimentary and often considered part of either the adductor pollicis or the flexor pollicis brevis. When present, it originates from the medial side of the palmar surface of metacarpal I and inserts into both the base of the proximal phalanx of the thumb and into the extensor hood. A sesamoid bone often occurs in the tendon attached to the base of the phalanx.

The second palmar interosseous muscle originates from the medial surface of metacarpal II and inserts into the medial side of the extensor hood of the index finger.

The third and fourth palmar interossei originate from the lateral surfaces of metacarpals IV and V and insert into the lateral sides of the respective extensor hoods.

Like the tendons of the dorsal interossei, the tendons of the palmar interossei pass dorsal to the deep transverse metacarpal ligaments.

The palmar interossei adduct the thumb, index, ring, and little fingers with respect to a long axis through the middle finger. The movements occur at the metacarpophalangeal joints. Because the muscles insert into the extensor hoods, they also produce complex flexion and extension movements of the digits (Table 7.15).

### Adductor pollicis

The **adductor pollicis** is a large triangular muscle anterior to the plane of the interossei that crosses the palm (Fig. 7.106). It originates as two heads:

- a **transverse head** from the anterior aspect of the shaft of metacarpal III, and
- an **oblique head**, from the capitate and adjacent bases of metacarpals II and III.

The two heads converge laterally to form a tendon, which often contains a sesamoid bone, that inserts into both the medial side of the base of the proximal phalanx of the thumb and into the extensor hood.

The radial artery passes anteriorly and medially between the two heads of the muscle to enter the deep plane of the palm and form the deep palmar arch.

The adductor pollicis is a powerful adductor of the thumb and opposes the thumb to the rest of the digits in gripping (Table 7.15).

### Thenar muscles

The three thenar muscles (the opponens pollicis, flexor pollicis brevis, and abductor pollicis brevis muscles) are associated with opposition of the thumb to the fingers and with delicate movements of the thumb (Fig. 7.107) and are responsible for the prominent swelling (**thenar eminence**) on the lateral side of the palm at the base of the thumb.

The thenar muscles are innervated by the recurrent branch of the median nerve.

#### Opponens pollicis

The **opponens pollicis** muscle is the largest of the thenar muscles and lies deep to the other two (Fig. 7.107). Originating from the tubercle of the trapezium and the adjacent flexor retinaculum, it inserts along the entire length of the lateral margin and adjacent lateral palmar surface of metacarpal I.

The opponens pollicis rotates and flexes metacarpal I on the trapezium, so bringing the pad of the thumb into a position facing the pads of the fingers (Table 7.15).

#### Abductor pollicis brevis

The **abductor pollicis brevis** muscle overlies the opponens pollicis and is proximal to the flexor pollicis brevis muscle (Fig. 7.107). It originates from the tubercles of the scaphoid and trapezium and from the adjacent flexor retinaculum, and inserts into the lateral side of the base of the proximal phalanx of the thumb and into the extensor hood.

The abductor pollicis brevis abducts the thumb, principally at the metacarpophalangeal joint. Its action is most apparent when the thumb is maximally abducted and the proximal phalanx is moved out of line with the long axis of the metacarpal bone (Table 7.15).

#### Flexor pollicis brevis

The **flexor pollicis brevis** muscle is distal to the abductor pollicis brevis (Fig. 7.107). It originates mainly from the tubercle of the trapezium and adjacent flexor retinaculum, but it may also have deeper attachments to other carpal bones and associated ligaments. It inserts into the lateral side of the base of the proximal phalanx of the thumb. The tendon often contains a sesamoid bone.

The flexor pollicis brevis flexes the metacarpophalangeal joint of the thumb (Table 7.15).

### Hypothenar muscles

The hypothenar muscles (the opponens digiti minimi, abductor digiti minimi, and flexor digiti minimi brevis) contribute to the swelling (**hypothenar eminence**) on the medial side of the palm at the base of the little finger (Fig. 7.107). The hypothenar muscles are similar to the thenar muscles in name and in organization.

Unlike the thenar muscles, the hypothenar muscles are innervated by the deep branch of the ulnar nerve and not by the recurrent branch of the median nerve.

#### Opponens digiti minimi

The **opponens digiti minimi** muscle lies deep to the other two hypothenar muscles (Fig. 7.107). It originates from the hook of the hamate and from the adjacent flexor retinaculum and it inserts into the medial margin and palmar surface of metacarpal V. Its base is penetrated by the deep branches of the ulnar nerve and ulnar artery.

The opponens digiti minimi rotates metacarpal V toward the palm; however, because of the simple shape of the carpometacarpal joint and the presence of a deep transverse metacarpal ligament, which attaches the head of metacarpal V to that of the ring finger, the movement is much less dramatic than that of the thumb (Table 7.15).

#### Abductor digiti minimi

The **abductor digiti minimi** muscle overlies the opponens digiti minimi (Fig. 7.107). It originates from the pisiform bone, the pisohamate ligament, and the tendon of the flexor carpi ulnaris, and inserts into the medial side of the base of the proximal phalanx of the little finger and into the extensor hood.

The abductor digiti minimi is the principal abductor of the little finger (Table 7.15).

#### Flexor digiti minimi brevis

The **flexor digiti minimi brevis** muscle is lateral to the abductor digiti minimi (Fig. 7.107). It originates from the hook of the hamate bone and the adjacent flexor retinaculum and inserts with the abductor digiti minimi muscle into the medial side of the base of the proximal phalanx of the little finger.

The flexor digiti minimi brevis flexes the metacarpophalangeal joint.

### Lumbrical muscles

There are four lumbrical (worm-like) muscles, each of which is associated with one of the fingers. The muscles originate from the tendons of the flexor digitorum profundus in the palm:

- The medial two lumbricals are bipennate and originate from the flexor digitorum profundus tendons associated with the middle and ring fingers and the ring and little fingers, respectively.
- The lateral two lumbricals are unipennate muscles, originating from the flexor digitorum profundus tendons associated with the index and middle fingers, respectively.

The lumbricals pass dorsally around the lateral side of each finger, and insert into the extensor hood (Fig. 7.108). The tendons of the muscles are anterior to the deep transverse metacarpal ligaments.

The lumbricals are unique because they link flexor tendons with extensor tendons. Through their insertion into the extensor hoods, they participate in flexing the metacarpophalangeal joints and extending the interphalangeal joints.

The medial two lumbricals are innervated by the deep branch of the ulnar nerve; the lateral two lumbricals are innervated by digital branches of the median nerve (Table 7.15).

## Arteries and veins

The blood supply to the hand is by the radial and ulnar arteries, which form two interconnected vascular arches (superficial and deep) in the palm (Fig. 7.109). Vessels to the digits, muscles, and joints originate from the two arches and the parent arteries:

- The radial artery contributes substantially to the supply of the thumb and the lateral side of the index finger.

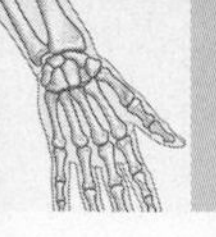

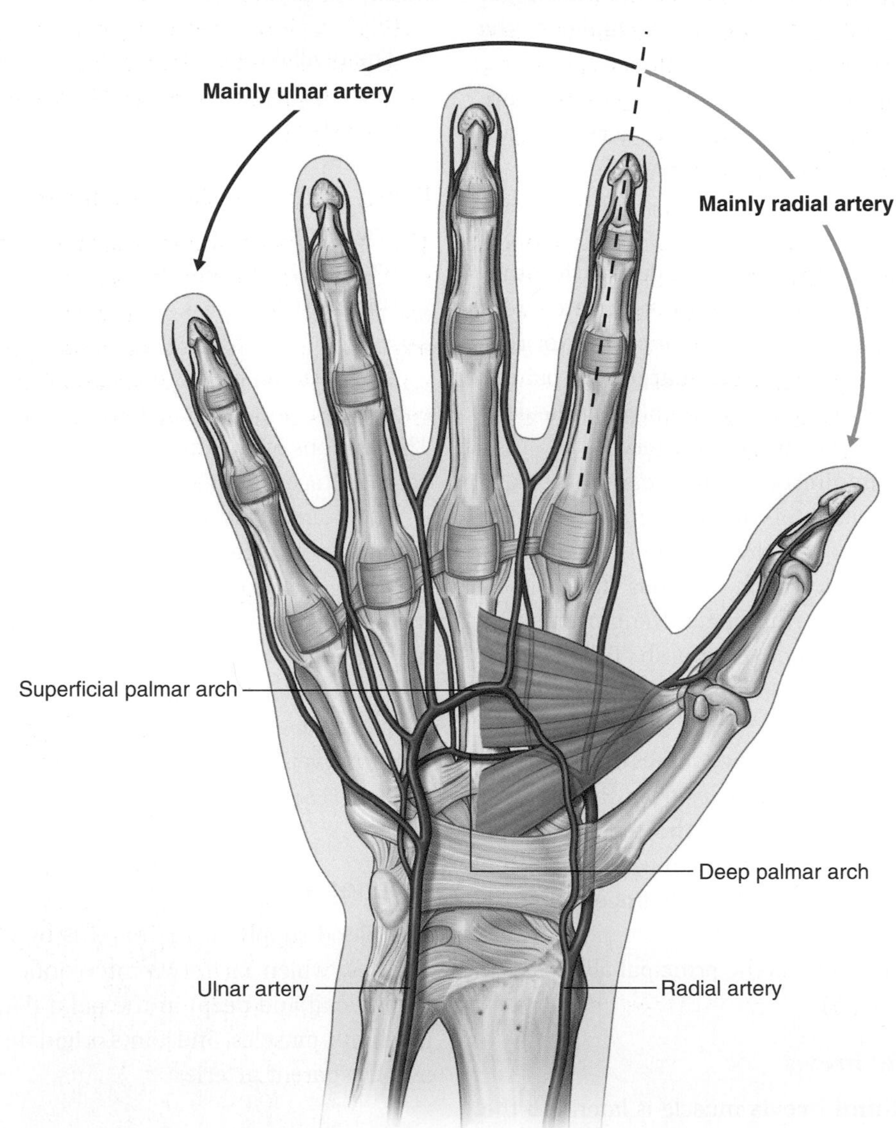


**Fig. 7.109** Arterial supply of the right hand.

- The remaining digits and the medial side of the index finger are supplied mainly by the ulnar artery.

### Ulnar artery and superficial palmar arch

The **ulnar artery** and ulnar nerve enter the hand on the medial side of the wrist (Fig. 7.110). The vessel lies between the palmaris brevis and the flexor retinaculum and is lateral to the ulnar nerve and the pisiform bone. Distally, the ulnar artery is medial to the hook of the hamate bone and then swings laterally across the palm, forming the **superficial palmar arch**, which is superficial to the long flexor tendons of the digits and just deep to the palmar aponeurosis. On the lateral side of the palm, the arch communicates with a palmar branch of the radial artery.

One branch of the ulnar artery in the hand is the **deep palmar branch** (Figs. 7.109 and 7.110), which arises from the medial aspect of the ulnar artery, just distal to the pisiform, and penetrates the origin of the hypothenar muscles. It curves medially around the hook of the hamate to access the deep plane of the palm and to anastomose with the deep palmar arch derived from the radial artery.

Branches from the superficial palmar arch include:

- a palmar digital artery to the medial side of the little finger, and
- three large, **common palmar digital arteries**, which ultimately provide the principal blood supply to the lateral side of the little finger, both sides of the ring and middle fingers, and the medial side of the index finger (Fig. 7.110); they are joined by palmar metacarpal arteries from the deep palmar arch before bifurcating into the **proper palmar digital arteries**, which enter the fingers.

### Radial artery and deep palmar arch

The **radial artery** curves around the lateral side of the wrist and passes over the floor of the anatomical snuffbox and into the deep plane of the palm by penetrating anteriorly through the back of the hand (Figs. 7.109 and 7.111). It passes between the two heads of the first dorsal interosseous muscle and then between the two heads of the

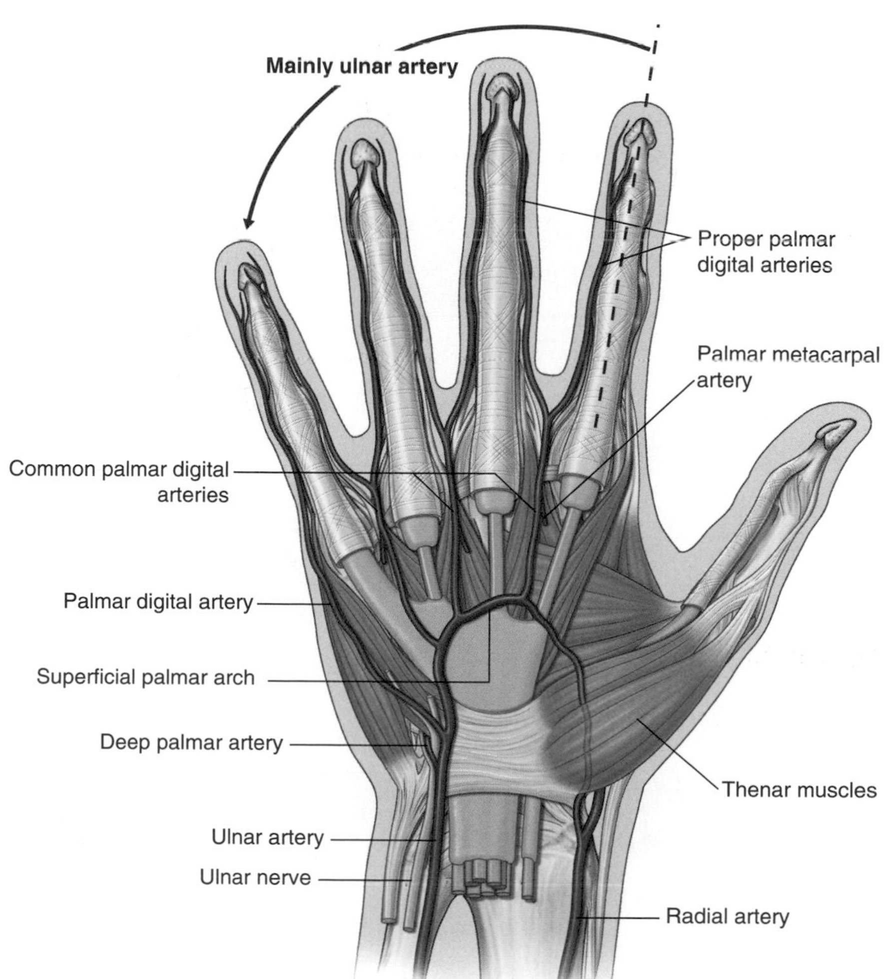


**Fig. 7.110** Superficial palmar arch, right hand.

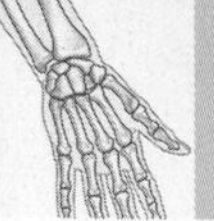



adductor pollicis to access the deep plane of the palm and form the deep palmar arch.

The **deep palmar arch** passes medially through the palm between the metacarpal bones and the long flexor tendons of the digits. On the medial side of the palm, it communicates with the deep palmar branch of the ulnar artery (Figs. 7.109 and 7.111).

Before penetrating the back of the hand, the radial artery gives rise to two vessels:

- a **dorsal carpal branch**, which passes medially as the **dorsal carpal arch**, across the wrist and gives rise to **three dorsal metacarpal arteries**, which subsequently divide to become small dorsal digital arteries, which enter the fingers; and
- the **first dorsal metacarpal artery**, which supplies adjacent sides of the index finger and thumb.

Two vessels, the **princeps pollicis artery** and the **radialis indicis artery**, arise from the radial artery in the plane between the first dorsal interosseous and adductor pollicis. The princeps pollicis artery is the major blood supply to the thumb, and the radialis indicis artery supplies the lateral side of the index finger.

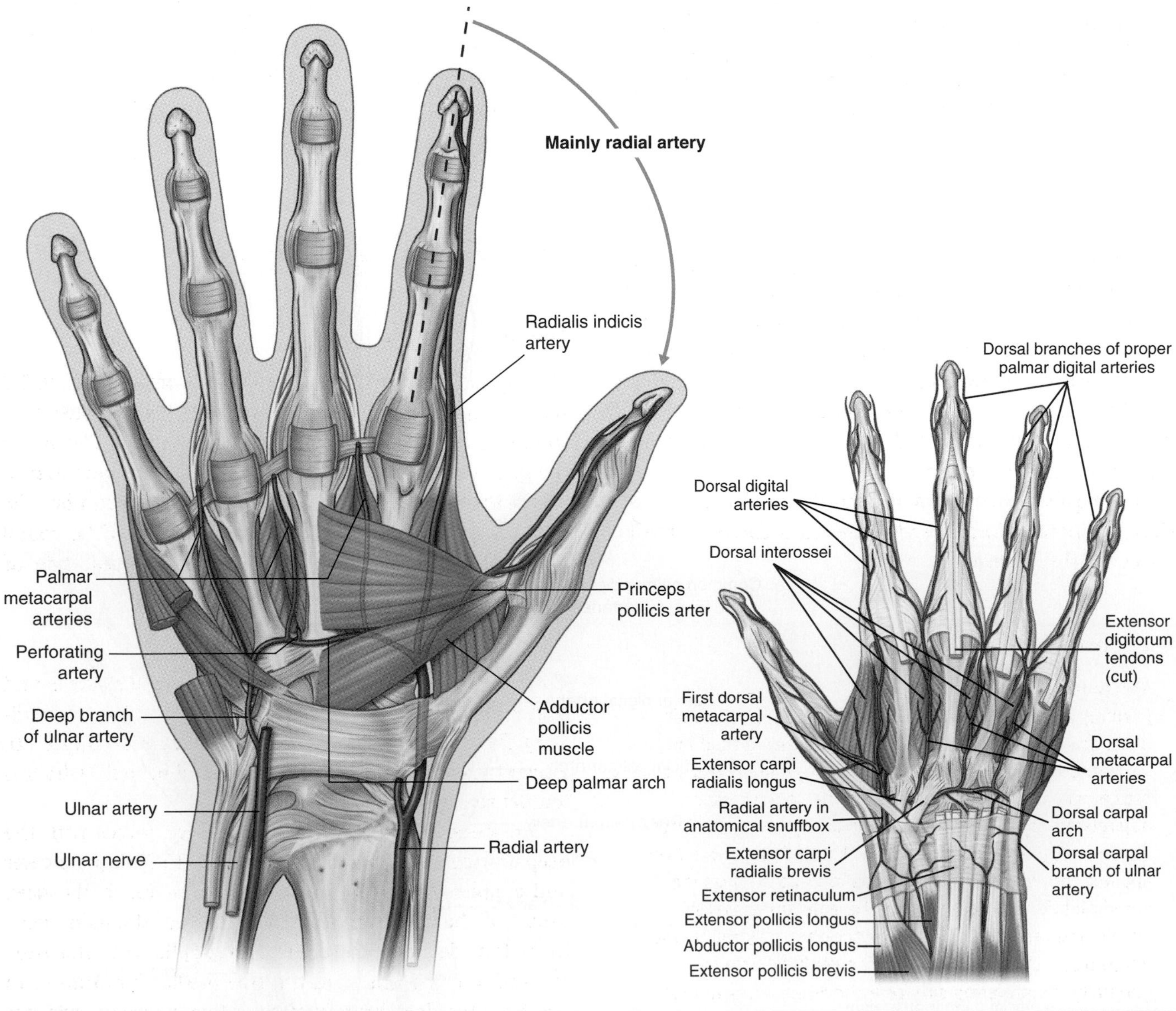

**Fig. 7.111** Deep palmar arch, right hand.

The deep palmar arch gives rise to:

- three **palmar metacarpal arteries**, which join the common palmar digital arteries from the superficial palmar arch; and
- three **perforating branches**, which pass posteriorly between the heads of origin of the dorsal interossei to anastomose with the dorsal metacarpal arteries from the dorsal carpal arch.

### In the clinic

**Allen's test**

To test for adequate anastomoses between the radial and ulnar arteries, compress both the radial and ulnar arteries at the wrist, then release pressure from one or the other, and determine the filling pattern of the hand. If there is little connection between the deep and superficial palmar arteries, only the thumb and lateral side of the index finger will fill with blood (become red) when pressure on the radial artery alone is released.

### Veins

As generally found in the upper limb, the hand contains interconnected networks of deep and superficial veins. The deep veins follow the arteries; the superficial veins drain into a dorsal venous network on the back of the hand over the metacarpal bones (Fig. 7.112).

The cephalic vein originates from the lateral side of the dorsal venous network and passes over the anatomical snuffbox into the forearm.

The basilic vein originates from the medial side of the dorsal venous network and passes into the dorsomedial aspect of the forearm.

### In the clinic

**Venipuncture**

In many patients, venous access is necessary for obtaining blood for laboratory testing and administering fluid and intravenous drugs. The ideal sites for venous access are typically in the cubital fossa and in the cephalic vein adjacent to the anatomical snuffbox. The veins are simply distended by use of a tourniquet. A tourniquet should be applied enough to allow the veins to become prominent. For straightforward blood tests the antecubital vein is usually the preferred site, and although it may not always be visible, it is easily palpated. The cephalic vein is generally the preferred site for a short-term intravenous cannula.

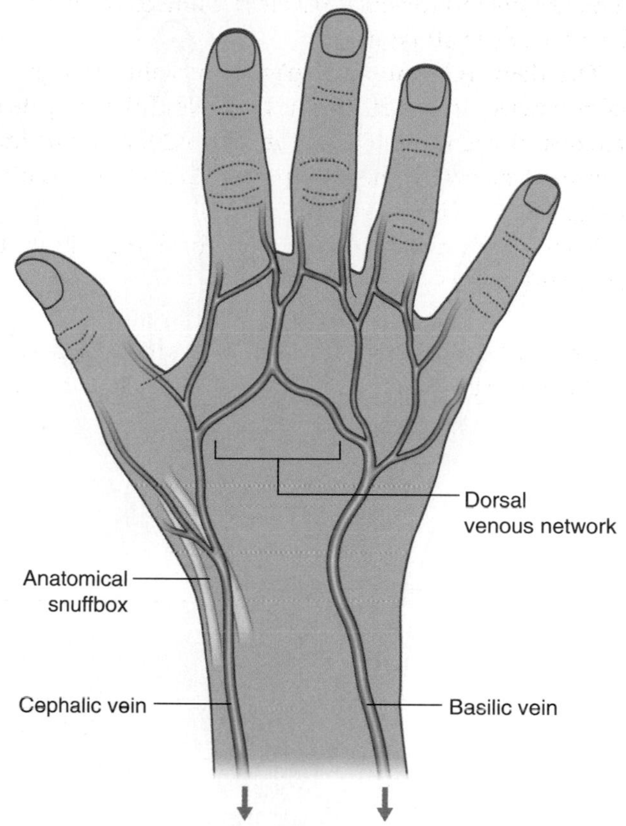


**Fig. 7.112** Dorsal venous arch of the right hand.

## Nerves

The hand is supplied by the ulnar, median, and radial nerves (Figs. 7.113 to 7.115). All three nerves contribute to cutaneous or general sensory innervation. The ulnar nerve innervates all intrinsic muscles of the hand except for the three thenar muscles and the two lateral lumbricals, which are innervated by the median nerve. The radial nerve only innervates skin on the dorsolateral side of the hand.

### Ulnar nerve

The ulnar nerve enters the hand lateral to the pisiform and posteromedially to the ulnar artery (Fig. 7.113). Immediately distal to the pisiform, it divides into a deep branch, which is mainly motor, and a superficial branch, which is mainly sensory.

The **deep branch** of the ulnar nerve passes with the deep branch of the ulnar artery (Fig. 7.113). It penetrates and supplies the hypothenar muscles to reach the deep aspect of the palm, arches laterally across the palm, deep to the long flexors of the digits, and supplies the interossei, the adductor pollicis, and the two medial lumbricals. In addition, the deep branch of the ulnar nerve contributes small articular branches to the wrist joint.

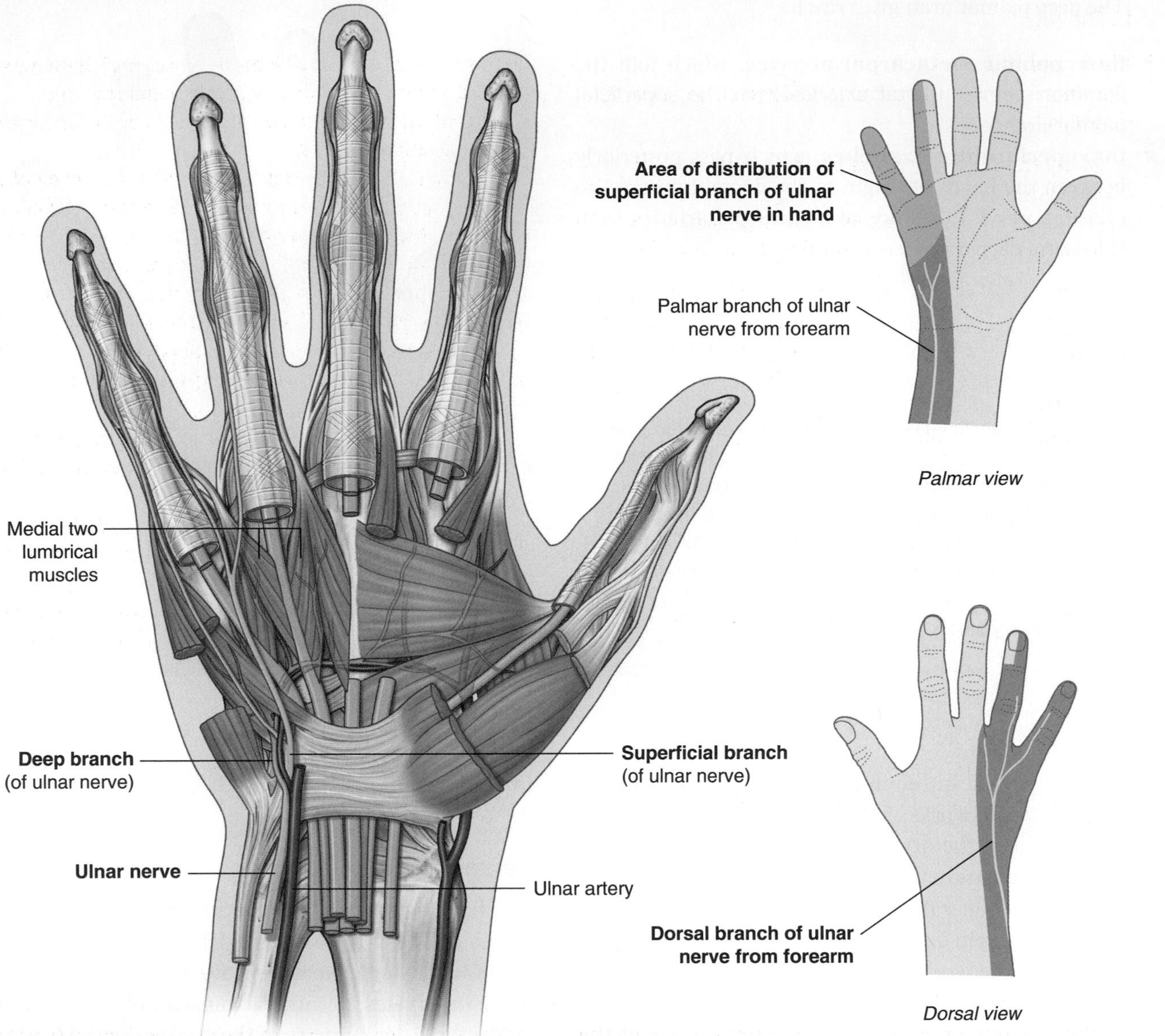

**Fig. 7.113** Ulnar nerve in the right hand.

## In the clinic

### Ulnar nerve injury

The ulnar nerve is most commonly injured at two sites: the elbow and the wrist.

- At the elbow, the nerve lies posterior to the medial epicondyle.
- At the wrist, the ulnar nerve passes superficial to the flexor retinaculum and lies lateral to the pisiform bone.

Ulnar nerve lesions are characterized by "clawing" of the hand, in which the metacarpophalangeal joints of the fingers are hyperextended and the interphalangeal joints are flexed because the function of most of the intrinsic muscles of the hand is lost (Fig. 7.114).

Clawing is most pronounced in the medial fingers because the function of all intrinsic muscles of these digits is lost while in the lateral two digits, the lumbricals are innervated by the median nerve. Function of the adductor pollicis muscle is also lost.

In lesions of the ulnar nerve at the elbow, function of the flexor carpi ulnaris muscle and flexor digitorum profundus to the medial two digits is lost as well. Clawing of the hand, particularly of the little and ring fingers, is worse with lesions of the ulnar nerve at the wrist than at the elbow because interruption of the nerve at the elbow paralyzes the ulnar half of the flexor digitorum profundus, which leads to lack of flexion at the distal interphalangeal joints in these fingers.

Ulnar nerve lesions at the elbow and wrist result in impaired sensory innervation on the palmar aspect of the medial one and one-half digits.

Damage to the ulnar nerve at the wrist or at a site proximal to the wrist can be distinguished by evaluating the status of function of the **dorsal branch** (cutaneous) of the ulnar nerve, which originates in distal regions of the forearm. This branch innervates skin over the dorsal surface of the hand on the medial side.

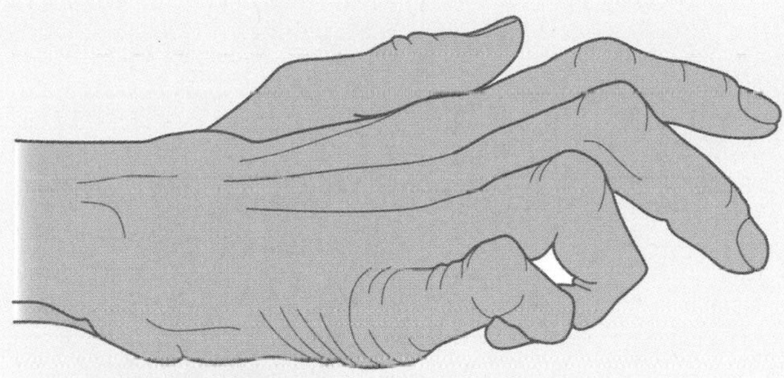

**Fig. 7.114** Typical appearance of a "clawed hand" due to a lesion of the ulnar nerve.

As the deep branch of the ulnar nerve passes across the palm, it lies in a fibro-osseous tunnel (Guyon's canal) between the hook of the hamate and the flexor tendons. Occasionally, small outpouchings of synovial membrane (ganglia) from the joints of the carpus compress the nerve within this canal, producing sensory and motor symptoms.

The superficial branch of the ulnar nerve innervates the palmaris brevis muscle and continues across the palm to supply skin on the palmar surface of the little finger and the medial half of the ring finger (Fig. 7.113).

## Median nerve

The median nerve is the most important sensory nerve in the hand because it innervates skin on the thumb, index and middle fingers, and lateral side of the ring finger (Fig. 7.115). The nervous system, using touch, gathers information about the environment from this area, particularly from the skin on the thumb and index finger. In addition, sensory information from the lateral three and one-half digits enables the fingers to be positioned with the appropriate amount of force when using precision grip.

The median nerve also innervates the thenar muscles that are responsible for opposition of the thumb to the other digits.

The median nerve enters the hand by passing through the carpal tunnel and divides into a recurrent branch and palmar digital branches (Fig. 7.115).

The **recurrent branch** of the median nerve innervates the three thenar muscles. Originating from the lateral side of the median nerve near the distal margin of the flexor retinaculum, it curves around the margin of the retinaculum and passes proximally over the flexor pollicis brevis muscle. The recurrent branch then passes between the flexor pollicis brevis and abductor pollicis brevis to end in the opponens pollicis.

The **palmar digital nerves** cross the palm deep to the palmar aponeurosis and the superficial palmar arch and enter the digits. They innervate skin on the palmar surfaces of the lateral three and one-half digits and cutaneous regions over the dorsal aspects of the distal phalanges (nail beds) of the same digits. In addition to skin, the digital nerves supply the lateral two lumbrical muscles.

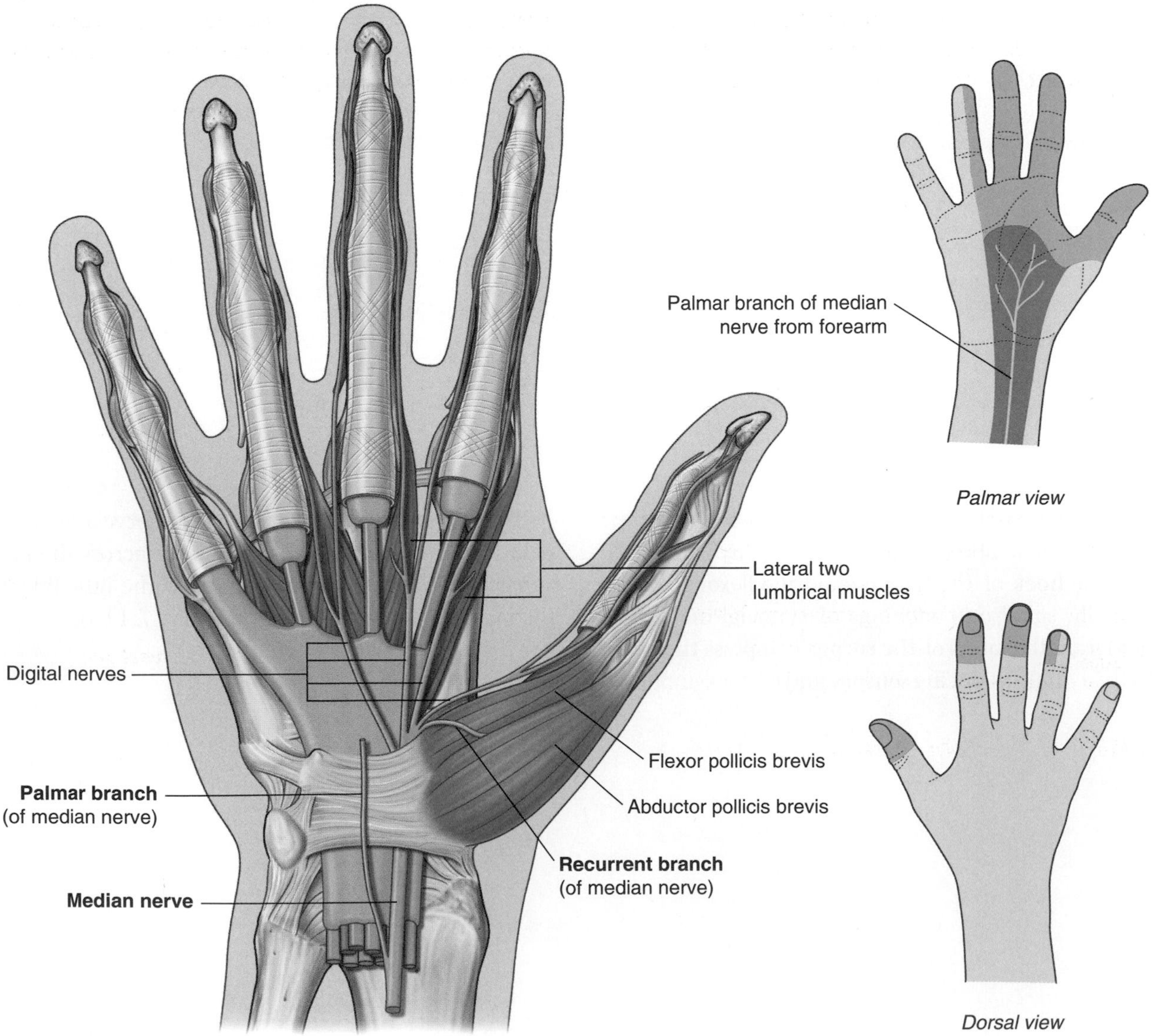

**Fig. 7.115** Median nerve in the right hand.

## Superficial branch of the radial nerve

The only part of the radial nerve that enters the hand is the superficial branch (Fig. 7.116). It enters the hand by passing over the anatomical snuffbox on the dorsolateral side of the wrist. Terminal branches of the nerve can be palpated or "rolled" against the tendon of the extensor pollicis longus as they cross the anatomical snuffbox.

The superficial branch of the radial nerve innervates skin over the dorsolateral aspect of the palm and the dorsal aspects of the lateral three and one-half digits distally to approximately the terminal interphalangeal joints.

### In the clinic

**Radial nerve injury**

Around the elbow joint the radial nerve divides into its two terminal branches—the superficial branch and the deep branch.

The most common radial nerve injury is damage to the nerve in the radial groove of the humerus, which produces a global paralysis of the muscles of the posterior compartment, resulting in wrist drop. Radial nerve damage can result from fracture of the shaft of the humerus as the radial nerve spirals around in the radial groove. The typical injury produces reduction of sensation in the cutaneous distribution, predominantly over the posterior aspect of the hand. Severing the posterior interosseous nerve (continuation of deep branch of radial nerve) may paralyze the muscles of the posterior compartment of the forearm, but the nerve supply is variable. Typically, the patient may not be able to extend the fingers.

The distal branches of the superficial branch of the radial nerve can be readily palpated as "cords" passing over the tendon of the extensor pollicis longus in the anatomical snuffbox. Damage to these branches is of little consequence because they supply only a small area of skin.

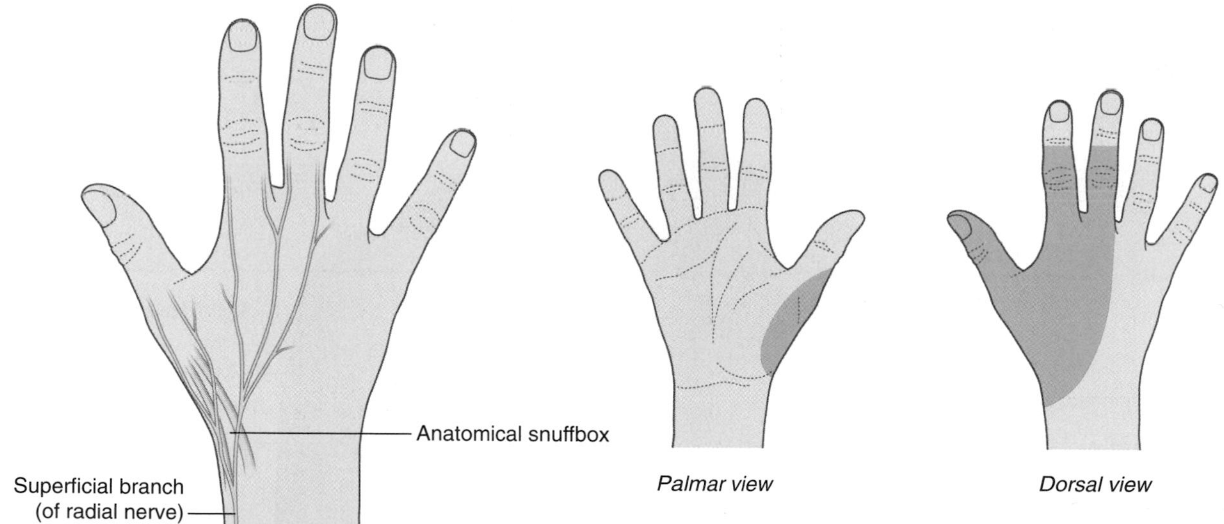


**Fig. 7.116** Radial nerve in the right hand.

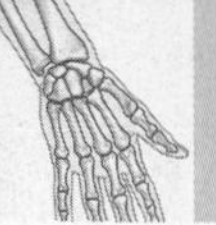



# Surface anatomy

## Upper limb surface anatomy

Tendons, muscles, and bony landmarks in the upper limb are used to locate major arteries, veins, and nerves. Asking patients to maneuver their upper limbs in specific ways is essential for performing neurological examinations.

- Tendons are used to test reflexes associated with specific spinal cord segments.
- Vessels are used clinically as points of entry into the vascular system (for collecting blood and administering pharmaceuticals and nutrients), and for taking blood pressure and pulses.
- Nerves can become entrapped or be damaged in regions where they are related to bone or pass through confined spaces.

### Bony landmarks and muscles of the posterior scapular region

The medial border, inferior angle, and part of the lateral border of the scapula can be palpated on a patient, as can the spine and acromion. The superior border and angle of the scapula are deep to soft tissue and are not readily palpable. The supraspinatus and infraspinatus muscles can be palpated above and below the spine, respectively (Fig. 7.117).

The trapezius muscle is responsible for the smooth contour of the lateral side of the neck and over the superior aspect of the shoulder.

The deltoid muscles form the muscular eminence inferior to the acromion and around the glenohumeral joint. The axillary nerve passes posteriorly around the surgical neck of the humerus deep to the deltoid muscle.

The latissimus dorsi muscle forms much of the muscle mass underlying the posterior axillary skin fold extending obliquely upward from the trunk to the arm. The teres major muscle passes from the inferior angle of the scapula to the upper humerus and contributes to this posterior axillary skin fold laterally.

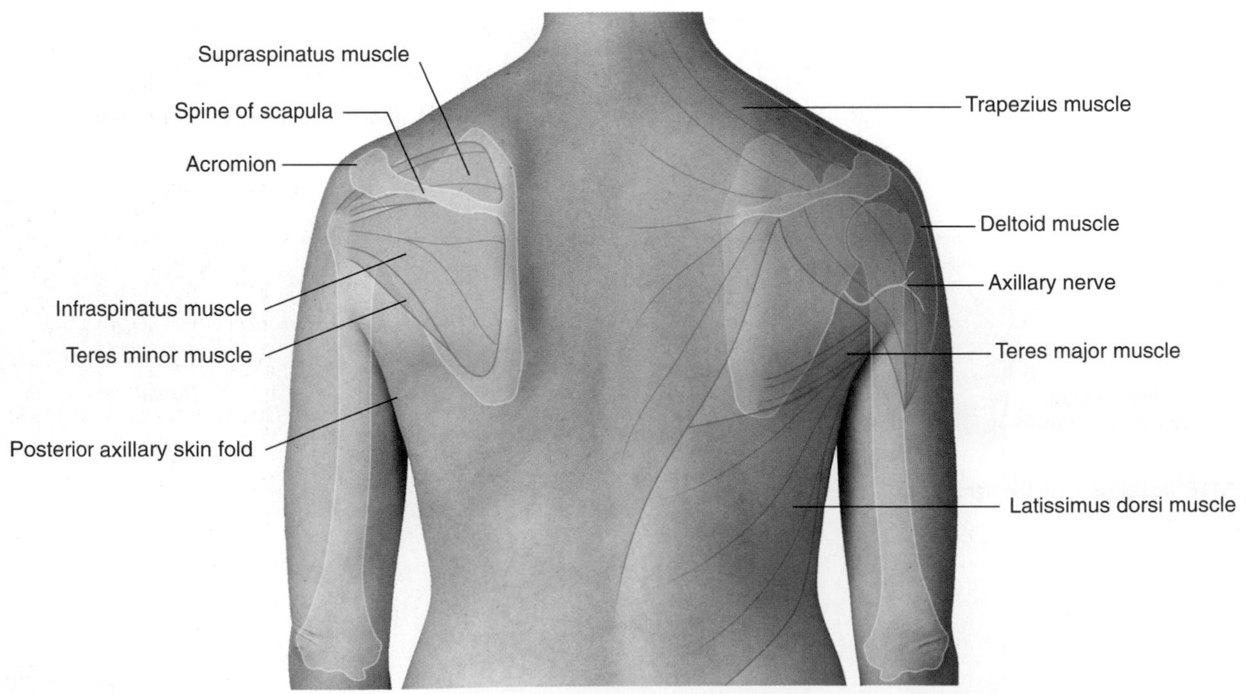


**Fig. 7.117** Bony landmarks and muscles of the posterior scapular region. Posterior view of shoulder and back.

## Visualizing the axilla and locating contents and related structures

The axillary inlet and outlet and walls of the axilla can be established using skin folds and palpable bony landmarks (Fig. 7.118).

- The anterior margin of the axillary inlet is the clavicle, which can be palpated along its entire length. The lateral limit of the axillary inlet is approximated by the tip of the coracoid process, which is palpable immediately below the lateral third of the clavicle and deep to the medial margin of the deltoid muscle.
- The inferior margin of the anterior axillary wall is the anterior axillary skin fold, which overlies the lower margin of the pectoralis major muscle.
- The inferior margin of the posterior axillary wall is the posterior axillary skin fold, which overlies the margins of the teres major muscle laterally and latissimus dorsi muscle medially.
- The medial wall of the axilla is the upper part of the serratus anterior muscle overlying the thoracic wall.

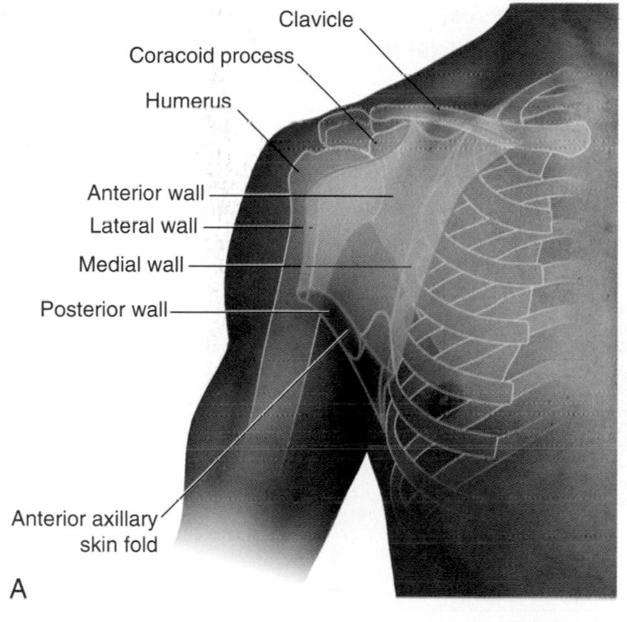


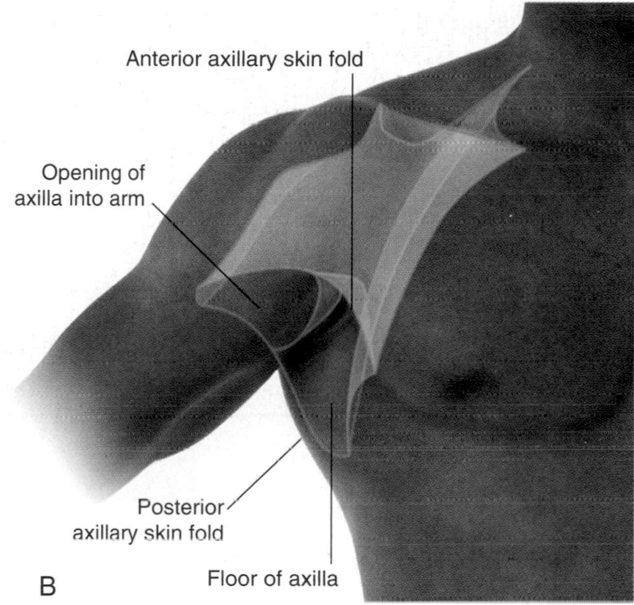


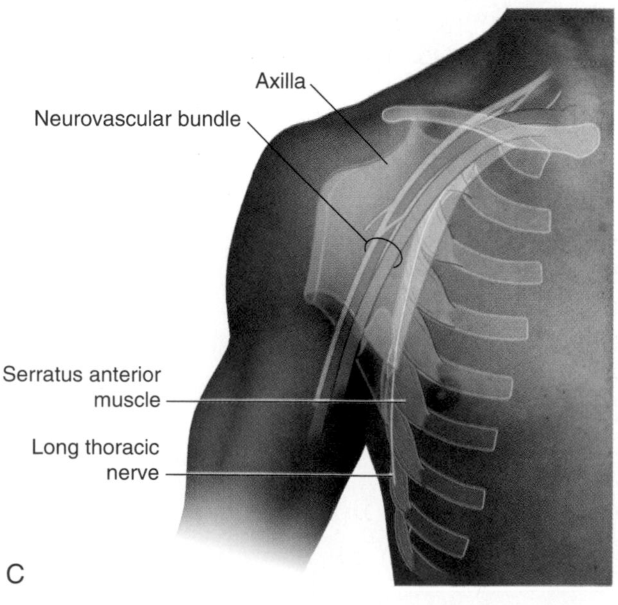


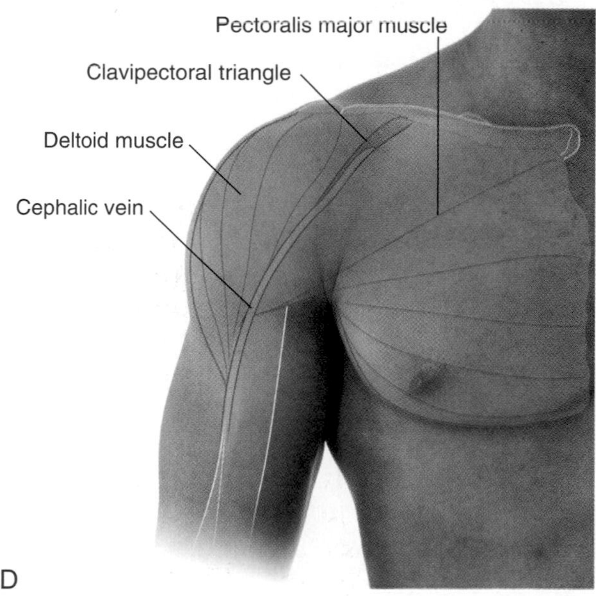


**Fig. 7.118** Visualizing the axilla and locating its contents and related structures. **A.** Anterior shoulder showing folds and walls of the axilla. **B.** Anterior shoulder showing outlet and floor of the axilla. **C.** Anterior view showing the axillary neurovascular bundle and long thoracic nerve. **D.** Anterior view of the shoulder showing the clavipectoral triangle with the cephalic vein.

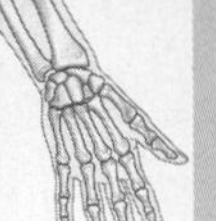

The long thoracic nerve passes vertically out of the axilla and down the lateral surface of the serratus anterior muscle in a position just anterior to the posterior axillary skin fold.

- The lateral boundary of the axilla is the humerus.
- The floor of the axilla is the dome of skin between the posterior and anterior axillary skin folds.

Major vessels, nerves, and lymphatics travel between the upper limb and the trunk by passing through the axilla.

The axillary artery, axillary vein, and components of the brachial plexus pass through the axilla and into the arm by traveling lateral to the dome of skin that forms the floor. This neurovascular bundle can be palpated by placing a hand into this dome of skin and pressing laterally against the humerus.

The cephalic vein travels in superficial fascia in the cleft between the deltoid muscle and the pectoralis major muscle and penetrates deep fascia in the clavipectoral triangle to join with the axillary vein.

### Locating the brachial artery in the arm

The brachial artery is on the medial side of the arm in the cleft between the biceps brachii and triceps brachii muscles (Fig. 7.119). The median nerve courses with the brachial artery, whereas the ulnar nerve deviates posteriorly from the vessel in distal regions.

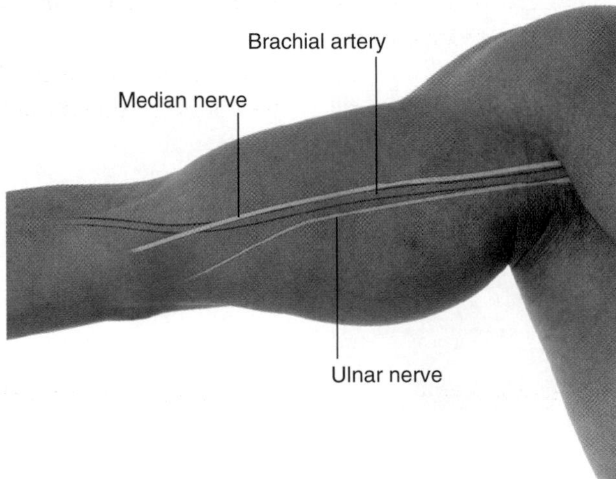


**Fig. 7.119** Locating the brachial artery in the right arm (medial view of arm with brachial artery, median nerve, and ulnar nerve).

### The triceps brachii tendon and position of the radial nerve

The triceps brachii muscle forms the soft tissue mass posterior to the humerus, and the tendon inserts onto the olecranon of the ulna, which is readily palpable and forms the bony protuberance at the "tip" of the elbow (Fig. 7.120).

The brachioradialis muscle is also visible as a muscular bulge on the lateral aspect of the arm. It is particularly prominent when the forearm is half pronated, flexed at the elbow against resistance, and viewed anteriorly.

The radial nerve in the distal arm emerges from behind the humerus to lie deep to the brachioradialis muscle.

### Cubital fossa (anterior view)

The cubital fossa lies anterior to the elbow joint and contains the biceps brachii tendon, the brachial artery, and the median nerve (Fig. 7.121).

The base of the cubital fossa is an imaginary line between the readily palpable medial and lateral epicondyles of the humerus. The lateral and medial borders are formed by the brachioradialis and pronator teres muscles, respectively. The margin of the brachioradialis can be found by asking a subject to flex the semipronated forearm against resistance. The margin of the pronator teres can be estimated by an oblique line extending between the medial epicondyle and the midpoint along the length of the lateral surface of the forearm. The approximate apex of the cubital fossa is where this line meets the margin of the brachioradialis muscle.

Contents of the cubital fossa, from lateral to medial, are the tendon of the biceps brachii, the brachial artery, and the median nerve. The tendon of the biceps brachii is easily palpable. Often the cephalic, basilic, and median cubital veins are visible in the subcutaneous fascia overlying the cubital fossa.

The ulnar nerve passes behind the medial epicondyle of the humerus and can be "rolled" here against the bone.

The radial nerve travels into the forearm deep to the margin of the brachioradialis muscle anterior to the elbow joint.

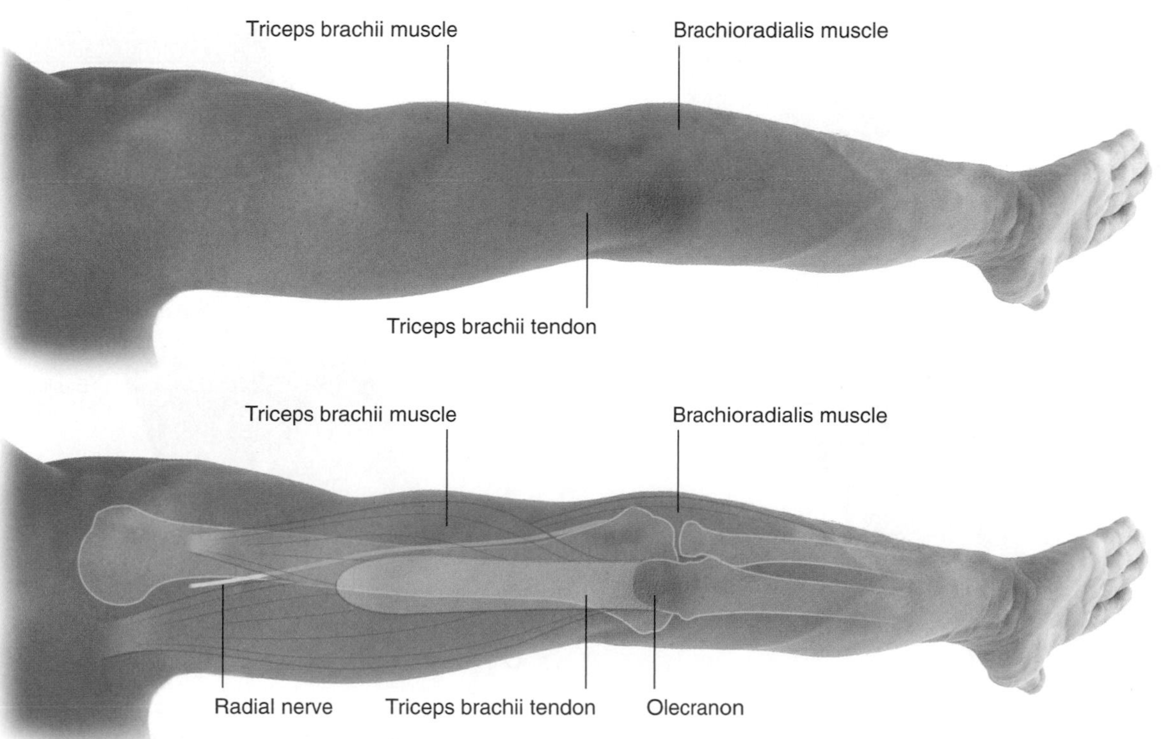


**Fig. 7.120** Triceps brachii tendon and position of the radial nerve (posterior view of right arm).

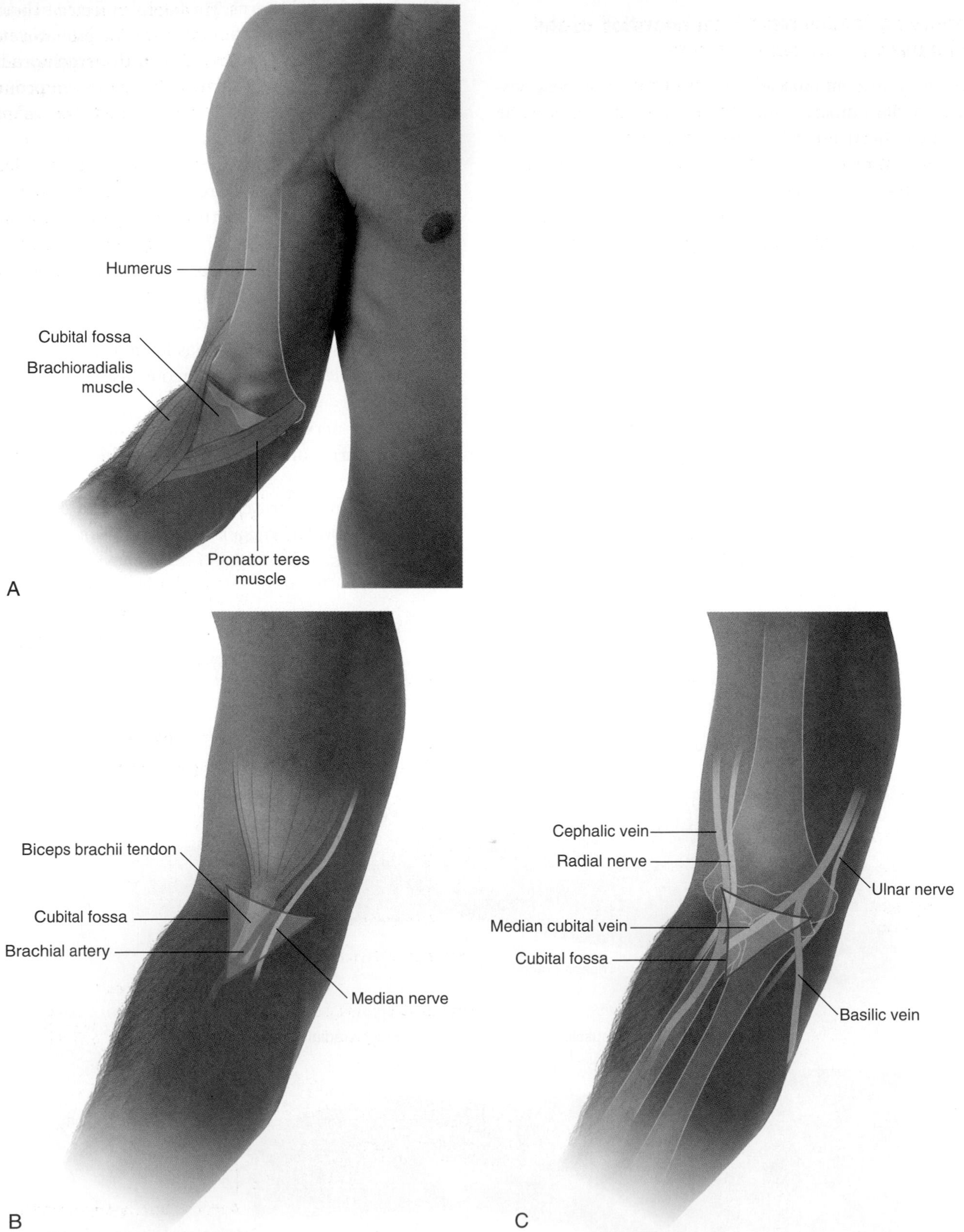

**Fig. 7.121** Cubital fossa (anterior view, right arm). **A.** Anterior view. **B.** Boundaries and contents. **C.** Showing radial and ulnar nerves, and veins.

## Identifying tendons and locating major vessels and nerves in the distal forearm

Tendons that pass from the forearm into the hand are readily visible in the distal forearm and can be used as landmarks to locate major vessels and nerves.

In the anterior aspect of the distal forearm, the tendons of the flexor carpi radialis, flexor carpi ulnaris, and palmaris longus muscles can be easily located either by palpating or by asking a patient to flex the wrist against resistance.

- The tendon of flexor carpi radialis is located approximately at the junction between the lateral and middle thirds of an imaginary line drawn transversely across the distal forearm. The radial artery is immediately lateral to this tendon and this site is used for taking a radial pulse (Fig. 7.122A).
- The tendon of the flexor carpi ulnaris is easily palpated along the medial margin of the forearm and inserts on the pisiform, which can also be palpated by following the tendon to the base of the hypothenar eminence of the hand. The ulnar artery and ulnar nerve travel

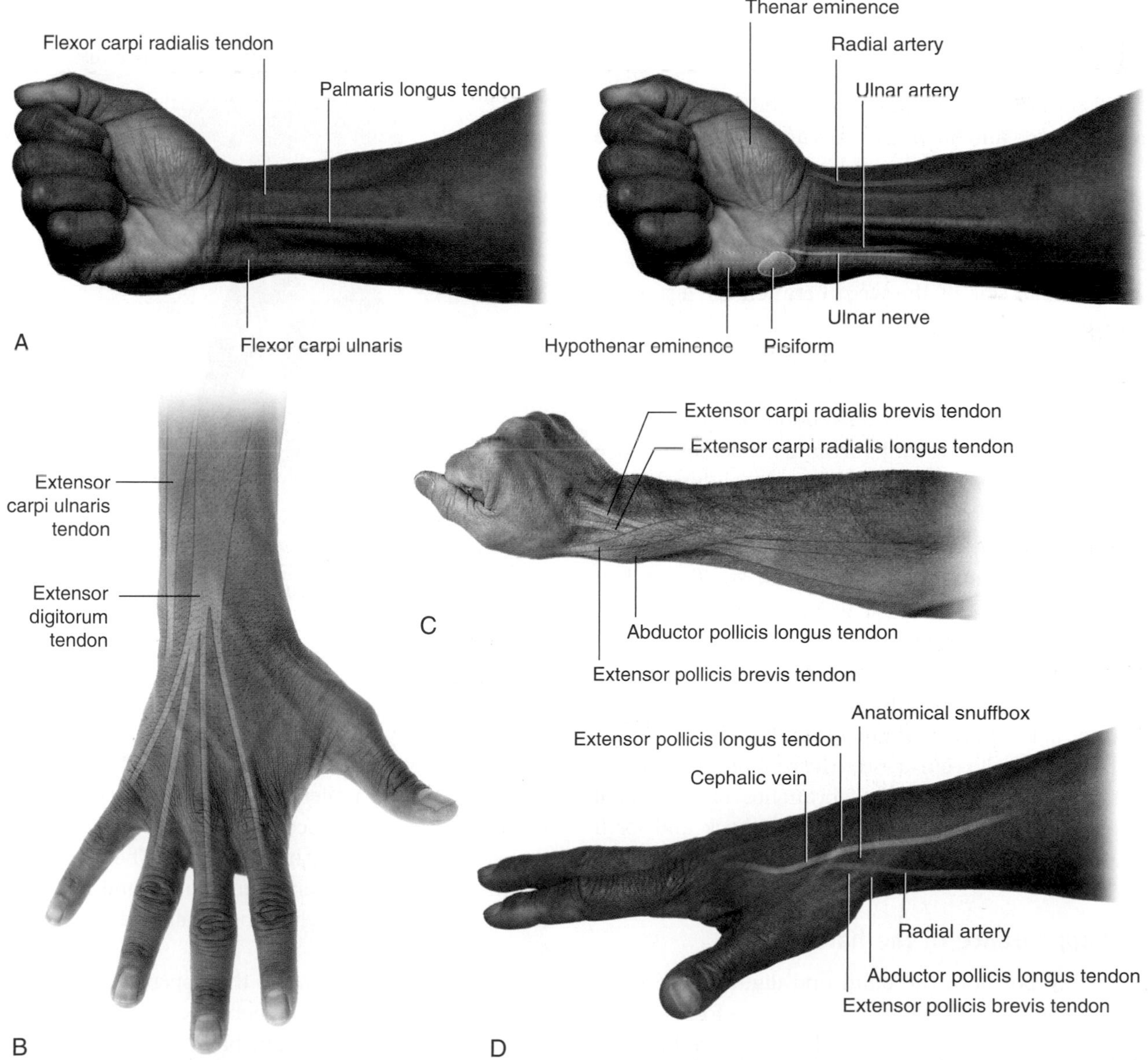


**Fig. 7.122** Identifying tendons and locating major vessels and nerves in the distal right forearm. **A.** Anterior distal forearm and wrist. **B.** Posterior distal forearm and wrist. **C.** Lateral view of posterior wrist and forearm. **D.** Anatomical snuffbox.

through the distal forearm and into the hand under the lateral lip of the flexor carpi ulnaris tendon and lateral to the pisiform.

- The palmaris longus tendon may be absent, but when present, lies medial to the flexor carpi radialis tendon and is particularly prominent when the wrist is flexed against resistance. The median nerve is also medial to the flexor carpi radialis tendon and lies under the palmaris longus tendon.
- The long tendons of the digits of the hand are deep to the median nerve and between the long flexors of the wrist. Their position can be visualized by rapidly and repeatedly flexing and extending the fingers from medial to lateral.
- In the posterior distal forearm and wrist, the tendons of the extensor digitorum (Fig. 7.122B) are in the midline and radiate into the index, middle, ring, and little fingers from the wrist.
- The distal ends of the tendons of the extensor carpi radialis longus and brevis muscles are on the lateral side of the wrist (Fig. 7.122C) and can be accentuated by making a tight fist and extending the wrist against resistance.
- The tendon of the extensor carpi ulnaris can be felt on the far medial side of the wrist between the distal end of the ulna and the wrist.
- Hyperextension and abduction of the thumb reveals the anatomical snuffbox (Fig. 7.122D). The medial margin of this triangular area is the tendon of the extensor pollicis longus, which swings around the dorsal tubercle of the radius and then travels into the thumb. The lateral margin is formed by the tendons of the extensor pollicis brevis and abductor pollicis longus. The radial artery passes through the anatomical snuffbox when traveling laterally around the wrist to reach the back of the hand and penetrate the base of the first dorsal interosseous muscle to access the deep aspect of the palm of the hand. The pulse of the radial artery can be felt in the floor of the anatomical snuffbox in the relaxed wrist. The cephalic vein crosses the roof of the anatomical snuffbox, and cutaneous branches of the radial nerve can be felt by moving a finger back and forth along the tendon of the extensor pollicis longus muscle.

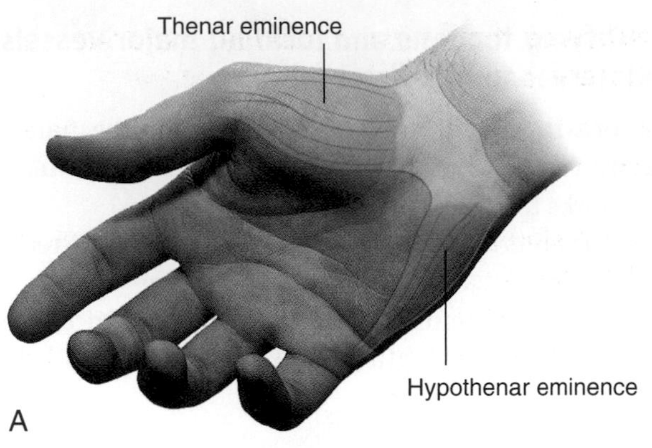


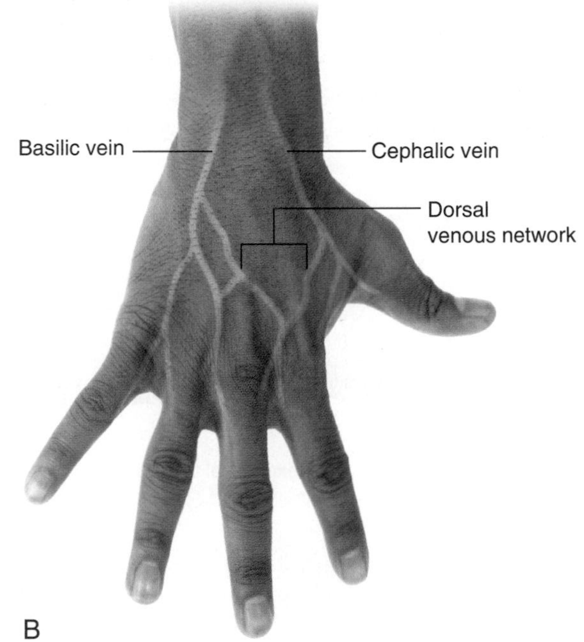


**Fig. 7.123** Normal appearances of the right hand. **A.** Palmar view with the thenar and hypothenar eminences and finger arcade. **B.** Dorsal view with dorsal venous network.

## Normal appearance of the hand

In the resting position, the palm and digits of the hand have a characteristic appearance. The fingers form a flexed arcade, with the little finger flexed the most and the index finger flexed the least (Fig. 7.123A). The pad of the thumb is positioned at a 90° angle to the pads of the fingers.

A thenar eminence occurs at the base of the thumb and is formed by the underlying thenar muscles. A similar hypothenar eminence occurs along the medial margin of the palm at the base of the little finger. The appearance of the thenar and hypothenar eminences, and the positions of the fingers change when the ulnar and median nerves are compromised.

Major superficial veins of the upper limb begin in the hand from a dorsal venous network (Fig. 7.123B), which overlies the metacarpals. The basilic vein originates from the medial side of the network and the cephalic vein originates from the lateral side.

## Position of the flexor retinaculum and the recurrent branch of the median nerve

The proximal margin of the flexor retinaculum can be determined using two bony landmarks.

- The pisiform bone is readily palpable at the distal end of the flexor carpi ulnaris tendon.
- The tubercle of the scaphoid can be palpated at the distal end of the flexor carpi radialis tendon as it enters the wrist (Fig. 7.124).

An imaginary line between these two points marks the proximal margin of the flexor retinaculum. The distal margin of the flexor retinaculum is approximately deep to the point where the anterior margin of the thenar eminence meets the hypothenar eminence near the base of the palm.

The recurrent branch of the median nerve lies deep to the skin and deep fascia overlying the anterior margin of the thenar eminence near the midline of the palm.

## Motor function of the median and ulnar nerves in the hand

The ability to flex the metacarpophalangeal joints while at the same time extending the interphalangeal joints of the fingers is entirely dependent on the intrinsic muscles of the hand (Fig. 7.125A). These muscles are mainly innervated by the deep branch of the ulnar nerve, which carries fibers from spinal cord level (C8)T1.

Adducting the fingers to grasp an object placed between them is caused by the palmar interossei muscles, which are innervated by the deep branch of the ulnar nerve carrying fibers from spinal cord level (C8)T1.

The ability to grasp an object between the pad of the thumb and the pad of one of the fingers depends on normal

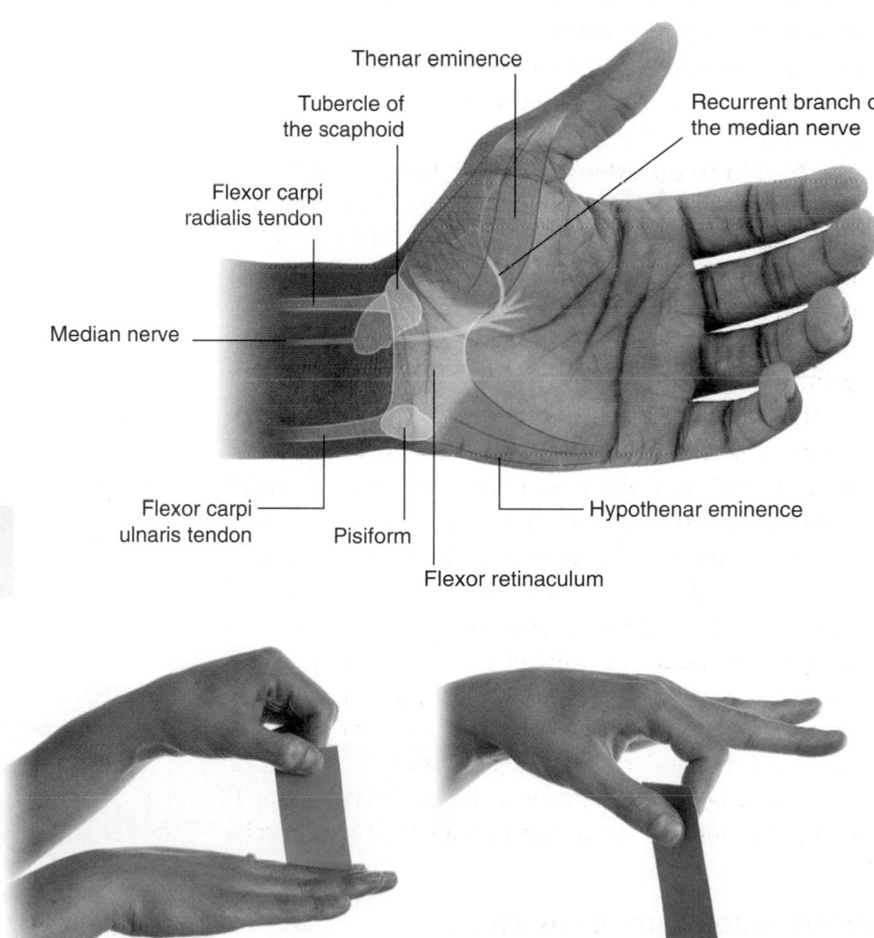


**Fig. 7.124** Anterior view of left hand to show the position of the flexor retinaculum and recurrent branch of the median nerve.

**Fig. 7.125** Motor function of the ulnar and median nerves in the hand. **A.** Flexing the metacarpophalangeal joints and extending the interphalangeal joints: the "ta-ta" position. **B.** Grasping an object between the fingers. **C.** Grasping an object between the pad of the thumb and pad of the index finger.

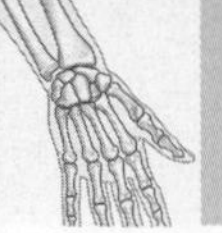



functioning of the thenar muscles, which are innervated by the recurrent branch of the median nerve carrying fibers from spinal cord level C8(T1).

## Visualizing the positions of the superficial and deep palmar arches

The positions of the superficial and deep palmar arches in the hand can be visualized using bony landmarks, muscle eminences, and skin creases (Fig. 7.126).

- The superficial palmar arch begins as a continuation of the ulnar artery, which lies lateral to the pisiform bone at the wrist. The arch curves laterally across the palm anterior to the long flexor tendons in the hand. The arch reaches as high as the proximal transverse skin crease of the palm and terminates laterally by joining a vessel of variable size, which crosses the thenar eminence from the radial artery in the distal forearm.
- The deep palmar arch originates on the lateral side of the palm deep to the long flexor tendons and between the proximal ends of metacarpals I and II. It arches medially across the palm and terminates by joining the deep branch of the ulnar artery, which passes through the base of the hypothenar muscles and between the pisiform and hook of the hamate. The deep palmar arch is more proximal in the hand than the superficial palmar arch and lies approximately one-half of the distance between the distal wrist crease and the proximal transverse skin crease of the palm.

## Pulse points

Peripheral pulses can be felt at six locations in the upper limb (Fig. 7.127).

- Axillary pulse: axillary artery in the axilla lateral to the apex of the dome of skin covering the floor of the axilla.
- Brachial pulse in midarm: brachial artery on the medial side of the arm in the cleft between the biceps brachii and triceps brachii muscles. This is the position where a blood pressure cuff is placed.
- Brachial pulse in the cubital fossa: brachial artery medial to the tendon of the biceps brachii muscle. This is the position where a stethoscope is placed to hear the pulse of the vessel when taking a blood pressure reading.
- Radial pulse in the distal forearm: radial artery immediately lateral to the tendon of the flexor carpi radialis muscle. This is the most common site for "taking a pulse."
- Ulnar pulse in the distal forearm: ulnar artery immediately under the lateral margin of the flexor carpi ulnaris tendon and proximal to the pisiform.
- Radial pulse in the anatomical snuffbox: radial artery as it crosses the lateral side of the wrist between the tendon of the extensor pollicis longus muscle and the tendons of the extensor pollicis brevis and abductor pollicis longus muscles.

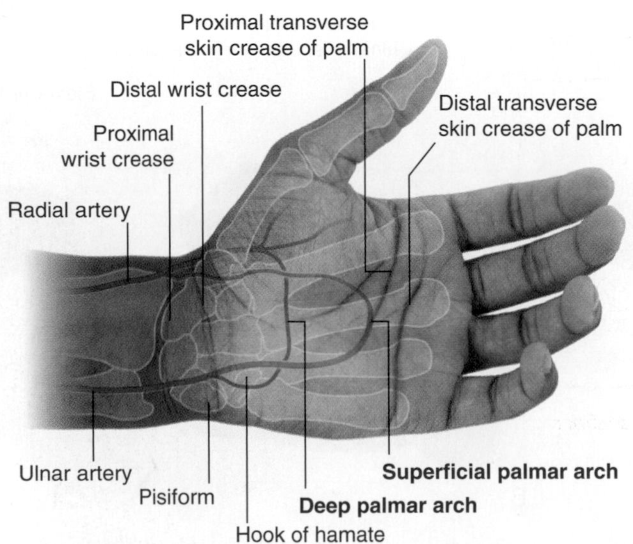


**Fig. 7.126** Visualizing the positions of the superficial and deep palmar arches, left hand. The proximal transverse skin crease of the palm and distal wrist crease are labeled and the superficial and deep palmar arches shown in overlay. This also shows the position of the pisiform and the hook of the hamate.

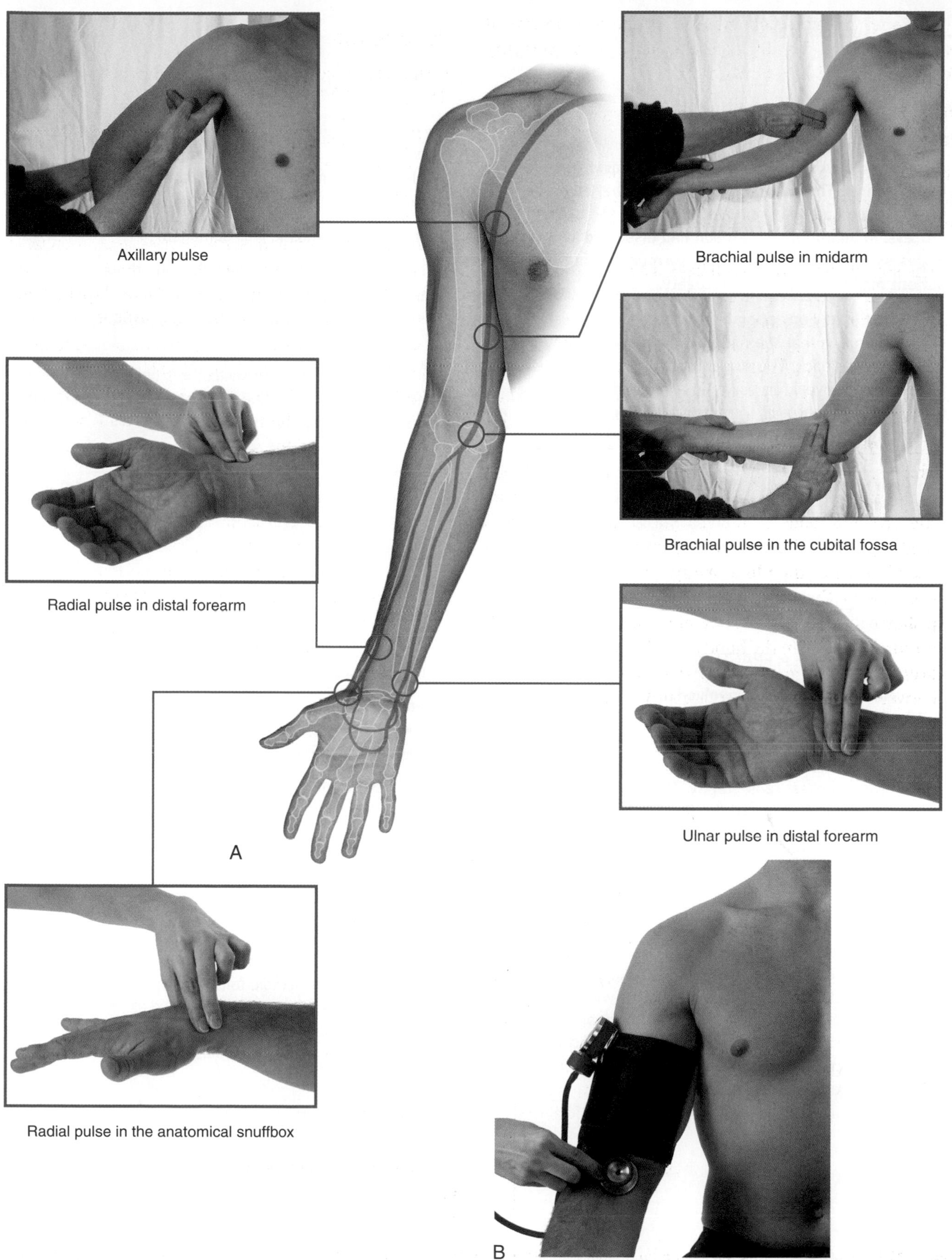


**Fig. 7.127** Where to take peripheral artery pulses in the upper limb. **A.** Pulse points. **B.** Placement of blood pressure cuff and stethoscope.

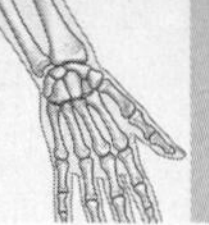



# Clinical cases

## Case 1

WINGED SCAPULA

**A 57-year-old woman underwent a right mastectomy for a breast cancer. The surgical note reported that all of the breast tissue had been removed, including the axillary process. In addition, the surgeon had dissected all lymph nodes within the axilla with their surrounding fat. The patient made an uneventful recovery.**

At the first follow-up appointment, the patient's husband told the surgeon that she had now developed a bony "spike" on her back. The surgeon was intrigued and asked the patient to reveal this spike. At examination, the spike was the inferior angle of the scapula, which appeared to be sticking out posteriorly ("winged"). Raising the arms accentuated this structure.

The medial border of the scapula was accentuated and it was noted that there was some loss of bulk of the serratus anterior muscle, which attaches to the tip of the scapula.

The nerve to this muscle was damaged.

During the surgery on the axilla, the long thoracic nerve was damaged as it passed down the lateral thoracic wall on the external surface of the serratus anterior, just deep to the skin and subcutaneous fascia.

Because the nerve was transected, it is unlikely that the patient will improve, but she was happy that she had an adequate explanation for the spike.

## Case 2

COMPLICATION OF A FRACTURED FIRST RIB

**A 25-year-old woman was involved in a motor vehicle accident and thrown from her motorcycle. When she was admitted to the emergency room, she was unconscious. A series of tests and investigations were performed, one of which included chest radiography. The attending physician noted a complex fracture of the first rib on the left.**

Many important structures that supply the upper limb pass over rib I.

It is important to test the nerves that supply the arm and hand, although this is extremely difficult to do in an unconscious patient. However, some muscle reflexes can be determined using a tendon hammer. Also, it may be possible to test for pain reflexes in patients with altered consciousness levels. Palpation of the axillary artery, brachial artery, radial artery, and ulnar artery pulses is necessary because a fracture of the first rib can sever and denude the subclavian artery, which passes over it.

A chest drain was immediately inserted because the lung had collapsed. The fractured first rib had damaged the visceral and parietal pleurae, allowing air from a torn lung to escape into the pleural cavity. The lung collapsed, and the pleural cavity filled with air, which impaired lung function.

A tube was inserted between the ribs, and the air was sucked out to re-inflate the lung.

The first rib is a deep structure at the base of the neck. It is not uncommon for ribs to be broken after minor injuries, including sports injuries. However, rib I, which lies at the base of the neck, is surrounded by muscles and soft tissues that provide it with considerable protection. Therefore a patient with a fracture of the first rib has undoubtedly been subjected to a considerable force, which usually occurs in a deceleration injury. Other injuries should always be sought and the patient should be managed with a high level of concern for deep neck and mediastinal injuries.

## Case 3

HOW TO EXAMINE THE HAND

**A resident was asked to carry out a clinical assessment of a patient's hand. He examined the following:**

**Musculoskeletal system**

The musculoskeletal system includes the bones, joints, muscles, and tendons. The resident looked for abnormalities and muscle wasting. Knowing which areas are wasted identifies the nerve that supplies them. She palpated the individual bones and palpated the scaphoid with the wrist in ulnar deviation. She examined the movement of joints because they may be restricted by joint disease or inability of muscular contraction.

**Circulation**

Palpation of both radial and ulnar pulses is necessary. The resident looked for capillary return to assess how well the hand was perfused.

**Examination of the nerves**

The three main nerves to the hand should be tested.

***Median nerve***

The median nerve innervates the skin on the palmar aspect of the lateral three and one-half digits, the dorsal aspect of the distal phalanx, half of the middle phalanges of the same fingers, and a variable amount on the radial side of the palm of the hand. Median nerve damage results in wasting of the thenar eminence, absence of abduction of the thumb, and absence of opposition of the thumb.

***Ulnar nerve***

The ulnar nerve innervates the skin of the anterior and posterior surfaces of the little finger and the ulnar side of the ring finger, the skin over the hypothenar eminence, and a similar strip of skin posteriorly. Sometimes the ulnar nerve innervates all the skin of the ring finger and the ulnar side of the middle finger.

An ulnar nerve palsy results in wasting of the hypothenar eminence, absent flexion of the distal interphalangeal joints of the little and ring fingers, and absent abduction and adduction of the fingers. Adduction of the thumb also is affected.

***Radial nerve***

The radial nerve innervates a small area of skin over the lateral aspect of metacarpal I and the back of the first web space.

The radial nerve also produces extension of the wrist and extension of the metacarpophalangeal and interphalangeal joints and of the digits.

A very simple examination would include tests for the median nerve by opposition of the thumb, for the ulnar nerve by abduction and adduction of the digits, and for the radial nerve by extension of the wrist and fingers and feeling on the back of the first web space.

# 8

# Head and Neck

**ADDITIONAL LEARNING RESOURCES for Chapter 8, *Head and Neck*, on STUDENT CONSULT (www.studentconsult.com):**

- Self-Assessment—National Board style multiple-choice questions, Chapter 8
- Short Questions—these are questions requiring short responses, Chapter 8
- Interactive Surface Anatomy—interactive surface animations, Chapter 8
- Clinical Cases, Chapter 8
  - Brainstem tumor
  - Branchial cyst
  - Complication of orbital fracture
  - Epiglottitis
  - Extradural hematoma
  - Multinodular goiter
  - Otosclerosis
  - Pancoast's tumor
  - Parotid duct calculus
  - Pituitary macroadenoma
  - Posterior communicating artery aneurysm
  - Recurrent epistaxis
  - Scalp laceration
  - Scaphocephaly
  - Stenosis of the internal carotid artery
  - Subclavian steal syndrome
  - Temporal arteritis
  - Tonsillitis

**FREE Online Anatomy and Embryology Self-Study Course**

- Anatomy modules 1 through 3 and 39 through 52
- Embryology modules 72 through 77

# Conceptual overview

## GENERAL DESCRIPTION

The head and neck are anatomically complex areas of the body.

### Head

#### Major compartments

The head is composed of a series of compartments, which are formed by bone and soft tissues. They are:

- the cranial cavity,
- two ears,
- two orbits,
- two nasal cavities, and
- an oral cavity (Fig. 8.1).

The **cranial cavity** is the largest compartment and contains the brain and associated membranes (meninges).

Most of the ear apparatus on each side is contained within one of the bones forming the floor of the cranial cavity. The external parts of the ears extend laterally from these regions.

The two **orbits** contain the eyes. They are cone-shaped chambers immediately inferior to the anterior aspect of the cranial cavity, and the apex of each cone is directed posteromedially. The walls of the orbits are bone, whereas the base of each conical chamber can be opened and closed by the eyelids.

The **nasal cavities** are the upper parts of the respiratory tract and are between the orbits. They have walls, floors, and ceilings, which are predominantly composed of bone and cartilage. The anterior openings to the nasal cavities are **nares** (**nostrils**), and the posterior openings are **choanae** (**posterior nasal apertures**).

Continuous with the nasal cavities are air-filled extensions (**paranasal sinuses**), which project laterally, superiorly, and posteriorly into surrounding bones. The largest, the **maxillary sinuses**, are inferior to the orbits.

The **oral cavity** is inferior to the nasal cavities, and separated from them by the **hard** and **soft palates**. The floor of the oral cavity is formed entirely of soft tissues.

The anterior opening to the oral cavity is the **oral fissure** (mouth), and the posterior opening is the **oropharyngeal isthmus**. Unlike the nares and choanae, which are continuously open, both the oral fissure and oropharyngeal isthmus can be opened and closed by surrounding soft tissues.

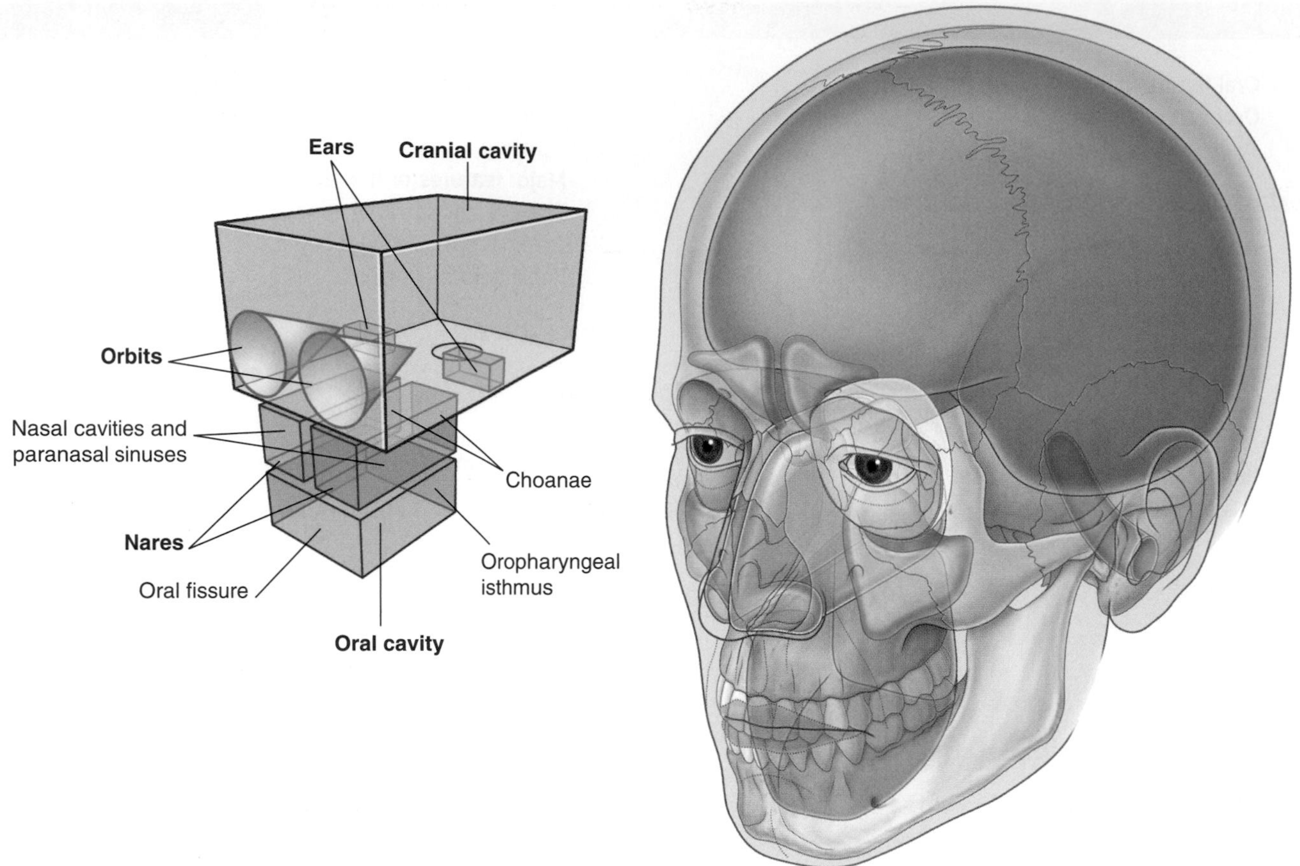


**Fig. 8.1** Major compartments of the head and neck.

## Other anatomically defined regions

In addition to the major compartments of the head, two other anatomically defined regions (infratemporal fossa and pterygopalatine fossa) of the head on each side are areas of transition from one compartment of the head to another (Fig. 8.2). The face and scalp also are anatomically defined areas of the head and are related to external surfaces.

The **infratemporal fossa** is an area between the posterior aspect (ramus) of the mandible and a flat region of bone (lateral plate of the pterygoid process) just posterior to the upper jaw (maxilla). This fossa, bounded by bone and soft tissues, is a conduit for one of the major cranial nerves—the mandibular nerve (the mandibular division of the trigeminal nerve [$V_3$]), which passes between the cranial and oral cavities.

The **pterygopalatine fossa** on each side is just posterior to the upper jaw. This small fossa communicates with the cranial cavity, the infratemporal fossa, the orbit, the nasal cavity, and the oral cavity. A major structure passing through the pterygopalatine fossa is the maxillary nerve (the maxillary division of the trigeminal nerve [$V_2$]).

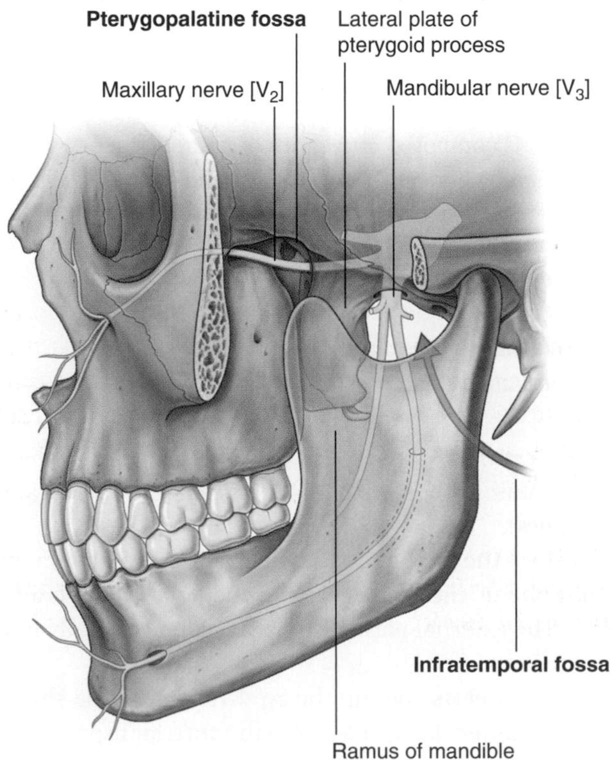


**Fig. 8.2** Areas of transition from one compartment of the head to another.

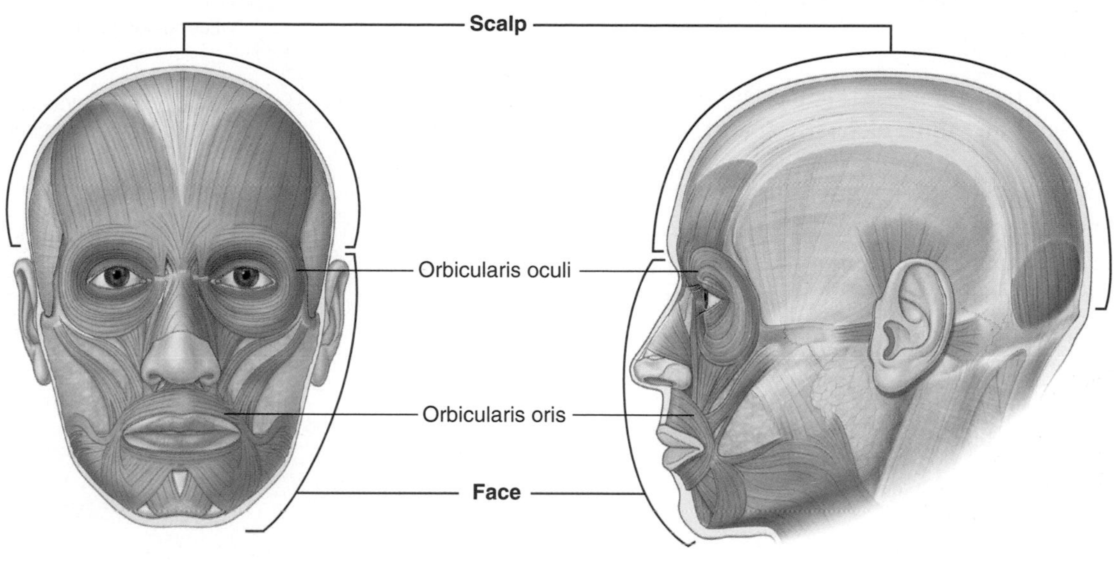


**Fig. 8.3** Muscles of the face.

The **face** is the anterior aspect of the head and contains a unique group of muscles that move the skin relative to underlying bone and control the anterior openings to the orbits and oral cavity (Fig. 8.3).

The **scalp** covers the superior, posterior, and lateral regions of the head (Fig. 8.3).

## Neck

The **neck** extends from the head above to the shoulders and thorax below (Fig. 8.4). Its superior boundary is along the inferior margins of the mandible and bone features on the posterior aspect of the skull. The posterior neck is higher than the anterior neck to connect cervical viscera with the posterior openings of the nasal and oral cavities.

The inferior boundary of the neck extends from the top of the sternum, along the clavicle, and onto the adjacent acromion, a bony projection of the scapula. Posteriorly, the inferior limit of the neck is less well defined, but can be approximated by a line between the acromion and the spinous process of vertebra CVII, which is prominent and easily palpable. The inferior border of the neck encloses the **base of the neck**.

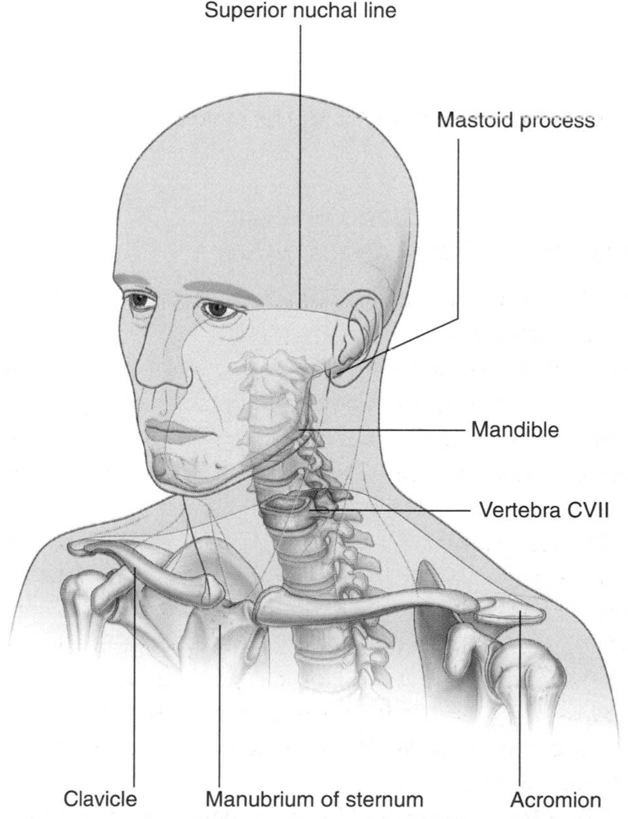


**Fig. 8.4** Boundaries of the neck.

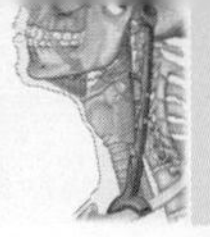

## Compartments

The neck has four major compartments (Fig. 8.5), which are enclosed by an outer musculofascial collar:

- The vertebral compartment contains the cervical vertebrae and associated postural muscles.
- The visceral compartment contains important glands (thyroid, parathyroid, and thymus), and parts of the respiratory and digestive tracts that pass between the head and thorax.
- The two vascular compartments, one on each side, contain the major blood vessels and the vagus nerve.

## Larynx and pharynx

The neck contains two specialized structures associated with the digestive and respiratory tracts—the larynx and pharynx.

The **larynx** (Fig. 8.6) is the upper part of the lower airway and is attached below to the top of the trachea and above, by a flexible membrane, to the hyoid bone, which in turn is attached to the floor of the oral cavity. A number of cartilages form a supportive framework for the larynx, which has a hollow central channel. The dimensions of

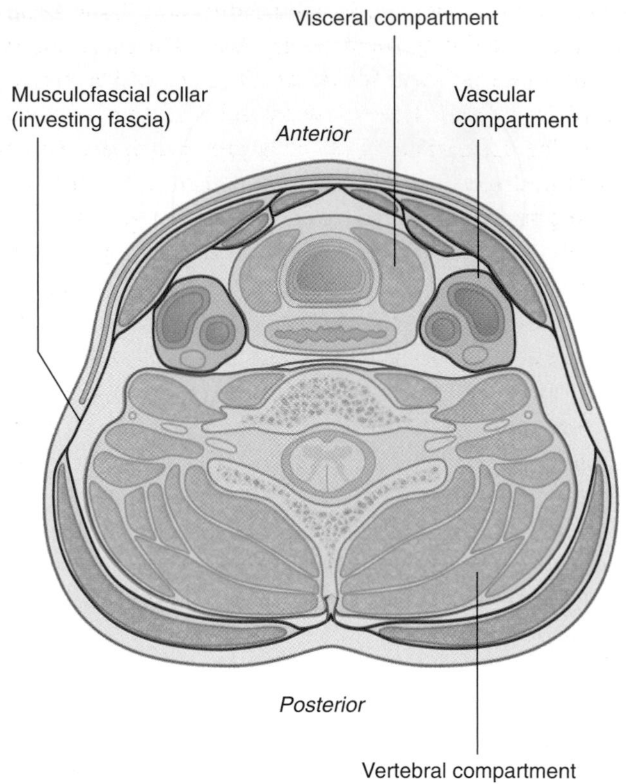


**Fig. 8.5** Major compartments of the neck.

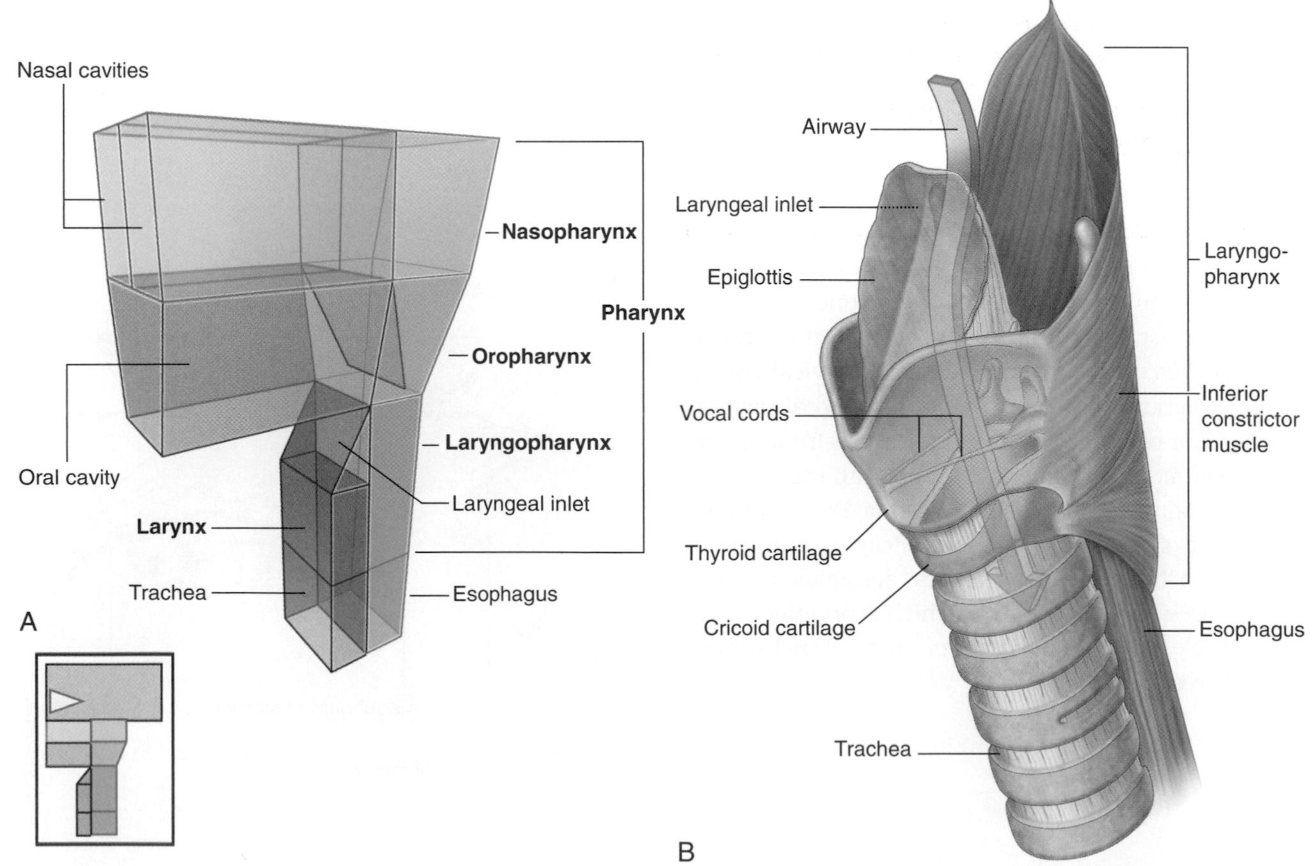


**Fig. 8.6** Specialized structures of the neck. **A.** Conceptual view. **B.** Anatomical view.

this central channel can be adjusted by soft tissue structures associated with the laryngeal wall. The most important of these are two lateral vocal folds, which project toward each other from adjacent sides of the laryngeal cavity. The upper opening of the larynx (**laryngeal inlet**) is tilted posteriorly, and is continuous with the pharynx.

The **pharynx** (Fig. 8.6) is a chamber in the shape of a half-cylinder with walls formed by muscles and fascia. Above, the walls are attached to the base of the skull, and below to the margins of the esophagus. On each side, the walls are attached to the lateral margins of the nasal cavities, the oral cavity, and the larynx. The two nasal cavities, the oral cavity, and the larynx therefore open into the anterior aspect of the pharynx, and the esophagus opens inferiorly.

The part of the pharynx posterior to the nasal cavities is the **nasopharynx**. Those parts posterior to the oral cavity and larynx are the **oropharynx** and **laryngopharynx**, respectively.

## FUNCTIONS

### Protection

The head houses and protects the brain and all the receptor systems associated with the special senses—the nasal cavities associated with smell, the orbits with vision, the ears with hearing and balance, and the oral cavity with taste.

### Contains upper parts of respiratory and digestive tracts

The head contains the upper parts of the respiratory and digestive systems—the nasal and oral cavities—which have structural features for modifying the air or food passing into each system.

### Communication

The head and neck are involved in communication. Sounds produced by the larynx are modified in the pharynx and oral cavity to produce speech. In addition, the muscles of facial expression adjust the contours of the face to relay nonverbal signals.

### Positioning the head

The neck supports and positions the head. Importantly, it enables an individual to position sensory systems in the head relative to environmental cues without moving the entire body.

### Connects the upper and lower respiratory and digestive tracts

The neck contains specialized structures (pharynx and larynx) that connect the upper parts of the digestive and respiratory tracts (nasal and oral cavities) in the head, with the esophagus and trachea, which begin relatively low in the neck and pass into the thorax.

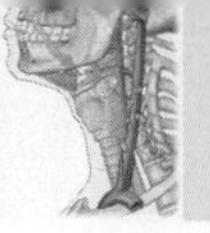



## COMPONENT PARTS

### Skull

The many bones of the head collectively form the skull (Fig. 8.7A). Most of these bones are interconnected by **sutures**, which are immovable fibrous joints (Fig. 8.7B).

In the fetus and newborn, large membranous and unossified gaps (**fontanelles**) between the bones of the skull, particularly between the large flat bones that cover the top of the cranial cavity (Fig. 8.7C), allow:

- the head to deform during its passage through the birth canal, and
- postnatal growth.

Most of the fontanelles close during the first year of life. Full ossification of the thin connective tissue ligaments separating the bones at the suture lines begins in the late twenties, and is normally completed in the fifth decade of life.

There are only three pairs of synovial joints on each side in the head. The largest are the temporomandibular joints between the lower jaw (mandible) and the temporal bone. The other two synovial joints are between the three tiny bones in the middle ear, the malleus, incus, and stapes.

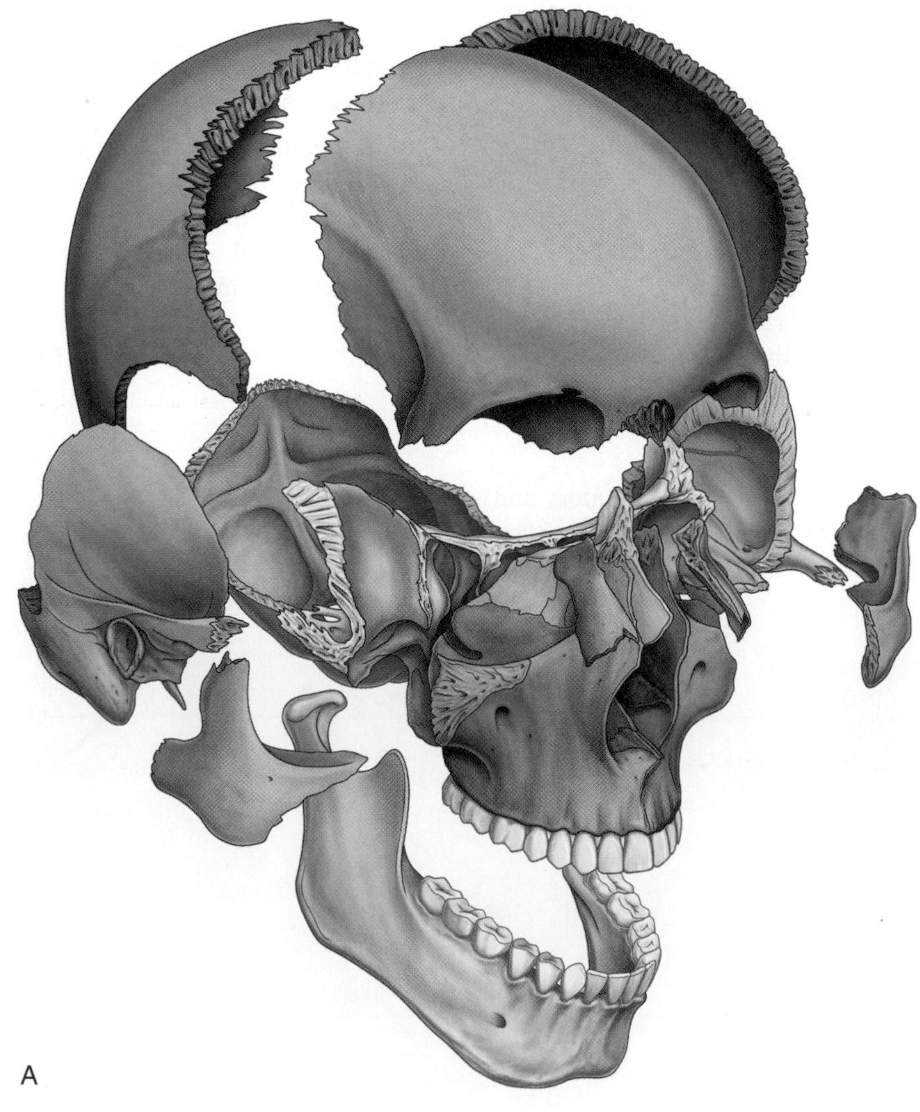


**Fig. 8.7** Skull. **A.** Bones.

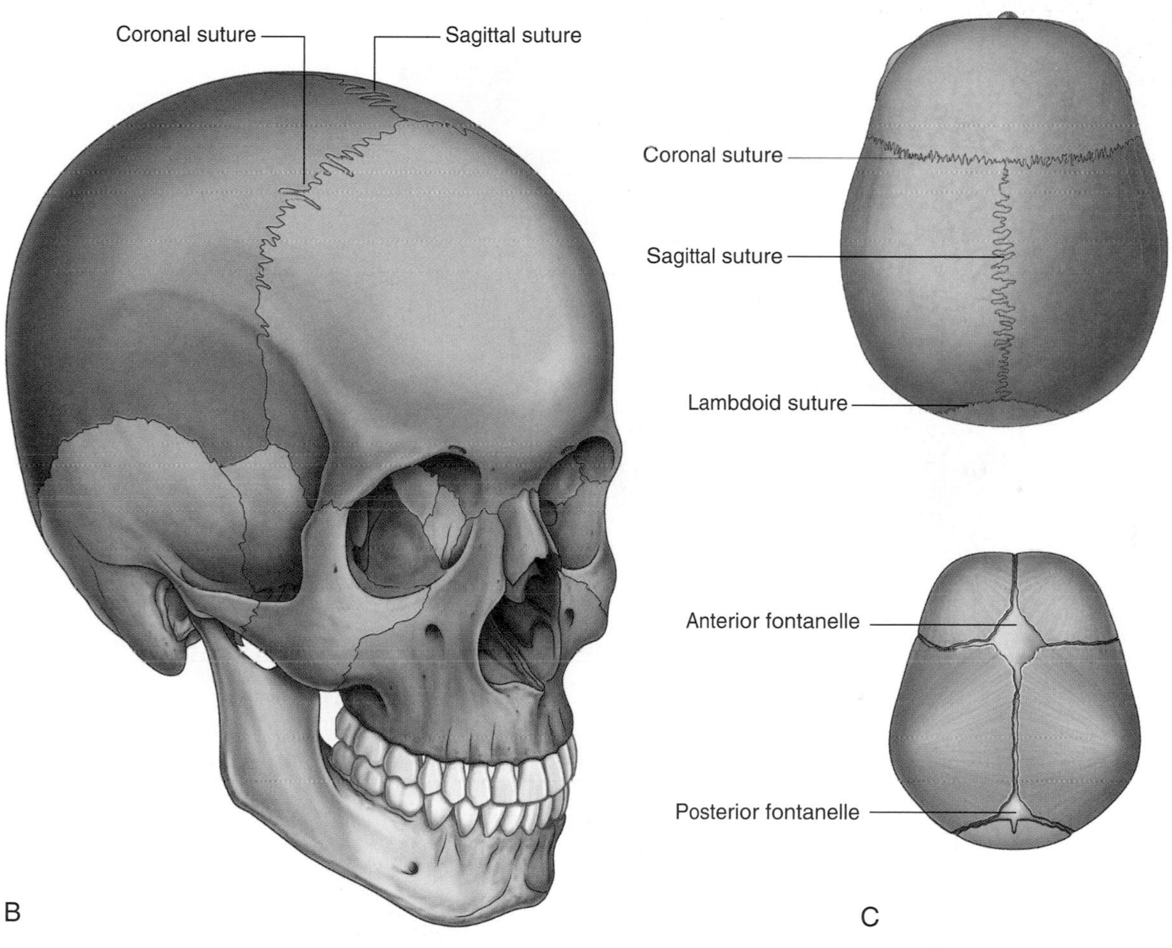


**Fig. 8.7, cont'd** Skull. **B.** Sutures. **C.** Fontanelles and lambdoid suture.

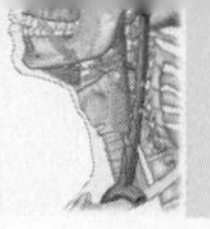



## Cervical vertebrae

The seven cervical vertebrae form the bony framework of the neck.

Cervical vertebrae (Fig. 8.8A) are characterized by:

- small bodies,
- bifid spinous processes, and
- transverse processes that contain a foramen (**foramen transversarium**).

Together the foramina transversaria form a longitudinal passage on each side of the cervical vertebral column for blood vessels (vertebral artery and veins) passing between the base of the neck and the cranial cavity.

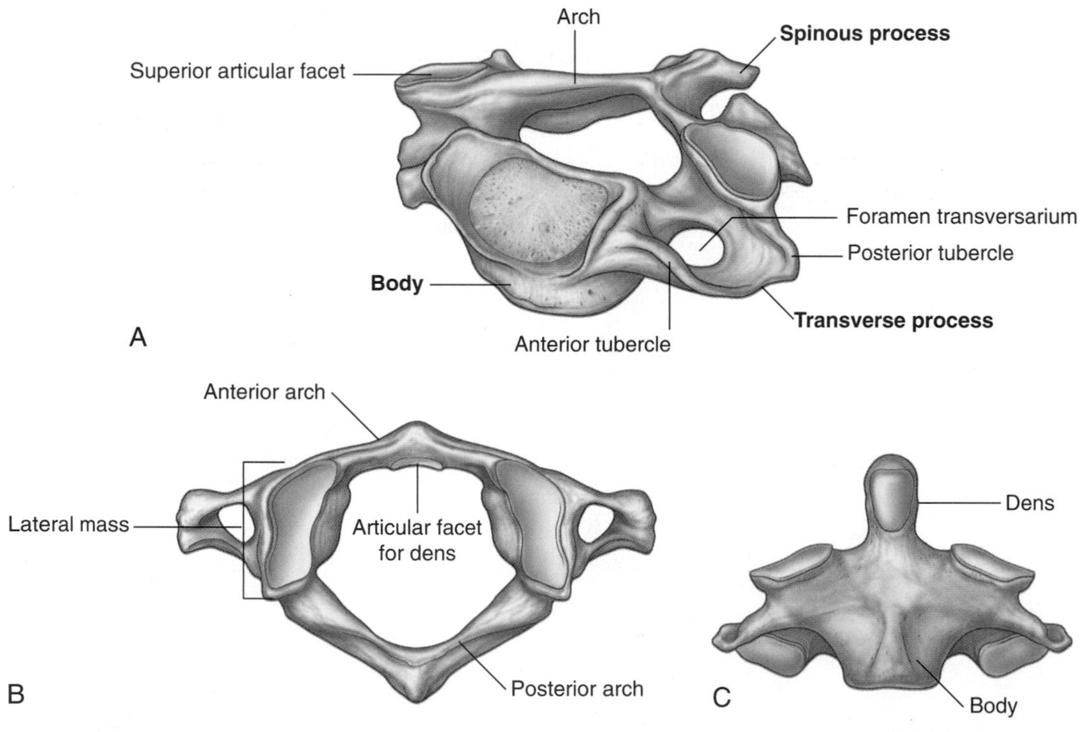


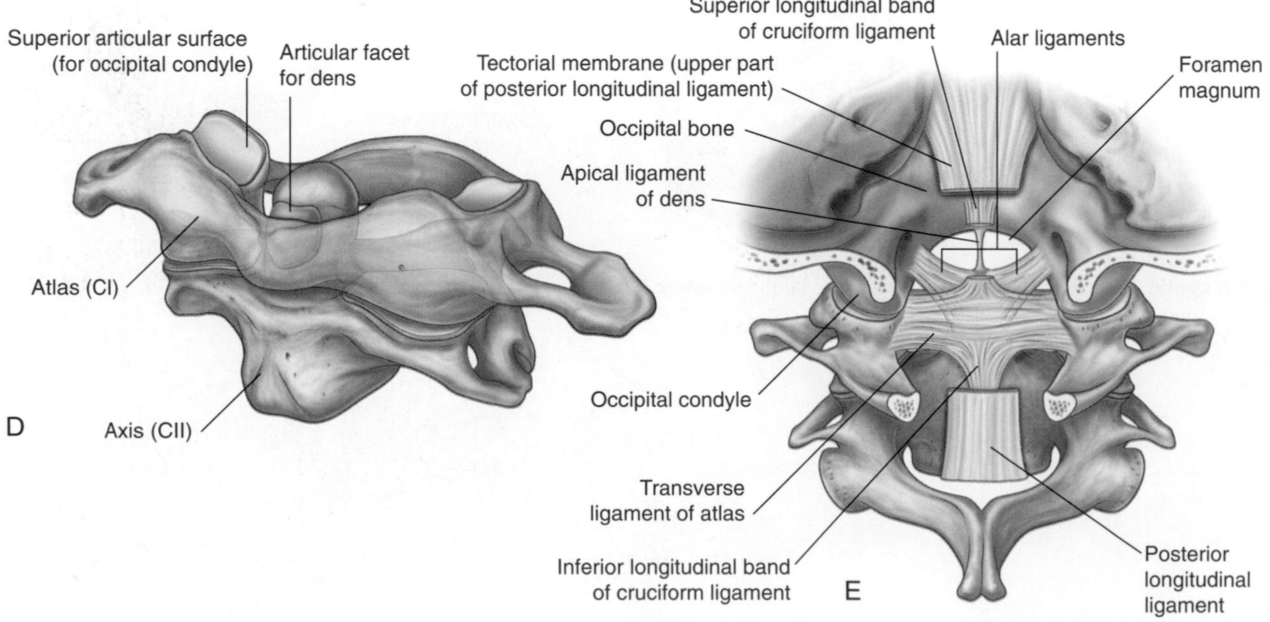


**Fig. 8.8** Cervical vertebrae. **A.** Typical features. **B.** Atlas—vertebra CI (superior view). **C.** Axis—vertebra CII (anterior view). **D.** Atlas and axis (anterolateral view). **E.** Atlanto-occipital joint (posterior view).

The typical transverse process of a cervical vertebra also has **anterior** and **posterior tubercles** for muscle attachment. The anterior tubercles are derived from the same embryological elements that give rise to ribs in the thoracic region. Occasionally, cervical ribs develop from these elements, particularly in association with the lower cervical vertebrae.

The upper two cervical vertebrae (CI and CII) are modified for moving the head (Fig. 8.8B–E; see also Chapter 2).

## Hyoid bone

The hyoid bone is a small U-shaped bone (Fig. 8.9A) oriented in the horizontal plane just superior to the larynx, where it can be palpated and moved from side to side.

- The **body of the hyoid bone** is anterior and forms the base of the U.
- The two arms of the U (**greater horns**) project posteriorly from the lateral ends of the body.

The hyoid bone does not articulate directly with any other skeletal elements in the head and neck.

The hyoid bone is a highly movable and strong bony anchor for a number of muscles and soft tissue structures in the head and neck. Significantly, it is at the interface between three dynamic compartments:

- Superiorly, it is attached to the floor of the oral cavity.
- Inferiorly, it is attached to the larynx.
- Posteriorly, it is attached to the pharynx (Fig. 8.9B).

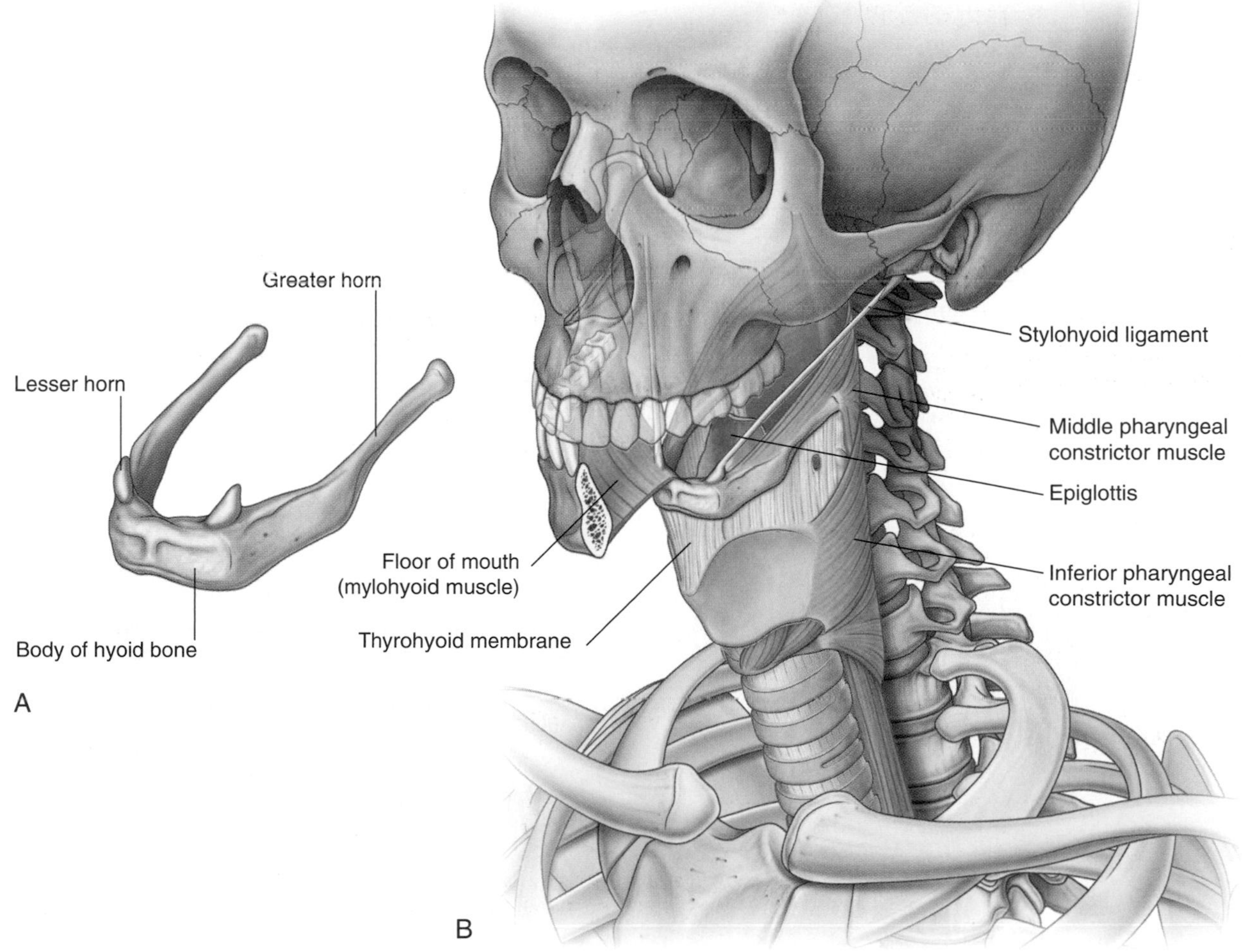


**Fig. 8.9** Hyoid. **A.** Bone. **B.** Attachments.

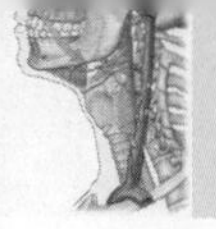



## Soft palate

The soft palate is a soft tissue flap-like structure "hinged" to the back of the hard palate (Fig. 8.10A) with a free posterior margin. It can be elevated and depressed by muscles (Fig. 8.10B).

The soft palate and associated structures can be clearly seen through an open mouth.

## Muscles

The skeletal muscles of the head and neck can be grouped on the basis of function, innervation, and embryological derivation.

### In the head

The muscle groups in the head include:

- the extra-ocular muscles (move the eyeball and open the upper eyelid),
- muscles of the middle ear (adjust the movement of the middle ear bones),
- muscles of facial expression (move the face),
- muscles of mastication (move the jaw—temporomandibular joint),
- muscles of the soft palate (elevate and depress the palate), and
- muscles of the tongue (move and change the contour of the tongue).

### In the neck

In the neck, major muscle groups include:

- muscles of the pharynx (constrict and elevate the pharynx),
- muscles of the larynx (adjust the dimensions of the air pathway),
- strap muscles (position the larynx and hyoid bone in the neck),
- muscles of the outer cervical collar (move the head and upper limb), and
- postural muscles in the muscular compartment of the neck (position the neck and head).

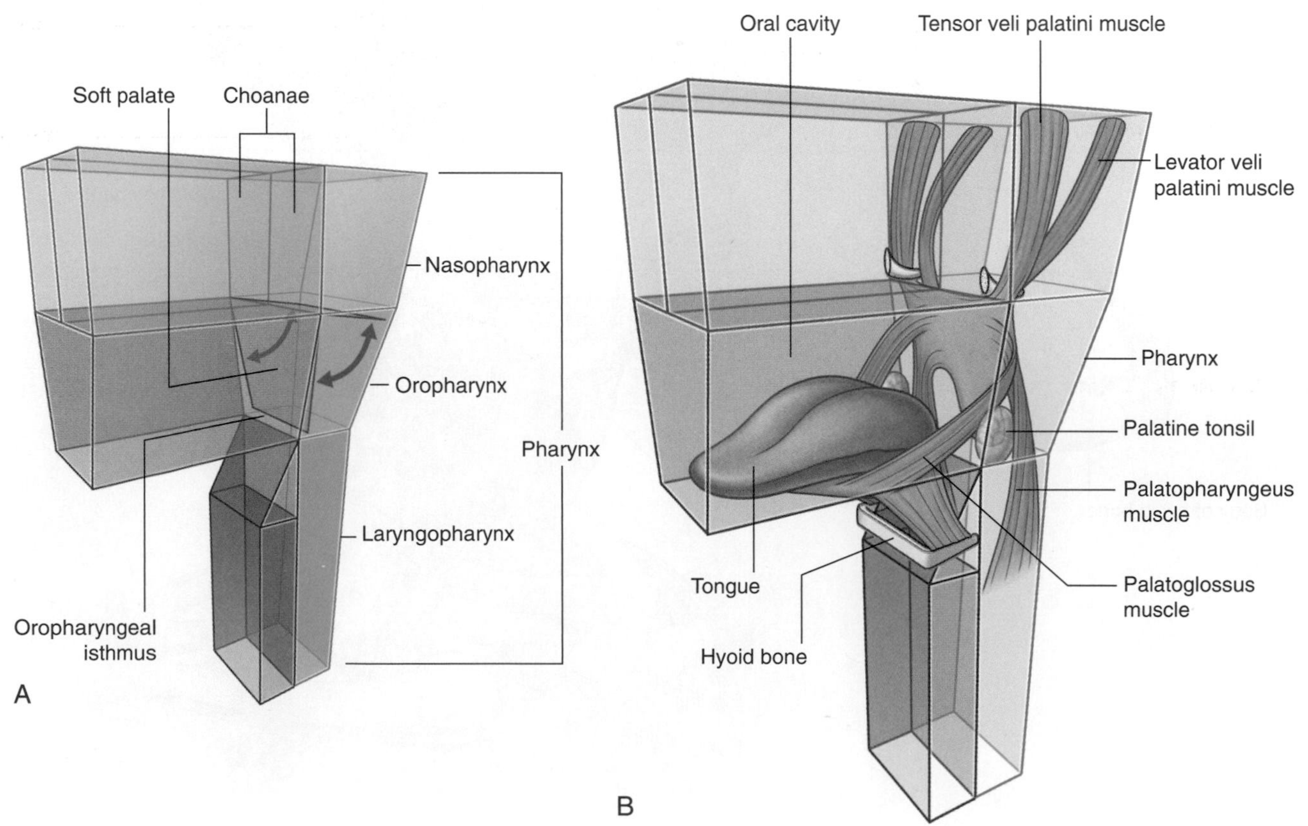


**Fig. 8.10** Soft palate. **A.** Position. **B.** Muscles.

## RELATIONSHIP TO OTHER REGIONS

### Thorax

The **superior thoracic aperture** (**thoracic inlet**) opens directly into the base of the neck (Fig. 8.11). Structures passing between the head and thorax pass up and down through the superior thoracic aperture and the visceral compartment of the neck. At the base of the neck, the trachea is immediately anterior to the esophagus, which is directly anterior to the vertebral column. There are major veins, arteries, and nerves anterior and lateral to the trachea.

### Upper limbs

There is an axillary inlet (gateway to the upper limb) on each side of the superior thoracic aperture at the base of the neck (Fig. 8.11):

- Structures such as blood vessels pass over rib I when passing between the axillary inlet and thorax.
- Cervical components of the brachial plexus pass directly from the neck through the axillary inlets to enter the upper limb.

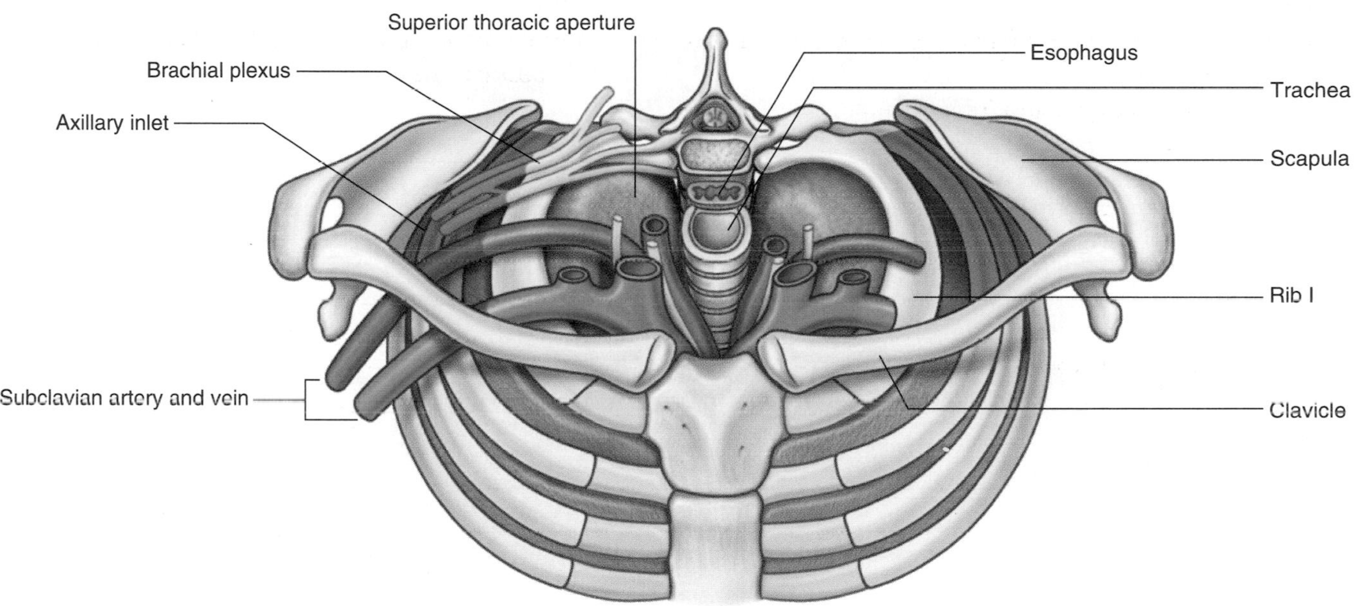


**Fig. 8.11** Superior thoracic aperture and axillary inlets.

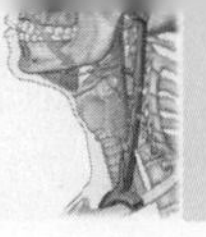

## KEY FEATURES

### Vertebral levels CIII/IV and CV/VI

In the neck, the two important vertebral levels (Fig. 8.12) are:

- between CIII and CIV, at approximately the superior border of the thyroid cartilage of the larynx (which can be palpated) and where the major artery on each side of the neck (the **common carotid artery**) bifurcates into internal and external carotid arteries; and
- between CV and CVI, which marks the lower limit of the pharynx and larynx, and the superior limit of the trachea and esophagus—the indentation between the cricoid cartilage of the larynx and the first tracheal ring can be palpated.

The internal carotid artery has no branches in the neck and ascends into the skull to supply much of the brain. It also supplies the eye and orbit. Other regions of the head and neck are supplied by branches of the external carotid artery.

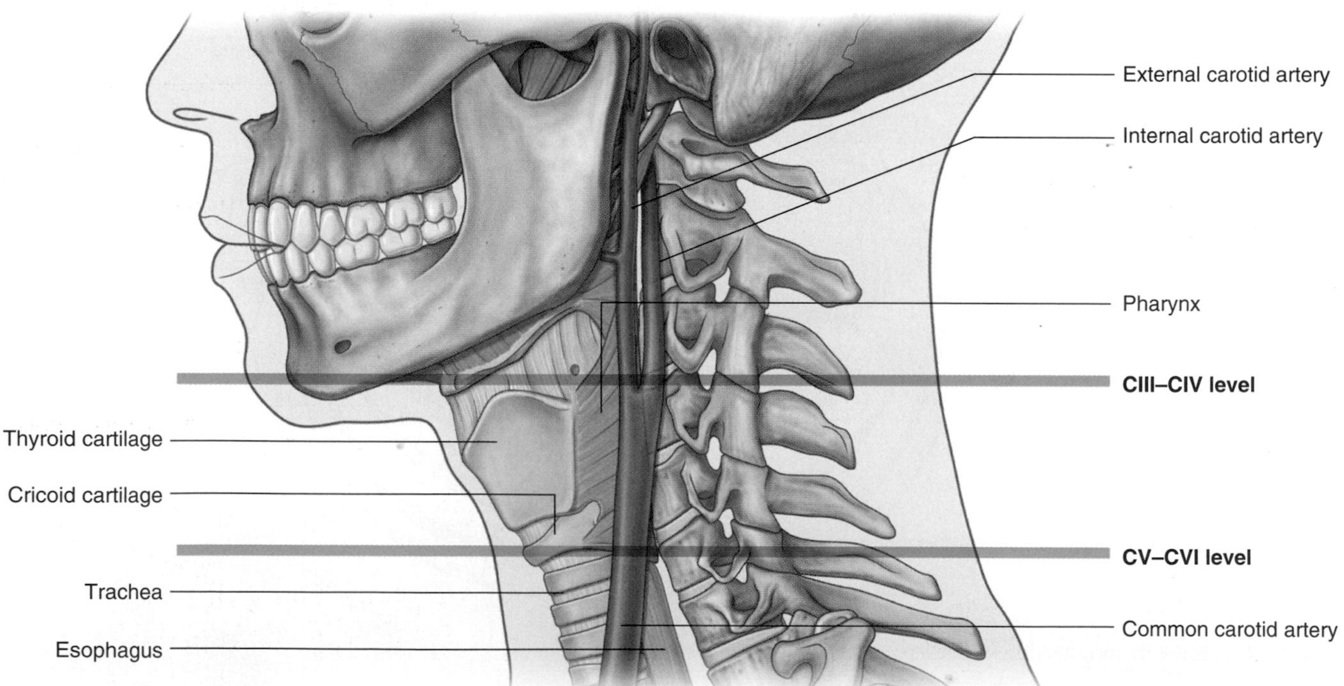


**Fig. 8.12** Important vertebral levels—CIII/CIV and CV/CVI.

## Airway in the neck

The larynx (Fig. 8.13) and the trachea are anterior to the digestive tract in the neck, and can be accessed directly when upper parts of the system are blocked. A **cricothyrotomy** makes use of the easiest route of access through the **cricothyroid ligament** (cricovocal membrane, cricothyroid membrane) between the cricoid and thyroid cartilages of the larynx. The ligament can be palpated in the midline, and usually there are only small blood vessels, connective tissue, and skin (though occasionally, a small lobe of the thyroid gland—pyramidal lobe) overlying it. At a lower level, the airway can be accessed surgically through the anterior wall of the trachea by **tracheostomy**. This route of entry is complicated because large veins and part of the thyroid gland overlie this region.

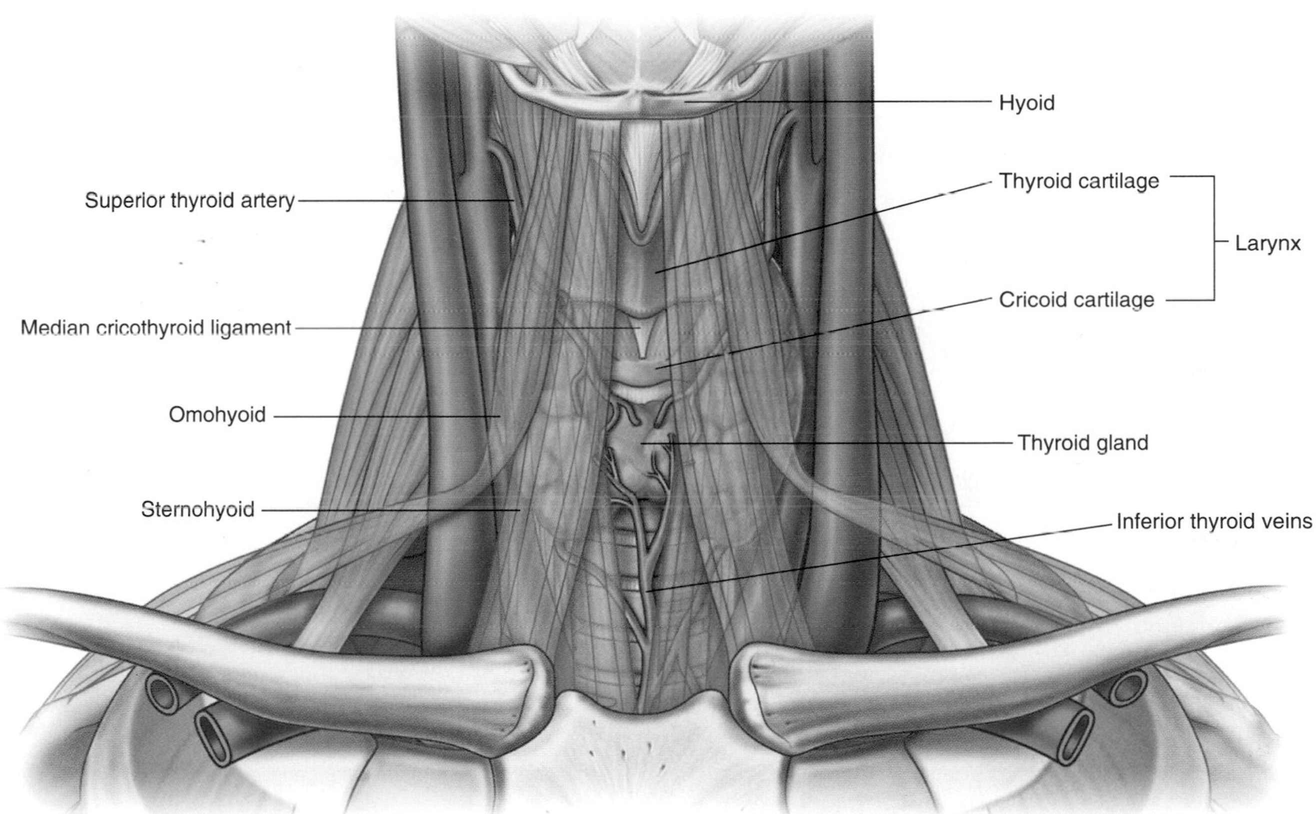


**Fig. 8.13** Larynx and associated structures in the neck.

## Cranial nerves

There are twelve pairs of cranial nerves and their defining feature is that they exit the cranial cavity through foramina or fissures.

All cranial nerves innervate structures in the head or neck. In addition, the **vagus nerve [X]** descends through the neck and into the thorax and abdomen where it innervates viscera.

Parasympathetic fibers in the head are carried out of the brain as part of four cranial nerves—the oculomotor nerve [III], the facial nerve [VII], the glossopharyngeal nerve [IX], and the vagus nerve [X] (Fig. 8.14). Parasympathetic fibers in the oculomotor nerve [III], the facial nerve [VII], and the glossopharyngeal nerve [IX] destined for target tissues in the head leave these nerves, and are distributed with branches of the trigeminal nerve [V].

The vagus nerve [X] leaves the head and neck to deliver parasympathetic fibers to the thoracic and abdominal viscera.

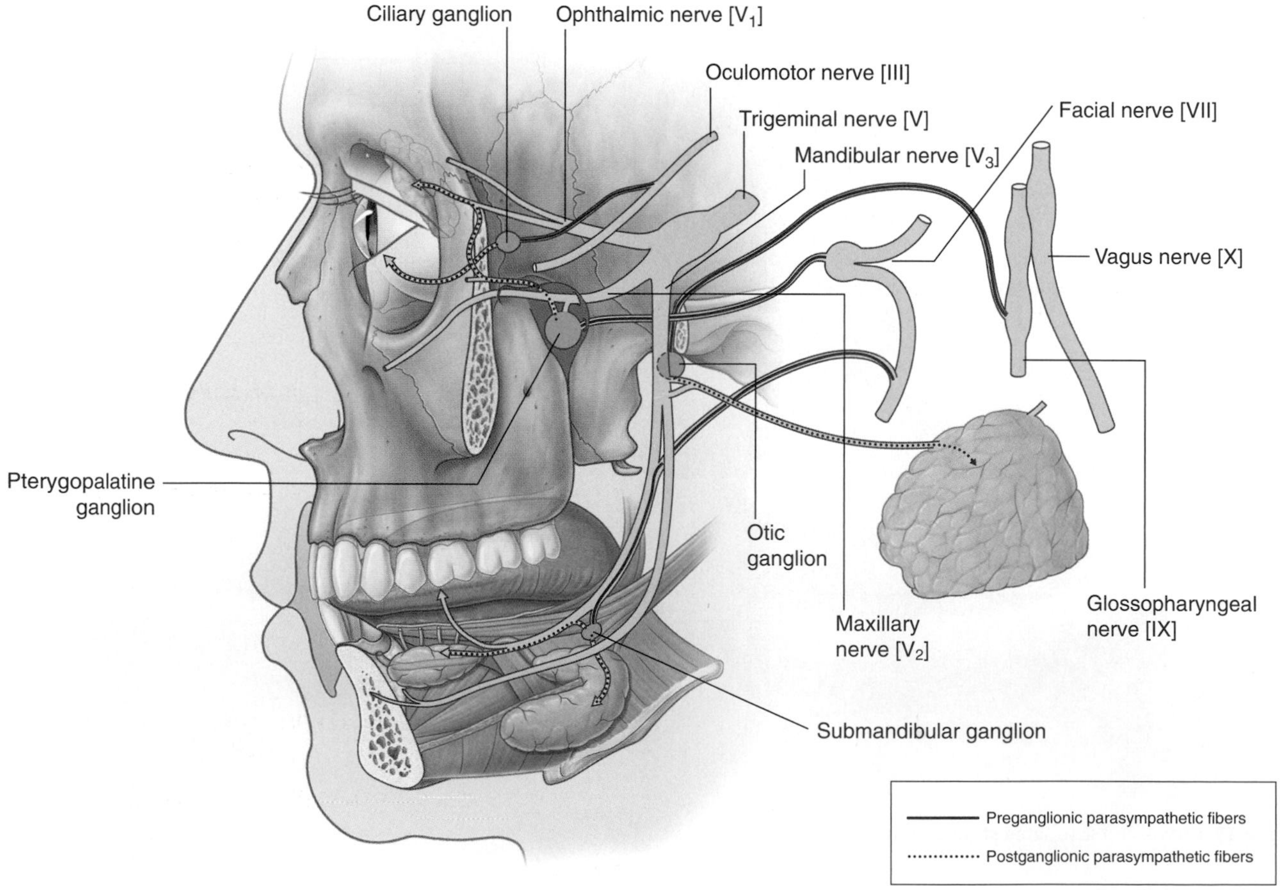

**Fig. 8.14** Cranial nerves and parasympathetic innervation.

## Cervical nerves

There are eight cervical nerves (C1 to C8):

- C1 to C7 emerge from the vertebral canal above their respective vertebrae.
- C8 emerges between vertebrae CVII and TI (Fig. 8.15A).

The anterior rami of C1 to C4 form the **cervical plexus**. The major branches from this plexus supply the strap muscles, the diaphragm (phrenic nerve), skin on the anterior and lateral parts of the neck, skin on the upper anterior thoracic wall, and skin on the inferior parts of the head (Fig. 8.15B).

The anterior rami of C5 to C8, together with a large component of the anterior ramus of T1, form the **brachial plexus**, which innervates the upper limb.

## Functional separation of the digestive and respiratory passages

The pharynx is a common chamber for the digestive and respiratory tracts. Consequently, breathing can take place through the mouth as well as through the nose, and

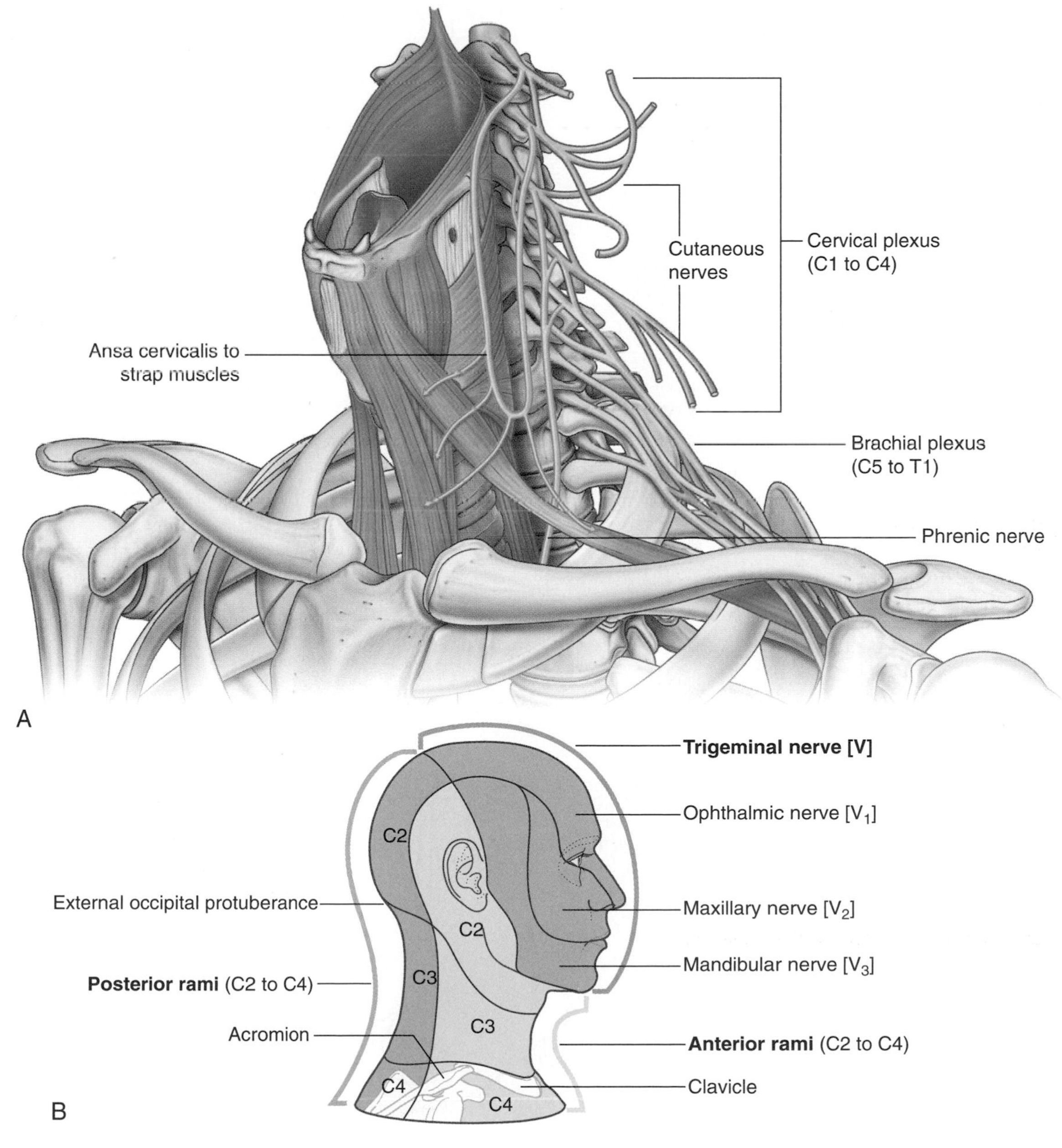


**Fig. 8.15** Cervical nerves. **A.** Structure. **B.** Dermatomes.

material from the oral cavity can potentially enter either the esophagus or the larynx. Importantly:

- The lower airway can be accessed through the oral cavity by intubation.
- The digestive tract (esophagus) can be accessed through the nasal cavity by feeding tubes.

Normally, the soft palate, epiglottis, and soft tissue structures within the larynx act as valves to prevent food and liquid from entering lower parts of the respiratory tract (Fig. 8.16A).

During normal breathing, the airway is open and air passes freely through the nasal cavities (or oral cavity), pharynx, larynx, and trachea (Fig. 8.16A). The lumen of

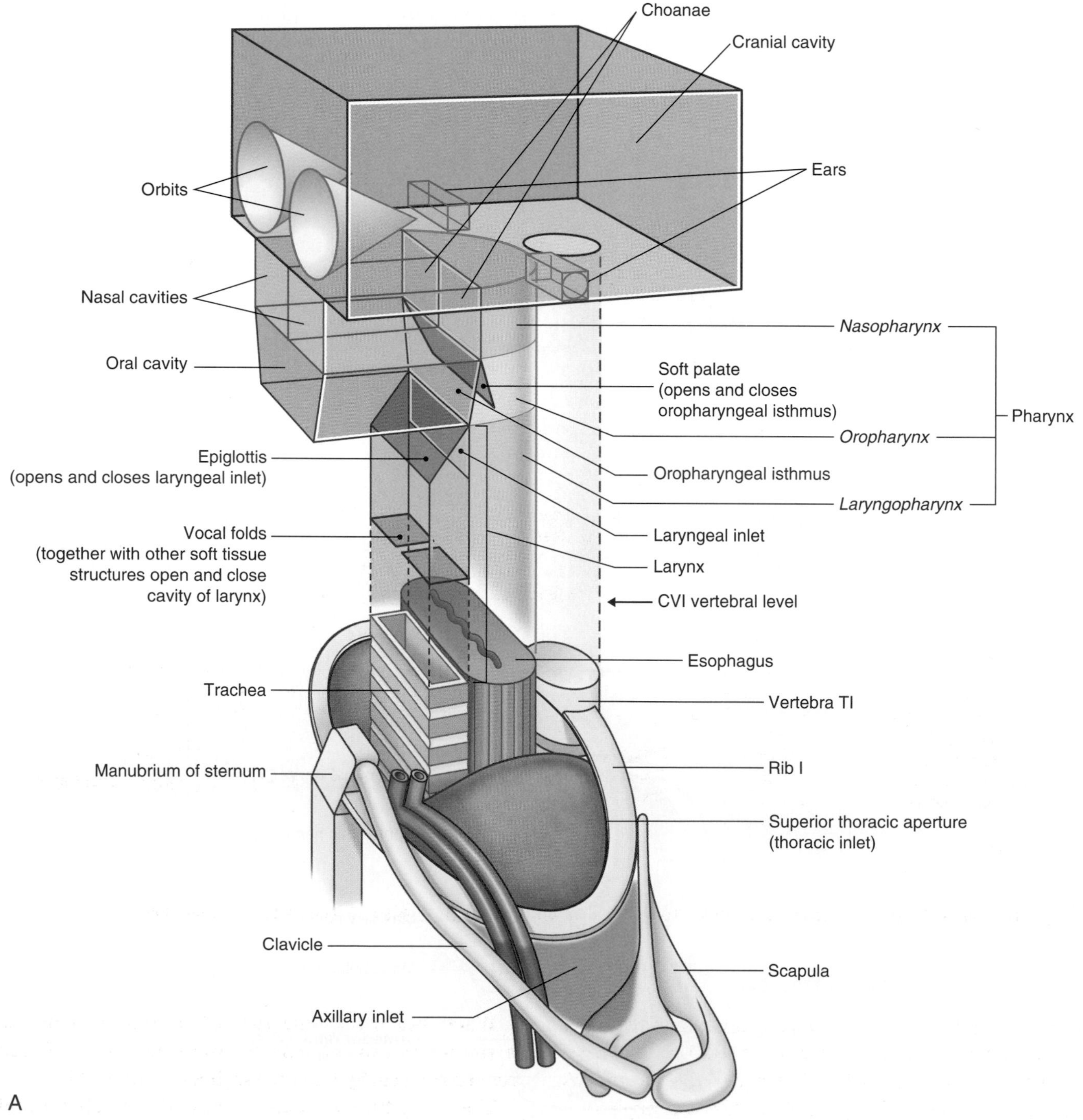


**Fig. 8.16** Larynx, soft palate, epiglottis, and oropharyngeal isthmus. **A.** Overall design.

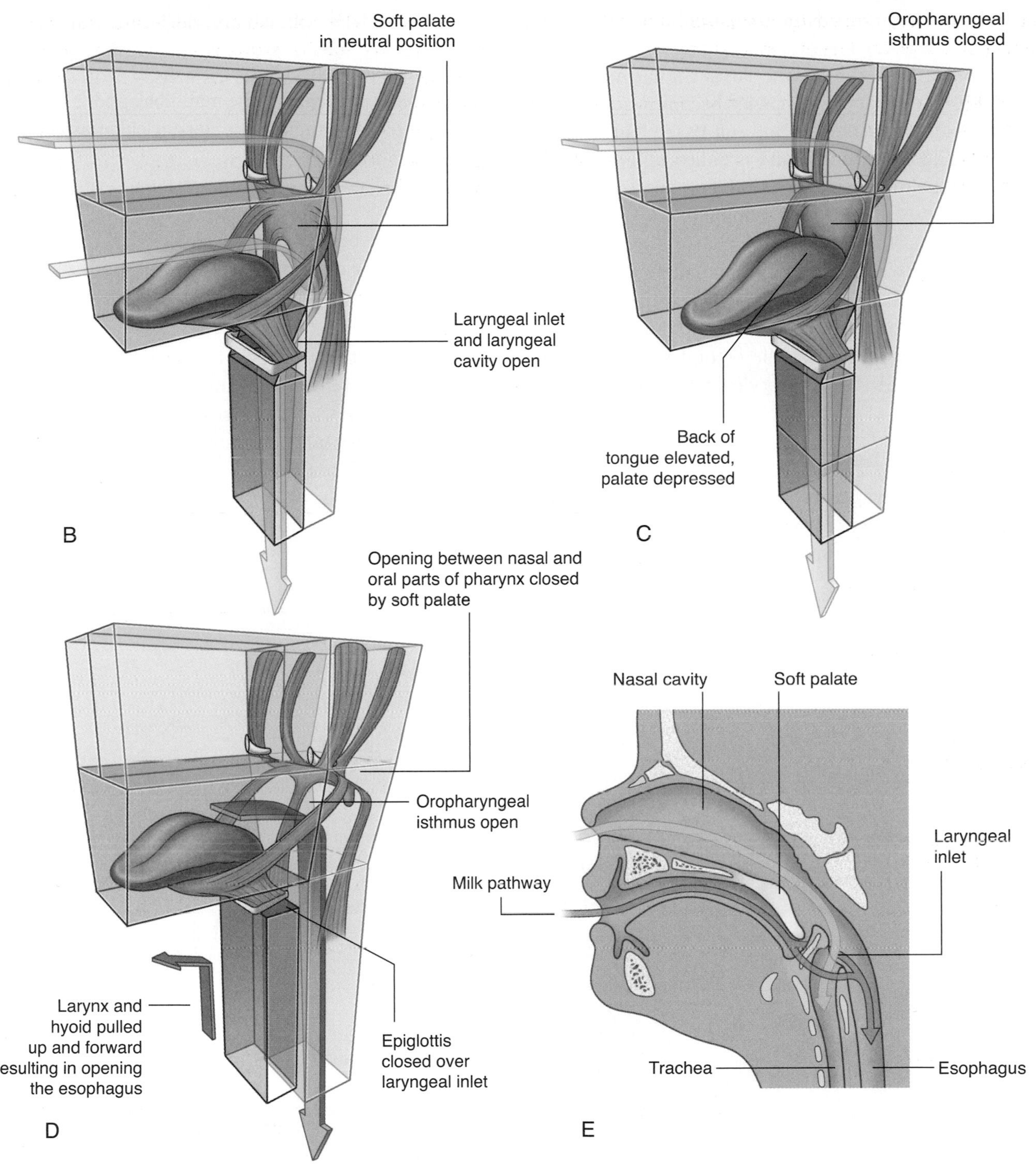


**Fig. 8.16, cont'd B.** Normal breathing. **C.** Breathing with food or liquid in the oral cavity. **D.** Swallowing. **E.** In a newborn child.

the esophagus is normally closed because, unlike the airway, it has no skeletal support structures to hold it open.

When the oral cavity is full of liquid or food, the soft palate is swung down (depressed) to close the oropharyngeal isthmus, thereby allowing manipulation of food and fluid in the oral cavity while breathing (Fig. 8.16C).

When swallowing, the soft palate and parts of the larynx act as valves to ensure proper movement of food from the oral cavity into the esophagus (Fig. 8.16D).

The soft palate elevates to open the oropharyngeal isthmus while at the same time sealing off the nasal part of the pharynx from the oral part. This prevents food and fluid from moving upward into the nasopharynx and nasal cavities.

The epiglottis of the larynx closes the laryngeal inlet and much of the laryngeal cavity becomes occluded by opposition of the vocal folds and soft tissue folds superior to them. In addition, the larynx is pulled up and forward to facilitate the moving of food and fluid over and around the closed larynx and into the esophagus.

In newborns, the larynx is high in the neck and the epiglottis is above the level of the soft palate (Fig. 8.16E). Babies can therefore suckle and breathe at the same time. Liquid flows around the larynx without any danger of entering the airway. During the second year of life, the larynx descends into the low cervical position characteristic of adults.

## Triangles of the neck

The two muscles (trapezius and sternocleidomastoid) that form part of the outer cervical collar divide the neck into anterior and posterior triangles on each side (Fig. 8.17).

The boundaries of each anterior triangle are:

- the median vertical line of the neck,
- the inferior margin of the mandible, and
- the anterior margin of the sternocleidomastoid muscle.

The posterior triangle is bounded by:

- the middle one-third of the clavicle,
- the anterior margin of the trapezius, and
- the posterior margin of the sternocleidomastoid.

Major structures that pass between the head and thorax can be accessed through the anterior triangle.

The posterior triangle in part lies over the axillary inlet, and is associated with structures (nerves and vessels) that pass into and out of the upper limb.

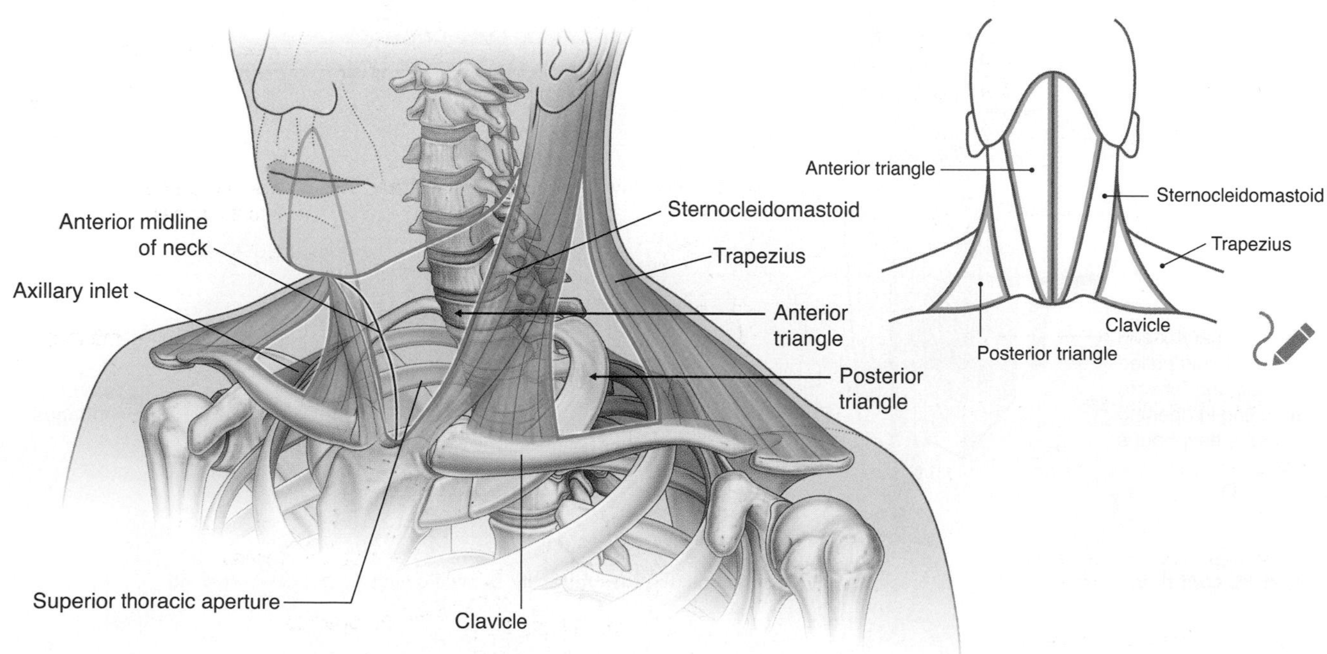


**Fig. 8.17** Anterior and posterior triangles of neck.

# Regional anatomy

## SKULL

The skull has 22 bones, excluding the ossicles of the ear. Except for the mandible, which forms the lower jaw, the bones of the skull are attached to each other by sutures, are immobile, and form the **cranium**.

The cranium can be subdivided into:

- an upper domed part (the **calvaria**), which covers the cranial cavity containing the brain,
- a base that consists of the floor of the cranial cavity, and
- a lower anterior part—the **facial skeleton** (**viscerocranium**).

The bones forming the calvaria are mainly the paired temporal and parietal bones, and parts of the unpaired frontal, sphenoid, and occipital bones.

The bones forming the base of the cranium are mainly parts of the sphenoid, temporal, and occipital bones.

The bones forming the facial skeleton are the paired nasal bones, palatine bones, lacrimal bones, zygomatic bones, maxillae and inferior nasal conchae and the unpaired vomer.

The mandible is not part of the cranium nor part of the facial skeleton.

### Anterior view

The anterior view of the skull includes the **forehead** superiorly, and, inferiorly, the orbits, the **nasal region**, the part of the face between the orbit and the upper jaw, the upper jaw, and the lower jaw (Fig. 8.18).

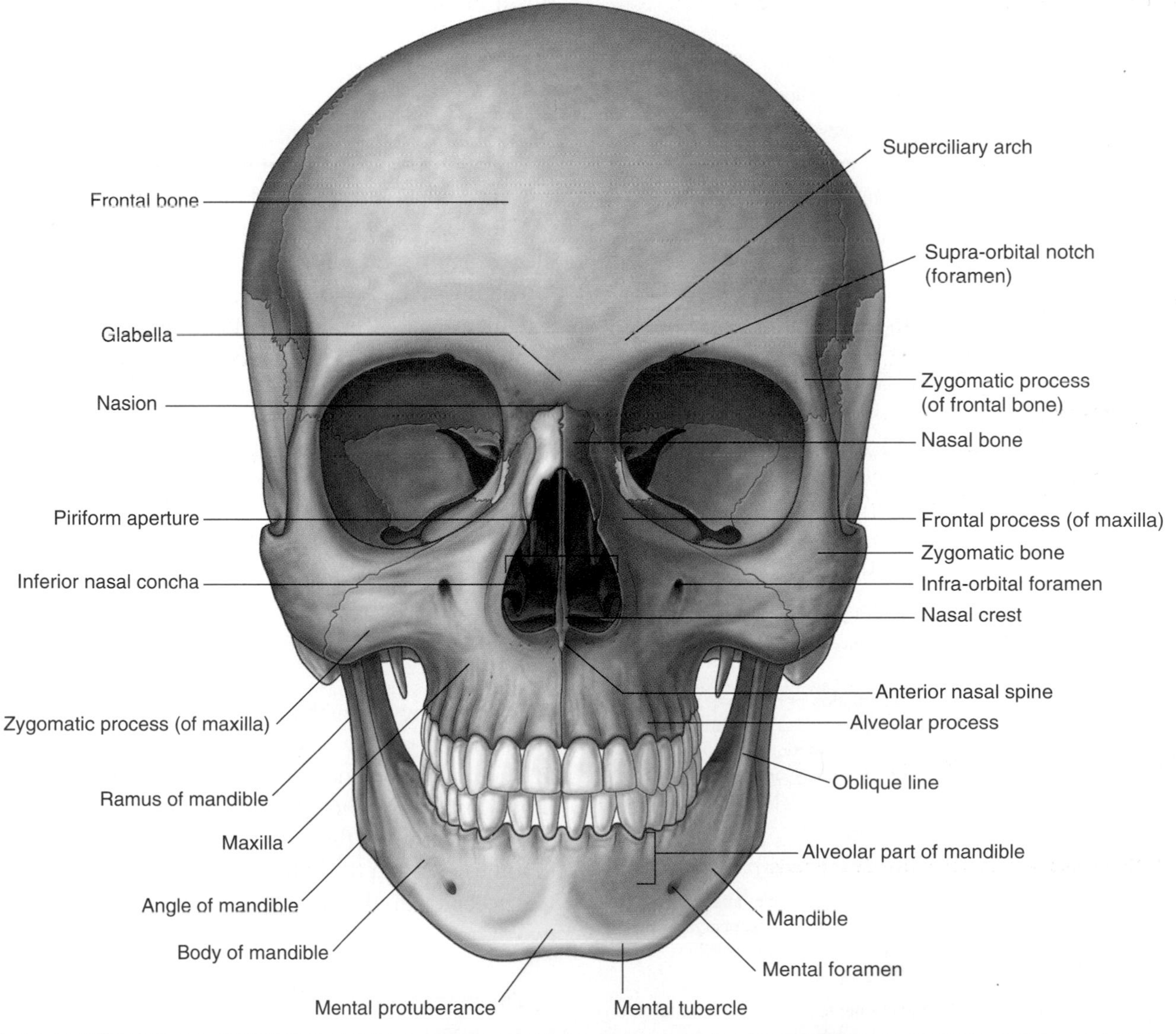


**Fig. 8.18** Anterior view of the skull.

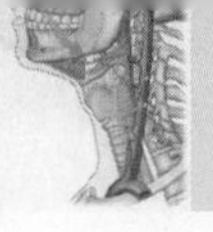



### Frontal bone

The forehead consists of the **frontal bone**, which also forms the superior part of the rim of each orbit (Fig. 8.18).

Just superior to the rim of the orbit on each side are the raised **superciliary arches**. These are more pronounced in men than in women. Between these arches is a small depression (the **glabella**).

Clearly visible in the medial part of the superior rim of each orbit is the **supra-orbital foramen** (**supra-orbital notch**; Table 8.1).

Medially, the frontal bone projects inferiorly forming a part of the medial rim of the orbit.

Laterally, the **zygomatic process** of the frontal bone projects inferiorly forming the upper lateral rim of the orbit. This process articulates with the **frontal process** of the zygomatic bone.

**Table 8.1** External foramina of the skull

| Foramen | Structures passing through foramen |
|---|---|
| **ANTERIOR VIEW** | |
| Supra-orbital foramen | Supra-orbital nerve and vessels |
| Infra-orbital foramen | Infra-orbital nerve and vessels |
| Mental foramen | Mental nerve and vessels |
| **LATERAL VIEW** | |
| Zygomaticofacial foramen | Zygomaticofacial nerve |
| **SUPERIOR VIEW** | |
| Parietal foramen | Emissary veins |
| **INFERIOR VIEW** | |
| Incisive foramen | Nasopalatine nerve; sphenopalatine vessels |
| Greater palatine foramen | Greater palatine nerve and vessels |
| Lesser palatine foramen | Lesser palatine nerves and vessels |
| Pterygoid canal | Nerve of pterygoid canal and vessels |
| Foramen ovale | Mandibular nerve [$V_3$]; lesser petrosal nerve |
| Foramen spinosum | Middle meningeal artery |
| Foramen lacerum | Filled with cartilage |
| Carotid canal | Internal carotid artery and nerve plexus |
| Foramen magnum | Continuation of brain and spinal cord; vertebral arteries and nerve plexuses; anterior spinal artery; posterior spinal arteries; roots of accessory nerve [XI]; meninges |
| Condylar canal | Emissary veins |
| Hypoglossal canal | Hypoglossal nerve [XII] and vessels |
| Jugular foramen | Internal jugular vein; inferior petrosal sinus; glossopharyngeal nerve [IX]; vagus nerve [X]; accessory nerve [XI] |
| Stylomastoid foramen | Facial nerve [VII] |

### Zygomatic and nasal bones

The lower lateral rim of the orbit, as well as the lateral part of the inferior rim of the orbit is formed by the **zygomatic bone** (the cheekbone).

Superiorly, in the nasal region the paired nasal bones articulate with each other in the midline, and with the frontal bone superiorly. The center of the **frontonasal suture** formed by the articulation of the nasal bones and the frontal bone is the **nasion**.

Laterally, each nasal bone articulates with the **frontal process** of each maxilla.

Inferiorly, the **piriform aperture** is the large opening in the nasal region and the anterior opening of the nasal cavity. It is bounded superiorly by the nasal bones and laterally and inferiorly by each maxilla.

Visible through the piriform aperture are the fused **nasal crests**, forming the lower part of the bony **nasal septum** and ending anteriorly as the **anterior nasal spine**, and the paired **inferior nasal conchae**.

### Maxillae

The part of the face between the orbit and the upper teeth and each upper jaw is formed by the paired maxillae.

Superiorly, each maxilla contributes to the inferior and medial rims of the orbit.

Laterally, the **zygomatic process** of each maxilla articulates with the zygomatic bone and medially, the frontal process of each maxilla articulates with the frontal bone.

Inferiorly, the part of each maxilla, lateral to the opening of the nasal cavity, is the **body of the maxilla**.

On the anterior surface of the body of the maxilla, just below the inferior rim of the orbit, is the **infra-orbital foramen** (Table 8.1).

Inferiorly, each maxilla ends as the **alveolar process**, which contains the teeth and forms the upper jaw.

### Mandible

The lower jaw (mandible) is the most inferior structure in the anterior view of the skull. It consists of the **body of the mandible** anteriorly and the **ramus of the mandible** posteriorly. These meet posteriorly at the **angle of the mandible**. All these parts of the mandible are visible, to some extent, in the anterior view.

The body of the mandible is arbitrarily divided into two parts:

- The lower part is the **base of the mandible**.
- The upper part is the **alveolar part of the mandible**.

The alveolar part of the mandible contains the teeth and is resorbed when the teeth are removed. The base of the mandible has a midline swelling (the **mental protuberance**) on its anterior surface where the two sides of the mandible come together. Just lateral to the mental protuberance, on either side, are slightly more pronounced bumps (**mental tubercles**).

Laterally, a **mental foramen** (Table 8.1) is visible halfway between the upper border of the alveolar part of the mandible and the lower border of the base of the mandible. Continuing past this foramen is a ridge (the **oblique line**) passing from the front of the ramus onto the body of the mandible. The oblique line is a point of attachment for muscles that depress the lower lip.

## Lateral view

The lateral view of the skull consists of the lateral wall of the cranium, which includes lateral portions of the calvaria and the facial skeleton, and half of the lower jaw (Fig. 8.19):

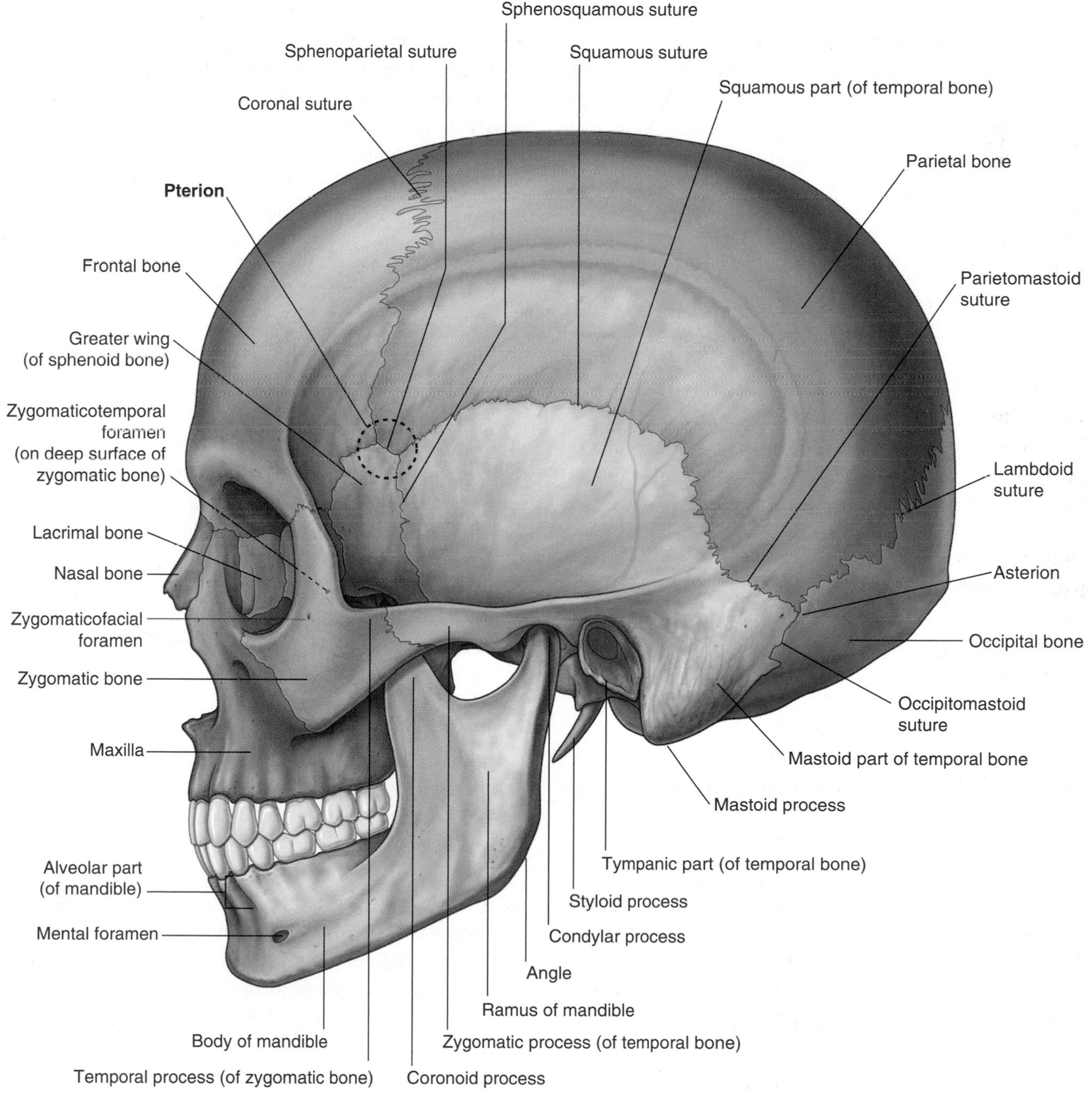


**Fig. 8.19** Lateral view of the skull.

- Bones forming the lateral portion of the calvaria include the frontal, parietal, occipital, sphenoid, and temporal bones.
- Bones forming the visible part of the facial skeleton include the nasal, maxilla, and zygomatic bones.
- The mandible forms the visible part of the lower jaw.

### Lateral portion of the calvaria

The lateral portion of the calvaria begins anteriorly with the frontal bone. In upper regions, the frontal bone articulates with the parietal bone at the **coronal suture**. The parietal bone then articulates with the occipital bone at the **lambdoid suture**.

In lower parts of the lateral portion of the calvaria, the frontal bone articulates with the **greater wing of the sphenoid bone** (Fig. 8.19), which then articulates with the parietal bone at the **sphenoparietal suture**, and with the anterior edge of the temporal bone at the **sphenosquamous suture**.

The junction where the frontal, parietal, sphenoid, and temporal bones are in close proximity is the **pterion**. The clinical consequences of a skull fracture in this area can be very serious. The bone in this area is particularly thin and overlies the anterior division of the middle meningeal artery, which can be torn by a skull fracture in this area, resulting in an extradural hematoma.

The final articulation across the lower part of the lateral portion of the calvaria is between the temporal bone and the occipital bone at the **occipitomastoid suture**.

#### Temporal bone

A major contributor to the lower portion of the lateral wall of the cranium is the temporal bone (Fig. 8.19), which consists of several parts:

- The **squamous part** has the appearance of a large flat plate, forms the anterior and superior parts of the temporal bone, contributes to the lateral wall of the cranium, and articulates anteriorly with the greater wing of the sphenoid bone at the sphenosquamous suture, and with the parietal bone superiorly at the squamous suture.
- The **zygomatic process** is an anterior bony projection from the lower surface of the squamous part of the temporal bone that initially projects laterally and then curves anteriorly to articulate with the temporal process of the zygomatic bone to form the **zygomatic arch**.
- Immediately below the origin of the zygomatic process from the squamous part of the temporal bone is the **tympanic part** of the temporal bone, and clearly visible on the surface of this part is the **external acoustic opening** leading to the **external acoustic meatus** (ear canal).
- The petromastoid part, which is usually separated into a **petrous part** and a **mastoid part** for descriptive purposes.

The mastoid part is the most posterior part of the temporal bone, and is the only part of the petromastoid part of the temporal bone seen on a lateral view of the skull. It is continuous with the squamous part of the temporal bone anteriorly, and articulates with the parietal bone superiorly at the **parietomastoid suture**, and with the occipital bone posteriorly at the occipitomastoid suture. These two sutures are continuous with each other, and the parietomastoid suture is continuous with the squamous suture.

Inferiorly, a large bony prominence (the **mastoid process**) projects from the inferior border of the mastoid part of the temporal bone. This is a point of attachment for several muscles.

Medial to the mastoid process, the **styloid process** projects from the lower border of the temporal bone.

### Visible part of the facial skeleton

The bones of the viscerocranium visible in a lateral view of the skull include the nasal, maxilla, and zygomatic bones (Fig. 8.19) as follows:

- A nasal bone anteriorly.
- The maxilla with its alveolar process containing teeth forming the upper jaw; anteriorly, it articulates with the nasal bone; superiorly, it contributes to the formation of the inferior and medial borders of the orbit; medially, its frontal process articulates with the frontal bone; laterally, its zygomatic process articulates with the zygomatic bone.
- The zygomatic bone, an irregularly shaped bone with a rounded lateral surface that forms the prominence of the cheek, is a visual centerpiece in this view—medially, it assists in the formation of the inferior rim of the orbit through its articulation with the zygomatic process of the maxilla; superiorly, its frontal process articulates with the zygomatic process of the frontal bone assisting in the formation of the lateral rim of the orbit; laterally, seen prominently in this view of the skull, the horizontal temporal process of the zygomatic bone projects backward to articulate with the zygomatic process of the temporal bone and so form the zygomatic arch.

Usually a small foramen (the **zygomaticofacial foramen**; Table 8.1) is visible on the lateral surface of the zygomatic bone. A **zygomaticotemporal foramen** is present on the medial deep surface of the bone.

### Mandible

The final bony structure visible in a lateral view of the skull is the mandible. Inferiorly in the anterior part of this view, it consists of the anterior body of the mandible, a posterior ramus of the mandible, and the angle of the mandible where the inferior margin of the mandible meets the posterior margin of the ramus (Fig. 8.19).

The teeth are in the alveolar part of the body of the mandible and the mental protuberance is visible in this view.

The mental foramen is on the lateral surface of the body, and on the superior part of the ramus **condylar** and **coronoid processes** extend upward.

The condylar process is involved in articulation of the mandible with the temporal bone, and the coronoid process is the point of attachment for the temporalis muscle.

## Posterior view

The occipital, parietal, and temporal bones are seen in the posterior view of the skull.

### Occipital bone

Centrally the flat or **squamous part of the occipital bone** is the main structure in this view of the skull (Fig. 8.20). It articulates superiorly with the paired parietal bones at the lambdoid suture and laterally with each temporal bone at the occipitomastoid sutures. Along the lambdoid suture small islands of bone (**sutural bones** or wormian bones) may be observed.

Several bony landmarks are visible on the occipital bone. There is a midline projection (the **external occipital protuberance**) with curved lines extending laterally from

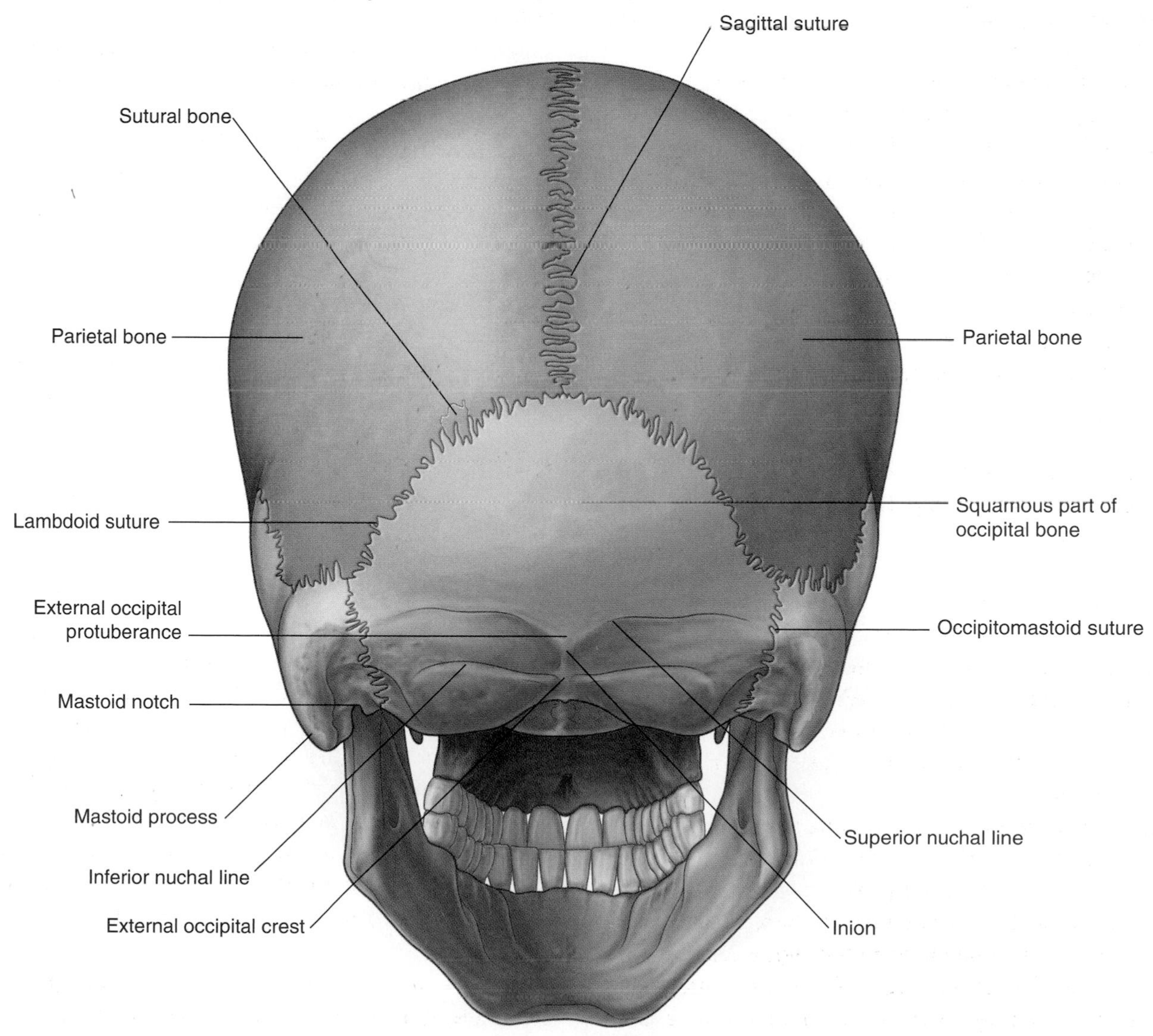


**Fig. 8.20** Posterior view of the skull.

it (**superior nuchal lines**). The most prominent point of the external occipital protuberance is the **inion**. About 1 inch (2.5 cm) below the superior nuchal lines two additional lines (the **inferior nuchal lines**) curve laterally. Extending downward from the external occipital protuberance is the **external occipital crest**.

### Temporal bones

Laterally, the temporal bones are visible in the posterior view of the skull, with the mastoid processes being the prominent feature (Fig. 8.20). On the inferomedial border of each mastoid process is a notch (the **mastoid notch**), which is a point of attachment for the posterior belly of the digastric muscle.

## Superior view

The frontal bone, parietal bones, and occipital bone are seen in a superior view of the skull (Fig. 8.21). These bones make up the superior part of the calvaria or the **calva** (skullcap).

In an anterior to posterior direction:

- The unpaired frontal bone articulates with the paired parietal bones at the coronal suture.
- The two parietal bones articulate with each other in the midline at the sagittal suture.
- The parietal bones articulate with the unpaired occipital bone at the lambdoid suture.

The junction of the sagittal and coronal sutures is the **bregma**, and the junction of the sagittal and lambdoid sutures is the **lambda**.

The only foramina visible in this view of the skull may be the paired parietal foramina, posteriorly, one on each parietal bone just lateral to the sagittal suture (Fig. 8.21).

The bones making up the calvaria (Fig. 8.22) are unique in their structure, consisting of dense internal and external tables of compact bone separated by a layer of spongy bone (the **diploë**).

## Inferior view

The base of the skull is seen in the inferior view and extends anteriorly from the middle incisor teeth posteriorly to the superior nuchal lines and laterally to the mastoid processes and zygomatic arches (Fig. 8.23).

For descriptive purposes the base of the skull is often divided into:

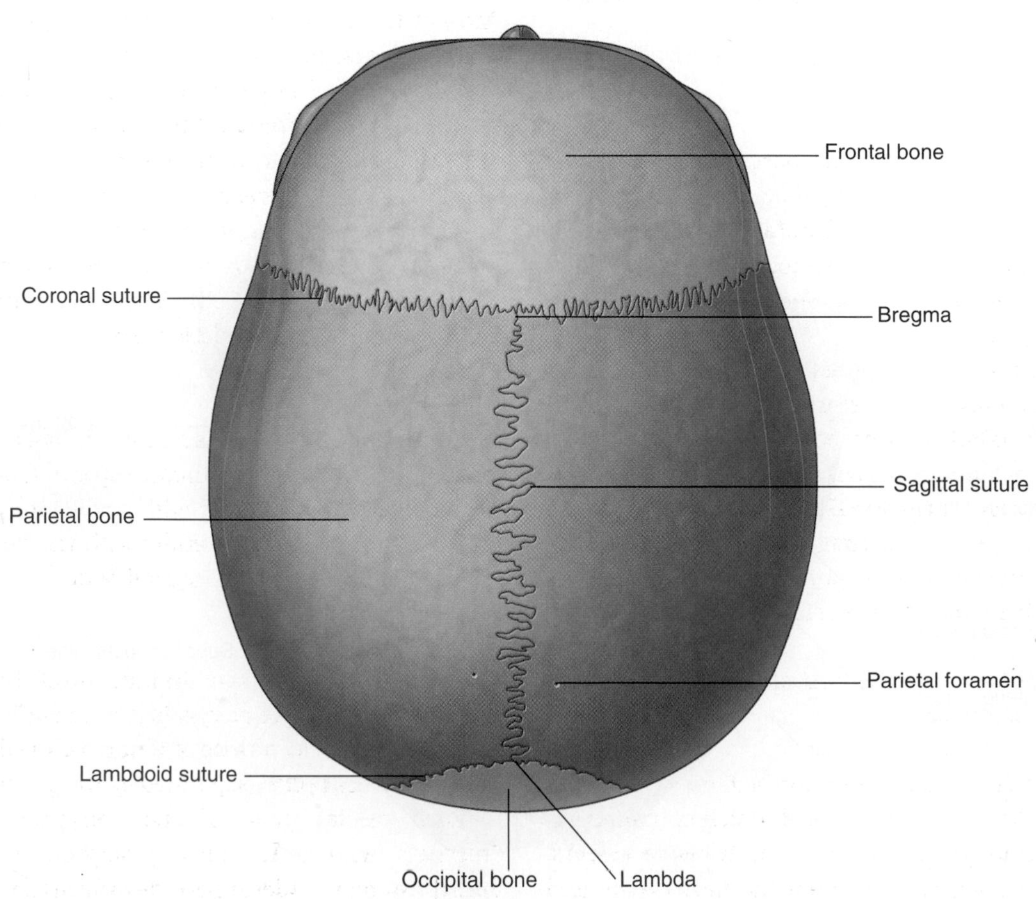


**Fig. 8.21** Superior view of the skull.

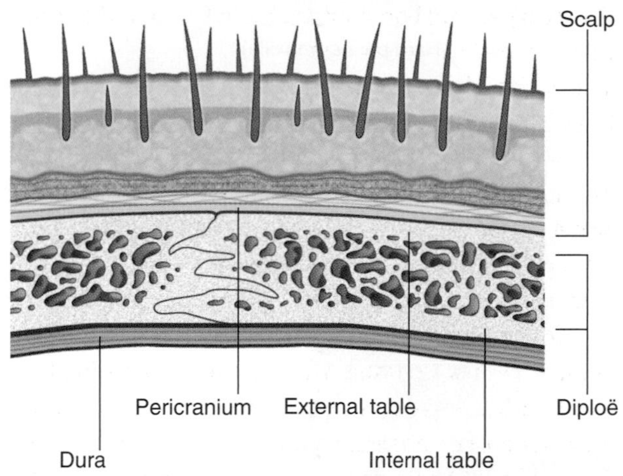


**Fig. 8.22** Calvaria.

- an anterior part, which includes the teeth and the hard palate,
- a middle part, which extends from behind the hard palate to the anterior margin of the foramen magnum, and
- a posterior part, which extends from the anterior edge of the foramen magnum to the superior nuchal lines.

### Anterior part

The main features of the anterior part of the base of the skull are the teeth and the hard palate.

The teeth project from the **alveolar processes** of the two maxillae. These processes are together arranged in a U-shaped alveolar arch that borders the hard palate on three sides (Fig. 8.23).

The **hard palate** is composed of the **palatine processes** of each maxilla anteriorly and the **horizontal plates** of each **palatine bone** posteriorly.

The paired palatine processes of each maxilla meet in the midline at the **intermaxillary suture**, the paired maxillae and the paired palatine bones meet at the **palatomaxillary suture**, and the paired horizontal plates of each palatine bone meet in the midline at the **interpalatine suture**.

Several additional features are also visible when the hard palate is examined:

- the **incisive fossa** in the anterior midline immediately posterior to the teeth, the walls of which contain **incisive foramina** (the openings of the **incisive canals**, which are passageways between the hard palate and nasal cavity);
- the **greater palatine foramina** near the posterolateral border of the hard palate on each side, which lead to **greater palatine canals**;
- just posterior to the greater palatine foramina, the **lesser palatine foramina** in the **pyramidal process** of each palatine bone, which lead to **lesser palatine canals**;
- a midline pointed projection (the **posterior nasal spine**) in the free posterior border of the hard palate.

### Middle part

The middle part of the base of the skull is complex:

- Forming the anterior half are the vomer and sphenoid bones.
- Forming the posterior half are the occipital and paired temporal bones.

#### Anterior half

##### Vomer

Anteriorly, the small vomer is in the midline, resting on the sphenoid bone (Fig. 8.23). It contributes to the formation of the bony nasal septum separating the two choanae.

##### Sphenoid

Most of the anterior part of the middle part of the base of the skull consists of the sphenoid bone.

The sphenoid bone is made up of a centrally placed **body**, paired **greater and lesser wings** projecting laterally from the body, and two downward projecting **pterygoid processes** immediately lateral to each choana.

Three parts of the sphenoid bone, the body, greater wings, and pterygoid processes, are seen in the inferior view of the skull (Fig. 8.23). The lesser wing of the sphenoid is not seen in the inferior view.

##### Body

The body of the sphenoid is a centrally placed cube of bone containing two large air sinuses separated by a septum.

It articulates anteriorly with the vomer, ethmoid, and palatine bones, posterolaterally with the temporal bones, and posteriorly with the occipital bone.

##### Pterygoid processes

Extending downward from the junction of the body and the greater wings are the pterygoid processes (Fig. 8.23). Each of these processes consists of a narrow **medial plate** and broader **lateral plate** separated by the **pterygoid fossa**.

Each medial plate of the pterygoid process ends inferiorly with a hook-like projection, the **pterygoid hamulus**, and divides superiorly to form the small, shallow **scaphoid fossa**.

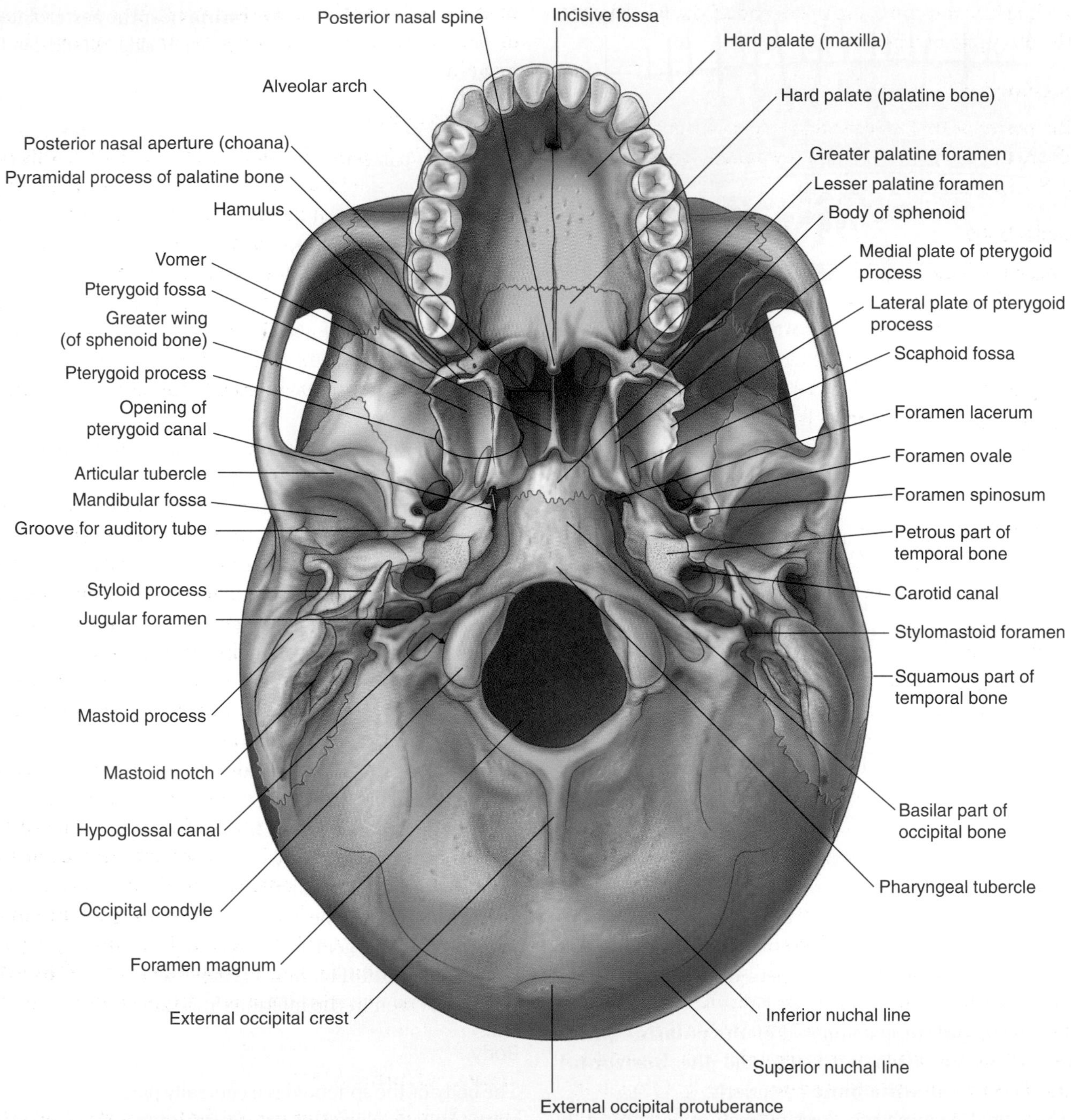

**Fig. 8.23** Inferior view of the skull.

Just superior to the scaphoid fossa, at the root of the medial plate of the pterygoid process is the opening of the **pterygoid canal**, which passes forward from near the anterior margin of the foramen lacerum.

### Greater wing

Lateral to the lateral plate of the pterygoid process is the greater wing of the sphenoid (Fig. 8.23), which not only forms a part of the base of the skull but also continues laterally to form part of the lateral wall of the skull. It articulates laterally and posteriorly with parts of the temporal bone.

Important features visible on the surface of the greater wing in an inferior view of the skull are the foramen ovale and the foramen spinosum on the posterolateral border extending outward from the upper end of the lateral plate of the pterygoid process.

#### Posterior half

In the posterior half of the middle part of the base of the skull are the occipital bone and the paired temporal bones (Fig. 8.23).

##### Occipital bone

The occipital bone, or more specifically its **basilar part**, is in the midline immediately posterior to the body of the sphenoid. It extends posteriorly to the **foramen magnum** and is bounded laterally by the temporal bones.

Prominent on the basilar part of the occipital bone is the **pharyngeal tubercle**, a bony protuberance for the attachment of parts of the pharynx to the base of the skull (Fig. 8.23).

##### Temporal bone

Immediately lateral to the basilar part of the occipital bone is the petrous part of the petromastoid part of each temporal bone.

Wedge-shaped in its appearance, with its **apex** anteromedial, the petrous part of the temporal bone is between the greater wing of the sphenoid anteriorly and the basilar part of the occipital bone posteriorly. The apex forms one of the boundaries of the **foramen lacerum**, an irregular opening filled in life with cartilage (Fig. 8.23).

The other boundaries of the foramen lacerum are the basilar part of the occipital bone medially and the body of the sphenoid anteriorly.

Posterolateral from the foramen lacerum along the petrous part of the temporal bone is the large circular opening for the **carotid canal**.

Between the petrous part of the temporal bone and the greater wing of the sphenoid is a groove for the cartilaginous part of the **pharyngotympanic tube** (**auditory tube**). This groove continues posterolaterally into a bony canal in the petrous part of the temporal bone for the pharyngotympanic tube.

Just lateral to the greater wing of the sphenoid is the squamous part of the temporal bone, which participates in the temporomandibular joint. It contains the **mandibular fossa**, which is a concavity where the head of the mandible articulates with the base of the skull. An important feature of this articulation is the prominent **articular tubercle**, which is the downward projection of the anterior border of the mandibular fossa (Fig. 8.23).

### Posterior part

The posterior part of the base of the skull extends from the anterior edge of the foramen magnum posteriorly to the superior nuchal lines (Fig. 8.23). It consists of parts of the occipital bone centrally and the temporal bones laterally.

#### Occipital bone

The occipital bone is the major bony element of this part of the base of the skull (Fig. 8.23). It has four parts organized around the foramen magnum, which is a prominent feature of this part of the base of the skull and through which the brain and spinal cord are continuous.

The parts of the occipital bone are the squamous part, which is posterior to the foramen magnum, the **lateral parts**, which are lateral to the foramen magnum, and the **basilar part**, which is anterior to the foramen magnum (Fig. 8.23).

The squamous and lateral parts are components of the posterior part of the base of the skull.

The most visible feature of the squamous part of the occipital bone when examining the inferior view of the skull is a ridge of bone (the external occipital crest), which extends downward from the external occipital protuberance toward the foramen magnum. The inferior nuchal lines arc laterally from the midpoint of the crest.

Immediately lateral to the foramen magnum are the lateral parts of the occipital bones, which contain numerous important structural features.

On each anterolateral border of the foramen magnum are the rounded **occipital condyles** (Fig. 8.23). These paired structures articulate with the atlas (vertebra CI). Posterior to each condyle is a depression (the **condylar fossa**) containing a **condylar canal**, and anterior and superior to each condyle is the large **hypoglossal canal**. Lateral to each hypoglossal canal is a large, irregular **jugular foramen** formed by opposition of the **jugular notch** of the occipital bone and **jugular notch** of the temporal bone.

#### Temporal bone

Laterally in the posterior part of the base of the skull is the temporal bone. The parts of the temporal bone seen in this location are the mastoid part of the petromastoid part and the styloid process (Fig. 8.23).

The lateral edge of the mastoid part is identified by the large cone-shaped mastoid process projecting from its inferior surface. This prominent bony structure is the point of attachment for several muscles. On the medial aspect of the mastoid process is the deep mastoid notch, which is also an attachment point for a muscle.

Anteromedial to the mastoid process is the needle-shaped styloid process projecting from the lower border of the temporal bone. The styloid process is also a point of attachment for numerous muscles and ligaments.

Finally, between the styloid process and the mastoid process is the stylomastoid foramen.

## CRANIAL CAVITY

The cranial cavity is the space within the cranium that contains the brain, meninges, proximal parts of the cranial nerves, blood vessels, and cranial venous sinuses.

### Roof

The calvaria is the dome-shaped roof that protects the superior aspect of the brain. It consists mainly of the frontal bone anteriorly, the paired parietal bones in the middle, and the occipital bone posteriorly (Fig. 8.24).

Sutures visible internally include:

- the coronal suture, between the frontal and parietal bones,
- the sagittal suture, between the paired parietal bones, and
- the lambdoid suture, between the parietal and occipital bones.

Visible junctions of these sutures are the bregma, where the coronal and sagittal sutures meet, and the lambda, where the lambdoid and sagittal sutures meet.

Other markings on the internal surface of the calva include bony ridges and numerous grooves and pits.

From anterior to posterior, features seen on the bony roof of the cranial cavity are:

- a midline ridge of bone extending from the surface of the frontal bone (the **frontal crest**), which is a point of attachment for the **falx cerebri** (a specialization of the dura mater that partially separates the two cerebral hemispheres);

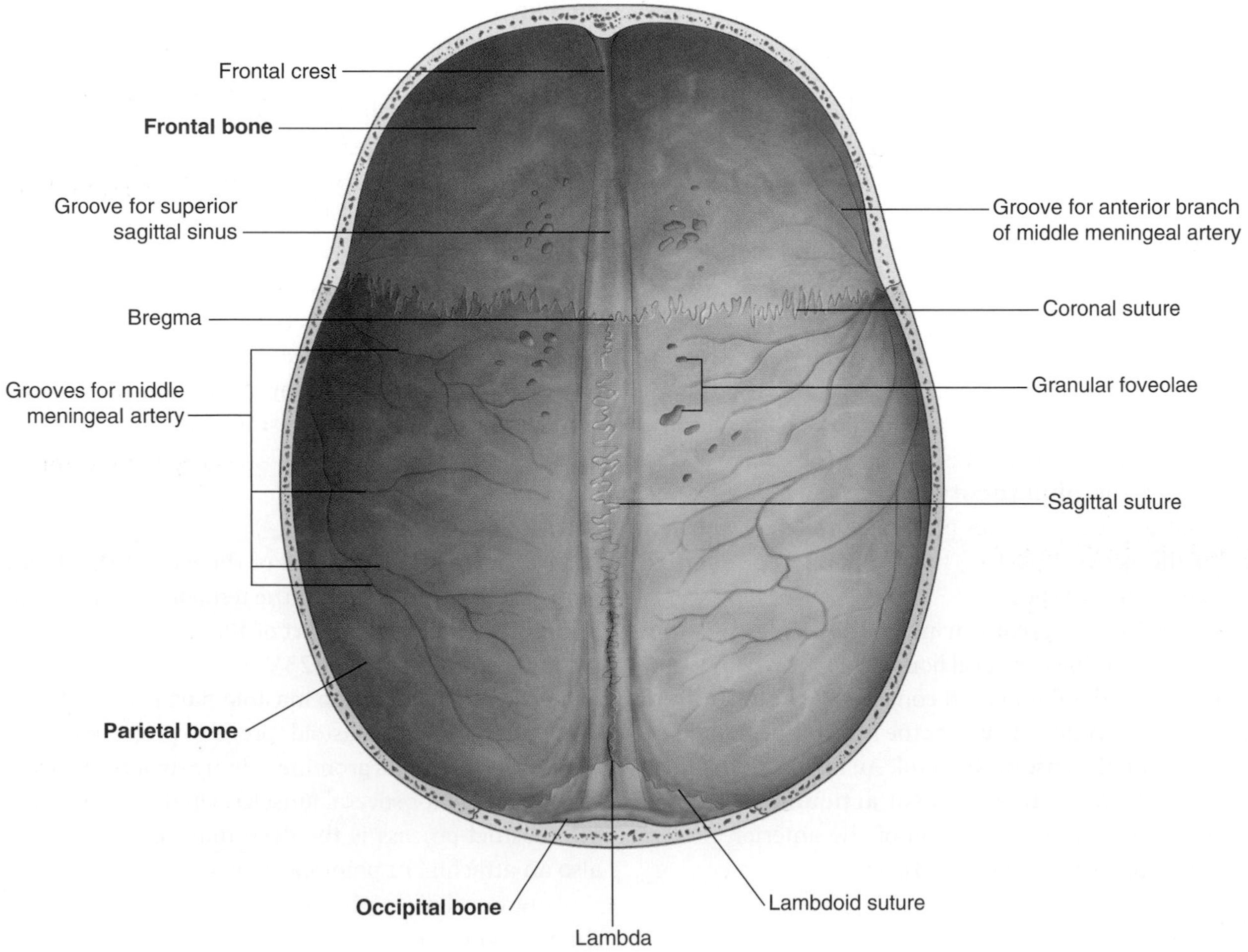

**Fig. 8.24** Roof of the cranial cavity.

- at the superior point of the termination of the frontal crest the beginning of the **groove for the superior sagittal sinus**, which widens and deepens posteriorly and marks the position of the superior sagittal sinus (an intradural venous structure);
- on either side of the groove for the superior sagittal sinus throughout its course, a small number of depressions and pits (the **granular foveolae**), which mark the location of arachnoid granulations (prominent structures readily identifiable when a brain with its meningeal coverings is examined; the arachnoid granulations are involved in the reabsorption of cerebrospinal fluid); and
- on the lateral aspects of the roof of the cranial cavity, smaller grooves created by various meningeal vessels.

## Floor

The floor of the cranial cavity is divided into anterior, middle, and posterior cranial fossae.

### Anterior cranial fossa

Parts of the frontal, ethmoid, and sphenoid bones form the anterior cranial fossa (Fig. 8.25). Its floor is composed of:

- frontal bone in the anterior and lateral direction,
- ethmoid bone in the midline, and
- two parts of the sphenoid bone posteriorly, the body (midline) and the lesser wings (laterally).

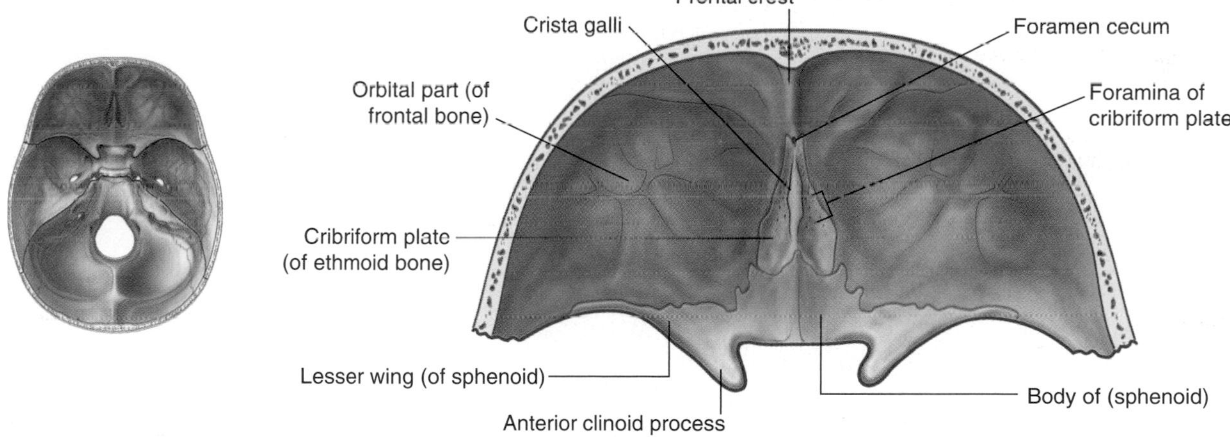


**Fig. 8.25** Anterior cranial fossa.

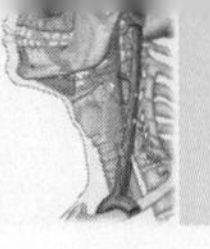



The anterior cranial fossa is above the nasal cavity and the orbits, and it is filled by the frontal lobes of the cerebral hemispheres.

Anteriorly, a small wedge-shaped midline crest of bone (the frontal crest) projects from the frontal bone. This is a point of attachment for the falx cerebri. Immediately posterior to the frontal crest is the **foramen cecum** (Table 8.2). This foramen between the frontal and ethmoid bones may transmit emissary veins connecting the nasal cavity with the superior sagittal sinus.

Posterior to the frontal crest is a prominent wedge of bone projecting superiorly from the **ethmoid** (the **crista galli**). This is another point of attachment for the falx cerebri, which is the vertical extension of dura mater partially separating the two cerebral hemispheres.

Lateral to the crista galli is the **cribriform plate** of the ethmoid bone (Fig. 8.25). This is a sieve-like structure, which allows small olfactory nerve fibers to pass through its foramina from the nasal mucosa to the olfactory bulb. The olfactory nerves are commonly referred to collectively as the olfactory nerve [I].

On each side of the ethmoid, the floor of the anterior cranial fossa is formed by relatively thin plates of frontal bone (the **orbital part** of the frontal bone), which also forms the roof of the orbit below. Posterior to both the frontal and ethmoid bones, the rest of the floor of the anterior cranial fossa is formed by the body and lesser wings of the sphenoid. In the midline, the body extends anteriorly between the orbital parts of the frontal bone to reach the ethmoid bone and posteriorly it extends into the middle cranial fossa.

The boundary between the anterior and middle cranial fossae in the midline is the anterior edge of the prechiasmatic sulcus, a smooth groove stretching between the optic canals across the body of the sphenoid.

### Lesser wings of the sphenoid

The two lesser wings of the sphenoid project laterally from the body of the sphenoid and form a distinct boundary between the lateral parts of the anterior and middle cranial fossae.

Overhanging the anterior part of the middle cranial fossae, each lesser wing ends laterally as a sharp point at the junction of the frontal bone and the greater wing of the sphenoid near the upper lateral edge of the superior orbital fissure that is formed between the greater and lesser wings.

Medially each lesser wing widens, curves posteriorly, and ends as a rounded **anterior clinoid process** (Fig. 8.25). These processes serve as the anterior point of

**Table 8.2** Internal foramina of the skull

| Foramen | Structures passing through foramen |
|---|---|
| **ANTERIOR CRANIAL FOSSA** | |
| Foramen cecum | Emissary veins to nasal cavity |
| Olfactory foramen in cribriform plate | Olfactory nerves [I] |
| **MIDDLE CRANIAL FOSSA** | |
| Optic canal | Optic nerve [II]; ophthalmic artery |
| Superior orbital fissure | Oculomotor nerve [III]; trochlear nerve [IV]; ophthalmic division of the trigeminal nerve [$V_1$]; abducent nerve [VI]; ophthalmic veins |
| Foramen rotundum | Maxillary division of the trigeminal nerve [$V_2$] |
| Foramen ovale | Mandibular division of the trigeminal nerve [$V_3$]; lesser petrosal nerve |
| Foramen spinosum | Middle meningeal artery |
| Hiatus for the greater petrosal nerve | Greater petrosal nerve |
| Hiatus for the lesser petrosal nerve | Lesser petrosal nerve |
| **POSTERIOR CRANIAL FOSSA** | |
| Foramen magnum | End of brainstem/beginning of spinal cord; vertebral arteries; spinal roots of the accessory nerve; meninges |
| Internal acoustic meatus | Facial nerve [VII]; vestibulocochlear nerve [VIII]; labyrinthine artery |
| Jugular foramen | Glossopharyngeal nerve [IX]; vagus nerve [X]; accessory nerve [XI]; inferior petrosal sinus, sigmoid sinus (forming internal jugular vein) |
| Hypoglossal canal | Hypoglossal nerve [XII]; meningeal branch of the ascending pharyngeal artery |
| Condylar canal | Emissary vein |

attachment for the **tentorium cerebelli**, which is a sheet of dura that separates the posterior part of the cerebral hemispheres from the cerebellum. Just anterior to each anterior clinoid process is a circular opening in the lesser wing of the sphenoid (the **optic canal**), through which the ophthalmic artery and optic nerve [II] pass as they exit the cranial cavity to enter the orbit. The optic canals are usually included in the middle cranial fossa.

## Middle cranial fossa

The middle cranial fossa consists of parts of the sphenoid and temporal bones (Fig. 8.26).

The boundary between the anterior and middle cranial fossae in the midline is the anterior edge of the prechiasmatic sulcus, which is a smooth groove stretching between the optic canals across the body of the sphenoid.

The posterior boundaries of the middle cranial fossa are formed by the anterior surface, as high as the superior border, of the petrous part of the petromastoid part of the temporal bone.

### Sphenoid

The floor in the midline of the middle cranial fossa is elevated and formed by the body of the sphenoid. Lateral to this are large depressions formed on either side by the greater wing of the sphenoid and the squamous part of the temporal bone. These depressions contain the temporal lobes of the brain.

#### Sella turcica

Just posterior to the chiasmatic sulcus is the uniquely modified remainder of the body of the sphenoid (the **sella turcica**), which consists of a deep central area (the **hypophyseal fossa**) containing the pituitary gland with anterior and posterior vertical walls of bone (Fig. 8.26).

The anterior wall of the sella is vertical in position with its superior extent visible as a slight elevation (the **tuberculum sellae**) at the posterior edge of the chiasmatic sulcus.

Lateral projections from the corners of the tuberculum sellae (the **middle clinoid processes**) are sometimes evident.

The posterior wall of the sella turcica is the **dorsum sellae**, a large ridge of bone projecting upward and forward. At the top of this bony ridge the lateral edges contain rounded projections (the **posterior clinoid processes**), which are points of attachment, like the anterior clinoid processes, for the tentorium cerebelli.

#### Fissures and foramina

Lateral to each side of the body of the sphenoid, the floor of the middle cranial fossa is formed on either side by the greater wing of the sphenoid (Fig. 8.26).

A diagonal gap, the **superior orbital fissure**, separates the greater wing of the sphenoid from the lesser wing and is a major passageway between the middle cranial fossa and the orbit. Passing through the fissure are the

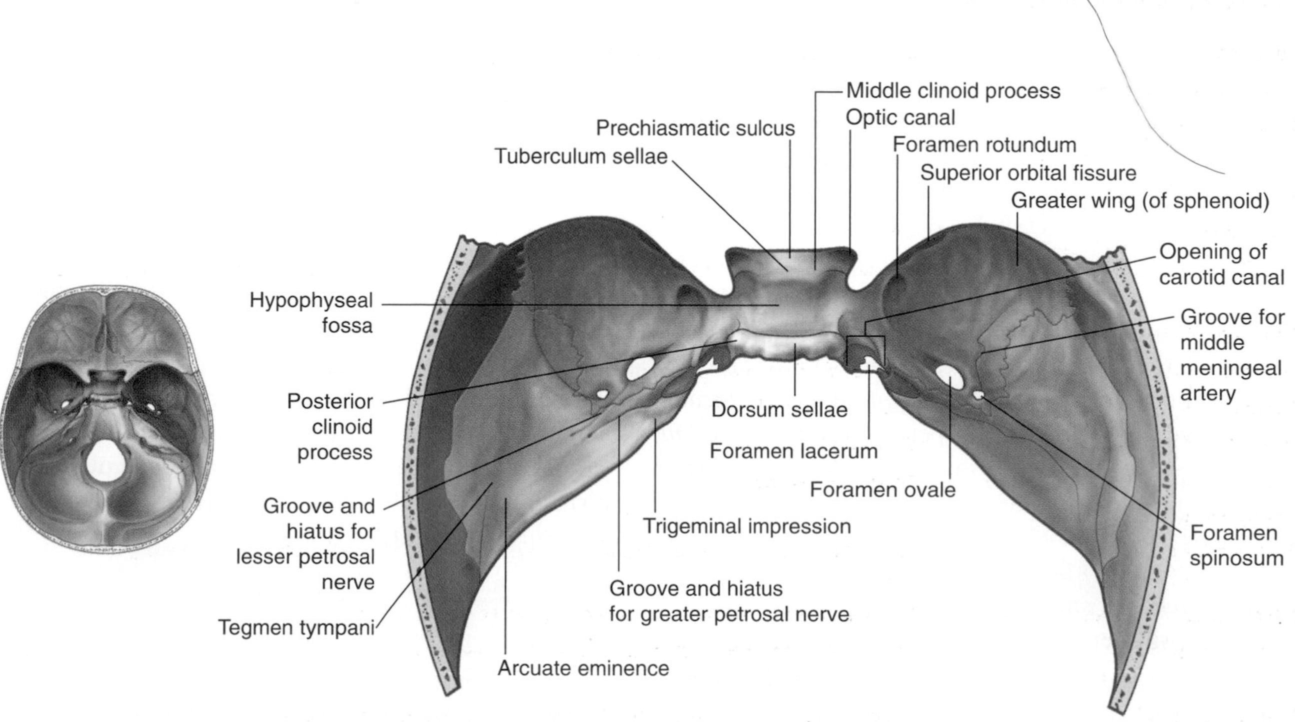


**Fig. 8.26** Middle cranial fossa.

oculomotor nerve [III], the trochlear nerve [IV], the ophthalmic nerve [$V_1$], the abducent nerve [VI], and ophthalmic veins.

Posterior to the medial end of the superior orbital fissure on the floor of the middle cranial fossa is a rounded foramen projecting in an anterior direction (the **foramen rotundum**), through which the maxillary nerve [$V_2$] passes from the middle cranial fossa to the pterygopalatine fossa.

Posterolateral to the **foramen rotundum** is a large oval opening (the **foramen ovale**), which allows structures to pass between the extracranial infratemporal fossa and the middle cranial fossa. The mandibular nerve [$V_3$], lesser petrosal nerve (carrying fibers from the tympanic plexus that originally came from the glossopharyngeal nerve [IX]) and, occasionally, a small vessel (the accessory middle meningeal artery), pass through this foramen.

Posterolateral from the foramen ovale is the small foramen spinosum (Fig. 8.26). This opening also connects the infratemporal fossa with the middle cranial fossa. The middle meningeal artery and its associated veins pass through this foramen and, once inside, the groove for the middle meningeal artery across the floor and lateral wall of the middle cranial fossa clearly marks their path.

Posteromedial to the foramen ovale is the **rounded intracranial opening** of the **carotid canal**. Directly inferior to this opening is an irregular foramen (the **foramen lacerum**) (Fig. 8.26). Clearly observed in the inferior view of the skull, the foramen lacerum is closed in life by a cartilaginous plug, and no structures pass through it completely.

### Temporal bone

The posterior boundary of the middle cranial fossa is formed by the anterior surface of the petrous part of the petromastoid part of the temporal bone.

Medially, there is a slight depression (**trigeminal impression**) in the anterior surface of the petrous part of the temporal bone (Fig. 8.26), which marks the location of the sensory ganglion for the trigeminal nerve [V].

Lateral to the trigeminal impression and on the anterior surface of the petrous part of the temporal bone is a small linear groove that passes in a superolateral direction and ends in a foramen (the **groove** and **hiatus for the greater petrosal nerve**). The greater petrosal nerve is a branch of the facial nerve [VII].

Anterolateral to the groove for the greater petrosal nerve is a second, smaller **groove** and **hiatus for the lesser petrosal nerve**, a branch from the tympanic plexus carrying fibers that originally came from the glossopharyngeal nerve [IX] (Fig. 8.26).

Above and lateral to the small openings for the greater and lesser petrosal nerves, near the superior ridge of the petrous part of the temporal bone, is a rounded protrusion of bone (the **arcuate eminence**) produced by the underlying anterior semicircular canal of the inner ear.

Just anterior and lateral to the arcuate eminence the anterior surface of the petrous part of the temporal bone is slightly depressed. This region is the **tegmen tympani**, and marks the thin bony roof of the middle ear cavity.

## Posterior cranial fossa

The posterior cranial fossa consists mostly of parts of the temporal and occipital bones, with small contributions from the sphenoid and parietal bones (Fig. 8.27). It is the largest and deepest of the three cranial fossae and contains the brainstem (midbrain, pons, and medulla) and the cerebellum.

### Boundaries

The anterior boundaries of the posterior cranial fossa in the midline are the dorsum sellae and the **clivus** (Fig. 8.27). The clivus is a slope of bone that extends upward from the foramen magnum. It is formed by contributions from the body of the sphenoid and from the basilar part of the occipital bone.

Laterally the anterior boundaries of the posterior cranial fossa are the superior border of the petrous part of the petromastoid part of the temporal bone.

Posteriorly the squamous part of the occipital bone to the level of the transverse groove is the major boundary, while laterally the petromastoid part of the temporal bone and small parts of the occipital and parietal bones border the fossa.

### Foramen magnum

Centrally, in the deepest part of the posterior cranial fossa, is the largest foramen in the skull, the foramen magnum. It is surrounded by the basilar part of the occipital bone anteriorly, the lateral parts of the occipital bone on either side, and the squamous part of the occipital bone posteriorly.

The spinal cord passes superiorly through the foramen magnum to continue as the brainstem.

Also passing through the foramen magnum are the vertebral arteries, the meninges, and the spinal roots of the accessory nerve [XI].

### Grooves and foramina

The clivus slopes upward from the foramen magnum. Lateral to the clivus is a **groove for the inferior petrosal sinus** between the basilar part of the occipital bone and the petrous part of the petromastoid part of the temporal bone (Fig. 8.27).

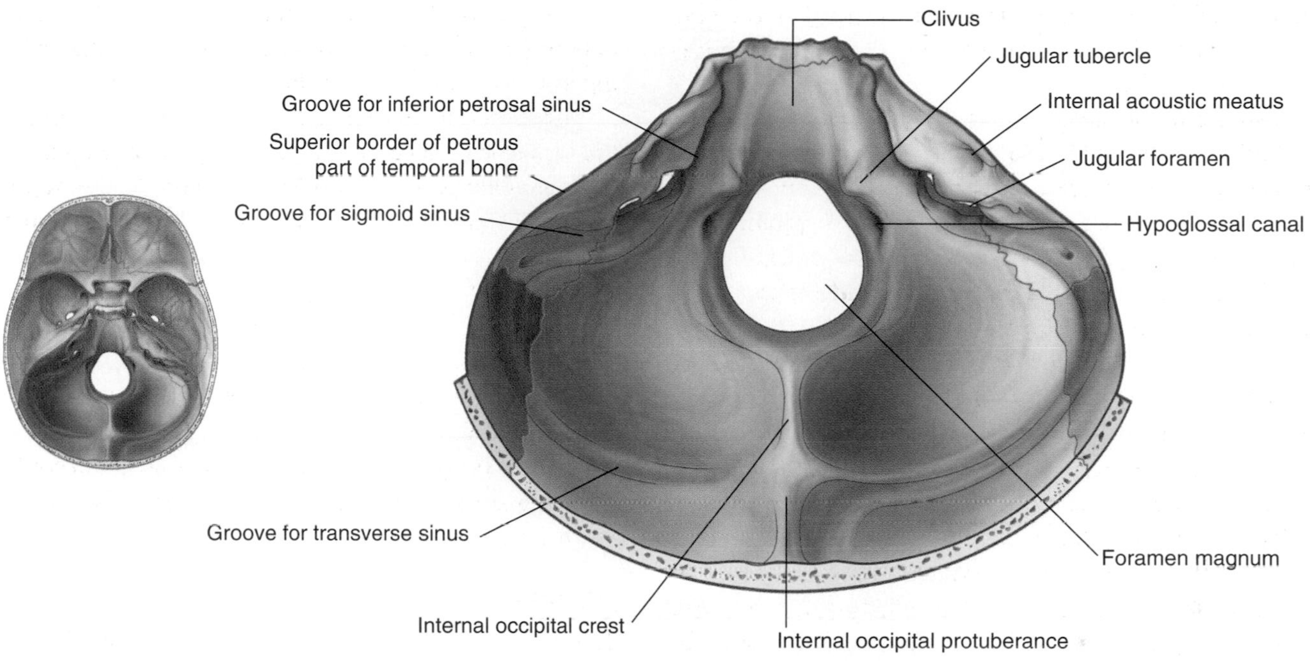


**Fig. 8.27** Posterior cranial fossa.

Laterally, across the upper half of the posterior surface of the petrous part of the temporal bone, is an oval foramen (the **internal acoustic meatus**). The facial [VII] and vestibulocochlear [VIII] nerves, and the labyrinthine artery pass through it.

Inferior to the internal acoustic meatus the temporal bone is separated from the occipital bone by the large jugular foramen (Fig. 8.27). Leading to this foramen from the medial side is the groove for the inferior petrosal sinus, and from the lateral side the **groove for the sigmoid sinus**.

The sigmoid sinus passes into the jugular foramen, and is continuous with the internal jugular vein, while the inferior petrosal sinus empties into the internal jugular vein in the area of the jugular foramen.

Also passing through the jugular foramen are the glossopharyngeal nerve [IX], the vagus nerve [X], and the accessory nerve [XI].

Medial to the jugular foramen is a large rounded mound of the occipital bone (the **jugular tubercle**). Just inferior to this, and superior to the foramen magnum, is the **hypoglossal canal**, through which the hypoglossal nerve [XII] leaves the posterior cranial fossa, and a meningeal branch of the ascending pharyngeal artery enters the posterior cranial fossa.

Just posterolateral to the hypoglossal canal is the small **condylar canal** that, when present, transmits an emissary vein.

### Squamous part of the occipital bone

The squamous part of the occipital bone has several prominent features (Fig. 8.27):

- Running upward in the midline from the foramen magnum is the internal occipital crest.
- On either side of the internal occipital crest, the floor of the posterior cranial fossa is concave to accommodate the cerebellar hemispheres.
- The internal occipital crest ends superiorly in a bony prominence (the internal occipital protuberance).
- Extending laterally from the internal occipital protuberance are grooves produced by the transverse sinuses, which continue laterally, eventually joining a groove for each sigmoid sinus—each of these grooves then turns inferiorly toward the jugular foramina.

The transverse and sigmoid sinuses are intradural venous sinuses.

### Foramina and fissures through which major structures enter and leave the cranial cavity

Foramina and fissures through which major structures pass between the cranial cavity and other regions are summarized in Fig. 8.28.

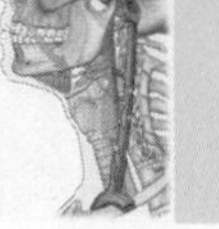

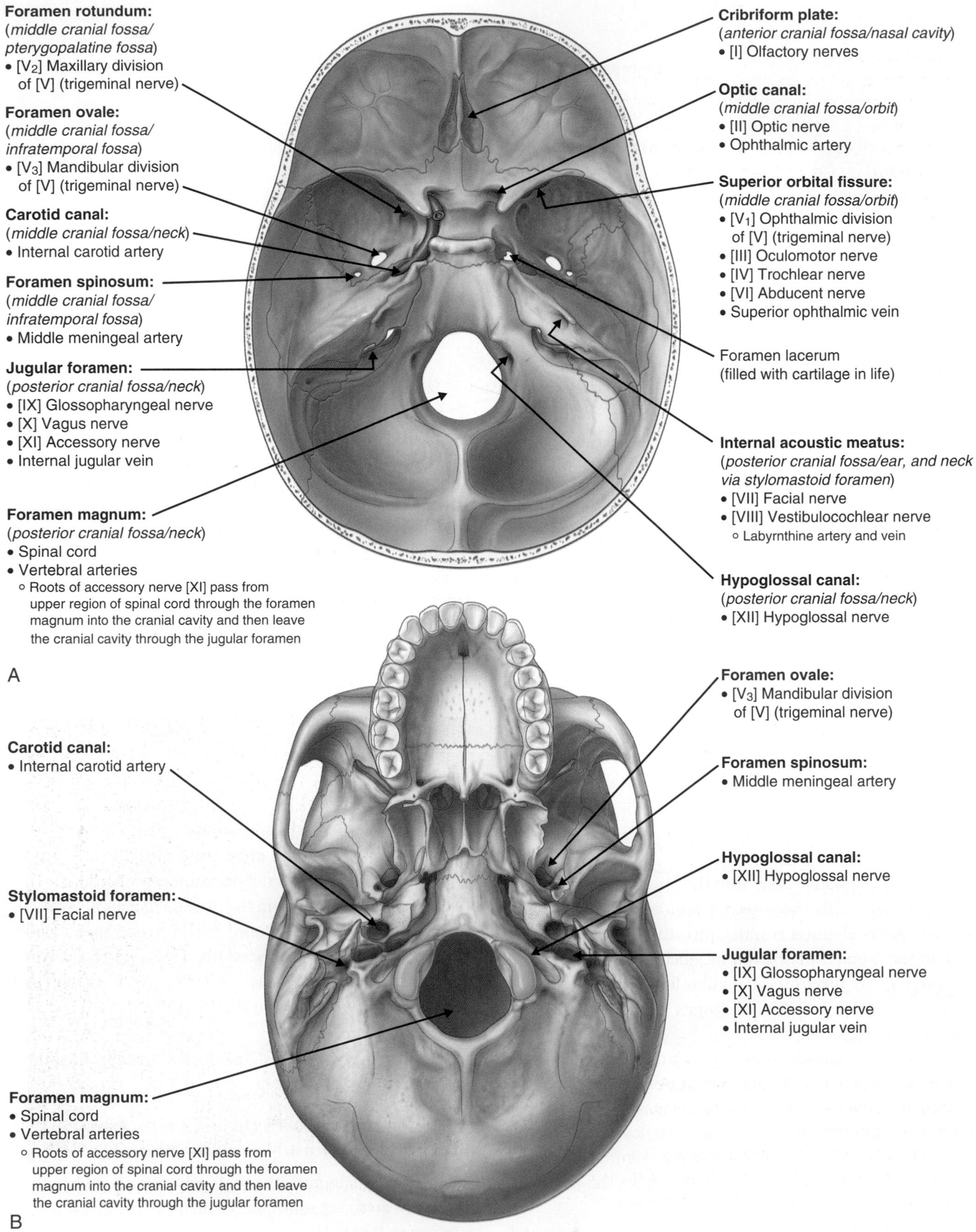

**Fig. 8.28** Summary of foramina and fissures through which major structures enter and leave the cranial cavity. **A.** Floor of cranial cavity. Also indicated are the regions between which each foramen or fissure communicates. **B.** Inferior aspect of cranium.

## In the clinic

**Craniosynostosis**

Some babies can be born with ossified fusion (synostosis) of one or more of the cranial sutures. This can result in an irregular head shape because the pattern and direction of skull growth are altered. In the majority of cases the cause is unknown, and in a minority of cases it may be caused by a genetic syndrome.

## In the clinic

**Medical imaging of the head**

***Radiography***

Until recently, the standard method of imaging the head was plain radiography. The radiographs are taken in three standard projections—the posteroanterior view, the lateral view, and the Towne's view (anteroposterior [AP] axial—head in anatomical position). Additional views are obtained to assess the foramina at the base of the skull and the facial bones. Currently, skull radiographs are used in cases of trauma, but such use is declining. Skull fractures are relatively easily detected (Fig. 8.29). The patient is assessed and treatment is based upon the underlying neurological or potential neurological complications.

***Computed tomography***

Since the development of computed tomography (CT), cerebral CT has become the "workhorse" of neuroradiological examination. It is ideally used for head injury because the brain and its coverings can be easily and quickly examined and blood is easily detected. By altering the mathematical algorithm of the data set the bones can also be demonstrated.

With intravenous contrast, CT angiography can be used to demonstrate the position and the size of an intracerebral aneurysm before endovascular treatment.

***Magnetic resonance imaging***

Magnetic resonance imaging (MRI) is unsurpassed by other imaging techniques in its ability for contrast resolution. The brain and its coverings, cerebrospinal fluid (CSF), and vertebral column can be easily and quickly examined. Newer imaging sequences permit CSF suppression to define periventricular lesions.

Magnetic resonance angiography has been extremely useful in determining the completeness of the intracranial vasculature (circle of Willis), which is necessary in some surgical conditions.

MRI is also a powerful tool in the assessment of carotid stenosis.

***Ultrasonography***

It is now possible to carry out intracranial Doppler studies, which enable a surgeon to detect whether a patient is experiencing cerebral embolization from a carotid plaque.

Extracranial ultrasound is extremely important in tumor staging and in assessing neck masses and the carotid bifurcation (Fig. 8.30).

Ultrasound is useful in children because they have an acoustic window through the fontanelles.

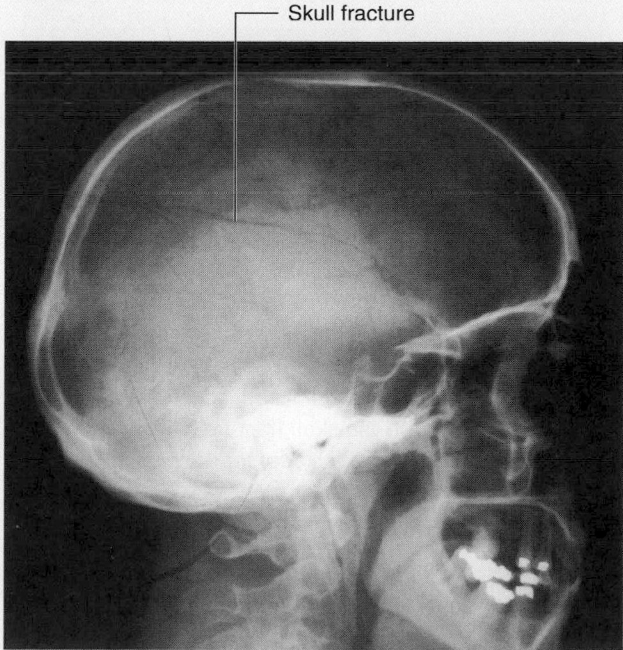


**Fig. 8.29** Skull fracture seen on a skull radiograph (patient in supine position).

*(continues)*

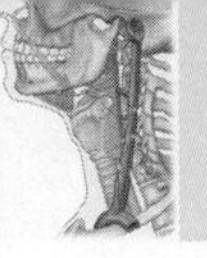



## In the clinic—cont'd

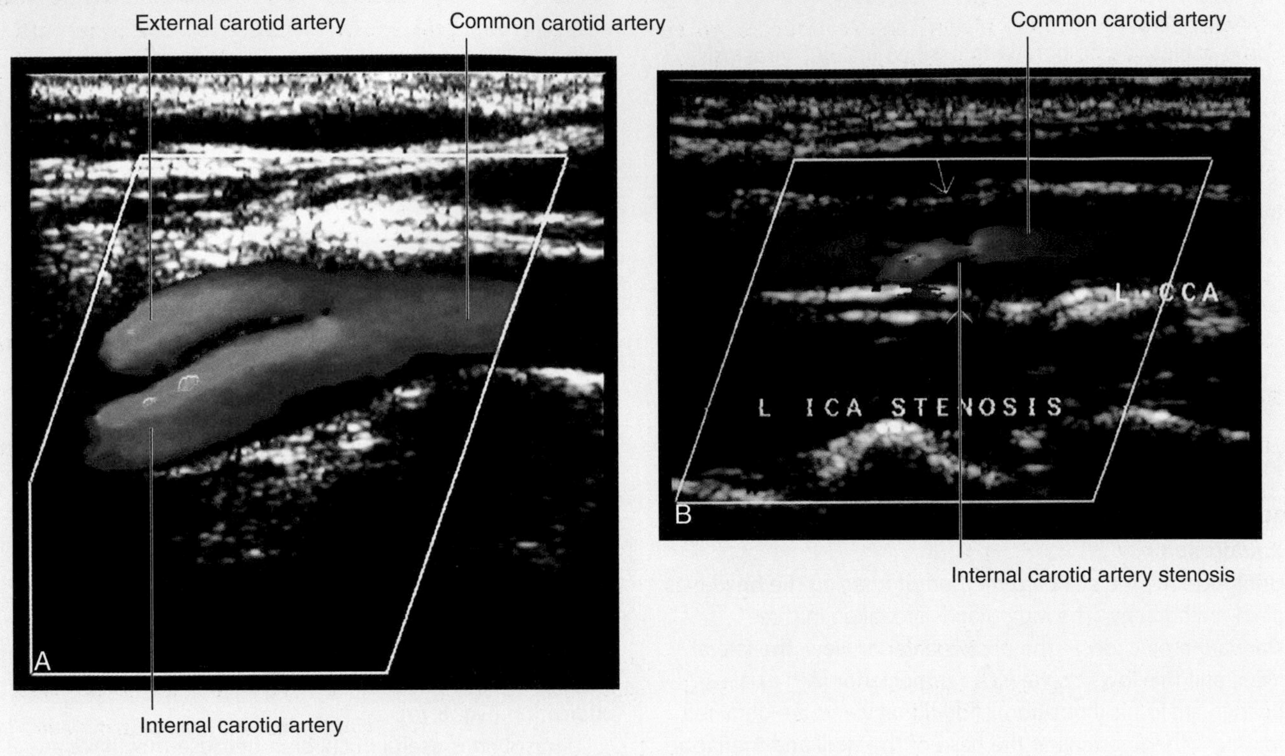


**Fig. 8.30** Ultrasound scans. **A.** Normal carotid bifurcation. **B.** Internal carotid artery stenosis.

## In the clinic

### Fractures of the skull vault

The skull vault is a remarkably strong structure because it protects our most vital organ, the brain. The shape of the skull vault is of critical importance and its biomechanics prevent fracture. From a clinical standpoint skull fractures alert clinicians to the nature and force of an injury and potential complications. The fracture itself is usually of little consequence (unlike, say, a fracture of the tibia). Of key importance is the need to minimize the extent of primary brain injury and to treat potential secondary complications, rather than focusing on the skull fracture. Skull fractures that have particular significance include depressed skull fractures, compound fractures, and pterion fractures.

#### *Depressed skull fractures*

In a depressed skull fracture a bony fragment is depressed below the normal skull convexity. This may lead to secondary arterial and venous damage with hematoma formation. A primary brain injury can also result from this type of fracture.

#### *Compound fractures*

In a compound fracture there is a fracture of the bone together with a breach of the skin, which may allow an infection to enter. Typically these fractures are associated with scalp lacerations and can usually be treated with antibiotics.

Important complications of compound fractures include meningitis, which may be fatal.

A more subtle type of compound fracture involves fractures across the sinuses. These may not be appreciated on first inspection, but are an important potential cause of morbidity and should be considered in patients who develop intracranial infections secondary to trauma.

#### *Pterion fractures*

The pterion is an important clinical point on the lateral aspect of the skull. At the pterion the frontal, parietal, greater wing of the sphenoid, and temporal bones come together. Importantly, deep to this structure is the middle meningeal artery. An injury to this point of the skull is extremely serious because damage to this vessel may produce a significant extradural hematoma, which can be fatal.

## MENINGES

The brain, as well as the spinal cord, is surrounded by three layers of membranes (the **meninges**, Fig. 8.31A)—a tough, outer layer (the **dura mater**), a delicate, middle layer (the **arachnoid mater**), and an inner layer firmly attached to the surface of the brain (the **pia mater**).

The cranial meninges are continuous with, and similar to, the spinal meninges through the foramen magnum, with one important distinction—the cranial dura mater consists of two layers, and only one of these is continuous through the foramen magnum (Fig. 8.31B).

### Cranial dura mater

The cranial dura mater is a thick, tough, outer covering of the brain. It consists of an outer periosteal layer and an inner meningeal layer (Fig. 8.31A):

- The outer **periosteal layer** is firmly attached to the skull, is the periosteum of the cranial cavity, contains the meningeal arteries, and is continuous with the periosteum on the outer surface of the skull at the foramen magnum and other intracranial foramina (Fig. 8.31B).
- The inner **meningeal layer** is in close contact with the arachnoid mater and is continuous with the spinal dura mater through the foramen magnum.

The two layers of dura separate from each other at numerous locations to form two unique types of structures (Fig. 8.31A):

- dural partitions, which project inward and incompletely separate parts of the brain, and
- intracranial venous structures.

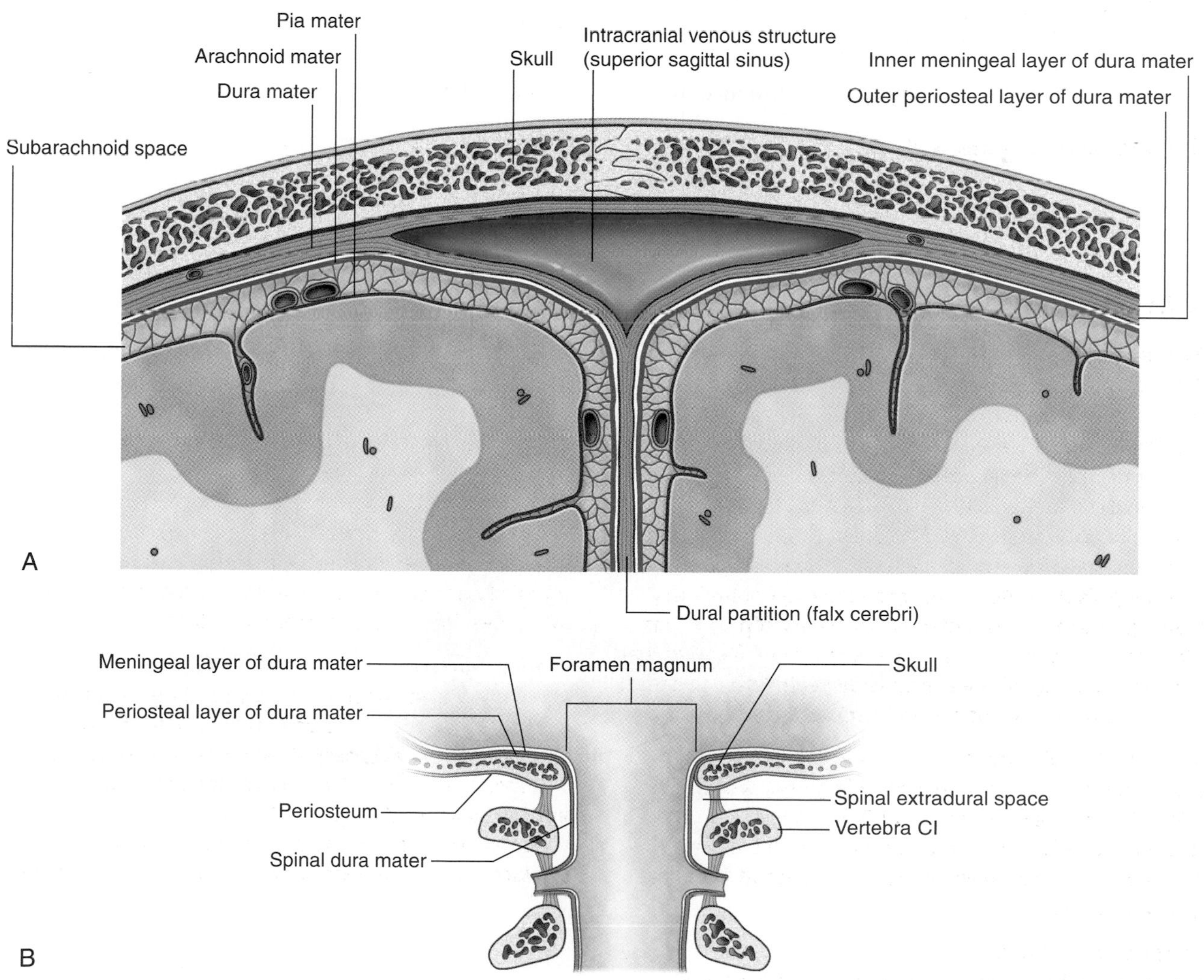


**Fig. 8.31** Cranial meninges. **A.** Superior coronal view. **B.** Continuity with the spinal meninges.

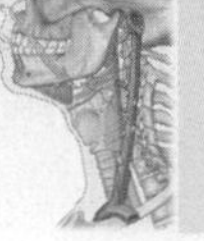

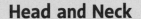

## Dural partitions

The dural partitions project into the cranial cavity and partially subdivide the cranial cavity. They include the falx cerebri, tentorium cerebelli, falx cerebelli, and diaphragma sellae.

### Falx cerebri

The falx cerebri (Fig. 8.32) is a crescent-shaped downward projection of meningeal dura mater from the dura lining the calva that passes between the two cerebral hemispheres. It is attached anteriorly to the crista galli of the ethmoid bone and frontal crest of the frontal bone. Posteriorly it is attached to and blends with the tentorium cerebelli.

### Tentorium cerebelli

The tentorium cerebelli (Fig. 8.32) is a horizontal projection of the meningeal dura mater that covers and separates the cerebellum in the posterior cranial fossa from the posterior parts of the cerebral hemispheres. It is attached posteriorly to the occipital bone along the grooves for the transverse sinuses. Laterally, it is attached to the superior border of the petrous part of the temporal bone, ending anteriorly at the anterior and posterior clinoid processes.

The anterior and medial borders of the tentorium cerebelli are free, forming an oval opening in the midline (the **tentorial notch**), through which the midbrain passes.

### Falx cerebelli

The falx cerebelli (Fig. 8.32) is a small midline projection of meningeal dura mater in the posterior cranial fossa. It is attached posteriorly to the internal occipital crest of the occipital bone and superiorly to the tentorium cerebelli. Its anterior edge is free and is between the two cerebellar hemispheres.

### Diaphragma sellae

The final dural projection is the diaphragma sellae (Fig. 8.32). This small horizontal shelf of meningeal dura mater covers the hypophyseal fossa in the sella turcica of the sphenoid bone. There is an opening in the center of the diaphragma sellae through which passes the **infundibulum**, connecting the pituitary gland with the base of the brain, and any accompanying blood vessels.

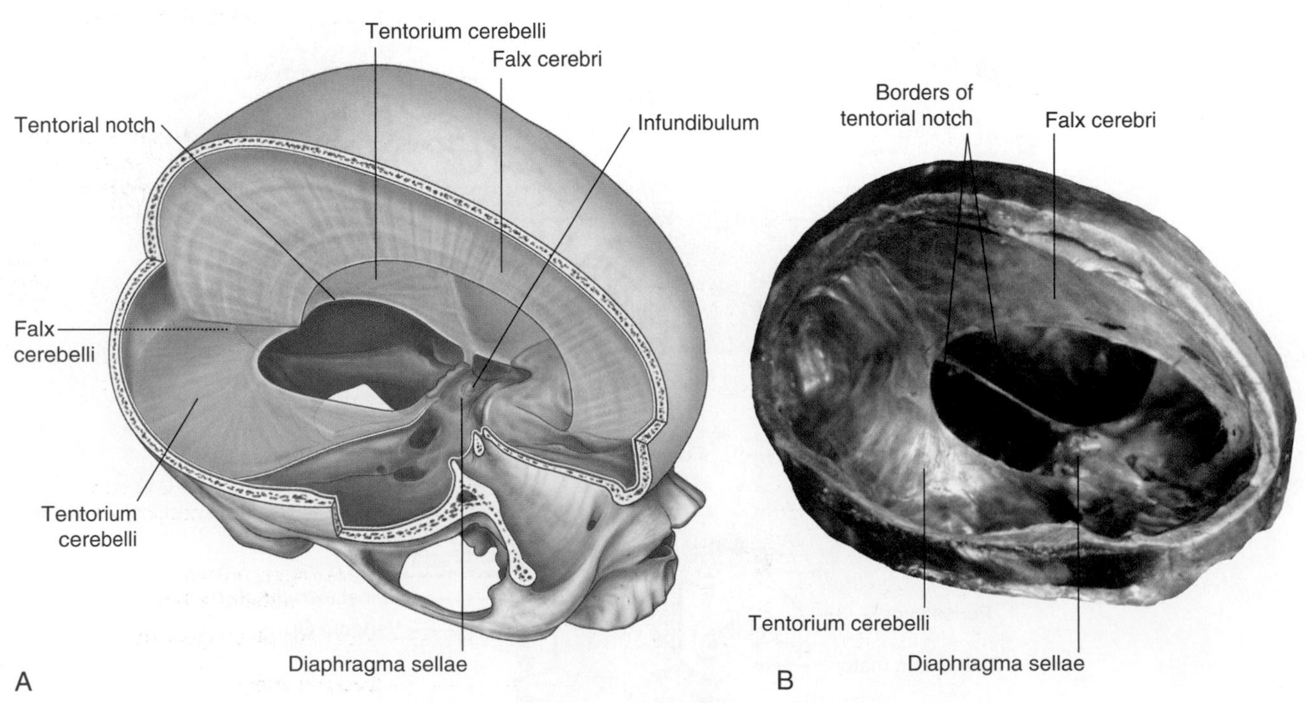


**Fig. 8.32** Dural partitions. **A.** Diagram. **B.** Dissection.

### Arterial supply

The arterial supply to the dura mater (Fig. 8.33) travels in the outer periosteal layer of the dura and consists of:

- **anterior meningeal arteries** in the anterior cranial fossa,
- the **middle** and **accessory meningeal arteries** in the middle cranial fossa, and
- the **posterior meningeal artery** and other meningeal branches in the posterior cranial fossa.

All are small arteries except for the middle meningeal artery, which is much larger and supplies the greatest part of the dura.

The anterior meningeal arteries are branches of the **ethmoidal arteries**.

The middle meningeal artery is a branch of the maxillary artery. It enters the middle cranial fossa through the foramen spinosum and divides into anterior and posterior branches:

- The anterior branch passes in an almost vertical direction to reach the vertex of the skull, crossing the pterion during its course.
- The posterior branch passes in a posterosuperior direction, supplying this region of the middle cranial fossa.

The accessory meningeal artery is usually a small branch of the maxillary artery that enters the middle cranial fossa through the foramen ovale and supplies areas medial to this foramen.

The posterior meningeal artery and other meningeal branches supplying the dura mater in the posterior cranial fossa come from several sources (Fig. 8.33):

- The posterior meningeal artery, the terminal branch of the **ascending pharyngeal artery**, enters the posterior cranial fossa through the jugular foramen.
- A meningeal branch from the ascending pharyngeal artery enters the posterior cranial fossa through the hypoglossal canal.
- Meningeal branches from the **occipital artery** enter the posterior cranial fossa through the jugular foramen and the mastoid foramen.
- A meningeal branch from the **vertebral artery** arises as the vertebral artery enters the posterior cranial fossa through the foramen magnum.

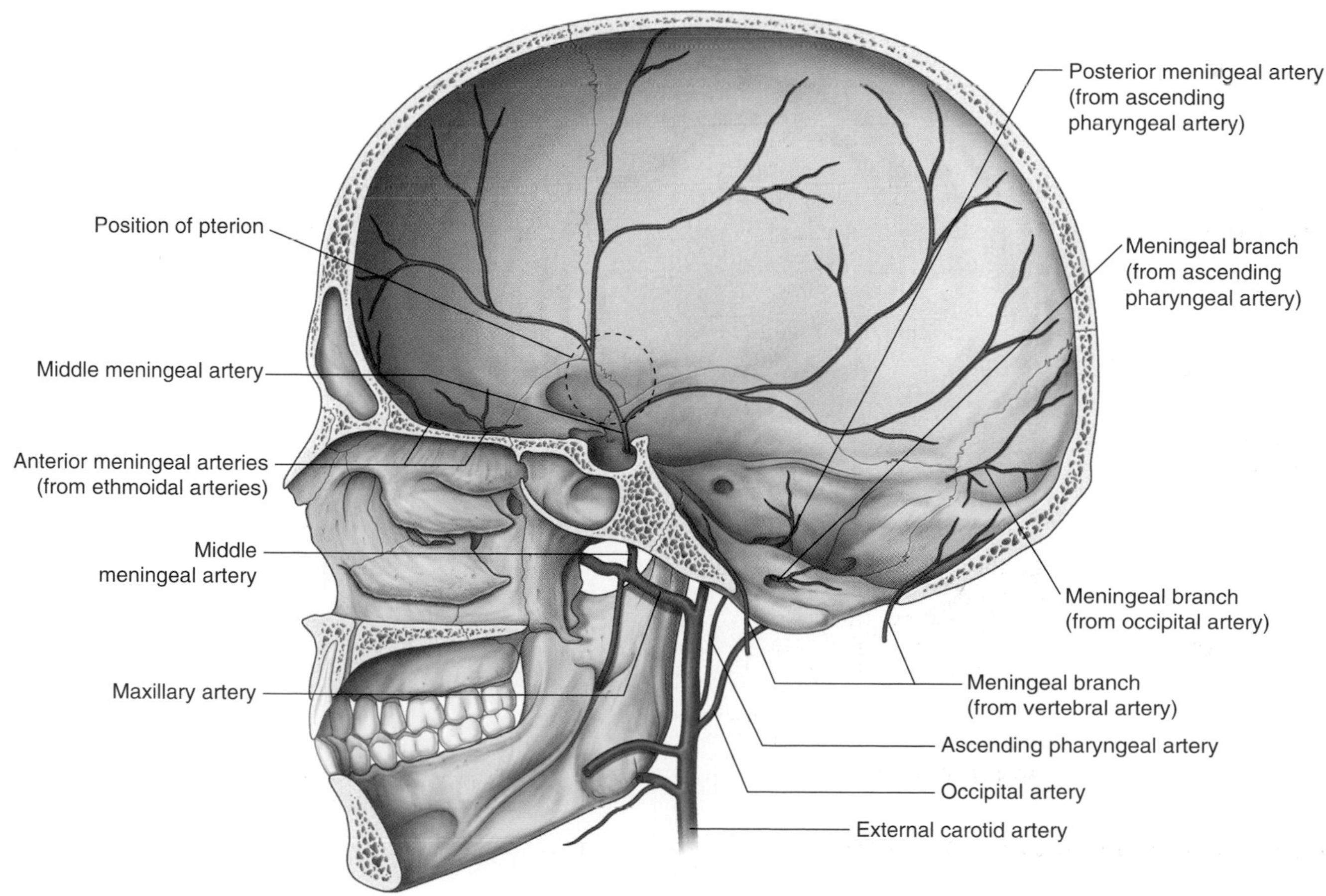


**Fig. 8.33** Dural arterial supply.

### Innervation

Innervation of the dura mater (Fig. 8.34) is by small meningeal branches of all three divisions of the trigeminal nerve [$V_1$, $V_2$, and $V_3$], the vagus nerve [X], and the first, second, and, sometimes, third cervical nerves. (Possible involvement of the glossopharyngeal [IX] and hypoglossal nerves [XII] in the posterior cranial fossa has also been reported.)

In the anterior cranial fossa meningeal branches from the ethmoidal nerves, which are branches of the ophthalmic nerve [$V_1$], supply the floor and the anterior part of the falx cerebri.

Additionally, a meningeal branch of the ophthalmic nerve [$V_1$] turns and runs posteriorly, supplying the tentorium cerebelli and the posterior part of the falx cerebri.

The middle cranial fossa is supplied medially by meningeal branches from the maxillary nerve [$V_2$] and laterally, along the distribution of the middle meningeal artery, by meningeal branches from the mandibular nerve [$V_3$].

The posterior cranial fossa is supplied by meningeal branches from the first, second, and, sometimes, third cervical nerves, which enter the fossa through the foramen magnum, the hypoglossal canal, and the jugular foramen. Meningeal branches of the vagus nerve [X] have also been described. (Possible contributions from the glossopharyngeal [IX] and hypoglossal [XII] nerves have also been reported.)

## Arachnoid mater

The arachnoid mater is a thin, avascular membrane that lines, but is not adherent to, the inner surface of the dura mater (Fig. 8.35). From its inner surface thin processes or trabeculae extend downward, cross the subarachnoid space, and become continuous with the pia mater.

Unlike the pia, the arachnoid does not enter the grooves or fissures of the brain, except for the longitudinal fissure between the two cerebral hemispheres.

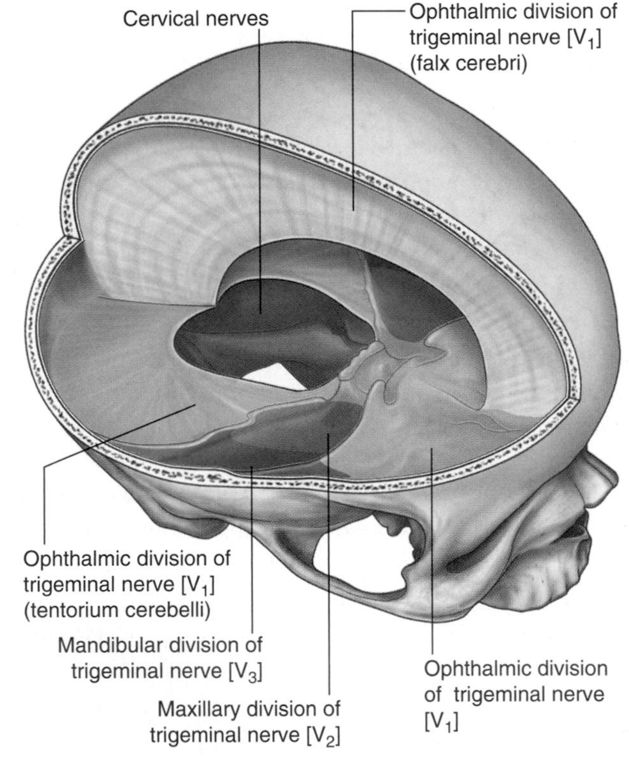


**Fig. 8.34** Dural innervation.

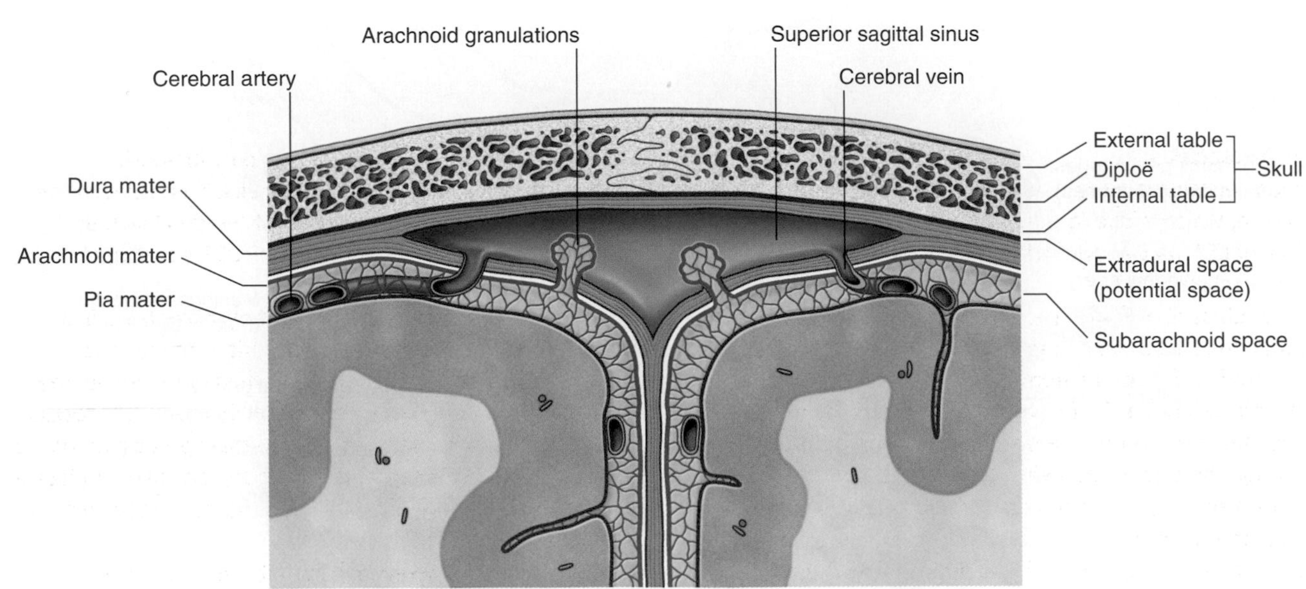


**Fig. 8.35** Arrangement of the meninges and spaces.

### Pia mater

The pia mater is a thin, delicate membrane that closely invests the surface of the brain (Fig. 8.35). It follows the contours of the brain, entering the grooves and fissures on its surface, and is closely applied to the roots of the cranial nerves at their origins.

### Arrangement of meninges and spaces

There is a unique arrangement of meninges coupled with real and potential spaces within the cranial cavity (Fig. 8.35).

A potential space is related to the dura mater, while a real space exists between the arachnoid mater and the pia mater.

#### Extradural space

The potential space between dura mater and bone is the **extradural space** (Fig. 8.35). Normally, the outer or periosteal layer of dura mater is firmly attached to the bones surrounding the cranial cavity.

This potential space between dura and bone can become a fluid-filled actual space when a traumatic event results in a vascular hemorrhage. Bleeding into the extradural space primarily due to rupture of a meningeal artery or less often from a torn dural venous sinus results in an extradural hematoma.

#### Subdural space

Anatomically, a true subdural space does not exist. Blood collecting in this region (subdural hematoma) due to injury represents a dissection of the dural border cell layer, which is the innermost lining of the meningeal dura. Dural border cells are flattened cells surrounded by extracellular spaces filled with amorphous material. While very infrequent, an occasional cell junction may be seen between these cells and the underlying arachnoid layer. Bleeding due to the tearing of a cerebral vein as it crosses through the dura to enter a dural venous sinus can result in a subdural hematoma.

#### Subarachnoid space

Deep to the arachnoid mater is the only normally occurring fluid-filled space associated with the meninges, the **subarachnoid space** (Fig. 8.35). It occurs because the arachnoid mater clings to the inner surface of the dura mater and does not follow the contour of the brain, while the pia mater, being against the surface of the brain, closely follows the grooves and fissures on the surface of the brain. The narrow subarachnoid space is therefore created between these two membranes (Fig. 8.35).

The subarachnoid space surrounds the brain and spinal cord and in certain locations it enlarges into expanded areas (subarachnoid **cisterns**). It contains cerebrospinal fluid (CSF) and blood vessels.

Cerebrospinal fluid is produced by the choroid plexus, primarily in the ventricles of the brain. It is a clear, colorless, cell-free fluid that circulates through the subarachnoid space surrounding the brain and spinal cord.

The CSF returns to the venous system through **arachnoid villi**. These project as clumps (**arachnoid granulations**) into the superior sagittal sinus, which is a dural venous sinus, and its lateral extensions, the **lateral lacunae** (Fig. 8.35).

### In the clinic

**Hydrocephalus**

Hydrocephalus is a dilation of the cerebral ventricular system, which is due to either an obstruction to the flow of CSF, an overproduction of CSF, or a failure of reabsorption of CSF.

Cerebrospinal fluid is secreted by the choroid plexus within the lateral, third, and fourth ventricles of the brain. As it is produced it passes from the lateral ventricles through the interventricular foramina (the foramina of Monro) to enter the third ventricle. From the third ventricle it passes through the cerebral aqueduct (aqueduct of Sylvius) into the fourth ventricle, and from here it passes into the subarachnoid space via the midline foramen or the two lateral foramina (foramen of Magendie and foramina of Luschka).

The CSF passes around the spinal cord inferiorly, envelops the brain superiorly, and is absorbed through the arachnoid granulations in the walls of the dural venous sinuses. In adults almost half a liter of CSF is produced per day.

In adults the commonest cause of hydrocephalus is an interruption of the normal CSF absorption through the arachnoid granulations. This occurs when blood enters the subarachnoid space after subarachnoid hemorrhage, passes over the brain, and interferes with normal CSF absorption. To prevent severe hydrocephalus it may be necessary to place a small catheter through the brain into the ventricular system to relieve the pressure.

Other causes of hydrocephalus include congenital obstruction of the aqueduct of Sylvius and a variety of

*(continues)*

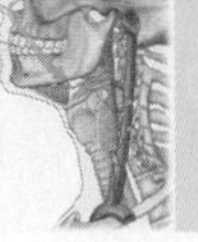



## In the clinic—cont'd

tumors (e.g., a midbrain tumor), where the mass obstructs the aqueduct. Rare causes include choroid plexus tumors that secrete CSF.

In children, hydrocephalus is always dramatic in its later stages. The hydrocephalus increases the size and dimensions of the ventricle, and as a result the brain enlarges. Because the skull sutures are not fused, the head expands. Cranial enlargement in utero may make a vaginal delivery impossible, and delivery then has to be by caesarean section.

Both CT and MRI enable a radiologist to determine the site of obstruction and in most cases the cause of the obstruction. A distinction must be made between ventricular enlargement due to hydrocephalus and that due to a variety of other causes (e.g., cerebral atrophy).

## In the clinic

### Cerebrospinal fluid leak

Leakage of CSF from the subarachnoid space may occur after any procedure in and around the brain, spinal cord, and meningeal membranes. These procedures include lumbar spine surgery, epidural injection, and CSF aspiration.

In "cerebrospinal fluid leak" syndrome, CSF leaks out of the subarachnoid space and through the dura mater for no apparent reason. The clinical consequences of this include dizziness, nausea, fatigue, and a metallic taste in the mouth. Other effects also include facial nerve weakness and double vision.

## In the clinic

### Meningitis

Meningitis is a rare infection of the leptomeninges (the **leptomeninges** are a combination of the arachnoid mater and the pia mater). Infection of the meninges typically occurs via a blood-borne route, though in some cases it may be by direct spread (e.g., trauma) or from the nasal cavities through the cribriform plate in the ethmoid bone.

Certain types of bacterial inflammation of the meninges are so virulent that overwhelming inflammation and sepsis with cerebral irritation can cause the patient to rapidly pass into a coma and die.

Meningitis is usually treatable with antibiotics.

Certain types of bacteria that produce meningitis produce other effects; for example, subcutaneous hemorrhage (ecchymoses) is a feature of meningococcal meningitis.

The typical history of meningitis is nonspecific at first. The patient may have mild headache, fever, drowsiness, and nausea. As the infection progresses, photophobia (light intolerance) and ecchymosis may ensue. Straight leg raising causes marked neck pain and discomfort (Kernig's sign) and an emergency hospital admission is warranted.

Immediate treatment consists of very-high-dose intravenous antibiotics and supportive management.

## In the clinic

### Brain tumors

Determination of the anatomical structure from which a tumor arises is of the utmost importance, particularly when it arises within the cranial vault. Misinterpretation of the location of a lesion and its site of origin may have devastating consequences for the patient.

When assessing any lesion in the brain, it is important to define whether it is intra-axial (within the brain) or extra-axial (outside the brain).

Typical extra-axial tumors include meningiomas (tumors of the meninges) and acoustic neuromas. Meningiomas typically arise from the meninges, with preferred sites including regions at and around the falx cerebri, the free edge of the tentorium cerebelli, and the anterior margin of the middle cranial fossa. Acoustic neuromas are typically at and around the vestibulocochlear nerve [VIII] and in the cerebellopontine angle.

Intra-axial lesions are either primary or secondary. By far the commonest type are the secondary brain lesions, which in most cases are metastatic tumor deposits.

Metastatic tumor lesions are typically found in patients with either breast carcinoma or lung carcinoma, though many other malignancies can give rise to cerebral metastases.

Primary brain lesions are rare and range from benign tumors to extremely aggressive lesions with a poor prognosis. These tumors arise from the different cell lines and include gliomas, oligodendrocytomas, and choroid plexus tumors. Primary brain tumors may occur at any age, though there is a small peak incidence in the first few years of life followed by a later peak in early to middle age.

## BRAIN AND ITS BLOOD SUPPLY

### Brain

The brain is a component of the central nervous system.

During development the brain can be divided into five continuous parts (Figs. 8.36 and 8.37). From rostral (or cranial) to caudal they are:

- The **telencephalon** (**cerebrum**) becomes the large cerebral hemispheres. The surface of these hemispheres consists of elevations (gyri) and depressions (sulci), and the hemispheres are partially separated by a deep longitudinal fissure. The cerebrum fills the area of the cranial cavity above the tentorium cerebelli and is subdivided into lobes based on position.
- The **diencephalon**, which is hidden from view in the adult brain by the cerebral hemispheres, consists of the thalamus, hypothalamus, and other related structures, and classically is considered to be the most rostral part of the brainstem. (However, in common usage today, the term brainstem usually refers to the midbrain, pons, and medulla.)
- The **mesencephalon** (**midbrain**), which is the first part of the brainstem seen when an intact adult brain is examined, spans the junction between the middle and posterior cranial fossae.
- The **metencephalon**, which gives rise to the cerebellum (consisting of two lateral hemispheres and a midline part in the posterior cranial fossa below the tentorium cerebelli) and the pons (anterior to the cerebellum, and is a bulging part of the brainstem in the most anterior part of the posterior cranial fossa against the clivus and dorsum sellae).
- The **myelencephalon** (**medulla oblongata**), the caudalmost part of the brainstem, ends at the foramen magnum or the uppermost rootlets of the first cervical nerve and to which cranial nerves VI to XII are attached.

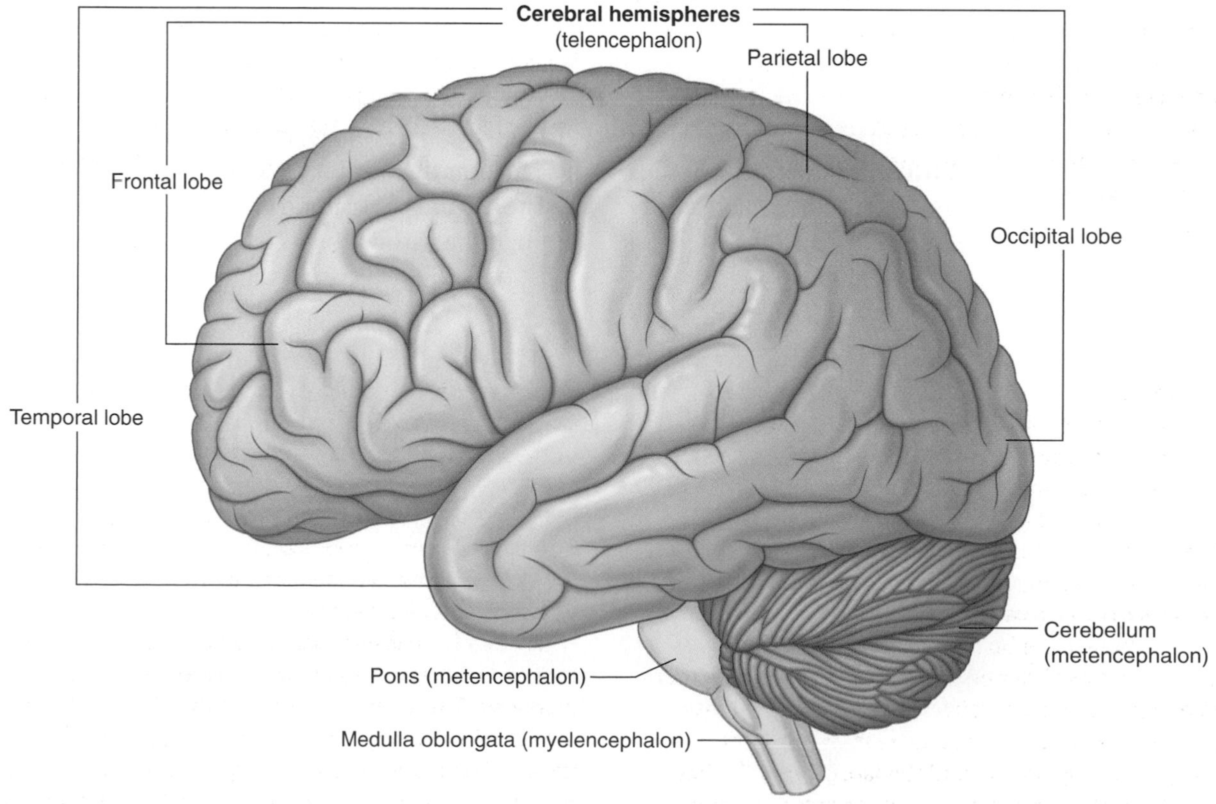


**Fig. 8.36** Lateral view of the brain.

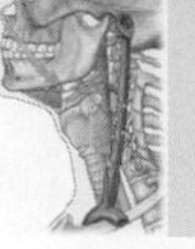


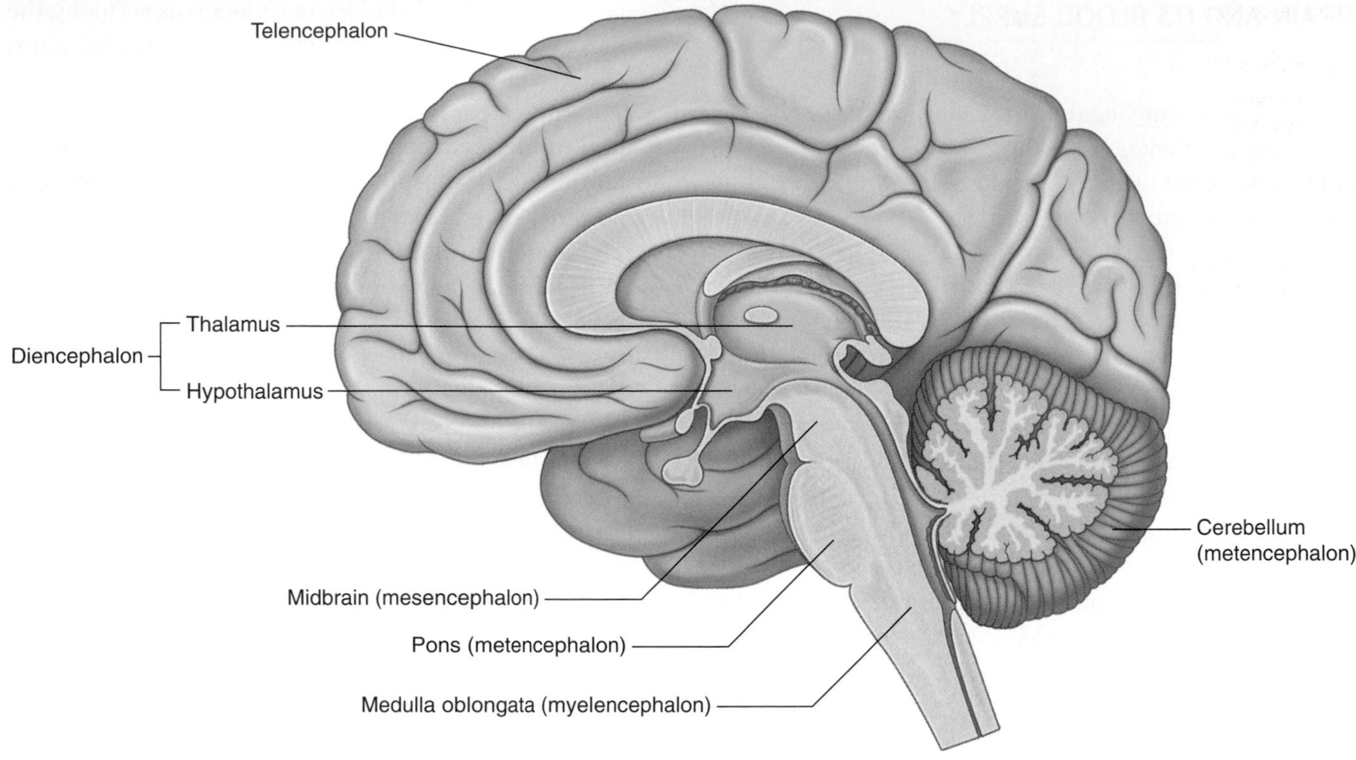


**Fig. 8.37** Sagittal section of the brain.

## Blood supply

The brain receives its arterial supply from two pairs of vessels, the **vertebral** and **internal carotid arteries** (Fig. 8.38), which are interconnected in the cranial cavity to produce a **cerebral arterial circle** (of Willis).

The two vertebral arteries enter the cranial cavity through the foramen magnum and just inferior to the pons fuse to form the **basilar artery**.

The two internal carotid arteries enter the cranial cavity through the carotid canals on either side.

**Fig. 8.38** Arterial supply to the brain. **A.** Diagram. **B.** Magnetic resonance angiogram showing normal carotid and vertebral arteries. **C.** Enhanced CT scan of carotid vessels.

Anterior communicating
Anterior cerebral
Cerebral arterial circle
Middle cerebral
Ophthalmic
Posterior communicating
Posterior cerebral
Basilar
Right internal carotid
Left internal carotid
Right common carotid
Left vertebral
Right vertebral
Right subclavian
Left subclavian
Brachiocephalic
Left common carotid
Aortic arch
A

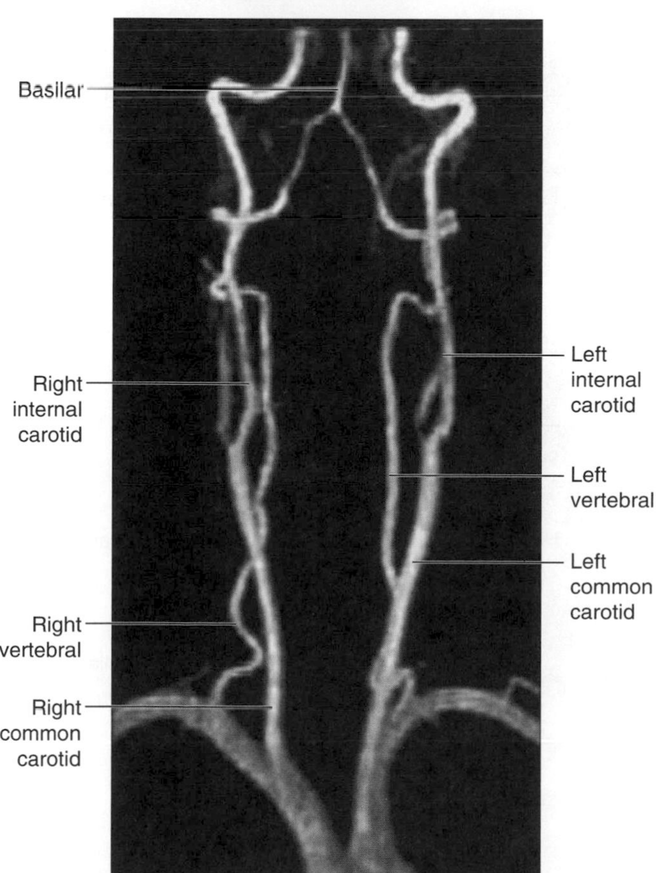


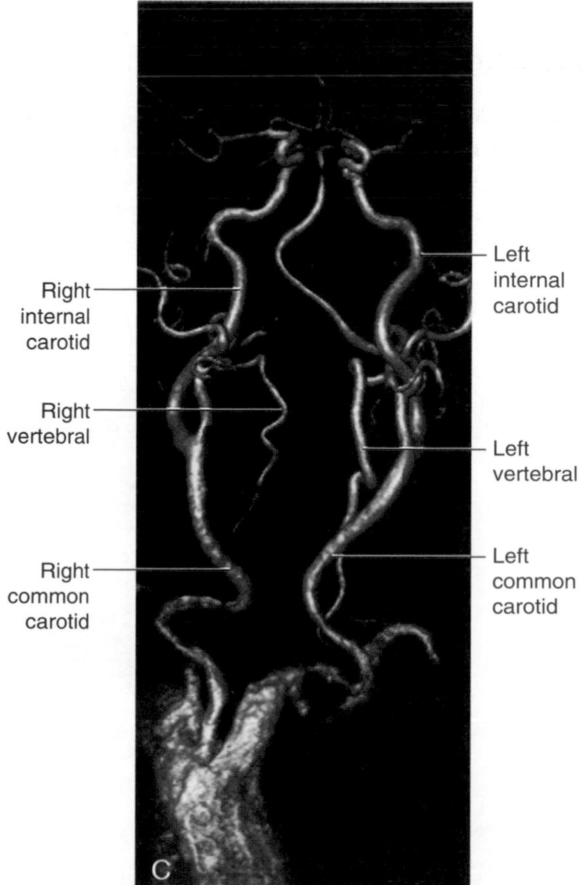

### Vertebral arteries

Each vertebral artery arises from the first part of each **subclavian artery** (Fig. 8.38) in the lower part of the neck, and passes superiorly through the foramen transversarium of the upper six cervical vertebrae. On entering the cranial cavity through the foramen magnum each vertebral artery gives off a small meningeal branch.

Continuing forward, the vertebral artery gives rise to three additional branches before joining with its companion vessel to form the basilar artery (Figs. 8.38 and 8.39):

- The first is a **posterior inferior cerebellar artery**.
- A second branch is the **posterior spinal artery**, which passes posteriorly around the medulla and then descends on the posterior surface of the spinal cord in the area of the attachment of the posterior roots—there are two posterior spinal arteries, one on each side (although the posterior spinal arteries can originate directly from the vertebral arteries, they more commonly branch from the posterior inferior cerebellar arteries).
- A third branch joins with its companion from the other side to form the single **anterior spinal artery**, which then descends in the anterior median fissure of the spinal cord.

The basilar artery travels in a rostral direction along the anterior aspect of the pons (Fig. 8.39). Its branches in a caudal to rostral direction include the **anterior inferior cerebellar arteries**, several small **pontine arteries**, and the **superior cerebellar arteries**. The basilar artery ends as a bifurcation, giving rise to two **posterior cerebral arteries**.

### Internal carotid arteries

The two internal carotid arteries arise as one of the two terminal branches of the common carotid arteries (Fig. 8.38). They proceed superiorly to the base of the skull where they enter the carotid canal.

Entering the cranial cavity each internal carotid artery gives off the **ophthalmic artery**, the **posterior communicating artery**, the **middle cerebral artery**, and the **anterior cerebral artery** (Fig. 8.39).

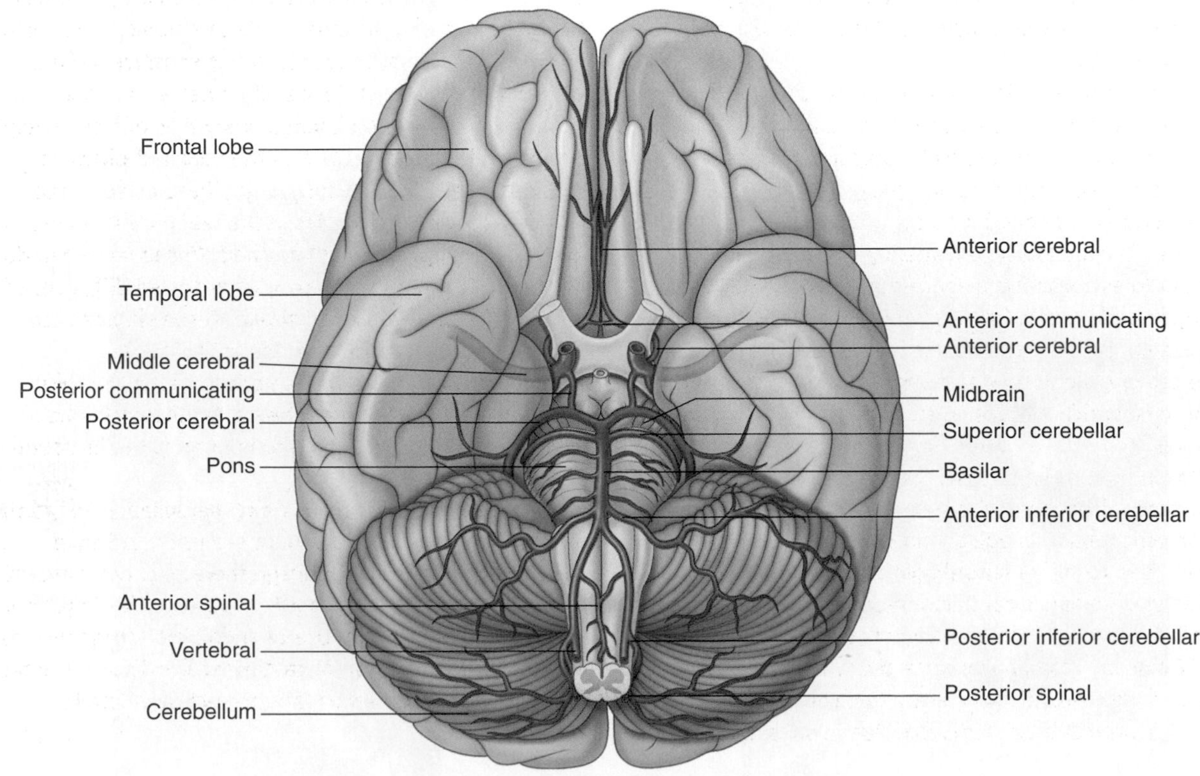


**Fig. 8.39** Arteries on the base of the brain.

### Cerebral arterial circle

The cerebral arterial circle (of Willis) is formed at the base of the brain by the interconnecting vertebrobasilar and internal carotid systems of vessels (Fig. 8.38). This anastomotic interconnection is accomplished by:

- an anterior communicating artery connecting the left and right anterior cerebral arteries to each other, and
- two posterior communicating arteries, one on each side, connecting the internal carotid artery with the posterior cerebral artery (Figs. 8.38 and 8.39).

## In the clinic

### Stroke

A stroke, or cerebrovascular accident (CVA), is defined as the interruption of blood flow to the brain or brainstem resulting in impaired neurological function lasting more than 24 hours. Neurological impairment resolving within 24 hours is known as a transient ischemic attack (TIA) or mini-stroke. Based on their etiology, strokes are broadly classified as either ischemic or hemorrhagic. Ischemic strokes are further divided into those caused by thrombotic or embolic phenomena. The latter is by far the commonest type of stroke and is often caused by emboli that originate from atherosclerotic plaques in the carotid arteries that migrate into and block smaller intracranial vessels. Hemorrhagic strokes are caused by rupture of blood vessels.

The risk factors for stroke are those of cardiovascular disease, such as diabetes, hypertension, and smoking. In younger patients underlying clotting disorders, use of oral contraceptives, and illicit substance abuse (such as cocaine) are additional causes.

The symptoms and signs of a stroke depend on the distribution of impaired brain perfusion. Common presentations include rapid-onset hemiparesis or hemisensory loss, visual field deficits, dysarthria, ataxia, and a decreased level of consciousness.

Stroke is a neurological emergency. It is therefore important to establish the diagnosis as early as possible so that urgent and potentially life-saving treatment can be administered. Potent thrombolytic (blood-thinning) drugs can restore cerebral blood flow and improved patient outcome if administered within 3 to 4.5 hours of onset of the patient's symptoms.

Following initial clinical history taking and neurological examination, all patients with suspected stroke should undergo urgent brain imaging with computed tomography (CT). This is to identify hemorrhagic strokes for which thrombolytic therapy is contraindicated and to exclude an alternative diagnosis such as malignancy. In ischemic stroke, early CT imaging may appear normal or can show a relatively darker area of low density that corresponds to the region of abnormal brain perfusion. Due to subsequent brain edema and swelling, the affected brain also loses its normal sulcal pattern (Fig. 8.40A). If thrombolysis is performed, a 24-hour follow-up CT scan is routinely carried out to evaluate for complications such as intracranial hemorrhage.

Additional diagnostic workup of stroke includes hematological and biochemical blood tests to identify causes such as hypoglycemia or underlying clotting disorders. A toxicology screen may be useful to identify substance intoxication, which can mimic stroke.

The full extent of neurological injury can be evaluated on subsequent magnetic resonance imaging (MRI) of the brain, which has better soft tissue resolution compared to CT. MRI is also useful for identifying strokes that may be too small to detect on a CT scan. MRI scans are produced by using complicated algorithms that create a series of images, also known as sequences. Various sequences can be obtained to assess different anatomical and physiological properties of the brain. A stroke, whether acute or chronic, will appear as a bright region on a sequence that is sensitive to fluid (T2 weighted) (Fig. 8.40B). To identify whether a stroke is acute, further sequences are obtained, known as diffusion-weighted imaging (DWI) (Fig. 8.40C) and the apparent diffusion coefficient (ADC) (Fig. 8.40D) map. These evaluate the diffusion of water molecules in the brain. If the region of abnormality appears bright on the DWI sequence and dark on the ADC map, this is known as restricted diffusion, which is compatible with an acute stroke. These changes can persist for up to a week after the initial insult.

Imaging of the carotid and vertebral arteries is also performed to assess for any treatable atherosclerotic changes and stenosis. This can be done with ultrasound, CT, or less frequently, MRI.

Management of a stroke is multidisciplinary. Supportive treatment to stabilize the patient is a priority. Stroke specialists, speech and language therapists, occupational therapists, and physiotherapists have key roles in patient rehabilitation. Long-term use of antiplatelet drugs such as aspirin and modification of cardiovascular disease risk factors are important in the secondary prevention of stroke.

*(continues)*

## In the clinic—cont'd

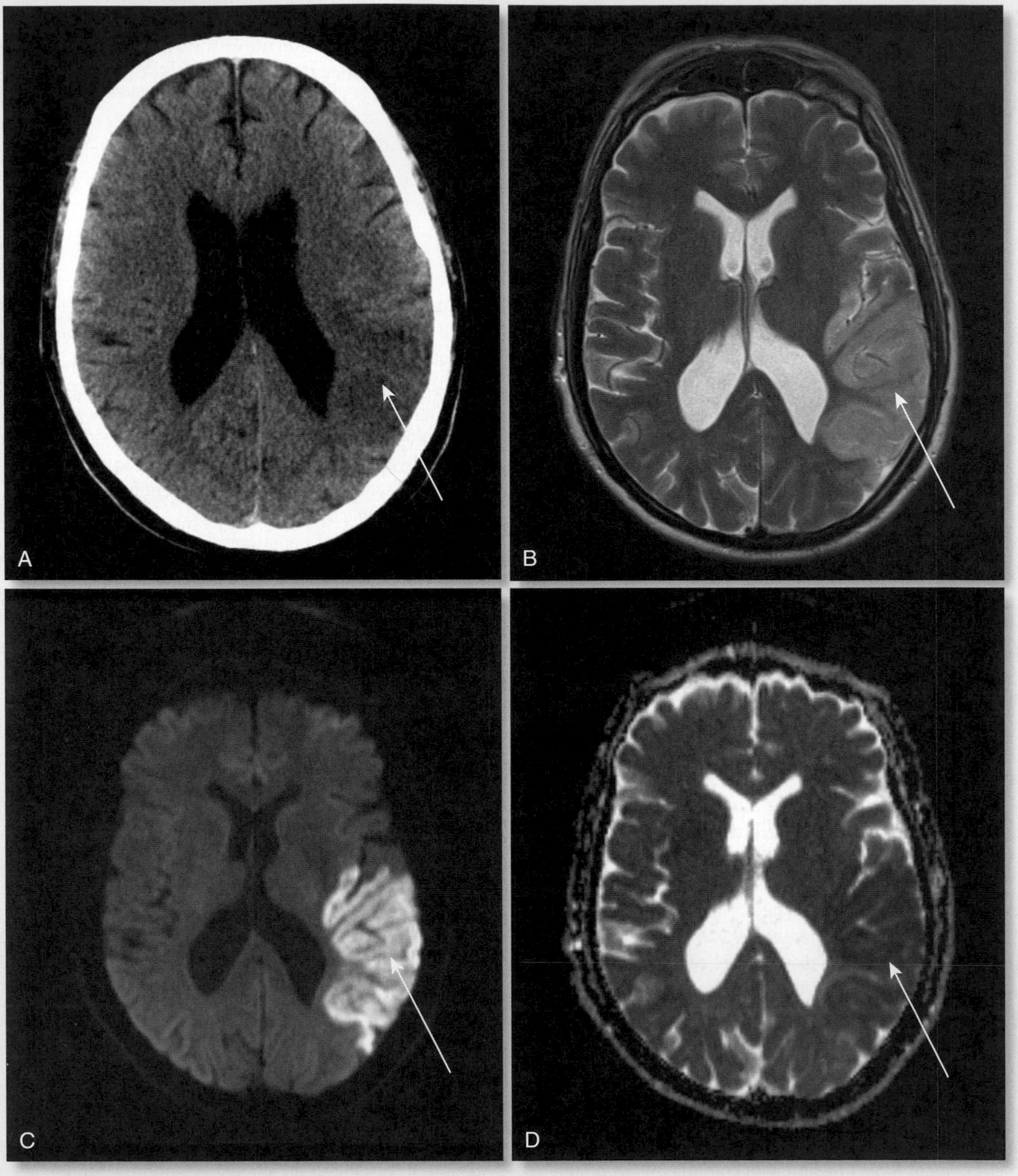

A

B

C

D

**Fig. 8.40** Different imaging modalities used to evaluate a stroke ***(arrows)***. **A.** CT scan. **B.** T2- weighted CT. **C.** Diffusion-weighted image (DWI). **D.** Apparent diffusion coefficient image (ADC).

## In the clinic

### Endarterectomy

Endarterectomy is a surgical procedure to remove atheromatous plaque from arteries.

Atheromatous plaques occur in the subendothelial layer of vessels and consist of lipid-laden macrophages and cholesterol debris. The developing plaque eventually accumulates fibrous connective tissue and calcifies. Plaque commonly occurs around vessel bifurcations, limiting blood flow, and may embolize to distal organs.

During endarterectomy, plaque is removed and the vessel reopened. In many instances a patch of material is sewn over the hole in the vessel, enabling improved flow and preventing narrowing from the suturing of the vessel.

## In the clinic

### Intracerebral aneurysms

Cerebral aneurysms arise from the vessels in and around the cerebral arterial circle (of Willis). They typically occur in and around the anterior communicating artery, the posterior communicating artery, the branches of the middle cerebral artery, the distal end of the basilar artery (Fig. 8.41), and the posterior inferior cerebellar artery.

As the aneurysms enlarge, they have a significant risk of rupture. Typically patients have no idea that there is anything wrong. As the aneurysm ruptures, the patient complains of a sudden-onset "thunderclap" headache that produces neck stiffness and may induce vomiting. In a number of patients death ensues, but many patients reach the hospital, where the diagnosis is established. An initial CT scan demonstrates blood within the subarachnoid space, and this may be associated with an intracerebral bleed. Further management usually includes cerebral angiography, which enables the radiologist to determine the site, size, and origin of the aneurysm.

Usually patients undergo complex surgery to ligate the neck of the aneurysm. More recently radiological intervention has superseded the management of some aneurysms in specific sites. This treatment involves cannulation of the femoral artery, and placement of a long catheter through the aorta into the carotid circulation and thence into the cerebral circulation. The tip of the catheter is placed within the aneurysm and is packed with fine microcoils (Fig. 8.42), which seals the rupture.

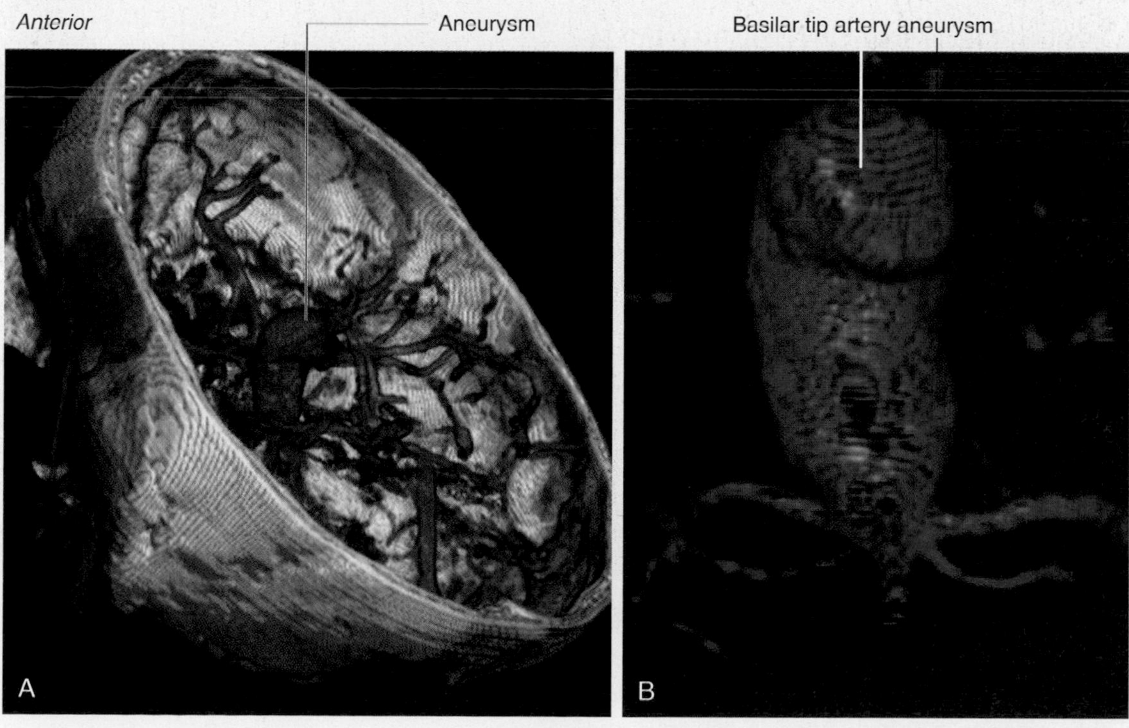


**Fig. 8.41** Basilar tip aneurysm. **A.** Three-dimensional cranial cutaway CT scan. **B.** Magnified view of aneurysm.

*(continues)*

## In the clinic—cont'd

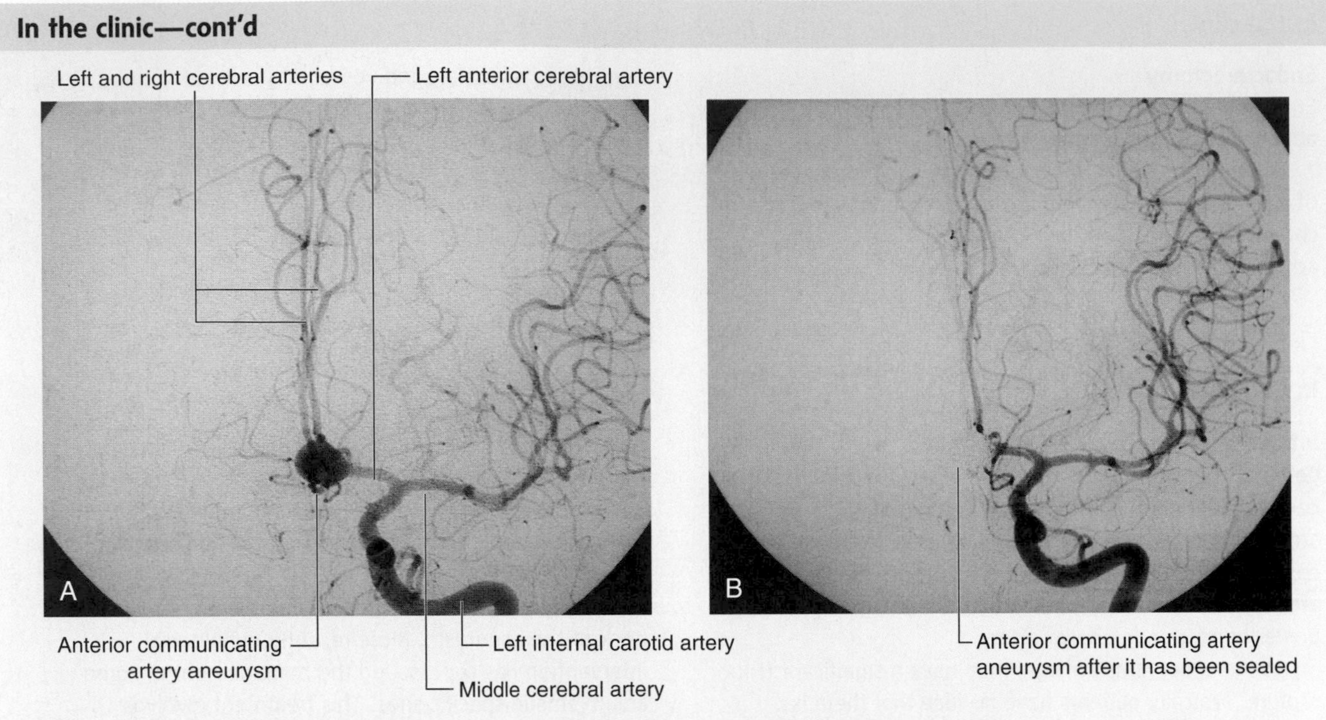


**Fig. 8.42** Anterior communicating aneurysm. **A.** Left carotid angiogram. **B.** Left carotid angiogram after embolization.

## Venous drainage

Venous drainage of the brain begins internally as networks of small venous channels lead to larger cerebral veins, cerebellar veins, and veins draining the brainstem, which eventually empty into **dural venous sinuses**. The dural venous sinuses are endothelial-lined spaces between the outer periosteal and the inner meningeal layers of the dura mater, and eventually lead to the **internal jugular veins**.

Also emptying into the dural venous sinuses are **diploic veins**, which run between the internal and external tables of compact bone in the roof of the cranial cavity, and **emissary veins**, which pass from outside the cranial cavity to the dural venous sinuses (Fig. 8.43).

The emissary veins are important clinically because they can be a conduit through which infections can enter the cranial cavity because they have no valves.

### Dural venous sinuses

The dural venous sinuses include the superior sagittal, inferior sagittal, straight, transverse, sigmoid, and occipital sinuses, the confluence of sinuses, and the cavernous, sphenoparietal, superior petrosal, inferior petrosal, and basilar sinuses (Fig. 8.44, Table 8.3).

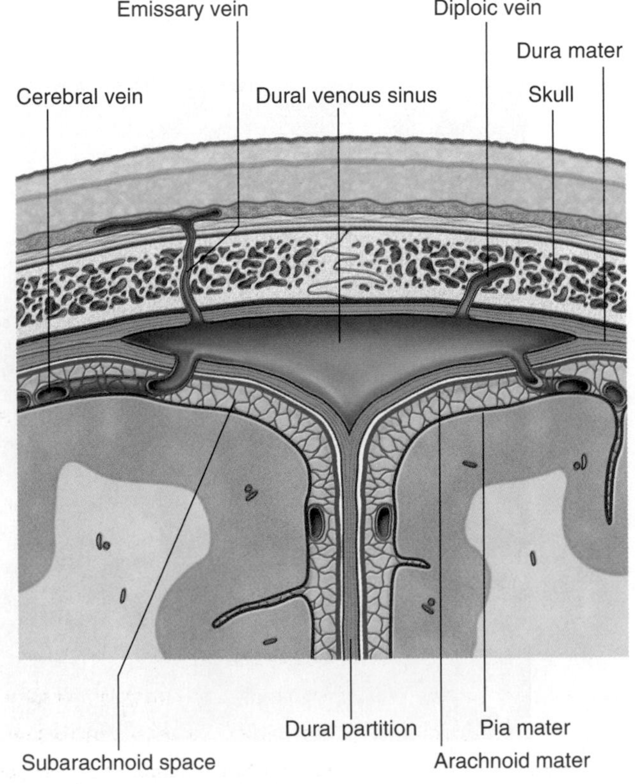


**Fig. 8.43** Dural venous sinuses.

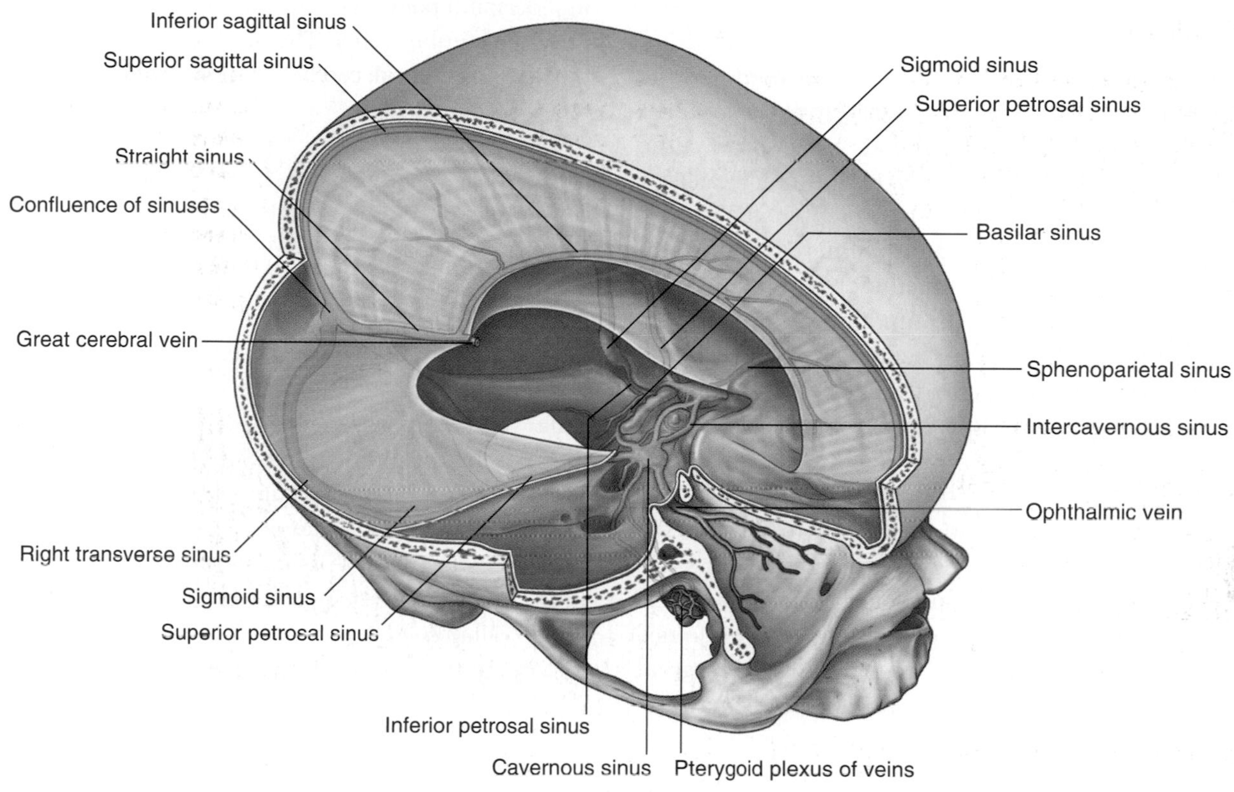


**Fig. 8.44** Veins, meninges, and dural venous sinuses.

**Table 8.3** Dural venous sinuses

| Dural sinus | Location | Receives |
|---|---|---|
| Superior sagittal | Superior border of falx cerebri | Superior cerebral, diploic, and emissary veins and CSF |
| Inferior sagittal | Inferior margin of falx cerebri | A few cerebral veins and veins from the falx cerebri |
| Straight | Junction of falx cerebri and tentorium cerebelli | Inferior sagittal sinus, great cerebral vein, posterior cerebral veins, superior cerebellar veins, and veins from the falx cerebri |
| Occipital | In falx cerebelli against occipital bone | Communicates inferiorly with vertebral plexus of veins |
| Confluence of sinuses | Dilated space at the internal occipital protuberance | Superior sagittal, straight, and occipital sinuses |
| Transverse (right and left) | Horizontal extensions from the confluence of sinuses along the posterior and lateral attachments of the tentorium cerebelli | Drainage from confluence of sinuses (right—transverse and usually superior sagittal sinuses; left—transverse and usually straight sinuses); also superior petrosal sinus, and inferior cerebral, cerebellar, diploic, and emissary veins |
| Sigmoid (right and left) | Continuation of transverse sinuses to internal jugular vein; groove of parietal, temporal, and occipital bones | Transverse sinuses, and cerebral, cerebellar, diploic, and emissary veins |
| Cavernous (paired) | Lateral aspect of body of sphenoid | Cerebral and ophthalmic veins, sphenoparietal sinuses, and emissary veins from pterygoid plexus of veins |
| Intercavernous | Crossing sella turcica | Interconnect cavernous sinuses |
| Sphenoparietal (paired) | Inferior surface of lesser wings of sphenoid | Diploic and meningeal veins |
| Superior petrosal (paired) | Superior margin of petrous part of temporal bone | Cavernous sinus, and cerebral and cerebellar veins |
| Inferior petrosal (paired) | Groove between petrous part of temporal bone and occipital bone ending in internal jugular vein | Cavernous sinus, cerebellar veins, and veins from the internal ear and brainstem |
| Basilar | Clivus, just posterior to sella turcica of sphenoid | Connect bilateral inferior petrosal sinuses and communicate with vertebral plexus of veins |

### Superior sagittal sinus

The superior sagittal sinus is in the superior border of the falx cerebri (Fig. 8.44). It begins anteriorly at the foramen cecum, where it may receive a small emissary vein from the nasal cavity, and ends posteriorly in the confluence of sinuses, usually bending to the right to empty into the right transverse sinus. The superior sagittal sinus communicates with lateral extensions (lateral lacunae) of the sinus containing numerous arachnoid granulations.

The superior sagittal sinus usually receives cerebral veins from the superior surface of the cerebral hemispheres, diploic and emissary veins, and veins from the falx cerebri.

### Inferior sagittal and straight sinuses

The inferior sagittal sinus is in the inferior margin of the falx cerebri (Fig. 8.44). It receives a few cerebral veins and veins from the falx cerebri, and ends posteriorly at the anterior edge of the tentorium cerebelli, where it is joined by the great cerebral vein and together with the great cerebral vein forms the straight sinus (Fig. 8.44).

The straight sinus continues posteriorly along the junction of the falx cerebri and the tentorium cerebelli and ends in the confluence of sinuses, usually bending to the left to empty into the left transverse sinus.

The straight sinus usually receives blood from the inferior sagittal sinus, cerebral veins (from the posterior part of the cerebral hemispheres), the great cerebral vein (draining deep areas of the cerebral hemispheres), superior cerebellar veins, and veins from the falx cerebri.

### Confluence of sinuses, transverse and sigmoid sinuses

The superior sagittal and straight sinuses, and the occipital sinus (in the falx cerebelli) empty into the confluence of sinuses, which is a dilated space at the internal occipital protuberance (Fig. 8.44) and is drained by the right and left transverse sinuses.

The paired transverse sinuses extend in horizontal directions from the confluence of sinuses where the tentorium cerebelli joins the lateral and posterior walls of the cranial cavity.

The right transverse sinus usually receives blood from the superior sagittal sinus and the left transverse sinus usually receives blood from the straight sinus.

The transverse sinuses also receive blood from the superior petrosal sinus, veins from the inferior parts of the cerebral hemispheres and the cerebellum, and diploic and emissary veins.

As the transverse sinuses leave the surface of the occipital bone, they become the sigmoid sinuses (Fig. 8.44), which turn inferiorly, grooving the parietal, temporal, and occipital bones, before ending at the beginning of the internal jugular veins. The sigmoid sinuses also receive blood from cerebral, cerebellar, diploic, and emissary veins.

### Cavernous sinuses

The paired cavernous sinuses are against the lateral aspect of the body of the sphenoid bone on either side of the sella turcica (Figs. 8.45 and 8.46). They are of great clinical importance because of their connections and the structures that pass through them.

The cavernous sinuses receive blood not only from cerebral veins but also from the ophthalmic veins (from the orbit) and emissary veins (from the pterygoid plexus of veins in the infratemporal fossa). These connections provide pathways for infections to pass from extracranial sites into intracranial locations. In addition, because structures pass through the cavernous sinuses and are located in the walls of these sinuses they are vulnerable to injury due to inflammation.

Structures passing through each cavernous sinus are:

- the internal carotid artery, and
- the abducent nerve [VI].

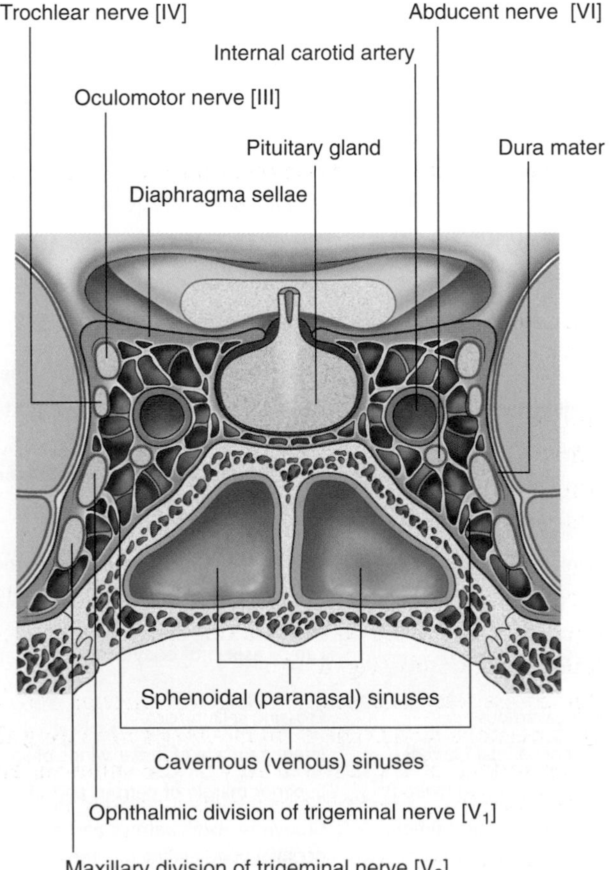


**Fig. 8.45** Cavernous sinuses.

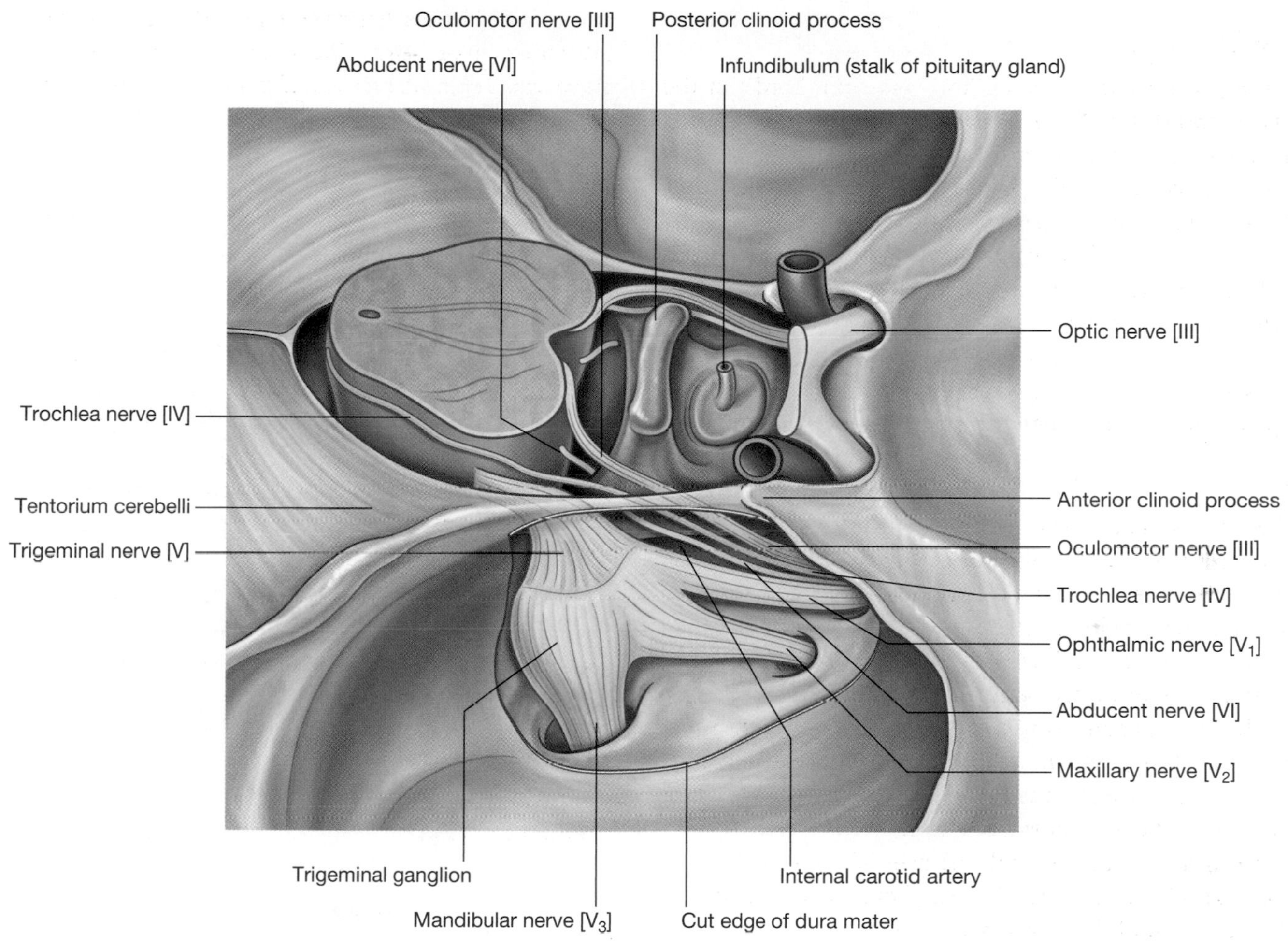


**Fig. 8.46** Lateral view of right cavernous sinus with meningeal layer of dura removed to show contents.

Structures in the lateral wall of each cavernous sinus are, from superior to inferior:

- the oculomotor nerve [III],
- the trochlear nerve [IV],
- the ophthalmic nerve [$V_1$], and
- the maxillary nerve [$V_2$].

Connecting the right and left cavernous sinuses are the intercavernous sinuses on the anterior and posterior sides of the pituitary stalk (Fig. 8.44).

Sphenoparietal sinuses drain into the anterior ends of each cavernous sinus. These small sinuses are along the inferior surface of the lesser wings of the sphenoid and receive blood from the diploic and meningeal veins.

### Superior and inferior petrosal sinuses

The superior petrosal sinuses drain the cavernous sinuses into the transverse sinuses. Each superior petrosal sinus begins at the posterior end of the cavernous sinus, passes posterolaterally along the superior margin of the petrous part of each temporal bone, and connects to the transverse sinus (Fig. 8.44). The superior petrosal sinuses also receive cerebral and cerebellar veins.

The inferior petrosal sinuses also begin at the posterior ends of the cavernous sinuses. These bilateral sinuses pass posteroinferiorly in a groove between the petrous part of the temporal bone and the basal part of the occipital bone, ending in the internal jugular veins. They assist in draining the cavernous sinuses and also receive blood from cerebellar veins and veins from the internal ear and brainstem.

Basilar sinuses connect the inferior petrosal sinuses to each other and to the vertebral plexus of veins. They are on the clivus, just posterior to the sella turcica of the sphenoid bone (Fig. 8.44).

## In the clinic

### Scalp and meninges

Summary of relationships and clinical significance of the scalp and meninges (Fig. 8.47).

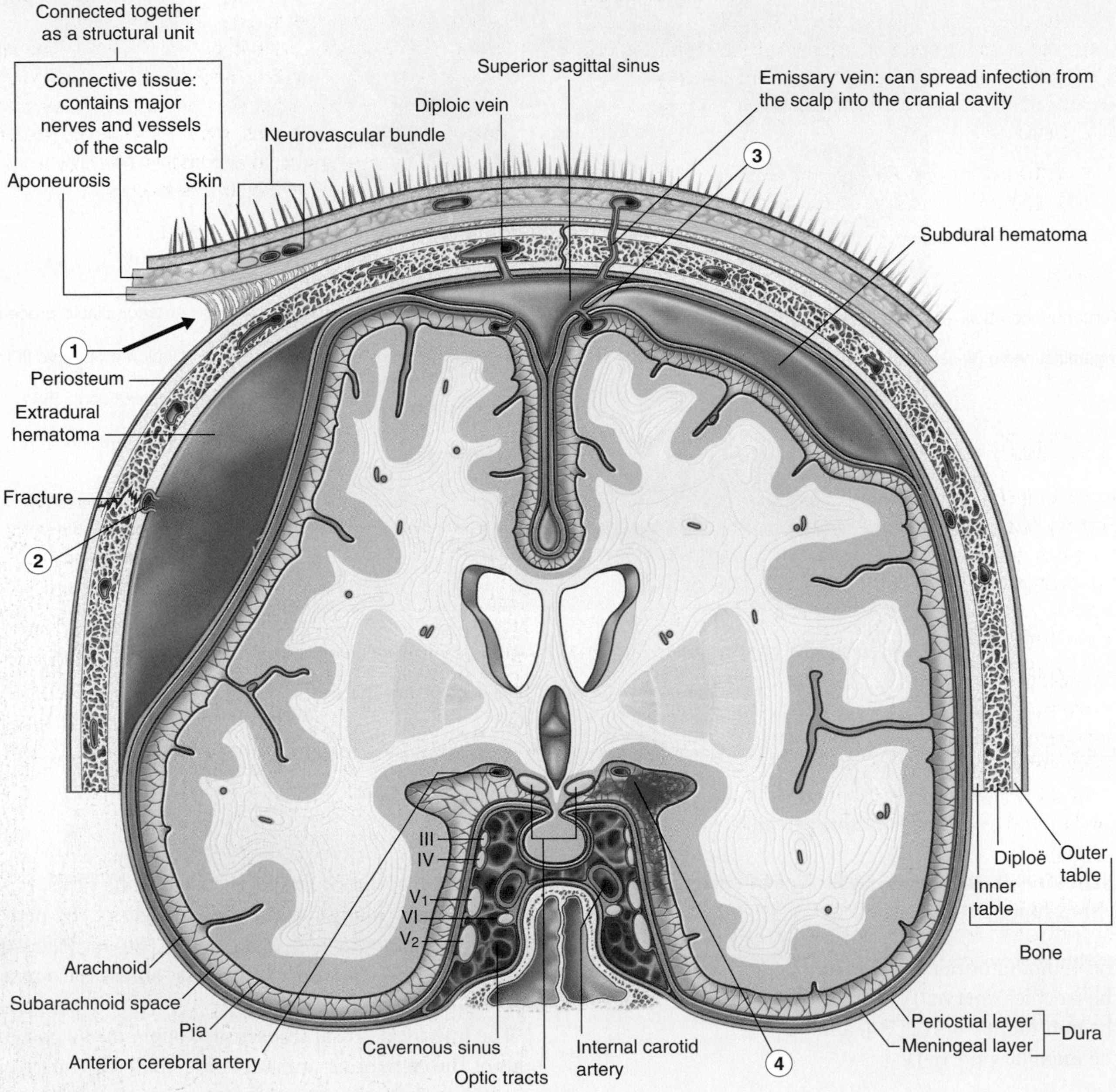

Connected together as a structural unit
Connective tissue: contains major nerves and vessels of the scalp
Aponeurosis
Skin
Neurovascular bundle
Diploic vein
Superior sagittal sinus
Emissary vein: can spread infection from the scalp into the cranial cavity
3
Subdural hematoma
1
Periosteum
Extradural hematoma
Fracture
2
III
IV
$V_1$
VI
$V_2$
Diploë
Outer table
Inner table
Bone
Arachnoid
Subarachnoid space
Pia
Anterior cerebral artery
Cavernous sinus
Optic tracts
Internal carotid artery
4
Periostial layer
Meningeal layer
Dura

(1) Loose connective tissue (danger area)
- In scalping injuries, this is the layer in which separation occurs.
- Infection can easily spread in this layer.
- Blunt trauma can result in hemorrhage in this layer (blood can spread forward into the face, resulting in "black eyes").

(2) Rupture of the middle meningeal artery (branches) by fracture of the inner table of bone results in extradural hematoma. Under pressure, the blood progressively separates dura from the bone.

(3) Tear to cerebral vein where it crosses dura to enter cranial venous sinus can result in subdural hematoma. The tear separates a thin layer of meningeal dura from that which remains attached to the periosteal layer. As a result, the hematoma is covered by an inner limiting membrane derived from part of the meningeal dura.

(4) Aneurysm
- Ruptured aneurysms of vessels of the cerebral arterial circle hemorrhage directly into the subarachnoid space and CSF.

**Fig. 8.47** Scalp and meninges.

## In the clinic

### Head injury

Head trauma is a common injury and is a significant cause of morbidity and death. Head injury may occur in isolation, but often the patient has other injuries; it should always be suspected in patients with multiple injuries. Among patients with multiple trauma, 50% die from the head injury.

At the time of the initial head injury two processes take place.

- First the primary brain injury may involve primary axonal and cellular damage, which results from the shearing deceleration forces within the brain. These injuries are generally not repairable. Further primary brain injuries include intracerebral hemorrhage and penetrating injuries, which may directly destroy gray and white matter.
- The secondary injuries are sequelae of the initial trauma. They include scalp laceration, fracture of the cranial vault, disruption of intracerebral arteries and veins, intracerebral edema, and infection. In most cases these can be treated if diagnosed early, and rapid and effective treatment will significantly improve the patient's recovery and prognosis.

## In the clinic

### Types of intracranial hemorrhage

#### *Primary brain hemorrhage*

The many causes of a primary brain hemorrhage include aneurysm rupture, hypertension (intracerebral hematoma secondary to high blood pressure), and bleeding after cerebral infarction.

#### *Extradural hemorrhage*

An extradural hemorrhage (Fig. 8.48) is caused by arterial damage and results from tearing of the branches of the middle meningeal artery, which typically occurs in the region of the pterion. Blood collects between the periosteal layer of the dura and the calvaria and under arterial pressure slowly expands.

The typical history is of a blow to the head (often during a sporting activity) that produces a minor loss of consciousness. Following the injury the patient usually regains consciousness and has a lucid interval for a period of hours. After this, rapid drowsiness and unconsciousness ensue, which may lead to death.

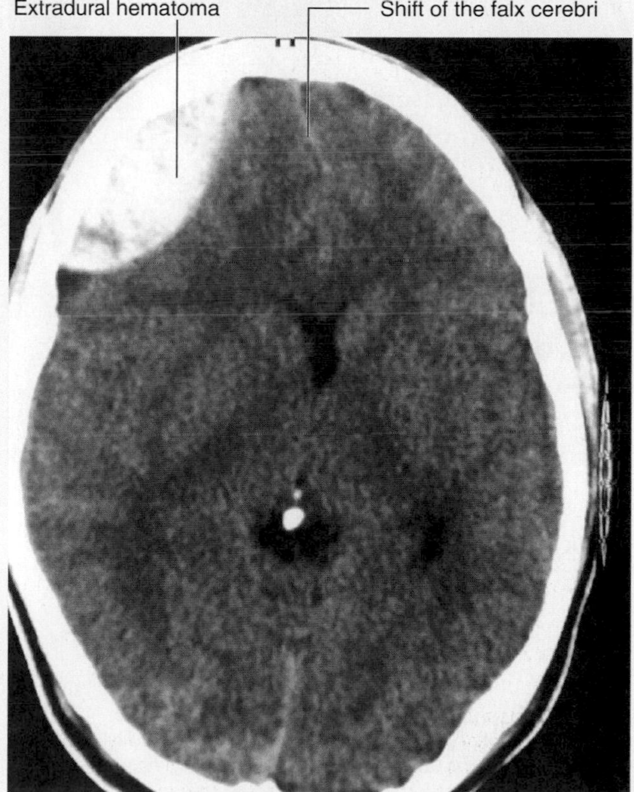


**Fig. 8.48** Extradural hematoma. Axial CT scan of brain.

*(continues)*

## In the clinic—cont'd

***Subdural hematoma***

A subdural hematoma (Fig. 8.49) results from venous bleeding, usually from torn cerebral veins where they enter the superior sagittal sinus. The tear and resulting seepage of blood separates the thin layer of dural border cells from the rest of the dura as the hematoma develops.

Patients at most risk of developing a subdural hematoma are the young and elderly. The increased CSF space in patients with cerebral atrophy results in a greater than normal stress on the cerebral veins entering the sagittal sinus. The clinical history usually includes a trivial injury followed by an insidious loss of consciousness or alteration of personality.

***Subarachnoid hemorrhage***

Subarachnoid hemorrhage (Fig. 8.50) may occur in patients who have undergone significant cerebral trauma, but typically it results from a ruptured intracerebral aneurysm arising from the vessels supplying and around the arterial circle (of Willis).

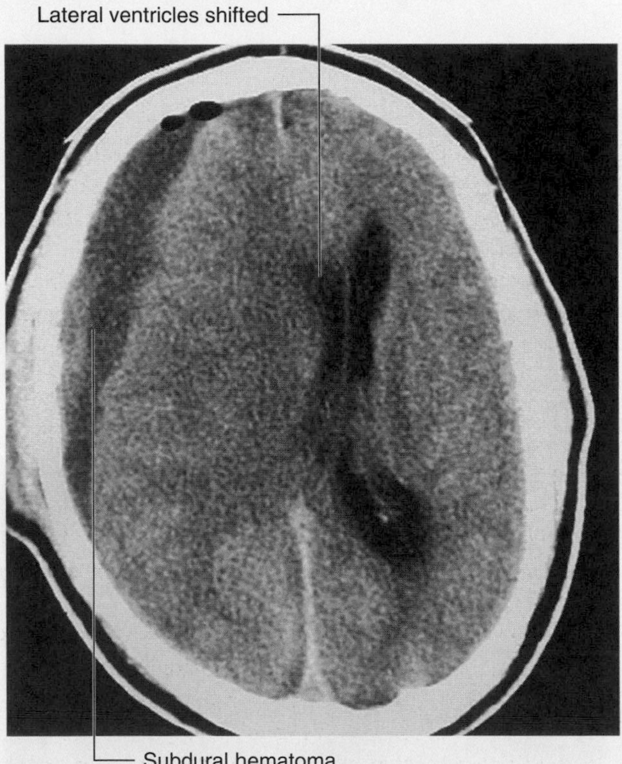


**Fig. 8.49** Chronic (low-density) subdural hematoma. Axial CT scan of brain.

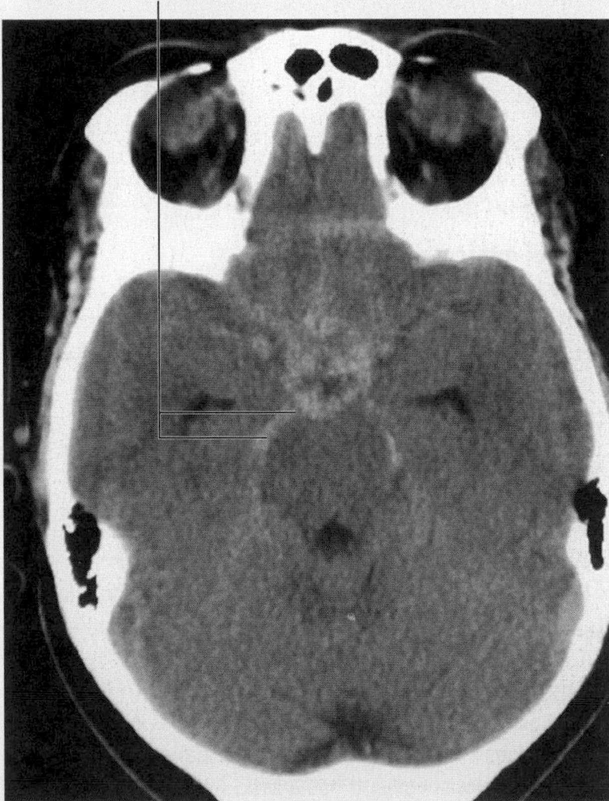


**Fig. 8.50** Subarachnoid hemorrhage. Axial CT scan of brain.

## In the clinic

### Tuberculosis of the central nervous system

Tuberculosis (TB) may invade the central nervous system, including the brain, spinal cord, and meninges (Fig. 8.51). Symptoms of brain TB include headache, neck stiffness, weight loss, and fever. Symptoms of spinal cord TB include leg weakness and fecal and urinary incontinence. Meningitis can cause altered mental status, fever, and seizures. Treatment usually requires a cocktail of drugs for 1 year, but treatment for brain TB can require 2 years.

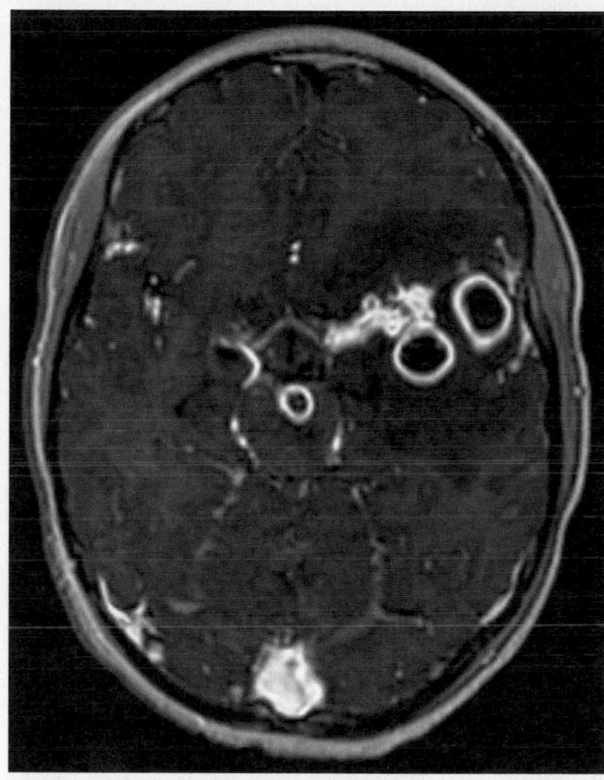

**Fig. 8.51** MRI of the brain shows peripherally enhancing tuberculosis lesions in the left temporal lobe and cerebral peduncle.

## In the clinic

### Emissary veins

Emissary veins connect extracranial veins with intracranial veins and are important clinically because they can be a conduit through which infections can enter the cranial cavity. Emissary veins lack valves, as do the majority of veins in the head and neck.

## In the clinic

### Concussion

Concussion (mild traumatic brain injury [MTBI]) is the most common type of traumatic brain injury. The injury typically results from a rapid deceleration of the head or by a rotation of the brain within the cranial cavity. General symptoms of MTBI can include posttraumatic amnesia, confusion, loss of consciousness, headache, dizziness, vomiting, lack of motor coordination, and light sensitivity. The diagnosis of concussion, MTBI, is based on the event, the current neurological status, and the state of consciousness of the patient.

## In the clinic

### Clinical assessment of patients with head injury

Clinical assessment of patients with head injury always appears relatively straightforward. In reality it is usually far from straightforward.

Patients may have a wide spectrum of modes of injury from a simple fall to complex multiple trauma. The age of the patient and ability to communicate about the injuries are important factors.

The circumstances in which the injury may have occurred should be documented because some head injuries result from a serious assault, and the physician may be required to give evidence to a court of law.

Determining the severity of head injury may be difficult because some injuries occur as a result of or in association with alcohol intoxication.

Even when the diagnosis has been made and the correct management has been instigated, the circumstances in which the injury occurred and the environment to which the patient will return after treatment need to be reviewed to prevent further injuries (e.g., an elderly person tripping on loose carpet on a staircase).

A thorough clinical examination includes all systems, but with a special focus on the central and peripheral nervous systems. The level of consciousness must also be assessed and accurately documented using the Glasgow Coma Scale, which allows clinicians to place a numerical value upon the level of consciousness so that any deterioration or improvement can be measured and quantified.

#### *Glasgow Coma Scale*

The Glasgow Coma Scale was proposed in 1974 and is now widely accepted throughout the world. There is a total score of 15 points, such that 15/15 indicates that the patient is alert and fully oriented, whereas 3/15 indicates a severe and deep coma. The points score comprises a best motor response (total of 6 points), best verbal response (total of 5 points), and best eye movement response (total of 4 points).

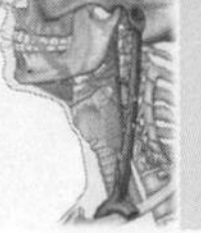


## In the clinic

### Treatment of head injury

Treatment of primary brain injury is extremely limited. Axonal disruption and cellular death are generally irrecoverable. Whenever the brain is injured, like most tissues, it swells. Because the brain is encased within a fixed space (the skull), swelling impairs cerebral function and has two other important effects.

- First, the swelling compresses the blood supply into the skull, resulting in a physiologically dramatic increase in blood pressure.
- Second, the cerebral swelling may be diffuse, eventually squeezing the brain and brainstem through the foramen magnum (**coning**). This compression and disruption of the brainstem may lead to a loss of basic cardiorespiratory function, and death will ensue. Focal cerebral edema may cause one side of the brain to herniate beneath the falx cerebri (**falcine herniation**).

Simple measures to prevent the swelling include hyperventilation (which alters the intracerebral acid–base balance and decreases swelling) and intravenous corticosteroids (though their action is often delayed).

Extracerebral hematoma may be removed surgically.

The outcome of patients with head injury depends on how the secondary injury is managed. Even with a severe primary injury, patients may recover to lead a normal life.

## In the clinic

### Increased intracranial pressure and coning

The skull is a closed bony compartment, and the brain and cerebrospinal fluid are maintained physiologically within a narrow intracranial pressure range. Any new space-occupying lesion, such as a hematoma, an injury that leads to brain swelling, or a brain tumor, can increase intracranial pressure and compress the brain. In severe cases, the brain may be squeezed down into the foramen magnum, giving it a cone shape, termed cerebral herniation, or "coning." This may in turn compress the brainstem and upper cervical spinal cord, which can be fatal.

Congenital herniation or coning of the cerebellar tonsils through the foramen magnum can also occur if the posterior fossa is too small, a condition known as Chiari I malformation (Fig. 8.52). This often causes no problems in childhood and may only start causing symptoms in adulthood.

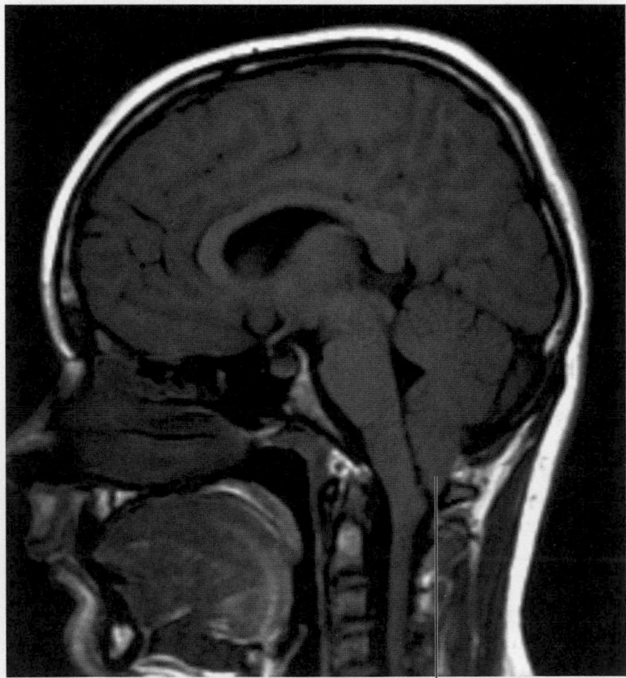


**Fig. 8.52** MRI of the brain reveals an incidental Chiari I malformation with herniation of the the cerebellar tonsils through the foramen magnum, giving rise to a cone shape.

## CRANIAL NERVES

The 12 pairs of cranial nerves are part of the peripheral nervous system (PNS) and pass through foramina or fissures in the cranial cavity. All nerves except one, the accessory nerve [XI], originate from the brain.

In addition to having somatic and visceral components similar to those of spinal nerves, some cranial nerves also contain special sensory and motor components (Tables 8.4 and 8.5).

The special sensory components are associated with hearing, seeing, smelling, balancing, and tasting.

Special motor components include those that innervate skeletal muscles derived embryologically from the pharyngeal arches and not from somites.

In human embryology, six pharyngeal arches are designated, but the fifth pharyngeal arch never develops. Each of the pharyngeal arches that does develop is associated with a developing cranial nerve or one of its branches.

**Table 8.4** Cranial nerve functional components

| Functional component | Abbreviation | General function | Cranial nerves containing component |
|---|---|---|---|
| General somatic afferent | GSA | Perception of touch, pain, temperature | Trigeminal nerve [V]; facial nerve [VII]; glossopharyngeal nerve [IX]; vagus nerve [X] |
| General visceral afferent | GVA | Sensory input from viscera | Glossopharyngeal nerve [IX]; vagus nerve [X] |
| Special afferent* | SA | Smell, taste, vision, hearing, and balance | Olfactory nerve [I]; optic nerve [II]; facial nerve [VII]; vestibulocochlear nerve [VIII]; glossopharyngeal nerve [IX]; vagus nerve [X] |
| General somatic efferent | GSE | Motor innervation to skeletal (voluntary) muscles | Oculomotor nerve [III]; trochlear nerve [IV]; abducent nerve [VI]; hypoglossal nerve [XII] |
| General visceral efferent | GVE | Motor innervation to smooth muscle, heart muscle, and glands | Oculomotor nerve [III]; facial nerve [VII]; glossopharyngeal nerve [IX]; vagus nerve [X] |
| Branchial efferent** | BE | Motor innervation to skeletal muscles derived from pharyngeal arch mesoderm | Trigeminal nerve [V]; facial nerve [VII]; glossopharyngeal nerve [IX]; vagus nerve [X]; accessory nerve [XI] (see Diogo R et al. *Nature* 2015;520:466–473) |

Other terminology used when describing functional components:
*Special sensory, or special visceral afferent (SVA): smell, taste. Special somatic afferent (SSA): vision, hearing, balance.
**Special visceral efferent (SVE) or branchial motor.

**Table 8.5** Cranial nerves (see Table 8.4 for abbreviations)

| | COMPONENT | | | |
|---|---|---|---|---|
| **Nerve** | **Afferent** | **Efferent** | **Exit from skull** | **Function** |
| Olfactory nerve [I] | SA | | Cribriform plate of ethmoid bone | Smell |
| Optic nerve [II] | SA | | Optic canal | Vision |
| Oculomotor nerve [III] | | GSE, GVE | Superior orbital fissure | GSE—innervates levator palpebrae superioris, superior rectus, inferior rectus, medial rectus, and inferior oblique muscles<br>GVE—innervates sphincter pupillae for pupillary constriction; ciliary muscles for accommodation of the lens for near vision |
| Trochlear nerve [IV] | | GSE | Superior orbital fissure | Innervates superior oblique muscle |
| Trigeminal nerve [V] | GSA | BE | Superior orbital fissure—ophthalmic division [$V_1$]<br>Foramen rotundum—maxillary nerve [$V_2$]<br>Foramen ovale—mandibular division [$V_3$] | GSA—sensory from: ophthalmic division [$V_1$]—eyes, conjunctiva, orbital contents, nasal cavity, frontal sinus, ethmoidal cells, upper eyelid, dorsum of nose, anterior part of scalp, dura in anterior cranial fossa, superior part of tentorium cerebelli; maxillary nerve [$V_2$]—dura in middle cranial fossa, nasopharynx, palate, nasal cavity, upper teeth, maxillary sinus, skin covering the side of the nose, lower eyelid, cheek, upper lip; mandibular division [$V_3$]—skin of lower face, cheek, lower lip, anterior part of external ear, part of external acoustic meatus, temporal fossa, anterior two-thirds of tongue, lower teeth, mastoid air cells, mucous membranes of cheek, mandible, dura in middle cranial fossa<br>BE—innervates temporalis, masseter, medial and lateral pterygoids, tensor tympani, tensor veli palatini, anterior belly of digastric, and mylohyoid muscles |



*Continued*

**Table 8.5** Cranial nerves (see Table 8.4 for abbreviations)—cont'd

| Nerve | COMPONENT Afferent | Efferent | Exit from skull | Function |
|---|---|---|---|---|
| Abducent nerve [VI] | | GSE | Superior orbital fissure | Innervates lateral rectus muscle |
| Facial nerve [VII] | GSA, SA | GVE, BE | Stylomastoid foramen (nerve leaves cranial cavity through internal acoustic meatus and gives rise to branches in the facial canal of the temporal bone prior to exiting through the stylomastoid foramen; these branches leave the skull through other fissures and canals.) | GSA—sensory from part of external acoustic meatus and deeper parts of auricle<br>SA—taste from anterior two-thirds of tongue<br>GVE—innervates lacrimal gland, submandibular and sublingual salivary glands, and mucous membranes of nasal cavity, hard and soft palates<br>BE—innervates muscles of face (muscles of facial expression) and scalp derived from the second pharyngeal arch, and stapedius, posterior belly of digastric, stylohyoid muscles |
| Vestibulocochlear nerve [VIII] | SA | | (Nerve leaves cranial cavity through internal acoustic meatus) | Vestibular division—balance<br>Cochlear division—hearing |
| Glossopharyngeal nerve [IX] | GVA, SA, GSA | GVE, BE | Jugular foramen | GVA—sensory from carotid body and sinus<br>GSA—posterior one-third of tongue, palatine tonsils, oropharynx, and mucosa of middle ear, pharyngotympanic tube, and mastoid air cells<br>SA—taste from posterior one-third of tongue<br>GVE—innervates parotid salivary gland<br>BE—innervates stylopharyngeus muscle |
| Vagus nerve [X] | GSA, GVA, SA | GVE, BE | Jugular foramen | GSA—sensory from larynx, laryngopharynx, deeper parts of auricle, part of external acoustic meatus, and dura in posterior cranial fossa<br>GVA—sensory from aortic body chemoreceptors and aortic arch baroreceptors, esophagus, bronchi, lungs, heart, and abdominal viscera of the foregut and midgut<br>SA—taste from the epiglottis and pharynx<br>GVE—innervates smooth muscle and glands in the pharynx, larynx, thoracic viscera, and abdominal viscera of the foregut and midgut<br>BE—innervates one tongue muscle (palatoglossus), muscles of soft palate (except tensor veli palatini), pharynx (except stylopharyngeus), and larynx |
| Accessory nerve [XI] | | BE | Jugular foramen | Innervates sternocleidomastoid and trapezius muscles [for classification as BE see Diogo R et al. *Nature* 2015;520:466–473.] |
| Hypoglossal nerve [XII] | | GSE | Hypoglossal canal | Innervates hyoglossus, genioglossus, and styloglossus muscles and all intrinsic muscles of the tongue |

These cranial nerves carry efferent fibers that innervate the musculature derived from the pharyngeal arch.

Innervation of the musculature derived from the five pharyngeal arches that do develop is as follows:

- first arch—trigeminal nerve [$V_3$],
- second arch—facial nerve [VII],
- third arch—glossopharyngeal nerve [IX],
- fourth arch—superior laryngeal branch of the vagus nerve [X],
- sixth arch—recurrent laryngeal branch of the vagus nerve [X],
- posterior arches—accessory nerve [XI].

## Olfactory nerve [I]

The **olfactory nerve [I]** carries special afferent (SA) fibers for the sense of smell. Its sensory neurons have:

- peripheral processes that act as receptors in the nasal mucosa, and
- central processes that return information to the brain.

The receptors are in the roof and upper parts of the nasal cavity, and the central processes, after joining into small bundles, enter the cranial cavity by passing through the cribriform plate of the ethmoid bone (Fig. 8.53). They terminate by synapsing with secondary neurons in the olfactory bulbs (Fig. 8.54).

## Optic nerve [II]

The **optic nerve [II]** carries SA fibers for vision. These fibers return information to the brain from photoreceptors in the retina. Neuronal processes leave the retinal

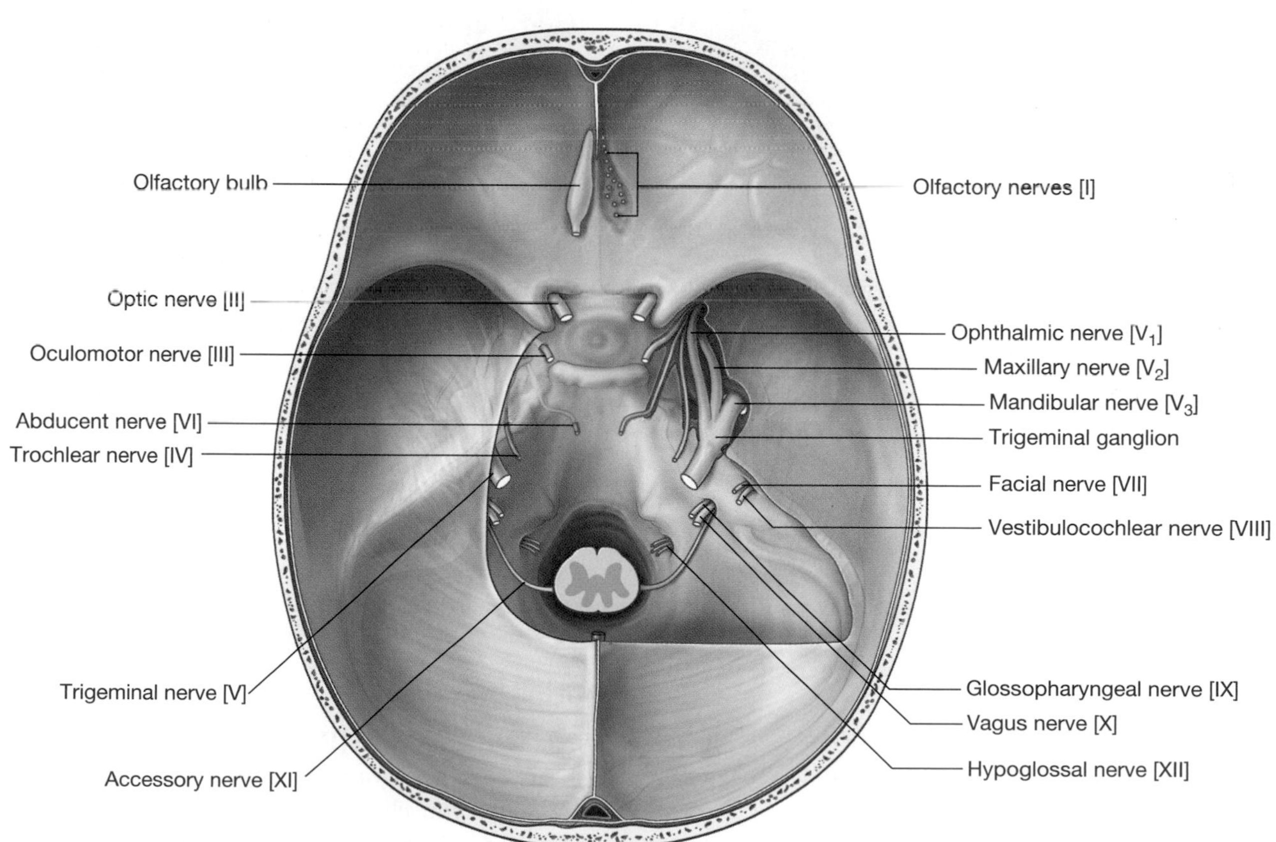


**Fig. 8.53** Cranial nerves exiting the cranial cavity.

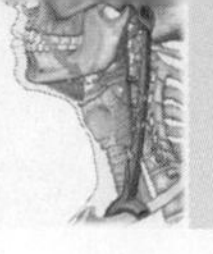

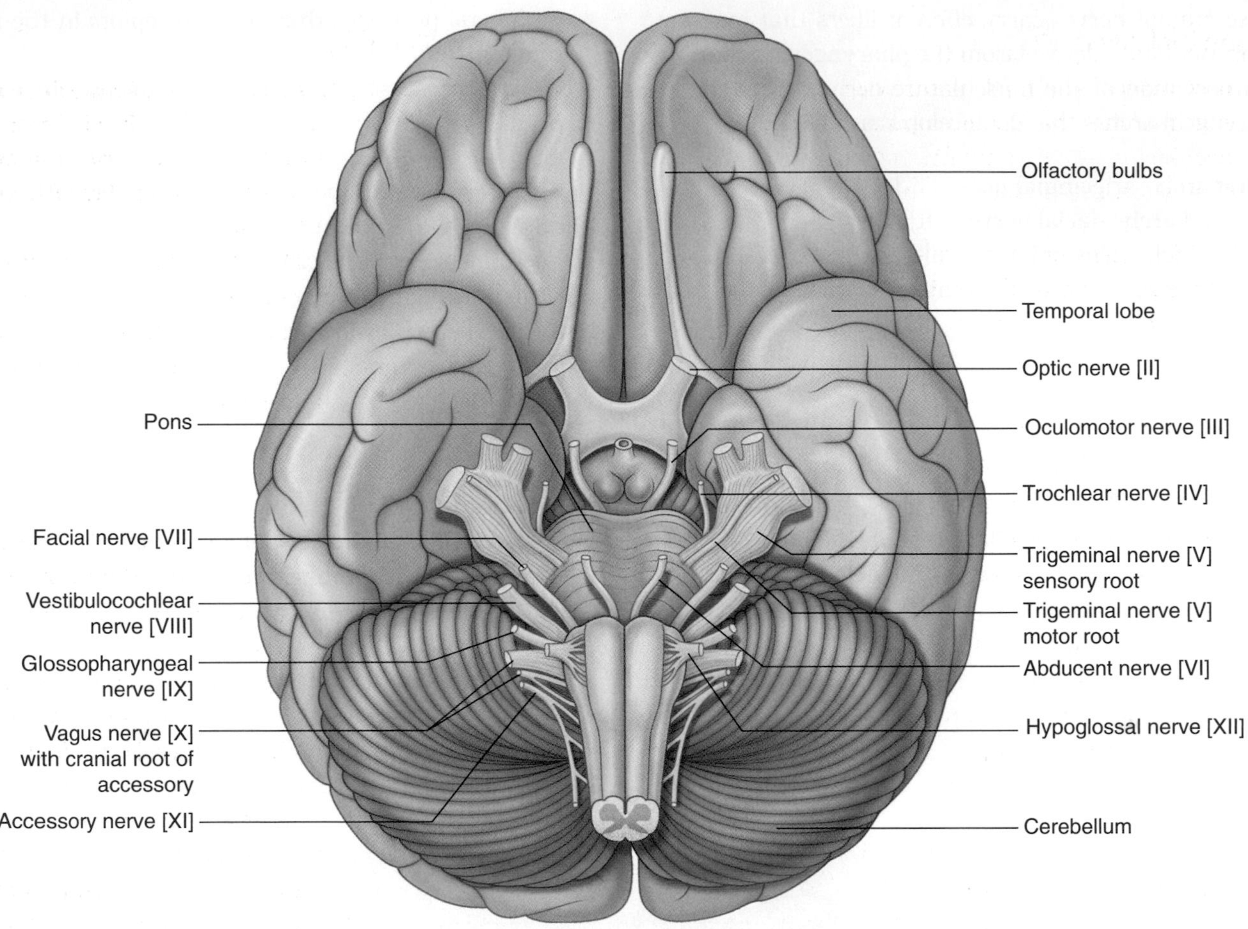


**Fig. 8.54** Cranial nerves on the base of the brain.

receptors, join into small bundles, and are carried by the optic nerves to other components of the visual system in the brain. The optic nerves enter the cranial cavity through the optic canals (Fig. 8.53).

## Oculomotor nerve [III]

The **oculomotor nerve [III]** carries two types of fibers:

- General somatic efferent (GSE) fibers innervate most of the extra-ocular muscles.
- General visceral efferent (GVE) fibers are part of the parasympathetic part of the autonomic division of the PNS.

The oculomotor nerve [III] leaves the anterior surface of the brainstem between the midbrain and the pons (Fig. 8.54). It enters the anterior edge of the tentorium cerebelli, continues in an anterior direction in the lateral wall of the cavernous sinus (Figs. 8.53 and 8.54; see Fig. 8.45), and leaves the cranial cavity through the superior orbital fissure.

In the orbit, the GSE fibers in the oculomotor nerve innervate levator palpebrae superioris, superior rectus, inferior rectus, medial rectus, and inferior oblique muscles.

The GVE fibers are preganglionic parasympathetic fibers that synapse in the ciliary ganglion and ultimately innervate the sphincter pupillae muscle, responsible for pupillary constriction, and the ciliary muscles, responsible for accommodation of the lens for near vision.

## Trochlear nerve [IV]

The **trochlear nerve [IV]** is a cranial nerve that carries GSE fibers to innervate the superior oblique muscle, an extra-ocular muscle in the orbit. It arises in the midbrain and is the only cranial nerve to exit from the posterior surface of the brainstem (Fig. 8.54). After curving around the midbrain, it enters the inferior surface of the free edge of the tentorium cerebelli, continues in an anterior direction in the lateral wall of the cavernous sinus (Figs. 8.53 and 8.54; see Fig. 8.45), and enters the orbit through the superior orbital fissure.

## Trigeminal nerve [V]

The **trigeminal nerve [V]** is the major general sensory nerve of the head and also innervates muscles that move the lower jaw. It carries general somatic afferent (GSA) and branchial efferent (BE) fibers:

- The GSA fibers provide sensory input from the face, anterior one-half of the scalp, mucous membranes of the oral and nasal cavities and the paranasal sinuses, the nasopharynx, part of the ear and external acoustic meatus, part of the tympanic membrane, the orbital contents and conjunctiva, and the dura mater in the anterior and middle cranial fossae.
- The BE fibers innervate the muscles of mastication; the tensor tympani, tensor veli palatini, and mylohyoid muscles; and the anterior belly of the digastric muscle.

The trigeminal nerve exits from the anterolateral surface of the pons as a large sensory root and a small motor root (Fig. 8.54). These roots continue forward out of the posterior cranial fossa and into the middle cranial fossa by passing over the medial tip of the petrous part of the temporal bone (Fig. 8.53).

In the middle cranial fossa the sensory root expands into the **trigeminal ganglion** (Fig. 8.53), which contains cell bodies for the sensory neurons in the trigeminal nerve and is comparable to a spinal ganglion. The ganglion is in a depression (the trigeminal depression) on the anterior surface of the petrous part of the temporal bone, in a dural cave (the **trigeminal cave**). The motor root is below and completely separate from the sensory root at this point.

Arising from the anterior border of the trigeminal ganglion are the three terminal divisions of the trigeminal nerve, which in descending order are:

- the **ophthalmic nerve (ophthalmic division [$V_1$])**,
- the **maxillary nerve (maxillary division [$V_2$])**, and
- the **mandibular nerve (mandibular division [$V_3$])**.

### Ophthalmic nerve [$V_1$]

The ophthalmic nerve [$V_1$] passes forward in the dura of the lateral wall of the cavernous sinus (see Fig. 8.45), leaves the cranial cavity, and enters the orbit through the superior orbital fissure (Fig. 8.53).

The ophthalmic nerve [$V_1$] carries sensory branches from the eyes, conjunctiva, and orbital contents, including the lacrimal gland. It also receives sensory branches from the nasal cavity, frontal sinus, ethmoidal cells, falx cerebri, dura in the anterior cranial fossa and superior parts of the tentorium cerebelli, upper eyelid, dorsum of the nose, and the anterior part of the scalp.

### Maxillary nerve [$V_2$]

The maxillary nerve [$V_2$] passes forward in the dura mater of the lateral wall of the cavernous sinus just inferior to the ophthalmic nerve [$V_1$] (see Fig. 8.45), leaves the cranial cavity through the foramen rotundum (Fig. 8.53), and enters the pterygopalatine fossa.

The maxillary nerve [$V_2$] receives sensory branches from the dura in the middle cranial fossa, the nasopharynx, the palate, the nasal cavity, teeth of the upper jaw, maxillary sinus, and skin covering the side of the nose, the lower eyelid, the cheek, and the upper lip.

### Mandibular nerve [$V_3$]

The mandibular nerve [$V_3$] leaves the inferior margin of the trigeminal ganglion and leaves the skull through the foramen ovale (Fig. 8.53), and enters the infratemporal fossa.

The motor root of the trigeminal nerve also passes through the foramen ovale and unites with the sensory component of the mandibular nerve [$V_3$] outside the skull. Thus the mandibular nerve [$V_3$] is the only division of the trigeminal nerve that contains a motor component.

Outside the skull the motor fibers innervate the four muscles of mastication (temporalis, masseter, and medial and lateral pterygoids), as well as the tensor tympani muscle, the tensor veli palatini muscle, the anterior belly of the digastric muscle, and the mylohyoid muscle.

The mandibular nerve [$V_3$] also receives sensory branches from the skin of the lower face, cheek, lower lip, anterior part of the external ear, part of the external acoustic meatus and the temporal region, the anterior two-thirds of the tongue, the teeth of the lower jaw, the mastoid air cells, the mucous membranes of the cheek, the mandible, and dura in the middle cranial fossa.

## Abducent nerve [VI]

The **abducent nerve [VI]** carries GSE fibers to innervate the lateral rectus muscle in the orbit. It arises from the brainstem between the pons and medulla and passes forward, piercing the dura covering the clivus (Figs. 8.53 and 8.54). Continuing upward in a dural canal, it crosses the superior edge of the petrous part of the temporal bone, enters and crosses the cavernous sinus (see Fig. 8.45) just inferolateral to the internal carotid artery, and enters the orbit through the superior orbital fissure.

## Facial nerve [VII]

The **facial nerve [VII]** carries GSA, SA, GVE, and BE fibers:

- The GSA fibers provide sensory input from part of the external acoustic meatus and deeper parts of the auricle.
- The SA fibers are for taste from the anterior two-thirds of the tongue.
- The GVE fibers are part of the parasympathetic part of the autonomic division of the PNS and stimulate secretomotor activity in the lacrimal gland, submandibular and sublingual salivary glands, and glands in the mucous membranes of the nasal cavity, and hard and soft palates.
- The BE fibers innervate the muscles of the face (muscles of facial expression) and scalp derived from the second pharyngeal arch, and the stapedius muscle, the posterior belly of the digastric muscle, and the stylohyoid muscle.

The facial nerve [VII] attaches to the lateral surface of the brainstem, between the pons and medulla oblongata (Fig. 8.54). It consists of a large motor root and a smaller sensory root (the **intermediate nerve**):

- The intermediate nerve contains the SA fibers for taste, the parasympathetic GVE fibers, and the GSA fibers.
- The larger motor root contains the BE fibers.

The motor and sensory roots cross the posterior cranial fossa and leave the cranial cavity through the internal acoustic meatus (Fig. 8.53). After entering the facial canal in the petrous part of the temporal bone, the two roots fuse and form the facial nerve [VII]. Near this point the nerve enlarges as the **geniculate ganglion**, which is similar to a spinal ganglion containing cell bodies for sensory neurons.

At the geniculate ganglion the facial nerve [VII] turns and gives off the **greater petrosal nerve**, which carries mainly preganglionic parasympathetic (GVE) fibers (Table 8.6).

The facial nerve [VII] continues along the bony canal, giving off the **nerve to the stapedius** and the **chorda tympani**, before exiting the skull through the stylomastoid foramen.

The chorda tympani carries taste (SA) fibers from the anterior two-thirds of the tongue and preganglionic parasympathetic (GVE) fibers destined for the submandibular ganglion (Table 8.6).

## Vestibulocochlear nerve [VIII]

The vestibulocochlear nerve [VIII] carries SA fibers for hearing and balance, and consists of two divisions:

- a vestibular component for balance, and
- a cochlear component for hearing.

The vestibulocochlear nerve [VIII] attaches to the lateral surface of the brainstem, between the pons and medulla, after emerging from the internal acoustic meatus and crossing the posterior cranial fossa (Figs. 8.53 and 8.54). The two divisions combine into the single nerve seen in the posterior cranial fossa within the substance of the petrous part of the temporal bone.

## Glossopharyngeal nerve [IX]

The glossopharyngeal nerve [IX] carries GVA, GSA, SA, GVE, and BE fibers:

- The GVA fibers provide sensory input from the carotid body and sinus.
- The GSA fibers provide sensory input from the posterior one-third of the tongue, palatine tonsils, oropharynx, and mucosa of the middle ear, pharyngotympanic tube, and mastoid air cells.
- The SA fibers are for taste from the posterior one-third of the tongue.
- The GVE fibers are part of the parasympathetic part of the autonomic division of the PNS and stimulate secretomotor activity in the parotid salivary gland.
- The BE fibers innervate the muscle derived from the third pharyngeal arch (the stylopharyngeus muscle).

**Table 8.6** Parasympathetic ganglia of the head

| Ganglion | Cranial nerve origin of preganglionic fibers | Branch supplying preganglionic fibers to ganglion | Function |
|---|---|---|---|
| Ciliary | Oculomotor nerve [III] | Branch to ciliary ganglion | Innervation of sphincter pupillae muscle for pupillary constriction, and ciliary muscles for accommodation of the lens for near vision |
| Pterygopalatine | Facial nerve [VII] | Greater petrosal nerve | Innervation of lacrimal gland, and mucous glands of nasal cavity, maxillary sinus, and palate |
| Otic | Glossopharyngeal nerve [IX] | Lesser petrosal nerve | Innervation of parotid gland |
| Submandibular | Facial nerve [VII] | Chorda tympani to lingual | Innervation of submandibular and sublingual glands |

## In the clinic

### Cranial nerve lesions

| Cranial Nerve | Clinical Findings | Example of Lesion |
|---|---|---|
| Olfactory nerve [I] | Loss of smell (anosmia) | Injury to the cribriform plate; congenital absence |
| Optic nerve [II] | Blindness/visual field abnormalities, loss of pupillary constriction | Direct trauma to the orbit; disruption of the optic pathway |
| Oculomotor nerve [III] | Dilated pupil, ptosis, loss of normal pupillary reflex, eye moves down inferiorly and laterally (down and out) | Pressure from an aneurysm arising from the posterior communicating, posterior cerebral, or superior cerebellar artery; pressure from a herniating cerebral uncus (false localizing sign); cavernous sinus mass or thrombosis |
| Trochlear nerve [IV] | Inability to look inferiorly when the eye is adducted (down and in) | Along the course of the nerve around the brainstem; orbital fracture |
| Trigeminal nerve [V] | Loss of sensation and pain in the region supplied by the three divisions of the nerve over the face; loss of motor function of the muscles of mastication on the side of the lesion | Typically, in the region of the trigeminal ganglion, though local masses around the foramina through which the divisions pass can produce symptoms |
| Abducent nerve [VI] | Inability of lateral eye movement | Brain lesion or cavernous sinus lesion extending onto the orbit |
| Facial nerve [VII] | Paralysis of facial muscles<br>Abnormal taste sensation from the anterior two-thirds of the tongue and dry conjunctivae<br>Paralysis of contralateral facial muscles below the eye | Damage to the branches within the parotid gland<br>Injury to temporal bone; viral inflammation of nerve<br>Brainstem injury |
| Vestibulocochlear nerve [VIII] | Progressive unilateral hearing loss and tinnitus (ringing in the ear) | Tumor at the cerebellopontine angle |
| Glossopharyngeal nerve [IX] | Loss of taste to the posterior one-third of the tongue and sensation of the soft palate | Brainstem lesion; penetrating neck injury |
| Vagus nerve [X] | Soft palate deviation with deviation of the uvula to the normal side; vocal cord paralysis | Brainstem lesion; penetrating neck injury |
| Accessory nerve [XI] | Paralysis of sternocleidomastoid and trapezius muscles | Penetrating injury to the posterior triangle of the neck |
| Hypoglossal nerve [XII] | Atrophy of ipsilateral muscles of the tongue and deviation toward the affected side; speech disturbance | Penetrating injury to the neck and skull base pathology |

## In the clinic

### Overview of cranial nerves

#### Cranial nerve reflexes

Corneal (blink) reflex

- Afferent—Trigeminal nerve (CN V)
- Efferent—Facial nerve (CN VII)

Gag reflex

- Afferent—Glossopharyngeal nerve (CN IX)
- Efferent—Vagus nerve (CN X)

Pupillary (light) reflex

- Afferent—optic nerve (CN II)
- Efferent—oculomotor nerve (CN III)

*(continues)*

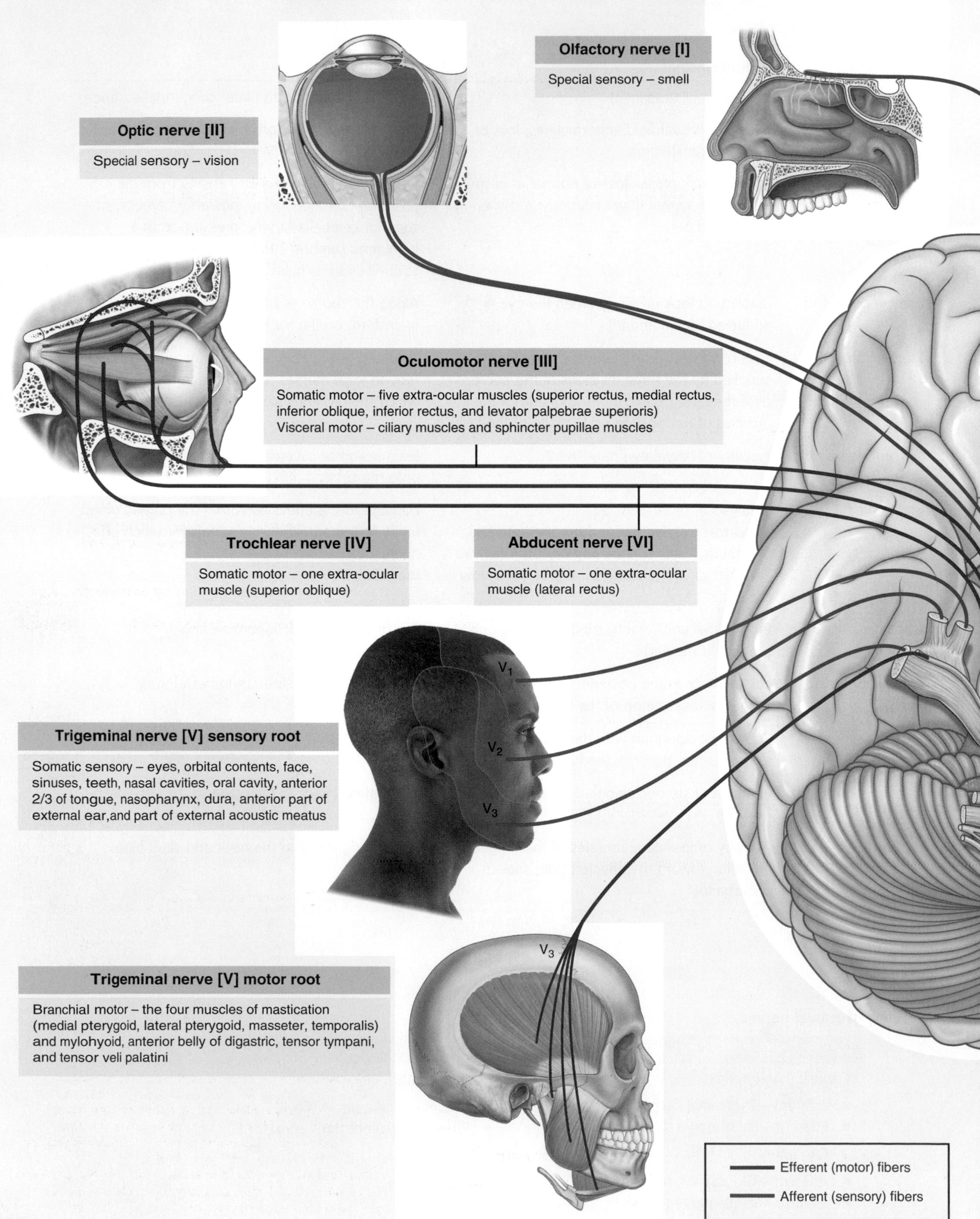

Olfactory nerve [I]
Special sensory – smell
Optic nerve [II]
Special sensory – vision
Oculomotor nerve [III]
Somatic motor – five extra-ocular muscles (superior rectus, medial rectus, inferior oblique, inferior rectus, and levator palpebrae superioris)
Visceral motor – ciliary muscles and sphincter pupillae muscles
Trochlear nerve [IV]
Somatic motor – one extra-ocular muscle (superior oblique)
Abducent nerve [VI]
Somatic motor – one extra-ocular muscle (lateral rectus)
V1
V2
V3
Trigeminal nerve [V] sensory root
Somatic sensory – eyes, orbital contents, face, sinuses, teeth, nasal cavities, oral cavity, anterior 2/3 of tongue, nasopharynx, dura, anterior part of external ear,and part of external acoustic meatus
V3
Trigeminal nerve [V] motor root
Branchial motor – the four muscles of mastication (medial pterygoid, lateral pterygoid, masseter, temporalis) and mylohyoid, anterior belly of digastric, tensor tympani, and tensor veli palatini
Efferent (motor) fibers
Afferent (sensory) fibers

**Fig. 8.55** Overview of cranial nerves.

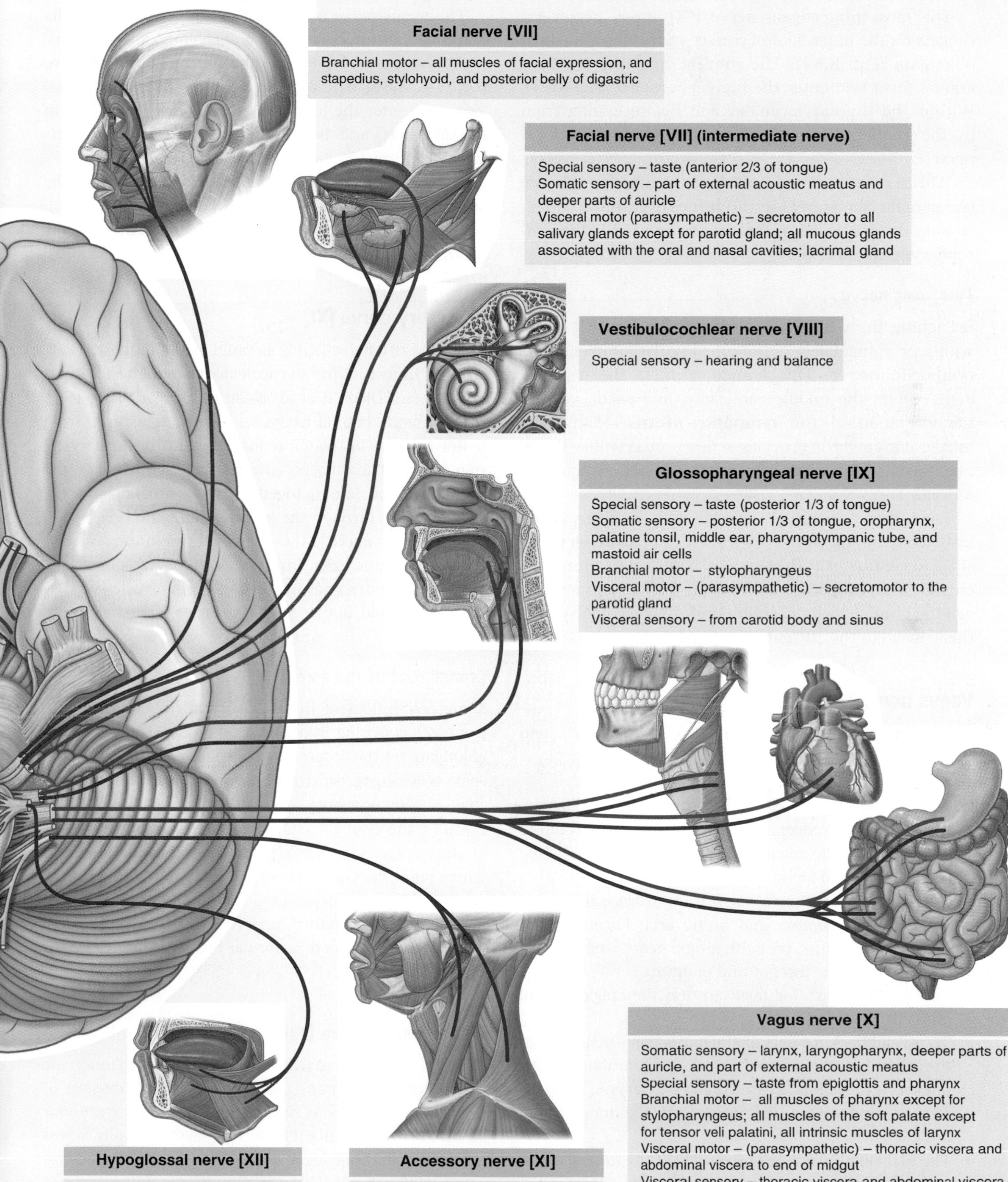
Facial nerve [VII]
Branchial motor – all muscles of facial expression, and stapedius, stylohyoid, and posterior belly of digastric
Facial nerve [VII] (intermediate nerve)
Special sensory – taste (anterior 2/3 of tongue)
Somatic sensory – part of external acoustic meatus and deeper parts of auricle
Visceral motor (parasympathetic) – secretomotor to all salivary glands except for parotid gland; all mucous glands associated with the oral and nasal cavities; lacrimal gland
Vestibulocochlear nerve [VIII]
Special sensory – hearing and balance
Glossopharyngeal nerve [IX]
Special sensory – taste (posterior 1/3 of tongue)
Somatic sensory – posterior 1/3 of tongue, oropharynx, palatine tonsil, middle ear, pharyngotympanic tube, and mastoid air cells
Branchial motor – stylopharyngeus
Visceral motor – (parasympathetic) – secretomotor to the parotid gland
Visceral sensory – from carotid body and sinus
Vagus nerve [X]
Somatic sensory – larynx, laryngopharynx, deeper parts of auricle, and part of external acoustic meatus
Special sensory – taste from epiglottis and pharynx
Branchial motor – all muscles of pharynx except for stylopharyngeus; all muscles of the soft palate except for tensor veli palatini, all intrinsic muscles of larynx
Visceral motor – (parasympathetic) – thoracic viscera and abdominal viscera to end of midgut
Visceral sensory – thoracic viscera and abdominal viscera to end of midgut, chemo- and baroreceptors (and in some cases carotid body)
Hypoglossal nerve [XII]
Somatic motor – all muscles of the tongue except palatoglossus
Accessory nerve [XI]
Branchial motor – sternocleidomastoid and trapezius

The glossopharyngeal nerve [IX] arises as several rootlets on the anterolateral surface of the upper medulla oblongata (Fig. 8.54). The rootlets cross the posterior cranial fossa and enter the jugular foramen (Fig. 8.53). Within the jugular foramen, and before exiting from it, the rootlets merge to form the glossopharyngeal nerve.

Within or immediately outside the jugular foramen are two ganglia (the **superior** and **inferior ganglia**), which contain the cell bodies of the sensory neurons in the glossopharyngeal nerve [IX].

#### Tympanic nerve

Branching from the glossopharyngeal nerve [IX] either within or immediately outside the jugular foramen is the **tympanic nerve**. This branch reenters the temporal bone, enters the middle ear cavity, and participates in the formation of the **tympanic plexus**. Within the middle ear cavity it provides sensory innervation to the mucosa of the cavity, pharyngotympanic tube, and mastoid air cells.

The tympanic nerve also contributes GVE fibers, which leave the tympanic plexus in the **lesser petrosal nerve**—a small nerve that exits the temporal bone, enters the middle cranial fossa, and descends through the foramen ovale to exit the cranial cavity carrying preganglionic parasympathetic fibers to the otic ganglion (Table 8.6).

### Vagus nerve [X]

The vagus nerve [X] carries GSA, GVA, SA, GVE, and BE fibers:

- The GSA fibers provide sensory input from the larynx, laryngopharynx, deeper parts of the auricle, part of the external acoustic meatus, and the dura mater in the posterior cranial fossa.
- The GVA fibers provide sensory input from the aortic body chemoreceptors and aortic arch baroreceptors, and the esophagus, bronchi, lungs, heart, and abdominal viscera in the foregut and midgut.
- The SA fibers are for taste around the epiglottis and pharynx.
- The GVE fibers are part of the parasympathetic part of the autonomic division of the PNS and stimulate smooth muscle and glands in the pharynx, larynx, thoracic viscera, and abdominal viscera of the foregut and midgut.
- The BE fibers innervate one muscle of the tongue (palatoglossus), the muscles of the soft palate (except the tensor veli palatini), pharynx (except the stylopharyngeus), and larynx.

The vagus nerve arises as a group of rootlets on the anterolateral surface of the medulla oblongata just inferior to the rootlets arising to form the glossopharyngeal nerve [IX] (Fig. 8.54). The rootlets cross the posterior cranial fossa and enter the jugular foramen (Fig. 8.53). Within this foramen, and before exiting from it, the rootlets merge to form the vagus nerve [X]. Within or immediately outside the jugular foramen are two ganglia, the **superior** (jugular) and **inferior** (nodose) **ganglia**, which contain the cell bodies of the sensory neurons in the vagus nerve [X].

### Accessory nerve [XI]

The accessory nerve [XI] is a cranial nerve that carries BE fibers to innervate the sternocleidomastoid and trapezius muscles (see Diogo R et al. *Nature* 2015;520:466–473). It is a unique cranial nerve because its roots arise from motor neurons in the upper five segments of the cervical spinal cord. These fibers leave the lateral surface of the spinal cord and, joining together as they ascend, enter the cranial cavity through the foramen magnum (Fig. 8.54). The accessory nerve [XI] continues through the posterior cranial fossa and exits through the jugular foramen (Fig. 8.53). It then descends in the neck to innervate the sternocleidomastoid and trapezius muscles from their deep surfaces.

#### Cranial root of the accessory nerve

Some descriptions of the accessory nerve [XI] refer to a few rootlets arising from the caudal part of the medulla oblongata on the anterolateral surface just inferior to the rootlets arising to form the vagus nerve [X] as the "cranial" root of the accessory nerve (Fig. 8.54). Leaving the medulla, the cranial roots course with the "spinal" roots of the accessory nerve [XI] into the jugular foramen, at which point the cranial roots join the vagus nerve [X]. As part of the vagus nerve [X], they are distributed to the pharyngeal musculature innervated by the vagus nerve [X] and are therefore described as being part of the vagus nerve [X].

### Hypoglossal nerve [XII]

The hypoglossal nerve [XII] carries GSE fibers to innervate all intrinsic muscles and most of the extrinsic muscles of the tongue. It arises as several rootlets from the anterior surface of the medulla (Fig. 8.54), passes laterally across the posterior cranial fossa, and exits through the hypoglossal canal (Fig. 8.53). This nerve innervates the hyoglossus, styloglossus, and genioglossus muscles and all intrinsic muscles of the tongue.

## FACE

A face-to-face meeting is an important initial contact between individuals. Part of this exchange is the use of facial expressions to convey emotions. In fact, a physician can gain important information about an individual's general health by observing a patient's face.

Thus an understanding of the unique organization of the various structures between the superciliary arches superiorly, the lower edge of the mandible inferiorly, and as far back as the ears on either side, the area defined as the face, is particularly useful in the practice of medicine.

### Muscles

The muscles of the face (Fig. 8.56) develop from the second pharyngeal arch and are innervated by branches of the

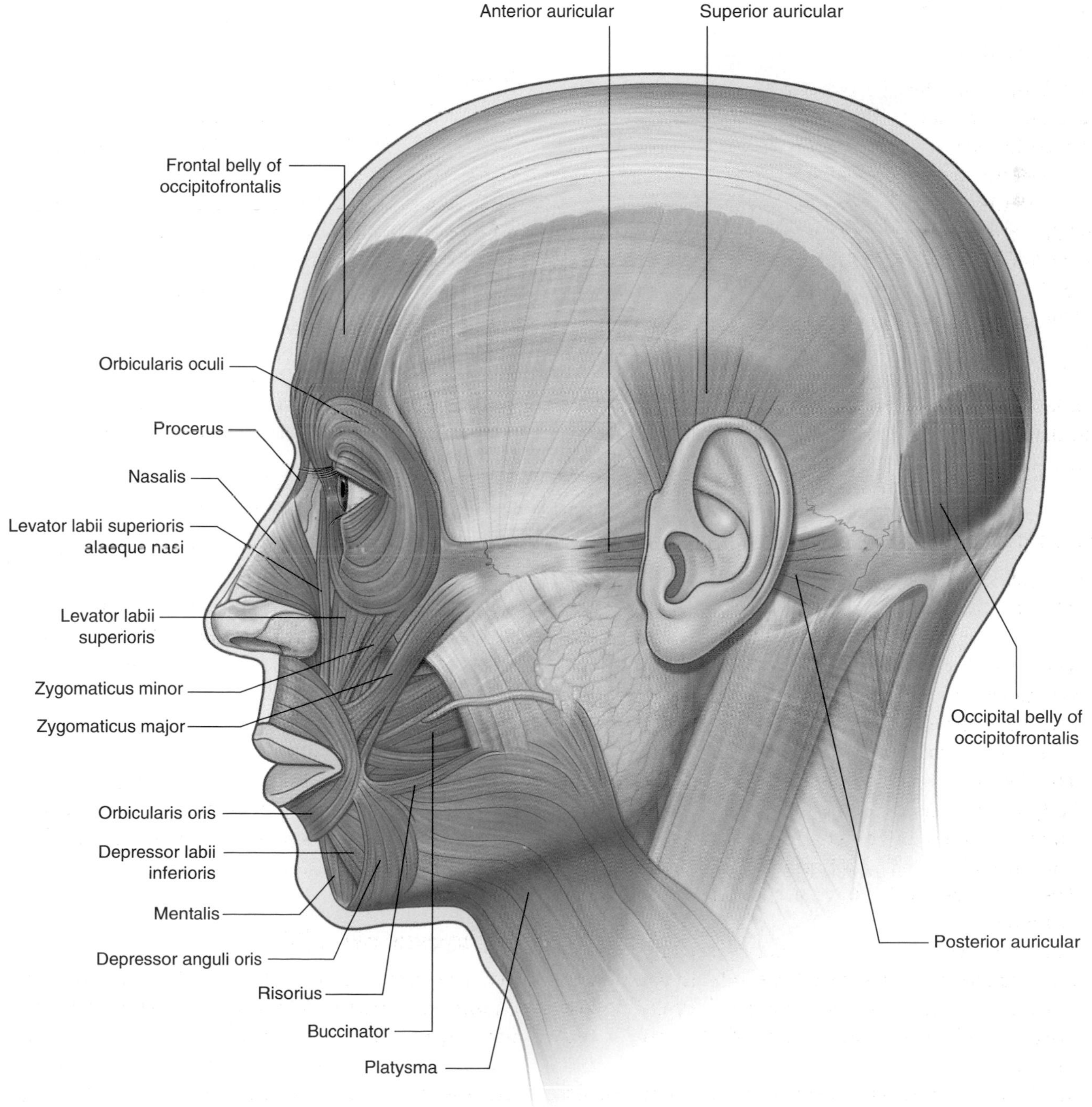


**Fig. 8.56** Facial muscles.

facial nerve [VII]. They are in the superficial fascia, with origins from either bone or fascia, and insertions into the skin.

Because these muscles control expressions of the face, they are sometimes referred to as muscles of "facial expression." They also act as sphincters and dilators of the orifices of the face (i.e., the orbits, nose, and mouth). This organizational arrangement into functional groups provides a logical approach to understanding these muscles (Table 8.7).

### Orbital group

Two muscles are associated with the orbital group—the orbicularis oculi and the corrugator supercilii.

**Table 8.7** Muscles of the face

| Muscle | Origin | Insertion | Innervation | Function |
|---|---|---|---|---|
| **ORBITAL GROUP** | | | | |
| Orbicularis oculi | | | | |
| —Palpebral part | Medial palpebral ligament | Lateral palpebral raphe | Facial nerve [VII] | Closes the eyelids gently |
| —Orbital part | Nasal part of frontal bone; frontal process of maxilla; medial palpebral ligament | Fibers form an uninterrupted ellipse around orbit | Facial nerve [VII] | Closes the eyelids forcefully |
| Corrugator supercilii | Medial end of the superciliary arch | Skin of the medial half of eyebrow | Facial nerve [VII] | Draws the eyebrows medially and downward |
| **NASAL GROUP** | | | | |
| Nasalis | | | | |
| —Transverse part | Maxilla just lateral to nose | Aponeurosis across dorsum of nose with muscle fibers from the other side | Facial nerve [VII] | Compresses nasal aperture |
| —Alar part | Maxilla over lateral incisor | Alar cartilage of nose | Facial nerve [VII] | Draws cartilage downward and laterally, opening nostril |
| Procerus | Nasal bone and upper part of lateral nasal cartilage | Skin of lower forehead between eyebrows | Facial nerve [VII] | Draws down medial angle of eyebrows, producing transverse wrinkles over bridge of nose |
| Depressor septi | Maxilla above medial incisor | Mobile part of the nasal septum | Facial nerve [VII] | Pulls nose inferiorly |
| **ORAL GROUP** | | | | |
| Depressor anguli oris | Oblique line of mandible below canine, premolar, and first molar teeth | Skin at the corner of mouth and blending with orbicularis oris | Facial nerve [VII] | Draws corner of mouth downward and laterally |
| Depressor labii inferioris | Anterior part of oblique line of mandible | Lower lip at midline; blends with muscle from opposite side | Facial nerve [VII] | Draws lower lip downward and laterally |
| Mentalis | Mandible inferior to incisor teeth | Skin of chin | Facial nerve [VII] | Raises and protrudes lower lip as it wrinkles skin on chin |

**Table 8.7** Muscles of the face—cont'd

| Muscle | Origin | Insertion | Innervation | Function |
|---|---|---|---|---|
| Risorius | Fascia over masseter muscle | Skin at the corner of the mouth | Facial nerve [VII] | Retracts corner of mouth |
| Zygomaticus major | Posterior part of lateral surface of zygomatic bone | Skin at the corner of the mouth | Facial nerve [VII] | Draws the corner of the mouth upward and laterally |
| Zygomaticus minor | Anterior part of lateral surface of zygomatic bone | Upper lip just medial to corner of mouth | Facial nerve [VII] | Draws the upper lip upward |
| Levator labii superioris | Infra-orbital margin of maxilla | Skin of upper lateral half of upper lip | Facial nerve [VII] | Raises upper lip; helps form nasolabial furrow |
| Levator labii superioris alaeque nasi | Frontal process of maxilla | Alar cartilage of nose and upper lip | Facial nerve [VII] | Raises upper lip and opens nostril |
| Levator anguli oris | Maxilla below infra-orbital foramen | Skin at the corner of mouth | Facial nerve [VII] | Raises corner of mouth; helps form nasolabial furrow |
| Orbicularis oris | From muscles in area; maxilla and mandible in midline | Forms ellipse around mouth | Facial nerve [VII] | Closes lips; protrudes lips |
| Buccinator | Posterior parts of maxilla and mandible; pterygomandibular raphe | Blends with orbicularis oris and into lips | Facial nerve [VII] | Presses the cheek against teeth; compresses distended cheeks |
| **OTHER MUSCLES OR GROUPS** | | | | |
| Anterior auricular | Anterior part of temporal fascia | Into helix of ear | Facial nerve [VII] | Draws ear upward and forward |
| Superior auricular | Epicranial aponeurosis on side of head | Upper part of auricle | Facial nerve [VII] | Elevates ear |
| Posterior auricular | Mastoid process of temporal bone | Convexity of concha of ear | Facial nerve [VII] | Draws ear upward and backward |
| Occipitofrontalis | | | | |
| —Frontal belly | Skin of eyebrows | Into galea aponeurotica | Facial nerve [VII] | Wrinkles forehead; raises eyebrows |
| —Occipital belly | Lateral part of superior nuchal line of occipital bone and mastoid process of temporal bone | Into galea aponeurotica | Facial nerve [VII] | Draws scalp backward |

### Orbicularis oculi

The **orbicularis oculi** is a large muscle that completely surrounds each orbital orifice and extends into each eyelid (Fig. 8.57). It closes the eyelids. It has two major parts:

- The outer **orbital part** is a broad ring that encircles the orbital orifice and extends outward beyond the orbital rim.
- The inner **palpebral part** is in the eyelids and consists of muscle fibers originating in the medial corner of the eye that arch across each lid to attach laterally.

The orbital and palpebral parts have specific roles to play during eyelid closure. The palpebral part closes the eye gently, whereas the orbital part closes the eye more forcefully and produces some wrinkling on the forehead.

An additional small lacrimal part of the orbicularis oculi muscle is deep, medial in position, and attaches to bone posterior to the lacrimal sac of the lacrimal apparatus in the orbit.

### Corrugator supercilii

The second muscle in the orbital group is the much smaller **corrugator supercilii** (Fig. 8.57), which is deep to the eyebrows and the orbicularis oculi muscle and is active when frowning. It arises from the medial end of the superciliary arch, passing upward and laterally to insert into the skin of the medial half of the eyebrow. It draws the eyebrows toward the midline, causing vertical wrinkles above the nose.

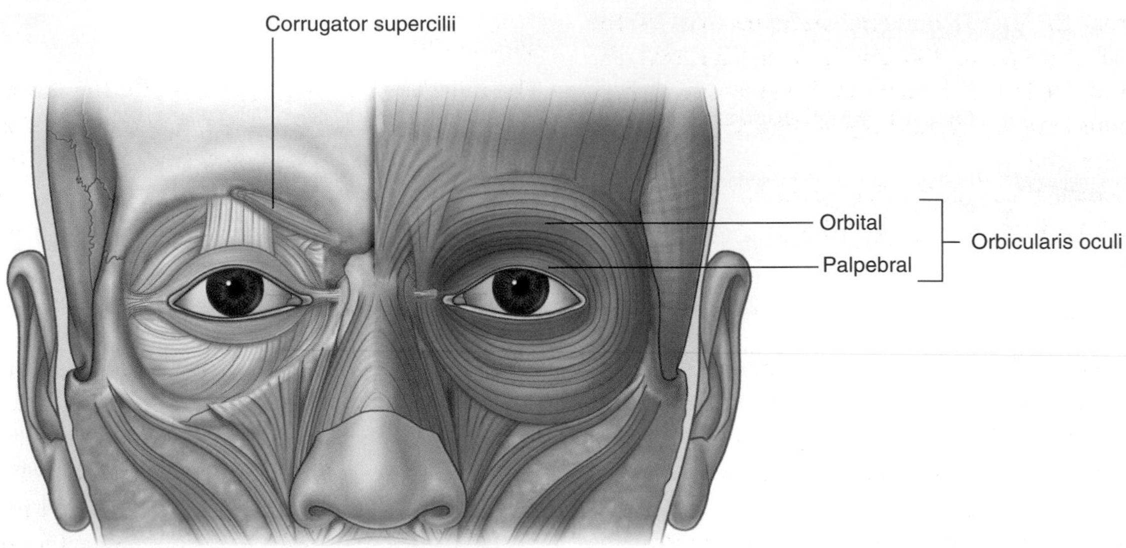


**Fig. 8.57** Orbital group of facial muscles.

## Nasal group

Three muscles are associated with the nasal group—the nasalis, the procerus, and the depressor septi nasi (Fig. 8.58).

### Nasalis

The largest and best developed of the muscles of the nasal group is the **nasalis**, which is active when the nares are flared (Fig. 8.58). It consists of a transverse part (the compressor naris) and an alar part (the dilator naris):

- The **transverse part** of the nasalis compresses the nares—it originates from the maxilla and its fibers pass upward and medially to insert, along with fibers from the same muscle on the opposite side, into an aponeurosis across the dorsum of the nose.
- The **alar part** of the nasalis draws the alar cartilages downward and laterally, so opening the nares—it originates from the maxilla, below and medial to the transverse part, and inserts into the alar cartilage.

### Procerus

The **procerus** is a small muscle superficial to the nasal bone and is active when an individual frowns (Fig. 8.58).

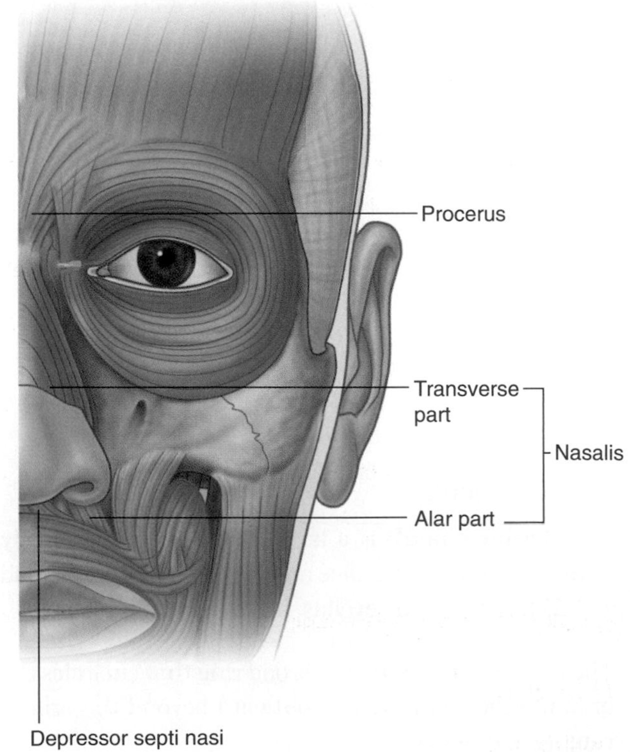


**Fig. 8.58** Nasal group of facial muscles.

It arises from the nasal bone and the upper part of the lateral nasal cartilage and inserts into the skin over the lower part of the forehead between the eyebrows. It may be continuous with the frontal belly of the occipitofrontalis muscle of the scalp.

The procerus draws the medial border of the eyebrows downward to produce transverse wrinkles over the bridge of the nose.

### Depressor septi nasi

The final muscle in the nasal group is the **depressor septi nasi**, another muscle that assists in widening the nares (Fig. 8.58). Its fibers arise from the maxilla above the central incisor tooth and ascend to insert into the lower part of the nasal septum.

The depressor septi nasi pulls the nose inferiorly, so assisting the alar part of the nasalis in opening the nares.

## Oral group

The muscles in the oral group move the lips and cheek. They include the orbicularis oris and buccinator muscles, and a lower and upper group of muscles (Fig. 8.59). Many of these muscles intersect just lateral to the corner of the mouth on each side at a structure termed the modiolus.

### Orbicularis oris

The **orbicularis oris** is a complex muscle consisting of fibers that completely encircle the mouth (Fig. 8.59). Its function is apparent when pursing the lips, as occurs during whistling. Some of its fibers originate near the midline from the maxilla superiorly and the mandible inferiorly, whereas other fibers are derived from both the buccinator, in the cheek, and the numerous other muscles

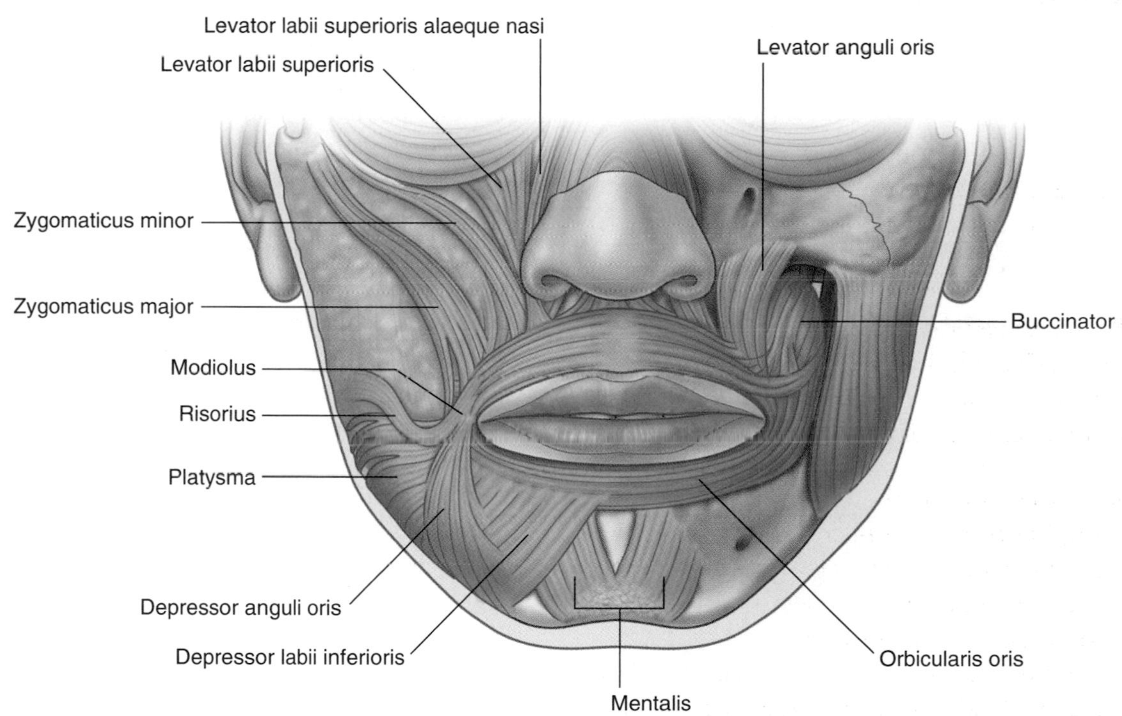


**Fig. 8.59** Oral group of facial muscles.

## In the clinic

### Facelifts and botox

Facelift surgery (rhytidectomy) aims to lift up and pull back the skin in the lower half of the face and neck to make the face more taught. Careful placement of the incisions is important to ensure there is no skin or facial distortion and to avoid hair loss. The commonest incisions are placed in the temporal region on each side, extending to the helices of the ears, then tracking behind the tragus, around the earlobes, and then to the occiput.

Botox is derived from the toxin produced by the bacterium *Clostridium botulinum*, which blocks neuromuscular junctions resulting in muscle relaxation. It is used in many therapies including strabismus (crossed eyes) where it is injected into extra-ocular muscles. Its injection is also used to treat uncontrolled blinking (blepharospasm), spastic muscle conditions, and overactive bladder disorders, as well as to relax facial muscles to improve the cosmetic appearances of lines and wrinkles and to treat patients with excessive sweating (hyperhidrosis).

acting on the lips. It inserts into the skin and mucous membrane of the lips, and into itself.

Contraction of the orbicularis oris narrows the mouth and closes the lips.

### Buccinator

The buccinator forms the muscular component of the cheek and is used every time air expanding the cheeks is forcefully expelled (Figs. 8.59 and 8.60). It is in the space between the mandible and the maxilla, deep to the other facial muscles in the area.

The buccinator arises from the posterior part of the maxilla and mandible opposite the molar teeth and the **pterygomandibular raphe**, which is a tendinous band between the pterygoid hamulus superiorly and the mandible inferiorly and is a point of attachment for the buccinator and superior pharyngeal constrictor muscles.

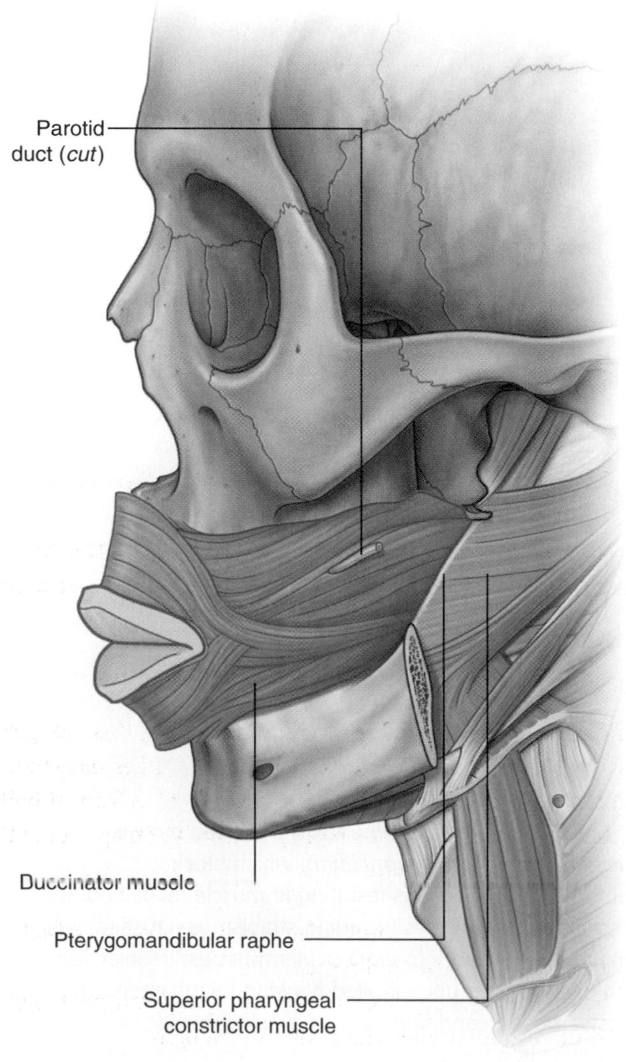


**Fig. 8.60** Buccinator muscle.

The fibers of the buccinator pass toward the corner of the mouth to insert into the lips, blending with fibers from the orbicularis oris in a unique fashion. Central fibers of the buccinator cross so that lower fibers enter the upper lip and upper fibers enter the lower lip (Fig. 8.60). The highest and lowest fibers of the buccinator do not cross and enter the upper and lower lips, respectively.

Contraction of the buccinator presses the cheek against the teeth. This keeps the cheek taut and aids in mastication by preventing food from accumulating between the teeth and the cheek. The muscle also assists in the forceful expulsion of air from the cheeks.

### Lower group of oral muscles

The muscles in the lower group consist of the depressor anguli oris, depressor labii inferioris. and mentalis (Fig. 8.59).

- The **depressor anguli oris** is active during frowning. It arises along the side of the mandible below the canine, premolar, and first molar teeth and inserts into skin and the upper part of the orbicularis oris near the corner of the mouth. It depresses the corner of the mouth.
- The **depressor labii inferioris** arises from the front of the mandible, deep to the depressor anguli oris. Its fibers move superiorly and medially, some merging with fibers from the same muscle on the opposite side and fibers from the orbicularis oris before inserting into the lower lip. It depresses the lower lip and moves it laterally.
- The **mentalis** helps position the lip when drinking from a cup or when pouting. It is the deepest muscle of the lower group arising from the mandible just inferior to the incisor teeth, with its fibers passing downward and medially to insert into the skin of the chin. It raises and protrudes the lower lip as it wrinkles the skin of the chin.

### Upper group of oral muscles

The muscles of the upper group of oral muscles consist of the risorius, zygomaticus major, zygomaticus minor, levator labii superioris, levator labii superioris alaeque nasi, and levator anguli oris (Fig. 8.59).

- The **risorius** helps produce a grin (Fig. 8.59). It is a thin, superficial muscle that extends laterally from the corner of the mouth in a slightly upward direction. Contraction of its fibers pulls the corner of the mouth laterally and upward.
- The **zygomaticus major** and **zygomaticus minor** help produce a smile (Fig. 8.59). The zygomaticus major

is a superficial muscle that arises deep to the orbicularis oculi along the posterior part of the lateral surface of the zygomatic bone, and passes downward and forward, blending with the orbicularis oris and inserting into skin at the corner of the mouth. The zygomaticus minor arises from the zygomatic bone anterior to the origin of the zygomaticus major, parallels the path of the zygomaticus major, and inserts into the upper lip medial to the corner of the mouth. Both zygomaticus muscles raise the corner of the mouth and move it laterally.

- The **levator labii superioris** deepens the furrow between the nose and the corner of the mouth during sadness (Fig. 8.59). It arises from the maxilla just superior to the infra-orbital foramen, and its fibers pass downward and medially to blend with the orbicularis oris and insert into the skin of the upper lip.
- The **levator labii superioris alaeque nasi** is medial to the levator labii superioris, arises from the maxilla next to the nose, and inserts into both the alar cartilage of the nose and skin of the upper lip (Fig. 8.59). It may assist in flaring the nares.
- The **levator anguli oris** is more deeply placed and covered by the other two levators and the zygomaticus muscles (Fig. 8.59). It arises from the maxilla, just inferior to the infra-orbital foramen and inserts into the skin at the corner of the mouth. It elevates the corner of the mouth and may help deepen the furrow between the nose and the corner of the mouth during sadness.

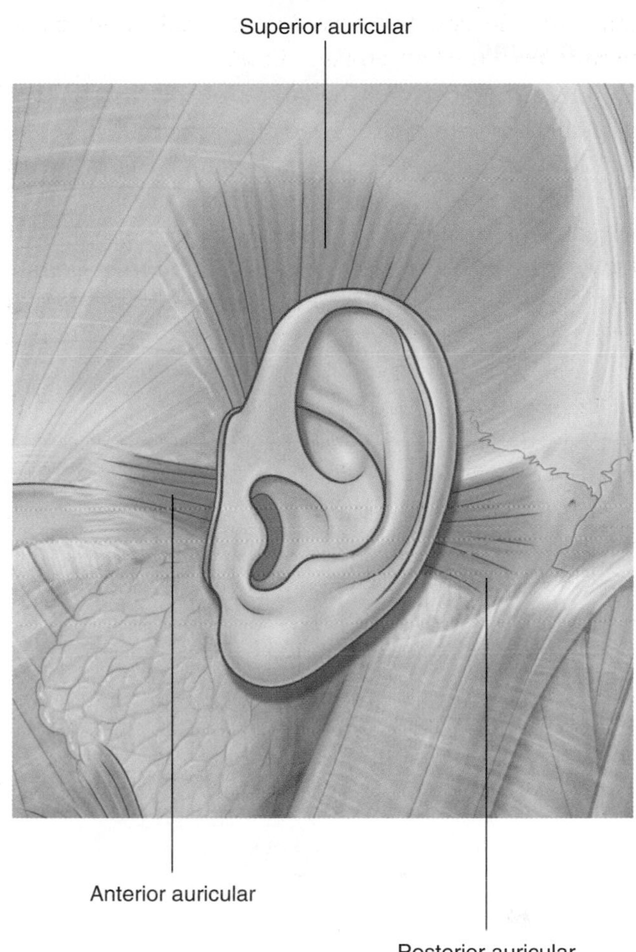


**Fig. 8.61** Auricular muscles.

### Other muscles or muscle groups

Several additional muscles or groups of muscles not in the area defined as the face, but derived from the second pharyngeal arch and innervated by the facial nerve [VII], are considered muscles of facial expression. They include the platysma, auricular, and occipitofrontalis muscles (see Fig. 8.56).

#### Platysma

The **platysma** is a large, thin sheet of muscle in the superficial fascia of the neck. It arises below the clavicle in the upper part of the thorax and ascends through the neck to the mandible. At this point, the more medial fibers insert on the mandible, whereas the lateral fibers join with muscles around the mouth.

The platysma tenses the skin of the neck and can move the lower lip and corners of the mouth down.

#### Auricular muscles

Three of these muscles, "other muscles of facial expression," are associated with the ear—the anterior, superior, and posterior **auricular muscles** (Fig. 8.61):

- The anterior muscle is anterolateral and pulls the ear upward and forward.
- The superior muscle is superior and elevates the ear.
- The posterior muscle is posterior and retracts and elevates the ear.

#### Occipitofrontalis

The **occipitofrontalis** is the final muscle in this category of "other muscles of facial expression" and is associated with the scalp (see Fig. 8.56). It consists of a frontal belly anteriorly and an occipital belly posteriorly. An aponeurotic tendon connects the two:

- The frontal belly covers the forehead and is attached to the skin of the eyebrows.
- The occipital belly arises from the posterior aspect of the skull and is smaller than the frontal belly.

The occipitofrontalis muscles move the scalp and wrinkle the forehead.

## Parotid gland

The **parotid glands** are the largest of the three pairs of main salivary glands in the head and numerous structures pass through them. They are anterior to and below the lower half of the ear, superficial, posterior, and deep to the ramus of the mandible (Fig. 8.62). They extend down to the lower border of the mandible and up to the zygomatic arch. Posteriorly they cover the anterior part of the sternocleidomastoid muscle and continue anteriorly to halfway across the masseter muscle.

The **parotid duct** leaves the anterior edge of the parotid gland midway between the zygomatic arch and the corner of the mouth (Fig. 8.62). It crosses the face in a transverse direction and, after crossing the medial border of the masseter muscle, turns deeply into the buccal fat pad and pierces the buccinator muscle. It opens into the oral cavity near the second upper molar tooth.

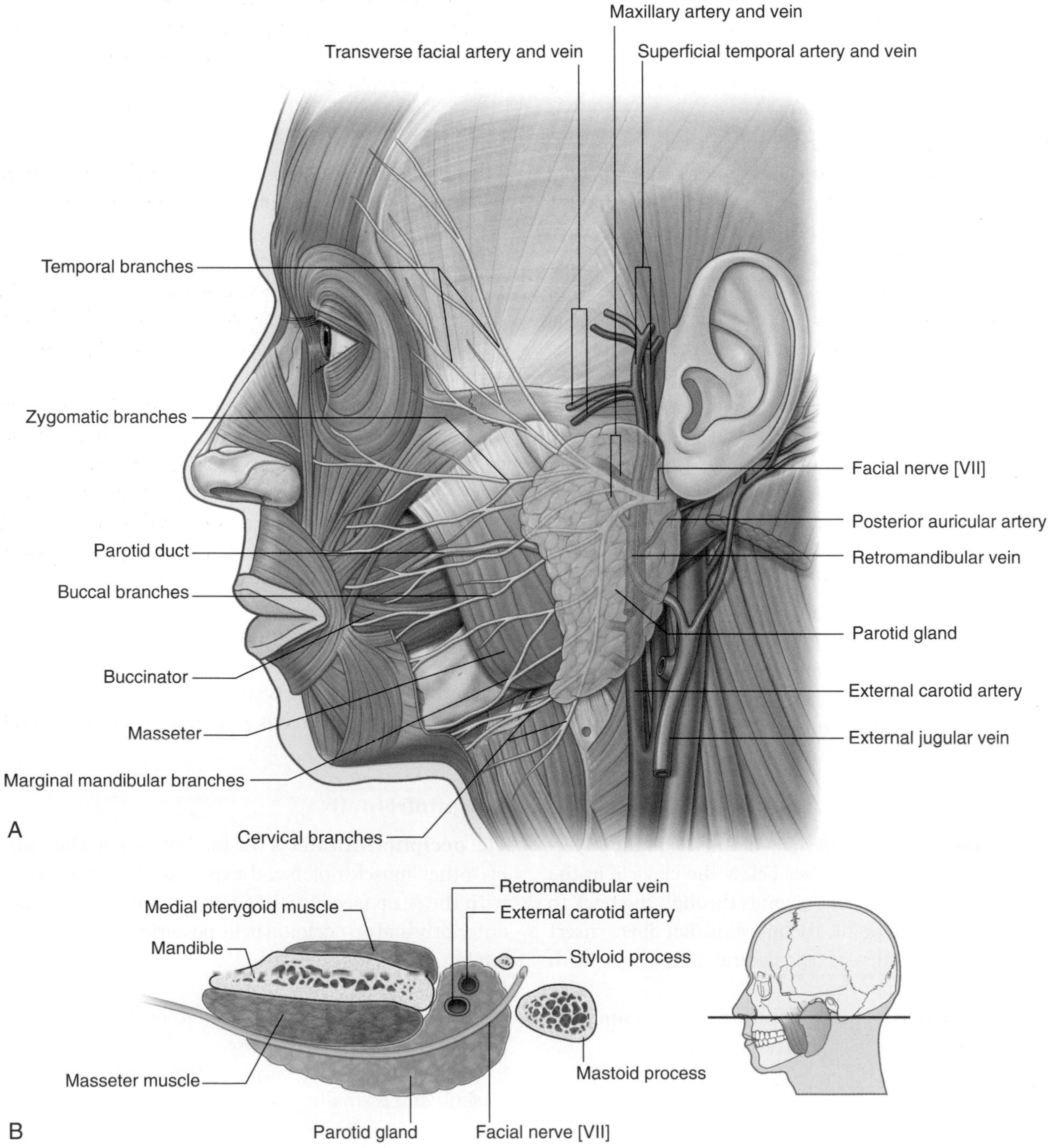

**Fig. 8.62** Parotid gland. **A.** Lateral view. **B.** Cross section.

### Important relationships

Several major structures enter and pass through or pass just deep to the parotid gland. These include the facial nerve [VII], the external carotid artery and its branches, and the retromandibular vein and its tributaries (Fig. 8.62).

#### Facial nerve

The facial nerve [VII] exits the skull through the stylomastoid foramen and then passes into the parotid gland, where it usually divides into upper and lower trunks. These pass through the substance of the parotid gland, where there may be further branching and anastomosing of the nerves.

Five terminal groups of branches of the facial nerve [VII]—the **temporal**, **zygomatic**, **buccal**, **marginal mandibular**, and **cervical branches**—emerge from the upper, anterior, and lower borders of the parotid gland (Fig. 8.62).

The intimate relationships between the facial nerve [VII] and the parotid gland mean that surgical removal of the parotid gland is a difficult dissection if all branches of the facial nerve [VII] are to be spared.

#### External carotid artery and its branches

The external carotid artery enters into or passes deep to the inferior border of the parotid gland (Fig. 8.62). As it continues in a superior direction, it gives off the **posterior auricular artery** before dividing into its two terminal branches (the **maxillary** and **superficial temporal arteries**) near the lower border of the ear:

- The maxillary artery passes horizontally, deep to the mandible.
- The superficial temporal artery continues in a superior direction and emerges from the upper border of the gland after giving off the **transverse facial artery**.

#### Retromandibular vein and its tributaries

The retromandibular vein is formed in the substance of the parotid gland when the **superficial temporal** and **maxillary veins** join together (Fig. 8.62), and passes inferiorly in the substance of the parotid gland. It usually divides into anterior and posterior branches just below the inferior border of the gland.

### Arterial supply

The parotid gland receives its arterial supply from the numerous arteries that pass through its substance.

### Innervation

Sensory innervation of the parotid gland is provided by the **auriculotemporal nerve**, which is a branch of the mandibular nerve [$V_3$]. This division of the trigeminal nerve exits the skull through the foramen ovale.

The auriculotemporal nerve also carries secretomotor fibers to the parotid gland. These postganglionic parasympathetic fibers have their origin in the otic ganglion associated with the mandibular nerve [$V_3$] and are just inferior to the foramen ovale. Preganglionic parasympathetic fibers to the otic ganglion come from the glossopharyngeal nerve [IX].

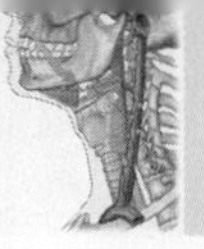



## In the clinic

### Parotid gland

The parotid gland is the largest of the paired salivary glands and is enclosed within the split investing layer of deep cervical fascia.

The parotid gland produces a watery saliva and salivary amylase, which are necessary for food bolus formation, oral digestion, and smooth passage of the bolus into the upper gastrointestinal tract.

### *Tumors of the parotid gland*

The commonest tumors of the parotid gland (Fig. 8.63) are benign and typically involve the more superficial part of the gland. These include pleomorphic adenoma and adenolymphoma. Their importance is in relation to their anatomical position. The relationship of any tumor to the branches of the facial nerve [VII] must be defined because resection of the tumor may damage the nerve.

### *Parotid gland stones*

It is not uncommon for stones to develop within the parotid gland. They typically occur within the main confluence of the ducts and within the main parotid duct. The patient usually complains of intense pain when salivating and tends to avoid foods that produce this symptom. The pain can be easily reproduced in the clinic by squirting lemon juice into the patient's mouth.

Surgery depends upon where the stone is. If it is within the anterior aspect of the duct, a simple incision in the buccal mucosa with a sphincterotomy may allow removal. If the stone is farther back within the main duct, complete gland excision may be necessary.

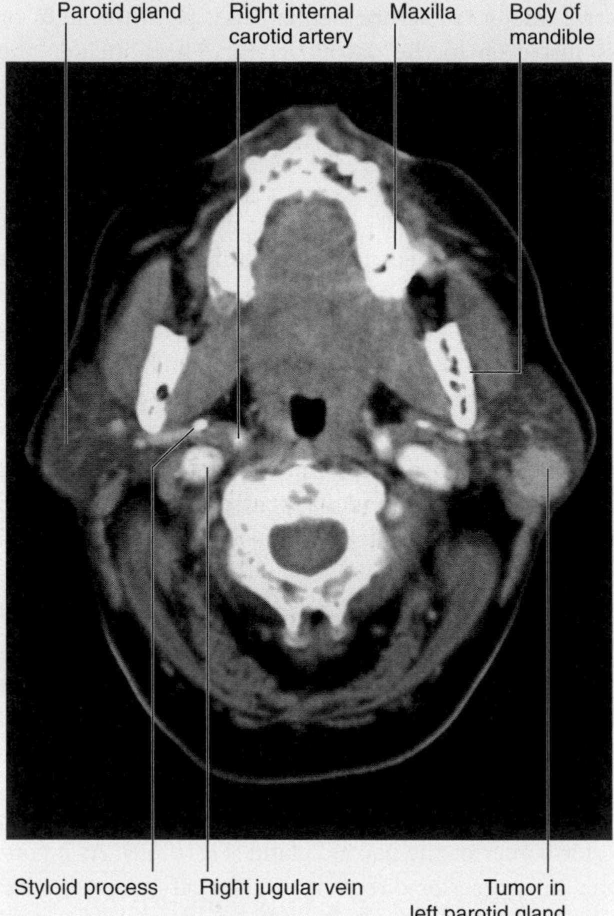


**Fig. 8.63** Tumor in parotid gland. Axial CT scan.

## Innervation

During development a cranial nerve becomes associated with each of the pharyngeal arches. Because the face is primarily derived from the first and second pharyngeal arches, innervation of neighboring facial structures is as follows:

- The trigeminal nerve [V] innervates facial structures derived from the first arch.
- The facial nerve [VII] innervates facial structures derived from the second arch.

### Sensory innervation

Because the face is derived developmentally from a number of structures originating from the first pharyngeal arch, cutaneous innervation of the face is by branches of the trigeminal nerve [V].

The trigeminal nerve [V] divides into three major divisions—the ophthalmic [$V_1$], maxillary [$V_2$], and mandibular [$V_3$]—before leaving the middle cranial fossa (Fig. 8.64). Each of these divisions passes out of the cranial cavity to innervate a part of the face, so most of the skin covering the face is innervated solely by branches of the trigeminal nerve [V]. The exception is a small area covering the angle and lower border of the ramus of the mandible and parts of the ear, where the facial [VII], vagus [X], and cervical nerves contribute to the innervation.

#### Ophthalmic nerve [$V_1$]

The ophthalmic nerve [$V_1$] exits the skull through the superior orbital fissure and enters the orbit. Its branches (Fig. 8.64) that innervate the face include:

- the **supra-orbital** and **supratrochlear nerves**, which leave the orbit superiorly and innervate the upper eyelid, forehead, and scalp;

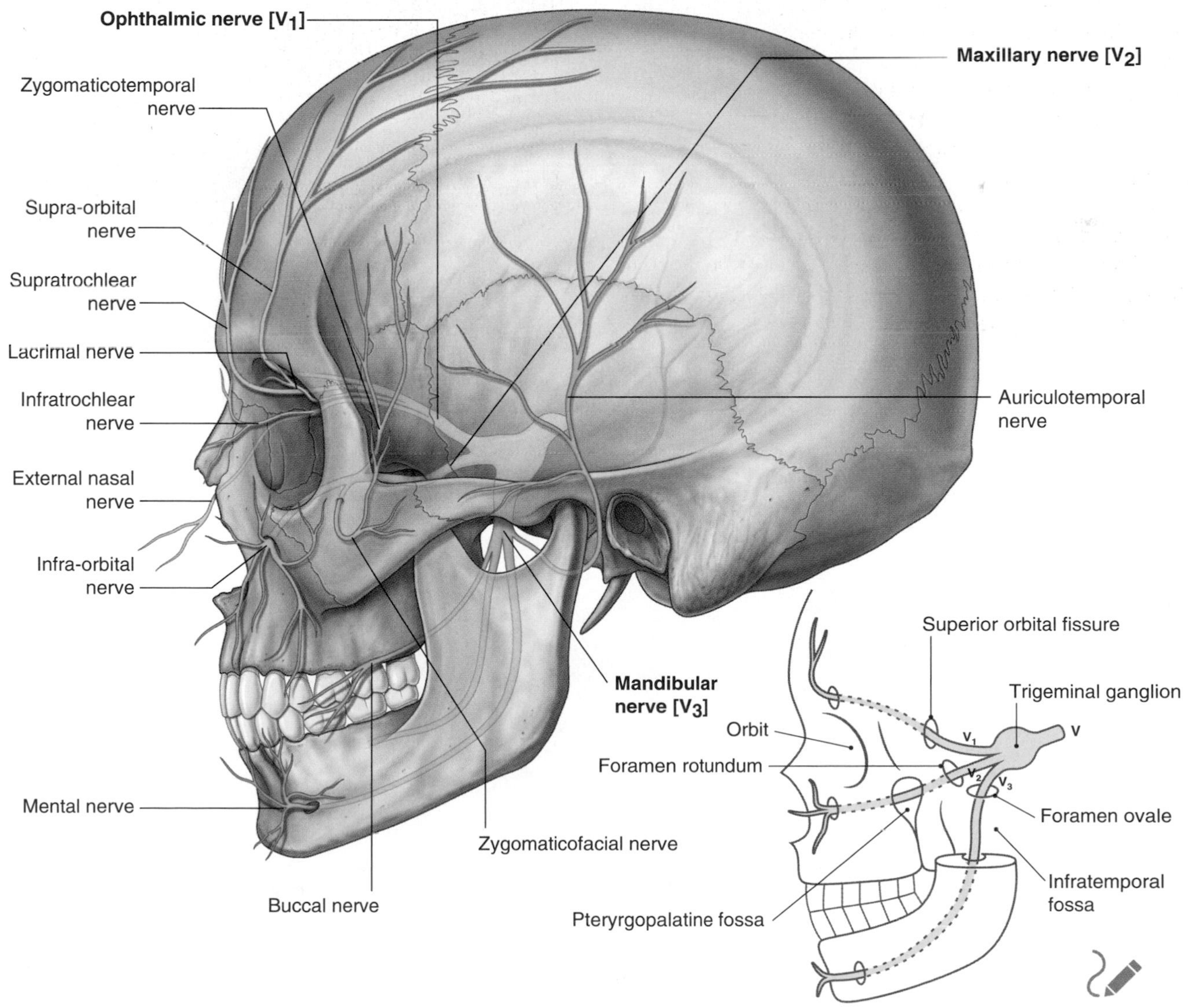


**Fig. 8.64** Trigeminal nerve [V] leaving the skull.

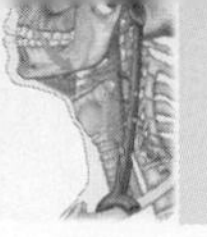
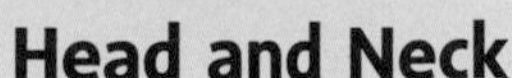


- the **infratrochlear nerve**, which exits the orbit in the medial angle to innervate the medial half of the upper eyelid, the skin in the area of the medial angle, and the side of the nose;
- the **lacrimal nerve**, which exits the orbit in the lateral angle to innervate the lateral half of the upper eyelid and the skin in the area of the lateral angle; and
- the **external nasal nerve**, which supplies the anterior part of the nose (Fig. 8.65).

### Maxillary nerve [$V_2$]

The maxillary nerve [$V_2$] exits the skull through the foramen rotundum. Branches (Fig. 8.64) that innervate the face include:

- a small **zygomaticotemporal branch**, which exits the zygomatic bone and supplies a small area of the anterior temple above the zygomatic arch;
- a small **zygomaticofacial branch**, which exits the zygomatic bone and supplies a small area of skin over the zygomatic bone; and
- the large **infra-orbital nerve**, which exits the maxilla through the infra-orbital foramen and immediately divides into multiple branches to supply the lower eyelid, cheek, side of the nose, and upper lip (Fig. 8.65).

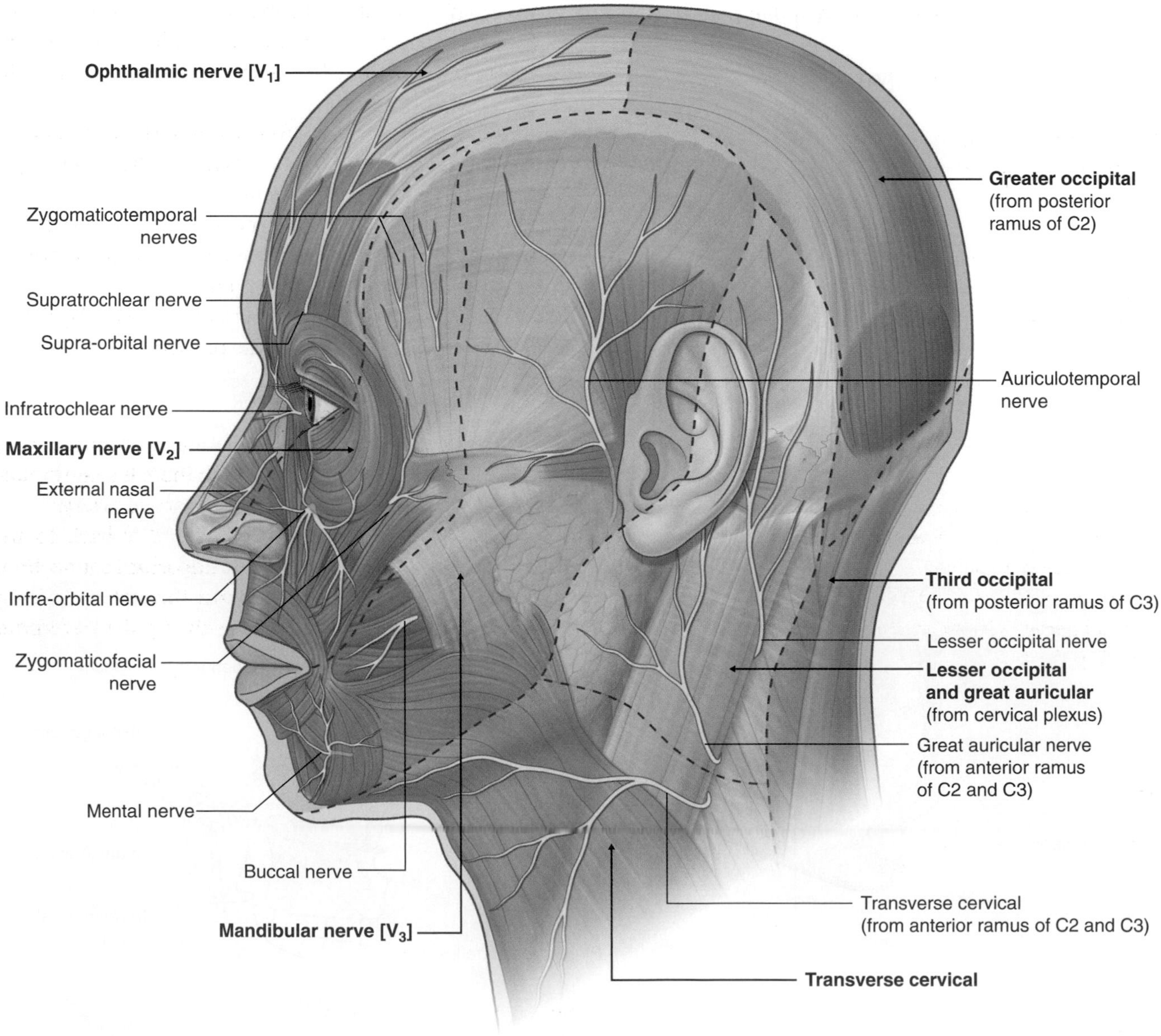

**Ophthalmic nerve [$V_1$]**
Zygomaticotemporal nerves
Supratrochlear nerve
Supra-orbital nerve
Infratrochlear nerve
**Maxillary nerve [$V_2$]**
External nasal nerve
Infra-orbital nerve
Zygomaticofacial nerve
Mental nerve
Buccal nerve
**Mandibular nerve [$V_3$]**
**Greater occipital** (from posterior ramus of C2)
Auriculotemporal nerve
**Third occipital** (from posterior ramus of C3)
Lesser occipital nerve
**Lesser occipital and great auricular** (from cervical plexus)
Great auricular nerve (from anterior ramus of C2 and C3)
Transverse cervical (from anterior ramus of C2 and C3)
**Transverse cervical**

**Fig. 8.65** Cutaneous distribution of the trigeminal nerve [V].

### Mandibular nerve [$V_3$]

The mandibular nerve [$V_3$] exits the skull through the foramen ovale. Branches (Fig. 8.65) innervating the face include:

- the **auriculotemporal nerve**, which enters the face just posterior to the temporomandibular joint, passes through the parotid gland, and ascends just anterior to the ear to supply the external acoustic meatus, the surface of the tympanic membrane (eardrum), and a large area of the temple;
- the **buccal nerve**, which is on the surface of the buccinator muscle supplying the cheek; and
- the **mental nerve**, which exits the mandible through the mental foramen and immediately divides into multiple branches to supply the skin and mucous membrane of the lower lip and skin of the chin (Fig. 8.65).

### Motor innervation

The muscles of the face, as well as those associated with the external ear and the scalp, are derived from the second pharyngeal arch. The cranial nerve associated with this arch is the facial nerve [VII] and therefore branches of the facial nerve [VII] innervate all these muscles.

The facial nerve [VII] exits the posterior cranial fossa through the internal acoustic meatus. It passes through the temporal bone, giving off several branches, and emerges from the base of the skull through the stylomastoid foramen (Fig. 8.66). At this point it gives off the **posterior auricular nerve**. This branch passes upward, behind the ear, to supply the occipital belly of the occipitofrontalis muscle of the scalp and the posterior auricular muscle of the ear.

The main stem of the facial nerve [VII] then gives off another branch, which innervates the posterior belly of the digastric muscle and the stylohyoid muscle. At this point, the facial nerve [VII] enters the deep surface of the parotid gland (Fig. 8.66B).

Once in the parotid gland, the main stem of the facial nerve [VII] usually divides into upper (temporofacial) and lower (cervicofacial) branches. As these branches pass through the substance of the parotid gland they may branch further or take part in an anastomotic network (the parotid plexus).

Whatever types of interconnections occur, five terminal groups of branches of the facial nerve [VII]—the temporal, zygomatic, buccal, marginal mandibular, and cervical branches—emerge from the parotid gland (Fig. 8.66A).

Although there are variations in the pattern of distribution of the five terminal groups of branches, the basic pattern is as follows:

- Temporal branches exit from the superior border of the parotid gland to supply muscles in the area of the temple, forehead, and supra-orbital area.
- Zygomatic branches emerge from the anterosuperior border of the parotid gland to supply muscles in the infra-orbital area, the lateral nasal area, and the upper lip.
- Buccal branches emerge from the anterior border of the parotid gland to supply muscles in the cheek, the upper lip, and the corner of the mouth.
- Marginal mandibular branches emerge from the anteroinferior border of the parotid gland to supply muscles of the lower lip and chin.
- Cervical branches emerge from the inferior border of the parotid gland to supply the platysma.

## Vessels

The arterial supply to the face is primarily from branches of the external carotid artery, though there is some limited supply from a branch of the internal carotid artery.

Similarly, most of the venous return is back to the internal jugular vein, though some important connections from the face result in venous return through a clinically relevant intracranial pathway involving the cavernous sinus.

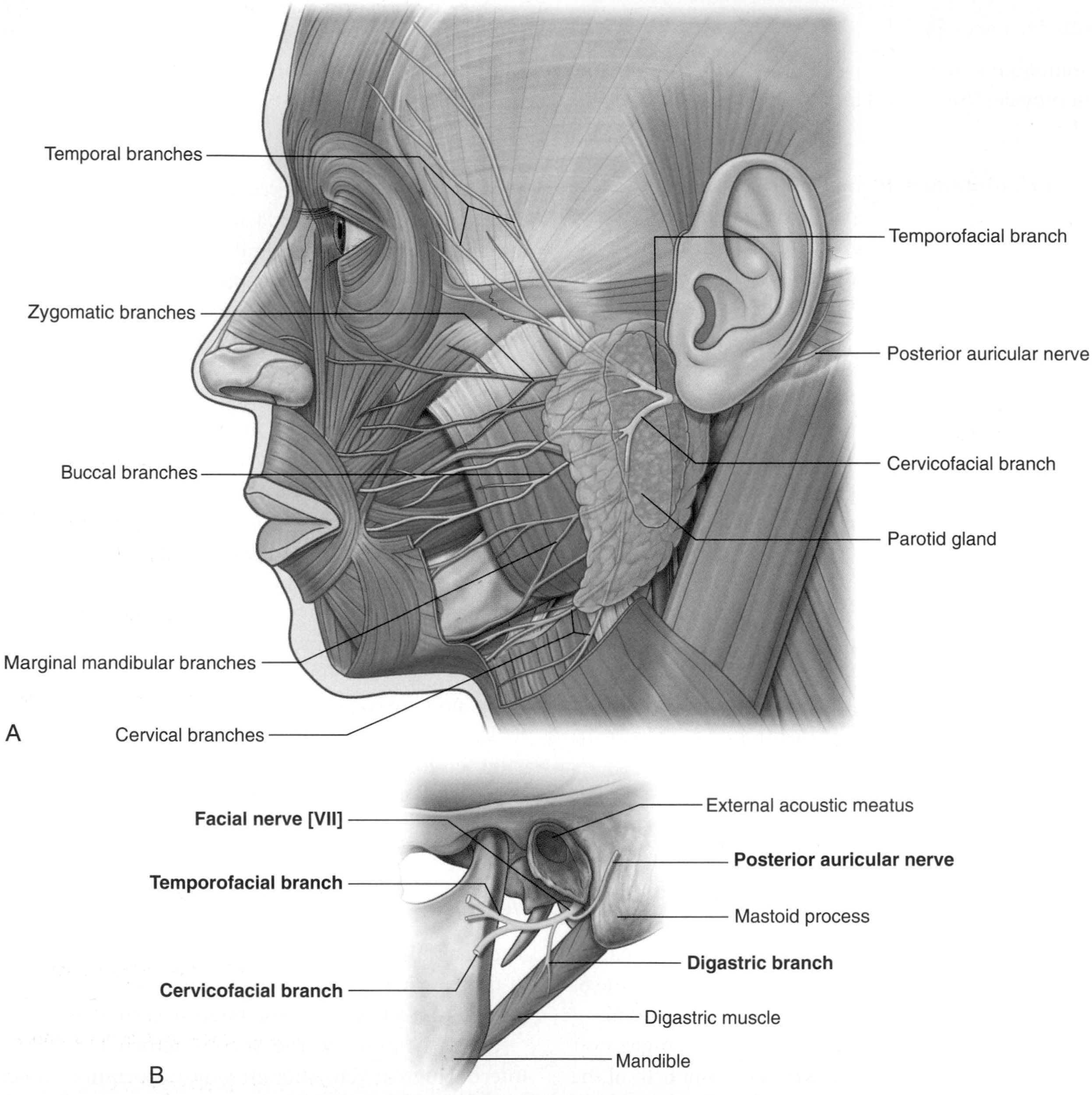

**Fig. 8.66** Facial nerve [VII] on the face. **A.** Terminal branches. **B.** Branches before entering the parotid gland.

## Arteries

### Facial artery

The facial artery is the major vessel supplying the face (Fig. 8.67). It branches from the anterior surface of the external carotid artery, passes up through the deep structures of the neck, and appears at the lower border of the mandible after passing posterior to the submandibular gland. Curving around the inferior border of the mandible just anterior to the masseter, where its pulse can be felt, the facial artery then enters the face. From this point the facial artery runs upward and medially in a tortuous course. It passes along the side of the nose and terminates as the **angular artery** at the medial corner of the eye.

Along its path the facial artery is deep to the platysma, risorius, and zygomaticus major and minor, superficial to the buccinator and levator anguli oris, and may pass superficially to or through the levator labii superioris.

Branches of the facial artery include the superior and inferior labial branches and the lateral nasal branch (Fig. 8.67).

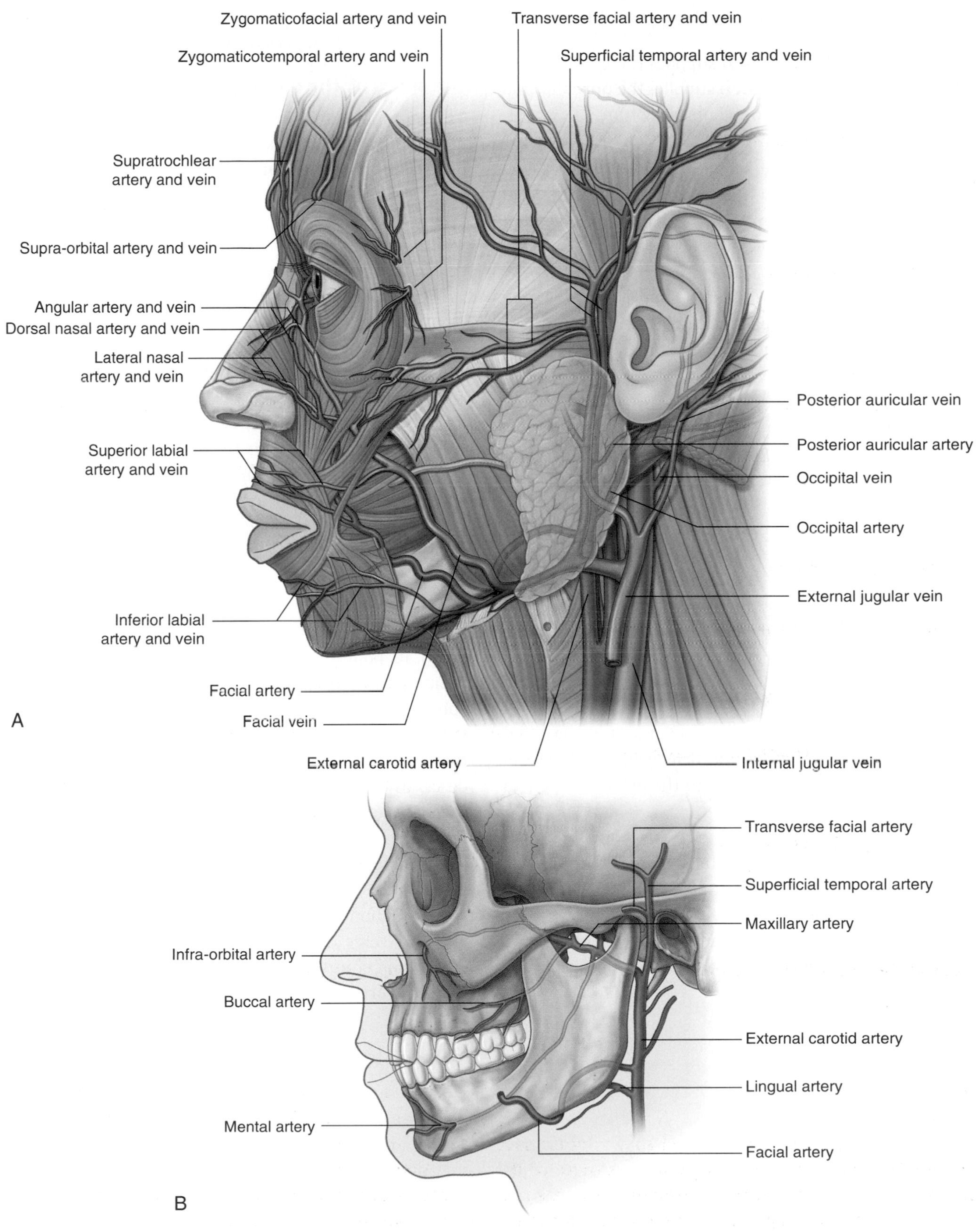
Zygomaticofacial artery and vein
Transverse facial artery and vein
Zygomaticotemporal artery and vein
Superficial temporal artery and vein
Supratrochlear artery and vein
Supra-orbital artery and vein
Angular artery and vein
Dorsal nasal artery and vein
Lateral nasal artery and vein
Superior labial artery and vein
Posterior auricular vein
Posterior auricular artery
Occipital vein
Occipital artery
External jugular vein
Inferior labial artery and vein
Facial artery
Facial vein
A
External carotid artery
Internal jugular vein
Transverse facial artery
Superficial temporal artery
Maxillary artery
Infra-orbital artery
Buccal artery
External carotid artery
Lingual artery
Mental artery
Facial artery
B

**Fig. 8.67** Vasculature of the face. **A.** Lateral view. **B.** Branches of the maxillary artery.

The labial branches arise near the corner of the mouth:

- The **inferior labial branch** supplies the lower lip.
- The **superior labial branch** supplies the upper lip, and also provides a branch to the nasal septum.

Near the midline, the superior and inferior labial branches anastomose with their companion arteries from the opposite side of the face. This provides an important connection between the facial arteries and the external carotid arteries of opposite sides.

The **lateral nasal branch** is a small branch arising from the facial artery as it passes along the side of the nose. It supplies the lateral surface and dorsum of the nose.

### Transverse facial artery

Another contributor to the vascular supply of the face is the transverse facial artery (Fig. 8.67), which is a branch of the superficial temporal artery (the smaller of the two terminal branches of the external carotid artery).

The transverse facial artery arises from the superficial temporal artery within the substance of the parotid gland, passes through the gland, and crosses the face in a transverse direction. Lying on the superficial surface of the masseter muscle, it is between the zygomatic arch and the parotid duct.

### Branches of the maxillary artery

The maxillary artery, the larger of the two terminal branches of the external carotid artery, gives off several small branches which contribute to the arterial supply to the face:

- The **infra-orbital artery** enters the face through the infra-orbital foramen and supplies the lower eyelid, upper lip, and the area between these structures.
- The **buccal artery** enters the face on the superficial surface of the buccinator muscle and supplies structures in this area.
- The **mental artery** enters the face through the mental foramen and supplies the chin.

### Branches of the ophthalmic artery

Three small arteries from the internal carotid artery also contribute to the arterial supply of the face. These vessels arise from the **ophthalmic artery**, a branch of the internal carotid artery, after the ophthalmic artery enters the orbit:

- The **zygomaticofacial** and **zygomaticotemporal arteries** come from the lacrimal branch of the ophthalmic artery (Fig. 8.67), enter the face through the zygomaticofacial and zygomaticotemporal foramina, and supply the area of the face over the zygomatic bone.
- The **dorsal nasal artery**, a terminal branch of the ophthalmic artery, exits the orbit in the medial corner, and supplies the dorsum of the nose.
- The **supraorbital and supratrochlear arteries** supply the anterior scalp.

## Veins

### Facial vein

The facial vein is the major vein draining the face (Fig. 8.67). Its point of origin is near the medial corner of the orbit as the **supratrochlear** and **supra-orbital veins** come together to form the **angular vein**. This vein becomes the facial vein as it proceeds inferiorly and assumes a position just posterior to the facial artery. The facial vein descends across the face with the facial artery until it reaches the inferior border of the mandible. Here the artery and vein part company and the facial vein passes superficial to the submandibular gland to enter the internal jugular vein.

Throughout its course the facial vein receives tributaries from veins draining the eyelids, external nose, lips, cheek, and chin that accompany the various branches of the facial artery.

### Transverse facial vein

The transverse facial vein is a small vein that accompanies the transverse facial artery in its journey across the face (Fig. 8.67). It empties into the superficial temporal vein within the substance of the parotid gland.

### Intracranial venous connections

As it crosses the face, the facial vein has numerous connections with venous channels passing into deeper regions of the head (Fig. 8.68):

- near the medial corner of the orbit, it communicates with ophthalmic veins;
- in the area of the cheek it communicates with veins passing into the infra-orbital foramen;
- it also communicates with veins passing into deeper regions of the face (i.e., the deep facial vein connecting with the pterygoid plexus of veins).

All these venous channels have interconnections with the intracranial cavernous sinus through emissary veins that connect intracranial with extracranial veins. There are no valves in the facial vein or any other venous channels in the head, so blood can move in any direction. Because of the interconnections between the veins, infections of the face, primarily above the mouth (i.e., the "danger area") should be handled with great care to prevent the dissemination of infectious material in an intracranial direction.

## Lymphatic drainage

Lymphatic drainage from the face primarily moves toward three groups of lymph nodes (Fig. 8.69):

- **submental nodes** inferior and posterior to the chin, which drain lymphatics from the medial part of the lower lip and chin bilaterally;
- **submandibular nodes** superficial to the submandibular gland and inferior to the body of the mandible, which drain the lymphatics from the medial corner of the orbit, most of the external nose the medial part of the cheek, the upper lip, and the lateral part of the lower lip that follows the course of the facial artery;
- **pre-auricular and parotid** nodes anterior to the ear, which drain lymphatics from most of the eyelids, a part of the external nose, and the lateral part of the cheek.

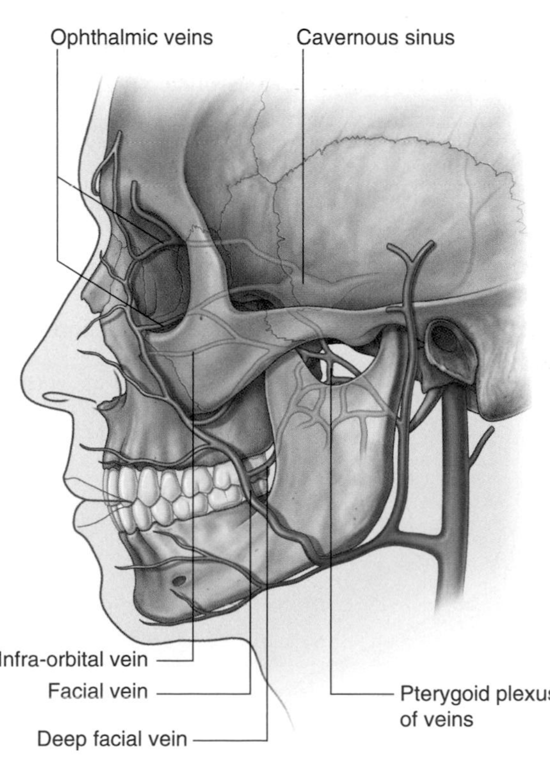


**Fig. 8.68** Intracranial venous connections.

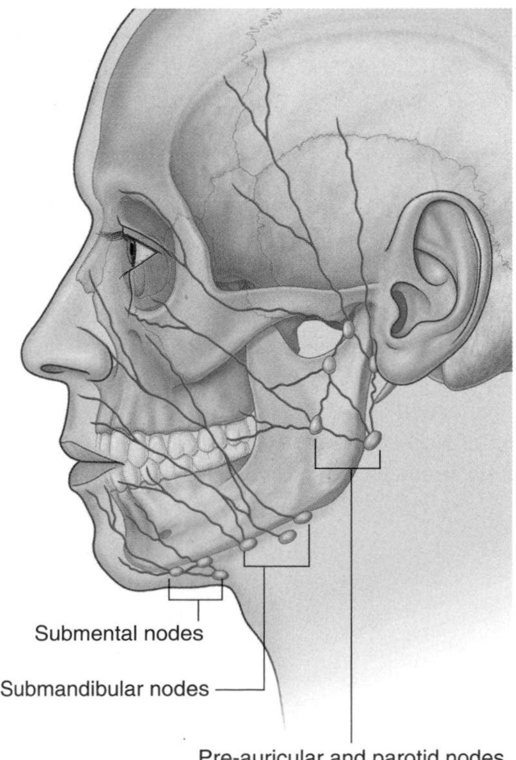


**Fig. 8.69** Lymphatic drainage of the face.

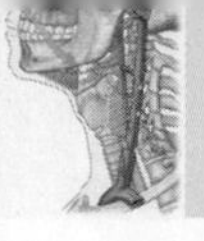



## In the clinic

### Facial nerve [VII] palsy (Bell's palsy)

The complexity of the facial nerve [VII] is demonstrated by the different pathological processes and sites at which these processes occur.

The facial nerve [VII] is formed from the nuclei within the brainstem emerging at the junction of the pons and the medulla. It enters the internal acoustic meatus, passes to the geniculate ganglion (which gives rise to further branches), and emerges from the skull base after a complex course within the temporal bone, leaving through the stylomastoid foramen. It enters the parotid gland and gives rise to five terminal groups of branches that supply muscles in the face and a number of additional branches that supply deeper or more posterior muscles. A series of lesions may affect the nerve along its course, and it is possible, with good clinical expertise, to determine the exact site of the lesion in relation to the course of the nerve.

#### *Central lesions*

A primary brainstem lesion affecting the motor nucleus of the facial nerve [VII] would lead to ipsilateral (same side) weakness of the whole face. However, because the upper part of the nucleus receives motor input from the left and right cerebral hemispheres a lesion occurring above the nucleus leads to contralateral lower facial weakness. In this example, motor innervation to the upper face is spared because the upper part of the nucleus receives input from both hemispheres. Preservation and loss of the special functions are determined by the extent of the lesion.

#### *Lesions at and around the geniculate ganglion*

Typically lesions at and around the geniculate ganglion are accompanied by loss of motor function on the whole of the ipsilateral (same) side of the face. Taste to the anterior two-thirds of the tongue, lacrimation, and some salivation also are likely to be affected because the lesion is proximal to the greater petrosal and chorda tympani branches of the nerve.

#### *Lesions at and around the stylomastoid foramen*

Lesions at and around the stylomastoid foramen are the commonest abnormality of the facial nerve [VII] and usually result from a viral inflammation of the nerve within the bony canal before exiting through the stylomastoid foramen. Typically the patient has an ipsilateral loss of motor function of the whole side of the face. Not only does this produce an unusual appearance, but it also complicates chewing of food. Lacrimation and taste may not be affected if the lesion remains distal to the greater petrosal and chorda tympani branches that originate deep in the temporal bone.

## In the clinic

### Trigeminal neuralgia

Trigeminal neuralgia (tic douloureux) is a complex sensory disorder of the sensory root of the trigeminal nerve. Typically the pain is in the region of the mandibular [$V_3$] and maxillary [$V_2$] nerves, and is usually of sudden onset, is excruciating in nature, and may be triggered by touching a sensitive region of skin.

The etiology of trigeminal neuralgia is unknown, although anomalous blood vessels lying adjacent to the sensory route of the maxillary [$V_2$] and mandibular [$V_3$] nerves may be involved.

If symptoms persist and are unresponsive to medical care, surgical exploration of the trigeminal nerve (which is not without risk) may be necessary to remove any aberrant vessels.

## SCALP

The scalp is the part of the head that extends from the superciliary arches anteriorly to the external occipital protuberance and superior nuchal lines posteriorly. Laterally it continues inferiorly to the zygomatic arch.

The scalp is a multilayered structure with layers that can be defined by the word itself:

- S—skin,
- C—connective tissue (dense),
- A—aponeurotic layer,
- L—loose connective tissue, and
- P—pericranium (Fig. 8.70).

### Layers

Examining the layers of the scalp reveals that the first three layers are tightly held together, forming a single unit. This unit is sometimes referred to as the scalp proper and is the tissue torn away during serious "scalping" injuries.

#### Skin

The skin is the outer layer of the scalp (Figs. 8.70 and 8.71). It is similar structurally to skin throughout the body with the exception that hair is present on a large amount of it.

#### Connective tissue (dense)

Deep to the skin is dense connective tissue. This layer anchors the skin to the third layer and contains the arteries, veins, and nerves supplying the scalp. When the scalp is cut, the dense connective tissue surrounding the vessels tends to hold cut vessels open. This results in profuse bleeding.

#### Aponeurotic layer

The deepest layer of the first three layers is the aponeurotic layer. Firmly attached to the skin by the dense connective

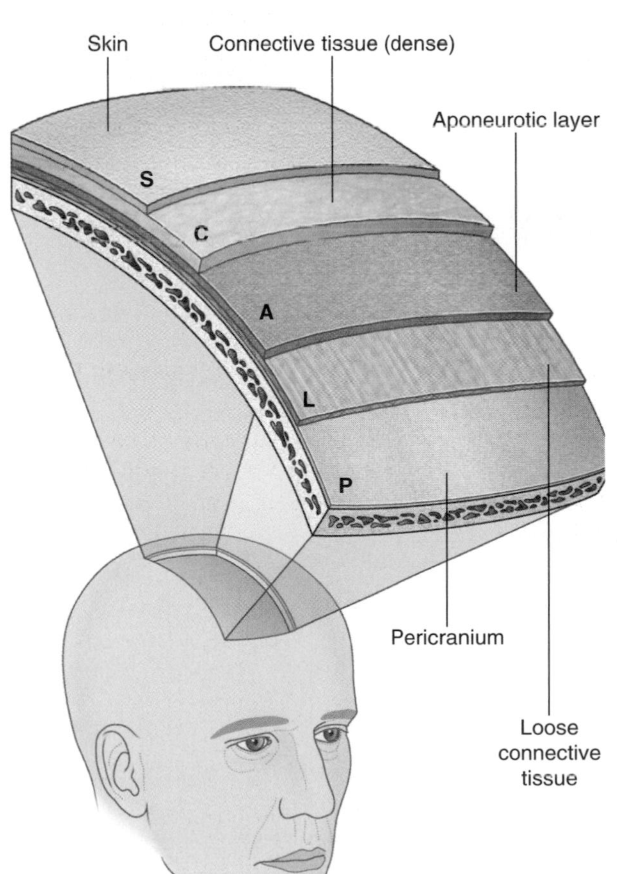


**Fig. 8.70** SCALP.

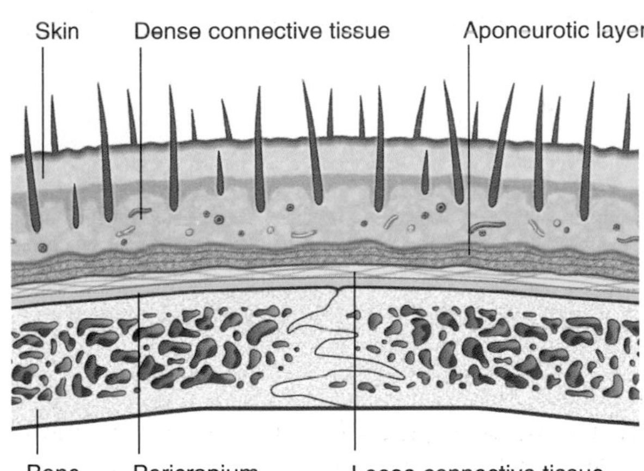


**Fig. 8.71** Layers of the scalp.

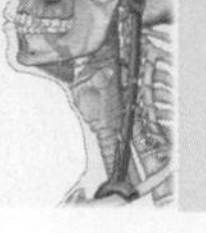



tissue of the second layer, this layer consists of the occipitofrontalis muscle, which has a frontal belly anteriorly, an occipital belly posteriorly, and an aponeurotic tendon—the **epicranial aponeurosis** (**galea aponeurotica**)—connecting the two (Fig. 8.72).

The frontal belly of the occipitofrontalis begins anteriorly where it is attached to the skin of the eyebrows. It passes upward, across the forehead, to become continuous with the aponeurotic tendon.

Posteriorly, each occipital belly of the occipitofrontalis arises from the lateral part of the superior nuchal line of the occipital bone and the mastoid process of the temporal bone. It also passes superiorly to attach to the aponeurotic tendon.

The occipitofrontalis muscles move the scalp, wrinkle the forehead, and raise the eyebrows. The frontal belly is innervated by temporal branches of the facial nerve [VII] and the posterior belly by the posterior auricular branch.

### Loose connective tissue

A layer of loose connective tissue separates the aponeurotic layer from the pericranium and facilitates movement of the scalp proper over the calvaria (Figs. 8.70 and 8.72). Because of its consistency, infections tend to localize and spread through the loose connective tissue (also see "In the clinic" on p. 878).

### Pericranium

The pericranium is the deepest layer of the scalp and is the periosteum on the outer surface of the calvaria. It is attached to the bones of the calvaria but is removable, except in the area of the sutures.

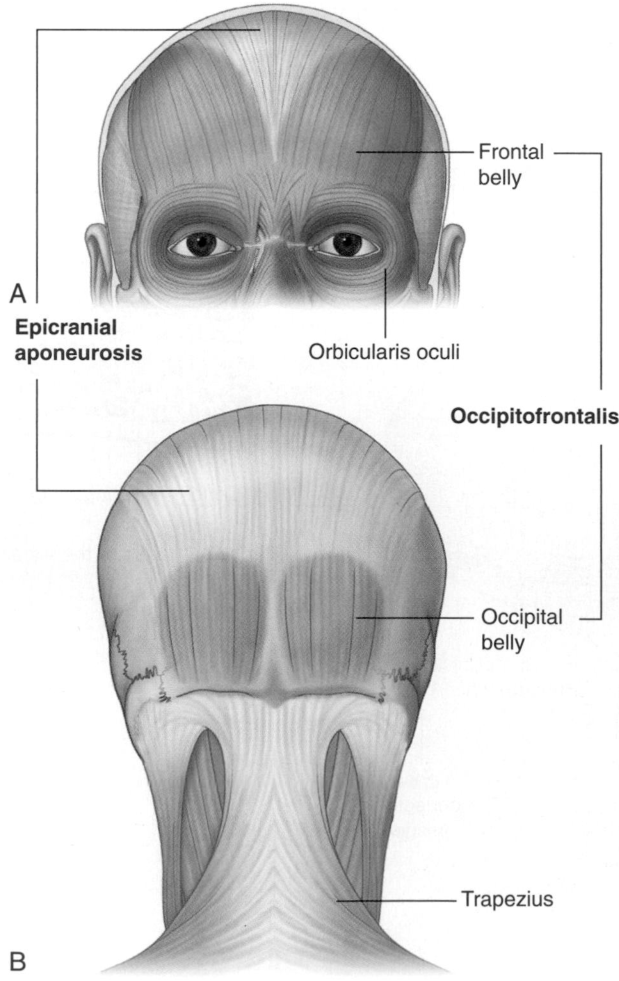


**Fig. 8.72** Occipitofrontalis muscle. **A.** Frontal belly. **B.** Occipital belly.

### Innervation

Sensory innervation of the scalp is from two major sources, cranial nerves or cervical nerves, depending on whether it is anterior or posterior to the ears and the vertex of the head (Fig. 8.73). The occipitofrontalis muscle is innervated by branches of the facial nerve [VII].

#### Anterior to the ears and the vertex

Branches of the trigeminal nerve [V] supply the scalp anterior to the ears and the vertex of the head (Fig. 8.73). These branches are the supratrochlear, supra-orbital, zygomaticotemporal, and auriculotemporal nerves:

- The **supratrochlear nerve** exits the orbit, passes through the frontalis muscle, continues superiorly across the front of the forehead, and supplies the front of the forehead near the midline.
- The **supra-orbital nerve** exits the orbit through the supra-orbital notch or foramen, passes through the frontalis muscle, and continues superiorly across the scalp as far back as the vertex of the head.
- The **zygomaticotemporal nerve** exits the skull through a foramen in the zygomatic bone and supplies the scalp over a small anterior area of the temple.
- The **auriculotemporal nerve** exits from the skull, deep to the parotid gland, passes just anterior to the ear, continues superiorly anterior to the ear until nearly reaching the vertex of the head, and supplies the scalp over the temporal region and anterior to the ear to near the vertex.

#### Posterior to the ears and the vertex

Posterior to the ears and vertex, sensory innervation of the scalp is by cervical nerves, specifically branches from spinal

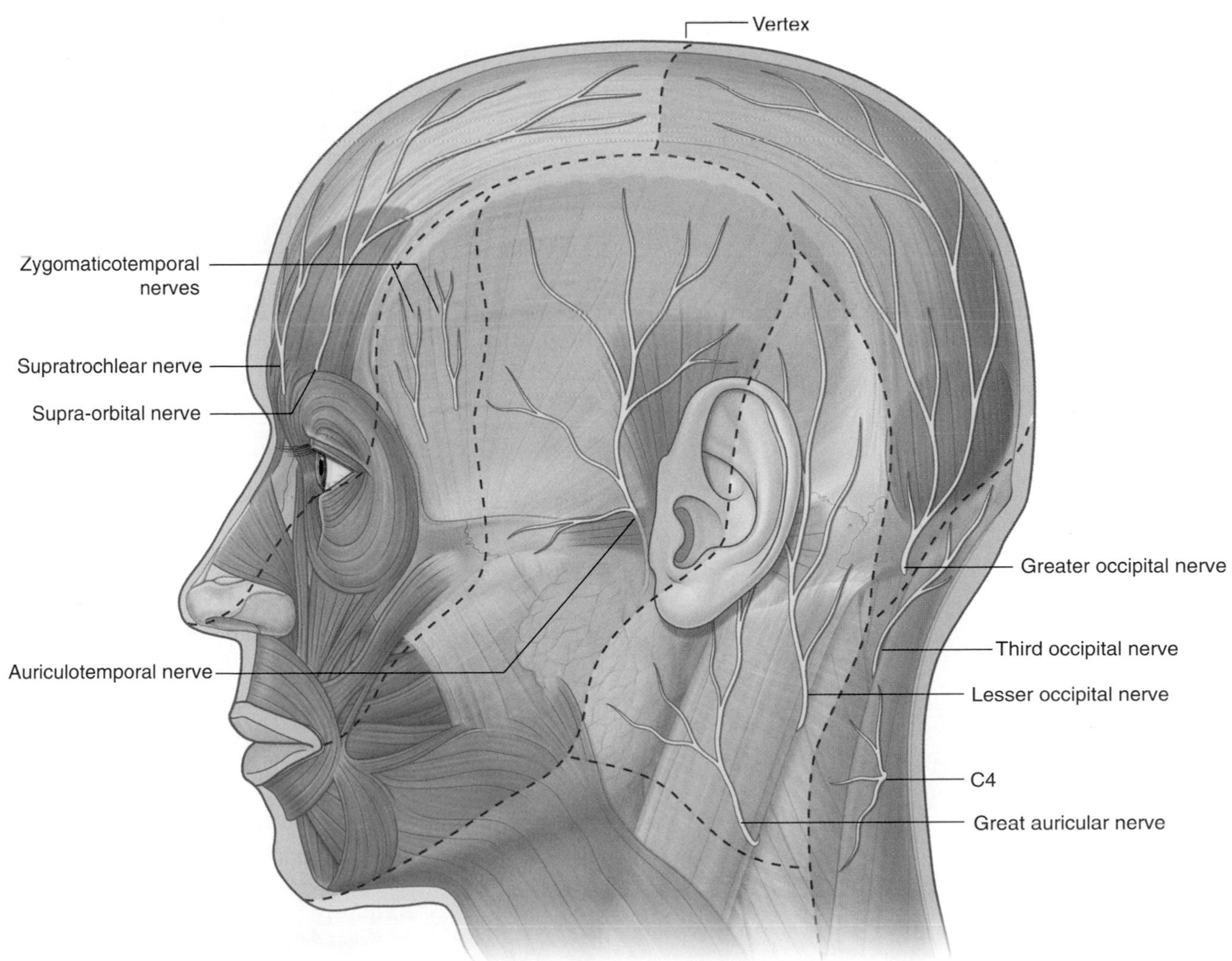


**Fig. 8.73** Innervation of the scalp.

cord levels C2 and C3 (Fig. 8.73). These branches are the great auricular, the lesser occipital, the greater occipital, and the third occipital nerves:

- The **great auricular nerve** is a branch of the cervical plexus, arises from the anterior rami of the C2 and C3 spinal nerves, ascends on the surface of the sternocleidomastoid muscle, and innervates a small area of the scalp just posterior to the ear.
- The **lesser occipital nerve** is also a branch of the cervical plexus, arises from the anterior ramus of the C2 spinal nerve, ascends on the posterior border of the sternocleidomastoid muscle, and supplies an area of the scalp posterior and superior to the ear.
- The **greater occipital nerve** is a branch of the posterior ramus of the C2 spinal nerve, emerges just inferior to the obliquus capitis inferior muscle, ascends superficial to the suboccipital triangle, pierces the semispinalis capitis and trapezius muscles, and then spreads out to supply a large part of the posterior scalp as far superiorly as the vertex.
- The **third occipital nerve** is a branch of the posterior ramus of the C3 spinal nerve, pierces the semispinalis capitis and trapezius muscles, and supplies a small area of the lower part of the scalp.

## Vessels

### Arteries

Arteries supplying the scalp (Fig. 8.74) are branches of either the external carotid artery or the ophthalmic artery, which is a branch of the internal carotid artery.

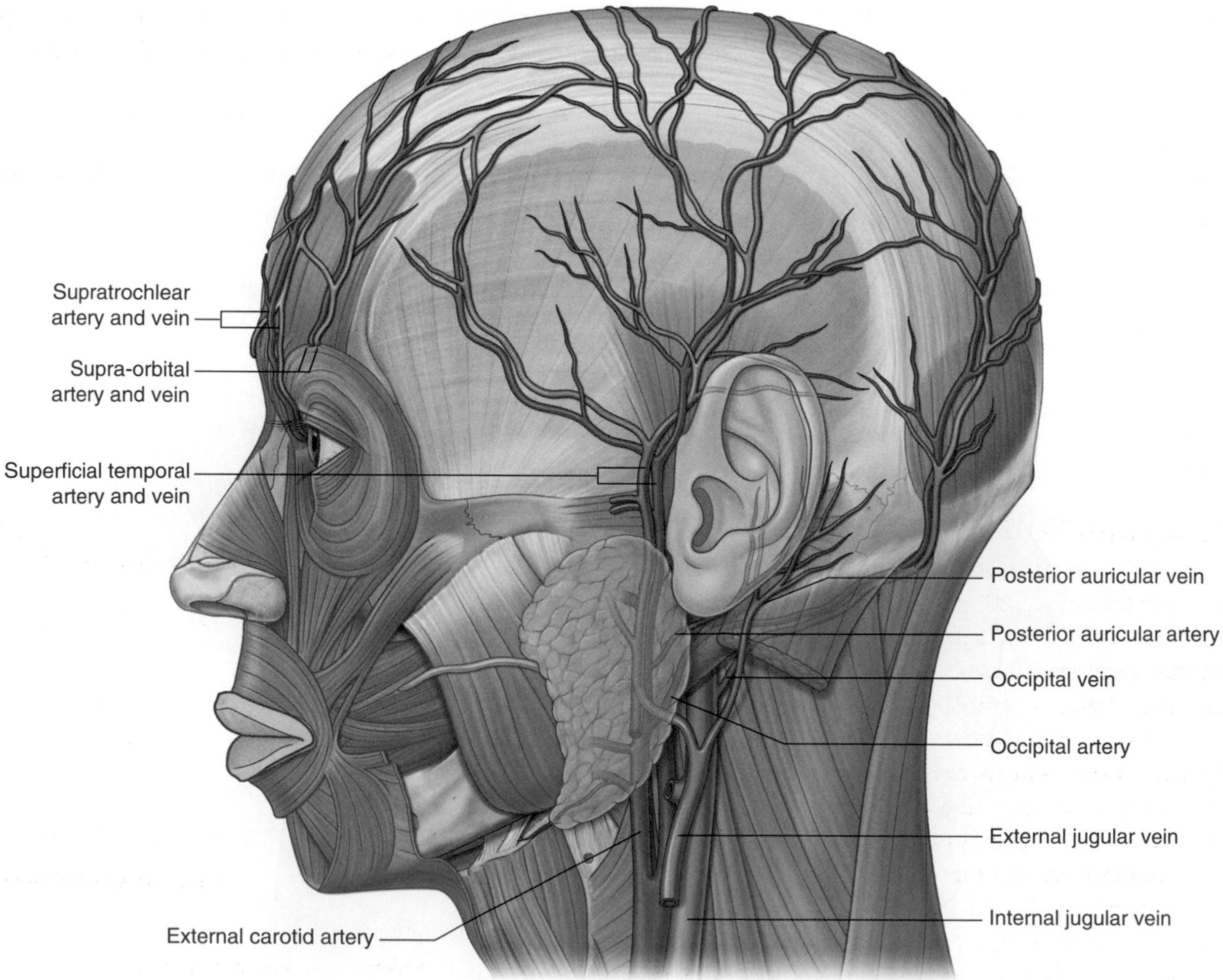

**Fig. 8.74** Vasculature of the scalp.

### Branches from the ophthalmic artery

The supratrochlear and supra-orbital arteries supply the anterior and superior aspects of the scalp. They branch from the ophthalmic artery while it is in the orbit, continue through the orbit, and exit onto the forehead in association with the supratrochlear and supra-orbital nerves. Like the nerves, the arteries ascend across the forehead to supply the scalp as far posteriorly as the vertex of the head.

### Branches from the external carotid artery

Three branches of the external carotid artery supply the largest part of the scalp—the superficial temporal, posterior auricular, and occipital arteries supply the lateral and posterior aspects of the scalp (Fig. 8.74):

- The smallest branch (the **posterior auricular artery**) leaves the posterior aspect of the external carotid artery, passes through deeper structures, and emerges to supply an area of the scalp posterior to the ear.
- Also arising from the posterior aspect of the external carotid artery is the **occipital artery**, which ascends in a posterior direction, passes through several layers of back musculature, and emerges to supply a large part of the posterior aspect of the scalp.
- The third arterial branch supplying the scalp is the **superficial temporal artery**, a terminal branch of the external carotid artery that passes superiorly, just anterior to the ear, divides into anterior and posterior branches, and supplies almost the entire lateral aspect of the scalp.

## Veins

Veins draining the scalp follow a pattern similar to the arteries:

- The **supratrochlear** and **supra-orbital veins** drain the anterior part of the scalp from the superciliary arches to the vertex of the head (Fig. 8.74), pass inferior to the superciliary arches, communicate with the ophthalmic veins in the orbit, and continue inferiorly to participate in the formation of the angular vein, which is the upper tributary to the facial vein.
- The **superficial temporal vein** drains the entire lateral area of the scalp before passing inferiorly to join in the formation of the retromandibular vein.
- The **posterior auricular vein** drains the area of the scalp posterior to the ear and eventually empties into a tributary of the retromandibular vein.
- The **occipital vein** drains the posterior aspect of the scalp from the external occipital protuberance and superior nuchal lines to the vertex of the head; deeper, it passes through the musculature in the posterior neck to join in the formation of the plexus of veins in the suboccipital triangle.

> **In the clinic**
>
> **Scalp laceration**
>
> The scalp has an extremely rich blood supply from the external carotid arteries, so lacerations of the scalp tend to bleed profusely. Importantly, scalp bleeding is predominantly arterial, because of two reasons. First, in the erect position the venous pressure is extremely low. Second, the vessels do not retract and close when lacerated because the connective tissue in which they are found holds them open.

## Lymphatic drainage

Lymphatic drainage of the scalp generally follows the pattern of arterial distribution.

The lymphatics in the occipital region initially drain to occipital nodes near the attachment of the trapezius muscle at the base of the skull (Fig. 8.75). Further along

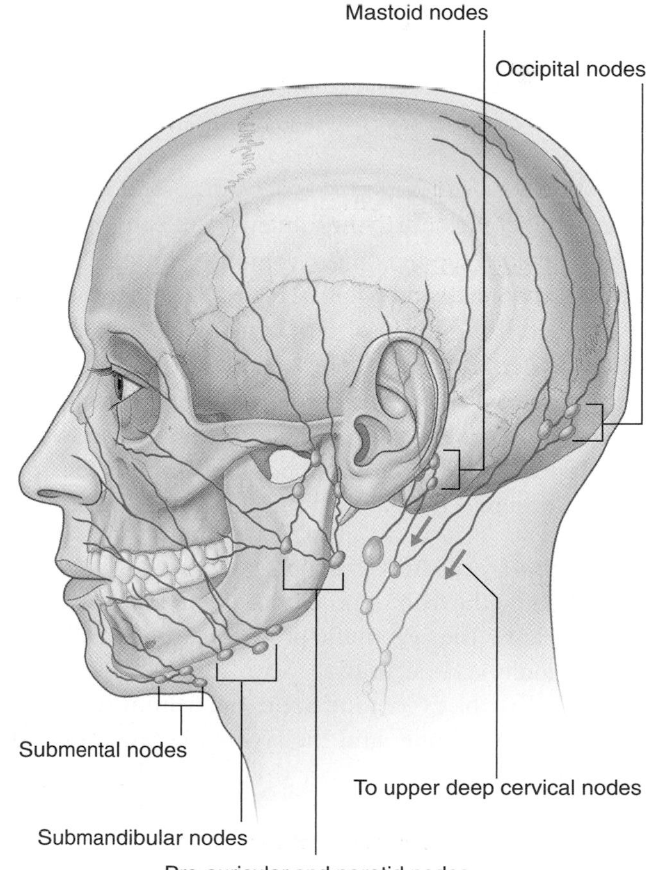


**Fig. 8.75** Lymphatic drainage of the scalp.

the pathway occipital nodes drain into upper deep cervical nodes. There is also some direct drainage to upper deep cervical nodes from this part of the scalp.

Lymphatics from the upper part of the scalp drain in two directions:

- Posterior to the vertex of the head they drain to **mastoid nodes** (retro-auricular/posterior auricular nodes) posterior to the ear near the mastoid process of the temporal bone, and efferent vessels from these nodes drain into upper deep cervical nodes.
- Anterior to the vertex of the head they drain to pre-auricular and parotid nodes anterior to the ear on the surface of the parotid gland.

Finally, there may be some lymphatic drainage from the forehead to the submandibular nodes through efferent vessels that follow the facial artery.

## ORBIT

The orbits are bilateral structures in the upper half of the face below the anterior cranial fossa and anterior to the middle cranial fossa that contain the eyeball, the optic nerve, the extra-ocular muscles, the lacrimal apparatus, adipose tissue, fascia, and the nerves and vessels that supply these structures.

### Bony orbit

Seven bones contribute to the framework of each orbit (Fig. 8.76). They are the maxilla, zygomatic, frontal, ethmoid, lacrimal, sphenoid, and palatine bones. Together they give the bony orbit the shape of a pyramid, with its wide base opening anteriorly onto the face and its apex extending in a posteromedial direction. Completing the pyramid configuration are medial, lateral, superior, and inferior walls.

The apex of the pyramid-shaped bony orbit is the optic foramen, whereas the base (the orbital rim) is formed:

- superiorly by the frontal bone,
- medially by the frontal process of the maxilla,
- inferiorly by the zygomatic process of the maxilla and the zygomatic bone, and
- laterally by the zygomatic bone, the frontal process of the zygomatic bone, and the zygomatic process of the frontal bone.

#### Roof

The **roof (superior wall)** of the bony orbit is made up of the orbital part of the frontal bone with a small contribution from the sphenoid bone (Fig. 8.76). This thin plate of

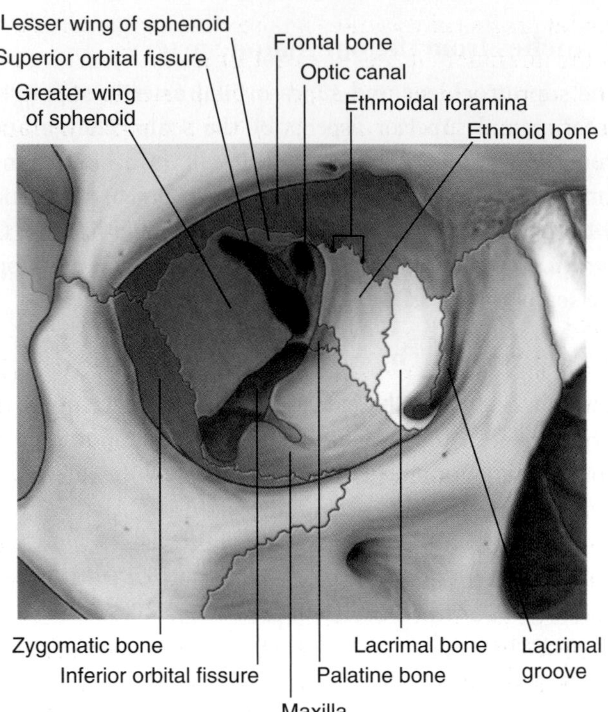


**Fig. 8.76** Bones of the orbit.

bone separates the contents of the orbit from the brain in the anterior cranial fossa.

Unique features of the superior wall include:

- anteromedially, the trochlear fovea, for the attachment of a pulley through which the superior oblique muscle passes, and the possible intrusion of part of the frontal sinus;
- anterolaterally, a depression (the lacrimal fossa) for the orbital part of the lacrimal gland.

Posteriorly, the lesser wing of the sphenoid bone completes the roof.

#### Medial wall

The **medial walls** of the paired bony orbits are parallel to each other and each consists of four bones—the maxilla, lacrimal, ethmoid, and sphenoid bones (Fig. 8.76).

The largest contributor to the medial wall is the orbital plate of the ethmoid bone. This part of the ethmoid bone contains collections of ethmoidal cells, which are clearly visible in a dried skull.

Also visible, at the junction between the roof and the medial wall, usually associated with the frontoethmoidal suture, are the **anterior** and **posterior ethmoidal foramina**. The anterior and posterior ethmoidal nerves and vessels leave the orbit through these openings.

Anterior to the ethmoid bone is the small lacrimal bone, and completing the anterior part of the medial wall is the frontal process of the maxilla. These two bones participate in the formation of the **lacrimal groove**, which contains the lacrimal sac and is bound by the **posterior lacrimal crest** (part of the lacrimal bone) and the **anterior lacrimal crest** (part of the maxilla).

Posterior to the ethmoid bone the medial wall is completed by a small part of the sphenoid bone, which forms a part of the medial wall of the optic canal.

### Floor

The **floor** (**inferior wall**) of the bony orbit, which is also the roof of the maxillary sinus, consists primarily of the orbital surface of the maxilla (Fig. 8.76), with small contributions from the zygomatic and palatine bones.

Beginning posteriorly and continuing along the lateral boundary of the floor of the bony orbit is the inferior orbital fissure. Beyond the anterior end of the fissure the zygomatic bone completes the floor of the bony orbit.

Posteriorly, the orbital process of the palatine bone makes a small contribution to the floor of the bony orbit near the junction of the maxilla, ethmoid, and sphenoid bones.

### Lateral wall

The **lateral wall** of the bony orbit consists of contributions from two bones—anteriorly, the zygomatic bone and posteriorly, the greater wing of the sphenoid bone (Fig. 8.76). The superior orbital fissure is between the greater wing of the sphenoid and the lesser wing of the sphenoid that forms part of the roof.

#### In the clinic

**Orbital fracture**

Fractures of the orbit are not uncommon and may involve the orbital margins with extension into the maxilla, frontal, and zygomatic bones. These fractures are often part of complex facial fractures. Fractures within the orbit frequently occur within the floor and the medial wall; however, superior and lateral wall fractures also occur. Inferior orbital floor fractures are one of the commonest types of injuries. These fractures may drag the inferior oblique muscle and associated tissues into the fracture line. In these instances, patients may have upward gaze failure (upward gaze diplopia) in the affected eye. Medial wall fractures characteristically show air within the orbit in radiographs. This is due to fracture of the ethmoidal labyrinth, permitting direct continuity between the orbit and the ethmoidal paranasal sinuses. Occasionally, patients feel a full sensation within the orbit when blowing the nose.

## Eyelids

The upper and lower eyelids are anterior structures that, when closed, protect the surface of the eyeball.

The space between the eyelids, when they are open, is the **palpebral fissure**.

The layers of the eyelids, from anterior to posterior, consist of skin, subcutaneous tissue, voluntary muscle, the orbital septum, the tarsus, and conjunctiva (Fig. 8.77).

The upper and lower eyelids are basically similar in structure except for the addition of two muscles in the upper eyelid.

### Skin and subcutaneous tissue

The skin of the eyelids is not particularly substantial, and only a thin layer of connective tissue separates the skin from the underlying voluntary muscle layer (Fig. 8.77). The thin layer of connective tissue and its loose arrangement account for the accumulation of fluid (blood) when an injury occurs.

### Orbicularis oculi

The muscle fibers encountered next in an anteroposterior direction through the eyelid belong to the **palpebral part** of the **orbicularis oculi** (Fig. 8.77). This muscle is part of the larger orbicularis oculi muscle, which consists primarily of two parts—an **orbital part**, which surrounds the orbit, and the palpebral part, which is in the eyelids. The orbicularis oculi is innervated by the facial nerve [VII] and closes the eyelids.

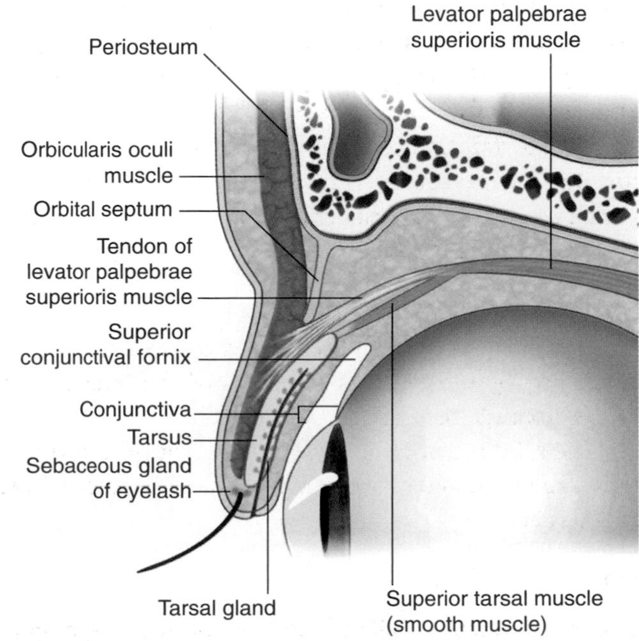


**Fig. 8.77** Eyelids.

The palpebral part is thin and anchored medially by the **medial palpebral ligament** (Fig. 8.78), which attaches to the anterior lacrimal crest and laterally blends with fibers from the muscle in the lower eyelid at the **lateral palpebral ligament** (Fig. 8.78).

A third part of the orbicularis oculi muscle that can be identified consists of fibers on the medial border, which pass deeply to attach to the posterior lacrimal crest. These fibers form the lacrimal part of the orbicularis oculi, which may be involved in the drainage of tears.

### Orbital septum

Deep to the palpebral part of the orbicularis oculi is an extension of periosteum into both the upper and lower eyelids from the margin of the orbit (Fig. 8.79). This is the **orbital septum**, which extends downward into the upper eyelid and upward into the lower eyelid and is continuous with the periosteum outside and inside the orbit (Fig. 8.79). The orbital septum attaches to the tendon of the levator palpebrae superioris muscle in the upper eyelid and attaches to the tarsus in the lower eyelid.

### Tarsus and levator palpebrae superioris

Providing major support for each eyelid is the tarsus (Fig. 8.80). There is a large **superior tarsus** in the upper eyelid and a smaller **inferior tarsus** in the lower eyelid (Fig. 8.80). These plates of dense connective tissue are attached medially to the anterior lacrimal crest of the maxilla by the medial palpebral ligament and laterally to the orbital tubercle on the zygomatic bone by the lateral palpebral ligament.

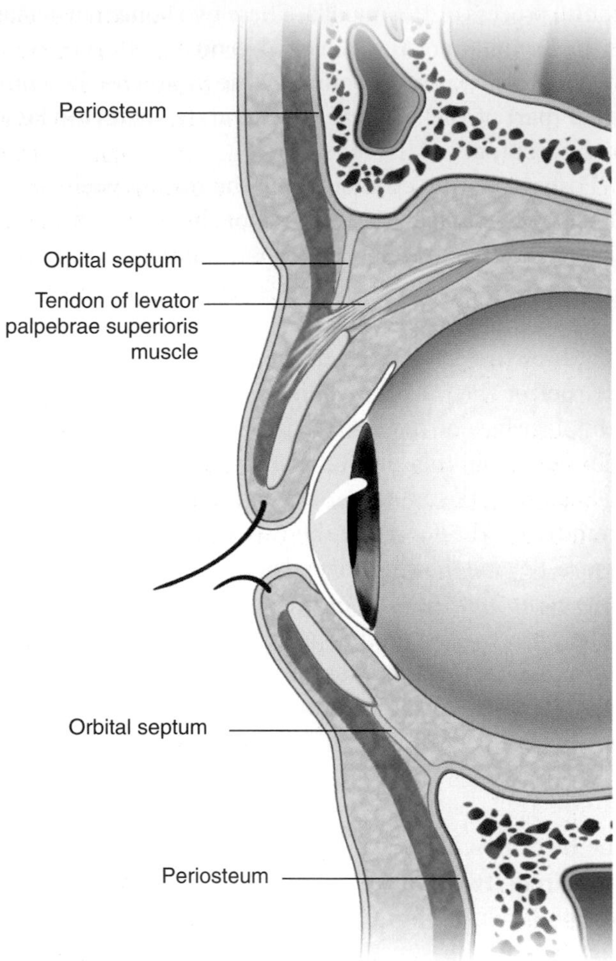

**Fig. 8.79** Orbital septum.

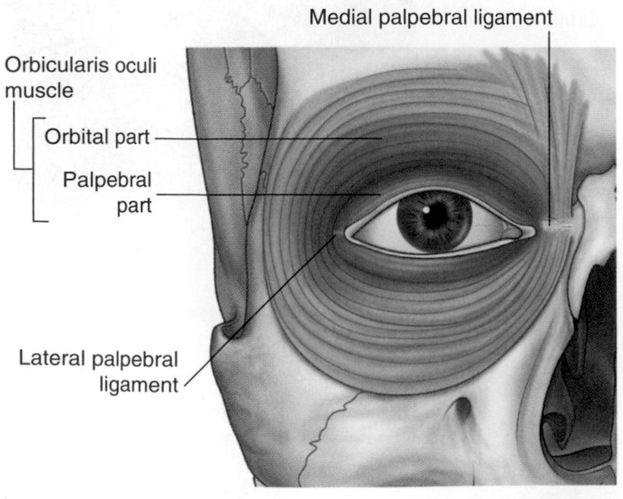


**Fig. 8.78** Orbicularis oculi muscle.

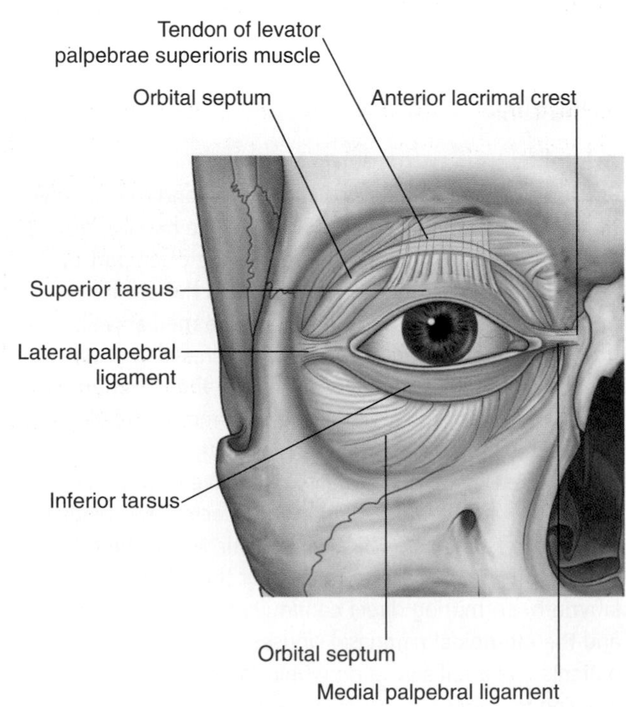


**Fig. 8.80** Tarsal plates.

Although the tarsal plates in the upper and lower eyelids are generally similar in structure and function, there is one unique difference. Associated with the tarsus in the upper eyelid is the **levator palpebrae superioris muscle** (Fig. 8.80), which raises the eyelid. Its origin is from the posterior part of the roof of the orbit, just superior to the optic foramen, and it inserts into the anterior surface of the superior tarsus, with the possibility of a few fibers attaching to the skin of the upper eyelid. It is innervated by the oculomotor nerve [III].

In companion with the levator palpebrae superioris muscle is a collection of smooth muscle fibers passing from the inferior surface of the levator to the upper edge of the superior tarsus (see Fig. 8.77). Innervated by postganglionic sympathetic fibers from the superior cervical ganglion, this muscle is the **superior tarsal muscle**.

Loss of function of either the levator palpebrae superioris muscle or the superior tarsal muscle results in a ptosis or drooping of the upper eyelid.

### Conjunctiva

The structure of the eyelid is completed by a thin membrane (the **conjunctiva**), which covers the posterior surface of each eyelid (see Fig. 8.77). This membrane covers the full extent of the posterior surface of each eyelid before reflecting onto the outer surface (**sclera**) of the eyeball. It attaches to the eyeball at the junction between the sclera and the cornea. With this membrane in place, a **conjunctival sac** is formed when the eyelids are closed, and the upper and lower extensions of this sac are the **superior** and **inferior conjunctival fornices** (Fig. 8.77).

### Glands

Embedded in the tarsal plates are tarsal glands (see Fig. 8.77), which empty onto the free margin of each eyelid. These glands are modified sebaceous glands and secrete an oily substance that increases the viscosity of the tears and decreases the rate of evaporation of tears from the surface of the eyeball. Blockage and inflammation of a tarsal gland is a **chalazion** and is on the inner surface of the eyelid.

The tarsal glands are not the only glands associated with the eyelids. Associated with the eyelash follicles are sebaceous and sweat glands (see Fig. 8.77). Blockage and inflammation of either of these is a **stye** and is on the edge of the eyelid.

### Vessels

The arterial supply to the eyelids is from the numerous vessels in the area (Fig. 8.81). They include:

- the supratrochlear, supra-orbital, lacrimal, and dorsal nasal arteries from the ophthalmic artery;
- the angular artery from the facial artery;
- the transverse facial artery from the superficial temporal artery; and
- branches from the superficial temporal artery itself.

Venous drainage follows an external pattern through veins associated with the various arteries and an internal pattern moving into the orbit through connections with the ophthalmic veins.

Lymphatic drainage is primarily to the parotid nodes, with some drainage from the medial corner of the eye along lymphatic vessels associated with the angular and facial arteries to the submandibular nodes.

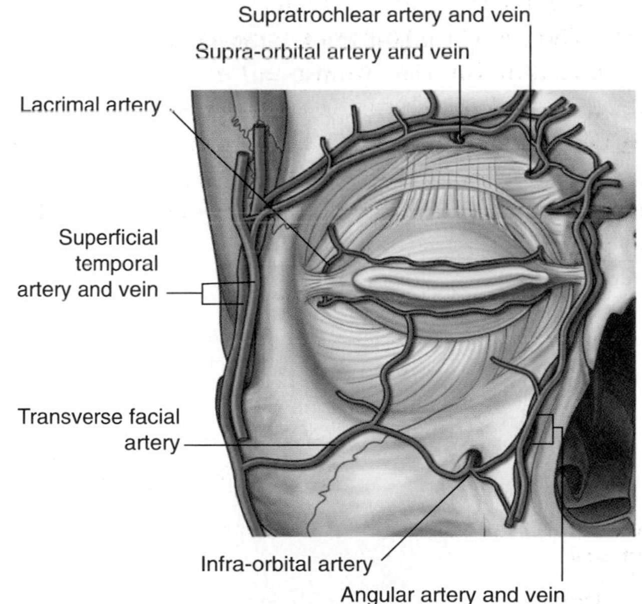


**Fig. 8.81** Vasculature of the eyelids.

## Innervation

Innervation of the eyelids includes both sensory and motor components.

The sensory nerves are all branches of the trigeminal nerve [V] (Fig. 8.82). Palpebral branches arise from:

- the supra-orbital, supratrochlear, infratrochlear, and lacrimal branches of the ophthalmic nerve [$V_1$]; and
- the infra-orbital branch of the maxillary nerve [$V_2$].

Motor innervation is from:

- the facial nerve [VII], which innervates the palpebral part of the orbicularis oculi;
- the oculomotor nerve [III], which innervates the levator palpebrae superioris; and
- sympathetic fibers, which innervate the superior tarsal muscle.

Loss of innervation of the orbicularis oculi by the facial nerve [VII] causes an inability to close the eyelids tightly and the lower eyelid droops away, resulting in a spillage of tears.

Loss of innervation of the levator palpebrae superioris by the oculomotor nerve causes an inability to open the superior eyelid voluntarily, producing a complete ptosis.

Loss of innervation of the superior tarsal muscle by sympathetic fibers causes a constant partial ptosis.

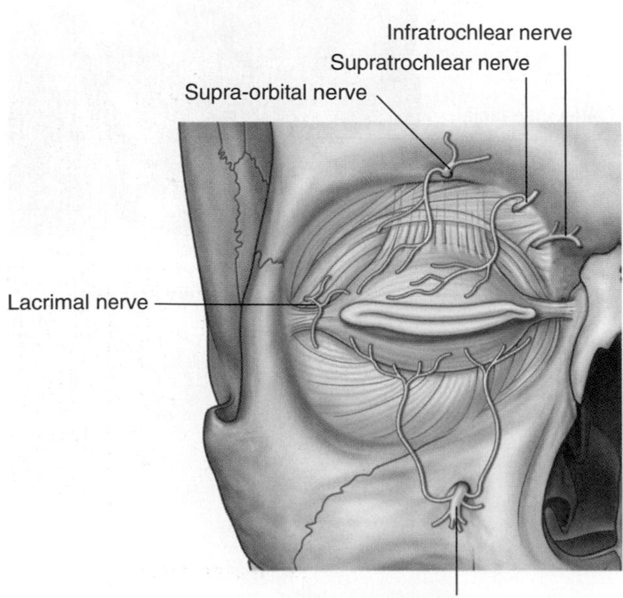


**Fig. 8.82** Innervation of the eyelids.

### In the clinic

**Horner's syndrome**

Horner's syndrome is caused by any lesion that leads to a loss of sympathetic function in the head. It is characterized by three typical features:

- pupillary constriction due to paralysis of the dilator pupillae muscle,
- partial ptosis (drooping of the upper eyelid) due to paralysis of the superior tarsal muscle, and
- absence of sweating on the ipsilateral side of the face and the neck due to absence of innervation of the sweat glands.

Secondary changes may also include:

- ipsilateral vasodilation due to loss of the normal sympathetic control of the subcutaneous blood vessels, and
- enophthalmos (sinking of the eye)—believed to result from paralysis of the orbitalis muscle, although this is an uncommon feature of Horner's syndrome.

The orbitalis muscle spans the inferior orbital fissure and helps maintain the forward position of orbital contents.

The commonest cause for Horner's syndrome is a tumor eroding the cervicothoracic ganglion, which is typically an apical lung tumor.

***Surgically induced Horner's syndrome***

A surgically induced Horner's syndrome may be necessary for patients who suffer severe hyperhidrosis (sweating). This often debilitating condition may be so severe that patients are confined to their home for fear of embarrassment. Treatment is relatively straightforward. The patient is anesthetized and a bifurcate endotracheal tube is placed into the left and right main bronchi. A small incision is made in the intercostal space on the appropriate side, and a surgically induced pneumothorax is created. The patient is ventilated through the contralateral lung.

Using an endoscope the apex of the thoracic cavity can be viewed from inside and the cervicothoracic ganglion readily identified. Obliterative techniques include thermocoagulation and surgical excision. After the ganglion has been destroyed, the endoscope is removed, the lung is reinflated, and the small hole is sutured.

## Lacrimal apparatus

The lacrimal apparatus is involved in the production, movement, and drainage of fluid from the surface of the eyeball. It is made up of the **lacrimal gland** and its ducts, the **lacrimal canaliculi**, the **lacrimal sac**, and the **nasolacrimal duct**.

The lacrimal gland is anterior in the superolateral region of the orbit (Fig. 8.83) and is divided into two parts by the levator palpebrae superioris (Fig. 8.84):

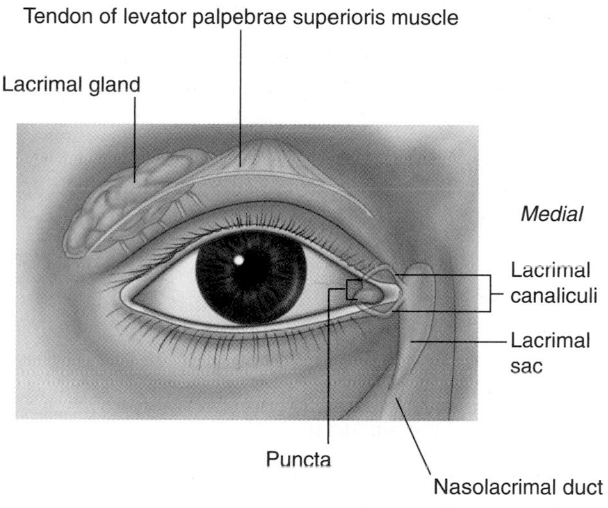


**Fig. 8.83** Lacrimal gland, anterior view.

- The larger **orbital part** is in a depression, the lacrimal fossa, in the frontal bone.
- The smaller **palpebral part** is inferior to the levator palpebrae superioris in the superolateral part of the eyelid.

Numerous ducts empty the glandular secretions into the lateral part of the superior fornix of the conjunctiva.

Fluid is continually being secreted by the lacrimal gland and moved across the surface of the eyeball from lateral to medial as the eyelids blink.

The fluid accumulates medially in the **lacrimal lake** and is drained from the lake by the lacrimal canaliculi, one canaliculus associated with each eyelid (Fig. 8.83). The **lacrimal punctum** is the opening through which fluid enters each canaliculus.

Passing medially, the lacrimal canaliculi eventually join the lacrimal sac between the anterior and posterior lacrimal crests, posterior to the medial palpebral ligament and anterior to the lacrimal part of the orbicularis oculi muscle (Figs. 8.85 and 8.86). When the orbicularis oculi muscle contracts during blinking, the small lacrimal part of the muscle may dilate the lacrimal sac and draw tears into it through the canaliculi from the conjunctival sac.

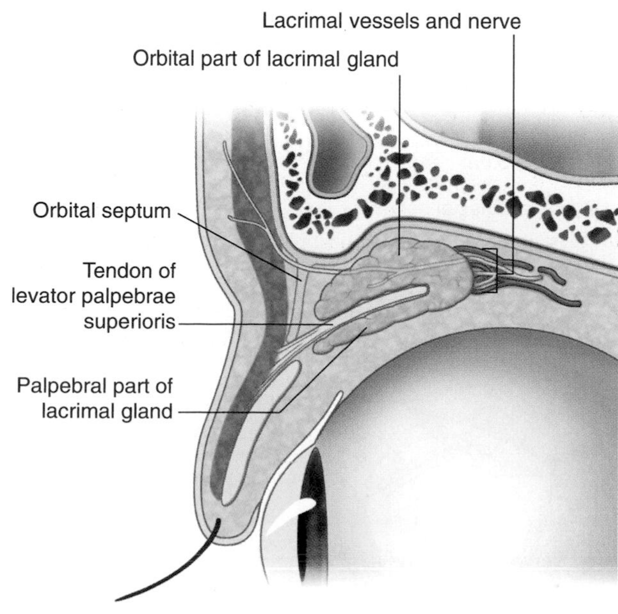


**Fig. 8.84** Lacrimal gland and levator palpebrae superioris.

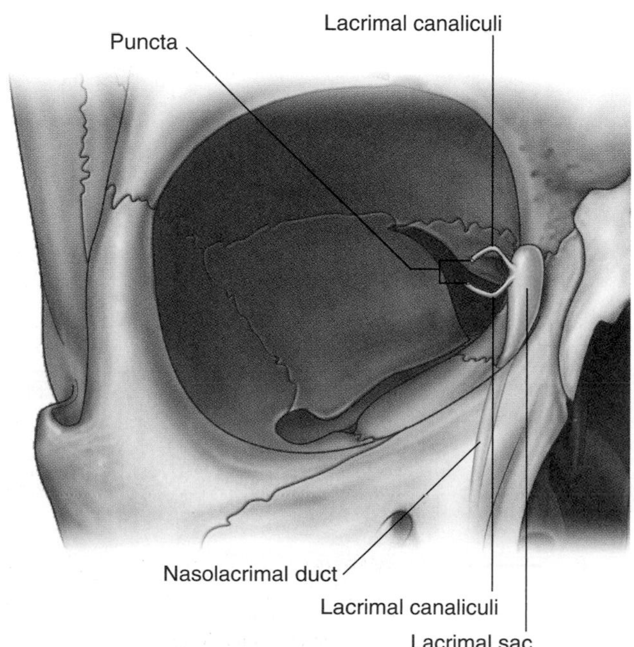


**Fig. 8.85** The lacrimal sac.

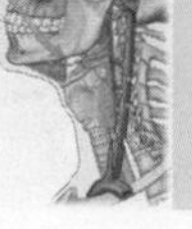



## Innervation

The innervation of the lacrimal gland involves three different components (Fig. 8.87).

### Sensory innervation

Sensory neurons from the lacrimal gland return to the CNS through the lacrimal branch of the ophthalmic nerve [$V_1$].

### Secretomotor (parasympathetic) innervation

Secretomotor fibers from the parasympathetic part of the autonomic division of the PNS stimulate fluid secretion from the lacrimal gland. These preganglionic parasympathetic neurons leave the CNS in the facial nerve [VII], enter the greater petrosal nerve (a branch of the facial nerve [VII]), and continue with this nerve until it becomes the **nerve of the pterygoid canal** (Fig. 8.87).

The nerve of the pterygoid canal eventually joins the pterygopalatine ganglion where the preganglionic parasympathetic neurons synapse on postganglionic parasympathetic neurons. The postganglionic neurons join the maxillary nerve [$V_2$] and continue with it until the zygomatic nerve branches from it, and travel with the zygomatic nerve until it gives off the zygomaticotemporal nerve, which eventually distributes postganglionic parasympathetic fibers in a small branch that joins the lacrimal nerve. The lacrimal nerve passes to the lacrimal gland.

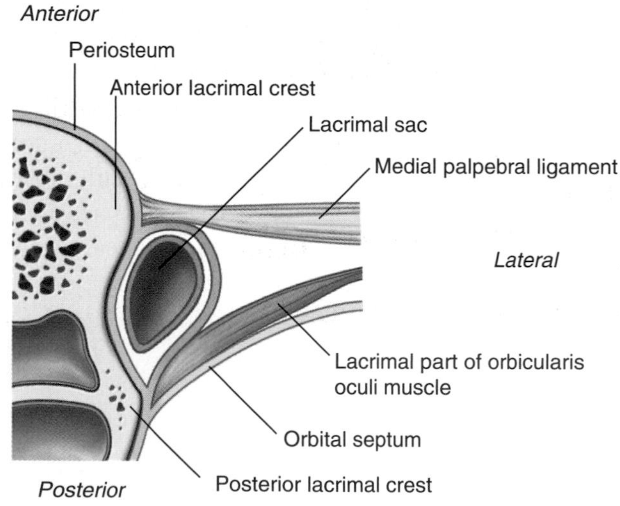


**Fig. 8.86** Position of lacrimal sac.

### Sympathetic innervation

Sympathetic innervation of the lacrimal gland follows a similar path as parasympathetic innervation. Postganglionic sympathetic fibers originating in the superior cervical ganglion travel along the plexus surrounding the internal carotid artery (Fig. 8.87). They leave this plexus as the deep petrosal nerve and join the parasympathetic fibers in the nerve of the pterygoid canal. Passing through the pterygopalatine ganglion, the sympathetic fibers from this point onward follow the same path as the parasympathetic fibers to the lacrimal gland.

## Vessels

The arterial supply to the lacrimal gland is by branches from the ophthalmic artery and venous drainage is through the ophthalmic veins.

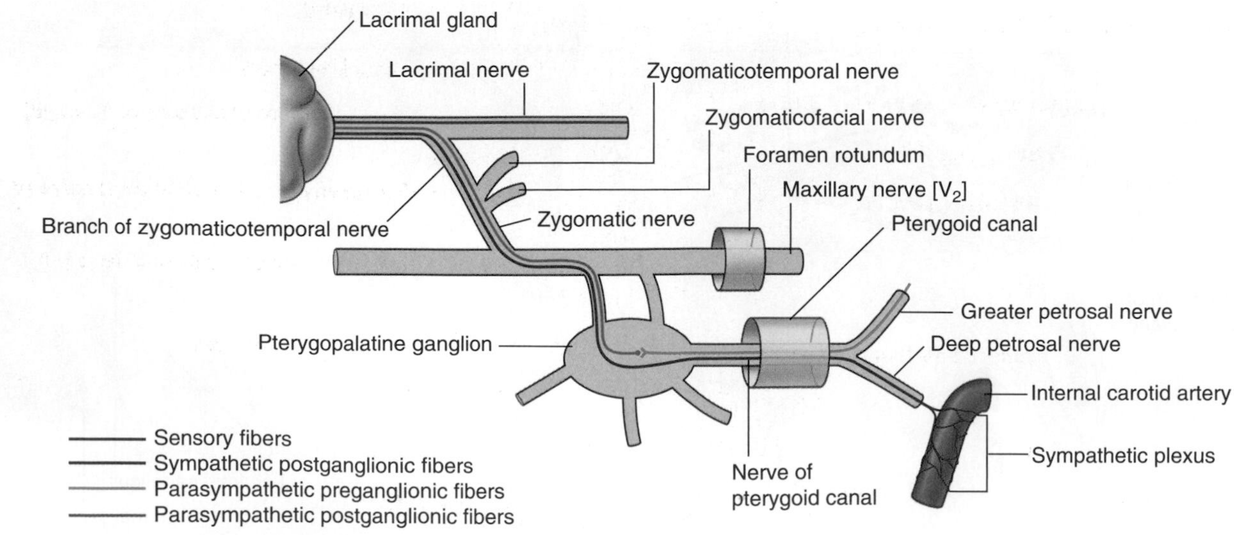


**Fig. 8.87** Innervation of the lacrimal gland.

## Fissures and foramina

Numerous structures enter and leave the orbit through a variety of openings (Fig. 8.88).

### Optic canal

When the bony orbit is viewed from an anterolateral position, the round opening at the apex of the pyramidal-shaped orbit is the optic canal, which opens into the middle cranial fossa and is bounded medially by the body of the sphenoid and laterally by the lesser wing of the sphenoid. Passing through the optic canal are the optic nerve and the ophthalmic artery (Fig. 8.89).

### Superior orbital fissure

Just lateral to the optic canal is a triangular-shaped gap between the roof and lateral wall of the bony orbit. This is the superior orbital fissure and allows structures to pass between the orbit and the middle cranial fossa (Fig. 8.88).

Passing through the superior orbital fissure are the superior and inferior branches of the oculomotor nerve [III], the trochlear nerve [IV], the abducent nerve [VI], the lacrimal, frontal, and nasociliary branches of the ophthalmic nerve [$V_1$], and the superior ophthalmic vein (Fig. 8.89).

### Inferior orbital fissure

Separating the lateral wall of the orbit from the floor of the orbit is a longitudinal opening, the inferior orbital fissure (Fig. 8.88). Its borders are the greater wing of the sphenoid

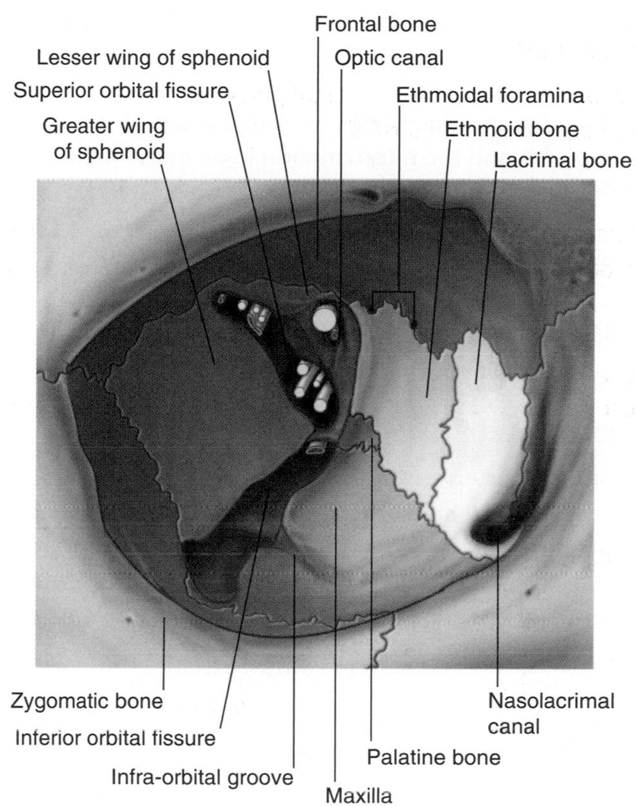


**Fig. 8.88** Openings into the bony orbit.

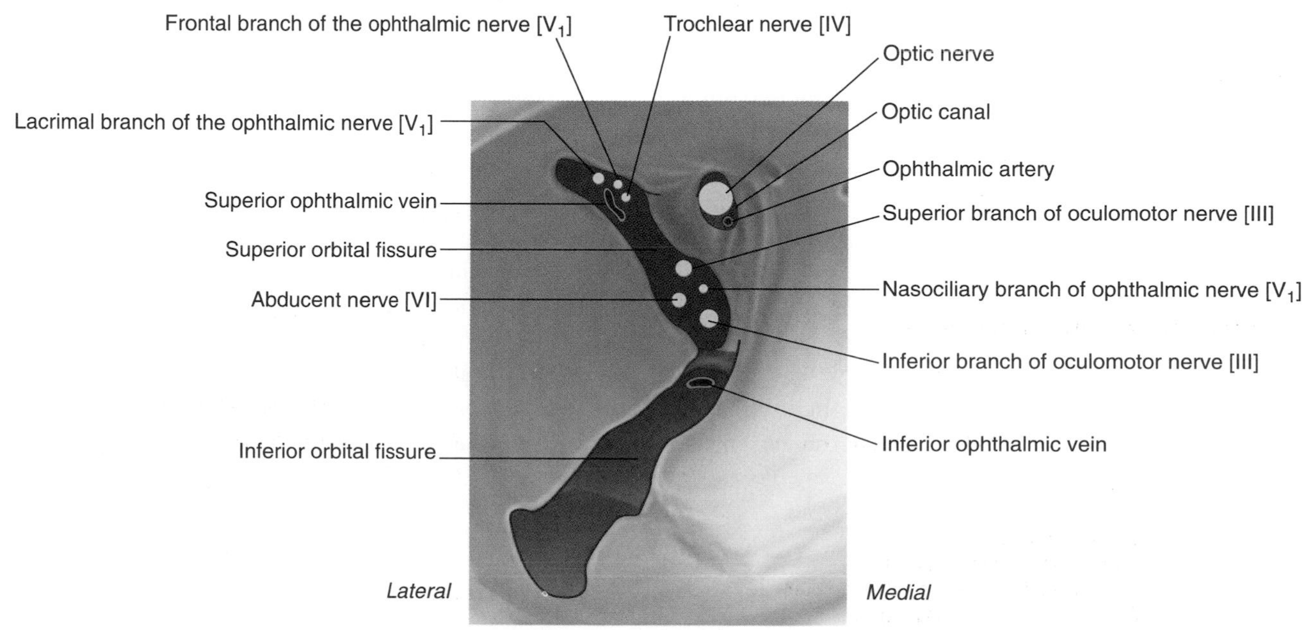


**Fig. 8.89** Optic canal and superior orbital fissure.

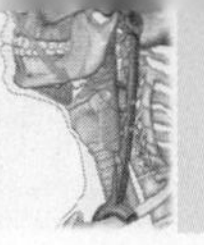

and the maxilla, palatine, and zygomatic bones. This long fissure allows communication between:

- the orbit and the pterygopalatine fossa posteriorly,
- the orbit and the infratemporal fossa in the middle, and
- the orbit and the temporal fossa posterolaterally.

Passing through the inferior orbital fissure are the maxillary nerve [$V_2$] and its zygomatic branch, the infra-orbital vessels, and a vein communicating with the pterygoid plexus of veins.

### Infra-orbital foramen

Beginning posteriorly and crossing about two-thirds of the inferior orbital fissure, a groove (the **infra-orbital groove**) is encountered, which continues anteriorly across the floor of the orbit (Fig. 8.88). This groove connects with the **infra-orbital canal** that opens onto the face at the **infra-orbital foramen**.

The infra-orbital nerve, part of the maxillary nerve [$V_2$], and vessels pass through this structure as they exit onto the face.

### Other openings

Associated with the medial wall of the bony orbit are several smaller openings (Fig. 8.88).

The **anterior** and **posterior ethmoidal foramina** are at the junction between the superior and medial walls. These openings provide exits from the orbit into the ethmoid bone for the anterior and posterior ethmoidal nerves and vessels.

Completing the openings on the medial wall is a canal in the lower part of the wall anteriorly. Clearly visible is the depression for the lacrimal sac formed by the lacrimal bone and the frontal process of the maxilla. This depression is continuous with the nasolacrimal canal, which leads to the inferior nasal meatus. Contained within the nasolacrimal canal is the nasolacrimal duct, a part of the lacrimal apparatus.

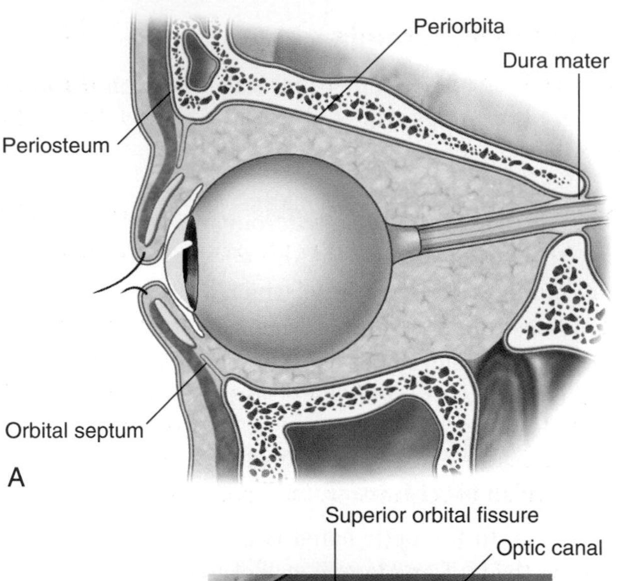


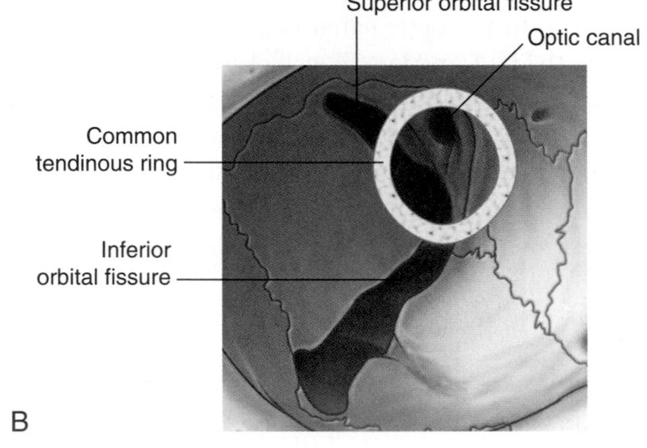


**Fig. 8.90** Periorbita. **A.** Lateral view. **B.** Common tendinous ring.

## Fascial specializations

### Periorbita

The periosteum lining the bones that form the orbit is the **periorbita** (Fig. 8.90A). It is continuous at the margins of the orbit with the periosteum on the outer surface of the skull and sends extensions into the upper and lower eyelids (the **orbital septa**).

At the various openings where the orbit communicates with the cranial cavity the periorbita is continuous with the periosteal layer of dura mater. In the posterior part of the orbit, the periorbita thickens around the optic canal and the central part of the superior orbital fissure. This is the point of origin of the four rectus muscles and is the **common tendinous ring**.

### Fascial sheath of the eyeball

The **fascial sheath of the eyeball** (bulbar sheath) is a layer of fascia that encloses a major part of the eyeball (Figs. 8.91 and 8.92):

- Posteriorly, it is firmly attached to the sclera (the white part of the eyeball) around the point of entrance of the optic nerve into the eyeball.
- Anteriorly, it is firmly attached to the sclera near the edge of the cornea (the clear part of the eyeball).
- Additionally, as the muscles approach the eyeball, the investing fascia surrounding each muscle blends with the fascial sheath of the eyeball as the muscles pass through and continue to their point of attachment.

A specialized lower part of the fascial sheath of the eyeball is the **suspensory ligament** (Figs. 8.91 and 8.92), which supports the eyeball. This "sling-like" structure is made up of the fascial sheath of the eyeball and contributions from the two inferior ocular muscles and the medial and lateral ocular muscles.

### Check ligaments of the medial and lateral rectus muscles

Other fascial specializations in the orbit are the check ligaments (Fig. 8.92). These are expansions of the investing fascia covering the medial and lateral rectus muscles, which attach to the medial and lateral walls of the bony orbit:

- The medial check ligament is an extension from the fascia covering the medial rectus muscle and attaches immediately posterior to the posterior lacrimal crest of the lacrimal bone.
- The lateral check ligament is an extension from the fascia covering the lateral rectus muscle and is attached to the orbital tubercle of the zygomatic bone.

Functionally, the positioning of these ligaments seems to restrict the medial and lateral rectus muscles, thus the names of the fascial specializations.

## Muscles

There are two groups of muscles within the orbit:

- **extrinsic muscles of eyeball (extra-ocular muscles)** involved in movements of the eyeball or raising upper eyelids, and
- intrinsic muscles within the eyeball, which control the shape of the lens and size of the pupil.

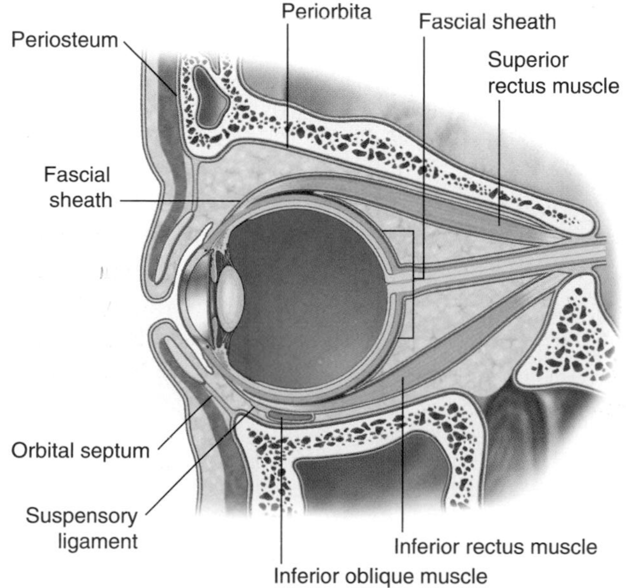


Fig. 8.91 Fascial sheath of the eyeball.

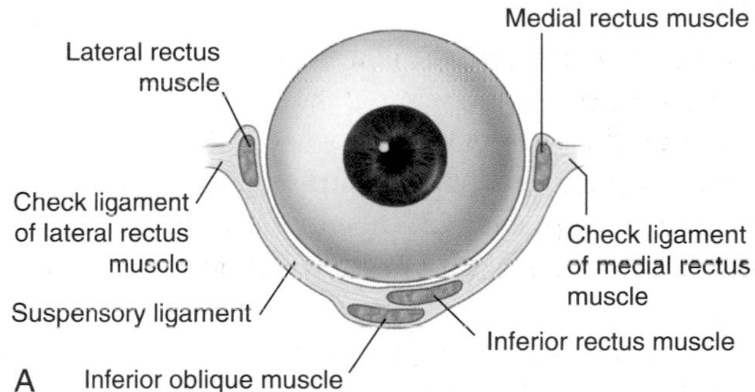


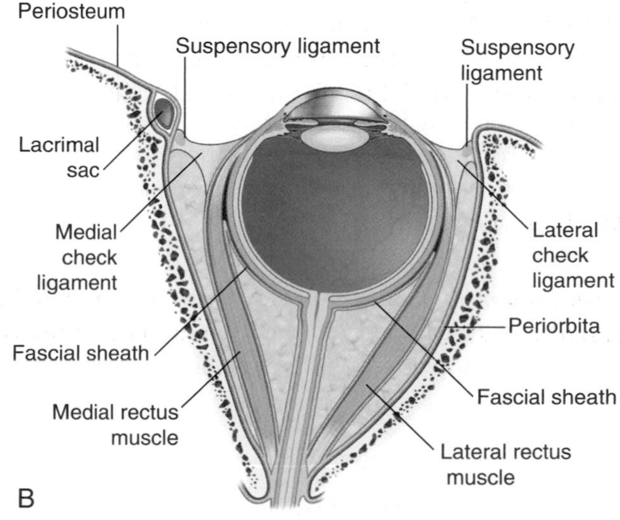


**Fig. 8.92** Check ligaments. **A.** Anterior view. **B.** Superior view.

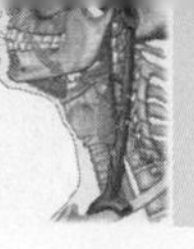



**Table 8.8** Extrinsic (extra-ocular) muscles

| Muscle | Origin | Insertion | Innervation | Function |
|---|---|---|---|---|
| Levator palpebrae superioris | Lesser wing of sphenoid anterior to optic canal | Anterior surface of tarsal plate; a few fibers to skin and superior conjunctival fornix | Oculomotor nerve [III]—superior branch | Elevation of upper eyelid |
| Superior rectus | Superior part of common tendinous ring | Anterior half of eyeball superiorly | Oculomotor nerve [III]—superior branch | Elevation, adduction, medial rotation of eyeball |
| Inferior rectus | Inferior part of common tendinous ring | Anterior half of eyeball inferiorly | Oculomotor nerve [III]—inferior branch | Depression, adduction, lateral rotation of eyeball |
| Medial rectus | Medial part of common tendinous ring | Anterior half of eyeball medially | Oculomotor nerve [III]—inferior branch | Adduction of eyeball |
| Lateral rectus | Lateral part of common tendinous ring | Anterior half of eyeball laterally | Abducent nerve [VI] | Abduction of eyeball |
| Superior oblique | Body of sphenoid, superior and medial to optic canal | Outer posterior quadrant of eyeball (superior surface) | Trochlear nerve [IV] | Depression, abduction, internal rotation of eyeball |
| Inferior oblique | Medial floor of orbit posterior to rim; maxilla lateral to nasolacrimal groove | Outer posterior quadrant of eyeball (inferior surface) | Oculomotor nerve [III]—inferior branch | Elevation, abduction, external rotation of eyeball |

The extrinsic muscles include the levator palpebrae superioris, superior rectus, inferior rectus, medial rectus, lateral rectus, superior oblique, and inferior oblique.

The intrinsic muscles include the ciliary muscle, the sphincter pupillae, and the dilator pupillae.

## Extrinsic muscles

Of the seven muscles in the extrinsic group of muscles, one raises the eyelids, whereas the other six move the eyeball itself (Table 8.8).

The movements of the eyeball, in three dimensions, (Fig. 8.93) are:

- elevation—moving the pupil superiorly,
- depression—moving the pupil inferiorly,
- abduction—moving the pupil laterally,
- adduction—moving the pupil medially,
- internal rotation (intorsion)—rotating the upper part of the pupil medially (or toward the nose), and
- external rotation (extorsion)—rotating the upper part of the pupil laterally (or toward the temple).

The axis of each orbit is directed slightly laterally from back to front, but each eyeball is directed anteriorly (Fig. 8.94). Therefore the pull of some muscles has multiple effects on the movement of the eyeball, whereas that of others has a single effect.

### Levator palpebrae superioris

Levator palpebrae superioris raises the upper eyelid (Table 8.8). It is the most superior muscle in the orbit, originating from the roof, just anterior to the optic canal on the inferior

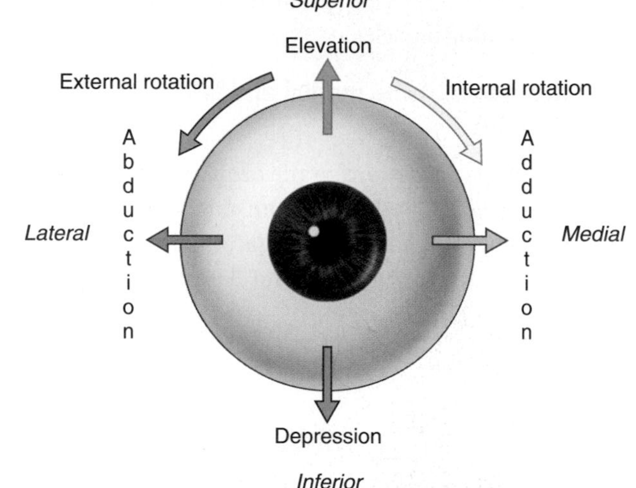


**Fig. 8.93** Movements of the eyeball.

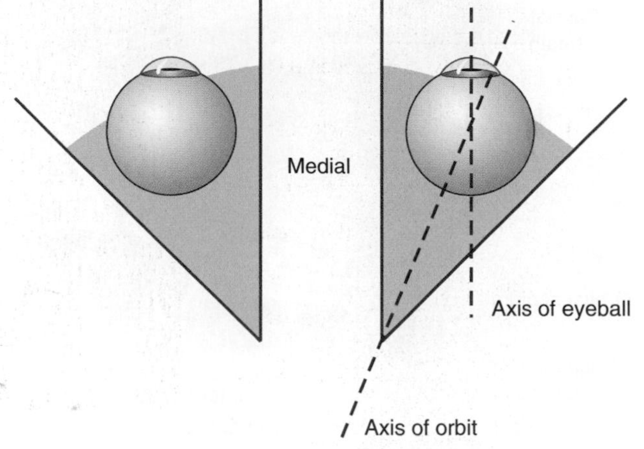


**Fig. 8.94** Axes of the eyeball and orbit.

surface of the lesser wing of the sphenoid (Fig. 8.95B). Its primary point of insertion is into the anterior surface of the superior tarsus, but a few fibers also attach to the skin of the upper eyelid and the superior conjunctival fornix.

Innervation is by the superior branch of the oculomotor nerve [III].

Contraction of the levator palpebrae superioris raises the upper eyelid.

A unique feature of the levator palpebrae superioris is that a collection of smooth muscle fibers passes from its inferior surface to the upper edge of the superior tarsus (see Fig. 8.77). This group of smooth muscle fibers (the superior tarsal muscle) help maintain eyelid elevation and are innervated by postganglionic sympathetic fibers from the superior cervical ganglion.

Loss of oculomotor nerve [III] function results in complete ptosis or drooping of the superior eyelid, whereas loss of sympathetic innervation to the superior tarsal muscle results in partial ptosis.

### Rectus muscles

Four rectus muscles occupy medial, lateral, inferior, and superior positions as they pass from their origins posteriorly to their points of attachment on the anterior half of the eyeball (Fig. 8.95 and Table 8.8). They originate as a group from a common tendinous ring at the apex of the orbit and form a cone of muscles as they pass forward to their attachment on the eyeball.

#### Superior and inferior rectus muscles

The superior and inferior rectus muscles have complicated actions because the apex of the orbit, where the muscles originate, is medial to the central axis of the eyeball when looking directly forward:

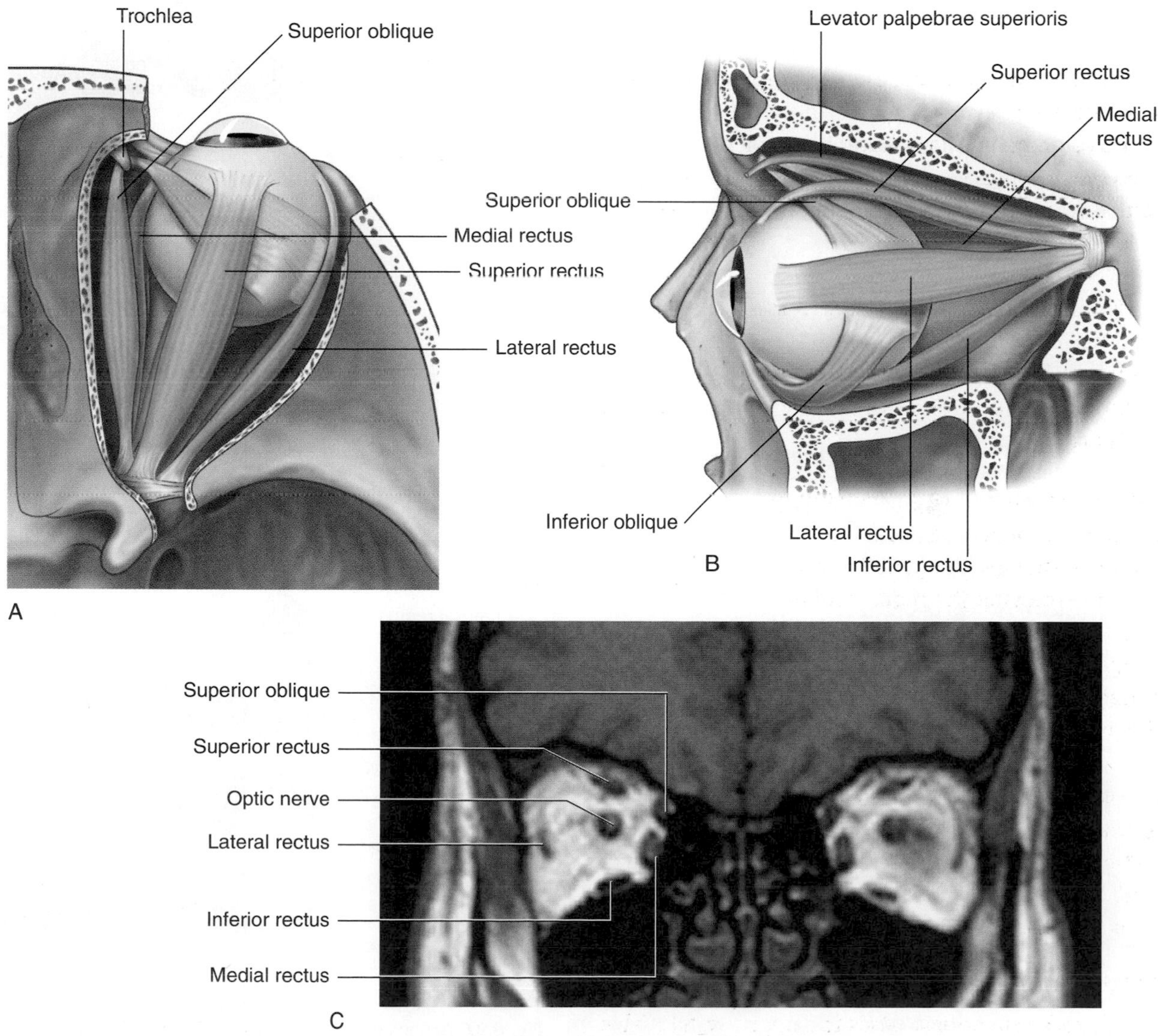


**Fig. 8.95** Muscles of the eyeball. **A.** Superior view. **B.** Lateral view. **C.** Coronal magnetic resonance image through the eye.

- The **superior rectus** originates from the superior part of the common tendinous ring above the optic canal.
- The **inferior rectus** originates from the inferior part of the common tendinous ring below the optic canal (Fig. 8.96).

As these muscles pass forward in the orbit to attach to the anterior half of the eyeball, they are also directed laterally (Fig. 8.95). Because of these orientations:

- Contraction of the superior rectus elevates, adducts, and internally rotates the eyeball (Fig. 8.97A).
- Contraction of the inferior rectus depresses, adducts, and externally rotates the eyeball (Fig. 8.97A).

The **superior branch** of the oculomotor nerve [III] innervates the superior rectus, and the **inferior branch** of the oculomotor nerve [III] innervates the inferior rectus.

To isolate the function of and to test the superior and inferior rectus muscles, a patient is asked to track a physician's finger laterally and then either upward or downward (Fig. 8.97B). The first movement brings the axis of the eyeball into alignment with the long axis of the superior and inferior rectus muscles. Moving the finger upward tests the superior rectus muscle and moving it downward tests the inferior rectus muscle (Fig. 8.97B).

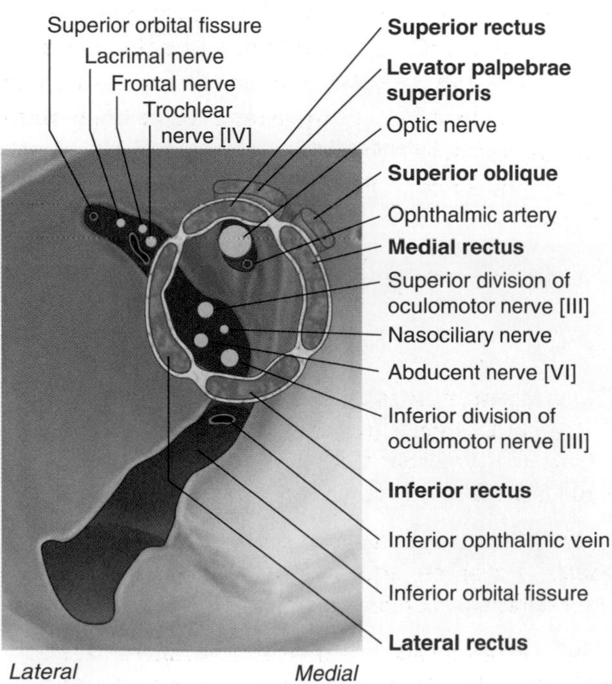


**Fig. 8.96** Origins of muscles of the eyeball, coronal view.

## Medial and lateral rectus muscles

The orientation and actions of the medial and lateral rectus muscles are more straightforward than those of the superior and inferior rectus muscles.

The **medial rectus** originates from the medial part of the common tendinous ring medial to and below the optic canal, whereas the **lateral rectus** originates from the lateral part of the common tendinous ring as the common tendinous ring bridges the superior orbital fissure (Fig. 8.96).

The medial and lateral rectus muscles pass forward and attach to the anterior half of the eyeball (Fig. 8.95).

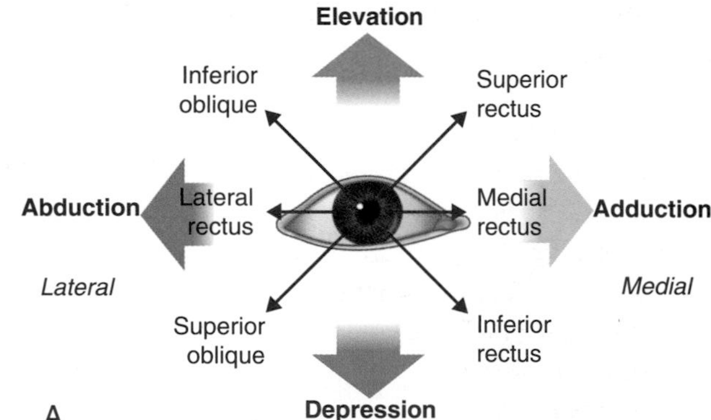


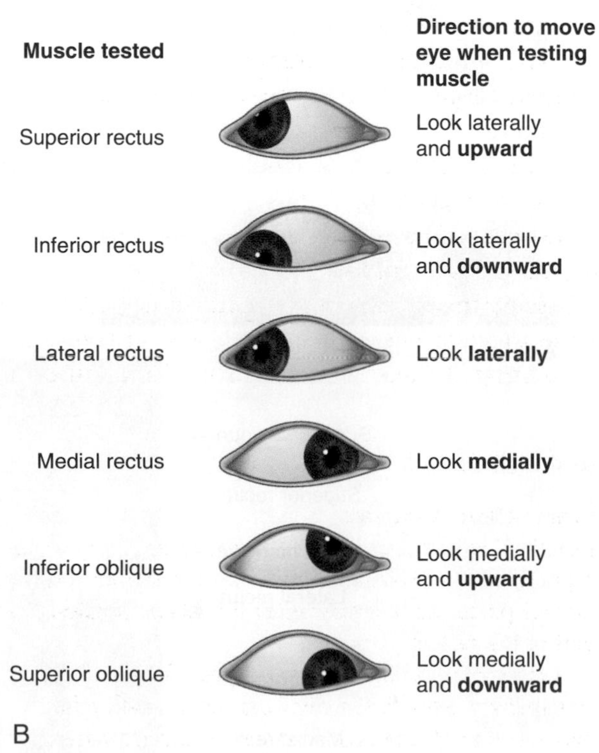


**Fig. 8.97** Actions of muscles of the eyeball. **A.** Action of individual muscles (anatomical action). **B.** Movement of eye when testing specific muscle (clinical testing).

Contraction of medial rectus adducts the eyeball, whereas contraction of lateral rectus abducts the eyeball (Fig. 8.97A).

The inferior branch of the oculomotor nerve [III] innervates the medial rectus, and the abducent nerve [VI] innervates the lateral rectus.

To isolate the function of and test the medial and lateral rectus muscles, a patient is asked to track a physician's finger medially and laterally, respectively, in the horizontal plane (Fig. 8.97B).

### Oblique muscles

The oblique muscles are in the superior and inferior parts of the orbit, do not originate from the common tendinous ring, are angular in their approaches to the eyeball, and, unlike the rectus muscles, attach to the posterior half of the eyeball (Table 8.8).

#### Superior oblique

The superior oblique arises from the body of the sphenoid, superior and medial to the optic canal and medial to the origin of the levator palpebrae superioris (Figs. 8.95 and 8.96). It passes forward, along the medial border of the roof of the orbit, until it reaches a fibrocartilaginous pulley (the **trochlea**), which is attached to the trochlear fovea of the frontal bone.

The tendon of the superior oblique passes through the trochlea and turns laterally to cross the eyeball in a posterolateral direction. It continues deep to the superior rectus muscle and inserts into the outer posterior quadrant of the eyeball.

Contraction of the superior oblique therefore directs the pupil down and out (Fig. 8.97A).

The trochlear nerve [IV] innervates the superior oblique along its superior surface.

To isolate the function of and to test the superior oblique muscle, a patient is asked to track a physician's finger medially to bring the axis of the tendon of the muscle into alignment with the axis of the eyeball, and then to look down, which tests the muscle (Fig. 8.97B).

#### Inferior oblique

The inferior oblique is the only extrinsic muscle that does not take origin from the posterior part of the orbit. It arises from the medial side of the floor of the orbit, just posterior to the orbital rim, and is attached to the orbital surface of the maxilla just lateral to the nasolacrimal groove (Fig. 8.95).

The inferior oblique crosses the floor of the orbit in a posterolateral direction between the inferior rectus and the floor of the orbit, before inserting into the outer posterior quadrant just under the lateral rectus.

Contraction of the inferior oblique directs the pupil up and out (Fig. 8.97A).

The inferior branch of the oculomotor nerve innervates the inferior oblique.

To isolate the function of and to test the inferior oblique muscle, a patient is asked to track a physician's finger medially to bring the axis of the eyeball into alignment with the axis of the muscle and then to look up, which tests the muscle (Fig. 8.97B).

### Extrinsic muscles and eyeball movements

Six of the seven extrinsic muscles of the orbit are directly involved in movements of the eyeball.

For each of the rectus muscles, the medial, lateral, inferior, and superior, and the superior and inferior obliques, a specific action or group of actions can be described (Table 8.8). However, these muscles do not act in isolation. They work as teams of muscles in the coordinated movement of the eyeball to position the pupil as needed.

For example, although the lateral rectus is the muscle primarily responsible for moving the eyeball laterally, it is assisted in this action by the superior and inferior oblique muscles.

### In the clinic

#### Examination of the eye

Examination of the eye includes assessment of the visual capabilities, the extrinsic musculature and its function, and disease processes that may affect the eye in isolation or as part of the systemic process.

Examination of the eye includes tests for visual acuity, astigmatism, visual fields, and color interpretation (to exclude color blindness) in a variety of circumstances. The physician also assesses the retina, the optic nerve and its coverings, the lens, and the cornea.

The extrinsic muscles are supplied by the abducent nerve [VI], the trochlear nerve [IV], and the oculomotor nerve [III].

The extrinsic muscles work synergistically to provide appropriate and conjugate eye movement:

- lateral rectus—abducent nerve [VI],
- superior oblique—trochlear nerve [IV], and
- remainder—oculomotor nerve [III].

The eye may be affected in systemic diseases. Diabetes mellitus typically affects the eye and may cause cataracts,

*(continues)*

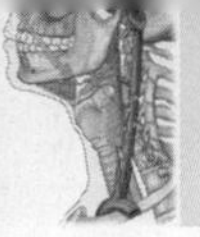

## In the clinic—cont'd

macular disease, and retinal hemorrhage, all impairing vision.

Occasionally unilateral paralysis of the extra-ocular muscles occurs and is due to brainstem injury or direct nerve injury, which may be associated with tumor compression or trauma. The paralysis of a muscle is easily demonstrated when the patient attempts to move the eye in the direction associated with normal action of that muscle. Typically the patient complains of double vision (diplopia).

***Loss of innervation of the muscles around the eye***

Loss of innervation of the orbicularis oculi by the facial nerve [VII] causes an inability to close the eyelids tightly, allowing the lower eyelid to droop away causing spillage of tears. This loss of tears allows drying of the conjunctiva, which may ulcerate, so allowing secondary infection.

Loss of innervation of the levator palpebrae superioris by oculomotor nerve [III] damage causes an inability of the superior eyelid to elevate, producing a complete ptosis. Usually, oculomotor nerve [III] damage is caused by severe head injury.

Loss of innervation of the superior tarsal muscle by sympathetic fibers causes a constant partial ptosis. Any lesion along the sympathetic trunk can induce this. An apical pulmonary malignancy should always be suspected because the ptosis may be part of Horner's syndrome (see "In the clinic" on p. 920).

## In the clinic

**The "H-test"**

A simple "formula" for remembering the nerves that innervate the extraocular muscles is "LR6SO4 and all the rest are 3" (lateral rectus [VI], superior oblique [IV], all the rest including levator palpebrae superioris are [III]).

The function of all extrinsic muscles and their nerves [III, IV, VI] that move the eyeball in both orbits can all easily be tested at the same time by having the patient track, without moving his or her head, an object such as the tip of a pen or a finger moved in an "H" pattern—starting from the midline between the two eyes (Fig. 8.98).

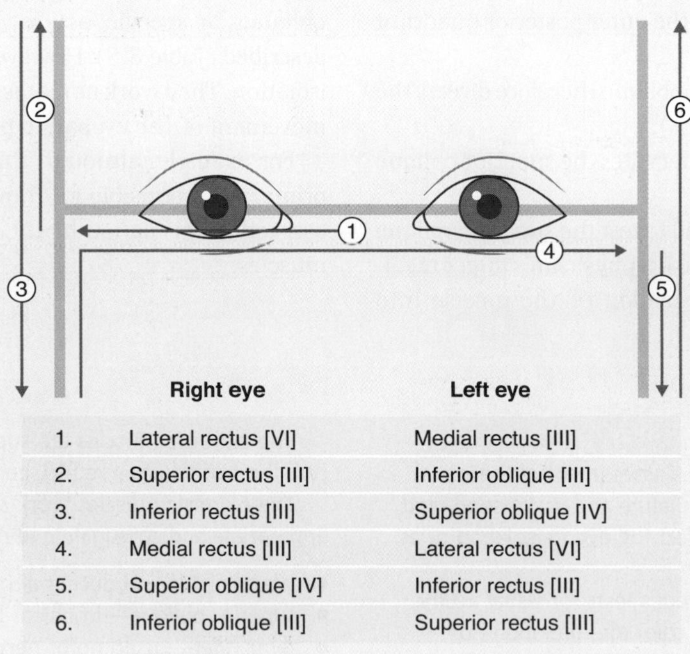


| | Right eye | Left eye |
|---|---|---|
| 1. | Lateral rectus [VI] | Medial rectus [III] |
| 2. | Superior rectus [III] | Inferior oblique [III] |
| 3. | Inferior rectus [III] | Superior oblique [IV] |
| 4. | Medial rectus [III] | Lateral rectus [VI] |
| 5. | Superior oblique [IV] | Inferior rectus [III] |
| 6. | Inferior oblique [III] | Superior rectus [III] |

**Fig. 8.98** The "H-test."

## Vessels

### Arteries

The arterial supply to the structures in the orbit, including the eyeball, is by the ophthalmic artery (Fig. 8.99). This vessel is a branch of the internal carotid artery, given off immediately after the internal carotid artery leaves the cavernous sinus. The ophthalmic artery passes into the orbit through the optic canal with the optic nerve.

In the orbit the ophthalmic artery initially lies inferior and lateral to the optic nerve (Fig. 8.99). As it passes forward in the orbit, it crosses superior to the optic nerve and proceeds anteriorly on the medial side of the orbit.

In the orbit the ophthalmic artery gives off numerous branches as follows:

- the **lacrimal artery**, which arises from the ophthalmic artery on the lateral side of the optic nerve, and passes anteriorly on the lateral side of the orbit, supplying the lacrimal gland, muscles, the anterior ciliary branch to the eyeball, and the lateral sides of the eyelid;
- the **central retinal artery**, which enters the optic nerve, proceeds down the center of the nerve to the retina, and is clearly seen when viewing the retina with an ophthalmoscope—occlusion of this vessel or of the parent artery leads to blindness;
- the **long** and **short posterior ciliary arteries**, which are branches that enter the eyeball posteriorly, piercing the sclera, and supplying structures inside the eyeball;
- the **muscular arteries**, which are branches supplying the intrinsic muscles of the eyeball;
- the **supra-orbital artery**, which usually arises from the ophthalmic artery immediately after it has crossed the optic nerve, proceeds anteriorly, and exits the orbit through the supra-orbital foramen with the supra-orbital nerve—it supplies the forehead and scalp as it passes across these areas to the vertex of the skull;
- the **posterior ethmoidal artery**, which exits the orbit through the posterior ethmoidal foramen to supply the ethmoidal cells and nasal cavity;
- the **anterior ethmoidal artery**, which exits the orbit through the anterior ethmoidal foramen, enters the cranial cavity giving off the anterior meningeal branch, and continues into the nasal cavity supplying the septum and lateral wall, and ending as the dorsal nasal artery;
- the **medial palpebral arteries**, which are small branches supplying the medial area of the upper and lower eyelids;
- the **dorsal nasal artery**, which is one of the two terminal branches of the ophthalmic artery, leaves the orbit to supply the upper surface of the nose; and
- the **supratrochlear artery**, which is the other terminal branch of the ophthalmic artery and leaves the orbit with the supratrochlear nerve, supplying the forehead as it passes across it in a superior direction.

### Veins

There are two venous channels in the orbit, the superior and inferior ophthalmic veins (Fig. 8.100).

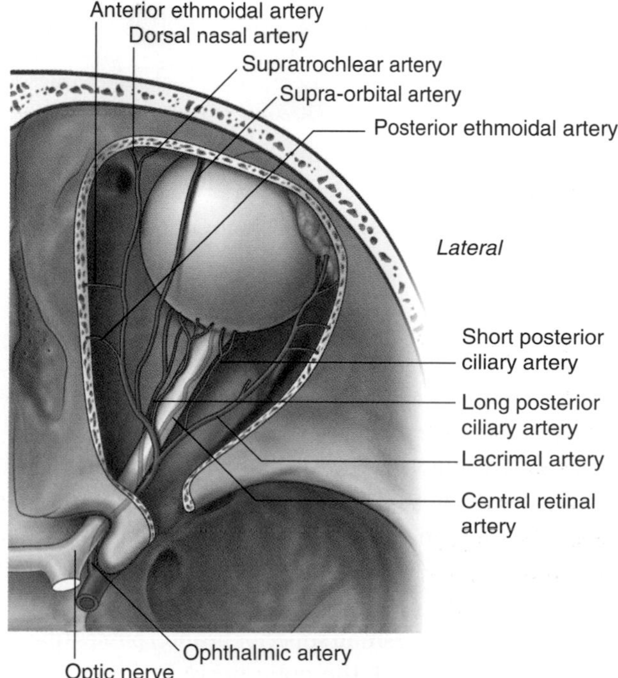


**Fig. 8.99** Arterial supply to the orbit and eyeball.

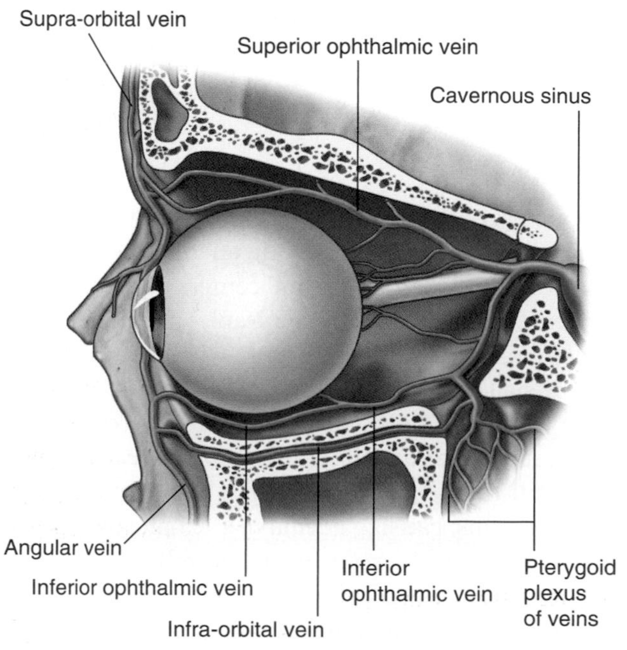


**Fig. 8.100** Venous drainage of the orbit and eyeball.

The **superior ophthalmic vein** begins in the anterior area of the orbit as connecting veins from the supra-orbital vein and the angular vein join together. It passes across the superior part of the orbit, receiving tributaries from the companion veins to the branches of the ophthalmic artery and veins draining the posterior part of the eyeball. Posteriorly, it leaves the orbit through the superior orbital fissure and enters the cavernous sinus.

The **inferior ophthalmic vein** is smaller than the superior ophthalmic vein, begins anteriorly, and passes across the inferior part of the orbit. It receives various tributaries from muscles and the posterior part of the eyeball as it crosses the orbit.

The inferior ophthalmic vein leaves the orbit posteriorly by:

- joining with the superior ophthalmic vein,
- passing through the superior orbital fissure on its own to join the cavernous sinus, or
- passing through the inferior orbital fissure to join with the pterygoid plexus of veins in the infratemporal fossa.

Because the ophthalmic veins communicate with the cavernous sinus, they act as a route by which infections can spread from outside to inside the cranial cavity.

## Innervation

Numerous nerves pass into the orbit and innervate structures within its bony walls. They include the optic nerve [II], the oculomotor nerve [III], the trochlear nerve [IV], the abducent nerve [VI], and autonomic nerves. Other nerves such as the ophthalmic nerve [$V_1$] innervate orbital structures and then travel out of the orbit to innervate other regions.

### Optic nerve

The optic nerve [II] is not a true cranial nerve, but rather an extension of the brain carrying afferent fibers from the retina of the eyeball to the visual centers of the brain. The optic nerve is surrounded by the cranial meninges, including the subarachnoid space, which extends as far forward as the eyeball.

Any increase in intracranial pressure therefore results in increased pressure in the subarachnoid space surrounding the optic nerve. This may impede venous return along the retinal veins, causing edema of the optic disc (papilledema), which can be seen when the retina is examined using an ophthalmoscope.

The optic nerve leaves the orbit through the optic canal (Fig. 8.101). It is accompanied in the optic canal by the ophthalmic artery.

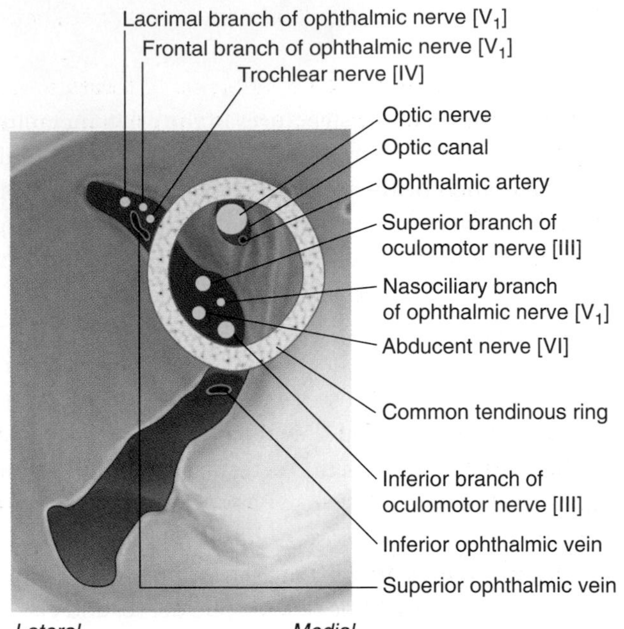


**Fig. 8.101** Innervation of the orbit and eyeball.

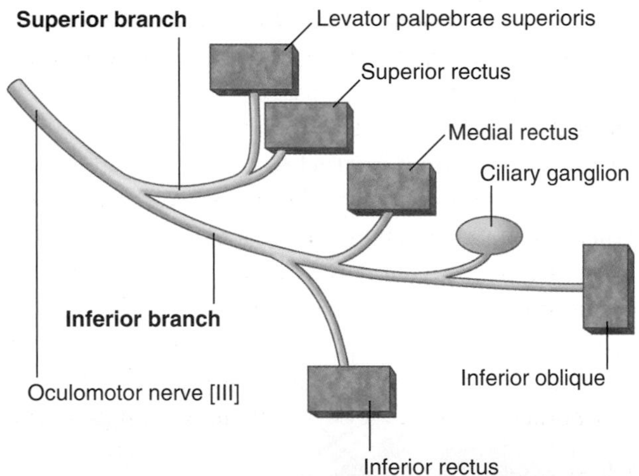


**Fig. 8.102** Oculomotor nerve [III] and its divisions.

### Oculomotor nerve

The oculomotor nerve [III] leaves the anterior surface of the brainstem between the midbrain and the pons. It passes forward in the lateral wall of the cavernous sinus.

Just before entering the orbit the oculomotor nerve [III] divides into superior and inferior branches (Fig. 8.102). These branches enter the orbit through the superior orbital fissure, lying within the common tendinous ring (Fig. 8.101).

Inside the orbit the small superior branch passes upward over the lateral side of the optic nerve to innervate the superior rectus and levator palpebrae superioris muscles (Fig. 8.102).

The large inferior branch divides into three branches:

- one passing below the optic nerve as it passes to the medial side of the orbit to innervate the medial rectus muscle,
- a second descending to innervate the inferior rectus muscle, and
- the third descending as it runs forward along the floor of the orbit to innervate the inferior oblique muscle (Fig. 8.102).

As the third branch descends, it gives off the **branch to the ciliary ganglion**. This is the parasympathetic root to the ciliary ganglion and carries preganglionic parasympathetic fibers that will synapse in the ciliary ganglion with postganglionic parasympathetic fibers. The postganglionic fibers are distributed to the eyeball through short ciliary nerves and innervate the sphincter pupillae and ciliary muscles.

### Trochlear nerve

The trochlear nerve [IV] arises from the posterior surface of the midbrain, and passes around the midbrain to enter the edge of the tentorium cerebelli. It continues on an intradural path arriving in and passing through the lateral wall of the cavernous sinus just below the oculomotor nerve [III].

Just before entering the orbit, the trochlear nerve ascends, passing across the oculomotor nerve [III] and entering the orbit through the superior orbital fissure above the common tendinous ring (Fig. 8.101). In the orbit the trochlear nerve [IV] ascends and turns medially, crossing above the levator palpebrae superioris muscle to enter the upper border of the superior oblique muscle (Fig. 8.103).

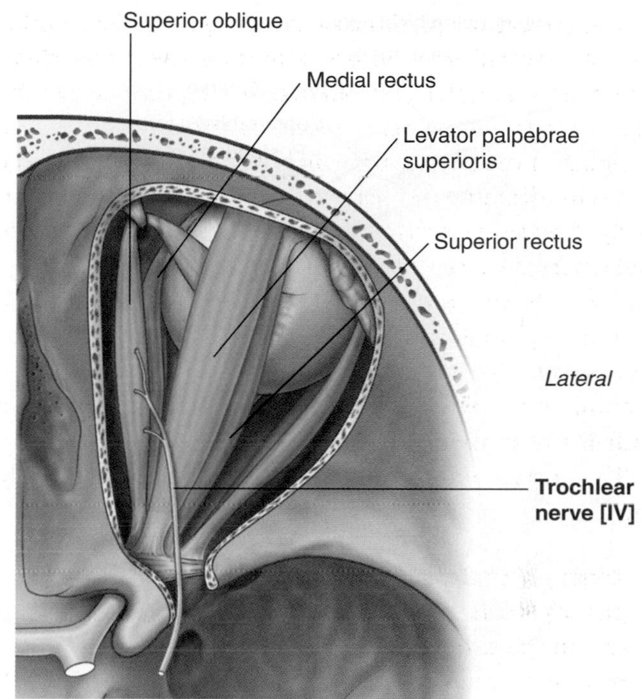


**Fig. 8.103** Trochlear nerve [IV] in the orbit.

### Abducent nerve

The abducent nerve [VI] arises from the brainstem between the pons and medulla. It enters the dura covering the clivus and continues in a dural canal until it reaches the cavernous sinus.

The abducent nerve enters the cavernous sinus and runs through the sinus lateral to the internal carotid artery. It passes out of the sinus and enters the orbit through the superior orbital fissure within the common tendinous ring (Fig. 8.101). Once in the orbit it courses laterally to supply the lateral rectus muscle.

### Postganglionic sympathetic fibers

Preganglionic sympathetic fibers arise from the upper segments of the thoracic spinal cord, mainly T1. They enter the sympathetic chain through white rami communicantes, and ascend to the **superior cervical ganglion** where they synapse with postganglionic sympathetic fibers.

The postganglionic fibers are distributed along the internal carotid artery and its branches.

The postganglionic sympathetic fibers destined for the orbit travel with the ophthalmic artery. Once in the orbit the fibers are distributed to the eyeball either by:

- passing through the ciliary ganglion, without synapsing, and joining the short ciliary nerves, which pass from the ganglion to the eyeball; or
- passing through long ciliary nerves to reach the eyeball.

In the eyeball postganglionic sympathetic fibers innervate the dilator pupillae muscle.

### Ophthalmic nerve [$V_1$]

The ophthalmic nerve [$V_1$] is the smallest and most superior of the three divisions of the trigeminal nerve. This purely sensory nerve receives input from structures in the orbit and from additional branches on the face and scalp.

Leaving the trigeminal ganglion, the ophthalmic nerve [$V_1$] passes forward in the lateral wall of the cavernous sinus inferior to the trochlear [IV] and oculomotor [III] nerves. Just before it enters the orbit it divides into three branches—the nasociliary, lacrimal, and frontal nerves

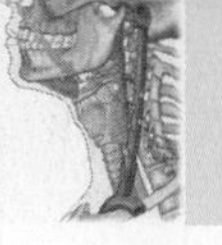



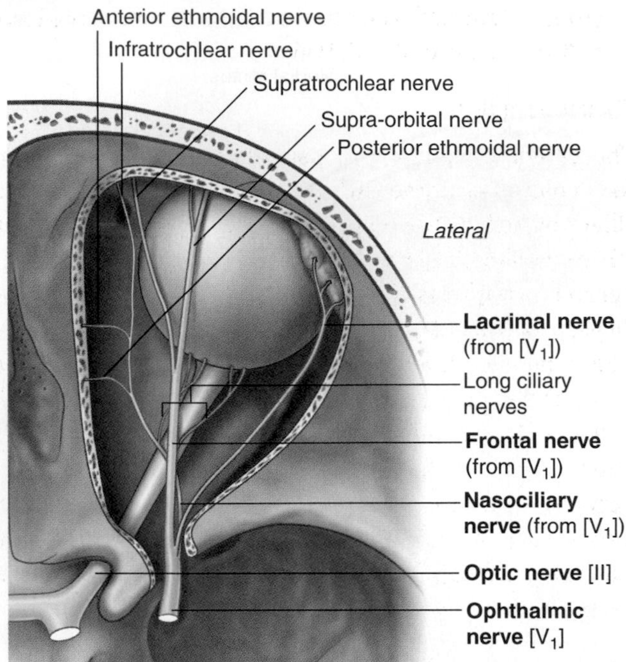


**Fig. 8.104** Ophthalmic nerve [$V_1$] and its divisions.

(Fig. 8.104). These branches enter the orbit through the superior orbital fissure with the frontal and lacrimal nerves outside the common tendinous ring, and the nasociliary nerve within the common tendinous ring (Fig. 8.101).

### Lacrimal nerve

The lacrimal nerve is the smallest of the three branches of the ophthalmic nerve [$V_1$]. Once in the orbit it passes forward along the upper border of the lateral rectus muscle (Fig. 8.105). It receives a branch from the zygomaticotemporal nerve, which carries parasympathetic and sympathetic postganglionic fibers for distribution to the lacrimal gland.

Reaching the anterolateral aspect of the orbit, the lacrimal nerve supplies the lacrimal gland, conjunctiva, and lateral part of the upper eyelid.

### Frontal nerve

The frontal nerve is the largest branch of the ophthalmic nerve [$V_1$] and receives sensory input from areas outside the orbit. Exiting the superior orbital fissure, this branch passes forward between the levator palpebrae superioris and the periorbita on the roof of the orbit (Fig. 8.101). About midway across the orbit it divides into its two terminal branches—the supra-orbital and supratrochlear nerves (Figs. 8.104 and 8.105):

- The **supratrochlear nerve** continues forward in an anteromedial direction, passing above the trochlea, exits the orbit medial to the supra-orbital foramen, and supplies the conjunctiva and skin of the upper eyelid and the skin on the lower medial part of the forehead.
- The **supra-orbital nerve** is the larger of the two branches, continues forward, passing between the levator palpebrae superioris muscle and the periorbita covering the roof of the orbit (Fig. 8.105), exits the orbit through the supra-orbital notch and ascends across the forehead and scalp, supplying the upper eyelid and conjunctiva, the forehead, and as far posteriorly as the middle of the scalp.

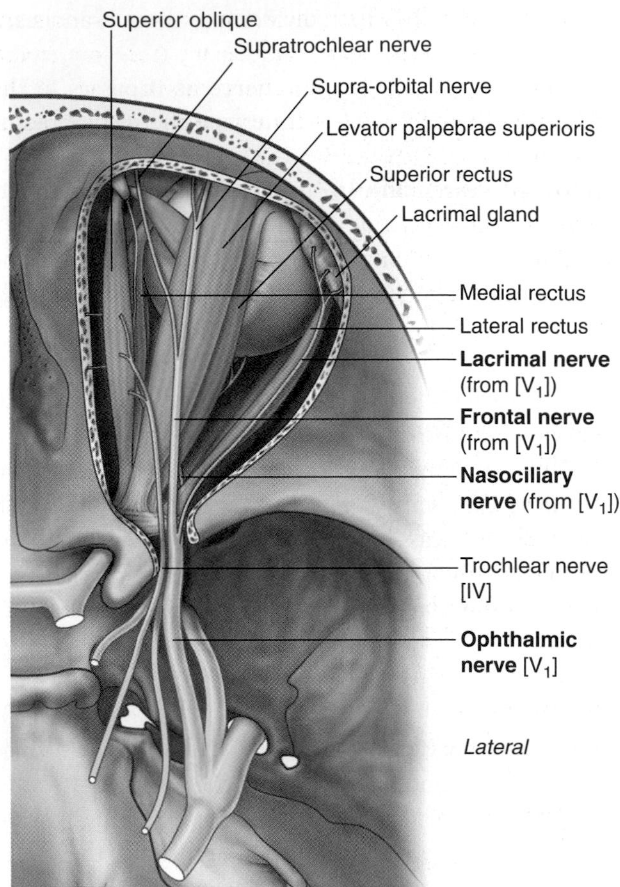


**Fig. 8.105** Relationship of the ophthalmic nerve [$V_1$] and its divisions to the muscles of the eyeball.

### Nasociliary nerve

The nasociliary nerve is intermediate in size between the frontal and lacrimal nerves and is usually the first branch from the ophthalmic nerve (Fig. 8.104). It is most deeply placed in the orbit, entering the area within the common tendinous ring between the superior and inferior branches of the oculomotor nerve [III] (see Fig. 8.101).

Once in the orbit, the nasociliary nerve crosses the superior surface of the optic nerve as it passes in a medial direction below the superior rectus muscle (Figs. 8.104 and 8.106). Its first branch, the **communicating branch with the ciliary ganglion** (**sensory root to the ciliary ganglion**), is given off early in its path through the orbit.

The nasociliary nerve continues forward along the medial wall of the orbit, between the superior oblique and the medial rectus muscles, giving off several branches (Fig. 8.106). These include:

- the **long ciliary nerves**, which are sensory to the eyeball but may also contain sympathetic fibers for pupillary dilation;
- the **posterior ethmoidal nerve**, which exits the orbit through the posterior ethmoidal foramen to supply posterior ethmoidal cells and the sphenoidal sinus;
- the **infratrochlear nerve**, which distributes to the medial part of the upper and lower eyelids, the lacrimal sac, and skin of the upper half of the nose; and
- the **anterior ethmoidal nerve**, which exits the orbit through the anterior ethmoidal foramen to supply the anterior cranial fossa, nasal cavity, and skin of the lower half of the nose (Fig. 8.106).

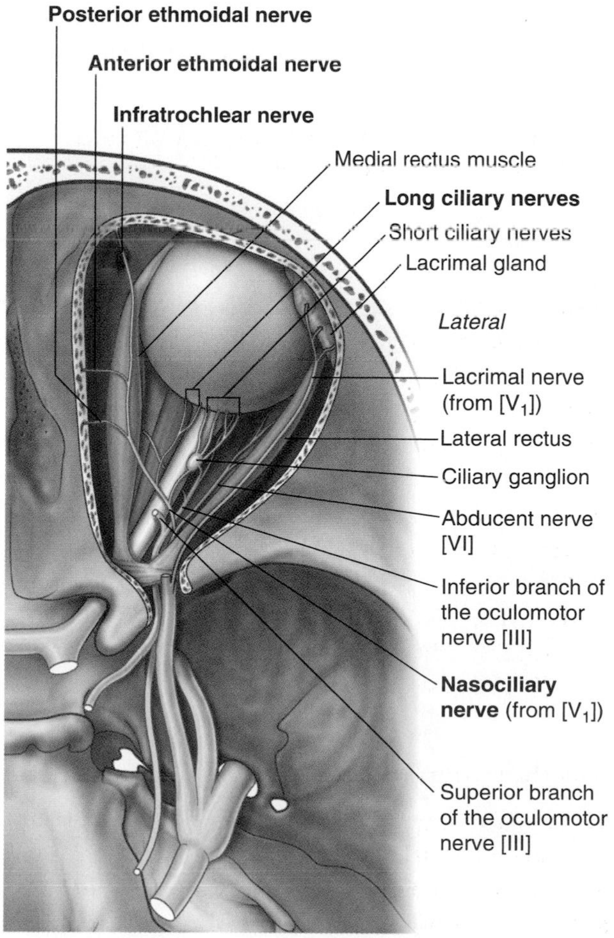

**Fig. 8.106** Course of the nasociliary nerve (from $[V_1]$) in the orbit.

## Ciliary ganglion

The ciliary ganglion is a parasympathetic ganglion of the oculomotor nerve [III]. It is associated with the nasociliary branch of the ophthalmic nerve $[V_1]$ and is the site where preganglionic and postganglionic parasympathetic neurons synapse as fibers from this part of the autonomic division of the PNS make their way to the eyeball. The ciliary ganglion is also traversed by postganglionic sympathetic fibers and sensory fibers as they travel to the eyeball.

The ciliary ganglion is a very small ganglion, in the posterior part of the orbit immediately lateral to the optic nerve and between the optic nerve and the lateral rectus muscle (Fig. 8.106). It is usually described as receiving at least two, and possibly three, branches or roots from other nerves in the orbit.

### Parasympathetic root

As the inferior branch of the oculomotor nerve [III] passes the area of the ciliary ganglion, it sends a branch to the ganglion (the parasympathetic root). The parasympathetic branch carries preganglionic parasympathetic fibers, which enter the ganglion and synapse with postganglionic parasympathetic fibers within the ganglion (Fig. 8.107).

The postganglionic parasympathetic fibers leave the ganglion through short ciliary nerves, which enter the posterior aspect of the eyeball around the optic nerve.

In the eyeball the parasympathetic fibers innervate:

- the **sphincter pupillae muscle**, responsible for pupillary constriction, and
- the **ciliary muscle**, responsible for accommodation of the lens of the eye for near vision.

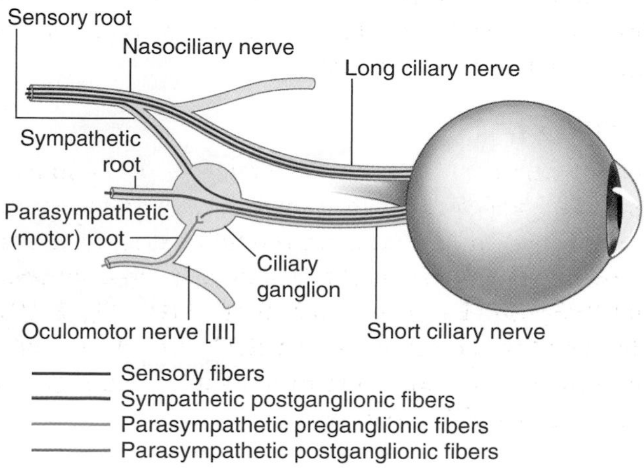


**Fig. 8.107** Ciliary ganglion.

### Sensory root

A second branch (the sensory root), passes from the nasociliary nerve to the ganglion (Fig. 8.107). This branch enters the posterosuperior aspect of the ganglion, and carries sensory fibers, which pass through the ganglion and continue along the short ciliary nerves to the eyeball. These fibers are responsible for sensory innervation to all parts of the eyeball; however, the sympathetic fibers also may take alternative routes to the eyeball.

### Sympathetic root

The third branch to the ciliary ganglion is the most variable. This branch, when present, is the sympathetic root and contains postganglionic sympathetic fibers from the superior cervical ganglion (Fig. 8.107). These fibers travel up the internal carotid artery, leave the plexus surrounding the artery in the cavernous sinus, and enter the orbit through the common tendinous ring. In the orbit they enter the posterior aspect of the ciliary ganglion, cross the ganglion, and continue along the short ciliary nerves to the eyeball; however, the sympathetic fibers also may take alternative routes to the eyeball.

Sympathetic fibers to the eyeball may not enter the ganglion as a separate sympathetic root. Rather, the postganglionic sympathetic fibers may leave the plexus associated with the internal carotid artery in the cavernous sinus, join the ophthalmic nerve [$V_1$], and course into the ciliary ganglion in the sensory root from the nasociliary nerve. In addition, the sympathetic fibers carried in the nasociliary nerve may not enter the ganglion at all and may course directly into the eyeball in the long ciliary nerves (Fig. 8.107). Whatever their path, postganglionic sympathetic fibers reach the eyeball and innervate the dilator pupillae muscle.

## Eyeball

The globe-shaped eyeball occupies the anterior part of the orbit. Its rounded shape is disrupted anteriorly, where it bulges outward. This outward projection represents about one-sixth of the total area of the eyeball and is the transparent cornea (Fig. 8.108).

Posterior to the cornea and in order from front to back are the anterior chamber, the iris and pupil, the posterior chamber, the lens, the postremal (vitreous) chamber, and the retina.

### Anterior and posterior chambers

The **anterior chamber** is the area directly posterior to the cornea and anterior to the colored part of the eye (**iris**). The central opening in the iris is the **pupil**. Posterior to the

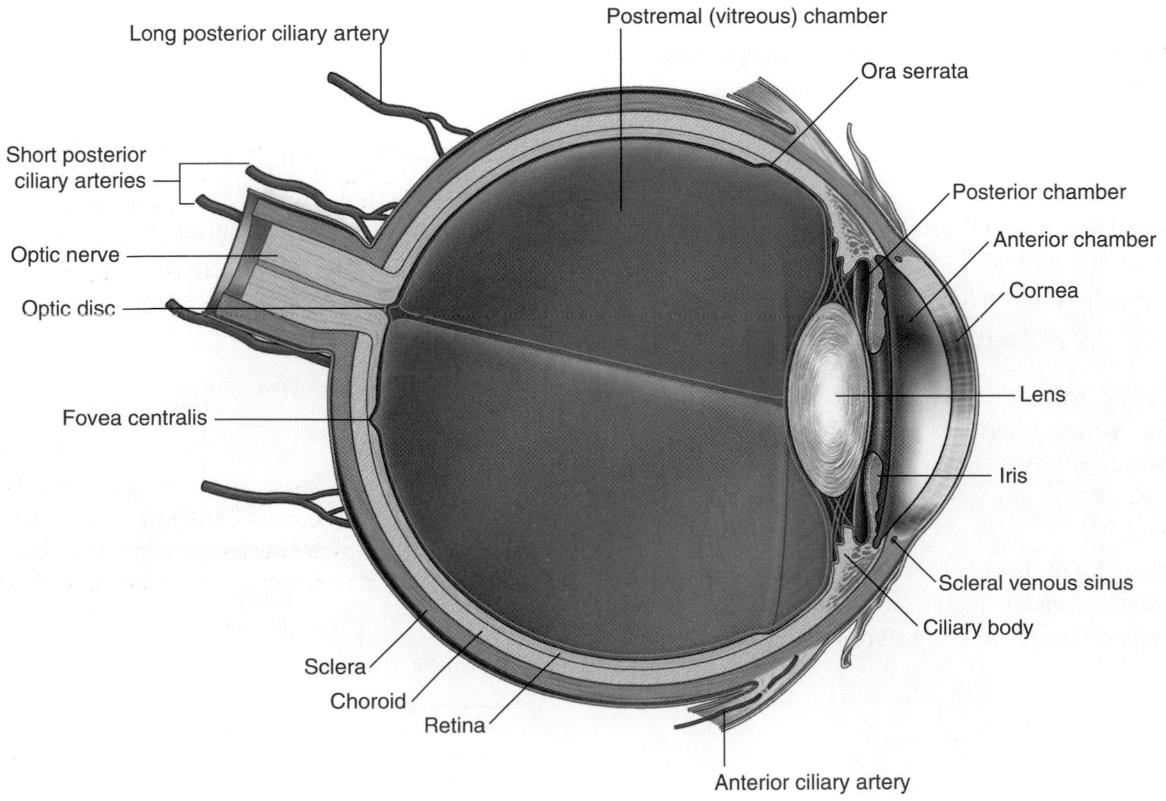


**Fig. 8.108** Eyeball.

iris and anterior to the lens is the smaller **posterior chamber**.

The anterior and posterior chambers are continuous with each other through the pupillary opening. They are filled with a fluid (**aqueous humor**), which is secreted into the posterior chamber, flows into the anterior chamber through the pupil, and is absorbed into the **scleral venous sinus** (the canal of Schlemm), which is a circular venous channel at the junction between the cornea and the iris (Fig. 8.108).

The aqueous humor supplies nutrients to the avascular cornea and lens and maintains the intra-ocular pressure. If the normal cycle of its production and absorption is disturbed so that the amount of fluid increases, intra-ocular pressure will increase. This condition (glaucoma) can lead to a variety of visual problems.

### Lens and vitreous humor

The **lens** separates the anterior one-fifth of the eyeball from the posterior four-fifths (Fig. 8.108). It is a transparent, biconvex elastic disc attached circumferentially to muscles associated with the outer wall of the eyeball. This lateral attachment provides the lens with the ability to change its refractive ability to maintain visual acuity. The clinical term for opacity of the lens is a cataract.

The posterior four-fifths of the eyeball, from the lens to the retina, is occupied by the postremal (vitreous) chamber (Fig. 8.108). This segment is filled with a transparent, gelatinous substance—the **vitreous body** (**vitreous humor**). This substance, unlike aqueous humor, cannot be replaced.

### Walls of the eyeball

Surrounding the internal components of the eyeball are the walls of the eyeball. They consist of three layers: an outer fibrous layer, a middle vascular layer, and an inner retinal layer (Fig. 8.108).

- The outer fibrous layer consists of the sclera posteriorly and the cornea anteriorly.
- The middle vascular layer consists of the **choroid** posteriorly and is continuous with the ciliary body and iris anteriorly.
- The inner layer consists of the optic part of the **retina** posteriorly and the nonvisual retina that covers the internal surface of the ciliary body and iris anteriorly.

### Vessels

#### Arterial supply

The arterial supply to the eyeball is from several sources:

- The short posterior ciliary arteries are branches from the ophthalmic artery that pierce the sclera around the optic nerve and enter the choroid layer (Fig. 8.108).
- The long posterior ciliary arteries, usually two, enter the sclera on the medial and lateral sides of the optic nerve and proceed anteriorly in the choroid layer to anastomose with the anterior ciliary arteries.
- The anterior ciliary arteries are branches of the arteries supplying the muscles (Fig. 8.108)—as the muscles attach to the sclera, these arteries pierce the sclera to anastomose with the long posterior ciliary arteries in the choroid layer.
- The central retinal artery that has traversed the optic nerve and enters the area of the retina at the optic disc.

#### Venous drainage

Venous drainage of the eyeball is primarily related to drainage of the choroid layer. Four large veins (the **vorticose veins**) are involved in this process. They exit through the sclera from each of the posterior quadrants of the eyeball and enter the superior and inferior ophthalmic veins. There is also a central retinal vein accompanying the central retinal artery.

### In the clinic

**Glaucoma**

Intraocular pressure will rise if the normal cycle of aqueous humor fluid production and absorption is disturbed so that the amount of fluid increases. This condition is glaucoma and can lead to a variety of visual problems including blindness, which results from compression of the retina and its blood supply.

### In the clinic

**Cataracts**

With increasing age and in certain disease states the lens of the eye becomes opaque. Increasing opacity results in increasing visual impairment. A common operation is excision of the cloudy lens and replacement with a new man-made lens.

## In the clinic

**Ophthalmoscopy**

Direct visualization of the postremal (vitreous) chamber of the eye is possible in most clinical settings. It is achieved using an ophthalmoscope, which is a small battery-operated light with a tiny lens that allows direct visualization of the postremal (vitreous) chamber and the posterior wall of the eye through the pupil and the lens. It is sometimes necessary to place a drug directly onto the eye to dilate the pupil for better visualization.

The optic nerve, observed as the optic disc, is easily seen. The typical four branches of the central retinal artery and the fovea are also seen.

Using ophthalmoscopy the physician can look for diseases of the optic nerve, vascular abnormalities, and changes within the retina (Fig. 8.109).

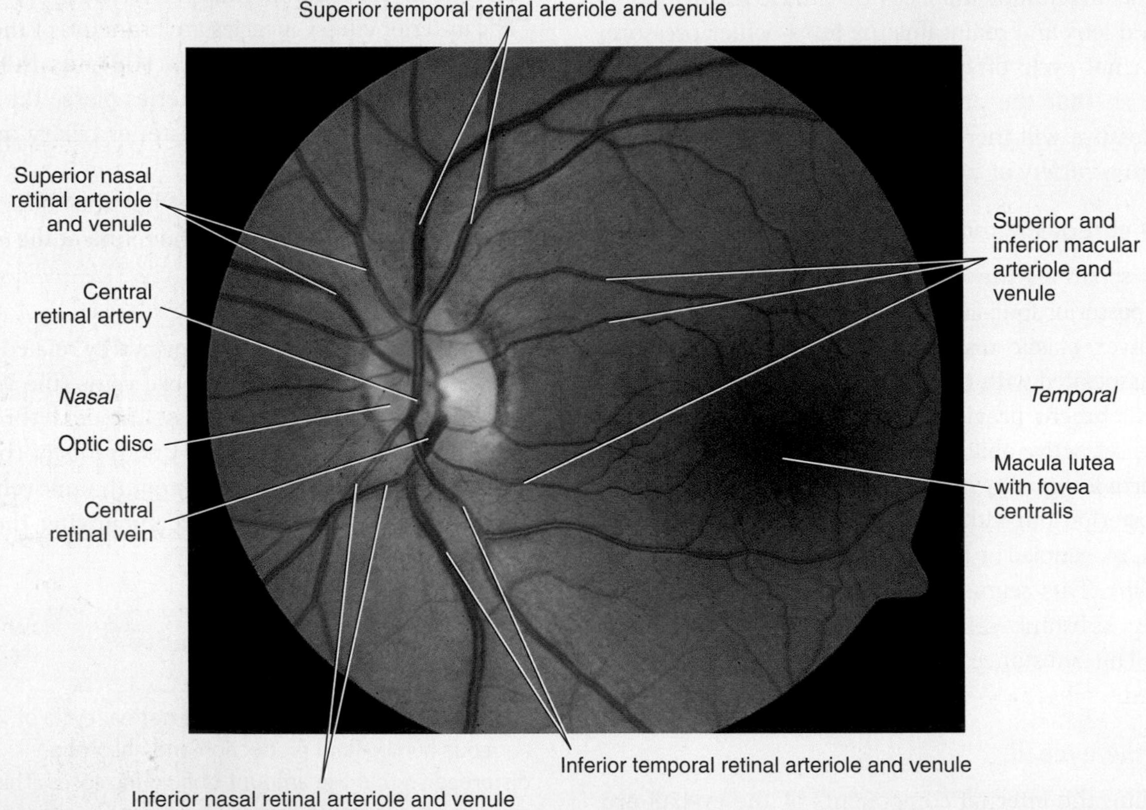


**Fig. 8.109** Ophthalmoscopic view of posterior chamber of the right eye.

## Fibrous layer of the eyeball

The fibrous layer of the eyeball consists of two components—the sclera covers the posterior and lateral parts of the eyeball, about five-sixths of the surface, and the cornea covers the anterior part (Fig. 8.108).

### Sclera

The sclera is an opaque layer of dense connective tissue that can be seen anteriorly through its conjunctival covering as the "white of the eye." It is pierced by numerous vessels and nerves, including the optic nerve posteriorly and provides attachment for the various muscles involved in eyeball movements.

The fascial sheath of the eyeball covers the surface of the sclera externally from the entrance of the optic nerve to the corneoscleral junction while internally the surface of the sclera is loosely attached to the choroid of the vascular layer.

### Cornea

Continuous with the sclera anteriorly is the transparent cornea. It covers the anterior one-sixth of the surface of the eyeball and, being transparent, allows light to enter the eyeball.

### Vascular layer of the eyeball

The vascular layer of the eyeball consists of three continuous parts—the choroid, the ciliary body, and the iris from posterior to anterior (Fig. 8.108).

#### Choroid

The choroid is posterior and represents approximately two-thirds of the vascular layer. It is a thin, highly vascular, pigmented layer consisting of smaller vessels adjacent to the retina and larger vessels more peripherally. It is firmly attached to the retina internally and loosely attached to the sclera externally.

#### Ciliary body

Extending from the anterior border of the choroid is the ciliary body (Fig. 8.108). This triangular-shaped structure, between the choroid and the iris, forms a complete ring around the eyeball. Its components include the ciliary muscle and the ciliary processes (Fig. 8.110).

The **ciliary muscle** consists of smooth muscle fibers arranged longitudinally, circularly, and radially. Controlled by parasympathetics traveling to the orbit in the oculomotor nerve [III], these muscle fibers, on contraction, decrease the size of the ring formed by the ciliary body.

The **ciliary processes** are longitudinal ridges projecting from the inner surface of the ciliary body (Fig. 8.110). Extending from them are **zonular fibers** attached to the lens of the eyeball, which suspend the lens in its proper position and collectively form the **suspensory ligament of the lens**.

Contraction of the ciliary muscle decreases the size of the ring formed by the ciliary body. This reduces tension on the suspensory ligament of the lens. The lens therefore becomes more rounded (relaxed) resulting in accommodation of the lens for near vision.

Ciliary processes also contribute to the formation of aqueous humor.

#### Iris

Completing the vascular layer of the eyeball anteriorly is the iris (Fig. 8.108). This circular structure, projecting outward from the ciliary body, is the colored part of the eye with a central opening (the pupil). Controlling the size of the pupil are smooth muscle fibers (sphincter pupillae) and myoepithelial cells (dilator pupillae) within the iris (Fig. 8.110):

- Fibers arranged in a circular pattern make up the **sphincter pupillae muscle** (Table 8.9), which is innervated by parasympathetics—contraction of its fibers decreases or constricts the pupillary opening.

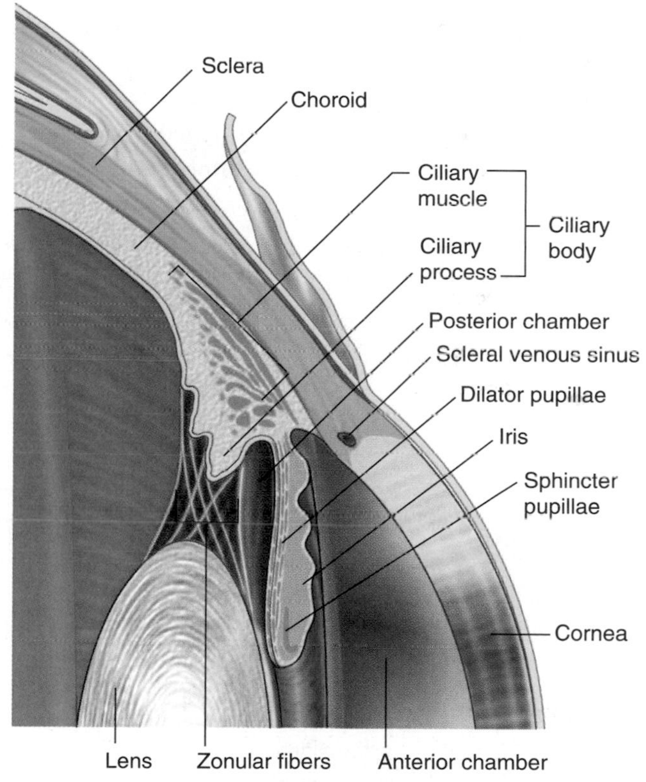


**Fig. 8.110** Ciliary body.

**Table 8.9** Intrinsic muscles of the eye

| Muscle | Location | Innervation | Function |
|---|---|---|---|
| Ciliary | Muscle fibers in the ciliary body | Parasympathetics from the oculomotor nerve [III] | Constricts ciliary body, relaxes tension on lens, lens becomes more rounded |
| Sphincter pupillae | Circularly arranged fibers in the iris | Parasympathetics from the oculomotor nerve [III] | Constricts pupil |
| Dilator pupillae | Radially arranged fibers in the iris | Sympathetics from the superior cervical ganglion (T1) | Dilates pupil |

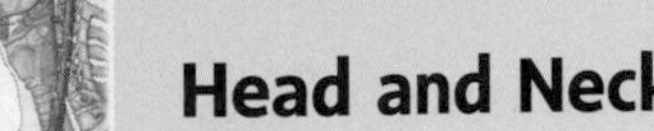

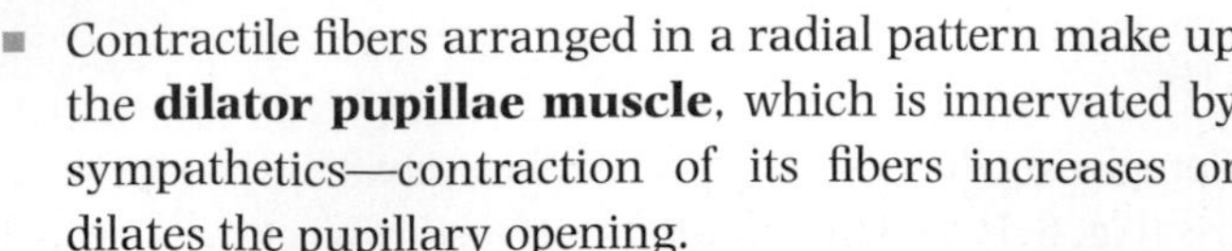



- Contractile fibers arranged in a radial pattern make up the **dilator pupillae muscle**, which is innervated by sympathetics—contraction of its fibers increases or dilates the pupillary opening.

### Inner layer of the eyeball

The inner layer of the eyeball is the retina (Fig. 8.108). It consists of two parts. Posteriorly and laterally is the **optic part of the retina**, which is sensitive to light, and anteriorly is the **nonvisual part**, which covers the internal surface of the ciliary body and the iris. The junction between these parts is an irregular line (the **ora serrata**).

#### Optic part of the retina

The optic part of the retina consists of two layers, an outer pigmented layer and an inner neural layer:

- The **pigmented layer** is firmly attached to the choroid and continues anteriorly over the internal surface of the ciliary body and iris.
- The **neural layer**, which can be further subdivided into its various neural components, is only attached to the pigmented layer around the optic nerve and at the ora serrata.

It is the neural layer that separates in the case of a detached retina.

Several obvious features are visible on the posterior surface of the optic part of the retina.

The **optic disc** is where the optic nerve leaves the retina (Fig. 8.109). It is lighter than the surrounding retina and branches of the central retinal artery spread from this point outward to supply the retina. As there are no light-sensitive receptor cells in the optic disc, it is referred to as a blind spot in the retina.

Lateral to the optic disc a small area with a hint of yellowish coloration is the **macula lutea** with its central depression, the **fovea centralis** (Fig. 8.109). This is the thinnest area of the retina and visual sensitivity here is higher than elsewhere in the retina because it has fewer **rods** (light-sensitive receptor cells that function in dim light and are insensitive to color) and more **cones** (light-sensitive receptor cells that respond to bright light and are sensitive to color).

#### In the clinic

**High-definition optical coherence tomography**

High-definition optical coherence tomography (HD-OCT) (Fig. 8.111) is a procedure used to obtain subsurface images of translucent or opaque materials. It is similar to ultrasound, except that it uses light instead of sound to produce high-resolution cross-sectional images. It is especially useful in the diagnosis and management of optic nerve and retinal diseases.

***Epiretinal membrane***

An epiretinal membrane (Fig. 8.112) is a thin sheet of fibrous tissue that develops on the surface of the retina in the area of the macula and can cause visual problems. If the visual problems are significant, surgical removal of the membrane may be necessary.

## In the clinic—cont'd

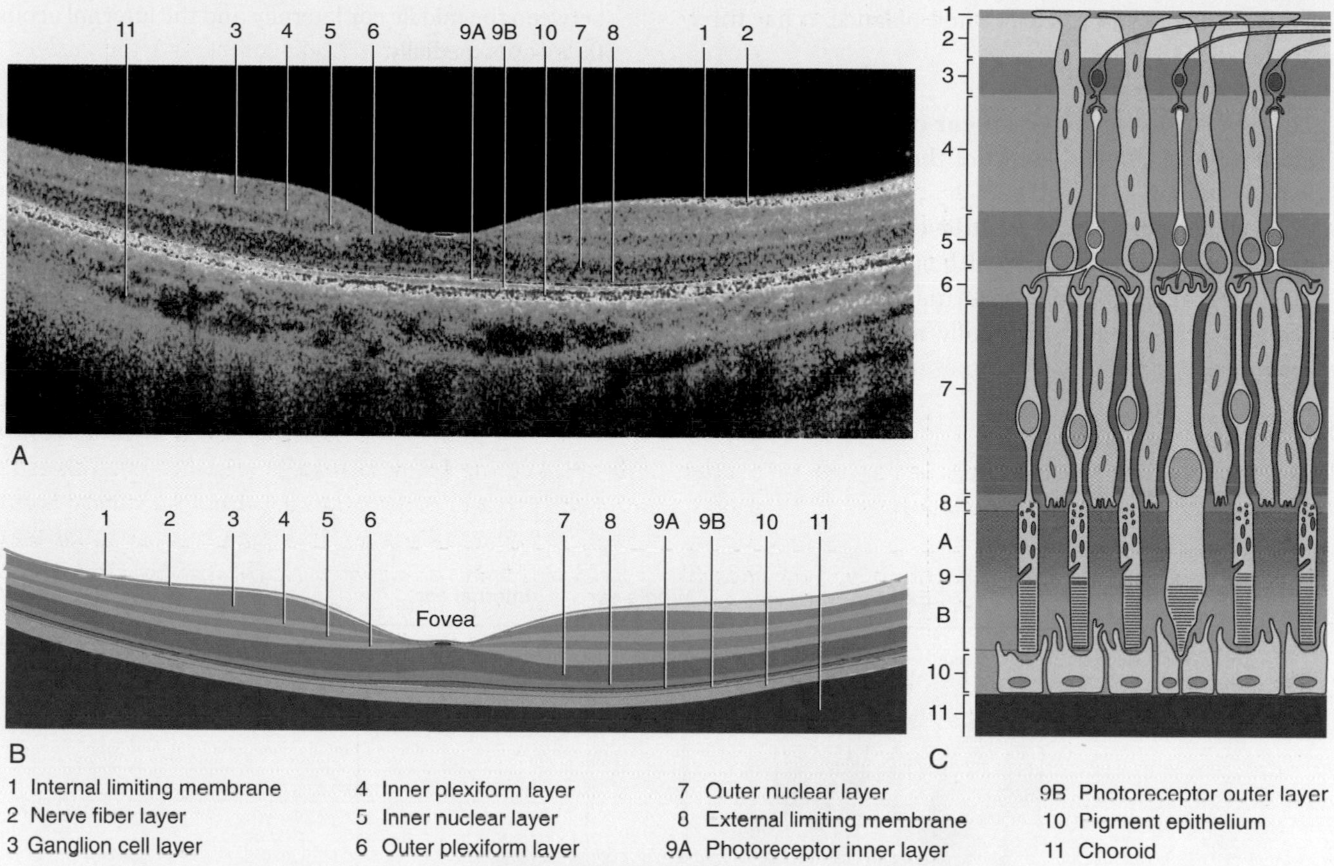

11
3
4
5
6
9A
9B
10
7
8
1
2
A
Fovea
B
C
1 Internal limiting membrane
2 Nerve fiber layer
3 Ganglion cell layer
4 Inner plexiform layer
5 Inner nuclear layer
6 Outer plexiform layer
7 Outer nuclear layer
8 External limiting membrane
9A Photoreceptor inner layer
9B Photoreceptor outer layer
10 Pigment epithelium
11 Choroid

**Fig. 8.111** Layers of the retina in a healthy eye. **A.** HD-OCT scan of a healthy eye. **B.** Schematic indicating the layers of the retina on an HD-OCT scan of a healthy eye. **C.** Diagram illustrating the layers of the retina.

Optic disc
Epiretinal membrane
N
T
A Left eye
Optic disc
Fovea
T
N
B Right eye

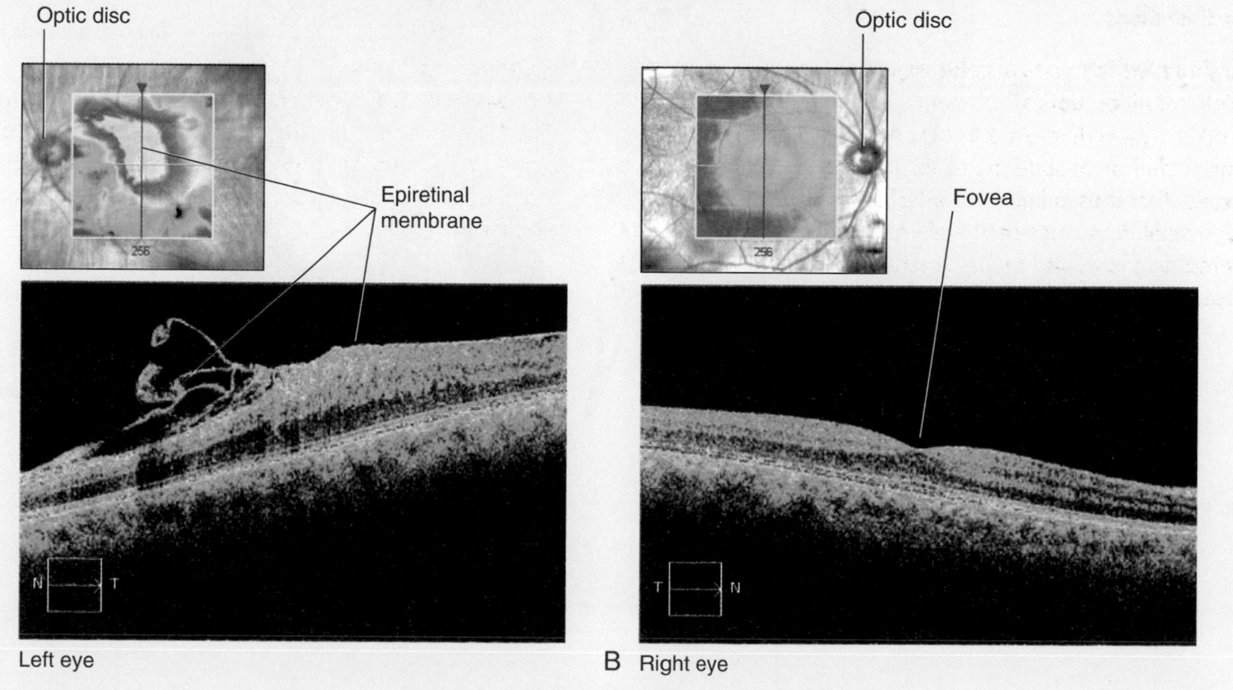

**Fig. 8.112** High-definition optical coherence tomography (HD-OCT). **A.** Diseased eye. **B.** Healthy eye.

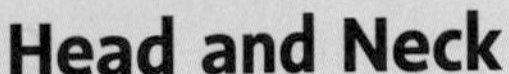


## EAR

The ear is the organ of hearing and balance. It has three parts (Fig. 8.113):

- The first part is the **external ear** consisting of the part attached to the lateral aspect of the head and the canal leading inward.
- The second part is the **middle ear**—a cavity in the petrous part of the temporal bone bounded laterally, and separated from the external canal, by a membrane and connected internally to the pharynx by a narrow tube.
- The third part is the **internal ear** consisting of a series of cavities within the petrous part of the temporal bone between the middle ear laterally and the internal acoustic meatus medially.

The internal ear converts the mechanical signals received from the middle ear, which start as sound captured by the external ear, into electrical signals to transfer information to the brain. The internal ear also contains receptors that detect motion and position.

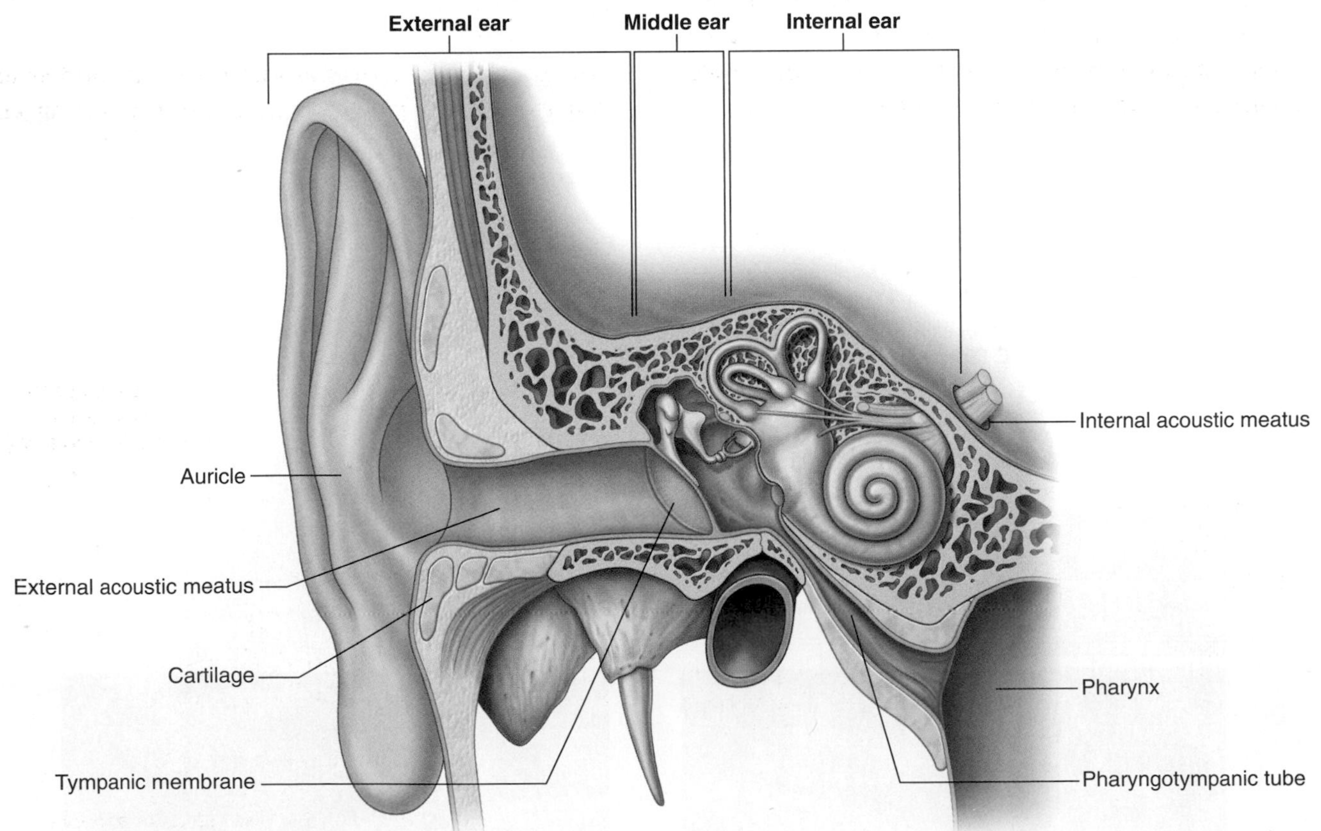


**Fig. 8.113** Right ear.

## External ear

The external ear consists of two parts. The part projecting from the side of the head is the **auricle** (**pinna**) and the canal leading inward is the **external acoustic meatus**.

### Auricle

The auricle is on the side of the head and assists in capturing sound. It consists of cartilage covered with skin and arranged in a pattern of various elevations and depressions (Fig. 8.114).

The large outside rim of the auricle is the **helix**. It ends inferiorly at the fleshy lobule, the only part of the auricle not supported by cartilage.

The hollow center of the auricle is the **concha of the auricle**. The external acoustic meatus leaves from the depths of this area.

Just anterior to the opening of the external acoustic meatus, in front of the concha, is an elevation (the **tragus**). Opposite the tragus, and above the fleshy **lobule**, is another elevation (the **antitragus**). A smaller curved rim, parallel and anterior to the helix, is the **antihelix**.

#### Muscles

Numerous intrinsic and extrinsic muscles are associated with the auricle:

- The intrinsic muscles pass between the cartilaginous parts of the auricle and may change the shape of the auricle.
- The extrinsic muscles, the anterior, superior, and posterior auricular muscles, pass from the scalp or skull to the auricle and may also play a role in positioning of the auricle (see Fig. 8.56).

Both groups of muscles are innervated by the facial nerve [VII].

#### Innervation

Sensory innervation of the auricle is from many sources (Fig. 8.115):

- The outer more superficial surfaces of the auricle are supplied by the great auricular nerve (anterior and

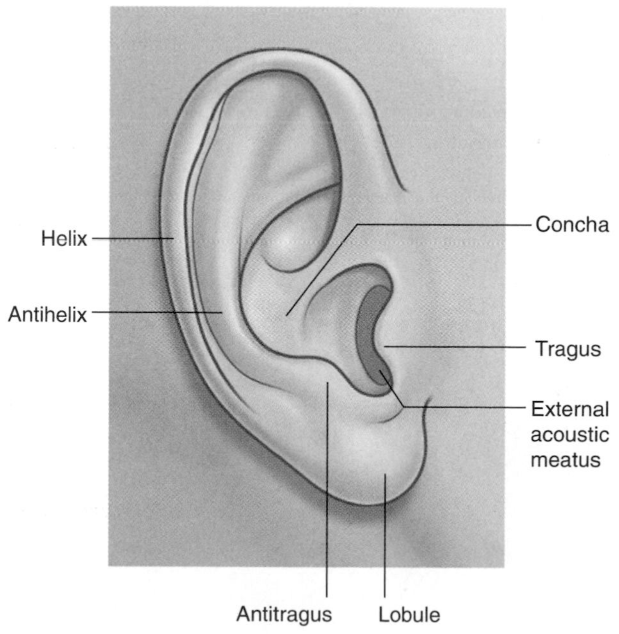


**Fig. 8.114** Auricle.

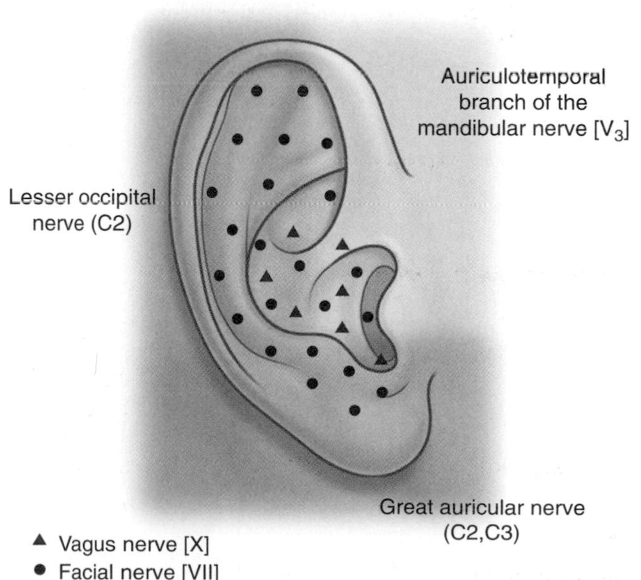


**Fig. 8.115** Sensory innervation of the auricle.

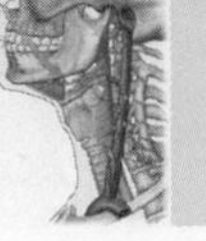



posterior inferior portions) and the lesser occipital nerve (posterosuperior portion) from the cervical plexus and the auriculotemporal branch of the mandibular nerve [$V_3$] (anterosuperior portion).

- The deeper parts of the auricle are supplied by the vagus nerve [X] (the auricular branch) and the facial nerve [VII] (which sends a branch to the auricular branch of the vagus nerve [X]).

### Vessels

The arterial supply to the auricle is from numerous sources. The external carotid artery supplies the posterior auricular artery, the superficial temporal artery supplies anterior auricular branches, and the occipital artery supplies a branch.

Venous drainage is through vessels following the arteries.

Lymphatic drainage of the auricle passes anteriorly into parotid nodes and posteriorly into mastoid nodes, and possibly into the upper deep cervical nodes.

## External acoustic meatus

The external acoustic meatus extends from the deepest part of the concha to the **tympanic membrane** (eardrum), a distance of approximately 1 inch (2.5 cm) (Fig. 8.116). Its walls consist of cartilage and bone. The lateral one-third is formed from cartilaginous extensions from some of the auricular cartilages and the medial two-thirds is a bony tunnel in the temporal bone.

Throughout its length the external acoustic meatus is covered with skin, some of which contains hair and modified sweat glands producing **cerumen** (earwax). Its diameter varies, being wider laterally and narrow medially.

The external acoustic meatus does not follow a straight course. From the external opening it passes upward in an anterior direction, then turns slightly posteriorly still passing upward, and finally, turns again in an anterior direction with a slight descent. For examination purposes, observation of the external acoustic meatus and tympanic membrane can be improved by pulling the ear superiorly, posteriorly, and slightly laterally.

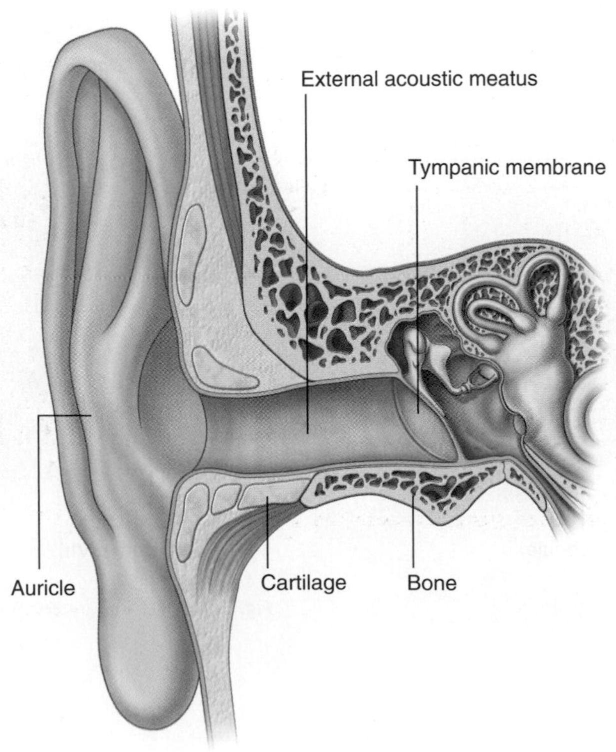


**Fig. 8.116** External acoustic meatus.

### Innervation

Sensory innervation of the external acoustic meatus is from several of the cranial nerves. The major sensory input travels through branches of the auriculotemporal nerve, a branch of the mandibular nerve [$V_3$] (anterior and superior walls), and in the auricular branch of the vagus nerve [X] (posterior and inferior walls). A minor sensory input may also come from a branch of the facial nerve [VII] to the auricular branch of the vagus nerve [X].

## Tympanic membrane

The tympanic membrane separates the external acoustic meatus from the middle ear (Figs. 8.117 and 8.118). It is at an angle, sloping medially from top to bottom and posteriorly to anteriorly. Its lateral surface therefore faces inferiorly and anteriorly. It consists of a connective tissue core lined with skin on the outside and mucous membrane on the inside.

Around the periphery of the tympanic membrane a **fibrocartilaginous ring** attaches it to the tympanic part of the temporal bone. At its center, a concavity is produced

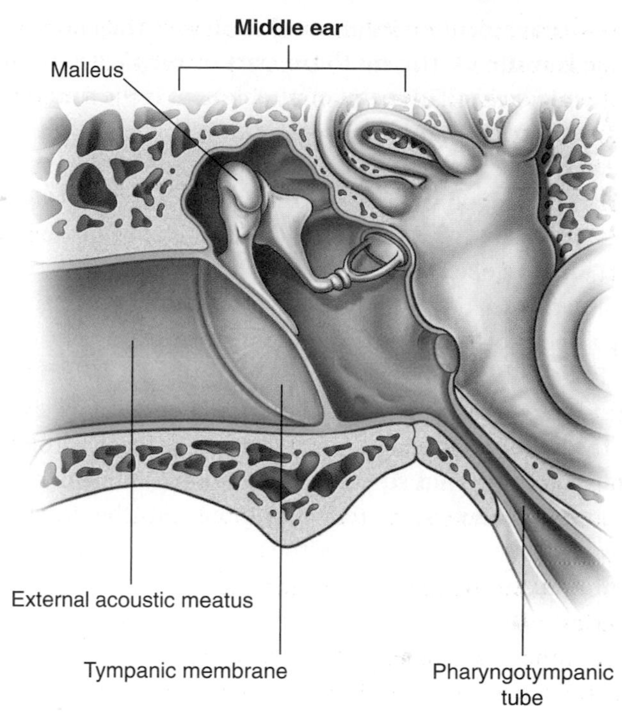


**Fig. 8.117** Middle ear.

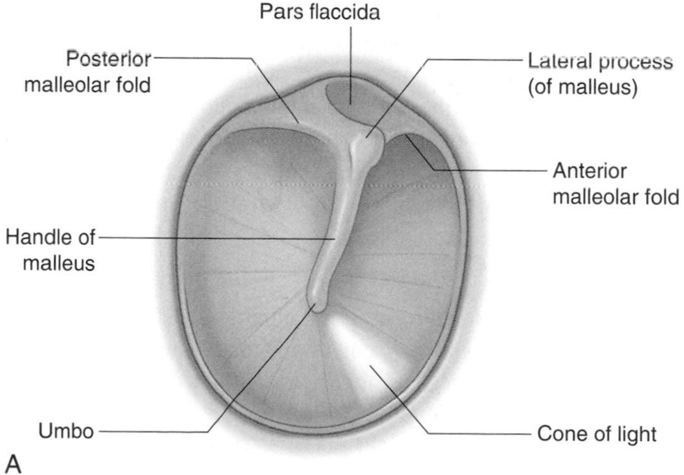


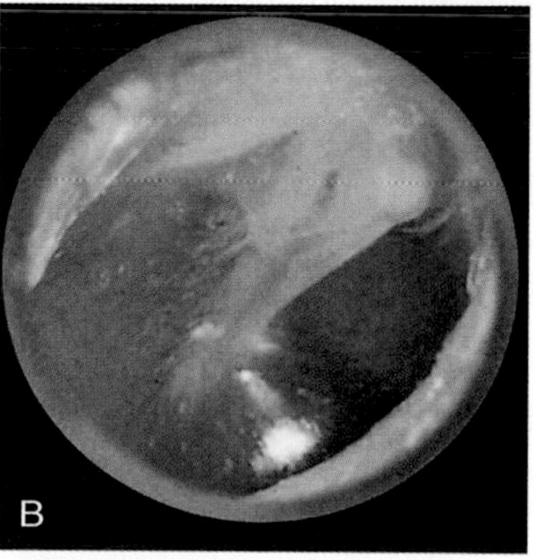


**Fig. 8.118** Tympanic membrane (right ear). **A.** Diagram. **B.** Otoscopic view.

by the attachment on its internal surface of the lower end of the **handle of the malleus**, part of the malleus bone in the middle ear. This point of attachment is the **umbo of the tympanic membrane**.

Anteroinferior to the umbo of the tympanic membrane a bright reflection of light, referred to as the cone of light, is usually visible when examining the tympanic membrane with an otoscope.

Superior to the umbo in an anterior direction is the attachment of the rest of the handle of the malleus (Fig. 8.118). At the most superior extent of this line of attachment a small bulge in the membrane marks the position of the **lateral process** of the malleus as it projects against the internal surface of the tympanic membrane. Extending away from this elevation, on the internal surface of the membrane, are the **anterior** and **posterior malleolar folds**. Superior to these folds the tympanic membrane is thin and slack (the **pars flaccida**), whereas the rest of the membrane is thick and taut (the **pars tensa**).

### Innervation

Innervation of the external and internal surfaces of the tympanic membrane is by several cranial nerves:

- Sensory innervation of the skin on the outer surface of the tympanic membrane is primarily by the auriculotemporal nerve, a branch of the mandibular nerve [$V_3$] with additional participation of the auricular branch of the vagus nerve [X], a small contribution by a branch of the facial nerve [VII] to the auricular branch of the vagus nerve [X], and possibly a contribution from the glossopharyngeal nerve [IX].
- Sensory innervation of the mucous membrane on the inner surface of the tympanic membrane is carried entirely by the glossopharyngeal [IX] nerve.

## In the clinic

### Otitis media

The eustachian tube links the middle ear and pharynx and balances the pressure between the outer and middle ear. Colds and allergies, particularly in children, can result in swelling of the lining of the eustachian tube, which can then impair normal drainage of fluid from the middle ear. The fluid then builds up behind the tympanic membrane, providing an attractive environment for bacteria and viruses to grow and cause otitis media. Left untreated, otitis media can lead to perforation of the tympanic membrane, hearing loss, meningitis, and brain abscess.

## In the clinic

### Examination of the ear

The ear comprises three components—the external, middle, and internal ear.

Clinical examination is carried out to assess hearing and balance. Further examination involves use of an otoscope or other imaging techniques.

#### *External ear*

The external ear is easily examined. The external acoustic meatus and the tympanic membrane require otoscopic examination (Fig. 8.118B). An otoscope is a device through which light can be shone and the image magnified to inspect the external acoustic meatus and the tympanic membrane.

The examination begins by grasping the posterosuperior aspect of the ear and gently retracting it to straighten the external auditory meatus. The normal tympanic membrane is relatively translucent and has a gray–reddish tinge. The handle of the malleus is visible near the center of the membrane. In the 5 o'clock position a cone of light is always demonstrated.

#### *Middle and inner ears*

The middle ear is investigated by CT and MRI to visualize the malleus, incus, and stapes. The relationship of these bones to the middle ear cavity is determined and any masses identified.

The inner ear is also assessed by CT and MRI.

## In the clinic

**Swimmer's ear**

Swimmer's ear, often called otitis externa, is a painful condition resulting from an infection in the external acoustic meatus. It frequently occurs in swimmers.

## In the clinic

**Surfer's ear**

Surfer's ear, which is prevalent among individuals who surf or swim in cold water, results from the development of a "bony lump" in the external acoustic meatus. Growth of the lump eventually constricts the meatus and reduces hearing in the affected ear.

## In the clinic

**Tympanic membrane perforation**

Although perforation of the tympanic membrane (eardrum) has many causes, trauma and infection are the most common.

Ruptures of the tympanic membrane tend to heal spontaneously, but surgical intervention may be necessary if the rupture is large.

Occasionally, it may be necessary to enter the middle ear through the tympanic membrane. Because the chorda tympani runs in the upper one-third of the tympanic membrane, incisions are always below this level. The richer blood supply to the posterior aspect of the tympanic membrane determines the standard surgical approach in the posteroinferior aspect.

Otitis media (infection of the middle ear) is common and can lead to perforation of the tympanic membrane. The infection can usually be treated with antibiotics. If the infection persists, the chronic inflammatory change may damage the ossicular chain and other structures within the middle ear to produce deafness.

## Middle ear

The middle ear is an air-filled, mucous membrane–lined space in the temporal bone between the tympanic membrane laterally and the lateral wall of the internal ear medially. It is described as consisting of two parts (Fig. 8.119):

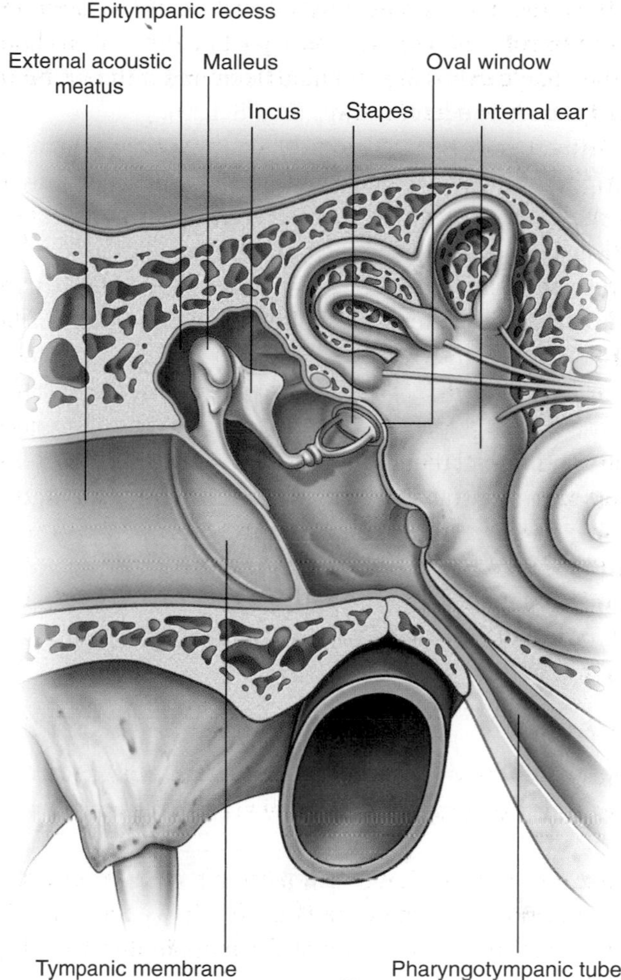


**Fig. 8.119** Parts of the middle ear.

- the **tympanic cavity** immediately adjacent to the tympanic membrane, and
- the **epitympanic recess** superiorly.

The middle ear communicates with the mastoid area posteriorly and the nasopharynx (via the pharyngotympanic tube) anteriorly. Its basic function is to transmit vibrations of the tympanic membrane across the cavity of the middle ear to the internal ear. It accomplishes this through three interconnected but movable bones that bridge the space between the tympanic membrane and the internal ear. These bones are the malleus (connected to the tympanic membrane), the incus (connected to the malleus by a synovial joint), and the stapes (connected to the incus by a synovial joint, and attached to the lateral wall of the internal ear at the oval window).

### Boundaries

The middle ear has a roof and a floor, and anterior, posterior, medial, and lateral walls (Fig. 8.120).

#### Tegmental wall

The tegmental wall (roof) of the middle ear consists of a thin layer of bone, which separates the middle ear from the middle cranial fossa. This layer of bone is the tegmen tympani on the anterior surface of the petrous part of the temporal bone.

#### Jugular wall

The jugular wall (floor) of the middle ear consists of a thin layer of bone that separates it from the internal jugular vein. Occasionally, the floor is thickened by the presence of mastoid air cells.

Near the medial border of the floor is a small aperture, through which the tympanic branch from the glossopharyngeal nerve [IX] enters the middle ear.

#### Membranous wall

The membranous (lateral) wall of the middle ear consists almost entirely of the tympanic membrane, but because the tympanic membrane does not extend superiorly into the epitympanic recess, the upper part of the membranous wall of the middle ear is the bony lateral wall of the epitympanic recess.

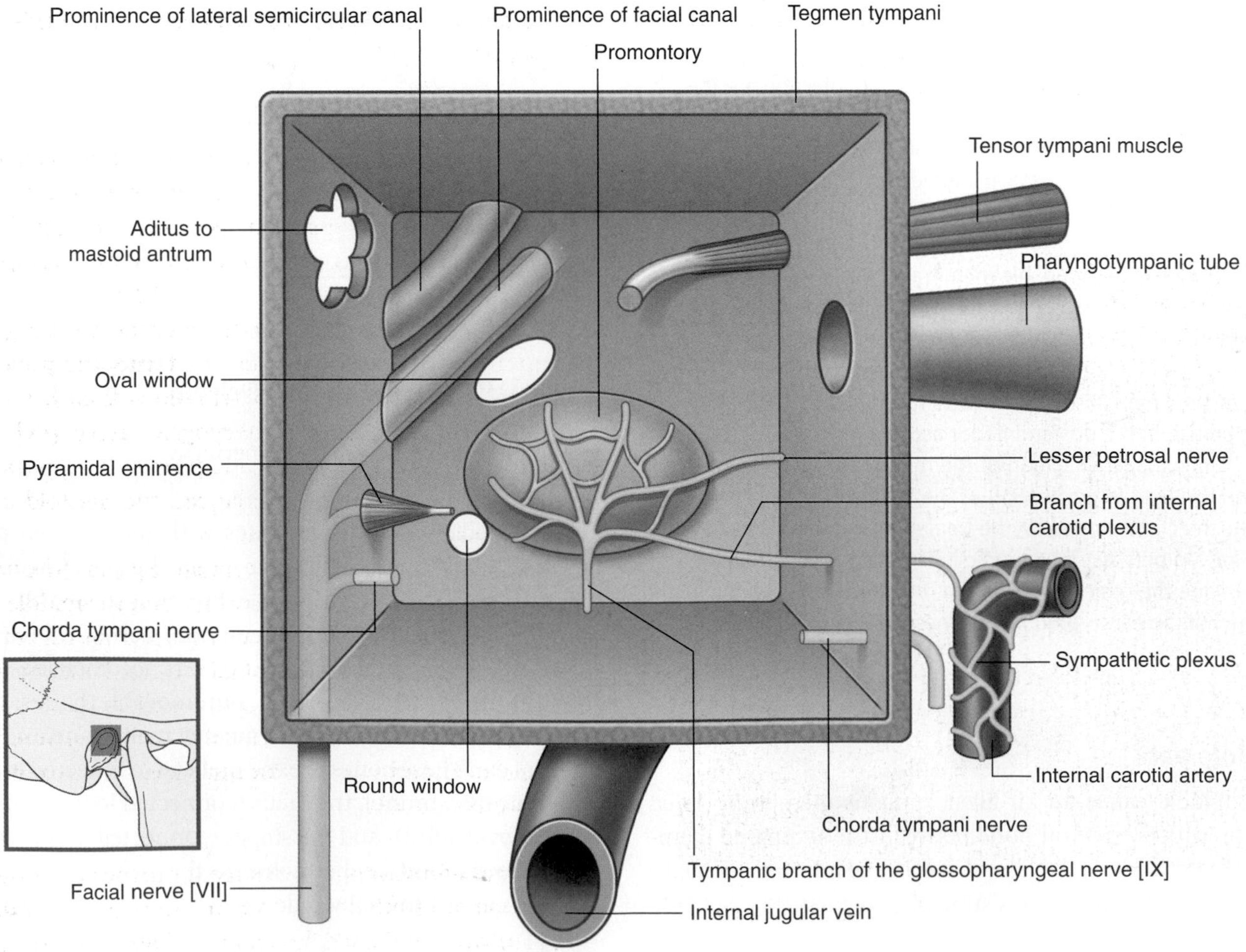

**Fig. 8.120** Boundaries of the right middle ear.

### Mastoid wall

The mastoid (posterior) wall of the middle ear is only partially complete. The lower part of this wall consists of a bony partition between the tympanic cavity and mastoid air cells. Superiorly, the epitympanic recess is continuous with the **aditus to the mastoid antrum** (Figs. 8.120 and 8.121).

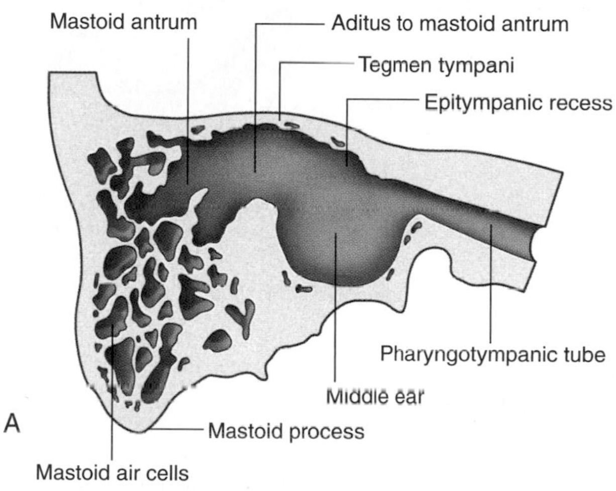


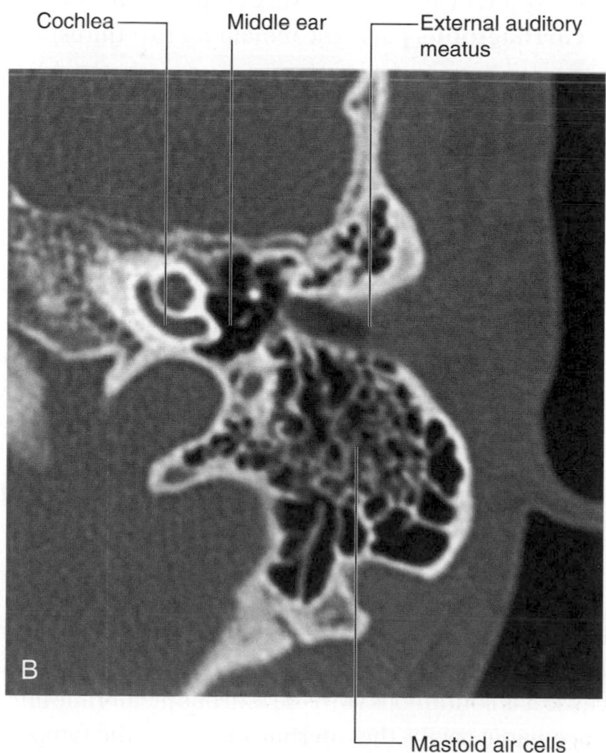


**Fig. 8.121** Mastoid antrum and surrounding bone. **A.** Diagram. **B.** High-resolution CT scan of left ear (petrous temporal bone).

Associated with the mastoid wall are:

- the pyramidal eminence, a small elevation through which the tendon of the stapedius muscle enters the middle ear; and
- the opening through which the chorda tympani nerve, a branch of the facial nerve [VII], enters the middle ear.

### Anterior wall

The anterior wall of the middle ear is only partially complete. The lower part consists of a thin layer of bone that separates the tympanic cavity from the internal carotid artery. Superiorly, the wall is deficient because of the presence of:

- a large opening for the entrance of the pharyngotympanic tube into the middle ear, and
- a smaller opening for the canal containing the tensor tympani muscle.

The foramen for the exit of the chorda tympani nerve from the middle ear is also associated with this wall (Fig. 8.120).

### Labyrinthine wall

The labyrinthine (medial) wall of the middle ear is also the lateral wall of the internal ear. A prominent structure on this wall is a rounded bulge (the **promontory**) produced by the basal coil of the **cochlea**, which is an internal ear structure involved with hearing (Fig. 8.120).

Associated with the mucous membrane covering the promontory is a plexus of nerves (the **tympanic plexus**), which consists primarily of contributions from the tympanic branch of the glossopharyngeal nerve [IX] and branches from the internal carotid plexus. It supplies the mucous membrane of the middle ear, the mastoid area, and the pharyngotympanic tube.

Additionally, a branch of the tympanic plexus (the lesser petrosal nerve) leaves the promontory and the middle ear, travels across the anterior surface of the petrous part of the temporal bone, and leaves the middle cranial fossa through the foramen ovale to enter the otic ganglion. Other structures associated with the labyrinthine wall are two openings, the oval and round windows, and two prominent elevations (Fig. 8.120):

- The **oval window** is posterosuperior to the promontory, is the point of attachment for the **base of the stapes** (**footplate**), and ends the chain of bones that transfer vibrations initiated by the tympanic membrane to the cochlea of the internal ear.

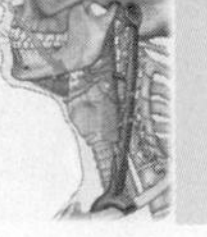



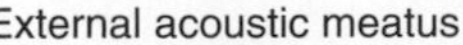

- The **round window** is posteroinferior to the promontory.
- Posterior and superior to the oval window on the medial wall is the **prominence of the facial canal**, which is a ridge of bone produced by the facial nerve [VII] in its canal as it passes through the temporal bone.
- Just above and posterior to the prominence of the facial canal is a broader ridge of bone (**prominence of the lateral semicircular canal**) produced by the lateral semicircular canal, which is a structure involved in detecting motion.

### Mastoid area

Posterior to the epitympanic recess of the middle ear is the aditus to the mastoid antrum, which is the opening to the mastoid antrum (Fig. 8.121).

The **mastoid antrum** is a cavity continuous with collections of air-filled spaces (the **mastoid cells**), throughout the mastoid part of the temporal bone, including the mastoid process. The mastoid antrum is separated from the middle cranial fossa above by only the thin tegmen tympani.

The mucous membrane lining the mastoid air cells is continuous with the mucous membrane throughout the middle ear. Therefore infections in the middle ear can easily spread into the mastoid area.

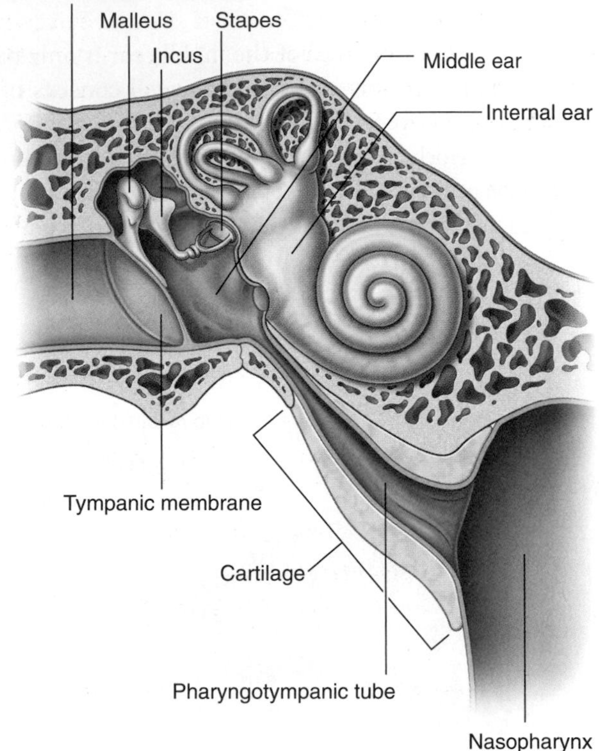


**Fig. 8.122** Pharyngotympanic tube.

#### In the clinic

**Mastoiditis**

Infection within the mastoid antrum and mastoid cells is usually secondary to infection in the middle ear. The mastoid cells provide an excellent culture medium for infection. Infection of the bone (osteomyelitis) may also develop, spreading into the middle cranial fossa.

Drainage of the pus within the mastoid air cells is necessary and there are numerous approaches for doing this. When undertaking this type of surgery, it is extremely important that care is taken not to damage the mastoid wall of the middle ear to prevent injury to the facial nerve [VII]. Any breach of the inner table of the cranial vault may allow bacteria to enter the cranial cavity and meningitis will ensue.

### Pharyngotympanic tube

The pharyngotympanic tube connects the middle ear with the nasopharynx (Fig. 8.122) and equalizes pressure on both sides of the tympanic membrane. Its opening in the middle ear is on the anterior wall, and from here it extends forward, medially, and downward to enter the nasopharynx just posterior to the inferior meatus of the nasal cavity. It consists of:

- a **bony part** (the one-third nearest the middle ear); and
- a **cartilaginous part** (the remaining two-thirds).

The opening of the bony part is clearly visible on the inferior surface of the skull at the junction of the squamous and petrous parts of the temporal bone immediately posterior to the foramen ovale and foramen spinosum.

#### Vessels

The arterial supply to the pharyngotympanic tube is from several sources. Branches arise from the **ascending pharyngeal artery** (a branch of the external carotid artery) and from two branches of the maxillary artery (the middle meningeal artery and the artery of the pterygoid canal).

Venous drainage of the pharyngotympanic tube is to the pterygoid plexus of veins in the infratemporal fossa.

#### Innervation

Innervation of the mucous membrane lining the pharyngotympanic tube is primarily from the tympanic plexus because it is continuous with the mucous membrane lining the tympanic cavity, the internal surface of the tympanic membrane, and the mastoid antrum and mastoid cells. This plexus receives its major contribution from the tympanic nerve, a branch of the glossopharyngeal nerve [IX].

### Auditory ossicles

The bones of the middle ear consist of the malleus, incus, and stapes. They form an osseous chain across the middle ear from the tympanic membrane to the oval window of the internal ear (Fig. 8.123).

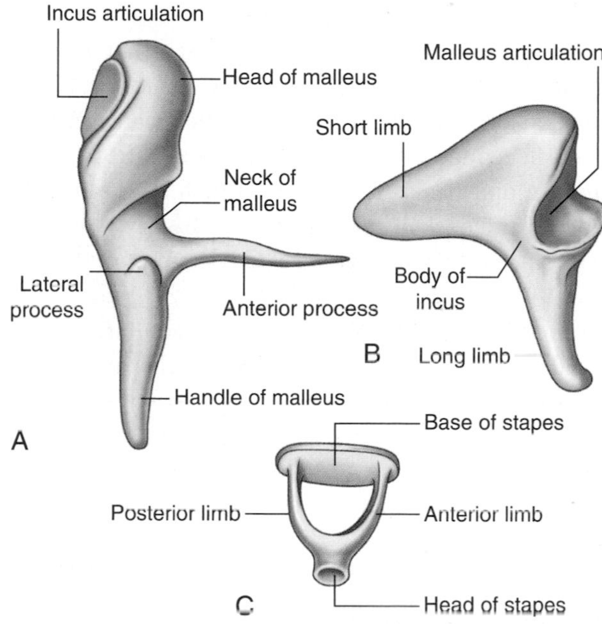


**Fig. 8.123** Auditory ossicles. **A.** Malleus. **B.** Incus. **C.** Stapes.

Muscles associated with the auditory ossicles modulate movement during the transmission of vibrations.

#### Malleus

The malleus is the largest of the auditory ossicles and is attached to the tympanic membrane. Identifiable parts include the **head of the malleus**, **neck of the malleus**, **anterior and lateral processes**, and **handle of the malleus** (Fig. 8.123). The head of the malleus is the rounded upper part of the malleus in the epitympanic recess. Its posterior surface articulates with the incus.

Inferior to the head of the malleus is the constricted neck of the malleus, and below this are the anterior and lateral processes:

- The anterior process is attached to the anterior wall of the middle ear by a ligament.
- The lateral process is attached to the anterior and posterior malleolar folds of the tympanic membrane.

The downward extension of the malleus, below the anterior and lateral processes, is the handle of the malleus, which is attached to the tympanic membrane.

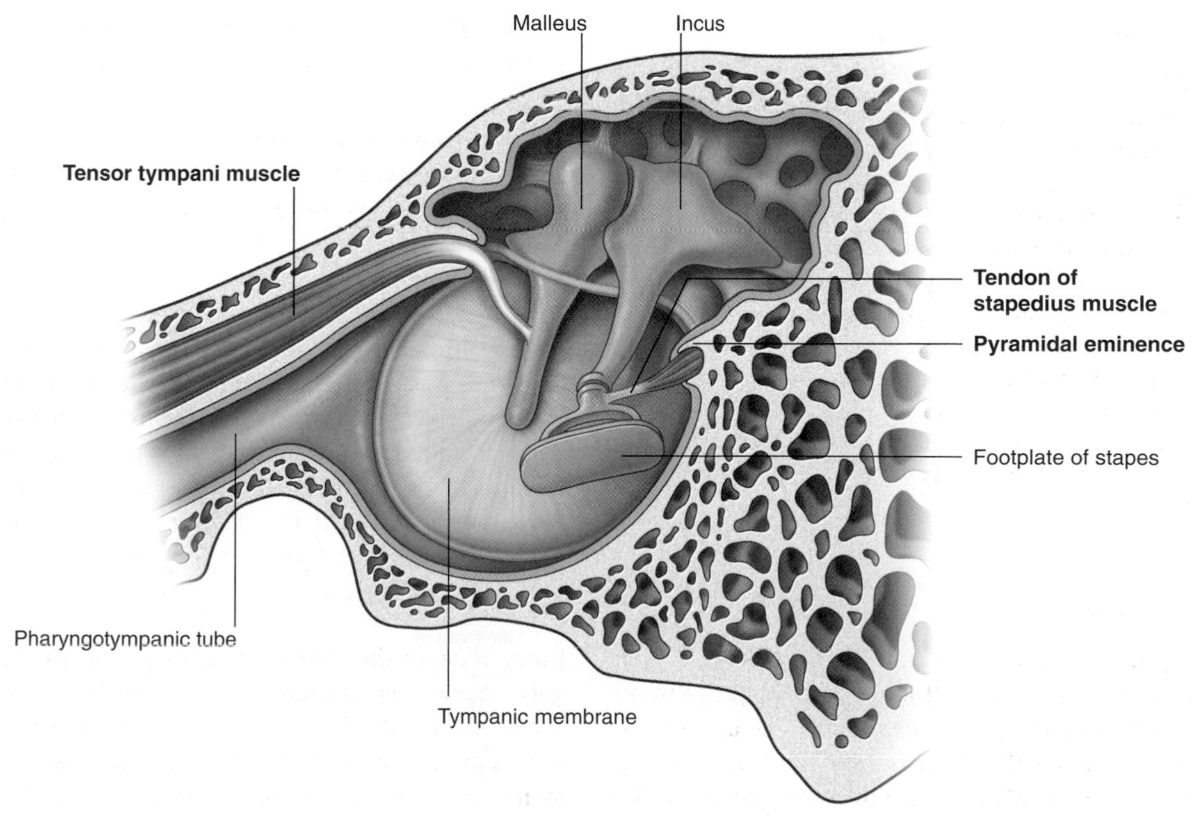


**Fig. 8.124** Muscles associated with the auditory ossicles (right ear).

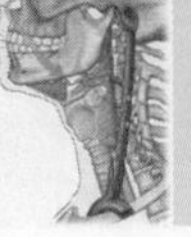



### Incus

The second bone in the series of auditory ossicles is the incus. It consists of the **body of the incus** and **long** and **short limbs** (Fig. 8.123):

- The enlarged body of the incus articulates with the head of the malleus and is in the epitympanic recess.
- The long limb extends downward from the body, paralleling the handle of the malleus, and ends by bending medially to articulate with the stapes.
- The short limb extends posteriorly and is attached by a ligament to the upper posterior wall of the middle ear.

### Stapes

The stapes is the most medial bone in the osseous chain and is attached to the oval window. It consists of the **head of the stapes**, **anterior** and **posterior limbs**, and the **base of the stapes** (Fig. 8.123):

- The head of the stapes is directed laterally and articulates with the long process of the incus.
- The two limbs separate from each other and attach to the oval base.
- The base of the stapes fits into the oval window on the labyrinthine wall of the middle ear.

### Muscles associated with the ossicles

Two muscles are associated with the bony ossicles of the middle ear—the tensor tympani and stapedius (Fig. 8.124 and Table 8.10).

#### Tensor tympani

The tensor tympani muscle lies in a bony canal above the pharyngotympanic tube. It originates from the cartilaginous part of the pharyngotympanic tube, the greater wing of the sphenoid, and its own bony canal, and passes through its canal in a posterior direction, ending in a rounded tendon that inserts into the upper part of the handle of the malleus.

Innervation of the tensor tympani is by a branch from the mandibular nerve [$V_3$].

Contraction of the tensor tympani pulls the handle of the malleus medially. This tenses the tympanic membrane, reducing the force of vibrations in response to loud noises.

#### Stapedius

The stapedius muscle is a very small muscle that originates from inside the pyramidal eminence, which is a small projection on the mastoid wall of the middle ear (Fig. 8.124). Its tendon emerges from the apex of the pyramidal eminence and passes forward to attach to the posterior surface of the neck of the stapes.

The stapedius is innervated by a branch from the facial nerve [VII].

Contraction of the stapedius muscle, usually in response to loud noises, pulls the stapes posteriorly and prevents excessive oscillation.

### Vessels

Numerous arteries supply the structures in the middle ear:

- the two largest branches are the **tympanic branch** of the maxillary artery and the **mastoid branch** of the occipital or posterior auricular arteries;
- smaller branches come from the middle meningeal artery, the ascending pharyngeal artery, the artery of the pterygoid canal, and tympanic branches from the internal carotid artery.

Venous drainage of the middle ear returns to the pterygoid plexus of veins and the superior petrosal sinus.

### Innervation

The tympanic plexus innervates the mucous membrane lining the walls and contents of the middle ear, which includes the mastoid area and the pharyngotympanic tube. It is formed by the **tympanic nerve**, a branch of the glossopharyngeal nerve [IX], and from branches of the internal carotid plexus. The tympanic plexus occurs in the

**Table 8.10** Muscles of the middle ear

| Muscle | Origin | Insertion | Innervation | Function |
|---|---|---|---|---|
| Tensor tympani | Cartilaginous part of pharyngotympanic tube, greater wing of sphenoid, its own bony canal | Upper part of handle of malleus | Branch from mandibular nerve [$V_3$] | Contraction pulls handle of malleus medially, tensing tympanic membrane |
| Stapedius | Attached to inside of pyramidal eminence | Neck of stapes | Branch of facial nerve [VII] | Contraction pulls stapes posteriorly, preventing excessive oscillation |

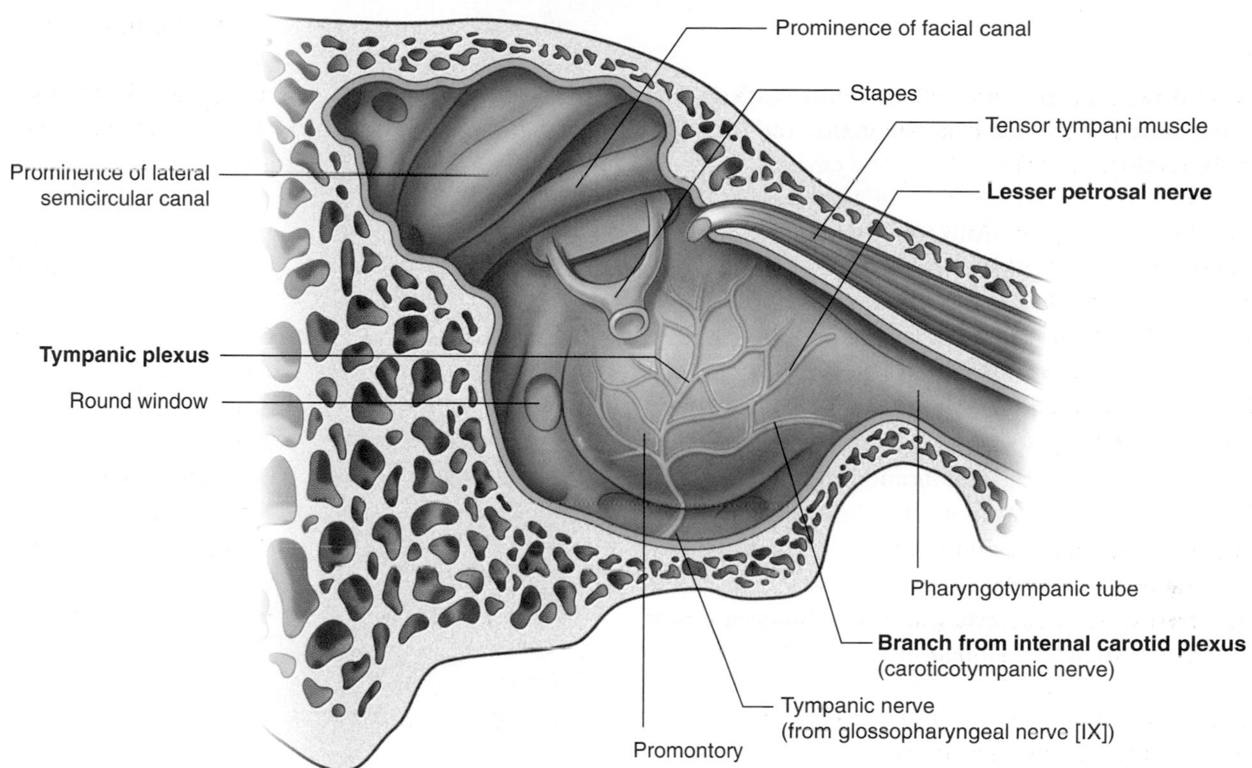


**Fig. 8.125** Innervation of the middle ear.

mucous membrane covering the promontory, which is the rounded bulge on the labyrinthine wall of the middle ear (Fig. 8.125).

As the glossopharyngeal nerve [IX] exits the skull through the jugular foramen, it gives off the tympanic nerve. This branch reenters the skull through a small foramen and passes through the bone to the middle ear.

Once in the middle ear, the tympanic nerve forms the **tympanic plexus**, along with branches from the plexus of nerves surrounding the internal carotid artery (**caroticotympanic nerves**). Branches from the tympanic plexus supply the mucous membranes of the middle ear, including the pharyngotympanic tube and the mastoid area.

The tympanic plexus also gives off a major branch (the lesser petrosal nerve), which supplies preganglionic parasympathetic fibers to the otic ganglion (Fig. 8.125).

The lesser petrosal nerve leaves the area of the promontory, exits the middle ear, travels through the petrous part of the temporal bone, and exits onto the anterior surface of the petrous part of the temporal bone through a hiatus just below the hiatus for the greater petrosal nerve (Fig. 8.126). It continues diagonally across the anterior surface of the temporal bone before exiting the middle cranial fossa through the foramen ovale. Once outside the skull it enters the otic ganglion.

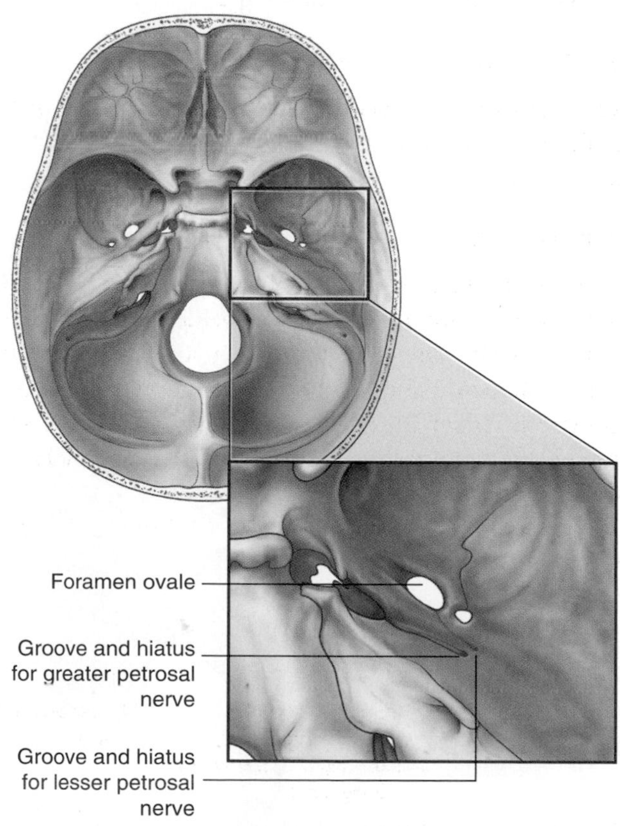


**Fig. 8.126** Grooves and hiatuses for the greater and lesser petrosal nerves.

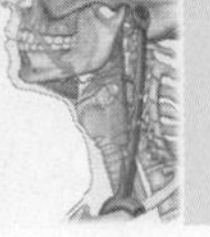



## Internal ear

The internal ear consists of a series of bony cavities (the **bony labyrinth**) and membranous ducts and sacs (the **membranous labyrinth**) within these cavities. All these structures are in the petrous part of the temporal bone between the middle ear laterally and the internal acoustic meatus medially (Figs. 8.127 and 8.128).

The bony labyrinth consists of the **vestibule**, three **semicircular canals**, and the **cochlea** (Fig. 8.128). These bony cavities are lined with periosteum and contain a clear fluid (the **perilymph**).

Suspended within the perilymph but not filling all spaces of the bony labyrinth is the membranous labyrinth, which consists of the **semicircular ducts**, the **cochlear duct**, and two sacs (the **utricle** and the **saccule**). These membranous spaces are filled with **endolymph**.

The structures in the internal ear convey information to the brain about balance and hearing:

- The cochlear duct is the organ of hearing.
- The semicircular ducts, utricle, and saccule are the organs of balance.

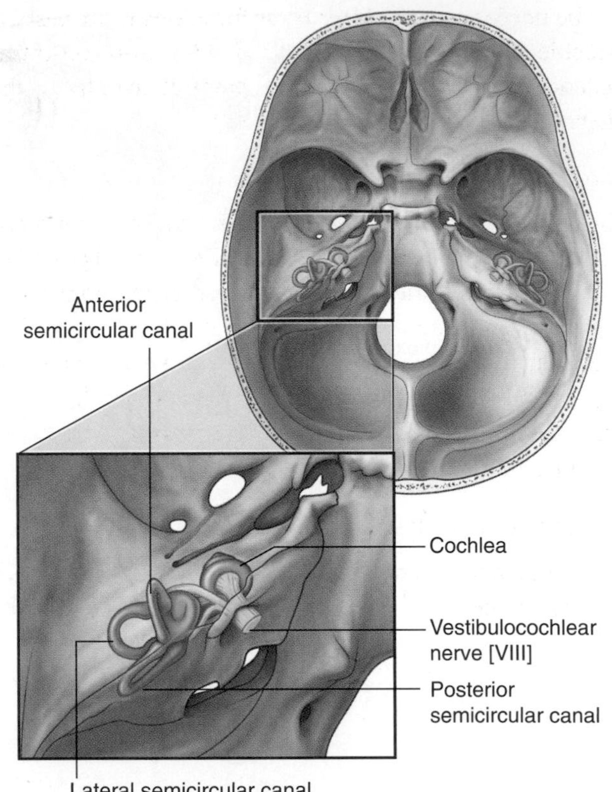


**Fig. 8.127** Location of the internal ear in temporal bone.





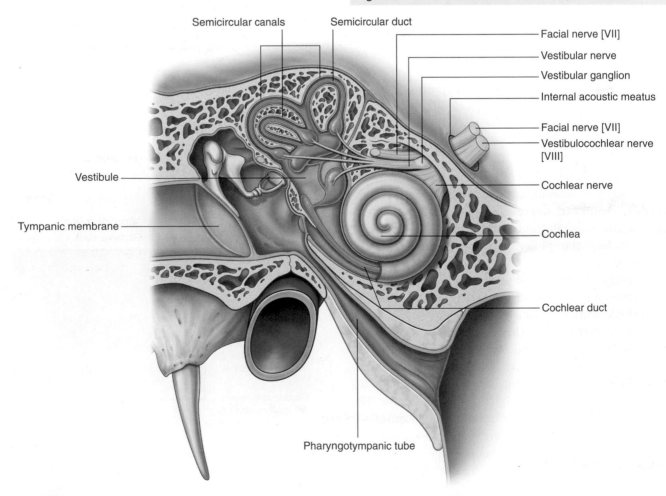

**Fig. 8.128** Internal ear.

The nerve responsible for these functions is the vestibulocochlear nerve [VIII], which divides into vestibular (balance) and cochlear (hearing) parts after entering the internal acoustic meatus (Fig. 8.128).

## Bony labyrinth

The vestibule, which contains the oval window in its lateral wall, is the central part of the bony labyrinth (Fig. 8.129). It communicates anteriorly with the cochlea and posterosuperiorly with the semicircular canals.

A narrow canal (the **vestibular aqueduct**) leaves the vestibule, and passes through the temporal bone to open on the posterior surface of the petrous part of the temporal bone.

### Semicircular canals

Projecting in a posterosuperior direction from the vestibule are the **anterior**, **posterior**, and **lateral semicircular canals** (Fig. 8.129). Each of these canals forms two-thirds of a circle connected at both ends to the vestibule and with one end dilated to form the **ampulla**. The canals are oriented so that each canal is at right angles to the other two.

### Cochlea

Projecting in an anterior direction from the vestibule is the cochlea, which is a bony structure that twists on itself two and one-half to two and three-quarter times around a central column of bone (the **modiolus**). This arrangement produces a cone-shaped structure with a **base of the cochlea** that faces posteromedially and an apex that faces anterolaterally (Fig. 8.130). This positions the wide base of the modiolus near the internal acoustic meatus, where it is entered by branches of the cochlear part of the vestibulocochlear nerve [VIII].

Extending laterally throughout the length of the modiolus is a thin lamina of bone (the **lamina of the modiolus**,

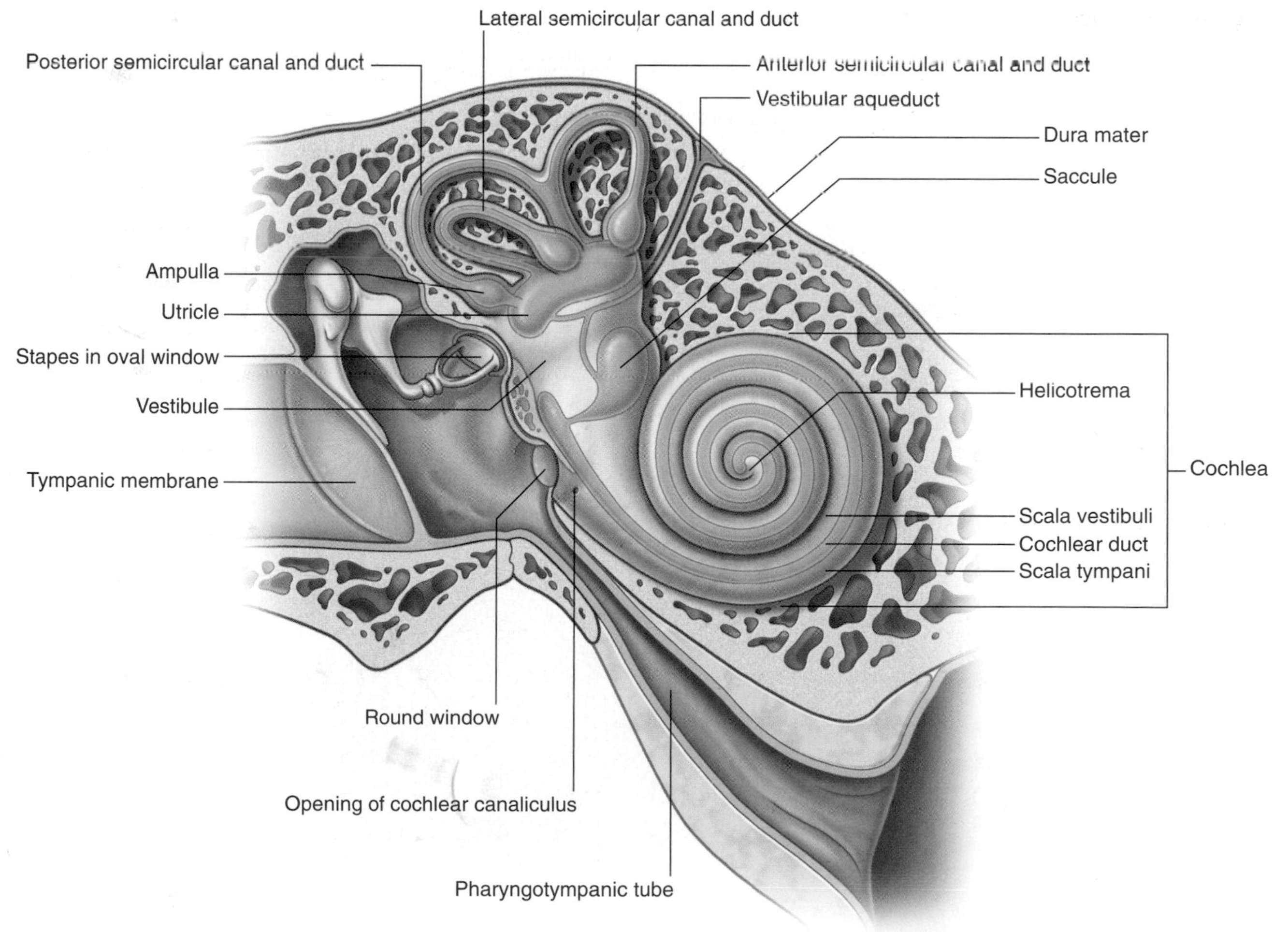


**Fig. 8.129** Bony labyrinth.

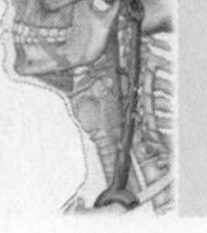

or **spiral lamina**). Circling around the modiolus, and held in a central position by its attachment to the lamina of the modiolus, is the cochlear duct, which is a component of the membranous labyrinth.

Attached peripherally to the outer wall of the cochlea, the cochlear duct creates two canals (the **scala vestibuli** and the **scala tympani**), which extend throughout the cochlea and are continuous with each other at the apex through a narrow slit (the **helicotrema**):

- The scala vestibuli is continuous with the vestibule.
- The scala tympani is separated from the middle ear by the secondary tympanic membrane covering the round window (Fig. 8.131).

Finally, near the round window is a small channel (the **cochlear canaliculus**), which passes through the temporal bone and opens on its inferior surface into the posterior cranial fossa. This provides a connection between the perilymph-containing cochlea and the subarachnoid space (Fig. 8.131).

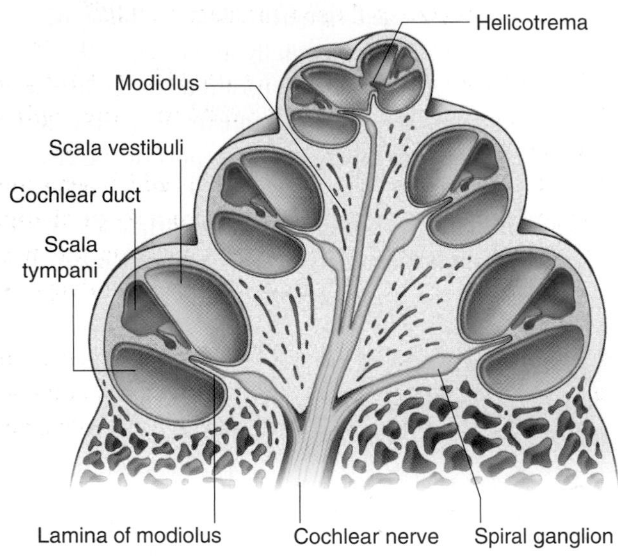


**Fig. 8.130** Cochlea.

## Membranous labyrinth

The membranous labyrinth is a continuous system of ducts and sacs within the bony labyrinth. It is filled with endolymph and separated from the periosteum that covers the walls of the bony labyrinth by perilymph.

Consisting of two sacs (the utricle and the saccule) and four ducts (the three semicircular ducts and the cochlear duct), the membranous labyrinth has unique functions related to balance and hearing:

- The utricle, saccule, and three semicircular ducts are part of the vestibular apparatus (i.e., organs of balance).
- The cochlear duct is the organ of hearing.

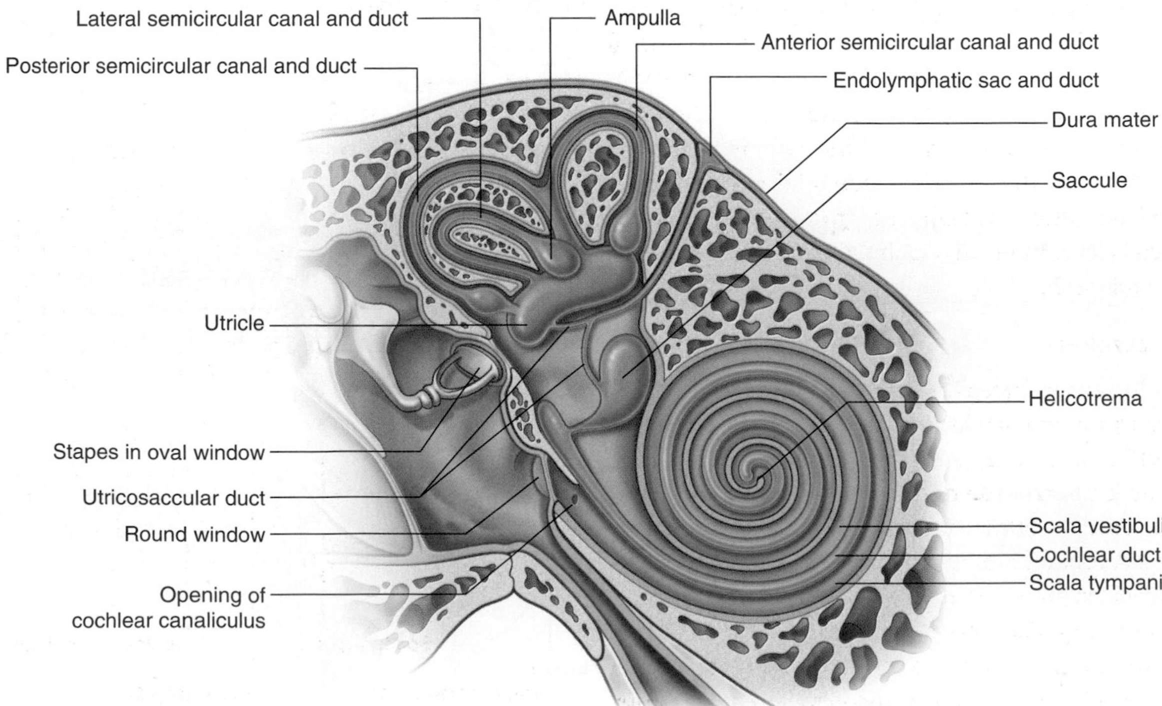


**Fig. 8.131** Membranous labyrinth.

The general organization of the parts of the membranous labyrinth (Fig. 8.131) places:

- the cochlear duct within the cochlea of the bony labyrinth, anteriorly,
- the three semicircular ducts within the three semicircular canals of the bony labyrinth, posteriorly, and
- the saccule and utricle within the vestibule of the bony labyrinth, in the middle.

## Organs of balance

Five of the six components of the membranous labyrinth are concerned with balance. These are the two sacs (the utricle and the saccule) and three ducts (the anterior, posterior, and lateral semicircular ducts).

### Utricle, saccule, and endolymphatic duct

The utricle is the larger of the two sacs. It is oval, elongated and irregular in shape and is in the posterosuperior part of the vestibule of the bony labyrinth.

The three semicircular ducts empty into the utricle. Each semicircular duct is similar in shape, including a dilated end forming the ampulla, to its complementary bony semicircular canal, only much smaller.

The saccule is a smaller, rounded sac lying in the anteroinferior part of the vestibule of the bony labyrinth (Fig. 8.131). The cochlear duct empties into it.

The utriculosaccular duct establishes continuity between all components of the membranous labyrinth and connects the utricle and saccule. Branching from this small duct is the **endolymphatic duct**, which enters the vestibular aqueduct (a channel through the temporal bone) to emerge onto the posterior surface of the petrous part of the temporal bone in the posterior cranial fossa. Here the endolymphatic duct expands into the **endolymphatic sac**, which is an extradural pouch that functions in resorption of endolymph.

### Sensory receptors

Functionally, sensory receptors for balance are organized into unique structures that are located in each of the components of the vestibular apparatus. In the utricle and saccule the sense organ is the **macula of the utricle** and the **macula of the saccule**, respectively, and in the ampulla of each of the three semicircular ducts it is the **crista**.

The utricle responds to linear acceleration in the horizontal plane and sideways head tilts, while the saccule responds to linear acceleration in the vertical plane, such as forward-backward and upward-downward movements. In contrast, the receptors in the three semicircular ducts respond to rotational movement in any direction.

## Organ of hearing

### Cochlear duct

The cochlear duct has a central position in the cochlea of the bony labyrinth dividing it into two canals (the scala vestibuli and the scala tympani). It is maintained in this position by being attached centrally to the lamina of the modiolus, which is a thin lamina of bone extending from the modiolus (the central bony core of the cochlea) and peripherally to the outer wall of the cochlea (Fig. 8.132).

Thus, the triangular-shaped cochlear duct has:

- an outer wall against the bony cochlea consisting of thickened, epithelial-lined periosteum (the **spiral ligament**),
- a roof (the **vestibular membrane**), which separates the endolymph in the cochlear duct from the perilymph in the scala vestibuli and consists of a membrane with a connective tissue core lined on either side with epithelium, and
- a floor, which separates the endolymph in the cochlear duct from the perilymph in the scala tympani and consists of the free edge of the lamina of the modiolus, and

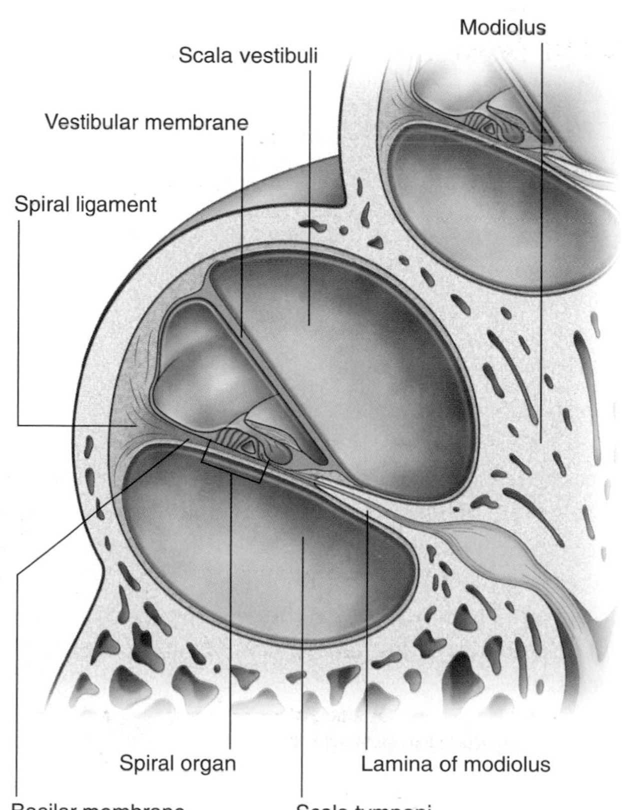


**Fig. 8.132** Membranous labyrinth, cross section.

a membrane (the **basilar membrane**) extending from this free edge of the lamina of the modiolus to an extension of the spiral ligament covering the outer wall of the cochlea.

The **spiral organ** is the organ of hearing, rests on the basilar membrane, and projects into the enclosed, endolymph-filled cochlear duct (Fig. 8.132).

### Vessels

The arterial supply to the internal ear is divided between vessels supplying the bony labyrinth and the membranous labyrinth.

The bony labyrinth is supplied by the same arteries that supply the surrounding temporal bone—these include an anterior tympanic branch from the maxillary artery, a stylomastoid branch from the posterior auricular artery, and a petrosal branch from the middle meningeal artery.

The membranous labyrinth is supplied by the **labyrinthine artery**, which either arises from the anteroinferior cerebellar artery or is a direct branch of the basilar artery—whatever its origin, it enters the internal acoustic meatus with the facial [VII] and vestibulocochlear [VIII] nerves and eventually divides into:

- a **cochlear branch**, which passes through the modiolus and supplies the cochlear duct; and
- one or two **vestibular branches**, which supply the vestibular apparatus.

Venous drainage of the membranous labyrinth is through vestibular veins and cochlear veins, which follow the arteries. These come together to form a **labyrinthine vein**, which eventually empties into either the inferior petrosal sinus or the sigmoid sinus.

### Innervation

The vestibulocochlear nerve [VIII] carries special afferent fibers for hearing (the cochlear component) and balance (the vestibular component). It enters the lateral surface of the brainstem, between the pons and medulla, after exiting the temporal bone through the internal acoustic meatus and crossing the posterior cranial fossa.

Inside the temporal bone, at the distal end of the internal acoustic meatus, the vestibulocochlear nerve divides to form:

- the **cochlear nerve**, and
- the **vestibular nerve**.

The vestibular nerve enlarges to form the **vestibular ganglion**, before dividing into **superior and inferior parts**, which distribute to the three semicircular ducts and the utricle and saccule (see Fig. 8.128).



The cochlear nerve enters the base of the cochlea and passes upward through the modiolus. The ganglion cells of the cochlear nerve are in the **spiral ganglion** at the base of the lamina of the modiolus as it winds around the modiolus (Fig. 8.130). Branches of the cochlear nerve pass through the lamina of the modiolus to innervate the receptors in the spiral organ.

## Facial nerve [VII] in the temporal bone

The facial nerve [VII] is closely associated with the vestibulocochlear nerve [VIII] as it enters the internal acoustic meatus of the temporal bone. Traveling through the temporal bone, its path and several of its branches are directly related to the internal and middle ears.

The facial nerve [VII] enters the internal acoustic meatus in the petrous part of the temporal bone (Fig. 8.133A). The vestibulocochlear nerve and the labyrinthine artery accompany it.

At the distal end of the internal acoustic meatus, the facial nerve [VII] enters the facial canal and continues laterally between the internal and middle ears. At this point the facial nerve [VII] enlarges and bends posteriorly and laterally. The enlargement is the sensory **geniculate ganglion**. As the facial canal continues, the facial nerve [VII] turns sharply downward, and running in an almost vertical direction, it exits the skull through the stylomastoid foramen (Fig. 8.133A).

### Branches

***Greater petrosal nerve.*** At the geniculate ganglion, the facial nerve [VII] gives off the greater petrosal nerve (Fig. 8.133A). This is the first branch of the facial nerve [VII]. The greater petrosal nerve leaves the geniculate ganglion, travels anteromedially through the temporal bone, and emerges through the hiatus for the greater petrosal nerve on the anterior surface of the petrous part of the temporal bone (see Fig. 8.126). The greater petrosal nerve carries preganglionic parasympathetic fibers to the pterygopalatine ganglion.

Continuing beyond the bend, the position of the facial nerve [VII] is indicated on the medial wall of the middle ear by a bulge (see Fig. 8.125).

***Nerve to stapedius and chorda tympani.*** Near the beginning of its vertical descent, the facial nerve [VII] gives off a small branch, the nerve to the stapedius (Fig. 8.133), which innervates the stapedius muscle, and just before it exits the skull the facial nerve [VII] gives off the chorda tympani nerve.

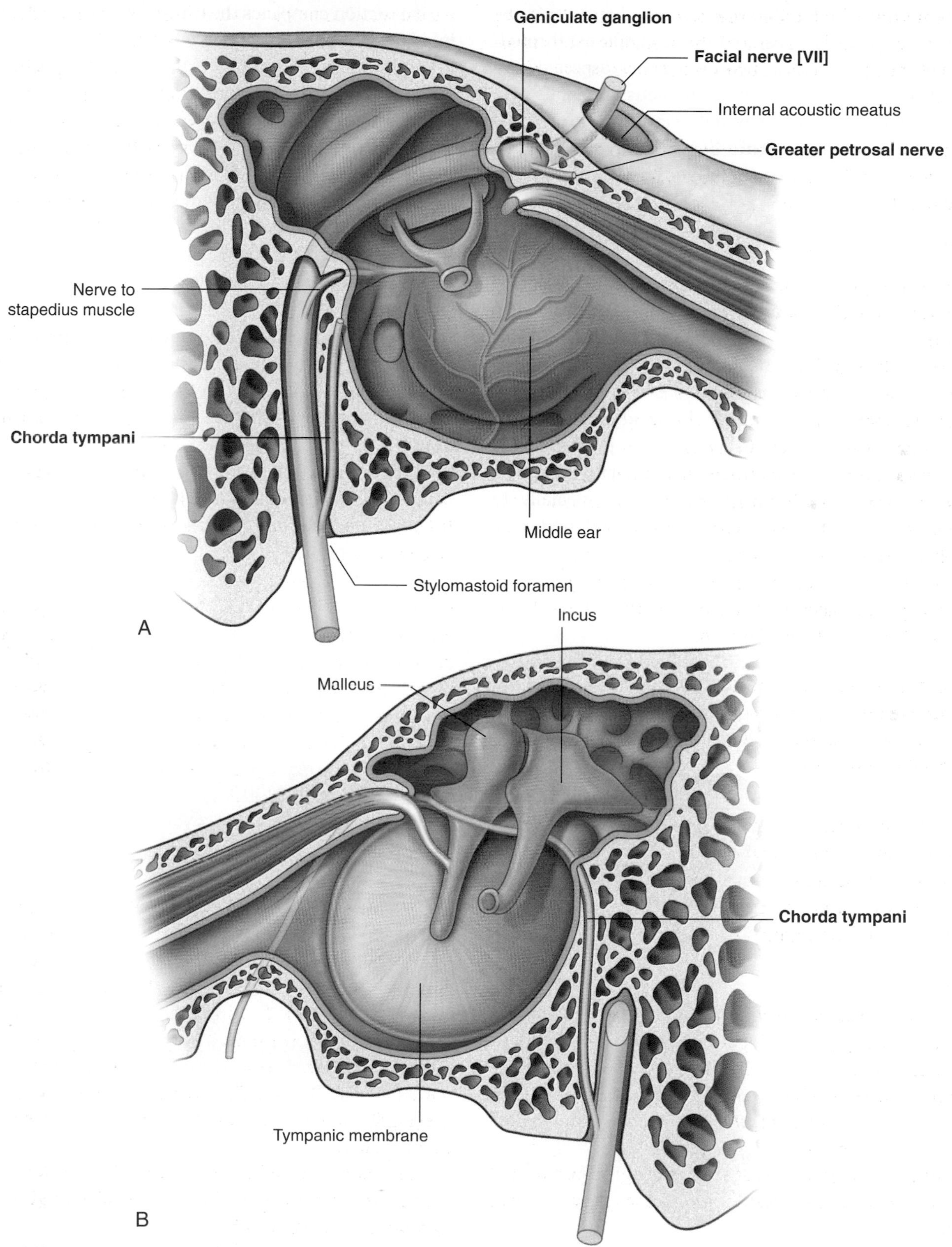


**Fig. 8.133 A.** Facial nerve in the temporal bone. **B.** Chorda tympani in the temporal bone.

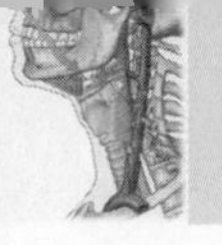



The chorda tympani does not immediately exit the temporal bone, but ascends to enter the middle ear through its posterior wall, passing near the upper aspect of the tympanic membrane between the malleus and incus (Fig. 8.133B). It then exits the middle ear through a canal leading to the **petrotympanic fissure** and exits the skull through this fissure to join the lingual nerve in the infratemporal fossa.

### Transmission of sound

A sound wave enters the external acoustic meatus and strikes the tympanic membrane moving it medially (Fig. 8.134). As the handle of the malleus is attached to this membrane, it also moves medially. This moves the head of the malleus laterally. Because the heads of the malleus and incus articulate with each other, the head of the incus is also moved laterally. This pushes the long process of the incus medially. The long process articulates with the stapes, so its movement causes the stapes to move medially. In turn, because the base of the stapes is attached to the oval window, the oval window is also moved medially.

This action completes the transfer of a large-amplitude, low-force, airborne wave that vibrates the tympanic membrane into a small-amplitude, high-force vibration of the oval window, which generates a wave in the fluid-filled scala vestibuli of the cochlea.

The wave established in the perilymph of the scala vestibuli moves through the cochlea and causes an outward bulging of the secondary tympanic membrane covering the round window at the lower end of the scala tympani (Fig. 8.134). This causes the basilar membrane to vibrate, which in turn leads to stimulation of receptor cells in the spiral organ.

The receptor cells send impulses back to the brain through the cochlear part of the vestibulocochlear nerve [VIII] where they are interpreted as sound.

If the sounds are too loud, causing excessive movement of the tympanic membrane, contraction of the tensor tympani muscle (attached to the malleus) and/or the stapedius muscle (attached to the stapes) dampens the vibrations of the ossicles and decreases the force of the vibrations reaching the oval window.

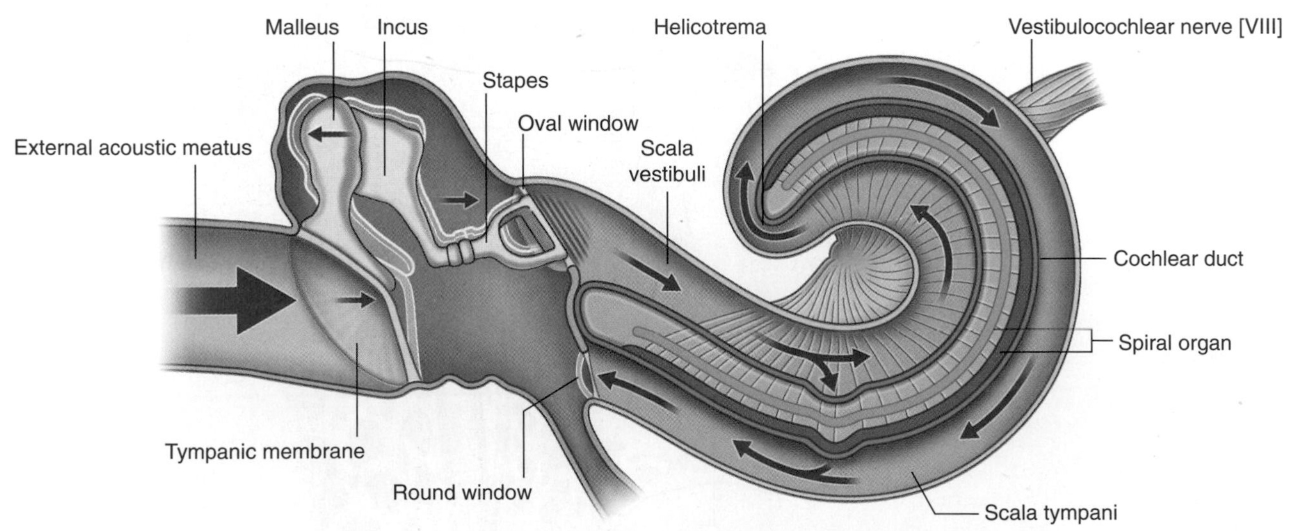


**Fig. 8.134** Transmission of sound.

## TEMPORAL AND INFRATEMPORAL FOSSAE

The temporal and infratemporal fossae are interconnected spaces on the lateral side of the head (Fig. 8.135). Their boundaries are formed by bone and soft tissues.

The temporal fossa is superior to the infratemporal fossa, above the zygomatic arch, and communicates with the infratemporal fossa below through the gap between the zygomatic arch and the more medial surface of the skull.

The infratemporal fossa is a wedge-shaped space deep to the masseter muscle and the underlying ramus of the mandible. Structures that travel between the cranial cavity, neck, pterygopalatine fossa, floor of the oral cavity, floor of the orbit, temporal fossa, and superficial regions of the head pass through it.

Of the four muscles of mastication (masseter, temporalis, medial pterygoid, and lateral pterygoid) that move the lower jaw at the temporomandibular joint, one (masseter) is lateral to the infratemporal fossa, two (medial and lateral pterygoid) are in the infratemporal fossa, and one fills the temporal fossa.

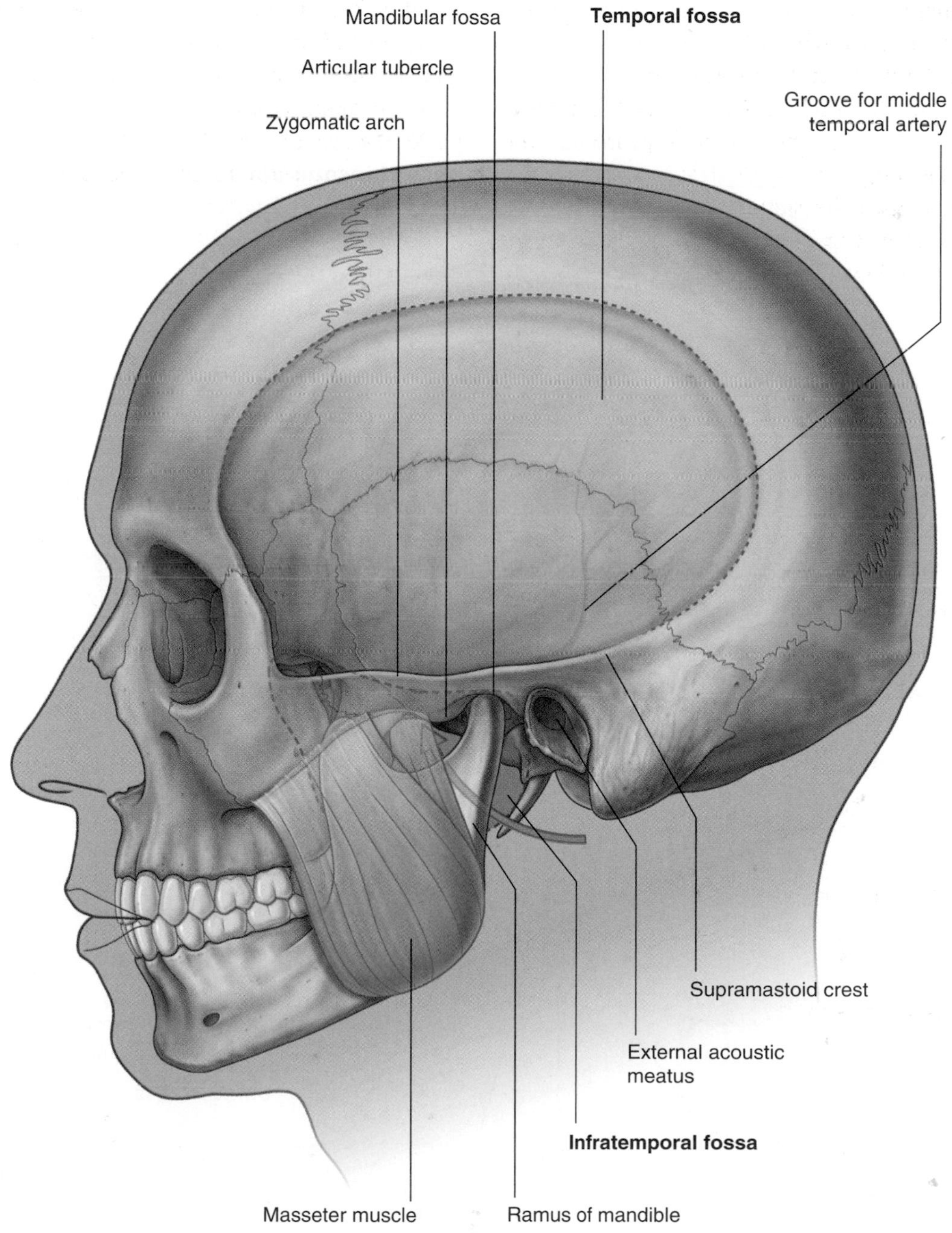


**Fig. 8.135** Temporal and infratemporal fossae.

## Bony framework

Bones that contribute significantly to the boundaries of the temporal and infratemporal fossae include the temporal, zygomatic, and sphenoid bones, and the maxilla and mandible (Figs. 8.136 and 8.137).

Parts of the frontal and parietal bones are also involved.

### Temporal bone

The squamous part of the temporal bone forms part of the bony framework of the temporal and infratemporal fossae.

The tympanic part of the temporal bone forms the posteromedial corner of the roof of the infratemporal fossa, and also articulates with the head of the mandible to form the temporomandibular joint.

The lateral surface of the squamous part of the temporal bone is marked by two surface features on the medial wall of the temporal fossa:

- a transversely oriented **supramastoid crest**, which extends posteriorly from the base of the zygomatic process and marks the posteroinferior border of the temporal fossa; and
- a vertically oriented **groove for the middle temporal artery**, a branch of the superficial temporal artery.

Two features that participate in forming the temporomandibular joint on the inferior aspect of the root of the zygomatic process are the articular tubercle and the mandibular fossa. Both are elongate from medial to lateral. Posterior to the mandibular fossa is the external acoustic meatus. The tympanic part of the temporal bone is a flat concave plate of bone that curves inferiorly from the back of the mandibular fossa and forms part of the wall of the external auditory meatus.

When viewed from inferiorly, there is a distinct **tympanosquamous fissure** between the tympanic and

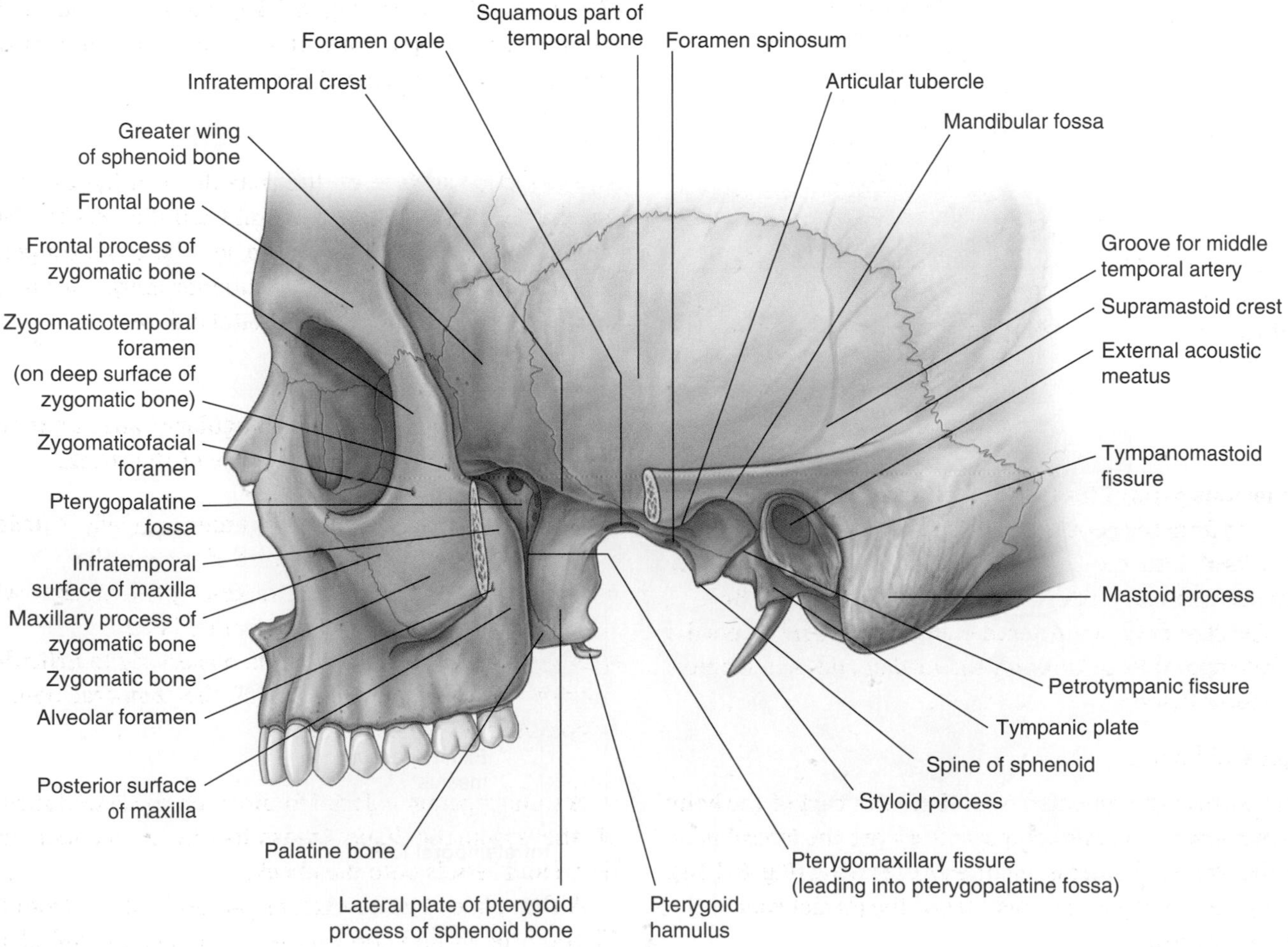

**Fig. 8.136** Bony features related to the temporal and infratemporal fossae.

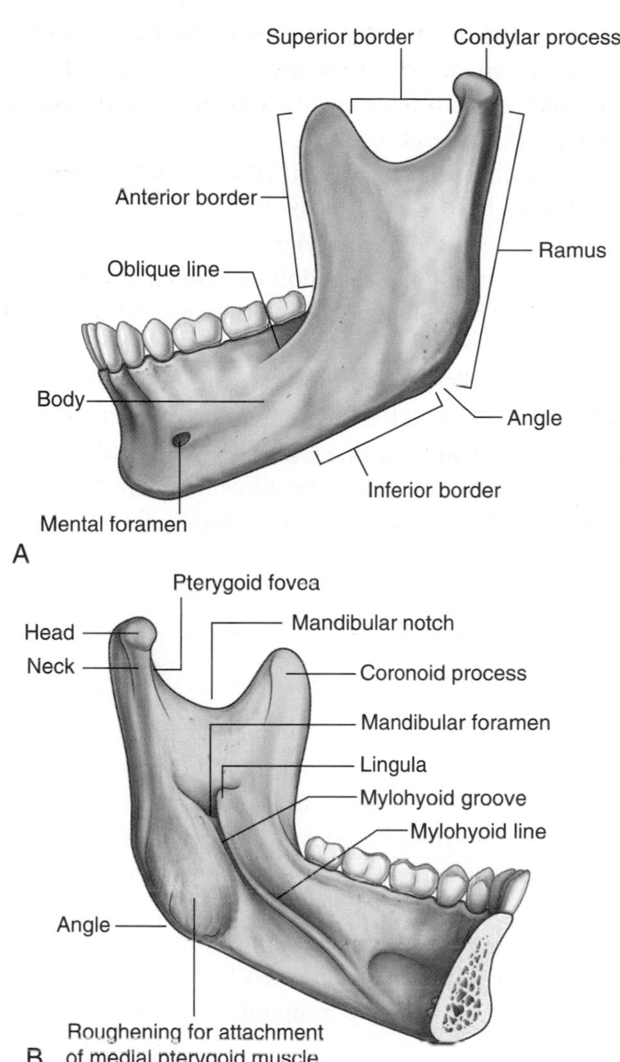


**Fig. 8.137** Mandible. **A.** Lateral view of left side. **B.** Medial view of left side.

squamous parts of the temporal bone. Medially, a small slip of bone from the petrous part of the temporal bone insinuates itself into the fissure and forms a **petrotympanic fissure** between it and the tympanic part (Fig. 8.136).

The chorda tympani nerve exits the skull and enters the infratemporal fossa through the medial end of the petrotympanic fissure.

### Sphenoid bone

The parts of the sphenoid bone that form part of the bony framework of the infratemporal fossa are the lateral plate of the pterygoid process and the greater wing (Fig. 8.136). The greater wing also forms part of the medial wall of the temporal fossa.

The greater wings extend one on each side from the body of the sphenoid. They project laterally from the body and curve superiorly. The inferior and lateral surfaces form the roof of the infratemporal fossa and the medial wall of the temporal fossa, respectively.

The sharply angled boundary between the lateral and inferior surfaces of the greater wing is the **infratemporal crest** (Fig. 8.136). Two apertures (the foramen ovale and the foramen spinosum) pass through the base of the greater wing and allow the mandibular nerve [$V_3$] and the middle meningeal artery, respectively, to pass between the middle cranial fossa and infratemporal fossa. In addition, one or more small sphenoidal emissary foramina penetrate the base of the greater wing anteromedial to the foramen ovale and allow emissary veins to pass between the pterygoid plexus of veins in the infratemporal fossa and the cavernous sinus in the middle cranial fossa.

Projecting vertically downward from the greater wing immediately medial to the foramen spinosum is the irregularly shaped **spine of the sphenoid**, which is the attachment site for the cranial end of the sphenomandibular ligament.

The lateral plate of the pterygoid process is a vertically oriented sheet of bone that projects posterolaterally from the pterygoid process (Fig. 8.136). Its lateral and medial surfaces provide attachment for the lateral and medial pterygoid muscles, respectively.

### Maxilla

The posterior surface of the maxilla contributes to the anterior wall of the infratemporal fossa (Fig. 8.136). This surface is marked by a foramen for the posterosuperior alveolar nerve and vessels. The superior margin forms the inferior border of the inferior orbital fissure.

### Zygomatic bone

The zygomatic bone is a quadrangular-shaped bone that forms the palpable bony prominence of the cheek:

- A **maxillary process** extends anteromedially to articulate with the zygomatic process of the maxilla.
- A **frontal process** extends superiorly to articulate with the zygomatic process of the frontal bone.
- A **temporal process** extends posteriorly to articulate with the zygomatic process of the temporal bone to complete the zygomatic arch.

A small zygomaticofacial foramen on the lateral surface of the zygomatic bone transmits the zygomaticofacial nerve and vessels onto the cheek.

A thin plate of bone extends posteromedially from the frontal process and contributes to the lateral wall of the orbit on one side and the anterior wall of the temporal fossa on the other. A zygomaticotemporal foramen on the

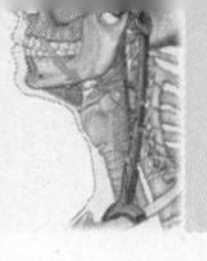

temporal fossa surface of the plate where it attaches to the frontal process is for the zygomaticotemporal nerve.

### Ramus of mandible

The **ramus of the mandible** is quadrangular in shape and has medial and lateral surfaces and condylar and coronoid processes (Fig. 8.137).

The lateral surface of the ramus of the mandible is generally smooth except for the presence of a few obliquely oriented ridges. Most of the lateral surface provides attachment for the masseter muscle.

The posterior and inferior borders of the ramus intersect to form the **angle of the mandible**, while the superior border is notched to form the **mandibular notch**. The anterior border is sharp and is continuous below with the **oblique line** on the body of the mandible.

The **coronoid process** extends superiorly from the junction of the anterior and superior borders of the ramus. It is a flat, triangular process that provides attachment for the temporalis muscle.

The **condylar process** extends superiorly from the posterior and superior borders of the ramus. It consists of:

- the **head of the mandible**, which is expanded medially and participates in forming the temporomandibular joint; and
- the **neck of the mandible**, which bears a shallow depression (the **pterygoid fovea**) on its anterior surface for attachment of the lateral pterygoid muscle.

The medial surface of the ramus of the mandible is the lateral wall of the infratemporal fossa (Fig. 8.137B). Its most distinctive feature is the **mandibular foramen**, which is the superior opening of the mandibular canal. The inferior alveolar nerve and vessels pass through this foramen.

Immediately anterosuperior to the mandibular foramen is a triangular elevation (the **lingula**) for attachment of the mandibular end of the sphenomandibular ligament.

An elongate groove (the **mylohyoid groove**) extends anteroinferiorly from the mandibular foramen. The **nerve to the mylohyoid** is in this groove.

Posteroinferior to the mylohyoid groove and mandibular foramen, the medial surface of the ramus of the mandible is roughened for attachment of the medial pterygoid muscle.

## Temporomandibular joints

The temporomandibular joints, one on each side, allow opening and closing of the mouth and complex chewing or side-to-side movements of the lower jaw.

Each joint is synovial and is formed between the head of the mandible and the articular fossa and articular tubercle of the temporal bone (Fig. 8.138A).

Unlike most other synovial joints where the articular surfaces of the bones are covered by a layer of hyaline cartilage, those of the temporomandibular joint are covered

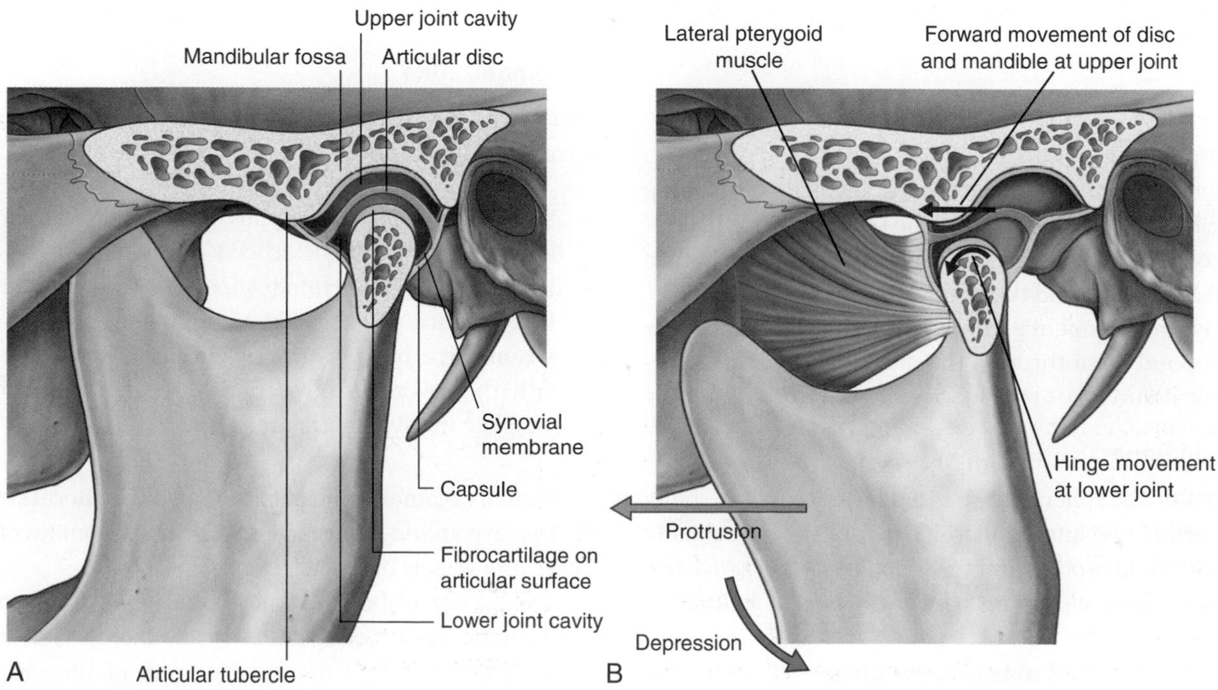


Fig. 8.138 Temporomandibular joint. **A.** Mouth closed. **B.** Mouth open.

by fibrocartilage. In addition, the joint is completely divided by a fibrous **articular disc** into two parts:

- The lower part of the joint allows mainly the hinge-like depression and elevation of the mandible.
- The upper part of the joint allows the head of the mandible to translocate forward (protrusion) onto the articular tubercle and backward (retraction) into the mandibular fossa.

Opening the mouth involves both depression and protrusion (Fig. 8.138B).

The forward or protrusive movement allows greater depression of the mandible by preventing backward movement of the angle of the mandible into structures in the neck.

### Joint capsule

The **synovial membrane** of the joint capsule lines all nonarticular surfaces of the upper and lower compartments of the joint and is attached to the margins of the articular disc.

The **fibrous membrane** of the joint capsule encloses the temporomandibular joint complex and is attached:

- above along the anterior margin of the articular tubercle,
- laterally and medially along the margins of the articular fossa,
- posteriorly to the region of the tympanosquamous suture, and
- below around the upper part of the neck of the mandible.

The articular disc attaches around its periphery to the inner aspect of the fibrous membrane.

### Extracapsular ligaments

Three extracapsular ligaments are associated with the temporomandibular joint—the lateral, sphenomandibular, and the stylomandibular ligaments (Fig. 8.139):

- The **lateral ligament** is closest to the joint, just lateral to the capsule, and runs diagonally backward from the margin of the articular tubercle to the neck of the mandible.
- The **sphenomandibular ligament** is medial to the temporomandibular joint, runs from the spine of the sphenoid bone at the base of the skull to the lingula on the medial side of the ramus of the mandible.
- The **stylomandibular ligament** passes from the styloid process of the temporal bone to the posterior margin and angle of the mandible.

### Movements of the mandible

A chewing or grinding motion occurs when the movements at the temporomandibular joint on one side are coordinated with a reciprocal set of movements at the joint on the other side. Movements of the mandible include depression, elevation, protrusion, and retraction (Fig. 8.140):

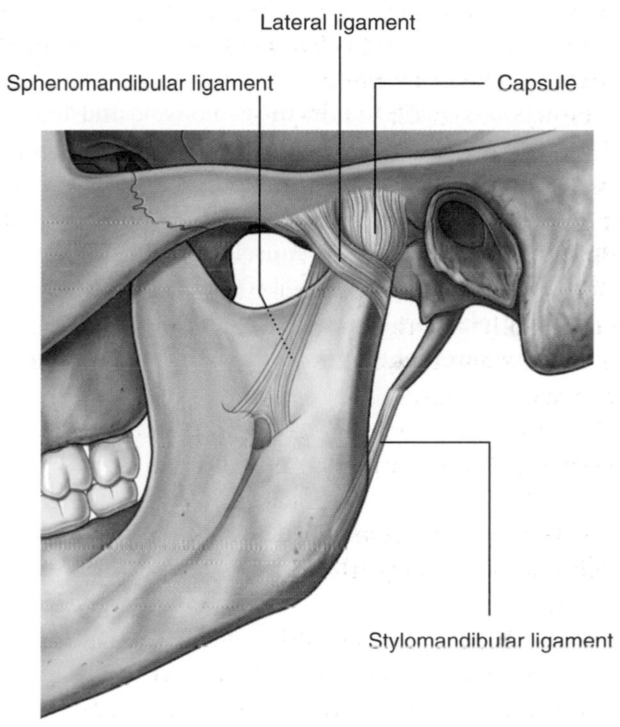


**Fig. 8.139** Ligaments associated with the temporomandibular joint.

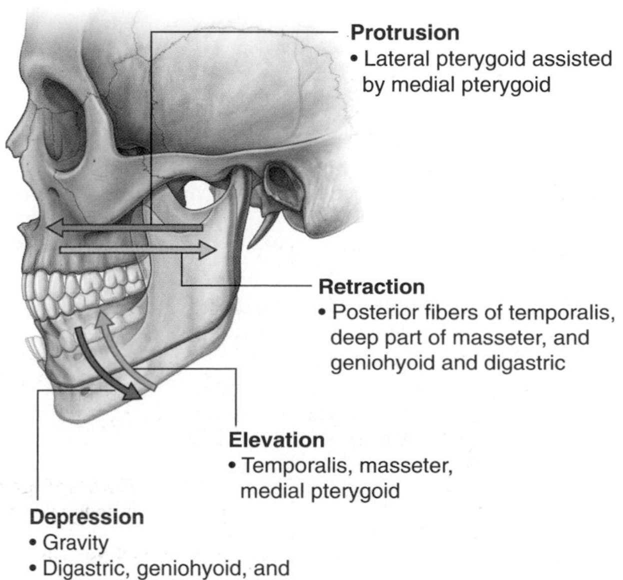


**Fig. 8.140** Movements of the temporomandibular joint.

- Depression is generated by the digastric, geniohyoid, and mylohyoid muscles on both sides, is normally assisted by gravity, and, because it involves forward movement of the head of the mandible onto the articular tubercle, the lateral pterygoid muscles are also involved.
- Elevation is a very powerful movement generated by the temporalis, masseter, and medial pterygoid muscles and also involves movement of the head of the mandible into the mandibular fossa.
- Protraction is mainly achieved by the lateral pterygoid muscle, with some assistance by the medial pterygoid.
- Retraction is carried out by the geniohyoid and digastric muscles, and by the posterior and deep fibers of the temporalis and masseter muscles, respectively.

Except for the geniohyoid muscle, which is innervated by the C1 spinal nerve, all muscles that move the temporomandibular joints are innervated by the mandibular nerve [$V_3$] by branches that originate in the infratemporal fossa.

## Masseter muscle

The **masseter** muscle is a powerful muscle of mastication that elevates the mandible (Fig. 8.141 and Table 8.11). It overlies the lateral surface of the ramus of the mandible.

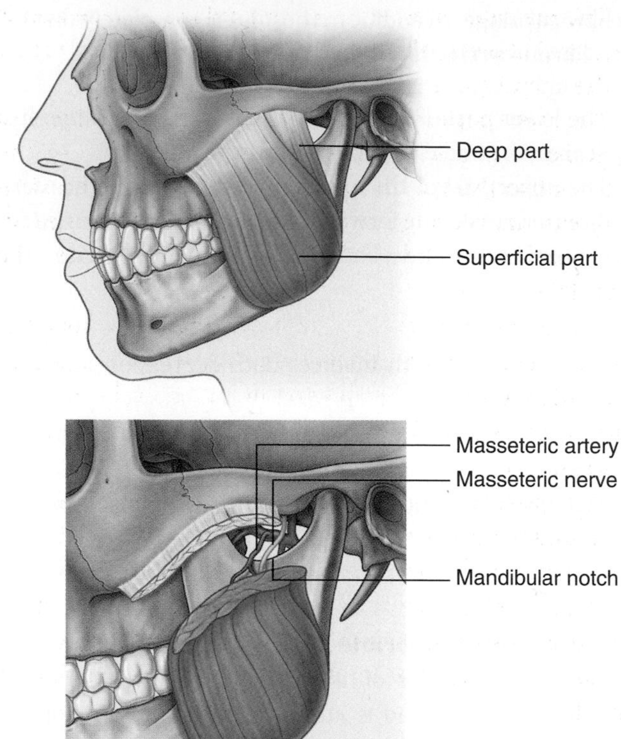


**Fig. 8.141** Masseter muscle.

**Table 8.11** Muscles of mastication

| Muscle | Origin | Insertion | Innervation | Function |
|---|---|---|---|---|
| Masseter | Zygomatic arch and maxillary process of the zygomatic bone | Lateral surface of ramus of mandible | Masseteric nerve from the anterior trunk of the mandibular nerve [$V_3$] | Elevation of mandible |
| Temporalis | Bone of temporal fossa and temporal fascia | Coronoid process of mandible and anterior margin of ramus of mandible almost to last molar tooth | Deep temporal nerves from the anterior trunk of the mandibular nerve [$V_3$] | Elevation and retraction of mandible |
| Medial pterygoid | Deep head—medial surface of lateral plate of pterygoid process and pyramidal process of palatine bone; superficial head—tuberosity of the maxilla and pyramidal process of palatine bone | Medial surface of mandible near angle | Nerve to medial pterygoid from the mandibular nerve [$V_3$] | Elevation and side-to-side movements of the mandible |
| Lateral pterygoid | Upper head—roof of infratemporal fossa; lower head—lateral surface of lateral plate of the pterygoid process | Capsule of temporomandibular joint in the region of attachment to the articular disc and to the pterygoid fovea on the neck of mandible | Nerve to lateral pterygoid directly from the anterior trunk of the mandibular nerve [$V_3$] or from the buccal branch | Protrusion and side-to-side movements of the mandible |

The masseter muscle is quadrangular in shape and is anchored above to the zygomatic arch and below to most of the lateral surface of the ramus of the mandible.

The more **superficial part** of the masseter originates from the maxillary process of the zygomatic bone and the anterior two-thirds of the zygomatic process of the maxilla. It inserts into the angle of the mandible and related posterior part of the lateral surface of the ramus of the mandible.

The **deep part** of the masseter originates from the medial aspect of the zygomatic arch and the posterior part of its inferior margin and inserts into the central and upper part of the ramus of the mandible as high as the coronoid process.

The masseter is innervated by the masseteric nerve from the mandibular nerve [$V_3$] and supplied with blood by the masseteric artery from the maxillary artery.

The masseteric nerve and artery originate in the infratemporal fossa and pass laterally over the margin of the mandibular notch to enter the deep surface of the masseter muscle.

## Temporal fossa

The temporal fossa is a narrow fan-shaped space that covers the lateral surface of the skull (Fig. 8.142A):

- Its upper margin is defined by a pair of temporal lines that arch across the skull from the zygomatic process of the frontal bone to the supramastoid crest of the temporal bone.
- It is limited laterally by the **temporal fascia**, which is a tough, fan-shaped aponeurosis overlying the temporalis muscle and attached by its outer margin to the superior temporal line and by its inferior margin to the zygomatic arch.
- Anteriorly, it is limited by the posterior surface of the frontal process of the zygomatic bone and the posterior surface of the zygomatic process of the frontal bone, which separate the temporal fossa behind from the orbit in front.

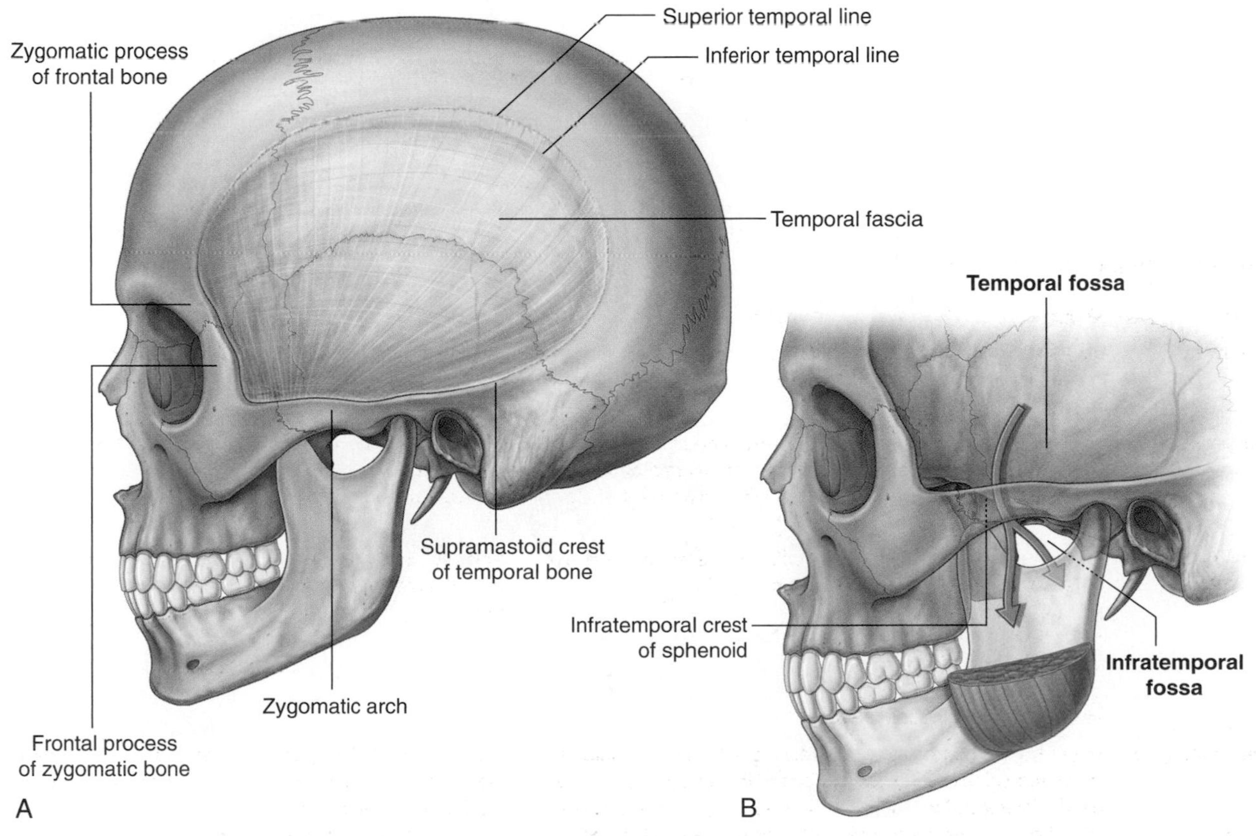


**Fig. 8.142** Temporal fossa. **A.** Lateral view. **B.** Lateral view showing the infratemporal fossa.

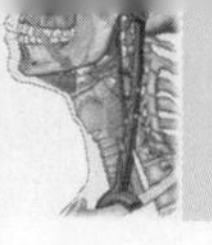



- Its inferior margin is marked by the zygomatic arch laterally and by the infratemporal crest of the greater wing of the sphenoid medially (Fig. 8.142B)—between these two features, the floor of the temporal fossa is open medially to the infratemporal fossa and laterally to the region containing the masseter muscle.

## Contents

The major structure in the temporal fossa is the temporalis muscle.

Also passing through the fossa is the zygomaticotemporal branch of the maxillary nerve [$V_2$], which enters the region through the zygomaticotemporal foramen on the temporal fossa surface of the zygomatic bone.

### Temporalis muscle

The **temporalis muscle** is a large, fan-shaped muscle that fills much of the temporal fossa (Fig. 8.143). It originates from the bony surfaces of the fossa superiorly to the inferior temporal line and is attached laterally to the surface of the temporal fascia. The more anterior fibers are oriented vertically while the more posterior fibers are oriented horizontally. The fibers converge inferiorly to form a tendon, which passes between the zygomatic arch and the infratemporal crest of the greater wing of the sphenoid to insert on the coronoid process of the mandible.

The temporalis muscle attaches down the anterior surface of the coronoid process and along the related margin of the ramus of the mandible, almost to the last molar tooth.

The temporalis is a powerful elevator of the mandible. Because this movement involves posterior translocation of the head of the mandible from the articular tubercle of the temporal bone and back into the mandibular fossa, the temporalis also retracts the mandible or pulls it posteriorly. In addition, the temporalis participates in side-to-side movements of the mandible.

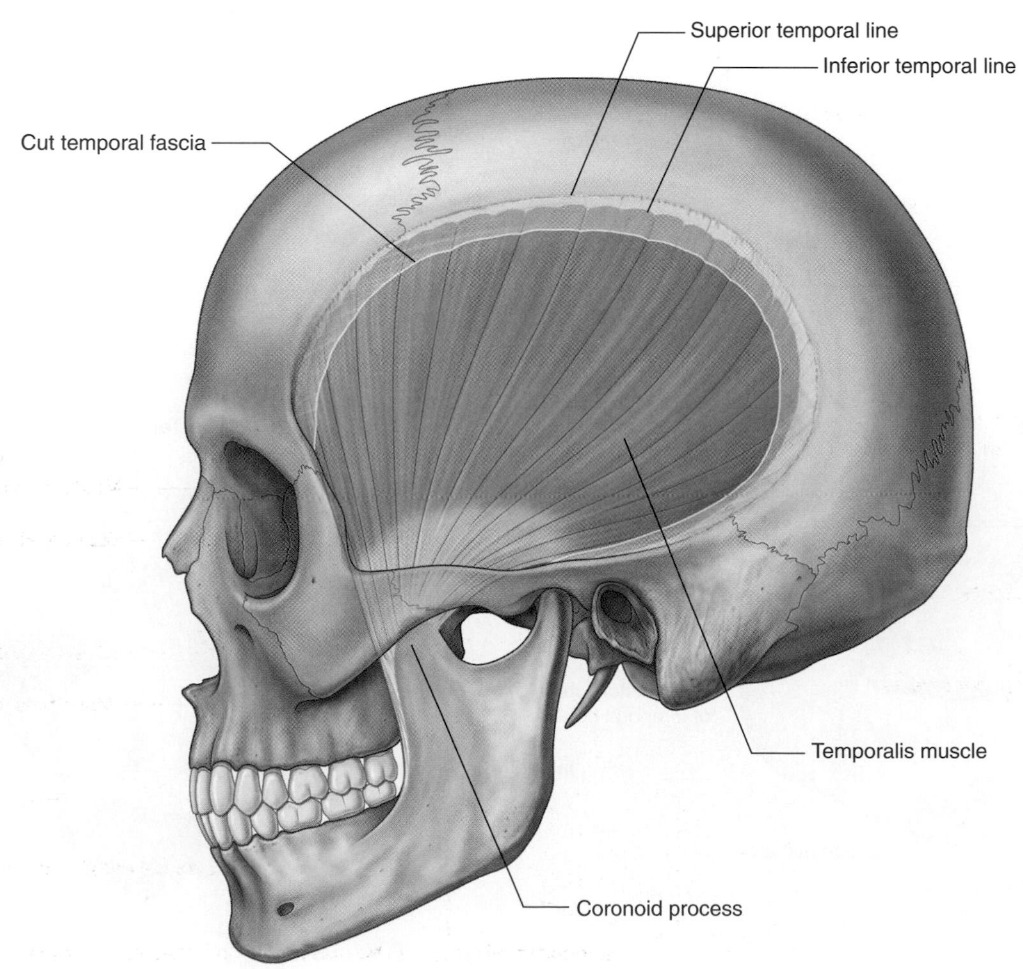


**Fig. 8.143** Temporalis muscle. Lateral view.

The temporalis is innervated by deep temporal nerves that originate from the mandibular nerve [$V_3$] in the infratemporal fossa and then pass into the temporal fossa.

Blood supply of the temporalis is by deep temporal arteries, which travel with the nerves, and the middle temporal artery, which penetrates the temporal fascia at the posterior end of the zygomatic arch.

### Deep temporal nerves

The deep temporal nerves, usually two in number, originate from the anterior trunk of the mandibular nerve [$V_3$] in the infratemporal fossa (Fig. 8.144). They pass superiorly and around the infratemporal crest of the greater wing of the sphenoid to enter the temporal fossa deep to the temporalis muscle, and supply the temporalis muscle.

### Zygomaticotemporal nerve

The zygomaticotemporal nerve is a branch of the zygomatic nerve (see Fig. 8.87, p. 922). The zygomatic nerve is a branch of the maxillary nerve [$V_2$], which originates in the pterygopalatine fossa and passes into the orbit.

The zygomaticotemporal nerve enters the temporal fossa through one or more small foramina on the temporal fossa surface of the zygomatic bone.

Branches of the zygomaticotemporal nerve pass superiorly between the bone and the temporalis muscle to penetrate the temporal fascia and supply the skin of the temple (Fig. 8.144).

### Deep temporal arteries

Normally two in number, these vessels originate from the maxillary artery in the infratemporal fossa and travel with the deep temporal nerves around the infratemporal crest of the greater wing of the sphenoid to supply the temporalis muscle (Fig. 8.144). They anastomose with branches of the middle temporal artery.

### Middle temporal artery

The middle temporal artery originates from the superficial temporal artery just superior to the root of the zygomatic arch between this structure and the external ear (Fig. 8.144). It penetrates the temporalis fascia, passes under

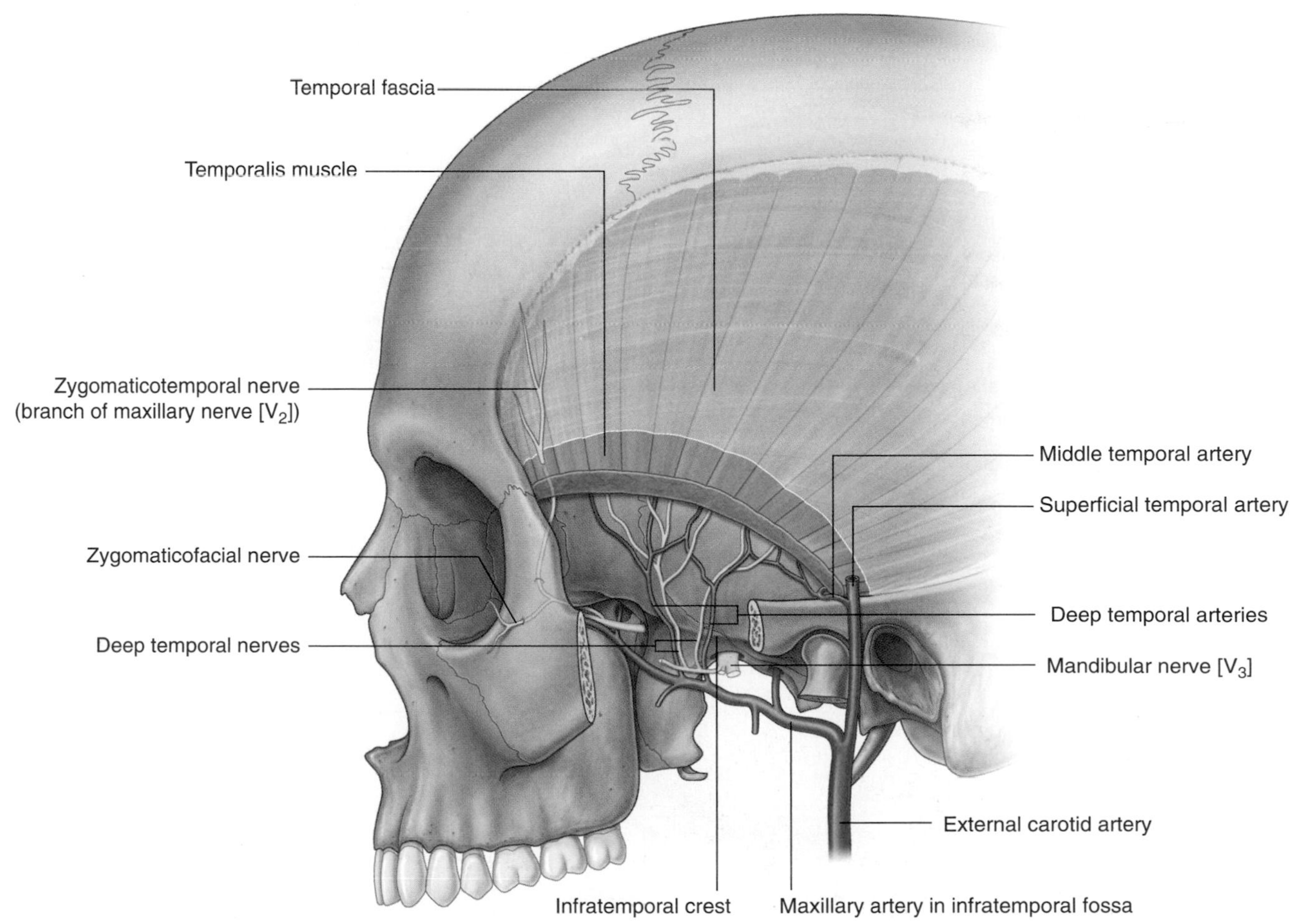


**Fig. 8.144** Nerves and arteries of the temporal fossa.

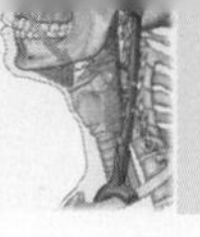



the margin of the temporalis muscle, and travels superiorly on the deep surface of the temporalis muscle.

The middle temporal artery supplies the temporalis and anastomoses with branches of the deep temporal arteries.

## Infratemporal fossa

The wedge-shaped infratemporal fossa is inferior to the temporal fossa and between the ramus of the mandible laterally and the wall of the pharynx medially. It has a roof, a lateral wall, and a medial wall, and is open to the neck posteroinferiorly (Fig. 8.145):

- The **roof** is formed by the inferior surfaces of the greater wing of the sphenoid and the temporal bone, contains the foramen spinosum, foramen ovale, and the petrotympanic fissure, and lateral to the infratemporal crest of the greater wing of the sphenoid, is open superiorly to the temporal fossa.
- The **lateral wall** is the medial surface of the ramus of the mandible, which contains the opening to the mandibular canal.
- The **medial wall** is formed anteriorly by the lateral plate of the pterygoid process and more posteriorly by the pharynx and by two muscles of the soft palate (tensor and levator veli palatini muscles), and contains the pterygomaxillary fissure anteriorly, which allows structures to pass between the infratemporal and pterygopalatine fossae.
- The **anterior wall** is formed by part of the posterior surface of the maxilla and contains the alveolar foramen, and the upper part opens as the inferior orbital fissure into the orbit.

### Contents

Major contents of the infratemporal fossa include the sphenomandibular ligament, medial and lateral pterygoid muscles (Table 8.11), the maxillary artery, the mandibular nerve [$V_3$], branches of the facial nerve [VII] and the glossopharyngeal nerve [IX], and the pterygoid plexus of veins.

#### Sphenomandibular ligament

The sphenomandibular ligament is an extracapsular ligament of the temporomandibular joint. It is attached

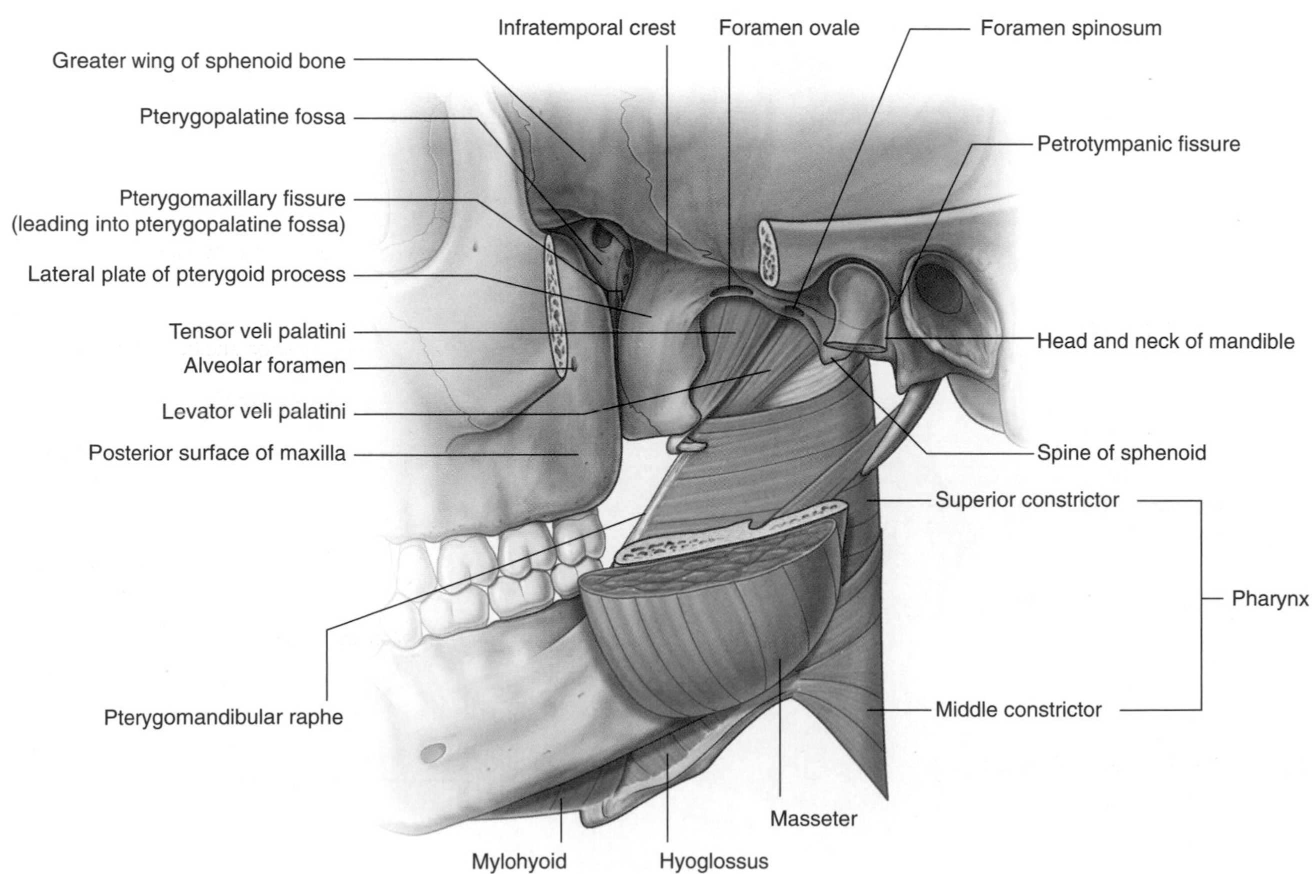


**Fig. 8.145** Borders of the infratemporal fossa.

superiorly to the spine of the sphenoid bone and expands inferiorly to attach to the lingula of the mandible and the posterior margin of the mandibular foramen (Fig. 8.146).

### Medial pterygoid

The **medial pterygoid** muscle is quadrangular in shape and has deep and superficial heads (Fig. 8.146):

- The **deep head** is attached above to the medial surface of the lateral plate of the pterygoid process and the associated surface of the pyramidal process of the palatine bone, and descends obliquely downward, medial to the sphenomandibular ligament, to attach to the roughened medial surface of the ramus of the mandible near the angle of the mandible.
- The **superficial head** originates from the tuberosity of the maxilla and adjacent pyramidal process of the palatine bone and joins with the deep head to insert on the mandible.

The medial pterygoid mainly elevates the mandible. Because it passes obliquely backward to insert into the mandible, it also assists the lateral pterygoid muscle in protruding the lower jaw.

The medial pterygoid is innervated by the nerve to the medial pterygoid from the mandibular nerve [$V_3$].

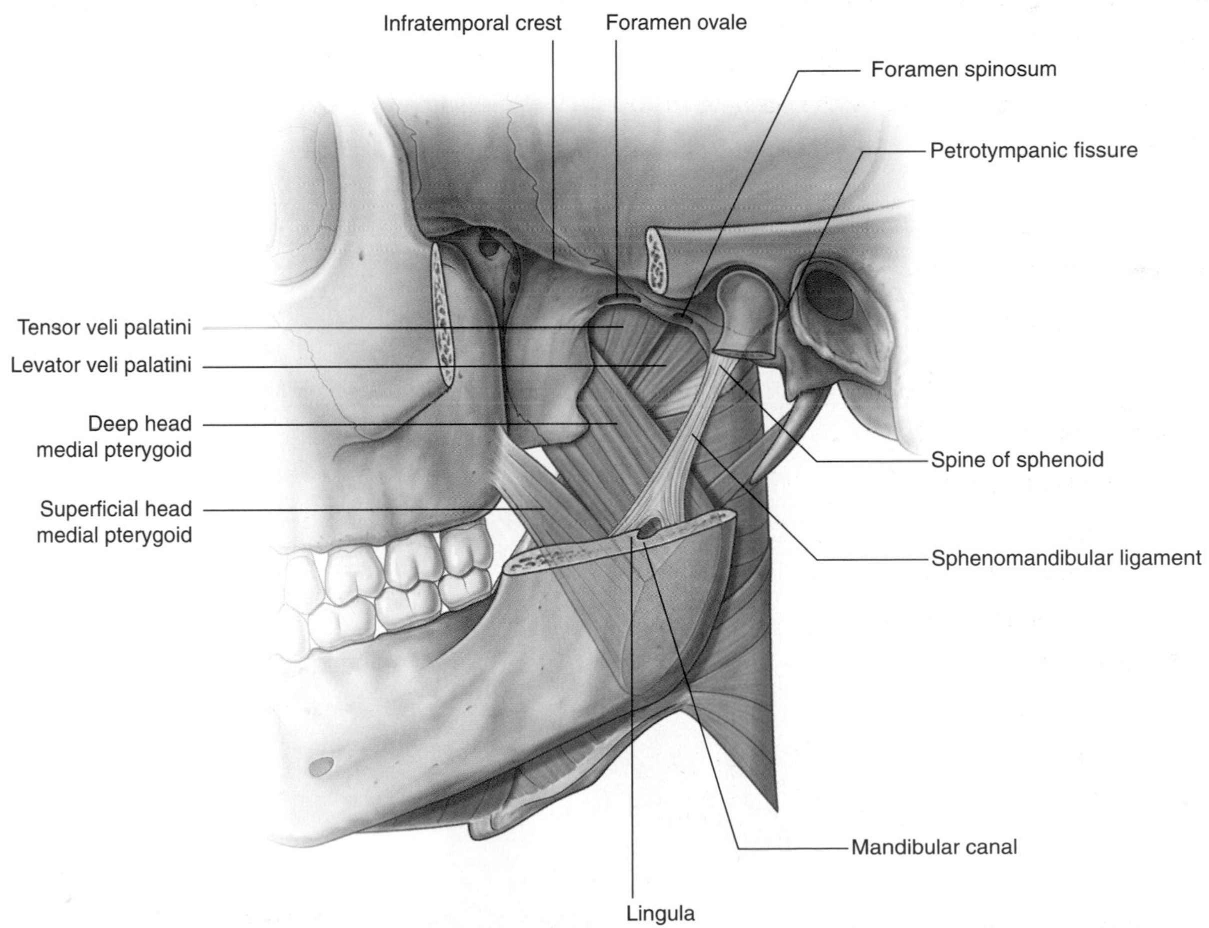


**Fig. 8.146** Medial pterygoid muscle.

### Lateral pterygoid

The lateral pterygoid is a thick triangular muscle and like the medial pterygoid muscle has two heads (Fig. 8.147):

- The **upper head** originates from the roof of the infratemporal fossa (inferior surface of the greater wing of the sphenoid and the infratemporal crest) lateral to the foramen ovale and foramen spinosum.
- The **lower head** is larger than the upper head and originates from the lateral surface of the lateral plate of the pterygoid process, and the inferior part insinuates itself between the cranial attachments of the two heads of the medial pterygoid.

The fibers from both heads of the lateral pterygoid muscle converge to insert into the pterygoid fovea of the neck of the mandible and into the capsule of the temporomandibular joint in the region where the capsule is attached internally to the articular disc.

Unlike the medial pterygoid muscle whose fibers tend to be oriented vertically, those of the lateral pterygoid are oriented almost horizontally. As a result, when the lateral pterygoid contracts it pulls the articular disc and head of the mandible forward onto the articular tubercle and is therefore the major protruder of the lower jaw.

The lateral pterygoid is innervated by the nerve to the lateral pterygoid from the mandibular nerve [$V_3$].

When the lateral and medial pterygoids contract on only one side, the chin moves to the opposite side. When opposite movements at the two temporomandibular joints are coordinated, a chewing movement results.

### Mandibular nerve [$V_3$]

The mandibular nerve [$V_3$] is the largest of the three divisions of the trigeminal nerve [V].

Unlike the ophthalmic [$V_1$] and maxillary [$V_2$] nerves, which are purely sensory, the mandibular nerve [$V_3$] is both motor and sensory.

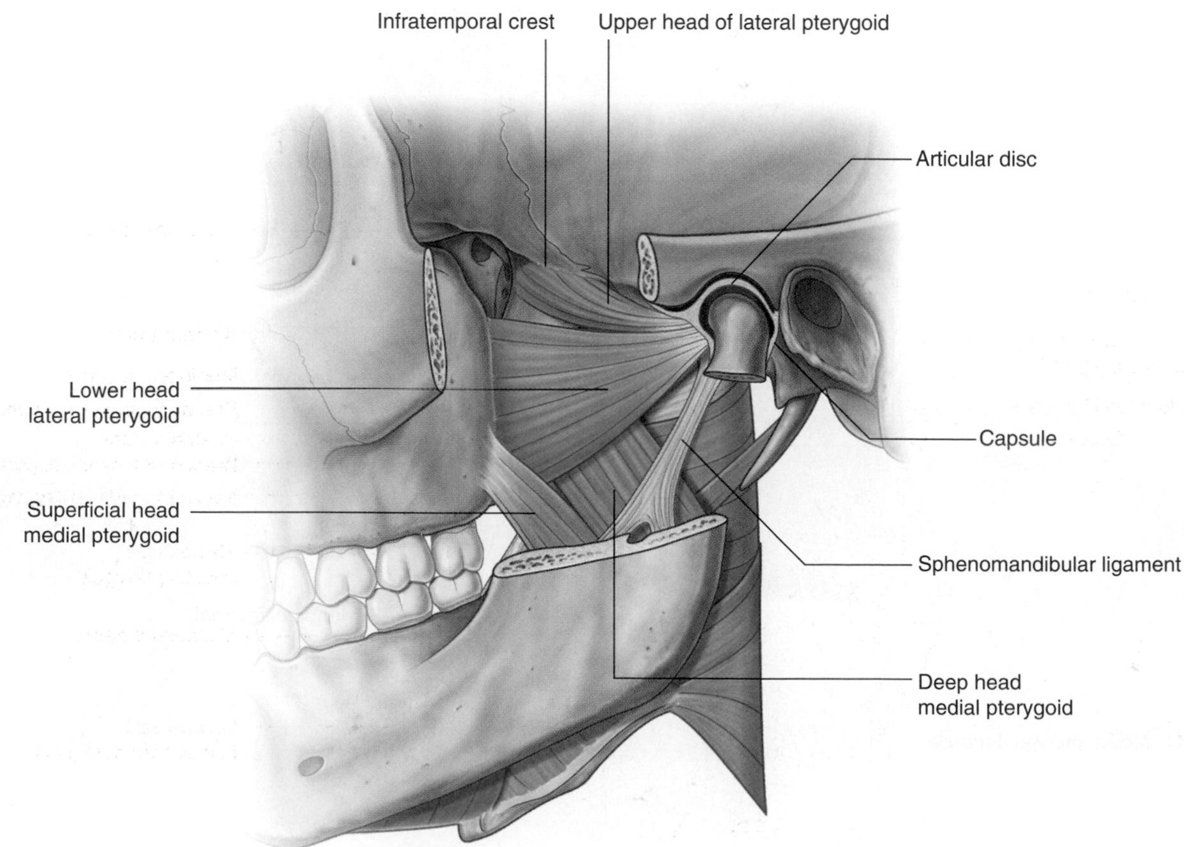


**Fig. 8.147** Lateral pterygoid muscle.

In addition to carrying general sensation from the teeth and gingivae of the mandible, the anterior two-thirds of the tongue, mucosa on the floor of the oral cavity, the lower lip, skin over the temple and lower face, and part of the cranial dura mater, the mandibular nerve [$V_3$] also carries motor innervation to most of the muscles that move the mandible, one of the muscles (tensor tympani) in the middle ear, and one of the muscles of the soft palate (tensor veli palatini).

All branches of the mandibular nerve [$V_3$] originate in the infratemporal fossa.

Like the ophthalmic [$V_1$] and maxillary [$V_2$] nerves, the sensory part of the mandibular nerve [$V_3$] originates from the trigeminal ganglion in the middle cranial fossa (Fig. 8.148):

- The sensory part of the mandibular nerve [$V_3$] drops vertically through the foramen ovale and enters the infratemporal fossa between the tensor veli palatini muscle and the upper head of the lateral pterygoid muscle.
- The small motor root of the trigeminal nerve [V] passes medial to the trigeminal ganglion in the cranial cavity, then passes through the foramen ovale and immediately joins the sensory part of the mandibular nerve [$V_3$].

### Branches

Soon after the sensory and motor roots join, the mandibular nerve [$V_3$] gives rise to a small meningeal branch and to the nerve to the medial pterygoid, and then divides into anterior and posterior trunks (Fig. 8.148):

- Branches from the anterior trunk are the buccal, masseteric, and deep temporal nerves, and the nerve to the lateral pterygoid, all of which, except the buccal nerve (which is predominantly sensory) are motor nerves.
- Branches from the posterior trunk are the auriculotemporal, lingual, and inferior alveolar nerves, all of which, except a small nerve (nerve to the mylohyoid) that branches from the inferior alveolar nerve, are sensory nerves.

### Meningeal branch

The meningeal branch originates from the medial side of the mandibular nerve [$V_3$] and ascends to leave the infratemporal fossa with the middle meningeal artery and reenter the cranial cavity through the foramen spinosum (Fig. 8.148). It is sensory for the dura mater, mainly of the middle cranial fossa, and also supplies the mastoid cells that communicate with the middle ear.

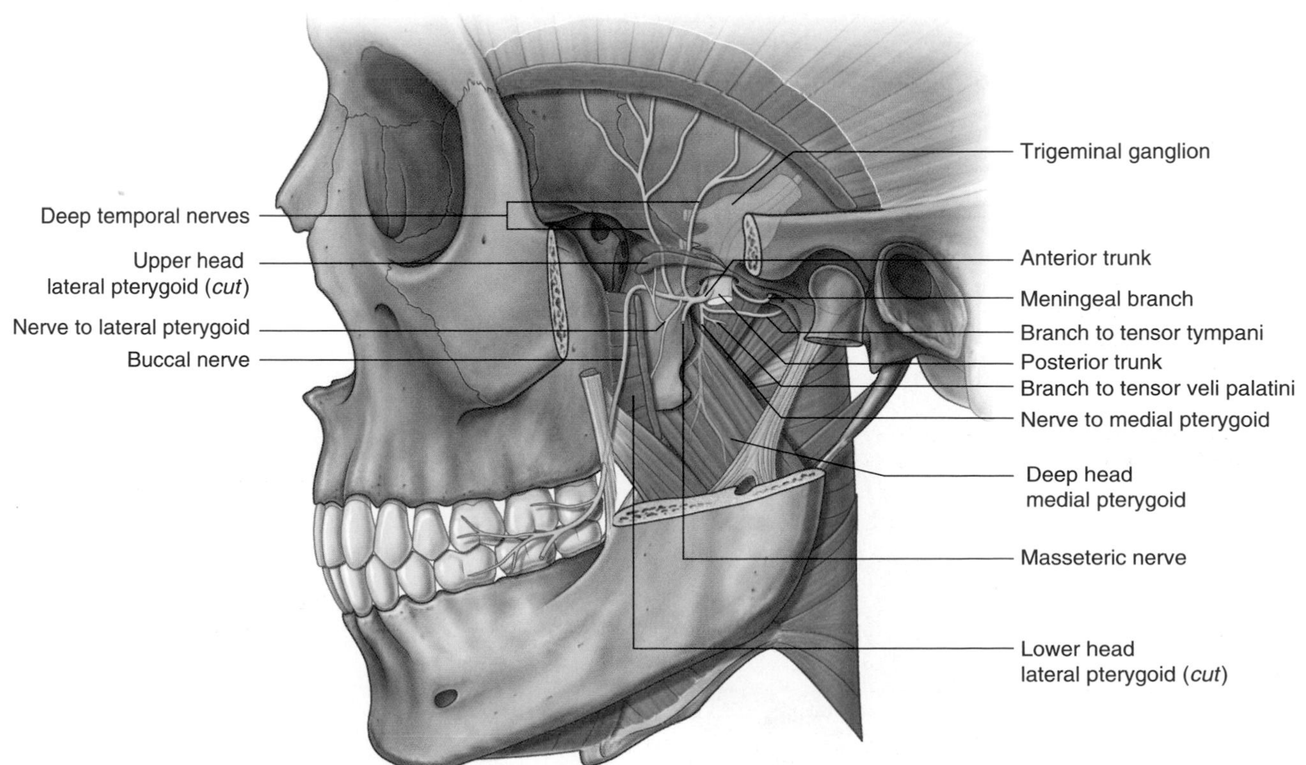


**Fig. 8.148** Mandibular nerve [$V_3$]—anterior trunk. Meningeal branch and nerve to medial pterygoid.

### Nerve to medial pterygoid

The nerve to the medial pterygoid also originates medially from the mandibular nerve [$V_3$] (Fig. 8.148). It descends to enter and supply the deep surface of the medial pterygoid muscle. Near its origin from the mandibular nerve [$V_3$], it has two small branches:

- One of these supplies the tensor veli palatini.
- The other ascends to supply the tensor tympani muscle, which occupies a small bony canal above and parallel to the pharyngotympanic tube in the temporal bone.

### Buccal nerve

The buccal nerve is a branch of the anterior trunk of the mandibular nerve [$V_3$] (Fig. 8.148). It is predominantly a sensory nerve, but may also carry the motor innervation to the lateral pterygoid muscle and to part of the temporalis muscle.

The buccal nerve passes laterally between the upper and lower heads of the lateral pterygoid and then descends around the anterior margin of the insertion of the temporalis muscle to the anterior margin of the ramus of the mandible, often slipping through the tendon of the temporalis. It continues into the cheek lateral to the buccinator muscle to supply general sensory nerves to the adjacent skin and oral mucosa and the buccal gingivae of the lower molars.

### Masseteric nerve

The masseteric nerve is a branch of the anterior trunk of the mandibular nerve [$V_3$] (Fig. 8.148; also see Fig. 8.141). It passes laterally over the lateral pterygoid muscle and through the mandibular notch to penetrate and supply the masseter muscle.

### Deep temporal nerves

The deep temporal nerves, usually two in number, originate from the anterior trunk of the mandibular nerve [$V_3$] (Fig. 8.148; also see Fig. 8.144). They pass laterally above the lateral pterygoid muscle and curve around the infratemporal crest to ascend in the temporal fossa and supply the temporalis muscle from its deep surface.

### Nerve to lateral pterygoid

The nerve to the lateral pterygoid may originate directly as a branch from the anterior trunk of the mandibular nerve [$V_3$] or from its buccal branch (Fig. 8.148). From its origin, it passes directly into the deep surface of the lateral pterygoid muscle.

### Auriculotemporal nerve

The auriculotemporal nerve is the first branch of the posterior trunk of the mandibular nerve [$V_3$] and originates as two roots, which pass posteriorly around the middle meningeal artery ascending from the maxillary artery to the foramen spinosum (Fig. 8.149).

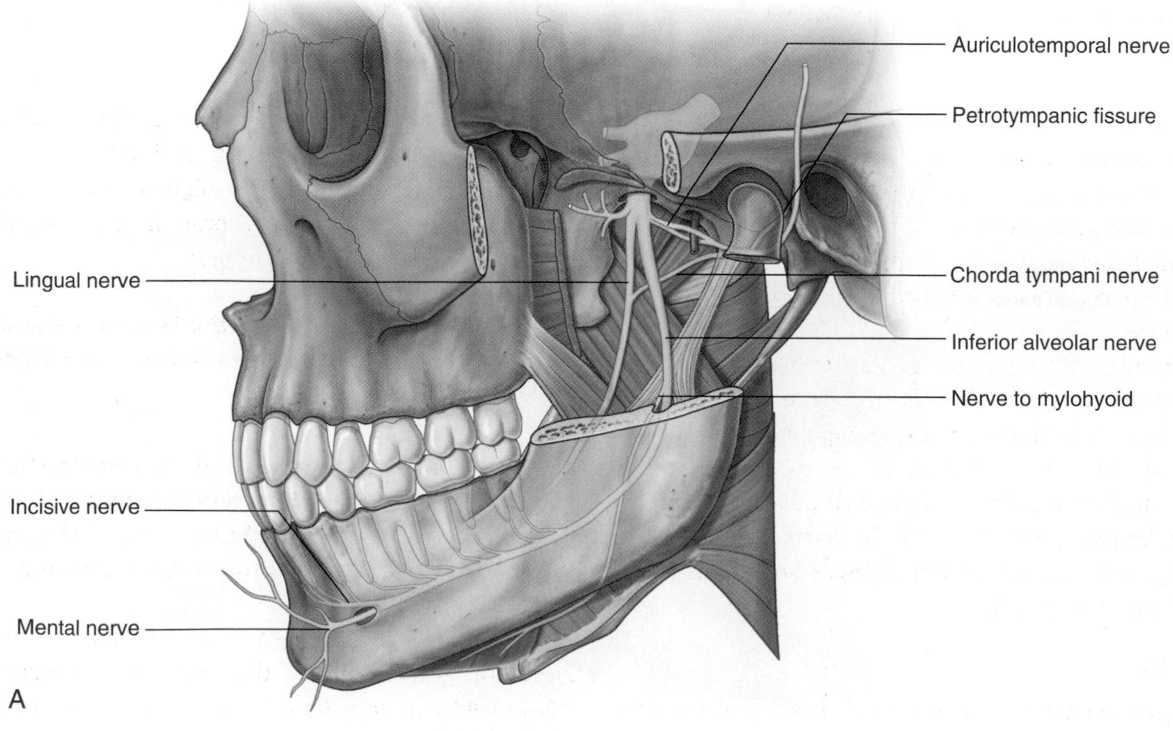


**Fig. 8.149** Mandibular nerve [$V_3$]—posterior trunk. **A.** Lateral view.

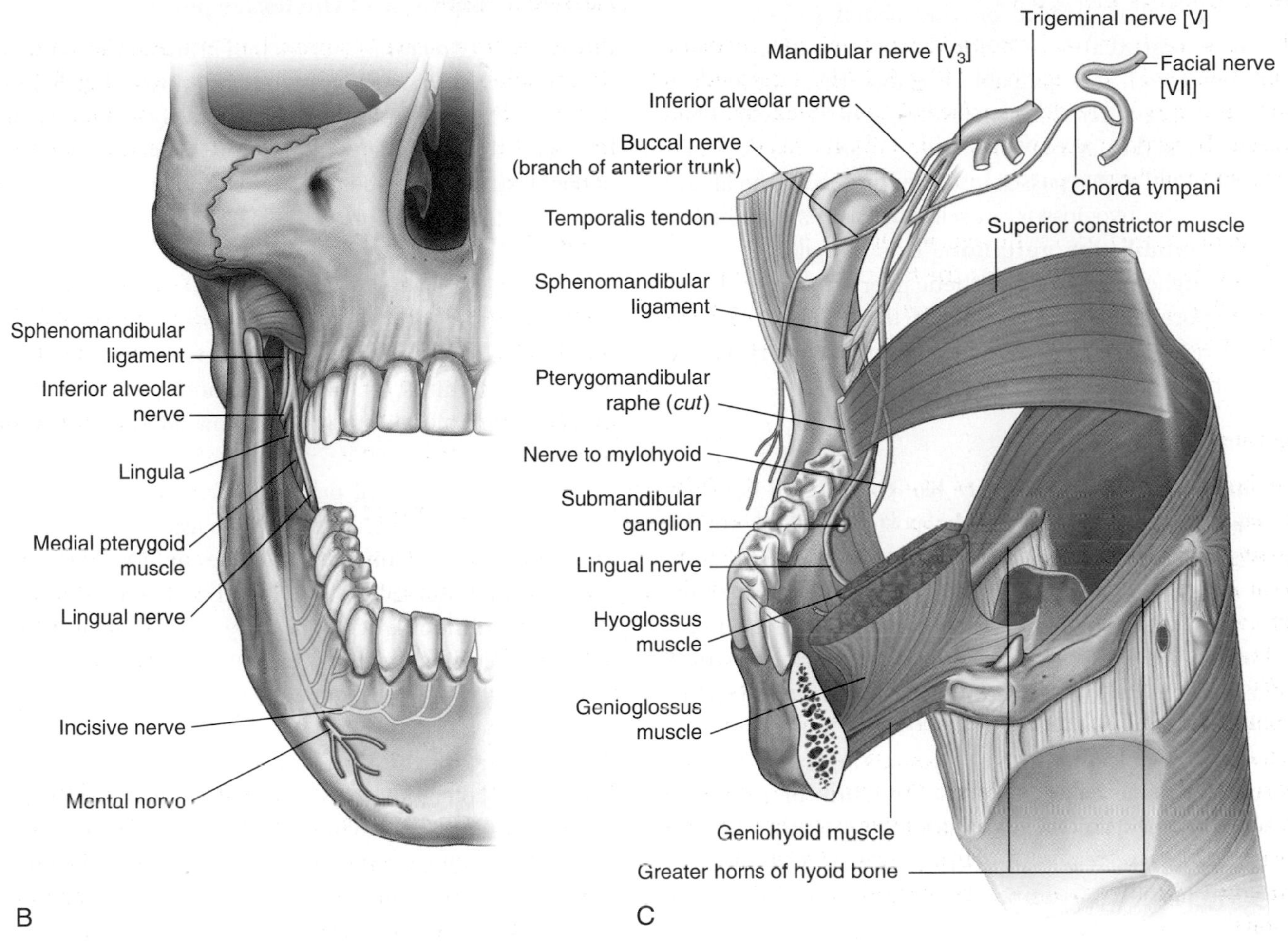


**Fig. 8.149, cont'd** B. Anterior view. C. Anteromedial view.

The auriculotemporal nerve passes first between the tensor veli palatini muscle and the upper head of the lateral pterygoid muscle, and then between the sphenomandibular ligament and the neck of the mandible. It curves laterally around the neck of the mandible and then ascends deep to the parotid gland between the temporomandibular joint and ear.

The terminal branches of the auriculotemporal nerve carry general sensation from skin over a large area of the temple. In addition, the auriculotemporal nerve contributes to sensory innervation of the external ear, the external auditory meatus, tympanic membrane, and temporomandibular joint. It also delivers postganglionic parasympathetic nerves from the glossopharyngeal nerve [IX] to the parotid gland.

## Lingual nerve

The **lingual nerve** is a major sensory branch of the posterior trunk of the mandibular nerve [$V_3$] (Fig. 8.149A,B). It carries general sensation from the anterior two-thirds of the tongue, oral mucosa on the floor of the oral cavity, and lingual gingivae associated with the lower teeth.

The lingual nerve is joined high in the infratemporal fossa by the chorda tympani branch of the facial nerve [VII] (Fig. 8.149C), which carries:

- taste from the anterior two-thirds of the tongue, and
- parasympathetic fibers to all salivary glands below the level of the oral fissure.

The lingual nerve first descends between the tensor veli palatini muscle and the lateral pterygoid muscle, where it is joined by the chorda tympani nerve, and then descends across the lateral surface of the medial pterygoid muscle to enter the oral cavity.

The lingual nerve enters the oral cavity between the posterior attachment of the mylohyoid muscle to the mylohyoid line and the attachment of the superior constrictor of the pharynx to the pterygomandibular raphe. As the lingual nerve enters the floor of the oral cavity,

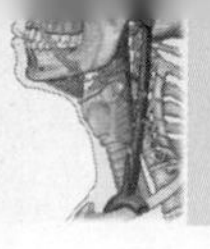



it is in a shallow groove on the medial surface of the mandible immediately inferior to the last molar tooth. In this position, it is palpable through the oral mucosa and in danger when one is operating on the molar teeth and gingivae (Fig. 8.149C).

The lingual nerve passes into the tongue on the lateral surface of the hyoglossus muscle where it is attached to the **submandibular ganglion**. This ganglion is where the preganglionic parasympathetic fibers carried from the infratemporal fossa into the floor of the oral cavity on the lingual nerve synapse with postganglionic parasympathetic fibers (see Fig. 8.150).

### Inferior alveolar nerve

The **inferior alveolar nerve**, like the lingual nerve, is a major sensory branch of the posterior trunk of the mandibular nerve [$V_3$] (Fig. 8.149A–C). In addition to innervating all lower teeth and much of the associated gingivae, it also supplies the mucosa and skin of the lower lip and skin of the chin. It has one motor branch, which innervates the mylohyoid muscle and the anterior belly of the digastric muscle.

The inferior alveolar nerve originates deep to the lateral pterygoid muscle from the posterior trunk of the mandibular nerve [$V_3$] in association with the lingual nerve. It descends on the lateral surface of the medial pterygoid muscle, passes between the sphenomandibular ligament and the ramus of the mandible, and then enters the mandibular canal through the mandibular foramen. Just before entering the mandibular foramen, it gives origin to the **nerve to the mylohyoid** (Fig. 8.149C), which lies in the mylohyoid groove inferior to the foramen and continues anteriorly below the floor of the oral cavity to innervate the mylohyoid muscle and the anterior belly of the digastric muscle.

The inferior alveolar nerve passes anteriorly within the mandibular canal of the lower jaw. The mandibular canal and its contents are inferior to the roots of the molar teeth, and the roots can sometimes curve around the canal making extraction of these teeth difficult.

The inferior alveolar nerve supplies branches to the three molar teeth and the second premolar tooth and associated labial gingivae, and then divides into its two terminal branches:

- the **incisive nerve**, which continues in the mandibular canal to supply the first premolar, incisor, and canine teeth, and related gingivae; and
- the **mental nerve**, which exits the mandible through the mental foramen and supplies the lower lip and chin (Fig. 8.149A,B). The mental nerve is palpable and sometimes visible through the oral mucosa adjacent to the roots of the premolar teeth.



## Chorda tympani and the lesser petrosal nerve

Branches of two cranial nerves join branches of the mandibular nerve [$V_3$] in the infratemporal fossa (Fig. 8.150). These are the chorda tympani branch of the facial nerve [VII] and the lesser petrosal nerve, a branch of the tympanic plexus in the middle ear, which had its origin from a branch of the glossopharyngeal nerve [IX] (see Fig. 8.125, p. 953).

### Chorda tympani

The chorda tympani (Fig. 8.150) carries taste from the anterior two-thirds of the tongue and parasympathetic innervation to all salivary glands below the level of the oral fissure.

The chorda tympani originates from the facial nerve [VII] within the temporal bone and in association with the mastoid wall of the middle ear, passes anteriorly through a small canal, and enters the lateral aspect of the middle ear. As it continues anterosuperiorly across the middle ear, it is separated from the tympanic membrane by the handle of the malleus. It leaves the middle ear through the medial end of the petrotympanic fissure, enters the infratemporal fossa, descends medial to the spine of the sphenoid and then to the lateral pterygoid muscle, and joins the lingual nerve.

Preganglionic parasympathetic fibers carried in the chorda tympani synapse with postganglionic parasympathetic fibers in the submandibular ganglion, which "hangs off" the lingual nerve in the floor of the oral cavity (Fig. 8.150).

Postganglionic parasympathetic fibers leave the submandibular ganglion and either:

- reenter the lingual nerve to travel with its terminal branches to reach target tissues, or
- pass directly from the submandibular ganglion into glands (Fig. 8.150).

The taste (SA) fibers do not pass through the ganglion and are distributed with terminal branches of the lingual nerve.

### In the clinic

#### Lingual nerve injury

A lingual nerve injury proximal to where the chorda tympani joins it in the infratemporal fossa will produce loss of general sensation from the anterior two-thirds of the tongue, oral mucosa, gingivae, the lower lip, and the chin.

If a lingual nerve lesion is distal to the site where it is joined by the chorda tympani, secretion from the salivary glands below the oral fissure and taste from the anterior two-thirds of the tongue will also be lost.

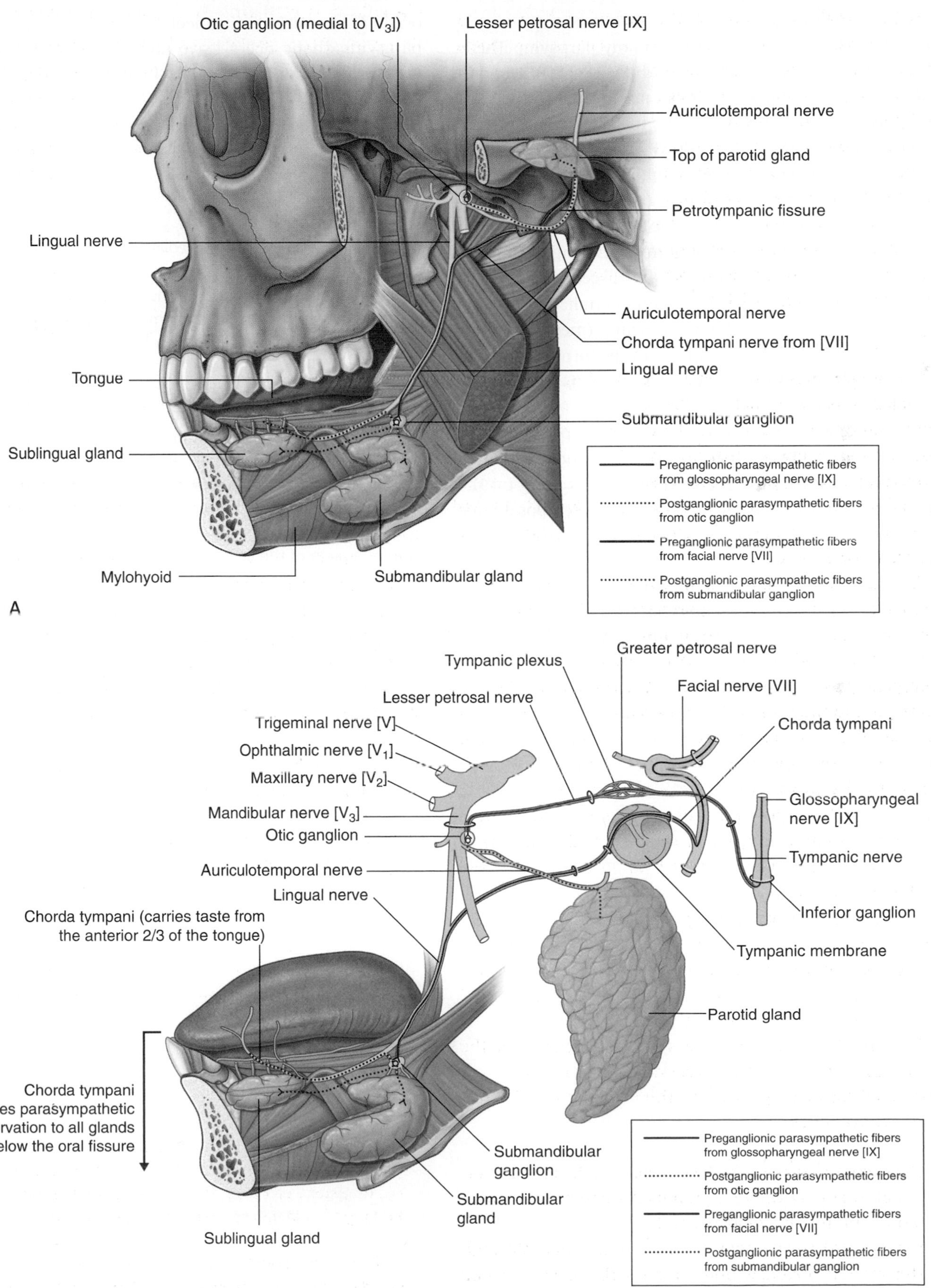
Otic ganglion (medial to [V3])
Lesser petrosal nerve [IX]
Auriculotemporal nerve
Top of parotid gland
Petrotympanic fissure
Lingual nerve
Auriculotemporal nerve
Chorda tympani nerve from [VII]
Lingual nerve
Tongue
Submandibular ganglion
Sublingual gland
Preganglionic parasympathetic fibers from glossopharyngeal nerve [IX]
Postganglionic parasympathetic fibers from otic ganglion
Preganglionic parasympathetic fibers from facial nerve [VII]
Postganglionic parasympathetic fibers from submandibular ganglion
Mylohyoid
Submandibular gland
A
Greater petrosal nerve
Tympanic plexus
Facial nerve [VII]
Lesser petrosal nerve
Trigeminal nerve [V]
Chorda tympani
Ophthalmic nerve [V1]
Maxillary nerve [V2]
Glossopharyngeal nerve [IX]
Mandibular nerve [V3]
Otic ganglion
Tympanic nerve
Auriculotemporal nerve
Lingual nerve
Inferior ganglion
Chorda tympani (carries taste from the anterior 2/3 of the tongue)
Tympanic membrane
Parotid gland
Chorda tympani carries parasympathetic innervation to all glands below the oral fissure
Submandibular ganglion
Submandibular gland
Sublingual gland
Preganglionic parasympathetic fibers from glossopharyngeal nerve [IX]
Postganglionic parasympathetic fibers from otic ganglion
Preganglionic parasympathetic fibers from facial nerve [VII]
Postganglionic parasympathetic fibers from submandibular ganglion
B

**Fig. 8.150** Chorda tympani and lesser petrosal nerves. **A.** Course after emerging from the skull. **B.** Course of parasympathetic fibers.

### Lesser petrosal nerve

The lesser petrosal nerve carries mainly parasympathetic fibers destined for the parotid gland (Fig. 8.150). The preganglionic parasympathetic fibers are located in the glossopharyngeal nerve [IX] as it exits the jugular foramen at the base of the skull. Branching from the glossopharyngeal nerve [IX] either within or immediately outside the jugular foramen is the tympanic nerve (Fig. 8.150B).

The tympanic nerve reenters the temporal bone through a small foramen on the ridge of bone separating the jugular foramen from the carotid canal and ascends through a small bony canal (inferior tympanic canaliculus) to the promontory located on the labyrinthine (medial) wall of the middle ear. Here it participates in the formation of the tympanic plexus. The lesser petrosal nerve is a branch of this plexus (Fig. 8.150B).

The lesser petrosal nerve contains mainly preganglionic parasympathetic fibers. It leaves the middle ear and enters the middle cranial fossa through a small opening on the anterior surface of the petrous part of the temporal bone just lateral and inferior to the opening for the greater petrosal nerve, a branch of the facial nerve [VII]. The lesser petrosal nerve then passes medially and descends through the foramen ovale with the mandibular nerve [$V_3$].

In the infratemporal fossa, the preganglionic parasympathetic fibers synapse with cell bodies of postganglionic parasympathetic fibers in the otic ganglion located on the medial side of the mandibular nerve [$V_3$] around the origin of the nerve to the medial pterygoid. Postganglionic parasympathetic fibers leave the otic ganglion and join the auriculotemporal nerve, which carries them to the parotid gland.

### In the clinic

#### Dental anesthesia

Anesthesia of the inferior alveolar nerve is widely practiced by most dentists. The inferior alveolar nerve is one of the largest branches of the mandibular nerve [$V_3$], carries the sensory branches from the teeth and mandible, and receives sensory information from the skin over most of the mandible.

The inferior alveolar nerve passes into the mandibular canal, courses through the body of the mandible, and eventually emerges through the mental foramen into the chin.

Dental procedures require perineuronal infiltration of the inferior alveolar nerve by local anesthetic. To anesthetize this nerve the needle is placed lateral to the anterior arch of the fauces (palatoglossal arch) in the oral cavity and is advanced along the medial border around the inferior third of the ramus of the mandible so that anesthetic can be deposited in this region.

It is also possible to anesthetize the infra-orbital and buccal nerves, depending on where the anesthesia is needed.

### Maxillary artery

The maxillary artery is the largest branch of the external carotid artery in the neck and is a major source of blood supply for the nasal cavity, the lateral wall and roof of the oral cavity, all teeth, and the dura mater in the cranial cavity. It passes through and supplies the infratemporal fossa and then enters the pterygopalatine fossa, where it gives origin to terminal branches (Fig. 8.151).

The maxillary artery originates within the substance of the parotid gland and then passes forward, between the neck of the mandible and sphenomandibular ligament, into the infratemporal fossa. It ascends obliquely through the infratemporal fossa to enter the pterygopalatine fossa by passing through the pterygomaxillary fissure. This part of the vessel may pass either lateral or medial to the lower head of the lateral pterygoid. If it passes medial to the lower head, the maxillary artery then loops laterally between the upper and lower heads of the lateral pterygoid to access the **pterygomaxillary fissure**.

#### Branches

Branches of the maxillary artery are as follows (Fig. 8.151):

- The first part of the maxillary artery (the part between the neck of the mandible and the sphenomandibular ligament) gives origin to two major branches (the middle meningeal and inferior alveolar arteries) and a number of smaller branches (deep auricular, anterior tympanic, and accessory meningeal).
- The second part of the maxillary artery (the part related to the lateral pterygoid muscle) gives origin to deep temporal, masseteric, buccal, and pterygoid branches, which course with branches of the mandibular nerve [$V_3$].
- The third part of the maxillary artery is in the pterygopalatine fossa (see Fig. 8.158)

#### Middle meningeal artery

The middle meningeal artery ascends vertically from the maxillary artery and passes through the foramen spinosum to enter the cranial cavity (Fig. 8.151). In the infratemporal fossa, it passes superiorly between the sphenomandibular ligament on the medial side and the lateral pterygoid muscle on the lateral side. Just inferior to the foramen spinosum, it passes between the two roots of the auriculotemporal nerve at their origin from the mandibular nerve [$V_3$] (Fig. 8.151).

The middle meningeal artery is the largest of the meningeal vessels and supplies much of the dura mater, bone, and related bone marrow of the cranial cavity walls.

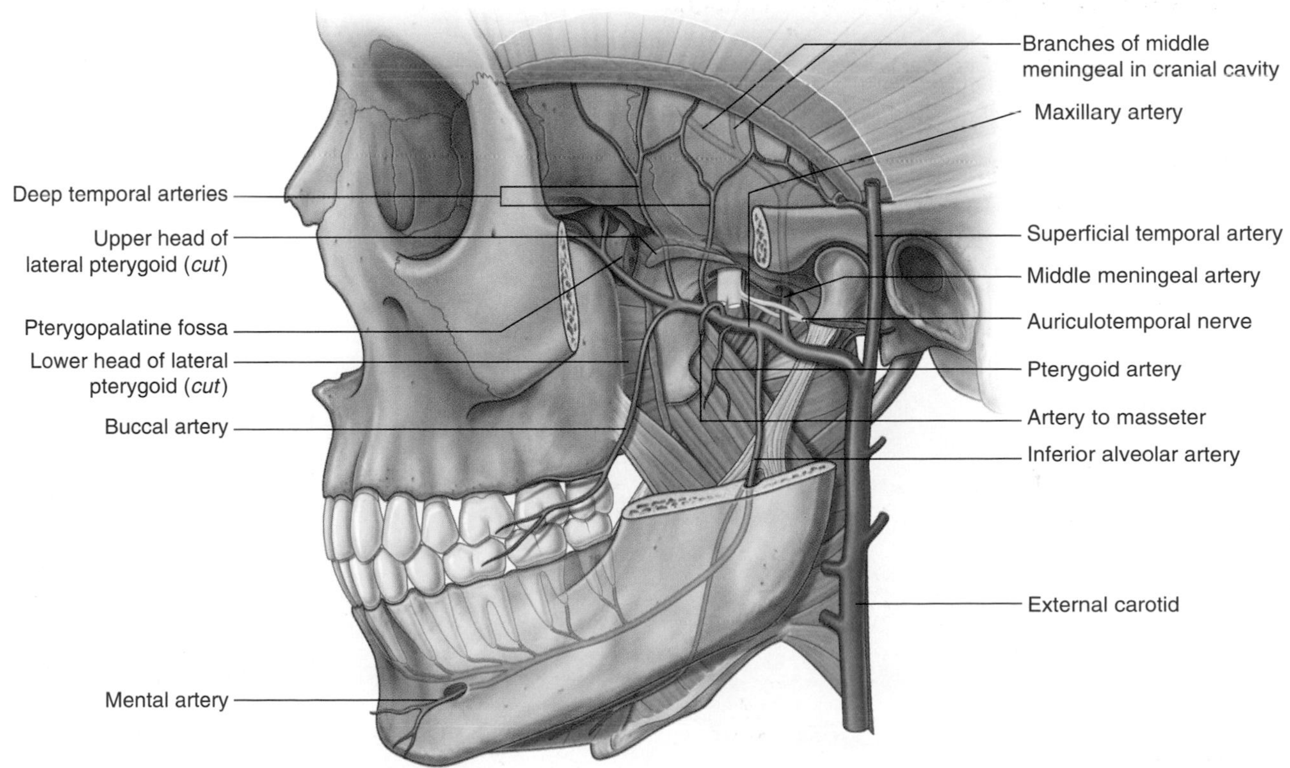


**Fig. 8.151** Maxillary artery.

Within the cranial cavity, the middle meningeal artery and its branches travel in the periosteal (outer) layer of dura mater, which is tightly adherent to the bony walls. As major branches of the middle meningeal artery pass superiorly up the walls of the cranial cavity, they can be damaged by lateral blows to the head. When the vessels are torn, the leaking blood, which is under arterial pressure, slowly separates the dura mater from its attachment to the bone, resulting in an extradural hematoma.

### Inferior alveolar artery

The inferior alveolar artery descends from the maxillary artery to enter the mandibular foramen and canal with the inferior alveolar nerve (Fig. 8.151). It is distributed with the inferior alveolar nerve and supplies all lower teeth, and contributes to the supply of the buccal gingivae, chin, and lower lip.

Before entering the mandible, the inferior alveolar artery gives origin to a small mylohyoid branch, which accompanies the nerve to the mylohyoid.

### Deep auricular, anterior tympanic, and accessory meningeal arteries

The deep auricular, anterior tympanic, and accessory meningeal arteries are small branches from the first part of the maxillary artery and contribute to the blood supply of the external acoustic meatus, deep surface of the tympanic membrane, and cranial dura mater, respectively.

The accessory meningeal branch also contributes small branches to surrounding muscles in the infratemporal fossa before ascending through the foramen ovale into the cranial cavity to supply the dura mater.

### Branches from the second part

Deep temporal arteries, usually two in number, originate from the second part of the maxillary artery and travel with the deep temporal nerves to supply the temporalis muscle in the temporal fossa (Fig. 8.151).

Numerous pterygoid arteries also originate from the second part of the maxillary artery and supply the pterygoid muscles.

The masseteric artery, also from the second part of the maxillary artery, accompanies the masseteric nerve laterally through the mandibular notch to supply the masseter muscle.

The buccal artery is distributed with the buccal nerve and supplies skin, muscle, and oral mucosa of the cheek.

## Pterygoid plexus

The **pterygoid plexus** is a network of veins between the medial and lateral pterygoid muscles, and between the lateral pterygoid and temporalis muscles (Fig. 8.152).

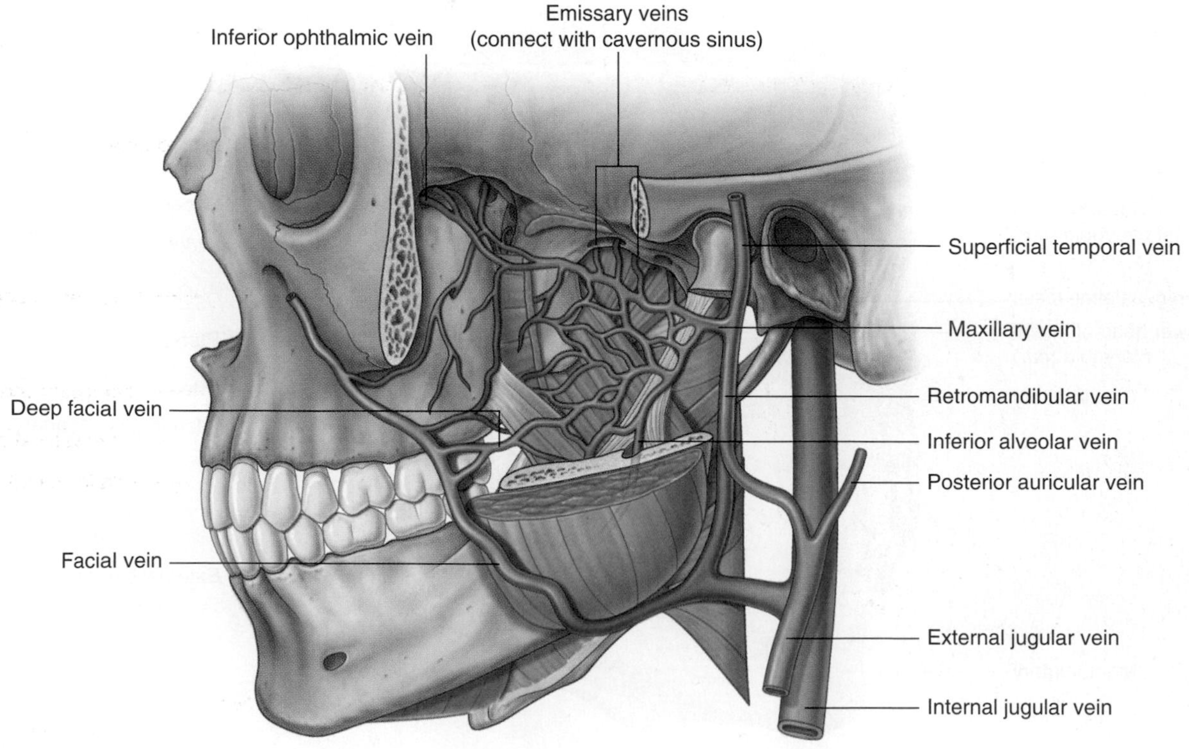


**Fig. 8.152** Pterygoid plexus of veins.

Veins that drain regions supplied by arteries branching from the maxillary artery in the infratemporal fossa and pterygopalatine fossa connect with the pterygoid plexus. These tributary veins include those that drain the nasal cavity, roof and lateral wall of the oral cavity, all teeth, muscles of the infratemporal fossa, paranasal sinuses, and nasopharynx. In addition, the inferior ophthalmic vein from the orbit can drain through the inferior orbital fissure into the pterygoid plexus.

Significantly, small emissary veins often connect the pterygoid plexus in the infratemporal fossa to the cavernous sinus in the cranial cavity. These emissary veins, which pass through the foramen ovale, through the cartilage that fills the foramen lacerum, and through a small sphenoidal foramen on the medial side of the lateral plate of the pterygoid process at the base of the skull, are a route by which infections can spread into the cranial cavity from structures, such as the teeth, that are drained by the pterygoid plexus. Also, because there are no valves in veins of the head and neck, anesthetic inadvertently injected under pressure into veins of the pterygoid plexus can backflow into tissues or into the cranial cavity.

The pterygoid plexus connects:

- posteriorly, via a short maxillary vein, with the retromandibular vein in the neck; and
- anteriorly, via a deep facial vein, with the facial vein on the face.

## PTERYGOPALATINE FOSSA

The pterygopalatine fossa is an inverted teardrop-shaped space between bones on the lateral side of the skull immediately posterior to the maxilla (Fig. 8.153).

Although small in size, the pterygopalatine fossa communicates via fissures and foramina in its walls with the:

- middle cranial fossa,
- infratemporal fossa,
- floor of the orbit,
- lateral wall of the nasal cavity,
- oropharynx, and
- roof of the oral cavity.

Because of its strategic location, the pterygopalatine fossa is a major site of distribution for the maxillary nerve [$V_2$] and for the terminal part of the maxillary artery. It also contains the pterygopalatine ganglion where preganglionic parasympathetic fibers originating in the facial nerve [VII] synapse with postganglionic parasympathetic fibers and these fibers, along with sympathetic fibers originating from the T1 spinal cord level join branches of the maxillary nerve [$V_2$].

All the upper teeth receive their innervation and blood supply from the maxillary nerve [$V_2$] and the terminal part of the maxillary artery, respectively, that pass through the pterygopalatine fossa.

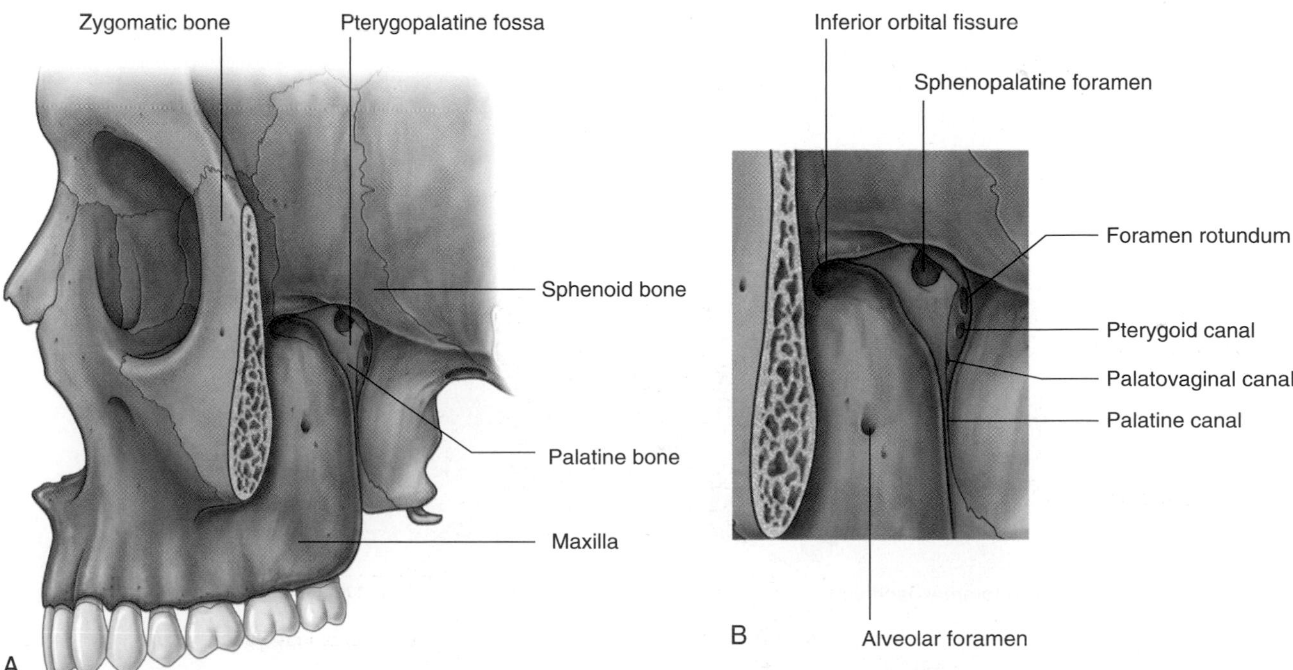


**Fig. 8.153** Pterygopalatine fossa. **A.** Anterolateral view. **B.** Lateral view.

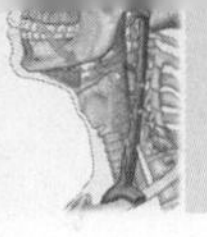



## Skeletal framework

The walls of the pterygopalatine fossa are formed by parts of the palatine, maxilla, and sphenoid bones (Fig. 8.153):

- The anterior wall is formed by the posterior surface of the maxilla.
- The medial wall is formed by the lateral surface of the palatine bone.
- The posterior wall and roof are formed by parts of the sphenoid bone.

### Sphenoid bone

The part of the sphenoid bone that contributes to the formation of the pterygopalatine fossa is the anterosuperior surface of the pterygoid process (Fig. 8.154). Opening onto this surface are two large foramina:

- The maxillary nerve [$V_2$] passes through the most lateral and superior of these—**the foramen rotundum**—which communicates posteriorly with the middle cranial fossa (Fig. 8.154B).
- The greater petrosal nerve from the facial nerve [VII] and sympathetic fibers from the internal carotid plexus join to form the nerve of the pterygoid canal that passes forward into the pterygopalatine fossa through the more medial and inferior foramen—**the anterior opening of the pterygoid canal**.

#### Pterygoid canal

The **pterygoid canal** (Fig. 8.154A) is a bony canal running horizontally through the root of the pterygoid process of the sphenoid bone. It opens anteriorly into the pterygopalatine fossa. Posteriorly it continues through the cartilage filling the foramen lacerum and opens into

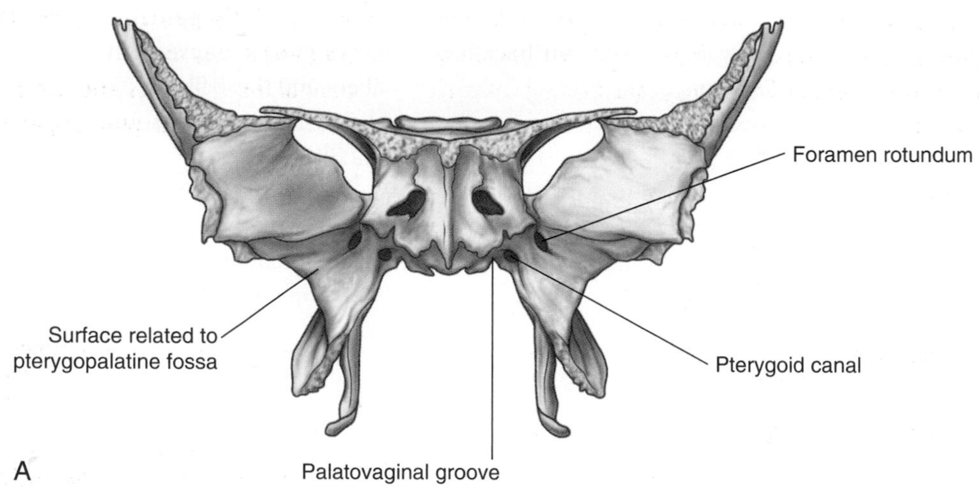


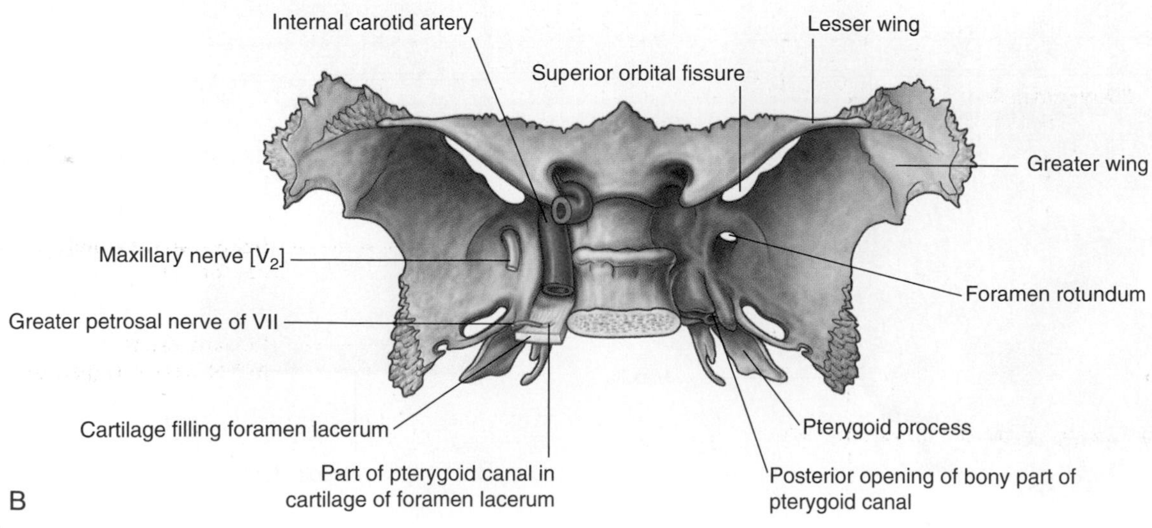


**Fig. 8.154** Sphenoid bone. **A.** Anterior view. **B.** Posterosuperior view.

the middle cranial fossa just anteroinferior to the internal carotid artery as the vessel enters the cranial cavity through the carotid canal (Fig. 8.154B).

## Gateways

Seven foramina and fissures provide apertures through which structures enter and leave the pterygopalatine fossa (Fig. 8.155):

- The foramen rotundum and pterygoid canal communicate with the middle cranial fossa and open onto the posterior wall.
- A small **palatovaginal canal** opens onto the posterior wall and leads to the nasopharynx.
- The palatine canal leads to the roof of the oral cavity (hard palate) and opens inferiorly.
- The sphenopalatine foramen opens onto the lateral wall of the nasal cavity and is in the medial wall.
- The lateral aspect of the pterygopalatine fossa is continuous with the infratemporal fossa via a large gap (the **pterygomaxillary fissure**) between the posterior surface of the maxilla and pterygoid process of the sphenoid bone.
- The superior aspect of the anterior wall of the fossa opens into the floor of the orbit via the inferior orbital fissure.

## Contents

The maxillary nerve [$V_2$] and terminal part of the maxillary artery enter and branch within the pterygopalatine fossa. In addition, the nerve of the pterygoid canal enters the fossa carrying:

- preganglionic parasympathetic fibers from the greater petrosal branch of the facial nerve [VII], and
- postganglionic sympathetic fibers from the deep petrosal branch of the carotid plexus.

The preganglionic parasympathetic fibers synapse in the pterygopalatine ganglion and both the sympathetic and postganglionic parasympathetic fibers pass with branches of the maxillary nerve [$V_2$] out of the fossa and into adjacent regions.

In addition to nerves and arteries, veins and lymphatics also pass through the pterygopalatine fossa.

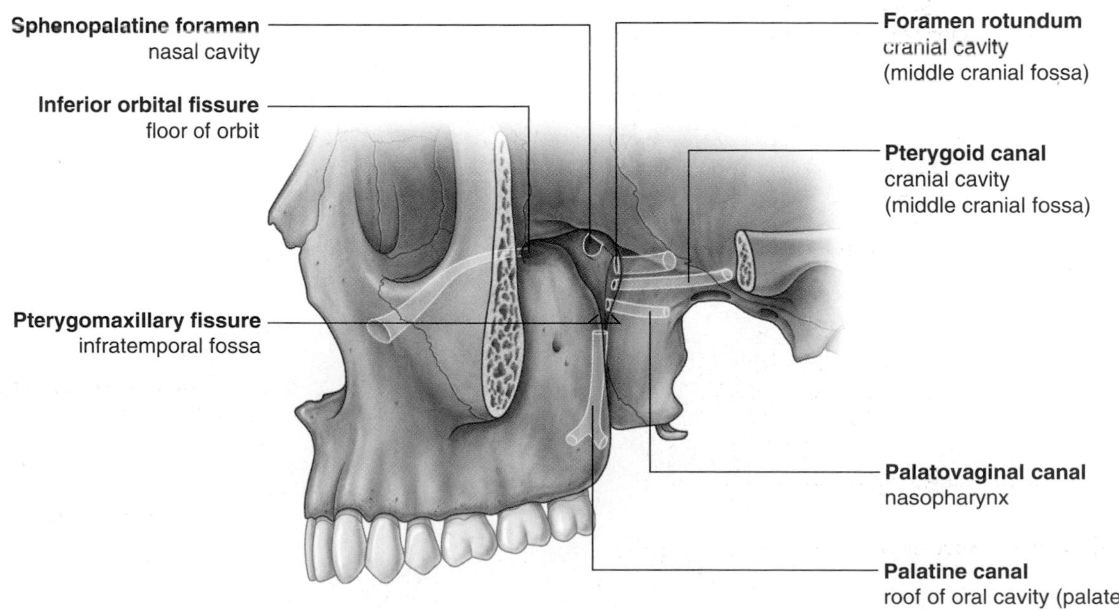


**Fig. 8.155** Gateways of the pterygopalatine fossa.

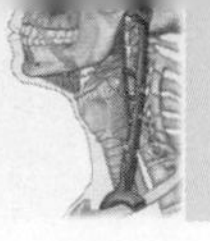



## Maxillary nerve [$V_2$]

The maxillary nerve [$V_2$] is purely sensory. It originates from the trigeminal ganglion in the cranial cavity, exits the middle cranial fossa, and enters the pterygopalatine fossa through the foramen rotundum (Fig. 8.156). It passes anteriorly through the fossa and exits as the infra-orbital nerve through the inferior orbital fissure.

While passing through the pterygopalatine fossa, the maxillary nerve [$V_2$] gives rise to the zygomatic nerve, the posterior superior alveolar nerve, and two ganglionic branches (Fig. 8.156). The two ganglionic branches originate from its inferior surface and pass through the pterygopalatine ganglion.

Postganglionic parasympathetic fibers, arising in the pterygopalatine ganglion, join the general sensory branches of the maxillary nerve [$V_2$] in the pterygopalatine ganglion, as do postganglionic sympathetic fibers from the carotid plexus. The three types of fibers leave the ganglion as orbital, palatine, nasal, and pharyngeal branches.

### Branches

***Orbital branches.*** The **orbital branches** are small and pass through the inferior orbital fissure to contribute to the supply of the orbital wall and of the sphenoidal and ethmoidal sinuses.

***Greater and lesser palatine nerves.*** The **greater** and **lesser palatine nerves** (Fig. 8.156) pass inferiorly from the pterygopalatine ganglion, enter and pass through the palatine canal, and enter the oral surface of the palate through the greater and lesser palatine foramina.

The greater palatine nerve passes forward on the roof of the oral cavity to innervate mucosa and glands of the hard palate and the adjacent gingiva, almost as far forward as the incisor teeth.

In the palatine canal, the greater palatine nerve gives origin to **posterior inferior nasal nerves**, which pass medially through small foramina in the perpendicular plate of the palatine bone and contribute to the innervation of the lateral nasal wall.

After passing through the lesser palatine foramen, the lesser palatine nerve passes posteriorly to supply the soft palate.

***Nasal nerves.*** The nasal nerves (Fig. 8.156), approximately seven in number, pass medially through the sphenopalatine foramen to enter the nasal cavity. Most pass anteriorly to supply the lateral wall of the nasal cavity, while others pass across the roof to supply the medial wall.

One of the nerves passing across the roof to supply the medial wall of the nasal cavity (the **nasopalatine nerve**) is the largest of the nasal nerves and passes anteriorly

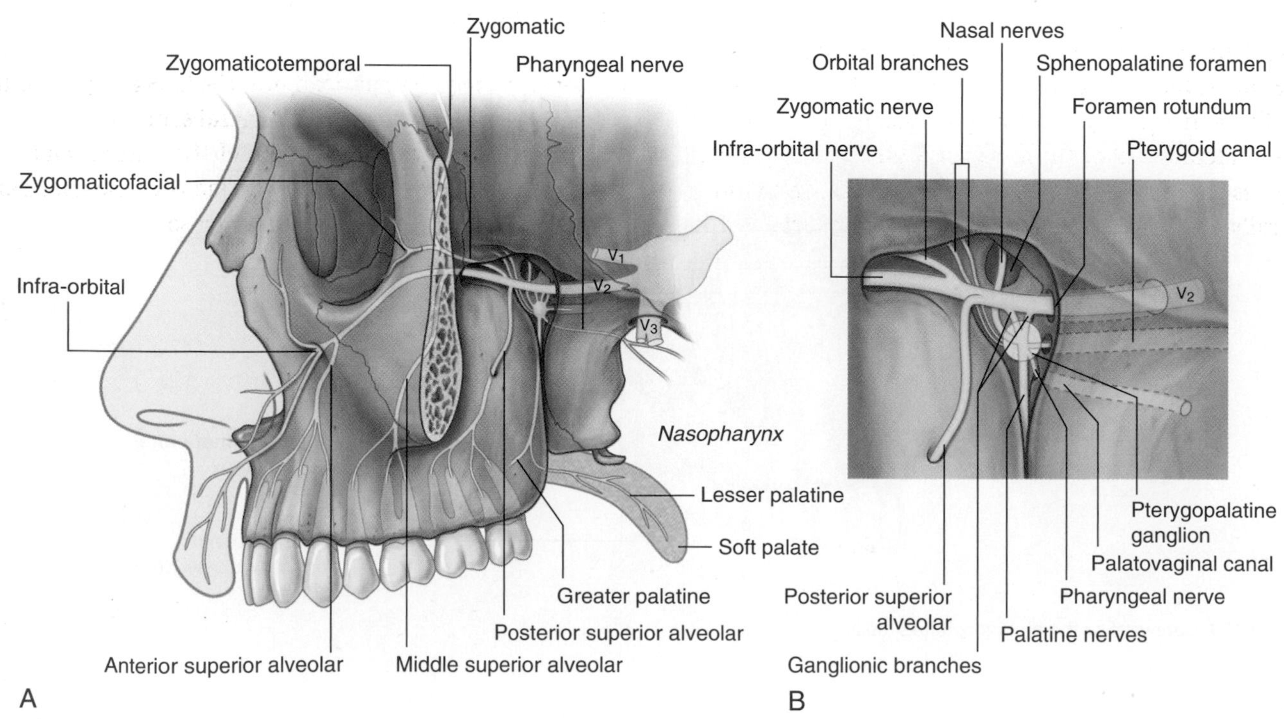


**Fig. 8.156** Maxillary nerve [$V_2$]. **A.** Terminal branches. **B.** In relationship to the pterygopalatine ganglion.

down the nasal septum, through the incisive canal and fossa in the hard palate to enter the roof of the oral cavity and supply mucosa, gingiva, and glands adjacent to the incisor teeth.

***Pharyngeal nerve.*** The **pharyngeal nerve** (Fig. 8.156) passes posteriorly from the pterygopalatine ganglion, and leaves the fossa through the palatovaginal canal, which it then exits to supply the mucosa and glands of the nasopharynx.

***Zygomatic nerve.*** The **zygomatic nerve** (Fig. 8.156) originates directly from the maxillary nerve [$V_2$] in the pterygopalatine fossa, which it leaves to enter the orbit through the inferior orbital fissure. It passes forward on the lateral orbital wall and divides into zygomaticotemporal and zygomaticofacial branches:

- The **zygomaticotemporal branch** continues forward at the base of the lateral orbital wall, passes through a small bony canal in the zygomatic bone to enter the temporal fossa through a small foramen in the lateral orbital margin on the posterior surface of the frontal process of the zygomatic bone, and passes superficially to supply skin over the temple.
- The **zygomaticofacial branch** also passes forward at the base of the lateral orbital wall and leaves through a small bony canal, in the orbital margin, which opens via multiple small foramina on the anterolateral surface of the zygomatic bone, and its branches supply the adjacent skin.

***Posterior superior alveolar nerve.*** The **posterior superior alveolar nerve** (Fig. 8.156) originates from the maxillary nerve [$V_2$] in the pterygopalatine fossa and passes laterally out of the fossa through the pterygomaxillary fissure to enter the infratemporal fossa. It continues laterally and inferiorly to enter the posterior surface of the maxilla through a small alveolar foramen approximately midway between the last molar tooth and the inferior orbital fissure. It then passes inferiorly just deep to the mucosa of the maxillary sinus to join the **superior dental plexus**.

The posterior superior alveolar nerve supplies the molar teeth and adjacent buccal gingivae, and contributes to the supply of the maxillary sinus.

***Infra-orbital nerve.*** The infra-orbital nerve (Fig. 8.156) is the anterior continuation of the maxillary nerve [$V_2$] that leaves the pterygopalatine fossa through the inferior orbital fissure. It lies first in the infra-orbital groove in the floor of the orbit and then continues forward in the infra-orbital canal.

While in the infra-orbital groove and canal, the infra-orbital nerve gives origin to **middle** and **anterior superior alveolar nerves**, respectively, which ultimately join the **superior alveolar plexus** to supply the upper teeth:

- The middle superior alveolar nerve also supplies the maxillary sinus.
- The anterior superior alveolar nerve also gives origin to a small nasal branch, which passes medially through the lateral wall of the nasal cavity to supply parts of the areas of the nasal floor and walls.

The infra-orbital nerve exits the infra-orbital canal through the infra-orbital foramen inferior to the orbital margin and divides into nasal, palpebral, and superior labial branches:

- Nasal branches supply skin over the lateral aspect of the external nose and part of the nasal septum.
- Palpebral branches supply skin of the lower eyelid.
- Superior labial branches supply skin over the cheek and upper lip, and the related oral mucosa.

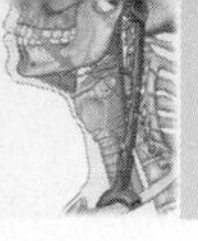



### Nerve of the pterygoid canal and the pterygopalatine ganglion

The nerve of the pterygoid canal (Fig. 8.157) is formed in the middle cranial fossa by the union of:

- the greater petrosal nerve (a branch of the facial nerve [VII]), and
- the deep petrosal nerve (a branch of the internal carotid plexus).

The nerve of the pterygoid canal passes into the pterygopalatine fossa and joins the pterygopalatine ganglion. It carries mainly preganglionic parasympathetic and postganglionic sympathetic fibers.

#### Greater petrosal nerve

The greater petrosal nerve, which originates from the geniculate ganglion of the facial nerve [VII] in the temporal bone, exits the temporal bone through a small canal that opens via a fissure onto the anterior surface of the petrous part of the temporal bone. It passes anteromedially along the posterior margin of the middle cranial fossa and then under the internal carotid artery to reach the superior surface of the cartilage filling the foramen lacerum.

As the greater petrosal nerve passes under the internal carotid artery, it is joined by the deep petrosal nerve to form the nerve of the pterygoid canal.

The greater petrosal nerve carries parasympathetic innervation to all glands above the oral fissure, including:

- mucous glands in the nasal cavity,
- salivary glands in the upper half of the oral cavity, and
- the lacrimal gland in the orbit.

The greater petrosal nerve also carries some taste (SA) fibers from the soft palate in the lesser palatine nerve.

#### Deep petrosal nerve

The **deep petrosal nerve** is formed by postganglionic sympathetic fibers that originate in the **superior cervical sympathetic ganglion** in the neck and leave the ganglion as the **internal carotid nerve**.

Preganglionic fibers that synapse in the ganglion are from the T1 spinal nerve.

The internal carotid nerve forms the internal carotid plexus around the internal carotid artery as the internal carotid artery passes through the skull and into the cranial cavity. Some of the fibers from the internal carotid plexus converge to form the deep petrosal nerve, which leaves the internal carotid plexus in the middle cranial fossa and joins the greater petrosal branch of the facial nerve [VII].

The deep petrosal nerve carries postganglionic sympathetic fibers destined mainly for blood vessels.

#### Pterygopalatine ganglion

The nerve of the pterygoid canal enters the superior surface of the cartilage that fills the foramen lacerum and passes anteriorly through the cartilage to enter the pterygoid canal in the root of the pterygoid process of the sphenoid bone. It passes through the canal and into the

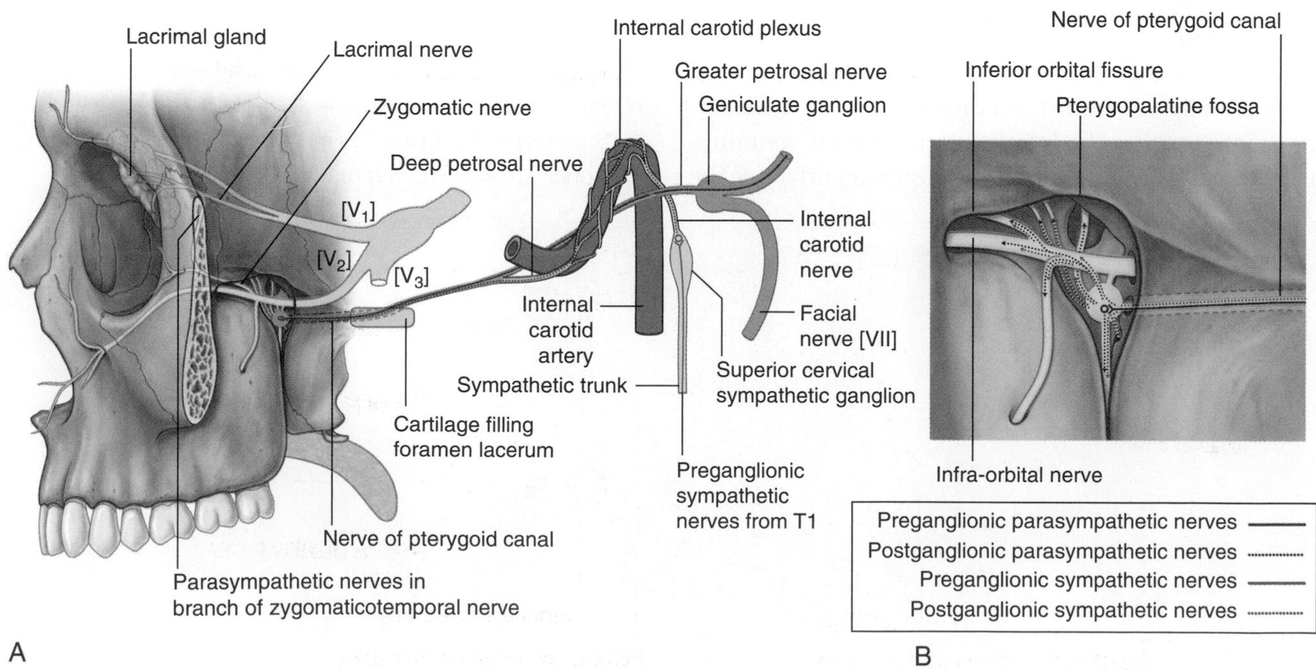


**Fig. 8.157** Nerve of the pterygoid canal. **A.** Overview. **B.** In relationship to the pterygopalatine ganglion.

pterygopalatine fossa where it joins the pterygopalatine ganglion formed around the branches of the maxillary nerve [$V_2$] (Fig. 8.157).

The **pterygopalatine ganglion** is the largest of the four parasympathetic ganglia in the head and is formed by the cell bodies of the postganglionic neurons associated with preganglionic parasympathetic fibers of the facial nerve [VII] carried by the greater petrosal nerve and the nerve of the pterygoid canal.

The postganglionic parasympathetic fibers that originate in the pterygopalatine ganglion, together with postganglionic sympathetic fibers passing through the ganglion, join fibers from the ganglionic branches of the maxillary nerve [$V_2$] to form orbital, palatine, nasal, and pharyngeal branches, which leave the ganglion.

Other postganglionic parasympathetic and sympathetic fibers pass superiorly through the ganglionic branches of the maxillary nerve [$V_2$] to enter the main trunk of the maxillary nerve and be distributed with the zygomatic, posterior superior alveolar, and infra-orbital nerves. Of these, the postganglionic parasympathetic and sympathetic fibers that pass into the orbit with the zygomatic nerve are particularly important because they ultimately innervate the lacrimal gland.

### Innervation of the lacrimal gland

Approximately midway along the orbital wall, the postganglionic parasympathetic and sympathetic fibers leave the zygomaticotemporal branch of the zygomatic nerve and form a special autonomic nerve, which travels up the lateral orbital wall to join the lacrimal nerve (Fig. 8.157; also see Fig. 8.87).

The lacrimal nerve is a major general sensory branch of the ophthalmic nerve [$V_1$], which passes forward in the orbit at the margin between the lateral wall and roof.

The postganglionic parasympathetic and sympathetic fibers pass with the lacrimal nerve to the lacrimal gland.

A lesion anywhere along the course of parasympathetic fibers that leave the brain as part of the facial nerve [VII] and are ultimately carried to the lacrimal gland along branches of the ophthalmic nerve [$V_1$] results in "dry eye" and can eventually lead to loss of vision in the affected eye.

## Maxillary artery

The maxillary artery is a major branch of the external carotid artery in the neck. It originates adjacent to the neck of the mandible, passes forward through the infratemporal fossa, and then enters the pterygopalatine fossa through the pterygomaxillary fissure (Fig. 8.158).

The part of the maxillary artery in the pterygopalatine fossa (the third part) is anterior to the pterygopalatine ganglion and gives origin to branches that accompany branches of the maxillary nerve [$V_2$] and the pterygopalatine ganglion.

Branches of the maxillary artery include the posterior superior alveolar, infra-orbital, greater palatine, pharyngeal, and sphenopalatine arteries, and the artery of the pterygoid canal (Fig. 8.158). Collectively, these branches supply much of the nasal cavity, the roof of the oral cavity,

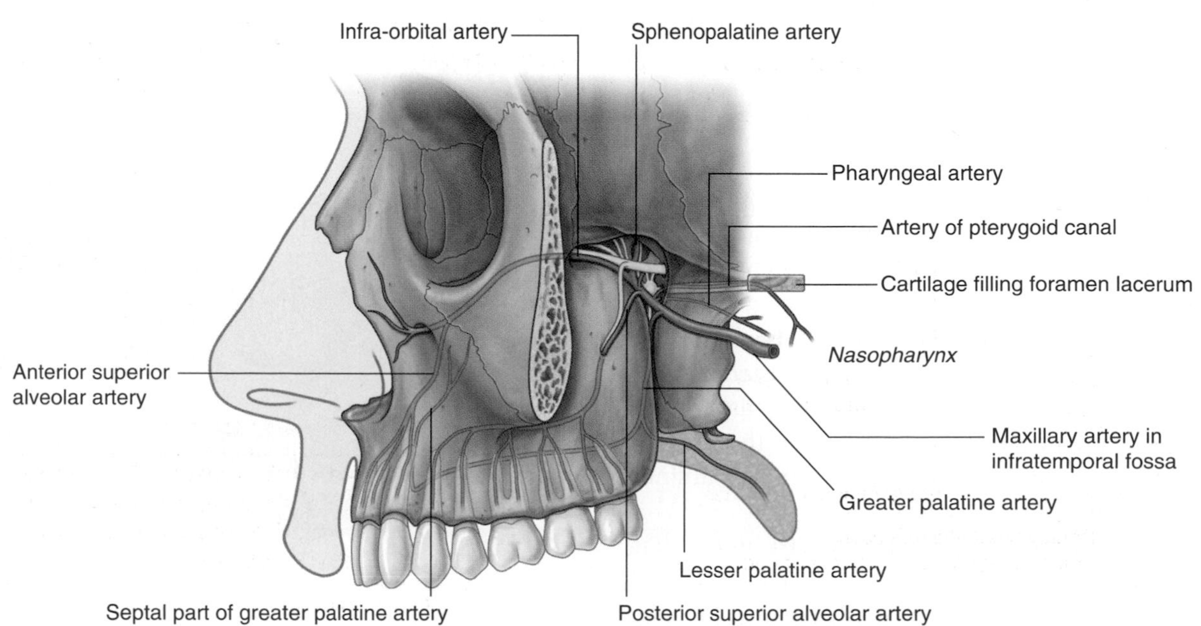


**Fig. 8.158** Maxillary artery in the pterygopalatine fossa.

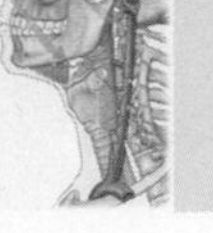



and all upper teeth. In addition, they contribute to the blood supply of the sinuses, oropharynx, and floor of the orbit.

### Branches

***Posterior superior alveolar artery.*** The **posterior superior alveolar artery** (Fig. 8.158) originates from the maxillary artery as it passes through the pterygomaxillary fissure. It meets the posterior superior alveolar nerve, accompanies it through the alveolar foramen on the infratemporal surface of the maxilla, and supplies the molar and premolar teeth, adjacent gingiva, and the maxillary sinus.

***Infra-orbital artery.*** The infra-orbital artery (Fig. 8.158) passes forward with the infra-orbital nerve and leaves the pterygopalatine fossa through the inferior orbital fissure. With the infra-orbital nerve, it lies in the infra-orbital groove and infra-orbital canal, and emerges through the infra-orbital foramen to supply parts of the face.

Within the infra-orbital canal, the infra-orbital artery gives origin to:

- branches that contribute to the blood supply of structures near the floor of the orbit—the inferior rectus and inferior oblique muscles, and the lacrimal sac; and
- **anterior superior alveolar arteries** (Fig. 8.158), which supply the incisor and canine teeth and the maxillary sinus.

***Greater palatine artery.*** The **greater palatine artery** (Fig. 8.158) passes inferiorly with the palatine nerves into the palatine canal. It gives origin to a **lesser palatine branch** (Fig. 8.158), which passes through the lesser palatine foramen to supply the soft palate, and then continues through the greater palatine foramen to supply the hard palate. The latter vessel passes forward on the inferior surface of the palate to enter the incisive fossa and pass superiorly through the incisive canal to supply the anterior aspect of the septal wall of the nasal cavity.

***Pharyngeal branch.*** The **pharyngeal branch** (Fig. 8.158) of the maxillary artery travels posteriorly and leaves the pterygopalatine fossa through the palatovaginal canal with the pharyngeal nerve. It supplies the posterior aspect of the roof of the nasal cavity, the sphenoidal sinus, and the pharyngotympanic tube.

***Sphenopalatine artery.*** The **sphenopalatine artery** (Fig. 8.158) is the terminal branch of the maxillary artery. It leaves the pterygopalatine fossa medially through the sphenopalatine foramen and accompanies the nasal nerves, giving off:

- posterior lateral nasal arteries, which supply the lateral wall of the nasal cavity and contribute to the supply of the paranasal sinuses; and
- posterior septal branches, which travel medially across the roof to supply the nasal septum—the largest of these branches passes anteriorly down the septum to anastomose with the end of the greater palatine artery.

***Artery of pterygoid canal.*** The **artery of the pterygoid canal** passes posteriorly into the pterygoid canal. It supplies surrounding tissues and terminates, after passing inferiorly through cartilage filling the foramen lacerum, in the mucosa of the nasopharynx.

## Veins

Veins that drain areas supplied by branches of the terminal part of the maxillary artery generally travel with these branches back into the pterygopalatine fossa.

The veins coalesce in the pterygopalatine fossa and then pass laterally through the pterygomaxillary fissure to join the pterygoid plexus of veins in the infratemporal fossa (Fig. 8.159).

The infra-orbital vein, which drains the inferior aspect of the orbit, may pass directly into the infratemporal fossa through the lateral aspect of the inferior orbital fissure, so bypassing the pterygopalatine fossa.

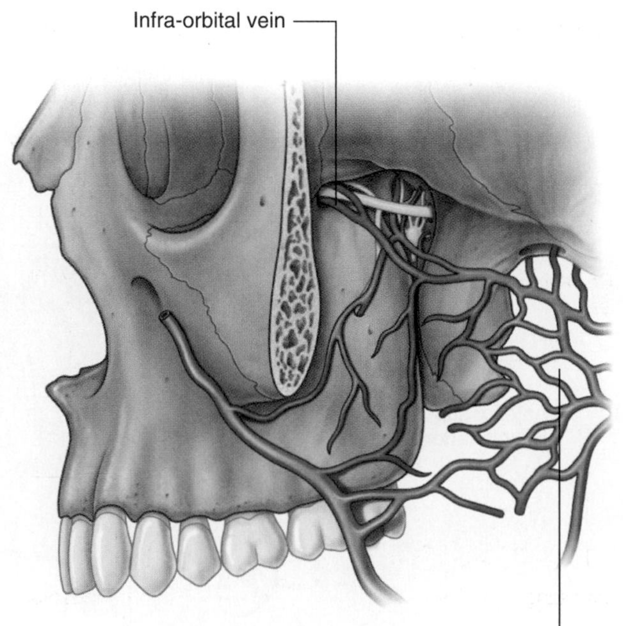


**Fig. 8.159** Veins of the pterygopalatine fossa.

## NECK

The neck is a tube providing continuity from the head to the trunk. It extends anteriorly from the lower border of the mandible to the upper surface of the manubrium of the sternum, and posteriorly from the superior nuchal line on the occipital bone of the skull to the intervertebral disc between the CVII and TI vertebrae. Within the tube, four compartments provide longitudinal organization (Fig. 8.160):

- The visceral compartment is anterior and contains parts of the digestive and respiratory systems, and several endocrine glands.
- The vertebral compartment is posterior and contains the cervical vertebrae, spinal cord, cervical nerves, and muscles associated with the vertebral column.
- The two vascular compartments, one on each side, are lateral and contain the major blood vessels and the vagus nerve [X].

All these compartments are contained within unique layers of cervical fascia.

For descriptive purposes the neck is divided into anterior and posterior triangles (Fig. 8.161):

- The boundaries of the **anterior triangle** are the anterior border of the sternocleidomastoid muscle, the inferior border of the mandible, and the midline of the neck.
- The boundaries of the **posterior triangle** are the posterior border of the sternocleidomastoid muscle, the anterior border of the trapezius muscle, and the middle one-third of the clavicle.

### Fascia

The fascia of the neck has a number of unique features.

The **superficial fascia** in the neck contains a thin sheet of muscle (the **platysma**), which begins in the superficial fascia of the thorax, runs upward to attach to the mandible and blend with the muscles on the face, is innervated by the cervical branch of the facial nerve [VII], and is only found in this location.

Deep to the superficial fascia, the deep cervical fascia is organized into several distinct layers (Fig. 8.160). These include:

- an investing layer, which surrounds all structures in the neck;
- the prevertebral layer, which surrounds the vertebral column and the deep muscles associated with the back;
- the pretracheal layer, which encloses the viscera of the neck; and
- the carotid sheaths, which receive a contribution from the other three fascial layers and surround the two major neurovascular bundles on either side of the neck.

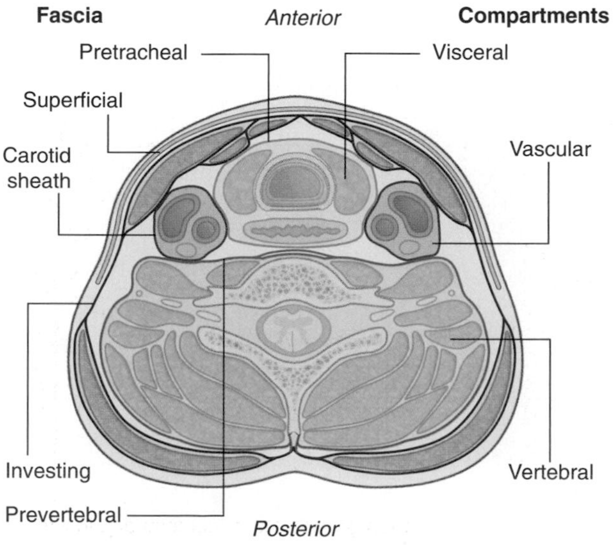

**Fig. 8.160** Compartments of the neck.

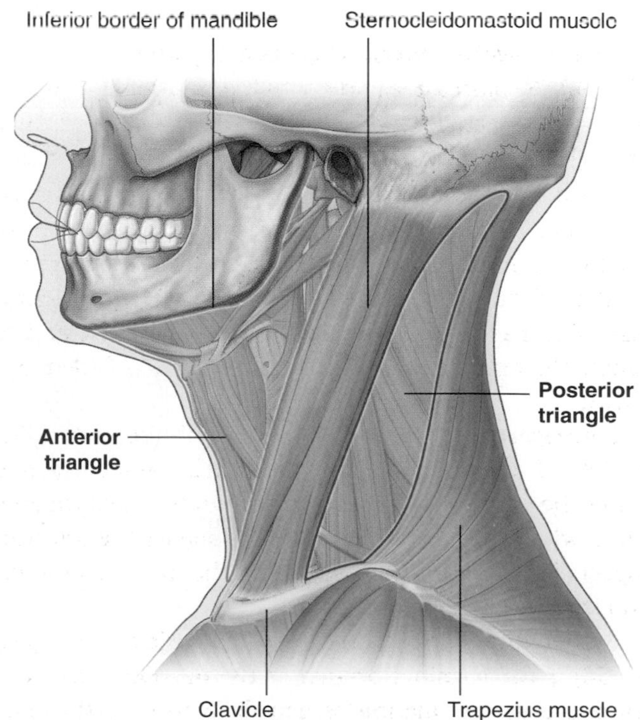


**Fig. 8.161** Anterior and posterior triangles of the neck.

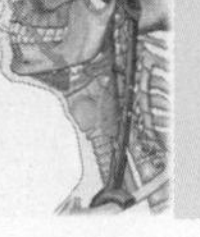



### Investing layer

The **investing layer** completely surrounds the neck and encloses the trapezius and sternocleidomastoid muscles (Fig. 8.162).

Attaching posteriorly to the ligamentum nuchae and the spinous process of the CVII vertebra, this fascial layer splits as it passes forward to enclose the trapezius muscle, reunites into a single layer as it forms the roof of the posterior triangle, splits again to surround the sternocleidomastoid muscle, and reunites again to join its twin from the other side.

Anteriorly, the investing fascia merges with fascia surrounding the infrahyoid muscles.

The investing fascia is attached:

- superiorly to the external occipital protuberance and the superior nuchal line,
- laterally to the mastoid process and zygomatic arch, and
- inferiorly to the spine of the scapula, the acromion, the clavicle, and the manubrium of the sternum.

The external and anterior jugular veins, and the lesser occipital, great auricular, transverse cervical, and supraclavicular nerves, all branches of the cervical plexus, pierce the investing fascia.

### Prevertebral layer

The prevertebral layer is a cylindrical layer of fascia that surrounds the vertebral column and the muscles associated with it (Fig. 8.162). Muscles in this group include the prevertebral muscles, the anterior, middle, and posterior scalene muscles, and the deep muscles of the back.

The prevertebral fascia is attached posteriorly along the length of the ligamentum nuchae, and superiorly forms a continuous circular line attaching to the base of the skull. This circle begins:

- anteriorly as the fascia attaches to the basilar part of the occipital bone, the area of the jugular foramen, and the carotid canal;
- continues laterally, attaching to the mastoid process; and
- continues posteriorly along the superior nuchal line ending at the external occipital protuberance, where it associates with its partner from the opposite side.

Anteriorly, the prevertebral fascia is attached to the anterior surfaces of the transverse processes and bodies of vertebrae CI to CVII.

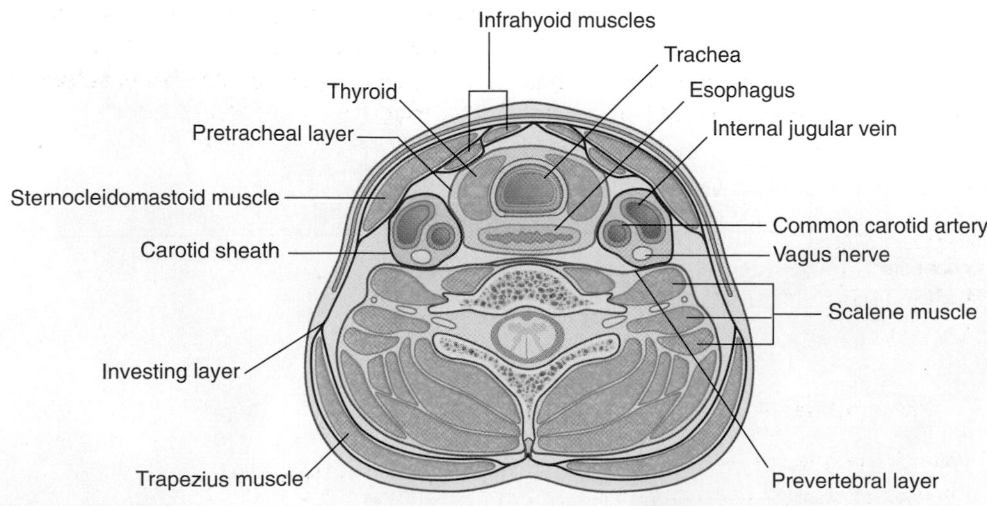


**Fig. 8.162** Fascia of neck, transverse view.

The prevertebral fascia passing between the attachment points on the transverse processes is unique. In this location, it splits into two layers, creating a longitudinal fascial space containing loose connective tissue that extends from the base of the skull through the thorax (Figs. 8.162 and 8.163).

There is one additional specialization of the prevertebral fascia in the lower region of the neck. The prevertebral fascia in an anterolateral position extends from the anterior and middle scalene muscles to surround the brachial plexus and subclavian artery as these structures pass into the axilla. This fascial extension is the **axillary sheath**.

### Pretracheal layer

The **pretracheal layer** consists of a collection of fascias that surround the trachea, esophagus, and thyroid gland (Fig. 8.162). Anteriorly, it consists of a pretracheal fascia that crosses the neck and encloses the infrahyoid muscles, and covers the trachea and the thyroid gland. The pretracheal fascia begins superiorly at the hyoid bone and ends inferiorly in the upper thoracic cavity. Laterally, this fascia encloses the thyroid gland and more posteriorly is continuous with fascia that surrounds the esophagus.

Posterior to the pharynx, the pretracheal layer is referred to as the buccopharyngeal fascia and separates the pharynx from the prevertebral layer (Fig. 8.163).

The buccopharyngeal fascia begins superiorly at the base of the skull and merges with fascia covering the esophagus that then continues inferiorly into the thoracic cavity.

### Carotid sheath

Each **carotid sheath** is a column of fascia that surrounds the common carotid artery, the internal carotid artery, the internal jugular vein, and the vagus nerve as these structures pass through the neck (Fig. 8.162).

It receives contributions from the investing, prevertebral, and pretracheal layers, though the extent of each component's contribution varies.

### Fascial compartments

The arrangement of the various layers of cervical fascia organizes the neck into four longitudinal compartments (Fig. 8.160):

- The first compartment is the largest, includes the other three, and consists of the area surrounded by the investing layer.
- The second compartment consists of the vertebral column and the deep muscles associated with this structure, and is the area contained within the prevertebral layer.
- The third compartment (the visceral compartment) contains the pharynx, the trachea, the esophagus, and the thyroid gland, which are surrounded by the pretracheal layer.
- Finally, there is a compartment (the carotid sheath) consisting of the neurovascular structures that pass from the base of the skull to the thoracic cavity, and the sheath enclosing these structures receives contributions from the other cervical fascias.

### Fascial spaces

Between the fascial layers in the neck are spaces that may provide a conduit for the spread of infections from the neck to the mediastinum.

Three spaces could be involved in this process (Fig. 8.163):

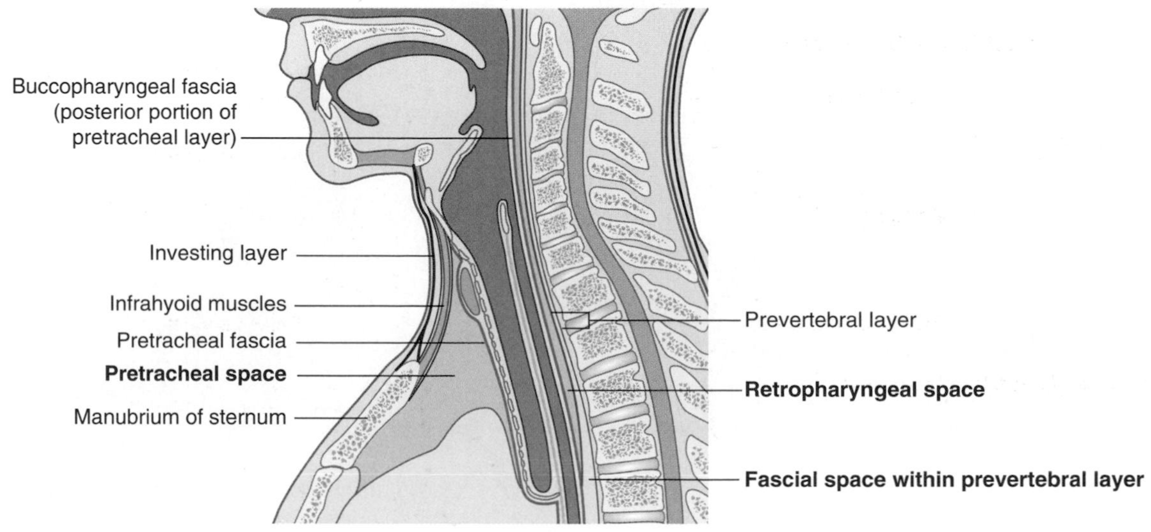


**Fig. 8.163** Fascia of the neck, sagittal view.

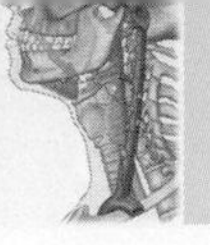



- The first is the **pretracheal space** between the investing layer of cervical fascia covering the posterior surface of the infrahyoid muscles and the pretracheal fascia (covering the anterior surface of the trachea and the thyroid gland), which passes between the neck and the anterior part of the superior mediastinum.
- The second is the **retropharyngeal space** between the buccopharyngeal fascia (on the posterior surface of the pharynx and esophagus) and the prevertebral fascia (on the anterior surface of the transverse processes and bodies of the cervical vertebrae), which extends from the base of the skull to the upper part of the posterior mediastinum.
- The **third space** is within the prevertebral layer covering the anterior surface of the transverse processes and bodies of the cervical vertebrae. This layer splits into two laminae to create a fascial space that begins at the base of the skull and extends through the posterior mediastinum to the diaphragm.

## Superficial venous drainage

The external jugular and anterior jugular veins are the primary venous channels for superficial venous drainage of the neck (Fig. 8.164).

### External jugular veins

The external jugular vein is formed posterior to the angle of the mandible as the **posterior auricular vein** and the **retromandibular vein** join:

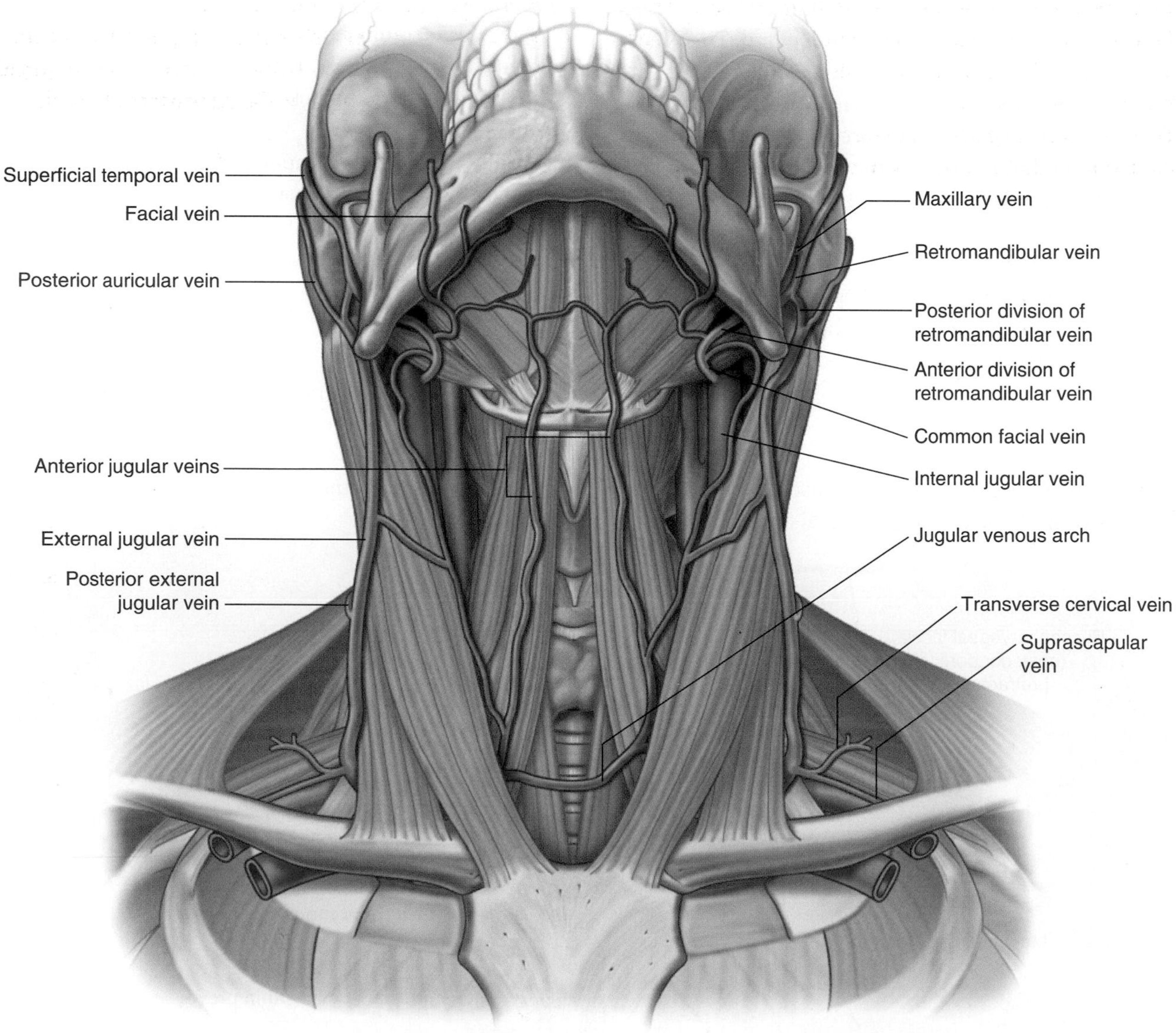

**Fig. 8.164** Superficial veins of neck.

- The posterior auricular vein drains the scalp behind and above the ear.
- The retromandibular vein is formed when the **superficial temporal** and **maxillary veins** join in the substance of the parotid gland and it descends to the angle of mandible, where it divides into an anterior and a posterior division (Fig. 8.164)—the posterior division joins the posterior auricular vein to form the external jugular vein, and the anterior division joins the **facial vein** to form the common facial vein, which passes deep and becomes a tributary to the internal jugular vein.

Once formed, the external jugular vein passes straight down the neck in the superficial fascia and is superficial to the sternocleidomastoid muscle throughout its course, crossing it diagonally as it descends.

Reaching the lower part of the neck, just superior to the clavicle and immediately posterior to the sternocleidomastoid muscle, the external jugular vein pierces the investing layer of cervical fascia, passes deep to the clavicle, and enters the **subclavian vein**.

Tributaries received by the external jugular vein along its course include the **posterior external jugular vein** (draining superficial areas of the back of the neck) and the **transverse cervical** and **suprascapular veins** (draining the posterior scapular region).

### Anterior jugular veins

The **anterior jugular veins**, although variable and inconsistent, are usually described as draining the anterior aspect of the neck (Fig. 8.164). These paired venous channels, which begin as small veins, come together at or just superior to the hyoid bone. Once formed, each anterior jugular vein descends on either side of the midline of the neck.

Inferiorly, near the medial attachment of the sternocleidomastoid muscle, each anterior jugular vein pierces the investing layer of cervical fascia to enter the subclavian vein. Occasionally, the anterior jugular vein may enter the external jugular vein immediately before the external jugular vein enters the subclavian vein.

Often, the right and left anterior jugular veins communicate with each other, being connected by a **jugular venous arch** in the area of the suprasternal notch.

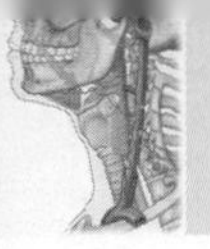



## In the clinic

### Central venous access

In most instances, access to peripheral veins of the arm and the leg will suffice for administering intravenous drugs and fluids and for obtaining blood for analysis; however, in certain circumstances it is necessary to place larger-bore catheters in the central veins, for example, for dialysis, parenteral nutrition, or the administration of drugs that have a tendency to produce phlebitis.

"Blind puncture" of the subclavian and jugular veins to obtain central venous access used to be standard practice. However, subclavian vein puncture is not without complications. As the subclavian vein passes inferiorly, posterior to the clavicle, it passes over the apex of the lung. Any misplacement of a needle into or through this structure may puncture the apical pleura, producing a pneumothorax. Inadvertent arterial puncture and vein laceration may also produce a hemopneumothorax.

A puncture of the internal jugular vein (Fig. 8.165) carries fewer risks, but local hematoma and damage to the carotid artery are again important complications.

Current practice is to identify major vessels using ultrasound and to obtain central venous access under direct vision to avoid any significant complication.

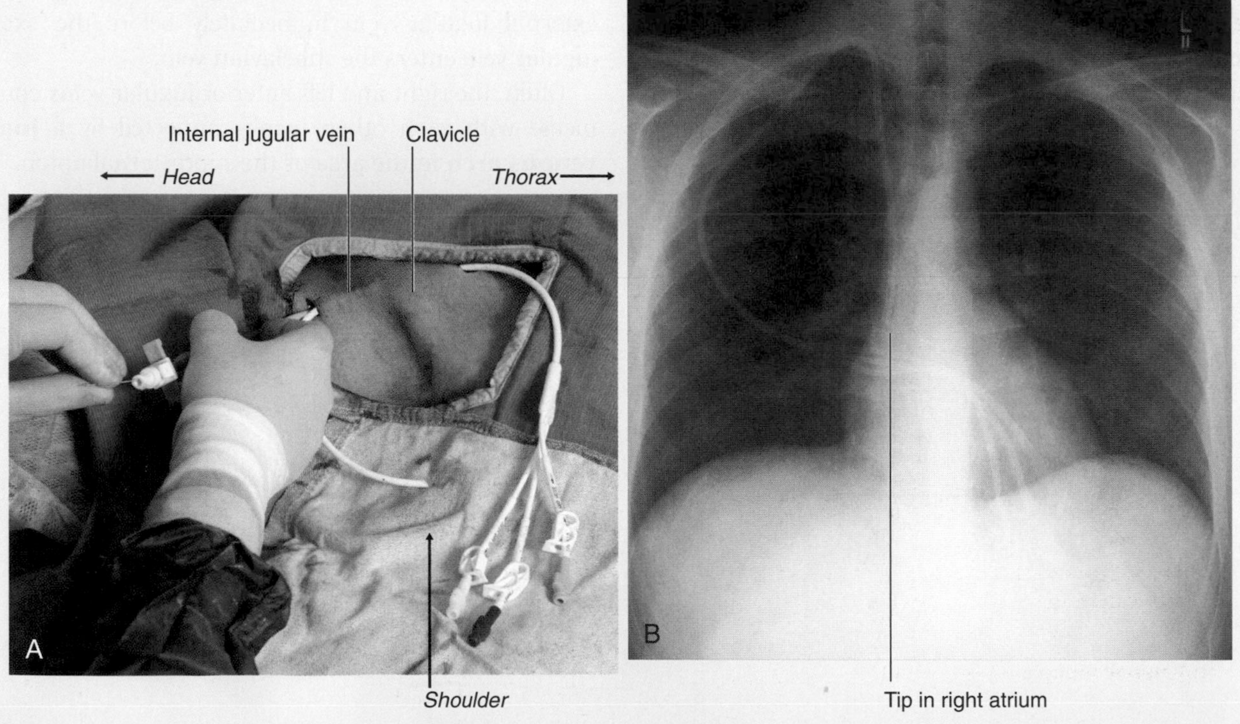


**Fig. 8.165** Placing a central venous catheter in the neck. **A.** Clinical procedure. **B.** Chest radiograph showing that the tip of the catheter is in the origin of the right atrium.

## Anterior triangle of the neck

The anterior triangle of the neck is outlined by the anterior border of the sternocleidomastoid muscle laterally, the inferior border of the mandible superiorly, and the midline of the neck medially (Fig. 8.166). It is further subdivided into several smaller triangles as follows:

- The **submandibular triangle** is outlined by the inferior border of the mandible superiorly and the anterior and posterior bellies of the digastric muscle inferiorly.
- The **submental triangle** is outlined by the hyoid bone inferiorly, the anterior belly of the digastric muscle laterally, and the midline.
- The **muscular triangle** is outlined by the hyoid bone superiorly, the superior belly of the omohyoid muscle, and the anterior border of the sternocleidomastoid muscle laterally, and the midline.
- The **carotid triangle** is outlined by the superior belly of the omohyoid muscle anteroinferiorly, the stylohyoid muscle and posterior belly of the digastric superiorly, and the anterior border of the sternocleidomastoid muscle posteriorly.

Each of these triangles contains numerous structures that can be identified as being within a specific triangle, passing into a specific triangle from outside the area, originating in one triangle and passing to another triangle, or passing through several triangles while passing through the region.

A discussion of the anterior triangle of the neck must therefore combine a systemic approach, describing the muscles, vessels, and nerves in the area, with a regional approach, describing the contents of each triangle.

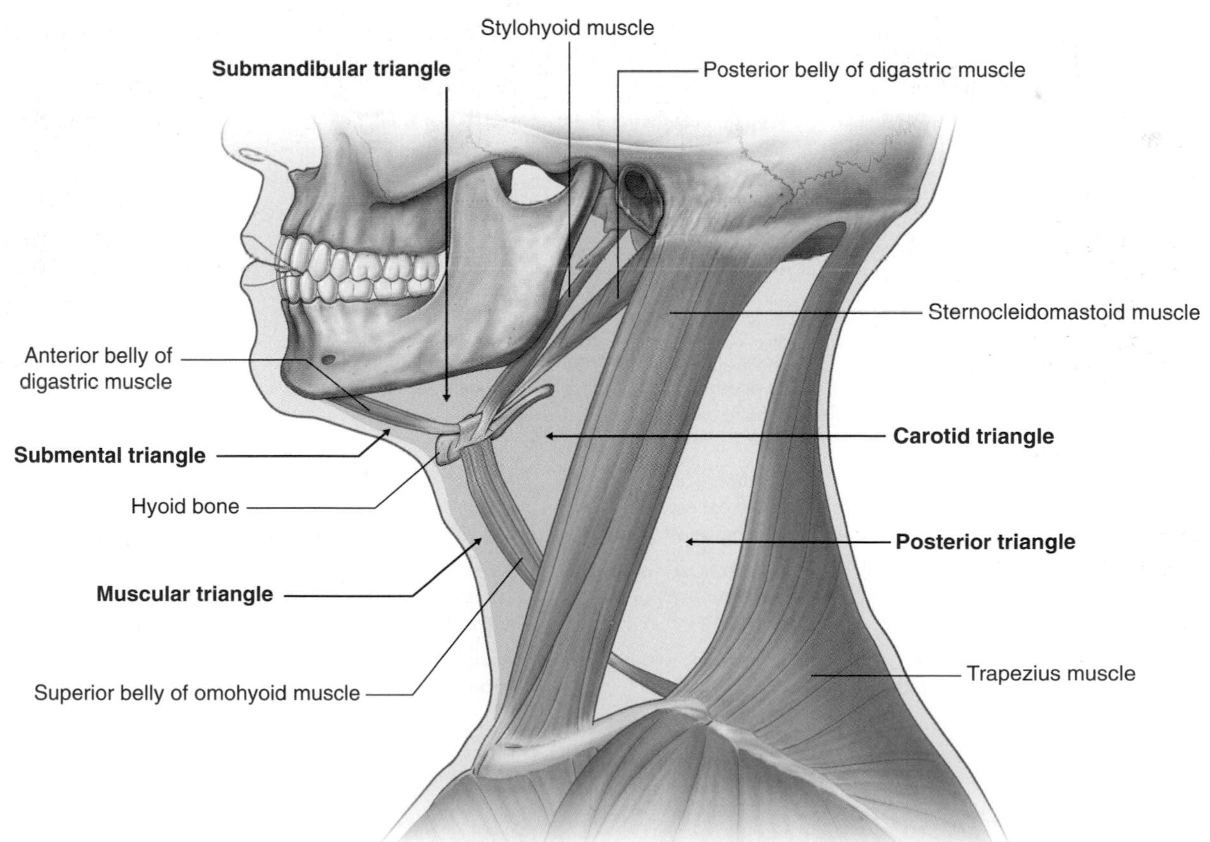


**Fig. 8.166** Borders and subdivisions of the anterior triangle of the neck.

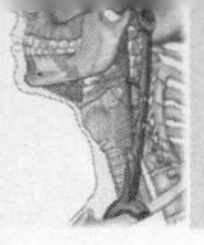
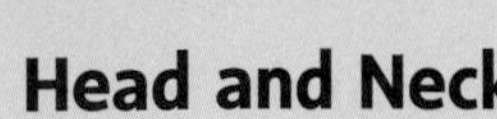



## Muscles

The muscles in the anterior triangle of the neck (Table 8.12) can be grouped according to their location relative to the hyoid bone:

- Muscles superior to the hyoid are classified as **suprahyoid muscles** and include the stylohyoid, digastric, mylohyoid, and geniohyoid.
- Muscles inferior to the hyoid are **infrahyoid muscles** and include the omohyoid, sternohyoid, thyrohyoid, and sternothyroid.

### Suprahyoid muscles

The four pairs of suprahyoid muscles are related to the submental and submandibular triangles (Fig. 8.166). They pass in a superior direction from the hyoid bone to the skull or mandible and raise the hyoid, as occurs during swallowing.

**Table 8.12** Anterior triangle of neck (suprahyoid and infrahyoid muscles)

| Muscle | Origin | Insertion | Innervation | Function |
|---|---|---|---|---|
| Stylohyoid | Base of styloid process | Lateral area of body of hyoid bone | Facial nerve [VII] | Pulls hyoid bone upward in a posterosuperior direction |
| Digastric | | | | |
| —Anterior belly | Digastric fossa on lower inside of mandible | Attachment of tendon between two bellies to body of hyoid bone | Mylohyoid nerve from inferior alveolar branch of mandibular nerve [$V_3$] | Opens mouth by lowering mandible; raises hyoid bone |
| —Posterior belly | Mastoid notch on medial side of mastoid process of temporal bone | Same as anterior belly | Facial nerve [VII] | Pulls hyoid bone upward and back |
| Mylohyoid | Mylohyoid line on mandible | Body of hyoid bone and fibers from muscle on opposite side | Mylohyoid nerve from inferior alveolar branch of mandibular nerve [$V_3$] | Support and elevation of floor of mouth; elevation of hyoid |
| Geniohyoid | Inferior mental spine on inner surface of mandible | Anterior surface of body of hyoid bone | Branch from anterior ramus of C1 (carried along the hypoglossal nerve [XII]) | Fixed mandible elevates and pulls hyoid bone forward; fixed hyoid bone pulls mandible downward and inward |
| Sternohyoid | Posterior aspect of sternoclavicular joint and adjacent manubrium of sternum | Body of hyoid bone medial to attachment of omohyoid muscle | Anterior rami of C1 to C3 through the ansa cervicalis | Depresses hyoid bone after swallowing |
| Omohyoid | Superior border of scapula medial to suprascapular notch | Lower border of body of hyoid bone just lateral to attachment of sternohyoid | Anterior rami of C1 to C3 through the ansa cervicalis | Depresses and fixes hyoid bone |
| Thyrohyoid | Oblique line on lamina of thyroid cartilage | Greater horn and adjacent aspect of body of hyoid bone | Fibers from anterior ramus of C1 carried along hypoglossal nerve [XII] | Depresses hyoid bone, but when hyoid bone is fixed raises larynx |
| Sternothyroid | Posterior surface of manubrium of sternum | Oblique line on lamina of thyroid cartilage | Anterior rami of C1 to C3 through the ansa cervicalis | Draws larynx (thyroid cartilage) downward |

### Stylohyoid

The **stylohyoid muscle** arises from the base of the styloid process and passes anteroinferiorly to attach to the lateral area of the body of the hyoid bone (Fig. 8.167). During swallowing it pulls the hyoid bone posterosuperiorly and it is innervated by the facial nerve [VII].

### Digastric

The **digastric muscle** has two bellies connected by a tendon, which attaches to the body of the hyoid bone (Fig. 8.167):

- The **posterior belly** arises from the mastoid notch on the medial side of the mastoid process of the temporal bone.
- The **anterior belly** arises from the digastric fossa on the lower inside of the mandible.

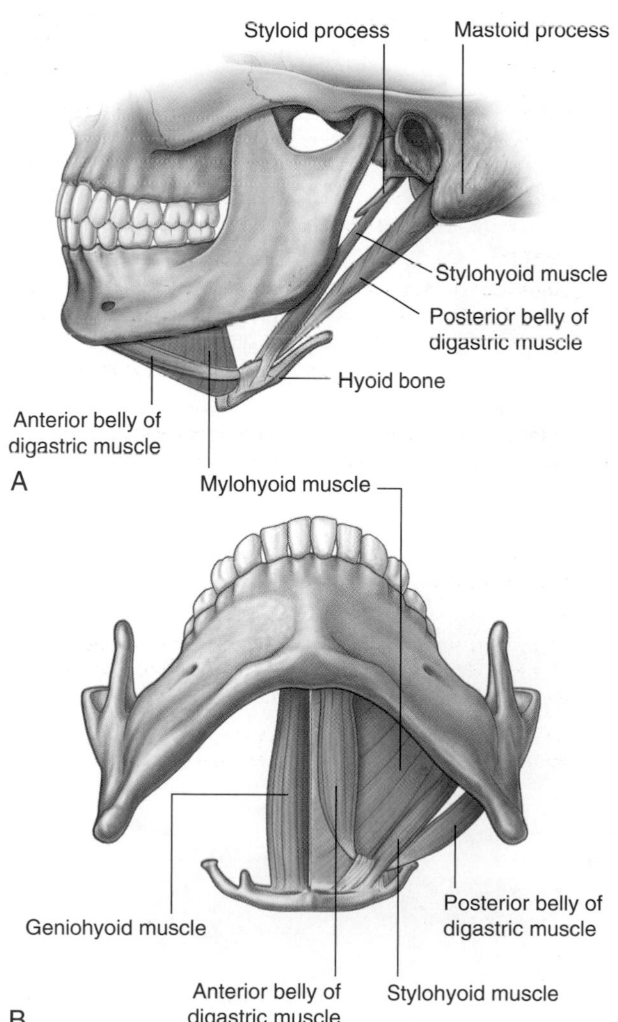


**Fig. 8.167** Suprahyoid muscles. **A.** Lateral view. **B.** Inferior view.

The tendon between the two bellies, which is attached to the body of the hyoid bone, is the point of insertion of both bellies. Because of this arrangement, the muscle has multiple actions depending on which bone is fixed:

- When the mandible is fixed, the digastric muscle raises the hyoid bone.
- When the hyoid bone is fixed, the digastric muscle opens the mouth by lowering the mandible.

Innervation of the digastric muscle is from two different cranial nerves.

The innervation of the posterior belly of the digastric muscle is by the facial nerve [VII], whereas the anterior belly of the muscle is innervated by the mandibular division [$V_3$] of the trigeminal nerve [V].

### Mylohyoid

The **mylohyoid muscle** is superior to the anterior belly of the digastric and, with its partner from the opposite side, forms the floor of the mouth (Fig. 8.167). It originates from the mylohyoid line on the medial surface of the body of the mandible and inserts into the hyoid bone and also blends with the mylohyoid muscle from the opposite side.

This mylohyoid muscle supports and elevates the floor of the mouth and elevates the hyoid bone. It is innervated by the mandibular division [$V_3$] of the trigeminal nerve [V].

### Geniohyoid

The **geniohyoid muscle** is superior to the floor of the oral cavity and is not generally considered a muscle of the anterior triangle of the neck; however, it can be regarded as a suprahyoid muscle. It is the final muscle in the suprahyoid group (Fig. 8.167). A narrow muscle, it is superior to the medial part of each mylohyoid muscle. The muscles from each side are next to each other in the midline.

The geniohyoid arises from the inferior mental spine of the mandible and passes backward and downward to insert on the body of the hyoid bone.

It has two functions depending on which bone is fixed:

- Fixation of the mandible elevates and pulls the hyoid bone forward.
- Fixation of the hyoid bone pulls the mandible downward and inward.

The geniohyoid is innervated by a branch from the anterior ramus of C1 carried along the hypoglossal nerve [XII].

## Infrahyoid muscles

The four infrahyoid muscles are related to the muscular triangle (Fig. 8.166). They attach the hyoid bone to inferior structures and depress the hyoid bone. They also provide a stable point of attachment for the suprahyoid muscles. Because of their appearance, they are sometimes referred to as the **"strap muscles."**

### Sternohyoid

The sternohyoid muscle is a long, thin muscle originating from the posterior aspect of the sternoclavicular joint and adjacent manubrium of the sternum (Fig. 8.168). It ascends to insert onto the body of the hyoid bone. It depresses the hyoid bone and is innervated by the anterior rami of C1 to C3 through the ansa cervicalis.

### Omohyoid

Lateral to the sternohyoid muscle is the omohyoid muscle (Fig. 8.168). This muscle consists of two bellies with an intermediate tendon in both the posterior and anterior triangles of the neck:

- The **inferior belly** begins on the superior border of the scapula, medial to the suprascapular notch, and passes forward and upward across the posterior triangle ending at the intermediate tendon.
- The **superior belly** begins at the intermediate tendon and ascends to attach to the body of the hyoid bone just lateral to the attachment of the sternohyoid.
- The intermediate tendon is attached to the clavicle, near its medial end, by a fascial sling.

The omohyoid depresses and fixes the hyoid bone. It is innervated by the anterior rami of C1 to C3 through the ansa cervicalis.

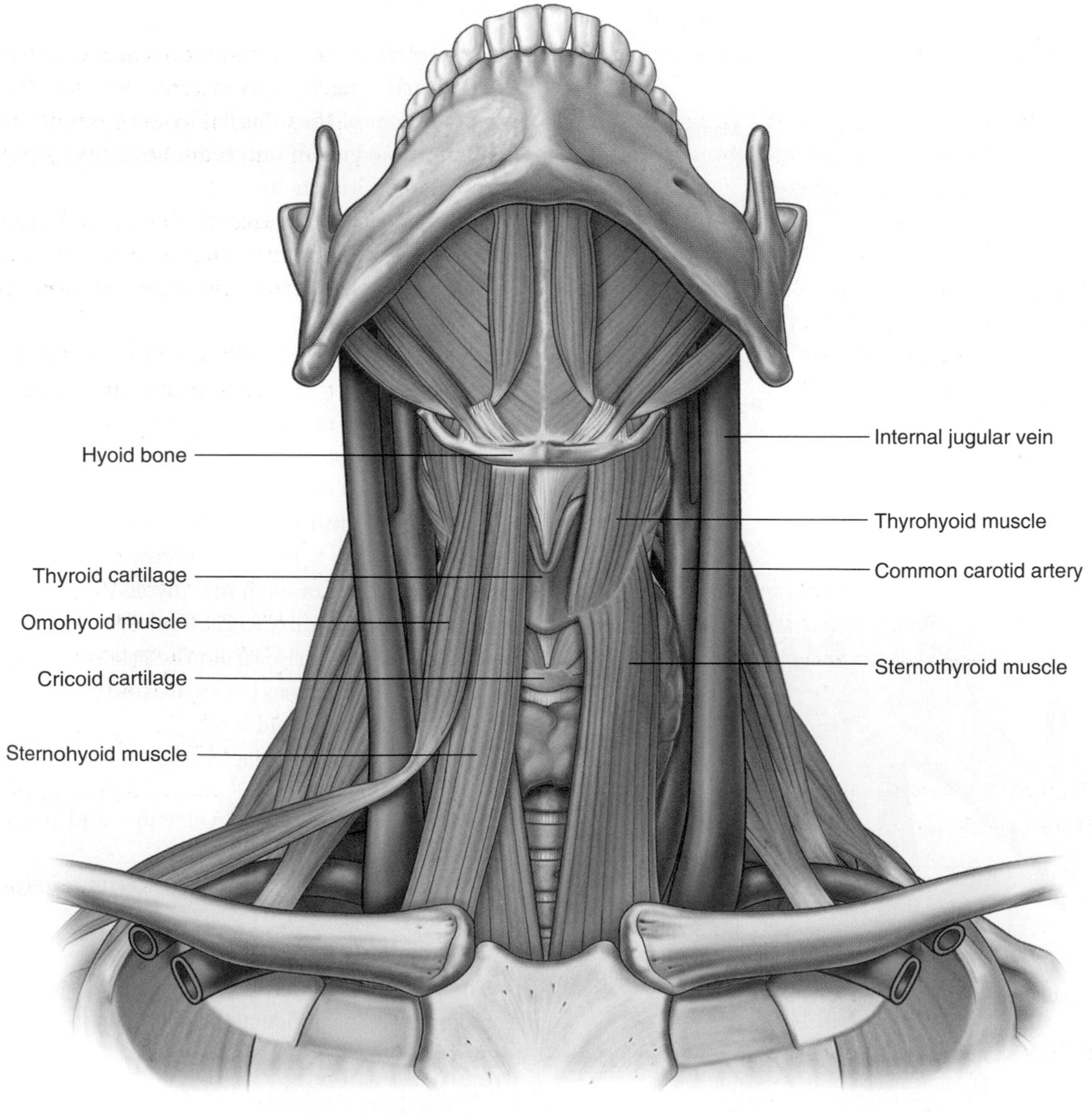

**Fig. 8.168** Infrahyoid muscles.

### Thyrohyoid

The thyrohyoid muscle is deep to the superior parts of the omohyoid and sternohyoid (Fig. 8.168). Originating at the oblique line on the lamina of the thyroid cartilage it passes upward to insert into the greater horn and adjacent aspect of the body of the hyoid bone.

The thyrohyoid muscle has variable functions depending on which bone is fixed. Generally, it depresses the hyoid, but when the hyoid is fixed it raises the larynx (e.g., when high notes are sung). It is innervated by fibers from the anterior ramus of C1 that travel with the hypoglossal nerve [XII].

### Sternothyroid

Lying beneath the sternohyoid and in continuity with the thyrohyoid, the sternothyroid is the last muscle in the infrahyoid group (Fig. 8.168). It arises from the posterior surface of the manubrium of the sternum and passes upward to attach to the oblique line on the lamina of the thyroid cartilage.

The sternothyroid muscle draws the larynx (thyroid cartilage) downward and is innervated by the anterior rami of C1 to C3 through the ansa cervicalis.

## Vessels

Passing through the anterior triangle of the neck are the common carotid arteries and their branches, the external and internal carotid arteries. These vessels supply all structures of the head and neck.

Associated with this arterial system are the internal jugular vein and its tributaries. These vessels receive blood from all structures of the head and neck.

## Carotid system

### Common carotid arteries

The **common carotid arteries** are the beginning of the carotid system (Fig. 8.169):

- The **right common carotid artery** originates from the brachiocephalic trunk immediately posterior to the right sternoclavicular joint and is entirely in the neck throughout its course.
- The **left common carotid artery** begins in the thorax as a direct branch of the arch of the aorta and passes superiorly to enter the neck near the left sternoclavicular joint.

Both right and left common carotid arteries ascend through the neck, just lateral to the trachea and esophagus, within a fascial compartment (the carotid sheath). They give off no branches as they pass through the neck.

Near the superior edge of the thyroid cartilage each common carotid artery divides into its two terminal branches—the **external** and **internal carotid arteries** (Fig. 8.170).

The superior part of each common carotid artery and its division into external and internal carotid arteries occurs

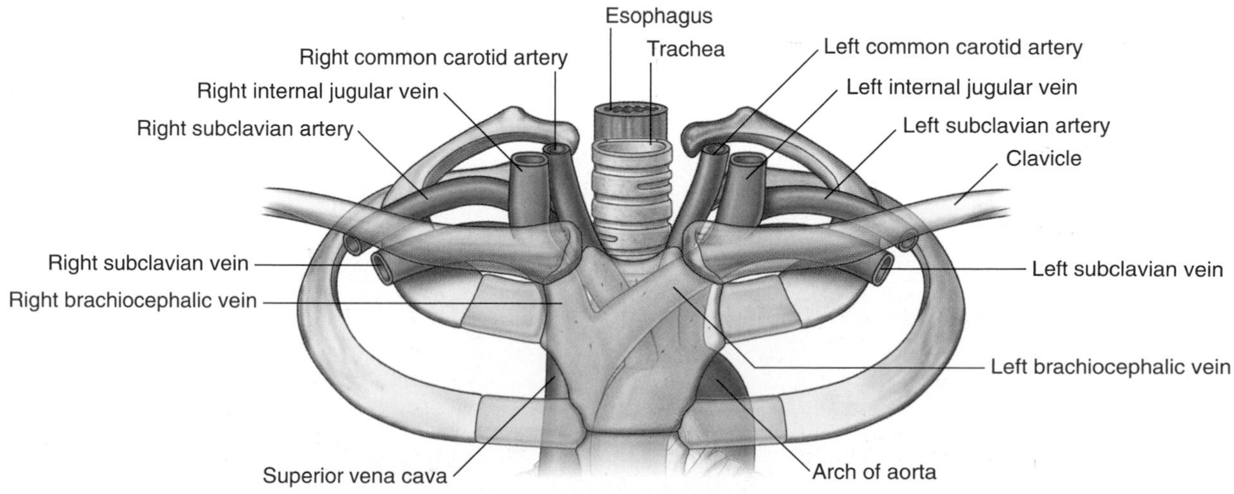


**Fig. 8.169** Origin of common carotid arteries.

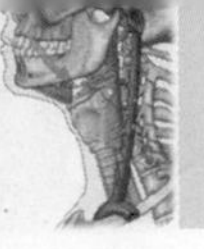



in the carotid triangle (Fig. 8.170), which is a subdivision of the anterior triangle of the neck (see Fig. 8.166).

At the bifurcation, the common carotid artery and the beginning of the internal carotid artery are dilated. This dilation is the **carotid sinus** (Fig. 8.171) and contains receptors that monitor changes in blood pressure and are innervated by a branch of the glossopharyngeal nerve [IX].

Another accumulation of receptors in the area of the bifurcation is responsible for detecting changes in blood chemistry, primarily oxygen content. This is the **carotid body** and is innervated by branches from both the glossopharyngeal [IX] and vagus [X] nerves.

## Internal carotid arteries

After its origin, the internal carotid artery ascends toward the base of the skull (Fig. 8.171). It gives off no branches in the neck and enters the cranial cavity through the carotid canal in the petrous part of the temporal bone.

The internal carotid arteries supply the cerebral hemispheres, the eyes and the contents of the orbits, and the forehead.

## External carotid arteries

The external carotid arteries begin giving off branches immediately after the bifurcation of the common carotid arteries (Fig. 8.171 and Table 8.13) as follows:

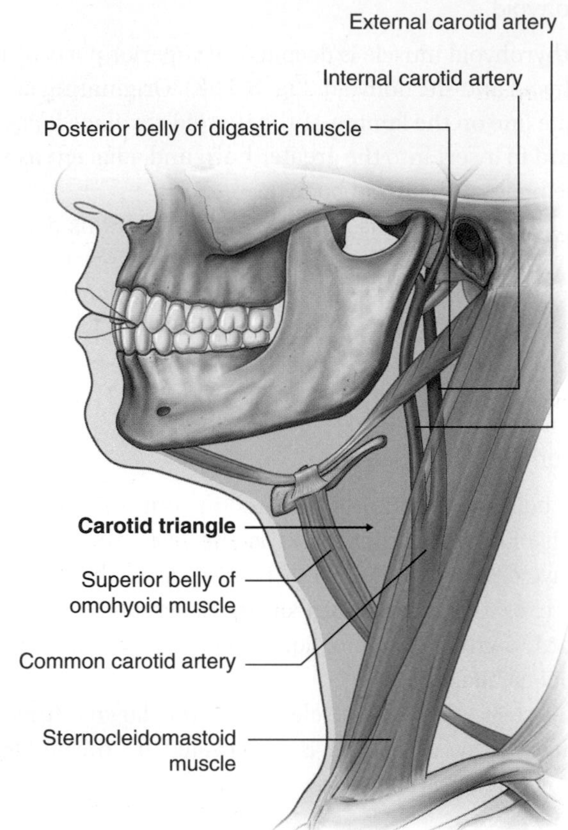


**Fig. 8.170** Carotid triangle.

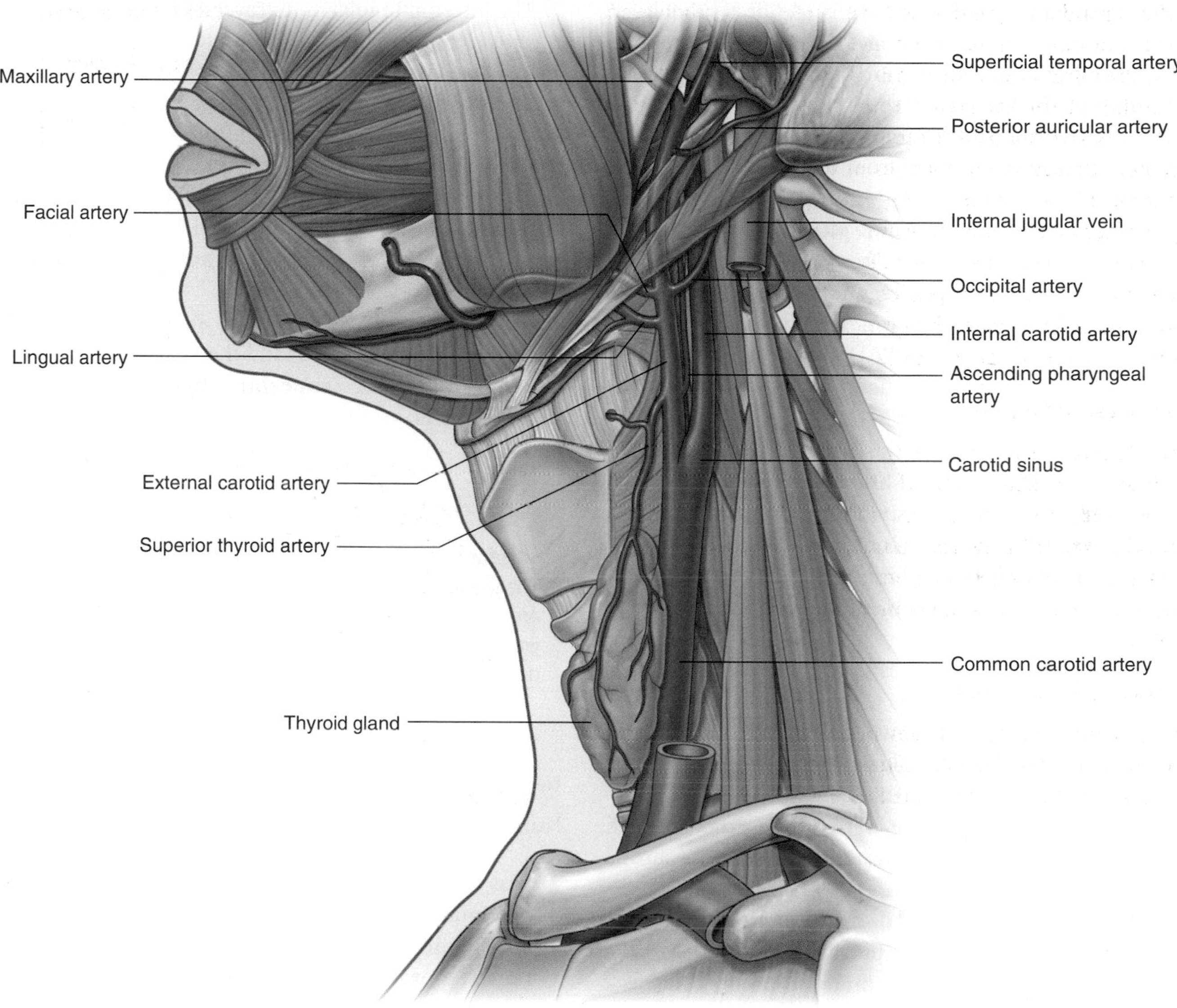


**Fig. 8.171** Carotid system.

**Table 8.13** Branches of the external carotid artery

| Branch | Supplies |
|---|---|
| Superior thyroid artery | Thyrohyoid muscle, internal structures of the larynx, sternocleidomastoid and cricothyroid muscles, thyroid gland |
| Ascending pharyngeal artery | Pharyngeal constrictors and stylopharyngeus muscle, palate, palatine tonsil, pharyngotympanic tube, meninges in posterior cranial fossa |
| Lingual artery | Muscles of the tongue, palatine tonsil, soft palate, epiglottis, floor of mouth, sublingual gland |
| Facial artery | All structures in the face from the inferior border of the mandible anterior to the masseter muscle to the medial corner of the eye, the soft palate, palatine tonsil, pharyngotympanic tube, submandibular gland |
| Occipital artery | Sternocleidomastoid muscle, meninges in posterior cranial fossa, mastoid cells, deep muscles of the back, posterior scalp |
| Posterior auricular artery | Parotid gland and nearby muscles, external ear and scalp posterior to ear, middle and inner ear structures |
| Superficial temporal artery | Parotid gland and duct, masseter muscle, lateral face, anterior part of external ear, temporalis muscle, parietal and temporal fossae |
| Maxillary artery | External acoustic meatus, lateral and medial surface of tympanic membrane, temporomandibular joint, dura mater on lateral wall of skull and inner table of cranial bones, trigeminal ganglion and dura in vicinity, mylohyoid muscle, mandibular teeth, skin on chin, temporalis muscle, outer table of bones of skull in temporal fossa, structures in infratemporal fossa, maxillary sinus, upper teeth and gingivae, infra-orbital skin, palate, roof of pharynx, nasal cavity |

- The **superior thyroid artery** is the first branch—it arises from the anterior surface near or at the bifurcation and passes in a downward and forward direction to reach the superior pole of the thyroid gland.
- The **ascending pharyngeal artery** is the second and smallest branch—it arises from the posterior aspect of the external carotid artery and ascends between the internal carotid artery and the pharynx.
- The **lingual artery** arises from the anterior surface of the external carotid artery just above the superior thyroid artery at the level of the hyoid bone, passes deep to the hypoglossal nerve [XII], and passes between the middle constrictor of the pharynx and hyoglossus muscles.
- The **facial artery** is the third anterior branch of the external carotid artery—it arises just above the lingual artery, passes deep to the stylohyoid and posterior belly of the digastric muscles, continues deep between the submandibular gland and mandible, and emerges over the edge of the mandible just anterior to the masseter muscle, to enter the face.
- The **occipital artery** arises from the posterior surface of the external carotid artery, near the level of origin of the facial artery, passes upward and posteriorly deep to the posterior belly of the digastric muscle, and emerges on the posterior aspect of the scalp.
- The **posterior auricular artery** is a small branch arising from the posterior surface of the external carotid artery and passes upward and posteriorly.
- The **superficial temporal artery** is one of the terminal branches and appears as an upward continuation of the external carotid artery—beginning posterior to the neck of the mandible, it passes anterior to the ear, crosses the zygomatic process of the temporal bone, and above this point divides into anterior and posterior branches.
- The **maxillary artery** is the larger of the two terminal branches of the external carotid artery—arising posterior to the neck of the mandible, it passes through the parotid gland, continues medial to the neck of the mandible and into the infratemporal fossa, and continues through this area into the pterygopalatine fossa.

### Veins

Collecting blood from the skull, brain, superficial face, and parts of the neck, the **internal jugular vein** begins as a dilated continuation of the **sigmoid sinus**, which is a dural venous sinus. This initial dilated part is referred to as the **superior bulb of jugular vein** and receives another dural venous sinus (the **inferior petrosal sinus**) soon after it is formed. It exits the skull through the jugular foramen associated with the glossopharyngeal [IX], vagus [X], and accessory [XI] nerves, and enters the carotid sheath.



The internal jugular vein traverses the neck within the carotid sheath, initially posterior to the internal carotid artery, but passes to a more lateral position farther down. It remains lateral to the common carotid artery through the rest of the neck with the vagus nerve [X] posterior and partially between the two vessels.

The paired internal jugular veins join with the subclavian veins posterior to the sternal end of the clavicle to form the right and left **brachiocephalic veins** (Fig. 8.169).

Tributaries to each internal jugular vein include the inferior petrosal sinus, and the **facial**, **lingual**, **pharyngeal**, **occipital**, **superior thyroid**, and **middle thyroid veins**.

#### In the clinic

**Jugular venous pulse**

The jugular venous pulse is an important clinical sign that enables the physician to assess the venous pressure and waveform and is a reflection of the functioning of the right side of the heart.

### Nerves

Numerous cranial and peripheral nerves:

- pass through the anterior triangle of the neck as they continue to their final destination,
- send branches to structures in or forming boundaries of the anterior triangle of the neck, and
- while in the anterior triangle of the neck, send branches to nearby structures.

The cranial nerves in these categories include the facial [VII], glossopharyngeal [IX], vagus [X], accessory [XI], and hypoglossal [XII].

Branches of spinal nerves in these categories include the transverse cervical nerve from the cervical plexus and the upper and lower roots of the ansa cervicalis.

#### Facial nerve [VII]

After emerging from the stylomastoid foramen, the facial nerve [VII] gives off branches that innervate two muscles associated with the anterior triangle of the neck:

- the posterior belly of the digastric, and
- the stylohyoid.

The facial nerve [VII] also innervates the platysma muscle that overlies the anterior triangle and part of the posterior triangle of the neck.

### Glossopharyngeal nerve [IX]

The glossopharyngeal nerve [IX] leaves the cranial cavity through the jugular foramen. It begins its descent between the internal carotid artery and the internal jugular vein, lying deep to the styloid process and the muscles associated with the styloid process. As the glossopharyngeal nerve [IX] completes its descent, it passes forward between the internal and external carotid arteries, and curves around the lateral border of the stylopharyngeus muscle (Fig. 8.172). At this point, it continues in an anterior direction, deep to the hyoglossus muscle, to reach the base of the tongue and the area of the palatine tonsil.

As the glossopharyngeal nerve [IX] passes through the area of the anterior triangle of the neck it innervates the stylopharyngeus muscle, sends a branch to the carotid sinus, and supplies sensory branches to the pharynx.

### Vagus nerve [X]

The vagus nerve [X] exits the cranial cavity through the jugular foramen between the glossopharyngeal [IX] and accessory [XI] nerves.

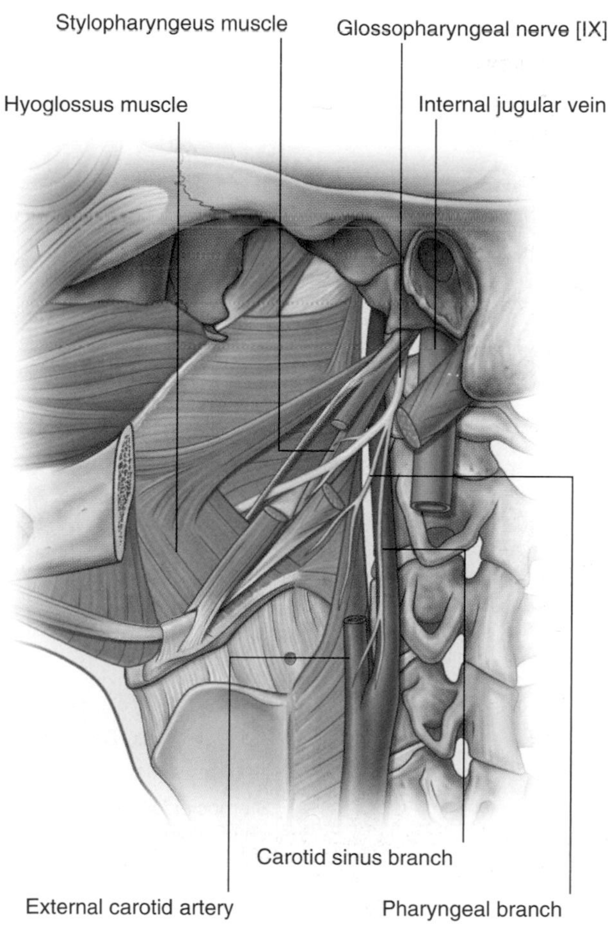


**Fig. 8.172** Glossopharyngeal nerve [IX] in the anterior triangle of the neck.

Outside the skull the vagus nerve [X] enters the carotid sheath and descends through the neck enclosed in this structure medial to the internal jugular vein and posterior to the internal carotid and common carotid arteries (Fig. 8.173).

Branches of the vagus nerve [X] as it passes through the anterior triangle of the neck include a motor branch to the pharynx, a branch to the carotid body, the superior laryngeal nerve (which divides into external and internal laryngeal branches), and possibly a cardiac branch.

### Accessory nerve [XI]

The accessory nerve [XI] is the most posterior of the three cranial nerves exiting the cranial cavity through the jugular foramen. It begins its descent medial to the internal jugular vein, emerging from between the internal jugular vein and internal carotid artery to cross the lateral surface

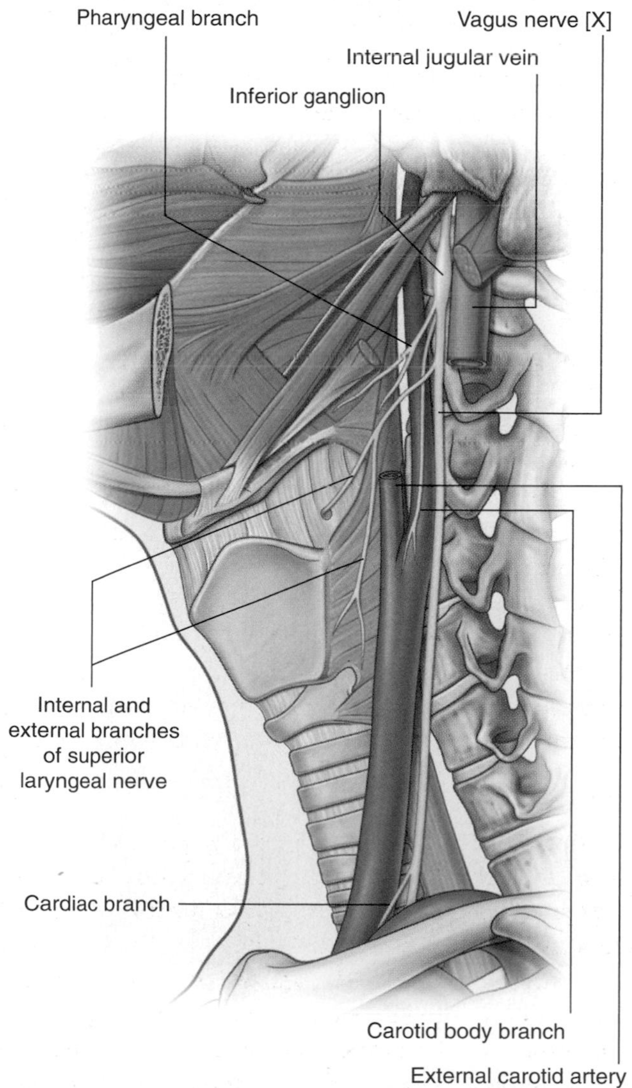


**Fig. 8.173** Vagus nerve [X] in the anterior triangle of the neck.

of the internal jugular vein as it passes downward and backward to disappear either into or beneath the anterior border of the sternocleidomastoid muscle (Fig. 8.174).

The accessory nerve gives off no branches as it passes through the anterior triangle of the neck.

### Hypoglossal nerve [XII]

The hypoglossal nerve [XII] leaves the cranial cavity through the hypoglossal canal and is medial to the internal jugular vein and internal carotid artery immediately outside the skull. As it descends, it passes outward between the internal jugular vein and internal carotid artery (Fig. 8.175). At this point it passes forward, hooking around the occipital artery, across the lateral surfaces of the internal and external carotid arteries and the lingual artery, and then continues deep to the posterior belly of the digastric and stylohyoid muscles. It passes over the surface of the hyoglossus muscle and disappears deep to the mylohyoid muscle.

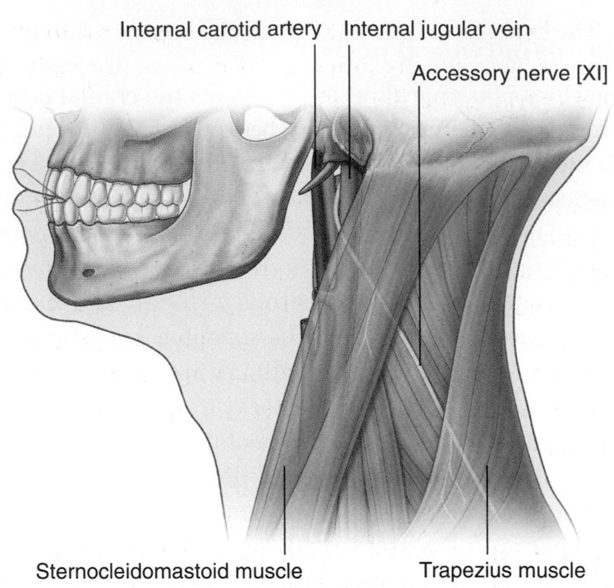


**Fig. 8.174** Accessory nerve [XI] in the posterior triangle of the neck.

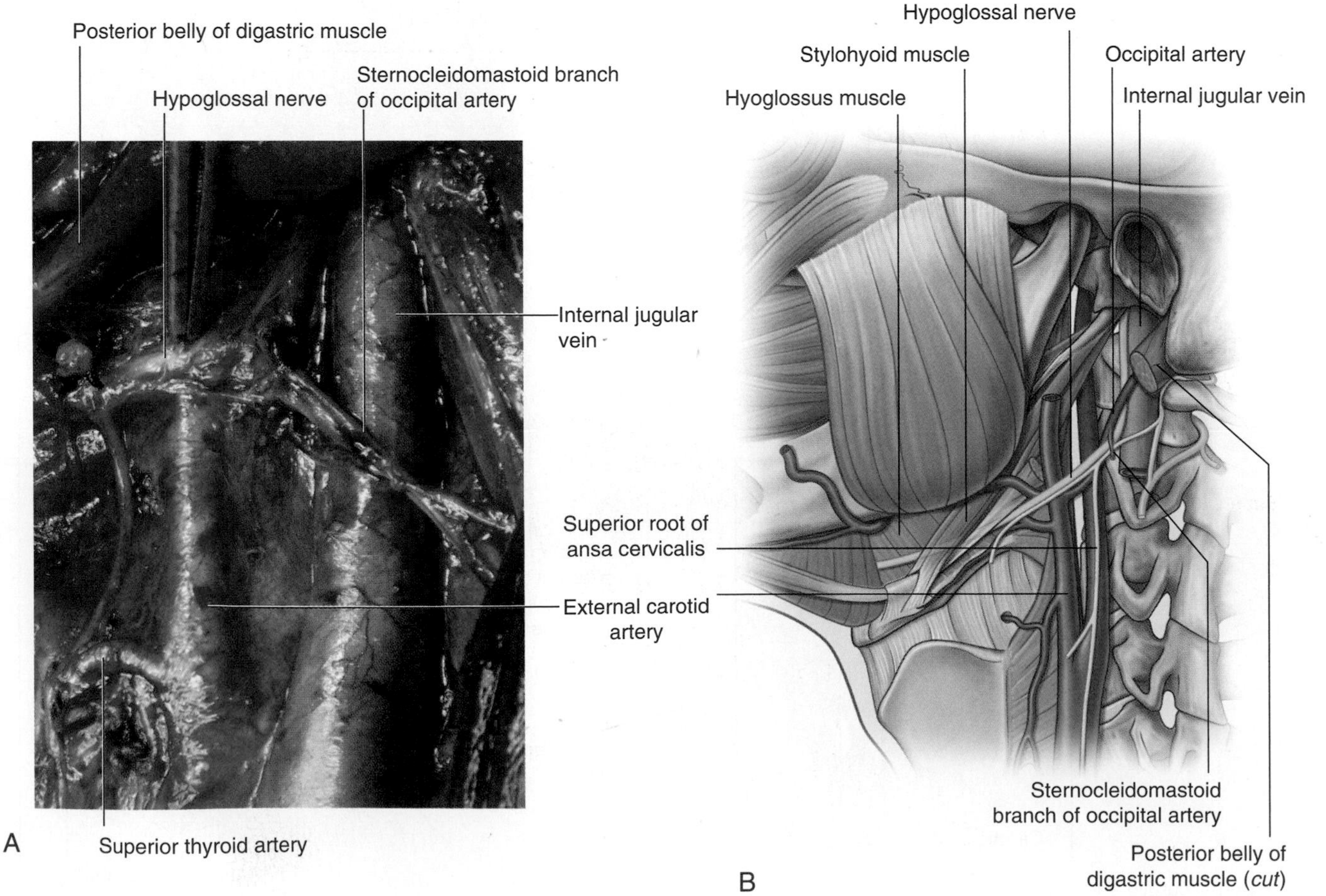

**Fig. 8.175** Hypoglossal nerve [XII]. **A.** Surgical view of hypoglossal nerve in anterior triangle of the neck. **B.** Diagram.

The hypoglossal nerve [XII], which supplies the tongue, does not give off any branches as it passes through the anterior triangle of the neck.

### Transverse cervical nerve

The transverse cervical nerve is a branch of the cervical plexus arising from the anterior rami of cervical nerves C2 and C3. It emerges from beneath the posterior border of the sternocleidomastoid muscle, near the middle of the muscle, and loops around the sternocleidomastoid to cross its anterior surface in a transverse direction (Fig. 8.176). It continues across the neck and provides cutaneous innervation to this area.

### Ansa cervicalis

The ansa cervicalis is a loop of nerve fibers from cervical nerves C1 to C3 that innervate the "strap muscles" in the anterior triangle of the neck (Fig. 8.177). It begins as

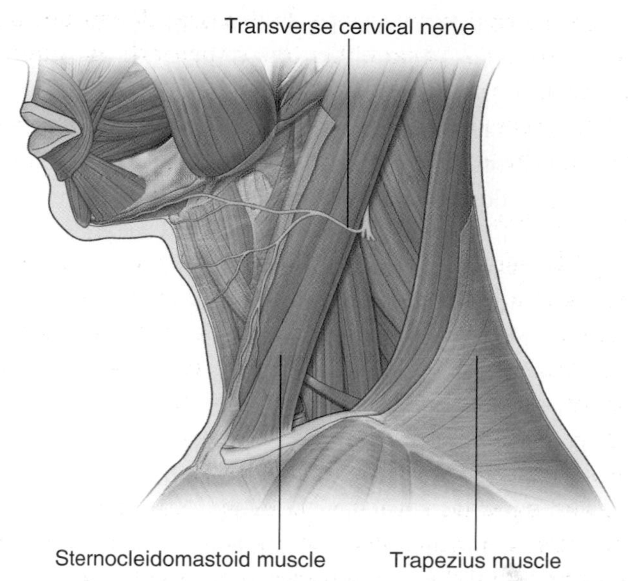


**Fig. 8.176** Transverse cervical nerve in the anterior triangle of the neck.

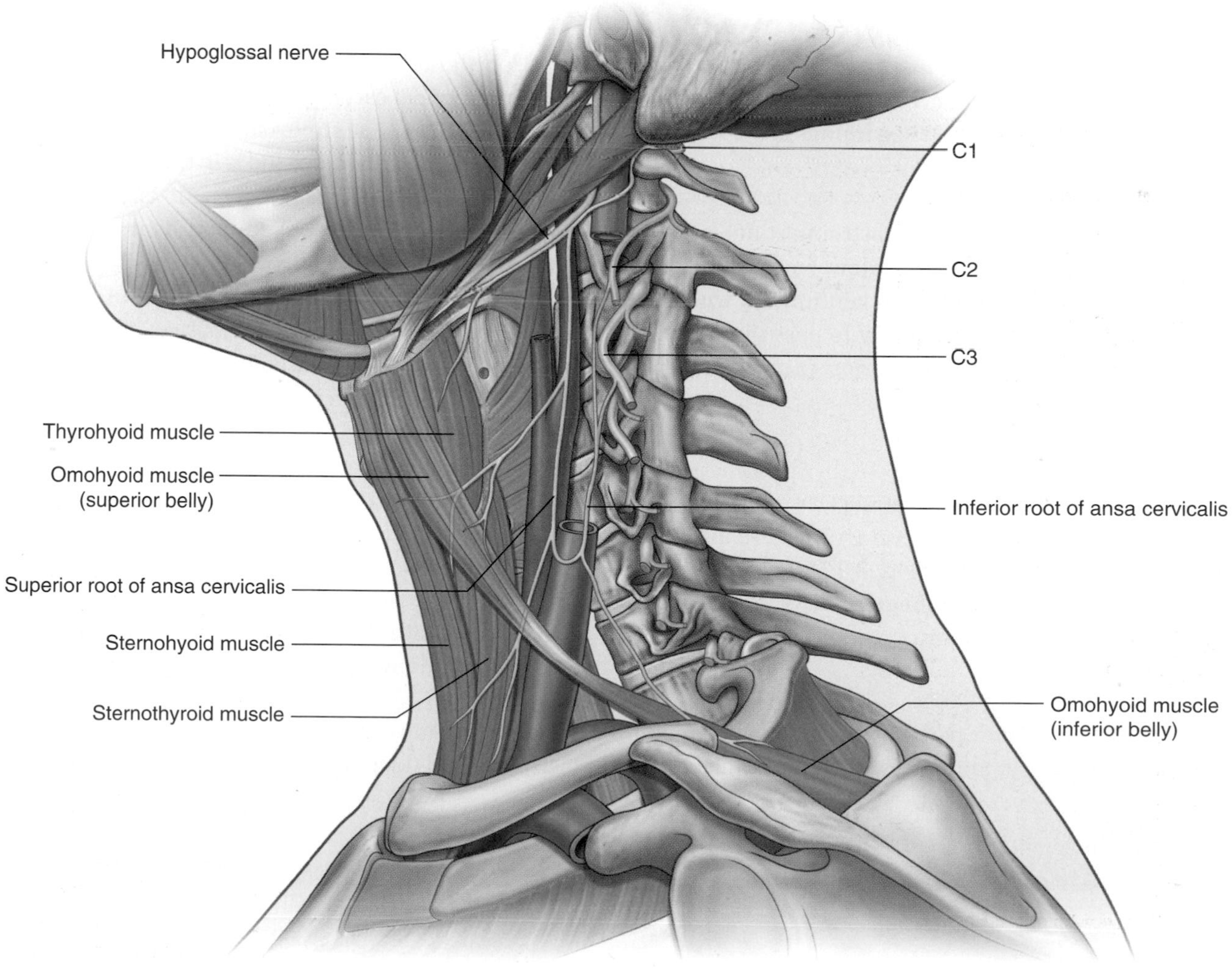

**Fig. 8.177** Ansa cervicalis.

branches from the cervical nerve C1 join the hypoglossal nerve [XII] soon after it leaves the skull.

As the hypoglossal nerve [XII] completes its descent and begins to pass forward across the internal and external carotid arteries, some of the cervical nerve fibers leave it and descend between the internal jugular vein and the internal, and then common, carotid arteries. These nerve fibers are the **superior root** of the ansa cervicalis and innervate the superior belly of the omohyoid muscle, and the upper parts of the sternohyoid and sternothyroid muscles.

Completing the loop is a direct branch from the cervical plexus containing nerve fibers from the second and third cervical nerves C2 and C3 (Fig. 8.177). This is the **inferior root** of the ansa cervicalis. It descends either medial or lateral to the internal jugular vein before turning medially to join the superior root. At this location, the ansa cervicalis gives off branches that innervate the inferior belly of the omohyoid, and the lower parts of the sternohyoid and sternothyroid muscles.

## Elements of the gastrointestinal and respiratory systems

The esophagus, trachea, pharynx, and larynx lie in the neck and are related to the anterior triangles.

### Esophagus

The esophagus is part of the gastrointestinal system and has only a short course in the lower neck. It begins at vertebral level CVI, where it is continuous with the pharynx above and courses inferiorly to pass through the thoracic inlet. It lies directly anterior to the vertebral column (Fig. 8.178B).

### Trachea

The trachea is part of the lower airway and, like the esophagus, begins at vertebral level CVI, where it is continuous with the larynx above (Fig. 8.178B). The trachea lies directly anterior to the esophagus and passes inferiorly in the midline to enter the thorax.

### Pharynx and larynx

The pharynx is a common pathway for air and food, and it connects respiratory and digestive compartments in the head with similar compartments in the lower neck (see pp. 1029–1041).

The larynx is the upper end of the lower airway. It is continuous with the trachea below and the pharynx posterosuperiorly (see pp. 1041–1058).

## Thyroid and parathyroid glands

The thyroid and parathyroid glands are endocrine glands positioned anteriorly in the neck.

Both glands begin as pharyngeal outgrowths that migrate caudally to their final positions as development continues.

The thyroid gland is a large, unpaired gland, while the parathyroid glands, usually four in number, are small and are on the posterior surface of the thyroid gland.

### Thyroid gland

The thyroid gland is anterior in the neck below and lateral to the thyroid cartilage (Fig. 8.178). It consists of two lateral **lobes** (which cover the anterolateral surfaces of the trachea, the cricoid cartilage, and the lower part of the thyroid cartilage) with an **isthmus** that connects the lateral lobes and crosses the anterior surfaces of the second and third tracheal cartilages.

Lying deep to the sternohyoid, sternothyroid, and omohyoid muscles, the thyroid gland is in the visceral compartment of the neck. This compartment also includes the pharynx, trachea, and esophagus and is surrounded by the pretracheal layers of fascia.

The thyroid gland arises as a median outgrowth from the floor of the pharynx near the base of the tongue. The foramen cecum of the tongue indicates the site of origin and the thyroglossal duct marks the path of migration of the thyroid gland to its final adult location. The thyroglossal duct usually disappears early in development, but remnants may persist as a cyst or as a connection to the foramen cecum (i.e., a fistula).

There may also be functional thyroid gland:

- associated with the tongue (a lingual thyroid),
- anywhere along the path of migration of the thyroid gland, or
- extending upward from the gland along the path of the thyroglossal duct (a pyramidal lobe).

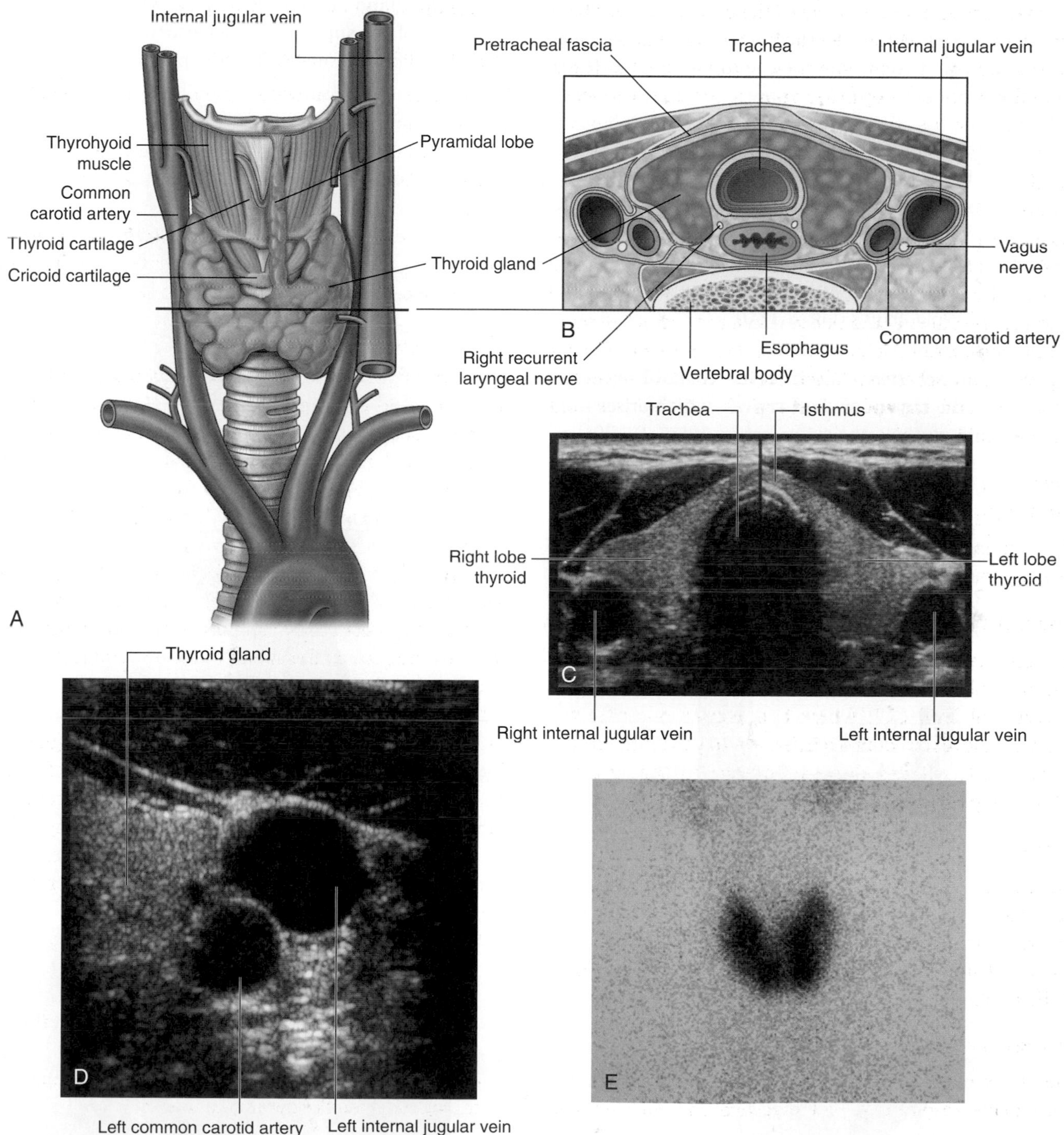


**Fig. 8.178** Thyroid gland in the anterior triangle of neck. **A.** Anterior view. **B.** Transverse view. **C.** Ultrasound scan—compound axial view of the neck. **D.** Ultrasound scan—axial view of the neck. **E.** Nuclear medicine scan—normal thyroid uptake of pertechnetate in the neck.

### Arterial supply

Two major arteries supply the thyroid gland.

***Superior thyroid artery.*** The superior thyroid artery is the first branch of the external carotid artery (Fig. 8.179). It descends, passing along the lateral margin of the thyrohyoid muscle, to reach the superior pole of the lateral lobe of the gland where it divides into anterior and posterior glandular branches:

- The **anterior glandular branch** passes along the superior border of the thyroid gland and anastomoses with its twin from the opposite side across the isthmus (Fig. 8.179).
- The **posterior glandular branch** passes to the posterior side of the gland and may anastomose with the inferior thyroid artery (Fig. 8.180).

***Inferior thyroid artery.*** The **inferior thyroid artery** is a branch of the **thyrocervical trunk**, which arises from the first part of the subclavian artery (Figs. 8.179 and 8.180). It ascends along the medial edge of the anterior scalene muscle, passes posteriorly to the carotid sheath, and reaches the inferior pole of the lateral lobe of the thyroid gland.

At the thyroid gland the inferior thyroid artery divides into an:

- inferior branch, which supplies the lower part of the thyroid gland and anastomoses with the posterior branch of the superior thyroid artery, and
- an ascending branch, which supplies the parathyroid glands.

Fig. 8.179 Vasculature of the thyroid: anterior view.

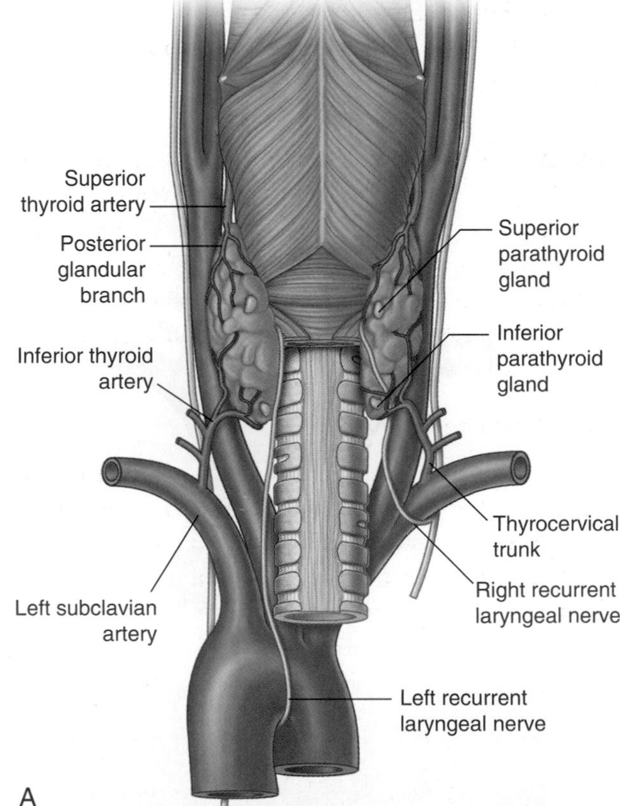


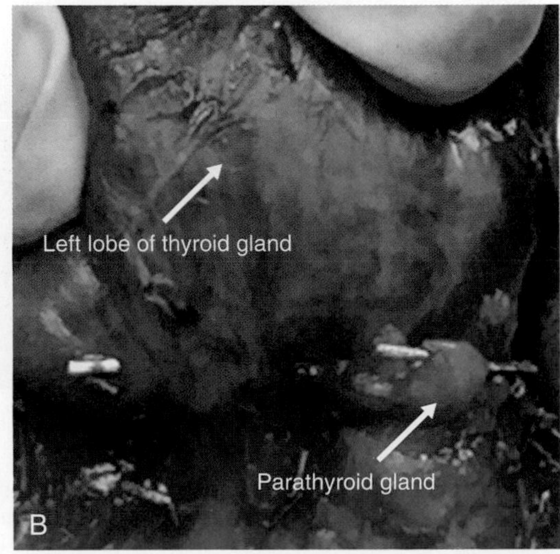


**Fig. 8.180** Superior and inferior thyroid arteries and left and right recurrent laryngeal nerves and thyroid and parathyroid glands. **A.** Posterior view. **B.** Surgical (anterolateral) view of parathyroid gland with left lobe of thyroid retracted.

Occasionally, a small **thyroid ima artery** arises from the brachiocephalic trunk or the arch of the aorta and ascends on the anterior surface of the trachea to supply the thyroid gland.

### Venous and lymphatic drainage

Three veins drain the thyroid gland (Fig. 8.179):

- The **superior thyroid vein** primarily drains the area supplied by the superior thyroid artery.
- The **middle** and **inferior thyroid veins** drain the rest of the thyroid gland.

The superior and middle thyroid veins drain into the internal jugular vein, and the inferior thyroid veins empty into the right and left brachiocephalic veins, respectively.

Lymphatic drainage of the thyroid gland is to nodes beside the trachea (paratracheal nodes) and to deep cervical nodes inferior to the omohyoid muscle along the internal jugular vein.

### Recurrent laryngeal nerves

The thyroid gland is closely related to the recurrent laryngeal nerves. After branching from the vagus nerve [X] and looping around the subclavian artery on the right and the arch of the aorta on the left, the **recurrent laryngeal nerves** ascend in a groove between the trachea and esophagus (Fig. 8.180). They pass deep to the posteromedial surface of the lateral lobes of the thyroid gland and enter the larynx by passing deep to the lower margin of the inferior constrictor of the pharynx.

Together with branches of the inferior thyroid arteries, the recurrent laryngeal nerves are clearly related to, and may pass through ligaments, one on each side, that bind the thyroid gland to the trachea and to the cricoid cartilage of the larynx. These relationships need to be considered when surgically removing or manipulating the thyroid gland.

## Parathyroid glands

The parathyroid glands are two pairs of small, ovoid, yellowish structures on the deep surface of the lateral lobes of the thyroid gland. They are designated as the superior and inferior parathyroid glands (Fig. 8.180). However, their position is quite variable and they may be anywhere from the carotid bifurcation superiorly to the mediastinum inferiorly.

Derived from the third (the inferior parathyroid glands) and fourth (the superior parathyroid glands) pharyngeal pouches, these paired structures migrate to their final adult positions and are named accordingly.

The arteries supplying the parathyroid glands are the inferior thyroid arteries, and venous and lymphatic drainage follows that described for the thyroid gland.

### In the clinic

#### Thyroid gland

The thyroid gland develops from a small region of tissue near the base of the tongue. This tissue descends as the thyroglossal duct from the foramen cecum in the posterior aspect of the tongue to pass adjacent to the anterior aspect of the middle of the hyoid bone. The thyroid tissue continues to migrate inferiorly and eventually comes to rest at the anterior aspect of the trachea in the root of the neck.

Consequently, the migration of thyroid tissue may be arrested anywhere along the embryological descent of the gland. Ectopic thyroid tissue is relatively rare. More frequently seen is the cystic change that arises from the thyroglossal duct. The usual symptom of a thyroglossal duct cyst is a midline mass. Ultrasound easily demonstrates its nature and position, and treatment is by surgical excision. The whole of the duct as well as a small part of the anterior aspect of the hyoid bone must be excised to prevent recurrence.

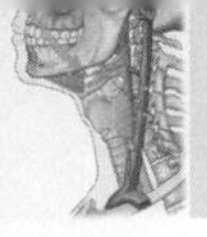



## In the clinic

### Thyroidectomy

A thyroidectomy is a common surgical procedure. In most cases it involves excision of part or most of the thyroid gland. This surgical procedure is usually carried out for benign diseases, such as multinodular goiter and thyroid cancer.

Given the location of the thyroid gland, there is a possibility of damaging other structures when carrying out a thyroidectomy, namely the parathyroid glands and the recurrent laryngeal nerve (Fig. 8.181). Assessment of the vocal folds is necessary before and after thyroid surgery because the recurrent laryngeal nerves are closely related to ligaments that bind the gland to the larynx and can be easily traumatized during surgical procedures.

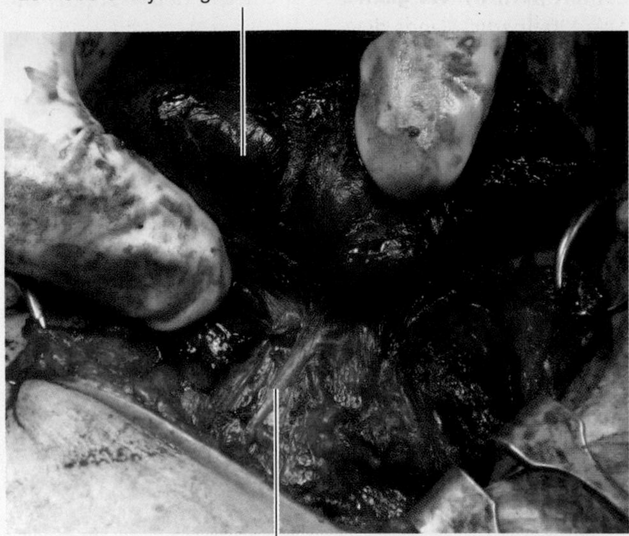


**Fig. 8.181** Surgical view of left lobe of enlarged thyroid (goiter) retracted to show close association with recurrent laryngeal nerve.

## In the clinic

### Thyroid gland pathology

Thyroid gland pathology is extremely complex. In essence, thyroid gland pathology should be assessed from two points of view. First, the thyroid gland may be diffusely or focally enlarged, for which there are numerous causes. Second, the thyroid gland may undersecrete or oversecrete the hormone thyroxine.

One of the commonest disorders of the thyroid gland is a **multinodular goiter,** which is a diffuse irregular enlargement of the thyroid gland with areas of thyroid hypertrophy and colloid cyst formation. Most patients are euthyroid (i.e., have normal serum thyroxine levels). The typical symptom is a diffuse mass in the neck, which may be managed medically or may need surgical excision if the mass is large enough to affect the patient's life or cause respiratory problems.

Isolated nodules in the thyroid gland may be a dominant nodule in a multinodular gland or possibly an isolated tumor of the thyroid gland. Isolated tumors may or may not secrete thyroxine depending on their cellular morphology. Treatment is usually by excision.

Immunological diseases may affect the thyroid gland and may overstimulate it to produce excessive thyroxine. These diseases may be associated with other extrathyroid manifestations, which include exophthalmos, pretibial myxedema, and nail changes. Other causes of diffuse thyroid stimulation include viral thyroiditis. Some diseases may cause atrophy of the thyroid gland, leading to undersecretion of thyroxine (**myxedema**).

## In the clinic

### Ectopic parathyroid glands

The parathyroid glands develop from the third and fourth pharyngeal pouches and translocate to their more adult locations during development. The position of the glands can be highly variable, sometimes being situated high in the neck or in the thorax. Tumors develop in any of these locations (Fig. 8.182).

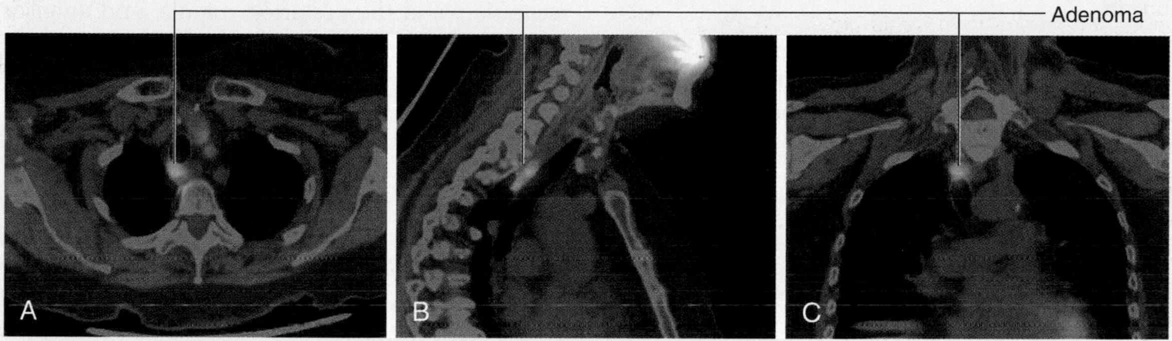


**Fig. 8.182** Ectopic parathyroid adenoma in superior mediastinum. Noncontrast hybrid single photon emission computed tomography/computed tomography (SPECT/CT). **A.** Transverse view. **B.** Sagittal view. **C.** Coronal view.

## Location of structures in different regions of the anterior triangle of the neck

The regional location of major structures in the anterior triangle of the neck is summarized in Table 8.14. Structures can be identified as being within a specific subdivision, passing into a specific subdivision from outside the area, originating in one subdivision and passing to another subdivision, or passing through several subdivisions while traversing the region.

**Table 8.14** Subdivisions of the anterior triangle of the neck—a regional approach

| Subdivision | Boundaries | Contents |
|---|---|---|
| Submental triangle (unpaired) | Mandibular symphysis; anterior belly of digastric muscle; body of hyoid bone | Submental lymph nodes; tributaries forming the anterior jugular vein |
| Submandibular triangle (paired) | Lower border of mandible; anterior belly of digastric muscle; posterior belly of digastric muscle | Submandibular gland; submandibular lymph nodes; hypoglossal nerve [XII]; mylohyoid nerve; facial artery and vein |
| Carotid triangle (paired) | Posterior belly of digastric muscle; superior belly of omohyoid muscle; anterior border of sternocleidomastoid muscle | Tributaries to common facial vein; cervical branch of facial nerve [VII]; common carotid artery; external and internal carotid arteries; superior thyroid; ascending pharyngeal; lingual, facial, and occipital arteries; internal jugular vein; vagus [X], accessory [XI], and hypoglossal [XII] nerves; superior and inferior roots of ansa cervicalis; transverse cervical nerve |
| Muscular triangle (paired) | Midline of neck; superior belly of omohyoid muscle; anterior border of sternocleidomastoid muscle | Sternohyoid, omohyoid, sternohyoid, and thyrohyoid muscles; thyroid and parathyroid glands; pharynx |

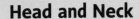

## Posterior triangle of the neck

The posterior triangle of the neck is on the lateral aspect of the neck in direct continuity with the upper limb (Fig. 8.183). It is bordered:

- anteriorly by the posterior edge of the sternocleidomastoid muscle,
- posteriorly by the anterior edge of the trapezius muscle,
- basally by the middle one-third of the clavicle, and
- apically by the occipital bone just posterior to the mastoid process where the attachments of the trapezius and sternocleidomastoid come together.

The roof of the posterior triangle consists of an investing layer of cervical fascia that surrounds the sternocleidomastoid and trapezius muscles as it passes through the region.

The muscular floor of the posterior triangle is covered by the prevertebral layer of cervical fascia; and from superior to inferior consists of the splenius capitis, levator scapulae, and the posterior, middle, and anterior scalene muscles.

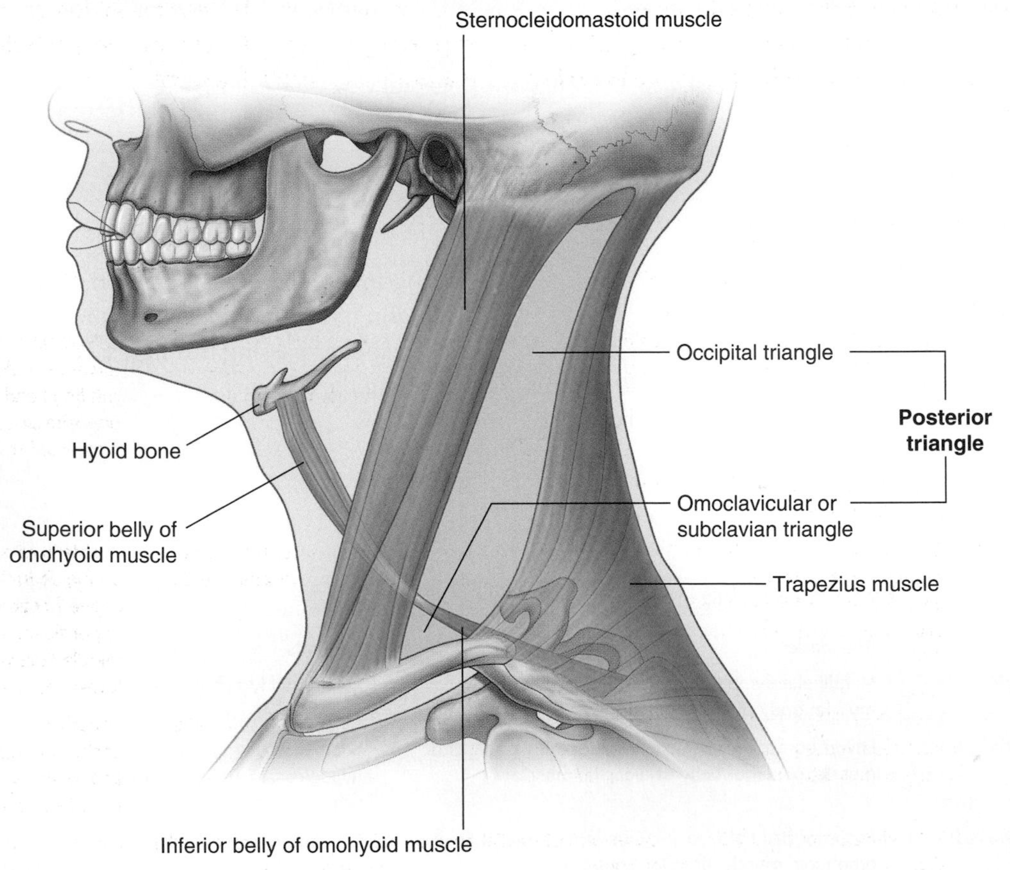


**Fig. 8.183** Borders of the posterior triangle of the neck.

### Muscles

Numerous muscles participate in forming the borders and floor of the posterior triangle of the neck (Table 8.15).

In addition, the **omohyoid** muscle passes across the inferior part of the posterior triangle before disappearing under the sternocleidomastoid muscle and emerging in the anterior triangle (Fig. 8.184). It is enclosed in the investing layer of cervical fascia and crosses the posterior triangle from lateral to medial as it continues in a superior direction. It originates on the superior border of the scapula, just medial to the scapular notch and eventually inserts into the inferior border of the body of the hyoid bone. It has two bellies connected by a tendon, which is anchored by a fascial sling to the clavicle:

- The **superior belly** is in the anterior triangle.
- The **inferior belly** crosses the posterior triangle, subdividing it into a small, **omoclavicular or subclavian triangle** inferiorly and a much larger **occipital triangle** superiorly.

The omohyoid is innervated by branches of the ansa cervicalis (anterior rami from C1 to C3) and it depresses the hyoid bone.

### Vessels

#### External jugular vein

One of the most superficial structures passing through the posterior triangle of the neck is the external jugular vein (Fig. 8.185). This large vein forms near the angle of the mandible, when the posterior branch of the retromandibular and posterior auricular veins join, and descends through the neck in the superficial fascia.

After crossing the sternocleidomastoid muscle, the external jugular vein enters the posterior triangle and continues its vertical descent.

**Table 8.15** Muscles associated with the posterior triangle of the neck; parentheses indicate possible involvement

| Muscle | Origin | Insertion | Innervation | Function |
|---|---|---|---|---|
| Sternocleidomastoid | | | | |
| —Sternal head | Upper part of anterior surface of manubrium of sternum | Lateral one-half of superior nuchal line | Accessory nerve [XI] and branches from anterior rami of C2 to C3 (C4) | Individually—will tilt head toward shoulder on same side rotating head to turn face to opposite side; acting together, draw head forward |
| —Clavicular head | Superior surface of medial one-third of clavicle | Lateral surface of mastoid process | | |
| Trapezius | Superior nuchal line; external occipital protuberance; ligamentum nuchae; spinous processes of vertebrae CVII to TXII | Lateral one-third of clavicle; acromion; spine of scapula | Motor—accessory nerve [XI]; proprioception—C3 and C4 | Assists in rotating the scapula during abduction of humerus above horizontal; upper fibers—elevate, middle fibers—adduct, lower fibers—depress scapula |
| Splenius capitis | Lower half of ligamentum nuchae; spinous processes of vertebrae CVII to TIV | Mastoid process, skull below lateral one-third of superior nuchal line | Posterior rami of middle cervical nerves | Together, draw head backward; individually, draw and rotate head to one side (turn face to same side) |
| Levator scapulae | Transverse processes of CI to CIV | Upper part of medial border of scapula | C3, C4; and dorsal scapular nerve (C4, C5) | Elevates scapula |
| Posterior scalene | Posterior tubercles of transverse processes of vertebrae CIV to CVI | Upper surface of rib II | Anterior rami of C5 to C7 | Elevation of rib II |
| Middle scalene | Transverse processes of vertebrae CII to CVII | Upper surface of rib I posterior to the groove for the subclavian artery | Anterior rami of C3 to C7 | Elevation of rib I |
| Anterior scalene | Anterior tubercles of the transverse processes of vertebrae CIII to CVI | Scalene tubercle and upper surface of rib I | Anterior rami of C4 to C7 | Elevation of rib I |
| Omohyoid | Superior border of scapula medial to scapular notch | Inferior border of body of hyoid bone | Ansa cervicalis; anterior rami of C1 to C3 | Depress the hyoid bone |

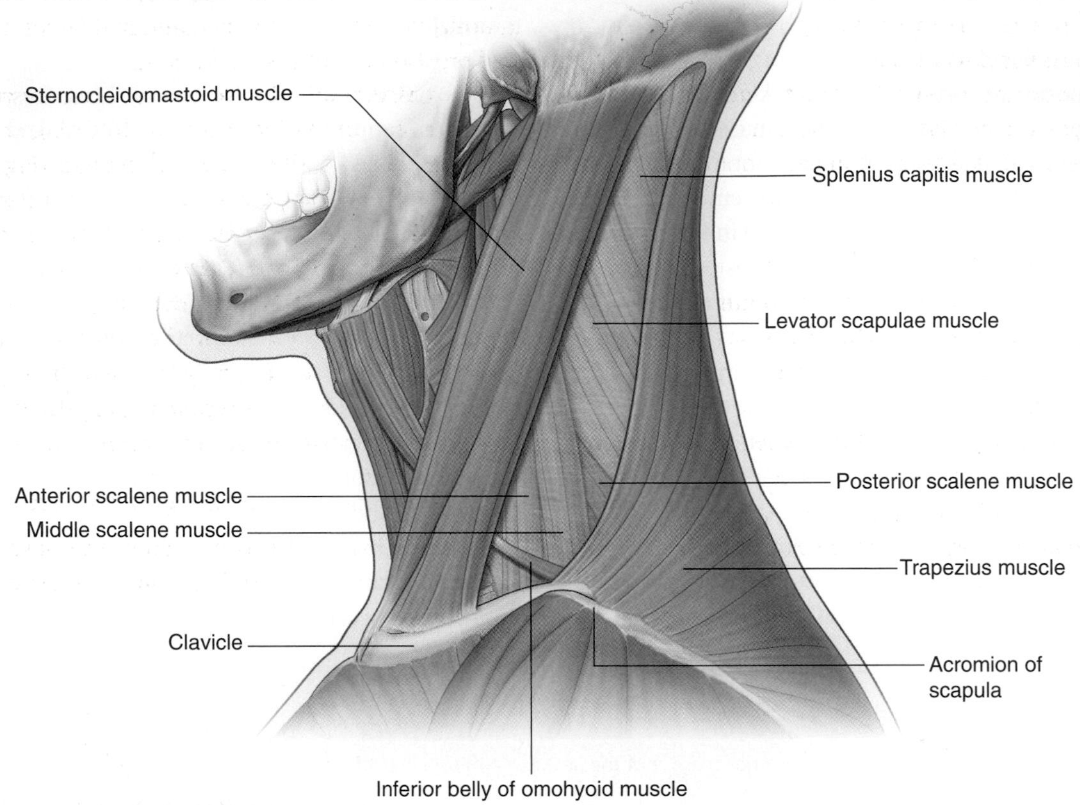


**Fig. 8.184** Muscles of the posterior triangle of the neck.

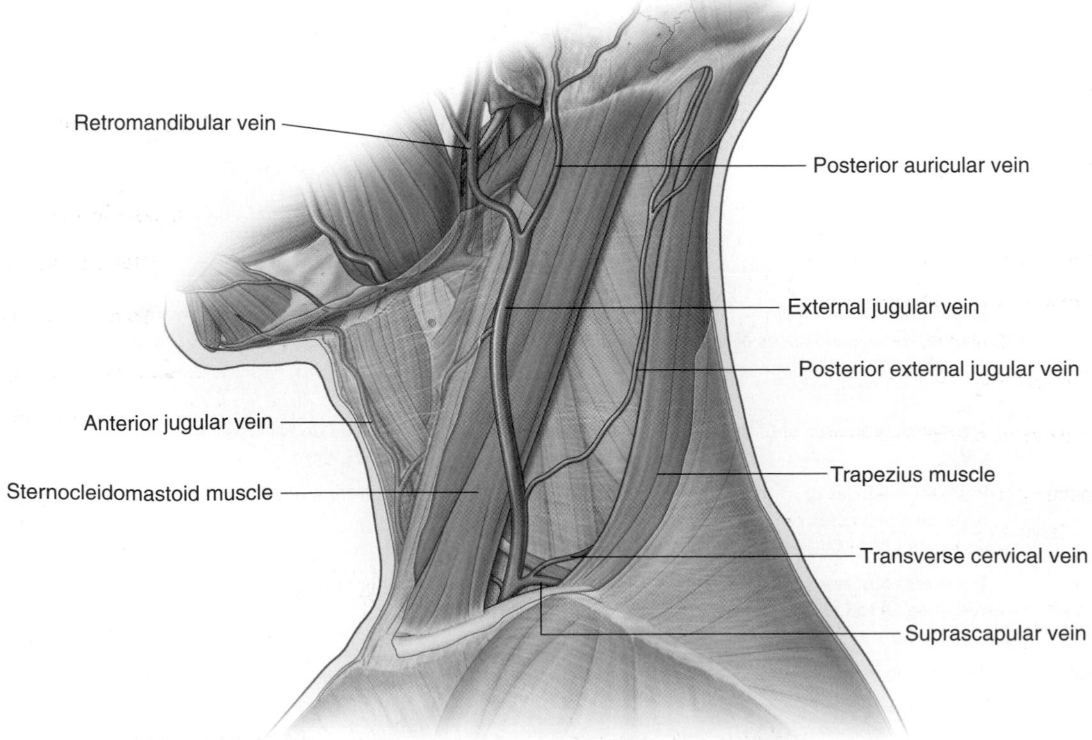


**Fig. 8.185** External jugular vein in the posterior triangle of the neck.

In the lower part of the posterior triangle, the external jugular vein pierces the investing layer of cervical fascia and ends in the subclavian vein.

Tributaries to the external jugular vein while it traverses the posterior triangle of the neck include the transverse cervical, suprascapular, and anterior jugular veins.

### Subclavian artery and its branches

Several arteries are found within the boundaries of the posterior triangle of the neck. The largest is the third part of the subclavian artery as it crosses the base of the posterior triangle (Fig. 8.186).

The **first part of the subclavian artery** ascends to the medial border of the anterior scalene muscle from either the brachiocephalic trunk on the right side or directly from the arch of the aorta on the left side. It has numerous branches.

The **second part of the subclavian artery** passes laterally between the anterior and middle scalene muscles, and one branch may arise from it.

The **third part of the subclavian artery** emerges from between the anterior and middle scalene muscles to cross the base of the posterior triangle (Fig. 8.186). It extends from the lateral border of the anterior scalene muscle to the lateral border of rib I where it becomes the **axillary artery** and continues into the upper limb.

A single branch (the **dorsal scapular artery**) may arise from the third part of the subclavian artery. This branch passes posterolaterally to reach the superior angle of the scapula where it descends along the medial border of the scapula posterior to the rhomboid muscles.

### Transverse cervical and suprascapular arteries

Two other small arteries also cross the base of the posterior triangle. These are the transverse cervical and the

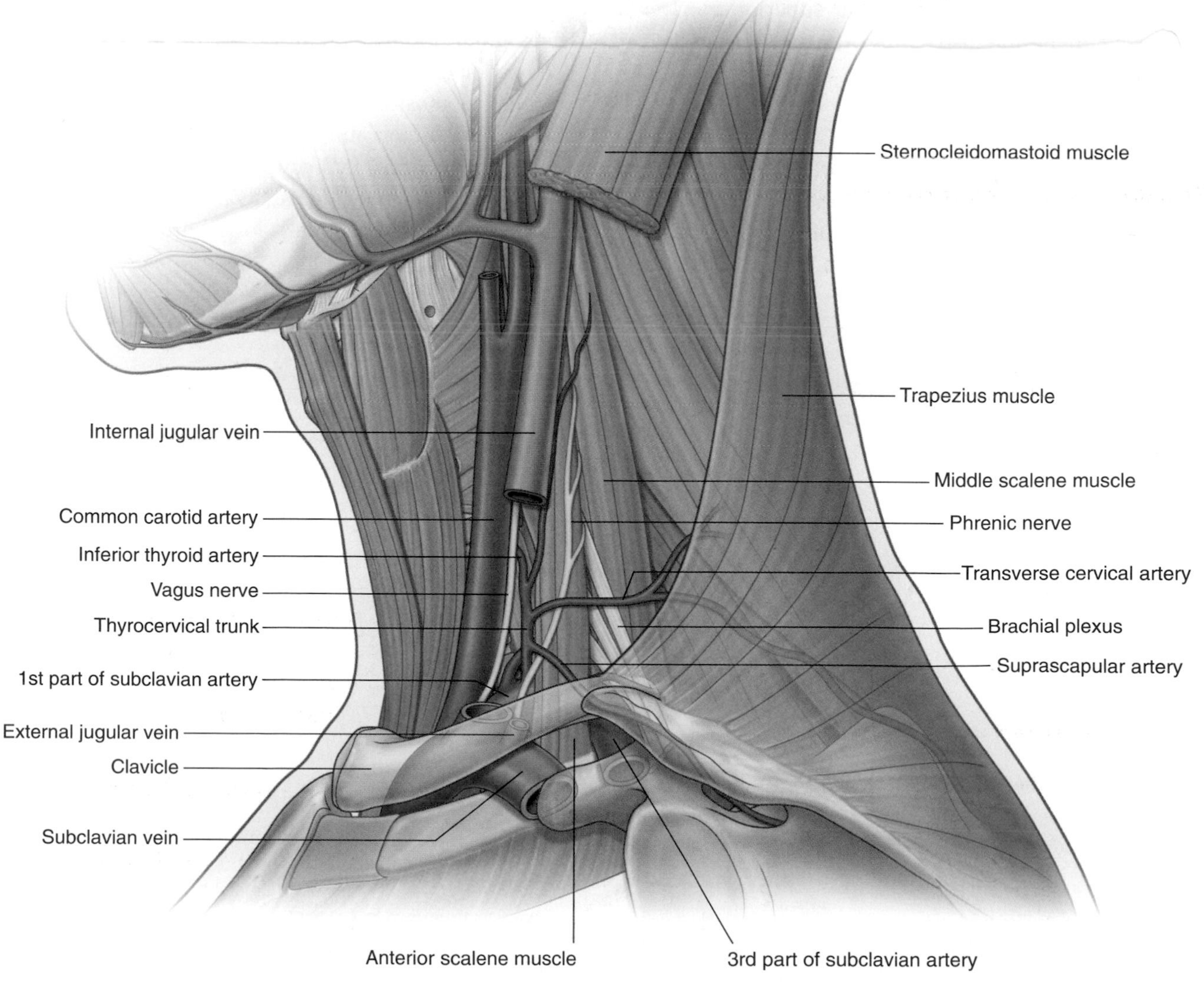


**Fig. 8.186** Arteries in the posterior triangle of the neck.

suprascapular arteries (Fig. 8.186). They are both branches of the thyrocervical trunk, which arises from the first part of the subclavian artery.

After branching from the thyrocervical trunk, the **transverse cervical artery** passes laterally and slightly posteriorly across the base of the posterior triangle anterior to the anterior scalene muscle and the brachial plexus. Reaching the deep surface of the trapezius muscle, it divides into superficial and deep branches:

- The **superficial branch** continues on the deep surface of the trapezius muscle.
- The **deep branch** continues on the deep surface of the rhomboid muscles near the medial border of the scapula.

The **suprascapular artery**, also a branch of the thyrocervical trunk, passes laterally, in a slightly downward direction across the lowest part of the posterior triangle, and ends up posterior to the clavicle (Fig. 8.186). Approaching the scapula, it passes over the superior transverse scapular ligament and distributes branches to muscles on the posterior surface of the scapula.

### Veins

Veins accompany all the arteries described previously.

The **subclavian vein** is a continuation of the axillary vein and begins at the lateral border of rib I. As it crosses the base of the posterior triangle, the external jugular, and, possibly, the suprascapular and transverse cervical veins enter it (Fig. 8.185). It ends by joining with the internal jugular vein to form the brachiocephalic vein near the sternoclavicular joint. In the posterior triangle it is anterior to, and slightly lower than, the subclavian artery and passes anterior to the anterior scalene muscle.

Transverse cervical and suprascapular veins travel with each of the similarly named arteries. These veins become tributaries to either the external jugular vein or the initial part of the subclavian vein.

## Nerves

A variety of nerves pass through or are within the posterior triangle. These include the accessory nerve [XI], branches of the cervical plexus, components forming the brachial plexus, and branches of the brachial plexus.

### Accessory nerve

The accessory nerve [XI] exits the cranial cavity through the jugular foramen. It descends through the neck in a posterior direction, to reach the anterior border of the sternocleidomastoid muscle. Passing either deep to or through and innervating the sternocleidomastoid muscle, the accessory nerve [XI] continues to descend and enters the posterior triangle (Fig. 8.187). It crosses the posterior triangle, still in an obliquely downward direction, within

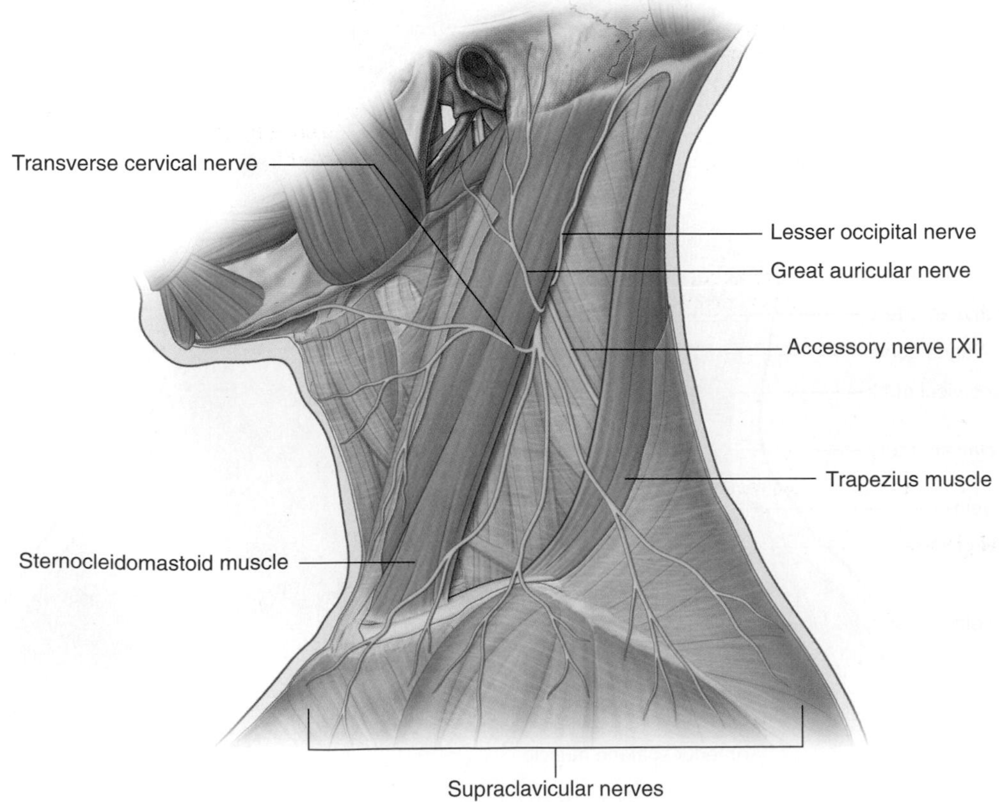


**Fig. 8.187** Accessory nerve and cutaneous branches of the cervical plexus in the posterior triangle of the neck.

the investing layer of cervical fascia as this fascia crosses between the sternocleidomastoid and trapezius muscles. When the accessory nerve [XI] reaches the anterior border of the trapezius muscle, it continues on the deep surface of the trapezius and innervates it. The superficial location of the accessory nerve as it crosses the posterior triangle makes it susceptible to injury.

## Cervical plexus

The cervical plexus is formed by the anterior rami of cervical nerves C1 to C4 (Fig. 8.188).

The cervical plexus forms in the substance of the muscles making up the floor of the posterior triangle within the prevertebral layer of cervical fascia, and consists of:

- muscular (or deep) branches, and
- cutaneous (or superficial) branches.

The cutaneous branches are visible in the posterior triangle emerging from beneath the posterior border of the sternocleidomastoid muscle (Fig. 8.187).

### Muscular branches

Muscular (deep) branches of the cervical plexus distribute to several groups of muscles. A major branch is the **phrenic nerve**, which supplies the diaphragm with both sensory and motor innervation (Fig. 8.188). It arises from the anterior rami of cervical nerves C3 to C5. Hooking around the upper lateral border of the anterior scalene muscle, the nerve continues inferiorly across the anterior surface of the anterior scalene within the prevertebral fascia to enter the thorax (Fig. 8.189). As the nerve descends in the neck, it is "pinned" to the anterior scalene muscle by the transverse cervical and suprascapular arteries.

Several muscular branches of the cervical plexus supply prevertebral and lateral vertebral muscles, including the rectus capitis anterior, rectus capitis lateralis, longus colli, and longus capitis (Fig. 8.189 and Table 8.16).

The cervical plexus also contributes to the formation of the superior and inferior roots of the ansa cervicalis (Fig. 8.188). This loop of nerves receives contributions from the anterior rami of the cervical nerves C1 to C3 and innervates the infrahyoid muscles.

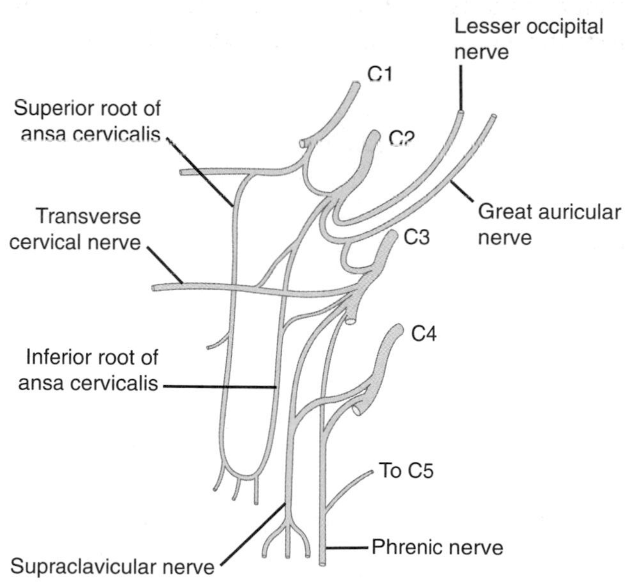


**Fig. 8.188** Cervical plexus.

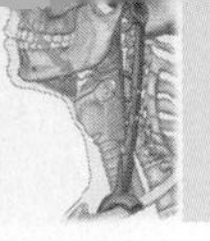





**Fig. 8.189** Prevertebral and lateral vertebral muscles supplied by cervical plexus.

**Table 8.16** Prevertebral and lateral vertebral muscles

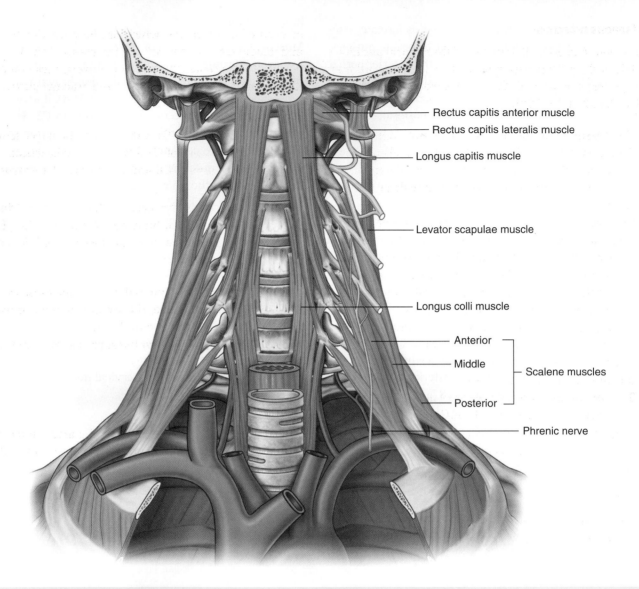

| Muscle | Origin | Insertion | Innervation | Function |
|---|---|---|---|---|
| Rectus capitis anterior | Anterior surface of lateral part of atlas and its transverse process | Inferior surface of basilar part of occipital bone | Branches from anterior rami of C1, C2 | Flexes head at atlanto-occipital joint |
| Rectus capitis lateralis | Superior surface of transverse process of atlas | Inferior surface of jugular process of occipital bone | Branches from anterior rami of C1, C2 | Flexes head laterally to same side |
| Longus colli | | | | |
| —Superior oblique part | Anterior tubercles of transverse processes of vertebrae CIII to CV | Tubercle of anterior arch of atlas | Branches from anterior rami of C2 to C6 | Flexes neck anteriorly and laterally and slight rotation to opposite side |
| —Inferior oblique part | Anterior surface of bodies of vertebrae TI, TII, and maybe TIII | Anterior tubercles of transverse processes of vertebrae CV and CVI | | |
| —Vertical part | Anterior surface of bodies of TI to TIII and CV to CVII | Anterior surface of bodies of vertebrae CII to CIV | | |
| Longus capitis | Tendinous slips to transverse processes of vertebrae CIII to CVI | Inferior surface of basilar part of occipital bone | Branches from anterior rami of C1 to C3 | Flexes the head |

### Cutaneous branches

Cutaneous (superficial) branches of the cervical plexus are visible in the posterior triangle as they pass outward from the posterior border of the sternocleidomastoid muscle (Figs. 8.187 and 8.188):

- The **lesser occipital nerve** consists of contributions from cervical nerve C2 (Fig. 8.188), ascends along the posterior border of the sternocleidomastoid muscle, and distributes to the skin of the neck and scalp posterior to the ear.
- The **great auricular nerve** consists of branches from cervical nerves C2 and C3, emerges from the posterior border of the sternocleidomastoid muscle, and ascends across the muscle to the base of the ear, supplying the skin of the parotid region, the ear, and the mastoid area.
- The **transverse cervical nerve** consists of branches from the cervical nerves C2 and C3, passes around the midpart of the sternocleidomastoid muscle, and continues horizontally across the muscle to supply the lateral and anterior parts of the neck.
- The **supraclavicular nerves** are a group of cutaneous nerves from cervical nerves C3 and C4 that, after emerging from beneath the posterior border of the sternocleidomastoid muscle, descend and supply the skin over the clavicle and shoulder as far inferiorly as rib II.

### Brachial plexus

The brachial plexus forms from the anterior rami of cervical nerves C5 to C8 and thoracic nerve T1. The contributions of each of these nerves, which are between the anterior and middle scalene muscles, are the **roots** of the brachial plexus. As the roots emerge from between these muscles, they form the next component of the brachial plexus (the **trunks**) as follows:

- the anterior rami of C5 and C6 form the upper trunk,
- the anterior ramus of C7 forms the middle trunk,
- the anterior rami of C8 and T1 form the lower trunk.

The trunks cross the base of the posterior triangle (see Fig. 8.186). Several branches of the brachial plexus may be visible in the posterior triangle (see Fig. 7.54 on pg. 730). These include the:

- **dorsal scapular nerve** to the rhomboid muscles,
- **long thoracic nerve** to the serratus anterior muscle,
- nerve to the subclavius muscle, and
- **suprascapular nerve** to the supraspinatus and infraspinatus muscles.

## Root of the neck

The root of the neck (Fig. 8.190) is the area immediately superior to the superior thoracic aperture and axillary inlets. It is bounded by:

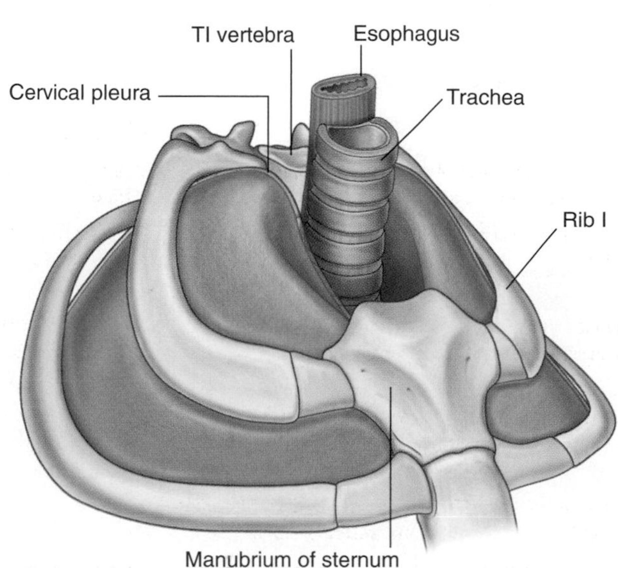


**Fig. 8.190** Root of the neck.

- the top of the manubrium of the sternum and superior margin of the clavicle anteriorly, and
- the top of the thoracic vertebra TI and the superior margin of the scapula to the coracoid process posteriorly.

It contains structures passing between the neck, thorax, and upper limb. There is also an extension of the thoracic cavity projecting into the root of the neck (Fig. 8.190). This consists of an upward projection of the pleural cavity, on both sides, and includes the cervical part of the parietal pleura (cupula), and the apical part of the superior lobe of each lung.

Anteriorly, the pleural cavity extends above the top of the manubrium of the sternum and superior border of rib I, while posteriorly, due to the downward slope of the superior thoracic aperture, the pleural cavity remains below the top of vertebra TI.

## Vessels

### Subclavian arteries

The subclavian arteries on both sides arch upward out of the thorax to enter the root of the neck (Fig. 8.191).

The **right subclavian artery** begins posterior to the sternoclavicular joint as one of two terminal branches of the brachiocephalic trunk. It arches superiorly and laterally to pass anterior to the extension of the pleural cavity in the root of the neck and posterior to the anterior scalene muscle. Continuing laterally across rib I, it becomes the **axillary artery** as it crosses its lateral border.

The **left subclavian artery** begins lower in the thorax than the right subclavian artery as a direct branch of the arch of the aorta. Lying posterior to the left common carotid artery and lateral to the trachea, it ascends and arches laterally, passing anterior to the extension of the pleural cavity and posterior to the anterior scalene muscle.

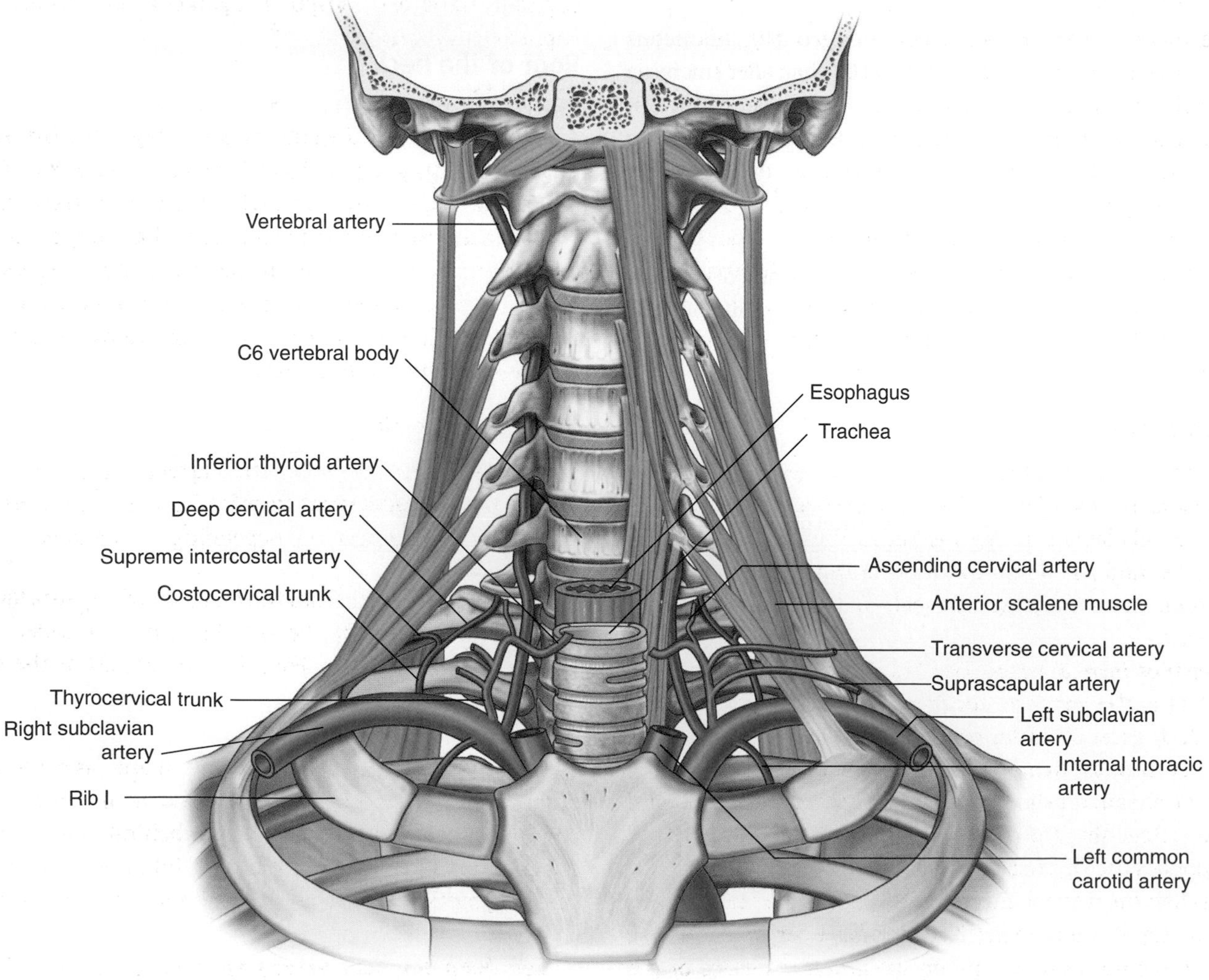

**Fig. 8.191** Vasculature of the root of the neck.

It continues laterally over rib I, and becomes the axillary artery as it crosses the lateral border of rib I.

Both subclavian arteries are divided into three parts by the anterior scalene muscle (Fig. 8.191):

- The first part extends from the origin of the artery to the anterior scalene muscle.
- The second part is the part of the artery posterior to the anterior scalene muscle.
- The third part is the part lateral to the anterior scalene muscle before the artery reaches the lateral border of rib I.

All branches from the right and left subclavian arteries arise from the first part of the artery, except in the case of one branch (the costocervical trunk) on the right side (Fig. 8.191). The branches include the vertebral artery, the thyrocervical trunk, the internal thoracic artery, and the costocervical trunk.

### Vertebral artery

The **vertebral artery** is the first branch of the subclavian artery as it enters the root of the neck (Fig. 8.191). A large branch, arising from the first part of the subclavian artery medial to the anterior scalene muscle, it ascends and enters the foramen in the transverse process of vertebra CVI. Continuing to pass superiorly, the vertebral artery passes through the foramina of vertebrae CV to CI. At the superior border of vertebra CI, the artery turns medially and crosses the posterior arch of vertebra CI. From here it passes through the foramen magnum to enter the posterior cranial fossa.

### Thyrocervical trunk

The second branch of the subclavian artery is the **thyrocervical trunk** (Fig. 8.191). It arises from the first part of the subclavian artery medial to the anterior scalene muscle, and divides into three branches—the inferior thyroid, the transverse cervical, and the suprascapular arteries.

***Inferior thyroid artery.*** The inferior thyroid artery (Fig. 8.191) is the superior continuation of the thyrocervical trunk. It ascends, anterior to the anterior scalene muscle, and eventually turns medially, crossing posterior to the carotid sheath and its contents and anterior to the vertebral artery. Reaching the posterior surface of the thyroid gland it supplies the thyroid gland.

When the inferior thyroid artery turns medially, it gives off an important branch (the **ascending cervical artery**), which continues to ascend on the anterior surface of the prevertebral muscles, supplying these muscles and sending branches to the spinal cord.

***Transverse cervical artery.*** The middle branch of the thyrocervical trunk is the **transverse cervical artery** (Fig. 8.191). This branch passes laterally, across the anterior surface of the anterior scalene muscle and the phrenic nerve, and enters and crosses the base of the posterior triangle of the neck. It continues to the deep surface of the trapezius muscle, where it divides into superficial and deep branches:

- The **superficial branch** continues on the deep surface of the trapezius muscle.
- The **deep branch** continues on the deep surface of the rhomboid muscles near the medial border of the scapula.

***Suprascapular artery.*** The lowest branch of the thyrocervical trunk is the **suprascapular artery** (Fig. 8.191). This branch passes laterally, crossing anterior to the anterior scalene muscle, the phrenic nerve, the third part of the subclavian artery, and the trunks of the brachial plexus. At the superior border of the scapula, it crosses over the superior transverse scapular ligament and enters the supraspinatus fossa.

### Internal thoracic artery

The third branch of the subclavian artery is the **internal thoracic artery** (Fig. 8.191). This artery branches from the inferior edge of the subclavian artery and descends.

It passes posterior to the clavicle and the large veins in the region and anterior to the pleural cavity. It enters the thoracic cavity posterior to the ribs and anterior to the transversus thoracis muscle and continues to descend giving off numerous branches.

### Costocervical trunk

The final branch of the subclavian artery in the root of the neck is the **costocervical trunk** (Fig. 8.191). It arises in a slightly different position, depending on the side:

- On the left, it arises from the first part of the subclavian artery, just medial to the anterior scalene muscle.
- On the right, it arises from the second part of the subclavian artery.

On both sides, the costocervical trunk ascends and passes posteriorly over the dome of the pleural cavity and continues in a posterior direction behind the anterior scalene muscle. Eventually it divides into two branches—the deep cervical and the supreme intercostal arteries:

- The **deep cervical artery** ascends in the back of the neck and anastomoses with the descending branch of the occipital artery.

- The **supreme intercostal artery** descends anterior to rib I and divides to form the posterior intercostal arteries for the first two intercostal spaces.

### Veins

Numerous veins pass through the root of the neck. Small veins accompany each of the arteries described above, and large veins form major drainage channels.

The **subclavian veins** begin at the lateral margin of rib I as continuations of the **axillary veins**. Passing medially on each side, just anterior to the anterior scalene muscles, each subclavian vein is joined by the internal jugular vein to form the brachiocephalic veins.

The only tributary to each subclavian vein is an external jugular vein.

The veins accompanying the numerous arteries in this region empty into other veins.

### Nerves

Several nerves and components of the nervous system pass through the root of the neck.

#### Phrenic nerves

The phrenic nerves are branches of the cervical plexus and arise on each side as contributions from the anterior rami of cervical nerves C3 to C5 come together. Passing around the upper lateral border of each anterior scalene muscle, the phrenic nerves continue inferiorly across the anterior surface of each anterior scalene muscle within the prevertebral layer of cervical fascia (Fig. 8.192). Leaving the lower edge of the anterior scalene muscle each phrenic nerve passes between the subclavian vein and artery to enter the thorax and continue to the diaphragm.

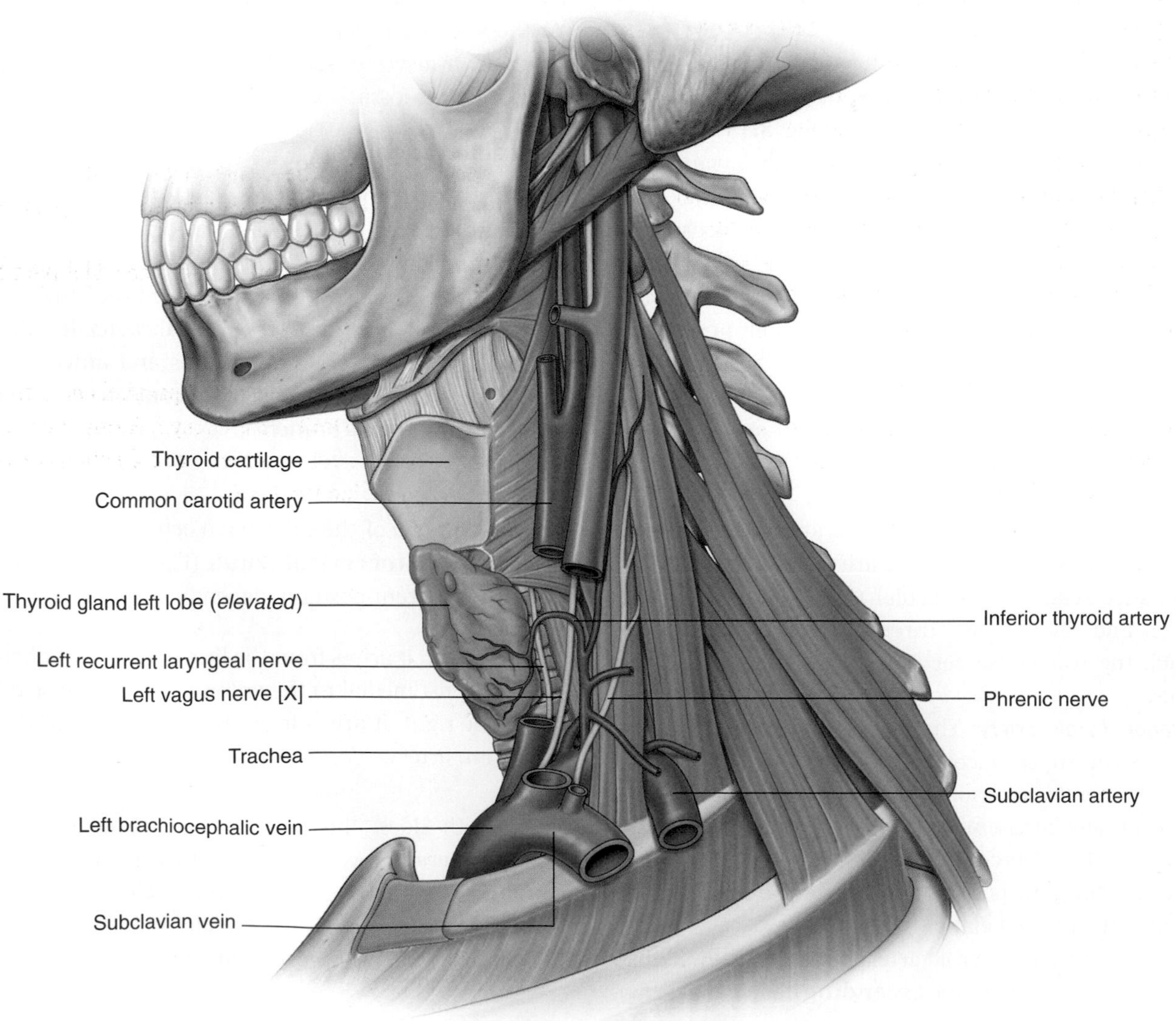

**Fig. 8.192** Nerves in the root of the neck.

### Vagus nerves [X]

The vagus nerves [X] descend through the neck within the carotid sheath, posterior to and just between the common carotid artery and the internal jugular vein.

In the lower part of the neck, the vagus nerves [X] give off cardiac branches, which continue downward and medially, passing posterior to the subclavian arteries to disappear into the thorax.

In the root of the neck, each vagus nerve [X] passes anterior to the subclavian artery and posterior to the subclavian vein as it enters the thorax (Fig. 8.192).

### Recurrent laryngeal nerves

The right and left recurrent laryngeal nerves are visible as they originate in (the right recurrent laryngeal nerve), or pass through (the left recurrent laryngeal nerve), the root of the neck.

The **right recurrent laryngeal nerve** is a branch of the right vagus nerve [X] as it reaches the lower edge of the first part of the subclavian artery in the root of the neck. It passes around the subclavian artery and upward and medially in a groove between the trachea and the esophagus as it heads to the larynx.

The **left recurrent laryngeal nerve** is a branch of the left vagus nerve [X] as it crosses the arch of the aorta in the superior mediastinum. It passes below and behind the arch of the aorta and ascends beside the trachea to the larynx (Fig. 8.192).

### Sympathetic nervous system

Various components of the sympathetic nervous system are visible as they pass through the root of the neck (Fig. 8.193). These include:

- the cervical part of the sympathetic trunk,
- the ganglia associated with the cervical part of the sympathetic trunk, and
- cardiac nerves branching from the cervical part of the sympathetic trunk.

The sympathetic trunks are two parallel cords that run from the base of the skull to the coccyx. Along the way they are punctuated by ganglia, which are collections of neuronal cell bodies outside the CNS.

#### In the clinic

**Recurrent laryngeal nerve palsy**

Damage to either the right or left recurrent laryngeal nerve may lead initially to a hoarse voice and finally to an inability to speak. Recurrent laryngeal nerve palsy can occur from disruption of the nerves anywhere along their course. Furthermore, interruption of the vagus nerves before the division of the recurrent laryngeal nerves can also produce vocal symptoms.

Lung cancer in the apex of the right lung can affect the right recurrent laryngeal nerve, whereas cancers that infiltrate into the area between the pulmonary artery and aorta, an area known clinically as the "aortopulmonary window," can affect the left recurrent laryngeal nerve. Thyroid surgery also can traumatize the recurrent laryngeal nerves.

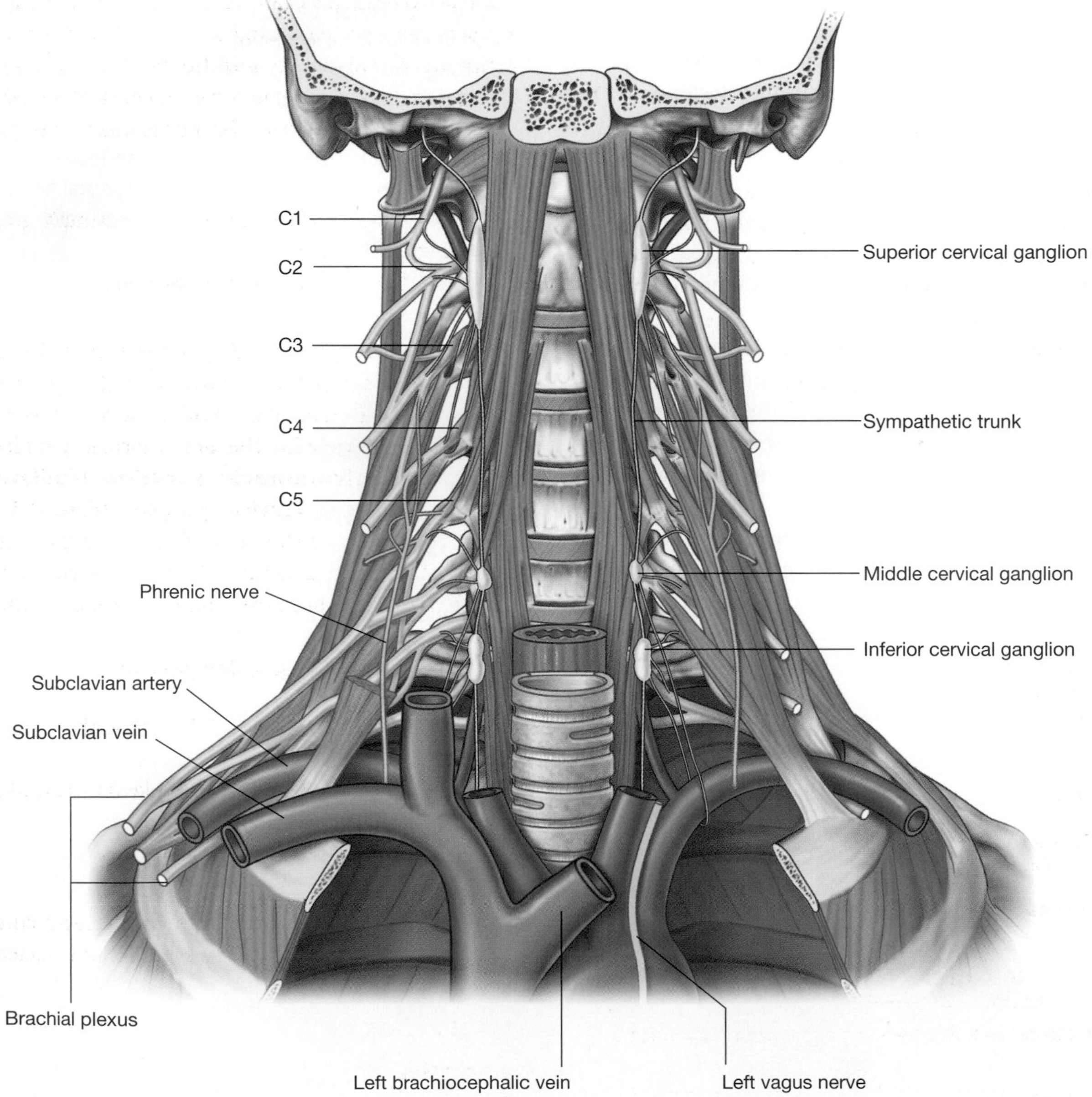

**Fig. 8.193** Components of the sympathetic nervous system in the root of the neck.

### Cervical part of the sympathetic trunk

The **cervical part of the sympathetic trunk** is anterior to the longus colli and longus capitis muscles, and posterior to the common carotid artery in the carotid sheath and the internal carotid artery. It is connected to each cervical spinal nerve by a gray ramus communicans (Fig. 8.194). There are no white rami communicantes in the cervical region.

### Ganglia

Three ganglia are usually described along the course of the sympathetic trunk in the cervical region, and in these ganglia ascending preganglionic sympathetic fibers from upper thoracic spinal cord levels synapse with postganglionic sympathetic fibers. The postganglionic sympathetic fibers are distributed in branches from these ganglia.

***Superior cervical ganglion.*** A very large **superior cervical ganglion** in the area of cervical vertebrae CI and CII marks the superior extent of the sympathetic trunk (Figs. 8.193 and 8.194). Its branches pass to:

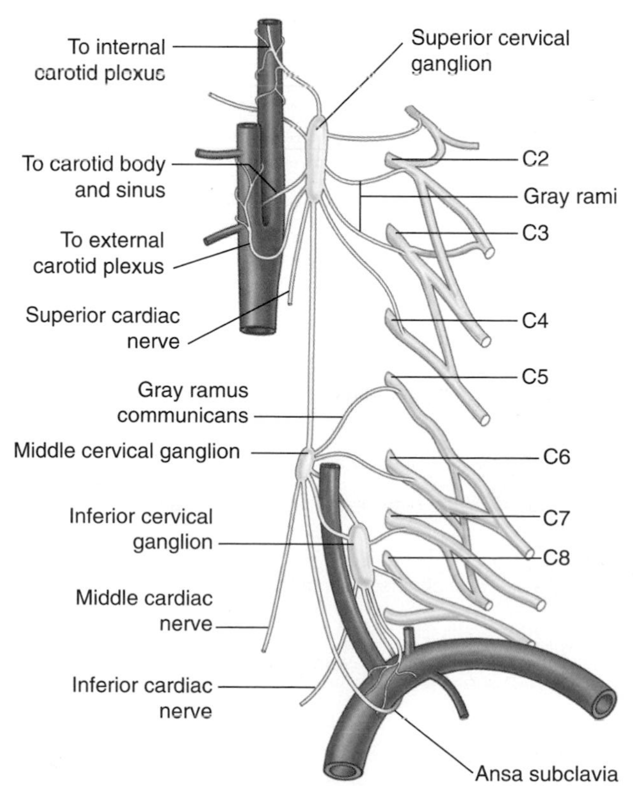


**Fig. 8.194** Cervical part of the sympathetic trunk.

- the internal carotid and external carotid arteries, forming plexuses around these vessels,
- cervical spinal nerves C1 to C4 through gray rami communicantes,
- the pharynx, and
- the heart as **superior cardiac nerves**.

***Middle cervical ganglion.*** A second ganglion inferior to the superior cervical ganglion along the course of the sympathetic trunk (the **middle cervical ganglion**) is encountered at about the level of cervical vertebra CVI (Figs. 8.193 and 8.194). Branches from this ganglion pass to:

- cervical spinal nerves C5 and C6 through gray rami communicantes, and
- the heart as **middle cardiac nerves**.

***Inferior cervical ganglion.*** At the lower end of the cervical part of the sympathetic trunk is another ganglion (the inferior cervical ganglion), which becomes very large when it combines with the first thoracic ganglion and forms the **cervicothoracic ganglion** (**stellate ganglion**). The inferior cervical ganglion (Figs. 8.193 and 8.194) is anterior to the neck of rib I and the transverse process of cervical vertebra CVII, and posterior to the first part of the subclavian artery and the origin of the vertebral artery.

Branches from this ganglion pass to:

- spinal nerves C7 to T1 through gray rami communicantes,
- the vertebral artery, forming a plexus associated with this vessel, and
- the heart as **inferior cardiac nerves**.

This ganglion may also receive white rami communicantes from thoracic spinal nerve T1, and occasionally, from T2.

## Lymphatics

### Thoracic duct

The **thoracic duct** is a major lymphatic channel that begins in the abdomen, passes superiorly through the thorax, and ends in the venous channels in the neck. It passes through the lower thoracic cavity in the midline with:

- the thoracic aorta on the left,
- the azygos vein on the right, and
- the esophagus anteriorly.

At about the level of thoracic vertebra TV the thoracic duct passes to the left and continues to ascend just to the left of the esophagus. It passes through the superior mediastinum and enters the root of the neck to the left of the esophagus (Fig. 8.195). Arching laterally, it passes posterior to the carotid sheath and turns inferiorly in front of the thyrocervical trunk, the phrenic nerve, and the vertebral artery.

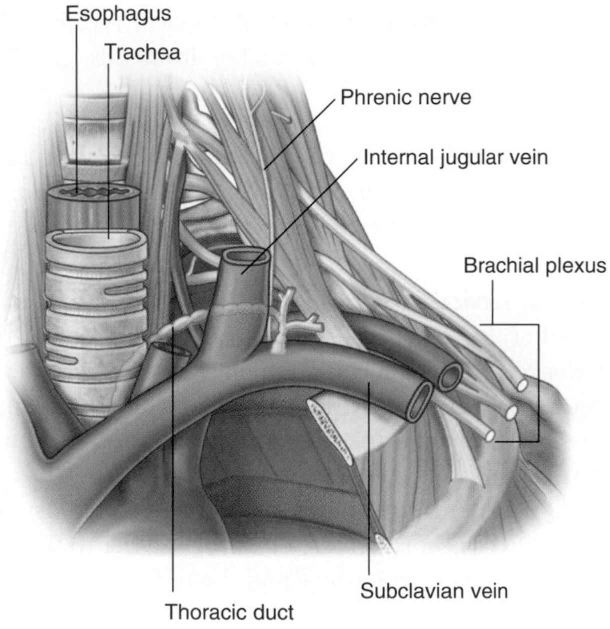


**Fig. 8.195** Thoracic duct in the root of the neck.

The thoracic duct terminates in the junction between the left internal jugular and the left subclavian veins (Fig. 8.195). Near its junction with the venous system it is joined by:

- the **left jugular trunk**, which drains lymph from the left side of the head and neck,
- the **left subclavian trunk**, which drains lymph from the left upper limb, and
- occasionally, the **left bronchomediastinal trunk**, which drains lymph from the left half of the thoracic structures (Fig. 8.196).

A similar confluence of three lymphatic trunks occurs on the right side of the body. Emptying into the junction between the right internal jugular and right subclavian veins are:

- the **right jugular trunk** from the head and neck,
- the **right subclavian trunk** from the right upper limb, and
- occasionally, the **right bronchomediastinal trunk** carrying lymph from the structures in the right half of the thoracic cavity and the right upper intercostal spaces (Fig. 8.196).

There is variability in how these trunks enter the veins. They may combine into a single right lymphatic duct to enter the venous system or enter as three separate trunks.

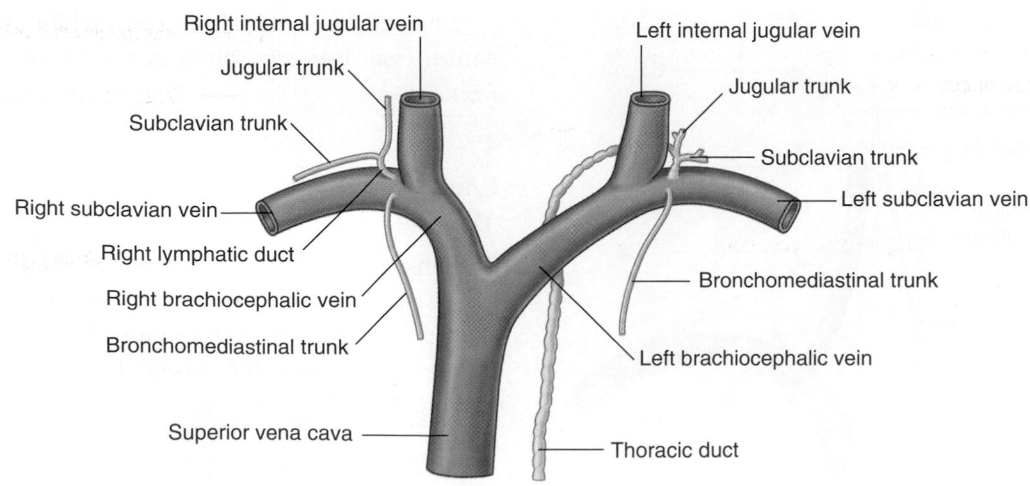


**Fig. 8.196** Termination of lymphatic trunks in the root of the neck.

## Lymphatics of the neck

A description of the organization of the lymphatic system in the neck becomes a summary of the lymphatic system in the head and neck. It is impossible to separate the two regions. The components of this system include superficial nodes around the head, superficial cervical nodes along the external jugular vein, and deep cervical nodes forming a chain along the internal jugular vein (Fig. 8.197).

The basic pattern of drainage is for superficial lymphatic vessels to drain to the superficial nodes. Some of these drain to the superficial cervical nodes on their way to the deep cervical nodes and others drain directly to the deep cervical nodes.

### Superficial lymph nodes

Five groups of superficial lymph nodes form a ring around the head and are primarily responsible for the lymphatic

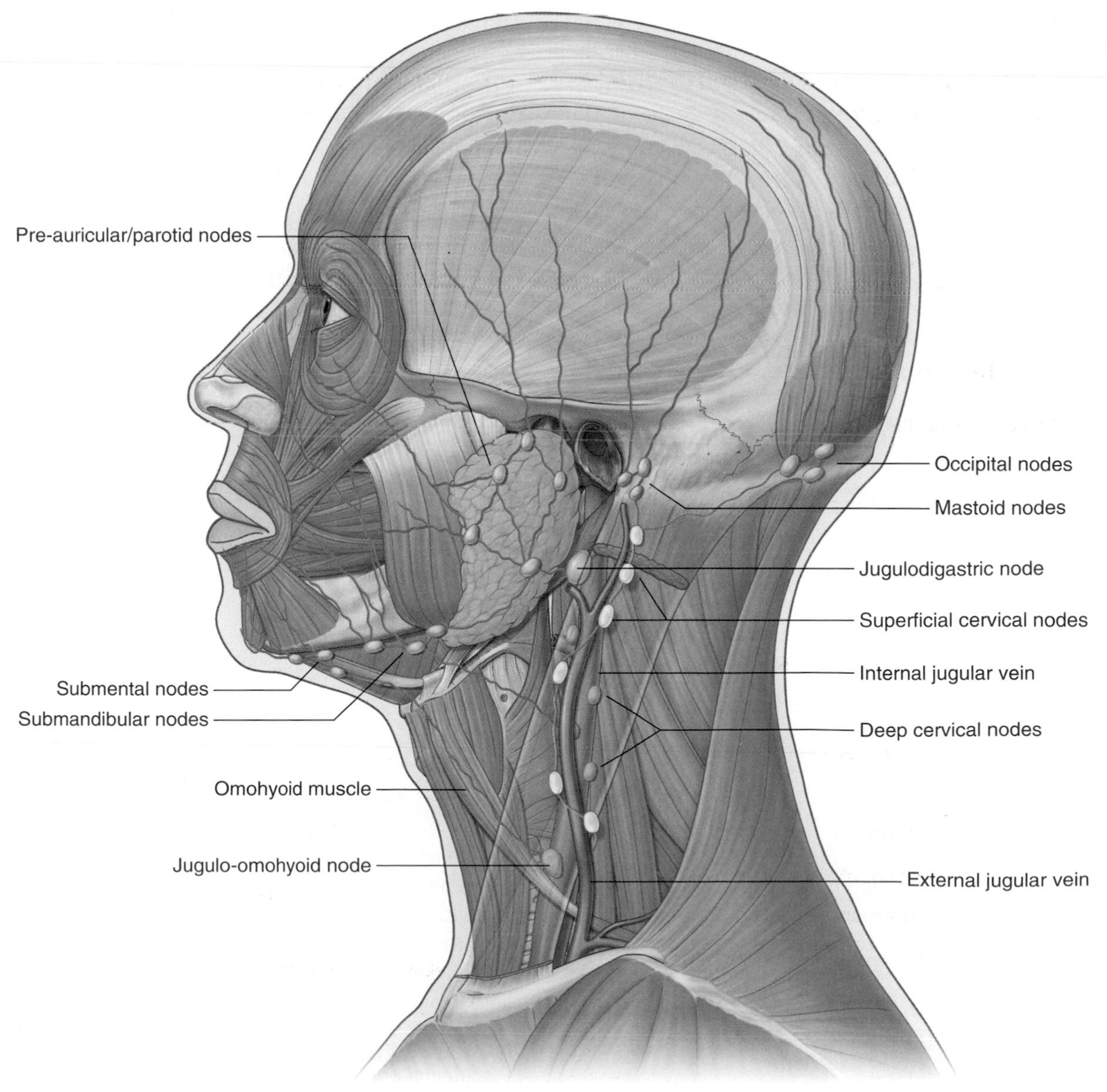


**Fig. 8.197** Lymphatic system in the neck.

drainage of the face and scalp. Their pattern of drainage is very similar to the area of distribution of the arteries near their location.

Beginning posteriorly these groups (Fig. 8.197) are:

- **occipital nodes** near the attachment of the trapezius muscle to the skull and associated with the occipital artery—lymphatic drainage is from the posterior scalp and neck;
- **mastoid nodes** (**retro-auricular/posterior auricular nodes**) posterior to the ear near the attachment of the sternocleidomastoid muscle and associated with the posterior auricular artery—lymphatic drainage is from the posterolateral half of the scalp;
- **pre-auricular and parotid nodes** anterior to the ear and associated with the superficial temporal and transverse facial arteries—lymphatic drainage is from the anterior surface of the auricle, the anterolateral scalp, the upper half of the face, the eyelids, and the cheeks;
- **submandibular nodes** inferior to the body of the mandible and associated with the facial artery—lymphatic drainage is from structures along the path of the facial artery as high as the forehead, as well as the gingivae, the teeth, and the tongue;
- **submental nodes** inferior and posterior to the chin—lymphatic drainage is from the center part of the lower lip, the chin, the floor of the mouth, the tip of the tongue, and the lower incisor teeth.

Lymphatic flow from these superficial lymph nodes passes in several directions:

- Drainage from the occipital and mastoid nodes passes to the superficial cervical nodes along the external jugular vein.
- Drainage from the pre-auricular and parotid nodes, the submandibular nodes, and the submental nodes passes to the deep cervical nodes.

### Superficial cervical lymph nodes

The **superficial cervical nodes** are a collection of lymph nodes along the external jugular vein on the superficial surface of the sternocleidomastoid muscle (Fig. 8.197). They primarily receive lymphatic drainage from the posterior and posterolateral regions of the scalp through the occipital and mastoid nodes, and send lymphatic vessels in the direction of the deep cervical nodes.

### Deep cervical lymph nodes

The **deep cervical nodes** are a collection of lymph nodes that form a chain along the internal jugular vein (Fig. 8.197). They are divided into upper and lower groups where the intermediate tendon of the omohyoid muscle crosses the common carotid artery and the internal jugular vein.

The most superior node in the upper deep cervical group is the **jugulodigastric node** (Fig. 8.197). This large node is where the posterior belly of the digastric muscle crosses the internal jugular vein and receives lymphatic drainage from the tonsils and tonsillar region.

Another large node, usually associated with the lower deep cervical group because it is at or just inferior to the intermediate tendon of the omohyoid muscle, is the **jugulo-omohyoid node** (Fig. 8.197). This node receives lymphatic drainage from the tongue.

The deep cervical nodes eventually receive all lymphatic drainage from the head and neck either directly or through regional groups of nodes.

From the deep cervical nodes, lymphatic vessels form the right and left jugular trunks, which empty into the right lymphatic duct on the right side or the thoracic duct on the left side.

## In the clinic

**Clinical lymphatic drainage of the head and neck**

Enlargement of the neck lymph nodes (cervical lymphadenopathy) is a common manifestation of disease processes that occur in the head and neck. It is also a common manifestation of diffuse diseases of the body, which include lymphoma, sarcoidosis, and certain types of viral infection such as glandular fever and human immunodeficiency virus (HIV) infection.

Evaluation of cervical lymph nodes is extremely important in determining the nature and etiology of the primary disease process that has produced nodal enlargement.

Clinical evaluation includes a general health assessment, particularly relating to symptoms from the head and neck. Examination of the nodes themselves often gives the clinician a clue as to the nature of the pathological process.

- Soft, tender, and inflamed lymph nodes suggest an acute inflammatory process, which is most likely to be infective.

### In the clinic—cont'd

- Firm multinodular large-volume rubbery nodes often suggest a diagnosis of lymphoma.

Examination should also include careful assessment of other nodal regions, including the supraclavicular fossae, the axillae, the retroperitoneum, and the inguinal regions.

Further examination may include digestive tract endoscopy, chest radiography, and body CT scanning.

Most cervical lymph nodes are easily palpable and suitable for biopsy to establish a tissue diagnosis. Biopsy can be performed using ultrasound for guidance and good samples of lymph nodes may be obtained.

The lymphatic drainage of the neck is somewhat complex, clinically. A relatively simple "level" system of nodal enlargement has been designed that is extremely helpful in evaluating lymph node spread of primary head and neck tumors. Once the number of levels of nodes are determined, and the size of the lymph nodes, the best mode of treatment can be instituted. This may include surgery, radiotherapy, and chemotherapy. The lymph node level also enables a prognosis to be made. The levels are as follows (Fig. 8.199):

- Level I—from the midline of the submental triangle up to the level of the submandibular gland.
- Level II—from the skull base to the level of the hyoid bone anteriorly from the posterior border of the sternocleidomastoid muscle.
- Level III—the inferior aspect of the hyoid bone to the bottom cricoid arch and anterior to the posterior border of the sternocleidomastoid up to the midline.
- Level IV—from the inferior aspect of the cricoid to the top of the manubrium of the sternum and anterior to the posterior border of the sternocleidomastoid muscle.

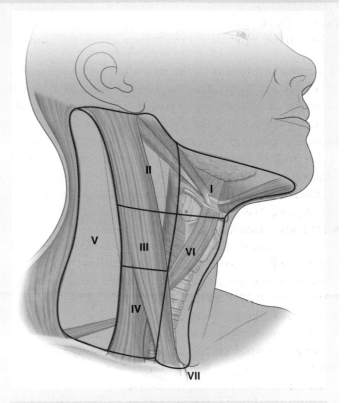

**Fig. 8.199** Neck regions (levels) that are used clinically to evaluate lymph nodes.

- Level V—posterior to the sternocleidomastoid muscle and anterior to the trapezius muscle above the level of the clavicle.
- Level VI—below the hyoid bone and above the jugular (sternal) notch in the midline.
- Level VII—below the level of the jugular (sternal) notch.

## PHARYNX

The pharynx is a musculofascial half-cylinder that links the oral and nasal cavities in the head to the larynx and esophagus in the neck (Fig. 8.198). The pharyngeal cavity is a common pathway for air and food.

The pharynx is attached above to the base of the skull and is continuous below, approximately at the level of vertebra CVI, with the top of the esophagus. The walls of the pharynx are attached anteriorly to the margins of the nasal cavities, oral cavity, and larynx. Based on these anterior relationships the pharynx is subdivided into three regions, the nasopharynx, oropharynx, and laryngopharynx:

- The posterior apertures (choanae) of the nasal cavities open into the nasopharynx.

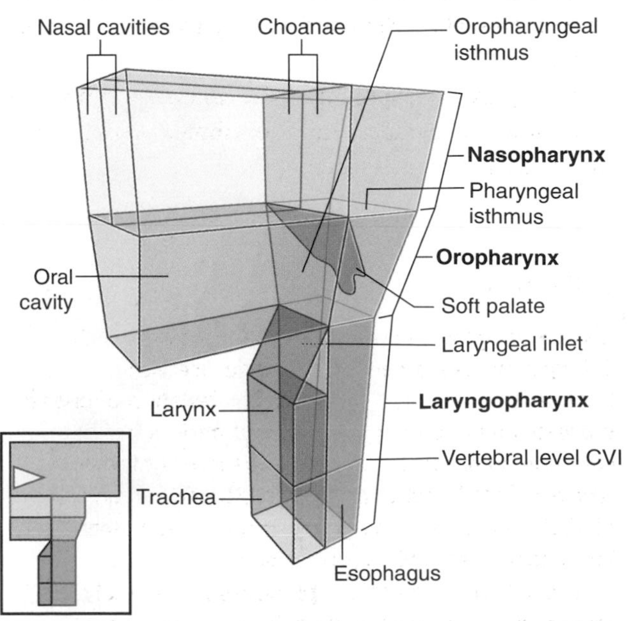


**Fig. 8.198** Pharynx.

- The posterior opening of the oral cavity (oropharyngeal isthmus) opens into the oropharynx.
- The superior aperture of the larynx (laryngeal inlet) opens into the laryngopharynx.

In addition to these openings, the pharyngeal cavity is related anteriorly to the posterior one-third of the tongue and to the posterior aspect of the larynx. The pharyngotympanic tubes open into the lateral walls of the nasopharynx.

Lingual, pharyngeal, and palatine tonsils are on the deep surface of the pharyngeal walls.

The pharynx is separated from the posteriorly positioned vertebral column by a thin retropharyngeal space containing loose connective tissue.

Although the soft palate is generally considered as part of the roof of the oral cavity, it is also related to the pharynx. The soft palate is attached to the posterior margin of the hard palate and is a type of "flutter valve" that can:

- swing up (elevate) to close the pharyngeal isthmus, and seal off the nasopharynx from the oropharynx, and
- swing down (depress) to close the oropharyngeal isthmus and seal off the oral cavity from the oropharynx.

## Skeletal framework

The superior and anterior margins of the pharyngeal wall are attached to bone and cartilage, and to ligaments. The two sides of the pharyngeal wall are welded together posteriorly in the midline by a vertically oriented cord-like ligament (the pharyngeal raphe). This connective tissue structure descends from the pharyngeal tubercle on the base of the skull to the level of cervical vertebra CVI where the raphe blends with connective tissue in the posterior wall of the esophagus.

There is an irregular C-shaped line of pharyngeal wall attachment on the base of the skull (Fig. 8.200). The open part of the C faces the nasal cavities. Each arm of the C begins at the posterior margin of the medial plate of the pterygoid process of the sphenoid bone, just inferior to the cartilaginous part of the pharyngotympanic tube. The line crosses inferior to the pharyngotympanic tube and then

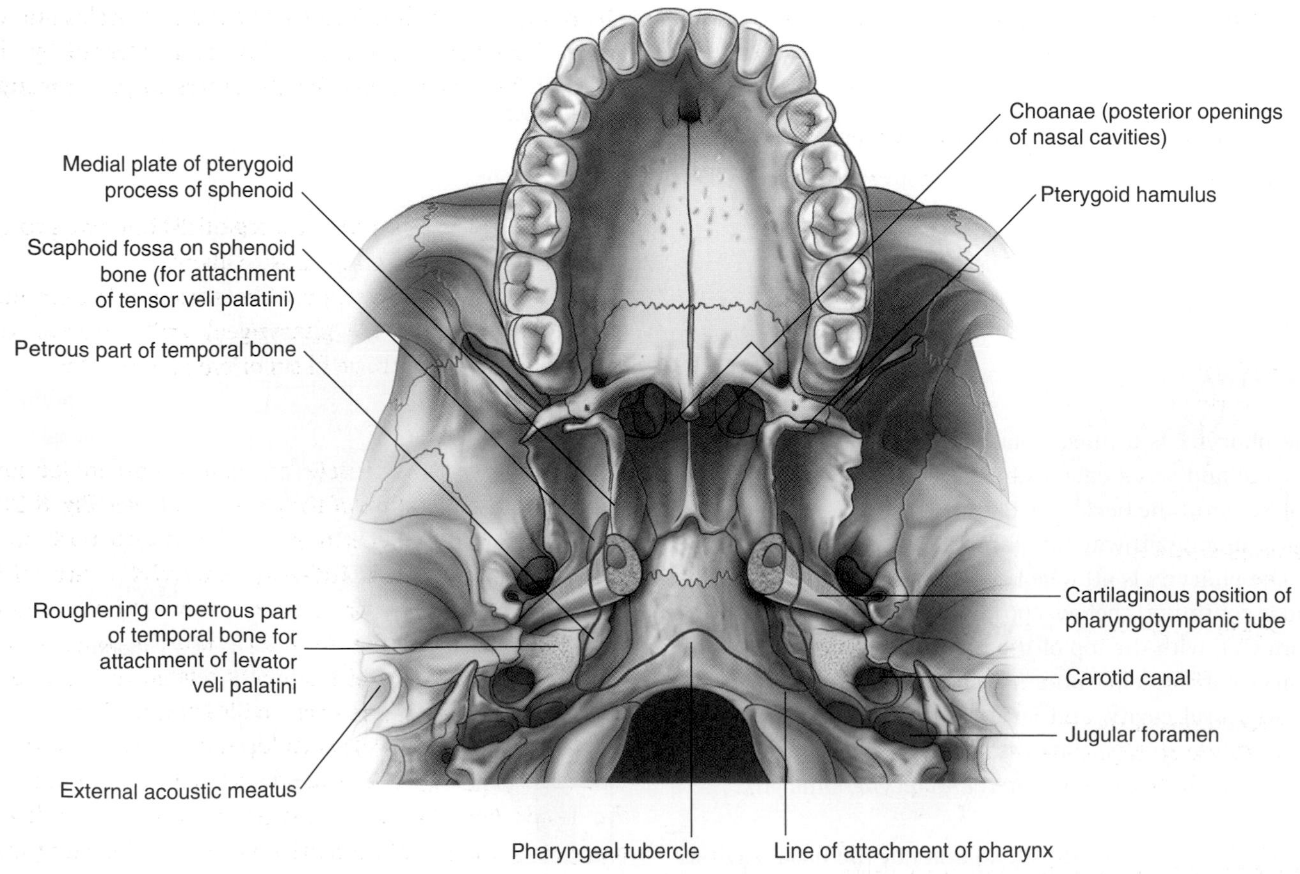




**Fig. 8.200** Line of attachment of the pharynx to the base of the skull.

passes onto the petrous part of the temporal bone where it is just medial to the roughening for the attachment of one of the muscles (levator veli palatini) of the soft palate. From here, the line swings medially onto the occipital bone and joins the line from the other side at a prominent elevation of bone in the midline (the pharyngeal tubercle).

## Anterior vertical line of attachment for the lateral pharyngeal walls

The vertical line of attachment for the lateral pharyngeal walls to structures related to the nasal and oral cavities and larynx is discontinuous and in three parts (Fig. 8.201).

### First part

On each side, the anterior line of attachment of the lateral pharyngeal wall begins superiorly on the posterior edge of the medial pterygoid plate of the sphenoid bone just inferior to where the pharyngotympanic tube lies against this plate. It continues inferiorly along the edge of the medial plate of the pterygoid process and onto the pterygoid hamulus. From this point, the line descends along the pterygomandibular raphe to the mandible where this part of the line terminates.

The **pterygomandibular raphe** is a linear cord-like connective tissue ligament that spans the distance between the tip of the pterygoid hamulus and a triangular roughening immediately posterior to the third molar on the mandible. It joins a muscle of the lateral pharyngeal wall (superior constrictor) with a muscle of the lateral wall of the oral cavity (buccinator).

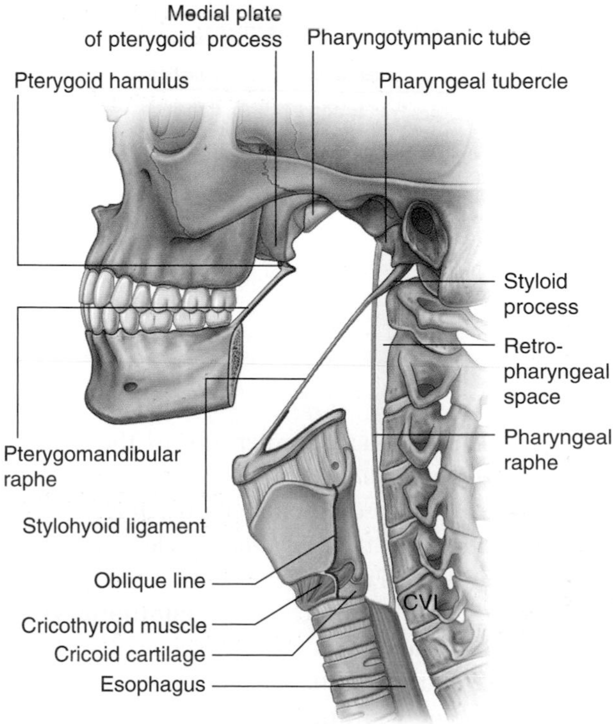


**Fig. 8.201** Attachments of the lateral pharyngeal wall.

### Second part

The second part of the line of attachment of the lateral pharyngeal wall is related to the hyoid bone. It begins on the lower aspect of the stylohyoid ligament, which connects the tip of the styloid process of the temporal bone to the lesser horn of the hyoid bone. The line continues onto the lesser horn and then turns and runs posteriorly along the entire upper surface of the greater horn of the hyoid where it terminates.

### Third part

The most inferior and third part of the line of attachment of the lateral pharyngeal wall begins superiorly on the superior tubercle of the thyroid cartilage, and descends along the oblique line to the inferior tubercle.

From the inferior tubercle, the line of attachment continues over the cricothyroid muscle along a tendinous thickening of fascia to the cricoid cartilage where it terminates.

## Pharyngeal wall

The pharyngeal wall is formed by skeletal muscles and by fascia. Gaps between the muscles are reinforced by the fascia and provide routes for structures to pass through the wall.

### Muscles

The muscles of the pharynx are organized into two groups based on the orientation of muscle fibers.

The constrictor muscles have fibers oriented in a circular direction relative to the pharyngeal wall, whereas the longitudinal muscles have fibers oriented vertically.

#### Constrictor muscles

The three constrictor muscles on each side are major contributors to the structure of the pharyngeal wall (Fig. 8.202 and Table 8.17) and their names indicate their position—**superior, middle,** and **inferior constrictor muscles.** Posteriorly, the muscles from each side are joined together by the pharyngeal raphe. Anteriorly, these muscles attach to bones, cartilages, and ligaments related to the lateral margins of the nasal and oral cavities and the larynx.

The constrictor muscles overlap each other in a fashion resembling the walls of three flower pots stacked one on the other. The inferior constrictors overlap the lower margins of the middle constrictors and, in the same way, the middle constrictors overlap the superior constrictors.

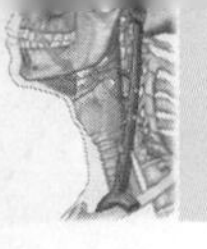

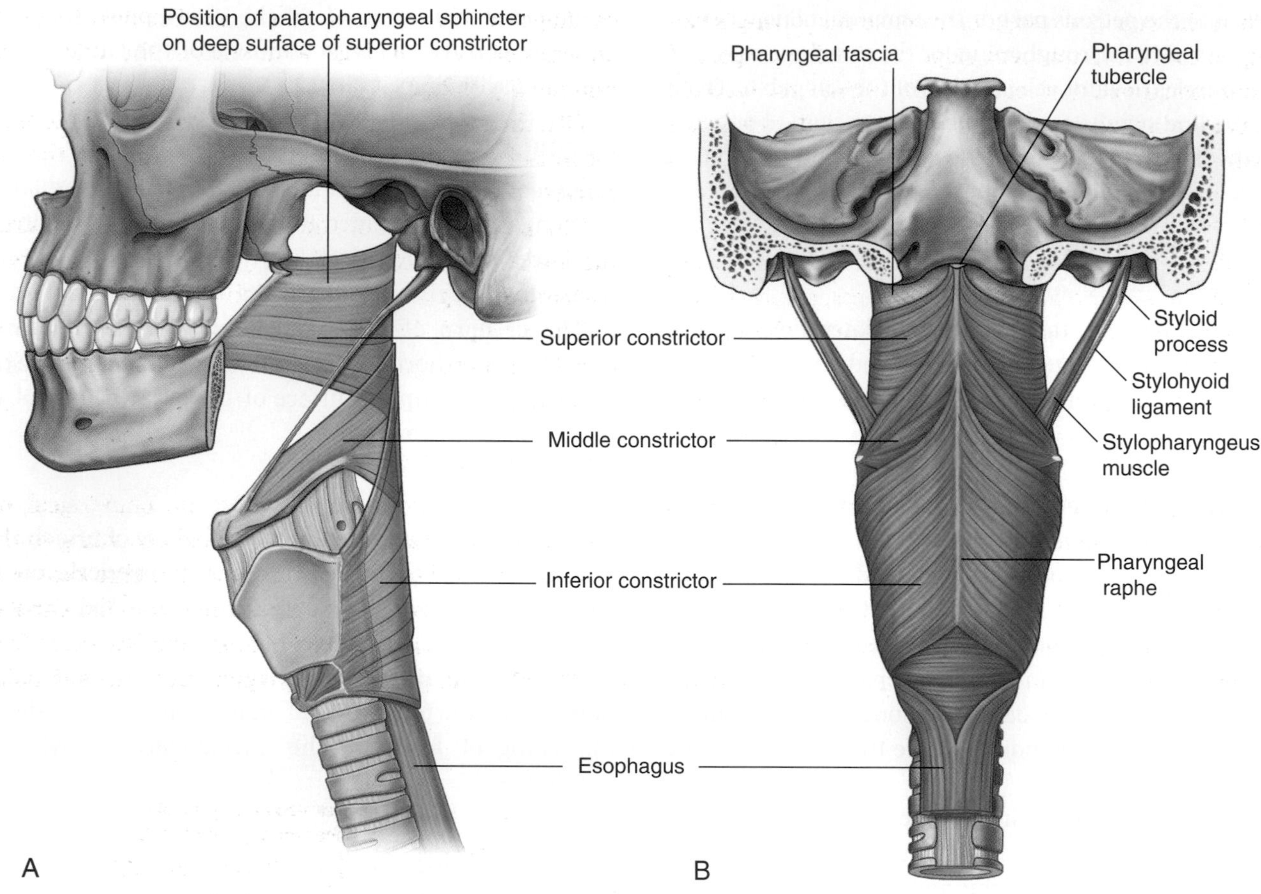


**Fig. 8.202** Constrictor muscles of the pharynx. **A.** Lateral view. **B.** Posterior view.

**Table 8.17** Constrictor muscles of the pharynx

| Muscle | Posterior attachment | Anterior attachment | Innervation | Function |
|---|---|---|---|---|
| Superior constrictor | Pharyngeal raphe | Pterygomandibular raphe and adjacent bone on the mandible and pterygoid hamulus | Vagus nerve [X] | Constriction of pharynx |
| Middle constrictor | Pharyngeal raphe | Upper margin of greater horn of hyoid bone and adjacent margins of lesser horn and stylohyoid ligament | Vagus nerve [X] | Constriction of pharynx |
| Inferior constrictor | Pharyngeal raphe | Cricoid cartilage, oblique line of thyroid cartilage, and a ligament that spans between these attachments and crosses the cricothyroid muscle | Vagus nerve [X] | Constriction of pharynx |

Collectively, the muscles constrict or narrow the pharyngeal cavity.

When the constrictor muscles contract sequentially from top to bottom, as in swallowing, they move a bolus of food through the pharynx and into the esophagus.

All of the constrictors are innervated by the pharyngeal branch of the vagus nerve [X].

### Superior constrictors

The superior constrictor muscles together bracket the upper part of the pharyngeal cavity (Fig. 8.202).

Each muscle is attached anteriorly to the pterygoid hamulus, pterygomandibular raphe, and adjacent bone of the mandible. From these attachments, the muscle fans out posteriorly and joins with its partner muscle from the other side at the pharyngeal raphe.

A special band of muscle (the **palatopharyngeal sphincter**) originates from the anterolateral surface of the soft palate and circles the inner aspect of the pharyngeal wall, blending with the inner aspect of the superior constrictor.

When the superior constrictor constricts during swallowing, it forms a prominent ridge on the deep aspect of the pharyngeal wall that catches the margin of the elevated soft palate, which then seals closed the pharyngeal isthmus between the nasopharynx and oropharynx.

### Middle constrictors

The middle constrictor muscles are attached to the lower aspect of the stylohyoid ligament, the lesser horn of the hyoid bone, and the entire upper surface of the greater horn of the hyoid (Fig. 8.202).

Like the superior constrictors, the middle constrictor muscles fan out posteriorly and attach to the pharyngeal raphe.

The posterior part of the middle constrictors overlaps the superior constrictors.

### Inferior constrictors

The inferior constrictor muscles attach anteriorly to the oblique line of the thyroid cartilage, the cricoid cartilage, and a ligament that spans between these two attachments to cartilage and crosses the cricothyroid muscle (Fig. 8.202).

Like the other constrictor muscles, the inferior constrictor muscles spread out posteriorly and attach to the pharyngeal raphe.

The posterior part of the inferior constrictors overlaps the middle constrictors. Inferiorly, the muscle fibers blend with and attach into the wall of the esophagus.

The parts of the inferior constrictors attached to the cricoid cartilage bracket the narrowest part of the pharyngeal cavity.

## Longitudinal muscles

The three longitudinal muscles of the pharyngeal wall (Fig. 8.203 and Table 8.18) are named according to their origins—**stylopharyngeus** from the styloid process of the temporal bone, **salpingopharyngeus** from the cartilaginous part of the pharyngotympanic tube (*salpinx* is Greek for "tube"), and **palatopharyngeus** from the soft palate.

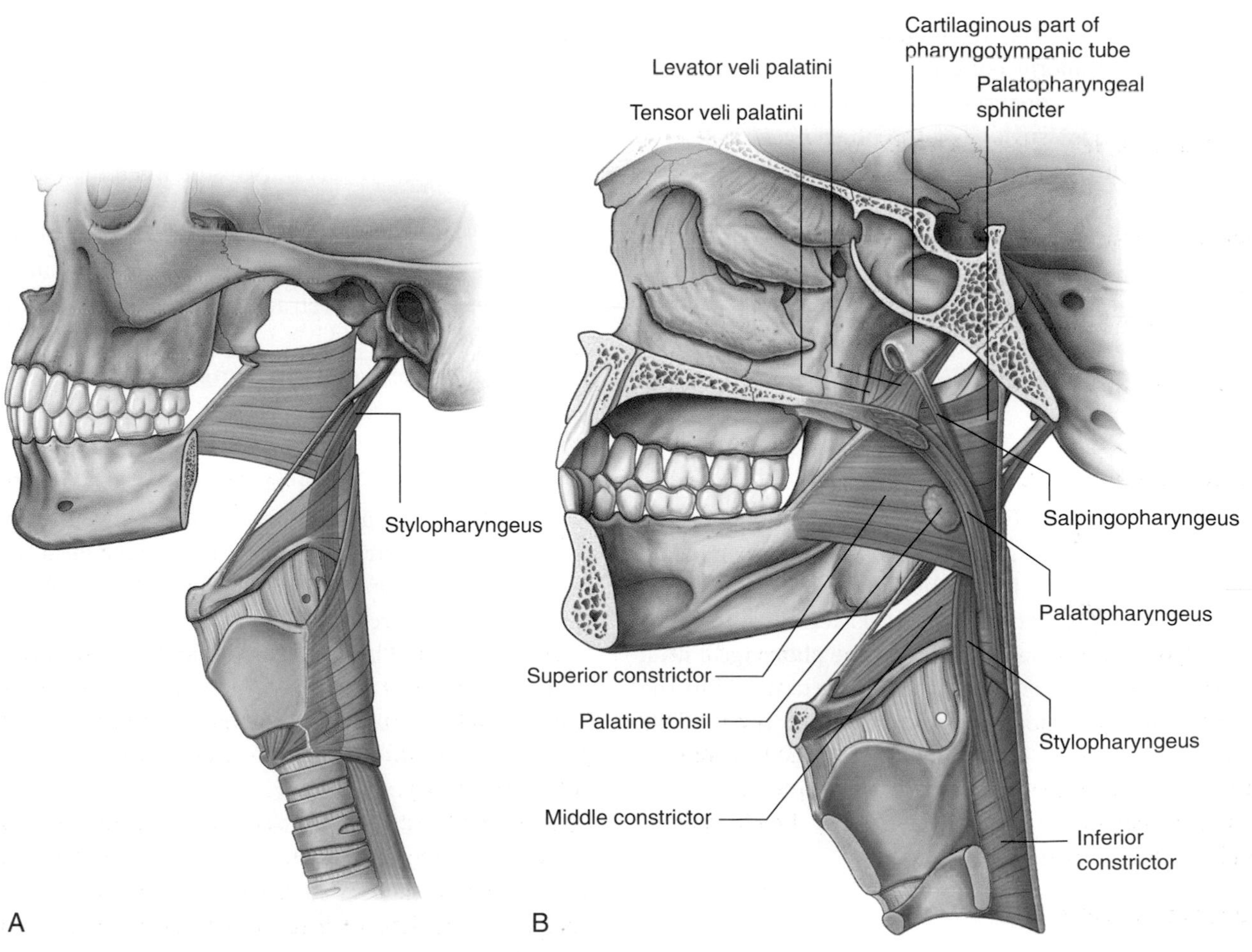


**Fig. 8.203** Longitudinal muscles of the pharynx. **A.** Stylopharyngeus muscle. **B.** Medial view.

Table 8.18 Longitudinal muscles of the pharynx

| Muscle | Origin | Insertion | Innervation | Function |
|---|---|---|---|---|
| Stylopharyngeus | Medial side of base of styloid process | Pharyngeal wall | Glossopharyngeal nerve [IX] | Elevation of the pharynx |
| Salpingopharyngeus | Inferior aspect of pharyngeal end of pharyngotympanic tube | Pharyngeal wall | Vagus nerve [X] | Elevation of the pharynx |
| Palatopharyngeus | Upper surface of palatine aponeurosis | Pharyngeal wall | Vagus nerve [X] | Elevation of the pharynx; closure of the oropharyngeal isthmus |

From their sites of origin, these muscles descend and attach into the pharyngeal wall.

The longitudinal muscles elevate the pharyngeal wall, or during swallowing, pull the pharyngeal wall up and over a bolus of food being moved through the pharynx and into the esophagus.

### Stylopharyngeus

The cylindrical stylopharyngeus muscle (Fig. 8.203A) originates from the base of the medial surface of the styloid process of the temporal bone and descends between the superior and middle constrictor muscles to fan out on, and blend with, the deep surface of the pharyngeal wall. It is innervated by the glossopharyngeal nerve [IX].

### Salpingopharyngeus

The salpingopharyngeus (Fig. 8.203B) is a small muscle originating from the inferior aspect of the pharyngotympanic tube, descending on, and blending into, the deep surface of the pharyngeal wall. It is innervated by the vagus nerve [X].

### Palatopharyngeus

The palatopharyngeus (Fig. 8.203B), in addition to being a muscle of the pharynx, is also a muscle of the soft palate (see pp. 1098–1099). It is attached to the upper surface of the palatine aponeurosis, and passes posteriorly and inferiorly to blend with the deep surface of the pharyngeal wall.

The palatopharyngeus forms an important fold in the overlying mucosa (the **palatopharyngeal arch**). This arch is visible through the oral cavity and is a landmark for finding the **palatine tonsil**, which is immediately anterior to it on the oropharyngeal wall.

In addition to elevating the pharynx, the palatopharyngeus participates in closing the oropharyngeal isthmus by depressing the palate and moving the palatopharyngeal fold toward the midline.

The palatopharyngeus is innervated by the vagus nerve [X].

## Fascia

The pharyngeal fascia is separated into two layers, which sandwich the pharyngeal muscles between them:

- A thin layer (**buccopharyngeal fascia**) coats the outside of the muscular part of the wall and is a component of the pretracheal layer of cervical fascia (see p. 991).
- A much thicker layer (**pharyngobasilar fascia**) lines the inner surface.

The fascia reinforces the pharyngeal wall where muscle is deficient. This is particularly evident above the level of the superior constrictor where the pharyngeal wall is formed almost entirely of fascia (Fig. 8.203). This part of the wall is reinforced externally by muscles of the soft palate (tensor and levator veli palatini).

## Gaps in the pharyngeal wall and structures passing through them

Gaps between muscles of the pharyngeal wall provide important routes for muscles and neurovascular tissues (Fig. 8.204).

Above the margin of the superior constrictor, the pharyngeal wall is deficient in muscle and completed by pharyngeal fascia.

The tensor and levator veli palatini muscles of the soft palate initially descend from the base of the skull and are lateral to the pharyngeal fascia. In this position, they reinforce the pharyngeal wall:

- The levator veli palatini passes through the pharyngeal fascia inferior to the pharyngotympanic tube and enters the soft palate.
- The tendon of the tensor veli palatini turns medially around the pterygoid hamulus and passes through the origin of the buccinator muscle to enter the soft palate.

One of the largest and most important apertures in the pharyngeal wall is between the superior and middle constrictor muscles of the pharynx and the posterior border of the mylohyoid muscle, which forms the floor of the mouth (Fig. 8.204). This triangular-shaped gap (**oropharyngeal triangle**) not only enables the stylopharyngeus to slip into the pharyngeal wall, but also allows muscles, nerves, and vessels to pass between regions lateral to the pharyngeal wall and the oral cavity, particularly to the tongue.

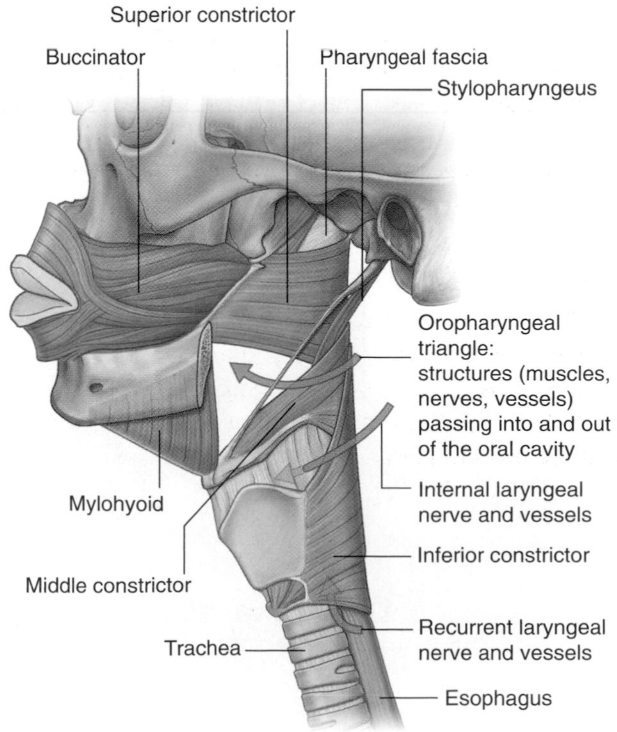


**Fig. 8.204** Gaps between muscles in the pharyngeal wall.

The gap between the middle and inferior constrictor muscles allows the internal laryngeal vessels and nerve access to the aperture in the thyrohyoid membrane to enter the larynx.

The recurrent laryngeal nerves and accompanying inferior laryngeal vessels enter the larynx posterior to the inferior horn of the thyroid cartilage deep to the inferior margin of the inferior constrictor muscle.

## Nasopharynx

The nasopharynx is behind the posterior apertures (choanae) of the nasal cavities and above the level of the soft palate (Fig. 8.205). Its ceiling is formed by the sloping base of the skull and consists of the posterior part of the body of the sphenoid bone and the basal part of the occipital bone. The ceiling and lateral walls of the nasopharynx form a domed vault at the top of the pharyngeal cavity that is always open.

The cavity of the nasopharynx is continuous below with the cavity of the oropharynx at the pharyngeal isthmus. The position of the pharyngeal isthmus is marked on the pharyngeal wall by a mucosal fold caused by the underlying palatopharyngeal sphincter, which is part of the superior constrictor muscle.

Elevation of the soft palate and constriction of the palatopharyngeal sphincter closes the pharyngeal isthmus during swallowing and separates the nasopharynx from the oropharynx.

There is a large collection of lymphoid tissue (the **pharyngeal tonsil**) in the mucosa covering the roof of the nasopharynx. Enlargement of this tonsil, known then as adenoids, can occlude the nasopharynx so that breathing is only possible through the oral cavity (Fig. 8.205A).

The most prominent features on each lateral wall of the nasopharynx are:

- the pharyngeal opening of the pharyngotympanic tube, and
- mucosal elevations and folds covering the end of the pharyngotympanic tube and the adjacent muscles.

The opening of the pharyngotympanic tube is posterior to and slightly above the level of the hard palate, and lateral to the top of the soft palate (Fig. 8.205A).

Because the pharyngotympanic tube projects into the nasopharynx from a posterolateral direction, its posterior

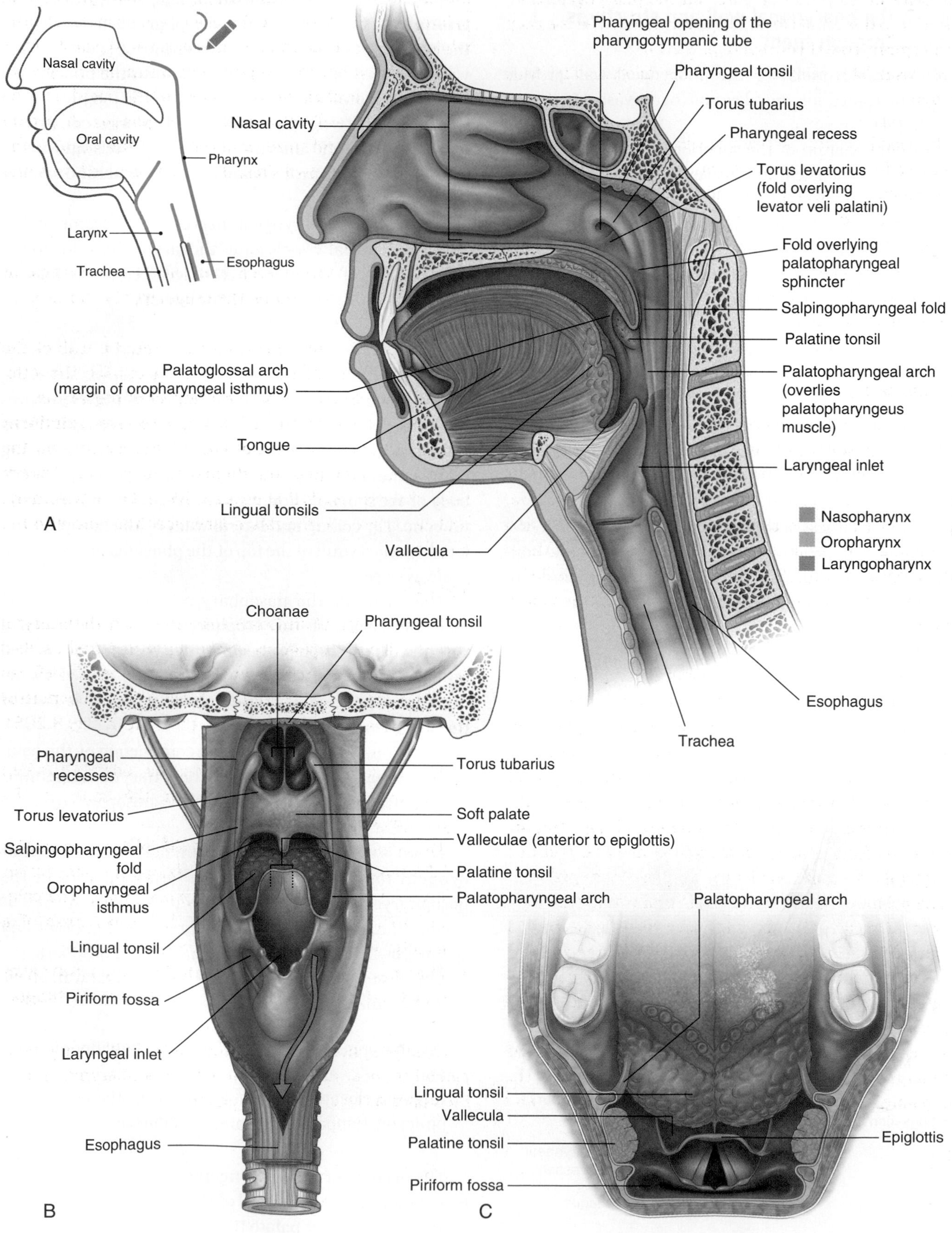

**Fig. 8.205** Mucosal features of the pharynx. **A.** Lateral view. **B.** Posterior view with the pharyngeal wall opened. **C.** Superior view.

rim forms an elevation or bulge on the pharyngeal wall. Posterior to this tubal elevation (**torus tubarius**) is a deep recess (**pharyngeal recess**) (Fig. 8.205A).

Mucosal folds related to the pharyngotympanic tube include:

- the small vertical **salpingopharyngeal fold**, which descends from the tubal elevation and overlies the salpingopharyngeus muscle, and
- a broad fold or elevation (**torus levatorius**) that appears to emerge from just under the opening of the pharyngotympanic tube, continues medially onto the upper surface of the soft palate, and overlies the levator veli palatini muscle.

## Oropharynx

The oropharynx is posterior to the oral cavity, inferior to the level of the soft palate, and superior to the upper margin of the epiglottis (Fig. 8.205). The palatoglossal folds (arches), one on each side, that cover the palatoglossal muscles, mark the boundary between the oral cavity and the oropharynx. The arched opening between the two folds is the oropharyngeal isthmus. Just posterior and medial to these folds are another pair of folds (arches), the palatopharyngeal folds, one on each side, that overlie the palatopharyngeus muscles.

The anterior wall of the oropharynx inferior to the oropharyngeal isthmus is formed by the upper part of the posterior one-third or pharyngeal part of the tongue. Large collections of lymphoid tissue (the lingual tonsils) are in the mucosa covering this part of the tongue. A pair of mucosal pouches (**valleculae**), one on each side of the midline, between the base of the tongue and epiglottis, are depressions formed between a midline mucosal fold and two lateral folds that connect the tongue to the epiglottis.

The palatine tonsils are on the lateral walls of the oropharynx. On each side, there is a large ovoid collection of lymphoid tissue in the mucosa lining the superior constrictor muscle and between the palatoglossal and palatopharyngeal arches. The palatine tonsils are visible through the oral cavity just posterior to the palatoglossal folds.

When holding liquid or solids in the oral cavity, the oropharyngeal isthmus is closed by depression of the soft palate, elevation of the back of the tongue, and movement toward the midline of the palatoglossal and palatopharyngeal folds. This allows a person to breathe while chewing or manipulating material in the oral cavity.

On swallowing, the oropharyngeal isthmus is opened, the palate is elevated, the laryngeal cavity is closed, and the food or liquid is directed into the esophagus. A person cannot breathe and swallow at the same time because the airway is closed at two sites, the pharyngeal isthmus and the larynx.

## Laryngopharynx

The laryngopharynx extends from the superior margin of the epiglottis to the top of the esophagus at the level of vertebra CVI (Fig. 8.205).

The laryngeal inlet opens into the anterior wall of the laryngopharynx. Inferior to the laryngeal inlet, the anterior wall consists of the posterior aspect of the larynx.

There is another pair of mucosal recesses (**piriform fossae**) between the central part of the larynx and the more lateral lamina of the thyroid cartilage. The piriform fossae form channels that direct solids and liquids from the oral cavity around the raised laryngeal inlet and into the esophagus.

## Tonsils

Collections of lymphoid tissue in the mucosa of the pharynx surrounding the openings of the nasal and oral cavities (Waldeyer's tonsillar ring) are part of the body's defense system. The largest of these collections form distinct masses (**tonsils**). Tonsils occur mainly in three areas (Fig. 8.205):

- The pharyngeal tonsil, known as adenoids when enlarged, is in the midline on the roof of the nasopharynx.
- The palatine tonsils are on each side of the oropharynx between the palatoglossal and palatopharyngeal arches just posterior to the oropharyngeal isthmus. (The palatine tonsils are visible through the open mouth of a patient when the tongue is depressed.)
- The lingual tonsils refer collectively to numerous lymphoid nodules on the posterior one-third of the tongue.

Small lymphoid nodules also occur in the pharyngotympanic tube near its opening into the nasopharynx, and on the upper surface of the soft palate.

## Vessels

### Arteries

Numerous vessels supply the pharyngeal wall (Fig. 8.206).

Arteries that supply upper parts of the pharynx include:

- the ascending pharyngeal artery,
- the ascending palatine and tonsillar branches of the facial artery, and
- numerous branches of the maxillary and the lingual arteries.

All these vessels are from the external carotid artery.

Arteries that supply the lower parts of the pharynx include pharyngeal branches from the inferior thyroid artery, which originates from the thyrocervical trunk of the subclavian artery.

The major blood supply to the palatine tonsil is from the tonsillar branch of the facial artery, which penetrates the superior constrictor muscle.



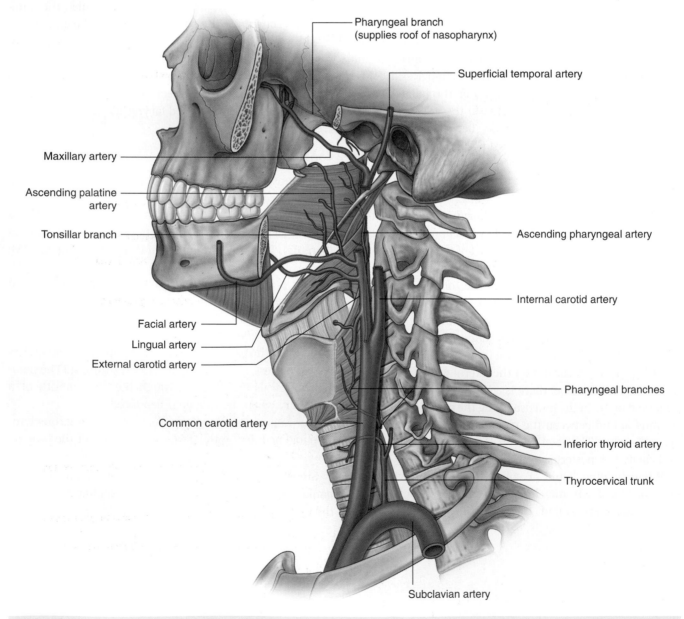

**Fig. 8.206** Arterial supply of the pharynx.

### Veins

Veins of the pharynx form a plexus, which drains superiorly into the pterygoid plexus in the infratemporal fossa, and inferiorly into the facial and internal jugular veins (Fig. 8.207).

### Lymphatics

Lymphatic vessels from the pharynx drain into the deep cervical nodes and include **retropharyngeal** (between the nasopharynx and vertebral column), **paratracheal**, and **infrahyoid nodes** (Fig. 8.207).

The palatine tonsils drain through the pharyngeal wall into the jugulodigastric nodes in the region where the facial vein drains into the internal jugular vein (and inferior to the posterior belly of the digastric muscle).

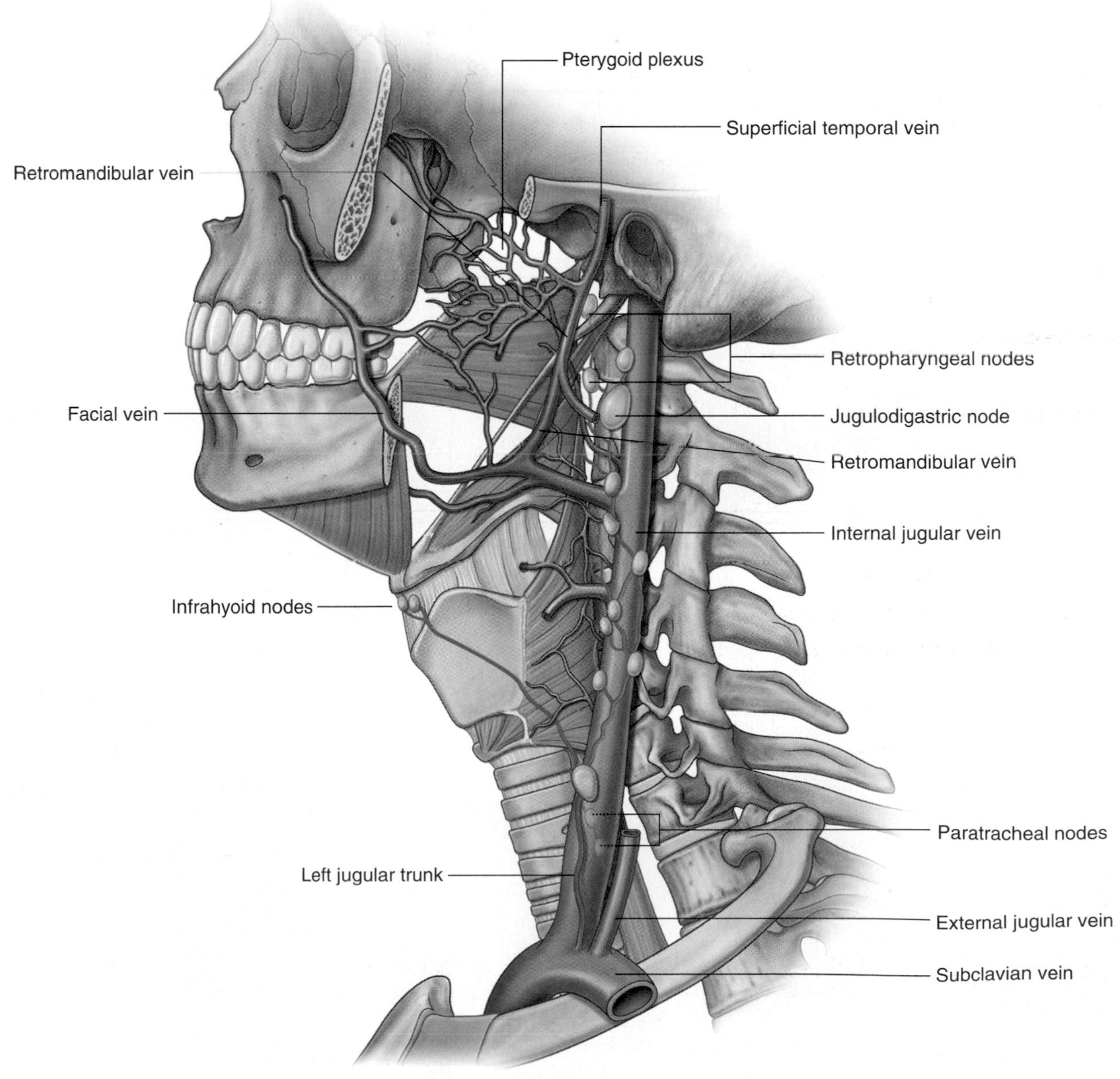


**Fig. 8.207** Venous and lymphatic drainage of the pharynx.

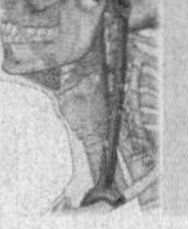



## Nerves

Motor and most sensory innervation (except for the nasal region) of the pharynx is mainly through branches of the vagus [X] and glossopharyngeal [IX] nerves, which form a plexus in the outer fascia of the pharyngeal wall (Fig. 8.208A).

The **pharyngeal plexus** is formed by:

- the pharyngeal branch of the vagus nerve [X],
- branches from the **external laryngeal nerve** from the **superior laryngeal branch** of the vagus nerve [X], and
- pharyngeal branches of the glossopharyngeal nerve [IX].

The **pharyngeal branch of the vagus nerve [X]** originates from the upper part of its **inferior ganglion** above the origin of the superior laryngeal nerve and is the major motor nerve of the pharynx.

All muscles of the pharynx are innervated by the vagus nerve [X] mainly through the pharyngeal plexus, except for the stylopharyngeus, which is innervated directly by a branch of the glossopharyngeal nerve [IX] (Fig. 8.208B).

Each subdivision of the pharynx has a different sensory innervation:

- The nasopharynx is innervated by a pharyngeal branch of the maxillary nerve [$V_2$] that originates in the pterygopalatine fossa and passes through the palatovaginal canal in the sphenoid bone to reach the roof of the pharynx.
- The oropharynx is innervated by the glossopharyngeal nerve [IX] via the pharyngeal plexus.
- The laryngopharynx is innervated by the vagus nerve [X] via the internal branch of the superior laryngeal nerve.

### Glossopharyngeal nerve [IX]

The glossopharyngeal nerve [IX] is related to the pharynx throughout most of its course outside the cranial cavity.

After exiting the skull through the jugular foramen, the glossopharyngeal nerve [IX] descends on the posterior surface of the stylopharyngeus muscle (Fig. 8.208B),

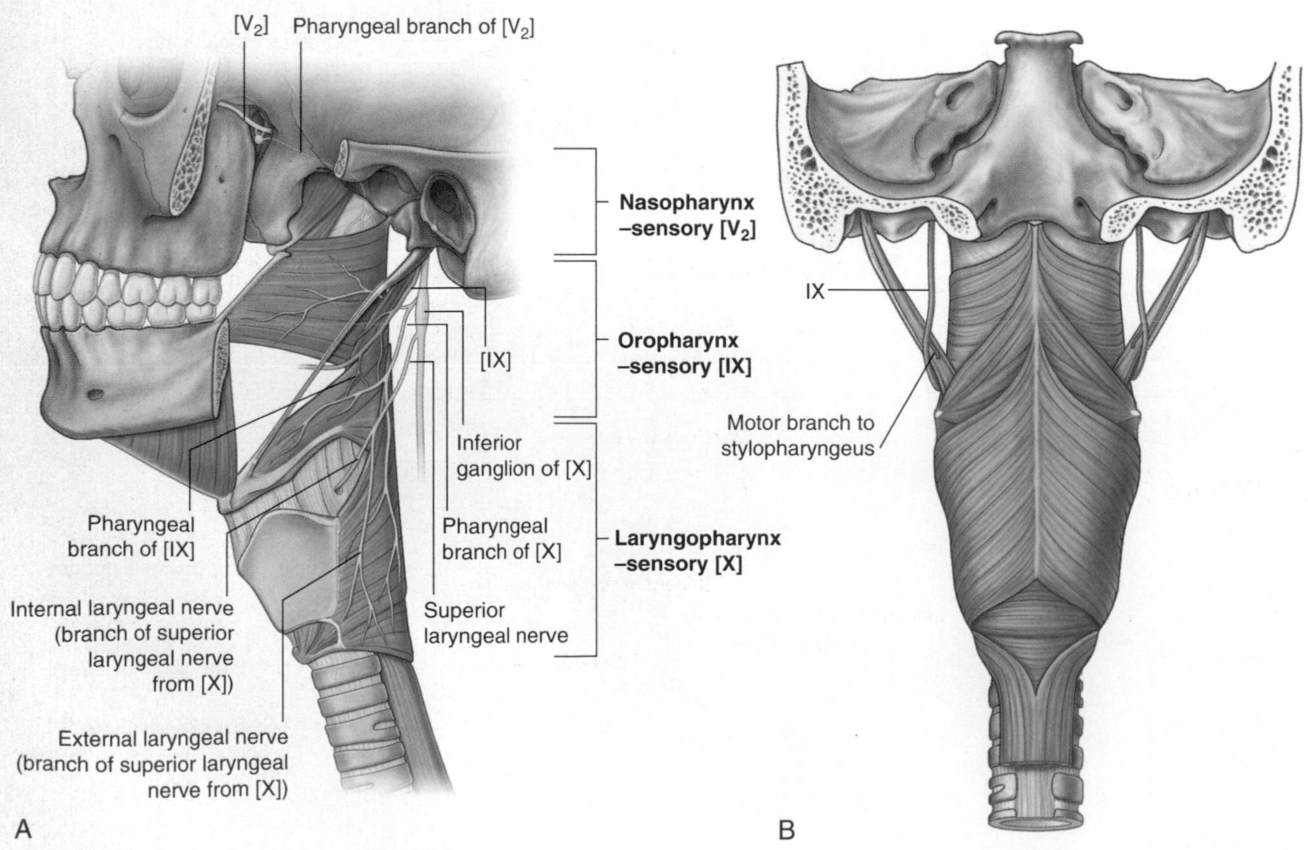


**Fig. 8.208** Innervation of the pharynx. **A.** Lateral view. **B.** Posterior view showing innervation of stylopharyngeus muscle.

passes onto the lateral surface of the stylopharyngeus, and then passes anteriorly through the gap (oropharyngeal triangle) between the superior constrictor, middle constrictor, and mylohyoid muscles to eventually reach the posterior aspect of the tongue.

As the glossopharyngeal nerve [IX] passes under the free edge of the superior constrictor, it is just inferior to the palatine tonsil lying on the deep surface of the superior constrictor.

Pharyngeal branches to the pharyngeal plexus and a motor branch to the stylopharyngeus muscle are among branches that originate from the glossopharyngeal nerve [IX] in the neck. Because sensory innervation of the oropharynx is by the glossopharyngeal nerve [IX], this nerve carries sensory innervation from the palatine tonsil and is also the afferent limb of the gag reflex (see "In the clinic" on p. 889).

## LARYNX

The larynx is a hollow musculoligamentous structure with a cartilaginous framework that caps the lower respiratory tract.

The cavity of the larynx is continuous below with the trachea, and above opens into the pharynx immediately posterior and slightly inferior to the tongue and the posterior opening (oropharyngeal isthmus) of the oral cavity (Fig. 8.209A,B).

The larynx is both a valve (or sphincter) to close the lower respiratory tract, and an instrument to produce sound. It is composed of:

- three large unpaired cartilages (cricoid, thyroid, and epiglottis),

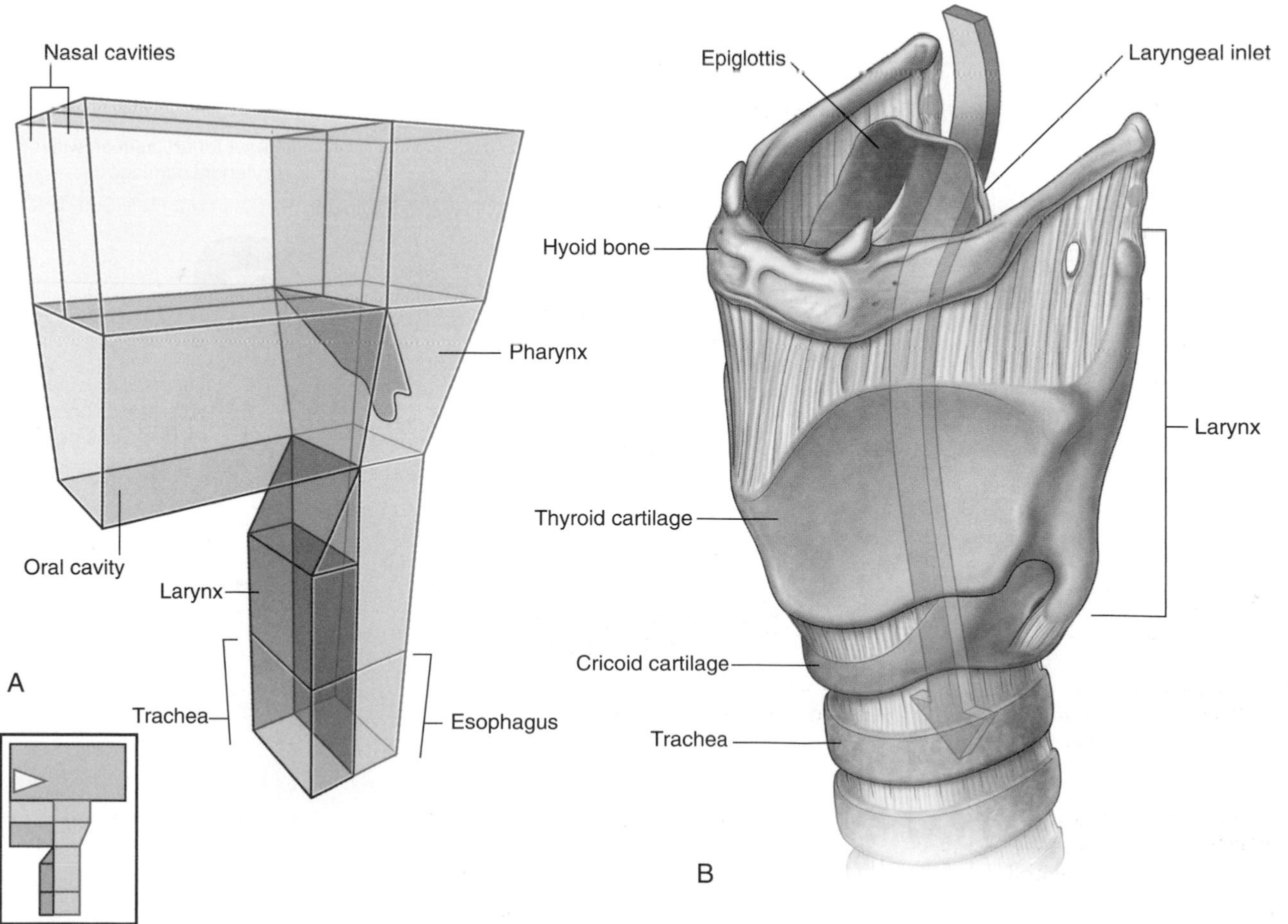


**Fig. 8.209** Larynx. **A.** Relationship to other cavities. **B.** Lateral view.

- three pairs of smaller cartilages (arytenoid, corniculate, and cuneiform), and
- a fibro-elastic membrane and numerous intrinsic muscles.

The larynx is suspended from the hyoid bone above and attached to the trachea below by membranes and ligaments. It is highly mobile in the neck and can be moved up and down and forward and backward by the action of extrinsic muscles that attach either to the larynx itself or to the hyoid bone.

During swallowing, the dramatic upward and forward movements of the larynx facilitate closing the laryngeal inlet and opening the esophagus.

Motor and sensory innervation of the larynx is provided by the vagus nerve [X].

## Laryngeal cartilages

### Cricoid cartilage

The cricoid cartilage is the most inferior of the laryngeal cartilages and completely encircles the airway (Fig. 8.210). It is shaped like a signet ring with a broad **lamina of cricoid cartilage** posterior to the airway and a much narrower **arch of cricoid cartilage** circling anteriorly.

The posterior surface of the lamina is characterized by two shallow oval depressions separated by a vertical ridge. The esophagus is attached to the ridge and the depressions are for attachment of the posterior crico-arytenoid muscles.

The cricoid cartilage has two articular facets on each side for articulation with other laryngeal cartilages:

- One facet is on the sloping superolateral surface of the lamina and articulates with the base of an arytenoid cartilage.
- The other facet is on the lateral surface of the lamina near its base and is for articulation with the medial surface of the inferior horn of the thyroid cartilage.

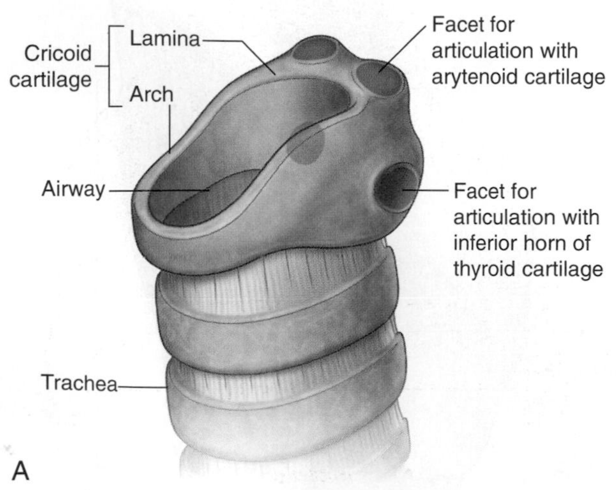


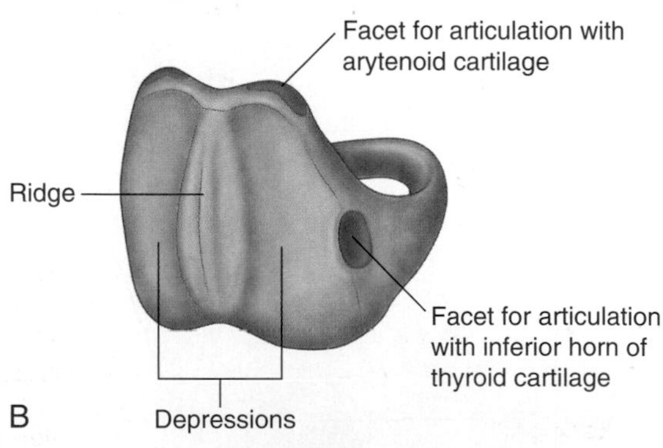


**Fig. 8.210** Cricoid cartilage. **A.** Anterolateral view. **B.** Posterior view.

### Thyroid cartilage

The thyroid cartilage (Fig. 8.211) is the largest of the laryngeal cartilages. It is formed by a right and a left lamina, which are widely separated posteriorly, but converge and join anteriorly. The most superior point of the site of fusion between the two broad flat laminae projects forward as the **laryngeal prominence** (Adam's apple). The angle between the two laminae is more acute in men (90°) than in women (120°) so the laryngeal prominence is more apparent in men than women.

Just superior to the laryngeal prominence, the **superior thyroid notch** separates the two laminae as they diverge laterally. Both the superior thyroid notch and the laryngeal prominence are palpable landmarks in the neck. There is a less distinct **inferior thyroid notch** in the midline along the base of the thyroid cartilage.

The posterior margin of each lamina of the thyroid cartilage is elongated to form a **superior horn** and an **inferior horn**:

- The medial surface of the inferior horn has a facet for articulation with the cricoid cartilage.
- The superior horn is connected by a **lateral thyrohyoid ligament** to the posterior end of the greater horn of the hyoid bone.

The lateral surface of each thyroid lamina is marked by a ridge (the **oblique line**), which curves anteriorly from the base of the superior horn to a little short of midway along the inferior margin of the lamina.

The ends of the oblique line are expanded to form **superior** and **inferior thyroid tubercles**. The oblique line is a site of attachment for the extrinsic muscles of the larynx (sternothyroid, thyrohyoid, and inferior constrictor).

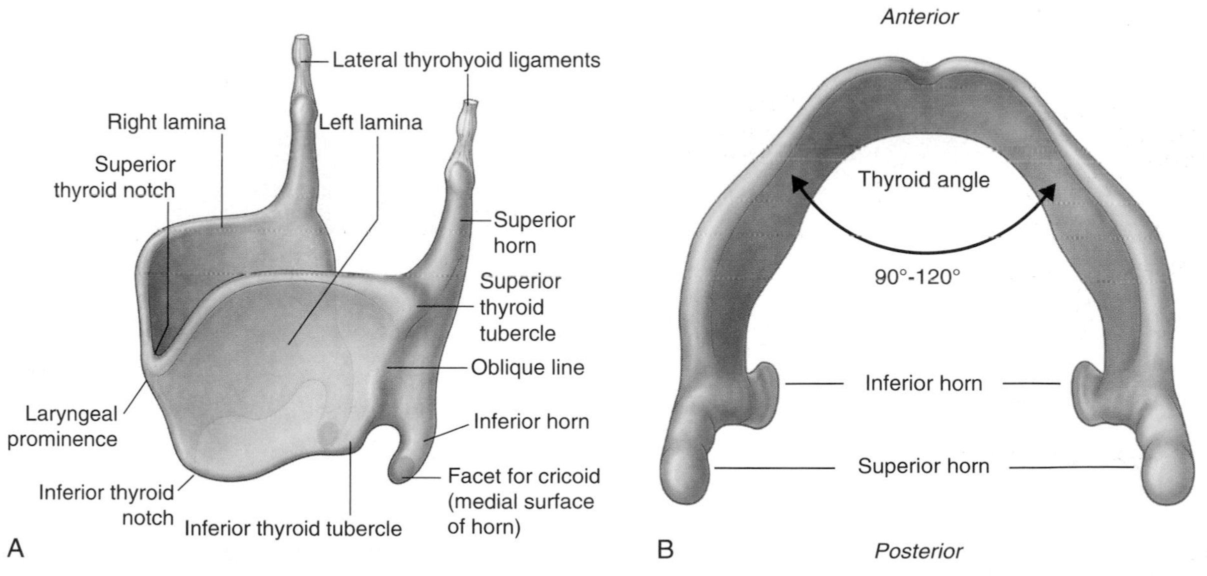


**Fig. 8.211** Thyroid cartilage. **A.** Anterolateral view. **B.** Superior view.

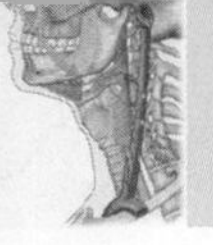



## Epiglottis

The epiglottis is a leaf-shaped cartilage attached by its stem to the posterior aspect of the thyroid cartilage at the angle (Fig. 8.212) and projects posterosuperiorly from its attachment to the thyroid cartilage. The attachment is via the **thyro-epiglottic ligament** in the midline approximately midway between the laryngeal prominence and the inferior thyroid notch. The upper margin of the epiglottis is behind the pharyngeal part of the tongue.

The inferior half of the posterior surface of the epiglottis is raised slightly to form an epiglottic tubercle.

## Arytenoid cartilages

The two arytenoid cartilages are pyramid-shaped cartilages with three surfaces, a **base of arytenoid cartilage** and an **apex of arytenoid cartilage** (Fig. 8.213):

- The base is concave and articulates with the sloping articular facet on the superolateral surface of the lamina of cricoid cartilage.
- The apex articulates with a corniculate cartilage.
- The **medial surface** of each cartilage faces the other.
- The **anterolateral surface** has two depressions, separated by a ridge, for muscle (vocalis) and ligament (vestibular ligament) attachment.
- The **posterior surface** is covered by the transverse arytenoid muscle (see Fig. 8.223).

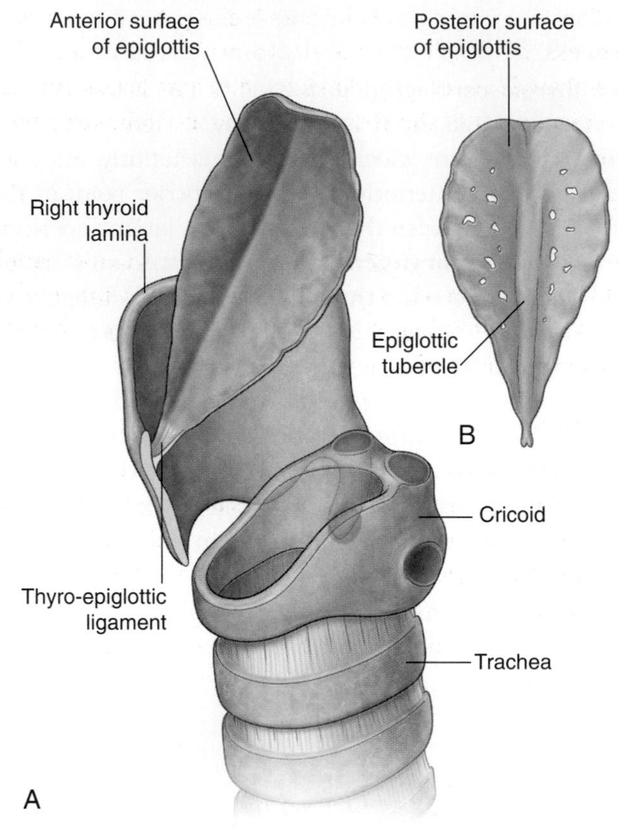


**Fig. 8.212** Epiglottis. **A.** Anterolateral view. **B.** Posterior surface.

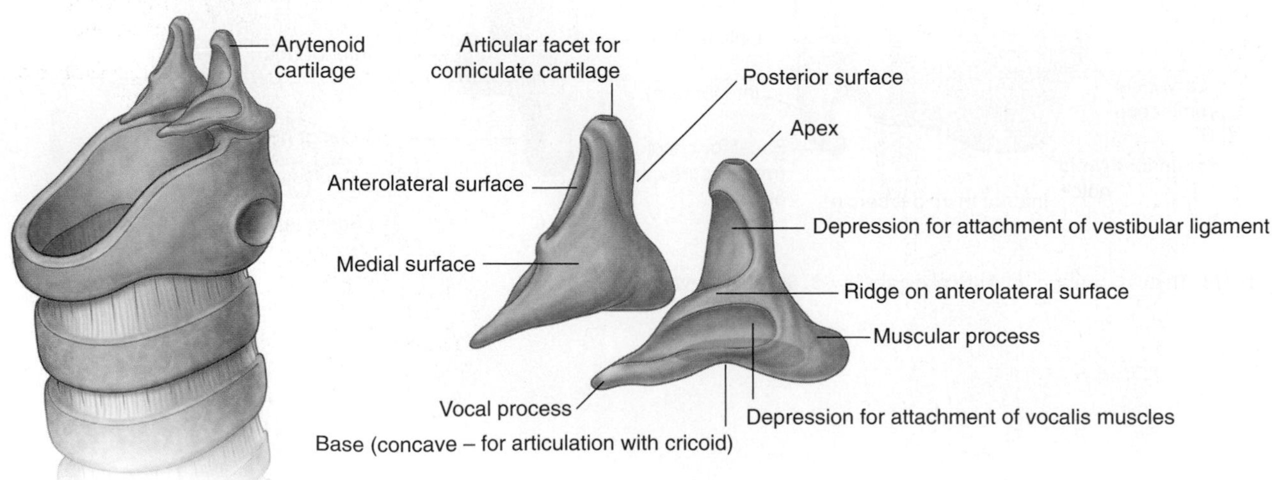


**Fig. 8.213** Arytenoid cartilages.

The anterior angle of the base is elongated into a **vocal process** to which the vocal ligament is attached. The lateral angle is similarly elongated into a muscular process for attachment of the posterior and lateral crico-arytenoid muscles.

### Corniculate

The corniculate cartilages (Fig. 8.214) are two small conical cartilages whose bases articulate with the apices of the arytenoid cartilages. Their apices project posteromedially toward each other.

### Cuneiform

These two small club-shaped cartilages (Fig. 8.214) lie anterior to the corniculate cartilages and are suspended in the part of the fibro-elastic membrane of the larynx that attaches the arytenoid cartilages to the lateral margin of the epiglottis.

## Extrinsic ligaments

### Thyrohyoid membrane

The thyrohyoid membrane is a tough fibro-elastic ligament that spans between the superior margin of the thyroid cartilage below and the hyoid bone above (Fig. 8.215). It is attached to the superior margin of the thyroid laminae and adjacent anterior margins of the superior horns, and ascends medial to the greater horns and posterior to the body of the hyoid bone to attach to the superior margins of these structures.

An aperture in the lateral part of the thyrohyoid membrane on each side is for the superior laryngeal artery, the internal branch of the superior laryngeal nerve, and lymphatics.

The posterior borders of the thyrohyoid membrane are thickened to form the **lateral thyrohyoid ligaments**. The membrane is also thickened anteriorly in the midline to form the **median thyrohyoid ligament**.

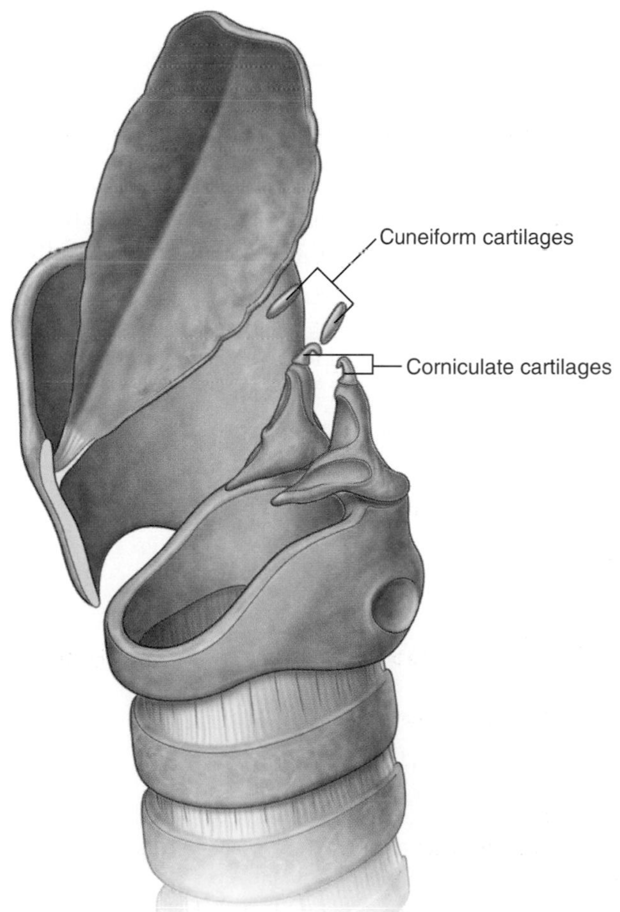


**Fig. 8.214** Corniculate and cuneiform cartilages.

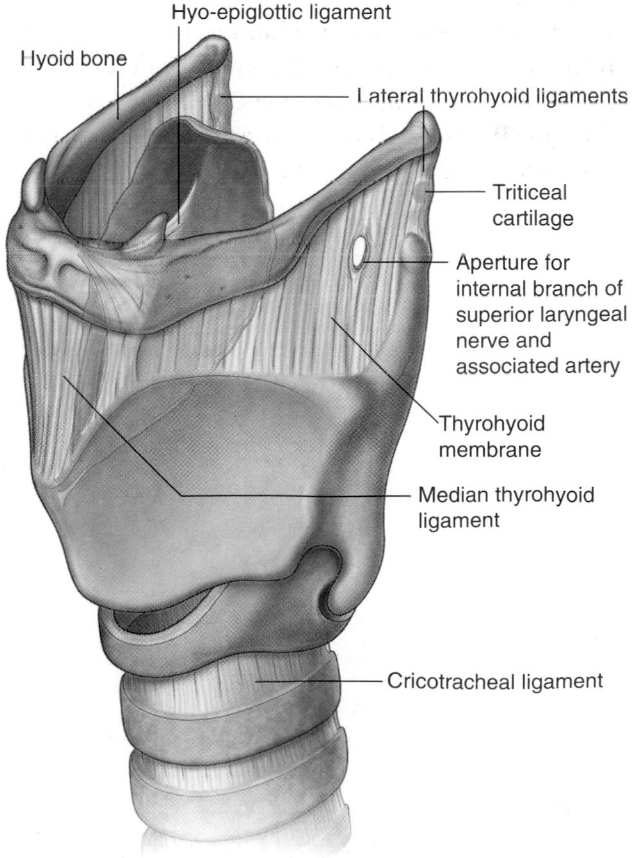


**Fig. 8.215** Extrinsic ligaments of the larynx.

Occasionally, there is a small cartilage (**triticeal cartilage**) in each lateral thyrohyoid ligament.

### Hyo-epiglottic ligament

The hyo-epiglottic ligament (Fig. 8.215) extends from the midline of the epiglottis, anterosuperiorly to the body of the hyoid bone.

### Cricotracheal ligament

The cricotracheal ligament (Fig. 8.215) runs from the lower border of the cricoid cartilage to the adjacent upper border of the first tracheal cartilage.

## Intrinsic ligaments

### Fibro-elastic membrane of the larynx

The fibro-elastic membrane of the larynx links together the laryngeal cartilages and completes the architectural framework of the laryngeal cavity. It is composed of two parts—a lower conus elasticus and an upper quadrangular membrane.

#### Conus elasticus (cricovocal membrane)

The conus elasticus (Fig. 8.216) is attached to the arch of cricoid cartilage and extends superiorly to end in a free upper margin within the space enclosed by the thyroid cartilage. On each side, this upper free margin attaches:

- anteriorly to the thyroid cartilage, and
- posteriorly to the vocal processes of the arytenoid cartilages.

The free margin between these two points of attachment is thickened to form the **vocal ligament**, which is under the **vocal fold** (**true vocal cord**) of the larynx.

The conus elasticus is also thickened anteriorly in the midline to form a distinct **median cricothyroid ligament**, which spans the distance between the arch of cricoid cartilage and the inferior thyroid notch and adjacent deep surface of the thyroid cartilage up to the attachment of the vocal ligaments.

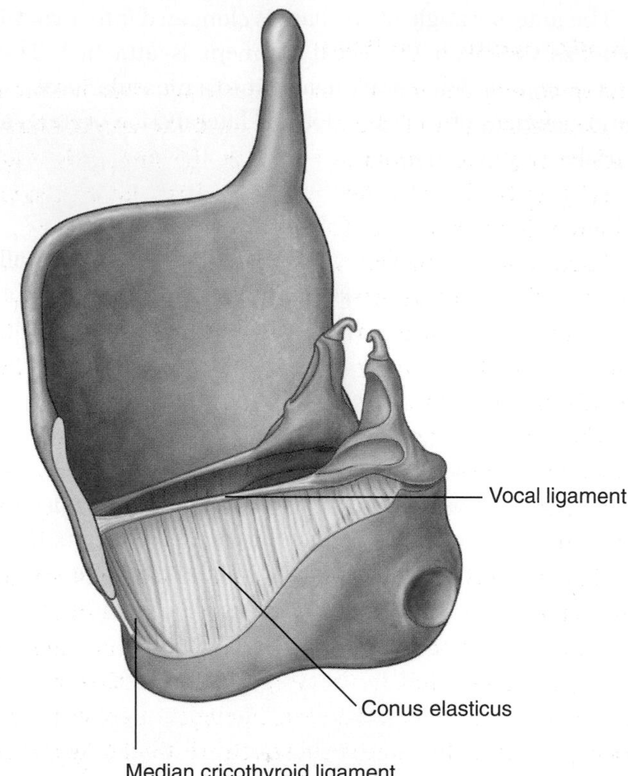


**Fig. 8.216** Cricothyroid ligament.

In emergency situations, when the airway is blocked above the level of the vocal folds, the median cricothyroid ligament can be perforated to establish an airway. Except for small vessels and the occasional presence of a pyramidal lobe of the thyroid gland, normally there are few structures between the median cricothyroid ligament and skin.

### Quadrangular membrane

The quadrangular membrane on each side runs between the lateral margin of the epiglottis and the anterolateral surface of the arytenoid cartilage on the same side (Fig. 8.217). It is also attached to the corniculate cartilage, which articulates with the apex of arytenoid cartilage.

Each quadrangular membrane has a free upper margin, between the top of the epiglottis and the corniculate cartilage, and a free lower margin. The free lower margin is thickened to form the **vestibular ligament** under the **vestibular fold** (**false vocal cord**) of the larynx.

The vestibular ligament is attached posteriorly to the superior depression on the anterolateral surface of the arytenoid cartilage and anteriorly to the thyroid angle just superior to the attachment of the vocal ligament.

On each side, the vestibular ligament of the quadrangular membrane is separated from the vocal ligament of the cricothyroid ligament below by a gap. Because the vestibular ligament attaches to the anterolateral surface of the arytenoid cartilage and the vocal ligament attaches to the vocal process of the same cartilage, the vestibular ligament is lateral to the vocal ligament when viewed from above (Fig. 8.218).

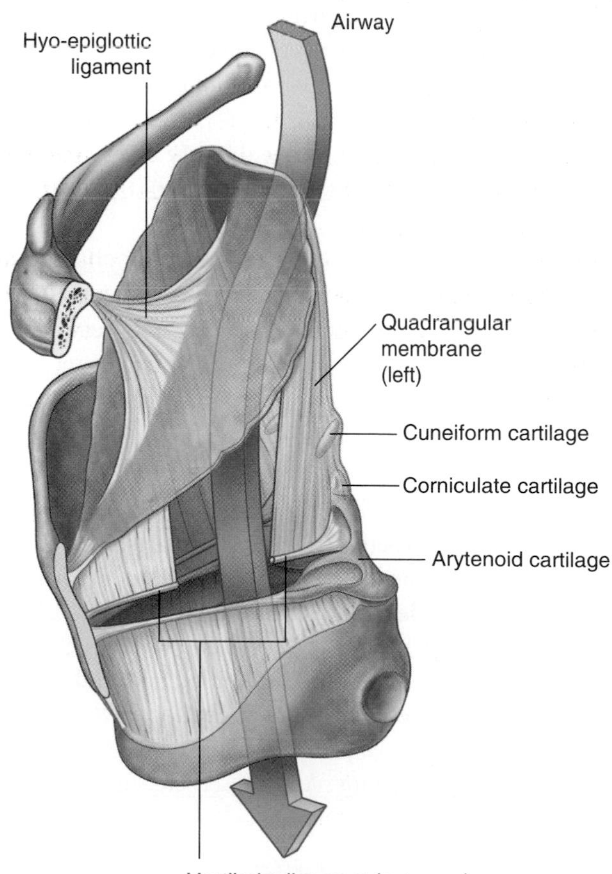


**Fig. 8.217** Quadrangular membrane.

## Laryngeal joints

### Cricothyroid joints

The joints between the inferior horns of the thyroid cartilage and the cricoid cartilage, and between the cricoid cartilage and arytenoid cartilages are synovial. Each is surrounded by a capsule and is reinforced by associated ligaments. The cricothyroid joints enable the thyroid cartilage to move forward and tilt downward on the cricoid cartilage (Fig. 8.219).

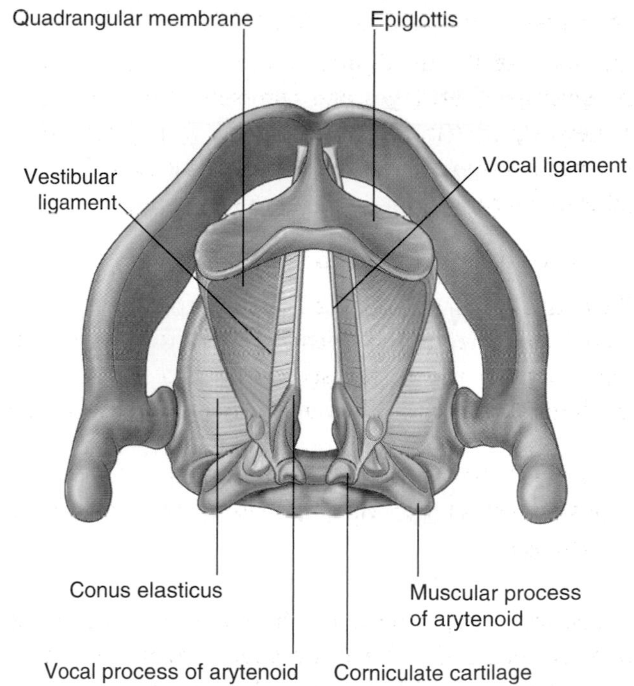


**Fig. 8.218** Fibro-elastic membrane of the larynx (superior view).

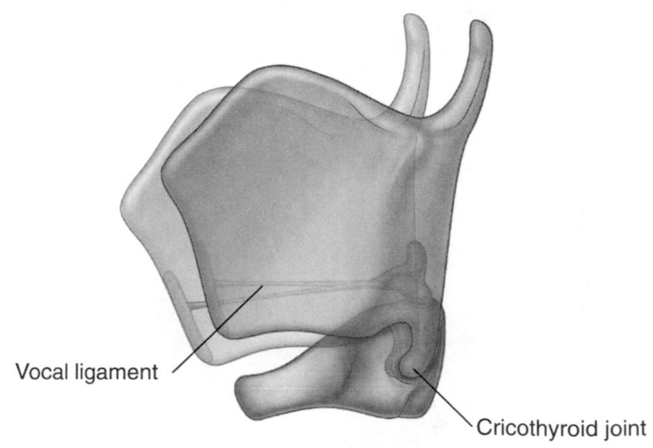


**Fig. 8.219** Movements of the cricothyroid joints.

Because the vocal ligaments pass between the posterior aspect of the thyroid angle and the arytenoid cartilages that sit on the lamina of cricoid cartilage, forward movement and downward rotation of the thyroid cartilage on the cricoid cartilage effectively lengthens and puts tension on the vocal ligaments.

### Crico-arytenoid joints

The crico-arytenoid joints between articular facets on the superolateral surfaces of the cricoid cartilage and the bases of the arytenoid cartilages enable the arytenoid cartilages to slide away or toward each other and to rotate so that the vocal processes pivot either toward or away from the midline. These movements abduct and adduct the vocal ligaments (Fig. 8.220).

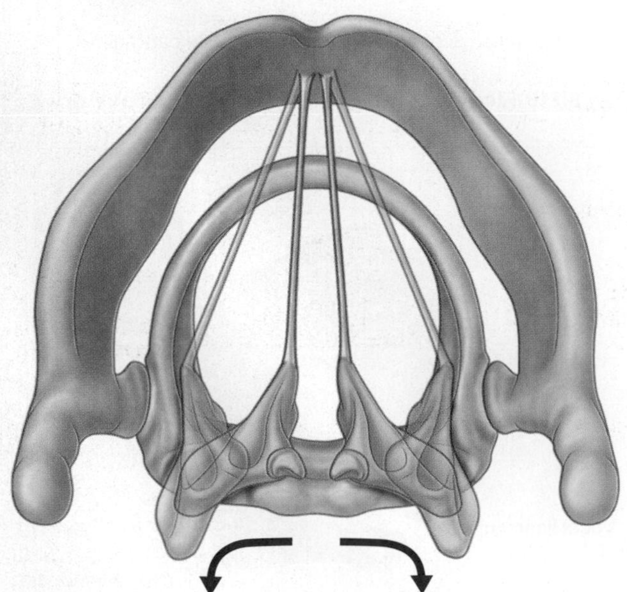

Fig. 8.220 Movements of the crico-arytenoid joints.

## Cavity of the larynx

The central cavity of the larynx (Fig. 8.221) is tubular and lined by mucosa. Its architectural support is provided by the fibro-elastic membrane of the larynx and by the laryngeal cartilages to which it is attached.

The superior aperture of the cavity (laryngeal inlet) opens into the anterior aspect of the pharynx just below and posterior to the tongue (Fig. 8.221A):

- Its anterior border is formed by mucosa covering the superior margin of the epiglottis.
- Its lateral borders are formed by mucosal folds (**ary-epiglottic folds**), which enclose the superior margins of the quadrangular membranes and adjacent soft tissues, and two tubercles on the more posterolateral margin of the laryngeal inlet on each side mark the positions of the underlying cuneiform and corniculate cartilages.
- Its posterior border in the midline is formed by a mucosal fold that forms a depression (**interarytenoid notch**) between the two corniculate tubercles.

The inferior opening of the laryngeal cavity is continuous with the lumen of the trachea, is completely encircled by the cricoid cartilage, and is horizontal in position unlike the laryngeal inlet, which is oblique and points posterosuperiorly into the pharynx. In addition, the inferior opening is continuously open, whereas the laryngeal inlet can be closed by downward movement of the epiglottis.

### Division into three major regions

Two pairs of mucosal folds, the vestibular and vocal folds, which project medially from the lateral walls of the laryngeal cavity, constrict it and divide it into three major regions—the vestibule, a middle chamber, and the infraglottic cavity (Fig. 8.221B):

- The **vestibule** is the upper chamber of the laryngeal cavity between the laryngeal inlet and the vestibular folds, which encloses the vestibular ligaments and associated soft tissues.
- The middle part of the laryngeal cavity is very thin and is between the vestibular folds above and the vocal folds below.
- The **infraglottic space** is the most inferior chamber of the laryngeal cavity and is between the vocal folds (which encloses the vocal ligaments and related soft tissues) and the inferior opening of the larynx.

### Laryngeal ventricles and saccules

On each side, the mucosa of the middle cavity bulges laterally through the gap between the vestibular and vocal ligaments to produce an expanded trough-shaped space (a **laryngeal ventricle**) (Fig. 8.221A). An elongate tubular extension of each ventricle (laryngeal saccule) projects anterosuperiorly between the vestibular fold and thyroid cartilage and may reach as high as the top of the thyroid cartilage. Within the walls of these laryngeal saccules are numerous mucous glands. Mucus secreted into the saccules lubricates the vocal folds.

### Rima vestibuli and rima glottidis

When viewed from above (Fig. 8.221C,D), there is a triangular opening (the **rima vestibuli**) between the two

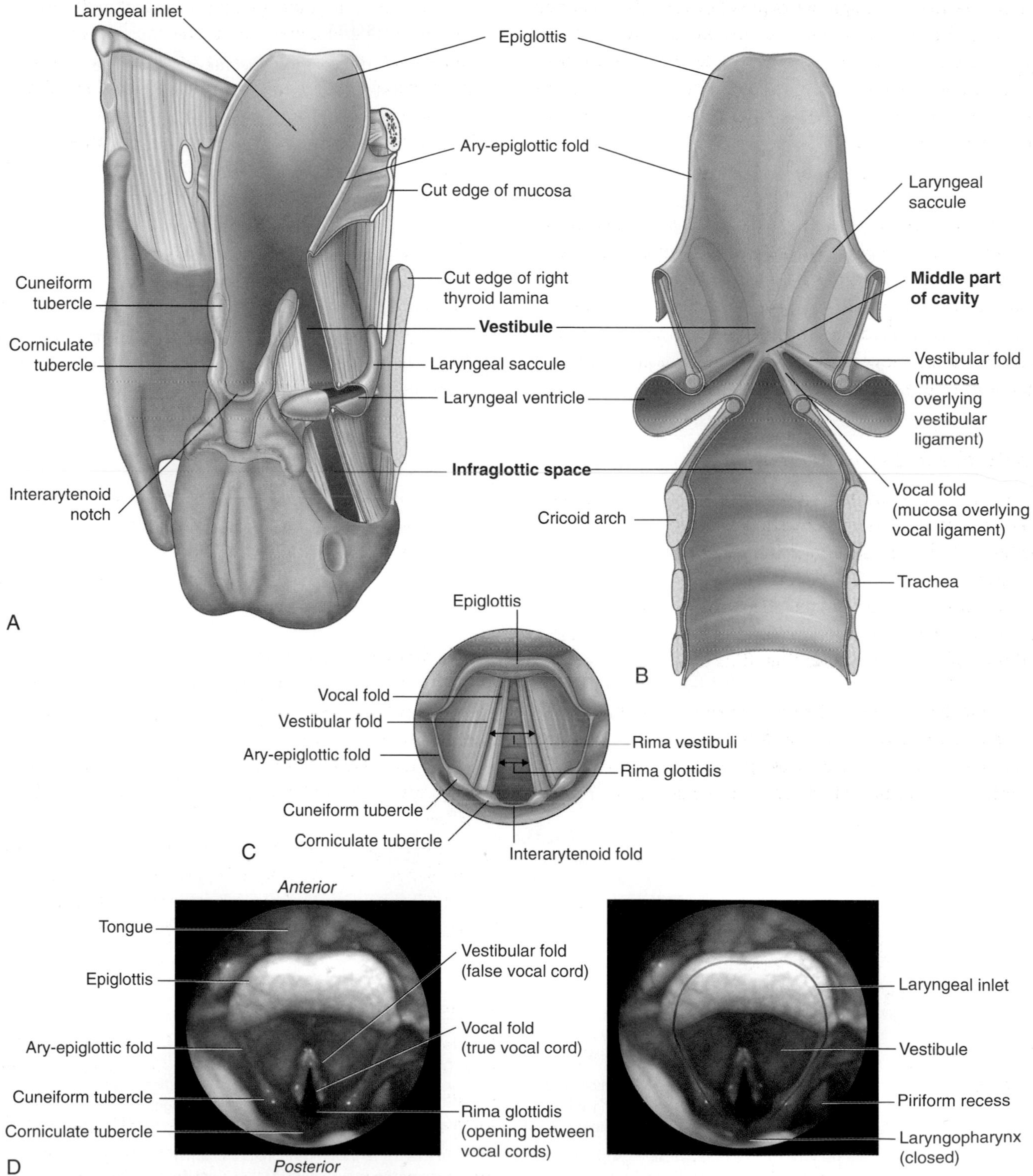


**Fig. 8.221** Laryngeal cavity. **A.** Posterolateral view. **B.** Posterior view (cut away). **C.** Superior view through the laryngeal inlet. **D.** Labeled photograph of the larynx, superior view.

adjacent vestibular folds at the entrance to the middle chamber of the laryngeal cavity. The apex of the opening is anterior and its base is formed by the posterior wall of the laryngeal cavity.

Inferior to the vestibular folds, the vocal folds (true vocal cords) and adjacent mucosa-covered parts of the arytenoid cartilages form the lateral walls of a similar, but narrower, triangular opening (the **rima glottidis** between the two adjacent vocal folds). This opening separates the middle chamber above from the infraglottic cavity below. The base of this triangular opening is formed by the fold of mucosa (**interarytenoid fold**) at the bottom of the interarytenoid notch.

Both the rima glottidis and the rima vestibuli can be opened and closed by movement of the arytenoid cartilages and associated fibro-elastic membranes.

## Intrinsic muscles

The intrinsic muscles of the larynx (Table 8.19) adjust tension in the vocal ligaments, open and close the rima glottidis, control the inner dimensions of the vestibule, close the rima vestibuli, and facilitate closing of the laryngeal inlet. They do this mainly by:

- acting on the cricothyroid and crico-arytenoid joints,
- adjusting the distance between the epiglottis and arytenoid cartilages,
- pulling directly on the vocal ligaments, and
- forcing soft tissues associated with the quadrangular membranes and vestibular ligaments toward the midline.

**Table 8.19** Intrinsic muscles of the larynx

| Muscle | Origin | Insertion | Innervation | Function |
|---|---|---|---|---|
| Cricothyroid | Anterolateral aspect of arch of cricoid cartilage | Oblique part—inferior horn of the thyroid cartilage; straight part—inferior margin of thyroid cartilage | External branch of superior laryngeal nerve from the vagus nerve [X] | Forward and downward rotation of the thyroid cartilage at the cricothyroid joint |
| Posterior crico-arytenoid | Oval depression on posterior surface of lamina of cricoid cartilage | Posterior surface of muscular process of arytenoid cartilage | Recurrent laryngeal branch of the vagus nerve [X] | Abduction and external rotation of the arytenoid cartilage. The posterior crico-arytenoid muscles are the primary abductors of the vocal folds. In other words, they are the primary openers of the rima glottidis. |
| Lateral crico-arytenoid | Superior surface of arch of cricoid cartilage | Anterior surface of muscular process of arytenoid cartilage | Recurrent laryngeal branch of the vagus nerve [X] | Internal rotation of the arytenoid cartilage and adduction of vocal folds |
| Transverse arytenoid | Lateral border of posterior surface of arytenoid cartilage | Lateral border of posterior surface of opposite arytenoid cartilage | Recurrent laryngeal branch of the vagus nerve [X] | Adduction of arytenoid cartilages |
| Oblique arytenoid | Posterior surface of muscular process of arytenoid cartilage | Posterior surface of apex of adjacent arytenoid cartilage; extends into ary-epiglottic fold | Recurrent laryngeal branch of the vagus nerve [X] | Sphincter of the laryngeal inlet |
| Thyro-arytenoid | Thyroid angle and adjacent cricothyroid ligament | Anterolateral surface of arytenoid cartilage; some fibers continue in ary-epiglottic folds to the lateral margin of the epiglottis | Recurrent laryngeal branch of the vagus nerve [X] | Sphincter of vestibule and of laryngeal inlet |
| Vocalis | Lateral surface of vocal process of arytenoid cartilage | Vocal ligament and thyroid angle | Recurrent laryngeal branch of the vagus nerve [X] | Adjusts tension in vocal folds |

### Cricothyroid muscles

The fan-shaped **cricothyroid muscles** are attached to the anterolateral surfaces of the arch of the cricoid cartilage and expand superiorly and posteriorly to attach to the thyroid cartilage (Fig. 8.222).

Each muscle has an oblique part and a straight part:

- The **oblique part** runs in a posterior direction from the arch of the cricoid cartilage to the inferior horn of the thyroid cartilage.
- The **straight part** runs more vertically from the arch of the cricoid cartilage to the posteroinferior margin of the thyroid lamina.

The cricothyroid muscles move the cricothyroid joints. They pull the thyroid cartilage forward and rotate it down relative to the cricoid cartilage. These actions lengthen the vocal folds.

The cricothyroid muscles are the only intrinsic muscles of the larynx innervated by the superior laryngeal branches of the vagus nerves [X]. All other intrinsic muscles are innervated by the recurrent laryngeal branches of the vagus nerves [X].

### Posterior crico-arytenoid muscles

There is a right and a left **posterior crico-arytenoid muscle** (Fig. 8.223). The fibers of each muscle originate from a large shallow depression on the posterior surface of the lamina of the cricoid cartilage, and run superiorly and laterally to converge on the muscular processes of the arytenoid cartilage.

The posterior crico-arytenoid muscles abduct and externally (laterally) rotate the arytenoid cartilages, thereby opening the rima glottidis. These muscles are the primary abductors of the vocal folds. They are innervated by the recurrent laryngeal branches of the vagus nerves [X].

### Lateral crico-arytenoid muscles

The **lateral crico-arytenoid muscle** on each side originates from the upper surface of the arch of the cricoid cartilage, and runs posteriorly and superiorly to insert on the muscular process of the arytenoid cartilage (Fig. 8.223).

The lateral crico-arytenoid muscles internally rotate the arytenoid cartilages. These movements result in adducted (closed) vocal folds.

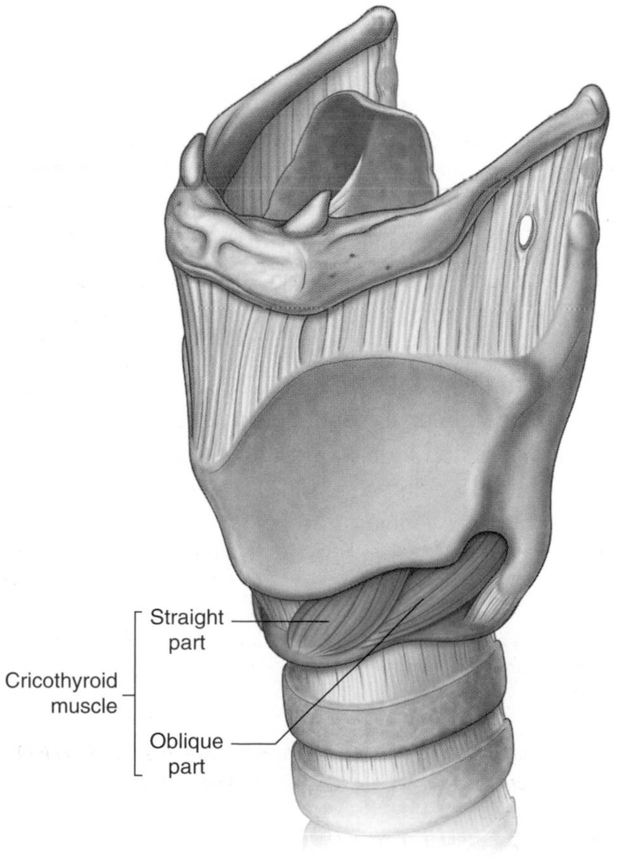

**Fig. 8.222** Cricothyroid muscle.

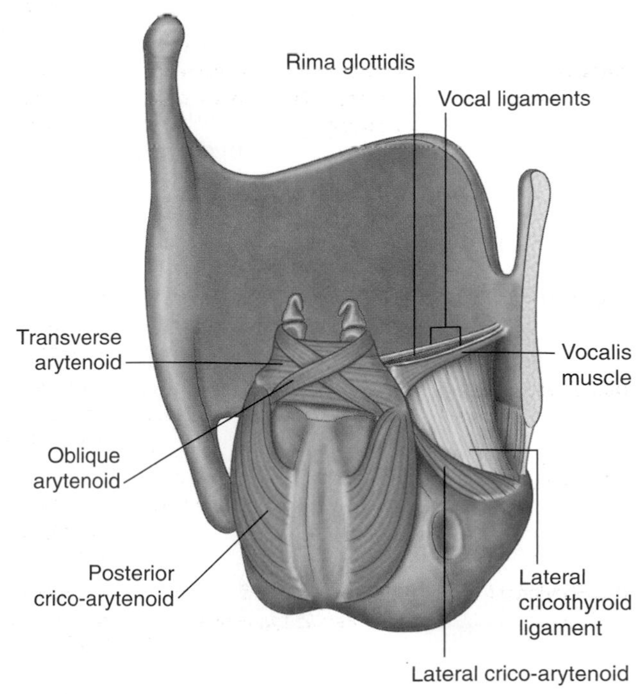


**Fig. 8.223** Crico-arytenoid, oblique and transverse arytenoid, and vocalis muscles.

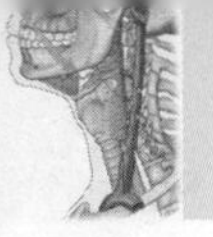



The lateral crico-arytenoids are innervated by the recurrent laryngeal branches of the vagus nerves [X].

### Transverse arytenoid muscle

The single **transverse arytenoid muscle** spans the distance between adjacent lateral margins of the arytenoid cartilages and covers the posterior surfaces of these cartilages (Fig. 8.223). It adducts the arytenoid cartilages and is innervated by the recurrent laryngeal branches of the vagus nerves [X].

### Oblique arytenoid muscles

Each of the two **oblique arytenoid muscles** runs from the posterior surface of the muscular process of one arytenoid cartilage to the apex of the arytenoid cartilage on the other side (Fig. 8.223). Some fibers of the muscle continue laterally around the margin of the arytenoid cartilage and into the ary-epiglottic fold where they continue as the **ary-epiglottic part** of the muscle (Fig. 8.224).

The oblique arytenoids can narrow the laryngeal inlet by constricting the distance between the arytenoid cartilages and the epiglottis. They are innervated by the recurrent laryngeal branches of the vagus nerves [X].

### Vocalis

The **vocalis muscles** are elongate muscles lateral to and running parallel with each vocal ligament (Fig. 8.223). The fibers in each muscle are attached posteriorly to the lateral surface of the vocal process and adjacent depression on the anterolateral surface of the arytenoid cartilage, and anteriorly insert along the length of the vocal ligament to the thyroid angle.

The vocalis muscles adjust tension in the vocal folds and are innervated by the recurrent laryngeal branches of the vagus nerves [X].

### Thyro-arytenoid muscles

The two **thyro-arytenoid muscles** are broad flat muscles lateral to the fibro-elastic membrane of the larynx and the laryngeal ventricles and saccules (Fig. 8.224). Each muscle runs from a vertical line of origin on the lower half of the thyroid angle and adjacent external surface of the crico-thyroid ligament to the anterolateral surface of the arytenoid cartilage. Some of the fibers may continue into the ary-epiglottic fold and reach the margin of the epiglottis. These fibers are the **thyro-epiglottic part** of the muscle.

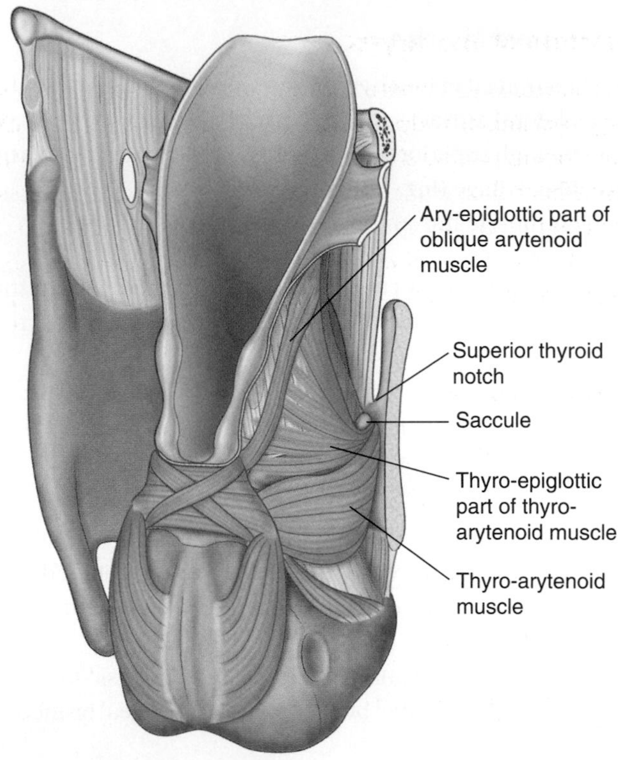


**Fig. 8.224** Thyro-arytenoid muscle.

Because the thyro-arytenoid muscles are broad and lateral to the quadrangular membrane, they act as a sphincter of the vestibule by pushing soft tissues medial to the muscles toward the midline. The muscles also narrow the laryngeal inlet by pulling the arytenoid cartilages forward while simultaneously pulling the epiglottis toward the arytenoid cartilages.

The thyro-arytenoid muscles are innervated by the recurrent laryngeal branches of the vagus nerves [X].

## Function of the larynx

The larynx is an elaborate sphincter for the lower respiratory tract and provides a mechanism for producing sounds. Adjustments of the size of the central cavity of the larynx result from changes in the dimensions of the rima glottidis, the rima vestibuli, the vestibule, and the laryngeal inlet (Fig. 8.225). These changes result from muscle actions and laryngeal mechanics.

### Respiration

During quiet respiration, the laryngeal inlet, vestibule, rima vestibuli, and rima glottidis are open. The arytenoid cartilages are abducted and the rima glottidis is triangular shaped (Fig. 8.225A). During forced inspiration (Fig. 8.225B), the arytenoid cartilages are rotated laterally, mainly by the action of the posterior crico-arytenoid muscles. As a result, the vocal folds are abducted and the

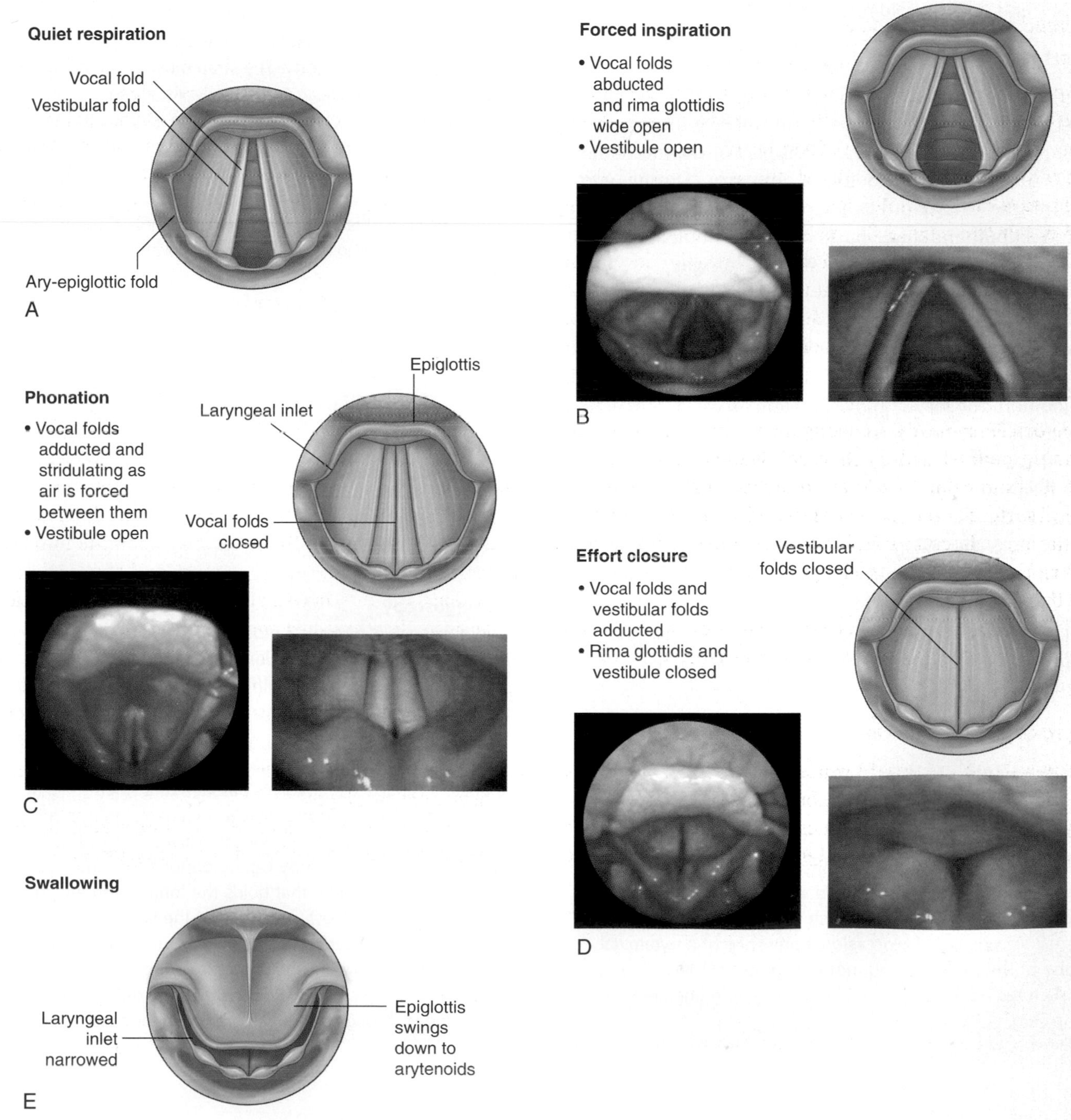

**Fig. 8.225** Laryngeal function. **A.** Quiet respiration. **B.** Forced inspiration. **C.** Phonation. **D.** Effort closure. **E.** Swallowing.

rima glottidis widens into a rhomboid shape, which effectively increases the diameter of the laryngeal airway.

### Phonation

When phonating, the arytenoid cartilages and vocal folds are adducted and air is forced through the closed rima glottidis (Fig. 8.225C). This action causes the vocal folds to vibrate against each other and produce sounds, which can then be modified by the upper parts of the airway and oral cavity. Tension in the vocal folds can be adjusted by the vocalis and cricothyroid muscles.

### Effort closure

Effort closure of the larynx (Fig. 8.225D) occurs when air is retained in the thoracic cavity to stabilize the trunk, for example during heavy lifting, or as part of the mechanism for increasing intra-abdominal pressure. During effort closure, the rima glottidis is completely closed, as is the rima vestibuli and lower parts of the vestibule. The result is to completely and forcefully shut the airway.

### Swallowing

During swallowing, the rima glottidis, rima vestibuli, and vestibule are closed and the laryngeal inlet is narrowed. In addition, the larynx moves up and forward. This action causes the epiglottis to swing downward toward the arytenoid cartilages and to effectively narrow or close the laryngeal inlet (Fig. 8.225E). The up and forward movement of the larynx also opens the esophagus, which is attached to the posterior aspect of the lamina of the cricoid cartilage. All these actions together prevent solids and liquids from entry into the airway and facilitate their movement through the piriform fossae into the esophagus.

#### In the clinic

**Cricothyrotomy**

In emergency situations, when the airway is blocked above the level of the vocal folds, the median cricothyroid ligament can be perforated and a small tube inserted through the incision to establish an airway. Except for small vessels and the occasional presence of a pyramidal lobe of the thyroid gland, normally there are few structures between the median cricothyroid ligament and the skin.

#### In the clinic

**Tracheostomy**

A tracheostomy is a surgical procedure in which a hole is made in the trachea and a tube is inserted to enable ventilation.

A tracheostomy is typically performed when there is obstruction to the larynx as a result of inhalation of a foreign body, severe edema secondary to anaphylactic reaction, or severe head and neck trauma.

The typical situation in which a tracheostomy is performed is in the calm atmosphere of an operating theater. A small transverse incision is placed in the lower third of the neck anteriorly. The strap muscles are deviated laterally and the trachea can be easily visualized. Occasionally it is necessary to divide the isthmus of the thyroid gland. An incision is made in the second and third tracheal rings and a small tracheostomy tube inserted.

After the tracheostomy has been in situ for the required length of time, it is simply removed. The hole through which it was inserted almost inevitably closes without any intervention.

Patients with long-term tracheostomies are unable to vocalize because no air is passing through the vocal cords.

#### In the clinic

**Laryngoscopy**

Laryngoscopy is a medical procedure that is used to inspect the larynx. The functions of laryngoscopy include the evaluation of patients with difficulty swallowing, assessment of the vocal cords, and assessment of the larynx for tumors, masses, and weak voice.

The larynx is typically visualized using two methods. Indirect laryngoscopy involves passage of a small rod-mounted mirror (not dissimilar to a dental mirror) into the oropharynx permitting indirect visualization of the larynx. Direct laryngoscopy can be performed using a device with a curved metal tip that holds the tongue and epiglottis forward, allowing direct inspection of the larynx. This procedure can be performed only in the unconscious patient or in a patient in whom the gag reflex is not intact. Other methods of inspection include the passage of fiberoptic endoscopes through either the oral cavity or nasal cavity.

## Vessels

### Arteries

The major blood supply to the larynx is by the superior and inferior laryngeal arteries (Fig. 8.226):

- The **superior laryngeal artery** originates near the upper margin of the thyroid cartilage from the superior thyroid branch of the external carotid artery, and accompanies the internal branch of the superior laryngeal nerve through the thyrohyoid membrane to reach the larynx.
- The **inferior laryngeal artery** originates from the inferior thyroid branch of the thyrocervical trunk of the subclavian artery low in the neck and, together with the recurrent laryngeal nerve, ascends in the groove between the esophagus and trachea—it enters the larynx by passing deep to the margin of the inferior constrictor muscle of the pharynx.

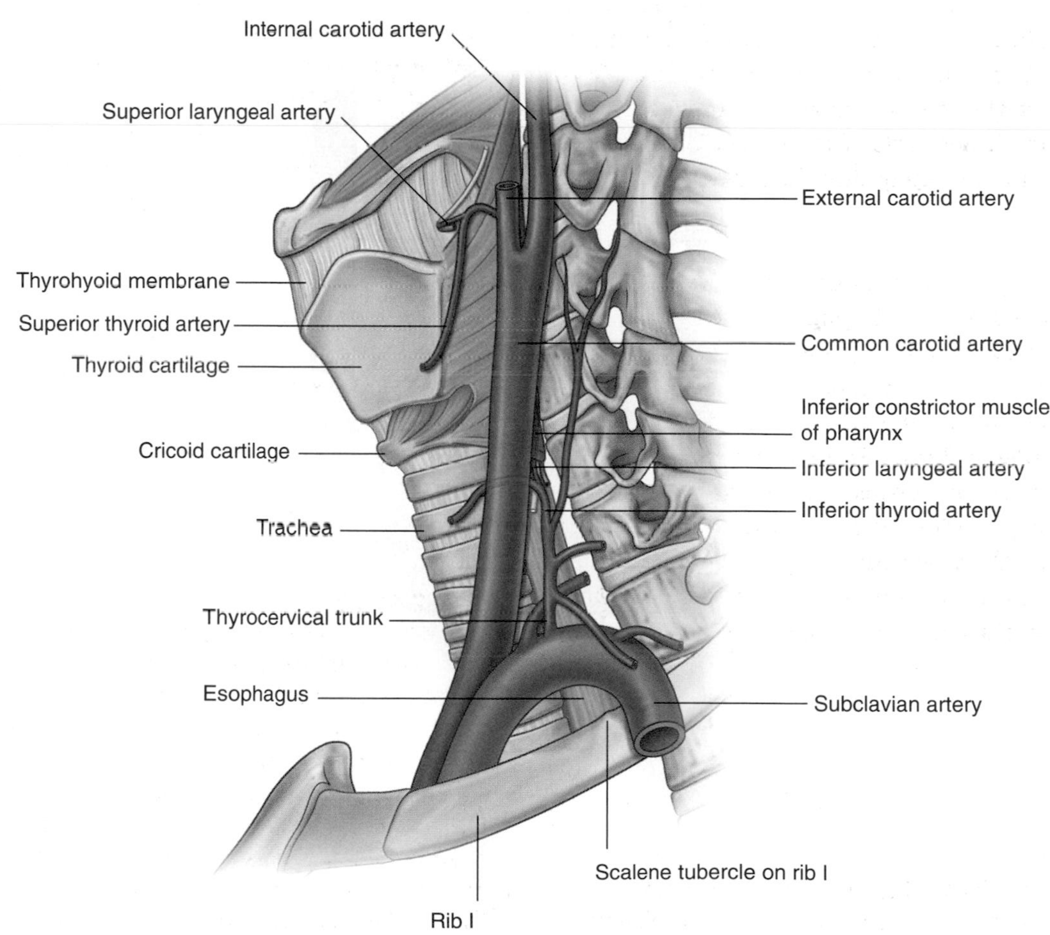


**Fig. 8.226** Arterial supply of the larynx, left lateral view.

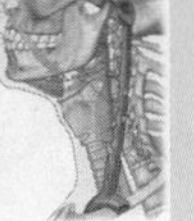



### Veins

Veins draining the larynx accompany the arteries:

- **Superior laryngeal veins** drain into superior thyroid veins, which in turn drain into the internal jugular veins (Fig. 8.227).
- **Inferior laryngeal veins** drain into inferior thyroid veins, which drain into the left brachiocephalic vein.

### Lymphatics

Lymphatics drain regions above and below the vocal folds:

- Those above the vocal folds follow the superior laryngeal artery and terminate in deep cervical nodes associated with the bifurcation of the common carotid artery.
- Those below the vocal folds drain into deep nodes associated with the inferior thyroid artery or with nodes associated with the front of the cricothyroid ligament or upper trachea.

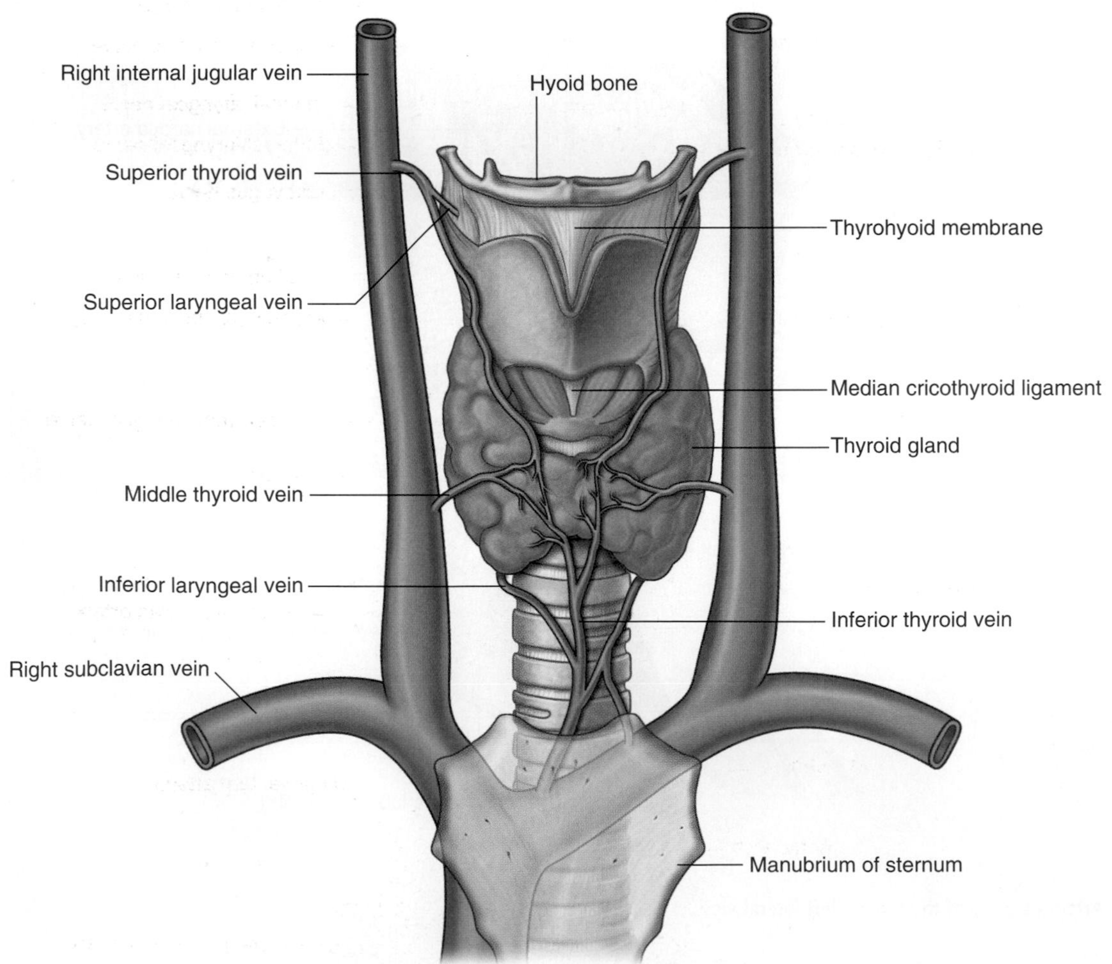


**Fig. 8.227** Venous drainage of the larynx, anterior view.

## Nerves

Sensory and motor innervation of the larynx is by two branches of the vagus nerves [X]—the superior laryngeal nerves and the recurrent laryngeal nerves (Fig. 8.228).

### Superior laryngeal nerves

The **superior laryngeal nerves** originate from the inferior vagal ganglia high in the neck (Fig. 8.228). On each side, the nerve descends medial to the internal carotid artery and divides into internal and external branches just above the level of the superior horn of the hyoid bone:

- The external branch (**external laryngeal nerve**) descends along the lateral wall of the pharynx to supply and penetrate the inferior constrictor of the pharynx and ends by supplying the cricothyroid muscle.
- The internal branch (**internal laryngeal nerve**) passes anteroinferiorly to penetrate the thyrohyoid membrane—it is mainly sensory and supplies the laryngeal cavity down to the level of the vocal folds.

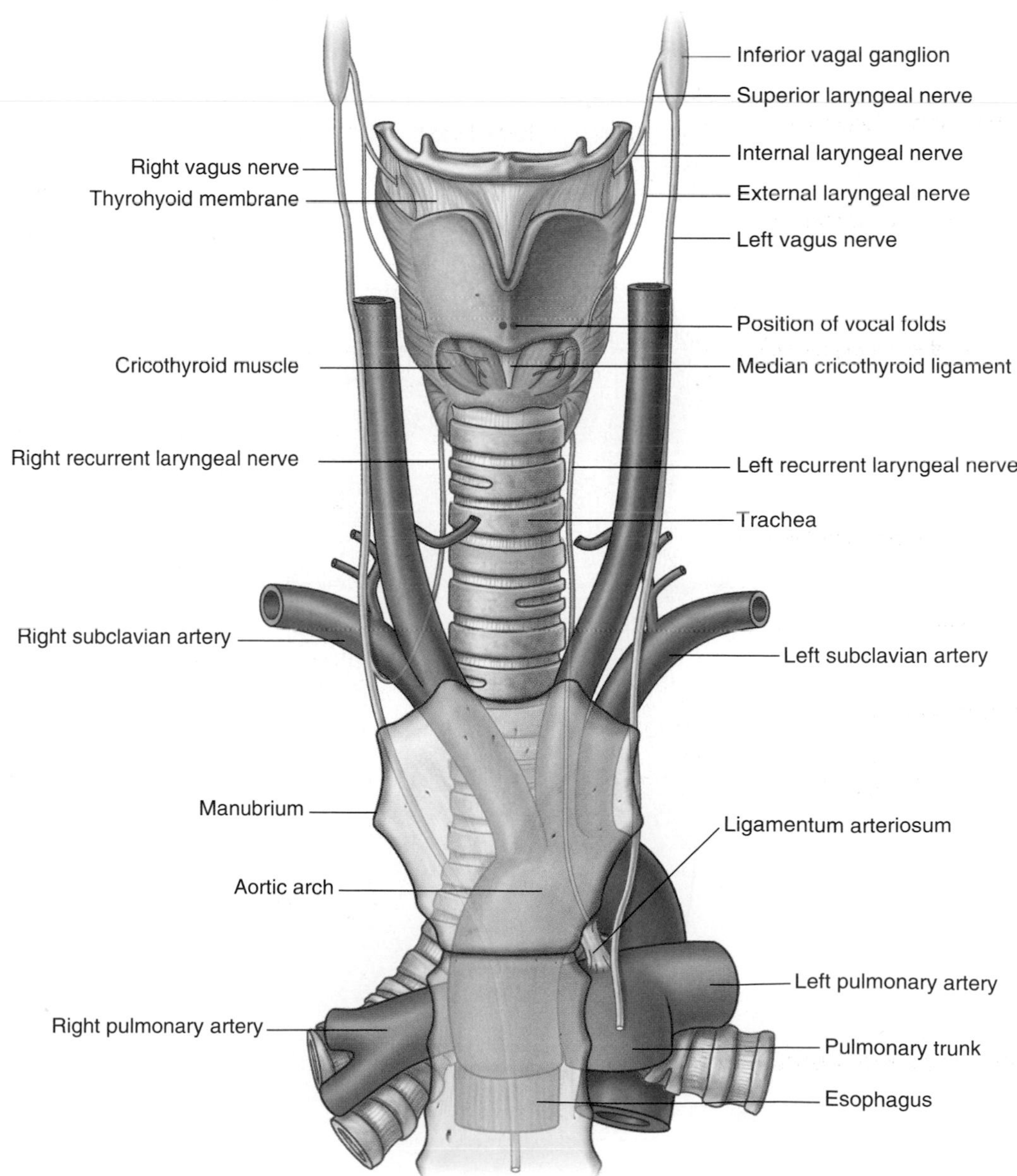


**Fig. 8.228** Innervation of the larynx.

### Recurrent laryngeal nerves

The recurrent laryngeal nerves are (Fig. 8.228):

- sensory to the laryngeal cavity below the level of the vocal folds, and
- motor to all intrinsic muscles of the larynx except for the cricothyroid.

The left recurrent laryngeal nerve originates in the thorax, whereas the right recurrent laryngeal nerve originates in the root of the neck. Both nerves generally ascend in the neck in the groove between the esophagus and trachea and enter the larynx deep to the margin of the inferior constrictor. They may pass medial to, lateral to, or through the lateral ligament of the thyroid gland, which attaches the thyroid gland to the trachea and lower part of the cricoid cartilage on each side.

## NASAL CAVITIES

The two nasal cavities are the uppermost parts of the respiratory tract and contain the olfactory receptors. They are elongated wedge-shaped spaces with a large inferior base and a narrow superior apex (Figs. 8.229 and 8.230) and are held open by a skeletal framework consisting mainly of bone and cartilage.

The smaller anterior regions of the cavities are enclosed by the external nose, whereas the larger posterior regions are more central within the skull. The anterior apertures of the nasal cavities are the nares, which open onto the inferior surface of the nose. The posterior apertures are the choanae, which open into the nasopharynx.

The nasal cavities are separated:

- from each other by a midline nasal septum,
- from the oral cavity below by the hard palate, and

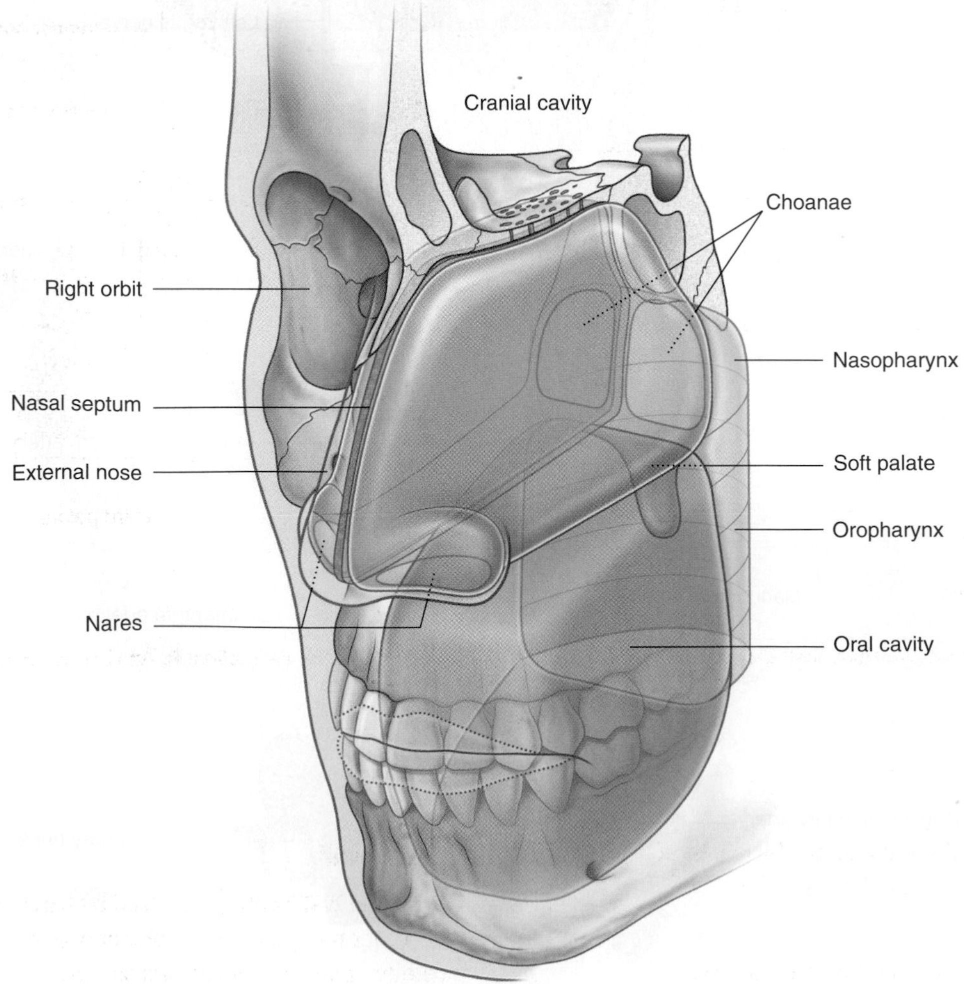


**Fig. 8.229** Nasal cavities (anterolateral view). Relationship to other cavities.

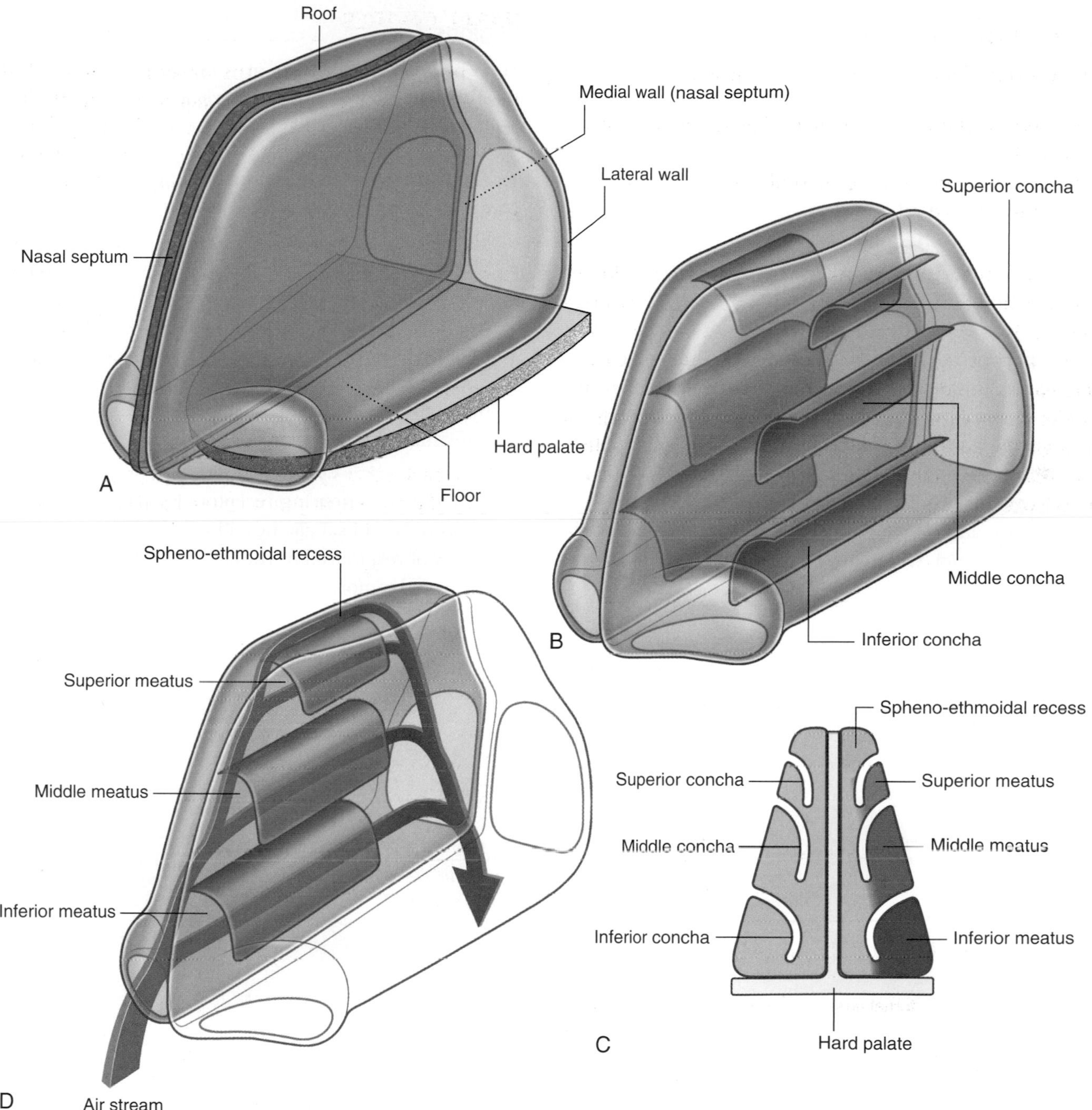


**Fig. 8.230** Nasal cavities. **A.** Floor, roof, and lateral walls. **B.** Conchae on lateral walls. **C.** Coronal section. **D.** Air channels in right nasal cavity.

- from the cranial cavity above by parts of the frontal, ethmoid, and sphenoid bones.

Lateral to the nasal cavities are the orbits.

Each nasal cavity has a floor, roof, medial wall, and lateral wall (Fig. 8.230A).

## Lateral wall

The lateral wall is characterized by three curved shelves of bone (conchae), which are one above the other and project medially and inferiorly across the nasal cavity (Fig. 8.230B). The medial, anterior, and posterior margins of the conchae are free.

The conchae divide each nasal cavity into four air channels (Fig. 8.230C,D):

- an **inferior nasal meatus** between the **inferior concha** and the nasal floor,
- a **middle nasal meatus** between the inferior and **middle concha**,
- a **superior nasal meatus** between the middle and **superior concha**, and
- a **spheno-ethmoidal recess** between the superior concha and the nasal roof.

These conchae increase the surface area of contact between tissues of the lateral wall and the respired air.

The openings of the paranasal sinuses, which are extensions of the nasal cavity that erode into the surrounding bones during childhood and early adulthood, are on the lateral wall and roof of the nasal cavities (Fig. 8.231). In addition, the lateral wall also contains the opening of the nasolacrimal duct, which drains tears from the eye into the nasal cavity.

## Regions

Each nasal cavity consists of three general regions—the nasal vestibule, the respiratory region, and the olfactory region (Fig. 8.232):

- The **nasal vestibule** is a small dilated space just internal to the naris that is lined by skin and contains hair follicles.
- The **respiratory region** is the largest part of the nasal cavity, has a rich neurovascular supply, and is lined by respiratory epithelium composed mainly of ciliated and mucous cells.
- The **olfactory region** is small, is at the apex of each nasal cavity, is lined by olfactory epithelium, and contains the olfactory receptors.

In addition to housing receptors for the sense of smell (olfaction), the nasal cavities adjust the temperature and humidity of respired air by the action of a rich blood supply, and trap and remove particulate matter from the airway by

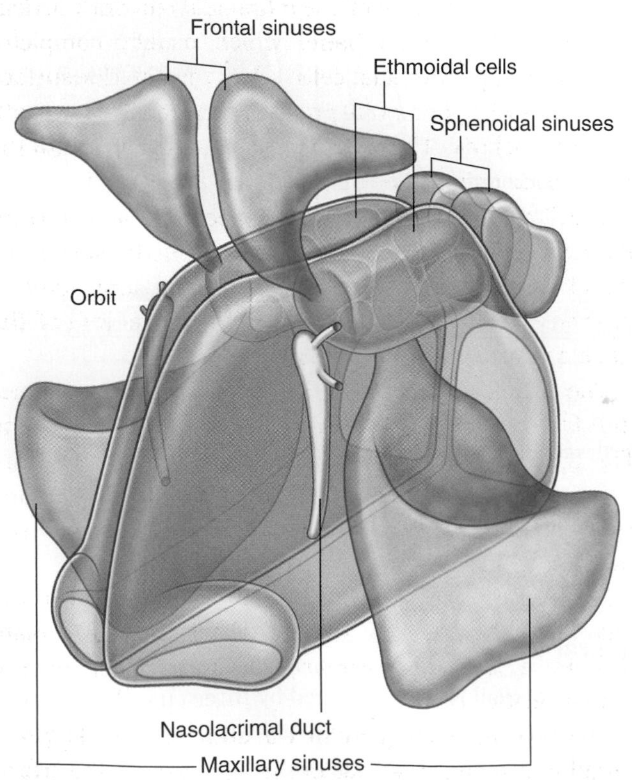


**Fig. 8.231** Paranasal sinuses and nasolacrimal duct.

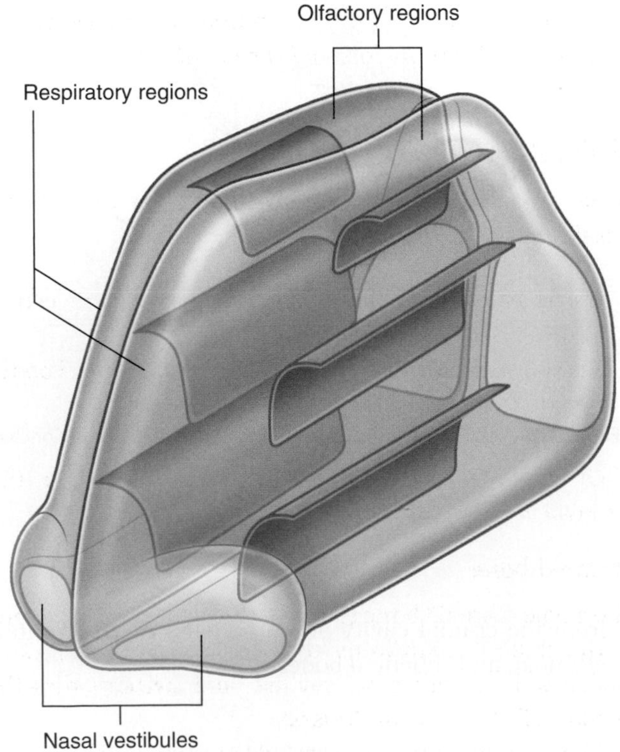


**Fig. 8.232** Regions of the nasal cavities.

filtering the air through hair in the vestibule and by capturing foreign material in abundant mucus. The mucus normally is moved posteriorly by cilia on epithelial cells in the nasal cavities and is swallowed.

## Innervation and blood supply

Innervation of the nasal cavities is by three cranial nerves:

- Olfaction is carried by the olfactory nerve [I].
- General sensation is carried by the trigeminal nerve [V], the anterior region by the ophthalmic nerve [$V_1$], and the posterior region by the maxillary nerve [$V_2$].
- All glands are innervated by parasympathetic fibers in the facial nerve [VII] (greater petrosal nerve), which join branches of the maxillary nerve [$V_2$] in the pterygopalatine fossa.

Sympathetic fibers are ultimately derived from the T1 spinal cord level. They synapse mainly in the superior cervical sympathetic ganglion, and postganglionic fibers reach the nasal cavities along blood vessels, or by joining branches of the maxillary nerve [$V_2$] in the pterygopalatine fossa.

Blood supply to the nasal cavities is by:

- terminal branches of the maxillary and facial arteries, which originate from the external carotid artery, and
- ethmoidal branches of the ophthalmic artery, which originates from the internal carotid artery.

## Skeletal framework

Bones that contribute to the skeletal framework of the nasal cavities include:

- the unpaired ethmoid, sphenoid, frontal, and vomer bones, and
- the paired nasal, maxillary, palatine, and lacrimal bones and inferior conchae.

Of all the bones associated with the nasal cavities, the ethmoid is a key element.

### Ethmoid bone

The single ethmoid bone is one of the most complex bones in the skull. It contributes to the roof, lateral wall, and medial wall of both nasal cavities, and contains the ethmoidal cells (ethmoidal sinuses).

The ethmoid bone is cuboidal in overall shape (Fig. 8.233A) and is composed of two rectangular box-shaped **ethmoidal labyrinths**, one on each side, united superiorly across the midline by a perforated sheet of bone (the **cribriform plate**). A second sheet of bone (the **perpendicular plate**) descends vertically in the median sagittal plane from the cribriform plate to form part of the nasal septum.

Each ethmoidal labyrinth is composed of two delicate sheets of bone, which sandwich between them the ethmoidal cells.

- The lateral sheet of bone (the **orbital plate**) is flat and forms part of the medial wall of the orbit.
- The medial sheet of bone forms the upper part of the lateral wall of the nasal cavity and is characterized by two processes and a swelling (Fig. 8.233B)—the two processes are curved shelves of bone (the superior and middle conchae), which project across the nasal cavity and curve downward ending in free medial margins, while inferior to the origin of the middle concha, the middle ethmoidal cells form a prominent bulge (the **ethmoidal bulla**), on the medial wall of the labyrinth.

Extending anterosuperiorly from just under the bulla is a groove (the **ethmoidal infundibulum**), which continues upward, and narrows to form a channel that penetrates the ethmoidal labyrinth and opens into the frontal sinus. This channel is for the frontonasal duct, which drains the frontal sinus.

The superior surface of the ethmoidal labyrinth articulates with the frontal bone, which usually completes the roof of the ethmoidal cells, while the anterior surface articulates with the frontal process of the maxilla and with the lacrimal bone. The inferior surface articulates with the upper medial margin of the maxilla.

A delicate irregularly shaped projection (the **uncinate process**) on the anterior aspect of the inferior surface of the ethmoidal labyrinth extends posteroinferiorly across a large defect (**maxillary hiatus**) in the medial wall of the maxilla to articulate with the inferior concha.

The cribriform plate is at the apex of the nasal cavities and fills the **ethmoidal notch** in the frontal bone (Fig. 8.233) and separates the nasal cavities below from the cranial cavity above. Small perforations in the bone allow the fibers of the olfactory nerve [I] to pass between the two regions.

A large triangular process (the **crista galli**) at the midline on the superior surface of the cribriform plate anchors a fold (falx cerebri) of dura mater in the cranial cavity.

The perpendicular plate of the ethmoid bone is quadrangular in shape, descends in the midline from the cribriform plate, and forms the upper part of the median nasal septum (Fig. 8.233). It articulates:

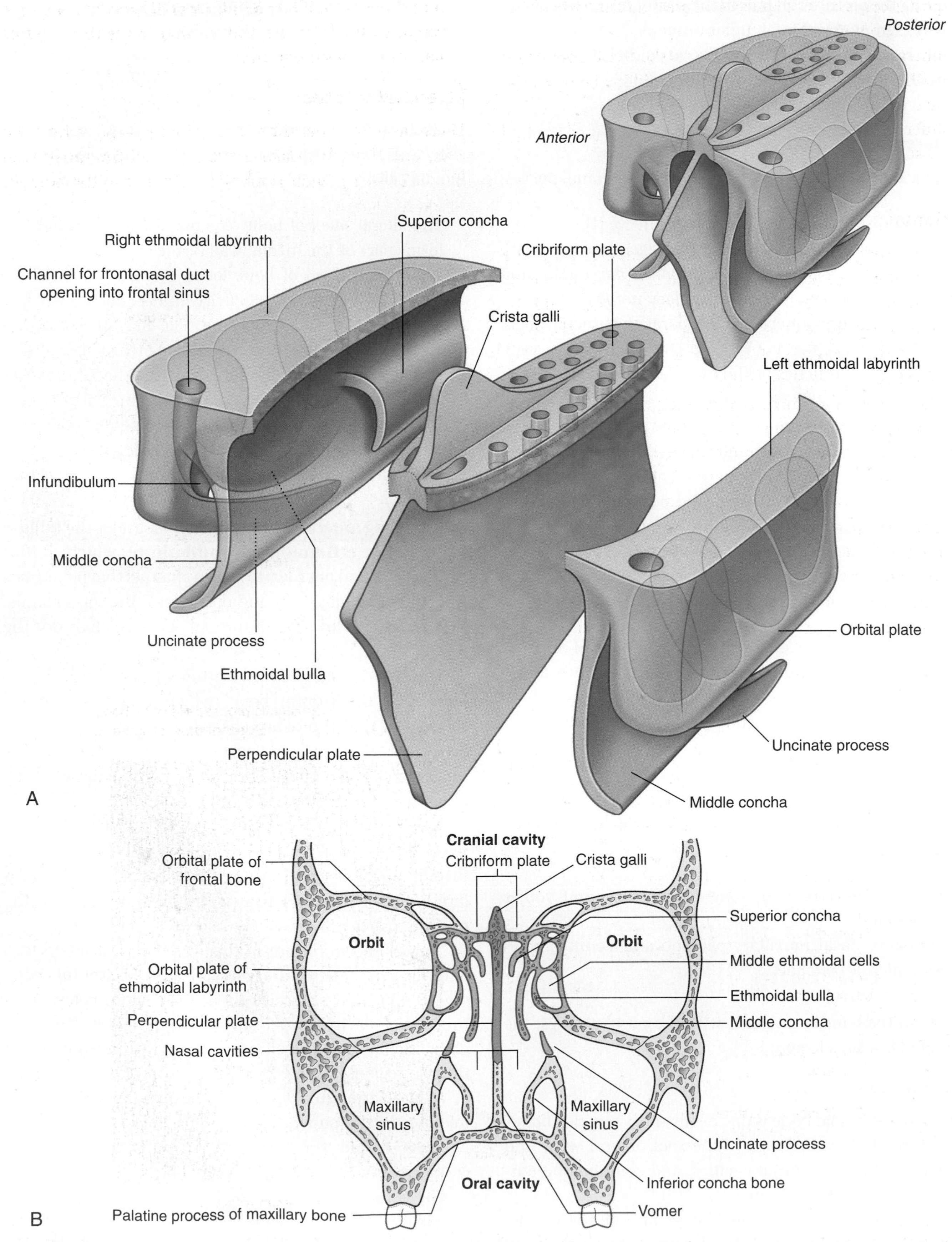
Posterior
Anterior
Superior concha
Right ethmoidal labyrinth
Cribriform plate
Channel for frontonasal duct opening into frontal sinus
Crista galli
Left ethmoidal labyrinth
Infundibulum
Middle concha
Orbital plate
Uncinate process
Ethmoidal bulla
Uncinate process
Perpendicular plate
A
Middle concha
Cranial cavity
Cribriform plate
Crista galli
Orbital plate of frontal bone
Superior concha
Orbit
Orbit
Orbital plate of ethmoidal labyrinth
Middle ethmoidal cells
Ethmoidal bulla
Perpendicular plate
Middle concha
Nasal cavities
Maxillary sinus
Maxillary sinus
Uncinate process
Oral cavity
Inferior concha bone
B
Palatine process of maxillary bone
Vomer

**Fig. 8.233** Ethmoid bone. **A.** Overall shape. **B.** Coronal section through skull.

- posteriorly with the sphenoidal crest on the body of the sphenoid bone,
- anteriorly with the nasal spine on the frontal bone and with the site of articulation at the midline between the two nasal bones, and
- inferiorly and anteriorly with the septal cartilage and posteriorly with the vomer.

## External nose

The external nose extends the nasal cavities onto the front of the face and positions the nares so that they point downward (Fig. 8.234). It is pyramidal in shape with its apex anterior in position. The upper angle of the nose between the openings of the orbits is continuous with the forehead.

Like posterior regions, the anterior parts of the nasal cavities found within the nose are held open by a skeletal framework, which is composed partly of bone and mainly of cartilage:

- The bony parts are where the nose is continuous with the skull—here the nasal bones and parts of the maxillae and frontal bones provide support.
- Anteriorly, and on each side, support is provided by **lateral processes** of the septal cartilage, **major alar** and three or four **minor alar cartilages**, and a single septal cartilage in the midline that forms the anterior part of the nasal septum.

## Paranasal sinuses

There are four paranasal air sinuses—the ethmoidal cells, and the sphenoidal, maxillary, and frontal sinuses (Fig. 8.235A,B). Each is named according to the bone in which it is found.

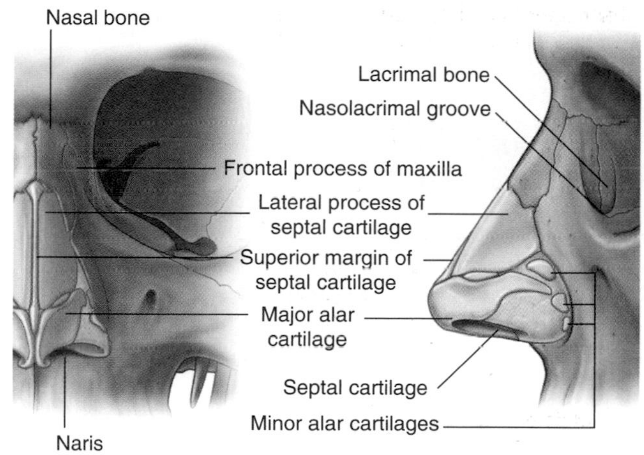


**Fig. 8.234** External nose.

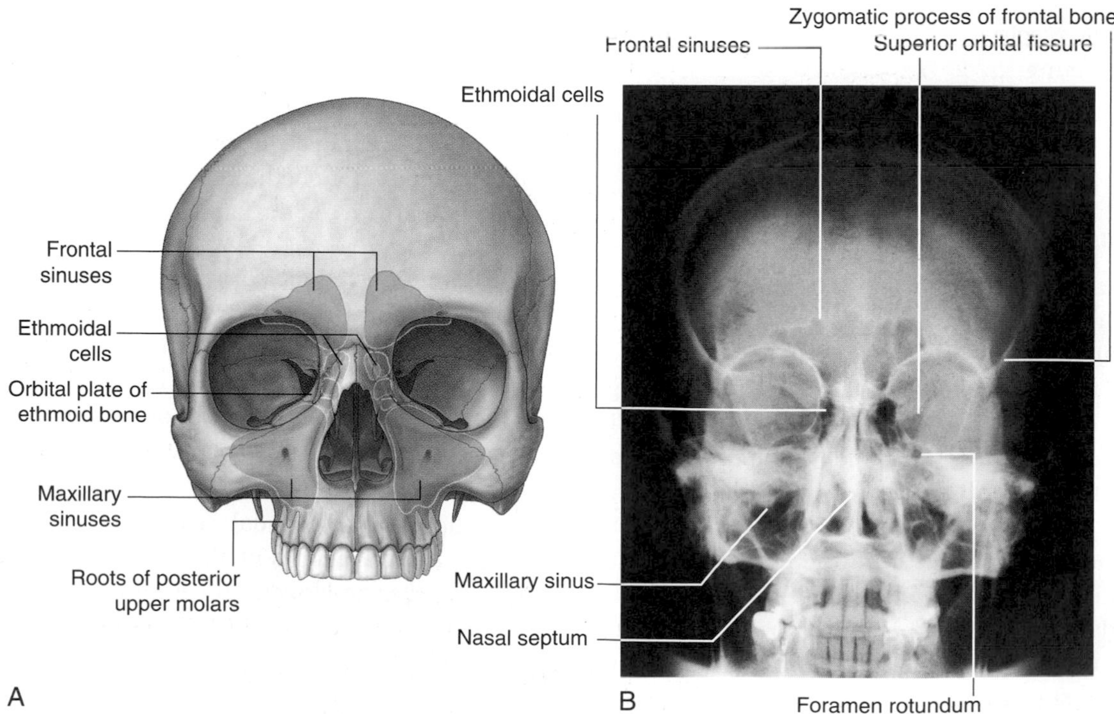


**Fig. 8.235** Paranasal sinuses. **A.** Anterior view. **B.** Posteroanterior skull radiograph. *Continued*

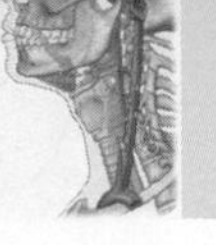



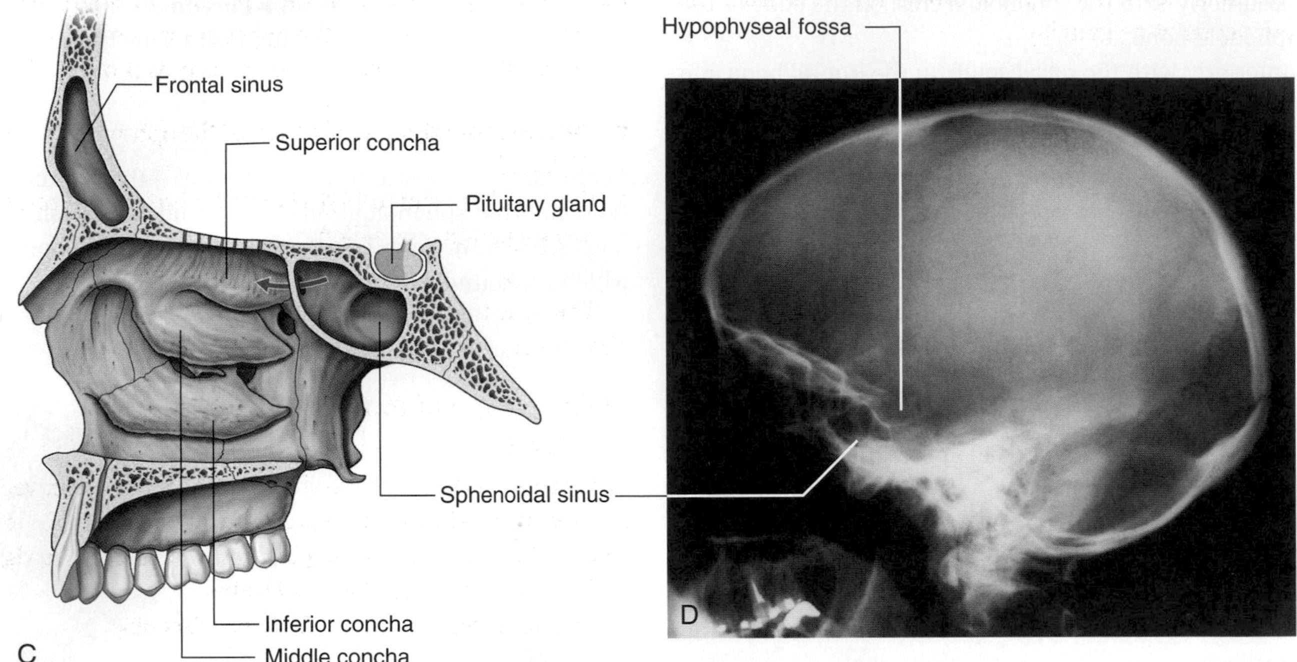


**Fig. 8.235, cont'd** C. Paramedian view of right nasal cavity. **D.** Lateral skull radiograph.

The paranasal sinuses develop as outgrowths from the nasal cavities and erode into the surrounding bones. All of the paranasal sinuses:

- are lined by respiratory mucosa, which is ciliated and mucus secreting,
- open into the nasal cavities, and
- are innervated by branches of the trigeminal nerve [V].

### Frontal sinuses

The frontal sinuses, one on each side, are variable in size and are the most superior of the sinuses (Fig. 8.235A–C). Each is triangular in shape and is in the part of the frontal bone under the forehead. The base of each triangular sinus is oriented vertically in the bone at the midline above the bridge of the nose and the apex is laterally approximately one-third of the way along the upper margin of the orbit.

Each frontal sinus drains onto the lateral wall of the middle meatus via the frontonasal duct, which penetrates the ethmoidal labyrinth and continues as the ethmoidal infundibulum at the front end of the **semilunar hiatus**.

The frontal sinuses are innervated by branches of the supra-orbital nerve from the ophthalmic nerve [$V_1$]. Their blood supply is from branches of the anterior ethmoidal arteries.

### Ethmoidal cells

The ethmoidal cells on each side fill the ethmoidal labyrinth (Fig. 8.235A,B). Each cluster of cells is separated from the orbit by the thin orbital plate of the ethmoidal labyrinth, and from the nasal cavity by the medial wall of the ethmoidal labyrinth.

The ethmoidal cells are formed by a variable number of individual air chambers, which are divided into anterior, middle, and posterior ethmoidal cells based on the location of their apertures on the lateral wall of the nasal cavity:

- The anterior ethmoidal cells open into the ethmoidal infundibulum or the frontonasal duct.
- The middle ethmoidal cells open onto the ethmoidal bulla, or onto the lateral wall just above this structure.
- The posterior ethmoidal cells open onto the lateral wall of the superior nasal meatus.

Because the ethmoidal cells often erode into bones beyond the boundaries of the ethmoidal labyrinth, their walls may be completed by the frontal, maxillary, lacrimal, sphenoid, and palatine bones.

The ethmoidal cells are innervated by:

- the **anterior** and **posterior ethmoidal branches** of the nasociliary nerve from the ophthalmic nerve [$V_1$], and

- the maxillary nerve [$V_2$] via orbital branches from the pterygopalatine ganglion.

The ethmoidal cells receive their blood supply through branches of the anterior and posterior ethmoidal arteries.

### Maxillary sinuses

The maxillary sinuses, one on each side, are the largest of the paranasal sinuses and completely fill the bodies of the maxillae (Fig. 8.235A,B). Each is pyramidal in shape with the apex directed laterally and the base deep to the lateral wall of the adjacent nasal cavity. The medial wall or base of the maxillary sinus is formed by the maxilla, and by parts of the inferior concha and palatine bone that overlie the maxillary hiatus.

The opening of the maxillary sinus is near the top of the base, in the center of the semilunar hiatus, which grooves the lateral wall of the middle nasal meatus.

Relationships of the maxillary sinus are as follows:

- The superolateral surface (roof) is related above to the orbit.
- The anterolateral surface is related below to the roots of the upper molar and premolar teeth and in front to the face.
- The posterior wall is related behind to the infratemporal fossa.

The maxillary sinuses are innervated by infra-orbital and alveolar branches of the maxillary nerve [$V_2$], and receive their blood through branches from the infra-orbital and superior alveolar branches of the maxillary arteries.

### Sphenoidal sinuses

The sphenoidal sinuses, one on either side within the body of the sphenoid, open into the roof of the nasal cavity via apertures on the posterior wall of the spheno-ethmoidal recess (Fig. 8.235C,D). The apertures are high on the anterior walls of the sphenoid sinuses.

The sphenoidal sinuses are related:

- above to the cranial cavity, particularly to the pituitary gland and to the optic chiasm,
- laterally, to the cranial cavity, particularly to the cavernous sinuses, and
- below and in front, to the nasal cavities.

Because only thin shelves of bone separate the sphenoidal sinuses from the nasal cavities below and hypophyseal fossa above, the pituitary gland can be surgically approached through the roof of the nasal cavities by passing first through the anteroinferior aspect of the sphenoid bone and into the sphenoidal sinuses and then through the top of the sphenoid bone into the hypophyseal fossa.

Innervation of the sphenoidal sinuses is provided by:

- the posterior ethmoidal branch of the ophthalmic nerve [$V_1$], and
- the maxillary nerve [$V_2$] via orbital branches from the pterygopalatine ganglion.

The sphenoidal sinuses are supplied by branches of the pharyngeal arteries from the maxillary arteries.

## Walls, floor, and roof

### Medial wall

The medial wall of each nasal cavity is the mucosa-covered surface of the thin nasal septum, which is oriented vertically in the median sagittal plane and separates the right and left nasal cavities from each other.

The nasal septum (Fig. 8.236) consists of:

- the **septal nasal cartilage** anteriorly,
- posteriorly, mainly the vomer and the perpendicular plate of the ethmoid bone,
- small contributions by the nasal bones where they meet in the midline, and the nasal spine of the frontal bone, and
- contributions by the nasal crests of the maxillary and palatine bones, rostrum of the sphenoid bone, and the incisor crest of the maxilla.

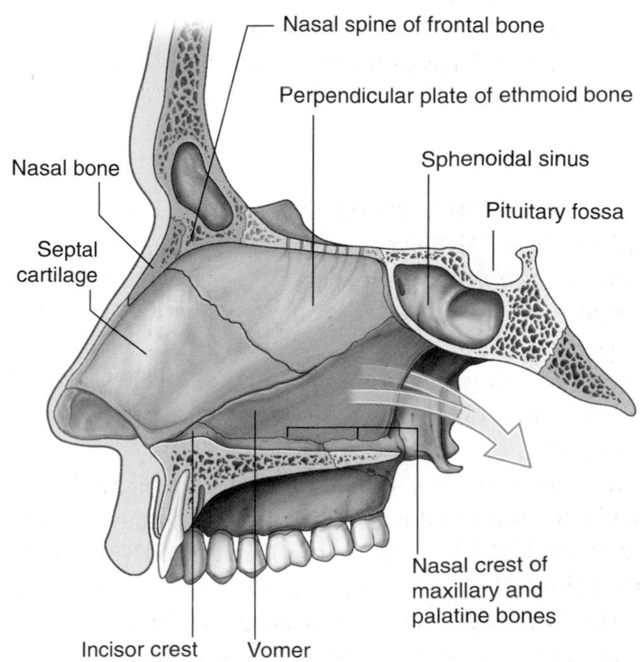


**Fig. 8.236** Medial wall of the nasal cavity—the nasal septum.

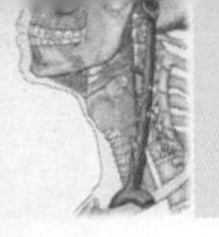



### In the clinic

**Deviated nasal septum**

The nasal septum is typically situated in the midline; however, septal deviation to one side or the other is not uncommon, and in many cases is secondary to direct trauma. Extreme septal deviation can produce nasal occlusion. The deviation can be corrected surgically.

## Floor

The floor of each nasal cavity (Fig. 8.237) is smooth, concave, and much wider than the roof. It consists of:

- soft tissues of the external nose, and
- the upper surface of the palatine process of the maxilla and the horizontal plate of the palatine bone, which together form the hard palate.

The naris opens anteriorly into the floor, and the superior aperture of the incisive canal is deep to the mucosa immediately lateral to the nasal septum near the front of the hard palate.

## Roof

The roof of the nasal cavity is narrow and is highest in central regions where it is formed by the cribriform plate of the ethmoid bone (Fig. 8.238).

Anterior to the cribriform plate the roof slopes inferiorly to the nares and is formed by:

- the nasal spine of the frontal bone and the nasal bones, and
- the lateral processes of the septal cartilage and major alar cartilages of the external nose.

Posteriorly, the roof of each cavity slopes inferiorly to the choana and is formed by:

- the anterior surface of the sphenoid bone,
- the ala of the vomer and adjacent sphenoidal process of the palatine bone, and
- the vaginal process of the medial plate of the pterygoid process.

Underlying the mucosa, the roof is perforated superiorly by openings in the cribriform plate, and anterior to these openings by a separate foramen for the anterior ethmoidal nerve and vessels.

The opening between the sphenoidal sinus and the spheno-ethmoidal recess is on the posterior slope of the roof.

## Lateral wall

The lateral wall of each nasal cavity is complex and is formed by bone, cartilage, and soft tissues.

**Fig. 8.237** Floor of the nasal cavity (superior view).

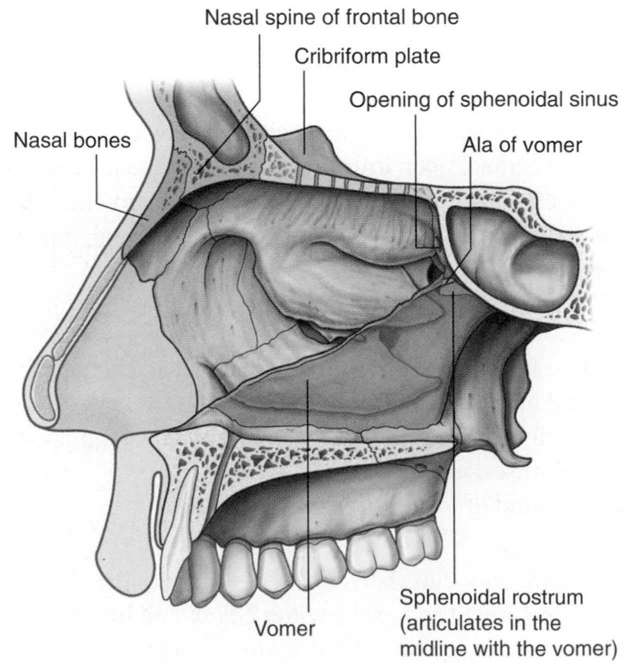


**Fig. 8.238** Roof of the nasal cavity.

Bony support for the lateral wall (Fig. 8.239A) is provided by:

- the ethmoidal labyrinth, superior concha, middle concha and uncinate process,
- the perpendicular plate of the palatine bone,
- the medial pterygoid plate of the sphenoid bone,
- the medial surfaces of the lacrimal bones and maxillae, and
- the inferior concha.

In the external nose, the lateral wall of the cavity is supported by cartilage (lateral process of the septal cartilage and major and minor alar cartilages) and by soft tissues. The surface of the lateral wall is irregular in contour and is interrupted by the three nasal conchae.

The inferior, middle, and superior conchae (Fig. 8.239B) extend medially across the nasal cavity, separating it into four air channels, an inferior, middle, and superior meatus and a spheno-ethmoidal recess. The conchae do not extend forward into the external nose. The anterior end of each concha curves inferiorly to form a lip that overlies the end of the related meatus.

Immediately inferior to the attachment of the middle concha and just anterior to the midpoint of the concha, the lateral wall of the middle meatus elevates to form the dome-shaped ethmoidal bulla (Fig. 8.239C). This is formed by the underlying middle ethmoidal cells, which expand the medial wall of the ethmoidal labyrinth.

Inferior to the ethmoidal bulla is a curved gutter (the semilunar hiatus), which is formed by the mucosa covering the lateral wall as it spans a defect in the bony wall between the ethmoidal bulla above and the uncinate process below.

The anterior end of the semilunar hiatus forms a channel (the ethmoidal infundibulum), which curves upward and continues as the frontonasal duct through the anterior part of the ethmoidal labyrinth to open into the frontal sinus.

The nasolacrimal duct and most of the paranasal sinuses open onto the lateral wall of the nasal cavity (Fig. 8.239C):

- The nasolacrimal duct opens onto the lateral wall of the inferior nasal meatus under the anterior lip of the inferior concha—it drains tears from the conjunctival sac of the eye into the nasal cavity and originates at the inferior end of the lacrimal sac on the anteromedial wall of the orbit.
- The frontal sinus drains via the frontonasal duct and ethmoidal infundibulum into the anterior end of the semilunar hiatus on the lateral wall of the middle nasal meatus—the anterior ethmoidal cells drain into the frontonasal duct or ethmoidal infundibulum (in some cases, the frontal sinus drains directly into the anterior

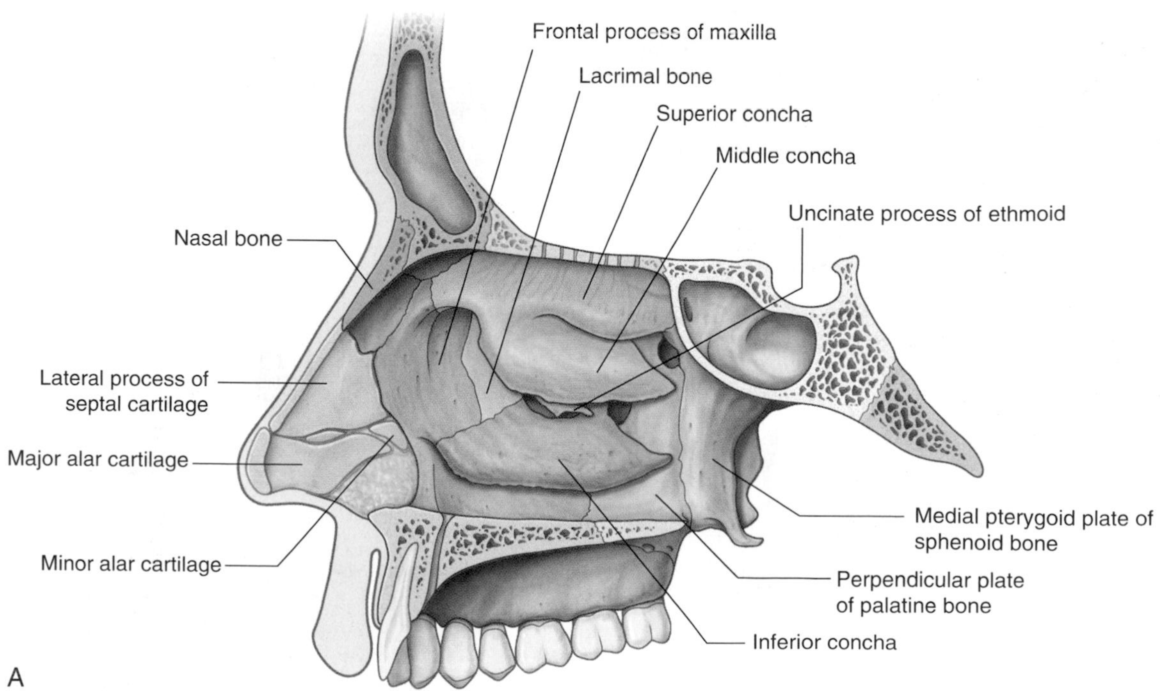


**Fig. 8.239** Lateral wall of the nasal cavity. **A.** Bones.

*Continued*

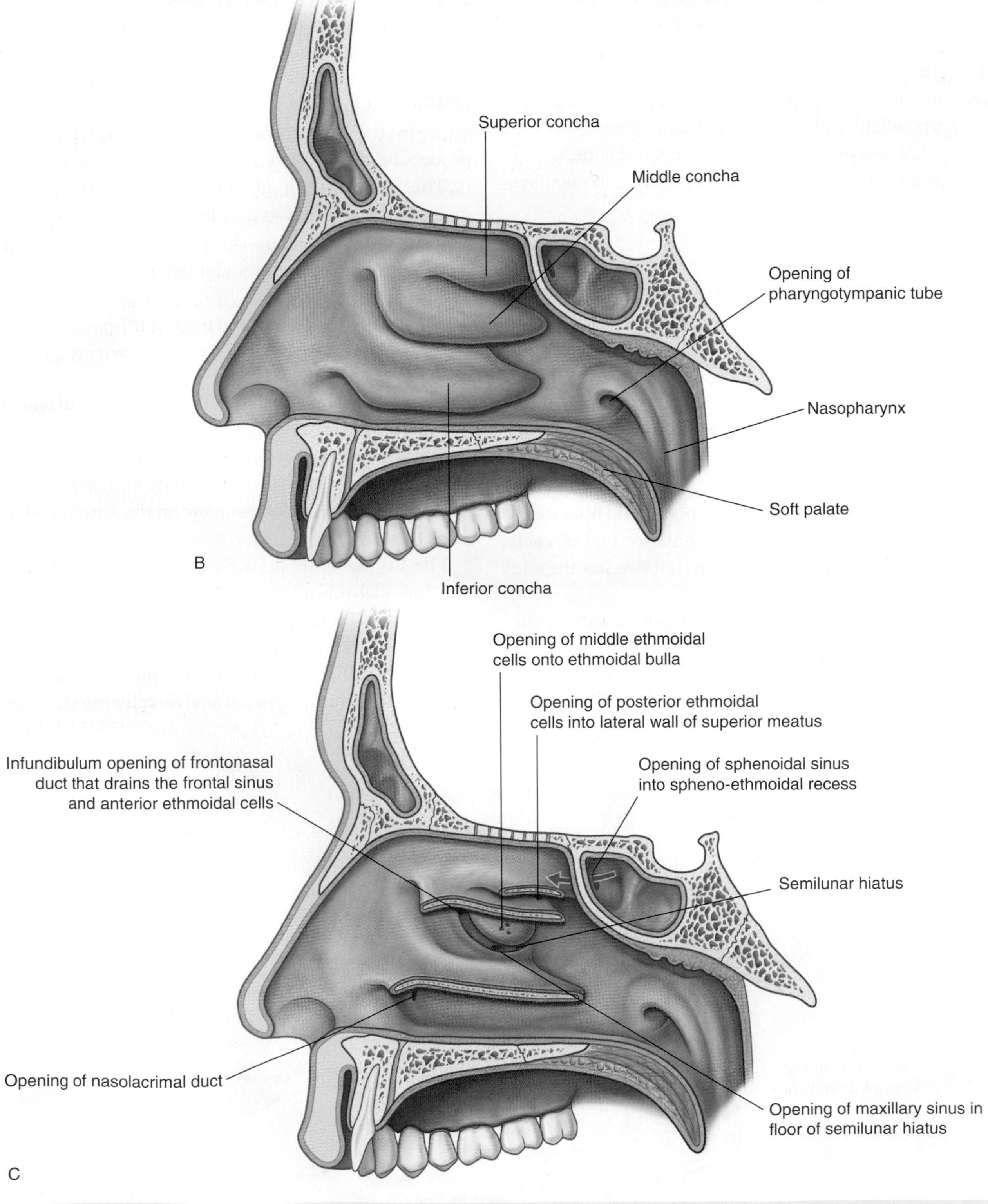

**Fig. 8.239, cont'd** **B.** Covered with mucosa. **C.** Conchae broken away at attachment to lateral wall.

end of the middle nasal meatus and the frontonasal duct ends blindly in the anterior ethmoidal cells).

- The middle ethmoidal cells open onto or just above the ethmoidal bulla.
- The posterior ethmoidal cells usually open onto the lateral wall of the superior nasal meatus.
- The large maxillary sinus opens into the semilunar hiatus, usually just inferior to the center of the ethmoidal bulla—this opening is near the roof of the maxillary sinus.

The only paranasal sinus that does not drain onto the lateral wall of the nasal cavity is the sphenoidal sinus, which usually opens onto the sloping posterior roof of the nasal cavity.

## Nares

The nares are oval apertures on the inferior aspect of the external nose and are the anterior openings of the nasal cavities (Fig. 8.240A). They are held open by the surrounding alar cartilages and septal cartilage, and by the inferior nasal spine and adjacent margins of the maxillae.

Although the nares are continuously open, they can be widened further by the action of the related muscles of

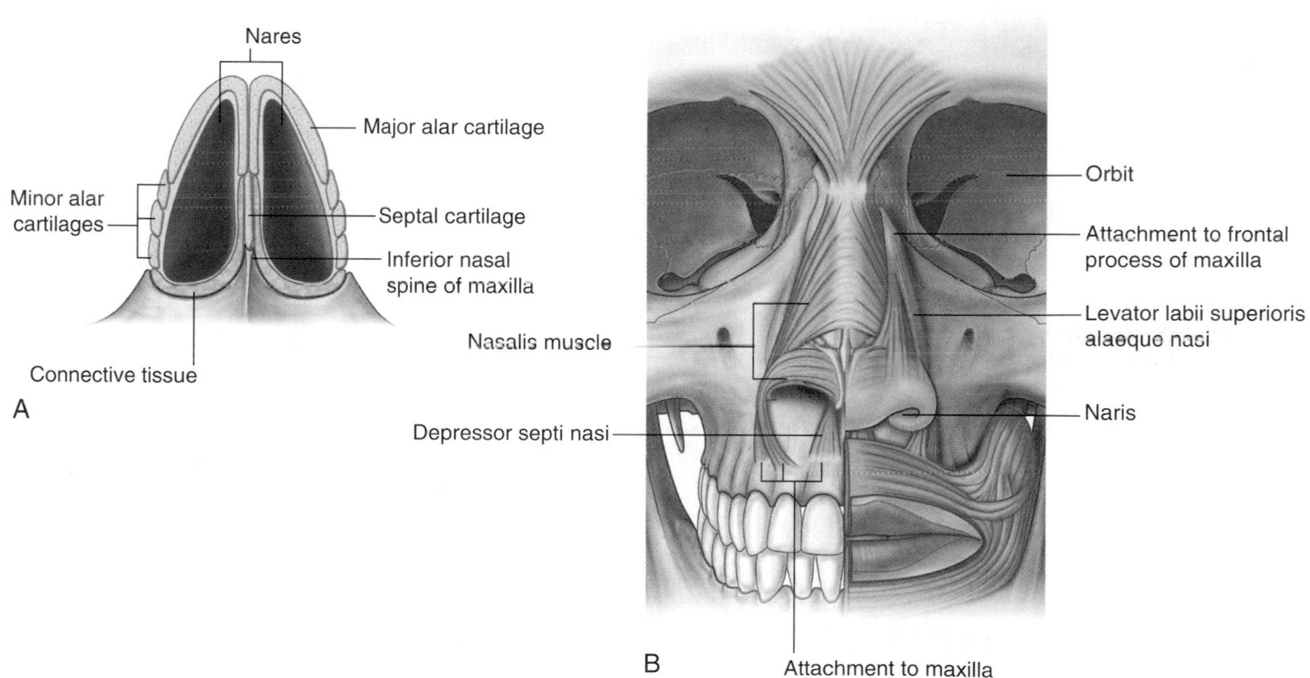


**Fig. 8.240** Nares. **A.** Inferior view. **B.** Associated muscles.

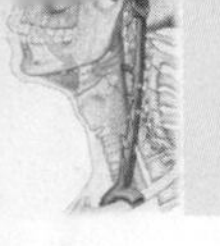



facial expression (nasalis, depressor septi nasi, and levator labii superioris alaeque nasi muscles; Fig. 8.240B).

## Choanae

The choanae are the oval-shaped openings between the nasal cavities and the nasopharynx (Fig. 8.241). Unlike the nares, which have flexible borders of cartilage and soft tissues, the choanae are rigid openings completely surrounded by bone, and their margins are formed:

- inferiorly by the posterior border of the horizontal plate of the palatine bone,
- laterally by the posterior margin of the medial plate of the pterygoid process, and
- medially by the posterior border of the vomer.

The roof of the choanae is formed:

- anteriorly by the ala of the vomer and the vaginal process of the medial plate of the pterygoid process, and
- posteriorly by the body of the sphenoid bone.

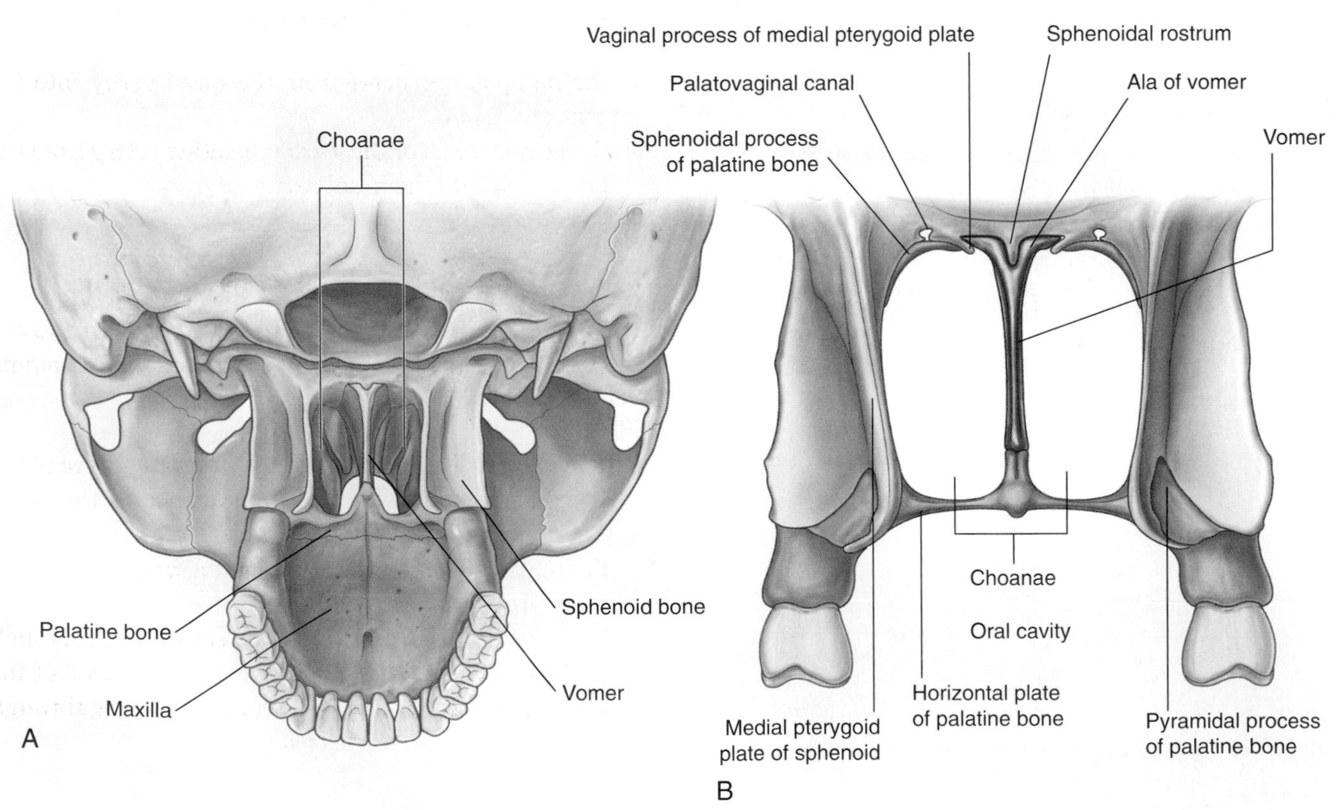


**Fig. 8.241** Choanae (posterior view). **A.** Overview. **B.** Magnified view.

## Gateways

There are a number of routes by which nerves and vessels enter and leave the soft tissues lining each nasal cavity (Fig. 8.242), and these include the cribriform plate, sphenopalatine foramen, incisive canal, and small foramina in the lateral wall, and around the margin of the nares.

### Cribriform plate

The fibers of the olfactory nerve [I] exit the nasal cavity and enter the cranial cavity through perforations in the cribriform plate. In addition, small foramina between the cribriform plate and surrounding bone allow the anterior ethmoidal nerve, a branch of the ophthalmic nerve [$V_1$], and accompanying vessels to pass from the orbit into the cranial cavity and then down into the nasal cavity.

In addition, there is a connection in some individuals between nasal veins and the superior sagittal sinus of the cranial cavity through a prominent foramen (the foramen cecum) in the midline between the crista galli and frontal bone.

### Sphenopalatine foramen

One of the most important routes by which nerves and vessels enter and leave the nasal cavity is the sphenopalatine foramen in the posterolateral wall of the superior nasal meatus. This foramen is just superior to the attachment of the posterior end of the middle nasal concha and is formed by the sphenopalatine notch in the palatine bone and the body of the sphenoid bone.

The sphenopalatine foramen is a route of communication between the nasal cavity and the pterygopalatine fossa. Major structures passing through the foramen are:

- the sphenopalatine branch of the maxillary artery,
- the nasopalatine branch of the maxillary nerve [$V_2$], and
- superior nasal branches of the maxillary nerve [$V_2$].

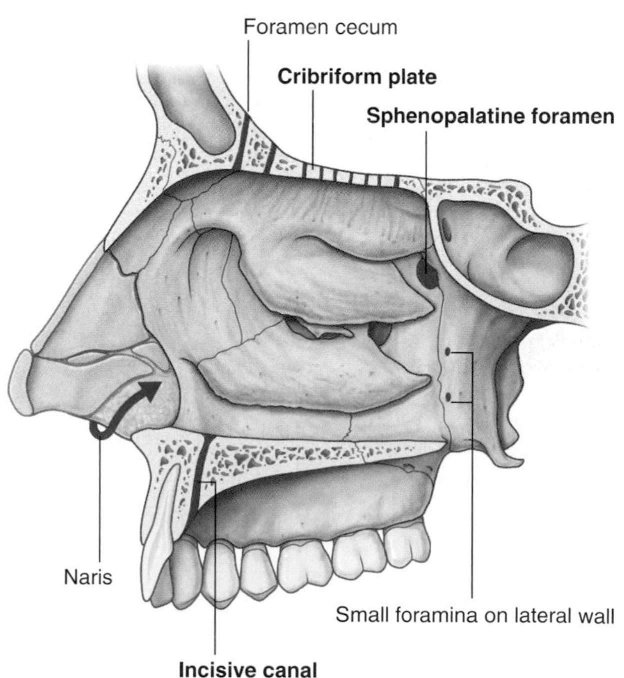


**Fig. 8.242** Gateways to the nasal cavities.

### Incisive canal

Another route by which structures enter and leave the nasal cavities is through the **incisive canal** in the floor of each nasal cavity. This canal is immediately lateral to the nasal septum and just posterosuperior to the root of the central incisor in the maxilla. The two incisive canals, one on each side, both open into the single unpaired incisive fossa in the roof of the oral cavity and transmit:

- the nasopalatine nerve from the nasal cavity into the oral cavity, and
- the terminal end of the greater palatine artery from the oral cavity into the nasal cavity.

### Small foramina in the lateral wall

Other routes by which vessels and nerves get into and out of the nasal cavity include the nares and small foramina in the lateral wall:

- Internal nasal branches of the infra-orbital nerve of the maxillary nerve [$V_2$] and alar branches of the nasal artery from the facial artery loop around the margin of the naris to gain entry to the lateral wall of the nasal cavity from the face.
- Inferior nasal branches from the greater palatine branch of the maxillary nerve [$V_2$] enter the lateral wall of the nasal cavity from the palatine canal by passing through small foramina on the lateral wall.

## Vessels

The nasal cavities have a rich vascular supply for altering the humidity and temperature of respired air. In fact, the submucosa of the respiratory region, particularly that related to the conchae and septum, is often described as "erectile" or "cavernous" because the tissue enlarges or shrinks depending on the amount of blood flowing into the system.

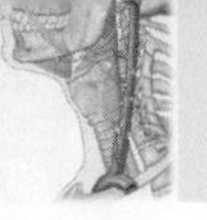



### Arteries

Arteries that supply the nasal cavity include vessels that originate from both the internal and external carotid arteries (Fig. 8.243):

- Vessels that originate from branches of the external carotid artery include the sphenopalatine, greater palatine, superior labial, and lateral nasal arteries.
- Vessels that originate from branches of the internal carotid artery are the anterior and posterior ethmoidal arteries.

#### Sphenopalatine artery

The largest vessel supplying the nasal cavity is the **sphenopalatine artery** (Fig. 8.243), which is the terminal branch of the maxillary artery in the pterygopalatine fossa. It leaves the pterygopalatine fossa and enters the nasal cavity by passing medially through the sphenopalatine foramen and onto the lateral wall of the nasal cavity.

**Posterior lateral nasal branches** supply a large part of the lateral wall and anastomose anteriorly with branches from the anterior and posterior ethmoidal arteries, and with lateral nasal branches of the facial artery.

**Posterior septal branches** of the sphenopalatine artery pass over the roof of the cavity and onto the nasal septum where they contribute to the blood supply of the medial wall. One of these latter branches continues forward down the nasal septum to anastomose with the terminal end of the greater palatine artery and septal branches of the superior labial artery.

#### Greater palatine artery

The terminal end of the **greater palatine artery** enters the anterior aspect of the floor of the nasal cavity by passing up through the incisive canal from the roof of the oral cavity (Fig. 8.243).

Like the sphenopalatine artery, the greater palatine artery arises in the pterygopalatine fossa as a branch of the maxillary artery. It passes first onto the roof of the oral cavity by passing down through the palatine canal and greater palatine foramen to the posterior aspect of the palate, then passes forward on the undersurface of the palate, and up through the incisive fossa and canal to reach the floor of the nasal cavity. The greater palatine artery supplies anterior regions of the medial wall and adjacent floor of the nasal cavity, and anastomoses with the septal branch of the sphenopalatine artery.

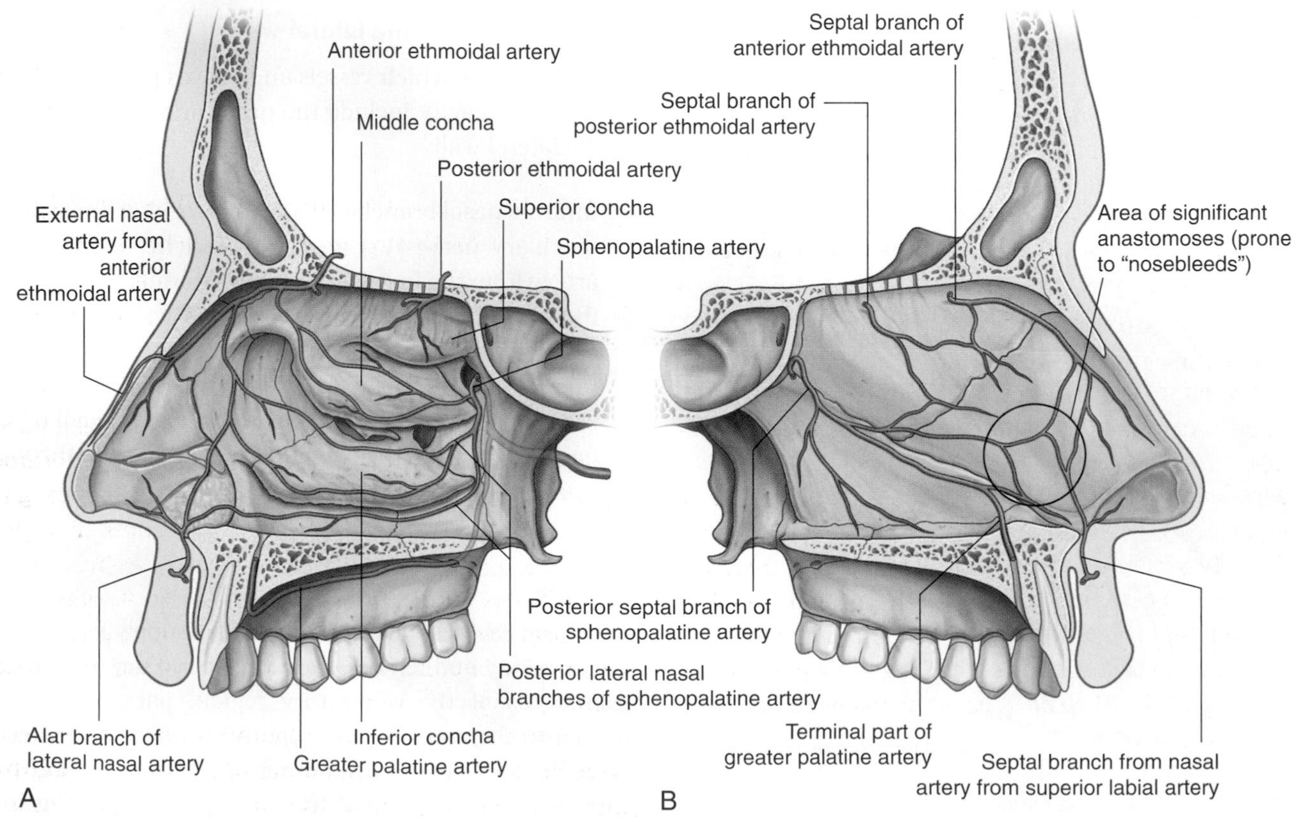


**Fig. 8.243** Arterial supply of the nasal cavities. **A.** Lateral wall of the right nasal cavity. **B.** Septum (medial wall of right nasal cavity).

### Superior labial and lateral nasal arteries

The superior labial artery and the lateral nasal artery originate from the facial artery on the front of the face.

The **superior labial artery** originates from the facial artery near the lateral end of the oral fissure and passes medially in the lip, supplying the lip and giving rise to branches that supply the nose and nasal cavity. An alar branch supplies the region around the lateral aspect of the naris and a septal branch passes into the nasal cavity and supplies anterior regions of the nasal septum.

The **lateral nasal artery** originates from the facial artery in association with the margin of the external nose and contributes to the blood supply of the external nose. Alar branches pass around the lateral margin of the naris and supply the nasal vestibule.

### Anterior and posterior ethmoidal arteries

The anterior and posterior ethmoidal arteries (Fig. 8.243) originate in the orbit from the ophthalmic artery, which originates in the cranial cavity as a major branch of the internal carotid artery. They pass through canals in the medial wall of the orbit between the ethmoidal labyrinth and frontal bone, supply the adjacent paranasal sinuses, and then enter the cranial cavity immediately lateral and superior to the cribriform plate.

The **posterior ethmoidal artery** descends into the nasal cavity through the cribriform plate and has branches to the upper parts of the medial and lateral walls.

The **anterior ethmoidal artery** passes forward, with the accompanying anterior ethmoidal nerve, in a groove on the cribriform plate and enters the nasal cavity by descending through a slit-like foramen immediately lateral to the crista galli. It gives rise to branches that supply the medial (septal) and lateral wall of the nasal cavity and then continues forward on the deep surface of the nasal bone, and terminates by passing between the nasal bone and lateral nasal cartilage to emerge on the external nose as the external nasal branch to supply skin and adjacent tissues.

Vessels that supply the nasal cavities form extensive anastomoses with each other. This is particularly evident in the anterior region of the medial wall where there are anastomoses between branches of the greater palatine, sphenopalatine, superior labial, and anterior ethmoidal arteries, and where the vessels are relatively close to the surface (Fig. 8.243B). This area is the major site of nosebleeds, or epistaxis.

## Veins

Veins draining the nasal cavities generally follow the arteries (Fig. 8.244):

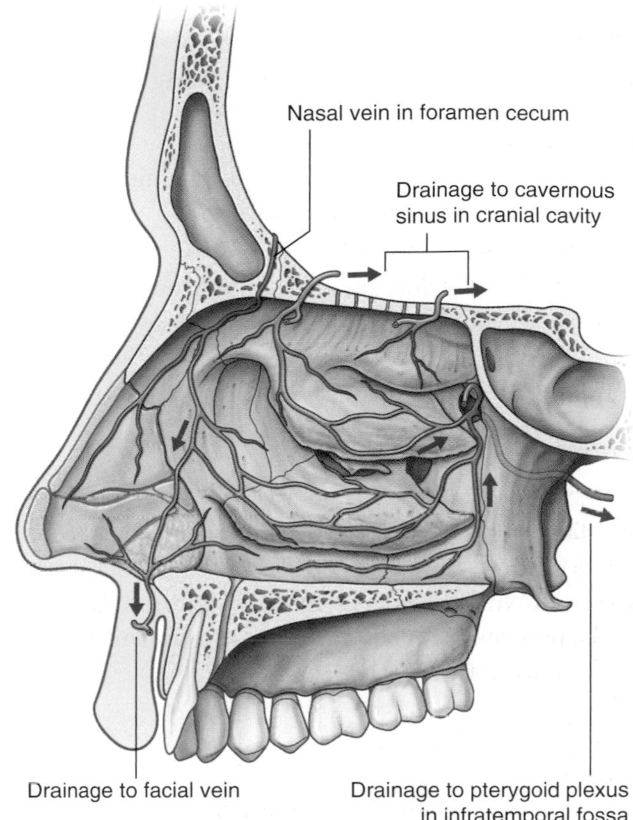


**Fig. 8.244** Venous drainage of the nasal cavities.

- Veins that pass with branches that ultimately originate from the maxillary artery drain into the pterygoid plexus of veins in the infratemporal fossa.
- Veins from anterior regions of the nasal cavities join the facial vein.

In some individuals, an additional nasal vein passes superiorly through a midline aperture (the foramen cecum), in the frontal bone anterior to the crista galli, and joins with the anterior end of the superior sagittal sinus. Because this nasal vein connects an intracranial venous sinus with extracranial veins, it is classified as an emissary vein. Emissary veins in general are routes by which infections can track from peripheral regions into the cranial cavity.

Veins that accompany the anterior and posterior ethmoidal arteries are tributaries of the superior ophthalmic vein, which is one of the largest emissary veins and drains into the cavernous sinus on either side of the hypophyseal fossa.

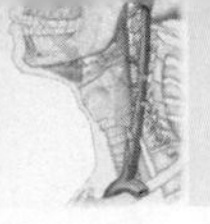



## Innervation

Nerves that innervate the nasal cavities (Fig. 8.245) are:

- the olfactory nerve [I] for olfaction, and
- branches of the ophthalmic [$V_1$] and maxillary [$V_2$] nerves for general sensation.

Secretomotor innervation of mucous glands in the nasal cavities and paranasal sinuses is by parasympathetic fibers from the facial nerve [VII], which mainly join branches of the maxillary nerve [$V_2$] in the pterygopalatine fossa.

### Olfactory nerve [I]

The olfactory nerve [I] is composed of axons from receptors in the olfactory epithelium at the top of each nasal cavity. Bundles of these axons pass superiorly through perforations in the cribriform plate to synapse with neurons in the olfactory bulb of the brain.

### Branches from the ophthalmic nerve [$V_1$]

Branches from the ophthalmic nerve [$V_1$] that innervate the nasal cavity are the anterior and posterior ethmoidal nerves, which originate from the nasociliary nerve in the orbit.

#### Anterior and posterior ethmoidal nerves

The anterior ethmoidal nerve (Fig. 8.245) travels with the anterior ethmoidal artery and leaves the orbit through a canal between the ethmoidal labyrinth and the frontal bone. It passes through and supplies the adjacent ethmoidal cells and frontal sinus, and then enters the cranial cavity immediately lateral and superior to the cribriform plate. It then travels forward in a groove on the cribriform plate and enters the nasal cavity by descending through a slit-like foramen immediately lateral to the crista galli. It has branches to the medial and lateral wall of the nasal cavity and then continues forward on the undersurface of the nasal bone. It passes onto the external surface

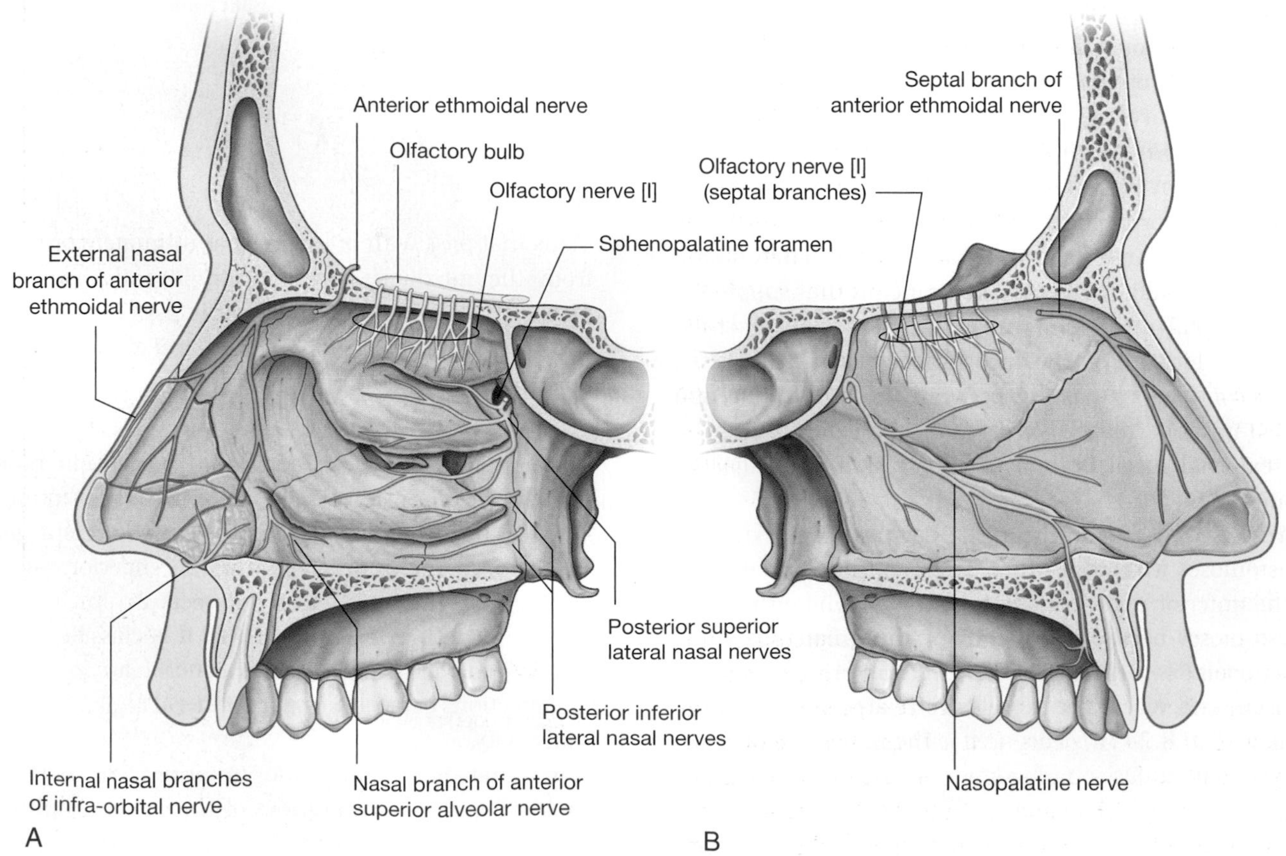


**Fig. 8.245** Innervation of the nasal cavities. **A.** Lateral wall of right nasal cavity. **B.** Medial wall of right nasal cavity.

of the nose by traveling between the nasal bone and lateral nasal cartilage, and then terminates as the **external nasal nerve**, which supplies skin around the naris, in the nasal vestibule, and on the tip of the nose.

Like the anterior ethmoidal nerve, the posterior ethmoidal nerve leaves the orbit through a similar canal in the medial wall of the orbit. It terminates by supplying the mucosa of the ethmoidal cells and sphenoidal sinus and normally does not extend into the nasal cavity itself.

### Branches from the maxillary nerve [$V_2$]

A number of nasal branches from the maxillary nerve [$V_2$] innervate the nasal cavity. Many of these nasal branches (Fig. 8.245) originate in the pterygopalatine fossa, which is just lateral to the lateral wall of the nasal cavity, and leave the fossa to enter the nasal cavity by passing medially through the sphenopalatine foramen or through smaller foramina in the lateral wall:

- A number of these nerves (**posterior superior lateral nasal nerves**) pass forward on and supply the lateral wall of the nasal cavity.
- Others (**posterior superior medial nasal nerves**) cross the roof to the nasal septum and supply both these regions.
- The largest of these nerves is the **nasopalatine nerve**, which passes forward and down the medial wall of the nasal cavity to pass through the incisive canal onto the roof of the oral cavity, and terminates by supplying the oral mucosa posterior to the incisor teeth.
- Other nasal nerves (**posterior inferior nasal nerves**) originate from the greater palatine nerve, descending from the pterygopalatine fossa in the palatine canal just lateral to the nasal cavity, and pass through small bony foramina to innervate the lateral wall of the nasal cavity.
- A small nasal nerve also originates from the anterior superior alveolar branch of the infra-orbital nerve and passes medially through the maxilla to supply the lateral wall near the anterior end of the inferior concha.

### Parasympathetic innervation

Secretomotor innervation of glands in the mucosa of the nasal cavity and paranasal sinuses is by preganglionic parasympathetic fibers carried in the greater petrosal branch of the facial nerve [VII]. These fibers enter the pterygopalatine fossa and synapse in the pterygopalatine ganglion (see Fig. 8.157 and pp. 986–987). Postganglionic parasympathetic fibers then join branches of the maxillary nerve [$V_2$] to leave the fossa and ultimately reach target glands.

### Sympathetic innervation

Sympathetic innervation, mainly involved with regulating blood flow in the nasal mucosa, is from spinal cord level T1. Preganglionic sympathetic fibers enter the sympathetic trunk and ascend to synapse in the superior cervical sympathetic ganglion. Postganglionic sympathetic fibers pass onto the internal carotid artery, enter the cranial cavity, and then leave the internal carotid artery to form the deep petrosal nerve, which joins the greater petrosal nerve of the facial nerve [VII] and enters the pterygopalatine fossa (see Figs. 8.156 and 8.157 and pp. 984–986).

Like the parasympathetic fibers, the sympathetic fibers follow branches of the maxillary nerve [$V_2$] into the nasal cavity.

### Lymphatics

Lymph from anterior regions of the nasal cavities drains forward onto the face by passing around the margins of the nares (Fig. 8.246). These lymphatics ultimately connect with the submandibular nodes.

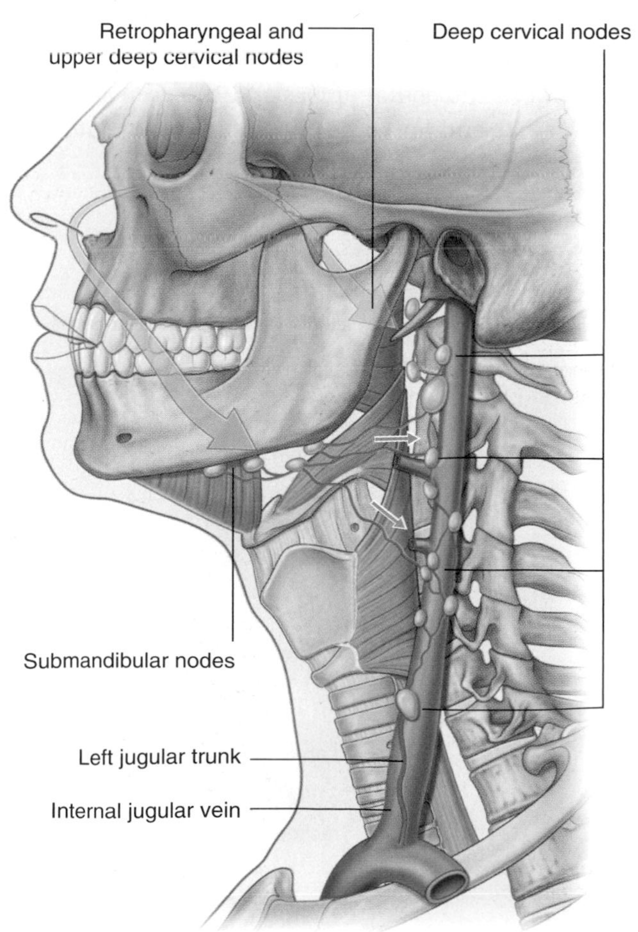


**Fig. 8.246** Lymphatic drainage of the nasal cavities.

Lymph from posterior regions of the nasal cavity and the paranasal sinuses drains into upper deep cervical nodes. Some of this lymph passes first through the retropharyngeal nodes.

## ORAL CAVITY

The oral cavity is inferior to the nasal cavities (Fig. 8.247A). It has a roof and floor and lateral walls, opens onto the face through the oral fissure, and is continuous with the cavity of the pharynx at the oropharyngeal isthmus.

The roof of the oral cavity consists of the hard and soft palates. The floor is formed mainly of soft tissues, which include a muscular diaphragm and the tongue. The lateral walls (cheeks) are muscular and merge anteriorly with the lips surrounding the **oral fissure** (the anterior opening of the oral cavity).

The posterior aperture of the oral cavity is the oropharyngeal isthmus, which opens into the oral part of the pharynx.

The oral cavity is separated into two regions by the upper and lower dental arches consisting of the teeth and alveolar bone that supports them (Fig. 8.247B):

- The outer **oral vestibule**, which is horseshoe shaped, is between the dental arches and the deep surfaces of the cheeks and lips—the oral fissure opens into it and can be opened and closed by muscles of facial expression, and by movements of the lower jaw.
- The inner **oral cavity proper** is enclosed by the dental arches.

The degree of separation between the upper and lower arches is established by elevating or depressing the lower jaw (mandible) at the temporomandibular joint.

The oropharyngeal isthmus at the back of the oral cavity proper can be opened and closed by surrounding soft tissues, which include the soft palate and tongue.

The oral cavity has multiple functions:

- It is the inlet for the digestive system involved with the initial processing of food, which is aided by secretions from salivary glands.
- It manipulates sounds produced by the larynx and one outcome of this is speech.

It can be used for breathing because it opens into the pharynx, which is a common pathway for food and air. For this reason, the oral cavity can be used by physicians to access the lower airway, and dentists use "rubber dams" to prevent debris such as tooth fragments from passing through the oropharyngeal isthmus and pharynx into either the esophagus or the lower airway.

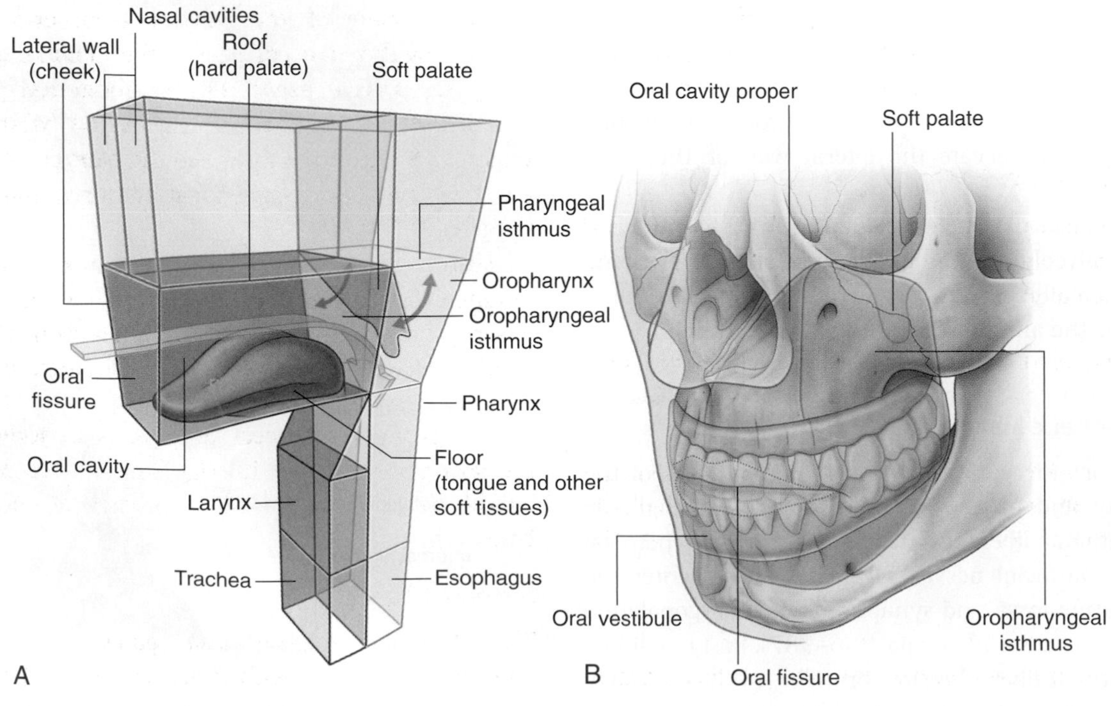


**Fig. 8.247** Oral cavity. **A.** Relationship to other cavities. **B.** Oral vestibule and oral cavity proper.

## Multiple nerves innervate the oral cavity

General sensory innervation is carried predominantly by branches of the trigeminal nerve [V]:

- The upper parts of the cavity, including the palate and the upper teeth, are innervated by branches of the maxillary nerve [$V_2$].
- The lower parts, including the teeth and oral part of the tongue, are innervated by branches of the mandibular nerve [$V_3$].
- Taste (special afferent [SA]) from the oral part or anterior two-thirds of the tongue is carried by branches of the facial nerve [VII], which join and are distributed with branches of the trigeminal nerve [V].
- Parasympathetic fibers to the glands within the oral cavity are also carried by branches of the facial nerve [VII], which are distributed with branches of the trigeminal nerve [V].
- Sympathetic fibers in the oral cavity ultimately come from spinal cord level T1, synapse in the superior cervical sympathetic ganglion, and are eventually distributed to the oral cavity along branches of the trigeminal nerve [V] or directly along blood vessels.

All muscles of the tongue are innervated by the hypoglossal nerve [XII], except the palatoglossus, which is innervated by the vagus nerve [X].

All muscles of the soft palate are innervated by the vagus nerve [X], except for the tensor veli palatini, which is innervated by a branch from the mandibular nerve [$V_3$]. The muscle (mylohyoid) that forms the floor of the oral cavity is also innervated by the mandibular nerve [$V_3$].

## Skeletal framework

Bones that contribute to the skeletal framework of the oral cavity or are related to the anatomy of structures in the oral cavity include:

- the paired maxillae, palatine, and temporal bones, and
- the unpaired mandible, sphenoid, and hyoid bones.

In addition, the cartilaginous parts of the pharyngotympanic tubes on the inferior aspect of the base of the skull are related to the attachment of muscles of the soft palate.

### Maxillae

The two maxillae contribute substantially to the architecture of the roof of the oral cavity. The parts involved are the alveolar and palatine processes (Fig. 8.248A).

The palatine process is a horizontal shelf that projects from the medial surface of each maxilla. It originates just superior to the medial aspect of the alveolar process and extends to the midline where it is joined, at a suture, with the palatine process from the other side. Together, the two palatine processes form the anterior two-thirds of the hard palate.

In the midline on the inferior surface of the hard palate and at the anterior end of the intermaxillary suture is a single small fossa (incisive fossa) just behind the incisor teeth. Two incisive canals, one on each side, extend posterosuperiorly from the roof of this fossa to open onto the floor of the nasal cavity. The canals and fossae allow passage of the greater palatine vessels and the nasopalatine nerves.

### Palatine bones

The parts of each L-shaped palatine bone that contribute to the roof of the oral cavity are the horizontal plate and the pyramidal process (Fig. 8.248A).

The horizontal plate projects medially from the inferior aspect of the palatine bone and is joined by sutures to its partner in the midline and, on the same side, with the palatine process of the maxilla anteriorly.

A single **posterior nasal spine** is formed at the midline where the two horizontal plates join and projects backward from the margin of the hard palate. The posterior margin of the horizontal plates and the posterior nasal spine are associated with attachment of the soft palate.

The greater palatine foramen, formed mainly by the horizontal plate of the palatine bone and completed laterally by the adjacent part of the maxilla, opens onto the posterolateral aspect of the horizontal plate. This foramen is the inferior opening of the palatine canal, which continues superiorly into the pterygopalatine fossa and transmits the greater palatine nerve and vessels to the palate.

Also opening onto the palatine bone is the lesser palatine foramen. This foramen is the inferior opening of the short lesser palatine canal, which branches from the greater palatine canal and transmits the lesser palatine nerve and vessels to the soft palate.

The pyramidal process projects posteriorly and fills the space between the inferior ends of the medial and lateral plates of the pterygoid process of the sphenoid bone.

### Sphenoid bone

The pterygoid processes and spines of the sphenoid bone are associated with structures related to the soft palate, which forms part of the roof of the oral cavity (Fig. 8.248A).

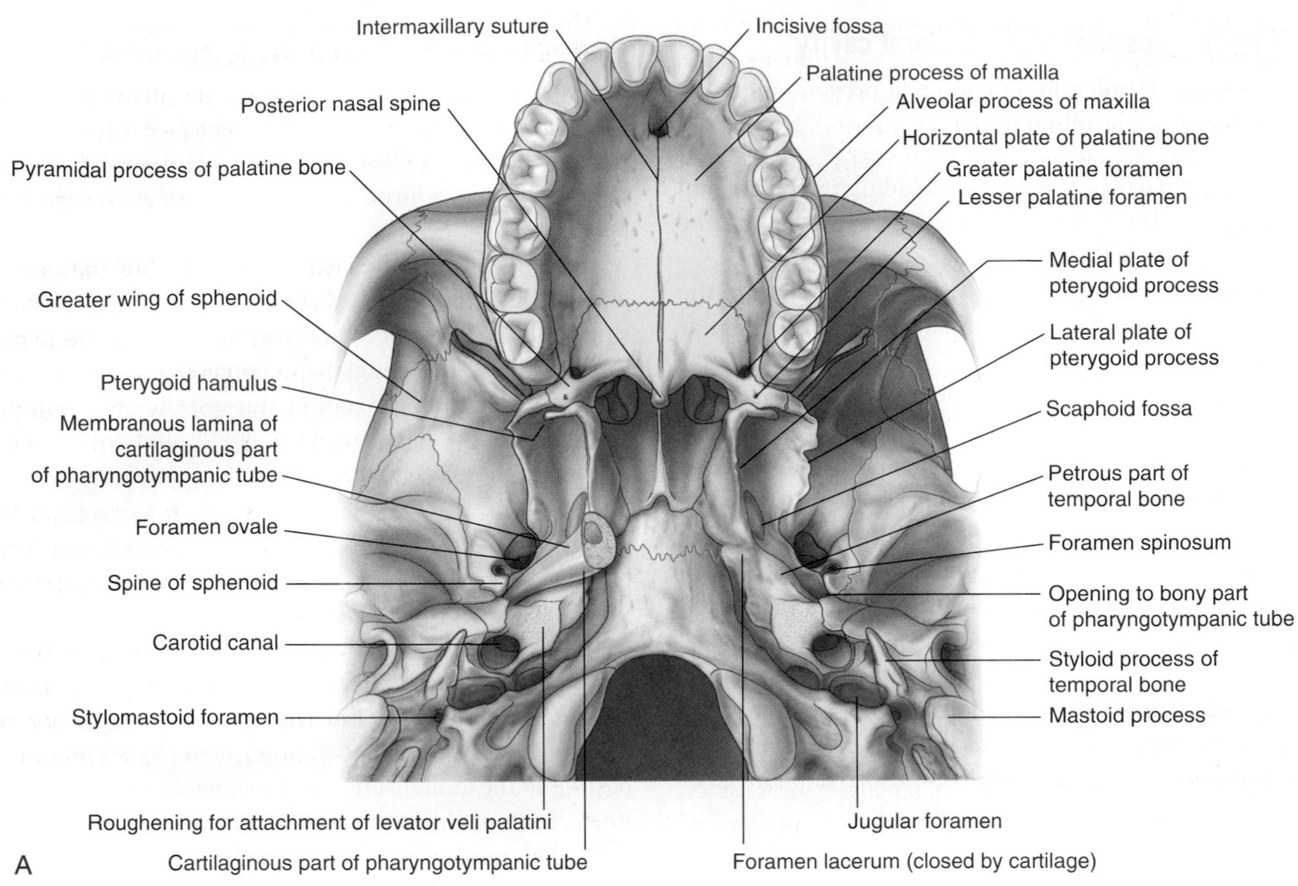


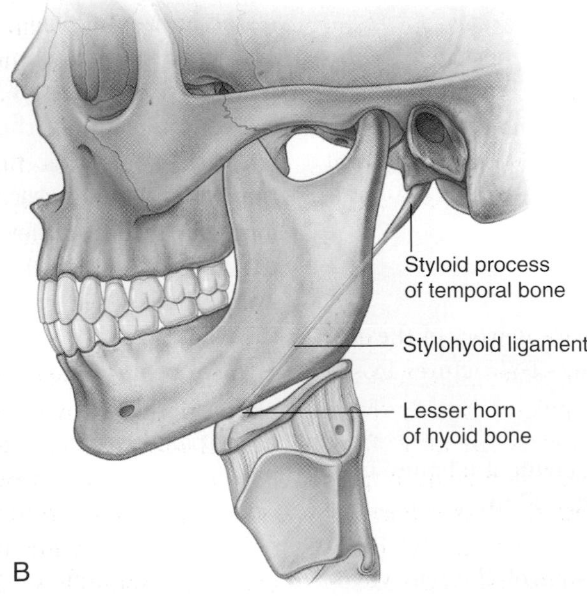


**Fig. 8.248** Base and lateral aspects of the skull. **A.** Features in the base of the skull related to structures associated with the oral cavity. **B.** Styloid process of the temporal bone.

The pterygoid processes descend, one on each side, from the lateral aspect of the body of the sphenoid bone. Each process has a medial and a lateral plate. These two vertically oriented plates project from the posterior aspect of the process. The V-shaped gap that occurs inferiorly between the two plates is filled by the pyramidal process of the palatine bone.

Projecting posterolaterally from the inferior margin of the medial plate of the pterygoid process is an elongate hook-shaped structure (the pterygoid hamulus). This hamulus is immediately behind the alveolar arch and inferior to the posterior margin of the hard palate. It is:

- a "pulley" for one of the muscles (tensor veli palatini) of the soft palate, and
- the attachment site for the upper end of the pterygomandibular raphe, which is attached below to the mandible and joins together the superior constrictor of the pharynx and the buccinator muscle of the cheek.

At the root of the medial plate of the pterygoid process on the base of the skull is a small canoe-shaped fossa (**scaphoid fossa**), which begins just medial to the foramen ovale and descends anteriorly and medially to the root of the medial plate of the pterygoid process (Fig. 8.248A). This fossa is for the attachment of one of the muscles of the soft palate (tensor veli palatini).

The spines of the sphenoid, one on each side, are vertical projections from the inferior surfaces of the greater wings of the sphenoid bone (Fig. 8.248A). Each spine is immediately posteromedial to the foramen spinosum.

The medial aspect of the spine provides attachment for the most lateral part of the tensor veli palatini muscle of the soft palate.

### Temporal bone

The styloid process and inferior aspect of the petrous part of the temporal bone provide attachment for muscles associated with the tongue and soft palate, respectively.

The styloid process projects anteroinferiorly from the underside of the temporal bone. It can be as long as 1 inch (2.5 cm) and points toward the lesser horn of the hyoid bone to which it is attached by the stylohyoid ligament (Fig. 8.248B). The root of the styloid process is immediately anterior to the stylomastoid foramen and lateral to the jugular foramen. The styloglossus muscle of the tongue attaches to the anterolateral surface of the styloid process.

The inferior aspect of the temporal bone has a triangular roughened area immediately anteromedial to the opening of the carotid canal (Fig. 8.248A). The levator veli palatini muscle of the soft palate is attached here.

### Cartilaginous part of the pharyngotympanic tube

The trumpet-shaped cartilaginous part of the pharyngotympanic tube is in a groove between the anterior margin of the petrous part of the temporal bone and the posterior margin of the greater wing of the sphenoid (Fig. 8.248A).

The medial and lateral walls of the cartilaginous part of the pharyngotympanic tube are formed mainly of cartilage, whereas the more inferolateral wall is more fibrous and is known as the **membranous lamina**.

The apex of the cartilaginous part of the pharyngotympanic tube connects laterally to the opening of the bony part in the temporal bone.

The expanded medial end of the cartilaginous part of the pharyngotympanic tube is immediately posterior to the upper margin of the medial plate of the pterygoid process and opens into the nasopharynx.

The cartilaginous part of the pharyngotympanic tube is lateral to the attachment of the levator veli palatini muscle and medial to the spine of the sphenoid. The tensor veli palatini muscle is attached, in part, to the membranous lamina.

### Mandible

The mandible is the bone of the lower jaw (Fig. 8.249). It consists of a body of right and left parts, which are fused anteriorly in the midline (**mandibular symphysis**), and two rami. The site of fusion is particularly visible on the external surface of the bone as a small vertical ridge in the midline.

The upper surface of the body of the mandible bears the alveolar arch (Fig. 8.249B), which anchors the lower teeth, and on its external surface on each side is a small mental foramen (Fig. 8.249B).

Posterior to the mandibular symphysis on the internal surface of the mandible are two pairs of small spines, one pair immediately above the other pair. These are the **superior** and **inferior mental spines** (**superior** and **inferior genial spines**) (Fig. 8.249A,C), and are attachment sites for a pair of muscles that pass into the tongue and a pair of muscles that connect the mandible to the hyoid bone.

Extending from the midline and originating inferior to the mental spines is a raised line or ridge (the **mylohyoid line**) (Fig. 8.249C), which runs posteriorly and superiorly along the internal surface of each side of the body of the mandible to end just below the level of the last molar tooth.

Above the anterior one-third of the mylohyoid line is a shallow depression (the **sublingual fossa**) (Fig. 8.249C), and below the posterior two-thirds of the mylohyoid line is another depression (the **submandibular fossa**) (Fig. 8.249C).

Between the last molar tooth and the mylohyoid line is a shallow groove for the lingual nerve.

Immediately posterior to the last molar tooth on the medial upper surface of the body of the mandible is a small triangular depression (**retromolar triangle**) (Fig. 8.249A,C). The pterygomandibular raphe attaches just medial to the apex of this triangle and extends from here to the tip of the pterygoid hamulus above.

The ramus of the mandible, one on each side, is quadrangular shaped and oriented in the sagittal plane. On the medial surface of the ramus is a large **mandibular foramen** for transmission of the inferior alveolar nerve and vessels (Fig. 8.249C).

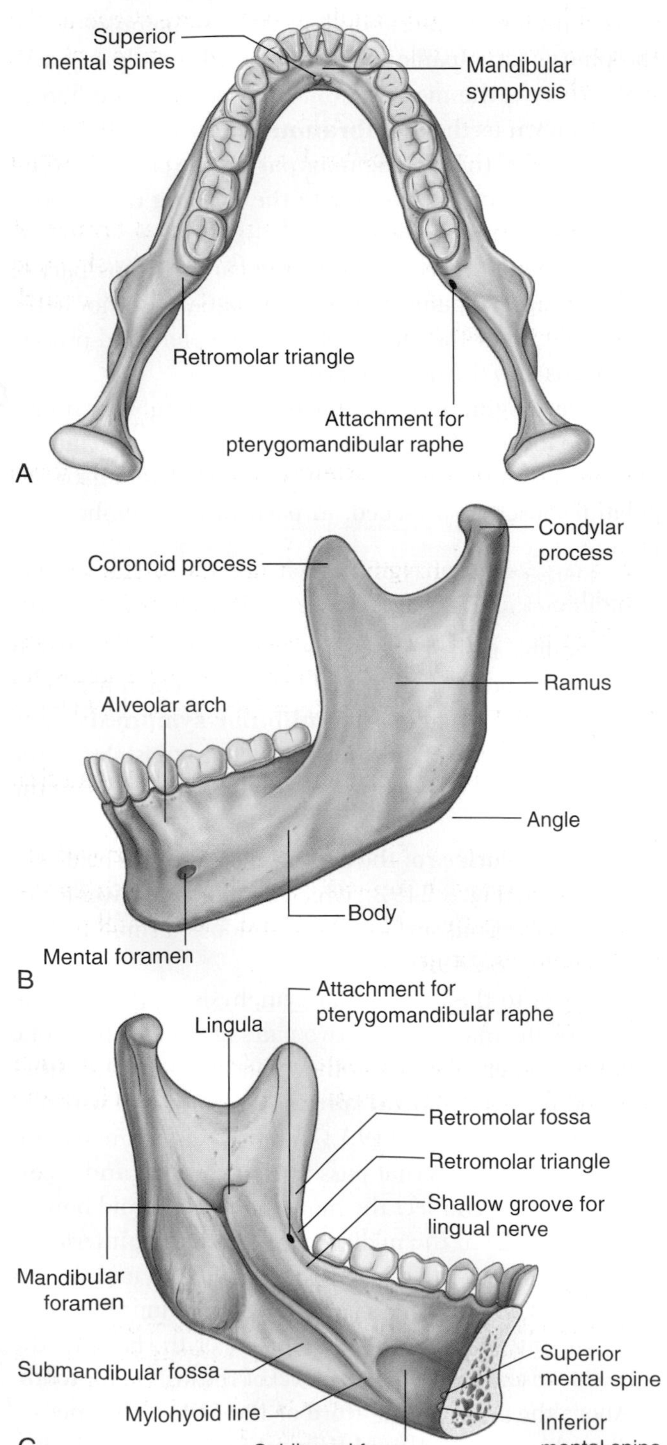


**Fig. 8.249** Mandible. **A.** Superior view. **B.** Lateral view. **C.** Medial view.

### Hyoid bone

The hyoid bone is a small U-shaped bone in the neck between the larynx and the mandible. It has an anterior body of hyoid bone and two large greater horns, one on each side, which project posteriorly and superiorly from the body (Fig. 8.250). There are two small conical lesser horns on the superior surface where the greater horns join with the body. The stylohyoid ligaments attach to the apices of the lesser horns.

The hyoid bone is a key bone in the neck because it connects the floor of the oral cavity in front with the pharynx behind and the larynx below.

## Walls: the cheeks

The walls of the oral cavity are formed by the cheeks.

Each cheek consists of fascia and a layer of skeletal muscle sandwiched between skin externally and oral mucosa internally. The thin layer of skeletal muscle within the cheeks is principally the buccinator muscle.

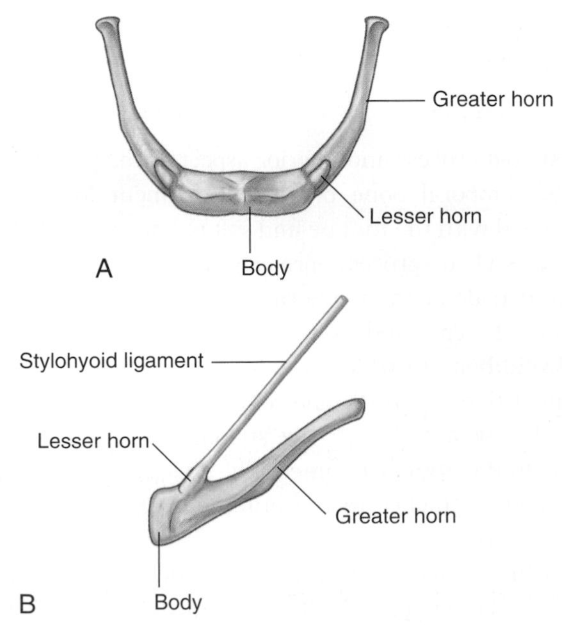


**Fig. 8.250** Hyoid bone. **A.** Anterior view. **B.** Lateral view.

### Buccinator

The buccinator muscle is one of the muscles of facial expression (Fig. 8.251). It is in the same plane as the superior constrictor muscle of the pharynx. In fact, the posterior margin of the buccinator muscle is joined to the anterior margin of the superior constrictor muscle by the pterygomandibular raphe, which runs between the tip of the pterygoid hamulus of the sphenoid bone above and a roughened area of bone immediately behind the last molar tooth on the mandible below.

The buccinator and superior constrictor muscles therefore provide continuity between the walls of the oral and pharyngeal cavities.

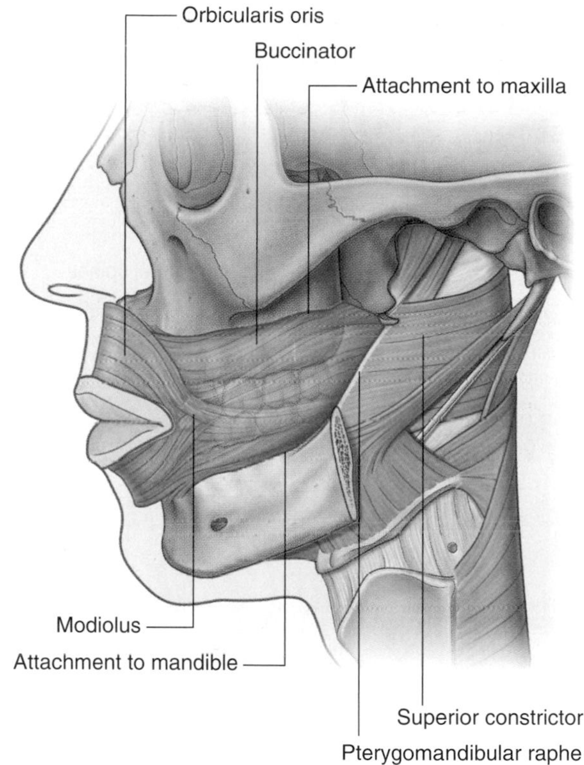


**Fig. 8.251** Buccinator muscle.

The buccinator muscle, in addition to originating from the pterygomandibular raphe, also originates directly from the alveolar part of the mandible and alveolar process of the maxilla.

From its three sites of origin, the muscle fibers of the buccinator run forward to blend with those of the orbicularis oris muscle and to insert into the modiolus, which is a small button-shaped nodule of connective tissue at the interface between the muscles of the lips and cheeks on each side.

The buccinator muscle holds the cheeks against the alveolar arches and keeps food between the teeth when chewing.

The buccinator is innervated by the buccal branch of the facial nerve [VII]. General sensation from the skin and oral mucosa of the cheeks is carried by the buccal branch of the mandibular nerve [$V_3$].

## Floor

The floor of the oral cavity proper is formed mainly by three structures:

- a muscular diaphragm, which fills the U-shaped gap between the left and right sides of the body of the mandible and is composed of the paired mylohyoid muscles;
- two cord-like geniohyoid muscles above the diaphragm, which run from the mandible in front to the hyoid bone behind; and
- the tongue, which is superior to the geniohyoid muscles.

Also present in the floor of the oral cavity proper are salivary glands and their ducts. The largest of these glands, on each side, are the sublingual gland and the oral part of the submandibular gland.

### Mylohyoid muscles

The two thin mylohyoid muscles (Table 8.20), one on each side, together form a muscular diaphragm that defines the

**Table 8.20** Muscles in the floor of the oral cavity

| Muscle | Origin | Insertion | Innervation | Function |
|---|---|---|---|---|
| Mylohyoid | Mylohyoid line of mandible | Median fibrous raphe and adjacent part of hyoid bone | Nerve to mylohyoid from the inferior alveolar branch of mandibular nerve [$V_3$] | Supports and elevates floor of oral cavity; depresses mandible when hyoid is fixed; elevates and pulls hyoid forward when mandible is fixed |
| Geniohyoid | Inferior mental spines of mandible | Body of hyoid bone | C1 | Elevates and pulls hyoid bone forward; depresses mandible when hyoid is fixed |

inferior limit of the floor of the oral cavity (Fig. 8.252A). Each muscle is triangular in shape with its apex pointed forward.

The lateral margin of each triangular muscle is attached to the mylohyoid line on the medial side of the body of the mandible. From here, the muscle fibers run slightly downward to the medial margin at the midline where the fibers are joined together with those of their partner muscle on the other side by a raphe. The raphe extends from the posterior aspect of the mandibular symphysis in front to the body of the hyoid bone behind.

The posterior margin of each mylohyoid muscle is free except for a small medial attachment to the hyoid bone.

The mylohyoid muscles:

- contribute structural support to the floor of the oral cavity,
- participate in elevating and pulling forward the hyoid bone, and therefore the attached larynx, during the initial stages of swallowing, and
- when the hyoid bone is fixed in position, depress the mandible and open the mouth.

Like the muscles of mastication, the mylohyoid muscles are innervated by the mandibular nerve [$V_3$]. The specific branch that innervates the mylohyoid muscles is the nerve to the mylohyoid from the inferior alveolar nerve.

### Geniohyoid muscles

The geniohyoid muscles (Table 8.20) are paired cord-like muscles that run, one on either side of the midline, from the inferior mental spines on the posterior surface of the mandibular symphysis to the anterior surface of the body of the hyoid bone (Fig. 8.252B,C). They are immediately

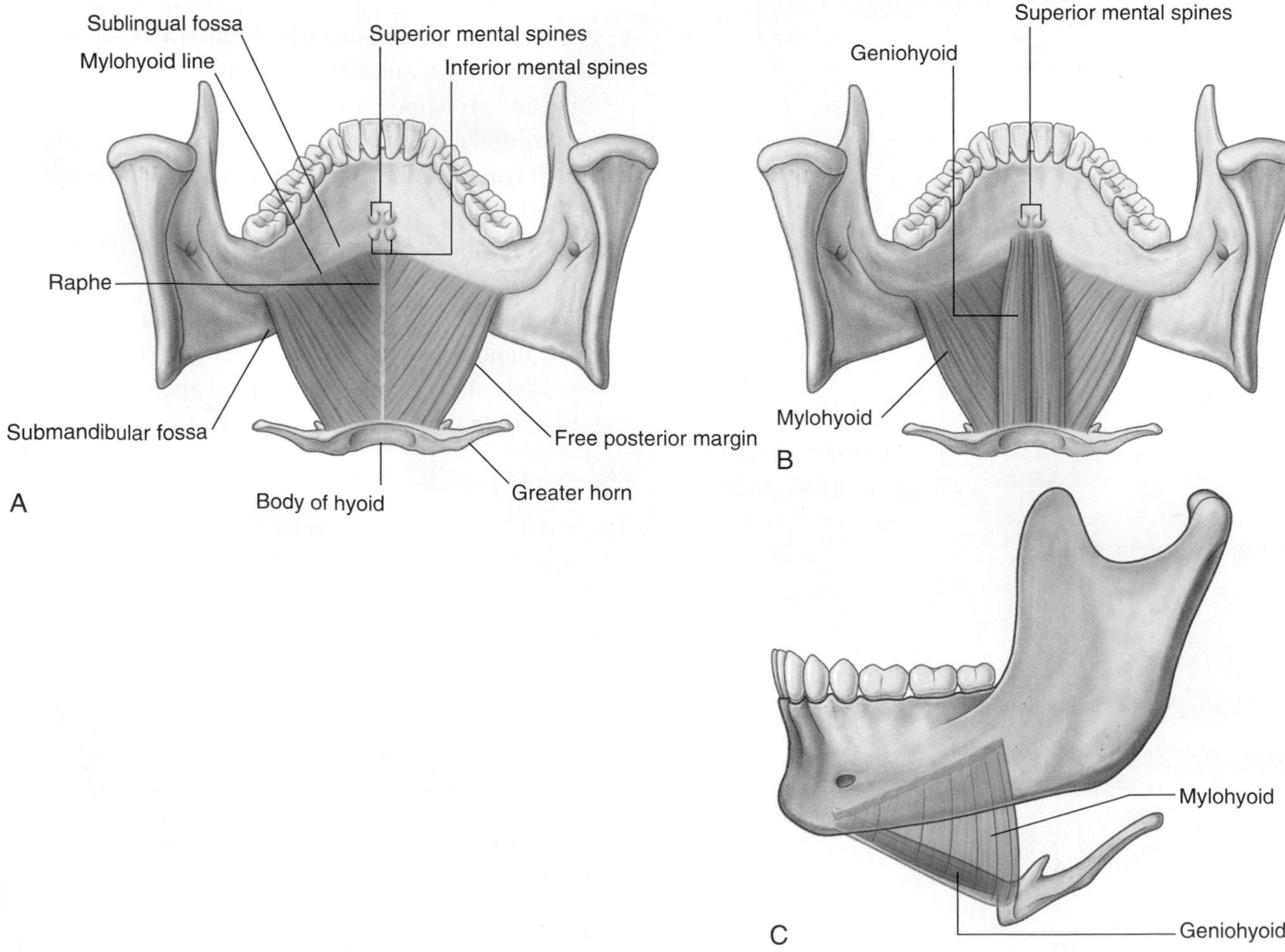

**Fig. 8.252** **A.** Mylohyoid muscles. **B.** Geniohyoid muscles. **C.** Lateral view.

superior to the mylohyoid muscles in the floor of the mouth and inferior to the genioglossus muscles that form part of the root of the tongue.

The geniohyoid muscles:

- mainly pull the hyoid bone, and therefore the attached larynx, up and forward during swallowing; and
- because they pass posteroinferiorly from the mandible to the hyoid bone, when the hyoid bone is fixed, they can act with the mylohyoid muscles to depress the mandible and open the mouth.

Unlike other muscles that move the mandible at the temporomandibular joint, the geniohyoid muscles are innervated by a branch of cervical nerve C1, which "hitchhikes" from the neck along the hypoglossal nerve [XII] into the floor of the oral cavity.

### Gateway into the floor of the oral cavity

In addition to defining the lower limit of the floor of the oral cavity, the free posterior border of the mylohyoid muscle on each side forms one of the three margins of a large triangular aperture **(oropharyngeal triangle)**, which is a major route by which structures in the upper neck and infratemporal fossa of the head pass to and from structures in the floor of the oral cavity (Fig. 8.253). The other two muscles that complete the margins of the aperture are the superior and middle constrictor muscles of the pharynx.

Most structures that pass through the aperture are associated with the tongue and include muscles (hyoglossus, styloglossus), vessels (lingual artery and vein), nerves (lingual, hypoglossal [XII], glossopharyngeal [IX]), and lymphatics.

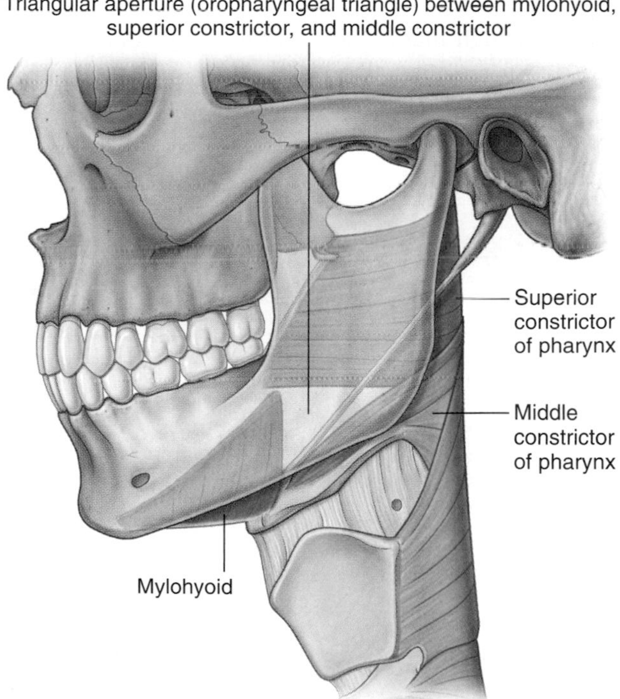


**Fig. 8.253** Gateway into the floor of the oral cavity.

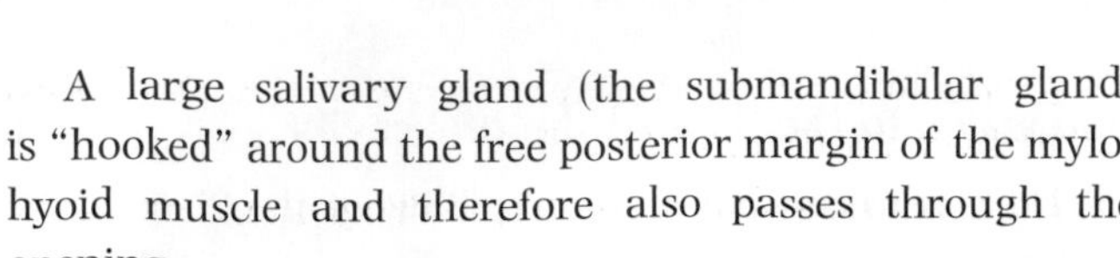


A large salivary gland (the submandibular gland) is "hooked" around the free posterior margin of the mylohyoid muscle and therefore also passes through the opening.

## Tongue

The tongue is a muscular structure that forms part of the floor of the oral cavity and part of the anterior wall of the oropharynx (Fig. 8.254A). Its anterior part is in the oral cavity and is somewhat triangular in shape with a blunt **apex of the tongue**. The apex is directed anteriorly and sits immediately behind the incisor teeth. The **root of the tongue** is attached to the mandible and the hyoid bone.

The superior surface of the oral or anterior two-thirds of the tongue is oriented in the horizontal plane.

The pharyngeal surface or posterior one-third of the tongue curves inferiorly and becomes oriented more in the vertical plane. The oral and pharyngeal surfaces are separated by a V-shaped **terminal sulcus of the tongue**. This terminal sulcus forms the inferior margin of the oropharyngeal isthmus between the oral and pharyngeal cavities. At the apex of the V-shaped sulcus is a small depression (the **foramen cecum of the tongue**), which marks the site in the embryo where the epithelium invaginated to form the thyroid gland. In some people a thyroglossal duct persists and connects the foramen cecum on the tongue with the thyroid gland in the neck.

### Papillae

The superior surface of the oral part of the tongue is covered by hundreds of papillae (Fig. 8.254B):

- **Filiform papillae** are small cone-shaped projections of the mucosa that end in one or more points.
- **Fungiform papillae** are rounder in shape and larger than the filiform papillae, and tend to be concentrated along the margins of the tongue.
- The largest of the papillae are the vallate papillae, which are blunt-ended cylindrical papillae invaginations in the tongue's surface—there are only about 8 to 12 vallate papillae in a single V-shaped line immediately anterior to the terminal sulcus of the tongue.
- **Foliate papillae** are linear folds of mucosa on the sides of the tongue near the terminal sulcus of tongue.

The papillae in general increase the area of contact between the surface of the tongue and the contents of the oral cavity. All except the filiform papillae have taste buds on their surfaces.

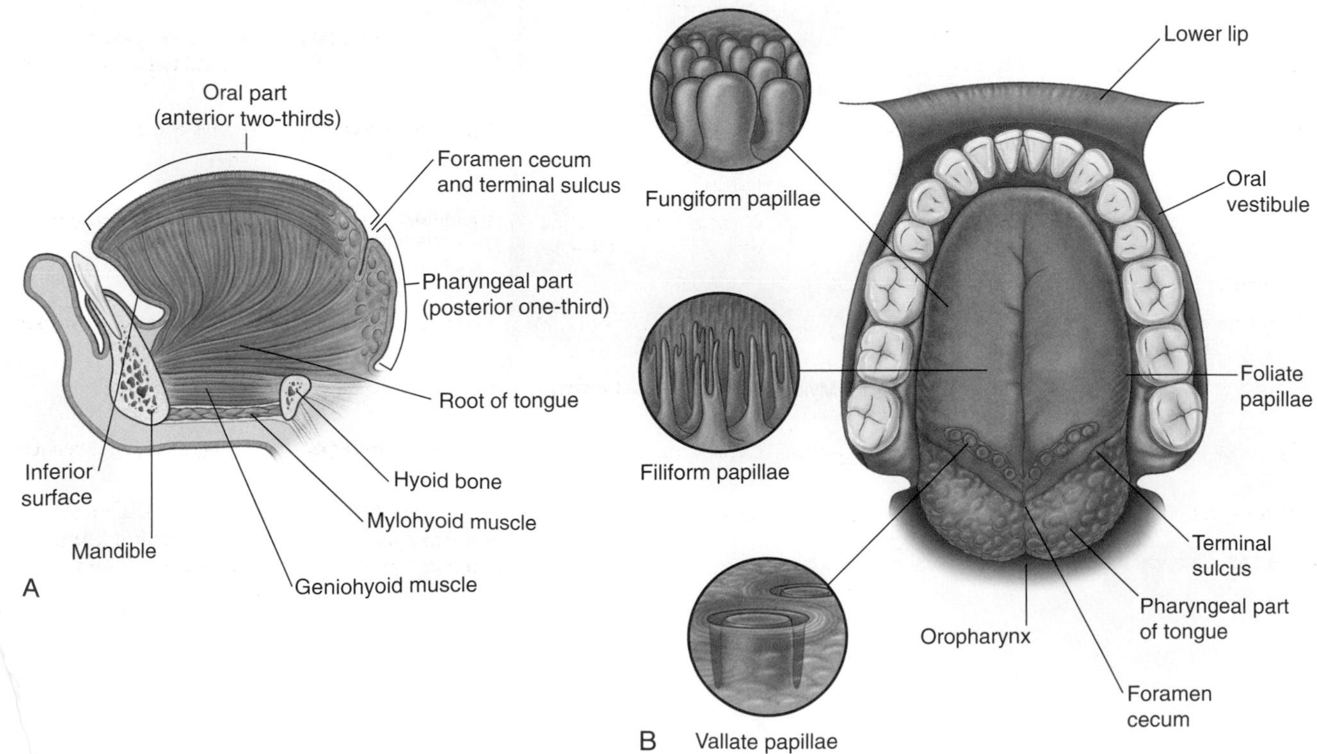


**Fig. 8.254** Tongue. **A.** Paramedian sagittal section. **B.** Superior view.

### Inferior surface of tongue

The undersurface of the oral part of the tongue lacks papillae, but does have a number of linear mucosal folds (see Fig. 8.265). A single median fold (the **frenulum of the tongue**) is continuous with the mucosa covering the floor of the oral cavity, and overlies the lower margin of a midline sagittal septum, which internally separates the right and left sides of the tongue. On each side of the frenulum is a lingual vein, and lateral to each vein is a rough **fimbriated fold**.

### Pharyngeal surface

The mucosa covering the pharyngeal surface of the tongue is irregular in contour because of the many small nodules of lymphoid tissue in the submucosa. These nodules are collectively the **lingual tonsil**.

There are no papillae on the pharyngeal surface.

### Muscles

The bulk of the tongue is composed of muscle (Fig. 8.254 and Table 8.21).

The tongue is completely divided into left and right halves by a median sagittal septum composed of connective tissue. This means that all muscles of the tongue are paired. There are intrinsic and extrinsic lingual muscles.

Except for the palatoglossus, which is innervated by the vagus nerve [X], all muscles of the tongue are innervated by the hypoglossal nerve [XII].

**Table 8.21** Muscles of the tongue

| Muscle | Origin | Insertion | Innervation | Function |
|---|---|---|---|---|
| **Intrinsic** | | | | |
| Superior longitudinal (just deep to surface of tongue) | Submucosal connective tissue at the back of the tongue and from the median septum of the tongue | Muscle fibers pass forward and obliquely to submucosal connective tissue and mucosa on margins of tongue | Hypoglossal nerve [XII] | Shortens tongue; curls apex and sides of tongue |
| Inferior longitudinal (between genioglossus and hyoglossus muscles) | Root of tongue (some fibers from hyoid) | Apex of tongue | Hypoglossal nerve [XII] | Shortens tongue; uncurls apex and turns it downward |
| Transverse | Median septum of the tongue | Submucosal connective tissue on lateral margins of tongue | Hypoglossal nerve [XII] | Narrows and elongates tongue |
| Vertical | Submucosal connective tissue on dorsum of tongue | Connective tissue in more ventral regions of tongue | Hypoglossal nerve [XII] | Flattens and widens tongue |
| **Extrinsic** | | | | |
| Genioglossus | Superior mental spines | Body of hyoid; entire length of tongue | Hypoglossal nerve [XII] | Protrudes tongue; depresses center of tongue |
| Hyoglossus | Greater horn and adjacent part of body of hyoid bone | Lateral surface of tongue | Hypoglossal nerve [XII] | Depresses tongue |
| Styloglossus | Styloid process (anterolateral surface) | Lateral surface of tongue | Hypoglossal nerve [XII] | Elevates and retracts tongue |
| Palatoglossus | Inferior surface of palatine aponeurosis | Lateral margin of tongue | Vagus nerve [X] (via pharyngeal branch to pharyngeal plexus) | Depresses palate; moves palatoglossal fold toward midline; elevates back of the tongue |

### Intrinsic muscles

The intrinsic muscles of the tongue (Fig. 8.255) originate and insert within the substance of the tongue. They are divided into **superior longitudinal**, **inferior longitudinal**, **transverse**, and **vertical muscles**, and they alter the shape of the tongue by:

- lengthening and shortening it,
- curling and uncurling its apex and edges, and
- flattening and rounding its surface.

Working in pairs or one side at a time the intrinsic muscles of the tongue contribute to precision movements of the tongue required for speech, eating, and swallowing.

### Extrinsic muscles

Extrinsic muscles of the tongue (Fig. 8.255 and Table 8.21) originate from structures outside the tongue and insert into the tongue. There are four major extrinsic muscles on each side, the genioglossus, hyoglossus, styloglossus, and palatoglossus. These muscles protrude, retract, depress, and elevate the tongue.

#### Genioglossus

The thick fan-shaped **genioglossus muscles** make a substantial contribution to the structure of the tongue. They occur on either side of the midline septum that separates left and right halves of the tongue.

The genioglossus muscles originate from the superior mental spines on the posterior surface of the mandibular

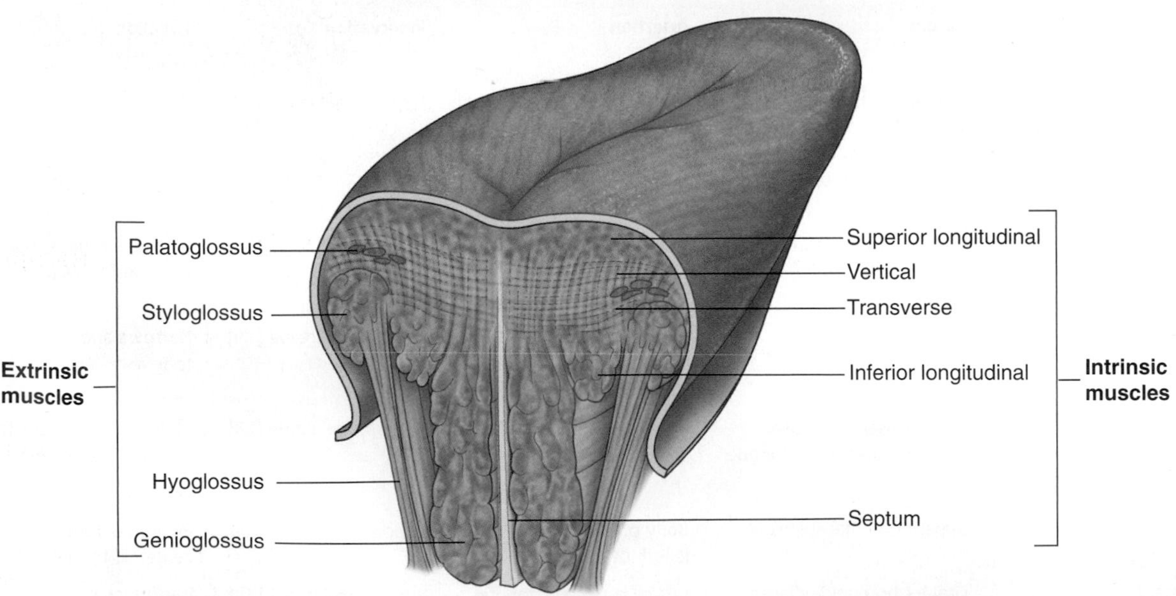


**Fig. 8.255** Muscles of the tongue.

symphysis immediately superior to the origin of the geniohyoid muscles from the inferior mental spines (Fig. 8.256). From this small site of origin, each muscle expands posteriorly and superiorly. The most inferior fibers attach to the hyoid bone. The remaining fibers spread out superiorly to blend with the intrinsic muscles along virtually the entire length of the tongue.

The genioglossus muscles:

- depress the central part of the tongue, and
- protrude the anterior part of the tongue out of the oral fissure (i.e., stick the tongue out).

Like most muscles of the tongue, the genioglossus muscles are innervated by the hypoglossal nerves [XII].

Asking a patient to "stick your tongue out" can be used as a test for the hypoglossal nerves [XII]. If the nerves are functioning normally, the tongue should protrude evenly in the midline. If the nerve on one side is not fully functional, the tip of the tongue will point to that side.

### Hyoglossus

The hyoglossus muscles are thin quadrangular muscles lateral to the genioglossus muscles (Fig. 8.257).

Each hyoglossus muscle originates from the entire length of the greater horn and the adjacent part of the body of the hyoid bone. At its origin from the hyoid bone, the hyoglossus muscle is lateral to the attachment of the middle constrictor muscle of the pharynx. The muscle passes superiorly and anteriorly through the gap

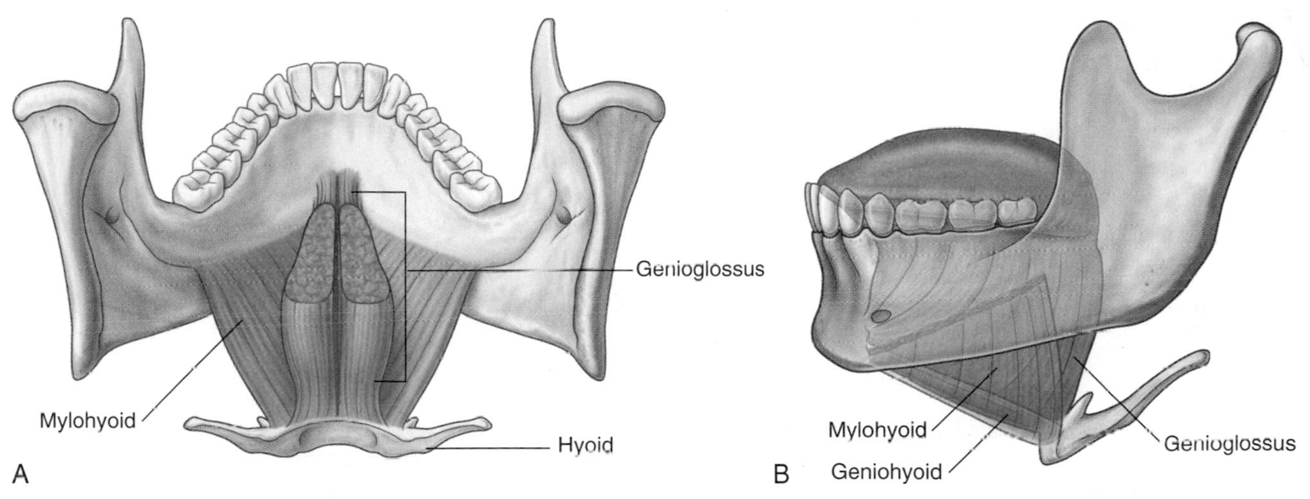


**Fig. 8.256** Genioglossus muscles. **A.** Posterior view. **B.** Lateral (left) view.

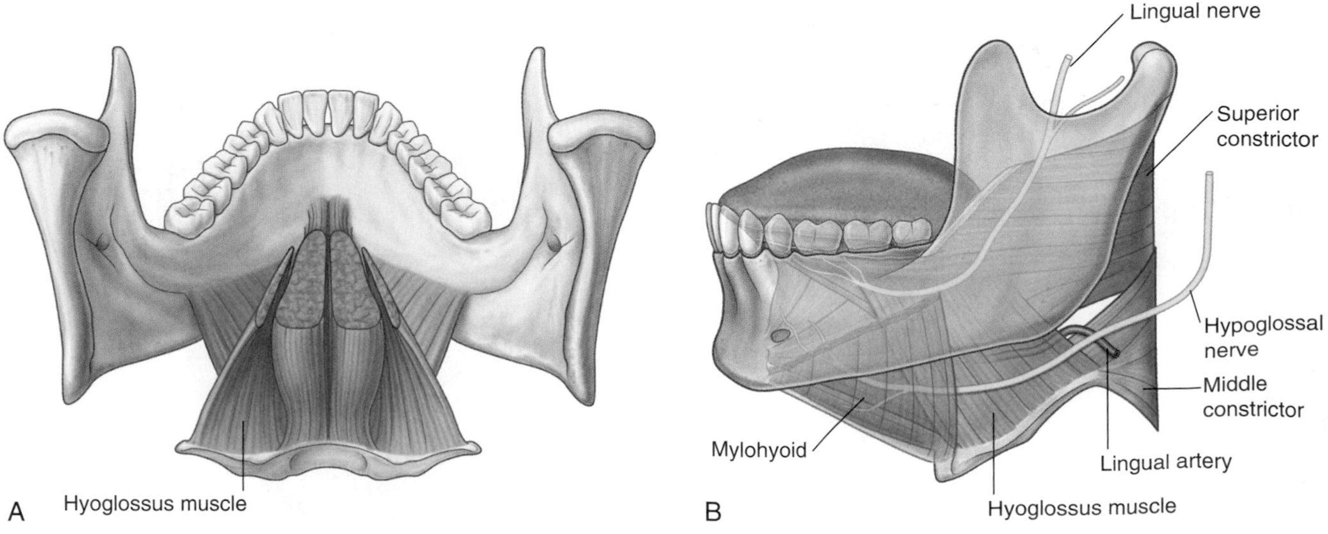


**Fig. 8.257** Hyoglossus muscles. **A.** Posterior view. **B.** Lateral (left) view.

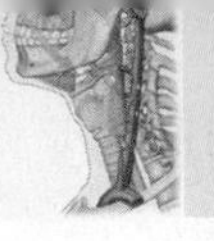

(oropharyngeal triangle) between the superior constrictor, middle constrictor, and mylohyoid to insert into the tongue lateral to the genioglossus and medial to the styloglossus.

The hyoglossus muscle depresses the tongue and is innervated by the hypoglossal nerve [XII].

***An important landmark.*** The hyoglossus muscle is an important landmark in the floor of the oral cavity:

- The lingual artery from the external carotid artery in the neck enters the tongue deep to the hyoglossus, between the hyoglossus and genioglossus.
- The hypoglossal nerve [XII] and lingual nerve (branch of the mandibular nerve [$V_3$]), from the neck and infratemporal fossa of the head, respectively, enter the tongue on the external surface of the hyoglossus.

### Styloglossus

The styloglossus muscles originate from the anterior surface of the styloid processes of the temporal bones. From here, each muscle passes inferiorly and medially through the gap (oropharyngeal triangle) between the middle constrictor, superior constrictor, and mylohyoid muscles to enter the lateral surface of the tongue where they blend with the superior margin of the hyoglossus and with the intrinsic muscles (Fig. 8.258).

The styloglossus muscles retract the tongue and pull the back of the tongue superiorly. They are innervated by the hypoglossal nerves [XII].

### Palatoglossus

The palatoglossus muscles are muscles of the soft palate and the tongue. Each originates from the undersurface of the palatine aponeurosis and passes anteroinferiorly to the lateral side of the tongue (Fig. 8.259).

The palatoglossus muscles:

- elevate the back of the tongue,
- move the palatoglossal arches of mucosa toward the midline, and
- depress the soft palate.

These movements facilitate closing of the oropharyngeal isthmus and as a result separate the oral cavity from the oropharynx.

Unlike other muscles of the tongue, but similar to most other muscles of the soft palate, the palatoglossus muscles are innervated by the vagus nerves [X].

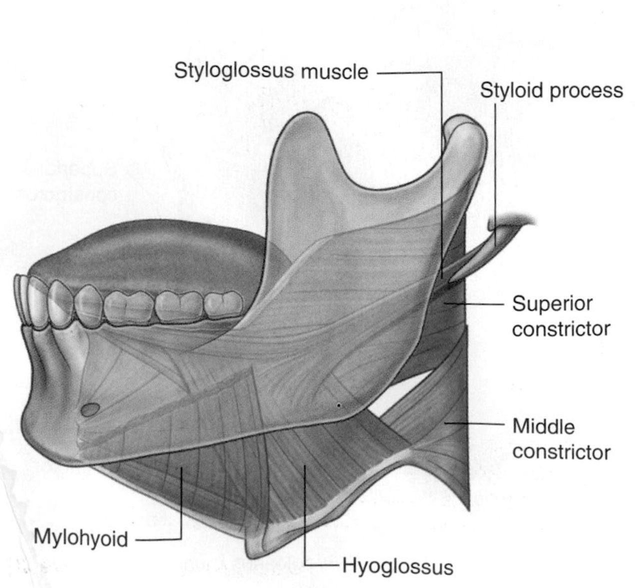


**Fig. 8.258** Styloglossus muscles.

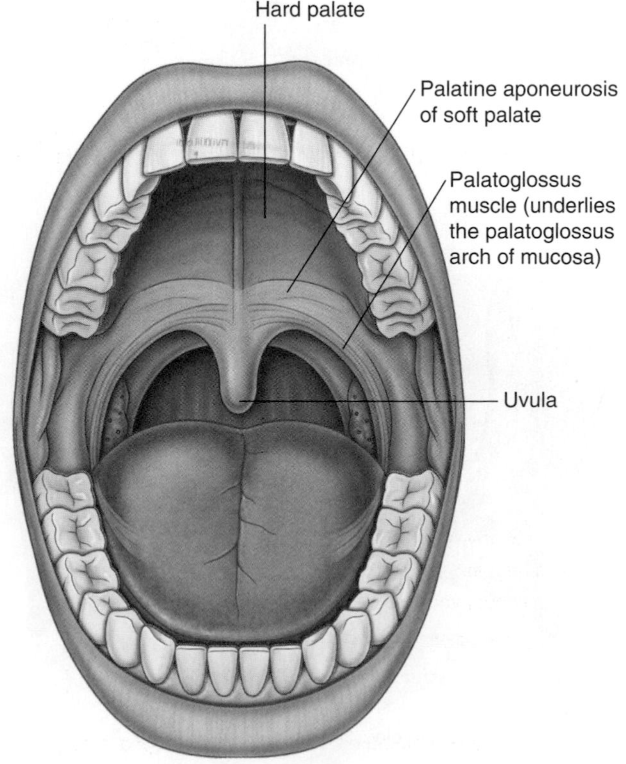


**Fig. 8.259** Palatoglossus muscles.

## Vessels

### Arteries

The major artery of the tongue is the **lingual artery** (Fig. 8.260).

On each side, the lingual artery originates from the external carotid artery in the neck adjacent to the tip of the greater horn of the hyoid bone. It forms an upward bend and then loops downward and forward to pass deep to the hyoglossus muscle, and accompanies the muscle through the aperture (oropharyngeal triangle) formed by the margins of the mylohyoid, superior constrictor, and middle constrictor muscles, and enters the floor of the oral cavity.

The lingual artery then travels forward in the plane between the hyoglossus and genioglossus muscles to the apex of the tongue.

In addition to the tongue, the lingual artery supplies the sublingual gland, gingiva, and oral mucosa in the floor of the oral cavity.

### Veins

The tongue is drained by dorsal lingual and deep lingual veins (Fig. 8.260).

The **deep lingual veins** are visible through the mucosa on the undersurface of the tongue. Although they accompany the lingual arteries in anterior parts of the tongue, they become separated from the arteries posteriorly by the hyoglossus muscles. On each side, the deep lingual vein travels with the hypoglossal nerve [XII] on the external surface of the hyoglossus muscle and passes out of the floor of the oral cavity through the aperture (oropharyngeal triangle) formed by the margins of the mylohyoid, superior constrictor, and middle constrictor muscles. It joins the internal jugular vein in the neck.

The **dorsal lingual vein** follows the lingual artery between the hyoglossus and genioglossus muscles and, like the deep lingual vein, drains into the internal jugular vein in the neck.

## Innervation

Innervation of the tongue is complex and involves a number of nerves (Figs. 8.260 and 8.261).

### Glossopharyngeal nerve [IX]

Taste (SA) and general sensation from the pharyngeal part of the tongue are carried by the glossopharyngeal nerve [IX].

The glossopharyngeal nerve [IX] leaves the skull through the jugular foramen and descends along the

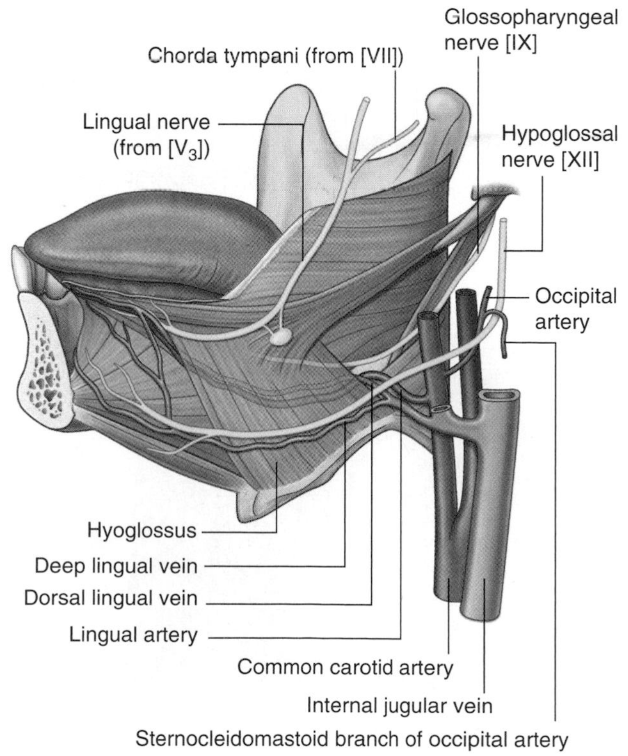

**Fig. 8.260** Arteries, veins, and nerves of the tongue.

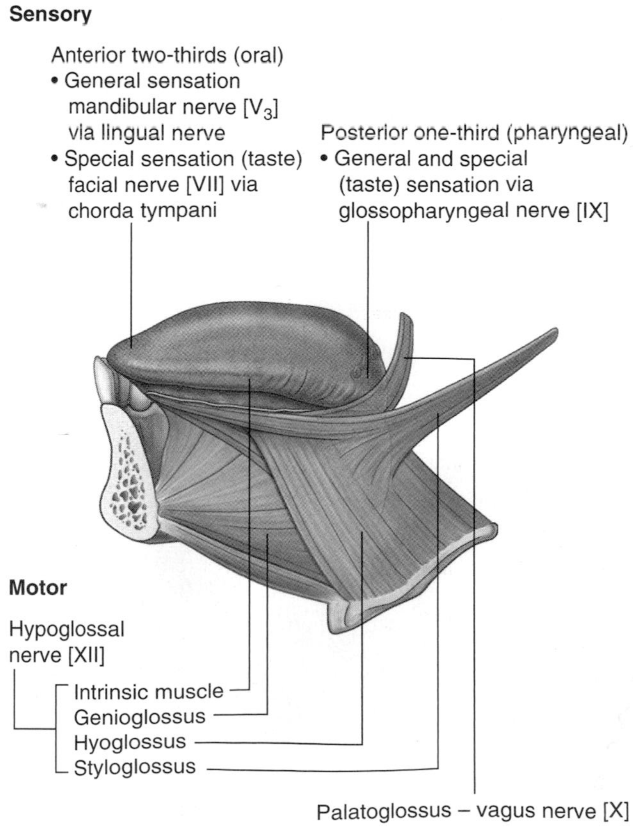

**Fig. 8.261** Innervation of the tongue.

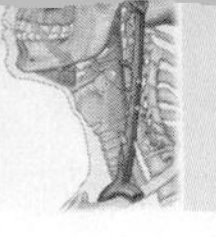

posterior surface of the stylopharyngeus muscle. It passes around the lateral surface of the stylopharyngeus and then slips through the posterior aspect of the gap (oropharyngeal triangle) between the superior constrictor, middle constrictor, and mylohyoid muscles. The nerve then passes forward on the oropharyngeal wall just below the inferior pole of the palatine tonsil and enters the pharyngeal part of the tongue deep to the styloglossus and hyoglossus muscles. In addition to taste and general sensation on the posterior one-third of the tongue, branches creep anterior to the terminal sulcus of the tongue to carry taste (SA) and general sensation from the vallate papillae.

### Lingual nerve

General sensory innervation from the anterior two-thirds or oral part of the tongue is carried by the **lingual nerve**, which is a major branch of the mandibular nerve [$V_3$]. It originates in the infratemporal fossa and passes anteriorly into the floor of the oral cavity by passing through the gap (oropharyngeal triangle) between the mylohyoid, superior constrictor, and middle constrictor muscles (Fig. 8.262). As it travels through the gap, it passes immediately inferior to the attachment of the superior constrictor to the mandible and continues forward on the medial surface of the mandible adjacent to the last molar tooth and deep to the gingiva. In this position, the nerve can be palpated against the bone by placing a finger into the oral cavity.

The lingual nerve then continues anteromedially across the floor of the oral cavity, loops under the submandibular duct, and ascends into the tongue on the external and superior surface of the hyoglossus muscle.

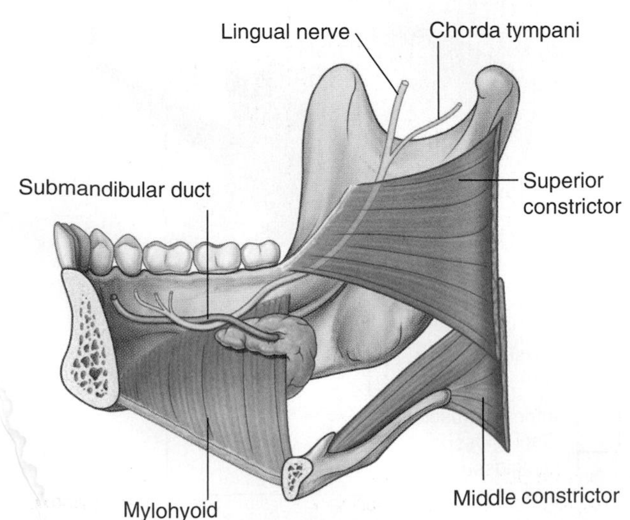


**Fig. 8.262** Lingual nerve in the floor of the oral cavity (medial view).

In addition to general sensation from the oral part of the tongue, the lingual nerve also carries general sensation from the mucosa on the floor of the oral cavity and gingiva associated with the lower teeth. The lingual nerve also carries parasympathetic and taste fibers from the oral part of the tongue that are part of the facial nerve [VII].

### Facial nerve [VII]

Taste (SA) from the oral part of the tongue is carried into the central nervous system by the facial nerve [VII]. Special sensory (SA) fibers of the facial nerve [VII] leave the tongue and oral cavity as part of the lingual nerve. The fibers then enter the chorda tympani nerve, which is a branch of the facial nerve [VII] that joins the lingual nerve in the infratemporal fossa (Fig. 8.262; also see p. 976).

### Hypoglossal nerve [XII]

All muscles of the tongue are innervated by the hypoglossal nerve [XII] except for the palatoglossus muscle, which is innervated by the vagus nerve [X].

The hypoglossal nerve [XII] leaves the skull through the hypoglossal canal and descends almost vertically in the neck to a level just below the angle of the mandible (Fig. 8.263). Here it angles sharply forward around the

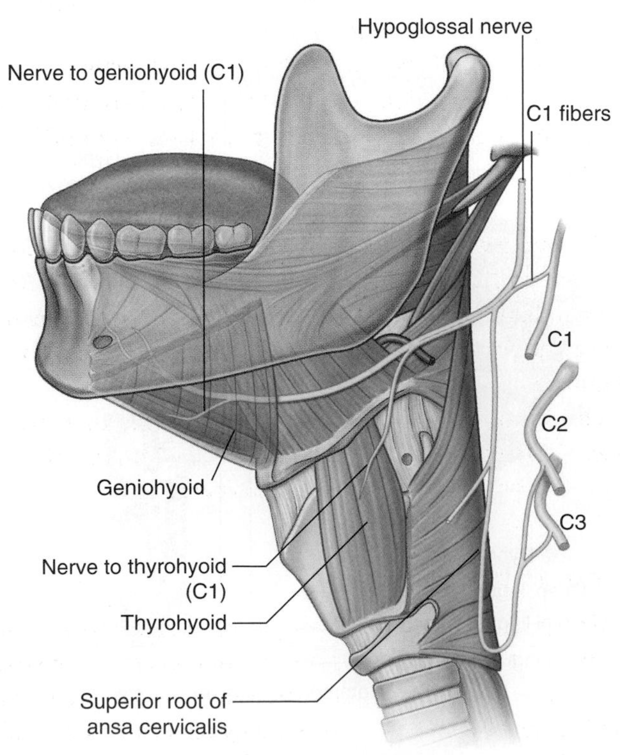


**Fig. 8.263** Hypoglossal nerve and C1 fibers.

sternocleidomastoid branch of the occipital artery, crosses the external carotid artery, and continues forward, crossing the loop of the lingual artery, to reach the external surface of the lower one-third of the hyoglossus muscle.

The hypoglossal nerve [XII] follows the hyoglossus muscle through the gap (oropharyngeal triangle) between the superior constrictor, middle constrictor, and mylohyoid muscles to reach the tongue.

In the upper neck, a branch from the anterior ramus of C1 joins the hypoglossal nerve [XII]. Most of these C1 fibers leave the hypoglossal nerve [XII] as the superior root of the ansa cervicalis (Fig. 8.263). Near the posterior border of the hyoglossus muscle, the remaining fibers leave the hypoglossal nerve [XII] and form two nerves:

- the thyrohyoid branch, which remains in the neck to innervate the thyrohyoid muscle, and
- the branch to the geniohyoid, which passes into the floor of the oral cavity to innervate the geniohyoid.

### Lymphatics

All lymphatic vessels from the tongue ultimately drain into the deep cervical chain of nodes along the internal jugular vein:

- The pharyngeal part of the tongue drains through the pharyngeal wall directly into mainly the jugulodigastric node of the deep cervical chain.
- The oral part of the tongue drains both directly into the deep cervical nodes, and indirectly into these nodes by passing first through the mylohyoid muscle and into submental and submandibular nodes.

The submental nodes are inferior to the mylohyoid muscles and between the digastric muscles, while the submandibular nodes are below the floor of the oral cavity along the inner aspect of the inferior margins of the mandible.

The tip of the tongue drains through the mylohyoid muscle into the submental nodes and then into mainly the jugulo-omohyoid node of the deep cervical chain.

## Salivary glands

Salivary glands are glands that open or secrete into the oral cavity. Most are small glands in the submucosa or mucosa of the oral epithelium lining the tongue, palate, cheeks, and lips, and open into the oral cavity directly or via small ducts. In addition to these small glands are much larger glands, which include the paired parotid, submandibular, and sublingual glands.

### Parotid gland

The parotid gland (see pp. 900–901) on each side is entirely outside the boundaries of the oral cavity in a shallow triangular-shaped trench (Fig. 8.264) formed by:

- the sternocleidomastoid muscle behind,
- the ramus of the mandible in front, and
- superiorly, the base of the trench is formed by the external acoustic meatus and the posterior aspect of the zygomatic arch.

The gland normally extends anteriorly over the masseter muscle, and inferiorly over the posterior belly of the digastric muscle.

The parotid duct passes anteriorly across the external surface of the masseter muscle and then turns medially to penetrate the buccinator muscle of the cheek and open into the oral cavity adjacent to the crown of the second upper molar tooth.

The parotid gland encloses the external carotid artery, the retromandibular vein, and the origin of the extracranial part of the facial nerve [VII].

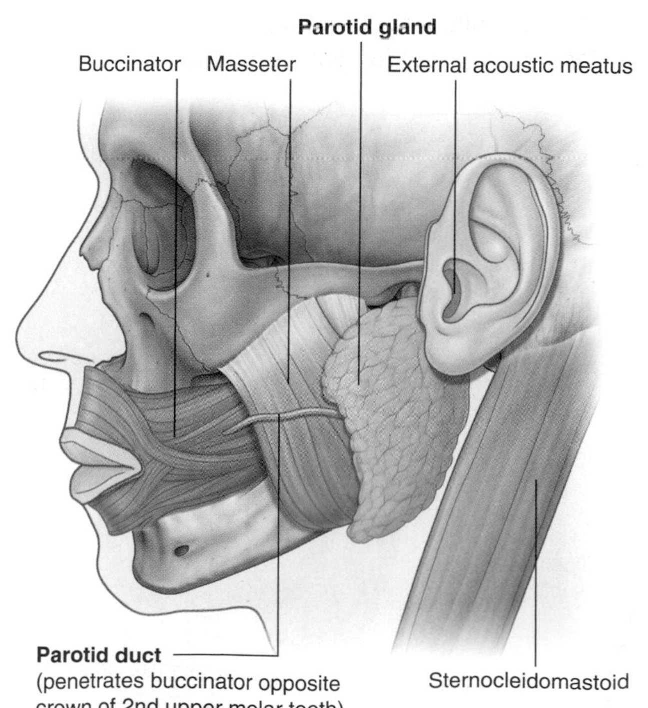


**Fig. 8.264** Parotid gland.

### Submandibular glands

The elongate **submandibular glands** are smaller than the parotid glands but larger than the sublingual glands. Each is hook shaped (Fig. 8.265A,B):

- The larger arm of the hook is directed forward in the horizontal plane below the mylohyoid muscle and is therefore outside the boundaries of the oral cavity—this larger superficial part of the gland is directly against a shallow impression on the medial side of the mandible (submandibular fossa) inferior to the mylohyoid line.
- The smaller arm of the hook (or deep part) of the gland loops around the posterior margin of the mylohyoid muscle to enter and lie within the floor of the oral cavity where it is lateral to the root of the tongue on the lateral surface of the hyoglossus muscle.

The **submandibular duct** emerges from the medial side of the deep part of the gland in the oral cavity and passes forward to open on the summit of a small **sublingual caruncle** (papilla) beside the base of the frenulum of the tongue (Fig. 8.265C,D).

The lingual nerve loops under the submandibular duct, crossing first the lateral side and then the medial side of the duct, as the nerve descends anteromedially through the floor of the oral cavity and then ascends into the tongue.

### Sublingual glands

The sublingual glands are the smallest of the three major paired salivary glands. Each is almond shaped and is immediately lateral to the submandibular duct and associated lingual nerve in the floor of the oral cavity (Fig. 8.265).

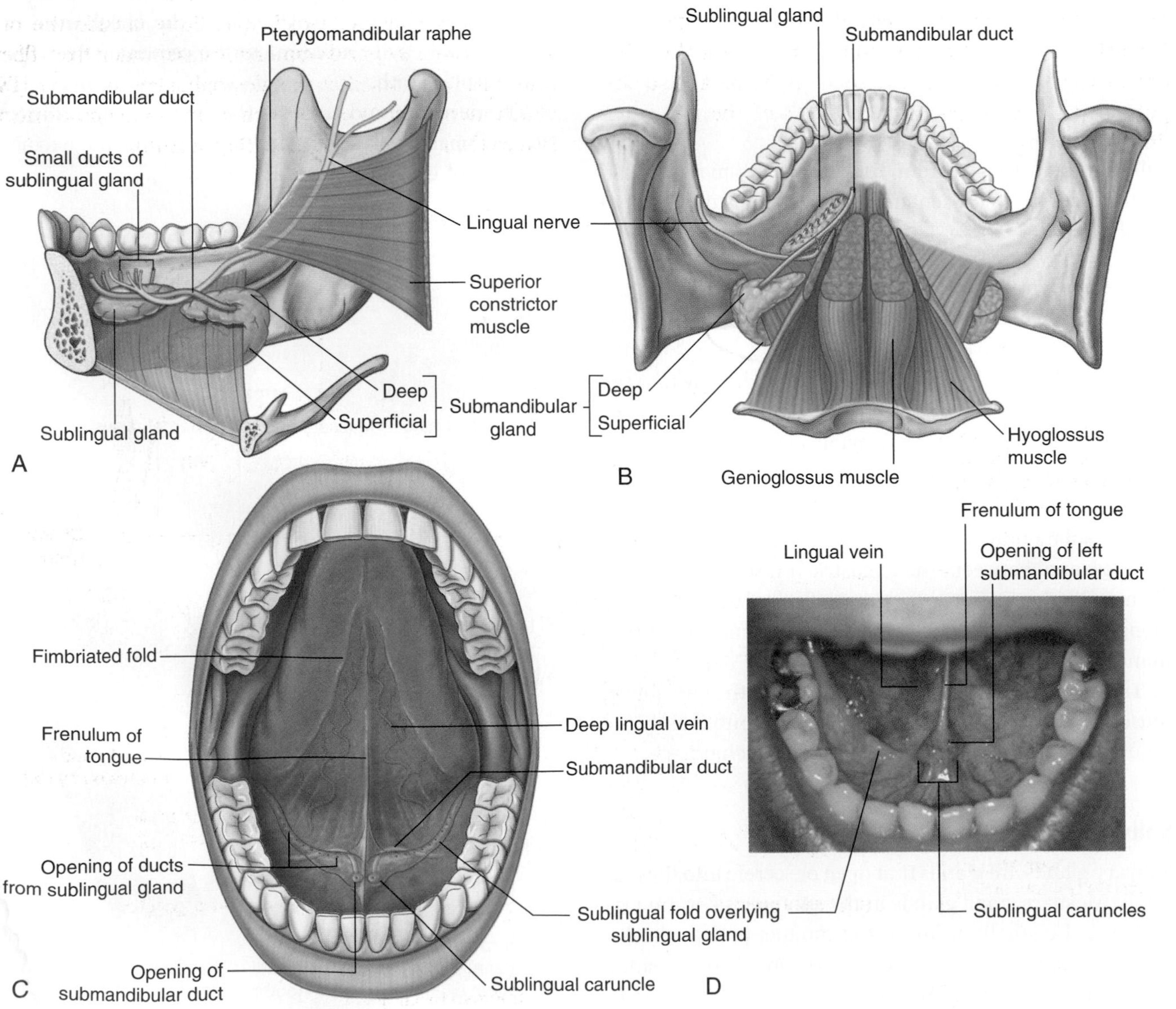

**Fig. 8.265** Submandibular and sublingual glands. **A.** Medial view. **B.** Posterior view. **C.** Anterior view. **D.** Anterosuperior view.

Each sublingual gland lies directly against the medial surface of the mandible where it forms a shallow groove (sublingual fossa) superior to the anterior one-third of the mylohyoid line.

The superior margin of the sublingual gland raises an elongate fold of mucosa (**sublingual fold**), which extends from the posterolateral aspect of the floor of the oral cavity to the sublingual papilla beside the base of the frenulum of the tongue at the midline anteriorly (Fig. 8.265D).

The sublingual gland drains into the oral cavity via numerous small ducts (minor sublingual ducts), which open onto the crest of the sublingual fold. Occasionally, the more anterior part of the gland is drained by a duct (major sublingual duct) that opens together with the submandibular duct on the sublingual caruncle.

### Vessels

Vessels that supply the parotid gland originate from the external carotid artery and from its branches that are adjacent to the gland. The submandibular and sublingual glands are supplied by branches of the facial and lingual arteries.

Veins from the parotid gland drain into the external jugular vein, and those from the submandibular and sublingual glands drain into lingual and facial veins.

Lymphatic vessels from the parotid gland drain into nodes that are on or in the gland. These parotid nodes then drain into superficial and deep cervical nodes.

Lymphatics from the submandibular and sublingual glands drain mainly into submandibular nodes and then into deep cervical nodes, particularly the jugulo-omohyoid node.

### Innervation

#### Parasympathetic

Parasympathetic innervation to all salivary glands in the oral cavity is by branches of the facial nerve [VII], which join branches of the maxillary [$V_2$] and mandibular [$V_3$] nerves to reach their target destinations.

The parotid gland, which is entirely outside the oral cavity, receives its parasympathetic innervation from fibers that initially traveled in the glossopharyngeal nerve [IX], which eventually join a branch of the mandibular nerve [$V_3$] in the infratemporal fossa (Fig. 8.266).

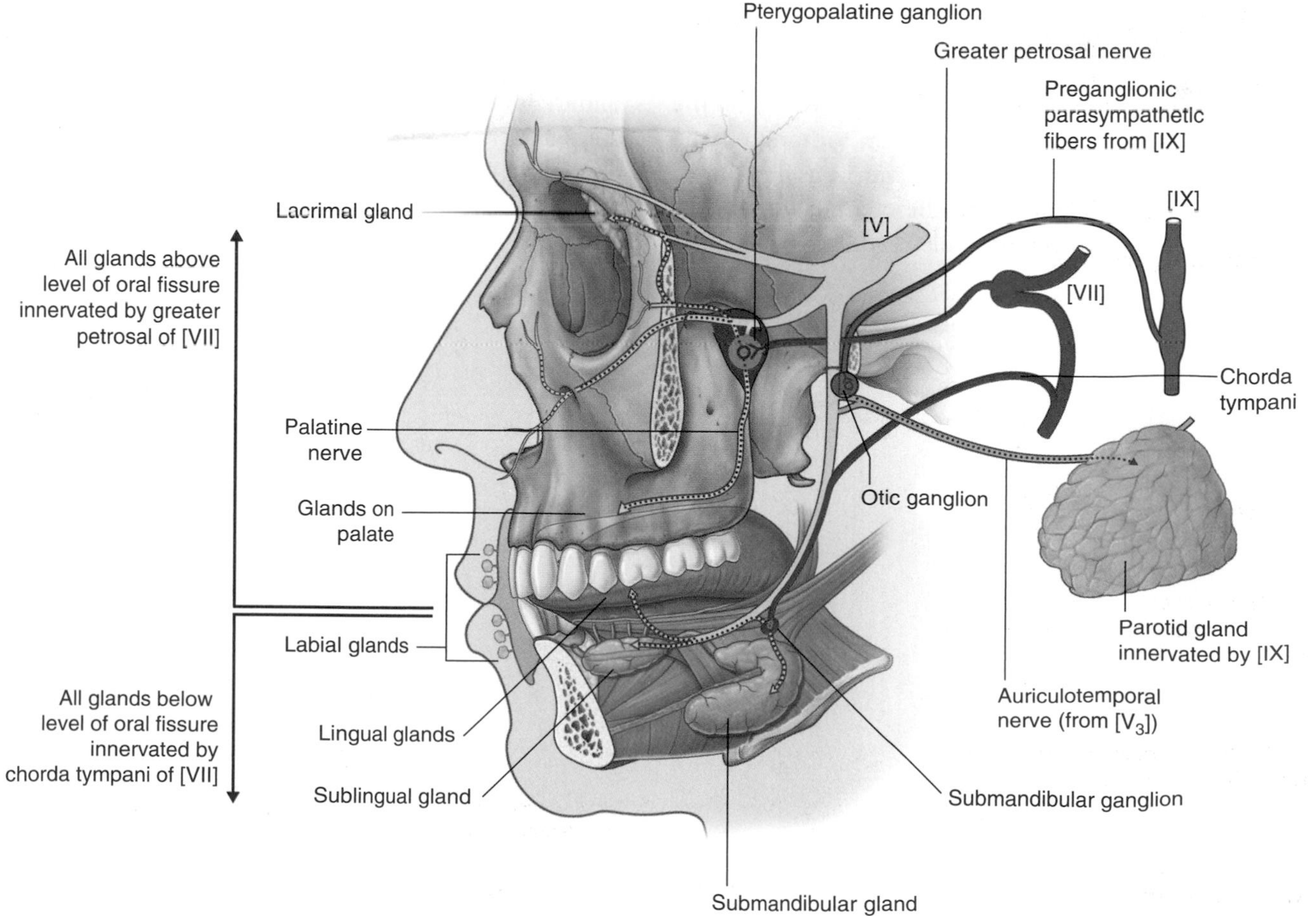


**Fig. 8.266** Summary of parasympathetic (secretomotor) innervation of glands in the head.

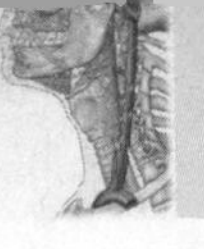



### Greater petrosal nerve

All salivary glands above the level of the oral fissure, as well as all mucus glands in the nose and the lacrimal gland in the orbit, are innervated by parasympathetic fibers carried in the greater petrosal branch of the facial nerve [VII] (Fig. 8.266). Preganglionic parasympathetic fibers carried in this nerve enter the pterygopalatine fossa and synapse with postganglionic parasympathetic fibers in the pterygopalatine ganglion formed around branches of the maxillary nerve [$V_2$]. Postganglionic parasympathetic fibers join general sensory branches of the maxillary nerve, such as the palatine nerves, destined for the roof of the oral cavity, to reach their target glands.

### Chorda tympani

All glands below the level of the oral fissure, which include those small glands in the floor of the oral cavity, in the lower lip, and in the tongue, and the larger submandibular and sublingual glands, are innervated by parasympathetic fibers carried in the chorda tympani branch of the facial nerve [VII] (Fig. 8.266).

The chorda tympani joins the lingual branch of the mandibular nerve [$V_3$] in the infratemporal fossa and passes with it into the oral cavity. On the external surface of the hyoglossus muscle, preganglionic parasympathetic fibers leave the inferior aspect of the lingual nerve to synapse with postganglionic parasympathetic fibers in the submandibular ganglion, which appears to hang off the lingual nerve (Fig. 8.267). Postganglionic parasympathetic fibers leave the ganglion and pass directly to the submandibular and sublingual glands while others hop back onto the lingual nerve and travel with branches of the lingual nerve to target glands.

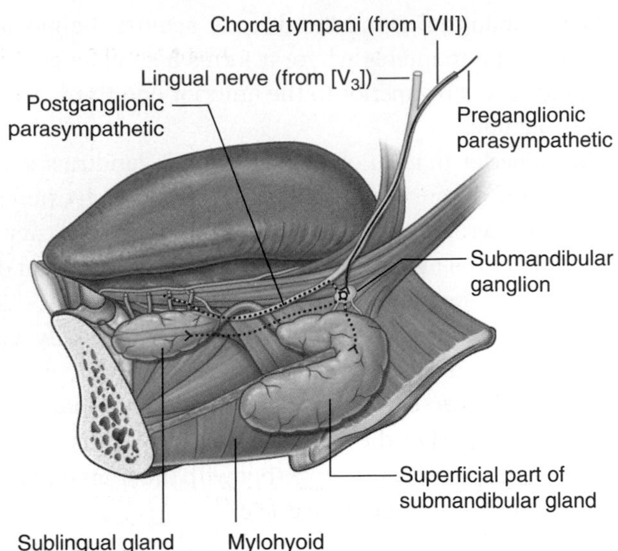


**Fig. 8.267** Course of parasympathetic fibers carried in the chorda tympani nerve.

### Sympathetic

Sympathetic innervation to the salivary glands is from spinal cord level T1. Preganglionic sympathetic fibers enter the sympathetic trunk and ascend to synapse in the superior cervical sympathetic ganglion (Fig. 8.268). Postganglionic fibers hop onto adjacent blood vessels and nerves to reach the glands.

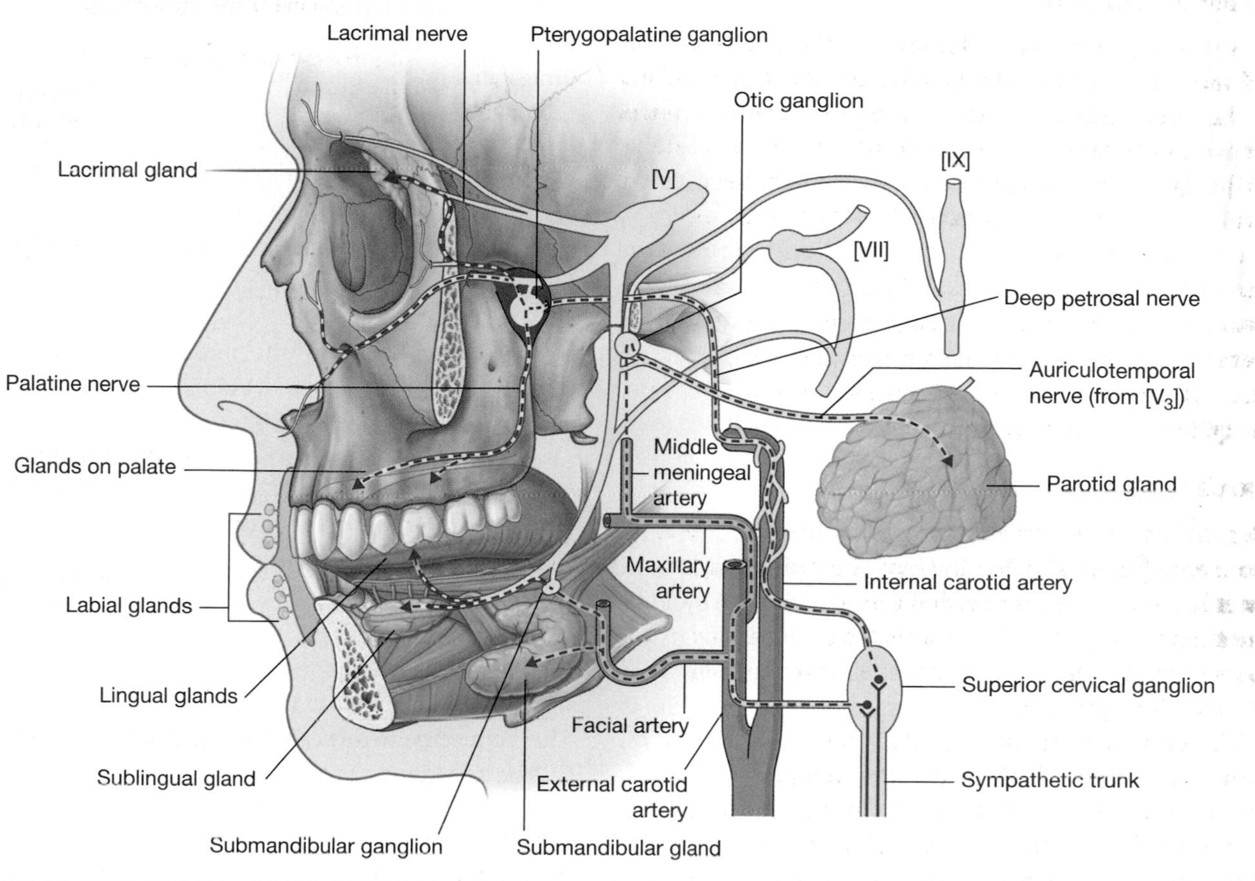


**Fig. 8.268** Summary of sympathetic innervation of glands in the head.

## Roof—palate

The roof of the oral cavity consists of the palate, which has two parts—an anterior hard palate and a posterior soft palate (Fig. 8.269).

### Hard palate

The hard palate separates the oral cavity from the nasal cavities. It consists of a bony plate covered above and below by mucosa:

- Above, it is covered by respiratory mucosa and forms the floor of the nasal cavities.
- Below, it is covered by a tightly bound layer of oral mucosa and forms much of the roof of the oral cavity (Fig. 8.269).

The palatine processes of the maxillae form the anterior three-quarters of the hard palate. The horizontal plates of the palatine bones form the posterior one-quarter. In the oral cavity, the upper alveolar arch borders the hard palate anteriorly and laterally. Posteriorly, the hard palate is continuous with the soft palate.

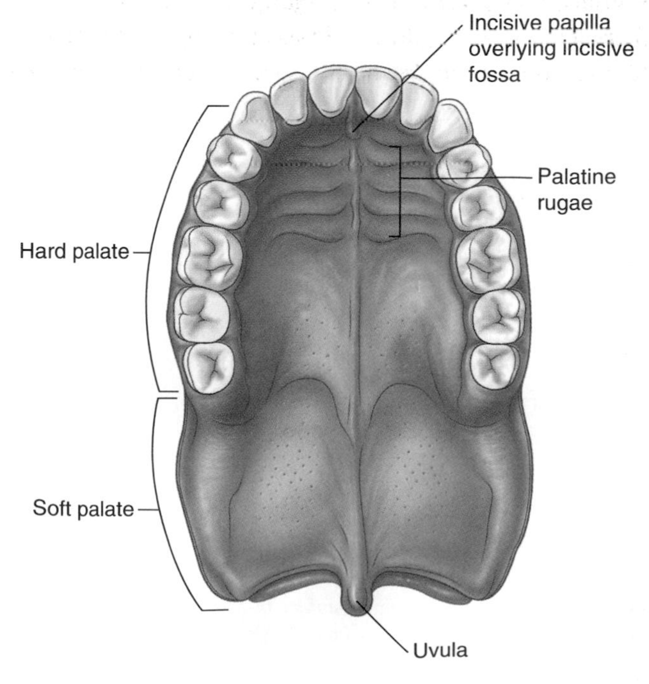


**Fig. 8.269** Palate.

The mucosa of the hard palate in the oral cavity possesses numerous **transverse palatine folds** (**palatine rugae**) and a median longitudinal ridge (**palatine raphe**), which ends anteriorly in a small oval elevation (**incisive papilla**). The incisive papilla (Fig. 8.269) overlies the incisive fossa formed between the horizontal plates of the maxillae immediately behind the incisor teeth.

## Soft palate

The soft palate (Fig. 8.269) continues posteriorly from the hard palate and acts as a valve that can be:

- depressed to help close the oropharyngeal isthmus, and
- elevated to separate the nasopharynx from the oropharynx.

The soft palate is formed and moved by four muscles and is covered by mucosa that is continuous with the mucosa lining the pharynx and oral and nasal cavities.

The small tear-shaped muscular projection that hangs from the posterior free margin of the soft palate is the **uvula**.

### Muscles of the soft palate

Five muscles (Table 8.22) on each side contribute to the formation and movement of the soft palate. Two of these, the tensor veli palatini and levator veli palatini, descend into the palate from the base of the skull. Two others, the palatoglossus and palatopharyngeus, ascend into the palate from the tongue and pharynx, respectively. The last muscle, the musculus uvulae, is associated with the uvula.

All muscles of the palate are innervated by the vagus nerve [X], except for the tensor veli palatini, which is innervated by the mandibular nerve [$V_3$] (via the nerve to the medial pterygoid).

#### Tensor veli palatini and the palatine aponeurosis

The **tensor veli palatini muscle** is composed of two parts—a vertical muscular part and a more horizontal fibrous part, which forms the palatine aponeurosis (Fig. 8.270A).

The vertical part of the tensor veli palatini is thin and triangular in shape with its base attached to the skull and its apex pointed inferiorly. The base is attached along an oblique line that begins medially at the scaphoid fossa near the root of the pterygoid process of the sphenoid bone and continues laterally along the membranous part of the pharyngotympanic tube to the spine of the sphenoid bone.

The tensor veli palatini descends vertically along the lateral surface of the medial plate of the pterygoid process and pharyngeal wall to the pterygoid hamulus where the fibers converge to form a small tendon (Fig. 8.270A).

The tendon loops 90° medially around the pterygoid hamulus, penetrating the origin of the buccinator muscle as it does, and expands like a fan to form the fibrous horizontal part of the muscle. This fibrous part is continuous across the midline with its partner on the other side to form the palatine aponeurosis.

The **palatine aponeurosis** is attached anteriorly to the margin of the hard palate, but is unattached posteriorly where it ends in a free margin. This expansive aponeurosis

**Table 8.22** Muscles of the soft palate

| Muscle | Origin | Insertion | Innervation | Function |
|---|---|---|---|---|
| Tensor veli palatini | Scaphoid fossa of sphenoid bone; fibrous part of pharyngotympanic tube; spine of sphenoid | Palatine aponeurosis | Mandibular nerve [$V_3$] via the branch to medial pterygoid muscle | Tenses the soft palate; opens the pharyngotympanic tube |
| Levator veli palatini | Petrous part of temporal bone anterior to opening for carotid canal | Superior surface of palatine aponeurosis | Vagus nerve [X] via pharyngeal branch to pharyngeal plexus | Only muscle to elevate the soft palate above the neutral position |
| Palatopharyngeus | Superior surface of palatine aponeurosis | Pharyngeal wall | Vagus nerve [X] via pharyngeal branch to pharyngeal plexus | Depresses soft palate; moves palatopharyngeal arch toward midline; elevates pharynx |
| Palatoglossus | Inferior surface of palatine aponeurosis | Lateral margin of tongue | Vagus nerve [X] via pharyngeal branch to pharyngeal plexus | Depresses palate; moves palatoglossal arch toward midline; elevates back of the tongue |
| Musculus uvulae | Posterior nasal spine of hard palate | Connective tissue of uvula | Vagus nerve [X] via pharyngeal branch to pharyngeal plexus | Elevates and retracts uvula; thickens central region of soft palate |

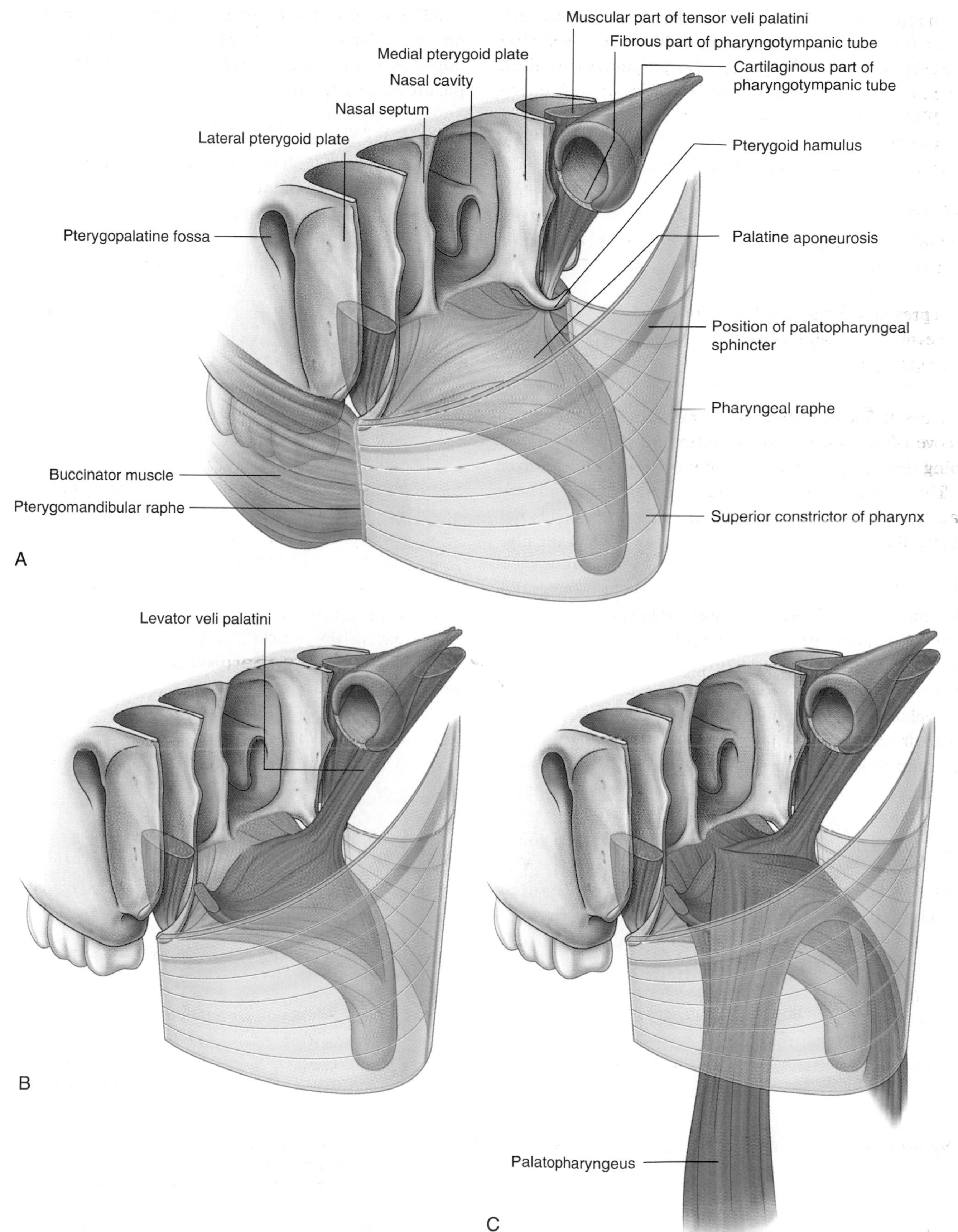


**Fig. 8.270** **A.** Tensor veli palatini muscles and the palatine aponeurosis. **B.** Levator veli palatini muscles. **C.** Palatopharyngeus muscles.

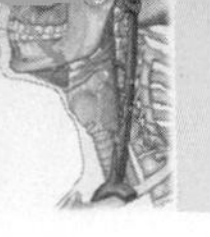



is the major structural element of the soft palate to which the other muscles of the palate attach.

The tensor veli palatini:

- tenses (makes firm) the soft palate so that the other muscles attached to the palate can work more effectively, and
- opens the pharyngotympanic tube when the palate moves during yawning and swallowing as a result of its attachment superiorly to the membranous part of the pharyngotympanic tube.

The tensor veli palatini is innervated by the nerve to the medial pterygoid from the mandibular nerve [$V_3$].

### Levator veli palatini

The levator veli palatini muscle originates from the base of the skull and descends to the upper surface of the palatine aponeurosis (Fig. 8.270B). On the skull, it originates from a roughened area on the petrous part of the temporal bone immediately anterior to the opening of the carotid canal. Some fibers also originate from adjacent parts of the pharyngotympanic tube.

The levator veli palatini passes anteroinferiorly through fascia of the pharyngeal wall, passes medial to the pharyngotympanic tube, and inserts onto the palatine aponeurosis (Fig. 8.270B). Its fibers interlace at the midline with those of the levator veli palatini on the other side.

Unlike the tensor veli palatini muscles, the levator veli palatini muscles do not pass around each pterygoid hamulus, but course directly from the base of the skull to the upper surface of the palatine aponeurosis. Therefore, they are the only muscles that can elevate the palate above the neutral position and close the pharyngeal isthmus between the nasopharynx and oropharynx.

The levator veli palatini is innervated by the vagus nerve [X] through the pharyngeal branch to the pharyngeal plexus. Clinically, the levator veli palatini can be tested by asking a patient to say "ah." If the muscle on each side is functioning normally, the palate elevates evenly in the midline. If one side is not functioning, the palate deviates away from the abnormal side.

### Palatopharyngeus

The palatopharyngeus muscle originates from the superior surface of the palatine aponeurosis and passes posterolaterally over its margin to descend and become one of the longitudinal muscles of the pharyngeal wall (Fig. 8.270C). It is attached to the palatine aponeurosis by two flat lamellae separated by the levator veli palatini muscle. The more anterior and lateral of these two lamellae is attached to the posterior margin of the hard palate as well as to the palatine aponeurosis.

The two palatopharyngeus muscles, one on each side, underlie the **palatopharyngeal arches** on the oropharyngeal wall. The palatopharyngeal arches lie posterior and medial to the **palatoglossal arches** when viewed anteriorly through the oral cavity (Fig. 8.271).

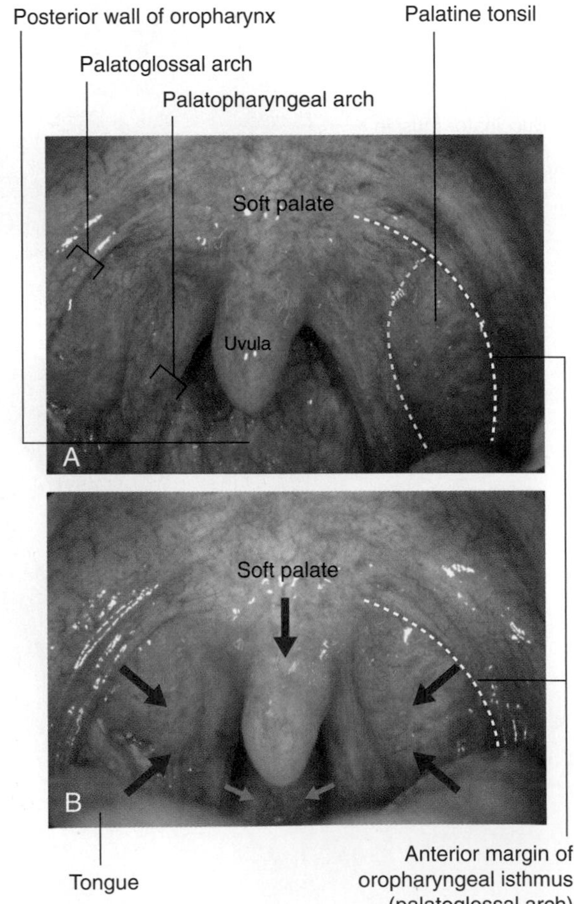


**Fig. 8.271** Open mouth with soft palate. **A.** Oropharyngeal isthmus opened. **B.** Oropharyngeal isthmus closed.

On each side, the palatine tonsil is between the palatopharyngeal and palatoglossal arches on the lateral oropharyngeal wall (Fig. 8.271A).

The palatopharyngeus muscles:

- depress the palate and move the palatopharyngeal arches toward the midline like curtains—both these actions help close the oropharyngeal isthmus; and
- elevate the pharynx during swallowing.

The palatopharyngeus is innervated by the vagus nerve [X] through the pharyngeal branch to the pharyngeal plexus.

## Palatoglossus

The palatoglossus muscle attaches to the inferior (oral) surface of the palatine aponeurosis and passes inferiorly and anteriorly into the lateral surface of the tongue (Fig. 8.272).

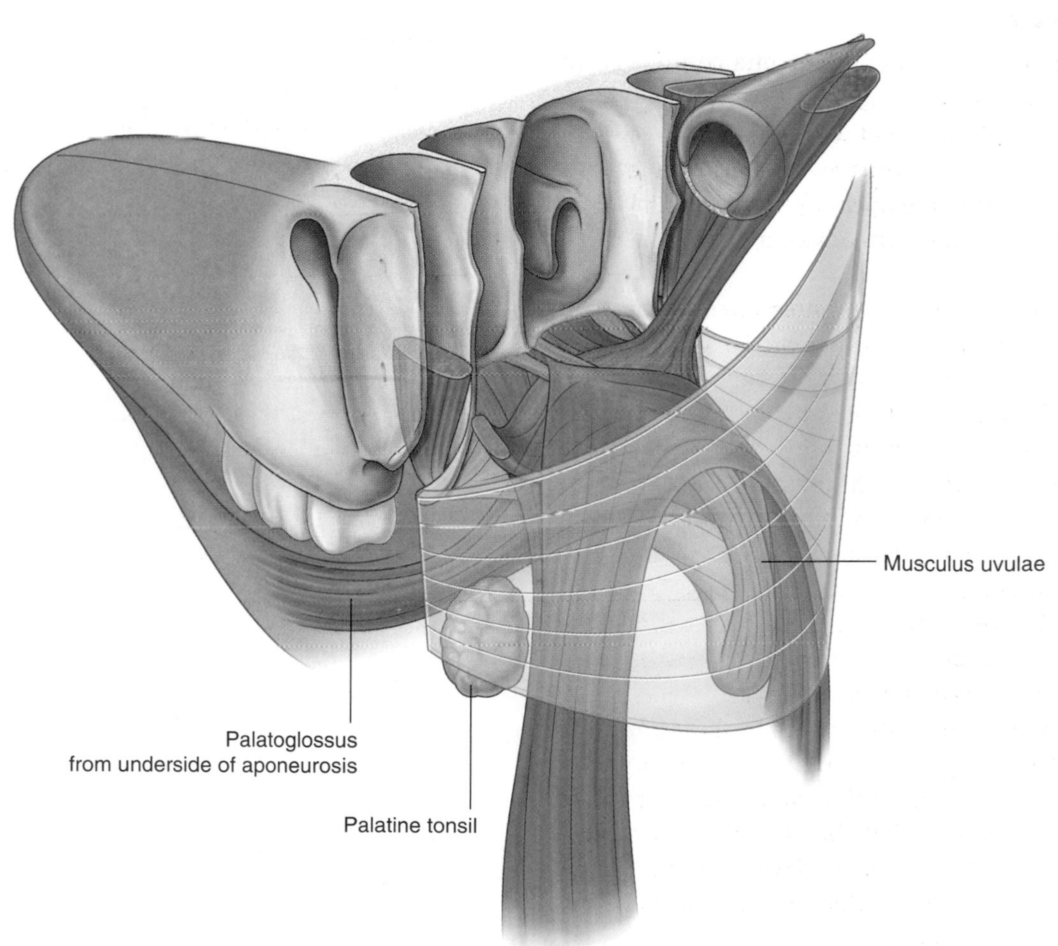


**Fig. 8.272** Palatoglossus muscles and musculus uvulae.

The palatoglossus muscle underlies a fold of mucosa that arches from the soft palate to the tongue. These palatoglossal arches, one on each side, are lateral and anterior to the palatopharyngeal arches and define the lateral margins of the oropharyngeal isthmus (Fig. 8.271A).

The palatine tonsil is between the palatoglossal and palatopharyngeal arches on the lateral oropharyngeal wall (Figs. 8.271 and 8.272).

The palatoglossus muscles depress the palate, move the palatoglossal arches toward the midline like curtains, and elevate the back of the tongue. These actions help close the oropharyngeal isthmus.

The palatoglossus is innervated by the vagus nerve [X] through the pharyngeal branch to the pharyngeal plexus.

#### Musculus uvulae

The musculus uvulae originates from the posterior nasal spine on the posterior margin of the hard palate and passes directly posteriorly over the dorsal aspect of the palatine aponeurosis to insert into connective tissue underlying the mucosa of the uvula (Fig. 8.272). It passes between the two lamellae of the palatopharyngeus superior to the attachment of the levator veli palatini. Along the midline, the musculus uvulae blends with its partner on the other side.

The musculus uvulae elevates and retracts the uvula. This action thickens the central part of the soft palate and helps the levator veli palatini muscles close the pharyngeal isthmus between the nasopharynx and oropharynx.

The musculus uvulae is innervated by the vagus nerve [X] through the pharyngeal branch to the pharyngeal plexus.

## Vessels

### Arteries

Arteries of the palate include the greater palatine branch of the maxillary artery, the ascending palatine branch of the facial artery, and the palatine branch of the ascending pharyngeal artery. The maxillary, facial, and ascending pharyngeal arteries are all branches that arise in the neck from the external carotid artery (Fig. 8.273).

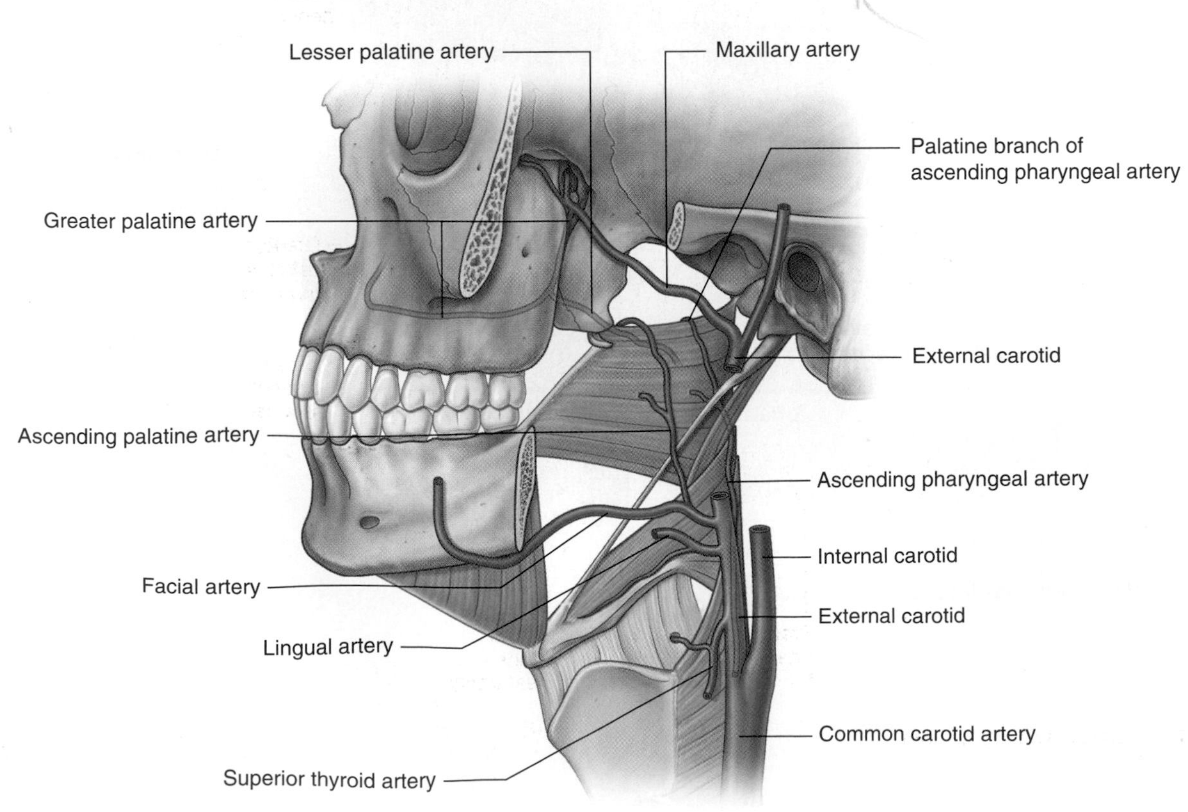


**Fig. 8.273** Arteries of the palate.

### Ascending palatine artery and palatine branch

The **ascending palatine artery** of the facial artery ascends along the external surface of the pharynx. The palatine branch loops medially over the top of the superior constrictor muscle of the pharynx to penetrate the pharyngeal fascia with the levator veli palatini muscle and follow the levator veli palatini to the soft palate.

The **palatine branch** of the ascending pharyngeal artery follows the same course as the palatine branch of the ascending palatine artery from the facial artery and may replace the vessel.

### Greater palatine artery

The **greater palatine artery** originates from the maxillary artery in the pterygopalatine fossa. It descends into the palatine canal where it gives origin to a small **lesser palatine branch**, and then continues through the greater palatine foramen onto the inferior surface of the hard palate (Fig. 8.274). The greater palatine artery passes forward on the hard palate and then leaves the palate superiorly through the incisive canal to enter the medial wall of the nasal cavity where it terminates. The greater palatine artery is the major artery of the hard palate. It also supplies palatal gingiva. The lesser palatine branch passes through the lesser palatine foramen just posterior to the greater palatine foramen, and contributes to the vascular supply of the soft palate.

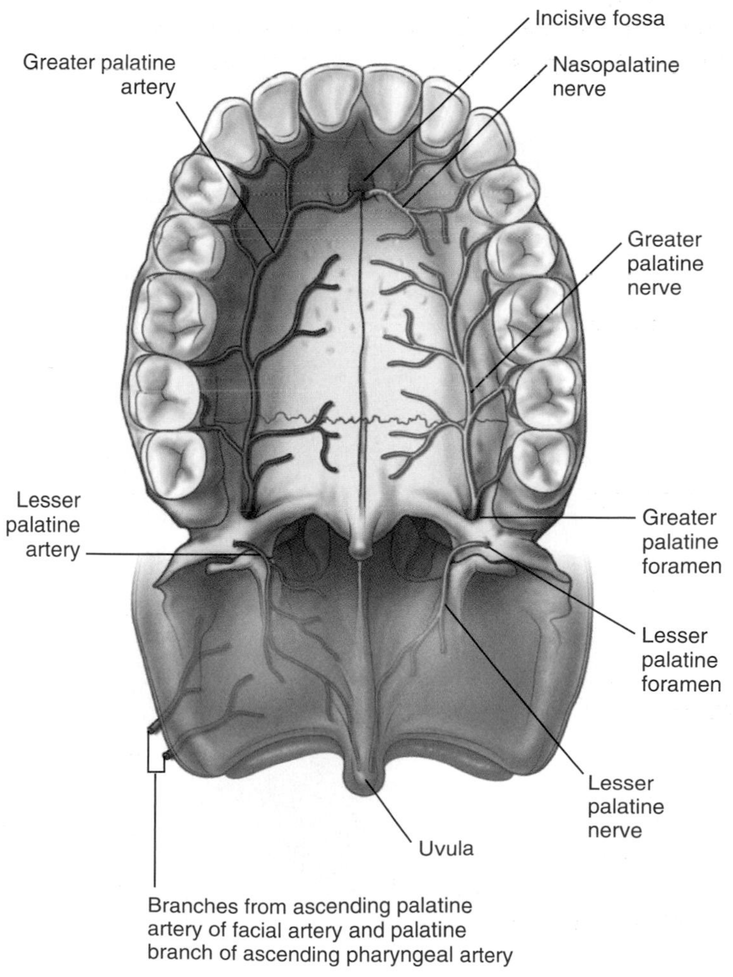


**Fig. 8.274** Palatine nerves and arteries.

### Veins

Veins from the palate generally follow the arteries and ultimately drain into the pterygoid plexus of veins in the infratemporal fossa (Fig. 8.275; also see pp. 980–981), or into a network of veins associated with the palatine tonsil, which drain into the pharyngeal plexus of veins or directly into the facial vein.

### Lymphatics

Lymphatic vessels from the palate drain into deep cervical nodes (Fig. 8.275).

## Innervation

The palate is supplied by the greater and lesser palatine nerves and the nasopalatine nerve (Figs. 8.274 and 8.276).

General sensory fibers carried in all these nerves originate in the pterygopalatine fossa from the maxillary nerve [$V_2$].

Parasympathetic (to glands) and SA (taste on soft palate) fibers from a branch of the facial nerve [VII] join the nerves in the pterygopalatine fossa, as do the sympathetics (mainly to blood vessels) ultimately derived from the T1 spinal cord level.

### Greater and lesser palatine nerves

The greater and lesser palatine nerves descend through the pterygopalatine fossa and palatine canal to reach the palate (Fig. 8.276):

- The greater palatine nerve travels through the greater palatine foramen and turns anteriorly to supply the hard palate and gingiva as far as the first premolar.

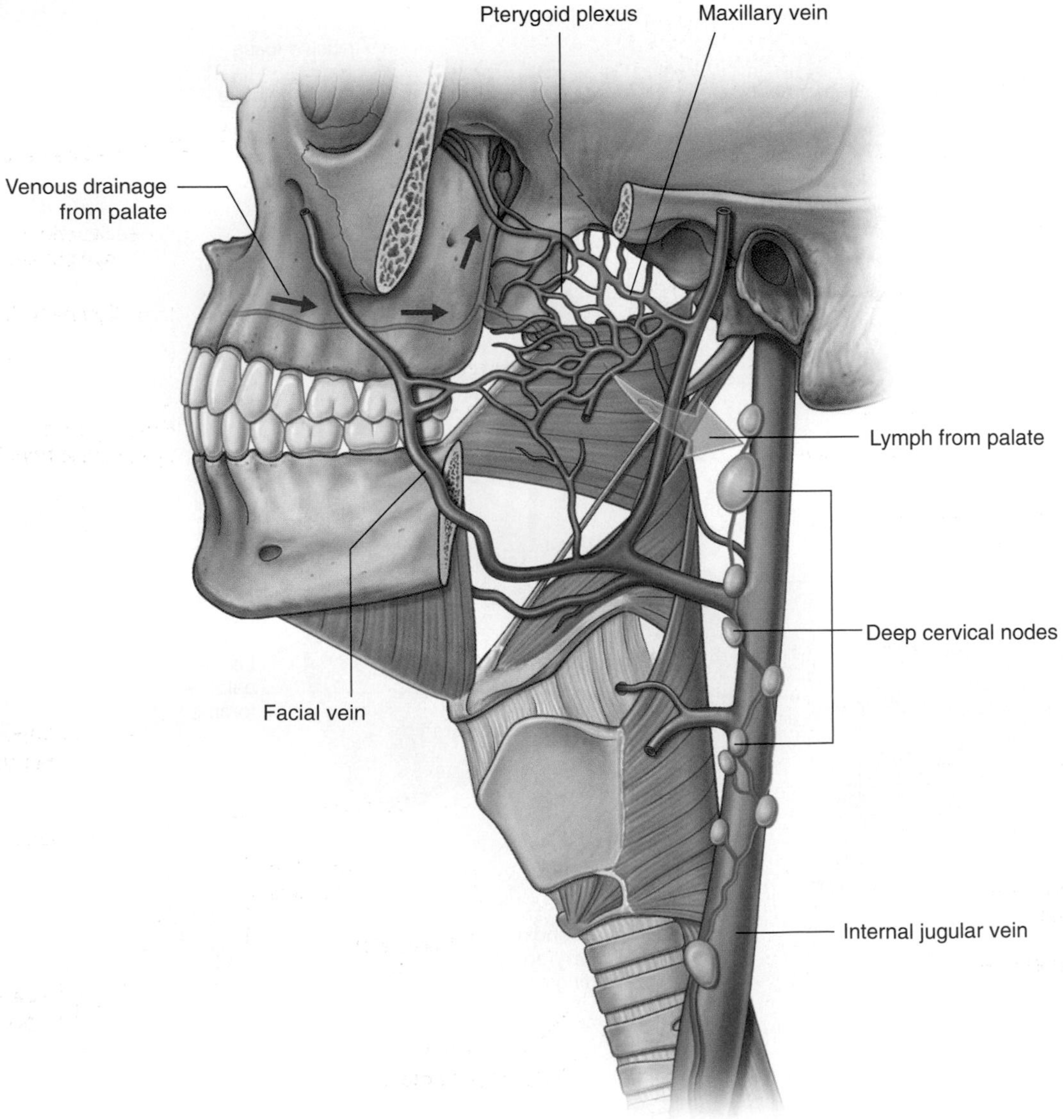

**Fig. 8.275** Venous and lymphatic drainage of the palate.

- The lesser palatine nerve passes posteromedially to supply the soft palate.

### Nasopalatine nerve

The nasopalatine nerve also originates in the pterygopalatine fossa, but passes medially into the nasal cavity. It continues medially over the roof of the nasal cavity to reach the medial wall, then anteriorly and obliquely down the wall to reach the incisive canal in the anterior floor, and descends through the incisive canal and fossa to reach the inferior surface of the hard palate (Fig. 8.276).

The nasopalatine nerve supplies gingiva and mucosa adjacent to the incisors and canine.

## Oral fissure and lips

The oral fissure is the slit-like opening between the lips that connects the oral vestibule to the outside (Fig. 8.277). It

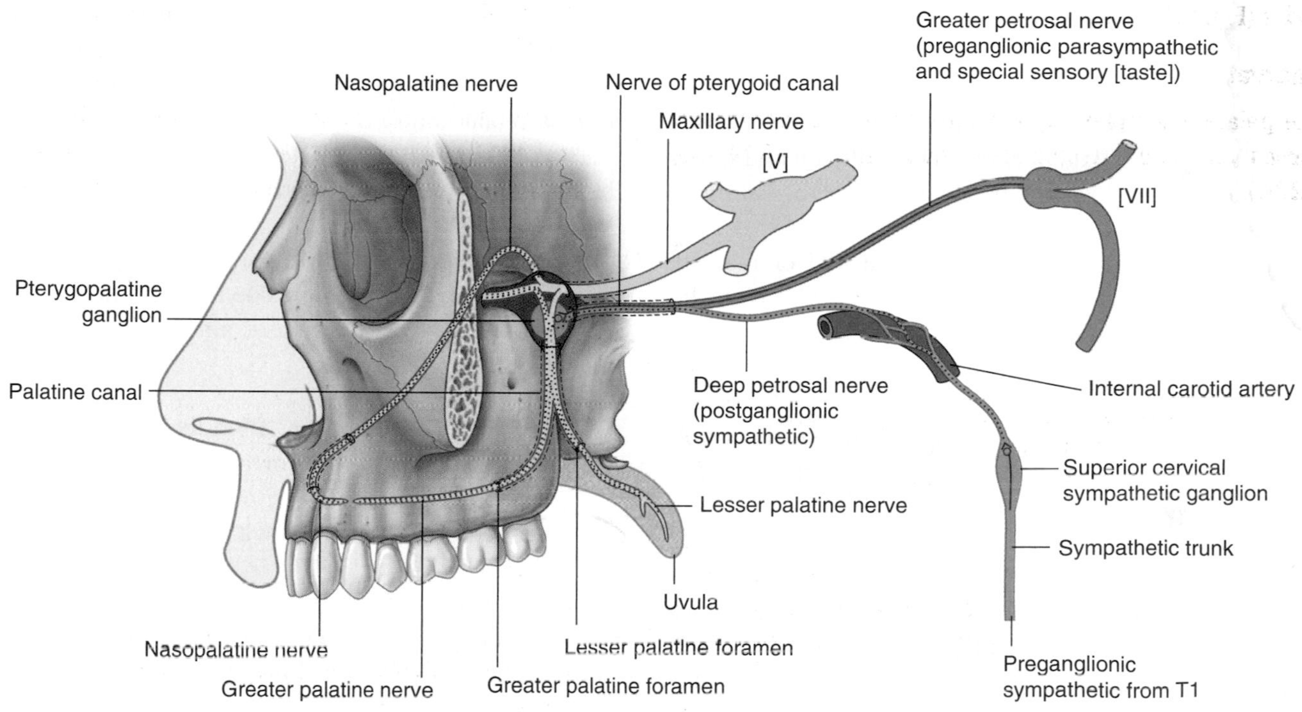


**Fig. 8.276** Innervation of the palate.

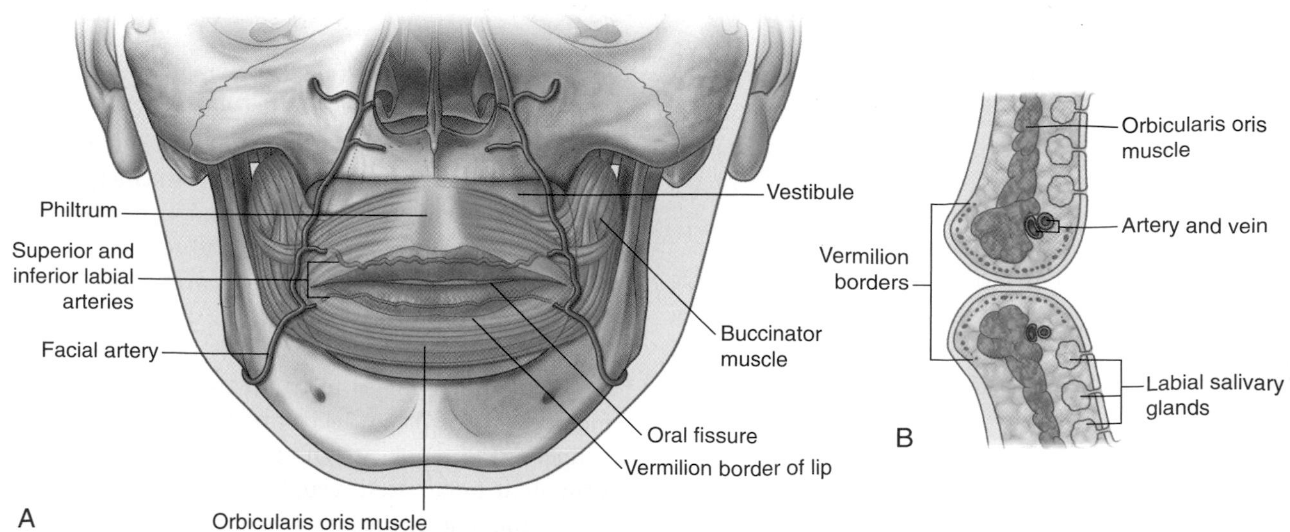


**Fig. 8.277** Oral fissure and lips. **A.** Anterior view. **B.** Sagittal section.

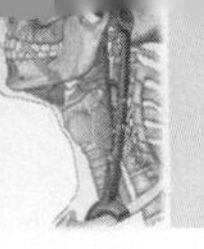



can be opened and closed, and altered in shape by the movements of the muscles of facial expression associated with the lips and surrounding regions, and by movements of the lower jaw (mandible).

The **lips** are entirely composed of soft tissues (Fig. 8.277B). They are lined internally by oral mucosa and covered externally by skin. Externally, there is an area of transition from the thicker skin that covers the face to the thinner skin that overlies the margins of the lips and continues as oral mucosa onto the deep surfaces of the lips.

Blood vessels are closer to the surface in areas where the skin is thin and as a consequence there is a vermilion border that covers the margins of the lips.

The upper lip has a shallow vertical groove on its external surface (the **philtrum**) sandwiched between two elevated ridges of skin (Fig. 8.277A). The philtrum and ridges are formed embryologically by fusion of the medial nasal processes.

On the inner surface of both lips, a fold of mucosa (the **median labial frenulum**) connects the lip to the adjacent gum.

The lips enclose the orbicularis oris muscle, neurovascular tissues, and labial glands (Fig. 8.277B). The small pea-shaped labial glands are between the muscle tissue and the oral mucosa and open into the oral vestibule.

A number of muscles of facial expression control the shape and size of the oral fissure. The most important of these is the orbicularis oris muscle, which encircles the orifice and acts as a sphincter. A number of other muscles of facial expression blend into the orbicularis oris or other tissues of the lips and open or adjust the contours of the oral fissure. These include the buccinator, levator labii superioris, zygomaticus major and minor, levator anguli oris, depressor labii inferioris, depressor anguli oris, and platysma (see pp. 897–899).

## Oropharyngeal isthmus

The oropharyngeal isthmus is the opening between the oral cavity and the oropharynx (see Fig. 8.271). It is formed:

- laterally by the palatoglossal arches;
- superiorly by the soft palate; and
- inferiorly by the sulcus terminalis of the tongue that divides the oral surface of the tongue (anterior two-thirds) from the pharyngeal surface (posterior one-third).

The oropharyngeal isthmus can be closed by elevation of the posterior aspect of the tongue, depression of the palate, and medial movement of the palatoglossal arches toward the midline.

Medial movement of the palatopharyngeal arches medial and posterior to the palatoglossal arches is also involved in closing the oropharyngeal isthmus. By closing the oropharyngeal isthmus, food or liquid can be held in the oral cavity while breathing.

## Teeth and gingivae

The **teeth** are attached to sockets (alveoli) in two elevated arches of bone on the mandible below and the maxillae above (alveolar arches). If the teeth are removed, the alveolar bone is resorbed and the arches disappear.

The **gingivae** (**gums**) are specialized regions of the oral mucosa that surround the teeth and cover adjacent regions of the alveolar bone.

The different types of teeth are distinguished on the basis of morphology, position, and function (Fig. 8.278A).

In adults, there are 32 teeth, 16 in the upper jaw and 16 in the lower jaw. On each side in both maxillary and mandibular arches are two incisor, one canine, two premolar, and three molar teeth.

- The **incisor teeth** are the "front teeth" and have one root and a chisel-shaped crown, which "cuts."
- The **canine teeth** are posterior to the incisors, are the longest teeth, have a crown with a single pointed cusp, and "grasp."
- The **premolar teeth** (bicuspids) have a crown with two pointed cusps, one on the buccal (cheek) side of the tooth and the other on the lingual (tongue) or palatal (palate) side, generally have one root (but the upper first premolar next to the canine may have two), and "grind."
- The **molar teeth** are behind the premolar teeth, have three roots and crowns with three to five cusps, and "grind."

Two successive sets of teeth develop in humans, deciduous teeth ("baby" teeth) (Fig. 8.278B) and permanent teeth ("adult" teeth). The deciduous teeth emerge from the gingivae at between six months and two years of age. Permanent teeth begin to emerge and replace the deciduous teeth at around age six years, and can continue to emerge into adulthood.

The 20 deciduous teeth consist of two incisor, one canine, and two molar teeth on each side of the upper and lower jaws. These teeth are replaced by the incisor, canine, and premolar teeth of the permanent teeth. The permanent molar teeth erupt posterior to the deciduous molars and require the jaws to elongate forward to accommodate them.

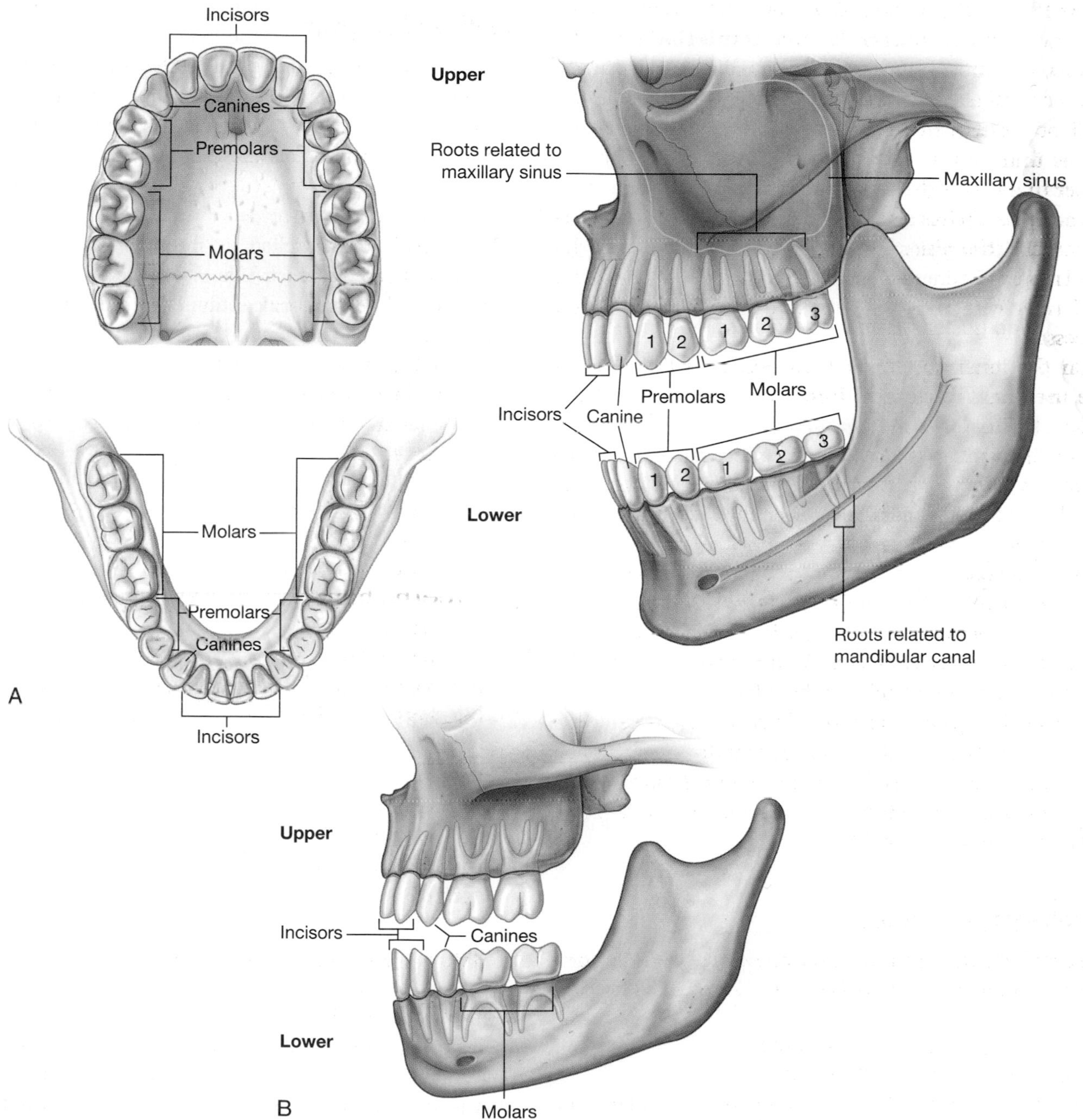


**Fig. 8.278** Teeth. **A.** Adult upper and lower permanent teeth. **B.** Deciduous ("baby") teeth.

## Vessels

### Arteries

All teeth are supplied by vessels that branch either directly or indirectly from the maxillary artery (Fig. 8.279).

#### Inferior alveolar artery

All lower teeth are supplied by the **inferior alveolar artery**, which originates from the maxillary artery in the infratemporal fossa. The vessel enters the mandibular canal of the mandible, passes anteriorly in bone supplying vessels to the more posterior teeth, and divides opposite the first premolar into **incisor** and **mental branches**. The mental branch leaves the mental foramen to supply the chin, while the incisor branch continues in bone to supply the anterior teeth and adjacent structures.

#### Anterior and posterior superior alveolar arteries

All upper teeth are supplied by anterior and posterior superior alveolar arteries.

The **posterior superior alveolar artery** originates from the maxillary artery just after the maxillary artery enters the pterygopalatine fossa and it leaves the fossa through the pterygomaxillary fissure. It descends on the posterolateral surface of the maxilla, branches, and enters small canals in the bone to supply the molar and premolar teeth.

The **anterior superior alveolar artery** originates from the infra-orbital artery, which arises from the maxillary artery in the pterygopalatine fossa. The infra-orbital artery leaves the pterygopalatine fossa through the inferior orbital fissure and enters the inferior orbital groove and canal in the floor of the orbit. The anterior superior alveolar artery originates from the infra-orbital artery in the infra-orbital canal. It passes through bone and branches to supply the incisor and canine teeth.

#### Gingival supply

The gingivae are supplied by multiple vessels and the source depends on which side of each tooth the gingiva is—the side facing the oral vestibule or cheek (vestibular or buccal side), or the side facing the tongue or palate (lingual or palatal side):

- Buccal gingiva of the lower teeth is supplied by branches from the inferior alveolar artery, whereas the lingual side is supplied by branches from the lingual artery of the tongue.

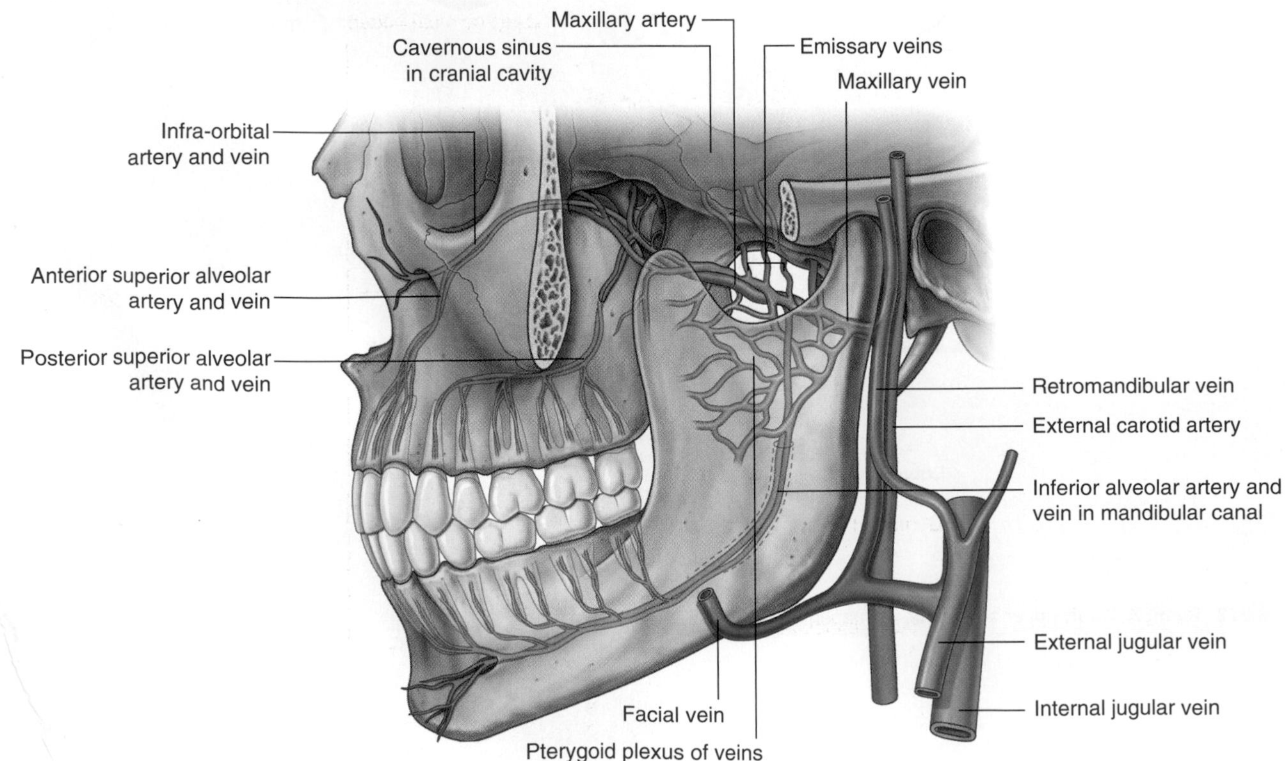


**Fig. 8.279** Arteries and veins of the teeth.

- Buccal gingiva of the upper teeth is supplied by branches of the anterior and posterior superior alveolar arteries.
- Palatal gingiva is supplied by branches from the nasopalatine (incisor and canine teeth) and greater palatine (premolar and molar teeth) arteries.

### Veins

Veins from the upper and lower teeth generally follow the arteries (Fig. 8.279).

Inferior alveolar veins from the lower teeth, and superior alveolar veins from the upper teeth drain mainly into the pterygoid plexus of veins in the infratemporal fossa, although some drainage from the anterior teeth may be via tributaries of the facial vein.

The pterygoid plexus drains mainly into the maxillary vein and ultimately into the retromandibular vein and jugular system of veins. In addition, small communicating vessels pass superiorly, from the plexus, and pass through small emissary foramina in the base of the skull to connect with the cavernous sinus in the cranial cavity. Infection originating in the teeth can track into the cranial cavity through these small emissary veins.

Venous drainage from the teeth can also be via vessels that pass through the mental foramen to connect with the facial vein.

Veins from the gingivae also follow the arteries and ultimately drain into the facial vein or into the pterygoid plexus of veins.

### Lymphatics

Lymphatic vessels from the teeth and gingivae drain mainly into submandibular, submental, and deep cervical nodes (Fig. 8.280).

## Innervation

All nerves that innervate the teeth and gingivae are branches of the trigeminal nerve [V] (Figs. 8.281 and 8.282).

### Inferior alveolar nerve

The lower teeth are all innervated by branches from the inferior alveolar nerve, which originates in the infratemporal fossa from the mandibular nerve [$V_3$] (Figs. 8.281 and 8.282). The inferior alveolar nerve and its accompanying vessels enter the mandibular foramen on the medial surface

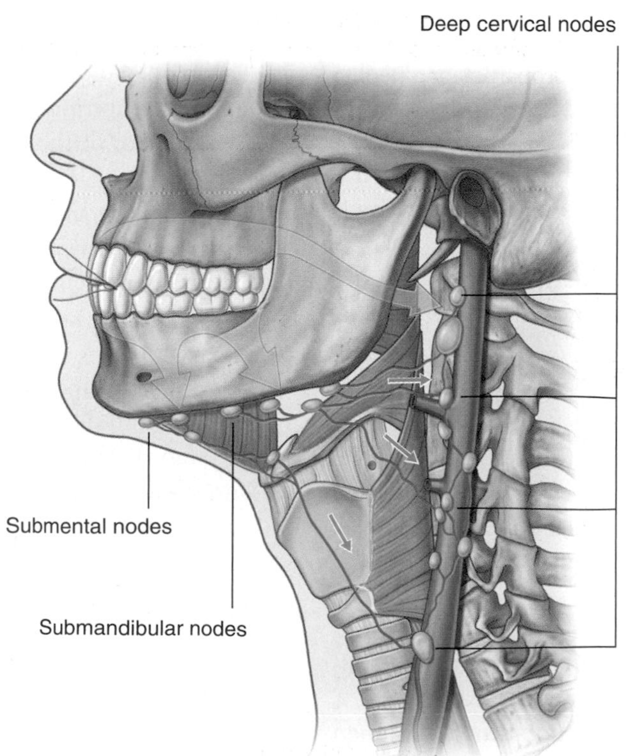


**Fig. 8.280** Lymphatic drainage of the teeth and gums.

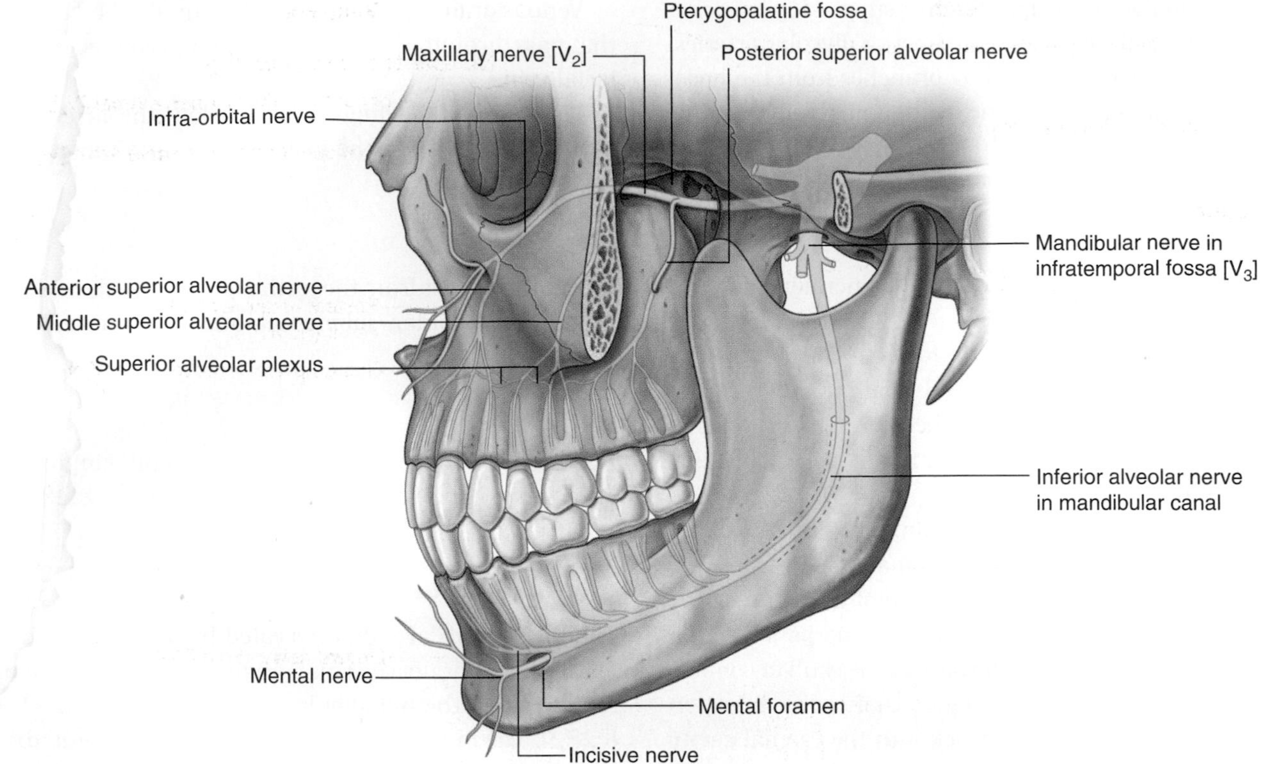


**Fig. 8.281** Innervation of the teeth.

of the ramus of the mandible and travel anteriorly through the bone in the mandibular canal. Branches to the back teeth originate directly from the inferior alveolar nerve.

Adjacent to the first premolar tooth, the inferior alveolar nerve divides into incisive and mental branches:

- The **incisive branch** innervates the first premolar, the canine, and the incisor teeth, together with the associated vestibular (buccal) gingiva.
- The **mental nerve** exits the mandible through the mental foramen and innervates the chin and lower lip.

### Anterior, middle, and posterior superior alveolar nerves

All upper teeth are innervated by the anterior, middle, and posterior superior alveolar nerves, which originate directly or indirectly from the maxillary nerve [$V_2$] (Figs. 8.281 and 8.282).

The posterior superior alveolar nerve originates directly from the maxillary nerve [$V_2$] in the pterygopalatine fossa, exits the pterygopalatine fossa through the pterygomaxillary fissure, and descends on the posterolateral surface of the maxilla. It enters the maxilla through a small foramen approximately midway between the pterygomaxillary fissure and the last molar tooth, and passes through the bone in the wall of the maxillary sinus. The posterior superior alveolar nerve then innervates the molar teeth through the superior alveolar plexus formed by the posterior, middle, and anterior alveolar nerves.

The middle and anterior superior alveolar nerves originate from the infra-orbital branch of the maxillary nerve [$V_2$] in the floor of the orbit:

- The middle superior alveolar nerve arises from the infra-orbital nerve in the infra-orbital groove, passes through the bone in the lateral wall of the maxillary sinus, and innervates the premolar teeth via the superior alveolar plexus.
- The anterior superior alveolar nerve originates from the infra-orbital nerve in the infra-orbital canal, passes through the maxilla in the anterior wall of the maxillary sinus, and via the superior alveolar plexus, supplies the canine and incisor teeth.

### Innervation of gingivae

Like the teeth, the gingivae are innervated by nerves that ultimately originate from the trigeminal nerve [V] (Fig. 8.282):

- Gingiva associated with the upper teeth is innervated by branches derived from the maxillary nerve [$V_2$].

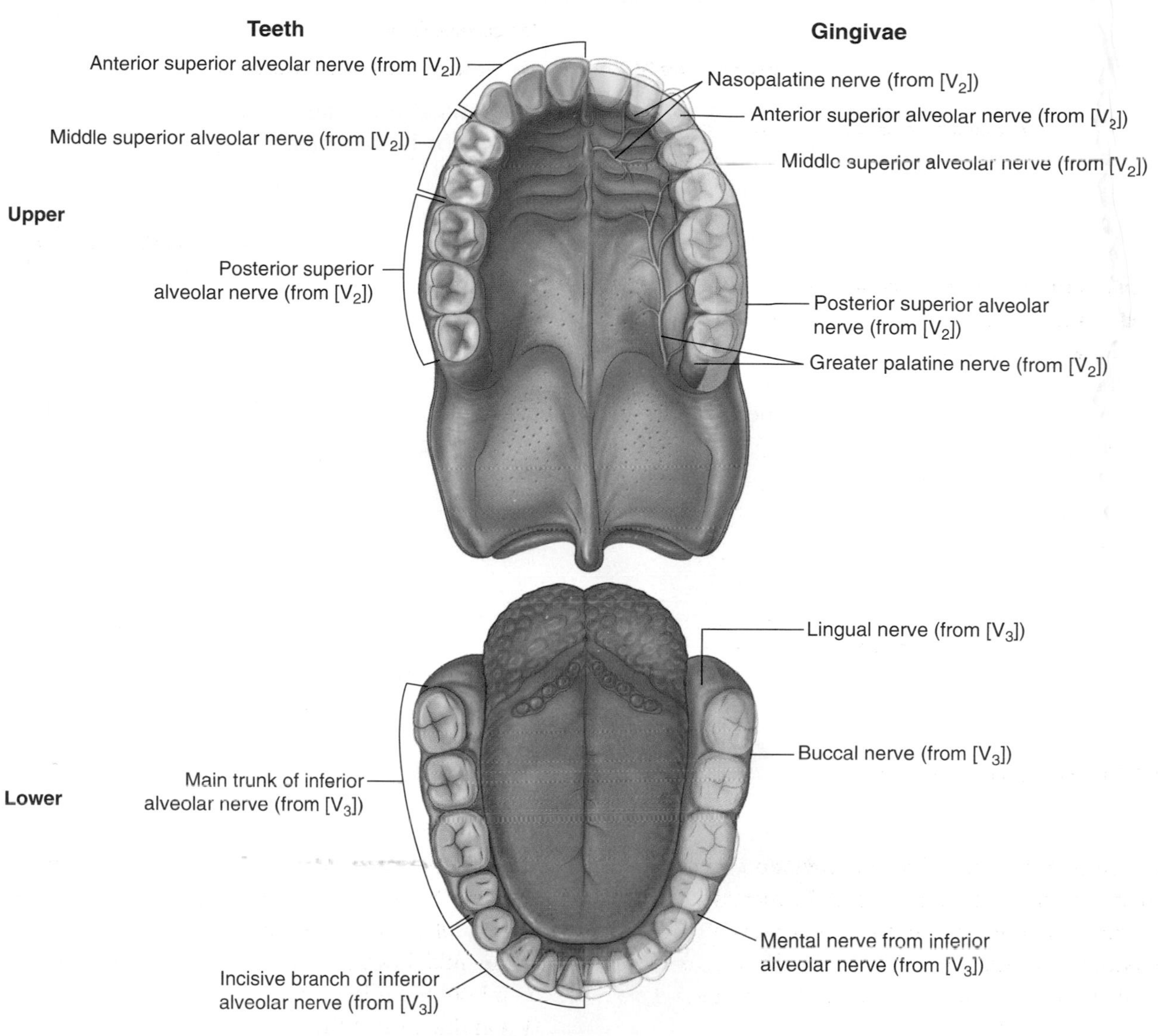


**Fig. 8.282** Innervation of the teeth and gums.

- Gingiva associated with the lower teeth is innervated by branches of the mandibular nerve [$V_3$].

The gingiva on the buccal side of the upper teeth is innervated by the anterior, middle, and superior alveolar nerves, which also innervate the adjacent teeth. Gingiva on the palatal (lingual) side of the same teeth is innervated by the nasopalatine and the greater palatine nerves:

- The nasopalatine nerve innervates gingiva associated with the incisor and canine teeth.
- The greater palatine nerve supplies gingiva associated with the remaining teeth.

The gingiva associated with the (buccal) side of the mandibular incisor, canine, and premolar teeth is innervated by the mental branch of the inferior alveolar nerve. Gingiva on the buccal side of the mandibular molar teeth is innervated by the buccal nerve, which originates in the infratemporal fossa from the mandibular nerve [$V_3$]. Gingiva adjacent to the lingual surface of all lower teeth is innervated by the lingual nerve.

## In the clinic

### Head and neck cancer

Most cancers of the oral cavity, oropharynx, nasopharynx, larynx, sinuses, and salivary glands arise from the epithelial cells that line them, resulting in squamous cell carcinoma. The majority of these are related to cell damage caused by smoking and alcohol use. Certain viruses are also related to cancers in the head and neck, including human papillomavirus (HPV) and Epstein-Barr virus (EBV).

# Surface anatomy

## Head and neck surface anatomy

Skeletal landmarks in the head and neck are used for locating major blood vessels, glands, and muscles, and for locating points of access to the airway.

Neurological examination of the cranial and upper cervical nerves is carried out by assessing function in the head and neck.

In addition, information about the general status of body health can often be obtained by evaluating surface features, the eye and the oral cavity, and the characteristics of speech.

## Anatomical position of the head and major landmarks

The head is in the anatomical position when the inferior margins of the bony orbits and the superior margins of the external acoustic meatuses are in the same horizontal plane (Frankfort plane).

In addition to the external acoustic meatus and the bony margin of the orbit, other features that are palpable include the head of the mandible, zygomatic arch, zygomatic bone, mastoid process, and external occipital protuberance (Fig. 8.283).

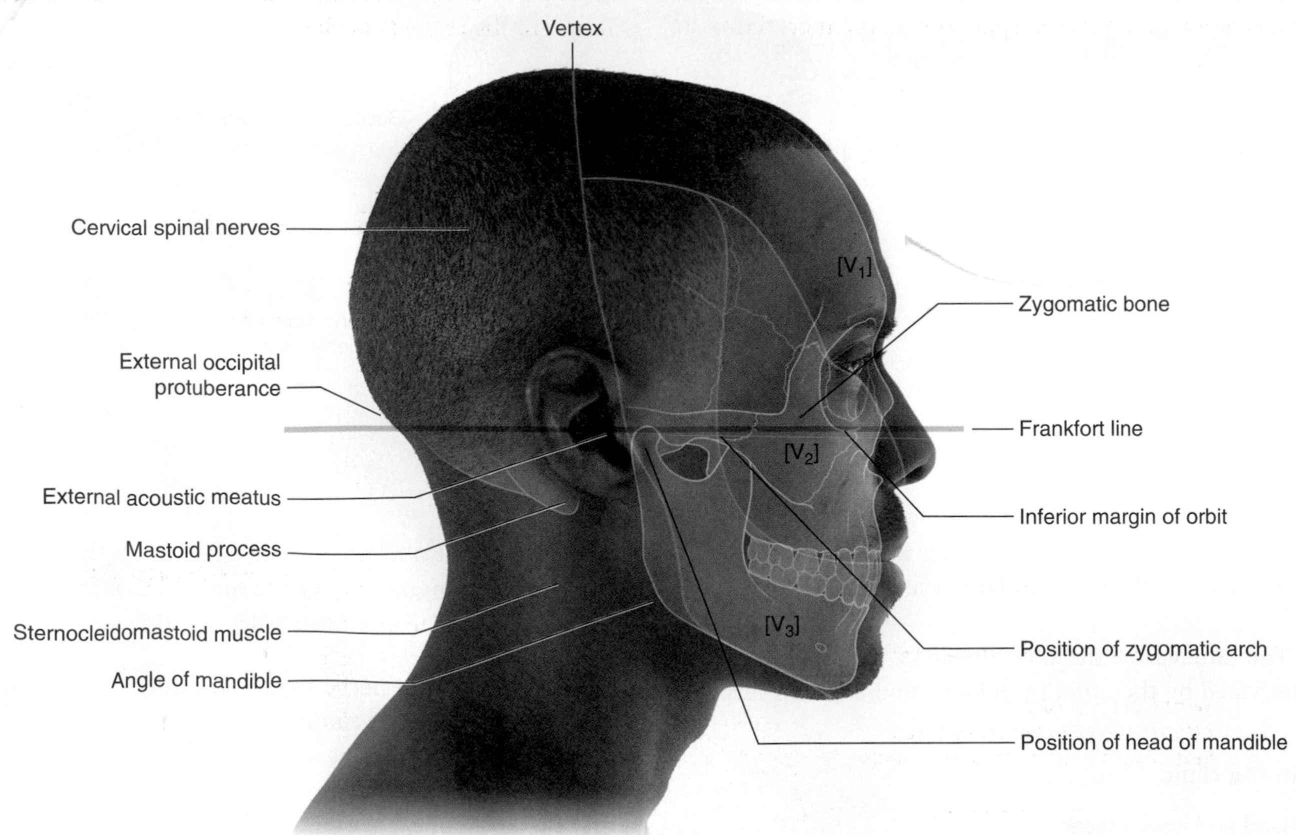


**Fig. 8.283** Anatomical position of the head and major landmarks. Lateral head and neck of a man.

The head of the mandible is anterior to the external ear and behind and inferior to the posterior end of the zygomatic arch. It is best found by opening and closing the jaw and palpating the head of the mandible as it moves forward onto the articular tubercle and then back into the mandibular fossa, respectively.

The zygomatic arch extends forward from the region of the temporomandibular joint to the zygomatic bone, which forms a bony prominence lateral to the inferior margin of the anterior opening of the orbit.

The mastoid process is a large bony protuberance that is easily palpable posterior to the inferior aspect of the external acoustic meatus. The superior end of the sternocleidomastoid muscle attaches to the mastoid process.

The external occipital protuberance is palpable in the midline posteriorly where the contour of the skull curves sharply forward. This landmark marks the point superficially where the back of the neck joins the head.

Another clinically useful feature of the head is the vertex. This is the highest point of the head in the anatomical position and marks the approximate point on the scalp where there is a transition from cervical to cranial innervation of the scalp. Anterior to the vertex, the scalp and face are innervated by the trigeminal nerve [V]. Posterior to the vertex, the scalp is innervated by branches from cervical spinal nerves.

## Visualizing structures at the CIII/CIV and CVI vertebral levels

Two vertebral levels in the neck are associated with important anatomical features (Fig. 8.284).

The intervertebral disc between the CIII and CIV vertebrae is in the same horizontal plane as the bifurcation of the common carotid artery into the internal and external carotid arteries. This level is approximately at the upper margin of the thyroid cartilage.

Vertebral level CVI marks the transition from pharynx to esophagus and larynx to trachea. The CVI vertebral level therefore marks the superior ends of the esophagus and trachea and is approximately at the level of the inferior margin of the cricoid cartilage.

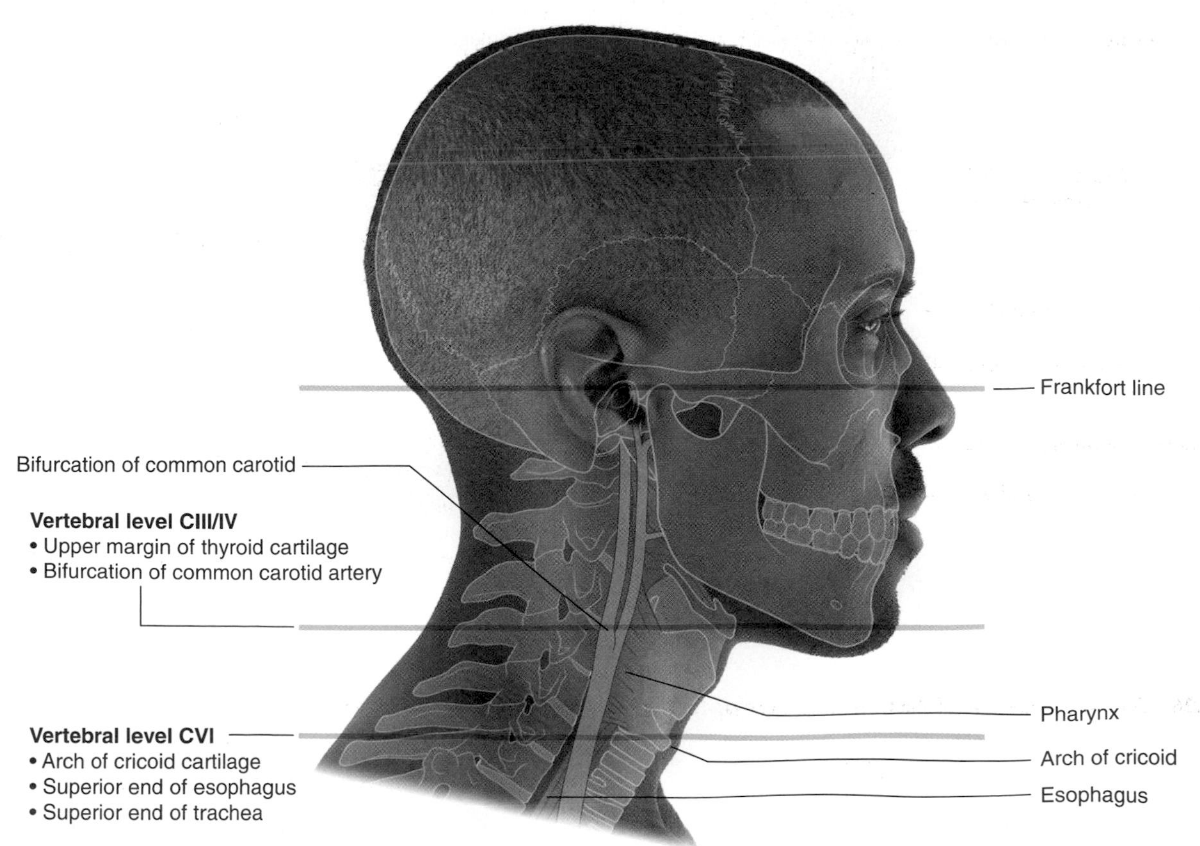


**Fig. 8.284** Visualizing structures at the CIII/IV and CVI vertebral levels. Lateral head and neck of a man.

## How to outline the anterior and posterior triangles of the neck

The boundaries of the anterior and posterior triangles on each side of the neck are easily established using readily visible bony and muscular landmarks (Fig. 8.285).

The base of each anterior triangle is the inferior margin of the mandible, the anterior margin is the midline of the neck, and the posterior margin is the anterior border of the sternocleidomastoid muscle. The apex of each anterior triangle points inferiorly and is at the suprasternal notch.

The anterior triangles are associated with structures such as the airway and digestive tract, and nerves and vessels that pass between the thorax and head. They are also associated with the thyroid and parathyroid glands.

The base of each posterior triangle is the middle one-third of the clavicle. The medial margin is the posterior border of the sternocleidomastoid muscle, and the lateral margin is the anterior border of the trapezius muscle. The apex points superiorly and is immediately posteroinferior to the mastoid process.

The posterior triangles are associated with nerves and vessels that pass into and out of the upper limbs.

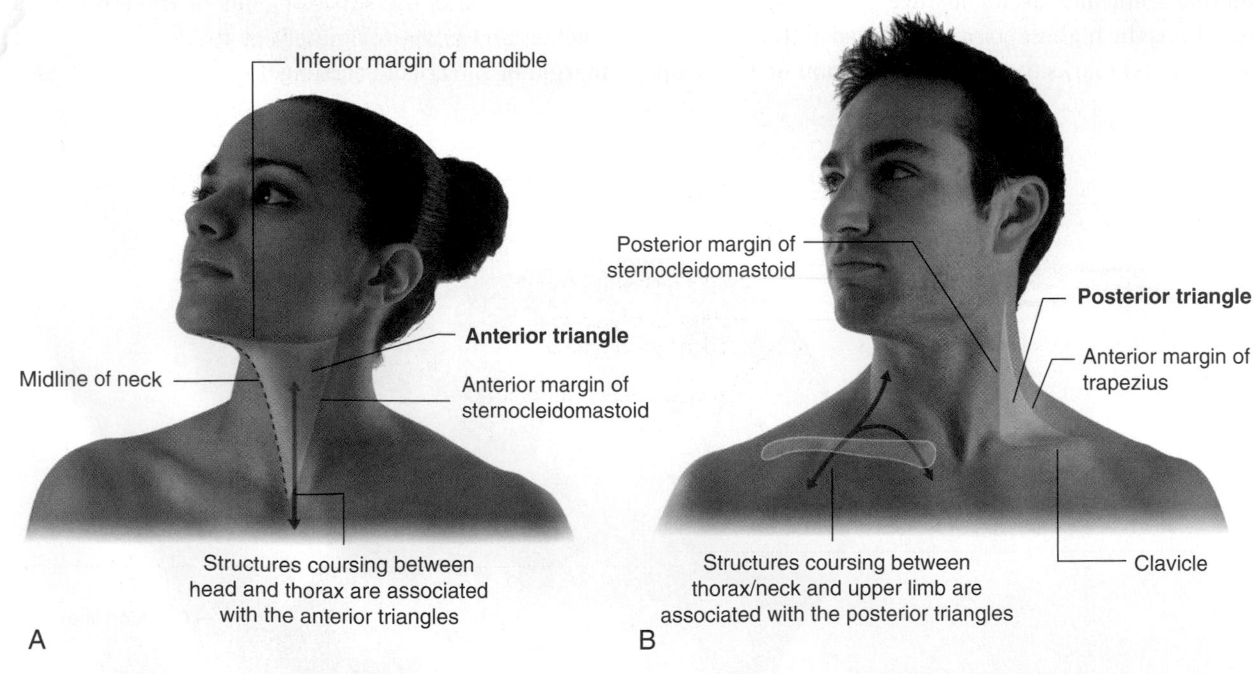


**Fig. 8.285** How to outline the anterior and posterior triangles of the neck. **A.** In a woman, anterolateral view. The left anterior triangle is indicated. **B.** In a man, anterior view of the posterior triangle.

## How to locate the cricothyroid ligament

An important structure to locate in the neck is the median cricothyroid ligament (Fig. 8.286) because artificial penetration of this membrane in emergency situations can provide access to the lower airway when the upper airway above the level of the vocal folds is blocked.

The ligament can be easily found using palpable features of the larynx as landmarks.

Using a finger to gently feel laryngeal structures in the midline, first find the thyroid notch in the superior margin of the thyroid cartilage and then move the finger inferiorly over the laryngeal prominence and down the anterior surface of the thyroid angle. As the finger crosses the inferior margin of the thyroid cartilage in the midline, a soft depression is felt before the finger slides onto the arch of the cricoid cartilage, which is hard.

The soft depression between the lower margin of the thyroid cartilage and the arch of the cricoid is the position of the median cricothyroid ligament.

A tube passed through the median cricothyroid ligament enters the airway just inferior to the position of the vocal folds of the larynx.

Structures that may occur in or cross the midline between the skin and the median cricothyroid ligament include the pyramidal lobe of the thyroid gland and small vessels, respectively.

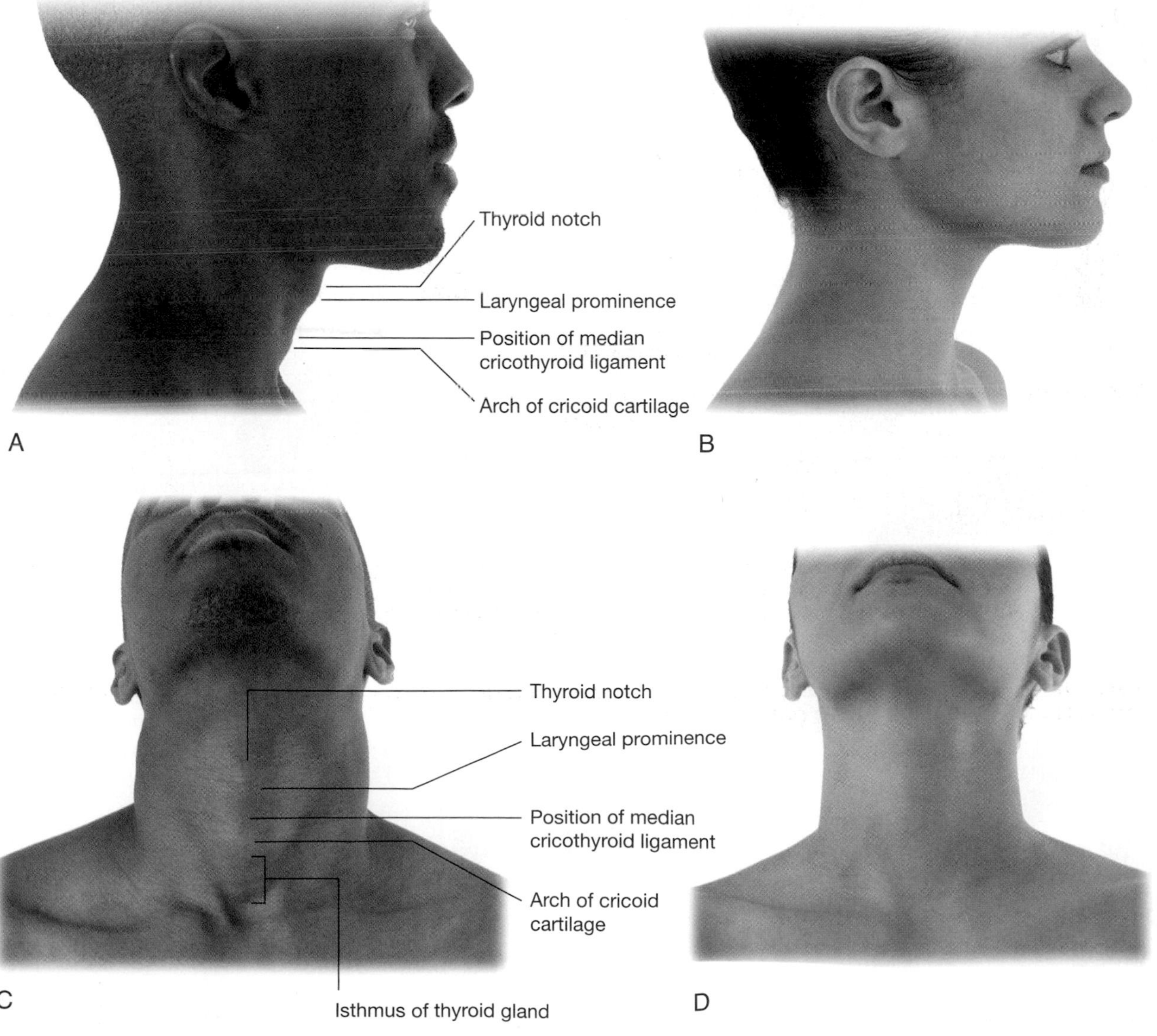


**Fig. 8.286** How to locate the median cricothyroid ligament. **A.** In a man, lateral view of head and neck. **B.** In a woman, lateral view of head and neck. **C.** In a man, anterior neck with the chin elevated. **D.** In a woman, anterior neck with the chin elevated.

Inferior to the cricoid cartilage, the upper cartilage of the larynx can sometimes be palpated above the level of the isthmus of the thyroid gland that crosses the trachea anteriorly.

The landmarks used for finding the cricothyroid ligament are similar in men and women; however, because the laminae of the thyroid cartilage meet at a more acute angle in men, the structures are more prominent in men than in women.

## How to find the thyroid gland

The left and right lobes of the thyroid gland are in the anterior triangles in the lower neck on either side of the airway and digestive tract inferior to the position of the oblique line of the thyroid cartilage (Fig. 8.287). In fact, the sternothyroid muscles, which attach superiorly to the oblique lines, lie anterior to the lobes of the thyroid gland and prevent the lobes from moving upward in the neck.

The lobes of the thyroid gland can be most easily palpated by finding the thyroid prominence and arch of the cricoid cartilage and then feeling posterolateral to the larynx.

The isthmus of the thyroid gland crosses anterior to the upper end of the trachea and can be easily palpated in the midline inferior to the arch of the cricoid.

The presence of the isthmus of the thyroid gland makes palpating the tracheal cartilages difficult in the neck. Also, the presence of the isthmus of the thyroid gland and the associated vessels found in and crossing the midline makes it difficult to artificially enter the airway anteriorly through the trachea. This procedure, a tracheostomy, is a surgical procedure.

## Estimating the position of the middle meningeal artery

The middle meningeal artery (Fig. 8.288) is a branch of the maxillary artery in the infratemporal fossa. It enters the skull through the foramen spinosum and is within the dura mater lining the cranial cavity.

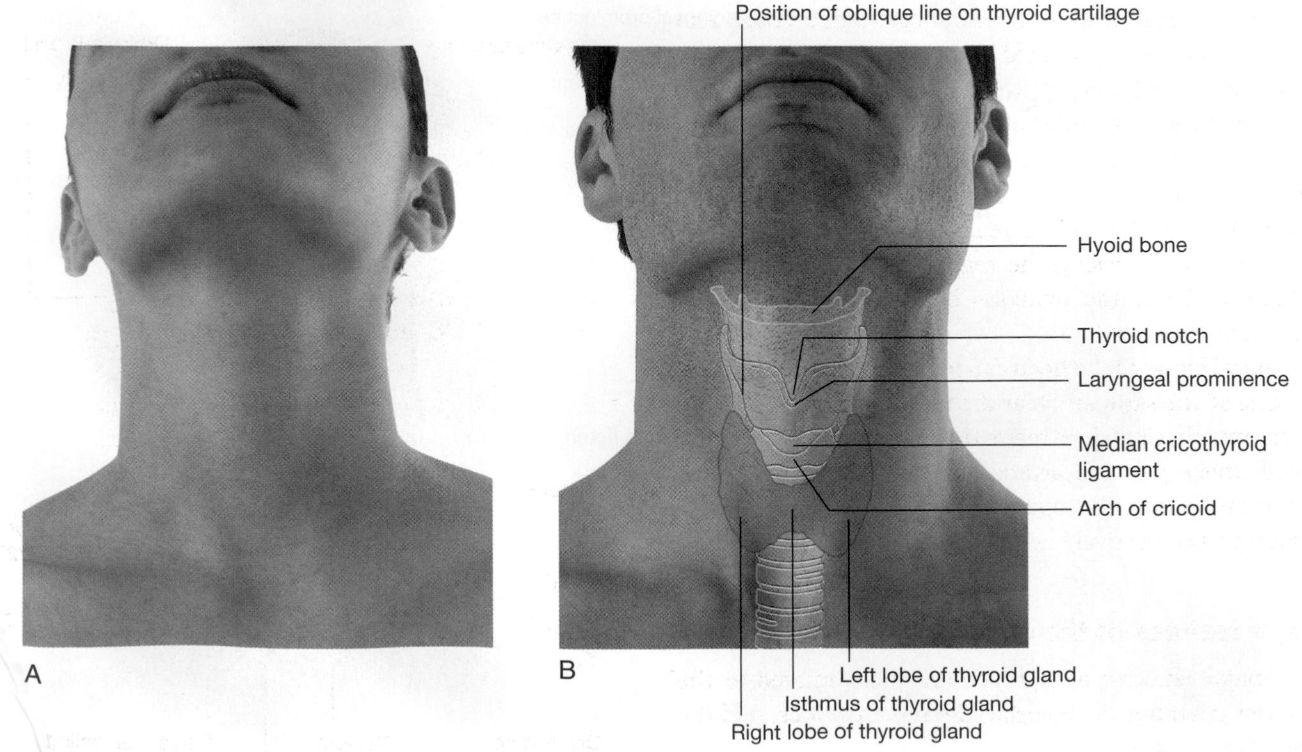


**Fig. 8.287** How to find the thyroid gland. **A.** In a woman, anterior view of neck. **B.** In a man, anterior view of neck.

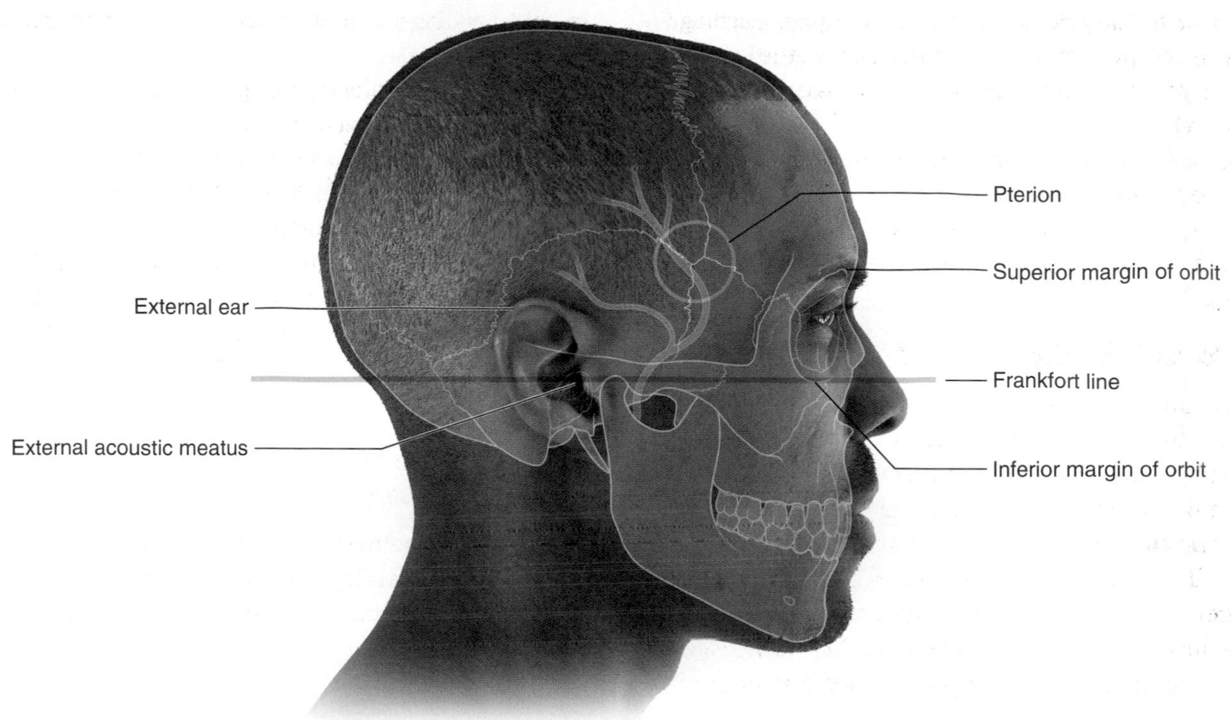


**Fig. 8.288** Estimating the position of the middle meningeal artery. Lateral head and neck of a man.

In lateral blows to the head the middle meningeal artery can be ruptured, leading to extradural hemorrhage and eventual death if not treated.

The anterior branch of the middle meningeal artery is the part of the vessel most often torn. This branch is in the temple region of the head, approximately midway between the superior margin of the orbit and the upper part of the external ear in the pterion region. The pterion is a small circular area enclosing the region where the sphenoid, frontal, parietal, and temporal bones of the skull come together.

Lateral blows to the head can fracture the internal table of bone of the skull and tear the middle meningeal artery in the outer layer of dura mater that is fused to the cranium. Blood under pulsatile arterial pressure leaks out of the vessel and gradually separates the dura from the bone, forming a progressively larger extradural hematoma.

## Major features of the face

The major features of the face are those related to the anterior openings of the orbit, the nasal cavities, and the oral cavity (Fig. 8.289).

The palpebral fissures are between the upper and lower eyelids and can be opened and closed. The oral fissure is the gap between the upper and lower lips and can also be opened and closed.

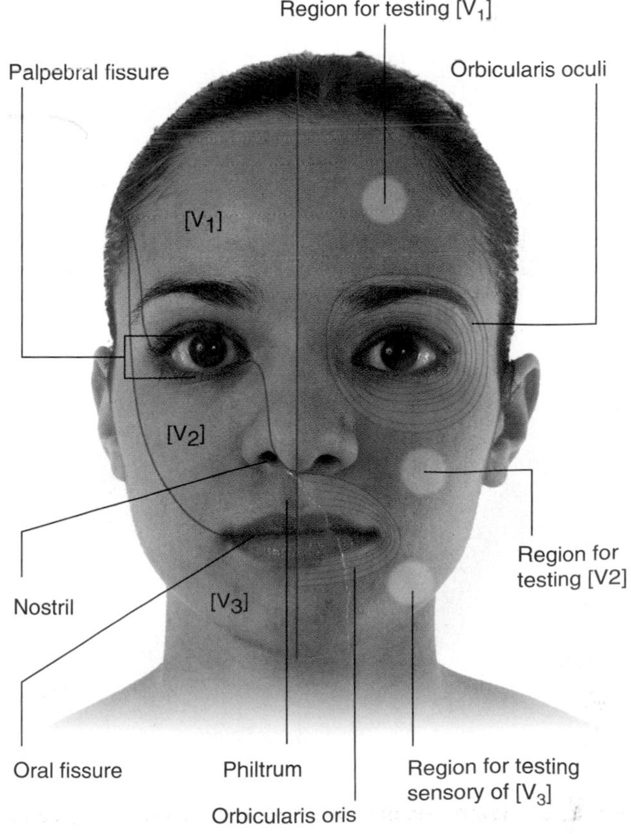


**Fig. 8.289** Major features of the face. Anterior head and neck of a woman.

The sphincter muscles of the oral and palpebral fissures are the orbicularis oris and orbicularis oculi muscles, respectively. These muscles are innervated by the facial nerve [VII].

The nares are the anterior apertures of the nasal cavities and are continuously open.

The vertical groove in the midline between the external nose and the upper lip is the philtrum.

Sensory innervation of the face is carried by the trigeminal nerve [V]. The three divisions of this nerve are represented on the face and can be tested by touching the forehead (the ophthalmic nerve [$V_1$]), the anterior cheek (the maxillary nerve [$V_2$]), and skin over the anterior body of the mandible (the mandibular nerve [$V_3$]).

## The eye and lacrimal apparatus

Major features of the eye include the sclera, cornea, iris, and pupil (Fig. 8.290). The cornea is continuous with the sclera and is the clear circular region of the external covering of the eye through which the pupil and iris are visible. The sclera is not transparent and is normally white.

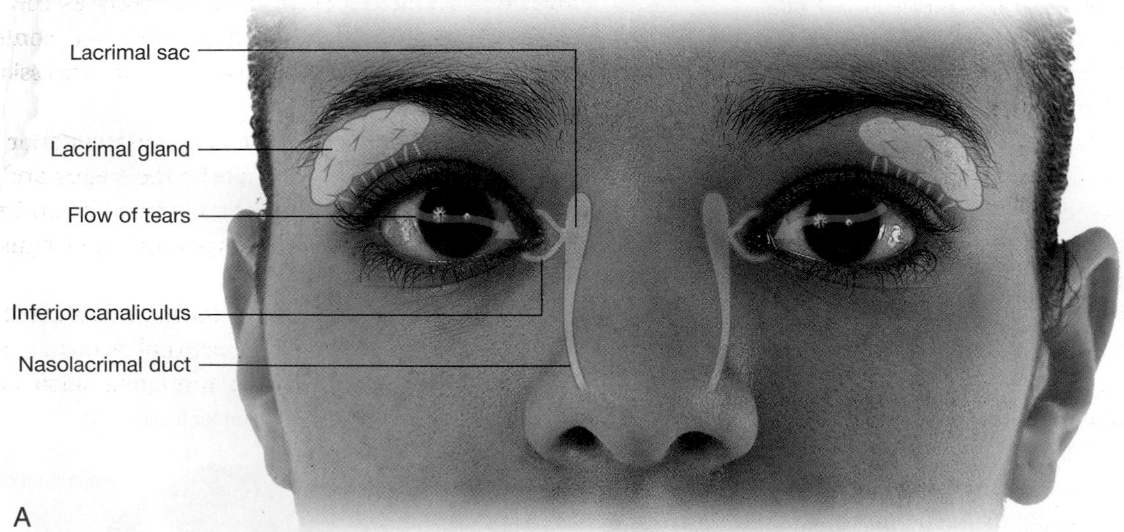


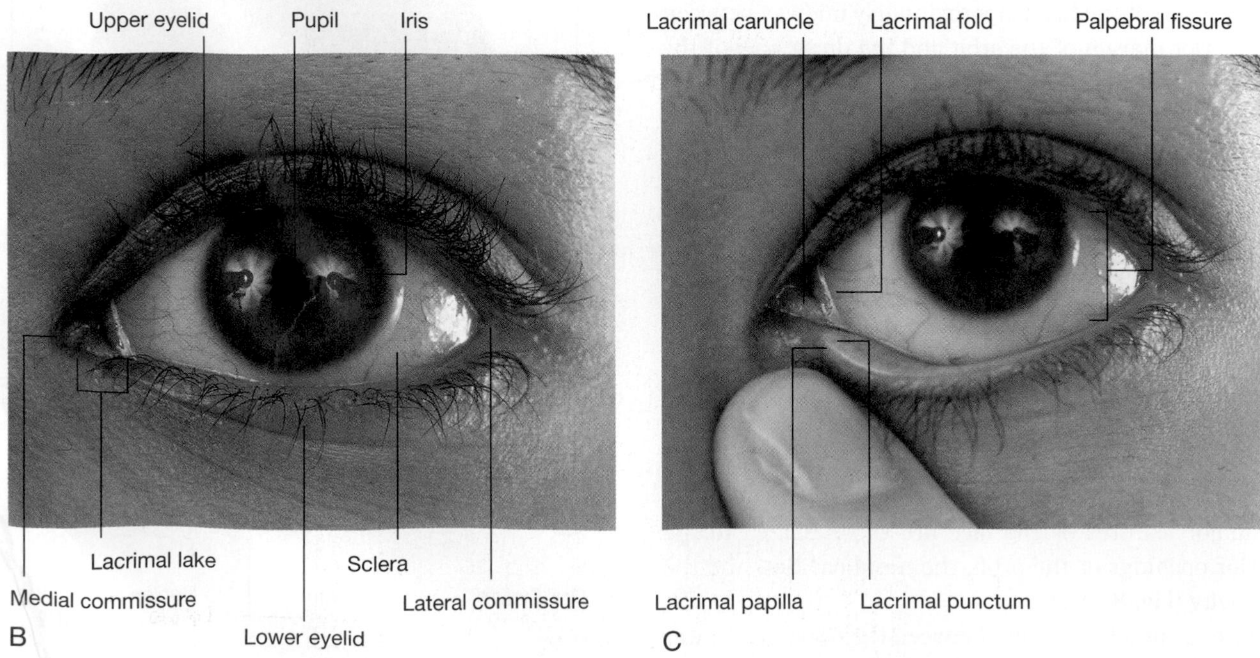


**Fig. 8.290** Eye and lacrimal apparatus. **A.** Face of a woman. Lacrimal apparatus and the flow of tears are indicated. **B.** Left eye and surrounding structures. **C.** Left eye and surrounding structures with lower eyelid pulled down to reveal the lacrimal papilla and lacrimal punctum.

The upper and lower eyelids of each eye enclose between them the palpebral fissure. The eyelids come together at the medial and lateral palpebral commissures on either side of each eye.

At the medial side of the palpebral fissure and lateral to the medial palpebral commissure is a small triangular soft tissue structure (the lacrimal lake).

The elevated mound of tissue on the medial side of the lacrimal lake is the lacrimal caruncle, and the lateral margin overlying the sclera is the lacrimal fold.

The lacrimal apparatus consists of the lacrimal gland and the system of ducts and channels that collects the tears and drain them into the nasal cavity. Tears hydrate and maintain the transparency of the cornea.

The lacrimal gland is associated with the upper eyelid and is in a small depression in the lateral roof of the orbit just posterior to the orbital margin. The multiple small ducts of the gland open into the upper margin of the conjunctival sac, which is the thin gap between the deep surface of the eyelid and the cornea.

Tears are swept medially over the eye by blinking and are collected in small openings (lacrimal puncta), one on each of the upper and lower eyelids near the lacrimal lake.

Each punctum is on a small raised mound of tissue (a lacrimal papilla), and is the opening of a small canal (lacrimal canaliculus) that connects with the lacrimal sac.

The lacrimal sac is in the lacrimal fossa on the medial side of the orbit. From the lacrimal sac, tears drain via the nasolacrimal duct into the nasal cavity.

## External ear

The external ear (Fig. 8.291) consists of the auricle and the external acoustic meatus. The auricle is supported by cartilage and is covered by skin. The external acoustic meatus is near the anterior margin of the auricle.

The auricle is characterized by a number of depressions, eminences, and folds. The folded outer margin of the auricle is the helix, which ends inferiorly as the lobule. A smaller fold (the antihelix) parallels the contour of the helix and is separated from it by a depression (the scaphoid fossa).

The tragus is a small eminence anteroinferior to the external acoustic meatus. Opposite the tragus and at the end of the antihelix is another eminence (the antitragus). The depression between the tragus and antitragus is the intertragic incisure.

The deepest depression (the concha) is bracketed by the antihelix and leads into the external acoustic meatus. Other depressions include the triangular fossa and the cymba conchae.

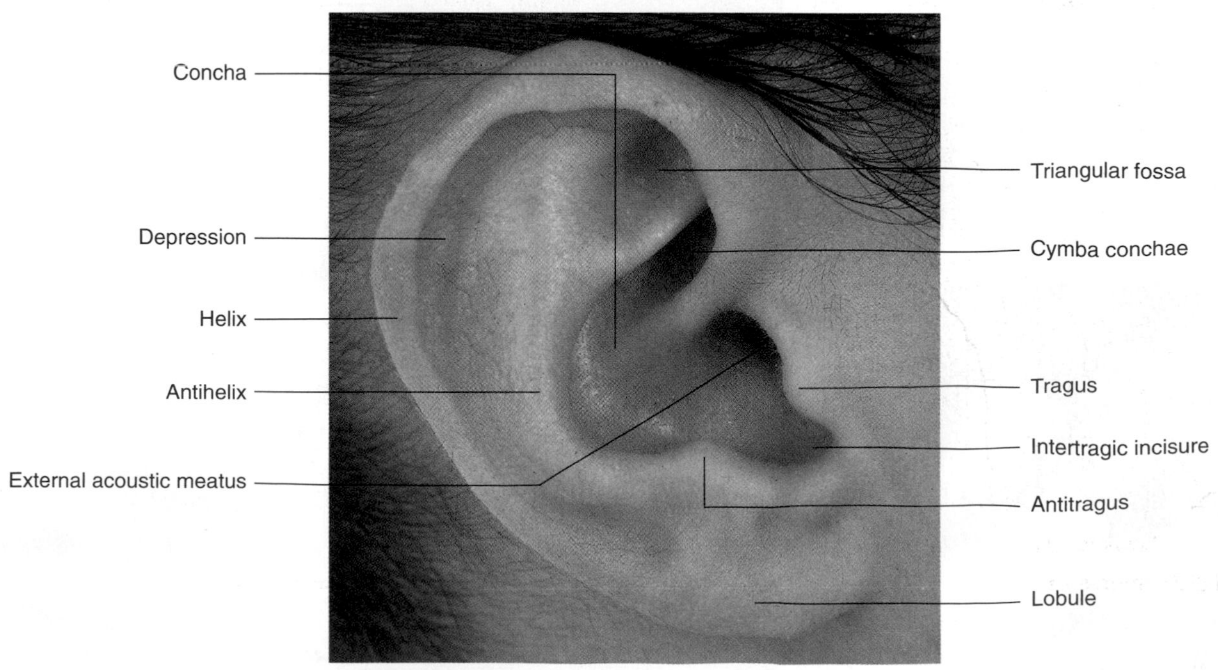


**Fig. 8.291** External ear. Lateral view of the right ear of a woman.

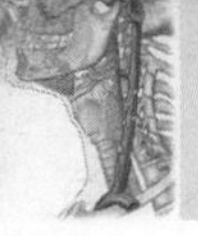



## Pulse points

Arterial pulses can be felt at four locations in the head and neck (Fig. 8.292).

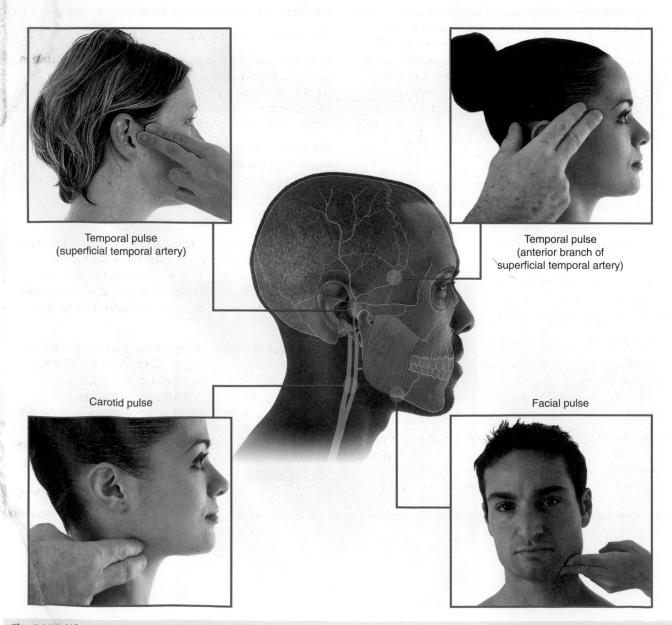

- Carotid pulse—the common or external carotid artery can be palpated in the anterior triangle of the neck. This is one of the strongest pulses in the body. The pulse can be obtained by palpating either the common carotid artery posterolateral to the larynx or the external carotid artery immediately lateral to the pharynx midway between the superior margin of the thyroid cartilage below and the greater horn of the hyoid bone above.
- Facial pulse—the facial artery can be palpated as it crosses the inferior border of the mandible immediately adjacent to the anterior margin of the masseter muscle.
- Temporal pulse—the superficial temporal artery can be palpated anterior to the ear and immediately posterosuperior to the position of the temporomandibular joint.
- Temporal pulse—the anterior branch of the superficial temporal artery can be palpated posterior to the zygomatic process of the frontal bone as it passes lateral to the temporal fascia and into anterolateral regions of the scalp. In some individuals pulsations of the superficial temporal artery can be seen through the skin.



**Fig. 8.292** Where to take arterial pulses in the head and neck.

# Clinical cases

## Case 1

MULTINODULAR GOITER

**A 50-year-old overweight woman came to the doctor complaining of hoarseness of voice and noisy breathing. She was also concerned at the increase in size of her neck. On examination she had a slow pulse rate (45 beats per minute). She also had an irregular knobby mass in the anterior aspect of the lower neck, which deviated the trachea to the right.**

A clinical diagnosis of a multinodular goiter and hypothyroidism was made.

Enlargement of the thyroid gland is due to increased secretion of thyroid-stimulating hormone, which is usually secondary to diminished output of thyroid hormones. The thyroid undergoes periods of activity and regression, which can lead to the formation of nodules, some of which are solid and some of which are partially cystic (colloid cysts). This nodule formation is compounded by areas of fibrosis within the gland. Other causes of multinodular goiter include iodine deficiency and in certain circumstances, drugs that interfere with the metabolism and production of thyroxine. The typical symptom of a goiter is a painless swelling of the thyroid gland. It may be smooth or nodular, and occasionally it may extend into the superior mediastinum as a retrosternal goiter.

The trachea was deviated.

The enlargement of the thyroid gland due to a multinodular goiter may not be symmetrical. In this case there was significant asymmetrical enlargement of the left lobe of the thyroid deviating the trachea to the right.

The patient had a hoarse voice and noisy breathing.

If the thyroid gland enlargement is significant it can compress the trachea, narrowing it to such an extent that a "crowing sound" is heard during inspiration (stridor).

Other possible causes for hoarseness include paralysis of the vocal cord due to compression of the left recurrent laryngeal nerve from the goiter. Of concern is the possibility of malignant change within the goiter directly invading the recurrent laryngeal nerve. Fortunately, malignant change is rare within the thyroid gland.

When patients have a relatively low production of thyroxine such that the basal metabolic rate is reduced they become more susceptible to infection, including throat and upper respiratory tract infections.

On examination the thyroid gland moved during swallowing.

Characteristically, an enlarged thyroid gland is evident as a neck mass arising on one or both sides of the trachea. The enlarged thyroid gland moves on swallowing because it is attached to the larynx by the pretracheal fascia.

The patient was hypothyroid.

Hypothyroidism refers to the clinical and biochemical state in which the thyroid gland is underactive (hyperthyroidism refers to an overactive thyroid gland). Some patients have thyroid masses and no clinical or biochemical abnormalities—these patients are euthyroid.

The hormone thyroxine controls the basal metabolic rate; therefore, low levels of thyroxine affect the resting pulse rate and may produce other changes, including weight gain, and in some cases depression.

The patient was insistent upon surgery.

After discussion about the risks and complications, a subtotal thyroidectomy was performed. After the procedure the patient complained of tingling in her hands and feet and around her mouth, and carpopedal spasm. These symptoms are typical of tetany and are caused by low serum calcium levels.

The etiology of the low serum calcium level was trauma and bruising of the four parathyroid glands left in situ after the operation. Undoubtedly the trauma of removal of such a large thyroid gland produced a change within the parathyroid gland, which failed to function appropriately. The secretion of parathyroid hormone rapidly decreased over the next 24 hours, resulting in increased excitability of peripheral nerves, manifest by carpopedal spasm and orofacial tingling. Muscle spasms can also be elicited by tapping the facial

*(continues)*

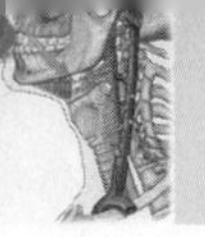



## Case 1—cont'd

nerve [VII] as it emerges from the parotid gland to produce twitching of the facial muscles (Chvostek's sign).

The patient recovered from these symptoms due to a low calcium level over the next 24 hours.

At her return to the clinic the patient was placed on supplementary oral thyroxine, which is necessary after removal of the thyroid gland.

The patient also complained of a hoarse voice.

The etiology of her hoarse voice was damage to the recurrent laryngeal nerve.

The recurrent laryngeal nerve lies close to the thyroid gland. It may be damaged in difficult surgical procedures, and this may produce unilateral spasm of the ipsilateral vocal cord to produce a hoarse voice.

Since the thyroidectomy and institution of thyroxine treatment, the patient has lost weight and has no further complaints.

## Case 2

EXTRADURAL HEMATOMA

**A 33-year-old man was playing cricket for his local Sunday team. As the new bowler pitched the ball short, it bounced higher than he anticipated and hit him on the side of his head. He immediately fell to the ground unconscious, but after about 30 seconds he was helped to his feet and felt otherwise well. It was noted he had some bruising around his temple. He decided not to continue playing and went to watch the match from the side. Over the next hour he became extremely sleepy and was eventually unrousable. He was rushed to hospital.**

When he was admitted to hospital, the patient's breathing was shallow and irregular and it was necessary to intubate him. A skull radiograph demonstrated a fracture in the region of the pterion. No other abnormality was demonstrated other than minor soft tissue bruising over the left temporal fossa.

A CT scan was performed.

The CT scan demonstrated a lentiform area of high density within the left cranial fossa.

A diagnosis of extradural hemorrhage was made.

Fractures in the region of the pterion are extremely dangerous. A division of the middle meningeal artery passes deep to this structure and is subject to laceration and disruption, especially in conjunction with a skull injury in this region. In this case the middle meningeal artery was torn and started to bleed, producing a large extradural clot.

The patient's blood pressure began to increase.

Within the skull there is a fixed volume and clearly what goes in must come out (e.g., blood, cerebrospinal fluid). If there is a space-occupying lesion, such as an extradural hematoma, there is no space into which it can decompress. As the lesion expands, the brain becomes compressed and the intracranial pressure increases. This pressure compresses vessels, so lowering the cerebral perfusion pressure. To combat this the homeostatic mechanisms of the body increase the blood pressure to overcome the increase in intracerebral pressure. Unfortunately, the increase in intracranial pressure is compounded by the cerebral edema that occurs at and after the initial insult.

An urgent surgical procedure was performed.

Burr holes were placed around the region of the hematoma and it was evacuated. The small branch of the middle meningeal artery was ligated and the patient spent a few days in the intensive care unit. Fortunately the patient made an uneventful recovery.

## Case 3

COMPLICATION OF ORBITAL FRACTURE

**A 35-year-old man was involved in a fight and sustained a punch to the right orbit. He came to the emergency department with double vision.**

The double vision was only in one plane.

Examination of the orbits revealed that when the patient was asked to look upward the right eye was unable to move superiorly when adducted. There was some limitation in general eye movement. Assessment of the lateral rectus muscle (abducent nerve [VI]), superior oblique muscle (trochlear nerve [IV]), and the rest of the eye muscles (oculomotor nerve [III]) was otherwise unremarkable.

The patient underwent a CT scan.

A CT scan of the facial bones demonstrated a fracture through the floor of the orbit (Fig. 8.293).

A careful review of this CT scan demonstrated that the inferior oblique muscle had been pulled inferiorly with the fragment of bone in the fracture. This produced a tethering effect, so when the patient was asked to gaze in the upward direction, the left eye was able to do so but the right eye was unable to because of the tethered inferior oblique muscle.

The patient underwent surgical exploration to elevate the small bony fragment and return the inferior oblique to its appropriate position. On follow-up the patient had no complications.

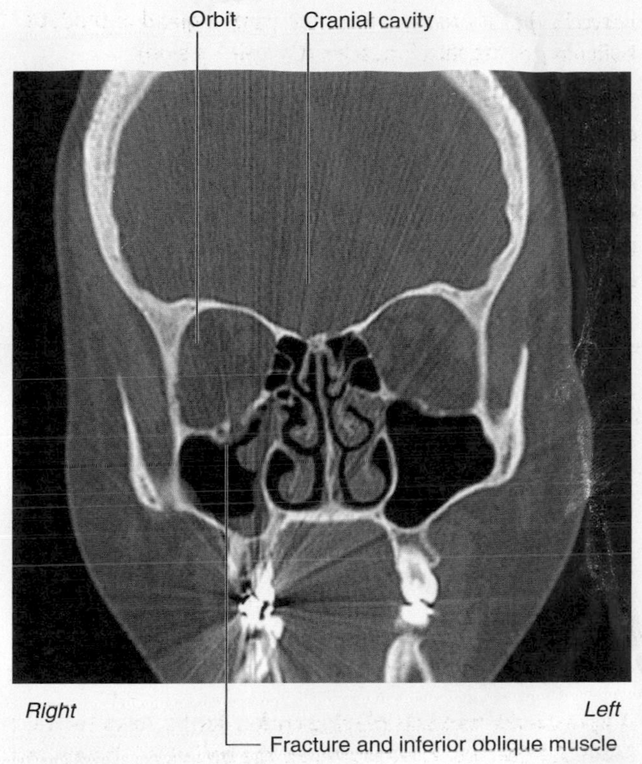


**Fig. 8.293** Coronal CT scan demonstrating an orbital blowout fracture.

# Index



Page numbers followed by "*f*" indicate figures, "*t*" indicate tables, "*b*" indicate boxes, and "*e*" indicate online content.

# Index

## C

# Index



## M

## O



## P

## S